# Fruit Flies (Tephritida

## Phylogeny and Evolution of Behavior

# Fruit Flies (Tephritida

## Phylogeny and Evolution of Behavior

Edited by

Martín Aluja, Ph.D.
Allen L. Norrbom, Ph.D.

CRC Press
Taylor & Francis Group
Boca Raton London New York

CRC Press is an imprint of the
Taylor & Francis Group, an **informa** business

**Cover Photographs: (Top to Bottom)** (1) Regurgitated droplets deposited by *Anastrepha serpentina* female. These droplets are then reingested by the same female. (Photograph by R. Wilson.) (2) Mating pair of *Anastrepha serpentina*. (Photograph by R. Wilson.) (3) Feeding in damaged guava by females of *Anastrepha ludens* (head in hole) and *A. fraterculus* (entire fly in hole). (Photograph by E. Piedra.) (4) Male of *Ceratitis capitata* releasing pheromone during calling bout. (Photograph by K. Kaneshiro.)

CRC Press
Taylor & Francis Group
6000 Broken Sound Parkway NW, Suite 300
Boca Raton, FL 33487-2742

First issued in paperback 2019



ISBN-13: 978-0-8493-1275-5 (hbk)
ISBN-13: 978-0-367-39910-8 (pbk)



**Library of Congress Cataloging-in-Publication Data**

Fruit flies (Tephritidae) : phylogeny and evolution of behavior / edited by Martin Aluja and Allen L. Norrbom.
p. cm.
Includes bibliographical references and index.
ISBN 0-8493-1275-2 (alk. paper)
1. Tephritidae--Phylogeny. 2. Tephritidae--Behavior. I. Aluja, Martin. II. Norrbom, Allen L. (Allen Lee), 1957-

QL537.T42 F775 1999
595.77'4--dc21

99-045290
CIP

Library of Congress Card Number 99-045290

**Visit the Taylor & Francis Web site at**
**http://www.taylorandfrancis.com**

**and the CRC Press Web site at**
**http://www.crcpress.com**

# Preface

The main purpose of this book is to discuss and promote the study of the evolution of fruit fly (Diptera: Tephritidae) behavior. As a basis for this discussion, we review the current state of knowledge and publish considerable new findings on various aspects of fruit fly behavior, phylogeny, and related subjects. Fruit flies are considered one of the most economically important groups of insect pests worldwide, and study of their behavior plays a critical role in their regulation and control. To give only one example, the annual cost of the establishment of the Mediterranean fruit fly in California has been estimated at approximately $1.2 billion by the California Department of Food and Agriculture. We note that the success of one of the most environmentally friendly control measures, the Sterile Insect Technique, hinges on a deep understanding of behavioral mechanisms (i.e., sexual selection) which will be thoroughly discussed in this book.

Given the tremendous economic importance of many species of Tephritidae, the study of fruit flies has been traditionally biased toward applied aspects (e.g., management, monitoring, mass rearing). Nevertheless, their ecological and behavioral plasticity render them ideal study objects to address basic biological and evolutionary questions of interest to a wide audience. Fruit flies have been used as models for the development of general theories on, for example, speciation processes (Bush 1975), mating behavior (Höglund and Alatalo 1995; Eberhard 1996; Shelly and Whittier 1997) and demography (Carey 1993). In the past 15 years, a number of broad-ranging books on fruit fly biology and/or management have been published: Cavalloro (1983; 1986), Mangel et al. (1986), Economopoulos (1987), Robinson and Hooper (1989), Vijaysegaran and Ibrahim (1991), White and Elson-Harris (1992), Aluja (1993), Aluja and Liedo (1993), Calkins et al. (1994), and McPheron and Steck (1996). In these books, one can find isolated chapters (in some cases very short reviews) dealing with behavior of particular genera or groups of fruit fly species, but there is a strong bias toward economically important species. The topics of phylogeny and evolution of behavior are barely addressed and, when so, only superficially. In this book we attempt a more comprehensive and thorough approach, covering all behaviors in a broad range of tephritids, incorporating phylogenies as much as possible. We are therefore confident that this book will bridge an important information gap in a highly visible group of insects and also serve as a blueprint for basic and applied behavioral research on fruit flies and other organisms in the coming years.

Hypotheses of phylogenetic relationships are valuable tools to understand the evolution of any trait in a group of organisms, and their use is becoming widely accepted in the study of animal behavior (see Martins 1996). Promoting their use by students of fruit fly behavior is one of the major goals of this book. Despite the economic importance of fruit flies, few rigorous, comprehensive phylogenies have been published for tephritid groups (see review by Norrbom et al. 1999a). Even for the economically important genera *Anastrepha*, *Bactrocera*, *Ceratitis*, *Dacus*, and *Toxotrypana* there are no, or at most partial, phylogenies available. On the other hand, in the last decade analyses of phylogenetic relationships have intensified in systematic work on Tephritidae, and nearly all genera and species of fruit flies worldwide have just been cataloged (Norrbom et al. 1999b). Thus, we felt the time was ripe to ask the world's leading tephritid systematists to tackle this phylogenetic problem. Their response has exceeded our greatest expectations.

Although we must warn the reader that the level of methodology varies from the latest cladistic techniques to "scenario writing," and that some results are still preliminary (but kindly included by the authors at our request), the phylogenetic chapters of this book include analyses to at least a minimal level for nearly all groups of fruit flies whose behavior has been studied to a significant degree. Notably, they include analyses of relationships among the families most closely related to fruit flies and among the higher groups (subfamilies and tribes) within Tephritidae (Korneyev); a preliminary analysis of higher tephritid relationships based on mitochondrial DNA data (Han and McPheron); a review of relationships within the subtribe Carpomyina, which includes *Rhagoletis*, by far the most intensively studied fruit fly genus in terms of diversity of phylogenetic methods (Smith and Bush); the first comprehensive morphological analysis and the first molecular analysis of *Anastrepha* and *Toxotrypana* (Norrbom et al. and McPheron et al.); the first, although preliminary, cladistic analyses of *Ceratitis* (De Meyer) and *Bactrocera* and *Dacus* (White); and the first cladistic analyses of the tribe Trypetini (Han), the subfamily Blepharoneurinae (Norrbom and Condon), and the tephritine tribe Myopitini (Freidberg and Norrbom) and *Tephritis* group of genera (Merz).

The book is intended to become a general reference. It therefore contains reviews of past and present work but also indicates potential future directions of research. We have asked authors not only to review all past work, but also to make an effort to present novel, unpublished results and to try to foster the conceptual advancement of the field.

This volume is divided into eight sections. Section 1 provides a general framework for the rest of the book. It contains an overview of the phylogeny of the Tephritoidea (Korneyev), a review of the behaviors exhibited by flies in the families most closely related to the Tephritidae (Sivinski), and a historical review of studies on tephritid behavior (Díaz-Fleischer and Aluja).

Each of the next five sections covers the phylogeny and behavior of important groups of tephritids — Blepharoneurinae (Norrbom and Condon; Condon and Norrbom), Phytalmiinae (Dodson), Toxotrypanini (Norrbom et al.; McPheron et al.; Aluja et al.; and Landolt), Carpomyina (Smith and Bush; Prokopy and Papaj), Dacinae (De Meyer; Eberhard; Yuval and Hendrichs; White; Drew and Hancock; Drew and Romig), Trypetini (Han), and Tephritinae (Korneyev; Freidberg and Norrbom; Merz; Headrick and Goeden). The classification of these groups follows that by Korneyev (Chapter 4). A detailed list of genera, following a slightly different classification, is included by Norrbom et al. (1999a). In the exceptional case of the genus *Ceratitis* there are two separate behavior chapters: one dealing specifically with the sexual behavior of *C. capitata* (Eberhard), and the other reviewing the other behaviors of this species and all behavior of the other species of *Ceratitis* (Yuval and Hendrichs). This is because *C. capitata* is by far the best-studied fruit fly, and a review of only its sexual behavior turned out to be as long as a review of all the remaining behaviors for the whole genus.

The next section of the book deals with specific topics, but on a broader scale than in previous chapters. We asked the authors of most of these chapters to analyze a specific topic based on all relevant data from across the Tephritidae and, where appropriate, to examine it from a phylogenetic viewpoint. The chapters on the evolution of feeding, mating, and oviposition behavior (Drew and Yuval, Sivinski et al., and Díaz-Fleischer et al., respectively) are among the most relevant to the primary goal of the book. They are products of teams of authors, and we wish to recognize here the monumental task of the lead authors in coordinating these chapters. Other chapters in this section include a review of tephritid population structure by Berlocher, which clearly explains a complex subject, and another by Cayol on the changes in sexual behavior and some life history traits generated by inadvertent artificial selection during mass rearing of certain economically important species (e.g., *C. capitata*, *B. cucurbitae*, *B. dorsalis*, *A. ludens*). This chapter was of particular interest to us because it addresses a highly applied aspect of fruit fly behavior and ecology. Framing such findings within an evolutionary perspective will be of great value to action programs, whose technical personnel may lack the proper theoretical background to understand the underlying mechanisms behind inadvertent artificial selection during mass rearing, and to allow them to design proper schemes to monitor and avoid it.

Since we wanted to end the book with a bang, we invited Robert Heath and Kenneth Kaneshiro to address topics of great applied and basic interest: Sexual pheromones in tephritid flies and the behavior of a distantly related group of flies (*Drosophila*). Heath and his collaborators have been generating very interesting information on the chemical characterization of the sexual pheromones of several species of *Anastrepha*. Such information, if investigated in additional species and genera, would be a novel and potentially useful data set for analysis of phylogenetic relationships in fruit flies. The fact that this information will be published here for the first time will no doubt make this chapter one of the most cited pieces of work on fruit fly chemical ecology. Ending this section with an overview of behavior in *Drosophila*, a group in which phylogenetic relationships are better resolved and the evolution of behavior fairly well understood, allows us both to honor the important contributions of Kenneth Kaneshiro to the study of fruit flies, and to draw interesting parallels between drosophilids and tephritids.

This book clearly has a hybrid nature and many readers are likely to be specialists in one field who are unfamiliar with the jargon used in others. Despite the efforts to review the meaning of some of these terms at the symposium in Xalapa, there was still some loss of communication between taxonomists and behaviorists. The final section of the book therefore contains a glossary of terminology in the areas of systematics, morphology, phylogeny reconstruction, ecology, and behavior (White et al.). We sincerely hope that such information will assist researchers throughout the world and will be of great value in standardizing the terminology used in publications of this nature (a much-needed effort).

Before ending, we would like to highlight some important aspects related to the genesis of this book and to our desire to recognize the pioneering work of D. Elmo Hardy and Ronald J. Prokopy. The idea of putting together a volume like this came when one of us (M.A.) was chairman of the Working Group on Fruit Flies of the International Organization for Biological Control of Noxious Plants and Animals (IOBC). This working group has a long and successful history, thanks largely to the efforts of the distinguished Swiss entomologist, Ernst Boller. M.A. felt the strong urge, as chairman of this group, to contribute something substantial to the field and, at the same time, to honor the rich history of the group. A book on a challenging topic, such as the evolution of fruit fly behavior, seemed like the best solution. As a result of fruitful discussions with A.L.N. and key researchers in the field, it was decided to organize a symposium during which all speakers would bring drafts of their prospective book chapters. The idea was to give everybody a chance to present and discuss the contents of their manuscripts with all other authors and to allow the editors to start working directly with authors during the early stages of the writing process. Fortunately, our hopes were realized and 35 scientists from 14 countries met in Xalapa, Veracruz, Mexico from February 16 to 21, 1998. The symposium, attended by nearly all of the world's top experts on fruit fly taxonomy and behavior, turned out to be a fantastic experience for everybody involved and, as expected, allowed all authors to get a clear idea of what was missing in their chapters. Also, it allowed all participants to discuss in great detail the topic of the symposium and the future book. This was particularly important to those participating in the collaborative effort of writing chapters on the evolution of feeding, mating, and oviposition behaviors. Among many other reasons, we believe that the symposium was worthwhile just to promote collaboration among the many participants who had never met and particularly to promote interaction between the taxonomists and behaviorists. For example, none of us except a few of the taxonomists had ever met Valery Korneyev, a Ukranian who is the most knowledgeable scientist in the area of fruit fly phylogeny and whose work until recently had been published exclusively in Russian.

The fact that this book is dedicated to D. Elmo Hardy and Ronald J. Prokopy should be easily understood. We believe, as do many others, that these two distinguished scientists have had a major impact in the fields of fruit fly behavior and systematics. Recognizing their fundamental contributions seemed to us both an obligation and a natural consequence of the deep respect we have for their work.

We sincerely hope that this volume will motivate many to share our enthusiasm and admiration for this wonderful group of insects, who zealously guard the secrets of their lives. Attempting to unravel these secrets is both motivating and humbling. ¡Vivan las moscas de la fruta! Long live the fruit flies!

**Martín Aluja**
*Xalapa, Veracruz, Mexico*

**Allen L. Norrbom**
*Washington, D.C., U.S.A.*

## REFERENCES


Aluja, M. 1993. *Manejo Integrado de las Moscas de la Fruta*. Trillas, Mexico D.F. 257 pp.

Aluja, M. and P. Liedo (eds.). 1993. *Fruit Flies: Biology and Management, Proceedings of the International Symposium on Fruit Flies of Economic Importance, 1990, Antigua Guatemala, Guatemala, October 14–20, 1990*. Springer-Verlag, New York. 492 pp.

Bush, G. 1975. Modes of Animal Speciation. *Annu. Rev. Ecol. Syst.* 6: 339–364.

Calkins, C.O., W. Klassen, and P. Liedo (eds.). 1994. *Fruit Flies and the Sterile Insect Technique*. CRC Press, Boca Raton. 258 pp.

Carey, J. 1993. *Applied Demography for Biologists*. Oxford University Press, New York. 206 pp.

Cavalloro, R. (ed.). 1983. *Fruit Flies of Economic Importance. Proceedings of the CEC/IOBC International Symposium, Athens, Greece, November 16–19, 1982*. A.A. Balkema, Rotterdam. 642 pp.

Cavalloro, R. (ed.). 1986. *Fruit Flies of Economic Importance 84, Proceedings of the CEC/IOBC "ad-hoc meeting,"* Hamburg, August 1994. A.A. Balkema, Rotterdam. 221 pp.

Eberhard, W. 1996. *Female Control: Sexual Selection by Cryptic Female Choice*. Princeton University Press, Princeton. 501 pp.

Economopoulos, A.P. (ed.). 1987. *Fruit Flies. Proceedings of the Second International Symposium, September 16–21, 1986, Colymbari, Crete, Greece*. Elsevier Science Publishers, Amsterdam. 590 pp.

Höglund, J. and R.V. Alatalo. 1995. *Leks*. Princeton University Press, Princeton. 248 pp.

Mangel, M., J.R. Carey and R.E. Plant (eds.). 1986. *Pest Control: Operations and Systems Analysis in Fruit Fly Management*. NATO Advanced Science Institutes Series G: Ecological Sciences 11. Springer-Verlag, Berlin. 465 pp.

Martins, E.P. (ed.). 1996. *Phylogenies and the Comparative Method in Animal Behavior*. Oxford University Press, New York. 415 pp.

McPheron, B.A. and G.J. Steck (eds.). 1996. *Fruit Fly Pests: A World Assessment of Their Biology and Management*. St. Lucie Press, Delray Beach. 586 pp.

Norrbom, A.L., L.E. Carroll, and A. Freidberg. 1999a. Status of knowledge. In *Fruit Fly Expert Identification System and Systematic Information Database* (F.C. Thomson, ed.), pp. 9–47. *Myia* (1998) 9, 524 pp.

Norrbom, A.L., L.E. Carroll, F.C. Thompson, I.M. White, A. Freidberg. 1999b. Systematic database of names. In *Fruit Fly Expert Identification System and Systematic Information Database* (F.C. Thomson, ed.), pp. 65–251. *Myia* (1998) 9, 524 pp.

Robinson, A.S. and G. Hooper (eds.). 1989. *Fruit Flies: Their Biology, Natural Enemies and Control*. In *World Crop Pests* (W. Helle, ed.), Vol. 3A, 372 pp. and Vol. 3B, 447 pp. Elsevier Science Publishers, Amsterdam.

Shelly, T.E. and T.S Whittier. 1997. Lek behavior of insects. In *The Evolution of Mating Systems in Insects and Arachnids* (J.C. Choe and B.J. Crespi, eds.), pp. 273–293. Cambridge University Press, Cambridge. 387 pp.

Vijaysegaran, S. and A.G. Ibrahim. 1991. *First International Symposium on Fruit Flies in the Tropics, Kuala Lumpur, 1988*. Malaysian Agricultural Research and Development Institute & Malaysian Plant Protection Society, Kuala Lumpur. 430 pp.

White, I.M. and M.M. Elson-Harris. 1992. *Fruit Flies of Economic Significance: Their Identification and Bionomics*. CAB International, Wallingford. 601 pp.

# Contributors

**Martín Aluja, Ph.D.**
Department of Animal Behavior
Instituto de Ecología, A.C.
Km 2.5 Antigua Carretera a Coatepec
Apartado Postal 63
91000 Xalapa, Veracruz, México
alujam@ecologia.edu.mx

**Stewart H. Berlocher, Ph.D.**
Department of Entomology
University of Illinois at Urbana-Champaign
320 Morrill Hall
505 S. Goodwin Avenue
Urbana, IL 61801, U.S.A.
stewartb@uiuc.edu

**Guy L. Bush, Ph.D.**
Department of Zoology
Michigan State University
203 Natural Sciences Bldg.
East Lansing, MI 48824-1115, U.S.A.
bushfly@pilot.msu.edu

**Lynn E. Carroll, Ph.D.**
c/o Smithsonian Institution MRC 168
Washington, D.C. 20560, U.S.A.
lcarroll@sel.barc.usda.gov

**Jean-Pierre Cayol, Ph.D.**
Entomology Unit
FAO/IAEA Agriculture and
Biotechnology Laboratory
A-2444 Seibersdorf, Austria

Current address:
CIRAD-FLHOR Guyane Campus
Agronomique
BP 701 F-97387
Kourou Cedex, French Guiana
cayol_j@kourou.cirad.fr

**William G. Eberhard, Ph.D.**
Escuela de Biología
Universidad de Costa Rica
Ciudad Universitaria "Rodrigo Facio"
San José, Costa Rica
weberhar@cariari.ucr.ac.cr

**Marty A. Condon, Ph.D.**
Cornell College
600 First Street West
Mount Vernon, IA 52314-1098, U.S.A.
mcondon@cornell-iowa.edu

**Marc De Meyer, Ph.D.**
Entomology Section
Royal Museum for Central Africa
Leuvensesteenweg 13
B-3080 Tervuren, Belgium
demeyer@africamuseum.be

**Francisco Díaz-Fleischer, B.Sc.**
Campaña Nacional contra las Moscas
de la Fruta
Desarrollo de Métodos
2a. Avenida Sur 5 - Altos
37000 Tapachula, Chiapas, México
fleische@ecologia.edu.mx

**Gary N. Dodson, Ph.D.**
Biology Department
Ball State University
Muncie, IN 47306, U.S.A.
gdodson@bsu.edu

**Richard A.I. Drew, D.Sc.**
Australian School of Environmental Studies
Griffith University, Nathan Campus
Queensland 4111, Australia
D.Drew@mailbox.gu.edu.au

**Barbara D. Dueben, Ph.D.**
Center for Medical, Agricultural and Veterinary
Entomology, Agricultural Research Service
U.S. Department of Agriculture
1600 SW 23rd. Drive, P.O. Box 14565
Gainesville, FL 32604, U.S.A.
bdueben@gainesville.usda.ufl.edu

**Nancy D. Epsky, Ph.D.**
Center for Medical, Agricultural
and Veterinary Entomology
Agricultural Research Service
U.S. Department of Agriculture - ARS
1600 SW 23rd Drive, P.O. Box 14565
Gainesville, FL 32604, U.S.A.
nepsky@gainesville.usda.ufl.edu

**Amnon Freidberg, Ph.D.**
Department of Zoology
George S. Wise Faculty of Life Sciences
Tel Aviv University
Ramat Aviv, 699798 Tel Aviv, Israel
afdipter@post.tau.ac.il

**Richard D. Goeden, Ph.D.**
Department of Entomology
University of California
Riverside, CA 92521, U.S.A.
rgoeden@ucrac1.ucr.edu

**Ho-Yeon Han, Ph.D.**
Department of Life Science
Yonsei University
234 Maeji-ri
Wonju-shi Kangwon-do 220-710, Korea
hyhan@dragon.yonsei.ac.kr

**David L. Hancock, Ph.D.**
P.O. Box 2464
Cairns, Queensland 4870, Australia
hancoc@prose.dpi.qld.gov.au

**David H. Headrick, Ph.D.**
Crop Science Department
California Polytechnic State University
San Luis Obispo, CA 93407, U.S.A.
dheadric@polymail.cpunix.calpoly.edu

**Robert R. Heath, Ph.D.**
Center for Medical, Agricultural
and Veterinary Entomology
Agricultural Research Service
U.S. Department of Agriculture
1600 SW 23rd Drive, P.O. Box 14565
Gainesville, FL 32604, U.S.A.
Bheath@gainesville.usda.ufl.edu
or miarh@ars-grin.gov

**Jorge Hendrichs, Ph.D.**
Insect & Pest Control Section
International Atomic Energy Agency
Wagramerstrasse 5, P.O. Box 100
A-1400 Vienna, Austria
J.Hendrichs@iaea.org

**Vicente Hernández-Ortiz, M.S.**
Department of Entomology
Instituto de Ecología, A.C.
Apartado Postal 63
91000 Xalapa, Veracruz, México
hernanvi@ecologia.edu.mx

**Isabel Jácome, B.Sc.**
Department of Animal Behavior
Instituto de Ecología, A.C.
Km 2.5 Antigua Carretera a Coatepec
Apartado Postal 63
91000 Xalapa, Veracruz, México
isabelj@ecologia.edu.mx

**Kenneth Y. Kaneshiro, Ph.D.**
Center for Conservation Research and Training
University of Hawaii
Honolulu, HI 96822, U.S.A.
kykanesh@hawaii.edu

**Valery A. Korneyev, Ph.D.**
Schmalhausen Institute of Zoology
Ukranian Academy of Sciences
B. Khmelnitski Street 15
252601 Kiev 30, Ukraine
korneyev@entom.freenet.kiev.ua

**Peter J. Landolt, Ph.D.**
Yakima Agricultural Research Laboratory
Agricultural Research Service
U.S. Department of Agriculture
5230 Konnowac Pass Road
Wapato, WA 98951, U.S.A.
landolt@yarl.gov

**Bruce A. McPheron, Ph.D.**
Department of Entomology
The Pennsylvania State University
501 A.S.I. Bldg.
University Park, PA 16802, U.S.A.
bam10@psu.edu

**Bernhard Merz, Ph.D.**
Muséum d'Histoire Naturelle
Département d'Entomologie
C.P. 6434, CH-1211 Genève, Switzerland
bernhard.merz@mhn.ville-ge.ch

**Allen L. Norrbom, Ph.D.**
Systematic Entomology Laboratory
PSI, ARS, USDA
c/o Smithsonian Institution, MRC 168
Washington, D.C. 20560-0168, U.S.A.
anorrbom@sel.barc.usda.gov

**Daniel R. Papaj, Ph.D.**
Department of Ecology
and Evolutionary Biology
University of Arizona
Tucson, AZ 85721, U.S.A.
papaj@u.arizona.edu

**Jaime C. Piñero, B.Sc.**
Department of Animal Behavior
Instituto de Ecología, A.C.
Km 2.5 Antigua Carretera a Coatepec
Apartado Postal 63
91000 Xalapa, Veracruz, México
pineroja@ecologia.edu.mx

**Ronald J. Prokopy, Ph.D.**
Department of Entomology
Fernand Hall
University of Massachusetts
Amherst, MA 01003, U.S.A.
prokopy@ent.umass.edu

**David C. Robacker, Ph.D.**
Subtropical Agricultural Research Center
Agricultural Research Service
U.S. Department of Agriculture
2301 S. International Blvd
Weslaco, TX 78596, U.S.A.
robacker@pop.tamu.edu

**Meredith C. Romig, Diploma of Fine Arts**
Australian School of Environmental Studies
Griffith University Nathan Campus
Queensland 4111, Australia
M.Romig@mailbox.gu.edu.au

**Janisete G. Silva, Ph.D.**
Department of Entomology
The Pennsylvannia State University
501 A.S.I. Bldg.
University Park, PA 16802, U.S.A.
jgs10@psu.edu

**John Sivinski, Ph.D.**
Center for Medical, Agricultural
and Veterinary Entomology
Agricultural Research Service
U.S. Department of Agriculture
1600 SW 23rd. Drive, P.O. Box 14565
Gainesville, FL 32604, U.S.A.
jsivinski@gainesville.usda.ufl.edu

**James J. Smith, Ph.D.**
Department of Zoology
Michigan State University
203 Natural Sciences Bldg.
East Lansing, MI 48824-1115, U.S.A.
jimsmith@pilot.msu.edu

**Ian M. White, Ph.D.**
Department of Entomology
The Natural History Museum
Cromwell Road, London, SW7 5BD
United Kingdom
imw@nhm.ac.uk

**Boaz Yuval, Ph.D.**
Department of Entomology
Hebrew University of Jerusalem
P.O. Box 12, Rehovot 76100, Israel
yuval@agri.huji.ac.il

**Roberto A. Zucchi, Ph.D.**
Departamento de Entomología
Escola Superior de Agricultura
"Luiz de Queiroz"
Universidade de São Paulo
Caixa Postal 9
CEP 13418-900, Piracicaba, São Paulo, Brazil
razucchi@carpa.ciagri.usp.br

**Symposium participants.** Back row, left to right (standing): Jorge Hendrichs, Larissa Guillén, Martín Aluja, Lynn E. Carroll, Amnon Freidberg, James J. Smith, Gary N. Dodson, David H. Headrick. Second row, left to right (standing): Jaime Piñero, Mark De Meyer, Isabel Jácome, Bernhard Merz, Cecilia García Viesca, Richard A.I. Drew, Peter J. Landolt, Daniel R. Papaj, Meredith C. Romig, Ronald J. Prokopy, Richard D. Goeden, Stewart H. Berlocher, Boaz Yuval, Losalini Leweniqila, Bruce A. McPheron, Jean-Pierre Cayol, Ian M. White (face covered), Raquel Cervantes, John Sivinski, Gary Steck. Third row, left to right (kneeling): Francisco Díaz-Fleischer, José Arredondo, Salvador Meza, Kenneth Y. Kaneshiro, Roberto A. Zucchi, Robert R. Heath, Allen L. Norrbom, Ho-Yeon Han. Fourth row, left to right (seated): Valery A. Korneyev, Guy L. Bush, Vicente Hernández-Ortiz and Jesús Reyes-Flores.

# Acknowledgments

All chapters in this book were peer-reviewed by at least one other symposium participant, by one or more outside reviewers, and by at least one editor. We are extremely grateful to our following colleagues for sharing their valuable time and expertise: Alessandra Baptista, Edward M. Barrows, Elizabeth Bernays, Ernst Boller, Brian V. Brown, John W. Brown, Guy L. Bush, Carrol O. Calkins, Martha A. Condon, Marc De Meyer, Gary N. Dodson, Richard A. I. Drew, William G. Eberhard, Alejandro Espinoza de los Monteros, Amnon Freidberg, Douglas J. Futuyma, Raymond J. Gagné, Ho-Yeon Han, David L. Hancock, David H. Headrick, Jorge Hendrichs, Thomas S. Hoffmeister, Kenneth Y. Kaneshiro, Waldemar Klassen, Alexander S. Konstantinov, Valery A. Korneyev, Marian Kotrba, Peter J. Landolt, Steven W. Lingafelter, James E. Lloyd, Rogelio Macías-Ordoñez, Stephen A. Marshall, Wayne N. Mathis, David K. McAlpine, Stuart H. McKamey, Bruce A. McPheron, Bernhard Merz, Takahisa Miyatake, James Nation, Francisco Ornelas, Daniel R. Papaj, Thomas Pape, Ronald J. Prokopy, David C. Robacker, Sonja J. Scheffer, John Sivinski, James J. Smith, Gary Steck, F. Christian Thompson, James Tumlinson, Terry A. Wheeler, Brian M. Wiegmann, Norman E. Woodley, and Boaz Yuval. We wish to sincerely thank all of the authors of the book chapters and participants at the symposium for their contributions. We are grateful for your enthusiastic participation and appreciate your patience and cooperation during the editorial process. In particular, we wish to acknowledge Francisco Díaz-Fleischer, Daniel R. Papaj, Richard Drew, and John Sivinski for their efforts as lead authors of the evolution chapters. We thank Jesús Reyes (Director, Campaña Nacional contra las Moscas de la Fruta, Erich Wajnberg (General Secretary of the International Organization for Biological Control of Noxious Animals and Plants, IOBC) and Sergio Guevara Sada (Instituto de Ecología, A.C.) for constant support and encouragement. Thanks are also due to Alejandra Meyenberg (Subdirección de Asuntos Bilaterales del Consejo Nacional de Ciencia y Tecnología, CONACyT), Claudia Gómez (Academia Mexicana de Ciencias), Mauricio Fortes (Coordinador del Programa de Visitas de Profesores Distinguidos de la Academia Mexicana de Ciencias y la Fundación México Estados-Unidos para la Ciencia), and Gloria East and L. Whetten Reed (Research & Scientific Exchanges Division, ICD, FAS, USDA) for helping finance the symposium. We also gratefully acknowledge the important support of John Sulzycki and Christine Andreasen (CRC Press). Working with you, John and Chris, was always a great pleasure. We thank Bianca Delfosse for her critical contributions to the symposium and for her constant encouragement throughout this project. We also thank Larissa Guillén, Raquel Cervantes, Isabel Jácome, Gloria Lagunes, Rogelio Macías-Ordoñez, Alfonso Díaz, Carlos Leal-Melgar, Enrique Bombela, Fernando Martínez-Lasso, Fernando Gallegos, Patricia Moreno, Juan Chávez, and Guadalupe López for important logistical support during the symposium. The help of Larissa Guillén, Lucrecia Rodriguez, Jaime Piñero, Gloria Lagunes, Isabel Jácome, and Diana Pérez-Staples during the editorial process was invaluable. Finally, we thank our wives for their patience and understanding during the many hours we spent editing this work.

The symposium was made possible through the generous financial support of the following agencies: Campaña Nacional contra las Moscas de la Fruta; Research & Scientific Exchanges Division (ICD, FAS, USDA); International Organization for Biological Control of Noxious Animals and Plants (IOBC); Instituto de Ecología, A.C.; Consejo Nacional de Ciencia y Tecnología (CONACyT); Academia Mexicana de Ciencias; Fundación México Estados-Unidos para la Ciencia; and Comité Local de Sanidad Vegetal (Veracruz, México). We also acknowledge an important donation by the Campaña Nacional contra las Moscas de la Fruta to defray partially the publication costs of this book.

*To Ronald J. Prokopy and D. Elmo Hardy*

*Pioneers in the study of*
*fruit fly behavior and systematics*

# Contents

## *SECTION IV — SUBFAMILY TRYPETINAE*



## *SECTION V — SUBFAMILY DACINAE*

# Section I

## General Framework

# 1 Phylogenetic Relationships among the Families of the Superfamily Tephritoidea

*Valery A. Korneyev*

## CONTENTS



## 1.1 INTRODUCTION

The superfamily Tephritoidea includes the families Lonchaeidae, Piophilidae, Pallopteridae, Richardiidae, Ulidiidae (= Otitidae, = Pterocallidae), Platystomatidae, Pyrgotidae, and Tephritidae. The phylogenetic relationships of the superfamily, including the ground plan characters and proofs of monophyly of the included families, were discussed by J. F. McAlpine (1976; 1977; 1981; 1989). There is some disagreement among contemporary taxonomists (e.g., Colless and McAlpine 1970; Griffiths 1972; J. F. McAlpine 1989) on the relationships of the families within the group. Among the problems unsolved by those authors, the relationships among the Tephritidae and allied families (Platystomatidae, Pyrgotidae, and Ulidiidae) have never been thoroughly analyzed.

0-8493-1275-2/00/$0.00+$.50

The mosaic distribution of numerous key characters in the families of Tephritoidea is well recognized, although no attempts to reevaluate the phylogenetic definitions of the main suprageneric groupings have been made since Hennig (1958), Griffiths (1972), and J. F. McAlpine (1989) discussed the syn- and autapomorphies of the included families. Recently, this problem was considered by Freidberg (1994), who noted the need to clarify the definitions of the Tephritidae and allied families, and by Korneyev (1994), who gave a short review of hypothesized ground plans and derived conditions of several characters of key importance within the Pyrgotidae, Platystomatidae, and Tephritidae.

This chapter is an outgrowth of an attempt to analyze the polarity of several characters used for analysis within the Tephritidae, and it emphasizes the relationships among that family and those most closely related to it. In this study, many characters used by J. F. McAlpine (1977; 1989) were reexamined for many genera of Tephritoidea, and some valuable new characters were found. The following abbreviations are used: A = apomorphy, AA = autapomorphy, P = plesiomorphy, SA = synapomorphy, SP = symplesiomorphy. The Tachiniscinae, previously recognized as a family, here is considered a subfamily of Tephritidae.

## 1.2 ANALYSIS OF RELATIONSHIPS AMONG TAXA OF TEPHRITOIDEA

### 1.2.1 Monophyly and Ground Plan of Tephritoidea

J. F. McAlpine (1976; 1977; 1981; 1989) discussed the phylogeny of the Tephritoidea and suggested most of the following character states as the ground plan of the common ancestor of the families included here. This ground plan, derived mostly from his analysis of character polarity within the Tephritoidea, shows certain similarity with the ground plan of Tanypezidae, s. lat. (including Strongylophthalmyiidae), that Steyskal (1987a) compared for the Strongylophthalmyiidae and Tanypezidae, s. str., which he considered as two separate families. The position and concept of Tanypezidae are rather controversial. Hennig (1958) placed them as the families Tanypezidae and Strongylophthalmyiidae in the superfamily Nothyboidea that generally corresponds to the superfamily Diopsoidea in the sense of J. F. McAlpine (1989). Colless and D. K. McAlpine (1970), Griffiths (1972), and, recently, D. K. McAlpine (1997) considered them as one family Tanypezidae. Moreover, Griffiths (1972) and D. K. McAlpine (1997) excluded them from the Nothyboidea. According to D. K. McAlpine (1997: 175), Tanypezidae share some symplesiomorphies and apparently some synapomorphies with Heleomyzidae, Nerioidea, and Tephritoidea, but actually are not close to Diopsoidea.

Indeed, both Tanypezidae and the families of Nerioidea (Cypselosomatidae, Micropezidae, and Neriidae) share three fronto-orbital setae in the ground plan (P? SA?), vein $A_2$ completely absent (SA of Nerioidea, Tanypezidae, and Diopsoidea?), tergum and sternum 7 of female at least partially fused and forming oviscape (SA of Nerioidea, Tanypezidae, and Tephritoidea?), tergum and sternum 8 of female elongate (SA of Nerioidea, Tanypezidae, and Tephritoidea?), epiphallus well-developed (P?), male tergum 6 well-developed (P); phallus long, anteriorly directed (SA of Nerioidea, Tanypezidae, and Tephritoidea?), phallapodeme extremely long (P?; SA of Helomyzoidea, Nerioidea, and Tanypezidae?), and no thickened setae (prensisetae) on surstylus (P).

#### 1.2.1.1 Comparison of the Tephritoidea and Tanypezidae Ground Plans

The following character states, which are plesiomorphic or of uncertain polarity, appear to be present in the ground plans of both groups: Compound eyes closer together in male than in female (P); frons without secondary frontal setae (P); postvertical setae divergent (P); anepisternum hairy or short setulose (polarity unclear); katepisternum with some setae (polarity unclear); Sc free from $R_1$ (P?), complete (P) (in Strongylophthalmyiinae broken at apex, AA); pterostigmal section rather short (P); $R_{2+3}$, $R_{4+5}$, and M relatively straight (P); cell bcu present, elongate but (probably) without an angular prolongation at posterior apex (P?); legs simple, rather uniformly hairy, without strong setae, except for posterodorsal setae on fore femur (P); male with abdominal sternite 6 rather symmetrical, setose, completely developed (P), with abdominal tergite 7 and 8 greatly reduced and shifted toward left side of abdomen (P), and with abdominal sternite 7 asymmetrical, partially fused and shifted toward left side of abdomen (P); female with seven pairs of functional spiracles between the sternites and tergites (P).

The following additional ground plan characters are worthy of comparison:

| **Tephritoidea** | **Tanypezidae** |
| --- | --- |
| Body rather hairy; setae weakly differentiated (polarity unclear) (P?; AA of Lonchaeidae?) | Body not hairy; setae strongly differentiated (polarity unclear) |
| Fronto-orbital plate restricted to posterior 0.3 to 0.5 of frons, anterior portion largely desclerotized and bearing no frontal setae (SA?, or SA with Conopidae?) | Frons with broadly sclerotized lateral margins (A? compared with most Calyptratae) reaching anterior margin and bearing frontal setae (P) |
| Frons with two pairs of orbital setae only (SA?, or SA with Conopidae?) | Frons with three pairs of fronto-orbital setae (P?) |
| Oral vibrissa absent (SA?), except in Piophilidae (P? or AA?) | Oral vibrissa absent (A? or SA with Tephritoidea?) |
| Anepisternum with linear internal phragma anterior to anepisternal setae (SA?; known also in some Conopidae, Sciomyzidae, and Heleomyzidae) | Anepisternum without internal phragma anterior to anepisternal setae (P?) |
| Anepimeron bare (polarity unclear) | Anepimeron with some setae (polarity unclear) |
| Costa at least with weakening near humeral vein (P?) | Costa never broken or weak near humeral cross vein (A?) |
| Costa unbroken near apex of Sc (polarity unknown), but perhaps with weakening (P?) | Costa weakened (Tanypezinae, P?) or broken (Strongylophthalmyiinae, AA) or near apex of Sc |
| $R_1$ setulose or microtrichose, but not setose above (P?) | $R_1$ setulose (Strongylophthalmyiinae) or setose above (Tanypezinae, AA?) |
| Anal vein extending to wing margin (P) | Anal vein not extending to wing margin (A) |
| Vein $A_2$ expressed at least as a fold (P) | Vein $A_2$ completely absent (A) |
| Male abdominal tergite 6 greatly reduced (absent except in some *Dasiops* species) (SA) | Male abdominal tergite 6 large, setose, complete (P) |
| Male with seven pairs of functional abdominal spiracles (P) | Male with six pairs of functional abdominal spiracles (A) |
| Male gonostylus small, plate-like (A) | Male gonostylus small, plate-like (Tanypezinae, A) or rather long (?), rod-like (Strongylophthalmyiinae, P) |
| Male surstylus with teeth-like modified setae (prensisetae) (SA) | Male surstylus without prensisetae (P) |
| Male distiphallus elongate, convoluted (A?) and directed anteroventrally (P) (but in all (?) Lonchaeidae long, bowed posteriorly (AA?) | Male distiphallus rather long, not convoluted, directed anteroventrally (P) |
| Female sterna 4 to 6 with long rodlike apodemes anteriorly (A) | Female sterna without rodlike apodemes anteriorly (P) |
| Female tergum and sternum 6 transverse, not forming an additional joint of oviscape (P) | Female tergum and sternum 6 long, not forming an additional joint of oviscape (P) |
| Female with seventh sternite and tergite separate in ground plan of Pallopteridae + Piophilidae (P) and fused forming a stout tube or oviscape in other families (A) | Female with seventh sternite and tergite anterolaterally fused forming a stout tube or oviscape (in Strongylophthalmiinae) (SA? with Nerioidea and Diopsoidea), or with seventh sternite and tergite divided into two pairs of sclerotized longitudinal strips (Tanypezinae) (AA) |
| Female with seventh sternite and tergite posteriorly forming two pairs of sclerotized strips (taeniae) divided by desclerotized areas (SA) | Female with seventh sternite and tergite either divided over all its length, forming 2 pairs of sclerotized longitudinal strips (Tanypezinae, AA) or with uniformly semimembranous eversible membrane (Strongylophthalmyiinae, AA) |

| **Tephritoidea** | **Tanypezidae** |
|---|---|
| Female with the taeniae of eversible membrane bearing setulae (see McAlpine 1987a: 793, Fig. 62.21; 1987 b: 841, Fig. 68.6) (P) | [Does not exist] |
| Female tergite 8 and sternite 8 divided longitudinally into paired struts that form the shaft of an aculeus (SA with Tanypezidae?) | Female tergite 8 and sternite 8 divided longitudinally into paired struts that form the shaft of an aculeus (SA with Tephritoidea?) |
| Female hypoproct fused with cerci, only remainders of it can be traced anteriorly (best in Lonchaeidae) (SA) | Female hypoproct well-developed as a separate plate anterior of cerci (P) |
| Female epiproct absent (SA?) | Female epiproct well-developed (P) |
| Female cerci dorsally fused, ventrally with distinct notch (SA with Tanypezidae?) | Female cerci dorsally fused, ventrally with distinct notch (SA with Tephritoidea?) |
| Female with three sclerotized spermathecae (P) | Female with two or one sclerotized spermathecae (A) |
| Female with two spermathecal ducts: two of three spermathecae on common branched duct (P) (SA of Acalyptratae) | Female with two spermathecal ducts (P) |
| Males with aerial swarming habit, and initial stages of mating taking place on the wing (in Lonchaeidae) (P) | Unknown |
| Immature stages occurring in dead wood or other decaying plant material (P) | Immature stages probably occurring in decaying plant material (Strongylophthalmyiinae under the bark of dead trees) |

Therefore, Tanypezidae apparently is one of the best candidates to be the sister group of Tephritoidea because of possessing at least similarly modified female tergum and sternum 8 and cerci. An alternative sister group of the Tephritoidea, according to J. F. McAlpine (1989), is the family Conopidae. Among other characters, conopids have strongly reduced fronto-orbital plates that are rather similar to those of the Tephritoidea (SA?), anepisternal phragma (short, but present, at least in Myopinae) (SA?), phallapodeme small (but different in shape from those in both Lonchaeidae and Tanypezidae), female tergum and sternum 6 free and transverse (SP with Tephritoidea), female tergum and sternum 7 fused (SA with Nerioidea, Diopsoidea, Tanypezidae, and Tephritoidea?), but female cerci separate and not forming apical caplike cercal unit (P, compared with the synapomorphic state of this character in Tanypezidae and Tephritoidea). Their relationship needs further consideration.

The monophyly of the superfamily Tephritoidea is supported by the following synapomorphies: (1) male tergum 6 strongly reduced or absent; (2) surstylus, or medial surstylus if there are two, bearing toothlike prensisetae (in Piophilidae only in one genus; prensisetae also occur in Drosophilidae, due to obvious homoplasy); (3) female sterna 4 to 6 with anterior rodlike apodemes; and (4) female tergosternum 7 consisting of two portions, the anterior one that forms a tubular "oviscape," and the posterior comprising two pairs of longitudinal taeniae. The other characters that J. F. McAlpine (1989: 1438) listed as autapomorphic ground plan features are probably synapomorphies with Tanypezidae.

### 1.2.2 Cladistic Parsimony Analysis of Tephritoidea

A cladistic analysis was performed on the major subgroups of the superfamily Tephritoidea and some allied families, using the program Hennig86 (Farris 1988). The analyzed character state matrix (see Table 1.2) included 33 taxa (including hypothetical outgroup) and 44 characters, which are listed in Table 1.1. All the characters that have three or more states were considered additive. Autapomorphies were not included. Species representing several outgroups and multiple species of most of the ingroup taxa were included to determine general patterns of distribution of characters suspected to be homoplastic.

## TABLE 1.1
## Characters Used in the Cladistic Parsimony Analysis

1. Dummy character
2. Compound eyes of males closer together than those of females: 0, yes; 1, no
3. Fronto-orbital plate: 0, complete from posterior to anterior margin of frons; 1, restricted to the posterior 0.3 to 0.5 of frons, anterior portion largely desclerotized
4. Fronto-orbital plates: 0, with three pairs of fronto-orbital setae (or more); 1, with two setae on posterior half only
5. Secondary frontal setae: 0, not developed; 1, developed at anterior frons margin
6. Oral vibrissa: 0, developed; 1, not developed
7. Dorsal cleft or notch of pedicel: 0, developed; 1, lacking
8. Ocelli: 0, well-developed; 1, anterior ocellus reduced; 2, all ocelli lacking
9. Ocellar seta: 0, well-developed; 1, reduced or lacking
10. Ocellar seta, when present: 0, lateroclinate; 1, proclinate
11. Presutural supra-alar seta: 0, lacking; 1, developed
12. Intrapostalar seta: 0, absent; 1, present
13. Anepimeron: 0, bare; 1, with some setae
14. Anepimeral tubercle: 0, absent; 1, developed
15. Costa at humeral vein: 0, not weakened; 1, weakened
16. Subcostal weakening of costa: 0, absent; 1, well-developed
17. Costal spurs at apex of subcostal vein: 0, absent; 1, developed
18. Vein Sc: 0, entire; 1, apically broken
19. Vein $R_1$: 0, bare; 1, setulose on stigmatal portion only; 2, completely setulose
20. Vein $R_{4+5}$: 0, bare; 1, setulose
21. Cell bcu: 0, closed by inwardly curved vein without extension; 1, with posteroapical extension; 2, closed by straight, perpendicular vein (nonadditive?)
22. Vein $A_2$: 0, developed; 1, lacking
23. Male tergum 6: 0, absent; 1, rudimentary; 2, well-developed
24. Male sternum 6: 0, more or less symmetrical, moderately long, setulose; 1, asymmetrical, narrow, bare
25. Male abdominal spiracle 6: 0, absent; 1, present
26. Male abdominal spiracle 7: 0, absent; 1, present
27. Gonopods: 0, well-developed; 1, buttonlike, small; 2, completely lacking
28. Phallapodeme: 0, longer than hypandrium, bacilliform; 1, short
29. Phallus: 0, shorter than hypandrium; 1, longer than hypandrium
30. Phallus: 0, straight, noncoiled apically; 1, coiled
31. Phallus: 0, directed anteroventrally; 1, directed posteroventrally
32. Phallus: 0, bare or trichose; 1, heavily spinulose
33. Prensisetae: 0, absent; 1, developed
34. Prenisetae: 0, not surmounted on lobe; 1, if present, surmounted on mesal lobe (medial surstylus)
35. Female tergum and sternum 7: 0, free; 1, fused
36. Oviscape: 0, neither much longer than aculeus, nor downward curved; 1, long, massive and downward curved
37. Eversible membrane: 0, without taeniae; 1, with two pairs of taeniae
38. Eversible membrane or taeniae, if present: 0, setulose; 1, bare
39. Female last abdominal segments: 0, not forming tactile or cutting aculeus; 1, consisting of paired sclerites of tergum and sternum 8 and cercal unit
40. Aculeus when everted: 0, directed posteriorly; 1, directed dorsally
41. Aculeus: 0, not short and stiletto-like; 1, short and stiletto-like
42. Female: apodemes of sterna 5 and 6: 0, absent; 1, developed
43. Number of spermathecae: 0, three; 1, two; 2, four
44. Number of spermathecal ducts: 0, three; 1, two

**TABLE 1.2**
**Matrix of Character States Used in Cladistic Parsimony Analysis (character numbers refer to Table 1.1)**

| | Character Numbers | | | | | | | | | | | | | | | | | | | | | | | | | | | | | | | | | | | | | | | | | | |
|---|---|---|---|---|---|---|---|---|---|---|---|---|---|---|---|---|---|---|---|---|---|---|---|---|---|---|---|---|---|---|---|---|---|---|---|---|---|---|---|---|---|---|---|
| | 2 | 3 | 4 | 5 | 6 | 7 | 8 | 9 | 10 | 11 | 12 | 13 | 14 | 15 | 16 | 17 | 18 | 19 | 20 | 21 | 22 | 23 | 24 | 25 | 26 | 27 | 28 | 29 | 30 | 31 | 32 | 33 | 34 | 35 | 36 | 37 | 38 | 39 | 40 | 41 | 42 | 43 | 44 |
| Outgroup | 0 | 0 | 0 | 0 | 0 | 0 | 0 | 0 | 1 | 1 | 0 | 0 | 0 | 0 | ? | 0 | 0 | ? | 0 | 0 | 0 | 2 | 0 | 1 | 1 | 0 | ? | 0 | 0 | ? | 0 | 0 | ? | 0 | 0 | ? | ? | 0 | 0 | 0 | 0 | 0 | 0 |
| *Cordylura* | 0 | 0 | 0 | 0 | 0 | 0 | 0 | 0 | 1 | 1 | 1 | 0 | 0 | 1 | 1 | 0 | 0 | 1 | 0 | 0 | 0 | 2 | 0 | 1 | 1 | 0 | ? | 0 | 0 | 0 | 0 | 0 | ? | 0 | 0 | ? | ? | 0 | 0 | 0 | 0 | 0 | 0 |
| Conopidae | 1 | 1 | 1 | 0 | 1 | 0 | 0 | 0 | 1 | 0 | 1 | 0 | 0 | 0 | 0 | 0 | 0 | 0 | 0 | 0 | 0 | 1 | 0 | 1 | 1 | 0 | ? | 1 | 1 | 0 | 0 | 0 | ? | ? | 0 | ? | 0 | 0 | 0 | 0 | 0 | 0 | 1 |
| *Raineria* | 1 | 0 | 0 | 0 | 1 | 0 | 0 | 0 | ? | 0 | 0 | 0 | 0 | 0 | 0 | 0 | 0 | 0 | 0 | 0 | 1 | 2 | 0 | 1 | 1 | 0 | 0 | 1 | 0 | 0 | 0 | 0 | ? | 1 | 0 | ? | 0 | 0 | 0 | 0 | 0 | 0 | 1 |
| *Tanypeza* | 0 | 0 | 0 | 0 | 1 | 0 | 0 | 0 | 1 | 0 | 1 | 2 | 0 | 0 | 1 | 0 | 0 | 2 | 0 | 0 | 1 | 2 | 0 | 0 | 0 | 0 | 0 | 1 | 0 | 0 | 0 | 0 | ? | 1 | 0 | 1 | 0 | 1 | 0 | 0 | 0 | 0 | 1 |
| *Strongylophthalmyia* | 1 | 0 | 0 | 0 | 1 | 0 | 0 | 0 | 1 | 0 | 1 | 2 | 0 | 0 | 1 | 0 | 1 | ? | 0 | 0 | 1 | 2 | 0 | 0 | 0 | 0 | 0 | 1 | 0 | 0 | 0 | 0 | ? | 1 | 0 | 0 | 0 | 1 | 0 | 0 | 0 | 0 | 1 |
| *Ramuliseta* | 0 | ? | 1 | 0 | 1 | 0 | 0 | 0 | ? | 0 | ? | 0 | 0 | 0 | 1 | 0 | 1 | 2 | 0 | 0 | 0 | 1 | 0 | 0 | 0 | ? | 0 | 0 | 0 | ? | 0 | 0 | ? | ? | 0 | 0 | ? | 0 | 0 | 0 | 0 | 0 | 1 |
| *Neottiophilum* | ? | ? | 1 | 0 | 0 | 0 | 0 | 0 | 1 | 1 | 1 | 0 | 0 | 1 | 1 | 0 | 0 | 2 | 0 | 0 | 0 | 0 | 1 | ? | ? | 0 | ? | 1 | ? | 0 | 0 | ? | 0 | 0 | 0 | 1 | 0 | 1 | 0 | 0 | 0 | 1 | 1 |
| *Mycetaulus* | 1 | ? | 1 | 0 | 0 | 0 | 0 | 0 | 1 | 1 | 1 | 0 | 0 | 1 | 1 | 0 | 0 | 0 | 0 | 0 | 0 | 0 | 1 | 0 | 0 | 0 | 1 | 1 | ? | 0 | 0 | ? | 0 | 0 | 0 | 1 | 0 | 1 | 0 | 0 | 0 | 1 | 1 |
| Dasiopinae | 0 | 1 | 1 | 0 | 1 | 0 | 0 | 0 | 1 | 1 | 1 | 0 | 0 | 1 | 1 | 0 | 0 | 0 | 0 | 0 | 0 | 1 | 0 | 1 | 1 | 1 | 1 | 0 | 0 | 1 | 0 | 1 | 0 | 1 | 0 | 1 | 0 | 1 | 0 | 0 | 1 | 0 | 1 |
| Lonchaeinae | 0 | 1 | 1 | 0 | 1 | 0 | 0 | 0 | 1 | 1 | 1 | 0 | 0 | 1 | 1 | 0 | 0 | 0 | 0 | 0 | 0 | 0 | 0 | 1 | 1 | 1 | 1 | 0 | 0 | 1 | 0 | 1 | 0 | 1 | 0 | 1 | 0 | 1 | 0 | 0 | 1 | 0 | 1 |
| Pallopteridae | 1 | 1 | 1 | 0 | 1 | 0 | 0 | 0 | 1 | 1 | 1 | 0 | 0 | 1 | 1 | 0 | 0 | 0 | 0 | 0 | 0 | 0 | 0 | 1 | 0 | 1 | 0 | 1 | 1 | 0 | 0 | 1 | 0 | 0 | 0 | 1 | 0 | 1 | 0 | 0 | 1 | 1 | 1 |
| *Omomyia* | 1 | 1 | 1 | 0 | 1 | 0 | 0 | 0 | 1 | 1 | 1 | 0 | 0 | 1 | 1 | 0 | 0 | 0 | 0 | 0 | 0 | 0 | 0 | 1 | 1 | 1 | 1 | 2 | 1 | 0 | 0 | 1 | 1 | 1 | 0 | 1 | 1 | 1 | 0 | 0 | 0 | 0 | 1 |
| *Richardia* | 1 | 1 | 1 | 0 | 1 | 0 | 0 | 0 | 1 | 1 | 1 | 0 | 0 | 0 | 1 | 0 | 1 | 0 | 0 | 0 | 0 | 0 | 0 | 1 | 1 | 1 | 1 | 1 | 1 | 0 | 0 | 1 | 1 | 1 | 0 | 1 | 1 | 1 | 0 | 0 | ? | 0 | 1 |
| *Seioptera* | 1 | 1 | 1 | 0 | 1 | 1 | 0 | 0 | 0 | 0 | 1 | 0 | 0 | 1 | 1 | 0 | 0 | 0 | 0 | 1 | 0 | 0 | 1 | 0 | 0 | 1 | 1 | 1 | 1 | 0 | 0 | 1 | 1 | 1 | 0 | 1 | 1 | 1 | 0 | 0 | 0 | 2 | 1 |
| *Physiphora* | 1 | 1 | 1 | 0 | 1 | 1 | 0 | 0 | 0 | 0 | 1 | 0 | 0 | 1 | 1 | 0 | 0 | 0 | 0 | 1 | 0 | 0 | 1 | 0 | 0 | 1 | 1 | 1 | 1 | 0 | 0 | 1 | 1 | 1 | 0 | 1 | 1 | 1 | 0 | 0 | 0 | 0 | 1 |
| *Pterocalla* | 1 | 1 | 1 | 0 | 1 | 1 | 0 | 0 | 0 | 0 | 1 | 0 | 0 | 1 | 1 | 0 | 0 | 1 | 0 | 1 | 0 | 0 | 1 | 0 | 0 | 1 | 1 | 1 | 1 | 0 | 0 | 1 | 1 | 1 | 0 | 1 | 1 | 1 | 0 | 0 | 0 | 0 | 1 |
| *Euxesta* | 1 | 1 | 1 | 0 | 1 | 1 | 0 | 0 | 0 | 0 | 1 | 0 | 0 | 1 | 1 | 0 | 0 | 0 | 0 | 1 | 0 | 0 | 1 | 0 | 0 | 1 | 1 | 1 | 1 | 0 | 0 | 1 | 1 | 1 | 0 | 1 | 1 | 1 | 0 | 0 | 0 | 1 | 1 |
| *Delphinia* | 1 | 1 | 1 | 0 | 1 | 1 | 0 | 0 | 0 | 0 | 1 | 0 | 0 | 1 | 1 | 0 | 0 | 1 | 0 | 1 | 0 | 0 | 1 | 0 | 0 | 1 | 1 | 1 | 1 | 0 | 1 | 1 | 1 | 1 | 0 | 1 | 1 | 1 | 0 | 0 | 1 | 0 | 1 |
| *Myennis* | 1 | 1 | 1 | 0 | 1 | 1 | 0 | 0 | 0 | 0 | 1 | 0 | 0 | 1 | 1 | 0 | 0 | 1 | 0 | 1 | 0 | 0 | 1 | 0 | 0 | 1 | 1 | 1 | 1 | 0 | 1 | 1 | 1 | 1 | 0 | 1 | 1 | 1 | 0 | 0 | 1 | 0 | 1 |
| *Otites* | 1 | 1 | 1 | 0 | 1 | 1 | 0 | 0 | 0 | 0 | 1 | 0 | 0 | 1 | 1 | 0 | 0 | 1 | 0 | 1 | 0 | 0 | 1 | 0 | 0 | 1 | 1 | 1 | 1 | 0 | 1 | 1 | 1 | 1 | 0 | 1 | 1 | 1 | 0 | 0 | 1 | 0 | 1 |
| *Poecilotraphera* | 1 | 1 | 1 | 1 | 1 | 0 | 0 | 0 | 0 | 0 | 1 | 1 | 0 | 1 | 1 | 0 | 1 | 2 | 1 | 2 | 0 | 0 | 1 | 0 | 0 | ? | 1 | 1 | 1 | 0 | 0 | 1 | 1 | 1 | 0 | 1 | 1 | 1 | 0 | 0 | 0 | 1 | 1 |
| *Angitula* | 1 | 1 | 1 | 0 | 1 | 0 | 0 | 1 | ? | 0 | 0 | 1 | 0 | 1 | 1 | 0 | 0 | 2 | 1 | 2 | 0 | 0 | 1 | 0 | 0 | 1 | 1 | 1 | 1 | 0 | 0 | 1 | 1 | 1 | 0 | 1 | 1 | 1 | 0 | 0 | 0 | 0 | 1 |
| *Rivellia* | 1 | 1 | 1 | 0 | 1 | 0 | 0 | 0 | 0 | 1 | 0 | 1 | 0 | 1 | 1 | 0 | 0 | 2 | 1 | 2 | 0 | 0 | 1 | 0 | 0 | 1 | 1 | 1 | 1 | 0 | 0 | 1 | 1 | 1 | 0 | 1 | 1 | 1 | 0 | 0 | 0 | 0 | 1 |
| *Scholastes* | 1 | 1 | 1 | 0 | 1 | 0 | 0 | 1 | ? | 0 | 1 | 1 | 0 | 1 | 1 | 0 | 0 | 2 | 1 | 2 | 0 | 0 | 1 | 0 | 0 | 1 | 1 | 1 | 1 | 0 | 0 | 1 | 1 | 1 | 0 | 1 | 1 | 1 | 0 | 0 | 0 | 0 | 1 |
| *Adapsilia* | 1 | 1 | 1 | 0 | 1 | 1 | 2 | 0 | 1 | 0 | 1 | 1 | 0 | 1 | 1 | 0 | 1 | 2 | 1 | 1 | 0 | 0 | 1 | 0 | 0 | 1 | 1 | 1 | 1 | 0 | 0 | 0 | 1 | 1 | 1 | 0 | ? | 1 | 0 | 0 | 0 | 0 | 0 |
| *Prodalmannia* | 1 | 1 | 1 | 0 | 1 | 0 | 0 | 0 | 1 | 1 | ? | 1 | 0 | 1 | 1 | 0 | 1 | 2 | 1 | 1 | 0 | 0 | 1 | 0 | 0 | ? | 1 | 1 | 1 | 0 | 0 | ? | ? | 1 | 1 | ? | ? | 1 | 0 | 0 | 0 | 0 | 0 |
| *Ortalotrypeta* | 1 | 1 | 1 | 1 | 1 | 0 | 0 | 0 | 1 | 1 | 1 | 1 | 0 | 1 | 1 | 1 | 1 | 2 | 1 | 1 | 0 | 0 | 1 | 0 | 0 | 2 | 1 | 1 | 1 | 0 | 0 | 1 | 1 | 1 | 0 | ? | ? | 1 | 1 | 1 | 0 | 0 | 0 |
| *Tachinisca* | 1 | 1 | 1 | 1 | 1 | 1 | 1 | 1 | ? | 1 | 1 | 1 | 1 | 1 | 1 | 1 | 0 | 2 | 1 | 1 | 0 | 0 | 1 | 0 | 0 | 1 | 1 | 1 | 1 | 0 | 0 | 1 | 1 | 1 | 0 | ? | ? | 1 | 1 | 1 | 0 | 0 | 0 |
| *Bibundia* | 1 | 1 | 1 | 0 | 1 | 1 | 2 | 1 | ? | 1 | 1 | 1 | 1 | 1 | 1 | ? | 1 | 2 | 1 | 1 | 0 | 0 | 1 | 0 | 0 | ? | ? | 1 | 1 | 0 | 0 | 1 | ? | ? | 0 | ? | ? | ? | ? | ? | ? | ? | ? |
| *Blepharoneura* | 1 | 1 | 1 | 1 | 1 | 0 | 0 | 0 | 1 | 1 | 1 | 1 | 0 | 1 | 1 | 1 | 1 | 2 | 1 | 1 | 0 | 0 | 1 | 0 | 0 | 2 | 1 | 1 | 1 | 0 | 0 | 1 | 1 | 1 | 0 | 1 | 1 | 1 | 0 | 0 | ? | 0 | 0 |
| *Acanthonevra* | 1 | 1 | 1 | 1 | 1 | 0 | 0 | 1 | ? | 1 | 1 | 1 | 0 | 1 | 1 | 1 | 1 | 2 | 1 | 1 | 0 | 0 | 1 | 0 | 0 | 2 | 1 | 1 | 1 | 0 | 0 | 1 | 1 | 1 | 0 | 1 | 1 | 1 | 0 | 0 | 0 | 0 | 0 |
| *Terellia* | 1 | 1 | 1 | 1 | 1 | 0 | 0 | 0 | 1 | 1 | 0 | 1 | 0 | 1 | 1 | 1 | 1 | 2 | 0 | 1 | 0 | 0 | 1 | 0 | 0 | 2 | 1 | 1 | 1 | 0 | 0 | 1 | 1 | 1 | 0 | 1 | 1 | 1 | 0 | 0 | 1 | 1 | 0 |

The matrix (Table 1.2) was analyzed using the mhennig* option combined with bb* (branch swapping) option. The strict consensus tree obtained using command "nelsen" from the set of resulting trees is shown in Figure 1.1A. The tree length = 112 steps, consistency index = 0.44, retention index = 0.46. The second tree (Figure 1.1B) was obtained using the mhennig* and bb* options followed by a series of successive character weightings (xs w command followed by mhennig* bb* until no further change occurred). The tree length = 321, consistency index = 0.63, retention index = 0.88.

Both trees show the distribution of certain characters considered important synapomorphies of the taxa within the superfamily Tephritoidea. Due to the small sample of characters and taxa, the trees do not show reliable relationships outside the superfamily, and must be considered with caution.

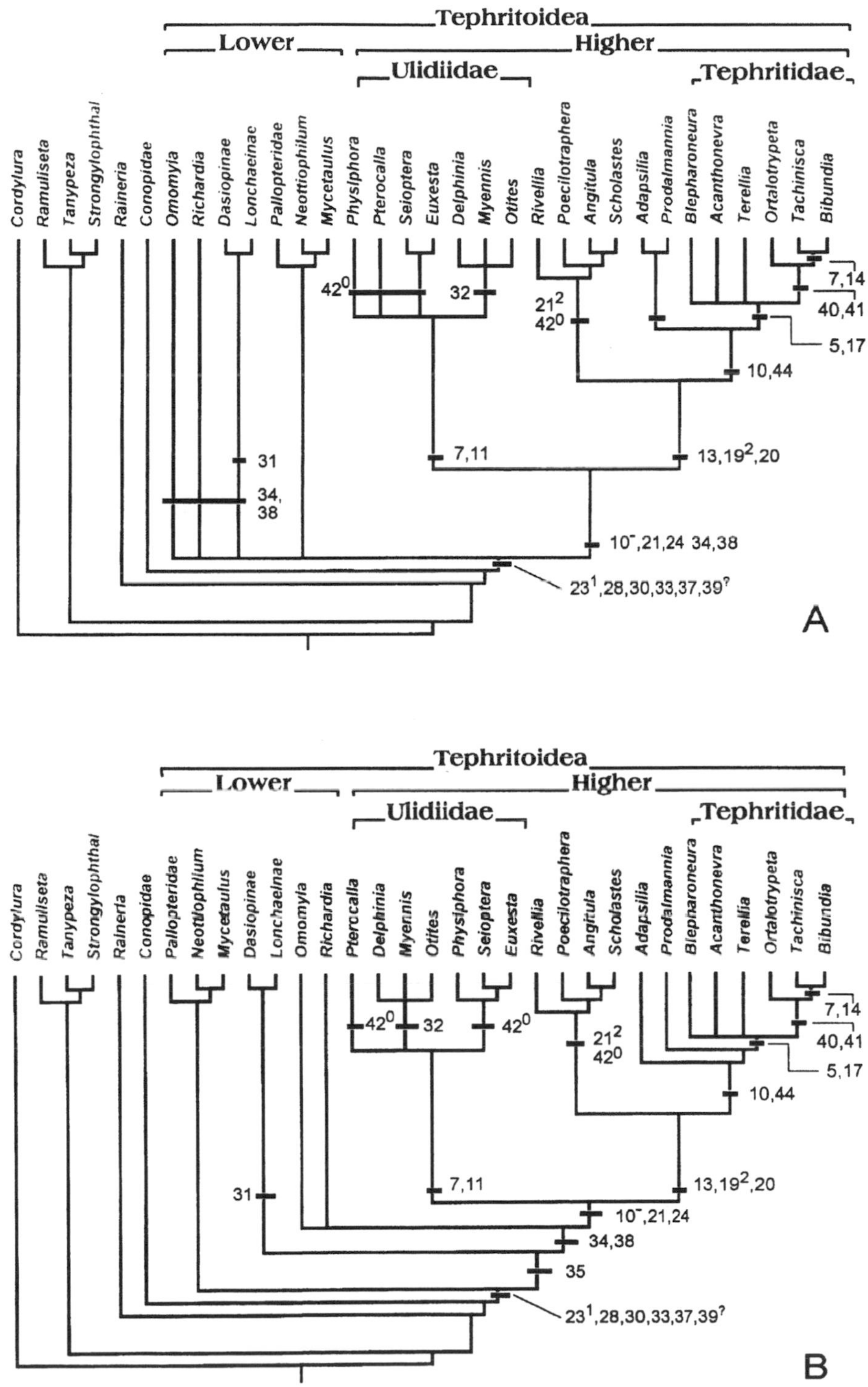


**FIGURE 1.1** Possible phylogenetic relationships among families of Tephritoidea. (A) Consensus tree, unweighted analysis; (B) Analysis using successive character weighting.

For convenience, the term *Lower Tephritoidea* is used hereafter to refer to the Lonchaeidae, Piophilidae, Pallopteridae, and Richardiidae together, and the term *Higher Tephritoidea* for the Ulidiidae (= Otitidae, = Pterocallidae), Platystomatidae, Pyrgotidae, and Tephritidae. The Higher Tephritoidea is clearly a monophyletic cluster which will be discussed in the following sections. Detailed analysis of the phylogeny of the Lower Tephritoidea, which may be paraphyletic (Figure 1.1B), is beyond the scope of this chapter, although the families may be related as follows. The relationships among these families are not well resolved (Figure 1.1A), but the results of this analysis are different in some respects from those of J. F. McAlpine (1989), who hypothesized that the Lonchaeidae are the sister group of all other Tephritoidea, and that the Richardiidae are the sister group of Piophilidae + Pallopteridae. Pallopteridae does appear to be the sister group of Piophilidae, but the basal branching of the superfamily is unresolved, with the Piophilidae + Pallopteridae, Lonchaeidae, Richardiidae, and the Higher Tephritoidea polytomic in the unweighted analysis (Figure 1.1A). The relationship of Piophilidae + Pallopteridae and other Tephritoidea except Lonchaeidae is supported by one apomorphy, the soft, elongate and setulose phallus, whereas the monophyly of all of the families except Lonchaeidae is supported by another, female tergum and sternum 7 fused to form the oviscape. The latter hypothesis is favored in the weighted analysis (Figure 1.1B). In the weighted analysis, Richardiidae (including Epiplateinae that may deserve family status) is hypothesized to be the sister group of the Higher Tephritoidea (Ulidiidae + Platystomatidae + Pyrgotidae + Tephritidae) on the basis of bare taenia in the female ovipositor and the presence of the medial surstylus in the male.

### 1.2.3 Monophyly and Ground Plan of the Higher Tephritoidea

J. F. McAlpine (1989: 1440) suggested the following autapomorphic characters of the tephritoid subgroup (= Higher Tephritoidea). "Males with spiracles 6 and 7 lost." "Sternite 6 reduced and asymmetrical, fused with sternites 7 and 8." It must be added that sternite 6 is bare, and this may be considered an additional synapomorphy. Among the Lower Tephritoidea, it is bare in Piophilidae, but setose in its sister group, Pallopteridae. "Gonopods reduced" and "parameres lost or greatly reduced." Actually the gonocoxites (= gonites, gonopods) and gonostyli (= parameres) in Ulidiidae are as well developed as in Tanypezidae, Lonchaeidae, and Epiplateinae, only being somewhat less setulose. These last two characters certainly are not synapomorphies belonging to the ground plan of the Higher Tephritoidea, nor is the following character. "Females with tergite and sternite of segment 7 fully fused." There always remains a poorly sclerotized fold between tergum and sternum 7 in Lonchaeidae, Richardiinae, Epiplateinae, Ulidiidae, most Platystomatidae, and many Tephritidae (e.g., Phytalmiinae). The completely integrated oviscape can be found only in Pyrgotidae and most Tephritidae (including Tachiniscinae).

There are some additional synapomorphies that must belong to the ground plan of the Higher Tephritoidea:

*Vein $R_1$ setulose dorsally* — All the species of Tephritidae, Pyrgotidae, and Platystomatidae possess this character, and most ulidiid tribes (except for Ulidiini itself) include genera where $R_1$ is at least partially setulose. The genera with $R_1$ setulose are extremely rare among other tephritoid families. This character occasionally can be found in Tanypezidae and *Automola* Loew of Epiplateinae, and *Neottiophilum* Frauenfeld and *Actenoptera* Czerny of the Piophilidae. Its mosaic distribution among genera and tribes of Ulidiidae does not preclude that the absence of setae was the ancestral state, but recent analysis has shown that it is more parsimonious to hypothesize the setulose $R_1$ as plesiomorphic (Kameneva and Korneyev, in prep.).

*Cell bcu with a posteroapical extension* — Most of the Higher Tephritoidea have an angular extension of this cell, whereas the Lower Tephritoidea have it closed by an arcuate, recurrent vein $Cu_2$. I consider this character a synapomorphy of the Higher Tephritoidea. There are considerable exceptions among them, like all Platystomatidae and a few species of Seiopterini, Cephalini, and Pterocallini (Ulidiidae), and some Tephritidae, but in most of these cases the shape of cell bcu certainly differs from that in the Lower Tephritoidea and usually can be proved to be the result of character reversal.

There are some plesiomorphic characters belonging to the ground plan of the Higher Tephritoidea that deserve mention, such as having a proepisternal seta (in all the Lower Tephritoidea and most Ulidiidae), and presence of the two costal weakenings, humeral and subcostal (in Neottiophilinae, most Epiplateinae, Ulidiidae, Pyrgotidae, and all Tephritidae, including Tachiniscinae).

Furthermore, the ground plan of the genitalic characters is as follows:

*Male protandrium* — All the Lower Tephritoidea have male sternum 6 setose, and the male sixth (and often also the seventh) spiracle present, whereas the higher tephritoid families share a highly reduced and bare male sternum 6 and abdominal spiracles 6 and 7 absent. Both characters are therefore synapomorphies of the latter group of families (J. F. McAlpine 1989).

*Hypandrium* — In the Tephritoidea it is always C-shaped, without a posterior bridge, and usually somewhat broadened anteriorly. Its posterior arms usually touch the epandrium and anterolateral arms of the subepandrial sclerite. The general plan is common for Lonchaeidae, Pallopteridae, Epiplateinae, and Ulidiidae, and is modified in various ways in the remaining families.

*Gonocoxites* (*gonites*) — In the ground plan of Tephritoidea they are very low, flattened, usually setulose sclerites well separated from the hypandrium. They form a rather shallow and broad phallic guide, joined to the phallapodeme. The so-called fultelliform apodeme first described in Tephritidae, and then attributed to some Lower Tephritoidea, e.g., Lonchaeidae, Epiplateinae, and Piophilidae (J. F. McAlpine 1976), actually is an innovation of Platystomatidae, Pyrgotidae, and Tephritidae that exists neither in the lower tephritids, nor in Ulidiidae. Gonostyli (parameres) are always small, buttonlike sclerites with four to eight setulae (trichoid or campaniform sensilla?), except in Piophilidae, where there are more produced, and most Platystomatidae and Tephritidae, where they very often completely disappear.

*Phallus* (*aedeagus*) — It consists of a proximal sclerotized tube or ring, called the basiphallus, that in some Lower Tephritoidea and many other Cyclorrhapha has an unpaired posteroventral process, the epiphallus, and is basally articulated with the paired posterior end of the phallapodeme. The remaining part is a flexible, long distiphallus. In the Lower Tephritoidea and Tanypezidae, it includes two sclerotized taeniae and is covered with various cuticular structures — microtrichia, scales, or spurs. Either the bare or haired distiphallus may belong to the ground plan of the Higher Tephritoidea. In Tanypezidae, Piophilidae, Richardiidae, and Ulidiinae the apical part of the distiphallus occasionally is differentiated by a constriction or has a modified arrangement of the cuticular structures, but this is not homologous with the discrete structure, termed the glans by D. K. McAlpine (1973) and Korneyev (1984), found in Platystomatidae, Pyrgotidae, and Tephritidae. The glans, which is separated from the rest of the distiphallus by deep folds and flexion or torsion, is composed of two or three sclerites of the acrophallus, guarding the gonopore, and lateral laps overlapping the acrophallus. It is certainly different from the similar structures of some Lower Tephritoidea (also called "glans" by J. F. McAlpine, 1989) and some Ulidiinae, and does not belong to the ground plan of all the Higher Tephritoidea.

*Epandrium* — The surstylus (or lateral surstylus) in the ground plan of the Tephritoidea (see Norrbom and McAlpine 1997) and in many Ulidiidae is not fused to the epandrium, and has two articulations with it: to the anterolateral and posterolateral margin of the epandrium. The subepandrial sclerite is commonly H-shaped in the ground plan of the Higher Tephritoidea, with two shorter arms laterally wrapping the basiphallus, and the longer arms associated with the outer surstyli and forming the medial surstyli, which are usually fingerlike, and in the ground plan bear subapically two to three thickened, clasping setae, called prensisetae.

*Female preabdomen* — Female sternites 4 to 6 bear anterior rodlike apodemes in Lonchaeidae, Pallopteridae, Ulidiidae: Otitinae, and most (but not all) Tephritinae. This is the more common character state than the absence of the apodemes, and therefore is believed to belong to the ground plan of the superfamily in whole and the Higher Tephritoidea in particular.

*Female postabdomen* — In the ground plan, tergum and the sternum 7 are free from each other (this condition is known in Pallopteridae and Piophilidae), whereas in the Lonchaeidae, Richardiidae, and the Higher Tephritoidea they are fused and form a tergosternum 7, or oviscape, that is completely sclerotized, with two lateral folds. Taeniae of eversible membrane are never setulose in the Higher Tephritoidea (contrary to Lonchaeidae, Pallopteridae, and Piophilidae, but similar to both Richardiinae and Epiplateinae), always separated from the oviscape by a narrow unsclerotized fold, and small triangular lobes of the oviscape between the taeniae both dorsally and ventrally. The eversible membrane between the taeniae is usually impregnated by tiny sclerotized patches that may form monodentate scales on the middle portion of the tube, and three to five dentate scales on the posterior portion; these structures correspond to the microtrichia of the membrane in Lonchaeidae, and I consider the presence of such structures to belong to the Tephritoidea ground plan. Completely bare and heavily toothed membranes are both derived states found in Ulidiidae, Platystomatidae, and Tephritidae. A weakly sclerotized aculeus with two pairs of longitudinal sclerotized areas covered by setulae, and the cercal unit with a ventral notch and six to eight long setulae, similar to that of *Lonchaea* Fallén and *Palloptera* Fallén, both are in the ground plan of the Higher Tephritoidea.

*Spermathecae* — Three (1 + 2) spermathecae, with the paired spermathecae on a common spermathecal duct (like in most Lonchaeidae, Ulidiidae, and Platystomatidae), is certainly the ground plan state of the character.

## 1.2.4 Relationships among Taxa of the Higher Tephritoidea

### 1.2.4.1 Ulidiidae (= Pterocallidae, Otitidae, Ortalidae)

J. F. McAlpine (1989: 1440) discussed general characteristics and synapomorphies of this family, but our recent studies (Kameneva and Korneyev, in prep.) show that his statements need thorough reconsideration. Our most recent investigations show that the family must be subdivided into two subfamilies, Ulidiinae and Otitinae, but they must be completely redefined. The subfamily Ulidiinae includes the tribes Seiopterini, Pterocallini, Lipsanini (= Euxestini), and Ulidiini that share the complete reduction of the apodemes of sterna 4 to 6 in females (SA) and shortly setulose or bare phallus of males (P). The subfamily Otitinae includes the tribes Cephalini, Otitini, and an undescribed tribe that have the well-developed apodemes of sternites 4 to 6 in females (P) and long spinose phallus of males (SA).

As indicated by McAlpine, "pedicel with dorsal notch reduced" and "presutural supra-alar bristle absent" are doubtless synapomorphies that well support the monophyly of Ulidiidae, but the following characters, which McAlpine considered synapomorphies of Ulidiidae, actually are well represented in Lonchaeidae and Richardiidae: Epiplateinae (or also in some genera of Platystomatidae and Tephritidae) and therefore rather belong to the ground plan of the Higher Tephritoidea:

"Proepisternal, anepisternal, and katepisternal bristles present," and "postsutural acrostichal and intra-alar bristles present" — All these setae are present in Lonchaeidae, Richardiidae (including Epiplateinae), and in the ground plan of Ulidiidae.

"Anepimeron bare" — It is bare also in all the Lower Tephritoidea families, and therefore belongs to the superfamily ground plan, and is not a synapomorphy of ulidiids. To the contrary, the presence of anepimeral setae is a synapomorphy of Platystomatidae + Pyrgotidae + Tephritidae (see below).

"Costa with a subcostal break" — The presence of the subcostal weakening really belongs to the ground plan of the whole superfamily, but there are no strong subcostal breaks in ulidiids like those observed in Tephritidae, and some Platystomatidae and Pyrgotidae. The expression of this character is strongly dependent on territorial and mating behavior: the species of Tephritoidea using wing displays involving frontal torsion and folding of the wing disk (Platystomatidae: Trapherinae, some Scholastinae; Tephritidae) usually have this break well expressed.

"Crossvein sc-r and pterostigma absent" — There is no true vein sc-r in Tephritoidea, but some species of Tephritidae, Pyrgotidae, and Platystomatidae may have a very short spur vein on $R_1$ at the level of the subcostal vein apex. However, it is not present in most representatives of these families, nor in any Lower Tephritoidea, and its absence is not an autapomorphy of Ulidiidae.

"Aedeagus tightly coiled, and stored in right, ventrolateral side of abdomen" — Actually, the way the phallus of most Ulidiidae (e.g., Seiopterini and Pterocallini) is stored does not differ from that of Richardiidae: Epiplateinae (and apparently of Richardiinae). Some Otitinae (but not all Ulidiidae) have the phallus coiled very tightly.

### 1.2.4.2 Platystomatidae + Pyrgotidae + Tephritidae (including Tachiniscinae)

J. F. McAlpine (1989) proposed the following synapomorphies for these three families:

"Anepisternal phragma strong and complete" — This character, indeed, occurs in all members of these families, but a complete phragma occurs also in Cephalini (Ulidiidae: Otitinae) that does not show close relationship to this family complex. Anyway, this character must be used with caution, as in all these cases it may be the result of homoplasy.

"Aedeagus elongate and looped, and bearing a complex apical glans that is stored more or less dorsally under tergite 5" — The complex glans of the phallus occurring in Platystomatidae + Pyrgotidae + Tephritidae is a novel character. Although many representatives of the Pallopteridae, Piophilidae, Richardiidae, and Ulidiidae have the apical portion of the distiphallus modified or covered with spines or microtrichia different from those covering the basal portion, these modifications are not homologous with the glans of Platystomatidae + Pyrgotidae + Tephritidae. The glans is presumed to align the gonopore and spermathecal duct entrances during copulation (Solinas and Nuzzaci 1984). The dorsal position of the phallic pouch is another synapomorphy of these families. Neither the homology nor polarity of the structures of the glans are understood well in many cases within this complex, and especially between Platystomatidae and Tephritidae.

"Aedeagal apodeme fultelliform, i.e., extensively fused with hypandrium" — The essence of the fultelliform phallapodeme is the following. The anterior margins of the gonocoxites are strongly sclerotized, forming two rodlike thickenings called "arms" or "vanes of fultella," firmly fused to the medial portion of the phallapodeme, and lateroventrally articulated with the striplike lateral sclerites (lateral remainders of gonocoxites). Posteriorly, the phallapodeme may be undivided or forked, and articulates with the anterior margin of the basiphallus. The surface of the triangle between the posterior processes, the vanes, and the lateral sclerites corresponds to the gonocoxites (= gonites) of Tanypezidae, the Lower Tephritoidea and Ulidiidae. In Platystomatidae + Pyrgotidae + Tephritidae, the gonocoxites do not bear any setae, which is another synapomorphy of these families. Most Platystomatidae, Pyrgotidae, and generalized genera of Tephritidae (including Tachiniscinae) have the vanes of the phallapodeme rather long, strong, and approximated to form a deep and narrow structure. This type of phallapodeme is hypothesized to be the initial condition of the character, whereas the shallow and broad shape with rather weak vanes in most Tephritidae is presumed to be the derived state.

The characters above are certainly synapomorphies that support monophyly of this complex of families. The following characters, also proposed by McAlpine as synapomorphies of these three families, are of doubtful status:

"Pedicel with an elongate dorsal seam" — Most Tephritidae and Platystomatidae have such a seam that differentiates them from both Ulidiidae and most Pyrgotidae, but I cannot confirm that these seams are longer or shorter than in Lonchaeidae or in Richardiidae: Epiplateinae. Some primitive pyrgotids (e.g., *Prodalmannia* Bezzi and some other Toxurini) also have such a seam. Therefore, the presence of the seam is certainly a plesiomorphy, whereas its absence is an apomorphy, probably a synapomorphy of many Pyrgotidae and also of Tephritidae: Tachiniscinae: Tachiniscini.

"Greater ampulla more or less developed" — The greater ampulla of Tephritoidea is rather variable in size. This character needs special revision that is beyond the scope of the current work.

"Lower lobe of calypter frequently broadened" — This character occurs in some Platystomatinae, but certainly is not a synapomorphy of the whole complex.

"Posterior notopleural bristle surmounting a tubercle (notopleural callus)" — This callus is rather well-developed in Ulidiidae, and may characterize the Higher Tephritoidea as a whole, rather than just the Platystomatidae + Pyrgotidae + Tephritidae.

"Surstylus consisting of an outer and an inner lobe (as opposed to an anterior and a posterior lobe, respectively), the latter bearing several prensisetae" — Prensisetae are present in at least some taxa of all the families of Tephritoidea, and the medial surstylus is present in Richardiidae and the Higher Tephritoidea, thus neither character is a synapomorphy at this level.

The following characters were not included by J. F. McAlpine (1989) in his list of synapomorphies of the complex, but apparently are synapomorphies of Platystomatidae + Pyrgotidae + Tephritidae:

Proepisternal seta absent — J. F. McAlpine (1989) suggested that this condition belongs to the ground plan of the Higher Tephritoidea, but, as most Ulidiidae and Lower Tephritoidea do have this seta, the reverse polarity is likely. The proepisternal seta is absent in some genera of Cephalini (Ulidiidae: Otitinae).

Anepimeral seta present — J. F. McAlpine (1989) suggested that the absence of the anepimeral seta is a synapomorphy of Ulidiidae (= Otitidae), but actually this character state belongs to the ground plan of the superfamily; the anepimeral seta is a novelty of the Platystomatidae + Pyrgotidae + Tephritidae cluster.

Basal portion of vein $R_1$ setulose dorsally; vein $R_{4+5}$ setulose dorsally — Most Lower Tephritoidea and most Ulidiidae have the section of $R_1$ bare between the levels of the humeral vein and the apex of Sc (setulose occasionally in certain genera or groups of species), and $R_{4+5}$ always bare. All the Platystomatidae, Pyrgotidae, and Tephritidae (including Tachiniscinae) have $R_1$ completely setulose above, and generalized genera in the main lineages inside these families have $R_{4+5}$ setulose. For this reason both characters are presumed to be synapomorphies of these families, belonging to their ground plan, as is the following character.

Scutellum with three pairs of setae — Outside of these three families, the "extra" scutellar setae occur occasionally in a few species of three or four very distantly related genera of Lonchaeidae and Ulidiidae, but occur in most Platystomatidae, some Pyrgotidae, and in the most generalized genera of four different lineages of Tephritidae (Blepharoneurinae, most Tachiniscinae, most Phytalmiinae, except Phytalmiini, and in a few genera of Trypetinae).

Certain characters should be mentioned as important plesiomorphies belonging to the ground plan of Platystomatidae + Pyrgotidae + Tephritidae:

Ocellar setae lateroclinate — This state is common in Ulidiidae, and is found also in *Rivellia* Robineau-Desvoidy and *Lule* Speiser (Platystomatidae); most Platystomatidae lack ocellar setae, and in the Pyrgotidae and Tephritidae they are proclinate when present.

Gonostyli present, buttonlike — Rudiments of gonostyli are present in several genera of Platystomatidae, Pyrgotidae, and Tachiniscinae, but not in most of Tephritidae.

Aculeus with the cercal unit separate from tergum 8 — Generalized genera of Platystomatidae and Tephritidae have such a type of aculeus, very similar to that of most Lonchaeidae and Ulidiidae.

Two right spermathecae have common spermathecal duct that bifurcates at apical one-sixth to two-thirds of its length — Platystomatidae have the same state of this character as the Lower Tephritoidea and Ulidiidae.

### 1.2.5 Relationships among the Families of the Tephritid Complex of Families

J. F. McAlpine (1989: 1440) suggested that there is a sister group relationship between the Platystomatidae and the Tephritidae + Pyrgotidae, but did not present any convincing evidence of monophyly of the latter branch. Korneyev (1992) presumed Platystomatidae and Tephritidae to be sister taxa, misinterpreting the absence of gonostyli as a synapomorphy. Actually, many platystomatids (see Hara 1987; 1989; 1992; 1993) and some Tachiniscinae of the Tephritidae possess gonostyli.

### 1.2.5.1 Platystomatidae

J. F. McAlpine (1989: 1440) presumed that the Platystomatidae have three autapomorphies that prove their monophyly, but none of them can be accepted without reservation:

"Pedicel with a long dorsal seam" — Not only Platystomatidae, but most Tephritidae and Lower Tephritoidea have a more or less developed seam (see comments above).

"Postocellar bristles weak or absent" — At least species of *Lule* (Trapherinae) and *Rivellia* (Platystomatinae), the two genera with the most setae developed, have postocellar setae, although they are short. This suggests that the postocellars may have disappeared independently in different lineages.

"Female with abdominal tergite 6 reduced or absent" — Females of *Poecilotraphera* Hendel (Trapherinae), and *Giraffomyia* Sharp (Angitulinae) have tergum 6 short, but completely developed, with spiracles 5 and 6 situated ventrolaterally of them (not on the dorsal side). *R. syngenesiae* (Fabricius) has tergite 6 short, but with a few setae (Hara 1989: 796, Figure 8), and most other species of Platystomatidae examined thus far have tergum 6 strongly reduced, and spiracles 5 and 6 displaced dorsally.

Instead, the following characters apparently are real synapomorphies of the family:

Cell bcu is closed by a straight vein $Cu_2$ that is perpendicular to Cu (neither curved inward, nor bent, with the posterior portion directed outward) — Compared with ulidiids, Platystomatidae never have a lobelike extension of the basal cubital cell. Most platystomatids have the posteroapical corner of this cell forming a right angle. There are some cases of a similarly closed cell bcu in various distantly related Tephritidae that are of homoplastic origin. Some species in different groups of platystomatids have this cell closed by an arcuate or inwardly curved vein; in all these cases these species are closely related to others that have bcu closed by a straight vein, showing that this character state appeared due to secondary modification. In Angitulinae, where this vein may be directed outward, it remains almost straight; otherwise, the three genera of this subfamily are very close to Platystomatinae, having male genitalia similar to most members of that subfamily, as well as the following character.

Anterior apodemes of sternites 4 to 6 lacking — Sternal apodemes are lacking in all Ulidiinae (Kameneva and Korneyev, in prep.), all Platystomatidae (A. Whittington, unpublished data), all Pyrgotidae, and in some lineages of Tephritidae. Once lacking in generalized genera, they never are found to reappear in advanced genera of the same cluster. For this reason they are hypothesized to be a character that does not reverse. As the presence of the apodemes appears to belong to the ground plans of Ulidiidae + Platystomatidae + Pyrgotidae + Tephritidae, as well as the cluster Pyrgotidae + Tephritidae that is the sister group (see below) of Platystomatidae, their lack is presumed to be completely independent loss in Platystomatidae. It is interesting that in both Ulidiinae and Platystomatidae the reduction of female tergum 6 happens where the sternal apodemes are lacking (and apparently follows their reduction). This correlation of characters is believed to explain numerous cases of abdominal tergite loss in Trapherinae, Plastotephritinae, Scholastinae, and Platystomatinae.

### 1.2.5.2 Pyrgotidae + Tephritidae (including Tachiniscinae)

J. F. McAlpine (1989: 1441) has not given any positive proof of the monophyly of this cluster. He proposed the presence of a posteroapical lobe on cell bcu as a synapomorphy, but this character state is also present in most Ulidiidae, and is probably a synapomorphy for the Higher Tephritoidea.

McAlpine also suggested that "the shift in larval feeding habits from saprophytic to living plants or insects is a significant synapotypic feature of these three families." However, the Tephritidae are not uniformly phytophagous (Hardy 1986; Dodson and Daniels 1988), and the ground plan of this family, and that of the Pyrgotidae + Tephritidae, includes saprophagy or saproxylophagy. The most primitive larval environment of the whole cluster very probably is decaying wood, as in Lonchaeidae, Pallopteridae, and Ulidiidae (Seiopterini). Current analysis shows three characters that are possible synapomorphies of these two families:

Ocellar setae proclinate — In both Ulidiidae and in the few genera of Platystomatidae that have ocellar setae they are rather lateroclinate, although the Lower Tephritoidea, the Pyrgotidae, and the Tephritidae have them proclinate. The independent origin of lateroclinate ocellar setae in Ulidiidae and Platystomatidae is equally likely as the evolution of that character state in the ground plan of the Higher Tephritoidea, with reversal to proclinate setae in Pyrgotidae + Tephritidae; thus, the status of this character as a synapomorphy for the latter clade is unresolved. Both *Tachinisca* Kertész and *Bibundia* Bischof (Tachiniscinae: Tachiniscini) have no ocellar setae at all, but they are proclinate in *Tachiniscidia* Malloch (Tachiniscinae: Tachiniscini) (see Barraclough 1995, Figure 20) and in the Ortalotrypetini.

Subcostal vein with sharp anterior bend, section apical to bend weak or reduced to a fold — Usually considered an autapomorphy of Tephritidae, this character is occasionally present in Platystomatidae (*Traphera* Loew, *Piara* Loew), is common in Pyrgotidae, and is present in almost all Tephritidae (except some Tachiniscinae, Phytalmiini, and *Matsumurania* Shiraki, probably due to reversal). This is equally likely to be a synapomorphy of Pyrgotidae + Tephritidae as to have evolved independently in each family.

Three full spermathecal ducts developed — Contrary to all other Tephritoidea, all the Pyrgotidae and Tephritidae (including Tachiniscinae) have three ducts, at least in the ground plan.

The relationships among some subfamilies assigned to the Pyrgotidae and Tephritidae are uncertain, and therefore the monophyly and relationships of both families may be resolved only tentatively.

### 1.2.5.3 Pyrgotidae

Monophyly of this family is supported by the following apomorphies:

Proepisternum forms a ridge, bearing strong setae — This character was first proposed by Aczél (1956a: 164) to differentiate the Pyrgotidae from Tephritidae. I have not examined enough material throughout Pyrgotidae to confirm that it is present everywhere, but it is well expressed in *Cardiacera* Macquart, *Parageloemyia* Hendel, and *Adapsilia* Waga. But I cannot find strong differences between this character state in Pyrgotidae, and what we see in Tachiniscinae and other Tephritidae (e.g., in *Acanthonevra* Macquart).

Abdominal sternites of female lacking anterior apodemes and overlapping to form conspicuous "pockets" between them — Thus far, this character (see Steyskal 1987b, Figure 65.8) seems to be present in all the genera assigned to Pyrgotidae (and also in Ctenostylidae, which with certainty were proved not to belong in Pyrgotidae nor the Higher Tephritoidea!), and surprisingly also in *Matsumurania*, attributed to Tephritidae: Phytalmiini. The relationships of Ctenostylidae and *Matsumurania* will be discussed below.

Female oviscape very long, usually exceeding joint length of abdominal tergites — The long, stout, conical oviscape is represented in all the genera of Pyrgotinae and Toxurinae (including *Descoleia* Aczél). *Matsumurania* has the oviscape very long but rather soft, like in Platystomatidae and Phytalmiinae (Tephritidae).

The cercal unit completely integrated into the aculeus — The aculeus is always stiletto-like, with an acute point in most Pyrgotinae and Toxurinae, except for that of *Descoleia*, which seems to have a rather needlelike aculeus (Hardy 1954, Figure 1e, f; Aczél 1956a, b, c; Barraclough 1994).

Larvae are internal parasitoids on adult beetles — Actually this synapotypy has not been proved throughout the family, but is rather evident from the two peculiar morphological characters (last two synapomorphies above). According to Pitkin (1989), some Hymenoptera are also hosts of Pyrgotidae.

The following characters are apparently useful for phylogenetic analysis within Pyrgotidae, but their distribution and origin need further study. Aczél (1956a) divided the Pyrgotidae into three subfamilies, Toxurinae (the subcostal vein with a broad break that forms a perpendicular fold very similar to that of Tephritinae, ocelli usually present, pedicular notch deep); Pyrgotinae (Teretrurini + Pyrgotini); (the subcostal vein ends close to costa or reaches it, and does not form the broad perpendicular section, ocelli absent, pedicular notch not expressed); and Lochmostyliinae (proboscis reduced, male with ramulose arista and narrow frons). D. K. McAlpine (1989) proposed a different arrangement into subfamilies, and removed Ctenostylidae (= Lochmostyliinae) from the Tephritoidea. According to him, the two major groups to be distinguished in the Pyrgotidae are the Teretrurinae, which have sternites 1 and 2 fully developed and free (plesiomorphy), and the Pyrgotinae (including Toxurini), which have sternites 1 and 2 fused into one synsternite (apomorphy). This subdivision needs to be further reconsidered, but that is beyond the scope of this chapter.

The third subfamily, Ctenostylinae (= Lochmostyliinae), recently was removed from Pyrgotidae (D. K. McAlpine 1989; Barraclough 1994) because it lacks several synapomorphies of Tephritoidea. It differs in the following significant characters: male sternum and tergum 6 rather well-developed; synsternite 7 + 8 absent; phallus extremely short; female sternum 1 largely desclerotized; aculeus short and blunt, including two rudimentary cerci that bear setae.

Of the other acalyptrate superfamilies, the Ctenostylidae may be related to the Tanypezidae which share the following characters: frons sexually dimorphic (P); cell bcu closed by arcuate vein (P); legs slender (polarity uncertain); male tergite 6 present (P); tubular oviscape (= tergosternum 7) (SA?); and paired female cerci (P). An alternative hypothesis is the relationship of Ctenostylidae with Conopidae, which share similarly shortened thorax (SA?), scutellum with six setae (SA?), vein $A_2$ present (P), male sternum 6 well developed (P), female ovipositor consisting of stout oviscape and eversible membrane (SA?) with separate setulose cerci and rudimentary tergum 9 (P) (see Smith and Peterson 1987, Figures 54.15, 54.18; D. K. McAlpine 1989, Figures 2 to 4).

### 1.2.5.4 Arguments for the Pyrgotidae + Tachiniscinae Sister Group Relationship

J. F. McAlpine (1989) presumed that the Pyrgotidae and Tachiniscinae are sister groups, based mostly on the following characters:

Larvae endophagous parasites — Both Pyrgotidae and Tachiniscinae are known to be parasites, but members of the former family infest adult Scarabaeidae and Hymenoptera (Aczél 1956a; Pitkin 1989), whereas the only known host of any species of Tachiniscinae is a saturniid caterpillar (Roberts 1969). These hosts are rather different, and parasitism may have arisen independently.

"Pedicel with dorsal notch weak or absent" — Both Pyrgotidae and Tachiniscinae include species that have a dorsal notch or cleft of the pedicel and others that have the notch smoothed, which shows this character is subject to homoplasy.

Other characters that might appear to support this relationship are as follows:

Ocelli absent — Most Teretrurinae and Toxurini (Pyrgotidae: Pyrgotinae, according to D. K. McAlpine 1989) have ocelli well developed, whereas the majority of Pyrgotini do not. Among Tachiniscinae, *Tachinisca* have the anterior ocellus conspicuously smaller than the other two, and *Bibundia* lack all the ocelli. But other genera herein assigned to Tachiniscinae have all the ocelli well developed (see below), thus there has been parallel reduction of ocelli in both taxa and this is not a synapomorphy of Pyrgotidae + Tachiniscinae.

Abdominal sternites of females lacking anterior rodlike apodemes — A widespread character, often appearing due to obvious homoplasy (e.g., in Ulidiinae and Phytalmiinae).

Cercal unit completely integrated into acutely pointed aculeus — A cutting and piercing type of aculeus is an adaptive feature that may be either homoplasy or a synapomorphy of these two taxa. The aculeus shape and characters of associated structures (oviscape, eversible membrane) are very different in the two taxa.

Because all of these characters are homoplastic or of uncertain homology, they hardly can be accepted as undoubted evidence of a sister group relationship between the Pyrgotidae and Tachiniscinae. I believe that the following characters better support the closer relationship of the Tachiniscinae and other Tephritidae.

#### 1.2.5.5 Arguments for the Tachiniscinae + Other Tephritidae Sister Group Relationship

Costa deeply broken before the apex of subcosta, and with two or more "costal spurs" (enlarged, stout setulae) present at subcostal break — The main evidence for the sister group relationship of Tachiniscinae and other Tephritidae is the presence of costal spurs in Tachiniscinae (at least in *Tachinisca cyaneiventris* Kertész and all Ortalotrypetini except perhaps *Neortalotrypeta*) along with the subcostal break of the costa, as in the ground plan of all the subfamilies of Tephritidae. This character does not occur elsewhere in the Tephritoidea, and hardly can be considered likely homoplasy.

Frontal setae stout and inclinate — All Tachiniscinae and nearly all other Tephritidae have frontal setae well developed. The only few cases where a similar state of this character occurs in other Tephritoidea are *Chaetopsis* (Ulidiidae: Ulidiinae: Lipsanini), *Poecilotraphera* (Platystomatidae: Trapherinae), and *Toxopyrgota* (Pyrgotidae: Pyrgotinae: Pyrgotini), which are not closely related.

#### 1.2.5.6 Status of the Tachiniscinae

Dissection of the female terminalia of the neotropical *Tachinisca cyaneiventris* (Figure 1.2) showed that it is extremely similar to the highly derived ovipositor structure described and figured for Tephritidae: Ortalotrypetini by Norrbom (1994, Figure 4C). The oviscape is short, with a posterodorsal opening. The eversible membrane is densely spinulose ventrally, finely microtrichose or bare dorsally; its ventral side 1.5 to 2 times longer than the dorsal one. The aculeus is rather short, acute, without any traces of the cercal unit, and when everted is directed dorsally. An identical ovipositor structure was reported for the four genera included in the Ortalotrypetini (*Cyaforma* Wang, *Neortalotrypeta* Norrbom, *Ortalotrypeta* Hendel, and the fossil *Protortalotrypeta* Norrbom) (Norrbom 1994). The posterodorsal opening of the oviscape now also is known for *O. idana* Hendel (Korneyev, unpublished data). The strongly similar chaetotaxy and other characters show continuous variation between the Tachiniscini and Ortalotrypetini, and they eventually may be considered subjective synonyms. Most species of Tachiniscinae fit very well the diagnosis of the Tephritidae and possess the main autapomorphies of that family. The differences of the remaining genera (*Tachinisca*, *Bibundia*, and *Tachiniscidia* Malloch) are of autapomorphic origin, and I prefer to consider this taxon as a whole unit within the Tephritidae (see above). This action maintains the integrity of the family Tephritidae that otherwise cannot be easily defined.

## 1.3 CONCLUSIONS AND FUTURE RESEARCH NEEDS

The relationships among the families of the superfamily Tephritoidea are reanalyzed, based on new characters and those used by J. F. McAlpine (1989). The reconstructed ground plan of the superfamily is based both on comparison with outgroups (Calyptratae, Conopidae, Heleomyzidae, Nerioidea, Diopsoidea) and on the distribution in the ingroup.

The monophyly of the superfamily Tephritoidea is shown to be supported by (1) male tergum 6 strongly reduced or absent; (2) the surstylus, or medial surstylus if there are two, bearing toothlike prensisetae; (3) female sterna 4 to 6 with anterior rodlike apodemes (except for Piophilidae?);

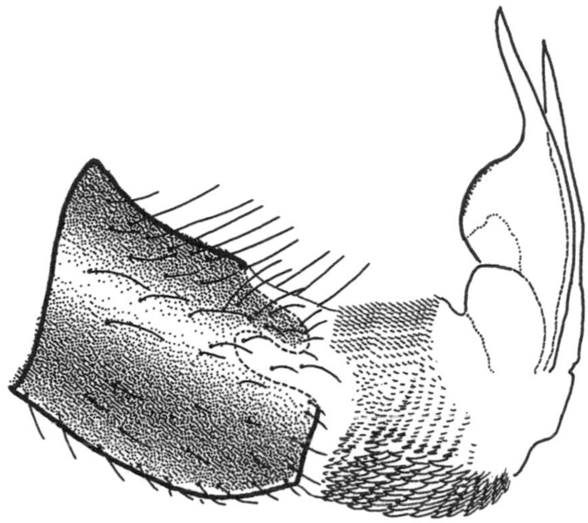

**FIGURE 1.2** Female terminalia, lateral view, *Tachinisca cyaneiventris* Kertész.

(4) female tergosternum 7 consisting of two portions, the anterior one that forms a tubular "oviscape" and the posterior comprising two pairs of longitudinal taeniae. Alternative hypotheses of sister group relationship of Tephritoidea with Tanypezidae or Conopidae are briefly discussed. Both taxa have characters that partially contradict these hypotheses.

The monophyly of the Higher Tephritoidea (Ulidiidae + Platystomatidae + Pyrgotidae + Tephritidae) and of the Platystomatidae + Pyrgotidae + Tephritidae cluster is well supported by several characters (see Figure 1.1). Ulidiidae are characterized by rather numerous plesiomorphic characters and two synapomorphies (the pedicel with dorsal notch reduced, and the presutural supra-alar bristle absent), both of very low weight and strongly subject to homoplasy.

The Platystomatidae (with some reservation concerning Trapherinae and Plastotephritinae, which were not thoroughly examined during this study) are a monophyletic group characterized by the combination of some plesiomorphies (aculeus with separate cercal unit, two spermathecal ducts with three spermathecae, etc.) and two synapomorphies, both of low value (cell bcu is closed by a straight vein $Cu_2$ that is perpendicular to Cu, and anterior apodemes of sternites 4 to 6 lacking).

The Pyrgotidae and Tephritidae (including Tachiniscinae) share at least one synapomorphy, the presence of three spermathecal ducts. The relationships between Pyrgotidae and Tephritidae are discussed. These families can be defined with some precaution, as no genera of Teretrurinae and Toxurini (that include rather primitive Pyrgotidae) were available for examination of genitalic characters. Tachiniscini, formerly considered to be a separate family, were found to be closely related to Ortalotrypetini of the Tephritidae and are included in that family rather than associated with Pyrgotidae.

There are certain gaps in our knowledge of Tephritoidea phylogeny that we need to fill in the future: (1) the problem of polytomy in the cluster Piophilidae/Pallopteridae + Lonchaeidae + other families; (2) the monophyly of Richardiidae (including Epiplateinae) is not proved, and the status of Epiplateinae needs further consideration; (3) the monophyly of Ulidiidae is supported by several synapomorphies of low weight, and additional proofs, both morphological and biochemical are needed; (4) the status of the subfamilies of the Platystomatidae must be reconsidered based on additional characters and cladistic analysis of relationships in the family; and (5) further analysis is needed of the morphology and phylogenetic relationships in the "Lower Pyrgotidae" (Teretrurinae, Pyrgotinae: Toxurini) to test the limits and diagnoses of both Pyrgotidae and its sister group, Tephritidae. At least two genera, *Descoleia* Aczél and *Matsumurania* Shiraki, deserve additional study, as they do not fit the current concepts of either Pyrgotidae or Tephritidae.

## ACKNOWLEDGMENTS

The three chapters presented in this book (Chapters 1, 4, and 22) have appeared as a result of long-term activity that was encouraged by Prof. Yu. G. Verves, my advisor at the very early stages of scientific work, and by my colleagues who supplied me with specimens of non-Palearctic flies important for this study: A. L. Norrbom (USDA SEL, Washington, D.C.), A. Freidberg (Tel-Aviv University), and B. Merz (ETH Zürich). Many Australasian taxa were available for analysis due to enormous work done by D. E. Hardy (University of Hawaii) and his assistants who illustrated his papers with numerous figures, especially those of male and female terminalia. Recent papers by R. A. I. Drew, D. L. Hancock, A. L. Norrbom, and H.-Y. Han were essential sources of morphological information. I thank R. Contreras-Lichtenberg for the great opportunity to study specimens in the collection of the Naturhistorisches Museum Wien and for arranging a NHMW visitor's grant in 1995; numerous USNM specimens were studied during my visit supported by a grant from the Smithsonian Institution. A. Whittington (Edinburgh) kindly read early proofs of this chapter and shared his comments on the relationships of Platystomatidae. A. L. Norrbom, A. Freidberg, and one anonymous reviewer offered valuable comments on the manuscript. Some new phylogenetic conclusions were the result of fruitful discussions with A. Freidberg, H.-Y. Han, E. P. Kameneva, B. Merz, A. L. Norrbom, and A. Whittington. I wish to thank M. Aluja and A. L. Norrbom for their enormous efforts to organize the symposium and for inviting me to attend that extraordinary meeting of the best experts in the taxonomy, phylogeny, and behavior of the Tephritidae. Financial support from Consejo Nacional de Ciencia y Tecnología (CONACyT) — Subdirección de Asuntos Bilaterales is greatly appreciated. I acknowledge the support from Campaña Nacional contra las Moscas de la Fruta (SAGAR-IICA) and thank Lic. A. Meyenberg (CONACyT), Ing. J. Reyes (SAGAR), C. Leal (IE), V. Domínguez (IE) and E. Bombela (IE) for arranging my trip to Xalapa. My warmest and sincere thanks are due to B. Delfosse (IE), and the team members: L. Guillén, I. Jácome, R. Cervantes, and J. Piñero.

I also thank Elena and Severin for their patience. They inspired this work every minute of the day for many years.

## REFERENCES

Aczél, M. 1956a. Revisión parcial de las Pyrgotidae Neotropicales y Antarcticas, con sinópsis de los géneros y especies (Diptera, Acalyptratae). *Rev. Bras. Entomol.* 4: 161–184.

Aczél, M. 1956b. Revisión parcial de las Pyrgotidae Neotropicales y Antarcticas, con sinópsis de los géneros y especies (Diptera, Acalyptratae). *Rev. Bras. Entomol.* 5: 1–70.

Aczél, M. 1956c. Revisión parcial de las Pyrgotidae Neotropicales y Antarcticas, con sinópsis de los géneros y especies (Diptera, Acalyptratae). *Rev. Bras. Entomol.* 6: 161–184.

Barraclough, D.A. 1994. A review of Afrotropical Ctenostylidae (Diptera: Schizophora: ?Tephritoidea), with redescription of *Ramuliseta lindneri* Keiser, 1952. *Ann. Natal Mus.* 35: 5–14.

Barraclough, D.A. 1995. An illustrated identification key to the acalyptrate fly families (Diptera: Schizophora) occurring in southern Africa. *Ann. Natal Mus.* 36: 97–133.

Colless, D.H. and D.K. McAlpine. 1970. Diptera (flies). In *The Insects of Australia*, pp. 656–740. Melbourne University Press, Carlton.

Dodson, G. and G. Daniels. 1988. Diptera reared from *Dysoxylum gaudichaudianum* (Juss.) Miq. at Iron Range, northern Queensland. *Aust. Entomol. Mag.* 15: 77–79.

Farris, J.S. 1988. *Hennig86, version 1.5.* Published by the author, Port Jefferson Station.

Freidberg, A. 1994. Is the Tephritidae a good family? In *Third International Congress of Dipterology. Abstract volume* (J.F. O'Hara, ed.), pp. 16–17. University of Guelph, Guelph.

Griffiths, G.C.D. 1972. *The Phylogenetic Classification of the Diptera Schizophora with Special Reference to the Structure of the Male Postabdomen.* W. Junk, The Hague, 340 pp.

Hara, H. 1987. A revision of the genus *Prosthiochaeta* (Diptera, Platystomatidae). *Kontyû* (Tokyo) 55: 684–695.

Hara, H. 1989. Identity of *Rivellia fusca* (Thomson) and description of a new allied species, with special reference to the structure of abdomen of *Rivellia* (Diptera, Platystomatidae). *Jpn. J. Entomol.* 57: 793–802.

Hara, H. 1992. Description of a new species of *Rivellia* (Diptera, Platystomatidae) from Japan. *Jpn. J. Entomol.* 60: 427–431.

Hara, H. 1993. *Rivellia basilaris* (Wiedemann) (Diptera, Platystomatidae) and its allied species in east Asia. I. *Jpn. J. Entomol.* 61: 819–831.

Hardy, D.E. 1954. Notes and descriptions of Australian fruit flies (Diptera: Tephritidae). *Proc. Hawaii. Entomol. Soc.* 15: 327–333.

Hardy, D.E. 1986. Fruit flies of the subtribe Acanthonevrina of Indonesia, New Guinea, and the Bismarck and Solomon Islands (Diptera: Tephritidae: Trypetinae: Acanthonevrini). *Pac. Insects Monogr.* 42: 1–191.

Hennig, W. 1958. Die Familien der Diptera Schizophora und ihre phylogenetischen Verwandtschaftsbeziehungen. *Beitr. Entomol.* 8: 505–688.

Korneyev, V.A. 1984. Comparative morphology of aedeagus of Tephritidae (Diptera). In *Diptera (Insecta), Their Systematics, Geographic Destribution [sic] and Ecology, Proc. of the 3rd All-Union Dipterological Symposium,* Kiev/Belaya Tserkov, Sept. 15–17, 1982 (O.A. Scarlato, ed.), pp. 67–72. Zoological Institute, Leningrad [in Russian].

Korneyev, V.A. 1992. Reclassification of the Palaearctic Tephritidae (Diptera). Communication 1. *Vestn. Zool.* 1992(4): 31–38. [in Russian].

Korneyev, V.A. 1994. Monophyly, groundplan and the sister-groups in the families Pyrgotidae, Platystomatidae and Tephritidae. In *Third International Congress of Dipterology. Abstract Volume* (J.F. O'Hara, ed.), pp. 112–113. University of Guelph, Guelph.

McAlpine, D.K. 1973. The Australian Platystomatidae (Diptera, Schizophora) with a revision of five genera. *Mem. Aust. Mus.* (1972) 15: 1–256.

McAlpine, D.K. 1989. The taxonomic position of the Ctenostylidae (= Lochmostyliinae; Diptera: Schizophora). *Mem. Inst. Oswaldo Cruz,* 84, Suppl. 4: 365–371.

McAlpine, D.K. 1997. Gobryidae, a new family of acalyptrate flies (Diptera: Diopsoidea), and a discussion of relationships of the diopsoid families. *Rec. Aust. Mus.* 49: 167–194.

McAlpine, J.F. 1976. Systematic position of the genus *Omomyia* and its transference to the Richardiidae (Diptera). *Can. Entomol.* 108: 849–853.

McAlpine, J.F. 1977. A revised classification of the Piophilidae, including 'Neottiophilidae' and 'Thyreophoridae.' *Mem. Entomol. Soc. Can.* 103, 66 pp.

McAlpine, J.F. 1981. Morphology and terminology. In *Manual of Nearctic Diptera.* Vol. 1 (J.F. McAlpine et al., coord.), pp. 9–63. Monograph of the Biosystematics Research Institute, No. 27. Agriculture Canada, Ottawa.

McAlpine, J.F. 1989. Phylogeny and classification of the Muscomorpha. In *Manual of Nearctic Diptera.* Vol. 3 (J.F. McAlpine, ed.), pp. 1397–1518. Monograph of the Biosystematics Research Centre, No. 32. Agriculture Canada, Ottawa.

Norrbom, A.L. 1994. New genera of Tephritidae (Diptera) from Brazil and Dominican Amber, with phylogenetic analysis of the tribe Ortalotrypetini. *Insecta Mundi* 8: 1–15.

Norrbom, A.L. and J.F. McAlpine. 1997. A revision of the Neotropical species of *Dasiops* Rondani (Diptera: Lonchaeidae) attacking *Passiflora* (Passifloraceae). *Mem. Entomol. Soc. Wash.* (1996) 18: 189–211.

Pitkin, B.R. 1989. Family Pyrgotidae. In *Catalog of the Diptera of the Australasian and Oceanian Regions* (N.L. Evenhuis, ed.), pp. 498–501. Bishop Museum Press, Honolulu and E.J. Brill, Leiden.

Roberts, H. 1969. Forest insects of Nigeria with notes on their biology and distribution. *Inst. Pap. For. Inst. Oxf.* 44: 1–206.

Smith, K.G.V. and B.V. Peterson. 1987. Conopidae. In *Manual of Nearctic Diptera.* Vol. 2 (J.F. McAlpine, ed.), pp. 749–756. Monograph of the Biosystematics Research Centre, No. 28. Agriculture Canada, Ottawa.

Solinas, M. and G. Nuzzaci. 1984. Functional anatomy of *Dacus oleae* Gmel. female genitalia in relation to insemination and fertilization processes. *Entomologica* (Bari) 19: 135–165.

Steyskal, G.C. 1987a. Tanypezidae. In *Manual of Nearctic Diptera.* Vol. 2 (J.F. McAlpine, ed.), pp. 791–797. Monograph of the Biosystematics Research Centre, No. 28. Agriculture Canada, Ottawa.

Steyskal, G.C. 1987b. Pyrgotididae. In *Manual of Nearctic Diptera.* Vol. 2 (J.F. McAlpine, ed.), pp. 813–816. Monograph of the Biosystematics Research Centre, No. 28. Agriculture Canada, Ottawa.

## APPENDIX 1.1: TAXA EXAMINED

### OUTGROUPS

Scatophagidae: *Cordylura ciliata* (Meigen), *Norellia spinipes* (Meigen).
Conopidae: *Dalmannia aculeata* (L.), *Sicus ferrugineus* (L.).
Micropezidae: *Raineria calceata* (Fallén).
Tanypezidae: *Tanypeza longimana* Fallén, *Strongylophthalmyia ustulata* (Zetterstedt).
Ctenostylidae: *Ramuliseta* sp. (female: Vietnam).

### INGROUPS

Piophilidae: *Neottiophilum praeustum* (Meigen), *Piophila casei* (L.), *Mycetaulus asiaticus* Gregor.

Pallopteridae: *Palloptera ustulata* Fallén, *Temnosira saltuum* (L.), *Toxoneura quinquemaculata* (Macquart).

Lonchaeidae: *Lonchaea chorea* (Fabricius).

Richardiidae: *Omomyia hirsuta* Coquillett, *Odontomera nitens* (Schiner), *Richardia podagrica* (Fabricius).

Ulidiidae: many species of all Palearctic genera (except for *Ulidiopsis* Hennig), *Pterocalla quadrata* Wulp, *Pterocalla* spp., *Paragorgopsis maculata* Giglio-Tos, *Xanthacrona bipustulata* Wulp, *Notogramma cimiciforme* Loew, *Eumetopiella* spp., *Euxesta notata* (Wiedemann), *Chaetopsis fulvifrons* (Macquart), *Perissoneura diversipennis* Malloch, *Oedopa capito* Loew, *Callopistromyia annulipes* (Macquart), *Delphinia picta* (Fabricius).

Platystomatidae: *Traphera azurea* Hendel, *Piara cyanea* Hendel, *Poecilotraphera hainanensis* Steyskal, *P. gamma* Hendel, *Lule* sp., *Conopariella tibialis* (Hendel), *Scholastes* spp., *Naupoda platessa* Osten Sacken, *N. contracta* Hendel, *Pterogenia* sp. cf. *flavopicta* Hennig, *Trigonosoma tropida* Hendel, *Parardelio pilosus* Hendel, *Angitula cyanea* Guérin-Méneville, *Giraffomyia* sp. cf. *regularis* Malloch, *Bromophila caffra* Macquart, *Steyskaliella tuberculifrons* Soós, many Palearctic species of *Platystoma* Robineau-Desvoidy and *Rivellia* Robineau-Desvoidy.

Pyrgotidae: *Adapsilia alini* Hering, *A. coarctata* Waga, *A. wagai* (Bigot), *Cardiacera microcera* (Portschinsky), *Parageloemyia* sp., *Teliophleps mandschurica* Hering.

Tephritidae: *Tachinisca cyaneiventris* Kertész, *Bibundia hermanni* Bischof, *Ortalotrypeta idana* Hendel, *Cyaforma shenonica* Wang (Tachiniscinae), and numerous other genera from all subfamilies and most tribes.

# 2 Breeding Habits and Sex in Families Closely Related to the Tephritidae: Opportunities for Comparative Studies of the Evolution of Fruit Fly Behavior

*John Sivinski*

**CONTENTS**



## 2.1 INTRODUCTION

There are reasons to address the tephrit*oid* flies other than fruit flies, i.e., the Lonchaeidae, Ulidiidae (= Otitidae), Platystomatidae, Pyrgotidae, Tachiniscidae, Richardiidae, Pallopteridae, and Piophilidae, in a book devoted to the Tephritidae. Foremost is that comparative studies of tephritid behavior might be improved by the larger, more various data set created by inclusion of nontephritid flies. The potential advantages become clearer following a consideration of the nature of the comparative method itself.

Correlations between the variations in a particular type of behavior and the different ecological circumstances in which the variations are performed are important means of suggesting how the behaviors evolved (e.g., Thornhill 1984). Such comparative studies have long played an important role in the study of fruit flies, a classic example being the relationship between the spatial distribution of larval foods and the types of places males forage for mates and the investments they are likely to make in courtship signals (Emlen and Oring 1977; Prokopy 1980; Burk 1981; see following section on Lonchaeidae).

While correlations of niches and behaviors can be very revealing, there are two types of errors that may occur in comparisons of organisms, and the likelihood of these errors is influenced by the scope of the comparison. The first is mistakenly ascribing similar traits to convergence and dismissing the importance of differences in particular selective contexts. This can result from comparing closely related insects which may share a trait simply because of recent common descent. There may have been insufficient opportunity, perhaps due to lack of time or shared genomic coadaptations, for natural selection to result in divergence. The second error is to compare distantly related organisms and underestimate the role of phylogeny in restricting convergence; for example, a hypothesis that large wings will evolve in a certain habitat should not be weakened because a wingless centipede was collected along with broad-winged flies from some particular site.

Thus, the justification for the consideration of other tephritoid families is that they may provide material for comparisons that do not suffer from being made among insects that are either "too closely" or "too distantly" related. I offer no formula that predicts when such comparisons would be particularly useful, only urge fruit fly researchers to keep the Ulidiidae, Lonchaeidae, and other related families in mind and be aware of the opportunities they may afford to identify convergent evolutionary patterns and their ecological correlates. To that end, I briefly review the phylogeny and breeding habits of the nontephritid tephritoids, and comment on what they might reveal about mating behaviors in fruit flies.

## 2.2 TEPHRITOID FAMILIES

### 2.2.1 Lonchaeidae

Within the superfamily Tephritoidea there are three monophyletic subgroups (McAlpine 1989; Figure 2.1; but also see Korneyev, Chapter 1). One, consisting of the Lonchaeidae alone, is the sister group to the other subgroups and is distinguished by, among other things, its unpatterned wings and aerial-swarm mating systems (pigmented wings do occur in *Dasiops gracilis* Norrbom and McAlpine (1997) and various other lonchaeid species, but the patterns are usually faint and diffuse).

The function of wing patterning in tephritoids is obscure, although when the folded wings of some fruit flies are seen from behind, the bands appear to mimic a salticid spider. This resemblance can deter spider attacks (e.g., Greene et al. 1987). While both sexes typically have similarly patterned wings (Foote et al. 1993), it is possible that bands and spots on slowly moving wings serve as sexual or agonistic signals of some sort (Sivinski and Webb 1986). If wing patterns are signals, then they would be of lesser value in insect swarms, such as formed by lonchaeids, since the rapidly moving wings would obscure the pattern (Sivinski and Petersson 1997). Bold patterns on the wings of the largely nocturnal tephritoid family Pyrgotidae (Steyskal 1978) may be evidence against a universal intraspecific signaling role for the markings. In some tephritids, for example, *Trupanea* spp., male wing markings are fainter than those of females and some features may be interrupted or missing (Foote et al. 1993). This might also be inconsistent with sexual signaling since males typically have more elaborate displays.

Swarms are rarely found among the cyclorrhaphous Diptera, especially in acalyptrate families, but the lonchaeids are exceptions. For example, 12 species were discovered swarming at the same time of year on a hilltop in Quebec (McAlpine and Munroe 1968). Several to ten aggregated males engaged in rapid, spiral-like, zigzag flights over forest paths, typically in a beam of sunlight. Similar swarms of males, some much larger (>100 individuals) and others with females found crawling on adjacent branch tips, have been observed in different Canadian locations and in Australia (citations in McAlpine and Munroe 1968).

While lonchaeid swarms are unique in the Tephritoidea, male aggregations in the form of leks are common in the Tephritidae (Aluja et al., Chapter 15; Eberhard, Chapter 18). Could a comparison of the two families reveal similarities and differences that led to the evolution of two different group-based mating systems? The distribution of larval feeding sites seems to be correlated to the

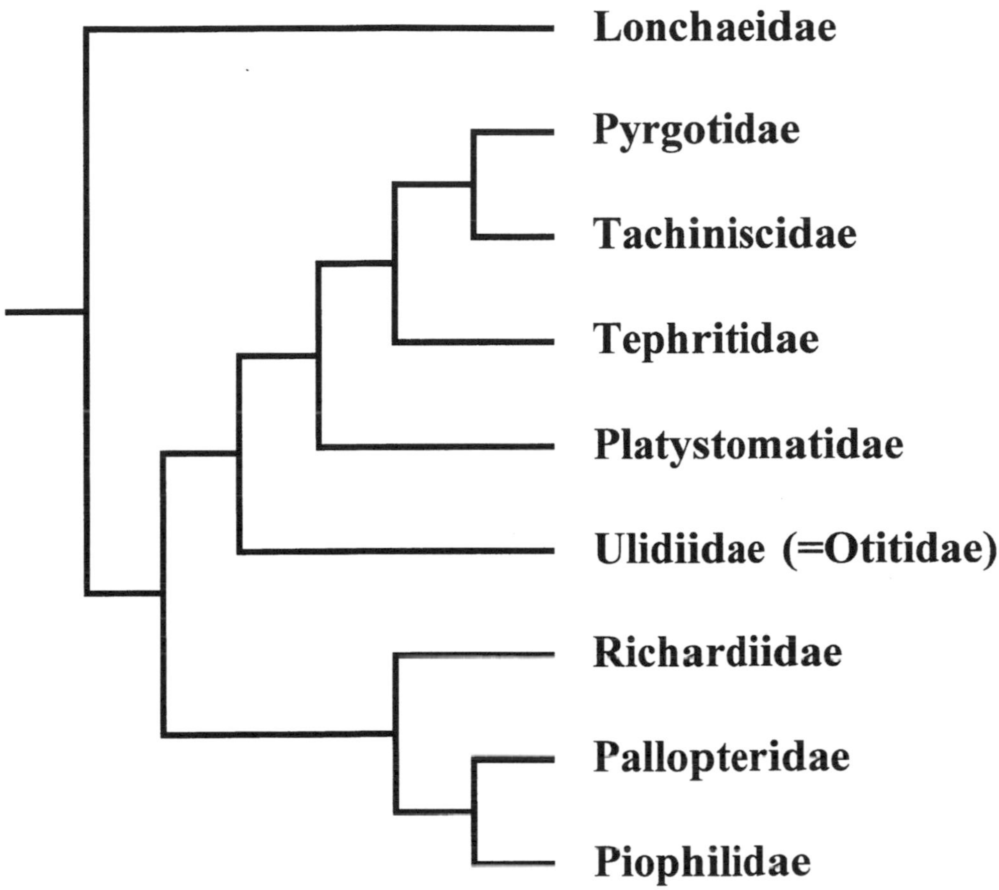


**FIGURE 2.1** A cladogram representing the derivation of the tephritoid families. (Modified from McAlpine J.F., *Manual of Nearctic Diptera,* Vol. 3, Agriculture and Agrifood Canada, 1989. Reproduced with the permission of the Minister of Public Works and Government Services Canada.)

occurrence of certain kinds of insect mating systems. Prokopy (1980) and Burk (1981) brought to the attention of tephritidists that there are circumstances where female fruit flies cannot readily exercise precopulatory mate choice, for example, when oviposition sites are discrete, scattered, and relatively rare (see Emlen and Oring 1977). Males can wait by these resources and attempt to copulate with arriving females. Under these conditions it might benefit a female to mate immediately rather than expend time and energy choosing a particular male, all the while distracted from exploiting the resource. Where there is little opportunity for females to choose, there is little advantage for males to invest in courtship signals. However, where oviposition sites are relatively common and homogeneously distributed, females are not concentrated and their locations become unpredictable. Males have little chance to control access to resources and females are free to choose mates. Males may then compete through signals (Sivinski 1997), and perhaps aggregate (lek) away from oviposition sites (Alcock 1987).

Theories concerning the role of oviposition site distribution in the evolution of swarming, like those that address lekking, first suppose that males find it difficult to predict the presence of females near resources, and so participate in mating systems away from resources, where the sexes meet by "convention" (Sivinski and Petersson 1997). Swarm sites, hilltops, sunbeams, branch tips, etc. might originally have been useful as navigation markers (Sullivan 1981). Since moving insects pass over or near these markers, their vicinities would have contained unusually high densities of otherwise rarely encountered females (Parker 1978; Alcock 1987); in the terminology of lek evolutionary theory such sites are "hot spots" (Bradbury 1979). One difference between leks, occurring as they do on a substrate, and aerial swarms is that males in leks have a platform from which to produce visual, acoustic, and/or chemical signals, all of which can presumably be tracked back to the emitter. This allows females to compare males and make informed mate choices. These signals are more difficult to track in aerial swarms, and appear to be rare in swarming species of flies (Sivinski and Petersson 1997). Thus, while leks and swarms may derive from similar conditions of female unpredictability, subsequent sexual selection may follow different paths.

What is the spatial-temporal distribution of resources used by lonchaeid females and are there any parallels with the resources used by lekking tephritids? In general, the Lonchaeidae are saprophages, with two main feeding types (Ferrar 1987). The first consume rotting fruit and vegetable material, and many follow attacks by tephritid fruit flies. A few are primary attackers of fruits and pine cones; for example, *D. alveofrons* McAlpine oviposits in apricots in a manner similar to that of tephritids (Moffitt and Yaruss 1961; see also figures of *Silba adipata* McAlpine in Katsoyannos 1983). The host ranges of Brazilian *Neosilba* spp. closely parallel those of local populations of *Anastrepha* spp. and *Ceratitis capitata* (Wiedemann) (as *Silba* spp.; Malavasi and Morgante 1980; Souza et al. 1983). The second feeding type is found under bark, in close association with wood-boring Coleoptera. Maggots consume weakened and dead beetle larvae and pupae (Ferrar 1987), although some can complete development on beetle frass alone. The large genus *Lonchaea* contains most of the larvae of this type.

Swarming species have been collected from genera that both secondarily attack fruit (e.g., *Silba horriomedia* McAlpine) and feed upon beetles under bark (numerous *Lonchaea* spp.). The first set of resources has obvious parallels with those exploited by fruit flies, and presumably generates similar selection pressures. It would be particularly interesting to compare the ecology and mating behavior of lonchaeids that oviposit into fruits previously attacked by lekking tephritids, such as the Brazilian *Neosilba* spp. mentioned above. Do these lonchaeids swarm? How similar are the larval resource distributions; for example, are the lonchaeid host ranges narrower? What factors might be responsible for the evolution of lekking in the one and swarming in the other? To what extent has the divergence between swarming and lekking influenced precopulatory courtship signals?

### 2.2.2 Ulidiidae

The Tephritoid subgroup containing the Tephritidae also encompasses the Ulidiidae, Platystomatidae, Pyrgotidae, and Tachiniscidae (Figure 2.1; see, however, Korneyev, Chapters 1 and 4, who subsumes the Tachiniscidae into the Tephritidae). All are characterized by patterned wings and sexual behaviors entirely completed while standing on a substrate (McAlpine 1989; see, however, rare aerial courtship components in species such as *A. robusta* Greene [Aluja 1993]). The Ulidiidae are separated from the remaining families in the subgroup largely on the basis of genital morphology (McAlpine 1989).

The ulidiids develop in a variety of breeding sites, the most common being rotting fruits and vegetables, and other decaying media such as pond muck, refuse, and dung (Ferrar 1987). However, there is a strong tendency to attack living plant tissue (e.g., *Euxesta stigmatias* Loew on sweet corn, *Zea mays* L.; Seal and Jansson 1989; Seal et al. 1995). These species tend to have narrow host ranges. Unlike the Lonchaeidae, those species that oviposit under bark appear to develop on beetle frass, rather than upon the beetles themselves.

Ulidiid behavior is diverse and sometimes spectacular. As in the Tephritidae, wing movements by individuals of both sexes are common. *Callopistromyia annulipes* (Macquart) "struts" with the wings upraised like a peacock's tail. The attitude so struck the well-known dipterist J. M. Aldrich that he mounted his specimens in the display position (Steyskal 1979). Wing waving by *Tritoxa incurva* Loew was thought by Allen and Foote (1975) to play some role in courtship, as does perhaps the frequent expulsion, and subsequent holding of, droplets in the mouth, a behavior which includes expansion of an orange-colored oral membrane.

While there are some behavioral similarities with fruit flies, other courtship behaviors in the Ulidiidae differ from those typical of tephritids. For example, in *E. stigmatias* females in the presence of males extend their aculeus (ovipositors), which stretch nearly as long as the body. Males then rub their labella down its length (Seal and Jansson 1989; see also similar activities reported by Perez (1911) as cited in Richards (1924); note that females of the tephritid *A. striata* Schiner may touch their ovipositors to the males' heads (Aluja et al. 1993)). Such a behavior might reflect different sensory systems in the two families (e.g., different chemosensillia on the ulidiid ovipositor) and/or a more saliva-borne pheromone in the male. The remarkable courtship of *Physiphora demandata* (Fabricius) consists of highly variable sequences of displays presented in territories held on twigs and grass stems. It includes (1) male drumming of the female's head and thorax with his forelegs; (2) vibration of the male's body; (3) wing waving; (4) wing flicking (supination?); (5) foreleg lift and wave; (6) midleg raise (abduction); and (7) the male quickly backing up several centimeters and then hurrying forward (Alcock and Pyle 1979). Female behavior is also complex and includes a bizarre episode where she places her extended proboscis on the male's back and appears to pull him backward in a spiraling course for several centimeters (Figure 2.2). Following a successful courtship, pairs remain coupled for over 2 h, which prevents females from mating again during the afternoon breeding period. The relationship between the opportunity to remate and copulation duration might likewise be profitably examined in the Tephritidae where pairings among species can be highly variable (see Sivinski et al., Chapter 28).

Larvae of *P. demandata* develop in dung and rotting vegetation, and Alcock and Pyle (1979) note that male dung-breeding flies are typically able to control access to oviposition resources and so do not advertise their suitability to females with elaborate/expensive signals. They suggest that the apparent paradox of an off-resource mating system containing elaborate male signals having evolved in a species exploiting a discrete and relatively rare oviposition site (dung) may be resolved by the flies' additional use of rotting vegetation. Such resources may be "... too widely distributed to be easily monitored or defended."

In addition to complex courtships, some ulidiid males have their eyes on stalks, which may be used by males to defend oviposition sites from sexual rivals (Wilkinson and Dodson 1997). Thus, sexual selection has generated a variety of adaptations used presumably to convince choosy females of male sexual suitability and/or deter rival males from occupying resources. The wide range of ulidiid behaviors, many of which resemble those of tephritids, and breeding sites, most of which are dissimilar to those of tephritids, seem to make them a particularly interesting group for comparison and contrast with fruit flies.

### 2.2.3 Platystomatidae

The Platystomatidae is the sister group of the remainder of the tephritid-containing subgroup (Tephritidae, Pyrgotidae, and Tachiniscidae), and differs from them by the largely saprophtyic feeding habits of its larvae (Figure 2.1; McAlpine 1989). Platystomatids breed in rotting tree trunks, bulbs, roots and fruit, dried flowers and dead grass stems, dung and fungus (Ferrar 1987). Mass graves dug in World War II sometimes produced prodigious numbers of *Platystoma lugubre* Robineau-Desvoidy (Hennig 1945, as cited in Ferrar 1987). Unlike ulidiids, a few species are predaceous; *Elassogaster linearis* (Walker) larvae are important predators of locust eggs in the Philippines (Greathead 1963) and *Euprosopia megastigma* McAlpine was found eating a scarab grub (McAlpine 1973a). Species of the cosmopolitan genus *Rivellia* typically attack the nitrogen-producing root nodules of legumes, occasionally reaching pest status (e.g., the soybean nodule fly *R. quadrifasciata* (Macquart); Koethe and Van Duyn 1988).

A striking component of platystomatid sexual behavior is the frequently encountered passing of fluids from the male's mouth to the female's (trophallaxis, Figure 2.3). The contents of these droplets are unknown. Alcock and Pyle (1979) suggest that the salivary substances passed in certain tephritids constitute a "food present," and that large investments of this sort would be presented in

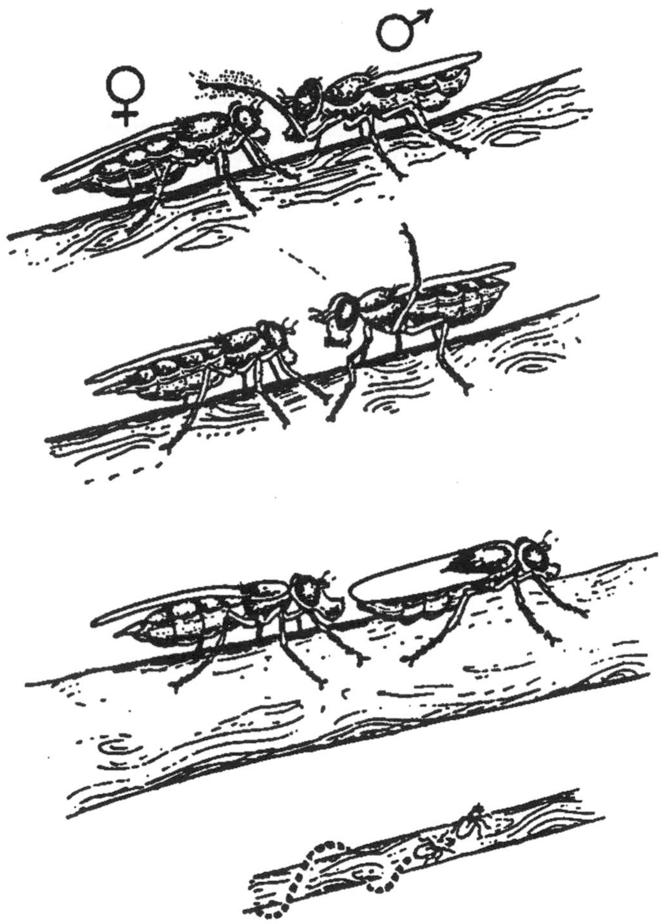

**FIGURE 2.2** The courtship of the ulidiid *Physiphora demandata* is particularly complex, and contains such unusual features as the signals involving midlegs, and the female touching the male's abdomen with her extended proboscis and then "pulling" him in a spiral course backward along a twig. (From Alcock, J. and D.W. Pyle, *Z. Tierpsychol.* 49: 352–362, 1979. With permission.)

a "... courtship free from elaborate displays"; that is, a relatively large male investment would be valuable to the female and she would be selected to obtain the resource rather than waste opportunities by making fine precopulatory distinctions about the genetic qualities of potential mates. For example, male *Stenopa vulnerata* (Loew) stand next to froth mass deposits on leaves making the stylized wing movements similar to many (most?) Tephritidae (Novak and Foote 1975). Females are mounted as they arrive to feed on the male's cache. Similar behaviors are exhibited by *Icterica seriata* (Loew) (Foote 1967) and *Dirioxa pornia* (Walker) (Pritchard 1967, as *Rioxa pornia*).

However, platystomatid courtships containing trophallaxis often appear to be relatively complex, as is also the case in some more recently studied Tephritidae (see Headrick and Goeden 1994; Sivinski et al., Chapter 28). In *Rivella melliginis* (Fabricius) males fan their wings, sway their bodies, and rotate. After the arrival of the female they form the droplet and continue to wing fan. This is followed by foretarsal tapping, circling, and eventual mounting (McMichael et al. 1990). Females of *R. boscii* Robineau-Desvoidy run about in circles on leaves with the males in close pursuit (Piersol 1907). After mounting, male *R. boscii* produce multiple clear droplets which the female consumes, and in some cases he dismounts and run in circles around his mate while she consumes the regurgitant. He then remounts and produces another droplet (see also Michelmore's

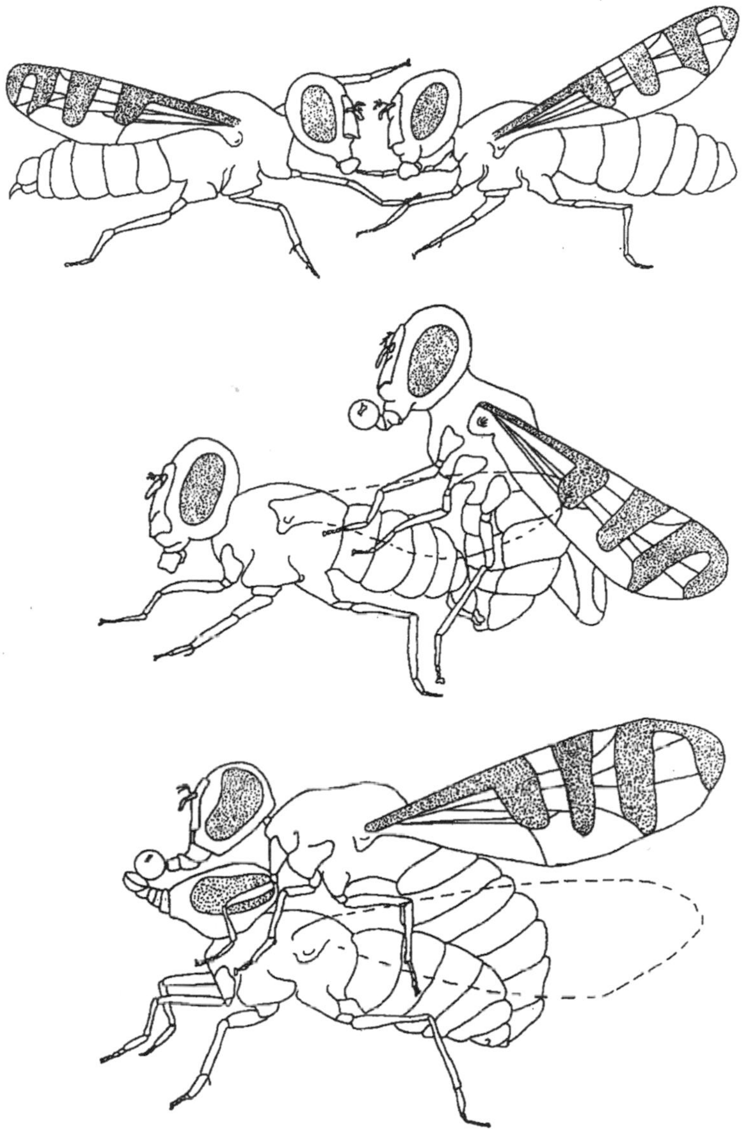

**FIGURE 2.3** Trophallaxis occurs during mounting in *Rivellia melliginis*; transfers of fluids at this point in mating appear to be relatively common in the Platystomatidae. (From McMichael, B. et al., *Ann. Entomol. Soc. Am.* 83: 967–974, 1990. With permission.)

1928 description of trophallaxis in *Platystoma seminationis* (Fabricius)). McAlpine (1973b) observed three species of Australian *Euprosopia,* one of which presented the female with a regurgitant. One stage of courtship — the mounted stage prior to copulation where males of the other species stroke, tap, and comb the female — was indeed absent in this species, although the premounting stage seems to be well developed, with male displays of blackened foretarsi and tapping by the male's proboscis on the female's wings and abdomen.

A comparison of the regurgitant contents and courtship complexities in the "simple" tephritids and the Platystomatidae and "complex" Tephritidae might reveal different functions for trophallaxis. Those provided by some male tephritids seem to have a higher concentration of solids than those produced by Platystomatidae. In some *Eutreta* spp. the regurgitant is a frothy mass deposited on leaves, which collapses into a viscid mucus when probed with a needle (Stoltzfus and Foote 1965).

**FIGURE 2.4** The eyes of a male *Achias* sp. from Papua, New Guinea occur on the ends of stalks, each of which is longer than the insect's body. (From Sivinski, J., *Fla. Entomol.* 80: 144–164, 1997. With permission.)

Froths are also produced by males of *S. vulnerata* and *I. seriata* (Novak and Foote 1975; Foote 1967). Foam ball "mating lures" are made by male *Afrocneros mundus* (Loew) (Oldroyd 1964), and male *Spathulina tristis* (Loew) transfer a viscous milky-white fluid to females *following* copulation (Freidberg 1982). While some tephritids produce a platystomatid-like clear droplet that is passed from labella to labella, this type of trophallaxis is relatively uncommon in fruit flies (Aluja et al. 1993; Headrick and Goeden 1994; see Sivinski et al., Chapter 28). If such viscous offerings are nutritionally more substantial than those of platystomatids, they could represent a greater reward to females which, given the validity of Alcock and Pyle's (1979) hypothesis, might ultimately lead to the relative simplification of courtships. There was no evidence in this case that the material influenced female longevity or fecundity, although the possible effects on the success of the offspring were not examined. However, trophallaxis increased female longevity and was associated with greater fecundity in the tephritid *A. striata* (Sánchez-Martínez 1998; Aluja et al., Chapter 15).

Other "wet" courtship activities in the Platystomatidae include males regurgitating on the thorax of mounted females and then imbibing their own fluids (*Euprosopia tenuicornis* Macquart), and the peculiar female production of an anal fluid which is ingested by the male prior to mounting (*E. anostigma* McAlpine) (McAlpine 1973b). In *E. anostigma* the males have a remarkable projection on the hind trochanter that is used to comb the "soft, downy pubescence" on tergite 3 of the female abdomen. The appearance of females that have been so combed "suggests that some liquid secretion contacts it during the combing process." Fluids that are not provided by males or not imbibed by females might be some sort of chemical or tactile signal, and may serve a different function than those that are consumed.

A number of platystomatid males have modification of their heads that are used in agonistic interactions with sexual rivals. These vary from broadening of the face into a surface used to push against the face of another male (McAlpine 1975), to extremely well-developed stalk eyes (e.g., *Achias* spp.; McAlpine 1979; Figure 2.4). The latter serve a similar role to the antlers projecting from the cheeks of tephritids in the genus *Phytalmia* (Wilkinson and Dodson 1997; Dodson, Chapter 8), a genus whose unusual breeding habits, oviposition into fallen timber rather than living plant tissue, is more typical of platystomatids. The parallel development of antlers and stalk eyes could be ascribed to similar opportunities to defend rare, discrete resources.

### 2.2.4 Pyrgotidae and Tachiniscidae

The pyrgotids are endoparasitoids of adult scarab beetles (or rarely Hymenoptera). They are further separated from the Tephritidae on the basis of their generally crepuscular or nocturnal habits and specialized ovipositors, both apparent adaptations for finding and then successfully penetrating the defenses of their largely nocturnal and armored hosts (exceptions include the diurnal *Peltodasia flaviseta* (Aldrich); Clausen et al. 1933, as *Adapsilia flaviseta*). Typically, female flies wait in the vicinity of feeding beetles for the host to take flight. The fly then pounces on the scarab, pierces the soft dorsal surface of the abdomen, and falls with the host to the ground where it lays one to several eggs (Forbes 1907). *Maenomenus ensifer* Bezzi oviposits in the anus of feeding hosts (Paramonov 1958). There are conflicting claims as to the effects of pyrgotid parasitism on populations of turf-feeding Scarabaeidae (e.g., Crocker et al. 1996), but the flies are, at least on some occasions, locally abundant (e.g., Clausen et al. 1933). Male beetles are twice as likely as females to be parasitized by *P. flaviseta* (Clausen et al. 1933), a feature that might minimize their impact on pest populations.

There are few accounts of pyrgotid behavior, other than those involved with egg-laying. In general flies are captured at lights (Lago 1981), and actively hunting females are typically near or on the host plants of their prey (Clausen et al. 1933). As noted earlier, the presence of boldly patterned wings in a largely nocturnal family (Steyskal 1978) might not support the notion that the markings serve as intraspecific signals.

The Tachiniscidae are even more poorly known than the Pyrgotidae. Korneyev (Chapters 1 and 4) includes them in the Tephritidae, as the subfamily Tachiniscinae (this came to my attention only as this book was going to press). This is a tiny group. Only three genera with only four described species had been included (Ferrar 1987), although Korneyev (Chapter 4) includes five more genera. McAlpine (1989) also raised some doubt as to its familial status, but suggested that in the future it may be reclassified as a supergeneric taxon of the Pyrgotidae. Tachiniscids occur in South America, eastern Asia, and Africa. *Tachinisca* and *Bibundia* are burly, hairy flies that resemble tachinids, while *Tachiniscidia* mimic vespid wasps. The only host record is for a *Bibundia* sp. which emerged from the pupae of two species of African Saturniidae (Roberts 1969, as cited in Ferrar 1987). Apparently, nothing is known of the sexual behaviors of any tachiniscid genera.

### 2.2.5 Richardiidae, Pallopteridae, and Piophilidae

The third monophyletic group within the Tephritoidea consists of the Richardiidae, Pallopteridae, and Piophilidae (Figure 2.1; but also see Korneyev, Chapter 1, regarding the Richardiidae). The Richardiidae is considered the most generalized family within the subgroup, and is distinguished by heavily spinose hind femora, strong bristles on the second abdominal tergite, reductions in features of the male genitalia (the phallapodeme, gonopods, and parameres), and the presence of only two spermathecae (McAlpine 1989). The few breeding records suggest a larval diet of rotting vegetable matter, rotting fruit, decaying trunks, and flowers (Ferrar 1987). The ~30 genera are essentially restricted to the neotropics, with only eight species extending as far north as the United States.

Recently, interest has focused on the possibility of male genitalia having a communicative role; that is, the various projections, flanges, and enlargements of the aedeagus or phallus in certain insects may have evolved as tactile displays, perhaps through "Fisherian" runaway sexual selection (e.g., Eberhard 1996). A condition in females that would promote this kind of male display is the postcopulatory ability to choose which ejaculates will be used to fertilize eggs. Thus, a genitalic display provided during copulation would still be useful to a choosy female sampling various potential sires for her offspring. One means of doing this is to store sperm from different males in different spermathecae and preferentially use those from certain locations. For example, in the yellow dung fly, *Scathophaga stercoraria* (L.), sperm from larger males are stored separately and used more often to fertilize eggs (Ward 1993). Reductions in the Richardiidae of both phallic structures and the number of spermathecae (from three to two) may reflect less postcopulatory female choice and diminished male genitalic displays. Comparisons of sperm usage, genital complexity, and courtship in richardiids with some of the better-endowed tephritids may be particularly revealing. Fruit flies have a wide range of reproductive morphologies; for example, females of various *Rhagoletis* species have two or three spermathecae and *Oedicarena* species have four (Foote et al. 1993).

The Richardiidae are yet another tephritoid family containing stalk-eyed species (Wilkinson and Dodson 1997), and again offer opportunities to correlate breeding habits with the presence of these extraordinary male adaptations. A considerable sexual dimorphism in size in *Omomyia hirsuta* Coquillett is described by Barber (1908): "To these moist spots [on wood] came flies in large numbers, apparently two species, one of which was large, 6 mm in length, yellowish, covered with long yellowish hairs. The other was much smaller, black, shiny, with a distinct dark spot near the tip of the wing. The large wooly one appeared very aggressive, alighting often upon the black, shiny one and with his wooly legs outspread so as to hide his captive completely he would run about on the moist wood as if he were a single specimen. But at last I saw a very small wooly specimen alight upon a large shiny black one and in this case saw copulation take place. Then the true state of affairs dawned upon me." Other richardiids (e.g., *Beebeomyia* sp.) also mate on or near host plants (Seifert and Seifert 1976).

The Pallopteridae are a small family of 52 species (+ 3 fossil species from Baltic amber). They occur mainly in the Palaearctic and Nearctic Regions with a scattering of species as far disjunct as the Falkland Islands, New Zealand, and Israel (Ferrar 1987). Unlike the Richardiidae, the parameres of male pallopterids are large and strongly sclerotized. In general, they infest shoots and stems of herbaceous plants, or live under bark, often in association with wood-boring Coleoptera. As with the Lonchaeidae, there is some question about the actual diet of the larvae living under bark. Some are clearly predaceous (Morge 1958, as cited in Ferrar 1987), but others have been collected in tree trunks with no evidence of beetles being present (Morge 1956, as cited in Ferrar 1987). Larvae of *Palloptera umbellatarum* (Fabricius) occupy the galls produced by tephritids in composite flowerheads, where they may be secondary consumers of the gall tissue (Niblett 1946). If this results in similar distributions of oviposition opportunities for both the tephritid and the pallopterid females, there may be a chance to examine further the relationship between resource distributions and mating system evolution (see previous section on Lonchaeidae). McAlpine (1989) notes that in the Pallopteridae the "habit of vibrating wings during sexual excitement [is] strongly developed," a characteristic they share with many tephritids (Sivinski et al. 1984; Aluja et al., Chapter 15; Sivinski et al., Chapter 28).

The Piophilidae, consisting of 71 species in 23 genera, is another relatively small family, and is the sister group of the Pallopteridae (Figure 2.1). The subfamily Neottiophilinae is composed of three species, one of which, *Neottiophilum praeustum* (Meigen), lives in birds' nests where larvae feed ectoparasitically on the blood of nestlings (Hutson 1978). Piophilinae develop in a variety of decaying materials, ranging from rotting leaves to the bones of whales to human corpses (including pupae found in a 2000-year-old Egyptian mummy; Cockburn et al. 1975). Because of their feeding habits and the ability of the larvae to flip themselves considerable distances, the maggots are sometimes called "bone-skippers." *Piophila casei* (L.), the cosmopolitan "cheese-skipper," has long been a major problem in human foods. Larvae feed deep inside such things as hams and cheeses with little outward sign of infestation. Not surprisingly, *P. casei* is a leading cause of human myiasis, and is the most common insect found in the human intestine (James 1947).

Species of *Protopiophila* occasionally occur together on a single piece of cervid carrion, where males of *P. litigata* Bonduriansky and *P. latipes* (Meigen) will unsuccessfully attempt heterospecific copulations (Bonduriansky 1995). However, male distributions and sexual behaviors in the two species are generally quite different. *Protopiophila litigata* larvae develop in the porous matrix of discarded cervid antlers. Some males defend aggregated territories on the upward surface of moose antlers, and remain coupled to females who later move toward oviposition sites following insemination. Other males position themselves near oviposition sites where they attempt to dislodge coupled males. Males of *P. latipes* fight with one another in and on corpses where females come to mate and lay eggs. They do not engage in mate guarding. Comparison of resource distributions in the two species may reveal the different selective pressures that have resulted in their behavioral divergence. (Is it important that discarded antlers are apt to be more abundant, persistent, smaller, and seasonally occurring than corpses? See similar patterns in Tephritidae of distinct mating behaviors on different parts of host plants in Headrick and Goeden, Chapter 25.)

Males of *Centrophlebomyia furcata* (Fabricius) are considerably larger than females, an atypical condition in the Tephritoidea and one shared with the previously mentioned richardiid *Omomyia hirsuta* (see Dodson, Chapter 8). Neither species engages in elaborate courtships, but both are described as "covering" their mates, presumably to prevent access to other males (Freidberg 1981). These mating systems may have evolved in response to the predictable nature of females at rare discrete resources (pockets of decay in wood and carcasses). The sexual dimorphism of *Amphipogon* spp. suggests interesting signaling/agonistic/courtship behaviors. Males have a long "beard" of

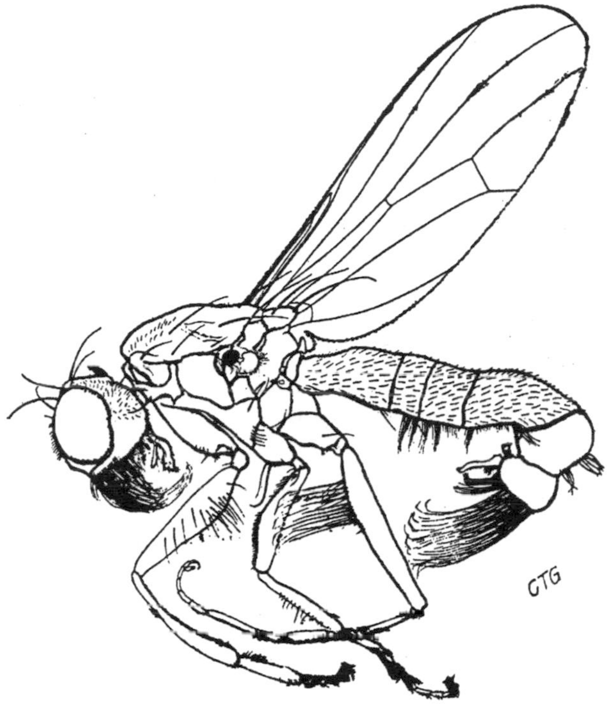


**FIGURE 2.5** The males of the "bearded flies," *Amphipogon* spp. (Piophilidae), are considerably more hirsute than the females. (From Cole, F., *The Flies of Western North America,* University of California Press, Berkeley, 1969. With permission.)

curved and crinkly bristles, and swollen femora "adorned with clumps of setula, bristles and hairs" (McAlpine 1977; Figure 2.5). The hind leg is particularly elaborate, with additional spines, distortions, and setae. Larvae develop in fungi, but the role of the peculiar male pelage is unknown.

## 2.3 CONCLUSIONS AND FUTURE RESEARCH NEEDS

### 2.3.1 Sexual Selection in the Tephritoidea

There are 64 families of acalyptrate Diptera; in eight of these, some species (particularly the males of some species), bear eye-stalks and in five families males have "antlers" (Wilkinson and Dodson 1997). Four of the eight families with eye-stalks and three of the five with antlers are tephritoids, which suggests a particularly rich history of intrasexual selection within the superfamily. Ornaments used in courtships are likewise abundant in the Tephritoidea (Sivinski 1997). What might account for this concentration of elaborate (and presumably expensive) communication devices? Some characteristics that might bear examination are:

1. The relatively large size of many tephritoids relative to many other acalyptrates. Larger size might mean an increased ability for males to control access by females and sexual rivals to a resource. All other things being equal, large size reduces the numbers of offspring that can develop in a particular resource and so limits the number of opportunities for females to oviposit into it. This, in turn, would tend to make the resource more valuable to foraging females, and this could have consequences for the practicality of female mate choice.

2. The appearance of behavioral richness is due to chance. Perhaps any particular adaptation is more likely to occur in larger families, such as the Tephritidae with 4000+ species (although the Richardiidae with only ~30 genera have males with both stalk eyes or antlers).

3. Fisherian runaway sexual selection may be more likely to occur where females have the time to sample male displays and the capacity to remember the range of variation in displays. Long life and good memory allows females to choose the most extreme examples within the pool of potential mates. Alexander et al. (1997) have argued that these characteristics are rare in insects and that complex courtships and ornaments would more likely be due to "arms races" between choosy females and males advertising actual qualities than to Fisherian selection. The evidence supporting this argument is debatable (Eberhard 1997); however, it may be that long-lived insects with good memories are particularly "preadapted" to bouts of runaway sexual selection, and that extravagant signals resulting from Fisherian selection could accumulate in such taxa. Certain tephritids can be extremely long lived for flies; for example, *Anastrepha suspensa* (Loew) lives up to 1 year in the laboratory (Sivinski 1994), but at present little is known about the relative quality of their memories.

### 2.3.2 Opportunities for Comparative Studies

The similarities and differences among Tephritoidea provide an extensive pool of subjects for use in comparative studies that center on tephritid behaviors. Some of the problems in the evolution of fruit fly behavior that might be addressed by a consideration of related families are reviewed below.

1. Wing patterns and wing movements are common in the tephritids, but except for some cases of spider or wasp mimicry, their purpose, particularly their sexual function (if any), is obscure. The absence of wing markings in the Lonchaeidae, the only family that regularly forms aerial mating swarms, the presence of wing markings in the largely nocturnal Pyrgotidae, and the prevalence of wing vibrations in sexually excited Pallopteridae may offer opportunities to form hypotheses about the evolution of this striking phenomenon.

2. There has been considerable interest in the evolution of mating aggregations in the Tephritidae, and how males and females perform within leks. Swarming by the Lonchaeidae offers an alternative aggregated mating system to compare and contrast with fruit fly leks. Investigations into lonchaeid larval resource distributions, which in some instances seem to parallel that of lekking tephritids, might offer insights into why males aggregate, and once aggregated, why some swarm and others lek.

3. The Ulidiidae have many complex intersexual and intrasexual signals reminiscent of the Tephritidae. However, their breeding habits are very different, as they largely oviposit in decaying plant material rather than live plants. Resource distribution over space and time has been implicated in the evolution of complexity in courtships and agonistic encounters. These different resource sets would permit comparative tests that are more independent of the nature of the resource, and allow more confidence to be placed on the distributional characteristics of the resource.

4. While the Ulidiidae, and other tephritoid families, share some courtship behaviors with Tephritidae, there are also differences, such as the male mouthing of the extended aculeus. Do these differences reflect different sensory systems, different types of information passed by advertising males to choosy females, or are they evidence of Fisherian sexual selection leading to an arbitrary distribution of displays within and between taxa?

5. Platystomatid courtships often include trophallaxis, but the materials passed from males to females seems to be of a different nature than those transferred by many tephritids; they are described as clearer and less viscous than the froths commonly produced by tephritids. In addition, the courtships of platystomatids that include trophallaxis may be more complex than those performed by males of some (but not all) tephritids with oral secretions. This correlation may represent a greater investment by these male tephritids in their fluids and foams.

6. Male genitalia may have a role in communication as well as being organs of insemination; that is, they may be tactile displays that are judged by females who may make decisions about the paternity of their offspring after sampling multiple sexual partners. One means by which females

could exercise control of different ejaculates would be to sort them into different spermathecae. In the Richardiidae there is both a simplification of male genitalia and a reduction in the number of spermathecae. Does this loss of complexity reflect a lesser role for male genitalic displays and female sperm control? If so, can simplification in the Richardiidae help illuminate the evolution of the broad range of phallic complexities and spermathecal numbers in the Tephritidae?

## ACKNOWLEDGMENTS

James Lloyd, Denise Johanowicz, Martín Aluja, and Ken Kaneshiro suggested many useful improvements to the manuscript. Gina Posey and Valerie Malcolm were invaluable in obtaining references and formatting the chapter. Partial funding to attend the conference in Xalapa was obtained from the following donors: Campaña Nacional contra las Moscas de la Fruta (Mexico), International Organization for Biological Control of Noxious Animals and Plants (IOBC), and Instituto de Ecología, A.C. (Mexico).

## REFERENCES

Alcock, J. 1987. Leks and hilltopping in insects. *J. Nat. Hist.* 21: 319–328.

Alcock, J. and D.W. Pyle. 1979. The complex courtship of *Physiphora demandata* (F.) (Diptera: Ulidiidae). *Z. Tierpsychol.* 49: 352–362.

Alexander, R.D., D.C. Marshall, and J.R. Cooley. 1997. Evolutionary perspectives on insect mating. In *The Evolution of Mating Systems in Insects and Arachnids* (J.C. Choe and B.J. Crespi, eds.), pp. 4–31. Cambridge University Press, Cambridge. 387 pp.

Allen, E.J. and B. Foote. 1967. Biology and immature stages of three species of Otitidae (Diptera) which have saprophagous larvae. *Ann. Entomol. Soc. Am.* 60: 826–836.

Allen, E.J. and B. Foote. 1975. Biology and immature stages of *Tritoxa incurva* (Diptera: Otitidae). *Proc. Entomol. Soc. Wash.* 77: 246–257.

Aluja, M. 1993. Unusual calling behavior of *Anastrepha robusta* (Diptera: Tephritidae) flies in nature. *Fla. Entomol.* 76: 391–395.

Aluja, M., I. Jácome, A. Birke, N. Lozado, and G. Quintero. 1993. Basic patterns of behavior in wild *Anastrepha striata* (Diptera: Tephritidae) flies under field-cage conditions. *Ann. Entomol. Soc. Am.* 86: 776–793.

Barber, H.S. 1908. Note on *Omomyia hirsuta* Coquillett (Diptera, Phycodromidae). *Proc. Entomol. Soc. Wash.* 9: 28–29.

Bonduriansky, R. 1995. A new Nearctic species of *Protopiophila* Duda (Diptera: Piophilidae) with notes on its behaviour and comparison with *P. latipes* (Meigen). *Can. Entomol.* 127: 859–863.

Bradbury, J.W. 1979. The evolution of leks. In *Natural Selection and Social Behavior* (R.D. Alexander and D.W. Tinkle, eds.), pp. 138–169. Blackwell Scientific Publications, Oxford. 532 pp.

Burk, T. 1981. Signaling and sex in acalyptrate flies. *Fla. Entomol.* 64: 30–43.

Clausen, C.P., H. Jaynes, and T.R. Gardner. 1933. Further investigations of the parasites of *Popillia japonica* in the Far East. *U.S. Dep. Agric. Tech. Bull.* No. 366.

Cockburn, A., R. Barraco, T. Reyman, and W. Peck. 1975. Autopsy of an Egyptian mummy. *Science* 187: 1155–1160.

Cole, F. 1969. *The Flies of Western North America.* University of California Press, Berkeley. 693 pp.

Crocker, R.L., L. Rodriguez-del-Bosque, W.T. Nailon, Jr., and X. Wei. 1996. Flight periods in Texas of three parasites (Diptera: Pyrgotidae) of adult *Phyllophaga* spp. (Coleoptera: Scarabaeidae), and egg production by *Pyrgota undata. Southwest. Entomol.* 21: 317–324.

Eberhard, W.G. 1996. *Female Control: Sexual Selection by Cryptic Female Choice.* Princeton University Press, Princeton. 501 pp.

Eberhard, W.G. 1997. Cryptic female choice. In *The Evolution of Mating Systems in Insects and Arachnids* (J.C. Choe and B.J. Crespi, eds.), pp. 32–57. Cambridge University Press, Cambridge. 387 pp.

Emlen, S.T. and L.W. Oring. 1977. Ecology, sexual selection, and the evolution of mating systems. *Science* 197: 215–223.

Ferrar, P. 1987. *A Guide to the Breeding Habits of Immature Stages of Diptera Cyclorrapha (Part 1: text).* E.J. Brill/Scandinavian Science Press, Leiden. 478 pp.

Foote, B.A. 1967. Biology and immature stages of fruit flies: the genus *Icterica. Ann. Entomol. Soc. Am.* 60: 1295–1305.

Foote, R.H., F.L. Blanc, and A.L. Norrbom. 1993. *Handbook of the Fruit Flies (Diptera: Tephritidae) of America North of Mexico.* Comstock Publishing Associates, Ithaca. 571 pp.

Forbes, S.A. 1907. On the life history, habits, and economic relations of the white-grubs and May-beetles. *Univ. Illinois Agric. Exp. Sta. Bull.* No. 116.

Freidberg, A. 1981. Taxonomy, natural history and immature stages of the bone-skipper, *Centrophlebomyia furcata* (Fabricius) (Diptera: Piophilidae, Thyreophorina). *Entomol. Scand.* 12: 320–326.

Freidberg, A. 1982. Courtship and post-mating behaviour of the fleabane gall fly, *Spathulina tristis* (Diptera: Tephritidae). *Entomol. Gene.* 7: 273–285.

Greathead, D.J. 1963. A review of the insect enemies of Acridoidea (Orthoptera). *Trans. R. Entomol. Soc. London* 114: 437–517.

Greene, E., L. Orsak, and D. Whitman. 1987. A tephritid fly mimics the territorial displays of its jumping spider predators. *Science* 366: 310–312.

Headrick, D.H. and R.D. Goeden. 1994. Reproductive behavior of California fruit flies and the classification and evolution of Tephritidae (Diptera) mating systems. *Stud. Dipterol.* 1: 194–252.

Hutson, A.M. 1978. Associations with other animals and micro-organisms. In *A Dipterist*'s *Handbook* (A. Stubbs and P. Chandler, eds.), pp. 143–151. The Amateur Entomologists Society, Hanworth. 255 pp.

James, M.T. 1947. The flies that cause myiasis in man. *Misc. Publ. U.S. Dep. Agric.* 631: 1–175.

Katsoyannos, B.I. 1983. Field observations on the biology and behavior of the black fig fly *Silba adipata* McAlpine (Diptera, Lonchaeidae), and trapping experiments. *Z. Angew. Entomol.* 95: 471–476.

Koethe, J.K. and J.W. Van Duyn. 1988. Influence of soil surface conditions and host plant on soybean nodule fly *Rivellia quadrifasciata* (Macquart) (Diptera: Platystomatidae) oviposition. *J. Entomol. Sci.* 23: 251–256.

Lago, P.K. 1981. Records of Pyrgotidae from Mississippi, with additional notes on the distribution of *Sphecomyiella valida* (Harris) and *Pyrgota undata* Weidemann (Diptera). *Entomol. News* 92: 115–118.

Malavasi, A. and J.S. Morgante. 1980. Biologia de "moscas-das-frutas" (Diptera, Tephritidae). II: Indices de infestação em differentes hospedeiros e localidades. *Rev. Bras. Biol.* 40: 17–24.

McAlpine, D.K. 1973a. The Australian Platystomatidae (Diptera, Schizophora) with a revision of five genera. *Mem. Aust. Mus.* 15: 1–256.

McAlpine, D.K. 1973b. Observations on sexual behavior in some Australian Platystomatidae (Diptera, Schizophora). *Rec. Aust. Mus.* 29: 1–10.

McAlpine, D.K. 1975. Combat between males of *Pogonortalis doclea* (Diptera, Platystomatidae) and its relation to structural modification. *Aust. Entomol. Mag.* 2: 104–107.

McAlpine, D.K. 1979. Agonistic behavior in *Achias australis* (Diptera, Platystomatidae) and the significance of eye stalks. In *Sexual Selection and Reproduction (*M. Blum and N. Blum, eds.), pp. 221–230. Academic Press, New York. 463 pp.

McAlpine, J.F. 1977. A revised classification of the Piophilidae, including 'Neottiophilidae' and 'Thyreophoridae' (Diptera: Schizophora). *Mem. Entomol. Soc. Can.* 103, 66 p.

McAlpine, J.F. 1989. Phylogeny and classification of the Muscomorpha. In *Manual of Nearctic Diptera,* Vol. 3. (J.F. McAlpine, ed.), pp. 1397–1518. Monograph of the Biosystematics Research Centre, No. 32. Agriculture Canada, Ottawa. 1581 pp.

McAlpine, J.F. and D. Munroe. 1968. Swarming of lonchaeid flies and other insects, with descriptions of four new species of Lonchaeidae (Diptera). *Can. Entomol.* 100: 1154–1178.

McMichael, B., B. Foote, and B. Bowker. 1990. Biology of *Rivellia melliginis* (Diptera: Platystomatidae): a consumer of the nitrogen-fixing root nodules of black locust (Leguminosae). *Ann. Entomol. Soc. Am.* 83: 967–974.

Michaelmore, A.P.G. 1928. A mating habit of *Platystoma seminationis* F. (Diptera, Ortalidae). *Entomologist* 61: 241–243.

Moffitt, H. and F. Yaruss. 1961. *Dasiops alveofrons*, a new pest of apricots in California. *J. Econ. Entomol.* 54: 504–505.

Niblett, M. 1946. Diptera bred from flower-heads of Compositae. *Entomol. Rec. J. Var.* 58: 121–123.

Norrbom, A.L. and J.F. McAlpine. 1997. A revision of the neotropical species of *Dasiops* Rondani (Diptera: Lonchaeidae) attacking *Passiflora* (Passifloraceae). *Mem. Entomol. Soc. Wash.* 18: 189–211.

Novak, J.A. and B. Foote. 1975. Biology and immature stages of fruit flies: the genus *Stenopa*. *J. Kans. Entomol. Soc.* 48: 42–52.

Oldroyd, H. 1964. *The Natural History of Flies.* W.W. Norton, New York. 324 pp.

Paramonov, S.J. 1958. A review of the Australian Pyrgotidae (Diptera). *Aust. J. Zool.* 6: 89–137.

Parker, G.A. 1978. Evolution of competitive mate searching. *Annu. Rev. Entomol.* 23: 173–196.

Piersol, W. 1907. The curious mating habit of *Rivellia boscii*. *Am. Nat.* 41: 465–467.

Pritchard, G. 1967. Laboratory observations on the mating behavior of the island fruit fly *Rioxa pornia* (Diptera: Tephritidae). *J. Aust. Entomol. Soc.* 6: 127–132.

Prokopy, R. 1980. Mating behavior of frugivorous Tephritidae in nature. In *Proceedings of a Symposium on Fruit Fly Problems*, XVI Intern. Cong. Entomol. (J. Koyama, ed.), pp. 37–46. Kyoto.

Richards, O.W. 1924. Sexual selections and allied problems in the insects. *Biol. Rev.* 2: 298–360.

Sánchez-Martínez, A.S. 1998. Efecto de la Dieta y Tamaño de Machos de *Anastrepha ludens* y *A. striata* Schiner (Diptera: Tephritidae) en su Competitividad Sexual. B.Sc. thesis, Universidad Veracruzana, Xalapa, Veracruz, Mexico.

Seal, D. and R. Jansson. 1989. Biology and management of corn-silk fly, *Euxesta stigmatias* Loew (Diptera: Otitidae), on sweet corn in southern Florida. *Proc. Fla. State Hortic. Soc.* 102: 370–373.

Seal, D., R. Jansson, and K. Bondari. 1995. Bionomics of *Euxesta stigmatias* (Diptera: Otitidae) on sweet corn. *Environ. Entomol.* 24: 917–922.

Seifert, R. and F. Seifert. 1976. Natural history of insects living in inflorescences of two species of *Heliconia*. *J. N.Y. Entomol. Soc.* 84: 233–242.

Sivinski, J. 1994. Longevity in the Caribbean fruit fly; effects of sex, strain and sexual experience. *Fla. Entomol.* 76: 635–644.

Sivinski, J. 1997. Ornaments in the Diptera. *Fla. Entomol.* 80: 144–164.

Sivinski, J. and E. Petersson. 1997. Mate choice and species isolation in swarming insects. In *The Evolution of Mating Systems in Insects and Arachnids* (J.C. Choe and B.J. Crespi, eds.), pp. 294–309. Cambridge University Press, Cambridge. 387 pp.

Sivinski, J. and J.C. Webb. 1986. Changes in a Caribbean fruit fly acoustic signal with social situation (Diptera: Tephritidae). *Ann. Entomol. Soc. Am.* 79: 146–149.

Sivinski, J., T. Burk, and J.C. Webb. 1984. Acoustic courtship signals in the caribfly, *Anastrepha suspensa*. *Anim. Behav.* 32: 1011–1016.

Souza, H., M. Cytrynowicz, J. Morgante, and O. Pava. 1983. Occurrence of *Anastrepha fracterculus* (Weid.), *Ceratitis capitata* (Weid.) (Diptera: Tephritidae) and *Silba* spp. (Diptera: Lonchaeidae) eggs in oviposition bores on three host fruits. *Rev. Bras. Entomol.* 27: 191–195.

Steyskal, G.C. 1978. Synopsis of the North American Pyrgotidae. *Proc. Entomol. Soc. Wash.* 80: 149–155.

Steyskal, G.C. 1979. Biological, anatomical, and distribution notes on the genus *Callopistromyia* (Diptera: Otitidae). *Proc. Entomol. Soc. Wash.* 81: 450–455.

Stoltzfus, W. and B. Foote. 1965. The use of froth masses in courtship of *Eutreta* (Diptera: Tephritidae). *Proc. Entomol. Soc. Wash.* 67: 263–264.

Sullivan, R. 1981. Insect swarming and mating. *Fla. Entomol.* 64: 44–65.

Thornhill, R. 1984. Scientific methodology in entomology. *Fla. Entomol.* 67: 74–96.

Ward, P.I. 1993. Females influence sperm storage and use in the yellow dung fly *Scathophaga stercoraria* (L.). *Behav. Ecol. Sociobiol.* 32: 313–319.

Wilkinson, G. and G. Dodson. 1997. Function and evolution of antlers and eye-stalks in flies. In *The Evolution of Mating Systems in Insects and Arachnids* (J.C. Choe and B.J. Crespi, eds.), pp. 310–328. Cambridge University Press, Cambridge. 387 pp.

# 3 Behavior of Tephritid Flies: A Historical Perspective

*Francisco Díaz-Fleischer and Martín Aluja*

## CONTENTS



## 3.1 INTRODUCTION

Fruit flies (Diptera: Tephritidae) represent, in our opinion, one of the best study systems any researcher can find. These insects are well suited for interspecific comparisons, field and laboratory studies, and studies under seminatural conditions. Since adults of many species are relatively easy to obtain, there is no difficulty in quantifying their behavior and performing long-term studies on marked individuals. In addition, some species in the subfamilies Dacinae and Trypetinae are notorious pests. This combination of features has motivated entomologists worldwide to study this group for more than a century. Species such as *Rhagoletis pomonella* (Walsh), *R. cerasi* (L.), *Ceratitis capitata* (Wiedemann), *Bactrocera dorsalis* (Hendel), *B. oleae* (Rossi), *B. cucurbitae* (Coquillett), *B. tryoni* (Froggatt), *Anastrepha ludens* (Loew), *A. obliqua* (Macquart), and *A. suspensa* (Loew) have been the subject of important studies on basic biology, ecology, evolution of animal–plant interactions, sexual selection, and speciation (work reviewed by Baker et al. 1944; Christenson and Foote 1960; Bateman 1972; Bush 1974; Bateman et al. 1976; Boller and Prokopy 1976; Boller and Chambers 1977b; Burk 1981; Prokopy 1977; 1980; 1982; Zwölfer 1983; 1988; Freidberg 1984a; Fletcher 1987; Drew 1987; Aluja 1994; Headrick and Goeden 1998; also see reviews in this volume by Aluja et al., Chapter 15; Drew and Romig, Chapter 21; Eberhard, Chapter 18; Headrick and Goeden, Chapter 25; Prokopy and Papaj, Chapter 10; and Yuval and Hendrichs, Chapter 17).

Studies on fruit fly behavior can be divided into those that describe patterns (usually under natural conditions) and those that experimentally seek to decipher the mechanisms regulating a particular behavior (usually under laboratory or seminatural conditions). Although valuable descriptive work

0-8493-1275-2/00/$0.00+$.50

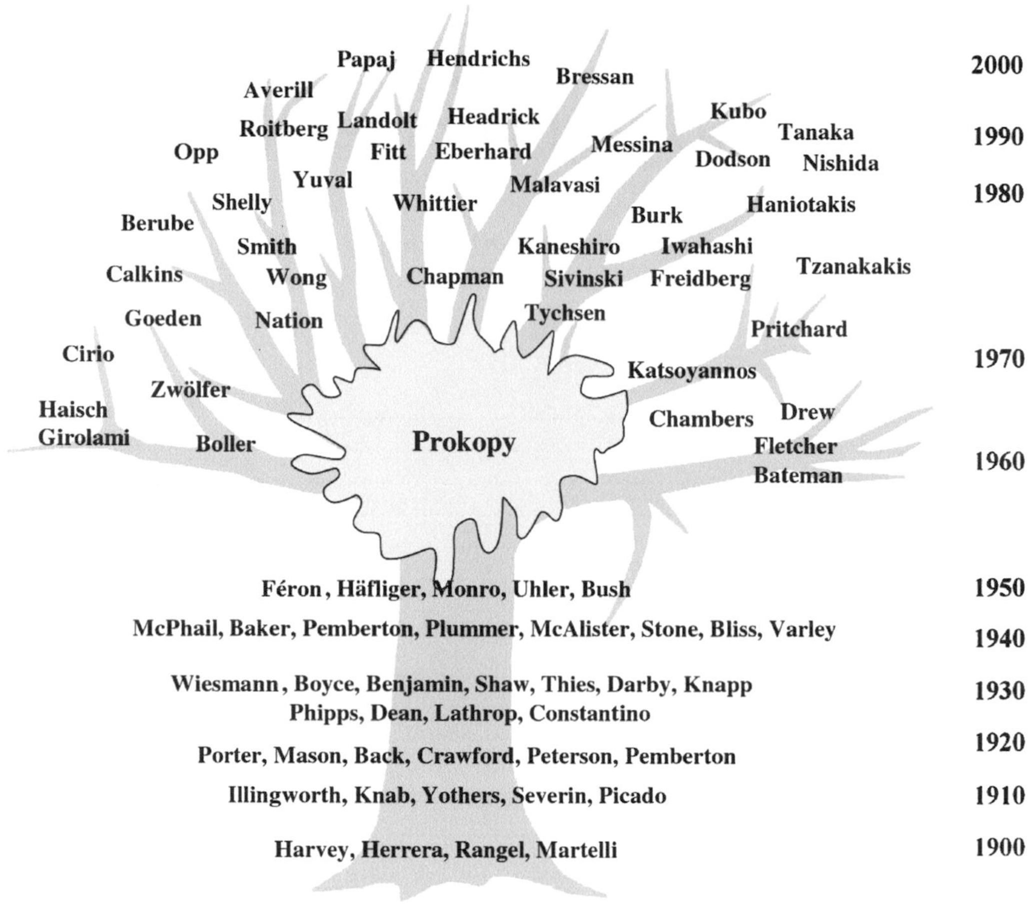


**FIGURE 3.1** Historical tree of behavioral studies on fruit flies, indicating the division between the first descriptive period and the experimental period mainly triggered by Ronald J. Prokopy and his students and collaborators.

is still carried out, the approach to the study of tephritid behavior has recently changed. Research efforts are currently aimed at understanding behavior in a more integrated fashion, taking into account environmental factors as well as the physiological state and experience of individuals. The pioneering work of R. J. Prokopy (Figure 3.1) has undoubtedly had the strongest influence on the direction of fruit fly experimental research. Over the past 28 years, this researcher and his collaborators have studied and deciphered many of the mechanisms regulating fruit fly mating (e.g., Prokopy and Bush 1973a; Prokopy and Hendrichs 1979) and oviposition behavior (e.g., Prokopy and Boller 1971; Prokopy 1972; Prokopy and Bush 1973b), visual and olfactory orientation (e.g., Prokopy 1968), feeding (e.g., Prokopy 1976) and learning (Prokopy et al. 1986). We also feel that part of the current boom in fruit fly behavioral research is intimately related to the seminal studies on sympatric speciation and behavioral reinforcement carried out by another pioneer in fruit fly research, G. L. Bush (1969; 1974).

In this chapter we review, from a historical perspective, the work that has been done on fruit fly behavior during the past 100 years. We place special emphasis on the pioneering work by many forgotten scientists who have not received the credit they deserve for having unraveled many of the secrets fruit flies kept for themselves during millennia. We begin by citing a series of anecdotal but highly intuitive and revealing studies from the late 19th century and conclude with a review of

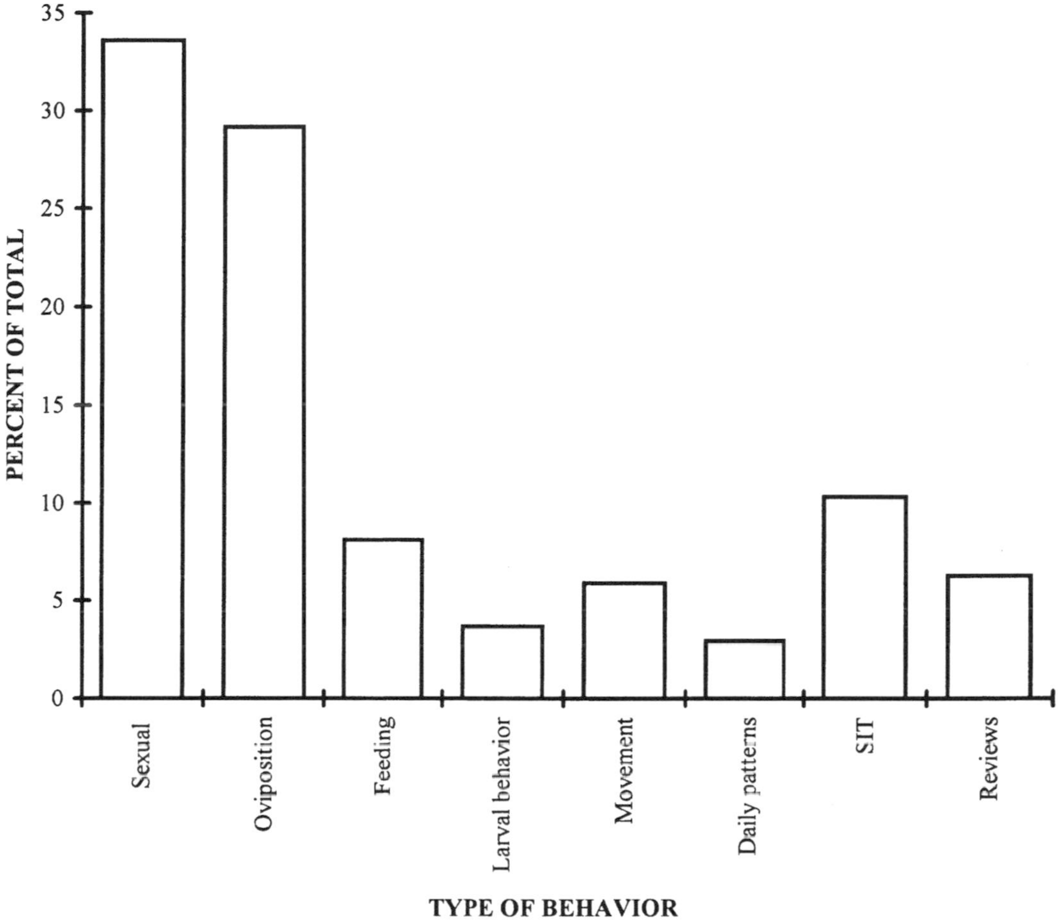


**FIGURE 3.2** Distribution by topic (expressed as percent of total) of behavioral studies conducted on fruit flies.

a wide variety of topics related to current research on tephritids. We also present a brief overview of those research topics and geographic locales that have received the most attention in the literature, in order to highlight those that have a potential for future research (Figures 3.2 and 3.3).

## 3.2 HISTORY OF TEPHRITID RESEARCH

### 3.2.1 Early Period

We have, somewhat arbitrarily, defined the early period as extending from the late 1800s to the end of the 1960s. This section includes, for the most part, anecdotal observations and some very insightful and detailed descriptions of fruit fly behavioral patterns. The economic impact of tephritids seems to have triggered research on behavior, since most of the pioneering work was done on species that are fruit-crop pests. The Mediterranean fruit fly, *C. capitata*, was the subject of many studies, some of which can be considered jewels of the fruit fly literature. One is the paper by the Italian entomologist G. Martelli entitled "Alcune note intorno ai costumi ed danni della mosca delle arance (*C. capitata*)" [Notes on the habits and damage of the orange fly (*C. capitata*)] (Martelli 1909). This paper appears to be the first one describing fruit fly host marking behavior and oviposition on previously occupied hosts.

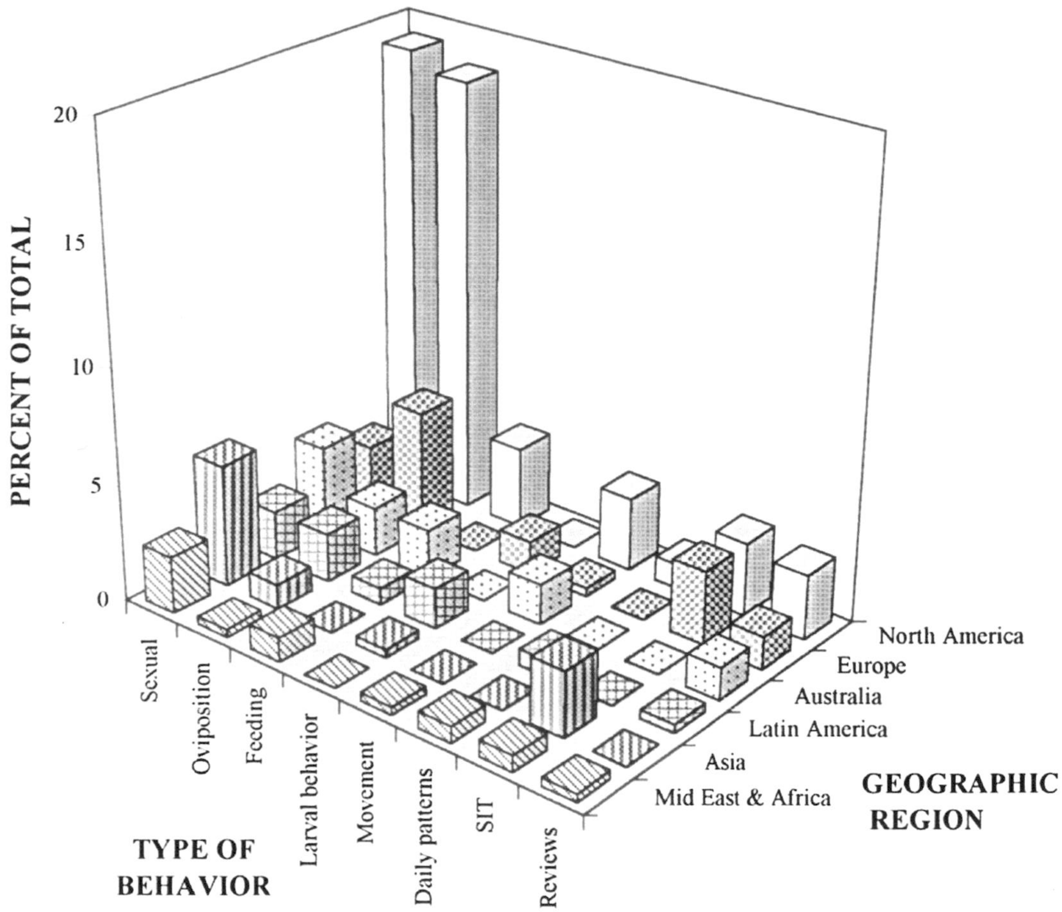


**FIGURE 3.3** Distribution of behavioral studies on fruit flies by geographic region, indicating distribution by topic. Note the great emphasis placed on the sexual and egg-laying aspects of behavior in studies carried out in North America.

E. A. Back, C. E. Pemberton, G. Constantino, and M. Féron should be recognized as the pioneers in the areas of *C. capitata* mating and oviposition behavior (e.g., courtship, calling, mating success, clutch size), responses to olfactory stimuli, and basic studies on demography (Back and Pemberton 1915; Féron 1957; 1958; 1959; 1960a; 1962). Constantino (1930) published, among morphological descriptions of adults and larvae, very accurate descriptions of sexual development, mating, oviposition, clutch size, and the use of previous oviposition holes. He also tested host odors as attractants for traps. Silvestri (1914) also described the oviposition behavior of *C. capitata* and the use of fruit wounds as oviposition substrates. Using marked flies and traps, Severin and Hartung (1912) carried out interesting work on Mediterranean fruit fly flight capacity and movement ability. The first studies on the relationships between natural food quality and reproduction were performed by Hanna (1947).

The early research history on Mediterranean fruit fly attractants is very interesting. It started with a housewife in Australia in the early part of the century who placed some freshly baked pies on garden fence posts to cool. She treated the bases of the posts with kerosene to prevent ants from attacking the pies, and her husband discovered that the kerosene attracted Mediterranean fruit flies. Subsequently, kerosene was used for years in Australia and Hawaii to "control" *C. capitata*. It was not until Severin and Severin (1913; 1914) discovered that kerosene only attracted males that its use was discontinued. The next widely used attractant was angelica seed oil. The Mediterranean fruit fly infestation in Florida in 1956 consumed the world's supply of angelica seed oil. A breakdown and bioassay of the oil components resulted in the synthesis of medlure. Later, through chlorination of medlure, the widely used Mediterranean fruit fly attractant trimedlure was obtained (C.O. Calkins, personal communication).

For the genus *Rhagoletis*, we find very detailed descriptions of oviposition behavior, adult movement, and larval emergence behavior of the apple maggot fly, *R. pomonella*, in Porter (1928). Oviposition, mating, and diapause in this species were described by Illingworth (1912) and Farleman (1933). The remarkable work of Boyce (1934) describes patterns of oviposition, mating, territoriality, combat between males, and feeding behaviors as well as some aspects of biotic mortality in *R. completa* Cresson on walnuts (*Juglans* sp.). Brooks (1921) describes how *R. suavis* (Loew) males defend territories near sites of mechanical injury on walnuts. Fluke and Allen (1931), working on diets, found that protein was necessary for reproduction. There were also early studies on the relationship between environmental factors and mating periods in Severin (1917), on adult movement and host distribution in Phipps and Dirks (1933), on host odor attractiveness in Hodson (1943), and on natural food sources in Neilson and Wood (1966). Distribution of eggs in apple fruits and trees was studied by Leroux and Mukerji (1963). Finally, Dean and Chapman (1973) carried out important studies on the bionomics of *R. pomonella.*

In the case of the European cherry fly, *R. cerasi*, detailed descriptions and ingenious experiments on host searching, responses to host shape and color, oviposition, and host-marking behaviors were reported by Wiesmann (1937). In an earlier study, the same author (Wiesmann 1933) observed *R. cerasi* flies feeding on extrafloral nectaries of cherry trees. Important statistical tests on egg-distribution patterns were carried out by Häfliger (1953), who concluded that the uniform distribution of eggs when abundant uninfested cherries were available was probably influenced by the elaborate fruit-marking procedure performed by the cherry fly when it drags its ovipositor over the fruit surface.

In the genus *Anastrepha*, the most prominent pioneering work is that of A. L. Herrera, L. de la Barrera, and A. F. Rangel in Mexico (for details see Aluja et al., Chapter 15), that of researchers from the U.S. Department of Agriculture working in Mexico and the southern United States, and that of C. Picado in Costa Rica. Thus, we find interesting descriptions of larval behavior inside fruits as well as oviposition and mating behavior in Herrera et al. (1901), Picado (1920), McPhail and Berry (1936), and Plummer et al. (1941). McPhail and Bliss (1933) reported that in *A. ludens,* mating occurred at dusk and lasted from 20 min to 3 h. They also observed that pairs maintained in cages mated repeatedly at intervals throughout their lives. In the same article, these two authors also reported the effect of fruit ripeness on feeding and oviposition preference. Searching for refuge, protection from environmental factors, and resting behavior were studied by Crawford (1927). Shaw (1947) found that the distribution of *A. serpentina* (Wiedemann) adults was restricted to the vicinities of their hosts. The effect of color on attraction was studied by McPhail (1937; 1939) in an attempt to increase trap efficiency. Darby and Knapp (1934) studied some aspects of pupation behavior in *A. ludens*. They discovered that pupae were found practically equidistant from the center of a fallen fruit and only on rare occasions could be found underneath a fruit. To demonstrate that the Mexican *A. fraterculus* (Wiedemann) was different from its South American "sister," Baker (1945) carried out experiments on oviposition preference and larval development in many host fruits, including oranges, a common host of the South American *A. fraterculus*. His work showed that Mexican flies did not accept oranges or mangoes as hosts.

The behavior of flies in the genus *Bactrocera* (formerly *Dacus*) was also studied from a wide perspective during this early period. There are reports on acoustic signals and their importance in mating behavior in *B. tryoni* in Myers (1952) and Monro (1953). Féron (1960b) studied acoustic signaling and mating behavior of *B. oleae* males. Martin et al. (1953) examined the effect of male sexual maturity on female attraction and mating behavior, as well as methods of control and rearing in *B. oleae*. Rearing in the laboratory started with a series of studies on reproductive behavior. Allmann (1938; 1940; 1941) carried out studies on the use of previous oviposition sites in fruits, patterns of oviposition behavior, and immature development of *B. tryoni* in the laboratory. Oviposition patterns and mating behavior were also studied under laboratory conditions in *B. tryoni* and *B. cacuminata* Hering by Myers (1952). Field and laboratory studies on the mating behavior of *B. cucurbitae* were carried out by Back and Pemberton (1917). The limiting effect of environmental factors such as light and temperature on the mating behavior of *B. dorsalis* was explored by Davis (1954) and Roan et al. (1954). Severin et al. (1914) summarized the results of a study on the habits of *B. cucurbitae*, including oviposition, feeding, and manifestation of "fear" by adults, as well as feeding and jumping behavior by larvae. Maher (1957) described the feeding and oviposition behavior of *Dacus ciliatus* Loew.

Notes on oviposition and larval behavior of *Toxotrypana curvicauda* Gerstaecker can be found in Knab and Yothers (1914). In 1922, 8 years later, Mason published an account on the life history and seasonality of this insect, together with descriptions of adult movement patterns, copulation, and oviposition behaviors.

The genus *Zonosemata*, for which six species have been described in North America (including Mexico) and one in South America, was the focus of early work by Peterson (1923). This author described host searching and mating behavior in the pepper maggot fly, *Z. electa* (Say). Benjamin (1934) and Burdette (1935) studied the life history, oviposition, and mating behavior of the same species.

Severin's study (1917) on the life history, habits, natural enemies, and methods of control of the currant or gooseberry fruit fly, *Euphranta canadensis* (Loew), is an outstanding reference with demographic information and data on emergence, feeding, oviposition, and mating. This work also includes the feeding and jumping habits of larvae.

During this early period one can also find outstanding studies on the behavior of nonfrugivorous tephritids in the literature. The work of Varley (1947) on oviposition behavior and the effect of feeding and mating on the fecundity and survival of females of *Urophora jaceana* (Hering) is one of the most valuable studies on the biology and ecology of any tephritid. Uhler's (1951) study on *Eurosta solidaginis* (Fitch) is also noteworthy for its accurate and detailed descriptions of the biology, life history, and habits of both adults and larvae, including emergence, oviposition, and mating. Demographic parameters such as survival and fecundity are also quantified in this study. Finally, the general studies on reproduction by Tauber and Toschi, which include descriptions of the biology, mating, and oviposition behaviors of *Euleia fratria* (Loew) (Tauber and Toschi 1965a), *Tephritis stigmatica* (Coquillett) (Tauber and Toschi 1965b), and *Aciurina ferruginea* (Doane) (Tauber and Tauber 1967) are among the last pioneer efforts of this early period.

### 3.2.2 Contemporary Period

Recent approaches to the study of tephritid behavior are diverse, ranging from simple observations to complex experiments attempting to identify causes and effects. Fruit fly behavior can be classified into five categories: mating, oviposition, feeding, trivial movement and migration, and larval behavior. Most studies during the contemporary period address aspects such as foraging behavior, communication, orientation, learning, mate choice, and territoriality. In addition, studies on daily behavioral patterns were aimed at locating each of these behaviors in time and space. There is a clear inclination toward studies on reproductive behavior. Courtship, mating, and oviposition strategies have been analyzed from many perspectives. Food foraging and feeding behavior, including factors that stimulate adults to search for and locate food, have received attention in recent times. The development of the sterile insect technique (SIT) deserves special attention. SIT-related behavioral studies are based on the manipulation of oviposition and mating behaviors. Almost all research has concentrated on developing methods to assess male competitiveness adequately under laboratory and field conditions.

In the following section, we present a summary of the contemporary work that we consider to be relevant to the description and understanding of mechanisms underlying the above-mentioned categories of fruit fly behavior. We include some studies from the 1960s because of their experimental nature and holistic approach.

#### 3.2.2.1 Sexual Behavior

The numerous publications cited in many chapters in this volume make it obvious that sexual behavior in fruit flies is a highly complex and diverse process. A theoretical framework for its study was provided by Bateman (1972). His division of tephritids into r-selected and K-selected species based on environmental factors was later used by Prokopy (1980) and Burk (1981) to characterize mating systems of tropical and temperate species. Sivinski and Burk (1989) reviewed sexual selection in tephritids with an emphasis on communication mechanisms and relations between courtship strategies and the distribution of oviposition sites. Many ecological aspects of the reproductive behavior were studied by Burk (1983). Without doubt, one of the most complete studies on tephritid reproductive behavior is that of Headrick and Goeden (1994), who analyzed the mating behavior of 48 nonfrugivorous tephritids under laboratory conditions, and generated a glossary of terms to standardize behavioral descriptions of courtship, as well as a database to develop evolutionary and ecological hypotheses on mating behavior.

Sexual behavior in fruit flies has been analyzed stage by stage, resulting in four main areas of research (Sivinski et al., Chapter 28): (1) precopulatory behavior, which includes mechanisms involved in bringing the sexes together for mating; (2) mating behavior, including courtship; (3) sexual selection; and (4) mating success, including cryptic female choice.

Precopulatory behavior includes different stages, most of them related to the mating system. Precopulatory behavioral studies are mainly related to location of the mating site (e.g., Prokopy et al. 1971; 1972; 1996; Zwölfer 1974; Smith and Prokopy 1980; Drew 1987) and intraspecific communication among individuals via visual, chemical, and acoustic signals.

Most of the work on visual signals refers to the distinctive wing patterns and the characteristic wing movements used by males to attract females or during aggressive displays for defending territories or resources (Dodson 1978; 1982; Aluja et al. 1983; Burk 1983). With the help of video, Headrick and Goeden (1994) were able to catalog the display behaviors of 48 nonfrugivorous species. They also categorized wing movements and discussed the function and evolution of wing patterns.

Chemical signals have been studied from different perspectives, such as the way flies store and release sex pheromones (Fletcher 1968) and how these infochemicals attract females and influence mating success (Nation 1972; Perdomo 1974; Katsoyannos 1976; Landolt et al. 1985; Landolt and Averill 1999). Researchers have explored questions such as why virgin females are more attracted by sex pheromones than mated ones (Prokopy 1975a; Katsoyannos 1982), and how pheromone quality and pheromone release behavior can be affected by the quality and quantity of the food that male flies ingest (Epsky and Heath 1993; Yuval et al. 1998). Reviews of sex pheromone release behavior were written by Koyama (1989), Nation (1979), and Kuba (1991). An interesting review on the use of pheromones and parapheromones in the control of tephritids was written by Sivinski and Calkins (1986).

Male wing vibrations can occur during pheromone emission, fighting, or after mounting a female. During these rapid wing movements, acoustic signals are produced. Behavioral effects of tephritid songs have been extensively studied at the USDA Insect Attractants, Behavior and Basic Biology Laboratory (currently USDA-ARS Center for Medical, Agricultural, and Veterinary Entomology) in Gainesville, Florida (Webb et al. 1976; 1983a; 1983b; 1984; 1985; Burk and Webb 1983; Sivinski et al. 1984; Sivinski and Webb 1985a, b; 1986). Studies on male-calling patterns in relation to time of day, food, and host fruit presence were carried out by Landolt and Sivinski (1992).

The lek mating system of most tropical fruit fly species has represented a fascinating area of research. The work of Tychsen (1977) on *B. tryoni* is specially valuable because of the accurate description of lek formation and how wind direction influences lek position in a tree. Interestingly,

Tychsen described the leks he observed as "settled swarms" because of their similarity to aggregations of flying males observed in other dipterans. We note that Tychsen's work was published during the same year that Emlen and Oring (1977) published their seminal theoretical work that classified mating systems and used the term *lek* in relation to invertebrates. Lekking studies in fruit flies include aspects such as interspecific lek formation, lek size, and interactions between two species of flies on the same host tree (Aluja et al. 1983), male–male aggressive behavior, and the importance of body size in territory defense and acoustic signaling (Burk 1984; Dodson 1986; Iwahashi and Majima 1986). Hendrichs (1986) discussed behavioral patterns exhibited by males while in a lek, and lek size, territoriality and male success in relation to lek position and body size, and male territory-invading strategies. Arita and Kaneshiro (1989) studied lek behavior, defensive strategies to preserve territories, and courtship behavior. Further studies have explored the distribution of sexes in relation to lek area and male success in relation to male position within a lek (Sivinski 1989). Sivinski et al. (1994) also studied territory-marking behavior, which elicited male site fidelity and female preference for marked leaves. Other authors have tried to decipher the effect of tree architecture on lek distribution. These investigations suggest that tree volume influences lek formation (Shelly and Whittier 1994a). Studies on lek behavior and its relationship to ecological factors were reviewed and discussed by Prokopy (1980) and Burk (1983; 1991).

Not all tropical tephritids use leks as mating strategies. *Toxotrypana curvicauda* males are solitary, territorial, and use pheromones to call females from host fruits (Landolt and Hendrichs 1983). Interestingly, the combination of fruit odor and male pheromone enhances attractiveness to females (Landolt et al. 1992; Landolt, Chapter 14).

The resource-defense mating system of temperate species has also provided a good study model. Descriptions of displays during agonistic interactions on fruit in *Rhagoletis* flies can be found in Biggs (1972) and Ali Niazee (1974), as well as interspecific interactions in Messina and Subler (1995). Oviposition site guarding by male walnut flies and its consequences for mating success were studied in detail by Papaj (1994a). Fruit marking with sex pheromone by *R. boycei* males and its function as an oviposition stimulant was studied by Papaj et al. (1996).

Interspecific recognition of sexual pheromones among fruit fly species is an important topic from an evolutionary perspective. It has been poorly studied and thus represents an interesting area for future research. The few laboratory bioassay studies carried out to date have produced confusing results. Better bioassays must be developed in the future. This is because it appears that the specificity of response by various species may depend on components which operate over longer distances and thus wind tunnel studies provide inadequate experimental conditions to study this phenomenon (Kobayashi et al. 1978; Fitt 1981a).

With respect to mating behavior, studies such as those by Prokopy and Hendrichs (1979) describe the sexual activities of Mediterranean fruit flies throughout the diel activity period. They observed that males use leks and oviposition resource defense to acquire mates and that fruit marked with host-marking pheromone arrests males. They also suggested that males mate with virgin females on vegetation and with mated females on fruit. Pritchard's work (1967) describes how *Dirioxa pornia* (Walker) males secrete a volatile sex attractant and produce a mound of foam that acts as a nuptial gift on which females feed while copulating. Other studies on nuptial gifts in nonfrugivorous species were performed by Stoltzfus and Foote (1965), Foote (1967), and Novak and Foote (1975). There are also studies on mating behavior indicating that trophallaxis is an important component of the courtship ritual and female choice in both nonfrugivorous (Batra 1979; Freidberg 1981; 1984b; 1986; 1997) and frugivorous species (Aluja et al. 1993). A curious form of trophallaxis behavior after copulation in nonfrugivorous flies is reported and its function discussed by Freidberg (1982). Another interesting aspect of mating behavior is reported by Alonso-Pimentel and Papaj (1996a) who worked with *R. juglandis* Cresson. These authors found that a change in operational sex ratio influences copulation duration. A male-biased ratio increases copulation duration probably because males seek to avoid sperm competition. A female-biased ratio reduces copulation duration and increases the number of copulations of males. The risk of sperm competition also affects ejaculate size. Gage (1991) suggested that male Mediterranean fruit flies respond to an increased threat of sperm competition by inseminating the female with larger sperm loads.

Detailed descriptions of courtship and mating have been published for different fruit fly species including temperate species (Prokopy and Bush 1973a; Smith and Prokopy 1982), tropical lekking species (Arakaki et al. 1984; Dodson 1987; Whittier et al. 1992) and nonfrugivorous species (Freidberg 1981). Using video recordings of *C. capitata* courtship, Briceño et al. (1996) described several hitherto unreported details of calling and mounting behaviors. Hendrichs and Reyes (1987) described reproductive behavior, including mating, oviposition, and postmating male guarding, in *Dacus longistylus* Wiedemann. Eberhard and Pereira (1993) discovered that courtship in *C. capitata* sometimes occurs after the male has mounted the female but before he has achieved intromission. They reported that the male nips the female at the tip of her abdomen with the surstyli of the genitalia.

Copulatory frequency has been another interesting area of research. The early work of Neilson and McAllan (1965) reported that mating frequency increases female fertility but not fecundity. Studies on mating frequency range from simple observations (Tzanakakis et al. 1968; Nakagawa et al. 1971; Robacker et al. 1985) to experiments that address the complex effects of multiple matings on female fecundity and survival (Opp and Prokopy 1986; Sivinski and Heath 1988; Saul and McCombs 1993; 1994; Whittier and Shelly 1993; Sivinski 1993; Landolt 1994; Mangan 1997; Trujillo 1998).

Some mating behavior studies have used a wide-ranging approach. For example, the study on *A. fraterculus* carried out by De Lima et al. (1994) includes the analysis of factors affecting sexual maturation and patterns of courtship, mating, and oviposition. Behavioral observations of *Paraceratitella eurycephala* Hardy indicated that females use host-marking pheromones to reduce larval overcrowding in the host (Fitt 1981b). In the same study, Fitt reports that *P. eurycephala* do not exhibit slow wing movements and posturing during courtship. Mating, however, involves wing vibration and possible pheromone release by the male to attract females.

A third area of studies on sexual behavior deals with the relation between sexual selection, male mating success, and female fitness. The work of Whittier et al. (1994) and Whittier and Kaneshiro (1991; 1995) is particularly relevant because of their attempt to decipher the factors that influence female choice. Sexual selection studies also include aspects that relate male mating behavior with mating success (Robacker et al. 1991). The relationship between previous sexual experience and male reproductive success has also been an active area of research, mainly for tropical lekking species. Questions like why females prefer some males over others are still in the air (Sivinski 1984; Shelly and Whittier 1993b). Recently, Hunt et al. (1998) reported that fluctuating asymmetry of the male anterior orbital seta is correlated to *C. capitata* male mating success.

There is a fourth category of sexual behavior studies. These describe factors underlying seasonal and daily rhythms of sexual activity (Causse et al. 1966; Tychsen and Fletcher 1971; Tychsen 1975; Loher and Zervas 1979; Boller et al. 1981; Katsoyannos 1982; Warburg and Yuval 1997a; Aluja et al., Chapter 15). Photoperiod as a regulator of circadian rhythms was studied by Tychsen (1978). Smith (1979) examined the genetic mechanism that controls the timing of mating and responses to light intensity. He found that sexual behavior is influenced by the interaction of light intensity and the circadian clock. Studies of the effect of environmental factors on mating behavior include the work of Myburgh (1962), who found that Mediterranean fruit fly copulation was initiated above ~2000 lux, at a temperature of 19 to 31°C, and a relative humidity of 30 to 95%.

Host race formation and sympatric speciation related to host fidelity, mating behavior, and the hybridization of sibling species constitute special categories in the study of sexual behavior in fruit flies (Bush 1969; Bierbaum and Bush 1988; Feder and Bush 1989). These studies provided the groundwork for studies on the evolution of behavior.

### 3.2.2.2 Oviposition Behavior

This aspect of fruit fly behavior has been extensively studied from different perspectives mainly due to the economic impact of ovipositions in commercial fruit. Studies can be divided into two categories: prealighting and postalighting behavior on host fruits. The former includes host search behaviors as well as the responses to visual and olfactory clues flies use to find hosts. Studies on postalighting behavior refer to the activities of flies once they are on host fruit (Díaz-Fleischer et al., Chapter 30).

Visual stimuli have been studied from the perspective of color and shape. Foliage color, tree shape, and tree size all have been shown to play a role in influencing fly arrival to trees (Sanders 1968a; Moericke et al. 1975; Moericke 1976; Katsoyannos et al. 1986). Host color has an important effect on fly searching behavior, and background characteristics have a substantial influence on the ability of flies to distinguish colors (Prokopy 1968; Owens and Prokopy 1986). Cytrynowicz et al. (1982) found that two different fruit fly species use different visual cues when seeking fruit. Studies of visual stimuli also include applied research on host and trap models, aimed at developing better control methods.

The effect of host odor on fly search behavior has been the subject of detailed studies, which include work on attraction of flies to odor sources (Reissig 1974; Katsoyannos et al. 1997) and the study of precise intra- and intertree cues used by flies to find hosts (Aluja and Prokopy 1993). An accurate method for the study of intratree host finding behavior was developed by Aluja et al. (1989). Studies of host odor and visual stimulus interactions during intra-tree host searching indicated that flies use odor stimuli to move among trees and visual stimuli when searching in a given tree (Aluja and Prokopy 1992). These types of studies have been reviewed by Prokopy (1986) and Fletcher and Prokopy (1991).

Roitberg and Prokopy (1982; 1984) and Roitberg (1985) have studied the dynamics of foraging behavior in relation to the nearness of neighboring host trees and have found that females spend less time in a tree when female density is high. Intratree foraging behavior in relation to host fruit density and quality was studied by Prokopy et al. (1989b, c), who showed that females spend more time in trees that have more and better quality host fruit. Studies on orientation, searching, and host-location behavior were reviewed by Prokopy and Roitberg (1983; 1989) and Prokopy (1993).

Postalighting behavior has been studied from many angles. For example, Pritchard (1969) carried out an interesting study on the oviposition behavior of *B. tryoni.* He analyzed the patterns of egg distribution within and between fruits and how female flies used oviposition punctures. The same author also reported aggressive interactions among females while on a fruit. Postalighting behavioral studies have addressed questions such as the selection of oviposition sites (Sanders 1962; 1968b; 1969a, b; Katsoyannos et al. 1986; Fernandes-Da-Silva and Zucoloto 1993; Papaj and Aluja 1993), differences in oviposition behavior among species and the implications of oviposition choices for speciation (Diehl and Prokopy 1986; Bierbaum and Bush 1988; Feder and Bush 1989; Boller et al. 1998), general descriptions of oviposition behavior (Cirio 1971; Barros et al. 1983), the relationship of fruit ripeness to clutch size (Abasa 1972; Papaj and Messing 1996), and female preference for a certain degree of fruit ripeness (Jang and Light 1991). Messina and Jones (1990) discussed how fruit maturation might account for the pattern of infestation in three fruit species. Berrigan et al. (1988), working in the laboratory with artificial hosts, studied the effect of fly age and density, host size, host density and color on clutch size in *A. ludens*. Messina (1990) reported that physical components of host choice, such as color and size, were important for host preference in two *Rhagoletis* species.

In other studies of oviposition behavior, special attention has been paid to substances that deter oviposition. Host marking with epideictic pheromones that deter conspecific females from ovipositing has been reported in many species of frugivorous Trypetinae (Cirio 1972; Prokopy 1972; 1975b; 1981a, b; Katsoyannos 1975; Prokopy et al. 1977; 1978; 1982b; Boller 1981; Averill and

Prokopy 1981; 1982; 1987; Mumtaz and Ali Niazee 1983; Boller and Hurter 1985; Boller et al. 1987; Landolt and Averill 1999; Aluja et al., Chapter 15) and also in some nonfrugivorous species (Straw 1989; Pitarra and Katsoyannos 1990; Lalonde and Roitberg 1992). Host fruit juices can also act as deterrents to oviposition in *B. oleae* (Girolami et al. 1981). Host-marking pheromones (HMPs) have been reported to arrest male flies in species with a resource defense mating system (Katsoyannos 1975; Prokopy and Hendrichs 1979). The functional significance of HMPs on egg allocation and its effects on host infestation was studied by Papaj et al. (1989). Roitberg and Prokopy (1981) established that, for *R. pomonella* females, previous experience with HMP is necessary in order for them to recognize it. These authors also determined that host deprivation alters female response to HMPs (Roitberg and Prokopy 1983). Research on marking includes the effect of factors such as age and diet, both of which influence the release of HMPs (Averill and Prokopy 1988). The study of interspecific recognition of HMPs in *Rhagoletis* flies has also helped elucidate relationships between species (Prokopy et al. 1976; Prokopy 1981a). Katsoyannos and Boller (1976; 1980) carried out field tests with raw extracts of cherry fruit fly HMP to determine if fruit infestation was reduced. At a later date, Boller (1981) and Boller and Hurter (1998) discussed the use of HMPs in controlling the cherry fruit fly, *R. cerasi.* A synthetic cherry fruit fly HMP (Ernst and Wagner 1989) was successfully tested 12 years later as an infestation-reducing agent in a commercial cherry orchard and in field-cage experiments (Aluja and Boller 1992a, b; Boller and Aluja 1992; Boller and Hurter 1998). Averill and Prokopy (1989b) studied the within-canopy distribution of *R. pomonella* eggs, among fruit sampled in hawthorn trees (*Crataegus mollis* (Torr. and A. Gray) Scheele). They found that the egg-distribution pattern changed over time. At the beginning of the season, there was an aggregated pattern of egg distribution, which changed to a random pattern and finally ended with an even pattern at the end of the season. These authors suggested that fruit phenology affected the availability of oviposition sites and indicated that when suitable fruits were available, deployment of an HMP reduced intraspecific competition. Remund et al. (1980) found that the distribution of *R. cerasi* eggs under field conditions did not follow a Poisson distribution because of the innate tendency of cherry fruit flies to oviposit only one egg per visit on a fruit and also due to the deterring effect of the HMP. Papaj (1994b) performed an interesting analysis on the evolutionary implications of marking, while Roitberg and Mangel (1988) and Mangel and Roitberg (1989) developed evolutionary models related to host-marking behavior and to dynamic information associated with host acceptance. Furthermore, Roitberg et al. (1982) developed a foraging behavior model in relation to oviposition site selection. Papaj et al. (1990) studied clutch size in relation to host quality (HMP marked fruits). Papaj and Messing (1996) took the latter study one step further and explored the effect of fruit ripeness and quality on oviposition behavior. They found that *Ceratitis capitata* females modulate oviposition behavior based on the degree of fruit ripeness, fruit size, and degree of infestation (infested vs. uninfested). Prokopy and Koyama (1982) studied oviposition site partitioning mechanisms in the nonmarking *B. cucurbitae*. For a general overview on host-marking behavior, see Averill and Prokopy (1989a).

Research on oviposition behavior also includes studies on superparasitism as a time-saving strategy (Papaj and Alonso-Pimentel 1997), the effects of egg load on searching and host preference behavior (Prokopy et al. 1994), and the distribution of the sexes in relation to egg load (Alonso-Pimentel and Papaj 1996b). Resident advantage in competition for a host fruit as a function of the physiological state of flies was recently tested by Papaj and Messing (1998). Another recent study by Freese and Zwölfer (1996) reported that females of *Urophora cardui* (L.) adjust clutch size according to bud quality. Lalonde and Roitberg (1994) studied the effect of clutch size and offspring fitness in relation to pollen availability in male and female flowers. Fitt (1986a) reported that oviposition behavior was one of the most important limiting factors in host use by five species of *Bactrocera*, and that the rate of ovary maturation was different between generalist and specialist species (Fitt 1986b; 1990). Oviposition behavior in relation to inter- and intraspecific interactions among flies on hosts was studied by Fitt (1984). The same author (1989) reviewed species competition for hosts and factors that limit species distribution.

Research on special issues relating to oviposition behavior include the influence of learning on host location success rates, oviposition site selection, and host acceptance (Prokopy et al. 1982a; 1986; 1989a; 1993a; 1994a; Cooley et al. 1986; Boller 1998). Prokopy and Papaj (1989) studied the capacity of fruit flies to learn to recognize different stages of fruit ripeness, while Papaj and Prokopy (1988) explored how flies use fruit phytochemicals to identify the type of host they prefer for oviposition. Other studies on oviposition behavior have described how grouped females oviposit more than solitary ones (Prokopy and Bush 1973c), and how social facilitation behavior enhances oviposition "drive" (Prokopy and Duan 1998; Prokopy and Reynolds 1998).

### 3.2.2.3 Feeding Behavior

Feeding, an essential activity for survival and reproduction, has also been widely studied. Research has focused on three main areas: the search for feeding sites, foraging in general terms, and feeding behavior in relation to food quality and its effect on fitness (Drew and Yuval, Chapter 27). The first category includes observations of searching behavior to locate feeding sites and systematic descriptions of fly movements on host and nonhost trees while flies are searching for protein and carbohydrates (Hendrichs and Hendrichs 1990; Hendrichs and Prokopy 1990; Hendrichs et al. 1990).

Studies on foraging have investigated the effect of ingested food quality on food-seeking behavior. These studies report that tree residence time is influenced by food quality, and that handling and processing times are longer than foraging time (Hendrichs et al. 1990; 1993). Prokopy et al. (1993b) studied the influence of physiological states on foraging behavior and found that hungry females exhibited a higher propensity to discover food than well-fed females. Using a different approach, Cangussu and Zucoloto (1994) explored physiological aspects of feeding behavior in relation to food quality and food selection in *C. capitata*.

The third category includes studies of feeding behavior around natural food sources. Prokopy and collaborators (1993c) carried out research on animal feces as a natural source of protein and found that the age and origin of feces were important factors in determining their degree of attractiveness. The role of bacteria as a food source for adult fruit flies and how they affect their behavior and fitness have been studied extensively by Drew and collaborators (Drew et al. 1983; Drew and Lloyd 1987; Drew 1987) and by Fitt and O'Brien (1985), who reported that some bacteria found on ripening fruits also exist in the digestive tract of flies and that females transmit these bacteria to their offspring during oviposition. Reviews on the importance of bacteria as a food source were written by Drew and Lloyd (1989; 1990).

Hendrichs et al. (1992) studied postfeeding behavior in relation to oral evaporation of excess water in food and found that the time required to evaporate excess water was directly related to the amount of water the ingested food contained. A review of food foraging behavior was recently written by Hendrichs and Prokopy (1994).

Another group of studies in this category explored the effect of feeding on reproduction. Hagen (1953; 1958) was the first to correlate experimentally the importance of protein and carbohydrates with fly reproduction. Recently, Yuval and collaborators have explored the effect of protein on mating behavior. For example, the work of Blay and Yuval (1997) confirmed that the diet of male Mediterranean fruit flies affects their ability to gain copulations with virgin females, and also affects receptivity of females to further copulations. In laboratory studies, they also found that protein-fed Mediterranean fruit flies expend more energy in sexual signaling than protein-deprived individuals (Warburg and Yuval 1996; 1997b). Jácome et al. (1995; 1999) found that lipogenesis capability is influenced by the adult diet in *Anastrepha serpentina*, and that when given a choice between a high-quality food (an open fruit) and a low-quality one (dry crystals of sucrose), females of the same species prefer the low-quality one. These authors called this phenomenon a "junk food syndrome."

Applied research on feeding includes comparative studies of behavioral responses to bait sprays and natural foods. The work of Prokopy et al. (1992) showed that *C. capitata* flies prefer bird feces over proteinaceous baits used in combination with an insecticide. Drew and Fay (1987), when comparing the relative attractiveness of *B. tryoni* to proteinaceous suspensions of ammonia and bacteria, found that bacterial odors and ammonia were strong attractants. Research has also been done on the importance of specific substances such as ammonia as a cue to food location and their role as feeding stimulants (Bateman and Morton 1981; Heath et al. 1994).

### 3.2.2.4 Larval Behavior

Larval behavior has been studied less intensively than adult behavior. Site preference inside a fruit or artificial host, competition between larvae, and the relationship between larval and adult performance have been the main topics of study in this category. Rajamannar (1962) used a simple "choice chamber" technique to determine the preference of larvae for a wide variety of fruits. He found little correlation between the attractiveness of a host and its suitability as larval food. Later, Debouzie (1977a, b; 1981) studied the effect of initial population size and competition among larvae in *C. capitata.* This work demonstrated the importance of competition in larval development. In-depth studies by Zucoloto (1988; 1993a, b) and Fernandes-Da-Silva and Zucoloto (1993) showed that fruit site selection for oviposition by adult females is related to carbohydrate contents and larval performance. Zucoloto (1990) also observed that larvae can select feeding sites inside a fruit, moving to areas with the highest levels of carbohydrates. Fruit maturity has a great influence on the development rate and movement of larvae within the fruit (Calkins and Ashley 1988). Jumping behavior of *C. capitata* larvae, rhythms, and photoperiod regulation of pupation were studied by Causse (1974).

### 3.2.2.5 Movement

For a perspective of studies on adult movement we follow the division of Bateman (1972), separating studies related to resource foraging behavior into those in which there is a physiological basis for movement (resulting in a tendency to remain in areas with ample feeding and oviposition sources), from studies that explore flight capacity related to long-range movements. Fruit fly movements related to foraging behavior have been studied from several perspectives, including the effect of the physiological state (i.e., egg load and nutritional state) (Malavasi and Prokopy 1992; Prokopy et al. 1994b; 1995). These studies relate female feeding and oviposition history to resource availability in different trees. They show that mature (high egg load) protein deprived females spend more time on trees having proteinaceous food and fruit than on trees that only have fruit. Prior experience and host tree distance also influence female movement (Roitberg et al. 1982; Roitberg and Prokopy 1983; 1984; Roitberg 1985). The effects of environmental factors such as wind speed and direction on intertree foraging behavior have been studied by Aluja and Prokopy (1992) and Aluja et al. (1989; 1993). Using marked flies, Opp and Prokopy (1987) monitored seasonal changes and trivial movements in *R. pomonella.* Roitberg (1985) and Roitberg et al. (1982) developed a model of foraging behavior taking egg load and fly experience into account. The work of Hendrichs et al. (1991) is especially noteworthy because of its integral point of view on sexual and nutritional foraging. Reviews on foraging behavior were compiled by Prokopy and Roitberg (1983; 1989).

Research on tephritid long-range movements include studies to assess capacity for colonizing new areas or orchards (Maxwell 1968; Haisch 1970; Boller et al. 1971; Neilson 1971). Tropical fruit fly species are strong fliers and have a considerable capacity for long-distance displacements (Christenson and Foote 1960; Bateman 1979; Al-Zaghal and Mustafa 1986; MacFarlane et al. 1987). Wong et al. (1982) found that most *C. capitata* flies were caught downwind from the release point in trapping studies. Baker et al. (1986) found the same patterns in *C. capitata* and *A. ludens.* Especially relevant is the work of Fletcher (1973; 1974), who studied patterns of short- and long-range displacement in *B. tryoni* and was able to provide an overview of this behavior. Studies on movement have also been used to monitor the establishment of released species used for weed control (Story and Anderson 1978). A study on *B. oleae* movement as a function of seasonal change was conducted by Fletcher and Kapatos (1981). Correlations between movement and wing shape have been proposed by Sivinski and Dodson (1992).

Fruit flies have also been tethered in an attempt to determine flight capacity (flight mill studies). Nevertheless, the results of this kind of work have been difficult to extrapolate to field conditions (Maxwell and Parson 1968; Remund and Boller 1975; Sharp et al. 1975; Sharp 1978; Chapman 1982; Nakamori and Simizu 1983).

Research on fruit fly movement has been reviewed to varying degrees by Christenson and Foote (1960), Bateman (1972), Bateman et al. (1976), Fletcher (1987; 1989) and more recently by Aluja (1993), who placed great emphasis on properly defining terms and on avoiding unsubstantiated reports of, for example, migration.

#### 3.2.2.6 Daily Activity Patterns

Studies in this area have tried to pinpoint behavior in time and space, and have also tried to understand the relationship of individuals with their habitat. A review of this can be found in Aluja and Birke (1993) and Aluja et al. (1993). Pioneering studies in this area were carried out by Prokopy et al. (1971; 1972) on *R. pomonella* and on *R. fausta* (Osten Sacken) (Prokopy 1976). Further research on diel patterns in temperate species was carried out on *R. mendax* Curran (Smith and Prokopy 1981) and *R. cingulata* (Loew) (Smith 1984). For tropical and subtropical species, studies on *A. fraterculus* (Malavasi et al. 1983), *C. capitata* (Hendrichs and Hendrichs 1990; Warburg and Yuval 1997a), *A. obliqua* (Aluja and Birke 1993), *A. striata* Schiner (Aluja et al. 1993), and *T. curvicauda* (Aluja et al. 1997) are of particular interest because they consider both orchards and neighboring vegetation in an effort to obtain a more integrated ecological view of daily activity patterns.

#### 3.2.2.7 Fly Behavior Related to Mass Rearing and Control Operations

Behavioral research on mass-reared insects has received considerable attention in the last 20 years, since the SIT started to be widely used for the eradication or control of tephritid pests. Researchers have attempted to develop proper methodologies to compare wild fly behavior with that of artificially reared ones, and this has resulted in the establishment of procedures for quality control. Extensive reviews of behavioral studies and quality control methods for mass-reared insects were published by Boller and Chambers (1977a, b), Chambers (1977), Burk and Calkins (1983), Calkins (1989), and Calkins et al. (1996) (also see Cayol, Chapter 31).

Most of the work done in this field has focused on sexual competitiveness and flight ability of sterile males. Studies on sexual competitiveness have been, for the most part, comparative experiments of irradiated vs. unirradiated flies aimed at assessing the effect of irradiation on this important trait (Holbrook and Fujimoto 1970; Hooper 1971; 1972; Boller et al. 1975). Laboratory-reared and wild males have been compared to assess the degree of competitiveness and compatibility of both lineages (Susuki and Koyama 1980; Kuba and Koyama 1982; Iwahashi et al. 1983; Liimatainen et al. 1997). A mating propensity index for comparing wild and laboratory flies under laboratory conditions was developed by Boller et al. (1975; 1977) for the European cherry fruit fly and later used in other fruit flies. Boller et al. (1970; 1971) had previously studied the behavioral and ecological aspects of *R. cerasi* flies to develop methods for releasing irradiated flies. Studies by Rössler (1975a, b) revealed sexual isolation between laboratory-reared flies and wild-type Mediterranean fruit flies. Discoveries like Rössler's demonstrated the need for studies on sexual compatibility between laboratory-reared and wild flies.

Based on an analysis of the importance of behavior in mass-rearing operations, a general concept for quality control was established (Boller 1972). This incorporated the adoption of quality control methodology from industry to the needs of insect-rearing facilities (Boller and Chambers 1977a, b; Chambers 1977). The combined efforts of independently operating quality control groups in Europe, the United States, and Japan resulted in a first compilation of potential methods to measure the quality of mass-reared fruit flies (Boller and Chambers 1977b). A standardized set of quality tests for routine testing of Mediterranean fruit fly strains with laboratory methods was assembled in a "Rapid Quality Control System" (Boller et al. 1981). The system includes a test for calibrating pupae size, flight ability, irritability, pheromone response, and mating propensity, and was validated by an international team of experts convened in Spain in 1979. This laboratory system was complemented by an analogous set of field tests (Boller et al. 1977; Chambers et al. 1983), and procedures to improve the mating performance of Mediterranean fruit flies by selection (Boller and Calkins 1984).

Calkins (1984) discussed the importance of understanding fruit fly mating behavior when trying to produce highly competitive males in sterile male release programs. Calkins (1989; 1991) analyzed lek behavior and its implications for the assessment of quality in mass-reared flies and also discussed behavioral changes caused by mass rearing. Shelly et al. (1994) found that sterile male *C. capitata* are incapable of acquiring mates even when they participate in natural leks. Walder and Calkins (1993) detected that the mating behavior of *A. suspensa* males was reduced by increasing the radiation dose. A study by McInnis et al. (1996) reported possible behavioral resistance of wild types to SIT. The improvement of the mating propensity of *C. capitata* at different temperatures by selection was reported by Boller and Calkins (1984).

There is an important body of research on the flight ability of mass-reared insects which includes field studies comparing wild with sterile mass-reared strains. De Murtas et al. (1972a, b) studied movement capability in relation to irradiation dose and feeding, and found a negative correlation between ability to fly and irradiation dose. Hamada (1980) did the same kind of study with *B. cucurbitae*. Studies such as those reported by Boller et al. (1970; 1981) and Boller and Remund (1973) explored species behavioral attributes to evaluate methods of quality control. Sterile *B. oleae* movement was studied by comparing flight capacity and assessing long-distance displacements (Fletcher and Economopoulos 1976; Economopoulos et al. 1978). Kawai et al. (1978) studied the movement patterns of sterile flies among islets close to Kume Island, the point of release. MacFarlane et al. (1987), also evaluated the flight ability of irradiated *B. tryoni*. Studies by Sharp (1976; 1980) helped implement flight propensity tests in mass-rearing systems and tethered flight experiments. Sharp et al. (1983) reported procedures to improve flight propensity traits by selection. An interesting report is the one by Ozaki and Kobayashi (1981), who found that pupal handling influences flight ability in mass-reared tephritids.

## 3.3 CONCLUSIONS

This overview of research done on fruit fly behavior has revealed a rich and interesting history. We have outlined different study areas and lines of research in an attempt to create awareness of those topics which have received the most attention and those on which further research is most needed. A number of factors make fruit flies model organisms for studies on behavior. For example, their behavior can often be observed directly under natural and seminatural conditions and fitness components can be measured directly. The potential for future investigation is almost as daunting as it is enticing. Uncovering the secrets that these insects have closely guarded over the thousands of years they have interacted with humans, via the study of mechanisms controlling their behavior, is a challenge we must face, not only to develop efficient and environmentally friendly methods of control, but also to understand the ecological role of behavior more accurately.

## ACKNOWLEDGMENTS

The authors are grateful to Rogelio Macías-Ordóñez, Diana Pérez-Staples, Bianca Delfosse, and Warren Hide for helping us write this chapter. We thank Ron Prokopy, Carrol Calkins, Richard

Drew, Ernst Boller, and John Sivinski for reviewing and adding interesting and important information to this chapter. The story on the Australian housewife discovering that kerosene attracted Mediterranean fruit flies was shared with us by C. Calkins. We also acknowledge the invaluable support of John Sivinski and Gina Posey (U.S. Department of Agriculture, Gainesville, Florida), Bruno Patrian (Swiss Federal Research Station, Wädenswil), and Juan Rull (University of Massachusetts) for finding and sending us many of the old and hard-to-get references cited here. This chapter was written thanks to the financial support provided by the Campaña Nacional contra las Moscas de la Fruta (SAGAR-IICA) (Mexico), the International Organization for Biological Control of Noxious Animals and Plants (IOBC), and the Instituto de Ecología, A.C.

## REFERENCES

Abasa, R.O. 1972. The Mediterranean fruit fly, *Ceratitis capitata* Wied. Laboratory investigations of its reproductive behaviour in *Coffea arabica* in Kenya. *East Afr. Agric. For. J.* 37: 181–184.

Al-Zaghal, K. and T. Mustafa. 1986. Flight activity of olive fruit fly (*Dacus oleae* Gmel.) (Diptera: Tephritidae) in Jordan. *J. Appl. Entomol.* 103: 452–456.

Ali Niazee, M.T. 1974. The western cherry fruit fly, *Rhagoletis fausta*. II. Aggressive behaviour. *Can. Entomol.* 106: 1021–1024.

Allmann, S.L. 1938. Breeding experiments with the Queensland fruit-fly (*Strumeta tryoni* Frogg.). *J. Aust. Inst. Agric. Sci.* 4: 204–205.

Allmann, S.L. 1940. The Queensland fruit-fly. Observations on breeding and development. *N.S.W. Agric. Gaz.* 50: 499–501; 547–549.

Allmann, S.L. 1941. Observations on various species of fruit flies. *J. Aust. Inst. Agric. Sci.* 7: 155–156.

Alonso-Pimentel, H. and D.R. Papaj. 1996a. Operational sex ratio vs. gender density as determinants of copulation duration in the walnut fly, *Rhagoletis juglandis* (Diptera: Tephritidae). *Behav. Ecol. Sociobiol.* 39: 171–180.

Alonso-Pimentel, H. and D.R. Papaj. 1996b. Patterns of egg load in the walnut fly *Rhagoletis juglandis* (Diptera: Tephritidae) in nature and their possible significance for distribution of sexes. *Ann. Entomol. Soc. Am.* 89: 875–882.

Aluja, M. 1993. The study of movement in tephritid flies: review of concepts and recent advances. In *Fruit Flies: Biology and Management* (M. Aluja and P. Liedo, eds.), pp. 105–113. Springer-Verlag, New York.

Aluja, M. 1994. Bionomics and management of *Anastrepha*. *Annu. Rev. Entomol.* 39: 155–178.

Aluja, M. and A. Birke. 1993. Habitat use by *Anastrepha obliqua* flies (Diptera: Tephritidae) in a mixed mango and tropical plum orchard. *Ann. Entomol. Soc. Am.* 86: 799–812.

Aluja, M. and E.F. Boller. 1992a. Host marking pheromone of *Rhagoletis cerasi:* foraging behavior in response to synthetic pheromonal isomers. *J. Chem. Ecol.* 18: 1299–1311.

Aluja, M. and E.F. Boller. 1992b. Host marking pheromone of *Rhagoletis cerasi*: field deployment of synthetic pheromone as a novel cherry fruit fly management strategy. *Entomol. Exp. Appl.* 65: 141–147.

Aluja, M. and R.J. Prokopy. 1992. Host search behaviour by *Rhagoletis pomonella* flies: inter-tree movement patterns in response to wind-borne fruit volatiles under field conditions. *Physiol. Entomol.* 17: 1–8.

Aluja, M. and R.J. Prokopy. 1993. Host odor and visual stimulus interaction during intratree host finding behavior of *Rhagoletis pomonella* flies. *J. Chem. Ecol.* 19: 2671–2695.

Aluja, M., M. Cabrera, and J. Hendrichs. 1983. General behavior and interactions between *Anastrepha ludens* and *A. obliqua* under seminatural conditions. I. Lekking behavior and male territoriality. In *Fruit Flies of Economic Importance* (R. Cavalloro, ed.), pp. 122–132. A.A. Balkema, Rotterdam. 642 pp.

Aluja, M., R.J. Prokopy, J.S. Elkinton, and F. Laurence. 1989. Novel approach for tracking and quantifying the movement patterns of insects in three dimensions under seminatural conditions. *Environ. Entomol.* 18: 1–7.

Aluja, M., R.J. Prokopy, J.P. Buonaccorsi, and R.T. Cardé. 1993a. Wind tunnel assays of olfactory responses of *Rhagoletis pomonella* flies to apple volatiles. *Entomol. Exp. Appl.* 68: 99–108.

Aluja, M., I. Jácome, A. Birke, N. Lozada, and G. Quintero. 1993b. Basic patterns of behavior in wild *Anastrepha striata* (Diptera: Tephritidae) flies under field cage conditions. *Ann. Entomol. Soc. Am.* 86: 766–783.

Aluja, M., A. Jiménez, J. Piñero, M. Camino, L. Aldana, M.E. Valdés, V. Castrejón, I. Jácome, A.B. Dávila, and R. Figueroa. 1997. Daily activity patterns and within-field distribution of papaya fruit flies (Diptera: Tephritidae) in Morelos and Veracruz, Mexico. *Ann. Entomol. Soc. Am.* 90: 505–520.

Arakaki, N., H. Kuba, and H. Soemori. 1984. Mating behavior of the Oriental fruit fly, *Dacus dorsalis* Hendel (Diptera: Tephritidae). *Appl. Entomol. Zool.* 19: 42–51.

Arita, L.H. and K.Y. Kaneshiro. 1989. Sexual selection and lek behavior in the Mediterranean fruit fly *Ceratitis capitata* (Wied.) Tephritidae. *Pac. Sci.* 43: 135–143.

Averill, A.L. and R.J. Prokopy. 1981. Oviposition-deterring fruit marking pheromone in *Rhagoletis basiola. Fla. Entomol.* 64: 221–226.

Averill, A.L. and R.J. Prokopy. 1982. Oviposition-deterring fruit marking pheromone in *Rhagoletis zephyria. J. Ga. Entomol. Soc.* 17: 315–319.

Averill, A.L. and R.J. Prokopy. 1987. Intraspecific competition in the tephritid fruit fly *Rhagoletis pomonella. Ecology* 68: 878–886.

Averill, A.L. and R.J. Prokopy. 1988. Factors influencing release of host marking pheromone by *Rhagoletis pomonella* flies. *J. Chem. Ecol.* 14: 95–111.

Averill, A.L. and R.J. Prokopy. 1989a. Host marking pheromones. In *Fruit Flies: Their Biology, Natural Enemies and Control* (A.S. Robinson and G. Hooper, eds.), pp. 207–219. In *World Crop Pests* (W. Helle, ed.), Vol. 3A. Elsevier Science Publishers, Amsterdam.

Averill, A.L. and R.J. Prokopy. 1989b. Distribution patterns of *Rhagoletis pomonella* (Diptera: Tephritidae) eggs in hawthorn. *Ann. Entomol. Soc. Am.* 82: 38–44.

Back, E.A. and C.E. Pemberton. 1915. Susceptibility of citrus fruits to the attack of the Mediterranean fruit fly. *J. Agric. Res. Wash.* III (4): 311–330.

Back, E.A. and C.E. Pemberton. 1917. The melon fly in Hawaii. *U.S. Dep. Agric. Bull.* 491: 1–64.

Baker, E.W. 1945. Studies on the Mexican fruitfly known as *Anastrepha fraterculus. J. Econ. Entomol.* 38: 95–100.

Baker, E.W., W.E. Stone, C.C. Plummer, and M. McPhail. 1944. A review of studies on the Mexican fruitfly and related species. *U.S. Dep. Agric. Misc. Pub.* No. 531.

Baker, P.S., A.S.T. Chan, and M.A. Jimeno-Zavala. 1986. Dispersal and orientation of sterile *Ceratitis capitata* and *Anastrepha ludens* (Tephritidae) in Chiapas, Mexico. *J. Appl. Ecol.* 23: 27–38.

Barros, M.D., M. Novaes, and A. Malavasi. 1983. Estudos do comportamento de oviposicao de *Anastrepha fraterculus* (Wiedmann [sic], 1830) (Diptera, Tephritidae) em condicoes naturais e de laboratorio. *An. Soc. Entomol. Bras.* 12: 243–247.

Bateman, M.A. 1972. The ecology of fruit flies. *Annu. Rev. Entomol.* 17: 493–518.

Bateman, M.A. 1979. Dispersal and species interaction as factors in the establishment and success of tropical fruit flies in new areas. *Proc. Ecol. Soc. Aust.* 10: 106–112.

Bateman, M.A. and T.C. Morton. 1981. The importance of ammonia in proteinaceous attractants on fruit flies (Tephritidae). *Aust. J. Agric. Res.* 32: 883–903.

Bateman, M.A., E.F. Boller, G.L Bush, D.L. Chambers, A.P Economopoulos, B.S. Fletcher, M.D. Huettel, V. Moericke, and R.J. Prokopy. 1976. Fruit flies. In *Studies in Biological Control* (V.L. Delucchi, ed.), pp. 11–49. International Biological Programme 9. Cambridge University Press, Cambridge. 304 pp.

Batra, S.W.T. 1979. Reproductive behavior of *Euaresta bella* and *E. festiva* (Diptera: Tephritidae), potential agents for the biological control of adventive North American ragweeds (*Abrosia* spp.) in Eurasia. *J. N.Y. Entomol. Soc.* 87: 118–125.

Benjamin, F.H. 1934. Descriptions of some native trypetid flies with notes on their habits. *U.S. Dep. Agric. Tech. Bull.* No. 401, 95 pp.

Berrigan, D.A., J.R. Carey, J. Guillen, and H. Celedonio. 1988. Age and host effects on clutch size in the Mexican fruit fly, *Anastrepha ludens. Entomol. Exp. Appl.* 47: 73–80.

Bierbaum, T.J. and G.L. Bush. 1988. Divergence in key host examining and acceptance behaviors of the sibling species *Rhagoletis mendax* and *R. pomonella* (Diptera: Tephritidae). In *Ecology and Management of Economically Important Fruit Flies* (M.T. Ali-Niazee, ed.), pp. 26–55. Agricultural Experiment Station, Oregon State University, Corvallis. Special Report 830.

Biggs, J.D. 1972. Aggressive behaviour in the adult of apple maggot (Diptera: Tephritidae). *Can. Entomol.* 104: 349–353.

Blay, S. and B. Yuval. 1997. Nutritional correlates of reproductive success of male Mediterranean fruit flies (Diptera: Tephritidae). *Anim. Behav.* 54: 59–66.

Boller, E.F. 1972. Behavioral aspects of mass rearing of insects. *Entomophaga* 17: 9–25.

Boller, E.F. 1981. Oviposition deterring pheromone of the European cherry fruit fly, *Rhagoletis cerasi L.*; Status of research and potential applications. In *Management of Insect Pests with Semiochemicals* (E.R. Mitchell, ed.), pp. 457–462. Plenum Press, New York. 514 pp.

Boller, E.F. and M. Aluja. 1992. Oviposition deterring pheromone of *Rhagoletis cerasi:* biological activity of 4 synthetic isomers and HMP discrimination of two host races as measured by an improved bioassay. *J. Appl. Entomol.* 113: 113–119.

Boller, E.F. and C.O. Calkins. 1984. Measuring, monitoring and improving the quality of mass-reared Mediterranean fruit flies, *Ceratitis capitata* (Wied.). 3. Improvement of quality by selection. *J. Appl. Entomol.* 98: 1–15.

Boller, E.F. and D.L. Chambers. 1977a. Quality of mass-reared insects. In *Biological Control of Insects by Augmentation of Natural Enemies* (R.L. Ridgway and S.B. Vinson, eds.), pp. 219–236. Plenum Press, New York.

Boller, E.F. and D.L. Chambers. 1977b. Quality Control — An Idea Book for Fruit Fly Workers. *IOBC/WPRS Bull.* 162 pp.

Boller, E.F. and J. Hurter. 1985. Oviposition deterring pheromone in *Rhagoletis cerasi*: Behavioral laboratory test to measure pheromone activity. *Entomol. Exp. Appl.* 39: 163–169.

Boller, E.F. and J. Hurter. 1998. The marking pheromone of the cherry fruit fly: a novel non-toxic and ecologically safe technique to protect cherries against cherry fruit fly infestations. In *Book of Abstracts, 2nd International Symposium on Insect Pheromones, 20 March–3 April, Wageningen*, pp. 99–101. Wageningen.

Boller, E.F. and R.J. Prokopy. 1976. Bionomics and management of *Rhagoletis*. *Annu. Rev. Entomol.* 21: 223–246.

Boller, E.F. and U. Remund. 1973. The flight characteristics of the olive fly (*Dacus oleae* Gmel.) under various conditions observed in connection with the sterile insects release method in Greece. Tech. Rep. FAO contract No. SF/GRE 25-1/AGP. 23 pp.

Boller, E.F., A. Haisch, and R.J. Prokopy. 1970. Ecological and behavioral studies preparing the application of the sterile-insect-release-method (SIRM) against *Rhagoletis cerasi.* In *International Atomic Energy Agency Symposium Sterility Principle for Insect Control or Eradication*, pp. 77–86. International Atomic Energy Agency, Athens.

Boller, E.F., A. Haisch, and R.J. Prokopy. 1971. Sterile-insect release method against *Rhagoletis cerasi* L. Preparatory ecological and behavioural studies. In *Sterility Principle for Insect Control or Eradication.* International Atomic Energy Agency.

Boller, E.F., U. Remund, and W. Zehner. 1975. Sterilization and its influence on the quality of the European cherry fruit fly, *Rhagoletis cerasi* L. In *Proc. Panel on Sterility Principle for Insect Control 1974,* pp. 179–189. International Atomic Energy Agency.

Boller, E.F., U. Remund, B. Katsoyannos, and W. Berchtold. 1977. Quality control in European cherry fruit fly: Evaluation of mating activity in laboratory and field-cage tests. *J. Appl. Entomol.* 83: 182–201.

Boller, E.F., B.I. Katsoyannos, U. Remund, and D.L. Chambers. 1981. Measuring, monitoring and improving the quality of mass-reared Mediterranean fruit flies (*Ceratitis capitata* Wied.). The RAPID quality control system for early warning. *J. Appl. Entomol.* 92: 67–83.

Boller, E.F., R. Schöni, and G.L. Bush. 1987. Oviposition deterring pheromone in *Rhagoletis cerasi*: biological activity of a pure single compound verified in semi-field test. *Entomol Exp. Appl.* 45: 17–22.

Boller, E.F., B.I. Katsoyannos, and C. Hippe. 1998. Host races of *Rhagoletis cerasi* L.: Effect of prior adult experience on oviposition site preference. *J. Appl. Entomol.* 122: 231–237.

Boyce, A.M. 1934. Bionomics of the walnut husk fly, *Rhagoletis completa. Hilgardia* 8: 363–579.

Briceño, R.D., D. Ramos, and W.G. Eberhard. 1996. Courtship behavior of *Ceratitis capitata* (Diptera: Tephritidae) in captivity. *Fla. Entomol.* 79: 130–143.

Brooks, F.E. 1921. Walnut husk maggot. *U.S. Dep. Agric. Bull.* No. 992, 8 pp.

Burdette, R.C. 1935. The biology and control of the pepper maggot *Zonosemata electa* Say. Trypetidae. *N.J. Agric. Exp. Sta. Bull.* 585, 30 pp.

Burk, T. 1981. Signaling and sex in acalyptrate flies. *Fla. Entomol.* 64: 30–43.

Burk, T. 1983. Behavioral ecology of mating in the Caribbean fruit fly, *Anastrepha suspensa* (Loew) (Diptera: Tephritidae). *Fla. Entomol.* 66: 330–344.

Burk, T. 1984. Male–male interactions in Caribbean fruit flies, *Anastrepha suspensa* (Loew) (Diptera: Tephritidae): territorial fights and signaling stimulation. *Fla. Entomol.* 67: 542–548.

Burk, T. 1991. Sex in leks: an overview of sexual behavior in *Anastrepha* fruit flies. In *Proceedings of the International Symposium on the Biology and Control of Fruit Flies* (K. Kawasaki, O. Iwahashi, and K.Y. Kaneshiro, eds.), pp. 2–4. Ginowan, Okinawa.

Burk, T. and C.O. Calkins. 1983. Medfly mating behavior and control strategies. *Fla. Entomol.* 66: 3–18.

Burk, T. and J.C. Webb. 1983. Effect of male size on calling propensity song parameters, and mating success in the Caribbean fruit flies, *Anastrepha suspensa* (Loew). *Ann. Entomol. Soc. Am.* 76: 678–682.

Bush, G.L. 1969. Mating behavior, host specificity, and the ecological significance of sibling species in frugivorous flies of the genus *Rhagoletis* (Diptera: Tephritidae). *Am. Nat.* 103: 669–672.

Bush, G.L. 1974. The mechanism of sympatric host race formation of the true fruit flies (Tephritidae). In *Evolutionary Patterns and Processes* (M.J.D. White, ed.), pp. 3–23. Academic Press, London.

Calkins, C.O. 1984. The importance of understanding fruit fly mating behavior in sterile male release programs (Diptera: Tephritidae). *Folia Entomol. Mex.* 61: 205–213.

Calkins, C.O. 1989. Quality control. In *Fruit Flies: Their Biology, Natural Enemies and Control* (A.S. Robinson and G. Hooper, eds.), pp. 153–165. In *World Crop Pests* (W. Helle, ed.), Vol. 3B. Elsevier Science Publishers, Amsterdam.

Calkins, C.O. 1991. The effect of mass rearing on mating behavior of Mediterranean fruit flies. In *Proceedings of the International Symposium on the Biology and Control of Fruit Flies* (K. Kawasaki, O. Iwahashi, and K.Y. Kaneshiro, eds.), pp. 153–160. Ginowan, Okinawa.

Calkins, C.O. and T.R. Ashley. 1988. Temporal and seasonal differences in movement of the Caribbean fruit fly larvae in grapefruit and the relationship to detection by acoustics. *Fla. Entomol.* 71: 409–416.

Calkins, C.O., T.R. Ashley, and D.L. Chambers. 1996. Implementation of technical and managerial system for quality control in Mediterranean fruit fly (*Ceratitis capitata*) sterile release program. In *Fruit Fly Pests: A World Assessment of Their Biology and Management* (B.A. McPheron and G.J. Steck, eds.), pp. 399–404. St. Lucie Press, Delray Beach.

Cangussu, J.A. and F.S. Zucoloto. 1994. Self-selection and perception threshold in adult females of *Ceratitis capitata* (Diptera: Tephritidae). *J. Insect Physiol.* 41: 223–227.

Causse, R. 1974. Étude d'un rhythme circadien du comportement de prénymphose chez *Ceratitis capitata* Wiedemann (Diptères: Trypetidae). *Ann. Zool. Ecol. Anim.* 6: 475–498.

Causse, R., M. Féron, and M.M. Serment. 1966. Rhythmes nycthéméraux d'activité sexualle inverses l'un de l'autre chez deux Diptères Trypetidae, *Dacus oleae* Gmelin, et *Ceratitis capitata* Wiedemann. *C. R. Seances Acad. Sci. Paris* 262: 1558–1560.

Chambers, D.L. 1977. Quality control in mass rearing. *Annu. Rev. Entomol.* 22: 289–308.

Chambers, D.L., C.O. Calkins, E.F. Boller, Y. Ito, and R.T. Cunningham. 1983. Measuring, monitoring and improving the quality of mass-reared Mediterranean fruit flies. 2. Field tests for confirming and extending laboratory results. *J. Appl. Entomol.* 96: 285–303.

Chapman, M.G. 1982. Experimental analysis of the patterns of tethered flight in the Queensland fruit fly, *Dacus tryoni*. *Physiol. Entomol.* 7: 143–150.

Christenson, L.D. and R.H. Foote. 1960. Biology of fruit flies. *Annu. Rev. Entomol.* 5: 171–192.

Cirio, U. 1971. Reperti sul meccanismo stimolo-risposta nell'ovideposizione del *Dacus oleae* Gmelin (Diptera: Trypetidae). *Redia* 52: 577–600.

Cirio, U. 1972. Osservazioni sul comportamento di ovideposizione della *Rhagoletis completa* Cresson (Diptera: Trypetidae) in laboratorio. *Atti IX Congr. Naz. Ital. Entomol.* 99–117.

Constantino, G. 1930. Contributo alla conocenza della mosca delle frutta (*Ceratitis capitata* Wied.) (Diptera: Trypaneidae). *Boll. Lab. Zool. Sen. Agric. Portici.* 23: 237–322.

Cooley, S.S., R.J. Prokopy, P.T. McDonald, and T.T.Y. Wong. 1986. Learning in oviposition site selection by *Ceratitis capitata* flies. *Entomol. Exp. Appl.* 40: 47–51.

Crawford, D.L. 1927. Investigation of Mexican fruit fly (*Anastrepha ludens* Loew) in Mexico. *Calif. Dep. Agric. Mon. Bull.* 16: 422–445.

Cytrynowicz, M., J.S. Morgante, and H.M.L. De Souza. 1982. Visual responses of South American fruit flies, *Anastrepha fraterculus* and Mediterranean fruit flies, *Ceratitis capitata*, to colored rectangles and spheres. *Environ. Entomol.* 11: 1202–1210.

Darby, H.H. and E.M. Knapp. 1934. Studies of the Mexican fruit fly *Anastrepha ludens* (Loew). *U.S. Dep. Agric. Bull.* 444, 20 pp.

Davis, C.J. 1954. Light intensity and temperature as factors limiting the mating of the Oriental fruit-fly. *Ann. Entomol. Soc. Am.* 47: 593–594.

Dean R.W. and P.J. Chapman. 1973. Bionomics of the apple maggot in eastern New York. *Search Agric. Entomol.* 3: 1–62.

De Lima, I.S., P.E. Howse, and L.A.B. Salles. 1994. Reproductive behaviour of the South American fruit fly *Anastrepha fraterculus* (Diptera: Tephritidae): laboratory and field studies. *Physiol. Entomol.* 19: 271–277.

De Murtas, I.D., U. Cirio, and D. Enkerlin. 1972a. Dispersal of *Ceratitis capitata* on Procida Island. I. Distribution of sterilized Mediterranean fruit fly on Procida and its relation to irradiated doses and feeding. *OEPP/EPPO Bull.* 6: 63–68.

De Murtas, I.D., U. Cirio, and D. Enkerlin. 1972b. Dispersal of *Ceratitis capitata* on Procida Island. II. Movement studies. *OEPP/EPPO Bull.* 6: 69–76.

Debouzie, D. 1977a. Étude de la compétition larvaire chez *Ceratitis capitata* (Diptera: Trypetidae). *Arch. Zool. Exp. Gen.* 315–334.

Debouzie, D. 1977b. Effect of initial population size on *Ceratitis* productivity under limited food conditions. *Ann. Zool. Ecol. Anim.* 9: 367–368.

Debouzie, D. 1981. Analyse experimentale de l'utilisation des resources dans un système simplifé formé d'une banane attaquée par la mouche mediterraneenne des fruits *Ceratitis capitata. Acta Oecol. Gener.* 2: 371–374.

Diehl, S.R. and R.J. Prokopy 1986. Host-selection behavioral differences between the fruit fly sibling species *Rhagoletis pomonella* and *R. mendax* (Diptera: Tephritidae). *Ann. Entomol. Soc. Am.* 79: 266–271.

Dodson, G. 1978. Behavioral, Anatomical and Physiological Aspects of Reproduction in the Caribbean Fruit Fly, *Anastrepha suspensa* (Loew). M.S. thesis, University of Florida, Gainesville, 68 pp.

Dodson, G. 1982. Mating and territoriality in wild *Anastrepha suspensa* (Diptera: Tephritidae) in field cages. *J. Ga. Entomol. Soc.* 17: 189–200.

Dodson, G. 1986. Lek mating system and large male aggressive advantage in a gall forming tephritid fly (Diptera: Tephritidae). *Ethology* 72: 99–108.

Dodson, G. 1987. Biological observations on *Aciurina trixa* and *Valentibulla dodsoni* (Diptera: Tephritidae) in New Mexico. *Ann. Entomol. Soc. Am.* 80: 494–500.

Drew, R.A.I. 1987. Behavioral strategies of fruit flies species of genus *Dacus* (Diptera: Tephritidae) significant in mating and host-plant relationship. *Bull. Entomol. Res.* 77: 73–81.

Drew, R.A.I. and H.A. Fay 1987. Comparison of the roles of ammonia and bacteria in the attraction of *Dacus tryoni* (Froggatt) (Queensland fruit fly) to proteinaceous suspensions. *J. Plant Prot. Trop.* 5: 127–130.

Drew, R.A.I. and A.C. Lloyd. 1987. Relationship of fruit flies (Diptera: Tephritidae) and their bacteria to host plants. *Ann. Entomol. Soc. Am.* 80: 629–636.

Drew, R.A.I. and A.C. Lloyd. 1989. Bacteria associated with fruit flies and their host plants. In *Fruit Flies: Their Biology, Natural Enemies and Control* (A.S. Robinson and G. Hooper, eds.), pp. 131–140. In *World Crop Pests* (W. Helle, ed.), Vol. 3A. Elsevier Science Publishers, Amsterdam.

Drew, R.A.I. and A.C. Lloyd. 1990. The role of bacteria in the life cycle of tephritid fruit flies. In *Microbial Mediation of Plant Herbivore Interactions* (P. Barbosa, V. Krischik, and C.L. Jones, eds.). Wiley, New York.

Drew, R.A.I., A.C. Courtice, and D.S. Teacke. 1983. Bacteria as a natural source of food for adult fruit flies (Diptera: Tephritidae). *Oecologia.* 60: 279–284.

Eberhard, W.G. and F. Pereira. 1993. Functions of the male genitalic surstyli in the Mediterranean fruit fly, *Ceratitis capitata* (Diptera: Tephritidae). *J. Kans. Entomol. Soc.* 66: 427–433.

Economopoulos, A.P., G.E. Haniotakis, J. Mathioudis, L. Missis, and P. Kinigakis. 1978. Long distance flight of wild and artificially-reared *Dacus oleae* (Gmelin) (Diptera: Tephritidae). *J. Appl. Entomol.* 87: 101–108.

Emlen, S.T. and L.W. Oring. 1977. Ecology, sexual selection and the evolution of mating systems. *Science* 197: 215–223.

Epsky, N.D. and R.R. Heath. 1993. Food availability and pheromone production by males of *Anastrepha suspensa* (Diptera: Tephritidae). *Environ. Entomol.* 22: 942–947.

Ernst, B. and B. Wagner. 1989. Synthesis of the oviposition-deterring pheromone (ODP) in *Rhagoletis cerasi* L. *Helv. Chim. Acta* 72: 165–171.

Farleman, M.G. 1933. Observations on the habits of flies belonging to the genus *Rhagoletis*. *J. Econ. Entomol.* 26: 825–828.

Feder, J.L. and G.L. Bush. 1989. A field test of differential host-plant usage between two sibling species of *Rhagoletis pomonella* fruit flies (Diptera: Tephritidae) and its consequences for sympatric models of speciation. *Evolution* 43: 1813–1819.

Fernandes-Da-Silva, P.G. and F.S. Zucoloto. 1993. The influence of host nutritive value on the performance and food selection in *Ceratitis capitata* (Diptera: Tephritidae). *J. Insect Physiol.* 10: 883–810.

Féron, M. 1957. Le comportement de ponte de *Ceratitis capitata* Wied. influencé de la lumière. *Rev. Pathol. Veg. Entomol. Agric. Fr.* 36(3): 127–143.

Féron, M. 1958. Mise en évidence d'un stimulus significatif dans le comportement de *Ceratitis capitata* Wied. (Diptera: Trypetidae). *C.R. Acad. Sci.* 246: 1590–1592.

Féron, M. 1959. Attraction chimique du male de *Ceratitis capitata* Wied. (Diptera: Trypetidae) pour la femelle. *C.R. Acad. Sci.* 248: 2403–2404.

Féron, M. 1960a. Bipotentiale de comportement mâle et female chez un insecte, *Ceratitis capitata* Wied. (Diptera: Trypetidae). *C.R. Acad. Sci.* 250: 2067–2069.

Féron, M. 1960b. l'Appel sonore du mâle dans le comportement sexuel de *Dacus oleae* Gmel. *Bull. Soc. Entomol. Fr.* 65: 139–143.

Féron, M. 1962. l'Instinct de réproduction chez la mouche méditerranéenne des fruits *Ceratitis capitata* Wied. (Dipt. Trypetidae). Comportement sexuel. Comportement de ponte. *Rev. Pathol. Veg. Entomol. Agric. Fr.* 41: 3–129.

Fitt, G.P. 1981a. Inter- and intraspecific responses to sex pheromones in laboratory bioassays by females of three species of tephritid fruit flies from Northern Australia. *Entomol. Exp. Appl.* 30: 40–44.

Fitt, G.P. 1981b. Observations on the biology and behaviour of *Paraceratitella eurycephala* (Diptera: Tephritidae) in Northern Australia. *J. Aust. Entomol. Soc.* 20: 1–7.

Fitt, G.P. 1984. Oviposition behaviour of two tephritid fruit flies, *Dacus tryoni* and *Dacus jarvisi*, as influenced by the presence of larvae in the host fruit. *Oecologia* 62: 37–46.

Fitt, G.P. 1986a. The roles of adult and larval specialization in limiting the occurrence of five species of *Dacus* (Diptera: Tephritidae) in cultivated fruits. *Oecologia* 69: 101–109.

Fitt, G.P. 1986b. The influence of shortage of hosts on the specificity of oviposition behaviour in species of *Dacus* (Diptera: Tephritidae). *Physiol. Entomol.* 11: 133–143.

Fitt, G.P. 1989. The role of interspecific interactions in the dynamics of tephritid population. In *Fruit Flies: Their Biology, Natural Enemies and Control* (A.S. Robinson and G. Hooper, eds.), pp. 281–300. In *World Crop Pests* (W. Helle, ed.), Vol. 3B. Elsevier Science Publishers, Amsterdam.

Fitt, G.P. 1990. Variation in ovariole number and egg size of species of *Dacus* (Diptera: Tephritidae) and their relation to host specialization. *Ecol. Entomol.* 15: 255–264.

Fitt, G.P. and R.W. O'Brien, 1985. Bacteria associated with four species of *Dacus* (Diptera: Tephritidae) and their role in the nutrition of the larvae. *Oecologia* 67: 447–454.

Fletcher, B.S. 1968. Storage and release of sex pheromone by the Queensland fruit fly *Dacus tryoni* (Diptera: Trypetidae). *Nature* 219: 631–2.

Fletcher, B.S. 1973. The ecology of natural populations of the Queensland fruit fly, *Dacus tryoni*. IV. The immigration and emigration of adults. *Aust. J. Zool.* 21: 541–565.

Fletcher, B.S. 1974. The ecology of natural population of the Queensland fruit fly, *Dacus tryoni*. V. The dispersal of adults. *Aust. J. Zool.* 22: 189–202.

Fletcher, B.S. 1987. The biology of dacine fruit flies. *Annu. Rev. Entomol.* 32: 115–144.

Fletcher, B.S. 1989. Movements of tephritid fruit flies. In *Fruit Flies: Their Biology, Natural Enemies and Control* (A.S. Robinson and G. Hooper, eds.), pp. 209–219. In *World Crop Pests* (W. Helle, ed.), Vol. 3B. Elsevier Science Publishers, Amsterdam.

Fletcher, B.S. and A.P. Economopoulos. 1976. Dispersal of normal and irradiated laboratory strains of the olive fly in an olive grove. *Entomol. Exp. Appl.* 20: 183–194.

Fletcher, B.S. and E.T. Kapatos 1981. Dispersal of the olive fruit fly, *Dacus oleae* during the summer period on Corfu. *Entomol. Exp. Appl.* 29: 1–8.

Fletcher, B.S. and R.J. Prokopy. 1991. Host location and oviposition in tephritid fruit flies. In *Reproductive Behavior of Insects. Individuals and Populations* (W.J. Bailey and J. Ridsdill-Smith, eds.), pp. 139–171. Chapman & Hall, London.

Fluke, C.L., Jr. and T.C. Allen. 1931. The role of yeast in life history studies of the apple maggot, *Rhagoletis pomonella* Walsh. *J. Econ. Entomol.* 24: 77–80.

Foote, B.A. 1967. Biology and immature stages of fruit flies: the genus *Icterica* (Diptera: Tephritidae). *Ann. Entomol. Soc. Am.* 60: 1295–1305.

Freese, G. and H. Zwölfer. 1996. The problem of the optimal clutch size in a tritrophic system: the oviposition strategy of the thistle gallfly *Urophora cardui* (Diptera: Tephritidae). *Oecologia* 108: 293–302.

Freidberg, A. 1981. Mating behaviour of *Scistopterum moebiusi* Becker (Diptera: Tephritidae). *Isr. J. Entomol.* 15: 89–95.

Freidberg, A. 1982. Courtship post-mating behavior of the fleabane gall fly *Spathulina tristis* (Diptera: Tephritidae). *Entomol. Gen.* 7: 273–285.

Freidberg, A. 1984a. Gall Tephritidae (Diptera). In *Biology of Gall Insects* (T.N. Ananthakrishnan, ed.), pp. 129–167, Oxford and IBH Publishing Co., New Delhi.

Freidberg, A. 1984b. The mating behavior of *Asteia elegantula* with biological notes of some others Asteiidae (Diptera). *Entomol. Gen.* 7: 217–224.

Freidberg, A. 1986. Mating trophallaxis in Diptera. First Congress of Dipterology, Budapest. 1–2.

Freidberg, A. 1997. Mating trophallaxis in *Metasphenisca negeviana* (Freidberg) (Diptera: Tephritidae). *Isr. J. Entomol.* 31: 199–203.

Gage, M.J.G. 1991. Risk of sperm competition directly affects ejaculate size in the Mediterranean fruit fly. *Anim. Behav.* 42: 1036–1037.

Girolami, V.A., A. Vianello, A. Strapazzon, E. Ragazzi, and E. Veronese. 1981. Ovipositional deterrents in *Dacus oleae*. *Entomol. Exp. Appl.* 29: 177–188.

Häfliger, E. 1953. Das Auswahlvermögen der Kirschenfliege bei der Eiablage (Eine statistische Studie). *Mitt. Schweiz. Entomol. Ges.* 26: 258–264.

Hagen, K.S. 1953. Influence of adult nutrition upon the reproduction of three fruit flies species. *Joint Legislative Committee on Agriculture and Livestock Problems, Third Special Report on Oriental Fruit Fly,* pp. 72–76. Senate of State of California.

Hagen, K.S. 1958. Honeydew as an adult fruit fly diet affecting reproduction. In *Proc. Tenth Int. Congr. Entomol.*, Montreal, 1956, Vol. 3, p. 25.

Haisch, A. 1970. Observations on the flying behaviour of the European cherry fruit fly (*Rhagoletis cerasi* L.). IAEA-SM 186/7: 191–199.

Hamada, R. 1980. Studies on the dispersal behavior of melon flies, *Dacus cucurbitae* Coquillet (Diptera: Tephritidae) and the influence of gamma-irradiation on dispersal. *Appl. Entomol. Zool.* 15: 363–371.

Hanna, A.D. 1947. Studies on the Mediterranean fruit-fly *Ceratitis capitata* Wied. *Bull. Soc. Fouad* 1er. 31: 251–279.

Headrick, D.H. and R.D. Goeden. 1994. Reproductive behavior of California fruit flies and the classification and evolution of Tephritidae (Diptera) mating systems. *Stud. Dipterol.* 1: 194–252.

Headrick, D.H. and R.D. Goeden. 1998. The biology of nonfrugivorous tephritid fruit flies. *Annu. Rev. Entomol.* 43: 217–241.

Heath, R.R., N.D. Epsky, S. Bloem, K. Bloem, F. Acajabon, A. Guzman, and D.L. Chambers. 1994. PH effect on the attractiveness of a corn hydrolysate to the Mediterranean fruit fly and several *Anastrepha* species (Diptera: Tephritidae) *J. Econ. Entomol.* 87: 1008–1013.

Hendrichs, J. 1986. Sexual Selection in Wild and Sterile Caribbean Fruit Flies *Anastrepha suspensa* (Loew). M.Sc. thesis, University of Florida, Gainesville.

Hendrichs, J. and M.A. Hendrichs. 1990. Mediterranean fruit fly, *Ceratitis capitata* (Diptera: Tephritidae) in nature: location and diel pattern of feeding and other activities on fruiting and non-fruiting host and nonhosts. *Ann. Entomol. Soc. Am.* 83: 632–641.

Hendrichs, J. and R.J. Prokopy. 1990. Where do apple maggot flies find fruit in nature? *Mass. Fruit Notes* 55: 1–3.

Hendrichs, J. and J. Reyes. 1987. Reproductive behavior and post-mating female guarding in the monophagous multivoltine *Dacus longistylus* (Wied.) (Diptera: Tephritidae) in southern Egypt. In *Fruit Flies: Proceedings of the Second International Symposium, 16–21 September 1986, Colymbari, Crete, Greece* (A.P. Economopoulos, ed.), pp. 303–313. Elsevier Science Publishers, Amsterdam.

Hendrichs, J. and R.J. Prokopy. 1994. Food foraging behavior of frugivorous fruit flies. In *Fruit Flies and the Sterile Insect Technique* (C.O. Calkins, W. Klassen, and P. Liedo, eds.), pp. 37–55. CRC Press, Boca Raton.

Hendrichs, J., C. Lauzon, S. Cooley, and R.J. Prokopy. 1990. What kind of fruit do the apple maggot flies need for survival and reproduction? *Mass. Fruit Notes* 55: 9–11.

Hendrichs, J., B.I. Katsoyannos, D.R. Papaj, and R.J. Prokopy. 1991. Sex differences in movement between natural feeding and mating sites and tradeoffs food consumption, mating success and predator evasion in Mediterranean fruit flies (Diptera: Tephritidae). *Oecologia* 86: 223–231.

Hendrichs, J., S.S. Cooley, and R.J. Prokopy. 1992. Post-feeding bubbling behavior in fluid feeding Diptera concentration of crop contents by oral evaporation of excess water. *Physiol. Entomol.* 17: 153–161.

Hendrichs, J., B.S. Fletcher, and R.J. Prokopy. 1993. Feeding behavior of *Rhagoletis pomonella* flies (Diptera: Tephritidae): effect of initial food quantity and quality on food foraging, handling cost, and bubbling. *J. Insect Behav.* 6: 43–64.

Herrera, A.L., A.F. Rangel, and L. De la Barrera. 1901. El gusano de la fruta (*Instrypeta ludens* I.D.B.). Mex. *Com. Parasit. Agric. Bot.* 1: 5–44, 76–86, 115–124, 170, 184–196, 291.

Hodson, A.C. 1943. Lures attractive to the apple maggot. *J. Econ. Entomol.* 36: 545–548.

Holbrook, F.R. and M.S. Fujimoto. 1970. Mating competitiveness of unirradiated and irradiated Mediterranean fruit flies. *J. Econ. Entomol.* 63: 1175–1176.

Hooper, G.H.S. 1971. Competitiveness of gamma sterilized males of the Mediterranean fruit fly: effect of irradiating pupae in nitrogen. *J. Econ. Entomol.* 64: 1364–1368.

Hooper, G.H.S. 1972. Sterilization of the Mediterranean fruit fly with gamma radiation: effect on male competitiveness and change in fertility of females alternately mated with irradiated and untreated males. *J. Econ. Entomol.* 65: 1–6.

Hunt, M.K., C.S. Crean, R.J. Wood, and A.S. Gilburn. 1998. Fluctuating asymmetry and sexual selection in the Mediterranean fruitfly (Diptera: Tephritidae). *Biol. J. Linn. Soc.* 64: 385–396.

Illingworth, J.F. 1912. A study of the biology of the apple maggot (*Rhagoletis pomonella)* together with an investigation of methods of control. *Cornell Univ. Agric. Exp. Stn. Bull.* 324: 125–188.

Iwahashi, O. and T. Majima. 1986. Lek formation and male–male competition in the melon fly, *Dacus cucurbitae* Coquillett (Diptera: Tephritidae). *Appl. Entomol. Zool.* 21: 70–75.

Iwahashi, O., Y. Ito, and M. Shiyomi. 1983. A field evaluation of the sexual competitiveness of sterile melon flies, *Dacus (Zegudacus) cucurbitae*. *Ecol. Entomol.* 8: 43–48.

Jácome, I., M. Aluja, P. Liedo, and D. Nestel. 1995. The influence of adult diet and age on lipid reserves in the tropical fruit fly *Anastrepha serpentina* (Diptera: Tephritidae). *J. Insect. Physiol.* 41: 1075–1086.

Jácome, I., M. Aluja, and P. Liedo. 1999. Impact of adult diet on demographic and population parameters in the tropical fruit fly *Anastrepha serpentina* (Diptera: Tephritidae). *Bull. Entomol. Res.* 89: 165–175.

Jang, E.B. and D.M. Light. 1991. Behavioral responses of female Oriental fruit flies to the odor of papayas at three ripeness stages in laboratory flight tunnel (Diptera: Tephritidae). *J. Insect Behav.* 6: 751–762.

Katsoyannos, B.I. 1975. Oviposition-deterring, male-arresting, fruit-marking pheromone in *Rhagoletis cerasi*. *Environ. Entomol.* 4: 801–807.

Katsoyannos, B.I. 1976. Female attraction to males in *Rhagoletis cerasi*. *Environ. Entomol.* 5: 474–476.

Katsoyannos, B.I. 1982. Male sex pheromone of *Rhagoletis cerasi* L. (Diptera: Tephritidae): factors affecting release and response and its role in the mating behavior. *J. Appl. Entomol.* 94: 187–198.

Katsoyannos, B.I. and E.F. Boller. 1976. First field application of oviposition-deterring marking pheromone of the European cherry fruit fly. *Environ. Entomol.* 5: 151–152.

Katsoyannos, B.I. and E.F. Boller. 1980. Second field application of oviposition-deterring pheromone of the European cherry fruit fly, *Rhagoletis cerasi* L. (Diptera: Tephritidae). *J. Appl. Entomol.* 89: 278–281.

Katsoyannos, B.I., K. Panagiotidou, and I. Kechagia. 1986. Effect of color properties on the selection of oviposition site by *Ceratitis capitata*. *Entomol. Exp. Appl.* 42: 187–193.

Katsoyannos, B.I., N.A. Kouloussis, and N.T. Papadopoulos. 1997. Response of *Ceratitis capitata* to citrus chemicals under semi-natural conditions. *Entomol. Exp. Appl.* 82: 181–188.

Kawai, A., O. Iwahashi, and Y. Ito. 1978. Movement of the sterilized melon fly from Kume Island to adjacent islets. *Appl. Entomol. Zool.* 13: 314–315.

Knab, F. and W.W. Yothers. 1914. Papaya fruit fly. *J. Agric. Res.* 6: 447–456.

Kobayashi, R.M., K. Ohinata, D.L. Chambers, and M.S. Fujimoto. 1978. Sex pheromones of the Oriental fruit fly and the melon fruit fly: mating behavior, bioassay method, and attraction of females by live males and by suspected pheromone glands of males. *Environ. Entomol.* 7: 107–112.

Koyama, J. 1989. Mating pheromones: tropical dacines. In *Fruit Flies: Their Biology, Natural Enemies and Control* (A.S. Robinson and G. Hooper, eds.), pp. 165–168. In *World Crop Pests* (W. Helle, ed.), Vol. 3A. Elsevier Science Publishers, Amsterdam.

Kuba, H. 1991. Sex pheromones and mating behavior in dacines. In *Proceedings of the International Symposium on the Biology and Control of Fruit Flies* (K. Kawasaki, O. Iwahashi, and K.Y. Kaneshiro, eds.), pp. 223–232. Ginowan, Okinawa.

Kuba, H. and J. Koyama. 1982. Mating behavior of the melon fly, *Dacus cucurbitae* Coquillett (Diptera: Tephritidae): comparative studies of one wild and two laboratory strains. *Appl. Entomol. Zool.* 17: 559–568.

Lalonde, R.G. and B.D. Roitberg. 1992. Host selection behavior of the thistle-feeding fly: choices and consequences. *Oecologia* 90: 534–539.

Lalonde, R.G. and B.D. Roitberg. 1994. Pollen availability, seed production and seed predator clutch size in tephritid-thistle system. *Evol. Ecol.* 8: 188–195.

Landolt, P.J. 1994. Mating frequency of the papaya fruit fly (Diptera: Tephritidae) with and without host fruit. *Fla. Entomol.* 77: 305–312.

Landolt, P.J. and A.L. Averill. Fruit flies. In *Pheromones on Non-Lepidopteran Insects Associated with Agricultural Plants* (A. Minks and J. Hardie, eds.). CAB INTERNATIONAL, in press.

Landolt, P.J. and J. Hendrichs. 1983. Reproductive behavior of the papaya fruit fly, *Toxotrypana curvicauda* Gerstaecker (Diptera: Tephritidae). *Ann. Entomol. Soc. Am.* 76: 413–417.

Landolt, P.J. and J. Sivinski. 1992. Effects of time of day, adult food and host fruit on the incidence of calling by male Caribbean fruit flies (Diptera: Tephritidae). *Environ. Entomol.* 21: 382–387.

Landolt, P.J., R.R. Heath, and J.R. King. 1985. Behavioral responses of female papaya fruit flies, *Toxotrypana curvicauda* (Diptera: Tephritidae), to male-produced sex pheromone. *Ann. Entomol. Soc. Am.* 78: 751–755.

Landolt, P.J., R.R. Heath, and D.L. Chambers. 1992. Oriented flight responses of females Mediterranean fruit flies to calling males, odor of calling males, and a synthetic pheromone blend. *Entomol. Exp. Appl.* 65: 259–266.

Leroux, E.J. and M.K. Mukerji. 1963. Notes on the distribution of immature stages of the apple maggot *Rhagoletis pomonella* (Walsh) (Diptera: Trypetidae) on apple in Quebec. *Ann. Entomol. Soc. Que.* 8: 60–70.

Liimatainen, J.O., A. Hikkala, and T. Shelly. 1997. Courtship behavior in the Mediterranean fruit fly *Ceratitis capitata* (Diptera: Tephritidae): a comparison of wild and mass reared males. *Ann. Entomol. Soc. Am.* 90: 836–843.

Loher, W. and G. Zervas. 1979. The mating rhythm of the olive fruitfly, *Dacus oleae* Gmelin. *J. Appl. Entomol.* 88: 425–435.

MacFarlane, J.R., R.W. East, R.A.I. Drew, and G.A. Betlinski. 1987. The dispersal of irradiated Queensland fruit fly *Dacus tryoni* (Froggatt) (Diptera: Tephritidae) in South-eastern Australia. *Aust. J. Zool.* 35: 275–282.

Maher, A. 1957. On the bionomics of *Dacus ciliatus* Loew. *Bull. Soc. Entomol. Egypte* 41: 527–533.

Malavasi, A. and R.J. Prokopy. 1992. Effect of food deprivation on the foraging behavior of *Rhagoletis pomonella* females for food and host hawthorn fruit. *J. Entomol. Sci.* 27: 185–193.

Malavasi, A., J.S. Morgante, and R.J. Prokopy. 1983. Distribution and activities of *Anastrepha fraterculus* (Diptera: Tephritidae) flies on host and nonhost trees. *Ann. Entomol. Soc. Am.* 76: 286–292.

Mangan, R.L. 1997. Effects of strain and access to males on female longevity, lifetime oviposition rate, and egg fertility of the Mexican fruit fly (Diptera: Tephritidae). *J. Econ. Entomol.* 90: 945–954.

Mangel, M. and B.D. Roitberg. 1989. Dynamic information and host acceptance by a tephritid fruit fly. *Ecol. Entomol.* 14: 181–189.

Martelli, G. 1909. Alcune note intorno ai costumi ed ai danni della mosca delle arance (*Ceratitis capitata* Wied.). *Boll. Lab. Zool. Sen. Agric. Portici.* 4: 120–127.

Martin, H., J.R. Geigy, and S.A. Bâle. 1953. Observations biologiques et essais de traitements contre la Mouche de l'olive (*Dacus oleae* Rossi) dans la province de Tarragone (Espagne) de 1946 à 1948. *Rev. Pathol. Veg. Entomol. Agric. Fr.* 21: 361–402.

Mason, A.C. 1922. Biology of the papaya fruit fly *Toxotrypana curvicauda*, in Florida. *U.S. Dep. Agric. Bull.* No. 1081.

Maxwell, C.W. 1968. Apple maggot adult dispersion in a New Brunswick apple orchard. *J. Econ. Entomol.* 61: 103–106.

Maxwell, C.W. and E.C. Parson. 1968. Tethered flight of apple maggot flies. *J. Econ. Entomol.* 61: 1157–1159.

McInnis, D.O., D.R. Lane, and C.G. Jackson. 1996. Behavioral resistance to the sterile insect technique by Mediterranean fruit fly (Diptera: Tephritidae) in Hawaii. *Ann. Entomol. Soc. Am.* 89: 739–744.

McPhail, M. 1937. Relation of time of day, temperature and evaporation to attractiveness of fermenting sugar solutions to the Mexican fruitfly. *J. Econ. Entomol.* 30: 793–799.

McPhail, M. 1939. Protein lures for fruit flies. *J. Econ. Entomol.* 32: 758–761.

McPhail, M. and N.O. Berry. 1936. Observations on *Anastrepha pallens* (Coq.) reared from wild fruits in the lower Rio Grande Valley of Texas during the spring of 1932. *J. Econ. Entomol.* 29: 405–410.

McPhail, M. and C.I. Bliss. 1933. Observations on the Mexican fruit fly and some related species in Cuernavaca, Mexico, in 1928 and 1929. *U.S. Dep. Agric. Circ.* 255.

Messina, F.J. 1990. Components of host choice by two *Rhagoletis* species (Diptera: Tephritidae) in Utah. *J. Kans. Entomol. Soc.* 63: 80–87.

Messina, F.J. and V.P. Jones. 1990. Relationship between fruit phenology and infestation by the apple maggot (Diptera: Tephritidae) in Utah. *Ann. Entomol. Soc. Am.* 83: 742–752.

Messina, F.J. and J.K. Subler. 1995. Conspecific and heterospecific interactions of male *Rhagoletis* flies (Diptera: Tephritidae) on a shared host. *J. Kans. Entomol. Soc.* 68: 206–213.

Moericke, V. 1976. Responses of tephritid flies to color stimuli. In *Studies in Biological Control* (V.L. Delucchi, ed.), pp. 167–174. International Biological Programme, 9. Cambridge University Press, Cambridge. 304 pp.

Moericke, V., R.J. Prokopy, S. Berlocher, and G.L. Bush. 1975. Visual stimuli associated with attraction of *Rhagoletis pomonella* flies to trees. *Entomol. Exp. Appl.* 18: 497–507.

Monro, J. 1953. Stridulation in the Queensland fruit-fly (*Dacus (Strumeta) tryoni* Frogg.). *Aust. J. Sci.* 16: 60–62.

Mumtaz, M.M. and M.T. Ali Niazee. 1983. The oviposition deterring pheromone in the western cherry fruit fly, *Rhagoletis indifferens* Curran (Dipt. Tephritidae). *J. Appl. Entomol.* 96: 83–93.

Myburgh, A.C. 1962. Mating habits of the fruit flies *Ceratitis capitata* (Wied.) and *Pterandrus rosa* (Ksh.). *S. Afr. J. Agric. Sci.* 5: 547–464.

Myers, K. 1952. Oviposition and mating behavior of the Queensland fruit fly (*Dacus* (*Strumeta*) *tryoni* Frogg.) and the solanum fruit fly (*Dacus* (*Strumeta*) *cacuminata* (Hering)). *Aust. J. Biol. Sci.* 5: 264–281.

Nakagawa, S., G.J. Farias, D. Suda, R.T. Cunningham, and D.L. Chambers. 1971. Reproduction of the Mediterranean fruit fly: frequency of mating in the laboratory. *Ann. Entomol. Soc. Am.* 64: 949 950.

Nakamori, H and K. Simizu. 1983. Comparison of flight ability between wild and mass-reared melon fly, *Dacus cucurbitae* Coquillet (Diptera: Tephritidae) using a flight mill. *Appl. Entomol. Zool.* 18: 371–378.

Nation, J.L. 1972. Courtship behavior and evidence for sex attractant in the male Caribbean fruit fly, *Anastrepha suspensa*. *Ann. Entomol. Soc. Am.* 65: 1364–1367.

Nation, J.L. 1979. The role of pheromones in the mating system of *Anastrepha* fruit flies. In *Fruit Flies: Their Biology, Natural Enemies and Control* (A.S. Robinson and G. Hooper, eds.), pp. 189–206. In *World Crop Pests* (W. Helle, ed.), Vol. 3A. Elsevier Science Publishers, Amsterdam.

Neilson, W.T.A. 1971. Dispersal studies of a natural population of the apple maggot adults. *J. Econ. Entomol.* 64: 648–653.

Neilson, W.T.A. and J.W. McAllan. 1965. Effects of mating and fecundity of the apple maggot *Rhagoletis pomonella* (Walsh). *Can. Entomol.* 97: 276–279.

Neilson, W.T.A. and F.A. Wood. 1966. Natural food source of the apple maggot. *J. Econ. Entomol.* 59: 997–998.

Novak, J.A. and B.A. Foote. 1975. Biology and immature stages of fruit flies: the genus *Stenopa* (Diptera: Tephritidae) *J. Kans. Entomol. Soc.* 1: 42–52.

Opp, S.B. and R.J. Prokopy. 1986. Variation in laboratory oviposition by *Rhagoletis pomonella* (Diptera: Tephritidae) in relation to mating status. *Ann. Entomol. Soc. Am.* 79: 705–710.

Opp, S.B. and R.J. Prokopy. 1987. Seasonal changes in resighting of marked wild *Rhagoletis pomonella* (Diptera: Tephritidae) flies in nature. *Fla. Entomol.* 70: 450–457.

Owens, E.D. and R.J. Prokopy. 1986. Relationship between reflectance spectra of host plant surfaces and visual detection of host fruit by *Rhagoletis pomonella* flies. *Physiol. Entomol.* 11: 297–307.

Ozaki, E.T. and R.M. Kobayashi. 1981. Effects of pupal handling during laboratory rearing on adult eclosion and flight capability in three tephritid species. *J. Econ. Entomol.* 74: 520–525.

Papaj, D.R. 1994a. Oviposition site guarding by male walnut flies and its possible consequences of mating success. *Behav. Ecol. Sociobiol.* 34: 187–195.

Papaj, D.R. 1994b. Use and avoidance of occupied host as a dynamic process in tephritid flies. In *Insects-Plants Interactions* Vol. V (E.A. Bernays, ed.), pp. 25–46. CRC Press, Boca Raton.

Papaj, D.R. and H. Alonso-Pimentel. 1997. Why walnut flies superparasitize: time savings as a possible explanation. *Oecologia* 109: 166–174.

Papaj, D.R. and M. Aluja. 1993. Temporal dynamics of host-marking in the tropical tephritid fly *Anastrepha ludens*. *Physiol. Entomol.* 18: 279–284.

Papaj, D.R. and R.H. Messing. 1996. Functional shifts in the use of parasitized host by a tephritid fly: the role of host quality. *Behav. Ecol.* 7: 235–242.

Papaj, D.R., and R.H. Messing. 1998. Asymmetries in physiological state as a possible cause of resident advantage in contest. *Behaviour* 135: 1013–1030.

Papaj, D.R. and R.J. Prokopy. 1988. Phytochemical basis of learning in *Rhagoletis pomonella* and other herbivorous insects. *J. Chem. Ecol.* 12: 1125–1143.

Papaj, D.R., B.D. Roitberg, and S.B. Opp. 1989. Serial effects of host infestation on egg allocation by the Mediterranean fruit fly: a rule of a thumb and its functional significance. *J. Anim. Ecol.* 58: 955–970.

Papaj, D.R., B.D. Roitberg, S.B. Opp, M. Aluja, R.J. Prokopy, and T.T.Y. Wong. 1990. Effect of marking pheromone on clutch size in the Mediterranean fruit fly. *Physiol. Entomol.* 15: 463–468.

Papaj, D.R., J.M. Garcia, and H. Alonso-Pimentel. 1996. Marking host fruit by male *Rhagoletis boycei* Cresson flies (Diptera: Tephritidae) and its effect on egg-laying. *J. Insect Behav.* 9: 585–597.

Perdomo, A.J. 1974. Sex and Aggregation Pheromone Bioassays and Mating Observations in the Caribbean Fruit Fly *Anastrepha suspensa* (Loew), under Field Conditions. Ph.D. dissertation, University of Florida, Gainesville.

Peterson, A. 1923. The pepper maggot, a new pest of peppers and eggplants. *N.J. Agric. Exp. Stn. Bull.* 373: 1–23.

Phipps, C.R. and C.O. Dirks. 1933. Dispersal of the apple maggot 1932 studies. *J. Econ. Entomol.* 26: 344–349.

Picado, C. 1920. Historia del gusano de la guayaba. *Colegio de Señoritas Publ.* Serie A No. 2. 28 pp.

Pitarra, K. and B.I. Katsoyannos. 1990. Evidence for a host-marking pheromone in *Chaetorellia australis. Entomol. Exp. Appl.* 54: 287–295.

Plummer, C.C., M. McPhail, and J.W. Monk. 1941. The yellow chapote: a native host of the Mexican fruitfly. *U.S. Dep. Agric. Tech. Bull.* 755, 12 pp.

Porter, B.A. 1928. The apple maggot. *U.S. Dep. Agric. Tech. Bull.* 66: 1–47.

Pritchard, G. 1967. Laboratory observations of the mating behaviour of the island fruit fly *Rioxa pomia* (Diptera: Tephritidae) *J. Aust. Entomol. Soc.* 6: 127–132.

Pritchard, G. 1969. The ecology of the natural population of the Queensland fruit fly, *Dacus tryoni*. II. The distribution of eggs in relation to behavior. *Aust. J. Zool.* 17: 923–311.

Prokopy, R.J. 1968. Visual responses of apple maggot flies *Rhagoletis pomonella* (Diptera: Tephritidae): orchard studies. *Entomol. Exp. Appl.* 11: 403–422.

Prokopy, R.J. 1972. Evidence for a marking pheromone deterring repeated oviposition in apple maggot flies. *Environ. Entomol.* 1: 326–332.

Prokopy, R.J. 1975a. Mating behavior in *Rhagoletis pomonella* (Diptera Tephritidae). V. virgin females attraction to male odor. *Can. Entomol.* 107: 905–908.

Prokopy, R.J. 1975b. Oviposition-deterring fruit marking pheromone in *Rhagoletis fausta. Environ. Entomol.* 4: 298–300.

Prokopy, R.J. 1976. Feeding, mating and oviposition activities of *Rhagoletis fausta* flies in nature. *Ann. Entomol. Soc. Am.* 69: 899–904.

Prokopy, R.J. 1977. Stimuli influencing trophic relations in Tephritidae. *Colloq. Int. C.N.R.S.* 265: 305–336.

Prokopy, R.J. 1980. Mating behavior in frugivorous Tephritidae in nature. In *Proceedings of a Symposium on Fruit Fly Problems* (J. Koyama, ed.), pp. 37–45. XVI International Congress of Entomology. Kyoto, Japan.

Prokopy, R.J. 1981a. Oviposition deterring pheromone system of apple maggot flies. In *Management of Insect Pest with Semiochemicals* (E.R. Mitchell, ed.), pp. 477–494. Plenum, New York.

Prokopy, R.J. 1981b. Epideitic pheromones that influence spacing patterns of phytophagous insects, In *Semiochemicals: Their Role in Pest Control* (D.A. Nordlund, R.L. Jones, and W.J. Lewis, eds.), pp. 181–213. Wiley, New York.

Prokopy, R.J. 1982. Tephritid relationship with plants. In *Fruit Flies of Economic Importance* (R. Cavalloro, ed.), pp. 230–239. A.A. Balkema, Rotterdam.

Prokopy, R.J. 1986. Visual and olfactory stimulus interaction in resource finding by insects. In *Mechanisms in Insects Olfaction* (T.L. Payne, M.C. Birch, and C.E.J. Kennedy, eds.), pp. 81–89. Oxford University Press, Oxford.

Prokopy, R.J. 1993. Levels of quantitative investigation of tephritid fly foraging behavior. In *Fruit Flies: Biology and Management* (M. Aluja and P. Liedo, eds.), pp. 165–171. Springer-Verlag, New York.

Prokopy, R.J. and E.F. Boller. 1971. Stimuli eliciting oviposition of the European cherry fruit flies, *Rhagoletis cerasi* (Diptera: Tephritidae), into inanimate objets. *Entomol. Exp. Appl.* 14: 1–14.

Prokopy, R.J. and G.L. Bush. 1973a. Mating behavior in *Rhagoletis pomonella* (Diptera: Tephritidae). IV. Courtship. *Can. Entomol.* 105: 873–891.

Prokopy, R.J. and G.L. Bush. 1973b. Oviposition responses to different sizes of artificial fruit by flies of *Rhagoletis pomonella* species group. *Ann. Entomol. Soc. Am.* 66: 927–929.

Prokopy, R.J. and G.L. Bush. 1973c. Oviposition by grouped and isolated apple maggot flies. *Ann. Entomol. Soc. Am.* 66: 1197–1200.

Prokopy, R.J. and J.J. Duan. 1998. Socially facilitated egglaying behavior in Mediterranean fruit flies. *Behav. Ecol. Sociobiol.* 42: 117–122.

Prokopy, R.J. and J. Hendrichs. 1979. Mating behavior of *Ceratitis capitata* on a field caged host tree. *Ann. Entomol. Soc. Am.* 72: 642–648.

Prokopy, R.J. and J. Koyama. 1982. Oviposition site partitioning in *Dacus cucurbitae*. *Entomol. Exp. Appl.* 31: 428–432.

Prokopy, R.J. and D.R. Papaj. 1989. Can ovipositing *Rhagoletis pomonella* females (Diptera: Tephritidae) learn to discriminate among different ripeness stages of the same host biotype? *Fla. Entomol.* 72: 489–494.

Prokopy, R.J. and A.H. Reynolds. 1998. Ovipositional enhancement through social facilitation behavior in *Rhagoletis pomonella* flies. *Entomol. Exp. Appl.* 86: 281–286.

Prokopy, R.J. and B.D. Roitberg. 1983. Foraging behavior of true fruit flies. *Am. Sci.* 72: 41–49.

Prokopy, R.J., and B.D. Roitberg. 1989. Fruit fly foraging behavior. In *Fruit Flies: Their Biology, Natural Enemies and Control* (A.S. Robinson and G. Hooper, eds.), pp. 293–306. In *World Crop Pests* (W. Helle, ed.), Vol. 3A. Elsevier Science Publishers, Amsterdam.

Prokopy, R.J., E.W. Bennet, and G.L. Bush. 1971. Mating behavior in *Rhagoletis pomonella* (Diptera: Tephritidae) I. Site of assembly. *Can. Entomol.* 112: 1319–1320.

Prokopy, R.J., E.W. Bennet, and G.L. Bush. 1972. Mating behavior in *Rhagoletis pomonella* (Diptera: Tephritidae) II. Temporal organization. *Can. Entomol.* 104: 97–104.

Prokopy, R.J., W.H. Reissig, and V. Moericke. 1976. Marking pheromone deterring repeated oviposition in *Rhagoletis* flies. *Entomol. Exp. Appl.* 20: 170–178.

Prokopy, R.J., P.D. Greany, and D.L. Chambers. 1977. Oviposition-deterring pheromone in *Anastrepha suspensa*. *Environ. Entomol.* 6: 463–465.

Prokopy, R.J., J.R. Ziegler, and T.T.Y. Wong. 1978. Deterrence of repeated oviposition by fruit marking pheromone in *Ceratitis capitata* (Diptera: Tephritidae). *J. Chem. Ecol.* 4: 55–63.

Prokopy, R.J., A.L. Averill, S.S. Cooley, and C.A. Roitberg. 1982a. Associative learning in egglaying site selection by apple maggot flies. *Science* 218: 76–77.

Prokopy, R.J., A. Malavasi, and J.S. Morgante. 1982b. Oviposition-deterring pheromone in *Anastrepha fraterculus* flies. *J. Chem. Ecol.* 8: 763–771.

Prokopy, R.J., D.R. Papaj, S.S. Cooley, and C. Kallet. 1986. On the nature of learning in oviposition site acceptance by apple maggot flies. *Anim. Behav.* 34: 98–107.

Prokopy, R.J., T.A. Green, and T.T.Y. Wong. 1989a. Learning to find host fruit in *Ceratitis capitata* flies. *Entomol. Exp. Appl.* 45: 65–72.

Prokopy, R.J., D.R. Papaj, S.B. Opp, and T.T.Y. Wong. 1989b. Intra-tree foraging behavior of *Ceratitis capitata* flies in relation to host fruit density and quality. *Entomol. Exp. Appl.* 45: 251–258.

Prokopy, R.J., M. Aluja, and T.T.Y. Wong. 1989c. Foraging behavior of laboratory cultured Mediterranean fruit flies on field-cage host trees. *Proc. Hawaii. Entomol. Soc.* 29: 103–109.

Prokopy, R.J., D.R. Papaj, J. Hendrichs, and T.T.Y. Wong. 1992. Behavioral responses of *Ceratitis capitata* flies to bait spray droplets and natural food. *Entomol. Exp. Appl.* 64: 247–257.

Prokopy, R.J., S.S. Cooley, and D.R. Papaj. 1993a. How well can naive *Rhagoletis* flies learn to discriminate fruit for oviposition? *J. Insect Behav.* 6: 167–176.

Prokopy, R.J., S.S. Cooley, J.J. Prokopy, Q. Quan, and J.P. Buonaccorsi. 1993b. Interactive effects of resource abundance and state of adults on residence of apple maggot (Diptera: Tephritidae) flies in host tree patches. *Environ. Entomol.* 2: 304–315.

Prokopy, R.J., L.H. Chiou, and R.I. Vargas. 1993c. Effect of source and condition of animal excrement on attractiveness to adults of *Ceratitis capitata* (Diptera: Tephritidae). *Environ. Entomol.* 22: 453–458.

Prokopy, R.J., C. Bergweiler, L. Galarza, and J. Schwerin. 1994a. Prior experience affects the visual ability of *Rhagoletis pomonella* flies (Diptera: Tephritidae) to find host fruit. *J. Insect Behav.* 7: 663–677.

Prokopy, R.J., B.D. Roitberg, and R.I. Vargas. 1994b. Effect of egg load on finding and acceptance of host fruit in *Ceratitis capitata* flies. *Physiol. Entomol.* 19: 124–132.

Prokopy, R.J., S.S. Cooley, I. Luna, and J.J. Duan. 1995. Combined influence of protein hunger and egg load on the resource foraging behavior of *Rhagoletis pomonella* flies (Diptera: Tephritidae). *Eur. J. Entomol.* 92: 655–666.

Prokopy, R.J., R. Poramarcom, M. Sutantawong, R. Dokmaihom, and J. Hendrichs. 1996. Localization of mating behavior of released *Bactrocera dorsalis* flies on host fruit in an orchard. *J. Insect Behav.* 9: 133–142.

Rajamannar, N. 1962. Growth, orientation and feeding behavior of the larvae of the melon *Dacus cucurbitae* Coq. on various plants. *Proc. Natl. Inst. Sci. India* 28: 133–142.

Reissig, W.H. 1974. Field test of the response of the *Rhagoletis pomonella* to apples. *Environ. Entomol.* 3: 733–736.

Remund, U. and E.F. Boller. 1975. Qualitätskontrolle bei Insekten: Messung von Flugparametern. *J. Appl. Entomol.* 78: 113–126.

Remund, U., B.I. Katsoyannos, E.F. Boller, and W. Berchtold. 1980. Zur Eiverteilung der Kirschenfliege, *Rhagoletis cerasi* L. (Dipt. Tephritidae), im Freiland. *Mitt. Schweiz. Entomol. Ges.* 53: 401–405.

Roan, C.C., N.E. Flitters, and C.J. Davis. 1954. Light intensity and temperature as factors limiting the mating of the Oriental fruit-fly. *Ann. Entomol. Soc. Am.* 47: 593–594.

Robacker, D.C., S.J. Ingleand, and W.G. Hart. 1985. Mating frequency and response to male-produced pheromone by virgin and mated females of the Mexican fruit fly. *Southwest. Entomol.* 10: 215–221.

Robacker, D.C., R.L. Mangan, D.S. Moreno, and A.M. Tarshis-Moreno. 1991. Mating behavior and male mating success in wild *Anastrepha ludens* (Diptera: Tephritidae) on a field cage host tree. *J. Insect Behav.* 4: 471–487.

Roitberg, B.D. 1985. Search dynamics in fruit parasitic insects. *J. Insect Physiol.* 31: 865–872.

Roitberg, B.D. and M. Mangel. 1988. On the evolutionary ecology of marking pheromones. *Evol. Ecol.* 2: 289–315.

Roitberg, B.D. and R.J. Prokopy. 1981. Experience required for pheromone recognition in the apple maggot fly. *Nature* 292: 540–541.

Roitberg, B.D. and R.J. Prokopy. 1982. Influence of inter-tree distance on foraging behaviour of *Rhagoletis pomonella* in the field. *Ecol. Entomol.* 7: 437–442.

Roitberg, B.D. and R.J. Prokopy. 1983. Hosts deprivation influence on response of *Rhagoletis pomonella* to its oviposition deterring pheromone. *Physiol. Entomol.* 8: 69–72.

Roitberg, B.D. and R.J. Prokopy. 1984. Host visitation as a determinant of search persistence in fruit parasitic tephritid flies. *Oecologia* 62: 7–12.

Roitberg, B.D., J.C. Van Lenteren, J.J.M. Van Alphen, F. Galis, and R.J. Prokopy. 1982. Foraging behaviour of *Rhagoletis pomonella* a parasite of hawthorn (*Crataegus viridis*), in nature. *J. Anim. Ecol.* 51: 307–325.

Rössler, Y. 1975a. The ability to inseminate: a comparison between a laboratory reared and field collected populations of the Mediterranean fruitfly, *Ceratitis capitata. Entomol. Exp. Appl.* 18: 255–257.

Rössler, Y. 1975b. Reproductive differences between laboratory-reared and field populations of the Mediterranean fruitfly, *Ceratitis capitata. Ann. Entomol. Soc. Am.* 68: 987–991.

Sanders, W. 1962. Das Verhalten der Mittelmeerfruchtfliege *Ceratitis capitata* Wied. bei der Eiablage. *Z. Tierpsychol.* 19: 1–28.

Sanders, W. 1968a. Die Eiablagehandlung der Mittelmeerfruchtfliege, *Ceratitis capitata* Wied. Ihre Anhängigkeit von Farbe und Gliederung des Umfeldes. *Z. Tierpsychol.* 25: 588–60.

Sanders, W. 1968b. Die Eiablagehandlung der Mittelmeerfruchtfliege, *Ceratitis capitata* Wied. Ihre Anhängigkeit von Grösse und Dichte der Früchte. *Z. Tierpsychol.* 25: 1–23.

Sanders, W. 1969a. Die Eiablagehandlung der Mittelmeerfruchtfliege, *Ceratitis capitata* Wied. Ihre Anhängigkeit von der Oberflächen- und Innenfeuchte der Früchte. *Z. Tierpsychol.* 26: 236–242.

Sanders, W. 1969b. Die Eiablagehandlung der Mittelmeerfruchtfliege, *Ceratitis capitata* Wied. Der Einfluss der Zwischenflüge auf die Wahl des Eiablageortes. *Z. Tierpsychol.* 26: 853–865.

Saul, S.H. and S.D. McCombs. 1993. Dynamics of sperm use in the Mediterranean fruit fly (Diptera: Tephritidae). Reproductive fitness of multiple-mated females and sequentially mated males. *Ann. Entomol. Soc. Am.* 86: 198–202.

Saul, S.H. and S.D. McCombs. 1994. Increased remating frequency in sex ratio distorted lines of the Mediterranean fruit flies (Diptera: Tephritidae). *Ann. Entomol. Soc. Am.* 86: 631–637.

Severin, H.H.P. 1917. The currant fruit fly. *Maine Agric. Exp. Stn. Bull.* 264: 177–247.

Severin, H.H.P. and W.J. Hartung. 1912. The flight of two thousand marked male Mediterranean fruit flies (*Ceratitis capitata* Wied.). *Ann. Entomol. Soc. Am.* 5: 400–407.

Severin, H.H.P. and H.C. Severin. 1913. A historical account on the use of kerosene to trap the Mediterranean fruit fly (*Ceratitis capitata* Wied.). *J. Econ. Entomol.* 6: 347–351.

Severin, H.H.P. and H.C. Severin. 1914. Behavior of the Mediterranean fruit fly (*Ceratitis capitata* Wied.) toward kerosene. *J. Anim. Behav.* 4: 223–227.

Severin, H.H.P., H.C. Severin, and W.J. Hartung. 1914. The ravages, life history, weights of stages, natural enemies and methods of control of the melon fly (*Dacus cucurbitae* Coq.). *Ann. Entomol. Soc. Am.* 3: 178–207.

Sharp, J.L. 1976. Comparison of flight ability of wild-type and laboratory-reared Caribbean fruit flies on a flight mill. *J. Ga. Entomol. Soc.* 11: 255–258.

Sharp, J.L. 1978. Tethered flight of apple maggot flies. *Fla. Entomol.* 61: 199–200.

Sharp, J.L. 1980. Flight propensity of *Anastrepha suspensa*. *J. Econ. Entomol.* 73: 631–633.

Sharp, J.L., D.L. Chambers, and F.H. Haromoto. 1975. Flight mill and stroboscope studies of oriental fruit flies and melon flies including observations on Mediterranean fruit flies. *Proc. Hawaii. Entomol. Soc.* 22: 137–144.

Sharp, J.L., E.F. Boller, and D.L. Chambers. 1983. Selection for flight propensity of laboratory and wild strains of *Anastrepha suspensa* and *Ceratitis capitata*. (Diptera: Tephritidae). *J. Econ. Entomol.* 76: 302–305.

Shaw, J.G. 1947. Hosts and distribution of *Anastrepha serpentina* in northeastern Mexico. *J. Econ. Entomol.* 40: 34–40.

Shelly, T.E. and T.S. Whittier. 1994a. Lek distribution in the Mediterranean fruit fly (Diptera: Tephritidae): influence of tree size, foliage density and neighborhood. *Proc. Hawaii. Entomol. Soc.* 32: 113–122.

Shelly, T.E. and T.S. Whittier. 1994b. Effect of sexual experience on the mating success of males of the Mediterranean fruit fly *Ceratitis capitata* Wied. (Diptera: Tephritidae). *Proc. Hawaii. Entomol. Soc.* 32: 91–94.

Shelly, T.E., T.S. Whittier, and K. Kaneshiro. 1994. Sterile insect release and the natural mating system of the Mediterranean fruit fly, *Ceratitis capitata* (Diptera: Tephritidae). *Ann. Entomol. Soc. Am.* 87: 470–481.

Silvestri, F. 1914. Report of an expedition to Africa in search of the natural enemies of fruit flies (Trypaneidae), with descriptions, observations and biological notes. *Hawaii Board Agric. For. Div. Entomol. Bull.* 3, 176 pp.

Sivinski, J. 1984. Effect of sexual experience on male mating success in a lek forming tephritid *Anastrepha suspensa* (Loew). *Fla. Entomol.* 67: 126–130.

Sivinski, J. 1989. Lekking and the small-scale distribution of the sexes in the Caribbean fruit fly *Anastrepha suspensa* (Loew). *J. Insect Behav.* 2: 3–13.

Sivinski, J. 1993. Longevity and fecundity in the Caribbean fruit fly (Diptera: Tephritidae): effect of mating, strain and body size. *Fla. Entomol.* 76: 635–644.

Sivinski, J. and T. Burk. 1989. Reproductive and mating behaviour. In *Fruit Flies: Their Biology, Natural Enemies and Control* (A.S. Robinson and G. Hooper, eds.), pp. 343–351. In *World Crop Pests* (W. Helle, ed.), Vol. 3A. Elsevier Science Publishers, Amsterdam.

Sivinski, J. and C.O. Calkins. 1986. Pheromones and parapheromones in the control of tephritids. *Fla. Entomol.* 69: 157–168.

Sivinski, J. and G. Dodson. 1992. Sexual dimorphism in *Anastrepha suspensa* (Loew) and other tephritid fruit flies: possible roles of developmental rate, fecundity and dispersal. *J. Insect Behav.* 5: 491–506.

Sivinski, J. and R.R. Heath. 1988. Effects of oviposition on remating, response to pheromones and longevity in female Caribbean fruit fly, *Anastrepha suspensa* (Diptera: Tephritidae). *Ann. Entomol. Soc. Am.* 81: 1021–1024.

Sivinski, J. and J.C. Webb. 1985a. Sound production and reception in the Caribfly, *Anastrepha suspensa*. *Fla. Entomol.* 68: 273–278.

Sivinski, J. and J.C. Webb. 1985b. The form and function of acoustic courtship signals of the papaya fruit fly *Toxotrypana curvicauda*. *Fla. Entomol.* 68: 634–641.

Sivinski, J. and J.C. Webb. 1986. Changes in Caribbean fruit fly acoustic signal with social situation (Diptera: Tephritidae). *Ann. Entomol. Soc. Am.* 79: 146–149.

Sivinski, J., T. Burk, and J.C. Webb. 1984. Acoustic courtship signals in the Caribbean fruit fly, *Anastrepha suspensa* (Loew). *Anim. Behav.* 32: 1011–1016.

Sivinski, J., N. Epsky, and R.R. Heath. 1994. Pheromone deposition on leaf territories by male Caribbean fruit flies, *Anastrepha suspensa* (Loew) (Diptera: Tephritidae). *J. Insect Behav.* 7: 43–51.

Smith, P.H. 1979. Genetic manipulation of the circadian clock's timing of sexual behavior in the Queensland fruit fly, *Dacus tryoni* and *Dacus neohumeralis*. *Physiol. Entomol.* 4: 71–78.

Smith, D.C. 1984. Feeding, mating, and oviposition by *Rhagoletis cingulata* (Diptera: Tephritidae) flies in nature. *Ann. Entomol. Soc. Am.* 77: 702–704.

Smith, D.C. and R.J. Prokopy. 1980. Mating behavior of *Rhagoletis pomonella* (Diptera: Tephritidae). VI. Site of early-season encounters. *Can. Entomol.* 112: 585–590.

Smith, D.C. and R.J. Prokopy. 1981. Seasonal and diurnal activity of *Rhagoletis mendax* flies in nature. *Ann. Entomol. Soc. Am.* 74: 462–466.

Smith, D.C. and R.J. Prokopy. 1982. Mating behavior of *Rhagoletis mendax* (Diptera: Tephritidae) flies in nature. *Ann. Entomol. Soc. Am.* 75: 388–392.

Stoltzfus, W.B. and B.A. Foote. 1965. The use of froth masses in courtship of *Eutetra* (Diptera: Tephritidae). *Proc. Entomol. Soc. Wash.* 67: 263–264.

Story, J.M. and N.L. Anderson. 1978. Release and establishment of *Urophora affinis* (Diptera: Tephritidae) on spotted knapweed in western Montana. *Environ. Entomol.* 7: 445–448.

Straw, N.A. 1989. Evidence for an oviposition-deterring pheromone in *Tephritis bardanae* (Schrank) (Diptera: Tephritidae). *Oecologia* 78: 121–130.

Susuki, Y. and J. Koyama. 1980. Temporal aspects of mating behavior of the melon fly, *Dacus cucurbitae* Coquillett (Diptera: Tephritidae): A comparison between laboratory and wild strains. *Appl. Entomol. Zool.* 15: 215–224.

Tauber, M.J. and C.A. Tauber. 1967. Reproductive behavior and biology of the gall-former *Aciurina ferruginea* (Doane) (Diptera: Tephritidae). *Can. J. Zool.* 45: 907–913.

Tauber, M.J. and C.A. Toschi. 1965a. Bionomics of *Euleia fratria* (Loew) (Diptera: Tephritidae). I. Life history and mating behavior. *Can. J. Zool.* 43: 369–379.

Tauber, M.J. and C.A. Toschi. 1965b. Life history and mating behavior of *Tephritis stigmatica* (Coquillett) (Diptera: Tephritidae). *Pan-Pac. Entomol.* 41: 73–79.

Trujillo, G. 1998. Efecto de la Dieta, Tamaño de los Adultos, Presencia de Hospedero y Condición Fértil o Estéril de los Machos en el Número de Apareamientos y Periodo Refractorio de Hembras de *A. ludens* (Loew) y *A. obliqua* (Macquart) (Diptera: Tephritidae). B.Sc. thesis, Universidad Veracruzana, Veracruz, México.

Tychsen, P.H. 1975. Circadian control of sexual drive level in *Dacus tryoni* (Diptera: Tephritidae). *Behaviour* 54: 111–114.

Tychsen, P.H. 1977. Mating behavior of the Queensland fruit fly *Dacus tryoni*, in field cages. *J. Aust. Entomol. Soc.* 16: 459–465.

Tychsen, P.H. 1978. The effect of photoperiod on the circadian rhythm of mating responsiveness in the fruit fly, *Dacus tryoni*. *Physiol. Entomol.* 3: 65–69.

Tychsen, P.H. and B.S. Fletcher. 1971. Studies on the rhythm of mating in the Queensland fruit fly, *Dacus tryoni* (Frogg). *J. Insect Physiol.* 17: 2139–2156.

Tzanakakis, M.E., J.A. Tsitsipis, and M.E. Economopoulos. 1968. Frequency of mating in females of the olive fruit fly under laboratory conditions. *J. Econ. Entomol.* 61: 1309–1312.

Uhler, L.D. 1951. Biology and ecology of the goldenrod gall fly, *Eurosta solidaginis* (Fitch). *Cornell Univ. Agric. Exp. Stn. Mem.* 300: 1–51.

Varley, G.C. 1947. The natural control of population balance in the knapweed gall-fly (*Urophora jaceana*). *J. Anim. Ecol.* 16: 139–187.

Walder, J.M.M. and C.O. Calkins. 1993. Effects of gamma radiation on the sterility and behavioral quality of the Caribbean fruit fly, *Anastrepha suspensa* (Loew) (Diptera: Tephritidae). *Sci. Agric. Piracicaba* 50: 157–165.

Warburg, M.S. and B. Yuval. 1996. Effects of diet and activity on lipid levels of adult Mediterranean fruit flies. *Physiol Entomol.* 21: 151–158.

Warburg, M.S. and B. Yuval. 1997a. Circadian patterns of feeding and reproductive activities of Mediterranean fruit flies (Diptera: Tephritidae) on various hosts in Israel. *Ann. Entomol. Soc. Am.* 90: 487–495.

Warburg, M.S. and B. Yuval. 1997b. Effects of energetic reserves on behavioral patterns of Mediterranean fruit flies (Diptera: Tephritidae). *Oecologia* 112: 314–319.

Webb, J.C., J.L. Sharp, D.L. Chambers, J.J. McDow, and J.C. Benner. 1976. The analysis and identification of sound produced by the male Caribbean fruit fly, *Anastrepha suspensa* (Loew). *Ann. Entomol. Soc. Am.* 65: 415–420.

Webb, J.C., C.O. Calkins, D.L. Chambers, W. Schwienbacher, and K. Russ. 1983a. Acoustical aspects of behavior of Mediterranean fruit fly, *Ceratitis capitata*: analysis and identification of courtship sounds. *Entomol. Exp. Appl.* 33: 1–8.

Webb, J.C., T. Burk, and J. Sivinski. 1983b. Attraction of female Caribbean fruit flies (*Anastrepha suspensa* (Loew)) (Diptera: Tephritidae) to males and male-produce stimuli in field cages. *Ann. Entomol. Soc. Am.* 76: 996–998.

Webb, J.C., J. Sivinski, and C. Litzkow. 1984. Acoustical behavior and sexual success in Caribbean fruit fly, *Anastrepha suspensa* (Loew) (Diptera: Tephritidae). *Environ. Entomol.* 13: 650–656.

Webb, J.C., J. Sivinski, and B. Smittle. B. 1985. Acoustical signals courtship and sexual success in irradiated Caribfly *Anastrepha suspensa* (Loew) (Diptera: Tephritidae). *Fla. Entomol.* 70: 103–109.

Whittier, T.S. and K.Y. Kaneshiro. 1991. Male mating success and female fitness in the Mediterranean fruit fly (Diptera: Tephritidae). *Ann. Entomol. Soc. Am.* 84: 608–611.

Whittier, T.S. and K.Y. Kaneshiro. 1995. Intersexual selection in the Mediterranean fruit fly: does female choice enhance fitness? *Evolution* 49: 990–996.

Whittier, T.S. and T.E. Shelly. 1993. Productivity of singly vs. multiply mated females Mediterranean fruit flies, *Ceratitis capitata* (Diptera: Tephritidae). *J. Kans. Entomol. Soc.* 66: 200–209.

Whittier, T.S., K.Y. Kaneshiro, and L.D. Prescott. 1992. Mating behavior of Mediterranean fruit flies (Diptera: Tephritidae) in a natural environment. *Ann. Entomol. Soc. Am.* 85: 214–218.

Whittier, T.S., F.Y. Nam, T.E. Shelly, and K.Y. Kaneshiro. 1994. Male courtship success and female discrimination in the Mediterranean fruit fly (Diptera: Tephritidae). *J. Insect. Behav.* 7: 159–170.

Wiesmann, R. 1933. Untersuchungen über die Lebensgeschichte und Bekämpfung der Kirschenfliege *Rhagoletis cerasi* Linné. *I. Mitt. Landw. Jahrb. Schweiz.* 1933: 711–760.

Wiesmann, R. 1937. Die Orientierung der Kirschenfliege *Rhagoletis cerasi* L. *Mitt. Landw. Jahrb. Schweiz.* 51: 1080–1109.

Wong, T.T.Y., L. Whitehand, R.M. Kobayashi, K. Ohinata, N. Tanaka, and E.J. Harris 1982. Mediterranean fruit fly: dispersal of wild and irradiated and untreated laboratory-reared males. *Environ. Entomol.* 11: 339–343.

Yuval, B., R. Kaspi, S. Shloush, and M.S. Warburg. 1998. Nutritional reserves regulate male participation in Mediterranean fruit fly leks. *Ecol. Entomol.* 23: 211–215.

Zucoloto, F.S. 1988. Qualitative and quantitative competition for food in *Ceratitis capitata* (Diptera: Tephritidae). *Rev. Bras. Biol.* 48: 523–526.

Zucoloto, F.S. 1990. Effects of flavour and nutritional value on diet selection by *Ceratitis capitata* larvae (Diptera: Tephritidae). *J. Physiol. Entomol.* 37: 21–25.

Zucoloto, F.S. 1993a. Acceptability of different Brazilian fruits to *Ceratitis capitata* (Diptera: Tephritidae) and fly performance on each species. *Braz. J. Med. Biol. Res.* 26: 291–298.

Zucoloto, F.S. 1993b. Nutritive value and selection of diets containing different carbohydrates by larvae of *Ceratitis capitata* (Diptera: Tephritidae). *Rev. Bras. Biol.* 53: 611–618.

Zwölfer, H. 1974. Das Treffpunkt-Prinzip als Kommunikationsstrategie und Isolationsmechanismus bei Bohrfliegen (Diptera: Trypetidae). *Entomol. Germ.* 1: 11–20.

Zwölfer, H. 1983. Life systems and strategies of resource exploitation in tephritids. In *Fruit Flies of Economic Importance* (R. Cavalloro, ed.), pp. 16–30. A.A. Balkema, Rotterdam.

Zwölfer, H. 1988. Evolutionary and ecological relationships of the insect fauna of thistles. *Ann. Rev. Entomol.* 33: 103–122.

# Section II

## Higher Relationships of Tephritidae

# 4 Phylogenetic Relationships among Higher Groups of Tephritidae

*Valery A. Korneyev*

## CONTENTS



0-8493-1275-2/00/$0.00+$.50

## 4.1 INTRODUCTION

A previous chapter (Korneyev, Chapter 1) presented the ground plan of the superfamily Tephritoidea and its main phylogenetic clusters, including the family Tephritidae. It was hypothesized that the family Pyrgotidae is the sister group of the Tephritidae, and that the latter family must be expanded to incorporate the Tachiniscidae which actually are aberrant members of the group previously known as Ortalotrypetini.

This chapter attempts to trace the phylogeny of the family Tephritidae and its main taxonomic groupings based mainly on morphological evidence. It focuses on the relationships among the subfamilies and within those other than the Tephritinae, which are analyzed in detail by Korneyev (Chapter 22).

My attempts to apply computer analysis for the reconstruction of the phylogeny of the Tephritidae were frustrated by large gaps in data on genital and larval morphology, especially in tropical groups. Many taxa were scored only from descriptions or illustrations in the literature (those examined personally are listed in Appendix 4.1), and in numerous cases critical characters have not been reported or I was unsure of the meaning of the descriptions. Missing data are an even more difficult obstacle than the numerous cases of homoplasy that appear to have occurred within the family. The data matrices and results of the analyses are reported, mainly to indicate the characters analyzed and show the distributions of their states. Because of the above-mentioned problems and computer limitations that restricted a complete analysis, I have low confidence in some of the cladistic results, and the putative phylogenetic hypotheses discussed below often do not conform to the cladograms.

The cladistic analysis was performed on the major subgroups of the family Tephritidae using the program Hennig86 (Farris 1988). A character state matrix consisting of 149 terminal taxa and 112 characters (including dummy outgroup and character) (Table 4.1) was used. All the characters except #56 were considered additive. Autapomorphies were not included. The hypothesized ground plan character set for the family was used as outgroup, and representative species or genera of all subfamilies and most tribes were included. Only a few taxa of the large subfamily Tephritinae, which is analyzed in detail by Korneyev (Chapter 22), are included. An analysis of the full matrix was not possible with my current computer (IBM PC/Pentium II Celeron 266/64 mb RAM), so the matrix was divided into two partially overlapping portions. Table 4.2 includes 105 taxa, predominantly of the Lower Tephritidae, and Table 4.3 includes 89, mostly of the Higher Tephritidae. Each matrix was analyzed by the mhennig* option combined with bb* (branch swapping) option in an unweighted analysis. Each was then analyzed using the mhennig bb* option combined with a series of successive character weightings. A Nelson consensus tree (option nelsen) was then obtained for the set of most parsimonious trees resulting from each of these analyses (Figures 4.1 to 4.4). In all of the analyses there was overflow, meaning that the set of resulting trees may not have included the shortest possible trees or additional equally parsimonious trees. These limitations also reduced my confidence in these strict cladistic results.

## TABLE 4.1
## Characters Used in the Cladistic Parsimony Analysis (plesiomorphic state is coded 0 unless stated otherwise)

1. A dummy character
2. Larval mask: 0, without placoid structures; 1, with placoid structures
3. Stomal sensory organ: 0, without preoral teeth; 1, with preoral teeth (synapomorphy of some Carpomyini)
4. Oral ridges of larva: 0, rather numerous (>7); 1, not numerous (<8)
5. Dental sclerite of larva: 0, conspicuous; 1, reduced or absent
6. Anal segment of larva: 0, without sclerotized process; 1, with sclerotized pronglike process (synapomorphy of Zaceratini?)
7. Dorsal cleft or notch of pedicel: 0, developed; 1, lacking
8. Medial lobe of pedicel: 0, normal; 1, enlarged (synapomorphy of Epacrocerini)
9. First flagellomere: 0, not pointed; 1, pointed dorsoapically (synapomorphy of Carpomyini)
10. First flagellomere: 0, not apically acute; 1, apically acute (synapomorphy of some Gastrozonini)
11. Pubescence of arista: 0, long (plumose); 1, moderately developed; 2, short (or bare) (polarity uncertain)
12. Ventral row of hairs of arista: 0, dorsal and ventral rows equally long or short; 1, absent (when the dorsal row is present) (synapomorphy of Epacrocerini?)
13. Fronto-orbital plates: 0, extending onto posterior half of frons; 1, not reaching posterior half of frons
14. Frontal setae: 0, none; 1, one to three pairs present; 2, more than three pairs present
15. Number of frontal setae: 2, two (plesiomorphy?); 1, one; 0, zero; 3, three; 4, four or more
16. Vertical plates: 0, rather short; 1, reaching middle of frons length
17. Ocelli: 0, all ocelli well developed; 1, anterior ocellus reduced; 2, all ocelli lacking
18. Ocellar seta: 0, well developed; 1, reduced or absent
19. Postocellar seta: 0, absent; 1, setula-like; 2, moderate; 3, very long
20. Lateral vertical seta: 0, longer than 0.5 length of medial vertical seta; 1, 0.5 of medial vertical seta length or shorter
21. Paravertical seta: 0, shorter than postocular setae; 1, as long as postocular setae; 2, much longer than postocular setae
22. Laterodorsal area of occiput (medial of postocular row): 0, with numerous setulae (or second row); 1, bare
23. Genal expansion or process: 0, absent; 1, well developed
24. Upper proepisternal setae (at the stigma level): 0, weak and light setulae only; 1, one black seta among 6 to 10 yellow setulae (synapomorphy of Xarnutini and Hexachaetini)
25. Postpronotal setae: 1, one (plesiomorphy); 0, none; 2, two; 3, three
26. Presutural supra-alar seta: 0, present; 1, absent
27. Postsutural supra-alar setae: 0, one pair; 1, two pairs
28. Intrapostalar seta: 0, present; 1, absent
29. Yellow medial vitta on scutum: 0, absent; 1, present
30. Scutellar setae: 0, three or more pairs; 1, two or fewer pairs
31. Scutellar setae: 0, three or fewer pairs; 1, more than three pairs
32. Medial (intercalary) scutellar seta: 0, indistinct or absent; 1, 0.1 to 0.3 length of basal scutellar seta; 2, 0.4 or more length of basal scutellar seta (plesiomorphy)
33. Scutellum disk: 0, bare; 1, setulose only laterally (plesiomorphy); 2, setulose medially
34. Anterior notopleural seta: 0, as long as or longer than posterior notopleural seta; 1, shorter than posterior notopleural seta; 2, absent
35. Prephragmal anepisternal seta: 0, absent; 1, present
36. Anepisternum at medioventral margin: 0, without additional setae; 1, with additional seta (synapomorphy of *Hexacinia* and *Erectovena*?)
37. Katepisternal setae: 0, present; 1, lacking or rudimentary
38. Anepimeral seta: 0, absent; 1, weak; 2, well developed (plesiomorphy)
39. Anepimeral setae: 0, one (or none); 1, two to four
40. Anepimeral tubercle: 0, absent; 1, developed (synapomorphy of some Tachiniscini)
41. Anatergite: 0, bare or microtrichose; 1, long trichose
42. Postcoxal metathoracic bridge: 0, membranous; 1, completely sclerotized
43. Costal spine at subcostal break: 0, present; 1, absent
44. Subcostal vein at apex: 0, not constricted; 1, constricted at very apex; 2, interrupted at 0.3 cell width

**TABLE 4.1 (continued)**
**Characters Used in the Cladistic Parsimony Analysis (plesiomorphic state is coded 0 unless stated otherwise)**

45. Subcostal cell: 0, not very long; 1, very long; 2, extremely long
46. Stem vein before humeral cross vein: 0, bare; 1, setulose
47. Basal dilation of cell $r_1$ proximal of the Rs node: 0, without isolated dark spot; 1, with an isolated dark spot (synapomorphy of some Ceratitidini)
48. Radial sector: 0, bare; 1, setulose
49. Vein $R_{2+3}$ (in Acanthonevrini): 0 straight; 1, sinuous
50. Vein $R_{2+3}$: 0, without spur veins; 1, sinuous with three spur veins (posterior, anterior, posterior) (synapomorphy of *Polyara* group)
51. Vein $R_{4+5}$: 0, setulose; 1, bare distal to R-M; 2, completely bare or with a few setulae basally
52. Vein $R_{4+5}$: 0, bare or with 1 row of setulae; 1, with two to three rows of setulae (synapomorphy of Tachiniscini)
53. Vein M before R-M: 0, straight; 1, sinuous (synapomorphy of Zaceratini)
54. Number of hyaline spots at margin of cell m: 0, zero; 1, one; 2, two; 3, three; 4, four
55. Apical section of M: 0, not bent anteriorly; 1, V-shaped, bent anteriorly (synapomorphy of *Polyara* group)
56. Cell bcu: 0, with short posteroapical extension; 1, with very long extension; 2, closed by straight or bowed vein, without any extension (nonadditive)
57. Vein $Cu_1$ distally: 0, bare; 1, setulose
58. Vein $Cu_2$: 0, bare; 1, setulose
59. Subapical portion of cell $R_{4+5}$: 0, without oblique dark stripe; 1, with oblique dark stripe
60. Apical portion of wing: 0, without large round spot; 1, with large round spot
61. Crossvein DM-Cu in relation to posteroapical wing margin: 0, not parallel or subparallel; 1, subparallel; 2, parallel
62. Vein M smoothly curved anteriorly into costa: 0, no; 1, yes
63. Fore femur: 0, nearly as long as midfemur; 1, half as long as midfemur (synapomorphy of *Soita* group of Adramini)
64. Fore femur in male ventrally densely setose ("feathered"): 0, no; 1, yes (synapomorphy of *Ptilona* subgroup)
65. Veins C and M not reaching posteroapical corner of cell $R_{4+5}$: 0, no; 1, yes (synapomorphy of Tachiniscini)
66. Mid femur with rows of short spurs: 0, no; 1, with one; 2, with two (synapomorphy of some Adramini)
67. Anterodorsal seta(e) on midtibia: 0, absent; 1, present
68. Posterodorsal setae on midtibia: 0, absent; 1, present
69. Second, shorter spur on midtibia: 0, shorter than 0.3 length of longest spur; 1, longer than 0.3 length of longest spur; 2, longer, than 0.8 length of longest spur (polarity unresolved)
70. First abdominal tergum: 0, without strong lateral seta; 1, with strong lateral seta
71. At least small, buttonlike, rudimentary gonopods: 0, present; 1, lacking (in *Tachinisca* only)
72. Vanes of phallapodeme: 0, separate at bases; 1, fused basally
73. Phallic guide: 0, narrow and deep; 1, broad and shallow
74. Epandrium: 0, short elliptic in outline, with the outer surstylus not much narrower than the length of epandrium in profile (epandrium height 1.5 to 3.5 times surstylus width); 1, very high in profile, four to seven times higher than long (synapomorphy of Phytalmiinae)
75. Epandrium: 0, short elliptic in outline, with the lateral surstylus not much narrower than the length of epandrium in profile; 1, "P-shaped," i.e., produced behind narrow surstylus (synapomorphy of Dacini, of Paraterelliini, and some other groups of genera)
76. Epandrium in posterior view: 0, elongate; 1, oval
77. Lateral and medial surstyli: 0, not long and narrow; 1, long and narrow (synapomorphy of *Clusiosoma* and allied genera)
78. Medial prensiseta: 0, not on ridge; 1, situated apically on a ridge, blunt and directed laterally, to the outer surstylus (synapomorphy of Paraterelliina)
79. Male proctiger: 0, not enlarged; 1, very large, exceeding surstyli (apomorphy of Gastrozonini)
80. Rectal glands: 0, inconspicuous or absent; 1, very large (apomorphy of some Gastrozonini)
81. Basal lobe of glans: 0, absent or entirely membranous; 1, single, well-developed, covered with papillae or sclerotized
82. Glans of phallus: 0, without median granulate sclerite; 1, with median granulate sclerite
83. Glans, dorsal sclerite: 0, absent or not covered with cells; 1, covered with hexagonal or fusiform cells
84. Apex of subapical lobe of glans (at least in ground plan of group): 0, not trumpet-shaped; 1, trumpet-shaped
85. Medial apodeme at anterior margin of female sternum 6 (and usually also 5): 0, present; 1, absent

**TABLE 4.1 (continued)**
**Characters Used in the Cladistic Parsimony Analysis**
**(plesiomorphic state is coded 0 unless stated otherwise)**

86. T-shaped desclerotized area on oviscape apex: 0, absent; 1, present (synapomorphy of Carpomyina)
87. Apicomedial lobe of oviscape: 0, simple, sclerotized in the middle; 1, bilobate, desclerotized in the middle (synapomorphy of Acanthonevrini); 2, absent
88. Taeniae of eversible membrane (ventrally): 0, absent; 1, one third membrane length; 2, half membrane length; 3, very long (3 is plesiomorphic state)
89. Sclerotized monodentate scales on eversible membrane: 0, absent; 1, present
90. Aculeus when everted: 0, directed posteriorly; 1, directed dorsally
91. Aculeus: 0, not short stiletto-like; 1, short stiletto-like
92. Aculeus length: 0, more than half length of oviscape; 1, half as long as oviscape
93. Aculeus: 0, not modified as in state 1; 1, tergite 8 approximately two to three times broader than cercal unit, blunt, not forming cutting blade (synapomorphy of Epacrocerini)
94. Aculeus: 0, not modified as in state 1; 1, tergite 8 expanded posterolaterally, bladelike flattened, serrate
95. Cercal unit with conspicuously thickened medial portion: 0, no; 1, yes (synapomorphy of Phascini)
96. Cercal unit tapered posteriorly of preapical setae: 0, no; 1, yes (synapomorphy of Phascini)
97. Cercal unit apically: 0, not incised; 1, v-shaped incised
98. Basal setulae on cercal unit (ventral and dorsal): 0, well developed; 1, reduced
99. Number of lateral setae of cercal unit: 0, four (or one short + two long); 1, two (or three very short)
100. Subapical (posterior) of the four lateral setae on female cercal unit: 0, long; 1, short, less than 2.5 times longer than anterior setae in the row
101. Spermatheca shape: 0, spherical or hemispherical; 1, elongate
102. Spermathecae: 0, smooth; 1, papillose
103. Spermathecae: 0, smooth; 1, wrinkled
104. Spermathecae: 0, without apical nipple, 1, with apical nipple (like in *Phasca*)
105. Number of spermathecae: 0, three; 1, two
106. Number of spermathecal ducts: 0, three; 1, two
107. Subapical portion of spermathecal duct: 0, narrow; 1, dilated considerable distance
108. Larvae: 0, associated with dead wood (saprophagous or predaceous); 1, phytophagous; 2, parasitoids; 3, not associated with dead wood, saprophagous or zoophagous
109. Immature stages occurring in living plant material: 0, no; 1, yes
110. Larvae bore asparagus stems: 0, no; 1, yes
111. Larvae feeding in flowerheads of Asteraceae: 0, no; 1, yes
112. Larvae presumed zoophagous: 0, no; 1, yes

From the unweighted analysis of the Lower Tephritidae using the matrix of Table 4.2, 917 trees (with overflow) resulted (length = 541, consistency index (ci) = 0.24, retention index (ri) = 0.73). The consensus tree (length = 610, ci = 0.21, ri = 0.69) is shown in Figure 4.1. From the weighted analysis of this matrix, 917 trees (with overflow) resulted (length = 682, ci = 0.63, ri = 0.90). The consensus tree (length = 859, ci = 0.50, ri = 0.84) is shown in Figure 4.2. From the unweighted analysis of the Higher Tephritidae using the second matrix (Table 4.3), 1079 trees (with overflow) resulted (length = 575, ci = 0.24, ri = 0.69). The consensus tree (length = 608, ci = 0.22, ri = 0.66) is shown in Figure 4.3. And from the weighted analysis of this matrix, 1079 trees (with overflow) resulted (length = 643, ci = 0.65, ri = 0.88). The consensus tree (length = 664, ci = 0.62, ri = 0.87) is shown in Figure 4.4.

The current analysis does not resolve polytomy inside two main clusters (corresponding to the Lower and the Higher Tephritidae) nor the relationships among the tribes of Phytalmiinae and Higher Tephritidae. However, many subfamilies and tribes are well supported as monophyletic by synapomorphies, as indicated in the figures.

**TABLE 4.2**
**Character State Matrix for Cladistic Analysis of Lower Tephritidae (character numbers refer to Table 4.1.)**

| | Character Numbers | | | | | | | | | | | |
|---|---|---|---|---|---|---|---|---|---|---|---|---|
| | | 1111111112 | 2222222223 | 3333333334 | 4444444445 | 5555555556 | 6666666667 | 7777777778 | 8888888889 | 9999999990 | 0000000001 | 11 |
| | 1234567890 | 1234567890 | 1234567890 | 1234567890 | 1234567890 | 1234567890 | 1234567890 | 1234567890 | 1234567890 | 1234567890 | 1234567890 | 12 |
| Tephritidae ground plan | 0000000000 | 0000??0000 | 0?00100000 | 0210000200 | 0000000000 | 000?000000 | 00000000?0 | 0000000000 | 0000000200 | 0000000000 | 0000000000 | 0? |
| *Tachiniscidia* | 000???1000 | 1?01310010 | 0?02201000 | 0220100211 | 0??1000000 | ?10?000000 | 000010?0?0 | ?????????? | ?????0?0?1 | ?????????? | ?????????0 | 01 |
| *Tachinisca* | 000???1000 | 1?01301100 | 0?02201000 | 1220100211 | 0001000000 | 1101000000 | 0001010?0 | 000010?010 | 0000?02001 | 1000010 1?1 | 1000000??0 | 01 |
| *Bibundia* | 000???1000 | 1?01302100 | 0?02201000 | 1210100211 | 0001000000 | 1101000000 | 0001010?0 | ?????????? | ?????0?0?1 | ?????????? | ?????????0 | 01 |
| *Protortalotrypeta* | 000???1000 | 1?02400110 | 0?0?201100 | 0210000200 | 0001000000 | 1001000000 | 000010?0?0 | ?????????? | ?????0?0?1 | ?????????? | ?????????0 | 0? |
| *Ortalotrypeta idana* | 000???0000 | 1?11310010 | 2001201000 | 0220000200 | 0001000000 | 0001000000 | 000001020 | 1010000000 | 0000?02001 | 10000101?1 | 0100000??0 | 01 |
| *O. singula* | 000???0000 | 1?11310010 | 2001101?00 | 02?0000200 | 0001000000 | 0001000000 | 000000?0?0 | 1010000000 | 0000?0?0?1 | ?0000?0??? | ?000000??0 | 01 |
| *O. isshikii* | 000???0000 | 1?11310010 | 200?300000 | 0220000200 | 0001000000 | 0001000000 | 000000?0?0 | 1010000000 | 0000?02001 | 10000101?1 | ?000000??0 | 01 |
| *Cyaforma macula* | 000???0000 | 1?11310010 | 2002301000 | 02?0000200 | 0001000000 | 0001000000 | 000000?0?0 | 1010000000 | 0000?020?1 | ?0000?0??? | ?000000??0 | 01 |
| *C. shenonica* | 000???0000 | 1?11310010 | 2001100100 | 02?0000200 | 0001000000 | 0001000000 | 000000?0?0 | 1010000000 | 0000?0?0?1 | 10000101?1 | ?000000??0 | 01 |
| *Neortalotrypeta* | 000???0000 | ??11100100 | 0?0?100101 | 02?0000200 | 00?1000000 | 0001000000 | 000000?0?0 | ?????00000 | 0000?020?1 | ?0??0101?1 | ?000??0??0 | 01 |
| *Ischyropteron* | 000????000 | ????????00 | 0?0?100101 | 0210000200 | 0001000000 | 0001000000 | 000000?0?0 | ?????00000 | 0000?02??1 | ?0??0?0??? | ?000??0??0 | 01 |
| *Blepharoneura* A | 000???0000 | 1?11210030 | 0100200000 | 0200100200 | 0002000000 | 0003011100 | 0000000000 | 1010000000 | 0000000110 | 0000010101 | 0000000110 | 00 |
| *Hexaptilona* | 000???0000 | 1?11210030 | 0100200000 | 01?0100200 | 0002000000 | 000300?100 | 0000000000 | 1010000000 | 0000000110 | 0000010100 | 0000000110 | 00 |
| *Baryglossa* | 000???0000 | 1?11310030 | 0100200000 | 01?0100200 | 0002000000 | 000200?100 | 0000000000 | 1010000000 | 0000000110 | 000001010? | 0000000110 | 00 |
| *Ceratodacus* | 000???1000 | 2?11210020 | 0000200100 | 0??0100200 | 0001000000 | 000100?000 | 00000000?0 | 1010000000 | 00000001?0 | 00000?0?0? | 0000000110 | 00 |
| *Matsumurania* | 000???0000 | 2?02400100 | 0000110101 | 0000001100 | 0010000000 | 0001020000 | 0000000010 | ?????????0? | ?00?100000 | 00000001?1 | 1000000??0 | 01 |
| *Acanthonevra fuscipennis* | 000???0000 | 0011110121 | 0100100100 | 0100000200 | 0001100110 | 0001000000 | 0000000100 | 1001000000 | 0000101300 | 0000000000 | 0000000000 | 00 |
| *A. dunlopi* | 000???0000 | 0011110121 | 0100100100 | 01?0000?00 | 0001100?00 | 0001000000 | 0001000?00 | 1001000000 | 0000101300 | 0000000000 | 0000000000 | 00 |
| *A. desperata* | 000???0000 | 0011210121 | 0100100100 | 0200000200 | 0001100000 | 0001000000 | 0000000100 | 1001000000 | 0000101300 | 0000000000 | 0000000000 | 00 |
| *A. pteropleuralis* | 0000000000 | 0011110121 | 0100100100 | 0100000100 | 0001100110 | 0001000000 | 0001000000 | 1001000000 | 0000101300 | 0000000000 | 0000000000 | 00 |
| *A. ochropleura* | 000???0000 | 0011110121 | 0100100100 | 01?0000?00 | 0001100?10 | 0001000000 | 0001000?00 | 1001000000 | 0000101300 | 0000000000 | 0000000000 | 00 |
| *Lenitovena trigona* | 0000000000 | 0011110121 | 0100100100 | 0100000100 | 0001100110 | 0001000000 | 0001000000 | 1001000000 | 0000101300 | 0000000000 | 0000000000 | 00 |
| *Rioxoptilona vaga* | 000???0000 | 0011210121 | 0100100000 | 0210000100 | 0001100000 | 0001000000 | 0000000110 | 1001000000 | 1000101300 | 0000000000 | 0000000000 | 00 |
| *Erectovena amurensis* | 000???0000 | 0011210121 | 0100100100 | 0200010200 | 0001100000 | 0001000000 | 0000000?00 | 1001000000 | 0000101300 | 0000000000 | 0000000000 | 00 |
| *Hexacinia punctifera* | 000???0000 | 0011210121 | 0100100000 | 0210010?00 | 0001000000 | 0002000000 | 0000001100 | 1001000000 | 0000101300 | 0000000000 | 0000000000 | 00 |
| *Paedohexacinia flavithorax* | 000???0000 | 0111210121 | 0100100000 | 02?0010200 | 0011000000 | 0000000000 | 0000000110 | 1?01000000 | 000?10??00 | 0010010100 | 000000??0 | 00 |
| *Pseudacanthoneura sexgutta* | 000???0000 | 0011210121 | 0100100000 | 0210000200 | 0001000010 | 0003000100 | 0000000?10 | 1001000000 | 000?10??00 | 0000000000 | 00000?0?00 | 00 |
| *Pseudoneothemara exul* | 000???0000 | 0011210120 | 0100100000 | 0210000200 | 0001100010 | 0003000110 | 0000000?10 | 1001000000 | 000?10???0 | 0000000000 | 0001000??0 | 00 |
| *Neothemara formosipennis* | 000???0000 | 0011210121 | 0100100000 | 0210000200 | 0001100110 | 0003000000 | 0000000?10 | 1001000000 | 0000101300 | 0000000000 | 0000000000 | 00 |
| *Hexaresta multistriga* | 000???0000 | 0011210121 | 0100100000 | 0200000?00 | 0001000110 | 0003000000 | 0000000?00 | 1001000000 | 0000101300 | 0000000000 | 0000000000 | 00 |
| *Themaroides quadrifera* | 000???0000 | 0011210120 | 0100101000 | 1220000200 | 0001000000 | 0001000000 | 0000000120 | 1101000000 | 0000101?00 | 0000000000 | 000000??0 | 00 |
| *Themarohystrix alpina* | 000???0000 | 0011210111 | 0100201000 | 0220000200 | 0001000000 | 0000001100 | 0000000?20 | 1?0?000000 | 0000101?00 | 0000010000 | 000000??0 | 00 |
| *Themaroidopsis* | 000???0000 | 0011210120 | 0100101000 | 0220000200 | 0001100000 | 0001000000 | 0000000120 | 1101000000 | 0000101?00 | 0000000000 | 000000??0 | 00 |

Character Numbers

| | | 1111111112 | 2222222223 | 3333333334 | 4444444445 | 5555555556 | 6666666667 | 7777777778 | 8888888889 | 9999999990 | 0000000001 | 11 |
|---|---|---|---|---|---|---|---|---|---|---|---|---|
| | 1234567890 | 1234567890 | 1234567890 | 1234567890 | 1234567890 | 1234567890 | 1234567890 | 1234567890 | 1234567890 | 1234567890 | 1234567890 | 12 |
| *Trypanocentra* | 000???0000 | 0011210120 | 0100100000 | 0200000000 | 0001000000 | 0001001100 | 0000000110 | 1101000000 | 0000101?00 | 0000000000 | 0000000??0 | 00 |
| *Rabaulia* | 000???0000 | 0011210120 | 0100100000 | 0200000000 | 0001000000 | 0001001100 | 0000000110 | 1101000000 | 0000101?00 | 0000000000 | 0000000??0 | 00 |
| *Rabauliomorpha* | 000???0000 | 0011210120 | 0100100000 | 0200000000 | 0001000000 | 0001001100 | 0000000110 | 1101000000 | 0000101?00 | 0000000000 | 0000000??0 | 00 |
| *Cheesmanomyia* | 000???0000 | 0011210120 | 0100100000 | 0200000000 | 0001200000 | 0001001100 | 0000000100 | 1101001000 | 0000101?00 | 0000000000 | 0000000??0 | 00 |
| *Clusiosoma* | 000???0000 | 0011210120 | 0100100000 | 0200000000 | 0001000000 | 0001001100 | 00000001?0 | 1101001000 | 0000101?00 | 0000000000 | 0000000??0 | 00 |
| *Paraclusiosoma* | 000???0000 | 0011210120 | 0100100000 | 0200000000 | 0001000000 | 0001001100 | 00000001?0 | 1101001000 | 0000101?00 | 0000000000 | 0000000??0 | 00 |
| *Cribrorioxa* | 000???0000 | 0011210130 | 0?02101?00 | 01?0000100 | 0002100000 | 0000000000 | 000000??00 | ?????00000 | 0000?0??0? | ?0??000000 | 0000??0??0 | 00 |
| *Enoplopteron* | 000???0000 | 0011210121 | 0100101000 | 0210000100 | 0001000000 | 0000000000 | 000000?120 | 1101000000 | 0000100?00 | 0000000000 | 0000000??0 | 00 |
| *Bululoa* | 000???0000 | 0011210120 | 0102101000 | 0220000100 | 000?000000 | 0001001100 | 000000??20 | 1???000000 | 0000?0??00 | 0000000000 | 0000??0??0 | 00 |
| *Diarrhegma* | 000???0000 | 0011210121 | 0100100000 | 0200000?00 | 0001000000 | 0001000000 | 000000??20 | 1001000000 | 0000101200 | 0000000000 | 0000000000 | 00 |
| *Rioxa* | 000???0000 | 0011210121 | 0100100100 | 0100000?00 | 0001200000 | 0001000000 | 000000??00 | 1101000000 | 0000101?00 | 0000000000 | 0000000??0 | 00 |
| *Termitorioxa* | 000???0000 | 1111100121 | 010?100000 | 0210000200 | 0001000000 | 0001000000 | 000000?110 | 1101000000 | 0000101300 | 0000000000 | 0000000000 | 00 |
| *Themara* | 000???0000 | 0011110120 | 0100100100 | 01?0000100 | 0001000110 | 0001001000 | 0001000?00 | 1?01000000 | 0000101?00 | 0000000000 | 0000000??0 | 00 |
| *Homoiothemara* | 000???0000 | 0011110110 | 0100100101 | 0000000100 | 0011000000 | 0001000010 | 000100??00 | ?????0?000 | 0000?0??00 | 0000000000 | 0000000??0 | 00 |
| *Orienticaelum femoratum* | 000???0000 | 0011210121 | 0100100100 | 01?0000?00 | 0001000?00 | 0001000000 | 0000000?00 | 1001000000 | 0000101300 | 0000000000 | 0000000000 | 00 |
| *Chaetomerella nigrifacies* | 000???0000 | 0011210121 | 0100110100 | 01?0000?00 | 0001100?00 | 0001000000 | 0000000?00 | 1001000000 | 0000101300 | 0000000000 | 0000000000 | 00 |
| *Micronevrina apicalis* | 0?????0000 | 0011210120 | 0100110001 | 0000000000 | 0001000000 | 0001000000 | 0000000000 | 1?0??00000 | 000?101110 | 0001001111 | 000000011? | ?0 |
| *Stymbara concisa* | 000???0000 | 00111101?0 | 010?110101 | 0020001000 | 0011100010 | 0001000000 | 0000000?10 | 1001000000 | 000010???? | 0000??0?0? | ??????0??0 | 00 |
| *Acanthonevroides* | 000???0000 | 0011110120 | 0100100000 | 0220000200 | 0001000000 | 0001000000 | 0000000110 | 1??1000000 | 000010???0 | 0000000000 | 0000000??0 | 00 |
| *Emheringia longiplaga* | 000???0000 | 0011110120 | 0100100000 | 0?20000200 | 0001000010 | 0001000010 | 100?0???00 | ?????0?00? | ?00?10???0 | 0000000000 | ??????0??0 | 00 |
| *Aridonevra cunnamullae* | 000???0000 | 0011110120 | 0100100001 | 0120000200 | 0001000010 | 0001000010 | 100000??20 | ?????0?00? | ?00?10???0 | 0000000000 | ??????0??0 | 00 |
| *Austronevra* | 000???0000 | 0011110120 | 0100100100 | 0100000200 | 0001100000 | 0001000000 | 0000000?10 | 1?01000000 | 000010??00 | 0100000000 | 0000000000 | 00 |
| *Austrorioxa acidiomorpha* | 000???0000 | 0011110120 | 0100100100 | 0100000200 | 0001200000 | 0001000010 | 0000000?00 | 1????0?00? | ?00?10??00 | 0100000000 | 0000000000 | 00 |
| *Lumirioxa aurauariae* | 000???0000 | 1?11110120 | 0102100000 | 0110000200 | 0001000000 | 0001000010 | 0001000100 | 1?01000000 | 0000101?00 | 0?00000000 | 0001000000 | 00 |
| *Dirioxa pornia* | 0000000000 | 0111110120 | 0100100000 | 0200000200 | 0001000000 | 1001000000 | 0000000?00 | 1?01000000 | 0000101200 | 0000000000 | 00010001?0 | 00 |
| *Taeniorioxa quinaria* | 000???0000 | 0011210121 | 0100100100 | 0210000200 | 0001000000 | 0001000010 | 0000000120 | 1?01000000 | 0000101?00 | 0000000000 | 0001000??0 | 00 |
| *Ptilona confinis* | 000???0000 | 0011110121 | 0100110101 | 0110000100 | 1001100000 | 0001000000 | 0000000000 | 1001000000 | 0000101300 | 0000000000 | 0000000000 | 00 |
| *Colobostroter* | 000???0000 | 0011100111 | 0100000101 | 00210010?0 | 0011100010 | ?000000000 | 0000000010 | ?00??00000 | 0000?0???0 | 0000000000 | 0001000??0 | 00 |
| *Copiolepis* | 000???0000 | 0011210121 | 0102100101 | 0000000200 | 00111000?0 | 0000000000 | 0000000010 | 100??0?000 | ?00??0???0 | 0000000?00 | 0000000??0 | 00 |
| *Sophira limbata* | 000???0000 | 0011110111 | 01101?0100 | 0020001100 | 0011100000 | 0001000000 | 0000000010 | 1001000000 | 0000101200 | 0000000000 | 0000000??0 | 00 |
| *S. signifera* | 000???0000 | 0011110111 | 01001?0100 | 0020001100 | 0011100000 | 0001000000 | 1000000010 | 1001000000 | 0000101200 | 0000000000 | 0000000??0 | 00 |
| *Dacopsis flava* | 000???0000 | 0011210111 | 0100110100 | 0100001100 | 0011200000 | 0001000000 | 1000000000 | 1?0?000000 | ?00??0?200 | 0000000000 | 1001000000 | 00 |
| *Loriomyia guttipennis* | 000???0000 | 0111100111 | 0100110?01 | 00?0001000 | 0011000010 | ?001000000 | 0000000000 | 1001000000 | 0000101200 | 0000000000 | 0000000??0 | 00 |
| *Terastiomyia distorta* | 000???0000 | 0110000111 | 0100000101 | 00?1001000 | 0011200010 | 0000100000 | 2000000000 | 1001000000 | 0000?0?2?0 | 0000000000 | 000?000??0 | 00 |
| *Phasca maculifacies* | 000???0000 | 0011210120 | 0100100000 | 0100000200 | 0001000000 | 0001000010 | 0000000010 | 1101000000 | 0000100100 | 0000110101 | 0001000110 | 00 |
| *P. ortalioides* | 000???0000 | 0011210120 | 0100100000 | 0100000200 | 0001000000 | 0001000010 | 0000000010 | 1101000000 | 0000100100 | 0000110101 | 0001000110 | 00 |
| *Paraphasca taenifera* | 000???0000 | 0011210120 | 0100100000 | 0100000200 | 0001000000 | 0001000010 | 0000000010 | 1101000000 | 0000100100 | 0000110000 | 0001000110 | 00 |
| *Othniocera pictipennis* | 000???0000 | 0011210120 | 0100100000 | 0100000200 | 0001000000 | 0001000010 | 0000000010 | 1101000000 | 0000100100 | 0000110000 | 0001000110 | 00 |
| *Diarrhegmoides* | 000???0000 | 0011210120 | 0100100000 | 0100000200 | 0001000000 | 0001000010 | 0000000010 | 1101000000 | 0000100100 | 0000110101 | 0000000110 | 00 |

**TABLE 4.2 (continued)**
**Character State Matrix for Cladistic Analysis of Lower Tephritidae (character numbers refer to Table 4.1.)**

| | Character Numbers | | | | | | | | | | | |
|---|---|---|---|---|---|---|---|---|---|---|---|---|
| | | 1111111112 | 2222222223 | 3333333334 | 4444444445 | 5555555556 | 6666666667 | 7777777778 | 8888888889 | 9999999990 | 0000000001 | 11 |
| | 1234567890 | 1234567890 | 1234567890 | 1234567890 | 1234567890 | 1234567890 | 1234567890 | 1234567890 | 1234567890 | 1234567890 | 1234567890 | 12 |
| *Stigmatomyia arcuata* | 000???0000 | 0011210120 | 0100100000 | 0100000200 | 0001000000 | 0001000010 | 0000000010 | 1101000000 | 0000100100 | 0000110000 | 0001000110 | 00 |
| *Xenosophira* | 000???0000 | 0011110120 | 0100100001 | 0000000200 | 0001000000 | 0001000010 | 0000000100 | ??01000000 | 0000100100 | 0000110000 | 0001000??0 | 00 |
| *Polyara* | 000???0000 | 0000100100 | 0100100000 | 0100000200 | 0001000011 | 0001100000 | 0000000120 | 1101000000 | 0000100000 | 0000010001 | ??????0110 | 00 |
| *Polyaroidea* | 000???0000 | 0000100100 | 0100100000 | 0100000200 | 0001000011 | 0001100000 | 0000000120 | 1101000000 | 0000100000 | 0000010001 | ??????0110 | 00 |
| *Pseudacrotoxa* | 000???0000 | 0000100100 | 0100100000 | 0100000200 | 0001000011 | 0001100000 | 0000000120 | 1101000000 | 0000100000 | 0000010001 | ??????0110 | 00 |
| *Phytalmia mouldsi* | 000???0000 | 0000100101 | 0120010101 | 0002001000 | 0110000000 | 0001020000 | 0000000010 | 1101000000 | 0000100000 | 0000000000 | 0000000000 | 00 |
| *Diplochorda* | 000???0000 | 0000100101 | 0110010101 | 0002001000 | 0110200000 | 0001020000 | 0000000010 | 1101000000 | 0000100000 | 0000000000 | 0000000000 | 00 |
| *Hexachaeta eximia* | 0?????0000 | 2?01300020 | 0101100100 | 0220000200 | 0001000000 | 0001000110 | 0000000111 | 1100100000 | 000?000110 | 0000101?1 | 1100000110 | 00 |
| *Callistomyia horni* | 0?????0000 | 2?01300121 | 0101100101 | 0010000200 | 0001000000 | 0000000001 | 0000020111 | 1??0100000 | 000?000110 | 0000010111 | 0100000110 | 00 |
| *Xarnuta leucotela* | 0?????0000 | 2?01300120 | 0102100000 | 1220000200 | 0001010000 | 0000000000 | 0000000120 | 1000100000 | 100?000110 | 0000010111 | 0100000110 | 00 |
| *X. sabahensis* | 0????00000 | 2?01300120 | 0102100000 | 1220000200 | 0001010000 | 0002000000 | 0000000120 | 1000100000 | 100?000110 | 0000101?1 | 0000000110 | 00 |
| *X. stellaris* | 0?????0000 | 2?01300120 | 0102100000 | 1220000200 | 0001010000 | 0002000010 | 0000000120 | 1000100000 | 1001000110 | 00010011?1 | 1010000110 | 00 |
| *Paracanthonevra boettcheri* | 0?????0000 | 2?013001?? | 010?100?01 | 00?0000200 | 0002000000 | 0000010000 | 0000000?0? | 1????0?00? | ?00??00??0 | 000?0?1??1 | 0100000110 | 00 |
| *Griphomyia* | 0?????0000 | 2?01100020 | 010?100000 | 0100000200 | 0001000000 | 0000000000 | 0000000100 | 1000000000 | 000??00??0 | 0000010001 | 1011000110 | 00 |
| *Anastrepha fraterculus* | 0000000000 | 2?02400120 | 0100100111 | 0020000200 | 0001000100 | 1002010010 | 0100000000 | 1010000010 | 1001000110 | 0000010111 | 1100000110 | 00 |
| *Celidodacus obnubilis* | 0?????0000 | 0001300121 | 0100100101 | 0010000200 | 1002000000 | 1000000000 | 0000020110 | 1000000000 | ?001000210 | 0100000000 | 01?0000110 | 00 |
| *Adrama* | 000?100000 | 1?11200101 | 010?010111 | 0000001?00 | 11?2000000 | 1000000000 | 0000020110 | 1000000000 | ?00?000210 | 0100110101 | 1110000110 | 00 |
| *Parastenopa limata* | 0?????0000 | 2?01300120 | 010?100101 | 0010000200 | 0002000000 | 1002000010 | 0000000000 | 10100000?0 | 0?11000300 | 0000010111 | 01?0000110 | 00 |
| *Myoleja lucida* | 0101100000 | 2?01300020 | 0102100101 | 0010000200 | 0002000000 | 1002000010 | 0000000000 | 10100000?0 | 0111000300 | 0000010111 | 01?0000110 | 00 |
| *Platyparea discoidea* | 0?????0000 | 2?01400030 | 0002100101 | 001000?200 | 0012000000 | 1001000010 | 0000000010 | 10100000?0 | 01?1000110 | 0000010111 | 11?0000110 | 00 |
| *Strauzia perfecta* | 0101100000 | 2?01300020 | 0102100101 | 0010000200 | 0002000000 | 1001000010 | 0000000000 | 10100000?0 | 0111000110 | 0000010111 | 01?0000110 | 00 |
| *Zonosemata vittigera* | 00?1?00010 | 2?01300120 | 0100100111 | 0010000200 | 0001000000 | 00010000?0 | 0000000110 | 1010000000 | 0?0?010110 | 0000010111 | 01?0000110 | 00 |
| *Rhagoletis alternata* | 0000100010 | 2?01300020 | 0100100101 | 0010000200 | 0002000000 | 20010000?0 | 0000000100 | 1010000000 | 000?010110 | 0000010111 | 11?0110110 | 00 |
| *Plioreocepta* | 01011100?0 | 2?01300020 | 0102100101 | 0010000200 | 0002000000 | 10110000?0 | 0000000000 | 1010010000 | ?00?000110 | 0000010111 | 1110110111 | 00 |
| *Zacerata* | 0????10000 | 2?01300020 | 010?100101 | 0000000200 | 0002000000 | 10110200?0 | 0000000?00 | 1010010000 | ?001000110 | 0000010111 | 11?0???111 | 00 |
| *Terellia* A | 0101100000 | 2?01300020 | 0100100101 | 00?0000200 | 0002000000 | 20010000?0 | 0000000?00 | 1010010000 | 1001000110 | 0000010?11 | 1110111110 | 10 |
| *Enicoptera palawanica* | 0?????0000 | 0001310130 | 010?100101 | 0020000200 | 0?12200?00 | 0012110010 | 2000000??0 | 1??0000000 | 100?000200 | 0000010100 | 010011?110 | 00 |
| *Monacrostichus citricola* | 0?????0000 | 2100010100 | 010?010101 | 00?0001100 | 0112200000 | 0011010000 | 1000000000 | 1010000000 | ?00?000?10 | 0000010100 | 1??011?110 | 00 |
| *Taeniostola apicata* | 0?????0000 | 0001310020 | 0100100101 | 00?0000200 | 000?000000 | 0001010010 | 0000000110 | 1010000000 | 100?000200 | 0000000000 | 1010110110 | 00 |
| *T. limbata* | 0?????0000 | 0001310020 | 0100100101 | 00?0000200 | 000?000000 | 0001010010 | 0000000110 | 1010000000 | 100?000200 | 0000000000 | 1010110110 | 00 |
| *Ceratitis capitata* | 0000000000 | 1?11210020 | 1100100101 | 0010000200 | 0001001000 | 00010000?0 | 0000000110 | 1010100010 | 1001000110 | 0000110?01 | 1110111110 | 00 |
| *Bactrocera dorsalis* | 0000000000 | 2101200100 | 0100010101 | 00200010?0 | 0112100000 | 010?0100?0 | 0000000010 | 101010001? | 1001?001?0 | 0000?10100 | 1??011?110 | 00 |

**TABLE 4.3**
**Character State Matrix for Cladistic Analysis of Higher Tephritidae (character numbers refer to Table 4.1)**

| | Character Numbers | | | | | | | | | | |
|---|---|---|---|---|---|---|---|---|---|---|---|
| | | 1111111112 | 2222222223 | 3333333334 | 4444444445 | 5555555556 | 6666666667 | 7777777778 | 8888888889 | 9999999990 | 0000000001 | 11 |
| | 1234567890 | 1234567890 | 1234567890 | 1234567890 | 1234567890 | 1234567890 | 1234567890 | 1234567890 | 1234567890 | 1234567890 | 1234567890 | 12 |
| Tephritidae ground plan | 0000000000 | 0000??0000 | 0?00100000 | 0210000200 | 0000000000 | 000?000000 | 00000000?0 | 0000000000 | 0000000200 | 0000000000 | 0000000000 | 0? |
| *Tachinisca* | 000???1000 | 1?01301100 | 0?02201000 | 1220100211 | 0001000000 | 1101000000 | 00001010?0 | 000010?010 | 0000?02001 | 10000101?1 | 1000000??0 | 01 |
| *Ortalotrypeta idana* | 000???0000 | 1?11310010 | 2001201000 | 0220000200 | 0001000000 | 0001000000 | 0000001020 | 1010000000 | 0000?02001 | 10000101?1 | 0100000??0 | 01 |
| *Hexaptilona* | 000???0000 | 1?11210030 | 0100200000 | 01?0100200 | 0002000000 | 000300?100 | 0000000000 | 1010000000 | 0000000110 | 0000010100 | 000000110 | 00 |
| *Matsumurania* | 000???0000 | 2?02400100 | 0000110101 | 0000001100 | 0010000000 | 0001020000 | 0000000010 | ????????0? | ?00?100000 | 00000001?1 | 1000000??0 | 01 |
| *Acanthonevra pteropleuralis* | 0000000000 | 0011110121 | 0100100100 | 0100000100 | 0001100110 | 0001000000 | 0001000000 | 1001000000 | 0000101300 | 0000000000 | 0000000000 | 00 |
| *Themaroides quadrifera* | 000???0000 | 0011210120 | 0100101000 | 1220000200 | 0001000000 | 0001000000 | 0000000120 | 1101000000 | 0000101?00 | 0000000000 | 0000000??0 | 00 |
| *Clusiosoma* | 000???0000 | 0011210120 | 0100100000 | 0200000000 | 0001000000 | 0001001100 | 00000001?0 | 1101001000 | 0000101?00 | 0000000000 | 0000000??0 | 00 |
| *Themara* | 000???0000 | 0011110120 | 0100100100 | 01?0000100 | 0001000110 | 0001001000 | 0001000?00 | 1?01000000 | 0000101?00 | 0000000000 | 0000000??0 | 00 |
| *Homoiothemara* | 000???0000 | 0011110110 | 0100100101 | 0000000100 | 0011000000 | 0001000010 | 000100??00 | ?????0?000 | 0000?0??00 | 0000000000 | 0000000??0 | 00 |
| *Chaetomerella nigrifacies* | 000???0000 | 0011210121 | 0100110100 | 01?0000?00 | 0001100?00 | 0001000000 | 0000000?00 | 1001000000 | 0000101300 | 0000000000 | 0000000000 | 00 |
| *Micronevrina apicalis* | 0?????0000 | 0011210120 | 0100110001 | 0000000000 | 0001000000 | 0001000000 | 0000000000 | 1?0??00000 | 000?101110 | 0001001111 | 00000011? | ?0 |
| *Dirioxa pornia* | 0000000000 | 0111110120 | 0100100000 | 0200000200 | 0001000000 | 1001000000 | 0000000?00 | 1?01000000 | 0000101200 | 0000000000 | 00010001?0 | 00 |
| *Ptilona confinis* | 000???0000 | 0011110121 | 0100110101 | 0110000100 | 1001100000 | 0001000000 | 0000000000 | 1001000000 | 0000101300 | 0000000000 | 0000000000 | 00 |
| *Sophira limbata* | 000???0000 | 0011110111 | 01101?0100 | 0020001100 | 0011100000 | 0001000000 | 0000000010 | 1001000000 | 0000101200 | 0000000000 | 0000000??0 | 00 |
| *Terastiomyia distorta* | 000???0000 | 0110000111 | 0100000101 | 00?1001000 | 0011200010 | 0000100000 | 2000000000 | 1001000000 | 0000?0?2?0 | 0000000000 | 000?000??0 | 00 |
| *Phasca maculifacies* | 000???0000 | 0011210120 | 0100100000 | 0100000200 | 0001000000 | 0001000010 | 0000000010 | 1101000000 | 0000100100 | 0000110101 | 0001000110 | 00 |
| *Xenosophira* | 000???0000 | 0011110120 | 0100100001 | 0000000200 | 0001000000 | 0001000010 | 0000000100 | ??01000000 | 0000100100 | 0000110000 | 0001000??0 | 00 |
| *Polyara* | 000???0000 | 0000100100 | 0100100000 | 0100000200 | 0001000011 | 0001100000 | 0000000120 | 1101000000 | 0000100000 | 0000010001 | ??????0110 | 00 |
| *Phytalmia mouldsi* | 000???0000 | 0000100101 | 0120010101 | 0002001000 | 0110000000 | 0001020000 | 0000000010 | 1101000000 | 0000100000 | 0000000000 | 0000000000 | 00 |
| *Hexachaeta eximia* | 0?????0000 | 2?01300020 | 0101100100 | 0220000200 | 0001000000 | 0001000110 | 0000000111 | 1100100000 | 000?000110 | 00000101?1 | 1100000110 | 00 |
| *Alincocallistomyia* | 0?????0000 | 2?013001?1 | 0100100100 | 0210000200 | 0011000000 | 0000000001 | 000000011? | 1????0?00? | ?00??00?10 | 0000000??? | ??00??0110 | 00 |
| *Callistomyia horni* | 0?????0000 | 2?01300121 | 0101100101 | 0010000200 | 0001000000 | 0000000001 | 0000020111 | 1??0100000 | 000?000110 | 0000010111 | 0100000110 | 00 |
| *Xarnuta leucotela* | 0?????0000 | 2?01300120 | 0102100000 | 1220000200 | 0001010000 | 0000000000 | 0000000120 | 1000100000 | 100?000110 | 0000010111 | 0100000110 | 00 |
| *X. sabahensis* | 0????00000 | 2?01300120 | 0102100000 | 1220000200 | 0001010000 | 0002000000 | 0000000120 | 1000100000 | 100?000110 | 00000101?1 | 0000000110 | 00 |
| *X. stellaris* | 0?????0000 | 2?01300120 | 0102100000 | 1220000200 | 0001010000 | 0002000010 | 0000000120 | 1000100000 | 1001000110 | 00010011?1 | 1010000110 | 00 |
| *Paracanthonevra boettcheri* | 0?????0000 | 2?013001?? | 010?100?01 | 00?0000200 | 0002000000 | 0000010000 | 0000000?0? | 1????0?00? | ?00??00??0 | 000?0?1??1 | 0100000110 | 00 |
| *Griphomyia* | 0?????0000 | 2?01100020 | 010?100000 | 0100000200 | 0001000000 | 0000000000 | 0000000100 | 1000000000 | 000??00??0 | 0000010001 | 1011000110 | 00 |
| *Anastrepha fraterculus* | 0000000000 | 2?02400120 | 0100100111 | 0020000200 | 0001000100 | 1002010010 | 0100000000 | 1010000010 | 1001000110 | 0000010111 | 1100000110 | 00 |
| *A. serpentina* | 0000000000 | 2?02400120 | 0100100111 | 0010001200 | 0001000100 | 1001010000 | 0100000000 | 1010000010 | 1001000110 | 0000010111 | 1100000110 | 00 |
| *Pseudomyoleja nigricrus* | 0?????0000 | 2?01300111 | 0102100101 | 00?0000200 | 000?000000 | 10010000?0 | 000000??20 | 10?0000000 | ?001000010 | 01101101?1 | 100000???0 | 0? |
| *Rivelliomima* | 0?????0000 | 2?01300111 | 010?100101 | 00?0000200 | 000?000000 | 2011000010 | 00000????0 | 10?0000000 | ?00?000110 | 0010110111 | 1000001110 | 00 |

**TABLE 4.3 (continued)**
**Character State Matrix for Cladistic Analysis of Higher Tephritidae (character numbers refer to Table 4.1)**

| | Character Numbers | | | | | | | | | | | |
|---|---|---|---|---|---|---|---|---|---|---|---|---|
| | | 1111111112 | 2222222223 | 3333333334 | 4444444445 | 5555555556 | 6666666667 | 7777777778 | 8888888889 | 9999999990 | 0000000001 | 11 |
| | 1234567890 | 1234567890 | 1234567890 | 1234567890 | 1234567890 | 1234567890 | 1234567890 | 1234567890 | 1234567890 | 1234567890 | 1234567890 | 12 |
| *Sosiopsila* | 0?????0000 | 0001200111 | 0100110101 | 0020001100 | 0?12000000 | 0000020000 | 0000000110 | 1??0000000 | ?001100110 | 0000?10111 | 0110001110 | 00 |
| *Celidodacus obnubilis* | 0?????0000 | 0001300121 | 0100100101 | 0010000200 | 1002000000 | 1000000000 | 0000020110 | 1000000000 | ?001000210 | 0100000000 | 01?0000110 | 00 |
| *Piestometopon* | 0?????0000 | 1?01300121 | 0100110101 | 0010000200 | 1?02000000 | 2002000010 | 0000020110 | 1????0?000 | ?00?000??0 | 0100??0??? | ??????110 | 00 |
| *Cyclopsia univittata* | 0?????0000 | 1?11200101 | 0100110101 | 0010000?00 | 1?02000000 | ?001000000 | 0000000?00 | 1??0000000 | ?00??002?0 | 0100100100 | ??11320??0 | 00 |
| *Dimeringophrys* | 000???0000 | 0011100101 | 010?110101 | 0010000200 | 1002000000 | 100?000000 | 00000?0?10 | 1??0000000 | ?001000310 | 0110100100 | 1??0001??0 | 00 |
| *Adrama* | 000?100000 | 1?11200101 | 010?010111 | 0000001?00 | 11?2000000 | 1000000000 | 0000020110 | 1000000000 | ?00?000210 | 0100110101 | 1110000110 | 00 |
| *Euphranta (E.) ocellata* | 000???0000 | 0001300121 | 010?110101 | 0010000200 | 1002010000 | 2001000001 | 0000000110 | 1100000000 | ?001000210 | 0110?10101 | ?1?000?110 | 00 |
| *E. (E.) connexa* | 0000100000 | 1?01300121 | 0100110101 | 0010000200 | 1002010000 | 2001000000 | 0000000110 | 1100000000 | ?001000210 | 0100?10101 | 11?000?110 | 00 |
| *E. (Rhacochlaena) bischofi* | 000???0000 | 1?01300121 | 010?110101 | 0010000200 | 1002000000 | 1001000000 | 0000000110 | 1100000000 | ?001000210 | 0110?10101 | 11?0000110 | 00 |
| *E. (Rhacochlaena) toxoneura* | 0000100000 | 2?01300121 | 0102110101 | 0010000200 | 1002000000 | 1001000000 | 0000000110 | 1100000000 | ?001000210 | 0100?10101 | 111000?110 | 00 |
| *Coelopacidia strigata* | 0?????0000 | 2?11110110 | 0100110101 | 000?000200 | 1111000000 | 0001001100 | 0010000110 | 1????0?00? | ?00??00??0 | 0?010011?1 | 1110000110 | 00 |
| *Trypanophion gigas* | 0?????0000 | 2?11110110 | 0100110101 | 000?000200 | 1111000000 | 0001001100 | 0010000110 | 1????0?00? | ?00??00??0 | 0?010011?1 | 1110000110 | 00 |
| *Soita cylindrica* | 0?????0000 | 2?11110110 | 0100110101 | 000?000200 | 1111000000 | 0001001100 | 0010000110 | 1????0?00? | ?00??00??0 | 0?010011?1 | 1110000110 | 00 |
| *Acidoxantha* | 0?????0000 | 2?01200020 | 0100100101 | 0000000200 | 0002000000 | 2001000000 | 0000000100 | 1010000010 | 11?1000300 | 0000010111 | 01?0100110 | 00 |
| *Parastenopa limata* | 0?????0000 | 2?01300120 | 010?100101 | 0010000200 | 0002000000 | 1002000010 | 0000000000 | 1010000?0 | 0?11000300 | 0000010111 | 01?0000110 | 00 |
| *Paramyiolia nigricornis* | 0?????0000 | 2?01300020 | 010?100101 | 0010000200 | 0002000000 | 1002000010 | 0000000000 | 1010000?0 | 0?11000300 | 0000010111 | 01?0000110 | 00 |
| *Myoleja lucida* | 0101100000 | 2?01300020 | 0102100101 | 0010000200 | 0002000000 | 1002000010 | 0000000000 | 1010000?0 | 0111000300 | 0000010111 | 01?0000110 | 00 |
| *Platyparea discoidea* | 0?????0000 | 2?01400030 | 0002100101 | 001000?200 | 0012000000 | 1001000010 | 0000000010 | 1010000?0 | 01?1000110 | 0000010111 | 11?0000110 | 00 |
| *Strauzia perfecta* | 0101100000 | 2?01300020 | 0102100101 | 0010000200 | 0002000000 | 1001000010 | 0000000000 | 1010000?0 | 0111000110 | 0000010111 | 01?0000110 | 00 |
| *Zonosemata vittigera* | 00?1?00010 | 2?01300120 | 0100100111 | 0010000200 | 0001000000 | 00010000?0 | 0000000110 | 1010000000 | 0?0?010110 | 0000010111 | 01?0000110 | 00 |
| *Rhagoletis cerasi* | 0011100010 | 2?01300020 | 0100100101 | 0010000200 | 0002000000 | 20010000?0 | 0000000100 | 1010000000 | 0101010110 | 0000010111 | 11?0000110 | 00 |
| *R. alternata* | 0000100010 | 2?01300020 | 0100100101 | 0010000200 | 0002000000 | 20010000?0 | 0000000100 | 1010000000 | 000?010110 | 0000010111 | 11?0110110 | 00 |
| *Rhagoletis* sp. (Mexico) | 00????0010 | 2?01300020 | 0100100101 | 0010000200 | 0002000000 | 1002000010 | 0000000110 | 1010000000 | 0?0?010110 | 0000010111 | 01?0110110 | 00 |
| *R. striatella* | 0011100010 | 1?01300020 | 0100100101 | 0010000200 | 0002000000 | 0002000010 | 0000000110 | 1010000000 | 0?0?010110 | 0000010111 | 11?0110110 | 00 |
| *R. turpiniae* | 001???0010 | 2?01300020 | 0100100101 | 0010000200 | 0002000000 | 2002000010 | 0000000100 | 1010000000 | 0?01010110 | 0000010111 | 11?0100110 | 00 |
| *Paraterellia* A | 0?????0000 | 2?01300020 | 0100100101 | 0010000200 | 0002000000 | 10010000?0 | 0000000010 | 1010100100 | 1101000110 | 0000010111 | 11?0000110 | 00 |
| *Paraterellia* B | 0?????0000 | 2?01300020 | 0100100101 | 0010000200 | 0002000000 | 10010000?0 | 0000000010 | 1010100100 | 1?0?000110 | 0000010111 | 11?0000110 | 00 |
| *Oedicarena* A | 0????00000 | 2?01300030 | 0102100101 | 0010000200 | 0002000000 | 21010000?0 | 0000010000 | 1010100100 | 1?0?000110 | 0000010111 | 11?0210110 | 00 |
| *Oedicarena* B | 0????00000 | 2?01300030 | 0102100101 | 0010000200 | 0002000000 | 21010000?0 | 00000?0000 | 1010100100 | ??0?000110 | 0000010111 | 11?0210110 | 00 |
| *Notomma* A | 0?????0000 | 2?01300020 | 010?100101 | 0010000200 | 000?000000 | 10110000?0 | 0000000010 | 1010100000 | ??0?000110 | 0000010111 | 11?0000110 | 00 |
| *Notomma* B | 0?????0000 | 2?01300020 | 010?100101 | 0010000200 | 000?000000 | 10110000?0 | 0000000?20 | 1010100000 | ??0?000110 | 0000010111 | 11?0000110 | 00 |
| *Plioreocepta* | 01011100?0 | 2?01300020 | 0102100101 | 0010000200 | 0002000000 | 10110000?0 | 0000000000 | 1010010000 | ?00?000110 | 0000010111 | 1110110111 | 00 |

| | Character Numbers | | | | | | | | | | | |
|---|---|---|---|---|---|---|---|---|---|---|---|---|
| | | 1111111112 | 2222222223 | 3333333334 | 4444444445 | 5555555556 | 6666666667 | 7777777778 | 8888888889 | 9999999990 | 0000000001 | 11 |
| | 1234567890 | 1234567890 | 1234567890 | 1234567890 | 1234567890 | 1234567890 | 1234567890 | 1234567890 | 1234567890 | 1234567890 | 1234567890 | 12 |
| *Zacerata* | 0????10000 | 2?01300020 | 010?100101 | 0000000200 | 0002000000 | 10110200?0 | 0000000?00 | 1010010000 | ?001000110 | 0000010111 | 11?0???111 | 00 |
| *Terellia* A | 0101100000 | 2?01300020 | 0100100101 | 00?0000200 | 0002000000 | 2001000?0 | 0000000?00 | 1010010000 | 1001000110 | 000010?11 | 1110111110 | 10 |
| *Terellia* B | 01011?0000 | 2?01300020 | 0100100101 | 0010000200 | 0002000000 | 2001000?0 | 0000000?00 | 1010010000 | 1001000110 | 000010?11 | 1110111110 | 10 |
| *Tomoplagia* | 0101100000 | 2?01300020 | 0100100101 | 0010000200 | 0002000000 | 0002000?0 | 0000000100 | 1010010000 | 100?000110 | 0000010111 | 1110111110 | 10 |
| *Acanthiophilus* | 0101100000 | 2?01300021 | 0100100101 | 0010000200 | 0002000000 | 2003000?0 | 0000000?00 | 1010010000 | ?00?000110 | 0000010111 | 11?0111110 | 10 |
| *Enicoptera palawanica* | 0?????0000 | 0001310130 | 010?100101 | 0020000200 | 0?12200?00 | 0012110010 | 2000000??0 | 1??0000000 | 100?000200 | 0000010100 | 010011?110 | 00 |
| *Monacrostichus citricola* | 0?????0000 | 2100010100 | 010?010101 | 00?0001100 | 0112200000 | 0011010000 | 1000000000 | 1010000000 | ?00?000?10 | 0000010100 | 1??011?110 | 00 |
| *Ichneumonopsis burmensis* | 0?????0000 | 0000010100 | 010?010101 | 00?0001100 | 0?1?000000 | 0001010000 | 0000000000 | 1010000000 | ?00?0002?0 | 0000010000 | 0010110110 | 00 |
| *Taeniostola apicata* | 0?????0000 | 0001310020 | 0100100101 | 00?0000200 | 000?000000 | 0001010010 | 0000000110 | 1010000000 | 100?000200 | 0000000000 | 1010110110 | 00 |
| *T. limbata* | 0?????0000 | 0001310020 | 0100100101 | 00?0000200 | 000?000000 | 0001010010 | 0000000110 | 1010000000 | 100?000200 | 0000000000 | 1010110110 | 00 |
| *Paraxarnuta bambusae* | 0?????0001 | 0011210030 | 1100100101 | 0020000200 | 0001000000 | 000?0100?0 | 0000010110 | 101010001? | 000?000??0 | 0000100000 | 1010110110 | 00 |
| *Phaeospilodes fritilla* | 0?????0001 | 0001210020 | 1100100101 | 0010000200 | 0001000000 | 00020000?0 | 0000000110 | 1010100011 | 100?000??0 | 000010?00 | 1010110110 | 00 |
| *Acrotaeniostola quadrifasciata* | 0?????0001 | 1001210020 | 1100100101 | 0010000200 | 0001000000 | 0001010000 | 0000000110 | 1010?00010 | ?00?000??0 | 000010?00 | 001?110110 | 00 |
| *Acroceratitis histrionica* | 0?????0001 | 0001210020 | 0100100101 | 0010000200 | 0001000000 | 0001010010 | 0000000110 | 1010?00010 | 100?0002?0 | 0000010000 | 1010111110 | 00 |
| *Gastrozona fasciventris* | 00?0?00000 | 0001310120 | 0100100101 | 00?0000200 | 0001000000 | 0002000010 | 0000000?00 | 1010100011 | ?00?0002?0 | 0000110000 | ????11?110 | 00 |
| *Paragastrozona japonica* | 0?????0000 | 0011310020 | 0102100101 | 0010000200 | 0001000000 | 0002000010 | 0000000100 | 1010000000 | ?00?000200 | 00001101?1 | 001?110110 | 00 |
| *Anoplomus flexuosus* | 0?????0001 | 0001200120 | 01000?0101 | 00?0000?00 | 0001000000 | 0001010010 | 0000000?20 | 1010000010 | ?00?0002?0 | 0000110000 | 111111?110 | 00 |
| *Pardalaspinus laqueata* | 0?????0000 | 0001300020 | 0100100101 | 00?0000200 | 000?000000 | 0001010010 | 0000000??0 | 1010?0001? | ?00?000??0 | 0000?10??0 | ????11?110 | 00 |
| *Carpophthorella capillata* | 0?????0000 | 0001400120 | 0100100101 | 0020000200 | 0001000000 | 0002010010 | 0000000??0 | 1010100011 | ?00?000??0 | 0001110?00 | 0011111110 | 00 |
| *Ceratitis capitata* | 0000000000 | 1?11210020 | 1100100101 | 0010000200 | 0001001000 | 0001000?0 | 0000000110 | 1010100010 | 1001000110 | 0000110?01 | 1110111110 | 00 |
| *Neoceratitis asiatica* | 0?????0000 | 2?11210020 | 1100100101 | 00?0000200 | 0001001000 | 0002000010 | 0000000110 | 1010100010 | 1001000110 | 0000?10?11 | ??????110 | 00 |
| *Trirhithrum coffeae* | 0000000000 | 0011210020 | 1100100101 | 00?0000200 | 0001001000 | 0001000?0 | 0000000110 | 1010100010 | 1001000110 | 0000?10?01 | ????11?110 | 00 |
| *Capparimyia savastani* | 0?????0000 | 2?11210020 | 0100100101 | 0010000200 | 0001001000 | 0001000?0 | 0000000110 | 1010100010 | 1001000110 | 0000?10??1 | ????11?110 | 00 |
| *Dacus armatus* | 0000000000 | 2101200100 | 0100010111 | 0020001070 | 0112100000 | 000?0100?0 | 0000000000 | 101010001? | 1001?001?0 | 0000?10100 | 1?1011?110 | 00 |
| *Bactrocera dorsalis* | 0000000000 | 2101200100 | 0100010101 | 00200010?0 | 0112100000 | 010?0100?0 | 0000000010 | 101010001? | 1001?001?0 | 0000?10100 | 1??011?110 | 00 |

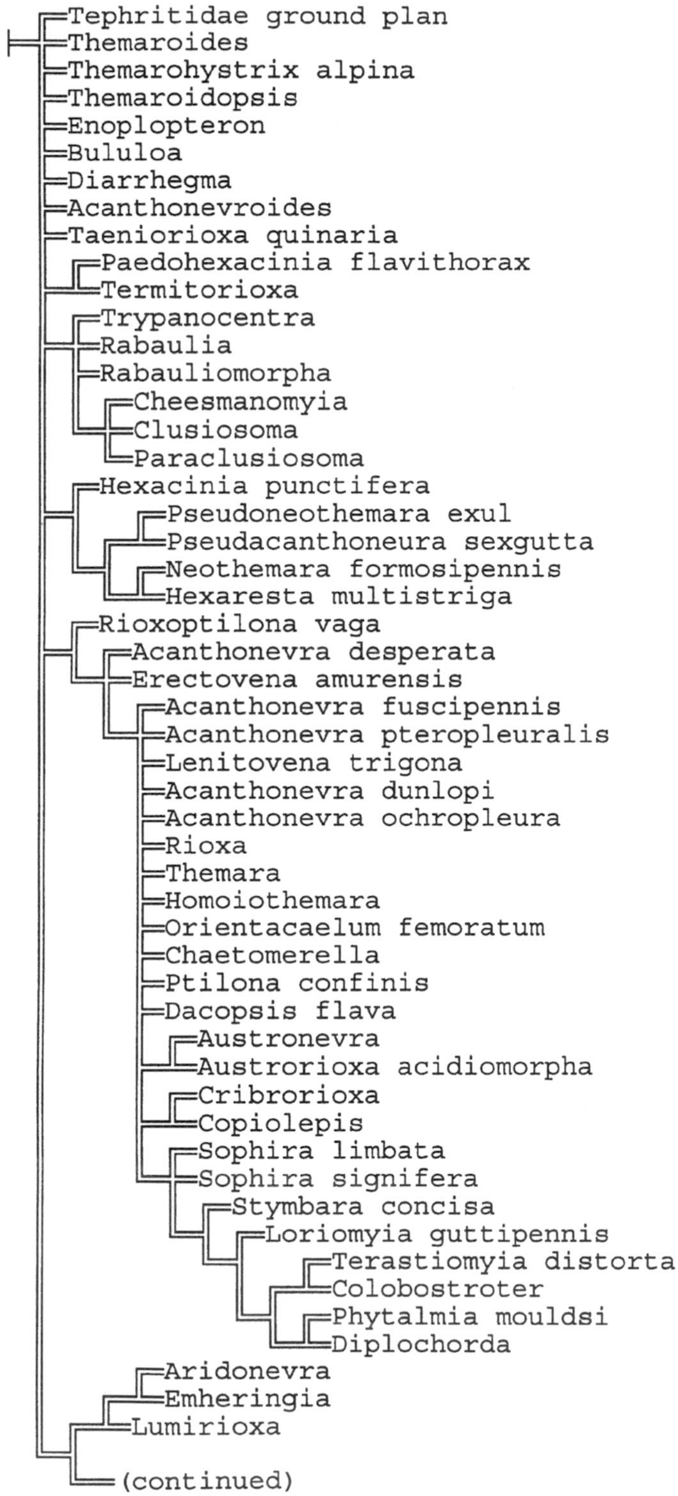


**FIGURE 4.1** Results of unweighted cladistic analysis of Lower Tephritinae based on taxa and characters in Table 4.2. Consensus tree of trees resulting from analysis using Hennig86 (options mhennig*; bb*; Nelsen).

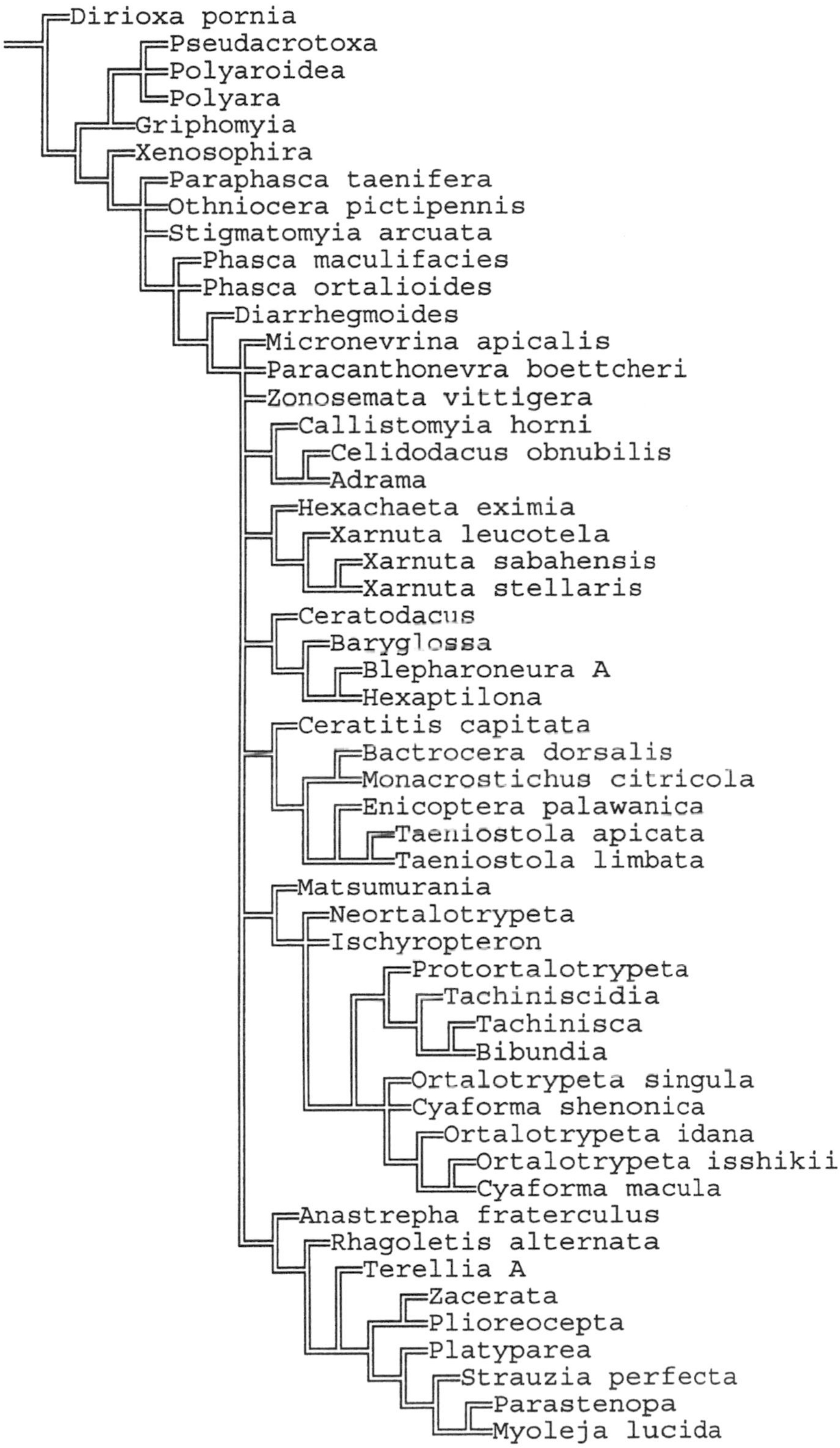


**FIGURE 4.1** (continued)

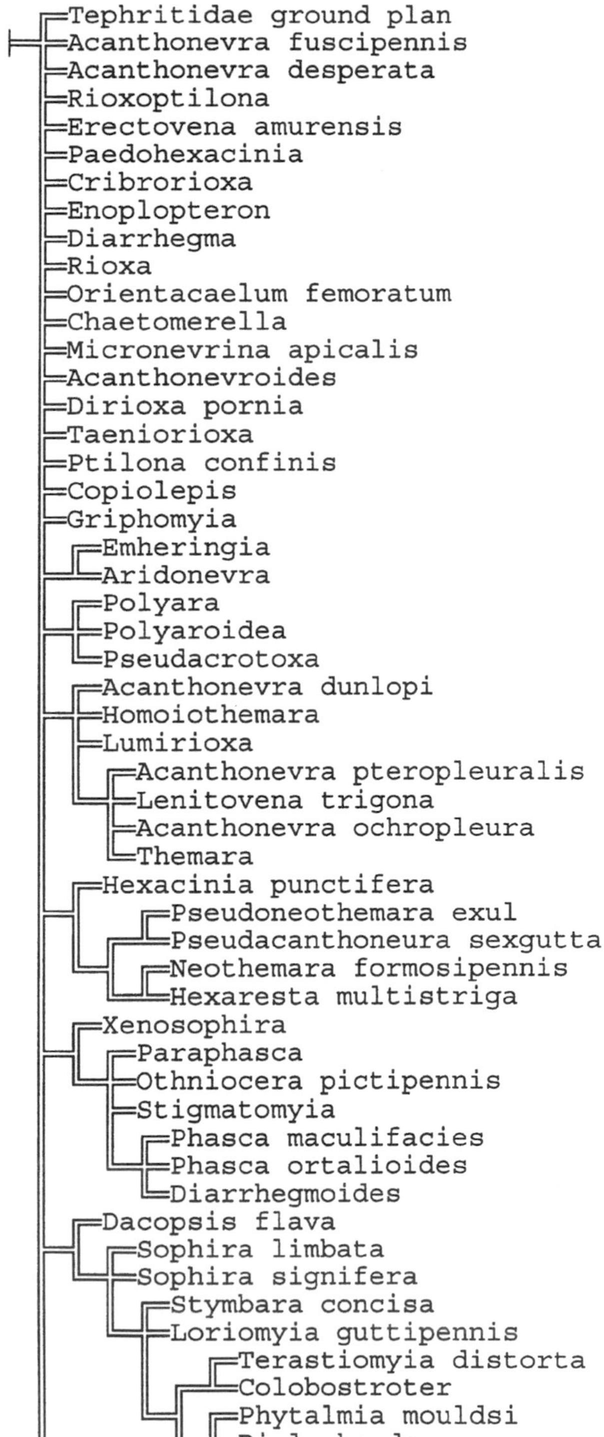


(continued)

**FIGURE 4.2** Results of weighted cladistic analysis of Lower Tephritinae based on taxa and characters in Table 4.2. Consensus tree resulting from analysis using Hennig86 (options mhennig*; bb*; followed by successive weighting).

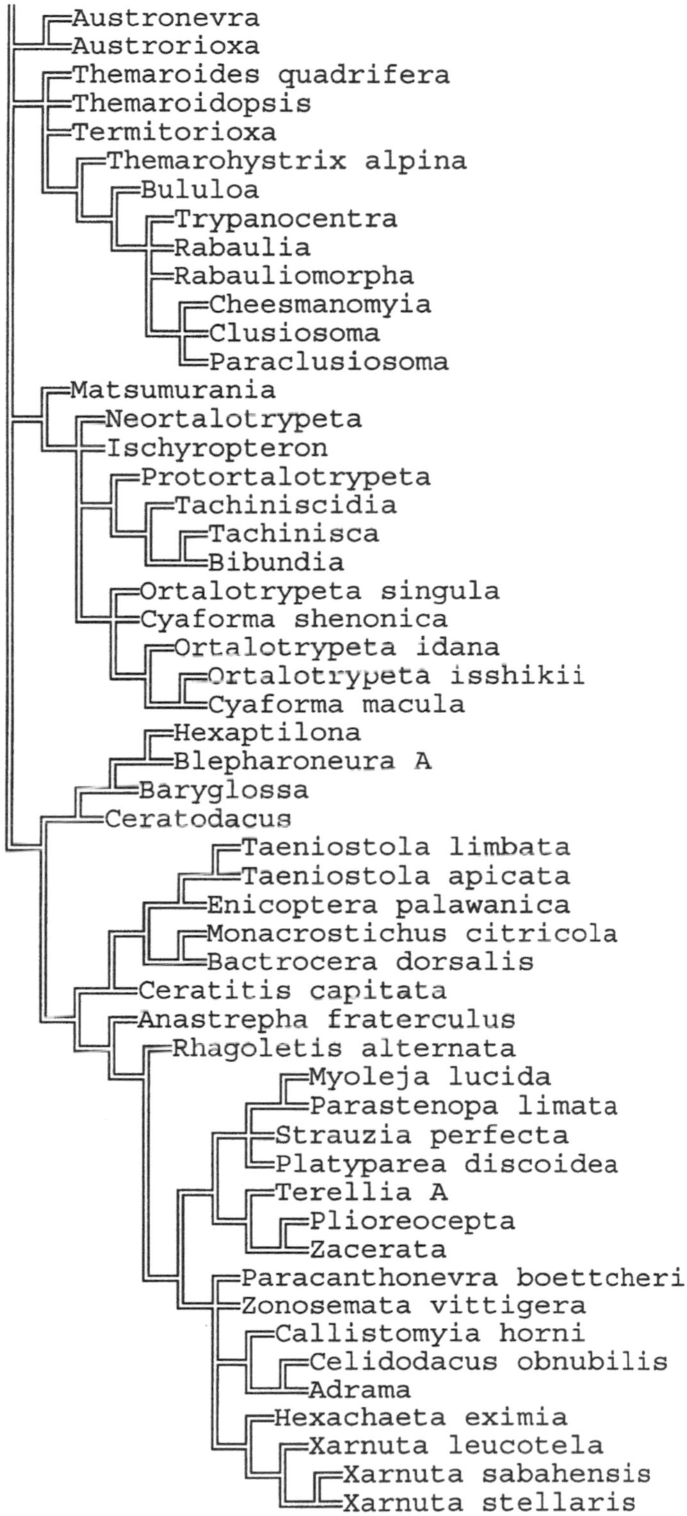


**FIGURE 4.2 (continued)**

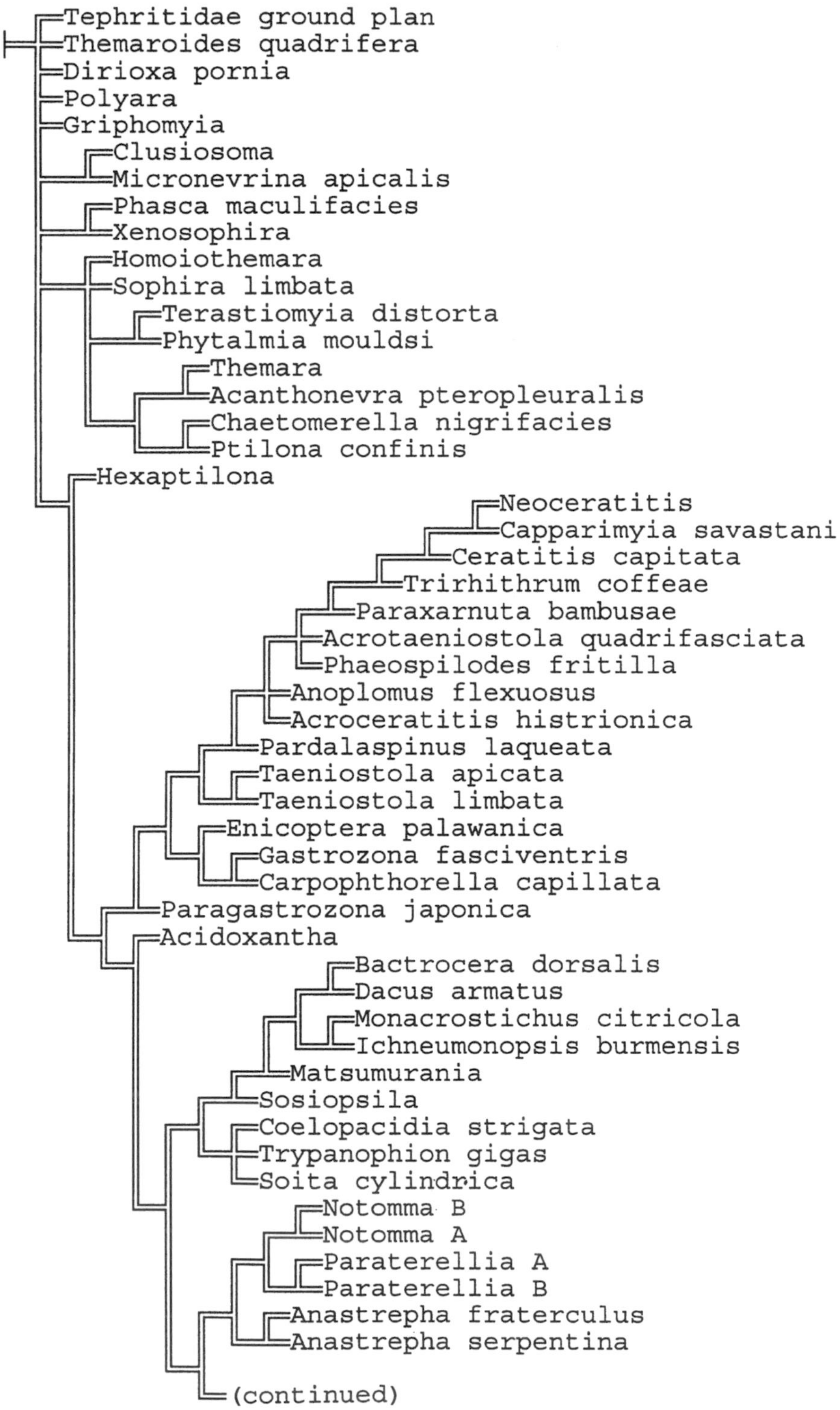


**FIGURE 4.3** Results of unweighted cladistic analysis of Higher Tephritinae based on taxa and characters in Table 4.3. Consensus tree of trees resulting from analysis using Hennig86 (options mhennig*; bb*; Nelsen).

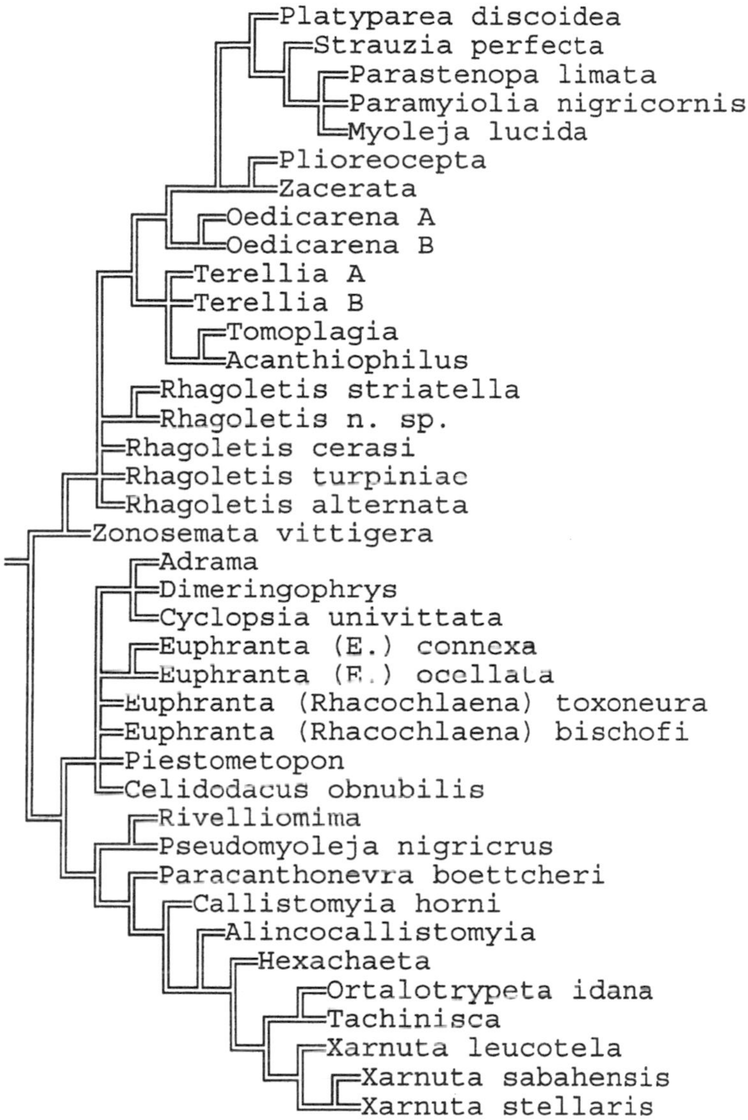


**FIGURE 4.3** (continued)

## 4.2 DEFINITION AND MONOPHYLY OF TEPHRITIDAE (= TACHINISCIDAE, N. SYN.)

The Tephritidae are not uniformly phytophagous; the family includes many saprophagous (including saproxylophagous) species and also occasional zoophages, gall inquilines, and nonspecialized phytophages (Krivosheina 1982; Hardy 1986b; Dodson and Daniels 1988; Permkam and Hancock 1995). Inclusion of the species formerly treated as the family Tachiniscidae (that are believed to be parasitoids) therefore does not destroy the ecological homogenity of Tephritidae because the family has diverse habits of larval feeding. On the contrary, keeping Tachiniscidae as a separate family (that would include the genera assigned to Ortalotrypetini) would not permit the recognition of the Tephritidae as a monophyletic family.

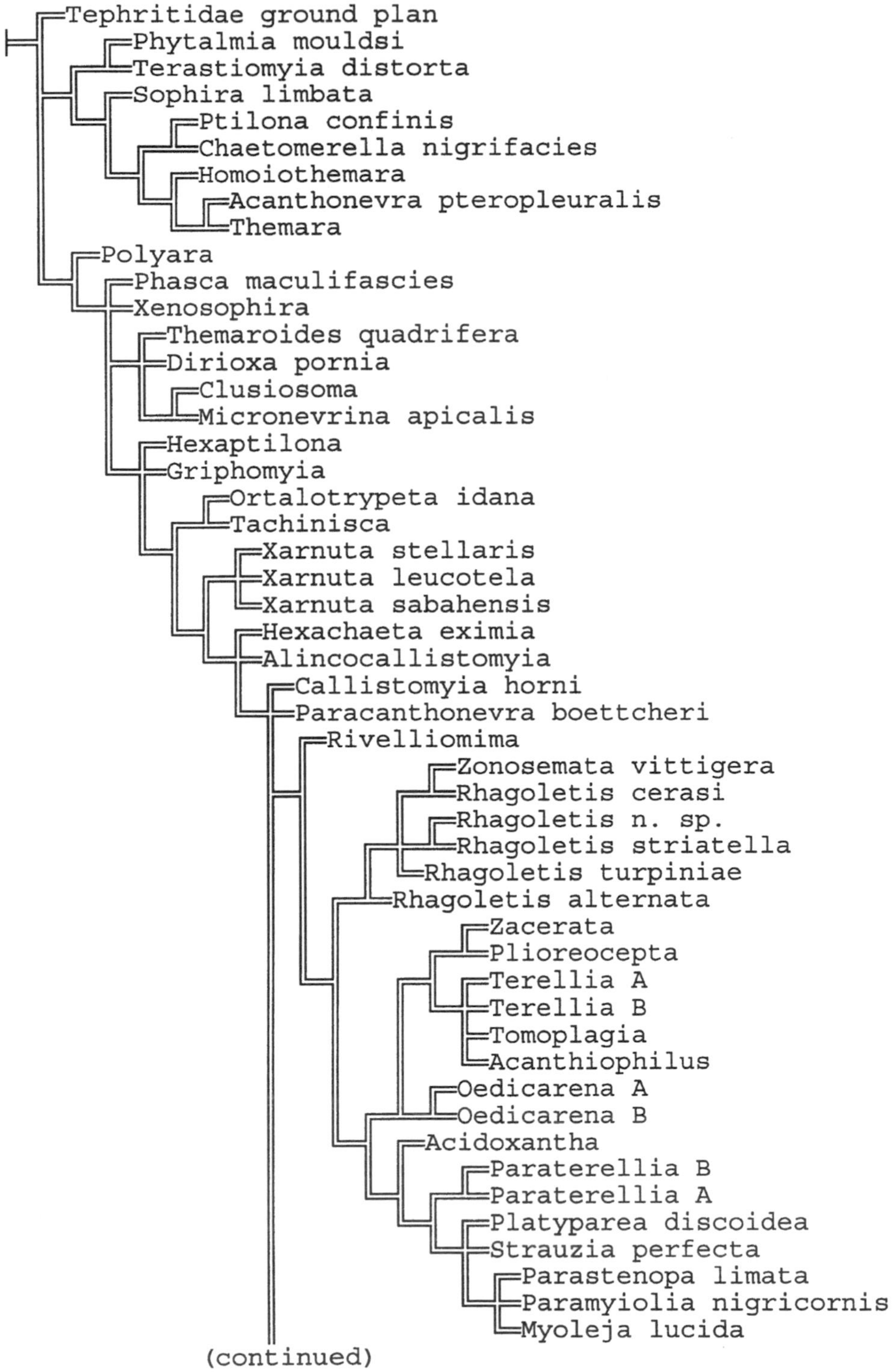


**FIGURE 4.4** Results of weighted cladistic analysis of Higher Tephritinae based on taxa and characters in Table 4.3. Consensus tree of trees resulting from analysis using Hennig86 (options mhennig*; bb*; followed by successive weighting).

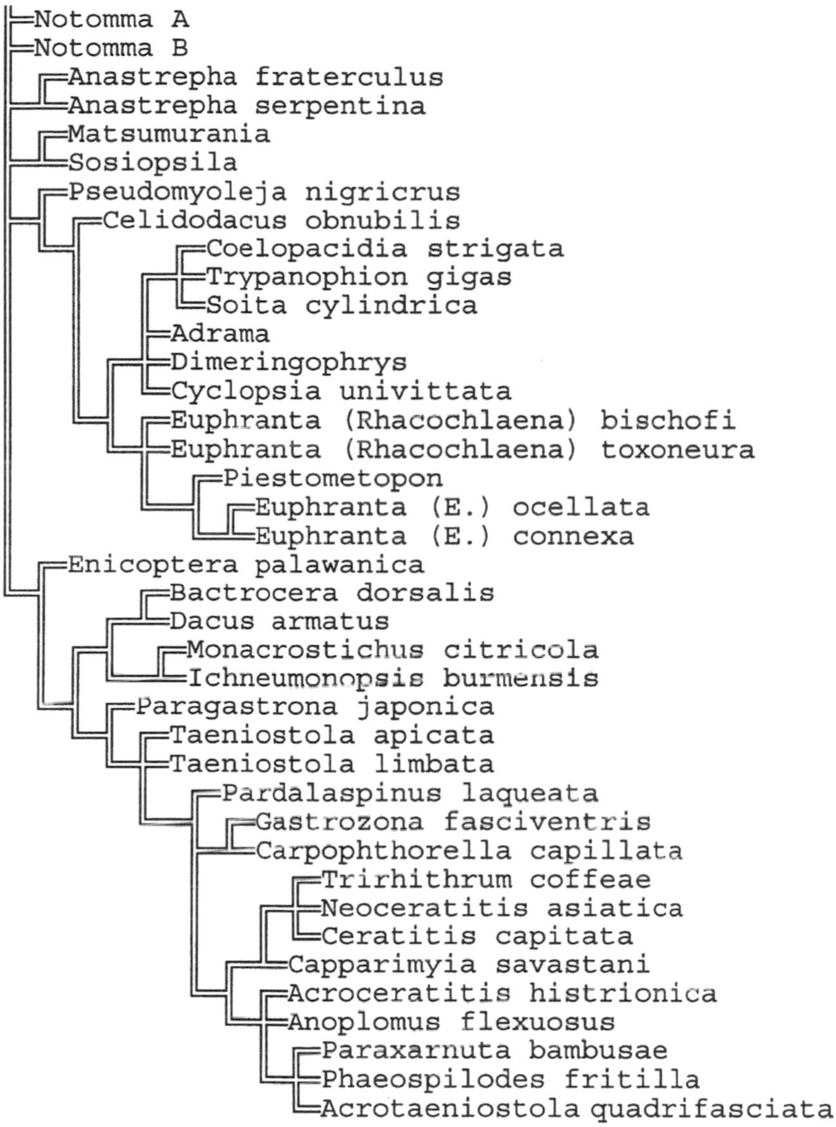


**FIGURE 4.4** (continued)

The hypothesis that the Tephritidae (including Tachiniscidae) are a monophyletic group is supported by the following characters that appear to be autapomorphies, despite their occasional reversal within the family or rare occurrence elsewhere:

1. *Frontal plates developed* — Most Tephritoidea have the fronto-orbital plate restricted to the posterior half of the frons (the orbital plate). In the Higher Tephritoidea secondarily sclerotized anterior corners of the frons occasionally occur (in *Poecilotraphera* Hendel, Platystomatidae: Trapherinae), and perhaps in *Chaetopsis* Loew (Ulidiidae: Lipsanini); this is probably homoplasy. It is difficult to find any frontal plate in Pyrgotidae. In the Tephritidae, the frontal plate appears to be a prolongation of the parafacial onto the frontal surface. It can be seen rather clearly, even if there are no frontal setae (*Terastiomyia* Bigot, *Ichneumonopsis* Hardy, *Monacrostichus* Bezzi, etc.). A shorter frontal plate (like in Acanthonevrini) is believed to be plesiomorphic vs. the longer frontal plate (like in Trypetini), but homoplasy probably occurs in different lineages of Tephritidae.

2. *Frontal setae, always much longer than surrounding setulae, developed on the frontal plate* — As noted in a previous chapter (Korneyev, Chapter 1), there are a few genera in allied families that also show a similar state of this character. They either belong to rather advanced branches within the Ulidiidae and Pyrgotidae (*Chaetopsis* and *Toxopyrgota* Hendel) and therefore cannot be closely related to Tephritidae, showing certain homoplasy in this character, or their position needs further clarification to prove or disprove the homoplastic origin of the character (*Poecilotraphera*).

3. *Costal vein with a deep constriction or break before the apex of the subcostal vein* — The subcostal break, present in the ground plan of the Higher Tephritoidea, is very distinctive throughout the Tephritidae. It is believed to be involved in wing displays during territorial or courtship behavior that include frontal (dorsoventral) torsion of spread wings.

4. *Two or three costal spines (enlarged and thickened setae) guarding such a break* — This character does not occur in the related families and appears to be a synapomorphy of the Tephritidae. The costal spines are indistinct in a few genera, most of which have a wasplike, slender appearance (*Matsumurania,* most Phytalmiini, Toxotrypanini, Dacini), but usually can be found in related taxa. It shows that secondary reduction of the costal spines associated with the subcostal break happens in different lineages of Tephritidae.

Among the characters autapomorphic for Tephritidae, McAlpine (1989) pointed out that they have the greater ampulla relatively strongly developed. Indeed, this character is usually well developed in all Tephritidae, but was not found by me nor reported by previous investigators in Pyrgotidae. It is absent in the Ulidiidae, and hardly developed in the Platystomatidae, where ampulla-like sclerites can be found in some genera, but are indistinct in most representatives. The greater ampulla is weakly developed in Tachiniscinae and many Dacinae species, but well developed in Blepharoneurinae, *Xarnuta* Walker (Norrbom, personal communication), *Matsumurania* Shiraki, most Phytalmiinae, Trypetinae, and Tephritinae (Korneyev, personal observation). This character needs further survey.

The basic branching of the Tephritidae into Tachiniscinae + Blepharoneurinae + Phytalmiinae + the Higher Tephritidae is an unresolved polytomy.

## 4.3 SUBFAMILY TACHINISCINAE, N. STAT.

Monophyly of this subfamily is proved by the following characters: (1) arista short pubescent (autapomorphy?, also in Blepharoneurinae and some Higher Tephritidae, due to homoplasy or synapomorphy; polarity is presumed from the presence of the long pubescent arista in the ground plans of most subfamilies of Platystomatidae and of Phytalmiinae); (2) two postpronotal setae in the ground plan (autapomorphy or synapomorphy with Blepharoneurinae) (one pair in *Ischyropteron* Bigot and *Neortalotrypeta* Norrbom); (3) two postsutural supra-alar setae (autapomorphy; one pair in some *Cyaforma* Wang, *Neortalotrypeta,* and *Ischyropteron*); (4) oviscape opened posterodorsally (autapomorphy); (5) eversible membrane without taeniae but with basoventral area of dense, very dark, larger scales (autapomorphy); and (6) aculeus short, stiletto-like (autapomorphy). The characteristic aculeus suggests that all the representatives of the subfamily are parasitoids like *Bibundia* Bischoff.

The very early separation of this subfamily from the general trunk of the Tephritidae is seen from possession of the rudimentary gonostyli in *Tachinisca* (completely lost in other Tephritidae, including two examined Ortalotrypetini species, apparently independently from other tephritids). Like in the Blepharoneurinae (see below), the current circumtropical distribution of Tachiniscinae may show that they existed before the isolation of the southerly members of the Eocene geoflora into Western and Eastern Hemisphere elements since the Miocene (Bush 1966; Norrbom et al. 1999).

### 4.3.1 TRIBE TACHINISCINI, N. STAT.

The monophyly of this tribe, corresponding to the family Tachiniscidae *auct.*, is supported by (1) lack of the pedicellar notch; (2) one to three anterior (prephragmal) anepisternal setae developed (autapomophy, one seta present in the Blepharoneurinae, apparently due to homoplasy); (3) presence of two or three anepimeral setae that are (4) mounted on a tubercle (small in *Tachiniscidia* Malloch); and (5) usually more than three pairs of scutellar setae. It is interesting that *Tachinisca cyaneiventris* Kértesz possesses a unique combination of plesiomorphies among the Tephritidae: (1) vanes of phallapodeme narrowly approximated, but (2) joined to fully developed lateral sclerites; (3) gonostylus rudimentary, but clearly present; and (4) glans with a sclerotized subapical lobe ending in a tail-like membranous apex ("flag"), not a trumpetlike dilation.

The tribe includes the Neotropical *Tachinisca* and Afrotropical *Bibundia* and *Tachiniscidia* Malloch. A species of *Bibundia* was reared from saturniid caterpillars (Roberts 1969). The fossil *Protortalotrypeta* Norrbom shares with them the head relatively narrow (short) in lateral view, long frontal plates with more than two frontal setae, short orbital plates (except in *Tachiniscidia* where the orbital plates are moderately long) and short ocellar, postocellar, and postvertical (paravertical) setae (symplesiomorphies or reversal?), posterior position of dorsocentral seta (aligned with intra-alar or postalar seta), the costa not reaching medial vein, and a very shallow or absent pedicellar notch (synapomorphies?). Otherwise, it is much smaller (3.7 mm long compared with 12 to 15 mm of *Tachinisca* and *Bibundia*).

### 4.3.2 Tribe Ortalotrypetini

Norrbom (1994a) gave a definition of this tribe and considered the relationships among the genera included here: eastern Palearctic and Oriental *Ortalotrypeta* Hendel and *Cyaforma* Wang, and Neotropical *Neortalotrypeta* Norrbom and *Protortalotrypeta* (fossil, Dominican amber). The Neotropical *Ischyropteron* Bigot is another genus that belongs in this tribe (Norrbom et al. 1999).

This tribe differs from the Tachiniscini mostly by possessing plesiomorphic characters (pedicellar notch present, ocellar setae long, etc.), and may be paraphyletic. *Ortalotrypeta idana* Hendel and *C. shenonica* Wang have the arms of the phallapodeme widely separated (synapomorphy; males of the other three genera unknown), and most species (except perhaps *I. nigricaudatum* Bigot, in which the head of the holotype is missing) have the gena high and the postgena bulging. Fossil *P. grimaldii* Norrbom possibly belongs to the Tachiniscini (see discussion above) or may be the sister group of that tribe.

### 4.3.3 *Matsumurania*, Unplaced Genus

*Matsumurania* has often been placed in Phytalmiinae. It has a similar elongate appearance and no apodemes on female sternum 6 (synapomorphy with Phytalmiinae or Tachiniscinae?), but the cercal unit is completely integrated into a long, broad, and blunt aculeus (apomorphy), the arista is short setulose (apomorphy), there are three (to four) frontal setae (plesiomorphy), one orbital seta (apomorphy), and one well-developed postpronotal seta (plesiomorphy), and the anterior notopleural seta is not shortened (plesiomorphy). *Matsumurania* has three completely separate spermathecal ducts, and certainly belongs to the Pyrgotidae + Tephritidae lineage. When dissecting a female of *M. sapporensis* (Matsumura), I found the abdomen to be full of first instar larvae, which means the species is viviparous and, thus, probably zoophagous. Therefore, its characters sharply disagree with the concept of the Phytalmiini. *Matsumurania* shares no significant synapomorphies with Blepharoneurinae, nor with the Higher Tephritidae. From the characteristics of the postocular rows of setae and probable zoophagy it may be somehow related to Tachiniscinae, but it differs by the oviscape, eversible membrane, and aculeus being unmodified. *Matsumurania* and *Ortalotrypeta* share the postocular area of the occiput with two or three rows of postocular setae instead of one (symplesiomorphy?, also known in some Ulidiidae), whereas in other Tephritidae the area posterior to the postocular row is bare (synapomorphy?). Since males are unknown, I cannot place this genus more precisely.

## 4.4 SUBFAMILY BLEPHARONEURINAE

Relationships within this group are analyzed by Norrbom and Condon (Chapter 6). Together with Tachiniscinae, Blepharoneurinae is apparently another group that separated from the common trunk of the Tephritidae at an early stage of its phylogenesis (Han and McPheron 1997). It possesses a rare combination of plesiomorphies, like well-developed ocellar seta, three pairs of scutellar setae, and female tergum 6 with anterior apodeme, with numerous autapomorphies. The monophyly of this subfamily is supported by the following characters: (1) arista short pubescent (autapomorphy or synapomorphy with Tachiniscinae, Trypetinae, and Tephritinae); (2) two postpronotal setae in the ground plan (autapomorphy or synapomorphy with Tachiniscinae); (3) anterior (prephragmal) anepisternal seta present on the upper portion of sclerite (autapomorphy or synapomorphy with Tachiniscini); (4) vanes of the phallapodeme broadly separate (autapomorphy or synapomorphy with Tachiniscinae: Ortalotrypetini and Higher Tephritidae); (5) scales on anterior half of eversible membrane well developed, but mostly tri- or multidentate with a large central tooth (Korneyev 1994, Figures 3.6 and 3.7), corresponding to the serrate scales of the posterior section of the other Tephritoidea (autapomorphy, partially homoplastic with some other phytophagous Tephritidae that have the scales on anterior portion of membrane mostly monodentate; not known for *Ceratodacus* Hendel); and (6) aculeus cutting, lobed or serrate (autapomorphy), in *Hexaptilona* Hering and *Blepharoneura femoralis* group with traces of fusion of cercal unit to tergum 8 (in fossil *Ceratodacus priscus* Norrbom and Condon fused, entire, but long and densely setose).

Condon and Norrbom (1994), Norrbom et al. (1999), and Norrbom and Condon (Chapter 6) included in this group *Blepharoneura* Wulp, *Baryglossa* Bezzi, *Hexaptilona*, *Problepharoneura* Norrbom and Condon, and, tentatively, *Ceratodacus*. Recently, these flies were classified either as a separate subfamily (Korneyev 1994) or as a tribe within the Phytalmiinae (Norrbom et al. 1999).

Some genera of Tachiniscinae and Blepharoneurinae share certain apomorphies and it is tempting to infer that these subfamilies are sister groups. But most of these characters are well-developed supernumerary setae that very probably are homoplasies in the case of the heavily setose Tachiniscini. Thus far the data are incomplete and this question needs further study.

Larval phytophagy and having the "piercing" aculeus and broadly separated vanes of the phallapodeme suggest that the Blepharoneurinae may be somehow related to the Higher Tephritidae, but both these characters appear highly subject to homoplasy. Two plesiomorphic characters of the Blepharoneurinae, the smooth mushroomlike spermathecae and the presence of a subapical lobe on the glans, which is, however, membranous or sclerotized and "flaglike" (Norrbom and Condon, Chapter 6), instead of a trumpetlike dilation, neither support their inclusion into the Higher Tephritidae, nor contradict a possible sister group relationship with them.

## 4.5 SUBFAMILY PHYTALMIINAE

The main groups of genera included here were recently redefined by Korneyev (1994) and then grouped with Blepharoneurini by Norrbom et al. (1999) as the subfamily Phytalmiinae that generally corresponds to the Acanthonevrina *sensu* Hardy (1986b) plus the Phytalmiini. Phytalmiinae is the oldest name applicable to this group.

I have discovered at most three characters that support the monophyly of this group comprising predominantly species that mostly have saprophagous (or very generalized phytophagous) larvae, and a very primitive, "tactile" (noncutting) type of aculeus that does not differ from that in the Lonchaeidae, Ulidiidae, and Platystomatidae: (1) in profile, epandrium elongated in the dorsoventral direction, often "barlike," with the lateral surstylus usually not narrower than the epandrium itself; there are no mesally directed lobes of the lateral surstylus against the prensisetae, and the medial surstylus is usually narrow and weak (autapomorphy; in the *Clusiosoma* subgroup different, but apparently due to secondary modification); (2) hypandrium with lateral sclerites rudimentary; vanes of the phallapodeme articulated with the anterior end of the hypandrium (autapomorphy; also in some Adramini, probably as a result of homoplasy); and (3) abdominal sterna 4 to 6 without anterior apodeme (autapomorphy; also in Ulidiidae: Ulidiinae and Platystomatidae).

All the Phytalmiinae have the ocellar setae small or rudimentary (polarity uncertain), no enlarged dentiform scales on the eversible membrane in the ground plan (plesiomorphy), and three spermathecae with smooth surface (plesiomorphy) (Korneyev 1994). The vanes of the phallapodeme are narrowly separated in the ground plan (plesiomorphy), but may be partially or completely fused, showing multiple cases of homoplasy.

The subfamily includes species breeding under the bark of fallen trees (mostly on early stages of decay) (Krivosheina 1982; Hardy 1986b; Dodson and Daniels 1988), termite galleries (Hill 1921), and occasionally in different tissues of dead or living plants, usually young large monocotyledonous grasses (bamboos, etc.) or in fruits.

There are four main lineages in the Phytalmiinae presumed to be monophyletic, but their relationships are still unresolved. These are the Acanthonevrini, Phytalmiini, Phascini, and Epacrocerini. The latter tribe is assigned to Phytalmiinae tentatively. There also are additional genera still unplaced at the tribal level.

### 4.5.1 Tribe Acanthonevrini

Korneyev (1994) defined this group by a single, but unique synapomorphy that supports its monophyly: the sclerotized medioapical lobes of the oviscape (dorsally and ventrally between the bases of the taeniae) are desclerotized medially and have a somewhat W-shaped appearance. This character is not developed in *Enoplopteron* Meijere and has not been examined in most species of *Sophira* Walker (except for *S. limbata* Enderlein), a genus that may be polyphyletic, and some other genera. The Australasian species of Acanthonevrini were revised by Hardy (1986b) and Permkam and Hancock (1995), but some genera assigned to this group because of having six scutellar setae (plesiomorphy) actually belong to Phascini (Phytalmiinae).

Relationships among the genera of Acanthonevrini cannot be successfully resolved without analysis of genital characters (e.g., structure of the vanes of phallapodeme, inner structures of the glans, etc.). Thus far, in most species they have not been examined. No external characters (presence/number of frontal and orbital setae, "intrapostalar" (posterior dorsocentral?), medial scutellar, and anepimeral setae, midtibial spurs, shape and setation of veins) show stability to be a reliable synapomorphy for large groups of genera within the tribe. Most of these characters have a mosaic distribution within the Acanthonevrini that cannot be reliably resolved in the most parsimonious trees.

#### 4.5.1.1 *Themaroides* Group of Genera

This group includes taxa characterized by the presence of the intrapostalar seta (plesiomorphy), rather long medial scutellar seta, two equal or subequal midtibial spurs, and other plesiomorphic features in its ground plan (two frontal and two orbital setae, separate vanes of the phallapodeme). Some smaller subgroups of genera within this group (see below) share apomorphic characters and can be characterized as monophyletic, but other genera are ungrouped, and their relationships are not resolved.

##### *4.5.1.1.1* Clusiosoma *Subgroup*

Genera that belong here include *Cheesmanomyia* Malloch, *Clusiosoma* Malloch, *Clusiosomina* Malloch, *Hemiclusiosoma* Hardy, *Nothoclusiosoma* Hardy, *Rabaulia* Malloch, *Rabauliomorpha* Hardy, and *Trypanocentra* Hendel. They share the anepimeral seta rudimentary or lacking (autapomorphy?), scutellum bare, wing mostly dark, without hyaline wedges or spots (autapomorphy or synapomorphy with *Paedohexacinia* Hardy), Cu setulose (autapomorphy), vanes of the phallapodeme fused (autapomorphy?), in combination with two pairs of frontal setae (plesiomorphy), the intrapostalar seta well developed (plesiomorphy), and the cercal unit of the aculeus bluntly truncate (plesiomorphy).

*Paedohexacinia* shares a similar wing pattern, the bare scutellum (synapomorphy), two frontal setae and two midtibial spurs (plesiomorphy), and may be the sister group of this subgroup; it also has more or less thickened male fore femur (in one species with strong setae, like in *Clusiosoma* and allied genera). Its cercal unit is tapered apically (autapomorphy) and vein Cu is bare (plesiomorphy or reversal). *Alloeomyia* Hardy also may fit in this subgroup because of the rudimentary anepimeral seta and setulose vein M (synapomorphy with this group?), but differs by its wing pattern type and partially setulose scutellum (plesiomorphy).

#### *4.5.1.1.2* Themaroides *Subgroup*

Genera that belong here include *Themaroides* Hendel, *Themarohystrix* Hendel, *Enoplopteron* Meijere, *Themaroidopsis* Hering, and *Bululoa* Hardy. They are large, heavily setulose flies with a second pair of postsutural supra-alar setae (autapomorphy?; also in some Tachiniscinae), two very strong midtibial spurs (apomorphy?), wing mostly dark without marginal indentations (synapomorphy with the *Clusiosoma* group or autapomorphy of the *Themaroides* group except for some species of *Themaroides* and *Themaroidopsis*?). Veins M and Cu are setulose in *Bululoa, Enoplopteron, Themarohystrix,* and some *Themaroidopsis* (synapomorphy with the *Clusiosoma* group?), the vanes of the phallapodeme are fused in *Themarohystrix* (apomorphy) but separated in *Enoplopteron* (plesiomorphy), in combination with the two pairs of frontal setae in the ground plan (plesiomorphy), and body chaetotaxy complete (plesiomorphy). *Bululoa* and *Enoplopteron* have a similarly setose costal vein (see Hardy 1986b, Figures 16, 35d, 36, and 37c), but Hardy suggests that they are not related.

The monophyly of this subgroup is tentatively supported. Its relationship with the *Clusiosoma* subgroup is supported by similarity in wing pattern and setulosity of veins M and Cu, but thus far is not supported by genitalic characters, which have been examined extremely unevenly. *Themaroides quadrifera* Walker (dissected in this study) has the apicomedial lobe of the oviscape clearly bilobate and fits well the concept of the tribe proposed by Korneyev (1994), but *E. hieroglyphicum* Meijere has this sclerite uniformly sclerotized (plesiomorphy or reversal?).

Permkam and Hancock (1995) assigned several species of Acanthonevrini with two long unequal midtibial spurs and a second (small) postsutural supra-alar seta to *Termitorioxa* Bezzi, and other species without such setae to *Acanthonevroides* Permkam and Hancock and *Taeniorioxa* Permkam and Hancock; they all share a similar shape of spermathecae with *Themarohystrix* and *Themaroides* and may be somehow related to this group. I consider them below as unplaced genera.

#### *4.5.1.1.3* Neothemara *Subgroup*

Genera belonging here include *Neothemara* Malloch, *Pseudacanthonevra* Malloch, *Hexaresta* Hering, *Lyronotum* Hering, and *Pseudoneothemara* Hardy. These genera have the wing dark, with one to three spots in cell $r_1$ and three marginal hyaline spots in the medial cell (autapomorphy?). Cu is often setulose in *Pseudacanthoneura* and *Pseudoneothemara* (synapomorphy), the vanes of the phallapodeme are separate in *Neothemara* and *Hexaresta* (plesiomorphy), in combination with the two pairs of frontal setae (plesiomorphy), and body chaetotaxy complete (plesiomorphy).

The wing pattern in this subgroup is uncommon among the Acanthonevrini and is presumed to be a synapomorphy of the taxa placed here; some *Hexacinia* share the presence of the three marginal spots in cell m, but this genus belongs to a different cluster within the tribe. *Neothemara* and *Pseudacanthoneura* both have vein $R_{2+3}$ undulate, but it is not clear if they are closely related.

### 4.5.1.2 *Diarrhegma* Group of Genera

*Diarrhegma* Bezzi, *Termitorioxa* Bezzi, *Acanthonevroides* Permkam and Hancock, *Aridonevra* Permkam and Hancock, and *Taeniorioxa* Permkam and Hancock share two midtibial spurs (plesiomorphy?) plus the intrapostalar seta well developed (plesiomorphy) without any peculiar characters of external morphology that have been reported that could be considered as a synapomorphy with the other groups within the tribe. Further study is needed to determine if this group is monophyletic.

*Diarrhegma* can be easily distinguished by its white and bare scutellum (autapomorphy), but otherwise has a set of characters very close to the ground plan of the tribe. *Termitorioxa*, *Acanthonevroides*, and *Taeniorioxa* have a very similar spermathecal shape: spherical, with a small apical nipple and a spindlelike expanded neck (see Permkam and Hancock 1995, Figures 29, 179, 184, and 204), but this structure cannot be considered as a synapomorphy of these three genera; at least, the genera of the *Themaroides* subgroup have the spermathecae rather similar (Hardy 1986b, Figures 93a, 99c, and 100e). Further analysis using additional characters is necessary to decide if they really are related.

*Aridonevra* is an Australian genus from an arid area, which is uncommon for Acanthonevrini; most occur in humid areas (rain forests, etc.). It shares with the Moluccan genus *Emheringia* Hardy two apomorphic characters: the dorsocentral seta on a level with the intra-alar seta and vein DM-Cu slanted, with the posterior end proximal to the anterior end (so the ultimate section of M is longer than the penultimate one); they have very similar wing patterns. Nevertheless, they differ in many chaetotaxy characters that contradict their close relationship.

### 4.5.1.3 *Dirioxa* Group of Genera

Among the genera with the intrapostalar seta present (plesiomorphy), *Dirioxa* Hendel, *Emheringia* Hardy, and *Lumirioxa* Permkam and Hancock differ by having only one long midtibial spur (apomorphy?) that may be a synapomorphy with the *Acanthonevra* group, but this hypothesis needs more evidence.

*Dirioxa* fits the ground plan of the *Acanthonevra* group (two frontal and two orbital setae, $R_{2+3}$ straight, medial scutellar seta relatively long) except in having the arista ventrally bare (autapomorphy?; also in Epacrocerini) and larvae associated with decaying fruits rather than with dead wood. *Emheringia* shares several apomorphies (wing venation and pattern, posterior position of dorsocentral seta) with *Aridonevra* (see above), but other characters apparently contradict their close relationship. *Lumirioxa* is a monotypic genus that fits well in the following cluster (and especially the *Ptilona* subgroup), sharing a single pair of frontal setae, male foreleg densely setulose on the ventral surface?, and one midtibial spur. It differs by the intrapostalar and anepimeral setae well developed.

### 4.5.1.4 *Acanthonevra* Group of Genera

This large complex is believed to be monophyletic because of several plesiomorphic features in its ground plan (two frontal setae, vanes of the phallapodeme separate) combined with the absence of the intrapostalar seta (autapomorphy compared to Tachiniscinae, Blepharoneurinae, and the other genus groups of Acanthonevrini, but common for most other Tephritidae, apparently due to homoplasy) and short secondary spurs of the midtibia (apomorphy?). Some groups of thoracic setae are reduced in more than half of the genera included in this group. Some genera of the *Diarrhegma* group, which may be paraphyletic (see above), may be related to this group.

The *Acanthonevra* group also must be considered provisional, as many species assigned to *Sophira* Walker, *Terastiomyia* Bigot, *Exallosophira* Hardy, and some other genera were not examined for the presence of the double medioapical lobe of the oviscape, a synapomorphy of the Acanthonevrini. Some of them may actually belong in other tribes.

This group includes a small number of generalized genera with a well-developed anepimeral seta (plesiomorphy), and a larger, probably monophyletic cluster of subgroups that share rudimentary or lacking anepimeral seta (apomorphy?; also in *Clusiosoma* group apparently due to homoplasy).

#### *4.5.1.4.1* Acanthonevra *Subgroup*

This subgroup includes genera sharing strong anepimeral and two orbital setae (plesiomorphy), and may be paraphyletic. Relationships of the taxa here included are not resolved. Some of them

(*A. fuscipennis* Macquart, *Anchiacanthonevra* Hardy, *Austronevra* Permkam and Hancock) have only one frontal seta (autapomorphy or synapomorphy with the genera of the *Ptilona* subgroup), whereas the others usually have two. These others include some species recently assigned to *Acanthonevra* — *desperata* Hering, *incerta* Hering, *shinonagai* Hardy, *Erectovena amurensis* (Portschinsky) (= *Acanthonevra speciosa* Hendel), and some others — *Hexacinia* Macquart, *Gressittidium* Hardy, and *Copiolepis* Hardy. The two genera *Erectovena* Ito and *Hexacinia* both have an additional, anteroventral anepisternal seta, but there is no further evidence of their close relationship. The position of *A. fuscipennis,* the type species of *Acanthonevra*, is uncertain within this complex, because the other species assigned to that genus are believed to belong elsewhere. Nevertheless, I leave the concept and the limits of the genus pending until the genitalic characters of more species are examined in detail.

*Rioxa* Walker and *Ectopomyia* Hardy, known to me from descriptions only, apparently belong to a monophyletic group also including *Hexacinia*. Whether all of their species have the anepimeral seta is unknown to me. According to Norrbom (personal communication), *R. lucifer* Hering and *R. sexmaculata* Wulp have it strongly. All three genera share long and broad surstyli (synapomorphy), very long and wrinkled preglans area of distiphallus (synapomorphy?), and oval spermathecae with long and bulbous necks and apical nipples (synapomorphy). They also have some yellow head setae (frontal, orbital, and postocellar setae in *Hexacinia*, posterior orbital seta in *Ectopomyia*, and postocellar seta in *Rioxa*). The sister group relationship of *Ectopomyia* and at least some *Rioxa* species is evident from their extremely long surstyli, but the monophyly of *Rioxa*, which is rather heterogeneous, needs further confirmation.

*Chaetomerella* Meijere and *Orienticaelum* Ito seem to fit in the *Ptilona* subgroup because of the rudimentary anepimeral seta (synapomorphy?). On the other hand, they have a short setulose arista (synapomorphy) and two frontal setae (plesiomorphy), the latter not known in the genera of that subgroup. The absence of the anepimeral seta occurs here along with the general reduction of thoracic setae (presutural, dorsocentral, and acrostichal) and therefore may be a case of homoplasy. I have analyzed no additional characters that solve whether to place these two genera in the *Acanthonevra* subgroup or the *Ptilona* subgroup.

*Micronevrina* Permkam and Hancock superficially resembles this group because of having two frontal setae, anepimeral seta rudimentary or absent, and presutural and medial scutellar setae absent; the intrapostalar seta is weak in one species and absent in the others. *Micronevrina* differs from other Acanthonevrini by having the aculeus highly modified (cercal unit fused with tergum 8, posterolateral margins of the latter flattened and serrate) and eversible membrane densely covered with dentiform (mostly monodentate) scales (Permkam and Hancock 1995, Figures 116, 135, and 144), showing strong resemblance to *Xarnuta* Walker, some *Hexachaeta* Loew, and *Soita* Walker of the Higher Tephritidae. Nevertheless, they have the medioapical lobes of the oviscape desclerotized medially and no apodemes of female sterna 5 and 6, and therefore are aberrant members of Acanthonevrini, clearly showing that such characters of the aculeus and eversible membrane are homoplasies of *Micronevrina*.

#### *4.5.1.4.2* Ptilona *Subgroup*

This subgroup includes genera that have the anepimeral seta rudimentary or absent (apomorphy) and is presumed to be monophyletic. Most of the included genera have at most one frontal seta (autapomorphy, synapomorphy, or homoplasy with several species assigned to the *Acanthonevra* subgroup; see above).

*Rioxoptilona* Hendel and *Austrorioxa* Permkam and Hancock have the anepimeral seta present, but thin, and none of the synapomorphies of the following groups. Both genera are presumed to be generalized members of the *Ptilona* subgroup, and actually share no additional synapomorphies. *Rioxoptilona vaga* Wulp was considered to be a member of *Acanthonevra* (see Hardy 1973; 1977; 1986b), but as far as species assigned to the latter genus are discussed here under different lineages, I prefer to consider *Rioxoptilona* as a separate genus to point out its disagreement with the concepts of both groups. *Austrorioxa* also does not show further relationship to any other genera of this subgroup.

*Themara* Walker, *Ptilona* Wulp, *Lenitovena* Ito, and *Homoiothemara* Hardy appear to be a complex of closely related genera within the *Ptilona* subgroup. They are moderate-sized flies with hyaline spots and wedges on dark wings, usually associated with rotting trees, and sharing, along with the intrapostalar seta lacking (synapomorphy of the whole complex), either the setulose Rs (except in *Ptilona* and *Homoiothemara*), or/and heavily setulose ventral surface of male fore femur and fore tibia (except in some *Themara* species), and medial scutellar seta very weak to absent (in *Ptilona* and *Homoiothemara*) (autapomorphy). *Lenitovena* was synonymized with *Acanthonevra*, but its status actually depends on the concept of the latter genus and of *Themara*. *Ptilona* and *Homoiothemara* share straight vein $R_{2+3}$ and bare vein Rs (plesiomorphies), single pair of orbital setae (synapomorphy?), and the medial scutellar seta lacking (synapomorphy?), but otherwise they are quite unlike, and their close relationship is questionable. Both *Themara* and *Lenitovena* have $R_{2+3}$ undulate, which seems to be a synapomorphy, although this character is quite widespread among Phytalmiinae, including the type species of *Acanthonevra*. Species assigned to *Themara* have veins M and Cu setulose (synapomorphy?). The type species of *Themara* does not have stalked eyes nor "feathered" forelegs, and has a normally developed anepimeral seta (as does the type species of *Acanthonevra*). It does not fit the diagnosis of the *Ptilona* subgroup at all. Some species recently assigned to *Acanthonevra* and *Themara* obviously need reconsideration of their generic position. In most species assigned to *Themara* and a few species assigned to *Acanthonevra* — *trigona* Matsumura (the type-species of *Lenitovena* Ito), *pteropleuralis* Hendel, *ochropleura* Hering — Rs is setulose, and males of these species and of *T. hirsuta* Perkins have the foreleg "feathered"; both characters, as well as the setulosity of veins M and Cu have a mosaic distribution. Setulae occasionally occur on M in some specimens of *P. confinis* Walker and *A. ochropleura*. Apparently, the sex-associated setulosity of the foreleg and the size of the anepimeral seta show reversal in some cases. Study of additional, especially genitalic characters is needed to solve relationships within this group. It is very likely that *Themara* and *Lenitovena* are synonyms, but I leave this pending, until more information is available.

*Sophira* Walker, *Soosina* Hering, *Kambangania* Meijere, *Loriomyia* Kértesz, *Felderimyia* Hendel, *Antisophira* Hardy, *Exallosophira* Hardy, *Langatia* Hancock and Drew, *Cleitamiphanes* Hering, and possibly *Colobostroter* Enderlein, *Terastiomyia* Bigot, and *Pseudosophira* Malloch, appear to be another complex of related genera. The members of this complex are large species with long, often variously modified wings, and reduced chaetotaxy. The costal spines are indistinct and minute (synapomorphy). All the genera share the medial scutellar seta rudimentary or completely lacking (also in *Ptilona*, see above, but apparently due to homoplasy). *Kambangania* and *Soosina* have been considered subgenera of *Sophira* (see Hardy 1980), but their relationships need further consideration. Some species assigned to *Themara* (e.g., *T. maculipennis* Westwood) fit the diagnosis of this complex of genera better than the preceding. The distributions of all the genera (except for *Loriomyia* from New Guinea and *Exallosophira* from the Solomons), is restricted to Southeast Asia (Thailand, Vietnam, Malaysia, Philippines, and Indonesia west of Weber's Line).

I tentatively consider this cluster the sister group of the *Ptilona* complex of genera. They share the medial scutellar and the anepimeral setae are very minute or completely absent, and each group differs from the other by a single autapomorphy (the "feathered" foreleg in the *Ptilona* complex and the lack of the costal spines in the *Sophira* complex). If several species of *Themara* lack the "feathering" of the foreleg due to reversal of this sex-dependent character, then at least some members of the *Sophira* complex may be derived from the *Ptiolona* complex rather than belonging to its sister group. *Loriomyia guttipennis* Kértesz shares an apically long setose proctiger, general shape of surstyli, semi-globose spermathecae and thickened costa (all apomorphies) with *Themara maculipennis* and *hirtipes* Rondani. Also, species of *Felderimyia* and *Langatia* share vein M setulose with *Themara* species (synapomorphies?).

Some taxa here considered to belong to this group were previously assigned to the Phytalmiini (Korneyev 1994). Both absence of the costal spines and reduction of head and thoracic setae may be either a synapomorphy or a result of homoplasy. Thus far, the monophyly of this cluster is not supported by any genitalic characters, but they have not been studied for most included taxa. Also, there is no good evidence of the monophyly of the *Sophira* group of genera *sensu* Hardy (1980), i.e., the yellowish flies with the katepisternal seta lacking. The species that Hardy assigned here have rather different appearance, and their close relationship has not been supported by additional synapomorphies (for instance, by genital characters).

Also, thus far, only *S. limbata* Enderlein has been examined and found to have the bilobate medioapical lobe of the oviscape; other species have not been checked. Some of the representatives of *Sophira*, like *S. limbata borneensis* Hering, possess male genal processes (Hardy 1988, Figure 34c) similar to some *Terastiomyia* and some Phytalmiini.

*Cleitamiphanes heinrichi* Hering (see Hardy 1988, Figure 14) appears to be similar to *S.* (*s. str.*) *plagifera* Walker and *S.* (*s. str.*) *spectabilis* Hardy in wing pattern and other characters (two orbital setae, thoracic chaetotaxy, body coloration) and may be congeneric with them.

*Terastiomyia, Pseudosophira,* and *Colobostroter* were recently assigned to Phytalmiini (Korneyev 1994) because of having the anterior notopleural seta short or lacking. They may belong either to that tribe or to this group of Acanthonevrini. The structure of the medioapical sclerite of the oviscape, like in most *Sophira*, has not been examined. *Pseudosophira* and *Terastiomyia* share similar structure of male terminalia (lateral sclerites of hypandrium lacking — synapomorphy?; phallapodeme short — synapomorphy) (Hardy 1974, Figure 62g; 1986b, Figure 12e) and head chaetotaxy (frontal, posterior orbital, and lateral vertical setae lacking — synapomorphy), and may be sister groups or synonyms. Relationships of the other included taxa are not resolved.

### 4.5.2 Tribe Phytalmiini

This tribe was strictly defined by McAlpine and Schneider (1978) to include the three most aberrant genera with long narrow wings, petiolate abdomen, and the metathoracic postcoxal bridge completely sclerotized (autapomorphy): *Phytalmia* Gerstaecker, *Diplochorda* Osten Sacken, and *Sessilina* McAlpine and Schneider. They certainly form a monophyletic group that share shortened anterior notopleural seta (autapomorphy or synapomorphy with *Terastiomyia* and allied genera of Acanthonevrini), elongate wing (cell bm 1.8 to 2.5 times longer than DM-Cu), chaetotaxy strongly reduced (one frontal, one orbital, ocellar, lateral vertical, postpronotal, presutural, and dorsocentral setae lacking), the secondary midtibial spur longer than half of the longest one (polarity uncertain), the apicomedial lobe of the oviscape not bilobate (plesiomorphy), and the aculeus half as long as the oviscape (autapomorphy; also in some Adramini).

There are several genera that share reduced chaetotaxy with the Phytalmiini but have a non-sclerotized metathoracic bridge (plesiomorphy) (*Antisophira*, *Cleitamiphanes*, *Colobostroter*, *Ortaloptera* Edwards, *Terastiomyia*, *Pseudosophira*). Of them, only *Ortaloptera* is believed to be a related group to the remaining Phytalmiini. The others are considered under Acanthonevrini (see above) and apparently do not form a monophyletic cluster with the Phytalmiini *sensu* McAlpine and Schneider (1978).

*Ortaloptera* and the other three genera of Phytalmiini share the aculeus half as long as the oviscape and the vanes of the phallapodeme fused. It also has a Papuan distribution and similar modifications of the gena, as discussed and figured by McAlpine and Schneider (1978) and Hardy (1988).

### 4.5.3 Tribe Phascini

This tribe was defined by Korneyev (1994) as a subfamily, but was recently inferred to belong to the same clade as Phytalmiini and Acanthonevrini. Its monophyly is supported by two synapomorphies: (1) wing pattern with an inverted V-shaped mark (autapomorphy?; similar condition in

Trypetinae: Toxotrypanini and Dacinae: Gastrozonini); (2) spermathecae bare, with nipplelike apex (autapomorphy?; also in many different genera of Phytalmiinae). These two characters are combined with the fused vanes of the phallapodeme and its anterior part short, ridgelike (synapomorphy with Phytalmiini and *Polyara* group?) and characteristic plesiomorphic features of acanthonevrine chaetotaxy (intrapostalar seta developed, midtibia with two strong subequal spurs). All Phascini are apparently confined to the Papuan Region.

Phascini (except for *Paraphasca* Hardy) have the aculeus of the typically "piercing" type (the cercal unit dorsoventrally compressed and tapered distal to the subapical setulae; these setulae are directed laterally and are all subequal in length rather than the two distal setae being much longer than the two anterior; basal setulae of cercal unit short or rudimentary) and are presumed to have phytophagous larvae.

### 4.5.4 Tribe Epacrocerini

Epacrocerini (see Hardy 1982, *Epacrocerus* group; Korneyev 1994, Epacrocerinae) are tentatively retained in the Phytalmiinae because of head setae arrangement, intrapostalar seta present, aculeus tactile and spermathecae bare (all plesiomorphies); but no other characters to support their relationship to this subfamily have been examined. Like in Phascini, the eversible membrane is impregnated by sclerotized, apparently dentiform, structures (Hardy 1982, Figure 4d), but the structures of male genitalia seem to be very different (Hardy 1982, Figure 5d). Monophyly of the tribe is supported by two synapomorphies: the similarly modified medial lobe of the pedicel (Hardy 1982, Figures 1b and 5b) and tergum and sternum 8 forming strongly dilated aculeus shaft (Hardy 1982, Figures 2a and 6d).

### 4.5.5 Unplaced Genera Presumed To Be Associated with Phytalmiinae

The group of species allied to *Polyara* Walker (see Hardy 1986b; 1988; Korneyev 1994) are placed into Phytalmiinae tentatively because the presence or absence of the apodemes of the female abdominal sternites has not been examined for them. Otherwise they possess the arrangement of frontal and orbital setae characteristic for Phascini and Acanthonevrini. The two midtibial spurs (plesiomorphy?) and the structure of male genitalia in *Polyara* (Hardy 1986b, Figure 67) and *Pseudacrotoxa* Hering (Hardy 1988, Figure 33c) (fused vanes of phallapodeme: synapomorphy?) are similar to those in *Phasca*, so this group may be somehow related to Phascini. The moderately to well-developed ocellar seta is uncommon in the Phytalmiinae and is rather characteristic for the Trypetinae + Dacinae + Tephritinae cluster.

*Robertsomyia* Hardy superficially resembles Phytalmiini — most setae lacking, wing basally narrowed (autapomorphy or synapomorphy with Phytalmiini), male epandrium and surstyli similar to those in other Phytalmiinae — but the phallapodeme has broadly separated vanes (polarity unresolved) and apparently there are no anterior apodemes of female sterna 4 to 6 (synapomorphy with Phytalmiinae?) (see Hardy 1983a). The phylogenetic relationship of this genus is unclear.

## 4.6 RELATIONSHIPS AMONG THE TAXA OF THE HIGHER TEPHRITIDAE (TRYPETINAE *S. LAT.* + TEPHRITINAE + DACINAE)

Most included taxa share the female sternite 6 with the anterior apodeme (plesiomorphy that differentiates them from the Phytalmiinae). To our current knowledge, larvae of all these flies are obligatorily (not occasionally) phytophagous, and females usually have similar structures of the eversible membrane (monodentate scales) and aculeus that arise due to adaptation for oviposition into intact plant tissues. However, in some cases these ovipositor structures apparently have arisen in a parallel way and therefore may not be "good" synapomorphies.

The monophyly of Trypetinae *s. lat.* + Tephritinae + Dacinae is not supported by any unique, totally consistent synapomorphies. Indeed, every character of the five presumed synapomorphies has a mosaic distribution, proving that all these characters either can develop reversibly or be homoplastic. These putative synapomorphies are (1) glans with one membranous but sculptured (with minute tubercles or denticles) or even partially sclerotized (in Tephritinae: Terelliini) basal lobe (most Blepharoneurinae; see Norrbom and Condon, Chapter 6) and a few Phytalmiinae have a similar lobe, but without any sclerotization); (2) glans with trumpet-shaped subapical lobe; (3) aculeus of piercing or cutting type, with tip fused, not "tactile"; (4) eversible membrane covered by dentiform scales; and (5) spermathecae papillose or spinulose.

The primary branching of the cluster changes under different options of computer analysis. In some trees, the Dacinae appear to be the sister group of the Trypetinae *s. lat.* + Tephritinae, whereas in others the Trypetinae *s. lat.* is a nonmonophyletic aggregation, with both Dacinae and Tephritinae derived from it. Evidence for both hypotheses is equally poor, and all trees are poorly resolved because many character states (especially larval and genitalic) are missing in the matrix.

The mosaic distribution of some plesiomorphies often contradicts hypotheses of monophyly of lineages based on other apomorphies. For example, the absence of the intrapostalar seta in the tribes other than Xarnutini (apomorphy) does not necessarily support the monophyly of that cluster. It contradicts the distribution of the monodentate scales of the eversible membrane (apomorphy) that are present in Xarnutini and Hexachaetini, but absent in Nitrariomyiini of Trypetinae and Gastrozonini of Dacinae that both have no intrapostalar seta.

The presence of the medial scutellar seta (plesiomorphy — in *Hexachaeta, Alincocallistomyia,* and *Xarnuta*) and two midtibial spurs (plesiomorphy — in *Hexachaeta, Alincocallistomyia,* and *Callistomyia* of Hexachaetini and *Xarnuta* of the Xarnutini), and blunt female cercal unit with long subapical lateral setae (plesiomorphy — in *Celidodacus* Hendel and *Cyclopsia* Malloch of the Adramini) makes evident that apomorphic characters corresponding to these mosaically distributed plesiomorphies may arise independently in different lineages (Toxotrypanini, which is apparently related to Hexachaetini and Xarnutini, and remaining genera of Adramini).

The early separation of the Dacinae is supported by the presence of several plesiomorphies that are absent in the other Higher Tephritidae. In the ground plan of the Dacinae (in some Gastrozonini) there are no dentiform scales on the eversible membrane, four relatively long lateral setulae on the female cercal unit are present, with the subapical one much longer than the two basal setulae, and the unit is generally blunt and fused to the tergum 8 with faint remainders of the suture (Hardy 1988, Figures 31a and 38a). The monodentate scales of the eversible membrane in the Dacinae: Ceratitidini are absent in Gastrozonini and in the ground plan of the subfamily and presumably arose to arise independently from Trypetinae and Tephritinae.

The complete absence of monodentate scales and presence of long taeniae on the eversible membrane in Nitrariomyiina (see below) is presumed to be a plesiomorphy; thus far, no additional cases which show that the last character may be reversal are known. If so, the monodentate scales in Trypetini also appear independently from the other tribes of this cluster.

The two taxa Hexachaetini and Nitrariomyiina have disjunctive amphi-Pacific distributions, with the centers of diversity in Southeast Asia and the Neotropical Region that may additionally support the hypotheses of their early separation from the common stem (in Miocene or earlier).

There are four main lineages whose members are presumed to retain archaic features and therefore separated at an early stage of tephritid evolution. These are Hexachaetini (possibly with Toxotrypanini as a derived group), Adramini, Nitrariomyiina (possibly forming a monophyletic cluster with Trypetina, Zaceratini, and Tephritinae), and Dacinae. If further study supports this hypothesis, the Trypetinae *s. lat.* are a para- or even polyphyletic cluster that needs further taxonomic and nomenclatural improvements.

## 4.7 SUBFAMILY TRYPETINAE *S. LAT.*

This is a possibly poly- or paraphyletic cluster that remains after exclusion of the monophyletic Dacinae and Tephritinae. Monophyly can be proved for most tribes and groups of genera here included, but the relationships among them are poorly resolved.

### 4.7.1 Tribes Xarnutini and Hexachaetini

This is an aggregation of taxa which retain three or more pairs of scutellar setae in the ground plan and two midtibial spurs (symplesiomorphies with Lower Tephritidae) and tergum 6 of female with anterior apodeme (symplesiomorphy with the Higher Tephritidae and Blepharoneurinae) in combination with well-developed black proepisternal seta (synapomorphy of Xarnutini + Hexachaetini?; also in some Tachiniscinae and Pyrgotidae due to homoplasy; apparently absent in *Alincocallistomyia* Hardy and some *Hexachaeta* Loew), the anterior surface of the midfemur with a row of setae or spines (synapomorphy), eversible membrane covered by monodentate scales (synapomorphy with other Trypetinae *s. lat.* or autapomorphy), spermathecae more or less papillose (apomorphy of the Higher Tephritidae; except for one group of species in *Xarnuta*), aculeus of cutting type (synapomorphy?), sometimes with apically bifurcate cercal unit (synapomorphy?; also in *Micronevrina* of Acanthonevrini and *Soita* Walker of Adramini). Most of its members share also high vertical eyes (synapomorphy?), elongate first flagellomere (synapomorphy), three frontal setae on lower two-thirds of frons, two orbital setae on upper quarter, and ocellar seta very short.

#### 4.7.1.1 Tribe Xarnutini

Autapomorphies of this tribe are the enlarged and flattened scutellum with two or more additional marginal setae and densely setulose disk. It includes two closely related Australasian and Oriental genera, *Xarnuta* Walker and *Platystomopsis* Hering (Korneyev 1994).

*Xarnuta* includes three groups of species characterized by different aculeus and spermathecae structure. (1) Species of the *leucotela* group (*X. leucotela* Walker, *confusa* Malloch, and *inopinata* Hering) have the aculeus laterally compressed (synapomorphy) and the spermathecae simple, oval, and conspicuously papillose (see Hardy 1986b, Figures 115a and 115d; Korneyev 1994, Figure 5; Permkam and Hancock 1995, Figure 218), and the wing pattern lacking regular spots or marginal incisions. (2) In the *stellaris* group (*X. stellaris* Hardy, and apparently *X. lativentris* Walker and *fenestellata* Hering; genitalia not dissected) the aculeus is dorsoventrally compressed with the tip serrate and bifurcate (autapomorphy?; also in some *Hexachaeta*), the spermathecae consist of two parts and apparently are nonpapillose, but rather tuberculose, and the wing pattern has regular marginal incisions (see Hardy 1986b, Figure 121). (3) *Xarnuta sabahensis* Hardy is an isolated species that has the aculeus slightly compressed dorsoventrally with two pairs of subapical steps (corresponding to the two ventrally directed projections of the *leucotela* group) and the spermathecae spherical, smooth, like in Phytalmiinae and Blepharoneurinae, with a conspicuous neck, but without the bulbous dilation characteristic for the *stellaris* group (plesiomorphy or autapomorphy due to reversal), and the wing pattern with regular marginal incisions, and the abdomen black spotted (see Hardy 1986b, Figure 120). The polarity of the main characters is unclear, and the polytomy of the three monophyletic groups is unresolved. The distribution of the genus is predominantly Australasian.

The status of the two species described from the Caspian area of the Palearctic region ("Talysch-Gebiet") is questionable; I was unable to find any conspicuous differences between *X. stellaris* and *X. fenestellata* (type examined) and *X. leucotela* and *X. inopinata* (type examined). They have never been rediscovered from the Talysh region (Richter, personal communication) and apparently were based upon mislabeled specimens presumably from the Philippines or from the western part of Indonesia, where both *X. leucotela* and *X. stellata* occur.

#### 4.7.1.2 Tribe Hexachaetini

The genera included in this tribe possess all the characters common for the Xarnutini + Hexachaetini cluster as discussed above, and also share lack of intrapostalar seta. Compared to the ground plan of the Lower Tephritidae and to Xarnutini, the latter character is a possible synapomorphy of Hexachaetini, but this character is common to all the other Trypetinae *s. lat.*, Tephritinae, and Dacinae, so the monophyly of this tribe based on this single apomorphy is questionable.

The tribe includes the Neotropical *Hexachaeta*, and Australasian *Callistomyia* Bezzi and *Alincocallistomyia*. *Callistomyia* differs from the monotypic *Alincocallistomyia* by two apomorphies: acrostichal seta and the third (medial) scutellar seta absent. Similarly, *Hexachaeta* includes one species lacking the medial scutellar seta (described as *Neohexachaeta* Lima, now considered a synonym of *Hexachaeta*).

All these genera were examined rather incompletely, and a relevant analysis cannot be provided to clarify relationships among these genera without additional study of genitalic and larval characters.

### 4.7.2 Tribe Toxotrypanini

This tribe was defined by Hancock (1986) to include *Anastrepha* Schiner and *Toxotrypana* Gerstaecker. It has been well defined both morphologically and biochemically by Norrbom (1985), Hancock (1986), and other authors (see Norrbom et al., Chapter 12). *Anastrepha* and Hexachaetini share similar head shape and chaetotaxy (see above), but the polarity and value of these characters is unclear. Han and McPheron (1997) postulated a sister group relationship between *Hexachaeta* and the Toxotrypanini, based on DNA sequence data, and Norrbom et al. (1999) included *Hexachaeta* in that tribe on this basis.

### 4.7.3 Tribe Rivelliomimini

This tribe was established by Hancock (1986) for one Oriental and two Afrotropical genera, *Ornithoschema* Meijere, *Rivelliomima* Bezzi, and *Xanthanomoea* Bezzi. The monophyly of the tribe is supported by having last terga of male and female preabdomen with shining black swollen spots, similar chaetotaxy, and similar structure of female terminalia. Relationships within the tribe, as well as its position within the subfamily Trypetinae, are not resolved. Differences between the genera are limited to the position of crossveins R-M and DM-Cu and minute details of the wing pattern, and they may rather be treated as one genus.

### 4.7.4 Tribe Adramini

This group was reviewed recently by Hardy (1973; 1974; 1983b; 1986a), Korneyev (1994), Hancock and Drew (1995), and Permkam and Hancock (1995). Monophyly of the tribe is supported by the presence of the long, fine, erect trichia on the anatergite (apomorphic), but recently the genus *Ptilona* which has similar trichia was found to be very distantly related to *Adrama* and *Euphranta* (it belongs in the Phytalmiinae: Acanthonevrini), so the value of this character as a synapomorphy, without other supporting apomorphic characters, is less convincing than has been supposed. Some genera of Adramini have a long setulose arista, complete chaetotaxy, long taeniae, and blunt aculeus with long setulae on the tip (plesiomorphies); this suggests that the corresponding apomorphies appear in the Adramini independently from those in the other tribes, due to homoplasy.

Within this tribe, one group of genera includes *Adrama* Walker, *Conradtina* Enderlein, *Piestometopon* Meijere (= *Elleipsa* Hardy), *Meracanthomyia* Hendel, *Munromyia* Bezzi, and *Celidodacus* Hendel. It can be distinguished by the combination of ventral spurs on the mid and hind femora and the aculeus short, one-third to two-thirds as long as the oviscape (synapomorphies). *Hardyadrama* Lee also belongs to this cluster because of having ventral spurs on the mid and hind femora; although its aculeus is long and needlelike, this is apparently due to secondary change.

The Afrotropical genus *Celidodacus* differs from other Adramini by its complete chaetotaxy, similar to that in outgroups (for instance, Carpomyini), and its tactile aculeus. It is believed to be the most generalized among the genera allied to *Adrama*. At least *Adrama* and *Celidodacus* share pubescent arista and same structure of epandrium; both characters, therefore, are believed to belong to the ground plan of this group of genera as a whole.

*Soita*, the genus of strange, slender-bodied Oriental flies with trichose anatergites, fits the diagnosis of Adramini well, but also shows strong affinities of the serrate aculeus with *Micronevrina* (Acanthonevrini), *Xarnuta* (Xarnutini), and some *Hexachaeta* (Hexachaetini); *Soita* and *Micronevrina* also share the anterior position of the anterior orbital seta that is atypical for Adramini. Nevertheless, the cephalopharyngeal skeleton of *S. cylindrica* (Hendel) larva is characteristic for Adramini, showing such apomorphic features as the parastomal sclerite fused to the hypopharyngeal sclerite at the middle of its length (rather than to anterior part of pharyngeal sclerite) and the labial sclerite as long as the hypopharyngeal sclerite (Kagesawa 1998, Figure 15; in that paper the labial sclerite is misinterpreted as the dental sclerite). *Soita* share several probable synapomorphies (two frontal and one orbital setae, anteriorly produced frons, short setulose arista, and mostly hyaline wings) with the other slender-bodied Adramini that have the mid and hind femora without ventral spines (symplesiomorphy). These are the Afrotropical *Coelopacidia* Enderlein, *Trypanophion* Bezzi, and Australasian *Ichneumonosoma* Meijere. *Soita* and *Coelopacidia* also have two (or often three) midtibial spurs (symplesiomorphy?).

The third group consists of genera that have the arista mostly long pubescent, wing with a distinctive pattern, mid and hind femora nonspinose, and the body not very slender (all plesiomorphies). It includes the more generalized genera *Euphranta* Loew (including the subgenus *Rhacochlaena* Loew), *Dimeringophrys* Enderlein (= *Tetrameringophrys* Hardy), *Coelotrypes* Bezzi, and *Cyclopsia* Malloch. This group has no apparent synapomorphies, and therefore may not be monophyletic.

The type species of *Euphranta, Rhacochlaena, Epochra,* and *Macrotrypeta* (the latter three currently considered to be synonyms of the first genus) are known to have long surstyli and proctiger and the phallapodeme vanes fused; both characters undoubtedly are apomorphic and may delimit much, or in the case of the latter character state, all of the genus *Euphranta* as monophyletic. Little is known about the phallapodeme condition in most non-Holarctic species, but at least some members of *Euphranta* (*E. flavoscutellata* Hardy, *E. ocellata* Hardy, *E. canangae* Hardy) are known to have short surstyli. *Rhacochlaena*, the largest subgenus of *Euphranta*, is very heterogeneous in wing pattern, head chaetotaxy, and the shape of male epandrium and female aculeus and needs further study of male genitalia to confirm its monophyly.

*Scolocolus* Hardy has long surstyli and proctiger, and the phallapodeme vanes probably fused (see Hardy 1970, Figure 11e) and fits well the concept of the *Euphranta* group of genera, although it has the midfemur spinulose, and reduced postpronotal and katepisternal setae.

At least some species of both *Adrama* and *Euphranta* share also the very long posterior arm of the phallapodeme, which might be a synapomorphy of the two genus groups discussed above. However, this character was not examined in *Celidodacus, Conradtina,* and some other generalized taxa in this tribe.

*Pelmatops* Enderlein, *Pseudopelmatops* Shiraki, *Paraeuphranta* Hardy, *Crinitosophira* Hardy, and *Brandtomyia* Hardy cannot be referred to the genus groups above: they have no mid or hind femoral spurs, and their genitalic characters were not examined.

*Adramoides* Hardy is clearly an *Adrama*-like genus, with the midfemur spinulose and the aculeus short, with the cercal unit integrated into it, similar to that in *Adrama*. Despite these synapomorphies and general affinities in the pubescence of arista, and wing venation and pattern, these two genera differ in a character that was believed to be of crucial importance as an autapomorphy of the tribe; *Adramoides* has the anatergite bare. This shows that at least some members of Adramini may have the "main" synapomorphy lacking secondarily, and that the concept of the tribe requires further revision.

The definition of Adramini is based on a single apomorphy; in the case of character reversal, an adramine species without anatergal trichia cannot be identified as a member of this tribe. Descriptions and figures of adramine larvae (Phillips 1946; Kandybina 1977; White and Elson-Harris 1992; Kagesawa 1998) show that Adramini apparently can be defined well by some larval characters.

### 4.7.5 Unplaced Genera, Presumed To Be Allied to the Adramini

The two Afrotropical genera *Sosiopsila* Bezzi and *Pseudomyoleja* Han and Freidberg have a trumpet-shaped subapical lobe of the glans, the eversible membrane with dentiform scales, and the aculeus of the piercing type, and belong to the Higher Tephritidae. Both genera appear to be closer related to Adramini than to any other tribe of the Higher Tephritidae, although they have the anatergite bare or only microtrichose.

*Sosiopsila* and some adramines share long pubescent arista (plesiomorphy), three frontal and one orbital setae, rudimentary ocellar seta, reduced thoracic chaetotaxy (presutural supra-alar, dorsocentral, acrostichal, and apical scutellar setae lacking) (synapomorphy?) and slender body (synapomorphy?) (see Munro 1984, Figure 139). All of these putative synapomorphies are of low value, however, as they occur convergently in other Tephritidae. Both the hypandrium and epandrium of *Sosiopsila* do not possess any apomorphic features of Adramini, and are similar to those of Toxotrypanini, Nitrariomyiini, Trypetini, Carpomyini, and *Pseudomyoleja.*

*Pseudomyoleja* has a short aculeus (Han and Freidberg 1994, Figure 8) that may be considered a synapomorphy with the Adramini. Most features that characterize adramines (see above), in *Pseudomyoleja* are in plesiomorphic condition, except the arista bare (apomorphy) and the eversible membrane microtrichose (autapomorphy) and with strongly reduced and modified taeniae (autapomorphy). With caution, it could be considered the sister group of the Adramini.

### 4.7.6 Tribe Carpomyini

The relationship of this tribe within the Trypetinae is uncertain. The included genera have an acute piercing aculeus with at most three small equal preapical setae, like in most Trypetini, Nitrariomyiini, or Tephritinae, but, otherwise, have rather generalized habitus and structure of terminalia, at least in the ground plan.

#### 4.7.6.1 Subtribes Notommatina and Paraterelliina

Two small subtribes, Notommatina (*Notomma* Bezzi and *Malica* Richter) and Paraterelliina (*Oedicarena* Loew and *Paraterellia* Foote) were proposed by Korneyev (1996) in the tribe Carpomyini. Actually, such a placement is putative, as the homologies or polarities of several significant morphological characteristics of both taxa are unresolved. Based on some molecular data, *Oedicarena* shows only distant relationship to the Carpomyini and appears closer to Dacinae (Han and McPheron 1997), but the trees of Berlocher and Bush (1982) and Smith and Bush (1997) show *Oedicarena* within Carpomyini. Further study, including biochemical analysis, is required to clarify the position of these two subtribes.

#### 4.7.6.2 Subtribe Carpomyina

The general characteristics of this tribe (including phylogeny) were given briefly by Jenkins (1996), McPheron and Han (1997), Smith and Bush (1997), and Norrbom et al. (1999). The phylogenetic relationships within it are discussed in Chapter 9 by Smith and Bush. These works discuss considerable adult morphological and DNA sequence data, but most omit characters of larval morphology.

Jenkins (1996) hypothesized that the reduction of the trumpet-shaped, sclerotized subapical lobe of the glans to a nonsclerotized lobe is an autapomorphy of the Carpomyini. Norrbom (1989) considered the desclerotized apical area of the oviscape as a synapomorphy for Carpomyina. All the examined representatives of the subtribe have it, including *Zonosemata* Benjamin (Norrbom, personal communication).

## 4.7.7 The Tribes Possibly Allied to the Tephritinae

A possible cluster with four main branches, including the tribes Trypetini, Nitrariomyiini (= Acidoxanthini, Chetostomatina), and Zaceratini, plus the subfamily Tephritinae, is supported by larval morphology. In at least Trypetini, Nitrariomyiini, and Tephritinae, there is an unpaired lobe, termed the "median oral lobe" in the Tephritinae by Headrick and Goeden (1990) (autapomorphy), attached posteriorly to the hypopharyngeal sclerite, that has not been detected in members of any other tephritid tribe (Carroll, Headrick, and Steck, personal communication). It was well illustrated for *Acidia cognata* (Wiedemann), *Anomoia purmunda* (Harris), and *Acanthiophilus helianthi* (Rossi) by Belcari (1989a, Figure 3-4; 1989b, Figures 4-5, 42, and 46, as "lobo labiale"). The area of the cephalic segment between the antennae is covered by placoid stuctures, and the surface lateral of the oral ridges has similar placoid structures, often with serrate ventral flaps, with a net of channels between them. This "placoid mask" is present in the ground plan of all four taxa, and is considered to be another synapomorphy.

From comparison with Phytalmiinae: Acanthonevrini, Dacinae: Gastrozonini, and Trypetinae: Toxotrypanini as alternative outgroups, the ground plan of this cluster is inferred to include the arista short pubescent (autapomorphy of this cluster or synapomorphy with some other Trypetinae), three frontal setae on anterior two-fifths of frons length (autapomorphy of this cluster or synapomorphy with some other Trypetinae), and two orbital setae (position of uncertain polarity), $R_{4+5}$ setulose apicad of the DM-Cu level (plesiomorphy), surstyli thick and short (polarity uncertain; highly variable; Toxotrypanini, most primitive Adramini, and the majority of Nitrariomyiini and Tephritinae have short surstyli), glans with trumpetlike subapical lobe (synapomorphy of the Higher Tephritidae?), sculptured inner surface of the praeputium (symplesiomorphy), two semitubular lobes of the acrophallus (symplesiomorphy), glans with basal lobe (synapomorphy of the Higher Tephritidae), aculeus not much shorter than the oviscape (plesiomorphy), cercal unit completely integrated into aculeus (synapomorphy of Higher Tephritidae?), and three papillose spermathecae (plesiomorphy). It is not clear if the presence of the monodentate scales on the eversible membrane belongs to the ground plan or arose in three taxa of the cluster independently from Toxotrypanini, Adramini, and Carpomyini. At least all the Nitrariomyiini have the membrane devoid of these scales (see discussion below).

### 4.7.7.1 Tribe Nitrariomyiini

This tribe was established by Korneyev (1996) for two Palearctic genera, *Nitrariomyia* Rohdendorf and *Kerzhnerella* Richter, but recent study suggests that these genera belong to a larger monophyletic cluster (Korneyev, unpublished data) that includes the genera assigned to the subtribe Acidoxanthina (*Acidoxantha* Hendel, *Craspedoxanthitea*) and the *Anomoia-Chetostoma* group of genera (Han 1992; Korneyev 1996). Han (Chapter 11) considers the latter group the subtribe Chetostomatina of the Trypetini, but suspects that it belongs to this broader group for which the senior available name would be Nitrariomyiina. The names Acidoxanthina and Chetostomatina may need to be formally synonymized in the future, but such an action needs additional study that would involve also larval characters and biochemical data.

Monophyly of Nitrariomyiini in this broad sense is supported by the following characters: (1) the aculeus is narrow and needlelike, and is usually much longer than the oviscape, when retracted always with the tip exposed (autapomorphy); (2) the eversible membrane has very long taeniae (plesiomorphy?); and (3) no monodentate scales (plesiomorphy or autapomorphy?). The glans has both the subapical lobe (plesiomorphy) and hexagonal sculpture of the "dorsal sclerite" (*sensu* Han 1992, and Chapter 11) in the Chetostomatina, but in *Acidoxantha* it is slender and lacks any sclerotization of the praeputium except for the "medial sclerite" (*sensu* Han), whereas in *Nitrariomyia* both characters are completely lacking. Larvae breed mainly in fleshy fruits or sometimes as inquilines in galls; *Acidoxantha* are associated with flower buds.

#### 4.7.7.2 Tribe Trypetini

See Chapter 11 by Han.

#### 4.7.7.3 Zaceratini and Tephritinae

The relationship of these two taxa is discussed by Korneyev (Chapter 22). That Zaceratini is the sister group of Tephritinae is supported by a single synapomorphy: the epandrium is oval in posterior view in *Zacerata* Coquillett and *Plioreocepta* Korneyev, the only genera of Zaceratini, and in all or most genera of most tribes of Tephritinae. Both groups share a very similar placoid structure of the third instar larval mask (autapomorphy of Zaceratini + Tephritinae or synapomorphy with Trypetini and Nitrariomyiini). The median oral lobe that is reported to be well developed in larvae of Nitrariomyiini, Trypetini, and Tephritini has not yet been reported in Zaceratini.

## 4.8 SUBFAMILY DACINAE

This taxon consists of three tribes, Gastrozononini, Dacini, and Ceratitidini (ranked as subtribes of the tribe Dacini in the subfamily Trypetinae by Norrbom et al. 1999). Phylogenetic relationships within the Dacini and of the genus *Ceratitis* MacLeay (Ceratitidini) are considered by Drew and Hancock (Chapter 19) and De Meyer (Chapter 16), respectively.

Monophyly of Dacinae is tentatively based on the large size of the male proctiger (synapomorphy), which usually is larger than the epandrium, and on having only two spermathecae (synapomorphy?; also in Tephritinae and some Trypetinae). Larvae of Dacini and Ceratitidini possess many plesiomorphies, including the habitus of the maggot, numerous oral ridges, long sicklelike mouthhooks, and well-developed creeping welts. The ridge on the tubercles below the posterior spiracles is very likely a synapomorphy of at least Dacini and Ceratitidini (Carroll, personal communication); larvae of Gastrozonini are undescribed.

Adults of Gastrozonini and Ceratitidini often have the anterior orbital seta situated close to the middle of frons length (plesiomorphy rarely occurring in other Higher Tephritidae: in *Soita* of Adramini and *Noeeta* Robineau-Desvoidy of Tephritinae, but apparently due to character reversal).

Gastrozonini is a poorly studied group that may not be monophyletic. It consists of various genera that do not have the autapomorphies of the Dacini or of the Ceratitidini. Relationships within Gastrozonini are not well understood, mainly because of missing data of larval and adult (genital) morphology. The tribe is largely restricted to the Oriental Region, with some genera reaching the Palearctic and Afrotropical Regions.

*Enicoptera* Macquart and *Ichneumonopsis* Hardy, two Oriental genera with a long first flagellomere, rudimentary ocellar seta, spotted head, and black and yellow marked body, apparently are related to the Dacini. They differ by having the arista plumose, as well as some other plesiomorphies, but otherwise seem to belong to the same monophyletic cluster with the Dacini. *Enicoptera* have a rather long extension of the basal cubital cell and the venation generally shifted to the anterior margin, like in most dacines. It also has the venation of the posterior half of the wing (especially shape of the discal medial cell) very similar to that in *Monacrostichus*.

The genera of the *Acroceratitis* group (Oriental *Acroceratitis* Hendel, *Paraxarnuta* Hardy, *Phaeospilodes* Hering, Afrotropical *Bistrispinaria* Munro, and apparently Oriental *Spilocosmia* Bezzi) share long, wrinkled, and twisted spermathecae (apomorphy) and long, apically pointed first flagellomere (synapomorphy) and may be somehow related. Both the spermatheca shape and long first flagellomere also are similar to those in Dacini (except for the flagellomere pointed), and the two groups may be closely related.

There are a few small monophyletic groups in the Gastrozonini that can be recognized. *Gastrozona* Macquart and *Carpophthorella* Hendel share ocellar seta rudimentary (polarity unresolved), epandrium and surstyli very slender (synapomorphy), male proctiger with very large rectal glands (synapomorphy), and a narrow, piercing or cutting aculeus (synapomorphy) with or without steps but with two short and two rather long subapical setulae (plesiomorphy). *Chaetellipsis* Bezzi also have the slender epandrium, shape of aculeus and spermathecae like in *Carpophthorella*, but no enormous rectal glands.

Hancock and Drew (1994) redefined the concept of the *Anoplomus* group to include seven genera arranged into two subgroups. According to them, it is characterized by the presence of a dark body (apomorphy?), swollen scutellum (synapomorphy with Ceratitidini), white or grayish pubescence on the scutum and abdomen (synapomorphy with Ceratitidini?), and the basal dark area in the wing broken into isolated spots and streaks (synapomorphy with Ceratitidini?).

The *Anoplomus* subgroup includes *Anoplomus* Bezzi and *Sinanoplomus* Zia which share two apical midtibial spurs (symplesiomorphy?) and *Proanoplomus* Shiraki which have one spur, but wing pattern details similar (see Hancock and Drew 1994: 869–870). In *Anoplomus* and *Proanoplomus* the spermathecae are of the same shape, oval, wrinkled, apparently not covered by papillae or denticles, with short apical nipple (synapomorphy?) (see Hardy 1973, Figures 128 and 131; 1974, Figure 90). Palearctic *Paragastrozona* Shiraki has the same wing pattern type as in *Proanoplomus*; also, they have only one midtibial spur (apomorphy?), and share the presence of a black apical spot on the swollen scutellum (both apomorphies of the *Anoplomus* group). *Paragastrozona* differs by numerous autapomorphies (arista short pubescent, mesonotal yellow vitta and lateral scutellar spots lacking, male femora swollen and claws and pulvillae enlarged, spermathecae globose rather than oval, oviscape with a subbasal swelling, etc.), but otherwise belongs to one cluster along with the Chinese species of *Proanoplomus* that have the scutellum yellow (plesiomorphy?).

Apparently, the appearance of simple dentiform scales on the eversible membrane associated with fruit-feeding is an apomorphy that separates the tribes Ceratitidini and Gastrozonini. This character was reported at least for *Ceratitella* Malloch (Hardy 1987, Figure 6d and e) and *Neoceratitis* Hendel (Korneyev 1994) of the *Ceratitella* subgroup *sensu* Hancock and Drew (1994).

The Ceratitidini is a monophyletic group defined by several autapomorphies: (1) R-M crossvein positioned very basally, at or proximad of the $R_1$ apex; (2) eversible membrane covered with monodentate scales; (3) spermathecae oval or pear-shaped, papillose or spinulose (the two latter characters are similarly developed in Trypetinae, but probably arose independently due to homoplasy). They retain a pubescent arista, light setulose scutum, and swollen scutellum like in the *Anoplomus* subgroup that apparently is closely allied to the Ceratitidini. I prefer to consider it as a complex within Gastrozonini, and to include the *Ceratitella* subgroup in the Ceratitidini.

## 4.9 CONCLUSIONS AND FUTURE RESEARCH NEEDS

The relationships among the subfamilies and tribes of Tephritidae has not been satisfactorily resolved by this study. Many characters were found to be strongly subject to homoplasy, and larval and genitalic characters have not been studied for more than half of the species used in the analysis.

The basic branching of Tephritidae is not resolved, although the subfamilies Tachiniscinae, Blepharoneurinae, and Tephritinae, and most tribes considered in this study appear to be well-supported monophyletic groups. The subfamilies Phytalmiinae and Dacinae, and the cluster called Higher Tephritidae (Trypetinae + Dacinae + Tephritinae) are very probably monophyletic, but this needs further confirmation. The monophyly of the subfamily Trypetinae is not supported by any characters; it is very possibly a paraphyletic remainder of the Higher Tephritidae minus Tephritinae and Dacinae, with the relationships among its tribes mostly unresolved. In particular, the Zaceratini, and perhaps the Trypetini and Nitrariomyiini, appear to be more closely related to the Tephritinae than to other Trypetinae.

The distribution of Tachiniscinae, Blepharoneurinae, and, with some reservations, of Hexachaetini and Nitrariomyiini, which occur in subtropical and tropical regions of the Old and New World, is probably the result of the separation of these taxa at early stages of the evolution of the family.

It is expected that various structures of the phallus, hypandrium, and female terminalia, when thoroughly examined and depicted, could add resolution to the analysis of Phytalmiinae and many other taxa of Tephritidae. The taxa included in this study were those available for study or described in the literature; thus there may be errors due to incomplete sampling. When species that possess the most primitive features of each lineage are lacking, homoplasies may be misinterpreted as synapomorphies, leading to false hypotheses of relationship. Furthermore, the technique of analysis used in this study always allowed character reversal, even when it is unlikely (for instance, tactile aculeus → cutting aculeus → tactile aculeus). To achieve better resolution in the future, the current data set must be completed and reanalyzed using more powerful computer equipment that can handle such a large matrix of characters.

## ACKNOWLEDGMENTS

I am indebted to Allen Norrbom (SEL, USDA, Washington, D.C.), Amnon Freidberg (Tel Aviv University), Bernhard Merz (Muséum d'Histoire Naturelle, Genève), Ho-Yeon Han (Yonsei University, Korea), and Lynn Carroll (formerly USDA, SEL) for fruitful discussions and dissemination of new facts and data they discovered, as well as the specimens they shared with me. Their ideas and criticism were very beneficial to this work.

Papers recently written by R. A. I. Drew, A. Freidberg, D. L. Hancock, H.-Y. Han, and A. L. Norrbom were essential sources of morphological information. I thank R.-E. Contreras-Lichtenberg for the great opportunity to study specimens in the collection of the Naturhistorisches Museum Wien and for arranging a visitor's grant from the NMW in 1995. Support generously provided by the Smithsonian Institution through a Visitor Grant in January 1993 greatly aided this investigation and is acknowledged. A. L. Norrbom and D. L. Hancock offered valuable comments on the manuscript. D. Barraclough (Natal Museum) kindly supplied me with necessary data on morphology of *Tachiniscidia*.

I wish to thank M. Aluja, Instituto de Ecologia, Xalapa, Mexico and A. L. Norrbom for their enormous efforts to organize the symposium and to invite me to attend this extraordinary meeting of the best experts in the taxonomy, phylogeny, and behavior of the Tephritidae. Financial support of Consejo Nacional de Ciencia y Tecnología (CONACyT)–Subdirección de Asuntos Bilaterales is greatly appreciated. I acknowledge the support from Campaña Nacional contra las Moscas de la Fruta (SAGAR-IICA) and thank Lic. A. Meyenberg (CONACyT), Ing. J. Reyes (SAGAR), C. Leal (IE), V. Domínguez (IE), and E. Bombela (IE) for arranging my trip to Xalapa. My warmest and sincere thanks are due to B. Delfosse (IE) and the team members: L. Guillén, I. Jácome, R. Cervantes, and J. Piñero.

I wish to express my sincere thanks to my colleagues from the Schmalhausen Institute of Zoology and "Vestnik Zoologii" editorial board for their patience. They kindly allowed me to stay away from my duties while preparing this chapter.

I also thank Elena and Severin. They were with me and inspired this work every minute of the day for many years.

## REFERENCES


Belcari, A. 1989a. Contributi alla conoscenza dei ditteri tefritidi. III. Descrizzione della larva matura di *Phagocarpus permundus* (Harris) (Diptera Tephritidae). *Frustula Entomol. (N. S.)*. (1986) 9: 77–106.

Belcari, A. 1989b. Contributi alla conoscenza dei ditteri tefritidi. IV. Descrizione della larva di terza eta di *Acanthiophilus helianthi* (Rossi), *Dacus oleae* (Gmel.), *Ceratitis capitata* (Wied.), *Acidia cognata* Wied. e considerazioni preliminari sulle differenziazioni morfologiche legate al diverso trofismo. *Frustula Entomol. (N. S.)*. (1987) 10: 82–125.

Berlocher, S.H. and G.L. Bush 1982. An electrophoretic analysis of *Rhagoletis* (Diptera: Tephritidae) phylogeny. *Syst. Zool.* 31: 136–155.

Bush, G.L. 1966. The taxonomy, cytology, and evolution of the genus *Rhagoletis* in North America (Diptera, Tephritidae). *Bull. Mus. Comp. Zool.* 134: 431–562.

Condon, M.A. and A.L. Norrbom. 1994. Three sympatric species of *Blepharoneura* (Dipterra: Tephritidae) on a single species of host (*Gurania spinulosa*, Cucurbitaceae): new species and new taxonomic methods. *Syst. Entomol.* 19: 279–304.

Dodson, G. and G. Daniels 1988. Diptera reared from *Dysoxylum gaudichaudianum* (Juss.) Miq. at Iron Range, northern Queensland. *Aust. Entomol. Mag.* 15: 77–79.

Farris, J.S. 1988. Hennig86. Version 1.5. Published by the author, Port Jefferson Station.

Han, H.-Y. 1992. Classification of the Tribe Trypetini (Diptera: Tephritidae: Trypetinae). Ph.D. dissertation, Pennsylvania State University, University Park. 275 pp.

Han, H.-Y. and A. Freidberg. 1994. *Pseudomyoleja*, a new Afrotropical genus of Tephritidae (Diptera). *J. Afr. Zool.* 108: 547–554.

Han, H.-Y. and B.A. McPheron. 1997. Molecular phylogenetic study of Tephritidae (Insecta: Diptera) using sequences of the mitochondrial 16S ribosomal DNA. *Mol. Phylogenet. Evol.* 7(1): 17–32.

Hancock, D.L. 1986. Classification of the Trypetinae (Diptera: Tephritidae), with a discussion of the Afrotropical fauna. *J. Entomol. Soc. South. Afr.* 49: 275–305.

Hancock, D.L. and R.A.I. Drew. 1994. Notes on *Anoplomus* Bezzi and related genera (Diptera: Tephritidae) in Southeast Asia and Africa. *Raffles Bull. Zool.* 42: 869–883.

Hancock, D.L. and R.A.I. Drew. 1995. New genus, species and synonyms of Asian Trypetinae (Diptera: Tephritidae). *Malays. J. Sci.* 16A: 45–59.

Hardy, D.E. 1970. Tephritidae (Diptera) collected by the Noona Dan expedition in the Philippine and Bismarck Islands. *Entomol. Medd.* 38: 71–136.

Hardy, D.E. 1973. The fruit flies (Tephritidae — Diptera) of Thailand and bordering countries. *Pac. Insects Monogr.* 31: 1–353.

Hardy, D.E. 1974. The fruit flies of the Philippines (Diptera: Tephritidae). *Pac. Insects Monogr.* 32: 1–266.

Hardy, D.E. 1977. Family Tephritidae (Trypetidae, Trupaneidae). In *A Catalog of the Diptera of the Oriental Region. Vol. III, Suborder Cyclorrhapha (excluding Division Aschiza)* (M.D. Delfinado and D.E. Hardy, eds.), pp. 44–134. University of Hawaii Press, Honolulu. 854 pp.

Hardy, D.E. 1980. The *Sophira* group of fruit fly genera (Diptera: Tephritidae). *Pac. Insects* 22: 123–161.

Hardy, D.E. 1982. The *Epacrocerus* complex of genera in New Guinea (Diptera: Tephritidae: Acanthonevrini). *Mem. Entomol. Soc. Wash.* 10: 78–92.

Hardy, D.E. 1983a. *Robertsomyia* an aberrant new genus of Phytalmiini from Papua New Guinea (Tephritidae: Diptera). *Int. J. Entomol.* 25: 152–205.

Hardy, D.E. 1983b. Fruit flies of the tribe Euphrantini of Indonesia, New Guinea and adjacent islands (Tephritidae: Diptera). *Int. J. Entomol.* 25: 152–205.

Hardy, D.E. 1986a. The Adramini of Indonesia, New Guinea and adjacent islands (Diptera: Tephritidae: Trypetinae). *Proc. Hawaii. Entomol. Soc.* 27: 53–78.

Hardy, D.E. 1986b. Fruit flies of the subtribe Acanthonevrina of Indonesia, New Guinea, and the Bismarck and Solomon Islands (Diptera: Tephritidae: Trypetinae: Acanthonevrini). *Pac. Insects Monogr.* 42: 1–191.

Hardy, D.E. 1987. The Trypetini, Aciurini and Ceratitini of Indonesia, New Guinea and adjacent islands of the Bismarcks and Solomons (Diptera: Tephritidae: Trypetinae). *Entomography* 5: 247–374.

Hardy, D.E. 1988. Fruit flies of the subtribe Gastrozonina of Indonesia, New Guinea and the Bismarck and Solomon islands (Diptera, Tephritidae, Trypetinae, Acanthonevrini). *Zool. Scr.* 17: 77–121.

Headrick, D.H. and R.D. Goeden. 1990. Description of the immature stages of *Paracantha gentilis* (Diptera: Tephritidae). *Ann. Entomol. Soc. Am.* 83: 220–229.

Hill, G.F. 1921. Notes on some Diptera found in association with Termites. *Proc. Linn. Soc. N.S.W.* 46: 216–220.

Jenkins, J. 1996. Systematic Study of *Rhagoletis* and Related Genera (Diptera: Tephritidae). Ph.D. dissertation, Michigan State University, East Lansing. 184 pp.

Kagesawa, N. 1998. Notes on *Soita cylindrica* (Hendel) from Japan (Diptera: Tephritidae). *Entomol. Sci.* 1: 241–248.

Kandybina, M.N. 1977. Lichinki plodovykh mukh-pestrokrylok (Diptera: Tephritidae) [Larvae of fruit-infesting fruit flies (Diptera: Tephritidae)]. *Opred. Faune SSSR* 114: 1–212.

Korneyev, V.A. 1994. Reclassification of Palearctic Tephritidae (Diptera). Communication 2. *Vestn. Zool.* [27](1): 3–17.

Korneyev, V.A. 1996. Reclassification of Palearctic Tephritidae (Diptera). Communication 3. *Vestn. Zool.* (1995) [28](5–6): 25–48.

Krivosheina, N.P. 1982. Unusual environment of the larvae of Diptera, Trypetidae. *Nauchn. Dokl. Vyssh. Shk. Biol. Nauki* 1982(2): 29–33.

McAlpine, D.K. and M.A. Schneider. 1978. A systematic study of *Phytalmia* (Diptera: Tephritidae), with description of a new genus. *Syst. Entomol.* 3: 159–175.

McAlpine, J.F. 1989. Phylogeny and classification of the Muscomorpha. In *Manual of Nearctic Diptera. Vol. 3* (J.F. McAlpine, ed.), pp. 1397–1518. Monograph of the Biosystematics Research Centre, No. 32. Agriculture Canada, Ottawa.

McPheron, B.A. and H.-Y. Han. 1997. Phylogenetic analysis of North American *Rhagoletis* (Diptera: Tephritidae) and related genera using mitochondrial DNA sequence data. *Mol. Phylogenet. Evol.* 7: 1–16.

Munro, H.K. 1984. A taxonomic treatise on the Dacidae (Tephritoidea, Diptera) of Africa. *Entomol. Mem. S. Afr. Dept. Agric.* 61, 313 pp.

Norrbom, A.L. 1985. Phylogenetic Analysis and Taxonomy of the *Cryptostrepha, Daciformis, Robusta* and *Schausi* Species Groups of *Anastrepha* Schiner (Diptera: Tephritidae). Ph.D. dissertation, Pennsylvania State University, University Park. 355 pp.

Norrbom, A.L. 1989. The status of *Urophora acuticornis* and *U. sabroskyi* (Diptera: Tephritidae). *Entomol. News* 100: 59–66.

Norrbom, A.L. 1994a. New genera of Tephritidae (Diptera) from Brazil and Dominican amber, with phylogenetic analysis of the tribe Ortalotrypetini. *Insecta Mundi* 8: 1–15.

Norrbom, A.L. 1994b. New species and phylogenetic analysis of *Cryptodacus, Haywardina,* and *Rhagoletotrypeta* (Diptera: Tephritidae). *Insecta Mundi* 8: 37–65.

Norrbom, A.L., L.E. Carroll, and A. Freidberg. 1999. Status of knowledge. In *Fruit Fly Expert Identification System and Systematic Information Database* (F.C. Thompson, ed.), pp. 9–47. *Myia* (1998) 9, 524 pp.

Permkam, S. and D.L. Hancock. 1995. Australian Trypetinae (Diptera: Tephritidae). *Invertebr. Taxon.* 9: 1047–1209.

Phillips, V.T. 1946. The biology and identification of trypetid larvae. *Mem. Am. Entomol. Soc.* 12: 1–161.

Roberts, H. 1969. Forest Insects of Nigeria with notes on their biology and distribution. *Inst. Pap. For. Inst. Oxf.* 44: 1–206.

Smith, J.J. and G.L. Bush. 1997. Phylogeny of the genus *Rhagoletis* (Diptera: Tephritidae) inferred from DNA sequences of mitochondrial cytochrom oxidase II. *Mol. Phylogenet. Evol.* 7: 33–43.

White, I.M. and M.M. Elson-Harris. 1992. *Fruit Flies of Economic Significance: Their Identification and Bionomics.* CAB International, Wallingford. 601 pp.

## APPENDIX 4.1: TAXA EXAMINED

Tachiniscini: *Tachinisca cyaneiventris* Kertész, *Bibundia hermanni* Bischof.

Ortalotrypetini: *Ortalotrypeta idana* Hendel, *Cyaforma shenonica* Wang.

Lower Tephritidae *incertae sedis*: *Matsumurania sapporensis* (Matsumura).

Blepharoneurinae: *Ceratodacus longicornis* Hendel, *C. priscus* Norrbom and Condon, *Hexaptilona hexacinioides* (Hering), *H. palpata* (Hendel), *Blepharoneura* spp.

Acanthonevrini: *Acanthonevra fuscipennis* Macquart, *A. desperata* (Hering), *Colobostroter pulchralis* Enderlein, *Diplochorda trineata* Meijere, *Enoplopteron hieroglyphicum* Meijere, *Erectovena amurensis* (Portschinsky), *Hexacinia pellucens* Hardy, *H. radiosa* (Rondani), *Lenitovena trigona* (Matsumura), *Loriomyia guttipennis* Kertész, *Micronevrina montana* Permkam & Hancock, *Ortaloptera cleitamina* Edwards, *Ptilona confinis* (Walker), *Rioxa sexmaculata* (Wulp), *Rioxoptilona vaga* (Wiedemann), *Robertsomyia paradoxa* Hardy, *Sophira limbata* Enderlein, *Terastiomyia lobifera* Bigot, *Themaroides quadrifer* (Walker).

Phascini: *Xenosophira invibrissata* Hardy.

Xarnutini: *Xarnuta fenestellata* Hering, *X. inopinata* Hering, *X. leucotela* Walker, *X. sabahensis* Hardy.

Hexachaetini: *Callistomyia horni* Hendel, *Hexachaeta eximia* (Wiedemann), *H. valida* Lima.

Toxotrypanini: *Anastrepha fraterculus* (Wiedemann), *A. ludens* (Loew), *A. obliqua* (Macquart), *A. serpentina* (Wiedemann), *Toxotrypana curvicauda* Gerstaecker.

Adramini: *Adrama determinata* (Walker), *Celidodacus obnubilus* (Karsch), *Coelopacidia strigata* Bezzi, *Euphranta* (*s. str.*) *connexa* (Fabricius), *E.* (*Rhacochlaena*) *canadensis* (Loew), *E.* (*Rhacochlaena*) *ortalidina* (Portschinsky), *E.* (*Rhacochlaena*) *toxoneura* (Loew), *Pelmatops ichneumoneus* (Walker), *Pseudopelmatops continentalis* Zia and Chen, *Sosiopsila rotunda* Munro, *Trypanophion gigas* Bezzi.

Carpomyina: most Holarctic species of *Carpomya* Costa and *Rhagoletis* Loew, *Rhagoletis striatella* Wulp, *R. turpiniae* Hernández, *Zonosemata vittigera* (Coquillett).

Notommatina: *Malica caraganae* Richter, *Notomma bioculatum* Bezzi, n. sp. aff. *mutilum* Bezzi.

Paraterelliina: *Oedicarena tetanops* (Loew), *Paraterellia immaculata* Blanc.

Nitrariomyiini: most Palearctic species; *Acidoxantha totoflava* Hardy.

Trypetini: most Palearctic species; *Strauzia* sp.

Zaceratini: *Plioreocepta poeciloptera* (Schrank), *Zacerata asparagi* Coquillett.

Gastrozonini: *Acrotaeniostola sexvittata* Hendel, *Paragastrozona japonica* (Miyake), *Paraxarnuta bambusae* Hardy, *Phaeospilodes torquata* Hering.

Dacini: *Bactrocera* (*s. str.*) *dorsalis* (Hendel), *B.* (*B.*) *zonata* (Saunders), *B.* (*Daculus*) *oleae* (Rossi), *Dacus* (*s. str.*) *bivittatus* (Bigot), *D.* (*s. str.*) *demmerezi* (Bezzi), *D.* (*Didacus*) *ciliatus* Loew, *D.* (*Didacus*) *vertebratus* Bezzi, *D.* (*Leptoxyda*) *persicus* Hendel.

Ceratitidini: *Ceratitis capitata* (Wiedemann), *C. catoirii* Guérin-Méneville, *C. rosa* Karsch, *Neoceratitis asiatica* (Becker), *N. cyanescens* (Bezzi), *Capparimyia savastani* (Martelli), *Trirhithrum coffeae* Bezzi.

# 5 Nucleotide Sequence Data as a Tool to Test Phylogenetic Relationships among Higher Groups of Tephritidae: A Case Study Using Mitochondrial Ribosomal DNA

*Ho-Yeon Han and Bruce A. McPheron*

## CONTENTS



## 5.1 INTRODUCTION

Members of the fly family Tephritidae have been the subjects of extensive biological investigation. The more than 4000 species of tephritids alone represent a significant evolutionary and economic fauna. Tephritid research has contributed to our general understanding of basic biological problems as well as pest control. Despite the importance of the family, the higher classification of the Tephritidae is in an unsatisfactory state (Freidberg 1984; Hancock 1986; Foote et al. 1993; Norrbom et al. 1999; Korneyev, Chapter 4). Years of work by many dipterists using morphological characters have yielded poor resolution of higher relationships within the Tephritidae. Various subfamilies and tribes have been defined, but the limits of many of them are uncertain, the relationships among them are largely unresolved, and the status of many higher groups as monophyletic taxa needs to

0-8493-1275-2/00/$0.00+$.50

be tested. This situation may be the combined outcome of the large size of the group, that previous systematic studies were largely regionally biased and focused on species description, and, especially, the fact that many morphological characters intergrade between higher taxa (Freidberg 1984).

The recent development of molecular systematics using nucleotide sequence data has provided new possibilities for tephritid classification. Sufficient DNA copies for sequencing can be easily obtained from a single specimen using the polymerase chain reaction (PCR) (Mullis et al. 1986). This technique replaces the more time-consuming gene cloning process to obtain a usable amount of DNA. Furthermore, PCR allows analysis of DNA from specimens that would not yield clonable nucleic acids (Thomas et al. 1989; De Salle et al. 1992). We have successfully amplified and sequenced DNA from alcohol-preserved and pinned specimens of Tephritidae and other taxa (Han and McPheron 1997). Molecular data will not only illuminate the problems from a new direction, but also provide a tool to test existing classifications. As an initial step, we have applied mitochondrial ribosomal DNA data toward resolving problems of the higher classification of the Tephritidae (Han and McPheron 1997; McPheron and Han 1997).

## 5.2 LABORATORY TECHNIQUES

Nucleic acid extractions follow a standard protocol optimized for single individual flies (Sheppard et al. 1992). For pinned or alcohol specimens, we further modified the technique by adding a step involving initial incubation at 55°C using proteinase-K (Han and McPheron 1997). Using this extraction technique, we have successfully amplified an approximately 600-base-pair-long rDNA region from up to 5-year-old pinned specimens and 20-year-old alcohol-preserved specimens. For a relatively fresh specimen, a single leg would provide a sufficient amount of good-quality DNA, and, thus, most of the body could be saved for other purposes.

The region to be analyzed is amplified using standard PCR approaches (e.g., Kocher et al. 1989; Simon et al. 1991). The double-stranded amplification product (generally, 40 amplification cycles) is gel purified by isolating the desired band using agarose gel electrophoresis. This product is reamplified asymmetrically, using one of the PCR primers or an internal primer as a limiting primer (1: 25~100 ratio). Single-stranded DNA is concentrated in a Millipore Ultrafree-MC 30,000 MW filter and used as template for the sequencing reaction. The single-stranded DNA is sequenced by the dideoxy, chain-termination method (Sanger et al. 1977) using Sequenase (Amersham Co.). Both strands should be sequenced to minimize errors.

Based on the alignment between the published mosquito and drosophilid sequences (Clary and Wolstenholme 1985; Beard et al. 1993), we have designed a number of primers for PCR and sequencing. We included this alignment in Appendix 5.1 with three additional tephritid species for which we sequenced nearly the entire 16S rDNA and a portion of 12S rDNA. The primers we used for PCR and sequencing are listed in Table 5.1, and Appendix 5.1 shows their positions. All of the primers have worked well with the tephritid species we have studied so far. Additional primers, as necessary, can be easily designed from the alignment provided in Appendix 5.1. We have used a portion of the 16S rDNA flanked by primers LR-J-12883 and LR-N-13770 for our recent studies (Han and McPheron 1997; McPheron and Han 1997), and the rest of the 16S and 12S genes are also currently being sequenced in a continuing investigation. As shown in Figure 5.3, the additional rDNA regions have similar nucleotide composition and variability, and, thus, should improve the resolving power of the analysis by increasing the number of characters twofold.

## 5.3 PHYLOGENETIC INFERENCE USING MITOCHONDRIAL RIBOSOMAL DNA

### 5.3.1 Mitochondrial DNA in Insect Systematics

Sequences of animal mitochondrial DNA (mtDNA) have been used extensively in phylogenetic studies at a wide variety of taxonomic levels. For insect systematics, various protein coding genes,

**TABLE 5.1**
**Sequence Information for Oligonucleotide Primers for 12/16S Ribosomal DNA Sequencing and PCR**

| Gene | Primer Name | Abbreviation[a] | Sequence |
|---|---|---|---|
| 12S | SR-N-14877 | A12D | 5′-ATGTAAATTTTTGTGTGAAT-3′ |
| | SR-N-14588 | A12C | 5′-CTAGGATTAGATACCCTATTAT-3′ |
| | SR-J-14554 | S12E | 5′-TTAAGTTTCAAGAACATAAC-3′ |
| | SR-J-14176 | S12A | 5′-CATTCTAGATACACTTTCCAGT-3′ |
| 16S | TV-N-14112 | A16D | 5′-AGCATTTCATTTACATTGAA-3′ |
| | LR-N-13770 | A16C | 5′-AGAAATGAAATGTTATTCGT-3′ |
| | LR-J-13677 | S16B | 5′-AGCTTATCCCATAAAATATT-3′ |
| | LR-N-13398 | A16F | 5′-CGCCTGTTTATCAAAAACAT-3′ |
| | LR-J-13323 | S16A | 5′-ACTAATGATTATGCTACCTT-3′ |
| | LR-N-13182 | A16X | 5′-TTAAAAGACGAGAAGACCCTA-3′ |
| | LR-J-13021 | S16M | 5′-ACGCTGTTATCCCTAAAGTA-3′ |
| | LR-J-12883 | S16R | 5′-CTCCGGTTTGAACTCAGATC-3′ |

[a] See Figure 5.1 and Appendix 5.1.

transfer RNA genes, and ribosomal RNA genes have been used to assess phylogenetic relationships at population to ordinal levels. Phylogenetic studies using mtDNA sequences were summarized in detail by Simon et al. (1994).

Animal mtDNA possesses certain properties that make it particularly useful in phylogenetic study. First, numerous molecules of circular mitochondrial DNA exist within a single cell, and the chance of finding intact portions of the molecule from a degraded specimen is much higher than for a single-copy nuclear gene. This is especially important for tephritid systematics because many species are only available as pinned or alcohol-preserved specimens. Individuals commonly are homoplasmic or nearly so, with a single mtDNA sequence predominating in all tissues, probably because of bottlenecks in mtDNA numbers in intermediate germ cell generation (Avise 1994). Second, different regions of the mtDNA change at different rates. For example, the control region changes very rapidly, both within and between species, whereas the rRNA genes evolve more slowly, with some parts retaining nearly complete sequence identity with homologous portions of prokaryotic and eukaryotic cytoplasmic rRNAs (Moritz et al. 1987). This rate heterogeneity is the reason why various mtDNA genes have been used in the systematics of both populations and higher taxa. Finally, mtDNA is transmitted predominantly through maternal lines (Avise and Vrijenhoek 1987). Genotypes for mtDNA thus represent nonrecombining characters, and their inferred evolutionary interrelationships may be interpreted as estimates of matriarchal phylogeny (Avise 1994). This property, by providing more simplified interpretation of genetic markers, will be particularly useful when our study is expanded to sibling species or populations of Tephritidae. Complete mtDNA sequences of three dipteran species, *Drosophila yakuba* Burla (Clary and Wolstenholme 1985), *Anopheles quadrimaculatus* Say (Cockburn et al. 1990), and *A. gambiae* Giles (Beard et al. 1993), are currently known, so preliminary predictions of relative evolutionary rates for various genes and the design of PCR and sequencing primers for dipteran taxa are feasible (as in Appendix 5.1). In addition, Simon et al. (1994) compiled many conserved PCR/sequencing primers tested for different animal groups, especially insects.

Among the mitochondrial genes, the two ribosomal RNA genes (mt 16S and 12S rDNAs) display very low intraspecific variation (McPheron and Han 1997; Han and McPheron 1997), and thus show great promise for the systematic investigations of species and higher taxa. We therefore selected these two genes for our initial study of Tephritidae using molecular systematics, particularly for establishing the limits of tephritid higher taxa and interrelationships among them.

### 5.3.2 Phylogenetic Constraints of Fly Mitochondrial Ribosomal DNA

It is crucial to understand the general properties of the gene under investigation since most phylogenetic methods rely on a set of assumptions defined by the properties of the character set. We are able to make some generalizations about the portion of the 16S gene we analyzed for tephritid taxa.

A-T bias in arthropod mtDNA is well known (Simon et al. 1994). For the 925 bp aligned segment of tephritid 16S rDNA, the average ratio of A:T:C:G is 37:43:7:13 (Han and McPheron 1997). This ratio is highly consistent, with a small standard error, for over 100 tephritid species we have sequenced for this gene region, and is similar to that in other dipteran taxa (Figure 5.1A). This ratio may result from a consistent asymmetrical substitution bias within certain taxa. The number of nucleotide substitutions calculated over the inferred phylogenetic tree illustrates this well (Figure 5.2A). Phylogenetic weights calculated using MacClade (Maddison and Maddison 1992) show substitution frequencies that are severely disproportional between different substitution types (Figure 5.2B). When transition/transversion (ts/tv) ratios were plotted with proportional distances (P-distance) based on all pair-wise comparisons, the ratios were initially high (>2.0) but decreased to about 0.5 when P-distance reached 0.05, showing saturation by multiple substitutions (Figure 5.3D). The cause of this observation is evident when each substitution type is individually plotted (Figure 5.3A to C). For example, the frequency of transition A ↔ G is initially high but appears to be obscured by A ↔ T and T ↔ G transversions (Figure 5.3A to C). However, the phylogenetic information content of the 16S gene is still high, even when the ts/tv ratio is beyond the saturation point (Figures 5.4 and 5.5A). We believe that this is true because, except for A ↔ G transitions, all other substitutions increase almost linearly up to a P-distance of 0.12 (Figure 5.3A to C).

Different portions of the ribosomal DNA are known to evolve at different rates (Hillis and Dixon 1991; Simon et al. 1994). For example, in both the 16S and 12S genes, the 5′ half (domains 1 and 2), on average, is less conserved than the 3′ half (domains 3 and 4). In tephritids, when we plot the 16S data sets on the inferred phylogenetic trees, fairly consistent variability profiles over all character sites are observed (Figure 5.2C, D). A similar pattern is found in the variability plot even between *Drosophila* and the mosquitoes (Figure 5.1B). These observations suggest that different rDNA portions are evolving at different rates in our taxa.

### 5.3.3 Comparison of Tree-Building Methods

As discussed in the above section, the tephritid mt rDNA appears to evolve under a variety of constraints. Therefore, it is important to select phylogenetic methods either designed for a similar evolutionary model or at least insensitive to violation of the assumptions of constant evolutionary rate both in different lineages and different characters. Over 100 different tree-building methods are available, and choosing the most suitable method for a certain data set could be a daunting task (Huelsenbeck and Hillis 1993). Fortunately, determining the relative efficiency of different methods has become a hot issue, and a number of articles on this topic have been published (e.g., Nei 1991; Hasegawa and Fujiwara 1993; Huelsenbeck and Hillis 1993; Kim et al. 1993; Kuhner and Felsenstein 1994; Tateno et al. 1994). An interesting consensus based on the simulation tests conducted by these authors is that the simple neighbor-joining (NJ) method is efficient in most cases of rate heterogeneity, while other popular techniques, such as maximum parsimony, are often sensitive to violation of the assumption of constant evolutionary rate. Therefore, for our rDNA data set, where such heterogeneity is highly suspected, the NJ method seems to be a reasonable choice for a phylogenetic analysis. In our experience with several tephritid data sets analyzed using UPGMA, maximum parsimony (MP), and NJ methods, the NJ method also was the most efficient for recovering supraspecific taxa previously well established by morphological data (Han and McPheron 1997; McPheron and Han 1997). For distance analyses, progressively more realistic genetic distance measures have been devised to correct for multiple substitutions (Jukes and Cantor

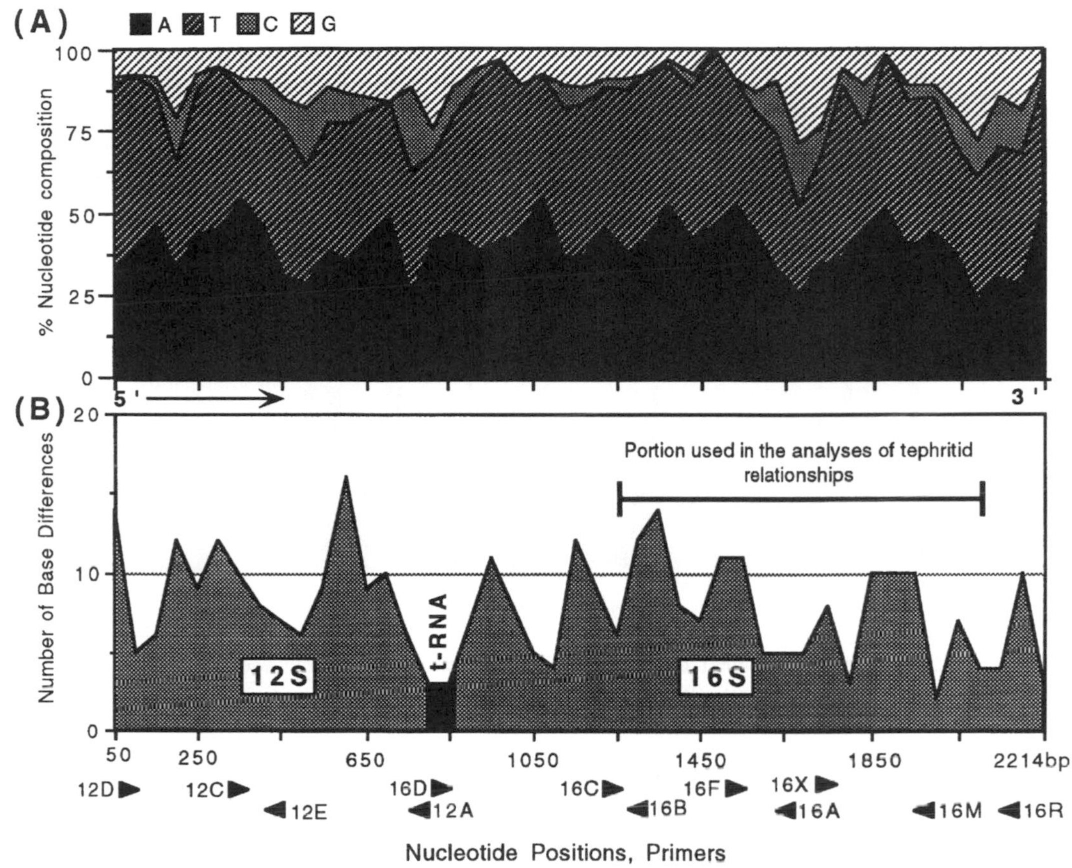


**FIGURE 5.1** (A) Average percentage nucleotide composition and (B) number of base differences between *Drosophila yakuba* and *Anopheles gambiae* for the entire 12S/t-RNA/16S genes (50-base nonoverlapping sliding window). Primer positions (abbreviations matching Table 5.1 and Appendix 5.1 except for first letter) are marked by arrows under the plot.

1969; Tajima and Nei 1984; Kimura 1980; Tamura 1992; Tamura and Nei 1993). Among them, the Tamura distance (Tamura 1992) was specifically designed for A+T bias cases in Diptera. For the two data sets we have published (Han and McPheron 1997; McPheron and Han 1997), however, several different distance measures produced identical NJ trees (empirical guidelines for selecting an appropriate distance measure were suggested by Nei 1991 and Kumar et al. 1993).

There is a widely held notion that agreement among trees estimated by different methods lends greater credibility to the phylogenetic estimates (Avise 1994). Kim (1993) has shown that a simulation test actually supports this notion and developed an index that can serve as a measure of the reliability of the joint estimate. Even though we do not hastily accept this unorthodox index as a measure of reliability, it is worthwhile to generate several different topologies by different methods; any topological differences might provide clues to determine the causes of this variation.

We believe that MP results may be improved by adopting various weighting schemes. Characters can be weighted by sites (Van de Peer et al. 1993; Goloboff 1993) or by substitution type (Wheeler 1990; Williams and Fitch 1990; Knight and Mindell 1993; Collins et al. 1994). We did try weighted parsimony methods using step matrices similar to Figure 5.2B. This, however, did not improve the topology, probably because of a severe violation of the triangle inequality (Wheeler 1993). For example, a T $\rightarrow$ G substitution (cost of 20 steps) is more costly than the sum of two intermediate substitutions (T $\rightarrow$ A + A $\rightarrow$ G = 4 steps). We can also weight each nucleotide position based upon

(A) Number of Substitutions

| from \ to | A | T | C | G |
|---|---|---|---|---|
| A | | 439 | 14 | 297 |
| T | 547 | | 103 | 107 |
| C | 9 | 55 | | 1 |
| G | 93 | 43 | 1 | |

(B) Phylogenetic Weights

| from \ to | A | T | C | G |
|---|---|---|---|---|
| A | | 1 | 20 | 3 |
| T | 1 | | 20 | 20 |
| C | 9 | 1 | | 20 |
| G | 1 | 4 | 20 | |

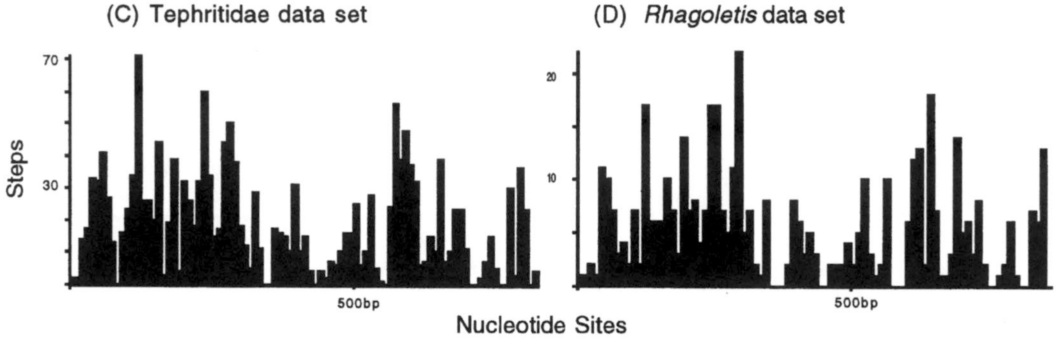


**FIGURE 5.2** (A) Average number of nucleotide substitutions reconstructed over the tephritid neighbor-joining tree. (B) Phylogenetic weights calculated as the reciprocal of substitution frequencies (scaled to a range of 1 to 20). (C) Number of phylogenetic steps for base positions, reconstructed over the NJ tree. (D) Same, based on the *Rhagoletis* data set. (A to C) Either modified from or based on the data presented in Han and McPheron (1997); (D) based on the data from McPheron and Han (1997).

our empirical data (Figure 5.2C, D). Van de Peer et al. (1993) proposed such a data-based weighting method, which can be applied to both parsimony and distance methods. Using distance analysis of 18S nuclear ribosomal RNA sequences, they were able to reconstruct an inferred vertebrate phylogeny highly consistent with the consensus view of paleontologists, while unweighted methods failed to recognize established relationships.

### 5.3.4 Statistical Tests

In the NJ tree, confidence probability values (Pc) from the standard error test (Rzhetsky and Nei 1992) tend to be higher than bootstrap probability values (Pb) (Felsenstein 1985), especially at the deeper branches defining the higher taxa above subtribe (Figure 5.4). This either implies that Pc overestimates the confidence of the branches or Pb underestimates them. Recent studies indicate that the latter may be the case (Sitnikova et al. 1995; Sitnikova 1996). They showed that Pc was, in fact, the complement of the P-value used in standard statistical tests, but Pb was not. In simulation tests using four and six taxa data sets, Sitnikova et al. (1995) found that Pb usually underestimated the extent of statistical support of species clusters, especially when the true tree was starlike and the number of sequences in the tree increased. Our data set might be a similar case because we used a relatively large number of sequences, and some branch lengths of the unknown "true phylogeny" are likely to be close to zero (= starlike topology) due to the rapid adaptive radiation of this family, resulting in over 4000 contemporary species in 40 to 70 million years (Rohdendorf 1964; Prokopy and Roitberg 1984).

A comparison of two NJ trees based, respectively, on 16S rDNA and mitochondrial cytochrome oxidase subunit II (COII) genes provided empirical evidence to support the reliability of the standard error test (Figure 5.5A; see also Section 5.4.2). Between these two trees, none of the corresponding branches supported by over 80% Pc show conflicting relationships. Further, branches supported by

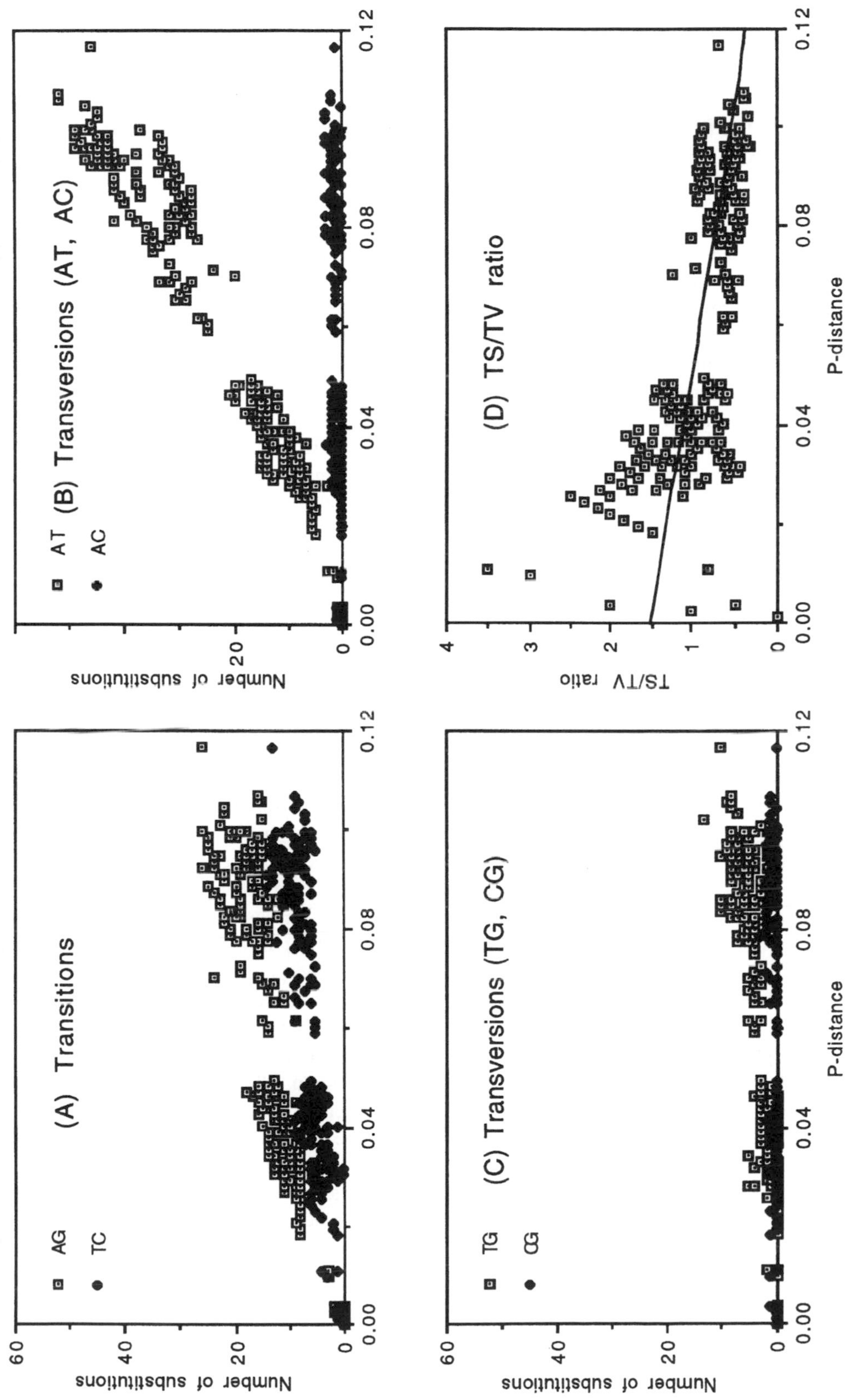


**FIGURE 5.3** Number of substitutions (A to C) and transition/transversion ratio (D) plotted over proportional distances based on all pair-wise comparisons of the *Rhagoletis* 16S rDNA data set of McPheron and Han (1997).

over 95% Pc in one tree are also recognized in the other tree by similarly high Pc values. Therefore, we can conclude that the branches supported by over 80% Pc are worthy of attention and those supported by over 95% Pc are highly likely to reflect true phylogeny (see also Sitnikova et al. 1995).

## 5.4 TEPHRITID RELATIONSHIPS INFERRED FROM MITOCHONDRIAL RIBOSOMAL DNA

### 5.4.1 Generic to Subfamily Relationships

Our recent study (Han and McPheron 1997) showed that an inferred phylogeny based on partial 16S rDNA sequences was not only highly congruent with morphological classification, but also suggested several previously unknown relationships (Figure 5.4). The following is a summary of tephritid relationships significantly supported by the interior branch test using the NJ method (modified from Han and McPheron 1997).

1. Excluding *Parastenopa limata* (Coquillett), the monophyly of Trypetini was strongly supported at 97% Pc. Within the Trypetini, the presence of two subtribes (Trypetina and Chetostomatina) was also consistent with morphological studies (Han 1992 and Chapter 11). The only disagreement was the position of *P. limata*. However, an expanded analysis based on a data set including 350 additional base pairs and seven more species of Trypetini supports the inclusion of *Parastenopa* as the most basal lineage within the subtribe Chetostomatina (Han, in press). Considering that the genus *Parastenopa* is chiefly Neotropical while the majority of the Chetostomatina have an Old World distribution, *Parastenopa* may possibly be a remnant of the early evolution of the Chetostomatina (see also Han, Chapter 11).

2. The monotypic genus *Plioreocepta* has been traditionally classified in the subfamily Trypetinae, but Korneyev (1987; 1996; Chapter 22) has placed it in the tribe Zaceratini and suggested that this tribe may be the sister group of the subfamily Tephritinae based on larval and genitalic characters. Our data strongly support (at 99% Pc) the monophyly of *P. poeciloptera* (Schrank) plus three species representing the subfamily Tephritinae. This may mean that *Plioreocepta* is either an aberrant member of, or the sister group to, the Tephritinae. The latter interpretation seems more likely for the following reasons. Tephritinae is a large and diverse subfamily of over 100 genera, but they are believed to be a monophyletic group including all of the flower- and stem-infesting and gall-forming taxa that breed mainly in Asteraceae, Acanthaceae, Verbenaceae, and Lamiaceae. A majority of the Tephritinae can also be defined morphologically by characters that are common but not consistently present, such as dense thoracic microtrichia, various setae or setulae lanceolate and pale, and scapular setae poorly differentiated or absent (Foote et al. 1993; Norrbom et al. 1999; Korneyev, Chapter 22). *Plioreocepta* lacks such biological and morphological synapomorphies for the subfamily. This hypothesized relationship requires more rigorous testing based on an increased number of taxa representing all the major lineages within the subfamily Tephritinae.

3. Monophyly of *Anastrepha* and *Toxotrypana* is clearly supported by previous work based on morphological (Norrbom and Foote 1989), immunological (Kitto 1983; Sarma et al. 1987), and DNA sequence data (Han and McPheron 1994). These two genera are placed in the tribe Toxotrypanini (Foote et al. 1993). Our study not only reconfirmed the monophyly of these two genera but also suggested the genus *Hexachaeta* as their possible sister group at high Pc (97%). This relationship was never previously suggested, but it makes good zoogeographic and taxonomic sense. All these genera are mostly Neotropical, and *Hexachaeta* can be placed at least in the same subfamily (Trypetinae) based on morphology. None of other New World genera of Trypetinae show any indication of being the sister group of *Anastrepha* and *Toxotrypana* based on our study and morphological evidence (Norrbom, personal communication). *Hexachaeta* is now included in the tribe Toxotrypanini based on our molecular study (Norrbom et al. 1999).

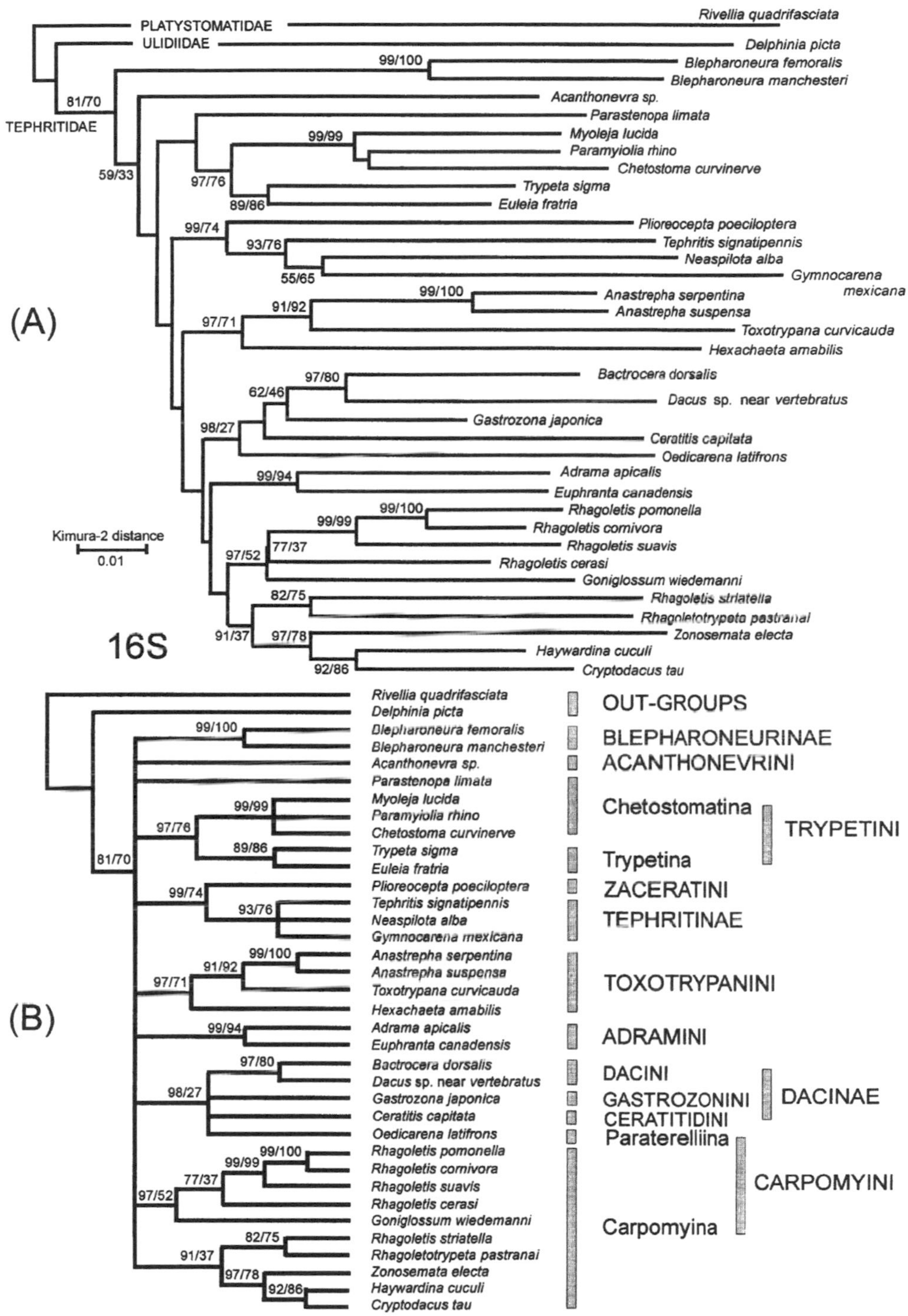


**FIGURE 5.4** (A) Tephritid relationships inferred from a neighbor-joining tree based on Kimura two-parameter distances with pair-wise deletion of gaps and missing data, using partial sequences of the mitochondrial 16S rDNA. Numbers on each interior branch are the confidence probability (Pc) value followed by the bootstrap probability (Pb) value in percentage (Pc values higher than 50% are indicated). (B) Same topology after collapsing the branches with Pc values lower than 75%. (Modified from Han and McPheron 1997.)

4. The tribe Ceratitidini had been included in the heterogeneous subfamily Trypetinae until Hancock (1986) suggested its close relationship with Dacini based on morphological evidence. He suggested subfamily status for each taxon, but White and Elson-Harris (1992) and Foote et al. (1993) formally recognized the larger subfamily Dacinae including the tribes Dacini and Ceratitidini. Our study also supports the close relationship between these two tribes, but neither supports nor rejects the monophyly of Ceratitidini. One interesting new hypothesis proposed by our study is a close relationship between the New World genus *Oedicarena* and the Dacinae (at 98% Pc). Since all the current members of Dacinae have Old World distributions, the ancestor of *Oedicarena* might have diverged from the stem group of the subfamily. Considering the enormous diversity of the subfamily (at least 40 genera and 1000 species, all from the Old World), we will need sequence data for a number of additional dacine taxa to test the phylogenetic position of *Oedicarena*. Contrary to our hypothesis, Korneyev (1996) placed *Oedicarena* in the subtribe Paraterelliina of the Capomyini on the basis of morphology. In a study using nucleotide sequences of subunit II of the mitochondrial cytochrome oxydase gene, Smith and Bush (1997) also suggested a close relationship between *O. latifrons* (Wulp) and *Rhagoletis striatella* Wulp based on a weighted parsimony analysis (posteriori weighting based on rescaled consistency indices, RC) whose utility is not currently well understood.

5. Although we only examined two species, monophyly of the tribe Adramini is supported at 99% Pc. *Adrama apicalis* Shiraki and *Euphranta canadensis* (Loew) had been placed in different tribes until Korneyev (1994) synonymized Euphrantini with Adramini. Our result not only supports Korneyev's synonymy but also is congruent with previous morphological studies, in which Adramini and Euphrantini were considered closely related (Hardy 1973; 1974; Hancock 1986, Foote et al. 1993).

6. In our NJ tree, the monophyly of the subtribe Carpomyina was suggested topologically but the statistical support for this relationship is weak. Instead, two monophyletic groups within the Carpomyina were recognized at relatively high Pc values (97 and 91%, respectively). One group includes the Palearctic genus *Goniglossum* plus four *Rhagoletis* species, and the other group includes four Neotropical genera plus *R. striatella*. Therefore, our study strongly suggests the nonmonophyletic nature of the diverse genus *Rhagoletis*. Indeed, closer relationships of Solanaceae-breeding *Rhagoletis* species, including *R. striatella*, to *Zonosemata* and other Neotropical carpomyine genera have been suggested by previous studies (Bush 1965; Berlocher and Bush 1982). In a recent phylogenetic study of several genera of Carpomyina based on morphological characters, Norrbom (1994) indicated that the New World genera *Cryptodacus*, *Haywardina*, *Rhagoletotrypeta*, and *Zonosemata* differed from other Carpomyina in having a white medial scutal stripe or spot. This character state might well be a morphological synapomorphy defining a group composed of these four genera plus *R. striatella* and probably some other Neotropical *Rhagoletis* spp., in which a character reversal might have occurred. Our study also suggests that *Zonosemata*, *Haywardina*, and *Cryptodacus* form a monophyletic group, although relationships among this group, *Rhagoletotrypeta* and *Rhagoletis striatella* are not well resolved. Relationships among North American *Rhagoletis* species are extensively discussed in McPheron and Han (1997) and Smith and Bush (1997) (see also next section).

### 5.4.2 Comparison of 16S rDNA and Cytochrome Oxidase II

We have also studied interspecific relationships within the genus *Rhagoletis* based on the same 16S rDNA region used for the above analysis (McPheron and Han 1997). Coincidentally, at the same time, Smith and Bush (1997) published a phylogenetic analysis of *Rhagoletis* based on the mitochondrial cytochrome oxidase subunit II (COII) gene sequences. Because both studies included many of the same species, we have a great opportunity to compare two data sets. We therefore reanalyzed the sequence data for the taxa common to both studies: 21 species of *Rhagoletis*, *Zonosemata electa* (Say), *Rhagoletotrypeta pastranai* Aczél, *Oedicarena latifrons* (Wulp),

*Euphranta canadensis* (Loew), and *Ceratitis capitata* (Wiedemann). For the latter species, we used a 16S rDNA sequence from Han and McPheron (1997). Even though the monophyly and sister group relationship of Carpomyina were not clear (Figure 5.4), we selected *C. capitata* as the outgroup to make our inferred phylogenetic trees directly comparable with those of Smith and Bush (1997). The 16S rDNA sequences, in which sequence length variation was observed, were aligned using CLUSTAL W software (version 1.7, Thompson et al. 1997).

Interestingly, both analyses produced similar NJ trees (Figure 5.5A). Furthermore, branches supported by Pc values greater than 95% are completely identical. This is consistent with predictions from previous simulation tests and our empirical observations (see Section 5.3.4). It is also interesting to note that the COII tree is slightly better defined in the shallower branches, while the close relationship between *R. conversa* (Brethes) and *R. striatella* is only resolved in the deeper branches of the 16S tree (Figure 5.5A). This suggests that the COII gene may have evolved faster, at least within the Carpomyina. This rate difference, however, may have arbitrarily resulted from the elimination of nucleotide positions with gaps and ambiguous characters from the 16S data set (we used 822 base pairs after eliminating 36 sites). Nevertheless, the congruence test of two data sets (Figure 5.5A) and the combined analysis (Figure 5.5B) have provided better insight into *Rhagoletis* phylogeny. Even though no additional relationships are suggested by this approach, the *Rhagoletis* relationships discussed in previous works (McPheron and Han 1997; Smith and Bush 1997) are much more confidently supported. For this reason, we highly recommend sequencing at least two genes to resolve tephritid higher relationships. We are currently trying to compare the mitochondrial 16S and 12S rDNAs. Because the molecular evolutionary constraints on these two ribosomal RNA genes are similar (Figure 5.1), the same tree-building and statistical methods can be applied for both separate and combined analyses.

## 5.5 CONCLUSIONS AND FUTURE RESEARCH NEEDS

Tephritid higher classification and phylogeny have been poorly resolved in the past. Many tephritid subfamilies and tribes are weakly defined by morphological characters, the relationships among many higher taxa are undefined, and classification schemes often do not serve as predictors of evolutionary relationships. We do believe that tephritid higher classification can be further improved by a phylogenetic approach based on comparative morphological study of global taxa. However, the morphological approach alone is not only severely limited by the set of usable characters, but also hindered by frequent cases of homoplasy (Han, Chapter 11). Therefore, we believe that recent developments in molecular systematics will greatly contribute to our knowledge of tephritid phylogenetic history by providing an almost unlimited source of taxonomic characters. Analysis of relationships among tephritid lineages based on molecular data can be used both to test morphological hypotheses and to erect new hypotheses that may direct further investigation.

Recent studies using mitochondrial DNA sequences (Han and McPheron 1997; McPheron and Han 1997; Smith and Bush 1997) clearly demonstrated the value of molecular data to improve tephritid higher classification. Inferred relationships from these studies not only are largely congruent with the well-established portions of the morphological classification, but also suggest previously unknown relationships. In fact, they already have contributed toward a sound classification of Tephritidae. For example, based on our molecular study (Han and McPheron 1997), the morphological hypothesis of the newly defined monophyletic tribe Trypetini was positively supported (Han, Chapter 11), and the genus *Hexachaeta* was included in the tribe Toxotrypanini (Norrbom et al. 1999). We believe that such cases will sharply increase as molecular sequences of more tephritid taxa are explored.

We are now expanding our analysis in terms of the number of taxa and characters. Currently, an additional 300 base pairs of the 16S gene plus about 700 base pairs of the 12S gene are being added to our sequence database. This approach will improve the resolving power of our analysis by increasing the number of characters twofold and by providing two inferred phylogenies to

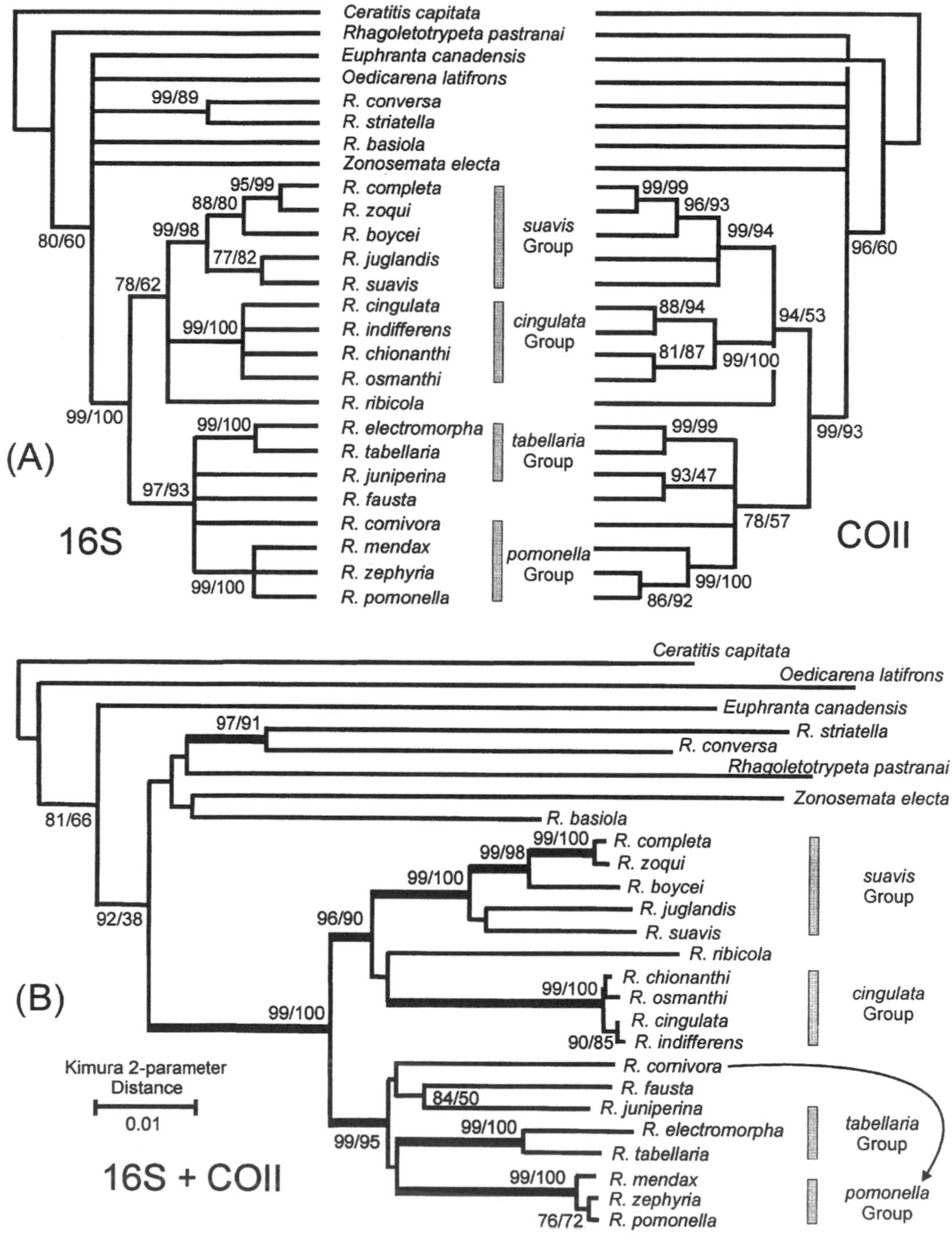


**FIGURE 5.5** (A) Comparison of two *Rhagoletis* neighbor-joining topologies based on the mitochondrial 16S rDNA and COII genes. Kimura two-parameter distances with complete deletion of gaps and missing data are used. *Ceratitis capitata* was used as outgroup. Numbers on each interior branch are the confidence probability (Pc) value followed by the bootstrap probability (Pb) value in percentage. All the branches with Pc values lower than 75% were collapsed. (B) NJ trees based on the combined anaysis of 16S rDNA and COII genes. Thicker lines represent branches supported by over 95% Pc. (Based on the data from Han and McPheron 1997; McPheron and Han 1997; and Smith and Bush 1997.)

compare. Considering the phylogenetic constraints of the mt rDNA (see Section 5.3.2), we also believe that additional nuclear genes should be explored to obtain additional characters and to generate alternative hypotheses of relationships. For this reason, we are currently sequencing a portion of the nuclear 28S rDNA, which already showed a great potential value to resolve relationships among the fly families of the infraorder Culicomorpha (Pawlowski et al. 1996). We are hoping that our approach will eventually provide answers to longstanding questions in tephritid systematics such as the basal phylogeny of Tephritidae in conjunction with the origin of phytophagy and relationships among families of the superfamily Tephritoidea.

## ACKNOWLEDGMENTS

We thank Stewart Berlocher, Nancy Bowers, Marty Condon, Jeff Feder, Amnon Freidberg, Daniel Frias, John Jenkins, Bernhard Merz, Al Norrbom, Dan Papaj, Hugh Robertson, John Sivinski, and Gary Steck for providing some of the taxa analyzed. Al Norrbom and an anonymous reviewer kindly reviewed this chapter and made helpful suggestions for its improvement. Support for this research came from the U.S. Department of Agriculture (Agreement 95-37302-1808) to BAM and Korea Science and Engineering Foundation (Project No. 961-0508-066-2) to HYH. We also thank the Campaña Nacional contra las Moscas de la Fruta (Mexico), International Organization for Biological Control of Animals and Plants, Instituto de Ecología, A.C. (Mexico), and Consejo Nacional de Ciencia y Tecnología (Mexico) for their financial support of the symposium, and USDA-ICD-RSED for partial funding of BAM's travel to participate in it.

## REFERENCES

Avise, J.C. 1994. *Molecular Markers, Natural History and Evolution*. Chapman & Hall, New York. 511 pp.

Avise, J.C. and R.C. Vrijenhoek. 1987. Mode of inheritance and variation of mitochondrial DNA in hybridogenetic fishes of the genus *Poeciliopsis*. *Mol. Biol. Evol.* 4: 514–525.

Beard, C.B., D.M. Hamm, F.H. Collins. 1993. The mitochondrial genome of the mosquito *Anopheles gambiae*: DNA sequence, genome organization, and comparisons with mitochondrial sequences of other insects. *Insect Mol. Biol.* 2: 103–124.

Berlocher, S.H. and G.L. Bush. 1982. An electrophoretic analysis of *Rhagoletis* (Diptera: Tephritidae) phylogeny. *Syst. Zool.* 31: 136–155.

Bush, G.L. 1965. The genus *Zonosemata*, with notes on the cytology of two species (Diptera-Tephritidae). *Psyche* 72: 307–323.

Clary, D.O. and D.R. Wolstenholme. 1985. The mitochondrial DNA molecule of *Drosophila yakuba*: nucleotide sequence, gene organization, and genetic code. *J. Mol. Evol.* 22: 252–271.

Cockburn, A.F., S.E. Mitchell, and J.A. Seawright. 1990. Cloning of the mitochondrial genome of *Anopheles quadrimaculatus*. *Arch. Insect Biochem. Physiol.* 14: 31–36.

Collins, T.M., F. Kraus, and G. Estabrook. 1994. Compositional effects and weighting of nucleotide sequences for phylogenetic analysis. *Syst. Biol.* 43: 449–459.

De Salle, R., J. Gatesy, W. Wheeler, and D. Grimaldi. 1992. DNA sequence from a fossil termite in Oligo-Miocene amber and their phylogenetic implications. *Science* 257: 1933–1936.

Felsenstein, J. 1985. Confidence limits on phylogenies: an approach using the bootstrap. *Evolution* 39: 783–791.

Foote, R.H., F.L. Blanc, and A.L. Norrbom. 1993. *Handbook of the Fruit Flies (Diptera: Tephritidae) of America North of Mexico*. Comstock Publishing Associates, Ithaca. 571 pp.

Freidberg, A. 1984. Gall Tephritidae (Diptera). In *Biology of Gall Insects* (T.N. Ananthakrishnan, ed.), pp. 129–167. Oxford and IBH Publishing Co., New Delhi.

Goloboff, P.A. 1993. Estimating character weights during tree-search. *Cladistics* 9: 83–91.

Han, H.-Y. 1992. Classification of the Tribe Trypetini (Diptera: Tephritidae: Trypetinae). Ph.D. dissertation, Pennsylvania State University, University Park. 274 pp.

Han, H.-Y. Molecular phylogenetic study of the tribe Trypetini (Diptera: Tephritidae) using mitochondrial 16S DNA sequences. *Biochem. Syst. Ecol.*, in press.

Han, H.-Y. and B.A. McPheron. 1994. Phylogenetic study of selected tephritid flies (Insecta: Diptera: Tephritidae) using partial sequences of the nuclear 18S ribosomal DNA. *Biochem. Syst. Ecol.* 22: 447–457.

Han, H.-Y. and B.A. McPheron. 1997. Molecular phylogenetic study of Tephritidae (Insecta: Diptera) using partial sequences of the mitochondrial 16S ribosomal DNA. *Mol. Phyl. Evol.* 7: 17–32.

Hancock, D.L. 1986. Classification of the Trypetinae (Diptera: Tephritidae), with a discussion of the Afrotropical fauna. *J. Entomol. Soc. South. Afr.* 49: 275–305.

Hardy, D.E. 1973. The fruit flies (Tephritidae-Diptera) of Thailand and bordering countries. *Pac. Insects Monogr.* 31: 1–353.

Hardy, D.E. 1974. The fruit flies of the Philippines (Diptera: Tephritidae). *Pac. Insects Monogr.* 32: 1–266.

Hasegawa, M. and M. Fujiwara. 1993. Relative efficiencies of the maximum likelihood, maximum parsimony, and neighbor-joining methods for estimating protein phylogeny. *Mol. Phyl. Evol.* 2: 1–5.

Hillis, D.M. and M.T. Dixon. 1991. Ribosomal DNA: molecular evolution and phylogenetic inference. *Q. Rev. Biol.* 66: 441–453.

Huelsenbeck, J.P. and D.M. Hillis. 1993. Success of phylogenetic methods in the four-taxon case. *Syst. Biol.* 42: 247–264.

Jukes, T.H. and C.R. Cantor. 1969. Evolution of protein molecules. In *Mammalian Protein Metabolism* (H.N. Munro, ed.), pp. 21–132. Academic Press, New York.

Kim, J. 1993. Improving the accuracy of phylogenetic estimation by combining different methods. *Syst. Biol.* 42: 331–340.

Kim, J., F.J. Rohlf, and R.R. Sokal. 1993. The accuracy of phylogenetic estimation using the neighbor-joining method. *Evolution* 47: 471–486.

Kimura, M. 1980. A simple method for estimating evolutionary rate of base substitutions through comparative studies of nucleotide sequences. *J. Mol. Evol.* 16: 111–120.

Kitto, G.B. 1983. An immunological approach to the phylogeny of the Tephritidae. In *Fruit Flies of Economic Importance. Proceedings of the CEC/IOBC International Symposium, Athens, 1982* (R. Cavalloro, ed.), pp. 203–211. A.A. Balkema, Rotterdam. 642 pp.

Knight, A. and D.P. Mindell. 1993. Substitution bias, weighting of DNA sequence evolution, and the phylogenetic position of Fea's viper. *Syst. Biol.* 42: 18–31.

Kocher, T. D, W.K. Thomas, A. Meyer, S.V. Edwards, S. Pääbo, F.X. Villablanca, and A.C. Wilson. 1989. Dynamics of mitochondrial DNA sequence evolution in animals: amplification and sequencing with conserved primers. *Proc. Natl. Acad. Sci. U.S.A.* 86: 6196–6200.

Korneyev, V.A. 1987. The asparagus fly and its taxonomic position in the family Tephritidae (Diptera). *Vestn. Zool.* 1987(1): 39–44 [in Russian].

Korneyev, V.A. 1994. Reclassification of the Palaearctic Tephritidae (Diptera). Communication 2. *Vestn. Zool.* 1994 (1): 3–17 [in Russian].

Korneyev, V.A. 1996. Reclassification of Palaearctic Tephritidae (Diptera). Communication 3. *Vestn. Zool.* 1995 (5–6): 25–48.

Kuhner, M.K. and J. Felsenstein. 1994. A simulation comparison of phylogeny algorithms under equal and unequal evolutionary rates. *Mol. Biol. Evol.* 11: 459–468.

Kumar, S., K. Tamura, and M. Nei. 1993. *MEGA: Molecular Evolutionary Genetics Analysis*, version 1.0. Institute of Molecular Evolutionary Genetics, Pennsylvania State University, University Park.

Maddison, W.P. and D.R. Maddison. 1992. *MacClade*, version 3. Sinauer Associates, Inc., Sunderland, MA.

McPheron, B.A. and H.-Y. Han. 1997. Phylogenetic analysis of North American *Rhagoletis* (Diptera: Tephritidae) and related genera using mitochondrial DNA sequence data. *Mol. Phyl. Evol.* 7: 1–16.

Moritz, C., T.E. Dowling, and W.M. Brown. 1987. Evolution of animal mitochondrial DNA: relevance for population biology and systematics. *Annu. Rev. Ecol. Syst.* 18: 269–292.

Mullis, K., F. Faloona, S. Scharf, R.K. Saiki, G. Horn, and H. Erlich. 1986. Specific enzymatic amplification of DNA *in vitro*: the polymerase chain reaction. *Cold Spring Harbor Symp. Quant. Biol.* 51: 263–273.

Nei, M. 1991. Relative efficiencies of different tree-making methods for molecular data. In *Phylogenetic Analysis of DNA Sequences* (M.M. Miyamoto and J. Cracraft, eds.), pp. 90–128. Oxford University Press, New York. 358 pp.

Norrbom, A.L. 1994. New species and phylogenetic analysis of *Cryptodacus*, *Haywardina*, and *Rhagoletotrypeta* (Diptera: Tephritidae). *Insecta Mundi* 8: 37–65.

Norrbom, A.L. and R.H. Foote. 1989. The taxonomy and zoogeography of the genus *Anastrepha* (Diptera: Tephritidae). In *Fruit Flies: Their Biology, Natural Enemies and Control* (A.S. Robinson and G. Hooper, eds.), pp. 15–26. In *World Crop Pests* (W. Helle, ed.), Vol. 3A. Elsevier Science Publishers, Amsterdam. 372 pp.

Norrbom, A.L., L.E. Carroll, and A. Freidberg. 1999. Status of knowledge. In *Fruit Fly Expert Identification System and Systematic Information Database* (F.C. Thompson, ed.), pp. 9–47. *Myia* (1998) 9, 524 pp.

Pawlowski, J., R. Szadziewski, D. Kmieciak, J. Fahrni, and G. Bittar. 1996. Phylogeny of the infraorder Culicomorpha (Diptera: Nematocera) based on 28S RNA gene sequences. *Syst. Entomol.* 21: 167–178.

Prokopy, R.J. and B.D. Roitberg. 1984. Foraging behavior of true fruit flies. *Am. Sci.* 72: 41–49.

Rohdendorf, B.B. 1964. The historical development of two-winged insects. *Tr. Paleont. Inst. Akad. Nauk SSSR* 100: 1–311 [in Russian, English translation published by University of Alberta Press, 1974].

Rzhetsky, A. and M. Nei. 1992. A simple method for estimating and testing minimum-evolution trees. *Mol. Biol. Evol.* 9: 945–967.

Sanger, F., S. Nickelen, and A.R. Coulson. 1977. DNA sequencing with chain-terminating inhibitors. *Proc. Natl. Acad. Sci. U.S.A.* 74: 5463–5467.

Sarma, R., G.B. Kitto, S. Berlocher, and G.L. Bush. 1987. Biochemical and immunological studies on an α-glycerophosphate dehydrogenase from the tephritid fly, *Anastrepha suspensa*. *Arch. Insect Biochem. Physiol.* 4: 271–286.

Sheppard, W.S., G.J. Steck, and B.A. McPheron. 1992. Geographic populations of the medfly may be differentiated by mitochondrial DNA variation. *Experientia* 49: 1010–1013.

Simon, C., A. Franke, and A. Martin. 1991. The polymerase chain reaction: DNA extraction and amplification. In *Molecular Techniques in Taxonomy* (G.M. Hewitt, A.W.B. Johnston, and J.P.W. Young, eds.), pp. 329–355. NATO Advanced Studies Institute, H57. Springer-Verlag, Berlin.

Simon, C., F. Frati, A. Beckenback, B. Crespi, H. Liu, and P. Flook. 1994. Evolution, weighting, and phylogenetic utility of mitochondrial gene sequences and a compilation of conserved polymerase chain reaction primers. *Ann. Entomol. Soc. Am.* 87: 651–701.

Sitnikova, T. 1996. Bootstrap method of interior-branch test for phylogenetic trees. *Mol. Biol. Evol.* 13: 605–611.

Sitnikova, T., A. Rzhetsky, and M. Nei. 1995. Interior-branch and bootstrap tests of phylogenetic trees. *Mol. Biol. Evol.* 12: 1–15.

Smith, J.J. and G.L. Bush. 1997. Phylogeny of the genus *Rhagoletis* (Diptera: Tephritidae) inferred from DNA sequences of mitochondrial cytochrome oxidase II. *Mol. Phyl. Evol.* 7: 33–43.

Tajima, F. and M. Nei. 1984. Estimation of evolutionary distance between nucleotide sequences. *Mol. Biol. Evol.* 1: 269–285.

Tamura, K. 1992. Estimation of the number of nucleotide substitutions when there are strong transition-transversion and G+C content biases. *Mol. Biol. Evol.* 9: 678–687.

Tamura, K. and M. Nei. 1993. Estimation of the number of nucleotide substitutions in the control region of mitochondrial DNA in humans and chimpanzees. *Mol. Biol. Evol.* 10: 512–526.

Tateno, Y., N. Takezaki, and M. Nei. 1994. Relative efficiencies of the maximum-likelihood, neighbor-joining, and maximum-parsimony methods when substitution rate varies with site. *Mol. Biol. Evol.* 11: 261–277.

Thomas, R.H., W. Schaffner, A.C. Wilson, and S. Pääbo. 1989. DNA phylogeny of the extinct marsupial wolf. *Nature* 340: 465–467.

Thompson, J.D., D.G. Higgins, and T.J. Gibson 1997. CLUSTAL W: improving the sensitivity of progressive multiple sequence alignment through sequence weighting, position specific gap penalties and weight matrix choice. *Nucl. Acids Res.* 22: 4673–4680.

Van de Peer, Y., J.-M. Neefs, P. De Rijk, and R. De Wachter. 1993. Reconstructing evolution from eukaryotic small-ribosomal-subunit RNA sequences: calibration of the molecular clock. *J. Mol. Evol.* 37: 221–232.

Wheeler, W.C. 1990. Combinatiorial weights in phylogenetic analysis: a statistical parsimony procedure. *Cladistics* 6: 269–275.

Wheeler, W.C. 1993. The triangle inequality and character analysis. *Mol. Biol. Evol.* 10: 707–712.

White, I.M. and M.M. Elson-Harris. 1992. *Fruit Flies of Economic Significance: Their Identification and Bionomics*. CAB International, Wallingford. 601 pp.

Williams, P.L. and W.M. Fitch. 1990. Phylogeny determination using dynamically weighted parsimony methods. *Methods Enzymol.* 183: 615–626.

## APPENDIX 5.1: ALIGNMENT OF 12S-16S RDNA SEQUENCES

Gaps inserted to improve alignment are indicated by a hyphen, and missing or ambiguous characters are indicated by a question mark. Dots indicate identity with the top sequence, and blank spaces indicate unavailability of sequence information. Abbreviations: AnopGamb = *Anopheles gambiae*; DrosYaku = *Drosophila yakuba*; RhagPomo = *Rhagoletis pomonella*; CeraCapi = *Ceratitis capitata*; EuleFrat = *Euleia fratria*. They represent Culicidae, Drosophilidae, and three species of Tephritidae. Underlined sequences indicate primer positions.

```
         >12S rRNA>                                                             80
AnopGamb AATAAGATTATTTTATTCTAGTTAAATATTTTATTATTATTTTATTTTACATGTAAATTTTTGTGTGAATTTTTATTAAT A12D
DrosYaku --.T.A.--G.......T.G.C.T..A.A...G......G...GA....T.......................A....T..
RhagPomo
CeraCapi
EuleFrat
                                                                                       160
AnopGamb TTTAAAAATTAATA-ATTTT---TAATTTATTCGCAGTAATTAATATTAATTATAAAAGAAATTTTGAATTAGTAATAAT
DrosYaku ..A......A....T...A.AAA.T.......................T.A..T..........A...A...C....T.
RhagPomo
CeraCapi
EuleFrat
                                                                                       240
AnopGamb ATATAGTATTGGTAAAATTTGTGCCAGCTACTGCGGTTATACAAATGATGCAAATAAAAATTTTTAGTATTAGTTAAATT
DrosYaku .A.A.......ACC.....G........AGTC..........C...A..A........TT..........G.A.......
RhagPomo
CeraCapi
EuleFrat
                                                                                       320
AnopGamb GTTTATAATTATTTAATTTATATATAAATTTATTAGGTGAAATTTTTAAATTTATTTATTATTAAAATGTAGATTAATTT
DrosYaku .A..-..T...---..AA..A.AT.....A.....A..........AT.T..A.A.....-...T..AA..A--.....G
RhagPomo
CeraCapi
EuleFrat
                                                                                       400
AnopGamb AAGCTATAAAAATTTTATAATAAACTAGGATTAGATACCCTATTATTAAAATTAAATATATAAGAATACTTAAGTAGTAT A12C
DrosYaku ...T..A....T...A....A........................T...--.TG..A....-.T.G..A........A
RhagPomo ..A...TT..C...A.G.......
CeraCapi                                                         ..A.A.TA.GC.....G.......
EuleFrat                                                         ..AT.T.A..C.T.A.G......A
                                                                                       480
AnopGamb TAGTTATATTCTTAAAATTTAAAGAATTTGGCGGTGTTTTAGTCTATTTAGAGGAATCTGTTCTGTAATTGATAATCCAC S12E
DrosYaku .......G.....G...C.....A............A...........CC.......C.....T......C..........
RhagPomo .......G.....G...C.....A.....??....A............C.......C.......................
CeraCapi .......G.....G...C.....A.....??....A............C.......C....C..A...............
EuleFrat .......G.....G...C.....A.....??....A............C.......C.................-.....
                                                                                       560
AnopGamb GTTGGACCTCACTTAATTTTGTTTT-CAATTTGTATATCGCCGTCATCAGAATATATTATAAGATTAATAATTTTCTTGA
DrosYaku .A.......T......A.....AA.-..G...A....C..T...T..........T........A........A...AAT.
RhagPomo .A..T....T......A........T..G...A....C..T...T..A.......T........A............CAT.
CeraCapi .A..T...........A........T..G...A....C..T...T..........T....T...A............AAT.
EuleFrat .A..T...........A........T..G...A....C..T...T..........T....T...A.............GT.
                                                                                       640
AnopGamb TATTTCATTAAATAATATGTCAGGTCAAGGTGCAGTTTATGGTTAAGTAGAAATGGATTACAATAAATATATTTATACGG
DrosYaku AT...A..A...ATT...A....A........T..C....AT.......AT.....G...........T......A....
RhagPomo AT...T......ATT...A....A........T..-....AT..............G..........AT......AGT.A
CeraCapi AT..AGT...T.ATT...A....A........T..C....AT.....A...G....G...........T......A....
EuleFrat AT...T......ATT...A....A........T..C....AT..............G...........T......A.T.A
                                                                                       720
AnopGamb ATAATTTTTTGAAATAAAAATTTGAAGGTGGATTTAATAGTAATATAAAATAGATTATTTATATGATTATAGCTCTAAAA
DrosYaku ....AA..A.....A.-.TT...............GG.......A..T.T.A......A.A..T.....T..........
RhagPomo ..TTAAA.........-TTTAA....AA.......G........A..TTT........AAA..T.....T..........
CeraCapi ..TTAA....T.....CTTTAA.............G........A.ATT..A.......AGT.T.....T..........
EuleFrat ..CTAAA...TC...G-TTTAG..TTT........G........A..TTT.A......AAGG.T.....T..........
                                                                                       800
AnopGamb CATGCACACATCGCCCGTCGCTCTCATTTTTAAAATGAGATAAGTCGTAACATAGTAGATGTACTGGAAAGTGTATCTAG S12A
DrosYaku T...T...................T...A....GG.A...........................................
RhagPomo T.......................T...C....G..A..................
CeraCapi T.................A.....T.C.A....GG.A..................
EuleFrat T.......................T...A....GG.A..................
             >t-RNA Val>                                                               880
```

```
AnopGamb AATGACAATTTAAAGCTTAATTAGTAAAGTATTTCATTTACATTGAAAAGAAATTTGTGCAAATCAATTTAAATTGA--- A16D
DrosYaku ...................T.A.......C.....................TT...............A........---
RhagPomo                                                         ............A........---
CeraCapi                                                         ............A........---
EuleFra                                                          ............A........TAA
>16S rRNA>                                                                            960
AnopGamb TAATA-ATTATTTATTAATTAT-TTTTTTTATTTATAATTATTAATAAAAATAATTTTATTTTTTATAGTTTTAGTAAT
DrosYaku .T...TT............T.AA..A......A..A..A..---..G.....C.A..AA...A.A.............T.
RhagPomo .T.ATAT............TATA..GA...TA..TAT.A.T.ATTA....TA.T.A.ATAA...-.AT............
CeraCapi .T.ATAT............TATA..GA...TA..TAT.A.T.ATTA....TA.T.A.ATAA...-.AT............
EuleFrat AT.A.AT............T.T.AA...A....-TAT.A.T.ATTA....TA.T.A.A......-.AT............
                                                                                     1040
AnopGamb TTATAATGAAATAATAATTTTAATTATAGTGTATTA-GTATTTTAAAAGAATATTGAAATAATTTGAAAAATTTTTAATT
DrosYaku G.T...A...A.............A.....A.....-.....G........A...................A.....T..
RhagPomo ..G..G.....A......A.....A.....T.....A.....G.G.....TA.....................A...T..
CeraCapi ..G..G.....A......A.....A.....T.....A.....G.G.....TA.....................A...T..
EuleFrat ..T........A.....AA..T..A.....TA....T.....G.G................A...............T..
                                                                                     1120
AnopGamb TTTAAGAAAATTTAATTTATTGTACCTTGTGTATCAGGGTTTATTAAATAATAAATTATTATAATAATTTTTCTCGAATT
DrosYaku .AA..................................C.............A...A.--...T.T............T..
RhagPomo AAA..........T..............................C......T...A.--T..T.T...AA.......T..
CeraCapi AAA..........T..............................C......T...A.--T..T.T...AA.......T..
EuleFrat AAA................................................T...A.--G..T.T...AA.......T..
                                                                                     1200
AnopGamb TTAAAGATTTAATTATATATAAAAGTTATTGTGGAATAACTATTTTAAATATGTAATTAGAAATGAAATGTTAATCGTTT A16C
DrosYaku .A.....G.....ATA...TT.......A...AC.A...T....AT.....TAT...................T......
RhagPomo .AT....G........T...T.......A...TC.A...T.....A....GA.....................T......
CeraCapi .AT....G........T...T.......A...TC.A...T.....A....GA.....................T......
EuleFrat .AT....G.....A..T...G.......A...TC.A...T..........AA.T...................T......
                                                                                     1280
AnopGamb TAAAATATATCTAGTTTTTTAAGAAATGAATTTAATTTAG--CTTATTTATTTATTAAGTTAATTTTTTAATTTAATAA
DrosYaku .T...GG...................A............AAA....-.A.A......T.A..T..A........A.....
RhagPomo AT...GG.............T.....A............CAAG..T..A.....T.GTTTA.TTA..AAA-..AA..A..
CeraCapi AT...GG.............T.....A............CAAG..T..A.....T.GTTTA.TTA..AAA-..AA..A..
EuleFrat AT...GG.............T.....A...A........CAAA..T..A.....TCATTTAATT.A..AA-G.GA..A..
                                                                                     1360
AnopGamb TTAAATAAAGTAATATTTTAAGGGATGAGCTTTAAAATAAAATTTTATATTTTTTATAATTTTTAAATAAATATAAGCTT S16B
DrosYaku ..T.TA.TTT..........T.....A....A.........T.....A.AA.AA.A.TAAA..T...........T....
RhagPomo ...TTA.TTTG...............A..............T.A...A.AA.GAGTA.T...AAT.TA....G..T.AA.
CeraCapi ...TTA.TTTG...............A..............T.A...A.AA.GAGTA.T...AAT.TA....G..T.AA.
EuleFrat ...TT..TTTG...............A..............T.A...A.AA..AATA.T.A..ATTTA....G..T.TG.
                                                                                     1440
AnopGamb AAAAATAGCTATTATTAATAAATTTGTTATAATTTATTTTTTATAAAA-AATTATTTAATTTAAATTAAATTATTTATTA
DrosYaku .G..T....A........A....G................A..A.TT.-........-..A..T..T.............
RhagPomo .G..T.TT.A..C...TG..............A.......A..ATTTT-T........C.A.T.T..T..G.T.....A.
CeraCapi .G..T.TT.A..C...TG..............A.......A..ATTTT-T........C.A.T.T..T..G.T.....A.
EuleFrat .G..T....A..C...T..G.G..........A......A..TA.TTTGT........T.AAT.T..T....T.....A.
                                                                                     1520
AnopGamb AAATTTAAATTTTAATAATAA-AAATTTAGTAATTATGATAAAATTAGTATATAAATTTATATAAAGTAATTAATTT---
DrosYaku ....A.T..........T.T.-....A......A.......G............TA..G.TA...TA.....TT.A.---
RhagPomo ......TTTCA.A...TT.TT-.TGAAA.T...A...................TTTA.AT.G..C.TAT.A.TT.A.TT-
CeraCapi ......TTTCA.A...TT.TT-.TGAAA.T...A...................TTTA.AT.G..C.TAT.A.TT.A.TT-
EuleFrat ....A.TTT.........A.TT.T.AAA.AA..A...................T-T..AT.G.TT.TAT.A.TTAA.TAT
                                                                                     1600
AnopGamb GATAGGTTTTAATGAAGAATTCGGCAAATTAAATATATTCACCTGTTTAACAAAAACATGTCTTTTTGTATTTTATTTAA A16F
DrosYaku ..A.A....A...A.............--..TA..G...G............................A...A...A...
RhagPomo ....A....AT..A.....C.........T.T-...C..G............................AG..A.T.....
CeraCapi ....A....AT..A.....C.........T.T-...C..G............................AG..A.T.....
EuleFrat ....AT...ATG.A...............TTT....C..G............................AG..A.T.....
                                                                                     1680
AnopGamb AGTCTAGCCTGCCCACTG---AGTTTTAAAGGGCCGCGGTATTTTGACCGTGCGAAGGTAGCATAATCAATAGTCTTTTA S16A
DrosYaku ......A...........AA-.A.......T......A..........T....A...............T..........
RhagPomo .....GA...........AAT.A....G..T......A.......A..T....A...............T..........
CeraCapi .....GA...........AAT.A....G..T......A.......A..T....A...............T..........
EuleFrat ......A...........AAA.A.......T......A.......A..T....A...............T..........
                                                                                     1760

AnopGamb ATTGAAGGCTGGTATGAATGGTTGAATGAGATATATACTGTTTTTTTAAAATTT-ATATAGAACTTTATTTTTTAGTTAA
DrosYaku ............A...........G.C..A....TA.......CA..T......A.A.....T..............C..
RhagPomo ........................G.C..AG...AT.......CA.A.......AT.G....T.....A...........
CeraCapi ........................G.C..AG...AT.......CA.A.......AT.G....T.....A...........
EuleFrat ............A.....C.....G.C..AG...AG.......CA.AT......ATA.....T.....A...........
                                                                                     1840
```

```
AnopGamb AAAGCTAAAATTTAATTAAAAGACGAGAAGACCCTATAGATCTTTAT-TTTTATAAATTATAAATTATAAAGAATTTTAA A16X
DrosYaku ............A.T.......................A........A......TT...-.T...........T.AA.TT
RhagPomo ......................................A........A...A...T.A.T.T....T.TTT..TG.A.TT
CeraCapi ......................................A........A...A...T.A.T.T....T.TTT..TG.A.TT
EuleFrat .............T........................A........A........TA.T......T..TG..TA.A.TT
                                                                                    1920
AnopGamb AATTTATATTTTAAATAAAATTTTACTGGGGTGGTATTAAAATTAAATAAACTTTTATTATTTATTTA-CATTGATTTAT
DrosYaku .....TA..AAATT.A..T......T.......A..........T..A.........A.T..A.AAA.-....A......
RhagPomo T....TA......TTA.TT.....GT.......A.G........T...G........A.TA...AAA.T....A......
CeraCapi T....TA......TTA.TT.....GT.......A.G........T...G........A.TA...AAA.T....A......
EuleFrat T....T...AC..TTA.TT.....GT.......A..........T...T........A.T..AT..A.T....A......
                                                                                    2000
AnopGamb GAATAAAAGATCCTGTTTTATGGATTAAAAATTTAAGTTACCTTAGGGATAACAGCGTAATTTTTTTAGAGAGTTCATAT S16M
DrosYaku ......TT.....AT.AA..AT.........A.........T.........................G............
RhagPomo ..T.T.TT.....GT.AA..AC.....T..GA.........T.........................G........T...
CeraCapi ..T.T.TT.....GT.AA..AC.....T..GA.........T.........................G........T...
EuleFrat ..T.T.TT.....GT.AA..AC.....T..GA.........T.........................G............
                                                                                    2080
AnopGamb CGATAAAAAAGATTGCGACCTCGATGTTGGATTAAGAGTTATTTTTAGGTGTAGAAGTTTAAAGTTTAGGTCTGTTCGAC
DrosYaku .....................................TA..A....G.......CC...C...T....A...........
RhagPomo T..........T.........................TA...............CC.C.....TA...A...........
CeraCapi T..........T.........................TA...............CC.C.....TA...A...........
EuleFrat ...........T.........................TA...............CT.C.....CA...A...........
                                                                                    2160
AnopGamb CTTTGAATTCTTACATGATCTGAGTTCAAACCGGCGTAAGCCAGGTTGGTTTCTATCTTTAATAAATTATTATATTGTAG S16R
DrosYaku T...A.............................T...........................A.......A.....T...
RhagPomo T...A..
CeraCapi T...A..
EuleFrat T...A..
                                                                    2224
AnopGamb TACGAAAGGACCTAATATAAAAAATATAATTTTATT-TAAATGAAAATTATTAA--AATAATTT
DrosYaku ............A.....T....TA..T..A..T..A..T.A...T........TAT.....AA
RhagPomo
CeraCapi
EuleFrat
```

# *Section III*

## *Subfamilies Blepharoneurinae and Phytalmiinae*

# 6 Phylogeny of the Subfamily Blepharoneurinae

*Allen L. Norrbom and Marty A. Condon*

## CONTENTS



## 6.1 INTRODUCTION

The purpose of this chapter is to analyze the phylogenetic relationships among the genera of the subfamily Blepharoneurinae. This group appears to be one of the oldest lineages within the Tephritidae (Korneyev, Chapter 4), and understanding the relationships among its genera, particularly the enigmatic genus *Ceratodacus* Hendel, may help resolve the relationships among the basal groups of Tephritidae. Of more direct importance to us, however, is the value of this phylogenetic information in the study of evolutionary questions within the group, such as how the fascinating pattern of host usage within the genus *Blepharoneura* Loew evolved.

The Blepharoneurinae are a mostly tropical group of Tephritidae that includes five genera: *Ceratodacus*, here tentatively included in the subfamily, and known from one extant and one fossil Neotropical species; *Problepharoneura*, described here from a fossil species in Dominican amber; *Blepharoneura*, known from 22 currently recognized Neotropical species, although there may be more than 200 (Condon 1994); *Baryglossa* Bezzi, known from seven Afrotropical species; and *Hexaptilona* Hering, with two known species from the Oriental and eastern Palearctic Regions (see Norrbom et al. 1999b for a full list of species and their distributions). Although the biology of the species of *Ceratodacus* and *Problepharoneura* is unknown, the species of *Blepharoneura*, *Baryglossa,* and *Hexaptilona* are known, or suspected, to breed in plants of the family Cucurbitaceae. We have numerous rearing records of *Blepharoneura* species from male or female flowers, fruit, seeds, or stems of a variety of cucurbit genera and species, although individual species are highly

**FIGURE 6.1** Scanning electron micrograph of ventral side of labella, *Blepharoneura* sp. D, showing modified pseudotracheal ring tips. (From Driscoll, C.A. and M.A. Condon, *Ann. Entomol. Soc. Am.* 87: 448–453, 1994, Figure 3a. With permission of Entomological Society of America.)

host and tissue specific. There is only a single rearing record for *Baryglossa* (*B. tersa* was reared from "flowers of a cucurbitous plant"; Munro 1957), and although *Hexaptilona* has not been reared, Amnon Freidberg (personal communication) has collected *H. hexacinioides* on a species of Cucurbitaceae in Taiwan, suggesting that this genus also breeds in cucurbits.

Observations of the fascinating biology of *Blepharoneura* (see Condon and Norrbom, Chapter 7) led to our taxonomic study of the group. The adults of *Blepharoneura, Baryglossa*, and *Hexaptilona* have rows of spinelike, modified pseudotracheal ringtips on the labella (Figures 6.1 and 2A), which are used, at least in *Blepharoneura*, for rasping leaves or flowers of their hosts (Driscoll and Condon 1994; Condon and Norrbom 1994). This is the only group of Tephritidae in which the adults are known to feed by damaging plant tissue with their mouthparts, although others will feed on sap exuding from oviposition holes or other wounds. Individual species of *Blepharoneura* are narrowly host and tissue specific, but together they attack a broad range of cucurbit species and tissues. How this host-usage pattern evolved is a major question motivating our research (Condon and Steck 1997).

Blepharoneurinae can be recognized from other Tephritidae by the presence of a single large seta just anterior to the phragma on the anepisternum (Figure 6.2C,D) (Condon and Norrbom 1994). The presence of three pairs of scutellar setae and often two or more postpronotal setae are also useful diagnostic characters for the subfamily, as is the wing pattern, which except in *Ceratodacus* and *Problepharoneura* is mostly dark with hyaline spots or incisions (Figure 6.3).

The valid family group name for the Blepharoneurinae was proposed by Korneyev (1994). The earlier name, Blepharoneuridae Wolcott (1936), proposed without a diagnosis, is a *nomen nudum* and therefore unavailable (Sabrosky 1999). The group also has been treated as a tribe within the Phytalmiinae (Norrbom et al. 1999a).

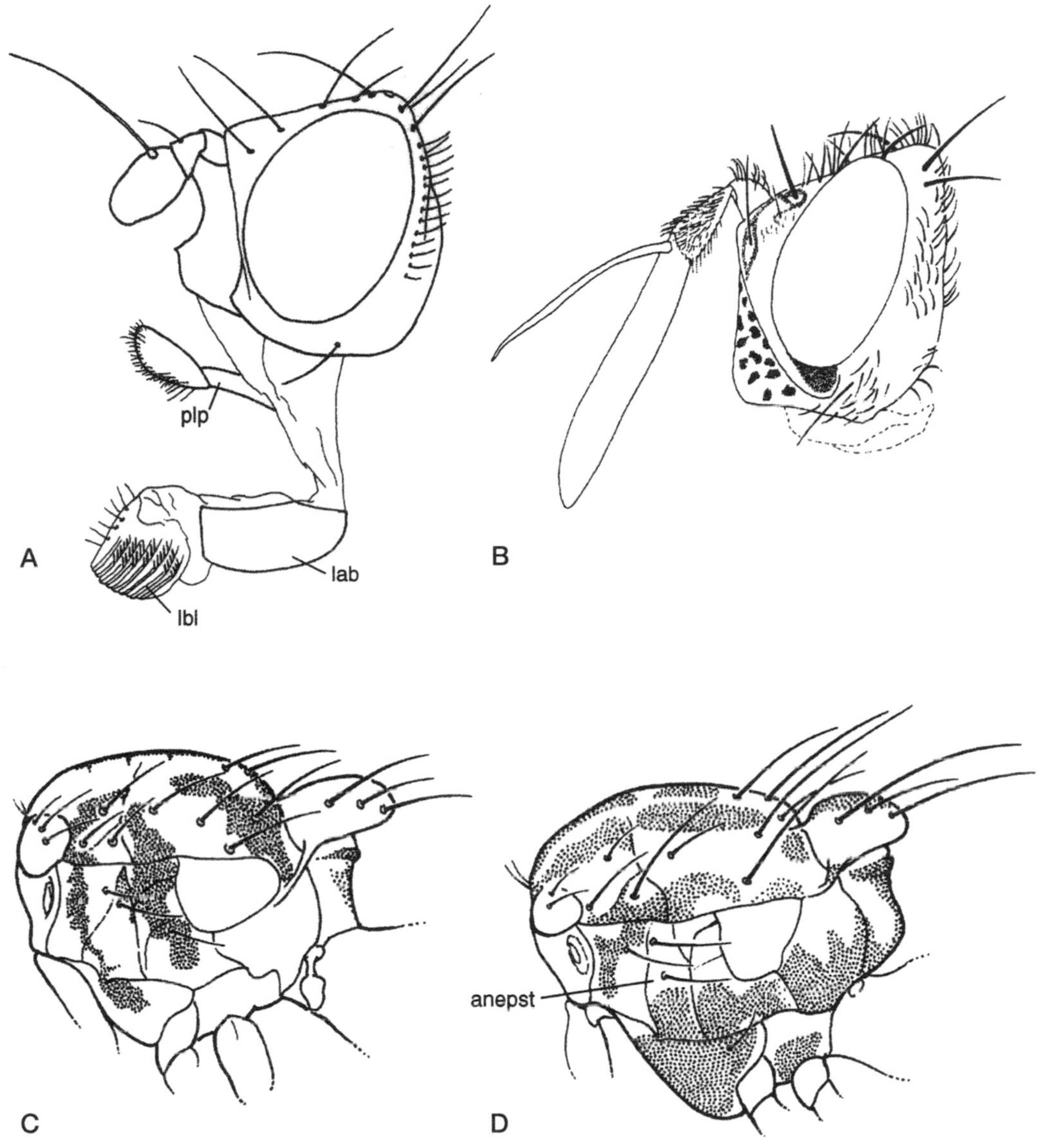


**FIGURE 6.2** (A,B) Heads, lateral view. (C,D) Thorax, lateral view (setulae not shown). (A) *Baryglossa trulla*, Bet Gherhia, Eritrea; (B) *Ceratodacus longicornis* (From Foote, R.H., *U.S. Dep. Agric. Tech. Bull.* No. 1600, 1980, Figure 19); (C) *Ceratodacus longicornis*, Pakitza, Peru; (D) *Blepharoneura femoralis*. anepst = anepisternum; lab = labium; lbl = labella; plp = palpus.

## 6.2 METHODS AND MATERIALS

Acronyms for depositories of specimens examined are as follows: BMNH = Natural History Museum, London; DEBUG = Department of Environmental Biology, University of Guelph; NMW = Naturhistorisches Museum, Wien; SANC = South African National Collection of Insects, Pretoria; USNM = National Museum of Natural History, Smithsonian Institution, Washington, D.C.; USP = Museum of Zoology, Universidade de São Paulo; ZMHU = Zoologisches Museum der Humboldt Universität. Methods used in the phylogenetic analysis are explained in Section 6.4.

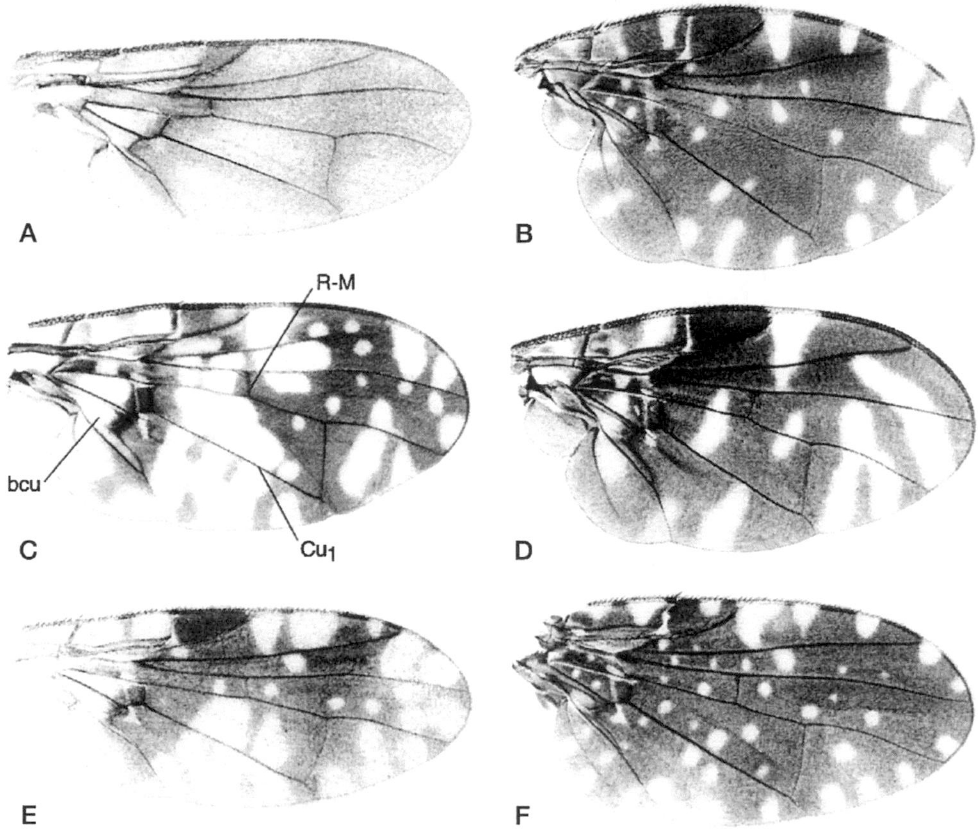


**FIGURE 6.3** Wings: (A) *Ceratodacus longicornis*; (B) *Blepharoneura parva* (*poecilosoma* group); (C) *Blepharoneura* n. sp. 43 (*femoralis* group); (D) *Blepharoneura* n. sp. 41 (*femoralis* group); (E) *Baryglossa trulla*; (F) *Hexaptilona hexacinioides*. bcu = basal cubital cell; $Cu_1$ = vein $Cu_1$; R-M = radial-medial crossvein.

The examined specimens of *Ceratodacus* and *Problepharoneura* are listed in Section 6.3. Those of *Baryglossa* and *Hexaptilona* are as follows: *B. mimella* — KENYA: Nyeri (S.), Oct 1948, van Someren, ♀ allotype (SANC); W. Ruwenzori, 8000 to 9000 ft, Jul 1945, van Someren, 1 ♂ paratype (SANC). *B. trulla* — ERITREA: Asmara, Bet Gherghis, 1 May 1950, G. De Lotto, no. 859, 1♀ (SANC). UGANDA: Ruwenzori Range, Kilembe, 4500 ft., Dec 1934 to Jan. 1935, F.W. Edwards, 2♂ paratypes (BMNH, SANC); Fort Portal, 4 Dec. 1934, F.W. Edwards, 2♀ paratypes (BMNH, SANC). *H. hexacinioides* Hering — TAIWAN: Toyenmongai bei Tainan, May 1910, Rolle 1♀ (ZMHU); Center: Wushe (Jenai), 1000 m, 45 km E Taichung, Rt. 14, 5 Oct. 1993, F. Kaplan and A. Freidberg, 1♂1♀ (USNM). *H. palpata* (Hendel) — CHINA: 23 Aug. 1957, 1♂1♀ (USNM). The examined specimens of *Blepharoneura atomaria* were listed by Condon and Norrbom (1994). Those of the other species are too numerous to list here and will be indicated in future publications.

## 6.3 TAXONOMY

### 6.3.1 Genus *Ceratodacus* Hendel

*Ceratodacus* Hendel 1914a: 81 (Type species, *longicornis* Hendel, by original designation); see Foote 1980: 23, for additional references.

*Ceratodacus* Hendel 1914b: 10 (Type species, *longicornis* Hendel, by original designation), preoccupied by Hendel 1914a.

*Diagnosis* — This genus can be recognized from all other Blepharoneurinae by the following characters which are unique within the subfamily: antenna elongate, extending at least to ventral margin of face; crossvein R-M at or proximal to basal two-fifths of cell dm and covered by subcostal crossband (the band is faint in *C. longicornis*) (Figure 6.5D); anepisternum without a large posterior seta dorsal to level of anterior seta (Figure 6.2C); and two or more postsutural supra-alar setae present.

*Description* — Body generally yellow to pale orange or brown, sometimes with dark brown markings, 4.3 to 5.75 mm long. Setae dark brown to black. Head: In lateral view (Figure 6.2B and 5A), anterior margin of face not strongly receding ventrally. Margin of parafacial receding ventrally. Postgena not bulging. Frons broad, setulose. Anterior ocellus sometimes absent. two to four frontal setae. Two orbital setae, posterior seta subequal to anterior. Ocellar seta slightly smaller than posterior orbital seta, strongly proclinate. Postocellar seta present. Paravertical seta absent (only setulae in this area). Postocular setae acuminate. Gena without brown spot below eye. Face with broadly rounded carina. Antenna with first flagellomere very long, extending to or beyond ventral margin of face. Arista short pubescent or densely pilose. Palpus not constricted medially. Labium small, not strongly convex. Labella without spinules or sclerotized ridges. Thorax: Mesonotum with following large setae: two to three postpronotal, two notopleural, one dorsocentral aligned with postalar (unknown for *C. priscus*), one intra-alar, one presutural, and two (rarely three) postsutural supra-alar, one postalar, and three scutellar. Acrostichal and intrapostalar setae absent (unknown for *C. priscus*). Anepisternum (Figure 6.5A) with one large seta just anterior to phragma; large posterior seta(e) ventral to level of anterior seta. Katepisternal seta present or absent. Anepimeral seta well developed. Greater ampulla well developed (at least in *C. longicornis*). Wing: Mostly hyaline, with brown subcostal, subapical, and anterior apical bands (Figure 6.5D) or with faint diffuse yellowish pattern with slightly darker subcostal band (Figure 6.3A). Crossvein R-M at basal one- to two-fifths of cell dm, covered by subcostal band. Cell bcu with relatively large posteroapical lobe (unknown for *C. priscus*). Dorsally, vein $R_1$ completely setulose; vein $R_{4+5}$ setulose to or beyond level of DM-Cu; vein $Cu_1$ nonsetulose. *Abdomen*: See species descriptions; the male is known only for *C. longicornis*, and the female only for *C. priscus*.

## 6.3.2 *Ceratodacus longicornis* Hendel

(Figures 6.2B, C, 6.3A, 6.4A to C)

*Ceratodacus longicornis* Hendel 1914a: 81 (lectotype ♂ (NMW), here designated, PERU); see Foote 1980: 23, for additional references.
*Ceratodacus longicornis* Hendel 1914b: 11 (holotype ♂ (NMW), PERU [unknown locality], Staudinger and Bang-Haas), preoccupied by Hendel 1914a; Hardy 1968: 111 [type data].
*Ceratodacus* sp.: Foote 1980: 23.

*Diagnosis* — This species, or at least the male which is the only known sex, is easily recognized by the following characters: anterior ocellus absent; postocular setae poorly differentiated from postgenal setulae; face yellow with numerous black spots; arista swollen, white, and densely pilose; acrostichal and katepisternal setae absent; R-M at basal one-fifth to one-quarter of cell dm; abdomen swollen, tergites not telescoped, very broad and extended to ventral side of abdomen; glans with large spinelike sclerite; and medial surstylus without prensisetae.

*Description* — Body generally yellow or pale orange-brown, with dark brown markings, 5.10 to 5.75 mm long. Setae black. Head: Shiny, without microtrichia except on antenna. In lateral view (Figure 6.2B), anterior margin of face slightly convex. Frons 1.5 times as broad as eye, with numerous black setulae laterally and extending across middle third, and usually with dark spots. Only two ocelli (anterior ocellus absent). Two (rarely three) frontal setae. Two orbital setae, posterior seta subequal to anterior. Ocellar seta moderately well developed, almost as long as posterior orbital seta, but slightly weaker, proclinate to slightly lateroclinate (in posterodorsal view, at 45 to 90° to other ocellar seta). Postocellar seta as large as ocellar seta, sometimes as large as posterior orbital seta. Postocular setae acuminate, weak and poorly differentiated from the numerous postgenal setulae. Gena with large brown spot below eye. Face with numerous large dark brown spots. Antennal grooves deep, carina strongly produced but evenly rounded. Antenna, especially first flagellomere, extremely long, extending well beyond ventral margin of face; pedicel with dorsal seam weak; arista stout, white, densely short pilose. Palpus with short microtrichia apically. Thorax: Shiny, without microtrichia except on anepisternum anteroventrally, greater ampulla, and anatergite; katatergite on posterior two-thirds densely covered with very small white setulae (or perhaps they are relatively large microtrichia). Pleura (Figure 6.2C) with two dark brown bands; one extending from just anterior to presutural supra-alar seta, across notopleuron and middle of anepisternum to katepisternum dorsally; one extending from just anterior to anterior postsutural supra-alar seta to middle of anepimeron. Scutum with triangular posteromedial dark spot, and postsutural, sublateral dark stripe or spot. Mesonotum 2.41 to 2.58 mm long, with following large setae: two (rarely three) postpronotal, two notopleural, one dorsocentral aligned with postalar (Peru ♂ with second, slightly more anterior dorsocentral), one intra-alar, one presutural, and two postsutural supra-alar (three postsuturals in one Brazil ♂), one postalar, and three scutellar. Lateral scapular seta present, medial absent. Acrostichal, intrapostalar, and katepisternal setae absent. Scutellum entirely finely setulose. Anepisternum (Figure 6.2C) with only one large posterior seta that is ventral to level of anterior seta. Greater ampulla large, produced. Legs: Tibiae, especially of hindleg, with apical half swollen. Midtibia with stout apical setae all relatively small or with ventral seta slightly larger than others; also with row of small posterior setae. Wing (Figure 6.3A): Pattern without distinct margins; mostly diffuse, faint brown, fading to hyaline in costal cell, basal medial cell, anal cells, cells $cu_1$ and m posteriorly, cell $r_1$ anteromedially, and cells $R_{2+3}$ and $R_{4+5}$ apicomedially; with slightly darker subcostal band from pterostigma to lobe of cell bcu, covering crossveins R-M and BM-Cu (this is more obvious with low or no magnification). R-M at basal one-fifth to one-quarter of cell dm. Cell bcu with moderately large posteroapical lobe 1.4 times as long as broadest width of cell. Vein $R_{4+5}$ densely setulose dorsally to beyond DM-Cu. Male abdomen: Swollen. Tergites fully extended (not telescoped), very broad and extended to ventral side of abdomen; with submedial stripe or row of dark brown spots. Hypandrium broad and flattened apically. Lateral sclerites absent. Surstyli (Figure 6.4A, B) short. Medial surstylus without prensisetae, but bilobed, and mesal lobe spinelike and projected. Subepandrial sclerite with only one bridge, ventral bridge absent. Proctiger very short, ringlike, not bilobed. Phallapodeme with arms broadly fused basally. Phallus 0.40 to 0.44 mm long, 0.17 times as long as mesonotum. Glans (Figure 6.4C) small and mostly membranous except for large, curved, spinelike lateral sclerite.

*Remarks* — Because both *Ceratodacus longicornis* Hendel (1914a) and *C. longicornis* Hendel (1914b) were indicated as "n. sp." and neither description refers to the other, technically both names are available and are homonyms. The former name was described from an unstated number of specimens, the latter from a single male, the holotype by monotypy. This specimen is in Cabinet 21, drawer 1 of the exotic Tephritidae in the NMW collection; it bears Hendel's determination label and was labeled as "type ♂" by Hardy (1968). To avoid ambiguity, the holotype of *longicornis* Hendel (1914b) is here designated as lectotype of *longicornis* Hendel (1914a). This species is known from only four collection series: the lectotype, and the specimens listed below from Brazil (reported by Foote 1980 as *Ceratodacus* sp.), Guyana, and Peru. Hendel's (1914b) description of this species was accurate except that the middle of the wing is not so hyaline as shown in his Taf. 1, Figure 6.1. The wing pattern is more extensive and diffuse.

*Specimens Examined* — BRAZIL: Mato Grosso: Jacaré, Fund. Bras. Cent., XII.1947, Sick, 4♂ (USP). GUYANA: Rupununi Dist., 200 ft, Kurupukari, W. side Essequibo R., 1° rain forest, Malaise ROM905064, 8-16 Oct. 1990, B. Hubley and L. Coote, 1♂ (DEBUG). PERU: Madre de Dios: Manu, Rio Manu, Pakitza, 250 m, 9 to 23.IX.1988, A. Freidberg, 1♂ (USNM).

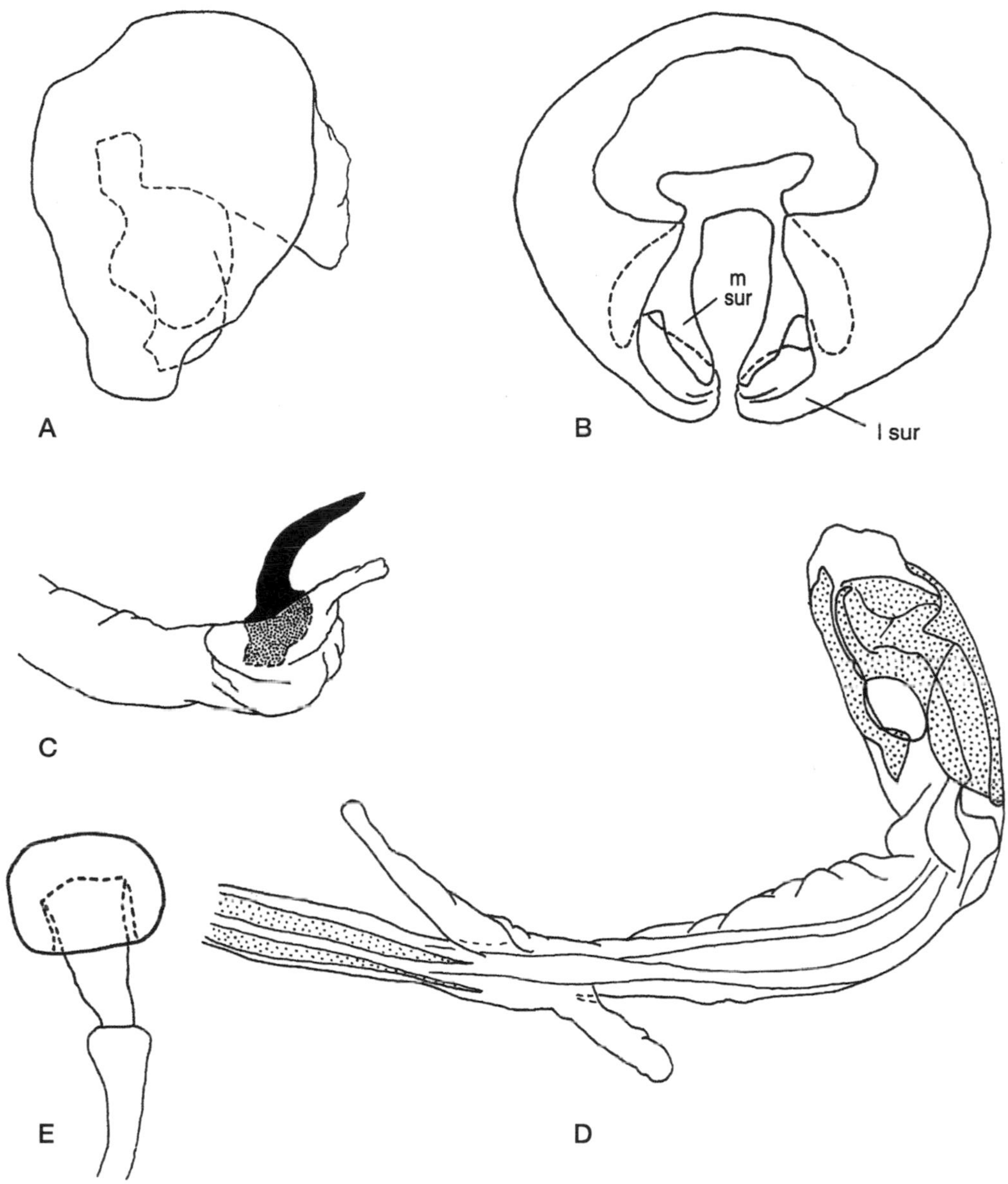


**FIGURE 6.4** Male and female genitalia: (A to C) *C. longicornis*, Pakitza, Peru; (A) epandrium, surstyli, and proctiger, lateral view (setae not shown); (B) epandrium and surstyli, posterior view (proctiger and setae not shown); (C) glans; (D) glans, *Baryglossa mimella*, paratype; (E) spermatheca (only one of three shown), *Blepharoneura femoralis*. l sur = lateral surstylus; m sur = medial surstylus.

### 6.3.3 *Ceratodacus priscus* Norrbom and Condon, n. sp.

(Figure 6.5A to D)

*Diagnosis* — This species differs from other Blepharoneurinae except *Problepharoneura antiqua* in having the wings narrowly and distinctly banded, and from all other members of the subfamily in having three postpronotal setae (*C. longicornis* rarely has three). See the diagnoses of *Problepharoneura* and *C. longicornis* for additional characters that differentiate them from this

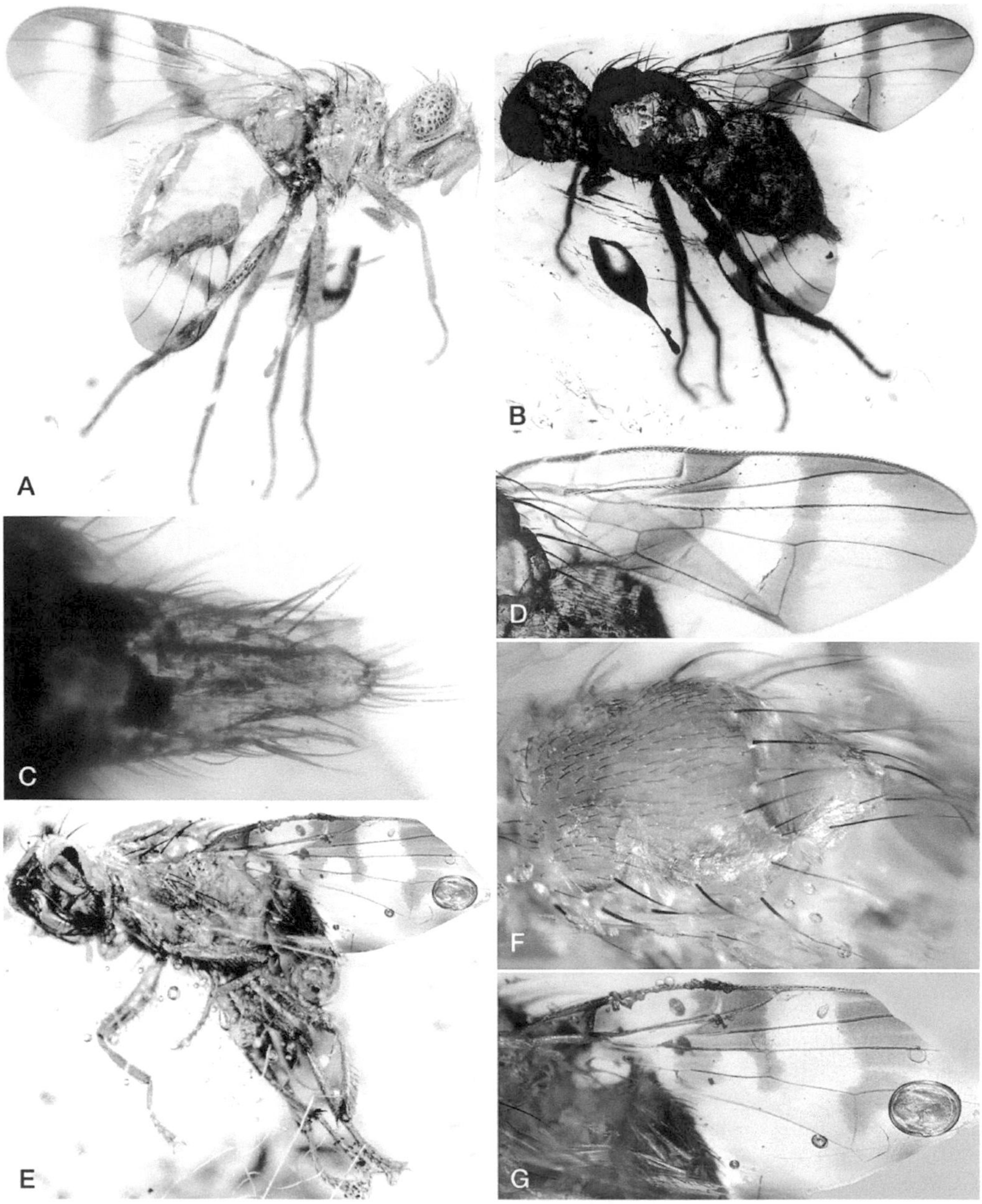


**FIGURE 6.5** (A to D) *Ceratodacus priscus*, holotype: (A) ventrolateral habitus (courtesy G. O. Poinar, Jr.); (B) dorsal habitus; (C) aculeus; (D) right wing. (E to G) *Problepharoneura antiqua*, holotype: (E) habitus; (F) mesonotum; (G) left wing.

species. It differs from other Tephritidae, except perhaps *C. longicornis* or *Problepharoneura antiqua* (both unknown from ♀), in having the aculeus tip with numerous long setulae.

*Description* — Body generally pale orange, 4.3 mm long. Setae dark brown. Head (Figure 6.5A): Frons without dark spots, broad, with row of five to six inclinate interfrontal setae slightly smaller than frontal setae and some additional setulae medially. Ocelli not clearly visible, right posterior and anterior ocelli appear present, area where left posterior ocellus woud be is damaged. Four frontal setae. Two orbital setae, posterior seta slightly smaller than anterior. Ocellar seta moderately well developed, slightly smaller than posterior orbital seta, strongly proclinate, subparallel to other ocellar seta. Postocellar seta small. Paravertical seta absent (several setulae in this area); occiput mostly bare. Postocular setae acuminate, well differentiated from the few postgenal setulae. Gena without brown spot below eye. Face without brown spots, with moderately produced, broadly rounded carina. Antenna with first flagellomere very long, extending to ventral margin of face. Arista short pubescent. Thorax: Entirely pale orange. Mesonotum 1.73 mm long, with following large setae: three postpronotal, two notopleural, one intra-alar, one presutural and two postsutural supra-alar, one postalar, and three scutellar (dorsocentral, acrostichal, and scapular areas missing). Anepisternum (Figure 6.5A) with one large and one smaller posterior seta ventral to level of anterior seta. Katepisternal seta well developed. Greater ampulla obscured. Legs: Tibiae not swollen apically. Midtibia with group of stout apical setae, one ventral seta distinctly larger; also with several posterior setae near midlength. Wing (Figure 6.5D): Mostly hyaline, with three moderate brown, transverse bands, including narrow subcostal band, from pterostigma to base of cell $cu_1$, covering crossveins R-M and BM- Cu; narrow, straight subapical band, extended from anterior wing margin midway between apices of veins $R_1$ and $R_{2+3}$, across DM-Cu to posterior wing margin, broadened slightly at posterior end; and anterior apical band, connected to subapical band anteriorly, filling all of apical part of cell $r_1$, and extended posteriorly beyond apex of vein M, becoming broad in apices of cells $R_{2+3}$ and $R_{4+5}$. Crossvein R-M at basal two-fifths of cell dm. Cell bcu not visible because of fold in wing. Vein $R_{4+5}$ dorsally with 15 setulae extended to level of DM-Cu. *Female abdomen*: Tergites slightly overlapping and not extending to ventral side of abdomen; without dark markings. Oviscape 0.4 mm long. Aculeus (Figure 6.5C) 0.34 mm long, apex bluntly rounded, with seven to eight pairs of long setulae.

*Holotype* — ♀ (Poinar collection, no. D 7-58C, at Oregon State University, for eventual deposit at California Academy of Sciences), DOMINICAN REPUBLIC: in amber piece from mines in Cordillera Septentrional, El Mamey Formation.

*Etymology* — The name of this species is a Latin adjective meaning ancient.

*Remarks* — The holotype is preserved in a piece of Dominican amber from the El Mamey Formation, which is of Upper Eocene age (Eberle et al. 1980; Poinar, personal communication), although the age of the amber is somewhat uncertain; it is usually considered to be of Miocene age (Grimaldi 1995), although estimates have ranged from 15 to 20 million years (Iturralde-Vincent and MacPhee 1996) to 30 to 45 million years (Cepek in Schlee 1990). The holotype is in good condition, except that a small part of the left side of the head, the left foreleg, most of the basal half of the left wing, and most of the mesonotum are missing, although the right edge of the scutum and scutellum are present so that most of the mesonotal chaetotaxy can be observed.

### 6.3.4 *Problepharoneura* Norrbom and Condon, n. gen.

*Type species* — *Problepharoneura antiqua* Norrbom and Condon, n. sp.

*Diagnosis* — See diagnosis for *P. antiqua*.

*Description* — Body generally yellow or pale brown, 4.5 mm long. Setae black. Head: Postgena not bulging. Frons broad, sparsely setulose. All three ocelli present. one frontal seta. two orbital setae, posterior seta subequal to anterior. Ocellar seta as long as posterior orbital seta, proclinate. Postocellar and paravertical setae present. Postocular setae acuminate, slender. Antenna short, not extending to ventral margin of face. Arista slender, short pubescent. Palpus not constricted medially. Labium small, not strongly convex. Thorax: Mesonotum (Figure 6.5F) with following large setae: two postpronotal (the medial one weaker), two notopleural, one dorsocentral slightly anterior to level of postalar seta, one acrostichal, two small intrapostalar, one intra-alar, one presutural and one postsutural supra-alar, one postalar, and three scutellar. Anepisternum with one large seta just anterior to phragma, and with row of setae near posterior margin, of which largest seta is dorsal to level of anterior seta. Katepisternal and anepimeral setae well developed. Greater ampulla obscured but appears well developed. Wing (Figure 6.5G): Mostly hyaline, with brown humeral, subcostal, discal, and subapical bands. Crossvein R-M at midlength of cell diameter, covered by discal band. Cell bcu with very small, acute posteroapical lobe. Dorsally, vein $R_1$ completely setulose; vein $R_{4+5}$ with at least several setulae proximal to level of R-M; vein $Cu_1$ nonsetulose. Male abdomen: Lateral surstylus short, slightly posteriorly curved. Medial surstylus with prensisetae.

*Etymology* — The name of this genus is derived from the Greek prefix *pro-* (before) plus *Blepharoneura*, in reference to its age and hypothesized relationship. Its gender is feminine.

### 6.3.5 *Problepharoneura antiqua* Norrbom and Condon, n. sp.

(Figure 6.5E to G)

*Diagnosis* — This species differs from other Blepharoneurinae except *Ceratodacus priscus* in having the wings with narrow and distinct bands. It differs from *C. priscus* in the location of crossvein R-M (at midlength of cell dm and not covered by subcostal band), the location of the largest posterior seta on the anepisternum (dorsal, not ventral, to level of anterior seta), the number of postsutural supra-alar setae (one, not two), and in wing pattern (discal band present, apical band absent).

*Description* — Body (Figure 6.5E) generally yellow or pale brown, 4.5 mm long. Setae black. Head: Anterior and lateral views of face distorted by debris and shape of amber piece. Frons without dark spots, approximately 1.5 times as broad as eye, with sparse, slender black setulae throughout. Only one well-developed frontal seta, near anterior margin of frons. Two orbital setae, posterior seta subequal to anterior. Ocellar seta moderately well developed, as long as posterior orbital seta, proclinate, subparallel to other ocellar seta. Postocellar seta almost as large as ocellar seta. Paravertical seta small. Gena large, without brown markings; with several large setae near middle. Row of four to five well-developed subvibrissal setae present. Face without brown spots, shape of carina obscured. Antenna short, extending two-thirds of distance to ventral margin of face. Labella underside not visible. Thorax: Without dark markings. Mesonotum (Figure 6.5F) 2 mm long. Legs: Tibiae not strongly swollen. Midtibia with two to three stout ventroapical setae, middle one largest. Wing (Figure 6.5G): Mostly hyaline, with four moderate brown, transverse bands, including humeral band from costal margin, covering crossvein H and extending to vein M; narrow subcostal band, from apex of cell c and pterostigma to base of cell $cu_1$, covering crossvein BM-Cu; discal band covering crossvein R-M and ending just beyond vein M in cell dm, also broadly connected to subcostal band to cover pterostigma and all of cell $R_{2+3}$ proximal to level of R-M; and irregular subapical band broadly covering apical two-thirds of marginal part of cell $r_1$, narrowing posteriorly, bending slightly proximally on vein M, then covering anterior three-quarters of crossvein DM-Cu. Wing apex hyaline, without apical band. Vein $R_{4+5}$ dorsally with at least several setulae proximal to level of R-M. Male abdomen: Tergites slightly overlapping and not extending to ventral side of abdomen; without dark markings. Genitalia partially exposed. Lateral surstylus short, slightly posteriorly curved. Medial surstylus somewhat obscured, but apparently with two prensisetae.

*Holotype* — ♂ (USNM, barcode ENT 00028651), DOMINICAN REPUBLIC: in amber.

*Etymology* — The name of this species is the Latin adjective meaning old or ancient.

*Remarks* — The holotype is in good condition. The tip of the left wing is missing, but that of the right wing, which is folded under the body, can be seen. The face is obscured by impurities and the curvature of the amber piece.

### 6.3.6 Key to the Genera of Blepharoneurinae

1. Wing reticulate, usually mostly dark, without bands or at most with bands on apical third (Figure 6.3B to F). Labium elongate and strongly convex (Figure 6.2A). Labella with rows of spinules (Figures 6.1 and 6.2A) .................. 2

- Wing mostly hyaline, with three to four narrow distinct bands (Figure 6.5D, G), or with diffuse yellowish pattern (Figure 6.3A). Labium small and not strongly convex. Labella without rows of spinules. . . . . . . . . . . . . . . . 4

2. Vein $Cu_1$ setulose dorsally. Palpus not constricted near midlength. New World, mostly Neotropical. . . . . . . . . . . . . . . . . . . . . . . . . . . . . . . . . *Blepharoneura* Loew

- Vein $Cu_1$ not setulose dorsally. Palpus constricted near midlength, often appearing two-segmented (Figure 6.2A). Old World. . . . . . . . . . . . . . . . . . . . . . . . . . . . 3

3. Facial carina keel-like, very narrow but strongly produced (Figure 6.2A). Arista bare. Dorsocentral seta aligned with or only slightly anterior to postalar seta. Glans with two membranous basal lateral lobes (Figure 6.4D). Afrotropical. . . . . . . . . . . . . . . . . . . . . . . . . . . . . . . . . . . . . . . . . . . . . . *Baryglossa* Bezzi

- Facial carina weak. Arista pubescent. Dorsocentral seta aligned approximately midway between postalar seta and postsutural supra-alar seta. Glans with a single membranous basal lateral lobe. Eastern Palearctic, Oriental. . . . . . . . . . . . . . . . . . . . . . . . . . . . . . . . . . . . . . *Hexaptilona* Hering

4. Antenna, especially first flagellomere, elongate, extending at least to ventral margin of face (Figure 6.3A, 6.5A). Crossvein R-M at or proximal to basal two-fifths of cell dm and covered by subcostal crossband (the band is faint in *C. longicornis*) (Figure 6.3A and 5D). Anepisternum without a large posterior seta dorsal to level of anterior seta (Figure 6.2C). Two or more postsutural supra-alar setae present. Neotropical, extant or Dominican amber. . . . . . . . . . . . . . . . . . . . . . . . . . . . . . . . . . . . . . *Ceratodacus* Hendel

- Antenna not elongate, not extending to ventral margin of face. Crossvein R-M near middle of cell diameter and covered by discal band, not by subcostal band (Figure 6.5G). Anepisternum with a large posterior seta dorsal to level of anterior seta (similar to Figure 6.2D). One postsutural supra-alar setae present. Dominican amber. . . . . . . *Problepharoneura*, n. gen.

## 6.4 PHYLOGENETIC RELATIONSHIPS

The genera of Blepharoneurinae sometimes have been included in the Acanthonevrini (Hardy 1977; Cogan and Munro 1980), which at one time comprised all genera, like the Blepharoneurinae, with three pairs of scutellar setae (see Hardy 1986). Hancock (1986) and Korneyev (1994) have shown that not all of these genera belong in the Acanthonevrini, which now includes some genera with only two pairs of scutellar setae. In an analysis of relationships among the higher groups of Tephritidae, Korneyev (Chapter 4) considers the Blepharoneurinae to be one of the basal lineages of Tephritidae; the exact relationships among the Blepharoneurinae, Tachiniscinae, Phytalmiinae, and the clade including the remaining Tephritidae are unresolved.

In some characters, such as the fused aculeus, the Blepharoneurinae appear most closely related to the Trypetinae–Tephritinae clade. These taxa also have the taenia, sclerotized strips at the base of the eversible membrane, broad and close to the midline, whereas in the Phytalmiinae, as in genera of the related families Ulidiidae and Platystomatidae, the taenia are narrow and widely separated (taeniae are absent in Tachiniscinae). Both characters could be related to phytophagy, however, and could have evolved together multiple times. There are incomplete sutures (perhaps the lines of fusion of the cercal unit and tergite 8) in the aculeus of many Blepharoneurinae (Figure 6.6A, C), which may indicate that the fusion of the aculeus occurred independently in the Blepharoneurinae (see Table 6.1, character #21).

The three spermathecae, at least in *Blepharoneura*, *Baryglossa,* and *Hexaptilona*, are broad, subspherical, without an apical lobe, with an elongate base, and the surface without minute denticles (Figure 6.4E). They are very similar to those of Phytalmiinae, but this resemblance was interpreted

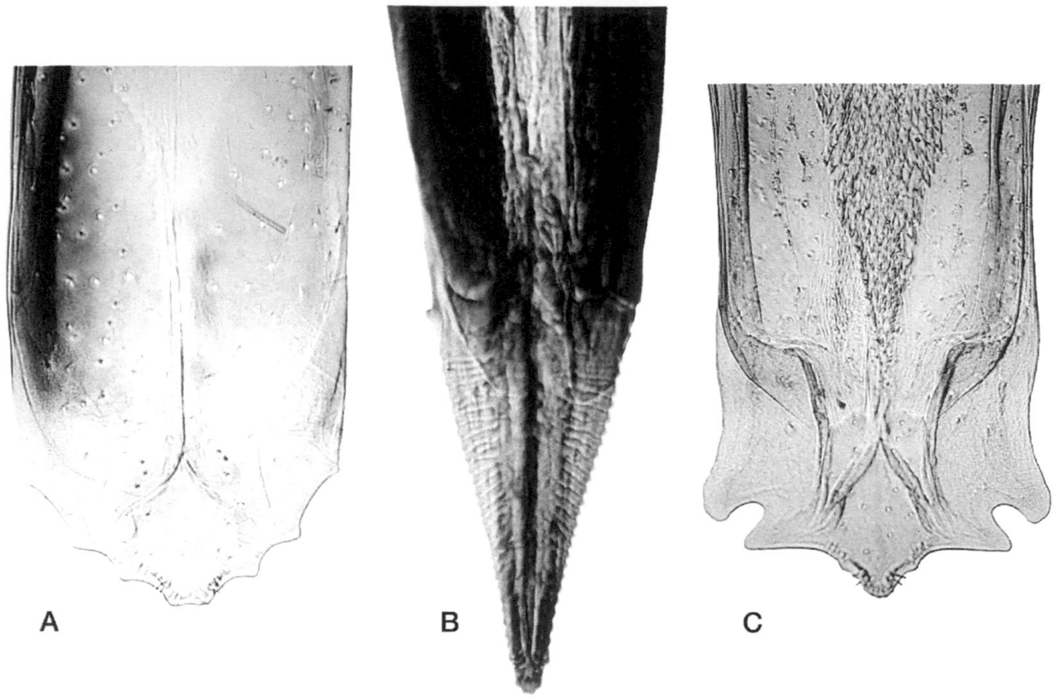


**FIGURE 6.6** Aculeus tips of *Blepharoneura*: (A) *Blepharoneura* sp. 8 (*femoralis* group); (B) *B. manchesteri* (*poecilosoma* group); (C) *Blepharoneura* sp. 9 (*femoralis* group).

as symplesiomorphy by Korneyev (1994, and this volume, Chapter 4). Tachiniscinae and most species of the Trypetinae–Tephritinae clade have denticles on the spermathecae.

Most Tachiniscinae and Blepharoneurinae have more than one postpronotal seta, which is perhaps a synapomorphy (Korneyev, Chapter 4); however, the former are generally setose flies and have additional setae in many areas. The two genera of Tachiniscinae that are less setose, *Neortalotrypeta* Norrbom and *Ischyropteron* Bigot, do not have more than one postpronotal seta (Norrbom 1994, and personal observation). We therefore do not consider this character to be strong evidence of relationship between these subfamilies.

Cladistic analyses of the Blepharoneurinae were conducted based on the 22 characters listed in Table 6.1. The character state distributions are shown in Table 6.2. Both species of *Ceratodacus* were included on the assumption that they are Tephritidae, although as noted below this needs further confirmation. The character matrix also included the single species of *Problepharoneura* and both species of *Hexaptilona*, and although only two species each of *Baryglossa* and the two species groups of *Blepharoneura* were included for simplicity, we have examined additional species that are consistent with the coding of the species in the matrix. For *Baryglossa*, these included males of *B. tersa* and examination of external characters for *B. emorsa*, *histrio*, and *oldroydi* (examined by ALN during visit to Natural History Museum, London), and for *Blepharoneura*, all other described species and many undescribed ones. Several analyses, with different taxa as the outgroup, were conducted on matrices derived from Table 6.2, all using the implicit enumeration option (ie*) of Hennig86 (Farris 1988). Because the sister group of the Blepharoneurinae is unknown, multiple analyses were run using different outgroup taxa chosen as representatives of possible sister taxa. The outgroups were the tribes Ortalotrypetini (Tachiniscinae) and Phytalmiini (Phytalmiinae), species representing three genera of Acanthonevrini (Phytalmiinae), and one species of Trypetinae. Characters were scored for the Ortalotrypetini based on examined specimens of the species listed by Norrbom (1994), and for the Phytalmiini based on examined specimens of

## TABLE 6.1
## Characters and States Used in Phylogenetic Analysis of Genera of Blepharoneurinae

1. Facial carina: 0, weak or if produced, broad and rounded; 1, keel-like, very narrow but strongly produced.
2. Frontal setae: 0, three or more; 1, 2. State 1 occurs in the Acanthonevrini and some other Trypetinae and Tephritinae, so this character appears highly subject to homoplasy and its polarity is uncertain. Only one seta is present in *Ptilona confinis* and many Phytalmiini; they were scored state 1.
3. Antenna, especially first flagellomere, elongate, extended at least to ventral margin of face: 0, no; 1, yes.
4. Labium: 0, small and/or not strongly convex; 1, elongate and strongly convex. This sclerite is relatively large in *Ptilona confinis*, but not as convex or elongate as in *Blepharoneura, Baryglossa*, and *Hexaptilona*.
5. Labella with rows of spinules (modified pseudotracheal ring tips): 0, no; 1, yes.
6. Palpus constricted near midlength, often appearing two-segmented: 0, no; 1, yes.
7. Arista: 0, pubescent or plumose; 1, bare. The Ortalotrypetini are variable for this character (Norrbom 1994). For this analysis they were coded state 0, the state in *Cyaforma* and *Ortalotrypeta*, and the more common state in other Tephritidae.
8. Postpronotal setae: 0, one; 1, two to three. *Blepharoneura, Baryglossa*, and *Hexaptilona* vary, often intraspecifically, from state 0 to 1. They are coded as state 1 for the analysis. No setae are present in *Diplochorda trineata*.
9. Dorsocentral seta: 0, aligned with or only slightly anterior to postalar seta; 1, approximately midway between postalar seta and postsutural supra-alar seta (if two postsutural supra-alars, the more posterior one). The dorsocentral seta is absent in most Phytalmiini, but is present, near the level of the postalar setae, in *Sessilina horida* (McAlpine and Schneider 1978), so this character was scored state 0 for the Phytalmiini. The position of the dorsocentral seta is variable in the Ortalotrypetini. For this analysis, they were coded state 0, the state in *Protortalotrypeta*, the most basal genus.
10. Intrapostalar seta: 0, absent; 1, present. This seta is located near the posterior margin of the scutum, slightly lateral to the dorsocentral line. State 1 occurs in some Acanthonevrini.
11. Postsutural supra-alar setae: 0, one; 1, two.
12. Number of scutellar setae: 0, two pairs; 1, three pairs. The number of pairs of setae varies in the Phytalmiini from one to two; it was scored state 0.
13. Anepisternum with a large seta just anterior to phragma: 0, no; 1, yes.
14. Anepisternal row of posterior setae: 0, dorsal seta present; 1, dorsal seta absent. The Phytalmiini was coded state 0; all of the setae are absent in *Phytalmia*, but the dorsal seta is present in *Diplochorda*.
15. Vein $Cu_1$ dorsally setulose: 0, no; 1, yes. State 1 occurs in several genera of Acanthonevrini.
16. Crossvein R-M: 0, near or apical to middle of cell dm, wing pattern variable, but R-M not in subcostal band; 1, at or basal to basal two-fifths of cell dm, in subcostal crossband.
17. Wing pattern: 0, mostly dark or reticulate; 1, mostly hyaline with narrow transverse bands. The Phytalmiini, which have wasp mimicry patterns, were tentatively coded state 1.
18. Lateral surstylus length: 0, medium length to long; 1, relatively short.
19. Glans: 0, without membranous basal lateral lobe; 1, with one membranous basal lateral lobe; 2, with two membranous basal lateral lobes. The polarity of this character is uncertain because the presence of basal lobes is highly variable in other Tephritidae. In the Blepharoneurinae, the lobes do not have minute spicules as do similar lobes, which are possibly homologous, in most Dacina, Ceratitidina, and Toxotrypanini. Few other Tephritidae have been reported to have basal lobes, although there are exceptions, e.g., three of five species of *Oedicarena* Loew, at least two species of *Rioxa* Walker (see Hardy 1986, Figure 80), and *Pseudophorellia* Lima (Norrbom, personal observation), and a careful survey for it has not been conducted.
20. Glans with small hooklike sclerite near middle of apex of sclerotized area: 0, no; 1, yes.
21. Aculeus tip: 0, not fused to tergite 8; 1, tip fused to tergite 8, but with incomplete sutures (perhaps remnants of sutures between tip and tergite 8); 2, tip completely fused to tergite 8.

**TABLE 6.1**
**Characters and States Used in Phylogenetic Analysis of Genera of Blepharoneurinae**

22 Aculeus tip: 0, with few or no internal channels running to marginal sensilla, margin entire (finely serrate in Ortalotrypetini); 1, with relatively few internal channels running to marginal sensilla, margin with several broadly spaced teeth or lobes; 2, with numerous channels running to marginal sensilla, margin with numerous fine serrations.

*Note*: State 0 is hypothesized as plesiomorphic (although for some characters this varies, depending upon the outgroup), and multistate transformation series are considered ordered.

**TABLE 6.2**
**Character State Distributions in Genera of Blepharoneurinae and Taxa Used as Outgroups**

| | Character numbers | | | | | | | | | | | | | | | | | | | | | |
|---|---|---|---|---|---|---|---|---|---|---|---|---|---|---|---|---|---|---|---|---|---|---|
| | 1 | 2 | 3 | 4 | 5 | 6 | 7 | 8 | 9 | 10 | 11 | 12 | 13 | 14 | 15 | 16 | 17 | 18 | 19 | 20 | 21 | 22 |
| Outgroups | | | | | | | | | | | | | | | | | | | | | | |
| Ortalotrypetini | 0 | 0 | 0 | 0 | 0 | 0 | 0 | 1 | 0 | 1 | 1 | 1 | 0 | 0 | 0 | 0 | 0 | 0 | 0 | 0 | 2 | 0 |
| Phytalmiini | 0 | 1 | 0 | 0 | 0 | 0 | 0 | 0 | 0 | 0 | 0 | 0 | 0 | 0 | 0 | 0 | 1 | 0 | 0 | 0 | 0 | 0 |
| Acanthonevrini | | | | | | | | | | | | | | | | | | | | | | |
| *Dirioxa pornia* | 0 | 1 | 0 | 0 | 0 | 0 | 0 | 0 | 1 | 1 | 0 | 1 | 0 | 0 | 0 | 0 | 0 | 0 | 1 | 0 | 0 | 0 |
| *Clusiosoma pleurale* | 0 | 1 | 0 | 0 | 0 | 0 | 0 | 0 | 1 | 1 | 1 | 1 | 0 | 0 | 1 | 0 | 0 | 0 | 0 | 0 | 0 | 0 |
| *Ptilona confinis* | 0 | 1 | 0 | 0 | 0 | 0 | 0 | 0 | 0 | 0 | 0 | 0 | 0 | 0 | 0 | 0 | 0 | 0 | 0 | 0 | 0 | 0 |
| Trypetinae | | | | | | | | | | | | | | | | | | | | | | |
| *Rhagoletis pomonella* | 0 | 0 | 0 | 0 | 0 | 0 | 0 | 0 | 1 | 0 | 0 | 0 | 0 | 0 | 0 | 0 | 1 | 0 | 0 | 0 | 2 | 0 |
| Ingroup | | | | | | | | | | | | | | | | | | | | | | |
| *Ceratodacus longicornis* | 0 | 1 | 1 | 0 | 0 | 0 | 0 | 1 | 0 | 0 | 1 | 1 | 1 | 1 | 0 | 1 | 1 | 1 | 0 | 0 | ? | ? |
| *C. priscus* | 0 | 0 | 1 | 0 | 0 | 0 | 0 | 1 | ? | ? | 1 | 1 | 1 | 1 | 0 | 1 | 1 | ? | ? | ? | 2 | 0 |
| *Problepharoneura antiqua* | 0 | 1 | 0 | 0 | 0 | 0 | 0 | 1 | 0 | 1 | 0 | 1 | 1 | 0 | 0 | 0 | 1 | 1 | ? | ? | ? | ? |
| *Blepharoneura rupta* | 0 | 1 | 0 | 1 | 1 | 0 | 0 | 1 | 0 | 1 | 0 | 1 | 1 | 0 | 1 | 0 | 0 | 1 | 1 | 0 | 1 | 1 |
| *B. femoralis* | 0 | 1 | 0 | 1 | 1 | 0 | 0 | 1 | 0 | 1 | 0 | 1 | 1 | 0 | 1 | 0 | 0 | 1 | 1 | 0 | 1 | 1 |
| *B. poecilosoma* | 0 | 1 | 0 | 1 | 1 | 0 | 0 | 1 | 0 | 1 | 0 | 1 | 1 | 0 | 1 | 0 | 0 | 1 | 1 | 0 | 2 | 2 |
| *B. atomaria* | 0 | 1 | 0 | 1 | 1 | 0 | 0 | 1 | 0 | 1 | 0 | 1 | 1 | 0 | 1 | 0 | 0 | 1 | 1 | 0 | 2 | 2 |
| *Hexaptilona palpata* | 0 | 1 | 0 | 1 | 1 | 1 | 0 | 1 | 1 | 1 | 0 | 1 | 1 | 0 | 0 | 0 | 0 | 1 | 1 | 0 | 1 | 1 |
| *H. hexacinioides* | 0 | 1 | 0 | 1 | 1 | 1 | 0 | 1 | 1 | 1 | 0 | 1 | 1 | 0 | 0 | 0 | 0 | 1 | 1 | 0 | 1 | 1 |
| *Baryglossa mimella* | 1 | 1 | 0 | 1 | 1 | 1 | 1 | 1 | 0 | 1 | 0 | 1 | 1 | 0 | 0 | 0 | 0 | 1 | 2 | 1 | 1 | 1 |
| *B. trulla* | 1 | 1 | 0 | 1 | 1 | 1 | 1 | 1 | 0 | 1 | 0 | 1 | 1 | 0 | 0 | 0 | 0 | 1 | 2 | 1 | 1 | 1 |

*Diplochorda trineata* and several species of *Phytalmia* and on published data for *Sessilina* (McAlpine and Schneider 1978).

Cladograms showing the possible phylogenetic relationships among the genera of Blepharoneurinae are presented in Figures 6.7 and 6.8. These two equally parsimonious trees (length 27 steps, consistency index (ci) = 0.92, retention index (ri) = 0.93) resulted with *Ptilona confinis* as the outgroup. They differ concerning the placement of *Problepharoneura*; whether it is the sister group of *Ceratodacus* or of the clade *Blepharoneura* + *Baryglossa* + *Hexaptilona*. A tree or trees matching at least one of these two trees in topology, although not always having the same length or character transformations, resulted with every other outgroup. Only one tree (length 26, ci = 0.96, ri = 0.96), with the same topology as Figure 6.7, resulted when the outgroup was the Phytalmiini, and only one tree (length 26, ci = 0.92, ri = 0.92), with the same topology as Figure 6.8, resulted when the outgroup was *Dirioxa pornia*. With *Clusiosoma pleurale*, *Rhagoletis pomonella,* or the Ortalotrypetini as outgroup, there were additional equally parsimonious trees in which the monophyly of *Blepharoneura* was unresolved or contradicted, but we consider them less likely to represent the true phylogeny (see discussion of *Blepharoneura* below). With *C. pleurale* as outgroup, there were four trees (length 28, ci = 0.85, ri = 0.87); two matched Figures 6.7 and 6.8 in topology, and the other two were similar to Figure 6.8, except with the monophyly of *Blepharoneura* unresolved or contradicted. With *Rhagoletis pomonella* as outgroup, there were four trees (length 27, ci = 0.88, ri = 0.90); one matched Figure 6.7, and the others were similar except concerning the relationships of the *Blepharoneura*. And with the Ortalotrypetini as outgroup, there were eight trees (length 25, ci = 0.88, ri = 0.90); two with the same topology as Figures 6.7 and 6.8, and the others similar to one of them except concerning the relationships of the *Blepharoneura*.

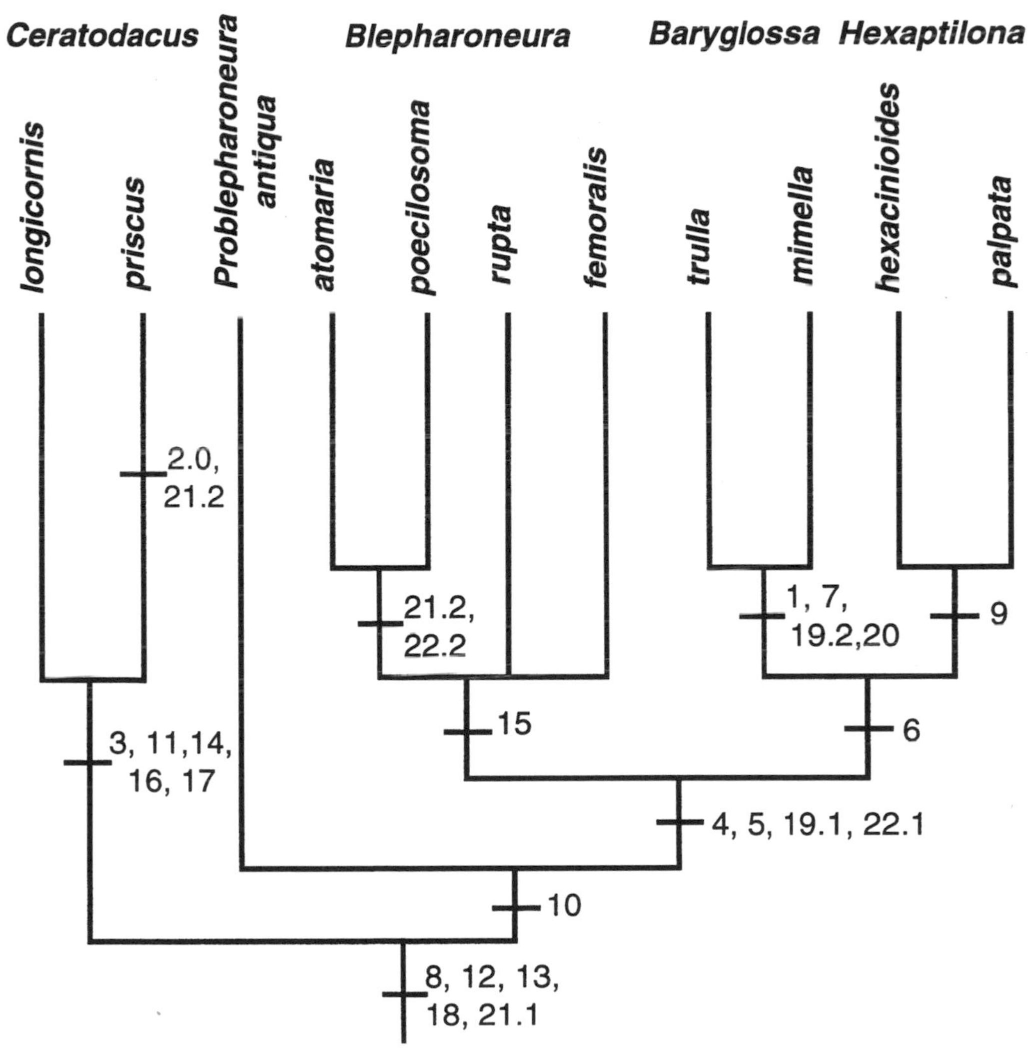


**FIGURE 6.7** Possible phylogenetic relationships of genera of Blepharoneurinae. One of two equally parsimonious trees resulting with *Ptilona confinis* as outgroup. Numbers, separated by commas, refer to characters listed in Tables 6.1 and 6.2; they represent state 1 of the character unless followed by ".0" or ".2," which indicate state 0 or 2, respectively. Alternate character evolutions are possible for characters 18, 20, 21, and 22.

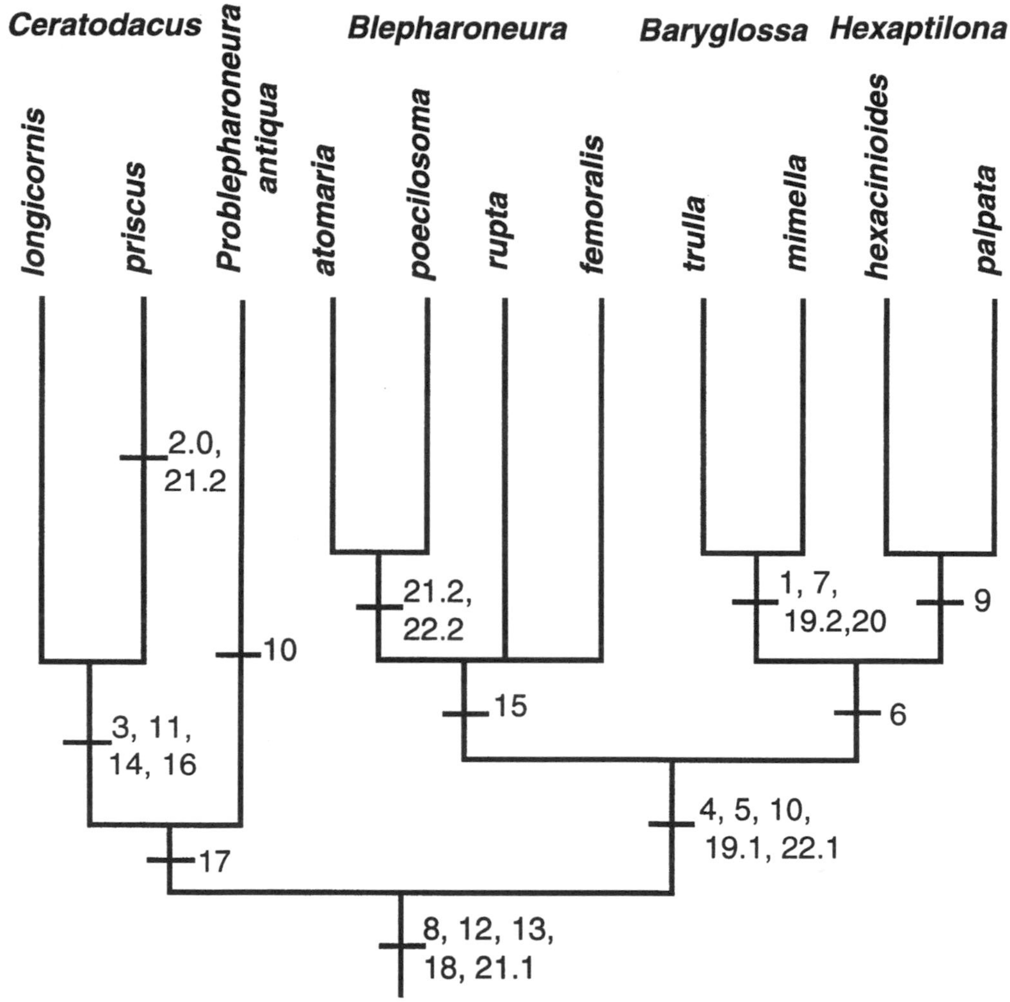


**FIGURE 6.8** Possible phylogenetic relationships of genera of Blepharoneurinae. One of two equally parsimonious trees resulting with *Ptilona confinis* as outgroup. Numbers, separated by commas, refer to characters listed in Tables 6.1 and 6.2; they represent state 1 of the character unless followed by ".0" or ".2," which indicate state 0 or 2, respectively. Alternate character evolutions are possible for characters 10 and 20.

The monophyly of the Blepharoneurinae is supported by two synapomorphies: (1) the presence of a large anepisternal seta slightly anterior to the phragma (Figure 6.2C,D) (character #13), to our knowledge, not present in Pyrgotidae, Platystomatidae, Ulidiidae, or other Tephritidae, except some Tachiniscinae, such as *Tachinsica cyaneiventris* Kertész, that are bristly in many areas and have several setae present in this area; and (2) more than one postpronotal seta present (#8); zero to one are present in Pyrgotidae, at least in taxa in the USNM, whereas in Tephritidae more than one is present only in some Tachiniscinae (Norrbom 1994; as Ortalotrypetini); this is a doubtful synapomorphy of Tachiniscinae + Blepharoneurinae (see above). The relatively short outer surstylus (#18), the presence of a third pair of scutellar setae (#12), and/or the fused aculeus tip (#21) might also be synapomorphies, but these characters vary in other Tephritidae, including among the outgroups used here, and their polarity is uncertain. Three pairs of scutellar setae occur in many Acanthonevrini and Tachiniscinae, as well as some Platystomatidae and Pyrgotidae, so this state may be plesiomorphic for the Tephritidae.

The only evidence contradicting the monophyly of the Blepharoneurinae involves *Ceratodacus*. There is little doubt that *Blepharoneura*, *Baryglossa*, and *Hexaptilona* form a monophyletic group, but the relationships of *Ceratodacus* and to a lesser extent *Problepharoneura* are less clear. Study of the relationships of the latter two genera is partly limited by lack of full genitalic data because each of the three species is known from a single sex. The only extant species, *C. longicornis*, is known only from males, whereas the fossil species *C. priscus* and *P. antiqua* are known from a single female and single male, respectively. *Problepharoneura* has mostly plesiomorphic or unknown character states compared with the other genera (see below). *Ceratodacus* appears to belong in the Tephritidae, and if so would certainly belong in the Blepharoneurinae, but it is possible that it is a primitive genus of Pyrgotidae. Part of the problem in determining its relationships is that most of the pertinent characters are equivocal and can be interpreted in different ways. Another problem is that the Pyrgotidae are not a well-defined group (Korneyev, Chapter 1).

Several characters suggest the possible relationship of *Ceratodacus* with the Pyrgotidae. (1) The prensisetae on the medial surstylus arc absent in *C. longicornis* (Figure 6.4B); they are absent in Pyrgotidae (Steyskal 1987, p. 814), except *Descoleia* (Norrbom, personal observation). (2) The aculeus of *C. priscus* (Figure 6.5C) is short and somewhat stout like in most pyrgotids, but its tip has numerous long setulae, unlike any extant Pyrgotidae or Tephritidae. *Ceratodacus* also lacks one character that may be an autapomorphy of Tephritidae: There are no enlarged costal setae at the subcostal break. Korneyev (Chapter 1) hypothesized this as an autapomorphy of Tephritidae, although noting that it has been lost in some taxa. It is also absent in the holotype of *P. antiqua*, and it is variable, often intraspecifically, in *Blepharoneura*, *Baryglossa*, and *Hexaptilona*. The anterior ocellus is absent in *C. longicornis*, and many pyrgotids lack ocelli, but Korneyev (Chapter 1) considers all three to have been present in the ground plan of Pyrgotidae.

On the other hand, *Ceratodacus* lacks some putative autapomorphies of Pyrgotidae. (1) In *C. longicornis* the proctiger is not bilobed, although it is very short and the lobes may have been reduced. The bilobed proctiger was considered a diagnosic character for Pyrgotidae by Aczél (1956: 166); in *Descoleia* it is long and not bilobed, but the ventral side has a medial longitudinal depression (Norrbom, personal observation). (2) The oviscape is very short and unmodified in *C. priscus*, whereas in most (all?) pyrgotids it is much longer than the aculeus, is curved, or bears modified setae or sclerotized structures; McAlpine (1989) considered this an autapomorphy of the family.

Furthermore, *Ceratodacus* appears to possess some autapomorphies of Tephritidae. (1) Both species of *Ceratodacus* have well-developed, inclinate frontal setae — (less differentiated in *C. priscus* (Figure 6.5A,B), but very distinct in *C. longicornis* (Figure 6.2B)). (2) The greater ampulla is well developed in *C. longicornis* (this area not visible in *C. priscus*). The species of *Ceratodacus* also share the two characters listed above as synapomorphies of the Blepharoneurinae, which also suggests their relationship with the Tephritidae rather than Pyrgotidae. The subcostal vein in both species of *Ceratodacus* has the apex sharply anteriorly turned and weak beyond the bend (Figures 6.3A and 6.5D), which has been considered an autapomorphy of Tephritidae; however, Korneyev (Chapter 1) considers this a likely synapomorphy of Tephritidae + Pyrgotidae, with some reversal in both families.

Synapomorphies supporting the monophyly of *Ceratodacus* include: (1) antenna elongate (see Table 6.1, character #3; Figures 6.2B and 6.5A); (2) dorsal posterior anepisternal seta absent (#14; Figures 6.2C and 6.5A); (3) crossvein R-M at or basal to basal two-fifths of cell dm and covered by subcostal band (#16; Figures 6.3A and 6.5D); and probably (4) two postsutural supra-alar setae (#11; the polarity of this character depends upon the outgroup; state 1 occurs in some Tachiniscinae and Acanthonevrini). The broad face and broadly rounded facial carina may be an additional synapomorphy. If *Ceratodacus* are really primitive Pyrgotidae rather than Blepharoneurinae (Tephritidae), the anterior anepisternal seta (#13) and additional postpronotal seta (#8) also would be synapomorphies of this genus that have evolved convergently in Blepharoneurinae.

*Ceratodacus longicornis* is a highly derived species with numerous autapomorphies, including anterior ocellus absent; postocular setae poorly differentiated from postgenal setulae; face with numerous black spots; arista swollen, white, and densely pilose; acrostichal and katepisternal setae absent; tibiae swollen; midtibia with all ventroapical spurlike setae relatively small; wing pattern diffuse, margins not well delimited; R-M at basal one-fifth to one-quarter of cell dm; and abdomen swollen, tergites not telescoped, very broad and extending to ventral side of abdomen. Some body markings are probably also autapomorphies. The male terminalia of *C. longicornis* (Figure 6.4A to C) include a highly derived glans and medial surstylus without prensisetae, but in other characters, such as the shape of the lateral surstylus, the subepandrial sclerite with only one strongly sclerotized bridge, the arms of the phallapodeme fused basally, the hypandrium flattened apically, and the lateral sclerites small or absent, *C. longicornis* is similar to *Blepharoneura*, *Baryglossa*, and *Hexaptilona*.

*Problepharoneura* was hypothesized in our analyses to be the sister group of either *Ceratodacus* (Figure 6.8) or of the clade including *Blepharoneura*, *Baryglossa*, and *Hexaptilona* (Figure 6.7) depending upon which outgroup was chosen, but there is not strong support for either relationship. There are few characters supporting either relationship, and all of them are relatively homoplastic (e.g., they vary among the outgroup taxa).

The hypothesis that *Problepharoneura* is more closely related to *Blepharoneura*, *Baryglossa*, and *Hexaptilona* is supported by the presence of an intrapostalar seta (character #10) in analyses with Phytalmiini, *Ptilona confinis,* or *Rhagoletis pomonella* as outgroup, or by loss of the second postsutural supra-alar seta (#11.0) in analyses with the Ortalotrypetini or *Clusiosoma pleurale* as outgroup. The reduced number of frontal setae (#2) also can be interpreted as a synapomorphy (also occurring as homoplasy in *Ceratodacus longicornis*) in analyses with the Ortalotrypetini or *R. pomonella* as the outgroup, but it is equally parsimonious that state 1 arose at the base of the tree and reversed in *C. priscus*. Several genitalic characters (#19, 21, and/or 22), which are unknown in *Problepharoneura*, also could be synapomorphies for this clade if *Problepharoneura* possesses state 1 of any of them.

Support for the hypothesis that *Problepharoneura* is more closely related to *Ceratodacus* is based mainly on the similar banded wing patterns of *P. antiqua* and *C. priscus* (#17). This requires the interpretation that the diffuse pattern of *C. longicornis* is derived from this type of banded pattern, which is not unreasonable. The other Blepharoneurinae have mostly dark or reticulate wing patterns. If the sister group of the Blepharoneuinae had a mostly dark pattern as in some of the outgroup taxa used in this analysis (e.g., *Dirioxa* or *Ptilona*), the narrow-banded pattern would be a synapomorphy for *Problepharoneura* + *Ceratodacus*.

The relationship of *Blepharoneura*, *Baryglossa*, and *Hexaptilona* was first hypothesized by Munro (1957), who noted the similarity in the shape of their proboscis (#4; Figure 6.2A) and the rows of spinulelike structures on their labella (#5; Figures 6.1 and 6.2A). The shape of the aculeus tip (#22) is probably another synapomorphy, as may be one or all of the following characters that vary among the outgroup taxa: wing pattern generally dark (#17); presence of a membranous, nonspiculose, basal lateral lobe on the glans (#19; Figure 6.4D); and a completely fused aculeus tip (#21). Depending upon what is the true sister group of Blepharoneurinae, each of these character states could be a synapomorphy of these three genera, a symplesiomorphy, or a synapomorphy with other tephritid taxa. The latter two characters, which are unknown for *Problepharoneura*, could also be synapomorphies for it and *Blepharoneura*, *Baryglossa*, and *Hexaptilona* if present in the former genus.

Among *Blepharoneura*, *Baryglossa,* and *Hexaptilona*, the latter two genera appear to be the more closely related, as indicated by their derived palpus shape (#6; Figure 6.2A). The anteriorly displaced dorsocentral seta (#9) indicates the monophyly of *Hexaptilona*, and the keel-like facial carina (#1), the bare arista (#7; Figure 6.2A), the small hooklike sclerite on the glans (#20; Figure 6.4D), and the second basal lobe on the glans (#19, state 2; Figure 6.4D) are autapomorphies of *Baryglossa* (the latter two male genitalic characters were not examined in all species of the genus).

The setulose vein $Cu_1$ (#15) is the only synapomorphy we have discovered that supports the monophyly of *Blepharoneura*. This character state is present in a few genera of Acanthonevrini (e.g., *Themara*, *Clusiosoma*, and several related genera), and if one of these taxa (e.g., *C. pleurale*) is selected as the outgroup, two trees in which *Blepharoneura*'s monophyly is uncertain result in addition to the ones with the same topology as Figures 6.7 and 6.8: they are similar to Figure 6.8 except one has the two species of the *Blepharoneura femoralis* group (*femoralis* and *rupta*) an unresolved polytomy with the *B. poecilosoma* group, and *Baryglossa* + *Hexaptilona*; and the other has *Blepharoneura* paraphyletic (i.e., with the two species of the *femoralis* group more closely related to *Baryglossa* + *Hexaptilona*). In the analyses with *Rhagoletis pomonella* or the Ortalotrypetini as the outgroup, there are also additional trees suggesting these same relationships among the taxa of *Blepharoneura*, plus another with the *femoralis* group monophyletic and the sister group of the *poecilosoma* group. Korneyev (Chapter 4) considered a setulose vein $Cu_1$ as derived within the Acanthonevrini, and although further analysis of that group is necessary to confirm the polarity of this character, it appears more likely that the setulose vein $Cu_1$ is independently derived within the Blepharoneurinae and that it does support the monophyly of *Blepharoneura*. The trees that differ from Figure 6.7 or 6.8 regarding the relationships of *Blepharoneura* are equally parsimonious with Figure 6.7 or 6.8 only by hypothesizing that the aculeus tip has incomplete sutures secondarily (i.e., character #21, state 1 arose from state 2) as a synapomorphy (e.g., for the *femoralis* group + *Baryglossa* + *Hexaptilona*). We consider it doubtful that these sutures arose secondarily and that the polarity shown in Table 6.1 is more likely, and that the trees differing from Figures 6.7 and 6.8 in topology are less likely to be the true phylogeny of the Blepharoneurinae.

Within *Blepharoneura*, the structure of the aculeus tip, with numerous internal channels and small marginal serrations (#22, state 2; Figure 6.6B), is a synapomorphy for the *poecilosoma* species group, which includes at least 60 species (Condon and Norrbom 1994). The monophyly of the other species group, the *femoralis* group, which includes at least 40 species (Condon and Norrbom 1994), is uncertain. Many, but not all, of the species have sclerotized scales on the medial membrane of the aculeus (Figure 6.6C), a character state that is unique to this group. Within both species groups, relationships are not well understood. In the *femoralis* group, we have recognized several subgroupings based upon aculeus tip shape (Norrbom and Condon, personal observation), but there is discordance between this character and certain wing pattern characters. Within the *poecilosoma* group, preliminary analyses based on isozymes (Condon and Steck 1997), morphology, and DNA sequences have produced fairly consistent results. The presence of an apical "flaglike" sclerite on the glans may be a synapomorphy grouping a large number of species, but analysis of many more species is needed.

## 6.5 CONCLUSIONS AND FUTURE RESEARCH NEEDS

Even with the inclusion of our unpublished data, the Blepharoneurinae are a group that is poorly known, both taxonomically and biologically. We believe that there remains a great diversity of species and a wealth of biological information yet to be discovered. In this chapter we have analyzed the relationships among the genera, which we hope will provide a solid framework for taxonomic and phylogenetic studies within the individual genera.

Relationships among the largest genera, *Blepharoneura*, *Baryglossa,* and *Hexaptilona*, are well resolved. *Blepharoneura*, which is predominantly Neotropical, appears to be the sister group of the clade including the other two genera, which occur in the Old World.

The relationships of *Problepharoneura* and *Ceratodacus* are less certain. *Problepharoneura*, known from a single male in Dominican amber, possesses a number of plesiomorphic character states compared with the other genera. It arises near the base of the Blepharoneurinae phylogenetic tree, either as the sister group of *Ceratodacus* or of the clade including *Blepharoneura* + *Baryglossa* + *Hexaptilona*.

*Ceratodacus*, known from a single extant species from South America (only males so far known) and another species from Dominican amber, appears to belong in the Blepharoneurinae, but also shares several characters with the Pyrgotidae. Discovery of the female and the biology of the extant species, *C. longicornis*, may help to clarify the relationships of the genus.

## ACKNOWLEDGMENTS

We are grateful to G. O. Poinar, Jr. (Oregon State University) for the opportunity to study the holotype of *Ceratodacus priscus*, for information concerning its origin and age, and for allowing us to use his photograph (Figure 6.5A). We also thank F. do Val (USP) and S. A. Marshall (DEBUG) for the loan of specimens of *C. longicornis*, M. Mansell (SANC) and N. P. Wyatt (BMNH) for loans of *Baryglossa*, and H. Schummann (ZMHU), X.-J. Wang (Academica Sinica, Bejing), and A. Freidberg for the loan or gift of specimens of *Hexaptilona*. W. N. Mathis, V. A. Korneyev, J. W. Brown, and N. E. Woodley kindly reviewed the manuscript. L. Rodriguez produced the wing images, and T. B. Griswald produced the line drawings in Figure 6.2C and D. We acknowledge the Compaña Nacional contra las Moscas de la Fruta (Mexico), International Organization for Biological Control of Noxious Animals and Plants (IOBC), Instituto de Ecologíca, A.C. (Mexico), for support of the symposium, and USDA-ICD-RSED for funding Norrbom's travel.

## REFERENCES

Aczél, M. 1956. Revisión parcial de las Pyrgotidae neotropicales y antárcticas, con sinópsis de los géneros y especies (Diptera, Acalyptratae). *Rev. Bras. Entomol.* 4: 161–184.

Cogan, B.H. and H.K. Munro. 1980. 40. Family Tephritidae, pp. 518–554. In *Catalogue of the Diptera of the Afrotropical Region* (R.W. Crosskey, ed., B.H. Cogan, P. Freeman, A.C. Pont, K.G.V. Smith, and H. Oldroyd, assist. eds.), British Museum (Natural History), London. 1437 pp.

Condon, M.A. 1994. Tom Sawyer meets insects: how biodiversity opens science to the public. *Biodiversity Lett.* 2: 159–162.

Condon, M.A. and A.L. Norrbom. 1994. Three sympatric species of *Blepharoneura* (Diptera: Tephritidae) on a single species of host *Gurania spinulosa* (Cucurbitaceae): new species and new taxonomic methods. *Syst. Entomol.* 19: 279–304.

Condon, M.A. and G.J. Steck. 1997. Evolution of host use in fruit flies of the genus *Blepharoneura* (Diptera: Tephritidae): cryptic species on sexually dimorphic host plants. *Biol. J. Linn. Soc.* 60: 443–466.

Driscoll, C.A. and M.A. Condon. 1994. Labellar modifications of *Blepharoneura* (Diptera: Tephritidae): Neotropical fruit flies that damage and feed on plant surfaces. *Ann. Entomol. Soc. Am.* 87: 448–453.

Eberle, W., W. Hirdes, R. Muff, and M. Paleaz. 1980. The geology of the Cordillera Septentrional. In *Proceedings of the 9th Caribbean Geological Conference.* August 1980. Santo Domingo, pp. 619–632.

Farris, J.S. 1988. *Hennig86. Version 1.5.* Published by the author, Port Jefferson Station.

Foote, R.H. 1980. Fruit fly genera south of the United States (Diptera: Tephritidae). *U.S. Dep. Agric. Tech. Bull.* No. 1600, 79 pp.

Grimaldi, D.A. 1995. The age of Dominican amber. In *Amber, Resinite, and Fossil Resins* (K.B. Anderson and J.C. Crelling, eds.), pp. 203–217. American Chemical Society Symposium Series, No. 617, Washington, D.C.

Hancock, D.L. 1986. Classification of the Trypetinae (Diptera: Tephritidae), with a discussion of the Afrotropical fauna. *J. Entomol. Soc. South.* Afr. 49: 275–305.

Hardy, D.E. 1968. The fruit fly types in the Naturhistorisches Museum, Wien (Tephritidae — Diptera). *Ann. Naturhist. Mus. Wien* 72: 107–155.

Hardy, D.E. 1977. Family Tephritidae (Trypetidae, Trupaneidae). In *A Catalog of the Diptera of the Oriental Region,* Vol. 3, *Suborder Cyclorrhapha (excluding Division Aschiza)* (M.D. Delfinado and D.E. Hardy, eds.), pp. 44–134. University of Hawaii Press, Honolulu. 854 pp.

Hardy, D.E. 1986. Fruit flies of the subtribe Acanthonevrina of Indonesia, New Guinea, and the Bismarck and Solomon Islands (Diptera: Tephritidae: Trypetinae: Acanthonevrini). *Pac. Insects Monogr.* 42: 1–191.

Hendel, F. 1914a. Die Gattungen der Bohrfliegen. *Wien. Entomol. Ztg.* 33: 73–98.

Hendel, F. 1914b. Die Bohrfliegen Südamerikas. *Abh. Ber. Königlich Zool. Anthropol. Ethnograph. Mus. Dresden* (1912) 14: 1–84.

Iturralde-Vincent, M.A. and R.D.E. MacPhee. 1996. Age and paleogeographical origin of Dominican amber. *Science* 273: 1850–1852.

Korneyev, V.A. 1994. Reclassification of the Palaearctic Tephritidae (Diptera), Communication 2. *Vestnik Zool.* 1994 (1): 3–17.

McAlpine, D.K. and M.A. Schneider. 1978. A systematic study of *Phytalmia* (Diptera, Tephritidae) with description of a new genus. *Syst. Entomol.* 3: 159–175.

McAlpine, J.F. 1989. Phylogeny and classification of the Muscomorpha. In *Manual of Nearctic Diptera.* Vol. 3 (J.F. McAlpine, ed.), pp. 1397–1518. Monograph of the Biosystematic Research Centre, No. 32. Agriculture Canada, Ottawa.

Munro, H.K. 1957. Trypetidae. In *Ruwenzori Expedition, 1934–35*, Vol. 2, no. 9, pp. 853–1054. British Museum of Natural History, London.

Norrbom, A.L. 1994. New genera of Tephritidae (Diptera) from Brazil and Dominican amber, with phylogenetic analysis of the tribe Ortalotrypetini. *Insecta Mundi* 8: 1–15.

Norrbom, A.L., L.E. Carroll, and A. Freidberg. 1999a. Status of knowledge. In *Fruit Fly Expert Identification System and Biosystematic Information Database* (F.C. Thompson, ed.), pp. 9–48. *Myia* (1998) 9, 524 pp.

Norrbom, A.L., L.E. Carroll, F.C. Thompson, I.M. White, and A. Freidberg. 1999b. Systematic database of names. In *Fruit Fly Expert Identification System and Biosystematic Information Database* (F.C. Thompson, ed.), pp. 65–251. *Myia* (1998) 9, 524 pp.

Sabrosky, C.W. 1999. Family-Group Names in Diptera. An Annotated Catalog. *Myia* 10, 576 pp.

Schlee, D. 1990. Das Bernstein-Kabinett. *Stutt. Beit. Naturk.*, Ser. C., 28: 1–100.

Steyskal, G.C. 1987. Pyrgotidae. In *Manual of Nearctic Diptera,* Vol. 2 (J.F. McAlpine, ed.), pp. 813–816. Monograph of the Biosystematics Research Institute, No. 28. Agriculture Canada, Ottawa.

Wolcott, G.N. 1936. Insectae Borinquenses. *J. Agric. Univ. Puerto Rico* 20: 1–600.

# 7 Behavior of Flies in the Genus *Blepharoneura* (Blepharoneurinae)

*Marty A. Condon and Allen L. Norrbom*

**CONTENTS**

## 7.1 INTRODUCTION

This chapter begins with an overview of the natural history of *Blepharoneura*, a species-rich Neotropical genus of highly host-specific species (most of which are undescribed), and its evolutionary and ecological puzzles. To highlight the intriguing problems presented by patterns of host use in *Blepharoneura*, our introduction reviews the hypothesis that host shifts from one part of a host plant to a different part of the same species of host could lead to speciation. Our chapter provides a brief overview of our current knowledge of larval host plants, as well as field and laboratory observations of behaviors of the larvae and adults. We describe wing displays in detail to provide a basis for comparison with other tephritids and other flies. After describing behaviors, we offer an outline for future research. Because *Blepharoneura* occupies a key basal position in the phylogeny of the Tephritidae, we include many recent unpublished observations of newly discovered undescribed taxa with the hope that our observations will facilitate more comparative work on tephritid behavior.

*Blepharoneura* is a member of one of the most basal clades of the Tephritidae (Condon and Norrbom 1994; Han and McPheron 1994; Korneyev, Chapter 4; Norrbom and Condon, Chapter 6). Although such a basal position alone might justify study of *Blepharoneura*, highly specific patterns of host use make *Blepharoneura* fascinating subjects for ecological and behavioral studies. All known hosts of *Blepharoneura* (as well as of other members of the Blepharoneurinae) are plants in the family Cucurbitaceae, one of the few families of flowering plants that characteristically has unisexual flowers. Many species of cucurbits are hosts to two or more sympatric species of *Blepharoneura* (Table 7.1), each of which infests a different part of the host plant (e.g., male flowers, female flowers, seeds, stems) (Condon and Norrbom 1994; Condon and Steck 1997). The repeated occurrence of such extreme specificity among morphologically similar sympatric species raises an intriguing question: Do sister species of *Blepharoneura* infest the same host? Phylogenetic analyses reveal one such relationship between species infesting a single host (*Gurania costaricensis* Cogn.): one species infests female flowers and another (sympatric) species infests male flowers (Condon and Steck 1997). How did such species diverge? To answer that question, we are studying the courtship behaviors of *Blepharoneura.*

Could host shifts from one plant part to another plant part result in divergence of populations of flies feeding on the same species of host plant? If courtship and mating take place in different locations, such divergence could take place (Bush 1969; 1994). Divergence may be particularly likely among populations of flies infesting functionally dioecious or sexually dimorphic hosts. On functionally dioecious hosts, male and female flowers rarely occur on the same individual plants at the same time, so flowers of different sexes can be as spatially (or temporally) isolated as different species of hosts (Condon and Gilbert 1988; 1990). If flies respond to visual cues associated with plant shape, morphological differences between male phase and female phase plants may further isolate "host-part-specific" populations of flies. Thus, host shifts from male parts to female parts of the same host species could interrupt gene flow between the host-part-specific populations of flies. We have collaborated with Gary Steck of the Florida Department of Agriculture, Dorothy Pumo of Hofstra University, and many students to analyze the phylogeny of *Blepharoneura* and to document courtship behaviors and patterns of host use. This chapter reports results from our ongoing field and laboratory studies.

## 7.2 DIVERSITY OF *BLEPHARONEURA*

Although Foote (1967) conservatively treated *Blepharoneura* as a genus of 13 valid species (20 described), his recognition of subtle variation in wing patterns among specimens reared from *G. spinulosa* Cogn. (Foote, unpublished data) pointed to the tremendous diversity of the genus. Condon (1994) estimated that *Blepharoneura* may include as many as 208 species, which can be divided into two main species groups: the *poecilosoma* group (~167 spp.), and the *femoralis* group

(~41 spp.). Those estimates of diversity are extrapolations from numbers of reared taxa per host taxon and are conservative. The estimates are conservative because not all potentially infested parts of plants have been collected (e.g., male flowers are common, but female flowers and fruit are rare in some collections), and levels of infestation of some host parts and species are low (e.g., stems of various species). To show how conservative our estimates are, we will give an example from records of *Blepharoneura* reared from *Gurania.*

To estimate the number of species of *Blepharoneura* that feed on *Gurania*, Condon divided the number of species of *Blepharoneura* reared from *Gurania* ($N$ = 16) by the total number of species of *Gurania* sampled ($N$ = 10), and then multiplied the number of species in the genus *Gurania* ($N$ = 40) by 1.6. The method yields an estimate of 64 species of *Blepharoneura* on *Gurania*. Those 16 species of *Blepharoneura* were reared from only 7 species of *Gurania*, from samples that do not include all potentially infested parts of the plants. One of the three sampled species of *Gurania* considered to be "nonhosts" actually is a host: recent collections in Ecuador yielded an additional species of *Blepharoneura*, and thus raise the percent species of *Gurania* that are infested. Thus, collections from some hosts do yield false negatives, which result in still lower estimates of *Blepharoneura* diversity.

Extrapolation from known hosts to potential hosts greatly underestimates diversity in the *femoralis* group (Condon 1994). Extrapolation suggests that the *femoralis* group has only 12 species; however, examination of specimens in collections reveals at least 41 morphologically distinct species (11 described). If species in this group are as cryptic morphologically as those in the *poecilosoma* group (Condon and Norrbom 1994; Condon and Steck 1997), then *Blepharoneura* may include many more than 208 species.

## 7.3 HOST SPECIFICITY

Current host records include plants in three tribes of the Cucurbitaceae (Table 7.1). Only two species of *Blepharoneura* have been reared from more than one species of plant. Most host records for *Blepharoneura* are from two functionally dioecious and sexually dimorphic genera, *Gurania* and *Psiguria*, which are known for their importance as components of the *Heliconius* butterfly community (Gilbert 1975; Condon and Gilbert 1988; 1990). Rarely are individuals of the same species of *Blepharoneura* reared from different parts of the same species of plant. For example, *B. perkinsi* Condon and Norrbom has only been reared from female flowers of *G. spinulosa*, never from male flowers ($N$ = 42). A sympatric species, *B. atomaria* (Fabricius), feeds almost exclusively on male flowers of *G. spinulosa*: of 60 specimens, 57 were reared from male flowers, and 3 from female flowers (Condon and Norrbom 1994; Condon and Steck 1997).

## 7.4 BEHAVIOR OF LARVAE

### 7.4.1 Flower Feeders

All flies reared from flowers are members of the *poecilosoma* species group and appear to be specific to flowers of a particular sex. In all cases in which flies have been reared from both male and female flowers of a particular species of cucurbit, flies reared from male flowers differ from those reared from female flowers (Table 7.1). One of the more puzzling aspects of the flower sex specificity of *Blepharoneura* is that larvae usually feed on the same tissue (calyx tissue) in both sex flowers instead of feeding on "sex-specific" parts of flowers. In male flowers of *Gurania* and *Psiguria*, larvae do not usually feed on anthers until after the flowers have fallen from the plant; however, in open male flowers of *Cucurbita pepo* L., larvae of *B. diva* Giglio-Tos can be found feeding within the anther columns. Occasionally, larvae in female flowers of *Gurania* and *Psiguria* feed on ovary tissue. Flowers with damaged ovaries invariably fall from the plant to the ground, where larvae continue to feed inside the fallen flower, just as larvae feed on fallen male flowers. Ovaries infested by larvae exude copious amounts of gelatinous sap, which may be conspicuous

**TABLE 7.1**
**Hosts and Collection Sites for Reared Specimens of 34 Species of *Blepharoneura***

| Host Taxa, Tribe and Species | Host Tissues and Collection Sites | | | |
|---|---|---|---|---|
| | Male Flower | Female Flower | Seed | Stem |
| **Tribe: Melothrieae** | | | | |
| *Psiguria racemosa* C. Jeffrey | *B.* sp. 4, Miranda, Venezuela | ? | ? | ? |
| *Anguria* (= *Psiguria*) *tabascensis* Donn. Sm. | *B.* sp. 18, Veracruz, Mexico | ? | ? | ? |
| *P. ternata* Roem. | *B.* sp. 25, Tambopata, Peru | ? | ? | ? |
| *P. triphylla* Miq. | *B.* sp. 31, Trinidad; Napo, Ecuador | ? | *B. manchesteri* Condon and Norrbom, Miranda, Venezuela | ? |
| *P. warscewiczii* Hook. | *B.* sp. 23, Sarapiqui, Costa Rica | ? | ? | ? |
| *P.* sp. A | *B.* sp. 26, Tambopata, Peru | ? | ? | ? |
| *P.* sp. B | ? | *B.* sp. 27, Tambopata, Peru | ? | ? |
| *Gurania acuminata* Cogn. | 0, Miranda, Venezuela | 0, Miranda, Venezuela | *B. manchesteri*, Miranda, Venezuela | *B. hirsuta* Bates, Miranda, Venezuela |
| *G. bignoniacea* (Poepp. and Endl.) C. Jeffrey | ? | *B.* sp. 24, Tambopata, Peru | ? | ? |
| *G. cissoides* (Benth.) Cogn. | *B.* sp. 28, Carajas, Brazil | ? | ? | ? |
| *G. costaricensis* Cogn. (<600 m elevation) | *B.* sp. 13, Sarapiqui, Costa Rica | *B.* sp. 14, Sarapiqui, Costa Rica | ? | ? |
| *G. costaricensis* (>600 m elevation) | *B.* sp. 11, Sarapiqui, Costa Rica | *B.* sp. 12, Sarapiqui, Costa Rica | ? | ? |
| *G. eggersii* Sprague and Hutchinson | 0, Rio Palenque, Ecuador | *B.* sp. 32, Rio Palenque, Ecuador | ? | ? |
| *G. eriantha* (Poepp. and Endl.) Cogn. | *B.* sp. 7, 8, Napo, Ecuador | *B.* sp. 9, Napo, Ecuador | ? | ? |
| *G. insolita* Cogn. | *B.* sp. 20, Tambopata, Peru | ? | ? | ? |
| *G. makoyana* (Lemaire) Cogn. | *B.* sp. 15, 16, Sarapiqui, Costa Rica | *B.* sp. 33, Sarapiqui, Costa Rica | ? | ? |
| *G. spinulosa* (Poepp. and Endl.) Cogn. | *B. atomaria* (Fabricius), Miranda, Venezuela; Trinidad | *B. perkinsi* Condon and Norrbom, Miranda, Venezuela; Trinidad | *B. manchesteri*, Miranda, Venezuela | 0, Miranda, Venezuela |
| *G. tubulosa* Cogn. (= *G. megistantha* Donn. Sm.) | *B.* sp. 17, Braulio Carillo, Costa Rica | ? | ? | ? |

**TABLE 7.1 (continued)**
**Hosts and Collection Sites for Reared Specimens of 34 Species of *Blepharoneura***

| Host Taxa, Tribe and Species | Host Tissues and Collection Sites | | | |
|---|---|---|---|---|
| | Male Flower | Female Flower | Seed | Stem |
| **Tribe: Cucurbiteae** | | | | |
| *Cucurbita pepo* L. | *B. diva* Giglio-Tos; various sites from Mexico, Costa Rica, Venezuela | ? | ? | ? |
| *Polyclathra cucumerina* Bertol. | *B.* sp. 22; Oaxaca, Mexico | ? | ? | ? |
| *Selysia prunifera* (Poepp. and Endl.) Cogn. | *B.* sp. 30; Miranda, Venezuela | ? | ? | ? |
| **Tribe: Sicyeae** | | | | |
| *Cyclanthera* sp. | ? | ? | *B. femoralis* Wulp, Zempoala, Mexico | Larvae found (not reared); San Gerardo de Dota, Costa Rica |
| *Microsechium helleri* (Peyr.) Cogn. | ? | ? | *B.* sp. 22, Zempoala, Mexico | ? |
| *Rytidostylis carthaginensis* (Jacq.) O. Ktze. | 0, Rio Palenque, Ecuador | ? | *B.* sp. 34, Rio Palenque, Ecuador; *B. poecilosoma* (Schiner), Guarapo, Venezuela | ? |
| *Rytidostylis gracilis* Hook. and Arn. | ? | ? | *B. poecilosoma*, Sacatepequez, Guatemala | ? |
| *Sechium* sp. | ? | ? | *B.* sp. 35, San Gerardo de Dota, Costa Rica | Larvae found (not reared); San Gerardo de Dota, Costa Rica |

Undescribed species are identified with a number. Zeros indicate plant tissues that have been sampled, but that have not been infested. Hosts that have not been sampled are designated by a "?".

to predators such as wasps and staphylinid beetles, which remove larvae from the oozing wounds (Color Figure 21C*). If larvae that feed on ovaries are exposed to increased predation, selection should favor larvae that remain inconspicuous by feeding only on calyx tissue.

In contrast to larvae in ovaries, larvae in calyces are very difficult to detect. Some pollinators, however, respond to the presence of larvae in flowers: *Heliconius* butterflies return repeatedly to uninfested male flowers of *P. warscewiczii* Hook., but after a single visit will not return to an infested flower (Murawski 1987). How (or why) larvae deter visitation by butterflies is unknown. The only visible signs of infestation are scars left on the calyx by the aculeus as it punctures the calyx tissue.

Environmental conditions that could affect larval development and vulnerability to predators also vary with the gender of flowers. Male flowers usually fall from the plant the day after anthesis, while female flowers with larvae in calyces (not ovaries) are retained on the plant throughout fruit development. Larvae fall to the ground with male flowers; however, larvae in calyces of female flowers must

* Color Figures follow p. 204.

drop from the calyx to the ground to pupariate. One exception to this pattern occurs in *G. eriantha* (Poepp. and Endl.) Cogn., a particularly "woolly" species that retains flowers of both sexes after anthesis. At least one of the three species of *Blepharoneura* known to infest flowers of *G. eriantha* in Ecuador pupariates among the older flowers on the inflorescence and does not drop to the ground.

### 7.4.2 Seed and Fruit Feeders

In contrast to flower feeders, which are all members of the *poecilosoma* group of *Blepharoneura*, members of the "seed and/or fruit feeding guild" belong to both the *femoralis* and *poecilosoma* groups of *Blepharoneura*. Only two species in the *poecilosoma* group have been reared from seeds or fruit: *B. poecilosoma* (Schiner) and *B. manchesteri* Condon and Norrbom. Little is known about *B. poecilosoma* except that it feeds on seeds of the explosive fruit of *Rytidostylis* (Table 7.1). *Blepharoneura manchesteri* infests seeds of the fleshy indehiscent fruit of *G. acuminata* Cogn., *G. spinulosa*, and *P. triphylla* Miq. Larvae of *B. manchesteri* feed on seeds and remain inside them until fruit ripen. When fruit ripen and the crisp bitter pulp surrounding seeds turns soft, juicy, and sweet, larvae burrow into the soft fruit rind where they remain until fruit are dispersed. Large bats (*Phyllostomus hastatus* (Pallas) and *P. discolor* Wagner) disperse larvae of *Blepharoneura* along with fruit and seeds of *Gurania* and *Psiguria*. Such long-distance dispersal of larvae may result in rather large effective population size of *B. manchesteri*, which might help explain why seed feeders are generalists that have diversified less than flower feeders.

Like members of the *B. poecilosoma* group, species in the *B. femoralis* group feed on either explosive (e.g., *Cyclanthera*) or indehiscent fruit (e.g., *Microsechium*). Unlike species in the *poecilosoma* group, which have been collected at lowland sites, *B. femoralis* group seed feeders have all been reared from hosts collected at high elevations (2820 m in Mexico, and 2400 m in Costa Rica). Other than host records, nothing is known about the natural histories of these species.

### 7.4.3 Stem Feeders

No members of the *poecilosoma* group are known to feed on stems, but at least one — and probably several — species in the *femoralis* group feed as stem borers on cucurbits. *Blepharoneura hirsuta* Bates infests young shoots of *G. acuminata* in the understory of very humid premontane forests at relatively low elevations (~600 m). Larvae tunnel through and destroy as much as a meter of actively growing shoots. Two other species in the *femoralis* group probably infest stems of a wild species of *Sechium*. At a high elevation (>2000 m) site in Costa Rica, we caught numerous adults of a *femoralis* group species in patches of wild *Sechium*. At the same site, Norrbom and INBio parataxonomists later found tephritid larvae in stems, one mining and the other feeding at the leaf nodes. Larvae were also found in stems of a *Cyclanthera* at that highland site.

## 7.5 BEHAVIOR OF ADULT FLIES

Whenever possible, we try to observe, record, and videotape behavior of flies in the field. Although flies' behaviors tend to be highly site specific in the field, the flies show a wide range of behaviors in the laboratory in the absence of any cues from plants. In the following we describe feeding behaviors and distribution of flies within habitats, leks, interactions, and wing displays.

### 7.5.1 Feeding

Adults of *Blepharoneura* rasp and damage plant surfaces — leaves, flowers, and fruit. Labella of species of *Blepharoneura*, as well as those of species of the related genera *Baryglossa* and *Hexaptilona*, bear robust spinules, which are bladelike pseudotracheal ring tips. The blades are braced by highly modified pseudotracheal rings. The blade-bearing rings are collapsed, displaced from the pseudotracheae, and fused to the edge of the blades (Driscoll and Condon 1994). Mouthparts of at least one species (*B. manchesteri*) are sexually dimorphic: females' labella are larger and bear a higher proportion of modified pseudotracheal rings than males (Condon et al. 1997).

Several species in the *poecilosoma* group rasp the surfaces of leaves of their host plants: on some hosts, a lacy network of tiny holes forms on the surface of the leaf as the rasped leaf expands. Such characteristic lacy damage provides indirect evidence that adults of the *femoralis* group (which also have labellar "spinules") also feed on leaves: stems of species infested by larvae of the *femoralis* group bear leaves with holes typical of those left by adult *Blepharoneura*. Not all adult *Blepharoneura* feed on leaves. Some (e.g., *B. perkinsi*) rasp the tips of pedicels from which female flowers have fallen. Others (e.g., *B. manchesteri*) feed on the surfaces of host fruit, leaving conspicuous scars on the rind of the fruit. Still others (e.g., *B. diva*) feed on the surfaces of petals of their host.

As *Blepharoneura* feed on the surfaces of plants, the flies' gut contents take on the color of the tissue the fly is abrading: abdomens of flies eating green parts of plants turn green, and abdomens of flies feeding on yellow petals turn yellow. The substances the flies are imbibing, ingesting, storing, or digesting are unknown. Flies are probably puncturing plant cells and mopping up released substances. Cucurbits contain very distinctive secondary compounds (cucurbitacins), which may be taken up by the flies; however, some species of *Blepharoneura* (e.g., *B. diva*, which feeds on petals of *Cucurbita*) feed on plant structures that can contain very low amounts of cucurbitacins (Andersen and Metcalf 1987). If adults do imbibe and sequester secondary compounds from plant tissues, the compounds do not appear to be particularly effective in defending flies against predators such as spiders. Crab spiders (Color Figure 21A*) and jumping spiders (Color Figure 21B* and D) both prey upon adults of *Blepharoneura* that feed on *G. spinulosa*.

### 7.5.2 Spatial Distribution of Adults

Our current understanding of the spatial distribution of adults is based on intensive study in northern Venezuela, several weeks of observations at two sites in Ecuador, and several days of observations made in Costa Rica. Determining whether sympatric species court and mate in the same places on a host is critical to the evaluation of the likelihood of gene flow and the evolution of reproductive isolation mechanisms. Many models of diversification in response to host shifts depend on the assumption that shifts in sites used for courtship accompany shifts in host use (Bush 1994). Evaluation of interactions among sympatric species can also help reveal ecological factors that could affect patterns of diversification. Thus, we focus on species in sympatry, and discuss behaviors observed within particular localities.

#### 7.5.2.1 Northern Venezuelan Sites

In June 1992, Condon and seven students attempted to find out if the courtship displays of three sympatric species of flies that infest *G. spinulosa* court and mate at the same or different locations — either on or off the host plant. We established a 140-m transect through roadside vegetation between Agua Blanca and Santa Crucita in Parque Nacional Guatopo, where we found and marked 45 leafy vegetative (not flowering) shoots of *G. spinulosa*. The transect did not contain female branches (either in fruit or in flower). We worked in two groups: one group surveyed "nonhost" vegetation, and another group surveyed "hosts." If a group of flies was discovered, one person stayed with the flies and recorded their behaviors while the rest of the group finished surveying the transect. The transect was checked once an hour from 0600 until 1800 on June 4 to 6, 8, and 11 to 13. Using wing patterns to identify the three *Blepharoneura* species known to use this host (Condon and Norrbom 1994), we recorded the positions, numbers, identities, and behaviors of all *Blepharoneura* we encountered on the transect.

All flies recorded in the transect were *B. atomaria*, the species specific to male flowers of *G. spinulosa*. We never observed *B. perkinsi* (which oviposit on female flowers of *G. spinulosa*) or *B. manchesteri* (which oviposit in seeds of *G. spinulosa*) in the transect; however, outside of the transect, we did observe females of *B. perkinsi* on female flowers of *G. spinulosa* and we

* Color Figures follow p. 204.

observed individuals of both sexes of *B. manchesteri* on fruit of *G. spinulosa*. Flies were observed on only three plants other than *G. spinulosa*: two of those plants were touching a branch of *G. spinulosa*, and the third was a fern. The fly found on the fern was teneral.

Over the course of the study, flies were observed on 20 of the 45 marked shoots of *G. spinulosa*. On any given day, flies were observed on no more than ten plants. During any given hour, no more than four plants had flies on them. Of 85 sightings of flies on single leaves of *G. spinulosa*, 50 were of single flies, 17 were of pairs of flies, and the remainder were groups of 3 to 9 flies. Flies were found on the transect from 0800 until 1800. Activity peaked between 1100 and 1400, when as many as ten flies were observed during an hour on the transect. Groups of flies included both males and females, usually aggregated on the lower surface of young actively expanding leaves. Both males and females feed on the surfaces of leaves. When not feeding, males actively pursued females (see description of courtship behaviors below).

#### 7.5.2.2 Ecuadorian Sites

In both eastern and western Ecuador (at Jatun Sacha Biological Station and Rio Palenque Biological Station, respectively), as in northern Venezuela, *B. atomaria* aggregate and feed under leaves of *G. spinulosa*. At both sites we found as many as five flies on a leaf (C. Thunberg and M. Condon, unpublished data). We did not find flies on nonhost species of plants in Ecuador.

#### 7.5.2.3 Costa Rican Sites

In August 1995, near Las Alturas Biological Station, we discovered numerous individuals (mainly males) of an undescribed species of *Blepharoneura* in two patches of a large leafed species of *Jessea*, possibly *cooperi* (Greenm.) H. Rob. and Cuatr. (Compositae: Senecionae). One patch had about 150 shoots of the plant and the other patch had about 25 shoots. One to three flies occupied leaves of many shoots. Because finding *Blepharoneura* on a composite was so surprising, we looked for flies under leaves of all plants within a 100 m radius of the two patches of *Jessea*. We found no other *Blepharoneura*. On the day we discovered the flies, we collected all the flies we could catch ($N > 13$). We returned to the site at the same time the next day and found a new set of flies occupying "posts" on the undersides of leaves in the same patch of *Jessea*. Again, we collected flies ($N = 19$), returned 2 days later, and found new flies occupying the leaves. We were struck by the abundance and distribution of flies in the patch of *Jessea*. This is the first observation of a consistent association of *Blepharoneura* with a noncucurbitaceous plant.

### 7.5.3 Wing Displays

Our studies of wing displays of *Blepharoneura* have focused on cryptic or taxonomically difficult groups for which behavior is a potentially valuable source of diagnostic field characters. Our descriptions of wing displays are based on field observations and on laboratory observations of flies that are at least 5 days old and are considered to be conspecific. Usually we place two or more males together with one or more females. We use that combination for two reasons: (1) some species appear to court in aggregations and may not exhibit full ranges of behaviors in the presence of only one other fly; (2) inclusion of two males permits observation of male–male interactions as well as male–female interactions. Wing displays of *Blepharoneura* vary qualitatively and quantitatively among species.

#### 7.5.3.1 Terminology

When applicable, we use the terminology of Headrick and Goeden (1994) to describe wing displays; however, the repertoire of *Blepharoneura* includes additional displays for which formal terminology has not been introduced. We use the word *display* to refer to wing positions or motions exhibited by the flies whether or not another fly is present. As such, the behaviors may not function as "displays." Some of the behaviors we have observed in *Blepharoneura* that have not been described by Headrick and Goeden (1994) do, however, include components of displays described by Headrick and Goeden (1994). We do not intend to introduce new formal terms for wing displays of *Blepharoneura* that have not been reported for other tephritids. We suggest that terminology be developed to address at least two different aspects of wing displays: position and motion. Because motion involves movement from one wing position to another, terms for specific wing positions would be particularly useful. An artificial (but effective) method might be to label positions as "Positions A, B, C" or "Positions 1, 2, 3," just as positions in classical ballet are labeled (e.g., first position, second position, etc.). By using terminology like that developed for choreography, descriptions could describe the direction and tempo of movements from position to position. Because tephritid wings can move and twist along so many axes, descriptions of the geometry of wing positions in three dimensions can be confusing — even to those of us who have observed the behaviors. To avoid such confusion and to help others visualize the behaviors of *Blepharoneura*, we use similes (e.g., "like a butterfly at rest") in our descriptions. Our goal is to develop a system of terms and methods for describing displays that will lead to a system of characters that can be interpreted in terms of homology.

### 7.5.3.2 Wing Displays of *Blepharoneura* and Other Tephritids

In the following sections, we list behaviors described by Headrick and Goeden (1994) that are shown by various species of *Blepharoneura*. We describe the context in which we observe the wing movements. We also describe additional behaviors (e.g., sidestepping) that often accompany the wing movements.

#### *7.5.3.2.1 Asynchronous Supination*

Both males and females of all species of *Blepharoneura* spend a considerable amount of time moving their wings asynchronously in a set of motions described by Headrick and Goeden (1994) as "asynchronous supination." Flies usually move one wing at a time, usually alternating wings. Wings are moved forward from a resting position and twisted from a position roughly parallel to the substrate to a position perpendicular to the substrate. While one wing is moved forward so that its long axis is roughly perpendicular to the long axis (= sagittal plane) of the body, the other wing stays at rest position. Flies routinely perform asynchronous supination in the absence of other flies. In the presence of another fly, flies will sometimes modify the rhythm with which the wings are moved. While facing another fly, a fly will hold the extended wing outstretched while taking a number of steps sideways. "Slow signal" and "wing waving," two behaviors of *Aciurina mexicana* (Aczél) described by Jenkins (1990), appear to be examples of asynchronous supination performed at either a slow pace ("slow signal") or a rather rapid tempo ("wing waving").

#### *7.5.3.2.2 Enantion (horizontal outstretch) and Synchronous Supination (vertical outstretch)*

We include both enantion and synchronous supination in the same section because both displays involve a position in which both wings are outstretched simultaneously and because *Blepharoneura* tend to exhibit them in similar situations (Figure 7.1A, positions 2 and 3; Figure 7.2B and C). Both males and females of some species of *Blepharoneura* hold both wings outstretched (with the long axis of the wing at a 90° angle to the sagittal plane of the body) simultaneously (Figure 7.1A). In this outstretched position, the surface of the wing can be horizontal in a "dragonfly-like" position, with its surface parallel to the ground (Figure 7.1A, position 2; *sensu* enantion, Headrick and Goeden 1994), or the surface of the wing can be vertical and perpendicular to the ground (Figure 7.1A, position 3; *sensu* synchronous supination, Headrick and Goeden 1994). These displays are exhibited in some species when members of the same sex approach each other in aggressive encounters (Figure 7.2B and C), but are also exhibited by males that appear to be courting females. Some species combine this wing display with a "thrusting" motion in which the fly's body jerks

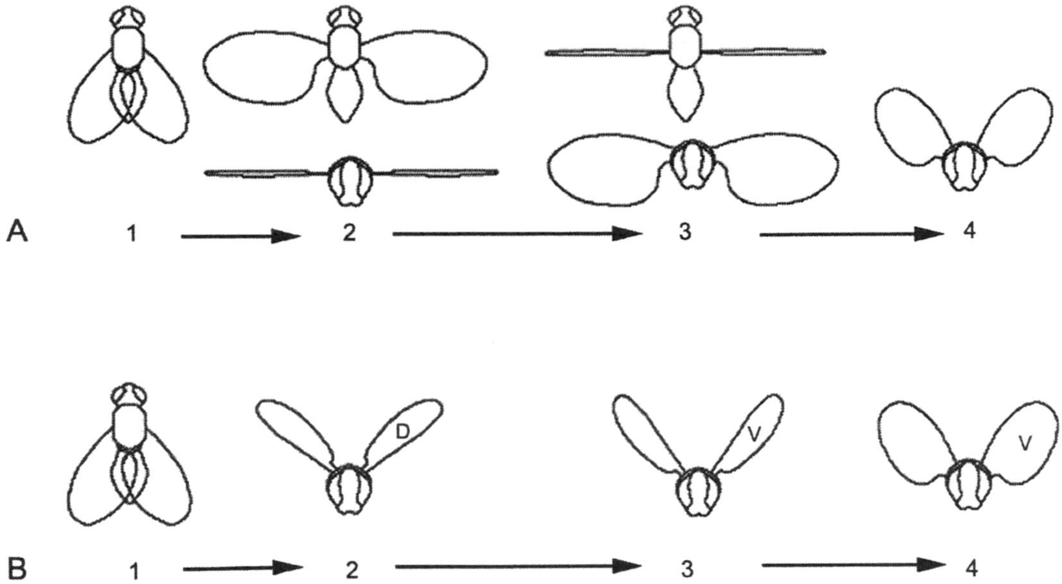


**FIGURE 7.1** Two pathways (A and B) to attaining a supinated full loft. Numbers denote different wing positions and the sequence in which they appear during a wing display. (A1) Position of fly at rest. (A2) Horizontal outstretched position (a component of enantion). Top = dorsal view of fly. Bottom = anterior view of fly. (A3) Vertical outstretched position (a component of synchronous supination). Top = dorsal view. Bottom = anterior view. (A4) Wings in a fully supinated loft in a position reminiscent of mouse ears. (B1) Position of fly at rest. (B2) Wings partially supinated and lofted. The posterior edge of the wing is proximal to the midline of the fly. (B3) Wings lofted and further supinated so that the anterior edge of the wing is proximal to the midline of the fly and the ventral surface of the wing faces forward. (B4) Wings completely supinated with the ventral (V) surface of the wing facing forward so the display achieves a position reminiscent of mouse ears.

forward (while its legs remain stationary), or with some degree of "body swaying" (as described by Headrick and Goeden 1994). Jenkins (1990) describes a vertical outstretched thrust in *A. mexicana* as a "wing thrust."

Species of *Blepharoneura* differ significantly in the frequency with which the "outstretched" positions are exhibited. Although Headrick and Goeden (1994) observed that species typically displaying enantion do not display asynchronous supination, asynchronous supination is performed by all species of *Blepharoneura*, even those that exhibit enantion.

#### *7.5.3.2.3 Hamation*

While keeping the surface of their wings parallel to the substrate, flies move their wings from side-to-side simultaneously as if the wings were both attached to the pendulum of a clock. Although Headrick and Goeden (1994) state that this is the most common display observed in species they studied, it is rarely shown by females of *Blepharoneura* and is usually seen only in species of *Blepharoneura* in which males perform a behavior we term *shiver* (see Section 5.3.3.4) Males usually perform hamation immediately before performing a "shiver" and generally show both behaviors while facing a female, usually directly in front of her, but sometimes facing her side.

### 7.5.3.3 Displays of *Blepharoneura* Not Reported for Other Tephritids

Few descriptions of tephritids' behaviors include descriptions as explicit as those of Headrick and Goeden (1994). For example, in comparing our observations with observations of *Anastrepha* Schiner, we were unable to determine exactly which wing positions and motions are involved in

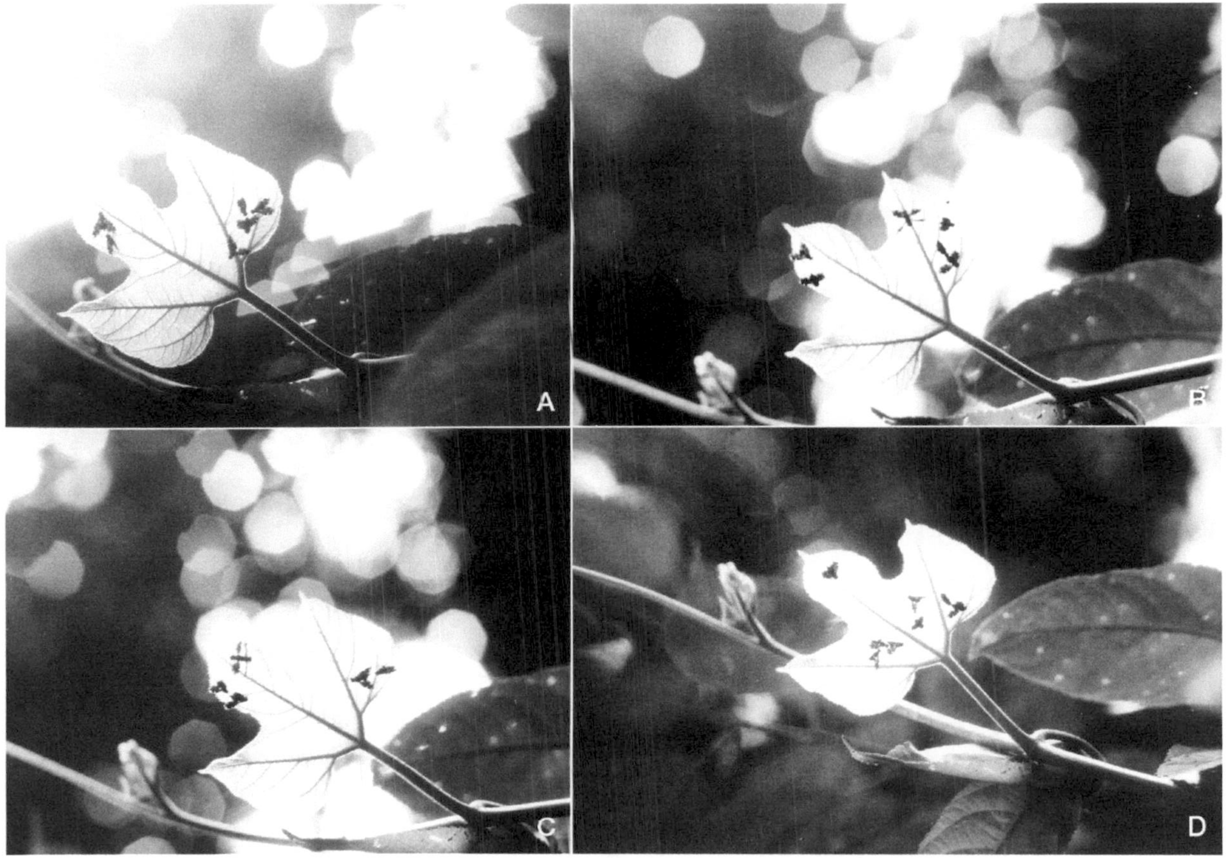


**FIGURE 7.2** Courtship arena of *B. atomaria*. The four photographs show assemblages of flies observed in sequence during 1 day on the same young leaf. (A) Six flies in arena: On the right lobe of the leaf are two males in pursuit of a female. The male lined up directly behind the female holds his wings in the "butterfly" position. Male flies behind females do not display agonistic behavior. On the middle lobe of the leaf a male with wings held aloft, but not fully aloft, pursues a female. Note the many small holes on the leaf caused by flies' feeding on the leaf. (B) Seven flies in arena: At the top of the right lobe are two males in a head-to-head display with their wings in an "outstretched" position. At the base of the right lobe is a male in "butterfly" position in line behind a female. (C) Six flies in arena: On the right lobe of the leaf, a male in "butterfly" position, with his abdomen slight curled, is still in line behind a female. On the center lobe of the leaf, two males are head-to-head with their wings outstretched, and next to them a male initiates pursuit of a female. (D) Eight flies in arena: On the right lobe, a male in "butterfly" position behind a female, touches his mouthparts to the female's oviscape. Just to the left of the center of the leaf are two males in pursuit of a female. The abdomen of the male directly behind the female is curled toward the female.

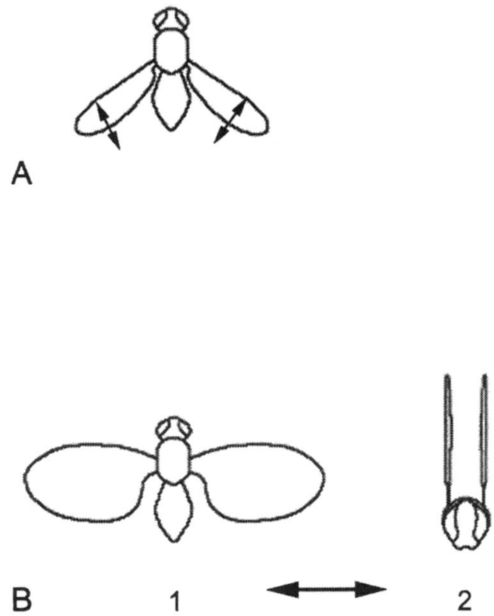


**FIGURE 7.3** Clapping (A) and flapping (B) displays of *Blepharoneura*. (A) Clapping: wings are tilted and the posterior edge is very rapidly moved downward and away from the midline and then returned at the same very rapid speed. The motion gives the impression that the dorsal edges of the wings are propelled toward each other in an aborted claplike movement. The motion is so rapid that the surfaces of the wings are blurred in frame-by-frame video playback but not during playback in real time. (B) Flapping: flies flap their wings by raising and lowering their wings repeatedly from an outstretched horizontal position to a full loft (a position that resembles the position of a butterfly at rest).

"wing fanning" and "semaphoring" (*sensu* Aluja et al., Chapter 15) or "wing buzzing" (Sivinski and Webb 1985). Thus, we do not necessarily consider the behaviors described below as autapomorphic for *Blepharoneura*. Instead, the behaviors we list in this section are behaviors we cannot match precisely with descriptions of behaviors of other flies.

#### *7.5.3.3.1 Flapping and Related Postures*

"Flapping" is a display (Figure 7.3B) that Headrick (personal communication) has not observed in the flies he has studied. Flapping is usually performed when two flies are facing each other. Flapping involves movement of wings up and down from a "horizontal outstretched" position (Figure 7.3B, position 1) to a "maximum loft" position (Figure 7.3B, position 2), and back down again, repeatedly. Wings held at maximum loft above the thorax resemble the position of the wings of a butterfly at rest: wings are held together, parallel to each other, so that the costal edge of the wing forms a 90° angle with the dorsal surface (= coronal plane) of the thorax.

Although we describe the "butterfly position" as wings held at "maximum loft," Headrick (personal communication) does not describe the up-and-down flapping motion as "lofting." Instead, Headrick and Goeden (1994) describe "lofting" as a display that involves "supinating the wings and raising them above the body in a line closely parallel with the long axis of the body." In other words, lofting begins with wings held parallel to each other in a position resembling wings of a damselfly at rest (Headrick, personal communication). Lofting occurs when the wings are held parallel to each other and are raised from the "damselfly position" to the "butterfly-at-rest" position. Headrick and Goeden (1994) note that relatively few species they studied exhibited lofting.

When in pursuit of females, males of at least one species of *Blepharoneura* (Venezuelan, but not Ecuadorian populations of *B. atomaria*) hold their wings in a "butterfly-at-rest" position without flapping (Figure 7.2B through D). The flies arrive at that position by raising their wings from a horizontal outstretched position (Figure 7.1A, position 2), not by "lofting" (*sensu* Headrick and Goeden 1994). We have never observed a species of *Blepharoneura* holding its wings in a "damselfly position."

Another display that involves raising wings above the thorax is a display that could be called "supinated full loft" (Figure 7.1, position 4). When flies exhibit this wing display their wings look like mouse ears when the fly is viewed head-on. This position is achieved through two quite different motions (Figure 7.1A and B). Some species that perform synchronous supination (vertical outstretch) can raise their extended wings to achieve the "supinated full loft" position (Figure 7.1A). The same species also can achieve the position via a different route (Figure 7.1B) by lifting their wings upward from the resting position while supinating them until the ventral surface of the wing is facing forward and the wings frame the fly's head like the ears of a mouse. The latter motion (Figure 7.1B) is similar to "lofting" (*sensu* Headrick and Goeden 1994) but without the wings being held parallel to each other. In "lofting" the wings are supinated to achieve the "damselfly" position, and then the parallel wings are lifted. In "supinated full loft," the wings are lifted and then supinated in the opposite direction (dorsal surfaces of the wings are moved away from each other instead of toward each other). Flies assuming the supinated full loft position often rock from side to side or thrust their bodies forward while keeping their legs stationary. The display is typical of male flies reared from highland populations of male flowers of *G. costaricensis*.

#### *7.5.3.3.2 Clapping*

This motion is exceedingly difficult to describe. It is not a true clap because wings never make contact with each other. Instead, the motion resembles an aborted clap: the dorsal surfaces of the wings are brought toward each other, but do not make contact. To perform this display, flies hold their wings in a tilted position (Figure 7.3A) midway between the resting position (Figure 7.1A, position 1) and a fully outstretched position (Figure 7.1A, positions 2 and 3). The wings are extended partially forward, and are tilted slightly (costal edge up and posterior edge angled downward and toward the abdomen) so that the plane of the wing forms a 45° angle relative to the substrate. To perform the display, the fly appears to move the posterior margins of the wings quickly toward each other (toward the midline of the body) and then back again, while holding the anterior margins of the wings relatively stationary (Figure 7.3A). This motion is extremely fast. When viewed in slow motion via frame-by-frame video playback, the surface of the wing is a blur; however, in real time the motion appears crisp and no blurring is visible. Recordings should be made to characterize the sound made by this motion. At least two species of *Blepharoneura* exhibit this display: *B. perkinsi* in Venezuela and the undescribed species that infests highland populations of male flowers of *G. costaricensis*. Neither species that infests the parapatric lowland populations of flowers of *G. costaricensis* exhibits this behavior.

#### *7.5.3.3.3 Wing Shivering*

This display, which is usually preceded by hamation, involves enantion, but wings are not extended to a full "outstretched" position. To produce the "wing shivering" effect, flies return horizontally extended wings rapidly to a position in which the posterior margins of the wings come very close together and may overlap. The motion is so rapid that the wings are blurred even during frame-by-frame video playback. During "wing shivering," males bob their abdomen upward so that it contacts the wings, apparently enhancing the wings' vibration. Males of several species of *Blepharoneura* engage in this behavior repeatedly when they are in the vicinity of females.

#### *7.5.3.3.4 Scissors*

In this behavior (which is a modified form of enantion, *sensu* Headrick and Goeden 1994), the wings are held with their surfaces parallel to the substrate and are repeatedly extended forward (away from each other) to about 60° from the costal edge of the wing to the longitudinal midline (with the angle measured relative to the posterior end of the fly), and then returned to resting position (like scissors opening and closing). Freidberg (1981) uses the term scissoring to describe such wing displays of the tephritid *Schistopterum moebiusi* Becker.

### 7.5.4 LEKS

We use the term *lek* to refer to aggregations in which courtship displays occur. *Blepharoneura* form two kinds of leklike aggregations. In one type of lek, individuals aggregate in a very small area. Within that area, females feed; however, their feeding need not be restricted to the resources in the area occupied by the aggregated individuals. For example, *B. atomaria* aggregate in "courtship arenas" on young leaves of the host plant where males actively court and pursue females (see Section 5.2; Figure 7.2). Although both males and females feed on the surfaces of the leaves, resources are not limited or defended: most young leaves are not occupied by flies. Males do not appear to be defending territories within the arena; however, males occasionally engage in head-to-head confrontations (Figure 7.2B and C) that usually result in one fly leaving the area. Male–male interactions in the courtship arenas are not always conspicuously aggressive. For example, more than one male can follow closely behind a female, and one can displace the other without any conspicuous signs of conflict (Figure 7.2A and D). Because both males and females actively feed on leaves that form the sites of these interactions, these courtship arenas do not fit definitions of "lek" that require that the aggregations are strictly limited to mating activities (Burk 1981; Headrick and Goeden 1994; Höglund and Alatalo 1995).

In contrast to such aggregations on leaves of host plants, the aggregations of flies found in stands of *Jessea* in Costa Rica more closely fit restrictive definitions of a lek. We observed single males holding territory-like positions on leaves. Resident males held their wings in a "mouse-ears" position (see Section 5.3.3.1) and chased other males off their leaves. Norrbom also observed a nonantagonistic display of asynchronous supination with sidestepping by a male toward a female on a leaf: the male faced the female and walked laterally to the right while extending his left wing, and then reversed the movements, walking to the left. The male repeated this display several times. The female faced the male but did nothing. The male eventually stopped the display and walked away from the female. We never observed adults rasp, penetrate, or consume leaf tissue and saw no signs of oviposition or other kinds of damage to the plants (which were not in flower). Norrbom did, however, observe both males and females with proboscises extended to the surface of *Jessea* leaves, apparently feeding on secretions on the surface of the leaf (as is common among many tephritids).

### 7.5.5 Interactions among Flies

#### 7.5.5.1 Female–Female Interactions

Female–female interactions occur less frequently than male–male, or female–male interactions. In some species, females at feeding sites will charge each other, holding their wings outstretched (either vertically or horizontally). In Ecuador, we have seen females of *B. atomaria* on male inflorescences of *G. spinulosa* chase other females off of inflorescences, but we have never seen such aggressive encounters among females ovipositing on fruit.

#### 7.5.5.2 Male–Male Interactions

Male–male interactions occur frequently in leklike aggregations (see Section 7.5.4). Some of the more common interactions involve males charging each other with their wings held outstretched. Males occasionally use their forelegs to touch each other, but such interactions do not escalate to behaviors such as the boxing or stilting bouts observed in other tephritids (Papaj 1994; Prokopy and Papaj, Chapter 10; Dodson 1997 and Chapter 8).

#### 7.5.5.3 Male–Female Interactions

In several species of *Blepharoneura*, males in pursuit of females remain in a position directly in line with a female's body and move synchronously with the female, keeping a straight line between the two. While following females, males touch their mouthparts to the tip of the female's oviscape and frequently curl their abdomens toward the female's oviscape (Figure 7.2D). Males of some species also use their legs to touch the face or ovipositor of females. Some species of *Blepharoneura* exhibit such behaviors more frequently than others. For example, males reared from lowland populations of female flowers of *G. costaricensis* touch the oviscape of females significantly more often than males of the two populations associated with male flowers of *G. costaricensis*.

Species differ conspicuously in the readiness with which they copulate. For example, five of seven pairs of Costa Rican flies reared from male flowers of *G. makoyana* (Lemaire) Cogn. copulated within the first 2 h of contact, while none of the 15 pairs of flies reared from lowland populations of male flowers of *G. costaricensis* copulated during trials lasting 2 h per pair. When pairs of the latter species were allowed to remain together for at least 6 h, two of three pairs copulated during the sixth hour of observation. Copulation usually lasts at least 20 min but less than 2 h. In the field, location of copulation relative to courtship arenas varies. For example, site of copulations appears to differ between geographically isolated populations of *B. atomaria*. In Venezuela, flies *in copula* were never observed by seven students working 12 h a day for 8 days, but in Ecuador, four pairs of flies were observed *in copula* by only five students working only 4 h a day. In Ecuador, copulation was initiated at courtship arenas (young leaves of the host plant). Pairs *in copula* sometimes moved to a different location either on or near the host plant.

We have not yet observed internal courtship or sexual selection at the level of genitalia, but the glans may be involved in elaborate internal "displays" (Eberhard 1990). The glans of many species of *Blepharoneura* bears a flaglike appendage at its tip. This flaglike structure, which is attached to the vesica (a membranous flexible area at the apex of the glans), can rotate and change its orientation. Some species have additional structures protruding from the vesica: *B. manchesteri* has a long, thin, flexible structure covered with hairlike protuberances (see Figure 4e in Condon and Norrbom 1994); and the vesica of flies reared from highland populations of female flowers of *G. costaricensis* bears a pendulous cyclindrical structure covered with chitinous papillae (see Figure 3 in Condon and Steck 1997). In some, but not all species, the inner surface of the sclerotized part of the glans bears spines (Condon and Norrbom 1994) or other conspicuous modifications of the surface (e.g., the inner surface of the glans of one species bears several wavy ridges). In all species of *Blepharoneura* we have examined, the glans has an opening or gap that may slip over a structure (as yet unidentified) within the female.

## 7.6 EVOLUTION OF BEHAVIOR

Preliminary evidence suggests that courtship behaviors may evolve more rapidly than host use in *Blepharoneura*. Preliminary results of phylogenetic analyses using mtDNA (cytb, and a region between COI and COII) suggest that fidelity to host part and host tissue has been highly conserved in *Blepharoneura*. Populations of flies infesting male flowers of *G. spinulosa* from northern Venezuela (*B. atomaria*) show less than 2% sequence divergence (COI, 249 base pairs) and no amino acid sequence divergence ($N = 62$ amino acid sequence of COI and 67 amino acid sequence of cytb) from flies reared from the same hosts in the Napo (Amazonian) region of Ecuador (Condon, Pucci, Rienzi, and Pumo, unpublished data); yet, patterns of courtship and mating differ conspicuously. In northern Venezuela, males in pursuit of females in courtship arenas on the surfaces of *G. spinulosa* leaves frequently hold their wings at maximum loft (the "butterfly" position) while in pursuit of females (Figure 7.2B through D); in contrast, males in Ecuadorian populations have never been observed with their wings aloft in the "butterfly position." In the Napo, males exhibit the "scissors" behavior, which was never observed in northern Venezuela. Venezuelan and Ecuadorian populations also differed markedly in the frequency with which pairs *in copula* were observed (see Section 7.5.5.3).

Surprisingly, behaviors of those allopatric populations of *B. atomaria* appear to be more divergent than the behaviors of sympatric populations of closely related cryptic species in Costa

Rica. Population genetic and phylogenetic analyses of allozyme data (Condon and Steck 1997) revealed a sister group formed by two lowland populations infesting flowers of *G. costaricensis*: one population infests male flowers and the other population infests female flowers. In contrast to the qualitative differences in behavior observed in allopatric conspecific populations, these sympatric Costa Rican populations show quantitative but not qualitative differences in behavior. They do, however, have qualitative morphological differences in their glans. If the sympatric species in Costa Rica are sibling species, they are particularly interesting because their use of the same host species suggests that host shifts from one plant part to another could lead to diversification. To determine the factors that limit gene flow between these populations, Condon and Cornell College students plan to return to Costa Rica to find out if these sympatric species court and mate in distinctive sites on or away from the host.

To learn more about the origins of behaviors in *Blepharoneura*, we need to learn more about other members of its clade (i.e., *Hexaptilona* and *Baryglossa*) and we need more information about groups basal to the Blepharoneurinae. Certainly, synapomorphies such as the labellar spinules suggest that some behaviors (e.g., phytophagy by adults) of *Blepharoneura* and its relatives may be unique to the Blepharoneurinae.

## 7.7 CONCLUSIONS AND FUTURE RESEARCH NEEDS

### 7.7.1 Conclusions

Much more is known about *Blepharoneura* than was known 10 years ago; however, much remains to be learned about their evolution and behavior. We continue to collect host records, rear flies, and record their behaviors. We continue to find new undescribed species. Because many of the species are so difficult to distinguish from each other on the basis of external morphology, our progress depends upon a combination of careful genetic and morphological work. Understanding the relationships between rates of morphological, behavioral, ecological, and molecular evolution depends upon continued collaboration among morphologists, ecologists, and molecular biologists.

### 7.7.2 Outline for Future Research

I. Systematics
  A. Rear specimens and document natural histories and behaviors of poorly known taxa
    1. Outgroups
      a. *Hexaptilona* (Asia)
      b. *Baryglossa* (Africa)
    2. *Blepharoneura femoralis* species group
    3. Taxa from hosts and host parts that have not yet been collected (i.e., stems and male flowers if only female flowers and fruit have been collected)
  B. Describe new taxa
  C. Phylogenetic analysis based on molecular, morphological, and behavioral data
    1. Determine homologies in elements of wing patterns, genitalia, behavior, and other taxonomically useful traits
    2. Determine how patterns of host use evolve (i.e., map host associations on phylogenies)
    3. Look for trends in diversification (e.g., are "flower clades" more diverse than "seed clades"?)

II. Biology
  A. What is the adaptive significance of adults' feeding behaviors?
    1. What do flies ingest?
    2. Do ingested compounds contribute to defense or nutrition or both?

B. Larval behavior
   1. Is there conflict (interference competition?) among larvae in a single flower?
   2. What determines whether larvae will
      a. Remain in seeds or migrate to fruit rind?
      b. Feed on the calyx or the ovary of a female flower?
C. Functional morphology
   1. Wings: What are the mechanical constraints to wing motion?
   2. Mouthparts: How do shapes and dimensions of labellar blades correspond to plant surfaces rasped?
   3. Genitalia: How does the glans work? Does it "bite"? Where does the sperm exit (from the gap at the tip of the glans or from the hole near the membranous basal lobe)?
D. Sexual selection
   1. Is sexual dimorphism (e.g., incipient evolution of "hammer-headedness") more pronounced in species that engage in male–male head-butting displays?
   2. How do courtship displays affect female choice?
E. Population structure: Are populations of flies dispersed by bats less subdivided than populations of flower-infesting flies?
F. Third trophic level interactions and host use: Does a shift from one host part to another decrease mortality due to parasitoids?

## ACKNOWLEDGMENTS

For help with allozymes and molecular analyses, we thank Gary Steck (Florida Department of Agriculture) and Dorothy Pumo (Hofstra University) and her students. For analyses of behaviors, we thank Cornell College students Bernadine Bailey, Antonio Centeno, and Cheyann Thunberg. We thank Cornell College staff and students, Beverly Garcia, Megan Kupko, Sarah Oppelt, Erica Osmundson, Heather Secrist, and David Wagenheim for help with illustrations and editing. We thank Manuel Zumbado and parataxonomists at INBio (Instituto Nacional de Biodiversidad) for their help in Costa Rica, and Javier Castrejón (UNAM) for his help in Mexico. We are very grateful for the collaboration of many students from the University of North Carolina, the University of Maryland, Hofstra University, and Cornell College. We thank Harold Robbins and Steve Smith (Smithsonian Institution), and J. Gonzalez (INBio) for providing determinations of specimens of plants. Condon was supported by a Smithsonian Postdoctoral Fellowship, the National Science Foundation (NSF HRD91-03322), Hofstra University, and Cornell College. We gratefully acknowledge partial funding for this chapter from the Campaña Nacional contra las Moscas de la Fruta (Mexico), International Organization for Biological Control of Noxious Animals and Plants (IOBC), Instituto de Ecología, A.C. (Mexico), and USDA-ICD-RSED.

## REFERENCES

Andersen, J.F. and R.L. Metcalf. 1987. Factors influencing distribution of *Diabrotica* species in blossoms of cultivated *Cucurbita* species. *J. Chem. Ecol.* 13: 681–699.

Burk, T. 1981. Signaling and sex in acalyptrate flies. *Fla. Entomol.* 64: 30–43.

Bush, G.L. 1969. Sympatric host race formation and speciation in frugivorous flies of the genus *Rhagoletis* (Diptera: Tephritidae). *Evolution* 23: 237–251.

Bush, G.L. 1994. Sympatric speciation in animals: new wine in old bottles. *Trends Ecol. Evol.* 9: 285–288.

Condon, M.A. 1994. Tom Sawyer meets insects: how biodiversity opens science to the public. *Biodiversity Lett.* 2: 159–162.

Condon, M.A. and L.E. Gilbert. 1988. Sex expression of *Gurania* and *Psiguria* (Guraniinae: Cucurbitaceae): neotropical vines that change sex. *Am. J. Bot.* 75: 875–884.

Condon, M.A. and L.E. Gilbert. 1990. Reproductive biology, demography, and natural history of neotropical vines *Gurania* and *Psiguria*. In *Biology and Chemistry of the Cucurbitaceae* (D.M. Bates, R.W. Robinson, and C. Jeffrey, eds.), pp. 150–166. Cornell University Press, Ithaca.

Condon, M.A. and A.L. Norrbom. 1994. Three sympatric species of *Blepharoneura* (Diptera: Tephritidae) on a single species of host (*Gurania spinulosa,* Cucurbitaceae): new species and new taxonomic methods. *Syst. Entomol.* 19: 279–304.

Condon, M.A. and G.J. Steck. 1997. Evolution of host use in fruit flies of the genus *Blepharoneura* (Diptera: Tephritidae): cryptic species on sexually dimorphic host plants. *Biol. J. Linn. Soc.* 60: 443–466.

Condon, M.A., A. Truelove, and L. Mathew. 1997. Sexual dimorphism in mouthparts of *Blepharoneura* Loew (Diptera: Tephritidae). *Proc. Entomol. Soc. Wash.* 99: 676–680.

Dodson, G.N. 1997. Resource defense mating system in antlered flies, *Phytalmia* spp. (Diptera: Tephritidae). *Ann. Entomol. Soc. Am.* 90: 496–504.

Driscoll, C.A. and M.A. Condon. 1994. Labellar modifications of *Blepharoneura* (Diptera: Tephritidae): neotropical fruit flies that damage and feed on plant surfaces. *Ann. Entomol. Soc. Am.* 87: 448–453.

Eberhard, W. 1990. Animal genitalia and female choice. *Am. Sci.* 78: 134–141.

Foote, R.H. 1967. Family Tephritidae (Trypetidae, Trupaneidae). In *A Catalogue of the Diptera of the Americas South of the United States* (N. Papavero, ed.), Departmento de Zoologia, Secretaria da Agricultura, São Paulo, Fasc. 57, 91 pp.

Freidberg, A. 1981. Mating behavior of *Schistopterum moebiusi* Becker (Diptera: Tephritidae). *Isr. J. Entomol.* 15: 89–95.

Gilbert, L.E. 1975. Ecological consequences of a coevolved mutualism between butterflies and plants. In *Coevolution of Plants and Animals* (L.E. Gilbert and P.R. Raven, eds.), pp. 210–240. University of Texas Press, Austin.

Han, H-Y. and B.A. McPheron. 1994. Phylogenetic study of selected tephritid fruit flies (Insecta: Diptera: Tephritidae) using partial sequences of the nuclear 18S ribosomal DNA. *Biochem. Syst. Ecol.* 22: 447–457.

Headrick, D.H. and R.D. Goeden. 1994. Reproductive behavior of California fruit flies and the classification and evolution of Tephritidae (Diptera) mating systems. *Stud. Dipterol.* 1: 195–252.

Höglund, J. and R.V. Alatalo. 1995. *Leks.* Princeton University Press, Princeton. 248 pp.

Jenkins, J. 1990. Mating behavior of *Aciurnia mexicana* (Aczél) (Diptera: Tephritidae). *Proc. Entomol. Soc. Wash.* 92: 66–75.

Murawski, D. 1987. Floral resource variation, pollinator response, and potential pollen flow in *Psiguria warscewiczii*: a comparison of *Heliconius* butterflies and hummingbirds. *Oecologia* (Berlin) 68: 161–167.

Papaj, D.R. 1994. Oviposition site guarding by male walnut flies and its possible consequences for mating success. *Behav. Ecol. Sociobiol.* 34: 187–195.

Sivinski, J. and J.C. Webb. 1985. The form and function of acoustic courtship signals of the papaya fruit fly, *Toxotrypana curvicauda* (Tephritidae). *Fla. Entomol.* 68: 634–641.

# 8 Behavior of the Phytalmiinae and the Evolution of Antlers in Tephritid Flies

*Gary N. Dodson*

## CONTENTS



## 8.1 INTRODUCTION

> "But we are chiefly concerned with structures by which one male is enabled to conquer another, either in battle or courtship, through his strength, pugnacity, ornaments, or music." (Darwin 1871)

There are many images in nature that excite the human mind both visually and by their embodiment of some associated action. Among these are traits of animals that have evolved via the process that Darwin (1871) termed sexual selection. We find beauty in the colors and shapes of animal sexual adornments. At the same time, we are fascinated by the large canines, shaggy manes, and massive horns that often characterize the males of a species. For better or worse, few things get the adrenaline flowing like the prospect of a fight. Our reaction to sexually dimorphic traits such as those shown in Plates 17 and 18 is probably influenced by our vertebrate-biased perspective. Invertebrates with large and elaborate structures that, to our minds, suggest weaponry, are a great source of fascination and speculation. Some of the most striking ornaments in the animal kingdom are found among fruit flies, and the ultimate among these are the antlerlike head modifications found within the subfamily Phytalmiinae.

Sexually dimorphic modifications of the head capsule exhibit tremendous variety within the Tephritidae. The species reviewed in this chapter have structures that bear a strong resemblance to the familiar antlers of cervid mammals. The resemblance is, of course, superficial since these processes are not shed and regrown as are mammalian antlers. The structures under review are paired, cuticular projections arising from the lateral margins of the head without causing displacement of the eyes.

0-8493-1275-2/00/$0.00+$.50

Given this delineation, these cheek projections do not include the eye stalks found in the tephritid genera *Themara* (Hardy 1973; 1974), *Pelmatops* and *Pseudopelmatops* (Hardy 1986a), or the wide heads (rudimentary eye stalks?) of *Homoiothemara* (Hardy 1988). Also excluded are highly modified, hornlike bristles such as those found between the eyes of various Trypetini, such as *Vidalia* spp. (e.g., Munro 1938) and *Paramyiolia rhino* (Steyskal) (Steyskal 1972) (see Han, Chapter 11).

This chapter reviews the limited information available on the life history and behavior of species in the subfamily Phytalmiinae. Attention is then focused on the species with antlerlike processes including a discussion of their phylogenetic placement, geographic distribution, and comparative morphology. Finally, I explain what is known regarding the function of these structures and make suggestions about the type of future work that would greatly benefit our understanding of these remarkable flies.

## 8.2 LIFE HISTORY AND BEHAVIOR OF PHYTALMIINAE

### 8.2.1 Larval Biology and Substrate (Host Plants)

Alfred Russell Wallace was the first scientist to examine species of Phytalmiinae, observing "deer-flies" during his visit to New Guinea (Wallace 1869). When collecting *Phytalmia alcicornis* (Saunders), *P. cervicornis* Gerstaecker, *P. megalotis* Gerstaecker, and *Diplochorda brevicornis* (Saunders), he noted that they "settled on fallen trees and decaying trunks." Recent evidence now suggests that decayed plant material is a primary larval food source across this subfamily as it is currently defined (Hardy 1986b; Dodson and Daniels 1988; Hancock and Drew 1995; Dodson 1997; Norrbom et al. 1999; but note that some of these authors treated the pertinent fly genera under the subfamily Trypetinae). Especially prominent among the known hosts are species of *Bambusa* (Hardy 1986b; Hancock and Drew 1995).

The only antlered tephritids for which there are definitive plant associations are species of *Phytalmia*. Four of the seven species have been reared from the decaying wood of rain forest trees (Dodson 1997; H. Roberts, personal communication). There is evidence that at least three of these species are extremely restricted in their host species acceptance (Dodson 1997). Decaying wood of *Dysoxylum gaudichaudianum* (Adr. Juss.) Miq. is apparently the only larval substrate for both *P. mouldsi* McAlpine and Schneider and *P. alcicornis*. It is also one of the tree species that attracts *P. cervicornis* (Dodson 1997). At least two nonantlered genera of Phytalmiinae have been reared from decaying *D. gaudichaudianum* as well (Hardy 1986b; Dodson and Daniels 1988).

Last instar larvae of *P. mouldsi* are capable of jumping astounding distances when outside the decaying wood (G. Dodson, unpublished data). The mechanism appeared to be the same or similar to that described for *Ceratitis capitata* (Wiedemann) (Maitland 1992) and is assumed to function in reducing the risk of predation when moving to a site for pupariation.

### 8.2.2 Mating Behavior

We know very little about the behavior of members of the Phytalmiinae. Two species are known to exhibit mating trophallaxis (see Sivinski et al., Chapter 28). *Afrocneros mundus* (Loew) produces a mound of foam that is used to attract females (Oldroyd 1964), but no details were provided. Pritchard (1967) studied the mating behavior of *Dirioxa pornia* (Walker) in the laboratory and described pleural distension, pheromone release, and slow wing movements associated with the production of a foam mound on which females fed during and after mating.

The only phytalmiines whose mating behavior has been studied in nature belong to the antlered genus *Phytalmia* (Dodson 1997). All *Phytalmia* species observed thus far exhibit a resource defense mating system, which is described in detail in Section 8.5.

**TABLE 8.1**
**Taxonomic Placement of the Antlered Species of Tephritidae, Their Gross Distributions, and the Number of Antlerless Congeners**

| Subfamily and Tribe | Genus | Species with Antlers | No. Congeners Lacking Antlers | Known Distribution |
|---|---|---|---|---|
| **Phytalmiinae** | | | | |
| Phytalmiini | *Diplochorda* | *aneura* Malloch | 6 | New Guinea |
| | | *australis* Permkam and Hancock | | New Guinea/Australia[a] |
| | | *brevicornis* (Saunders) | | New Guinea |
| | | *myrmex* Osten Sacken | | New Guinea |
| | *Ortaloptera* | *callistomyia* Hering | 1 | New Guinea |
| | *Phytalmia* | *alcicornis* (Saunders) | 0 | New Guinea[b] |
| | | *antilocapra* McAlpine and Schneider | | New Guinea |
| | | *biarmata* Malloch | | New Guinea |
| | | *cervicornis* Gerstaecker | | New Guinea[b] |
| | | *megalotis* Gerstaecker | | New Guinea[b] |
| | | *mouldsi* McAlpine and Schneider | | Australia[a] |
| | | *robertsi* Schneider | | New Guinea |
| | *Sessilina* | *nigrilinea* (Walker) | 2 | New Guinea |
| Acanthonevrini | *Terastiomyia* | *lobifera* Bigot | 2 | Sulawesi |
| | *Sophira* | *limbata* Hering | 17 | Borneo |

[a] Restricted to one area of the Cape York Peninsula.
[b] Also one adjacent island.

## 8.3 PHYLOGENETIC RELATIONSHIPS AND DISTRIBUTIONS OF ANTLERED FLIES

Flies with antlerlike head projections are known from five families of acalyptrate flies, three of which are members of the superfamily Tephritoidea. Wilkinson and Dodson (1997) determined that the taxonomic distribution of these families was statistically nonrandom. Within the Tephritidae, the phylogenetic influence is even more dramatic as the six genera known to exhibit antlerlike processes (Table 8.1) all belong to the subfamily Phytalmiinae (Korneyev, Chapter 4) and four of these genera belong to the tribe Phytalmiini. It is unclear if cheek projections are an apomorphic character for this tribe as currently defined (Korneyev, Chapter 4). All four genera in the tribe include species with antlers, but only one genus (*Phytalmia*) consists solely of species with the projections. They occur in 50% or less of the known species in each of the other five genera with antlered species (Table 8.1). This distribution within the subfamily together with their variation in form suggests that cheek projections may have evolved more than once within the Phytalmiinae rather than having been lost multiple times. However, neither interpretation can be satisfactorily tested until a thorough phylogenetic analysis of the subfamily is completed.

All of the antlered species of tephritids occur between the island of Borneo to the west and the Cape York Peninsula of Australia to the southeast, with the majority occurring on the island of New Guinea (Table 8.1). All of the species for which there is habitat information are found in rain forest. A good indicator of the variation in the distributional range of individual species can be found by considering the genus *Phytalmia*. Three species (*P. alcicornis, P. cervicornis, P. megalotis*) have been collected from virtually the full length of New Guinea (the world's second largest island) as well as one small adjacent island for each species (McAlpine and Schneider 1978). By contrast, *P. mouldsi* is restricted to a single, isolated rain forest area on the northeast coast of the Cape York

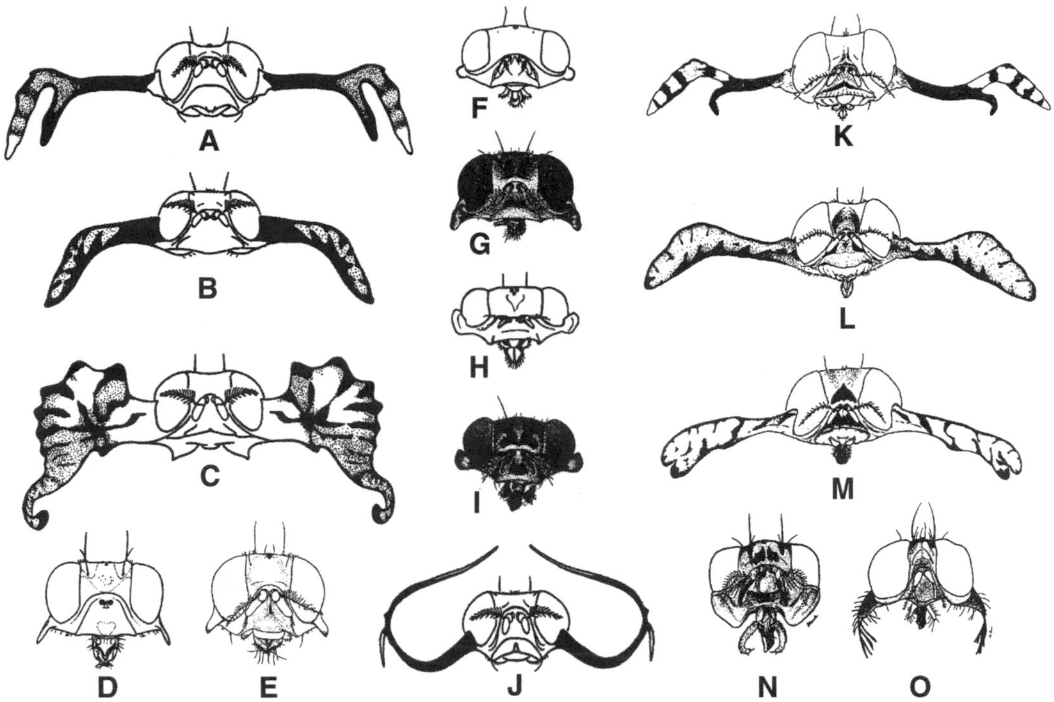


**FIGURE 8.1** Heads of males of Tephritidae with antlerlike cheek projections. Heads have been proportionately scaled so that they are of equal width across the outer eye margins. (A) *Phytalmia antilocapra*; (B) *P. mouldsi*; (C) *P. alcicornis*; (D) *Terastiomyia lobifera*; (E) *Sessilina nigrilinea*; (F) *Diplochorda aneura*; (G) *D. australis*; (H) *D. brevicornis*; (I) *D. myrmex*; (J) *P. cervicornis*; (K) *P. biarmata*; (L) *P. megalotis*; (M) *P. robertsi*; (N) *Ortaloptera callistomyia*; (O) *Sophira limbata boreensis.* (A, B, C, and J, from Wilkinson, G. and G. N. Dodson, in *The Evolution of Mating Systems in Insects and Arachinids,* J. Choe and B. Crespi, eds., Cambridge University Press, Cambridge, 310–328, 1997; d, from Hardy, D.E., *J. Kans. Entomol. Soc*. 31: 76–81, 1958; E, from McAlpine, D.K. and M.A. Schneider, *Syst. Entomol.* 3: 159–175, 1978; f, from Malloch, J.R., *Proc. Linn. Soc. N.S.W.* 64: 169–180, 1939; H, from Saunders, W.W., *Trans. Entomol. Soc. London* (new ser.) 5(mem.): 413–417, 1861; N and O, from Hardy, D.E., *Zool. Scr.* 17: 77–121, 1988. All with permission.)

Peninsula (McAlpine and Schneider 1978). The only species listed in Table 8.1 that occurs on more than one large "island" is *Diplochorda australis* Permkam and Hancock, found on New Guinea and on the Cape York Peninsula of Australia.

## 8.4 COMPARATIVE MORPHOLOGY

### 8.4.1 Structure of the Cheek Processes

Antlerlike cheek processes in tephritids occur only on males and vary from the small, disk-shaped protrusions on the lower half of the face of *Ortaloptera callistomyia* Hering to the massive, sometimes branched structures of *Phytalmia* spp. (Figure 8.1). These cuticular projections arise from distinct notches at the margins of the eyes in *Diplochorda*, *Phytalmia*, *Sessilina,* and *Terastiomyia*. In contrast, the eye margins are entire in *O. callistomyia* specimens and a side-view illustration of the head of *Sophira limbata borneensis* Hering shows the eye to be unnotched as well (Hardy 1988). In some species with notched eyes (e.g., *Phytalmia* spp. and *Sessilina nigrilinea* (Walker)), there is a correlated sinuation in the eye margin of females, whereas in at least one species (*Terastiomyia lobifera* Bigot) the sinuated eye margin is restricted to males.

Colors associated with *Phytalmia* antlers are quite vivid in living specimens (for examples, see Color Figures 17* and 18; Dodson 1989; Moffett 1997). Most have bold streaks or areas where strongly contrasting colors come in contact. The surfaces of the processes are mostly bare, but minute setae do occur in limited areas on all species of *Phytalmia* as well as *D. myrmex* Osten Sacken (G. Dodson, personal observation). No setae were observed on the antlers of two male *Sessilina* specimens (G. Dodson, personal observation). *Sophira limbata borneensis* is the only species with large bristles over the surface of the cheek projections, including the remarkably showy plumes at the apex (Figure 8.1d).

*Phytalmia* puparia have no external indications of the cheek projections, which expand following eclosion as hemolymph is pumped into them. Despite their robust appearance, the projections are quite flexible and slight pressure against them results in displacement (G. Dodson, personal observation). Antler size is tightly correlated with body size (Wilkinson and Dodson 1997) and the smallest males have little or no trace of the processes (McAlpine and Schneider 1978). Hardy (1986a) provided figures of male *T. lobifera* with processes and without processes and it seems likely that this is a result of the same phenomenon. The five male specimens of *T. lobifera* in the Natural History Museum collection conform to the pattern of largest antlers on the largest specimens (I. White, personal communication). Likewise, the single specimen that was available for the description and illustration of a male *S. nigrilinea* (Figure 8.1; McAlpine and Schneider 1978) may be a small male with correspondingly small cheek processes. I have seen specimens of *Sessilina* with significantly longer antlers in the National Forest Insect Collection of Papua New Guinea (now housed in Lae, Morobe Province, PNG) that were either *S. nigrilinea* or an undescribed species.

### 8.4.2 Associated Morphological Features

Three additional characters may be useful for predicting the behavior of phytalmiine species not yet studied. Spines on the fore femora, the shape of the epistomal margin of the face, and the ratio of foreleg to middle and hindleg length are all sexually dimorphic in *Phytalmia* spp. and are all associated with the mating system. All species of *Phytalmia* have large spines on the posteroventral surface of male fore femora that are absent from females (Saunders 1861; Malloch 1939; McAlpine and Schneider 1978; Schneider 1993). These spines are employed in the copulatory and, in certain species, the postcopulatory mounting position of male on female (see Section 8.5). It is noteworthy that the species with the minimum number of spines (*P. cervicornis*) differs from the others in that males do not remain mounted on females following copulation. Antlered species of *Diplochorda* have posteroventral femoral spines (there is some question as to their exact location on males of *D. brevicornis*; see Malloch 1939), and in at least one of these species the spines are absent or smaller on females (Malloch 1939). Species of *Diplochorda* without antlers lack femoral spines (Malloch 1939). They are present on males but not females of *O. callistomyia* (Hardy 1988). They are "stouter" on males compared with females of *S. nigrilinea* (McAlpine and Schneider 1978). The nonantlered *S. horrida* McAlpine and Schneider has bristles along the ventrum of the fore femur, but they vary in size and are not restricted to the posterior portion (McAlpine and Schneider 1978). Spines of the fore femora are not described for *Sophira limbata borneensis* or *T. lobifera*. The description of these femoral spines leads me to predict that males of species with robust spines will exhibit the "wing lock" mounting position observed for *Phytalmia* (see Section 8.5).

The epistomal margin of the face is a variable character across *Phytalmia*. In some species it is strongly produced (see Figure 8.2 in McAlpine and Schneider 1978); usually more so in males than females (e.g., *P. mouldsi* and *P. robertsi* Schneider). But this margin is not produced forward so greatly in either sex of *P. alcicornis* or *P. cervicornis*. Behavioral observations have revealed differences in male intrasexual interactions that are related to the form of the epistomal margin (see Section 8.5). Species with the more protruding epistome use this surface as a point of contact in pushing contests. We might therefore expect that non-*Phytalmia* species with strongly produced

---

* Color Figures follow p. 204.

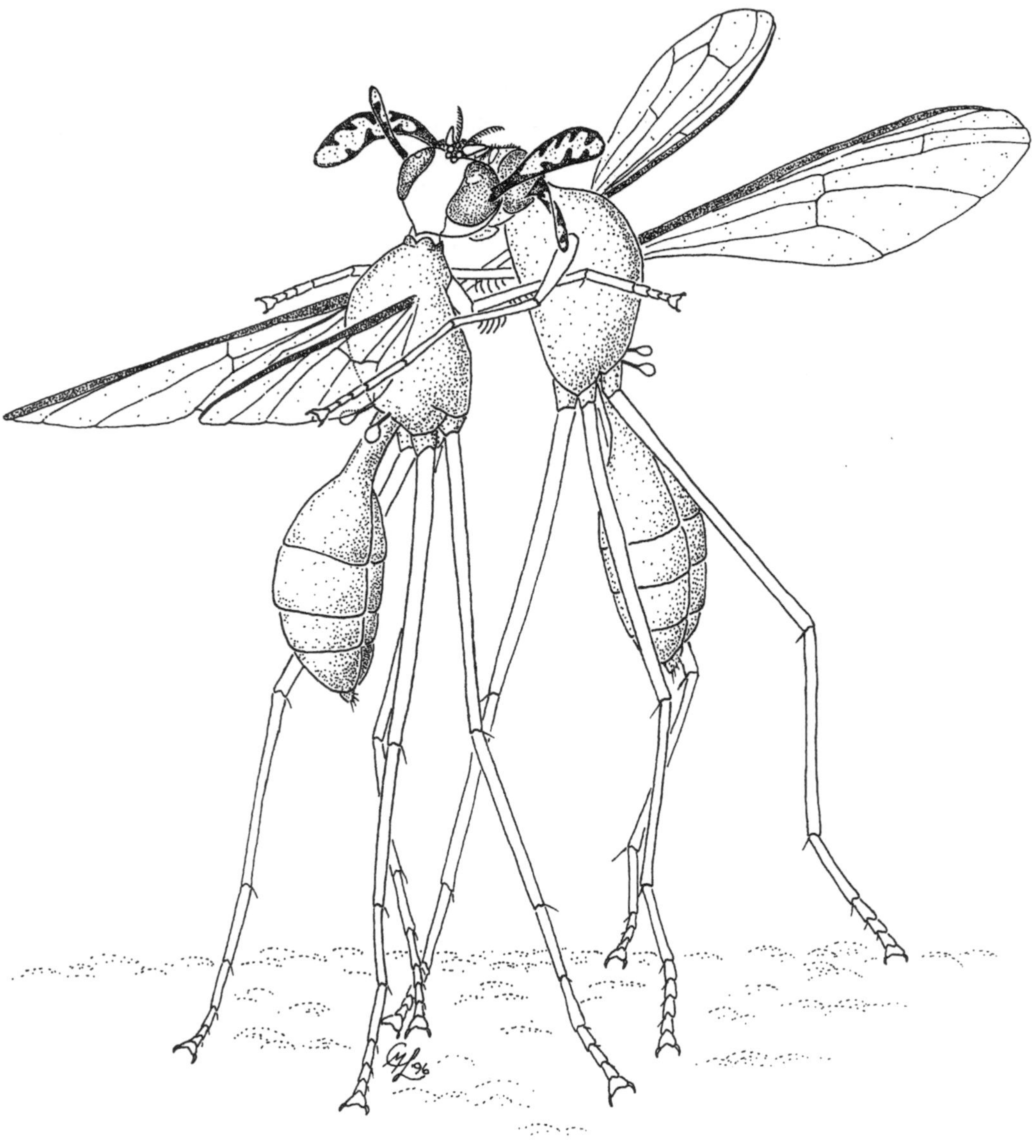

**FIGURE 8.2** Two similarly sized male *Phytalmia mouldsi* engaged in the vertically oriented pushing contest as part of their resource defense mating system. (From Dodson, G.N., *Ann. Entomol. Soc. Am.* 90: 496–504, 1997. With permission.)

epistomes on males will also employ them in agonistic encounters. Examination of specimens will be necessary to determine which species fit this prediction because the typical head-on facial illustrations make it difficult to determine if the epistomal margin projects forward.

Finally, leg length appears to be subject to sexual selection in some species of *Phytalmia*. *Phytalmia mouldsi* male intrasexual encounters involve pushing contests while raised up on the middle and hindlegs (Figure 8.2 and Section 8.5). The ratio of middle or hindleg length to foreleg length is greater for males than for females in this species (G. Dodson, unpublished data). By contrast this ratio does not differ between male and female *P. alcicornis*, a species that does not stilt up on its hindlegs during escalated male–male interactions. Sexual differences in leg length ratios should be examined across antlered species as a possible correlated character with vertically oriented male contests.

## 8.5 MATING SYSTEMS AND THE FUNCTION OF THE CHEEK PROCESSES

*Phytalmia* spp. are also the only antlered tephritids that have been studied with regard to mating behaviors. Moulds (1977) first described the behavior of *P. mouldsi*, revealing male defense of territories on decaying *D. gaudichaudianum* (identified at the time as *D. decandrum*). Individual males attempt to maintain exclusive possession of oviposition sites along the tree surface to which females will be attracted (Moulds 1977; Dodson 1997). Larger males have a distinct advantage in male–male agonistic encounters. Contests between males that differ noticeably in size are usually decided with little or no physical contact, while the most evenly matched individuals exhibit spectacular pushing contests to determine winners (Figure 8.2) (Dodson 1997). Possession of a territory is critical for optimal mating success as females must copulate with a male before they are able to oviposit at any guarded site (Dodson 1997).

*Phytalmia mouldsi* males react to an arriving female with a synchronous supination of the wings while blocking her path to the oviposition location (Moulds 1977). Successful males mount receptive females and clasp the bases of the female's wing between their own femur-tibia joint and a set of strongly recurved spines on the posteroventral surface of the femur (the "wing lock").

This mounted, wing-locked position is maintained following copulation as the female lays eggs at the site guarded by the male. Male–male interactions often continue between the mounted male and intruders (Moulds 1977). Both males and females mate multiple times and individual males can successfully defend a site for consecutive days at the same resource (Dodson 1997; G. Dodson, unpublished data). Mating system theory predicts that such resource defense mating systems should evolve only where the resource defended is rare and/or has a patchy distribution (Emlen and Oring 1977), and this expectation is met for *Phytalmia* in all tests thus far (Dodson 1997).

Three other species of *Phytalmia* are known to exhibit resource defense mating systems with interesting similarities and contrasts to *P. mouldsi* (Dodson 1997). Limited observations of *P. biarmata* Malloch suggest a virtually identical mating system, including the stilting position of males in escalated contests. Both *P. alcicornis* and *P. cervicornis* exhibit the same basic behaviors as *P. mouldsi* with the following notable exceptions. Males of both species hold their wings back (not supinated) when females approach guarded oviposition sites. Escalated contests between *P. alcicornis* males involve repeated forward thrusts (lunging) with the body axis remaining mostly parallel with the substrate, rather than upright stilting. Escalated contests between opponents in *P. cervicornis* are also less vertical and without the prolonged stilting of *P. mouldsi*. But perhaps the most surprising difference is that *P. cervicornis* males do not guard females as they oviposit following copulation.

While it is fairly certain that tephritid antlers evolved within the general context of male intrasexual competition, their specific function is less clear. Initial contact between opponents is made with the antlers in the four species of *Phytalmia* for which fighting has been observed. However, the role of the antlers differs among these species when contests continue beyond momentary contact. As noted above, escalated contests in *P. alcicornis* and *P. cervicornis* involve repeated, reciprocal thrusts with the anteroproximal portions of the antlers serving as the primary point of contact (Dodson 1977). By contrast, the stilting matches of *P. mouldsi* and *P. biarmata* utilize the protruding epistomal margin of the face for transmitting force rather than the antlers. Correspondingly, the epistome is not produced strongly forward on the faces of *P. alcicornis* and *P. cervicornis*.

When the antlers of *P. mouldsi* were shortened or removed experimentally, the winning percentage of the manipulated males decreased. However, large males still defeated small males in at least 75% of their encounters (G. Dodson, unpublished data). The deantlered males paid a cost in terms of the amount of energy expended in deciding the outcome of interactions. The percentage of escalated interactions increased after antler shortening, indicating that the bigger males lost the "luxury" of winning some encounters strictly on the basis of the visual cue of their size. Somewhat surprisingly, antler-shortening experiments with *P. alcicornis* yielded similar results. Relative body size remained the primary determinant of contest outcome even when their immense antlers were removed (G. Dodson, unpublished data). These results are consistent with the hypothesis that all lateral head projections on flies have evolved not as weapons per se, but as honest signals of size and, thus, the fighting capacity of their bearers (Wilkinson and Dodson 1997).

## 8.6 CONCLUSIONS AND FUTURE RESEARCH NEEDS

It is clear that the tephritids with antlerlike cheek projections are phylogenetically and geographically clustered. It is not at all clear *why* this is the case. Why do cheek projections not occur in other tephritid subfamilies or in other parts of the world? The latter part of this question may be made moot by the first part. Given that antlers are restricted to the Phytalmiinae, species possessing them are likely to be found only in the Australasian and Oriental Regions by default. Only 15% of the genera and less than 20% of the species in this subfamily are found outside of these regions (Norrbom et al. 1999). Whatever the reason that these structures are present only on phytalmiines, they are thereby limited to a small region.

The possibility that the restricted distribution of these antlered tephritids is attributable to a limited distribution of their larval substrate is not supported by the evidence at hand. *Phytalmia* species have been reared from trees belonging to three families (Dodson 1997; H. Roberts, personal communication) that together are widespread biogeographically. While it is true that the single tree genus *Dysoxylum* and more specifically *D. gaudichaudianum* serves as a host for multiple species within the subfamily Phytalmiinae, *Dysoxylum* is the most widely distributed Indopacific genus in the family Meliaceae (from India to New Zealand) and *D. gaudichaudianum* is "one of the most widely distributed [species] in the genus" (Mabberley et al. 1995). Therefore, having *Dysoxylum* as a host would not limit these flies to their present range, all else being equal. Whether the fly distributions are an historical remnant of a formerly more-limited host tree distribution remains to be tested.

Are cheek projections a genetic possibility only in the Phytalmiinae? The answer must be "no" since apparently morphologically equivalent structures occur in other acalyptrate fly families (Wilkinson and Dodson 1997). But there is obviously a genetic *propensity* for this phenomenon within a small segment of this subfamily since it has arisen repeatedly within one tribe. There is also a recognizable phylogenetic effect at a higher systematic level, given that cheek projections occur disproportionately often within the tephritoid families. Currently, it seems impossible to explain this heightened chance for the evolution of cheek projections within the developmental bauplan of phytalmiines and other tephritoids. However, the astounding rate at which our knowledge of genetics is increasing makes it reasonable to assume that we will soon be able to identify the genes responsible for this developmental potential.

Does the scarcity of this phenomenon have an ecological explanation? McAlpine (1982) hypothesized that all head modifications of this sort would be associated with male–male agonistic interactions. Wilkinson and Dodson (1997) further predicted that all lateral head projections would be found to function in the context of mutual assessment by opponents. Whereas these ideas may help explain why these particular species have head projections, they do not explain why other species in seemingly similar circumstances do not have them. In other words, it is reasonable to expect species with resource defense mating systems to be under selection for effective conflict resolution. This should lead to the development of traits that enhance the efficiency of determining the winner of conflicts, that is, the evolution of honest signals of dominance. But there are several other tephritid species in which males defend oviposition sites using similar fighting styles, and yet have no head projections (Sivinski et al., Chapter 28).

It is difficult to rule out the parsimonious explanation that antlered flies are found in the tropics simply because of the greater species richness of the region. If the evolution of a trait is not attributable to habitat-limited selection pressures, then it is more likely to evolve wherever more species are undergoing selection. More rare traits of any sort will occur by chance where there are more species. Alternatively, Wilkinson and Dodson (1997) hypothesized that longer average life spans are more likely in the tropics and that this life history trait could select, in part, for lower costs of contests among resource-defending males. If antlers are structures that function as honest signals of fighting prowess, then their presence saves all but the most evenly matched opponents from having an escalated encounter. With the potential for hundreds of contests over a long life span, the small savings per contest would add up and may be the trade-off that explains the evolution of such elaborate structures.

These tentative conclusions on the evolution of antlerlike projections in tephritids make the most worthwhile areas for future research fairly obvious. A knowledge of the phylogenetic relationships among tephritid species will allow us to assess the homology and evolution of the morphological and life history traits discussed in this chapter. The phylogenetic issue is currently being investigated with both morphological and molecular analyses (e.g., Chapters 4 and 5).

We know so little about the life histories of these species, any efforts to study them in their natural habitats is guaranteed to pay big dividends. What are the identities of their host plants and how specific are their relationships with them? Are oviposition sites a limiting resource for species other than *Phytalmia*? Are these flies long-lived as adults compared with other tephritids with similar mating ecologies?

Observations of mating behaviors will allow for testing the other predictions described above which are explicitly summarized here.

1. Species with cheek projections will have resource defense mating systems.
2. Males of species with robust posteroventral spines on the fore femora will exhibit the "wing lock" mounting position.
3. Species with the most strongly produced epistomal margins will utilize them in male–male contests.
4. Species with sexual differences in leg length ratios will engage in vertically oriented contests.

More than 130 years have passed since A. R. Wallace brought these remarkable flies to the attention of the scientific community. What we have learned in the past 22 years regarding their behavior has only added to the fascination and desire to know more about the flies that remind us of cervids. Because we do not hunt them and eat them, they go about their lives with little notice even from the people who live in the same forests. As the existence of these very forests is threatened by modern man's incredible consumption of natural resources, it is imperative that we learn more about them quickly. Antlered flies seem to be good candidates for charismatic symbols of the neglected invertebrate fauna that dominates the world's biota. Their continued existence and the implications for our own species' well-being deserves attention.

## ACKNOWLEDGMENTS

My wife, Jill Jereb, has been a great partner through all of these studies. David Yeates provided very helpful advice and comments on the manuscript. Ian White and David McAlpine generously supplied information on museum specimens. Will McClintock was a great help and a great friend in the field. The Resource and Conservation Foundation of Papua New Guinea facilitated visits to the Crater Mountain Wildlife Management Area. Debra Wright and Andy Mack built it and now people come! Cheryl LeBlanc drew Figure 8.2 and the heads labeled k, l, and m in Figure 8.1. S. P. Kim drew heads g and i in Figure 8.1. The composition of Figure 8.1 required Denise Jones' graphic design skills. We owe thanks to Hywel Roberts for his dedication to the collection of flies during his time in New Guinea. I thank Martín Aluja and Allen Norrbom for encouraging me to be part of the Xalapa meeting and the Campaña Nacional contra las Moscas de la Fruta, International Organization for Biological Control of Noxious Animals and Plants, and the Instituto de Ecología A.C. (Mexico) for partial funding. Financial support at various times during my studies of these flies has been provided by Ball State University,

University of Queensland, Indiana Academy of Science, *National Geographic Magazine,* Television New Zealand, and the Wildlife Conservation Society.

## REFERENCES


Darwin, C. 1871. *The Descent of Man and Selection in Relation to Sex.* Princeton University Press, Princeton.

Dodson, G. 1989. The horny antics of antlered flies. *Aust. Nat. Hist.* 12: 604–611.

Dodson, G.N. 1997. Resource defense mating system in antlered flies, *Phytalmia* spp. (Diptera: Tephritidae). *Ann. Entomol. Soc. Am.* 90: 496–504.

Dodson, G. and G. Daniels. 1988. Diptera reared from *Dysoxylum gaudichaudianum* (Juss.) Miq. at Iron Range, Northern Queensland. *Aust. Entomol. Mag.* 15: 77–79.

Emlen, S.T. and L.W. Oring. 1977. Ecology, sexual selection and the evolution of mating systems. *Science* 197: 215–222.

Hancock, D.L. and R.A.I. Drew 1995. Observations of the genus *Acanthonevra* Macquart in Thailand and Malaysia (Diptera: Tephritidae: Trypetinae). *The Entomologist* 114: 99–103.

Hardy, D.E. 1958. A review of the genus *Neosophira* Hendel (Diptera, Tephritidae). *J. Kans. Entomol. Soc.* 31: 76–81.

Hardy, D.E. 1973. The fruit flies (Tephritidae — Diptera) of Thailand and bordering countries. *Pac. Insects Monogr.* 31: 1–353.

Hardy, D.E. 1974. The fruit flies of the Philippines (Diptera: Tephritidae). *Pac. Insects Monogr.* 32: 1–266.

Hardy, D.E. 1986a. The Adramini of Indonesia, New Guinea and adjacent islands (Diptera: Tephritidae: Trypetinae). *Proc. Hawaii. Entomol. Soc.* 27: 53–78.

Hardy, D.E. 1986b. Fruit flies of the subtribe Acanthonevrina of Indonesia, New Guinea and the Bismarck and Solomon Islands (Diptera: Tephritidae: Trypetinae: Acanthronevrini). *Pac. Insects Monogr* 42: 1–42.

Hardy, D.E. 1988. Fruit flies of the subtribe Gastrozonina of Indonesia, New Guinea and the Bismarck and Solomon Islands (Diptera, Tephritidae, Trypetinae, Acanthronevrini). *Zool. Scr.* 17: 77–121.

Mabberley, D.J., C.M. Pannell, and A.M. Sing. 1995. *Flora Malesiana.* Series I, *Spermatophyta.* Vol. 12, Pt. 1, *Meliaceae,* pp. 1–407. Rijksherbarium/Hortus Botanicus, Leiden University, The Netherlands.

Maitland, D.P. 1992. Locomotion by jumping in the Mediterranean fruit fly larva *Ceratitis capitata. Nature* 355: 159–161.

Malloch, J.R. 1939. The Diptera of the Territory of New Guinea. IX Family Phytalmiidae. *Proc. Linn. Soc. N.S.W.* 64: 169–180.

McAlpine, D.K. 1982. The acalyptrate Diptera with special reference to the Platystomatidae. In *Biogeography and Ecology of New Guinea* (J.L. Gressitt, ed.), pp. 659–673. Dr. W. Junk Publishers, The Hague.

McAlpine, D.K. and M.A. Schneider. 1978. A systematic study of *Phytalmia* (Diptera: Tephritidae) with description of a new genus. *Syst. Entomol.* 3: 159–175.

Moffett, M. 1997. Flies that fight. *Nat. Geogr. Mag.* 192 (5): 68–77.

Moulds, M.S. 1977. Field observations on behaviour of a North Queensland species of *Phytalmia* (Diptera, Tephritidae). *J. Aust. Entomol. Soc.* 16: 347–352.

Munro, H.K. 1938. Studies on the Indian Trypetidae (Diptera). *Rec. Indian Mus.* 40: 21–37.

Norrbom, A.L., L.E. Carroll, and A. Freidberg. 1999. Status of knowledge. In *Fruit Fly Expert Identification System and Systematic Information Database* (F.C. Thompson, ed.), pp. 9–47. *Myia* (1998) 9, 524 pp.

Oldroyd, H. 1964. *The Natural History of Flies.* W.W. Norton, New York.

Pritchard, G. 1967. Laboratory observations on the mating behaviour of the island fruit fly *Rioxa pornia* (Diptera: Tephritidae). *J. Aust. Entomol. Soc.* 6: 127–132.

Saunders, W.W. 1861. On *Elaphomyia*, a genus of remarkable insects of the order Diptera. *Trans. Entomol. Soc. London* (new ser.) 5 (mem.): 413–417, Plates 12–13.

Schneider, M.A. 1993. A new species of *Phytalmia* (Diptera: Tephritidae) from Papua New Guinea. *Aust. Entomol.* 20: 3–8.

Steyskal, G.C. 1972. A new species of *Myoleja* with a key to North American species (Diptera: Tephritidae). *Fla. Entomol.* 55: 207–211.

Wallace, A.R. 1869. *The Malay Archipelago.* Macmillan, London.

Wilkinson, G. and G.N. Dodson. 1997. Function and evolution of antlers and eye-stalks in flies. In *The Evolution of Mating Systems in Insects and Arachnids* (J.C. Choe and B.J. Crespi, eds.), pp. 310–328. Cambridge University Press, Cambridge. 387 pp.

# Section IV

## Subfamily Trypetinae

# 9 Phylogeny of the Subtribe Carpomyina (Trypetinae), Emphasizing Relationships of the Genus *Rhagoletis*

*James J. Smith and Guy L. Bush*

## CONTENTS



## 9.1 INTRODUCTION

In this chapter we examine the phylogenetic relationships of flies in the tephritid subtribe Carpomyina. Our primary goal is to establish the current state of knowledge by consolidating information from various phylogenies that have been generated using morphology (Bush 1966; Norrbom 1994; 1997; Jenkins 1996), allozymes (Berlocher and Bush 1982; Berlocher et al. 1993) and DNA sequences (McPheron and Han 1997; Smith and Bush 1997).

We have chosen to take a geographic approach, discussing separately the fauna within each biogeographic region. It is our thesis that it will be difficult to understand the evolutionary history of the Carpomyina without considering the biological and geographic contexts in which the flies are found and have evolved. For example, the present-day geographic distributions of taxa, examined in the context of their genetic relationships and host plant use characteristics, may provide clues as to the mode of evolutionary divergence (Bush and Smith 1998). Geography is particularly important in understanding evolutionary relationships within Carpomyina given the fact that some groups apparently speciate sympatrically while others diverge allopatrically. Distribution data also provide clues that may allow us to identify progenitors and areas of endemism.

0-8493-1275-2/00/$0.00+$.50

A primary focus of this tour of the carpomyine taxa is whether the large genus *Rhagoletis* is monophyletic. It appears that some *Rhagoletis* species are more closely related to other carpomyine genera than they are to other *Rhagoletis*. We will also show how our understanding of the evolutionary relationships of Carpomyina has been hindered by both undersampling of taxa and characters.

Similarities and differences between morphological and molecular character evolution in Carpomyina are also discussed. Morphological features within different carpomyine lineages apparently are evolving at different rates, and the same appears to be true for DNA sequences. We describe two examples within *Rhagoletis* where the rate of morphological divergence has apparently been low despite considerable DNA divergence. We contrast these situations with those in which morphological and molecular divergence appear to be more highly correlated. These different patterns of morphological and molecular divergence observed in carpomyine flies are examined in the context of geographic distribution and host plant use. Using this information, we can study the relative importance of different processes driving divergence, such as natural selection in response to differential host use and genetic drift due to the geographic isolation of populations.

## 9.2 PHYLOGENETIC RELATIONSHIPS WITHIN THE SUBTRIBE CARPOMYINA

### 9.2.1 Taxonomic Status of the Subtribe Carpomyina

We focus much attention below on the morphological cladistic analysis of Jenkins (1996), in which the classification of Foote et al. (1993) was used as the basis of the sample design. These authors divided the Tephritidae into three subfamilies: Trypetinae, Dacinae, and Tephritinae (Figure 9.1). They recognized the Carpomyina, the subjects of this chapter, as a subtribe of the Trypetini, one of the tribes within Trypetinae. Recent work by several authors (e.g., Han and McPheron 1997; Norrbom et al. 1999a; Korneyev, Chapter 4; Han, Chapter 11) has led to the reclassification of some tephritid subfamilies, tribes, and subtribes, including the transfer of the Carpomyina from the Trypetini to the tribe Carpomyini, which also includes two other small subtribes. However, the component genera of Carpomyina have not changed significantly. Thus, we refer the reader to Korneyev's Chapter 4 for information on the current taxonomic status of Carpomyina and its phylogenetic relationships with other higher groups of Tephritidae.

There are eight to ten genera currently recognized in the Carpomyina. Most specialists consider the following to be valid genera in the subtribe: *Carpomya*, *Haywardina*, *Rhagoletis*, *Rhagoletotrypeta*, *Stoneola,* and *Zonosemata*. Norrbom (1997) synonymized *Myiopardalis* and *Goniglossum* into *Carpomya*, but we continue to recognize them as separate genera (see below). Recently, Norrbom et al. (1999a) also included *Scleropithus* in the Carpomyina, but we have been unable to study this monotypic Afrotropical genus. The constituent species of each of these genera, along with their host plants and geographic distributions, are given in Table 9.1. Additional nomenclatural data for them were listed by Norrbom et al. (1999b). The distributions of the genera are also shown below in Figures 9.6 to 9.9. *Cryptodacus*, *Haywardina,* and *Stoneola* are mostly or exclusively Neotropical genera, *Rhagoletotrypeta* and *Zonosemata* are Nearctic and Neotropical, and *Carpomya*, *Goniglossum,* and *Myiopardalis* are Palearctic except for one Neotropical *Carpomya* species that has recently been described (Norrbom 1997). *Rhagoletis* occurs in the Palearctic, Oriental, Nearctic, and Neotropical Regions.

Morphological data indicate that the subtribe Carpomyina is a monophyletic group. In each member of Carpomyina the subapical lobe of the glans is an elongate lobe or flap (Jenkins 1996). This appears to represent a synapomorphy for the subtribe, with the ancestral condition being a trumpet-shaped lobe (Jenkins 1996). However, Korneyev (personal communication) cautions that the derived character state involves reduction and is therefore subject to homoplasy.

Other proposed morphological synapomorphies for Carpomyina include a desclerotized area at the apex of the female oviscape and the presence of preoral teeth, usually distinctly sclerotized, in the larva (Norrbom 1989; Carroll 1992). Jenkins (1996) found the desclerotized area of the

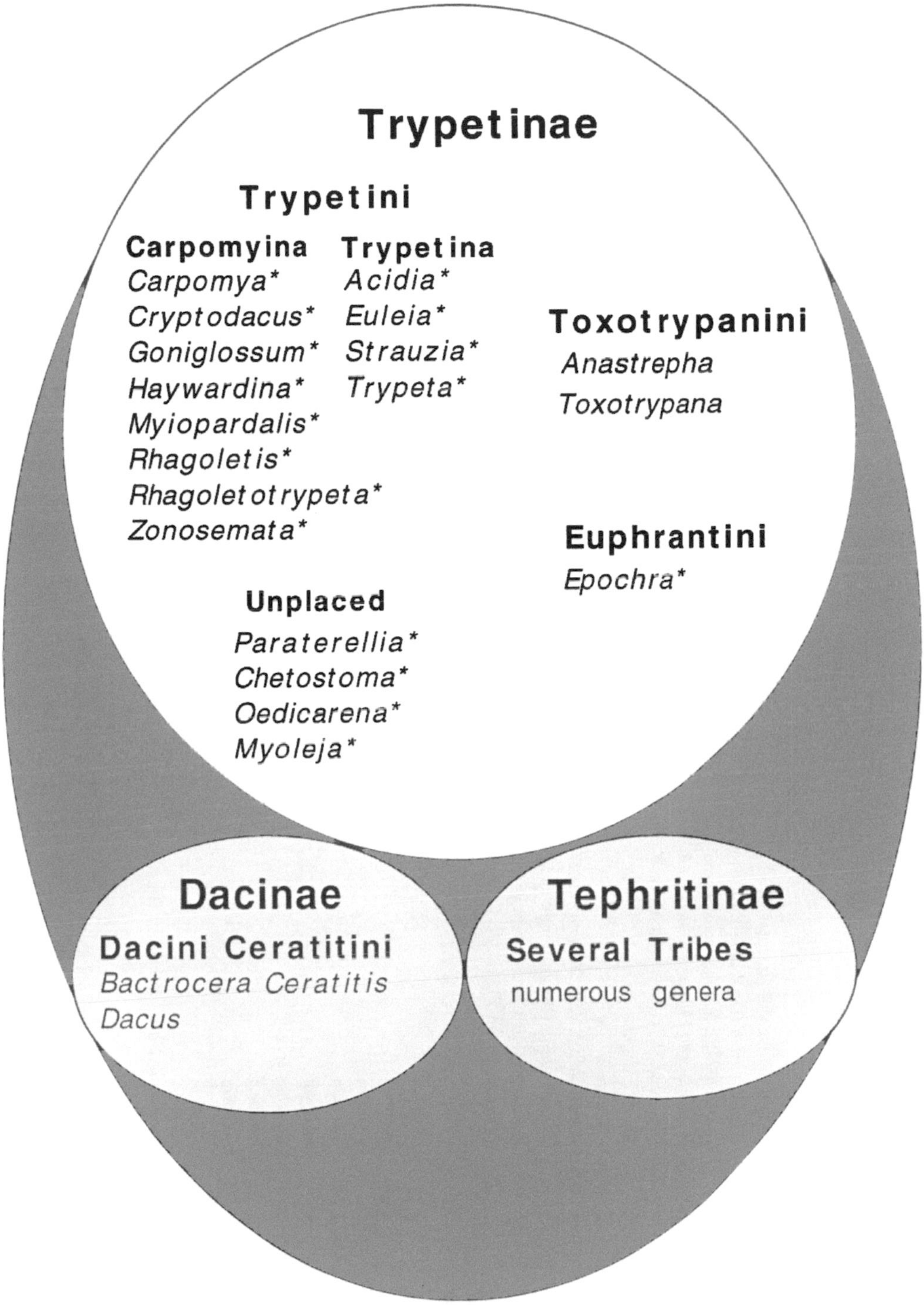


**FIGURE 9.1** Taxonomic classification of the Nearctic genera of Tephritidae plus *Acidia* following Foote et al. (1993). Subfamilies are shown, as well as selected tribes and subtribes. Starred genera were included in the morphological analysis of Jenkins (1996).

**TABLE 9.1**
**Classification, Hosts, and Geographic Ranges of the Carpomyina**

| Genus/Species | Status | Host Plant(s) | HP Family | Distribution | Reference(s) |
|---|---|---|---|---|---|
| ***Carpomya* Costa** | | | | | |
| *C. incompleta* (Becker) | s | *Ziziphus jujuba* Mill., *Z. lotus* (L.) Lam, *Z spina-christi* (L.) Willd., *Z. sativus* Gaertn. | Rhamnaceae | PA: S. Europe, Middle East, Egypt, Sudan, Ethiopia | White and Elson-Harris 1992, Kandybina 1977 |
| *C. schineri* (Loew) | s | *Rosa canina* L., *R. beggerana* Schrenk, *R. damascena* Mill., *R. gallica* L., *R. kokanica* (Rgl.) Juz., *R. pulverulenta* M. B., *R. rugosa* Thunb., *R. spinosissima* L., *R. villosa* L., *Rosa* sp. | Rosaceae | PA: Cent. Europe to Kazakstan and Israel | White and Elson-Harris 1992, Kandybina 1977 |
| *C. tica* Norrbom | s | Unknown | | NT: Costa Rica | Norrbom 1997 |
| *C. vesuviana* Costa | s | *Ziziphus jujuba* Mill., *Z. mauritiana* Lam., *Z. nummularia* (Burm.) Wight and Arn., *Z. rotundifolius* Lam., *Z. sativus* Gaertn., {*Psidium guajava* L.} | Rhamnaceae {Myrtaceae} | PA-OR: Italy, Bosnia, Caucasus, Cent. Asia, Pakistan, India, Thailand | Kandybina 1977 |
| ***Cryptodacus* Hendel** | | | | | |
| *C. lopezi* Norrbom | s | Unknown | | NT: Guatemala | Norrbom 1994 |
| *C. obliquus* Hendel | s | Unknown | | NT: Bolivia, Peru | Norrbom 1994 |
| *C. ornatus* Norrbom | s | Unknown | | NT: Colombia, N. Brazil | Norrbom 1994 |
| *C. parkeri* Norrbom | s | Unknown | | NT: Costa Rica | Norrbom 1994 |
| *C. quirozi* Norrbom | s | Unknown | | NT: Mexico (Veracruz) | Norrbom 1994 |
| *C. silvai* Lima | s | *Loranthus* sp. | Loranthaceae | NT: S. Brazil | Norrbom 1994 |
| *C. tau* (Foote) | s | Unknown | | NT: Mexico, Guatemala | Norrbom 1994 |
| *C. tigreroi* Norrbom | s | Unknown | | NT: Ecuador | Norrbom 1994 |
| ***Goniglossum* Rondani** | | | | | |
| *G. wiedemanni* (Meigen) | s | *Bryonia cretica* L., *B. syriaca* Boiss. | Cucurbitaceae | PA: Europe, Israel | Freidberg and Kugler 1989 |
| ***Haywardina* Aczél** | | | | | |
| *H. cuculi* (Hendel) | s | *Solanum trichoneuron* Lillo, {*S. argentium* Bitter and Lillo} | Solanaceae | NT: N. Argentina | Norrbom 1994 |
| *H. cuculiformis* (Aczél) | s | {*Solanum* sp.} | {Solanaceae} | NT: Peru | Norrbom 1994 |
| *H. obscura* Norrbom | s | Unknown | | NT: Argentina | Norrbom 1994 |
| *H. bimaculata* Norrbom | s | Unknown | | NT: Peru, Ecuador | Norrbom 1994 |

| | | | | | |
|---|---|---|---|---|---|
| ***Myiopardalis* Bezzi** | | | | | |
| *M. pardalina* (Bezzi) | s | *Cucumis melo* L. (prob. *Melo* of auth.), *C. sativus* L., *Ecballium elaterium* (L.) A. Rich., "Wild species of Cucurbitaceae"; [*Citrullus lanatus* (Thumb.) Matsum. and Nakai (= *C. vulgaris* Schrad.), *Cucurbita* spp.] | Cucurbitaceae | PA-OR: Caucasus, Turkey, Cyprus, and Egypt to W. India | Kandybina 1977, Freidberg and Kugler 1989, White and Elson-Harris 1992 |
| ***Rhagoletis* Loew (by species group)[a]** | | | | | |
| ***R. alternata* Group** | | | | | |
| *R. alternata* Fallén | s | *Rosa acicularis* Lindl., *R. alberti* Rgl. | Rosaceae | PA: Europe, Altai, S. Siberia | Kandybina 1977 |
| | | *R. canina* L., *R. rugosa* Thumb. | | PA: Far E. Russia, Japan | |
| | | *R. spinosissima* L., *R. villosa* L. | | PA: S.E. Kazakstan, N.E. Kirghizia | |
| | hr/us? | *Lonicera xylosteum* L. | Caprifoliaceae | | |
| *R. basiola* (Osten Sacken) | s | Many *Rosa* spp. (see Balduf 1959) | Rosaceae | NA: W. to E. North America | Balduf 1959 |
| *R. turanica* (Rohdendorf) | s | *Rosa kokanica* (Rgl.) Juz., *R. canina* L., *R. spinosissima* L. | Rosaceae | PA: Kazakstan, S. Kirghizia | Kandybina 1977, Korneyev and Merz 1997 |
| ***R. cerasi* Group** | | | | | |
| *R. almatensis* Rohdendorf | s | *Lonicera stenantha* Pojark. | Caprifoliaceae | PA: S.E. Kazakstan to N. Kirghizia | Kandybina 1977 |
| *R. berberidis* Jermy | s | *Berberis vulgaris* L. | Berberidaceae | PA: Cent. to E. Europe | Kandybina 1977 |
| *R. cerasi* (L.) | s | *Lonicera tartarica* L., *L. xylosteum* L. | Caprifoliaceae | PA: Europe | Kandybina 1977 |
| | | {*Lycium barbarum* L.} | {Solanaceae} | PA: Europe | White and Elson-Harris 1992 |
| | | {*Vaccinium myrtillus* L.} | {Ericaceae} | | White and Elson-Harris 1992 |
| *R.* nr. *cerasi* | hr/us? | *Prunus avium* (L.) L., *P. cerasus* L., *P. mahaleb* L., [*P. serotina* Ehrh.] | Rosaceae | PA: Cent. Europe | Kandybina 1977 |
| *R. faciata* Rohdendorf | hr/s? | *Lonicera* sp. (probably *L. xylosteum* L.) | Caprifoliaceae | PA: N.W. to E. Kazakstan | Korneyev and Merz 1997 |
| *R. nigripes* Rohdendorf | s | *Lonicera* sp. | Caprifoliaceae | PA: Tadjikistan, Kirghizia, N.E. Uzbekistan | Korneyev and Merz 1997 |
| ***R. cingulata* Group** | | | | | |
| *R. chionanthi* Bush | s | *Chionanthus virginicus* L. | Oleaceae | NA: S.E. | Bush 1966 |
| *R. cingulata* (Loew) | s | *Prunus serotina* Ehrh. *P. pensylvanica* L. (rare), *P. virginiana* L. (rare) [*P. avium* (L.) L., *P. cerasus* L., *P. mahaleb* L.] | Rosaceae | NA: N.E. N. America, N.C. Mexico | Bush 1966 |

**TABLE 9.1 (continued)**
**Classification, Hosts, and Geographic Ranges of the Carpomyina**

| Genus/Species | Status | Host Plant(s) | HP Family | Distribution | Reference(s) |
|---|---|---|---|---|---|
| *R. turpiniae* Hernández-Ortiz | s | *Turpinia insignis* (H.B. and K.) Tulane, *T. occidentalis breviflora* Croat | Staphyleaceae | NT: Mexico (Cent. to S. Veracruz) | Hernández-Ortiz 1993 |
| *R. indifferens* Curran | s | [*Prunus avium* (L.) L.], *P. emarginata* (Dougl.) Dietr., [*P. salicina* Lindl.], *P. subcordata* Benth., *P. virginiana* var. *demissa* (Nut.) Torrey | Rosaceae | NA: S.W. Canada to N.W. U.S. | Bush 1966 |
| *R. osmanthi* Bush | s | *Osmanthus americanus* (L.) Benth. and Hook. | Oleaceae | NA: S.E. U.S. | Bush 1966 |
| ***R. ferruginea*** **Group** | | | | | |
| *R. adusta* Foote | s | Unknown | | NT: S.E. Brazil (São Paulo) | Foote 1981 |
| *R. blanchardi* Aczél | s | *Lycopersicon esculentum* Mill., {*Citrus aurantium* L. (?)} | Solanaceae {Rutaceae} | NT: S. Cent. Brazil, Bolivia, Argentina | Foote 1981, Bush, pers. obs. |
| *R. ferruginea* Hendel | s | *Solanum* sp. (Joá) | Solanaceae | NT: S. Brazil to N. Argentina | Foote 1981 |
| | | *Lycopersicon esculentum* Mill. | Solanaceae | NT: S. Brazil to N. Argentina | De Santos, pers. comm. |
| | | {*Citrus sinensis* (L.) Osbeck (?)} | {Rutaceae} | NT: Brazil | Foote 1981 |
| ***R. flavicincta*** **Group (appears closely related to the Nearctic** ***R. ribicola*** **group)** | | | | | |
| *R. flavicincta* (Loew) | s | *Lonicera korolkowii* Stapf, *L. stenantha* Pojark., *L. nummularifolia* Jaub. and Spach, *Lonicera sp.* | Caprifoliaceae | PA: Europe, Russia, Kazakstan, Kirghizia, W. Uzbekistan | Kandybina 1977, Korneyev and Merz 1997 |
| *R. obsoleta* Hering | s? | *Lonicera* sp. | Caprifoliaceae | PA: W. Europe | Kandybina 1977 |
| *R. reducta* Hering | s? | *Lonicera gibbiflora* (Rupr.) Dipp., *L. ruprechtiana* Rgl., *L. xylosteum* L. | Caprifoliaceae | PA: Russia (Khabarovsk, Vladivostock), N. China | Kandybina 1977 |
| *R. rumpomaculata Hardy* | *s* | Unknown | | OR: Nepal | Hardy 1964 |
| *R. scutellata* Zia | s | *Lonicera ruprechtiana* Rgl. | Caprifoliaceae | PA: N.E. China (Gansu) | Kandybina 1977 |
| ***R. meigenii*** **Group (appears closely related to** ***Carpomya, Goniglossum,*** **and** ***Myiopardalis*****)** | | | | | |
| *R. caucasica* Kandybina and Richter | s | *Berberis* sp. | Berberidaceae | PA: Russia (Caucasus) | Kandybina 1977 |
| *R. chumsanica* (Rohdendorf) | s | *Berberis heteropoda* Schrenk | Berberidaceae | PA: S. Kazakstan | Kandybina 1977 |
| *R. kurentsovi* (Rohdendorf) | s | *Berberis amurensis* Rupr., *B. ruprechtii* Kom. | Berberidaceae | PA: Russia (Amurskaya) | Kandybina 1977 |
| *R. meigenii* (Loew) | s | *Berberis vulgaris* L. | Berberidaceae | PA: Europe, Caucasus, [N.E. U.S.] | Kandybina 1977 |
| *R. rohdendorfi* Korneyev and Merz | s | Swept from *Berberis* sp. | {Berberidaceae} | PA: Kirghizia | Korneyev and Merz 1997 |
| *R. samojlovitshae* (Rohdendorf) | s | *Berberis heteropoda* Schrenk | Berberidaceae | PA: S. Kazakstan, N. C. Kirghizia | Kandybina 1977 |

| | | | | | |
|---|---|---|---|---|---|
| ***R. nova* Group** | | | | | |
| *R. conversa* (Brèthes) | s | *Solanum tomatillo* Rémy. (= *S. ligustrinum* Lodd.) | Solanaceae | NT: Cent. Chile | Frías et al. 1987 |
| *R.* nr. *conversa* | hr/us? | *S. nigrum* L. | Solanaceae | NT: Cent. Chile | Frías, personal observation |
| *R. lycopersella* Smyth | s | *Lycopersicon pimpinellifolium* (L.) Mill., *L. esculentum* Mill. | Solanaceae | NT: Peru | Foote 1981 |
| *R. nova* (Schiner) | s | *Solanum muricatum* Ait. | Solanaceae | NT: Cent. Chile | Foote 1981 |
| *R. penela* Foote | s | Unknown | | NT: Cent. Chile | Foote 1981 |
| *R. tomatis* Foote | s | *Lycopersicon esculentum* Mill. | Solanaceae | NT: Peru-N. Chile | Foote 1981 |
| *R. willinki* Aczél | s | Unknown | | NT: W. Cent. Argentina | |
| ***R. pomonella* Group** | | | | | |
| *R. cornivora* Bush | s | *Cornus obliqua* Raf., *C. amomum* Mill., {*C. stolonifera* Michx.} | Cornaceae | NA: U.S. to S.E. Canada | Bush 1966 |
| *R. mendax* Curran | s | *Vaccinium corymbosum* L., *V. stamineum* L., *V. angustifolium* Aiton (= *V. ashi* of auth.), *Gaylussacia bacatta* (Wangenh.), *G. frondosa* (L.) Torr. and A. Gray, *G. dumosa* (Andrews) Torr. and A. Gray, {*Gautheria procumbens* L.}, {*Oxycoccoides* spp.} | Ericaceae | NA: U.S. to S.E. Canada | Payne and Berlocher 1995a |
| *R.* nr. *mendax* | us | *Cornus florida* L. | Cornaceae | NA: E. U.S. | Bush 1966 |
| *R. pomonella* (Walsh) | s | *Crataegus brainerdii* Sarg., *C. brachyacantha* Engelm. and Sarg., *C. crus-galli* L., *C. flabellata* (Spach) Kirchn., *C. flava* Ait., *C. greggiana* Egglest., *C. holmesiana* Ashe, *C. invisa* Sarg., *C. macrosperma* Ashe, *C. marshalii* Egglest., *C. mollis* (T. and G.) Scheele, *C. pedicellata* Sarg., *C. pruinosa* (Wendel.) K. Koch, *C. punctata* Jacq., *C. viridis* L., *C. texana* Buck | Rosaceae | NA: E. U.S. to S.E. Canada | Berlocher and Enquist 1993, Berlocher and McPheron 1996, Reissig and Smith 1978, Smith 1988 |
| | | (*C. douglasii* Lindl, [*C. monogyna* Jacq.]) | Rosaceae | NA: N.W. U.S. | McPheron 1990 |
| | | Rare or unconfirmed records: | | | |
| | | *Amelanchier bartramiana* (Tausch) Roemer, *A. spicata* (prob. *A. sanguianea* (Pursh) D.C.), | Rosaceae | NA: E. U.S. | Bush 1966, Wasbauer 1972 |
| | | *Aronia arbutifolia* L., *A. melanocarpa* Michx., [*Cotoneaster* sp. or *Pyracantha* sp.], [*Prunus persica* (L.) Batsch] | Rosaceae | | |

**TABLE 9.1 (continued)**
**Classification, Hosts, and Geographic Ranges of the Carpomyina**

| Genus/Species | Status | Host Plant(s) | HP Family | Distribution | Reference(s) |
|---|---|---|---|---|---|
| *R.* nr. *pomonella* | hr/us? | [*Malus pumila* L., *M. baccata* L.] | Rosaceae | NA: [N.E. U.S. to S.E. Canada] | Bush 1966 |
| *R.* nr. *pomonella* | gr/us? | *Crataegus mexicana* (Loud.) Rehd. | Rosaceae | NA: N. Cent. Mexico | Berlocher and McPheron 1996 |
| *R.* nr. *pomonella* | hr/us? | [*Malus pumila* L.] | Rosaceae | NA: N. Cent. Mexico | Bush 1966 |
| *R.* nr. *pomonella* | hr? | [*Prunus cerasus* L.] | Rosaceae | NA: [N. Cent., W. North America] | Shervis et al. 1970 |
| *R.* nr. *pomonella* | hr/us? | *Prunus angustifolia* Marsh., *P. umbelata* Ell., [*P. domestica* L. ?] | Rosaceae | NA: S.E. U.S. | Bush 1966 |
| | | | | NA: [NE U.S., S.E. Canada] | Bush 1966 |
| *R.* nr. *pomonella* | hr? | [*Rosa rugosa* Thunb.] | Rosaceae | NA: [U.S.A., MA, Cape Cod] | Prokopy and Berlocher 1980 |
| *R.* nr. *pomonella* | us | *Vaccinium arboreum* Marshall | Ericaceae | NA: S.E. U.S. | Payne and Berlocher 1995b |
| *R. zephyria* Snow | s | *Symphoricarpos albus* (L.) Blake, *S. albus* (L.) var. *laevigatus* (Fern.) Blake, *S. occidentalis* Hooker | Caprifoliaceae | NA: N. Cent. to N.E. U.S. to S.E. Canada | Bush and Smith, pers. obs. |
| | | | | NA: N.W. U.S. to S.W. Canada (N.E. NA) | Bush and Smith, pers. obs. |
| | | | | NA: N. Cent. U.S., S. Cent. Canada | Bush and Smith, pers. obs. |
| ***R. psalida* Group** | | | | | |
| *R. metallica* Hendel | s | Unknown | | NT: Venezuela, Peru | Foote 1981 |
| *R. psalida* Hendel | s | *Solanum tuberosum* L. | Solanaceae | NT: Peru, Bolivia | Foote 1981 |
| *R. rhytida* Hendel | s | Unknown | | NT: Ecuador, Bolivia | Foote 1981 |
| ***R. ribicola* Group (appears close to Palearctic *flavicincta* group)** | | | | | |
| *R. berberis* Curran | s | *Mahonia nervosa* (Pursh) Nutt., *M. aquifolium* (Pursh) Nutt. | Berberidaceae | NA: N.W. U.S. to S.W. Canada | Bush 1966 |
| *R. ribicola* Doane | s | *Ribes aureum* Pursh, *R. grossularia* L., [*R. sativum* Syme (= *vulgare* of auth.)] | Saxifragaceae | NA: N.W. U.S. to S.W. Canada | Bush 1966 |
| ***R. striatella* Group** | | | | | |
| *R. jamaicensis* Foote | s | Unknown | | NT: Jamaica, Costa Rica, Venezuela | Foote 1981, Norrbom et al. 1999b |
| *R. macquartii* (Loew) | s | Unknown | | NT: Cent. Brazil | Foote 1981 |
| *R. striatella* Wulp | s | *Physalis heterophylla* Nees., *P. longifolia* Nutt., *P. ixocarpa* Best. | Solanaceae | NA: Cent. to E. U.S. to N. Mexico NT: Mexico | Jenkins 1996, Bush, pers. obs. |

| | | | | | |
|---|---|---|---|---|---|
| ***R. suavis* Group** | | | | | |
| *R. completa* Cresson | s | *Juglans nigra* L., *J. microcarpa* Berl., *J. hirsuta* Manning, *J. major* (Torr.) Heller, [*J. regia* L.], | Juglandaceae | NA: S. Cent. to N. Cent. U.S., N.E. Mexico | Berlocher and Bush, pers. obs. |
| | | (*J. californica* S. Watson, *J. hindii* Rehd.), [*J. regia* L.], | Juglandaceae | NA: U.S. (California) | Boyce 1934 |
| | | [*Prunus persica* (L.) Batsch (rare)] | Rosaceae | NA: [U.S. (California)] | Boyce 1934 |
| *R. boycei* Cresson | s | *Juglans major* (Torr.) Heller, [*J. regia* L.] | Juglandaceae | NA: S.W. U.S. to N. Cent. Mexico | Berlocher and Bush, pers. obs. |
| *R. juglandis* Cresson | s | *Juglans major* (Torr.) Heller, [*J. regia* L.] | Juglandaceae | NA: S.W. U.S. to N. Cent. Mexico | Berlocher and Bush, pers. obs. |
| *R. ramosae* Hernández-Ortiz | s | *Juglans mollis* Engl. (= *J. major* var. *glabrata*) | Juglandaceae | NT: W.C. Mexico (Guerr. and Mich.) | Hernández-Ortiz 1985 |
| *R. suavis* (Loew) | s | *Juglans nigra* L., *J. cinerea* L., [*J. regia* L.], [*J. sieboldiana* Maxim.], | Juglandaceae | NA: E. U.S. | Bush 1966 |
| | | [*Prunus persica* (L.) Batsch (rare)] | Rosaceae | NA: U.S. (New York) | Dean 1969 |
| *R. zoqui* Bush | s | *Juglans mollis* Engl., *J. pyriformis* Liebm. | Juglandaceae | NT: E. Cent. Mexico, E. Mexico (Veracruz) | Berlocher and Bush, pers. obs., M. Aluja, pers. obs. |
| ***R. tabellaria* Group** | | | | | |
| *R. ebbettsi* Bush | s | Unknown | | NA: U.S. (California) | Bush 1966 |
| *R. electromorpha* Berlocher | s | *Cornus drummondii* C. A. Mey., *C. racemosa* Lam. | Cornaceae | NA: N. Cent. U.S. | Berlocher 1984 |
| | | | Elaeagnaceae | | |
| *R.* nr. *electromorpha* | us | *Shepherdia argentea* (Pursh) Nutt., *S. canadensis* (L.) Nutt. | Elaeagnaceae | NA: N. U.S. to S. Canada | Bush and Smith, pers. obs. |
| *R. persimilis* Bush | s | *Disporum tachycarpum* (S. Wat.) Benth. and Hook. | Liliaceae | NA: N.W. U.S. to S.W. Canada | Bush and Smith, pers. obs. |
| *R. tabellaria* (Fitch) | s | *Cornus stolonifera* Michx. | Cornaceae | NA: N. U.S. to S. Canada | Bush 1966, Wasbauer 1972 |
| *R.* nr. *tabellaria* | hr/us? | *Vaccinium parvifolium* Smith, *V. ovalifolium* Smith | Ericaceae | NA: N.W. U.S. to S.W. Canada | Bush 1966 |
| ***R. zernyi* Group** | | | | | |
| *R. flavigenualis* Hering | s | *Juniperus excelsa M.* Bieb., | Cupressaceae | PA: Turkey, N.W. Caucasus | Kandybina 1977 |
| | | *J. seravschanica* Kom., | | PA: S. Kirghizia, Tadjikistan | |
| | | *J. semiglobosa* Rgl., *J. sabina* L., *J. turkestanica* Kom. | | PA: S. Turkmenistan (Kopet-Dagh) | |
| *R. zernyi* Hendel | s | *Juniperus thurifera* L. | Cupressaceae | PA: N. Spain | Merz and Blasco-Zumeta 1995 |
| **Unplaced *Rhagoletis*** | | | | | |
| *R. acuticornis* | s | *Lycium berlandieri* Dunal | Solanaceae | NA: S.W. U.S. | Norrbom 1989 |
| *R. batava* Hering | s | *Hippophae rhamnoides* L. | Rhamnaceae | PA: N. Cent. to E. Europe, N. Caucasus, Kirghizia | Kandybina 1977 |

**TABLE 9.1 (continued)**
**Classification, Hosts, and Geographic Ranges of the Carpomyina**

| Genus/Species | Status | Host Plant(s) | HP Family | Distribution | Reference(s) |
|---|---|---|---|---|---|
| *R. bezziana* Hendel | s | Unknown | | OR: India (Uttar Pradesh) | Hendel 1931 |
| *R. emiliae* Richter | | Unknown | | PA: N. Kirghizia, Tadjikistan | Korneyev and Merz 1997 |
| *R. fausta* (Osten Sacken) | s | *Prunus cerasus* L., *P. emarginata* (Dougl.), [*P. mahaleb* L. (rare)], *P. pensylvanica* L., *P. serotina* Ehrh., *P. virginiana* var. *demissa* (Nutt.) Torr. | Rosaceae | NA: N.W. to N.E. U.S. | Bush 1966 |
| *R. juniperina* Marcovitch | s | *Juniperus monosperma* (Engelm.) Sarg., *J. virginiana* L. | Cupressaceae | NA: N.C. to N.E. to S.C. to S.E. U.S. | Bush 1966 |
| *R. magniterebra* (Rohdendorf) | s | *Berberis heteropoda* Schrenk | Berberidaceae | PA: Kazakstan, Kirghizia, N. Tadjikistan | Kandybina 1977 |
| *R. mongolica* Kandybina | s | *Juniperus sabina* L., *Juniperus* sp. | Cupressaceae | PA: S. Mongolia, E. Kirghizia | Kandybina 1977 |
| ***Rhagoletotrypeta* Aczél** | | | | | |
| *Rh. annulata* Aczél | s | *Celtis pallida* Torr. (= “granjeo huasteco”) | Ulmaceae | NT: Mexico, Costa Rica | Norrbom 1994 |
| *Rh. argentinensis* (Aczél) | s | Unknown | | NT: Argentina | Norrbom 1994 |
| *Rh. intermedia* Norrbom | s | Unknown | | NT: Mexico | Norrbom 1994 |
| *Rh. morgantei* Norrbom | s | Unknown | | NT: S. Brazil | Norrbom 1994 |
| *Rh. parallela* Norrbom | s | *Celtis* sp. (= “tala”) | Ulmaceae | NT: Argentina | Norrbom 1994 |
| *Rh. pastranai* Aczél | s | *Celtis iguanaea* (Jacq.) Sarg., *C. tala* Gill. | Ulmaceae | NT: S. Brazil to N. Argentina | Norrbom 1994 |
| *Rh. rohweri* Foote | s | {(*C. laevigata* Willd., *C. occidentalis* L., or *C. tenuifolis*)} | Ulmaceae | NA: (N.E. and N. Cent. U.S.) | Norrbom 1994 |
| *Rh. uniformis* Steyskal | s | *Celtis laevigata* Willd. | Ulmaceae | NA: U.S. (New Mexico, Texas) | Norrbom 1994 |
| *Rh. xanthogastra* Aczél | s | *Celtis tala* Gill., {“Ciruelo” = *Spondias* sp.} | Ulmaceae {Anacardiaceae} | NT: Argentina | Norrbom 1994 |
| ***Scleropithus* Munro** | | | | | |
| *S. glaphyrochalyps* Munro | s | *Strychnos henningsii* Gilg | Loganiaceae | AF: South Africa (Durban) | Munro 1939 |
| ***Stoneola* Hering** | | | | | |
| *S. fuscobasalis* (Hering) | s | Unknown | | NT: Peru (Chanchamayo) | Hering 1941 |

| | | | | | |
|---|---|---|---|---|---|
| ***Zonosemata* Benjamin** | | | | | |
| *Z. cocoyoc* Bush | s | *Solanum* sp. | Solanaceae | NA to NT: Cent. Mexico | Bush 1965, Norrbom 1990 |
| *Z. electa* (Say) | s | *Capsicum annuum* L., *Solanum aculeatissimum* Jacq., *S. carolinense* L., *S. melanocerasum* All., S. *melongena* L., *Lycopersicon esculentum* Mill. (rare), *Physalis subglabrata* Mack. and Bush, {*Rosa* sp.} | Solanaceae {Rosaceae} | NA: E. U.S., S.E. Canada | Bush 1965, Norrbom 1990, Wasbauer 1972 |
| *Z. macgregori* Hernández-Ortiz | s | Unknown | | NA: Mexico (Baja California) | Hernández-Ortiz 1989 |
| *Z. minuta* Bush | s | {"At fruit" of *Solanum verbascifolium* L. (probably *S. erianthum* D. Don.)} | {Solanaceae} | NT: Jamaica | Norrbom 1990 |
| *Z. scutellata* (Hendel) | s | {"En" or "ex fruto de huevo de gato" (*Solanum hirtum* Vahl or *hyporrhodium* A. Braun)} | Solanaceae | NT: Venezuela–Brazil | Norrbom 1990 |
| *Z. vidrapennis* Bush | s | *Solanum* sp. | Solanaceae | NT: Mexico | Bush 1965 |
| *Z. vittigera* (Coquillett) | s | *Solanum eleagnifolium* Cav. | Solanaceae | NA to NT: S.W. U.S. | Norrbom 1990, Bush 1965 |

Abbreviations: gr = geographic race; h = host race; s = described species; us = undescribed species; [] = plant introduced; () = fly introduced; { } doubtful host, needs verification by rearing; ? = status uncertain; PA = Palearctic; NA = Nearctic; NT = Neotropical; OR = Oriental; AF = Afrotropical.

[a] *Note: Rhagoletis* species groups are tentative and based on molecular and morphological phylogenetic relationships.

oviscape to be readily apparent in darkly pigmented carpomyines, but difficult to see in lightly pigmented species, and not present at all in *Zonosemata* species. However, Norrbom (personal communication) argues that the desclerotized area is definitely present in *Zonosemata*, and that Jenkins overcleared the specimens that he examined. Kandybina (1977) showed that several Palearctic carpomyine species, formerly placed in *Zonosema*, *Megarhagoletis,* and *Microrhagoletis* (currently synonyms of *Rhagoletis*), have no preoral teeth.

The only molecular study to date that addresses the issue of Carpomyina monophyly is that of Han and McPheron (1997). Their neighbor-joining analysis of mitochondrial 16S ribosomal RNA genes from 34 tephritid species supported the monophyly of Carpomyina. However, statistical support for the relationship was weak.

## 9.2.2 Sources of Data

### 9.2.2.1 Jenkins' Analysis

One of the primary works that we used in this chapter to evaluate phylogenetic relationships among carpomyine taxa is that of Jenkins (1996). Jenkins (1996) analyzed morphological features from 87 tephritid taxa, designing his taxon sample on the basis of the classification of Tephritidae of Foote et al. (1993) described above. Jenkins' analysis included eight genera in the subtribe Carpomyina, four genera in the subtribe Trypetina (still classified in the Trypetini), four genera that were unplaced within Trypetini (two now placed in Trypetini, and two now in the Carpomyini, subtribe Paraterelliina), and *Euphranta canadensis* (as *Epochra canadensis*) which was used as the outgroup taxon (see Figure 9.1).

Jenkins (1996) based his analysis on 77 morphological features (76 phylogenetically informative) of the head, thorax, wings, legs, abdomen, and male and female genitalia. Only chitinized structures were examined, in some cases using an electron microscope. Jenkins also restricted his analysis to characters with discrete variation, avoiding those that varied continuously. Jenkins' phylogenetic analysis using PAUP 3.1.1 yielded 13,100 most parsimonious reconstructions (MPRs) before exhausting the 8 MB of RAM that he had available in the PowerMac 7100/66 used in the analysis. Thus, neither the actual number of MPRs was determined, nor was whether or not the 13,100 trees obtained actually represented the shortest trees.

Jenkins' (1996) strict consensus of these 13,100 trees was notable in that the relationships of many taxa were unresolved within it. For example, most of the Old World *Rhagoletis* were unresolved with respect to their relationships to species in other carpomyine genera and other *Rhagoletis* species. Since one of our goals is to obtain working hypotheses based on morphological data that we can test using molecular data, we repeated the PAUP search when new computer hardware (PowerMac 8600/200, 32 MB RAM for PAUP) became available to us.

This analysis yielded 32,700 MPRs before exhausting the tree file size capacity of PAUP (length 531, consistency index (ci) = 0.576, retention index (ri) = 0.789). Thus, we still do not know how many MPRs there are and if the trees we have found represent the shortest trees possible. A strict consensus of these 32,700 trees was similar in topology (with notable exceptions, see below) to Jenkins' (1996) consensus tree. Because the strict consensus tree did not provide many testable hypotheses of relationships for carpomyine taxa, we summarized the 32,700 trees as a 50% majority-rule consensus (Figure 9.2). We realize that the 50% majority-rule consensus might mislead some readers into considering a particular topology to be more strongly supported by data than another by virtue of its appearance in *most* of the trees. However, this is not our intention. We simply are trying to obtain testable hypotheses of relationship, and the 50% majority-rule consensus of the 32,700 trees provides interesting, and in some cases not unexpected, relationships for us to test.

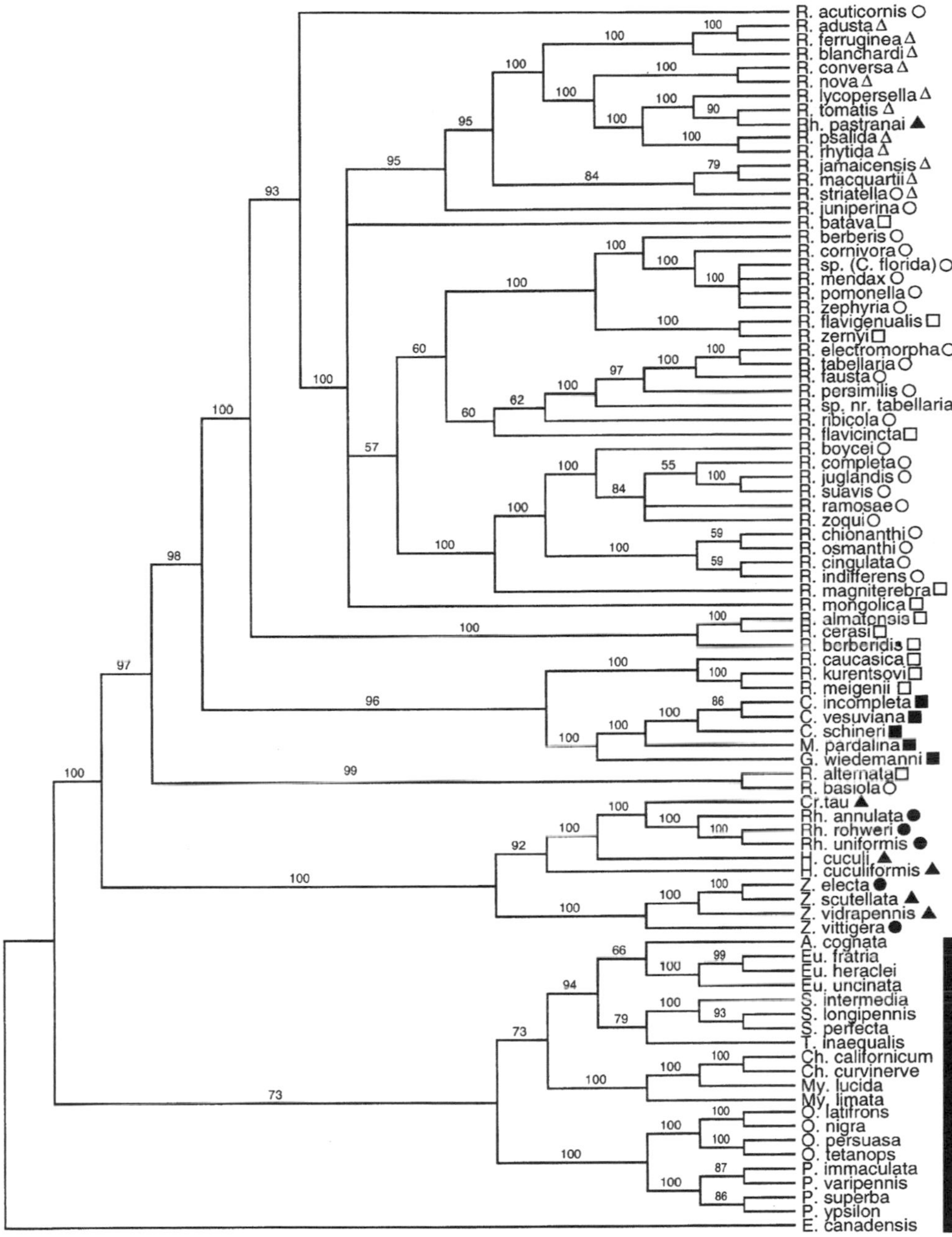


**FIGURE 9.2** Carpomyina phylogeny based on analysis of morphological characters. The phylogenetic tree was generated by reanalysis of the data set of Jenkins (1996) which included 87 taxa representing eight genera within Carpomyina (symbols; see below) and nine genera otherwise classified within the subfamily Trypetinae (solid bar). *Euphranta canadensis* was designated as the outgroup. Nearctic, Palearctic, and Neotropical *Rhagoletis* are designated with open circles (○), open squares (□), and open triangles (△), respectively, while other Nearctic, Palearctic, and Neotropical genera within the Carpomyina are indicated by closed circles (●), closed squares (■), and closed triangles (▲), respectively. Parsimony analysis using PAUP, based on 76 informative characters, yielded 32,700 MPRs of length 531 (ci = 0.576, ri = 0.789) before exhausting the 32 MB of RAM available on the Power Mac 8600/200 used for the analysis. The tree shown is a 50% majority-rule consensus of these 32,700 trees, and numbers on the branches indicate the percentage of the 32,700 trees that contain that particular clade.

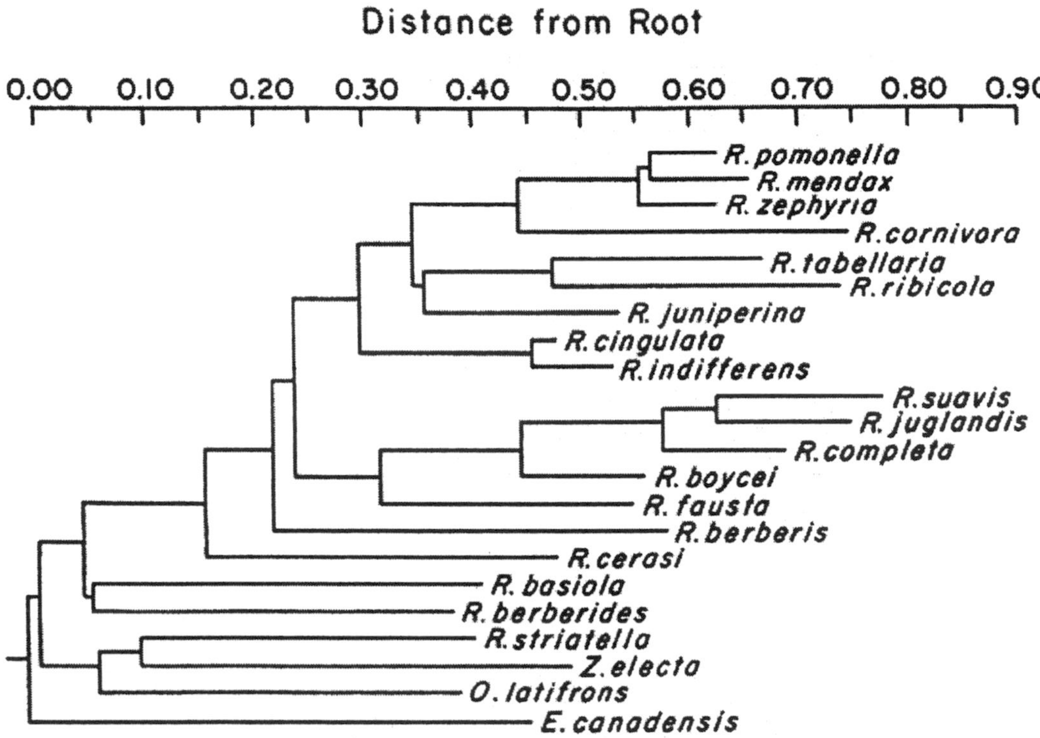


**FIGURE 9.3** Phylogeny of *Rhagoletis* based on analysis of allozyme variation. Wagner distance tree based on Manhattan distances. (From Berlocher, S.H. and Bush, G.L., *Syst. Zool.* 31: 136–155, 1982. With permission.)

#### 9.2.2.2 Other Data Sources

The discussion below also relies on many other data sources. Bush's (1966) monograph on the genus *Rhagoletis*, based on analysis of morphology, cytology, geographic distribution, and host use, provides the foundation for much of the taxonomy of *Rhagoletis*, in particular the Nearctic species. Berlocher and Bush (1982) analyzed phylogenetic relationships of *Rhagoletis* and related genera using allele frequency data at electrophoretic loci. One of the phylogenetic trees published in Berlocher and Bush (1982) is included here as Figure 9.3.

We have drawn heavily from Norrbom (1994) in our discussions of the relationships of the genera *Haywardina*, *Cryptodacus,* and *Rhagoletotrypeta*, and Norrbom (1997) for our discussion of the taxonomic status and relationships of *Carpomya*, *Goniglossum,* and *Myiopardalis*.

Phylogenetic relationships based on mitochondrial DNA sequences have been analyzed in papers by both McPheron and Han (1997) and Smith and Bush (1997). A phylogenetic tree representative of the former analysis can be found elsewhere in this volume (Han and McPheron, Chapter 5), while a cladogram similar to one published in Smith and Bush (1997) is included here as Figure 9.4.

### 9.2.3 Tour of the Taxa

In the following sections, we summarize the state of knowledge on the phylogenetic relationships of carpomyine species, discussing both inter- and intrageneric relationships. We will examine the phylogenetic relationships of the subtribe using the genus *Rhagoletis* as a frame of reference. *Rhagoletis* is by far the most speciose and widespread of the carpomyine genera, with 62 described

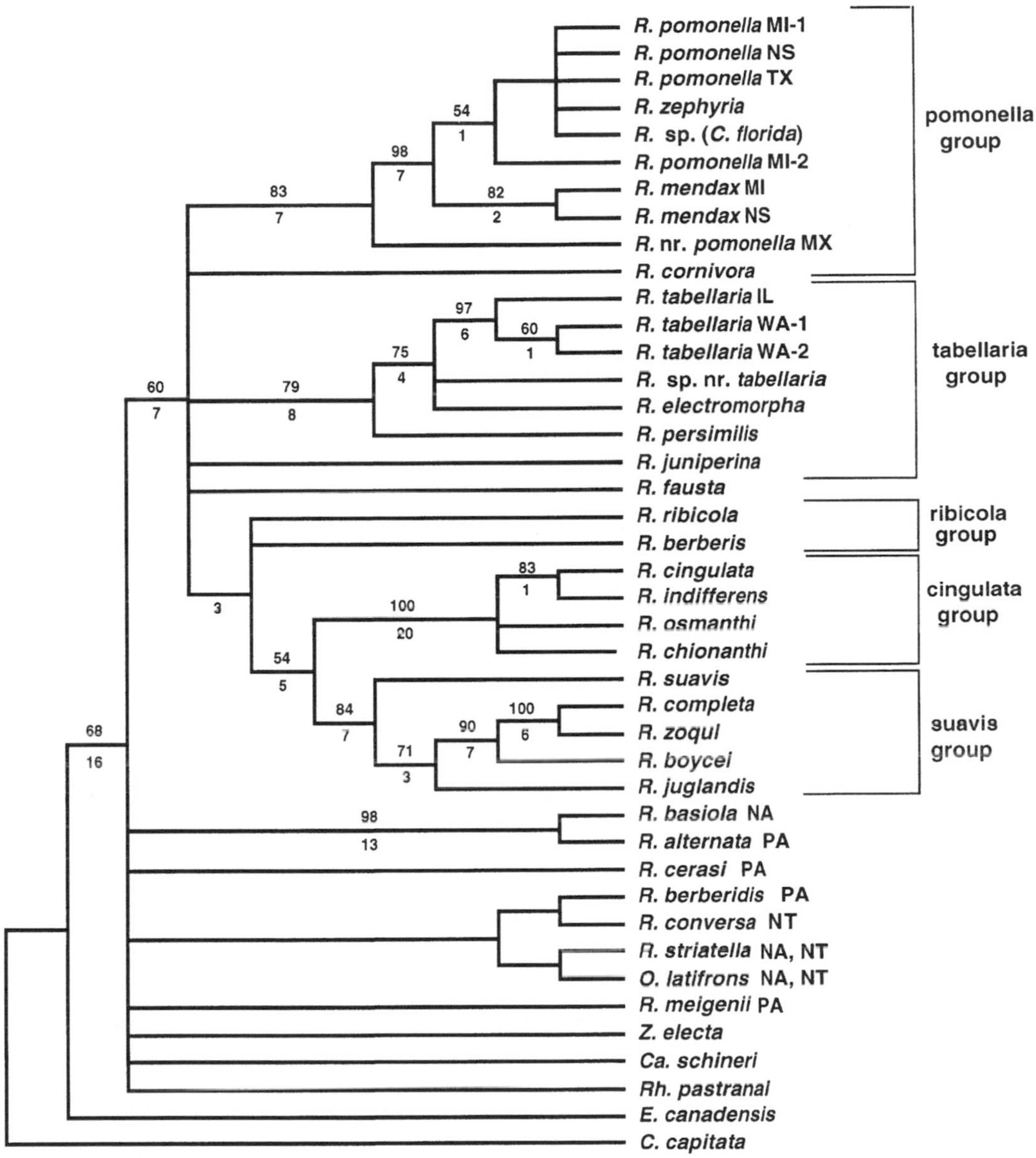


**FIGURE 9.4** Mitochondrial COII phylogeny of the genus *Rhagoletis*. Tree shown is a strict consensus of 200 MPRs (tree length = 840, ci = 0.377, ri = 0.537) obtained from analysis of nucleotide sequences of mitochondrial COII genes from 42 individuals in 26 *Rhagoletis* species and six other genera. Taxa in North American species groups are indicated, and distributions of other *Rhagoletis* species are shown following the taxon name. NA = Nearctic; PA = Palearctic; NT = Neotropical. Bootstrap values >50 are shown above branches (100 replicates), with branch lengths shown below branches.

valid species (not including new species or synonymies in Korneyev and Merz 1997) which occur in the Palearctic, Oriental, Nearctic, and Neotropical Regions (Table 9.1; Norrbom et al. 1999a,b). Data from several recent studies (Jenkins 1996; Han and McPheron 1997; McPheron and Han 1997; Smith and Bush 1997) indicate that the genus *Rhagoletis* as currently defined is not monophyletic. Seven of the other carpomyine genera have been implicated in some way with disruption of *Rhagoletis* monophyly.

#### 9.2.3.1 Nearctic Taxa

Our discussion of Nearctic taxa will focus exclusively on *Rhagoletis*. Although *Zonosemata* is primarily a Nearctic genus (see Table 9.1), it appears to be closely related to the Neotropical genera *Cryptodacus* and *Haywardina*, and the predominantly Neotropical genus *Rhagoletotrypeta*, and thus will be considered with them.

The most intensively studied flies in the subtribe Carpomyina are the North American *Rhagoletis* species. Bush (1966) established five taxonomic species groups (*pomonella*, *tabellaria*, *ribicola*, *cingulata*, and *suavis*) which contain the majority of the Nearctic *Rhagoletis* species (Table 9.1). Phylogenetic analysis of mitochondrial cytochrome oxidase II (COII) nucleotide sequences recovered a moderately well-supported monophyletic assemblage that included all members of these five North American species groups (Smith and Bush 1997). Monophyly of these five species groups was also supported by studies using mitochondrial 16S rDNA sequences (McPheron and Han 1997), electrophoretic enzyme alleles (Berlocher and Bush 1982), and cytogenetic data (Bush 1966). The latter study showed that only the *Rhagoletis* species in these five North American species groups have a pair of "dot" acrocentric chromosomes.

However, the observed monophyly may be a reflection of the fact that few Eurasian and Neotropical *Rhagoletis* species were included in these analyses. Jenkins' (1996) morphological cladistic analysis did not recover these five species groups as a monophyletic group and, in our reanalysis of Jenkins' data set, several Palearctic *Rhagoletis* species are grouped with the North American species (see Figure 9.2). For example, *R. batava* and *R. mongolica* are unresolved within a clade containing the combined Nearctic and Neotropical *Rhagoletis*. Thus, while it is clear that the five North American species groups are related, some other species not yet studied by molecular or cytological methods may actually belong to the same clade.

Mitochondrial DNA sequence data indicate that the combined *R. pomonella* and *R. tabellaria* species groups, plus the previously unplaced *R. fausta*, form a monophyletic assemblage (Smith and Bush 1997; McPheron and Han 1997). However, neither analyses of morphology (Bush 1966; Jenkins 1996) nor allozymes (Berlocher and Bush 1982) supported this clade. In addition, in our reanalysis of Jenkins' morphological data set (see Figure 9.2), the Palearctic species *R. flavigenualis* and *R. zernyi* form the sister group to the pomonella group (plus *R. berberis*), while *R. flavicinta* is the sister taxon to the *tabellaria* group. Unfortunately, none of the molecular studies conducted to date has included these Palearctic species, and thus sampling bias cannot be ruled out as a factor in the placement of these species.

Within the *pomonella* species group, it is clear that *R. pomonella*, *R. mendax,* and *R. zephyria* are the most closely related species. These three species are virtually indistinguishable on the basis of morphology — only slight differences in surstylus shape distinguish *R. pomonella* from *R. zephyria* (Bush 1966; Westcott 1982) and femur coloration alone distinguishes *R. pomonella* from *R. mendax* (Berlocher 1997). Although allozyme studies support the close relationships of *R. pomonella*, *R. mendax,* and *R. zephyria* (Berlocher 1981; Berlocher and Bush 1982; Berlocher et al. 1993), the exact relationships among the three species are not clear. These relationships are complicated further by the discovery of at least two undescribed species within the *pomonella* group — *R.* nr. *mendax* "*Cornus florida*," or flowering dogwood fly, which infests *Cornus florida* (Berlocher et al. 1993; "*R.* n. sp. A" of Smith and Bush 1997), and *R.* nr. *pomonella* "Vaccinium arboreum," or sparkleberry fly, which infests *Vaccinium arboreum* (Payne and Berlocher 1995b).

Figure 9.5 shows a mitochondrial COII haplotype network for sequences obtained from 18 individuals within the *pomonella* species group. In all, 11 different haplotypes were observed, and several haplotypes are shared by individuals representing different species. For example, *R. pomonella* and *R. zephyria* share haplotype 1, *R. pomonella* and the sparkleberry fly share haplotype 5, and *R. pomonella* and the flowering dogwood fly share haplotype 8. Given the relatively rapid rate of mtDNA divergence (Harrison 1989), these close genetic relationships indicate that divergence of *R. pomonella*, *R. zephyria,* and *R. mendax* (and the undescribed sparkleberry and

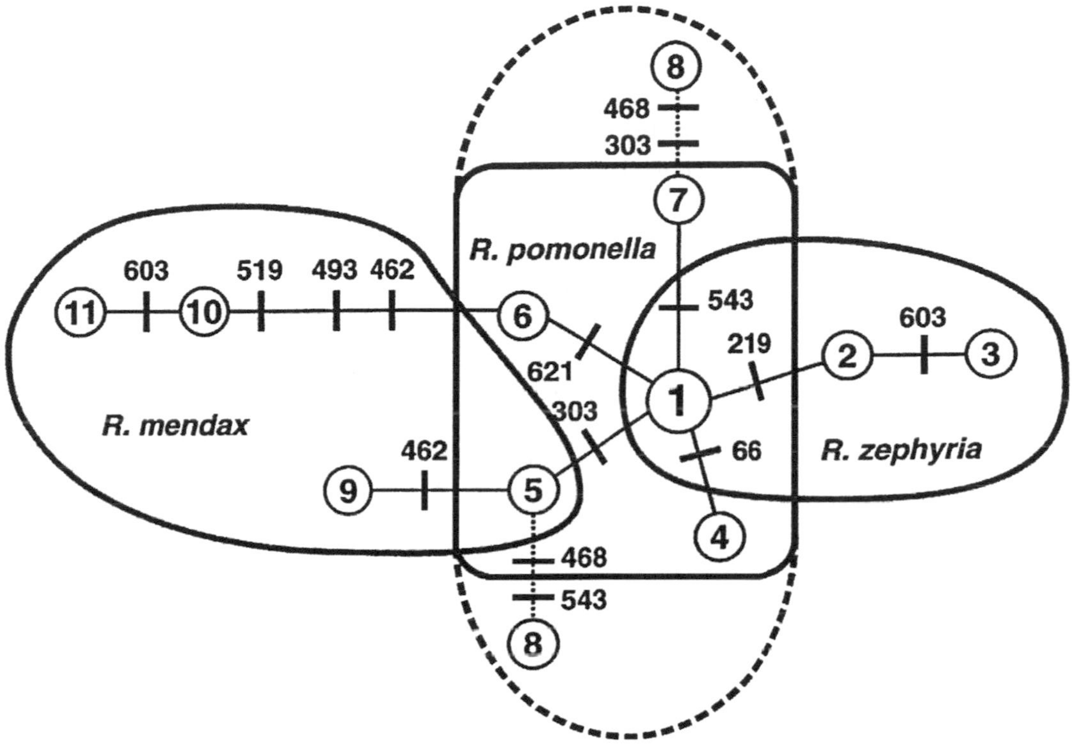


**FIGURE 9.5** Mitochondrial DNA haplotype network for the *R. pomonella* species group based on complete COII sequences (687 nt from 18 individuals representing *R. pomonella* (including the flowering dogwood fly), *R. zephyria*, and *R. mendax* (including the sparkleberry fly)). Numbers in circles represent the 11 different sequences (haplotypes) obtained from the 18 individuals. Numbered bars on lines represent single nucleotide changes between haplotypes, with numbers referring to nucleotide position in COII. Lines encircling haplotype numbers show which species contain which haplotypes. The network drawn is the single most parsimonious reconstruction for haplotype relationships, with the dotted lines encompassing haplotype 8 showing the two possibilities for its placement. Haplotype 1: *R. pomonella* NS (apple race), *R. zephyria* OR, *R. zephyria* MN1, *R. zephyria* MN2; haplotype 2: *R. zephyria* PA, *R. zephyria* ON2, *R. zephyria* MA; haplotype 3: *R. zephyria* ON1; haplotype 4: *R. pomonella* MN2 (haw race); haplotype 5: *R. pomonella* MI1 (haw race), sparkleberry fly; haplotype 6: *R. pomonella* MI2 (haw race); haplotype 7: *R. pomonella* MN1 (haw race), flowering dogwood fly; haplotype 8: *R. pomonella* TX; haplotype 9: *R. mendax* GA; haplotype 10: *R. mendax* NS; haplotype 11: *R. mendax* MI.

dogwood flies) has occurred relatively recently. Genetic distances indicate that *R. mendax* is more distantly related to *R. pomonella* than is *R. zephyria*. Further, while both *R. mendax* and *R. zephyria* share haplotypes with *R. pomonella*, there are no shared haplotypes between *R. mendax* and *R. zephyria* individuals.

The placement of an undescribed Mexican *Rhagoletis* taxon near *pomonella* (host *Crataegus mexicana*) as the sister taxon to the rest of the *pomonella* species group except *R. cornivora* is well supported by mitochondrial COII data (Smith and Bush 1997; Figure 9.4). The COII sequence from *R.* sp. nr. *pomonella* differs from the haplotype 1 sequence only by 22 transition substitutions (3.2% divergence, or $p$-distance = 0.032). Using a rate estimate for mtDNA of 2.1% sequence divergence per million years (Brower 1994) as a rough approximation, we calculate that *R.* sp. nr. *pomonella* from Mexico shared a common ancestor with its sister group approximately 1.5 million years ago. It would be particularly useful to establish the distribution and full host range of the *pomonella* group flies in Mexico. Several putative host plants other than *C. mexicana* are present in Mexico, for example, *Vaccinium stamineum* (Vander Kloet 1988), *Symphoricarpos microphyllus* (Jones 1940), and *Cornus florida*. Whether or not fruits of these plants in Mexico are infested is not known.

The monophyly of the *tabellaria* species group, except *R. juniperina*, is supported by both molecules and morphology (Jenkins 1996; Smith and Bush 1997), and the hypotheses of relationship among the included species are similar in these studies. Field work has resulted in the identification of *Disporum trachycarpum* (fairybells; Liliaceae) as a host plant for *R. persimilis*. According to the mitochondrial COII phylogeny (Smith and Bush 1997), *Rhagoletis persimilis* is the sister taxon to the remainder of the *R. tabellaria* species group (except *R. juniperina*). We have also identified a new *Rhagoletis* species ("*R.* n. sp. B" of Smith and Bush 1997), the buffaloberry fly, from *Shepherdia argentea* (thorny buffaloberry; Elaeagnaceae). Based on the COII data, the buffaloberry fly is related to the *Cornus*-infesting *R. electromorpha* and *R. tabellaria*. However, in Jenkins' morphological cladistic analysis (Jenkins 1996), the buffaloberry fly is the sister taxon to the remainder of the *R. tabellaria* species group (see Figure 9.2).

The report by Smith and Bush (1997) of *R. tabellaria* reared from *C. stolonifera* in Washington State expanded the known range of *R. tabellaria* on *C. stolonifera* to the Pacific coast. More extensive sampling will be required to determine the extent of infestation of *C. stolonifera* throughout its geographic range, which extends coast-to-coast throughout most of North America from Mexico to Alaska (Gleason and Cronquist 1963). In addition, *C. stolonifera* has a Eurasian sibling species, *C. occidentalis*, which to our knowledge has not been sampled and may harbor *R. tabellaria*-like flies.

*Rhagoletis* sp. nr. *tabellaria* (= *tabellaria* WA-2 of Smith and Bush 1997) infests *Vaccinium* spp. in the Pacific Northwest. The status of this taxon is unknown, and it may represent a host race of *R. tabellaria*, or a new species. The mitochondrial DNA data indicate that the *Cornus*-infesting *R. tabellaria* from Washington State may be more closely related to *R.* nr. *tabellaria* than they are to *Cornus*-infesting *R. tabellaria* from Illinois (Smith and Bush 1997). In either case, it appears that the evolution of *Vaccinium* infestation by *R.* nr. *tabellaria*, in the *R. tabellaria* species group, has occurred independently from the evolution of *Vaccinium* infestation by *R. mendax*, a member of the *R. pomonella* species group. One line of evidence supporting this hypothesis is that the *Vaccinium* spp. infested by *R. mendax* and *R.* nr. *tabellaria* have highly disjunct distributions (Vander Kloet 1988). Additionally, infestation of *Vaccinium* spp. is apparently a derived trait within *Rhagoletis*, with a host shift from *Cornus* to *Vaccinium* in the *R. tabellaria* group and a host shift from *Cornus* or *Crataegus* to *Vaccinium* in the *R. pomonella* group (Smith and Bush, personal observation).

The phylogenetic relationships of *R. cornivora*, *R. juniperina,* and *R. fausta*, both to each other and to the remainder of the *pomonella* and *tabellaria* species groups, are not resolved. The hypothesis that *R. cornivora* is the sister taxon to the remainder of the *pomonella* group is strongly supported by morphological and cytological characters (Bush 1966; Berlocher 1981; Jenkins 1996); morphological divergence of *R. cornivora* and the remainder of the *R. pomonella* group is relatively minor and they share the same distinctive karyotype. However, genetic divergence between *R. cornivora* and the other *R. pomonella* group species is significant, for both allozymes (Berlocher and Bush 1982; Figure 9.3) and mitochondrial COII sequences (Smith and Bush 1997; Figure 9.4). The phylogenetic placement of *R. cornivora* has been variable in allozyme trees (Berlocher 1981; Berlocher and Bush 1982), and is either poorly resolved or weakly supported in the mitochondrial DNA trees (McPheron and Han 1997; Smith and Bush 1997). In the mtDNA trees, *R. cornivora* is sometimes hypothesized to be more distantly related (e.g., sister group of *pomonella* + *tabellaria* groups) or in a clade with *R. juniperina* and *R. fausta*.

*Rhagoletis juniperina* was originally considered, on the basis of cytogenetic and morphological similarities, as a member of the *tabellaria* species group (Bush 1966). However, *R. juniperina* did not cluster with other members of the *tabellaria* group in analyses of morphology (Jenkins 1996), allozymes (Berlocher and Bush 1982), or mitochondrial DNA (McPheron and Han 1997; Smith

# *Color Figures*

**COLOR FIGURE 1** Feeding in damaged guava by females of *Anastrepha ludens* (head in hole) and *A. fraterculus* (entire fly in hole). (Photograph by E. Piedra.)

**COLOR FIGURE 2** Feeding on bird droppings by an unidentified species of Ulidiidae. (Photograph by R. Macías.)

**COLOR FIGURE 3** Regurgitated droplets deposited by *Anastrepha serpentina* female. These droplets are then reingested by the same female. (Photograph by R. Wilson.)

**COLOR FIGURE 4** "Bubbling" behavior exhibited by *Anastrepha obliqua*. (Photograph by R. Wilson.)

**COLOR FIGURE 5** Lek formed by five *Anastrepha obliqua* males in branch of citrus tree. Note the visiting female. All individuals have a colored dot on the mesonotum to facilitate quantification of individual fly behavior. (Photograph by J. Piñero.)

**COLOR FIGURE 6** Male of *Ceratitis capitata* releasing pheromone during calling bout. (Photograph by K. Kaneshiro.)

**COLOR FIGURE 7** Male of *Anastrepha hamata* in calling position. (A) Note pleural glands and evaginated anal membranes. (B) Male touching leaf pedicel with proctiger (marking territory with sexual pheromone). (Photographs by J. Piñero.)

**COLOR FIGURE 8** Mating pair of *Anastrepha serpentina*. (Photograph by R. Wilson.)

**COLOR FIGURE 9** Premating trophallaxis by *Anastrepha striata*. (Photograph by T. Hoffmeister.)

**FIGURE 10** Direct, stomodeal, postmating trophallaxis by *Spathulina sicula*. (Photograph by A. Freidberg.)

**COLOR FIGURE 11** *Ceratitis capitata* males fighting over leaf territory while another calls nearby. (Photograph by R. Macías.)

**COLOR FIGURE 12** *Toxotrypana curvicauda* female ovipositing in a papaya fruit. Note the drop of latex being released by the fruit at puncture site. (Photograph by E. Piedra.)

**COLOR FIGURE 13** *Taeniostola limbata* female. This fly lays its eggs in the oviposition holes bored by large weevils into bamboos. One or two individuals will spend hours on the elytra of the beetle waiting for it to complete its labors, then hop off to be the first to lay their eggs. For more information see Kovac, D. and I. Azarae, 1994, Depredations of a bamboo shoot weevil, *Nat. Malays.*, December, 115–122. (Photograph by D. Kovac.)

**COLOR FIGURE 14** (A) Mark (drop of latex) at oviposition wound made by *Anastrepha hamata* female. (B) Oviposited fruit with undeveloped seed (substrate on which eclosing larvae feeds — coin is 2 cm in diameter). (C) Mature fruit with two undamaged and one damaged seed (larvae of *A. hamata* feed on seed and exit fruit well before fruit ripens; they do so by biting their way through the hard pulp and jumping to the ground from heights of up to several meters; note callous tissue around exit hole). (Photograph by F. Díaz-Fleischer.)

**COLOR FIGURE 15** Sequence of events prior to, during, and after oviposition by *Anastrepha ludens*. (A) Insertion of aculeus into fruit. (B) Gradual extrusion of aculeus by female while exhibiting stereotyped cleaning behavior (with hind legs). (C) Entire aculeus extended while fly keeps cleaning it with hind legs. (D) Aculeus-dragging behavior exhibited by female inmediately after an oviposition bout (note the aculeus tip in contact with the fruit surface; a host marking pheromone is released during this process). (Photographs by R. Wilson.)

**FIGURE 16** Pheromone marks left by female of *Anastrepha spatulata* on fruit surface. (Photograph by M. López.)

**COLOR FIGURE 17** *Phytalmia mouldsi* males engaged in an escalated pushing contest over possession of a resource (an oviposition site) attractive to females. (Photograph by G. Dodson.)

**COLOR FIGURE 18** *Phytalmia alcicornis* males preparing to battle over a territory on a rotting rain forest tree trunk in Papua New Guinea. (Photograph by G. Dodson.)

**COLOR FIGURE 19** *Aciurina trixa* female in resting position. (Photograph by D. Headrick.)

**COLOR FIGURE 20** Close-up of brilliantly colored eye of *Myopites nigrescens*. (Photograph by A. Freidberg.)

**COLOR FIGURE 21** Predators of *Blepharoneura*: (A) Dead *Blepharoneura* dropped by a crab spider perched on a male flower of the host *Gurania spinulosa*. (B) Jumping spider, which resembles buds of host plant *G. spinulosa*, holds *Blepharoneura* at the base of a shoot bearing female flowers. (C) Staphylinid beetles preying upon larvae of *Blepharoneura* in ovaries of *G. spinulosa*. Beetle in upper left holds a larva, which was extracted from the oozing wound. (D) Close-up of the flower-mimicking jumping spider shown in (B). (Photographs by M. Condon.)

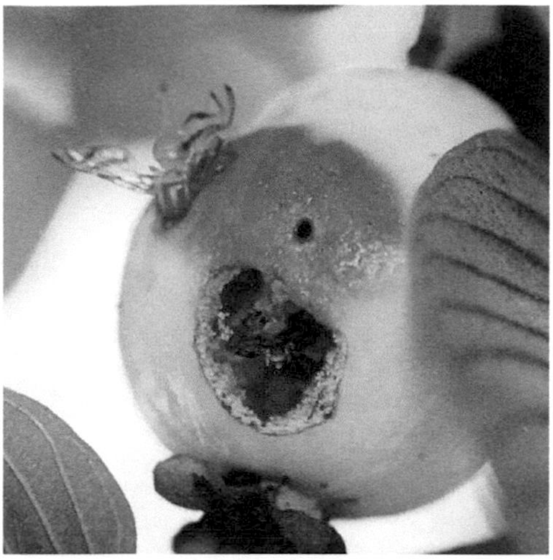

COLOR FIGURE 1

COLOR FIGURE 2

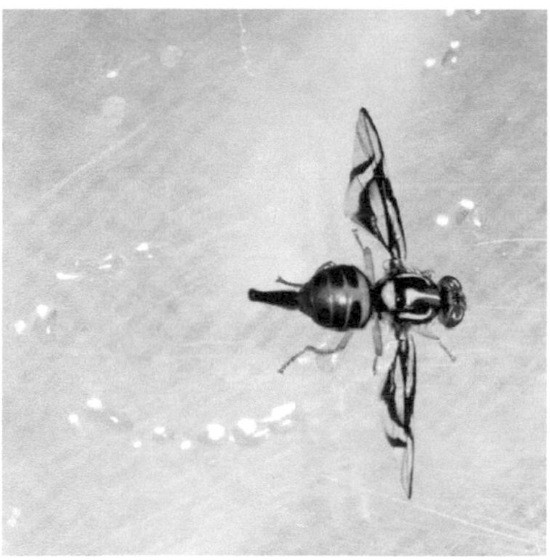

COLOR FIGURE 3

COLOR FIGURE 4

COLOR FIGURE 5

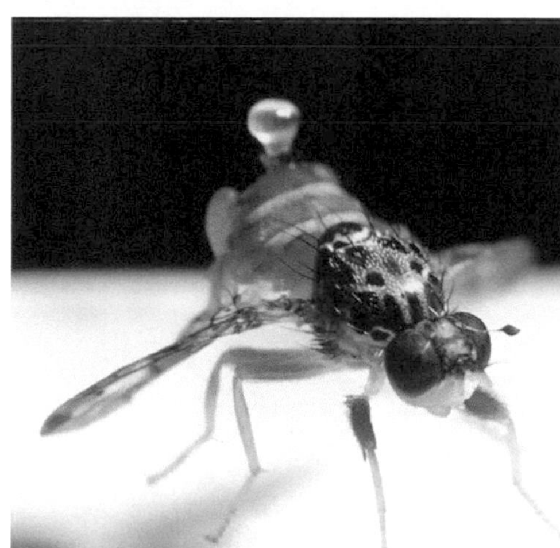

COLOR FIGURE 6

COLOR FIGURE 7A

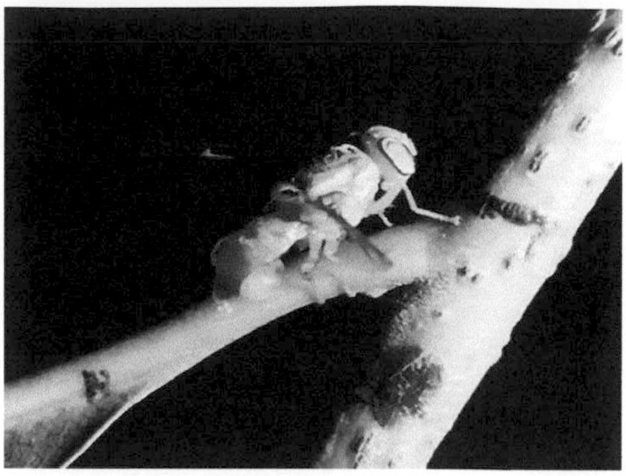

COLOR FIGURE 7B

COLOR FIGURE 8

COLOR FIGURE 9

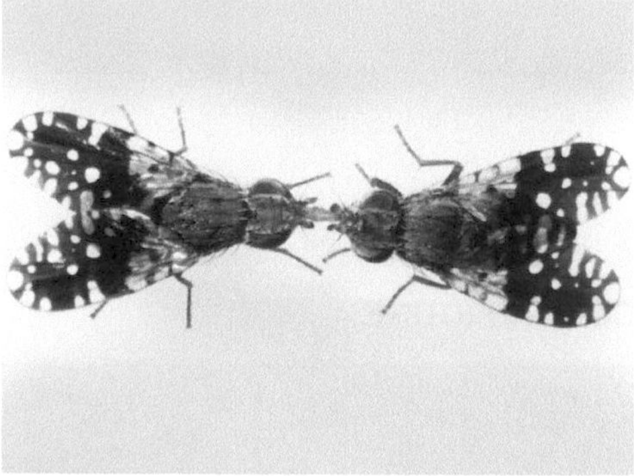

FIGURE 10

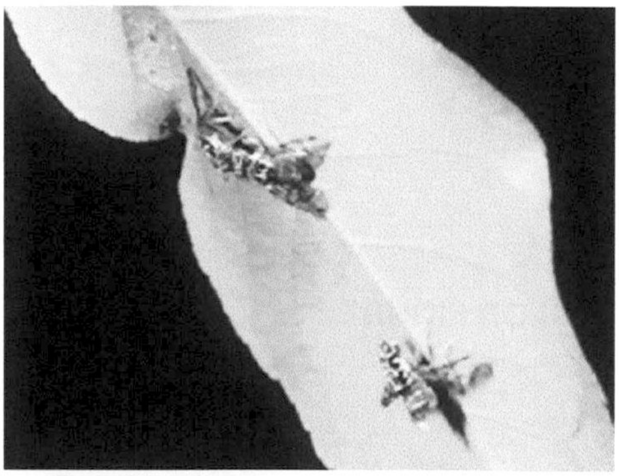

COLOR FIGURE 11

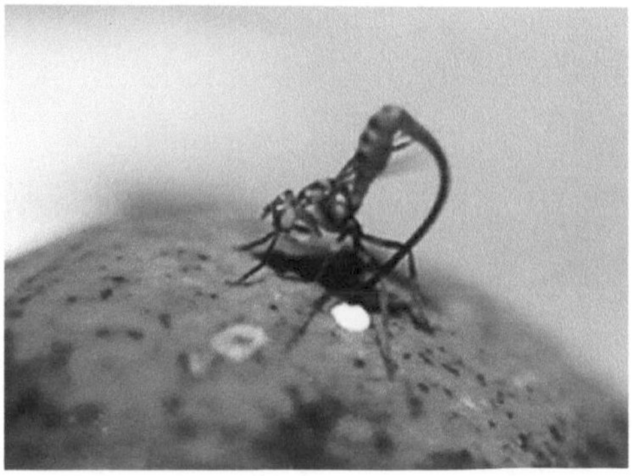

COLOR FIGURE 12

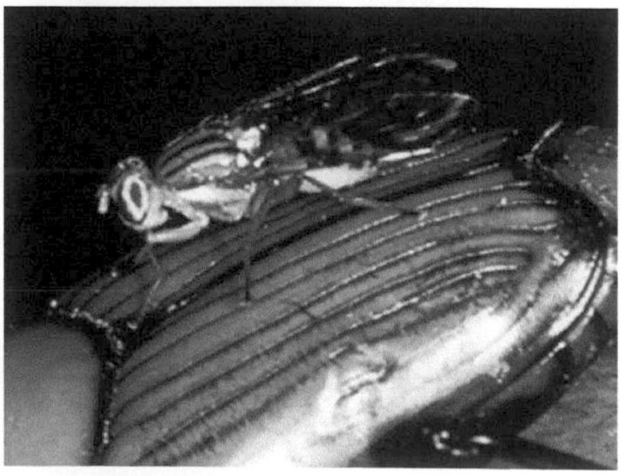

COLOR FIGURE 13

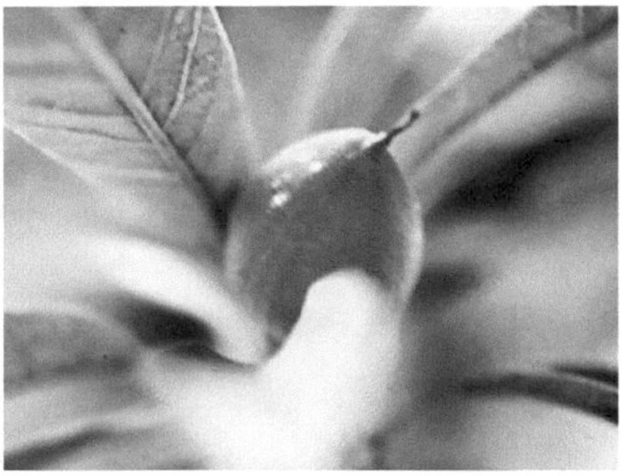

COLOR FIGURE 14A

COLOR FIGURE 14B

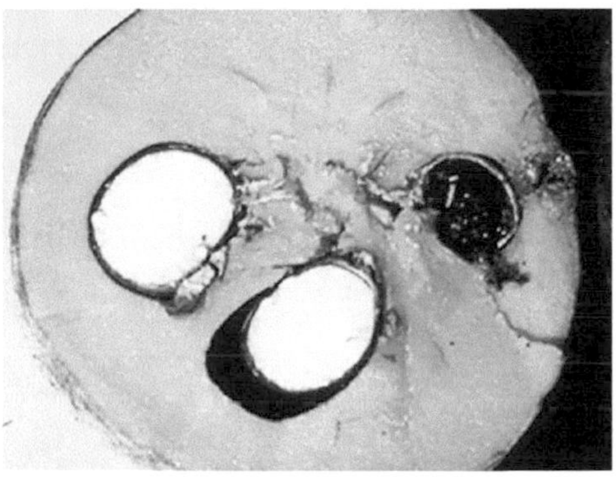

COLOR FIGURE 14C

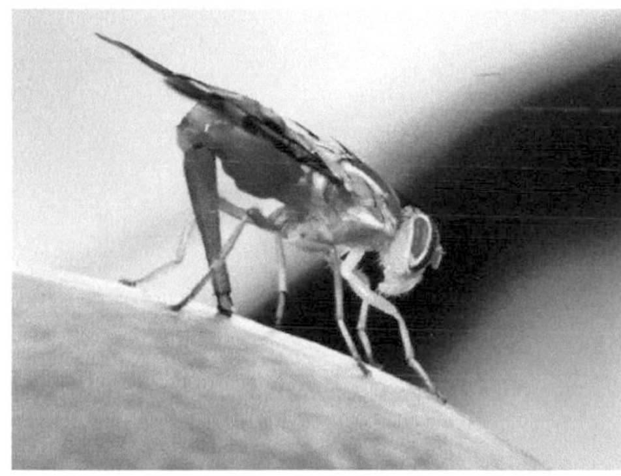

COLOR FIGURE 15A

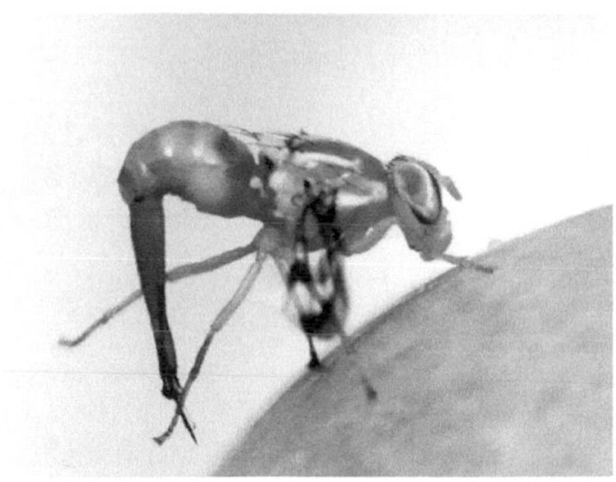

COLOR FIGURE 15B

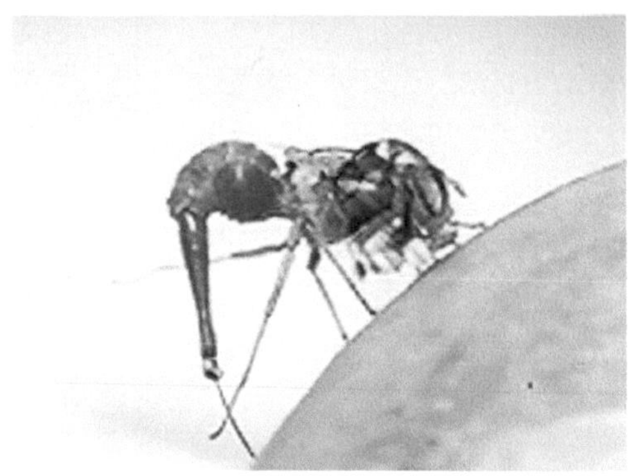

COLOR FIGURE 15C

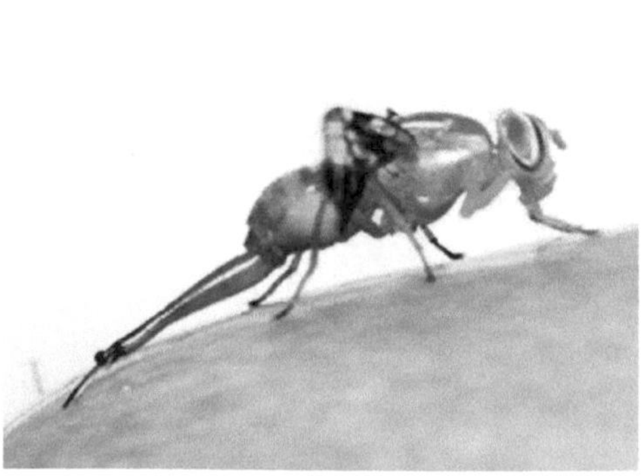

COLOR FIGURE 15D

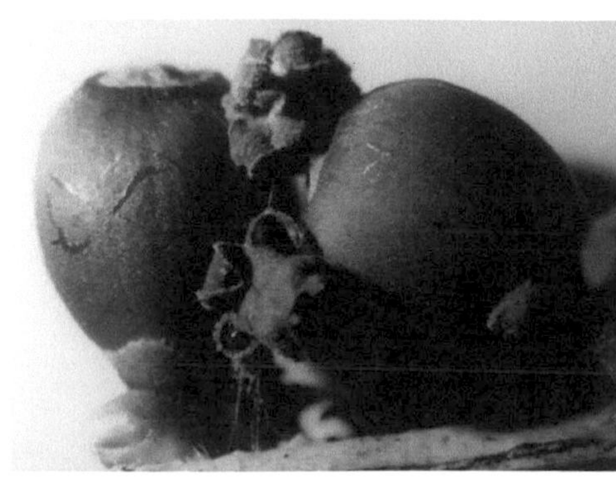

FIGURE 16

COLOR FIGURE 17

COLOR FIGURE 18

COLOR FIGURE 19

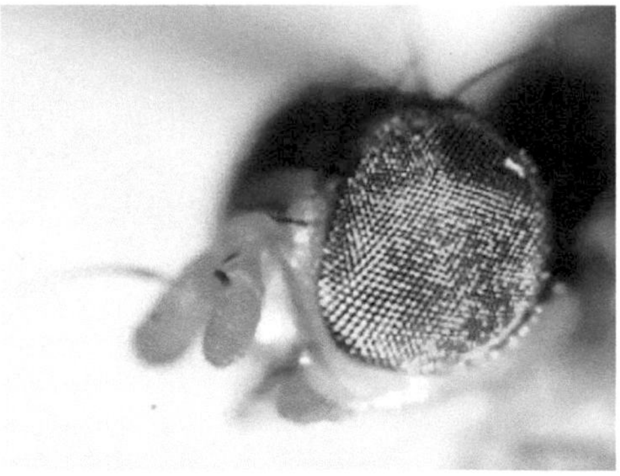

COLOR FIGURE 20

COLOR FIGURE 21A

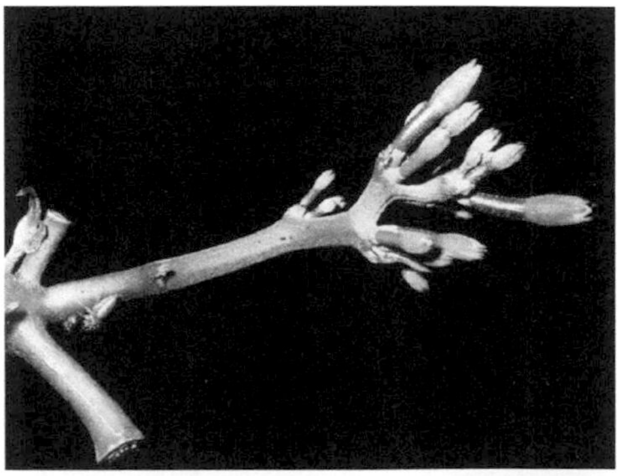

COLOR FIGURE 21B

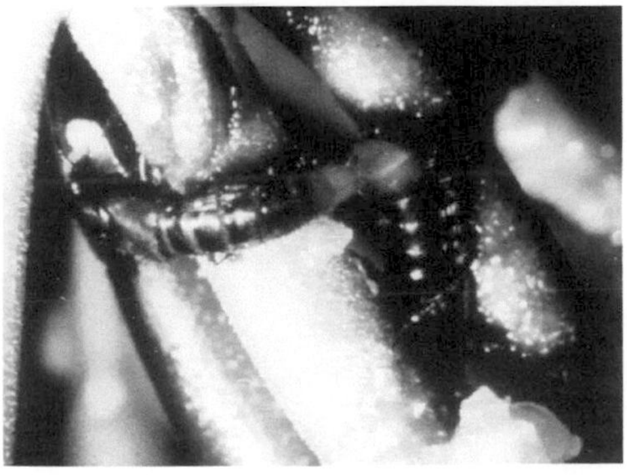

COLOR FIGURE 21C

COLOR FIGURE 21D

and Bush 1997). One notable hypothesis is that *R. juniperina* is the sister taxon to the South American *Rhagoletis*, a relationship suggested based on morphological data (Jenkins 1996; Figure 9.2). However, no characters in Jenkins' (1996) data set unambiguously support this relationship, and tree manipulations using MacClade (Maddison and Maddison 1992) show that several alternate placements of *R. juniperina* are equally parsimonious (Smith, unpublished data). Similarly, the phylogenetic relationship of *R. fausta* to the other *Rhagoletis* has been difficult to determine. It may be that many of these species diverged simultaneously during an adaptive radiation. In this case, phylogenetic resolution would not be expected (Hoelzer and Melnick 1994).

The mitochondrial COII data also indicate that the remaining three North American species groups (*ribicola, cingulata,* and *suavis*) may form a monophyletic group (Smith and Bush 1997). Support for this clade comes from Berlocher's (1993) observation that allozyme loci display similar sex-linkage patterns in the *ribicola, cingulata,* and *suavis* group species. Both molecules and morphology support a clade with the *R. cingulata* and *R. suavis* groups combined. Mitochondrial DNA data weakly support this clade (Smith and Bush 1997; Figure 9.4), and a single morphological feature, tufts of long setae at the tips of their surstyli, supports the group (Jenkins 1996). However, sampling bias may again be limiting our understanding of relationships. In the reanalysis of Jenkins' data set, the central Asian *R. magniterebra* is placed with the *cingulata* and *suavis* groups (see Figure 9.2).

### 9.2.3.2 Palearctic and Oriental Taxa

The Palearctic and Oriental carpomyine species are currently classified in four genera, *Rhagoletis, Carpomya, Goniglossum,* and *Myiopardalis*. The approximate geographic ranges of these genera are shown in Figures 9.6 and 9.7. Morphological cladistic analyses show that *Carpomya, Goniglossum,* and *Myiopardalis* form a monophyletic group (Jenkins 1996; Norrbom 1997) which disrupts the monophyly of *Rhagoletis* and may be the sister group of the *R. meigenii* species group (Jenkins 1996; Figure 9.2). Within this clade, *Carpomya* (*s. str.*) is apparently monophyletic. Jenkins (1996) hypothesized that the lateral surstylus having its setae short, stout, and proximally directed is a synapomorphy for *Carpomya*. In addition, he placed *Myiopardalis* as the sister taxon to *Carpomya* by virtue of the fact that the mediotergite in both *Carpomya* and *Myiopardalis* is pollinose (with dense vs. sparse microtrichia), a character state not shared by *Goniglossum* (or any other carpomyine species). Scutellum shape (almost semicircular in dorsal view, disk slightly to moderately convex) and the presence of two spermathecae may also be synapomorphies of *Carpomya* (*s. str.*), while katepisternum color (most of it darker than its dorsal margin) and epandrium color (mostly yellow with small dorsomedial brown spot) defines the monophyletic group including *Carpomya, Myiopardalis,* and *Goniglossum* (Norrbom 1997). Norrbom's (1997) analysis placed *M. pardalina* in a clade with *G. wiedemanni*, which together formed the sister group to *Carpomya s. str.*, with the spur vein on vein $R_{2+3}$ and use of hosts in the Cucurbitaceae hypothesized as synapomorphies.

*Carpomya tica* Norrbom (1997) represents the first report of *Carpomya* outside of the Palearctic or Oriental Regions. Phylogenetic analysis (Norrbom 1997) of *C. tica*, the three species of *Carpomya s. str., M. pardalina,* and *G. wiedemanni*, indicated that *C. tica* is the sister taxon to a group containing the other five species. In preference to creating an additional monotypic carpomyine genus, Norrbom (1997) synonymized *Goniglossum* and *Myiopardalis* with *Carpomya*, expanding the latter genus to include six species, including *C. tica*. *Carpomya* as thus expanded lacks a unique, consistent, morphological synapomorphy; it was defined on the basis of mesonotal coloration and microtrichia pattern with reversal postulated in *C. incompleta*, which lacks the dark markings and bare and black microtrichose areas of the other five species. However, *C. incompleta* was included based on other synapomorphies with the species exclusive of *C. tica*, and especially with *C. schineri* and *C. vesuviana* (Norrbom 1997).

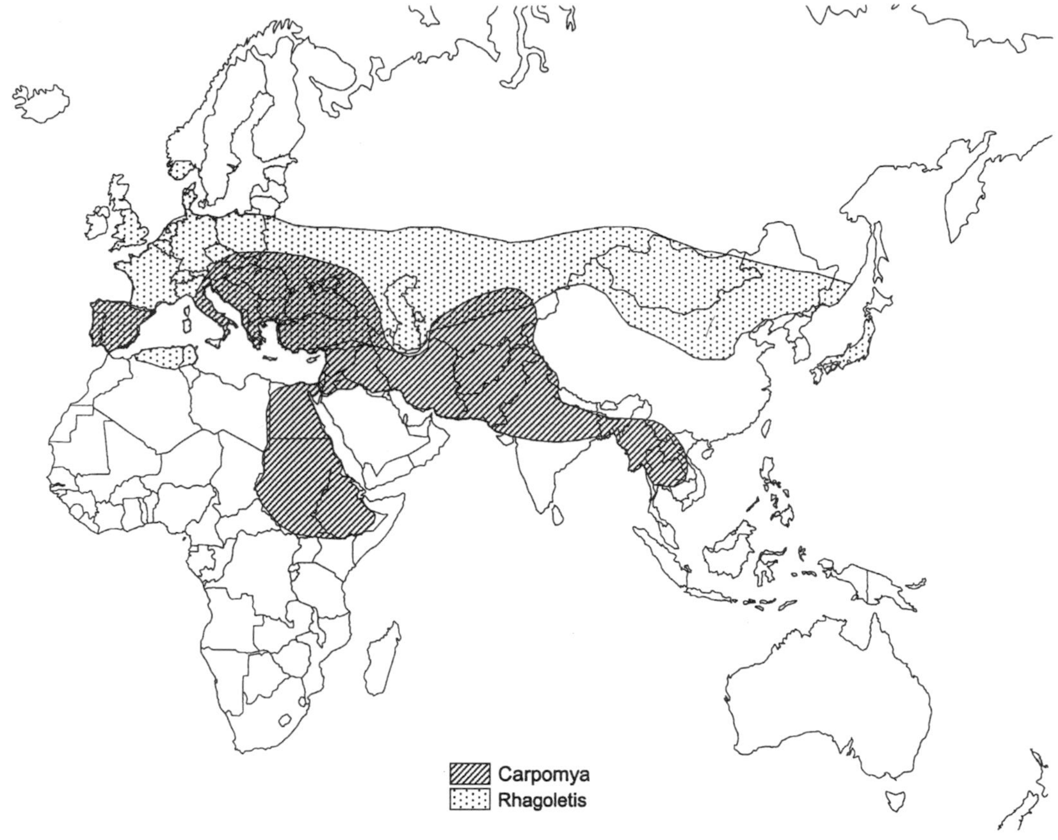


**FIGURE 9.6** Geographic distributions of *Rhagoletis* and *Carpomya* in the Old World. Ranges are approximate, and were compiled from literature sources. Keys are inset.

There are 24 described Palearctic and Oriental species of *Rhagoletis* which infest primarily *Rosa* spp., *Berberis* spp., *Juniperus* spp., and *Lonicera* spp. The phylogenetic relationships of the Old World *Rhagoletis* species are poorly understood, and they almost certainly do not represent a natural group. In Jenkins' study (1996), most of the Old World *Rhagoletis* were unresolved relative to other *Rhagoletis* species. However, reanalysis of Jenkins' morphological data set appears to provide insight into their phylogenetic affinities, and we have tentatively assigned some of the Old World *Rhagoletis* species into taxonomic species groups on the basis of this analysis, host use, and geographic distribution (see Table 9.1). However, we emphasize that the monophyly of each of these species groups remains to be tested.

In Jenkins' (1996) original analysis, he recovered the following clades containing Palearctic taxa: a monophyletic *cerasi* group containing *R. berberidis*, *R. almatensis*, and *R. cerasi*; *R. meigenii* and *R. kurentsovi*; and *R. flavigenualis* and *R. zernyi* as a monophyletic sister group to the Nearctic *R. pomonella* species group. The other Old World *Rhagoletis* in his study, *R. alternata*, *R. batava*, *R. caucasica*, *R. flavicinta*, *R. magniterebra*, and *R. mongolica*, formed an unresolved polytomy with the clades described above, a clade containing *Carpomya s. str.*, *M. pardalina,* and *G. wiedemanni*, and other clades containing the New World *Rhagoletis*.

In the 50% majority-rule consensus tree obtained upon reanalysis of Jenkins' data set (see Figure 9.2), *R. magniterebra* was recovered as the sister taxon of the Nearctic *R. cingulata* + *R. suavis* groups. In addition, *R. batava*, *R. flavicinta,* and *R. mongolica* belong to a large clade that is the sister group to the *R. cerasi* species group, which itself is monophyletic. Additionally, in

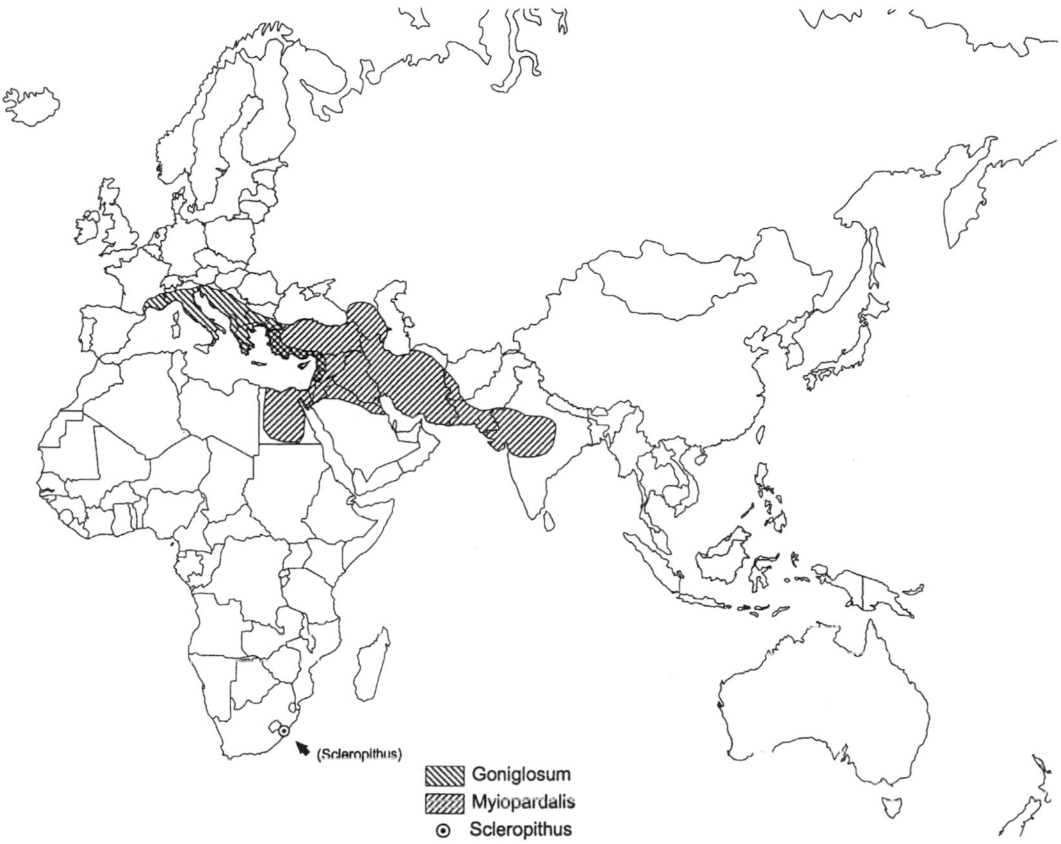


**FIGURE 9.7** Geographic distributions of *Goniglossum, Myiopardalis,* and *Scleropithus*. Ranges are approximate, and were compiled from literature sources. Keys are inset.

98% of the 32,700 trees, the *Rosa* infesting *R. basiola* (Nearctic) and *R. alternata* (Palearctic) emerge as the sister group to all other *Rhagoletis* and the Palearctic genera *Carpomya, Myiopardalis,* and *Goniglossum*. This raises the interesting possibility that resurrection of the genus *Zonosema* (Loew 1862) may be appropriate. Also, in 96% of the 32,700 trees, a clade consisting of the *Berberis*-infesting *R. caucasica, R. kurentsovi,* and *R. meigenii* appears as the sister group to the clade including *Carpomya, Myiopardalis,* and *Goniglossum*.

These hypotheses are not supported by DNA-based phylogenetic analyses of Palearctic *Rhagoletis* that have been carried out to date. Molecular phylogenetic studies provide some indication that *R. cerasi* may be closely related to *Goniglossum, Carpomya,* and *Myiopardalis*. Mitochondrial COII sequence analysis using successive approximations yielded bootstrap support for grouping *C. shineri* and *R. cerasi* (Smith and Bush 1997). The study by Smith and Bush (1997) also included individual flies of *R. meigenii* and *R. berberidis*, but the positions of these taxa in the phylogenies were not supported by the data and should be considered unresolved.

In none of the molecular studies to date has the taxonomic sample represented the full range of Palearctic taxa, and it will be helpful to test hypotheses including more of the pertinent taxa. Phylogenetic analyses tend to place several Palearctic Carpomyina genera and species in positions that are basal to the New World taxa. For example, in Han and McPheron's (1997) analysis of mitochondrial 16S rRNA genes, *G. wiedemanni* and *R. cerasi* were grouped basal to the North American *Rhagoletis*. This indicates that Palearctic carpomyines diverged from their Nearctic and Neotropical relatives fairly early. However, the basal position of *R. alternata* (Old World) and *R. basiola*

(New World) within the subtribe, and the placement of several Palearctic *Rhagoletis* species (e.g., *R. zernyi, R. flavigenualis, R. batava, R. magniterebra,* and *R. flavicinta*) within a predominantly New World clade (see Figure 9.2), suggest that the global divergence pattern of Carpomyina may not be a simple one.

### 9.2.3.3 Neotropical Taxa

The distributions of the Neotropical carpomyine genera are shown in Figures 9.8 and 9.9. There are four genera in the subtribe Carpomyina with predominantly Neotropical distributions, *Cryptodacus*, *Haywardina*, *Rhagoletotrypeta,* and *Stoneola*. *Stoneola* is a poorly known, monotypic genus whose relationships are uncertain and will not be discussed further. Various lines of evidence indicate that *Zonosemata* is closely related to *Cryptodacus*, *Haywardina,* and *Rhagoletotrypeta* and thus will be considered with them here. For example, Norrbom (1994) suggested the pale scutal stripe as a potential synapomorphy for these four genera, but the focus of his analysis was on the relationships within *Cryptodacus*, *Haywardina,* and *Rhagoletotrypeta*. Jenkins' (1996) analysis of morphology recovered a clade with *Cryptodacus*, *Haywardina*, *Rhagoletotrypeta,* and *Zonosemata* (excluding *Rh. pastranai*), although he identified no unambiguous synapomorphies for the group.

Analyses of neither morphological nor molecular characters have provided a clear picture of the relationships among *Cryptodacus*, *Haywardina*, *Rhagoletotrypeta,* and *Zonosemata*. For example, in Jenkins' (1996) morphological study, all species from *Cryptodacus*, *Haywardina*, *Rhagoletotrypeta,* and *Zonosemata* (except *Rh. pastranai*, see below) formed the sister group to the remainder of the Carpomyina. *Cryptodacus* was sister group to *Rhagoletotrypeta*, with *Haywardina* polyphyletic. On the other hand, Norrbom's (1994) cladistic analysis based on 50 morphological characters placed *Haywardina* as the sister genus to *Rhagoletotrypeta*, with *Cryptodacus* and then *Zonosemata* basal to this group, respectively (*Paraterellia* used as outgroup). However, Norrbom's topology was not stable, and he did not consider any hypothesis of relationship among these four genera to be well supported by his data. Inclusion of two Neotropical *Rhagoletis* species in the data matrix, *R. striatella* and *R. ferruginea*, changed the topology considerably, with *Cryptodacus* forming the sister group to five of the eight *Rhagoletotrypeta* species, and *Rhagoletotrypeta* and *Haywardina* becoming paraphyletic. Lack of monophyly of *Rhagoletotrypeta* was also observed in Jenkins' (1996) morphological study. *Rhagoletotrypeta pastranai* did not group with the other *Rhagoletotrypeta* in the analysis, *Rh. annulata*, *Rh. rohweri,* and *Rh. uniformis*, which did form a clade. Instead, *Rh. pastranai* grouped with *R. tomatis*, *R. lycopersella,* and other South American *Rhagoletis* species.

Analyses based on molecular characters also have not resolved the relationships among *Cryptodacus*, *Haywardina*, *Rhagoletotrypeta,* and *Zonosemata*. The study of McPheron and Han (1997), based on mitochondrial 16S rDNA sequences of a single individual from each genus, placed *Haywardina* as the sister genus to *Cryptodacus*, with *Zonosemata* and then *Rhagoletotrypeta* basal to this group, respectively. In Han and McPheron (1997), also based on mitochondrial 16S rDNA sequences, *Haywardina* was again placed as the sister genus to *Cryptodacus*, with *Zonosemata* sister to this group, and a clade consisting of *Rh. pastranai* and *R. striatella* basal to *Haywardina*, *Cryptodacus,* and *Zonosemata*. The study of Smith and Bush (1997), based on mitochondrial COII, included single individuals of *Z. electa* and *Rh. pastranai* whose relationships to each other and the other taxa in the study were unresolved.

Relationships within *Cryptodacus*, *Haywardina,* and *Rhagoletotrypeta* were examined in Norrbom's (1994) morphological study. Within *Cryptodacus*, *C. obliquus* appears to be the sister taxon to the remainder of the genus. *C. tigreroi* and *C. silvai* are a species pair, and *C. parkeri* and *C. quirozi* form another within a subgroup that also includes *C. ornatus*. The relationships of *C. tau* and *C. lopezi* are uncertain; they pair or arise as successive lineages between *C. obliquus* and the

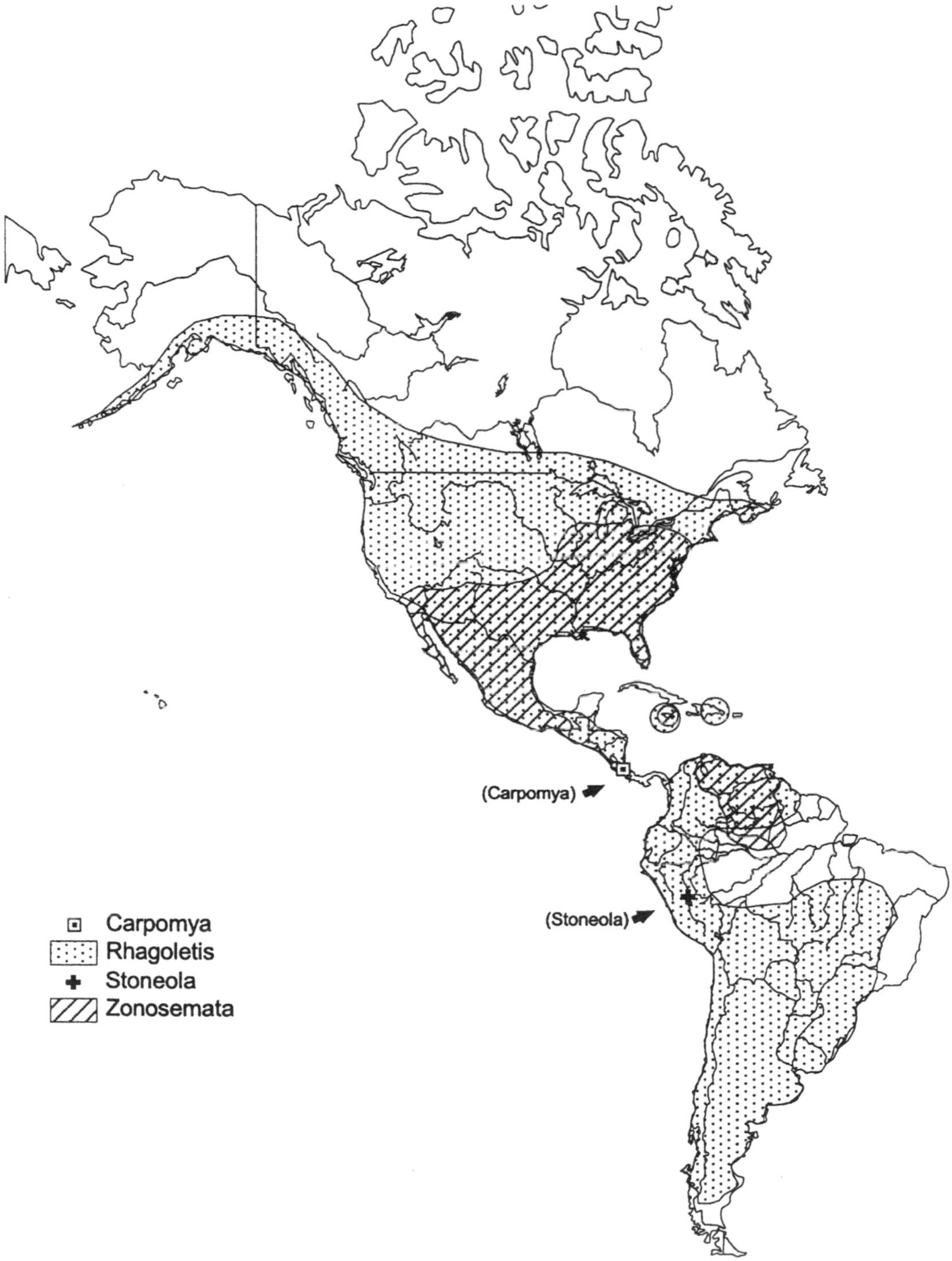


**FIGURE 9.8** Geographic distributions of *Rhagoletis*, *Zonosemata*, *Stoneola,* and *Carpomya* in the New World. Ranges are approximate, and were compiled from literature sources. Keys are inset.

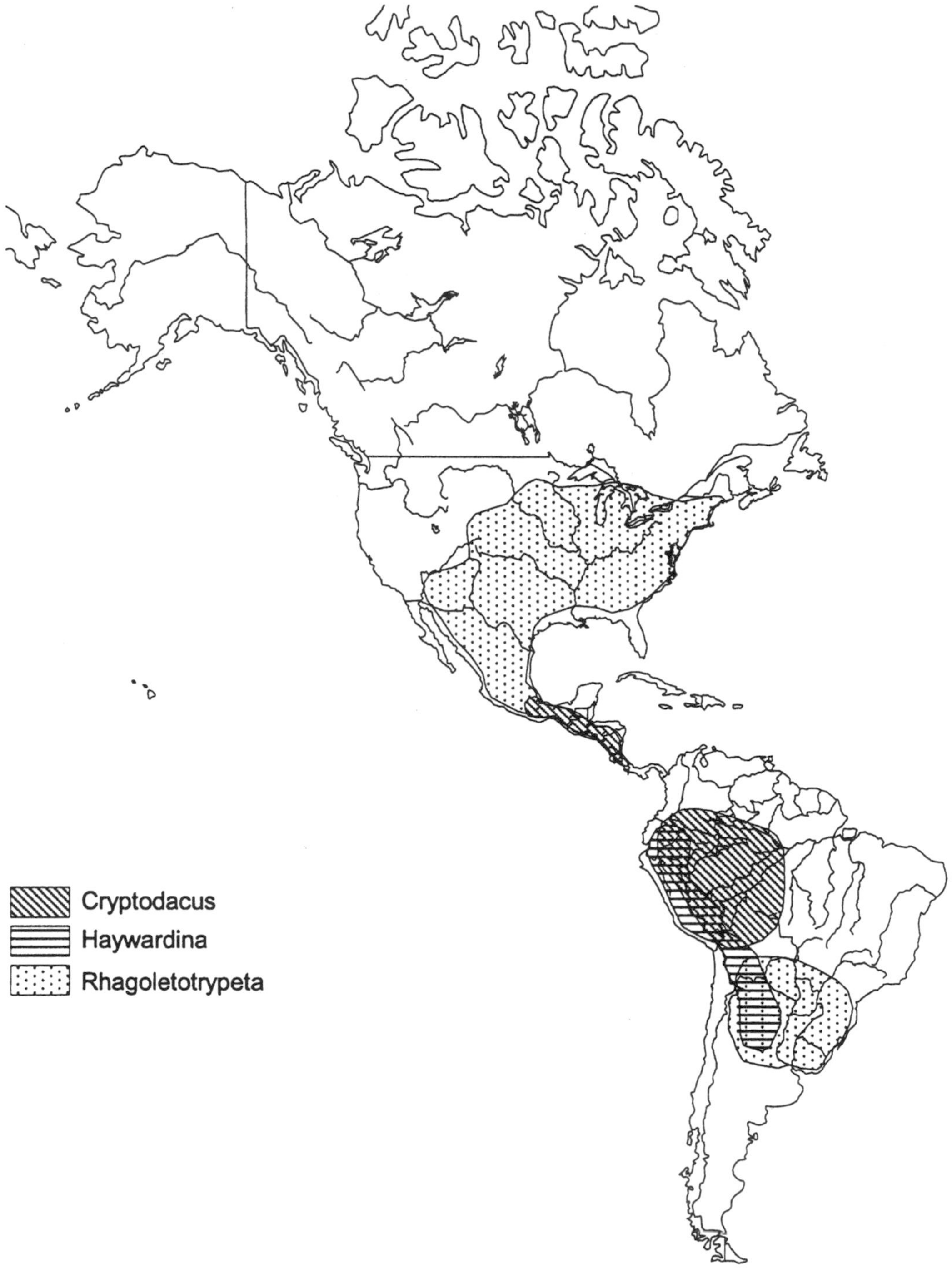


**FIGURE 9.9** Geographic distributions of *Cryptodacus, Haywardina,* and *Rhagoletotrypeta*. Ranges are approximate, and were compiled from literature sources. Keys are inset.

clade of the other five species. As mentioned above, *Haywardina* and *Rhagoletotrypeta* were not monophyletic in Norrbom's (1994) analysis when two Neotropical *Rhagoletis* species were included in the taxon sample. However, within *Haywardina, H. cuculiformis* was consistently the sister taxon of *H. cuculi*. *Rhagoletotrypeta* was divided into two species groups, which were each consistently monophyletic. The *xanthogastra* group contains *Rh. parallela*, *Rh. pastranai*, and *Rh. xanthogastra*, and the *annulata* group contains *Rh. annulata*, *Rh. argentinensis*, *Rh. morgantei*, *Rh. intermedia*, *Rh. uniformis,* and *Rh. rohweri*.

*Zonosemata* appears to be well supported as a monophyletic genus, although its intrageneric relationships have not been analyzed. One morphological synapomorphy for the genus is the shape and attachment points of the vesica of the glans (Jenkins 1996). The karyotype also appears to be distinct from *Rhagoletis* and probably all other carpomyine genera (Bush 1966). Also, Norrbom (1994) noted that the glans of *Zonosemata* species is relatively simple, with little distinct internal sclerotization, as opposed to the complex internal sclerotization occurring in other Carpomyina. Norrbom (1994) considered the large, elongate, and weakly sclerotized spermathecae to be a synapomorphy for *Zonosemata*, and Jenkins (1996) found the external surface of the spermathecal ducts to be annulated in *Zonosemata* species, yet smooth in all other Carpomyina.

Except for *R. turpiniae* of the *cingulata* group and *R. ramosae* of the *suavis* group, the Neotropical species of *Rhagoletis* belong to four species groups (*striatella*, *nova*, *psalida,* and *ferrugina*) which infest almost exclusively members of the plant family Solanaceae (Foote 1981). With the exception of *R. conversa*, which has been found infesting plums in Chile (D. Frías, personal communication; see Table 9.1), the non-Solanaceae host records are questionable. Phylogenetic relationships within and among these four *Rhagoletis* species groups are not well characterized. Jenkins' (1996) analysis of morphology indicated that the *striatella*, *nova*, *psalida,* and *ferruginia* species groups form a monophyletic group, with the notable exception of the inclusion of *R. pastranai* in this clade. The *R. striatella* group appears to have diverged earliest within this lineage, followed by the *R. ferruginea* group. The *R. psalida* group species appeared as the sister taxon to *R. lycopersella*, *R. tomatis,* and *Rh. pastranai*, creating a paraphyletic *R. nova* species group.

Molecular phylogenetic studies published to date have shed little light on relationships of Neotropical *Rhagoletis* as taxon sampling has been insufficient. The study of McPheron and Han (1997), based on mitochondrial 16S rDNA sequences, included only *R. striatella* and *R. conversa*, which formed a clade. The study of Smith and Bush (1997), based on mitochondrial COII, also included only *R. striatella* and *R. conversa* but could not resolve relationships between these taxa and other carpomyine genera.

Currently, in collaboration with Dr. Daniel Frías from the Instituto de Entomología, Universidad Metropolitana de Ciencias de la Educación, Santiago, Chile, and Dr. Priscilla Santos from the Instituto de Biociencias da Universidade de São Paulo, Brazil, we are producing a mitochondrial COII phylogeny of South American *Rhagoletis* based on a more representative sample. Preliminary analysis of individuals in the *R. nova* species group (M. Jaycox, J. Smith, D. Frías, and G. Bush, unpublished data) has yielded results that are congruent with the morphological analysis of Jenkins (1996). *Rhagoletis nova* and *R. conversa* are closely related, as suggested by Frías et al. (1987), and *R. lycopersella* appears as a more distant relative of these two species. Intraspecific variation was observed in *R. conversa*. Two haplotypes that differ by approximately 1% ($p$-distance = 0.01) have been sequenced from *R. conversa* flies infesting *Solanum nigrum*. One of the haplotypes is identical to a haplotype observed in a *R. conversa* fly infesting *S. tomatillo*.

## 9.3 MORPHOLOGICAL VS. MOLECULAR CHARACTER EVOLUTION IN THE CARPOMYINA

Our ultimate objective with respect to the systematics of Carpomyina is a stable phylogenetic classification. Reconstruction of the evolutionary history of Carpomyina will promote understanding of the evolutionary dynamics that gave rise to today's taxa. Completion of this objective will require not only analysis of morphological divergence, but also characterization of the genetic and biogeographic contexts in which morphological divergence has occurred.

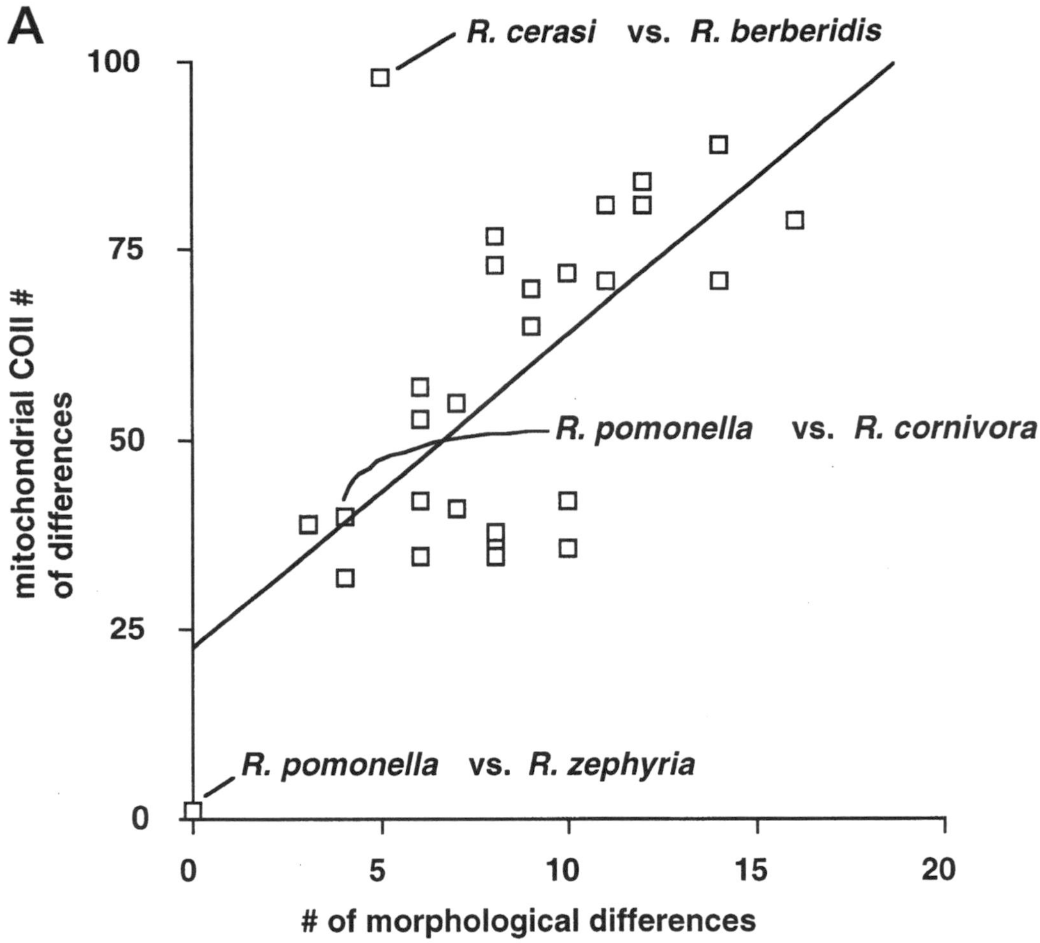


**FIGURE 9.10** Comparison of morphological vs. molecular divergence in *Rhagoletis* species. Pair-wise difference values (autapomorphic characters included) were generated for all possible comparisons of eight *Rhagoletis* species (*R. pomonella, R. zephyria, R. cornivora, R. juniperina, R. fausta, R. cingulata, R. cerasi,* and *R. berberidis*) using the morphological data set of Jenkins (1996), the allozyme data set of Berlocher and Bush (1982), and the mitochondrial DNA data set of Smith and Bush (1997). (A) mitochondrial DNA vs. morphology. For each species pair, the number of nucleotide differences in the COII gene was plotted as a function of number of morphological differences. (B) Allozymes vs. morphology. For each species pair, the Manhattan distance (allozyme) was plotted as a function of number of morphological differences.

Taxa are defined (for the most part) on morphological features. However, it is extremely difficult to make inferences about the time frame of divergence of taxa using morphological data, especially in the absence of a fossil record. Analysis of DNA sequences and the application of a molecular clock can lead to inferences about the time frame in which morphological change has occurred. One observation that has emerged from these studies is that morphological and molecular evolution are not always linked. Below, we point out some cases in which flies that are very close morphologically show disproportionately large genetic divergences.

As an illustration, we have compared morphological divergence with allozyme divergence and mtDNA divergence for eight *Rhagoletis* species (*R. pomonella, R. zephyria, R. cornivora, R. juniperina, R. fausta, R. cingulata, R. cerasi, and R. berberidis*). For the comparison, we used data from the morphological data set of Jenkins (1996), the allozyme data sets of Berlocher and Bush (1982) and Berlocher et al. (1993), and the mitochondrial DNA data set of Smith and Bush (1997).

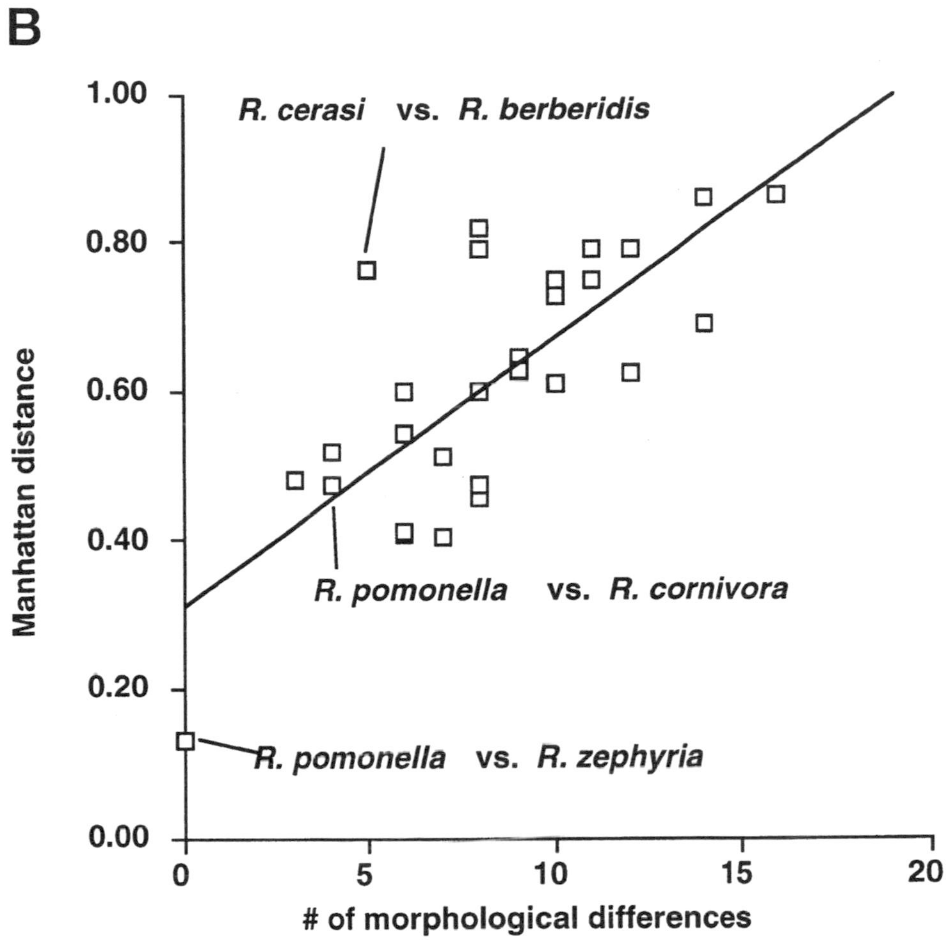


FIGURE 9.10 (continued)

The eight species chosen were common to all three studies, and the results (Figure 9.10) illustrate the differences in evolutionary rates of morphological and molecular characters.

For example, the closely related apple maggot fly, *R. pomonella*, and snowberry fly, *R. zephyria*, were scored identically in the morphological data set of Jenkins (1996). These species differ morphologically only in the shapes of their surstyli (Bush 1966; Westcott 1982), a character that was not scored by Jenkins, who restricted his analysis to characters with discrete variation. Analysis of allozyme variation shows that *R. pomonella* and *R. zephyria* share alleles at all loci except for a single fixed difference at the *Had* locus (Berlocher et al. 1993). Similarly, analysis of mitochondrial DNA shows that *R. zephyria* and *R. pomonella* share haplotypes (see Figure 9.5), with haplotypes unique to *R. zephyria* differing from their nearest *R. pomonella* counterparts by less than 0.5% (*p*-distance). Thus, comparison of morphological and genetic changes between *R. zephyria* and *R. pomonella* (see Figure 9.5) shows that these species are similar morphologically and genetically.

A second sibling species pair, *R. cornivora* and *R. pomonella*, are also quite similar morphologically, although there are four characterizable minor morphological differences (Jenkins 1996; shape of medial surstylus, microtrichia on epandrium, number of spermathecae, and denticles on eversible membrane). However, both allozymes and mtDNA have diverged to a greater extent (Figure 9.10). The extent of allozyme divergence has made placement of *R. cornivora* in allozyme phylogenies difficult (Berlocher and Bush 1982; Figure 9.3). In addition, mitochondrial COII divergence between *R. cor-*

*nivora* and its nearest *R. pomonella* haplotype is approximately 6% (*p*-distance). Thus, *R. cornivora* and *R. pomonella* appear similar morphologically yet are genetically highly divergent.

Finally, in a third species pair, *R. cerasi* and *R. berberidis*, morphological differences are again slight (five differences; Jenkins 1996). Indeed, *R. cerasi* and *R. berberidis* were considered to be the same species until Jermy's (1961) reclassification. However, in this species pair, genetic divergence is even more pronounced than in *R. cornivora* and *R. pomonella*. Allozyme differences are considerable (Manhattan distance = 0.78; Berlocher and Bush 1982) and the 16% difference (*p*-distance) in mitochondrial COII sequences was the largest of any *Rhagoletis* species pair observed by Smith and Bush (1997). Using Brower's (1994) estimate of 2.3% sequence divergence per million years (1.1 to 1.2% per lineage) for silent sites in arthropod mtDNA, we estimate that these two species (haplotypes) shared a common ancestor approximately 7 million years ago.

Molecular and morphological features may be subject to widely differing selection pressures that in some cases may obscure phylogenetic signal. Natural selection, whether stabilizing, directional, or diversifying, almost certainly plays a significant role in shaping morphological features that are used in mate choice or that confer a selective advantage in a given habitat (e.g., wing patterning in avoidance of predation). On the other hand, population genetic processes, such as the fixation of alternate alleles from a polymorphic ancestral population, may also lead to apparent disparities in evolutionary rates. The mitochondrial DNA difference observed between *R. pomonella* and *R. cornivora* (approximately 6%, *p*-distance) may have arisen in this manner.

Differences in evolutionary rate between different data sets provide important information about evolutionary history. When rates differ, such as in *R. cerasi* and *R. berberidis*, where morphological features have remained similar in the face of significant genetic change, we can ask whether or not there is something about the biology of these taxa that might explain reduced morphological variation. On the other hand, geographic distribution and host use data may indicate that an ancestral taxon was abundant and widespread, increasing the chances of fixation of alternate alleles. Also, it is clear that ecology plays an important role in the relationship between rates of morphological and genetic evolution, as emphasized by Bush and Smith (1998). Increased understanding of the mechanisms underlying different types of variation might allow more informed use of these data as phylogenetic characters.

## 9.4 CONCLUSIONS AND FUTURE RESEARCH NEEDS

The tephritid subtribe Carpomyina contains eight genera distributed in the Holarctic, Oriental, and Neotropical Regions. Much work has been done to describe the species within Carpomyina and define their geographic ranges. Inter- and intrageneric relationships within Carpomyina have been examined by a number of workers using both morphological and molecular characters. Although many relationships are not well resolved, they are coming into focus.

We should seek to obtain well-resolved phylogenies based on morphology for comparison with well-resolved phylogenies based on DNA sequences. The advent of molecular phylogenetic analyses provides us with a powerful tool with which we can test phylogenetic hypotheses proposed on the basis of morphological characters, life-history traits, and host relationships. This task is not simple. Morphological features provide few unambiguous clade-defining synapomorphies in Carpomyina. Establishing homologies and determining whether or not a particular character state is plesiomorphic or derived is time-consuming and difficult. On the other hand, molecular studies reveal that we need to include more taxa in our analyses and sequence more nucleotides from each taxon. Molecular studies are time-consuming, expensive, and often require destruction of specimens. This prohibits the inclusion of taxa for which only one or two individuals are known. In general, we have not yet sufficiently sampled molecular and morphological data for the Carpomina to arrive at a consensus on their relationships.

In summary, we need more complete taxon sampling and more characters from each taxon in order to obtain robust hypotheses of the relationships of the species belonging to the subtribe Carpomyina. Because most of the molecular phylogenetic analysis has been carried out using mtDNA, and the fact that mtDNA behaves differently than nuclear genes (Harrison 1989), one obvious need is for analysis of relationships based on nuclear gene sequences. Using nuclear gene data, much knowledge has been gained about relationships at all levels of phylogenetic resolution (Traut et al. 1992; Zheng et al. 1993; Cho et al. 1995, Estoup et al. 1996; Fang et al.1997). We hope these types of studies will be soon carried out within Carpomyina.

## ACKNOWLEDGMENTS

We thank Martín Aluja and everybody in Xalapa for making the symposium and, hence, this chapter possible. We thank John Jenkins for discussions regarding taxonomy of Carpomyina and for providing access to his data set for reanalysis, Marlene Cameron for assistance in preparing figures, Valery Korneyev for helpful discussions, and Vesna Gavrilovic and Matt Jaycox for critically reading the manuscript. We also thank two anonymous reviewers for their comments on an earlier version of the manuscript. Finally, we especially thank Al Norrbom for his comments, assistance, and patience with the editing of the final version of the manuscript.

We acknowledge the Campaña Nacional contra las Moscas de la Fruta (Mexico), International Organization for Biological Control of Noxious Animals and Plants (IOBC), Instituto de Ecología, A.C. (Mexico), and USDA-ICD-RSED for their financial support of the symposium, and we thank the Academia Mexicana de Ciencias and the Fundación Mexico-Estados Unidos para la Ciencia for providing travel funds for Dr. Bush.

## REFERENCES

Balduf, W.V. 1959. Obligatory and facultative insects in rose hips, their recognition and bionomics. *Ill. Biol. Monogr.* 26: 1–194.

Berlocher, S.H. 1981. A comparison of molecular and morphological data, and phenetic and cladistic methods, in the estimation of phylogeny in *Rhagoletis* (Diptera: Tephritidae). In *Application of Genetics and Cytology in Insect Systematics and Evolution* (M.W. Stock, ed.), p. 152. University of Idaho, Moscow.

Berlocher, S.H. 1984. A new North American species of *Rhagoletis* (Diptera: Tephritidae), with records of host plants of *Cornus*-infesting *Rhagoletis*. *J. Kans. Entomol. Soc.* 57: 237–242.

Berlocher, S.H. 1993. Gametic disequilibrium between allozyme loci and sex chromosomes in the genus *Rhagoletis*. *J. Hered.* 84: 431–437.

Berlocher, S.H. 1997. Can sympatric speciation be proven from phylogenetic and biogeographic evidence? In *Endless Forms: Species and Speciation* (D. Howard and S.H. Berlocher, eds.), pp. 99–113. Oxford University Press, Oxford.

Berlocher, S.H. and G.L. Bush. 1982. An electrophoretic analysis of *Rhagoletis* (Diptera: Tephritidae) phylogeny. *Syst. Zool.* 31: 136–155.

Berlocher, S.H. and M. Enquist. 1993. Distribution and host plants of the apple maggot fly, *Rhagoletis pomonella* (Diptera: Tephritidae) in Texas. *J. Kans. Entomol. Soc.* 66: 51–59.

Berlocher, S.H. and B.A. McPheron. 1996. Population structure of *Rhagoletis pomonella*, the apple maggot. *Heredity* 77: 83–99.

Berlocher, S.H., B.A. McPheron, J.L. Feder, and G.L. Bush. 1993. Genetic differentiation at allozyme loci in the *Rhagoletis pomonella* (Diptera: Tephritidae) species complex. *Ann. Entomol. Soc. Am.* 86: 1–11.

Boyce, A.M. 1934. Bionomics of the walnut husk fly, *Rhagoletis completa*. *Hilgardia* 8: 363–579.

Brower, A.V.Z. 1994. Rapid morphological radiation and convergence among races of the butterfly *Heliconius erato* inferred from patterns of mitochondrial DNA evolution. *Proc. Natl. Acad. Sci. U.S.A.* 91: 6491–6495.

Bush, G.L. 1965. The genus *Zonosemata*, with notes on the cytology of two species (Diptera-Tephritidae). *Psyche* 72: 307–323.

Bush, G.L. 1966. The taxonomy, cytology, and evolution of the genus *Rhagoletis* in North America (Diptera, Tephritidae). *Bull. Mus. Comp. Zool.* 134: 431–562.

Bush, G.L. and J.J. Smith. 1998. The genetics and ecology of sympatric speciation: a case study. *Res. Popul. Ecol.* 40: 175–187.

Carroll, L.E. 1992. Systematics of Fruit Fly Larvae (Diptera: Tephritidae). Ph.D. dissertation, Texas A&M University, College Station. 341 pp.

Cho, S., A. Mitchell, J.C. Regier, C. Mitter, R.W. Poole, T.P. Friedlander, and S. Zhao. 1995. A highly conserved nuclear gene for low-level phylogenetics: elongation factor-1α recovers morphology-based tree for heliothine moths. *Mol. Biol. Evol.* 12: 650–656.

Dean, R.W. 1969. Infestation of peaches by *Rhagoletis suavis*. *J. Econ. Entomol.* 62: 490–491.

Estoup, A., M. Solignac, J.M. Cornuet, J. Goudet, and A. Scholls. 1996. Genetic differentiation of continental and island populations of *Bombus terrestris* (Hymenoptera: Apidae) in Europe. *Mol. Ecol.* 5: 19–31.

Fang, Q.Q., S. Cho, J.C. Regier, C. Mitter, M. Matthews, R.W. Poole, T.P. Friedlander, and S. Zhao. 1997. A new nuclear gene for insect phylogenetics: dopa decarboxylase is informative of relationships within Heliothinae (Lepidoptera: Noctuidae). *Syst. Biol.* 46: 269–283.

Foote, R.H. 1981. The genus *Rhagoletis* Loew south of the United States (Diptera: Tephritidae). *U.S. Dep. Agric. Tech. Bull.* No. 1607, 75 pp.

Foote, R.H., F.L. Blanc, and A.L. Norrbom. 1993. *Handbook of the Fruit Flies (Diptera: Tephritidae) of America North of Mexico.* Comstock Publishing Associates, Ithaca. 571 pp.

Freidberg, A. and J. Kugler. 1989. Fauna Palaestina. Diptera: Tephritidae, Insecta IV. The Israel Academy of Sciences and Humanities, Jerusalem. 212 pp.

Frías L.D., M. Ibarra, and A.M. Llanca. 1987. Un nuevo diseño alar en *Rhagoletis conversa* (Brethes) (Diptera: Tephritidae). *Rev. Chil. Entomol.* 15: 21–26.

Gleason, H.A. and A. Cronquist. 1963. *Manual of Vascular Plants of Northeastern United States and Adjacent Canada.* D. Van Nostrand, New York. 810 pp.

Han, H.-Y. and B.A. McPheron. 1997. Molecular phylogenetic study of Tephritidae (Insecta: Diptera) using partial sequences of the mitochondrial 16S ribosomal DNA. *Mol. Phylogenet. Evol.* 7: 17–32.

Hardy, D.E. 1964. Diptera from Nepal: The fruit flies (Diptera: Tephritidae). *Bull. Br. Mus. (Nat. Hist.) Entomol.* 15: 145–169.

Harrison, R.G. 1989. Animal mitochondrial DNA as a genetic marker in population and evolutionary biology. *Trends Ecol. Evol.* 4: 6–11.

Hendel, F. 1931. Kritische und synonymische Bemerkungen über Dipteren. *Verh. Zool. Bot. Ges. Wien* 81: 4–19.

Hering, E.M. 1941. Trypetidae (Dipt.). In *Beiträge zur Fauna Perus* Bd. 1 (E. Titschack), pp. 121–176. Hamburg.

Hernández-Ortiz, V. 1985. Descripción de una nueva especie mexicana del género *Rhagoletis* Loew (Diptera: Tephritidae). *Folia Entomol. Mex.* 64: 73–79.

Hernández-Ortiz, V. 1989. Una especie nueva de *Zonosemata* (Diptera: Tephritidae) y clave de idenificación de las especies del género. *An. Inst. Biol. University Nac. Auton. Mex. Ser. Zool.* 60: 205–210.

Hernández-Ortiz, V. 1993. Description of a new *Rhagoletis* species from tropical Mexico (Diptera: Tephritidae). *Proc. Entomol. Soc. Wash.* 95: 418–424.

Hoelzer, G.A. and D.J. Melnick. 1994. Patterns of speciation and limits to phylogenetic resolution. *Trends Ecol. Evol.* 9: 104–107.

Jenkins, J. 1996. Systematic Studies of *Rhagoletis* and Related Genera (Diptera: Tephritidae). Ph.D. dissertation, Michigan State University, East Lansing. 184 pp.

Jermy, T. 1961. Eine neue *Rhagoletis*-art (Diptera: Tephritidae) aus den Fruchten von *Berberis vulgaris* L. *Acta Zool. Acad. Sci. Hung.* 8: 133–137.

Jones, G.N. 1940. A monograph of the genus *Symphoricarpos*. *J. Arnold Arbor.* 21: 201–252.

Kandybina, M.N. 1977. Lichinki plodovykh mukh-pestrokrylok (Diptera: Tephritidae) [Larvae of fruit-infesting fruit flies (Diptera: Tephritidae)]. *Opred. Faune S.S.S.R.* 114: 1–212.

Korneyev, V.A. and B. Merz. 1997. A new species of *Rhagoletis* Loew (Diptera: Tephritidae), with notes on central Asian species. *J. Ukr. Entomol. Soc.* 3: 55–64.

Loew, H. 1862. Die Europaeischen Bohrfliegen (Trypetidae) W. Junk, Wien.

Maddison, W.P. and D.R. Maddison. 1992. *MacClade: Analysis of Phylogeny and Character Evolution. Version 3.0.* Sinauer Associates, Sunderland.

McPheron, B.A. 1990. Genetic structure of apple maggot fly (Diptera: Tephritidae) populations. *Ann. Entomol. Soc. Am.* 83: 568–577.

McPheron, B.A. and H.-Y. Han. 1997. Phylogenetic analysis of North American *Rhagoletis* (Diptera: Tephritidae) and related genera using mitochondrial DNA sequence data. *Mol. Phylogenet. Evol.* 7: 1–16.

Merz, B. and J. Blasco-Zumeta. 1995. The fruit flies (Diptera, Tephritidae) of the Moregros region (Zaragosa, Spain), with the record of the host plant of *Rhagoletis zernyi* Hendel, 1927. *Zapateri Rev. Aragon. Entomol.* 5: 127–134.

Munro, H.K. 1939. Studies in African Trypetidae, with descriptions of new species. *J. Entomol. Soc. South. Afr.* 1: 26–46.

Norrbom, A.L. 1989. The status of *Urophora acuticornis* and *U. sabroskyi* (Diptera: Tephritidae). *Entomol. News* 100: 59–66.

Norrbom, A.L. 1990. Notes on *Zonosemata* Benjamin (Diptera: Tephritidae) and the status of *Cryptodacus scutellatus* Hendel (= *Z. ica* Steyskal syn. n.). *Ann. Naturhist. Mus. Wien* 91: 53–55.

Norrbom, A.L. 1994. New species and phylogenetic analysis of *Cryptodacus*, *Haywardina*, and *Rhagoletotrypeta* (Diptera: Tephritidae). *Insecta Mundi* 8: 37–65

Norrbom, A.L. 1997. The genus *Carpomya* Costa (Diptera: Tephritidae): new synonymy, description of first American species, and phylogenetic analysis. *Proc. Entomol. Soc. Wash.* 99: 338–347.

Norrbom, A.L., L.E. Carroll, and A. Freidberg. 1999a. Status of knowledge. In *Fruit Fly Expert Identification System and Systematic Information Database* (F.C. Thompson, ed.), pp. 9–48. *Myia* (1998) 9, 524 pp.

Norrbom, A.L., L.E. Carroll, F.C. Thompson, I. White, and A. Freidberg. 1999b. Systematic database of names. In *Fruit Fly Expert Identification System and Systematic Information Database* (F.C. Thompson, ed.), pp. 65–251. *Myia* (1998) 9, 524 pp.

Payne, J.A. and S.H. Berlocher. 1995a. Distribution and host plants of the blueberry maggot fly, *Rhagoletis mendax* (Diptera: Tephritidae) in southeastern North America. *J. Kans. Entomol. Soc.* 68: 133–142.

Payne, J.A. and S.H. Berlocher. 1995b. Phenological and electrophoretic evidence for a new blueberry-infesting species in the *Rhagoletis pomonella* sibling species complex. *Entomol. Exp. Appl.* 75: 183–187.

Prokopy, R.J. and S.H. Berlocher. 1980. Establishment of *Rhagoletis pomonella* (Diptera: Tephritidae) on rose hips in southern New England. *Can. Entomol.* 112: 1319–1320.

Reissig, W.H. and D.C. Smith. 1978. Bionomics of *Rhagoletis pomonella* in *Crataegus*. *Ann. Entomol. Soc. Am.* 71: 155–159.

Shervis, L.J., G.M. Boush, and C.F. Koval. 1970. Infestation of sour cherries by apple maggot: confirmation of a previously uncertain host. *J. Econ. Entomol.* 63: 294–295.

Smith, D.C. 1988. Reproductive differences between *Rhagoletis* (Diptera: Tephritidae) fruit parasites of *Cornus amomum* and *C. florida* (Cornaceae). *J. N.Y. Entomol. Soc.* 96: 327–331.

Smith, J.J. and G.L. Bush. 1997. Phylogeny of the genus *Rhagoletis* (Diptera: Tephritidae) inferred from DNA sequences of mitochondrial cytochrome oxidase II. *Mol. Phylogenet. Evol.* 7: 33–43.

Swofford, D.L. 1993. *PAUP: Phylogenetic Analysis Using Parsimony, Version 3.1.* Illinois Natural History Survey, Champaign.

Traut, W., J.T. Epplen, D. Weichenhan, and J. Rohwedel. 1992. Inheritance and mutation of hypervariable (GATA)n microsatellite loci in a moth, *Ephestia kuehnilla*. *Genome* 35: 659–666.

Vander Kloet, S.P. 1988. The genus *Vaccinium* in North America. Publication 1828, Research Branch Agriculture Canada, Ottawa.

Wasbauer, M.S. 1972. An Annotated Host Catalog of the Fruit Flies of America North of Mexico (Diptera: Tephritidae). *Occas. Pap. Calif. Dep. Agric. Bur. Entomol.* 19: 172 pp.

Westcott, R.J. 1982. Differentiating adults of apple maggot, *Rhagoletis pomonella* (Walsh) from snowberry maggot, *R. zephyria* Snow (Diptera: Tephritidae) in Oregon. *Pan-Pac. Entomol.* 58: 25–30.

White, I.M. and M.M. Elson-Harris. 1992. *Fruit Flies of Economic Significance: Their Identification and Bionomics.* CAB International, Wallingford. 601 pp.

Zheng, L., F.H. Collins, V. Kumar, and F.C. Kafatos. 1993. A detailed genetic map for the X chromosome of the malaria vector, *Anopheles gambiae*. *Science* 261: 605–608.

# 10 Behavior of Flies of the Genera *Rhagoletis, Zonosemata,* and *Carpomya* (Trypetinae: Carpomyina)

*Ronald J. Prokopy and Daniel R. Papaj*

## CONTENTS



## 10.1 INTRODUCTION

Understanding the origin and adaptive value of a behavioral trait requires not only knowledge of phylogenetic relationships of the organism in question but also consideration of the nature of the environment in which the trait may have arisen (and in which the trait persists) together with consideration of the internal state of the organism, both of which can affect patterns of response to the environment. For insects, the behaviors of greatest relevance to survival and reproductive success are probably those associated with acquisition of essential resources such as food, mates, and egg-laying sites and those associated with defense against natural enemies.

0-8493-1275-2/00/$0.00+$.50

In this chapter, our primary goal is to describe, discuss, and compare the behavior of adult and larval members of three genera in the tribe Carpomyini: *Rhagoletis, Zonosemata,* and *Carpomya.* Larvae of all species in this tribe feed in the pulp of developing fruit and several species are important economic pests. For these three genera, there exists a relatively substantial literature on the behavior of several species of *Rhagoletis* flies, aspects of which have been reviewed by Bateman (1972), Dean and Chapman (1973), Boller and Prokopy (1976), Prokopy (1977a), Prokopy and Roitberg (1984), Averill and Prokopy (1989), Fletcher (1989), Prokopy and Roitberg (1989), Fletcher and Prokopy (1991), and Papaj (1993). Literature describing behavioral traits of *Zonosemata* flies is limited to two species: *Zonosemata vittigera* (Coquillett) and *Z. electa* (Say). Unfortunately, little information exists on the behavior of any species in the genus *Carpomya.*

Comparisons of behavior of species in the genera *Rhagoletis, Zonosemata,* and *Carpomya* are best made where sufficient information exists to permit useful comparisons. For the most part, this will involve species of agricultural importance as it is these species that have received the greatest attention. This is less than ideal because absence of behavioral information on nonagriculturally important species not only is constraining but potentially misleading. We will focus on traits which might elucidate phylogenetic patterns of behavior and on recent information that is relevant to this intent. Information on the behavior of *Rhagoletis* flies published before 1970 is discussed by Díaz-Fleischer and Aluja (Chapter 3).

The chapter begins with consideration of activity patterns of adults in space and time, followed by description and discussion of adult behavior associated with foraging for food, mates, and ovipositional sites. We divide the section on oviposition-site foraging into several components: host plant finding, host fruit finding, examination and acceptance of foliage, examination and acceptance of fruit, influence of host-marking pheromones, and social facilitation of egg-laying behavior. Then follows a section dealing specifically with studies on nongenetic variables that may affect adult foraging behavior, including the state of the environment and the physiological and informational state of the forager. Explicit treatment of state variables affecting the foraging behavior of adults for essential resources helps to ensure that all proximate factors shaping adult foraging behavior are considered (Mangel and Clark 1986; Mangel and Roitberg 1989). The concluding section on adult behavior deals with escape from natural enemies. The chapter concludes with a section on larval behavior.

## 10.2 ACTIVITY PATTERNS OF ADULTS IN SPACE AND TIME IN NATURE

To our knowledge, no studies exist which quantify the organization of feeding, mating, and egg-laying activities of *Zonosemata* or *Carpomya* flies over time or space in nature. However, studies of this sort have been carried out for several *Rhagoletis* species including *R. pomonella* (Walsh) (Prokopy et al. 1971; 1972; Smith and Prokopy 1980; Opp and Prokopy 1987; Hendrichs and Prokopy 1990), *R. mendax* Curran (Smith and Prokopy 1981; 1982), *R. cornivora* Bush (Smith 1985), *R. zephyria* Snow (Tracewski and Brunner 1987), *R. juglandis* Cresson (Alonso-Pimentel and Papaj 1996b), *R. cingulata* (Loew) (Smith 1984), *R. fausta* (Osten Sacken) (Prokopy 1976), *R. conversa* (Brèthes) (Frías et al. 1984), and *R. turpiniae* Hernández-Ortiz (Lozada et al. 1999). Several themes common to all observed *Rhagoletis* species emerge from these studies.

During the early part of the fly season, adults of both sexes are likely to be seen on foliage of host plants and nearby nonhost plants. As the fly season progresses, there is an ever greater tendency for adults, particularly males, to be more concentrated on host plants, particularly on fruit. These tendencies reflect the propensity of both sexes of *Rhagoletis* species to search for food early in the season but to aggregate increasingly on host plants with onset of sexual maturity and increasing ripeness of fruit. Where observed, initiation of a copulation was confined exclusively to fruiting host plants. None has been observed on nonhost plants, even after lengthy observations on nonhost plants in several studies. Odor of ripening host fruit, attractive to both sexes (Prokopy et al. 1973), could be serving as a substitute for a long-range attractive pheromone (not known to occur in any *Rhagoletis* species) in drawing both sexes to locales where mating occurs.

With respect to time of day of activity, feeding appears to occur with about equal frequency during all hours of daylight provided temperature is sufficiently high to permit movement and rainfall is absent. Copulation initiation and oviposition occur from midmorning to late afternoon in more temperate climates but, like feeding, may occur early in the day in warm climates as exemplified in *R. conversa* (Frías et al. 1984). When midday temperatures are very warm, adults of several species have been observed resting on the undersides of leaves of broad-leaf nonhost plants in the vicinity of hosts. One sort of behavior observed by those who have watched different species of *Rhagoletis* flies move about on host plants is a strong tendency of adults of both sexes to fly upward toward tops of host plants or interplant space toward evening. At night, adults have been detected almost exclusively in upper portions of host or nearby nonhost plants, rarely in middle or lower portions. Such behavior may have been shaped as a response to escape predators foraging in lower parts of plant canopies or as a response to abiotic conditions (light and temperature) favoring food foraging activity until darkness.

As pointed out by Fletcher (1989) and others, the great majority of *Rhagoletis* species has but one full generation each year. The same appears to be true for most studied species of *Zonosemata* and *Carpomya* flies. Exceptions are *R. lycopersella* Smyth (Smyth 1960), *R. tomatis* Foote (Frías 1993) and possibly other neotropical *Rhagoletis* species whose hosts are in the Solanaceae as well as *C. vesuviana* Costa in India (which completes six to nine generations annually) (Lakra and Singh 1989). Following puparial diapause, adults emerge beneath host plants that had fruit the previous year and at a time roughly synchronous with fruit suitable for egg laying during the current year. In addition, most *Rhagoletis* species are thought to have been nearly monophagous (infesting fruit in only a single genus of plants) up until the introduction of new cultivated or wild hosts in recent centuries. The same appears to be true for *Z. electa* and *Z. vittigera* (Burdette 1935; Goeden and Ricker 1971). As a consequence of these life history traits, there seems to have been little selection pressure for strong flight capability that would permit adults, unaided by wind, to move away from potential interspecific tephritid competitors and toward distant alternative hosts. Indeed, *Rhagoletis* and *Zonosemata* adults seem to be able to acquire all essential resources during comparatively local movement. Their rather small body size compared with many other tephritids is consistent with the notion of comparatively short dispersal capability (Sivinski and Dodson 1992).

In sum, information available to date suggests strong similarity among *Rhagoletis* species in the organization of behavior in space and time. Whether on account of inhabiting a more temperate climate, having a more limited host range, experiencing fewer confamilial competitors, and/or having a more limited dispersal capability, most species of *Rhagoletis* flies appear to be less canalized in activity pattern in time and space than are several frugivorous species of *Anastrepha*, *Bactrocera*, *Ceratitis* and *Toxotrypana* Gerstaecker flies (Aluja et al., Chapter 15; Drew and Romig, Chapter 21; Yuval and Hendrichs, Chapter 17; and Landolt, Chapter 14).

## 10.3 FORAGING BY ADULTS FOR FOOD

Behaviors associated with the acquisition of food are among the least-studied traits in *Rhagoletis*, *Zonosemata,* and *Carpomya* flies. Except in *R. pomonella*, much current information is anecdotal (Hendrichs and Prokopy 1994). Even so, some among-species comparisons can be made.

Like other frugivorous tephritids, flies of these three genera need to ingest carbohydrate and water for survival and protein or amino acids for egg production (e.g., Neilson and McAllan 1965). The need for carbohydrate and water essentially is daily but for protein it is less frequent. Sources of food for at least some species of these genera have been identified as foliar leachate, insect honeydew, nectar from extrafloral nectaries, juice oozing from wounded fruit, and bird droppings (Hendrichs and Prokopy 1994). Very little juice emanates from wounded host fruit of *R. pomonella* and adults infrequently acquire nutrients from this source (Hendrichs et al. 1993a; Hendrichs and Prokopy 1994). In contrast, adults of *R. mendax* (Smith and Prokopy 1981), *R. indifferens* Curran (Frick et al. 1954), *R. cingulata* (Smith 1984), *R. cerasi* (L.) (Leski 1963), *R. berberis* Curran

(Mayes and Roitberg 1987), and *Z. vittigera* (Goeden and Ricker 1971), among others, have been observed to feed rather extensively on juice from wounded fruit. While data are sparse, variation among species in adult feeding on fruit appears not to have a phylogenetic basis. Available evidence suggests that *Rhagoletis, Zonosemata,* and *Carpomya* flies are opportunistic in feeding behavior, acquiring carbohydrate and proteinaceous resources wherever convenient or possible. This may involve short bouts of "grazing" on leaf surfaces or longer feeding bouts on honeydew, fruit juices, or bird droppings.

To our knowledge, there exists only one report of a stereotyped sort of feeding behavior in a species of these three genera. In a study of *R. fausta* flies in sour cherry trees, Prokopy (1976) observed a highly consistent pattern of fly movement associated with obtaining food from the two extrafloral nectaries located on the petiole of each leaf. Following arrival on a leaf (almost always on the bottom surface), a fly would crawl to the leaf margin, over the edge onto the top surface, toward the midrib and then down the midrib toward the petiole, all in a matter of a few seconds. It would halt at each nectary, feed for 1 to 3 s and then fly upward to the bottom surface of a leaf immediately above. This pattern would be repeated over and over again until the fly reached the uppermost leaves on a branch or tree, after which it usually would make a long flight downward to lower leaves and begin the entire process again. Wiesmann (1933) observed *R. cerasi* flies feeding on extrafloral nectaries of sweet cherry trees but did not report on the pattern of feeding. Whether the stereotyped pattern of feeding on extrafloral nectaries observed in *R. fausta* is characteristic of all tephritids whose hosts have extrafoliar nectaries associated with each leaf or is peculiar to *R. fausta* awaits evaluation.

Little is known about how flies of these three genera find distant food sources, whether by random movement, by response to physical characteristics of sites harboring possible food, or by response to odor associated with potentially high-quality food, such as insect honeydew or bird droppings. With regard to physical characteristics, *Rhagoletis* and other tephritids may be guided by attractive visual stimuli, such as yellow-colored foliage on distal parts of plants. Many species of *Rhagoletis* flies are strongly attracted to yellow color (reviewed by Katsoyannos 1989), either as a response to yellow as the supernormal equivalent of reflection by green leaves or possibly in part as a consequence of selective responses to sites most likely to harbor insects that secrete honeydew (e.g., aphids), which also are highly attracted to yellow color (e.g., yellowish foliage) (Kring 1970). With respect to odor, recently it has been found that *R. pomonella* flies are attracted by odor emanating from bird droppings and that addition of antibiotic to bird droppings significantly reduces attractiveness, suggesting that microorganisms such as bacteria may be involved in the production of attractants (Prokopy et al. 1993a). It turns out that at least one bacterial species common to bird droppings (*Enterobacter agglomerans* Beijerinck) (Lauzon 1991) does in fact produce odor attractive to *R. pomonella* (MacCollom et al. 1992). Very recent information indicates that *R. pomonella* flies are attracted to odor emitted by specific strains of *E. agglomerans* but not by odor emitted by other strains of *E. agglomerans* or other bacteria (e.g., *Bacillus cerasus* (Griffin) Holland, *Streptococcus* sp.) isolated from bird feces (Lauzon et al. 1998). Additional testing has revealed the presence of a protein (uricase) that is unique to certain isolates of *E. agglomerans* that are attractive to *R. pomonella* and is absent from unattractive isolates (Lauzon et al. 1999). Presence of uricase permits breakdown of uric acid in bird feces. Such uricase-producing bacterial isolates (probably not confined to *E. agglomerans*) may be actively sought and selectively retained by *R. pomonella* flies.

In some insect-bacterial endosymbiont associations, there is a close correspondence between the phylogeny of the insect and that of the bacteria (Moran and Telang 1998). In contrast, there does not appear to be a close correspondence between *Rhagoletis* phylogenies and those of their associated bacterial symbionts. A detailed study by Howard et al. (1985) in which bacterial associates of seven different *Rhagoletis* species were examined revealed no evidence for a symbiotic relationship between a particular species of fly and a particular species or species complex of bacteria. Their finding is consistent with the idea that species or strains of bacteria attractive to *Rhagoletis* and probably other genera of tephritid flies (Drew and Lloyd 1989; Jang and Nishijima 1990) are not specific to host plants but can be found in association with a wide variety of vegetation containing bird feces or other substances serving as nutrients of bacteria. Thus, it seems unlikely that, as once thought, the complex of bacterial associates of *Rhagoletis* and other tephritid species can illuminate our understanding of phylogenetic relationships among tephritids.

The proximity of quality food to quality ovipositional sites can have a strong impact on the extent to which potential ovipositional sites might be used. To illustrate, *R. pomonella* females were found to lay significantly more eggs in native host hawthorn fruit near to a source of high-quality carbohydrate and protein (i.e., on the same tree) compared with fruit more distant from high-quality food (i.e., on a distant tree) (Averill and Prokopy 1993). This finding suggested that attractive odor of abundant natural food (such as bird feces) could result in accumulation of flies of both sexes on species of plants where fruit emit no attractive odor and are not permanent hosts but which could receive eggs and support larval development of temporary populations of food-seeking flies. In fact, such has been shown possible in *R. pomonella* in an experiment where sour cherries (which do not emit odor attractive to *R. pomonella*; A.L. Averill, unpublished data) have been found to be significantly more infested when adjacent to attractive odor of protein than when distant from odor of protein (R. Prokopy, unpublished data). Of course, permanent expansion of host range onto such fruit would undoubtedly require accompanying changes in chemosensory capabilities of flies associated with the new host (if not also in larval survival capability). Some species of *Rhagoletis* flies (e.g., *R. pomonella*) have expanded their host range considerably in recent centuries, and one wonders if reliable presence of high-quality food in close physical association with new hosts and within cruising range of current hosts might have been a contributing factor, at least initially.

Besides potentially enhancing ovipositional attempts into nonhost fruit, proximity of feeding sites to ovipositional sites might also enhance rapid development of oogenesis in females. Indeed, as shown recently by Alonso-Pimentel et al. (1998), ovarian development in the first egg maturation cycle in *R. juglandis* females is enhanced both in rate and degree by exposure to host fruit. Immature females finding food frequently in the vicinity of fruit and remaining near fruit for much of their prereproductive life could gain a reproductive edge over other females through more rapid oogenesis, apart from the contribution of food itself to egg production. The extent to which this finding for *R. juglandis* flies might be characteristic of other *Rhagoletis* or related species remains to be determined.

Among species of *Rhagoletis, Zonosemata,* and *Carpomya*, the dynamics of fly intratree foraging behaviors as affected by availability of foods of varying quality and quantity have been examined quantitatively only in *R. pomonella*, where Hendrichs et al. (1993b) studied feeding and postfeeding behaviors of individual flies presented with sources of food of varying concentration, volume, and amount of solute. It turned out that patch residence time was less closely linked to food foraging time than to food-handling and -processing time. With increasing dilution or volume of food ingested, engorged flies exhibited increasing postfeeding quiescence, during which they engaged in oral extrusion of droplets of liquid crop contents so as to eliminate excess water. One wonders whether the pattern of postingestion food processing found in *R. pomonella* is peculiar to this species or is characteristic of related species and largely a product of the type, amount, and concentration of food ingested.

In sum, except in *R. pomonella*, existing information on behaviors associated with food foraging and post-ingestion food processing in *Rhagoletis*, *Zonosemata,* and *Carpomya* flies is largely qualitative in nature. Too few quantitative data are available to permit species to be compared in informative ways.

## 10.4 FORAGING BY ADULTS FOR MATES

As with other aspects of behavior, little is known about mating in any group within the Carpomyini except in the genus *Rhagoletis*. In this genus, there is abundant information about mating in members of the *pomonella* and *suavis* groups. In addition, something is known about mating in *R. fausta* and the European species, *R. cerasi.*

*Rhagoletis* species are characterized by polygynandry, wherein both males and females mate multiple times over the course of their lives (*R. pomonella*: Opp 1988; *R. completa* Cresson: Opp et al. 1996; *R. juglandis*: D. Papaj and H. Alonso-Pimentel, unpublished data). As noted earlier, initiation of copulations appears to occur exclusively on host plants. Early in the season, before flies move to fruit, males and females alike are found in loose aggregations in which mating takes place (*R. pomonella*: Prokopy et al. 1971; *R. fausta*: Prokopy 1976; Smith and Prokopy 1980; *R. mendax*: Smith and Prokopy 1981; *R. juglandis*: Alonso-Pimentel and Papaj 1996a). Studies of *R. pomonella* and *R. cerasi* indicate that males produce a pheromone that attracts females, although probably over only a short distance (Prokopy and Bush 1973a; Prokopy 1975a; Katsoyannos 1976; Katsoyannos et al. 1980; although see Prokopy and Bush 1972).

As the fly season progresses, mating activity moves to the fruit. As fruit ripen, first males and then females appear on fruit in increasing numbers (*R. pomonella*: Prokopy et al. 1971; Smith and Prokopy 1980; *R. fausta*: Prokopy 1976; *R. mendax*: Smith and Prokopy 1981; *R. juglandis*: Alonso-Pimentel and Papaj 1996a). After moving to fruit, males defend the fruit from other males through a variety of displays, including wing-waving, foreleg-kicking, and boxing (e.g., Messina and Subler 1995). *Rhagoletis pomonella* males are arrested by host-marking pheromone; in addition, males deposit an unknown arrestant on fruit that tends to aggregate males on fruit or fruit surrogates (Prokopy and Bush 1972).

In members of the *suavis* group, males defend not just fruit but oviposition punctures on fruit (*R. juglandis* and *R. boycei* Cresson: Papaj 1994; *R. completa*: Lalonde and Mangel 1994). Since females in this group frequently deposit eggs in oviposition cavities established by other females, oviposition site-guarding increases male mating success (Papaj 1994).

On-fruit mating has been characterized as "forced copulation" because males attempt copulation with a minimum of courtship display and because females often appear to resist matings physically. Characterization of on-fruit mating as forced is somewhat controversial, in part because it is impossible to prove that females are being forced to copulate (what appears, for example, to be resistance may actually be evaluation of male quality on the part of the female). Nevertheless, several observations reported by Prokopy and Bush (1973a) and Smith and Prokopy (1980; 1982) for *R. pomonella* and *R. mendax* are consistent with the notion that males on fruit force copulations on females. First, Smith and Prokopy (1982) noted, albeit without providing data, that the presence of *R. mendax* males on fruit deterred conspecific females from visiting fruit. Second, Smith and Prokopy (1980, 1982) reported that, on fruit, both *R. pomonella* and *R. mendax* males tend to approach females from behind; by contrast, during early-season encounters on foliage, males approach females primarily from the front (Figure 10.1A). Associated with the difference in approach is a difference in the efficacy of mating (Figure 10.1B). On foliage, frontal approaches result proportionately more often in mating; on fruit, by contrast, rear approaches result proportionately more often in mating.

The "sneaky" behavior of males in on-fruit encounters is consistent with the notion that males act as sexual predators, forcing copulations upon females. From a functional perspective, forced copulation presumably results from a conflict of interests between the two sexes. Females arrive at fruit with live sperm and mature eggs (Alonso-Pimentel and Papaj 1996a) and presumably benefit most by laying eggs in those fruit. Where costs of mating (for example, in terms of opportunities to oviposit; compare Papaj and Alonso-Pimentel 1997) exceed costs in terms of resistance, females may be expected to resist matings. Males, by contrast, benefit mainly by mating as often as possible. Where costs of using force are relatively low, males may be expected to force matings upon females.

Both resource defense and forced copulation are facilitated by the pattern of sperm competition in *Rhagoletis*. Allozyme studies of *R. pomonella* and *R. completa* indicate a pattern of last male sperm precedence in which the last male to mate with a female tends (although not absolutely) to fertilize the next eggs out (Opp et al. 1990; 1996). Another strategy of advantage in the context of last male sperm precedence is postcopulatory mate guarding. Where such guarding reduces the probability of a female mating with another male, it may increase the likelihood that males sire

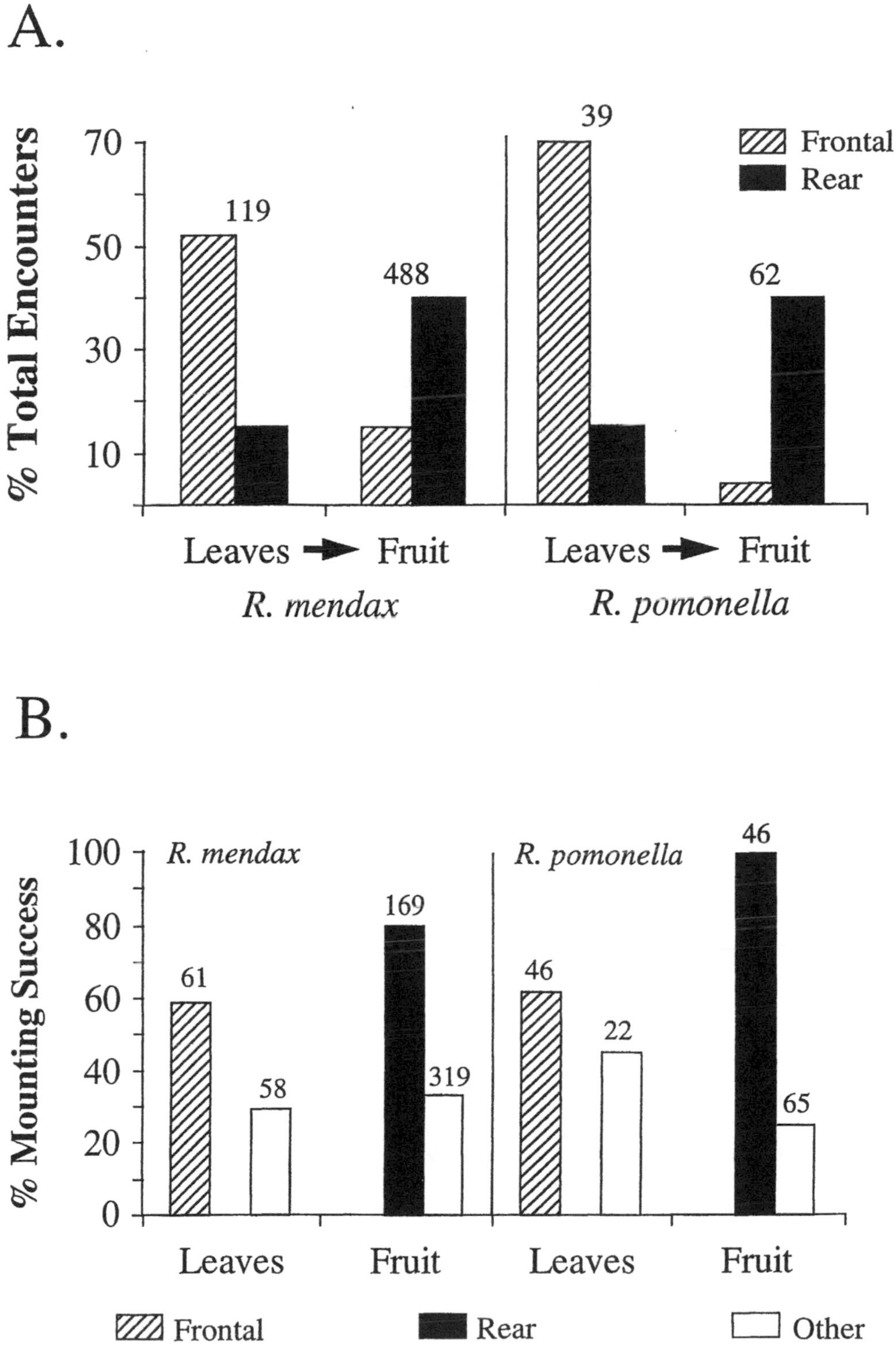


**FIGURE 10.1** (A) Shift in direction of approach by males according to location of mating encounters (fruit vs. foliage). (B) Percent mounting success for favored direction of approach vs. all other directions according to location of mating encounters (fruit vs. foliage). (Data rearranged from Smith and Prokopy 1980; 1982).

the next offspring produced by females. Opp et al. (1996) reported that *R. completa* males guard mates after copulation, while females engage in oviposition (although see Boyce 1934, who claimed that males engage in precopulatory guarding, standing in proximity to ovipositing females until oviposition is completed and then copulating with them). In Opp's studies, males engaged mainly in noncontact guarding in which the male was positioned in close proximity to the female. Whether guarding increases a male's fertilization success is not known, nor is it clear to what extent mate guarding is different in underlying mechanism from oviposition site guarding. Interestingly, less than 50% of copulations in *R. completa* are accompanied by guarding. Contact mate guarding has been observed also in *R. suavis* (Loew).

A form of contact guarding may occur in *R. juglandis*, where copulation duration has a distinctly bimodal distribution (Alonso-Pimentel and Papaj 1996b; Figure 10.2A). Short copulations tend to be 200 s or shorter; long copulations tend to be 400 s or longer. Copulations of intermediate duration are conspicuously absent. It is conceivable that sperm transfer occurs in the time period associated with short copulations and that the remainder of a long copulation constitutes a kind of guarding. It is equally plausible that each mode of copulation is associated with a particular pattern of sperm transfer and storage. At this point, data distinguishing between these possibilities are lacking. If long copulations do involve contact guarding, such guarding is highly dynamic. Both in the laboratory and the field, operational sex ratio has a dramatic effect on the proportion of males engaging in one or the other mode of copulation. In male-biased environments, copulations are predominantly long; in female-biased environments, copulations are predominantly short (Alonso-Pimentel and Papaj 1996b; Figure 10.2B). Similarly, the presence of fruit influences the dominant mode of copulation. In the presence of fruit, copulations tend to be short; in its absence, most copulations are long (Alonso-Pimentel and Papaj 1999). Both sets of results are consistent with sperm competition theory on male mating interests.

As might be expected where copulation is forced by "sneaky" males, evidence of *Rhagoletis* male courtship signaling on fruit is scant. Males in the *suavis* group appear to stand as exceptions to this pattern. *Rhagoletis completa* males engage in wing-waving displays toward females, in which males elevate their wings perpendicular to the long axis of their bodies (Opp et al. 1996; but see Boyce 1934, who describes on-fruit mating as distinctly male-forced). Their head-on approach is in contrast to the rear approach typical of members of the *pomonella* group. Male *R. boycei* mark fruit by dabbing the proctiger on the fruit surface (Papaj et al. 1996); marking occurs selectively at oviposition punctures which, in this species, are preferred by females for oviposition. In laboratory assays, females were more likely to attempt oviposition into male-exposed punctures than into unexposed punctures on the same fruit. Because on-fruit mating in this species, as in most *Rhagoletis* species, occurs in the context of oviposition, the mark serves to increase a male's mating opportunities. Whether the mark influences female receptivity in any additional way is not known.

Male *R. juglandis* vibrate their wings, a behavior produced selectively in the presence of females and associated with mating in both laboratory and field (H. Alonso-Pimentel, H. Spangler, R. Rogers, and D. Papaj, unpublished data). The vibration generates a low-frequency sound and may function to produce a courtship song, although visual and chemical functions have not been ruled out. Two other members of the *suavis* group, *R. suavis* and *R. zoqui* Bush, have yet to be studied in detail with respect to courtship signaling. *Rhagoletis zoqui* is closely related to *R. completa* (Smith and Bush, Chapter 9), and we might predict the occurrence of wing-waving displays in this species. *Rhagoletis suavis* males have also been observed to engage in wing-waving displays (D. Papaj, personal observation). Male–female interactions in *R. suavis* have not been studied intensively, but males have been reported to wing-flick at females (Bush 1966).

That members of the *R. suavis* group appear to express more in the way of courtship behavior than members of other groups within the genus *Rhagoletis* is consistent with an hypothesis put forward by Bush (1969). Bush contended that whereas most groups within North American *Rhagoletis* diversified through sympatric speciation by shifting from one host species to another, members of the *R. suavis* group, all of which use the same host genus, *Juglans*, speciated allopatrically. He

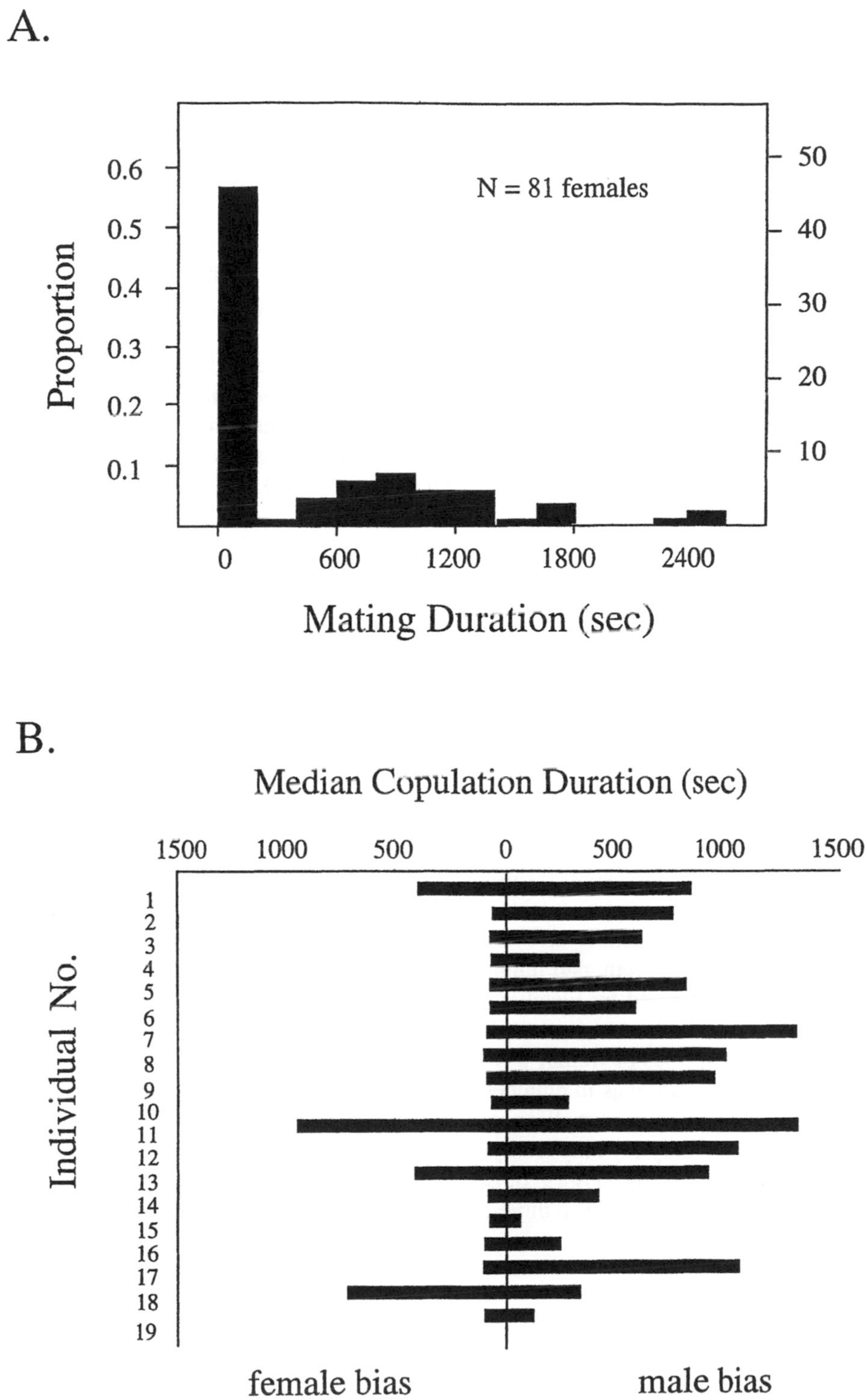


**FIGURE 10.2** (A) Frequency distribution of mating durations for *R. juglandis* in the field. (B) Median copulation duration (s) for individual female *R. juglandis* placed initially in female-biased laboratory cages and then in male-biased cages (or vice-versa). (Figure redrawn from Alonso-Pimentel and Papaj 1996a.)

further argued that use of different hosts by sympatrically speciating forms would generate immediate and complete reproductive isolation. By contrast, members of the *R. suavis* group likely came into secondary contact repeatedly. This difference in reproductive isolation was proposed to generate a difference in selection for species recognition mechanisms. Both the significant degree of sexual dimorphism in body markings among members of the *R. suavis* group reported by Bush (1966; 1969) and the developing database on courtship behavior within the genus are consistent with this prediction.

In summary, comparative data are again sparse and phylogenetic inferences consequently difficult to make. Unlike males of some *Anastrepha, Bactrocera,* and *Ceratitis* species (Aluja et al., Chapter 15; Drew and Romig, Chapter 21; and Yuval and Hendrichs, Chapter 17), *Rhagoletis* males do not form leks and do not engage in complex courtship signaling. Within the genus *Rhagoletis,* members of the *R. suavis* group differ from members of other species groups in several respects. First, males defend not just the fruit, as in other species groups, but oviposition sites on fruit. Second, males of species in the *R. suavis* group engage in a variety of species-typical courtship displays not observed in other members of the genus. These displays are not as elaborate as displays in lekking tephritid species, but are quite striking for members of the genus. Finally, mate guarding is reported only among members of the *R. suavis* group. These phylogenetic differences may reflect a difference in mode of speciation, as hypothesized by Bush (1969). Alternatively, it is conceivable that sexual selection within members of the *R. suavis* group has taken a different course than in members of other groups, perhaps because of ecological factors peculiar to that group. Noteworthy in this respect is the fact that species within the *R. suavis* group use only walnuts as hosts, a fruit not used by members of other groups. Could life on walnuts, a fruit which differs from native host fruit of other *Rhagoletis* species in key respects (e.g., fruit size and chemistry), have generated, through sexual selection, the species group differences in mating behavior?

## 10.5 FORAGING BY ADULTS FOR OVIPOSITION SITES

It is stating the obvious that a strong genetic component underlies the foraging behavior of *Rhagoletis, Zonosemata,* and *Carpomya* flies for host fruit as oviposition sites. If not, then these species would be less variable in host specificity than they are. Genetic endowment dictates not only chemical and physical sensory capabilities associated with host finding, examination, and acceptance but also sets limits on body size, ovipositor musculature and architecture, and duration of life, which in turn affect number of ovarioles, dispersal capability, ability to penetrate fruit of differing hardness and skin depth, and number of potential hosts of differing fruiting phenologies that can be used. As described by Fletcher (1989), there seems to exist limited variation among *Rhagoletis* species (if not also among members of all three genera) in these latter traits. Notable exceptions are the six species of walnut-infesting flies in the *R. suavis* species group. Most, if not all, species in this group lay eggs in clutches of ten or more eggs per ovipositional bout, whereas all other species of *Rhagoletis*, as well as species of *Zonosemata* and *Carpomya* for which information is available, normally lay only a single egg per ovipositional bout (Burdette 1935; Goeden and Ricker 1971; Boller and Prokopy 1976; Lakra and Singh 1983). In this section, we will examine the foraging behavior of flies in these three genera for host fruit.

### 10.5.1 Host Plant Finding

One can postulate that both odor and visual stimuli attract *Rhagoletis* and species of related genera to host plants bearing fruit suitable for oviposition. Unfortunately, except for studies on *R. pomonella* and to some extent *R. mendax*, information is lacking on this aspect of behavior in the genera considered here.

In *R. pomonella*, flies are able to detect odor of host fruit at distances of at least 20 m (and probably at much greater distances when odor amount or concentration is high) and detect visual stimuli of host trees at distances of a least 3 m (Green et al. 1994; in press). Components of host hawthorn and apple fruit odor attractive to *R. pomonella* have been identified and formulated into attractive blends (Averill et al. 1988). When adults detect odor from such a blend, they move upwind in a series of flights that eventually culminate in arrival at the odor source (Aluja and Prokopy 1992). Evidence to date suggests that green leaf volatiles play little if any role in the host plant finding behavior of *Rhagoletis* flies (Fletcher and Prokopy 1991).

Although to date no studies have been published on patterns of orientation of *R. mendax* adults to distant or within-plant host odor or visual stimuli, considerable information exists on comparative electroantennogram responses of *R. mendax* and *R. pomonella* flies (as well as hybrid adults) to extracts of host fruit volatiles of each species (Frey and Bush 1990; 1996; Frey et al. 1992). Antennae in each species are selectively tuned to odor components of host fruit of that species and are less responsive to components of host fruit of the sibling species, with responses of hybrids being inferior to those of either parental type. As hypothesized by these authors, modifications in host preference caused by minor genetic changes affecting the number, odor specificity, or ratio of specific antennal receptor cell types could be an important mechanism promoting host shifts in these species. Differences between species (or host races within species) in antennal sensitivity to host odor do not necessarily correspond to behavioral differences between species in response to distant host odor sources, and solid published information on the latter is lacking in comparisons between *R. pomonella* and *R. mendax*. Even so, an electroantennogram approach to comparing response profiles among *Rhagoletis* and related species to host odor might shed at least some light on degrees of phylogenetic relatedness among species. All available evidence (Prokopy and Owens 1983) suggests that responses to distant visual stimuli of host plants will probably be too similar to be useful as a tool in assessing relatedness among fly species.

If antennae in *R. pomonella* are selectively tuned to respond to components of fruit odor of current hosts, then from a chemosensory point of view, how has *R. pomonella* been able to expand its host range permanently to encompass new hosts? Apart from potential influences of aforementioned environmental factors on host range expansion, a key operative factor might be the degree of overlap in fruit volatile profile between a current host and a potential new host. Although there is comparatively little such overlap between volatile profiles of ripening hawthorn fruit (the native host of *R. pomonella*) and ripening cherry fruit (a newly acquired sporadic host of *R. pomonella*) (A. L. Averill, unpublished data), there is strong overlap between volatiles of ripening hawthorn fruit and volatiles of certain cultivars of ripening apples (newly acquired permanent hosts of *R. pomonella*) (Carle et al. 1987). Even so, subtle differences in electroantennogram responses between hawthorn-origin and apple-origin *R. pomonella* flies to one another's hosts have been detected (Frey and Bush 1990), suggesting that an electroantennogram approach might also be useful in determining degree of relatedness among host races within a species and assessing current nonhosts as candidates for future host range expansion. In this regard, one wonders just how similar volatile profiles of ripening stone fruits such as peaches and nectarines are to volatile profiles of walnuts. The latter are permanent hosts of several walnut-infesting *Rhagoletis* species, while the former are temporary if not incipient permanent hosts. If response to fruit volatiles is to be used as a vehicle for assessing relatedness among and within fly species, then careful attention must be paid to capturing volatiles only of growing or very fresh-picked fruit at a stage of ripeness conducive to substantial oviposition (Carle et al. 1987).

### 10.5.2 Host Fruit Finding

Once having arrived on a fruit-bearing host plant, tephritid adults need to find individual fruit for oviposition. The only published study to date conducted under natural conditions and characterizing, in depth, cues used by tephritid flies to find fruit within a plant canopy is in *R. pomonella*. Thus, Aluja and Prokopy (1993) found that *R. pomonella* flies use exclusively vision to find host fruit when host fruit are visually apparent, but use fruit odor together with visual stimuli to find visually unapparent fruit.

One of the most frequently studied behavioral patterns in *Rhagoletis* (although not in *Zonosemata* or *Carpomya*) concerns attraction to fruit-mimicking visual stimuli of different shapes, colors, and sizes (reviewed in Katsoyannos 1989; Fletcher and Prokopy 1991; Liburd et al. 1998). Studies on *R. pomonella*, *R. mendax*, *R. completa*, *R. cingulata*, *R. indifferens*, *R. fausta*, and *R. cerasi* indicate that adults of these species use the spherical shape of host fruit (a shape common to all host fruit of these species) as a dominant visual cue, with response greatest to sizes of spheres somewhat larger than native or acquired host fruit and to colors of spheres whose intensity contrasts most strongly against background of foliage and skylight (e.g., dark green or red spheres). Host fruit change color substantially as they ripen, and one would therefore expect few genetically based species-specific differences in response to fruit color, especially if response is based largely or exclusively on contrast of fruit against background (but see Riedl and Hislop 1985). An interesting exception is snowberry fruit, which is a host of *R. zephyria* and becomes white rather than dark toward maturity. One might expect a greater possibility for genetically based species-specific differences in response to fruit size, given that all known natural host fruit of some *Rhagoletis* and related species are small (less than 1 cm in diameter) while some host fruit of other *Rhagoletis* and related species are considerably larger (10 cm in diameter). However, evidence for genetically based differences among fly species in attraction to fruit of different sizes is weak (Prokopy 1977b), possibly because fly perception of fruit size varies according to distance of fly from fruit (Roitberg 1985).

### 10.5.3 Examination and Acceptance of Foliage

During the process of foraging for fruit within host plant canopies, tephritid flies make frequent landings on foliage. Are there intraspecific as well as interspecific differences in responses of flies to foliage of different plants? In *R. pomonella*, it turns out that foliage of certain kinds of nonhosts (e.g., pine trees, tomato plants) is much less conducive to fly exploration of the plant canopy than foliage of other kinds of nonhosts (e.g., birch) or hosts (hawthorn or apple) (Diehl et al. 1986). In fact, when on tomato foliage, *R. pomonella* flies exhibit signs of neurotoxicity, possibly from compounds released from plant hairs touched by foraging flies. Obviously, several neotropical species of *Rhagoletis* as well as *Zonosemata* have accommodated to characteristics of Solanaceae foliage. The only studies known to us on simultaneous comparisons of the foraging behavior of two tephritid species when on foliage of one another's host plants involve *R. pomonella* and *R. mendax* (Diehl and Prokopy 1986; Bierbaum and Bush 1988). Neither species exhibited a consistently greater tendency to forage among foliage of its own host plant (apple or blueberry) than among foliage of the plant of its sibling species. Except where foliar characteristics are radically different among hosts of *Rhagoletis* and related species (e.g., foliage of solanaceous vs. nonsolanaceaus plants), we anticipate fairly similar responses of flies to foliage and thus comparatively little value in using response to foliage as a means of assessing phylogenetic patterns in behavior.

### 10.5.4 Examination and Acceptance of Fruit

Postalighting examination of fruit by tephritids to determine whether a fruit is acceptable for oviposition involves touching the fruit surface with the antennae and mouthparts and probing the fruit skin with the ovipositor. In the only detailed study of fruit examining behavior known to us in *Rhagoletis* or related species, Bierbaum and Bush (1988) found that *R. pomonella* and *R. mendax* were likely to engage in these examining behaviors more extensively on their own host fruit than on host fruit of the sibling species. Video equipment with macrolens capability could aid greatly in providing detailed quantitative information on host fruit examining behaviors, which potentially could offer much insight into degree of genetic relatedness of studied species. In fact, this sort of approach could be considerably more useful than relying simply on whether an egg is or is not laid in a fruit.

One of the most extensively studied aspects of the behavior of *Rhagoletis* and related species involves characterization of stimuli that affect acceptance of fruit for oviposition. Stimuli include chemicals in surface waxes, physical attributes such as shape, color, and size of fruit, and the chemical composition and physical structure of the fruit flesh. Pertinent information on some *Rhagoletis* species has been reviewed in Fletcher and Prokopy (1991), but more information is now available.

With respect to chemical stimuli, Haisch and Levinson (1980) collected volatiles from ripening host and nonhost fruit of *R. cerasi*, impregnated ceresin wax models of cherry fruit with the volatiles, and measured ensuing amounts of egg laying by *R. cerasi* females in laboratory cages. Volatiles from two principal hosts, honeysuckle and sweet cherry, enhanced egg laying, whereas volatiles from three nonhosts decreased egg laying in relation to the amount of oviposition in untreated wax models. Similarly, Bierbaum and Bush (1990) used a solvent to extract chemicals from cuticular waxes of ripe apple or blueberry fruit, mixed each extract with ceresin wax, constructed hollow domes of such wax, and measured oviposition by *R. pomonella* and *R. mendax* females into the domes. Results paralleled ovipositional responses of each species of fly to apple and blueberry fruit. *Rhagoletis pomonella* flies laid more eggs than *R. mendax* flies in domes containing extract from apples, whereas the reverse was true for domes containing extract from blueberries. Although neither study attempted to determine whether the principal influence of fruit chemical stimuli was on alighting behavior as opposed to postalighting behavior, data from Diehl and Prokopy (1986) suggest that the major effect was probably enhancement of ovipositional propensity after alighting.

When comparing postalighting responses of different species or host races of flies to fruit chemical stimuli, one ought to be cognizant of the importance of ripeness of fruit to the outcome. To illustrate, Bierbaum and Bush (1990) showed a significantly greater ovipositional response of *R. pomonella* to extract of ripe compared with unripe apples, even though the latter were in fact stimulating. Degree of fruit ripeness has been implicated in fly response patterns to fruit not only in *R. pomonella* (Dean and Chapman 1973; Messina and Jones 1990; Murphy et al. 1991) but also in *R. indifferens* (Messina et al. 1991), and several *Rhagoletis* species that infest walnuts (Boyce 1934; Lalonde and Mangel 1994; Papaj and Alonso-Pimentel 1997).

More is known about physical stimuli than chemical stimuli eliciting boring attempts of *Rhagoletis* flies. In laboratory studies on *R. cerasi*, *R. pomonella,* and *R. completa* (Wiesmann 1937; Prokopy 1966; Prokopy and Boller 1971; Cirio 1972), inanimate models of spherical shape received more borings than models of other shapes that did not closely resemble the spherical shape of the host fruit. Assessment of boring propensity of *R. cerasi*, *R. pomonella*, *R. completa,* and *R. indifferens* into inanimate fruit models of various colors likewise has been carried out under laboratory conditions (Prokopy 1966; Prokopy and Boller 1971; Cirio 1972; Haisch and Levinson 1980; Messina 1990). Although in most of these studies at least moderate relationship was found between boring propensity into certain colors of models and known patterns of oviposition into differing colors of intact host fruit, the sources of laboratory light used did not mimic natural daylight sufficiently to permit valid detailed comparisons of species response patterns. Toughness of fruit skin in relation to penetrability by the aculeus also has been studied in *Rhagoletis* flies (e.g., Prokopy 1966; Prokopy and Boller 1971; Messina and Jones 1990; Messina et al. 1991; Papaj 1993; Lalonde and Mangel 1994). Because all known host fruit of *Rhagoletis* species are spherical in shape (or nearly so) and because fruit color and skin toughness often change rapidly as fruit ripen, none of these physical properties would appear to be useful for assessing degree of relatedness among *Rhagoletis* species. A possible exception to this generalization is the propensity of walnut flies (in the *R. suavis* species group) but no other species of *Rhagoletis* (or *Zonosemata* or *Carpomya*) studied to date to oviposit in existing punctures in host fruit rather than drill a new puncture for each ovipositional bout (Burdette 1935; Lakra and Singh 1983; Papaj 1993; Lalonde and Mangel 1994). As discussed by Papaj and Alonso-Pimentel (1997), it is possible that the propensity to use existing punctures for oviposition arose just once in *Rhagoletis* (in the ancestor of the *R. suavis* species group); because no other *Rhagoletis* species infest walnuts, it is not possible to be certain whether the trait depends critically on the taxon of fly species, on use of walnuts as hosts, or on some other habitat factor associated with specialization on walnuts.

In contrast to fly response to above physical characteristics of fruit, fly ovipositional response to fruit size has considerable potential for being useful in determining relationships among fly species. Studies have shown that in several species of *Rhagoletis* (e.g., *R. cerasi, R. pomonella, R. completa, R. indifferens, R. nova* (Schiner), and *R. tomatis*) as well as *Carpomya vesuviana*, females prefer a limited range of sizes of intact host fruit in which to oviposit (Wiesmann 1937; Cirio 1972; Lakra and Singh 1983; Frías 1986; 1995; Reissig et al. 1990; Messina and Jones 1990; Messina et al. 1991). More definitive studies involving use of inanimate models of host fruit have permitted separation of fruit size from other factors influencing oviposition. In one such study, *R. zephyria* and *R. cornivora*, whose natural host fruit range from about 4 to 10 mm in diameter, were found to deposit far more eggs in fruit models of 10 mm diameter than in 20, 40, or 70 mm models. In contrast, *R. mendax* (whose natural host fruit range from about 5 to 15 mm in diameter) deposited eggs in about equal numbers in 10 and 20 mm models, with few laid in 40 and 70 mm models, and *R. pomonella* (whose natural host fruit range from 15 to 70 mm in diameter) deposited eggs in about equal numbers in models of 20 and 40 mm diameter, with fewer laid in 10 or 70 mm models (Prokopy and Bush 1973b). Interestingly, this pattern of ovipositional propensity in *R. pomonella* was true for all three host races studied (hawthorn, cherry, and apple races), even though hawthorn and cherry fruit usually do not exceed 20 mm diameter when mature, whereas apples are usually larger than 50 mm diameter before being attacked by *R. pomonella.* In two other studies that dealt with postalighting responses of flies, *R. cerasi, R. indifferens,* and *R. pomonella* females were found to bore more often into fruit models that were similar to or slightly larger in size than respective native host fruits, and had lesser propensity to bore into models smaller than or much larger than native host fruit (Prokopy and Boller 1971; Papaj and Prokopy 1986; Messina 1990). Based upon the above evidence, it appears that in those *Rhagoletis* species whose native host fruits are small and which have not expanded their host range to include plants having larger fruits grown commercially, oviposition is rather narrowly restricted to sizes of fruit falling within the range of native hosts. In *Rhagoletis* species that have expanded their host range to embrace fruits of larger size, the range of fruit sizes acceptable for ovipostion is broader.

Once a female has bored through the skin of a fruit, chemical stimuli in the fruit flesh detected by sensilla on the aculeus of the ovipositor may determine whether an egg is laid or not (Stoffolano and Yin 1987). In both *R. completa* and *R. pomonella,* sugars (particularly glucose and fructose) introduced into the "flesh" of artificial fruit have been found to stimulate egg deposition, with malic acid and sodium chloride also being ovipositional stimulants in *R. pomonella* (Tsiropoulos and Hagen 1979; Girolami et al. 1986). The degree to which potential ovipositional stimulants and deterrents in flesh of fruit (at a stage conducive to oviposition) varies among host fruit of *Rhagoletis* and related genera remains to be determined. Lack of such information is not necessarily an impediment to exploring how different species of flies respond to different compositions of fruit flesh. One could excise spheres of appropriate-size flesh from a variety of host and nonhost fruit, dip them in ceresin wax, and compare amounts of egg laying into the flesh by different species or races of flies as a possible way of gaining better insight into degree of relatedness among fly species. Based on a number of studies, one should not expect, however, a tight fit between the degree to which a species of fruit is acceptable for oviposition and the degree to which it supports larvae to maturity (Glasgow 1933; Neilson 1967; Prokopy et al. 1988; Reissig et al. 1990).

### 10.5.5 Host-Marking Pheromones

Most aspects of oviposition behavior discussed above have been studied in so few species within the Carpomyini that meaningful phylogenetic inferences about the evolution of oviposition behavior simply cannot be drawn. This is not true for host-marking behavior. One of the best-studied elements of the behavior of *Rhagoletis* flies is that of response to host-marking pheromone deposited during dragging of the aculeus, the distal part of the ovipositor, on the fruit surface following egg deposition. To date, there is firm evidence for the existence of host-marking pheromone in at least 13 species of *Rhagoletis*: *R. completa* (Cirio 1972), *R. pomonella* (Prokopy 1972), *R. cerasi* (Katsoyannos 1975), *R. fausta* (Prokopy 1975b), *R. cingulata*, *R. cornivora*, *R. mendax* and *R. tabellaria* (Fitch) (Prokopy et al. 1976), *R. indifferens* (Prokopy et al. 1976; Mumtaz and Ali Niazee 1983), *R. basiola* (Osten Sacken) (Averill and Prokopy 1981), *R. zephyria* (Averill and Prokopy 1982), *R. alternata* (Fallén) (Zwölfer 1983), and *R. berberis* (Mayes and Roitberg 1987). In addition, evidence suggesting deposition of host-marking pheromone exists for *R. nova* (Frías 1986), *R. conversa* (Frías 1993), and *R. tomatis* (Frías 1995). Deposition of host marking pheromone following egg laying is a trait common to most species of *Anastrepha* and *Ceratitis* (Aluja et al., Chapter 15; Yuval and Hendrichs, Chapter 17), but apparently is absent in *Bactrocera* (Drew and Romig, Chapter 21).

Just as information on degree of behavioral recognition of one species' sex pheromone by another species can provide useful behavioral clues to degree of genetic relatedness between species, so also might degree of interspecific behavioral recognition of marking pheromone provide information bearing upon genetic relationships and phylogenies of *Rhagoletis*. Data in Table 10.1 represent a summary of degree of interspecific and interpopulation recognition of host-marking pheromone derived from behavioral assays of *Rhagoletis* species reported in Prokopy et al. (1976) and Averill and Prokopy (1981; 1982). The data clearly show that flies from different species groups (Smith and Bush, Chapter 9) do not recognize each other's marking pheromone, whereas different populations of the same species strongly recognize each other's marking pheromone. Within the same species group, by contrast, there is considerable variation in degree of heterospecific pheromone recognition. For example, *R. pomonella* showed strong recognition of marking pheromone deposited by *R. mendax*, *R. cornivora*, and *R. zephyria*, but each of these latter three species showed at best only intermediate recognition of marking pheromone of *R. pomonella*.

Of all information available to date on comparative behavioral response patterns of several different species of *Rhagoletis* to environmental stimuli, the above information on recognition of marking pheromone is perhaps the most valuable for application to investigations of genetic relatedness among *Rhagoletis* species. Population studies on *R. pomonella* and *R. cerasi* have shown that particular sensilla on the tarsi act as receptors for host-marking pheromone (e.g., Bowdan 1984; Stadler et al. 1994). Electrophysiological assays of the sort described by Stadler et al. (1994) could complement or even supplant behavioral assays in future studies of degree of marking pheromone recognition among different *Rhagoletis* species.

Considerable knowledge has been gained on site of production, chemical structure, mode of release, and site of detection of host-marking pheromone in certain *Rhagoletis* species, particularly *R. pomonella* and *R. cerasi* (reviewed in Averill and Prokopy 1989; also Aluja and Boller 1992; Boller and Aluja 1992; Stadler et al. 1994). Equivalent progress has been made in understanding the evolutionary ecology of host-marking pheromones, especially in *R. pomonella* and *R. basiola* (Roitberg and Prokopy 1987; Roitberg and Mangel 1988; Roitberg and Lalonde 1991; Papaj 1993; Papaj et al. 1996; Papaj and Alonso-Pimentel 1997; Hoffmeister and Roitberg 1997; 1998). Among the relationships or patterns uncovered or hypothesized throughout these studies on evolutionary ecology, perhaps the following are the most relevant to the purposes at hand. The value of host-marking pheromone is greatest to overall fitness of individuals when the resources for larval development are limited, as they are in small fruit, where older larvae usually win in competition with younger larvae. Host marking may have evolved primarily as a means whereby individuals avoid laying additional eggs in hosts into which they themselves have already oviposited and that additional benefits in terms of increased offspring survival accrued when this also resulted in avoidance of hosts marked by other individuals. *Rhagoletis* marking pheromones evolved as relatively short-lived water-soluble compounds because the main advantage of host marking is to the female that does the marking, with most of the advantage gained during the first 2 days in a patch of host fruit. Parasitoids use fly marking pheromones as cues for finding fly eggs or larvae in which

**TABLE 10.1**
**Degree to Which One Species or Population of *Rhagoletis* Flies Responded Behaviorally to Host-Marking Pheromone Deposited by the Same or a Different Species or Population of *Rhagoletis* Flies**

| Species Group | Species Responding to Pheromone | Host and Geographic Origin | Species Depositing Pheromone | | | |
|---|---|---|---|---|---|---|
| | | | *pomonella* | *cingulata* | *fausta* | |
| *pomonella* | *pomonella* | *Malus*, Wisconsin | S | N | N | |
| *cingulata* | *cingulata* | *Prunus*, Wisconsin | N | S | N | |
| Unplaced | *fausta* | *Prunus*, New York | N | N | S | |
| | | | *pomonella* | *basiola* | | |
| *pomonella* | *pomonella* | *Malus*, Massachusetts | S | NI | | |
| *alternata* | *basiola* | *Rosa*, Massachusetts | N | S | | |
| | | | *cornivora* | *tabellaria* | | |
| *pomonella* | *cornivora* | *Cornus*, Wisconsin | S | N | | |
| *tabellaria* | *tabellaria* | *Cornus*, Wisconsin | N | S | | |
| | | | *pomonella* | *mendax* | *cornivora* | *zephyria* |
| *pomonella* | *pomonella* | *Malus*, Wisconsin | S | S | S | S |
| *pomonella* | *mendax* | *Vaccinium*, Nova Scotia | I | S | S | — |
| *pomonella* | *cornivora* | *Cornus*, Wisconsin | N | I | S | — |
| *pomonella* | *zephyria* | *Symphoricarpos*, Connecticut | I | — | — | S |
| | | | *pomonella* (WI) | *pomonella* (TX) | | |
| *pomonella* | *pomonella* | *Malus*, Wisconsin | S | S | | |
| *pomonella* | *pomonella* | *Crataegus*, Texas | S | S | | |
| | | | *cornivora* (WI) | *cornivora* (TX) | | |
| *pomonella* | *cornivora* | *Cornus*, Wisconsin | S | S | | |
| *pomonella* | *cornivora* | *Cornus*, Texas | S | S | | |
| | | | *cingulata* (WI) | *cingulata* (NY) | | |
| *cingulata* | *cingulata* | *Prunus*, Wisconsin | S | S | | |
| *cingulata* | *cingulata* | *Prunus*, New York | S | S | | |

[a] S, I, and N = strong, intermediate, or no recognition of pheromone by responding females.

Indices of response are based on data presented in Prokopy et al. (1976) and Averill and Prokopy (1981; 1982).

to oviposit, thereby exerting selection pressure for short signal duration of marking pheromone. Selection pressure for production and deposition of marking pheromone may have been particularly weak in species where native fruit are difficult to penetrate and/or are characteristically large (e.g., walnut-infesting species) because the value of an existing puncture exceeds loss stemming from possible competition with other larvae and/or because susceptible fruit are always large enough to support many larvae to maturity.

In regard to the latter point, it is instructive that only two of the four walnut-infesting *Rhagoletis* species studied to date (*R. completa* and *R. juglandis*) have been shown to deposit marking pheromone which deters ovipostion (Cirio 1972; C. Nufio and D.R. Papaj, unpublished data). In

*R. boycei* and *R. suavis*, aculeus dragging is highly inconsistent in its occurrence (D.R. Papaj, personal observation). In *R. boycei*, males mark the fruit by repeatedly dipping the tip of the abdomen on the surface of a fruit in the vicinity of a puncture; rather than being deterred by male marking pheromone, females show enhanced propensity to oviposit in areas where males have marked (Papaj et al. 1996). As noted earlier, walnut-infesting species of *Rhagoletis* usually lay several eggs per clutch, whereas in all the other *Rhagoletis* species studied to date, normally only a single egg is laid per ovipositional bout. Unlike the small native host fruit characteristic of most *Rhagoletis* species, walnuts are comparatively large and capable of supporting many larvae to maturity (Papaj 1993). The possible reduction in expression of host-marking behavior within some members of the *R. suavis* group could reflect either (1) a common ancestor of the genus that exhibited reduced marking behavior or (2) given expression of host-marking behavior in the common ancestor of the genus, reduction of host-marking behavior arose within the *R. suavis* group itself. To distinguish between these equally parsimonious alternatives, it would be most helpful to know if species of taxa that are outgroups for *Rhagoletis*, for example, *Zonosemata* species, engage in host-marking behavior. Unfortunately, as yet there is no information on deposition of host-marking pheromone by *Zonosemata* or *Carpomya* flies. In *Z. electa* (Foote 1968), *Z. vittigera* (Goeden and Ricker 1971), and *C. vesuviana* (Lakra and Singh 1983), however, an observed tendency toward a single oviposition in small fruit hints at a mechanism, such as deployment of host-marking pheromone, influencing egg distribution.

In addition to host-marking pheromone on the fruit surface, chemical and/or physical signals from eggs or larvae in the fruit flesh could in principle furnish information on whether a fruit is occupied by conspecifics. To date, no study has shown that potential stimuli associated with eggs (e.g., possible odor of eggs emitted through oviposition punctures) elicit responses from *Rhagoletis*, *Zonosemata*, or *Carpomya* adults. However, there is evidence both in *R. completa* and *R. pomonella* that adults can in fact detect the presence of developing conspecific larvae in fruit, although it is not known whether this is done via chemical or physical signals emanating directly from larvae themselves or from chemical signals associated with decaying tissues of infested fruit (Cirio 1972; Averill and Prokopy 1987).

### 10.5.6 Social Facilitation of Oviposition Behavior

Socially facilitated behavior is considered to be behavior of an individual that is initiated or increased in frequency or intensity in the presence of other individuals engaged in the same behavior (Clayton 1978). It lies at one end of a continuum (or on one side of a fulcrum) of potential influence of conspecifics upon one another. The other end of the continuum (or the other side of the fulcrum) is behavior shaped by intraspecific competition for resources, which may take the form of physical contests between individuals. For example, Biggs (1972) observed that at least on a few occasions, *R. pomonella* females engaged in physical contests when jointly occupying the same resource. Response to marking pheromone deposited after egg laying also is an expression of behavior affected by competition.

Recently, it was found that *R. pomonella* flies in laboratory cages laid more eggs per female in host fruit when maintained in groups than when maintained singly and that the effect of grouping was a proximate one, largely coincident with the day females were placed in groups (Prokopy and Reynolds 1998). The most important factor contributing to the oviposition enhancing effect of grouping was shown to be presence of an ovipositing female on a fruit, which had a stimulating effect on the propensity of an arriving female to bore into a fruit provided the latter was not displaced by the resident female following a physical contest. Similarly, Robertson et al. (1995) observed that *R. basiola* females that foraged for hosts in the presence of conspecific females were less likely to reject fruit that received host-marking pheromone than females that foraged for hosts under solitary conditions.

The importance of social facilitation and its counterpart, physical displacement, in shaping ovipositional decisions and patterns in *R. pomonella* and *R. basiola* females in nature remains to be determined. Frequency of female–female encounters on the same fruit, even under high population conditions, may be too few to be of biological significance relative to other factors affecting oviposition behavior. Essentially nothing is known about possible effects of adult age, egg load, prior ovipositional experience, weather, or quality, availability, and distribution of hosts on the degree to which socially facilitated egg laying behavior is expressed in *R. pomonella* and *R. basiola*. As postulated by Robertson et al. (1995), through socially facilitated behavior, a host-foraging female might acquire information that a particular host is of acceptable quality for its offspring when it sees a conspecific female ovipositing in the same host. In *Rhagoletis*, ovipositing in a host simultaneously receiving eggs from another fly would give rise to similar-age larvae, which should have equal access to host resources.

Although not strictly within the realm of social facilitation as defined by Clayton (1978) and others, the aforementioned influence of marking by male *R. boycei* on a female's propensity to attempt oviposition into existing punctures on walnut fruits could perhaps be considered as a form of socially facilitated ovipositional behavior, albeit intersexual rather than intrasexual facilitation.

## 10.6 STATE VARIABLES AFFECTING ADULT FORAGING BEHAVIOR

Thus far, we have attempted to point out that response patterns of flies to stimuli associated with the finding, examination, or acceptance of food, mates, and oviposition sites may be affected by several variables that impact upon expressions of behaviors that are genetically based. Here, we will deal overtly with the principal types of state variables affecting the behavior of *Rhagoletis, Zonosemata,* and *Carpomya* flies: environmental state, physiological state, and informational state.

### 10.6.1 Environmental State

Among all components of the environment inhabited by tephritid flies, the abundance of quality resources, particularly food and egg-laying sites, is perhaps more important than any other in shaping fly behavior. Some aspects of the influence of resource abundance on acquisition of food were discussed earlier in the section on food foraging behavior. Here, we will focus on abundance of hosts.

Studies of host use patterns across several orders of insects show that plant species which are consistently abundant across space and time accumulate the most insect species using such plants as hosts (Bernays and Chapman 1994). This being a general pattern, it follows that fruiting plants that are at least moderately abundant in a locale year after year have a greater chance of becoming a permanent or temporary host of a *Rhagoletis* or related species than are fruiting plants that are sporadic in appearance across space and time.

Perhaps part of the answer to why some species of *Rhagoletis* have broader host ranges (albeit not exceeding oligophagy) than other *Rhagoletis* species lies in a continual annual abundance of potentially usable hosts within the normal cruising range of existing hosts. This line of thought might apply especially to introduced cultivated hosts that are grown in abundance as crops. For example, the fact that *R. pomonella* flies have been known to establish temporary but not truly permanent populations on fruiting sour cherry trees may be at least in part attributable to the proximity of colonized sour cherry trees to existing permanent hosts. The same may apply to temporary, if not permanent, expansion of *R. completa* onto peaches and nectarines as hosts that support larval development (Yokoyama and Miller 1994). Frequently, large orchards of these stone fruit grow in proximity to walnut trees, the original and favored host of *R. completa*. As mentioned earlier, abundance of quality food on potentially new hosts may also be a contributing factor to the process of host range expansion.

Abundance of host fruit can have a marked influence on the manner in which *Rhagoletis* flies forage for and utilize fruit within host plants and on the tendency to leave present plants and explore new ones. As in *R. pomonella* (Roitberg et al. 1982), there is probably a general tendency among *Rhagoletis* and related flies to search more intensively, oviposit more often, and remain longer in patches of host plants bearing many rather than few fruit. However, as shown in studies of *R. cornivora* (Borowicz and Juliano 1986), *R. indifferens* (Messina 1989), and *R. basiola* (Roitberg and Mangel 1997), various habitat-associated factors can complicate such relationships between host fruit abundance and fly searching behavior.

Besides resource abundance, natural enemies such as predators or parasitoids in the environment may have strong proximate effects on behaviors of *Rhagoletis* and related flies as well as more ultimate effects that shape host use patterns. These effects are dealt with in a succeeding section on escape from natural enemies.

As discussed earlier, intraspecific competition for fruit resources, mediated in part via deposition of and response to host-marking pheromone, can have profound effects on the behavior of *Rhagoletis* flies. But what about interspecific competitors? There exist at least two studies, one on *R. indifferens* (Messina 1989) and the other on *R. pomonella* (Feder et al. 1995), showing that *Rhagoletis* larvae do not develop as well in the presence of larvae of other orders of insects occupying the same fruit as in the absence of such. In *R. pomonella*, effects of interspecific competitors were much more pronounced on native host hawthorn than on recently acquired host apple fruit. One wonders whether ovipositing adults are able to detect the presence of interspecific competitors in fruit and alter egg-laying patterns accordingly.

In sum, several environmental factors described here can affect the way in which flies might behave in relation to existing and potentially new hosts. Differences in levels of these factors among past or present environments could either mask or exaggerate observed behavioral differences among species or host races currently occupying an environment. It is important to be aware of such potential environmental influences when considering degrees of relatedness among species based on observed behavior.

### 10.6.2 Physiological State

The physiological state of a fly can have a profound influence on the extent of its behavioral response to potential resources of food, mates, or egg-laying sites. Rather little is known about the influence of physiological state on responses to food or mates in *Rhagoletis* or related species. More is known about effects of physiological state on response to egg-laying sites.

To date, the only quantitative studies on the impact of varying degrees of fly hunger on food-foraging behavior in *Rhagoletis* and related genera have been on *R. pomonella* in individual host trees in field cages and in patches of host trees in open fields (Malavasi and Prokopy 1992; Prokopy et al. 1994a; 1995). Not surprisingly, middle-aged (14- to 19-day-old) females deprived of all food (carbohydrate as well as protein) for the 18 to 24 h preceding release, or of protein alone since eclosion, were much more likely to be found at sources of protein and carbohydrate within the first few hours after release than were nondeprived females or females deprived for an intermediate period of time. Whether or not adults of *Rhagoletis* and related genera experience a physiologically based need for carbohydrate and/or for protein at particular age-related stages during life remains to be determined.

How physiological state influences mating behavior is an open question. Certainly, one would predict such influences. Once-mated females with high egg load, for example, may resist subsequent matings to a greater extent than females with low egg load; similarly, copulation duration, where matings occur, may be relatively shorter when egg load is relatively high. However, data on these points are generally lacking for any member of the Carpomyini.

Investigation has been made of the impact of age, hunger, mating status, and egg load on host fruit-foraging behavior in *R. pomonella* and to a lesser extent in *R. basiola*. In *R. pomonella*, adults are unresponsive to olfactory or visual stimuli of host fruit until they reach an age that more or less coincides with earliest observable development of ovaries in the ovarioles, but response to such stimuli does occur before eggs are ready to be laid (Averill et al. 1988; Duan and Prokopy 1994). As mentioned earlier, Alonso-Pimentel et al. (1998) have shown that in *R. juglandis*, proximity of adults to physical stimuli of fruit prior to adult reproductive maturity can have a strong effect on the rate and degree of ovarian development. Among *R. pomonella* adults varying in degree of hunger for protein (and coincidentally varying in egg load), the hungriest were found to spend the most time in association with food and the least time in association with fruit in patches of host trees, whereas the reverse was true for the least hungry; those of intermediate hunger divided their time roughly equally between sites containing food and fruit (Prokopy et al. 1994a; 1995). Interestingly, stress or a "sense of malaise" resulting from being deprived of all food for 24 h did not cause either *R. pomonella* or *R. mendax* females to "dump" eggs in abnormally large numbers in either high-ranking or low-ranking host fruit (Prokopy et al. 1993b).

With respect to effects of mating status on oviposition-related behavior, Opp and Prokopy (1986) showed that in *R. pomonella*, females that mated once did not differ from virgin females in total fecundity, rate of egg-laying or egg laying longevity, whereas multiply mated females exceeded virgin and once-mated females in all three of these traits. Telang et al. (1996) found that holding *R. completa* females continuously with a mate (during which females had the opportunity to mate multiply) tended to promote oviposition and egg hatch, though trends were not statistically significant. Roitberg (1989) examined the influence of mating status (virgin vs. mated) together with the influence of ovipositional status (continuous availability of egg-laying sites vs. semideprivation of egg-laying sites) on *R. basiola* females and found no effect of mating status on survivorship (longevity) but a negative effect of continuous availability of oviposition sites on survivorship.

More in-depth attention has been devoted to effects of physiological state of *Rhagoletis* flies on production of and response to host-marking pheromone than to any other aspect related to fruit-foraging behavior. In *R. pomonella*, the amount of pheromone released on successively offered fruit (over a day or a week) does not change appreciably within a given fly nor does fly diet affect pheromone activity; but older flies release less active pheromone than younger flies and starvation reduces the amount of trail substance deposited (Averill and Prokopy 1988). Also in *R. pomonella*, it turns out that the greater the duration or denial of access of adults to host fruit for oviposition, the less the tendency to respond to presence of host-marking pheromone and the greater the tendency to superparasitize fruit (Roitberg and Prokopy 1983; Mangel and Roitberg 1989). In an elegant experiment on *R. basiola* flies, Randen and Roitberg (1996) manipulated egg load so that it was independent of other factors that could influence response to host fruit. They found that the greater the egg load, the greater the tendency to superparasitize pheromone-marked fruit. Finally, in *R. juglandis*, clutch size is strongly and directly related to egg load (D. Papaj and H. Alonso-Pimentel, personal observation).

Taken as a whole, the aforementioned studies dealing with different facets of the influence of adult physiological state on behavior suggest that no matter what behavioral trait is under investigation, it is unlikely that genetically based differences in fly response to resource stimuli can be affirmed without carefully controlling for potential effects of differences in physiological state.

### 10.6.3 Informational State

The behavior of an individual can be markedly affected by its informational state, or more specifically, information about the type, abundance, or distribution of resources or enemies encountered in the past and information about particular abiotic conditions. During the past 15 years or so, several studies have been carried out on effects of prior experience of *Rhagoletis* flies on different components of foraging behavior for food, mates, and egg-laying sites.

With respect to foraging for food, Averill et al. (1996) found that *R. mendax* females which had recent experience feeding for a short time on carbohydrate or protein searched more intensely in the canopy of a host plant than females lacking recent feeding experience, a finding consistent with the notion of "success-motivated search." The same study revealed that brief feeding experience of *R. mendax* as well as *R. pomonella* flies on bird feces (a source of protein) had no detectable effect on the finding of bird feces within a host plant soon after the initial brief bout of feeding had ended. Similar results for *R. pomonella* were reported by Prokopy et al. (1994a). Recently, Prokopy et al. (1998) evaluated whether *R. pomonella* flies were capable of learning to associate the presence of high-quality carbohydrate as food with the color of surrogate foliage on which food could be found. Response to yellowish surrogate foliage was always great and was not enhanced by lengthy prior experience of feeding on yellowish surrogate foliage having carbohydrate. However, response to green surrogate foliage was always weaker than to yellowish surrogate foliage, and was significantly enhanced (albeit only to a modest degree) by substantial prior experience feeding on green surrogate foliage containing carbohydrate. Conceivably, species of *Rhagoletis* or other tephritids whose habitats habitually contain very limited amounts of food distributed in a discrete fashion among plant types or plant parts might have acquired a better capability of learning to associate presence of food with a particular type or structure of plant than have species whose food sources have usually been more ubiquitous and reliable. This notion awaits evaluation. Much more in-depth research needs to be conducted before any firm conclusions can be drawn on effects of prior feeding experience on food foraging behavior in *Rhagoletis* and related genera.

The role of informational state in mating behavior is not well characterized. However, in work summarized above, Alonso-Pimentel and Papaj (1996b) demonstrated that *R. juglandis* responds dynamically at the level of the individual to changes in operational sex ratio, even when such changes occur over a relatively short time period (<1 day). For instance, flies placed in male-biased environments tended to copulate for relatively long periods of time; when placed in female-biased environments ~24 h later, flies tended to copulate for relatively short periods of time. As noted above, the pattern in duration is consistent with predictions made from sperm competition theory. Alonso-Pimentel and Papaj argue that such dynamic responses may be relevant to situations in nature, where local sex ratio in the vicinity of host walnut fruit (where most mating takes place) seems to vary unpredictably from one fruit to the next within the same tree.

These responses to local sex ratio are dynamic, but do not necessarily involve learning. To date, the only experimental information on effects of prior experience on mating behavior in *Rhagoletis* is from a study by Prokopy et al. (1989) on the propensity of *R. pomonella* males to reside on fruit as territories at which to acquire females. Males whose most recent territories were exclusively on hawthorn fruit took up residence on hawthorn fruit for a longer time than on apple fruit, whereas the reverse was true for males whose most recent territories were on apple fruit. If the predictions of Roitberg et al. (1993) (that learning about the availability of mates is more likely to occur where mating decisions are many and each has minor fitness consequences) can be generalized, then prior experience as a factor influencing components of mating behavior in *Rhagoletis* flies may be rather limited owing to a limited number of elements comprising courtship and mating behavior in *Rhagoletis* flies (although, as mentioned above, walnut-infesting species may prove an exception to this generalization).

Considerable knowledge has been gained on effects of prior egg-laying experience on ovipositional-site foraging behavior in *R. pomonella* and to a lesser extent in *R. mendax* and *R. suavis*. A single oviposition has been found to stimulate extensive subsequent search for fruit by *R. pomonella* and *R. mendax* flies released onto host plants, with search involving many more visits to leaves and a much higher probability of discovering fruit in comparison with flies lacking recent ovipositional experience (Roitberg et al. 1982; Averill et al. 1996). Up to three consecutive ovipositions by an *R. pomonella* female has no apparent effect on type of fruit sought during searching, but three consecutive days of oviposition (approximately nine ovipositions) into the same fruit type can lead to a significantly reduced propensity to alight on fruit of a different type, particularly when the familiar type is red and the unfamiliar type is green (Prokopy et al. 1994b). Apparently, relatively inconspicuous fruit, such as green fruit against a background of green foliage, are less readily detected by *R. pomonella* females accustomed to finding more conspicuous fruit, such as red fruit. Prokopy et al. (1994b) found no detectable effects of fruit size or fruit odor on ability of *R. pomonella* females to find familiar compared with unfamiliar fruit.

In *R. pomonella*, prior ovipositional experience has a greater impact on postalighting responses to fruit than on prealighting responses, as evidenced by the finding that three consecutive ovipositions into same-type fruit are in fact sufficient to generate significant rejection of unfamiliar fruit (Prokopy et al. 1982; 1994b), particularly fruit that differ in size and chemistry from familiar fruit (Papaj and Prokopy 1986; Prokopy et al. 1994b). Ovipositional familiarity with a particular color of fruit seems to have little impact on postalighting acceptance or rejection of fruit of the same or a different color (Prokopy and Papaj 1989; Prokopy et al. 1994b). Effects of prior ovipositional experience on subsequent egg-laying behavior in *R. pomonella* are not confined simply to propensity to find and accept the first familiar or unfamiliar fruit encountered following ovipositional experience (the principal criterion used in the above studies). As shown by Papaj and Prokopy (1988), effects of prior ovipositional experience in *R. pomonella* extend through a succession of egg-laying bouts of a female in a fruit-rich patch of host trees. Interestingly, the nature of ovipositional-site learning in *R. pomonella* and perhaps also in other tephritids appears to involve greater propensity to reject unfamiliar stimuli (even stimuli associated with an unfamiliar biotype of the same fruit species) rather than greater propensity to accept familiar stimuli as a consequence of increasing experience with familiar stimuli (Prokopy et al. 1986; Prokopy and Papaj 1988).

*Rhagoletis mendax* and *R. suavis* flies, like *R. pomonella* flies, also have been shown to learn characteristics of host fruit that affect propensity to accept or reject hosts after alighting (Prokopy et al. 1993c). These authors developed a "learning index" as an aid for determining whether relative specialists, such as *R. suavis* and *R. mendax*, might be less capable of learning fruit stimuli than relative generalists, such as *R. pomonella*, where host range is broader than in either of the other two species. Although the two relative specialists did, in most experiments, have lower-ranking learning indicies than *R. pomonella*, apparent differences in learning capability among these three species were postulated to depend as much or more upon degrees of difference in the physical and chemical nature of host fruit presented as upon real species' differences in capacities to learn.

Beside learning to recognize various fruit stimuli per se, *R. pomonella* females have also been shown to learn to recognize host-marking pheromone deposited on the fruit surface. Naive females are unable to recognize marking pheromone deposited by another female whereas females having had but a single encounter with conspecific marking pheromone are well able to recognize the pheromone (Roitberg and Prokopy 1981). The potential adaptive value of learning to recognize marking pheromone and other stimuli associated with foraging for ovipositional sites remains uncertain but has been discussed by Roitberg and Prokopy (1981) and Papaj and Prokopy (1989).

Although we have only begun to learn about the role of informational state as a component shaping resource-foraging behavior in *Rhagoletis* and related genera, it is clear from studies to date, particularly on *R. pomonella*, that the prior experience of an individual cannot be ignored when conducting behavioral studies bearing upon degree of genetic relatedness among species. In fact, the informational state of the individual may be closely connected to its physiological state and the state of the environment in a way that demands that all three of these factors be considered as interacting variables. Models illustrating interconnectiveness of these variables for *Rhagoletis* and other insects have been developed by Mangel and Roitberg (1989), Roitberg et al. (1990), Jaenike and Papaj (1992), and Roitberg et al. (1993).

## 10.7 ADULT ESCAPE FROM NATURAL ENEMIES

As mentioned briefly in an earlier section, natural enemies of *Rhagoletis* and related genera can have a substantial influence on proximate as well as ultimate causes of behavior in flies. Here, we will focus on some recent studies that show or suggest a direct or indirect impact of certain parasitoids and predators on the behavior of *Rhagoletis* and *Zonosemata* flies.

Several studies have shown that hymenopterous parasitoids of *Rhagoletis* eggs or larvae use cues such as visual and chemical stimuli of host fruit, fly marking pheromone deposited on the fruit surface, and movement of larvae within the fruit to locate fruit harboring potential hosts for egg laying (e.g., Prokopy and Webster 1978; Glas and Vet 1983; Roitberg and Lalonde 1991). The question arises as to whether flies adjust their foraging behavior for fruit in a way that confers escape from parasitoid attack on their progeny. Direct evidence that this may be so was found recently in *R. basiola* by Hoffmeister and Roitberg (1997). These flies were observed to recognize fruit searched recently by *Halticoptera rosae* Burks adults (parasitoids of fly eggs), possibly using marking pheromone deposited on fruit by parasitoid adults as the informational cue for presence of parasitoids in the habitat. Hoffmeister and Roitberg (1997) postulated that flies might alter the spatial distribution of their eggs in response to recognition of parasitoid presence. Earlier, Juliano and Borowicz (1987) observed increasing levels of parasitism of *R. cornivora* larvae by *Opius richmondi* (Gahan) adults with increasing density of host larvae, a pattern which led them to speculate that risk of greater parasitoid load may be one reason why, in their studies, *R. cornivora* females tended to abandon host plants before all or most fruit had been used for oviposition. In the same vein, Roitberg and Lalonde (1991) developed models suggesting that threat of increased parasitism by *H. rosae* adults of *R. basiola* eggs may be one reason underlying variation among *R. basiola* adults in quantity and quality of marking pheromone deposited after egg laying: flies that deposit less pheromone, although running greater risk of intraspecific competition among their larval progeny, may compensate for this disadvantage by reduced risk of attack by *H. rosae*, which use marking pheromone as a cue for finding host eggs.

Considerable evidence now exists that several species of Nearctic *Rhagoletis* (notably *R. pomonella*) as well as several species of Palearctic *Rhagoletis* have left behind many or most of the parasitoids attacking their egg and larval progeny when they expanded their host ranges to include new kinds of fruit (Maier 1981; Diehl 1984; Ali Niazee 1985; Hoffmeister 1992; Gut and Brunner 1994; Feder 1995). Decreased larval survival on newly acquired hosts may be offset by protection against parasitism and could be an important ingredient in initial establishment of new host races of *Rhagoletis* flies (Hoffmeister 1992; Feder 1995). How long such an advantage of protection against parasitism might persist remains to be determined, but parasitoids may be much more constrained than tephritid flies in adapting to cues from hosts.

In regard to predators, Monteith (1972) was among the first to elaborate upon the close morphological resemblance of *Rhagoletis* flies to predaceous jumping spiders (Araneae: Salticidae). He showed that jumping spiders seemed to avoid *R. pomonella* flies as potential prey but did not propose a mechanism. Later, Eisner (1984) described a striking physical resemblance of *Z. vittigera* flies to jumping spiders present in the habitat of that species. He proposed that imitation of jumping spiders could be a benefit to flies in escaping other sorts of potential predators that might shun jumping spiders on account of poisonous bites.

Subsequently, apparent mimicry of jumping spiders was studied in depth in *R. zephyria* (Mather and Roitberg 1987) and *Z. vittigera* (Greene et al. 1987). Through simple, elegant protocols that involved altering wing and body patterns of the flies, it was found in both species that fly wing patterns were effective mimics of leg patterns of jumping spiders, which reacted to flies (by retreat) in the same way they reacted to conspecific spiders. Retreat was especially pronounced in response to wing-waving flies compared with sedentary flies. Greene et al. (1987) went on to speculate that wing patterns in tephritid flies evolved initially to serve in intraspecific communication and only later came to serve in a defensive capacity.

Whitman et al. (1988) made special note of movement patterns of *Z. vittigera* flies, which characteristically are rather jerky and involve short flights that resemble spider jumps to nearby vegetation. As reported in Averill et al. (1996), different species of tephritid flies exhibit characteristically different patterns of movement when foraging within plants. For example, *R. mendax* adults are more prone than *R. pomonella* adults to be perturbed by sudden nearby movement and to engage in a sudden series of short hops or flights. One wonders about the extent to which differences in movement patterns (or wing and body patterns) among species of *Rhagoletis*, *Zonosemata,* and *Carpomya* might have evolved in response to differences in selection pressure associated with presence of different species complexes of jumping spiders (or other kinds of predators) in occupied habitats. It could be revealing to compare assemblages of predators on native host plants of flies with assemblages on newly acquired hosts and critically examine movement patterns of flies associated with each host type.

From the preceding, it seems likely that predators and/or parasitoids can indeed have an influence not only on morphological traits of adult *Rhagoletis* and related genera but also on behaviors as simple as the architecture of movement or as complex as those involved in sampling patches of hosts as potential egg-laying sites. Perhaps an understanding of phylogenetic relationships among species in this group could be improved by examining behaviors as they may have evolved under selection by particular assemblages of natural enemies identifiable with particular types of host plants.

## 10.8 BEHAVIOR OF LARVAE

With rare exception (McAlister 1932), *Rhagoletis*, *Zonosemata,* and *Carpomya* larvae confine their feeding activity to the fruit in which eggs have been deposited. Larvae of these genera face four sorts of challenges that may affect their behavior prior to forming puparia: acquiring nutrients that are conducive to growth; avoiding ingestion of potentially harmful chemicals; avoiding potentially harmful effects of intra- or interspecific competitors and natural enemies; and finding a suitable site for forming a puparium without perishing. Unfortunately, few studies of behavioral responses of larvae to these challenges have been conducted in *Rhagoletis* and related genera (in contrast to work done on other tephritid genera — see Aluja et al., Chapter 15; and Yuval and Hendrichs, Chapter 17).

Although artificial diets have been developed for larvae of *R. pomonella* (Neilson 1967) and *R. cerasi* (Haisch and Boller 1971), essentially nothing is known about specific sorts of nutrients required by *Rhagoletis* larvae and whether, following egg hatch, they move preferentially to specific parts of fruit to acquire such nutrients. Regarding allelochemicals potentially harmful to larvae, both Neilson (1967) and Reissig et al. (1990) found that larvae of *R. pomonella* were unable to complete development in certain species or biotypes of crab apples that may have contained compounds toxic to the larvae. Pree (1977) confirmed that "resistance" in certain crab apples to development of *R. pomonella* larvae was correlated with total phenol content and demonstrated that addition of 1000 ppm of any of several phenolic acids (gallic, tannic, or o-coumaric acids, quercetin, naringenin, *d*-catechin) to artificial diet prevented larval development. Nothing is known about movement of *R. pomonella* larvae within fruit in relation to potential intrafruit variation in content of phenolic acids. Quite possibly for *Rhagoletis* and related genera, host fruits are rather uniform (within fruit) in distribution of nutrients and allelochemicals and larval behavior is unaffected. In addition to nutrients and allelochemicals, fruit texture and firmness are thought to affect rates of development of *R. pomonella* larvae (Dean and Chapman 1973), possibly accounting for the repeated observation (Dean and Chapman 1973; Reissig 1979) that larvae of this species suffer much higher mortality in firm-flesh than in soft-flesh cultivars of apple and in apples remaining on a tree until autumn than in apples that have fallen and softened.

One of the more fascinating components of the biology of *Rhagoletis*, *Zonosemata,* and *Carpomya* flies involves use of fruits of Solanaceae as larval hosts in most studied Neotropical species of *Rhagoletis* (Smith and Bush, Chapter 9) and all studied species of *Zonosemata* (Burdette 1935; Foote 1968; Goeden and Ricker 1971). Solanaceous fruit contain a-tomatine and other allelochemicals that have strong detrimental effects on larvae of many species of tephritids (e.g., Chan and Tam 1985). For gaining insight into phylogenetic relationships, it could be revealing to compare physiologies and behaviors of larvae of species of *Rhagoletis* and *Zonosemata* in relation to response to a-tomatine and other allelochemicals common to solanaceous plants. Conceivably, certain species of bacteria associated closely with tephritid species whose larvae can develop in solanaceous fruit play a role in detoxifying allelochemicals in the fruit, as they do in detoxifying insecticide (Boush and Matsumura 1967) and harmful allelochemicals present in food ingested from foliar surfaces by *R. pomonella* flies (C. R. Lauzon and R. J. Prokopy, unpublished data). It is also conceivable that allelochemicals such as a-tomatine serve as essential "vitamins" for larvae of those tephritid species that use solonaceous plants as hosts (G.L. Bush, personal communication). It follows that larvae of these species might not be able to survive in nonsolonaceous hosts lacking a-tomatine.

Although limited, evidence to date suggests that larval competition for resources in species of *Rhagoletis* that lay eggs singly takes the form of interference or contest competition (as in *R. alternata* and *R. pomonella*) in which the activities of one larva directly or indirectly limit the access of rival larvae to a resource (Zwölfer 1983; Averill and Prokopy 1987). Such interference may occur through territorial behavior, physiological suppression, or cannibalism. Although it is not known which of these mechanisms comes into play in intraspecific competition in *R. alternata* and *R. pomonella*, cannibalism has been documented in *Anastrepha suspensa* larvae (Carroll 1986). If direct competitive interactions do occur in *Rhagoletis* larvae, size and consistency of host fruit could be important factors in the severity of such competition. As proposed by Fletcher (1989), the occurrence of contest rather than scramble competition in species of tephritids whose puparia must overwinter to ensure survival (e.g., most *Rhagoletis* flies) may ensure that competition does not result in undersized pupae that would suffer greater risk of winter mortality. The nature and degree of competition in *Rhagoletis* larvae is less well understood in the walnut-infesting species of the *suavis* group, where eggs are laid in clutches of approximately ten or more and where a single fruit usually supports many larvae to maturity (Boyce 1934). In these species, it is conceivable that larvae cooperate in utilizing the host resource; such cooperation would favor both deposition of clutches of more than a single egg as well as repeated oviposition in a fruit and even reuse of existing oviposition cavities (see Papaj 1993). All of these traits are typical of species within the *suavis* group but found nowhere else within the genus. Evidence as to the occurrence and nature of competition and possible cooperation in members of the *suavis* group is only beginning to emerge (C. Nufio, D. R. Papaj, and H. Alonso-Pimentel, unpublished data).

No direct information is available on whether larvae of *Rhagoletis* and related genera move to interior parts of fruit or engage in other action to evade attack by ovipositing parasitoids. Methodological difficulties inherent in making direct observations of larvae within intact fruit pose an impediment to studies of this sort. Indirect information from dissection of hawthorn fruit to determine location of *R. pomonella* larvae suggests that larvae reside closer to the fruit surface and are more susceptible to parasitoid attack in the presence than in the absence of interspecific competitors, such as larvae of moths and weevils that feed near the center of hawthorn fruit (Feder 1995). When mature, *Rhagoletis* and *Zonosemata* larvae bore holes through the skin of host fruit when preparing for emergence and drop from falling fruit to the ground early in the morning in search of sites favorable for pupation, with the onset of light and increasing temperature apparently being two of the stimuli triggering early-morning emergence (Goeden and Ricker 1971; Boller and Prokopy 1976). In at least some species of *Carpomya*, larvae tend to pupate within fallen fruit rather than drop to the ground and pupate in soil (e.g., Monastero 1970). One wonders if a tendency to pupate within fruit might be associated with a relative lack of vertebrate predation upon larval-infested fruit, a relatively high load of pupal predators in the soil, lack of need to find a suitable site for undergoing pupal diapause, or simply the physiological state of the fruit. In contrast to larvae of *Ceratitis* species (Yuval and Hendrichs, Chapter 17), larvae of members of the Carpomyini do not exhibit "jumping" behavior to avoid predators after exiting from fruit.

## 10.9 CONCLUSIONS AND FUTURE RESEARCH NEEDS

Our review of the body of work on behavior of *Rhagoletis*, *Zonosemata,* and *Carpomya* flies raises a number of issues relevant to the theme of this volume, specifically the evolution and phylogeny of behavior.

First, despite the tremendous effort invested in understanding the behavior of *Rhagoletis* flies, a genus of importance both to agriculture and to the study of speciation, we are nevertheless lacking relevant data for many, if not most, species within the genus. As stated eloquently by B. A. McPheron (personal communication), work by behavioral ecologists has tended to be vertical in nature, delving very deeply into the behavior of just one or a few species. We clearly need to move horizontally in our work, including more species from more groups. Moreover, although it is not surprising that work within the Carpomyini, as with work in other tephritid groups, has focused on species of economic importance, such a bias is clearly not conducive to a rigorous phylogenetic analysis of behavior. It both reduces the number of species for which data are available (i.e., our "phylogenetic sample size") and may skew that sample in ways that further impede meaningful analysis. At the same time, a comparative survey is potentially of significance to students of pest management. In other words, both students of phylogeny and students of pest management should be of like mind as to the value of comparative study, even when such study includes species of little or no economic importance.

Second, the need for inclusion of more species notwithstanding, the significant body of knowledge of behavior accumulated for a few species (especially *R. pomonella*) has much to offer those interested in phylogenetic patterns. For example, this substantial database sounds a cautionary note regarding how working on more than one species might trade-off against the depth of work on any one species. For instance, whereas the resource defense component of mating in *Rhagoletis* is very conspicuous and relatively easily observed, the foliage-based component early in the season is not nearly so obvious. A comparative study that included many species but that made less intensive study of any one species could easily miss the foliage-based component of mating; yet, inconspicuous or not, such matings may be of critical significance to the evolution of courtship or to speciation processes, and thus of significance to the phylogenetic biologist.

Likewise, the considerable evidence bearing on the dynamics of individual behavior in *Rhagoletis* species, including learning and motivational processes, reminds us that behavior poses certain challenges for phylogenetic biology that morphology may not. In particular, learning processes that adapt individuals to their environment can generate phylogenetic patterns that resemble patterns generated by evolution under natural selection. Suppose, for example, that the effects of experience on host acceptance noted for *R. pomonella* were common in *Rhagoletis* species generally, namely, that experience with host fruit of one type alters acceptance of host fruit of other types. In lieu of knowledge of fly learning, a student who conducted field assays of fruit preference for a variety of species might infer a stronger pattern of evolutionary divergence in host fruit responses than is actually the case. Only by working under controlled conditions and conducting assays with naive individuals can that student define a pattern of evolutionary divergence in fruit preference that is independent of the effects of experience. Our message is simple but general in scope: the dynamics of individual behavior simply cannot be ignored in any phylogenetic analysis of behavior. In this regard, the work ahead of us is imposing in magnitude. At the same time, our understanding of the dynamics themselves, and their underlying processes, will surely benefit from a consideration of fly phylogeny (compare Prokopy et al. 1993c). In this regard, the work, if imposing, may nevertheless be highly rewarding.

In addition to these broader challenges of longer-range concern, there are challenges of a somewhat more specific nature that arise from the account given in this chapter, which bear upon the phylogeny of behavior in the tribe Carpomyini and which can be addressed in the nearer term. We present some of these challenges here in the form of questions.

What are the underlying factors that confine all members of certain species groups to development on a narrow range of hosts (e.g., all members of the *R. suavis* species group are confined to species of *Juglans* as hosts; all members of the *R. nova* species group are confined to species of *Solanum* and *Lycopersicon* as hosts)? Why are members of other species groups able to infest a comparatively broad range of unrelated species as hosts (e.g., the *R. pomonella* and *R. tabellaria* species groups each infest hosts of at least four different plant families)? Are the host ranges of members of the *R. suavis* and *R. nova* species groups constrained by lack of particular bacterial associates that could, if present, permit expansion onto other kinds of fruit; are the host ranges constrained by presence of certain compounds in host fruit (e.g., juglone in *Juglans*, a-tomatine in *Solanum* and *Lycopersicon*) that serve as essential nutrients for larvae and are absent or in low amount in current nonhost fruit; or are the host ranges constrained by particular chemical and physical cues associated with host finding and acceptance that are present in current hosts but absent or in low amount in current nonhosts?

By the same token, what factors have precluded use of *Juglans* as hosts by *Rhagoletis* species other than those in the *R. suavis* species group and use of *Solanum* and *Lycopersicon* as hosts by *Rhagoletis* species other than those of the species groups of neotropical origin? Do differences in host use reflect total dissimilarity of chemical and physical cues used in host finding and acceptance between fruit of *Juglans* or *Solanum/Lycopersicon,* on the one hand, and fruit colonized by *Rhagoletis* species that are not members of the *suavis* and *nova* species groups, on the other hand; in the case of *Solanum* and *Lycopersicon*, do they reflect a failure of most *Rhagoletis* species outside the Neotropics to tolerate negative effects of toxins released by trichomes present on foliar surfaces of *Solanum* and *Lycopersium* plants and contacted by adults when foraging within plant canopies; or do host use differences reflect a lack of associated bacteria or lack of physiological/biochemical mechanisms capable of detoxifying harmful plant secondary compounds in the fruit?

Apart from contrasts between members and nonmembers of the *suavis* and *nova* species groups of *Rhagoletis*, how tightly linked is the relationship between chemical profiles and sizes of existing host fruit and degree of monophagy or stenophagy in *Rhagoletis*, *Zonosemata,* and *Carpomya* flies? Does tightness of this relationship constrain host range expansion?

Finally, to what extent are reported differences among species in behavior a reflection of intrinsic species differences vs. differences in the ecological contexts in which behavior was recorded? For example, does reuse of oviposition sites by members of the *R. suavis* group reflect an intrinsic difference from members of other groups? Or, might members of other groups behave similarly if provided with relatively large and impenetrable fruit? If individuals of *Rhagoletis* species that are not members of the *suavis* group were offered artificial fruit that possessed the same essential chemical stimuli as the species' natural host fruit but were large and harder to puncture, would these individuals use preexisting punctures for egg deposition and would they deposit more than one egg per clutch, as members of the *suavis* group do? Such knowledge could contribute toward understanding the origin of using existing punctures and laying multiple eggs per clutch by members of the *suavis* species group.

## ACKNOWLEDGMENTS

We are grateful to A.L. Averill, B.D. Roitberg, M. Aluja, and two anonymous reviewers for very constructive suggestions on an earlier version of this chapter, and to Jonathan Black for assistance in typing. Work on this chapter was supported by Campaña Nacional contra las Moscas de la Fruta (Mexico), the International Organization of Biological Control of Noxious Animals and Plants, Instituto de Ecología A.C. (Mexico), and USDA-ICD-RSED.

## REFERENCES


Ali Niazee, M.T. 1985. Opiine parasitoids of *Rhagoletis pomonella* and *R. zephyria* in the Willamette Valley, Oregon. *Can. Entomol.* 117: 163–166.

Alonso-Pimentel, H. and D.R. Papaj. 1996a. Operational sex ratio vs. gender density as determinants of copulation duration in the walnut fly, *Rhagoletis juglandis* (Diptera: Tephritidae). *Behav. Ecol. Sociobiol.* 39: 171–180.

Alonso-Pimentel, H. and D.R. Papaj. 1996b. Patterns of egg load in the walnut fly, *Rhagoletis juglandis* (Diptera: Tephritidae), in nature and their possible significance for distribution of sexes. *Ann. Entomol. Soc. Am.* 89: 875–882.

Alonso-Pimentel, H. and D.R. Papaj. 1999. Territory and sex ratio as determinants of copulation duration in the fly *Rhagoletis juglandis*. *Anim. Behav.* 57: 1063–1069.

Alonso-Pimentel, H., J.B. Korer, C. Nufio, and D.R. Papaj. 1998. Role of color and shape stimuli in host-enhanced oogenesis in the walnut fly *Rhagoletis juglandis*. *Physiol. Entomol.* 23: 97–104.

Aluja, M. and E.F. Boller. 1992. Host marking pheromone of *Rhagoletis cerasi*: foraging behavior in response to synthetic pheromonal isomers. *J. Chem. Ecol.* 18: 1299–1311.

Aluja, M. and R.J. Prokopy. 1992. Host search behavior of *Rhagoletis pomonella* flies: inter-tree movement patterns in response to wind-borne fruit volatiles under field conditions. *Physiol. Entomol.* 17: 1–8.

Aluja, M. and R.J. Prokopy. 1993. Host odor and visual stimuli interaction during intratree host finding behavior of *Rhagoletis pomonella* flies. *J. Chem. Ecol.* 19: 2671–2686.

Averill, A.L. and R.J. Prokopy. 1981. Oviposition-deterring fruit marking pheromone in *Rhagoletis basiola*. *Fla. Entomol.* 64: 221–226.

Averill, A.L. and R.J. Prokopy. 1982. Oviposition-deterring fruit marking pheromone in *Rhagoletis zephyria*. *J. Ga. Entomol. Soc.* 17: 315–319.

Averill, A.L. and R.J. Prokopy. 1987. Intraspecific competition in the tephritid fruit fly *Rhagoletis pomonella*. *Ecology* 68: 878–886.

Averill, A.L. and R.J. Prokopy. 1988. Factors influencing release of host marking pheromone by *Rhagoletis pomonella* flies. *J. Chem. Ecol.* 14: 95–111.

Averill, A.L. and R.J. Prokopy. 1989. Host-marking pheromones. In *Fruit Flies: Their Biology, Natural Enemies and Control* (A.S. Robinson and G. Hooper, eds.), pp. 207–219. In *World Crop Pests* (W. Helle, ed.), Vol. 3A. Elsevier Science Publishers, Amsterdam.

Averill, A.L. and R.J. Prokopy. 1993. Foraging of *Rhagoletis pomonella* flies in relation to interactive food and fruit resources. *Entomol. Exp. Appl.* 66: 179–185.

Averill, A.L., R.J. Prokopy, M.M. Sylvia, P.P. Connor, and T.T. Wong. 1996. Effects of recent experience on foraging in tephritid fruit flies. *J. Insect Behav.* 9: 571–583.

Averill, A.L., W.H. Reissig, and W.L. Roelofs. 1998. Specificity of olfactory responses in the tephritid fly *Rhagoletis pomonella*. *Entomol. Exp. Appl.* 47: 211–222.

Bateman, M.A. 1972. The ecology of fruit flies. *Annu. Rev. Entomol.* 17: 493–518.

Bernays, E.A. and R.F. Chapman. 1994. *Host-Plant Selection by Phytophagous Insects*. Chapman & Hall, New York. 312 pp.

Bierbaum, T.J. and G.L. Bush. 1988. Divergence in key host examining and acceptance behaviors of the sibling species *Rhagoletis mendax* and *R. pomonella*. *Oregon State Univ. Agric. Exp. Stn. Spec. Rep.* 830: 26–55.

Bierbaum, T.J. and G.L. Bush. 1990. Host fruit chemical stimuli eliciting distinct ovipositional responses from sibling species of *Rhagoletis* fruit flies. *Entomol. Exp. Appl.* 56: 165–177.

Biggs, J.D. 1972. Aggressive behavior in the adult apple maggot. *Can. Entomol.* 140: 349–353.

Boller, E.F. and M. Aluja. 1992. Ovipositon deterring pheromone in *Rhagoletis cerasi*. *J. Appl. Entomol.* 113: 113–119.

Boller, E.F. and R.J. Prokopy. 1976. Bionomics and management of *Rhagoletis*. *Annu. Rev. Entomol.* 21: 223–246.

Borowicz, V.A. and S.A. Juliano. 1986. Inverse density-dependent parasitism of *Cornus amomum* fruit by *Rhagoletis cornivora*. *Ecology* 67: 639–643.

Boush, G.M. and F. Matsumura. 1967. Insecticidal degradation by *Pseudomonas melophthora*, the bacterial symbiote of the apple maggot. *J. Econ. Entomol.* 60: 918–920.

Bowdan, E. 1984. Electrophysiological responses of tarsal contact chemoreceptors of the apple maggot fly *Rhagoletis pomonella* to salt, sucrose and oviposition-deterring pheromone. *J. Comp. Physiol.* (A) 154: 143–152.

Boyce, A.M. 1934. Bionomics of the walnut husk fly, *Rhagoletis completa*. *Hilgandia* 8: 363–579.

Burdette, R.C. 1935. The biology and control of the pepper maggot, *Zonosemata electa*. *N.J. Agric. Exp. Stn. Bull.* 585, pp. 1–24.

Bush, G.L. 1966. The taxonomy, cytology, and evolution of the genus *Rhagoletis* in North America (Diptera — Tephritidae). *Bull. Mus. Comp. Zool.* 134: 431–562.

Bush, G.L. 1969. Mating behavior, host specificity, and the ecological significance of sibling species in frugivorous flies of the genus *Rhagoletis* (Diptera — Tephritidae). *Am. Nat.* 103: 669–672.

Carle, S.A., A.L. Averill, G.S. Rule, W.H. Reissig, and W.L. Roelofs. 1987. Variation in host fruit volatiles attractive to the apple maggot fly. *J. Chem. Ecol.* 13: 795–805.

Carroll, J.F. 1986. Caribbean fruit flies, *Anastrepha suspensa*, reared from eggs to adults on cannabalistic diet. *Proc. Entomol. Soc. Wash.* 88: 253–256.

Chan, H.T. and S.Y.T. Tam. 1985. Toxicity of a-tomatine to larvae of the Mediterranean fruit fly. *J. Econ. Entomol.* 78: 305–307.

Cirio, U. 1972. Observazioni sul comportamento di ovideposizione della *Rhagoletis completa* in laboratorio. In *Proc. 9th Italian Cong. Entomol.*, Siena, pp. 99–117.

Clayton, D.A. 1978. Socially facilitated behavior. *Q. Rev. Biol.* 53: 373–392.

Dean, R.W. and P.J. Chapman. 1973. Bionomics of the apple maggot fly in Eastern New York. *Search Agric. Entomol.* (Geneva, NY) 3(10): 1–61.

Diehl, S.R. 1984. The Role of Host Plant Shifts in the Ecology and Speciation of *Rhagoletis* flies. Ph.D. dissertation, University of Texas, Austin.

Diehl, S.R. and R.J. Prokopy. 1986. Host selection behavior differences between the fruit fly sibling species *Rhagoletis pomonella* and *R. mendax*. *Ann. Entomol. Soc. Am.* 79: 266–271.

Diehl, S.R., R.J. Prokopy, and S. Henderson. 1986. The role of stimuli associated with branches and foliage in host selection by *Rhagoletis pomonella*. In *Fruit Flies of Economic Importance 84* (R. Cavalloro, ed.), pp. 191–196. A.A. Balkema, Rotterdam.

Drew, R.A.I. and A.C. Lloyd. 1989. Bacteria associated with fruit flies and their host plants. In *Fruit Flies: Their Biology, Natural Enemies and Control* (A.S. Robinson and G. Hooper, eds.), pp. 131–140. In *World Crop Pests* (W. Helle, ed.), Vol. 3A. Elsevier Science Publishers, Amsterdam.

Duan, J.J.D. and R.J. Prokopy. 1994. Apple maggot fly response to red sphere traps in relation to fly age and experience. *Entomol. Exp. Appl.* 73: 279–287.

Eisner, T. 1984. A fly that mimics jumping spiders. *Psyche* 92: 103–104.

Feder, J.L. 1995. The effects of parasitoids on sympatic host races of *Rhagoletis pomonella*. *Ecology* 76: 801–813.

Feder, J.L., K. Reynolds, W. Go, and E.C. Wang. 1995. Intra- and interspecific competition and host race formation in the apple maggot fly, *Rhagoletis pomonella*. *Oecologia* 101: 416–425.

Fletcher, B. 1989. Life history strategies of tephritid fruit flies. In *Fruit Flies: Their Biology, Natural Enemies and Control* (A.S. Robinson and G. Hooper, eds.), pp. 195–208. In *World Crop Pests* (W. Helle, ed.), Vol. 3A. Elsevier Science Publishers, Amsterdam.

Fletcher, B.S. and R.J. Prokopy. 1991. Host location and oviposition in tephritid fruit flies. In *Reproductive Behavior of Insects* (W.J. Bailey and J. Ridsdill-Smith, eds.), pp. 138–171. Chapman & Hall, New York.

Foote, W.H. 1968. The importance of *Solanum carolinense* as a host of the pepper maggot, *Zonosemata electa* (Say), in Southwestern Ontario. *Proc. Entomol. Soc. Ont.* 98: 16–18.

Frey, J.E. and G.L. Bush. 1990. *Rhagoletis* sibling species and host races differ in host odor recognition. *Entomol. Exp. Appl.* 57: 123–131.

Frey, J.E. and G.L. Bush. 1996. Impaired host odor reception in hybrids between the sibling species *Rhagoletis pomonella* and *R. mendax*. *Entomol. Exp. Appl.* 80: 163–165.

Frey, J.E., T.J. Bierbaum, and G.L. Bush. 1992. Differences among sibling species *Rhagoletis mendax* and *R. pomonella* in their antennal sensitivity to host fruit compounds. *J. Chem. Ecol.* 18: 2011–2024.

Frías, D. 1986. Biologia poblacionale de *Rhagoletis nova*. *Rev. Chil. Entomol.* 13: 75–84.

Frías, D. 1993. Evolutionary biology of certain Chilean *Rhagoletis* species. In *Fruit Flies: Biology and Management* (M. Aluja and P. Liedo, eds.), pp. 21–28. Springer-Verlag, New York.

Frías, D. 1995. Oviposition behavior of *Rhagoletis tomatis* in tomato. *Acta Entomol. Chil.* 19: 159–162.

Frías, D., A. Malavasi, J.S. Morgante. 1984. Field observations of distribution and activities of *Rhagoletis conversa* on two hosts in nature. *Ann. Entomol. Soc. Am.* 77: 548–551.

Frick, K.E., H.G. Simkover, and H.S. Telford. 1954. Bionomics of the cherry fruit fly in Eastern Washington. *Wash. Agric. Exp. Stn. Tech. Bull.* 3: 66 pp.

Girolami, V., A. Strapazzon, R. Crnjar, A.M. Anjioy, P. Pietra, J.G. Stoffolano, and R.J. Prokopy. 1986. Behavior and sensory physiology of *Rhagoletis pomonella* in relation to oviposition stimulants and deterrents in fruit. In *Fruit Flies of Economic Importance* (R. Cavalloro, ed.), pp. 183–190. A.A. Balkema, Rotterdam.

Glas, P.C. and L.E.M. Vet. 1983. Host habitat location and host location by *Diachasma alloeum*, a parasitoid of *Rhagoletis pomonella. Neth. J. Zool.* 33: 41–54.

Glasgow, H. 1933. The host relations of our cherry fruit flies. *J. Econ. Entomol.* 26: 431–438.

Goeden, R.D. and D.W. Ricker. 1971. Biology of *Zonosemata vittigera* relative to silverleaf nightshade. *J. Econ. Entomol.* 64: 417–421.

Green, T.A., R.J. Prokopy, and D.W. Hosmer. 1994. Distance of response to host tree models by female apple maggot flies, *Rhagoletis pomonella*: interaction of visual and olfactory stimuli. *J. Chem. Ecol.* 20: 2393–2413.

Green, T.A., R.J. Prokopy, and D.W. Hosmer. Distance of response to a synthetic component of host fruit odor by apple maggot flies. *J. Chem. Ecol.*, in press.

Greene, E., L.J. Orsazk, and D.W. Whitman. 1987. A tephritid fly mimics the territorial displays of its jumping spider predators. *Science* 236: 310–312.

Gut, L.J. and J.F. Brunner. 1994. Parasitision of the apple maggot, *Rhagoletis pomonella*, infesting hawthorns in Washington. *Entomophaga* 39: 41–49.

Haisch, A. and E.F. Boller. 1971. The genetic control of the European cherry fruit fly: progress report rearing and sterilization. Symposium on Sterility Principle for Insect Control or Eradication. I.A.E.A. Athens (1970): 67–76.

Haisch, A. and H.Z. Levinson. 1980. Influences of fruit volatiles and coloration on oviposition of the cherry fruit fly. *Naturwissenschaften* 67: 44–45.

Hendrichs, J. and R.J. Prokopy. 1990. Where do apple maggot flies find food in nature? *Mass. Fruit Notes* 55 (3): 1–3.

Hendrichs, J. and R.J. Prokopy. 1994. Food foraging behavior of frugivorous fruit flies. In *Fruit Flies and the Sterile Insect Technique* (C.A. Calkins, W. Klassen, P. Liedo, eds.), pp. 37–55. CRC Press, Boca Raton.

Hendrichs, J., C.R. Lauzon, S.S. Cooley, and R.J. Prokopy. 1993a. Contribution of natural food sources to adult longevity and fecundity in *Rhagoletis pomonella. Ann. Entomol. Soc. Am.* 86: 250–264.

Hendrichs, J., B.S. Fletcher, and R.J. Prokopy. 1993b. Feeding behavior of *Rhagoletis pomonella* flies: effect of initial food quantity and quality on food foraging, handling costs and bubbling. *J. Insect Behav.* 6: 43–64.

Hoffmeister, T. 1992. Factors determining the structure and diversity of parasitoid complexes in tephritid fruit flies. *Oecologia* 89: 288–297.

Hoffmeister, T. and B.D. Roitberg. 1997. Counterespionage in an insect herbivore-parasitoid system. *Naturwissenschaften* 84: 117–119.

Hoffmeister, T. and B.D. Roitberg. 1998. Evolution of signal persistence under parasitoid exploitation. *Ecoscience* 5: 312–320.

Howard, D.J., G.L. Bush, and J.A. Breznak. 1985. The evolutionary significance of bacteria associated with *Rhagoletis. Evolution* 39: 405–417.

Jaenike, J. and D.R. Papaj. 1992. Behavioral plasticity and patterns of host use by insects. In *Insect Chemical Ecology: An Evolutionary Approach* (B.D. Roitberg and M.B. Isman, eds.), pp. 245–264. Chapman & Hall, New York.

Jang, E.B. and K.A. Nishijima. 1990. Identification and attractancy of bacteria associated with *Dacus dorsalis. Environ. Entomol.* 19: 1726–1731.

Juliano, S.A. and V.A. Borowicz. 1987. Parasitision of a frugivorous fly, *Rhagoletis cornivora*, by the wasp *Opius richmondi*: relationships to fruit and host density. *Can. Zool.* 65: 1326–1330.

Katsoyannos, B.I. 1975. Oviposition-deterring, male-arresting fruit marking pheromone in *Rhagoletis cerasi. Environ. Entomol.* 4: 801–807.

Katsoyannos, B.I. 1976. Female attraction to males in *Rhagoletis cerasi. Environ. Entomol.* 5: 474–476.

Katsoyannos, B.I. 1989. Response to shape, size and color. In *Fruit Flies: Their Biology, Natural Enemies and Control* (A.S. Robinson and G. Hooper, eds.), pp. 307–324. In *World Crop Pests* (W. Helle, ed.), Vol. 3A. Elsevier Science Publishers, Amsterdam.

Katsoyannos, B.I., E.F. Boller, and U. Remund. 1980. A simple olfactometer for the investigation of sex pheromones and other olfactory arrestants in fruit flies and moths. *Z. Angew. Entomol.* 90: 105–112.

Kring, J.B. 1970. Flight behavior of aphids. *Annu. Rev. Entomol.* 17: 461–492.

Lakra, R.K. and Z. Singh. 1983. Oviposition behavior of ber fruitfly, *Carpomya vesuviana*, and relationship between its incidence and ruggedness in fruit in Haryana. *Indian J. Entomol.* 45: 48–59.

Lakra, R.K. and Z. Singh. 1989. Bionomics of *Zizyphus* fruitfly, *Carpomya vesuviana*, in Haryana. *Bull. Entomol.* 27: 13–27.

Lalonde, R.G. and M. Mangel. 1994. Seasonal effects on superparasitism by *Rhagoletis completa*. *J. Anim. Ecol.* 63: 583–588.

Lauzon, C.R. 1991. Microbial Ecology of a Fruit Fly Pest, *Rhagoletis pomonella*. Ph.D. dissertation, University of Vermont, Burlington.

Lauzon, C.R., R.E. Sjogren, S.E. Wright, and R.J. Prokopy. 1998. Attraction of *Rhagoletis pomonella* flies to odor of bacteria: apparent confinement to specialized members of Enterobacteriaceae. *Environ. Entomol.* 27: 853–857.

Lauzon, C.R., R.E. Sjogren, and R.J. Prokopy. Enzymatic capabilities of bacteria associated with apple maggot flies: a postulated role in attractancy. *J. Chem. Ecol.*, in press.

Leski, R. 1963. Studies on the biology and ecology of the cherry fruit fly, *Rhagoletis cerasi*. *Pol. Pismo Entomol.* 31/32: 154–240.

Liburd, O.E., S.R. Alm, R.A. Casagrande, S. Polavarapu. 1998. Effect of trap color, bait, shape and orientation in attraction of blueberry maggot flies. *J. Econ. Entomol.* 91: 243–249.

Lozada, N., M. Aluja, V. Hernández, A. Birke, and J. Piñero. Comportamento basico de *Rhagoletis turpiniae* bajo condiciones naturales, seminaturales y de laboratorio. *Acta Zool. Mex.*, in press.

MacCollom, G.B., C.R. Lauzon, R.W. Weires, and A.A Rutowski. 1992. Attraction of apple maggot to microbial isolates. *J. Econ. Entomol.* 85: 83–87.

Maier, C.T. 1981. Parasitoids emerging from puparia of *Rhagoletis pomonella* infesting hawthorn and apple in Connecticut. *Can. Entomol.* 113: 867–870.

Malavasi, A. and R.J. Prokopy. 1992. Effect of food deprivation on the foraging behavior of *Rhagoletis pomonella* females for food and host hawthorn fruit. *J. Entomol. Sci.* 27: 185–193.

Mangel, M. and C. Clark. 1986. Towards a unifying foraging theory. *Ecology* 67: 1127–1138.

Mangel, M. and B.D. Roitberg. 1989. Dynamic information and host acceptance by a tephritid fruit fly. *Ecol. Entomol.* 14: 181–189.

Mather, M.H. and B.D. Roitberg. 1987. A sheep in wolf's clothing: tephritid flies mimic spider predators. *Science* 236: 308–310.

Mayes, C.F. and B.D. Roitberg. 1987. Host discrimination in *Rhagoletis berberis*. *J. Entomol. Soc. Br. C.* 83: 39–43.

McAlister, L.C. 1932. An observation of a maggot, *Rhagoletis pomonella*, passing from one blueberry into an adjacent blueberry. *J. Econ. Entomol.* 25: 412–413.

Messina, F.J. 1989. Host preferences of hawthorn and cherry infesting flies of *Rhagoletis pomonella* in Utah. *Entomol. Exp. Appl.* 53: 89–92.

Messina, F.J. 1990. Components of host choice by two *Rhagoletis* species in Utah. *J. Kans. Entomol. Soc.* 63: 80–87.

Messina, F.J. and V.P. Jones. 1990. Relationship between fruit phenology and infestation and the apple maggot in Utah. *Ann. Entomol. Soc. Am.* 83: 742–752.

Messina, F.J. and J.K. Subler. 1995. Conspecific and heterospecific interactions of male *Rhagoletis* flies on a shared host. *J. Kans. Entomol. Soc.* 68: 206–213.

Messina, F.J., D.G. Alston, and V.P. Jones. 1991. Oviposition by the Western cherry fruit fly in relation to host development. *J. Kans. Entomol. Soc.* 64: 197–208.

Monastero, S. 1970. New observations on the biology of *Carpomya incompleta* injurious to jujube fruits. *Boll. Inst. Entomol. Agrar. Palermo* 7: 147–157.

Monteith, L.G. 1972. The status of the apple maggot, *Rhagoletis pomonella*, in Ontario. *Can. Entomol.* 103: 507–512.

Moran, N.A. and A. Telang. 1998. Bacteriocyte-associated symbionts of insects. *Bioscience* 48: 295–304.

Mumtaz, M.M. and M.T. Ali Niazee. 1983. The oviposition-deterring pheromone of the western cherry fruit fly, *Rhagoletis indifferens*. 1. Biological properties. *Z. Angew. Entomol.* 96: 83–93.

Murphy, B.C., L.T. Wilson, and R.V. Dowell. 1991. Quantifying apple maggot preference for apples to optimize the distribution of traps among trees. *Environ. Entomol.* 20: 981–987.

Neilson, W.T. A. 1967. Development and mortality of the apple maggot, *Rhagoletis pomonella*, in crab apples. *Can. Entomol.* 99: 217–219.

Neilson, W.T.A. 1969. Rearing larvae of the apple maggot on an artificial diet. *J. Econ. Entomol.* 62: 1028–1031.

Neilson, W.T.A. and J.W. McAllan. 1965. Artificial diets for the apple maggot. III. Improved, defined diets. *J. Econ. Entomol.* 58: 542–543.

Opp, S.B. 1988. Polygamous Mating System of a Tephritid Fruit Fly, *Rhagoletis pomonella*. Ph.D. dissertation, University of Massachusetts, Amherst.

Opp, S.B. and R.J. Prokopy. 1986. Variation in laboratory oviposition by *Rhagoletis pomonella* in relation to mating status. *Ann. Entomol. Soc. Am.* 79: 705–710.

Opp, S.B. and R.J. Prokopy. 1987. Seasonal changes in resightings of marked, wild *Rhagoletis pomonella* flies in nature. *Fla. Entomol.* 70: 449–457.

Opp, S.B., J. Ziegner, N. Bui, and R.J. Prokopy. 1990. Factors influencing estimates of sperm competition in *Rhagoletis pomonella* (Walsh) (Diptera: Tephritidae). *Ann. Entomol. Soc. Am.* 83: 521–526.

Opp, S.B., S.A. Spisak, A. Telang, and S.S. Hammond. 1996. In *Economic Fruit Flies: A World Assessment of Their Biology and Management* (B.A. McPheron and G.J. Steck, eds.), pp. 43–49. St. Lucie Press, Delray Beach.

Papaj, D.R. 1993. Use and avoidance of occupied hosts as a dynamic process in tephritid flies. In *Insect-Plant Interactions* Vol. 5 (E.A. Bernays, ed.), pp. 25–46. CRC Press, Boca Raton.

Papaj, D.R. 1994. Oviposition site guarding by male walnut flies and its possible consequences for mating success. *Behav. Ecol. Sociobiol.* 34: 187–195

Papaj, D.R. and H. Alonso-Pimentel. 1997. Why walnut flies superparasitize: time savings as a possible explanation. *Oecologia* 109: 166–174.

Papaj, D.R. and R.J. Prokopy. 1986. The phytochemical basis of learning in *Rhagoletis pomonella* and other herbivorous insects. *J. Chem. Ecol.* 12: 1125–1143.

Papaj, D.R. and R.J. Prokopy. 1988. The effect of prior experience on components of habitat preference in the apple maggot fly, *Rhagoletis pomonella. Oecologia* 76: 538–543.

Papaj, D.R. and R.J. Prokopy. 1989. Ecological and evolutionary aspects of learning in phytophagous insects. *Annu. Rev. Entomol.* 34: 315–350.

Papaj, D.R., J.M. Garcia, and H. Alonso-Pimentel. 1996. Marking of host fruit by male *Rhagoletis boycei* Cresson flies (Diptera: Tephritidae) and its effect on egg-laying. *J. Insect Behav.* 9: 585–598.

Pree, D.J. 1977. Resistance to development of larvae of the apple maggot in crab apples. *J. Econ. Entomol.* 70: 611–614.

Prokopy, R.J. 1966. Artificial oviposition devices for apple maggot. *J. Econ. Entomol.* 59: 384–387.

Prokopy, R.J. 1972. Evidence for a marking pheromone deterring repeated oviposition in apple maggot flies. *Environ. Entomol.* 1: 326–332.

Prokopy, R.J. 1975a. Mating behavior in *Rhagoletis pomonella*. V. Virgin female attraction to male odor. *Can. Entomol.* 107: 905–908.

Prokopy, R.J. 1975b. Oviposition deterring fruit marking pheromone in *Rhagoletis fausta. Environ. Entomol.* 4: 298–300.

Prokopy, R.J. 1976. Feeding, mating and oviposition activities of *Rhagoletis fausta* flies in nature. *Ann. Entomol. Soc. Am.* 69: 899–904.

Prokopy, R.J. 1977a. Stimuli influencing trophic relations in Tephritidae. *Colloq. Int. C.N.R.S.* 265: 305–336.

Prokopy, R.J. 1977b. Responses of *Rhagoletis pomonella* and *R. fausta* flies to different size spheres. *Can. Entomol.* 109: 593–596.

Prokopy, R.J. and E.F. Boller. 1971. Stimuli eliciting oviposition of *Rhagoletis cerasi* (Diptera: Tephritidae) into inanimate objects. *Entomol. Exp. Appl.* 14: 1–14.

Prokopy, R.J. and G.L. Bush. 1972. Mating behavior in *Rhagoletis pomonella*. III. Male aggregation in response to an arrestant. *Can. Entomol.* 104: 275–283.

Prokopy, R.J. and G.L. Bush. 1973a. Mating behavior in *Rhagoletis pomonella*. IV. Courtship. *Can. Entomol.* 195: 973–991.

Prokopy, R.J. and G.L. Bush. 1973b. Ovipositional response to different sizes of artificial fruit by flies of *Rhagoletis pomonella* species group. *Ann. Entomol. Soc. Am.* 66: 927–929.

Prokopy, R.J. and E.D. Owens. 1983. Visual basis of host plant selection by phytophagous insects. *Annu. Rev. Entomol.* 28: 337–364.

Prokopy, R.J. and D.R. Papaj. 1988. Learning of apple fruit biotypes by apple maggot flies. *J. Insect Behav.* 1: 67–74.

Prokopy, R.J. and D.R. Papaj. 1989. Can ovipositioning *Rhagoletis pomonella* females learn to discriminate among host fruit of the same biotype? *Fla. Entomol.* 72: 489–494.

Prokopy, R.J. and A.H. Reynolds. 1998. Ovipositional enhancement through socially facilitated behavior in *Rhagoletis pomonella* flies. *Entomol. Exp. Appl.* 86: 281–286

Prokopy, R.J. and B.D. Roitberg. 1984. Resource foraging behavior of true fruit flies. *Am. Sci.* 72: 41–49.

Prokopy, R.J. and B.D. Roitberg. 1989. Fruit fly foraging behavior. In *Fruit Flies: Their Biology, Natural Enemies and Control* (A.S. Robinson and G. Hooper, eds.), pp. 293–306. In *World Crop Pests* (W. Helle, ed.), Vol. 3A. Elsevier Science Publishers, Amsterdam.

Prokopy, R.J. and R. Webster. 1978. Oviposition-deterring pheromone of *Rhagoletis pomonella* as a kairomone for its parasitoid *Opius lectus*. *J. Chem. Ecol.* 4: 481–494.

Prokopy, R.J., E.W. Bennett, and G.L. Bush. 1971. Mating behavior in *Rhagoletis pomonella*. I. Site of assembly. *Can. Entomol.* 103: 1405–1409.

Prokopy, R.J., E.W. Bennett, and G.L. Bush. 1972. Mating behavior in *Rhagoletis pomonella*. II. Temporal organization. *Can. Entomol.* 104: 97–104.

Prokopy, R.J., V. Moericke, and G.L. Bush. 1973. Attraction of apple maggot flies to odor of apples. *Environ. Entomol.* 2: 743–749.

Prokopy, R.J., W.H. Reissig, and V. Moericke. 1976. Marking pheromones deterring repeated oviposition in *Rhagoletis* flies. *Entomol. Exp. Appl.* 20: 170–178.

Prokopy, R.J., A.L. Averill, S.S. Cooley, C.A. Roitberg, and C. Kallet. 1982. Associative learning in egglaying site selection by apple maggot flies. *Science* 218: 76–77.

Prokopy, R.J., S.S. Cooley, C. Kallet, and D. Papaj. 1986. On the nature of learning in oviposition site acceptance by apple maggot flies. *Anim. Behav.* 34: 98–107.

Prokopy, R.J., S.R. Diehl, and S.S. Cooley. 1988. Behavioral evidence for host races in *Rhagoletis pomonella* flies. *Oecologia* 76: 138–147.

Prokopy, R.J., S.S. Cooley, and S.B. Opp. 1989. Prior experience influences fruit residence of male apple maggot flies. *J. Insect Behav.* 2: 39–48.

Prokopy, R.J., S.S. Cooley, L. Galarza, C. Bergweiler, and C.R. Lauzon. 1993a. Bird droppings compete with bait sprays for *Rhagoletis pomonella* flies. *Can. Entomol.* 125: 413–422.

Prokopy, R.J., A.L. Averill, T.A. Green, and T.T.Y. Wong. 1993b. Does food shortage cause fruit flies to dump eggs? *Ann. Entomol. Soc. Am.* 86: 362–365.

Prokopy, R.J., S.S. Cooley, and D. Papaj. 1993c. How well can relative specialist *Rhagoletis* flies learn to discriminate fruit for oviposition? *J. Insect Behav.* 6: 167–176.

Prokopy, R.J., S.S. Cooley, J.J. Prokopy, Q. Quan, and J.F. Buonaccorsi. 1994a. Interactive effects of resource abudance and state of adults on residence of *Rhagoletis pomonella* flies in host tree patches. *Environ. Entomol.* 23: 304–315.

Prokopy, R.J., C. Bergweiler, L. Galarza, and J. Schwerin. 1994b. Prior experience affects visual ability of *Rhagoletis pomonella* flies to find host fruit. *J. Insect Behav.* 7: 663–677.

Prokopy, R.J., S.S. Cooley, I. Luna, and J.J. Duan. 1995. Combined influence of protein hunger and egg load on the resource foraging behavior of *Rhagoletis pomonella* flies. *Eur. J. Entomol.* 92: 655–666.

Prokopy, R.J., A.H. Reynolds, and L.J. van der Ent. 1998. Can *Rhagoletis pomonella* flies learn to associate presence of food on foliage with foliage color? *Eur. J. Entomol.* 95: 335–341.

Randen, E.J. van and B.D. Roitberg. 1996. The effect of egg load on superparasitism by the snowberry fly. *Entomol. Exp. Appl.* 79: 241–245.

Reissig, W.H. 1979. Survival of apple maggot larvae, *Rhagoletis pomonella*, in picked and unpicked apples. *Can. Entomol.* 111: 181–187.

Reissig, W.H., S.K. Brown, R.C. Lamb, and J.N. Cummins. 1990. Laboratory and field studies of resistance of crab apple clones to *Rhagoletis pomonella*. *Environ. Entomol.* 19: 565–572.

Riedl, H. and R. Hislop. 1985. Visual attraction of the walnut husk fly to colored rectangles and spheres. *Environ. Entomol.* 14: 810–814.

Robertson, I.C., B.D. Roitberg, I. Williams, and S.E. Senger. 1995. Contextual chemical ecology: an evolutionary approach to the chemical ecology of insects. *Am. Entomol.* 41: 237–239.

Roitberg, B.D. 1985. Search dynamics in fruit-parasitic insects. *J. Insect Physiol.* 31: 865–872.

Roitberg, B.D. 1989. The cost of reproduction in rosehip flies, *Rhagoletis basiola*: eggs are time. *Evol. Ecol.* 3: 183–188.

Roitberg, B.D. and R.G. Lalonde. 1991. Host marking enhances parasitism risk for a fruit-infesting fly *Rhagoletis basiola*. *Oikos* 61: 389–393.

Roitberg, B.D. and M. Mangel. 1988. On the evolutionary ecology of marking pheromones. *Evol. Ecol.* 2: 289–315.

Roitberg, B.D. and M. Mangel. 1997. Individuals on the landscape: behavior can mitigate landscape differences among habitats. *Oikos* 80: 234–240.

Roitberg, B.D. and R.J. Prokopy. 1981. Experience required for pheromone recognition by the apple maggot fly. *Nature* 292: 540–541.

Roitberg, B.D. and R.J. Prokopy. 1983. Host deprivation influence on response of *Rhagoletis pomonella* to its oviposition-deterring pheromone. *Physiol. Entomol.* 8: 69–72.

Roitberg, B.D. and R.J. Prokopy. 1987. Insects that mark host plants. *Bioscience* 37: 400–406.

Roitberg, B.D., J.C. van Lenteren, J.J.M. van Alphen, F. Galis, and R.J. Prokopy. 1982. Foraging behavior of *Rhagoletis pomonella*, a parasite of hawthorn. *J. Anim. Ecol.* 51: 307–325.

Roitberg, B.D., M. Mangel, and G. Tourigny. 1990. The density dependence of parasitism by tephritid fruit flies. *Ecology* 71: 1871–1885.

Roitberg, B.D., M.L. Reid, and C. Li. 1993. Choosing hosts and mates: the value of learning. In *Insect Learning: Ecological and Evolutionary Perspectives* (D.R. Papaj and A.C. Lewis, eds.), pp. 174–194. Chapman & Hall, New York.

Sivinski, J.M. and G. Dodson. 1992. Sexual dimorphism in *Anastrepha suspensa* and other tephritid fruit flies: possible roles of developmental rate, fecundity and dispersal. *J. Insect Behav.* 5: 491–506.

Smith, D.C. 1984. Feeding, mating and oviposition by *Rhagoletis cingulata* flies in nature. *Ann. Entomol. Soc. Am.* 77: 702–704.

Smith, D.C. 1985. General activity and reproductive behavior of *Rhagoletis cornivora* flies in nature. *J. N.Y. Entomol. Soc.* 93: 1052–1056.

Smith, D.C. and R.J. Prokopy. 1980. Mating behavior in *Rhagoletis pomonella* (Diptera: Tephritidae) VI. Site of early-season encounters. *Can. Entomol.* 112: 585–590.

Smith, D.C. and R.J. Prokopy. 1981. Seasonal and diurnal activity of *Rhagoletis mendax* flies in nature. *Ann. Entomol. Soc. Am.* 74: 462–466.

Smith, D.C. and R.J. Prokopy. 1982. Mating behavior of *Rhagoletis mendax* (Diptera: Tephritidae) flies in nature. *Ann. Entomol. Soc. Am.* 75: 388–392.

Smyth, E.G. 1960. A new tephritid fly injurious to tomatoes in Peru. *Calif. Dep. Agric. Bull.* 49: 16–22.

Stadler, E., B. Ernst, J. Hurter, and E. Boller. 1994. Tarsal contact chemoreceptors for the host marking pheromone of the cherry fruit fly, *Rhagoletis cerasi*: response to natural and synthetic compounds. *Physiol. Entomol.* 19: 139–151.

Stoffolano, J.G. and R.S. Yin. 1987. Structure and function of the ovipositor and associated sensilla of the apple maggot, *Rhagoletis pomonella*. *J. Insect Morph. Embryol.* 16: 41–69.

Telang, A., S.S. Hammond, and S.B. Opp. 1996. Effects of copulation frequencies on egglaying and egg hatch in the walnut husk fly, *Rhagoletis completa*. *Pan-Pac. Entomol.* 72: 235–237.

Tracewski, K.T. and J.F. Brunner. 1987. Seasonal and diurnal activity of *Rhagoletis zephyria*. *Melanderia* 45: 26–32.

Tsiropoulos, G.J. and K.S. Hagen. 1979. Oviposition behavior of laboratory-reared and wild Caribbean fruit flies: selected chemical influences. *Z. Angew. Entomol.* 89: 547–550.

Wiesmann, R. 1933. Untersuchungen über die Lebensgeschichte und Bekämpfung der Kirschfliege, *Rhagoletis cerasi*. *Landwirtsch. Jahrb. Schweiz* 1933: 711–760.

Wiesmann, R. 1937. Die Orientierung der Kirschfliege, *Rhagoletis cerasi*, bei der Eiablage. *Landwirtsch. Jahrb. Schweiz* 1937: 1080–1109.

Whitman, D.W., L. Orzak, and E. Greene. 1988. Spider mimicry in fruit flies (Diptera: Tephritidae): further experiments on the deterrence of jumping spiders by *Zonosemata vittigera* (Coquillett). *Ann. Entomol. Soc. Am.* 81: 532–536.

Yokoyama, V.Y. and G.T. Miller. 1994. Walnut husk fly pest-free and preovipositional periods and adult emergence for stone fruits exported to New Zealand. *J. Econ. Entomol.* 87: 747–751.

Zwölfer, H. 1983. Life systems and strategies of resource exploitation in tephritids. In *Fruit Flies of Economic Importance* (R. Cavalloro, ed.), pp. 16–30. A.A. Balkema, Rotterdam.

# 11 Phylogeny and Behavior of Flies in the Tribe Trypetini (Trypetinae)

*Ho-Yeon Han*

## CONTENTS



0-8493-1275-2/00/$0.00+$.50

## 11.1 INTRODUCTION

Based on a comparative morphological study of more than 250 tephritid species from all over the world, I here redefine the tribe Trypetini as a monophyletic group with two subtribes, Trypetina and Chetostomatina. I proposed this new classification in my Ph.D. dissertation (Han 1992), which is not an available publication for the purposes of zoological nomenclature. However, it has been adopted (sometimes with modifications) by several recent major tephritid publications (Hancock and Drew 1995; Permkam and Hancock 1995; Korneyev 1996; 1998; Wang 1996; Norrbom et al. 1999a, b), and, therefore, it is imperative to publish formally the justifications for this classification. Discussions in this chapter are largely based on my dissertation, with some modifications based on subsequent information.

Prior to the revised classification proposed in my dissertation, the tribe Trypetini was recognized as a heterogeneous assemblage of genera that lacked derived characters of other tribes in the subfamily Trypetinae. Many genera of uncertain relationships have been included in the Trypetini, and the demarcation of the tribe by different tephritid specialists varied. This confusion may have been a result of the following:

1. Most previous systematic studies were restricted to regional faunas (Hardy 1973; 1987; Ito 1983–1985; Hancock 1985; White 1988).
2. Useful taxonomic characters such as genitalic structures have not been adequately utilized for the higher classification, even though they were often used for species identifications (Hardy 1987; White 1988).
3. Until lately, rigorous phylogenetic logic was not used to infer relationships.

Therefore, my study aims at redefining the taxonomic limit of the tribe Trypetini and investigating the phylogenetic relationships among the genera within the tribe.

Most Trypetini are distributed in the Oriental and Palearctic Regions. Species of the subtribe Chetostomatina are mostly fruit-feeders, while the subtribe Trypetina includes all the known species of leaf-mining tephritids as well as some flies with different feeding behaviors, such as fruit-feeding and stem-mining. Currently, there are 285 valid species in 29 genera of Trypetina and 69 species in 7 genera of Chetostomatina, most of which were cataloged by Norrbom et al. (1999b).

## 11.2 STRATEGIES FOR HYPOTHESIS BUILDING AND TESTING

Since the previous concept of the tribe Trypetini was not defined on an explicit phylogenetic basis, initial hypotheses for a monophyletic demarcation of the tribe were built. As far as is known, species of the type genus *Trypeta* Meigen, including the type species *T. artemisiae* (Fabricius), are leaf miners. This leaf-mining behavior has been reported from less than 20 species of various tephritid genera scattered within the Trypetini, *sensu lato*. Because of its rarity, I postulated that leaf mining evolved only once within the Tephritidae. To search for potential synapomorphies of the leaf-mining group, I initially examined about ten known leaf miners plus some additional species that were congeners with the known leaf miners. As a result, the following was hypothesized: (1) leaf-mining tephritids form a monophyletic group; and (2) morphological synapomorphies define this leaf-mining group or a larger group including all the leaf-mining tephritids (see Section 11.3).

Since the initiation of this project, I have examined more than 250 species to test the above hypotheses. They include the rest of the known leaf-mining tephritids, most species placed in the same genera as the known leaf miners, members of the genera previously regarded to be related to the leaf miners, and some representative tephritids from other groups for comparison. Most genera involved in this study have been considered taxonomically difficult by previous workers because of lack of obvious diagnostic characters at the generic level and high variability of the characters often used to define genera (Hardy 1987). For example, the position of crossvein R-M, used to distinguish *Trypeta* and *Myoleja* (*s. lat.*), is actually variable within *Trypeta*, and extremely similar wing patterns occur in *Philophylla*, *Anomoia*, and even *Hexachaeta*, which belongs to the tribe Toxotrypanini. Similar examples are found in numerous other characters of wing pattern, venation, and chaetotaxy. For this reason, I have surveyed all the conventionally used morphological characters, including genitalia, some of which have proved useful at generic or higher levels.

I followed a traditional upward classification approach (Mayr and Ashlock 1991) to test the already established generic limits and to propose a more phylogenetically based generic classification within the Trypetini. Each genus was first delimited through an evaluation of similarity and differences. A hypothesis of monophyly was accepted if there were one or more possible synapomorphies. For example, *Acidiostigma* is supported as monophyletic based on its enlarged pterostigma (see Figure 11.5N, O), and *Stemonocera* by its uniquely modified male frontal plate (Figure 11.4K). Tests of monophyly were not always successful. No obvious synapomorphies were found for some genera that could be either monophyletic groups without currently recognized derived characters or nonmonophyletic groups (e.g., *Trypeta* and *Hemilea*).

The extent of the potential tribal synapomorphies was also surveyed within each genus and among genera in comparison with other tephritid tribes. As a result, a new monophyletic concept of the tribe Trypetini emerged with recognition of two subtribes. Numerous new combinations of scientific names were also proposed in my dissertation (Han 1992), and the majority of them have been adopted and formally published by me (Han et al. 1993; 1994a, b; Han 1996a, b; 1997a, b; Han and Wang 1997) or other tephritid workers (Hancock and Drew 1995; Permkam and Hancock 1995; Wang 1996; Korneyev 1998; Norrbom et al. 1999b).

## 11.3 MONOPHYLY AND SUBTRIBES OF THE TRYPETINI

A sister-group relationship between the subtribes Trypetina and Chetostomatina is supported based on their close morphological similarity, including a single potential synapomorphy: dorsal sclerite of glans with internal sculpture pattern of elongated granulation (Figures 11.7A to D vs. Figures 11.6H and I). This hypothesis, however, requires some reversals within the Chetostomatina and convergence elsewhere (i.e., in some Acanthonevrini). Therefore, a molecular analysis using mitochondrial 16S ribosomal DNA sequences was conducted to elucidate their relationships from a different perspective (Han and McPheron 1997; Chapter 5). This study supported their sister-group relationship (Figure 5.1), and a follow-up study based on additional taxa and more sequence characters convincingly supports this hypothesis (Han, in press).

Although the two subtribes of the Trypetini are relatively well defined (see the following sections), there is no single good diagnostic character to recognize the entire tribe. The included genera are better distinguished as Trypetini based on the combination of the diagnostic characters of either one of the subtribes.

### 11.3.1 SUBTRIBE TRYPETINA

From the initial stage of this project, the following three potential synapomorphies for a suprageneric taxon including *Trypeta* were recognized. The combination of these characters can now be used to diagnose the subtribe Trypetina: (1) median sclerite of glans with internal sculpture pattern of round granulation (Figures 11.2A to D and 11.7B to G); (2) dorsal sclerite of glans with internal sculpture pattern of elongated granulation (Figures 11.2A to D and 11.7A to F); and (3) aculeus broad, with lateral serration toward apex (Figures 11.2G and 11.10A to F). After extensive examination of more than 250 species, the granulated median sclerite has been consistently found within nearly all the leaf-mining genera and the rest of the genera recognized within the subtribe Trypetina. Secondary loss of this structure has been hypothesized only for the genus *Itosigo* (Figure 11.7A), and for a few species within *Acidiella* and *Vidalia* (Han et al. 1994b). Because of the relatively few cases of reversal within such a large group (285 species in 29 genera), I suggest that this structure is a robust synapomorphy defining the Trypetina. Character 2, on the other hand, is found within a few genera of the Chetostomatina, and, thus, is suggested as a synapomorphy of the two subtribes (see the previous section). Character 3 is also consistent within the Trypetina, but appears to be somewhat homoplastic because similar structures are sometimes found in nontrypetine taxa such as Toxotrypanini.

Since the completion of my dissertation (Han 1992), it has been indicated that the suggested synapomorphy for the Trypetina (the granulated median sclerite) was also observed in some other taxa such as *Oedicarena, Paraterellia, Euphranta, Acanthonevra,* and *Notomma* (Korneyev 1996; Jenkins 1996; see also Figure 11.8E to K). However, based upon the two most commonly used criteria to determine morphological homology (similarity of position and structure), I believe that the granulated median sclerite of the Trypetina is unlikely to be homologous with the granulated areas of those genera as evident from the comparison of Figures 11.7B to G and 11.8E to K. The granulated median sclerite in the Trypetina is almost always a clearly definable sclerotized area bound by a basal-to-dorsal furrow defining it from adjacent areas. Extra caution should be taken when determining homology between various structures in the tephritid glans, because rather highly heterogeneous structures are often found in many tephritid higher taxa.

*Description of Trypetina*: Generally yellow brown to dark brown with brown to dark brown setae. Head with one paired ocellar seta, one to two orbital setae, two to seven frontal setae (most commonly three pairs, as in Figure 11.4A, B); antennae closely approximated basally, arista bare to short pubescent; genal seta strong, yellow brown to dark brown; postgena moderately to strongly swollen, with long, fine setulae. Thorax (Figure 11.1A, C) with standard chaetotaxy of Tephritidae plus two pairs of scapular setae; position of dorsocentral seta varies considerably; scutellum with two pairs of marginal setae; anepisternum with one to two strong setae. Wing (Figures 11.1A and 11.5A to U) hyaline with yellow brown to dark brown pattern; extension of cell bcu slightly to moderately elongated. Male genitalia (Figures 11.1D, E and 11.2A to D) with lateral prensiseta from as long as to much shorter than medial prensiseta; median sclerite of glans mostly with internal sculpture pattern of round granulation; dorsal sclerite of glans mostly with internal sculpture pattern of elongated granulation; subapical lobe of glans usually present. Female postabdomen (Figures 11.2E to G) with strong teeth on eversible membrane; dorsal and ventral taeniae extended at most to midlength of membrane; aculeus usually broad, dorsoventrally flattened, with lateral serration toward apex; two to three spermathecae round to elliptic in outline, most often with spinular papillae. Eggs narrowly elliptic in outline with tiny knoblike micropylar end, but highly modified in a few taxa.

### 11.3.2 Chetostomatina, New Subtribe

It is not easy to differentiate the members of the subtribe Chetostomatina from the Trypetina based on external characters, and many trypetine species have been misplaced in the genus *Myoleja* of the Chetostomatina in the past. However, Chetostomatina can be readily distinguished by the following female postabdominal structures (Figures 11.9C, D and 11.10G to N): (1) eversible membrane nearly smooth, without strong spinules; (2) taeniae reaching apex of eversible membrane; and (3) aculeus long and slender (in most taxa laterally flattened and ventrally serrate). Of these characters, the first two are proposed as synapomorphies for the subtribe Chetostomatina. In addition, the laterally flattened and ventrally serrate aculeus tip convincingly supports the monophyly of five of the seven genera within the Chetostomatina (Figures 11.3 and 11.10I to N). In addition to the above characters, the egg of the Chetostomatina, so far as known, has a pointed apex (the opposite end from micropyle), which is considered apomorphic (plesiomorphy: rounded apex). Unfortunately, this character was examined only from a few species of *Anomoia, Chetostoma,* and *Paramyiolia* and the level at which it may be a synapomorphy is therefore uncertain.

The Nitrariomyiina and Acidoxanthina have been suggested as possible sister groups of the Chetostomatina because they have eversible membranes with long taeniae and reduced scales associated with their slender aculei (as in Figure 11.9E to G) (Korneyev 1996; and Chapter 4), but none of these taxa possesses these characters to the same extent as in the Chetostomatina. Similar elongation (or narrowing) of female terminalia appears to have occurred multiple times in tephritid evolution (e.g., in *Anastrephoides* within Trypetina).

*Description of Chetostomatina*: Generally yellow brown to dark brown with brown to dark brown setae. Head with one paired ocellar seta, two orbital setae, three frontal setae; antennae closely approximated basally, arista bare to short pubescent; genal seta strong, yellow brown to dark brown; postgena moderately to strongly swollen, with long, fine setulae. Thorax with standard chaetotaxy of Tephritidae with two pairs of scapular setae (as in Figure 11.1A, C); dorsocentral seta 0.4 to 1 distance from level of intra-alar seta to postsutural supra-alar seta; scutellum with two pairs of marginal setae; anepisternum with one to two strong setae. Wing hyaline with yellow brown to dark brown pattern; extension of cell bcu slightly to moderately elongated. Male genitalia (Figure 11.6L to O) with lateral prensiseta as long as or slightly shorter than medial prensiseta; glans (Figures 11.7H to J and 11.8A to D) without granulated median sclerite, dorsal sclerite often with internal sculpture pattern of numerous hexagonal cells; subapical lobe present. Female ovipositor (Figure 11.9C, D) usually with two ventral and two dorsal marginal setae on oviscape; eversible membrane cylindrical without any strong teeth; dorsal and ventral taeniae extending almost entire length of eversible membrane; aculeus long, slender; three round to elliptic spermathecae.

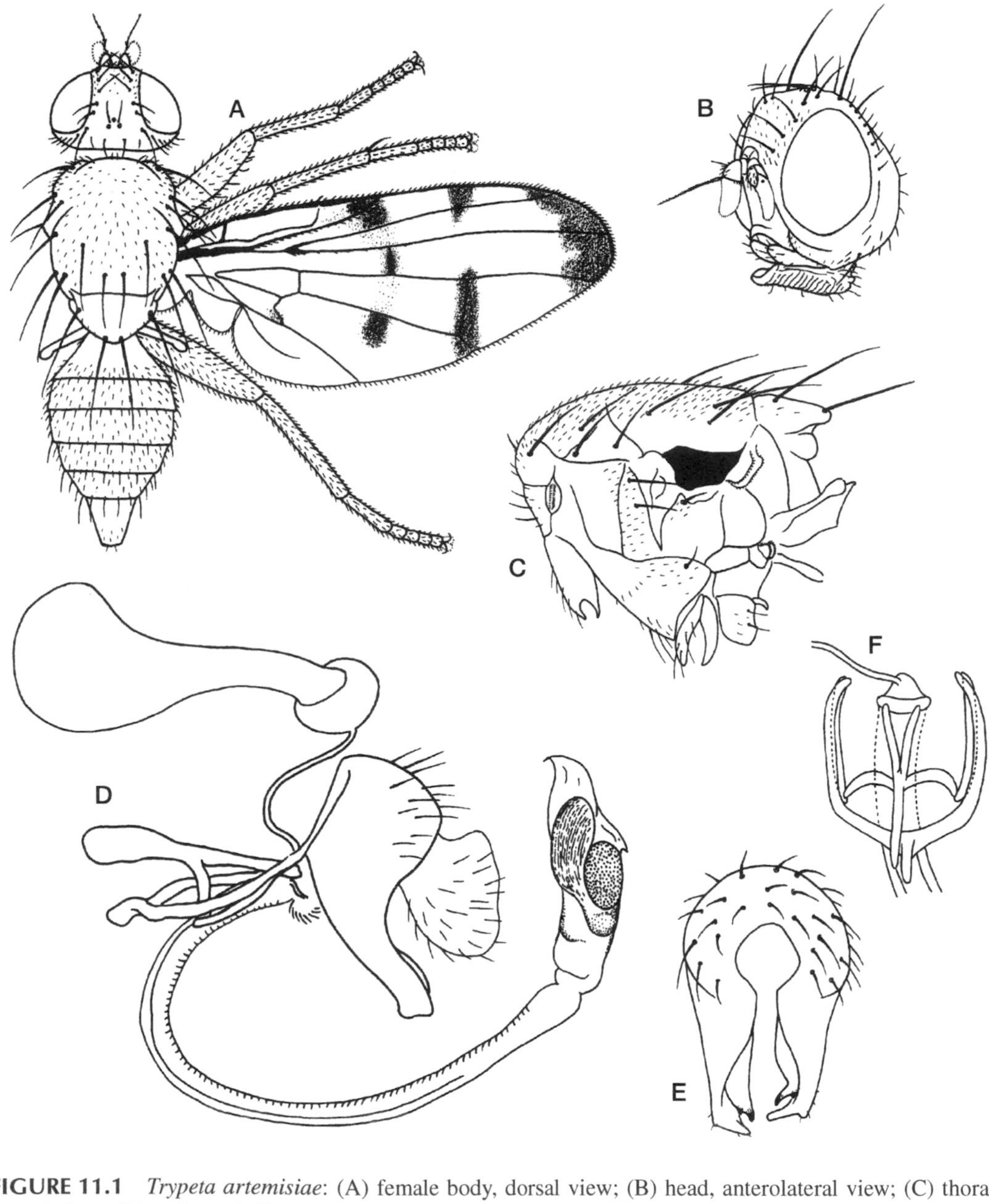


**FIGURE 11.1** *Trypeta artemisiae*: (A) female body, dorsal view; (B) head, anterolateral view; (C) thorax, lateral view; (D) male genitalia, lateral view; (E) epandrium and surstyli, posterior view (proctiger removed); (F) base of phallus.

Eggs elongated, broadly rounded on micropylar end with tiny knoblike structure bearing micropyle, tapering toward opposite end with pointed apex.

## 11.4 INTERGENERIC RELATIONSHIPS OF THE TRIBE TRYPETINI

### 11.4.1 Character Analysis

For intergeneric analysis to identify suprageneric groups within the Trypetini, only characters showing relatively small to no intrageneric variation were selected, excluding any autapomorphies

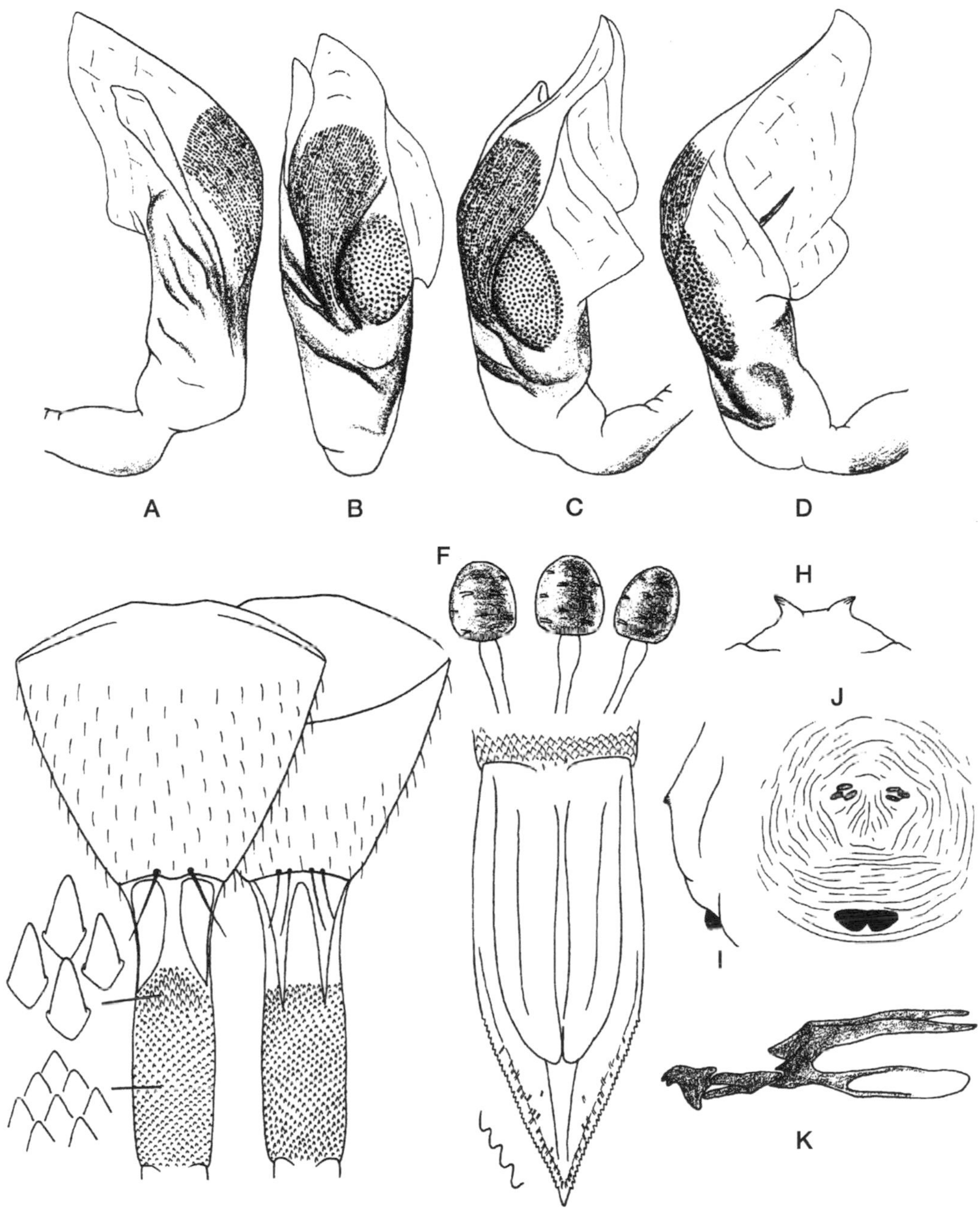


**FIGURE 11.2** *Trypeta artemisiae*: (A) glans, lateral view (left); (B) same, dorsal view; (C) same, dorsolateral view; (D) same, lateral view (right); (E) oviscape and eversible membrane, ventral and dorsal views (insets at 10× main figure); (F) spermathecae; (G) aculeus, ventral view; (H) posterior spiracle, dorsal view; (I) same, lateral view; (J) same, posterior view; (K) third instar cephalopharyngeal skeleton, lateral view.

of individual genera. Because the sister group of the Trypetini is not yet known, the polarity for each character was determined by surveying the character state distributions in other tephritid tribes. If a character state is more common in the other tephritid tribes, it is considered plesiomorphic, whereas its alternate state(s) is apomorphic (Wiley 1981; as one form of the outgroup criterion).

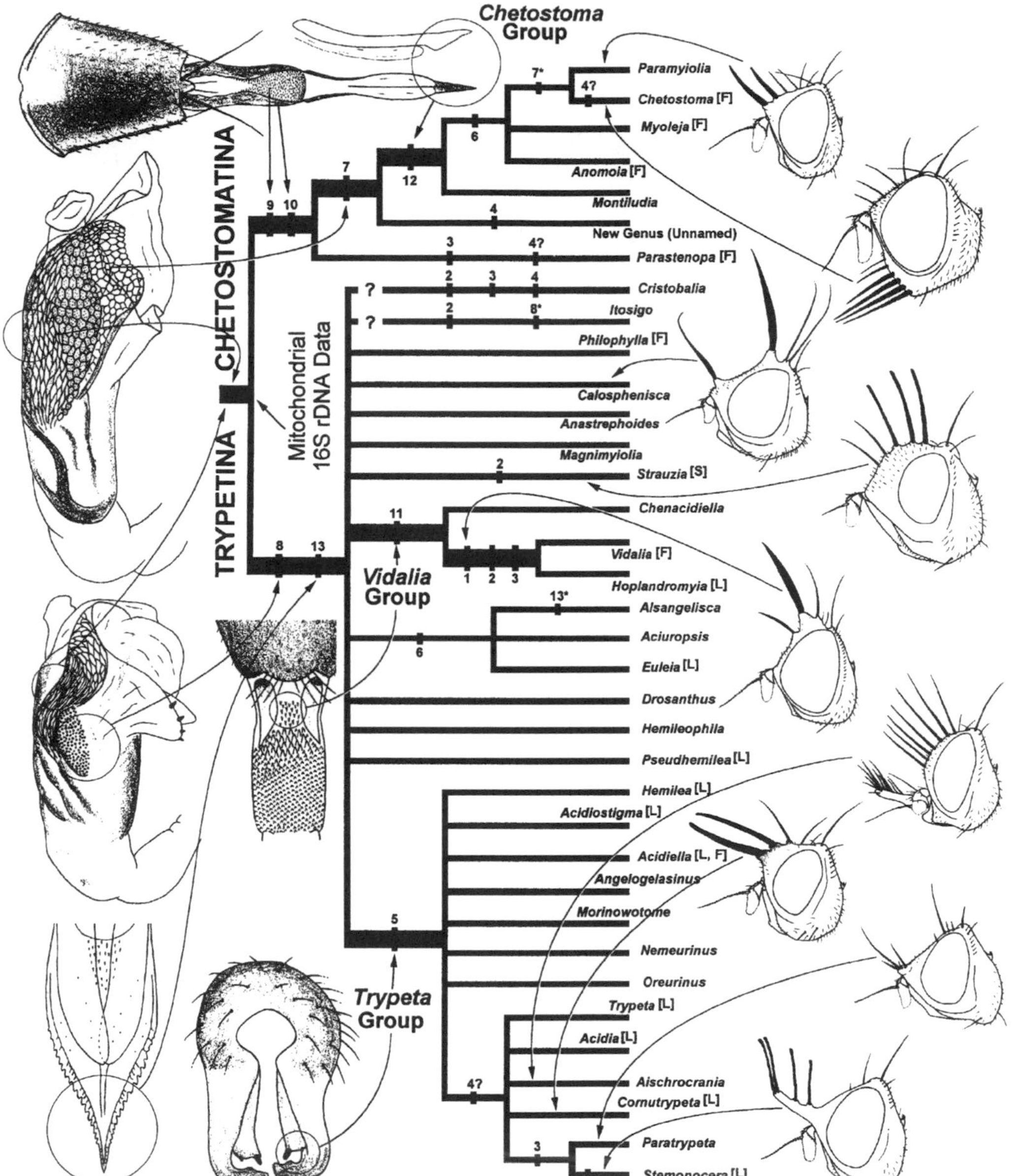


**FIGURE 11.3** Cladodogram of the tribe Trypetini. Numbers refer to apomorphies listed in the character analysis section (Section 11.4.1). * Indicates reversal. Graphic representations of some important synapomorphies are given on the left side. Autapomorphic occurrences of sexually dimorphic head structures are shown on the right side. Abbreviations: F = fruit feeder; L = leaf miner; S = stem borer.

The following 13 characters were selected after comparing the 33 genera of Trypetini. State 0 is considered plesiomorphic.

*Character 1.* Male frontal setae: 0, not modified, or if so, not as in state 1; 1, enlarged and flattened, with second one largest (Figure 11.4L, P). This peculiar sexually dimorphic structure is only found in *Hoplandromyia* and *Vidalia*. Enlargement of the male frontal setae also occurs in other genera such as *Paramyiolia, Calosphenisca, Strauzia, Stemonocera,* and *Paratrypeta,* but the shape of the setae, their arrangement, and even which particular setae are modified differs among these genera. Because of these differences, I believe that these enlargements of the male frontal setae evolved independently, probably due to similar selection pressure from certain sexual activities, such as male-to-male competitions (see Section 11.6). The other frontal setae modifications occur in single genera and are thus autapomorphies, the reason they are not included in this list.

*Character 2.* Number of orbital setae: 0, two; 1, one. A single orbital seta is considered apomorphic because having two setae is much more common in other tephritid taxa. Since both character states are also found in Ulidiidae, Platystomatidae, and Pyrgotidae, which are closely related to Tephritidae, this character appears to be highly subject to homoplasy.

*Character 3.* Ocellar seta: 0, at least 1.5 × as long as ocellar triangle; 1, reduced, about as long as or shorter than ocellar triangle. The reduction of the ocellar seta is considered apomorphic because it is less common in other tephritid taxa. Since both states also occur within related families of Tephritoidea, this character appears to have evolved independently numerous times.

*Character 4.* Distance along vein M between crossveins R-M and DM-Cu/distance between crossveins R-M and BM-Cu: 0, not more than 0.7; 1, greater than or slightly less than 1.0 (Figure 11.5A and T to V). The latter state is less common across the Tephritidae and is coded apomorphic, although this character appears to be highly subject to homoplasy.

*Character 5.* Lateral prensiseta: 0, as large as or only slightly smaller than medial prensiseta; 1, much smaller than medial prensiseta (Figures 11.1E and 11.6K). Although the apomorphic state at first may appear to be trivial, it is highly consistent within each genus.

*Character 6.* Dorsal sclerite of glans: 0, with internal sculpture pattern of elongated granulation (Figure 11.7A to D, F, H, and I); 1, without internal sculpture pattern of elongated granulation (Figures 11.7G, J and 11.8A to D). State 0 is hypothesized as a synapomorphy of the Trypetini, and its loss as apomorphic within the tribe.

*Character 7.* Dorsal sclerite of glans: 0, without apical pattern of hexagonal cells; 1, apically with pattern of hexagonal cells (Figures 11.7H to J and 11.8B). The apomorphic state is a unique condition not found in any other tephritid taxa nor other tephritoid families. This pattern is internal sculpture of the cuticle, and can be best observed under a light compound microscope using slide mounting.

*Character 8.* Median sclerite of glans: 0, not defined or without granulation; 1, with internal sculpture pattern of round granulation (Figure 11.7B to G). The median sclerite can be readily observed in Trypetini in laterodorsal view of the glans. The apomorphic condition is not found in any other tephritid taxa nor other tephritoid families.

*Character 9.* Eversible membrane: 0, with numerous strong denticles or scales (Figure 11.9A, B); 1, smooth or nearly so, without strong denticles or scales (Figure 11.9C, D). In most other taxa of the subfamily Trypetinae, variously shaped denticles are found on the eversible membrane. Thus, the reduction of these teeth is considered apomorphic. Most of the genera possessing the apomorphic state have a series of tiny scales on the eversible membrane, presumably homologous to the strong ones of other tephritids.

*Character 10.* Taeniae: 0, not more than 0.7 × as long as eversible membrane (Figure 11.9A, B); 1, reaching apex of eversible membrane (Figure 11.9C, D). The apomorphy is a unique state found only within Trypetini.

*Character 11.* Ventral side of eversible membrane: 0, basally without posteriorly directed spines; 1, basally with numerous posteriorly directed spines between taeniae (Figure 11.9B). The apomorphy is a unique state found only within Trypetini.

*Character 12.* Aculeus tip: 0, dorsoventrally flattened; 1, flattened in sagittal plane with ventral serration (Figures 11.9D and 11.10I to N). The apomorphy is a unique condition found only within Trypetini. Remotely similar modification has been observed only in the African genus *Notomma*, which does not belong to the Trypetini.

*Character 13.* Aculeus tip: 0, laterally without serration; 1, laterally with serration (Figure 11.10A to F). The apomorphic condition occurs consistently in many genera of Trypetini. It is considered apomorphic because similar conditions have been less commonly observed outside of the Trypetini.

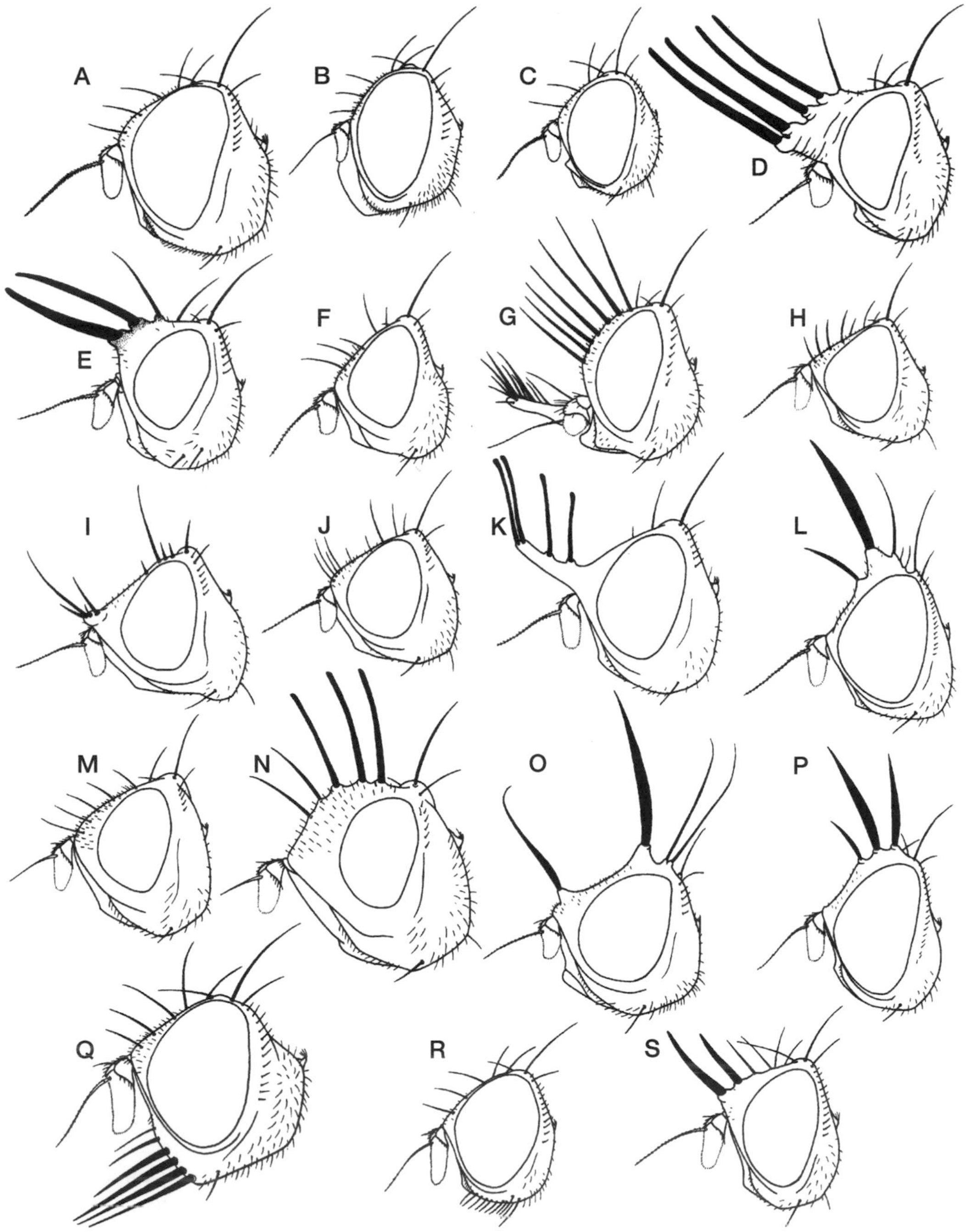


**FIGURE 11.4** Heads, lateral view (male except F and J female): (A) *Acidia cognata*; (B) *Euleia heraclei*; (C) *Alsangelisca takeuchii*; (D) *Cornutrypeta nigrifemur*; (E,F) *C. superciliata*; (G) *Aischrocrania aldrichi*; (H) *A. quadrisetata*; (I,J) *Paratrypeta appendiculata*; (K) *Stemonocera cornuta*; (L) *Vidalia bicolor*; (M) *Strauzia stoltzfusi*; (N) *S. intermedia*; (O) *Calosphenisca ensifera*; (P) *Hoplandromyia junodi*; (Q) *Chetostoma curvinerve*; (R) *Chetostoma californicum*; (S) *Paramyiolia takeuchii*.

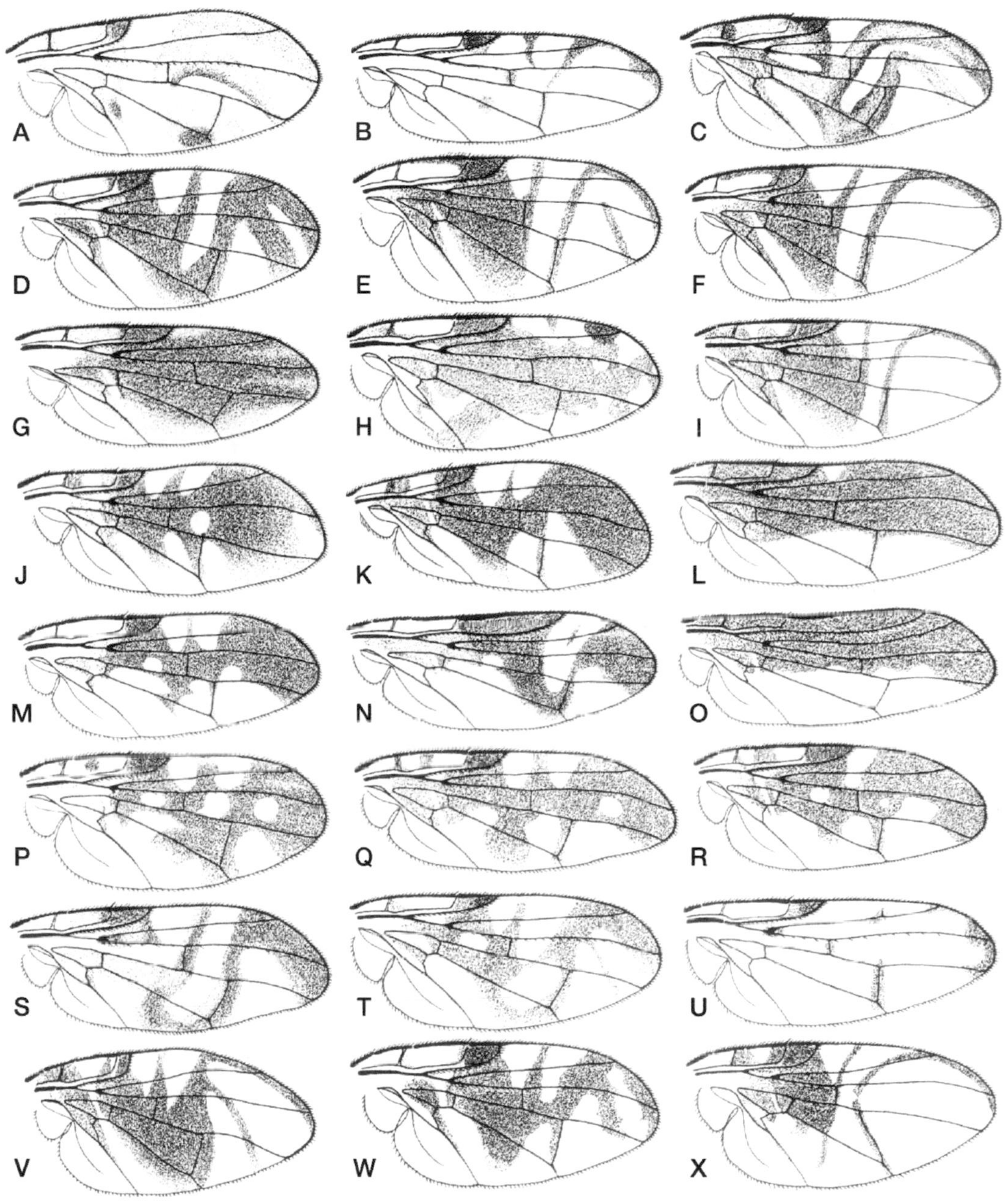


**FIGURE 11.5** Wings: (A) *Cristobalia lutea*; (B) *Itosigo bellus*; (C) *Anastrephoides matsumurai*; (D) *Philophylla caesio*; (E) *P. fossata*; (F) *P. taylori*; (G) *Magnimyiolia picea*; (H) *Chenacidiella purpureiseta*; (I) *Hoplandromyia junodi*; (J) *Aciuropsis pusio*; (K) *Euleia kovalevi*; (L) *Hemilea dimidiata*; (M) *Acidiella longipennis*; (N) *Acidiostigma longipennis*, male; (O) *A. nigritum,* male; (P) *Angelogelasinus naganoensis*; (Q) *Nemeurinus leucocelis*; (R) *Oreurinus cuspidatus*; (S) *Morinowotome egregia*; (T) *Acidia japonica*; (U) *Paratrypeta flavoscutata*; (V) *Parastenopa anastrephoides*; (W) *Montiludia nemorivaga*; (X) *Anomoia purmunda*.

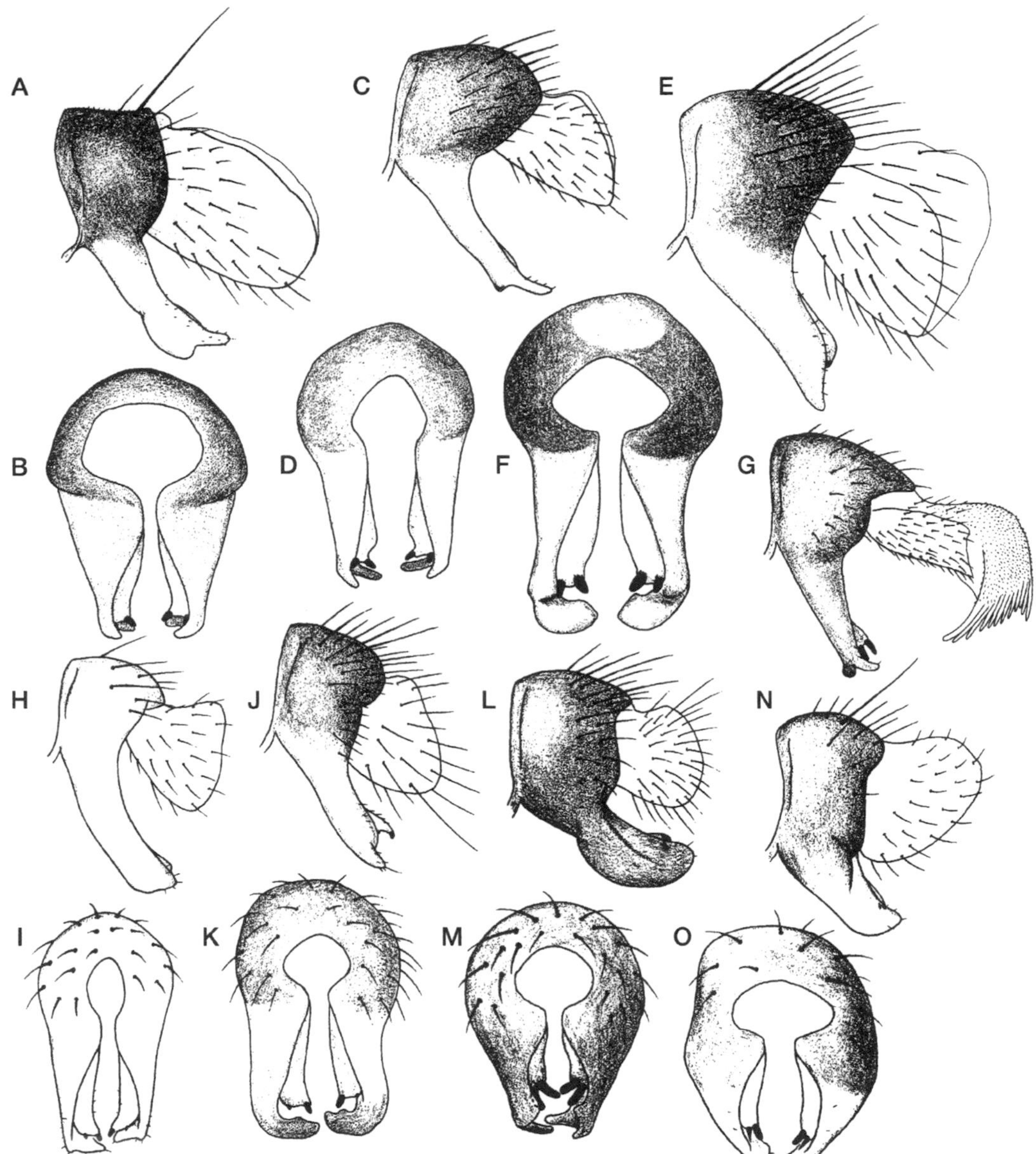


**FIGURE 11.6** Male genitalia: (A, C, E, G, H, J, L, N) epandrium, surstyli, and proctiger, lateral view; (B, D, F, I, K, M, O) epandrium and surstyli, posterior view (proctiger removed): (A,B) *Philophylla taylori*; (C,D) *Euleia heraclei*; (E,F) *Strauzia stoltzfusi*; (G) *Calosphenisca unicuneata*; (H,I) *Cornutrypeta spinifrons*; (J,K) *Acidiella longipennis*; (L,M) *Anomoia purmunda*; (N,O) *Paramyiolia takeuchii.*

### 11.4.2 Intergeneric Relationships

The cladogram (Figure 11.3) was constructed based on the 33 genera and 13 selected characters listed in Section 11.4.1 and Table 11.1. For characters 1, 7, 8, and 10 to 12, only a single forward change (ancestral to derived character state) was permitted, but with unlimited reversals allowed because they are considered as complex structures, which are unlikely to have evolved more than once but possibly could have been lost more than once. Therefore, the topology of the cladogram is heavily influenced by the characters with unique derived states. For the other characters, an

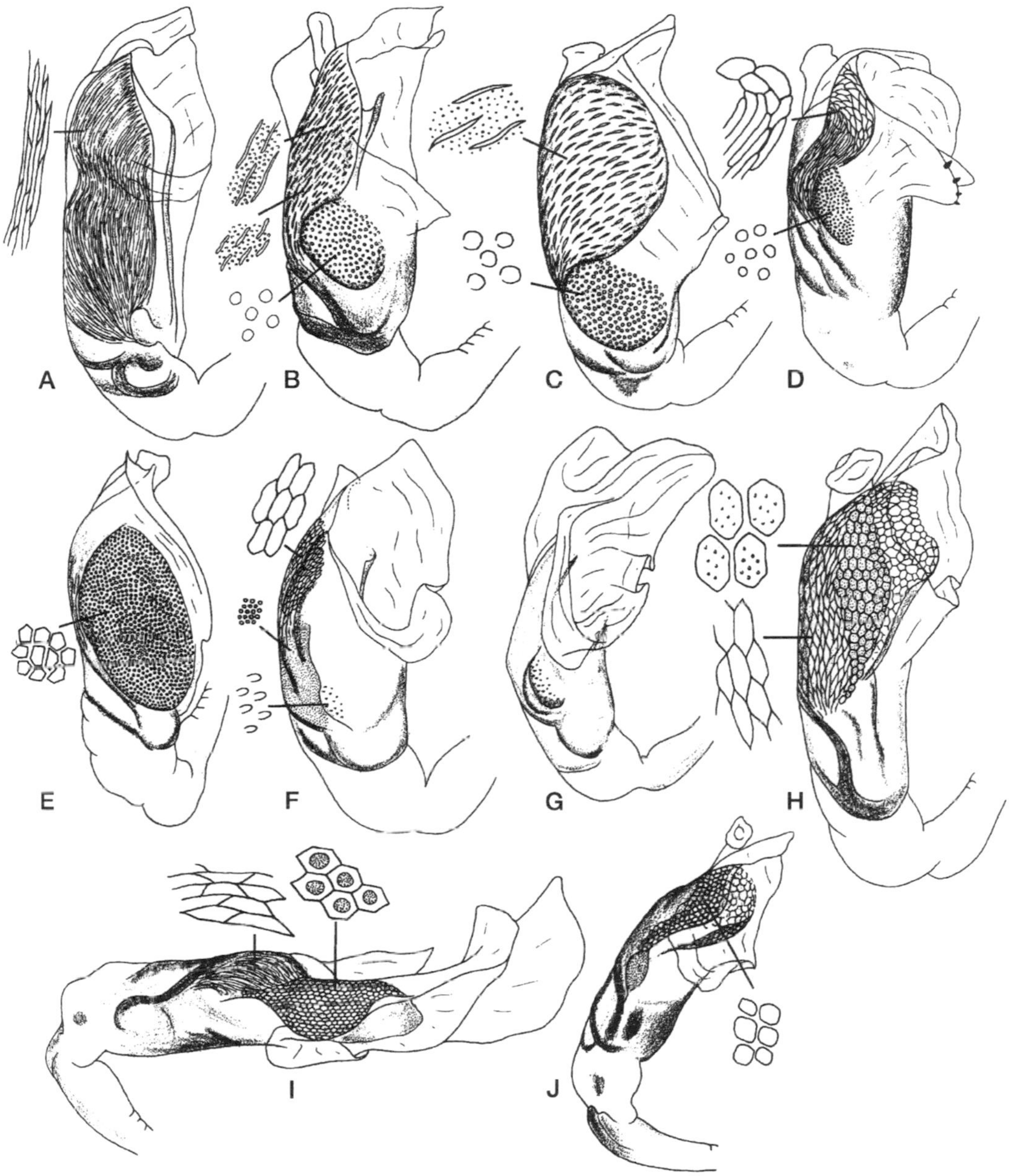


**FIGURE 11.7** Glans, dorsolateral view: (A) *Itosigo bellus*; (B) *Acidiostigma brevigaster*; (C) *Pseudhemilea accepta*; (D) *Acidiella longipennis*; (E) *Alsangelisca takeuchii*; (F) *Acidia cognata*; (G) *Hoplandromyia junodi*; (H) unnamed new genus; (I) *Montiludia nemorivaga*; (J) *Myoleja lucida.*

unlimited number of forward and backward changes were allowed as necessary. *Itosigo* and *Cristobalia,* whose membership in the Trypetini is uncertain (see Section 11.7.2), were excluded in the initial cladistic analysis, but later arbitrarily placed in their suspected positions with question marks (Figure 11.3). Because the number of characters (13) is much smaller than the number of taxa (33), this manually constructed cladogram is largely unresolved. Nevertheless, the following relationships within the Trypetini are hypothesized.

A large monophyletic clade, which is the newly recognized subtribe Trypetina, is characterized by the apomorphies of characters 8 and 13 (Figure 11.3; see also Section 11.3.1). Since the

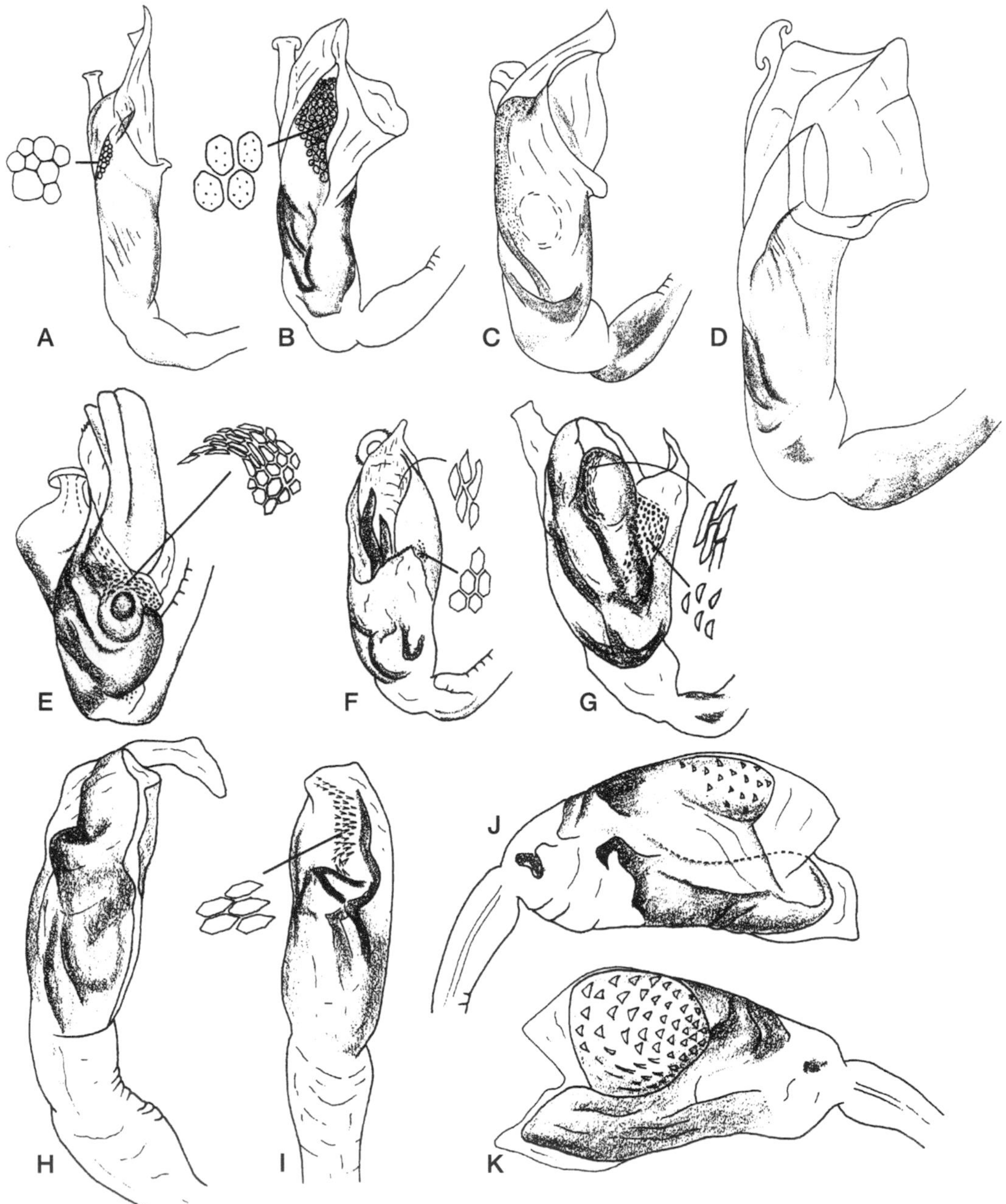


**FIGURE 11.8** Glans, dorsolateral view except as indicated: (A) *Parastenopa limata*; (B) *Anomoia purmunda*; (C) *Paramyiolia takeuchii*; (D) *Chetostoma curvinerve*; (E) *Euphranta canadensis*; (F) *Paraterellia immaculata*; (G) *Oedicarena tetanops*; (H) *Blepharoneura diva*; (I) same, dorsal view; (J) *Notomma galbanum*; (K) same, left lateral view.

apomorphy of character 8 is absent in *Itosigo*, we could treat this genus as the sister group of the rest of the Trypetina. However, I think it is also likely that this taxon secondarily lost the internal sculpture pattern of round granulation on its glans as occurred in a few species within *Acidiella* and *Vidalia* (Han et al. 1994b). A molecular study to test the taxonomic position of *Itosigo* is under way (Han, in preparation).

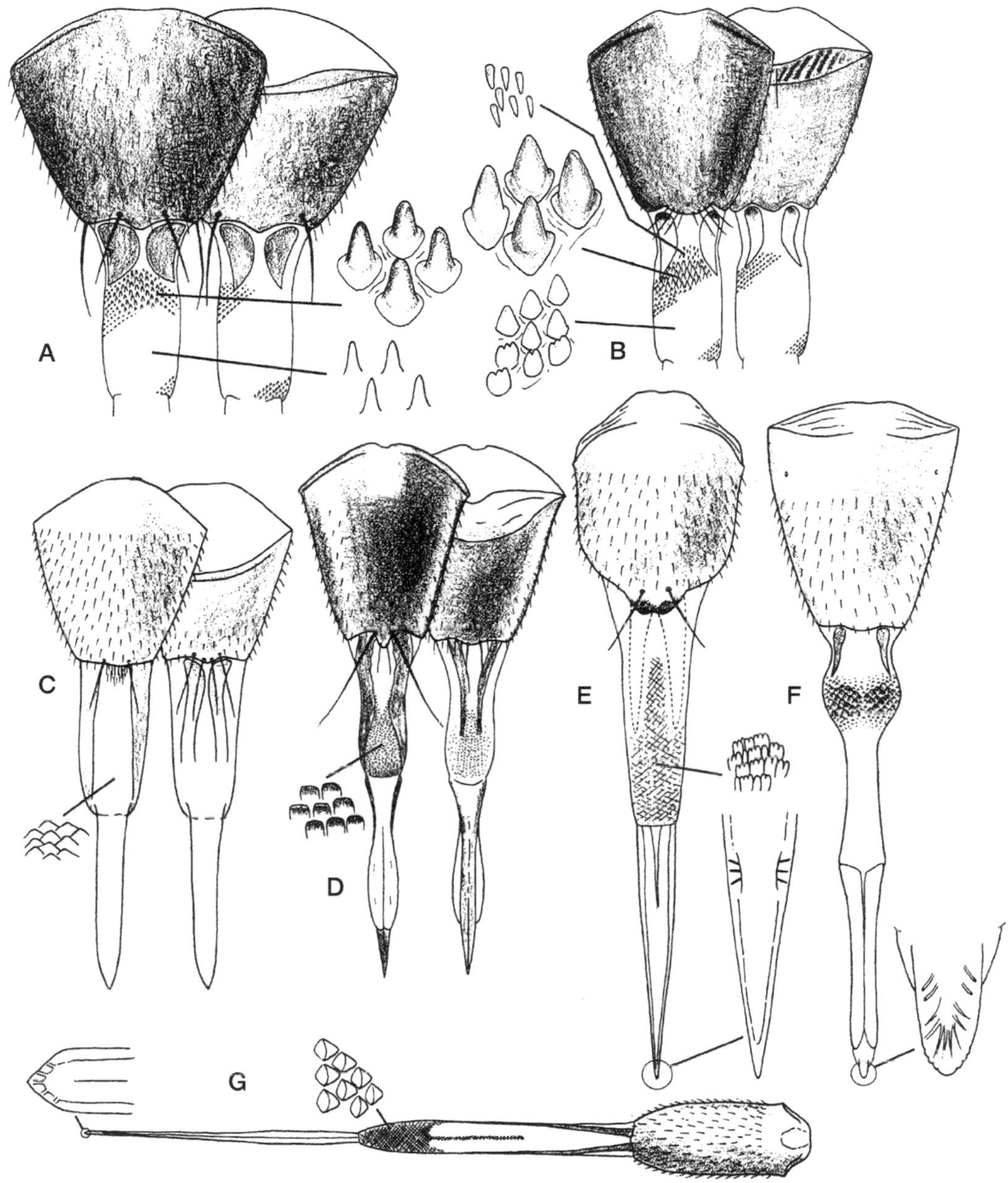


**FIGURE 11.9** Female postabdomen, ventral and dorsal views (insets at 10× main figures): (A) *Philophylla taylori*; (B) *Vidalia diffluata*; (C) unnamed new genus; (D) *Paramyiolia nigricornis*; (E) *Nitrariomyia lukjanovitshi*; (F) *Rivelliomima punctiventris*; (G) *Acidoxantha balabacensis*.

A robust monophyletic group including *Chenacidiella, Vidalia,* and *Hoplandromyia* (*Vidalia* group) is recognized based on their eversible membrane having numerous posteriorly directed spines ventrally between the taeniae (character 11). Such backward orientation of spines or scales has never been observed in other tephritids. In addition, a sister-group relationship between *Hoplandromyia* and *Vidalia* is supported by three synapomorphies, including the uniquely shaped frontal modification (Figure 11.4L, P).

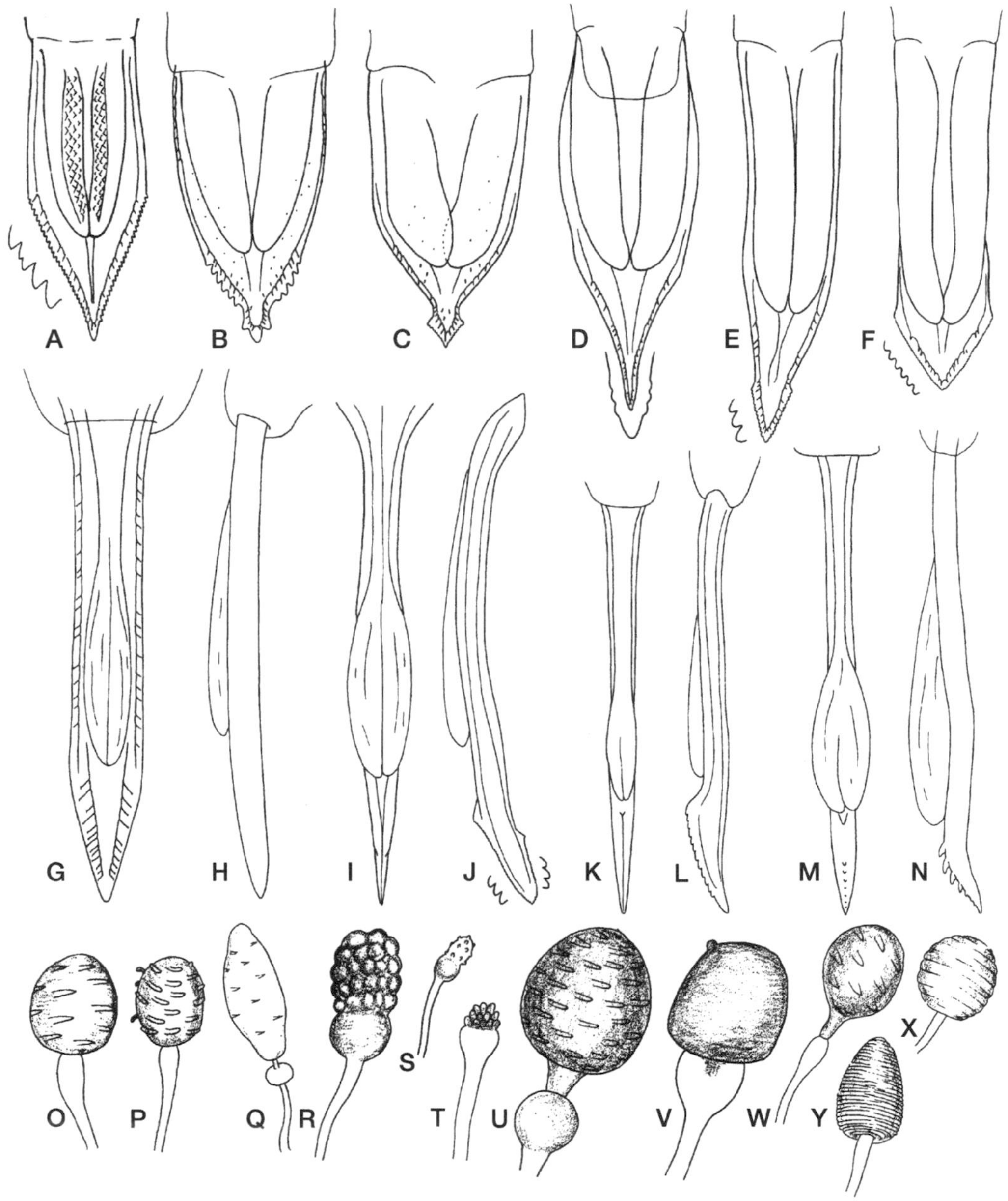


**FIGURE 11.10** Aculeus, ventral view except as indicated: (A) *Philophylla taylori*; (B) *Pseudhemilea longistigma*; (C) *P. acrotoxa*; (D) *Euleia kovalevi*; (E) *Magnimyiolia piceae*; (F) *Itosigo bellus*; (G) unnamed new genus; (H) same; lateral view; (I) *Chetostoma californicum*; (J) same, lateral view; (K) *Anomoia purmunda*; (L) same, lateral view; (M) *Myoleja lucida*; (N) same, lateral view. Spermathecae: (O) *Trypeta artemisiae*; (P) *Itosigo bellus*; (Q) *Philophylla connexa*; (R) *Euleia heraclei*; (S) *E. rotundiventris*; (T) *E. unifasciata*; (U) *Strauzia intermedia*; (V) *Acidia cognata*; (W) *Parastenopa anastrephoides*; (X) *Myoleja lucida*; (Y) *Paramyiolia takeuchii*.

The apomorphy of character 5 groups all the known leaf-mining genera except *Euleia*, *Hoplandromyia*, and *Pseudhemilea* (*Trypeta* group). As more host data become available, leaf-mining behavior may be found to be a synapomorphy for the *Trypeta* group plus a few more genera. In this case, the presence of leaf-mining behavior in *Hoplandromyia* and fruit-feeding behavior in

**TABLE 11.1**
**Character State Distribution of Characters Used in the Cladistic Analysis of the Tribe Trypetini (? = uncertain state)**

| | Character Numbers | | | | | | | | | | | | |
|---|---|---|---|---|---|---|---|---|---|---|---|---|---|
| | 1 | 2 | 3 | 4 | 5 | 6 | 7 | 8 | 9 | 10 | 11 | 12 | 13 |
| *Paramyiolia* | 0 | 0 | 0 | 0 | 0 | 1 | 0 | 0 | 1 | 1 | 0 | 1 | 0 |
| *Chetostoma* | 0 | 0 | 0 | 1/0 | 0 | 1 | 0 | 0 | 1 | 1 | 0 | 1 | 0 |
| *Myoleja* | 0 | 0 | 0 | 0 | 0 | 1 | 1 | 0 | 1 | 1 | 0 | 1 | 0 |
| *Anomoia* | 0 | 0 | 0 | 0 | 0 | 1 | 1 | 0 | 1 | 1 | 0 | 1 | 0 |
| *Montiludia* | 0 | 0 | 0 | 0 | 0 | 0 | 1 | 0 | 1 | 1 | 0 | 1 | 0 |
| New genus | 0 | 0 | 0 | 1 | 0 | 0 | 1 | 0 | 1 | 1 | 0 | 0 | 0 |
| *Parastenopa* | 0 | 0 | 1 | 1/0 | 0 | ? | ? | 0 | 1 | 1 | 0 | 0 | 0 |
| *Cristobalia* | 0 | 1 | 1 | 1 | ? | ? | ? | ? | 0 | 0 | 0 | 0 | 1 |
| *Itosigo* | 0 | 1 | 0 | 0 | 0 | 0 | 0 | 0 | 0 | 0 | 0 | 0 | 1 |
| *Philophylla* | 0 | 0 | 0 | 0 | 0 | 0 | 0 | 1 | 0 | 0 | 0 | 0 | 0/1 |
| *Calosphenisca* | 0 | 0 | 0 | 0 | 0 | 0 | 0 | 1 | 0 | 0 | 0 | 0 | 1 |
| *Anastrephoides* | 0 | 0 | 0 | 0 | 0 | 0 | 0 | 1 | 0 | 0 | 0 | 0 | 1 |
| *Magnimyiolia* | 0 | 0 | 0 | 0 | 0 | 0 | 0 | 1 | 0 | 0 | 0 | 0 | 1 |
| *Strauzia* | 0 | 1 | 0 | 0 | 0 | 0 | 0 | 1 | 0 | 0 | 0 | 0 | 1 |
| *Chenacidiella* | 0 | 0 | 0 | 0 | 0 | 0 | 0 | 1 | 0 | 0 | 1 | 0 | 1 |
| *Vidalia* | 1 | 1 | 1 | 0 | 0 | 0 | 0 | 1 | 0 | 0 | 1 | 0 | 1 |
| *Hoplandromyia* | 1 | 1 | 1 | 0 | 0 | 0 | 0 | 1 | 0 | 0 | 1 | 0 | 1 |
| *Alsangelisca* | 0 | 0 | 0 | 0 | 0 | 1 | 0 | 1 | 0 | 0 | 0 | 0 | 0 |
| *Aciuropsis* | 0 | 0 | 1 | 0 | 0 | 1 | 0 | 1 | 0 | 0 | 0 | 0 | 1 |
| *Euleia* | 0 | 0 | 0 | 0 | 0 | 1 | 0 | 1 | 0 | 0 | 0 | 0 | 1 |
| *Drosanthus* | 0 | 0 | 0 | 0 | 0 | 0 | 0 | 1 | ? | ? | 0 | 0 | ? |
| *Hemileophila* | 0 | 0 | 0 | 0 | 0 | 0 | 0 | 1 | 0 | 0 | 0 | 0 | 1 |
| *Pseudhemilea* | 0 | 0 | 0 | 0 | 0 | 0 | 0 | 1 | 0 | 0 | 0 | 0 | 1 |
| *Hemilea* | 0 | 0 | 0 | 0 | 1 | 0 | 0 | 1 | 0 | 0 | 0 | 0 | 1 |
| *Acidiostigma* | 0 | 0 | 0 | 0 | 1 | 0 | 0 | 1 | 0 | 0 | 0 | 0 | 1 |
| *Acidiella* | 0 | 0 | 0 | 0 | 1 | 0 | 0 | 1 | 0 | 0 | 0 | 0 | 1 |
| *Angelogelasinus* | 0 | 0 | 0 | 0 | 1 | 0 | 0 | 1 | 0 | 0 | 0 | 0 | 1 |
| *Morinowotome* | 0 | 0 | 0 | 0 | 1 | 0 | 0 | 1 | 0 | 0 | 0 | 0 | 1 |
| *Nemeurinus* | 0 | 0 | 1 | 0 | 1 | 0 | 0 | 1 | 0 | 0 | 0 | 0 | 1 |
| *Oreurinus* | 0 | 0 | 0 | 0 | 1 | 0 | 0 | 1 | 0 | 0 | 0 | 0 | 1 |
| *Trypeta* | 0 | 0 | 0 | 0/1 | 1 | 0 | 0 | 1 | 0 | 0 | 0 | 0 | 1 |
| *Acidia* | 0 | 0 | 0 | 1 | 1 | 0 | 0 | 1 | 0 | 0 | 0 | 0 | 1 |
| *Aischrocrania* | 0 | 0 | 0 | 0/1 | 1 | 0 | 0 | 1 | 0 | 0 | 0 | 0 | 1 |
| *Cornutrypeta* | 0 | 0 | 0 | 1 | 1 | 0 | 0 | 1 | 0 | 0 | 0 | 0 | 1 |
| *Paratrypeta* | 0 | 0 | 1 | 1 | 1 | 0 | 0 | 1 | 0 | 0 | 0 | 0 | 1 |
| *Stemonocera* | 0 | 1 | 1 | 0/1 | 1 | 0 | 0 | 1 | 0 | 0 | 0 | 0 | 1 |

its sister genus *Vidalia* has to be addressed, because either one of these behaviors may be a result of reversal or parallelism. This problem may be resolved when the host mode for *Chenacidiella* is discovered.

In addition to the above characters, the puparia of *Acidia cognata* (Wiedemann), *Trypeta artemisiae, T. concolor* (Wulp), *T. flaveola* Coquillett, *T. immaculata* (Macquart), *T. zoe* Meigen, and *Stemonocera cornuta* (Scopoli) have the middle opening of the hind spiracle projecting spinelike laterally (Figure 11.2H to J). This condition, so far as is known, is not found in any other Tephritidae nor in other tephritoid families. As the puparia of more species are known, the spinelike spiracle may turn out to be another synapomorphy of the *Trypeta* group.

The apomorphies of characters 9 and 10 unequivocally characterize a monophyletic group including seven genera, the newly recognized subtribe Chetostomatina (Figure 11.3; see also Section 11.3.2). Within this group, the monophyly of all the genera except *Parastenopa* is supported by the hexagonal pattern on the dorsal sclerite of their glans (character 7), with reversal hypothesized in *Paramyiolia* and *Chetostoma* (Figure 11.8C, D). The unusual laterally flattened and ventrally serrate aculeus tip (character 12) further characterizes *Montiludia, Anomoia, Chetostoma, Myoleja,* and *Paramyiolia* as a robust monophyletic group (*Chetostoma* group). Within the *Chetostoma* group, the apomorphy of character 6 groups four genera and places *Montiludia* at the most basal position. However, this topology should be interpreted with caution, because this character is variable even within some genera. A sister-group relationship between *Chetostoma* and *Paramyiolia* may be hypothesized based on the loss of hexagonal pattern of the glans (reversal of character 7), but it is also possible that they lost the pattern independently. Alternatively, similar spermathecal structure in *Paramyiolia* and some species of *Myoleja* may indicate their close relationship (Figure 11.10X, Y; see also Han 1996b). Since *Myoleja,* for now, includes the species lacking the derived characters of the other related genera, examination of more *Myoleja* species may further enhance the phylogenetic resolution.

One interesting outcome of the above phylogenetic approach is the recognition that certain characters, such as wing patterns and secondary sexual structures, are homoplastic in the Trypetini. For example, sexually dimorphic head structures (enlarged frontal setae, subvibrissal setae, or frontal plate) appear to have evolved at least nine times within the Trypetini (Figure 11.3). Similar selection pressure probably led to the development of these structures. So far as is known, trypetine males combat each other and often butt heads whether they have frontal horns or not (see Section 11.6).

## 11.5 DISTRIBUTION

The Trypetina largely have a Palearctic–Oriental distribution pattern. In the Western Hemisphere, *Euleia, Strauzia,* and *Trypeta* species occur mainly in North America, but a few *Trypeta* species extend their distribution to Mexico or Costa Rica (Han and Norrbom, in preparation). *Hoplandromyia,* the sister group of Palearctic–Oriental *Vidalia,* is largely endemic to the Afrotropical Region. The large genus *Philophylla* also extends its range to the Afrotropical Region. In the Australasian Region, most taxa occur in the Austro-Malayan Subregion, further indicating the strong Oriental center of diversity of the trypetine fauna. The number of Palearctic species is higher than Oriental species (Norrbom et al. 1999a). However, I believe that the Oriental trypetine fauna is much richer, because the number of undescribed species based on small numbers of specimens that I have seen suggests that many more Oriental species are still waiting to be discovered.

The Chetostomatina also occur mostly in the Palearctic and Oriental Regions with no Afrotropical representation. *Anomoia* and *Myoleja* extend their distributions to the Australasian Region (Austro-Malayan and Polynesian Subregions), but no species actually occurs on the Australian continent. In the Western Hemisphere, a small number of *Chetostoma* and *Paramyiolia* species are found in North America, but both genera probably originated in the Old World. Among the four known *Paramyiolia* species, two occur in Japan and the other two in the eastern United States. This disjunct pattern appears to be a typical Arcto-Tertiary relic distribution often found in higher plants (Cox and Moore 1980). *Paramyiolia* probably originated in the Tertiary and was once widely distributed in the Holarctic Region, and then might have only survived in the moist temperate forests in Japan and the Appalachians during the ice ages of the Quaternary.

*Parastenopa* is the only chetostomatine genus primarily from the Neotropical Region. It is interesting to note that this genus appears to be the basalmost lineage of the Chetostomatina (Figure 11.3). Further zoogeographic interpretation regarding this matter requires finer resolution of the basal tephritid phylogeny.

**TABLE 11.2**
**Host Plant Families and the Number of Trypetine Species Associated**

| | Chetostomatina | | Trypetina | |
|---|---|---|---|---|
| **Host Plant Family** | **Genus** | **No. Spp.** | **Genus** | **No. Spp.** |
| Apiaceae | | | *Euleia* | 3 lm |
| Aquifoliaceae | *Parastenopa* | 1 f, 1 gc? | | |
| | *Anomoia* | 1 sob | | |
| | | 1 f | | |
| Araliaceae | | | *Acidiella* | 4 f |
| | | | *Euleia* | 1 lm |
| | | | *Pseudhemilea* | 3 lm |
| | | | *Vidalia* | 2 f |
| | | | *? rohdendorfi* | 1 f |
| Asteraceae | | | *Acidia* | 1 lm |
| | | | *Cornutrypeta* | 1 lm |
| | | | *Euleia* | 2 lm |
| | | | *Hemilea* | 1 lm |
| | | | *Stemonocera* | 2 lm |
| | | | *Strauzia* | 10 sb |
| | | | *Trypeta* | 5 lm |
| Berberidaceae | *Anomoia* | 1 f | | |
| Caricaceae | | | *Philophylla* | 1 f |
| Caprifoliaceae | *Chetostoma* | 1 f | | |
| | *Myoleja* | 2 f | | |
| Lardizabalaceae | | | *Acidiella* | 1 lm |
| Melastomataceae | *Parastenopa* | 1 gc? | | |
| Polygonaceae | | | *Euleia* | 1 lm |
| Rosaceae | *Anomoia* | 1 f | | |
| Rubiaceae | *Anomoia* | 1 f | *Hoplandromyia* | 1 lm |
| Urticaceae | | | *Philophylla* | 1 lpm |
| Verbenaceae | | | *Acidiostigma* | 1 lm |
| | | | *Philophylla* | 5 f, 1 sd |

Abbreviations: f = in fruit; gc = gall commensal; lm = leaf miner; sb = stem borer; sd = in seed; sob = shoot borer.

## 11.6 BEHAVIOR

Host plants are known for only 52 species in 17 genera of Trypetini (Table 11.2), representing only one-sixth of the species and half of the genera of the tribe. More-detailed host plant data and source references are listed in Han (1998). The larvae of 22 species are known to feed on the plant family Asteraceae, but the others attack members of 13 additional plant families. It is still premature to discuss the evolution of their host associations, but their host data certainly are of great predictive value to understand their relationships.

Several different larval feeding strategies have been observed in the subtribe Trypetina. Larvae of five *Philophylla* species develop in the fruits of Verbenaceae and one species on Caricaceae (Malloch 1939; Hardy and Adachi 1956; Hardy 1973; 1987; Hancock and Drew 1994b). However, *P. caesio* (Harris) mines the leaf petiole of *Urtica dioica* (Urticaceae) (Beiger 1968) and *P. superflucta* (Enderlein) feeds on the seeds of *Clerodendrum inerme* (Verbenaceae) (Hardy 1987).

The larvae of *Vidalia bidens* Hendel and *V. thailandica* Hancock and Drew feed on the fruits of *Schefflera* spp. (Araliaceae) (Han et al. 1994b; Hancock and Drew 1994b). Species of the Nearctic genus *Strauzia* are the only known stem borers in the Trypetini, and their host plants are restricted to Asteraceae, especially various *Helianthus* species (Stoltzfus 1988).

All the known leaf-mining tephritids belong to the Trypetina, and they are largely associated with Asteraceae (Table 11.2). The species of *Euleia,* however, are known to mine the leaves of Apiaceae, and *Pseudhemilea* mine the leaves of Araliaceae. There are also four isolated records of additional plant families, each of which involves a single known leaf-mining species: Lardizabalaceae by *Acidiella kagoshimensis* (Miyake), Rubiaceae by *Hoplandromyia madagascarensis* Hancock, Polygonaceae by *Euleia heraclei* (Linnaeus), and Verbenaceae by *Acidiostigma polyfasciatum* (Frost 1924; Ito 1984; Hancock 1985).

The larvae of the Chetostomatina, so far as is known, develop in fruits, except for some species of the American genus *Parastenopa. Parastenopa* species show an interesting divergence of host-feeding strategies almost exclusively associated with the plant genus *Ilex* (Aquifoliaceae). *Parastenopa limata,* which is the only representative of the genus in North America, breeds in fruits of various *Ilex* species (Benjamin 1934; Phillips 1946). *Parastenopa ogloblini* mines young shoots of *I. paraguariensis,* and *P. elegans* (Blanchard) bores psyllid leaf galls on the same plant (Blanchard 1929). A fourth species, *P. marcetiae* Bezzi and Tavares, was reared from oval galls on *Marcetia* sp. (Melastomataceae), probably made by a cecidomyiid (Bezzi and Tavares 1916).

The behavior and life history of only a few species have been studied in some detail, so it is not easy to make generalizations. The following discussion is based mainly on studies of the relatively better known species, such as *Euleia heraclei, E. fratria* (Loew), *Myoleja lucida* (Fallén), *Trypeta flaveola* Coquillett, and *Strauzia* spp.

Two distinct patterns of adult behavior associated with their host plants are recognized. Both sexes of *Strauzia* species are usually found closely associated with their host plants (Stoltzfus 1988). Both feeding and mating occur on the host plant, where males usually patrol a leaf or rest on the undersurface. Similar behavior was observed also in *Trypeta concolor* and *Trypeta,* n. sp., in Mexico (Han and Norrbom, in preparation). In contrast, adults of *E. heraclei* and *T. flaveola* Coquillett (as *T. angustigena* Foote) have been seldom seen on their host plants despite their abundance (Frick 1971; Leroi 1977). Instead, feeding and mating sites for *E. heraclei* are in trees near their hosts (Leroi 1977). Only females visit the host plant and come back to the shelter tree immediately after ovipositing. Factors contributing to such behavior are not understood, but it may involve availability of adult food, such as honeydew or bird droppings, essential for survival and reproduction. In Tephritidae, olfactory attractants are also known to initiate, control, or modify various searching/foraging behaviors (Jang and Light 1996). However, these attractants have never been studied in depth for Trypetini, with the exception of a record of *Philophylla fossata* (Fabricius) responding to methyl eugenol in Malaysia and Thailand (Hancock and Drew 1994b).

In Korea, I have observed a large number of mating populations of *Itosigo bellus* Ito on the flowers of *Heracleum moellendorffi* Hance in August, and similarly many mating adults of *Trypeta trifasciata* Shiraki on the leaves of various trees on the top of Mt. Baegun (1087 m, Kangwon-do). Apparent mating leks were also discovered for several Korean trypetine species, but without actually observing copulation or even females. These species are *Oreurinus cuspidatus* Ito and *Hemileophila sibirica* (Portschinsky) on rocks along streams in deep woods, and *Acidiella issikii* (Shiraki), *A. circumvaga* (Ito), and *Acidiostigma s-nigrum* (Matsumura) on specific parts of trees (usually undersides of leaves) near the peak of Mt. Baegun. The aggregations of the latter three species consist of highly active males taking up territories on leaves, which they aggressively defend from incursions by other males. This behavior was observed in the afternoon, but I was not able to observe copulation or even the presence of females. In Dacinae, mating has been observed frequently at dusk (Fletcher 1987), which was normally the time I had to climb down the mountain.

Apparent territorial behavior of male *Strauzia* spp. has been documented. Intruders are chased away by wing movements or short forward thrusts of the body (Stoltzfus 1988). In the male-to-male encounter of *S. vittigera* (Loew), they would approach each other and soon one would dash at the other and crawl over him. Frequently, they use their sexually dimorphic hornlike frontal setae to butt against each other (Steyskal 1945). Such aggressive male butting behavior (with or without hornlike setae) has been consistently observed in all the known cases for the Trypetini: *Stemonocera mica* (Richter and Kandybina), *Acidiella issikii, A. circumvaga,* and *Acidiostigma s-nigrum* (Han, personal observation), *Chetostoma curvinerve* Rondani (Freidberg, personal communication), *Euleia fratria* (Tauber and Toschi 1964), and *Myoleja lucida* (Hoffmeister 1992). It is interesting to note that hornlike structures in the male have evolved at least nine times within the Trypetini according to my phylogenetic analysis (Figure 11.3). They are most frequently enlargements of frontal setae and often the associated part of the frons, but the subvibrissal setae in *Chetostoma* (Figure 11.4Q) and the second antennal segment in *Aischirocrania* (Figure 11.4G) are also modified to form such structures. I believe that this male butting behavior led to these remarkable parallel evolutions of sexual dimorphisms in many taxa within the Trypetini. In the past, many species with different types of enlarged frontal setae were placed in the genus *Vidalia, sensu lato* (Han et al. 1994b).

*Euleia fratria* displays an elaborate and persistent courtship behavior that lasts up to 19 min (Tauber and Toschi 1964). Stoltzfus (1988) also observed courtship of seven *Strauzia* species. Their courtship is short and may last only 2 to 3 s if the female is receptive. Mating averages 30 h in *E. heraclei* (Gardner 1921) and lasts up to 7 h in *Strauzia* species (Stoltzfus 1988).

The preoviposition period averages 7 days in *T. flaveola* and 60 h in *E. fratria* (Tauber and Toschi 1964; Frick 1971). Eggs are inserted under the epidermis and almost always laid on the lower surface of a horizontal leaf in both species. Stem-mining *Strauzia* species insert the eggs into the parenchyma of an internode sufficiently deeply for the egg to be completely protected (Stoltzfus 1988).

In leaf-mining species such as *E. heraclei, E. fratria,* and *T. concolor,* more than two larvae often develop in a single leaf or stem mine (Leroi 1974; Han and Norrbom, in preparation). In *Euleia* it is common to see second and third instar larvae leave an old mine and reenter a fresh portion of the leaf or a new leaf (Leroi 1974). The exit and entrance cuts are confined mainly to the dorsal surface. Most species pupate in soil, but some *Strauzia* species pupate in the stem or underground rhizome of the host plant (Stoltzfus 1988).

## 11.7 GENERA OF THE TRIBE TRYPETINI

### 11.7.1 Key to the Genera of Trypetini

1. Two pairs of frontal setae (Figure 11.4C) .......... *Alsangelisca* Ito
- Three or more pairs of frontal setae .......... 2
2. One pair of scutellar setae .......... 3
- Two pairs of scutellar setae .......... 4
3. Dorsocentral seta at same level as postsutural supra-alar seta .......... *Aciuropsis* Hardy
- Dorsocentral seta anterior to level of postsutural supra-alar seta .......... *Euleia rotundiventris* (Fallén)
4. Pterostigma very short or shorter than wide (Figure 11.5V) .......... *Parastenopa* Hendel [in part]
- Pterostigma longer than wide .......... 5
5. Katepisternal seta absent; dorsocentral seta anterior to level of intra-alar seta .......... *Acidia* Robineau-Desvoidy
- Single katepisternal seta present; dorsocentral seta posterior to level of postsutural supra-alar seta .......... 6
6. Lower margin of gena with several well-developed setae anterior to genal setae in both sexes but much more so in males (Figure 11.4Q, R) .......... *Chetostoma* Rondani
- Lower margin of gena without strong setae anterior to genal setae .......... 7

7. Anepisternum whitish on at least two-thirds of total area, contrasting well with darker adjacent area . . . . . . . . . . . . . . . . . . . . . . *Calosphenisca* Hendel
- Anepisternum without such large whitish area . . . . . . . . . . . . . . . . . . . . . . . . . . . . . . . . . . . . . . . 8
8. One long frontal seta situated near orbital setae and rest of frontal setae near antenna (Figure 11.4I, J); wing with vein $R_{2+3}$ anteriorly with short spurious vein (Figure 11.5U) . . . . . . . . . . . . . . . . . . . . *Paratrypeta* Han and Wang
- Frontal setae evenly spaced at least in female (as in Figure 11.4A, B); wing with vein $R_{2+3}$ without spurious vein . . . . . . . . . . . . . . . . . . . . . . . . . . . . . . . . . . . . . . . . . . 9
9. Single pair of orbital setae . . . . . . . . . . . . . . . . . . . . . . . . . . . . . . . . . . . . . . . . . . . . . . . . . . . . . . 10
- Two pairs of orbital setae . . . . . . . . . . . . . . . . . . . . . . . . . . . . . . . . . . . . . . . . . . . . . . . . . . . . . . 16
10. Lateral vertical seta shorter than one-third as long as medial vertical seta; wing with vein $R_{4+5}$ densely setulose beyond crossvein R-M (Figure 11.5A). . . . . . . . . . . . . . . . . . . . . . . . . . . . . . . . . . . . . . . . . . . . . *Cristobalia* Malloch
- Lateral vertical setae at least half as long as medial vertical seta; vein $R_{4+5}$ bare or nearly so beyond crossvein R-M. . . . . . . . . . . . . . . . . . . . . . . . . . . . . . . . 11
11. Ocellar seta about twice as long as ocellar triangle; males with or without sexually dimorphic frontal process. . . . . . . . . . . . . . . . . . . . . . . . . . . . . . . . . . . . . . 12
- Ocellar seta reduced and hairlike, shorter than 1.5 × as long as ocellar triangle; male always with sexually dimorphic frontal process . . . . . . . . . . . . . . . . . . . . . . . 14
12. Tarsomere 5 dark brown, contrasting well with yellow brown preceding tarsomeres; male with frontal setae enlarged (similar to Figure 11.4E) . . . . . . . . . . . . . . . . . . . . . *Cornutrypeta omeishana* Han and Wang
- Tarsomere 5 yellow brown, concolorous with preceding tarsomeres; male without enlarged frontal setae. . . . . . . . . . . . . . . . . . . . . . . . . . . . . . . . . . . . . *Itosigo* Ito
13. Anepisternum with single strong seta; additional seta, if present, less than half as long as upper seta; male frontal area produced to form antlerlike frontal horn with enlarged and blunt frontal setae (Figure 11.4K). . . . . . . . . . . . . . . . . . . . . . . . . . . . . . . . . . . . . . . . . . . . *Stemonocera* Rondani
- Anepisternum with two strong setae; if male frontal setae enlarged, then not on frontal horn (Figure 11.4L to N, P) . . . . . . . . . . . . . . . . . . . . . . . . . . . . . . . . . 14
14. First flagellomere not more than half length of face (Figure 11.4M, N). . . . . . . . . . . . . . . . . . . . . . . . . . . . . . . . . . *Strauzia* Robineau-Desvoidy
- First flagellomere more than half length of face (Figure 11.4L, P) . . . . . . . . . . . . . . . . . . . 15
15. Wing with narrow unbroken C-shaped band from DM-Cu to anterior margin, then to wing apex to slightly after $R_{4+5}$ (Figure 11.5I) . . . . . . . . . . . . . *Hoplandromyia* Bezzi
- Wing without distinct C-shaped band . . . . . . . . . . . . . . . . . . . . . . *Vidalia* Robineau-Desvoidy
16. Pterostigma at least 0.9× as long as cell c, often greatly elongated in males (Figure 11.5N, O); if pterostigma not as above, apical and subapical bands form distinct S-shaped band (*A. s-nigrum, A. yoshinoi*) . . . . . . . . . . . . . *Acidiostigma* Hendel
- Pterostigma at most 0.8× as long as cell c, without sexual dimorphism; without such S-shaped wing band . . . . . . . . . . . . . . . . . . . . . . . . . . . . . . . . . . . . . . . . . . . 17
17. Crossvein DM-Cu oblique with cell dm apically pointed (Figure 11.5X) . . . . . . . . . . . . . . . 18
- Crossvein DM-Cu not oblique . . . . . . . . . . . . . . . . . . . . . . . . . . . . . . . . . . . . . . . . . . . . . . . . . . 19
18. Wing with anal lobe largely hyaline (Figure 11.5X); eversible membrane not twisted; aculeus narrow, apex flattened laterally and ventrally serrate (Figure 11.10K, L) . . . . . . . . . . . . . . . . . . . . . . . . . . . . . . . . . . . . . . . . *Anomoia* Walker [in part]
- Wing with anal lobe largely dark brown; eversible membrane twisted; aculeus broad, dorsoventrally flattened, asymmetrical. . . . . . . . . . . . . . . . . . . . . . . . . . . . . . . . . . . . . . . . . . *Philophylla kraussi* (Hardy)

19. Head with at least five pairs of frontal setae of similar size (Figure 11.4G, H); pedicel usually elongated in males and strongly setulose in both sexes (Figure 11.4G) . . . . . . . . . . *Aischrocrania* Hendel
- Head with less than five pairs of frontal setae, or if five pairs, anterior four in male greatly enlarged (Figure 11.4D); male pedicel normal . . . . . . . . . . 20
20. Wing with narrow unbroken C- or inverted L-shaped band from DM-Cu to anterior margin and then to wing apex, ending before vein M (Figure 11.5F) . . . . . . . . . 21
- Wing without such C- or inverted L-shaped band. . . . . . . . . . 23
21. Dorsal sclerite of male glans with internal sculpture pattern of hexagonal cells (Figure 11.7H); female eversible membrane without strong teeth (Figure 11.9C, D); oviscape without lateral marginal setae . . . . . . . . . 22
- Dorsal sclerite of male glans without such pattern; female eversible membrane with strong teeth; oviscape with a pair of strong lateral marginal setae (Figure 11.9A) . . . . . . . . . . *Philophylla* Rondani [in part]
22. Aculeus with apex flattened in sagittal plane (Figure 11.10K, L) . . . . . . . . . . *Anomoia* Walker [in part]
- Aculeus with apex flattened dorsoventrally (Figure 11.10G, H). . . . . . . . Unnamed new genus
23. Wing apically with brown to dark brown F-shaped band (Figure 11.5D, E) . . . . . . . . . . *Philophylla* Rondani [in part]
- Wing apically without F-shaped band . . . . . . . . . . 24
24. Wing apically with two inverted L shaped bands (Figure 11.5C). . . . . . . . . . *Anastrephoides* Hendel
- Wing apically without two inverted L-shaped bands . . . . . . . . . . 25
25. Wing predominantly brown to dark brown anterior to vein M (Figure 11.5G, L). . . . . . . . . 26
- Wing with at least three hyaline markings anterior to vein M . . . . . . . . . . 30
26. Wing with anterior dark pattern broadly diffused with posterior hyaline area on cells $cu_1$ and m (Figure 11.5G). . . . . . . . . . *Magnimyiolia picea* Hering
- Wing with anterior dark pattern clearly separated from posterior hyaline area (Figure 11.5L) . . . . . . . . . . 27
27. Genal seta reduced . . . . . . . . . . *Pseudhemilea* Chen [in part]
- Genal seta clearly distinguished from nearby setulae . . . . . . . . . . 28
28. Supracervical setae black; usually four pairs of frontal setae. . . . . . . . . . *Hemileophila* Hering
- Supracervical setae pale yellow to yellow brown; three pairs of frontal setae . . . . . . . . . . 29
29. Dorsocentral seta closer to level of intra-alar seta than to level of postsutural supra-alar seta . . . . . . . . . . *Drosanthus* Hering
- Dorsocentral seta closer to level of postsutural supra-alar seta than to level of intra-alar seta. . . . . . . . . . *Hemilea* Loew
30. Cell $r_1$ largely dark brown, with single medial hyaline spot (Figure 11.5R). . . . . . . . . . *Oreurinus* Ito
- Cell $r_1$ with two hyaline spots or predominantly hyaline . . . . . . . . . . 31
31. Posterior portion of scutum and entire scutellum ivory white . . . . . . . . . . *Nemeurinus* Ito
- Scutum without such ivory white area. . . . . . . . . . 32
32. Apical scutellar seta at most four-fifths as long as basal scutellar seta; male with frontal setae greatly enlarged, with anterior one largest (Figure 11.4S); crossvein R-M at apical one-third to one-fourth of cell dm . . . . . . . . . . *Paramyiolia* Shiraki
- Apical scutellar seta slightly shorter than basal scutellar seta; if male with such enlarged frontal setae (in *Cornutrypeta*), then crossvein R-M close to middle of cell dm (at least not at apical one-third to one-fourth of cell dm; as in Figure 11.5U) . . . . . . . . . . 33

33. Vein $R_{4+5}$ dorsally with more than 15 setulae proximal to crossvein R-M (Figure 11.5G, W). . . . . 34
- Vein $R_{4+5}$ dorsally with less than 15 setulae proximal to crossvein R-M. . . . . 36

34. Dorsocentral seta about midway between levels of postsutural supra-alar seta and intra-alar seta . . . . . *Montiludia* Ito
- Dorsocentral seta at or slightly posterior to level of postsutural supra-alar seta . . . . . 35

35. Apical half of cell $r_1$ brown to dark brown . . . . . *Magnimyiolia* Shiraki [in part]
- Apical half of cell $r_1$ at least with some hyaline area (Figure 11.5H). . . . . *Chenacidiella* Shiraki

36. Arista almost bare (Figure 11.4B) . . . . . 37
- Arista short pubescent (as in Figure 11.4A) . . . . . 38

37. Ocellar seta reduced and hairlike, shorter than 1.5× as long as ocellar triangle . . . . . *Parastenopa* Hendel [in part]
- Ocellar setae at least twice as long as ocellar triangle (Figure 11.4B). . . . . *Euleia* Walker [in part]

38. Vein $R_{4+5}$ dorsally with at least ten setulae beyond crossvein R-M (Figure 11.5P) . . . . . 39
- Vein $R_{4+5}$ almost bare beyond crossvein R-M. . . . . 40

39. Mediotergite dark brown except narrow mesal longitudinal stripe. . . . . *Angelogelasinus* Ito
- Mediotergite yellow brown. . . . . *Myoleja boninensis* (Ito)

40. Anepisternum with single strong seta; additional seta, if present, less than half as long as upper seta . . . . . 41
- Anepisternum always with two setae, lower seta more than half as long as upper seta. . . . . 43

41. Crossvein R-M at apical two-thirds to three-fourths of cell dm (Figure 11.5S). . . . . *Morinowotome* Ito
- Crossvein R-M usually near middle of cell dm (Figure 11.1A). . . . . 42

42. Male with frontal setae greatly enlarged, with anterior one largest (Figure 11.4D, E, and S) . . . . . *Cornutrypeta* Han and Wang [in part]
- Male without enlarged frontal setae. . . . . *Trypeta* Meigen

43. Dorsal sclerite of glans with pattern of hexagonal cells (Figure 11.7J); aculeus slender, with apex flattened in sagittal plane (Figure 11.10M, N). . . . . *Myoleja* Rondani [in part]
- Dorsal sclerite of glans without such pattern; aculeus broad and dorsoventrally flattened . . . . . 44

44. Lateral surstylus with posterior lobe reduced and anterior lobe expanded inward (Figure 11.6K); lateral prensiseta much smaller than medial prensiseta; wing pattern variable but most species with patterns similar to Figure 11.5M, with two hyaline markings in cells $r_1$ and $r_{2+3}$, of which apical marking sometimes extends all the way to cell dm. . . . . *Acidiella* Hendel
- Lateral surstylus with both anterior and posterior lobes slightly elongated (as in Figure 11.6D); lateral prensiseta as long as or slightly shorter than medial prensiseta; wing pattern variable but many species with predominantly dark pattern (as in Figure 11.5L) . . . . . *Pseudhemilea* Chen [in part]

## 11.7.2 Genera of the Subtribe Trypetina

### 11.7.2.1 *Acidia* Robineau-Desvoidy

Distribution: Europe and Japan.
Biology: *Acidia cognata* (Wiedemann) is a leaf miner of various Asteraceae (Han 1998).

The generic name *Acidia* has been widely used for many species of similar appearance, but I have restricted this genus to *A. cognata* and *A. japonica* Shiraki (Han 1992). They can be readily distinguished from other Trypetini by their wing pattern with two hyaline incisions separated by a narrow yellowish brown band in posterior distal quarter of wing (Figure 11.5T), and the following synapomorphies: (1) katepisternal seta absent; (2) dorsocentral seta slightly anterior to the level of intra-alar seta; and (3) granulation of median sclerite of glans reduced or absent (Figure 11.7F).

### 11.7.2.2 *Acidiella* Hendel

Distribution: Eastern Palearctic, Oriental.
Biology: The larvae of *A. kagoshimensis* (Miyake) are known to mine the leaves of *Akebia quinata* (Lardizabalaceae) (Ito 1984). Larvae of *A. angustifascia* (Hering) and *A. echinopanacis* Kandybina are known to breed in the fruits of some Araliaceae (Kandybina 1966).

A monophyletic concept of *Acidiella* was proposed by Han (1992) and followed by Wang (1996) and Norrbom et al. (1999b). *Flaviludia* Ito was recently synonymized with *Acidiella* by Korneyev (1998). As true for many other genera of Trypetina, the generic limit of *Acidiella* has been much confused. Hardy (1987) considered *Acidiella* a synonym of *Myoleja*, which is only remotely related. The monophyly of *Acidiella* is supported by a single synapomorphy: posterior surstylar lobe reduced and anterior surstylar lobe expanded and bent inward (Figure 11.6J, K). A similar structure found in *Strauzia* (Figure 11.6E, F) is considered convergence, because none of the *Trypeta* group genera, which are more closely related to *Acidiella*, has the same structure. Most *Acidiella* species can be easily identified to genus without dissecting the male genitalia because of their similar wing patterns (as in Figure 11.5M with subapical and discal bands often separated). Many *Vidalia* spp. also have such wing patterns, but they can be differentiated from *Acidiella* by having only a single pair of orbital setae.

*Machaomyia* Hendel (1914; 1915) is a new synonym of *Acidiella*. *Acidiella caudata* (Hendel), n. comb., the type and only species ever included in *Machaomyia*, has a peculiar wing shape (only males known — posterior wing margin with a pointed projection), but this appears to be an autapomorphy. I believe that this species is closely related to *A. kagosimensis* and *A. maculata* (Shiraki) based on their similar genitalia (Han, personal observation). Furthermore, these two species also have sexually dimorphic wing shape.

*Acidiella dilutata* (Ito) was synonymized with *A. bipunctata* (Portschinsky) by Korneyev (1998), but a female specimen available to me differs in having five pairs of frontal setae and the apex of aculeus bent upward and laterally flattened. Although these two species closely resemble each other in general appearance, including very similar wing patterns, I believe they are distinct. Such aculeus structure otherwise has been found only in *A. malaisei* (Hering) from Burma (Han, in preparation). Korneyev (1998) also indicated that *A. spinifera* (Hering) is a presumed synonym of *A. bipunctata*. They do have very similar wing patterns and aculei, but comparison of their male genitalia is needed to confirm their conspecificity.

*Acidiella issikii*, the type species of *Pseudacidia* Shiraki, had been previously known from the unique holotype female from Korea, and was treated as a member of *Acidiella* based on its similar wing pattern (Han 1992). Korneyev (1998) recently examined both sexes of this species, and questioned generic placement of this species in *Acidiella*. I was also able to obtain a large number of both sexes of *A. issikii* and a closely related species, *A. circumvaga,* from Wonju-si, which is about 70 km southeast of the type locality. Preliminary examination of the male genitalic structure revealed that these two species do not have the single synapomorphy of *Acidiella*. At the moment, I do not know the phylogenetic position of these two species except that they clearly belong to the *Trypeta* group. A more careful study involving both morphological and molecular data may remove *Pseudacidia* from the synonymy of *Acidiella*.

#### 11.7.2.3 *Acidiostigma* Hendel

Distribution: Eastern Palearctic to Oriental.
Biology: The larvae of *A. polyfasciatum* (Miyake) mines the leaves of *Clerodendrum trichomum* (Verbenaceae) (Ito 1984).

Han and Wang (1997) proposed a new generic concept of *Acidiostigma*, including *Parahypenidium* Shiraki and *Shiracidia* Ito as synonyms. They also analyzed the relationships among the 16 recognized species. Most *Acidiostigma* species can be readily distinguished from other Trypetini by their predominantly dark wing pattern and elongated cell sc (Figure 11.5O). Similar wing patterns are also found in *Hemilea, Pseudhemilea,* and *Drosanthus,* which do not have cell sc elongated. Some species, including *A. longipennis* (Hendel) and *A. s-nigrum,* do not have this typical wing pattern, but they can still be distinguished by the elongated cell sc (Figure 11.5N), which is a synapomorphy indicating the monophyly of the genus (Han and Wang 1997).

#### 11.7.2.4 *Aciuropsis* Hardy

Distribution: Philippines, Papua New Guinea.
Biology: Unknown.

This monotypic genus, including only *A. pusio* Hardy, can be readily distinguished from other Trypetini by the following combination of characters: (1) wing length less than 2.5 mm; (2) with only single pair of scutellar setae; and (3) dorsocentral seta at same level as postsutural supra-alar seta. The relationship of this monotypic genus to other genera is not clear. Unfortunately, only male specimens were available to me, and the female character data are entirely based on Hardy (1987). Absence of apical scutellar setae and similarity in wing shape and pattern may indicate a close relationship to *Euleia rotundiventris* (Fallén) and *E. kovalevi* (Korneyev) (Figure 11.5J vs. 11.5K).

#### 11.7.2.5 *Aischrocrania* Hendel

Distribution: Eastern Palearctic to Oriental.
Biology: Unknown.

The eight species of *Aischrocrania* can be readily distinguished from other Trypetina by the following combination of characters: (1) at least five pairs of frontal setae and (2) antennal pedicel usually elongated in males (Figure 11.4G) and strongly setulose in both sexes except for *A. quadrisetata*. A remarkably similar modification in the male occurs in *Cerajocera* Rondani and in one species of *Polionota* Wulp (Norrbom 1988) (both belonging to the Tephritinae), undoubtedly as a result of convergence.

In addition to the five previously recognized *Aischrocrania* spp., Han (1992) included two more species for the following reasons. *Aischrocrahia quadrisetata* (Hering) does not have sexually dimorphic antennae, but was transferred here from *Vidalia* based on many additional characters common to both sexes including the five pairs of frontal setae. *Aischrocrania multipilosa* (Kwon) is also included for having the five pairs of frontal setae and strongly setulose pedicel in both sexes. *Neomyoleja* Tseng et al. (1992b) is here synonymized with *Aischrocrania*. No specimens of

*N. chowi* Tseng et al., n. comb., were available to me, but the original description clearly shows that it belongs in *Aischrocrania* by having five pairs of frontal setae and the pedicel with "blackish thick bristle-like hairs, several of them rather long." In addition, it has a typical wing pattern of the genus.

### 11.7.2.6 *Alsangelisca* Ito

Distribution: Japan, Russian Far East.
Biology: Unknown.

This monotypic genus, including only *A. takeuchii* Ito, is easily distinguished from other genera of the Trypetina by having only two pairs of frontal setae (Figure 11.4C). In addition, the following autapomorphies may facilitate its identification: (1) dorsal sclerite of glans reduced without any pattern (Figure 11.7E); (2) single seta on anepisternum; and (3) female aculeus without any apical serration.

### 11.7.2.7 *Anastrephoides* Hendel

Distribution: Eastern Palearctic.
Biology: Unknown.

The genus includes only two species, *A. gerckei* Hendel and *A. matsumurai* Shiraki. Based on its wing pattern (Figure 11.5C), which is remarkably similar to *Myoleja sinensis,* I tentatively placed *Anastrephoides* in synonymy of *Myoleja* (Han 1992). However, Korneyev (1996) treated it under the subtribe Trypetina. Recently, I was able to examine both sexes of *A. matsumurai,* and found that the female postabdomen was unusually long, but the aculeus tip apically serrate and the glans was typical of the Trypetina. A molecular study to elucidate its phylogenetic position within the Trypetini is under way (Han, in preparation).

### 11.7.2.8 *Angelogelasinus* Ito

Distribution: Japan, Russia.
Biology: Unknown.

The generic limit of *Angelogelasinus* is not resolved. Of the four recognized species (Ito 1984; Korneyev 1998), I examined only *A. naganoensis* (Shiraki) and *A. amuricola* (Hendel), but was not able to find any characters to define the genus other than their somewhat similar wing patterns (Figure 11.5P). No characters of generic significance have been found even after comparing Ito's descriptions of the other two species, *A. implicatus* Ito and *A. venustus* Ito.

### 11.7.2.9 *Calosphenisca* Hendel

Distribution: Eastern Palearctic to Oriental, and Australasian.
Biology: Unknown.

An expanded concept of this genus (as *Fusciludia*), including six species, was proposed by Han (1992) and followed by Hancock and Drew (1994a), Permkam and Hancock (1995), and Norrbom et al. (1999b). Korneyev (1998) synonymized *Fusciludia* under *Calosphenisca*, and, as a result, added two more species to the genus. The members of *Calosphenisca*, although previously placed in several different genera, can be readily distinguished by their almost entirely whitish anepisternum and scutellum, which can also be interpreted as a synapomorphy of the genus.

Among the eight recognized species, two show extreme morphological differentiation. The frons and associated setae have been drastically modified in males of *C. ensifera* Ito (Figure 11.4O).

This modification is somewhat similar to that of *Vidalia* (Figure 11.4L), which is probably why this species was originally described in *Vidalia*. In addition, it is the only species of *Calospheniscas* that has a single pair of orbital setae. This reduction might be genetically correlated with the occurrence of sexually dimorphic male frontal structure, because it is repeatedly observed in taxa having similar modifications (*Vidalia, Stemonocera,* and *Strauzia*). *Calospheniscas unicuneata* (Hardy) is also a highly apomorphic species, having unusual modifications in male abdominal structure: (1) pleural membrane of segments 3 to 4 with a pair of large, dark brown, blisterlike structures; and (2) proctiger dovetailed (Figure 11.6G).

### 11.7.2.10 *Chenacidiella* Shiraki

Distribution: Eastern Palearctic, Oriental.
Biology: Unknown.

The species of *Chenacidiella* are readily distinguished from other Trypetini by the following combination of characters: (1) female eversible membrane ventrally with numerous posteriorly directed small spines between taeniae (as in Figure 11.9B); (2) male frontal setae not modified; and (3) vein $R_{4+5}$ with at least 16 setulae between node and R-M (Figure 11.5H). Although the three species of *Chenacidiella* closely resemble each other, especially in wing pattern (Figure 11.5H), it is difficult to identify a synapomorphy for the genus. This genus forms the sister group of *Hoplandromyia* plus *Vidalia* based on a single synapomorphy (character 1, above) that is unique within Tephritidae (Han 1992; Han et al. 1994b).

### 11.7.2.11 *Cornutrypeta* Han and Wang

Distribution: Palearctic, Oriental.
Biology: *Cornutrypeta spinifrons* (Schroeder) mines the leaves of *Solidago virgaurea* (Niblett 1956).

The monophyly of *Cornutrypeta* is clearly supported by the unique modification of the male frontal setae: the anterior two to four frontal setae are enlarged and rod-shaped; the posterior frontal seta is always thinner, pointed, and erect (Figure 11.4D, E). The reduced number of dorsal setulae on vein $R_{4+5}$ (usually five or less but up to nine in two species) may be considered as another synapomorphy. Although the identity of *Cornutrypeta* males is relatively easy to determine by their characteristic frontal setae, it is virtually impossible to separate unassociated females from *Trypeta* females. See Han et al. (1993) for additional discussion of their relationships.

### 11.7.2.12 *Cristobalia* Malloch

Distribution: Solomon Islands.
Biology: Unknown.

This monotypic genus, including only *C. lutea* Malloch, can be readily distinguished from other Trypetini by the following combination of characters: (1) a single pair of orbital setae and (2) lateral vertical seta unusually short, not well distinguished from nearby postocular setae. It can be also identified based on its characteristic wing pattern (Figure 11.5A). Unfortunately, the male is unknown for this interesting genus. It is tentatively placed in the Trypetina based on its apically serrate aculeus. Presence of a single orbital seta may indicate a close relationship to *Itosigo,* but this character occurs in many remotely related genera within the Tephritidae, and this hypothesis is speculative until a male specimen is available.

### 11.7.2.13 *Drosanthus* Hering

Distribution: Known only from Java, Indonesia.
Biology: Unknown.

Because of its predominantly dark wing pattern, this monotypic genus, including only *D. melanopteryx* Hering, was previously placed in the synonymy of *Hemilea* (Han 1992; Norrbom et al. 1999b). It can be distinguished by having the following characters: (1) almost completely dark wing pattern with narrow hyaline area on posterior one-third of anal lobe and cell $cu_1$; and (2) dorsocentral seta one-third distance from level of intra-alar seta to postsutural supra-alar seta.

### 11.7.2.14 *Euleia* Walker

Distribution: Holarctic.
Biology: Larvae of three species are known to mine the leaves of Apiaceae spp. (Han 1998). One highly polyphagous species, *E. heraclei,* extends its host range to a few species of Asteraceae and Araliaceae.

The generic limit of *Euleia* has been highly confused in the past, but the status of this genus has been clarified by Korneyev (1991a, b) and Han (1992; 1996b). The species of *Euleia* may be distinguished by the following potential synapomorphies: (1) dorsocentral seta at same level as or anterior to postsutural supra-alar seta; (2) lateral serration of aculeus reduced and restricted to apex (Figure 11.10D); (3) median sclerite of glans with elongated granules; (4) subapical lobe of glans absent; and (5) spermatheca with a number of short and blunt tubercles on its surface (Figure 11.10R to T).

The relationships among the *Euleia* species are relatively clear — excluding *E. odnosumi* (Korneyev) and *E. scorpioides* (Richter and Kandybina) which I have not seen. I consider *E. unifasciata* (Blanc and Foote) the sister species of the rest of *Euleia,* which is characterized by the dorsal taeniae fused medially (in *Euleia unifasciata* and other genera of Trypetina, the dorsal taeniae are not fused). Within this group, there are two distinct monophyletic subgroups. *Euleia rotundiventris* and *E. kovalevi* can be characterized by the dorsocentral seta well anterior to the level of the postsutural supra-alar setae (in other *Euleia* species and other genera of Trypetina, the dorsocentral seta is at the same level as or posterior to the postsutural supra-alar seta). *Euleia fratria, E. heraclei, E. separata* (Becker), and *E. uncinata* (Coquillett) can be characterized by their increased spermatheca size. This particular synapomorphy is deduced from the following character polarity data within *Euleia*; none of the following character states is found in other genera of Trypetina: (1) reduced in size with numerous nipplelike structures (Figure 11.10T); (2) similar to state 1 but sclerotized lower lobe developed (Figure 11.10S); (3) increased considerably in size (Figure 11.10R). Character 1 is considered plesiomorphic because it is found in *E. unifasciata,* which is the sister species of the rest of the *Euleia* species.

### 11.7.2.15 *Hemilea* Loew

Distribution: Palearctic and Oriental to Austro-Malayan Subregion.
Biology: *Hemilea infuscata* Hering mines the leaves of *Lectuca laciniata* and *Taraxacum platycarpum,* both of which belong to the family Asteraceae (Sasakawa 1955; Kwon 1985).

An expanded concept of *Hemilea* was suggested by Han (1992), and followed by Norrbom et al. (1999b). However, this group of flies with predominantly dark wing patterns (as in Figure 11.5L) is obviously nonmonophyletic (Han 1992; Korneyev 1998). The type species, *H. dimidiata* (Costa), belongs to the *Trypeta* group by its reduced lateral prensiseta (as in Figure 11.1E), but some other previously synonymized taxa do not. Therefore, *Drosanthus, Hemileophila,* and *Pseudhemilea* are resurrected, and *Dryadodacryma* and *Hemileoides,* whose phylogenetic positions are uncertain, are moved to the unconfirmed generic list (Section 11.7.5). In this study, *Hemilea* is restricted to the species with the following combination of characters: (1) wing with extensive brown to dark brown pattern anterior to vein M (Figure 11.5L); (2) subcostal cell shorter than costal cell; and (3) lateral prensiseta much smaller than medial prensiseta (as in Figure 11.1E). Despite this reduced concept of *Hemilea*, there is no obvious synapomorphy defining this genus within the *Trypeta* group. The following species are included: *H. atrata* Hardy, *H. bipars* (Walker),* *H. clarilimbata* (Chen), *H. cnidella* Munro,* *H. dimidiata* (Costa), *H. infuscata* Hering, *H. lineomaculata* Hardy, *H. malgassa* Hancock,* and *H. praestans* Bezzi.*

### 11.7.2.16 *Hemileophila* Hering

Distribution: Russian Far East, China, Korea, Japan.
Biology: Unknown.

Because of its predominantly dark wing pattern, this monotypic genus was previously placed in the synonymy of *Hemilea* (Han 1992; Norrbom et al. 1999b), but it can be distinguished by having dark brown supracervical setae and four pairs of frontal setae in most specimens. Korneyev (1998) recently synonymized *H. alini* Hering and *H. undosa* Ito with *H. sibirica* (Portschinsky), which is the sole species of *Hemileophila.*

### 11.7.2.17 *Hoplandromyia* Bezzi

Distribution: Mainly Afrotropical, but two species Oriental.
Biology: *Hoplandromyia madagascariensis* Hancock mines the leaves of *Canthium humberti* (Rubiaceae) (Hancock 1985).

*Hoplandromyia* is a morphologically homogeneous genus readily distinguished by the following combination of characters: (1) three pairs of frontal setae at least in females, but highly modified in male with second seta greatly enlarged and flattened, third seta enlarged but smaller than second seta (Figure 11.4P); (2) wing with unbroken C-band from DM-Cu to anterior apical wing margin to slightly after $R_{4+5}$ (Figure 11.5I); (3) eversible membrane ventrally with numerous posteriorly directed small spines between taeniae (as in Figure 11.9B); (4) glans (Figure 11.7G) with median granulate sclerite largely reduced; (5) elongated granulation on dorsal sclerite of glans greatly reduced or absent; and (6) vesica of glans greatly enlarged. Of the above characters, Han et al. (1994b) hypothesized character 5 as a synapomorphy for *Hoplandromyia*. The relationships of this genus to *Vidalia* and *Chenacidiella*, which share character 3, was discussed by Han et al. (1994b).

### 11.7.2.18 *Itosigo* Ito

Distribution: Eastern Palearctic.
Biology: In Korea, I have observed a large number of mating adults of *I. bellus* on the flowers of *Heracleum moellendorffi* Hance in August. Larval host unknown.

This small genus includes only *I. bellus* Ito and *I. kuwayamai* (Shiraki). Both species are yellow brown flies that can be distinguished from other Trypetini by the following combination of characters: (1) one pair of orbital setae; (2) dorsal sclerite of glans relatively large with extensive internal sculpture pattern of elongated granulation (Figure 11.7A); and (3) glans without median granulate

* Genitalia not examined.

sclerite. Characters 1 and 2 can be considered synapomorphies of the genus but the absence of the median granulate sclerite may either be interpreted as a reversal or plesiomorphic condition in this taxon. If the latter is the case, *Itosigo* may be the sister group of the rest of the Trypetini. A molecular study showed that this species clearly belongs to the Trypetina, but did not resolve its position within the subtribe (Han, in press).

*Itosigo* may be congeneric with the monotypic genus *Carpophthoracidia* Shiraki, which also has only a pair of orbital setae but is not sexually dimorphic in frontal structure. Unfortunately, I was not able to obtain any specimens of *Carpophthoracidia* for this study. Another monotypic genus known only from females, *Cristobalia*, might also be related by sharing a single pair of orbital setae and similar aculeus structure, but I do not suggest any relationship at this time because of lack of male information that might be critical in resolving this matter.

### 11.7.2.19 *Magnimyiolia* Shiraki

Distribution: Eastern Palearctic and Oriental.
Biology: Unknown.

A new concept of *Magnimyiolia* was proposed by Han (1992), and followed by Norrbom et al. (1999b). The members of *Magnimyiolia* may be distinguished from other Trypetini by the following potential synapomorphies: (1) lateral serration on aculeus restricted to apical one-sixth or less (Figure 11.10E); and (2) micropylar end of egg greatly elongated, rod shaped. Unfortunately, seven of ten *Magnimyiolia* species are known only from males; I was able to examine females of two species, *M. picea* (Hering) and *M. fusca* (Ito). In addition to both sexes of these two species, the males of *M. animata* (Hering), *M. interrupta* Kwon, and *M. media* Ito were also examined.

*Magnimyiolia fusca* and *M. tumifrons* (Chen) are almost identical to each other, and may be shown to be conspecific as our understanding of their intraspecific variability improves. Within the genus, at least *M. picea, M. animata, M. interrupta,* and *M. media* appear to form a monophyletic group characterized by an unusually elongated vesica of the glans, but examination of the remaining species and, more importantly, female data are needed to clarify their relationships.

### 11.7.2.20 *Morinowotome* Ito

Distribution: Eastern Palearctic to Oriental (southern China and Taiwan).
Biology: Unknown.

Han (1992) recognized three species in *Morinowotome*, and a fourth, *itoi* Korneyev (1998), was subsequently described. Although these species resemble each other closely, no synapomorphy has been found to support their monophyly. Additional data, such as immature morphology, host records, and molecular sequence data, are needed to clarify their relationships. The following combination of characters can be used to distinguish *Morinowotome* from other members of the *Trypeta* genus group: (1) ocellar seta at least 2.5× as long as ocellar triangle; (2) dorsocentral seta at same level as or slightly posterior to postsutural supra-alar seta; (3) section of vein M between crossveins DM-Cu and R-M 0.5 to 0.6× as long as section between BM-Cu and R-M (Figure 11.5S).

### 11.7.2.21 *Nemeurinus* Ito

Distribution: Korea, Japan.
Biology: Unknown.

This monotypic genus, including only *N. leucocelis* Ito, can be readily distinguished from other genera of the Trypetini by its wing pattern, especially the large faint brownish spot on cell m

(Figure 11.5Q), and the ivory white postpronotal lobe, upper part of anepisternum, posterior part of scutum, and scutellum. The reduced lateral prensiseta on the male genitalia place it in the *Trypeta* group. No further characters indicate its relationships to other genera within the *Trypeta* group. A detailed redescription with discussion of the intraspecifically variable wing pattern was published recently (Han 1997b).

#### 11.7.2.22 *Oreurinus* Ito

Distribution: Korea, Japan.

Biology: Host unknown. Han (1997a) found a number of *O. cuspidatus* males sitting on rocks along a mountain stream in deep woods in Kangwon-do, Korea. A single female was swept from nearby vegetation.

Ito (1984) described this monotypic genus based on two female specimens of *O. cuspidatus* from Japan. Han (1997a) recently provided a detailed description of both sexes based on Korean specimens. The wing pattern of this species (Figure 11.5R) is similar to that of *Acidiella* spp. (Figure 11.5M), but it can be distinguished by having dark brown supracervical setae and almost entirely dark brown thoracic dorsum and abdomen.

#### 11.7.2.23 *Paratrypeta* Han and Wang

Distribution: China (Sichuan and Tibet).

Biology: Unknown.

The monophyly of two *Paratrypeta* species is supported by two synapomorphies (Han et al. 1994a): (1) frontal setae with long posteriormost one close to orbital setae and remaining two to four setae displaced anteriorly (Figure 11.4I, J); and (2) $R_{2+3}$ subapically with short, anteriorly directed spurious vein. See Han et al. (1994a) for additional discussion of their relationships.

#### 11.7.2.24 *Philophylla* Rondani

Distribution: Oriental, Australasian, Afrotropical, and Palearctic.

Biology: Six of the eight species with known host records breed in the fruits of Verbenaceae and Caricaceae (Table 11.2; Han 1998). There are some conflicting host records for *P. caesio,* but Beiger's (1968) record of leaf petiole-mining on *Urtica dioica* (Urticaceae), accompanied with an adult diagnosis, seems most reliable.

*Philophylla* is a large genus of more than 50 included species, most of which have been treated under the genus *Myoleja* in recent taxonomic studies (Hancock 1986; Hardy 1987; White 1988). As a result of my study (Han 1992), *Myoleja* is more narrowly defined as a small genus of Chetostomatina (see Section 11.7.3.4), and the majority of the 100 species formerly placed in *Myoleja* actually belong to various genera of Trypetina. *Philophylla desparata* (Hering), *P. kraussi* (Hardy), and *P. quadrata* (Malloch) are tentatively placed in *Philophylla* (Norrbom et al. 1999b), but may deserve placement in a new genus (see Han 1992 for further discussion).

*Philophylla* can be defined as a monophyletic group by the following unequivocal synapomorphies: (1) oviscape dorsally with a pair of strong lateral marginal setae (Figure 11.9A); and (2) lateral surstylus with anterior lobe broadly flattened and posterior lobe elongated (Figure 11.6A, B). Although dissection of genitalia is recommended to confirm generic placement, most species can be conveniently recognized as *Philophylla* by having one of three typical wing patterns (Figure 11.5D to F). Since wing patterns very similar to that shown in Figure 11.5F are also found in *Anomoia, Hoplandromyia,* and a few other genera, extra precaution is needed. More than half the included species have been confirmed by actual examination or data from the literature, but some unexamined species have been included here based on their similarity in wing pattern and other external characters to those confirmed. A careful revisionary study is needed to access the inter- and intrageneric relationships critically.

### 11.7.2.25 *Pseudhemilea* Chen

Distribution: Palearctic, Oriental.
Biology: Three species mine the leaves of Araliaceae spp. (Han 1998).

*Pseudhemilea* was previously placed in the synonymy of the heterogeneous genus *Hemilea* (Han 1992; Norrbom et al. 1999b). As Korneyev (1998) indicated, the status of this genus depends on examination of the type species, *P. nudiarista* (Chen), which was not available for this study. Nevertheless, I tentatively define the concept of this genus based on the fact that it is apparently closely related to *P. longistigma* (Shiraki) by sharing many characteristics including the reduced genal seta (Chen 1948; Ito 1984; Korneyev 1998), which can be hypothesized as a synapomorphy of these two species. The original description with color illustration of *P. longistigma* (Shiraki 1933) provides an unmistaken identity of this species except that his line drawing of a male head does have a distinct genal seta. For all the specimens that I have examined so far (13 males and 24 females from Korea, Manchuria, and Japan), the genal seta is clearly reduced in both sexes. Therefore, there might have been an error when Shiraki (1933) made this drawing. The demarcation of *Pseudhemilea* is equivalent to that of the *Hemilea longistigma* group (Han 1992). Some *Pseudhemilea* species have predominantly dark wing patterns as in *Hemilea*, but the other species have quite different wing patterns. They may be characterized by the following two potential synapomorphies: (1) dorsal sclerite of glans swollen (Figure 11.7C) and (2) female with two spermathecae. In addition to two species mentioned above, this genus includes *P. accepta* (Ito), *P. acrotoxa* (Hering), *P. freyi* (Hardy), *P. kalopanacis* (Ito), and *P. pilosa* (Ito), all of which are newly transferred (n. comb.). *Pseudhemilea araliae,* n. comb, another leaf-miner of Araliaceae, although not examined yet, is tentatively placed in this genus based on having the same host family.

### 11.7.2.26 *Stemonocera* Rondani

Distribution: Palearctic, except for two Indian species.
Biology: Two species, *S. cornuta* and *S. spinulosa,* mine the leaves of Asteraceae, including *Senecio* spp. (Han 1998).

A new concept of *Stemonocera* was suggested by Han (1992), and followed by Korneyev (1998) and Norrbom et al. (1999b). The ten species of *Stemonocera* can be readily distinguished from any other tephritids by the following combination of characters: (1) frons anteriorly produced to form a frontal horn in males (Figure 11.4K) and (2) with single pair of orbital setae. Most *Stemonocera* species previously have been included in *Vidalia* based on their sexually dimorphic frons and presence of only a single pair of orbital setae. Since the frontal modification of *Stemonocera* is structurally different from that of *Vidalia* (Figure 11.4L) and the reduction of orbital setae occurs in several remotely related genera, neither character justifies the inclusion of these species in *Vidalia*. According to the phylogenetic analysis, *Vidalia* is more closely related to *Chenacidiella* and *Hoplandromyia* than to *Stemonocera*, which belongs to the *Trypeta* genus group (Figure 11.3).

### 11.7.2.27 *Strauzia* Robineau-Desvoidy

Distribution: Nearctic.
Biology: Unlike other genera of the Trypetina, the host records of *Strauzia* species are well known (11 of the 12 known species) (Steyskal 1986; Stoltzfus 1988). They are, so far as is known, univoltine stem miners of various Asteraceae, especially *Helianthus* spp. Larvae

tunnel in the pith parenchyma of the host, moving up and down the stem (Brink 1923; Stoltzfus 1988). Pupariation occurs in the soil, upper root, or stem. The pupa forms in spring several weeks before adult emergence (Stoltzfus 1988). Adults are closely associated with their host plants where mating occurs.

The species of *Strauzia* can be readily distinguished from other Trypetini by the following apomorphies: (1) head with five or more frontal setae (Figure 11.4M, N); (2) with single or no orbital seta; (3) anterior surstylar lobe enlarged and bulbous, bent inward (Figure 11.6F); and (4) apical portion of spermathecal duct abruptly swollen to form a round, dark-colored chamber (Figure 11.10U). In addition, their exclusive stem-mining behavior on composite plants convincingly supports the monophyly of *Strauzia*. Within the 12 known *Strauzia* spp., the sexually dimorphic modification of the male frontal setae, which are enlarged and rod-shaped with blunt apices (Figure 11.4N), suggests the monophyly of ten species sharing this character. Despite recent taxonomic revisions (Steyskal 1986; Stoltzfus 1988), it is very difficult to identify some species because of extraordinarily high variability in key characters such as the color patterns of the wing and thorax. More careful study is needed to clarify the inter- and intraspecific variability of this genus.

### 11.7.2.28 *Trypeta* Meigen

Distribution: Oriental, Holarctic, and Neotropical south to Costa Rica.
Biology: Six *Trypeta* species are known to mine the leaves of the plant family Asteraceae (Han 1998).

*Trypeta* includes 36 described species, including *T. ambigua* (Shiraki), n. comb., and *T. retroflexa* (Wang), n. comb., which are here transferred from *Acidia* and *Sineuleia,* respectively. There are numerous undescribed species in the New World, including one from Costa Rica (Han and Norrbom, in preparation). Most *Trypeta* species may be distinguished from many genera of the *Trypeta* group by having crossvein R-M approximately at the middle of the cell dm (as in Figure 11.5U). *Acidia, Aischrocrania, Paratrypeta, Stemonocera,* and *Cornutrypeta* have similar wing venation, but are easily differentiated by their male secondary sexual characters and associated modifications in females. In other words, *Trypeta* is a taxon sharing a synapomorphy with these five genera, but lacking their other derived characters, and thus it is possibly paraphyletic. There is at least one large monophyletic group within *Trypeta*. It includes the type species, a few Palearctic species, and all the New World species, and is characterized by the reduction of the subapical lobe on the glans (Figure 11.2A-D).

### 11.7.2.29 *Vidalia* Robineau-Desvoidy

Distribution: Eastern Palearctic to Oriental and Australasian.
Biology: *Vidalia bidens* breeds in the fruits of *Schefflera subulata,* a member of Araliaceae, in West Malaysia (Han et al. 1994b).

There has been controversy about the name *Vidalia,* because the original description (Robineau-Desvoidy 1830) is inadequate, and the type specimen(s) from the East Indies (Indonesia) of the type species, *V. impressifrons* Robineau-Desvoidy, apparently has been lost (Munro 1938; Hardy 1987; Han et al. 1994b). Munro (1938) proposed *V. ceratophora* Bezzi as "neogenotype," but that was not a valid nomenclatural act. Since then, many tephritid species with enlarged male frontal setae have been placed in *Vidalia*. Han et al. (1993; 1994a) removed a number of species to three other genera (*Paratrypeta, Cornutrypeta,* and *Stemonocera*), and later (Han et al. 1994b) established a newly recognized monophyletic group under the resurrected name, *Pseudina* Malloch. However, Hancock and Drew (1995) synonymized *V. quadricornis* Meijere with *V. impressifrons,* resurrecting *Vidalia* for this taxon. Despite the fact that the original description of *V. impressifrons* is not adequate for positive identification, Hancock and Drew's treatment was followed by two recent major tephritid publications (Korneyev 1998; Norrbom et al. 1999b). After e-mail discussion involving seven tephritid taxonomists (Freidberg, Han, Hancock, Korneyev, Merz, Norrbom, and White), we agreed to keep the long-used name *Vidalia* by designating a neotype for the sake of nomenclatural stability.

Therefore, I am here designating the holotype of *V. quadricornis* as the neotype of *V. impressifrons*. Hardy (1987) provided a redescription of this specimen, which is sufficient for species identification. I also examined the neotype male, and found that it possessed the frontal modification and wing pattern typical of *Vidalia* (similar to Figures 11.4L and 11.5M), but could be easily distinguished from any other known *Vidalia* spp. by its predominantly dark coloration: (1) frons dark brown, contrasting with yellow brown occiput and gena; (2) scutum entirely shiny dark brown, contrasting well with the ivory white postpronotal lobes and scutellum; (3) thoracic pleura and legs yellow brown; and (4) abdominal T3-5 shiny dark brown, contrasting with yellow brown T1+2. Neotype data: Fort de Kock (Bukittinggi), Sumatra. Col. Date: written as "10.1913." The neotype male (= holotype of *V. quadricornis*) is in the Zoölogisch Museum, University of Amsterdam. The abdomen was dissected and kept in a genitalia vial.

The 16 species of *Vidalia* can be readily distinguished by the following combination of characters: (1) four pairs of frontal setae, at least in females; highly modified in male with second seta greatly enlarged and flattened, first seta sometimes enlarged and flattened but always much smaller than second seta, third seta usually short, fourth seta shortest, sometimes indistinguishable (Figure 11.4L); (2) female eversible membrane ventrally with numerous posteriorly directed small spines between taeniae (as in Figure 11.9B). Their monophyly and relationships to *Chenacidiella* and *Hoplandromyia*, which also possess character 2, are extensively discussed in Han et al. (1994b).

Korneyev (1996) suggested *V. rohdendorfi* Richter as the sister group of *Hoplandromyia,* and later (Korneyev 1998) indicated that it might be the sister group of the *Vidalia* group (a clade including *Chenacidiella, Hoplandromyia,* and *Vidalia*) or at least the latter two. However, I do not find any evidence that *V. rohdendorfi* is closely related to the *Vidalia* group. The shape and arrangement of the enlarged male frontal setae (Richter 1963) differs and is doubtfully homologous with those of typical *Vidalia* and *Hoplandromyia* (Figure 11.4L, P). More importantly, *V. rohdendorfi* does not have the posteriorly directed spines between the taeniae (Korneyev's 1996 drawing shows it has anteriorly directed spines instead). The C-shaped band and the reduced sculpture of the glans, which Korneyev (1996) discussed, are highly homoplastic, because they are repeatedly found in different lineages of Trypetini. In addition, the shape of the prensisetae apparently places it in the *Trypeta* group (as in Figure 11.6K), but a proper generic name cannot be given for it at this time. Based on the similar wing pattern and larval hosts (fruits of Araliaceae), *V. rohdendorfi* might be related to *Acidiella angustifascia* and *A. echinopanacis,* and all of them might have to be treated under the genus *Flaviludia*. More careful study is needed to resolve this problem.

## 11.7.3 Genera of the Chetostomatina, New Subtribe

### 11.7.3.1 *Anomoia* Walker

Distribution: Palearctic, Oriental, Australasian.

Biology: Host records are known for only two species. *Anomoia purmunda* (Harris) is probably one of the best biologically known Trypetini. It appears to be univoltine in Europe, and breeds in the fruits of many species of Berberidaceae and Rosaceae. After completing development, it pupates and overwinters in soil. Another species, *A. alboscutellata,* develops in green twigs of coffee (Rubiaceae) (Meijere 1911; Leefmans 1930; White and Elson-Harris 1992).

Most of the 32 species of *Anomoia* can be easily distinguished by their peculiar wing pattern and venation (Figure 11.5X): (1) a narrow C-shaped band from DM-Cu to anterior wing margin, extending apically to slightly beyond $R_{4+5}$, commonly interrupted in cell $r_{4+5}$; and (2) crossvein DM-Cu strongly oblique, making cell dm posteroapically pointed. *Anomoia alboscutellata* Wulp and *A. nigrithorax* Malloch have an unbroken C-shaped band as well as an almost perpendicular DM-Cu, which are characteristics of many *Philophylla* species (Figure 11.5F). In this case, only their sagittally flattened aculeus tip (synapomorphy of the *Chetostoma* group) indicates correct generic placement. Indeed, a few species that have been tentatively placed in *Philophylla* may turn out to belong in *Anomoia* as their female postabdomens are examined. The oblique position of DM-Cu (character 2 above) could be interpreted as a synapomorphy of *Anomoia* with reversal to the perpendicular position in *A. alboscutellata* and *A. nigrithorax*. Alternatively, we could consider these two species as the sister group of the rest of *Anomoia* and their DM-Cu orientation as plesiomorphic. Therefore, placement of them in *Anomoia* must be considered as tentative. More comprehensive study involving examination of many species of related genera is needed to clarify their relationships.

#### 11.7.3.2 *Chetostoma* Rondani

Distribution: Holarctic and Oriental.
Biology: Known only for *C. continuans* Zia, which develops in the fruits of honeysuckle (*Lonicera* spp.) (Kandybina 1966; 1977).

The species of *Chetostoma* can be readily distinguished by the following synapomorphies which are all unique within Tephritidae: (1) facial ridge anteroventrally with strong setae, especially pronounced in males (Figure 11.4Q, R); (2) subapical lobe of glans apically with two tiny lateral lobes (confirmed in five representative species; Figure 11.8D); and (3) aculeus with apicodorsal serration in addition to ventral serration (only three species examined: *C. curvinerve, C. californicum* Blanc, and *C. rubidum* (Coquillett); Figure 11.10J). Although *Chetostoma* clearly belongs to the *Chetostoma* group characterized by the sagittally flattened aculeus tip, its relationship within the clade is not well understood. Its wing pattern has some similarity to that of *Anomoia*, but whether this constitutes a synapomorphy is ambiguous. Absence of the honeycomb pattern on the glans may instead indicate relationship to *Paramyiolia,* but, again, this is not an unequivocal synapomorphy because loss or reduction of pattern on the glans is common in other genera of Trypetini.

#### 11.7.3.3 *Montiludia* Ito

Distribution: Korea, Japan.
Biology: Unknown.

The two species of *Montiludia* may be distinguished from other chetostomatines by the following combination of characters: (1) vein $R_{4+5}$ with at least 15 setulae between node and R-M (Figure 11.5W); (2) dorsal sclerite basally with pattern of fusiform cells and apically with pattern of extensive hexagonal cells (Figure 11.7I); and (3) vesica of glans enlarged (Figure 11.7I). Monophyly of this genus is hypothesized based on characters 1 and 3.

#### 11.7.3.4 *Myoleja* Rondani

Distribution: Palearctic and Papua New Guinea
Biology: Larvae of *M. lucida* and *M. sinensis* breed in the fruits of *Lonicera* spp. (Han 1998).

*Myoleja* includes four species: *M. boninensis* (Ito), *M. lucida* (Fallén), *M. megaloba* Hardy, and *M. sinensis* (Zia). These species may be distinguished from other chetostomatines by the following combination of characters: (1) apex of aculeus sagittally flattened with ventral serration

(Figure 11.10M, N); (2) head without sexually dimorphic enlargement of frontal setae or subvibrissal setae; (3) crossvein DM-Cu more or less perpendicular to vein M; and (4) body largely yellow brown. Before my Ph.D. study (Han 1992), many unrelated species were treated under *Myoleja,* and its concept was extremely inconsistent among different authors. According to the concept of some recent authors (Hardy 1973; 1974; 1987; White 1988), almost half the species of the Trypetini would belong in this taxon. Fortunately, their postabdominal structures provide several diagnostic characters that clarify the generic concept. Especially, the smooth and cylindrical eversible membrane and the sagittally flattened aculeus (as Figures 11.9D and 10I to N) clearly place *Myoleja* in the subtribe Chetostomatina. Therefore, the few true *Myoleja* species are more closely related to *Montiludia, Anomoia, Chetostoma,* and *Paramyiolia* than to most other species formerly placed in *Myoleja* and which belong in the Trypetina. Although all four recognized species of *Myoleja* resemble each other, this genus may not be monophyletic because it is defined on the basis of plesiomorphies; it includes those species not having derived characters of the other four related genera. *Anastrephoides,* which was tentatively treated as a junior synonym of *Myoleja* in my earlier work (Han 1992), is given a full generic status as suggested by Korneyev (1998). *Anastrephoides* has a wing pattern remarkably similar to *M. sinensis,* but actually belongs to the subtribe Trypetina (see Section 11.7.2.7). This again shows the homoplastic nature of tephritid wing patterns.

### 11.7.3.5 *Paramyiolia* Shiraki

Distribution: Holarctic: Japan and eastern North America.
Biology. Unknown.

A new concept of this genus was proposed by Han (1992; 1996b). *Paramyiolia* appears to be a robust monophyletic group supported by the following synapomorphies: (1) anterior two to three frontal setae greatly enlarged in males, with anteriormost seta largest (Figure 11.4S); and (2) lateral prensiseta sharply pointed (Figure 11.6O) (plesiomorphy: apically blunt). Further discussion of the phylogenetic relationships of the four included species can be found in Han (1996b).

### 11.7.3.6 *Parastenopa* Hendel

Distribution: Neotropical except for *P. limata* from North America.
Biology: Host records of four species are known (Han 1998). *Parastenopa limata* breeds in fruits of many *Ilex* species (Aquifoliaceae) in North America. The larvae of *P. ogloblini* mine tender shoots of *I. paraguariensis.* The type series of *P. elegans* was bred from the galls of *I. paraguariensis* caused by a psyllid (*Metaphalera spegassiniana* Lizer). Adults of *P. marcetiae* were reared from oval galls on *Marcetia* sp. (Melastomataceae), probably caused by cecidomyiid larvae on the axillary buds of the host plant.

An expanded concept of *Parastenopa* was proposed by Han (1992; 1996b). *Parastenopa* spp. can be readily distinguished by the following combination of characters: (1) ocellar seta reduced, at most as long as ocellar triangle; (2) single anepisternal seta; (3) eversible membrane smooth and cylindrical; and (4) aculeus slender with apex dorsoventrally flattened. I believe that characters 1 and 2 are synapomorphies because they are rare in Trypetina, which is the sister group of Chetostomatina. This genus is the only representative of Chetostomatina in the Neotropical Region. It is interesting to note that three of four *Parastenopa* species with known hosts are associated with *Ilex* species, although their feeding habits show substantial variation. More careful study in conjunction with the following unnamed Old World genus, which shares many characters, is needed to clarify their generic relationships.

#### 11.7.3.7 Unnamed New Genus

Distribution: Oriental.
Biology: Unknown.

This unnamed new genus is included here because it is essential for intergeneric analysis within the Chetostomatina. Illustrations of the male and female genitalia of the putative type species (new species) from Bhutan and Burma are provided (Figures 11.7H, 9C, 10G, H). Description of the new genus based on this new species plus *Euleia contemnens* Hering is under way (Han, in preparation).

### 11.7.4 Unconfirmed Genera of Trypetini

No specimens of the following genera were available for study. They are tentatively placed in the Trypetini based not on any observed synapomorphy but on evidence from the literature. Acquisition of these mostly monotypic genera to investigate their relationships is currently under way.

#### 11.7.4.1 *Apiculonia* Wang

Wang (1990) said that this monotypic genus was allied to *Trypeta* without indicating any synapomorphy. It appears somewhat similar to *Trypeta*, but crossvein R-M is situated far more apically. Examination of the genitalia is needed to resolve its relationships.

#### 11.7.4.2 *Carpophthoracidia* Shiraki

Hancock and Drew (1995) said that this monotypic genus was related to *Hemilea*, probably based on its predominantly dark wing pattern. If it indeed belongs to Trypetini, it is notable in having only a single pair of orbital setae without showing any male frontal modification. In the Trypetini, such loss of orbital setae usually occurs in association with male frontal modification (i.e., *Vidalia, Stemonocera, Strauzia,* etc.). The other trypetine genera with a single pair of orbital setae but without such frontal sexual dimorphism are *Itosigo, Cristobalia,* and an undescribed genus (Han, in preparation) related to *Vidalia*. The possible relationship of *Carpophthoracidia* to these taxa needs to be investigated in the future.

#### 11.7.4.3 *Cephalophysa* Hering

Korneyev (1996) placed this monotypic genus in the Trypetini and indicated that it might be close to *Platyparea* because of having similar shape of the female oviscape and male epandrium. Since *Platyparea* does not belong to the Trypetini, more-detailed data regarding *Cephalophysa*'s genitalia, especially the glans, are needed to resolve its relationship.

#### 11.7.4.4 *Cervarita* Tseng, Chu and Chen

This monotypic genus is based on a single male holotype with a modified frons (Tseng et al. 1992a). Its frontal horn (protrusion of frontal plate) is similar to those of *Stemonocera* and *Paratrypeta,* but *Cervarita* may be closer to the latter genus in having two pairs of orbital setae. Its genitalia need to be examined to determine its relationships.

#### 11.7.4.5 *Dryadodacryma* Ito

Ito (1984) included three species in *Dryadodacryma,* but Korneyev (1998) transferred *continuum* Ito to *Hemileoides* Rohdendorf. According to Korneyev's (1998) description of the female terminalia, this genus apparently belongs to Trypetina, but the male genitalia should be examined to confirm its taxonomic position and further determine its relationships. This genus was previously placed in the synonymy of the heterogeneous genus *Hemilea* (Han 1992; Norrbom et al. 1999b).

### 11.7.4.6 *Epinettyra* Permkam and Hancock

Permkam and Hancock (1995) placed this monotypic genus in the Trypetini, but indicated an isolated position within the tribe. Since it has only a single pair each of frontal and orbital setae, it appears closer to Acanthonevrini.

### 11.7.4.7 *Hemiristina* Permkam and Hancock

Permkam and Hancock (1995) said that this monotypic genus was referable to the subtribe Trypetina (*sensu* Han) without indicating any observed synapomorphy. The male genitalia should be examined to confirm its taxonomic position and further determine its relationships.

### 11.7.4.8 *Hemileoides* Rohdendorf

Korneyev (1998) recognized *H. theodori* Rohdendorf and *H. continuus* (Ito), under this genus. According to Korneyev's (1998) description of the female terminalia, this genus apparently belongs to Trypetina, but the male genitalia should be examined to confirm its taxonomic position and further determine its relationships. This genus was previously placed in the synonymy of the heterogeneous genus *Hemilea* (Han 1992; Norrbom et al. 1999b).

### 11.7.4.9 *Notommoides* Hancock

Freidberg (1994) placed this genus of two species in the Trypetini based on its having median granulation of the glans, but, according to the illustration provided, the homology of median granulation and the median granulated sclerite (*sensu* Han 1992) is in doubt (see Section 11.3.1).

### 11.7.4.10 *Paracanthonevra* Hardy

Hardy (1974) placed this genus of two species in the Acanthonevrini, but Korneyev (1996) moved it to the Trypetina without indicating any synapomorphy. According to the original description, it superficially resembles typical Trypetina, but genitalia should be examined to confirm its relationships.

### 11.7.4.11 *Paracristobalia* Hardy

Hardy (1987) erected *Paracristobalia* based on *P. polita* Hardy and two unnamed new species from New Guinea and adjacent areas. He placed this genus in a complex of genera nearest to *Cristobalia,* and Korneyev (1996) later placed it in Trypetina. However, Hardy (1987) showed that "*Paracristobalia* (n. sp.) B" had different aculeus tip shape than typical Trypetina in having a pair of subapical sensillae and a smooth lateral margin. Membership of this genus in Trypetini is in doubt.

### 11.7.4.12 *Prospheniscus* Shiraki

The original description of *P. miyakei* Shiraki (1933) shows superficial similarity to *Philophylla* spp., but its genitalia should be examined to confirm its relationships.

### 11.7.4.13 *Sinacidia* Chen

This genus of two species is likely to be a synonym of *Chetostoma* as also noted by Korneyev (1998). Both known species have wing patterns similar to that of *C. continuans* and a rather distinctly setulose subvibrissal area as in other *Chetostoma* spp.

Other than the above genera, *Acidoxantha* Hendel, *Breviculala* Ito, *Callistomyia* Bezzi, *Esacidia* Ito, *Nitrariomyia* Rohdendorf, *Platyparea* Loew, and *Plioreocepta* Korneyev have been either placed in or presumed as close relatives of the Trypetini (Permkam and Hancock 1995; Korneyev 1996; 1998; Norrbom et al. 1999a). However, detailed examination of the type species did not reveal any synapomorphies to link these genera to the Trypetini (Han, personal observation). More careful studies to elucidate their phylogenetic positions within Tephritidae are under way (Han, in preparation).

### 11.7.5 Unplaced Species of Trypetini

Based on the data currently available to me, I am not able to provide proper generic placements for the following species.

*Pseudacidia clotho* Korneyev — This fossil species is known only from a relatively well preserved wing in Miocene deposits from Caucasia (Korneyev 1982). Even though the pattern and venation are somewhat similar to *Philophylla casio* (Figure 11.5D), its placement in Trypetini is not certain.

*Hemilea malaisei* Hering — This species is superficially similar to *Hemilea* spp. for sharing the predominantly dark wing pattern (similar to Figure 11.5L). I was able to examine the holotype female from Miyanmar, which is the only known specimen of this species. Presence of the posteriorly directed spines between the taeniae indicates that this species might be a sister species of the *Vidalia* group (see Section 11.4.2). The male genitalia should be examined to confirm its taxonomic position and further determine its relationships.

*Acidia parallela* Meijere — I was able to examine the holotype female from Sumatra, which is the only known specimen of this species. Presence of the dorsoventrally flattened and serrate aculeus apparently places this species in the Trypetina, but male specimens are needed to find its proper generic placement.

*Myoleja quadrinota* Hardy — Hardy (1987) said that this species, known only by a single female from Java, Indonesia, fits in a complex of species with *freyi* (Hardy), which I placed in *Pseudhemilea* (Section 11.7.2). The male genitalia should be examined to determine its relationships.

*Myoleja reclusa* Hardy — I have not seen this species, which is known by a single male from the Solomon Islands (Hardy 1987). Based on the original description, I was not able to find a clue to relate this species to any known trypetine taxa. The genitalia need to be examined to determine its relationships.

*Vidalia rohdendorfi* Richter — A single paratype male of *V. furialis* Ito, a synonym of *rohdendorfi*, was available for this study, and Korneyev (1996) provided detailed illustrations of the genitalia of both sexes. Based on the data available, this species, from the Russian Far East and Japan, obviously belongs to the Trypetina, but its generic status is not clear (see Section 11.7.2.29). Ito and Tamaki (1995) synonymized *V. brevialis* with *V. rohdendorfi,* but, as Korneyev (1998) indicated, the original description of *V. brevialis* shows relatively distinct differences in male frontal structure and wing pattern, which are difficult to accept as intraspecific variation. A good series of specimens is needed to resolve this problem.

*Myoleja semipicta* Zia — I have not seen this species, which is known by a single male from Sichuan, China (Zia and Chen 1938). Based on the original description, I was not able to find a clue to relate this species to any known trypetine taxa. The genitalia need to be examined to determine its relationships.

## 11.8 CONCLUSIONS AND FUTURE RESEARCH NEEDS

The tribe Trypetini is narrowly delimited as a monophyletic group with two newly defined subtribes, Trypetina and Chetostomatina. A total of 285 species in 29 genera of Trypetina and 69 species in 7 genera of the Chetostomatina are recognized (most taxa listed in Norrbom et al. 1999b). The species of Trypetini have been previously placed in many loosely defined genera of the subfamily Trypetinae. This confusion has largely been clarified by this study through phylogenetic analysis of the majority of nominal taxa. A key to genera and monophyletic demarcations of supraspecific taxa are also provided.

In addition to the numerous nomenclatural changes proposed in my dissertation (Han 1992), and formally published by me (Han et al. 1993; 1994a, b; Han 1996a, b; 1997a, b; Han and Wang 1997) or other tephritid workers (Hancock and Drew 1995; Permkam and Hancock 1995; Wang 1996; Korneyev 1996; 1998; Norrbom et al. 1999b), I propose the following nomenclatural acts in this chapter: Chetostomatina Han, new subtribe; neotype designation of *Vidalia impressifrons* Robineau-Desvoidy; *Machaomyia* Hendel, n. syn. of *Acidiella* Hendel; *Neomyoleja* Tseng, Chu and Chen, n. syn. of *Aischrocrania* Hendel; *Acidiella caudata* (Hendel), n. comb. from *Machaomyia; Aischrocrania chowi* (Tseng, Chu and Chen), n. comb. from *Neomyoleja; Pseudhemilea accepta* (Ito), n. comb. from *Pseudacidia* Shiraki; *Pseudhemilea acrotoxa* (Hering), n. comb. from *Euleia* Walker; *P. freyi* (Hardy), n. comb. from *Acidiella; P. kalopanacis* (Ito), n. comb. from *Hyleurinus* Ito; *P. pilosa* (Ito), n. comb. from *Yamanowotome* Ito; *P. araliae* (Malloch), n. comb. from *Hemilea* Loew; *Trypeta ambigua* (Shiraki), n. comb. from *Acidia* Robineau-Desvoidy; *T. retroflexa* (Wang), n. comb. from *Sineuleia* Chen.

The subtribe Trypetina includes all the known leaf-mining tephritids and some other tephritids of different larval feeding behavior, including stem mining, fruit feeding, seed feeding, and leaf-petiole mining. I believe that the majority of genera are defined reasonably but the relationships among many of them are still unresolved. All the known leaf-mining genera except *Hoplandromyia, Euleia,* and *Pseudhemilea* are included within a single monophyletic group (the *Trypeta* group). Another monophyletic group includes *Chenacidiella, Vidalia,* and *Hoplandromyia* (the *Vidalia* group). The subtribe Chetostomatina includes seven genera, whose larvae are mostly fruit feeders. Except for the Neotropical genus *Parastenopa* and an unnamed new genus, the Chetostomatina is further characterized by an unusual sagittally flattened and ventrally serrate aculeus tip (Figure 11.10I to N).

Although this study clarified much taxonomic confusion, more work is needed to improve further the classification of the taxa treated here. There are two possible strategies toward this goal: (1) discovering more species and (2) exploiting additional taxonomic characters.

There is no doubt that many new species exist, especially in the Old World tropics, since many species are only known from small numbers of specimens, some only from the holotype. Indeed, the lack of information on both sexes has prevented the resolution of the relationships of many species. By better sampling, we should be able to obtain taxa critical to resolve some unanswered problems in the classification.

Use of previously little-used genitalic characters in this study has already significantly improved the classification. Most major lineages of Chetostomatina and Trypetina are characterized based on male and female genitalic structures, and many external characters, such as chaetotaxy and wing patterns, are homoplastic. Characters from immature stages, although not much used in this study, also have potential taxonomic value. The eggs of the Chetostomatina, so far as is known, are uniquely shaped, thus supporting the monophyly of this taxon (see Section 11.3.2). However, the importance of this character is uncertain because of the small sample size. The modified micropylar end, found in a few species of the Trypetina, may also have potential use as more data become available. In larvae and pupae, useful taxonomic characters may be found from the cephalopharyngeal skeleton (Figure 11.2K), anterior spiracle, posterior spiracle, and cuticle. Such characters already have been used to diagnose many tephritid species (e.g., Kandybina 1966; 1977).

Biological data have enormous potential value to resolve phylogenetic problems and improve the predictive power of the classification. For example, the new delimitation of the genus *Euleia* is convincingly supported by their exclusive leaf-mining behavior on Apiaceae. The monophyly of *Pseudhemilea* is also supported by their leaf-mining behavior in Araliaceae. Since half of the genera treated in this study do not have host records, additional biological data will have considerable impact on the interpretation of their phylogeny.

Other unconventional sources of taxonomic data may also be useful for the establishment of sound higher classification of Tephritidae when used in conjunction with conventional data. Molecular sequence data, for example, can provide many additional taxonomic characters useful in a wide range of taxonomic levels by providing both rapidly evolving and slowly evolving character sets. Some recent studies using molecular sequence data have already provided many interesting insights that could be used to improve tephritid higher classification (Han and McPheron 1994; 1997; McPheron and Han 1997; Smith and Bush 1997; see also Han and McPheron, Chapter 5).

## ACKNOWLEDGMENTS

A large portion of this study was done for my Ph.D. at Pennsylvania State University (Han 1992). I express my sincere appreciation to my advisor, Ke Chung Kim, for his guidance, patience, and personal concern during the study. I am also grateful to B. A. McPheron, J. R. Stauffer, Jr., K. R. Valley, and A. L. Wheeler for reviewing earlier drafts of the work as members of my doctoral committee. I especially thank B. A. McPheron for useful discussions about tephritid systematics and later providing me a great opportunity to conduct molecular systematics. I am greatly indebted to numerous curators for the loan of specimens (75 curators and institutions were listed in Han 1992). Without their cooperation, this study would not have been possible. I am most thankful to A. L. Norrbom, who initially suggested this project and provided me an unpublished list of possible members of the Trypetini. He has also given me valuable advice on a number of occasions. I also appreciate A. Freidberg, I. White, V. Korneyev, B. Merz, X.-J. Wang, and D. L. Hancock for sharing both their tephritid expertise and collections. A. L. Norrbom and two anonymous reviewers kindly commented on this chapter and made helpful suggestions for its improvement. Finally, I thank M. Aluja and his crew for organizing the symposium and making this chapter possible. This study was supported in part by Campaña Nacional contra las Moscas de la Fruta (Mexico), International Organization for Biological Control of Animals and Plants, Instituto de Ecología, A.C. (Mexico), Consejo Nacional de Ciencia y Tecnología (Mexico), Pennsylvania State University, and the Korea Science and Engineering Foundation (Project No. 961-0508-066-2).

## REFERENCES

Beiger, M. 1968. Notes on the Polish flies of the family Trypetidae, Diptera. *Fragm. Faun.* (Warsaw) 15: 45–49 [in Polish with English abstract].

Benjamin, F.H. 1934. Descriptions of some native trypetid flies with notes on their habits. *U.S. Dep. Agric. Tech. Bull.* 401, 95 pp.

Bezzi, M. and A.S. Tavares. 1916. Alguns muscideos cecidogenicos do Brazil. *Brotéria (Sér. Zool.)* 14: 155–170.

Blanchard, E.E. 1929. Descriptions of Argentine Diptera. *Physis* (Buenos Aires) 9: 458–465.

Brink, J.E. 1923. The sunflower maggot (*Straussia longipennis* Wied.). *Annu. Rep. Entomol. Soc. Ont.* 36: 72–74.

Chen, S.H. 1948. Notes on Chinese Trypetinae. *Sinensia* 18: 69–123.

Cox, C.B. and P.D. Moore. 1980. *Biogeography, an Ecological and Evolutionary Approach*, 3rd ed. Blackwell Science, Oxford. 234 pp.

Fletcher, B.S. 1987. The biology of Dacine fruit flies. *Annu. Rev. Entomol.* 32: 115–144.

Freidberg, A. 1994. A second species of *Notommoides* Hancock 1986, with a re-description of the type species (Diptera Tephritidae Trypetinae). *Trop. Zool.* 7: 333–341.

Frick, K.E. 1971. The biology of *Trypeta angustigena* Foote in central coastal California — host plants and notes. *J. Wash. Acad. Sci.* 61: 20–24.

Frost, S.W. 1924. A study of the leaf-mining Diptera of North America. *Cornell University Agric. Expt. Stn. Mem.* 78, 228 pp.

Gardner, J.C.M. 1921. The celery fly, life history, damage to plants and control. *Fruitgrower* 51: 829–831, 863–865, 896–897.

Han, H.-Y. 1992. Classification of the Tribe Trypetini (Diptera: Tephritidae: Trypetinae). Ph.D. dissertation, Pennsylvania State University, University Park, 274 pp.

Han, H.-Y. 1996a. A new *Cornutrypeta* species from Taiwan with notes on its phylogenetic relationships (Diptera: Tephritidae). *Insecta Koreana* 13: 113–119.

Han, H.-Y. 1996b. Taxonomic revision of *Paramyiolia* Shiraki (Diptera: Tephritidae: Trypetinae) with analyses of their phylogenetic relationships. *Entomol. Scand.* 27: 377–391.

Han, H.-Y. 1997a. Redescription of *Oreurinus cuspidatus* Ito (Diptera: Tephritidae). *Korean J. Biol. Sci.* 1: 259–263.

Han, H.-Y. 1997b. Redescription of *Nemerurinus leucocelis* Ito, the type species of the monotypic genus *Nemeurinus* Ito. *Korean J. Entomol.* 27: 117–122.

Han, H.-Y. 1998. A list of the reported host plants of the tribe Trypetini (Diptera: Tephritidae). *Korean J. Entomol.* 28: 355–368.

Han, H.-Y. Molecular phylogenetic study of the tribe Trypetini (Diptera: Tephritidae) using mitochondrial 16S DNA sequences. *Biochem. Syst. Ecol.*, in press.

Han, H.-Y. and B.A. McPheron. 1994. Phylogenetic study of selected tephritid flies (Insecta: Diptera: Tephritidae) using partial sequences of the nuclear 18S ribosomal DNA. *Biochem. Syst. Ecol.* 22: 447–457.

Han, H.-Y. and B.A. McPheron. 1997. Molecular phylogenetic study of Tephritidae (Insecta: Diptera) using partial sequences of the mitochondrial 16S ribosomal DNA. *Mol. Phylogenet. Evol.* 7: 17–32.

Han, H.-Y. and X.-J. Wang. 1997. A systematic revision of the genus *Acidiostigma* Hendel (Diptera: Tephritidae) from China, Korea, and Japan. *Insecta Koreana* 14: 81–118.

Han, H.-Y., X-J. Wang, and K.C. Kim. 1993. Revision of *Cornutrypeta* Han and Wang, a new tephritid genus proposed for Oriental and Palaearctic species (Diptera: Tephritidae). *Entomol. Scand.* 24: 167–184.

Han, H.-Y., X-J. Wang, K.C. Kim. 1994a. *Paratrypeta* Han and Wang, a new genus of Tephritidae (Diptera) from China. *Orient. Insects* 28: 49–56.

Han, H.-Y., X-J. Wang, and K.C. Kim. 1994b. Taxonomic review of *Pseudina* Malloch (Diptera: Tephritidae) with descriptions of two new species from China. *Orient. Insects* 28: 103–123.

Hancock, D.L. 1985. Trypetinae (Diptera: Tephritidae) from Madagascar. *J. Entomol. Soc. South. Afr.* 48: 283–301.

Hancock, D.L. 1986. Classification of the Trypetinae (Diptera: Tephritidae), with a discussion of the Afrotropical fauna. *J. Entomol. Soc. South. Afr.* 49: 275–305.

Hancock, D.L. and R.A.I. Drew. 1994a. Notes on some Pacific Island Trypetinae and Tephritinae (Diptera: Tephritidae). *Aust. Entomol.* 21: 21–30.

Hancock, D.L. and R.A.I. Drew. 1994b. New species and records of Asian Trypetinae (Diptera: Tephritidae). *Raffles Bull. Zool.* 42: 555–591.

Hancock, D.L. and R.A.I. Drew. 1995. New genus, species and synonyms of Asian Trypetinae (Diptera: Tephritidae). *Malay. J. Sci.* 16A: 45–59.

Hardy, D.E. 1973. The fruit flies (Tephritidae-Diptera) of Thailand and bordering countries. *Pac. Insects Monogr.* 31, 353 pp.

Hardy, D.E. 1974. The fruit flies of the Philippines (Diptera: Tephritidae). *Pac. Insects Monogr.* 32, 266 pp.

Hardy, D.E. 1987. The Trypetini, Aciurini and Ceratitini of Indonesia, New Guinea and adjacent islands of the Bismarcks and Solomons (Diptera: Tephritidae: Trypetinae). *Entomography* 5: 247–374.

Hardy, D.E. and M.S. Adachi. 1956. Diptera: Tephritidae. *Insects Micronesia* 14: 28 pp.

Hendel, F. 1914. Die Gattungen der Bohrfliegen. *Wein. Entomol. Ztg.* 33: 73–98.

Hendel, F. 1915. H. Sauter's Formosa-Ausbeute. Tephritinae. *Ann. Hist. Nat. Mus. Natl. Hung.* 13: 424–467.

Hoffmeister, T. 1992. Aspekte der Partnerfindung, Konkurrenz und Parasitierung frugivorer Bohrfliegen (Diptera: Tephritidae). Ph.D. dissertation, Christian Albrechts Universität, Kiel. 140 pp.

Ito, S. 1983–1985. *Die Japanischen Bohrfliegen.* Maruzen Co., Ltd., Osaka [pp. 1–48 published 1983; pp. 49–288 published 1984; pp. 289–352 published 1985].

Ito, S. and N. Tamaki. 1995. Fruitflies caught in fluorescent light trap on Mt. Akagi, Central Honshu, Japan (2). *Yosegaki* 74: 1787–1790 [in Japanese].

Jang, E.B. and D.M. Light. 1996. Olfactory semiochemicals of tephritids. In *Fruit Fly Pests: A World Assessment of Their Biology and Management* (B.A. McPheron and G.J. Steck, eds.), pp. 73–90. St. Lucie Press, Delray Beach.

Jenkins, J. 1996. Systematic Studies of *Rhagoletis* and Related Genera (Diptera: Tephritidae). Ph.D. dissertation, Michigan State University, East Lansing. 184 pp.

Kandybina, M.N. 1966. Contribution to the study of fruit flies (Tephritidae, Diptera) in the Far East of the USSR. *Entomol. Rev.* 45: 383–388 [English translation of *Entomol. Obozr.* 45: 677–687, 1996].

Kandybina, M.N. 1977. The larvae of fruit-flies (Diptera, Tephritidae). *Keys to the Fauna of the USSR* (114): 1–212 [in Russian].

Korneyev, V.A. 1982. A new fruit fly from the Miocene of the northern Caucasus. *Paleontol. J.* 16: 95–97 [English translation of *Paleontol. Zh.* 1982: 97–98].

Korneyev, V.A. 1991a. Tephritid flies of the genera allied to *Euleia* (Diptera, Tephritidae) of the USSR fauna. *Vestn. Zool.* 1991(3): 8–17 [in Russian].

Korneyev, V.A. 1991b. Tephritid flies of the genera allied to *Euleia* (Diptera, Tephritidae) of the USSR fauna. *Vestn. Zool.* 1991(3): 30–37 [in Russian].

Korneyev, V.A. 1996. Reclassification of Palaearctic Tephritidae (Diptera). Communication 3. *Vestn. Zool.* 1995(5–6): 25–48.

Korneyev, V.A. 1998. New data and nomenclatural notes on the Tephritidae (Diptera) of Far East Russia. II. *J. Ukr. Entomol. Soc.* 3: 5–48.

Kwon, Y.J. 1985. Classification of the fruitfly-pests from Korea. *Insecta Koreana* (Ser. 5): 49–112.

Leefmans, S. 1930. Ziekten en Plagen der Cultuurgewassen in Nederlandsch Oost-Indie in 1929. *Meded. Inst. Plziek.* 79: 1–100.

Leroi, B. 1974. A study of natural populations of the celery leaf-miner, *Philophylla heraclei* L. (Diptera, Tephritidae). II. Importance of changes of mines for larval populations. *Res. Popul. Ecol.* 15: 163–182 [in French].

Leroi, B. 1977. Relations biocoenotiques de la mouche du celeri, *Philophylla heraclei* L. (Diptere, Tephritidae): necessite de vegetaux complementaire pour les populations vivant sur celeri. In *Comportement des Insects et Milieu Ttrophique* (V. Labeyrie, ed.), pp. 443–454. Tours, France, Sept. 13–17, *Colloq. Int. C.N.R.S.* 265, 493 pp.

Malloch, J.R. 1939. Solomon Islands Trypetidae. XVI. *Ann. Mag. Nat. Hist.* ser. 11, 4: 228–278.

Mayr, E. and P.D. Ashlock. 1991. *Principles of Systematic Zoology,* 2nd ed.. McGraw-Hill, New York. 475 pp.

McPheron, B.A. and H.-Y. Han. 1997. Phylogenetic analysis of North American *Rhagoletis* (Diptera: Tephritidae) and related genera using mitochondrial DNA sequence data. *Mol. Phylogenet. Evol.* 7: 1–16.

Meijere, J.C.H. de. 1911. Studien über südostasiatische Dipteren, VI. *Tijdschr. Entomol.* 54: 258–432.

Munro, H.K. 1938. Studies on Indian Trypetidae (Diptera). *Mem. Indian Mus.* 40: 27–37.

Niblett, M. 1956. Notes on leaf-mining Diptera. *Proc. Trans. S. Lond. Entomol. Nat. Hist. Soc.* 1956: 151–153.

Norrbom, A.L. 1988. A revision of the neotropical genus *Polionota* Wulp (Diptera: Tephritidae). *Folia Entomol. Mex.* (1987) 73: 101–123.

Norrbom, A.L., L.E. Carroll, and A. Freidberg. 1999a. Status of knowledge. In *Fruit Fly Expert Identification System and Systematic Information Database* (F.C. Thompson, ed.), pp. 9–47. *Myia* (1998) 9, 524 pp.

Norrbom, A.L., L.E. Carroll, F.C. Thompson, I.M. White, and A. Freidberg. 1999b. Systematic database of names. In *Fruit Fly Expert Identification System and Systematic Information Database* (F.C. Thompson, ed.), pp. 65–251. *Myia* (1998) 9, 524 pp.

Permkam, S. and D.L. Hancock. 1995. Australian Trypetinae (Diptera: Tephritidae). *Invertebr. Taxon.* 9: 1047–1209.

Phillips, V.T. 1946. The biology and identification of trypetid larvae (Diptera: Trypetidae). *Mem. Am. Entomol. Soc.* 12: 161 pp.

Richter, V.A. 1963. A report on the fruit flies of Soviet Far East. *Zool. Zh.* 42: 770–772 [in Russian].

Robineau-Desvoidy, J.B. 1830. Essai sur les Myodaires. *Mém. Prés. Div. Sav. Acad. R. Sci. Inst. Fr.* [ser. 2] 2, 830 pp.

Sasakawa, M. 1955. Morphological and biological notes on the larvae of three leaf-mining Trypetid flies. *Kontyu* 22: 53–57 [in Japanese].

Shiraki, T. 1933. A systematic study of Trypetidae in the Japanese Empire. *Mem. Fac. Sci. Agric. Taihoku Imp. Univ.* 8 (Entomol. 2), 509 pp.

Smith, J.J. and G.L. Bush. 1997. Phylogeny of the genus *Rhagoletis* (Diptera: Tephritidae) inferred from DNA sequences of mitochondrial cytochrome oxidase II. *Mol. Phylogenet. Evol.* 7: 33–43.

Steyskal, G.C. 1945. Behavior of *Strauzia longipennis* var. *vittigera* Loew (Diptera, Trypetidae). *Bull. Brooklyn Entomol. Soc.* 39: 156.

Steyskal, G.C. 1986. Taxonomy of the adults of the genus *Strauzia* Robineau-Desvoidy (Diptera, Tephritidae). *Insecta Mundi* 1: 101–117.

Stoltzfus, W.B. 1988. The taxonomy and biology of *Strauzia* (Diptera: Tephritidae). *J. Iowa Acad. Sci.* 95: 117–126.

Tauber, M.J. and Toschi, C.A. 1965. Bionomics of *Euleia fratria*. *Can. J. Zool.* 43: 369–379.

Tseng, Y.-H., Y.-I. Chu, and C.-C. Chen. 1992a. A new genus *Cervarita* new genus of fruit flies from Taiwan (Diptera: Tephritidae: Trypetinae). *Chin. J. Entomol.* 12: 17–20.

Tseng, Y.-H., Y.-I. Chu, and C.-C. Chen. 1992b. A new genus *Neomyoleja* new genus of fruit flies from Taiwan (Diptera: Tephritidae: Trypetinae). *Plant Prot. Bull.* (Taichung) 34: 171–174.

Wang, X.-J. 1990. Notes on a new genus and species of Trypetinae from China. *Acta Zootaxon. Sin.* 15: 358–361 [in Chinese with English summary].

Wang, X.-J. 1996. The fruit flies (Diptera: Tephritidae) of the East Asia Region. *Acta Zootaxon. Sin.* (Suppl.), 338 pp.

White, I.M. 1988. Tephritid flies (Diptera: Tephritidae). *Handbk. Ident. Br. Insects* 10(5a): 134 pp.

White, I.M. and M.M. Elson-Harris. 1992. *Fruit Flies of Economic Significance: Their Identification and Bionomics.* CAB International, Wallingford. 601 pp.

Wiley, E.O. 1981. *Phylogenetics: The Theory and Practice of Phylogenetic Systematics.* John Wiley & Sons, New York. 439 pp.

Zia, Y. and Chen, S.H. 1938. Trypetidae of north China. *Sinensia* 9: 1–180.

# 12 Phylogeny of the Genera *Anastrepha* and *Toxotrypana* (Trypetinae: Toxotrypanini) Based on Morphology

*Allen L. Norrbom, Roberto A. Zucchi, and Vicente Hernández-Ortiz*

**CONTENTS**

## 12.1 INTRODUCTION

*Anastrepha* Schiner is the largest and most economically important genus of Tephritidae in the Americas, including major pest species such as the Mexican, Caribbean, and South American fruit flies. The closely related genus *Toxotrypana* Gerstaecker also contains economically important species, including the papaya fruit fly. Despite their economic status, little has been published about the phylogeny of *Anastrepha* and *Toxotrypana*, as is true for most groups of Tephritidae. In this chapter, we analyze the relationships among the species of these two genera based on morphological character data, including new as well as previously published information. We also summarize host plant data at the level of plant family and discuss several trends in host plant relationships.

Other than the phylogenetic analysis of the 13 species of the *A. daciformis* group by Norrbom (1998), there has been no rigorous cladistic analysis of *Anastrepha* or *Toxotrypana* using computer parsimony programs such as Hennig86 or PAUP. Although we have analyzed character polarities (see Table 12.2) and present some preliminary phylogenetic hypotheses in this chapter, we have not conducted a rigorous analysis because we doubt that there would be much resolution beyond the species group level at this time, and even some of the species groups would not be supported because of missing data or variable characters. Compared with the number of species, relatively few characters have been discovered that appear to be useful for phylogenetic analysis in these genera. Many of the taxonomically useful characters, such as aculeus length, intergrade to such an extent that dividing them into character states is problematic. For most of the morphological characters we have studied, the apomorphic state occurs in a relatively small number of species. We have found few apomorphies above this level. Thus our discussion focuses on the definition of the species groups. We indicate whether diagnostic character states are apomorphic, and if they are not unique, in what other species they are present. Some judgment regarding which character states are likely to be homoplastic is thus implied, but such hypotheses need to be further tested by more rigorous analysis and additional character data.

## 12.2 MORPHOLOGICAL CHARACTERS USEFUL IN *ANASTREPHA* AND *TOXOTRYPANA*

Color patterns on the body are useful characters in *Toxotrypana* and many *Anastrepha* species (Figures 12.1 and 12.2). The markings on the thorax are the most important taxonomically, although in some species there are markings on the head (e.g., some species of the *A. daciformis* group and males of some species of the *A. schausi* group) or the abdomen (e.g., *A. serpentina* group). *Anastrepha* species have a pair of sublateral stripes of the xanthine type, and frequently an unpaired medial one, although in generally pale-colored species they are often not well contrasted and may be difficult to see, especially in dried, preserved specimens. But they are obvious in generally dark-bodied species (Figure 12.1B, J). In the *A. fraterculus* group, in which most of the body is yellow or orange, the subscutellum and/or the mediotergite may have a characteristic dark brown lateral stripe or spot (Figure 12.2E, F). These markings vary intraspecifically in some species, but are nonetheless very useful taxonomically.

Microtrichia patterns, especially on the scutum, are often useful characters in *Anastrepha* (Norrbom 1985). As in most Tephritidae, the microtrichia are extremely fine and can be differentiated well only with a scanning electron microscope (see Figure 33.2C). Under the light microscope, where they are present the body surface looks duller or matte if they are very dense. Where there are no microtrichia, the cuticle appears shiny. The scutal microtrichia should not be confused with the scutal setulae, which are much larger and have alveoli, or sockets. In *Anastrepha* the scutum is frequently entirely microtrichose, but in many species it is bare, microtrichose only laterally and posteriorly, or sometimes (e.g., in the *striata* and *doryphoros* groups) there are nonmicrotrichose stripes. This character is subtle and difficult to see (particularly on specimens in alcohol or those from McPhail traps that often are covered with fine particulate matter), but is very

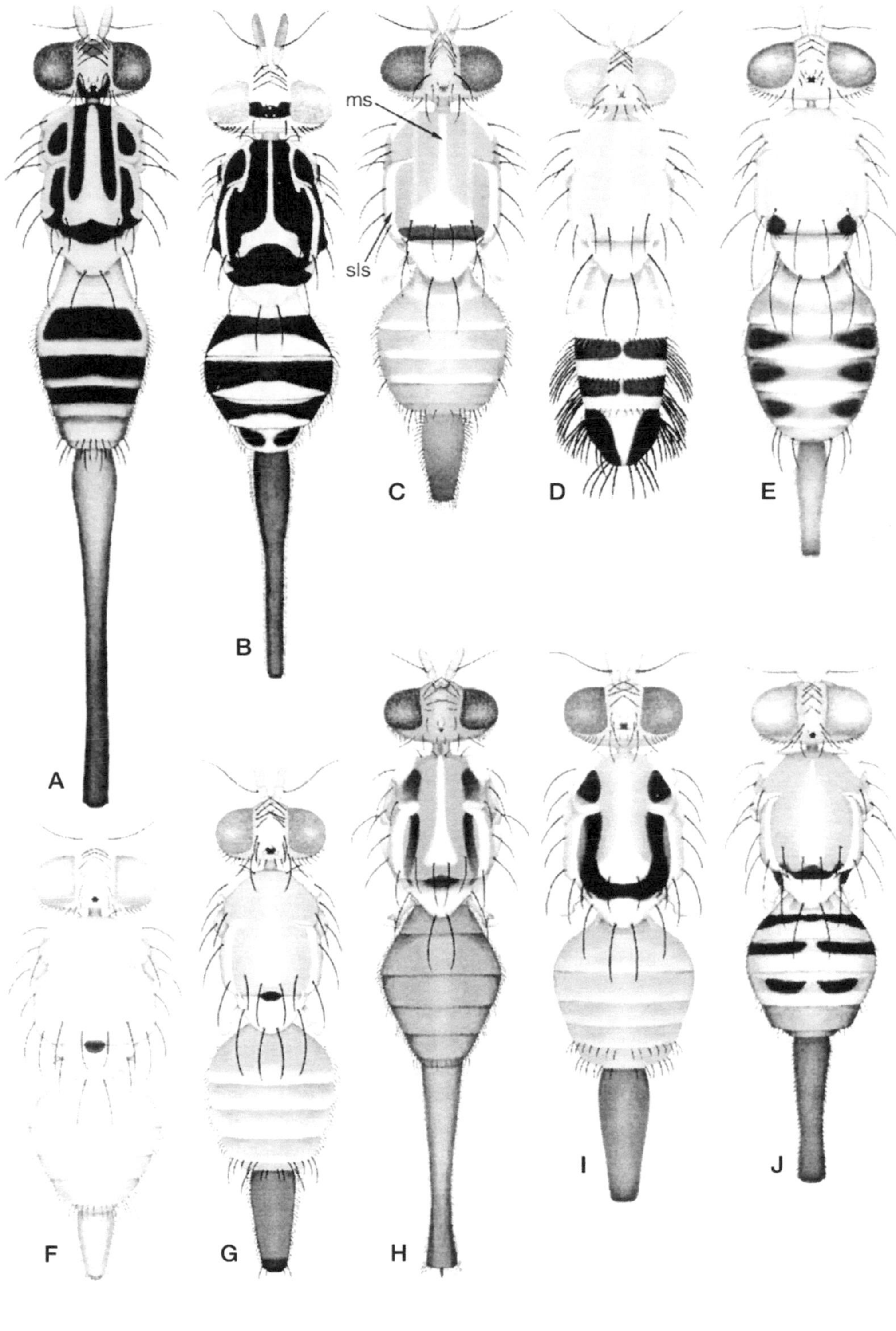


**FIGURE 12.1** Body color in *Anastrepha*: (A) *cordata*, *cryptostrepha* group; (B) *bicolor*, *daciformis* group; (C) *robusta*, *robusta* group; (D) *bellicauda* male, *schausi* group; (E) *punctata*, *punctata* group; (F) *spatulata*, *spatulata* group; (G) *suspensa*, *fraterculus* group; (H) *grandis*, *grandis* group; (I) *striata*, *striata* group; (J) *ocresia*, *serpentina* group; ms = medial stripe; sls = sublateral stripe.

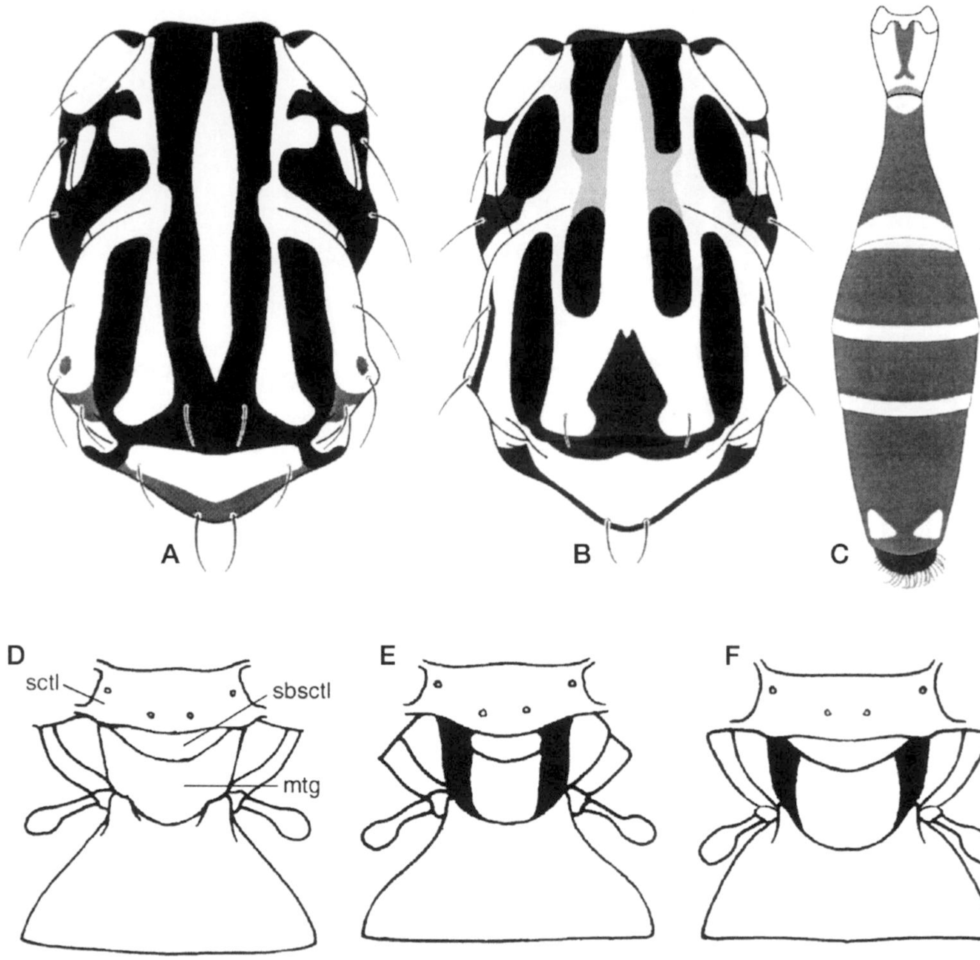


**FIGURE 12.2** Thorax and abdomen color in *Toxotrypana* and *Anastrepha*: (A,B) thorax, dorsal view; (C) abdomen; (D to F), subscutellum and mediotergite color patterns; (A) *T. littoralis*; (B,C) *T. australis*; (D) *A. manihoti*, *spatulata* group; (E) *A. bahiensis*, *fraterculus* group; (F) *A. obliqua*, *fraterculus* group; mtg = mediotergite; sctl = scutellum; sbsctl = subscutellum.

useful taxonomically. Its appearance may vary depending upon the angle of observation, and it is best viewed from an oblique anterodorsal angle.

The facial carina (= clypeal ridge of Stone 1942) is produced in some *Anastrepha* species (see Norrbom 1997). This is best seen in lateral view. In most species it is straight or concave in profile (Figure 12.4A, B), but the medial or dorsal parts may project giving a convex appearance (Figure 12.4C, E, F).

Most species of *Anastrepha* have a similar wing pattern (Figures 12.5E and 12.6A), consisting of three bands that have been termed the C-band, S-band, and V-band (Stone 1942; Steyskal 1977b). The C-band, or costal band, runs from the wing base along the anterior margin to the apex of vein $R_1$, filling cells bc, c, sc, and the bases of $r_1$, $r_{2+3}$, and br. It is often fainter in cells bc and c and/or is darker in the pterostigma. The S-band, which is somewhat S-shaped if viewed from the wing apex, runs from cell bcu, obliquely across R-M to the anterior wing margin, and then follows the margin to beyond the apex of vein $R_{4+5}$. The V-band forms an inverted V, with its apex on or near vein $R_{4+5}$; the subapical band (covering DM-Cu) forms the proximal arm, and the posterior apical band (crossing cell m) forms the distal arm. The wing patterns of about 95% of the species of

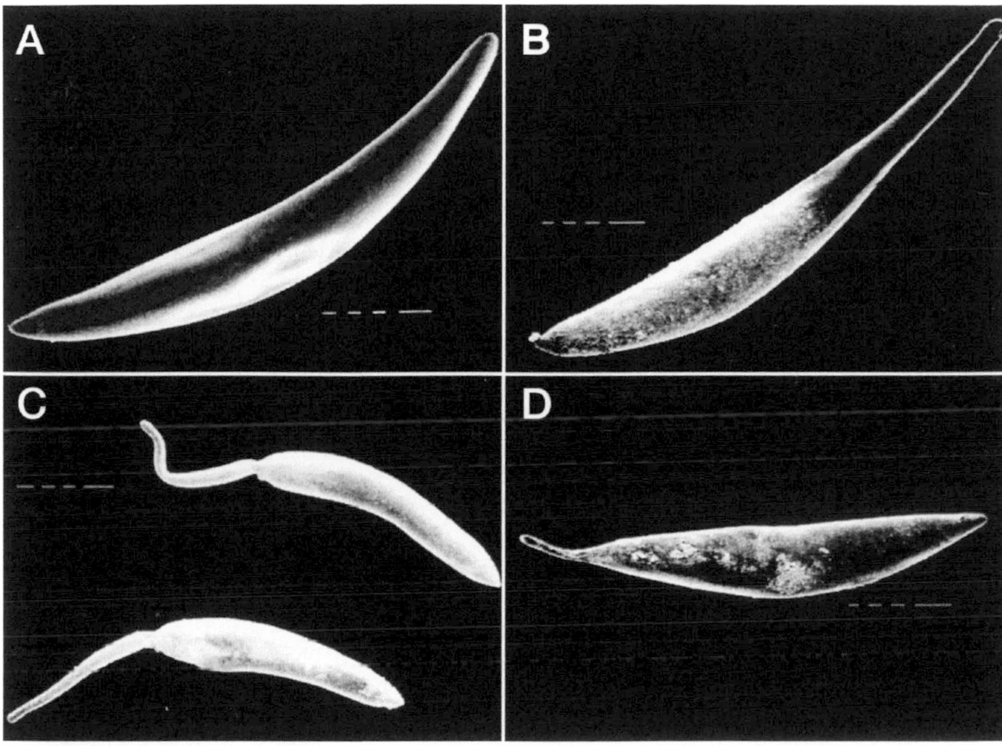


**FIGURE 12.3** *Anastrepha* eggs, scanning electron micrographs: (A) *leptozona*, *leptozona* group; (B) *ludens*, *fraterculus* group; (C) *pittieri*, *robusta* group; (D) *obliqua*, *fraterculus* group. (From Norrbom, A.L., Ph.D. dissertation, Pennsylvania State University, University Park, 1985.)

*Anastrepha* have some variation of this pattern or are easily derived from it. The three bands may be separated (Figure 12.6A) or connected (Figure 12.5E); usually they touch along vein $R_{4+5}$ when they are connected, although they occasionally connect elsewhere. Parts of some bands may be absent, especially the apex and/or the distal arm of the V-band.

A few species of *Anastrepha* have diffuse wing patterns, for example, *A. doryphoros* and *obscura*, but in most cases species closely related to them have typical or intermediate patterns. The male of *A. bellicauda* also has a very diffuse pattern (Figure 12.5J), but the recently discovered female has a more typical *Anastrepha* type pattern (Figure 12.5I). A few other species of *Anastrepha* and all of the species of *Toxotrypana* have a wasp mimicry wing pattern (Figure 12.5A, D), with only a long costal band that is not interrupted at the apex of vein $R_1$, and an infuscated area in cell bcu and along vein $A_1$+$Cu_2$. There are species with intermediate patterns in the *A. daciformis* and *grandis* groups (Figures 12.5C and 12.6C).

The main useful characters of the male genitalia (see Figure 12.8) in *Anastrepha* and *Toxotrypana* are the length of the phallus, which is correlated with the length of the female genitalia, and the shape of the surstyli, especially the lateral surstylus (Figures 12.8 through 12.10).

The female genitalia (Figure 12.12) of *Anastrepha* and *Toxotrypana* have an enlarged dorsobasal area of the eversible membrane that bears moderately to greatly enlarged scales or teeth (Figures 12.12A, C and 12.13 through 12.15). These large teeth have been called "the rasper" although there is no evidence regarding their function and use of that term is not recommended.

The shape, dentition, and length of the aculeus tip (Figures 12.16 and 12.17) are extremely variable in *Anastrepha* and along with aculeus length are among the most important taxonomic characters at the species level. The length of the aculeus tip has been defined as the distance from the apex of the inner margin of the sclerotized area on the ventral side of the aculeus (Figure 12.12D,

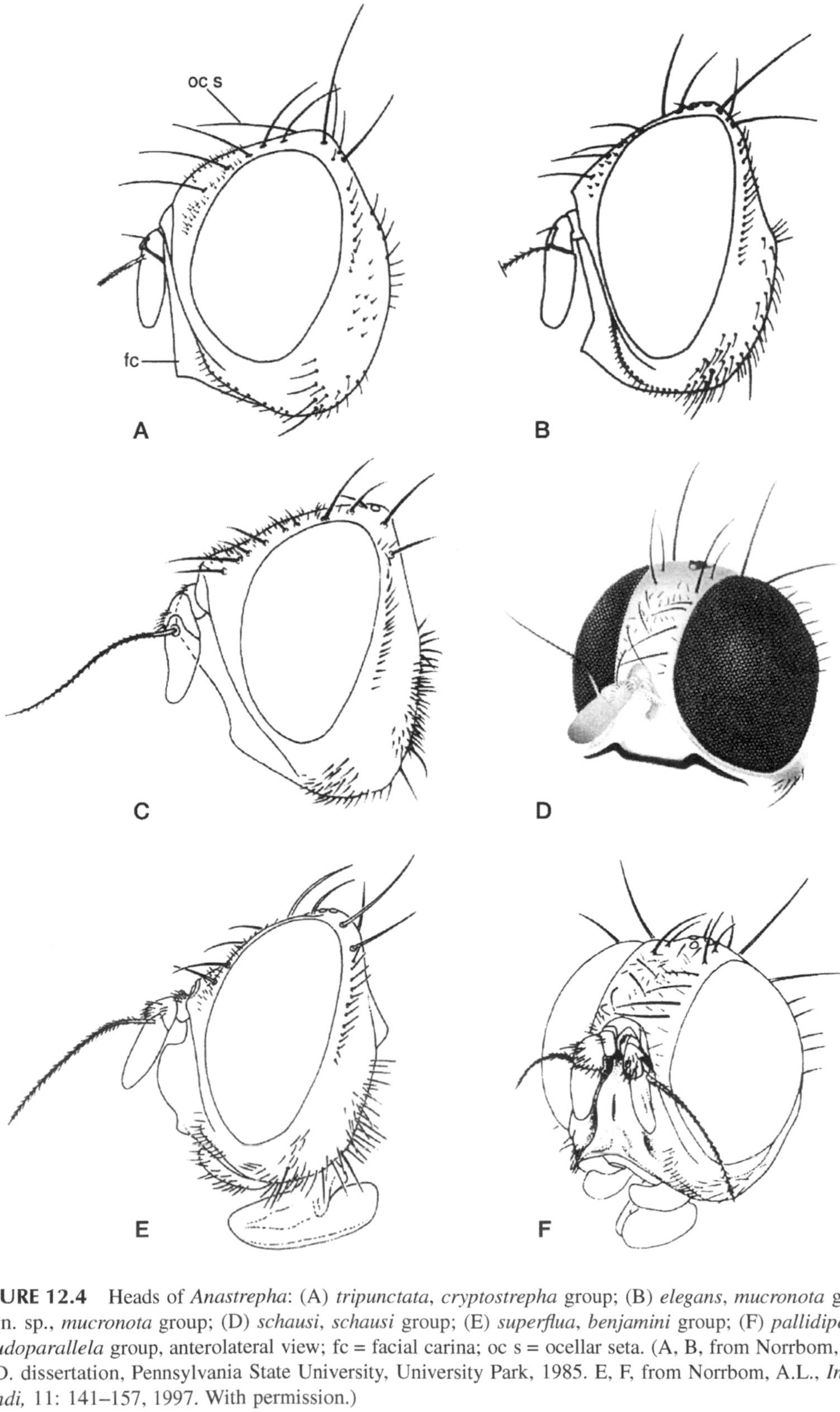


**FIGURE 12.4** Heads of *Anastrepha*: (A) *tripunctata*, *cryptostrepha* group; (B) *elegans*, *mucronota* group; (C) n. sp., *mucronota* group; (D) *schausi*, *schausi* group; (E) *superflua*, *benjamini* group; (F) *pallidipennis*, *pseudoparallela* group, anterolateral view; fc = facial carina; oc s = ocellar seta. (A, B, from Norrbom, A.L., Ph.D. dissertation, Pennsylvania State University, University Park, 1985. E, F, from Norrbom, A.L., *Insecta Mundi,* 11: 141–157, 1997. With permission.)

arrow) to the extreme apex. The former has sometimes been referred to as the opening of the cloaca or oviduct, but the cloacal opening is actually membranous and difficult to see, and is slightly more basal.

Regarding the immature stages, eggs have been described to some extent for *T. curvicauda* and 25 species of *Anastrepha* (Table 12.1). There are some phylogenetically useful characters in their gross morphology (Figure 12.3), particularly the presence of lobes in some species. Other characters, involving surface sculpture and vestiture, have been described in a few species, but need to be studied in greater detail (e.g., as by Murillo and Jirón 1994 and Selivon and Perondini 1999) and in more species. The lobes on the eggs, at least in *A. obliqua*, have a respiratory function (Seín 1933; Murillo and Jirón 1994). Their presence may be a relatively plastic character (i.e., easily evolved and subject to homoplasy). There are species with similar lobes (although often with the micropyle at the apex of the lobe) and other species without them within each of the genera *Aciurina*, *Chaetorellia*, and *Rhagoletis*, and species with lobes are also known in *Craspedoxantha* and *Paracantha* (Tauber and Tauber 1967; Freidberg 1985; White and Marquardt 1989; Headrick and Goeden 1990; 1993; Frías et al. 1993).

The larval stages of *Anastrepha* and *Toxotrypana* are also poorly known. First and/or second instars have been described only for *A. bistrigata*, *A. grandis*, and *A. ludens* (Steck and Malavasi 1988; Steck and Wharton 1988; Carroll and Wharton 1989). Third instars have been described to some extent for *T. curvicauda* and 16 species of *Anastrepha* (Norrbom 1985; Heppner 1986; Steck et al. 1990; White and Elson-Harris 1992). Most of the taxonomically useful characters are difficult to interpret phylogenetically. Most, such as the number of oral ridges, number of tubules of the anterior spiracle, or presence of spinules on the thorax and abdomen, are meristic and/or overlap considerably among the species and are difficult to divide into character states for phylogenetic analysis.

## 12.3 RELATIONSHIPS OF *ANASTREPHA* AND *TOXOTRYPANA*

*Anastrepha* and *Toxotrypana* belong to the tribe Toxotrypanini, which otherwise includes only the genus *Hexachaeta* Loew. All three genera are primarily Neotropical, although a few species of each genus extend slightly into the Nearctic Region.

*Hexachaeta*, which is currently being revised by Hernández-Ortiz, includes about 25 described and numerous undescribed species. These species comprise several groups that are well defined by morphological characters. *Hexachaeta*'s affinities have been enigmatic (Foote et al. 1993). It has been included in the Acanthonevrini on the basis of having three pairs of scutellar setae (Foote 1967), but Hancock (1986) and Korneyev (1994) rejected this hypothesis because the aculeus tip is completely fused and there are small denticles on the spermathecae. The latter author placed *Hexachaeta* in a monotypic tribe, Hexachaetini. *Hexachaeta* was tentatively included in the Toxotrypanini by Norrbom et al. (1999a) based mainly on results of a molecular study by Han and McPheron (1997). Although there is some morphological resemblance between *Hexachaeta* and the other two genera, to date no morphological synapomorphies have been proposed to support or contradict their relationship. Some *Hexachaeta* species, especially the *colombiana* group, somewhat resemble *Anastrepha* in body markings and wing pattern. Most species of *Hexachaeta* have the apical extension of cell bcu relatively large as in both *Anastrepha* and *Toxotrypana*, but this character state is not unique to the Toxotrypanini. The ocellar bristles are poorly developed in the species of the *H. socialis* and *amabilis* groups, as in all species of *Toxotrypana* and all but one or two species of *Anastrepha*. This is an apomorphic character state, but it occurs within various other higher groups of Tephritidae (Hernández-Ortiz, unpublished data). The male genitalia in *Hexachaeta* include very long, slender surstyli. In the female, the aculeus may be simple, multilobed, or lobed and serrate (Lima 1953a, b; 1954).

Although *Toxotrypana* was long placed in the Dacina (Loew 1873; Hardy 1955), there is strong morphological evidence that it forms a monophyletic group with *Anastrepha*. Snow (1895) was the

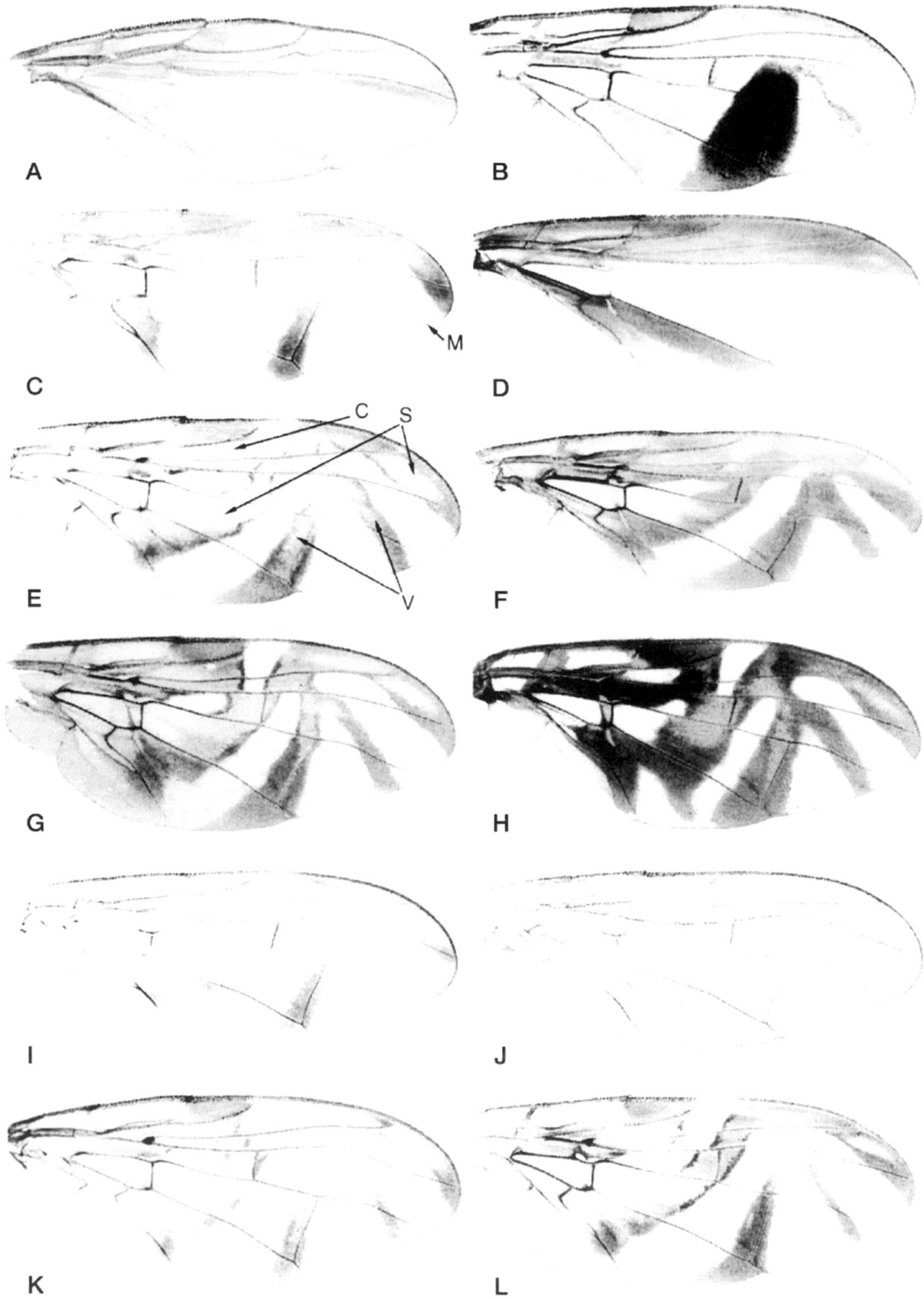


**FIGURE 12.5** Wing patterns of *Anastrepha* and *Toxotrypana*: (A) *T. nigra*; (B) *A. cordata*, *cryptostrepha* group; (C) *A. pallens*, *daciformis* group; (D) *A. zucchii*, *daciformis* group; (E) *A. dentata*, *dentata* group; (F) *A. gigantea*, *benjamini* group; (G) *A. robusta*, *robusta* group; (H) *A. fenestrata*, *robusta* group; (I) *A. bellicauda*, *schausi* group, female; (J) same, male; (K) *A. punctata*, *punctata* group; (L) *A. leptozona*, *leptozona* group; C = C-band; M = medial vein; S = S-band; V = V-band.

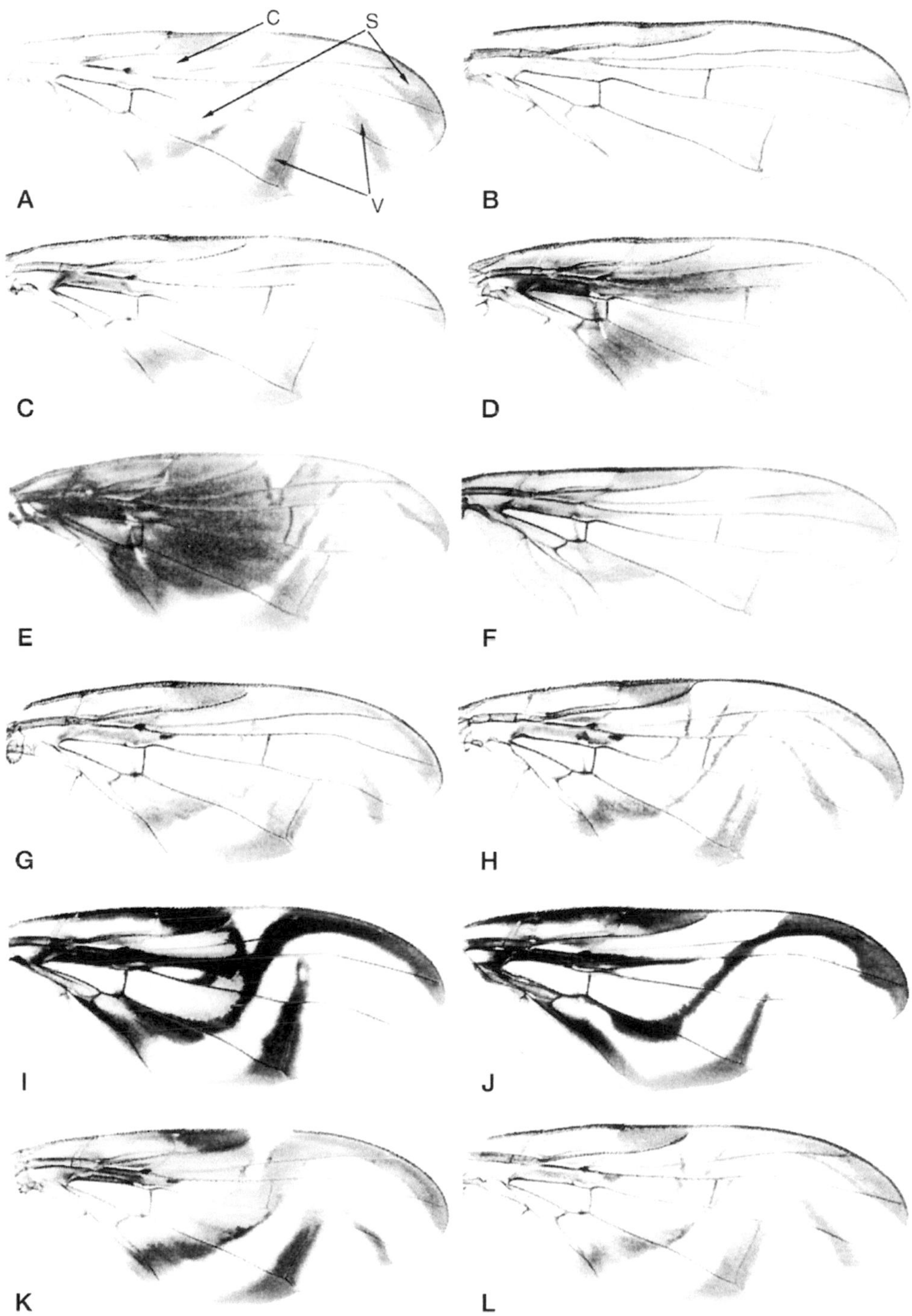


**FIGURE 12.6** Wing patterns of *Anastrepha*: (A) *nunezae*, *mucronota* group; (B) *bezzii*, *mucronota* group; (C) *grandicula*, *grandis* group; (D) *atrigona*, *grandis* group; (E) *fumipennis*, *grandis* group (F) *freidbergi*, *doryphoros* group; (G) *spatulata*, *spatulata* group; (H) *limae*, *pseudoparallela* group; (I) *anomala*, *serpentina* group; (J) *ornata*, *striata* group; (K) *striata*, *striata* group; (L) *canalis*, *fraterculus* group; C = C-band; S = S-band; V = V-band. (E, from Lima, A.M. da Costa, *Mem. Inst. Oswaldo Cruz,* 28:487–575, 1934.)

**TABLE 12.1**
**Dimensions (in mm) and Shape of Eggs of *Anastrepha* and *Toxotrypana* Species**

| Species | Source | No. | Length | Width | Lobe Length | General Shape | Sculpture |
|---|---|---|---|---|---|---|---|
| *T. curvidauda* | Knab and Yothers 1914[a] | ? | 2.55–2.75 | 0.18–0.20 | — | Strongly tapered | Absent |
| *A. cordata* | Norrbom 1985[a]: Holotype (Belize); and Panama (USNM) | 20 | 1.23–1.41 | 0.12–0.14 | — | Strongly tapered | Absent |
| *A. pallens* | Norrbom 1985[a]: Mexico: Reynosa, 19 Feb 1936, Hensley, no. 36-4132 (USNM) | 8 | 1.13–1.27 | 0.08–0.09 | — | Slender | Absent? |
| *A. atrox* | Norrbom 1985: Peru: Anchanchon, 18 Feb 1968, C. A. Korytkowski (USNM) | 2 | 1.41–1.60 | 0.13–0.15 | — | Strongly tapered | ? |
| *A. nigrifascia* | Norrbom 1985[a]: USA: Florida (USNM) | 1 | 1.53 | 0.13 | 0.70 | Slender, with long lobe | ? |
| *A. pittieri* | Norrbom 1985[a]: Venezuela: Rancho Grande, 15–16 Mar 1978, Heppner (USNM) | 10 | 0.85–0.90 | 0.15–0.18 | 0.52–0.60 | Stout, with long lobe | Absent |
| *A. leptozona* | Norrbom 1985[a]: Costa Rica: Golfito, 18 Aug 1957, A. Menke | 10 | 1.30–1.41 | 0.22–0.27 | — | Stout | Absent |
| *A. grandis* | Steck and Wharton 1988[a] | 10 | 2.06–2.25 | 0.20 | — | Strongly tapered | Absent |
| *A. shannoni* | Norrbom 1991[a] | 6 | 1.90–2.02 | 0.22–0.26 | — | Strongly tapered | Absent |
| *A. alveatoides* | Argentina: Cordoba, Jesus Maria, 6 Dec 1994 (USNM) | 2 | 1.35–1.41 | 0.22 | — | Stout | |
| *A. haywardi* | Argentina: Misiones (IML) | 11 | 2.20–2.41 | 0.27–0.33 | — | Tapered | |
| *A. manihoti* | Brazil: MG: Viçosa, 8 Mar 1933, E. J. Hambleton (USNM) | 5 | 1.54–1.71 | 0.25–0.29 | 0.62–0.69 | Stout, with long lobe | |
| *A. pickeli* | | 2 | Not measured | | | Stout, with long lobe | |
| *A. pastranai* | Panama (Z-4686) (USNM) | 5 | Not measured | | — | Stout | |
| *A. pseudoparallela* | Brazil: São Paulo; Oldenberg (USNM) | 6 | 1.04–1.09 | 0.16–0.19 | — | Stout | Absent |
| *A. serpentina* | Emmart 1933[a] | 50 | 1.657 ± 0.011 | 0.207 | — | Tapered | Absent |
| | Selivon and Perondini 1999[a] | | 1.66 ± 0.08 | 0.21 ± 0.01 | — | Tapered | Weak |
| *A. bistrigata* | Steck and Malavasi 1988[a] | 10 | 1.42–1.62 | 0.20–0.25 | — | Stout, apex acute | Strong |

| | | | | | | | |
|---|---|---|---|---|---|---|---|
| *A. striata* | Emmart 1933[a] | 50 | 1.40 ± 0.006 | 0.207 | — | Stout | Strong |
| *A. barbiellinii* | Brazil: Rio Grande do Sul; Stieglmayr (USNM) | 5 | 1.37–1.50 | 0.21–0.23 | 0.13–0.16 | Stout, with short lobe | |
| *A. obliqua* | Norrbom 1985[a]: Dominica: Clarke Hall, 12–18 Oct 1964, P.J. Spangler (USNM) | 10 | 1.15–1.30 | 0.19–0.23 | 0.14–0.16 | Stout, with short lobe | Moderate |
| | Seín 1933 (as *mombinpraeoptans*)[a] | ? | 1.4[b] | 0.4 | | Stout, with short lobe | Absent |
| | Emmart 1933 (as *fraterculus*)[a] | 31 | 1.433 ± 0.012[b] | 0.242 | | Stout, with short lobe | Moderate |
| | Murillo and Jirón 1994[a] | 30 | 1.181–1.584[b] | 0.202–0.259 | 0.086–0.245 | Stout, with short lobe | Moderate |
| *A. ludens* | eggs in alcohol (USNM), source unknown | 10 | 1.20–1.36 | 0.15–0.18 | — | Tapered | Moderate |
| | Carroll and Wharton 1989[a] | 30 | 1.37–1.60 | 0.18–0.21 | — | Tapered | Weak |
| | Emmart 1933[a] | 50 | 1.328 ± 0.07 | 0.20 | — | Tapered | Absent |
| *A. amita* | Trinidad: area XV (SE Caroni), 10 Jul 1990 (USNM47345) | 10 | 0.85–0.96 | 0.14–0.17 | — | Stout | Moderate? |
| *A. fraterculus* | Murillo and Jirón 1994[a] | | Not measured | | | Stout | Strong |
| *A. fraterculus* "I" | Selivon et al. 1997 | 40 | 1.35 ± 0.05 | | — | Stout | |
| *A. fraterculus* "II" | Selivon et al. 1997 | 40 | 1.42 ± 0.07 | | — | Stout | |
| *A. sororcula* | Selivon and Perondini 1999[a] | | 1.11 ± 0.12 | 0.19 ± 0.01 | — | Stout | Strong |
| *A. suspensa* | Seín 1933 (as *unipuncta*)[a] | ? | 1.00 | 0.30 | — | Stout | Weak–moderate |
| | Lawrence 1979 | ? | 1.20 ± 0.20 | 0.30 ± 0.01 | — | | |
| *A. tumida* | Costa Rica, Ala., 20 km S Upala, 21–30 Apr 1991, F.D. Parker | 6 | 1.27–1.35 | 0.22–0.24 | — | Stout | |

*Note*: For species with lobes, length indicates that of the main part of the egg, not including the lobe. Except for *A. ludens*, data for species from Norrbom (1985) or not cited from other publications are based on eggs dissected by Norrbom from abdomens of females.

[a] Includes illustration of egg.

[b] Total egg length; length of lobe not stated.

first to suggest this relationship, but his comments were apparently unknown to, or were ignored by, subsequent taxonomists who classified *Toxotrypana*. Only recently was the relationship of *Anastrepha* and *Toxotrypana* convincingly demonstrated. Kitto (1983) and Sarma et al. (1987) found *T. curvicauda* and *A. suspensa* (Loew) to be similar in an immunological analysis, and suggested that what had been considered strong morphological similarity of *Toxotrypana* and the Dacina was due to convergence. Reexaminations by Norrbom (1985), Hancock (1986), and Norrbom and Foote (1989) demonstrated that the resemblance of *Toxotrypana* and the Dacina actually is superficial (e.g., reduced chaetotaxy, but in different ways in each group). Morphology also strongly indicates the monophyly of *Toxotrypana* and *Anastrepha*, including the following synapomorphies: (1) the eversible membrane of the female is enlarged basally and bears a group of enlarged dorsal teeth (the teeth are secondarily reduced in some species of the *A. dentata* group, but still larger than in this area in other tephritids, and the basal area is still enlarged) (Figures 12.12 through 12.15); (2) the base of the oviscape of the female has lateral, flangelike lobes (smaller in *Toxotrypana*) (Figure 12.12A, B); (3) the male glans is weakly sclerotized medially and has a T-shaped, somewhat hooklike apical sclerite (Figure 12.11); (4) the surstyli are relatively short, the lateral surstylus without anterior or posterior lobes (Figures 12.8 through 12.10); and (5) vein M is anteriorly curved in the distal half of its last segment (distal to DM-Cu) (Figures 12.5 and 12.6). Other characters of uncertain polarity, such as the strongly sclerotized secondary connection of the subepandrial sclerite, the number, size, and arrangement of the sensilla of the aculeus tip (three pairs, none extended beyond lateral margin), and the posterior location of the dorsocentral seta, are similar in *Toxotrypana* and *Anastrepha*. These similarities do not contradict the hypothesis that *Anastrepha* and *Toxotrypana* form a monophyletic group, and some or all may be additional synapomorphies.

Autapomorphies indicating the monophyly of *Toxotrypana* include: (1) vein $R_{2+3}$ with three sharp bends, often with spur veins arising from them (Figure 12.5A); (2) male wing with costal setulae stout (sometimes intraspecifically variable) (Figure 12.5A); (3) wing pattern consisting of only broad complete costal band and faint streak over cell bcu (Figure 12.5A) (this also occurs, probably due to convergence, in some species of the *A. daciformis* and *A. grandis* species groups); (4) subapical bend in vein Sc weak; (5) scutum with a medial longitudinal depression; (6) many setae reduced (ocellar, frontal, orbital, postpronotal, acrostichal, dorsocentral, presutural supra-alar, anepisternal, katepisternal, and basal scutellar setae very small and weak or absent; each is at least sometimes present in at least one species); (7) abdomen petiolate (Figure 12.2C); and (8) body yellow with dark markings, or predominantly dark (Figure 12.2A to C).

Although there is strong morphological evidence for the monophyly of *Toxotrypana*, and for *Toxotrypana* + *Anastrepha*, the same cannot be said for *Anastrepha* alone. The apical curvature of the medial vein (e.g., Figures 12.5C and 12.6) was long considered a diagnostic character for the genus, but in a few species (e.g., Figure 12.5B, J, K) the curvature is weak and no stronger than what occurs within *Toxotrypana*. Within some species of *Toxotrypana* (e.g., *T. nigra*; Figure 12.5A), this vein meets the costa without an obvious angle, thus the range in this character overlaps in the two genera. The anterior bend in the distal half of the distal segment of the medial vein is a synapomorphy for both genera, but some *Anastrepha* species cannot be differentiated from *Toxotrypana* on the basis of having this vein more strongly curved. A possible synapomorphy for *Anastrepha* suggested by Norrbom (1985) is the typical wing pattern, consisting of C-, S-, and V-bands, and occurring in more than 90% of the species. This assumes that the atypical patterns found in a few species are secondary modifications, as appears to have occurred within the *A. daciformis* group in which there is a transition from the typical *Anastrepha* pattern to the wasp mimicry pattern (Norrbom 1998). However, because the wasp mimicry pattern has evolved at least twice within *Anastrepha* (in *A. aberrans* and within the *daciformis* group), the hypothesis that this pattern also could have evolved in *Toxotrypana* from the *Anastrepha* pattern should not be considered unlikely. Recent molecular studies (McPheron et al., Chapter 13) in fact suggest that some species of *Anastrepha* may be more closely related to *Toxotrypana*, or in other words, that *Anastrepha* is not monophyletic without including *Toxotrypana*. This would cause a nomenclatural problem, as the latter name has priority, but if these genera eventually need to be synonymized, *Anastrepha* should be used to conserve usage for the major pest species it includes. Because the monophyly of *Anastrepha* is in doubt, *Toxotrypana* should be included in any analysis of relationships within *Anastrepha*. We have used *Hexachaeta* as the outgroup for the analyses discussed in this chapter.

## 12.4 RELATIONSHIPS WITHIN *TOXOTRYPANA*

*Toxotrypana* includes seven described and at least six undescribed species (see Table 12.4; Norrbom and Zucchi, in preparation). There are two probably monophyletic species groups. The species of the *curvicauda* group have extremely long, strongly curved female terminalia, as in the best-known species, *T. curvicauda*. This is an apomorphic character, although it occurs, probably via convergence, in an undescribed *Anastrepha* species. The other species group within *Toxotrypana* includes at least four species, all undescribed, that have shorter, straight female terminalia, and the apical setae of male tergite 5 often short and stout.

Both species groups occur from Mexico to Central America, and following the Andes to Argentina and southern Brazil. *Toxotrypana curvicauda* also occurs in the Antilles. The known host plants belong to milky latex-bearing families: Asclepiadaceae, Caricaceae, and, possibly, Apocynaceae. The larvae feed on the developing seed tissues inside the thick husked fruits.

## 12.5 INTRAGENERIC CLASSIFICATION AND RELATIONSHIPS WITHIN *ANASTREPHA*

*Anastrepha* includes 197 currently recognized species (see Table 12.4). Of these valid names, about ten are poorly recognized and may be synonyms; they were insufficiently described and were based on lost, damaged, or male-type specimens. On the other hand, we know of nearly 50 additional undescribed species, and there are undoubtedly many more yet to be discovered. Some of what were thought to be widespread species, such as *A. fraterculus*, *A. hamata*, and *A. pickeli* (Norrbom 1985; Steck 1991; Canal 1997), are now known to be cryptic species complexes. Resolution of these complexes and the likelihood that other complexes occur within *Anastrepha* will further increase the total number of species.

The first groupings of species now classified within *Anastrepha* were proposed by Stone (1939a, b), who recognized two separate genera that were subsequently synonymized with *Anastrepha* by Steyskal (1977a): *Pseudodacus* (including four species now in the *daciformis* group) and *Lucumaphila* (including 11 species, 7 now in the *dentata* group). *Acrotoxa*, which was proposed by Loew (1873) without knowledge of Schiner's genus *Anastrepha*, and *Phobema*, proposed by Aldrich (1925) as a monotypic genus, are also synonyms of *Anastrepha*. Shaw (1962) first recognized the *spatulata* group for four species, two of which are now considered synonyms of *A. spatulata*. Korytkowski and Ojeda (1968) recognized ten numbered groups for 35 species they studied from Peru. Steyskal (1977b) recognized four additional species groups, the *benjamini*, *grandis*, *punctata*, and *serpentina* groups. Zucchi (1977) proposed the *fraterculus* species group, and Norrbom (1985) recognized an additional 12 species groups or subgroups, but only some of the results of these dissertations have been formally published and widely circulated. Norrbom and Kim (1988b) further modified the intrageneric classification and provided a checklist of the then recognized 180 *Anastrepha* species divided into 17 species groups, but they did not explain the basis for this classification. Norrbom (1991; 1998) revised the *grandis* and *daciformis* groups and added additional species. Norrbom (1997) suggested that the *benjamini* group (*sensu* Steyskal 1977b) was not monophyletic and transferred some species from it, as well as all of the species

## TABLE 12.2
## Morphological Characters Useful for Phylogenetic Analysis in *Toxotrypana* and *Anastrepha* (the plesiomorphic character state is coded 0)

1. Body color: 0, predominantly yellow or orange; 1, mostly dark orange to dark brown.
2. Facial carina: 0, not strongly produced medially or dorsally, concave or straight in lateral view; 1, strongly produced medially, convex in lateral view; 2, strongly produced dorsally; 3, weak, indistinct.
3. Ocellar seta: 0, well developed; 1, small and weak or absent.
4. Frontal, orbital, postpronotal, acrostichal, dorsocentral, presutural supra-alar, anepisternal, katepisternal, and basal scutellar setae: 0, well developed; 1, very small and weak or absent. Acrostichal setae are variable in *A. cordata* and absent in an undescribed species from Costa Rica.
5. Scutum with a medial longitudinal depression: 0, no; 1, yes.
6. Scutum: 0, without brown markings, or if present, in different pattern; 1, without brown markings except band on posterior margin; 2, with two pairs of dark brown stripes and band on posterior margin; 3, without brown markings except paired, circular spot near posterolateral corner. Additional types of scutal color patterns occur within *Hexachaeta*, and the *daciformis*, *dentata*, *grandis*, *serpentina*, and *striata* species groups; some species in the *daciformis* and *dentata* groups have irregular, often acute spots in the area of the circular spots of the *punctata* group.
7. Scutal microtrichia: 0, mostly or entirely microtrichose; 1, mostly or entirely bare of microtrichia; 2, microtrichose except dorsocentral bare stripe. The coding of this character is oversimplified, as there are additional patterns that could be defined as states.
8. Scutellum: 0, without dark markings, or if any present they are restricted to extreme base, apex, or semicircular area on disk; 1, bicolored, with at least basal third on sides and dorsum distinctly darker than apex.
9. Mediotergite and/or subscutellum with lateral dark brown stripe or spot and body otherwise without dark markings, except sometimes on scuto-scutellar suture: 0, no; 1, yes.
10. Costal setulae of male wing: 0, slender; 1, stout, often spur-like.
11. Vein $R_{2+3}$: 0, more or less straight; 1, sinuate; 2, with sharp bends, often with spur veins arising from them.
12. Vein M distal to DM-Cu: 0, straight or posteriorly curved; 1, anteriorly curved in distal half, but meeting costa at distinct angle; 2, more strongly curved apically, meeting costa in a smooth curve; 3, extremely strongly curved. Because there is nearly continuous variation in this character, it was difficult to divide into states and to code for many taxa; future refinement and rechecking of this character is needed.
13. Wing pattern: 0, not *Anastrepha*-type nor wasp mimic-type; 1, *Anastrepha*-type, with C-, S-, and V-bands and hyaline area at apex of vein $R_1$; 2, *Anastrepha*-type, but without hyaline area at apex of vein $R_1$ (with complete costal band as well as S-band); 3, wasp mimic-type, with only broad complete costal band and faint streak over cell bcu.
14. S-band with basal cleft: 0, weak or absent; 1, strong.
15. Abdomen: 0, not petiolate; 1, petiolate.
16. Abdomen: 0, unicolorous except paler posterior margins of tergites; 1, bicolored, partially dark brown. The dark markings occur in various patterns, some of which probably are not homologous.
17. Lateral surstylus length: 0, elongate, apically with anterior and posterior lobes; 1, short to medium length (shorter than epandrium height), without anterior or posterior lobes but not transversely flattened; 2, short to medium length, oriented obliquely; 3, short to medium length, apically transversely flattened; 4, extremely short, barely extended beyond prensisetae.
18. Lateral surstylus: 0, not bootshaped; 1, short and somewhat bootshaped, with laterally projecting apical lobe.
19. Proctiger: 0, weakly or not creased, sclerotized area usually continuous; 1, with pair of strong lateral creases, sclerotized area divided into three parts.
20. Phallus length: 0, more than 1.25 mm long, glans present; 1, short, less than 1.25 mm long, glans absent; 2, less than 0.30 mm long, glans absent.
21. Glans: 0, strongly sclerotized basally and medially, subapical lobe not T-shaped if present; 1, weakly sclerotized medially, subapical lobe T-shaped.
22. Glans: 0, without spines; 1, with minute spines.
23. Oviscape basally with lateral, flangelike lobes: 0, no; 1, yes.
24. Eversible membrane enlarged basally and bearing group of enlarged dorsal teeth: 0, no; 1, yes.
25. Eversible membrane dorsobasal teeth pattern: 0, usually triangular or semicircular, with teeth well sclerotized and gradually changing in size; 1, all teeth relatively small and weakly sclerotized; 2, small and weakly sclerotized except for medially interrupted apical row of large, strongly sclerotized, hooklike teeth; 3, teeth short and arranged in elongate pattern.

**TABLE 12.2 (continued)**
**Morphological Characters Useful for Phylogenetic Analysis in *Toxotrypana* and *Anastrepha* (the plesiomorphic character state is coded 0)**

26. Aculeus width: 0, greater than 0.05 mm; 1, less than 0.05 mm. Reduction in aculeus width has also occurred in some unplaced *Anastrepha* species.
27. Aculeus tip: 0, well defined, inner margin on ventral side distinct; 1, poorly defined, inner margin on ventral side indistinct.
28. Aculeus tip tapered, then parallel-sided, then tapered: 0, no; 1, yes.
29. Aculeus tip: 0, width < 0.18 mm, or if broad, not blunt; 1, extremely broad (width >0.18 mm), blunt.
30. Aculeus tip: 0, elongate and/or narrow; 1, short, broad, with numerous fine serrations extended beyond base; 2, short, broad, and triangular, with very large serrations extended to base.
31. Spermathecae: 0, moderately sclerotized; 1, weakly sclerotized; 2, membranous.
32. Third instar larva, hairs of posterior spiracle: 0, medium length to long, longer than width of spiracular opening; 1, very short, less than width of spiracular opening.
33. Egg, micropyle end: 0, without lobe; 1, with short lobe; 2, with long lobe.
34. Chromosome number (diploid): 0, 12 in both sexes; 1, 8 in both sexes; 2, female 12, male 11.

of the *chiclayae* group, to the *pseudoparallela* group. In a publication that came to our attention just prior to this chapter going to press, Tigrero (1998) grouped 31 species from Ecuador into 17 species groups. He did not explain the basis for this classification, however, and we have not had sufficient time to evaluate it thoroughly.

Other than that by Norrbom (1998) for the *daciformis* group, there has been no rigorous cladistic analysis of *Anastrepha* based on morphological characters using computer software. Although the relationships among most of the species groups are poorly understood, many of the species groups that have been recognized appear to be monophyletic (the status of others needs further analysis). These species groups and their included species are listed in Table 12.4. Nomenclatural and distributional data for most species are listed by Norrbom et al. (1999b). The limited previous analysis of relationships among the species groups of *Anastrepha* used *Toxotrypana* as the outgroup (Norrbom 1985), but new molecular data (McPheron et al., Chapter 13) suggest that *Anastrepha* may be paraphyletic without including *Toxotrypana*. Our present analysis therefore includes *Toxotrypana* as part of the ingroup, and uses *Hexachaeta* as the outgroup. Table 12.2 lists most of the characters that are discussed and indicates their polarities, and Table 12.3 shows the distributions of their states. A preliminary cladogram indicating possible relationships is shown in Figure 12.7.

### 12.5.1 Relationships among Species Groups of *Anastrepha*

Little morphological evidence has been discovered to indicate the relationships among the species groups of *Anastrepha* and *Toxotrypana*. The *A. cryptostrepha* group may be the most primitive clade (see discussion of that group). There is considerable evidence that the *A. daciformis* and *dentata* groups are sister groups, including the following synapomorphies: (1) phallus short, less than 1.25 mm long, and glans absent (Figure 12.9E); (2) aculeus extremely slender (Figure 12.16C, D), less than 0.05 mm wide except at base (reduction in aculeus width has occurred in certain other *Anastrepha* species, but whether this is convergence or a synapomorphy for some of these species and the *daciformis* + *dentata* groups remains uncertain); and (3) spermathecae weakly sclerotized. Norrbom (1998) suggested another possible synapomorphy — third instar larva with hairs of hind spiracle relatively short — but this character is known for only one species in each group, *A. pallens* and *A. sagittata* (Baker et al. 1944; Phillips 1946), and these hairs are also short in *Toxotrypana*, and relatively short in *A. interrupta* (*spatulata* group) and *A. limae* (*pseudoparallela* group) (Baker et al. 1944; Steck and Wharton 1988), so this character may be homoplasious or have evolved at a lower level. The *daciformis* and *dentata* groups share one additional apomorphy: Lateral surstylus very short and rounded, barely extended beyond the

**TABLE 12.3**
**Character State Distributions in *Toxotrypana* and Species Groups of *Anastrepha* (numbers of characters and states refer to Table 12.3)**

| | Character Numbers | | | | | | | | | | | | | | | | | | | | | | | | | | | | | | | | | |
|---|---|---|---|---|---|---|---|---|---|---|---|---|---|---|---|---|---|---|---|---|---|---|---|---|---|---|---|---|---|---|---|---|---|---|
| | 1 | 2 | 3 | 4 | 5 | 6 | 7 | 8 | 9 | 10 | 11 | 12 | 13 | 14 | 15 | 16 | 17 | 18 | 19 | 20 | 21 | 22 | 23 | 24 | 25 | 26 | 27 | 28 | 29 | 30 | 31 | 32 | 33 | 34 |
| *Hexachaeta* | 0 | 0 | 0,1 | 0 | 0 | 0 | 0 | 0 | 0 | 0 | 0 | 0 | 0 | 0 | 0 | 0,1 | 0 | 0 | 0 | 0 | 0 | 0 | 0 | 0 | ? | 0 | 0 | 0 | 0 | 0 | 0 | 0 | ? | ? |
| *Toxotrypana* | 0,1 | 0 | 1 | 1 | 1 | 2 | 1 | 0 | 0 | 1 | 2 | 1,2 | 3 | 0 | 1 | 0,1 | 4 | 0 | 0 | 0 | 1 | 0 | 1 | 1 | 0 | 0 | 0 | 0 | 0 | 0 | 0 | 1 | 0 | ? |
| *cryptostrepha* | 0 | 0 | 0,1 | 0 | 0 | 0,1,2 | 1 | 0 | 0 | 0 | 0 | 1,2 | 1,2 | 0 | 0 | 0,1 | 1 | 0 | 0 | 0 | 1 | 0 | 1 | 1 | 0 | 0,1 | 0 | 0 | 0 | 0 | 0 | ? | 0 | ? |
| *daciformis* | 0,1 | 0 | 1 | 0 | 0 | 0 | 0,1 | 1 | 0 | 0 | 0 | 2 | 1,2,3 | 0 | 0 | 0,1 | 4 | 0 | 0 | 2 | ? | ? | 1 | 1 | 2 | 1 | 0 | 0 | 0 | 0 | 2 | 1 | 0 | ? |
| *dentata* | 0 | 0 | 1 | 0 | 0 | 0,1 | 0 | 0 | 0 | 0 | 0 | 2 | 1 | 0 | 0 | 0 | 4 | 0 | 0 | 1 | ? | ? | 1 | 1 | 1 | 1 | 1 | 0 | 0 | 0 | 1 | 1 | ? | ? |
| *benjamini* | 0 | 1 | 1 | 0 | 0 | 0 | 0,1 | 0 | 0 | 0 | 0 | 2,1 | 1 | 1 | 0 | 0 | 3 | 0 | 0 | 0 | 1 | 0 | 1 | 1 | 0 | 0 | 0 | 0 | 0 | 0 | 0 | ? | ? | ? |
| *robusta* | 0 | 0 | 1 | 0 | 0 | 1,2 | 0 | 0 | 0 | 0 | 0 | 2,1 | 1 | 1 | 0 | 0 | 3,4 | 0 | 0 | 0 | 1 | 0 | 1 | 1 | 0 | 0 | 0 | 0 | 0 | 0 | 0 | ? | 2 | ? |
| *schausi* | 0 | 3 | 1 | 0 | 0 | 0 | 0,1 | 0 | 0 | 0 | 0 | 1,2 | 1 | 0 | 0 | 0,1 | 3 | 0 | 0 | 0 | 1 | 1 | 1 | 1 | 0 | 0 | 0 | 0 | 0 | 0 | 0 | ? | ? | ? |
| *punctata* | 0 | 0 | 1 | 0 | 0 | 3 | 0 | 0 | 0 | 0 | 0 | 1 | 1 | 0 | 0 | 1,0 | 2 | 0 | 0 | 0 | 1 | 0 | 1 | 1 | 0 | 1 | 0 | 0 | 0 | 0 | 0 | ? | ? | ? |
| *leptozona* | 0 | 0 | 1 | 0 | 0 | 0 | 0 | 0 | 0 | 0 | 0 | 3 | 1 | 0 | 0 | 0 | 3 | 1 | 0 | 0 | 1 | 0 | 1 | 1 | 0 | 0 | 0 | 0 | 0 | 0 | 0 | ? | 0 | 0 |
| *mucronota* | 0 | 0,2 | 1 | 0 | 0 | 0 | 0,1 | 0 | 0 | 0 | 0,1 | 2 | 1,2 | 0 | 0 | 0 | 3 | 0 | 0 | 0 | 1 | 0 | 1 | 1 | 0,3 | 0 | 0 | 0 | 0 | 0 | 0 | 0 | 0 | ? |
| *grandis* | 1 | 0,1 | 1 | 0 | 0 | 0 | 0 | 0 | 0 | 0 | 0 | 2 | 1,2,3 | 0 | 0 | 0,1 | 3 | 0 | 1 | 0 | 1 | 0 | 1 | 1 | 0 | 0 | 0 | 0 | 0 | 0 | 0 | 0 | 0 | ? |
| *doryphoros* | 0 | 0 | 1 | 0 | 0 | 0 | 2 | 0 | 0 | 0 | 0 | 2 | 1 | 0 | 0 | 0 | 3 | 0 | 1 | 0 | 1 | 0 | 1 | 1 | 3 | 0 | 0 | 1 | 0 | 0 | 0 | ? | ? | ? |
| *spatulata* | 0 | 0 | 1 | 0 | 0 | 0 | 0 | 0 | 0 | 0 | 0 | 2,3 | 1 | 0 | 0 | 0 | 3 | 0 | 1 | 0 | 1 | 0 | 1 | 1 | 0 | 0,1 | 0 | 0 | 0 | 0,1 | 0 | 1 | 0,2 | 1 |
| *ramosa* | 0 | 0 | 1 | 0 | 0 | 0 | 0 | 0 | 0 | 0 | 0 | 2 | 1 | 0 | 0 | 0 | 3 | 0 | 1 | 0 | 1 | 0 | 1 | 1 | 0 | 0 | 0 | 0 | 0 | 2 | 0 | ? | ? | ? |
| *pseudoparallela* | 0 | 0,1 | 1 | 0 | 0 | 0 | 0,1 | 0 | 0 | 0 | 0 | 2 | 1 | 0 | 0 | 0 | 3 | 0 | 1 | 0 | 1 | 0 | 1 | 1 | 0 | 0 | 0 | 0 | 0 | 0 | 0 | 1? | ? | 0 |
| *serpentina* | 1 | 0 | 1 | 0 | 0 | 0 | 0 | 0 | 0 | 0 | 0 | 2 | 1 | 0 | 0 | 1 | 3 | 0 | 1 | 0 | 1 | 0 | 1 | 1 | 0 | 0 | 0 | 0 | 0 | 0 | 0 | 0 | 0 | 2 |
| *striata* | 0 | 0 | 1 | 0 | 0 | 0 | 2 | 0 | 0 | 0 | 0 | 2 | 1 | 0 | 0 | 0,1 | 3 | 0 | 1 | 0 | 1 | 0 | 1 | 1 | 0 | 0 | 0 | 0 | 1 | 0 | 0 | 0 | 0 | 0,2 |
| *fraterculus* | 0 | 0 | 1 | 0 | 0 | 0 | 0 | 0 | 1 | 0 | 0 | 2 | 1 | 0 | 0 | 0 | 3 | 0 | 1 | 0 | 1 | 0 | 1 | 1 | 0 | 0 | 0 | 0 | 0 | 0 | 0 | 0 | 0,1 | 0 |

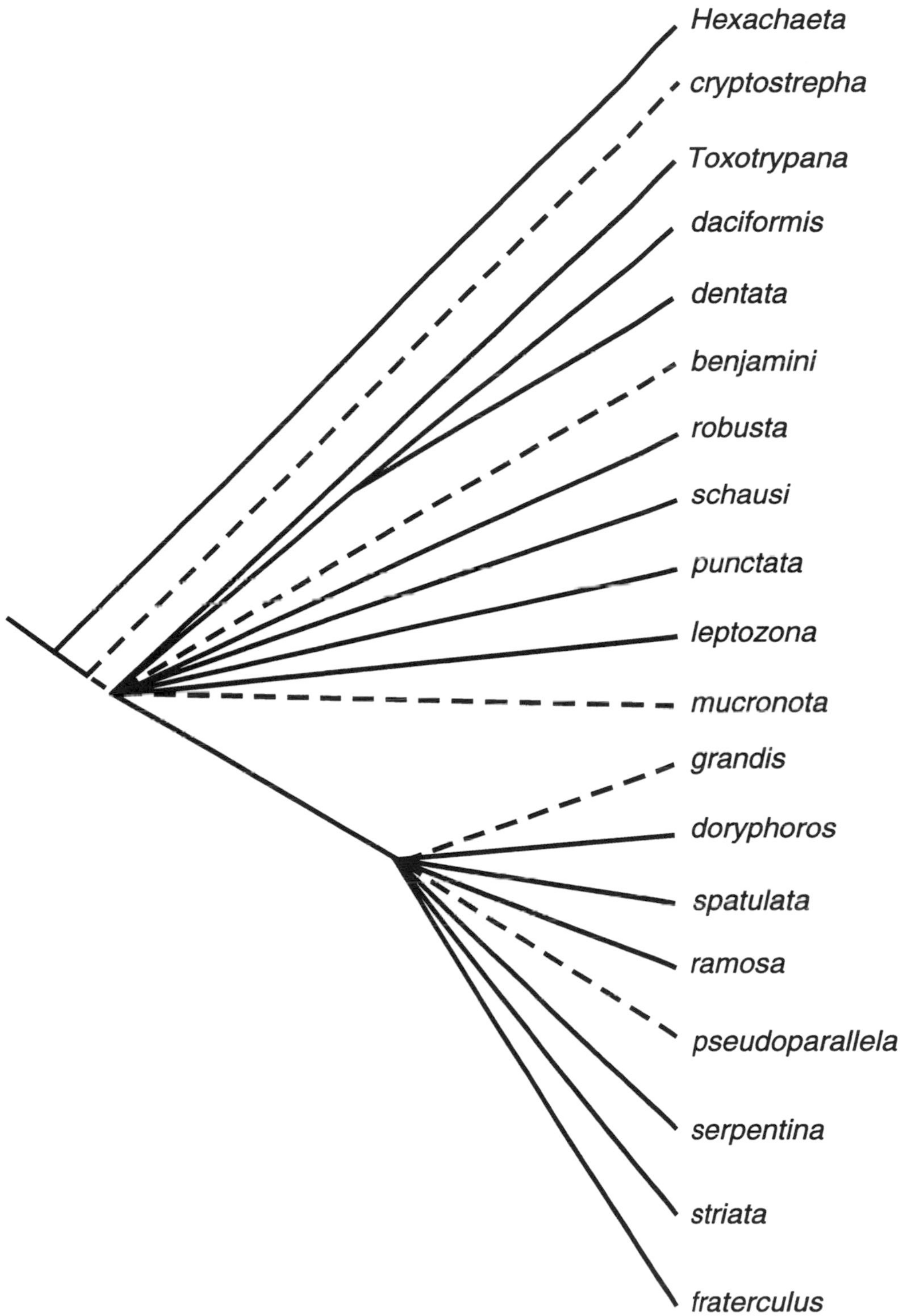


**FIGURE 12.7** Preliminary cladogram showing possible phylogenetic relationships among species groups of *Anastrepha* and *Toxotrypana*.

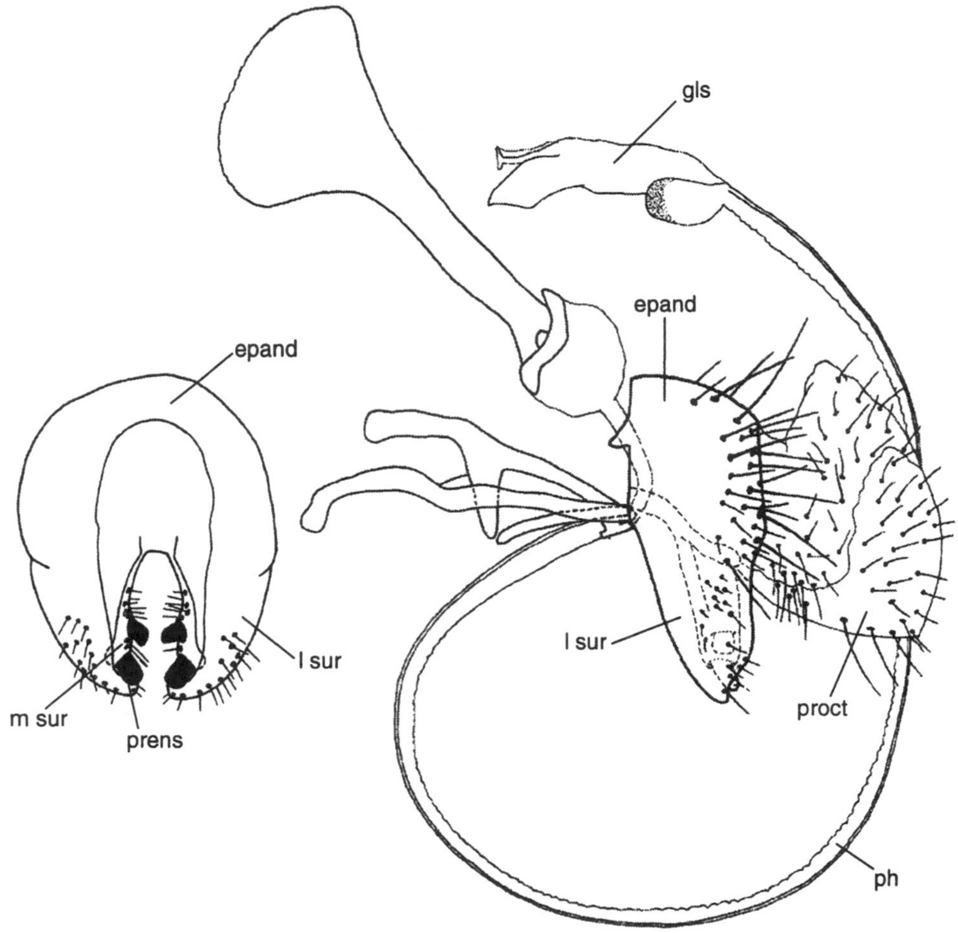


**FIGURE 12.8** Male genitalia, *Anastrepha nigrifascia*, *robusta* group: (A) epandrium and surstyli, posterior view; (B) complete genitalia, lateral view; epand = epandrium; gls = glans; l sur = lateral surstylus; m sur = medial surstylus; ph = phallus; proct = proctiger; prens = prensiseta. (From Norrbom A.L., Ph.D. dissertation, Pennsylvania State University, University Park, 1985.)

prensisetae (Figure 12.9F, H), and subepandrial sclerite relatively posterior in position, in lateral view its apex usually at posterior margin of epandrium (Figure 12.9E, G). But here again, the lateral surstylus is also rather short in *Toxotrypana*, the *A. robusta* group (especially in *A. binodosa*), and to a lesser extent in the *A. punctata*, *leptozona,* and *schausi* groups, so it could possibly be a synapomorphy for some or all of these groups in addition to the *A. daciformis* and *dentata* groups.

The *A. punctata* group resembles some species of the *A. cryptostrepha* and *leptozona* groups in lateral surstylus shape (Figure 12.9I, J), but the surstylus is oriented obliquely rather than being broadest in the longitudinal or transverse directions. Vein M is usually weakly curved apically (plesiomorphy) in the *punctata* and *schausi* groups, as in the *cryptostrepha* group (Figure 12.5B, I to K, but this state occurs in a few species in other species groups as well (e.g., *A. superflua* of the *benjamini* group).

A large clade may be formed by the *A. pseudoparallela* + *spatulata* + *ramosa* + *doryphoros* + *grandis* + *serpentina* + *striata* + *fraterculus* groups. In these groups the male proctiger has strong lateral creases, with the sclerotized area divided into three parts (Figure 12.10E, I, K, N). In other *Anastrepha* species groups and *Toxotrypana* it is weakly or not creased and the sclerotized area is usually continuous (Figures 12.8B and 12.9A, C, E, G).

### 12.5.2 Species Groups of *Anastrepha*

#### 12.5.2.1 The *cryptostrepha* Group

The following character states suggest that this group may be monophyletic: (1) male lateral surstylus very broad in lateral view (Figure 12.9C, D), not transversely flattened (polarity uncertain); (2) scutum mostly or entirely bare of microtrichia (apomorphic, but occurs sporadically outside this group); (3) aculeus tip short, nonserrate, and nearly round in cross section, approximately as broad in lateral view as in ventral view (Figure 12.16A, B) (apomorphic); and (4) vein M weakly curved apically (Figure 12.5B) (plesiomorphic, occurs occasionally in other groups). This may be the most primitive species group of *Anastrepha* if the first character is plesiomorphic (i.e., if the more transversely flattened condition found in other *Anastrepha* species groups is apomorphic), but the interpretation of this character is equivocal as there is no distinct anterior or posterior lobe on the lateral surstylus in the *cryptostrepha* group as there is in *Hexachaeta* and most Trypetinae.

Within the *cryptostrepha* group, *A. cryptostrepha* and *cordata* share a unique apomorphy of the male genitalia: The proctiger is sclerotized dorsally. The scutal markings of these species, although much less extensive in *A. cryptostrepha,* are another possible synapomorphy: *A. cryptostrepha* usually has a moderate brown band along the posterior margin (similar to Figure 12.1C), whereas *A. cordata* has extensive dark brown markings, including along the posterior margin (Figure 12.1A). *Anastrepha margarita, panamensis,* and *zeteki* appear to comprise another group of closely related species. They share the following probable synapomorphies: scutellum without microtrichia dorsally; aculeus very slender, less than 0.06 mm wide at midlength (reduction in aculeus width occurs in various other species groups); and lateral surstylus relatively long and acute.

Whether these two clades plus *A. tripunctata* form a monophyletic group is less certain (i.e., it is possible that the *cryptostrepha* group as a whole is paraphyletic). *Anastrepha tripunctata* possesses a possibly plesiomorphic character state, ocellar seta strong (Figure 12.4A), that suggests it could be the sister taxon to the rest of *Anastrepha* + *Toxotrypana*. The head shape of *A. tripunctata*, which is as long as high, was also interpreted as plesiomorphic by Norrbom (1985), but this is probably apomorphic based on outgroup comparison with *Hexachaeta*. There are also two undescribed species from Mexico that appear to be related to *A. tripunctata* based on their scutellar markings and wing pattern; at least one of them has a well-developed ocellar seta (C. Estrada, personal communication). Because the size of the ocellar seta is variable within *Hexachaeta*, the polarity of this character within *Anastrepha* is uncertain; the well-developed seta could be a synapomorphy for *A. tripunctata* and the new species from Mexico, or the reduced seta could be a synapomorphy for all other *Anastrepha* and *Toxotrypana* species (the latter hypothesis requires that there is homoplasy in characters 2 and 3 above, but given their variability across the genus this does not seem unrealistic). Study of the new Mexican species and analysis of *Hexachaeta* to determine the ground plan condition of this character in that genus may help to resolve the relationships of *A. tripunctata*.

The *cryptostrepha* group is predominantly distributed from Mexico to Venezuela, although *A. cryptostrepha* is known only from Ecuador and Peru. The known host plants of the group (for three of the eight spp.) are Apocynaceae and Sapotaceae.

#### 12.5.2.2 The *daciformis* Group

This is probably the most clearly monophyletic species group of *Anastrepha*, as indicated by the following unique apomorphies: (1) scutellum bicolored, with at least basal third on sides and dorsum distinctly darker than apex, and with basal scutellar seta within darker area (Figure 12.1B), except usually in *A. avispa* (scutellar markings are present on only a few other *Anastrepha* species, and in different patterns, suggesting they are not homologous); (2) eversible membrane with unique pattern of dorsobasal teeth (Figure 12.13B), small and weakly sclerotized except for medially interrupted apical row of large, strongly sclerotized, hooklike teeth; (3) spermathecae membranous;

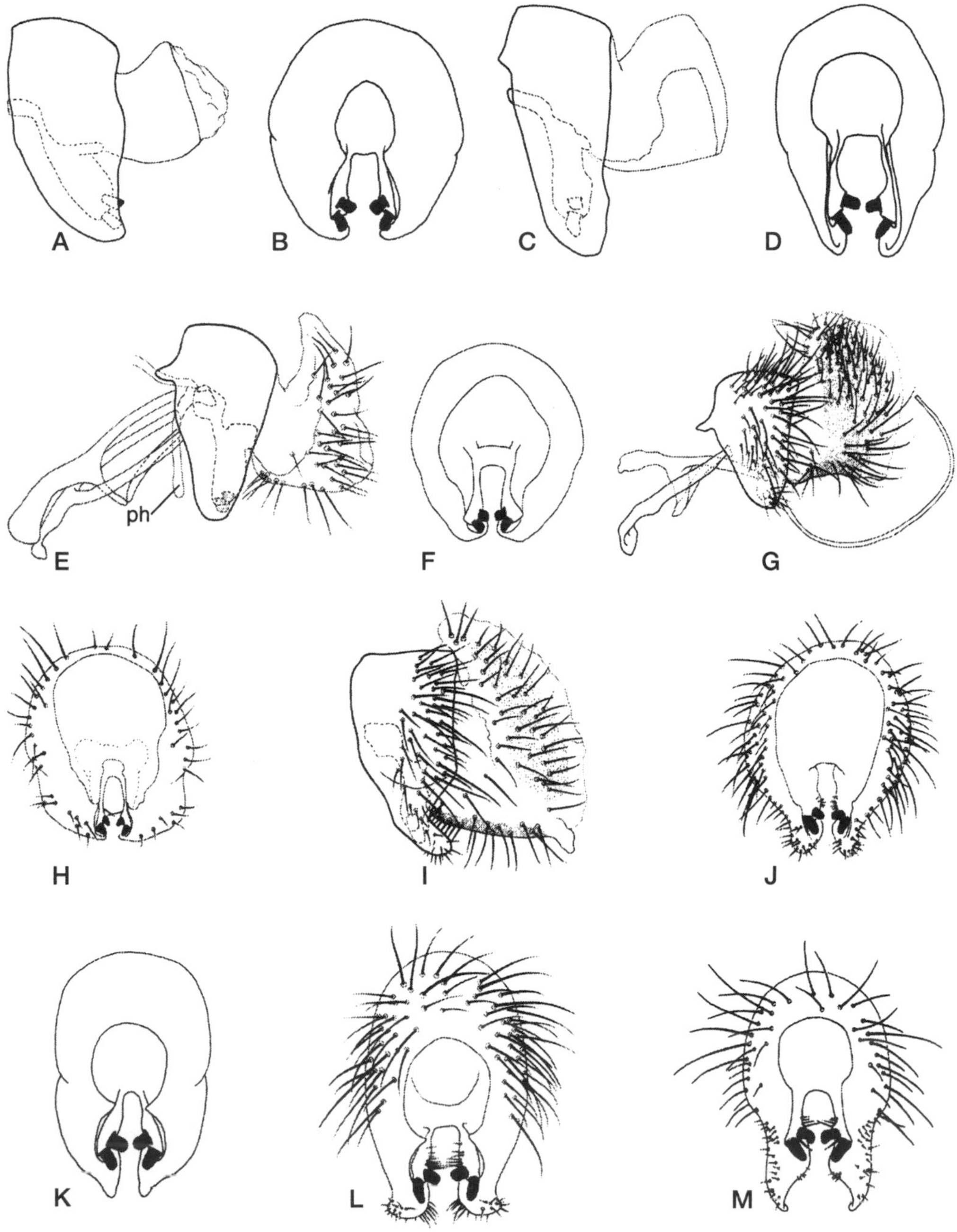


**FIGURE 12.9** Male genitalia of *Anastrepha* and *Toxotrypana*: (A, C, E, G, I) epandrium, surstyli and proctiger, lateral view; (B, D, F, H, J to M) epandrium and surstyli, posterior view; (A, B) *T. curvicauda*; (C, D) *A. tripunctata*, *cryptostrepha* group; (E, F) *A. maculata*, *daciformis* group; (G) *A. sagittata*, *dentata* group; (H) *A. hamata*, *dentata* group; (I, J) *A. aczeli*, *punctata* group; (K) *A. fernandezi*, *schausi* group, posterior view; (L) *A. barnesi*, *leptozona* group; (M) *A. nunezae*, *mucronota* group; ph = phallus. (A to D, from Norrbom A.L., Ph.D. dissertation, Pennsylvania State University, University Park, 1985. E, F, from Norrbom, A.L., *Proc Entomol. Soc. Wash.* 100: 160–192, 1998. K, from Norrbom, A.L. and Kim, K.C., *Ann. Entomol. Soc. Am.* 81: 164–173, 1988. With permission.)

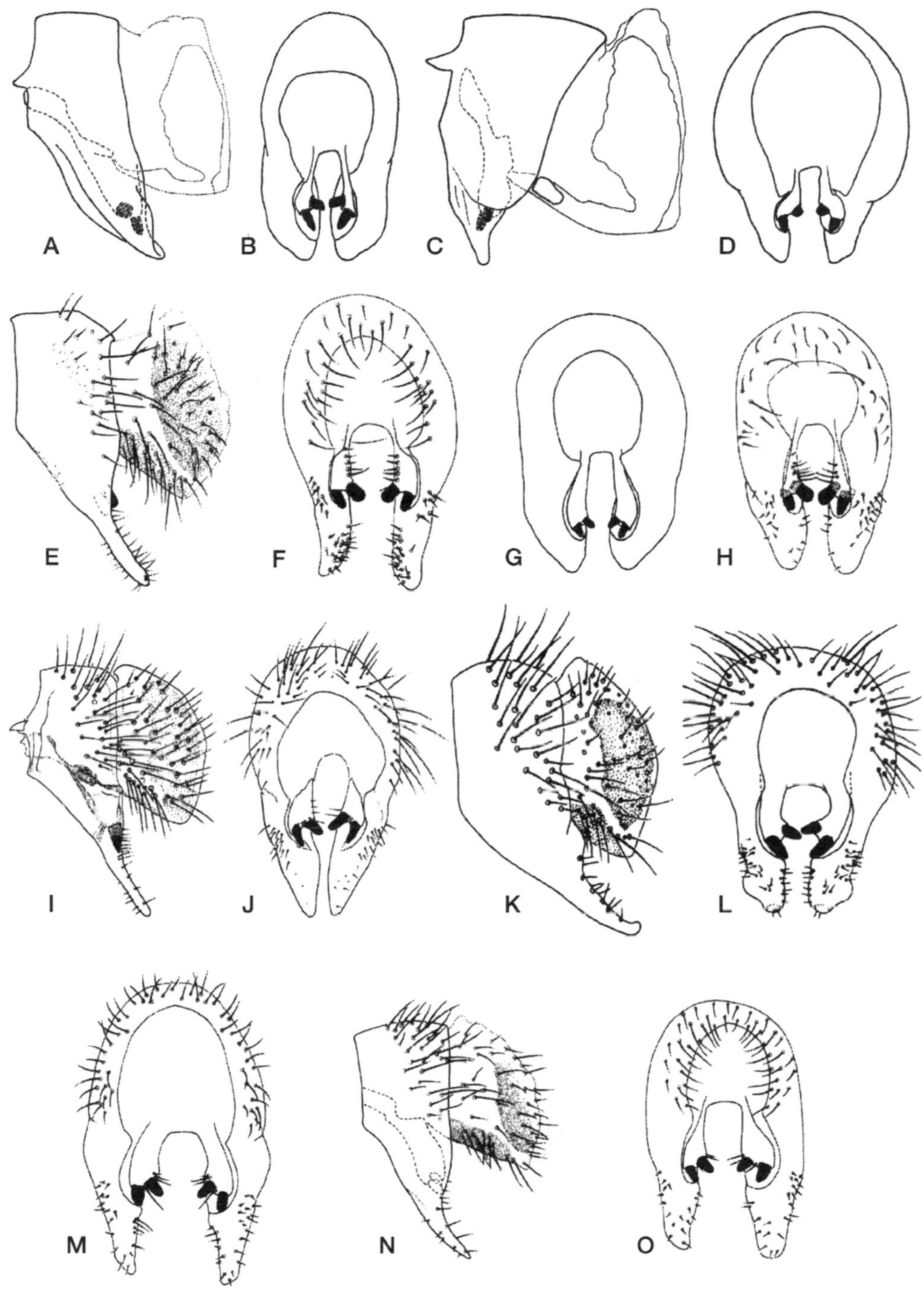


**FIGURE 12.10** Male genitalia of *Anastrepha*: (A, C, E, I, K, N) epandrium, surstyli and proctiger, lateral view; (B, D, F to H, J, L, M, O) epandrium and surstyli, posterior view; (A, B) *grandis*, *grandis* group; (C, D) *freidbergi*, *doryphoros* group; (E, F) *spatulata*, *spatulata* group; (G) *pallida*, *pseudoparallela* group; (H) *pastranai*, *pseudoparallela* group; (I, J) sp. nr. anomala, *serpentina* group; (K, L) *striata*, *striata* group; (M) *ludens*, *fraterculus* group; N, O, *fraterculus*, *fraterculus* group. (A, B, from Norrbom, A.L., *Proc. Entomol. Soc. Wash.* 93: 101–124, 1991. C, D, from Norrbom, A.L., *Proc. Entomol Soc. Wash.* 95: 52–58, 1993. G, from Norrbom, A.L., *Insecta Mundi* 11: 141–157, 1997. With permission.)

and (4) phallus extremely short, less than 0.30 mm long (Figure 12.9E). Relationships among the species of the *daciformis* group were analyzed by Norrbom (1998). The species with wasp mimicry wing patterns form a monophyletic group, including the *daciformis* complex (*A. castanea, daciformis*, and *katiyari*) and the *macrura* complex (*A. aquila, avispa, bicolor, macrura*, and *zucchii*), which together are the sister group of *A. pallens*. The relationships of the remaining four species (*A. antilliensis, murrayi, maculata*, and *stonei*) are less resolved. They either form a monophyletic group that is the sister group of the above group, or they arise as basal clades in the following order: *antilliensis, murrayi*, and *maculata + stonei*. *Anastrepha nambacoli* Tigrero (1992) is here considered a new synonym of *A. macrura*. Norrbom (1998) examined the holotype of *nambacoli* and reported it under *macrura*, not realizing the former name was available.

The *daciformis* group is widespread. *Anastrepha antilliensis, murrayi, maculata*, and *stonei* are endemic to the Antilles. *Anastrepha pallens* is Mesoamerican, the *daciformis* complex is restricted to South America, and the *macrura* complex includes three Mesoamerican and two South American species. The known host plants (for 3 of the 13 spp.) are Sapotaceae. The larvae feed on the seeds.

### 12.5.2.3 The *dentata* Group

Two synapomorphies indicate that this species group is monophyletic: (1) apex of aculeus poorly defined, extreme apex usually broader and sagittate (Figure 12.16D) (unique apomorphy); and (2) dorsobasal teeth of eversible membrane all small and weak (Figure 12.13C). Reduction in size and sclerotization of the dorsobasal teeth of the eversible membrane occurs in a few other species with slender aculei (e.g., *hamadryas*), but not as extensively as in this group; whether or not this is homoplasy or a synapomorphy of these species plus the *dentata* group is undetermined. This group is roughly equivalent to the genus *Lucumaphila* proposed by Stone (1939a), now considered a synonym of *Anastrepha*, although Steyskal (1977b) and Norrbom and Kim (1988b) removed four of the 11 species he included and added two others.

The *dentata* group is widespread, from southern Texas to Brazil, with one Antillean species. The known host plants (for only three spp.) are Sapotaceae. The larvae feed in the seeds only and tunnel out through the pulp of the fruit.

### 12.5.2.4 The *benjamini* Group

Two apomorphic character states, neither unique to this group, support its monophyly: (1) facial carina strongly produced medially and convex in lateral view (Figure 12.4E) (apomorphic but also occurs in the *pallidipennis* complex of the *pseudoparallela* group, and rarely in other species; a few other species, such as *A. atrox* and several new species of the *mucronota* group (Figure 12.4C), have the carina produced more dorsally (see Norrbom 1997)); and (2) female terminalia and male phallus very long, oviscape greater than 8.5 mm long (apomorphic, but occurs in other groups). This group may not be monophyletic, as the more compelling possible synapomorphy, the produced facial carina, apparently has evolved convergently within other species groups. At least *A. gigantea* and *magna* appear to be closely related, however (Norrbom 1997). *A. benjamini* has dense microtrichia around the lobe of cell bcu in both sexes, an unusual, autapomorphic character of unknown behavioral significance.

The *benjamini* group includes two species from Panama, one from Colombia and Venezuela, and one from southeastern Brazil. The only known host plant belongs to the Sapotaceae.

### 12.5.2.5 The *robusta* Group

The following character states appear to be synapomorphies of the *robusta* group: (1) scutum with dark brown band on posterior margin in most species (Figure 12.1C) (apomorphic, but this band is sometimes faint, and a similar one occurs in a few species in other species groups, e.g., *A. cryptostrepha*); (2) eversible membrane with relatively few, stout dorsobasal teeth (Figures 12.13D and 12.15C, D) (apomorphic, but numerous teeth are present in *A. speciosa*

(Figure 12.15E) and two related new species, possibly via reversal); and (3) the two species for which eggs have been examined (*A. nigrifascia* and *pittieri*) have a long lobe on the micropyle end (Figure 12.3C) (apomorphic, but this occurs, apparently by convergence, within the *spatulata* group and in other tephritid genera). Many of the species in this group (*fenestrata*, *lambda*, *pittieri*, *speciosa*, n. sp. near *lambda*, and two new species near *speciosa*) have a unique apomorphic wing character: There is an extension from the middle of the S-band that reaches the posterior wing margin in the middle of cell $cu_1$ (Figure 12.5H). This character further supports the inclusion of *speciosa* and the two new related species in the *robusta* group, despite their different eversible membranes. There are at least five undescribed species belonging to the *robusta* group (Norrbom, personal observation). Tigrero (1998) included *A. concava* and *A. montei* in his *fenestrata* group (*A. fenestrata* is here included in the *robusta* group). We place *A. montei* in the *spatulata* group. *Anastrepha concava*, which has a similar wing pattern to the *robusta* group (generally dark brown and with a distinct cleft in the base of the S-band) and undivided proctiger sclerotization, and the little studied species *A. connexa*, *hamadryas*, and *longicauda* might possibly be related to the *robusta* group or to the *benjamini* group, all of which have the basal cleft in the S-band. They differ from the *robusta* group in having long terminalia and relatively long lateral surstyli.

The *robusta* group is widespread, including species from Mexico to Brazil, with two species in the Antilles and Florida. The known host plants (for only two species) are Moraceae and Sapotaceae.

### 12.5.2.6 The *schausi* Group

The following unique apomorphies indicate that this group is monophyletic: (1) glans with minute spines (Figure 12.11F); (2) facial carina weak; and (3) lateral surstylus acute, lateral margin usually slightly concave (Figure 12.9K). All of the species appear to be sexually dimorphic, with the male face (Figure 12.4D), abdomen (Figure 12.1D), and/or wing having unusual markings or pattern, although not all of the species have the same dimorphic structures, and whether this can be considered a single homologous character is uncertain. The male of *A. bellicauda* has only some diffuse yellow markings on its wing (Figure 12.5J), but the recently discovered female has a more normal pattern in which all of the typical *Anastrepha* bands can be recognized (Figure 12.5I). This group was revised by Norrbom and Kim (1988a). The four included species are known from Costa Rica, Panama, Venezuela, Peru, and Bolivia. *Anastrepha bellicauda* is the only species to have been reared, but the identity of the host is uncertain, although it is probably a species of Sapotaceae (K. P. Katiyar, personal communication).

### 12.5.2.7 The *punctata* Group

Two unique apomorphies support this group as monophyletic: (1) scutum with pair of posterior brown spots (Figure 12.1E) (variable within an undescribed species near *A. luederwaldti*; brown markings in this area occur rarely in other species groups, e.g., in *A. dentata* and *pallens*, but differ in shape); and (2) lateral surstylus obliquely oriented, strongly curved and blunt apically (Figure 12.9I, J). This small group is known only from southern Brazil, Paraguay, and Argentina. The only known host plant belongs to the Myrtaceae.

### 12.5.2.8 The *leptozona* Group

Two synapomorphies indicate that this group is monophyletic: (1) lateral surstylus short and somewhat boot-shaped, with laterally projecting apical lobe (Figure 12.9L) (apomorphic); (2) vein M extremely strongly curved (Figure 12.5L) (occurs convergently in a few species in other species groups, e.g., *A. montei*). This small group is widely distributed in the mainland Neotropics. Two of the four species have known hosts, and both attack Sapotaceae, although *A. leptozona* breeds in a range of other plant families as well.

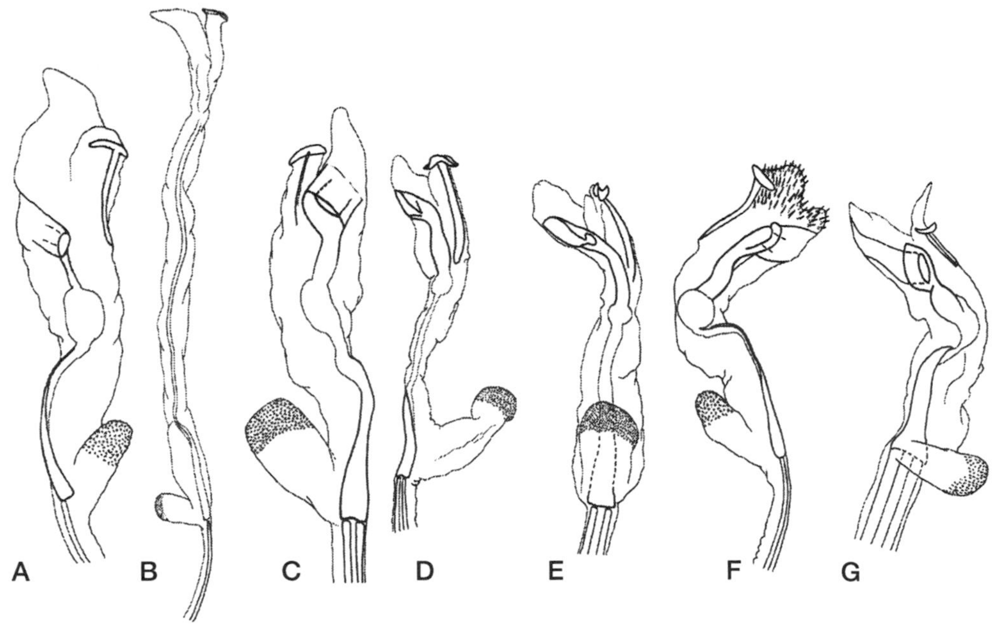


**FIGURE 12.11** Male genitalia of *Anastrepha* and *Toxotrypana*, glans: (A) *T. curvicauda*; (B) *A. zeteki*, *cryptostrepha* group; (C) *nigrifascia*, *robusta* group; (D, E) *bezzii*, *mucronota* group; (F) *fernandezi*, *schausi* group; (G) *grandis*, *grandis* group. (A to C from Norrbom, A.L., Ph.D. dissertation, Pennsylvania State University, University Park, 1985. D, E, G, from Norrbom, A.L., *Proc. Entomol. Soc. Wash.* 93: 101–124, 1991. F, from Norrbom, A.L. and Kim, K.C., *Ann Entomol. Soc. Am.* 81: 164–173, 1988. With permission.)

### 12.5.2.9 The *mucronota* Group

This is perhaps a paraphyletic group; it includes species with the male proctiger not strongly creased (plesiomorphic), but not fitting in any other species group above. But it might be monophyletic, and at least seems to include some groups of closely related species. The wing bands are usually well separated along vein $R_{4+5}$ (Figure 12.6A, B) (perhaps a synapomorphy, but this character is very variable across the genus and also occurs in the *leptozona*, *punctata,* and *schausi* groups), and vein $R_{2+3}$ tends to be sinuous (Figure 12.6B) (apomorphic, but varies, often intraspecifically). *Anastrepha minuta* has an unusual character of the eversible membrane that may have behavioral significance: On the ventral side, opposite the hooklike dorsobasal teeth, it has a large clump of fine, hair-like projections (Figure 12.13I). The *parallela* and *integra* groups of Tigrero (1998) fit here; he included *A. mucronota* in the former group.

The *mucronota* group is widespread, from Mexico to Argentina, with four species known from the Antilles. Five of the species, so far as is known, breed only in Bombacaceae. Six others have reported hosts belonging to the Sapotaceae (for four species), Sterculiaceae (for two species), Annonaceae (2 spp.), Passifloraceae and Rutaceae.

### 12.5.2.10 The *grandis* Group

This group is rather weakly supported by the following apomorphies and may not be monophyletic: (1) wing with complete marginal band, without hyaline mark at apex of vein $R_1$ (Figure 12.6C, D) (absent in two species; occurs convergently in *bezzii*, *cordata*, and within *daciformis* group); (2) distal arm of V-band absent (this also occurs within various other species groups); and (3) body color relatively dark (Figure 12.1H) (this also occurs in several other species groups). *Anastrepha bivittata* and *fumipennis* lack the first apomorphy (they have a hyaline marginal spot in cell $r_1$;

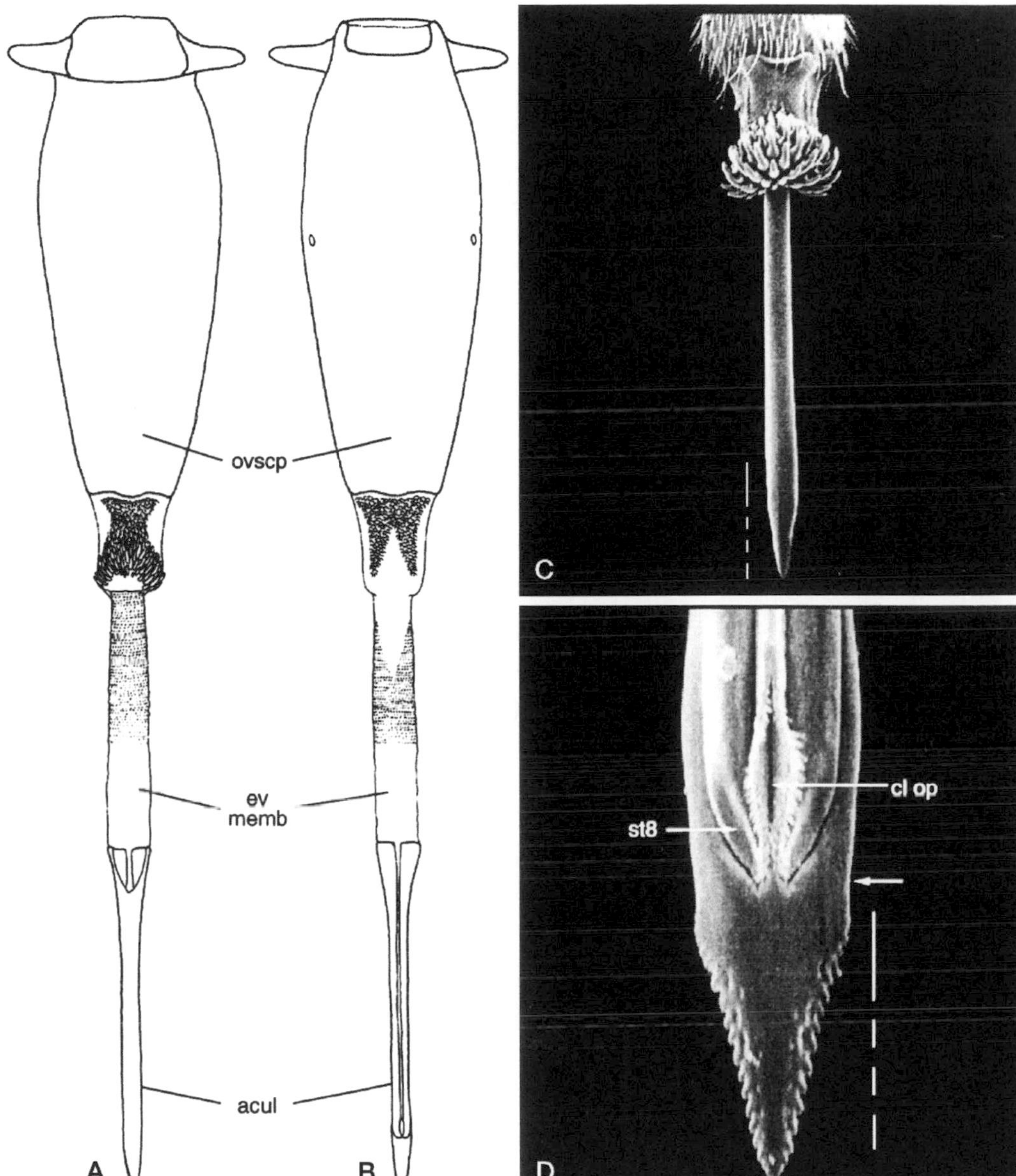


**FIGURE 12.12** Female genitalia of *Anastrepha*: (A) dorsal; (B) ventral; (C) base of eversible membrane and aculeus; (D) aculeus tip; (A, B) *nigrifascia*; (C) *ludens*; (D) *obliqua*; acul = aculeus; cl op = cloacal opening; ev memb = eversible membrane; ovscp = oviscape; st8 = sternite 8; arrow indicates proximal point of measurement of aculeus tip. (A, B, from Norrbom, A.L. and Kim, K.C., *Ann. Entomol. Soc. Am.* 81: 164–173, 1988. With permission.)

Figure 12.6E), but are included in this group because of other apomorphies shared with *A. atrigona*, including (1) cell bm infuscated (Figure 12.6D, E) (occurs rarely in other *Anastrepha*, e.g., *flavipennis*); (2) crossvein R-M relatively close to DM-Cu (Figure 12.6D, E); and (3) aculeus tip broadly rounded (Figure 12.16J; Norrbom 1991). The hypothesis that the presence of the hyaline spot in $r_1$ in *A. bivittata* and *fumipennis* is due to reversal (i.e., the *grandis* group is monophyletic) is equally parsimonious with the hypothesis that the complete marginal band (absence of a hyaline spot in $r_1$) in *A. atrigona* is due to homoplasy (i.e., the *grandis* group is not monophyletic), and

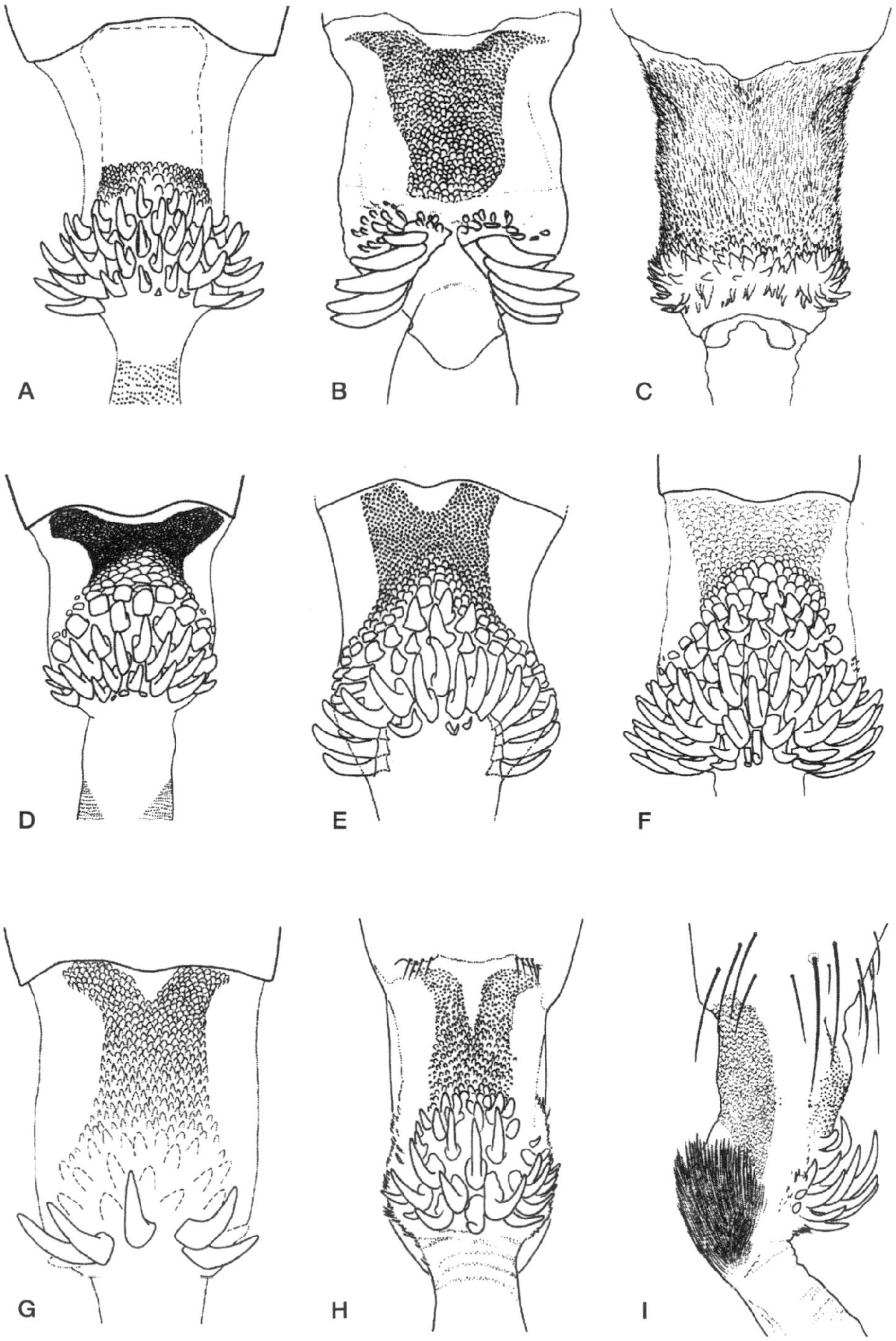


**FIGURE 12.13** Female genitalia of *Anastrepha*, base of eversible membrane, dorsal view except (I) lateral; (A) *tripunctata*, *cryptostrepha* group; (B) *macrura*, *daciformis* group; (C) *hamata*, *dentata* group; (D) *simulans*, *robusta* group; (E) *magna*, *benjamini* group; (F) *fernandezi*, *schausi* group; (G) *punctata*, *punctata* group; (H, I) *minuta*, *mucronota* group. (A, D, G, from Norrbom A.L., Ph.D. dissertation, Pennsylvania State University, University Park, 1985. B, from Norrbom, A.L., *Proc. Entomol. Soc. Wash.* 100: 160–192, 1998. E, from Norrbom, A.L., *Insecta Mundi* 11: 141–157, 1997. F, from Norrbom, A.L. and Kim, K.C., *Ann. Entomol. Soc. Am.* 81: 164–173, 1988. With permission.)

further testing of the monophyly of this group is needed. *Anastrepha aberrans* and a related undescribed species from Costa Rica lack the basal half of the S-band unlike the other species of the *grandis* group (apomorphic; occurs convergently within the *daciformis* group), and resemble *A. castilloi* in the pattern of the dorsobasal teeth of the eversible membrane. Steyskal (1977b) included *A. bezzii* in this group based on similarity in wing pattern, but Norrbom and Kim (1988b; see also Norrbom 1991) transferred it to the *mucronota* group based on genitalic and other characters. Except for the new species near *A. aberrans*, the *grandis* group is restricted to South America. Host plants are known only for *A. grandis*, which breeds in Cucurbitaceae.

### 12.5.2.11 The *doryphoros* Group

The following synapomorphies indicate that this group is monophyletic: (1) scutum with paired nonmicrotrichose submedial stripe (in most *Anastrepha* species, the scutum is almost entirely microtrichose or is mostly without microtrichia, although shorter, broader bare stripes occur in the *striata* group and some species of the *schausi* group); (2) wing with apical part of S-band broadly fused to V-band (Figure 12.6F), completely fused in cell $r_{2+3}$, and in cell $r_{4+5}$ separated at most by a hyaline area that does not extend to vein $R_{4+5}$, in cell m hyaline spot between arms of V-band not extended beyond vein M; (3) eversible membrane with dorsobasal teeth short and arranged in elongate pattern (Figure 12.14A) (*A. bezzii* (Figure 12.15F) has a somewhat similar elongate pattern, but the teeth are larger); and (4) aculeus tip tapered, then parallel-sided, then tapered (Figure 12.16L). This group includes a new species from Costa Rica in addition to *A. doryphoros* and *A. freidbergi*, which were revised by Norrbom (1993). *Anastrepha conjuncta*, known only from the male holotype, might be related to this group. As in the species of the *doryphoros* group, the medial scutal pale stripe is very narrow and not expanded posteriorly, and the genitalia are very long. The female of *conjuncta* is unknown, but the phallus of the holotype is over 15 mm long. This species lacks the other nongenitalic apomorphies of the *doryphoros* group, however. The scutum is entirely microtrichose, and, although the S-band is broadly fused to both the C-band and V-band, the wing pattern is typical for *Anastrepha*. Discovery of the female may help clarify the relationships of *A. conjuncta*. The *doryphoros* group occurs in Costa Rica, Panama, Colombia, Peru, and Bolivia. The host plants for the group are unknown.

### 12.5.2.12 The *spatulata* Group

The following two character states may be synapomorphies of this group: (1) aculeus tip short and broad (except in *montei* and *haywardi* in which the tip is extremely slender), with numerous fine serrations extended beyond the base (less extensively in *haywardi*; Figure 12.16M); and (2) in the two species whose karyotype has been studied (*montei* and *pickeli*), the diploid number (2n) is eight, whereas in most other *Anastrepha* species studied and in Tephritidae in general the diploid number is 12 (Solferini and Morgante 1987; Morgante et al. 1996). The status of the *spatulata* group needs further testing, as the former character state does not occur in all of the species (further evolution of this character in *A. montei* and *haywardi* must be postulated if the state described above is a synapomorphy for the entire species group), and the latter is unknown for most of the species, although it is probably a synapomophy for at least the *Manihot*-infesting species. *Anastrepha nascimentoi*, *rheediae*, and *tecta*, which have broad and finely serrate aculeus tips, may also belong in the *spatulata* group, although their tips are slightly longer and somewhat intermediate in shape between species of the *spatulata* and *pseudoparallela* groups. Tigrero (1998) placed *A. montei* (probably based on a misidentified female of the *A. pickeli* complex according to his figures of the aculeus and wing, especially the shape of vein M) in his *fenestrata* group, and *A. manihoti* in his *manihoti* subgroup of the *chiclayae* group. These three taxa, which breed in *Manihot* (Euphorbiaceae), and *A. haywardi*, whose host is unknown, appear instead to form a closely related group of species. *Anastrepha manihoti* and *pickeli* have elongate lobes on their eggs, but such lobes

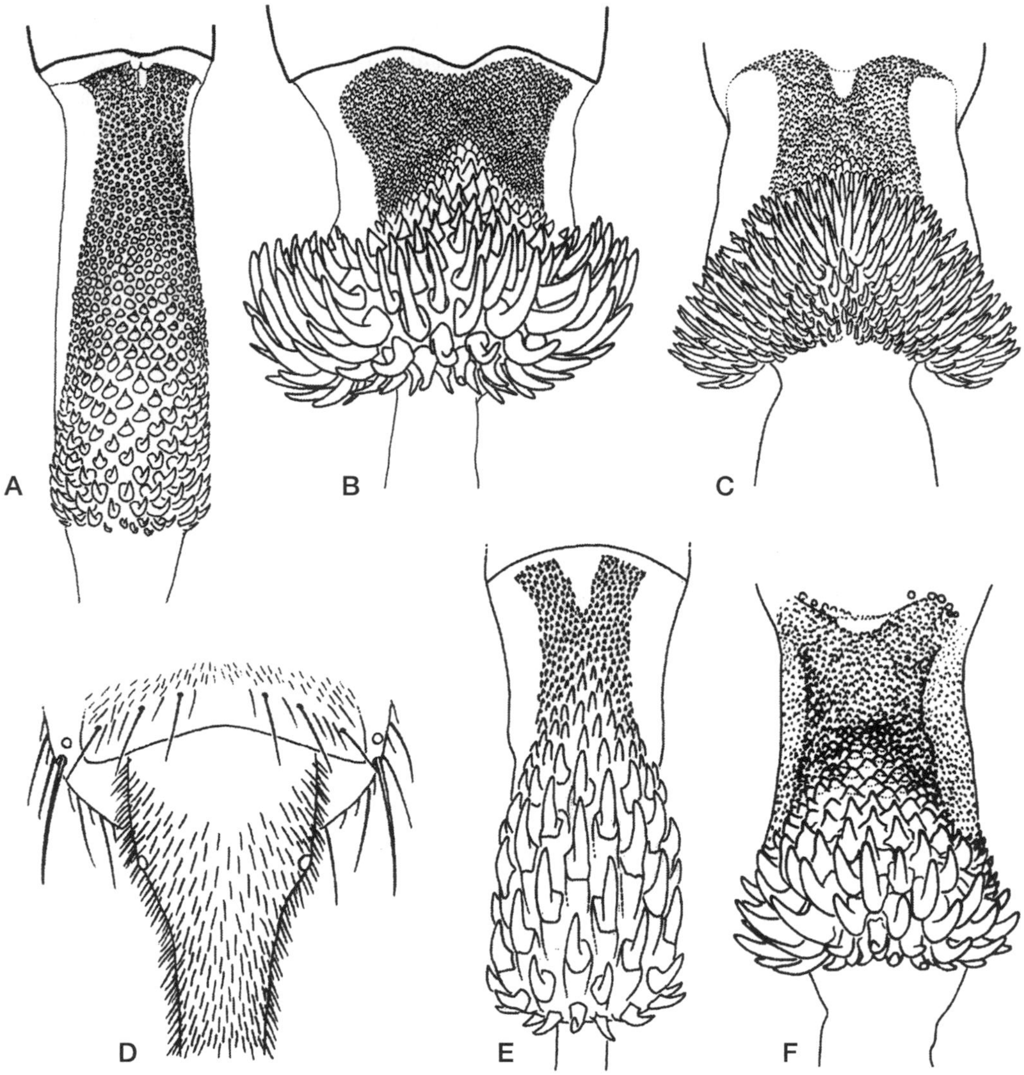


**FIGURE 12.14** Female genitalia of *Anastrepha*: (A to C, E to F) base of eversible membrane, dorsal view; (D) base of oviscape; (A) *doryphoros*, *doryphoros* group; (B) *grandis*, *grandis* group; (C) *consobrina*, *pseudoparallela* group; (D,E) *pallidipennis*, *pseudoparallela* group; (F) *anomala*, *serpentina* group. (A, from Norrbom, A.L., *Proc. Entomol. Soc. Wash.* 95: 52–58, 1993. B, from Norrbom, A.L., *Proc. Entomol. Soc. Wash.* 93: 101–124, 1991. C to E, from Norrbom, A.L., *Insecta Mundi* 11: 141–157, 1997. With permission.)

do not occur on the eggs of *A. alveatoides* and *haywardi*, so this is not a synapomorphy for the entire *spatulata* group.

The *spatulata* group occurs from Texas to Argentina, although it appears to be absent from most of Amazonia. One species occurs in the Antilles and Florida. Host plants are known for seven species, two of which breed in Olacaceae, the other five in Euphorbiaceae (one of these species also in Bombacaceae).

### 12.5.2.13 The *ramosa* Group

A single unique apomorphy indicates that this small species group is monophyletic: Aculeus tip short, broad, and triangular, with very large serrations extended to base (Figure 12.17F). This group includes only two species, from Central America and Panama, whose hosts are unknown.

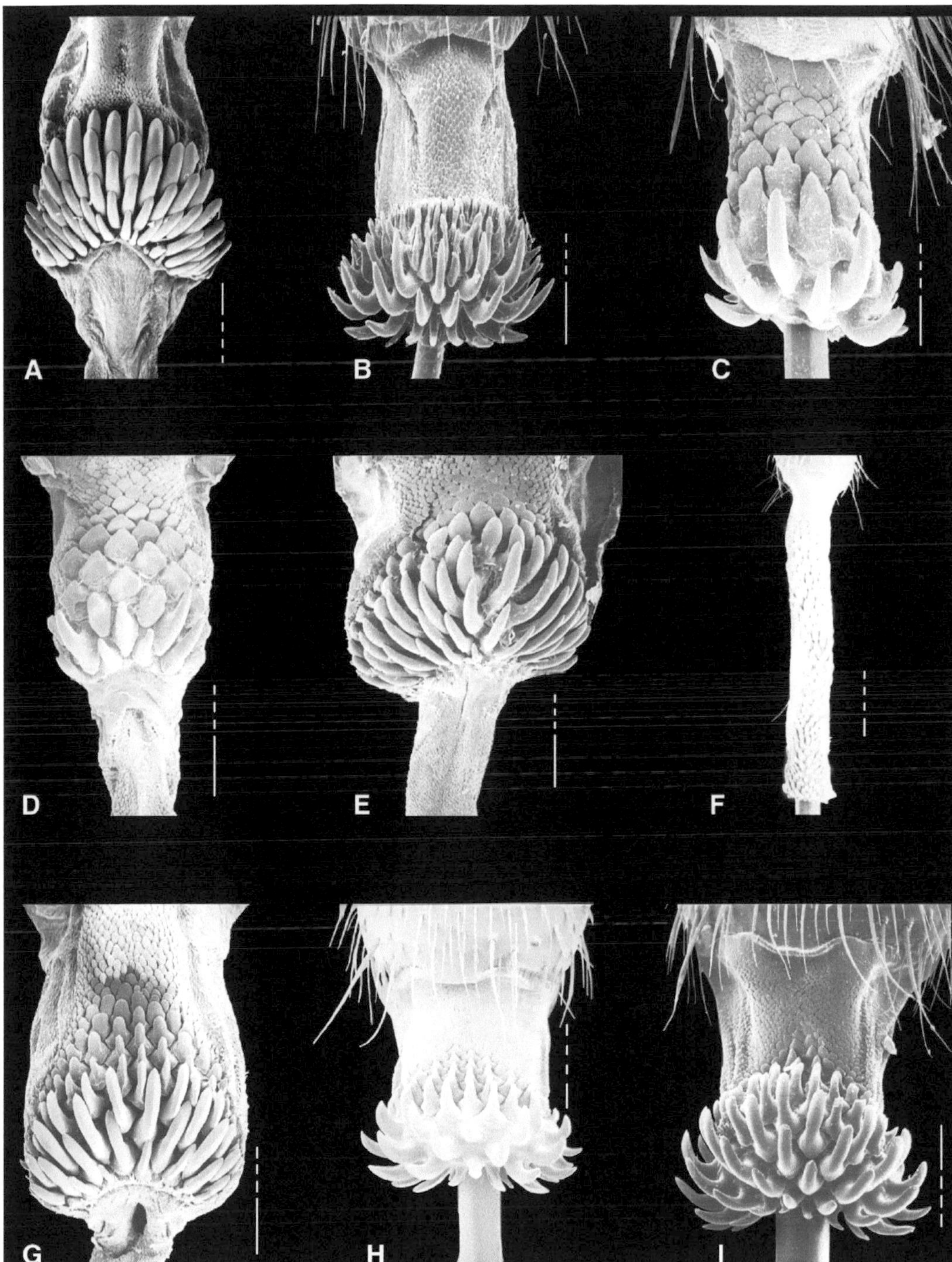


**FIGURE 12.15** Eversible membrane of *Anastrepha* and *Toxotrypana*, scanning electron micrographs: (A) *T. curvicauda*; (B) *A. zeteki*, *cryptostrepha* group; (C) *A. nigrifascia*, *robusta* group; (D) *A. pittieri*, *robusta* group; (E) *A. speciosa*, *robusta* group; (F) *A. bezzii*, *mucronota* group; (G) *A. atrox*, *mucronota* group; (H) *A. ramosa*, *ramosa* group; (I) *A. ludens*, *fraterculus* group. (A to C, from Norrbom, A.L., Ph.D. dissertation, Pennsylvania State University, University Park, 1985.)

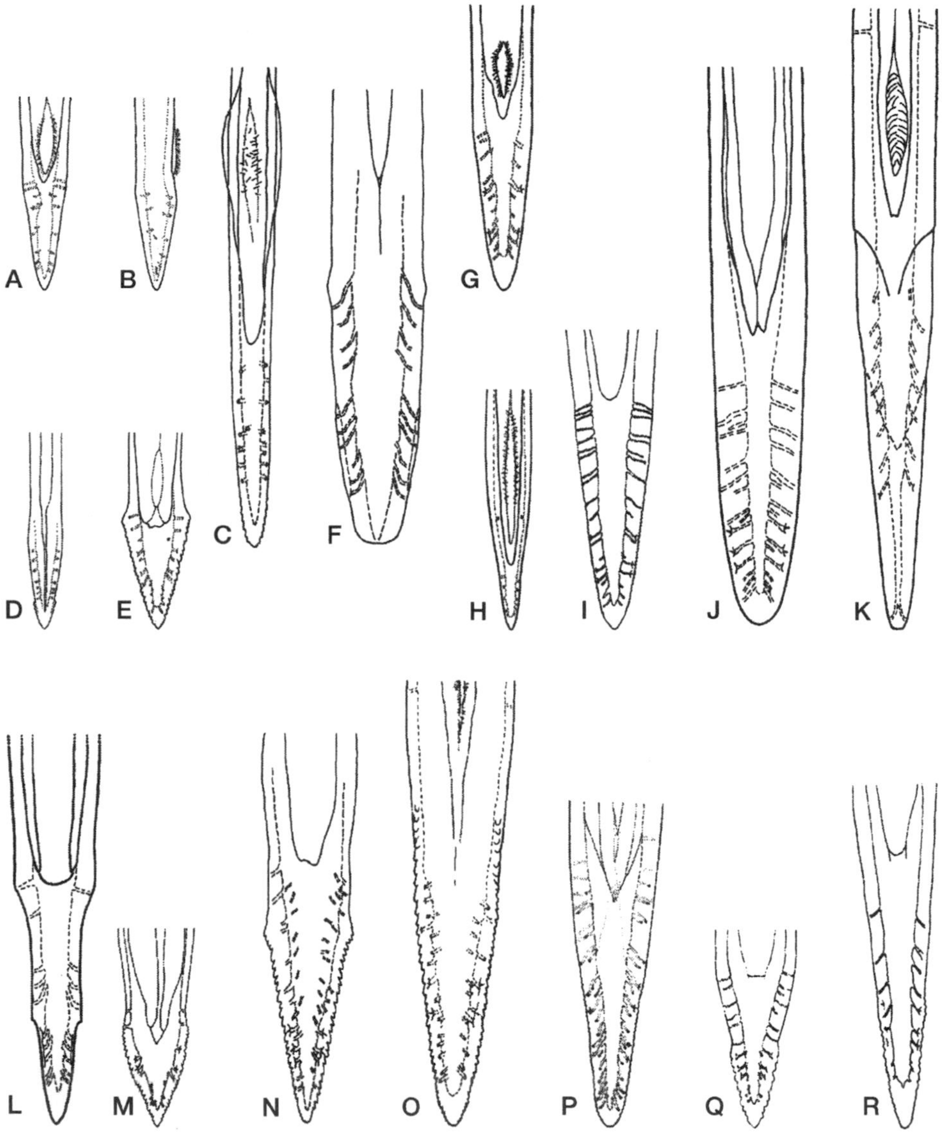


**FIGURE 12.16** Aculeus tip of *Anastrepha*: (A,B) *cordata*, *cryptostrepha* group; (C) *daciformis*, *daciformis* group; (D) *obscura*, *dentata* group; (E) *fenestrata*, *robusta* group; (F) *magna*, *benjamini* group; (G) *fernandezi*, *schausi* group; (H) *punctata*, *punctata* group; (I) *leptozona*, *leptozona* group; (J) *atrigona*, *grandis* group; (K) *grandis*, *grandis* group; (L) *doryphoros*, *doryphoros* group; (M) *pickeli*, *spatulata* group; (N) *pallidipennis*, *pseudoparallela* group; (O) *pseudoparallela*, *pseudoparallela* group; (P) *anomala*, *serpentina* group; (Q) *fraterculus*, *fraterculus* group; (R) *distincta*, *fraterculus* group. (A, B, D, E, H, from Norrbom, A.L., Ph.D. dissertation, Pennsylvania State University, University Park, 1985. C, from Norrbom, A.L., *Proc. Entomol. Soc. Wash.* 100: 160–192, 1998. F, from Norrbom, A.L., *Insecta Mundi* 11: 141–157, 1997. G, from Norrbom, A.L. and Kim, K.C., *Ann. Entomol. Soc. Am.* 81: 164–173, 1988. I, Q, from Stone, A. *U.S. Dep. Agric. Misc. Publ.* 439, 112 pp., 1942. J, K, from Norrbom, A.L., *Proc. Entomol. Soc. Wash.* 93: 101–124, 1991. L, from Norrbom, A.L., *Proc. Entomol. Soc. Wash.* 95: 52–58, 1993. R, from Hernández-Ortiz, V. El genero *Anastrepha* Schiner en Mexico (Diptera: Tephritidae). Taxonomia, distribucion y sus plantas huespedes, Instituto de Ecología and Sociedad Mexicana de Entomología, Xalapa, Mexico, 1992. With permission.)

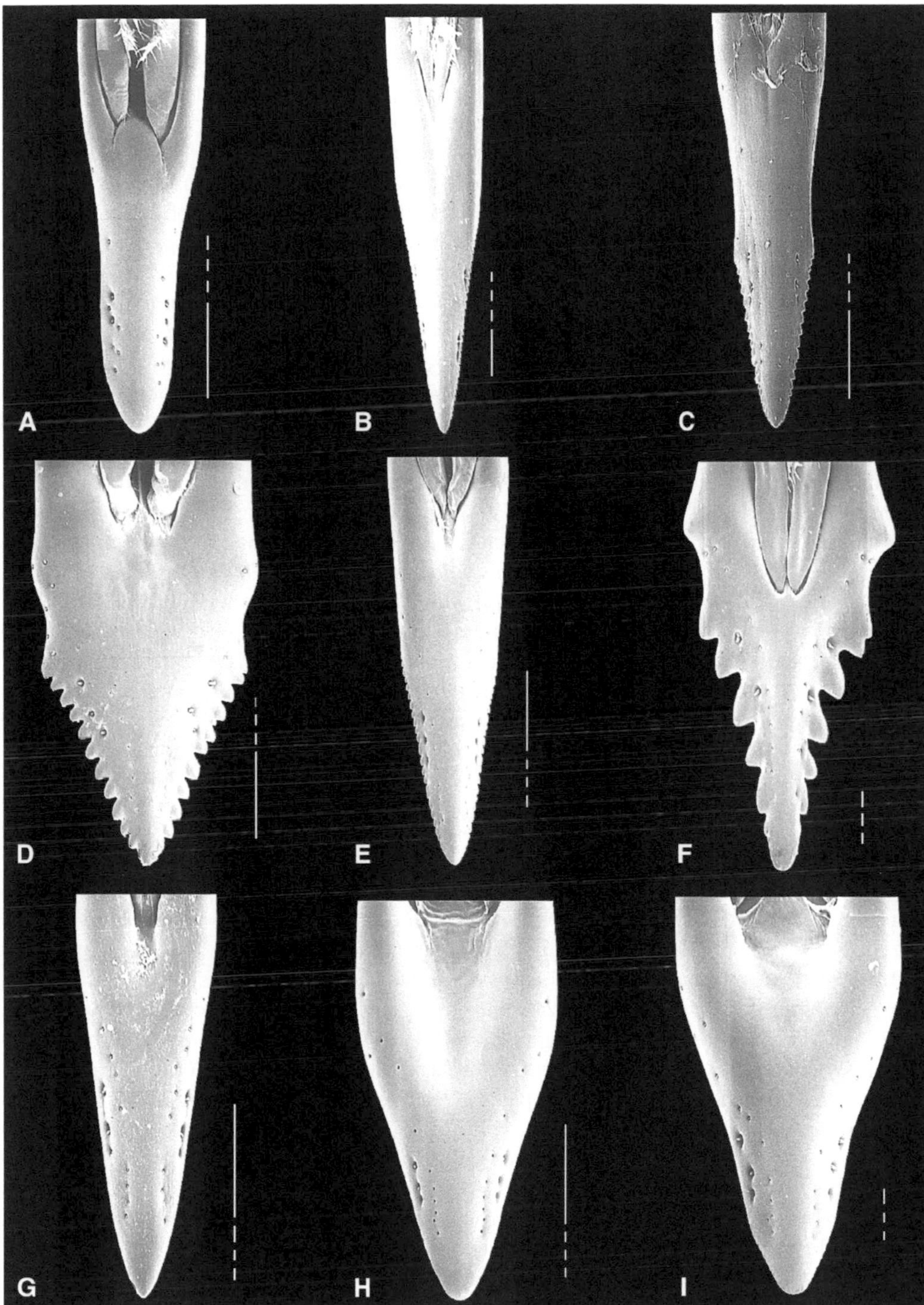


**FIGURE 12.17** Aculeus tip of *Anastrepha*, scanning electron micrographs: (A) *robusta*, *robusta* group; (B) *leptozona*, *leptozona* group; (C) *atrox*, *mucronota* group; (D) *elegans*, *mucronota* group; (E) *serpentina*, *serpentina* group; (F) *ramosa*, *ramosa* group; (G) *mucronota*, *mucronota* group; (H) *ornata*, *striata* group; (I) *striata*, *striata* group. (A, from Norrbom, A.L., Ph.D. dissertation, Pennsylvania State University, University Park, 1985.)

#### 12.5.2.14 The pseudoparallela Group

This group is recognized mainly on the basis of aculeus tip shape, but its status needs further testing, as the single putative synapomorphy here proposed is not present in all species. All but three species have the aculeus tip long, moderately broad, and mostly or entirely finely serrate (Figure 12.16N, O; except nonserrate in *velezi*, and much shorter in *passiflorae* and especially *xanthochaeta*), which is possibly a synapomorphy, although similar aculeus shapes occur in the *leptozona* and *serpentina* groups, so it may alternatively be a synapomorphy at a higher level. This group includes the species placed by Norrbom and Kim (1988b) in the *pseudoparallela* and *chiclayae* groups, which Norrbom (1997) considered to intergrade. *Anastrepha velezi* is included in this group because it has what Norrbom (1997) hypothesized to be a unique synapomorphy of the *pallidipennis* complex, which also includes *A. amnis*, *curitis*, *pallida*, and *pallidipennis*: The spiracle of the female oviscape is located very close to its base (Figure 12.14D). Some species of the *pallidipennis* complex resemble species of the *benjamini* group in facial carina shape (Figure 12.4F) and were placed in that group by Steyskal (1977b), but because this character occurs in a few other species that otherwise appear to belong to other species groups, Norrbom (1997) regarded it as more likely to be homoplastic than the oviscape character, and transferred the species of the *pallidipennis* complex to this species group. Based on host plant data, we have tentatively included *A. mburucuyae*, which was described from only males, as well as *A. xanthochaeta* and *passiflorae*, in the *pseudoparallela* group. These three species and the other 11 species (of 20 species) of the group for which host data are known breed in Passifloraceae (two species have reported hosts in other families in addition). Tigrero (1998) proposed the *palae* group for *A. townsendi*, which we include here, placed *A. pseudoparallela* in a subgroup of the *distincta* group (here considered part of the *fraterculus* group), and continued to recognize the *chiclayae* group, in which he also included *A. manihoti*. The *pseudoparallela* group is widespread, from Texas to Argentina, although only two species extend north of Panama and one into the Antilles.

#### 12.5.2.15 The *serpentina* Group

This group is supported by the following synapomorphies: (1) abdomen bicolored, partially dark brown (Figure 12.1J) (abdominal markings also occur in some species of the *cryptostrepha*, *schausi*, *punctata*, *grandis,* and *daciformis* groups, but differences in the patterns and/or other character state distributions (e.g., the synapomorphies of the *daciformis* group) suggest that this is homoplasy, except perhaps in *A. ornata* (see *striata* group)); (2) aculeus tip long, moderately broad, evenly tapered, and partially finely serrate (Figure 12.16P, 12.17E) (probably apomorphic, but this is an uncertain synapomorphy at this level because similar aculeus tip shapes occur in the *leptozona* and *pseudoparallela* groups); and (3) wing pattern relatively dark brown (Figure 12.6I) (occurs in occasional other *Anastrepha* species, e.g., within *robusta* group). Steyskal (1977b) also included *A. ornata, fenestrata,* and *phaeoptera* in this group. We include the former in the *striata* group, although it strongly resembles *A. serpentina* in wing pattern. Norrbom (1985) transferred *A. fenestrata* and *phaeoptera* to the *robusta* group, although the unique male holotype of the latter species should be restudied to confirm this action.

The *serpentina* group occurs mainly between Guatemala and Brazil, although *A. serpentina* extends to Texas and Argentina, and *A. ocresia* occurs in the Antilles and Florida. Host plants are known for five of the seven species, but vary considerably. Three species breed in Apocynaceae, and two in Sapotaceae (including *A. serpentina*, which has hosts in both families). The only native host known for *A. ocresia* belongs to the Myrtaceae, but it has been reared from an exotic species of Sapotaceae as well.

### 12.5.2.16 The *striata* group

The following apomorphies support this group as monophyletic: (1) aculeus tip extremely broad and gradually tapered to blunt point (Figure 12.17H,I) (unique apomorphy); and (2) scutum mostly microtrichose, with sublateral bare stripes (see discussion of this character under *doryphoros* group). The *striata* group includes *A. ornata*, from Colombia and Ecuador, and the disjunct pair, *A. striata*, from Mexico to Bolivia and northern Brazil, and *A. bistrigata*, from southeastern Brazil. The latter two species, which have very similar male genitalia (Figure 12.10K, L) and body markings (Figure 12.1I), are probably sister taxa. *Anastrepha ornata* is placed in this group based on the two hypothesized synapomorphies listed above, but it also resembles species of the *serpentina* group in abdominal color and wing pattern (Figure 12.6J). It may possibly belong in the *serpentina* group (Steyskal 1977b; Tigrero 1998), or these two species groups may be related, although there is homoplasy in at least some of the characters discussed above in any hypothesis of relationships among the species of the *serpentina* and *striata* groups. The single known native host plant of *A. ornata*, and most of those of *A. striata* and *bistrigata*, belong to the family Myrtaceae.

### 12.5.2.17 The *fraterculus* Group

The following possible synapomorphies suggest that this group is monophyletic: (1) mediotergite and/or subscutellum with lateral dark brown stripe or spot and body otherwise without dark markings, except sometimes on scuto-scutellar suture (Figure 12.2E, F) (apomorphic, but absent in two species and variable in a few others); (2) aculeus tip usually partially serrate and often with constriction just basal to serrate part (Figures 12.12D and 12.16Q,R) (neither attribute is unique to this group, but this aculeus shape may be a synapomorphy, depending upon what is the sister group of the *fraterculus* group); and (3) lateral surstylus usually parallel sided or slightly tapered and truncate apically (Figure 12.10M to O) (except in *schultzi*; apomorphic, but possibly not unique to the group). Norrbom (1985) attempted to divide the species placed here into two groups (*distincta* and *fraterculus* groups), but they have been found to intergrade. All of the included species except *A. barbiellinii* and *zuelaniae* at least sometimes have the lateral brown markings on the mediotergite or subscutellum. This includes *A. antunesi*, of which several specimens from Venezuela with lateral markings on the mediotergite have been observed (Norrbom, personal observation). Similar markings occur in various species of *Anastrepha* in which other areas of the body are dark (e.g., in the *daciformis*, *striata*, and *serpentina* groups), but such markings are otherwise rare in predominantly yellow species. An exception is *A. rheediae*, which might belong in the *spatulata* or *pseudoparallela* groups based on aculeus tip shape. *Anastrepha zuelaniae* is included here on the basis of its surstylus shape and aculeus tip shape. The egg of *A. barbiellinii* is very similar to that of *A. obliqua* (Figure 12.3D), having a short lobe on the micropyle end, and this species is tentatively included in the group on that basis. The *fraterculus* group is widespread, from Texas to Argentina, with two species occurring in the Antilles. The host plants are very diverse, and seven of the species are reported to have native hosts in three or more families.

### 12.5.2.18 Unplaced Species

Many of the 32 species which have not been placed in a species group are poorly studied and/or are known from only one sex. Tigrero (1998) recognized the *nigripalpis* group for *A. nigripalpis*, and the *bondari* group for *A. buscki*, but without further explanation. As noted above, *A. conjuncta* may be related to the *doryphoros* group, and *A. concava*, *connexa*, *hamadryas*, and *longicauda* share certain characters with the *robusta* and *benjamini* groups and may be related to one or both of them or form another species group. Species such as *A. acris*, *buscki*, and *palae*, which have distinctly creased proctigers in the male, belong to the large clade also including the *pseudoparallela*

+ *spatulata* + *ramosa* + *doryphoros* + *grandis* + *serpentina* + *striata* + *fraterculus* groups. Tigrero (1998) placed A. *tecta* and *rheediae* in his *rheediae* group, presumably based on their somewhat similar aculeus tips. These species are intermediate in this character between the *pseudoparallela* and *spatulata* groups, and we prefer to leave them unplaced at this time.

## 12.6 HOST PLANT RELATIONSHIPS

Published host plant records for *Anastrepha* and *Toxotrypana* have been compiled in a database (Norrbom, in press). Fields were included for the reliability of the record (based on Norrbom's subjective evaluation) and the origin of the plant (exotic or native, based on whether or not its original distribution coincided with that of the fly species). Only native hosts and reliable records were used to produce the host plant summary in Table 12.4, in which the host family or families are listed for each fly species. It was augmented by data from theses or unpublished sources contributed by Zucchi for 15 species from Brazil. Several trends in host relationships can be observed based on this data.

Most species of *Hexachaeta* and all species of *Toxotrypana* for which host plants are known attack latex-bearing plants. The hosts for the nine *Hexachaeta* species that have been reared belong to the Moraceae, except one species was reported from a Bignoniaceae species and another has been reported from fruit of one Verbenaceae and one Moraceae species (Silva et al. 1968; Foote et al. 1993; Hernández-Ortiz, unpublished data). The Bignoniaceae and Verbenaceae are not latex bearing. The known hosts of *Toxotrypana* species belong to the Caricaceae, Asclepiadaceae, and possibly Apocynaceae; at least two species attack hosts in both of the former families. *Toxotrypana* larvae feed on the developing seeds and ovaries within the chamber inside the fruit (the larvae are killed if they enter the flesh of papaya fruit before it is ripe; Knab and Yothers 1914; Landolt, Chapter 14), whereas those of at least one undescribed species of *Hexachaeta* feed inside the seeds within the fruit of their *Sorocea* host plant (Moraceae) (Norrbom, personal observation).

Many *Anastrepha* species, including the majority of those in the primitive clades, also breed in fruits of latex-bearing plants, especially Sapotaceae. Certain species groups are associated mainly with other plant families, such as the *spatulata* group on Euphorbiaceae and Olacaceae, the *pseudoparallela* group on Passifloraceae, the *grandis* group (or at least the one species of the group whose hosts are known) on Cucurbitaceae, the *striata* group on Myrtaceae, and some species of the *mucronota* group on Bombacaceae. In most published host plant records for *Anastrepha*, the part of the fruit attacked has not been mentioned. In future rearing studies, this would be useful for researchers to observe and report. At least in the case of *A. sagittata* (*dentata* group) and *A. katiyari* and *pallens* (*daciformis* group), the larva feeds exclusively in the seed (McPhail and Berry 1936; Baker et al. 1944; K. P. Katiyar, personal communication). Larvae of *A. pseudoparallela* (*pseudoparallela* group), and *A. montei* and *pickeli* (*spatulata* group) feed on developing seeds and associated tissues (Morgante et al. 1996; Stefani and Morgante 1996) similar to species of *Toxotrypana*. Larvae of *A. cordata* (*cryptostrepha* group) feed mainly on the seeds and later on the pulp of the fruit, apparently because immature fruits have large quantities of latex (Hernández-Ortiz and Pérez-Alonso 1993). Larvae of *A. crebra* (*mucronota* group) have been found feeding in the pulp and seeds at same time (Hernández-Ortiz and Pérez-Alonso 1993), and those of *A. steyskali* (*leptozona* group), and *A. anomala* and *serpentina* (*serpentina* group) reportedly can feed on both seeds and the fleshy mesocarp (Stone 1942; Korytkowski 1974). Conversely, in most of the generalist species, such as *A. obliqua*, *fraterculus*, *striata,* and *distincta*, feeding is primarily or exclusively on the mesocarp (Aluja 1994; but also see discussion of *A. ludens* in Chapter 15). The effects on behavior that may result from these different larval feeding modes may be worthy of study. For example, females of *Toxotrypana curvicauda* and *A. cordata*, which are seed feeders, do not need to forage for protein in order to produce eggs, unlike most of the pulp-feeding, pest species of *Anastrepha* (Landolt 1985 and Chapter 14; Aluja 1994).

**TABLE 12.4**
**Checklist of *Anastrepha* and *Toxotrypana* Species by Species Group with List of the Families of Their Native Host Plants (the numbers following host families are the number of genera/species of native hosts so far reported)**

| Taxa | Host Plant Families |
|---|---|
| *Anastrepha* | |
| *cryptostrepha* group | |
| *cordata* Aldrich | Apocynaceae 1/1 |
| *cryptostrepha* Hendel | |
| *margarita* Caraballo | |
| *panamensis* Greene | Sapotaceae 1/1 |
| *tripunctata* Wulp | |
| *zeteki* Greene | Sapotaceae 1/1 |
| two n.spp. nr. *tripunctata* | |
| *daciformis* group | |
| *antilliensis* Norrbom | |
| *aquila* Norrbom | |
| *avispa* Norrbom | |
| *bicolor* Stone | |
| *castanea* Norrbom | |
| *daciformis* Bezzi | |
| *katiyari* Norrbom | Sapotaceae 1/1 |
| *macrura* Hendel | Sapotaceae 1/1 |
| = *nambacoli* Tigrero, n. syn. | |
| *maculata* Norrbom | |
| *murrayi* Norrbom | |
| *pallens* Coquillett | Sapotaceae 1/2 |
| *stonei* Steyskal | |
| *zucchii* Norrbom | |
| *dentata* group | |
| *acidusa* (Walker) | |
| *dentata* (Stone) | |
| *flavissima* Hering | |
| *hamata* (Loew) | |
| *obscura* Aldrich | Sapotaceae 1/1 |
| *sagittata* (Stone) | Sapotaceae 1/1 |
| *sagittifera* Zucchi | |
| *urichi* Greene | |
| *zernyi* Lima | Sapotaceae 1/1 |
| n.sp. nr. *dentata* | |
| *benjamini* group | |
| *benjamini* Lima | Sapotaceae 1/1 |
| = *discessa* Stone | |
| *gigantea* Stone | |
| *magna* Norrbom | |
| *superflua* Stone | |
| *robusta* group | |
| *amaryllis* Tigrero | |
| *binodosa* Stone | |
| *fenestrata* Lutz and Lima | |
| *furcata* Lima | |
| *lambda* Hendel | |
| *nigrifascia* Stone | Sapotaceae 1/1 |

**TABLE 12.4 (continued)**
**Checklist of *Anastrepha* and *Toxotrypana* Species by Species Group with List of the Families of Their Native Host Plants (the numbers following host families are the number of genera/species of native hosts so far reported)**

| Taxa | Host Plant Families |
|---|---|
| *phaeoptera* Lima | Moraceae 1/1 |
| *pittieri* Caraballo | |
| *robusta* Greene | |
| *simulans* Zucchi | |
| *speciosa* Stone | |
| five to six new species | |
| *schausi* group | |
| *bellicauda* Norrbom | 1/1 Sapotaceae? |
| *fernandezi* Caraballo | |
| *hermosa* Norrbom | |
| *schausi* Aldrich | |
| *punctata* group | |
| *aczeli* Blanchard | |
| *luederwaldti* Lima | |
| *morvasi* Uramoto and Zucchi | |
| *punctata* Hendel | Myrtaceae 1/1 |
| = *dangeloi* Blanchard | |
| = *goldbachi* Blanchard | |
| = *hendeli* Greene | |
| = *minor* Lima | |
| = *pseudopunctata* Blanchard | |
| n. sp. nr. *luederwaldti* | |
| *leptozona* group | |
| *barnesi* Aldrich | Sapotaceae 1/1 |
| = *virescens* Lima | |
| *costalimai* Autuori | |
| *elongata* Fernández | |
| *leptozona* Hendel | Anacardiaceae 1/1, Icacinaceae 1/1, Myrtaceae 1/1, Quiinaceae 1/1, Rosaceae 1/1, Sapotaceae 2/6 |
| *steyskali* Korytkowski | |
| *mucronota* group | |
| *aphelocentema* Stone | Rutaceae 1/1, Sapotaceae 1/1 |
| *atrox* (Aldrich) | Annonaceae 1/1, Sapotaceae 1/1 |
| = *barandiaranae* Korytkowski and Ojeda | |
| *bezzii* Stone | Sterculiaceae 1/2 |
| = *balloui* Stone | |
| *borgmeieri* Lima | |
| *convoluta* Stone | |
| *crebra* Stone | Bombacaceae 1/2 |
| *debilis* Stone | |
| *edentata* Stone | |
| *elegans* Blanchard | |
| *galbina* Stone | |
| *greenei* Lima | Bombacaceae 1/2 |
| *hambletoni* Lima | |
| *hastata* Stone | |
| *insulae* Stone | |
| *integra* (Loew) | |

**TABLE 12.4 (continued)**
**Checklist of *Anastrepha* and *Toxotrypana* Species by Species Group with List of the Families of Their Native Host Plants (the numbers following host families are the number of genera/species of native hosts so far reported)**

| Taxa | Host Plant Families |
|---|---|
| *kuhlmanni* Lima | Passifloraceae 1/2 |
| *lanceola* Stone | |
| *loewi* Stone | |
| *megacantha* Zucchi | |
| *minuta* Stone | |
| *mucronota* Stone | Bombacaceae 1/1 |
| *nunezae* Steyskal | Bombacaceae 1/1 |
| *parallela* (Wiedemann) | Sapotaceae 1/2, Sterculiaceae 1/2 |
| *quararibeae* Lima | Bombacaceae 1/2 |
| *scobinae* Stone | |
| *similis* Greene | |
| *sinvali* Zucchi | |
| *soroana* Fernández and Rodríguez | |
| *submunda* Lima | Annonaceae 1/1, Sapotaceae 1/3 |
| *tubifera* (Walker) | |
| *undosa* Stone | |
| *grandis* group | |
| *aberrans* Norrbom | |
| *atrigona* Hendel | |
| *bivittata* (Macquart) | |
| *castilloi* Norrbom | |
| *fumipennis* Lima | |
| *grandicula* Norrbom | |
| *grandis* (Macquart) | Cucurbitaceae 1/2 |
| = *latifasciata* Hering | |
| = *schineri* Hendel | |
| *shannoni* Stone | |
| n. sp. (Brazil) | |
| n. sp. nr. *aberrans* | |
| *doryphoros* group | |
| *doryphoros* Stone | |
| *freidbergi* Norrbom | |
| n. sp. (Costa Rica) | |
| *spatulata* group | |
| *alveata* Stone | Olacaceae 1/1 |
| *alveatoides* Blanchard | |
| *distans* Hendel | |
| *haywardi* Blanchard | |
| *interrupta* Stone | Olacaceae 1/2 |
| *manihoti* Lima | Euphorbiaceae 1/1 |
| *montei* Lima | Euphorbiaceae 2/2 |
| = *procurvata* Blanchard | |
| *pickeli* Lima | Bombacaceae 1/1, Euphorbiaceae 1/1 |
| *spatulata* Stone | |
| = *infuscata* Shaw | |
| = *triangulata* Shaw | |
| *umbrosa* Blanchard | |

**TABLE 12.4 (continued)**
**Checklist of *Anastrepha* and *Toxotrypana* Species by Species Group with List of the Families of Their Native Host Plants (the numbers following host families are the number of genera/species of native hosts so far reported)**

| Taxa | Host Plant Families |
|---|---|
| two n. spp. nr. *pickeli* | Euphorbiaceae 1/2 |
| n. sp. (Guatemala) | |
| *ramosa* group | |
| *ramosa* Stone | |
| *subramosa* Stone | |
| *pseudoparallela* group | |
| *amnis* Stone | |
| *anduzei* Stone | |
| *chiclayae* Greene | Myrtaceae 1/1, Passifloraceae 1/2 |
| *consobrina* (Loew) | Passifloraceae 1/2 |
| = *zikani* Lima | |
| *curitis* Stone | Passifloraceae 1/2 |
| *dissimilis* Stone | Passifloraceae 1/1 |
| = *correntina* Blanchard | |
| *dryas* Stone | |
| *ethalea* (Walker) | Passifloraceae 1/2 |
| *limae* Stone | Passifloraceae 1/2 |
| *lutzi* Lima | Passifloraceae 1/1 |
| *mburucuyae* Blanchard | Passifloraceae 1/1 |
| *munda* Schiner | |
| *pallida* Norrbom | Passifloraceae 1/2 |
| *pallidipennis* Greene | Passifloraceae 1/1 |
| *passiflorae* Greene | Passifloraceae 1/1 |
| *pastranai* Blanchard | |
| *pseudoparallela* (Loew) | Annonaceae 1/1, Myrtaceae 1/1, Passifloraceae 1/5 |
| *townsendi* Greene | |
| *velezi* Norrbom | Passifloraceae 1/1 |
| *xanthochaeta* Hendel | Passifloraceae 1/1 |
| *serpentina* group | |
| *anomala* Stone | Apocynaceae 1/1, Theophrastaceae 1/1 |
| *ocresia* (Walker) | Myrtaceae 1/1 |
| = *tricincta* Loew | |
| *pulchra* Stone | Sapotaceae 1/1 |
| *serpentina* (Wiedemann)<br>= *vittithorax* Macquart | Anacardiaceae 1/1, Annonaceae 1/2, Apocynaceae 1/1, Clusiaceae 1/1, Ebenaceae 1/1, Euphorbiaceae 1/1, Hippocrateceae 1/1, Malpighiaceae 1/1, Myrtaceae 1/3, Rutaceae 2/2, Sapotaceae 5/14 |
| three new species | Apocynaceae 1/1 |
| *striata* group | |
| *bistrigata* Bezzi | Anacardiaceae 1/2, Myrtaceae 1/3, Sapotaceae 1/1 |
| *ornata* Aldrich | Myrtaceae 1/2 |
| *striata* Schiner<br>= *cancellaria* Fabricius | Anacardiaceae 1/2, Annonaceae 1/1, Ebenaceae 1/1, Fabaceae 1/1, Melastomataceae 1/1, Myrtaceae 4/14, Passifloraceae, Sapotaceae 2/2, Solanaceae 1/1 |
| *fraterculus* group | |
| *amita* Zucchi | Verbenaceae 1/1 |
| *ampliata* Hernández-Ortiz | |
| *antunesi* Lima | Anacardiaceae 2/3, Myrtaceae 2/2, Rubiaceae 1/1 |

**TABLE 12.4 (continued)**
**Checklist of *Anastrepha* and *Toxotrypana* Species by Species Group with List of the Families of Their Native Host Plants (the numbers following host families are the number of genera/species of native hosts so far reported)**

| Taxa | Host Plant Families |
|---|---|
| *bahiensis* Lima | Anacardiaceae 1/1, Juglandaceae 1/1, Moraceae 4/4, Myrtaceae 2/3 |
| *barbiellinii* Lima | Cactaceae 1/1 |
| *canalis* Stone | Staphylaceae 1/1 |
| *compressa* Stone | |
| *coronilli* Carrejo and Gonzalez | Melastomataceae 1/2 |
| *distincta* Greene<br>= *silvai* Lima | Anacardiaceae 1/1, Annonaceae 1/1, Caesalinaceae 1/1, Clusiaceae 1/1, Fabaceae 1/17, Juglandaceae 1/1, Myrtaceae 5/7, Sapotaceae 1/1 |
| *fischeri* Lima | |
| *fraterculus* (Wiedemann)<br>= *braziliensis* Greene<br>= *costarukmanii* Capoor<br>= *frutalis* Weyenbergh<br>= *lambayecae* Korytkowski and Ojeda<br>= *mellea* Walker<br>= *peruviana* Townsend<br>= *pseudofraterculus* Capoor<br>= *scholae* Capoor<br>= *soluta* Bezzi<br>= *unicolor* Loew | Anacardiaceae 2/3, Annonaceae 1/5, Bombacaceae 1/1, Caricaceae 1/1, Ebenaceae 1/1, Euphorbiaceae 1/1, Fabaceae 1/4, Hippocrateceae 1/1, Juglandaceae 1/1, Lauraceae 1/1, Malpighiaceae 1/1, Moraceae 1/2, Myrtaceae 8/26, Olacaceae 1/1, Oxalidaceae 1/1, Passifloraceae 1/1, Rosaceae 4/7, Sapindaceae 2/2, Sapotaceae 4/6, Solanaceae 2/2, Staphylaceae 1/1, Sterculiaceae 1/1 |
| *irradiata* Blanchard | |
| *irretita* Stone | |
| *ludens* (Loew)<br>= *lathana* Stone | Anacardiaceae 1/1, Annonaceae 1/3, Caricaceae 1/1, Fabaceae 1/2, Myrtaceae 1/1, Passifloraceae 1/1, Rutaceae 2/3, Sapotaceae 1/1 |
| *macra* Stone | |
| *matertela* Zucchi | |
| *minensis* Lima<br>= *extensa* Stone | Annonaceae 1/1, Myrtaceae 2/2 |
| *obliqua* (Macquart)<br>= *ligata* Lima<br>= *mombinpraeoptans* Seín<br>= *trinidadensis* Greene | Anacardiaceae 3/7, Apocynaceae 1/1, Bignoniaceae 1/1, Ebenaceae 1/1, Euphorbiaceae 1/1, Fabaceae 2/2, Malpighiaceae 2/2, Oxalidaceae 1/1, Melastomataceae 1/1, Myrtaceae 6/13, Passifloraceae 1/1, Rosaceae 3/3, Sapotaceae 2/4 |
| *perdita* Stone | Anacardiaceae 1/1 |
| *quiinae* Lima | Quiinaceae 1/1 |
| *reichardti* Zucchi | |
| *schultzi* Blanchard<br>= *inca* Stone<br>= *obliteratella* Blanchard | Myrtaceae 1/1 |
| *sororcula* Zucchi | Anacardiaceae 1/1, Fabaceae 1/1, Myrtaceae 5/11 |
| *suspensa* (Loew)<br>= *longimacula* Greene<br>= *unipuncta* Seín | Anacardiaceae 1/1, Annonaceace 1/2, Canellaceae 1/1, Caricaceae 1/1, Elaeocarpaceae 1/1, Euphorbiaceae 1/1, Malpighiaceae 1/1, Myrtaceae 4/7, Polygonaceae 1/1, Rosaceae 1/1, Rutaceae 1/1, Sapotaceae 2/3 |
| *tenella* Zucchi | |
| *turicai* Blanchard | |
| *turpiniae* Stone | Flacourtiaceae 1/1 |
| *zenildae* Zucchi | Anacardiaceae 1/1, Fabaceae 1/1, Melastomataceae 1/1, Myrtaceae 3/5, Rhamnaceae 1/1 |
| *zuelaniae* Stone | Flacourtiaceae 1/1 |

**TABLE 12.4 (continued)**
**Checklist of *Anastrepha* and *Toxotrypana* Species by Species Group with List of the Families of Their Native Host Plants (the numbers following host families are the number of genera/species of native hosts so far reported)**

| Taxa | Host Plant Families |
|---|---|
| Unplaced species | |
| *acris* Stone | Euphorbiaceae 1/1 |
| *barrettoi* Zucchi | |
| *belenensis* Zucchi | |
| *bondari* Lima | Clusiaceae 1/1, Moraceae 1/1, Mytaceae 1/1, Sapotaceae 1/1 |
| *buscki* Stone | |
| *caudata* Stone | |
| *concava* Greene | |
| *conjuncta* Hendel | |
| *connexa* Lima | |
| *cruzi* Lima | |
| *duckei* Lima | Flacourtiaceae 1/1 |
| *flavipennis* Greene | |
| *fractura* Stone | |
| *guianae* Stone | |
| *hamadryas* (Stone) | |
| *longicauda* Lima | |
| = *hendeliana* Lima | |
| *lutea* Stone | |
| *mixta* Zucchi | |
| *nascimentoi* Zucchi | |
| *nigripalpis* Hendel | |
| *pacifica* Hernández-Ortiz | |
| *palae* Stone | |
| *parishi* Stone | Myrtaceae 1/1 |
| *repanda* Blanchard | |
| *rheediae* Stone | Clusiaceae 1/3 |
| *rosilloi* Blanchard | |
| *sodalis* Stone | |
| *sylvicola* Knab | Capparidaceae 1/1 |
| *tecta* Zucchi | |
| *teli* Stone | |
| *teretis* Stone | |
| *tumida* Stone | |
| *Toxotrypana* | |
| *australis* Blanchard | Asclepiadaceae 1/1 |
| *curvicauda* Gerstaecker | Apocynaceae 1/1, Asclepiadaceae 1/3, Caricaceae 2/3 |
| *littoralis* Blanchard | Asclepiadaceae 1/1, Caricaceae 1/1 |
| *nigra* Blanchard | Asclepiadaceae 1/1 |
| = *pseudopicciola* Blanchard | |
| *picciola* Blanchard | |
| *proseni* Blanchard | Asclepiadaceae 1/1 |
| *recurcauda* Tigrero | |
| six to eight new species | Caricaceae 1/1 |

It is noteworthy that most of the generalist, pest species of *Anastrepha* belong to the *fraterculus* species group. Except for single species in the *serpentina*, *striata*, *leptozona,* and *pseudoparallela* groups, feeding on more than a few related hosts or one to two families by a single *Anastrepha* species is rare outside of the *fraterculus* group. The reasons for the preponderance of pest species in this one group may be varied, but there may be some connection to the larval feeding mode. Perhaps the specialization of the species of the *fraterculus* group toward feeding on a variety of pulpy, mature fruits, rather than on a narrower range of plants with particular toxic chemicals, has preadapted them for attacking cultivated fruits.

## 12.7 CONCLUSIONS AND FUTURE RESEARCH NEEDS

One way to assess our current understanding of the phylogeny of *Anastrepha* and *Toxotrypana* is to compare with other tephritid genera. In that sense, *Anastrepha* and *Toxotrypana* are better understood than most Tephritidae, but they have not been as well studied as *Rhagoletis*, the most intensely studied tephritid genus.

The monophyly of *Anastrepha* + *Toxotrypana* is well supported by morphological characters, as is that of *Toxotrypana*. Whether *Anastrepha* is monophyletic remains uncertain. Based mainly on morphological studies of the adult stage, 18 species groups, including 166 valid species, have been recognized within *Anastrepha* (another 32 species are unplaced). There is strong to fairly good character evidence (hypothesized synapomorphies), in roughly decreasing order, for the *A. daciformis*, *schausi*, *dentata*, *punctata*, *leptozona*, *doryphoros*, *ramosa*, *robusta*, *fraterculus*, *striata*, *serpentina*, and *spatulata* groups. The status of the *A. cryptostrepha*, *grandis*, *pseudoparallela*, *benjamini*, and *mucronota* groups is less certain, although each contains some species that are clearly closely related. Relationships among the species groups and within most of them are poorly resolved.

Among the most important questions remaining in the phylogeny of these two genera are the following. Is *Anastrepha* monophyletic or are some species more closely related to *Toxotrypana*? What are the relationships among the species groups of *Anastrepha*? For most of the groups, what are the relationships among the species? To answer these questions, much additional morphological study is needed to treat the many undescribed species remaining in both genera, and for many of the known species, to describe additional characters fully, such as the male genitalia. Sources of additional character data, such as larvae, eggs, and molecular characters need to be further explored. Determination of the natural host plants of the as yet biologically unknown species will be crucial to obtain the immature stages of these species and to provide other useful data, such as the part of the plant attacked (particularly, what tissue(s) within the fruit). Based on all these additional data and currently known information, a rigorous cladistic analysis is needed for the entire group.

## ACKNOWLEDGMENTS

We are grateful to the many scientists and institutions, too numerous to list here, who have loaned or sent us specimens for study. T. Litwak and L. Lawrence drew many of the illustrations, and L. Rodriguez produced the plates. We thank Sonja Scheffer, Marc De Meyer, and Norman Woodley for reviewing the manuscript. We acknowledge the Campaña Nacional contra las Moscas de la Fruta (Mexico), International Organization for Biological Control of Noxious Animals and Plants (IOBC), and Instituto de Ecología, A.C. (Mexico) for their financial support of the symposium, and USDA-ICD-RSED for funding Norrbom's travel.

## REFERENCES


Aldrich, J.M. 1925. New Diptera or two-winged flies in the United States National Museum. *Proc. U.S. Natl. Mus.* 66 (18): 36 pp. [No. 2555].

Aluja, M. 1994. Bionomics and management of *Anastrepha*. *Annu. Rev. Entomol.* 39: 155–178.

Baker, A.C., W.E. Stone, C.C. Plummer, and M. McPhail. 1944. A review of studies on the Mexican fruitfly and related Mexican species. *U.S. Dep. Agric. Misc. Publ.* 531, 155 pp.

Canal D., N.A. 1997. Levantamento, Flutuação Populacional e Análise Faunística das Espécies de Moscas-das-frutas (Dip., Tephritidae) en Quatro Municípios do Norte do Estado de Minas Gerais. Dissertation, Universidade de São Paulo, Piracicaba, 113 pp.

Carroll, L.E. and R.A. Wharton. 1989. Morphology of the immature stages of *Anastrepha ludens* (Diptera: Tephritidae). *Ann. Entomol. Soc. Am.* 82: 201–214.

Emmart, E.W. 1933. The eggs of four species of fruit flies of the genus *Anastrepha*. *Proc. Entomol. Soc. Wash.* 35: 184–191.

Foote, R.H. 1967. Family Tephritidae (Trypetidae, Trupaneidae). In *A Catalogue of the Diptera of the Americas South of the United States* (P.E. Vanzolini and N. Papavero, eds.), fasicle 57: 91 pp. Departamento de Zoologia, Secretaria da Agricultura, São Paulo.

Foote, R.H., F.L. Blanc, and A.L. Norrbom. 1993. *Handbook of the Fruit Flies (Diptera: Tephritidae) of America North of Mexico*. Comstock Publishing Associates, Ithaca. 571 pp.

Freidberg, A. 1985. The genus *Craspedoxantha* Bezzi (Diptera: Tephritidae: Terelliinae). *Ann. Natal Mus.* 27: 183–206.

Frías L., D., H. Martinez, and A. Alviña. 1993. Descripcion morfologica de los estados inmaduros de *Rhagoletis tomatis* Foote (Diptera: Tephritidae). *Acta Entomol. Chil.* 18: 31–40.

Han, H.Y. and B.A. McPheron. 1997. Molecular phylogenetic study of Tephritidae (Insecta: Diptera) using partial sequences of the mitochondrial 16S ribosomal DNA. *Mol. Phylogenet. Evol.* 7: 17–32.

Hancock, D.L. 1986. Classification of the Trypetinae (Diptera: Tephritidae), with a discussion of the Afro-tropical fauna. *J. Entomol. Soc. South. Afr.* 49: 275–305.

Hardy, D.E. 1955. A reclassification of the Dacini (Tephritidae-Diptera). *Ann. Entomol. Soc. Am.* 48: 425–437.

Headrick, D.H. and R.D. Goeden. 1990. Description of the immature stages of *Paracantha gentilis* (Diptera: Tephritidae). *Ann. Entomol. Soc. Am.* 83: 220–229.

Headrick, D.H. and R.D. Goeden. 1993. Life history and description of immature stages of *Aciurina thoracica* (Diptera: Tephritidae) on *Baccharis sarothroides* in southern California. *Ann. Entomol. Soc. Am.* 86: 68–79.

Heppner, J.B. 1986. Larvae of fruit flies. III. *Toxotrypana curvicauda* (Papaya fruit fly) (Diptera: Tephritidae). *Fla. Dept. Agric. Consum. Serv. Div. Plant Ind. Entomol. Circ.* 282, 2 pp.

Hernández-Ortiz, V. and R. Pérez-Alonso. 1993. The natural host plants of *Anastrepha* (Diptera: Tephritidae) in a tropical rain forest of Mexico. *Fla. Entomol.* 76: 447–460.

Kitto, G.B. 1983. An immunological approach to the phylogeny of the Tephritidae. In *Fruit Flies of Economic Importance. Proceedings of the CEC/IOBC International Symposium, Athens, Greece, November 16–19, 1982* (R. Cavalloro, ed.), pp. 203–211. A.A. Balkema, Rotterdam. 642 pp.

Knab, F. and W.W. Yothers. 1914. Papaya fruit fly. *J. Agric. Res.* 2: 447–453.

Korneyev, V.A. 1994. Reclassification of the Palaearctic Tephritidae (Diptera). Communication 2. *Vestn. Zool.* 1994 (1): 3–17.

Korytkowski G., C.A. 1974. Una nueva especie del género *Anastrepha* Schiner (Diptera: Tephritidae). *Rev. Peru. Entomol.* 17: 1–3.

Korytkowski G., C.A. and D. Ojeda P. 1968. Especies del género *Anastrepha* Schiner 1868 en el nor-oeste peruano. *Rev. Peru. Entomol.* 11: 32–70.

Landolt, P.J. 1985. Behavior of the papaya fruit fly *Toxotrypana curvicauda* Gerstaecker (Diptera: Tephritidae), in relation to its host plant, *Carica papaya* L. *Folia Entomol. Mex.* (1984) 61: 215–224.

Lawrence, P.O. 1979. Immature stages of the Caribbean fruit fly, *Anastrepha suspensa*. *Fla. Entomol.* 62: 214–219.

Lima, A.M. da Costa. 1953a. Moscas de frutas do U.S. National Museum (Smithsonian Institution) (II) (Diptera: Trypetidae). *An. Acad. Bras. Cienc.* 25: 153–155.

Lima, A.M. da Costa. 1953b. Moscas de frutas do U.S. National Museum (Smithsonian Institution) (III) (Diptera, Trypetidae). *An. Acad. Bras. Cienc.* 25: 557–566.

Lima, A.M. da Costa. 1954. Moscas de frutas do U.S. National Museum (Smithsonian Institution) (IV) (Diptera, Trypetidae). *An. Acad. Bras. Cienc.* 26: 277–282.

Loew, H. 1873. Monographs of the Diptera of North America. Part III. *Smithson. Misc. Collect.* 11 (3 [publ. 256]); 351 pp.

McPhail, M. and N.O. Berry. 1936. Observations on *Anastrepha pallens* (Coq.) reared from wild fruits in the lower Rio Grande Valley of Texas during the spring of 1932. *J. Econ. Entomol.* 29: 405–411.

Morgante, J.S., D. Selivon, V.N. Solferini, and A.S. do Nascimento. 1996. Genetic and morphological differentiation in the specialist species *Anastrepha pickeli* and *A. montei*. In *Fruit Fly Pests: A World Assessment of Their Biology and Management* (B.A. McPheron and G.J. Steck, eds.), pp. 259–276. St. Lucie Press, Delray Beach. 586 pp.

Murillo, T. and L.F. Jirón. 1994. Egg morphology of *Anastrepha obliqua* and some comparative aspects with eggs of *Anastrepha fraterculus* (Diptera: Tephritidae). *Fla. Entomol.* 77: 342–348.

Norrbom, A.L. 1985. Phylogenetic Analysis and Taxonomy of the *cryptostrepha*, *daciformis*, *robusta*, and *schausi* Species Groups of *Anastrepha* Schiner (Diptera: Tephritidae). Ph.D. dissertation, Pennsylvania State University, University Park. 354 pp.

Norrbom, A.L. 1991. The species of *Anastrepha* (Diptera: Tephritidae) with a *grandis*-type wing pattern. *Proc. Entomol. Soc. Wash.* 93: 101–124.

Norrbom, A.L. 1993. Two new species of *Anastrepha* (Diptera: Tephritidae) with atypical wing patterns. *Proc. Entomol. Soc. Wash.* 95: 52–58.

Norrbom, A.L. 1997. Revision of the *Anastrepha benjamini* species group and the *A. pallidipennis* complex (Diptera: Tephritidae). *Insecta Mundi* 11: 141–157.

Norrbom, A.L. 1998. A revision of the *Anastrepha daciformis* species group (Diptera: Tephritidae). *Proc. Entomol. Soc. Wash.* 100: 160–192.

Norrbom, A.L. Host plant database for *Anastrepha* and *Toxotrypana* (Diptera: Tephritidae: Toxotrypanini). *Diptera Data Dissemination Disk* 2, in press.

Norrbom, A.L. and R.H. Foote. 1989. The taxonomy and biogeography of the genus *Anastrepha* (Diptera: Tephritidae). In *Fruit Flies: Their Biology, Natural Enemies and Control* (A.S. Robinson and G. Hooper, eds.), pp. 15–26. In *World Crop Pests* (W. Helle, ed.), Vol. 3A. Elsevier Science Publishers, Amsterdam. 372 pp.

Norrbom, A.L. and K.C. Kim. 1988a. Revision of the *schausi* group of *Anastrepha* Schiner (Diptera: Tephritidae), with a discussion of the terminology of the female terminalia in the Tephritoidea. *Ann. Entomol. Soc. Am.* 81: 164–173.

Norrbom, A.L. and K.C. Kim. 1988b. A list of the reported host plants of the species of *Anastrepha* Schiner (Diptera: Tephritidae). U.S. Depy. Agric. Animal Plant Health Insp. Serv. No. 81–52, 114 pp.

Norrbom, A.L., L.E. Carroll, and A. Freidberg. 1999a. Status of knowledge. In *Fruit Fly Expert Identification System and Systematic Information Database* (F.C. Thompson, ed.), pp. 9–47. *Myia* (1998) 9, 524 pp.

Norrbom, A.L., L.E. Carroll, F.C. Thompson, I.M. White, and A. Freidberg. 1999b. Systematic database of names. In *Fruit Fly Expert Identification System and Systematic Information Database* (F.C. Thompson, ed.), pp. 65–251. *Myia* (1998) 9, 524 pp.

Phillips, V.T. 1946. The biology and identification of trypetid larvae (Diptera: Trypetidae). *Mem. Am. Entomol. Soc.* 12, 161 pp.

Sarma, R., G.B. Kitto, S.H. Berlocher, and G.L. Bush. 1987. Biochemical and immunological studies on an alpha-glycerophosphate dehydrogenase from the tephritid fly, *Anastrepha suspensa*. *Arch. Insect Biochem. Physiol.* 4: 271–286.

Seín, F., Jr. 1933. *Anastrepha* (Trypetydae [sic], Diptera) fruit flies in Puerto Rico. *J. Dept. Agric. P.R.* 17: 183–196.

Selivon, D. and A.L.P. Perondini. 1999. Description of *Anastrepha sororcula* and *A. serpentina* (Diptera: Tephritidae) eggs. *Fla. Entomol.* 82: 347–353.

Selivon, D., J.S. Morgante, and A.L.P. Perondini. 1997. Egg size, yolk mass extrusion and hatching behavior in two cryptic species of *Anastrepha fraterculus* (Wiedemann) (Diptera, Tephritidae). *Braz. J. Genet.* 91: 471–478.

Shaw, J.G. 1962. Species of the *spatulata* group of *Anastrepha* (Diptera: Tephritidae). *J. Kans. Entomol. Soc.* 35: 408–414.

Silva, A.G. d'Araujo e, C.R. Goncalves, D.M. Galvao, A.J.L. Goncalves, J. Gomes, M. do Nascimento Silva, and L. de Simoni. 1968. Quarto catalogo dos insetos que vivem nas plantas do Brasil. Seus parasitos e predadores. Parte II — 1.o Tomo. Insetos, hospedeiros, e inimigos naturais. Ministerio da Agricultura, Departamento de Defesa e Inspecao Agropecuaria, Servico de Defesa Sanitaria Vegetal, Laboratorio Central de Patologia Vegetal, Rio de Janeiro. 622 pp.

Snow, W.A. 1895. On *Toxotrypana* of Gerstaecker. *Kans. University Q.* 4: 117–119.

Solferini, V.N. and J.S. Morgante. 1987. Karyotype study of eight species of *Anastrepha* (Diptera: Tephritidae). *Caryologia* 40: 229–241.

Steck, G.J. 1991. Biochemical systematics and population genetic structure of *Anastrepha fraterculus* and related species (Diptera: Tephritidae). *Ann. Entomol. Soc. Am.* 84: 10–28.

Steck, G.J. and A. Malavasi. 1988. Description of immature stages of *Anastrepha bistrigata* (Diptera: Tephritidae). *Ann. Entomol. Soc. Am.* 81: 1004–1006.

Steck, G.J. and R.A. Wharton. 1988. Description of immature stages of *Anastrepha interrupta*, *A. limae*, and *A. grandis* (Diptera: Tephritidae). *Ann. Entomol. Soc. Am.* 81: 994–1003.

Steck, G.J., L.E. Carroll, H. Celedonio, and J.C. Guillen. 1990. Methods for identification of *Anastrepha* larvae (Diptera: Tephritidae), and key to 13 species. *Proc. Entomol. Soc. Wash.* 92: 333–346.

Stefani, R.N. and J.S. Morgante. 1996. Genetic variability in *Anastrepha pseudoparallela*: A specialist species. In *Fruit Fly Pests: A World Assessment of Their Biology and Management* (B.A. McPheron and G.J. Steck, eds.), pp. 259–276. St. Lucie Press, Delray Beach. 586 pp.

Steyskal, G.C. 1977a. Two new neotropical fruitflies of the genus *Anastrepha*, with notes on generic synonymy (Diptera, Tephritidae). *Proc. Entomol. Soc. Wash.* 79: 75–81.

Steyskal, G.C. 1977b. *Pictorial Key to Species of the Genus Anastrepha (Diptera: Tephritidae).* Entomological Society of Washington, Washington, D.C. 35 pp.

Stone, A. 1939a. A new genus of Trypetidae near *Anastrepha* (Diptera). *J. Wash. Acad. Sci.* 29: 340–350.

Stone, A. 1939b. A revision of the genus *Pseudodacus* Hendel (Dipt. Trypetidae). *Rev. Entomol. (Rio J.)* 10: 282–289.

Stone, A. 1942. The fruitflies of the genus *Anastrepha. U.S. Dep. Agric. Misc. Publ.* 439, 112 pp.

Tauber, M.J. and C.A. Tauber. 1967. Reproductive behavior and biology of the gall-former *Aciurina ferruginea* (Doane) (Diptera: Tephritidae). *Can. J. Zool.* 45: 907–913.

Tigrero, J. 1992. Descripción de dos nuevas especies de Tephritidae: Toxotrypaninae, presentes en Ecuador. *Rev. Rumipamba* 9: 102–112.

Tigrero, J. 1998. *Revisión de Especies de Moscas de la Fruta Presentes en el Ecuador.* Published by the author, Sangolquí, Ecuador, 55 pp.

White, I.M. and M.M. Elson-Harris. 1992. *Fruit Flies of Economic Significance: Their Identification and Bionomics.* CAB International, Wallingford. 601 pp.

White, I.M. and K. Marquardt. 1989. A revision of the genus *Chaetorellia* Hendel (Diptera: Tephritidae) including a new species associated with spotted knapweed, *Centaurea maculosa* Lam. (Asteraceae). *Bull. Entomol. Res.* 79: 453–487.

Zucchi, R.A. 1977. Taxonomia das Especies Brasileiras de *Anastrepha* Schiner, 1868 do Compexo *fraterculus* (Diptera, Tephritidae). Thesis, Universidade de São Paulo, Piracicaba. 63 pp.

# 13 Phylogeny of the Genera *Anastrepha* and *Toxotrypana* (Trypetinae: Toxotrypanini) Based upon 16S rRNA Mitochondrial DNA Sequences

*Bruce A. McPheron, Ho-Yeon Han, Janisete G. Silva, and Allen L. Norrbom*

**CONTENTS**



## 13.1 INTRODUCTION

The evolutionary history of the genus *Anastrepha* Schiner, an important taxon due to the economic effects of many of its 197 described species, has not been clarified by past studies. This is a critical lapse, as there are many interesting behaviors and ecologies within this group that would be best studied within the framework of a well-resolved phylogenetic hypothesis (e.g., see Chapter 15 by Aluja et al. and Chapter 29 by Heath et al.). Morphology and a few other characters, such as host plant use, have been used to group *Anastrepha* spp. into 18 species groups (the history of this effort and a new proposal are detailed in Chapter 12 by Norrbom et al.). This most recent effort is still a work in progress, and it is clear that much remains to be done in resolving the phylogeny of this group. Norrbom et al.

(Chapter 12) discuss the relationship of *Anastrepha* to other tephritid genera. It belongs in the tribe Toxotrypanini, along with the genera *Toxotrypana* Gerstaecker and *Hexachaeta* Loew.

In recent years, several phylogenetic studies of tephritid taxa have involved molecular methods. A number of studies of many tephritid groups have employed isozyme analysis. Morgante et al. (1980) and Steck (1991) both used isozymes to look at relationships among selected subsets of *Anastrepha* spp., but their studies were not designed to be comprehensive across the genus. More recently, phylogenetic studies using DNA sequence data have been employed to examine tephritid relationships (Han and McPheron 1994; 1997; and Chapter 5; Soto-Adames et al. 1994; McPheron and Han 1997; Smith and Bush 1997; and Chapter 9). These studies have initially focused on identifying the phylogenetic level at which certain molecules may be informative and, in some cases, discussing the relationships that are confidently supported by these sequence data. By far, most of the information from these molecular systematic studies to date has come from analysis of mitochondrial DNA sequences.

In this chapter we report the analysis of phylogenetic relationships of 40 species of *Anastrepha* belonging to 14 species groups. Also included are three *Toxotrypana* spp. and two *Hexachaeta* spp. The phylogeny of this group of species was inferred from a 930 base pair (bp) alignment of the large subunit ribosomal DNA (16S rDNA) of the mitochondrial DNA (mtDNA) using different techniques of phylogeny reconstruction. Our results are compared against the species group organization of Norrbom et al. (Chapter 12), which, although not a complete phylogenetic hypothesis, is the most comprehensive prediction of relationships for the group.

## 13.2 MATERIALS AND METHODS

We sequenced 40 species of *Anastrepha*, three species of *Toxotrypana*, and two species of *Hexachaeta*, representing the three genera in the tribe Toxotrypanini. The taxa used in this study, along with collection and preservation information, are listed in Table 13.1. Vouchers are deposited in the Frost Entomological Museum, Pennsylvania State University or the National Museum of Natural History, Smithsonian Institution.

Total nucleic acid extractions of individual flies followed the protocol previously described for frozen specimens (Sheppard et al. 1992). Pinned or alcohol-preserved specimens were extracted following a modification of this protocol described in Han and McPheron (1997), which involves an additional step of initial incubation at 55°C using proteinase K. DNA was extracted from single individuals of each taxon included in this study.

The polymerase chain reaction (PCR) was used to amplify a region comprising 930 bp within the mitochondrial 16S ribosomal DNA. The primers used for PCR and sequencing are displayed in Table 13.2. PCR was performed using primers LR-J-13323/LR-N-13770 and LR-J-12883/LR-N-13398 in 50 μl reaction volumes under the following conditions: 1X Promega reaction buffer, 250 μ*M* of each dATP, dCTP, dGTP, and dTTP, 1.25 μ*M* of each primer, 2.5 units of Promega *Taq* polymerase, and 1 μl of template DNA. For specimens with high DNA degradation, higher volumes of template (2 or 3 μl) were used. The cycle program consisted of 40 cycles of 1 min at 93°C, 1 min at 45°C, and 2 min at 72°C with a final cycle of 1 min at 94°C, 1 min at 50°C, and 9 min at 72°C.

Two approaches were used for sequencing. In the first approach, the double-stranded amplification product was gel purified in 1.5% agarose gels (1X TAE) following the procedure described in Han and McPheron (1997). The DNA obtained from the gel-purification process was then used as template for asymmetrical PCR in 60 μl volumes, under the same amplification conditions described above using one of the PCR primers or an internal primer as a limiting primer (1:25~100 ratio). Single-stranded DNA was washed (300 μl sterile water), concentrated three times in Millipore Ultrafree-MC 30,000 MW filters, and used as the template for sequencing reactions. Manual sequencing was performed using the dideoxy, chain-termination method according to the Sequenase 2.0 (Amersham Co.) protocol (as described in Han and McPheron 1997). In the second approach, the

**TABLE 13.1**
**Collection Data and Condition of Specimen for Taxa Sequenced in This Study (organized by species group as defined by Norrbom et al., Chapter 12)**

***Hexachaeta***

*amabilis* (Loew). Guatemala: Escuintala: Palín area, McPhail trap, 1992–1993, col. J. López; in ethanol

*fallax* Lima. Mexico: Nuevo Leon: Santiago, 24.II.1994, col. Esau; in ethanol

***Toxotrypana***

*australis* Blanchard. Argentina: Tucumán: Burruyacu, Taruca Pampa, Finca San Augustine, McPhail trap in citrus orchard, 14.V.92; pinned

*curvicauda* Gerstaecker. USA: Florida, 1990, col. J. Sivinski; frozen

*littoralis* Blanchard. Guatemala: El Portal, ex. *Gonolobus leianthus*, V.1994, col. J. López; in ethanol

***Anastrepha***

*cryptostrepha* group
- *cordata* Aldrich. Mexico: Veracruz: Los Tuxtlas: Coyame-Nanciyaga, ex. *Tabernaemontana alba*, 1994, col. M. Aluja et al.; frozen
- *panamensis* Greene. Panama, 30.III.1989, col. G. Tapia and C. Korytkowski; in ethanol

*daciformis* group
- *bicolor* (Stone). Mexico: Veracruz: Emiliano Zapata, La Jicayana, II.1996, col. P. Juárez; in ethanol
- *katiyari* Norrbom. Venezuela: Zulia: Mara, ex. *Sideroxylon obtusifolium*, 25.X.1995, col. K. Katiyar, Camacho, and J. Oroño; in ethanol
- *pallens* Coquillett. Mexico: Veracruz: Apazapan, 2.VI.1991, col. G. Quintero and L. Quiroz; in ethanol

*dentata* group
- *hamata* (Loew). Mexico: Veracruz: 1997, col. M. Aluja et al.; in ethanol

*robusta* group
- *nigrifascia* Stone. Bahamas: Abaco I., Bahama Star Grove, 9.VI.1994; in ethanol

*schausi* group
- *bellicauda* Norrbom. Venezuela: Trujillo: La Chirá, ex. cusco, col. 15.XII.95, col. K. Katiyar and J. Oroño; in ethanol

*punctata* group
- *punctata* Hendel. Argentina: Tucumán: Burruyacu, Taruca Pampa, Finca San Augustine, in McPhail trap in citrus orchard, 28.V.1989, col. ALN; pinned

*leptozona* group
- *leptozona* Hendel. Mexico: Chiapas: Tapachula, 27.III.94, ex. baricoco, col. M. Aluja et al.; frozen

*mucronota* group
- *aphelocentema* Stone. Mexico: Veracruz: Pozarica, em. 18.X.1997, col. M. Aluja et al.; frozen
- *bezzii* Lima. Venezuela: Aragua: Maracay, I.1991, reared ex. fruit of *Sterculia apetala*, col. J. Dedordy, reared by ALN; frozen
- *crebra* Stone. Mexico: Veracruz: Estación Biología Los Tuxtlas, ex. fruits of *Quararibea funebris*, coll. 22.VIII.1989, col. V. Hernández and ALN (891722)

*grandis* group
- *grandis* (Macquart). Brazil: São Paulo: Laranjal Paulista, ex. *Cucurbita pepo* fruit, col. 26.XI.1996, JGS; in ethanol

*spatulata* group
- *alveata* Stone. Mexico: Veracruz: Llano Grande Ravine, ex. *Ximenia americana*, 1994, col. M. Aluja et al.; frozen
- *manihoti* Lima. Venezuela: Zulia: Zona de Reserva de Burro Negro Munucupio, Langunilla, reared from *Manihot esculenta* stem, 22.VII.1996, col. K. Katiyar and J. Oroño; in ethanol
- *montei* Lima. Venezuela: Trujillo: El Helechal, McPhail trap baited with Staley, 15.IV.1996, col. K. Katiyar and J. Oroño; in ethanol
- *pickeli* Lima. Brazil: Bahia: Cruz das Almas. ex. *Manihot esculenta*, 1988, col. G. J. Steck; frozen.
- *spatulata* Stone. Tobago: Area II (St. David or St. George Par.), trap in mango, 27.VI.1989; in ethanol

**TABLE 13.1 (continued)**
**Collection Data and Condition of Specimen for Taxa Sequenced in This Study (organized by species group as defined by Norrbom et al., Chapter 12)**

*pseudoparallela* group
*dryas* Stone. Venezuela: Trujillo: El Helechal, McPhail trap, 9.IX.95, K. Katiyar and J. Oroño; in ethanol
*limae* Stone. Panama: Capira, 8.X.1989, col. G. Tapia and C. Korytkowski; in ethanol
*pallidipennis* Greene. Venezuela: Merida: Bachaquero (La Azulita), ex. *Passiflora quadrangularis*, col. 16.IX.95, Katiyar, Camacho, and Oroño; in ethanol
*pseudoparallela* (Loew). Brazil: São Paulo: Laranjal Paulista, ex. *Passiflora edulis* fruit, col. 2.I.1997, JGS; in ethanol
*serpentina* group
sp. nr. *anomala*. Venezuela: Marcillol, 7.V.1993, col. K. Katiyar; in ethanol
*serpentina* (Wiedemann). Venezuela: Aragua: Maracay, ex. *Manilkara zapota*, 1988, col. G. J. Steck; frozen
*striata* group
*bistrigata* Bezzi. Brazil: São Paulo: Louveira, from lab colony of J. S. Morgante and A. Malavasi, Dept. de Biol., USP, São Paulo, Brazil, 1990, G. J. Steck; frozen
*striata* Schiner. Venezuela: Merida: Merida, ex. *Psidium guajava*, 1988, col. G. J. Steck; frozen
*fraterculus* group
*amita* Zucchi. Trinidad: Victoria Par., X-XI.1991, K-80 and K-88; in ethanol
*bahiensis* Lima. Guatemala: Taxisco, ex. *Brosimum costaricum,* 1994, col. J. López; in ethanol
*barbiellinii* Lima. Brazil: Minas Gerais: Arceburgo, Fazenda Fortaleza, reared ex. *Pereskia aculeata* fruit, 1.VI.1991, col. ALN and R. A. Zucchi; pinned
*coronilli* Carrejo and González. Venezuela: Palmichal, 1.V.1993, col. K. Katiyar; in ethanol
*distincta* Greene. Brazil: Bahia: Cruz das Almas, ex. *Inga* sp., 1988, col. G. J. Steck; frozen
*fraterculus* (Wiedemann). Brazil: São Paulo: Bertioga, ex. *Terminalia catappa*, 1990, G. J. Steck; frozen
*fraterculus* (Wiedemann). Venezuela: Merida: Merida area, above 1600 m, ex. *Rubus glaucus*, V-VI.88, col. G. J. Steck; frozen. [these specimens are labeled *fraterculus** throughout the chapter]
*ludens* (Loew). Mexico: Veracruz: Actopan (Pueblo), ex. *Mangifera indica*, 20.V.1994, col. M. Aluja et al.; frozen
*obliqua* (Macquart). Mexico: Veracruz: Actopan (Pueblo), ex. *Spondias purpurea*, 20.IV.1994, col. M. Aluja et al.; frozen
*sororcula* Zucchi. Brazil: Bahia: Santo Antonio de Jesus. From lab colony of J. S. Morgante and A. Malavasi, 1990; frozen
*suspensa* (Loew). USA: From lab colony of Florida Dept. Agric., Gainesville, Florida; frozen
unplaced
*acris* Stone. Venezuela: Falcon: Boca del Tocuyo, reared from *Hippomane mancinella*, 7.V.1993, col. K. Katiyar and R. Matheus; in ethanol
*flavipennis* Greene. Panama: Capira, col. G. Tapia and C. Korytkowski, 26.IV.1989; in ethanol

ALN and JGS refer to collections made by the authors.

symmetrical PCR product was submitted to an enzymatic pretreatment (EPT) prior to sequencing using the PCR product presequencing kit (Amersham Co., catalog # U.S. 70995). EPT involves the use of two enzymes, shrimp alkaline phosphatase and exonuclease I, which remove the excess deoxynucleoside triphosphates (dNTPs), primers, and any extraneous single-stranded DNA produced by PCR amplification that would interfere with the sequencing reaction. We added 1 μl of each of the two enzymes to 5 μl of PCR product and incubated it at 37°C for 15 min followed by heating to 80°C for 15 min to inactivate the enzymes. Double-stranded DNA was then manually sequenced using the Thermo Sequenase radiolabeled terminator cycle sequencing kit (Amersham Co., catalog # U.S. 79750) protocol, which employs a single primer and repeated cycles of thermal denaturation, primer annealing, and polymerization. Reaction conditions consisted of 2 μl reaction buffer, 1 μl of 2 μ*M* sequencing primer, 14 μl MQ $H_2O$, 8 units of Thermo Sequenase DNA polymerase, 1 μl

**TABLE 13.2**
**Oligonucleotide Primers for PCR and Sequencing of the Mitochondrial 16S Ribosomal DNA**

| Primer | Sequence |
|---|---|
| LR-J-12883[a] | 5′-CTCCGGTTTGAACTCAGATC-3′ |
| LR-J-13021 | 5′-ACGCTGTTATCCCTAAAGTA-3′ |
| LR-N-13182 | 5′-TTAAAAGACGAGAAGACCCTA-3′ |
| LR-J-13323 | 5′-ACTAATGATTATGCTACCTT-3′ |
| LR-N-13398[a] | 5′-CGCCTGTTTATCAAAAACAT-3′ |
| LR-N-13770 | 5′-AGAAATGAAATGTTATTCGT-3′ |

[a] Primers from Xiong and Kocher (1991). All other primers were designed by eye (by HYH) based on an alignment between *Drosophila yakuba*, GenBank accession number X03240 (Clary and Wolstenholme 1985), and *Anopheles gambiae*, GenBank accession number L20934 (Beard et al. 1993).

EPT template, 1 μCi [α-$^{33}$P] ddNTPs (0.5 μl each of G, A, T, or C), and 2 μl termination master mix per sample. The cycle program consisted of 30 cycles of 30 s at 95°C, 30 s at 48°C, and 1 min at 72°C. We added 3.5 μl of stop solution to each tube before loading the sequencing gel. Sequences from both strands were obtained for each specimen.

Initial alignment of the sequences was conducted using SeqPup (Gilbert 1995), followed by manual refinement using MacClade 3.07 (Maddison and Maddison 1997). Phylogenetic relationship analyses were conducted using maximum parsimony (MP) and neighbor-joining (NJ) methods. The two *Hexachaeta* spp. were designated as the outgroup taxa in our analyses (Han and McPheron 1997; see below). We used PAUP*, version 4.0b1 (Swofford 1998), to perform MP analysis using the heuristic search procedure (tree-bisection-reconnection algorithm and the MULPARS option) to find the most parsimonious trees. All included characters were assigned equal weights in the MP analysis. We used 50 replications of random addition of taxa to evaluate possible bias due to input order. Bootstrapping (Felsenstein 1985) of the MP analysis (300 replicates) was conducted under the heuristic search procedure, with a maxtree setting of 100 trees. NJ analysis was performed using PAUP*, version 4.0b1 (Swofford 1998). An NJ tree was generated using the Jukes–Cantor distance (chosen based upon criteria in Kumar et al. 1993), but NJ topologies using other genetic distances, including proportional, Kimura 2-parameter, Tajima-Nei and Tamura-Nei, were also examined. We used bootstrapping (1000 replicates) to estimate the support for NJ topologies. Positions containing gaps or ambiguous alignments were excluded from both the MP and the NJ analyses. PAUP* was also used to calculate nucleotide composition of the 16S rDNA.

## 13.3 RESULTS AND DISCUSSION

### 13.3.1 Molecular Evolution of the 16S rDNA in *Anastrepha*

A total of 830 to 894 bp were sequenced from the 45 taxa included in this study, yielding a data matrix containing 930 bp when gaps were inserted to improve the alignment (Appendix 13.1). This region is homologous to that used in previous studies of the Tephritidae and *Rhagoletis* (Han and McPheron 1997; McPheron and Han 1997), encompassing positions 12,895–13,756 in the *Drosophila yakuba* complete mtDNA sequence (Clary and Wolstenholme 1985). As previously observed in other tephritid taxa, this is a highly A-T rich gene. Average nucleotide composition across the 45 taxa was 36.4% A, 43.9% T, 7.0% C, and 12.7% G. The average Jukes–Cantor distance among the 45 species included in the analysis was 0.083 ± 0.001; the level of sequence divergence ranged from a minimum of 0.003 to a maximum distance of 0.153.

Sequences reported in this chapter are available from GenBank under the accession numbers U39379 to U39382 and AF15049 to AF152091.

### 13.3.2 Phylogenetic Analyses

One of four most parsimonious trees recovered by MP analysis is shown in Figure 13.1. Of the 822 characters used in the analysis, 283 were variable and 175 were informative under parsimony. Bootstrap values higher than 50% are indicated above the appropriate branches. Topology of the four MP trees differed only in the position of terminal branches within the *fraterculus* species group (*barbiellinii* through *fraterculus** on Figure 13.1), so we are comfortable showing a single, resolved tree rather than a consensus of the four trees. The topology of the NJ tree is relatively similar, although not identical, in its assignment of species' membership in species groups and relative placement of the various species groups (Figure 13.2). Bootstrap values higher than 50% are again indicated on the figure. The NJ tree graphically shows that many branches are quite short (indicating a small genetic distance between the constituent species). No variation in tree topology was detected among NJ trees calculated from different distance measures. The NJ topology was notably longer than the MP tree, requiring 923 steps.

Many branches in both the MP and NJ trees have only low statistical support, which somewhat constrains the arguments we can make regarding the evolutionary history of this assemblage of taxa. We reduced the complexity of the data set by analyzing various combinations of taxa with only one or two exemplars of the different species groups (several combinations of taxa were analyzed, including an analysis using only the species for which the various groups were named; analyses not shown). Our hope was that this would increase the resolution at the base of the tree, but, using MP methods, there was no change in topology of the resulting trees.

### 13.3.3 Relationships within the Toxotrypanini

Previous DNA analysis (Han and McPheron 1997) placed the genus *Hexachaeta* as the sister taxon to *Anastrepha* + *Toxotrypana*, at that time the only two genera included in the tribe Toxotrypanini. On the strength of this relationship, Norrbom et al. (1999) tentatively included *Hexachaeta* as a member of this tribe. Further studies using a variety of characters are required to confirm this relationship, but we used *Hexachaeta* as the outgroup in our study, and we recovered a monophyletic ingroup (*Anastrepha* + *Toxotrypana*) in all analyses. To test the robustness of this assumption, we also conducted three additional MP analyses with *Rhagoletis pomonella*, *Oedicarena latifrons*, or *Ceratitis capitata* each designated separately as the outgroup taxon and a fourth analysis with all of those three species included as outgroups (results not shown). Han and McPheron (1997) found that *Hexachaeta*, *Toxotrypana*, and *Anastrepha* formed a monophyletic group within their analysis of 35 tephritid species representing many lineages in the family. Our study using three of those species with a much larger number of species representing *Anastrepha* and *Toxotrypana* reached the same conclusion. The Toxotrypanini as defined here was monophyletic, and *Anastrepha* + *Toxotrypana* remained monophyletic with respect to *Hexachaeta*. Thus, we are confident in using *Hexachaeta* as the outgroup to analyze the relationships between and within *Toxotrypana* and *Anastrepha*.

### 13.3.4 Relationships of *Toxotrypana* and *Anastrepha*

The three *Toxotrypana* spp. in our study were always strongly supported as a monophyletic taxon, in agreement with the morphological evidence (Norrbom et al., Chapter 12). As indicated by Norrbom et al., there is not strong morphological support for the monophyly of *Anastrepha* with respect to *Toxotrypana*, and our results suggest that the former indeed may not be monophyletic.

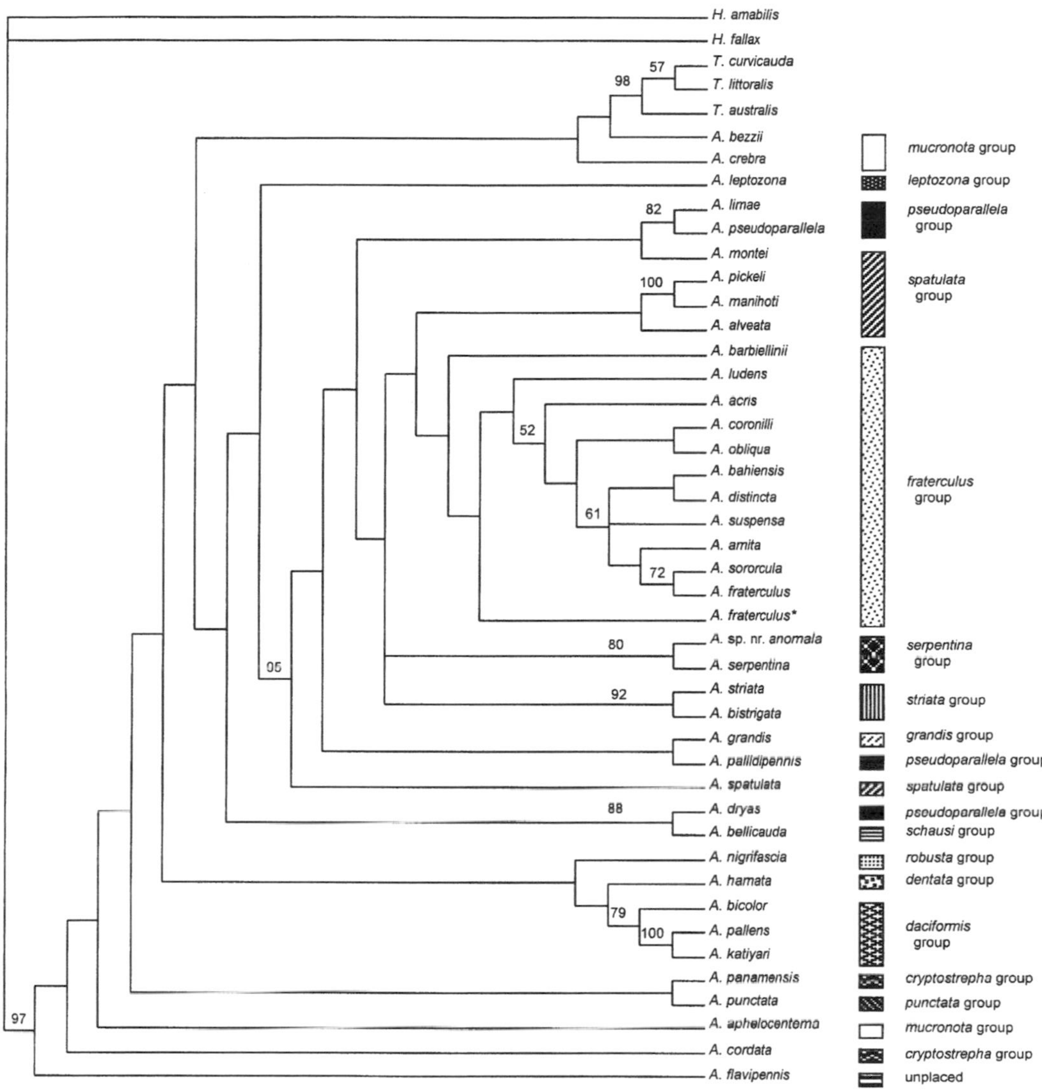


**FIGURE 13.1** Toxotrypanini phylogenetic relationships inferred from one of four most parsimonious trees. Numbers on the branches indicate bootstrap confidence limits higher than 50% (300 replications). Statistics of each most parsimonious tree: length = 892, consistency index with uninformative characters excluded = 0.44, retention index = 0.53, rescaled consistency index = 0.23. Taxon names correspond to the species listed in Table 13.1.

The basal relationships of *Anastrepha* + *Toxotrypana* were not well resolved in our analysis, so it is premature to declare that *Anastrepha* is paraphyletic with respect to *Toxotrypana*. However, our results, coupled with the uncertainty expressed by Norrbom et al. (Chapter 12), suggest that this matter deserves serious attention. In our analysis, members of many of the more ancestral *Anastrepha* species groups (e.g., *robusta*, *mucronota*, *cryptostrepha*, *punctata*, *daciformis*, and *dentata* groups) came out at the base of the tree along with *Toxotrypana*. Included among these was *A. flavipennis*, which is not currently assigned to a species group. This species may warrant further morphological examination.

**FIGURE 13.2** Toxotrypanini phylogenetic relationships inferred from a neighbor-joining tree based on Jukes–Cantor distances with complete deletion of gaps and missing data. Numbers on the branches indicate bootstrap values higher than 50% (1000 replications). Taxon names correspond to the species listed in Table 13.1.

### 13.3.5 Relationships within *Toxotrypana*

Our sampling of *Toxotrypana* was sparse, with only three species representing a single species group. The 16S data suggested a sister relationship between *T. curvicauda* and *T. littoralis*, although bootstrap support was not strong.

### 13.3.6 Relationships within *Anastrepha*

Of the 18 species groups delimited by Norrbom et al. (Chapter 12), we had exemplars from 14. In six of these groups we were able to obtain only a single species for this study, so there can be no discussion of the monophyly of those groups. We will discuss the relationships suggested by the

DNA sequence data in the same order as groups were discussed by Norrbom et al. to facilitate direct comparisons between molecules and morphology. Where we refer to a morphological hypothesis without citation, the hypothesis has been drawn from Norrbom et al. All of our results are presented graphically in Figures 13.1 and 13.2.

*cryptostrepha* group — This species group was hypothesized to be the most ancestral group in the genus. We included two species, *A. panamensis* and *A. cordata*. They did not group as sister taxa using either MP or NJ analytical methods, but both species arise near the base of the tree. This region of the tree is not well resolved (very low bootstrap values), so it is premature to draw strong conclusions. Norrbom et al. suggested that the *cryptostrepha* group as a whole might be paraphyletic, and they placed *A. cordata* and *A. panamensis* in separate clades within the group. Other taxa found consistently in this part of our MP tree were *A. punctata* (*punctata* group), *A. aphelocentema* (*mucronota* group), and *A. flavipennis* (unplaced). The two *cryptostrepha* group species are somewhat more widely separated in the NJ tree than in the MP tree but, again, there is very low resolution of this part of the tree.

*cryptostrepha* + *punctata* groups — A similarity in lateral surstylus shape was noted between members of these two groups, and the molecular data join *A. panamensis* (*cryptostrepha* group) and *A. punctata* (*punctata* group).

*daciformis* group — The *daciformis* species group is the only group within *Anastrepha* to receive a rigorous cladistic analysis based upon morphological characters (Norrbom 1998). We included three species, *A. bicolor*, *A. pallens*, and *A. katiyari*, and the group was recovered as a monophyletic cluster. DNA data join *A. pallens* and *A. katiyari* with high support, leaving *A. bicolor* as the sister group to this clade, which is somewhat at odds with Norrbom's (1998) results. However, many additional species in the group were not included, which could affect the recovered topology.

*daciformis* + *dentata* groups — These two groups are united by a number of morphological apomorphies, and *A. hamata* (*dentata* group) appears as predicted as the sister taxon to the *daciformis* group in our analysis. This topology is consistent in both MP and NJ analyses.

*mucronota* group — Two of the three species included in our study, *A. crebra* and *A. bezzii*, clustered together in our molecular analysis. In the MP analysis, these two species were placed on a branch with *Toxotrypana*. *Anastrepha aphelocentema* was well separated from the other two species. We must keep in mind that this portion of the tree is not well resolved, but our results suggest that Norrbom et al. may be correct in questioning the monophyly of this group.

*pseudoparallela* + *spatulata* + *ramosa* + *doryphoros* + *grandis* + *serpentina* + *striata* + *fraterculus* groups — A characteristic of the male genitalia (proctiger sclerotization) suggests the monophyly of this set of species groups, and, with a single exception to be discussed below, the DNA data support this hypothesis. We examined representatives of all but the *ramosa* and *doryphoros* groups, including a total of 26 taxa, and they clustered as a monophyletic group with the exception of *A. dryas* (*pseudoparallela* group), which joined *A. bellicauda* (*schausi* group) outside the large clade in both MP and NJ analyses. The average Jukes–Cantor distance among the members of this large group was $0.041 \pm 0.001$, one-half the average distance from the entire data set.

*spatulata* group — The five species from the *spatulata* group included in our study were not recovered as a monophyletic taxon. A sister taxon relationship was supported between *A. pickeli* and *A. manihoti*, with *A. alveata* joined to this clade. The position of *A. montei* depended upon the analysis; NJ analysis clustered this species with the preceding three, while MP analysis linked it with the *A. limae*–*A. pseudoparallela* clade. Even more variable was the position of *A. spatulata*. In the NJ analysis, it was the sister group of most of the *fraterculus* group, and in the MP analysis *A. spatulata* was the most basal branch within the larger *pseudoparallela*-through-*fraterculus* cluster defined above. In either case, *A. spatulata* was well removed from the remaining members of the *spatulata* group. The branch lengths in this entire cluster of species groups are relatively short, which leads us to suggest that the monophyly of the *spatulata* group should be further tested, not only by morphological analysis but also by other molecular tools with higher resolution potential.

*pseudoparallela* group — This species group was greatly rearranged by Norrbom (1997) and Norrbom et al. (Chatper 12) compared with previous classifications (e.g., Norrbom and Kim 1988). The current hypothesis, based upon a combination of morphological characters and host associations, is not supported by our results. Most puzzling is the placement of *A. dryas* as the sister taxon to *A. bellicauda*. As mentioned above, these taxa are outside the larger clade containing the other three members of the *pseudoparallela* group included in our study, a relationship well supported by the molecular data. Leaving the issue of *A. dryas* aside, the relationship *of A. limae* and *A. pseudoparallela* has good statistical support. These two species were formerly placed in the *chiclayae* and *pseudoparallela* groups, respectively (Norrbom and Kim 1988). *Anastrepha pallidipennis*, previously in the *benjamini* group (Norrbom and Kim 1988, Norrbom 1997), falls out near *A. grandis* in the molecular analyses, although this relationship lacks statistical support. Inclusion of more species from this large group could lead to some very interesting predictions from molecular analyses.

*serpentina* group — We included *A. serpentina* and a new species similar to *A. anomala* in this study. Their monophyly within the studied taxa was well supported.

*striata* group — Within the three-species *striata* group, *A. striata* and *A. bistrigata* are considered to be sister taxa, and the statistical support for their relationship was quite strong.

*fraterculus* group — This group has received more attention in recent years than most of the other groups, primarily because many of the pest species within *Anastrepha* belong to this group (see White and Elson-Harris 1992; Aluja 1994). It is a large group; only the *mucronota* group at present has more assigned species (Norrbom et al., Chapter 12). The number of species may increase; *A. fraterculus*, for example, is actually a complex of species (Steck 1991; Selivon 1996; see below). Relatively few hypotheses of interspecific relationships within the *fraterculus* group have been proposed. There are several species that are morphologically very similar to *A. fraterculus* itself (Araujo 1997), which may be indicative of a recent evolutionary history in this subset of the *fraterculus* group as a whole.

From our analysis we obtained relatively poor resolution among the 11 taxa included. It was within the *fraterculus* group that the relative positions of the species differed among the four most parsimonious MP trees. The average genetic distance among these species is $0.018 \pm 0.001$ (Jukes–Cantor distance), compared with the value of 0.083 reported above for the entire data set. The 16S rDNA has not evolved rapidly enough relative to the apparently young age of the group to provide a clear pattern of the divergence among constituent members. With that said, certain subsets of species were more similar than others on the basis of DNA data. A sister relationship was indicated between *A. fraterculus* and *A. sororcula*, consistent with similarities in morphology, karyotype, and isozymes (Selivon 1996). A group comprising *A. fraterculus*, *A. sororcula*, *A. amita*, *A. distincta*, *A. bahiensis*, and *A. suspensa* also had moderate statistical support (Figure 13.1). The previously unplaced species *A. acris* also appears to belong, on the basis of DNA data, in the *fraterculus* group. The membership of *A. barbiellinii*, whose inclusion in the group is only tentatively supported by morphological evidence, lies at the outer boundary of the *fraterculus* group in the MP analysis and is separated from the remaining members of the *fraterculus* group in the NJ analysis by the *striata* group and *A. spatulata*. Overall, the statistical support from the 16S data for monophyly of the *fraterculus* group is not strong, consistent with the low levels of divergence in this gene among these species. The positions of at least the *striata*, *serpentina*, and, perhaps, *spatulata* groups must be examined relative to the full *fraterculus* group as currently defined to resolve this issue.

*A. fraterculus*: single species vs. species complex — Steck (1991) and Selivon (1996), using isozyme data, clearly demonstrated that what had been called *A. fraterculus* was not a single species. We obtained specimens from Steck's study for our DNA-based analysis. Included in the present study were two *A. fraterculus* sequences, one from São Paulo state in Brazil (*fraterculus* in the figures) and the other from a high elevation site in Merida, Venezuela (*fraterculus** in the figures). It is the Brazilian *fraterculus* sequence that is united with *A. sororcula* and other members of the *fraterculus* group. The *fraterculus** sequence lies near the outside of the *fraterculus* group in the MP tree and is relatively deep in the branching pattern of the group in the NJ tree. The pair-wise distance between these two sequences, both from flies that were identified on the basis of morphology as *A. fraterculus*, is 0.022, which is greater than the average distance among members of the *fraterculus* group. We have sequenced portions of the 16S gene from 48 individuals (McPheron et al., unpublished data) from nine different locations (Mexico to Brazil). No intrasample (microgeographic) sequence variation has yet been observed; the Central American samples were all identical and were very similar to lowland Venezuela and Brazil, and, therefore, the high-elevation Venezuelan sample (represented by the *fraterculus** sequence), the only high-elevation sample studied so far, is quite distinct from all of these other *A. fraterculus* populations.

## 13.4 CONCLUSIONS AND FUTURE RESEARCH NEEDS

This is the first molecular study to attempt a broad look at the evolutionary relationships of *Anastrepha* and the Toxotrypanini. On the basis of this research, we are not proposing an extensive reorganization of the classification of the group, but our results provide the first independent test of the morphologically based classification. Clearly, further work is needed on both fronts.

The problems of membership in and relationships among species groups of *Anastrepha* have not been thoroughly resolved by this molecular study. The tribe and included genera pose some interesting questions at different degrees of divergence, and discrete analyses using different genes may be required to address these questions. Among the more interesting issues that may yield to subsequent molecular analysis is the question of *Toxotrypana* + *Anastrepha*. Our data suggest that this may not be a sister taxon relationship (i.e., *Anastrepha* may be paraphyletic), but the basal branches of our trees are not fully resolved.

The limits of some, but not all, of the morphologically defined *Anastrepha* species groups are supported by molecular data, but many of the species groups were defined very tentatively based upon morphology (Norrbom et al., Chapter 12). If specimens can be acquired for molecular analysis (a real constraint for many of the more ancestral species groups), DNA-based approaches could be used to help resolve these questions. From our results, and within the limits of our own taxon representation, we found support for the *daciformis* group, the *dentata* + *daciformis* groups, the *serpentina* group, the *striata* group, and, by and large, the *fraterculus* group. An assemblage of several of the more-derived species groups that had been predicted from morphology was also supported. Certain other groups were clearly not supported as monophyletic taxa, including the *cryptostrepha* group, the *mucronota* group, the *pseudoparallela* group, and, to some degree, the *spatulata* group. All of these observations should be treated as hypotheses. The Toxotrypanini, and particularly *Anastrepha*, is a case where morphology, ecology, behavior, biogeography, and genetics will be valuable as we generate and independently test phylogenetic hypotheses.

The molecular systematics of *Anastrepha* requires additional attention. We will examine some other regions of the mtDNA (perhaps cytochrome oxidase sequences; see Smith and Bush 1997; and Chapter 9; and Han and McPheron, Chapter 5) for recently evolved lineages. A thorough study of the *A. fraterculus* problem, including geographic variation across the entire range of the nominal species and comparison with closely related species, is clearly dictated. To resolve the basal branches of the phylogeny and determine the position of *Toxotrypana* and the more ancestral *Anastrepha* species groups, we believe that nuclear gene sequences may be more appropriate tools (e.g., Brower and DeSalle 1994; Mitchell et al. 1997). When a well-resolved phylogenetic framework does become available, we will be better able to conduct the fascinating task of analyzing the evolution of many characters, such as host plant use, host plant part (e.g., seed vs. fruit) use, chemical ecology, timing of behavior, and biogeographic patterns.

## ACKNOWLEDGMENTS

This chapter is based upon work supported by the Cooperative State Research, Education, and Extension Service, U.S. Department of Agriculture, under Agreement No. 95-37302-1808. We also are grateful for support from the Campaña Nacional contra las Moscas de la Fruta (Mexico), the International Organization for Biological Control of Noxious Animals and Plants, the Instituto de Ecología, A.C. (Mexico), and USDA-ICD-RSED. We thank the individuals cited in Table 13.1 who provided taxa for this study, particularly M. Aluja and members of his laboratory (special thanks to J. Piñero) for specimens from Mexico, D. Emmen for specimens from Panama, K. Katiyar and colleagues for specimens from Venezuela, J. López (APHIS-PPQ) for specimens from Guatemala, and G. Steck. M. Thomas and several reviewers provided comments on the manuscript.

## REFERENCES

Aluja, M. 1994. Bionomics and management of *Anastrepha*. *Annu. Rev. Entomol.* 39: 155–178.

Araujo, E.L. de. 1997. Estudo morfométrico no acúleo de cinco espécies de *Anastrepha* Schiner, 1868 (Diptera: Tephritidae) do grupo *fraterculus*. M.S. thesis, Universidade de São Paulo, Piracicaba.

Beard, C.B., D. Mills Hamm, and F.H. Collins. 1993. The mitochondrial genome of the mosquito *Anopheles gambiae*: DNA sequence, genome organization, and comparisons with mitochondrial sequences of other insects. *Insect Mol. Biol.* 2: 103–124.

Brower, A.V.Z. and R. DeSalle. 1994. Practical and theoretical considerations for choice of a DNA sequence region in insect molecular systematics, with a short review of published studies using nuclear gene regions. *Ann. Entomol. Soc. Am.* 87: 702–716.

Clary, D.O. and D.R. Wolstenholme. 1985. The mitochondrial DNA molecule of *Drosophila yakuba*: nucleotide sequence, gene organization, and genetic code. *J. Mol. Evol.* 22: 252–271.

Felsenstein, J. 1985. Confidence limits on phylogenies: an approach using the bootstrap. *Evolution* 39: 783–791.

Gilbert, D.G. 1995. *SeqPup: a biosequence editor and analysis application.* Version 0.6f.

Han, H.-Y. and B.A. McPheron. 1994. Molecular phylogenetic study of selected tephritid flies (Insecta: Diptera: Tephritidae) using partial sequences of the nuclear 18S ribosomal DNA. *Biochem. Syst. Ecol.* 22: 447–457.

Han, H.-Y. and B.A. McPheron. 1997. Molecular phylogenetic study of Tephritidae (Insecta: Diptera) using partial sequences of the mitochondrial 16S ribosomal DNA. *Mol. Phylogenet. Evol.* 7: 17–32.

Kumar, S., K. Tamura, and M. Nei. 1993. *MEGA: Molecular Evolutionary Genetics Analysis, version 1.0.* Institute of Molecular Evolutionary Genetics, Pennsylvania State University, University Park.

Maddison, W.P. and D.R. Maddison. 1997. *MacClade, version 3.07.* Sinauer Associates, Sunderland.

McPheron, B.A. and H.-Y. Han. 1997. Phylogenetic analysis of North American *Rhagoletis* (Diptera: Tephritidae) and related genera using mitochondrial DNA sequence data. *Mol. Phylogenet. Evol.* 7: 1–16.

Mitchell, A., S. Cho, J.C. Regier, C. Mitter, R.W. Poole, and M. Matthews. 1997. Phylogenetic utility of *elongation factor-1*$\alpha$ in Noctuoidea (Insecta: Lepidoptera): the limits of synonymous substitution. *Mol. Biol. Evol.* 14: 381–390.

Morgante, J.S., A. Malavasi, and G.L. Bush. 1980. Biochemical systematics and evolutionary relationships of Neotropical *Anastrepha*. *Ann. Entomol. Soc. Am.* 73: 622–630.

Norrbom, A.L. 1997. Revision of the *Anastrepha benjamini* species group and the *A. pallidipennis* complex (Diptera: Tephritidae). *Insecta Mundi* 11: 141–157.

Norrbom, A.L. 1998. A revision of the *Anastrepha daciformis* species group (Diptera: Tephritidae). *Proc. Entomol. Soc. Wash.* 100: 160–192.

Norrbom, A.L. and K.C. Kim. 1988. A list of the reported host plants of the species of *Anastrepha* Schiner (Diptera: Tephritidae). *U.S. Dep. Agric. Animal Plant Health Insp. Serv. (APHIS)*, No. 81–52, Hyattsville, 114 pp.

Norrbom, A.L., L.E. Carroll, and A. Freidberg. 1999. Status of knowledge. In *Fruit Fly Expert Identification System and Systematic Information Database* (F.C. Thompson, ed.), pp. 9–47. *Myia* (1998) 9, 524 pp.

Selivon, D. 1996. Estudo sobre a Diferenciação Populacional em *Anastrepha fraterculus* (Wiedemann) (Diptera, Tephritidae). Ph.D. thesis, Universidade de São Paulo, São Paulo.

Sheppard, W.S., G.J. Steck, and B.A. McPheron. 1992. Geographic populations of the medfly may be differentiated by mitochondrial DNA variation. *Experientia* 49: 1010–1013.

Smith, J.J. and G.L. Bush. 1997. Phylogeny of the genus *Rhagoletis* (Diptera: Tephritidae) inferred from DNA sequences of mitochondrial cytochrome oxidase II. *Mol. Phylogenet. Evol.* 7: 33–43.

Soto-Adames, F.N., H.M. Robertson, and S.H. Berlocher. 1994. Phylogenetic utility of partial sequences of $G_6pdh$ at different taxonomic levels in hexapod systematics with emphasis on Diptera. *Ann. Entomol. Soc. Am.* 87: 723–736.

Steck, G.J. 1991. Biochemical systematics and population structure of *Anastrepha fraterculus* and related species (Diptera: Tephritidae). *Ann. Entomol. Soc. Am.* 84: 10–28.

Swofford, D.L. 1998. *PAUP*: Phylogenetic Analysis Using Parsimony, version 4.0b1*. Sinauer Associates, Sunderland.

White, I.M. and M.M. Elson-Harris. 1992. *Fruit Flies of Economic Significance: Their Identification and Bionomics*. CAB International, Wallingford. 601 pp.

Xiong, B. and T.D. Kocher. 1991. Comparisons of mitochondrial DNA sequences of seven morphospecies of black flies (Diptera: Simuliidae). *Genome* 34: 306–311.

## APPENDIX 13.1: ALIGNMENT OF PARTIAL MITOCHONDRIAL 16S RDNA SEQUENCES FOR THE INCLUDED TAXA

An “N” refers to a character state that was ambiguous after sequencing, and a “?” means that the position was missing from our data collection. Gaps are indicated by “–” and a dot (.) means that the character state was identical to the state listed in the top row (*Hexachaeta amabilis*).

```
                                                                                                                1 1111111111 1111111111 1111111111 1111111111 1111111111 1111111111
                                        1 1111111112 2222222223 3333333334 4444444445 5555555556 6666666667 7777777778 8888888889 9999999990 0000000001 1111111112 2222222223 3333333334 4444444445 5555555556
                               1234567890 1234567890 1234567890 1234567890 1234567890 1234567890 1234567890 1234567890 1234567890 1234567890 1234567890 1234567890 1234567890 1234567890 1234567890 1234567890
Hexachaeta  amabilis           TAGTTTTTTT -ATAAATAAA TTTAATTTAC AAATTTTTGA TTTTTATTTT TTT-TAAGAA A---TTTAAA GAATTATTGA TTTGAATATT TTAAGGGATA AGCTTTAAAA TAAAG-TTTT AAAAA-TTAA TAATTATT-A TTTAAAA-TG TATGCTTATA
Hexachaeta  fallax             .......... ..G....... .......... ........T. .....TGA.. A.....TAT. ......A... A.......A. .........? C..T...... .......G.. ....T.AA.. .......... ......A... .A........ ........G.
Toxotrypana curvicauda         .......... ..G....... .......... ..G...C.A. ......A... A......T.. .......... A.....GAA. .......... .......... .......... ....A..A.. .......A.. A...A..... A......... ....T...A.
Toxotrypana australis          .......... ..G....... .......... ........A. .....TG... A....T.T.. .......... A.......A. .......... .......... ......T... .......A.. .......GTT A...A..... A......... ....T...A.
Toxotrypana littoralis         .......... ..G....... .......... ........A. .....TA... A....T.T.. .......... A.T.....A. .......... .......... .......... ....T..G.. .......ATT A...A..... A......... ....T...A.
Anastrepha  flavipennis        .......... ..G....... .......... ..T.....A. ......G... ...TA..A.. ....A..... A.T.....A. ........C. .......... .........G ....A..A.. .........T G...A..... A......A.. ....T...G.
Anastrepha  panamensis         .......... ..G....... .......... G.......A. ......A... A......T.. .......... A.T.....A. .......... .A........ ......TG.G ..G.T..A.. .......GT. A...A....G A......A.. .....A..G.
Anastrepha  nigrifascia        .......... ..G....... .......... ........A. ......GG.. .....TGA.. ....AA.... A.T.....A. .......... .......... .......... ....T.AA.. .......-TT A...A.G..T A......A.. ....TG..G.
Anastrepha  punctata           .......... ..G....... .......... ........A. ......A... A....G.A.. ....A..... A.......A. .........C .......... ........G. ....T..A.. .......ATT A...A....T A......A.. .....A..G.
Anastrepha  leptozona          .......... C.G....... .......... ........A. ...A...... ...T..TT.. ........T. A.T.....A. .......... .......... .......... ....AT.A.. ........TT A...A..... A......A.. ....TG..G.
Anastrepha  crebra             .......... ..G....... .......... ........A. ......A... A.....TT.. ....A..... A.......A. .......... .......... .......... .......A.. .......ATT A...A..... A......A.. ....TG..G.
Anastrepha  bezzii             .......... ..G....... .......... .G......A. ......A... A.....TT.. ....G..... A.T.....A. .......... .......... .......... ....A..A.. .......ATT A...A..... A......A.. ....TG..G.
Anastrepha  limae              .......... ..G....... .......... ........A. ......A... A....TTT.. .......... A.T.....A. .......... .......... .......... ....A..A.. .....T.ATT A...G..... A......A.. ....TG..G.
Anastrepha  pickeli            .......... ..G....... .......... ........A. ......A... A....TTT.. .......... A.T.....A. .....-.... .......... .......... ....A..A.. .....C.ATT G...A..... A..T...A.. .....A..G.
Anastrepha  sp. nr. anomala    .......... ..G....... .......... ........A. ......A... A....TTT.. ......C... A.T.....A. .......... .......... .......... ....T..A.. .....T.ATT A...A.A.AG A..T..CT.. ...?.A..G.
Anastrepha  serpentina         .......... ..G....... .......... ..G.....A. ......A... A....TTT.. .......... A.T.....A. .......... .......... .......... ....T..A.. .....T.ATT A...A.A..G A..T...A.. .....A..G.
Anastrepha  spatulata          .......... ..G....... .......... ........A. ......A... A..T.TTT.. .......T.. A.T.....A. .......... .......... ......G... ....A..A.. .....T.GTT G.G.A....G A..T...A.. ....TG..G.
Anastrepha  barbiellinii       .......... ..G....... .......... ........A. ......A... A....TTT.. .......... A.T.....A. .......... .......... .......... .......A.. .....C.ATT GG..A....G G..T...A.. .....A..G.
Anastrepha  striata            .......... ..G....... .......... ........A. ......A... A....TTT.. .......... A.T.....A. .......... .......... .......... ....A..A.. .....T.GTT G...A....G A..T...G.. .....G..G.
Anastrepha  bistrigata         .......... ..G....... .......... ........A. ......A... A....TTT.. .......... A.T.....A. .......... .......... .......... ....A..A.. .....T.ATT G...A....G A..T...A.. .....A..G.
Anastrepha  ludens             .......... ..G....... .......... ........A. ......A... .....TTT.. ....A..... A.T.....A. .......... .......... .......... ....A..A.. .....T.ATT G...A....T A..T...A.. ....TA..G.
Anastrepha  acris              .......... ..G....... .......... ........A. ......A... A....TTT.. .......... A.T.....A. .......... .......... .......... ....A..A.. .....TCATT G...A....T A..T...A.. ...-.A..G.
Anastrepha  coronilli          .......... ..G....... .......... ........A. ......A... A....TTTT. .......... A.T.....A. .......... .......... .......... ....A..A.. .....T.GTT G.G.A....T A..TG..A.. ....TA..G.
Anastrepha  obliqua            .......... ..G....... .......... ........A. ......A... A....TTTT. .......... A.T.....A. .......... .......... .......... ....A..A.. .....T.ATT G...A..C.T A..T...A.. .....A..G.
Anastrepha  bahiensis          .......... ..G....... .......... ........A. ......A... A..T.TTTT. .......... A.......A. .......... .......... .......... ....A..A.. .....TCATT G...A....T A..T...A.. .....A..G.
Anastrepha  distincta          .......... ..G....... .......... ........A. ......A... A....TTTT. ....A..... A.T.....A. .......... .......... .......... ....A..A.. .....TCATT G...A....T A..T...A.. .....A..G.
Anastrepha  suspensa           .......... ..G....... .......... ........A. ......A... A....TTTT. .......... A.T.....A. .......... .......... .......... ....A..A.. .....TCATT G...A....T A..T...A.. .....A..G.
Anastrepha  amita              .......... ..G....... .......... ........A. ......A... A....TTTT. .......... A.T.....A. .......... .......... .......... ....A..A.. .....TCATT G...A....T A..T...A.. .....A..G.
Anastrepha  sororcula          .......... ..G....... .......... ........A. ......A... A....TTTTT ...A...... A.T.....A. .......... .......... .......... ....A..A.. .....TCATT G...A....T A..T...A.. .....A..G.
Anastrepha  fraterculus        .......... ..G....... .......... ........A. .....GA... A....TTTTT ...A...... A.T.....A. .......... .......... .......... ....A..A.. .....TCATT G...A....T A..T...A.. .....A..G.
Anastrepha  fraterculus*       .......... ..G....... .......... ........A. ......A... A....TTT.. .......... A.T.....A. .......... .......... .......... ....A..A.. .....T.ATT G...A....T A..T...A.. .....A..G.
Anastrepha  dryas              .......... ..G....... .......... G.......A. ......A... A....TTAT. ....G..... A.T.....A. .......... .........G .......... ....A..A.. .....T..TT -...A....G -......A.. ....T...G.
Anastrepha  grandis            .......... ..G....... .......... G.......A. ......A... A....TTT.. .......G.. A.......A. .......... .......... ......G... ....A..A.G .....T.ATT G...A..A.G G......A.. .....A..G.
Anastrepha  pseudoparallela    ?????????? ..G....... .......... ........A. ......A... A....TTT.. .......... A.T.....A. .......... .......... .......... ....T..A.. .....T.ATT A...G..... A......A.. ....AGATG.
Anastrepha  montei             .......... ..G....... .......... ........A. ......AA.. A....TTT.. .......... A.T.....A. .......... .......... .......... ....A..A.. .....T.ATT G...G....T G..T...A.. ....TA..G.
Anastrepha  manihoti           .......... ..G....... .......... ........A. ......A... A....TTT.. .......... A.T.....A. .......... .......... .........G ....A..A.. .....T.ATT G.G.A..... A..T...A.. .....A..G.
Anastrepha  alveata            .......... ..G....... .......... ........A. ......A... A....TTT.. .......... A.T.....A. .......... .......... .......... ....A..A.. .....T.ATT G...A..A.G A..T...A.. .....A..G.
Anastrepha  cordata            ..?....... ..?....... .......... ........A. ......A... ......GT.. .......... A.T...AAA. .......... .......... ........G. ....T..... .....TA..G A.TAAT...T A...G..T.. ........G.
Anastrepha  bellicauda         ?......... ..G....... .......... ........A. ......A... A.....TT.. ....A..... A.......A. .......... .......... .......... ....A..A.. .......ATT A...A....C A......A.. ....TG..G.
Anastrepha  pallidipennis      .......... ..G....... .......... ........A. ......A... A....TTT.. .......... A.T.....A. .......... .......... .........G ....A.AG.. .....T.GTT G...AG...G A..T...A.. ....TA..G.
Anastrepha  hamata             .......... ..G....... .......... .......... ......A... .....TTT.T .AAAA..... A.......T. .......... ..T....... ........GG ....T..A.. ........TT G...A..... AG.T...A.. ....TG..G.
Anastrepha  bicolor            ?......... ..G....... .......... ........A. ......A... .....TTAG. ..AA...... A......AA. .......... .......... .......... ....T..A.. ....TTA.TT A...A..... A..T...A.. ....TG..G.
Anastrepha  aphelocentema      .......... .?G?.?...? ...??...?. ?.T.....A. ......A... .......A.. ....A..... A.T.....A. .......... .......... .......... ....A..A.. .......ATT AG..A..... A......A.. .....A..G.
Anastrepha  pallens            ?????????? .?........ ..A....... ........A. ......A... .....G.A.. ....A..... A.T.....A. .......... .......... .......... ....A..A.. ....TTA.TT ....A..... A..T...A.. ....TG..G.
Anastrepha  katiyari           .......... ..G....... ..A....... ........A. ......A... .....G.A.. ....G..... A.T.....A. .......... .......... .......... ....A..A.. ....TTA.TT ....A..... A..T...A.. ....TG..G.
```

| | | | | | | | | | | | | | | | | | |
|---|---|---|---|---|---|---|---|---|---|---|---|---|---|---|---|---|---|
| | | 1111111111 | 1111111111 | 1111111111 | 1111111112 | 2222222222 | 2222222222 | 2222222222 | 2222222222 | 2222222222 | 2222222222 | 2222222222 | 2222222222 | 2222222222 | 2222222223 | 3333333333 | 3333333333 |
| | | 6666666667 | 7777777778 | 8888888889 | 9999999990 | 0000000001 | 1111111112 | 2222222223 | 3333333334 | 4444444445 | 5555555556 | 6666666667 | 7777777778 | 8888888889 | 9999999990 | 0000000001 | 1111111112 |
| | | 1234567890 | 1234567890 | 1234567890 | 1234567890 | 1234567890 | 1234567890 | 1234567890 | 1234567890 | 1234567890 | 1234567890 | 1234567890 | 1234567890 | 1234567890 | 1234567890 | 1234567890 | 1234567890 |
| *Hexachaeta* | *amabilis* | ATTAGCAATC | ATTAATGATT | TTGTTATAAA | TTATTTTATA | A-TTTGT-AT | TATTTATTAA | TGTTTTAATT | TT--TTATAA | AAATG-TTTT | TATTAATAT- | TTATTGATAA | GTAATAATGA | TTAAATTAGT | ATA--TA-TT | T----TTTGT | TT-T-AATAA |
| *Hexachaeta* | *fallax* | .....T.... | ...T....A. | .......... | .......... | .....T.... | .........- | .A........ | .A.T...... | .......... | .G......A. | ...A.A.G.. | .......... | .AG....... | ......-... | .......... | ...A.G.... |
| *Toxotrypana* | *curvicauda* | .......... | ...T..A.G. | ..A....... | .....A..AG | .A...AA... | ........TT | AA........ | ....A...-. | ....TA.... | .......T.. | AA..AA.A.T | .......... | ..T....... | .......A-. | .....A.... | ...A.....T |
| *Toxotrypana* | *australis* | ....T..... | ...T..A.G. | ..A....... | .....A.... | .A...AAT.. | ......A.TT | .A........ | .......... | .....A.... | ..A-...TA. | .A..AA.A.T | .......... | ..T....... | .......AA. | .....A.... | ...A.....T |
| *Toxotrypana* | *littoralis* | .......... | ...T....G. | ..A....... | .....A..AT | TT..ATAT.. | ......A.TT | .A........ | ...T....-. | ....TA.... | .......TA. | .AT.AA.A.T | .C........ | ..T....... | .......GA. | -....A...- | ...A..T.TT |
| *Anastrepha* | *flavipennis* | .......... | ...T..A... | .......... | ......A..T | ..G..T.... | .........T | .A........ | AA.T...... | ....T..... | .......T.. | ...A.A.A.. | T......... | ..T....... | .......T.. | A....-.... | ..TA..G... |
| *Anastrepha* | *panamensis* | .......... | .....AA... | .......... | .......... | .....A.... | ......C.T. | .A........ | AA.T...... | ....A..... | .......T.. | ...A...G.. | .A.......G | .......... | ......TA.. | AAT..A.... | ...A.....T |
| *Anastrepha* | *nigrifascia* | .......... | ...T..A.A. | .......... | .......... | ..A.AT.... | ........T- | .......... | .A.T...... | G...T..... | .......T.. | .ACA...A.. | .......... | ..T....... | .......A.A | ACA.T..... | ......G..T |
| *Anastrepha* | *punctata* | .......... | ...T..A.A. | .......... | .....AATA. | G....T.... | ........T. | .A........ | AA.T...... | ....T..... | .......T.. | .A.A.A.A.. | T......... | .......... | .......T.. | .....-A... | .A.G.....T |
| *Anastrepha* | *leptozona* | ........C. | ...T.AA.G. | ..A....... | .......TAT | T.A..C...A | .........- | AA........ | A..T....G. | ....T..... | ...A...T.. | G.TA.A.A.. | ..G....... | .......... | .......AAA | .A...AA... | ...A.G.... |
| *Anastrepha* | *crebra* | .......... | ...TG.A.A. | .......... | .......... | ..AGGT.... | ......G.TT | .A........ | .A.T...... | ....T..... | .......T.. | ...A.A.A.. | A......... | ..T....... | .......TAA | AT...G.... | ...A..G..T |
| *Anastrepha* | *bezzii* | .......... | ......A.A. | ..A....... | .......... | ..AA.T.... | ........TT | .A........ | .A.T...... | ....T..... | .......T.. | ...A.A.A.. | .......... | ..T....... | .......A.. | -....G.... | ..TAT..... |
| *Anastrepha* | *limae* | .......... | ...T..A.G. | .......... | .......T.G | .....T.... | ......A.TT | G.....C... | GAAT...... | ....T..... | .......T.. | AA.A.A.A.. | .......... | ..G....... | ......TTAA | .AT..A.... | .A.A.....T |
| *Anastrepha* | *pickeli* | .......... | ...T..A... | .......... | ......A.AG | -....A.... | ......A.TT | ......C... | GAAT...... | ....T..... | .......T.. | GATA.A.A.. | .......... | .......... | ......TT.A | GT...G.... | .G.A.....T |
| *Anastrepha* | sp. nr. *anomala* | .......... | ...TG.A.G. | .......... | .......G.. | .......... | ......G.TT | .......... | .AAT...... | ....T..... | .......T.. | .ATA.A.A.. | .......... | .......... | ......TT.A | AT...A...- | .A.A.....T |
| *Anastrepha* | *serpentina* | .......... | ...TG.A.A. | .......... | .......G.G | .....A.... | ......G.TT | ......C... | GAAT...... | ....T..... | .......T.. | .ATA.A.A.. | .......... | .......... | ......TTAG | AT...A.... | .A.A.....T |
| *Anastrepha* | *spatulata* | .......... | ...TG.A.G. | .......... | .......... | .....A.... | ......A.TT | ......C... | GAAT...... | ....T..... | .......T.. | .AGA.A.A.. | .C....C... | .......... | ......TTAA | GAT..A.... | .A.A.....T |
| *Anastrepha* | *barbiellinii* | .......... | ...T..A.G. | .......... | .......... | .....T.... | ......A.TT | ......C... | GAAT...... | ....T..... | .T.....T.. | .AGA.A.A.. | .......... | .C........ | ......TT.A | AT...G.... | .A.A.....T |
| *Anastrepha* | *striata* | .......... | ...TG.A.G. | .......... | .......... | .....A.... | ......A.TT | ......C... | GAAT...... | ....T..... | .......T.. | .A.A.A.A.. | .......... | .......... | ......TTA. | AT...A.... | GA.A.G...T |
| *Anastrepha* | *bistrigata* | .......... | ...TG.A.GG | .......... | .......... | .....A.... | ......A.TT | ......C... | GAAT...... | ....T..... | .......T.. | AA.A.A.A.. | .......... | .......... | ......TTA. | -....A.... | GA.A.....T |
| *Anastrepha* | *ludens* | .......... | ...T..A.A. | .......... | .......... | .....A..G. | ......A.TT | ......C... | GAAT...... | ....T..... | .......T.. | AA.A.A.A.. | .......... | .......... | ...TA.TTAA | GT...A.... | .A.A.....T |
| *Anastrepha* | *acris* | .......... | ...T..A.G. | .......... | .......... | G....A..T. | ......A.TT | ......C... | GAAT...... | ....T..... | .......T.T | AA.A.A.G.. | .......... | .C.......? | ......TTAA | .T...A.... | .A.A.....T |
| *Anastrepha* | *coronilli* | .......... | ...T..A.G. | .......... | .......... | .....A.... | ......A.TT | ......C... | GAAT...... | ....T..... | .......T.T | AA.A.A.A.. | .......... | .C........ | ......TTAA | .T...A.... | .A.A.....T |
| *Anastrepha* | *obliqua* | .......... | ...T..A.G. | .......... | .......... | .....A.... | ......A.TT | .A....C... | GAAT...... | ....T..... | .......T.T | AA.A.A.A.. | .......... | .C........ | ......TTAA | .T...A.... | .A.A.....T |
| *Anastrepha* | *bahiensis* | .......... | ...T..A.G. | .......... | .......... | .....A.... | ......A.TT | ......C... | GAAT...... | ....T..... | .......T.T | AA.A.A.A.. | .......... | .C........ | ......TTAA | .TA.TA.... | .A.A.....T |
| *Anastrepha* | *distincta* | .......... | ...T..A.G. | .......... | .......... | G....A..G. | ......A.TT | ......C... | GAAT...... | ....T..... | .......T.T | AA.A.A.A.. | T......... | .......... | ......TTAA | .TA.TA.... | .A.A.....T |
| *Anastrepha* | *suspensa* | .......... | ...T..A.G. | .......... | .......... | .......... | ......A.TT | ......C... | GAAT...... | ....T..... | .......T.A | AA.A.A.A.. | .......... | .C........ | ......TTAA | .TG.TA.... | .A.A.....T |
| *Anastrepha* | *amita* | .......... | ...T..A.G. | .......... | .......... | .....A.... | ......G.TT | ......C... | GAAT...... | ....T..... | .......T.T | AA.A.A.A.. | .......... | .C........ | ......TTAA | .TA.TA.... | .A.A.....T |
| *Anastrepha* | *sororcula* | .......... | ...T..A.G. | .......... | .......... | .T...A.... | ......A.TT | .A....C... | GAAT...... | ....T..... | .......T.T | AA.A.A.A.. | .......... | .C........ | ......TTAA | .TA.TA.... | .A.A.....T |
| *Anastrepha* | *fraterculus* | .......... | ...T..A.G. | .......... | .......... | .T...A.... | ......A.TT | .A....C... | GAAT...... | ....T..... | .......T.T | AA.A.A.A.. | .......... | .C........ | ......TTAA | .TA.TA.... | .A.A.....T |
| *Anastrepha* | *fraterculus** | .......... | ...T..A.G. | .......... | .......... | .T..-A.... | ......A.TT | ......C... | GAAT...... | ....T..... | .......T.. | .A.A.A.A.. | .......... | .CG....... | ......TTAA | .TATTA.... | .A.A.....T |
| *Anastrepha* | *dryas* | .......... | ...TG.A.A. | .......... | .......... | .A...-.... | ......A.TT | AT...A.T.. | G-.T...... | ....T..... | ...G...... | ..TG.A.A.. | .......... | ..T....... | ....AATTAA | A...TA.... | ...A.G.... |
| *Anastrepha* | *grandis* | .......... | ...GG.A.G. | .........T | .......... | .T..A-.... | ......A.TT | ......C... | AAGT...... | ....T..... | ..AG...T.. | AA.....G.. | ..G....... | .......... | ....T.TAAA | .....A.... | ...A.T...T |
| *Anastrepha* | *pseudoparallela* | .......... | ...TG.A.A. | .......... | .......T.. | .G...T.... | ......A.TT | A.....C... | GAAT...... | ....T..... | .......T.. | .ACA.A.A.. | .......... | ..G....... | ....T.TAA. | A...TA.... | .A.A.....T |
| *Anastrepha* | *montei* | .......... | G..TG.A.G. | .......... | .......... | .....A.... | ....?T..T? | ?T...C?-.. | GAAT...... | ...AA..... | .TA....T.. | .ACA.A.A.. | .......... | .......... | ....T.TA.- | -...TA.... | .A.A.....T |
| *Anastrepha* | *manihoti* | .......... | ...T..A... | .......... | ......A.AG | T...ATA.T. | AT..A...TT | GT........ | GAAT...... | ....T..... | .......T.. | AACA.A.A.. | .......... | .......... | ....T.TTAA | -...TA.... | .G.......T |
| *Anastrepha* | *alveata* | .......... | ...T..A.G. | ..T....... | .........G | .....A.AT. | AT..AG..TT | GT...-C... | GAAT...... | ....T..... | .......T.. | AA.A.A.A.. | .......... | .......... | ....T.TAAA | A...TA.... | ...A.....T |
| *Anastrepha* | *cordata* | .......... | ...T.AA... | .......... | .......... | ..A..TA.T. | AT..A..AT- | GT...-.... | .-AT...... | ....T..... | ..AA...T.. | A.G..A.ATT | T......... | ..T....... | ....TAT.-- | -...TA.... | .G........ |
| *Anastrepha* | *bellicauda* | .......... | ...T..A.A. | .......... | .......... | .AA..TATTA | .T.A.TA.T. | -T...-.-.. | --AT...... | ....T..... | ..CA...T.. | ...A...A.. | A......... | ..T....... | ....TA.A-- | -...TA.... | ...A.....T |
| *Anastrepha* | *pallidipennis* | .......... | ...TG.A.G. | ..T....... | .......... | ..-..TAT.. | ......A.TT | G.....C... | GAAT...... | ....T..... | .......T.. | .A.A.A.A.. | .......... | .......... | ....T.TA.G | ATG.TG.... | ...A.....T |
| *Anastrepha* | *hamata* | .......... | ...G..A.A. | .......... | ......AT.T | .TGA.T.AT. | AT..AT.... | .T..A--... | A-AT...... | ....T..... | AT.....T.. | ..G..A.A.. | .......... | ..T....... | ....TA.AA- | -.T.TG.... | A..G.T...T |
| *Anastrepha* | *bicolor* | .......... | ......A.A. | .......... | .....AATAT | ..-A.T.AT. | AT..AT.ATT | .T..A--... | A-AT...... | ....T..... | .G.....T.. | ...A.A.A.. | .......... | .......... | ....TA.G.A | A...TA.... | .....G...T |
| *Anastrepha* | *aphelocentema* | .......... | ...T..A.A. | .......... | .......... | .T...TATTA | .T.A.T..T. | .T..A--... | .AAT...... | ....T..... | .......T.. | .....A.A.T | .......... | ..T....... | ....TA.GAA | C...TA.... | .G.AT..... |
| *Anastrepha* | *pallens* | .......... | ...TG.A.A. | .......... | .....AATAT | T.A..TATTA | .T.A.T.ATT | .T..A--... | -AAT....G. | ....T..... | .......T.. | .A.A.A.A.. | T......... | ..T....... | ....TA.G-- | -...TA.... | ?????-???? |
| *Anastrepha* | *katiyari* | .......... | ...TG.A.A. | .......... | .....AATAT | T.A..TATTA | .T.A.T.ATT | .T.-A--... | -AAT....G. | ....T..... | .......T.. | .A.A...A.. | T......... | ..T....... | ....TA.G-- | -...TA.... | ..TAA-TAT. |

| | | 3333333333<br>2222222223<br>1234567890 | 3333333333<br>3333333334<br>1234567890 | 3333333333<br>4444444445<br>1234567890 | 3333333333<br>5555555556<br>1234567890 | 3333333333<br>6666666667<br>1234567890 | 3333333333<br>7777777778<br>1234567890 | 3333333333<br>8888888889<br>1234567890 | 3333333334<br>9999999990<br>1234567890 | 4444444444<br>0000000001<br>1234567890 | 4444444444<br>1111111112<br>1234567890 | 4444444444<br>2222222223<br>1234567890 | 4444444444<br>3333333334<br>1234567890 | 4444444444<br>4444444445<br>1234567890 | 4444444444<br>5555555556<br>1234567890 | 4444444444<br>6666666667<br>1234567890 | 4444444444<br>7777777778<br>1234567890 |
|---|---|---|---|---|---|---|---|---|---|---|---|---|---|---|---|---|---|
| *Hexachaeta* | *amabilis* | TTTGATTTGA | TAAGTTTATA | TAAAGAACTC | GGCAAATTTT | G--------- | ---------- | ---TACTCG- | CCTGTTTAAC | AAAAACATGT | CTTTTTGAGT | TAATTTTAAA | GTCTGACCTG | CCCACTGAAA | TGT---TTTT | -AAATGGCCG | CAGTATTTTA |
| *Hexachaeta* | *fallax* | ...A...... | .......... | .......... | .......A.. | .......... | .......... | ....G..... | .......... | .......... | .......... | ..T....... | .......... | ........G- | .A-....... | .......... | .......C.G |
| *Toxotrypana* | *curvicauda* | ...TT...A. | A........T | ....T..... | .......... | AAA....... | .........T | TTA......G | .......... | .......... | .....A..A. | .TTA...... | ....A..... | .........- | .TA....... | .......... | .....A...G |
| *Toxotrypana* | *australis* | ...TG.A.A. | A......... | .......... | .........A | TAA......T | TAAATTTATG | TTA....... | .......... | .......... | ........A. | .TTA...... | ....A..... | .........- | ---....... | .......... | .....A...G |
| *Toxotrypana* | *littoralis* | ...T.AA.A. | A......... | ....T..... | .......... | AAA....... | .........T | TTA....... | .......... | .......... | ........A. | .TTA...... | ....A..... | .........- | .A........ | .G........ | .....A...G |
| *Anastrepha* | *flavipennis* | ...T...... | A......... | .......... | .........A | TATTGTATAT | A........T | TTA....... | .......... | .......... | ........AG | ..TA...... | .......... | ........T- | .AA....... | .......... | .........G |
| *Anastrepha* | *panamensis* | ...T.....G | A........G | .......... | .....T.... | T..TTTAAAA | AA........ | TTA....... | .......... | .......... | .........G | ..TA.A.... | ....AG.... | .........- | .AG...A... | .......... | .........G |
| *Anastrepha* | *nigrifascia* | A..TTG.... | A........G | .......... | .......--- | -......... | .........T | TTA....... | .......... | .......... | ........A. | .T.A...... | .......... | ........G- | ..A...A... | .......... | .........G |
| *Anastrepha* | *punctata* | ...T...... | A........G | .......... | .........A | T.......TT | ACAA...... | TTA....... | .......... | .......... | .......TA. | ..TA.AA... | ....A..... | .........- | .AA....... | .......... | .........G |
| *Anastrepha* | *leptozona* | A..AT....G | A......... | .......... | .........A | A.......TT | A.A......T | TTA....... | .......... | .......... | .......... | .GTA...... | .......... | .........- | ..AA..A... | .......... | .........G |
| *Anastrepha* | *crebra* | ...T...... | A......... | .......... | .........A | TAAA....TT | ATA..TA..T | TTA....... | .......... | .......... | ........AC | ..TA...... | ....A..... | .........- | .AA...A... | .......... | .........G |
| *Anastrepha* | *bezzii* | ...TTA...T | A......... | .......... | .........A | TAAT....TT | ATAATTATTT | TTA....... | .......... | .......... | ........AA | ..TA...... | ....A..... | ........G- | .AA....... | TG........ | .........G |
| *Anastrepha* | *limae* | ...ATA.... | A........G | .......... | .........- | A......... | .......... | TTA....... | .......... | .......... | .......... | ..TA...... | .......... | .........- | .AA....... | .......... | .........G |
| *Anastrepha* | *pickeli* | ...AT..... | G......... | .......... | .......... | A......... | .........T | TTA....... | .......... | .......... | ........A. | ..TA...... | .......... | .........- | .AA...G... | .......... | .........G |
| *Anastrepha* | sp. nr. *anomala* | ...ATA...- | G-.......T | ...-...... | .........- | A......... | .........T | TTA....... | .......... | .......... | .......... | .GTA...... | .......... | .........- | .AA....... | ........T. | .........G |
| *Anastrepha* | *serpentina* | ...ATA.... | G......... | .......... | .......... | A......... | .........T | TTA....... | .......... | .......... | ........A. | ..TA...... | .......... | .........- | .AA....... | .......... | .........G |
| *Anastrepha* | *spatulata* | ...ATG.... | .......... | .......... | .........- | A......... | .........T | TTA....... | .......... | .......... | ........A. | .GTA...... | .......... | .........- | .AATAA.... | .......... | .........G |
| *Anastrepha* | *barbiellinii* | ...ATAA... | G......... | .......... | .......... | A......... | .........T | TTA....... | .......... | .......... | ........A. | .GTA...... | .......... | .........- | .AA....... | .......... | .........G |
| *Anastrepha* | *striata* | ...ATAA... | A......... | .......... | .......... | A......... | .........T | TTA...C... | .......... | .......... | ........A. | ...A...... | .......... | .........- | .AA....... | .......... | .......C.G |
| *Anastrepha* | *bistrigata* | ...ATGA... | A........G | .......... | .......... | A......... | .........T | TTA....... | .......... | .......... | ........A. | ...A...... | .......... | .........- | .AA....... | .......... | .......C.G |
| *Anastrepha* | *ludens* | ...ATA.... | A......... | .......... | .......... | A......... | .........T | TTA....... | .......... | .......... | ........A. | ..TA...... | .......... | .........- | .AA....... | .......... | .........G |
| *Anastrepha* | *acris* | ...ATA.... | A......... | .......... | .........A | T......... | .........T | TTA....... | .......... | .......... | ........A. | ...A...... | .......... | .........- | .AA....... | .......... | .........G |
| *Anastrepha* | *coronilli* | ...ATA.... | A......... | .......... | .......... | A......... | .........T | TTA....... | .......... | .......... | ........A. | ...A...... | .......... | ........G- | .AA....... | .......... | .........G |
| *Anastrepha* | *obliqua* | ...ATA.... | A......... | .......... | .......... | A......... | .........T | TTA....... | ......-... | .......... | ........A. | ...A...... | .......... | .........- | .AA....... | .......... | .........G |
| *Anastrepha* | *bahiensis* | ...ATG.... | A......... | .......... | .........- | A......... | .........T | TTA....... | .......... | .......... | ........A. | ...A...... | .......... | .........- | .AA....... | .......... | .........G |
| *Anastrepha* | *distincta* | ...ATG.... | A......... | .......... | .......... | A......... | .........T | TTA....... | ......-... | .......... | ........A. | ...A...... | .......... | .........- | .AA....... | .......... | .........G |
| *Anastrepha* | *suspensa* | ...ATA.... | A......... | .......... | .......... | A......... | .........T | TTA....... | .......... | .......... | ........A. | ...A...... | .......... | .........- | .AA....... | .......... | .........G |
| *Anastrepha* | *amita* | ...ATA.... | G......... | .......... | .......... | A......... | .........T | TTA....... | .......... | .......... | ........A. | ...A...... | .......... | .........- | .AA....... | .......... | .........G |
| *Anastrepha* | *sororcula* | ...ATA.... | G......... | .......... | .......... | A......... | .........T | TTA....... | ......-... | .......... | ........A. | ...A...... | .......... | .........- | .AA....... | .......... | .........G |
| *Anastrepha* | *fraterculus* | ...ATA.... | G......... | .......... | .......... | A......... | .........T | TTA....... | .......... | .......... | ........A. | ...A...... | .......... | .........- | .AA....... | .......... | .........G |
| *Anastrepha* | *fraterculus** | ...ATA.... | A......... | .......... | .........A | TTT....... | .......... | ..A....... | .......... | .......... | ........A. | ..TA...... | .......... | .........- | .AA....... | .......... | .......C.G |
| *Anastrepha* | *dryas* | ...AT....T | A......... | .......... | .........A | TTTATTT... | .......... | ..A...C... | .......... | .......... | ........A. | .GTA...... | ....A..... | .........- | .AA...A... | .......... | .........G |
| *Anastrepha* | *grandis* | ...ATA...G | .......... | .......... | .......... | -......... | .......... | ..A....... | .......... | .......... | ........A. | .GTA.A.... | ....A..... | .........- | .AA....... | .......... | .........G |
| *Anastrepha* | *pseudoparallela* | ...ATA.... | G........G | .......... | ........AA | TT........ | .......... | ..A....... | .......... | ......TA.. | .......... | ..TA...... | .......... | .........- | .AA....... | .......... | .........G |
| *Anastrepha* | *montei* | ...ATA.... | G......... | .......... | .......... | ATTT...... | .......... | ..A....... | ...A...... | .......... | .......... | ..TA...... | .......... | .....A...- | .AA...A... | .G........ | .......C.G |
| *Anastrepha* | *manihoti* | ...AT..... | G......... | ....T..... | .......... | ATTT...... | .......... | ..A....... | .......... | .......... | ........A. | ..TA...... | .......... | .........- | .AA...G... | .......... | .........G |
| *Anastrepha* | *alveata* | ...ATA.... | G......... | .......... | .........- | ATTT...... | .......... | ..A....... | .......... | .......... | ........A. | ..TA...... | .......... | .........- | .AA....... | .......... | ...C.....G |
| *Anastrepha* | *cordata* | A..TTA...T | A......... | .......... | ........AA | AATTATTACT | .......... | ..A....... | .......... | .......... | .......G.A | .TTA...... | ....A..... | .........- | .T-...G... | .......... | .........G |
| *Anastrepha* | *bellicauda* | ...A.A.... | A......... | .......... | .........A | TAATTTATAT | T......... | ..A....... | .......... | .......... | ........A. | .GTA...... | ....A..... | .........- | .AA...A... | .......... | .........G |
| *Anastrepha* | *pallidipennis* | ...ATAC... | .........G | .......... | .......... | -......... | .......... | ..G....... | .......... | .......... | ........T. | ..TA...... | .......... | .........- | .AA....... | .......... | .........G |
| *Anastrepha* | *hamata* | A...T....G | A......... | .......... | .......AA. | TGGTATATTT | .......... | ..A....... | .......... | .......... | .......... | ...A...... | .......... | .........- | .AA...A... | T......... | .........G |
| *Anastrepha* | *bicolor* | ...AT..... | A......... | .......... | ........AA | TAAATAAGTT | ATTATTTAAA | TTA...C... | .......... | .......... | .......... | .......... | ....AG.... | .........- | .AA....... | T......... | .........G |
| *Anastrepha* | *aphelocentema* | ...AT....G | A........G | .......... | .......AAA | T......... | .......... | ..T....... | .......... | .......... | ........A. | ..TA...... | ....A..... | .........- | .TA....... | .......... | .........G |
| *Anastrepha* | *pallens* | ?????????? | ?????????? | ?????????? | ?????????A | TACATATATA | TAA....AAA | TTA....... | .......... | .......... | .......... | .G........ | ....A..... | .........- | .AA....... | T......... | .........G |
| *Anastrepha* | *katiyari* | .????????? | ?????????? | ?????????? | ?????????? | TATATGTTTA | TATATATAAA | TTA.-..... | .......... | .......... | .......... | .G........ | ....A..... | ...T.....- | .AA....... | TG........ | .........G |

| | | | | | | | | | | | | | | | | | |
|---|---|---|---|---|---|---|---|---|---|---|---|---|---|---|---|---|---|
| | | 4444444444 | 4444444445 | 5555555555 | 5555555555 | 5555555555 | 5555555555 | 5555555555 | 5555555555 | 5555555555 | 5555555555 | 5555555555 | 5555555556 | 6666666666 | 6666666666 | 6666666666 | 6666666666 |
| | | 8888888889 | 9999999990 | 0000000001 | 1111111112 | 2222222223 | 3333333334 | 4444444445 | 5555555556 | 6666666667 | 7777777778 | 8888888889 | 9999999990 | 0000000001 | 1111111112 | 2222222223 | 3333333334 |
| | | 1234567890 | 1234567890 | 1234567890 | 1234567890 | 1234567890 | 1234567890 | 1234567890 | 1234567890 | 1234567890 | 1234567890 | 1234567890 | 1234567890 | 1234567890 | 1234567890 | 1234567890 | 1234567890 |
| *Hexachaeta* | *amabilis* | ACTGTGCAAA | GGTAGCATAA | TCATTAGTCT | TTTAATTGAA | GGCTGGTATG | AACGGTTGGA | CGAAGTATAA | TC-TGTTTCA | TATAAAATTA | TAGTAAAATT | TTATATTTTA | GTAAAAAAGC | TAAAATAAAA | TTAAAAGACG | AGAAGACCCT | ATAAATCTTT |
| *Hexachaeta* | *fallax* | .......... | .......... | ?......... | .......... | .......... | ..T....... | .........T | ?......... | ......T... | ..A.TG.... | .......... | ..T....... | ......TT.T | .......... | .......... | .......... |
| *Toxotrypana* | *curvicauda* | .......... | .......... | .......... | .......... | ....T.A... | .......... | .......... | G......... | ......T... | ..T..G.... | .......... | .......... | ......TTT. | .......... | .......... | .......... |
| *Toxotrypana* | *australis* | .......... | .......... | ...G...... | .......... | ....T..... | .......... | .......... | .......... | ...T.TT... | .TT..G.... | .......... | ..T....... | ......TTT. | .......... | .......... | .......... |
| *Toxotrypana* | *littoralis* | .......... | .......... | .A........ | .......... | ....T.A... | ..T....... | .......... | .......... | ...?.?T... | .TT..G.... | .......... | .......... | ......TTT. | .......... | .......... | .......... |
| *Anastrepha* | *flavipennis* | .......... | .......... | ?......... | .......... | .......... | .......... | .A........ | A......... | ......T... | .TA..G.... | .......... | ..T....... | ......TTT. | .......... | .......... | .......... |
| *Anastrepha* | *panamensis* | .......... | .......... | ?......... | .......... | ......A... | ..T....... | .......... | .......... | ......T... | .TA..G.... | .......... | ..T....... | ......TTT. | .......... | .......... | .......... |
| *Anastrepha* | *nigrifascia* | .......... | .......... | ?....T.... | .......... | .......... | ..T....... | .......... | A......... | .......... | .TA..G.... | .......... | .......... | ......TTT. | .......... | .......... | .......... |
| *Anastrepha* | *punctata* | .......... | .......... | ?......... | .......... | .......... | .......... | .......... | .......... | .......... | .TA..G.... | .......... | ..T....... | ......TTT. | .......... | .......... | .......... |
| *Anastrepha* | *leptozona* | .......... | .......... | ........T. | .......... | .......... | ..T....... | .......... | G......... | .G....T..T | GTA..G.... | .......... | ..C....... | ......TTT. | .......... | .......... | .......... |
| *Anastrepha* | *crebra* | .......... | .......... | .......... | .......... | .......... | .......... | .......... | G......... | ......T... | .TA..G.... | .......... | ..T....... | ......TTT. | .......... | .......... | .......... |
| *Anastrepha* | *bezzii* | .......... | .......... | .......... | .......... | .......... | .......... | .......... | G......... | ......T... | .TA..G.... | .......... | ..T....... | ......TTT. | .......... | .......... | .......... |
| *Anastrepha* | *limae* | .......... | .......... | ?......... | .......... | .......... | .......... | .......... | A......... | ......T... | ....G..... | .......... | ..T....... | ......TTT. | .......... | .......... | .......... |
| *Anastrepha* | *pickeli* | .......... | .......... | .......... | .......... | .......... | ..T....... | .......... | G......... | ......T... | ..A..G.... | .......... | ..T....... | ......TTT. | .......... | .......... | .......... |
| *Anastrepha* | *sp. nr. anomala* | .......... | .......... | ?......... | .......... | .......... | .......... | .......... | ?......... | ......T... | ..A..G.... | .......... | ..T....... | ......TTT. | .......... | .......... | .......... |
| *Anastrepha* | *serpentina* | .......... | .......... | .......... | .......... | .......... | .......... | .......... | G......... | ......T... | ..A..G.... | .......... | ..T....... | ......TTT. | .......... | .......... | .......... |
| *Anastrepha* | *spatulata* | .......... | .......... | ?......... | .......... | .......... | .......... | .......... | ?......... | .G....T... | ..A..G.... | .......... | ..T....... | ......TTT. | .......... | .......... | .......... |
| *Anastrepha* | *barbiellinii* | .......... | .......... | .......... | .......... | .......... | .......... | .......... | A......... | ...T..T... | ..A..G.... | .......... | ..T....... | ......TTTT | .......... | .......... | .......... |
| *Anastrepha* | *striata* | .......... | .......... | .......... | .......... | .......... | .......... | ...G...... | G......... | ......T... | ..A..G.... | .......... | ..T....... | .......TT. | .......... | .......... | .......... |
| *Anastrepha* | *bistrigata* | .......... | .......... | .......... | .......... | .......... | .......... | ...G...... | G......... | ......T... | ..A..G.... | .A........ | ..T....... | .......TT. | .......... | .......... | .......... |
| *Anastrepha* | *ludens* | .......... | .......... | .......... | .......... | .......... | .......... | .......... | G......... | ......T... | ..A..G.... | .......... | ..T....... | ......TTT. | .......... | .......... | .......... |
| *Anastrepha* | *acris* | .......... | .......... | ?......... | .......... | .......... | .......... | .......... | ?......... | ......T... | ..A..G.... | .......... | ..T....... | ......TTT. | .......... | .......... | .......... |
| *Anastrepha* | *coronilli* | .......... | .......... | .......... | ........G. | .......... | .......... | .......... | A......... | ......T... | ..A..G.... | .......... | ..T....... | ......TTT. | .......... | .......... | .......... |
| *Anastrepha* | *obliqua* | .......... | .......... | .......... | ........G. | .......... | .......... | .......... | G......... | ......T... | ..A..G..C. | .......... | ..T....... | ......TTT. | .......... | .......... | .......... |
| *Anastrepha* | *bahiensis* | .......... | .......... | .......... | ........G. | .......... | .......... | .......... | A......... | ...T..T... | ..A..G.... | .......... | ..T....... | ......TTT. | .......... | .......... | .......... |
| *Anastrepha* | *distincta* | .......... | .......... | .......... | ........G. | .......... | .......... | .......... | G......... | ......T... | ..A..G.... | .......... | ..T....... | ......TTT. | .......... | .......... | .......... |
| *Anastrepha* | *suspensa* | .......... | .......... | .......... | ........G. | .......... | .......... | .......... | G......... | ......T... | ..A..G.... | .......... | ..T....... | ......TTT. | .......... | .......... | .......... |
| *Anastrepha* | *amita* | .......... | .......... | .......... | ........G. | .......... | .......... | .......... | A......... | ......T... | ..A..G.... | .......... | ..T....... | ......TTT. | .......... | .......... | .......... |
| *Anastrepha* | *sororcula* | .......... | .......... | .......... | ........G. | .......... | .......... | .......... | G......... | ......T... | ..A..G.... | .......... | ..T....... | ......TTT. | .......... | .......... | .......... |
| *Anastrepha* | *fraterculus* | .......... | .......... | .......... | ........G. | .......... | .......... | .......... | G......... | ......T... | ..A..G.... | .......... | ..T....... | ......TTT. | .......... | .......... | .......... |
| *Anastrepha* | *fraterculus** | .......... | .......... | .......... | .......... | .......... | .......... | .......... | G......... | ......T... | ..A..G.... | .......... | ..T....... | ......TTT. | .......... | .......... | .......... |
| *Anastrepha* | *dryas* | .......... | .......... | .......... | ........G. | .......... | ..T....... | ...G...... | G......... | ......T... | .TT..G.... | .......... | ..T....... | ......TTT. | .......... | .......... | .......... |
| *Anastrepha* | *grandis* | .......... | .......... | .......... | .......... | .......... | .......... | ...G...... | A......... | ......T... | ....G..... | .......... | ..T....... | ......TTT. | .......... | .......... | .......... |
| *Anastrepha* | *pseudoparallela* | .......... | .......... | .......... | .......... | .......... | .......... | .......... | G......... | ......T... | ..A..G.... | .......... | ..T....... | ......TTT. | .......... | .......... | .......... |
| *Anastrepha* | *montei* | .......... | .......... | .......... | .......... | .......... | ..T....... | .......... | G......... | ......T..G | ..A..G.... | .......... | ..T....... | ......TTT. | .......... | .......... | .......... |
| *Anastrepha* | *manihoti* | .......... | .......... | .......... | .......... | .......... | ..T....... | .......... | G......... | ......T... | .GA..G.... | .......... | ..T....... | ......TTT. | .......... | .......... | .......... |
| *Anastrepha* | *alveata* | .......... | .......... | .......... | .......... | .......... | .......... | .......... | G.G....... | ......T... | ..A..G.... | .......... | ..T....... | ......TTT. | .......... | .......... | .......... |
| *Anastrepha* | *cordata* | .......... | .......... | .......... | .......... | ......A... | ..T....... | .........G | G......... | ......T... | .TT..G.... | .......... | ..T....... | ......TTT. | .......... | .......... | .......... |
| *Anastrepha* | *bellicauda* | .......... | .......... | .......... | ........G. | .......... | ..T....... | ...G...... | G......... | ......T... | .TT..G.... | .......... | ..T....... | ......TTT. | .......... | .......... | .......... |
| *Anastrepha* | *pallidipennis* | .......... | .......... | .......... | ........G. | .......... | .......... | ...G...... | G......... | .T....T... | ....G..... | .......... | ..T....... | ......TT.G | .......... | .......... | .......... |
| *Anastrepha* | *hamata* | .......... | .......... | .......... | .......... | .......... | ..T....A.. | .......... | .......... | ...G..T... | ..A..G.... | ....T..... | .......... | ......TT.. | ..G....... | .......... | .......... |
| *Anastrepha* | *bicolor* | .......... | .......... | .......... | .......... | .......... | ..T....... | ...G.....T | A......... | ......T... | .TA..G.... | .......... | .......... | ......TTT. | .......... | .......... | .......... |
| *Anastrepha* | *aphelocentema* | .......... | .......... | .......... | .......... | .......... | ..T....... | ...G.....T | A......... | ......T... | .TA..G.... | .......... | ..T....... | ......TTT. | .......... | .......... | .......... |
| *Anastrepha* | *pallens* | .......... | .......... | .......... | .......... | .......... | ..T....... | .......... | G......... | ......T... | .TA..G.... | .......... | .......... | ......TTT. | .......... | .......... | .......... |
| *Anastrepha* | *katiyari* | .......... | .......... | .......... | .......... | .......... | ..T....... | .......... | G......... | ......T... | .TA..G.... | .......... | .......... | ......TTT. | .......... | .......... | .......... |

```
                              6666666666 6666666666 6666666666 6666666666 6666666666 6666666667 7777777777 7777777777 7777777777 7777777777 7777777777 7777777777 7777777777 7777777777 7777777777 7777777778
                              4444444445 5555555556 6666666667 7777777778 8888888889 9999999990 0000000001 1111111112 2222222223 3333333334 4444444445 5555555556 6666666667 7777777778 8888888889 9999999990
                              1234567890 1234567890 1234567890 1234567890 1234567890 1234567890 1234567890 1234567890 1234567890 1234567890 1234567890 1234567890 1234567890 1234567890 1234567890 1234567890
Hexachaeta   amabilis         -ATAATTAAT -AT-TAAATA AATTTTTTTG GTTTTTTTAT GTTTTTTTA- ATATTA-ATT ATTTTGTTGG GGTGATATTA AAATTTAATG AACTTTTAAT TTATTAAA-- TCATTAATTT ATGAATTATT GATCCATTGA TTAT-GATTA TAAGATTAAG
Hexachaeta   fallax           ....T...G. .....GTT.. ........G. A...G...TA -.....A.T. T.......A. .......... ......G... .........A .......... .....T..G. .......... ....T..G.. ........A. .A........ ..........
Toxotrypana  curvicauda       ....T..TT. ......TT.. .G.A....AA A...A....A -.....A... ...AA...A. .......... .......... .........T .......... ..T.G..... .......... .......... ........AG .AT....... ..........
Toxotrypana  australis        ....T..T.. .G....TT.. .G......GA T...A...TA -.....A... ..G.A.T.A. .......... ......G... .........T .......... ..T.....A. .......... .......... ........A. .AT....... ..........
Toxotrypana  littoralis       ....T..T.. ......TT.. ...A....AA A...A..A.A -.....A... ....G...A. .....A.... .......A.. .........T .......... ..T.A..... .......... .......... ........A. .AT....... ..........
Anastrepha   flavipennis      ....T..... .GA...TT.. ........G. A...A...TA -.....A... T.......A. .......... ......G... .......... .......... ..T.A..... .......... .......T.. ......A.A. .AT..T.... ..........
Anastrepha   panamensis       T...T..GG. .G....TT.. ........A. A..A...AGA -.....A... ........A. .......... .......... .........A .......... .AT.A...A. .......... ....T.G... ........AG .AT....... ..........
Anastrepha   nigrifascia      ....TC.T.. .T....TT.. .G......A. A...A...TA -.....A... T...G..GG. .......... ......G... .......... .......... AAT....GA. .......... .......G.. ........A. .AT....... ..........
Anastrepha   punctata         ....T..... ......TT.. ......A.A. A.AAA...TA -.....A... .......... .......... ......G... .........A .......... .AT.....A. .......... ....T..... ........A. .AT....... ..........
Anastrepha   leptozona        ........T. ......TTG. .T......G. A...A...TA -.....A... ...A...... .....A.... .......... .......... .......... ..T.G...A. .......... .......... ........A. .AT....... .....A....
Anastrepha   crebra           ....T..T.. .....GTT.. ........A. A...A...TA -.....A.T. ....A...A. .......... .......... .......... .......... .AT.....A. .......... ....G..... ........A. .AT....... ..........
Anastrepha   bezzii           ....T..T.. ......TT.. ........A. A...A....A -.....A... T...A..... .......... .......... .......... .......... .A.AA..... .......... .......... ........A. .AT....... ..........
Anastrepha   limae            .......... ...T...C.. .......... A..GAG..TA -......... .......... .......... ......G... .......... .......... ..T.....A. .......... .......... ........A. .A........ ..........
Anastrepha   pickeli          .......... .......C.. .......... A...G...TA -......... G......... .......... ......G... .......... .......... ..T.....A. .......... ......A... ........A. .A........ .....C....
Anastrepha   sp. nr. anomala  .......... TG.....C.. .......... A...A...TA -......... .....G.... .......... ......G... .......... .......... ..T.....A. .......... .......... ........A. .A........ ..........
Anastrepha   serpentina       .......G.. .G.....C.. .......... A...A...TA -......... .......... .......... ......G... .......... .......... ..T.....A. .......... ....G..... ........A. .A........ ..........
Anastrepha   spatulata        .......... .......T.. .......... A...A...TA -.......G. .......... .......... .......... .........A ........T. ..T....... .......... .......... ........AG .AT....... ..........
Anastrepha   barbiellinii     ....G..... .T.....C.. .......... A...A...TA -......... .......... .......... .......... .......... .......... ..T..T..AA .......... .......... ........A. .A........ ..........
Anastrepha   striata          .......... ..A....C.. .......... A...A...TA -......... T......... .......... ......G... .......... .......... ..T.....A. .......... .......... ........A. .A........ ..........
Anastrepha   bistrigata       .......... ..G....C.. .......... A...A...TA -......... T......... .......... ......G... .......... .......... ..T.....A. .......... .......... ........A. .A........ ..........
Anastrepha   ludens           ....T..... ....A..C.. .......... A...A...TA -.......TA .......... .......... .......... ......G... .......... ..T.A...A. .......... .......... ........A. .AG....... ..........
Anastrepha   acris            ....T..... .......C.. .......... ....A...TA -......... ...-...... .......... .......... .......... .......... ..T.....AA .......... .......... ........A. .AT....... ..........
Anastrepha   coronilli        ....T..... .......C.. .......... A...A...TA -......... .......... .......... .......... .......... .......... ..T.....A. .......... .......... ........A. .AT....... ..........
Anastrepha   obliqua          ....T..... .......C.. .......... A...A...TA -......... .......... .......... .......... .......... .......... ..T.....A. .......... .......... ........A. .AT....... ..........
Anastrepha   bahiensis        ....T..... ....A..C.. .......... A...A...TA -.......T. .......... .......... .......... .......... .......... ..T.....A. .......... .......... ........A. .AT....... ..........
Anastrepha   distincta        ....T..... ....A..C.. .......... A...A...TA -.......T. .......... .......... .......... .......... .......... ..T.....A. .......... .......... ........A. .AT....... ..........
Anastrepha   suspensa         ....T..... ....A..C.. .......... A...A...TA -.......T. .......... .......... .......... .......... .......... ..T.....A. .......... .......... ........A. .AT....... ..........
Anastrepha   amita            ....T..... ....A..C.. .......... A...A...TA -.......T. .......... .......... .......... .......... .......... ..T....... .......... .......... ........A. .AT....... ..........
Anastrepha   sororcula        ....T..... ....A..C.. .......... A...A...TA -.......T. .......... .......... .......... .......... .......... ..T....... .......... .......... ........A. .AT....... ..........
Anastrepha   fraterculus      ....T..... ....A..C.. .......... A...A...TA -.......T. .......... .......... .......... .......... .......... ..T....... .......... .......... ........A. .AT....... ..........
Anastrepha   fraterculus*     ....T..... .......C.. .......... A...A...TA -......... .......... .......... .......... .......... .......... ..T..T..AA .......... .......... ........AG .AT....... ..........
Anastrepha   dryas            ....T..... .T....TT.. .G........ A...G...TA -.....A... .......GA. .......... .......... .......... .......... .GT.....A. .......... .......... ........A. .AT....... ..........
Anastrepha   grandis          .......... .T.....C.. .......... A..GA...TA -.......G. .C........ .......... ......G... .........T .......... ..T...G.A. .......... ....T..G.. ......C.C. .AG....... ..........
Anastrepha   pseudoparallela  ........G. ...T...C.. .......... A..AAG..TA -......... .......... .......... ......G... .......... .......... ..T.....C. C......... .......... ........A. .A........ ..........
Anastrepha   montei           .......... .T.....C.. .......... A...A...TA -......... .......... .......... ......G... .......... .......... ..T.AG..AA .......... .......... ........A. .AG....... ..........
Anastrepha   manihoti         .......... .......C.. .......... A...G...TA -......... G.G....... .......... ......G... .......... .......... ..T.A...A. .......... ......AG.. ........A. .A........ .....C....
Anastrepha   alveata          .......T.. .......C.. .......... A...A...TA -......... ....A..... .......... ......G... .......... .......... ..T.....A. .......... ....G..... ........A. .AG....... ..........
Anastrepha   cordata          ....T..T.. ......TT.. .G......A. AA......TA -.....A... GA..A...A. .......... ......G... .........A .......... ..T.A..... .......... .......... ........A. .AT....... .T........
Anastrepha   bellicauda       ....T..... .T....TT.. .G........ A...G...TA -.....A... .......GA. .......... .......... .......... .......... .GT.....A. .......... .......... ........A. .AT....... ..........
Anastrepha   pallidipennis    ........G. .T...C.T.G .......... A...G...TA -.......T. TA.C...... .......... ......G... .......... .......... ..T.A..... .......... ....T..... ......G.A. .A........ ..........
Anastrepha   hamata           ....T..... ......TT.T ........G. A...A...TA -.....A... .......... .......... .......... .....A..C. .......... .GT.....GA .......... .......G.. ......A.AT ....T..... ..........
Anastrepha   bicolor          ....T..... .T....T... ........A. A.AAA...TA -.....A... ........A. .......... .......... .........T .......... .AT....GA. .......... ....T..G.. ......A.AG .AT....... ..........
Anastrepha   aphelocentema    ....T..GT. .T....TT.. ......A.A. A....G..TA -.....A... GA.A....A. .......... ......G... .......... .......... .AT....GAA .......... .......... ........A. .AT....... ..........
Anastrepha   pallens          ....T..... .T....TT.. .T........ A...A....A -....AA... .....G..G. .......... ......G... .........T .......... .A.AA...A. .......... ....T..... ......A.AG .AT....... ..........
Anastrepha   katiyari         ....T..... .T....TT.. .T........ A...A....A -....AA... ........G. .......... ......G... .........T .......... .A.AA..... .......... ....T.C... ......A.AG .AT....... ..........
```

| | | 8888888888 | 8888888888 | 8888888888 | 8888888888 | 8888888888 | 8888888888 | 8888888888 | 8888888888 | 8888888888 | 8888888889 | 9999999999 | 9999999999 | 9999999999 |
|---|---|---|---|---|---|---|---|---|---|---|---|---|---|---|
| | | 0000000001 | 1111111112 | 2222222223 | 3333333334 | 4444444445 | 5555555556 | 6666666667 | 7777777778 | 8888888889 | 9999999990 | 0000000001 | 1111111112 | 2222222223 |
| | | 1234567890 | 1234567890 | 1234567890 | 1234567890 | 1234567890 | 1234567890 | 1234567890 | 1234567890 | 1234567890 | 1234567890 | 1234567890 | 1234567890 | 1234567890 |
| *Hexachaeta* | *amabilis* | TTACTTTAGG | GATAACAGCG | TAATTTTTTT | GGAGAGTTCA | TATTGATAAA | AAAGTTTGCG | ACCTCGATGT | TGGATTAAGA | TAAAGA-TTA | GGTGTAGCTG | ?TT-AATTGG | TGAAGTCTGT | TCGACTTTTA |
| *Hexachaeta* | *fallax* | .......... | .......... | .......... | .......... | ...C...... | .......... | .......... | .......... | A.T.CC.... | ........C. | C.....G.AA | C-........ | .......... |
| *Toxotrypana* | *curvicauda* | .......... | .......... | .......... | A......... | ...C...... | .......... | .......... | .........G | ..T.A.T..G | .......... | -..C..A.AT | .-........ | .......... |
| *Toxotrypana* | *australis* | .......... | .......... | .......... | T......... | ...C...... | .......... | .......... | .........G | ..T.ATT..G | .......... | -..C..A.AT | .-........ | .......... |
| *Toxotrypana* | *littoralis* | .......... | .......... | .......... | T......... | ...C...... | .......... | .......... | .......... | ..T.TTT..G | .......... | -..C..A.AA | .-........ | .......... |
| *Anastrepha* | *flavipennis* | .......... | ........A. | .......... | T......... | ...C...... | .G........ | .......... | .........G | ..T.ATT..G | .......... | -..C..A.AT | .-........ | .......... |
| *Anastrepha* | *panamensis* | .......... | .......... | .......... | T........C | ...C..C... | .......... | .......... | .........G | ..T.ATT... | .......... | C.....ACAT | .-........ | .......... |
| *Anastrepha* | *nigrifascia* | .......... | .......... | .......... | T......... | ...C..C.G. | .......... | .......... | .........G | .GT.ATT..G | ........C. | -..C..G.AT | .-........ | .......... |
| *Anastrepha* | *punctata* | .......... | .......... | .......... | T......... | ...C..C... | .......... | .......... | .........G | ..T.ATT..G | .......T.. | -..C..A.AT | .-........ | .......... |
| *Anastrepha* | *leptozona* | .......... | .......... | ........C. | T......... | ...C..C.G. | .......... | .......... | .........G | ..T.ATT..G | .......... | -..C..A.AT | .-........ | .......... |
| *Anastrepha* | *crebra* | .......... | .......... | .......... | T......... | ...C..C... | .......... | .......... | .........G | ..T.ATT..G | .......... | -..C..A.AT | .-........ | .......... |
| *Anastrepha* | *bezzii* | .......... | .......... | .......... | T......... | ...C...... | .......... | .......... | .........G | ..T.ATT..G | ........C. | -..C..A.AT | .-........ | .......... |
| *Anastrepha* | *limae* | .......... | .......... | .......... | T......... | ...C..C... | .......... | .......... | .........G | ..T.ATT..G | .......... | -..C..AAAT | .-........ | .......... |
| *Anastrepha* | *pickeli* | .......... | .......... | .......... | T......... | ...C..C... | .......... | .......... | .........G | ..T..TT..G | .......... | -..C..AAAT | .-........ | .......... |
| *Anastrepha* | *sp. nr. anomala* | .......... | .......... | .......... | T......... | ...C..C... | .......... | .......... | .........G | ..T.ATT..G | ........C. | -..C..AAAT | .-........ | .......... |
| *Anastrepha* | *serpentina* | .......... | .......... | .......... | T......... | ...C..C... | .......... | .......... | .........G | ..T.ATT..G | ........C. | -..C..AAAT | .-........ | .......... |
| *Anastrepha* | *spatulata* | .......... | .......... | .......... | T......... | ...C..C... | .......... | .......... | .........G | ..T.ATT..G | .......... | -..C..AAAT | .-........ | .......... |
| *Anastrepha* | *barbiellinii* | .......... | .......... | .......... | T......... | ...C..C... | .......... | .......... | .........G | ..T.ATT..G | ........C. | -..C..AA.T | .-........ | .......... |
| *Anastrepha* | *striata* | .......... | .......... | .......... | T......... | ...C..C... | .......... | .......... | .........G | ..T.ATT..G | ........C. | -..C..AAAT | .-........ | .......... |
| *Anastrepha* | *bistrigata* | .......... | .......... | .......... | T......... | ...C..C... | .......... | .......... | .........G | ..T.ATT..G | ........C. | -..C..AAAT | .-........ | .......... |
| *Anastrepha* | *ludens* | .......... | .......... | .......... | T......... | ...C..C... | .......... | .......... | .........G | ..T.ATT..G | ........C. | -..C..AAAT | .-........ | .......... |
| *Anastrepha* | *acris* | .......... | .......... | .......... | TA........ | ...C..C... | .......... | .......... | .........G | ..T.ATT..G | .......TA. | -..C..AAAT | .-........ | .......... |
| *Anastrepha* | *coronilli* | .......... | .......... | .......... | T......... | ...C..C... | .......... | .......... | .........G | ..T.ATT..G | ........C. | -..C..AAAT | .-........ | .......... |
| *Anastrepha* | *obliqua* | .......... | .......... | .......... | T......... | ...C..C... | .......... | .......... | .........G | ..T.ATT..G | ........C. | -..C..AAAT | .-........ | .......... |
| *Anastrepha* | *bahiensis* | .......... | .......... | .......... | T......... | ...C..C... | .......... | .......... | .........G | ..T.ATT..G | ........C. | -..C..AAAT | .-........ | .......... |
| *Anastrepha* | *distincta* | .......... | .......... | .......... | T......... | ...C..C... | .......... | .......... | .........G | ..T.ATT..G | ........C. | -..C..AAAT | .-........ | .......... |
| *Anastrepha* | *suspensa* | .......... | .......... | .......... | T......... | ...C..C... | .......... | .......... | .........G | ..T.ATT..G | ........C. | -..C..AAAT | .-........ | .......... |
| *Anastrepha* | *amita* | .......... | .......... | .......... | T......... | ...C..C... | .......... | .......... | .........G | ..T.ATT..G | ........C. | -..C..AAAT | .-........ | .......... |
| *Anastrepha* | *sororcula* | .......... | .......... | .......... | T......... | ...C..C... | .......... | .......... | .........G | ..T.ATT..G | ........C. | -..C..AAAT | .-........ | .......... |
| *Anastrepha* | *fraterculus* | .......... | .......... | .......... | T...TA.... | ...C..C... | .......... | .......... | .........G | ..T.ATT..G | ........C. | -..C..AAAT | .-........ | .......... |
| *Anastrepha* | *fraterculus** | .......... | .......... | .......... | T......... | ...C..C... | .......... | .......... | .........G | ..T.ATT..G | ........C. | -..C..AAAT | .-........ | .........? |
| *Anastrepha* | *dryas* | .......... | .......... | .......... | T......... | ...C..C... | .......... | .......... | .......... | ..T.A.T..G | .......... | -..C?????? | ?????????? | ?????????? |
| *Anastrepha* | *grandis* | .......... | .......... | .......... | T......... | ...C..C..T | .......... | .......... | .........G | ..T.ATT..G | .......T.. | -..C..AAAT | .-........ | .........G |
| *Anastrepha* | *pseudoparallela* | .......... | .......... | .......... | T......... | ...C..C... | .......... | .......... | .........G | ..T.ATT..G | .......T.. | -..C..AAAT | .-........ | .......... |
| *Anastrepha* | *montei* | ...G...... | .......... | .......... | T......... | ...C..C... | .......... | .......... | .........G | ..T.ATT..G | .......... | -..C.????? | ?-???????? | ?????????? |
| *Anastrepha* | *manihoti* | .......... | .......... | .......... | T......... | ...C..C... | .......... | .......... | .........G | ..T..TT..G | .......... | -..C..AAAT | .-...ATA.. | .......... |
| *Anastrepha* | *alveata* | .......... | .......... | .......... | T......... | ...C..C... | .......... | .......... | .........G | ..T.ATT..G | ........C. | -..C..AAAT | .-........ | .......... |
| *Anastrepha* | *cordata* | .......... | .......... | .......... | T......... | ...C..C... | .......... | .......... | .......... | ..T.A.T..G | .......??? | -????????? | ?-???????? | ?????????? |
| *Anastrepha* | *bellicauda* | .......... | .......... | .......... | T......... | ...C..C... | .......... | .......... | .......... | ..T.A.T..G | .......... | -..C..A.AT | .-........ | .......... |
| *Anastrepha* | *pallidipennis* | .......... | .......... | .......... | T......... | ...C..C... | .......... | .......... | .........G | ..T.ATT..G | .......... | -..C..AAAT | .-........ | .......... |
| *Anastrepha* | *hamata* | .......... | .......... | .......... | T......... | ...C..C.T. | .......... | .......... | .........G | ..T.ATT..G | .......... | -..C..A.AT | .-........ | .........G |
| *Anastrepha* | *bicolor* | .......... | .......... | .......... | TA........ | ...C...... | .......... | .......... | .........G | ..T.ATT..G | .......... | -..C..AA-T | .-........ | .........G |
| *Anastrepha* | *aphelocentema* | .......... | .......... | .......... | T......... | ...C..C... | .......... | .......... | .........G | ..T..TT..G | .......... | -..C.TAAAC | .-........ | .......... |
| *Anastrepha* | *pallens* | .......... | .......... | .......... | TA........ | ...C..C... | .......... | .......... | .........G | ..T.ATT..G | .......... | -..C..A..T | .-........ | .......... |
| *Anastrepha* | *katiyari* | .......... | .......... | .......... | TA........ | ...C..C... | .......... | .......... | .........G | ..T.ATT..G | .......... | -..C..A..T | .-........ | .......... |

# 14 Behavior of Flies in the Genus *Toxotrypana* (Trypetinae: Toxotrypanini)

*Peter J. Landolt*

## CONTENTS



## 14.1 INTRODUCTION

The genus *Toxotrypana* is a small taxon of seven described and several undescribed species found exclusively in the New World (Blanchard 1959; Norrbom et al., Chapter 12). These species occur in tropical and subtropical areas of South, Central, and North America. *Toxotrypana* is closely aligned with, and may be synonymous with, the much larger Neotropical genus *Anastrepha* (see Chapters 12 and 13). Most information on *Toxotrypana* biology and behavior is on a single species, *T. curvicauda* Gerstaecker, also known as the papaya fruit fly. It has been studied principally because it is a pest of papaya fruit and is widely distributed from south Florida and south Texas of the United States through much of Central America and northern South America, including a number of Caribbean islands (Knab and Yothers 1914; Lawrence 1976).

Some aspects of the behavior and biology of *T. curvicauda* are included in two other reviews. Landolt and Heath (1996) summarized pheromone-based trapping systems for Florida fruit flies, including *T. curvicauda*. Landolt (1990) reviewed the chemical ecology of the papaya fruit fly, including the sex pheromone and evidence that host chemical cues are important to host finding. Readers are also referred to Chapter 29 by Heath et al. for information on the sex pheromone chemistry of *T. curvicauda*. This chapter compiles and summarizes information on all aspects of the behavior of flies in the genus *Toxotrypana*, with a particular emphasis on the papaya fruit fly and on sexual behavior, as is reflected in the published scientific literature.

## 14.2 LARVAL BEHAVIOR

Young papaya fruit fly larvae in papaya fruit feed exclusively on seeds (Knab and Yothers 1914; Landolt 1984a). Peña et al. (1986a) observed larval feeding on immature papaya seeds in the laboratory and found that they fed upon and preferred the testa and endosperm of the seeds and did not feed well on the sarcotesta (Fisher 1980). Nearly mature larvae may be found tunneling within and feeding on the thick pulp of papaya fruit (as evidenced by the abundant frass within the tunnels), but only when the fruit is maturing or senescing (Knab and Yothers 1914; Landolt 1984a).

The maturation of larvae and their exit from fruit is often coincident with ripening or senescence of the fruit. This change in fruit physiology is evident both in the consistency of the fruit pulp and in the color of fruit. Papaya fruit with mature larvae may be readily spotted on the tree because of a change in external coloration of the fruit from dark green to lighter green and yellow (Landolt 1984a). The pulp of the fruit at this time becomes softer and loses the ability to exude latex, which may (Knab and Yothers 1914) or may not (Mason 1922) be toxic to larvae. It is thought that these changes in the physiology of the fruit permit or facilitate mature larvae to burrow to the fruit surface and exit fruit (Knab and Yothers 1914). The premature senescence of papaya fruit infested with papaya fruit fly larvae may be caused by the presence of the larvae, but this has not been demonstrated.

Mature larvae exit papaya fruit and drop to the ground. After a hole is cut to the outside, multiple larvae may exit from the same hole (Mason 1922). Larval exiting may be prompted by vibration. When infested fruit are picked and are vibrated (such as when transported in a vehicle), many larvae will exit the fruit in a short period of time. It seems likely that larvae respond to the shock of a fruit hitting the ground by exiting the fruit. Larvae on the ground may burrow directly into the soil or may wander over its surface, possibly in relation to the soil texture (Knab and Yothers 1914; Lawrence 1976). On hard or stony ground, larvae are then exposed for a greater period of time and may be more prone to predation by ants (Mason 1922; Lawrence 1976).

## 14.3 ADULT BEHAVIOR

### 14.3.1 Feeding Behavior

Adult papaya fruit flies feed on sugar-rich materials. Mason (1922) found that papaya fruit flies could be readily killed with poisoned syrups on which they fed, and would feed on a variety of sweet materials and fruits placed with the flies in cages. Sharp and Landolt (1984) demonstrated stimulation of papaya fruit fly feeding by dark brown sugar or white refined sucrose in the laboratory. In the laboratory, adult papaya fruit flies that feed on sugar water solutions will imbibe sugar water until the abdomen is visibly distended and then may regurgitate droplets of water on the walls of the cage, often in lines. Similar behavior by apple maggot flies, *Rhagoletis pomonella* (Walsh) appears to be a means of concentrating the sugar (Hendrichs et al. 1992; 1993). In laboratory studies by Sharp and Landolt (1984), flies did not respond to formulations of protein hydrolysate materials (casein, yeast, soy, and torula yeast) that are commonly used as food or baits for frugivorous fruit flies (Steiner 1955), either in an arena assay or in an olfactometer. Papaya fruit fly females do not require protein feeding to develop oocytes (Landolt 1984b), as is necessary in many other tephritids (Bateman 1972). There is a dearth of information on papaya fruit fly feeding behavior in the field, although flies have been seen to feed on honeydew from whiteflies on papaya leaves and may feed on old latex from papaya wounds (Landolt and Hendrichs 1983).

Results of tests and surveys to trap fruit flies with food baits indicate poor attraction to such materials. Mason (1922) stated that papaya fruit flies do not appear to be attracted by any food. Generally, proteinaceous food baits commonly used for trapping tropical fruit flies are ineffective in capturing papaya fruit flies, in line with the lack of feeding on such materials in the laboratory.

However, papaya fruit flies were captured in traps baited with liver in Mexico (Enkerlin et al. 1989). Papaya fruit flies were captured in protein hydrolysate-baited traps when traps were placed in the higher branches of trees in Costa Rica (L. Jirón, personal communication) and in Florida. This observation indicates that a general lack of response of papaya fruit flies to protein hydrolysates in traps may be due in part to the location of traps in relation to where flies are present, in addition to a lack of feeding on proteinaceous materials.

### 14.3.2 Host Finding and Host Selection

Papaya fruit flies specialize on a small number of species of plants in the Caricaceae and Asclepiadaceae (Knab and Yothers 1914; Castrejón-Ayala 1987; Castrejón-Ayala and Camino-Lavín 1991; Landolt 1994a), and may be considered stenophagous, if not oligophagous. Other species of *Toxotrypana* also use fruits of plants in the Caricaceae and Asclepiadaceae as larval hosts (Blanchard 1959). There is a single record of *T. curvicauda* from mango fruit (Butcher 1952) which is here discounted as anomolous. The host plants of *Toxotrypana* are characterized by the production of latex upon injury and also by large fruit with a central seed cavity. These two characteristics of host fruit require particular behaviors by ovipositing females of *T. curvicauda*, and likely by other species of the genus as well.

Both sexes of the papaya fruit fly are active on papaya fruit in the field. Female papaya fruit flies become reproductively mature and sexually active after 5 to 6 days postemergence, and must locate suitable host trees for oviposition. Both males and females are active in papaya groves at specific times of the day and depart at the end of these activity periods (Landolt and Hendrichs 1983; Landolt 1984a; Castrejón-Ayala 1987; Landolt et al. 1991; Aluja et al. 1997a). Therefore, they must reorient to locate host fruit on a daily basis as long as they are reproductively active.

Host finding by the papaya fruit fly likely includes attraction to both host kairomones and the male-produced sex pheromone, as well as orientation to visual cues. Papaya fruit flies are attracted to odorants from papaya fruit and may arrive at hosts partly as a result of chemoanemotactic responses to host volatiles (Landolt and Reed 1990). In a flight tunnel, females are weakly attracted to immature papaya fruit, odor of fruit, and solvent washes of fruit. Female papaya fruit flies may also use male pheromone as a means of locating host fruit for oviposition. Females remain highly responsive to male pheromone following mating (Landolt and Heath 1988), and may orient to pheromone to locate hosts in addition to locating males that call from hosts. Odor of fresh green papaya fruit or solvent extract of papaya peel enhances female attraction to male pheromone (or male pheromone enhances female attraction to host odor) (Landolt et al. 1992).

Papaya fruit flies also use visual cues in their selection of hosts or oviposition sites. Flies oviposit in green immature fruit (Peña et al. 1986b) and are most sensitive to visible wavelengths in the blue (475 nm) and green (500 nm) ranges. Males, unmated females, and mated females all orient preferentially to green (Landolt et al. 1988). Females orient preferentially to green spheres over green panels, indicating significance of shape in addition to color. This information was used to contrive a trap for papaya fruit flies that is essentially a fruit mimic (Landolt et al. 1988). Peña et al. (1986b) demonstrated a color preference of ovipositing papaya fruit flies for green over yellow, indicating that responses to specific wavelengths of light play a role in female selection of fruit for oviposition.

Papaya fruit flies may also respond to host visual cues in their preference for fruit shape, fruit size, and to positions of fruit on a papaya tree. Flies orient to spheres more than rectangular panels (Landolt et al. 1988). Also, green sticky cylinders make effective papaya fruit fly traps (Heath et al. 1996), indicating that the flies may respond to a curved edge rather than, or in addition to, the overall shape of a fruit. Fruit flies may orient preferentially to fruit of a particular size, as shown for fruit flies of the *Rhagoletis pomonella* species group (Prokopy 1973). Reports of papaya fruit fly orientation to spheres and fruit of different sizes are not entirely consistent, however. In a flight tunnel, more unmated female flies landed on the largest green spheres tested (12.7 cm diameter) compared with smaller spheres (Landolt et al. 1988). Mason (1922) indicated that flies in the field oviposited into medium and larger fruit that were still immature, while Knab and Yothers (1914) reported oviposition in the field in large to mature fruit and not on very young fruit. However, Landolt (1985) found papaya fruit fly eggs in fruit ranging in size from 1.9 to 10. 9 cm diameter, with no eggs in fruit from 10.9 to 15 cm diameter. There has not yet been a quantitative assessment of rates of fly oviposition on fruits of different sizes and maturities on papaya trees. Also, the selection of fruit for oviposition is likely a result of female responses to both visual and chemical cues, even at a distance, confounding the relationship between fruit size and attractiveness.

### 14.3.3 Oviposition

Behaviors directly associated with oviposition of papaya fruit flies into papaya fruit were described by Landolt and Reed (1990), based on observations made in a laboratory setting. Females arriving on fruit exhibit a characteristic set of behaviors. First, females bob the head and thorax up and down while moving forward, contacting the fruit with the mouthparts at each bob. This may be chemosensory behavior, providing gustatory sampling of the fruit surface. Females then position the oviscape at a right angle to the fruit surface, with the abdomen oriented nearly vertically (Color Figure 12*). When the papaya fruit skin is penetrated with the tip of the aculeus, the distal part of the ovipositor, latex is exuded from the fruit skin. However, females often repeatedly assume this position without penetrating the fruit skin (no latex). It is not known if these repeated positionings are samplings for appropriate sensory cues received at the aculeus tip that may be necessary to initiate further drilling and penetration of the fruit, or if penetration of the fruit skin is difficult and an unsuccessful female may move to another spot on the fruit to attempt again. Females that penetrate the skin, causing the flow of latex, maintain this position for some time, presumably while moving the aculeus through the pulp of the fruit and depositing eggs in the fruit seed cavity. In the laboratory, mean oviposition times were 20.5 + 8.1 min and mean eggs deposited per bout was 29.0 + 18.1 (Landolt and Reed 1990), considerably higher than the 10.1 + 8.7 eggs found per field-collected fruit by Landolt (1985). Eggs are laid in strings or clusters (Mason 1922; Landolt 1985). Ovipositing females typically acquire a mound of solidifying latex around the oviscape but generally do not have difficulty removing the oviscape from this latex following the removal of the aculeus from the fruit. On rare occasions, however, dead females may be seen encased in latex on fruit (Mason 1922; Aluja et al. 1997a) or live females are seen with pieces of latex attached to the oviscape. Although Lawrence (1976) stated that female papaya fruit flies do not lay eggs in fruit that have already been infested, larvae of different sizes and presumably different ages may be found within the same fruit (Mason 1922; Landolt 1985). Multiple oviposition wounds, indicated by mounds of solidified latex, are often seen on papaya fruit in the field, but multiple batches of eggs directly associated with these wounds have not been documented. Following removal of the aculeus from the fruit, females then contact the surface of the fruit with the mouthparts again, but in a continuous fashion, rather than in the rhythmic bobbing pattern that occurs prior to oviposition. In a laboratory experiment, this behavior occurred for up to 6.5 min following oviposition. Because host-marking pheromones are produced and deposited by many species of frugivorous fruit flies (Averill and Prokopy 1989), this latter behavior may be a means of depositing such a pheromone to the papaya fruit surface. In other tephritids, host-marking pheromones are deposited from the tip of the aculeus during a behavior referred to as aculeus or ovipositor dragging (Porter 1928; Prokopy 1972). No such dragging of the aculeus has been observed in the papaya fruit fly and the hypothesis that *T. curvicauda* deposits a host-marking or oviposition-deterring pheromone has not been tested. The postoviposition behavior involving continuous contact between the mouthparts and fruit may be interpreted as feeding on fruit surfaces when observed in the field (see Landolt and Hendrichs 1983; Aluja et al. 1997a) and requires additional careful study.

In the field, oviposition occurs over the entire day (Mason 1922; Landolt and Hendrichs 1983) but may be focused during a narrow range of time of day. In south Florida, papaya fruit flies

* Color Figures follow p. 204.

oviposit in papaya primarily in late afternoon and evening (Knab and Yothers 1914; Mason 1922; Landolt and Hendrichs 1983). During cloudy weather, this activity may be more pronounced (Knab and Yothers 1914; Mason 1922) or may occur earlier in the day. In Buenas Aires, Puntarenas, Costa Rica, oviposition in a commercial papaya planting occurred primarily in late morning and early afternoon (Landolt et al. 1991). In Retalhuleu, Guatemala, females were active on fruit throughout much of the day except evening (Landolt et al. 1991). In Morelos and Veracruz, Mexico, the times of the day that females were present on papaya fruit shifted through the season, from afternoon in January, throughout the day in July, and during midday in August (Castrejón-Ayala 1987; Aluja et al. 1997a). The reasons for these differences in diel patterns, which are similar for other behavioral activities, are unclear, but may include temperature and humidity (Aluja et al. 1997a), sunlight, predation pressures, and genetic differences in papaya fruit flies in different geographic areas.

### 14.3.4 Sexual Behavior

As is typical with tephritid fruit flies (Burk 1981), *T. curvicauda* males and females exhibit a rich repertoire of sexual behaviors, including visual, chemical, and audio signaling during courtship and mating, as well as in associated station-taking, sex attraction, and intrasexual competition.

Female and male papaya fruit flies are attracted to a pheromone released by males. In a flight tunnel, it has been demonstrated that females are attracted to males placed in a small cage, flying upwind and landing on the cage (Landolt et al. 1985). Females were also attracted to airflow from over males that was piped into a flight tunnel. Papaya fruit fly attraction to male pheromone appears to involve a combination of chemoanemotaxis and self-steered counterturning as occurs in moth responses to sex pheromones (Baker 1989). These mechanisms were evident as upwind flights and both casting and hovering flight patterns that appeared to serve to keep flies within the odor plume and resulted in their arrival at the odor source (cage of males or pipe vent). Female chemotactic responses to odorants from males were differentiated from responses to visual displays and sounds produced by males by the piping of airflow from over males into the flight tunnel. Analyses of volatile chemicals present in headspace over males revealed the presence of 2,6-methylvinylpyrazine (Chuman et al. 1987).

The pheromonal role of this compound was confirmed and detailed in a series of laboratory experiments using a flight tunnel and in field tests evaluating 2,6-methylvinylpyrazine as a trap bait (Chuman et al. 1987; Landolt and Heath 1988; 1990; Landolt et al. 1991). In the laboratory the greatest percentage response of unmated mature females tested was to 1.65 μg of 2,6-methylvinylpyrazine per hour (Landolt and Heath 1988), while in the field the greatest numbers of papaya fruit flies were captured on traps baited with pheromone released at from 2 to about 12 male-equivalents (0.16 to nearly 1 μg per hour) (Landolt and Heath 1988; Landolt et al. 1988; 1991). In a field test, unmated females responded strongly to the pheromone and not to unbaited traps, while mated females were captured in good numbers on all traps, including unbaited traps, presumably in response to the visual aspects of the fruit mimic trap design. Females in the laboratory were not attracted to synthetic sex pheromone until they were reproductively mature (Landolt 1984b; Landolt and Heath 1988), with some flies responding when 5 days old and highest response rates occurring with 7- and 8-day-old flies. Females that were mated were still attracted to male pheromone in the laboratory (Landolt and Heath 1988), as they were in the field, although at a somewhat higher pheromone release rate than required to elicit a response from unmated females (Landolt and Heath 1990). The diel pattern to papaya fruit fly attraction to male pheromone was broad in the laboratory and more defined in the field. In the laboratory, mated females responded in late afternoon and evening, while unmated females responded from midmorning until late afternoon (Landolt and Heath 1988). In the field, males, unmated females and mated females were captured primarily during the last 4 h of light in south Florida (Landolt and Heath 1990). This difference in the diel pattern of activity of flies in the laboratory vs. in the field may be due to the low ambient light levels in laboratory situations, compared with the field.

Females and males also interact in courtship preceding mating. In an arena, sexually receptive males and females were variable in their interactions (Landolt et al. 1985). Females exhibited increased movement in the presence of a male (walking, running, turning) and in most cases (83%) approached the male directly. Most females also bobbed the abdomen and ovipositor in the presence of the male and tilted the abdomen and ovipositor up sharply while approaching the male. These behaviors appear to be visual signaling. Females also rotated (spun) directly in front the males and tapped the male on the vertex of the head with their fore tarsi. Males extended the abdominal pleural pouches, presumably as a means of releasing pheromone. Few males (17%) inititated encounters with females, although 28 of 64 males that paired had followed females after females initiated the encounters. Males also exhibited a side-to-side motion (rocking) while positioned directly in front of and facing females that may be some form of visual signaling. Males also assumed a crouching position (body close to the substrate) with wings close to the body, while facing females that may be visual signaling. Males lunged at females while vibrating the wings, sometimes followed by the male mounting the female. Most mountings began with the male placing his fore tarsi on the head of the female and then aligning his body somewhat parallel to the female before climbing onto her thorax. The male then moved posteriorly while on the torso of the female until contact was made between his genitalia and the tip of her oviscape. To bring the tip of the oviscape into position, the male grasped the distal part of the oviscape with his hind legs and pulled the oviscape forward while moving posteriorly on the female. The male phallus was inserted into the distal tip of the aculeus and was extended to the base of the ovipositor where sperm was deposited in the vicinity of the vagina, where the spermathecal ducts join the common oviduct. The location of the apical glans of the phallus at this time was confirmed by dissecting females in copula for various lengths of time, after freezing mating pairs on dry ice. The courtship sounds (acoustic signals) produced by male *T. curvicauda* were analyzed by Sivinski and Webb (1985). Two distinct sounds are produced, one is a song of one to five pulse trains produced just before mounting the female, and the other is a song of up to seven pulse trains produced while on the female and while the male attempts to insert his phallus into the ovipositor.

It is speculated that female papaya fruit flies may respond to male pheromone, or to male pheromone and host odors in combination, as a means of locating suitable oviposition sites (Landolt et al. 1992). Mature females remain responsive to male sex pheromone both shortly after a mating and up to several days following mating (Landolt and Heath 1988), indicating that their response is not entirely sexual. Similarly, mated females are captured on pheromone-baited traps in papaya orchards, although a proportion of these are likely responding to the visual aspects of the trap design (Landolt and Heath 1990). Female responses to immature papaya fruit odors and extracts are strongly enhanced by the presence of male pheromone, and such flies respond to male pheromone at low release rates in the presence of papaya odor (Landolt et al. 1992). Although papaya fruit fly mated females exhibited attraction responses to papaya odor in a flight tunnel, this response was weak when not in combination with male pheromone (Landolt and Reed 1990; Landolt et al. 1992). These findings suggest the possibility that mated females may respond to sex pheromone as a means of locating papaya fruit as oviposition sites.

Female papaya fruit flies generally mated once (six of ten), and some mated twice (four of ten) in a study of their behavior in a field cage enclosing a fruiting papaya tree (Landolt and Hendrichs 1983). This observation contributed to the speculation that the attraction of mated females to sex pheromone of males may be a host-finding strategy and not a mate-finding strategy (Landolt and Heath 1988; Landolt et al. 1992). In a subsequent study of mating frequency in the laboratory, females provided fruit for oviposition generally mated more than once (Landolt 1994b). Under those circumstances, remating appeared to be associated with access to papaya fruit for oviposition and does not appear to negate the assumption that newly mated females that have not oviposited would not be searching for another mate.

### 14.3.5 Male Station-Taking, Calling, and Territoriality

Male *T. curvicauda* on papaya trees exhibit station-taking, territoriality, and pheromonal calling on papaya fruit. Males occupy fruit for extended periods of time and defend the fruit from other males. Males may arrive on fruit already occupied by a male (or near fruit occupied by a male). This results in clearly agonistic interactions (Landolt and Hendrichs 1983). Males face each other, often with the head and thorax elevated from the fruit and the abdomen raised nearly vertically. While in this stance, they may alternately raise and lower the middle legs. More aggressive behavior includes striking (pawing) at the opponent with the forelegs, short forward rushes, striking or pushing the opponent directly with the forelegs, or striking the opponent with a wing while vibrating the wing. Males in these latter interactions often grapple, and may then fall from the fruit or perch. While on fruit, males also extrude abdominal pouches, presumably calling females with sex pheromone. In fruit flies of the genus *Anastrepha* and other Tephritidae, similar pouches are associated with columnar epidermal cells that appear to be exocrine in function (Nation 1974; 1981). These abdominal pouches in *T. curvicauda* are thought to be associated with exocrine glands and pheromone release, although this has not been supported experimentally.

Papaya fruit fly males, in Florida, typically arrive on papaya trees in late afternoon, or earlier in cloudy weather, and set up territories on papaya fruit (Landolt and Hendrichs 1983). In Mexico, male papaya fruit flies arrive at papaya trees from midmorning to late afternoon, depending on environmental conditions, and call from papaya fruit or leaf stems (Aluja et al. 1997a). Males, in both Florida and Mexico, generally arrive on average before females (Landolt and Hendrichs 1983; Castrejón-Ayala 1987). It is assumed that males on papaya are releasing sex pheromone to attract females to fruit, in order to obtain courting and mating opportunities. As discussed above, females may orient to male pheromone and host odors to arrive at host fruit, both for mating and oviposition.

### 14.3.6 Mimicry

The papaya fruit fly resembles a wasp (Knab and Yothers 1914; Mason 1922; Lawrence 1976) and is thought to be a mimic of certain species of Vespidae, such as *Polistes* spp. (Knab and Yothers 1914). It is commonly referred to as the papaya wasp in south Florida, where it is similar in coloration to brown paper wasps in the genus *Polistes*, such as *P. dorsalis* (Fabricius) and *P. exclamans* (Vierick), and *Mischocyttarus mexicanus* (Saussure), which are common in Florida and on many Caribbean islands. In Central America, the coloration of the papaya fruit fly is a brighter yellow, which matches the coloration of several social wasps prevalent there. The wings of the papaya fruit fly are darkened along the costal margin, resembling the appearance of the folded forewings of Vespidae (see Figure 12.5A). The long ovipositor of the female contributes to the misconception that it is a wasp and it is sometimes referred to as a stinger. Papaya fruit fly females also may mimic social wasps behaviorally. Females that are held with the fingers will curve the abdomen and orient the ovipositor around as if attempting to sting. It is not known if these resemblances to social wasps convey any advantage to the papaya fruit fly. Known predators of these flies include salticid jumping spiders (Mason 1922), lizards (*Anolis* species in Florida), and wasps (Aluja et al. 1997a). Females are particularly vulnerable to predation while ovipositing. This is evidenced by the ovipositors observed remaining in fruit after females are attacked and eaten, both in Florida (observations by the author and C. O. Calkins, personal communication) and in Mexico (Aluja et al. 1997a). The lizard *A. sagrei* occurs commonly in papaya trees, taking up residence among fruit clusters, and may prey heavily on papaya fruit flies. One female of this species held in captivity in Miami, Florida and fed papaya fruit flies subsequently died, however, possibly due to an inability to digest or pass the ovipositors of female flies. When the lizard was dissected after death, a mass of papaya fruit ovipositors was discovered in the intestine. If such lizards feed on females that are ovipositing, the ovipositor might often remain in the fruit and would not be ingested.

## 14.4 RELATIONSHIP TO BEHAVIOR OF OTHER SPECIES OF FRUIT FLIES

Most other species of frugivorous Tephritidae that have been studied in detail utilize fruit that are maturing, with larval feeding throughout the fruit pulp (but see Chapters 12 and 15). The papaya fruit fly is quite different in utilizing immature papaya fruit. This host utilization pattern may provide explanations for why young papaya fruit fly larvae feed on seeds and not on fruit pulp (to bypass defensive chemistry that is strong in immature fruit), why papaya fruit fly females have long ovipositors (to bypass the toxic pulp to oviposit in the seed cavity), why papaya fruit flies are wasp mimics (long ovipositors provide a predisposition and long oviposition times on exposed fruit exposes flies to predators), and why females do not need to feed on protein-rich materials to develop mature oocytes (young papaya fruit fly larvae feed on seeds which are more nutritious that fruit pulp).

Another interesting aspect of comparative behavior of frugivorous fruit flies is the relationships between host species (including host breadth) and mating behavior. Prokopy (1980) noted that males of species of *Rhagoletis* utilize the host plant as the mating encounter site, while males of *Anastrepha* species, *Dacus* species, and the Mediterranean fruit fly, *Ceratitis capitata* (Wiedemann), utilize nonhost mating encounter sites or leks. This difference in mating strategies was considered to be due to the predictability of host fruit availability in the univoltine temperate species considered and the relative unpredictability of host fruit availability in multivoltine tropical species discussed. The papaya fruit fly appears to exhibit mating behavior similar to that of *Rhagoletis* flies in that they utilize the host plant as a mating encounter site and are not known to form leks. However, unlike *Rhagoletis* flies, papaya fruit flies are tropical in distribution and multivoltine. The possible dichotomy in frugivorous fruit fly mating strategies (mating restricted to host plants vs. facultative use of leks) may relate more closely to host breadth rather than latitude or voltinism. Fruit flies with restricted host breadth are more closely adapted to their host plants (specialized) and may be more able to track fruiting cycles and to locate hosts in time and space. Polyphagous fruit flies shift hosts seasonally and geographically and do not adapt their emergence patterns and sexual behaviors to coincide with the fruiting cycles of one species of host. It seems that these consequences of polyphagy contribute to host fruit being less predictable for the several tropical pest species that have been intensively studied and that it is the specialization of monophagous and oligophagous flies on limited hosts that contributes to the relative "predictability" of those host fruit to those species of flies.

Predator avoidance may also impact sexual and host-related behaviors, and may be another significant selective force in the evolution of mate-finding strategies and their relationship to the host plant. Fruit flies may minimize time spent on fruit if predation by visually oriented predators is intense. This avoidance of predation may be another factor favoring nonresource-based aggregations of males (leks) and discouraging males from setting up territories on fruit and waiting for arriving females. The papaya fruit fly spends a great deal of time exposed on relatively large fruit, but may reduce the threat from predators through its possible mimicry of social wasps.

## 14.5 CONCLUSIONS AND FUTURE RESEARCH NEEDS

There are aspects of the behavior of *T. curvicauda* that are poorly understood. Research to date has focused heavily on fly behavior in the field when on papaya fruit and fly behavior in the laboratory that relates to sex pheromonal signaling and responses to pheromone. Little is known of their behavior in the field when they are not on papaya fruit. For example, papaya fruit flies depart from papaya trees at dusk in Florida or in midafternoon in Costa Rica, and it is assumed that they go to and remain in tall vegetation near papaya trees (Landolt 1984a; Aluja et al. 1997b). Also, it appears that immature female flies also are not on host plants. However, there have been no field studies of their behavior off of hosts. Studies are needed to determine more precisely where papaya fruit flies go when not on hosts and what their behavior is off of hosts. For example, does mating occur on nonhost plants? Do males release pheromone and attract mates? If so, what mate-finding and mating strategies are employed? Do papaya fruit flies feed when not on host plants?

The roles of host plant odors in host finding, host selection, and sexual behavior are poorly understood and host kairomones important to fly behavior are not characterized. It seems possible that fly chemotactic responses to papaya may be affected by such variables as damage to fruit (including latex release), fruit maturity, papaya varieties as well as host plant species, host-marking pheromone, and fruit diseases. Although it is evident that flies respond to papaya fruit odor, we have not identified any host chemicals that are attractive to papaya fruit flies or that enhance attraction responses to pheromone. It is also not known if host odors attractive to papaya fruit flies are exclusive to the fruit, or are produced by the remainder of the papaya tree. Additional chemical, as well as tactile, cues are likely involved in host selection and acceptance responses when females and males contact papaya fruit and during fruit puncturing by females.

There is no information for papaya fruit fly behavior in relation to host plants other than on papaya. It would be interesting to know if males perch on fruit of *Morrenia odorata* (Lindl.) and release pheromone to attract females, as they do on papaya, or if females sponge the surface of the fruit of *Gonolobus sorodius* A. Gray, as they do on papaya. If papaya fruit flies use host odors to locate or select papaya trees, do they also respond to similar or different chemicals from asclepiads? Since so many aspects of the biology of *T. curvicauda* relate to the behavior of the fly on the host plant, it would be quite valuable to have comparable information for *T. curvicauda* on other species of host plants.

The nearly complete lack of information on the behavior of species of *Toxotrypana* other than *T. curvicauda* is a serious handicap in comparing the behavior of the papaya fruit fly to other species of Tephritidae and to evaluating *Toxotrypana* behavior in comparison to what is known of *Anastrepha* species. Of particular interest is sexual behavior, particularly sexual signaling. Because of the close association between mating and host plants in the papaya fruit fly and the utilization of similar host plants throughout the other members of the genus (Asclepiadaceae and Caricaceae) (Blanchard 1959), comparative studies of sexual signaling might divulge how sympatric species of *Toxotrypana* are reproductively isolated.

Another area of interest is the hypothetical mimicry of social wasps by papaya fruit flies. At this time, it is attractive speculation but needs critical experimental testing. Well-designed and rigorous experimentation is needed to determine if papaya fruit flies do indeed benefit from looking like, acting like, or even smelling like a social wasp. Such experimentation might involve assays with likely predators, such as birds, large spiders, or lizards that avoid putative social wasp models (such as *Polistes* species) to determine if such predators that are experienced with models avoid papaya fruit flies.

To date, behavioral studies of papaya fruit flies, and fruit flies generally, are driven by economic (largely agricultural) needs. This certainly will continue to be the principal force determining what kinds of studies are conducted and what areas of research are promoted. To this end, there is a need to determine more accurately the roles of host odorants in female papaya fruit fly attraction to host fruit and to males, to characterize these odorants, and to determine their usefulness as attractants and as attractant synergists when combined with the male sex pheromone 2,6-methly-vinylpyrazine. It is possible that such a combination will provide a more powerful chemical attractant, to be used in pest control programs without the use of pesticides (Landolt et al. 1992; Heath et al. 1996; Aluja et al. 1997b).

## ACKNOWLEDGMENTS

Much of the research summarized in this chapter and ideas expressed are the results of extensive discussion and collaboration with other scientists. These include R. R. Heath, J. Hendrichs, J. Sivinski, and C. Calkins. Helpful comments to improve the manuscript were made by C. O. Calkins, J. Sivinski, and R. Zack. The writing of this chapter was made possible by the generous financial support of the International Organization for Biological Control of Noxious Animals and Plants (IOBC), the Campaña Nacional contra las Moscas de la Fruta (SAGAR-IICA) and USDA-ICD-RSED.

## REFERENCES


Aluja, M., A. Jiménez, J. Piñero, M. Camino, L. Aldana, M.E. Valdés, V. Castrejón, I. Jácome, A.B. Dávila, and R. Figueroa. 1997a. Daily activity patterns and within-field distribution of papaya fruit flies (Diptera: Tephritidae) in Morelos and Veracruz, Mexico. *Ann. Entomol. Soc. Am.* 90: 505–520.

Aluja, M., A. Jiménez, M. Camino, J. Piñero, L. Aldana, V. Castrejón, and M.E. Valdés. 1997b. Habitat manipulation to reduce papaya fruit fly (Diptera: Tephritidae) damage: orchard design, use of trap crops, and border trapping. *J. Econ. Entomol.* 90: 1567–1576.

Averill, A.L. and R.J. Prokopy. 1989. Host marking pheromones. In *Fruit Flies: Their Biology, Natural Enemies and Control* (A.S. Robinson and G. Hooper, eds.), pp. 207–219. In *World Crop Pests* (W. Helle, ed.), Vol. 3A. Elsevier Science Publishers, Amsterdam. 372 pp.

Baker, T.C. 1989. Sex pheromone communication in the Lepidoptera: new research progress. *Experientia* 45: 248–262.

Bateman, M.A. 1972. The ecology of fruit flies. *Annu. Rev. Entomol.* 17: 493–518.

Blanchard, E.F. 1959. El género *Toxotrypana* en la República Argentina (Diptera: Tephritidae). *Acta Zool. Lilloana* 17: 33–44.

Burk, T. 1981. Signaling and sex in acalyptrate flies. *Fla. Entomol.* 64: 30–43.

Butcher, F.G. 1952. The occurrence of papaya fruit fly in mango. *Proc. Fla. State Hortic. Soc.* 65: 196.

Castrejón-Ayala, F. 1987. Aspectos de la biología y hábitos de *Toxotrypana curvicauda* Gerst. (Diptera: Tephritidae) en condiciones de laboratorio y su distribución en una plantación de *Carica papaya* L. en Yuatepec, Morelos. B.Sc. dissertation. Instituto Politécnico Nacional, Mexico, D.F. 88 pp.

Castrejón-Ayala, F. and M. Camino-Lavín. 1991. New host plant record for *Toxotrypana curvicauda* (Diptera: Tephritidae). *Fla. Entomol.* 74: 466.

Chuman, T., P.J. Landolt, R.R. Heath, and J.H. Tumlinson. 1987. Isolation, identification, and synthesis of male-produced sex pheromone of papaya fruit fly, *Toxotrypana curvicauda* Gerstaecker (Diptera: Tephritidae). *J. Chem. Ecol.* 13: 1979–1992.

Enkerlin, D., L. Garcia R., and F. López M. 1989. Mexico, Central and South America. In *Fruit Flies: Their Biology, Natural Enemies and Control* (A.S. Robinson and G. Hooper, eds.), pp. 83–90. In *World Crop Pests* (W. Helle, ed.), Vol. 3A. Elsevier Science Publishers, Amsterdam. 372 pp.

Fisher, J.B. 1980. The vegetative and reproductive structure of papaya (*Carica papaya*). *Lyonia* 1:191–208.

Heath, R.R., N.D. Epsky, A. Jiménez, B.D. Dueben, P.J. Landolt, W.L. Meyer, M. Aluja, J. Rizzo, M. Camino, F. Jeronimo, and R.M. Baranowski. 1996. Improved pheromone-based trapping systems to monitor *Toxotrypana curvicauda* (Diptera: Tephritidae). *Fla. Entomol.* 79: 37–48.

Hendrichs, J., S.S. Cooley, and R.J. Prokopy. 1992. Postfeeding bubbling behavior in fluid feeding Diptera: concentration of crop contents by oral evaporation of excess water. *Physiol. Entomol.* 17: 153–161.

Hendrichs, J., B.S. Fletcher, and R.J. Prokopy. 1993. Feeding behavior of *Rhagoletis pomonella* flies (Diptera: Tephritidae): effect of initial food quantity and quality on food foraging, handling costs, and bubbling. *J. Insect Behav.* 6: 43–64.

Knab, F. and W.W. Yothers. 1914. Papaya fruit fly. *J. Agric. Res.* 2: 447–453.

Landolt, P.J. 1984a. Behavior of the papaya fruit fly *Toxotrypana curvicauda* Gerstaecker, in relation to its host plant, *Carica papaya* L. *Folia Entomol. Mex.* 61: 215–224.

Landolt, P.J. 1984b. Reproductive maturation and premating period of the papaya fruit fly, *Toxotrypana curvicauda* (Diptera: Tephritidae). *Fla. Entomol.* 67: 240–244.

Landolt, P.J. 1985. Papaya fruit fly eggs and larvae (Diptera: Tephritidae) in field-collected papaya fruit. *Fla. Entomol.* 68: 354–356.

Landolt, P.J. 1990. Chemical ecology of the papaya fruit fly. In *Fruit Flies: Biology and Management* (M. Aluja and P. Liedo, eds.), pp. 207–210. Springer-Verlag, New York. 492 pp.

Landolt, P.J. 1994a. Fruit of *Morrenia odorata* (Asclepiadaceae) as a host for the papaya fruit fly, *Toxotrypana curvicauda* (Diptera: Tephritidae). *Fla. Entomol.* 77: 287–288.

Landolt, P.J. 1994b. Mating frequency of the papaya fruit fly (Diptera: Tephritidae) with and without host fruit. *Fla. Entomol.* 77: 305–312.

Landolt, P.J. and R.R. Heath. 1988. Effects of age, mating, and time of day on behavioral responses of female papaya fruit flies, *Toxotrypana curvicauda* (Diptera: Tephritidae) to synthetic sex pheromone. *Environ. Entomol.* 17: 47–51.

Landolt, P.J. and R.R. Heath. 1990. Effects of pheromone release rate and time of day on catches of male and female papaya fruit flies (Diptera: Tephritidae) on fruit model traps baited with pheromone. *J. Econ. Entomol.* 83: 2040–2043.

Landolt, P.J. and R.R. Heath. 1996. Development of pheromone-based trapping systems for monitoring and controlling tephritid fruit flies in Florida. In *Pest Management in the Subtropics — A Florida Perspective* (D. Rosen and J. Capinera, eds.), pp. 197–297. Intercept. Ltd., Andover. 578 pp.

Landolt, P.J. and J. Hendrichs. 1983. Reproductive behavior of the papaya fruit fly, *Toxotrypana curvicauda* Gerstaecker (Diptera: Tephritidae). *Ann. Entomol. Soc. Am.* 76: 413–417.

Landolt, P.J. and H.C. Reed. 1990. Behavior of the papaya fruit fly (Diptera: Tephritidae): host finding and oviposition. *Environ. Entomol.* 19: 1305–1310.

Landolt, P.J., R.R. Heath, and J.R. King. 1985. Behavioral responses of female papaya fruit flies *Toxotrypana curvicauda* Gerstaecker (Diptera: Tephritidae) to male-produced sex pheromone. *Ann. Entomol. Soc. Am.* 78: 751–755.

Landolt, P.J., R.R. Heath, H.R. Agee, J.H. Tumlinson, and C.O. Calkins. 1988. Sex pheromone based trapping system for papaya fruit fly (Diptera: Tephritidae). *J. Econ. Entomol.* 81: 1163–1169.

Landolt, P.J., M. Gonzales, D.L. Chambers, and R.R. Heath. 1991. Comparison of field observations and trapping of papaya fruit fly in papaya plantings in Central America and Florida. *Fla. Entomol.* 74: 408–414.

Landolt, P.J., H.C. Reed, and R.R. Heath. 1992. Attraction of female papaya fruit fly (Diptera: Tephritidae) to male pheromone and host fruit. *Environ. Entomol.* 21: 1154–1159.

Lawrence, G.A. 1976. The papaya fruit fly. *J. Agric. Soc. Trinidad Tobago.* 76: 359–360.

Mason, A.C. 1922. Biology of the papaya fruit fly, *Toxotrypana curvicauda*, in Florida. *U.S. Dep. Agric. Bull.* No. 1081. 1–10.

Nation, J.L. 1974. The structure and development of two sex specific glands in male Caribbean fruit flies. *Ann. Entomol. Soc. Am.* 67: 732–734.

Nation, J.L. 1981. Sex specific glands in tephritid fruit flies of the genera *Anastrepha*, *Ceratitis*, *Dacus*, and *Rhagoletis* (Diptera: Tephritidae). *Int. J. Insect Morphol. and Embryol.* 10: 121–129.

Peña, J.E., D.F. Howard, and R.E. Litz. 1986a. Feeding behavior of *Toxotrypana curvicauda* (Diptera: Tephritidae) on young papaya seeds. *Fla. Entomol.* 69: 427–428.

Peña, J.E., R.M. Baranowski, and R.E. Litz. 1986b. Oviposition of the papaya fruit fly *Toxotrypana curvicauda* Gerstaecker as affected by fruit maturity. *Fla. Entomol.* 69: 344–348.

Porter, B.A. 1928. The apple maggot, *Rhagoletis pomonella* Walsh (Trypetidae). *U.S. Dep. Agric. Bull.* 66: 1–48.

Prokopy, R.J. 1972. Evidence for a marking pheromone deterring repeated oviposition in apple maggot flies. *Environ. Entomol.* 3: 326–332.

Prokopy, R.J. 1973. Ovipositional responses to different sizes of artificial fruit by flies of *Rhagoletis pomonella* species group. *Ann. Entomol. Soc. Am.* 66: 927–929.

Prokopy, R.J. 1980. Mating behavior of frugivorous Tephritidae in nature. *Proceedings of a Symposium on Fruit Fly Problems* (J. Koyama, ed.), pp. 37–46. XVI. Int. Congr. Entomol., Kyoto, Japan.

Sharp, J.L. and P.J. Landolt. 1984. Gustatory and olfactory behavior of the papaya fruit fly, *Toxotrypana curvicauda* Gerstaecker (Diptera: Tephritidae) in the laboratory with notes on longevity. *J. Ga. Entomol. Soc.* 19: 176–182.

Sivinski, J. and J.C. Webb. 1985. The form and function of acoustic courtship signals of the papaya fruit fly, *Toxotrypana curvicauda*. *Fla. Entomol.* 68: 634–641.

Steiner, L.F. 1955. Bait sprays for fruit fly control. *Agric. Chem.* 10: 32–34.

# 15 Behavior of Flies in the Genus *Anastrepha* (Trypetinae: Toxotrypanini)

*Martín Aluja, Jaime Piñero, Isabel Jácome, Francisco Díaz-Fleischer, and John Sivinski*

**CONTENTS**

## 15.1 INTRODUCTION

Flies in the genus *Anastrepha* Schiner offer a unique opportunity to study behavior using a comparative approach. The species that have been studied to date have revealed a remarkably plastic, variable, and complex behavioral repertoire. Male calling rhythms and behaviors performed during sexual encounters are good examples of this. Some species, like *A. robusta* Greene, perform elaborate in-flight loops during calling (wing fanning) bouts while other species, like *A. aphelocentema* Stone, *A. cordata* Aldrich, or the closely related species *Toxotrypana curvicauda* Gerstaecker, stand still and do not wing fan during the entire daily calling period (calling in these three species is discernible because males puff pleural glands and release a sexual pheromone). With respect to calling rhythms, some species call at dawn (*A. cordata*) while others do so at sunset (e.g., *A. spatulata* Stone). Another *Anastrepha* behavior where great variability is observed is oviposition. For example, species like *A. obliqua* (Macquart) lay strictly one egg/clutch, while others, like *A. grandis* (Macquart), are able to lay more than 100 eggs/clutch. Furthermore, there are a few species like *A. hamata* (Loew) that never mark a fruit, while many others deposit a host-marking pheromone after an oviposition bout, for example, *A. spatulata, A. leptozona* Hendel, *A. alveata* Stone, *A. ludens* (Loew), *A. grandis, A. fraterculus* (Wiedemann). This great variability in the behavioral repertoire, added to the fact that we now have a fairly well-supported phylogeny for the 197 known species (Norrbom et al., Chapter 12; and McPheron et al., Chapter 13), opens up the opportunity to build a behavioral phylogeny of unprecedented scope.

In this chapter, we will first provide a historical review of studies on *Anastrepha* behavior and summarize relevant information on their biology and natural history. Then, we will describe the few known facts about larval behavior, review known adult diel rhythms of activity, and address in detail each of the most important behaviors exhibited by *Anastrepha* adults: trivial movements, feeding, oviposition, mating, and shelter seeking and resting. In doing so, we will review all previously published work along with unpublished information recently generated on rare species such as *A. acris* Stone, *A. alveata*, *A. aphelocentema*, *A. bahiensis* Lima, *A. bezzii* Lima, *A. cordata*, *A. distincta* Greene, *A. leptozona*, *A. hamata,* and *A. spatulata.* Whenever possible, we will compare behavioral data on *Anastrepha* with relevant information on *T. curvicauda.* This species belongs to the genus that may be the sister group of *Anastrepha* or may even fall within *Anastrepha* (formal analysis of phylogenetic relationships of *Anastrepha* and *Toxotrypana* can be found in Norrbom et al., Chapter 12, and McPheron et al., Chapter 13). Our intent is also to lay down relevant facts for the formal discussion of the evolution of fruit fly behavior in the last part of this book (Drew and Yuval, Chapter 27; Díaz-Fleischer et al., Chapter 30; and Sivinski et al., Chapter 28).

## 15.2 HISTORY OF STUDIES ON *ANASTREPHA* BEHAVIOR

When reviewing the history of studies on *Anastrepha* behavior, several highlights emerge: (1) there is a series of highly insightful, but for the most part anecdotal, observations made by Mexican and American naturalists at the beginning of the 20th century in Mexico and Puerto Rico (L. de la Barrera and A. Rangel cited by Herrera 1905; Crawford 1918; 1927; Picado 1920; McPhail and Bliss 1933; McAlister et al. 1941; also see review by Baker et al. 1944). In our opinion, all of these authors deserve credit as pioneers in the study of the biology and behavior of *Anastrepha.* Between this period (1900 to 1944) and the burst of activity in the late 1970s to early 1980s, work on *Anastrepha* behavior was virtually halted. (2) Most in-depth studies on *Anastrepha* behavior are restricted to seven economically important species: *fraterculus*, *grandis*, *ludens*, *obliqua*, *serpentina* (Wiedemann), *striata* Schiner and *suspensa* (Loew) (e.g., Nation 1972; Perdomo 1974; Dodson 1982; Dickens et al. 1982; Aluja et al. 1983; Burk 1983; Malavasi et al. 1983; Morgante et al. 1983; Robacker and Hart 1985; Hendrichs 1986; Sivinski and Webb 1986; Sivinski 1988; 1989; Aluja et al. 1989; Robacker et al. 1991; Silva 1991; Silva and Malavasi 1993; Aluja and Birke 1993; Aluja et al. 1993; Sivinski et al. 1994). (3) Most studies on *Anastrepha* behavior stem mainly from three countries: Brazil, Mexico, and the United States (also one study from Costa Rica, Hedström 1991; in addition to references under item 2 above, Silva et al. 1985; Polloni and Silva 1987; Selivon 1991; Morgante et al. 1993; Aluja 1993a).

The first anecdotal observations made on aspects of behavior of an *Anastrepha* species are arguably those on *A. ludens*, described in 1873 by the German taxonomist H. Loew as *Trypeta ludens* and in early works often called the "orange worm." Alfonso Herrera (1900) compiled some observations by a series of U.S. naturalists working in Mexico between 1881 and 1897, indicating that *A. ludens* females deposited their eggs in the skin of oranges and that larvae exited the fruit and buried themselves in the ground to pupate. At a later date, Herrera (1905) cited a series of observations by Leopoldo de la Barrera and Amado Rangel (Mexican Ministry of Agriculture agents) on the "intelligence" of *A. ludens* adults and larvae. In the original text by Herrera (1905), Rangel and de la Barrera describe how larvae extracted from the pulp, "crawl day and night experiencing sensations of fear and malaise. If touched they contract and cease moving." Once development is completed "they frantically search for a dark site to pupate." While still feeding in the fruit "some larvae peep out through a hole drilled by them to breathe." On occasion, if the ripe fruit does not drop from the tree, some larvae "that cannot wait any longer" jump out of it from considerable heights. With respect to the adult fly, it is mentioned that "their senses are more perfect and their intellectual manifestations more complex." These senses "are mainly determined by sensations of fear, hunger and desire, especially a maternal desire that forces females to lay their eggs in a protected spot underneath the skin of an orange." These authors noted that females did not lay eggs in many parts of the orange, but instead they preferred one single site, perhaps "to economize time and labor." Rangel and de la Barrera were intrigued by the fact that females did not lay more than six to ten eggs per fruit and only in one slice of an orange. They attributed this to the "intelligence" of the female which "understood that if they deposited all the germs carried in their ovaries in one single fruit, there would not be enough food for all the progeny." In other words, "the *Trypeta* mother has sufficient insight to avoid putting on board all her progeny in one boat but instead distributes it in 8, 10 or more fruits." Rangel and de la Barrera also noted that guavas, a fruit smaller than oranges, hosted fewer larvae and that flies would only lay one egg in cherries because of the small size of this fruit.

When describing the general behavior of adults, Rangel and de la Barrera mentioned that the movements of flies are "very fast." If exposed to direct sunlight "flies move nervously." When on a leaf flies "exhibit excessive timidness, frequently turning graciously 360 degrees to face their enemy." Adults "like" to feed on juices oozing from fruit, especially oranges and guavas. In orchards, flies "like" to rest on the underside of leaves. Females prefer to sting mangoes in the "middle section" of the fruit. Oranges are always stung in the bottom part of the fruit. When a female is ready to oviposit, "it moves on the surface of the fruit in slow motion while searching for an appropriate spot. As soon as an ideal location is found it turns rapidly, inserts its ovipositor perpendicularly and stays motionless for up to two minutes. This procedure can be repeated 3 or 4 times before moving on to another fruit." Rangel and de la Barrera indicated, further, that females "prefer" to attack fruit that is in well-shaded parts of a tree and "avoid" fruit in branches that stick out from the crown (sun exposed). Fruits with a thick albedo are always "preferred" over those with a thin albedo.

All these insightful observations refer to phenomena such as clutch size regulation, oviposition site selection, enhanced fitness through efficient resource use, and physiological state (sensations of "fear," "hunger," "desire," "malaise"). We wanted to cite them specifically because few people know or have access to them. Other pioneering work on *Anastrepha* behavior is reviewed by Díaz-Fleischer and Aluja (Chapter 3).

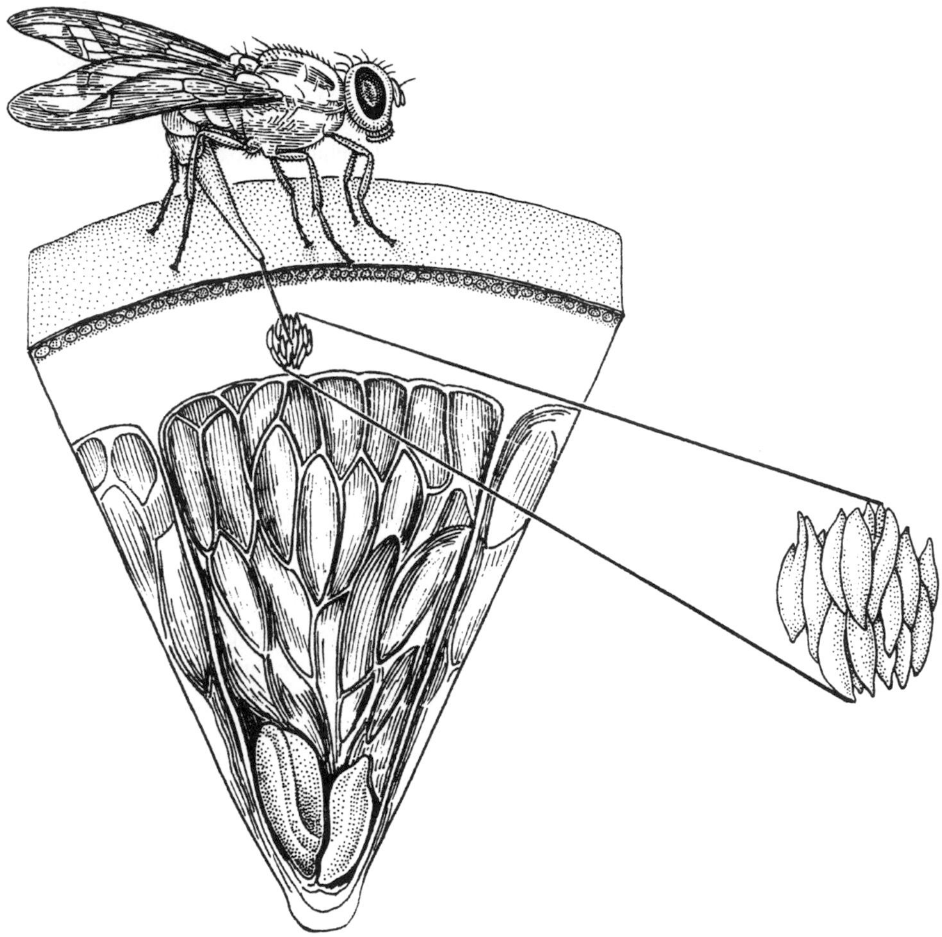

**FIGURE 15.1** *Anastrepha ludens* female depositing a clutch of eggs in the albedo of a grapefruit. (Courtesy of Andrea Birke.)

## 15.3 ASPECTS OF *ANASTREPHA* BIOLOGY AND NATURAL HISTORY RELEVANT TO BEHAVIOR

*Anastrepha* is endemic to the New World and is restricted to tropical and subtropical environments (Aluja 1994). Its range covers part of North America (states of Florida and Texas in the United States and all of Mexico), Central and South America (except Chile) and most of the Caribbean Islands (Hernández-Ortiz and Aluja 1993). The basic life cycle is similar among all *Anastrepha* species (see detailed review by Aluja 1994): females deposit their eggs in the epi- or mesocarp region of host fruit or, in some species such as *A. hamata*, in developing seeds (Color Figure 14*). Eggs are laid singly (e.g., *A. obliqua*) or in clutches (e.g., *A. ludens*) (Figure 15.1). The larvae of many species feed on the fruit pulp but it appears that the larvae of the primitive species (see Norrbom et al., Chapter 12, and McPheron et al., Chapter 13) feed on seeds (e.g., *A. cordata*, *A. hamata*, *A. bezzii*, *A. sagittata* (Stone), *A. pallens* Coquillett and *A. spatulata*; McPhail and Berry 1936; Baker et al. 1944; Santos et al. 1993; Hernández-Ortiz and Pérez-Alonso 1993; M. Aluja and M. López, unpublished data). Larvae of the close *Anastrepha* relative, *T. curvicauda,* also feed on

* Color Figures follow p. 204.

seeds (Peña et al. 1986). However, caution is warranted in drawing generalizations from this. *Anastrepha ludens*, for example, is well known for the damage it inflicts on commercial citrus groves. When attacking fruits in the genera *Citrus* or *Mangifera* (both introduced to the American Continent), larvae always feed on the albedo or the fruit pulp, but never on seeds. In contrast to this, in fruits of what are believed to be their ancestral host plants (*Sargentia greggii* S. Wats. and *Casimiroa edulis* Llave and Lex.), they basically feed on the seeds alone (*S. greggii*) or on both seeds and pulp (*C. edulis*) (Plummer et al. 1941; F. Díaz-Fleischer, unpublished data). Thus, widespread pulp feeding in *A. ludens* larvae could be a derived behavior that appeared recently as the result of the introduction of hosts such as *Citrus sinensis* L. (Osbeck) and *M. indica* L.

An interesting aspect of *Anastrepha* biology that bears directly on behavior is the phenomenon of two species partitioning the same host fruit or host tree. For example, larvae of *A. sagittata*, a close relative of *A. hamata*, feed on the seeds of the yellow zapote (*Pouteria campechiana* (Kunth) Baehni, while the pulp of the same fruit is used as a food source by larvae of *A. serpentina* (Baker et al. 1944). In the case of *A. striata* and *A. fraterculus*, two species that can also be found infesting the same fruit in the same tree (guava), there is evidence of a different type of resource partitioning. In this case, fruit in certain sectors of the tree crown are preferentially infested by one species, while the rest is infested by the other (M. Aluja, M. López, and J. Sivinski, unpublished data).

Other life history characteristics of *Anastrepha* species are highly variable. For example, life expectancy among *A. ludens*, *A. serpentina*, *A. crebra* Stone, and *A. bezzii* differs sharply. When adult *A. ludens* and *A. serpentina* were kept in 30-cc Plexiglas cages under laboratory conditions, 3.3 and 8.3%, respectively, were alive after 120 days (see Jácome, 1995, for detailed methodology). In contrast, when *A. crebra* and *A. bezzii* adults were maintained in exactly the same conditions, 60 and 46.6%, respectively, were alive after 120 days (M. Aluja, unpublished data). This may be related to the fact that both *A. crebra* and *A. bezzii* are monophagous, apparently univoltine species whose adults must survive for long periods to cope with the high environmental variability, which in turn determines the fruit production schedule of their host plants, *Quararibea funebris* (Llave) Vischer and *Sterculia apetala* (Jacq.) Karst. for *A. crebra* and *A. bezzii*, respectively. *Anastrepha ludens* and *A. serpentina*, on the other hand, are multivoltine, polyphagous species, which exploit a series of host species that appear in a progressive fashion throughout most of the year. Therefore, adults of multivoltine species do not need to survive for as long as univoltine species, whose hosts are available only once a year for a few weeks. Consistent with this, we note that the maximum longevity recorded for an adult *Anastrepha* individual, kept under laboratory conditions, is 431 days (M. Aluja and I. Jácome, unpublished data). This age was attained by a male of *A. alveata*, which is monophagous. An alternative strategy, allowing flies to survive periods during which host fruits are scarce or not available, would be entering diapause. Interestingly, diapause has never been documented in *Anastrepha*, even though the phenomenon appears to be widespread among the native parasitoids attacking many species in this genus (Aluja et al. 1998a).

## 15.4 DIEL RHYTHMS OF ACTIVITY

### 15.4.1 LARVAE

Diel larval activity patterns inside a fruit have been studied only in *A. suspensa*. Larvae of this species have been shown to feed within fruit continuously over a 24-h period (Webb and Landolt 1984). When larvae are ready to exit the fruit for pupation (after completing their development), they do so by following a distinct diel pattern. In the few species studied so far, this happens in late night or early morning hours. In *A. ludens*, 92% of all larvae exited before 0900 hours (McPhail and Bliss 1933). For *A. serpentina* and *A. striata*, the peak exit period was between 0600 and 0800 hours (M. Aluja, unpublished data). Emergence from fruit is stimulated by ambient temperature, rainfall, and by the impact of fruit falling from the tree (McPhail and Bliss 1933; Darby and Knapp 1934).

Exiting fruit during late night–early morning hours may be related both to danger of desiccation and predation and to ease of soil penetration. During that time, air humidity is close to 100% and there is no direct sunlight. Larvae crawling out of a fruit to bury themselves in the ground are thus at no risk of desiccation. Furthermore, because of damp air and soil conditions, burying procedures are greatly facilitated (reducing the risk of predation).

### 15.4.2 Adults

Despite the fact that there are clearly identifiable patterns of activity according to species, *Anastrepha* adults, in general, exhibit a high degree of plasticity in relation to most daily activities (e.g., time of day during which eggs are laid). Factors such as temperature, air humidity, or barometric pressure can influence the onset of a particular behavior. To illustrate this, *A. obliqua* adults observed in a mixed tropical plum (*Spondias purpurea* L.)/mango (*M. indica*) orchard in central Veracruz, Mexico oviposited during morning and afternoon hours, but never during midday hours (1100 to 1300). When individuals of the same population were forced to cope with different ecological conditions (i.e., cooler, wetter climate), females oviposited during the precise hours (1100 to 1500) they had avoided in a hotter and drier climate (details in Aluja and Birke 1993).

#### 15.4.2.1 Adult Emergence

There are very few studies reporting adult emergence patterns. It appears that emergence in most species takes place preferentially during morning hours. For example, 95.7 and 62% of all *A. ludens* and *A. striata* adults, respectively, emerged between 0600 to 1000 and 0900 to 1200 hours (McPhail and Bliss 1933; Aluja et al. 1993). In contrast, Santos et al. (1993) reported that most *A. bezzii* adults emerged between 1500 and 1800 hours. Exposure to sunlight and higher temperatures appears to stimulate emergence (McPhail and Bliss 1933). *Anastrepha striata* adults remained motionless immediately after emerging, until their wings expanded and dried out and they were able to fly. They then flew to a shaded area in a tree and remained motionless for most of the day on the underside of leaves (Aluja et al. 1993). In *A. serpentina*, flies emerging from under *Manilkara zapota* (L.) P. Royen trees flew to neighboring nonhost citrus trees, where they rested for the entire day (M. Aluja, unpublished data).

Daily calling, oviposition, feeding, and resting rhythms exhibited by the few species of *Anastrepha* studied to date will be described when each of these behaviors is analyzed in depth.

## 15.5 TRIVIAL MOVEMENTS

### 15.5.1 Larvae

Initial movements by newly hatched larvae are within a range of millimeters. As they grow, and food resources within the vicinity are depleted, larvae start to cover distances of centimeters. For example, when infesting citrus, *A. suspensa* larvae hatch in the flavedo (where eggs are deposited) and move to the albedo and pulp as they grow and the fruit ripens (Calkins and Webb 1988). The movement rate of larvae within the fruit can increase if the presence of parasitoids is detected. Larvae usually try to escape by wiggling movements or by burying themselves as deep as possible in the pulp (M. Aluja, personal observation). On occasion, third instar larvae drill a hole all the way to the surface of a fruit, extend the anterior of their bodies out of the fruit, and then crawl back into the fruit, either using the same gallery or a new one (A. Rangel and L. de la Barrera in Herrera 1905; Crawford 1918; 1927). The holes on the epidermis of the fruit are sometimes mistakenly used as evidence that the larvae have already exited the fruit, but often they are still feeding inside. It is possible that these galleries and holes are drilled to permit gas exchange. It is common to find rotting fruit where all larvae have died inside. Invariably, these fruits had no "respiration holes" and this may have resulted in noxious gases reaching lethal levels inside the fruit. It must be noted, however, that the larvae of some species do indeed exit the fruit while it is still hanging in the tree. For example, larvae of *A. hamata*, which feed exclusively on seeds, leave the fruit at a time when it is still unripe. They do so by biting their way through the hard pulp and jumping to the ground from heights of up to 10 m

(F. Díaz-Fleischer, personal observation). The only evidence of the former presence of the larvae is the exit hole around which callous tissue forms (Color Figure 14C*).

Once the fruit falls to the ground, larvae crawl out of it and bury themselves in the soil for pupation. Some do it right away, but some keep feeding and exit the fruit up to a month after it falls to the ground (Crawford 1918; 1927). Most commonly, larvae drill an exit hole in the part of the fruit touching the soil (where a humid microclimate forms because of the juices flowing). They usually pupate at a depth of 2 to 4 cm, depending on soil type and humidity (Bressan and Teles 1990; Salles and Carvalho 1993; Hennessey 1994; Hodgson et al. 1998). There are several factors that can modify this typical larval behavior. If the temperature rises to high levels in exposed fruit, larvae may crawl out and jump to the ground from heights of several meters (M. Aluja, personal observation). Despite the danger of desiccation (falling in an exposed spot with compacted soil) or predation (ants) the chances of such larvae surviving may be higher than if they remained in the fruit. Another factor that alters larval behavior is parasitism. Parasitized larvae bury themselves and pupate much faster (up to 12 h) than when unparasitized (Córdoba 1999).

### 15.5.2 Adults

As mentioned previously when discussing the behavior of teneral *A. striata* and *A. serpentina* individuals, when an adult emerges it flies to a densely foliated tree in a neighboring location and rests. Movements are circumscribed to a small area (e.g., natural patch of trees or section of orchard) and involve foraging for food and shelter/resting sites (Aluja 1993a). Mature males visit lek sites (e.g., *A. obliqua*) or patrol fruit (e.g., *A. bistrigata* Bezzi, in which males defend female resources). Females, after mating, search for suitable oviposition sites. Malavasi et al. (1983) reported high mobility of *A. fraterculus* adults between and within 11 host and nonhost trees in approximately 100 $m^2$ in Brazil. Male calling took place on nonhosts as well as hosts, but all observed mating pairs were located near the top of a tall nonhost tree. Feeding and oviposition occurred only in host trees. Resting took place on the bottom surface of leaves near the tops of host and nonhost trees. Flies moved toward the tops of trees at dusk, where they remained until the next day. When *A. obliqua* was studied in an orchard in which mango trees were surrounded by tropical plum trees (*S. purpurea*), Aluja and Birke (1993) reported that females used plum trees for feeding and oviposition, whereas males used them only for feeding. In contrast, females used mango trees principally for resting and feeding and males for sexual activities and resting. Flies moved back and forth between microhabitats, especially during early morning and late afternoon. Once all plums and mangoes had fallen to the ground or were harvested, *A. obliqua* adults started to show up in a neighboring chico zapote (*M. zapota*) orchard. There they used ripe fruit as food sources and some interspersed citrus trees as resting and male calling sites. A detailed analysis of the genetic structure of *A. obliqua* adults collected in plum, mango, chico zapote and citrus trees revealed no significant differences (B. McPheron, J. Piñero, and M. Aluja, unpublished data). This lends support to the hypothesis that members of a population actively move within an area using all available resources.

From the above, it is clear that *Anastrepha* adults readily move within a patch or orchard, but do not leave this patch if resources are plentiful. The contrary is true if the environment is unfavorable. In a series of release–recapture studies it was shown that the mobility of *A. ludens* and *A. fraterculus* was low if released in a place where vegetation, food, water, and oviposition substrates were plentiful (Plummer et al. 1941; Bressan and Teles 1991). In contrast, if flies of the same species are released in an unfavorable environment where conditions are dry, and there is a lack of host plants or adult food, they will quickly leave the release site (Shaw et al. 1967; Enkerlin 1987). Baker and Chan (1991a,b) and Baker et al. (1986) showed that wind affects the displacements of *A. ludens* and *A. obliqua* adults (i.e., mean fly movements were oriented in the direction of the prevailing wind). *Anastrepha ludens* has been reported to be able to move over remarkable distances.

---

* Color Figures follow p. 204.

For example, Christenson and Foote (1960) reported that individuals of this species flew approximately 135 km from breeding sites in Mexico to invade citrus groves in neighboring Texas. Shaw et al. (1967) trapped tepa-sterilized individuals of the same species up to 36 km from their release site. We believe that these long displacements were probably wind aided.

## 15.6 FEEDING BEHAVIOR

Feeding behavior in *Anastrepha* involves several modalities that are identical to those observed in other tephritid species such as *Rhagoletis pomonella* (Walsh) (Hendrichs et al. 1992; 1993) and *Ceratitis capitata* (Wiedemann) (Hendrichs et al. 1991): dabbing (grazing *sensu* Hendrichs et al. 1992), sucking, bubbling, and regurgitation. These behaviors were defined by Aluja et al. (1993). Dabbing is a "repetitive lowering of the proboscis to touch the surface on which the fly was feeding (usually a leaf) or while walking at increased rates of turning." This behavior has been studied in detail in *R. pomonella* by Hendrichs et al. (1992), who determined that adult flies could accrue small amounts of certain proteins and carbohydrates by grazing on leaf surfaces. Plants release such nutrients through leaching and guttation processes. Sucking "is the action of extending the proboscis to absorb liquids oozing from a fruit, water drops, or fresh bird feces" (Color Figure 1*). Bubbling is the "formation of a drop of liquid, of varying sizes, at the tip of the proboscis while the fly is sitting motionless" (Color Figure 4*). Regurgitation is the "deposition of a series of regurgitated drops on a leaf or fruit and reabsorption (reingestion) of those drops after varying intervals of time" (Color Figure 3*). Regurgitation behavior has been reported in *A. bistrigata*, *A. fraterculus*, *A. grandis*, *A. sororcula* Zucchi (Solferini 1990), *A. obliqua*, *A. serpentina*, *A. striata* (Aluja et al. 1989; 1993; Solferini 1990), *A. ludens* (Aluja et al. 1989), and *A. suspensa* (Hendrichs 1986). In the only study on *Anastrepha* where this behavior was quantified, Aluja et al. (1993), working with *A. striata*, determined that individuals deposit and reingest 23.5 drops within 12 min on the average. We believe that the number of drops depends on the type of food ingested. If it is high in water content, the number of drops should increase if flies are able to evaporate excess water from their food through regurgitation (Aluja et al. 1989). Alternatively, Drew et al. (1983) argued that through regurgitation, flies in the genus *Bactrocera* collect vital bacteria as an important source of protein. For a more wide-ranging outlook on fruit fly–bacteria relationships, see Drew and Lloyd (1991).

The few systematic studies carried out to date on feeding rhythms report varying patterns of feeding, depending on species and environmental conditions. For example, *A. suspensa* feeds throughout the day, but preferentially in the morning (Landolt and Davis-Hernández 1993). When *A. obliqua* was observed in an orchard where temperatures reached 45°C in unshaded areas, feeding followed a bimodal pattern with the most activity in the cooler afternoon hours (Aluja and Birke 1993). Feeding in the morning was mainly by females. Similar sexual differences in diel patterns of activities have been reported in other tephritids (e.g., *C. capitata*; Hendrichs and Hendrichs 1990; Hendrichs et al. 1991). Importantly, under a different set of environmental conditions (orchard in which temperature fluctuations were attenuated by densely foliated trees), *A. obliqua* fed principally between 1000 and 1500 hours (Aluja and Birke 1993). A similar pattern was reported by Malavasi et al. (1983) in *A. fraterculus*. Based on the above, we again underline the fact that when reporting feeding, or any other behavioral rhythm, all pertinent environmental variables should be quantified and described (e.g., barometric pressure, daily pattern of air temperature and humidity, light intensity, distribution and architecture of vegetation).

*Anastrepha serpentina* individuals fed for less than 5% of the 456 h of observation (M. Aluja and I. Jácome, unpublished data). Furthermore, an individual *A. serpentina* does not feed every day. In 39 days of uninterrupted observation, a typical *A. serpentina* individual provided with sugar, protein, and water *ad libitum* did not feed on 10 of the 39 days and fed minimally on 8 of the 39 days. Interestingly, if given a choice, most females tend to prefer sugar (sucrose) over protein or an open fruit. This "junk food syndrome" is described and discussed in detail in Jácome et al. (1999).

* Color Figures follow p. 204.

## 15.7 OVIPOSITION BEHAVIOR

Oviposition behavior of flies in the genus *Anastrepha* and in the closely related genus *Toxotrypana* follows a stereotyped pattern that includes the following steps: arrival on fruit (by flight or occasionally by walking), examination, and aculeus insertion (including superficial probing and actual skin puncture) (Barros et al. 1983; Landolt and Reed 1990). If an egg is laid, aculeus dragging invariably follows in species that mark. We note that this behavior (i.e., host marking) is only observed in *Anastrepha* (not in *Toxotrypana*). Examination involves moving in a straight or zigzag line on the surface of a fruit, while at the same time head-butting (bobbing *sensu* Landolt and Reed 1990) the surface. If a potential site for ovipositor insertion is detected, the speed of walking is reduced while head-butting periodicity is increased. Usually the turning angle becomes sharper, causing the fly to remain in the vicinity of the preselected site (movement in circular fashion). In *A. ludens*, the aculeus of the ovipositor is extruded at this stage, followed by aculeus insertion after movement has been completely halted (Color Figure 15*). The fly lifts its hind legs and abdomen, and inserts its aculeus at an angle of approximately 45° (Figure 15.1) (Color Figure 15A*). The probability of aculeus insertion after landing on a fruit depends on several factors: host type (primary vs. secondary), quality of fruit (e.g., degree of ripeness), and evidence of previous use by conspecifics (presence of host-marking pheromone). In *A. fraterculus*, the likelihood of a female making a puncture after landing on a fruit is 70% if it is a primary host like guava (Barros 1986). If it is a secondary host like apple, the likelihood drops to 51% (Sugayama et al. 1997). Flies invariably drag their aculeus after it is removed from the fruit flesh following release of eggs (Color Figure 15D*). An interesting facet of *Anastrepha* oviposition behavior is the fact that at least in *A. suspensa*, females are able to detect acoustical cues emitted by feeding larvae, and as a result, reject fruits that are infested (Sivinski 1987).

As noted before, ovipositional activities are greatly influenced by environmental conditions, especially ambient temperature. Despite this, overall patterns are still discernible. *Anastrepha ludens,* for example, prefers to oviposit between 1100 and 1400 hours (Birke 1995). In comparison, *A. obliqua* starts to oviposit much earlier (0700 hours) (Aluja and Birke 1993) and *A. serpentina* a little later (1200 hours) (M. Aluja, unpublished data). Oviposition behavior in *A. striata* follows a clear bell-shaped pattern beginning at 0800, ending at 1600, with a peak between 1200 and 1300 hours (Aluja et al. 1993). In the case of *A. hamata* and *A. fraterculus*, peak oviposition activity is observed at 1200 and between 1600 and 1700 hours, respectively (F. Díaz-Fleischer, unpublished data; Silva 1991; Sugayama et al. 1997).

Host-marking behavior (deposition of an oviposition-deterring pheromone through aculeus dragging after laying an egg; Figure 16*) has been reported in the following *Anastrepha* species: *bistrigata* (Selivon 1991), *fraterculus* (Prokopy et al. 1982), *grandis* (Silva 1991; Silva and Malavasi 1993), *ludens* (Papaj and Aluja 1993), *obliqua* (Aluja et al. 1998b), *pseudoparallela* (Loew) (Polloni and Silva 1987), *serpentina* (Aluja et al. 1998b), *sororcula* (Simoes et al. 1978), *striata* (Aluja et al. 1993), and *suspensa* (Prokopy et al. 1977). We have further evidence that *A. alveata*, *A. leptozona*, *A. acris,* and *A. spatulata* also exhibit host-marking behavior (M. Aluja, I. Jácome, C. Miguel, and M. López, unpublished data). Importantly, field and field-cage observations demonstrated that *A. cordata* and *A. hamata* do not exhibit host-marking behavior (M. Aluja and F. Díaz-Fleischer, unpublished data). These two species are considered primitive within *Anastrepha* (Norrbom et al., Chapter 12; McPheron et al., Chapter 13). In both cases, females lay eggs in fruit that release latex after being punctured (Color Figures 12 and 14A*). Interestingly, when observing *A. cordata* oviposition behavior in *Tabernaemontana alba* Mill. (Apocynaceae), we noted that females preferentially inserted their aculeus along the middle rib of the fruit. If the process of aculeus insertion/removal lasted less than 2 min, a drop of latex covered the site of insertion. If the oviposition bout lasted longer than 2 min, we never observed the formation of a latex drop at the site of ovipositor insertion. This led us to speculate that the female, while probing, is actually trying to circumvent latex channels in the fruit or is "milking" the fruit to force release of troublesome

---

* Color Figures and Figure 16 follow p. 204.

latex. Our hypothesis is based on the behavior exhibited by some caterpillars which bite leaves to cause latex flow and to facilitate ingestion of leaf material (Dillon et al. 1983). Because these two primitive *Anastrepha* species, together with the closely related species *T. curvicauda*, do not mark fruit after egg laying, we feel warranted to hypothesize that host marking could be a derived behavior in *Anastrepha* (for an in-depth analysis on the evolution of tephritid host-marking behavior, see Díaz-Fleischer et al., Chapter 30). Interestingly, all three species mentioned above lay their eggs in the developing seeds of the fruit or in the fruit cavity containing seeds (*T. curvicauda*), with larvae feeding on the seeds, not the pulp (Color Figures 14B, C*). The hosts of these species all release latex after being punctured. When *A. cordata* females remove their aculeus from the fruit, they immediately try to clean it. As part of the process, they rub the aculeus tip with their legs and occasionally "rub" it against the surface of the fruit (M. Aluja, personal observation). This would support the hypothesis of Fitt (1984) that aculeus dragging is an elaboration of cleaning behavior. During such a process feces and other digestive by-products are also deposited on the surface of the fruit. It is easy to envision how such a mechanism could have evolved into what we currently know as host-marking behavior. In this respect, it is interesting to note that *A. serpentina*, a species with larvae that occasionally feed on seeds and adults that exhibit host-marking behavior, uses hosts that release latex after puncture (several plants in the family Sapotaceae). It is possible that this species represents a transition in the evolution from seed feeding, nonhost marking to pulp feeding and host marking.

As discussed earlier, it is noteworthy that two species of *Anastrepha* can utilize the same host without competition. Larvae of *A. sagittata* and *A. serpentina* feed on the seeds and the pulp, respectively, of *Pouteria campechiana.* It may be that the mark of latex left by the female ovipositing deep in the pulp to reach the young seed is a sufficient signal to another conspecific female, whereas the female laying eggs in a maturing fruit with less latex must also mark with a pheromone to signal previous occupation. Also, seeds in an optimal stage for oviposition are highly ephemeral. In contrast, larvae in pulp are exposed to more competition. Since pulp is a less ephemeral resource, it needs to be protected for longer periods.

Finally, and as also mentioned before, some *Anastrepha* species compete for the same resource. For example, *A. fraterculus* and *A. striata* both infest *Psidium guajava* L. and individuals of both species can be found in a single fruit. The same can be true for *A. ludens* and *A. obliqua* infesting *Mangifera indica.* All of these species deposit a host-marking pheromone. In such direct competition, cross species recognition of the marking pheromone is critical for the survival of the larvae. Furthermore, simultaneous infestations of the same fruit by species in two genera (*A. fraterculus* and *C. capitata*) have been reported for peach, apple, and coffee in Brazil (Pavan and Souza 1979). *Ceratitis capitata* is an exotic species that was introduced in Brazil from either the Mediterranean area or the African continent in 1901 (Ihering 1901). Aluja et al. (1998b) have shown that *Anastrepha* individuals cannot recognize the host-marking pheromone (HMP) of *C. capitata* and, as a result, direct competition is unavoidable. It would be interesting to follow the probable development of intergeneric recognition of the HMP into the next century.

When comparing oviposition behavior among *Anastrepha* species, we have arbitrarily created four groups based on their clutch size: always one egg (e.g., *A. obliqua*; Celedonio-Hurtado et al. 1988), small (one to three eggs; e.g., *A. striata*; Aluja et al. 1993), medium-sized (1 to 40 eggs; e.g., *A. serpentina* and *A. ludens*; Jácome 1995 and Dávila 1995, respectively), and large clutches (>30 eggs; e.g., *A. grandis* and *A. bezzii*, Silva 1991 and M. Aluja and I. Jácome, unpublished data, respectively). Clutches of up to 110 eggs have been reported for *A. grandis* (Silva 1991). Importantly, even in clutch-laying species, ovipositing only one egg is also common. In studies aimed at determining the effect of adult diet on the basic demographic parameters of *A. serpentina* and *A. ludens*, it was determined that these species lay 1 to 19 and 1 to 23 eggs/clutch, respectively (Jácome 1995; Dávila 1995). Oviposition substrates used were 3-cm-diameter agar spheres wrapped in Parafilm (details of methodology in Boller 1968). *Anastrepha ludens* that were fed on sucrose and protein produced average clutch sizes of $3.96 \pm 0.3$ ($N = 264$; measurement of variance is S.E.; Dávila 1995).

* Color Figures follow p. 204.

In a much earlier study with the same species, McPhail and Bliss (1933) reported an average clutch of 5.4 ± 0.2 (range = 1 to 18), with flies spending 1 to 12 min per oviposition bout.

Berrigan et al. (1988) argued that clutch size in *A. ludens* was largely determined by host size and variation among females, ranging between 1 and 40 eggs/clutch. Host color, host density, fly density, and fly age did not affect clutch size. Average clutch size was 4.4 eggs and 12.7 eggs in 2-cm- and 11-cm-diameter artificial hosts, respectively. These authors report that in no-choice experiments, the increase in clutch size with increased host size appeared to peak at a host size of ~7 cm.

When analyzing the evolution of clutch size, one should take into account that most wild host fruits of *Anastrepha* tend to be small (3 to 30 g) (López et al. 1999) and that one or few eggs are laid in them. We thus believe that the large clutch sizes seen in some *Anastrepha* species of economic importance (e.g., *A. ludens*, *A. serpentina*, *A. grandis*) could be a recent development related to the appearance of artificially large, cultivated fruit. *Anastrepha ludens* coexisted for a long time with the host *S. greggii*, whose fruit weighs approximately 2 to 3 g and can only harbor one or two larvae (Plummer et al. 1941). In contrast, the commercial Marsh grapefruit commonly infested by *A. ludens* larvae weighs an average of 312 g (range = 150 to 800 g) (López et al. 1999) and can easily harbor more than 80 larvae. It would be interesting to ascertain clutch size in *A. ludens* populations that have not been in contact with cultivated plants (possibly in remote canyons in the states of Nuevo León and Sinaloa, Mexico).

## 15.8 MATING BEHAVIOR

The sexual behaviors of a handful of *Anastrepha*, mostly economically important species, have been examined in detail (Aluja 1994). Because some of these species (e.g., *A. suspensa*, *A. ludens*, and *A. obliqua*) have been mass-reared for sterile male releases, it has been necessary to understand male–male competitive interactions and female mate-choice criteria so that quality control measures could be instituted and trapping done more effectively (e.g., Burk 1986; Moreno et al. 1991; Mangan et al. 1992). Since pestiferous, polyphagous species have attracted a disproportionate amount of research, the behaviors reviewed below may not be representative of the genus as a whole.

In general, the known mating systems of *Anastrepha* species are complex and often include male territories (often aggregated in some degree to form leks), male pheromone emissions from pleural glands and evaginated anal membranes (Color Figure 7A*), pheromone depositions on leaves (Color Figure 7B*), wing fanning-acoustic signals (songs) produced both prior to and during coupling, wing motions (semaphoring) accompanied by "graceful" sideways-arching body movements, and extensions of the male mouthparts and other activities during copulation (see Sivinski et al. 1984; Sivinski and Burk 1989) (Color Figure 8*). Some species perform additional behaviors such as trophallaxis (the female consumption of fluids provided by the male; Aluja et al. 1993) (Color Figure 9*) and short looping flights over leaf territories (Aluja 1993b). Copulation durations vary with species and may serve as a means of preventing subsequent reinseminations by sexual rivals or as a means of transferring male material investments to mates or zygotes (see Alcock 1994). The evolution of sexual behavior, both in *Anastrepha* and other fruit flies, is considered in Sivinski et al. (Chapter 28).

### 15.8.1 Mate Acquisition Strategies

In broad terms, two mating strategies have been reported in *Anastrepha*: (1) resource defense by which males patrol and defend clumps of fruit that are attractive to receptive females (to date only reported in one monophagous species: *A. bistrigata*, Morgante et al. 1993) and (2) lek polygyny, which appears to be the norm in *Anastrepha*. Of the latter system there seem to be gradients as will be discussed later. In the case of leks, females visit these calling arenas, devoid of any resource, and actively choose a mating partner (see reviews by Burk 1983 and Hendrichs 1986).

Mating occurs on the leaves or leaf nodes (rarely on fruits) of host trees, or occasionally on the leaves of more hospitable tree canopies adjacent to hosts (e.g., Burk 1983; Malavasi et al. 1983;

* Color Figures follow p. 204.

Aluja and Birke 1993). Males typically maintain a territory on the underside of a leaf from which they emit pheromones and produce calling songs (see below).

Male behavior of *A. suspensa* in field cages and in the field has been described in detail by Dodson (1982), Burk (1983), and Hendrichs (1986). As the late afternoon period of sexual activity approaches, males become "alert," raising themselves off the surface, holding their wings at an acute angle, and stationing themselves at the leaf base, facing the stem and often astride the main leaf vein. As time passes, they begin to semaphore (see below) and eventually to call (see below).

Territories, at least those of polyphagous species, are often aggregated to form leks (e.g., Perdomo 1974; Dodson 1982) (Color Figure 5*), although the propensity to form leks varies with species. Even in species where aggregations are common, isolated males ("satellite males") may signal and obtain mates (e.g., Sivinski 1989; Robacker et al. 1991). There is no fixed definition in the literature of what constitutes a lek in *Anastrepha* species. Although two males on adjacent leaves probably represent an interacting group (see sections on male agonistic interactions and female mate-choice criteria), Aluja and Birke (1993) defined an *Anastrepha* lek as "an aggregation of at least 3 males calling simultaneously in a clearly defined area, usually adjacent leaves of a single branch" (Color Figure 5*). The largest *A. suspensa* lek observed by Sivinski (1989) contained nine calling males, none more than 15 cm from at least one other participant. Malavasi et al. (1983), working with *A. fraterculus*, noted a lek of five males within 80 cm of one another. In *A. suspensa*, leks form in areas where males and females are also likely to be found in nonsexual situations, presumably because of favorable microhabitats. Certain leaves may be occupied by signaling males for several consecutive days (Sivinski 1989). Aggregation locations in *A. obliqua* are also influenced by microhabitat (Aluja and Birke 1993) and calling territories of *A. ludens* tend to occur in the interior half of the tree canopy (Robacker et al. 1991). There is little apparent structure to leks, although smaller males of *A. suspensa* have a tendency to be located nearer the tips of branches. This could be due to no more than the combined tendencies of larger males to defeat smaller opponents and for flies in flight to move toward better lit areas (Sivinski 1989; see, however, Hendrichs 1986; and Section 15.8.11 on female mate-choice criteria).

Not all *Anastrepha* species form leks or form them as consistently as others. *Anastrepha striata* is characterized as an intermediate lekking species because of its propensity to call alone (Aluja et al. 1993). Furthermore, *A. bistrigata* males do not appear to lek at all and maintain territories on or near fruit-oviposition sites (Morgante et al. 1993).

### 15.8.2 Diel Periodicity

Daily rhythms in male calling vary sharply in *Anastrepha* (we only provide information on male calling, since mating is initiated at roughly the same time as calling activities take place). Its onset fluctuates from before sunrise to after sunset, depending on the species. To illustrate this, we have graphed the calling patterns exhibited by 20 *Anastrepha* and one *Toxotrypana* species (Figure 15.2).

Despite the fact that the daily pattern of calling rhythms appears to be one of the most hard-wired behaviors in *Anastrepha* (probably driven by a circadian clock; see Smith 1979 for an in-depth analysis of the phenomenon for other tephritid flies), it is also influenced by environmental conditions. As discussed by Aluja and Birke (1993), and references therein, "daily patterns of activity in insects have evolved in response to ecological factors and are strongly influenced by prevailing physical characteristics of the habitat." It is probable that selection acted upon genotypes that were more efficient at timing the broadcast of signals (e.g., pheromones, sounds) under situations such as a humid evergreen rain forest or a drier deciduous forest. For example, if temperatures are below or above the optimum (20 to 24°C), the onset of calling is delayed in *A. ludens* (M. Aluja and J. Piñero, unpublished data). A similar pattern was observed with *T. curvicauda*. Flies observed during January under cooler temperatures called much later than flies under warmer temperatures during August (see Figure 15.2; details of study in Aluja et al. 1997).

* Color Figures follow p. 204.

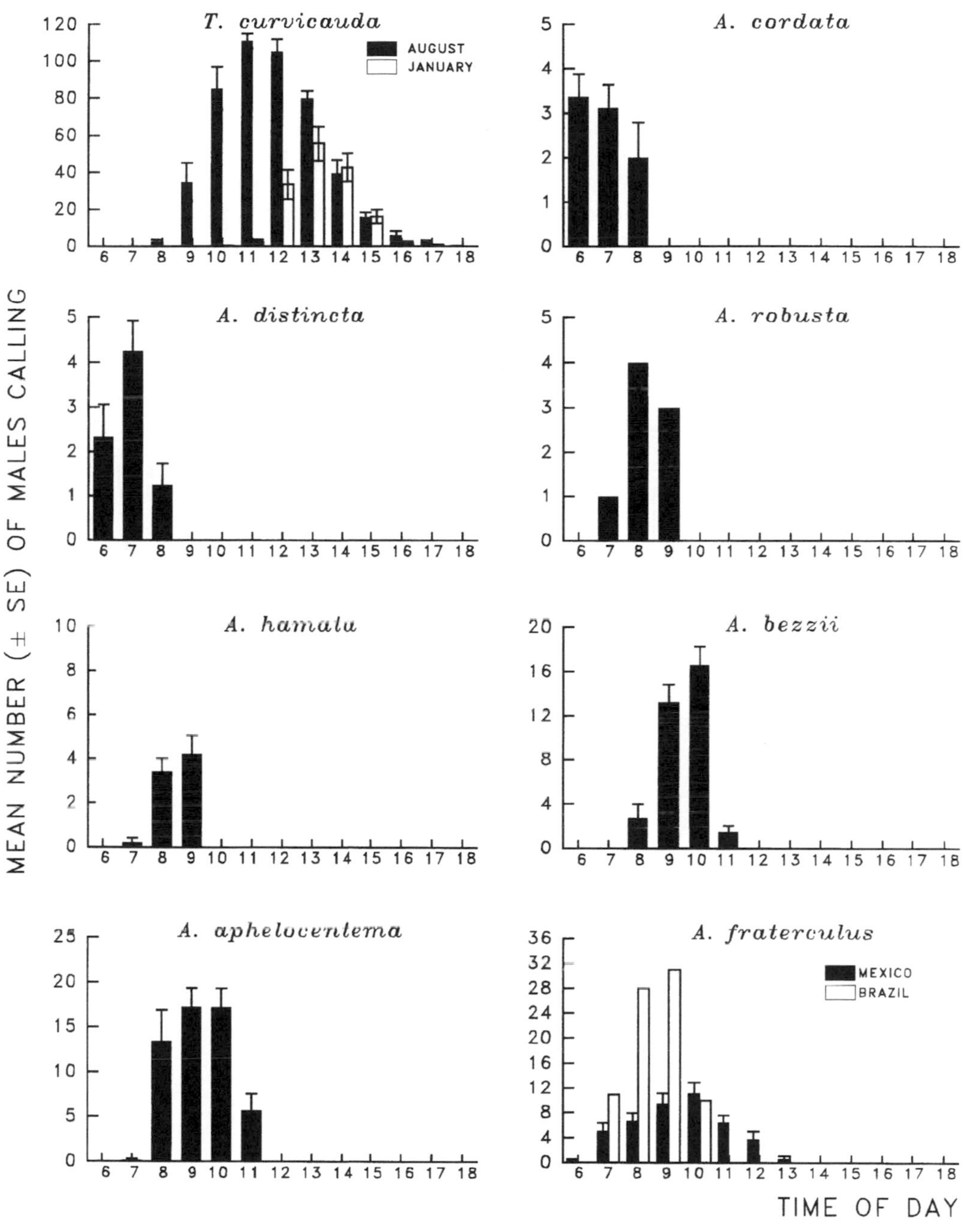


**FIGURE 15.2** Distribution of calling activities in 20 *Anastrepha* species and *Toxotrypana curvicauda*. Information on *A. acris*, *A. alveata*, *A. aphelocentema*, *A. bezzii*, *A. cordata*, *A. crebra*, *A. distincta*, *A. hamata*, *A. leptozona*, and *A. serpentina* comes from previously unpublished data by M. Aluja, I. Jácome, J. Piñero, and F. Díaz-Fleischer (methodology used to conduct these observations is described in detail in Aluja et al. 1993). Information on the remaining species comes from *A. bistrigata* (Selivon 1991), *A. fraterculus* (Malavasi et al. 1983; M. Aluja, J. Piñero, V. Hernández, and B. McPheron, unpublished data), *A. grandis* (Silva 1991), *A. ludens* (McPhail and Bliss 1933; Aluja et al. 1983), *A. obliqua* (Aluja and Birke 1993), *A. pseudoparallela* (Polloni and Silva 1987), *A. robusta* (Aluja 1993b), *A. spatulata* (M. Aluja and M. López, unpublished data), *A. suspensa* (Hendrichs 1986), *A. striata* (Aluja et al. 1993), and *T. curvicauda* (Aluja et al. 1997).

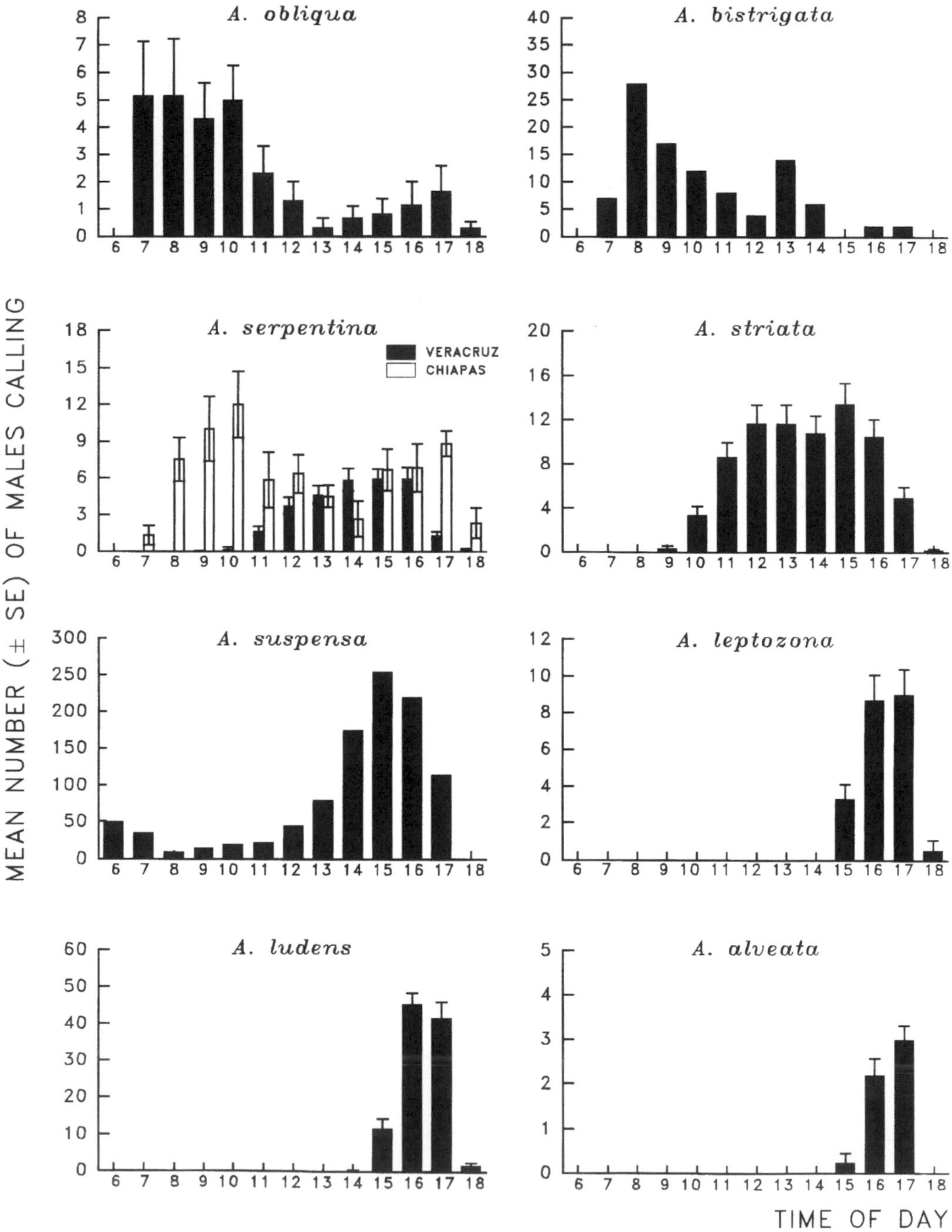


**FIGURE 15.2 (continued)**

Despite the fact that temperature can delay or push forward the onset of calling, we believe that the most critical extrinsic factor regulating calling (mating) behavior in *Anastrepha* is light intensity. F. López and R.H. Rhode (work reviewed by McFadden 1964) conducted a series of experiments on the effect of photoperiod and wavelength on the mating of *A. ludens*. These authors found that

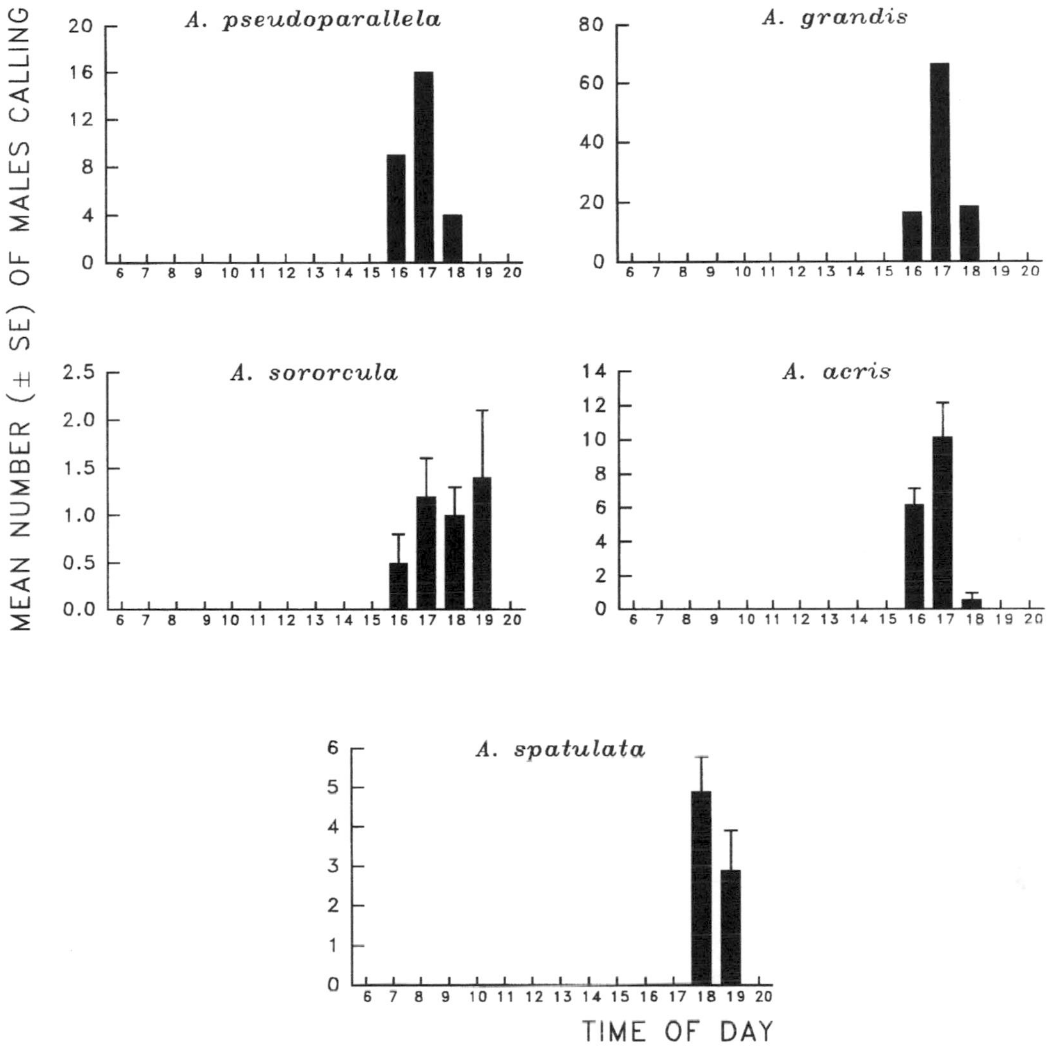


**FIGURE 15.2 (continued)**

under natural light conditions in the laboratory, caged *A. ludens* individuals began mating in midafternoon with a peak occurring between 1600 and 1800 hours. By varying this normal pattern under controlled conditions, flies could be induced to initiate mating during the midmorning hours and to reach a peak at 13:00 hours. In nature, *A. ludens* males only call in late afternoon (dusk) when light intensity is low (4 to 6 footcandles) (McPhail and Bliss 1933; Aluja et al. 1983).

Furthermore, there seem to be populational differences with respect to calling rhythms. In Figure 15.2, we describe the daily calling rhythm of *A. serpentina* cohorts stemming from two localities in Mexico separated by ~1200 km: the states of Veracruz (Apazapan) and Chiapas (Tapachula). Note that, unlike the Veracruz population, individuals from the Chiapas population had a much more extended calling period that included morning hours. A similar pattern is observed when analyzing calling rhythms of *A. fraterculus* populations in Mexico and Brazil (Figure 15.2). We currently lack enough information to interpret these variations in calling rhythms adequately. We could be dealing with biotypes adapted to local environmental conditions or with an unresolved species complex.

### 15.8.3 Male Agonistic Interactions

Lekking insects, such as many *Anastrepha* species, interact in three ways (Thornhill and Alcock 1983): (1) they are attracted to the sexual signals of other males (see Heath et al., Chapter 29); (2) they participate in signaling interactions such as mutual stimulation; and (3) they engage in aggressive interactions for the possession of territories (example of aggresive interaction in lek of *C. capitata* in Color Figure 11*).

*Anastrepha suspensa* males placed in the presence of other males begin pheromone signaling more quickly (have a shorter latency) than those maintained alone (Burk 1984). Male reactions to intruders consist of lowering the body to the leaf surface, pressing the wings against the side of the body, and extending the proboscis forward (the "arrowhead" posture; Dodson 1982; Burk 1983). Escalation of the confrontation leads to head butting and audible "aggression" songs produced by wing buzzing (similar behaviors have been observed in *A. ludens*; Robacker and Hart 1985; see below).

Resident males in wild populations of *A. ludens*, *A. obliqua,* and *A. suspensa* were usually able to defend their leaf territories against aggressive interlopers (Aluja et al. 1983; Burk 1983). In laboratory experiments with *A. suspensa*, prior residence on the territory and large size characterize males that win agonistic interactions (Burk 1984). The two factors interact so that large residents are very successful against smaller interlopers, and large residents are more successful than large interlopers, but small residents often lose their territory to large interlopers.

Given the apparent importance of large size to male *A. suspensa*, both in terms of intrasexual agonistic interactions and female mate choice, it is something of a mystery that males tend to be the smaller sex; i.e., they are not as large as a fruit fly can become in that particular niche. In a field-cage study of *A. ludens* sexual behavior, there was no relationship between male size and sexual success (Robacker et al. 1991; see section on female mate-choice criteria). Nonsexual explanations for small-male sexual dimorphism in the genus are provided by Sivinski and Dodson (1992).

### 15.8.4 Looping Flights at Territories

Male *Anastrepha* rarely leave and return to the same spot during their calling bouts. *Anastrepha robusta* and *A. leptozona* are exceptions (Aluja 1993b; M. Aluja, J. Piñero, and I. Jácome, unpublished data). *Anastrepha robusta* is a large and widely distributed species, but seldom observed. The males perform many of the behaviors common to other species of the genus, such as wing fanning (calling song), pheromone gland puffing, and touching the everted anal membrane to leaf surfaces (see relevant sections). An additional behavior is looping flights, performed repeatedly, 15 to 40 cm in distance, which depart from and return to the same leaf territory (Figure 15.3). More than 10 of these loops may be performed by an individual every minute (Aluja 1993b). In *A. leptozona*, the loops are not as wide but are also very common. Flying as a mating strategy is rare for tephritoids outside of the swarming Lonchaeidae (see Sivinski, Chapter 2).

### 15.8.5 Pheromones

As far as is known, male-produced pheromones are present in all species of *Anastrepha* (see Nation 1972, for an early report). Their production and composition is discussed by Epsky and Heath (1993) and Heath et al. (Chapter 29).

The males of many *Anastrepha* species touch their leaf territories with their evaginated anal membranes (i.e., proctiger), a structure associated with pheromone emission (e.g., Nation 1989; Aluja 1993b; Aluja and Birke 1993; Aluja et al. 1993). This behavior has been examined in *A. suspensa* (Sivinski et al. 1994). Bouts of anal touching occur while a second set of pheromone-related structures, the lateral abdominal pleural glands, are protruded. They also coincide with wing fanning, which may help disperse pheromones (see Section 15.8.6). Five of the eight components

* Color Figures follow p. 204.

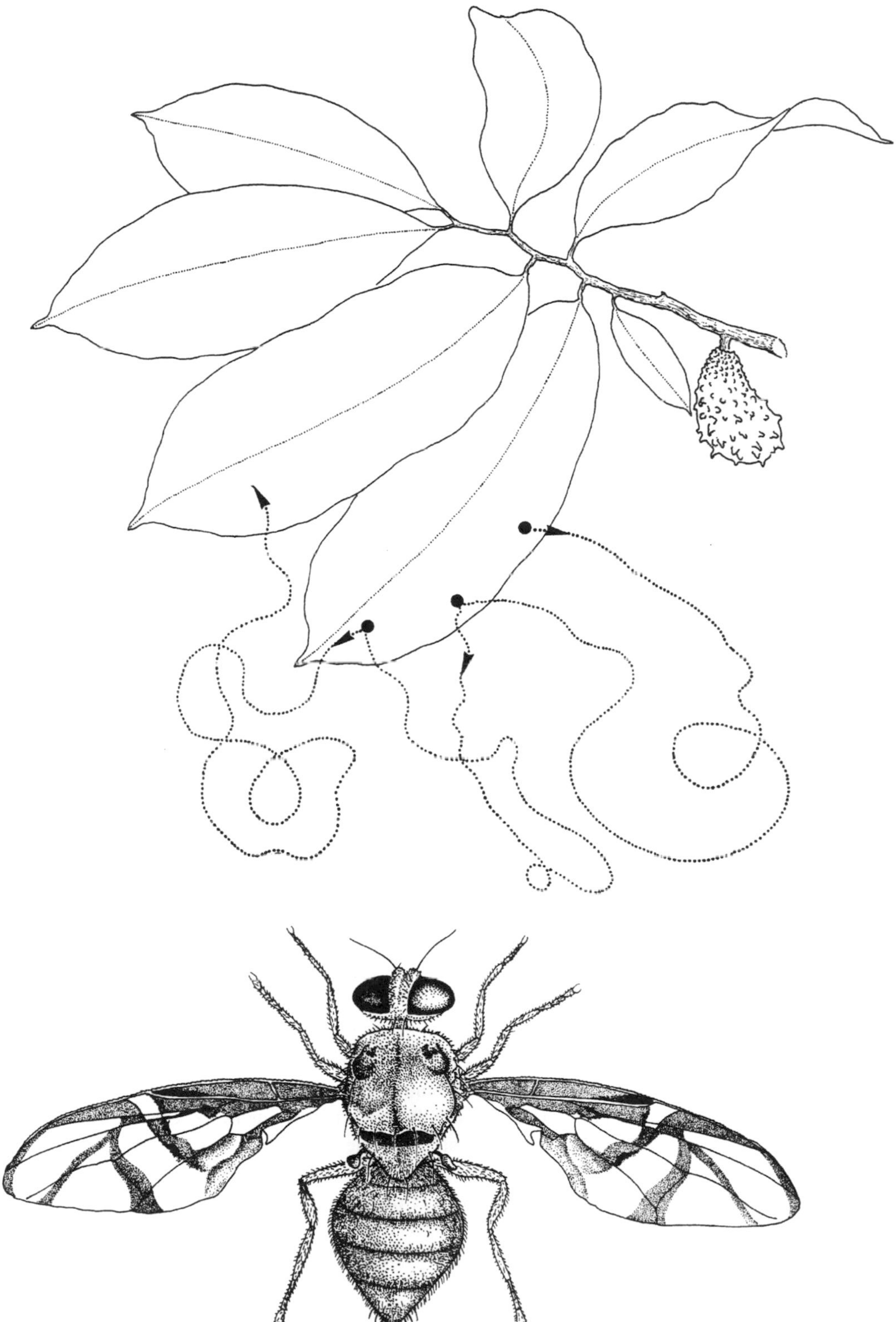

**FIGURE 15.3** Graphic representation of loops performed by *A. robusta* males while calling on an *Annona muricata* L. tree. (From Aluja, M., *Fla. Entomol.* 76: 391–395, 1993. With permission.)

of *A. suspensa*'s pheromone were recovered from leaves that had previously been held as territories by signaling males. Only one component was detected on leaves that had been held near, but not in contact with, signaling males. Thus, the bulk of the pheromone chemicals on the leaves were there due to anal touching. In the laboratory, mature virgin females were more likely to be in contact with leaves that had been anal-touched by males than with untouched control leaves. There was no difference in female contacts with leaves that had been kept with sexually immature males or females and their respective controls. The ability of leaf territories to hold pheromones and apparently enhance chemical signals may increase their value to territory holders. Pheromone deposits may be one reason why aggressive interlopers would want to expel territory holders and obtain their signaling platform (see section on mating systems and Sivinski et al., Chapter 28, on the evolution of mating behavior).

Related to the anal touching (dipping) behavior observed in males of *A. suspensa* and other *Anastrepha* species, is the anal dragging behavior exhibited by *A. hamata* males (M. Aluja, personal observation). In between wing-fanning bouts, males repeatedly drag their proctiger along leaf pedicels or branchlets (Color Figure 7B*). This seems to be an elaboration of the anal touching (dipping) behavior exhibited by other *Anastrepha* species. We have not been able to ascertain if males are depositing a pheromone while dragging the proctiger, but this seems likely.

### 15.8.6 Calling Songs

The wing fanning that accompanies anal membrane touching may help disperse pheromones deposited on leaf territories; however, it also generates an acoustic "calling song" that has sexual significance and has been examined in detail in *A. suspensa* (Webb et al. 1976; also see Sivinski et al., Chapter 28). Songs are produced by males on their calling stations both in the presence and absence of other flies, although the incidence of calling and the characteristics of the song vary under different social circumstances and presence/lack of adult food and host fruit (see below; Sivinski and Webb 1986; Landolt and Sivinski 1992). Calling songs in *A. suspensa* are largely produced by wing movements, with thoracic vibration making a lesser contribution (Sivinski and Webb 1985), and generally have a fundamental frequency of 140 to 150 Hz (Webb et al. 1984; Burk and Webb 1983). The typical form of *A. suspensa*'s song is repeated episodes of wing fanning (pulse trains), each ~0.5 s long separated by ~0.5-s-long pulse-train intervals. There is a sexual dimorphism in the shape of the wings in *Anastrepha*: those of the male are more oval (Sivinski and Webb 1985). It has been suggested that this shape is an adaptation for acoustic signaling; however, there are alternative explanations based on male flight abilities (Sivinski and Dodson 1992).

Female *A. suspensa* use male songs to locate calling sites. Sticky traps baited with recorded calling songs captured more virgin females than silent controls, but the range of attraction is unknown (Webb et al. 1983).

In addition to serving as an attractant, the calling song appears to represent an important courtship component. Certain song characteristics are correlated with male size (Burk and Webb 1983). For example, in *A. suspensa*, large males tend to have a greater propensity to sing, and have songs with shorter pulse-train intervals and lower fundamental frequencies. Since females given a choice of a large and a small male prefer to mate with the larger, it is possible that females could use the size-correlated song characteristics as criteria for choosing or rejecting a courting male.

In one laboratory study, *A. suspensa* males were divided into four categories of sexual success: (1) those that mated on their first attempt; (2) those that mated after repeated attempts; (3) those that mounted but were rejected; and (4) those that never mounted a female during the 30-min observation period (Webb et al. 1984). There was only one difference in their calling songs: the bandwidth of the fundamental frequency was broader in males that never mounted a female compared with males that copulated on their first attempt. Expanded bandwidth could reflect a

---

* Color Figures follow p. 204.

physical disability, and has been found in the acoustic signals of improperly irradiated *C. capitata* males suffering from malformed wings ("droopy-wing syndrome") (Little and Cunningham 1978).

However, in another study of *A. suspensa* female response to calling songs, there was evidence that the pulse-train interval was an important characteristic to females when choosing mates. Virgin females were more likely to react to recorded songs with short pulse-train intervals (a large-male characteristic) than to songs with long intervals (a small-male characteristic; Sivinski et al. 1984). There was no difference in the reactions of mated females, and males were most active in the periods of silence that separated the recordings.

Calling males in *A. suspensa* modify their songs in the presence of potential mates and sexual rivals by shortening the pulse-train interval when females are nearby and lengthening the pulse train when males are placed next to them (Sivinski and Webb 1986). Shortening the pulse-train interval near females is consistent with a short interval being a song characteristic that females find attractive in a mate.

The calling songs of the other species of *Anastrepha* that have been examined are similar in pulse train and pulse-train interval to those of *A. suspensa* (J. Sivinski, unpublished data). Mankin et al. (1996) compared the calling songs of *A. grandis*, *A. obliqua*, *A. sororcula*, and two populations of *A. fraterculus*. *Anastrepha grandis* had the longest pulse train and pulse-train interval, and lowest fundamental frequency. *Anastrepha fraterculus* had the shortest pulse train and pulse-train interval. However, there was no single characteristic that was unique in every species and could be used to identify a species or different populations of a species.

Importantly, there are some primitive *Anastrepha* species like *A. cordata* that do not produce any calling songs. The same is true in the case of *T. curvicauda*. In these species, males only raise their abdomen (with puffed pleural glands) but do not fan their wings. Wing fanning could be considered a derived behavior that may have evolved from aggressive male/male encounters during which individuals wing fan vigorously.

### 15.8.7 Precopulatory Songs

There is a second type of wing-fanning-acoustic signal performed by *Anastrepha* males, and which again has been examined in *A. suspensa*. As males mount females and attempt to insert their phallus into the ovipositor, they produce a "precopulatory song." This differs from the calling song by its continuous nature (no repeated pulse trains), its higher fundamental frequency (+19 Hz), greater energy (+199 mV under Fast Fourier Transform [FFT] curve), smaller waveform distortion (–35%), and lesser range of fundamental frequencies (–49 Hz) (Webb et al. 1984). In summary, it is a relatively pure, high, intense, and energetic sound that usually lasts until the male genitalia have begun to penetrate the female's uplifted ovipositor (~35 seconds), although particular precopulatory songs can last up to 15 min or more. Occasionally, a similar sound will be produced by copulating males in apparent response to female restlessness (i.e., females apparently do not cooperate completely with postintromission processes such as further penetration or transport of sperm). The energy invested by males in this final acoustic signal prior to insemination suggests that it is an important component of courtship that plays a critical role in maximizing male reproductive success (Eberhard 1994; 1996).

Mounted and singing males are frequently rejected by females before they can completely insert their phallus. Burk and Webb (1983) noted that males that copulated had songs that were ~10 dB more intense than those that did not copulate (dB, decibel, being a measure of sound pressure level) and were rejected, and that this difference occurred regardless of male size. Since an increase of 6 dB represents a doubling of sound output, the differences between successful and unsuccessful songs are dramatic. Webb et al. (1984) examined a larger sample of successful and unsuccessful songs and again found that successful songs had a significantly higher sound pressure level and were also more energetic, had narrower frequency bandwidths, and were less distorted. In experiments with recorded songs, muted (dealated) males were more likely to copulate if the correct form of the precopulatory song was broadcast in their cage at a high intensity than if the correct song was played at a low intensity or an incorrect calling song at a high intensity (Sivinski et al. 1984).

Precopulatory songs vary in different species. Descriptions of courtship in *A. sororcula* suggest a more energetic song than that produced by *A. obliqua* (Silva et al. 1985).

### 15.8.8 Wing Movements

Both sexes of many tephritids and species in related families move their patterned wings in a manner reminiscent of semaphoring (see Sivinski, Chapter 2, on the behaviors of related families; and Headrick and Goeden 1994, for detailed definitions of wing movement modalities). The wings may be brought forward either together (*enantion*; e.g., *A. ludens*) or alternately (*hamation*; e.g., *A. suspensa*, *A. fraterculus*; Robacker and Hart 1985; M. Aluja, unpublished data). The reason for these movements, produced by males and females, mature and immature individuals, and in both sexual and seemingly nonsexual situations, is a mystery. The energy that must go into the repeated motions suggests they have some important function, although their purpose is not necessarily communicative. For example, metabolic heat from flexing could keep flight muscles in optimal condition for escape. However, there is circumstantial evidence that at least some semaphoring has a sexual role. In *A. suspensa*, males greatly increase the rate of semaphoring in the presence of potential mates (Sivinski and Webb 1986). On the average, 35% of a courting male's time is spent in such wing movements, and this increased time investment coincides with a decline in the production of another signal, the calling song, with known sexual significance. Not all *Anastrepha* engage in semaphoring during courtship. When an *A. striata* male detects a female, he extends his proboscis and holds his wings close to his body ("arrowhead" posture described in Section 15.8.3 on male agonistic interactions; Aluja et al. 1993) (Figure 15.4).

### 15.8.9 Trophallaxis

While examples of trophallaxis (males providing an oral substance that is consumed by females) are relatively common in the nonfrugivorous Tephritidae, there is only a single instance described for *Anastrepha* (see Chapters 2 and 28 on the evolution of mating behavior and the behaviors of related families). *Anastrepha striata* females circle in front of courting males and periodically stop to touch their labella with the extended labella of the males (Figure 15.4; Aluja et al. 1993). During these touches a substance is offered by the male that is imbibed by the female. The only other described instance of male-produced substances transferred to mates occurs in *A. suspensa*, where radioactively labeled materials in the ejaculate have been subsequently discovered in unfertilized eggs and female tissue (Sivinski and Smittle 1987). However, there is no evidence that insemination enhances female longevity or that the ejaculates of particular males (i.e., large males) increase fecundity (Sivinski 1993).

### 15.8.10 Behaviors during Coupling

In *A. ludens*, the male palpitates the vertex of the female's head and dorsal anterior thorax, while the female palpitates the substrate. The position of the legs of male *A. ludens* is as follows: foretarsi along anterior thorax of female; mesotarsi along pleural region of midabdomen of female; metatarsi along oviscape or distal portion of female's wings (Dickens et al. 1982; see copulating pair of *A. serpentina* in Color Figure 8*). Male *A. bistrigata* perform more rapid and frequent palpitations of the female's head than do male *A. obliqua* (Silva et al. 1985). In *A. obliqua*, the mounted male releases an oral fluid onto the female's head (Silva et al. 1985).

* Color Figures follow p. 204.

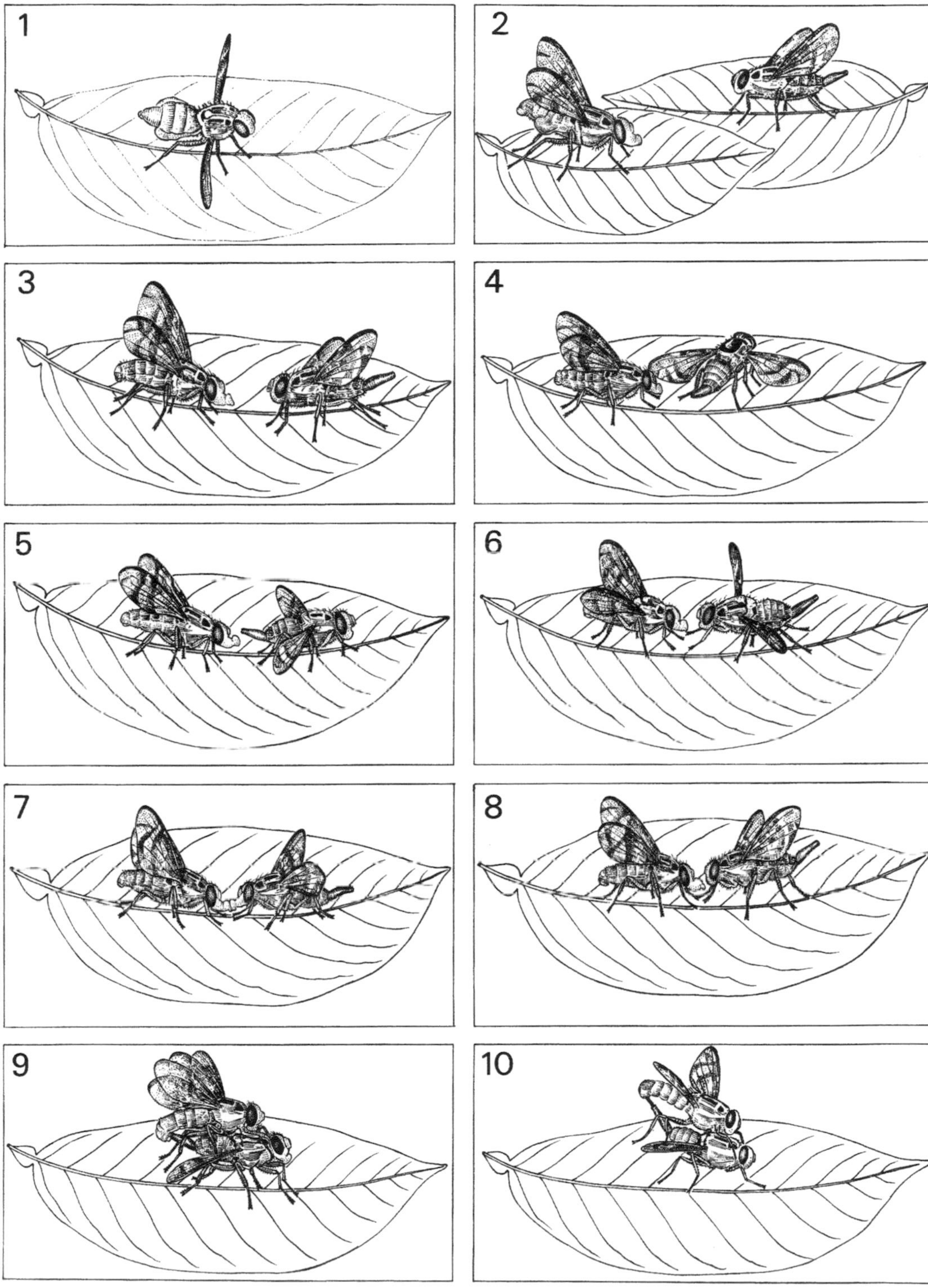


**FIGURE 15.4** Typical sequence of events before and during *A. striata* pair formation and prior to copulation. Note the labella-to-labella contacts (picture 7) and the stiff position of the male while the female circles in front of him. (From Aluja, M. et al., *Ann. Entomol. Soc. Am.* 86: 776–793, 1993. With permission.)

### 15.8.11 Further Criteria for Female Mate Choice

When given a choice between sexually experienced and virgin males, female *A. fraterculus* overwhelmingly preferred previously mated males (De Lima et al. 1994). In *A. suspensa*, large males that would have typically been chosen as mates were temporarily less attractive following mating (the effect dissipated after 2 h; Sivinski 1984). This occurred in spite of no decrease in calling song production. Courtship in *A. striata* includes repeated attempts (mean 13.3 ± 1.9, S.E.) by males to mount females (Aluja et al. 1993). Male persistence might be a quality solicited and then judged by choosy females (Thornhill and Alcock 1983).

In a field-cage study of *A. ludens*, four factors were positively correlated with male mating success: the number of days males survived, the tendency of a male to join a lek, the propensity to engage in fights with other males, and fighting ability (Robacker et al. 1991). All of these characteristics are related to the ability of the male to find and stay within a lek, and it is suggested that females search for areas of concentrated male activity containing highly contested leaves. By arriving at a contested leaf, females have a good chance of finding a male fit enough to have won or held its leaf during recent fighting. Most *A. ludens* matings occurred at the uppermost locations with leks (Robacker et al. 1991). Hendrichs (1986) examined the sexual behavior of *A. suspensa* in a field cage and found that males compete for leaves in the centers of aggregations and that females usually mate in the center as well. However, evidence that females make mate-choice decisions solely on the basis of male locations within leks rather than for the characteristics of particular males may be difficult to reconcile with laboratory studies that suggest individual qualities are important in mate choice.

Aluja et al. (1999) found that adult food quality affected male mating success in field-cage studies aimed at comparing the sexual behavior of *A. ludens*, *A. obliqua*, *A. serpentina,* and *A. striata.* Males that had fed on a protein-rich diet were more competitive than those that only had access to sugar. Furthermore, a poorly fed *A. striata* male that was able to copulate, considerably reduced its partner's life expectancy, fecundity, and the hatchability of the eggs produced by the mated female (Sánchez 1998). By contrast, in the case of *A. ludens*, adult diet had no effect on male mating success (Aluja et al. 1999).

### 15.8.12 Copulation Duration

Information on copula duration of 19 *Anastrepha* species is graphed in Figure 15.5. An interpretation of the tremendous range of copulation durations within this genus is offered by Sivinski et al. (Chapter 28).

### 15.8.13 Female Sexual Maturation Periods

In the few studied species, sexual maturation periods are: *A. serpentina* 14 days (Martínez et al. 1995), *A. striata* 14 to 15 days (Ramírez-Cruz et al. 1996), *A. fraterculus* 17 days (De Lima et al. 1994), *A. pseudoparallela* 18 days, *A. sororcula* 24 days and *A. bistrigata* 26 days (Silva et al. 1985), *A. obliqua* 7 to 19 days (Brazil; Bressan 1996) or 10 to 12 days (Mexico; M. Aluja, unpublished data), and *A. suspensa* 14 days (Dodson 1982).

### 15.8.14 Copulation Frequency

Comparing several *Anastrepha* species, Silva et al. (1985) found that *A. fraterculus*, *A. sororcula* and *A. pseudoparallela* females rarely mated more than once and, in one study, never more than three times. This is in contrast to *A. bistrigata* in which exceptional females mated up to 13 times (Silva et al. 1985). Laboratory-reared *A. fraterculus* females can remate up to eight times (De Lima et al. 1994). Female propensity to remate in *A. suspensa* is dependent on numerous oviposition opportunities being available, suggesting that sexual receptivity returns with the exhaustion of

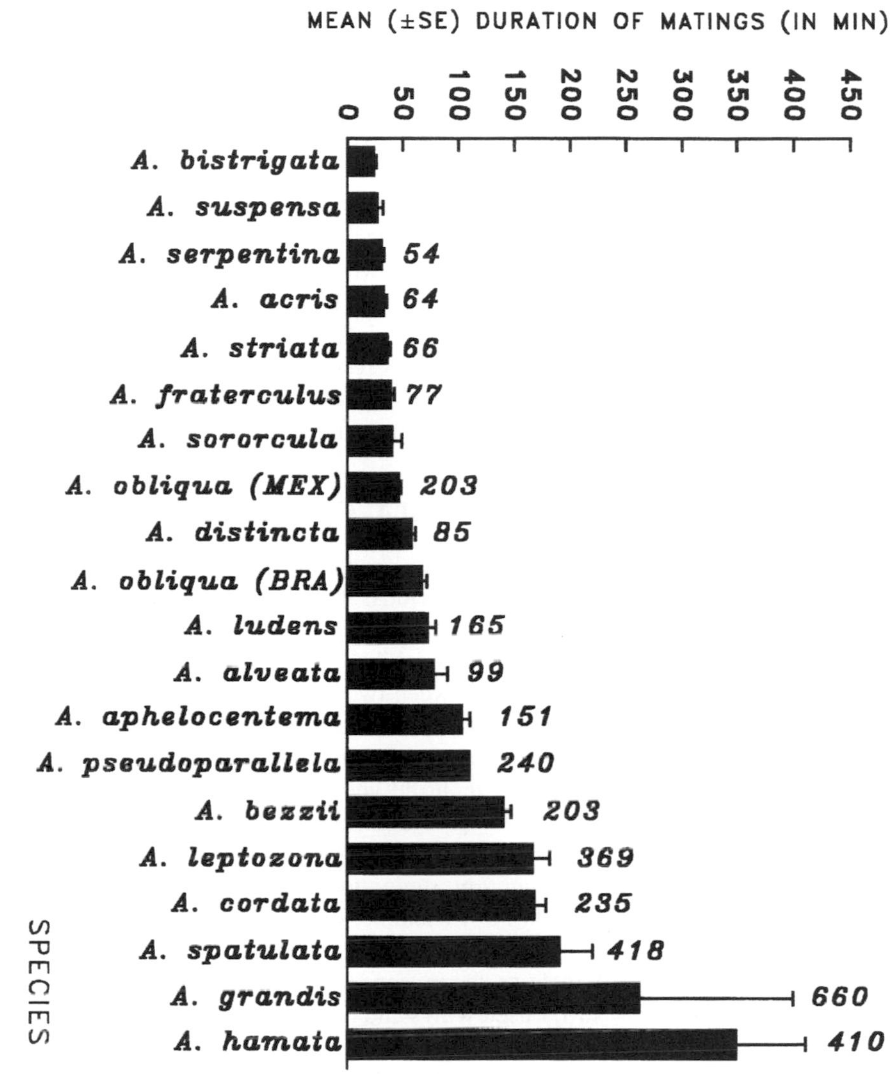


**FIGURE 15.5** Copulation duration of 19 *Anastrepha* species (numbers above bars represent maximum duration recorded). Information stems from the same sources cited in the Figure 15.2 caption.

sperm stores (Sivinski and Heath 1988). Recent work on *A. obliqua* and *A. ludens* revealed the female's refractory period to be regulated by mating status, size of male, feeding history of the male, fertility level of male (irradiated vs. unirradiated), and host availability (Robacker et al. 1985; Trujillo 1998). For example, 91% of the *A. ludens* females fed with protein and sugar mated at least once when in company of males fed with the same diet, whereas only 50% of the females mated when both sexes were fed on sucrose only. In the case of *A. obliqua*, 96% of females mated at least once, independently of the diet of either sex (Trujillo 1998).

## 15.8.15 Alternative Mating Strategies

Since in several species of *Anastrepha*, a minority of competitive/attractive males in leks obtain the majority of copulations, it may behoove the less competitive/attractive males to engage in alternative behaviors that give them a greater opportunity to mate (see Sivinski et al., Chapter 28). The examination of oviposition sites by male *A. suspensa* outside of the calling-lekking period might be such an alternative means to locate females, although unsolicited sexual advances on fruits are rarely successful (Burk 1983; Hendrichs 1986). Robacker et al. (1991) failed to find any evidence of a similar alternative male strategy in *A. ludens*.

## 15.9 SHELTER SEEKING AND RESTING BEHAVIOR

Crawford (1918; 1927) already observed that flies always seek shelter during the heat of the day in the shaded sides of leaves (bottom surface), branches, and fruit. Flies seek out densely foliated trees and stay there until temperatures drop. At least with *A. striata* and *A. obliqua*, both sexes usually rest at the same time (Aluja et al. 1993; Aluja and Birke 1993). Resting occurs throughout the day (Malavasi et al. 1983; Aluja et al. 1993), and is influenced by the physiological state of the individual (e.g., after feeding it is common to see flies resting), its age, and environmental conditions. As noted before, recently emerged individuals tend to rest continuously for several hours while sexually mature ones do so intermittently throughout the day.

## 15.10 CONCLUSIONS AND FUTURE RESEARCH NEEDS

Perusal of the chapters on the behavior of various groups of tephritids reveals the remarkable parallels between *Anastrepha* and many other tephritids. If we borrow a concept from the taxonomists, it appears that there is a general behavioral ground plan in tephritids. Based on this ground plan, at least one species in every tephritid genus exhibits "variations on the general theme." The same is even true when comparing behavior between two unrelated groups of flies: tephritids vs. drosophilids (Kaneshiro, Chapter 32). Here too, one is faced with the remarkable parallels in the behavioral repertoire between, for instance, some *Anastrepha* and *Drosophila* species. For example, when comparing the behavior of flies in the tephritid genera *Blepharoneura* and *Anastrepha*, we are struck by how similar certain body or wing movements are when encountering a conspecific or natural enemy or when alone. We specifically refer to holding raised wings in a stiff position ("arrowhead" posture), or to *hamation*, *enantion*, *supination* (synchronous and asynchronous), and *lofting* (*sensu* Headrick and Goeden 1994). For example, Condon and Norrbom (1994), reporting on the behavior of three sympatric species of *Blepharoneura*, show a picture in which a male of *B. atomaria* (Fabricius) pursues a female as she grazes on a leaf of a host plant. Remarkably, the position of the wings and body — stiff, raised wings and body in "attack position" (= "arrowhead" posture) — is almost identical to that observed in *A. striata* males when detecting a female in close proximity (see Figure 15.4; also see Aluja et al. 1993). To illustrate our point further, in *Anastrepha* there are species such as *A. fraterculus*, where males exhibit a neatly timed rhythm of *asynchronous supination* which is identical to the one exhibited by many flies in the subfamily Tephritinae (Headrick and Goeden 1994). For the evolutionary implications of the above, see Sivinski et al. (Chapter 28).

We would like to highlight the fact that our observations on the behavior of *Anastrepha* seem to mesh well with the phylogenies based on morphological characters and molecular data by Norrbom et al. (Chapter 12) and McPheron et al. (Chapter 13). For example, primitive *Anastrepha* species like *A. cordata* behave quite similarly to species in the probable sister group *Toxotrypana* (e.g., *T. curvicauda*). As discussed in this chapter and Chapter 12, it appears that the more primitive *Anastrepha* species attack latex-producing plants and feed preferentially on seeds. Seed feeding can be thus considered a primitive character. Interestingly, this character appears, on occasion, in derived species (e.g., *A. ludens* feeding on seeds of two of its native hosts). As more comparative information is accrued on the behavior of *Anastrepha*, it will be possible to map behavior onto the existing morphological and molecular phylogenies (Norrbom et al., Chapter 12; McPheron et al., Chapter 13). With this in mind, we recommend that any new study on the many species for which nothing is known consider a detailed quantification of the overall behavioral repertoire, and in particular oviposition and mating behaviors. In the case of oviposition it would be particularly useful to describe and quantify drilling behavior, host-marking behavior or lack thereof, aculeus-cleaning behavior immediately before or after an oviposition bout, clutch size, and host-selection and acceptance behaviors. In the case of mating behavior, the following would be useful information: daily calling and mating rhythms, length of matings, body postures, wing movements (such as *hamation*), pre- and postcopulatory songs or lack thereof (when possible, making recordings for pulse-train analysis), existence of trophallaxis or lack thereof, and pheromone release and deposition behaviors and mechanisms. Of interest here, too, is the mating system (e.g., lek polygyny vs. resource-based polygyny). Aside from behavior, it would also be very useful to record systematically (i.e., using the same methodology) in each species certain basic traits such as sexual maturation periods, survivorship schedules, and gross and net fecundities.

As pointed out in the introduction, *Anastrepha* offers the unique opportunity to compare behaviors among many species. The wide gamut of calling/mating rhythms described here provides impressive evidence of this. It would be highly desirable for future studies to utilize a standard methodology to facilitate comparison and interpretation of behavioral patterns. For example, it would be important to select study species along phylogenetic lines, ideally picking primitive and derived species and always stressing the comparative approach. It is also highly desirable that a greater effort be made to study behavior under natural conditions. This entails several challenges. For example, in tropical evergreen forests, trees are very tall and costly observation towers would be required. But if we are truly to understand behavior in flies of the genus *Anastrepha*, we first need to observe their behavior in nature, thoroughly trying to identify the most important environmental factors that influence it.

A further step we need to take is to start dissecting the genetic basis of behavior. Intraspecific variability in mating success is large and it seems there are grounds to believe that there are alternative mating strategies, such as delaying the onset of calling or participation in leks (M. Aluja, unpublished data). Interspecific crosses could also prove rewarding when trying to identify ancestral behavioral characters. We also need to dissect behavioral rhythms in terms of their genetic vs. environmental components. For example, there are rhythms such as timing of ovipositional activities that seem to be quite plastic and strongly influenced by environmental factors such as temperature. In contrast, there are others, such as calling rhythms, that appear to be mostly driven by internal clocks. Finally, we need to start analyzing behavior in terms of energetics and to address the fascinating topic of cryptic female choice. In these areas, the work on *Ceratitis capitata* by B. Yuval and W. Eberhard and their collaborators has elegantly set the stage for future studies on *Anastrepha* (see Eberhard 1996; Yuval and Hendrichs, Chapter 17).

With respect to the wide gamut of *Anastrepha* calling rhythms mentioned above, there are several selective forces that could have played a role in shaping them. First, temporal isolation may play an important role in reproductive isolation between *Anastrepha* species, especially those occurring in sympatry (Selivon and Morgante 1997; Sivinski et al., Chapter 28). When males of a given *Anastrepha* species are placed in a cage with females of the "wrong" species, they readily attempt mating. Furthermore, at peak hours of mating activity, it is common to observe males attempting to mate with other males. It thus appears that there is no early discrimination by males of potential mating partners. Such discrimination occurs during close-range interactions between the sexes (for a thorough analysis of this see Eberhard, Chapter 18). We thus believe that temporal isolation is a very effective mechanism allowing *Anastrepha* species living in sympatry to reduce energy waste and costly errors in choosing the appropriate mating partner.

To document this, Sivinski et al. (Chapter 28) compared calling rhythms of some *Anastrepha* species living in sympatry and in allopatry. A good example of the first case is represented by *A. suspensa* and *A. obliqua*. The center of evolution for these highly derived species is arguably the Greater Antilles, where both can be currently found. Close analysis of their daily patterns of calling reveals an almost perfect reversed mirror image (see Figure 15.2). While *A. obliqua* calls preferentially in the morning, *A. suspensa* does so in the afternoon hours. A similar pattern can be detected when comparing calling rhythms of *A. striata* and *A. fraterculus*, or by comparing those of *A. obliqua*, *A. ludens,* and *A. serpentina* (see Figure 15.2). In both cases, individuals of each species are likely to encounter each other because they either share a common host (*P. guajava* in the case of *A. striata* and *A. fraterculus*) or hosts, and resting or feeding sites (*M. indica* or mixed *M. zapota*, *Citrus* spp., and *M. indica* orchards in the case of *A. obliqua*, *A. ludens,* and *A. serpentina*). Note that the peak hour of calling never coincides. Another factor that could have played a role in shaping daily calling/mating rhythms is that many species never interact because of the fruiting phenology of their hosts. If such is the case, an overlap in calling rhythm would have no detrimental consequences.

Finally, a word on applied aspects of *Anastrepha* behavior. Based on the overwhelming behavioral diversity and complexity described in this chapter, it becomes clear that any attempt to mass rear *Anastrepha* needs to be coupled with highly sensitive quality control tests. These tests need to be oriented toward detecting consistency in a particular trait (e.g., mating success), as opposed to the narrow snapshots currently obtained with standard quality control tests. Our research clearly indicates that when studying a cohort of individuals over a period of time, what could be a competitive male on day 1 becomes uncompetitive 3 or 4 days later.

In conclusion, flies in the genus *Anastrepha* provide a unique, but to date not fully exploited, opportunity to study behavior using a comparative approach. First, there are potentially more than 200 species to be compared (Norrbom et al., Chapter 12). Second, the behavioral repertoire of those few species studied thus far is remarkably variable and complex. This could render the comparison of, for instance, mating or oviposition behaviors a highly rewarding endeavor. Comparing modalities of, for example, male calling or female aculeus dragging behavior between species could possibly allow the identification of apomorphic or plesiomorphic behaviors. Third, and as aptly discussed by Norrbom et al. (Chapter 12) and McPheron et al. (Chapter 13), the phylogeny of the group is now fairly well understood. This opens up the opportunity to compare formally behaviors between primitive and derived species and to attempt to construct a behavioral tree based on phylogeny. Fourth, and related to the latter, the behavior of *T. curvicauda*, a fly that belongs to the genus that may be the sister group of *Anastrepha* or may even fall within *Anastrepha*, is well studied. This allows interesting comparisons, and can provide clues to the possible evolutionary pathways of certain behaviors. Fifth, the genetics of behavior, a critical area when trying to unravel the evolution of behavior, can also be approached from a comparative perspective, since some species interbreed in the laboratory. This will allow us, at least in the near future, to determine the genetic mechanisms involved in controlling behavior. Sixth, both the sexual and host-marking pheromone systems of *Anastrepha* are currently being studied in depth (see Aluja et al. 1998; Heath et al., Chapter 29;). This will no doubt contribute to our knowledge on the phylogeny of the group, and will also allow us to understand how chemical barriers could have shaped interactions between species in nature.

## ACKNOWLEDGMENTS

Part of the work reported here was technically supported by Andrea Birke, Alejandro Vázquez, Anita Sánchez, Gloria Lagunes, Guadalupe Trujillo, and Everardo Bigurra. We thank Rogelio Macías-Ordóñez, Diana Pérez-Staples, Allen Norrbom, and three anonymous reviewers for their constructive comments on an earlier draft of this chapter. Original research reported here was financially supported by the following donors: Campaña Nacional contra las Moscas de la Fruta (SAGAR-IICA), International Foundation for Science (Project 051/93), Consejo Nacional de Ciencia y Tecnología (CONACyT) (projects D111-903537 and 0436P-N9506), Secretaría de Educación Pública (project DGICSA-902467), U.S. Department of Agriculture (USDA), Office of International

Cooperation and Development (ICD) (Project No. 198-23), USDA-ARS (Agricultural Research Service) (Agreement No. 58-6615-3-025), and Sistema de Investigación Regional del Golfo de México (SIGOLFO-CONACyT) (Project 96-01-003-V). The writing of this chapter was made possible by the generous financial support of the International Organization for Biological Control of Noxious Animals and Plants (IOBC) and the Campaña Nacional contra las Moscas de la Fruta (SAGAR-IICA, Mexico).

## REFERENCES

Alcock, J. 1994. Post-insemination associations between males and females in insects: the mate-guarding hypothesis. *Annu. Rev. Entomol.* 39: 1–21.

Aluja, M. 1993a. The study of movement in tephritid flies: review of concepts and recent advances. In *Fruit Flies: Biology and Management* (M. Aluja and P. Liedo, eds.), pp. 105–113. Springer-Verlag, New York.

Aluja, M. 1993b. Unusual calling behavior of *Anastrepha robusta* flies (Diptera: Tephritidae) in nature. *Fla. Entomol.* 76: 391–395.

Aluja, M. 1994. Bionomics and management of *Anastrepha*. *Annu. Rev. Entomol.* 39: 155–178.

Aluja, M. and A. Birke. 1993. Habitat use by *Anastrepha obliqua* flies (Diptera: Tephritidae) in a mixed mango and tropical plum orchard. *Ann. Entomol. Soc. Am.* 86: 799–812.

Aluja, M., J. Hendrichs, and M. Cabrera. 1983. Behavior and interactions between *Anastrepha ludens* (Loew) and *A. obliqua* (Macquart) on a field cage mango tree I. Lekking behavior and male territoriality. In *Fruit Flies of Economic Importance* (R. Cavalloro, ed.), pp. 122–133. A.A. Balkema, Rotterdam

Aluja, M., M. Cabrera, J. Guillén, and H. Celedonio. 1989. Behaviour of *Anastrepha ludens*, *A. obliqua* and *A. serpentina* (Diptera: Tephritidae) on a wild mango tree (*Mangifera indica*) harbouring three McPhail traps. *Insect Sci. Appl.* 10: 309–318.

Aluja, M., I. Jácome, A. Birke, N. Lozada, and G. Quintero. 1993. Basic patterns of behavior in wild *Anastrepha striata* (Diptera Tephritidae) flies under flied cage conditions. *Ann. Entomol. Soc. Am.* 86: 776–793.

Aluja, M., A. Jiménez, J. Piñero, M. Camino, L. Aldana, M.E. Valdés, V. Castrejón, I. Jácome, A.B. Dávila, and R. Figueroa. 1997. Daily activity patterns and within field distribution of papaya fruit flies (Diptera: Tephritidae) in Morelos and Veracruz, Mexico. *Ann. Entomol. Soc. Am.* 90: 505–520.

Aluja, M., M. López, and J. Sivinski. 1998a. Ecological evidence for diapause in four native and one exotic species of larval-pupal fruit fly (Diptera: Tephritidae) parasitoids in tropical environments. *Ann. Entomol. Soc. Am.* 91: 821–833.

Aluja, M., F. Díaz-Fleischer, A. J. F. Edmunds, and L. Hagmann. 1998b. Isolation, structural determination, synthesis, biological activity and application as pest management agent of the host marking pheromone (and derivatives thereof) of fruit flies (Diptera: Tephritidae) in the genus *Anastrepha*. Invention and Use Patent. Instituto Mexicano de la Propiedad Industrial (IMPI). Registration No. 988732. Mexico City.

Aluja, M., I. Jácome, and R. Macías. Effect of food quality on male mating success in four tropical fruit fly species of the genus *Anastrepha* (Diptera: Tephritidae). *Anim. Behav.*, in press.

Baker, A.C., W.E. Stone, C.C. Plummer, and M. McPhail. 1944. A review of studies on the Mexican fruit fly and related Mexican species. *U.S. Dep. Agric. Misc. Publ.* 531, 155 pp.

Baker, P.S. and A.S.T. Chan. 1991a. Quantification of tephritid fruit fly dispersal. Guidelines for a sterile release programme. *J. Appl. Entomol.* 112: 410–421.

Baker, P.S. and A.S.T. Chan. 1991b. Appetitive dispersal of sterile fruit flies: aspects of the methodology and analysis of trapping studies. *J. Appl. Entomol.* 112: 263–273.

Baker, P.S., A.S.T. Chan, and M.A. Jimeno Zavala. 1986. Dispersal and orientation of sterile *Ceratitis capitata* and *Anastrepha ludens* (Tephritidae) in Chiapas, México. *J. Appl. Ecol.* 23: 27–38.

Barros, M.D. 1986. Estudo da Estrategia de Oviposiçâo de Três Espécies de Tefritídeos (Diptera: Tephritidae) no estado de São Paulo. M.Sc. thesis, Universidade de São Paulo, Brazil, 134 pp.

Barros, M.D., M. Novaes, and A. Malavasi. 1983. Estudos do Comportamento de oviposiçâo de *Anastrepha fraterculus* (Wiedemann, 1830) (Diptera: Tephritidae) em condições naturais e de laboratório. *An. Soc. Entomol. Bras.* 12: 243–247.

Berrigan, D.A., J.R. Carey, J. Guillén, and H. Celedonio. 1988. Age and host effects on clutch size in the Mexican fruit fly, *Anastrepha ludens*. *Entomol. Exp. Appl.* 47: 73–80.

Birke, A. 1995. Comportamiento de Oviposición de la Mosca Mexicana de la Fruta *Anastrepha ludens* (Loew) y Uso de Ácido Giberélico para Disminuir la Susceptibilidad de Toronja *Citrus paradisi* al Ataque de Esta Plaga. B.Sc. thesis, Universidad Veracruzana, Veracruz, México. 84 pp.

Boller, E.F. 1968. An artificial egging device for the European cherry fruit fly *Rhagoletis cerasi*. *J. Econ. Entomol.* 61: 850–852.

Bressan, S. 1996. Desenvolvimiento e potencial reproductivo das femeas de *Anastrepha obliqua* (Macquart, 1835) (Diptera, Tephritidae) em condicoes naturais. *Rev. Bras. Entomol.* 40: 11–16.

Bressan, S. and M. da Costa Teles. 1990. Profundidade de pupacao de *Anastrepha obliqua* (Macquart, 1835) (Diptera: Tephritidae) em tres substratos. *An. Soc. Entomol. Bras.* 19: 471–479.

Bressan, S. and M. da Costa Teles. 1991. Recaptura de adultos marcados de *Anastrepha* spp. (Diptera: Tephritidae) liberados em apenas un ponto do pomar. *Rev. Bras. Entomol.* 35: 679–684.

Burk, T. 1983. Behavioral ecology of mating in the Caribbean fruit fly, *Anastrepha suspensa* (Loew) (Diptera: Tephritidae). *Fla. Entomol.* 66: 330–344.

Burk, T. 1984. Male–male interactions in Caribbean fruit flies, *Anastrepha suspensa* (Loew) (Diptera: Tephritidae): territorial fights and signaling stimulation. *Fla. Entomol.* 67: 542–547.

Burk, T. 1986. Caribbean fruit fly (Diptera: Tephritidae) mating propensity tests — effects of a "checkerboard" cage top. *Fla. Entomol.* 69: 428–430.

Burk, T. and J.C. Webb. 1983. Effect of male size on calling propensity, song parameters, and mating success in Caribbean fruit flies, *Anastrepha suspensa* (Loew) (Diptera: Tephritidae). *Ann. Entomol. Soc. Am.* 76: 678–682.

Calkins, C.O. and J.C. Webb. 1988. Temporal and seasonal differences in movement of Caribbean fruit fly larvae in grapefruit and the relationship to detection by acoustics. *Fla. Entomol.* 79: 409–416.

Celedonio-Hurtado, H., P. Liedo, M. Aluja, J. Guillen, D. Berrigan, and J.R. Carey. 1988. Demography of *Anastrepha ludens*, *A. obliqua* and *A. serpentina* (Diptera: Tephritidae) in Mexico. *Fla. Entomol.* 71: 111–120.

Christenson, L.E. and R.E. Foote. 1960. Biology of fruit flies. *Annu. Rev. Entomol.* 5: 171–192.

Condon, M.A. and A.L. Norrbom. 1994. Three sympatric species of *Blepharoneura* (Diptera: Tephritidae) on a single species of host (*Gurania spinulosa*, Cucurbitaeae): new species and new taxonomic methods. *Syst. Entomol.* 19: 279–304.

Córdoba, B.H. 1999. Efecto del Parasitismo Ejercido por *Doryctobracon crawfordi* (Viereck), *Diachasmimorpha longicaudata* (Ashmead) (Hymenoptera: Braconidae) y *Aganaspis pelleranoi* (Bréthes) (Hymenoptera: Eucoilidae) en la Velocidad y Profundidad de Pupación de Larvas de *Anastrepha ludens* (Loew) (Diptera: Tephritidae). B.Sc. thesis, Universidad Veracruzana, Veracruz, México. 56 pp.

Crawford, D.L. 1918. Insectos y otros invertebrados nocivos. La Mosca de la Naranja de México, *Anastrepha ludens*. *Rev. Agric. (Mexico)* 2: 458–462.

Crawford, D.L. 1927. Investigation of Mexican fruit fly (*Anastrepha ludens* Loew) in Mexico. *Calif. Dep. Agric. Monthly Bull.* 16: 422–445.

Darby, H.H. and E.M. Knapp. 1934. Studies of the Mexican fruit fly *Anastrepha ludens* (Loew). *U.S. Dep. Agric. Tech. Bull.* 444: 1–20.

Dávila, A. 1995. Estudio de Algunos Parámetros Demográficos de la Mosca Mexicana de la Fruta (*Anastrepha ludens* (Loew)) y su Relación con Diferentes Dietas. B.Sc. thesis, Universidad Veracruzana, Veracruz, México. 65 pp.

De Lima, I., P. Howse, and L. Salles. 1994. Reproductive behavior of the South America fruit fly *Anastrepha fraterculus* (Diptera: Tephritidae): laboratory and field studies. *Physiol. Entomol.* 19: 271–277.

Dickens, J.C., E. Solis, and W.G. Hart. 1982. Sexual development and mating behavior of the Mexican fruit fly, *Anastrepha ludens* (Loew). *Southwest. Entomol.* 7: 9–15.

Dillon, P.H., S. Lowrie, and D. McKey. 1983. Disarming the "evil woman"! Petiole constriction by a sphingid larva circumvents mechanical defenses of its host plant, *Cnidoscolus urnes* (Euphorbiaceae). *Biotropica* 15: 112–116.

Dodson, G. 1982. Mating and territoriality in wild *Anastrepha suspensa* (Diptera: Tephritidae) in field cages. *J. Ga. Entomol. Soc.* 17: 189–200.

Drew, R.A.I. and A.C. Lloyd. 1991. Bacteria in the life cycle of tephritid fruit flies. In *Microbial Mediation of Plant-Herbivore Interactions* (P. Barbosa, V.A. Krischnik, and C.G. Jones, eds.), pp. 441–465. John Wiley & Sons, New York.

Drew, R.A.I., A.C. Courtice, and D.S. Teakle. 1983. Bacteria as a natural source of food for adult fruit flies (Diptera: Tephritidae). *Oecologia* 60: 279–284.

Eberhard, W.G. 1994. Evidence for widespread courtship during copulation in 131 species of insects and spiders, and implications for cryptic female choice. *Evolution* 48: 711–733.

Eberhard, W.G. 1996. *Female Control: Sexual Selection by Cryptic Female Choice.* Princeton University Press, Princeton. 501 pp.

Enkerlin, W.R. 1987. Orientación y Dispersión de Poblaciones de la Mosca Mexicana de la Fruta (*Anastrepha ludens*, Loew), Estériles y Silvestres, en el Municipio de Allende, N.L., en el Periodo de Septiembre de 1985 a Agosto de 1986. M.Sc. thesis. Instituto Technológica de Estudios Superiores de Monterrey, Monterrey. 76 pp.

Epsky, N.D. and R.R. Heath. 1993. Food availability and pheromone productions by males of *Anastrepha suspensa* (Diptera: Tephritidae). *Environ. Entomol.* 22: 942–947.

Fitt, G.P. 1984. Oviposition behaviour of two tephritid fruit flies, *Dacus tryoni* and *Dacus jarvisi*, as influenced by the presence of larvae in the host fruit. *Oecologia* 62: 37–46.

Headrick, D.H. and R.D. Goeden. 1994. Reproductive behavior of California fruit flies and the classification and evolution of Tephritidae (Diptera) mating systemas. *Stud. Dipterol.* 1: 194–252.

Hedström, I. 1991. The Guava Fruit Fly, *Anastrepha striata* Schiner (Tephritidae) in Seasonal and Non-seasonal Neotropical Forest Environments. Ph.D. dissertation, Uppsala University, Uppsala.

Hendrichs, J. 1986. Sexual Selection in Wild and Sterile Caribbean Fruit Flies, *Anastrepha suspensa* (Loew) (Diptera: Tephritidae). M.Sc. thesis, University of Florida, Gainesville.

Hendrichs, J. and M.A. Hendrichs. 1990. Mediterranean fruit fly, *Ceratitis capitata* (Diptera: Tephritidae) in nature: location and diel pattern of feeding and other activities on fruiting and non-fruiting host and nonhosts. *Ann. Entomol. Soc. Am.* 83: 632–641.

Hendrichs, J., B.I. Katsoyannos, D.R. Papaj, and R.J. Prokopy. 1991. Sex differences in movement between natural feeding and mating sites and tradeoffs food consumption, mating success and predator evasion in Mediterranean fruit flies (Diptera: Tephritidae). *Oecologia* 86: 223–231.

Hendrichs, J., S. Cooley, and R.J. Prokopy. 1992. Post-feeding bubbling behavior in fluid feeding Diptera: concentration of crop contents by oral evaporation of excess water. *Physiol. Entomol.* 17: 153–161.

Hendrichs, J., S. Cooley, and R.J. Prokopy. 1993. Uptake of plant leachates by apple maggot flies. In *Fruit Flies: Biology and Management* (M. Aluja and P. Liedo, eds.), pp. 173–175. Springer-Verlag, New York. 492 pp.

Hennessey, M.K. 1994. Depth of pupation of Caribbean fruit fly (Diptera: Tephritidae) in soils in the laboratory. *Environ. Entomol.* 23: 1119–1123.

Hernández-Ortiz, V. and M. Aluja. 1993. Lista preliminar de especies del género Neotropical *Anastrepha* (Diptera: Tephritidae) con notas sobre su distribución y plantas hospederas. *Folia Entomol. Mex.* 88: 89–105.

Hernández-Ortiz, V. and R. Pérez-Alonso. 1993. The natural host plants of *Anastrepha* (Diptera: Tephritidae) in a tropical rain forest in Mexico. *Fla. Entomol.* 76: 447–460.

Herrera, A.L. 1900. El gusano de la naranja. *Bol. Soc. Agríc. Mex.* 14: 61–69.

Herrera, A.L. 1905. El gusano de la naranja. *Bol. Com. Parasitol. Agríc.* (México) 2: 307–415.

Hodgson, P.J., J. Sivinski, G. Quintero, and M. Aluja. 1998. Depth of pupation and survival of fruit fly (*Anastrepha* spp.: Tephritidae) pupae in a range of agricultural habitats. *Environ. Entomol.* 27: 1310–1314.

Ihering, H. von. 1901. Laranjas bichadas. *Rev. Agric.* 6: 179.

Jácome, I., 1995. Comportamiento de alimentación de la mosca del zapote *Anastrepha serpentina* (Wiedemann) (Diptera: Tephritidae). B.Sc. thesis, Universidad Veracruzana, Veracruz, México. 91 pp.

Jácome, I., M. Aluja, and P. Liedo. 1999. Impact of adult diet on demographic and population parameters of the tropical fruit fly *Anastrepha serpentina* (Diptera: Tephritidae). *Bull. Entomol. Res.* 89: 165–175.

Landolt, P.J. and K.M. Davis-Hernández. 1993. Temporal patterns of feeding by Caribbean fruit flies (Diptera: Tephritidae) on sucrose and hydrolyzed yeast. *Ann. Entomol. Soc. Am.* 86: 749–755.

Landolt, P.J. and H.C. Reed. 1990. Behavior of the papaya fruit fly (Diptera: Tephritidae): host finding and oviposition. *Environ. Entomol.* 19: 1305–1310.

Landolt, P.J. and J. Sivinski. 1992. Effects of time of day, adult food, and host fruit on incidence of calling by male Caribbean fruit flies (Diptera: Tephritidae). *Environ. Entomol.* 21: 382–387.

Little, H.F. and R.T. Cunningham. 1978. Missing indirect flight muscles in the Mediterranean fruit fly with droopy wing syndrome. *Ann. Entomol. Soc. Am.* 71: 517–518.

López, M., M. Aluja, and J. Sivinski. 1999. Hymenopterous larval-pupal and pupal parasitoids of *Anastrepha* flies (Diptera: Tephritidae) in Mexico. *Biol. Control* 15: 119–129.

Malavasi, A., J.S. Morgante, and R.J. Prokopy. 1983. Distribution and activities of *Anastrepha fraterculus* (Diptera: Tephritidae) flies on host and nonhost trees. *Ann. Entomol. Soc. Am.* 76: 286–292.

Mangan, R.L., D.S. Moreno, and M. Sanchez. 1992. Interaction of wild males and laboratory-adapted females of the Mexican fruit fly (Diptera: Tephritidae) in natural habitats. *Environ. Entomol.* 21: 294–300.

Mankin, R.W., A. Malavasi, and C. Aquino. 1996. Acoustical comparisons of calling songs from *Anastrepha* species in Brazil. In *Fruit Fly Pests: A World Assessment of Their Biology and Management* (B.A. McPheron and G.J. Steck, eds.), pp. 37–42. St. Lucie Press, Delray Beach. 586 pp.

Martínez, I., V. Hernández-Ortiz, and R. Luna-L. 1995. Desarrollo y maduración sexual en *Anastrepha serpentina* (Wiedemann) (Diptera: Tephritidae). *Acta Zool. Mex.* 65: 75–88.

McAlister, L.C., Jr., W.A. McCubbin, G.A. Pfaffman, W.T. Owrey, H.G. Taylor, and I.W. Berryhill. 1941. A Study of the Adult Populations of the West Indian Fruit Fly in Citrus Plantings in Puerto Rico. *Puerto Rico Exp. Station, U.S. Dep. Agric. Bull.* No. 41, 16 pp.

McFadden, M.W. 1964. Recent developments in our knowledge of the biology and ecology of the Mexican fruit fly. In Unpublished Minutes of the Planning Conference for Research on Tropical Fruit Flies in Mexico and Central America. Dec. 16–17, Mexico City, Mexico.

McPhail, M. and N.O. Berry. 1936. Observations on *Anastrepha pallens* (Coq.) reared from wild fruit in the lower Rio Grande Valley of Texas during the spring of 1932. *J. Econ. Entomol.* 29: 405–410.

McPhail, M. and C.I. Bliss. 1933. Observations on the Mexican fruit fly and some related species in Cuernavaca, Mexico in 1928 and 1929. *U.S. Dep. Agric. Circ.* 255, 1–24.

Moreno, D., M. Sánchez, D. Robacker, and J. Worely. 1991. Mating competitiveness of irradiated Mexican fruit fly (Diptera: Tephritidae). *J. Econ. Entomol.* 84: 1227–1234.

Morgante, J.S., A. Malavasi, and R.J. Prokopy. 1983. Mating behavior of wild *Anastrepha fraterculus* (Diptera: Tephritidae) on a caged host tree. *Fla. Entomol.* 66: 234–241.

Morgante, J.S., D. Selivon, V.N. Solferini, and S.R. Matioli. 1993. Evolutionary patterns in specialist and generalist species of *Anastrepha*. In *Fruit Flies: Biology and Management* (M. Aluja and P. Liedo, eds.), pp. 16–20. Springer-Verlag, New York. 492 pp.

Nation, J.L. 1972. Courtship behavior and evidence for a sex attractant in the male Caribbean fruit fly, *Anastrepha suspensa. Ann. Entomol. Soc. Am.* 65: 1364–1367.

Nation, J.L. 1989. The role of pheromones in the mating system of *Anastrepha* fruit flies. In *Fruit Flies: Their Biology, Natural Enemies and Control* (A.S. Robinson and G. Hooper, eds.), pp. 189–205. In *World Crop Pests* (W. Helle, ed.), Vol. 3A. Elsevier Science Publishers, Amsterdam. 369 pp.

Papaj, D.R. and M. Aluja. 1993. Temporal dynamics of host-marking in the tropical tephritid fly, *Anastrepha ludens. Physiol. Entomol.* 18: 279–284.

Pavan, O.H.O. and H.M.L. Souza. 1979. Competition between *Ceratitis capitata* and *Anastrepha fraterculus* in fruit crops. In *Proceedings IX International Congress of Plant Protection* (T. Kommendahl, ed.), pp. 158–162. Washington, D.C.

Peña, J.E., D.E. Howard, and R.E. Litz. 1986. Feeding behavior of *Toxotrypana curvicauda* (Diptera: Tephritidae) on young papaya seeds. *Fla. Entomol.* 69: 427–428.

Perdomo, A.J. 1974. Sex and Aggregation Pheromone Biossays and Mating Observations of the Caribbean Fruit Fly *Anastrepha suspensa* (Loew) under Field Conditions. Ph.D. dissertation, University of Florida, Gainesville. 127 pp.

Picado, C. 1920. La mosca de la guayaba. Publicaciones Entomológicas de Costa Rica. Colegio de Señoritas. Serie 2.

Plummer, C.C., M. McPhail, and J.W. Monk. 1941. The yellow chapote, a native host of the Mexican fruit fly. *U.S. Dep. Agric. Tech. Bull.* 775, 1–12.

Polloni, Y.J. and M.T. Da Silva. 1987. Considerations on the reproductive behavior of *Anastrepha pseudoparallela*, Loew 1873 (Diptera, Tephritidae). In *Fruit Flies: Proceedings of the Second International Symposium, 16–21 September 1986, Colymbari, Crete, Greece* (A.P. Economopoulos, ed.), pp. 295–301. Elsevier Science Publishers, Amsterdam. 590 pp.

Prokopy, R.J., P.D. Greany, and D.L. Chambers. 1977. Oviposition-deterring pheromone in *Anastrepha suspensa. Environ. Entomol.* 6: 463–465.

Prokopy, R.J., A. Malavasi, and J.S. Morgante. 1982. Oviposition deterring pheromone in *Anastrepha fraterculus* flies. *J. Chem. Ecol.* 8: 763–771.

Ramírez-Cruz, A., V. Hernández-Ortiz, and I. Martínez. 1996. Maduración ovárica en la "mosca de la Guayaba" *Anastrepha striata* Schiner (Diptera: Tephritidae). *Acta Zool. Mex.* 69: 105–116.

Robacker, D.C. and W. Hart. 1985. Courtship and territoriality of laboratory-reared Mexican fruit flies, *Anastrepha ludens* (Diptera: Tephritidae), in cages containing host and nonhost trees. *Ann. Entomol. Soc. Am.* 78: 488–494.

Robacker, D.C., S.J. Ingle, and W. Hart. 1985. Mating frequency and response to male-produced pheromone by virgin and mated females of the Mexican fruit fly. *Southwest. Entomol.* 10: 215–221.

Robacker, D.C., R.L. Mangan, D.S. Moreno, and A.M. Tarshis-Moreno. 1991. Mating behavior and male mating success in wild *Anastrepha ludens* (Diptera: Tephritidae) on a field caged host tree. *J. Insect Behav.* 4: 471–487.

Salles, L.A.B. and F.L.C. Carvalho. 1993. Profundidade da localizacao da puparia de *Anastrepha fraterculus* (Wied.) (Diptera: Tephritidae) em diferentes condicoes do solo. *An. Soc. Entomol. Bras.* 22: 299–305.

Sánchez, A. 1998. Efecto de la Dieta y Tamaño de Machos de *Anastrepha ludens* (Loew) y *A. striata* (Schiner) (Diptera: Tephritidae) en su Competitividad Sexual. B.Sc. thesis. Universidad Veracruzana, Veracruz, México.

Santos G.P., N. Anjos, J.C. Zanuncio, and S.L. Asis, Jr. 1993. Danos e aspectos biológicos de *Anastrepha bezzii* Lima, 1934 (Diptera: Tephritidae) em sementes de *Stercula chicha* St. Hill. (Sterculiaceae). *Rev. Bras. Entomol.* 37: 15–18.

Selivon, D. 1991. Algunos Aspectos do Comportamento de *Anastrepha striata* Schiner e *Anastrepha bistrigata Bezzii* (Diptera: Tephritidae). M.Sc. thesis, Instituto da Biociencias, Universidade de São Paulo, São Paulo.

Selivon, D. and J.S. Morgante. 1997. Reproductive isolation between *Anastrepha bistrigata* and *A. striata* (Diptera: Tephritidae). *Braz. J. Genet.* 20: 583–585.

Shaw, J.G., M. Sánchez-Rivello, L.M. Spishakoff, G. Trujillo, and F. López. 1967. Dispersal and migration of tepa-sterilized Mexican fruit flies. *J. Econ. Entomol.* 60: 992–994.

Silva, J.G. da. 1991. Biologia e Comportamento de *Anastrepha grandis* (Macquart, 1846) (Diptera: Tephritidae). M.Sc. thesis, Instituto da Biociencias da Universidade de São Paulo, São Paulo.

Silva, J.G. da and A. Malavasi. 1993. Mating and oviposition behavior of *Anastrepha grandis* under laboratory conditions. In *Fruit Flies: Biology and Management* (M. Aluja and P. Liedo, eds.), pp. 181–184. Springer-Verlag, New York. 492 pp.

Silva, M.T., Y.J. Polloni, and S. Bressan. 1985. Mating behavior of some fruit flies of the genus *Anastrepha* Schiner, 1868 (Diptera; Tephritidae) in the laboratory. *Rev. Bras. Entomol.* 29: 155–164.

Simoes, M.H., Y.J. Polloni, and M.A. Paludetti. 1978. Biologia de algumas especies de *Anastrepha* (Diptera: Tephritidae) em laboratorio. 3rd Latin-American Entomology Congress, Ilheus-Bahia, Brasil.

Sivinski, J. 1984. Effect of sexual experience on male mating success in a lek forming tephritid *Anastrepha suspensa* (Loew). *Fla. Entomol.* 67: 126–130.

Sivinski, J. 1987. Acoustical oviposition cues in the Caribbean fruit fly, *Anastrepha suspensa* (Diptera: Tephritidae). *Fla. Entomol.* 70: 171–172.

Sivinski, J. 1988. What do fruit fly songs mean? *Fla. Entomol.* 71: 463–466.

Sivinski, J. 1989. Lekking and the small-scale distribution of the sexes in the Caribbean fruit fly, *Anastrepha suspensa* (Loew). *J. Insect Behav.* 2: 3–13.

Sivinski, J. 1993. Longevity and fecundity in the Caribbean fruit fly (Diptera: Tephritidae): effects of mating, strain and body size. *Fla. Entomol.* 76: 635–644.

Sivinski, J. and T. Burk. 1989. Reproductive and mating behaviour. In *Fruit Flies: Their Biology, Natural Enemies and Control* (A.S. Robinson and G. Hooper, eds.), pp. 343–351. In *World Crop Pests* (W. Helle, ed.), Vol. 3A. Elsevier Science Publishers, Amsterdam. 369 pp.

Sivinski, J. and G. Dodson. 1992. Sexual dimorphism in *Anastrepha suspensa* (Loew) and other tephritid flies: possible roles of developmental rate, fecundity and dispersal. *J. Insect Behav.* 5: 491–506.

Sivinski, J. and R. Heath. 1988. Effects of oviposition on remating, response to pheromones, and longevity in female Caribbean fruit fly, *Anastrepha suspensa* (Diptera: Tephritidae). *Ann. Entomol. Soc. Am.* 811: 1021–1024.

Sivinski, J. and B. Smittle. 1987. Male transfer of materials to mates in the Caribbean fruit fly, *Anastrepha suspensa* (Diptera: Tephritidae). *Fla. Entomol.* 70: 233–238.

Sivinski, J. and J.C. Webb. 1985. Sound production and reception in the Caribfly, *Anastrepha suspensa* (Diptera: Tephritidae). *Fla. Entomol.* 68: 273–278.

Sivinski, J. and J.C. Webb. 1986. Changes in Caribbean fruit fly acoustic signal with social situation (Diptera: Tephritidae). *Ann. Entomol. Soc. Am.* 79: 146–149.

Sivinski, J., T. Burk, and J.C. Webb. 1984. Acoustic courtship signals in the Caribbean fruit fly, *Anastrepha suspensa* (Loew). *Anim. Behav.* 32: 1011–1016.

Sivinski, J.M., N. Epsky, and R.R. Heath. 1994. Pheromone deposition on leaf territories by male Caribbean fruit flies, *Anastrepha suspensa* (Loew) (Diptera: Tephritidae). *J. Insect Behav.* 7: 43–51.

Smith, P.H. 1979. Genetic manipulation of the circadian clock's timing of sexual behaviour in the Queensland fruit flies, *Dacus tryoni* and *Dacus neohumeralis*. *Physiol. Entomol.* 4: 71–78.

Solferini, V.N. 1990. Interacoes entre Bactérias e *Anastrepha* (Diptera: Tephritidae). Ph.D. dissertation, Universidade de São Paulo, Saõ Paulo.

Sugayama, R.L., E.S. Branco, A. Malavasi, A. Kovaleski, and I. Nora. 1997. Oviposition behavior of *Anastrepha fraterculus* in apple and diel pattern of activities in an apple orchard in Brazil. *Entomol. Exp. Appl.* 83: 239–245.

Thornhill, R. and J. Alcock. 1983. *The Evolution of Insect Mating Systems*. Harvard University Press, Cambridge. 547 pp.

Trujillo, G. 1998. Efecto de la Dieta, Tamaño de los Adultos, Presencia de Hospedero y Condición Fértil o Estéril de los Machos en el Número de Apareamientos y Periodo Refractorio de Hembras de *A. ludens* (Loew) y *A. obliqua* (Macquart) (Diptera: Tephritidae). B.Sc. thesis, Universidad Veracruzana, Veracruz, México.

Webb, J.C. and P.J. Landolt. 1984. Detecting insect larvae in fruit by vibrations produced. *J. Environ. Sci. Health A19* 3: 367–375.

Webb, J.C., J.L. Sharp, D.L. Chambers, J.J. McDow, and J.C. Benner. 1976. Analysis and identification of sounds produced by the male Caribbean fruit fly, *Anastrepha suspensa*. *Ann. Entomol. Soc. Am.* 69: 415–420.

Webb, J.C., T. Burk, and J. Sivinski. 1983. Attraction of female Caribbean fruit flies, *Anastrepha suspensa* (Diptera: Tephritidae), to the presence of males and male-produced stimuli in field cages. *Ann. Entomol. Soc. Am.* 76: 996–998.

Webb, J.C., J. Sivinski, and C. Litzkow. 1984. Acoustical behavior and sexual success in the Caribbean fruit fly, *Anastrepha suspensa* (Loew) (Diptera: Tephritidae). *Environ. Entomol.* 13: 650–656.

# Section V

## Subfamily Dacinae

# 16 Phylogeny of the Genus *Ceratitis* (Dacinae: Ceratitidini)

*Marc De Meyer*

## CONTENTS



## 16.1 INTRODUCTION

The systematic position, species composition of, and relationships within the genus *Ceratitis* MacLeay are confusing topics. Despite the fact that the most-studied fruit fly, *C. capitata* (Wiedemann), the Mediterranean fruit fly, belongs to this genus, our knowledge of other species within this higher taxon is very limited and it has never been the subject of a comprehensive review (White 1989). The only recent taxonomic studies were limited to the faunas of particular areas like the Malagasy Subregion (Hancock 1984) and Zimbabwe (Hancock 1987), or revised subgeneric positions and identification keys (Freidberg 1991; Hancock 1991; Hancock and White 1997). The genus *Ceratitis*, as it stands today, is a composite of six subgenera, comprising 78 species. The exact delimitation of the subgenera and which species belong to what subgenus has been and still is subject to continuous change. *Ceratitis* is native to the Afrotropics, although a few species are adventive elsewhere: *C.* (*Ceratitis*) *capitata* nearly worldwide, and *C.* (*Pterandrus*) *rosa* Karsch in Mauritius and Réunion. However, considering the growing production and trade of fruit, especially in African countries, there is great danger of further introductions with grave consequences for domestic fruit-growing industries.

A proper systematic revision of *Ceratitis* was therefore deemed necessary; however, this is an ongoing study. So far, the species traditionally placed in the subgenera *Pardalaspis* Bezzi and *Ceratalaspis* Hancock have been revised (De Meyer 1996; 1998). At the moment the subgenera *Ceratitis* (*s. str.*) and *Pterandrus* Bezzi are being studied, but only some species normally placed in these subgenera have been investigated. Therefore, the conclusions presented here are preliminary and definite taxonomic changes will only be made after the whole revision is finished. Here, I will point out some phylogenetic lineages, delimit species groups, and discuss the possible monophyly of some recognized taxa.

0-8493-1275-2/00/$0.00+$.50

## 16.2 RELATIONSHIPS WITHIN CERATITIDINI

The genus *Ceratitis* is usually classified in the subfamily Dacinae (or in the Dacini where this higher taxon is treated as a tribe of Trypetinae). The exact position of this higher taxon within the family Tephritidae is discussed in detail elsewhere in this book (Korneyev, Chapter 4; Han and McPheron, Chapter 5). Within the Dacinae, three subgroups are usually recognized: Dacini (or Dacina), Ceratitidini (or Ceratitidina), and Gastrozonini (or Gastrozonina). Different classifications exist for these subgroups, although without much evidence of phylogeny or cladistic analysis. The Dacini is usually considered a separate group, whereas Ceratitidini and Gastrozonini are either grouped together as an equivalent group to Dacini, or as two separate entities at the same taxonomic level with Dacini (throughout this chapter, all three groups are treated as separate entities). In general, however, the Gastrozonini are ranked closer to the Ceratitidini, both being morphologically very similar and sharing the following characters (after White and Elson-Harris 1992): scutum with dorsocentral setae, usually placed well forward; katepisternum usually with a seta; scutellum usually swollen dorsally; wing with basal medial cell (bm) about as wide as basal cubital cell (bcu); and vein $Cu_2$ along anterior side of basal cubital cell extension usually sinuous and only reaching about a quarter of the way to the wing margin from the basal part of the basal cubital cell. No unambiguous characters seem to diagnose both groups. White and Elson-Harris (1992) give as a differentiating character for the two groups the first flagellomere of the antenna being evenly rounded apically in Ceratitidini, but with a dorsoapical point in Gastrozonini. This character does not hold for several of the Asian Gastrozonini (Hancock, personal communication). Gastrozonini have in general a longer aristal plumosity (although some Gastrozonini have shorter, more pubescent hairs) (Hancock, personal communication), whereas Ceratitidini have an almost bare to short plumose arista (except for some *Ceratalaspis* species). Hancock (1999) mentions a biological difference, with Ceratitidini larvae having fruits or flower buds (from a wide spectrum of plant families) as hosts, while Gastrozonini larvae live in grasses (the Asian representatives being found in bamboo species; African species are reported from *Panicum*, *Sorghum*, *Zea*, and possibly *Saccharum*).

Hancock and White (1997) recognized the following nine genera as belonging to the *Ceratitis* genus group in Africa: *Capparimyia* Bezzi, *Carpophthoromyia* Austen, *Ceratitis* MacLeay, *Eumictoxenus* Munro, *Neoceratitis* Hendel, *Nippia* Munro, *Perilampsis* Bezzi, *Trirhithrum* Bezzi, and *Xanthorrachista* Hendel. Other additional African genera usually listed under Ceratitidini are *Ceratitoides* Hendel, *Clinotaenia* Bezzi, and *Leucotaeniella* Bezzi. Hancock (1999) discusses their systematic position. *Clinotaenia* and *Leucotaeniella* are probably related to *Bistrispinaria*, suggesting they all breed in grasses and belong to the Gastrozonini. The position of *Ceratitoides* is uncertain.

I was able to study representatives of all of these genera (except *Neoceratitis* and *Nippia*). The monophyly of and relationships within this group are not well established and could be a large topic of study on their own. I can, however, indicate some subgroups and point out the closest relatives of *Ceratitis*. The genera *Perilampsis*, *Trirhithrum,* and *Carpophthoromyia* seem to form a separate subgroup supported by the following possible synapomorphy: the dorsum and abdomen mainly shining black (primitive state not shining, dusted, with brownish ground color). According to Hancock (personal communication), *Nippia* probably also belongs here. Within this subgroup *Perilampsis* and *Carpophthoromyia* can be grouped together based on the synapomorphy of the presence of an oblique, white line running from the postpronotal lobe over the anepisternum. *Eumictoxenus* has the abdominal tergites predominately shining black and the scutellum completely white (as in *Perilampsis* and *Carpophthoromyia*), and is possibly related to this subgroup. The exact relationship of the genus *Xanthorrachista* to the other genera cannot be immediately established.

The genera *Capparimyia* and *Ceratitis* seem to be related. As synapomorphies the following are suggested: (1) basic wing pattern with bands showing a varying degree of brown and yellow patches, costal (marginal) band, discal and subapical crossbands (cubital band) present with variations on this pattern (primitive state: wing base without patches and without typical wing banding of costal, discal bands and subapical crossbands); and (2) scutellum yellow white, with two or three apical spots (joined or separate) (primitive state: scutellum unicolorous).

These groupings are only preliminary, however, and a thorough study of character distributions in outgroups and character state polarity is needed before a more rigorous analysis of relationships can be presented.

## 16.3 THE GENUS *CERATITIS* AS AN ENTITY

The genus *Ceratitis* as an entity can be differentiated by the synapomorphy of having three apical scutellar spots (albeit with a number of exceptions, which arguably could be considered evolved character states from the ground plan), whereas *Capparimyia* has only two apical scutellar spots. The attraction to terpinyl acetate could probably also be a synapomorphy for *Ceratitis* but has not been checked for all subgenera yet (Hancock, personal communication). In addition, there is the biological difference that *Capparimyia* species attack buds (solely of Capparidaceae), while *Ceratitis* species are found in fruits of a wide spectrum of plant families.

As mentioned earlier, a number of subgenera are included in *Ceratitis*. Most of them were originally described as separate genera but later incorporated into *Ceratitis*. A short overview of the taxonomic history is given below.

*Ceratitis* was described by MacLeay in 1829 with type species *Ceratitis citriperda* MacLeay, a name now synonymized with *C. capitata*. Subsequently, a subgenus, *Pinacochaeta*, was described by Munro (1933) with *C. pinax* Munro as type species. The genus *Pardalaspis* was described by Bezzi in 1918 with type species *Tephritis punctata* Wiedemann. In the same publication, Bezzi also described the genus *Pterandrus*, with *C. rosa* as type species. The first species now placed in *Ceratitis* was described by Wiedemann (1824): *Tephritis capitata* from 'India orient' (possibly referring to the East Indies, but see Pont 1995 for discussion on the exact position of this type locality). Further early descriptions were by Guérin-Méneville (1843), Walker (1849; 1853), Karsch (1887), Bigot (1891), and Graham (1908). Most *Ceratitis* species were described by Bezzi or Munro in various papers between 1909 and 1957.

The three genera mentioned above were considered as distinct genera and listed as such in the Tephritidae chapter of the *Catalogue of the Diptera of the Afrotropical Region* (Cogan and Munro 1980). All (sub)generic changes subsequent to Cogan and Munro (1980) are summarized in Table 16.1. The first major changes were suggested by Hancock (1984), who placed *Petrandrus* and *Pardalaspis* as subgenera of *Ceratitis*, and proposed a new subgenus, *Ceratalaspis*, to accommodate several species formerly included in *Pardalaspis*. He also synonymized the subgenus *Pinacochaeta* with *Ceratitis* (*s. str.*) and transferred a number of species to different subgenera.

Hancock (1987) placed *Hoplolophomyia* Bezzi (1926) (replacement name for *Hoplolopha* Bezzi 1920), as a subgenus of *Ceratitis* and put all species belonging to the former, except the type species *Hoplolophomyia cristata* Bezzi, under *Ceratalaspis*. *Hoplolophomyia* remains a monotypic subgenus. Freidberg (1991) followed Hancock (1984) except for changing the subgeneric placement of a few species. Recently, Hancock and White (1997) placed the monotypic genus *Acropteromma* Bezzi (1926) as a subgenus of *Ceratitis* and proposed a number of additional subgeneric species transfers.

Thus far, the genus *Ceratitis* comprises 78 species in six subgenera. The subgenus *Ceratalaspis* is the largest with 34 species. The subgenus *Pterandrus* has 24 species, *Pardalaspis* 10, *Ceratitis* (*s. str.*) 8, and the subgenera *Hoplolophomyia* and *Acropteromma* one species each. All species indicated with an asterisk (*) in Table 16.1 have not changed subgeneric classification since Cogan and Munro (1980), not taking into account the taxonomic ranking change from genus to subgenus. These number only 39, 14 of which were described after 1980, so that only 25 species are still in the same subgeneric combination as in 1980. This reflects the unstable classification within the genus.

**TABLE 16.1**
**List of All Species Included in the Genus *Ceratitis*, with Indication of Generic or Subgeneric Changes from 1980 to Present Day**

| | 80 | 84 | 85 | 87 | 91 | 96 | 97 | 98 |
|---|---|---|---|---|---|---|---|---|
| ***Ceratalaspis*** | | | | | | | | |
| *aliena* | Pl | Cl | | | | | | |
| **andranotobaka* | | Cl | | | | | | |
| *antistictica* | Pl | Ct | | | Cl | | | |
| *argenteobrunnea* | Pl | Cl | | | | | | |
| *brucei* | Pl | Cl | | | | | | |
| *connexa* | Hp | | | Cl | | | | |
| *contramedia* | Pl | Cl | | | | | | |
| *cosyra* | Pl | Cl | | | | | | |
| *discussa* | Pl | Cl | | | | | | |
| *divaricata* | Hp | | | Cl | | | | |
| *dumeti* | Pl | Cl | | | | | | |
| *epixantha* | Pl | Cl | | | | | | |
| *grahami* | Pl | Cl | | | | | | |
| *guttiformis* | Pl | Cl | | | | | | |
| *lentigera* | Pl | Cl | | | | | | |
| *lineata* | Pl | Cl | | | | | | |
| *lunata* | Pl | Cl | | | | | | |
| *marriotti* | Pl | Cl | | | | | | |
| *morstatti* | Pl | Cl | | | | | | |
| *nana* | Pl | Cl | | | | | | |
| *ovalis* | Pl | Cl | | | | | | |
| *quinaria* | Pl | Cl | | | | | | |
| *scaevolae* | Pl | Cl | | | | | | |
| *silvestrii* | Pl | Cl | | | | | | |
| *simi* | Pl | Cl | | | | | | |
| *stictica* | Pl | Cl | | | | | | |
| *striatella* | Pl | Cl | | | | | | |
| *turneri* | Pl | Cl | | | | | | |
| *venusta* | Hp | | | Cl | | | | |
| **neostictica* | | | | | | | | Cl |
| **hancocki* | | | | | | | | Cl |
| **mlimaensis* | | | | | | | | Cl |
| **paradumeti* | | | | | | | | Cl |
| **sucini* | | | | | | | | Cl |
| ***Ceratitis*** | | | | | | | | |
| **brachychaeta* | | | | | Ct | | | |
| **caetrata* | Ct | | | | | | | |
| **capitata* | Ct | | | | | | | |
| **catoirii* | Ct | | | | | | | |
| *cornuta* | Pt | Ct | | | | | | |
| **malgassa* | Ct | | | | | | | |
| **manjakatompo* | | Ct | | | | | | |
| **pinax* | Ct | | | | | | | |

| | 80 | 84 | 85 | 87 | 91 | 96 | 97 | Unpl |
|---|---|---|---|---|---|---|---|---|
| ***Pardalaspis*** | | | | | | | | |
| **bremii* | Pl | | | | | | | |
| **cuthbertsoni* | Pl | | | | | | | |
| **ditissima* | Pl | | | | | | | |
| **edwardsi* | Pl | | | | | | | |
| **punctata* | Pl | | | | | | | |
| **hamata* | | | | | | Pl | | |
| **serrata* | | | | | | Pl | | |
| **munroi* | | | | | | Pl | | |
| **zairensis* | | | | | | Pl | | |
| **semipunctata* | | | | | | Pl | | |
| ***Pterandrus*** | | | | | | | | |
| **acicularis* | Pt | | | | | | | Pt (A) |
| **anonae* | Pt | | | | | | | Pt (A) |
| **colae* | Pt | | | | | | | Pt (A) |
| **flexuosa* | Pt | | | | | | | Pt (A) |
| **fulicoides* | Pt | | | | | | | Pt (A) |
| **lepida* | Pt | | | | | | | Pt (A) |
| *melanopus* | Pl | Cl | | | | | Pt | Pt (A) |
| **penicillata* | Pt | | | | | | | Pt (A) |
| **pinnatifemur* | Pt | | | | | | | Pt (A) |
| **rosa* | Pt | | | | | | | Pt (A) |
| **rubivora* | Pt | | | | | | | Pt (A) |
| **tripteris* | Pt | | | | | | | Pt (A) |
| *bicincta* | Tm | | | | | | Pt | Pt (B) |
| *chirinda* | | | Ta | | | | Pt | Pt (B) |
| **curvata* | Pt | | | | | | | Pt (B) |
| *faceta* | Tm | | | | | | Pt | Pt (B) |
| *gravinotata* | Pt | Ta | | | Pt | | | Pt (B) |
| *inauratipes* | Tm | | | | | | Pt | Pt (B) |
| *lobata* | Pl | Pt | | | | | | Pt (B) |
| *pedestris* | Pl | Pt | | | | | | Pt (B) |
| **podocarpi* | Pt | | | | | | | Pt (B) |
| *querita* | Tm | | | | | | Pt | Pt (B) |
| *roubaudi* | Pl | Cl | | | | | Pt | Pt (B) |
| **tananarivana* | | Pt | | | | | | Pt (?) |
| ***Hoplolophomyia*** | | | | | | | | |
| **cristata* | Hp | | | | | | | |
| ***Acropteromma*** | | | | | | | | |
| **munroanum* | Ac | | | | | | | |
| ***Non-Ceratitis*** | | | | | | | | |
| *cyanescens* | Pl | Ta | | | | | | |

*Note*: Species indicated with * have not changed in classification since Cogan and Munro (1980). Literature references: 80 = Cogan and Munro 1980; 84 = Hancock 1984; 85 = Hancock 1985; 87 = Hancock 1987; 91 = Freidberg 1991; 96 = De Meyer 1996; 97 = Hancock and White 1997; 98 = De Meyer 1998; Unpl = Hancock, unpublished data. Ct = *Ceratitis* (*s. str.*), Cl = *Ceratalaspis*, Pl = *Pardalaspis*, Pt = *Pterandrus*, Hp = *Hoplolophomyia*, Ac = *Acropteromma*, Tm = *Trirhithrum*, Ta = *Trirhithromyia*.

## 16.4 SUBGENERIC CHARACTERS

The main problem in the study of *Ceratitis* classification has been to find characters to delimit the higher taxa (i.e., the subgenera) appropriately and accurately. A number of such characters have been proposed in the recent literature (Hancock 1984; Freidberg 1991; Hancock and White 1997). The emphasis in this chapter will be on the subgenera *Ceratitis*, *Pterandrus*, *Ceratalaspis,* and *Pardalaspis*, since they comprise the majority of the species. The subgenera *Hoplolophomyia* and *Acropteromma* are only briefly discussed below.

The following characters were used by Hancock (1984) to differentiate the subgenera when he incorporated *Pterandrus* and *Pardalaspis* into *Ceratitis* and proposed the new subgenus *Ceratalaspis*:

*Ceratitis* (*s. str.*): Head with modifications of anterior orbital bristle (not in all species), sometimes with prominent spatulate terminal process. Scutum red brown to gray with distinct black patches. Wing broad, with yellow areas in banding extensive. Males attracted to trimedlure [no mention of feathering of legs].

*Pterandrus*: Head with orbital bristles not modified. Scutum brown to gray without distinct patches. Wing narrow; markings dark, except in *C.* (*Pterandrus*) *tripteris* (Munro) and yellow areas much reduced. Males attracted to trimedlure. Legs with black feathering, ochraceous in *C.* (*Pterandrus*) *fulicoides* (Munro) or with black/white pattern.

*Pardalaspis*: Head with orbital bristles not modified, but frons with silvery areas. Scutum brown to gray, without distinct patches; postpronotal lobe uniformly brown. Wing with markings dark, yellow areas much reduced [no mention of wing size]. Males attracted to methyl eugenol [trimedlure in text is error]. Legs without feathering. Abdomen with spots. In general larger in size.

*Ceratalaspis*: Head with orbital bristles not modified. Scutum orange brown to red brown, with distinct black patches (reduced in some species). Wing narrow; with yellow areas extensive. Legs without feathering. Abdomen without spots (except in two species that have a different pattern than in *Pardalaspis*). No response to trimedlure or methyl-eugenol.

Freidberg (1991) elaborated the distinction between the subgenera *Pterandrus* and *Ceratitis* slightly:

*Ceratitis*: Head with anterior orbital bristle not black and acuminate, often blunt or spatulate; arista with short rays mainly dorsobasally but also a few present ventrobasally. Scutum with black and yellow pattern; including round or triangular black spot at mesal end of transverse suture. Wing broad, with banding yellow, bordered by brown. Fore femur with yellow bushy ornamentation; other legs neither feathered nor bushy.

*Pterandrus*: Head with orbital bristles black, not modified; arista with moderately long rays dorsally as well as ventrally. Scutum dull gray or brown; in most species with poorly defined, poorly contrasting spots on mesal end of transverse suture. Wing markings brown to black [no mention of wing size]. Legs with feathering usually black; midleg usually feathered or otherwise ornamented, foreleg and hindleg sometimes ornamented.

Freidberg used the presence of modified orbital setae as a diagnostic character for *Ceratitis* (*s. str.*) and therefore excluded *C. antistictica* Bezzi. However, he pointed out that some species do not fall completely within the limits of either of these two subgenera indicating intergradation.

Hancock and White (1997) and Hancock (unpublished data) differentiated two subgroups within the subgenus *Pterandrus*: (1) typical *Pterandrus* (group A of Hancock, unpublished data) with postpronotal lobe yellow, without black spot; and scutellum broadly yellow basally; (2) group B (Hancock, unpublished data) with postpronotal lobe with black spot; scutellum with extensive, often broadly trilobed black area, or entirely black; wing without posterior apical crossband (medial band), except in *C.* (*Pterandrus*) *gravinotata* (Munro); scutum black, with extensive gray tomentosity and two yellow or white prescutellar spots. Hancock and White (1997) listed the species belonging to group B but not the others. Hancock (unpublished data) lists a number of species for group A as indicated in Table 16.1. According to Hancock (personal communication), *C.* (*Pterandrus*) *tananarivana* Hancock appears not to belong in either group.

De Meyer (1998) revised the subgenus *Ceratalaspis* and differentiated a number of species groups, without assigning any formal taxonomic ranking to them or making any subgeneric changes:

1. *aliena* (Bezzi) group: Wing with costal band not joined to discal crossband, subapical crossband and posterior apical crossband joining costal band; scutellum with apical spots restricted to apical margin; postpronotal lobe unspotted; arista with short hairs; aculeus tip dentate, with one or two sets of subapical protuberances.
2. *cosyra* (Walker) group: Postpronotal lobe spotted; scutum with small spots present on mesal end of transverse suture, around base of dorsocentral and/or prescutellar acrostichal bristles; aculeus tip simple, acute.
3. *andranotobaka* Hancock group: Wing with subapical crossband free, posterior apical crossband attached to costal band; postpronotal lobe unspotted; prescutellar acrostichal spots large, connected to dorsocentral spots; aculeus tip with broad and rounded indentations.
4. *melanopus* (Hering) group: Wing with dark banding; scutal pattern dark; aculeus tip simply pointed. After this article was accepted for publication, Hancock and White (1997) moved *melanopus* and *roubaudi* (Bezzi) (forming the *melanopus* group) to *Pterandrus*.
5. *stictica* Bezzi group: Arista with medium long to distinctly plumose hairs; scutal pattern with distinct and extensive spots, presutural spot reaching as far as outer scapular seta, along median line, prescutellar acrostichal spots reaching dorsocentral bristles; aculeus tip simple, acute.
6. *dumeti* Munro group: Scutal pattern with presutural spots not reaching postpronotal lobe, prescutellar acrostichal spots not reaching dorsocentral bristles; aculeus tip with small lobes.
7. *guttiformis* Munro group: Spots extensive and surrounding postpronotum and along medial line; aculeus tip very broad, with heavy indentations.
8. *morstatti* Bezzi group: Scutal pattern without extensive spotting; postpronotal lobe unspotted; aculeus tip simple, acute.

A number of *Ceratalaspis* species could not be placed in any of these groups.

*Hoplolophomyia* is now considered to be a monotypic subgenus that includes the single species *C.* (*Hoplolophomyia*) *cristata*. In earlier classifications (like Cogan and Munro 1980) other species were included, but they have all been transferred to *Ceratalaspis*. *Ceratitis* (*Hoplolophomyia*) *cristata* is a large species; the wing pattern shows the normal banding except for the presence of a posterior apical crossband connected to the costal band, and subapical crossband connected to discal crossband, as in *C.* (*Ceratalaspis*) *connexa* (Bezzi) and *C.* (*Ceratalaspis*) *argenteobrunnea* Munro. The abdomen is unspotted, showing the pattern found in *Pterandrus*, with dark bands. An autapomorphic character is the cristate frons in the male. The relationship of this species within *Ceratitis* is uncertain, but it seems to truly belong in the genus.

*Acropteromma* is also a monotypic subgenus, including only the aberrant *C.* (*Acropteromma*) *munroanum* (Bezzi). It shows a number of character states also found in other *Ceratitis* species, like scutum without distinct patches; postpronotal lobe not distinctly spotted but only with slightly darker coloration; and scutellum with three apical spots. The abdominal banding is yellow with silvery bands. The wing pattern is almost completely lacking; only a distinct apical spot in the male wing is present (indistinct in females). Most striking, however, is the presence of bushy yellow orange feathering on the posterior side of the forefemur. This character is similar to that found in *Ceratitis* (*s. str.*), but the hairs are much shorter. In general appearance *C.* (*Acropteromma*) *munroanum* resembles *C.* (*Ceratalaspis*) *divaricata* (Munro) (which is itself an aberrant species among *Ceratalaspis* species). Hancock and White (1997) proposed it to be a derived species close to the *Pterandrus* group A, but this needs further study. At the moment this seems questionable because of the lack of dark abdominal bands, the absence of distinct spots on the scutellum, and the difference in the shape of the feathering of the forefemur.

## 16.5 PRELIMINARY CLADISTIC ANALYSIS

A preliminary cladistic analysis was performed on a number of representatives of the different subgenera. In total, 53 species were included in the analysis, including *C.* (*Hoplolophomyia*) *cristata*. An hypothetical outgroup (based on the presumed plesiomorphic character states of the characters involved) was used since the actual sister group of *Ceratitis* is not established. *Capparimyia* is a likely candidate (see above) but could also be an apomorphic group within *Ceratitis*. A cladistic revision of all taxa within the Ceratitidini is required to confirm this. Not all species from the main subgenera (*Ceratitis s.str.*, *Pterandrus*, *Pardalaspis*, and *Ceratalaspis*) were included because a number of them have not yet been studied. Others were excluded because of missing character data (e.g., only female known) or having an identical character coding as other species, for example, *C.* (*Pardalaspis*) *ditissima* (Munro) and *C.* (*Pardalaspis*) *edwardsi* (Munro) were excluded because they have the identical character states as *C.* (*Pardalaspis*) *punctata*. It is not the intention to give a detailed analysis for all species but merely to find the main species groups and to see how they are related to each other. Therefore, unique apomorphic characters (autapomorphies) were excluded.

The cladistic analysis was performed with Hennig86 (Farris 1988), using the options mh*; bb*. Mhennig (mh*) constructs several trees, each by a single pass, adding the terminals in several different sequences, and retaining the shortest trees found. Then it applies branch-swapping to each of the initial trees, retaining no more than one tree for each initial one. Option bb* applies extended branch-swapping to the trees in the current file, producing a new tree file and thereby generating all trees it can find. Hennig86 allows other tree-constructing methods that ensure finding all most parsimonious trees, but they proved to be too time-consuming because of the large amount of incongruity in the data matrix. Farris (1988) suggests in such cases the mh*, bb* combination. The produced profile was then imported in COMPONENT version 2.0 (Page 1993) to construct strict consensus and majority-rule consensus trees.

In a preliminary analysis only those characters were included that were mentioned in earlier publications to differentiate subgenera (especially Hancock 1984; Freidberg 1991; Hancock and White 1997). In total, 17 characters were used (Table 16.2, indicated with *). The matrix of character states found in all species studied is shown in Table 16.3. The mh* option produced nine shortest trees with length 64, consistency index (ci) 0.35, and retention index (ri) 0.80. The bb* option produced an overflow profile with 1746 trees of length 63, ci 0.36, and ri 0.80. The strict consensus tree option produced a largely unresolved tree (Figure 16.1). A strict consensus tree only includes those clades that are found in all trees present in the profile (in this case in all 1754 trees). A number of clades are hypothesized in this tree: (1) *Pardalaspis* + *C.* (*Ceratalaspis*) *divaricata*; (2) *Pterandrus* A group (except *C.* (*Pterandrus*) *fulicoides*) + *C.* (*Ceratalaspis*) *morstatti*; (3) *C.* (*Ceratalaspis*) *andranotobaka* group + *C.* (*Ceratalaspis*) *aliena* group + *C.* (*Ceratalaspis*) *connexa* + *C.* (*Hoplolophomyia*) *cristata*; (4) *Ceratitis* (*s. str.*) + *Pterandrus* B group + *C.* (*Ceratalaspis*) *stictica* group + *C.* (*Ceratalaspis*) *cosyra* group + *C.* (*Ceratalaspis*) *simi* Munro + *C.* (*Ceratalaspis*) *guttiformis*. At the base of the tree there is an unresolved trichotomy with *Pardalaspis* + *C.* (*Ceratalaspis*) *divaricata*, *C.* (*Pterandrus*) *melanopus*, and a clade including all other species. The large clade includes groups 2, 3, and 4 above in an unresolved bush also including *C.* (*Pterandrus*) *fulicoides*, *C.* (*Ceratalaspis*) *hancocki* De Meyer, *C.* (*Ceratalaspis*) *contramedia* (Munro),

**TABLE 16.2**
**Characters Used in Analysis of *Ceratitis* Phylogeny**

1. Ocellar bristle: 0, well developed, at least twice length of ocellar triangle; 1, poorly developed
2. Frontal bristles: 0, 2; 1, number varying, but not 2
*3. Male anterior orbital bristle: 0, simple; 1, modified; 2, spatulate
4. Frons: 0, longer than wide; 1, at most as long as wide
5. Frons: 0, convex; 1, not convex, either flat or concave
6. Frons: 0, not swollen; 1, swollen
*7. Male frons: 0, dull colored, not silvery; 1, with silvery patches
8. Male face: 0, without conspicuous banding, ground color pale white; 1, with conspicuous and contrasting banding or color (orange red)
*9. Arista: 0, with short hairs or bare; 1, with medium long hairs to plumose
10. Frons in lateral view: 0, not projecting at antennal implant; 1, slightly projecting; 2, strongly projecting
*11. Postpronotal lobe: 0, not spotted; 1, with black or brownish spot
*12. Anepisternal bristles: 0, one; 1, two to three
13. Subscutellum: 0, completely dark colored; 1, not completely dark, partly pale colored
14. Subscutellum: 0, without three spots; 1, with three dark spots, placed at same position as scutellar apical spots
*15. Scutellum ground color: 0, yellow white; 1, grayish; 2, grayish and apical spots reaching only halfway to base of disc
16. Scutellum: 0, with yellow-white ground color, either with well-developed apical spots reaching at least halfway to base, or with isolated medial spots; 1, ground color white, apical spots restricted to apical three-fifths
17. Scutellum basal spots: 0, absent; 1, present
*18. Scutellum apical spots: 0, not touching; 1, touching
*19. Mesonotum: 0, without well-defined, contrasting spots; 1, with well-developed and contrasting spots
*20. Mesonotal pubescent area: 0, pale, at most with orange tinge; 1, ash-gray
*21. Mesonotal spots: 0, not extensive; 1, extensive, including area behind postpronotal lobe, along median line, prescutellar spots reaching to dorsocentral setae
22. Costal band of wing: 0, not interrupted; 1, interrupted
*23. Wing bands: 0, mainly yellow; 1, yellow-black; 2, mainly dark
24. Discal crossband of wing: 0, joining costal band; 1, joining costal band, but interrupted in discal cell; 2, free
25. Subapical crossband: 0, free; 1, joining discal crossband; 2, joining costal band
*26. Posterior apical crossband of wing: 0, absent; 1, present and free; 2, present and joining costal band
*27. Forefemur: 0, without orange feathering; 1, with bushy orange feathering on posterior side
*28. Midtibia: 0, without orange feathering; 1, with orange feathering
*29. Midtibia: 0, without black feathering; 1, with black feathering
30. Forefemur: 0, without row of white hairs; 1, with row of white hairs
*31. Abdominal tergites: 0, without spots; 1, with dark spots
32. Abdominal tergites: 0, yellow, tergites 2 and 4 with silvery shine posteriorly; 1, partly with dark banding, at least tergite 3 with dark band
33. Aculeus tip: 0, not dentate; 1, dentate, with one or two sets of subapical protuberances
34. Aculeus tip: 0, not broad nor with rounded indentations; 1, broad, with rounded indentations
35. Aculeus tip: 0, not broad, nor with strong indentations; 1, broad, with strong indentations
*36. Wing: 0, slender (length/width ratio more than 1.8,; 1, slightly broad (ratio 1.7 to 1.8); 2, very broad (ratio distinctly less than 1.7)

*Note*: All characters are additive except characters 23 to 25. Characters indicated with * were used in first analysis; all characters were used in second analysis (see text for further explanation).

*C.* (*Ceratalaspis*) *quinaria* (Bezzi), *C.* (*Ceratalaspis*) *scaevolae* (Munro), *C.* (*Ceratalaspis*) *silvestrii* Bezzi, *C.* (*Ceratalaspis*) *dumeti,* and *C.* (*Ceratalaspis*) *epixantha* (Hering).

This first analysis indicates that the characters used in earlier publications do not allow unambiguous recognition of the groups. The subgenus *Ceratalaspis* does not form a monophyletic group,

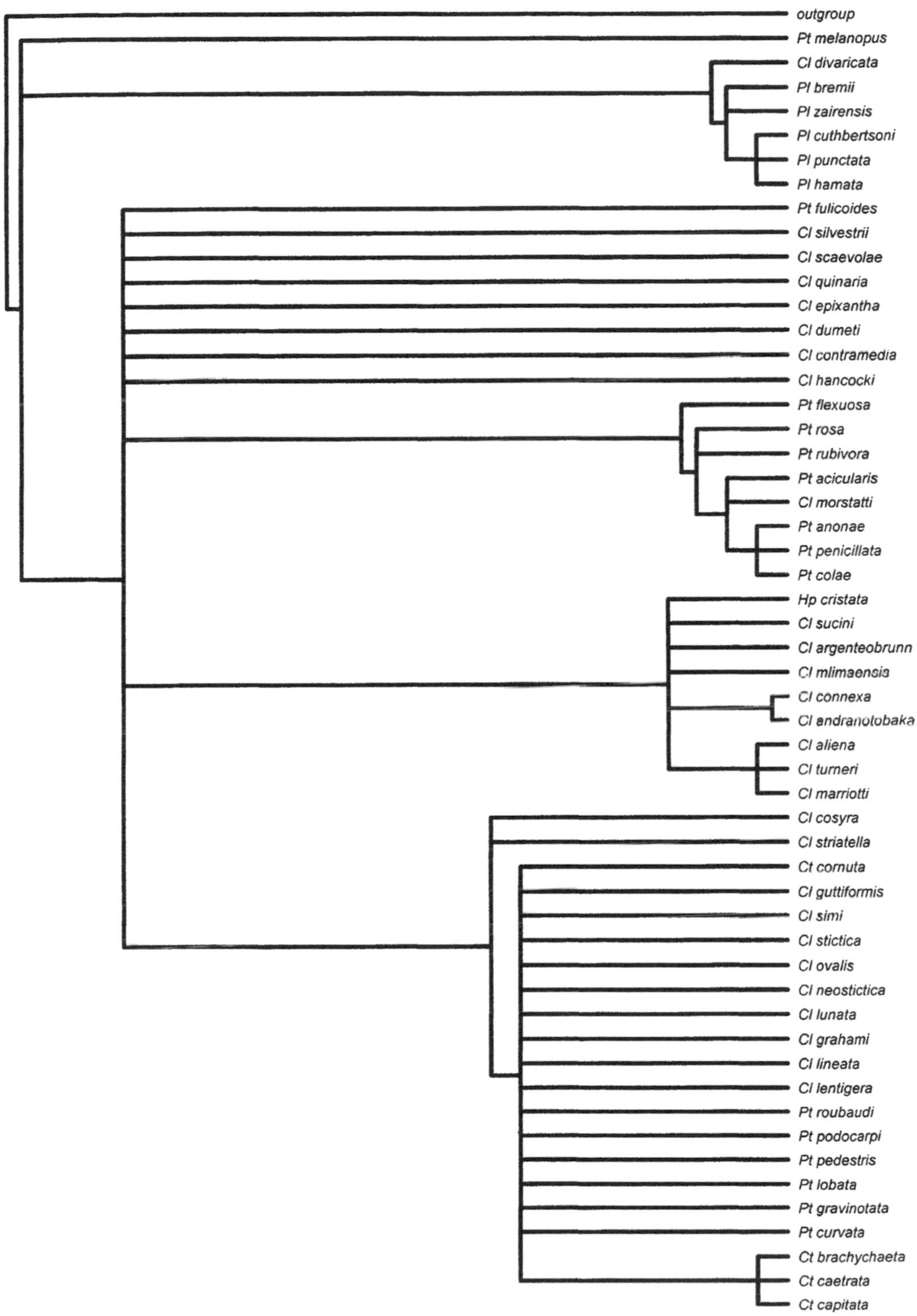


**FIGURE 16.1** Strict consensus tree from a profile of 1746 most parsimonious trees, based on an analysis of 17 morphological characters in *Ceratitis* species. Subgeneric affiliation indicated with abbreviations as in Table 16.1.

**TABLE 16.3**
**Character State Data Matrix for *Ceratitis* Species in Cladistic Analysis**

| Species | 1 | 2 | 3 | 4 | 5 | 6 | 7 | 8 | 9 | 10 | 11 | 12 | 13 | 14 | 15 | 16 | 17 | 18 | 19 | 20 | 21 | 22 | 23 | 24 | 25 | 26 | 27 | 28 | 29 | 30 | 31 | 32 | 33 | 34 | 35 | 36 |
|---|---|---|---|---|---|---|---|---|---|---|---|---|---|---|---|---|---|---|---|---|---|---|---|---|---|---|---|---|---|---|---|---|---|---|---|---|
| Outgroup | 0 | 0 | 0 | 0 | 0 | 0 | 0 | 0 | ? | 0 | 0 | 0 | 0 | 0 | 0 | 0 | ? | 0 | 0 | 0 | 0 | 0 | ? | 0 | 0 | 0 | 0 | 0 | 0 | 0 | 0 | 0 | 0 | 0 | 0 | 0 |
| 1 *Hp. cristata* | 0 | 0 | 0 | 0 | 0 | 0 | 0 | 0 | 0 | 0 | 0 | 0 | 0 | 0 | 0 | 0 | 0 | 0 | 0 | 0 | 0 | 0 | 1 | 0 | 0 | 2 | 0 | 0 | 0 | 0 | 0 | 1 | ? | ? | ? | 0 |
| 2 *Pl. bremii* | 0 | 0 | 0 | 1 | 1 | 0 | 1 | 1 | 0 | 2 | 0 | 1 | 1 | 0 | 0 | 0 | 0 | 0 | 0 | 0 | 0 | 0 | 0 | 0 | 0 | 0 | 0 | 0 | 0 | 0 | 1 | 0 | 0 | 0 | 0 | 0 |
| 3 *Pl. cuthbertsoni* | 0 | 0 | 0 | 1 | 1 | 0 | 1 | 1 | 0 | 2 | 0 | 1 | 0 | 0 | 0 | 0 | 0 | 0 | 0 | 0 | 0 | 1 | 2 | 1 | 0 | 0 | 0 | 0 | 0 | 0 | 1 | 0 | 0 | 0 | 0 | 0 |
| 4 *Pl. hamata* | 0 | 0 | 0 | 1 | 1 | 0 | 1 | 1 | 0 | 2 | 0 | 1 | 0 | 0 | 0 | 0 | 0 | 0 | 0 | 0 | 0 | 0 | 2 | 0 | 0 | 0 | 0 | 0 | 0 | 0 | 1 | 0 | 0 | 0 | 0 | 0 |
| 5 *Pl. punctata* | 0 | 0 | 0 | 1 | 1 | 0 | 1 | 1 | 0 | 2 | 0 | 1 | 0 | 0 | 0 | 0 | 1 | 0 | 0 | 0 | 0 | 0 | 2 | 0 | 0 | 0 | 0 | 0 | 0 | 0 | 1 | 0 | 0 | 0 | 0 | 0 |
| 6 *Pl. zairensis* | 0 | 0 | 0 | 1 | 1 | 0 | 1 | 1 | 0 | 2 | 0 | 1 | 0 | 0 | 0 | 0 | 0 | 0 | 0 | 0 | 0 | 0 | 1 | 0 | 0 | 0 | 0 | 0 | 0 | 0 | 1 | 0 | 0 | 0 | 0 | 0 |
| 7 *Ct. cornuta* | 0 | 0 | 0 | 0 | 0 | 0 | 0 | 0 | ? | 0 | 1 | 0 | 0 | 0 | 0 | 0 | 1 | 1 | 1 | ? | 1 | 0 | 0 | 0 | 0 | 1 | 1 | 0 | 0 | 0 | 0 | 0 | 0 | 0 | 0 | 2 |
| 8 *Ct. brachychaeta* | 0 | 0 | 1 | 0 | 0 | 1 | 0 | 0 | 0 | 0 | 1 | 0 | 0 | 0 | 0 | 0 | 1 | 1 | 1 | 1 | 1 | 1 | 0 | 0 | 0 | 0 | 1 | 0 | 0 | 0 | 0 | 0 | 0 | 0 | 0 | 2 |
| 9 *Ct. capitata* | 0 | 0 | 2 | 0 | 0 | 0 | 0 | 0 | 0 | 0 | 1 | 0 | 0 | 0 | 0 | 0 | 1 | 1 | 1 | 1 | 1 | 1 | 0 | 0 | 0 | 0 | 1 | 0 | 0 | 0 | 0 | 0 | 0 | 0 | 0 | 2 |
| 10 *Ct. caetrata* | 0 | 1 | 2 | 0 | 0 | 1 | 0 | 0 | 0 | 0 | 1 | 0 | 0 | 0 | 0 | 0 | 1 | 1 | 1 | 0 | 1 | 1 | 0 | 0 | 0 | 0 | 1 | 0 | 0 | 0 | 0 | 0 | 0 | 0 | 0 | 2 |
| 11 *Pt. curvata* | 0 | 0 | 0 | 0 | 0 | 0 | 0 | 0 | 1 | 0 | 1 | 0 | 0 | 0 | 0 | 0 | 1 | 1 | 1 | 1 | 1 | 0 | 2 | 0 | 0 | 0 | 0 | 0 | 1 | 1 | 0 | 1 | 0 | 0 | 0 | 1 |
| 12 *Pt. gravinotata* | 0 | 0 | 0 | 0 | 0 | 0 | 0 | 0 | 0 | 0 | 1 | 0 | 0 | 0 | 0 | 0 | 1 | 1 | 1 | 1 | 1 | 0 | 2 | 0 | 0 | 2 | 0 | 0 | 1 | 0 | 0 | 1 | 0 | 0 | 0 | 0 |
| 13 *Pt. lobata* | 0 | 0 | 0 | 0 | 1 | 0 | 0 | 0 | 1 | 0 | 1 | 0 | 0 | 0 | 0 | 0 | 1 | 1 | 1 | 1 | 1 | 0 | 2 | 0 | 0 | 0 | 0 | 0 | 0 | 1 | 0 | 1 | 0 | 0 | 0 | 2 |
| 14 *Pt. pedestris* | 0 | 0 | 0 | 0 | 0 | 0 | 0 | 0 | 0 | 0 | 1 | 0 | 0 | 0 | 0 | 0 | 1 | 0 | 1 | 1 | 1 | 0 | 2 | 0 | 0 | 0 | 0 | 0 | 0 | 0 | 0 | 1 | 0 | 0 | 0 | 1 |
| 15 *Pt. podocarpi* | 0 | 0 | 0 | 0 | 0 | 0 | 0 | 0 | 0 | 0 | 1 | 0 | 0 | 0 | 0 | 0 | 1 | 0 | 1 | 1 | 1 | ? | 2 | 0 | 0 | 0 | 0 | 0 | 1 | 0 | 0 | 1 | 0 | 0 | 0 | 0 |
| 16 *Pt. roubaudi* | 0 | 0 | 0 | 0 | 0 | 0 | 0 | 0 | 0 | 0 | 1 | 0 | 0 | 0 | 0 | 0 | 1 | 0 | 1 | 1 | 1 | 0 | 2 | 0 | 0 | 0 | 0 | 0 | 0 | 0 | 0 | 1 | 0 | 0 | 0 | ? |
| 17 *Pt. acicularis* | 0 | 0 | 0 | 0 | 0 | 0 | 0 | 0 | 1 | 0 | 0 | 0 | 0 | 0 | 2 | 0 | 0 | 0 | 0 | 0 | 0 | 1 | 1 | 0 | 0 | 0 | 0 | 0 | 1 | 0 | 0 | 1 | 0 | 0 | 0 | 0 |
| 18 *Pt. anonae* | 0 | 0 | 0 | 0 | 0 | 0 | 0 | 0 | 1 | 0 | 0 | 0 | 0 | 0 | 2 | 0 | 0 | 0 | 0 | 0 | 0 | 1 | 2 | 0 | 0 | 0 | 0 | 0 | 1 | 0 | 0 | 1 | 0 | 0 | 0 | 0 |
| 19 *Pt. colae* | 0 | 0 | 0 | 0 | 0 | 0 | 0 | 0 | 1 | 0 | 0 | 0 | 0 | 0 | 2 | 0 | 0 | 0 | 0 | 0 | 0 | 1 | 2 | 0 | 0 | 0 | 0 | 0 | 1 | 0 | 0 | 1 | 0 | 0 | 0 | 0 |
| 20 *Pt. flexuosa* | 0 | 0 | 0 | 0 | 0 | 0 | 0 | 0 | 1 | 0 | 0 | 0 | 0 | 0 | 0 | 0 | 0 | 0 | 0 | 0 | 0 | 1 | 2 | 2 | 0 | 0 | 0 | 0 | 1 | 0 | 0 | 1 | 0 | 0 | 0 | 0 |
| 21 *Pt. fulicoides* | 0 | 1 | 0 | 0 | 0 | 0 | 0 | 0 | 1 | 0 | 0 | 0 | 0 | 0 | 0 | 0 | 0 | 0 | 0 | 0 | 0 | 1 | 1 | 0 | 0 | 1 | 1 | 1 | 0 | 0 | 0 | 1 | 0 | 0 | 0 | 0 |
| 22 *Cl. morstatti* | 0 | 0 | 0 | 0 | 0 | 0 | 0 | 0 | 1 | 0 | 0 | 0 | 0 | 0 | 2 | 0 | 0 | 0 | 0 | 0 | 0 | 1 | 1 | 0 | 0 | 0 | 0 | 0 | 0 | 0 | 0 | 1 | 0 | 0 | 0 | 0 |
| 23 *Pt. penicillata* | 0 | 0 | 0 | 0 | 0 | 0 | 0 | 0 | 1 | 0 | 0 | 0 | 0 | 0 | 2 | 0 | 0 | 0 | 0 | 0 | 0 | 0 | 2 | 0 | 0 | 0 | 0 | 0 | 1 | 0 | 0 | 1 | 0 | 0 | 0 | 1 |
| 24 *Pt. rosa* | 0 | 0 | 0 | 0 | 0 | 0 | 0 | 0 | 1 | 0 | 0 | 0 | 0 | 0 | 1 | 0 | 0 | 0 | 0 | 0 | 0 | 1 | 1 | 0 | 0 | 0 | 0 | 0 | 1 | 0 | 0 | 1 | 0 | 0 | 0 | 0 |
| 25 *Pt. rubivora* | 0 | 0 | 0 | 0 | 0 | 0 | 0 | 0 | 1 | 0 | 0 | 0 | 0 | 0 | 1 | 0 | 0 | 0 | 0 | 0 | 0 | 1 | 1 | 0 | 0 | 1 | 0 | 0 | 1 | 0 | 0 | 1 | 0 | 0 | 0 | 1 |

| | | | | | | | | | | | | | | | | | | | | | | | | | | | | | | | | | | | | |
|---|---|---|---|---|---|---|---|---|---|---|---|---|---|---|---|---|---|---|---|---|---|---|---|---|---|---|---|---|---|---|---|---|---|---|---|---|
| 26 *Cl. cosyra* | 0 | 0 | 0 | 0 | 1 | 0 | 0 | 0 | 0 | 1 | 1 | 0 | 1 | 1 | 0 | 0 | 1 | 0 | 1 | 0 | 0 | 0 | 0 | 0 | 0 | 0 | 0 | 0 | 0 | 0 | 0 | 0 | 0 | 0 | 0 | 0 |
| 27 *Cl. lentigera* | 0 | 0 | 0 | 0 | 1 | 0 | 0 | 0 | 1 | 0 | 1 | 0 | 1 | 1 | 0 | 0 | 1 | 0 | 1 | 0 | 0 | 0 | 0 | 0 | 0 | 0 | 0 | 0 | 0 | 0 | 0 | 0 | 0 | 0 | 0 | 1 |
| 28 *Cl. lineata* | 0 | 0 | 0 | 0 | 1 | 0 | 0 | 0 | 0 | 0 | 1 | 0 | 1 | 1 | 0 | 0 | 1 | 0 | 1 | 0 | 0 | 0 | 0 | 0 | 0 | 0 | 0 | 0 | 0 | 0 | 0 | 0 | 0 | 0 | 0 | 1 |
| 29 *Cl. striatella* | 0 | 0 | 0 | 0 | 1 | 0 | 0 | 0 | 0 | 0 | 1 | ? | 1 | 1 | 0 | 0 | 1 | 0 | 1 | 0 | 0 | 0 | 0 | 0 | 1 | 0 | 0 | 0 | 0 | 0 | 0 | 0 | 0 | 0 | 0 | 0 |
| 30 *Cl. grahami* | 0 | 1 | 0 | 0 | 0 | 0 | 0 | 0 | 1 | 0 | 1 | 0 | 0 | 0 | 0 | 0 | 0 | 0 | 1 | 0 | 1 | 0 | 0 | 0 | 0 | 0 | 1 | 0 | 0 | 0 | 0 | 0 | 0 | 0 | 0 | 2 |
| 31 *Cl. lunata* | 0 | 0 | 0 | 0 | 0 | 0 | 0 | 0 | 1 | 0 | 1 | 0 | 1 | 1 | 0 | 0 | 1 | 0 | 1 | 0 | 1 | 0 | 0 | 0 | 0 | 0 | 0 | 0 | 0 | 0 | 0 | 0 | 0 | 0 | 0 | 2 |
| 32 *Cl. neostictica* | 0 | 0 | 0 | 0 | 0 | 0 | 0 | 0 | 1 | 0 | 1 | 0 | 0 | 0 | 0 | 0 | 1 | 0 | 1 | 0 | 1 | 0 | 0 | 0 | 0 | 1 | 1 | 1 | 0 | 0 | 0 | 0 | 0 | 0 | 0 | 2 |
| 33 *Cl. ovalis* | 0 | 0 | 0 | 0 | 0 | 0 | 0 | 0 | 1 | 0 | 1 | 0 | 0 | 0 | 0 | 0 | 1 | 0 | 1 | 0 | 1 | 0 | 0 | 0 | 0 | 0 | 0 | 0 | 0 | 0 | 0 | 1 | 0 | 0 | 0 | 2 |
| 34 *Cl. stictica* | 0 | 0 | 0 | 0 | 0 | 0 | 0 | 0 | 1 | 0 | 1 | 0 | 1 | 1 | 0 | 0 | 1 | 0 | 1 | 0 | 1 | 0 | 0 | 0 | 1 | ? | 0 | 0 | 0 | 0 | 0 | 0 | 0 | 0 | 0 | 2 |
| 35 *Cl. simi* | 0 | 0 | 0 | 0 | 1 | 0 | 0 | 0 | 0 | 1 | 1 | ? | 1 | 1 | 0 | 0 | ? | 0 | 1 | 0 | 0 | 0 | 0 | 0 | 1 | 1 | 0 | 0 | 0 | 0 | 0 | 0 | 0 | 0 | 0 | 1 |
| 36 *Cl. hancocki* | 0 | 0 | 0 | 0 | 0 | 0 | 0 | 0 | 1 | 0 | 1 | 0 | 1 | 0 | 0 | 0 | 1 | 0 | 1 | 0 | 0 | 0 | 1 | 0 | 1 | 0 | 0 | 0 | 0 | 0 | 0 | 0 | 0 | 0 | 1 | 0 |
| 37 *Cl. guttiformis* | 0 | 0 | 0 | 0 | 1 | 0 | 0 | 0 | 1 | 0 | 1 | 0 | 1 | 0 | 0 | 0 | 1 | 0 | 1 | 0 | 1 | 0 | 0 | 0 | ? | 0 | 0 | 0 | 0 | 0 | 0 | 0 | 0 | 0 | 1 | 0 |
| 38 *Cl. connexa* | 0 | 1 | 0 | 0 | 0 | 0 | 1 | 0 | 0 | 1 | 0 | 1 | 1 | 1 | 0 | 0 | 0 | 0 | 1 | 0 | 0 | 0 | 0 | 0 | 1 | 2 | 0 | 0 | 0 | 0 | 0 | 0 | 0 | 0 | 0 | 0 |
| 39 *Cl. contramedia* | 1 | 0 | 0 | 0 | 1 | 0 | 0 | 0 | 0 | 2 | 0 | 0 | 1 | 0 | 0 | 0 | 0 | 1 | 1 | 0 | 0 | 0 | 1 | 0 | 0 | 0 | 0 | 0 | 0 | 0 | 0 | 1 | 0 | 0 | 0 | 0 |
| 40 *Cl. divaricata* | 0 | 0 | 0 | 0 | 0 | 0 | 0 | 0 | 0 | 0 | 0 | 1 | 1 | 0 | 0 | 0 | 1 | 0 | 0 | 0 | 0 | 0 | 0 | 0 | 0 | 0 | 0 | 0 | 0 | 0 | 0 | 0 | 0 | 0 | 0 | 0 |
| 41 *Cl. dumeti* | 0 | 0 | 0 | 0 | 0 | 0 | 0 | 0 | 0 | 0 | 0 | ? | 1 | 1 | 0 | 0 | 0 | 0 | 1 | 0 | 0 | 0 | 0 | 0 | 0 | 0 | 0 | 0 | 0 | 0 | 0 | 0 | 0 | 0 | 0 | 0 |
| 42 *Cl. epixantha* | 0 | 0 | 0 | 0 | 0 | 0 | 0 | 0 | 0 | 0 | 0 | 1 | 1 | 1 | 0 | 0 | 0 | 0 | 1 | 0 | 0 | 0 | 1 | 0 | 0 | 0 | 0 | 0 | 0 | 0 | 0 | 1 | 0 | 0 | 0 | 0 |
| 43 *Cl. quinaria* | 0 | 0 | 0 | 0 | 1 | 0 | 0 | 0 | 0 | 0 | 0 | 0 | 1 | 0 | 0 | 0 | 0 | 0 | 1 | 0 | 0 | 0 | 0 | 0 | 0 | 0 | 0 | 0 | 0 | 0 | 0 | 0 | 0 | 0 | 0 | 0 |
| 44 *Cl. scaevolae* | 0 | 0 | 0 | 0 | 0 | 0 | 0 | 0 | 0 | 1 | 0 | 0 | 0 | 0 | 0 | 0 | 0 | 1 | 1 | 0 | 1 | 0 | 0 | 0 | 1 | 0 | 0 | 0 | 0 | 0 | 0 | 0 | 0 | 0 | 0 | 0 |
| 45 *Cl. silvestrii* | 0 | 0 | 0 | 0 | 0 | 0 | 0 | 0 | 0 | 0 | 0 | 0 | 0 | 0 | 0 | 0 | 0 | 0 | 1 | 0 | 0 | 0 | 0 | 0 | 0 | 0 | 0 | 0 | 0 | 0 | 0 | 0 | 0 | 0 | 0 | 0 |
| 46 *Cl. mlimaensis* | 0 | 0 | 0 | 0 | 0 | 0 | 0 | 0 | 0 | 0 | 0 | ? | 1 | 1 | 0 | 0 | 0 | 0 | 1 | 0 | 0 | 0 | 1 | 0 | 0 | 2 | 0 | 0 | 0 | 0 | 0 | 0 | 0 | 1 | 0 | 0 |
| 47 *Cl. andranotobaka* | 0 | 0 | 0 | 0 | 0 | 0 | 0 | 0 | 0 | 0 | 0 | 1 | 1 | 1 | 0 | 0 | 1 | 0 | 1 | 0 | 0 | 0 | 0 | 0 | 0 | 2 | 0 | 0 | 0 | 0 | 0 | 0 | 0 | 1 | 0 | 0 |
| 48 *Cl. argenteobrunnea* | 0 | 0 | 0 | 0 | 0 | 0 | 0 | 0 | 0 | 0 | 0 | 0 | 1 | 1 | 0 | 0 | 1 | 0 | 1 | 0 | 0 | 0 | 0 | 0 | 1 | 2 | 0 | 0 | 0 | 0 | 0 | 0 | 0 | 1 | 0 | 0 |
| 49 *Cl. sucini* | 0 | 0 | 0 | 0 | 0 | 0 | 0 | 0 | 0 | 0 | 0 | 0 | 1 | 1 | 0 | 0 | 0 | 0 | 1 | 0 | 0 | 0 | 0 | 0 | 0 | 2 | 0 | 0 | 0 | 0 | 0 | 0 | 0 | 1 | 0 | 0 |
| 50 *Cl. aliena* | 0 | 0 | 0 | 0 | 0 | 0 | 0 | 0 | 0 | 0 | 0 | 0 | 0 | 0 | 0 | 1 | 0 | 0 | 1 | 0 | 0 | 1 | 1 | 0 | 2 | 2 | 0 | 0 | 0 | 0 | 0 | 0 | 1 | 0 | 0 | 0 |
| 51 *Cl. marriotti* | 0 | 0 | 0 | 0 | 0 | 0 | 0 | 0 | 0 | 0 | 0 | 0 | 0 | 0 | 0 | 1 | 0 | 0 | 1 | 0 | 0 | 1 | 1 | 0 | 0 | 2 | 0 | 0 | 0 | 0 | 0 | 0 | 1 | 0 | 0 | 0 |
| 52 *Cl. turneri* | 0 | 0 | 0 | 0 | 0 | 0 | 0 | 0 | 0 | 0 | 0 | 0 | 0 | 0 | 0 | 1 | 0 | 0 | 1 | 0 | 0 | 1 | 2 | 0 | 2 | 2 | 0 | 0 | 0 | 0 | 0 | 0 | 1 | 0 | 0 | 0 |
| 53 *Pt. melanopus* | 0 | 0 | 0 | 0 | 0 | 0 | 0 | 0 | 0 | 0 | 0 | 0 | 0 | 0 | 0 | 0 | 0 | 0 | 0 | 0 | 0 | 1 | 2 | 0 | 0 | 0 | 0 | 0 | 0 | 0 | 0 | 1 | 0 | 0 | 0 | ? |

*Note*: 0 = plesiomorphic state; 1 and 2 = autapomorphic states; ? = unknown or more than one state present.

nor does *Pterandrus*, with its two species groups being placed at different positions within the tree. Only the subgenus *Pardalaspis* forms a monophyletic group.

In a second analysis, 36 characters were included (Table 16.2). These included the previous 17, plus an additional 19 that seemed informative to indicate relationships (including those used by De Meyer (1998) to differentiate the species groups of *Ceratalaspis*). The option mh* produced a single tree with length 125, ci 0.35, and ri 0.76. The bb* option again produced an overflow profile of 1754 trees. The strict consensus tree (Figure 16.2) of this profile supported the following groups: (1) *C.* (*Ceratalaspis*) *aliena* group; (2) *C.* (*Ceratalaspis*) *andranotobaka* group + *C.* (*Ceratalaspis*) *dumeti* group + *C.* (*Ceratalaspis*) *connexa* + *C.* (*Ceratalaspis*) *divaricata*; (3) *Pardalaspis* + *C.* (*Ceratalaspis*) *contramedia*; (4) *Ceratitis* (*s .str.*) + *Pterandrus* B group + *C.* (*Ceratalaspis*) *stictica* group + *C.* (*Ceratalaspis*) *cosyra* group + *C.* (*Ceratalaspis*) *guttiformis* group + *C.* (*Ceratalaspis*) *simi*. Again, a number of species were not placed, including all species of the *Pterandrus* A group and *C.* (*Hoplolophomyia*) *cristata*. Because of the overflow, ten replicates of the data matrix with random reordering of the taxa were analyzed similarly. All gave the same strict consensus tree, indicating that probably no shorter trees were missed. A majority-rule consensus tree was produced, to see whether the *Pterandrus* A group was not found in a large subset within the tree profile. A majority-rule consensus tree will only show those clusters that are found in the majority (more than 50%) of all trees in the profile. The majority-rule consensus tree (Figure 16.3) shows the same clusters as the strict consensus with the following change: the species in the unresolved bush at the base of the tree in Figure 16.2 form a monophyletic group (found in 1588 trees or 90% of the trees in the profile) in which *Pterandrus* A group, including *C.* (*Pterandrus*) *fulicoides* and *C.* (*Ceratalaspis*) *morstatti*, form the sister group of a cluster containing the *C.* (*Ceratalaspis*) *aliena* group + *C.* (*Hoplolophomyia*) *cristata*.

Caution must be taken to interpret these trees. First of all, 36 characters are far too few to hope for a fully resolved tree. Also, not all species of *Ceratitis* are included, in particular only half of the species of *Pterandrus* and *Ceratitis* (*s. str.*). Therefore, this is not a definitive phylogeny but merely an exercise to see whether the currently recognized subgenera hold up. The tree is especially weak in explaining the interrelationships between the recognized clusters.

The first analysis shows that the present classification is inadequate. The characters currently used to differentiate the subgenera do not define these same taxa as monophyletic, except for *Pardalaspis* (albeit with a number of related species which are doubtful) and the *Pterandrus* A group, including *C.* (*Ceratalaspis*) *morstatti*. *Ceratitis* (*s. str.*) and *Pterandrus* B group seem to be more closely related to each other and to some of the *Ceratalaspis* subgroups than *Pterandrus* B group is to *Pterandrus* A group.

The second analysis confirms these groups in most respects but provides further refinement. *Pterandrus* group A, including *C.* (*Pterandrus*) *fulicoides*, *C.* (*Pterandrus*) *melanopus*, and *C.* (*Ceratalaspis*) *morstatti*, is found in a majority of the trees in the profile. The *C.* (*Ceratalaspis*) *aliena* group is recognized as a separate cluster.

The character fits to the strict consensus tree of Figure 16.2 were analyzed with option xsteps c in Hennig86. This analysis showed that the wing characters used are very homoplastic, with characters 23, 25, 26, and 36 in Table 16.2 all needing nine steps or more and with a consistency index of 0.22 or less. They therefore do not seem to be useful characters for interpreting the relationships within *Ceratitis*. Most characters of the female aculeus tip (characters 33 to 35) and cephalic characters (characters 1 to 8) on the other hand are generally less homoplastic (1 to 4 steps needed). An exception is the plumosity of the arista (character 9). Characters of the thorax and of the male legs (feathering) are of intermediate usefulness. In this respect it is also important to note that characters used in earlier classifications of *Ceratitis* (those marked with * in Table 16.2) generally show higher homoplasy than the additional characters. This was to be expected since they did not unambiguously delimit the recognized groups as shown by the first analysis.

Those characters that show less homoplasy also strongly support a number of branches in the strict consensus tree, in particular, the subgenera *Pardalaspis* and *Ceratitis*, and the *C.* (*Ceratalaspis*)

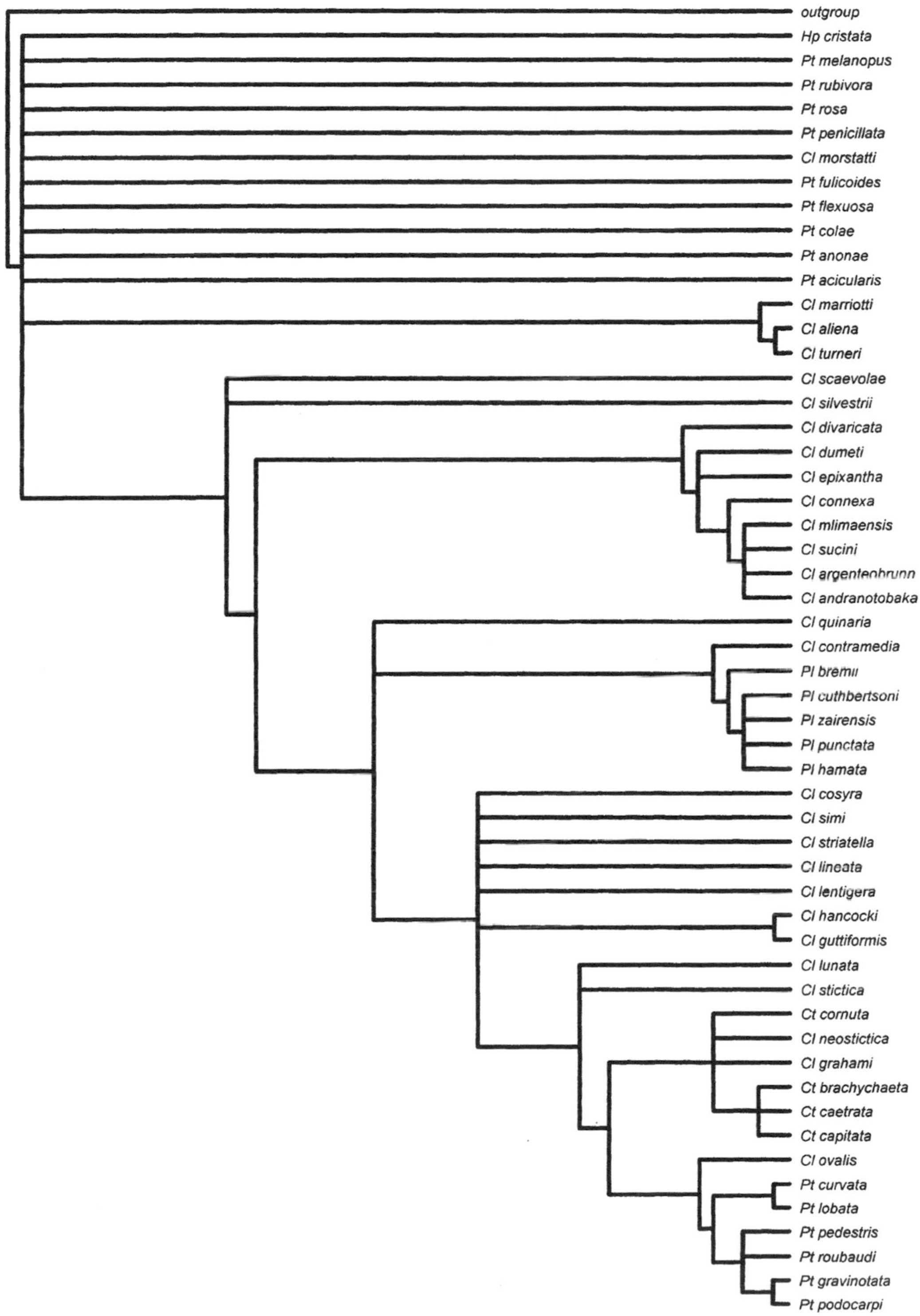


**FIGURE 16.2** Strict consensus tree from a profile of 1754 most parsimonious trees, based on an analysis of 36 morphological characters in *Ceratitis* species. Subgeneric affiliation indicated with abbreviations as in Table 16.1.

*aliena*, *guttiformis*, and *andranotobaka* groups. However, they do not support higher branches in the strict consensus tree. There is some indication that the mesonotal and scutellar characters might provide some further resolution in this respect. Male leg feathering also could provide some indications of higher relationships in the genus. However, there is a need for a more rigorous coding and a thorough investigation of the homology of certain characters.

## 16.6 CONCLUSIONS AND FUTURE RESEARCH NEEDS

In general, it can be concluded that there is a need for further detailed systematic revision of the genus *Ceratitis*. The only distinctly monophyletic group established thus far is *Pardalaspis*, but its exact position within the phylogenetic tree is not clear. *Pterandrus* group A also appears to be monophyletic. There is an indication of a close relationship between the *C.* (*Ceratalaspis*) *stictica* group, *Pterandrus* group B, and *Ceratitis* (*s. str.*). The latter seems to be largely a monophyletic group, but merely as a derived cluster within a subgroup of *Ceratalaspis*. The subgenus *Ceratalaspis* is not a monophyletic group, but a number of its species groups, as recognized by De Meyer (1998), are monophyletic. The position of the subgenus *Hoplolophomyia* is not certain, and the subgenus *Acropteromma* was not included in the analysis.

As mentioned above, certain morphological structures could prove to be useful in a further analysis, pending more rigorous coding and homology interpretation. There is also a need for further study to detect additional characters that can be used to further resolve relationships within *Ceratitis*. Perhaps, with detailed revision of *Ceratitis* (*s. str.*) and *Pterandrus*, this might be possible. It is doubtful, however, that one will be able to completely or largely resolve the phylogeny based solely on adult morphological characters. As mentioned for other groups within the Tephritidae, homoplasy is very high for many morphological characters. This homoplasy is seen at different taxonomic levels within the family, in higher taxa (Korneyev, Chapter 4) as well as in other genera like *Anastrepha* Schiner (Norrbom et al., Chapter 12). For this analysis of *Ceratitis*, this is reflected in the low consistency and retention indices of the most parsimonious trees. I believe a major problem in *Ceratitis* phylogeny will be the resolution of the relationships among the recognized subgroups because of the limited morphological characters that vary at this level. Therefore, there is a great need to search for other kind of characters.

Hancock (personal communication) mentions lures as a potentially useful character for phylogenetic analysis. Table 16.4 combines all data known for lure attractants (after Hancock 1987 and Hancock, personal communication). The attraction to terpinyl acetate as a possible synapomorphy for the genus *Ceratitis* was mentioned earlier. From Hancock's preliminary findings, there seems to be some specificity to certain lures within *Ceratitis*. All of the species of *Ceratitis* (*s. str.*) and *Pterandrus* so far tested react to trimedlure, and those of *Pardalaspis* respond to methyl eugenol. Species in *Ceratalaspis* have not reacted to either of those two lures, only to terpinyl acetate. The fact that *Nippia*, *Perilampsis,* and many dacine species are attracted to methyl eugenol could indicate that this is a plesiomorphic trait, with reaction to trimedlure and "no lure" (or lure unknown) being derived. The results for the limited number of taxa tested are congruent with the proposed tree in Figure 16.3 except that reaction to trimedlure would have arisen twice. It would be interesting to test lure reaction for a larger number of species, especially *Ceratalaspis* species, since the data for this group are limited. Only four species have been tested (Table 16.4). In addition there are other attractants that could be tested.

Host plant selection is another character that shows potentially interesting application. Figure 16.4 shows the majority-rule consensus tree of Figure 16.3 with the names of the species replaced by their major family of host plants (excluding occasional or ambiguous records). The same problem as for the lures applies here, in that the amount of data is too limited since usually only a few species within a species group have known hosts. Still there are some indications of correlation between the phylogeny and host plant usage; there appears to be host plant specificity for at least some groups. The only known hosts for all four species of the *C.* (*Ceratalaspis*) *aliena*

**TABLE 16.4**
**List of Known Lure Attractant Reactions for *Ceratitis* Species**

| Species | terp. ac. | trimed. | met.-eug. |
|---|---|---|---|
| ***Ceratitis* (*s. str.*)** | | | |
| *C. capitata* | * | * | 0 |
| *C. catoirii* | ? | * | ? |
| *C. cornuta* | * | * | 0 |
| *C. pinax* | * | ? | ? |
| ***Pardalaspis*** | | | |
| *C. bremii* | * | 0 | * |
| *C. cuthbertsoni* | ? | 0 | * |
| *C. ditissima* | ? | ? | * |
| *C. punctata* | * | 0 | * |
| ***Pterandrus*** | | | |
| *C. chirinda* | ? | * | ? |
| *C. podocarpi* | * | ? | ? |
| *C. pedestris* | * | * | 0 |
| *C. lobata* | * | ? | ? |
| *C. rosa* | * | * | 0 |
| *C. rubivora* | * | * | 0 |
| ***Ceratalaspis*** | | | |
| *C. cosyra* | * | 0 | 0 |
| *C. divaricata* | * | ? | ? |
| *C. quinaria* | * | 0 | 0 |
| *C. stictica* | ? | 0 | ? |
| ***Acropteromma*** | | | |
| *C. munroanum* | * | ? | ? |

*Note*: * = positive reaction, 0 = no reaction, ? = not tested for particular species, terp. ac. = terpinyl acetate, trimed. = trimedlure, met.-eug. = methyl eugenol.

After Hancock 1987; Hancock, personal communication.

group are Solanaceae, excluding a doubtful record for *C.* (*Ceratalaspis*) *turneri* (Munro) from "composites" based on material in the National Museums of Kenya, Nairobi, probably due to erroneous labeling. These are also the only *Ceratalaspis* species so far known to breed in Solanaceae. *Pardalaspis* species are mainly found in Apocynaceae and Sapotaceae, as well as a few occasional records from commercial hosts, except for *C.* (*Pardalaspis*) *punctata*, which seems to be polyphagous. These two plant families are similar in that they produce milky latex. All known records for species of the *Pterandrus* B group are found in Podocarpaceae and Loganiaceae, whereas a number of *Pterandrus* A species are recorded from different *Cola* species (Sterculiaceae) besides other families. Again, further data are needed, but the host plants known so far seem to corroborate, or at least not to contradict, the monophyly of the above taxa.

As indicated in other chapters of this book (Han and McPheron, Chapter 5; McPheron et al., Chapter 13; and Smith and Bush, Chapter 9), molecular data like DNA sequences do not necessarily resolve all phylogenetic problems but can provide an additional character set. So far, only *C.*

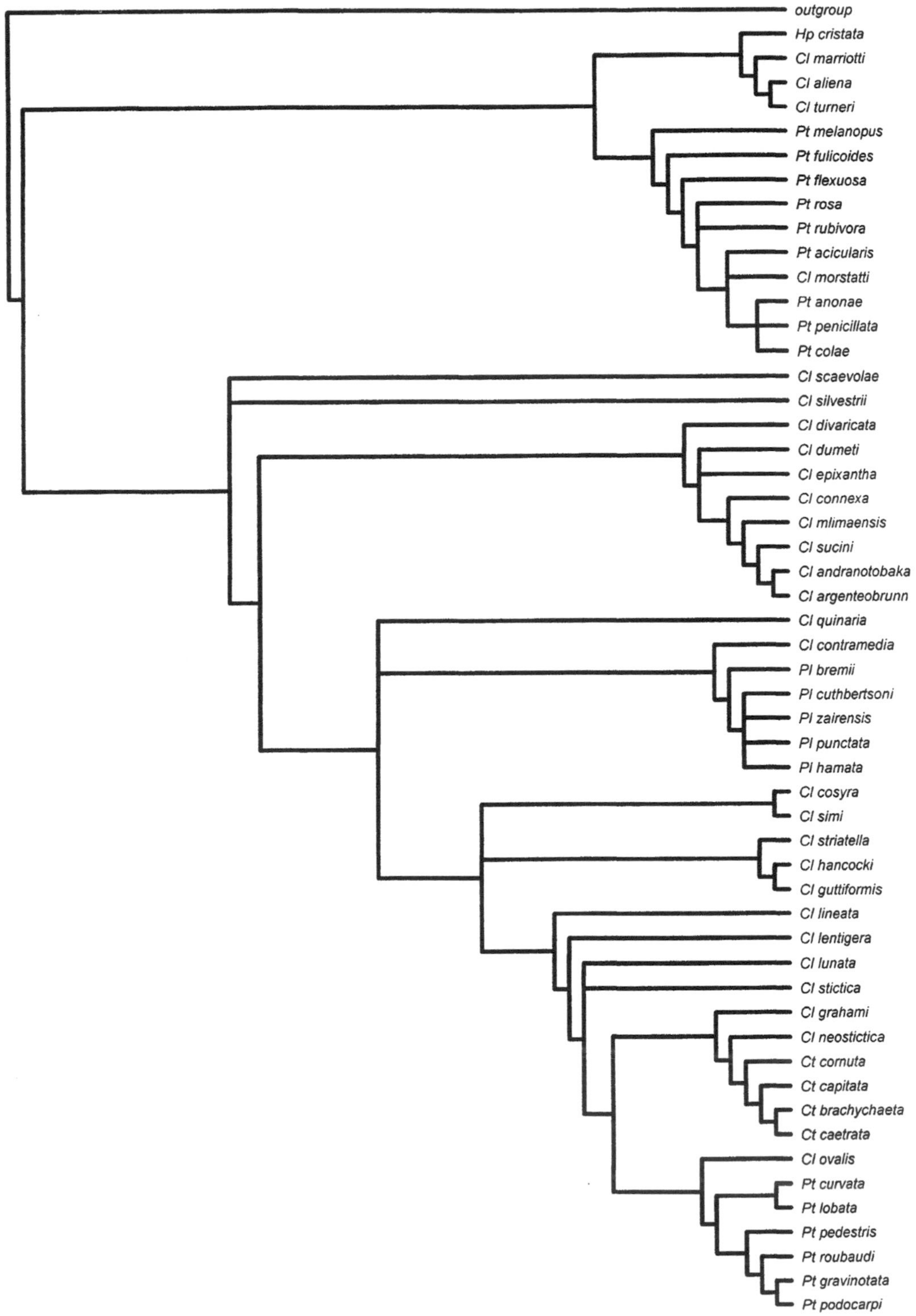


**FIGURE 16.3** Majority rule consensus tree from a profile of 1754 most parsimonious trees, based on an analysis of 36 morphological characters in *Ceratitis* species. Subgeneric affiliation indicated with abbreviations as in Table 16.1.

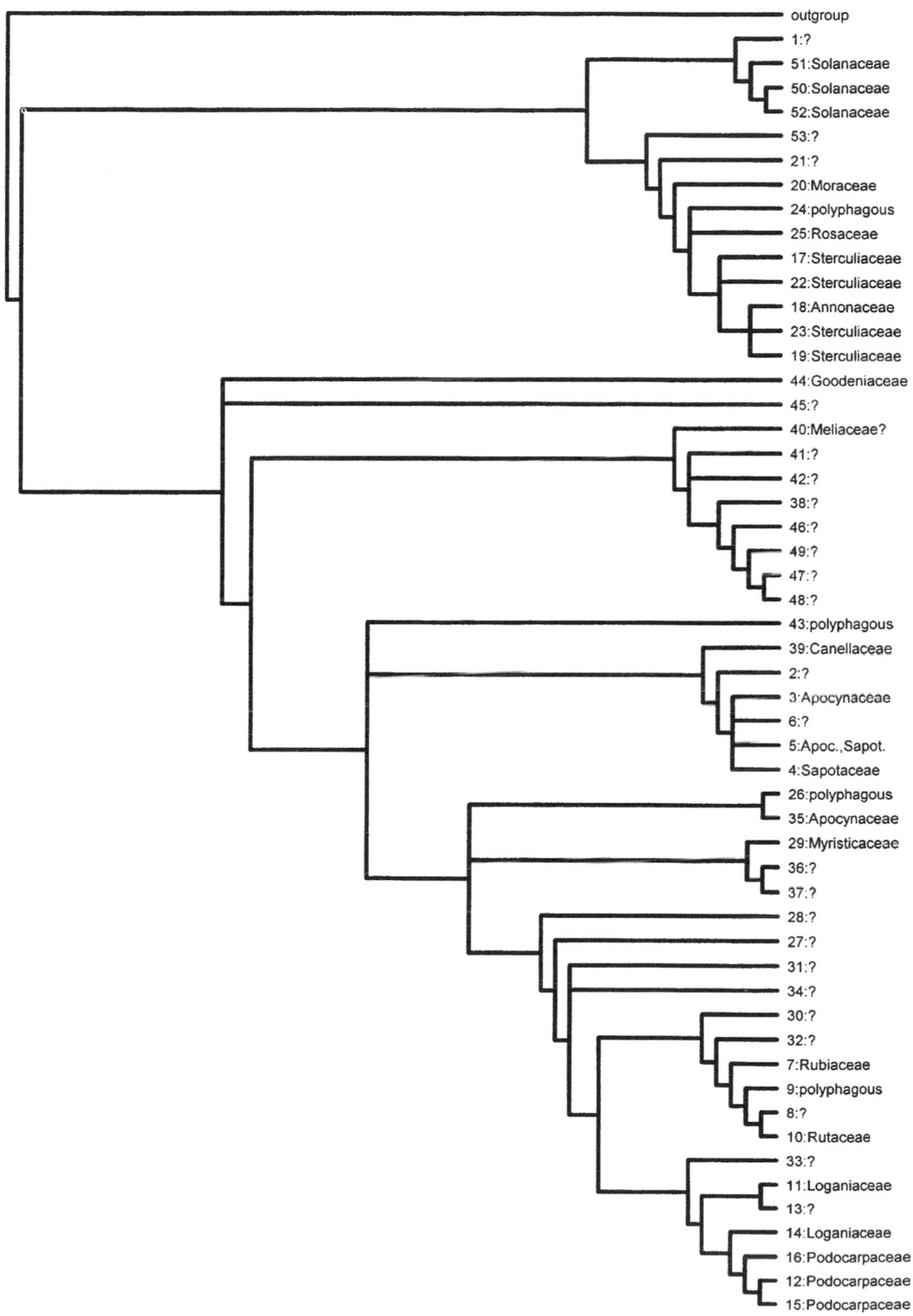


**FIGURE 16.4** Same as Figure 16.3 with indication of known host plants for *Ceratitis* species. Numbers refer to species names in Table 16.3.

*capitata* has been analyzed, within the framework of higher relationships among tephritids (Han and McPheron 1997; and Chapter 5). This species has also been studied at the population level and there have been various exploratory studies (see, for example, McPheron et al. 1994; Baruffi et al. 1995; Gasparich et al. 1995; Reyes and Ochando 1998), providing data on several regions of DNA that should be easy to study if samples of other species can be obtained. A molecular study on a number of selected species, representing species groups as resolved by the above analysis, could test the phylogenies produced by morphological characters and perhaps resolve the poorly understood regions.

Finally, widening the scope of behavioral studies from *C. capitata* to other *Ceratitis* species might provide us with phylogenetically useful data. An interesting aspect in this regard is the presence of secondary sexual characters in males. These characters, which have been used for subgenus recognition, may have specific functions in certain behaviors, deducing from similar studies in other tephritid (or other Diptera) groups. These characters include (1) modified orbital bristles and bushy feathering on the fore femur in *Ceratitis* (*s. str.*) (in Drosophilidae similar structures seem to play a role in pheromone distribution; Kaneshiro, Chapter 32); (2) black feathering on the midleg in *Pterandrus* species (similar structures are found in other Diptera, and in some tephritid groups the midleg is known to be used in displays during courtship); in addition, other types of feathering on the forelegs, sometimes in combination with silvery spots that are only visible from certain angles, are also found in species of this subgenus leading to the assumption that a wide array of different display patterns can be involved; and (3) frons silvery and contrasting with face in *Pardalaspis* species; although not reported as a display character, this could well have a function in courtship behavior given the fact that most of these interactions are performed face to face. No secondary sexual characters are reported for males of *Ceratalaspis* species, but this subgenus does show a larger variation in wing banding patterns and in female aculeus tip shapes. Wings have been shown to play a role in display behavior, and the variation in patterns found in this subgenus could have a role in recognition. The variation in aculeus tip shape has been a point of discussion during this symposium, and it has been stated that their variation cannot be explained solely by adaptation to different types of host plant texture (White, Chapter 20).

Unfortunately, the behavioral data published for *Ceratitis* species, except for the Mediterranean fruit fly (*C. capitata*), are scant (Yuval and Hendrichs, Chapter 17). In addition, for *C. capitata*, whose males have ornamented fore femora and spatulate anterior orbital bristles and whose behavior has been studied in some detail, there is no clear indication of the role these structures play in behavioral display. Limatainen et al. (1997) and Prokopy and Hendrichs (1979) list courtship behaviors in wild Mediterranean fruit fly populations, including only pheromone release (with abdominal raising), head rocking, and wing fanning and vibrating (see Eberhard, Chapter 18, for complete review of the sexual behavior of *C. capitata*). There is some evidence that the presence and symmetry of the bristles is related to the mating success of the males (Hunt et al. 1998; Mendez et al. 1998; Eberhard, personal communication). Again, there is a need for detailed comparative studies in other *Ceratitis* species. An in-depth study of the behavior of selected species could shed some light on the role of certain morphological features within the group.

## ACKNOWLEDGMENTS

I give special thanks to David Hancock (Australia) for his extensive comments on this manuscript and for kindly allowing me to cite from his unpublished manuscripts and observations. Thanks also to Allen Norrbom and an anonymous referee for reviewing and editing the first draft of this chapter, and to Boaz Yuval and Bill Eberhard for pointing out additional references on medfly behavior. I further wish to thank all curators and institutions who kindly put material at my disposal for this study: American Museum of Natural History, New York (D. Grimaldi); Natural History Museum, London (I. M. White, N. P. Wyatt); Bernice P. Bishop Museum, Honolulu (N. Evenhuis); Centre de Coopération Internationale en Recherche Agronomique pour le Développement, Montpellier (J. F. Vayssieres); Entomologischen Sammlung, Eidgenössische Hochschule, Zurich (B. Merz); Hungarian Natural History Museum, Budapest (A. Dely-Draskovits); International Centre for Insect Physiology and Ecology, Nairobi (S. Lux); Koninklijk Belgisch Instituut voor Natuurwetenschappen, Brussels (P. Grootaert); Museum für Naturkunde der Humboldt-Universität, Berlin (H. Wendt, M. Kotrba); Natural History Museum of Zimbabwe, Bulawayo (K. Donnan, M. Mawanza); National Museums of Kenya, Nairobi (M. Mungai); Natal Museum, Pietermaritzburg (D. Barraclough); Plant Protection Research Institute, Pretoria (M. Mansell); Tel Aviv University, Tel Aviv (A. Freidberg); National Museum of Natural History, Washington, D.C. (A. Norrbom). My participation in the Symposium on phylogeny and evolution of behavior of fruit flies held in Xalapa in 1998 was made possible through financial support from the Campaña Nacional contra las Moscas de la Fruta (Mexico), the International Organization for Biological Control of Noxious Animals and Plants (IOBC), the Instituto de Ecología, A.C. (Mexico), and the Fonds voor Wetenschappelijk Onderzoek – Vlaanderen (Belgium). Finally, I wish to thank Martín Aluja and Allen Norrbom for inviting me to the Xalapa symposium.

## REFERENCES

Baruffi, L., G. Damiani, C.R. Gulielmino, C. Mandi, A.R. Malacrida, and G. Casperi. 1995. Polymorphism within and between populations of *Ceratitis capitata*: comparison between RAPD and multi-locus enzyme electrophoresis data. *Heredity* 74: 425–437.

Bezzi, M. 1918. Notes on the Ethiopian fruit-flies of the family Trypaneidae, other than *Dacus* (s.l.), with descriptions of new genera and species (Dipt.) — I. *Bull. Entomol. Res.* 9: 13–46.

Bezzi, M. 1920. Notes on the Ethiopian fruit-flies of the family Trypaneidae, other than *Dacus* — III. *Bull. Entomol. Res.* 10: 211–272.

Bezzi, M. 1926. Nuove specie di Tripaneidi (Dipt.) dell'Africa del Sud. *Boll. Lab. Zool. Gen. Agric. R. Scuola Agric. Portici* 18: 276–300.

Bigot, J.M.F. 1891. Voyage de M. Ch. Alluaud dans le territoire d'Assinie 8e mémoire (Afrique occidentale) en juillet et août 1886. Diptères. *Ann. Soc. Entomol. Fr.* 60: 365–386.

Cogan, B.H. and H.K. Munro. 1980. Family Tephritidae. In *Catalogue of the Diptera of the Afrotropical Region* (R.W. Crosskey, ed.), pp. 518–554. British Museum (Natural History), London.

De Meyer, M. 1996. Revision of the subgenus *Ceratitis* (*Pardalaspis*) Bezzi 1918 (Diptera, Tephritidae, Ceratitini). *Syst. Entomol.* 21: 15–26.

De Meyer, M. 1998. Revision of the subgenus *Ceratitis* (*Ceratalaspis*) Hancock, 1984 (Diptera, Tephritidae). *Bull. Entomol. Res.* 88: 257–290.

Farris, J. 1988. *Hennig86, Version 1.5*. Published by the author, Port Jefferson Station.

Freidberg, A. 1991. A new species of *Ceratitis* (*Ceratitis*) (Diptera: Tephritidae), key to species of subgenera *Ceratitis* and *Pterandrus*, and record of *Pterandrus* fossil. *Bishop Mus. Occas. Pap.* 31: 166–173.

Gasparich, G.E., W.S. Sheppard, H.-Y. Han, B.A. McPheron, and G.J. Steck. 1995. Analysis of mitochondrial DNA and development of PCR-based diagnostic molecular markers for Mediterranean fruit fly (*Ceratitis capitata*) populations. *Insect. Mol. Biol.* 4: 61–67.

Graham, W.M. 1908. Some new and undescribed insect pests affecting cocoa in West Africa. *J. Econ. Biol.* 3: 113–117.

Guérin-Méneville, F.E. 1843. Monographie d'un genre de Muscides, nommé *Ceratitis*. *Rev. Zool. Soc. Cuvier* 1843: 194–201.

Han, H.-Y. and B.A. McPheron. 1997. Molecular phylogenetic study of Tephritidae (Insecta: Diptera) using partial sequences of the mitochondrial 16S ribosomal DNA. *Mol. Phylogenet. Evol.* 7: 17–32.

Hancock, D.L. 1984. Ceratitinae (Diptera: Tephritidae) from the Malagasy subregion. *J. Entomol. Soc. South. Afr.* 47: 277–301.

Hancock, D.L. 1987. Notes on some African Ceratitinae (Diptera: Tephritidae), with special reference to the Zimbabwean fauna. *Trans. Zimb. Sci. Assoc.* 63: 47–57.

Hancock, D.L. 1991. Revised tribal classification of various genera of Trypetinae and Ceratitinae, and the description of a new species of *Taomyia* Bezzi (Diptera: Tephritidae). *J. Entomol. Soc. South. Afr.* 54: 9–14.

Hancock, D.L. 1999. Grass-breeding fruit flies and their allies of Africa and Asia (Diptera: Tephritidae: Ceratitidinae). *J. Nat. Hist.* 33: 911–948.

Hancock, D.L. and I.M. White. 1997. The identity of *Trirhithrum nigrum* (Graham) and some new combinations in *Ceratitis* MacLeay (Diptera: Tephritidae). *Entomologist* 116: 192–197.

Hunt, M.K., C.S. Crean, R.J. Wood, and A.S. Gilburn. 1998. Fluctuating asymmetry and sexual selection in the Mediterranean fruitfly (Diptera, Tephritidae). *Biol. J. Linn. Soc.* 64: 385–396.

Karsch, F.A. 1887. Dipterologisches von der Delagoabai. *Entomol. Nachr.* 13: 22–26.

Limatainen, J., A. Hoikkala, and T. Shelly. 1997. Courtship behavior in *Ceratitis capitata* (Diptera: Tephritidae): comparison of wild and mass-reared males. *Ann. Entomol. Soc. Am.* 90: 836–843.

MacLeay, W.S. 1829. Notice of *Ceratitis citriperda*, an insect very destructive to oranges. *Zool. J. Lond.* 4: 475–482.

McPheron, B.A., G.E. Gasparich, H.-Y. Han, G.J. Steck, and W.S. Sheppard. 1994. Mitochondrial DNA restriction map for the Mediterranean fruit fly, *Ceratitis capitata. Biochem. Genet.* 32: 25–33.

Mendez, V., R.D. Briceno, and W.G. Eberhard. 1998. Functional significance of the capitate supra-fronto-orbital bristles of male Medflies (*Ceratitis capitata*) (Diptera, Tephritidae). *J. Kans. Entomol. Soc.* 71: 164–174.

Munro, H.K. 1933. Records of South African fruit-flies (Trypetidae, Diptera) with descriptions of new species. *Entomol. Mem. South Afr. Dept. Agric.* (1932) (1) 8: 25–45.

Page, R.D.M. 1993. *COMPONENT, Version 2.0.* Natural History Museum, London.

Pont, A. 1995. The Dipterist C.R.W. Wiedemann (1770–1840). His life, work and collections. *Steenstrupia* 21: 125–154.

Prokopy, R.J. and J. Hendrichs. 1979. Mating behavior of *Ceratitis capitata* on a field caged host tree. *Ann. Entomol. Soc. Am.* 72: 642–648.

Reyes, A. and M.D. Ochando. 1998. Use of molecular markers for detecting the geographical origin of *Ceratitis capitata* (Diptera: Tephritidae) populations. *Ann. Entomol. Soc. Am.* 91: 222–227.

Walker, F. 1849. *List of the Specimens of Dipterous Insects in the Collection of the British Museum.* Part IV. British Museum (Natural History), London. pp. 688–1172.

Walker, F. 1853. Diptera (Part IV). In *Insecta Saundersiana: Or Characters of Undescribed Insects in the Collection of William Wilson Saunders, Esq., F.R.S., F.L.S., etc.* Vol. 1, pp. 253–414. John Van Voorst, London.

White, I.M. 1989. The state of fruit fly taxonomy and future research priorities. In *Fruit Flies of Economic Importance 87. Proceedings of the CEC/IOBC International Symposium, Rome, 7–10 April 1987* (R. Cavalloro, ed.), pp. 543–552. A.A. Balkema, Rotterdam. 626 pp.

White, I.M. and M.M. Elson-Harris. 1992. *Fruit Flies of Economic Significance: Their Identification and Bionomics.* CAB International, Wallingford. 601 pp.

Wiedemann, C.R.W. 1824. *Munus rectoris in Academia Christiana Albertina aditurus Analecta entomologica ex Museo Regio Havniensi maxime congesta profert iconibusque illustrat.* Kiliae [Kiel]. 60 pp.

# 17 Behavior of Flies in the Genus *Ceratitis* (Dacinae: Ceratitidini)

*Boaz Yuval and Jorge Hendrichs*

**CONTENTS**



## 17.1 INTRODUCTION

All organisms exhibit an array of discrete, specific behaviors. These behaviors may be seen as fine-tuned adaptations that, on the proximate level, form a bridge between the physiological needs of the organism, a variety of phylogenetic constraints, and the environment in which the organism lives. On an ultimate level, these behaviors are highly evolved patterns that withstand the acid test of continuous natural selection and optimize the reproductive success of the individual that enacts them. These principles are very well illustrated in analyses of tephritid behavior in general, and that of *Ceratitis* flies in particular.

The genus *Ceratitis* encompasses some 78 species, several of significant economic importance (Table 17.1) (Hancock 1984; White and Elson-Harris 1992; De Meyer, Chapter 16). Paramount among these, in economic importance, is *C. capitata* (Wiedemann), the Mediterranean fruit fly, which is distributed in most tropical and temperate regions of the world, and constantly threatens

0-8493-1275-2/00/$0.00+$.50

**TABLE 17.1**
**Host Affinities and Distribution of Economically Important Flies of the Genus *Ceratitis***

| Species | Common Name | Distribution | Host Range |
|---|---|---|---|
| *C. cosyra* (Walker) | Mango fruit fly | East, Central, and South Africa | Polyphagous |
| *C. quinaria* (Bezzi) | Five-spotted or Zimbabwean fruit fly | East and South Africa | Polyphagous |
| *C. capitata* (Wiedemann) | Mediterranean fruit fly | Most tropical and subtropical regions of the world | Polyphagous |
| *C. catoirii* Guérin-Méneville | Mascarene fruit fly | Islands of Indian Ocean | Polyphagous |
| *C. malgassa* Munro | Madagascan fruit fly | Madagascar | Polyphagous |
| *C. punctata* (Wiedemann) | — | Africa | Polyphagous |
| *C. anonae* Graham | — | West, Central, and East Africa | Polyphagous |
| *C. colae Silvestri* | — | West Africa | Oligophagous |
| *C. pedestris* (Bezzi) | Strychnos fruit fly | Southern Africa | Oligophagous |
| *C. rosa* Korsch | Natal fruit fly | Africa and islands of Indian Ocean | Polyphagous |
| *C. rubivora* Coquillett | Blackberry fruit fly | South, West, and East Africa | Oligophagous |
| *C. aliena* (Bezzi) | — | East and South Africa | Monophagous |
| *C. giffardi* Bezzi | — | East and West Africa | Monophagous |
| *C. morstatti* Bezzi | — | West Africa | Monophagous |
| *C. silvestri* Bezzi | — | West Africa | Monophagous |
| *C. turneri* (Munro) | — | East Africa | Monophagous |
| *C. flexuosa* (Walker) | — | West and Central Africa | Monophagous |
| *C. penicillata* Bigot | — | West and Central Africa | Monophagous |

to invade or reinvade new areas (Christenson and Foote 1960; Bateman 1972; Carey 1991). Other polyphagous flies of the genus have potential as invaders as well, and may become cosmopolitan pests in the future (Steyskal 1982; Chen and Tseng 1992). The impressive biological success of these flies is supported by numerous adaptations, morphological, physiological, and behavioral, which allow them to thrive in diverse habitats that exert a range of environmental pressures on individual flies. In this chapter we attempt to present a synthesis of the behaviors exhibited by these flies, focusing on *C. capitata*, the best-studied member of the genus.

Attempting a view that integrates the various hierarchies that govern the behavior of individuals at each stage of the life cycle, we treat larval and adult behavior separately. Larvae of *Ceratitis* flies live fairly uncomplicated lives, ensconced as they are in the host fruit selected by their female parent. Nevertheless, as we show below, they are endowed with behaviors that allow them to utilize the host optimally, avoid predators and parasites, and pupate in the soil. Adult behavior is extremely complex, punctuated by the catholic host preferences of these flies, and their need to move around in heterogeneous environments, seeking carbohydrate and proteinaceous nutrition, mates, and oviposition sites. We present adult behavior from a foraging perspective, as each sex must forage sequentially for these resources (nutrition, mates, and oviposition hosts), constantly making decisions with grave reproductive consequences (Prokopy and Roitberg 1989). Thus, the chapter includes sections on larval behavior, and feeding, sexual, and oviposition behavior of adults. Given the economic importance of *C. capitata* and the practical implications of a thorough knowledge of sexual behavior and sexual selection in this species, this particular topic is covered in great detail in a separate chapter (Eberhard, Chapter 18).

## 17.2 LARVAL BEHAVIOR

Females deposit their eggs on a suitable host (more below), and, thereafter, the offspring are left to fend for themselves. First instar larvae quickly make their way through the peel toward the nutritious pulp of the fruit (Malio 1979). This journey is associated with high mortality (Back and

Pemberton 1915; Bodenheimer 1951; Carey 1984), probably due to the combined dangers of predation, parasitism, and exposure to the toxic oils produced within the flavedo. Once beyond the dangers of the fruit surface, cushioned and protected by the host, larvae are free to burrow in the pulp and feed to their fill, with only two behavioral goals — to exploit the host, and to avoid predators and parasites.

### 17.2.1 Larval Feeding Behavior

While larval development rate is host dependent (Carey 1984), the decision regarding which host to develop in is made by the adult female. Female oviposition decisions largely reflect the suitability of the host for larval development (e.g., Bateman 1972; Zucoloto 1993). It remains to the larva to exploit to the full the host in which it finds itself. This exploitation, combined with the cardinal host selection decision previously made by the female, will ultimately affect adult reproductive success (Debouzie 1977; Krainacker et al. 1987; 1989; Zucoloto 1988).

A number of laboratory studies indicate that larvae are indeed capable of adjusting their behavior in a manner that will optimize their intake of food. Larvae are able to recognize the diets that are best from a nutritional viewpoint. When chemically defined diets were tested in the laboratory, larvae always preferred those containing the largest amount of protein (except for older larvae, which preferred protein-free diets) (Zucoloto 1987). Larval sensory capabilities are highly discriminative. In a threshold study, larvae were able to recognize as little as 0.1% g yeast and 1.5% casein in their diet. When faced with a low-quality diet they adaptively move on to a better diet, if available (Zucoloto 1991). Larvae actively migrate to the most nutritious parts of the fruit. In an analytical laboratory study monitoring the behavior of larvae in various fruits, it was found that larvae choose the area of the fruit richest in nutrients. In addition, when placed on a nutrient-poor area, they move to the lower area of the fruit, which contains more solutes and sugars (Fernandes Da Silva and Zucoloto 1993). Furthermore, in the laboratory, preferences may be induced. When fed for 3 days on one of two diets (brewer's yeast or milk powder), larvae subsequently preferred the one on which they were fed previously (Canato and Zucoloto 1993). This behavior seems to be an attribute of any organism with a sophisticated sensory and nervous system, and may not have an evolved adaptive value to larvae that develop within a homogenous host.

### 17.2.2 Evasion of Parasitoids and Predators

Protecting the eggs from egg parasitoids (Bautista and Harris 1996) is in the realm of behavior of the female, who must endeavor to oviposit in a manner unobtrusive to potential enemies. After hatching, larvae are vulnerable to a number of parasitoids adapted to seek them out and oviposit in them (reviews by Greathead 1972; Wharton 1989; Headrick and Goeden 1996; see also Haramoto and Bess 1969; Steck et al. 1986; Purcell et al. 1996; Ramadan and Wong 1990; Wong et al. 1990; Ramadan et al. 1995). Do *Ceratitis* larvae actively avoid parasitoids while inside the fruit? Not much is known about this, due to the difficulty of observing within-host behavior of larvae. *Biosteres* parasitoids locate hosts by detecting their movements (Lawrence 1981), while others such as *Aganaspis* seek out larvae by entering preexisting holes in the fruit (Ovruski 1994). One might speculate that an evolutionary arms race (*sensu* Dawkins and Krebs 1979) might ensue, where larvae learn to avoid predators as these become better at detecting larvae (see also Calkins and Webb 1988). Functionally, a negative response to a heat gradient within the fruit may achieve the desired result of distancing from a parasitoids ovipositor (Bateman 1972). Although it may be that the pressure exerted by natural enemies does not justify wasting precious feeding time in efforts to avoid them (e.g., Morse 1986), within-host larval behavior may be a rewarding avenue of research.

Much more is known of the behavior of larvae in the prepupal phase. Feeding completed, last-instar Mediterranean fruit fly larvae migrate to the skin of their host, bore a neat round hole and, throwing caution to the wind, jump out to pupate in the soil. Jumping increases the larva's speed to 0.5 m/s per jump, or by 200-fold over crawling (Maitland 1992). During this stage they are vulnerable to a variety of generalist predators such as ants and beetles, and several pupal parasitoids (Podoler and Mazor 1981; Eskafi and Kolbe 1990; Wong et al. 1990; Ovruski 1995).

Pupation takes place in moist soil in *C. cosyra* (Walker) (Malio 1979) or in cracked, dry soil (*C. capitata*) (Cavalloro and Delrio 1975). Pupation of *C. capitata* is affected not by the chemical composition of the soil but by its physical structure and moisture content at the time of entry by the mature larvae, which tend to penetrate more deeply (as deep as 5 cm from the surface) into dry, cracked soil than into wet soil. This depth is not enough to protect pupae from extremes of heat or cold or from predators, and these factors together are the main causes of pupal mortality (Cavalloro and Delrio 1975). The preferred depth of penetration for *C. capitata* pupae is 5.5 to 11 mm (54.0% of larvae), while 5% pupate at a depth between 0 and 5.5 mm and only 3.5% move down to 27.5 mm (Jackson et al. 1998). *Ceratitis capitata* is better adapted to dry conditions than are the *Bactrocera* species studied by Jackson et al. (1998). However, both pupation depth and survival were affected by the moisture content of the sand. Mortality of pupae was greatest in dry sand at 0 to 5.5 mm, and cumulative mortality was 50% greater in dry sand than in wet sand.

While jumping away from predators may be adaptive, judicious timing of emergence from fruit may avoid encounters with predators altogether (Myburgh 1963). Mediterranean fruit fly larvae exhibit a daily rhythm in their departure from the host, as activity is concentrated within a narrow window of time, peaking just before dawn (Causse 1974). The adaptivennes of this rhythm becomes clear when activity of predators is considered. Eskafi and Kolbe (1990) noted that predation by the ant, *Solenopsis geminata* (Fabricius), on falling Mediterranean fruit fly larvae in the field ranged from 7 to 25% in coffee and orange orchards. Although seasonal change did not affect ant predation on pupae, time of day had a significant effect on predation on larvae by the ants. Larvae were most susceptible to predation when their emergence coincided with the activity period of the ants.

The pupae, defenseless, are vulnerable to pupal parasitoids and generalist predators (Wong et al. 1990; Ovruski 1995). There is some evidence that larval performance may affect subsequent susceptibility to pupal parasitoids. When Mediterranean fruit fly pupae, reared at different larval population densities, were exposed to a solitary endoparasitoid of fruit fly pupae, *Coptera occidentalis* (Muesbeck), a significantly higher percentage of parasitization occurred in smaller pupae (originating from high larval density) (Kazimirova and Vallo 1992).

## 17.3 ADULT BEHAVIOR

Upon emergence from the puparium, teneral adults burrow to the surface, usually around dawn (Myburgh 1963), by peristaltic movements of the still soft integument and contractions of the ptilinum. Once released of the pupal environment, they commence their adult life. Longevity differs between the sexes and depends greatly on temperature, nutrition, and reproductive activity (Carey et al. 1986; 1992; 1998; Rivnay 1950).

Observations on Mediterranean fruit flies in controlled environments (Prokopy and Hendrichs 1979; Prokopy et al. 1987) or in the field (Hendrichs and Hendrichs 1990; Hendrichs et al. 1991; Warburg and Yuval 1997a) have established that males and females have different behavioral agendas. Females must nourish themselves and their developing eggs by feeding on carbohydrate and protein, copulate with the best male they can find, and locate a fruit suitable for oviposition. Males must also feed on carbohydrate and protein, and engage in one of two extremely time- and energy-consuming mating tactics — lekking, fruit guarding, or, at different times of the day, both (Prokopy and Hendrichs 1979; Baker and van der Valk 1992; Whittier et al. 1992; Warburg and Yuval 1997a, b; Yuval et al. 1998). All of these activities may be performed in different locations, entailing frequent movement and incurring the risk of encountering a predator. Later in this chapter we consider how these flies budget their time between these diverse behavioral tasks. First, we detail aspects of feeding, sexual, and oviposition behavior.

### 17.3.1 Adult Feeding Behavior

Most adult insects need considerable amounts of carbohydrate, protein, and lipid in order to perform the biological activities necessary for survival and reproduction. In addition, other nutrients such as minerals, vitamins, and sterols are essential for reproduction (e.g., Tsistsipis 1989). Their foraging behavior often reflects their physiological needs. Nevertheless, metabolic capabilities may frequently compensate for lack of adequate resources in the environment. Stores of essential nutrients may be carried over from the larval or pupal stages, or these nutrients may be synthesized *de novo* by the adults following ingestion of the relevant precursors (reviews by Downer and Mathews 1976; Waldbauer 1968; Waldbauer and Friedman 1991).

A number of laboratory studies have addressed the question of how *Ceratitis* flies optimize the intake of food. Adults eclose with varying amounts of nutrient reserves, which reflect the larval environment. Crowding and poor host quality will produce small adults with low nutrient reserves (Carey 1984; Krainacker et al. 1987; Krainacker et al. 1989). However, unlike many short-lived insect species that concentrate on reproduction in the adult stage of their lives, and feed exclusively as immatures, adult *Ceratitis* may improve their fitness by locating and utilizing food present in the environment. As they continue on their quest for reproductive success (a quest begun at the moment of fertilization), a large proportion of their time is spent feeding. Their diet consists of carbohydrate, which fuels both their daily activities and lipogenesis, and of protein, which is crucial for egg development in females (Christenson and Foote 1960; Bateman 1972; Teran 1977; Hendrichs et al. 1991; Cangussu and Zucoloto 1995; Warburg and Yuval 1996), and reproductive success of males (Blay and Yuval 1997; Yuval et al. 1998).

#### 17.3.1.1 Carbohydrates and Lipids

The function of sugar feeding is twofold. On the one hand, sugars are needed to fuel the daily flight, foraging, and courtship activities of the flies (Warburg and Yuval 1997b) and, on the other hand, they serve as a substrate for lipogenesis. The issue of lipogenesis in *Ceratitis* and other tephritids is interesting. Lipids are not available in the natural diet of the flies. Some studies, focused on teneral adults, suggested that lipids are synthesized only during preadult stages and that adults have no lipogenic capabilities (Langley et al. 1972; Municio et al. 1973; Garcia et al. 1980; Pagani et al. 1980). Furthermore, quantitative analytical studies that focused on nutritionally stressed flies documented an age-related decrease in stored lipids (Nestel et al. 1985). Recent experiments, focused on adult *Anastrepha serpentina* (Wiedemann) (Jácome et al. 1995) and *C. capitata* (Nestel et al. 1986; Warburg and Yuval 1996), indicate conclusively that adult tephritids, if well fed, are able to synthesize lipids throughout their life.

The ability of flies to discriminate between sugars of varying concentrations has been studied with electrophysiological methods by Gothilf et al. (1971). This ability allows flies to adjust their intake of sucrose solutions according to the concentration, imbibing greater volumes of low concentrations and small volumes of highly concentrated sucrose. For concentrations ranging between 0.8 to 8%, the duration of the meal and the volume consumed increases with sucrose concentration. For sucrose concentrations of 8 to 32%, the duration and volume are constant. At concentrations above 32%, both the duration of meal and volume ingested decrease slightly. Considering energy invested vs. energy gain, maximal gain is achieved by ingestion of 24% sucrose. This suggests that flies have a "caloric goal" and consume the volume needed to meet it (Nestel et al. 1986; Warburg and Galun 1992; Canato and Zuculoto 1998).

#### 17.3.1.2 Protein

Protein is essential for realizing the reproductive success of both females and males. Several studies have examined the relationship between protein feeding and egg production by females, and there is a consensus that when a protein source is ingested by females throughout adult life, egg production is increased (Christenson and Foote 1960; Bateman 1972; Teran 1977; Hendrichs et al. 1991; Cangussu and Zucoloto 1995; Warburg and Yuval 1996). Protein-feeding also enhances male reproductive success. Protein-fed males are better at gaining copulations than protein-deprived males, and females mated to protein-fed males are less likely to remate than females copulated previously with protein-deprived males (Blay and Yuval 1997). In addition, field studies have shown that male participation in leks is regulated by the reserves of protein (and carbohydrates) they possess (Yuval et al. 1998).

#### 17.3.1.3 Sources of Carbohydrate and Protein in Nature

Carbohydrates are obtained by feeding on several sources: fruit damaged by birds, or overripe or rotting fruit oozing juice, honeydew, and possibly nectar as well (Katsoyannos 1983; Hendrichs and Hendrichs 1990; Hendrichs et al. 1991; Warburg and Yuval 1997a). Protein or its precursors may be acquired by feeding on protein rich fruit (such as figs), bird feces (preferably fresh! see Prokopy et al. 1993a), honeydew, and colonies of bacteria found on the leaf surface or on decomposing fruit (Drew et al. 1983; Hendrichs and Hendrichs 1990; Hendrichs et al. 1991; Warburg and Yuval 1997a). The combination of amino acids and sugar, which may be found in decomposing fruit and honeydew, is synergistically phagostimulatory (Galun 1989).

The attraction of flies to protein sources may be a factor in their dispersal into new habitats (Prokopy et al. 1996a). Their findings suggest that odor of natural food could lure flies to plants whose fruit emit little or no attractive odor and are not permanent hosts, but which are nonetheless susceptible to egg laying and larval development, resulting in temporary expansion of host range. The attraction to olfactory cues associated with protein food is exploited in control operations, by offering these cues as bait combined with poisons or in traps (Ripley and Hepburn 1929; Steiner 1952; 1955; Mazor et al. 1987; Roessler 1989; Quilici and Trahais 1993; Buitendag and Naude 1994; Epsky et al. 1995; 1996; Heath et al. 1995; Katsoyannos and Hendrichs 1995; Gazit et al. 1998). Ammonia, certain amines, certain fatty acids, and other unidentified volatiles (possibly of bacterial origin), seem to be the major factors responsible for attractiveness of protein to tephritids (e.g., Morton and Bateman 1981). Amino acids, although phagostimulatory, are probably not involved in long-range attraction. However, the water in a proteinaceous food bait could itself be a principal agent attractive to flies in low-rainfall climates (Cunningham et al. 1978).

Because of its importance in reproduction for both sexes, determining precisely where protein is acquired and in what quantities remains a worthwhile target.

#### 17.3.1.4 Hierarchies of Feeding Decisions

The interactions between sugar feeding and protein feeding are intriguing. Studies on flies where protein feeding is highly specialized — blood feeders – provide an instructive comparison. Many blood-feeding dipterans have evolved specific behaviors, with precise hierarchies and rhythms, for acquiring protein (blood) and carbohydrate (usually nectar) (Roitberg and Friend 1992; Yee et al. 1992; Yuval 1992). Is protein feeding a discrete appetitive behavior in *Ceratitis* flies, different in its expression and regulation from sugar feeding? This question may be answered by seeking correlations between physiological state and specific hungers, and looking at temporal patterns of sugar and protein feeding in the field.

Laboratory experiments have shown that females are selective with respect to choice of food, and deprivation of a protein source lowers the perception threshold for protein (Cangussu and Zucoloto 1995). Differences between the sexes in sugar and protein consumption are also significant. In an experiment comparing the olfactory, aggregatory, and feeding responses of normal laboratory stocks of *C. capitata* to those of flies irradiated 2 days before eclosion, several important effects were noted. Females consumed greater quantities of protein hydrolysate solutions, entered protein hydrolysate-baited olfactory traps, and aggregated on agar plates containing protein hydrolysate in greater numbers than males of the same age and condition. However, males consumed more sucrose than did females of the same age and condition. Irradiation resulted in reduced olfactory response, reduced total food intake by flies of both sexes, and a significant reduction in aggregation on, and intake of, protein hydrolysate by females, and of sugar consumption by males (Galun et al. 1985).

There is a growing interest in the influence of individual physiological and experiential state on resource foraging behavior (Prokopy and Roitberg 1984; 1989; Bell 1990; Barton Browne 1993; Prokopy 1993). This approach has been brought to bear in a number of studies on Mediterranean fruit fly feeding behavior. As we detail below, it appears that while experience immediately prior to release does not influence feeding decisions, protein hunger increases with age and drives protein-directed behavior. Prokopy et al. (1992) found that physiological state had a significant effect on attraction of Mediterranean fruit flies to protein sources. Degree of attraction was considerably greater among flies continuously deprived of protein since eclosion than of flies that had continuous access to protein in the days preceding the experiment. In addition, the attraction to protein (of protein-deprived flies) increased with age. However, brief contact with protein before the experiment did not increase subsequent response to protein. Similarly, Averill and co-workers (1996) investigated how the foraging behavior of several species of tephritid fruit flies is modified by experience immediately prior to release on host plants. One of several stimuli was supplied prior to release on a host plant, including contact with 20% sucrose, or contact with a mixture of protein food (bird feces and sucrose), or contact with water. Residence on the host tree was the same following exposure to sugar, protein, and fruit stimuli. No significant effect of recent brief experience with any of the stimuli on subsequent attraction to protein food was found. In an experiment focused on trap-entering behavior of 3- and 12-day-old male and female Mediterranean fruit flies, protein-deprived females and males were significantly more inclined to enter protein-odor-baited traps than were protein-fed females and males. Conversely, there was little or no influence of physiological state on propensity of flies to enter trimedlure-baited traps (Prokopy et al. 1996b). In another set of field-cage studies, *C. capitata* females that were starved, sugar fed, or protein and sugar fed were allowed to choose between host trees containing either protein or sucrose. Starved flies showed no preference, while attraction to protein in the other experimental groups increased significantly as they aged (Cohen and Yuval, unpublished data).

Field observations complement these experiments. Feeding behavior of wild Mediterranean fruit flies has been observed in three different Mediterranean habitats (Hendrichs and Hendrichs 1990; Hendrichs et al. 1991; Warburg and Yuval 1997a). Feeding on bird droppings in the field was seen in all of these habitats. In the relatively xeric environment of Chios, females were observed to forage away from oviposition hosts, presumably to seek protein nutrition. However, despite the abundance of bird feces in agricultural habitats in coastal Israel, relatively few feeding events were recorded on them, compared with feeding on carbohydrate sources (Warburg and Yuval 1997a). Attempts to quantify the amount of feeding on different food sources (fruit, feces, and leaf surfaces) revealed that female feeding events on these substrates were evenly distributed throughout the day. Thus, no specific time was allocated to protein feeding. Finally, male feeding events on putative sources of protein were significantly fewer than female sightings on the same sources (Hendrichs and Hendrichs 1990; Hendrichs et al. 1991; Warburg and Yuval 1997a), confirming laboratory observations of reduced protein ingestion by males (Galun et al. 1985). More work is needed to establish how these flies solve the sugar–protein quandary.

The relationship between nutrient levels and the expression of various behaviors has also been studied in the field. In a quantitative study, correlates were sought among lipid, sugar, and glycogen contents of males and females trapped in the field, and their behavior (Warburg and Yuval 1997b). Despite circadian fluctuations in carbohydrates, lipids, and glycogen, female behavioral patterns were not associated significantly with the amounts of these compounds. Thus, if there are any behavioral thresholds associated with levels of lipid, sugar, or glycogen, all females examined exceeded them. This leads us to conclude that in the habitat studied, where sugar meals in the form of ripe and rotting fruit were ubiquitous and abundant, sugars (and their metabolic products) do not constrain female behavior. In more xeric or monocultural environments within the Mediterranean fruit fly range, where sugar may be harder to come by, its availability may play a greater role in determining female feeding and reproductive behaviors (Hendrichs et al. 1991).

To date, no information associating protein levels with behavioral patterns of wild females is available. A little more is known about males. While sugar and its metabolic products were not associated with female behavioral patterns, they significantly affect patterns of male behavior, as do protein levels. Males engaged in the alternative mating tactic of fruit guarding (Prokopy and Hendrichs 1979) have a higher sugar/lipid ratio than lekking males (Warburg and Yuval 1997b). Furthermore, males collected while lekking in the field differ from males resting at the same time. Although their size was not significantly different, and the amounts of glycogen and lipid in both groups were similar, lekking males contained significantly more protein and sugar than resting males (Yuval et al. 1998). Thus, foraging success is linked to reproductive success.

In conclusion, feeding on sugar and protein by adults is an important and complex behavior, providing the fuel and resources for the reproductive behavior of males and females, discussed in the next sections.

### 17.3.2 Sexual Behavior

To realize their reproductive potential, male insects must clear three hurdles, whose height, metaphorically, is controlled by female choice. These are copulation, insemination, and fertilization. To gain copulations, males must often compete with rival males for access to females, or court females effectively (reviewed by Thornhill and Alcock 1983). Insemination and fertilization do not inevitably follow copulation, as these are determined by a series of complex interactions between the sexes, who frequently have conflicting reproductive interests. Accordingly, it is important to distinguish between these events because sperm competition (Parker 1970) and female-regulated postcopulatory mechanisms (Eberhard 1991) may act after copulation and insemination to affect the fitness of copulating males (Alcock 1994).

In general, the reproductive sequence of tephritids contains the following elements: male advertisement and courtship (consisting of olfactory, auditory, and visual displays), copulation, insemination, fertilization, and oviposition (Sivinski and Burk 1989). In most species studied, females are receptive to multiple males, and may mate several times throughout their lives (e.g., Whittier and Shelly 1993). Sperm are stored in specialized organs, the spermathecae, where presumably they are nourished and mobilized when needed for fertilization.

For several genera of tephritids, such as *Anastrepha* Schiner and *Bactrocera*, a wealth of comparative information on sexual behavior of congenerics is available (e.g., Aluja 1994; Sivinski et al., Chapter 28), allowing us to make inferences on the evolution of sexual behavior, and the relative importance of phylogenetic and ecological constraints in shaping mating systems (e.g., Parker 1978; Yuval 1994). Unfortunately, this is not the case for the genus *Ceratitis*. Although a substantial body of knowledge has accumulated on various aspects of the sexual biology of *C. capitata*, with the exception of two reports on the sexual behavior of *C. rosa* Karsch (Myburgh 1962; Etienne 1973), nothing has been published on the numerous other species in the genus. In the following paragraphs we briefly review the sexual behavior of these two species, and refer readers seeking more detail to the comprehensive review of Mediterranean fruit fly sexual behavior by Eberhard in Chapter 18.

#### 17.3.2.1 Mating Systems

The mating system of the Mediterranean fruit fly hinges on leks (Prokopy and Hendrichs 1979; Arita and Kaneshiro 1985; 1989; Whittier et al. 1992; Shelly et al. 1993; 1994), which by definition are non-resource-based mating aggregations (Höglund and Alatalo 1995). An alternative tactic, of fruit guarding, is employed by some males in the population (Prokopy and Hendrichs 1979; Warburg and Yuval 1997a). In Mediterranean fruit fly leks, males aggregate during late morning and early afternoon in the canopies of host trees and defend individual leaves as their mating territory (Prokopy and Hendrichs 1979). From these leaves males produce a complex of olfactory (pheromone calling), auditory, and visual signals which attract females (and other males) to the lek area (reviewed by Shelly and Whittier 1997).

During the daily lekking period, not all males in the population participate in leks. In a recent study (Yuval et al. 1998), the size and weight of males captured either lekking or resting at the same time in the vicinity of leks were measured. In addition, precise amounts of sugar, glycogen, lipid, and protein in each individual were established quantitatively. There was no significant size difference between lekking and resting males. However, lekking males were significantly heavier and contained significantly more sugars and protein than resting males. Leks, therefore, are apparently exclusive, and only males with adequate nutritional reserves may join.

The question of lek site selection in the field has been addressed in a number of studies (Table 17.2). Males are very selective in their choice of lek site and locations of leks are very stable through time. For example, Whittier et al. (1992), conducted a systematic census of 118 trees in a mixed orchard in Hawaii. Ten of these trees accounted for 80% of lek sightings, while 79 trees had zero sightings. As a result of male selectivity, leks (and their product — copulations) occur on only a small subset of available trees. There is very little information on the characteristics of these trees. In the only study of its kind, Shelly and Whittier (1995) associated lek distribution among individual trees of a single species (persimmon, *Diospyros kaki* L.) with various physical and positional characteristics of the trees. Multiple regression was performed using lek number as the dependent variable, and tree height, tree volume, foliage density, average distance to two nearest neighbors, and total lek number on the two nearest neighbors as independent variables. Leks were found to be nonrandomly distributed among persimmon trees, owing primarily to the large number of trees that contained no leks at all. Also, lek location was stable over the 60-day study period. The regression analysis identified only two variables, tree volume and total lek number on the two nearest trees that were significantly associated with lek distribution. Leks were rarely found on the smallest trees at the study site (but were evenly distributed among medium and large trees). Furthermore, the leks themselves were generally clustered in space, this result deriving largely from the fact that the five most frequently used persimmon trees were aggregated.

Once a tree is chosen, factors important in site selection within the tree could be abiotic (temperature, relative humidity, light intensity; Prokopy and Hendrichs 1979) and biotic (such as predation risk; Hendrichs et al. 1994; Hendrichs and Hendrichs 1990; 1998), as well as proximity to nutritional resources and oviposition sites, the presence of other males, and leaf size and integrity (Kaspi and Yuval 1999a). The relative weight of each factor should be reflected by the three-dimensional location of a male within the canopy, which optimally should provide protection from predators (Hendrichs and Hendrichs 1990; Hendrichs et al. 1994), wind, direct sunlight, and water loss (Prokopy and Hendrichs 1979; Arita and Kaneshiro 1989), while providing an effective platform from which to broadcast courtship signals (Kaspi and Yuval, 1999b).

The alternative reproductive tactic of fruit guarding is seen infrequently, and usually after lekking activity has ceased (Prokopy and Hendrichs 1979; Warburg and Yuval 1997a). Males that employ this tactic establish a territory on a suitable oviposition host and accost females that arrive on this host to oviposit. It is not clear which males in the population choose to employ this tactic, whether some males only fruit-guard, or if males who do not succeed in copulating females at leks proceed to fruit-guard in an attempt to realize their reproductive potential. Warburg and Yuval (1997b) compared the energetic status of lekking to fruit-guarding males. The sugar-to-lipid ratio of the latter was significantly higher than that of lekking males. This could indicate that energetic status dictates to some extent what mating tactic a male will employ, or simply reflect sugar feeding by males immediately before fruit guarding.

**TABLE 17.2**
**Field Studies of Lekking by *C. capitata* Males**

| Authors | Study Site | Physical Properties | Host Species | Predators | Alternative Tactics |
|---|---|---|---|---|---|
| Arita and Kaneshiro 1989 | Hawaii (Maui) | Direct sunlight, underside of leaves | Lemon, coffee, plum | — | — |
| Hendrichs and Hendrichs 1990 | Egypt | Dense canopy, direct sunlight | Orange, other species (but not fig) | Mantids, odonates, wasps | Fruit guarding in late afternoon |
| Hendrichs et al. 1991 | Greece (Chios) | Dense canopy, direct sunlight | Orange (not fig) | Wasps | Fruit guarding in late afternoon |
| Whittier et al. 1992 | Maui | Direct sunlight, underside of leaves | Lemon, persimmon (very consistent) | — | — |
| Shelly et al. 1993 | Maui | — | Persimmon, loquat, plum, lemon | — | — |
| Shelly et al. 1994 | Maui | — | Nonrandom and consistent | — | — |
| Warburg and Yuval 1997a | Israel | Under leaves | Orange, pitanga (not fig or guava) | Wasps | Fruit guarding |
| Shelly and Whittier 1995 | Maui | See text | Persimmon | — | — |
| Kaspi and Yuval 1999b | Israel | Under shaded leaves, position correlated to sun's azimuth | Pitanga | — | — |

*Ceratitis rosa* also employs a lek mating system (S. Quilici, personal communication), where males establish territories on the undersides of leaves, from which they produce a variety of courtship signals. In contrast to the Mediterranean fruit fly, leks of this species form in the late afternoon, and last until nightfall (Myburgh 1962; Etienne 1973).

### 17.3.2.2 Courtship

Courtship of the Mediterranean fruit fly, first described in detail by Féron (1962), consists of a sequence of discrete and ritualized behaviors, leading to copulation (Figure 17.1). These include emission of pheromone from the eversible rectal ampulla to attract a female, vigorous wing enation and audible buzzing following the arrival of a female, and additional displays possibly combined with tactile stimuli before mounting (Briceño et al. 1996; Liimatainen et al. 1998; and Eberhard, Chapter 18 present detailed analyses of these behaviors). These behavioral elements are also present in *C. rosa,* with the addition of visual stimuli associated with the midtibia hair comb (S. Quilici, personal communication).

Courtship is usually terminated at some stage by the female, who decamps (Arita and Kaneshiro 1989). Arita and Kaneshiro (1985) found that 15% of males account for 46% of matings in leks. Whittier et al. (1992), observing leks in nature, found that male size and territory location do not correlate with copulatory success, and 6% of the copulations observed featured previously mated males. Who then are the lucky few, the males that satisfy female scrutiny and proceed to copulate? This question is of paramount importance for the success of control operations based on the sterile insect technique, and its resolution is elusive. In laboratory experiments Churchill-Stanland et al. (1986) and Orozco and López (1992) documented an advantage for large males. However, Arita and Kaneshiro (1988), comparing the mating success of males raised on two different host fruits and competing for mates in the same mating cage found that, although significantly smaller in size, males emerging from coffee copulated more frequently than males emerging from cherry. The energetic status of these populations may have differed, and could be the key to understanding differences in copulatory success. Whittier et al. (1994) documented that male copulatory success was significantly related to the level of sexual activity: number of courtships performed, number of attempted copulations, and number of different females courted. Furthermore, Blay and Yuval (1997) found that males fed no protein were significantly less likely to copulate than were protein-fed males. In both diet groups, size was significantly associated with copulatory success. Finally, symmetry of the supra-orbital setae on the males' head has been correlated to copulatory success (Hunt et al. 1998). A synthesis of these results, all robust in their own right, but nevertheless sometimes contradictory, would be timely. In addition, exactly how females choose mates, and which cues are important in this choice, remain unresolved problems, and should be studied.

### 17.3.2.3 Copulation and Sperm Transfer

If accepted by the female, the male mounts and threads his phallus through the female's ovipositor into her anterior vagina. Sperm are apparently released at this site and must travel up the spermathecal ducts to be stored in the spermathecae (Solinas and Nuzzaci 1984; Eberhard, Chapter 18). During copulation, males perform various thrusting and tapping movements, whose function may be related to sperm transfer, or serve as copulatory courtship cues (*sensu* Eberhard 1991).

Laboratory studies suggest that copula duration is positively associated with both the amount of sperm stored by females (Farrias et al. 1972; Seo et al. 1990) and female nonreceptivity to later males (Saul et al. 1988). Strain differences in copula duration appear to be widespread among Mediterranean fruit flies. Studies of Mediterranean fruit flies in nature have also reported dramatically disparate copulation durations: Wong et al. (1984) and Whittier et al. (1992) both observed copulations at natural leks in Hawaii, and report copula durations averaging 144 and 179 min, respectively. Copula duration is dependent on laboratory strain and size asymmetry of the mating

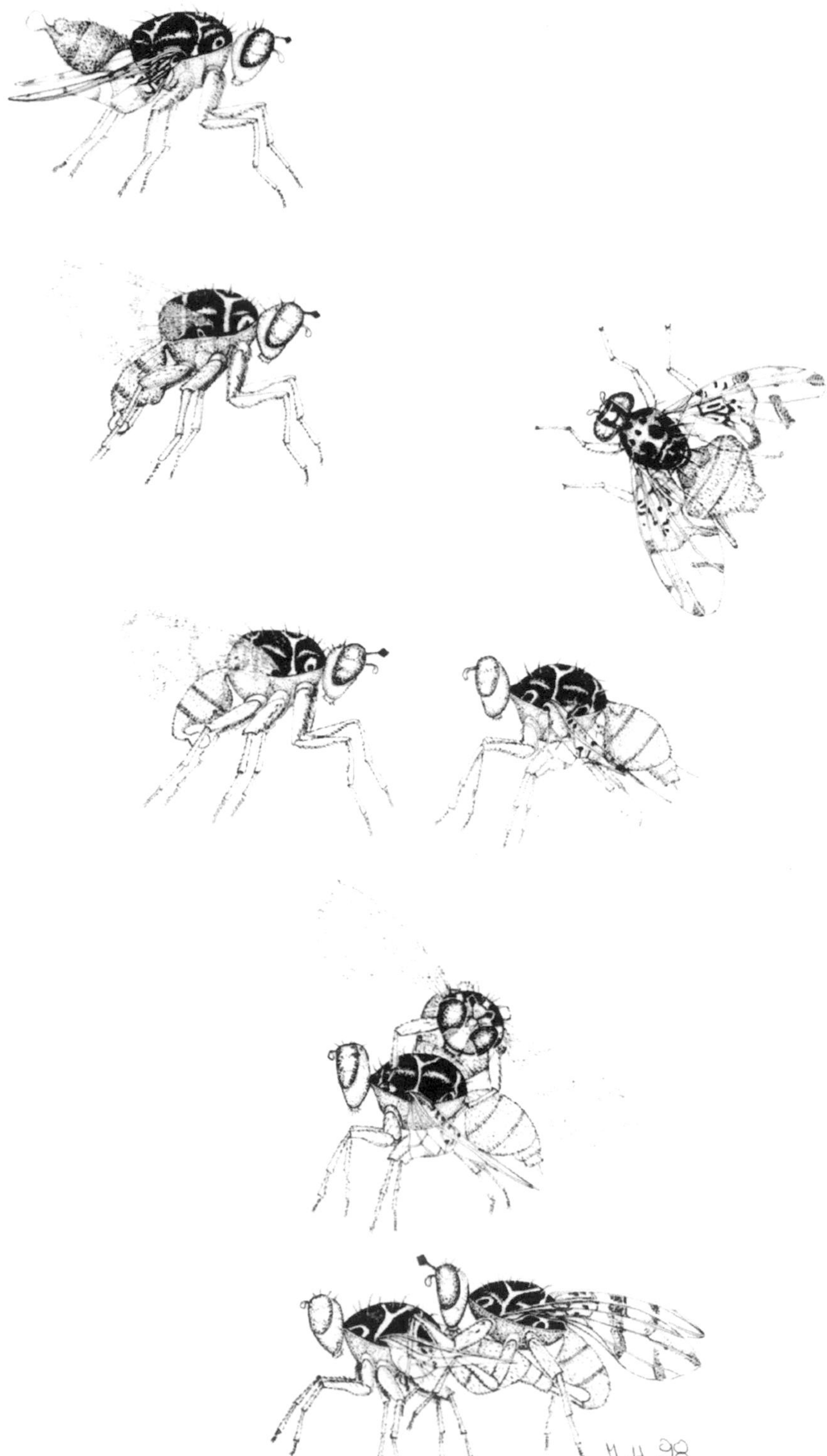

**FIGURE 17.1** Courtship and copulation in *Ceratitis capitata*. (Drawn by M. Yuval, after Féron 1962.)

pair: Flies of one laboratory strain copulated for longer than flies of another strain, and pairs in which males were smaller than females copulated for longer than pairs in which the flies were more evenly size-matched (Field et al., in press). Absolute female size also had a significant effect on copula duration, although less so than relative size, and no effect of absolute male size was detected.

Adult nutritional status also affects copula duration (Field and Yuval 1999). When males and females were either protein fed or deprived of protein, the effect of female nutrition on copula duration was stronger than that of male nutrition, suggesting that females exert substantial control over copulation. When both males and females were deprived of protein, copula duration was prolonged, both for the first mating and remating on the following day. The interests of protein-deprived males may be served by mating for longer, due to their lesser ability to obtain further matings. The benefit to females of mating for longer when protein deprived is less obvious, and may simply be due to a lesser ability to force the male to disengage. Another possibility is that females obtain a nutritional benefit from the male, in which case longer copulations would represent a confluence of male and female interests.

Copulations of *C. rosa* are of much longer duration than those of *C. capitata*. They commence at dusk and pairs remain in copula until the following morning (Myburgh 1962). The function of this prolonged mating has not been investigated.

Various aspects of sperm development (McInnis 1993) and transfer in Mediterranean fruit flies have been studied. Farrias et al. (1972) determined qualitatively that sperm accumulate in spermathecae during copula. Seo et al. (1990) found that after 2 h of copulation most (94%) sperm were in the spermathecae, and the remainder in other areas of the reproductive tract. Gage (1991) showed that risk of sperm competition affects ejaculate size in the Mediterranean fruit fly. Yuval et al. (1996) determined that male testes contained on average 34,000 sperm cells before copulation and approximately 14,000 after copulation. Intriguingly, most of this ejaculate never reaches the spermathecae. Spermathecae of once-mated laboratory-reared females contained 3212 sperm on average, similar to the amount found in field-collected females. Sperm allocation between the spermathecae, in both field and laboratory females, is significantly nonrandom; one organ (either left or right) contains significantly more sperm than expected by chance, suggesting that allocation of sperm is controlled, either by the male or by the female (Taylor and Yuval 1999).

Sperm is not the only substance transferred by males to females. Part of the ejaculate consists of male accessory gland secretions (MAGS), which contain many biologically active compounds that may affect the inseminated female in a number of ways (reviews by Leopold 1976; Chen 1984; Happ 1992; Eberhard and Cordero 1995). There is no precise information on the function of tephritid MAGS. However, patterns of female behavior (Jang 1995; Jang et al. 1998) are modulated by copulation or injections of MAGS — females thus treated are likely to avoid males and seek oviposition sites.

### 17.3.2.4 Female Remating and Sperm Precedence

Remating is common in laboratory strains of Mediterranean fruit flies. Saul et al. (1988) established second male sperm precedence in multiply mating females, and Saul and McCombs (1993a,b) discuss the adaptiveness and heritability of remating in Mediterranean fruit flies of various strains. Whittier and Shelly (1993) showed that multiple mating was adaptive, as multiply mated females have a significantly higher reproductive output than once-mated females. The proportion of females who remate varies in the various experimental settings. Hendrichs et al. (1996) found that in field cages 7% of females remate on the first day following the day of the initial mating, while in smaller laboratory cages as many as 50% (Whittier and Shelly 1993) or even 76% of females remate (Blay and Yuval 1997). In addition to these environmental effects (which may also reflect strain differences), physiological processes appear to regulate remating. There is a significant negative correlation between length of initial mating and proportion of Mediterranean fruit fly females that mate a second time

(Saul et al. 1988). Blay and Yuval (1997) found that the nutritional status of a female's first sexual partner affects her receptivity to further copulations. Females mated to protein-deprived males copulated again significantly more frequently than females whose first mate was protein fed. How female receptivity is regulated is unknown, and should prove to be a rewarding avenue for future research. In addition, following the fate, in the female reproductive tract, of ejaculates provided by different males presents an excellent model to study sexual selection at the male–female interface.

### 17.3.3 Oviposition Behavior

The genus *Ceratitis* encompasses, in terms of host use, a broad range from generalists to specialists, including polyphagous, oligophagous, and monophagous species (see Table 17.1). This impressive diversity in habitats is accompanied by numerous adaptations in ovipositor, especially aculeus, morphology (De Meyer, Chapter 16), reproductive physiology and oviposition behavior, as well as life strategies adapted to the phenology of the different hosts. Due to the very limited information on the oviposition behavior of other *Ceratitis* species (with the exception of *C. rosa*, the Natal fruit fly, see Barnes 1976; Quilici and Rivry 1996), here, too, we focus on *C. capitata*, the best-studied species of the group.

A distinctive characteristic of *C. capitata* is the great plasticity of its oviposition behavior (Prokopy et al. 1984; Katsoyannos 1989), which is reflected in the ability to colonize diverse habitats and to accept as hosts even fruit that are unable to support larval development (Carey 1984). Unlike other polyphagous tephritids of economic importance, including *C. rosa,* this plasticity of *C. capitata* results in highly nonadaptive behavior, such as egg laying in nonconvex screen surfaces with no color, tactile, or odor stimuli reminiscent of any fruit (Schwarz et al. 1981).

The oviposition behavior of *C. capitata*, and to a much lesser degree of other frugivorous *Ceratitis* species with a broad host range such as *C. rosa*, has been the object of many studies and various reviews (see Katsoyannos 1989; Prokopy and Roitberg 1989; Prokopy and Papaj, Chapter 10). The study of host selection and egg laying is of interest in these species because of their important pest status and significance of the findings for trapping purposes. Furthermore, oviposition is readily observable in nature and can be quantified and manipulated in the laboratory and field to answer questions on fly foraging and learning behaviors, as well as on fly physiology, ecology, and host-marking pheromones.

As detailed above, before being able to initiate egg laying, *C. capitata* females have to undergo a sexual maturation period after emergence during which they require a regular supply of amino acids and other nutrients to fuel oogenesis. This anautogeny, typical of most frugivorous tephritids, is the result of a rapid larval development in short-lived fruit, generally poor in proteins (Hendrichs and Prokopy 1994). *Ceratitis capitata* females have an average of 28 polytrophic ovarioles in each ovary (Hanna 1938) and can produce 300 to 1000 eggs throughout life, laying eggs every day (Christenson and Foote 1960; Bateman 1972). For this large reproductive capacity to be fully expressed they require these nutrients, generally on a daily basis, to continue producing oocytes in all successive stages of maturation.

Females visit fruit infrequently before becoming gravid, and then mostly to feed on fruit juices. However, after mating and successful reception of sperm and accessory gland products, female behavior changes from mate-foraging to host-foraging behavior (Levinson et al. 1990). Jang (1995), by injecting accessory gland extracts into mature virgin *C. capitata* females, demonstrated a dramatic and nearly immediate switch from male-pheromone to host-oriented olfactory behavior, even though these females had never mated or received sperm. In addition, Jang et al. (1998) showed strain differences in this response, with a progressive reduction in the ability to induce such a switch in behavior from laboratory-reared normal males to sterile males to wild males.

Already during the premating maturation period, females disperse more than males in search of food (Hendrichs et al. 1991), and sometimes also emigrate to new locations, thereby colonizing new host patches. Females with mature ovaries tend to remain on or very near fruiting host plants as long as fruit is available for egg laying. When fruit in an area becomes scarce and competition among females for remaining fruit is intense, females disperse rapidly, sometimes over long distances, in search of new host plants with acceptable fruit (Prokopy and Roitberg 1984; Hendrichs and Hendrichs 1990).

The major long-range stimuli guiding mature *C. capitata* and *C. rosa* to patches with host plants are volatile components related to a fruiting host tree consisting probably of a combination of ripening fruit odors (Sanders 1968; Guerin et al. 1983; Levinson et al. 1990; Ripley and Hepburn 1929), together with other odors such as green leaf volatiles and male pheromone odors (Light et al. 1988; Dickens et al. 1990), droppings of frugivorous birds (Prokopy et al. 1992), and fermenting fruit on the ground.

Host detection within the host habitat is a sequence of combined olfactory (Schoonhoven 1983; Levinson et al. 1990) and visual responses (Sanders 1968; Katsoyannos 1989). Upon arrival on host plants, visual characteristics play a major role in detecting and selecting fruits. To locate an individual fruit, gravid females employ a variety of fruit characters such as shape, size, and color. They prefer spheres over other shapes on the basis of the circular outline, as well as large fruit size (attraction increases as the size of spheres increased from 1.8 to 18 cm), whereas color properties of fruit and background are secondary to the convex shape and fruit size (Katsoyannos 1989).

Upon landing on a fruit, females assess whether it is acceptable for egg laying based on physical characteristics (surface structure and fruit quality and condition) perceptible by tactile stimuli and surface chemistry characteristics (both short-range fruit odors and the presence or absence of host-marking pheromone) perceptible by olfactory stimuli (Schoonhoven 1983; McDonald and McInnis 1985; Averill and Prokopy 1989; Prokopy and Roitberg 1989). As can be expected from polyphagous *C. capitata*, there is considerable intra- and interpopulation plasticity as well as wild and laboratory-reared strain effects in these behaviors and preferences (Prokopy and Roitberg 1984; Papaj et al. 1987; 1989a; Katsoyannos 1989).

The fate of fruit fly offspring is largely determined by the decisions made by female fruit flies of where to deposit their eggs. This is because larvae have no option to leave the host in which they were deposited by their mother to escape either low-quality fruit, intra- and interspecific competitors, or natural enemies. Female behavior involved in the choice of oviposition site is therefore of great consequence to offspring fitness. For example, even though over 200 species of fruit and vegetables have been confirmed to sustain *C. capitata* larval development (Liquido et al. 1991), egg depositions are not uncommon on hosts that are unable to sustain larvae to maturity (Carey 1984; Krainacker 1986). Furthermore, oviposition attempts are frequent on unripe fruit.

By ovipositing as early as possible in still green and hard maturing fruit (a typical characteristic of polyphagous frugivorous tephritid fruit flies, and which makes them such formidable commercial pests), females increase the probability that at least some of their offspring will complete larval development and leave the fruit before the fruit is consumed by frugivorous mammals or birds (Drew 1987), or before the mature fruit drops and is attacked by additional interspecific competitors such as ground-dwelling insects or vertebrates.

Most important from the point of view of offspring given the often intense intraspecific competition for oviposition sites is that females detect fruit early before it has been found by intraspecific competitors (Vargas et al. 1995). As a result of larval competition, survival and size of survivors decrease progressively with increasing larval density (Papaj et al. 1989b). Furthermore, as a result of the decreased size of *C. capitata* adults with increasing larval density in fruit, the fecundity of female progeny as well as the mating success of male progeny decreases (Papaj et al. 1989b; Orozco and López 1992). Interspecific competition has also been shown by Keiser et al. (1974) to play (at least in Hawaii), a major role, with *C. capitata* larvae being suppressed by *Bactrocera dorsalis* (Hendel) larvae when found together in the same fruit (but see Vargas et al. 1995).

As a result of the significantly lower fitness of progeny developing in heavily infested fruit, *C. capitata* females have evolved the capacity to protect their offspring both by depositing a host-marking pheromone and by being able to detect the pheromone deposited by competitors, to discriminate between uninfested and already infested fruit (Prokopy et al. 1978). The female drags the aculeus of her ovipositor after egg laying and deposits marking pheromone first locally around the oviposition site and eventually over the surface of the fruit. Averill and Prokopy (1989) have related the evolution of such host-marking pheromones in frugivorous tephritids to monophagous or oligophagous habits. They link these pheromones to enhanced likelihood of frequent or severe intraspecific larval competition. Its presence in a generalist such as *C. capitata,* however, indicates that this ecological characteristic is not exclusive of monophagous or oligophagous species. Actually, an international comparison of *C. capitata* host-marking pheromone (Boller et al. 1994) showed that Mediterranean fruit fly populations from all parts of its geographic distribution responded similarly to it, and that laboratory strains with extremely long rearing histories have still retained a reduced capability to perceive and react to its deterrent properties.

*Ceratitis capitata* females lay their eggs in clutches of one to ten eggs, with fruit size influencing fruit acceptability as an oviposition site by female flies. McDonald and McInnis (1985) reported average numbers of eggs per oviposition increasing by about one egg with every 10 mm increase in fruit diameter. Papaj (1990), however, showed that clutch size is independent of either fruit acceptability or size, but that host-marking pheromone accumulates more rapidly on small fruit in relation to fruit surface area, affecting the number of clutches deposited. Papaj et al. (1990) demonstrated that clutch size was significantly lower on uninfested fruit artificially marked with host-marking pheromone compared with uninfested and untreated fruit.

*Ceratitis capitata* females frequently deposit eggs in fruit cracks or wounds or preexisting oviposition punctures in infested fruit (Silvestri 1914; Back and Pemberton 1918; Papaj et al. 1989c; 1992). Despite the deterrence of host-marking pheromone present on infested fruit, the use of existing sites holds a number of advantages for female Mediterranean fruit flies. First, the proportion of successful egg-laying attempts in such sites reduces the wear on a female's aculeus (Jones et al. 1993). Second, the peel of unripe citrus fruit is particularly toxic to larvae (Back and Pemberton 1918; Carey 1984; Krainacker 1987; Katsoyannos et al. 1997). By preferring wounds that penetrate the peel to shallow ones (Papaj et al. 1989c), the survival of the female's progeny is increased as larvae are not exposed to the allelopathic essential oils of the flavedo of citrus fruit. Third, time to deposit a clutch decreased at least by half compared with oviposition at a new site. Fourth, as females are more vulnerable to attacks by yellow jacket wasps (*Vespula germanica* (L.)) and other predators while ovipositing than when engaged in other activities (Papaj et al. 1989c; Hendrichs and Hendrichs 1990; 1998), use of old oviposition punctures reduces significantly the period during which a female is especially vulnerable to predation.

On the other hand, as a result of a limited availability of preexisting sites and a relatively narrow afternoon window during which most oviposition takes place, egg-laying females are often interrupted or even displaced by other females while ovipositing in preexisting sites (Papaj et al. 1989c). The time and energy invested in intraspecific competition when exploiting an existing site at least partially counterbalance the disadvantages of pioneering a new puncture. Recently, Papaj and Messing (1996) assessed the dynamics of a female's response to egg-occupied fruit in relation to fruit size and surface penetrability. Females tended to avoid occupied fruit on both large and small fruit; however, as expected, avoidance was greater on small fruit. Females reused sites almost exclusively when fruit were unripe, whereas they tended to avoid occupied fruit, when fruit were ripe. When oviposition sites are contested, a resident advantage is apparent, and is associated with female age and behavior — older females, and those engaged in oviposition, are significantly more likely to win contests when residents (Papaj and Messing 1998).

Prokopy and Duan (1998) recently examined the effect of the presence of an ovipositing female on oviposition site selection. The behavior of individual mature female Mediterranean fruit flies, transferred from a holding cage without fruit to a clean host kumquat fruit already occupied by another Mediterranean fruit fly female engaged in oviposition, was monitored. A significantly greater proportion of ovipositionally naive females initiated ovipositor boring into a fruit in the presence than in the absence of an occupying Mediterranean fruit fly. These results suggest that some degree of social facilitation occurs in the interaction between naive and ovipositing females. The inexperienced females may benefit from acquisition of a cue demonstrating the acceptability of a host for oviposition. What benefits accrue to the female who first occupies the oviposition site, and how this facilitation is regulated, remain to be established.

Finally, learning also plays a role in host fruit acceptance. *Ceratitis capitata* females that oviposit several times in a given fruit type are subsequently highly prone to reject other equally acceptable fruit types (Cooley et al. 1986). This conditioning to hosts is reversible, with memory of ovipositional experience on a given fruit type lasting several hours or days (Papaj et al. 1989a). Fruit size is the principal character learned and used in finding familiar fruit, while color and odor appear to be of little or no importance in this regard (Prokopy et al. 1989). This seems to be the case in other polyphagous ceratitine species, as Quilici and Rivry (1996) found that *C. rosa* could not be conditioned to accept oviposition sites based on color alone. Learning also plays a role in laboratory flies, as mass-reared females with egg-laying experience or wild females have lower propensity to accept artificial oviposition substrates than naive females of an old laboratory strain (Prokopy et al. 1990).

In conclusion, considerable information on *C. capitata* oviposition behavior is now available, and some limited knowledge is being gained on other polyphagous *Ceratitis* species such as *C. catoirii* Guérin-Méneville, *C. cosyra,* and *C. rosa* (Quilici and Rivry 1996). However, no information is available on other species, some of which although not polyphagous, may also be of economic importance. A comparative study of behaviors involved in oviposition, specifically comparisons between monophagous and oligophagous species may pave the way to understand the evolution of this behavior in this genus. Thus, a comparison of the learning abilities of monophagous and polyphagous species in the genus could illuminate the question of the adaptiveness of learning. Does an increased learning ability go hand in hand with polyphagy, or is the learning ability of these flies a phylogenetic attribute?

### 17.3.4 Spatial and Temporal Patterns of Behavior

The partitioning of time between reproductive and feeding activities represents an evolutionary problem for which each species of insect finds a particular solution. This solution reflects the ecological imperatives of the insect's environment, and specific physiological and behavioral constraints (Roitberg 1985; Stearns 1989; Bell 1990). Some species solve this problem on a seasonal timescale, concentrating feeding and growth in one developmental period and reproduction in another (e.g., ephemeropterans and many lepidopterans). On the other hand, the reproductive success of numerous other species hinges on their ability, as adults, to forage efficiently for various nutritional resources. These resources must then be budgeted in such a manner as to satisfy the somatic and reproductive needs of the individual (e.g., Raubenheimer and Simpson 1995). Furthermore, these adults must decide when to forage for nutritional resources, when to invest time on reproduction (seek mates, copulate, oviposit), and when to desist altogether (Bell 1990). These decisions are frequently triggered by physiological thresholds, which dictate or regulate the expression of a specific behavior.

Discrete circadian behavioral patterns have been observed repeatedly in Mediterranean fruit fly populations in laboratory (Causse and Féron 1967; Smith 1989), seminatural (Prokopy and Hendrichs 1979; Prokopy et al. 1996b) and natural environments (Cayol 1996; Hendrichs and Hendrichs 1990; Hendrichs et al. 1991; Whittier et al. 1992; Warburg and Yuval 1997a). The observed activity patterns indicate a partitioning of the day between nutritional and reproductive activities. In general, male activities are regimented and discrete with a separation between feeding and reproductive activities, whereas females engage in different behaviors throughout the day. Males congregate in leks from midmorning to early afternoon, and rarely engage in other activities during this period. Fruit guarding, an alternative tactic, occurs mainly in the late afternoon (Prokopy and Hendrichs 1979). Feeding by both sexes takes place primarily in the evening on ripe and rotten fruit, leaf

**TABLE 17.3**
**Factors Affecting Spatial and Circadian Patterns of *C. capitata* Activity**

| Factor | Effect | Reference |
|---|---|---|
| **a. Environmental** | | |
| Temperature | Elevated temperatures inhibit flight and sexual activities. | Baker and Van der Valk 1992; Cayol 1996; Hendrichs and Hendrichs 1990; Prokopy et al. 1987 |
| Oviposition hosts | Lack of suitable hosts hastens female emmigration; host ripeness affects oviposition decisions. | Prokopy et al. 1987; 1993b |
| Nutritional resources | Female search for protein extends to nonhost patches. | Hendrichs and Hendrichs 1990; Hendrichs et al. 1991; Prokopy et al. 1996a |
| Lek sites | Specificity of lek sites causes patchy distribution of males during hours of sexual activity. | Shelly and Whittier 1995; Warburg and Yuval 1997a; Whittier et al. 1992a |
| Predators | Affect choice of oviposition and lek sites, and timing of feeding activities. | Hendrichs and Hendrichs 1990; Hendrichs et al. 1991; Hendrichs et al. 1994; 1998; Warburg and Yuval 1997a |
| **b. Physiological** | | |
| Age | Young flies do not participate in reproductive activities. | Bateman 1972; Christenson and Foote 1960; Féron 1962 |
| Egg load | Female responses to hosts increase. | Prokopy 1984; Prokopy et al. 1993b |
| Nutritional reserves | Protein hunger increases attraction to protein sources. | Averill et al. 1996; Warburg and Yuval 1997b |
| | Low sugar and protein levels proscribe males from participating in leks. | Yuval et al. 1998 |
| Mating status | Recent mating diverts females to oviposition sites. | Jang 1995; Jang et al. 1998 |
| Health | ??? | |

surfaces and fresh bird feces, although bursts of feeding occur in the morning as well, and females occasionally feed throughout the day (Hendrichs and Hendrichs 1990; Hendrichs et al. 1991; Warburg and Yuval 1997a). Oviposition, although taking place sporadically at all hours, is concentrated primarily in the second half of the day (Hendrichs and Hendrichs 1990; Hendrichs et al. 1991; Warburg and Yuval 1997a).

By comparing the results observed in Mediterranean habitats and New World habitats (Baker and Van der Valk 1992; Whittier et al. 1992), we can attempt to identify the hierarchies that govern Mediterranean fruit fly behavior (Table 17.3). These may be divided into environmental and internal (or physiological) factors, and we treat them separately below.

### 17.3.4.1 Environmental Factors Regulating Behavior Patterns

The major environmental factor regulating behavior is apparently temperature. In Egypt, when relatively high temperatures prevailed (Hendrichs and Hendrichs 1990), very little activity was seen during the middle of the day, resulting in a bimodal pattern of lekking by males, and oviposition and feeding by females. Similarly, a bimodal activity pattern was seen in Tunisia, when temperatures exceeded 30° C (Cayol 1996). In addition, rising temperatures drive flies toward the cooler center of the tree they inhabit (Hendrichs et al. 1990; Baker and Van der Valk 1992; Kaspi and Yuval, in press). In Chios, Hawaii, Mexico, and in Israel, temperatures were milder (at the time published observations were made), imposing no restrictions on behavior, resulting in similar, unimodal, circadian activity patterns.

Within the constraint imposed by temperature, other factors are important as well. Hendrichs et al. (1991) established that flies adjust their food-foraging activities in response to spatial, temporal, and seasonal distribution of food resources. Dispersal and activity patterns indicate that females go farther in search of protein than do males, a habit which, combined with encountering novel oviposition hosts, enables them to colonize new habitats (Hendrichs et al. 1991; Prokopy et al. 1996a).

The availability of suitable oviposition hosts is another factor that affects spatial behavior of females. Prokopy et al. (1987) investigated the intratree foraging behavior of individually released, wild-population Mediterranean fruit flies on trees which varied in the density and quality of fruit suitable for oviposition. Females emigrated within a few minutes after release on trees devoid of fruit. With increasing density of fruit, females tended to remain longer in trees, visit more fruit before leaving, oviposit more often, accept a proportionately smaller number of fruit visited, and emigrate sooner after the last egg was laid. Furthermore, females spent much less time and oviposited much less often on trees harboring pheromone-marked fruit than noninfested fruit.

The availability of lek sites may limit and shape spatial patterns of male behavior. As detailed above, males are very selective in their choice of lek site and locations of leks are very stable through time. As a result of male selectivity, leks (and their product — copulations) occur on only a small subset of available trees.

Predator avoidance also has a role in determining activity patterns. Predators of Mediterranean fruit flies in the field include odonates, vespid wasps, mantids, and hunting spiders (Hendrichs and Hendrichs 1990; 1998; Hendrichs et al. 1991; 1994; R. Kaspi and B. Yuval, unpublished data). In exposed locations, such as open foliage or on fruit, flies are often attacked or ambushed. The wing pattern of Mediterranean fruit flies (and many other tephritids) is considered to be an antipredatory adaptation, inasmuch as they mimic the shape of jumping spiders (Mather and Roitberg 1987). Overall, flies seem to be able to evade predation. For example, of every 12 ambush attempts by praying mantids, on average only one was successful; however, the odds were worse for ovipositing females, where one of every four attacks was successful (Hendrichs and Hendrichs 1990). Libellulid dragonflies, waiting on perches, seem to specialize on flies flying into or out of the foliage. Mantids specialize in ambushing flies on the foliage in the mornings and near fruits during the oviposition period. The activity of these predators may have significant effects on spatial and temporal patterns of Mediterranean fruit fly behavior. Vespid wasps are capable of homing in on males calling in leks and attacking the males congregating there, as well as pairs copulating nearby (Hendrichs et al. 1994; Hendrichs and Hendrichs 1998). Apparently, males are able to identify visually some of these predators and choose leks sites accordingly (R. Kaspi and B. Yuval, unpublished data). Thus one factor influencing lek site choice may be the probability of detection by predators, and may explain why leks are rarely found on fig trees (Hendrichs et al. 1994). Similarly, it has been suggested that oviposition site selection is affected by the probability of predation (Hendrichs et al. 1991). The intense peak of feeding by flies of both sexes may also be driven by predator activity. In Israel flies begin their period of intense feeding on the same fruits where wasps feed before, after wasp activity has ceased. There is evidence from other insects that activity patterns of predators may, to some extent, regulate the activity of their prey (e.g., Burk 1982; Yuval and Bouskila 1993, and references therein). It may be that where nutrients are not a limited resource, time to consume them without the threat of predation limits and regulates feeding.

#### 17.3.4.2 Internal Factors Regulating Behavior Patterns

Chief among the internal factors that govern female behavior appear to be protein hunger and search for oviposition sites motivated by egg load. Females forage farther than males to satisfy these needs, and, although concentrating these activities in the latter part of the day, they are much more flexible than males, temperature permitting. It may be that feeding and oviposition by females during the "off hours" of midmorning are motivated by hunger and heavy egg loads, respectively. This hypothesis remains to be critically tested in the field. Conversely, male feeding is constrained by an internal clock — they rarely feed during the hours allocated to sexual activities (Warburg and Yuval 1997a), even when their nutritional status denies them the ability to participate in leks (Yuval et al. 1998).

Onset of sexual behavior in males and sexual receptivity in females is age related. Thus, young flies spend their active time feeding, entering the sexual population when they are 7 to 10 days old.

Mating status critically affects behavior patterns. Following copulation and transfer of sperm and male accessory gland products, female behavior becomes host oriented (Jang 1995; Jang et al. 1998). Whether males lek on consecutive days in the field and if copulatory success affects subsequent sexual activity is not clear (but see Whittier et al. 1994).

Another factor that may affect individual patterns of behavior is health. Infection with pathogens has been shown to alter the behavior of insects of various species (Zuk 1987; 1988; Simmons and Zuk 1994; Polak and Markow 1995; Polak 1996). This question has not been addressed in *Ceratitis*.

## 17.4 CONCLUSIONS AND FUTURE RESEARCH NEEDS

Considerable progress has been made in understanding the behavior of at least one species of the genus *Ceratitis*. Nevertheless, the more we learn about the behavior of these extraordinary flies, the more questions that remain unanswered. Future research on flies of this genus should attempt to increase our basic knowledge of other species of *Ceratitis,* in addition to the intensively studied *C. capitata*. Specifically, there is room for a comparative approach analyzing the same behaviors in sympatric species of the genus. Such information will establish the relative weight of ecological and phylogenetic constraints in determining patterns of feeding and oviposition as well as the evolution of mating systems. Furthermore, interspecific competition of sympatric vs. nonsympatric species both at the larval and the adult stage should be scrutinized.

In addition to bringing other ceratitines into the fold of knowledge, much remains to be understood about the basic biology of the Mediterranean fruit fly. Understanding fly behavior in its original habitat in the presence of its original guild of parasitoids and other natural enemies (including birds and mammals) should provide novel insights. Larval behavior within the host is an enigma, as are the responses of larvae to natural enemies. The importance of bacteria in both larval and adult diets is another issue deserving further study. Feeding behavior of adults, which is intimately linked to reproductive success and pivotal in many control schemes, is still poorly understood, in particular the hierarchies between protein and sugar feeding, the sources of protein in the field, and the interactions between adult feeding and other patterns of adult behavior.

Oviposition behavior remains a frontier, which will yield insights on the trade-offs between selfish and cooperative behavior, as well as providing a relatively easy system in which to study the evolution of decision making in insects. Large gaps still exist in our understanding of the sexual biology of these flies. In particular, questions on determinants of male copulatory and reproductive success remain unanswered. In addition, the related but distinct question of how females choose males remains to be resolved.

Time flies, but flies wait for no one. There is still a lot of work ahead.

## ACKNOWLEDGMENTS

We thank S. Quilici for sharing unpublished information; W. Eberhard and S. Field for illuminating comments on parts of a preliminary draft; and D. Papaj, M. Aluja, and an anonymous reader for critical reviews of the whole chapter. M. Yuval drew Figure 17.1. B.Y.'s research was supported by the Israel Science Foundation, BARD, and the California Department of Food and Agriculture, and benefited immensely from the contributions of S. Blay, H. Cohen, S. Field, R. Kaspi, Y. Perry, P. Taylor, S. Shloush, and M. Warburg. J.H.' s work was supported by the Joint FAO/IAEA Division of Nuclear Techniques in Food and Agriculture in Vienna, with major contributions by M. Hendrichs, B. Katsoyannos, N. Kouloussis, N. Papadopoulos, R.J. Prokopy, and V. Wornoayporn. This chapter was prepared thanks to support from the Campaña Nacional contra las Moscas de la Fruta (Mexico), International Organization for Biological Control of Noxious Animals and Plants (IOBC), and Instituto de Ecología, A.C. (Mexico).

## REFERENCES

Alcock, J. 1994. Postinsemination associations between males and females in insects: the mate-guarding hypothesis. *Annu. Rev. Entomol.* 39: 1–21.

Aluja, M. 1994. Bionomics and management of *Anastrepha*. *Annu. Rev. Entomol.* 39: 155–178.

Arita, L.H. and K.Y. Kaneshiro. 1985. The dynamics of the lek system and mating success in males of the Mediterranean fruit fly, *Ceratitis capitata* (Wiedemann). *Proc. Hawaii Entomol. Soc.* 25: 39–48.

Arita, L. and K.Y. Kaneshiro. 1988. Body size and differential mating success between males of two populations of the Mediterranean fruit fly. *Pac. Sci.* 42: 173–177.

Arita, L.H. and K.Y. Kaneshiro. 1989. Sexual selection and lek behavior in the Mediterranean fruit fly *Ceratitis capitata* (Diptera: Tephritidae). *Pac. Sci.* 43: 135–143.

Averill A.L. and R.J. Prokopy. 1989. Host marking pheromones. In *Fruit Flies:Their Biology, Natural Enemies and Control* (A.S. Robinson and G. Hooper, eds.), pp. 207–219. In *World Crop Pests* (W. Helle, ed.), Vol. 3A. Elsevier Science Publishers, Amsterdam.

Averill, A.L., R.J. Prokopy, M.M. Sylvia, P.P. Connor, and T.T.Y. Wong. 1996. Effects of recent experience on foraging in tephritid fruit flies. *J. Insect Behav.* 9: 571–583.

Back, E.A. and C.E. Pemberton. 1915. Susceptibility of citrus fruit to the attack of the Mediterranean fruit fly. *J. Agric. Res.* 3: 311–330.

Baker, P.S. and H. Van der Valk. 1992. Distribution and behaviour of sterile Mediterranean fruit flies in a host tree. *J. Appl. Entomol.* 114: 67–76.

Barnes, B.N. 1976. Mass rearing of the Natal fruit fly *Pterandrus rosa* (Ksh.) Diptera: Tephritidae. *J. Entomol. Soc. South. Afr.* 39: 121–124.

Barton Browne, L. 1993. Physiologically induced changes in host oriented behavior. *Annu. Rev. Entomol.* 38: 1–25.

Bateman, M.A. 1972. The ecology of fruit flies. *Annu. Rev. Entomol.* 17: 493–518.

Bautista, R.C. and E.J. Harris. 1996. Effect of fruit substrates on parasitization of tephritid fruit flies (Diptera) by the parasitoid *Biosteres arisanus* (Hymenoptera: Braconidae). *Environ. Entomol.* 25: 470–475.

Bell, W.J. 1990. Searching behavior patterns in insects. *Annu. Rev. Entomol.* 35: 447–467.

Blay, S. and B. Yuval. 1997. Nutritional correlates to reproductive success of male Mediterranean fruit flies. *Anim. Behav.* 54: 59–66.

Bodenheimer, F.S. 1951. *Citrus Entomology in the Middle East*. W. Junk, The Hague. 661 pp.

Boller, E.F., C. Hippe, R.J. Prokopy, W. Enkerlin, B.I. Katsoyannos, J.S. Morgante, S. Quilici, D. Crespo de Stilinovic, and M. Zapater. 1994. Response of wild and laboratory-reared *Ceratitis capitata* Wied. (Diptera: Tephritidae) flies from different geographic origins to a standard host marking pheromone solution. *J. Appl. Entomol.* 118: 84–91.

Briceño, R.D., D. Ramos, and W.G. Eberhard. 1996. Courtship behavior of male *Ceratitis capitata* (Diptera: Tephritidae) in captivity. *Fla. Entomol.* 79: 130–143.

Buitendag, C.H. and W. Naude. 1994. Fruit-fly control: development of a new fruit-fly attractant and correct bait administration. *Citrus J.* 4: 22–25.

Burk, T. 1982. Evolutionary significance of predation on sexually signaling males. *Fla. Entomol.* 65: 90–104.

Calkins, C.O. and J.C. Webb. 1988. Temporal and seasonal differences in movement of the Caribbean fruit fly larvae in grapefruit and the relationship to detection by acoustics. *Fla. Entomol.* 71: 409–416.

Canato, C.M. and F.S. Zucoloto. 1993. Diet selection by *Ceratitis capitata* larvae (Diptera, Tephritidae): influence of the rearing diet and genetic factors. *J. Insect Physiol.* 39: 981–985.

Canato, C.M. and F.S. Zucoloto. 1998. Feeding behavior of *Ceratitis capitata* (Diptera: Tephritidae): influence of carbohydrate ingestion. *J. Insect Physiol.* 44: 149–155.

Cangussu, J.A. and F.S. Zucoloto. 1995. Self-selection and perception threshold in adult females of *Ceratitis capitata* (Diptera, Tephritidae). *J. Insect Physiol.* 41: 223–227.

Carey, J.R. 1984. Host-specific demographic studies of the Mediterranean fruit fly *Ceratitis capitata. Ecol. Entomol.* 9: 261–270.

Carey, J.R. 1991. Establishment of the Mediterranean fruit fly in California. *Science* 253: 1369–1373.

Carey, J.R., D.A. Krainacker, and R.I. Vargas. 1986. Life history response of female Mediterranean fruit flies, *Ceratitis capitata,* to periods of host deprivation. *Entomol. Exp. Appl.* 42: 159–167.

Carey, J.R., P. Liedo, D. Orozco, and J.W. Vaupel. 1992. Slowing of mortality rates at older ages in large medfly cohorts. *Science* 258: 457–461.

Carey, J.R., P. Liedo, H. Muller, J. Wang, and J. Vaupel. 1998. Dual modes of aging in Mediterranean fruit fly females. *Science* 281: 996–998.

Causse, R. 1974. Etude d'un rythme circadien du comportment de prenymphse chez *Ceratitis capitata* Wiedemann (Diptere Trypetidae). *Ann. Zool. Ecol. Anim.* 6: 475–498.

Causse, R. and M. Féron. 1967. Influence du rythme photoperiodique sur l'activite sexuelle de la mouche Mediterranee des fruits: *Ceratitis capitata* Wiedemann (Diptere Trypetidae). *Ann. Epiphyt.* 18: 175–192.

Cavalloro, R. and G. Delrio. 1975. Soil factors influencing the pupation of *Ceratitis capitata* Wiedemann. *Boll. Lab. Entomol. Agric. Filippo Silvestri* 32: 190–195.

Cayol, J.P. 1996. Box thorn, key early season host of the Mediterranean fruit fly. *Int. J. Pest. Manage.* 42: 325–329.

Chen, C.C. and Y.H. Tseng. 1992. Monitoring and survey of insect pests with potential to invade the Republic of China. Plant Quarantine in Asia and the Pacific: Report of an APO Study Meeting. pp. 42–52.

Chen, P.S. 1984. The functional morphology and biochemistry of insect male accessory glands and their secretions. *Annu. Rev. Entomol.* 29: 233–255.

Christenson, L.D. and R.H. Foote. 1960. Biology of fruit flies. *Annu. Rev. Entomol.* 5: 171–192.

Churchill Stanland, C., R. Stanland, T.T.Y. Wong, N. Tanaka, D.O. McInnis, and R.V. Dowell. 1986. Size as a factor in the mating propensity of Mediterranean fruit flies, *Ceratitis capitata* (Diptera: Tephritidae) in the laboratory. *J. Econ. Entomol.* 79: 614–619.

Cooley, S.S., R.J. Prokopy, P.T. McDonald, and T.T.Y. Wong. 1986. Learning in oviposition site selection by *Ceratitis capitata* flies. *Entomol. Exp. Appl.* 40: 47–51.

Cunningham, R.T., S. Nakagawa, D.Y. Suda, and T. Urago. 1978. Tephritid fruit fly trapping: liquid food baits in high and low rainfall climates. *J. Econ. Entomol.* 71: 762–763.

Dawkins, R. and J.R. Krebs. 1979. Arms races between and within species. *Proc. R. Soc. London B* 205: 489–511.

Debouzie, D. 1977. Effect of initial population size on *Ceratitis* productivity under limited food conditions. *Ann. Zool. Ecol. Anim.* 9: 367–368.

Dickens, J.C., E.B. Jang, D.M. Light, and A. Alford. 1990. Enhancement of insect pheromone responses by green leaf volatiles. *Naturwissenschaften* 77: 29–31.

Downer, R.G.H. and J.R. Mathews. 1976. Patterns of lipid distribution and utilization in insects. *Am. Zool.* 16: 733–745.

Drew, R.A.I. 1987. Reduction in fruit fly (Tephritidae: Dacini) populations in their endemic rainforest habitat by frugivorous vertebrates. *Aust. J. Zool.* 35: 283.

Drew, R.A.I., A.C. Courtice and D.S. Teakle. 1983. Bacteria as a natural source of food for adult fruit flies (Diptera: Tephritidae). *Oecologia* 60: 279–284.

Eberhard, W.G. 1991. Copulatory courtship and cryptic female choice in insects. *Biol. Rev.* 66: 1–31.

Eberhard, W.G. and C. Cordero. 1995. Sexual selection by cryptic female choice on male seminal products — a new bridge between sexual selection and reproductive physiology. *Trends Ecol. Evol.* 10: 493–496.

Epsky, N.D., R.R. Heath, A. Guzman, and W.L. Meyer. 1995. Visual cue and chemical cue interactions in a dry trap with food-based synthetic attractant for *Ceratitis capitata* and *Anastrepha ludens* (Diptera: Tephritidae). *Environ. Entomol.* 24: 1387–1395.

Epsky, N.D., R.R. Heath, G. Uchida, A. Guzman, J. Rizzo, R. Vargas, and F. Jeronimo. 1996. Capture of Mediterranean fruit flies (Diptera: Tephritidae) using color inserts in trimedlure-baited Jackson traps. *Environ. Entomol.* 25: 256–260.

Eskafi, F.M. and M.M. Kolbe. 1990. Predation on larval and pupal *Ceratitis capitata* (Diptera: Tephritidae) by the ant *Solenopsis geminata* (Hymenoptera: Formicidae) and other predators in Guatemala. *Environ. Entomol.* 19: 148–153.

Etienne, J. 1973. The artificial conditions necessary for the mass-rearing of *Ceratitis rosa* (Diptera: Trypetidae). *Entomol. Exp. App.* 16: 380–388.

Farrias, G.T., R.T. Cunningham, and S. Nakagawa. 1972. Reproduction in the Mediterranean fruit fly: abundance of stored sperm affected by duration of copulation and affecting egg hatch. *J. Econ. Entomol.* 65: 914–915.

Fernandes Da Silva, P.G. and F.S. Zucoloto. 1993. The influence of host nutritive value on the performance and food selection in *Ceratitis capitata* (Diptera, Tephritidae). *J. Insect Physiol.* 39: 883–887.

Féron, M. 1962. L'instinct de reproduction chez la mouche Mediterranee des fruits *Ceratitis capitata* Wied. (Dipt. Trypetidae). Comportement sexuel — comportment de ponte. *Rev. Pathol. Veg. Entomol. Agric. Fr.* 41: 1–129.

Field, S. and B. Yuval. 1999. Nutritional status affects copula duration in the Mediterranean fruit fly, *Ceratitis capitata* (Insecta: Tephritidae). *Ethol. Ecol. Evol.* 11: 61–70.

Field, S.A., P.W. Taylor, and B. Yuval. Sources of variability in copula duration of Mediterranean fruit flies. *Entomol. Exp. Appl.*, in press.

Gage, M.J.G. 1991. Risk of sperm competition directly affects ejaculate size in the Mediterranean fruit fly. *Anim. Behav.* 42: 1036–1037.

Galun, R. 1989. Phagostimulation of the Mediterranean fruit fly, *Ceratitis capitata* by ribonucleotides and related compounds. *Entomol. Exp. Appl.* 50: 133–140.

Galun, R., S. Gothilf, S. Blondheim, J.L. Sharp, M. Mazor, and A. Lachman. 1985. Comparison of aggregation and feeding responses by normal and irradiated fruit flies, *Ceratitis capitata* and *Anastrepha suspensa* (Diptera: Tephritidae). *Environ. Entomol.* 14: 726–732.

Garcia, R., A. Megias, and A.M. Municio. 1980. Biosynthesis of neutral lipids by mitochondria and microsomes during development of insects. *Comp. Biochem. Physiol.* 65: 13–23.

Gazit, Y., Y. Rossler, N.D. Epsky, and R.R. Heath. 1998. Trapping females of the Mediterranean fruit fly (Diptera: Tephritidae) in Israel: comparison of lures and trap type. *J. Econ. Entomol.* 91: 1355–1359

Gothilf, S., R. Galun, and M. Bar-Zeev. 1971. Taste reception in the Mediterranean fruit fly: electrophysiological and behavioural studies. *J. Insect Physiol.* 17: 1371–1384.

Greathead, D.J. 1972. Notes on coffee fruit-flies and their parasites at Kawanda (Uganda). *Tech. Bull. Comm. Inst. Biol. Control* 15: 11–18.

Guerin, P.M., U. Remund, E.F. Boller, B.I. Katsoyannos, and G. Delrio. 1983. Fruit fly electroantennogram and behavior responses to some generally occurring fruit volatiles. In *Fruit Flies of Economic Importance* (R. Cavalloro, ed.), pp. 248–251. A.A. Balkema, Rotterdam.

Hancock, D.L. 1984. Ceratitinae (Diptera: Tephritidae) from the Malagasy subregion. *J. Entomol. Soc. South. Afr.* 47: 277–301.

Hanna, A.D. 1938. Studies on the Mediterranean fruit fly: *Ceratitis capitata* Wied. I. The structure and operation of the reproductive organs. *Bull. Soc. Foread Ier. Entomol.* 22: 39–52.

Happ, G.M. 1992. Maturation of the male reproductive system and its endocrine regulation. *Annu. Rev. Entomol.* 37: 303–320.

Haramoto, F.H. and H.A. Bess. 1969. Recent studies on the abundance of the Oriental and Mediterranean fruit flies an the status of their parasites. *Proc. Hawaii. Entomol. Soc.* 20: 551–566.

Headrick, D.H. and R.D. Goeden. 1996. Issues concerning the eradication or establishment and biological control of the Mediterranean fruit fly, *Ceratitis capitata* (Wiedemann) (Diptera: Tephritidae), in California. *Biol. Control* 6: 412–421.

Heath, R.R., N.D. Epsky, A. Guzman, B. Dueben, A. Manukian, and W.L. Meyer. 1995. Development of a dry plastic insect trap with food-based synthetic attractant for the Mediterranean and Mexican fruit flies (Diptera: Tephritidae). *J. Econ. Entomol.* 88: 1307–1315.

Hendrichs, J. and M.A. Hendrichs. 1990. Mediterranean fruit fly (Diptera: Tephritidae) in nature: location and diel pattern of feeding and other activities on fruiting and nonfruiting hosts and nonhosts. *Ann. Entomol. Soc. Am.* 83: 632–641.

Hendrichs, M.A. and J. Hendrichs. 1998. Perfumed to be killed: interception of Mediterranean fruit fly (Diptera: Tephritidae) sexual signaling by predatory foraging wasps (Hymenoptera: Vespidae). *Ann. Entomol. Soc. Am.* 91: 228–234

Hendrichs, J. and R.J. Prokopy. 1994. Food foraging behavior of frugivorous fruit flies. In *Fruit Flies and the Sterile Insect Technique* (C.O. Calkins, W. Klassen, and P. Liedo, eds.), pp. 37–55. CRC Press, Boca Raton.

Hendrichs, J., B.I. Katsoyannos, D.R. Papaj, and R.J. Prokopy. 1991. Sex differences in movement between natural feeding and mating sites and tradeoffs between food consumption, mating success and predator evasion in Mediterranean fruit flies (Diptera: Tephritidae). *Oecologia* 86: 223–231.

Hendrichs, J., B.I. Katsoyannos, V. Wornoayporn, and M.A. Hendrichs. 1994. Odour-mediated foraging by yellowjacket wasps (Hymenoptera: Vespidae): predation on leks of pheromone-calling Mediterranean fruit fly males (Diptera: Tephritidae). *Oecologia* 99: 88–94.

Hendrichs, J., B. Katsoyannos, K. Gaggl, and V. Wornoayporn. 1996. Competitive behavior of males of Mediterranean fruit fly, *Ceratitis capitata*, genetic sexing strain *Vienna-42*. In *Fruit Fly Pests: A World Assessment of Their Biology and Management* (B.A. McPheron and G.J. Steck, eds.), pp. 405–414. St. Lucie Press, Delray Beach.

Höglund, J. and R.V. Alatalo. 1995. *Leks*. Princeton University Press, Princeton. 248 pp.

Hunt, M.K., C.S. Crean, R.J. Wood, and A.S. Gilburn. 1998. Fluctuating asymmetry and sexual selection in the Mediterranean fruitfly (Diptera, Tephritidae). *Biol. J. Linn. Soc.* 64: 385–396.

Jackson, C.G., J.P. Long, and L.M. Klungness. 1998. Depth of pupation in four species of fruit flies (Diptera: Tephritidae) in sand with and without moisture. *J. Econ. Entomol.* 91: 138–142.

Jácome, I., M. Aluja, P. Liedo, and D. Nestel. 1995. The influence of adult diet and age on lipid reserves in the tropical fruit fly *Anastrepha serpentina* (Diptera: Tephritidae). *J. Insect Physiol.* 41: 1079–1086.

Jang, E.B. 1995. Effect of mating and accessory gland injections on olfactory mediated behavior in the female Mediterranean fruit fly, *Ceratitis capitata*. *J. Insect Physiol.* 41: 705–710.

Jang, E.B., D.O. McInnis, D.R. Lance, and L.A. Carvalho. 1998. Mating-induced changes in olfactory-mediated behavior of laboratory-reared normal, sterile and wild female Mediterranean fruit flies (Diptera: Tephritidae) mated to conspecific males. *Ann. Entomol. Soc. Am.* 91: 139–144.

Jones, S.R., M.C. Zapater, and K.C. Kim. 1993. Morphological adaptation to different artificial oviposition substrates in the aculeus of *Ceratitis capitata* (Diptera: Tephritidae). *Ann. Entomol. Soc. Am.* 86: 153–157.

Kaspi, R. and B. Yuval. 1999a. Lek site selection by male Mediterranean fruit flies. *J. Insect Behav.* 12: 267–276.

Kaspi, R. and B. Yuval. 1999b. Mediterranean fruit fly leks: factors affecting male location. *Funct. Ecol.* 13: 539–545.

Katsoyannos, B.I. 1983. Captures of *Ceratitis capitata* and *Dacus oleae* flies (Diptera: Tephritidae) by McPhail traps and Rebell color traps suspended on citrus, fig and olive trees on Chios, Greece. In *Fruit Flies of Economic Importance* (R. Cavalloro, ed.), pp. 451–456. A.A. Balkema, Rotterdam.

Katsoyannos, B.I. 1989. Responses to shape, size and color. In *Fruit Flies, Their Biology, Natural Enemies and Control* (A.S. Robinson and G. Hooper, eds.), pp. 307–324. In *World Crop Pests* (W. Helle, ed.), Vol. 3A. Elsevier Science Publishers, Amsterdam.

Katsoyannos, B.I. and J. Hendrichs. 1995. Food bait enhancement of fruit mimics to attract Mediterranean fruit fly females. *J. Appl. Entomol.* 119: 211–213.

Katsoyannos, B.I., N.A. Kouloussis, and N.T. Papadopoulos. 1997. Response of *Ceratitis capitata* to citrus chemicals under semi-natural conditions. *Entomol. Exp. Appl.* 82: 181–188.

Kazimirova, M. and V. Vallo. 1992. Influence of larval density of Mediterranean fruit fly (*Ceratitis capitata*, Diptera, Tephritidae) on parasitization by a pupal parasitoid, *Coptera occidentalis* (Hymenoptera, Proctotrupoidea, Diapriidae). *Acta Entomol. Bohemoslov.* 89: 179–185.

Keiser, I., R.M. Kobayashi, D.H. Miyashita, E.J. Harris, E.L. Schneider, and D.L. Chambers. 1974. Suppression of Mediterranean fruit flies by Oriental fruit flies in mixed infestations in guava. *J. Econ. Entomol.* 67: 355–360.

Krainacker, D.A. 1986. Demography of the Mediterranean Fruit Fly: Larval Host Effects. M.S. dissertation, University of California, Davis. 360 pp.

Krainacker, D.A., J.R. Carey, and R.I. Vargas. 1987. Effect of larval host on life history traits of the Mediterranean fruit fly, *Ceratitis capitata*. *Oecologia* 73: 583–590.

Krainacker, D.A., J.R. Carey, and R.I. Vargas. 1989. Size-specific survival and fecundity for laboratory strains of two tephritid (Diptera: Tephritidae) species: implications for mass rearing. *J. Econ. Entomol.* 82: 104–108.

Langley, P.A., H. Maly, and F. Ruhm. 1972. Application of the sterility principle for the control of the Mediterranean fruit fly (*Ceratitis capitata*): pupal metabolism in relation to mass rearing techniques. *Entomol. Exp. Appl.* 15: 23–34.

Lawrence, P.O. 1981. Host vibration — a cue to host location by the parasite *Biosteres longicaudatus*. *Oecologia* 48: 249–251.

Leopold, R.A. 1976. The role of male accessory glands in insect reproduction. *Annu. Rev. Entomol.* 21: 199–221.

Levinson, H.Z., A.R. Levinson, and K. Mueller. 1990. Influence of some olfactory and optical properties of fruits on host location by the Mediterranean fruit fly (*Ceratitis capitata* Wied.). *J. Appl. Entomol.* 109: 44–54.

Light, D.M., E.B. Jang, and J.C. Dickens. 1988. Electroantennogram responses of the Mediterranean fruit fly, *Ceratitis capitata*, to a spectrum of plant volatiles. *J. Chem. Ecol.* 14: 159.

Liimatainen, J., A. Hoikkala, and T. Shelly. 1997. Courtship behavior in *Ceratitis capitata* (Diptera: Tephritidae): comparison of wild and mass-reared males. *Ann. Entomol. Soc. Am.* 90: 836–843.

Liquido, N., L.A. Shinoda and R.T. Cunningham. 1991. *Host Plants of the Mediterranean Fruit Fly (Diptera: Tephritidae): An Annotated Review.* Entomological Society of America, Lanham.

Maitland, D.P. 1992. Locomotion by jumping in the Mediterranean fruit-fly larva *Ceratitis capitata*. *Nature* 355: 159–161.

Malio, E. 1979. Observations on the mango fruit fly *Ceratitis cosyra* in the Coast Province, Kenya. *Kenya Entomol. Newslett.* 7.

Mather, M.H. and B.D. Roitberg. 1987. A sheep in wolf's clothing: tephritid flies mimic spider predators. *Science* 236: 308–310.

Mazor, M., S. Gothilf, and R. Galun. 1987. The role of ammonia in the attraction of females of the Mediterranean fruit fly to protein hydrolysate baits. *Entomol. Exp. Appl.* 43: 25–29.

McDonald, P.T. and D.O. McInnis. 1985. *Ceratitis capitata*: effect of host fruit size on the number of eggs per clutch. *Entomol. Exp. Appl.* 37: 207–211.

McInnis, D.O. 1993. Size differences between normal and irradiated sperm heads in mated female Mediterranean fruit flies (Diptera: Tephritidae). *Ann. Entomol. Soc. Am.* 86: 305–308.

Morse, D.H. 1986. Predatory risk to insects foraging at flowers. *Oikos* 46: 223–228.

Morton, T.C. and M.A. Bateman. 1981. Chemical studies on proteinacous attractants for fruit flies, including the identification of volatile constituents. *Aust. J. Agric. Res.* 32: 905–916.

Municio, A.M., J.M. Odriozola, A. Pineiro, and A. Ribera. 1973. *In vitro* and *in vivo* [14C] acetate incorporation during development of insects. *Insect Biochem.* 3: 19–29.

Myburgh, A.C. 1962. Mating habits of the fruit flies *Ceratitis capitata* (Wied.) and *Pterandus rosa* (Ksh.). *S. Afr. J. Agric. Sci.* 5: 457–464.

Myburgh, A.C. 1963. Diurnal rhythms in emergence of mature larvae from fruit and eclosion of adult *Pterandrus rosa* and *Ceratitis capitata*. *S. Afr. J. Agric. Sci.* 6: 41–46.

Nestel, D., R. Galun, and S. Friedman. 1985. Long-term regulation of sucrose intake by the adult Mediterranean fruit fly *Ceratitis capitata* (Wiedemann). *J. Insect. Physiol.* 31: 533–536.

Nestel, D., R. Galun, and S. Friedman. 1986. Balance energético en el adulto irradiado de *Ceratitits capitata* (Wied.) (Diptera: Tephritidae). *Folia Entomol. Mex.* 70: 75–85.

Orozco, D. and R.O. López. 1992. Mating competitiveness of wild and laboratory mass-reared medflies: effect of male size. In *Fruit Flies: Biology and Management* (M. Aluja and P. Liedo, eds.), pp. 185–188. Springer-Verlag, New York.

Ovruski, S.M. 1994. Comportamiento en la detección del húesped de *Aganaspis pelleranoi* (Hymenoptera: Eucoilidae), parasitoide de larvas de *Ceratitis capitata* (Diptera: Tephritidae). *Rev. Soc. Entomol. Argent.* 53: 121–127.

Ovruski, S.M. 1995. Pupal and larval-pupal parasitoids (Hymenoptera) obtained from *Anastrepha* spp. and *Ceratitis capitata* (Dipt.: Tephritidae) pupae collected in four localities of Tucuman Province, Argentina. *Entomophaga* 40: 367–370.

Pagani, R., A. Suarez, and A.M. Municio. 1980. Fatty acid patterns of the major lipid classes during development of *Ceratitis capitata*. *Comp. Biochem. Physiol.* 67: 511–518.

Papaj, D.R. 1990. Fruit size and clutch size in wild *Ceratitis capitata*. *Entomol. Exp. Appl.* 54: 195–198.

Papaj, D.R. and R.H. Messing. 1996. Functional shifts in the use of parasitized hosts by a tephritid fly: the role of host quality. *Behav. Ecol.* 7: 235–242.

Papaj, D.R., and R.H. Messing. 1998. Asymmetries in physiological state as a possible cause of resident advantage in contests. *Behaviour* 135: 1013–1030.

Papaj, D.R., R.J. Prokopy, P.T. McDonald, and T.T.Y. Wong. 1987. Differences in learning between wild and laboratory *Ceratitis capitata* flies. *Entomol. Exp. Appl.* 45: 65–72.

Papaj, D.R., S.B. Opp, R.J. Prokopy, and T.T.Y. Wong. 1989a. Cross-induction of fruit acceptance by the medfly *Ceratitis capitata*: the role of fruit size and chemistry. *J. Insect Behav.* 2: 241–251.

Papaj, D.R., B.D. Roitberg, and S.B. Opp. 1989b. Serial effects of host infestation on egg allocation by the Mediterranean fruit fly: a rule of thumb and its functional significance. *J. Anim. Ecol.* 58: 955–970.

Papaj, D.R., B.I. Katsoyannos, and J. Hendrichs. 1989c. Use of fruit wounds in oviposition by Mediterranean fruit flies. *Entomol. Exp. Appl.* 53: 203–209.

Papaj, D.R., B.D. Roitberg, S.B. Opp, M. Aluja, R. J. Prokopy, and T.T.Y. Wong. 1990. Effect of marking pheromone on clutch size in the Mediterranean fruit fly. *Physiol. Entomol.* 15: 463–468.

Papaj, D.R., R.J. Prokopy, and T.T.Y. Wong. 1992. Host-marking pheromone and use of previously-established oviposition sites by the Mediterranean fruit fly. *J. Insect Behav.* 5: 583–598.

Parker, G.A. 1970. Sperm competition and its evolutionary consequences in the insects. *Biol. Rev.* 45: 525–567.

Parker, G.A. 1978. The evolution of competitive mate searching. *Annu. Rev. Entomol.* 23: 176–196.

Podoler, H. and M. Mazor. 1981. *Dirhinus giffardii* Silvestri (Hym.: Chalcididae) as a parasite of the Mediterranean fruit fly, *Ceratitis capitata* (Wiedemann) (Dip.: Tephritidae). 2. Analysis of parasite responses. *Acta Oecol. Oecol. Appl.* 2: 299–309.

Polak, M. 1996. Ectoparasitic effects on host survival and reproduction: the *Drosophila-Macrocheles* association. *Ecology* 77: 1379–1389.

Polak, M. and T.A. Markow. 1995. Effect of ectoparasitic mites on sexual selection in a Sonoran desert fruit fly. *Evolution* 49: 660–669.

Prokopy, R. 1993. Levels of quantitative investigation of fruit fly foraging behavior. In *Fruit Flies: Biology and Management* (M. Aluja and P. Liedo, eds.), pp. 165–172. Springer-Verlag, New York.

Prokopy, R.J. and J.J. Duan. 1998. Socially facilitated egglaying behavior in Mediterranean fruit flies. *Behav. Ecol. Sociobiol.* 42: 117–122.

Prokopy, R.J. and J. Hendrichs. 1979. Mating behavior of *Ceratitis capitata* on a field-caged host tree. *Ann. Entomol. Soc. Am.* 72: 642–648.

Prokopy R.J. and B.D. Roitberg. 1984. Foraging behavior of true fruit flies. *Am. Sci.* 72: 41–49.

Prokopy, R.J. and B. Roitberg, D. 1989. Fruit fly foraging behavior. In *Fruit Flies: Their Biology, Natural Enemies and Control* (G. Robinson and A. Hooper, eds.), pp. 291–306. In *World Crop Pests* (W. Helle, ed.), Vol. 3A. Elsevier Science Publishers, Amsterdam.

Prokopy, R.J., J.R. Ziegler, and T.T. Wong. 1978. Deterrence of repeated oviposition by fruit-marking pheromone in *Ceratitis capitata* (Diptera: Tephitidae). *J. Chem. Ecol.* 4: 55–63.

Prokopy, R.J., P.T. McDonald, and T.T.Y. Wong. 1984. Inter-population variation among *Ceratitis capitata* flies in host acceptance pattern. *Entomol. Exp. Appl.* 35: 65–69.

Prokopy, R.J., D.R. Papaj, S.B. Opp, and T.T.Y. Wong. 1987. Intra-tree foraging behavior of *Ceratitis capitata* flies in relation to host fruit density and quality. *Entomol. Exp. Appl.* 45: 251–258.

Prokopy, R.J., T.A. Green, and T.T. Wong. 1989. Learning to find fruit in *Ceratitis capitata* flies. *Entomol. Exp. Appl.* 53: 65–72.

Prokopy, R.J., T.A. Green, T.T.Y. Wong, and D. McInnis. 1990. Influence of experience on acceptance of artificial oviposition substrates in *Ceratitis capitata* (Wiedemann). *Proc. Hawaii. Entomol. Soc.* 30: 91–95.

Prokopy, R.J., D.R. Papaj, J. Hendrichs, and T.T.Y. Wong. 1992. Behavioral responses of *Ceratitis capitata* flies to bait spray droplets and natural food. *Entomol. Exp. Appl.* 64: 247–257.

Prokopy, R.J., C. Hsu, R.I. Vargas, and C.L. Hsu. 1993a. Effect of source and condition of animal excrement on attractiveness to adults of *Ceratitis capitata* (Diptera: Tephritidae). *Environ. Entomol.* 22: 453–458.

Prokopy, R.J., A.L. Averill, T.A. Green, and T.T.Y. Wong. 1993b. Does food shortage cause fruit flies (Diptera: Tephritidae) to "dump" eggs? *Ann. Entomol. Soc. Am.* 86: 362–365.

Prokopy, R.J., J.J. Duan, and R.I. Vargas. 1996a. Potential for host range expansion in *Ceratitis capitata* flies: impact of proximity of adult food to egg laying sites. *Ecol. Entomol.* 21: 295–299.

Prokopy, R.J., S.S. Resilva, and R.I. Vargas. 1996b. Post-alighting behavior of *Ceratitis capitata* (Diptera: Tephritidae) on odor-baited traps. *Fla. Entomol.* 79: 422–428.

Purcell, M.F., A. Van Nieuwenhoven, and M.A. Batchelor. 1996. Bionomics of *Tetrastichus giffardianus* (Hymenoptera: Eulophidae): an endoparasitoid of tephritid fruit flies. *Environ. Entomol.* 25: 198–206.

Quilici, S. and L. Rivry. 1996. Influence of some visual stimuli on the selection of oviposition site by *Ceratitis (Pterandus) rosa*. In *Fruit Flies: A World Assessment of Their Biology and Management* (B.A. McPheron and G.J. Steck, eds.), pp. 59–65. St. Lucie Press, Delray Beach.

Quilici, S. and B. Trahais. 1993. Improving fruit fly trapping systems in Reunion Island. In *Fruit Flies: Biology and Management* (M. Aluja and P. Liedo, eds.), pp. 235–240. Springer-Verlag, New York.

Ramadan, M.M. and T.T.Y. Wong. 1990. Biological observations on *Tetrastichus giffardianus* (Hymenoptera: Eulophidae), a gregarious endoparasitoid of the Mediterranean fruit fly and the Oriental fruit fly (Diptera: Tephritidae). *Proc. Hawaii. Entomol. Soc.* 30: 59–62.

Ramadan, M.M., T.T.Y. Wong, and R.H. Messing. 1995. Reproductive Biology of *Biosteres vandenboschi* (Hymenoptera: Braconidae), a parasitoid of early-instar Oriental fruit fly. *Ann. Entomol. Soc. Am.* 88: 189–195.

Raubenheimer, D. and S.J. Simpson. 1995. Constructing nutrient budgets. *Entomol. Exp. Appl.* 77: 99–104.

Ripley, L.B. and G.A. Hepburn. 1929. Olfactory and visual reactions of the Natal fruit fly, *Pterandrus rosa*, as applied to control. *S. Afr. J. Sci.* 24: 449–458.

Rivnay, E. 1950. The Mediterranean fruit fly in Israel. *Bull. Entomol. Res.* 31: 321–341.

Roessler, Y. 1989. Insecticidal bait and cover sprays. In *Fruit Flies, Their Biology, Natural Enemies and Control* (A.S. Robinson and G. Hooper eds.), pp. 329–335. In *World Crop Pests* (W. Helle, ed.), Vol. 3A. Elsevier Science Publishers, Amsterdam.

Roitberg, B.D. 1985. Search dynamics in fruit-parasitic insects. *J. Insect Physiol.* 31: 865–872.

Roitberg, B.D. and W.G. Friend. 1992. A general theory for host seeking decisions in mosquitoes. *Bull. Math. Biol.* 54: 401–412.

Sanders, W. 1968. Die Eiablage der Mittelmeerfruchtfliege *Ceratitis capitata.* Ihre Abhängigkeit von Farbe und Gliederung des Umfeldes. *Z. Tierpsychol.* 25: 588–607.

Saul, S.H. and S.D. McCombs. 1993a. Dynamics of sperm use in the Mediterranean fruit fly (Diptera: Tephritidae): reproductive fitness of multiple-mated females and sequentially mated males. *Ann. Entomol. Soc. Am.* 86: 198–202.

Saul, S.S. and S.D. McCombs. 1993b. Increased remating frequency in sex ratio distorted lines of the Mediterranean fruit fly (Diptera: Tephritidae). *Ann. Entomol. Soc. Am.* 86: 631–637.

Saul, S.H., S.Y.T. Tam, and D.O. McInnis. 1988. Relationship between sperm competition and copulation duration in the Mediterranean fruit fly (Diptera: Tephritidae). *Ann. Entomol. Soc. Am.* 81: 498–502.

Schoonhoven, L.M. 1983. The role of chemoreception in hostplant finding and oviposition in phytophagous Diptera. In *Fruit Flies of Economic Importance* (R. Cavalloro, ed.), pp. 240–247. A.A. Balkema, Rotterdam.

Schwarz, A.J., A. Zambada, D.H.S. Orozco, J.L. Zavala, and C.O. Calkins. 1981. Mass production of the Mediterranean fruit fly at Metapa, Mexico. *Fla. Entomol.* 68: 467–477.

Seo, S.T., R.I. Vargas, J.E. Gilmore, R.S. Kurashima, and M.S. Fujimoto. 1990. Sperm transfer in normal and gamma-irradiated, laboratory-reared Mediterranean fruit flies (Diptera: Tephritidae). *J. Econ. Entomol.* 83: 1949–1953.

Shelly, T.E. and T.S. Whittier. 1995. Lek distribution in the Mediterranean fruit fly: influence of tree size, foliage density and neighborhood. *Proc. Hawaii. Entomol. Soc.* 32: 113–121.

Shelly, T.E. and T.S. Whittier. 1997. Lek behavior of insects. In *Mating Systems in Insects and Arachnids* (J.C. Choe and B.J. Crespi, eds.), pp. 273–293. Cambridge University Press, Cambridge.

Shelly, T.E., T.S. Whittier, and K.Y. Kaneshiro. 1993. Behavioral responses of Mediterranean fruit flies (Diptera: Tephritidae) to trimedlure baits: can leks be created artificially? *Ann. Entomol. Soc. Am.* 86: 341–351.

Shelly, T.E., T.S. Whittier, and K.Y. Kaneshiro. 1994. Sterile insect release and the natural mating system of the Mediterranean fruit fly, *Ceratitis capitata* (Diptera: Tephritidae). *Ann. Entomol. Soc. Am.* 87: 470–481.

Silvestri, F. 1914. Report of an expedition to Africa in search of the natural enemies of fruit flies (Trypaneidae), with descriptions, observations and biological notes. *Terr. Hawaii Div. Entomol. Bull.* No. 3: 176 pp.

Simmons, L.W. and M. Zuk. 1994. Age structure of parasitized and unparasitized populations of the field cricket *Teleogryllus oceanicus. Ethology* 98: 3–4.

Sivinski, J. and T. Burk. 1989. Reproductive and mating behavior. In *Fruit Flies: Their Biology, Natural Enemies and Control* (A.S. Robinson and G. Hooper, eds.), pp. 343–352. In *World Crop Pests* (W. Helle, ed.), Vol. 3A. Elsevier Science Publishers, Amsterdam.

Smith, P.H. 1989. Behavioral partitioning of the day and circadian rhythmicity. In *Fruit Flies: Their Biology, Natural Enemies and Control* (A.S. Robinson and G. Hooper, eds.), pp. 325–341. In *World Crop Pests* (W. Helle, ed.), Vol. 3A. Elsevier Science Publishers, Amsterdam.

Solinas, M. and G. Nuzzaci. 1984. Functional anatomy of *Dacus oleae* Gmel. female genitalia in relation to insemination and fertilization processes. *Entomologica* 19: 135–165.

Stearns, S.C. 1989. Trade-offs in life-history evolution. *Funct. Ecol.* 3: 259–268.

Steck, G.J., F.E. Gilstrap, R.A. Wharton, and W.G. Hart. 1986. Braconid parasitoids of Tephritidae (Diptera) infesting coffee and other fruits in West-Central Africa. *Entomophaga* 31: 59–67.

Steiner, L.F. 1952. Fruit fly control in Hawaii with poison bait spray containing protein hydrolysate. *J. Econ. Entomol.* 45: 838–843.

Steiner, L.F. 1955. Bait sprays for fruit fly control. *Agric. Chem.* 10: 32–34.

Steyskal, G.C. 1982. A second species of *Ceratitis* (Diptera: Tephritidae) adventive in the New World. *Proc. Entomol. Soc. Wash.* 84: 165–166.

Taylor, P.W. and B. Yuval. 1999. Post-copulatory sexual selection in the Mediterranean fruit fly: advantages for large and protein fed males. *Anim. Behav.* 58: 247–254.

Teran, H.R. 1977. Comportamiento alimentario y su corelacion a la reproduccion en hembras de *Ceratitis capitata* (Wied.) (Diptera, Trypetidae). *Rev. Agron. N.O. Argent.* 14: 17–34.

Thornhill, R. and J. Alcock 1983. *The Evolution of Insect Mating Systems.* Harvard University Press, Cambridge. 547 pp.

Tsistsipis, J.A. 1989. Nutrition: requirements. In *Fruit Flies: Their Biology, Natural Enemies and Control* (A.S. Robinson and G. Hooper, eds.), pp. 103–120. In *World Crop Pests* (W. Helle, ed.), Vol. 3A. Elsevier Science Publishers, Amsterdam.

Vargas, R.I., W.A. Walsh, and T. Nishida. 1995. Colonization of newly planted coffee fields: dominance of Mediterranean fruit fly over Oriental fruit fly (Diptera: Tephritidae). *J. Econ. Entomol.* 88: 620–627.

Waldbauer, G.P. 1968. The consumption and utilization of food by insects. *Adv. Insect Physiol.* 5: 229–288.

Waldbauer, G.P. and S. Friedman. 1991. Self-selection of optimal diets by insects. *Annu. Rev. Entomol.* 36: 43–63.

Warburg, I. and R. Galun. 1992. Ingestion of sucrose solutions by the Mediterranean fruit fly *Ceratitis capitata* (Wied). *J. Insect Physiol.* 38: 969–972.

Warburg, M.S. and B. Yuval. 1996. Effects of diet and activity on lipid levels of adult Mediterranean fruit flies. *Physiol. Entomol.* 21: 151–158.

Warburg, M.S. and B. Yuval. 1997a. Circadian patterns of feeding and reproductive activities of Mediterranean fruit flies (Diptera: Tephritidae) on various hosts in Israel. *Ann. Entomol. Soc. Am.* 90: 487–495.

Warburg, M.S. and B. Yuval. 1997b. Effects of energetic reserves on behavioral patterns of Mediterranean fruit flies (Diptera, Tephritidae). *Oecologia* 112 (3): 314–319.

Wharton, R.A. 1989. Classical biological control of fruit infesting Tephritidae. In *Fruit Flies: Their Biology, Natural Enemies and Control* (A.S. Robinson and G. Hooper, eds.), pp. 303–314. In *World Crop Pests* (W. Helle, ed.), Vol. 3A. Elsevier Science Publishers, Amsterdam.

White, I.M., and M.M. Elson-Harris. 1992. *Fruit Flies of Economic Significance: Their Identification and Bionomics.* CAB International, Wallingford. 601 pp.

Whittier, T.S. and T.E. Shelly. 1993. Productivity of singly vs. multiply mated female Mediterranean fruit flies, *Ceratitis capitata* (Diptera: Tephritidae). *J. Kans. Entomol. Soc.* 66: 200–209.

Whittier, T.S., K.Y. Kaneshiro, and L.D. Prescott. 1992. Mating behavior of Mediterranean fruit flies (Diptera: Tephritidae) in a natural environment. *Ann. Entomol. Soc. Am.* 85: 214–218.

Whittier, T.S., F.Y. Nam, T.E. Shelly, and K.Y. Kaneshiro. 1994. Male courtship success and female discrimination in the Mediterranean fruit fly (Diptera: Tephritidae). *J. Insect Behav.* 7: 159–170.

Wong, T.T.Y., R.M. Kobayashi, L.C. Whitehand, D.G. Henry, D.A. Zadig, and C.L. Denny. 1984. Mediterranean fruit fly (Diptera: Tephritidae): mating choices of irradiated laboratory-reared and untreated wild flies of California in laboratory cages. *J. Econ. Entomol.* 77: 58–62.

Wong, T.T.Y., M.M. Ramadan, D.O. McInnis, and N. Mochizuki. 1990. Influence of cohort age and host age on oviposition activity and offspring sex ratio of *Biosteres tryoni* (Hymenoptera: Braconidae), a larval parasitoid of *Ceratitis capitata* (Diptera: Tephritidae). *J. Econ. Entomol.* 83: 779–783.

Yee, W.L., W.A. Foster, M.J. Howe, and R.G. Hancock. 1992. Simultaneous field comparison of evening temporal distributions of nectar and blood feeding by *Aedes vexans* and *Ae. trivittatus* (Diptera: Culicidae) in Ohio. *J. Med. Entomol.* 29: 356–360.

Yuval, B. 1992. The other habit — sugar feeding by mosquitoes. *Bull. Soc. Vector Ecol.* 17: 150–156.

Yuval, B. 1994. The vertebrate host as mating encounter site for its ectoparasites: ecological and evolutionary considerations. *Bull. Soc. Vector Ecol.* 19: 115–120.

Yuval, B. and A. Bouskila. 1993. Temporal dynamics of mating and predation in mosquito swarms. *Oecologia* 95: 65–69.

Yuval, B., S. Blay, and R. Kaspi. 1996. Sperm transfer and storage in the Mediterranean fruit fly. *Ann. Entomol. Soc. Am.* 89: 486–492.

Yuval, B., R. Kaspi, S. Shloush, and M. Warburg. 1998. Nutritional reserves regulate male participation in Mediterranean fruit fly leks. *Ecol. Entomol.* 23: 211–215.

Zucoloto, F.S. 1987. Feeding habits of *Ceratitis capitata* (Diptera: Tephritidae): can larvae recognize a nutritionally effective diet? *J. Insect Physiol.* 33: 349–353.

Zucoloto, F.S. 1988. Qualitative and quantitative competition for food in *Ceratitis capitata* (Diptera, Tephritidae). *Rev. Bras. Biol.* 48: 523–526.

Zucoloto, F.S. 1991. Effects of flavour and nutritional value on diet selection by *Ceratitis capitata* larvae (Diptera, Tephritidae). *J. Insect Physiol.* 37: 21–25.

Zucoloto, F.S. 1993. Acceptability of different Brazilian fruits to *Ceratitis capitata* (Diptera, Tephritidae) and fly performance on each species. *Braz. J. Med. Biol. Res.* 26: 291–298.

Zuk, M. 1987. The effects of gregarine parasites, body size, and time of day on spermatophore production and sexual selection in field crickets. *Behav. Ecol. Sociobiol.* 21: 65–72.

Zuk, M. 1988. Parasite load, body size, and age of wild-caught male field crickets (Orthoptera: Gryllidae): effects on sexual selection. *Evolution* 42: 969–976.

# 18 Sexual Behavior and Sexual Selection in the Mediterranean Fruit Fly, *Ceratitis capitata* (Dacinae: Ceratitidini)

*William G. Eberhard*

**CONTENTS**



0-8493-1275-2/00/$0.00+$.50

## 18.1 INTRODUCTION

Despite recent dramatic increases in information on the sexual behavior of the Mediterranean fruit fly, *Ceratitis capitata* (Wiedemann), little is known about what determines the success or failure of a given courtship. This is surprising because the ultimate payoff from the great amounts of money that are spent annually to control Mediterranean fruit fly populations with sterile males depends entirely on the success or failure of courtships by these males in the wild (e.g., Calkins 1984). Recent studies indicate that an important limitation of the attempts to combat wild populations with sterile mass-reared males is the relatively low effectiveness of their courtship behavior (Shelly et al. 1994; Shelly and Whittier 1996; Hendrichs et al. 1996; Liimatainen et al. 1997). The technique, which is extremely benign in its ecological effects, has therefore become controversial (e.g., Burk and Calkins 1983). Nevertheless, basic facts, such as details of premount courtship movements by males, female responses to courting males that are necessary for mounts to occur, and male courtship behavior during copulation, have only recently been discovered. There is clearly a pressing need for further studies of the sexual behavior of Mediterranean fruit flies.

The following account is the first attempt to summarize the scattered literature on sex in the Mediterranean fruit fly. It concentrates almost exclusively on behavioral aspects of sexual interactions (for other facets of Mediterranean fruit fly biology and data on other species of *Ceratitis*, see Yuval and Hendrichs, Chapter 17). I have avoided simply listing previous studies concerning particular topics, and have instead tried to evaluate critically what these studies have and what they have not demonstrated. If the chapter seems a litany of "there are no data on …," it is because of our remarkable ignorance, despite more than three decades of concerted research.

## 18.2 LEKS

Male Mediterranean fruit flies release attractant pheromones in small leks. The number of males/lek averaged about 3 to 4 males under natural conditions and 8 to 16 following releases of mass-reared flies in Hawaii (Shelly et al. 1994; Shelly and Whittier 1996); 4 males after mass releases in Mexico (Baker and Van der Valk 1992); 3 to 6 males in a field cage in Guatemala (Prokopy and Hendrichs 1979); and 3 to 4 in a field cage in Costa Rica (Ramos 1991). Presumably, lek size is small because of lower reproductive payoffs/male in larger groups, but there are apparently no data on this point.

Male aggregations contain no resources required by females (e.g., Shelly and Whittier 1994a), and females have ample opportunity to select a mate within an aggregation. Each male in a lek occupies a separate leaf. Mating opportunities in nature apparently are not concentrated on particular leaves; for example, all but 2 of 71 copulations observed in the study of Whittier et al. (1992) occurred on different leaves. Males in leks move readily to nearby leaves (Zapién et al. 1983; Whittier et al. 1992; Shelly et al. 1993; 1994).

Both in a field cage in Guatemala (Prokopy and Hendrichs 1979) and in unrestrained populations in Hawaii (Arita and Kaneshiro 1989) and Israel (Warburg and Yuval 1997a), males congregated at leks from midmorning to early afternoon. Lekking activity in a hotter environment in Egypt peaked earlier in the morning and showed a temporary midday decline (Hendrichs and Hendrichs 1990). Males in captivity are more sexually active earlier in the day (Causse and Féron 1967). Participation in leks is correlated with nutritional state (Warburg and Yuval 1997b; see also Yuval and Hendrichs, Chapter 17). Males that have not fed recently are less likely to emit long-distance attractant pheromones (Landolt et al. 1992), and males found in leks are relatively heavy and have more protein and sugar in their bodies (Yuval et al. 1998). Substantial numbers of males occur outside leks even at peak hours of lek activity (Hendrichs and Hendrichs 1990; Shelly et al. 1993; Yuval et al. 1998). Presumably at least some of these males lack sufficient reserves to fuel lek activities (Warburg and Yuval 1997b), but other possible factors such as sexual immaturity, defeats in aggressive interactions, recent copulation, and migration have not been investigated. A direct experimental demonstration that reduced feeding results in reduced lek attendance remains to be performed. At least at high population densities, attendance at leks can be dangerous, as flies there are especially likely to be captured by the predatory yellow jacket, *Vespula germanica* (L.) (Hendrichs et al. 1994; Hendrichs and Hendrichs 1998).

Leks consistently occur on the sunny sides of particular trees (Whittier et al. 1992; Shelly et al. 1994). Direct sunlight may be important because it allows the male to see and respond to the female's shadow when she lands on the top of the leaf (K. Y. Kaneshiro, personal communication; Color Figure 6*). Creation of artificial leks with synthetic pheromone (e.g., Villeda et al. 1988; Hendrichs et al. 1989) has shown that the lek locations chosen by male flies are, for reasons not understood, superior to artificial leks for attracting females (Shelly et al. 1993). The reasons particular trees and particular sites within trees are preferred are not clear, and may be quite complex. In a group of persimmon trees, the trees which harbored leks were relatively clumped, and smaller trees were less likely to have leks (Shelly and Whittier 1994a). Other factors that have been mentioned as promoting lek formation include exposure to the morning sun, dense foliage, undamaged leaves, leaf size, the windward side of the tree, tree species (with preferences for citrus and persimmon), and visual and olfactory stimuli from fruit (Villeda et al. 1988; Arita and Kaneshiro 1989; Hendrichs and Hendrichs 1990; Whittier et al. 1992; Shelly and Whittier 1994a; 1996; see also Yuval and Hendrichs, Chapter 17). Kaspi and Yuval (1999) experimentally tested several factors in field cages.

The fact that females show such clear preferences for particular sites makes it uncertain whether the common use of field cages, which are *not* placed at natural lek sites (e.g., Calkins et al. 1996), can accurately evaluate the relative abilities of different types of males (typically mass-reared vs. wild) to mate with wild females (see Millar 1995, for a similar discussion of the possible limitations of bioassays of pheromone compounds). Mating trials in field cages may be especially stiff tests of male courtship abilities, because courtships in cages probably at least sometimes involve females that are less sexually receptive than those which arrive at leks in nature. Field-cage tests performed in the general area of natural leks might help resolve this question.

The location of a lek can shift within a tree over the course of a day (Hendrichs and Hendrichs 1990; Kaspi and Yuval 1999). The fluid nature of leks makes it difficult to define their borders in a biologically meaningful way. Some workers have used an arbitrary number of flies and the distance between them to define a "lek" (e.g., at least three calling males less than 10 to 15 cm apart; Hendrichs and Hendrichs 1990), while others have used the equally arbitrary, conservative tactic of considering all calling males in a given tree as members of a single lek (Shelly et al. 1994; Shelly and Whittier 1994a) (See Aluja et al., Chapter 15, for a discussion of similarly disparate distinctions in *Anastrepha* Schiner). More-detailed studies of male movements and possible within-lek coordination of behavior, and of female movements between males might help resolve this question in a more biologically reasonable manner.

### 18.2.1 Male Aggressive Behavior

Direct observations have yielded brief descriptions of males defending territories by lunging at and pushing intruders, "slashing" with their wings, or by more passively touching heads ("facing off") for up to 5 min (average, 93 s) without exerting "visible force" (Arita and Kaneshiro 1989). Video analyses of flies in captivity (Briceño et al. 1998) revealed a more complex array of least nine different aggressive behavior patterns, including (following the terminology of Headrick and Goeden 1994) wing arching, synchronous wing supination, short wing vibrations, wing enantion, butting, wing strikes, pawing and pounding with the front legs, and head pushing. The male produces sounds during wing vibration, wing strikes, and butting that differ from those produced in other contexts (Briceño et al. 1998). Females also aggressively butt other flies, and produce a short aggressive sound by vibrating their wings (Russ and Schwienbacher 1982).

* Color Figures follow p. 204.

The biological significance of male aggressive behavior is not clear. Territorial defense, which is crucial to male mating success in many lekking species (Höglund and Alatalo 1995), also has been thought to be important in Mediterranean fruit flies (e.g., Arita and Kaneshiro 1985; 1988). Territory defense in nature is, however, sometimes surprisingly weak. Intruders won 69% of 144 contests in nature in a wild Hawaiian population (Whittier et al. 1992). In contrast, intruders won only 18% of 43 interactions in leks of wild Costa Rican flies in a field cage (Ramos 1991; Briceño et al. 1998). One study in captivity showed that aggressive male interactions were more common on certain leaves of a plant in a cage, but receptive females in the same cage were not especially attracted to these leaves (Arita and Kaneshiro 1985). The individual males of a young Hawaiian mass-reared strain that were more successful in mating with females in captivity were not more successful at winning aggressive interactions with other males than were nonmaters (Whittier et al. 1994). In captivity, a "dominance value" that combined both a male's tendency to engage in aggressive interactions and his ability to win them showed substantial variation, even among the same three males from one day to the next (Arita and Kaneshiro 1985).

The variety of aggressive behavior patterns, most of which are shared with other tephritids (e.g., Headrick and Goeden 1994; Aluja et al., Chapter 15), suggests that aggression is probably selectively important for Mediterranean fruit flies in some context or contexts, although not necessarily in leks. The fact that males of a mass-reared strain, in which aggressive behavior is presumably less adaptive because of crowded breeding conditions, were less likely to defend their leaves against invading flies in a field cage (Briceño et al. 1998) is in accord with the idea that aggressive behavior in leks is selectively advantageous in the wild. In Hawaii, however, wild and mass-reared males were equally successful in defending leaves (T. E. Shelly, personal communication). The puzzle of the poor success rates of male Mediterranean fruit flies defending their leaves is emphasized by the fact that males of other tephritids often engage in quite elaborate fights, and show the expected pattern of resident males usually winning fights at leks (e.g., Dodson, Chapter 8; Aluja et al., Chapter 15; Headrick and Goeden, Chapter 25). Perhaps part of this difference stems from the greater value of particular leaves in some other groups, such as *Anastrepha*, in which male pheromone is deposited directly on the leaf and accumulates there (Aluja et al., Chapter 15).

### 18.2.2 Alternative Male Tactics and Their Importance

Flies also mate on fruit where feeding and oviposition occur (Prokopy and Hendrichs 1979), and which males defend aggressively (Warberg and Yuval 1997b). Males generally occur on fruit in the afternoon (Prokopy and Hendrichs 1979). Due to the difficulty of standardizing survey techniques in the field (see, for example, the greater proportion of flies observed on the upper surfaces of leaves when observation towers were used instead of the usual technique of observing flies from the ground; Baker and Van der Valk 1992), the relative frequencies of mating in nature in the contexts of leks and oviposition sites have not yet been convincingly determined. It appears, however, that leks are more important. The percentage of matings on fruit in a field cage in Guatemala was low — 15% of 46 copulations observed directly, and 7% of 214 pairs found copulating (Prokopy and Hendrichs 1979). It was even lower in other studies: 0% of 71 and 0% of 409 copulating pairs in the field in Hawaii (Whittier et al. 1992; Shelly et al. 1994), and 1 to 2% of 267 pairs found copulating in a wild population in Egypt (Hendrichs and Hendrichs 1990). Cool, rainy weather in the afternoon at the Hawaiian site may have been responsible for the lack of mating activity away from leks there (T. E. Shelly, personal communication), but this cannot be the explanation in the hot, dry site in Egypt. The sharp peak in the frequency with which copulating pairs were observed in the field occurs during the hours of the day when lekking is more common and male visits to fruit are rare (e.g., Hendrichs and Hendrichs 1990; Baker and Van der Valk 1992; Whittier et al. 1992), suggesting that lek matings are much more common.

## 18.3 MALE–FEMALE INTERACTIONS

### 18.3.1 Precopulatory Behavior

#### 18.3.1.1 Premount Courtship

The basic sequence of courtship is well known. While in leks and, less often, when alone (Hendrichs and Hendrichs 1990; Shelly et al. 1994), males release a pheromone that attracts both other males and females from a distance (Féron 1962; Ohinata et al. 1977; Shelly et al. 1993; see Sections 18.4.3.1 and 18.4.4.1). Male courtship begins after a female lands on the upper surface of the leaf and then walks over the edge and approaches the male on the underside. Féron (1962) was the first to describe carefully the sequence of behavior patterns in courtship.

1. Sexually active "calling" males rest immobile, with the abdominal pleura expanded, the tip of the abdomen directed dorsally, and the balloonlike rectal epithelial sac everted (Arita and Kaneshiro 1986; Color Figure 6*), thus presumably releasing pheromone (stage I of Féron; termed *pheromonal calling* in the descriptions below). The male sometimes touches the tip of his abdomen to the leaf, presumably depositing pheromone (Prokopy and Hendrichs 1979; see, however, Jones 1989 and Section 18.2.1 for evidence against this interpretation). Receptive females respond to male odor by flying upwind, zigzagging back and forth within the plume of pheromone in wind tunnel tests (Landolt et al. 1992).

2. When a female approaches, or if she lands on the top of the leaf and the male sees her shadow, the male's first response is usually to turn and face toward her (or her shadow). Usually as soon as a male orients toward the female he immediately deflects his abdomen ventrally and starts to vibrate his wings (stage II of Féron; wing vibration called *calling song* by Sivinski et al. 1989; the behavior will be termed *continuous wing vibration* in the descriptions below to contrast with the subsequent intermittent wing buzzing). The male's rectal pheromone sac and abdominal pleura both remain everted, and a plume of pheromone is probably wafted toward the female (Arita and Kaneshiro 1989). If the female is on the upper side of the leaf, she may tend to walk to the side toward which the male is facing as he vibrates his wings (Kaneshiro et al. 1995; Kaneshiro, Chapter 32; no data given). Video analysis shows that the abdominal pleura often pulse during continuous wing vibration (R. D. Briceño and W. Eberhard, unpublished data).

Presumably the same pheromone is emitted during calling and continuous wing vibration, though there are no studies of pheromone composition at different stages of courtship, nor have the air movements near courting pairs ever been studied directly. Arita and Kaneshiro (1989) speculated that the inflated pleural regions of the abdomen form a physical channel through which the male directs air currents carrying the pheromone, but the similar pleural expansion during calling, when the wings are not vibrated (above), argues against this explanation. It seems more likely that pleural expansion functions to expose glandular products that exude onto the abdomen's surface, as in other tephritids (Headrick and Goeden 1994; Headrick and Goeden, Chapter 25). The fact that wing removal reduced male copulation success (only 37% of females kept for 3 days with males were inseminated, as opposed to 78% of females kept with control males; Keiser et al. 1973) is in accord with the idea that pheromonal stimulation via air movements may be important, although it is certainly not definitive proof, because of possible changes in visual and auditory stimuli (Keiser et al. 1973).

Sounds are produced during continuous wing vibration (Rolli 1976; Webb et al. 1983; Sivinski et al. 1989; R. D. Briceño and W. Eberhard, unpublished data), but their significance as stimuli to the female has never been studied. Continuous wing vibration lasts up to 15 s or more, and may induce receptive females to turn to face and move toward the male (Section 18.4.4.2). There are no reports that males modulate either this display or the next on the basis of female proximity (as occurs in *Anastrepha suspensa* (Loew); Sivinski and Webb 1986), but other than a temporal analysis of intermittent buzzing that found no changes (Briceño et al. 1996), no careful checks have been performed.

---

* Color Figures follow p. 204.

3. After a variable amount of continuous wing vibration, the male switches abruptly to a second type of wing movement in which the wings are moved rhythmically forward and back while continuing to vibrate rapidly (Lux and Gaggl 1996) (stage III of Féron; wing buzzing called *approach song* by Sivinski et al. 1989; this behavior will be termed *intermittent wing buzzing* in the descriptions below). Video analyses showed that the sac formed by the rectal epithelium is partially or more often completely withdrawn into the male's body during intermittent wing buzzing (Briceño et al. 1996), so presumably pheromone from the anal gland is not directed toward the female during this behavior, as is sometimes stated (e.g., Arita and Kaneshiro 1986). The pleural regions of the abdomen remain inflated, however, during intermittent wing buzzing (R. D. Briceño and W. Eberhard, unpublished data).

Distinctive, more intense sounds are produced each time the male moves his wings anteriorly during intermittent wing buzzing (Rolli 1976; Webb et al. 1983; Sivinski et al. 1989; R. D. Briceño and W. Eberhard, unpublished data). The sound produced during continuous wing vibration also continues during the intervals between buzzes. Presumably the female is stimulated aurally by intermittent wing buzzing. Again, however, there are no experimental tests of the importance of sound as, for instance, was done in *A. suspensa* by Sivinski et al. (1984). Other possible but untested stimuli include vision, olfaction, and wind currents (see Section 18.4.4.2). The edges of the female's wings are often deflected rearward with each forward buzz of the male's wings (Kaneshiro et al. 1995), demonstrating that the behavior does produce air movements toward the female.

Rapid movements of the head ("head rocking") are performed in bursts during intermittent wing buzzing. The sexually dimorphic, capitate, anterior orbital bristles on the anterior surface of the male's head may be displayed visually to the female during head rocking (below). Similarly, the sexually dimorphic colors of the male's eyes and the highly reflective white area of pubescence on the anterior surface of his head (Holbrook et al. 1970; Landolt et al. 1992) may also be displayed to the female during head rocking. In these cases, however, it has not even been demonstrated that these colors have any effect on the female.

#### 18.3.1.2 Postmount Behavior

After courting the female, the male leaps onto her dorsum. He generally lands facing more or less toward her rear (the average angle with the posterior tip of the female in 20 taped sequences was $47 \pm 18°$; R. D. Briceño, personal communication). Often the female immediately falls or flies, and the male is dislodged both in captivity and also in the field (T. E. Shelly, personal communication). If she remains still, the male briefly buzzes his wings within an average of $0.4 \pm 0.3$ s after mounting (Russ and Schwienbacher 1982; Zapién et al. 1983; Arita and Kaneshiro 1989) and rocks his body rapidly forward and backward for an average of $1.6 \pm 0.2$ s (Briceño et al. 1996). Such buzzing and rocking movements are not a normal part of landing behavior, and were never observed in flies landing on the walls of rearing cages. Possibly this behavior produces favorable responses in females (inhibits female dislodgement behavior?), but this has never been tested. Arita and Kaneshiro (1989) reported that the male "clasps the female near the base of her wings" with his hind legs. Videotapes of mounting show, however, that the male's hind legs are lowered to press on the lateral or ventral surface of one side of the female's body, while his front legs grasp the lateral or ventral surface of the other side (R. D. Briceño and W. Eberhard, unpublished data).

After buzzing and rocking briefly, the male releases his grasp on the female and turns to align himself to face in the same direction, and attempts to copulate. As in other tephritids (Headrick and Goeden 1994), intromission can apparently only occur if the female's ovipositor is raised (Zapién et al. 1983) and if she partially everts her aculeus from within the tubular oviscape and eversible membrane, thereby giving the male a chance to clamp the tip of the aculeus with his genitalic surstyli (Eberhard and Pereira 1993). In some pairs, the female does not evert her aculeus, and the mounted male repeatedly nips her on the posterior edge of her eversible membrane with the posterior lobes of his lateral surstyli. Nips are often followed by partial eversions of the aculeus.

If the aculeus emerges far enough, the male clamps it securely, pressing it with four different portions of his surstyli (Eberhard and Pereira 1993), and then introduces his distiphallus into her vagina via the cloacal opening. In some pairs, the female never everts her aculeus enough for the male to grasp, and the male eventually leaves.

The male crosses his hind legs under the posterior portion of the female's abdomen and raises her oviscape while attempting to intromit. This may result in mechanical facilitation of intromission. In addition, he rubs the ventral surface of her abdomen periodically with his hind tarsi and tibiae if he is having difficulty achieving intromission in an apparent attempt to stimulate her to raise her abdomen (Briceño et al. 1996; see Headrick and Goeden 1994, for similar behavior in other tephritids).

## 18.3.2 Copulation

### 18.3.2.1 External Events (Copulatory Courtship)

On the outside of the female the male performs a number of movements during copulation (Eberhard and Pereira 1993; 1998), including several long series of "thrusting" movements of various rhythms and magnitudes, and rubbing movements with his middle and (less often) his hind legs on the ventral surface of her abdomen that probably represent copulatory courtship (Eberhard 1991). If this interpretation is correct, it implies that cryptic female choice of some sort occurs in this species (Eberhard 1994). Careful observations of leg and body positions showed that at least some smaller "thrusts" are brief rhythmic *pulls* on the aculeus, followed in each case by a slight retraction of the aculeus by the female into her eversible membrane (W. Eberhard, unpublished data). In addition, the male occasionally makes brief, forceful pulls on the aculeus that extend the female's ovipositor considerably. Pulls of this sort are common during the first few minutes of copulation (Eberhard and Pereira 1998).

### 18.3.2.2 Internal Events and Possible Sperm Competition and Cryptic Female Choice

Female Mediterranean fruit flies mate one or a few times (e.g., Katiyar and Valerio 1965; Whittier and Shelly 1993; Saul and McCombs 1993a; Yuval et al. 1996), so sperm competition as well as cryptic female choice may occur. Female receptivity to second matings has been reported to correlate negatively with female age (Chapman et al. 1998). Only a fraction of the sperm that leave the male's reproductive tract are eventually stored in the female (about 15 to 20% in flies in Israel; Yuval et al. 1996). The number of sperm transferred to the female's two spermathecae when a virgin female copulates averaged about 3000 to 4000 in mass-reared and mixed wild-mass-reared flies in Israel (Yuval et al. 1996), 1400 to 3500 in a mass-reared strain from Kenya and Egypt (Gage 1991), and only about 500 in mass-reared flies in Costa Rica (Camacho 1989). Female storage capacities are apparently substantially larger: numbers of sperm in the spermathecae of flies from Hawaii and Israel ranged up to estimated maxima of, respectively, 16,000 (Seo et al. 1990), and more than 8000 (Yuval et al. 1996).

Copulation usually lasts 90 to 195 min (Katiyar and Ramírez 1973; Wong et al. 1984; Camacho 1989; Seo et al. 1990; Whittier and Shelly 1993; Field et al., unpublished data), but some copulations last 15 min. or less (Camacho 1989; Seo et al. 1990). (The possibility that some "short copulations" were actually failed intromission attempts was apparently not checked in some studies.) Short copulations (<15 min) do not result in sperm transfer (Seo et al. 1990). Sperm transfer to the female's reproductive tract and arrival in her spermathecae is apparently gradual over the first 90 min. of copulation, and may (Seo et al. 1990) or may not (Farias et al. 1972) increase thereafter. A complete lack of sperm transfer after 100 min has also been reported (P. W. Taylor and B. Yuval, unpublished data, cited in Field et al., unpublished data). It is not known when during copulation the male accessory gland products, which affect female postcopulatory behavior and physiology, are transferred (in some flies these products are transferred separately from sperm; e.g., Merrett 1989 on *Lucilia cuprina* (Wiedemann)).

Copulation duration was somewhat shorter in an old mass-reared strain in Israel, when both male and female were smaller, and when the male was relatively large compared with the female (Field et al., unpublished data). There are apparently differences between mass-reared strains in the frequency of short sterile matings: 15% <30 min in Hawaii (Seo et al. 1990); about 5% in <100 min in Israel (Field et al., unpublished data); 7% in <50 min in Costa Rica (Camacho 1989). Due to the lack of reciprocal cross-strain pairings, it is not known whether males or females (or both) are responsible for failures of sperm transfer.

There are apparently no data regarding the possible presence of sperm in the ventral receptacle; this is an important storage site in some other acalyptrate flies (e.g., Fowler 1973; Kotrba 1995), and the ventral receptacle is clearly developed in Mediterranean fruit flies (e.g., Mungira et al. 1983; Eberhard and Pereira 1998). There are also no data on numbers of sperm transferred and moved into storage in copulations with nonvirgin females. Such transfer does occur, as multiply mated females in Israel had more stored sperm than did singly mated females (Yuval et al. 1996).

The last male to mate with a female tends to sire the majority of her progeny (on average 60 to 70%) in doubly mated females (Katiyar and Ramírez 1970; Saul et al. 1988; Saul and McCombs 1993b). The degree of sperm precedence varies in different genetic lines (Saul and McCombs 1993b). The proportions of offspring sired by first and second males do not change over a doubly mated female's lifetime (Saul and McComb 1993b), or with different times between matings (Katiyar and Ramírez 1970), suggesting that sperm mix freely in female storage organs. However, the fact that sperm precedence remained unchanged when the time between first and second copulations varied from 7 to 18 days (Katiyar and Ramírez 1970) argues against this simple interpretation, because the numbers of sperm from the first male would presumably be reduced by leakage and fertilization of eggs. Sperm numbers in the spermathecae decline over 6 to 7 weeks after mating, even when the female is not laying eggs (Cunningham et al. 1971; Nakagawa et al. 1971; see also Yuval et al. 1996).

Males that had been maintained previously in groups, and were then mated in the presence of other males transferred larger numbers of sperm than males kept and mated isolated from other males (Gage 1991). The sperm are generally distributed in significantly unequal numbers in the female's two spermathecae, suggesting possible control of sperm transport by either the female or the male (rather than filling based only on sperm movements; Yuval et al. 1996). Direct male control seems less likely because the male genitalia have no direct access beyond the mouths of the spermathecal ducts (Eberhard and Pereira 1998). Several kinds of data argue against Gage's idea that males compete via ejaculate size (Yuval et al. 1996): the lack of direct male access to storage sites; the apparently rapid disappearance of extra-spermathecal sperm from the female's reproductive tract (Seo et al. 1990); and the apparent rarity of immediate female remating, deduced from direct observations in captivity by Saul and McCombs (1993), and in the wild from the low numbers of copulations observed by Shelly et al. (1994), Whittier et al. (1992), and Shelly and Whittier (1996).

Contrary to expectations, several studies have found that even normal-length copulations sometimes result in the transfer of few or no sperm to the spermathecae. The highest frequencies of failure were found by Camacho (1989) in Costa Rican mass-reared flies: females had not received any sperm following 29% of 193 copulations; in an additional 5% only one of her two spermathecae had sperm. Unsuccessful copulations were only somewhat shorter (mean 108 ± 44, $N = 56$, vs. 130 ± 53, $N = 295$). Seo et al. (1990) found no sperm in the spermathecae of Hawaiian flies following 10% of 134 copulations that lasted 120 min. A possible female effect on sperm arrival in the spermathecae was suggested by two additional flies in which there was sperm in the bursa but not in the spermathecae. Of 114 pairs in this same study that were allowed to separate spontaneously (after 4 to 254 min of copulation), no sperm was transferred to the spermathecae in 25%. Using the less direct measure of offspring surviving to pupate, Saul and McCombs (1993b) found that 22% of 49 copulations of Hawaiian flies that lasted >15 min were failures.

Other studies have found less frequent failures. Wong et al. (1984) found that sperm was absent from the spermathecae 2 h after the end of copulation in only 2 to 5% of copulations of California flies. Whittier and Kaneshiro (1995), also working with Hawaiian flies, found that 7% of males that copulated failed to produce offspring, while Taylor and Yuval (1999) found a 6% failure rate in mass-reared flies in Israel. Mass-reared males in Israel were less effective inseminators than wild males (Rössler 1975a), and fertilization rates were reduced dramatically in wild females (from 90 to 24%) when mating occurred with males of a mass-reared strain instead of with wild males (Rössler 1975b). Females mated to larger and better-fed males were more likely to store sperm, and to store sperm in greater abundance than those mated to small and protein-starved males (Taylor and Yuval 1999).

In sum, it is clear that sustained mounts do not necessarily lead to intromission, that intromission does not necessarily lead to insemination, and that the frequency of failure to transfer sperm and have them arrive in the female's spermathecae is sometimes substantial (probably as a result of lack of female transport — see below). Even if sperm have arrived in the spermathecae, they sometimes fail to fertilize eggs. There are as yet no studies to determine why intromission, insemination, and fertilization failures occur, and why their frequencies vary. A basic understanding of the process of copulation is a necessary first step. Although movements of the male genitalia within the female during copulation are not normally included in descriptions of behavior, they are in effect "internal behavior." They are complex, and vary to some extent in Mediterranean fruit flies, and their obviously critical nature in the reproductive biology of this species make it important to understand them.

The events within the female are as yet only partially understood. Hanna (1938) gave a detailed description of copulation, but it is certain that he did not see some of the events he reported (Eberhard and Pereira 1998), thus casting doubt on his entire account. The process of intromission (reviewed in Eberhard and Pereira 1998) is an impressive feat, as the male's short, complex glans (distiphallus *sensu* Eberhard and Pereira 1998) is the tip of the distiphallus, the rest of which is thin, wirelike, and extremely long (about 40% the length of his body); the female's vagina is also long, and with the ovipositor in repose, it is folded sharply into an "S." Intromission commences when the glans, folded back onto the rest of the distiphallus (= basiphallus *sensu* Eberhard and Pereira 1998), enters the aculeus, but the very first stages of intromission have not yet been clarified. After the glans has passed through the aculeus, but before it reaches the sclerites that mark the inner portion of the vagina and the lower end of the bursa (the "stigma"), the distiphallus is unfolded so that the distal tip of the glans is deepest within the female. The distiphallus usually reaches this depth within the first 2 to 10 min of copulation, but there is substantial variation. The mechanism by which unfolding of the distiphallus is brought about has not been confirmed, but probably involves some of the long pulls on the female aculeus mentioned in the preceding section.

Once the distiphallus has been unfolded, it is still far short of the anterior end of the bursa where the spermathecal and ventral receptacle ducts open (Dallai et al. 1993), and where sperm transfer apparently occurs. The distiphallus gradually moves deeper into the female, due at least in part to inflations of the more basal of the two inflatable sacs on the glans. This sac produces a rearward pushing motion each time it is inflated. Rhythmic stiffening of the slender, main part of the distiphallus and thrusting or pushing movements of the male (above) may also help achieve deeper penetration.

When the glans has reached the inner end of the female bursa, a process on the glans (the "genital rod" of Eberhard and Pereira 1995; the "endophallus" of Solinas and Nuzzaci 1984) is at least sometimes inserted into the duct of the ventral receptacle, perhaps aided by rhythmic inflations of a second, larger distal sac on the glans (Dallai et al. 1993). This rod does not bear any opening, and may serve to orient the male genitalia (and perhaps stimulate the female) rather than to transfer sperm directly.

The details of how and where sperm and male accessory gland products are transferred are not known. Judging by the morphology of *Bactrocera oleae* (Rossi) (Solinas and Nuzzaci 1984), the tip of the male's ejaculatory duct is probably near the base of the rod, and thus may be brought into close approximation with the openings of the spermathecal ducts near the mouth of the ventral receptacle duct. Subsequent movement of sperm up the spermathecal ducts to the spermathecae is probably influenced by the female. H. Camacho (personal communication) has seen sperm moving up a spermathecal duct in a large, coherent mass, suggesting that contractions of the muscles of the duct are involved (see also Yuval et al. 1996). The arrangement of spermathecal duct muscles in *B. oleae* is such that their contraction will produce a partial vacuum within the duct (Solinas and Nuzzaci 1984).

### 18.3.3 Postcopulatory Behavior and Effects of Male Accessory Gland Products

It is not clear which sex is responsible for terminating copulation. This point is potentially important because the duration of copulation is correlated with the number of sperm transferred (preceding section), and it may also correlate with amounts of male accessory gland products transferred (no data available). Certainly the final withdrawal of the male's genitalia cannot occur without the male's cooperation, as he always dismounts and tugs repeatedly before he is able to pull his phallus out of the female. This leaves unexplained, however, observations like those of Seo et al. (1990) in which 15% of 114 males ended copulation after <30 min without having transferred any sperm. The possibility of internal female resistance during copulation (e.g., Eberhard and Kariko 1996) has never been tested. The fact that female nutritional status has a stronger effect on copulation duration than male nutrition (Field and Yuval 1999) hints at a female influence.

Direct male–female interactions end once the male withdraws his genitalia, but the effects of copulation on female behavior and physiology persist after the flies separate, and sometimes vary. Mating triggers a reduction in the female's attraction to male odor, and an increase in her attraction to the odor of fruit (Del Rio and Cavalloro 1979; Jang et al. 1998), evidently due to the effects of male accessory gland products transferred in the semen (Jang 1995).

The increased female tendencies to remate after copulating for shorter periods noted by Nakagawa et al. (1971; no data given) and Saul et al. (1988), after copulating with smaller males (S. Bloem et al. 1993), or with protein-starved males (Blay and Yuval 1997) may result from differences in quality or quantity of this male product, as indicated by unpublished data (Miyatake et al., cited in Chapman et al. 1998) on the effects of "transfer" (? = injection) of seminal fluid. Chapman (unpublished, cited in Chapman et al. 1998) found no effect on egg production (presumably oviposition) of injecting male accessory gland extracts in females. The possibility that other aspects of male precopulatory or copulatory courtship and copulation also inhibit female remating, or that they stimulate oviposition, sperm transport, or other female reproductive processes has never been tested. There are male effects on female refractory period (K. Bloem et al. 1993; S. Bloem et al. 1993) that may be due to male accessory gland products, as occurs in other tephritids (e.g., Kuba and Itô 1993).

Mating reduced a female's life expectancy, even though it had no effect on the numbers of eggs that she laid in one study (Chapman et al. 1998), but in another there was a positive correlation between remating and female longevity (Whittier and Shelly 1993). Whittier and Kaneshiro (1991) found that a male's mating history (virgin vs. nonvirgin) affected egg production by his mate.

Different strains may differ in some of these traits, as changes in attraction of females of an old Hawaiian mass-reared strain to oviposition cues as opposed to male-produced pheromone were apparently more marked after they mated with males of the same strain than after they mated with wild males (Jang et al. 1998). In the absence of reciprocal crosses, it is not clear whether these changes were due to changes in male stimuli or female responsiveness. If female bias in these responses consistently favors certain types of males, it may exercise sexual selection by cryptic female choice (Eberhard and Cordero 1995; Eberhard 1996). Rapid divergence, as appears to have occurred, is common in such sexually selected traits. Tests of the other postcopulatory female responses to male accessory gland products that often occur in other insects (Chen 1984) (induction of oviposition, maturation of eggs, and rejection of courting males) merit further study in Mediterranean fruit flies, as does the possibility of divergence in different strains of male triggering abilities and female responsiveness.

## 18.4 VARIATION IN MATING BEHAVIOR

### 18.4.1 "Abnormal" Variation

There are a number of "unnatural" behavior patterns seen in mass-reared flies whose significance in nature, if any, is not known. Old virgin females that have been kept in isolation from males execute male courtship movements that include continuous wing vibration and intermittent wing buzzing (Féron 1962; Arita and Kaneshiro 1983), and produce songs similar to those of courting males (Rolli 1976; Webb et al. 1983). Females can be induced by exposure to certain chemicals to perform continuous wing vibration and elevate their abdomens like calling males (T. E. Shelly, personal communication). The possibility that female preferences in different strains for different types of male courtship behavior might be reflected in how they perform these behaviors themselves has never been tested. Females exposed to blends of male odors fan their wings slowly, bob their abdomens, and butt heads (Jang et al. 1989a).

Males in mass-rearing cages sometimes mount other males and spend many minutes apparently attempting to copulate. Additional males sometimes mount such pairs (Féron 1962 reports chains of up to 10 males!). This homosexual behavior can be induced by exposure to several different compounds (McInnis and Warthen 1988). Homosexual mounting by wild males also occurs under field conditions (Prokopy and Hendrichs 1979). Mass-reared males also sometimes mount and attempt to mate with dead females on the bottom of rearing cages (W. Eberhard, unpublished data). Perhaps some of these deviant behaviors in captivity are due to pheromones with short-range aphrodesiac effects (Howse and Knapp 1996).

### 18.4.2 "Normal" Variation

The largely typological descriptions given in Section 18.3 do not do justice to the variability of Mediterranean fruit fly behavior. Many aspects of both male and female behavior vary widely. For instance, the approximate proportion of courtship time that was spent by Hawaiian flies in calling, continuous wing vibration, and intermittent wing buzzing had coefficients of variation that averaged 108% and ranged from 49 to 214% (calculated from Fig. 1 of Liimatainen et al. 1997). The songs of different males are distinguishable individually (Rolli 1976). The amount of head rocking varies among strains in both Hawaii (Lance et al., unpublished data) and Costa Rica, and is reduced (and absent in many courtships) in a mass-reared strain from Egypt (Briceño and Eberhard 1998). Since females may prefer less common male phenotypes (Rössler 1980), variation per se may be biologically important. An extreme case of variation in male behavior occurred in a field cage in Guatemala where sometimes (15% of 127 male–female interactions) males omitted all of the courtship behavior described above. They neither released pheromone nor buzzed their wings, but instead "slowly walked up to the front, side, or rear of a female and stroked her with the forelegs or mouthparts. Sometimes the female reciprocated these actions" (Prokopy and Hendrichs 1979); some of these interactions ended with attempted copulation. This male behavior has never been reported in any other study of courtship (most of which have been performed under laboratory conditions).

The order in which male behavior patterns occur also varies widely (Briceño et al. 1996; Liimatainen et al. 1997). Comparisons of the courtships of successful and unsuccessful males in two different strains (Liimatainen et al. 1997) consistently showed that the transition from continuous vibration to intermittent buzzing was less common in unsuccessful courtships. One surprise in the pattern of transitions was that females with successful males were more likely to approach the male during a break in courtship by the male. This could imply a short-term memory of male signals in the female, or perhaps a potent, lingering pheromone.

Female acceptance of copulation attempts also varies, and is far from automatic. The female allowed the male to mount in less than half of all courtships in field cages (50 and 24%, respectively, were successful on leaves and fruit; Prokopy and Hendrichs 1979); 82.5% succeeded in the field-cage study of Zapién et al. (1983); and 58% in the laboratory with virgin females (Whittier et al. 1994). Mounting attempts also frequently fail to result in copulations; 26 and 77%, respectively, were successful on leaves and fruit in a field cage (Prokopy and Hendrichs 1979) and 30% in the study of Zapién et al. (1983); 49% succeeded with virgin females in the laboratory (Whittier et al. 1994). Overall, only about 7% of male courtships of virgin females in the laboratory resulted in copulation (Whittier and Kaneshiro 1995). Laboratory mating trials under conditions that standardized light, temperature, site, time of day, rearing, and holding conditions, as well as strain and age of flies nevertheless showed highly significant day-to-day differences in mating frequency (Saul and McCombs 1993a), and mating frequency also varies considerably in field cages (McInnis et al. 1996).

Eventual understanding of Mediterranean fruit fly sexual behavior will depend largely on understanding why variations in male and female behavior occur, both in terms of the stimulus–response conditions that elicit them, and in their effects on the animals' reproductive success. Most of the rest of this chapter concerns the influence of different factors on variation in male and female behavior, and on the decisions that they make during sexual interactions.

## 18.4.3 Stimuli That Affect Triggering of Male Courtship and Mounting

### 18.4.3.1 Male Odors

Male Mediterranean fruit flies have long been known to be attracted from a distance by a variety of compounds (see reviews of the large literature by Millar 1995; Jang and Light 1996a), and lek formation probably depends on chemical attraction among males (e.g., Villeda et al. 1988; Hendrichs et al. 1989; Shelly et al. 1993). Olfactory pheromone stimuli from the artificial attractant trimedlure apparently trigger short-term increases in calling behavior by males in captivity (Shelly et al. 1996) (possible effects on other aspects of courtship behavior were not tested), but olfactory stimuli from other males performing pheromonal calling do not (McDonald 1987). Males also release pheromones in the absence of pheromones from other males, and removal of antennae from males, and thus presumed loss of olfaction (and also hearing), did not reduce the number of matings with females in small containers (Nakagawa et al. 1973). Overall, the picture of chemical attraction and activation of males with respect to the basic sexual biology of Mediterranean fruit flies remains confused.

Male pheromones are complex. Each male has sex-specific anal, pleural, and salivary glands that could produce chemical stimuli (Nation 1981). The total of sex-specific compounds isolated to date from air that has passed over courting and noncourting males and from the anal glands is in the range of 80 to 90 compounds (Landolt et al. 1992). The anal gland product itself is chemically complex (Jacobson et al. 1973; Ohinata et al. 1977; Jones 1989). Of the 58 compounds identified by Jang et al. (1989a) from mass-reared flies fed only on sugar, water, and hydrolyzed yeast, 24 also occurred in the odor of ripe nectarines (Light et al. 1992), and there is also a substantial overlap with citrus peel volatiles and a weaker overlap with odors from other fruit (Howse and Knapp 1996).

Surprisingly, compounds that are especially effective at attracting males (e.g., the sesquiterpene α-copaene, methyl eugenol, and trimedlure; see also Howse and Knapp 1996) are relatively ineffective at drawing females (Millar 1995; Howse and Knapp 1996). And some compounds that cause dramatic male aggregation responses and attempted mating in the laboratory have little attractant effect in the field (Howse and Knapp 1996). Further confusing the issue, no compound similar in structure to the three potent male attractants just mentioned occurs in male-produced odors (Millar 1995) (two unidentified sesquiterpenes occurred as traces; their effects on antennal responses were not tested; Jang et al. 1989a). Presumably some other, as yet unidentified, components of male pheromone attract males from a distance to leks.

There is contradictory evidence regarding the possibility that the attractant effects of trimedlure and α-copaene on males occur because males feed on them and possibly use them as precursors for pheromone compounds. McInnis and Warthen (1988) saw males feeding on both of these substances, and Nadel and Peleg (1965) found that attraction to trimedlure increased when males were starved. But the careful behavioral observations of Shelly et al. (1996) showed that although males were activated by exposure to trimedlure, they did not feed on it.

Possible close-range effects of odors on male sexual behavior have been less studied, and may be less confusing. Although it has apparently not been previously noted, three of the four compounds mentioned by Howse and Knapp (1996) as being effective attractants in the laboratory but not in the field (i.e., possible short-range activators) also occur in the odor of living males. These are geranyl acetate, linalool, and limonene, and are, respectively, major, intermediate, and trace components (Jang et al. 1989a).

The electrical responses of the chemical sense organs on the male antennae (EAG) were significantly greater than female EAG responses for only one compound in male odor, the "minor" component 6-methylhept-5-en-2-one. EAG responses are thought to measure the overall population of receptors that respond to a particular stimulus, and are presumably a subset of the abundant sensilla trichodea on the funiculus (Levinson et al. 1987). Of course, differences or lack of differences in the population of receptors constitute only weak evidence regarding male or female use of a particular male product, due to the likelihood of further, higher-order processing in the fly's central nervous system. No one, for instance, would claim that identical abilities of the eyes of men and women to sense curved shapes would constitute convincing evidence that such visual stimuli play no role in sexual attractiveness in humans.

The male attractant may differ from the compounds that attract females, but unfortunately the only field test of female attraction did not examine effects on males (Heath et al. 1991; J. Sivinski, personal communication). For it to be advantageous to a lekking male to produce a substance that attracts other males but not females, the additional males would have to produce enough increase in the arrivals of receptive females at the lek (via their own production of female attractant pheromone) to more than compensate for the increased competition for females that these males would represent. The fact that males may increase the time spent calling when another male is also calling nearby is presumably a result of selection resulting from this type of competition (McDonald 1987).

The only published data to test the prediction that there are more copulations/male (or at least more females/male) in larger leks come from studies of females arriving at artificial leks (Shelly et al. 1996). As far as they go, the data do not support the prediction of lower payoffs to males in larger leks. The number of females sighted near cups containing several males was significantly greater if more males in a cup called, but the number of females/male was the same (2.12 in leks averaging 4.8 males calling; 2.13 in cups with a mean of 3.1 males calling). When males were separated in different cups that were grouped in clusters of six on different trees, the ratio of frequency of observations of calling in two groups (2.38:1) was nearly equal to that of the numbers of females seen near these groups (2.42:1). At a slightly larger scale, leks that occurred in a cluster of five trees that had leks averaged significantly more females/male (0.79 + 0.51) than others in the same orchard (0.55 + 0.05) (Shelly and Whittier 1994a). The difficulty of delimiting individual leks in a biologically meaningful way in nature (Section 18.2.1), the possibility that the readiness of females to mate when they arrive at different-sized leks, and the fact that mating pairs sometimes leave leks because of harassment from other males (Hendrichs and Hendrichs 1990) will make it difficult to perform such tests and interpret their results. The generally small sizes of Mediterranean fruit fly leks suggest that the payoff/male decreases in especially large leks; Whittier et al. (1992) hint that this may be the case. This expected result would in turn seem to make a male pheromone product that attracts other males but not females paradoxical.

In summary, confusion still reigns with respect to male attraction and activation via chemical signals. Males attract other males from a distance and may also sexually activate them via airborne compounds. These compounds are not the same as those which attract females. The most potent known male long-distance attractants (at least one of which also activates males) are substances that do not even appear as traces among the 80-odd compounds collected from calling males. Other compounds, some of which *are* present in male pheromone, appear to have only a short-range activating effect on males. To date only a few of the more common compounds in male odor have been tested as long-distance attractants in field and wind tunnel tests. They attracted only females, but not in large numbers (Heath et al. 1991; Section 18.4.4.1).

### 18.4.3.2 Stimuli from Females

Visual stimulation from other nearby males apparently induces pheromonal calling behavior, since males exposed to their own mirror images in small vials performed calling behavior more often than similar males without mirrors (McDonald 1987). It may be that visual stimuli need to be combined with acoustic stimuli to be effective, however, since separating pairs of males by glass plates eliminated the stimulatory effect that occurred when they were separated only by screen (McDonald 1987). Removal of male antennae (presumably eliminating both olfaction and hearing) did not, however, reduce the number of matings with females in small containers (Nakagawa et al. 1973), so induction of calling may be more complex.

At least two types of visual stimuli trigger continuous male wing vibration. Féron (1962) saw calling males inside a transparent container turn and begin continuous wing vibration toward nearby flies (males or females) on the outside of the container. Arita and Kaneshiro (1989) saw similar responses in the field to the shadow of a female that had landed on the upper surface of the male's leaf. There are no data, however, regarding which aspects of visual stimuli from other flies trigger this male response. Consistent failure to elicit responses from calling males using models and dead females moved with magnets (W. Eberhard, unpublished data) suggest that the males may use relatively subtle visual cues.

Howse and Knapp (1996) proposed that males respond to certain chemical stimuli from females that the females acquire while feeding on host fruit before visiting the lek. This invocation of contamination to explain the presence of these particular stimuli at the lek may be unnecessary, however, since male calling pheromone contains several of the compounds that these authors argue may be brought to the lek as feeding contaminants (previous section).

Males make several decisions during courtship, and stimuli from the female during courtship are probably used as cues. Briceño et al. (1996) presented correlations regarding the decision by the male to attempt to mount the female. Females were closer, and were facing more directly toward the male in courtships that ended in mounting attempts. Detailed analysis of the final 6 s of courtships that ended in mounts showed that neither the distance to the female nor the angle she faced changed during the period immediately preceeding the male's leap. In fact, the female tended to be immobile. Distance and alignment factors may thus predispose the male to mount, but they are not the immediate cues that trigger his leap. Further, experimental rather than correlational studies are needed to test these ideas.

As is typical in tephritids (Headrick and Goeden 1994), female Mediterranean fruit fly responses to courting males are subtle and difficult to detect. Lux and Gaggl (1996) showed that some female activities at close range (touch legs, touch head, vibrate) may occur more often in pairs that eventually copulate (no tests of statistical significance were given). It was not determined whether these behavior patterns indicated female "approval" of the particular male (Lux and Gaggl 1996), or simply greater general receptivity of the female, independent of the male; nor was it shown that the male actually responded to these female behavior patterns.

Briceño and Eberhard (1998) found three female behavior patterns during courtship (lunge briefly toward the male, raise front legs and tap the male's front legs, and lean slowly rearward) that were significantly correlated with the likelihood that a courtship would end in a mounting attempt rather than being broken off by the male (their "tapping" behavior may be the same as the touch legs and touch head behavior patterns of Lux and Gaggl 1996). Nevertheless, males apparently did not use any of these female behavior patterns as cues to trigger mounting attempts: courtships in which these female behavior patterns occurred were not shorter than other successful courtships in which they did not occur.

Liimatainen et al. (1997) listed all female behavior patterns that occurred while a male was courting as "courtship elements," including walking, approaching, and standing, and Levinson et al. 1987 also mentioned female approach as "courtship." But the implication that these female behavior patterns affected and were affected by the male's behavior needs substantiation. As just noted, there is reason to think that several additional aspects of these behavior patterns, including the distance at which approaching and standing occur, the direction the female walks with respect to the male, the angle that the female's body makes with that of the male as she stands, and the direction in which she faces must all be included before it is possible to evaluate their significance as signals to courting males. The statistical analyses performed by Liimatainen et al. (1997) to demonstrate differences in transitions between behavior patterns in the two races were based on the assumption that all transitions were equally likely (J. Liimatainen, personal communication), which seems unlikely, since some behavior elements were much more common than others. Simple female failure to respond to the male could explain several of the changes in male behavior in successful and unsuccessful courtships in Table 4 of Liimantainen et al. (1997), and needs to be distinguished from possible female courtship. The conclusion that courtship by mass-reared males is characterized by "poorly integrated sequences of male–female communication" (Liimatainen et al. 1997) needs further study. In sum, it may well be true that a "dialogue" occurs between male and female during courtship, but its existence has yet to be demonstrated.

#### 18.4.3.3 Crowding

One particularly variable aspect of courtship is the durations of different behavior patterns and the numbers of times that the male executes repeated behavior patterns such as head rocks and wing buzzes. The coefficients of variation in durations of different male behavior patterns are substantial, generally on the order of 100% (Briceño and Eberhard 1998). Part of this variation is due to facultative reduction of duration as the density of nearby flies increases. Total courtship duration before mounting attempts was reduced by about half when courtships occurred at densities similar to those in mass-rearing cages, as compared with durations in isolated pairs of the same strain of flies (Briceño and Eberhard 1998). Reduced courtship duration reduces the chances of interruption by other flies under crowded conditions. Adjustments in duration occurred in both wild flies from Costa Rica and in mass-reared flies that had been derived from this wild stock. Perhaps the ability to make such adjustments evolved in the context of mating on host fruit rather than in leks, where crowding is minimal.

### 18.4.4 Stimuli Triggering Female Responses

#### 18.4.4.1 Male Odors

It seems certain that females in the field are originally attracted from long range to males in a lek by chemical stimuli (e.g., Féron 1962; Nakagawa et al. 1981; Jang et al. 1989a; Baker et al. 1990; Heath et al. 1991). At short range the female's orientation toward the male and her immobilization in front of him may also result from chemical, as well as visual and perhaps auditory stimuli. But there are no direct demonstrations of which aspects of male chemical stimuli are important for which female responses. The question whether there are any components of male pheromone that specifically attract only females or only males in the field seems to be still unresolved, although the relative attractiveness for males and females clearly varies with different compounds (Howse and Knapp 1996). Nakagawa et al. (1973) found that removal of one female antenna partially reduced copulation, and (as later confirmed by Levinson et al. 1987) that removal of both antennae nearly completely eliminated copulation. This effect was assumed to be due to loss of olfaction, but the possible effect of loss of hearing from amputation was not tested. The experimental protocols of these studies suggest that they detected short-range rather than long-range effects.

Most of the large literature on Mediterranean fruit fly pheromones involves compounds that attract males rather than females (above; see review of Millar 1995), and some studies of female responses have serious limitations. Some otherwise sophisticated chemical analyses have used only crude behavioral classifications of both male and female flies. As noted by Landolt et al. (1992), the laboratory studies of female responses that failed to include airflow and female responses to airflow may be of only limited significance. Female responses to male odors may also be modulated by simultaneous presentation of typical volatiles from green leaves, to which both male and female antennae are especially responsive (Light et al. 1992; Howse and Knapp 1996).

Males from which pheromones have been obtained have all been kept isolated from females, and no attempt has ever been made to obtain pheromones released during short-range interactions between the sexes. Even if no additional compounds are released during intermittent buzzing, the fact that the expanded rectal epithelium is usually withdrawn into the male's body during buzzing (Briceño et al. 1996) while the abdominal pleura remain expanded (R. D. Briceño and W. Eberhard, unpublished data) implies that the mix of pheromone components is changed. Interpretations are also complicated by failures to replicate both the presence of major pheromonal components and behavioral responses of females, and by the possible screening effects of other odors (e.g., Jacobson et al. 1973; Ohinata et al. 1973; Baker et al. 1985; Jang et al. 1989a; Howse and Knapp 1996).

Behavioral attraction of females in wind tunnels to a combination of three of the more common male substances were significant, but much weaker than responses to the odor of live males (Landolt et al. 1992). The same three compounds emitted from sticky black spheres attracted from about 1.5 to 2.8 times more females than spheres without pheromone compounds (estimated from figures in Heath et al. 1991). A blend of the five major identified components (which individually were only weakly attractive) was only somewhat less attractive in dual-choice flight tunnel bioassays than the odor of living males (Jang and Light 1996b). Mixtures of one to three male pheromone components (unspecified; apparently "linalool-based mixtures") probably competed poorly with males in the field, as many more females were attracted after the local population of males was reduced (Howse and Knapp 1996). Millar (1995) noted that to date only disappointingly small numbers of females have been attracted in the field using pheromone components. It should be kept in mind, however, that expectations for attraction of females as compared with males are quite different. On a given day only a limited proportion of the female population in the field are expected to be at the appropriate stage in their reproductive cycle to be attracted to lekking males; in contrast, males are probably attracted to leks during much of their adult lives (Burk and Calkins 1983).

EAG responses of female antennae to concentrations of male odors that were relatively high compared with the likely concentrations away from the immediate vicinity of a male were significantly greater than male responses for only 8 of 54 identified components of odors obtained from males (Jang et al. 1989a). But most of the major components of the male odor were *not* potent stimulators of female antennae, again emphasizing the limited value of EAG data in evaluating relative attractiveness.

The possible significance of the affinity between male odor and the odor of fruit (Light et al. 1992; Howse and Knapp 1996) remains a mystery, but it hints at the possibility that male pheromone compounds are sensory traps that exploit preexisting female sensitivities and attractions to oviposition sites (West-Eberhard 1984; Ryan 1990; Christy 1995). Arita and Kaneshiro (1988) suggested, on the basis of striking differences in male mating success, that the larval host plant may affect male pheromone composition, but the limited tests performed to date have not shown any effect (Levinson et al. 1992; Heath et al. 1994). It appears, however, that in some studies both male strain and larval rearing substrate varied simultaneously.

Some authors (Millar 1995; Howse and Knapp 1996) have remarked on the puzzling complexity of male pheromone composition in Mediterranean fruit flies as compared with the relatively simple composition typical of many long-range attractant pheromones that are produced by female Lepidoptera (Silberglied 1977; Phelan 1997). This difference is in fact predicted by sexual selection theory. Competitive signals under sexual selection (as may occur in Mediterranean fruit flies, although no direct demonstration of differential attraction to the pheromones of different males has ever been attempted) are expected to be more complex than noncompetitive signals (such as those from females) that are not under sexual selection (West-Eberhard 1984).

### 18.4.4.2 Male Behavior and Morphology

Many aspects of sexually dimorphic male courtship behavior and morphology presumably function to elicit crucial female responses that will enable the male to copulate with her and fertilize her eggs, such as inhibiting her aggressive responses to other flies, approaching a courting male, stopping nearby, facing him head-on, allowing him to mount, and transporting his sperm to her spermathecae. Specific data that would indicate which male cues elicit which female responses, however, are even more sparse than those just summarized for male responses. The only experimental data of which I am aware are the following. Féron (1962) observed that a female sometimes turned toward a male that was performing continuous wing vibration on the other side of a pane of glass, thus demonstrating the importance of visual stimuli. It was not clear which visual stimuli from the male induced this response. The apparently contradictory statement of Levinson et al. (1987) that "the sight of a calling male fails to induce sexual attraction" actually refers to copulation rather than attraction per se.

When the male's wings were removed, the percentage of females that were inseminated after 3 days with the males dropped by about half (Keiser et al. 1973); female behavior was not observed directly, however, and it was not clear whether visual, auditory, or chemical stimuli (via air currents laden with pheromone) were involved. Genetic alteration of male eye color (with the "apricot" allele) resulted in reduced male mating success (Rössler 1980), but the possibility that there are pleiotropic effects of this allele was not tested. Previous sexual experience (virgins vs. nonvirgins) did not affect mating success (Shelly and Whittier 1994b).

Elimination of all visual stimuli by keeping flies in darkness reduced insemination rates by a factor of more than 10 (Keiser et al. 1973), but it was not clear whether this was due to reduced courtship or reduced acceptance rates, or which of several possible visual stimuli may have been involved. When the capitate anterior orbital bristles of male flies from a mass-reared strain in Costa Rica were removed, there was an increased likelihood for females being courted in small mating chambers to attack males, and a decreased likelihood that mounting would occur (Mendez et al. 1998). Preliminary results of a similar experiment with a different mass-reared strain in field cages in Guatemala showed a similar effect on mating success (A. Gilburn, personal communication). The positions of the setae, the pattern of movements of the head during head rocking, and the lack of mobility of the setae argue against the hypothesis that the setae serve to increase the concentration of pheromone near the female (see Kaneshiro 1993; and Chapter 32). Their positions with respect to the sexually dimorphic color pattern on the male's eyes, their own pattern of colors (dark distal portion and light base; Mendez et al. 1998), and the white tips of homologous bristles in other *Ceratitis* species (S. Quilici, personal communication) suggest that instead they are visual signals, but further experimental work is needed to test this idea. Presumably visual stimuli from the capitate bristles are important during head rocking behavior rather than other stages of courtship, but there are no data on this point.

Hasson and Rossler (unpublished data) have argued that symmetry in bristle length or width may be used as a cue by females of an indicator of male vigor. Hunt et al. (1998) found that males with more symmetrical bristles had greater mating success in laboratory conditions, a result confirmed by more recent observations in field cages in Guatemala (A. Gilburn, personal communication). Variation in the orientation of the bristles with respect to the female during courtship makes it unlikely that females can judge the small differences in setal dimensions that are involved (Mendez et al. 1998), so it seems likely that the association between mating success and setal symmetry was an indirect consequence of female biases resulting from other male traits. It was not determined in any of these experiments which aspects of the bristles were important.

### 18.4.4.3 Male Size

The effects of male size on female receptivity are controversial. Churchill-Stanland et al. (1986) working with Hawaiian mass-reared flies, Blay and Yuval (1997) and Taylor and Yuval (1999) using mass-reared flies in Israel, and A. Gilburn and co-workers (A. Gilburn, personal communication) working with flies in Guatemala, found that larger or heavier males were either more successful in obtaining copulations or in copulating more rapidly. Orozco and López (1993) came to a similar conclusion using mass-reared flies in field cages in Mexico, but failed to test for statistical significance. The Hawaiian data suggest that the relative difference between male and female size, rather than the absolute size of the male may be most important; the smallest females mated equally well with males of all sizes.

In contrast, Arita and Kaneshiro (1988), Whittier et al. (1992; 1994), and Whittier and Kaneshiro (1995) in studies in both the field and in captivity with mass-reared flies in Hawaii, Calcagno et al. (unpublished data) with mass-reared flies in captivity in Argentina, Orozco and López (1993) with wild flies from Guatemala in field cages, and R. D. Briceño and R. Gonzalez (personal communication) with both wild and mass-reared flies in captivity in Costa Rica found that male copulatory success was not correlated with male size. Copulating males in the field in Hawaii averaged almost exactly the same size as solitary males trapped nearby using a trimedlure bait (Whittier et al. 1992); and slightly *smaller* males from one Hawaiian population were strongly preferred over males from another population by females from both populations in captivity (56 vs. 27 copulations when one male of each population was present) (Arita and Kaneshiro 1988).

The discrepancy in data on size biases, at least in Hawaiian flies, may stem from the fact that Churchill-Stanland et al. (1986) used an unusually large range of sizes, creating especially small flies of only 3 mg by crowding larvae. It is possible that there is no size advantage within the normal range of sizes in the field, but that abnormally small males that do not occur in the field are less successful. Unfortunately, the measures of size in the Churchill-Stanland et al. study did not include the measurements used in the other Hawaiian studies, so direct comparisons cannot be made. Hunt et al. (1998) did measure one of the dimensions used by Churchill-Stanland et al., and the sizes of their Guatemalan flies support this interpretation, as the total mean wing lengths (4.4 and 4.5 mm for mated and unmated males) correspond to flies in the intermediate to large size range of Churchill-Stanland et al. (1986). Blay and Yuval (1997) showed that males that had been protein starved as adults copulated less than normally fed males. In none of these studies was it determined whether any particular behavioral or morphological male trait associated with courtship was responsible for differential male success.

### 18.4.4.4 Differences among Males

Whatever the important stimuli may be, some individual males seem to be better at inducing females to mate than others. Lance and McInnis (1996) showed that the early study of Arita and Kaneshiro (1985) did not, as claimed, demonstrate statistically significant differences in mating abilities among males. But subsequent studies (Whittier et al. 1994; Whittier and Kaneshiro 1995; Lance et al., unpublished data) showed significant individual differences. Little is known about which traits make one male succeed and another fail. The most successful males in the study by Whittier et al. (1994) were especially persistent; they courted significantly more often, and attempted significantly more copulations. These correlations were only moderately strong, however, leading the authors to suggest that male activity was not the sole determinant of increased male copulatory success.

Success did not correlate with time spent calling, male size, proportion of courtships that led to mounting attempts, or success in aggressive interactions. There is indirect evidence that females compare males in the laboratory. When a female was courted by several males before mating, she ultimately copulated with the male that had the highest "copulation score" (based on mating success with other females) of the males that had courted her (Whittier et al. 1994). Father–son correlations and sibling analyses showed that female mate choice did not enhance offspring quality (Whittier and Kaneshiro 1995), although (as noted by the authors), the inability to show any heritable component to male size suggests a relatively insensitive experimental design.

#### 18.4.4.5 Necessity of Female Cooperation

Female cooperation in male mating attempts is clearly crucial at several stages. In general, a male will only get close enough to a female to court her if the female has been successfully attracted to the lek and to the male's leaf. As noted by Féron (1962), in close-range interactions that follow, the male is more or less immobile while he moves his wings and head during courtship, and it is generally the female that is attracted to and positions herself near the male (see also Whittier et al. 1992; Saul and McCombs 1993a; and Whittier and Kaneshiro 1995 on female choosiness). In addition, the mating site, the undersurface of a leaf, makes it possible for the female to break off an interaction both prior to mounting and immediately after mounting has occurred by simply dropping from the leaf (e.g., Whittier et al. 1994) (recently mounted pairs invariably separate in midair when the female drops from the leaf; Lance et al., unpublished data; T. E. Shelly, personal communication). Perhaps females choose these as mating sites precisely in order to be able to make this type of rejection possible (A. Gilburn, personal communication). Even males that have mounted and not been dislodged by the female sometimes have difficulties inducing the female to extend her aculeus, and abandon the female after genitalic contact lasting <15 min that probably does not involve intromission or sperm transfer (failures of this sort are apparently more common in one genetic race than another; Saul and McCombs 1993a, no data given).

It is important to remember that the female responses needed for copulation to take place usually do *not* occur. For instance, the large majority of 5- to 13-day-old virgin females in captivity were not attracted to extracts of the suite of odors produced by males (Jang et al. 1989a). Only about 7% of copulation attempts with virgin females in captivity were successful (Whittier and Kaneshiro 1995). Successful copulations with virgin females in field cages occurred in only 37% ($N = 105$) and 23% ($N = 1075$) of courtships (Prokopy and Hendrichs 1979; Lance et al., unpublished data). Even in leks in the field, where the only females that are present have already responded to male pheromones and are thus presumably sexually receptive, females often either walked away from a courting male and prevented mounting, or dislodged the male after he mounted (Whittier et al. 1994). Females can reliably dislodge newly mounted males by dropping from the leaf; they can prevent a mounted male from intromitting by simply failing to extend the aculeus; and they may be able to prevent arrival of sperm in the spermathecae by failing to transport them (Section 18.3.2.2).

Thus female cooperation at several stages of an interaction is important in determining a male's chances of mating, as in other tephritids (Headrick and Goeden 1994), and forceful intromission without female cooperation ("rape") (Prokopy and Hendrichs 1979; Prokopy and Papaj, Chapter 10) is unlikely if not impossible. None of the male attempts to "force copulation" with resisting females in captivity were successful (Whittier et al. 1994). It is imaginable that male persistence may make female resistance more costly than acceptance of copulation (Prokopy and Papaj, Chapter 10). This possibility seems weak, however, for Mediterranean fruit flies; the cost to the female cited by Prokopy and Papaj, reduced opportunities to oviposit, is almost certainly not in operation in Mediterranean fruit flies. Rejection of a male mating attempt may cost the female a few seconds of oviposition, but acceptance will cost her about 2 h, due to the long duration of copulation in this species.

## 18.5 COMPARISONS OF MASS-REARED AND WILD FLIES

### 18.5.1 MALES

There is general agreement that males of mass-reared strains of Mediterranean fruit flies are usually less successful at courting wild females under natural conditions (e.g., Rössler 1975b; Calkins 1984; Shelly et al. 1994; Shelly and Whittier 1996; McInnis et al. 1996; Hendrichs et al. 1996; Lance et al., unpublished data; see, however, Wong et al. 1983; also Cayol, Chapter 31). Initiation of mass-reared strains is difficult because wild flies often fail to reproduce under mass-rearing conditions (e.g., Rössler 1975b), so mass-reared flies are often from relatively old strains and the differences in sexual success are probably often associated with genetic differences. Differences in male success can be dramatic. For instance, in leks in the field in Hawaii with both wild and mass-reared males present, wild males obtained 93% of 41 matings with wild females, even though they represented only 20% of the males present at these leks (Shelly and Whittier 1996), and Lance et al. (unpublished data) showed, using large samples, that wild male courtships of wild females in field cages were 3.5 times more likely to result in copulation. The reasons are not clear, and a major remaining challenge in efforts to control Mediterranean fruit fly populations by releasing sterile males is to determine which aspects of a male and his behavior result in greater copulatory and fertilization success (Whittier and Kaneshiro 1995; McInnis et al. 1996). Mass-reared males in the field in Hawaii appeared to join, call, and remain in leks in normal fashion (Shelly et al. 1994), and Lance et al. (unpublished data) obtained similar results in field cages. One test of the temporal pattern of pheromone release by wild and mass-reared males in Guatemala failed to find differences between wild and mass-reared males, but did show significant differences in the ratios of three pheromonal components that may possibly be important in attracting females (Heath et al. 1991). Perhaps some inconsistencies in the pheromone components found in different studies (see Millar 1995) are due to variation of this sort.

There are differences between genetic strains in courtship duration. The crowded conditions in mass-rearing facilities probably result in stronger selection in mass-reared strains favoring shorter courtship (Calkins 1984), since interruption of courtship by other males is frequent in captivity (Briceño and Eberhard 1998). Interruption also occurs at least occasionally in nature (Hendrichs and Hendrichs 1990; Whittier et al. 1992). When flies of wild and mass-reared strains from Costa Rica were kept at the same density, males from the mass-reared strain executed significantly shorter courtships than males from the wild strain from which it had been derived (Briceño and Eberhard 1998). Two other mass-reared strains derived from Costa Rican and Egyptian flies also showed significantly lower courtship durations at comparable densities.

Thus variation in courtship duration is produced both by the fly's environment (shorter courtships when flies are more crowded; Section 18.4.3.3.) and by differences among strains. Head rocking may be especially susceptible to reduction. It showed particularly dramatic reductions, and was omitted entirely in 35 to 38% of successful courtships of Egyptian-derived flies. It should be noted, however, that the number of mass-reared strains studied to date is only three. There are additional data on duration of different types of courtship behavior in wild and mass-reared strains in Hawaii (Liimantainen et al. 1997), but they are presented in terms of the proportion of total courtship time rather than as absolute durations, and thus cannot be compared with the Costa Rican results. A similar reduction in courtship duration resulted in strains in which mass mating tests were carried out (Harris et al. 1988), and may have also resulted from mass-rearing in the tephritid *Bactrocera cucurbitae* (Coquillett) (Hibino and Iwahashi 1989; 1991).

Recent work (R. D. Briceño and W. Eberhard, unpublished data) with additional, larger samples confirmed that courtships of isolated pairs of wild Mediterranean fruit flies in Costa Rica lasted longer than those of isolated pairs of a mass-reared strain. Mass-reared males were also more likely to mount without courting the female previously, and initiated courtship (continuous wing vibration) at smaller male–female distances than did wild males (R. D. Briceño and W. Eberhard, unpublished data). All of these behavior patterns are probably selectively favorable under conditions of mass-rearing, although direct tests are only available for courtship duration.

Several alternative interpretations of male behavioral responses to crowding can be discarded. Mass-reared and wild flies did not show differences in activity levels in other nonsexual contexts, and the rates of execution of repetitive behavior patterns such as wing buzzes and head rocks were not different under different degrees of crowding or between wild and mass-reared flies (Briceño and Eberhard 1998). Possible differences in female acceptance signals were ruled out because males did not respond to the three female behavior patterns that were associated with mounting attempts (Section 18.4.3.2), and because the courtships of males of the mass-reared strain were equally short whether they courted mass-reared or wild females (Briceño and Eberhard 1998).

Thus inadvertent selection in mass-rearing cages appears to have resulted in evolutionary change in male courtship behavior. One important consequence of this interpretation is that it helps answer the important question posed by Lance and McInnis (1993) regarding the existence of genetic variation in male courtship abilities, and of the appropriateness of using sexual selection models in making decisions about attempts to control pest populations (e.g., Boake et al. 1996). Even in the mass-reared strains in Costa Rica, which were derived from a genetically relatively homogeneous population (Huettel et al. 1980; Fuerst 1988; see also Malacrida et al. 1991), and which may have suffered further loss of alleles as they came under domestication (see Cayol, Chapter 31), there was apparently enough variation to allow rapid response to selection under mass-rearing conditions.

### 18.5.2 Females

The opposite, female, side of the coin, has received less attention, but it obviously has important consequences, both for quality control in mass-reared strains, and for attempts to "refresh" mass-reared strains by introducing wild males (Rössler 1975b). Overall female receptivity to courtship has changed under mass-rearing, as females of mass-reared strains become receptive at a much earlier age (e.g., Rössler 1975b; Wong and Nakahara 1978), presumably due to selection in captivity that favors early reproduction. Females of one old (240 generation) mass-reared strain did not show the triggering effect on receptivity from exposure to guavas that was seen in wild females (Wong and Nakahara 1978), again possibly due to selection under mass-rearing. Artificial selection for slower mating in a moderately young mass-reared strain (about 25 generations old) resulted in females that were less responsive to male courtship (Harris et al. 1988). Rapid changes in female criteria apparently occurred in only 2 to 3 years in a wild population (also genetically impoverished) in Hawaii, in response to sustained releases of sterile males (McInnis et al. 1996).

When males and females of both wild and mass-reared strains are combined in mating trials in field cages, mass-reared females often have a stronger tendency to mate with mass-reared males than with wild males than do wild females (Guerra et al. 1986; Wong and Nakahara 1978; Wong et al. 1984); however, Lance et al. (unpublished data) found no preference in either long-range attraction to calling males or to male courtship in field cages. The trend to within-strain mating may occur in the field, but the only data available (Shelly et al. 1994) involve mass-raised males that were irradiated, a treatment that reduces the male's mating effectiveness and can make the relative differences between wild and mass-reared females less distinct (Wong et al. 1984).

A possible problem with many studies that has never been satisfactorily resolved is that mass-reared flies generally mature sexually more rapidly than wild flies (e.g., Wong and Nakahara 1978), so it is difficult to standardize age effects in comparisons with wild flies. Most studies choose an age for each strain at which males are clearly mature; whether these are really equivalent ages in all important respects has never been established.

Generally it is not known which female criteria are responsible for biases against mass-reared males, making it impossible to check whether changes in female preferences in mass-reared strains have evolved in concert in these strains with changes in male courtship behavior. Preliminary analyses suggest one possible case of such female–male coordination in a mass-reared strain in Costa Rica. While male courtship became shorter (Briceño and Eberhard 1998), female discrimination against shorter courtships appears to have diminished in a Fisherian type of process (R. D. Briceño and W. Eberhard, unpublished data).

It should be kept in mind that changes in mass-reared strains are not likely to be simple, indiscriminant "degeneration," but rather adaptive adjustments to the very different and in some respects especially competitive conditions in rearing cages. This means that changes i n certain directions are more likely than others in mass-reared strains. This expected pattern may simplify the search for the causes of the difficulties that mass-reared males experience when courting wild females (Section 18.6). Detailed observations of the selective regimes within mass-rearing cages could be useful in orienting future research.

It is also important to note that some of the altered behavior patterns in mass-reared strains may have already been present in wild flies and expressed in mating attempts at oviposition sites. For instance, the increased tendency for mass-reared males to mount females without courting them may also be more common on fruit than in leks (Prokopy and Hendrichs 1979). A similar trend may occur in shortened courtships (Briceño and Eberhard 1998). Further, more-detailed quantitative comparisons of male behavior at leks and fruit would be of interest.

## 18.6 COMPARISONS WITH OTHER TEPHRITIDS

As I will argue below with respect to studies of other species of *Ceratitis*, comparative data from other taxa can sometimes aid in understanding a particular species, particularly with respect to the differences between species. I have not made a careful search of the large literature on the sexual biology of other genera of tephritids, but it is possible to use chapters of this book (Dodson, Chapter 8; Prokopy and Papaj, Chapter 10; Landolt, Chapter 14; Aluja et al., Chapter 15; and Headrick and Goeden, Chapter 25) to gain a quick overview. My impression is that the general outlines of the sexual biology of the Mediterranean fruit fly are so typical of tephritids as a group that comparisons at this level can make only limited contributions to our understanding of the Mediterranean fruit fly.

The fact that during the reproductive season *Rhagoletis* males of a given species use alternate tactics for encountering and mating with females similar to those that occur in Mediterranean fruit fly on a diurnal timescale, from loose aggregations of males and females away from fruit early in the season to mating on fruits (sometimes without apparent courtship) later (Prokopy and Papaj, Chapter 10), and that similar differences occur both intra- and interspecifically in *Anastrepha* (Aluja et al., Chapter 15) support the appropriateness of the alternative tactic interpretation of Mediterranean fruit fly behavior (Prokopy and Hendrichs 1979; Section 18.2.2). The fact that *Toxotrypana curvicauda* Gerstaecker females are particularly attracted to combinations of odors from fruits and males (Landolt, Chapter 14) is in accord with the idea proposed here (Section 18.4.4.1) that the pheromones of male Mediterranean fruit flies act as sensory traps, but it is far from constituting proof. The fact that some male *A. suspensa* perform unmistakable courtship behavior after having introduced their genitalia into the female (the misnamed "precopulatory" songs; see Aluja et al., Chapter 15), and that male courtship behavior during copulation is widespread in other species of *Anastrepha* as well as in many other groups (e.g., Headrick and Goeden 1994; and Chapter 25) support the interpretation that the less dramatic rubbing movements of male Mediterranean fruit flies during copulation function as copulatory courtship (Eberhard and Periera 1998), and serve to induce as yet undetermined favorable postintromission female responses (Eberhard 1991; 1994). It appears that females of a variety of tephritids probably exercise cryptic female choice via paternity biases that are imposed after intromission has begun (Eberhard 1996). But again, more direct tests of this heretofore relatively neglected possibility will be needed to understand each particular case.

## 18.7 CONCLUSIONS AND FUTURE RESEARCH NEEDS

It seems appropriate to end this chapter by reemphasizing our degree of ignorance of what distinguishes successful from unsuccessful Mediterranean fruit fly courtships. The kinds of questions that have been asked to date are mostly comparatively crude, typically involving only total durations or the presence/absence of particular stimuli. Consider the following sample of the types of questions that still remain.

1. *Organization of behavior*: Do the patterns of transition between stages or the relative lengths of different stages of courtship matter for male copulation (Liimatainen et al. 1997) and fertilization success? Do males modulate their courtship on the basis of female stimuli (the dialogue idea of Lux and Gaggl 1996)? Are the different steps in female responses to males that lead to successful copulation influenced differently by different male signals?
2. *Songs*: Does the intensity of a male's song matter (as in another tephritid; Sivinski et al. 1984)? Do frequency, amplitude, or pattern of co-occurrence with buzzes matter? Do the durations of individual buzzes, or the periods between buzzes, or the relative durations of the two matter (as is the case in several species of *Drosophila*; Bennett-Clark and Ewing 1969; Tomaru and Oguma 1994)? Do the songs matter at all, or are they perhaps only incidental by-products of wing movements that function to produce visual and chemical displays (Kaneshiro, Chapter 32)? Are airborne songs accompanied by substrate-borne vibrations, and, if so, are these vibrations important?
3. *Visual stimuli*: Does the presence/absence or the pattern of head rocking movements matter? Are the sexually dimorphic male eye color and the silvery anterior portion of the head important? Do colors in the ultraviolet affect male success?
4. *Odors*: Do the relative concentrations of the different components of pheromones matter in long-distance attraction (Jones 1989) (as is the rule in Lepidoptera; e.g., Silberglied 1977)? Do they matter in short-range courtship? Are, in fact, the same pheromones used in long- and short-range interactions with females?
5. *Geographic variation*: Do female criteria for male traits vary within or between wild populations (e.g., Lance and McInnis 1993)? Do any aspects of male courtship vary geographically?
6. *Genetics*: Are sexually selected male morphological and behavioral traits all genetically independent, or are some genetically correlated with each other? Are female criteria genetically associated with particular male display traits, as predicted by Fisherian theory (e.g., Andersson 1994)?
7. *Female physiology*: Do differences in any male courtship stimuli result in female responses (or lack of responses) that could explain why some copulations fail to result in sperm transfer (e.g., Seo et al. 1990)? Are sperm numbers within the female related in any way to male quality (Yuval et al. 1996)? Is the transfer of male accessory gland products that affect female reproduction independent in any way from that of sperm, and if so, what are the differences? Do the effects of male accessory gland products on such female responses vary between strains, as suggested by other studies of cryptic female choice (Eberhard 1996)?
8. *Others*: Are reproductively significant postcopulatory responses by females such as oviposition and resistance to remating affected by precopulation courtship or by copulatory courtship? Are such processes affected by male genitalic size or form? Are there predictable differences between the male behavior or female criteria under field as opposed to field-cage or laboratory conditions? Do female criteria differ between virgin and mated females (e.g., Heath et al. 1991, on long-distance attraction)?

As is common in studies of possible sexual selection by female choice (e.g., Andersson 1994), male–female interactions confront us with a bewildering array of possible durations, intensities, sequences, and combinations of male stimuli that might possibly influence female receptivity. This problem is particularly acute in Mediterranean fruit flies because so many sensory modalities are probably involved. Most of these questions have great potential significance for improving the success of attempts to control pest populations of Mediterranean fruit flies with mass-reared flies. The most important recent progress is the studies in Hawaii (Shelly et al. 1994; Lance et al., unpublished data) that have now focused attention on male courtship per se, by showing that mass-reared males in the field perform about on par with wild males in events such as lek attendance and long-distance attraction of females (see, however, Cayol et al. 1999).

Three possible ways forward occur to me. Several simple experiments, along the lines of the removal of antennae (Nakagawa et al. 1973), wings (Sivinski et al. 1984), and the male capitate anterior orbital bristles (Mendez et al. 1998), could give important information, especially when combined with experimental replacement of particular stimuli (e.g., Sivinski et al. 1984), and with video analyses of the behavioral responses of females. A second tactic would be to determine the probable phylogeny of species in the genus *Ceratitis* (see De Meyer, Chapter 16), and combine this with information on courtship structures and behavior in these species. This could help determine which male traits in *C. capitata* are newly derived and are thus especially likely to be important in female choice. It will be important to keep in mind possible differences in the behavior of males and females from different geographic sites, as some differences of this sort are beginning to emerge (e.g., McInnis et al. 1996; Lance et al., unpublished data; Section 18.3.2.2).

Finally, it may be useful to study in more detail the types of selection that occur in mass-rearing cages (the "selection" process mentioned by Cayol, Chapter 31), in order to orient and focus searches for differences in the male courtship behavior (and female receptivity) of mass-reared strains. Traits that are probably markedly superior in these cages (e.g., shorter courtships, reduced aggression, perhaps mounting females without courtship) may be more likely to diverge rapidly in mass-reared strains. On the other hand, traits that are likely to be under little or no directional selection, and thus likely to diverge only by drift may change more slowly. For instance, visual stimuli seem less likely to diverge rapidly than courtship duration (unless artificial lighting has an effect on how courting males are perceived by females) (for a contrasting view that does not include this emphasis, see Cayol, Chapter 31).

The patterns in the fragmentary data available to date give preliminary confirmation that selection in cages has been especially important in producing divergence in Mediterranean fruit flies. Age of reproduction, courtship duration, mating without courtship, male–female distance when courtship is initiated, inhibition of receptivity without odors of plants, and male aggression are all likely to be under strong directional selection in mass-rearing cages, and all these traits differ between mass-reared and wild strains (respectively, Cayol, Chapter 31; Briceño and Eberhard 1998; R. D. Briceño and W. Eberhard, unpublished data; Wong and Nakahara 1978, Briceño et al. 1998). The fundamental frequencies of sounds from both continuous and intermittent wing vibration were also somewhat reduced in a mass-reared strain, contrary to predictions; but this may have been an effect of the larger male size of mass-reared flies (Sivinski et al. 1989); a longer history of mass-rearing resulted in a reduction rather than an accentuation of differences with wild flies. There are also differences in copulation duration (Section 18.3.2.2), but our lack of understanding of the reproductive significance of different durations prevents interpretation.

In contrast, effectiveness of pheromones to attract females, attendance at leks, durations of individual buzzes during intermittent buzzing, and times of the day when males are sexually active all seem less likely to be under strong selection in cages (there are no data, however, on these points), and all have diverged weakly, inconsistently, or not at all (Shelly et al. 1994; Lance et al., unpublished data; R. D. Briceño and W. Eberhard, unpublished data; see, however, Guerra et al. 1986; Wong and Nakahara 1978; and Cayol, Chapter 31). Mass-reared strains probably represent ongoing evolutionary experiments on the effects of selection due to altered environmental conditions.

Studies of the sexual behavior of *C. capitata* readily combine topics of both academic and practical interest. Aspects of basic biology that have immediate practical consequences include the amount of genetic variation in sexually selected characters, microevolution of sexual signals and female criteria under artificial selection, and the operation of an unusual lekking system in which male territoriality is relatively weak and male signals are numerous and extremely complex. Improved understanding on these fronts will be necessary to guide the search for ways to improve the mating behavior of mass-reared males, which is of high priority to ensure the success of pest control efforts throughout the world.

## ACKNOWLEDGMENTS

I am especially grateful to T. E. Shelly for insightful comments and unstinting help with references. I also thank M. Aluja, R. D. Briceño, A. Gilburn, K. Y. Kaneshiro, D. Roubik, M. J. West-Eberhard, B. Yuval, and an anonymous reviewer for useful comments on preliminary drafts; R. D. Briceño, A. Gilburn, D. Lance, T. Shelly, and B. Yuval for access to unpublished data; J.-P. Cayol for help with references; and J. Liimatainen and J. Sivinski for additional information. The Smithsonian Tropical Research Institute, the Vicerrectoría de Investigación of the Universidad de Costa Rica, and the International Atomic Energy Agency provided financial support. The writing of this chapter was made possible by the generous financial support of the International Organization for Biological Control of Noxious Animals and Plants (IOBC) and the Campaña Nacional contra las Moscas de la Fruta (SAGAR-IICA).

## REFERENCES

Andersson, M. 1994. *Sexual Selection*. Princeton University Press, Princeton.

Arita, L. and K.Y. Kaneshiro. 1983. Pseudomale courtship behavior of the female Mediterranean fruit fly, *Ceratitis capitata* (Wiedemann). *Proc. Hawaii. Entomol. Soc.* 24: 205–210.

Arita, L. and K.Y. Kaneshiro. 1985. The dynamics of the lek system and mating success in males of the Mediterranean fruit fly, *Ceratitis capitata* (Wiedemann). *Proc. Hawaii. Entomol. Soc.* 25: 39–48.

Arita, L. and K.Y. Kaneshiro. 1986. Structure and function of the rectal epithelium and anal glands during mating behavior in the Mediterranean fruit fly. *Proc. Hawaii. Entomol. Soc.* 26: 27–30.

Arita, L. and K.Y. Kaneshiro 1988. Body size and differential mating success between males of two populations of the Mediterranean fruit fly. *Pac. Sci.* 42: 173–177.

Arita, L. and K.Y. Kaneshiro. 1989. Sexual selection and lek behavior in the Mediterranean fruit fly, *Ceratitis capitata* (Diptera: Tephritidae). *Pac. Sci.* 43: 135–143.

Baker, P.S., P.E. Howse, R.N. Ondarza, and J. Reyes. 1990. Field trials of synthetic sex pheromone components of the male Mediterranean fruit fly (Diptera: Tephritidae) in Mexico. *J. Econ. Entomol.* 83: 2236–2245.

Baker, P.S. and H. Van der Valk. 1992. Distribution and behaviour of sterile Mediterranean fruit flies in a host tree. *J. Appl. Entomol.* 114: 67–76.

Baker, R., R.H. Herbert, and G.G. Grant. 1985. Isolation and identification of the sex pheromone of the Mediterranean fruit fly, *Ceratitis capitata* (Wied.). *J. Chem. Soc. Chem. Commun.* 824–825.

Bennett-Clark, H.C. and W.W. Ewing. 1969. Pulse interval as a critical parameter in the courtship song of *Drosophila melanogaster*. *Anim. Behav.* 17: 755–759.

Blay, S. and B. Yuval. 1997. Nutritional correlates of reproductive success of male Mediterranean fruit flies (Diptera: Tephritidae). *Anim. Behav.* 54: 5966.

Bloem, K., S. Bloem, N. Rizzo, and D.L. Chambers. 1993. Female medfly refractory period: effect of male reproductive status. In *Fruit Flies: Biology and Management* (M. Aluja and P. Liedo, eds.), pp.189–190. Springer-Verlag, New York.

Bloem, S., K. Bloem, N. Rizzo, and D.L. Chambers. 1993. Female medfly refractory period: effect of first mating with sterile males of different sizes. In *Fruit Flies: Biology and Management* (M. Aluja and P. Liedo, eds.), pp. 191–192. Springer-Verlag, New York.

Boake, C.R.B, T.E. Shelly, and K.Y. Kaneshiro. 1996. Sexual selection in relation to pest-management strategies. *Annu. Rev. Entomol.* 41: 211–229.

Briceño, R.D. and W.G. Eberhard. 1998. Medfly courtship duration: a sexually selected reaction norm changed by crowding. *Ethol. Ecol. Evol.* 10: 369–382.

Briceño, R.D., D. Ramos, and W.G. Eberhard. 1996. Courtship behavior of male medflies (*Ceratitis capitata*; Diptera: Tephritidae) in captivity. *Fla. Entomol.* 79: 1–15.

Briceño, R.D., D. Ramos, and W.G. Eberhard. 1998. Aggressive behavior in medflies (*Ceratitis capitata*) and the effects of mass-rearing (Diptera: Tephritidae). *J. Kans. Entomol. Soc.* 72: 17–27.

Burk, T. and C.O. Calkins. 1983. Medfly mating behavior and control strategies. *Fla. Entomol.* 66: 3–18.

Calkins, C.O. 1984. The importance of understanding fruit fly mating behavior in sterile male release programs (Diptera, Tephritidae). *Folia Entomol. Mex.* 61: 205–213.

Calkins, C.O., T.R. Ashley, and D.L. Chambers. 1996. Implementation of technical and managerial systems for quality control in Mediterranean fruit fly (*Ceratitis capitata*) sterile release programs. In *Fruit Fly Pests: A World Assessment of Their Biology and Management* (B.A. McPheron and G.J. Steck, eds.), pp. 399–404. St. Lucie Press, Delray Beach. 586 pp.

Camacho, H. 1989. Transferencia de Espermatozoides en la Mosca del Mediterráneo *Ceratitis capitata* Wied. (Diptera: Tephritidae). Master's thesis, Universidad de Costa Rica, San José.

Causse, R. and M. Féron. 1967. Influence du rythme photopériodique sur l'activité sexualle de la mouche Méditerranéenne des fruits: *Ceratitis capitata* Wiedemann (Diptère Trypetidae). *Ann. Epiphyt.* 18: 175–192.

Cayol, J.P., J. Vilardi, E. Rial, and M.T. Vera. 1999. New indices to measure the sexual compatibility and mating performance of medfly (Diptera: Tephritidae) laboratory reared strains under field cage conditions. *J. Econ. Entomol.* 92: 140–145.

Chapman, T., T. Miyatake, H.K. Smith, and L. Partridge. 1998. Interactions of mating, egg production and death rates in females of the Mediterranean fruit fly, *Ceratitis capitata*. *Proc. R. Soc. Lond. Ser. B* 265: 1879–1894.

Chen, P.S. 1984. The functional morphology and biochemistry of insect male accessory glands and their secretions. *Annu. Rev. Entomol.* 29: 233–255.

Christy, J. 1995. Mimicry, mate choice, and the sensory trap hypothesis. *Am. Nat.* 146: 171–181.

Churchill-Stanland, C., R. Stanland, T.T.Y. Wong, N. Tanaka, D.O. McInnis, and R.V. Dowell. 1986. Size as a factor in the mating propensity of Mediterranean fruit flies, *Ceratitis capitata* (Diptera: Tephritidae), in the laboratory. *J. Econ. Entomol.* 79: 614–619.

Cunningham, R.T., G.J. Farias, S. Nakagawa, and D.L. Chambers. 1971. Reproduction in the Mediterranean fruit fly: depletion of stored sperm in females. *Ann. Entomol. Soc. Am.* 64: 312–313.

Dallai, R., D. Marchini, and G. Del Bene. 1993. The ultrastructure of the sphermateca in *Ceratitis capitata* Wied. and *Dacus oleae* Gmel. (Diptera: Tephritidae). *Redia* 76: 147–167.

Del Rio, G. and R. Cavalloro. 1979. Influenza dell'accoppiamento sulla recettività sessuale e sull'ovideposizione in femmine di *Ceratitis capitata* Wiedemann. *Entomologica (Bari)* 15: 127–143.

Eberhard, W.G. 1991. Copulatory courtship and cryptic female choice in insects. *Biol. Rev.* 66: 1–31.

Eberhard, W.G. 1994. Evidence for widespread courtship during copulation in 131 species of insects and spiders, and implications for cryptic female choice. *Evolution* 48: 711–733.

Eberhard, W.G. 1996. *Female Control: Sexual Selection by Cryptic Female Choice*. Princeton University Press, Princeton. 501 pp.

Eberhard, W.G. and C. Cordero. 1995. Sexual selection by cryptic female choice on male seminal products — a new bridge between sexual selection and reproductive physiology. *Trends Ecol. Evol.* 10: 493–496.

Eberhard, W.G. and S. Kariko. 1996. Copulation behavior inside and outside the beetle *Macrohaltica jamaicensis* (Coleoptera, Chrysomelidae). *J. Ethol.* 14: 59–72.

Eberhard, W.G. and F. Pereira. 1993. Functions of the male genitalic surstyli in the Mediterranean fruit fly, *Ceratitis capitata* (Diptera: Tephritidae). *J. Kans. Entomol. Soc.* 66: 427–433.

Eberhard, W.G. and F. Pereira. 1998. The process of intromission in the Mediterranean fruit fly, *Ceratitis capitata* (Diptera: Tephritidae). *Psyche* (1995) 102: 99–120.

Farias, G.J., R.T. Cunningham, and S. Nakagawa. 1972. Reproduction in the Mediterranean fruit fly: abundance of stored sperm affected by duration of copulation, and affecting egg hatch. *J. Econ. Entomol.* 65: 914–915.

Féron, M. 1962. L'instinct de réproduction chez la mouche Mediterranéan des fruits *Ceratitis capitata* Wied. (Diptera: Trypetidae). Comportement sexuel. Comportement de ponte. *Rev. Pathol. Veg. Entomol. Agric. Fr.* 41: 1–129.

Field, S.A. and B. Yuval. 1999. Nutritional status affects copula duration in the Mediterranean fruit fly, *Ceratitis capitata* (Insecta: Tephritidae). *Ethol. Ecol. Evol.* 11: 61–70.

Fowler, 1973. Some aspects of the reproductive biology of *Drosophila*: sperm transfer, sperm storage, and sperm utilization. *Adv. Genet.* 17: 293–360.

Fuerst, P.A. 1988. Islands as models in population genetics. In *The Biogeography of the Island Region of Western Lake Erie* (J. Downhower, ed.), pp. 264–269. Ohio State University Press, Columbus.

Gage, M.J.G. 1991. Risk of sperm competition directly affects ejaculate size in the Mediterranean fruit fly. *Anim. Behav.* 42: 1036–1037.

Guerra, M., D. Orozco, A. Schwarz, and P. Liedo. 1986. Mating competitiveness of mass-reared and sterilized Med-flies compared with wild flies. In *Fruit Flies of Economic Importance 84* (R. Cavalloro, ed.), pp. 113–119. A.A. Balkema, Rotterdam.

Hanna, A.D. 1938. Studies of the Mediterranean fruit fly Wied. I. The structure and operation of the reproductive organs. *Bull. Soc. Fouad 1er Entomol.* 22: 39–48, plates I-V.

Harris, D.J., R.J. Wood, and S.E.R. Bailey. 1988. Two-way selection for mating activity in the Mediterranean fruit fly, *Ceratitis capitata. Entomol. Exp. Appl.* 60: 1–10.

Headrick, D.H. and R.D. Goeden. 1994. Reproductive behavior of California fruit flies and the classification and evolution of Tephritidae (Diptera) mating systems. *Stud. Dipterol.* 1: 194–252.

Heath, R.R., P.J. Landolt, J.H. Tomlinson, D.L. Chambers, R.E. Murphy, R.E. Doolittle, B.D. Dueban, J. Sivinski, and C.O. Calkins. 1991. Analysis, synthesis, formulation, and field testing of three major components of male Mediterranean fruit fly pheromone. *J. Chem. Ecol.* 17: 1925–1940.

Heath, R.R., N.D. Epsky, B.D. Duebeh, A. Guzman, and L.E. Andrade. 1994. Gamma radiation effects on production of four pheromonal components of male Mediterranean fruit fly (Diptera: Tephritidae). *J. Econ. Entomol.* 87: 904–909.

Hendrichs, J. and M.A. Hendrichs. 1990. Mediterranean fruit flies (Diptera: Tephritidae) in nature: location and diel pattern of feeding and other activities on fruiting and nonfruiting hosts and nonhosts. *Ann. Entomol. Soc. Am.* 83: 632–641.

Hendrichs, M.A. and J. Hendrichs. 1998. Perfumed to be killed: interception of Mediterranean fruit fly (Diptera: Tephritidae) sexual signaling by predatory foraging wasps (Hymenoptera: Vespidae). *Ann. Entomol. Soc. Am.* 91: 228–234.

Hendrichs, J., J. Reyes, and M. Aluja. 1989. Behaviour of female and male Mediterranean fruit flies, *Ceratitis capitata*, in and around Jackson traps placed on fruiting host trees. *Insect Sci. Appl.* 10: 285–294.

Hendrichs, J., I. Katsoyannos, D.R. Papaj, and R.J. Prokopy. 1991. Sex differences in movement between natural feeding and mating sites and tradeoffs between food consumption, mating success and predatory evasion in Mediterranean fruit flies (Diptera: Tephritidae). *Oecologia* 86: 223–231.

Hendrichs, J., B.I. Katsoyannos, V. Wornoayporn, and M.A. Hendrichs. 1994. Odour-mediated foraging by yellowjacket wasps (Hymenoptera: Vespidae): predation on leks of pheromone-calling Mediterranean fruit fly males (Diptera: Tephritidae). *Oecologia* 99: 88–94.

Hendrichs, J., B.I. Katsoyannos, K. Gaggl, and V. Wornoayporn. 1996. Competitive behavior of males of Mediterranean fruit flies, *Ceratitis capitata*, genetic sexing strain *Vienna-42*. In *Fruit Fly Pests: A World Assessment of Their Biology and Management* (B.A. McPheron and G.J. Steck, eds.), pp. 405–414. St. Lucie Press, Delray Beach. 586 pp.

Hibino, Y. and O. Iwahashi. 1989. Mating receptivity of wild type females for wild type males and mass-reared males in the melon fly, *Dacus cucurbitae* Coquillett (Diptera: Tephritidae). *Ann. Entomol. Zool.* 24: 152–154.

Hibino, Y. and O. Iwahashi. 1991. Appearance of wild females unreceptive to sterilized males on Okinawa Island in the erradication program of the melon fly, *Dacus curcurbitae* Coquillett (Diptera: Tephritidae). *Appl. Entomol. Zool.* 26: 265–270.

Höglund, J. and R.V. Alatalo. 1995. *Leks*. Princeton University Press, Princeton. 248 pp.

Holbrook, F.R., L.F. Steiner, and M.S. Fujimoto. 1970. Mating competitiveness of Mediterranean fruit flies marked with fluorescent powders. *J. Econ. Entomol.* 63: 454–455.

Howse, P.E. and J.J. Knapp. 1996. Pheromones of Mediterranean fruit flies: presumed mode of action and implications for improved trapping techniques. In *Fruit Fly Pests: A World Assessment of Their Biology and Management* (B.A. McPheron and G.J. Steck, eds.), pp. 91–99. St. Lucie Press, Delray Beach. 586 pp.

Huettel, M.P., A.P. Fuerst, T. Maruyama, and F. Chakraborty. 1980. Genetic effects of multiple bottlenecks in the Mediterranean fruit fly. *Genetics* 94: 547–548.

Hunt, M.K., C.S. Crean, R.J. Wood, and A.S. Gilburn. 1998. Fluctuating asymmetry and sexual selection in the Mediterranean fruit fly (Diptera, Tephritidae). *Biol. J. Linn. Soc.* 64: 385–396.

Jacobson, M., K. Ohinata, D.L. Chambers, W.A. Jones, and M. Fujimoto. 1973. Insect sex attractants. 13. Isolation, identification and synthesis of sex pheromones of the male Mediterranean fruit fly. *J. Med. Chem.* 16: 248–251.

Jang, E.B. 1995. Effects of mating and accessory gland injections on olfactory mediated behavior in the female Mediterranean fruit fly, *Ceratitis capitata*. *J. Insect Physiol.* 41: 705–710.

Jang, E.B. and D.M. Light. 1996a. Olfactory semiochemicals in tephritids. In *Fruit Fly Pests: A World Assessment of Their Biology and Management* (B.A. McPheron and G.J. Steck, eds.), pp. 73–90. St. Lucie Press, Delray Beach. 586 pp.

Jang, E.B. and D.M. Light. 1996b. Attraction of female Mediterranean fruit flies to identified components of the male-produced pheromone: qualitative aspects of major, intermediate and minor components. In *Fruit Fly Pests: A World Assessment of Their Biology and Management* (B.A. McPheron and G.J. Steck, eds.), pp. 115–121. St. Lucie Press, Delray Beach. 586 pp.

Jang, E.B., D.M. Light, R.A. Flath, J.T. Nagata, and T.R. Mon. 1989a. Electroantennogram responses of the Mediterranean fruit fly, *Ceratitis capitata* to identified volatile constituents from calling males. *Entomol. Exp. Appl.* 50: 7–19.

Jang, E.B., D.M. Light, J.C. Dickens, T.P. McGovern, and J.T. Nagata. 1989b. Electroantennogram responses of Mediterranean fruit fly, *Ceratitis capitata* (Diptera: Tephritidae) to trimedlure and its *trans* isomers. *J. Chem. Ecol.* 15: 2219–2231.

Jang, E.B., D.O. McInnis, D.R. Lance, and L.A. Carvalho. 1998. Mating-induced changes in olfactory-mediated behavior of laboratory-reared normal, sterile and wild female Mediterranean fruit flies (Diptera: Tephritidae) mated to conspecific males. *Ann. Entomol. Soc. Am.* 91: 139–144.

Jones, O.T. 1989. *Ceratitis capitata*. In *Fruit Flies: Their Biology, Natural Enemies and Control* (A.S. Robinson and G. Hooper, eds.), pp. 179–183. In *World Crop Pests* (W. Helle, ed.), Vol. 3A. Elsevier Science Publishers, Amsterdam. 372 pp.

Kaneshiro, K.Y. 1993. Introduction, colonization, and establishment of exotic insect populations: fruit flies in Hawaii and California. *Am. Entomol.* 39: 23–29.

Kaneshiro, K.Y., T.E. Shelly, and T.S. Whittier. 1995. Biology and control of the Mediterranean fruit fly: lessons from natural populations. In *Proceedings: The Mediterranean Fruit Fly in California: Defining Critical Research* (J.G. Morse, R.L. Metcalf, J.R. Carey, and R.V. Dowell, eds.), pp. 199–214. University of California, Riverside. 318 pp.

Kaspi, R. and B. Yuval. 1999. Lek site selection by male Mediterranean truit flies. *J. Insect Behav.* 12: 267–276.

Katiyar, K.P. and E. Ramírez. 1970. Mating frequency and fertility of Mediterranean fruit fly females alternately mated with normal and irradiated males. *J. Econ. Entomol.* 63: 1247–1250.

Katiyar, K.P. and E. Ramírez. 1973. Mating duration of gamma irradiated Mediterranean fruit fly males. *Turrialba* 23: 471–472.

Katiyar, K.P. and J. Valerio. 1965. El efecto de uno y varios apariemientos sobre la viabilidad de los huevos de la mosca del Mediterraneo (*Ceratitis capitata* Wied.). *Turrialba* 15: 248–249.

Keiser, I., R.M. Kobayashi, D.L. Chambers, and E.L. Schneider. 1973. Relation of sexual dimorphism in the wings, potential stridulation, and illumination to mating of oriental fruit fly, melon flies, and Mediterranean fruit fly in Hawaii. *Ann. Entomol. Soc. Am.* 66: 937–941.

Kotrba, M. 1995. The internal female genital organs of *Chaetodiopsis* and *Diasemopsis* (Diptera: Diopsidae) and their systematic relevance. *Ann. Natal Mus.* 36: 147–159.

Kuba, H. and Y. Itô. 1993. Remating inhibition in the melon fly, *Bactrocera* (= *Dacus*) *curcurbitae* (Diptera: Tephritidae): copulation with spermless males inhibits female remating. *J. Ethol.* 11: 23–28.

Lance, D.R. and D.O. McInnis. 1993. Misrepresented reality? *Am. Entomol.* 39: 196.

Lance, D.R., D.O. McInnis, P. Rendon, and C.G. Jackson. Courtship among sterile and wild Mediterranean fruit flies, *Ceratitis capitata* (Diptera: Tephritidae) in field cages in Hawaii and Guatemala. *J. Insect Behav.*, in press.

Landolt, P.J., R.R. Heath, and D.L. Chambers. 1992. Oriented flight responses of female Mediterranean fruit flies to calling males, odor of calling males, and a synthetic pheromone blend. *Entomol. Exp. Appl.* 65: 259–266.

Levinson, H.Z., A.R. Levinson, and K. Schäfer. 1987. Pheromone biology of the Mediterranean fruit fly (*Ceratitis capitata* Wied.) with emphasis on the functional anatomy of the pheromone glands and antennae as well as mating behaviour. *J. Appl. Entomol.* 107: 448–461.

Levinson, H.Z., A.R. Levinson, and K. Müller. 1992. Complexity of the sex pheromone of *Ceratitis capitata* Wied. (Dipt., Trypetidae). *J. Appl. Entomol.* 109: 156–162.

Light, D.M., E.B. Jang, and R.A. Flath. 1992. Electroantennogram responses of the Mediterranean fruit fly, *Ceratitis capitata*, to the volatile constituents of nectarines. *Entomol. Exp. Appl.* 63: 13–26.

Liimatainen, J., A. Hoikkala, and T.E. Shelly. 1997. Courtship behavior in *Ceratitis capitata* (Diptera: Tephritidae): comparison of wild and mass-reared males. *Ann. Entomol. Soc. Am.* 90: 836–843.

Lux, S.A. and K. Gaggl. 1996. Ethological analysis of medfly courtship: potential for quality control. In *Fruit Fly Pests: A World Assessment of Their Biology and Management* (B.A. McPheron and G.J. Steck, eds.), pp. 425–432. St. Lucie Press, Delray Beach. 586 pp.

Malacrida, A.R., C.R. Guglielmino, G. Gasperi, L. Baruffi, and R. Milani. 1991. Spatial and temporal differentiation in colonizing populations of *Ceratitis capitata*. *Heredity* 69: 101–111.

McDonald, P.T. 1987. Intragroup stimulation of pheromone release by male Mediterranean fruit flies. *Ann. Entomol. Soc. Am.* 80: 17–20.

McInnis, D.O. and J.D. Warthen. 1988. Mediterranean fruit fly (Diptera: Tephritidae): laboratory bioassay for attraction of males to leaf or stem substances from *Ficus* and *Litchi*. *J. Econ. Entomol.* 81: 1637–1640.

McInnis, D.O., D.R. Lance, and C.G. Jackson. 1996. Behavioral resistance to the sterile insect technique by Mediterranean fruit flies (Diptera: Tephritidae) in Hawaii. *Ann. Entomol. Soc. Am.* 89: 739–744.

Mendez, V., R.D. Briceño, and W.G. Eberhard. 1998. Functional significance of the capitate supra-fronto-orbital bristles of male medflies (*Ceratitis capitata*) (Diptera, Tephritidae). *J. Kans. Entomol. Soc.* 71: 164–174.

Merrett, D.J. 1989. The morphology of the phallosome and accessory gland material transfer during copulation in the blowfly, *Lucilia cuprina* (Insecta, Diptera). *Zoomorphology* 108: 359–366.

Millar, J.G. 1995. An overview of attractants for the Mediterranean fruit fly. In *Proceedings: The Mediterranean Fruit Fly in California: Defining Critical Research* (J.G. Morse, R.L. Metcalf, J.R. Carey, and R.V. Dowell, eds.), pp. 123–143. University of California, Riverside. 318 pp.

Mungira, M.L., F. Salom, and M. Muñiz. 1983. Estudio morfológico del aparato reproductor femenino de *Ceratitis capitata* Wied. (Dipt.:Trypetidae). *Biol. Serv. Plagas* 9: 31–44.

Nadel, D.J. and B.A. Peleg. 1965. The attraction of fed and starved males and females of the Mediterranean fruit fly, *Ceratitis capitata* Wied., to "trimedlure." *Isr. J. Agric. Res.* 15: 83–86.

Nakagawa, S., G.J. Farias, D. Suda, R.T. Cunningham, and D.L. Chambers. 1971. Reproduction of the Mediterranean fruit fly: frequency of mating in the laboratory. *Ann. Entomol. Soc. Am.* 64: 949–950.

Nakagawa, S., G.J. Farias, D. Suda, and D.L. Chambers. 1973. Mating behavior of the Mediterranean fruit fly following excision of the antennae. *Ann. Entomol. Soc. Am.* 66: 583–584.

Nakagawa, S., L.F. Steiner, and G.J. Farias. 1981. Response of virgin female Mediterranean fruit flies to live mature normal males, sterile males, and trimedlure in plastic traps. *J. Econ. Entomol.* 74: 566–567.

Nation, J.L. 1981. Sex-specific glands in tephritid fruit flies of the genera *Anastrepha*, *Ceratitis*, *Dacus*, and *Rhagoletis* (Diptera: Tephritidae). *Int. J. Insect Morph. Embryol.* 10: 121–129.

Ohinata, K., M.S. Fujimoto, D.L. Chambers, M. Jacobson, and D.C. Kamakahi. 1973. Mediterranean fruit fly: bioassay techniques for investigating sex pheromones. *J. Econ. Entomol.* 66: 812–814.

Ohinata, K., M. Jacobson, S. Nakagawa, M. Fujimoto, and H. Higa. 1977. Mediterranean fruit fly: laboratory and field evaluations of synthetic sex pheromones. *J. Environ. Sci. Health A* 12: 67–78.

Orozco, D. and R.O. López. 1993. Mating competitiveness of wild and laboratory mass-reared medflies: effect of male size. In *Fruit Flies: Biology and Management* (M. Aluja and P. Liedo, eds.), pp. 185–188. Springer-Verlag, New York. 492 pp.

Phelan, P.L. 1997. Evolution of mate-signaling in moths: phylogenetic considerations and predictions from the asymmetric tracking hypothesis. In *The Evolution of Mating Systems in Insects and Arachnids* (J.C. Choe and B.J. Crespi, eds.), pp. 240–256. Cambridge University Press, Cambridge. 387 pp.

Prokopy, R.J. and J. Hendrichs. 1979. Mating behavior of *Ceratitis capitata* on a field-caged host tree. *Ann. Entomol. Soc. Am.* 72: 642–648.

Ramos, D. 1991. Selección y Comportamiento de Apareamiento de la Mosca Mediterránea de la Fruta (*Ceratisis capitata*) en Laboratorio y la Comparación con una Cepa Silvestre Bajo Condiciones Seminaturales. Master's thesis, Universidad de Costa Rica, San José. 72 pp.

Rolli, V.K. 1976. Die akustischen Sexualsignale von *Ceratitis capitata* Wied. und *Dacus oleae* Gmel. *Z. Angew. Entomol.* 81: 219–223.

Rössler, Y. 1975a. The ability to inseminate: a comparison between laboratory-reared and field populations of the Mediterranean fruitfly (*Ceratitis capitata*). *Entomol. Exp. Appl.* 18: 255–260.

Rössler, Y. 1975b. Reproductive differences between laboratory-reared and field-collected populations of Mediterranean fruit fly, *Ceratitis capitata. Ann. Entomol. Soc. Am.* 68: 987–991.

Rössler, Y. 1980. Sexual competitiveness of males of the Mediterranean fruit fly, *Ceratitis capitata* (W.) (Diptera: Tephritidae), carrying a Y-chromosome translocation. *Bull. Entomol. Res.* 70: 649–656.

Russ, K. and W. Schwienbacher. 1982. Investigations on sound production of *Ceratitis capitata* L. In *Sterile Insect Technique and Radiation in Insect Control.* pp. 369–378. International Atomic Energy Agency, Vienna, 495 pp.

Ryan, M.J. 1990. Sexual selection, sensory systems and sensory exploitation. *Oxford Surv. Evol. Biol.* 7: 157–195.

Saul, S.H. and S.D. McCombs. 1993a. Increased remating frequency in sex ratio distorted lines of the Mediterranean fruit fly (Diptera: Tephritidae). *Ann. Entomol. Soc. Am.* 86: 631–637.

Saul, S.H. and S.D. McCombs. 1993b. Dynamics of sperm use in the Mediterranean fruit fly (Diptera: Tephritidae): reproductive fitness of multiple-mated females and sequentially mated males. *Ann. Entomol. Soc. Am.* 86: 198–202.

Saul, S.H., S.Y.T. Tam, and D.O. McInnis. 1988. Relationship between sperm competition and copulation duration in the Mediterranean fruit fly (Diptera: Tephritidae). *Ann. Entomol. Soc. Am.* 81: 498–502.

Seo, S.T., R.I. Vargas, J.E. Gilmore, R.S. Kurashima, and M.S. Fujimoto. 1990. Sperm transfer in normal and gamma-irradiated, laboratory-reared Mediterranean fruit flies (Diptera: Tephritidae). *J. Econ. Entomol.* 83: 1949–1953.

Shelly, T.E. and T.S. Whittier. 1994a. Lek distribution in the Mediterranean fruit fly (Diptera: Tephritidae): influence of tree size, foliage density, and neighborhood. *Proc. Hawaii Entomol. Soc.* 32: 113–121.

Shelly, T.E. and T.S. Whittier. 1994b. Effect of sexual experience on the mating success of males of the Mediterranean fruit fly, *Ceratitis capitata* W. (Diptera: Tephritidae). *Proc. Hawaii Entomol. Soc.* 32: 91–94.

Shelly, T.E. and T.S. Whittier. 1996. Mating competitiveness of sterile male Mediterranean fruit flies (Diptera: Tephritidae) in male-only releases. *Ann. Entomol. Soc. Am.* 89: 754–758.

Shelly, T.E., T.S. Whittier, and K.Y. Kaneshiro. 1993. Behavioral responses of Mediterranean fruit fly (Diptera: Tephritidae) to trimedlure baits: can leks be created artificially? *Ann. Entomol. Soc. Am.* 86: 341–351.

Shelly, T.E., T.S. Whittier, and K.Y. Kaneshiro. 1994. Sterile insect release and the natural mating system of the Mediterranean fruit fly, *Ceratitis capitata* (Diptera: Tephritidae). *Ann. Entomol. Soc. Am.* 87: 470–481.

Shelly, T.E., T.S. Whittier, and E. Villalobos. 1996. Trimedlure affects mating success and mate attraction in male Mediterranean fruit flies. *Entomol. Exp. Appl.* 78: 181–185.

Silberglied, R.E. 1977. Communication in the Lepidoptera. In *How Animals Communicate* (T.A. Sebeok, ed.), pp. 362–402. Indiana University Press, Bloomington.

Sivinski, J. and J.C. Webb. 1986. Changes in a Caribbean fruit fly acoustic signal with social situation (Diptera: Tephritidae). *Ann. Entomol. Soc. Am.* 79: 146–149.

Sivinski, J., T. Burk, and J.C. Webb. 1984. Acoustic courtship signals in the Caribbean fruit fly, *Anastrepha suspensa* (Loew). *Anim. Behav.* 32: 1011–1016.

Sivinski, J., C.O. Calkins, and J.C. Webb. 1989. Comparisons of acoustic courtship signals in wild and laboratory reared Mediterranean fruit fly *Ceratitis capitata. Fla. Entomol.* 72: 212–214.

Solinas, M. and G. Nuzzaci. 1984. Functional anatomy of *Dacus oleae* Gmel. female genitalia in relation to insemination and fertilization processes. *Entomologica* (*Bari*) 19: 135–165.

Taylor, P.W. and B. Yuval. 1999. Postcopulatory sexual selection in Mediterranean fruit flies, *Ceratitis capitata* (Tephritidae): advantages for large and protein-fed males. *Anim. Behav.* 58: 247–254.

Tomaru, M. and Y. Oguma. 1994. Differences in courtship song in the species of the *Drosophila auraria* complex. *Anim. Behav.* 47: 133–140.

Villeda, M.P., J. Hendrichs, M. Aluja, and J. Reyes. 1988. Mediterranean fruit fly *Ceratitis capitata*: behavior in nature in relation to different Jackson traps. *Fla. Entomol.* 71: 154–162.

Warburg, M.S. and B. Yuval. 1997a. Circadian patterns of feeding and reproductive activities of Mediterranean fruit flies (Diptera: Tephritidae) on various hosts in Israel. *Ann. Entomol. Soc. Am.* 90: 487–495.

Warburg, M.S. and B. Yuval. 1997b. Effects of energetic reserves on behavioral patterns of Mediterranean fruit flies (Diptera: Tephritidae). *Oecologia* 112: 314–319.

Webb, J.C., C.O. Calkins, D.L. Chambers, W. Schwienbacher, and K. Russ. 1983. Acoustical aspects of behavior of Mediterranean fruit flies, *Ceratitis capitata*: analysis and identification of courtship sounds. *Entomol. Exp. Appl.* 33: 1–8.

West-Eberhard, M.J. 1984. Sexual selection, competitive communication and species-specific signals in insects. In *Insect Communication* (T. Lewis, ed.), pp. 284–324. Academic Press, New York.

Whittier, T.S. and K.Y. Kaneshiro. 1991. Male mating success and female fitness in the Mediterranean fruit fly (Diptera: Tephritidae). *Ann. Entomol. Soc. Am.* 84: 608–611.

Whittier, T.S. and K.Y. Kaneshiro. 1995. Intersexual selection in the Mediterranean fruit fly: does female choice enhance fitness? *Evolution* 49: 990–996.

Whittier, T.S. and T.E. Shelly. 1993. Productivity of singly vs. multiply mated female Mediterranean fruit flies, *Ceratitis capitata* (Diptera: Tephritidae). *J. Kans. Entomol. Soc.* 66: 200–209.

Whittier, T.S., K.Y. Kaneshiro, and L.D. Prescott. 1992. Mating behavior of Mediterranean fruit flies (Diptera: Tephritidae) in a natural environment. *Ann. Entomol. Soc. Am.* 85: 214–218.

Whittier, T.S., F.Y. Nam, T.E. Shelly, and K.Y. Kaneshiro. 1994. Male courtship success and female discrimination in the Mediterranean fruit fly (Diptera: Tephritidae). *J. Insect Behav.* 7: 159–170.

Wong, T.T.Y. and L.M. Nakahara. 1978. Sexual development and mating response of laboratory-reared and native Mediterranean fruit flies. *Ann. Entomol. Soc. Am.* 71: 592–596.

Wong, T.Y.Y, J.I. Nishimoto, and H.M. Couey. 1983. Mediterranean fruit fly (Diptera: Tephritidae): further studies on selective mating response of wild and unirradiated and irradiated, laboratory-reared flies in field cages. *Ann. Entomol. Soc. Am.* 76: 51–55.

Wong, T.T.Y., R.M. Kobayashi, L.C. Whitehand, D.G. Henry, D.A. Zadig, and C.L. Denny. 1984. Mediterranean fruit fly (Diptera: Tephritidae): mating choices of irradiated laboratory-reared and untreated wild flies of California in laboratory cages. *J. Econ. Entomol.* 77: 58–62.

Yuval, B., S. Blay, and R. Kaspi. 1996. Sperm transfer and storage in the Mediterranean fruit fly (Diptera: Tephritidae). *Ann. Entomol. Soc. Am.* 89: 486–492.

Yuval, G., R. Kaspi, S. Shloush, and M.S. Warburg. 1998. Nutritional reserves regulate male participation in Mediterranean fruit fly leks. *Ecol. Entomol.* 23: 101–105.

Zapién, G., J. Hendrichs, P. Liedo, and A. Cisneros. 1983. Comparative mating behavior of wild and mass-reared sterile medfly *Ceratitis capitata* (Wied.) on a field caged host tree II. Female mate choice. In *Fruit Flies of Economic Importance* (R. Cavalloro, ed.), pp. 397–409. A.A. Balkema, Rotterdam.

# 19 Phylogeny of the Tribe Dacini (Dacinae) Based on Morphological, Distributional, and Biological Data

*Richard A. I. Drew and David L. Hancock*

## CONTENTS



## 19.1 INTRODUCTION

Taxonomic research on the tribe Dacini began in 1805 with the description of *Dacus armatus* Fabricius. From the mid-1880s to the present time, there has been a great deal of research particularly in the description of species and the definition of genera and subgenera. Little attention has been given to analyses of phylogenetic relationships within the group. This chapter concerns only the Dacini, treated here as a tribe in order to maintain uniformity within the present volume. The present authors do not consider current evidence sufficient to warrant reduction of the traditional subfamily status of this economically important group, or of the other taxa, the Ceratitidini and Gastrozonini, included in the broader subfamily Dacinae as used in this volume, despite their evident relationship.

A complete study of evolutionary relationships should take into consideration comparisons of morphological characters, geographic distributions, biological associations with host plants and genetic patterns (e.g., isozymes and DNA). In this chapter the phylogenies have been elucidated

0-8493-1275-2/00/$0.00+$.50

on the basis of comparative studies of morphological characters, biological associations, and geographic distributions. For the dacines of Southeast Asia and the South Pacific region, there is a considerable amount of morphological and biological data available. However, a full phylogenetic analysis, based on currently accepted principles, is not possible at the present time and must await a detailed analysis of the complicated Afrotropical fauna, together with a full revision of the Southeast Asian fauna, currently under way. Only then will it be possible to redefine the numerous subgenera as phylogenetically meaningful groupings. The following is offered as a "first-step" hypothesis of a possible phylogenetic scenario, based on the present authors' experience with the group.

As noted in the chapter on biology and behavior of the Dacini (Drew and Romig, Chapter 21), the close relationships of species with their host plants justifies the use of endemic host records in phylogenetic studies and the value of such comparisons will be evident in the data presented in this chapter. Indeed, the host records reflect changes in the reproductive biology between species, which in turn is fundamental to the processes of speciation.

## 19.2 SUPRASPECIFIC CLASSIFICATION OF DACINI

The Dacini are concentrated in two areas of the world, with significantly smaller numbers of species away from these centers. The two areas are the Afrotropical Region and from Southeast Asia to northeastern Australia. We recognize four genera in the tribe: *Bactrocera* Macquart, *Dacus* Fabricius, *Ichneumonopsis* Hardy, and *Monacrostichus* Bezzi (Drew and Hancock 1994; Drew et al. 1998). The tribe appears to consist mostly (or entirely) of wasp mimics and hence morphological similarities between them and other tephritid taxa may be misleading. The significant reduction in chaetotaxy of the head and thorax (e.g., postocellar, postpronotal, presutural, dorsocentral, and katepisternal bristles absent, ocellar and postvertical bristles absent or vestigial — postpronotal bristles are secondarily present in *B.* (*Notodacus*) Perkins) appears to be a synapomorphy for the tribe (convergent with Adramini). The majority of species belong in *Bactrocera* and *Dacus*, each of which is divided into a number of subgenera. In terms of studying evolutionary trends, it is most useful to consider the subgenera, as in the main these contain groups of apparently closely related species. The current division of *Bactrocera* and *Dacus* has morphological, biogeographic, and biological significance (Drew 1989). Characteristics of the four genera are listed in Table 19.1.

### 19.2.1 GENUS *ICHNEUMONOPSIS* HARDY

This is a monotypic genus possessing a number of characters (particularly the reduction in chaetotaxy and the shape of the scutellum) that link it to the Dacini (Drew and Hancock 1994). It differs from *Bactrocera* and *Dacus* in having the frontal bristles absent or vestigial and abdominal tergum VI in females short but distinct, resembling *Monacrostichus* in these characters.

This appears to be the most primitive genus of the Dacini (Drew and Hancock 1994) and possesses the following characters not seen elsewhere in the tribe: arista plumose, cell bm not distinctly expanded, vein $R_{4+5}$ setose, and spermathecae rounded. These characters also occur in the Gastrozonini and Ceratitidini.

Superficially, *Ichneumonopsis* resembles species of *Enicoptera* Macquart (in Gastrozonini) but the latter genus lacks forefemoral spines and has the postocellar, postpronotal, presutural and dorsocentral bristles present, three pairs of strong frontal bristles, distinct posterior abdominal bristles, two pairs of scutellar bristles and a different wing venation and scutellum shape (Hancock and Drew 1999).

### 19.2.2 GENUS *MONACROSTICHUS* BEZZI

This is a small genus with two known species, both with peculiar wing venation similar to that of *Enicoptera* (Gastrozonini). It differs from other Dacini in having transverse facial and scutal

**TABLE 19.1**
**Character States in the Genera of Dacini**

| Character | *Ichneumonopsis* | *Monacrostichus* | *Dacus* | *Bactrocera* |
|---|---|---|---|---|
| Abdomen shape | Elongate-oval | Elongate-oval | Elongate-oval or oval | Generally oval |
| Abdomen fusion | Terga not fused | Terga not fused | Terga fused ** | Terga not fused |
| Pecten on male tergum III | Absent | Absent | Present * or absent | Usually present * |
| Ceromata on tergum V | Absent | Absent | Present * | Usually present * |
| Female tergite VI | Short but distinct | Short but distinct | Vestigial, disassociated * | Vestigial, disassociated * |
| Oviscape | Very long | Short | Short or long | Short or long |
| Spermathecae | Rounded | Convoluted *^ | Convoluted *^ | Convoluted *^ |
| Posterior lobe of lateral surstylus | Long | Short | Long or short | Long or short |
| Posterior margin of male sternite V | With moderate concavity | With moderate concavity | With shallow to moderate concavity | With shallow to moderate or deep concavity |
| Wing cell sc | Short | Elongate and broad ** | Short | Short |
| Wing cell $r_1$ | With a transverse sclerotized line ** | Without sclerotized line | Without sclerotized line | Without sclerotized line |
| Wing cell bm | Narrow and not broadened | Narrow, basally broadened * | Broad * | Broad * |
| Wing cell dm | Broad | Basally narrow *** | Broad | Broad |
| Vein $R_{4+5}$ | Setose | Bare *^ | Bare *^ | Bare *^ |
| Vein $A_1$+$Cu_2$ | Ends at wing margin | Ends well before wing margin ** | Ends at wing margin | Ends at wing margin |
| Ventroapical spines on forefemur | Present | Present | Present or absent | Absent |
| Scutellum shape | Short and broad | Short and broad | Usually short and broad | Subtriangular |
| Prescutellar acrostichal bristles | Absent | Absent | Absent | Usually present *** |
| Inner postalar bristle | Vestigial ** | Present | Present | Present |
| Supra-alar bristle | Present | Present | Present or absent | Present or absent |
| Scutellar bristles | One pair (apical) | One pair (apical) | One pair (apical) | One or two pairs |
| Facial and scutal furrows | Absent | Present ** | Absent | Absent |
| Chaetotaxy of head and thorax | Much reduced *^ | Much reduced *^ | Much reduced *^ | Much reduced *^ |
| Frontal bristles | Absent or with 1 vestigial pair * | Absent * | Usually 2 pairs present | Usually 2 pairs present |
| Arista | Plumose | Bare *^ | Bare *^ | Bare *^ |
| Distribution | SE Asia | SE Asia | Primarily Africa and SE Asia | Primarily SE Asia–Pacific |
| Host plants | Unknown | Rutaceae (Citrus spp.) | Primarily Asclepiadaceae, Passifloraceae, and Cucurbitaceae | Primarily tropical and subtropical rain forest fruits |

*Note*: * = Shared apomorphy (^ = character state occurs outside tribe but presumed convergent); ** = apomorphy unique to genus; *** = possible apomorphy for genus within Dacini (character state also occurs outside tribe).

furrows, cell sc elongate and broad (elongate but narrow in *Enicoptera*), veins $R_{2+3}$ and M with pronounced curvatures, cell dm narrow basally (the latter two characters also seen in *Enicoptera*), cell bm narrow but somewhat expanded basally and vein $A_1$+$Cu_2$ ending well before wing margin (Drew and Hancock 1994). It differs from *Ichneumonopsis* in having an elongate antenna with bare arista, a bare vein $R_{4+5}$, short posterior lobe of lateral surstylus, short oviscape, and convoluted spermathecae (these characters very similar to those of *Dacus* subgenus *Callantra*).

### 19.2.3 GENUS *DACUS* FABRICIUS

Extensive studies are required on the Afrotropical fauna before the species of *Dacus* can be grouped into meaningful subgenera. Munro (1984) recognized a large number of African genera, but these are poorly defined on superficial characters and most represent no more than species groups. The overall homogeneity of the African fauna suggests a relatively recent derivation. At present, the following eight subgenera are recognizable: *Callantra* Walker, *Dacus* Fabricius, *Didacus* Collart, *Leptoxyda* Macquart, *Lophodacus* Collart, *Pionodacus* Munro, *Psilodacus* Collart, and *Tythocalama* Munro.

As recognized here, subgenus *Dacus* is equivalent to tribes Dacini plus Metidacini of Munro (1984). *Didacus* is equivalent to Munro's tribe Didacini and, if host data are as significant as is suspected, is probably polyphyletic. *Psilodacus* includes all species from Munro's tribes Psilodacini and Athlodacini that lack postsutural yellow vittae on the scutum (i.e., excluding *Leptoxyda* and *Lophodacus*). *Pionodacus* appears to be closely related to *Leptoxyda*.

*Paracallantra* Hendel cannot be defined because the type species *vespiformis* Hendel was poorly described and the type specimens are lost; however, it appears to be a synonym of *Callantra*. Subgenus *Semicallantra* Drew was originally placed in genus *Dacus* but the type species *aquilus* Drew has free abdominal terga and it belongs in *Bactrocera*. *Dacus* (*Asiadacus*) *aneuvittatus* Drew was originally described from New Caledonia as a species with free abdominal terga (Drew 1971). It was later transferred to *Bactrocera* (*Sinodacus*) (Drew 1989). Recently it was bred from *Tylophora* sp. (family Asclepiadaceae) (Amice and Sales 1997) and a study of the holotype has revealed that it has fused abdominal terga. Consequently, it is hereby transferred to *Dacus* (*Dacus*). *Neodacus* Perkins eventually may be resurrected from synonymy to accommodate those Asian and Australasian species currently included within *Dacus* and *Didacus*.

Species of subgenera *Callantra, Dacus,* and *Didacus* are recorded from Asclepiadaceae and Cucurbitaceae, *Leptoxyda, Pionodacus,* and *Psilodacus* from Asclepiadaceae, *Lophodacus* from the stamens of male flowers of Cucurbitaceae. Several species of African *Dacus* and *Didacus* have been reared from the fruit or flower buds of Passifloraceae and a few species of *Didacus* occur in the pods of Apocynaceae and Periplocaceae. It is significant that the Cucurbitaceae and Passifloraceae are considered to be closely related plant families (Cronquist 1981). Also the Asclepiadaceae, Periplocaceae, and Apocynaceae are closely related with some authors placing the Periplocaceae as a subfamily, Periplocoideae, of Asclepiadaceae (Heywood 1978). *Dacus* species that occur outside the Afrotropical Region all possess hosts in the Asclepiadaceae or Cucurbitaceae.

### 19.2.4 GENUS *BACTROCERA* MACQUART

This is a large genus with many species still to be described from many parts of Southeast Asia and from Papua New Guinea and the Solomon Islands. Consequently, when this work is completed there are likely to be some changes to our present definitions of subgenera, many of which currently are defined on the basis of presence or absence of a pecten on abdominal tergum III in males or supra-alar and prescutellar acrostichal bristles.

The following 30 subgenera are currently recognized from Africa, Southeast Asia, and the Pacific region: *Afrodacus* Bezzi, *Aglaodacus* Munro, *Apodacus* Perkins, *Asiadacus* Perkins, *Austrodacus* Perkins, *Bactrocera* Macquart, *Bulladacus* Drew and Hancock, *Daculus* Speiser, *Diplodacus* May,

*Gymnodacus* Munro, *Hemigymnodacus* Hardy, *Heminotodacus* Drew, *Hemiparatridacus* Drew, *Hemisurstylus* Drew, *Hemizeugodacus* Hardy, *Javadacus* Hardy, *Melanodacus* Perkins, *Nesodacus* Perkins, *Niuginidacus* Drew, *Notodacus* Perkins, *Papuodacus* Drew, *Paradacus* Perkins, *Paratridacus* Shiraki, *Parazeugodacus* Shiraki, *Queenslandacus* Drew, *Semicallantra* Drew, *Sinodacus* Zia, *Tetradacus* Miyake, *Trypetidacus* Drew, and *Zeugodacus* Hendel.

On the basis of morphological characters, such as length of lateral surstylus and shape of abdominal tergum V of male, the 28 Asian and Pacific subgenera have been divided into four groups as follows (Drew 1989):

| | |
|---|---|
| ***Bactrocera* group** | ***Zeugodacus* group** |
| *'Afrodacus'* | *Asiadacus* |
| *Apodacus* | *Austrodacus* |
| *Bactrocera* | *Diplodacus* |
| *Bulladacus* | *Hemigymnodacus* |
| *'Gymnodacus'* | *Heminotodacus* |
| *Notodacus* | *Hemiparatridacus* |
| *Semicallantra* | *Javadacus* |
| *Tetradacus* | *Nesodacus* |
| *Trypetidacus* | *Niuginidacus* |
| ***Melanodacus* group** | *Papuodacus* |
| *Hemisurstylus* | *Paradacus* |
| *Hemizeugodacus* | *Paratridacus* |
| *Melanodacus* | *Parazeugodacus* |
| ***Queenslandacus* group** | *Sinodacus* |
| *Queenslandacus* | *Zeugodacus* |

In the *Bactrocera* group the species placed in subgenera *'Afrodacus'* and *'Gymnodacus'* probably have no relationship to the true *Afrodacus* (including *Mauritidacus* Munro) and *Gymnodacus* of the Afrotropics which, together with *Daculus*, appear to belong in the *Melanodacus* group. The morphological similarities (based on unreliable bristle characters) between these two geographically distinct groups probably do not indicate their genetic separation, which may better be indicated by host plant differences. The Southeast Asian and Pacific species currently placed in *Afrodacus* and *Gymnodacus* probably should be placed in new subgenera but this cannot be done until some distinctive characters can be found and which will probably be based on isozyme and DNA analyses.

A fourth Afrotropical subgenus, *Aglaodacus* from Madagascar, is of uncertain affinity but has a wing shape similar to members of the *Zeugodacus* group and may represent an early offshoot of that group.

Within the *Zeugodacus* group, at least 50% of all known species are cucurbit feeders and are placed in at least eight subgenera: *Asiadacus, Austrodacus, Hemigymnodacus, Hemiparatridacus, Javadacus, Paradacus, Sinodacus,* and *Zeugodacus*. Several subgenera include non-cucurbit feeders, for example, *Diplodacus,* some *'Javadacus'* [*aberrans* group], *Paratridacus,* and *Parazeugodacus*, that suggest a link with the *Melanodacus* group.

## 19.3 BASIS FOR PHYLOGENETIC ASSESSMENT

For this study a comparative analysis has been made of the currently known dacine fauna on the basis of morphology, biology, biogeography, and responses to male lures. This is an attempt to combine taxonomic and biological information with the hope that it may provide, with reasonable accuracy, a picture of an historical event, the phylogeny of the tropical and subtropical Dacini. In this work there has been no attempt to compare a large number of taxa at species level, rather to study the key character states at the generic and subgeneric level. This approach provides a better understanding of the worldwide evolutionary patterns as a considerable amount of speciation has been localized within regions.

The outgroup for the Dacini appears to be the Ceratitidini + Gastrozonini (Ceratitinae of Hancock 1986), a view supported by rDNA studies (Han and McPheron 1997). In this group the arista is normally plumose or pubescent, the spermathecae are normally oval or rounded (but convoluted in some Gastrozonini), and there is a full (or almost full) complement of bristles. Host plants are primarily fleshy fruits or bamboo shoots. There are almost always two pairs of scutellar bristles, but the great reduction of bristles seen in the Dacini, plus the presence of only one pair of scutellar bristles in all but the *Melanodacus* and *Zeugodacus* groups of *Bactrocera*, suggest that the frequent occurrence of two pairs of scutellar bristles in the latter groups is a derived character. The usual presence of prescutellar acrostichal bristles in genus *Bactrocera* may also be a derived character within the Dacini and serves as a possible synapomorphy for that genus. Most members of the Dacini appear to be wasp mimics and this probably plays a part in wing pattern and abdominal shape characters, resulting in many cases of convergence.

### 19.3.1 Morphological Characters

An attempt has been made to determine which characters are primitive within the Dacini, based on their distribution within the genera (see Table 19.1). These characters are as follows. Head: frontal bristles reduced; combined lengths of antennal segments shorter than vertical length of head; arista plumose. Thorax: one pair of scutellar bristles; scutellum short and broad; presence of spines on apical ventral surfaces of forefemur; vein $R_{4+5}$ setose; cell bm narrowing basally. Abdomen: elongate-oval and club shaped; abdominal terga not fused (free); two rounded spermathecae.

It is likely that the original parent stock possessed a head with reduced bristles, particularly on the frons, short antenna (equal to or shorter than vertical length of head), plumose arista, thorax with one pair of scutellar bristles, a short broad scutellum, spines on the apicoventral surface of forefemur, abdomen elongate-oval and club shaped, terga free (not fused). *Ichneumonopsis* appears to be the first genus to have diverged and is probably the closest in appearance to the ancestor. It is unique among the Dacini in possessing a very long oviscape, cell bm narrowing basally, veins $R_{2+3}$ and M without pronounced curvatures, a peculiar, crossvein-like line of sclerotization across the middle of cell $r_1$, a plumose arista, and two rounded spermathecae. Convoluted spermathecae occur in most Dacini and also in many Gastrozonini, but appear to be secondarily derived in both groups. The next group to have diverged appears to be the genus *Monacrostichus*, which lies between *Ichneumonopsis* and the remaining Dacini in morphological characters. It is distinguished by having cell bm broader basally, cell sc elongate and broad, transverse furrows on the face and scutum, veins $R_{2+3}$ and M with pronounced curvatures so that cell dm is narrow over the basal two-thirds and with vein $A_1$+$Cu_2$ ending well before wing margin. The presence of both these primitive genera in Southeast Asia suggests that the Dacini originated in or close to that region.

The remainder of the Dacini appear to have split into two major groups on the basis of fusion or nonfusion of abdominal terga II to V. The fused terga group, genus *Dacus*, dispersed westward and speciated in the dry savannah habitats of Africa. The fused terga condition may be a water conservation mechanism for survival in drier environments. Pupation within the host fruiting body, particularly common in the Asclepiadaceae feeders, may also be an adaptation for survival in hot, dry habitats. Relatively few African species appear to have become adapted to rain forests. Two subgenera of African origin, *Didacus* and *Leptoxyda,* have radiated back to Asia as far as India/Sri Lanka where *D.* (*Didacus*) *keiseri*, *D.* (*Didacus*) *ciliatus* (both cucurbit feeders) and *D.* (*Leptoxyda*) *persicus* (in asclepiads) occur (Drew et al. 1998).

The original *Dacus* stock probably gave rise to the currently recognized Southeast Asian subgenera *Callantra*, '*Dacus*' and '*Didacus*.' *Callantra* are uniquely Southeast Asian/South Pacific and are identified by possessing fused terga, an elongate-oval and club-shaped abdomen, short broad scutellum, frontal bristles absent or reduced, some species with spines on the ventroapical surface of the forefemur, antenna elongate (combined lengths of all segments greater than vertical length of head), and abdominal sternum V with a slight concavity on posterior margin. The Asclepiadaceae feeding '*Dacus*' and '*Didacus*' of Asia and Australasia probably have only distant relationships to the African subgenera, although they are morphologically indistinguishable, and eventually may need to be placed in their own taxonomic grouping (for which the name *Neodacus* Perkins is available) to identify this distinction. A secondary radiation of the Southeast Asian group appears to have brought the subgenus *Tythocalama* to Madagascar, differing from *Callantra* in the shorter antenna.

The free terga group, genus *Bactrocera*, has undergone extensive speciation within Southeast Asia and the South Pacific and has split into two broad groupings. One possesses a deep "V" on the posterior margin of abdominal sternum V and one pair of scutellar bristles, the other has a shallow indentation or "V" on abdominal sternum V and usually two pairs of scutellar bristles. The first of these groups has divided into the monotypic *Queenslandacus* group with a long posterior lobes on the male surstyli and the *Bactrocera* group, with short posterior surstylus lobes, of eight recognizable subgenera which all possess oval-shaped abdomens, plus *Tetradacus* and *Semicallantra* which have elongate-oval abdomens and appear to be primitive subgenera within the *Bactrocera* group. Approximately 80% of all known species of Dacini occur within the *Bactrocera* group of subgenera and particularly in subgenus *Bactrocera*. The second group of genus *Bactrocera*, with a shallow indentation on the posterior margin of abdominal sternum V, has split into the *Zeugodacus* group of 15 known subgenera with long posterior lobes on the male surstyli and the *Melanodacus* group of three subgenera with short posterior surstylus lobes.

A small group of *Bactrocera* species occur in Africa and are placed in three closely related subgenera: *Afrodacus*, *Daculus,* and *Gymnodacus*. These appear to belong in the *Melanodacus* group and have probably resulted from a radiation westward from Southeast Asia. Although morphologically indistinguishable (except for the shallower indentation on sternum V) from similar groups of species in the latter region, host plant associations suggest that they are probably distant relatives. Consequently, the Southeast Asian and Pacific species placed in these subgenera may require new subgeneric names once most of the species have been described.

Finally, a single Madagascan species, placed in subgenus *Aglaodacus*, appears to be derived from an ancestral member of the *Zeugodacus* group. The male has a well-developed supernumerary lobe and broad anal streak on the wing. It presumably reached Madagascar from Southeast Asia via India.

### 19.3.2 Biology and Habitat Associations

As discussed by Drew and Romig (Chapter 21), the biology and behavior of the Dacini are such that species have a close-knit relationship with their host plants. In particular the reproductive biology of species is dependent upon factors within the host plant habitat. Consequently, understanding the patterns of distribution of their endemic host plants and speciation of their host plant ecosystems will contribute to elucidating the phylogenetic relationships of the Dacini. Because their endemic host plants belong primarily to the tropical and subtropical rain forests, an analysis of these is a suitable starting point.

Within the geographic range of the Dacini, the tropical and subtropical rain forests are distributed in West Africa, coastal East Africa, Madagascar and the Mascarene Islands, southwest India, Southeast Asia from Nepal to southern China in the north to the Indonesian islands in the south, Papua New Guinea, northeastern Australia, and some South Pacific islands. Endemic species of Dacini occur in all of these areas.

In general, the rain forests of Southeast Asia possess the greatest species richness, while those of Melanesia (e.g., Solomon Islands) are less rich and there is increasing floristic poverty eastward into the Pacific (Whitmore 1986). In contrast, there are no extremely floristically rich rain forests in Africa (e.g., those of Ghana, Nigeria, and Mauritius; Whitmore 1986). It is recognized that the factors that have contributed to the large number of plant species in the Indo-Malayan rain forests include considerable localized evolution of species, following the combining of the Laurasian and Gondwanan elements through continental drift (Whitmore 1986). Whitmore also stated that the long, stable history of tropical rain forests and their large number of ecological niches are conditions that, when plants and animals interact, result in coevolution.

In the rain forest plants, there are large numbers of sibling species best exhibited in almost all genera of Dipterocarpaceae, especially *Calophyllum* and *Garcinia* in the family Guttiferae and *Syzygium* in the Myrtaceae (Whitmore 1986). These are all very important hosts of species of the genus *Bactrocera*. Another characteristic feature of species-rich rain forests is the large number of endemic plant species that have localized distributions. It is of note that within the rain forests of Southeast Asia there are also a large number of endemic sibling species of Dacini with localized geographic distributions.

Evidence indicates that there has been considerable evolution of dacine species (particularly in the genus *Bactrocera*) in association with their rain forest host plant species. The distribution of numbers of species of Dacini and rain forest plants is similar when transects are taken from Southeast Asia westward to Africa and southeast from Papua New Guinea to the Pacific Islands. Species in genus *Bactrocera*, in particular, are mostly rain forest species. Also, there are a number of related dacine species and subgenera that utilize closely related plant families as their endemic hosts.

In terms of host plant associations, a likely line of evolution of dacine species appears to be as follows. The original parent stock appears to have arisen in what is now a part of Southeast Asia, possibly in India, associated with rain forest fruit. Early radiation from this stock produced *Ichneumonopsis* and *Monacrostichus*.

In a major division of the group, adaptation to dry habitats led to the evolution of the fused terga *Dacus*, its species largely occupying savannah or woodland habitats and breeding initially in the plant family Asclepiadaceae and, in Africa, the closely related families Apocynaceae and Periplocaceae. The existing subgenera *Dacus, Didacus, Leptoxyda, Pionodacus,* and *Psilodacus* are examples. Some feeding in Cucurbitaceae developed in the African *Dacus* as seen in the subgenera *Dacus*, *Didacus,* and *Lophodacus*. Also some species of African *Dacus* and *Didacus* feed on a closely related plant family, the Passifloraceae. The Asian *Dacus* subgenera *Callantra, 'Dacus,'* and *'Didacus'* appear to have developed from the same parental stock as the African fauna. The Asian species have been recorded from Asclepiadaceae and, in a few species of *Callantra*, Cucurbitaceae. An early offshoot from this Asian line appears to have reached Madagascar as subgenus *Tythocalama*. The Southeast Asian *Callantra*, *'Dacus'* and *'Didacus'* show strong evolutionary links to the African fauna through their common groups of endemic host plant families, the Asclepiadaceae, Apocynaceae, and Periplocaceae being one group and the Cucurbitaceae and Passifloraceae the other. Dispersal of the *Dacus* ancestor to Africa resulted in extensive speciation and a large number of included species. Host plants appear to be primitively Asclepiadaceae, later adapting also to Cucurbitaceae and Passifloraceae.

Ancestral *Bactrocera* underwent considerable speciation in a system of coevolution with rain forest plant species. Only one species, *B. alyxiae* (May), appears to be restricted to a host plant belonging to the *Dacus* line, namely, *Alyxia spicata* (family Apocynaceae); this and utilization of Apocynaceae by other species is presumably secondary. Extensive speciation in association with rain forest elements in the Australian and Papua New Guinea land areas has resulted in large numbers of species occurring in Papua New Guinea, northeastern Australia, and Pacific Islands as far east as the Society Islands. In this coevolutionary line, there has been a filtering effect with regard to numbers of both plant and fruit fly species, resulting in a continuing decline in numbers from Papua New Guinea to the south and east. This pattern is strongly exemplified in the *Bactrocera* group of subgenera.

Feeding in Cucurbitaceae developed strongly in the *Zeugodacus* group of subgenera, with approximately 60% of species having adapted to and evolved with this family, although the group also includes some soft fleshy fruit-feeding subgenera. The *Zeugodacus* group spread to Madagascar

**TABLE 19.2**
**Worldwide Geographic Distribution of Species of Dacini in Each of the Four Genera**

| | Total No. of Species | No. of Species of *Bactrocera* | No. of Species of *Dacus* | *Ichneumonopsis* | *Monacrostichus* |
|---|---|---|---|---|---|
| Africa (including Madagascar and Mascarene Islands) | 182 | 10 | 172 | 0 | 0 |
| Southeast Asia | 229 | 182 | 44 | 1 | 2 |
| Papua New Guinea | 168 | 155 | 13 | 0 | 0 |
| Australia | 87 | 75 | 12 | 0 | 0 |
| Solomons (including Bougainville) | 56 | 54 | 2 | 0 | 0 |
| Vanuatu | 13 | 12 | 1 | 0 | 0 |
| New Caledonia | 11 | 10 | 1 | 0 | 0 |
| Fiji | 4 | 4 | 0 | 0 | 0 |
| Tonga | 6 | 6 | 0 | 0 | 0 |
| Samoa | 7 | 7 | 0 | 0 | 0 |
| Niue | 2 | 2 | 0 | 0 | 0 |
| Cook Islands | 2 | 2 | 0 | 0 | 0 |
| Austral Islands | 2 | 2 | 0 | 0 | 0 |
| Society Islands | 2 | 2 | 0 | 0 | 0 |
| Marquesas Islands | 1 | 1 | 0 | 0 | 0 |
| Tuamotu Archipelago | 2 | 2 | 0 | 0 | 0 |
| Micronesia/N. Pacific | 2 | 2 | 0 | 0 | 0 |

as a single species in the endemic subgenus *Aglaodacus*. The *Melanodacus* group of *Bactrocera* spread to Africa and neighboring islands and gave rise to a small number of species in the rain forests and moist woodlands of that region, all referable to subgenera *Afrodacus, Daculus,* and *Gymnodacus*. Many species in the *Melanodacus* group (including all three African subgenera) have host plants within the Oleaceae. *Bactrocera* (*Daculus*) *oleae* subsequently spread to the Mediterranean and as far east as northwestern India together with cultivated olives. There were also small movements of Asclepiadaceae and Cucurbitaceae feeding *Dacus* back from Africa to India. These gave rise to two endemic Asian *Didacus* and *Leptoxyda* species. A more recent dispersal appears to have brought *D.* (*Didacus*) *ciliatus* to Asia.

### 19.3.3 Biogeography

Approximately 750 species of Dacini have been described worldwide. The distribution of species in each of the genera, *Bactrocera, Dacus, Ichneumonopsis,* and *Monacrostichus* is listed in Table 19.2. Approximately 68% belong to *Bactrocera* and 32% to *Dacus*. It is noteworthy that the greatest speciation in genus *Dacus* has occurred in Africa while prolific speciation in genus *Bactrocera* has occurred in Southeast Asia and Papua New Guinea.

In Africa, the proliferation of species has been associated with the dry savannah habitats where plants belonging to the families Asclepiadaceae, Cucurbitaceae, and Passifloraceae are common. The latter two are classed as closely related plant families (Cronquist 1981). The Asclepiadaceae are more common in Africa and less well represented in Southeast Asia and the Pacific and this pattern is similar to that for *Dacus* species. In Southeast Asia, the process of speciation has been associated with the Indo-Malayan rain forests and has involved primarily subgenus *Bactrocera*.

Whitmore (1986) described the distribution of the Indo-Malayan rain forests as a belt from Sumatra in the West to Papua New Guinea in the East and including the Malaysian Archipelago and the southern Thailand isthmus. In addition, there are substantial outlier areas in northwestern and southeastern Thailand, Indo-China, southern China, Philippines, Borneo, the Andaman Islands, southwest Sri Lanka, and the Western Ghats of India. Elements of the same tropical rain forests also extend into northeastern Australia, the Melanesian Archipelago (Solomon Islands, Vanuatu, Fiji, Samoa, and Tonga), and into Micronesia and Polynesia. The number of plant species and areas of land covered by these rain forests decline sharply with increasing distance south and east from Papua New Guinea. This pattern of Indo-Malayan rain forest distribution mirrors the distribution of *Bactrocera* species in the same region. For example, all land areas listed above contain endemic species of Dacini, particularly *Bactrocera* and the number of species declines in the same manner as for rain forest plant species.

The similarities in the biogeographic patterns of distribution of the dacine species and their known host groups provides strong evidence supporting the assertions by Drew and Romig, Chapter 21, on behavior and biology, that the fly species have a strong dependent relationship with their host plants. Certainly evolution of dacine species together with their host plant species is well supported.

### 19.3.4 Responses to Male Lures

The two lures most attractive to species of Dacini are cuelure and methyl eugenol. While the true biological relationships of these lures are open to question, Drew (1974) and Drew and Hooper (1981) showed that each fly species that responded did so to one only of these attractants, while some species did not respond to either. Also it was noted that most *Dacus* species responded to cuelure, some subgenera of *Bactrocera* were entirely nonresponding, and others entirely cuelure responding. This information plus the research of Drew (1987), which showed that the attractive component of the cuelure molecule was 2-butanone, which is produced by ripening fruit and some bacteria of the family Enterobacteriaceae, indicates that there is a taxonomic and evolutionary relationship between lure response and dacine species. The lure responses may reflect some evolutionary trends within the group.

The more primitive genera, *Ichneumonopsis* and *Monacrostichus,* do not respond to either lure. The lure responses evolved later with *Dacus* developing cuelure attraction and *Bactrocera* an attraction to either cuelure or methyl eugenol. Within the genus *Bactrocera*, the *Bactrocera* group of subgenera contain both cuelure and methyl eugenol–responding species, while the *Zeugodacus* group of subgenera possess either cuelure–responding species or species with no response to either lure. The *Melanodacus* group of subgenera show no response to lures and may have evolved from part of the *Zeugodacus* group that also possesses no response. Further, the small number of African *Bactrocera* (subgenera *Afrodacus, Daculus,* and *Gymnodacus*) also have no recorded responses to male lures and appear to have arisen from the *Melanodacus* group. This is also supported on morphological grounds.

Methyl eugenol is also a male lure for species of *Ceratitis* (subgenus *Pardalaspis* Bezzi) in Africa (Hancock 1987). Cuelure is known as an attractant only within the Dacini and may represent the "primitive" lure for both *Dacus* and *Bactrocera*, with methyl eugenol, which occurs naturally in various plant species, secondarily becoming a lure independently in various groups, particularly within *Bactrocera*, rarely so in *Dacus*. Vertlure, the only other male attractant known within the Dacini (Hancock 1985), is chemically very similar to cuelure.

## 19.4 EVOLUTION AND PHYLOGENY OF THE DACINI

Based on morphological, biological, and biogeographic information, the most likely scenario for the evolution and phylogenetic relationships of the Dacini is as follows: The parental stock appears to have originated in the Indian block of Gondwana as it drifted northward, with both *Ichneumonopsis* and *Monacrostichus* evolving there as early offshoots, the latter genus subsequently displaced to peripheral areas of the Malaysian Archipelago and the Philippines following unification with Laurasia. Genera *Bactrocera* and *Dacus* also may have differentiated during the Indian drifting phase, with the fused terga *Dacus* adapting to Asclepiadaceae in savannah woodlands, leaving the free terga *Bactrocera* associated with rain forest fruits.

Following unification of India with Laurasia, a dispersal took the savannah-adapted *Dacus* to Africa, leaving the ancestor of subgenera *Callantra*, '*Dacus*,' and '*Didacus*' in India, with an offshoot from there reaching Madagascar as *Tythocalama*. In Africa, a long period of isolation and independent speciation has resulted in a large fauna of genus *Dacus* associated largely with the dry savannah habitat of that continent and its vegetation.

Unification of India with Laurasia also enabled a long period of speciation in limited areas of rain forest in southern and eastern Asia, resulting in the development of a diverse dacine fauna based around the genus *Bactrocera*, including subgenus *Tetradacus* and ancestors of the *Zeugodacus* and *Melanodacus* groups of subgenera. The few known *Bactrocera* species in Africa, Madagascar, and the Mascarene Islands appear to have resulted from incursions from Asia of the ancestors of the *Melanodacus* and *Zeugodacus* groups, giving rise to subgenera *Afrodacus*, *Aglaodacus*, *Daculus*, and *Gymnodacus*. The presence of *Aglaodacus* in Madagascar suggests that the major groups of genus *Bactrocera*, namely, *Bactrocera*, *Melanodacus*, and *Zeugodacus* groups, had all differentiated before India became united with Laurasia.

Following the unification of Southeast Asian (present-day areas from southern China and Burma south to Malaysia and Indonesia) and Australasian Gondwanan land elements with Laurasia, *Bactrocera*, particularly the *Bactrocera* group of subgenera, speciated with the rich rain forest elements in these regions, spreading throughout the South Pacific. This was accompanied by a smaller radiation of *Dacus* species in drier areas. A subsequent radiation from Africa back into Asia resulted in the differentiation of *D.* (*Didacus*) *keiseri* and *D.* (*Leptoxyda*) *persicus* in India/Sri Lanka. A more recent dispersal, probably human-aided in historical times, brought *B.* (*Zeugodacus*) *cucurbitae* from Asia to Africa and *B.* (*Daculus*) *oleae* and *D.* (*Didacus*) *ciliatus* from Africa to western Asia and India.

The current concept of phylogenetic relationships within the Dacini, based on the above hypothesis, is given in Figure 19.1. This illustrates the primitive nature of *Ichneumonopsis* and *Monacrostichus*, the presumed Asian (Indian) origin of both *Dacus* and *Bactrocera*, the dispersal and subsequent radiation of *Dacus* within Africa, the breakup of *Bactrocera* into groups of subgenera, particularly in the Southeast Asian/Australasian region, plus the more recent dispersal of some *Dacus* groups into Southeast Asia and some *Bactrocera* groups into Africa.

For this scenario to be correct, the following must be true.

1. The Dacini are Gondwanan in origin. The absence of dacines (and their sister group the Ceratitidini + Gastrozonini) from South America, absence of primitive groups from Australasia and the Pacific, and presence of many primitive groups (genera and subgenera) in southern and eastern areas of Asia support the suggestion that the Indian plate (rather than the Southeast Asian/Australasian plate) was the center of evolution within the group.
2. Dispersal to Africa of the savannah-adapted *Dacus* occurred shortly after unification of Gondwanan elements (particularly India) with Laurasia. This is supported by the great species diversity in Africa but lack of distinctly differentiated higher taxa. The bulk of the Asian species of *Dacus*, although less numerous, show a diversity and geographic distribution pattern (into Australasia and the Pacific) that suggests a relatively long association with Southeast Asia, rather than a more recent dispersal from Africa. The suggestion that India is the center of origin is strengthened by the occurrence of the Asian-like taxa *Tythocalama* and *Aglaodacus* in Madagascar, faunal links apparently existing between these two countries around the time of Indian unification with Laurasia.

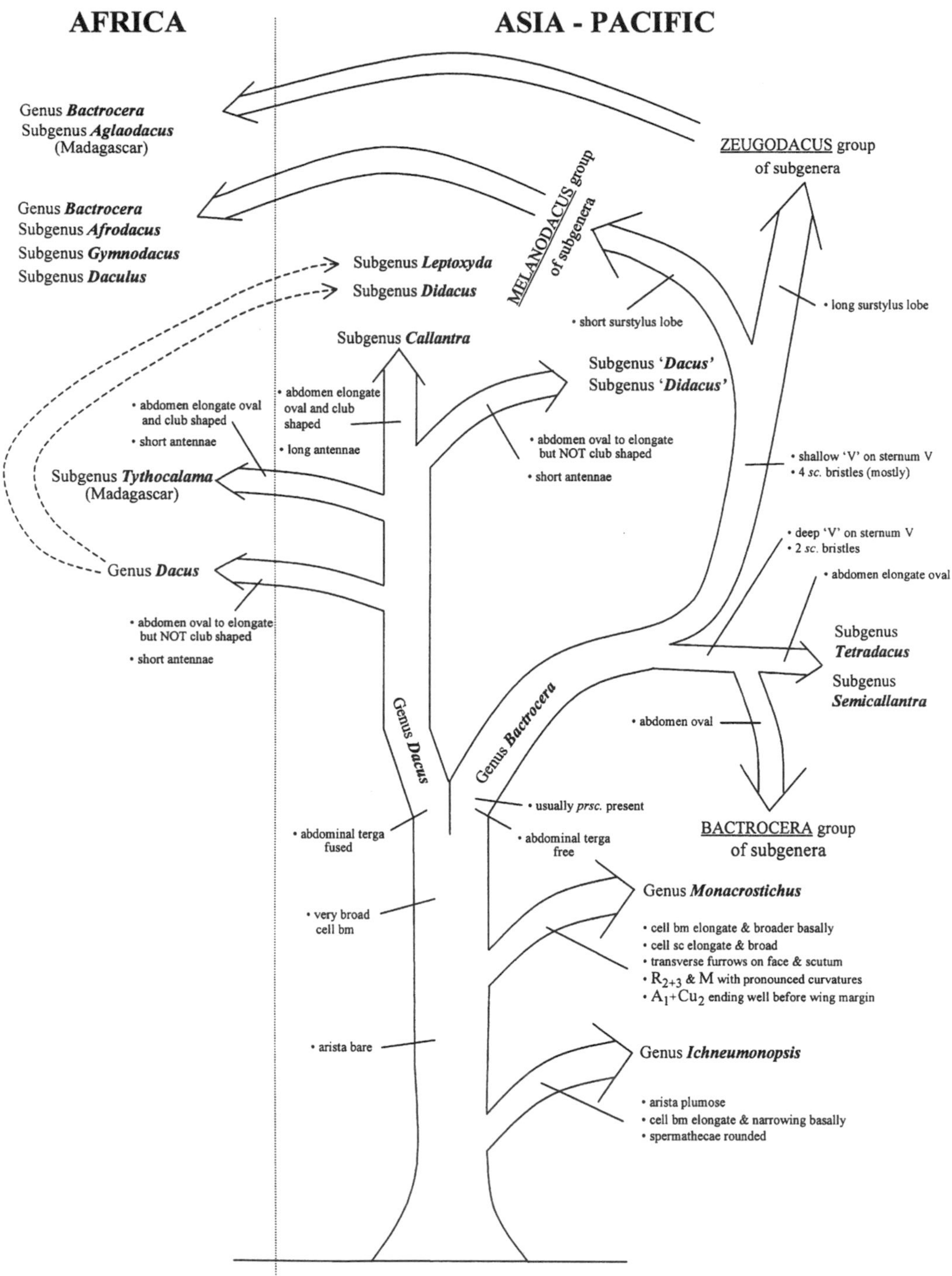


**FIGURE 19.1** Phylogenetic relationships of the supraspecific groups of the tribe Dacini.

3. Dispersal to Australasia and the South Pacific occurred after unification with Laurasia. This is supported by the relatively specialized nature of the fauna in these regions.
4. The events must be consistent with the geologic time frames and developments in the evolution of flowering plants in Southeast Asia and the Pacific.

## 19.5 GEOLOGIC TIME FRAME

The known endemic host plants of the Dacini that occur in the tropical/subtropical rain forests have been recorded in the late Cretaceous fossil beds (80 million years ago) through to the Oligocene (about 25 million years ago). For example, a common host family, the Myrtaceae (Allwood et al., in press), has pollen records in the upper Cretaceous and Tertiary deposits (Cronquist 1981). From pollen fossil records it has been deduced that Australia has been part of a primary center for the spread and evolution of flowering plants (Anonymous 1980). Further, it is believed that the climate in Gondwana between 200 and 60 million years ago favored subtropical flora and that the Gondwanan rain forest could have been the progenitor of the present known rain forests of Australia, India, Africa, and Madagascar (Anonymous 1980). The remarkably high number of rain forest plant genera (47) shared by the Indian Western Ghats and northeastern Australia supports the theory of a breakup of Gondwana from a major center of early evolution of tropical/subtropical flora. The Indian and Madagascan rain forests show similar close relationships.

Large gaps had developed between the major components of Gondwana from 80 million years ago. This time frame is consistent with the earliest fossil records of the endemic rain forest fruit fly host plants, supporting the concept that the fly species and their host plants have continued to coevolve over the Tertiary and Quaternary Periods. Following the breakup of Gondwana, there were extremely long periods of isolation that enabled the different landmasses to host extensive evolution of plants at the species level. In addition, the subsequent union of some of the landmasses resulted in intermingling of parts of their flora and fauna and further radiation of species. Thus, an ideal situation existed for rapid radiation of Dacini, particularly *Bactrocera*, into a rapidly increasing diversity of plant species in Southeast Asia, plus an open niche enabling rapid diversification into the Australasian and Pacific regions.

## 19.6 CONCLUSIONS AND FUTURE RESEARCH NEEDS

The Dacini are a major portion of the genuinely tropical/subtropical Tephritidae with Gondwanan origins, probably centered in India. Their current distribution is consistent with a pattern of dispersal and subsequent speciation from this original center, following unification with Laurasia. The current understanding of the morphology of the subfamily, together with data on their biology, habitat associations, and biogeography combine to present a picture of the phylogenetic relationships within the group, particularly at the generic and subgeneric level. Thorough analyses and revisions of the African and Southeast Asian faunas are required before a more-detailed phylogenetic analysis, based on definable subgenera, is possible. Similar studies are needed on the presumed outgroup, the Gastrozonini + Ceratitidini. Further studies at the species level and genetic research using tools such as DNA and isozymes will provide further light on the evolutionary history of the Dacini.

## ACKNOWLEDGMENTS

We thank the following donors for partial funding: Campaña Nacional contra las Moscas de la Fruta (Mexico), International Organization for Biological Control of Noxious Animals and Plants (IOBC), Instituto de Ecología, A.C. (Mexico), and Consejo Nacional de Ciencia y Tecnología (Mexico).

## REFERENCES

Allwood, A.J., A. Chinajariyawong, R.A.I. Drew, E.L. Hamacek, D.L. Hancock, C. Hengsawad, J.C. Jipanin, M. Jirasurat, C. Kong Krong, S. Kritsaneepaiboon, C.T.S. Leong, and S. Vijaysegaran. Host plant records for fruit flies (Diptera: Tephritidae) in Southeast Asia. *Raffles Bull. Zool.*, Suppl., in press.

Amice, R. and F. Sales. 1997. Fruit fly fauna in New Caledonia. In *Management of Fruit Flies in the Pacific* (A.J. Allwood and R.A.I. Drew, eds.), pp. 68–76. ACIAR Proceedings No. 76. ACIAR, Canberra.

Anonymous. 1980. New light on the origins of Australia's flora. *Ecos* 24: 3–9.

Cronquist, A. 1981. *An Integrated System of Classification of Flowering Plants*. Columbia University Press, New York.

Drew, R.A.I. 1971. New species of Dacinae (Diptera: Trypetidae) from the South Pacific area. *Queensl. J. Agric. Anim. Sci.* 28: 29–103.

Drew, R.A.I. 1974. The responses of fruit fly species (Diptera: Tephritidae) in the South Pacific area to male attractants. *J. Aust. Entomol. Soc.* 13: 267–270.

Drew, R.A.I. 1987. Behavioral strategies of fruit flies of the genus *Dacus* (Diptera: Tephritidae) significant in mating and host-plant relationships. *Bull. Entomol. Res.* 77: 73–81.

Drew, R.A.I. 1989. The tropical fruit flies (Diptera: Tephritidae: Dacinae) of the Australasian and Oceanian Regions. *Mem. Queensl. Mus.* 26: 1–521.

Drew, R.A.I. and D.L. Hancock. 1994. Revision of the tropical fruit flies (Diptera: Tephritidae: Dacinae) of South-east Asia. I. *Ichneumonopsis* Hardy and *Monacrostichus* Bezzi. *Invertebr. Taxon.* 8: 829–838.

Drew, R.A.I. and G.H.S. Hooper. 1981. The responses of fruit fly species (Diptera: Tephritidae) in Australia to various attractants. *J. Aust. Entomol. Soc.* 20: 201–205.

Drew, R.A.I., D.L. Hancock, and I.M. White. 1998. Revision of the tropical fruit flies (Diptera: Tephritidae: Dacinae) of South-east Asia. II. *Dacus* Fabricius. *Invertebr. Taxon.* 12: 567–654.

Han, H.-Y. and B.A. McPheron. 1997. Molecular phylogenetic study of Tephritidae (Insecta: Diptera) using partial sequences of the mitochondrial 16S ribosomal DNA. *Mol. Phylogenet. Evol.* 7: 17–32.

Hancock, D.L. 1985. New species and records of African Dacinae (Diptera: Tephritidae). *Arnoldia Zimb.* 9: 299–314.

Hancock, D.L. 1986. Classification of the Trypetinae (Diptera: Tephritidae), with a discussion of the Afrotropical fauna. *J. Entomol. Soc. South. Afr.* 49: 275–305.

Hancock, D.L. 1987. Notes on some African Ceratitinae (Diptera: Tephritidae), with special reference to the Zimbabwean fauna. *Trans. Zimb. Sci. Assoc.* 63: 47–57.

Hancock, D.L. and R.A.I. Drew. 1999. Bamboo-shoot fruit flies of Asia (Diptera: Tephritidae: Ceratitidinae). *J. Nat. Hist.* 33: 633–775.

Heywood, V.H. 1978. *Flowering Plants of the World*. Oxford University Press, Oxford. 336 pp.

Munro, H.K. 1984. A taxonomic treatise on the Dacidae (Tephritoidea, Diptera) of Africa. *Entomol. Mem. S. Afr. Dep. Agric.* 61: 1–313.

Whitmore, T.C. 1986. *Tropical Rainforests of the Far East*, 2nd ed. Oxford University Press, Oxford. 352 pp.

# 20 Morphological Features of the Tribe Dacini (Dacinae): Their Significance to Behavior and Classification

*Ian M. White*

## CONTENTS



## 20.1 INTRODUCTION

The completion of a data matrix (White and Hancock 1997) of coded descriptions of Indo-Australasian Dacini* provided an ideal opportunity to analyze character distributions and correlations. The purpose of this chapter is to discuss characters used for classification and identification of dacine fruit flies from the perspective of their behavior. In addition, a preliminary reanalysis of the classification is carried out and a number of previously undocumented observations of potential interest to the study of dacine behavior are reported. The small genera *Ichneumonopsis*** (one species) and *Monacrostichus* (two species) are not considered in detail here (see Drew and Hancock 1994b and Chapter 19), leaving only *Bactrocera* and *Dacus* with 447 and 57 species, respectively. White and Hancock (1997) covered all described and valid species of Dacini from the Oriental, Australasian, and Oceanic Regions, plus some species from the Afrotropical Region (all Mascarene species and African pest species), making a total of 507 species. A further 7 *Bactrocera* and 161 *Dacus* spp. are known from the Afrotropical Region but were not included by White and Hancock

* This group is here regarded as the tribe Dacini in the subfamily Dacinae (see Korneyev, Chapter 4), although Norrbom et al. (1999) rank it as the subtribe Dacina in the subfamily Trypetinae, tribe Dacini, and Drew (1989) and Drew and Hancock (1994b) rank it as the subfamily Dacinae.

** *Ichneumonopsis* is sometimes placed in the Gastrozonini (as Gastrozonina, Norrbom et al. 1999), and it has also been placed in the Adramini (Hardy 1973), a tribe of Trypetinae.

0-8493-1275-2/00/$0.00+$.50

(1997) and another 17 Asian *Dacus* spp. were recently described by Drew et al. (1998), bringing the totals to 454 *Bactrocera* spp. and 235 *Dacus* spp. They are therefore omitted from the analyses given here, although their inclusion would not be expected to alter the conclusions greatly since species of most subgenera of *Dacus* are included.

Many of the characters for which we have some understanding of function have been used in the construction of the subgeneric classification. The subgeneric distribution of these characters is therefore presented here as a basis for discussion (Table 20.1). The classification used by White and Hancock (1997) derives from Drew's (1989) monograph on the Australasian-Oceanic fauna, but with a few minor modifications to allow for inclusion of the Oriental fauna and newly discovered information (see Figure 20.6). Drew (1989) placed most Indo-Australasian species that had previously been classified as *Dacus* into the genus *Bactrocera* (a name formerly used for a subgenus, known as *Strumeta* in the older literature). This was based on the fact that most species with fused terga (those still called *Dacus*) were Afrotropical and mostly associated with just two plant families, Asclepiadaceae and Cucurbitaceae. The Indo-Australasian group formerly known as the genus *Callantra* have the same pattern of host association and fused terga, so they were placed as a subgenus of *Dacus*. The species with nonfused terga, that is, the genus *Bactrocera*, were subdivided into four groups of subgenera according to the length of the posterior lobe of the male lateral surstylus and the presence or absence of a notch in the hind margin of male sternite 5. Most species (all except eight), belong to two of those groups; the *Zeugodacus* group of subgenera are predominantly associated with Cucurbitaceae; the *Bactrocera* group of subgenera rarely attack cucurbits and include some species which attack a very wide range of plant families, for example, the Oriental fruit fly, *B. dorsalis* (Hendel).

One of the most outstanding features of the biology of the Dacinae (at least the Dacini and Ceratitidini) is that the males of many species are attracted to so-called male lures, that is, chemicals that are not pheromones but are nonetheless very potent male attractants of value for control through male suppression, for population monitoring and taxonomic survey. The chemicals that attract Dacini are known as methyl eugenol, cue-lure, and vert-lure, although the latter is only known to attract one species (see White and Elson-Harris 1994 for details of all major lures). Species in the *Bactrocera* group of subgenera respond to either methyl eugenol or cue-lure and in some cases lure response is the only tenable diagnostic feature for the identification of males. Conversely, in the *Zeugodacus* group and the genus *Dacus*, methyl eugenol response is rare. The pecten (row of long setae placed transversely, close to side of hind margin of tergite 3), plus a suite of correlated morphological features, appear to be involved in pheromone dispersal and it has been shown that male lures augment the effects of pheromones. Many of the same suite of characters are involved in sound production and the taxonomic distribution of these characters is discussed with respect to that behavior later in this chapter.

Some characters used by taxonomists to differentiate species may have evolved in response to, among other things, the need to be cryptic to potential predators, the need to be recognized by conspecifics, and/or the need to be seen as different by similar species sharing the same niche. Behavioral interactions with predators may have led to the evolution of wasp mimicry in the Dacini. The characters that may help fool a predator are discussed, although so little research has been carried out on mimicry in the group that little can be said. Similarly, the functional importance of some of the characters that help taxonomists distinguish similar species is generally little understood. However, recent work by the author and K. Mahmood, a postgraduate at the Natural History Museum, London (1993 to 1996), discovered some interesting differences between closely related species belonging to the Oriental fruit fly, *B.* (*Bactrocera*) *dorsalis* species complex which added to the details given in the recent revision by Drew and Hancock (1994a). These differences included the use of scutal microtrichia pattern and phallus length to differentiate some species pairs. They are summarized here and discussed with respect to possible resource partitioning by sympatric species.

**TABLE 20.1**
**Subgenera of *Bactrocera* and *Dacus* showing the Percentage of Species Known to Have Each Character of Interest**

| | tgf | sur | S5V | pec | bul | cer | prs | asp | wpd | tbp | lfg | syn | ltv | mdv | 4sc | *n* |
|---|---|---|---|---|---|---|---|---|---|---|---|---|---|---|---|---|
| *B.* (*Hemisurstylus*) | 0 | 0 | 0 | 0 | 0 | 100 | 0 | 100 | 0 | 0 | 0 | 0 | 100 | 0 | 100 | 1 |
| *B.* (*Melanodacus*) | 0 | 0 | 0 | 0 | 0 | 100 | 100 | 100 | 0 | 0 | 0 | 0 | 50 | 0 | 100 | 2 |
| *B.* (*Hemizeugodacus*) | 0 | 0 | 13 | 100 | 0 | 100 | 100 | 50 | 100 | 88 | 0 | 0 | 75 | 50 | 100 | 4 |
| *B.* (*Daculus*) | 0 | 0 | 50 | 100 | 0 | 100 | 0 | 0 | 100 | 100 | 0 | 0 | 0 | 0 | 0 | 1 |
| *B.* (*Bulladacus*) | 0 | 0 | 94 | 100 | 100 | 0 | 100 | 78 | 83 | 83 | 0 | 0 | 100 | 44 | 0 | 9 |
| *B.* (*Trypetidacus*) | 0 | 0 | 100 | 0 | 0 | 100 | 0 | 0 | 0 | 0 | 0 | 0 | 0 | 0 | 0 | 1 |
| *B.* (*Apodacus*) | 0 | 0 | 100 | 100 | 0 | (100) | 67 | 67 | 100 | 83 | 0 | 17 | 100 | 100 | 0 | 3 |
| *B.* (*Afrodacus*) | 0 | 0 | 100 | 100 | 0 | 100 | 100 | 7 | 100 | 93 | 0 | 0 | 57 | 0 | 7 | 7 |
| *B.* (*Notodacus*) | 0 | 0 | 100 | 100 | 0 | 100 | 100 | 100 | 100 | 100 | 0 | 0 | 100 | 100 | 0 | 2 |
| *B.* (*Bactrocera*) | 0 | 1 | 100 | 100 | 0 | 100 | 94 | 98 | 99 | 98 | 0 | 1 | 93 | 0 | 0 | 253 |
| *B.* (*Gymnodacus*) | 0 | 8 | 92 | 0 | 0 | 100 | 92 | 75 | 17 | 8 | 0 | 0 | 83 | 0 | 0 | 6 |
| *B.* (*Parazeugodacus*) | 0 | 10 | 0 | 100 | 0 | 100 | 100 | 100 | 100 | 100 | 0 | 0 | 80 | 0 | 100 | 5 |
| *B.* (*Tetradacus*) | 0 | 20 | 80 | 100 | 0 | 100 | 0 | 20 | 60 | 60 | 0 | 80 | 100 | 80 | 0 | 5 |
| *B.* (*Paratridacus*) | 0 | 83 | 8 | 0 | 0 | 100 | 100 | 100 | 0 | 0 | 0 | 0 | 100 | 50 | 100 | 6 |
| *B.* (*Semicallantra*) | 0 | 83 | 100 | 100 | 0 | 100 | 100 | 0 | 100 | 100 | 100 | 83 | 67 | 0 | 0 | 3 |
| *B.* (*Javadacus*) | 0 | 89 | 22 | 100 | 0 | 100 | 100 | 0 | 100 | 100 | 0 | 0 | 100 | 44 | 11 | 9 |
| *B.* (*Paradacus*) | 0 | 94 | 6 | 100 | 0 | 100 | 0 | 100 | 94 | 81 | 0 | 6 | 100 | 50 | 75 | 8 |
| *B.* (*Zeugodacus*) | 0 | 99 | 2 | 100 | 0 | 100 | 97 | 98 | 100 | 100 | 0 | 0 | 100 | 93 | 79 | 54 |
| *B.* (*Austrodacus*) | 0 | 100 | 0 | 0 | 0 | 100 | 0 | 0 | 0 | 0 | 0 | 0 | 100 | 100 | 100 | 1 |
| *B.* (*Hemiparatridacus*) | 0 | 100 | 0 | 0 | 0 | 100 | 0 | 100 | 0 | 0 | 0 | 0 | 100 | 100 | 100 | 1 |
| *B.* (*Nesodacus*) | 0 | 100 | 0 | 0 | 0 | 100 | 0 | 100 | 0 | 0 | 0 | 0 | 100 | 0 | 0 | 1 |
| *B.* (*Niuginidacus*) | 0 | 100 | 0 | 0 | 0 | 100 | 0 | 100 | 0 | 50 | 0 | 50 | 0 | 100 | 0 | 1 |
| *B.* (*Hemigymnodacus*) | 0 | 100 | 0 | 0 | 0 | 100 | 100 | 100 | 0 | 0 | 0 | 0 | 100 | 100 | 50 | 1 |
| *B.* (*Asiadacus*) | 0 | 100 | 0 | 100 | 0 | 100 | 0 | 0 | 75 | 75 | 0 | 25 | 100 | 88 | 0 | 4 |
| *B.* (*Diplodacus*) | 0 | 100 | 0 | 100 | 0 | 100 | 0 | 0 | 100 | 100 | 0 | 0 | 100 | 100 | 50 | 1 |
| *B.* (*Papuodacus*) | 0 | 100 | 0 | 100 | 0 | 100 | 50 | 0 | 100 | 100 | 0 | 0 | 100 | 100 | 100 | 1 |
| *B.* (*Heminotodacus*) | 0 | 100 | 0 | 100 | 0 | 100 | 100 | 0 | 100 | 100 | 0 | 0 | 100 | 0 | 100 | 1 |
| *B.* (*Queenslandacus*) | 0 | 100 | 100 | 0 | 0 | 100 | 0 | 0 | 0 | 50 | 0 | 0 | 0 | 0 | 0 | 1 |
| *B.* (*Sinodacus*) | 8 | 100 | 3 | 100 | 0 | 100 | 6 | 97 | 100 | 100 | 3 | 33 | 67 | 78 | 11 | 18 |
| *D.* (*Callantra*) | 100 | 35 | 4 | 88 | 0 | 100 | 0 | 83 | 81 | 60 | 100 | 98 | 8 | 33 | 0 | 24 |
| *D.* (*Dacus*) | 100 | 62 | 8 | 100 | 0 | 100 | 0 | 92 | 100 | 96 | 23 | 35 | 19 | 69 | 0 | 13 |
| *D.* (*Didacus*) | 100 | 63 | 8 | 100 | 0 | 100 | 0 | 4 | 88 | 88 | 13 | 33 | 8 | 38 | 0 | 12 |
| *D.* (*Leptoxyda*) | 100 | 100 | 0 | 0 | 0 | 100 | 0 | 0 | 0 | 0 | 0 | 0 | 50 | 100 | 0 | 1 |
| TOTAL % | 11 | 30 | 64 | 94 | 2 | 98 | 76 | 86 | 92 | 90 | 7 | 11 | 83 | 27 | 17 | 460 |

*Note:* The subgenera are ordered by the distribution of the first eight characters. Species for which the males were inadequately known were excluded. Species with variable, intermediate, or indeterminate states of any character were counted as a half (indeterminate states were commonly caused by observational difficulty). The characters were: tgf, tergites fused; sur, posterior lobe of male surstylus long; S5V, sternite 5 of male with a deep V-shaped notch in hind margin; pec, pecten of male present; bul, bulla of male next to cell bcu of wing; cer, ceromata present and oval (except in *Apodacus* in which they are narrow); prs, prescutellar acrostichal seta present; asp, postsutural supra-alar seta present; wpd, wing of male with a dense area of microtrichia around cell bcu; tbp, hind tibia of male with an apical microtrichose swelling; 1fg, first flagellomere very long; syn, syntergite 1+2 long and narrow (wasp-waisted); ltv, lateral postsutural yellow vitta present; mdv, medial postsutural yellow vitta present; 4sc, scutellum with four marginal setae. *n* = number of species considered.

The final sections of this chapter discuss other characters used in the subgeneric classification but for which we have little or no idea of function; then a preliminary cladistic analysis is presented, and the fit of nonmorphological data to that cladogram is discussed. The characters of unknown function that are discussed include the abdominal tergite fusion of *Dacus* spp., the male terminalia features which define the four groups of subgenera in *Bactrocera*, and differences in chaetotaxy which define subgenera within those groups. The current classification, which is highly dependent on these characters, is then tested by cladistic analysis.

## 20.2 PHEROMONE DISPERSAL AND MALE LURE RESPONSE

Male fruit flies in mass culture can be seen to emit a "smoke," which is known to contain male pheromone. At the time of "smoke" emission sexually mature males vibrate their wings. That is believed to help disperse the pheromone, and it also produces a sound. Kuba and Sokei (1988) studied this phenomenon in male melon flies, *B.* (*Zeugodacus*) *cucurbitae* (Coquillett), and they observed that pheromone droplets were taken from the rectal glands using the hind legs and then smeared onto the densely microtrichose area adjacent to cell bcu of the wings. The wings were then vibrated against the pecten (they called this the "targal bristles"). From Table 20.1 it can be seen that 94% of species examined by White and Hancock (1997) were found to have a pecten (pec). With very few exceptions those same species also have the dense microtrichose area on the wing (wpd) and a microtrichose swelling at the apex of the hind tibia (tbp), a feature not mentioned by Kuba and Sokei (1988). The only confirmed exceptions are *B.* (*Asiadacus*) *brachycera* (Bezzi), some *B.* (*Tetradacus*) spp. and some *D.* (*Callantra*) spp., all of which share the feature of a wasp-waisted (petiolate) abdomen (any other discrepancy in values shown in Table 20.1 between these three characters can be attributed to difficulties in deciding if the character was properly developed or not, missing body parts, or other difficulties in observing some species, which were therefore counted as 0.5 when summing up the totals). All three of these characters are peculiar to the males. Despite its prevalence in dacine species, few taxonomists have mentioned the hind tibial pad, a notable exception being Shiraki (1968). Furthermore, Kuba and Sokei (1988) referred to the hind tarsi being used to transfer pheromone and made no mention of the hind tibial pad. Studies carried out by K. Mahmood (unpublished work at the Natural History Museum, London, 1994) confirmed that there are no discernible sexual differences between the tarsi as might be expected if it really were the tarsi transferring the pheromone. It is therefore highly likely that it is the adjacent specialized swelling at the apex of the hind tibia (of the species which have a pecten), that is responsible for pheromone transfer. The question remains, however, what mechanism for mate attraction is used by those 6% of species whose males lack all these features?

Work by Todd Shelly and collaborators has shown that the tendency for males of *Bactrocera* spp. to be attracted to either methyl eugenol (ME; 4-allyl-1,2-dimethoxybenzene) or cue-lure (CL; 4(*p*-acetoxyphenyl)-2-butanone) is a related subject to that of pheromone dispersal. It has been shown that when exposed to ME, males of *B.* (*Bactrocera*) *dorsalis* (Hendel) (Shelly and Dewire 1994) and *B.* (*Bactrocera*) *philippinensis* Drew and Hancock (Shelly et al. 1996) increased their wing vibrating activity. Similarly, Shelly and Villalobos (1995) showed the same phenomenon when *B.* (*Zeugodacus*) *cucurbitae* males were exposed to CL. Furthermore, once fed, the flies show little interest in ever feeding on these substances again. Although male lures are not essential to pheromone production or mating success, it appears that both pheromone production (Fitt 1981) and mating success (Shelly et al. 1996) can be enhanced by males ingesting these chemicals. It seems likely that these chemicals are analogues of natural substances (Fletcher 1987) gathered by the flies to augment their pheromone production and mating success. For example, Shelly and Dewire (1994) found that over a series of separate experiments, around 70% of matings involved males that had fed on ME, and that even a 30-s feed still conferred a mating advantage 35 days later. However, it was uncertain if this was entirely due to pheromone differences, differences in wing fanning, or intermale differences in sounds produced (see next section).

ME is a very widely occurring chemical that has been found in many plant families and many plant parts, for example, Aristolochiaceae (*Asarum* spp.), Canellaceae (*Canella winterana* stems), Fabaceae (*Acacia farnesiana*), Lamiaceae (*Hyssopus officinalis*, *Ocimum* spp., *Rosmarinus officinalis*), Lauraceae (*Laurus nobilis* leaves, *Umbellularia californica*), Lecythidaceae (*Couroupita guianensis* flowers), Liliaceae (*Proiphys amboinensis*), Podocarpaceae (*Dacrydium franklinii*) (data from listed World Wide Web sources; Chuah et al. 1997; Yong 1990). Related chemicals such as eugenol and iso-eugenol may also attract the flies and these also occur in a wide range of plants, for example, Annonaceae (*Cananga odorata*) and Myrtaceae (*Syzygium aromaticum*) (data from listed World Wide Web sources). The latter is cloves whose oil has been used for the attraction and collection of these flies. Of all these plants only *C. odorata* has been reported as a host plant; it is attacked by *B.* (*Bactrocera*) *endiandrae* (Perkins and May) which is methyl eugenol attracted (Drew 1989). Furthermore, few of these plants occur within the natural geographic range of the Dacini, although related plants that do occur may be found to produce ME. *Ocimum* spp. are a notable example of ME-producing plants that do occur within the range, and the planting of *O. sanctum* (tulsi plant or holy basil) has been advocated as part of a control strategy (Shah and Patel 1976).

Shelly and Dewire (1994) questioned the generality of the hypothesis that naturally occurring male lures are important in the synthesis of male sex pheromone. They note that *Ceratitis capitata* (Wiedemann) does not ingest its male lures (e.g., trimedlure) but rests nearby. The author has noticed that when lures (CL or ME) are painted on a leaf or placed on a cotton wick hung in a tree (rather than contained in a trap), many *Bactrocera* spp. also rest nearby, especially the CL-attracted species. Drew (1974) suggested that for CL-attracted flies the real attraction was not the CL itself but its breakdown (hydroxy derivative) product, named Willison's lure or raspberry ketone (4-(*p*-hydroxyphenyl)-2-butanone). Cunningham (1989) was of the opinion that this could not be the case, since CL was considerably more attractive than Willison's lure. However, that misses the possibility that CL may constitute a superstimulus to attract flies into the area (or trap) while it may be condensates (perhaps Willison's lure or 2-butanone) on nearby surfaces that the flies actually ingest (the author has observed that the flies on the "nearby" surfaces do extend their proboscis and so appear to be feeding, although that could be for other reasons, such as bacterial feeding). Drew (1987), in a review of the importance of plant surface bacteria as adult food for dacines, noted that 2-butanone, a component of the CL molecule, was also a breakdown product of bacterial decomposition and itself a male attractant (although too volatile for normal use) of importance for bringing mature males into the feeding and oviposition site (host tree) of the developing females (the work of R.A.I. Drew and colleagues on bacterial feeding is reviewed in Chapters 21 and 27). It may be expected that this proposed manner of host attraction and the male lure phenomenon would be linked in some way, and there is a possibility that 2-butanone attraction represents an evolutionary precursor to the development of CL attraction (see Cunningham (1989) for a discussion of other common features of male lure chemistry).

Although not strictly relevant to the main theme of this chapter, it is worth mentioning that assemblages of both sexes of several species have sometimes been observed in trees, which were often not suitable hosts. These observations have been made by taxonomists (in Karnataka, India by the author; in Zimbabwe by D.L. Hancock, personal communication; and in Assam (India), Ethiopia, Kenya, Tanzania, Nigeria, by A. Freidberg, personal communication) and as they have not been reported in the literature before, an example is detailed here. Eight *Bactrocera* spp. were found in a single fig tree (*Ficus asperima*) in southern India (University of Agricultural Sciences, Bangalore, field station at Mudigere, Karnataka, India; 980 m). Flies were collected by visual observation and hand-netting by the author, D.L. Hancock, and S. Ramani, between 0630 and 0730 each day (20–22 May 1992). Traps (ME and CL) were placed nearby (about 10 and 100 m) but collected only a few flies, namely, *B.* (*Bactrocera*) *caryeae* (Kapoor) (ten males at ME), *B.* (*Bactrocera*) *dorsalis* (one male at ME), and *B.* (*Bactrocera*) *correcta* (Bezzi) (six males at ME), and most of these were from the trap farthest from the tree. No *B.* (*Bactrocera*) *caryeae* or *B.* (*Bactrocera*) *dorsalis* were found in the tree. Conversely, ME-attracted species such as *B.* (*Bactrocera*) *correcta* (one male),

*B.* (*Hemigymnodacus*) *diversa* (Coquillett) (two males, five females) and *B.* (*Bactrocera*) *versicolor* (Bezzi) (three females) were found in the tree, along with CL- attracted species *B.* (*Zeugodacus*) *cucurbitae* (two males) and a species related to *B.* (*Zeugodacus*) *tau* (Walker) (14 males, 36 females). In addition, *B.* (*Bactrocera*) *latifrons* (Hendel) (one male, one female) was found although it does not respond to either ME or CL, as were the following species of unknown lure response: *B.* (*Zeugodacus*) *duplicata* (Bezzi) (one male) and *B.* (*Sinodacus*) *watersi* (Hardy) (six males, five females). In each case the numbers relate only to the specimens in the collections at the Natural History Museum, London, which give an indication of species and sexual composition (it was not practical to total numbers sent to other collections). *Ficus* spp. are rarely hosts of Dacini and none was reared from the fruits collected off this tree. However, the ripe fruits had clearly attracted vertebrate feeders (birds and possibly monkeys) and many leaves had a rich covering of guano. The flies, which included females, were almost certainly attracted to odors of decomposition involving bacterial action on the surface of the guano-covered leaves. The almost complete non-concordance of the flies caught in lure traps and in the tree may simply be due to differences in the stage of sexual maturation at which flies might be attracted to each odor (lure trap or tree) but it may also be that for some species the tree represented a stronger stimulus. Regrettably this assemblage phenomenon has rarely been observed and always in places far from analytical facilities that would allow identification of the attractant mechanism involved. Given that male lures are the favored attractant for use for Dacini collection due their vastly greater attractancy compared with protein or ammonia baits, the chemistry that attracts these diverse mixed-sex assemblages cannot simply be dismissed as a mere ammonia source. R.A.I. Drew (personal communication 1998) has suggested that the flies in these assemblages are in their nonbreeding period. Observations have been made of the greater attraction of bird feces to *C. capitata* as compared with protein bait, especially when the flies had been starved of protein for several days (Prokopy et al. 1992).

There is some evidence that dacine flies may be pollinating agents of some male-attracting plants, and that other eugenol-type chemicals may also attract the flies. Yong (1993) reported males of *B.* (*Bactrocera*) *carambolae* Drew and Hancock and *B.* (*Bactrocera*) *papayae* Drew and Hancock on the spadix of *Spathiphyllum cannaefolium* (Araceae) in the mornings only, and their numbers peaking when the flowers had been opened about 5 days. Similarly, Lewis et al. (1988) reported that the heavy odor of this plant, and its attraction to *B.* (*Bactrocera*) *musae* (Tryon) in northern Queensland, declines as the pollen is lost. CL does not appear to be a naturally occurring chemical, but chemicals attracting *B.* (*Zeugodacus*) *cucurbitae* have been found to be emitted from orchids at the same time as the flies were most active (Toong and Tan 1994), and males of an unidentified species related to *B.* (*Bactrocera*) *dorsalis* have been found at unpollinated flowers of *Bulbophyllum cheiri* (Orchidaceae) in Malaysia (Toong and Tan 1994).* Nishida et al. (1993) identified Willison's lure** in the flowers of *Dendrobium superbum* (Orchidaceae), whose flowers were "licked" by males of *Bactrocera* (*Zeugodacus*) *cucurbitae*, and the chemical could be detected in their rectal glands from 6 h to 6 days after such a "feed." The possibility that at least some cases of attraction may be "payment" for pollination should be investigated, and orchids may be the true sources of natural attractants of CL responding flies. Taxonomists should be vigilant to observe any structures which may be orchid pollinia adhered to collected flies and not regard them as "dirt" to be cleaned off. The author has observed pollinia adhered to the scutum of a male of the *B.* (*Zeugodacus*) *tau* complex (Bogor Museum, Indonesia) and apparent pollinia (two pairs) placed either side of the front left leg of a female of the *B. emittens* (Walker) species complex (Natural History Museum, London). Species belonging to these complexes are only known to respond to CL; however, the reason for the presumed pollinia on a female is a little difficult to explain; perhaps because females

* Within the Oriental fruit fly (*B. dorsalis*) species complex 31 species are known to respond to CL and 26 to ME (Drew 1989; Drew and Hancock 1994a; White and Hancock 1997).

** Nishida et al. (1993) called this "benzylacetone (4-(4-hydroxyphenyl)-2-butanone)" but these are two different chemicals (D.J. Henshaw, personal communication, 1998); the latter name is an alternative for Willison's lure and that appears to be what they found.

do under some circumstances show a very low response to "male" lures. D.L. Hancock (personal communication) reported that ME-attracted flies, such as *B. endiandrae*, have been caught in ME traps, with two types of pollinia on them, one attached to the scutum and the other to the eyes. The author was unable to find specific data on the presence of eugenol-type chemicals in orchids but it is know that males of some South American orchid bees (Apidae, Euglossini, which also sequester male lures) are attracted to eugenol.* Furthermore, vanilla (traditionally an extract of orchid pods) is commercially synthesized from eugenol.**

Table 20.2 summarizes known lure response data. It can be seen that cue-lure is known to attract almost three times as many species as methyl eugenol. However, few conclusions can be drawn about the relative taxonomic distributions of the two major lure responses, except for the fact that methyl eugenol is a rare (and unconfirmed) response for subgenus *Zeugodacus* spp., uncommon in *Dacus* subgenus *Callantra* spp., and unknown in other *Dacus* spp., including all Afrotropical spp. However, some of the smaller subgenera regarded as close relatives of *Zeugodacus* (i.e., those with long lateral surstylus lobes) include species with a methyl eugenol response, namely, *Hemigymnodacus* and *Javadacus*.

The distribution of subgenera believed to lack a lure response or for which no lure response is yet known is also worthy of discussion. Presuming that the pecten has a function in helping to disperse pheromone and that the lure mimics a natural substance gathered in order to augment mating success, one might expect that those species that lacked a pecten used some as-yet-unknown alternative mate attraction system. Of the 13 subgenera that lack a pecten, 5 include species that still show a lure response, and many of the remainder may simply have an as-yet-undiscovered response, perhaps involving another chemical. Assuming that these species gather lure for the same reasons as species with a pecten, we must conclude that these species have alternative means of pheromone dispersal that have yet to be discovered. A reexamination of a sample of these species (those categorized as pests by White and Hancock 1997) showed that they all have a series of very thin, pale-colored setae in the "pecten area" of tergite 3, which are almost certainly homologous with a fully developed pecten (thick setae, normally dark in color). These species also have the smooth area between the pecten setae and the tergite margin, which Munro (1984) called the lamprine. These fine setae may of course act as a "soft" pecten and the fact that they can often be detected upon close examination suggests that in many (probably all) subgenera the lack of a distinct pecten is a derived condition. However, nearly all of these species lack both the areas of dense microtrichia around cell bcu and a swelling at the apex of the hind tibia.

One subgenus, namely, *Bulladacus*, has the unusual feature of a swelling on the anterior side of the extension of cell bcu, which appears to be caused by a cavity between the wing membranes. The author is not aware of this having been studied by scanning electron microscopy (SEM), but at the very least the swelling (bulla) would increase the surface area of microtrichia for holding pheromone and it may even have an opening, possibly ventrally. No *Bulladacus* spp. has been collected at either ME or CL, although three species have been bred sufficiently frequently as to suggest that they are fairly common and would have been found at known lures were they attracted to them (Yong 1994; Drew and Hancock 1995). This subgenus is also remarkable in that they are the only *Bactrocera* or *Dacus* spp. to lack ceromata. These flat oval structures on tergite 5 are areas of tiny wax-secreting pores (Munro 1984), but their function has not been studied.

## 20.3 ACOUSTICS

The wing vibration associated with pheromone dispersal also produces sound. Although it has been shown that sound production can be used as a measure of "health" in mass culture (e.g., Kanmiya et al. 1987), the importance of the sound generated by wing vibration is unclear. These sounds may

* http://www.neosoft.com/~worldc/bee.htm

** http://wwwchem.uwimona.edu.jm:1104/lectures/vanilla.html

**TABLE 20.2**
**Subgenera of *Bactrocera* and *Dacus* Showing the Percentage of Species Known to Respond to Each Synthetic Male Lure**

| | CL | ME | Vert lure | No lure | *n*1 | sur | pec | bul | cer | wpd | *n*2 |
|---|---|---|---|---|---|---|---|---|---|---|---|
| *B.* (*Hemisurstylus*) | ? | ? | ? | ? | 0 | 0 | 0 | 0 | 100 | 0 | 1 |
| *B.* (*Melanodacus*) | ? | ? | ? | ? | 0 | 0 | 0 | 0 | 100 | 0 | 2 |
| *B.* (*Hemizeugodacus*) | ? | ? | ? | ? | 0 | 0 | 100 | 0 | 100 | 100 | 4 |
| *B.* (*Daculus*) | 0 | 0 | 0 | 100 | 1 | 0 | 100 | 0 | 100 | 100 | 1 |
| *B.* (*Bulladacus*) | 0 | 0 | 0 | 100 | 3 | 0 | 100 | 100 | 0 | 83 | 9 |
| *B.* (*Trypetidacus*) | 0 | 100 | 0 | 0 | 1 | 0 | 0 | 0 | 100 | 0 | 1 |
| *B.* (*Apodacus*) | 0 | 100 | 0 | 0 | 3 | 0 | 100 | 0 | (100) | 100 | 3 |
| *B.* (*Afrodacus*) | 80 | 0 | 0 | 10 | 5 | 0 | 100 | 0 | 100 | 100 | 7 |
| *B.* (*Notodacus*) | 0 | 100 | 0 | 0 | 1 | 0 | 100 | 0 | 100 | 100 | 2 |
| *B.* (*Bactrocera*) | 65 | 33 | 0 | 2 | 193 | 1 | 100 | 0 | 100 | 99 | 253 |
| *B.* (*Gymnodacus*) | 50 | 0 | 0 | 50 | 2 | 8 | 0 | 0 | 100 | 17 | 6 |
| *B.* (*Parazeugodacus*) | ? | ? | ? | ? | 0 | 10 | 100 | 0 | 100 | 100 | 5 |
| *B.* (*Tetradacus*) | ? | ? | ? | 100 | 2 | 20 | 100 | 0 | 100 | 60 | 5 |
| *B.* (*Paratridacus*) | 0 | 40 | 0 | 60 | 5 | 83 | 0 | 0 | 100 | 0 | 6 |
| *B.* (*Semicallantra*) | 50 | 50 | 0 | 0 | 2 | 83 | 100 | 0 | 100 | 100 | 3 |
| *B.* (*Javadacus*) | 40 | 60 | 0 | 0 | 5 | 89 | 100 | 0 | 100 | 100 | 9 |
| *B.* (*Paradacus*) | 83 | 0 | 0 | 17 | 6 | 94 | 100 | 0 | 100 | 94 | 8 |
| *B.* (*Zeugodacus*) | 94 | 3 ? | 0 | 3 | 32 | 99 | 100 | 0 | 100 | 100 | 54 |
| *B.* (*Austrodacus*) | 0 | 0 | 0 | 100 | 1 | 100 | 0 | 0 | 100 | 0 | 1 |
| *B.* (*Hemiparatridacus*) | ? | ? | ? | ? | 0 | 100 | 0 | 0 | 100 | 0 | 1 |
| *B.* (*Nesodacus*) | ? | ? | ? | ? | 0 | 100 | 0 | 0 | 100 | 0 | 1 |
| *B.* (*Niuginidacus*) | 100 | 0 | 0 | 0 | 1 | 100 | 0 | 0 | 100 | 0 | 1 |
| *B.* (*Hemigymnodacus*) | 0 | 100 | 0 | 0 | 1 | 100 | 0 | 0 | 100 | 0 | 1 |
| *B.* (*Asiadacus*) | 100 | 0 | 0 | 0 | 3 | 100 | 100 | 0 | 100 | 75 | 4 |
| *B.* (*Diplodacus*) | ? | ? | ? | ? | 0 | 100 | 100 | 0 | 100 | 100 | 1 |
| *B.* (*Papuodacus*) | 100 | 0 | 0 | 0 | 1 | 100 | 100 | 0 | 100 | 100 | 1 |
| *B.* (*Heminotodacus*) | ? | ? | ? | ? | 0 | 100 | 100 | 0 | 100 | 100 | 1 |
| *B.* (*Queenslandacus*) | ? | ? | ? | ? | 0 | 100 | 0 | 0 | 100 | 0 | 1 |
| *B.* (*Sinodacus*) | 100 | 0 | 0 | 0 | 13 | 100 | 100 | 0 | 100 | 100 | 18 |
| *D.* (*Callantra*) | 82 | 18 | 0 | 0 | 11 | 35 | 88 | 0 | 100 | 81 | 24 |
| *D.* (*Dacus*) | 100 | 0 | 0 | 0 | 12 | 62 | 100 | 0 | 100 | 100 | 13 |
| *D.* (*Didacus*) | 72 | 0 | 14 | 14 | 7 | 63 | 100 | 0 | 100 | 88 | 12 |
| *D.* (*Leptoxyda*) | 0 | 0 | 0 | 100 | 1 | 100 | 0 | 0 | 100 | 0 | 1 |
| TOTAL % | 68 | 25 | <1 | 6 | 312 | 30 | 94 | 2 | 98 | 92 | 460 |

*Note:* The percentage of species having each character that may in some way be associated with the lure response phenomenon is shown in the right section. The subgeneric order follows Table 20.1. The characters were: sur, posterior lobe of male surstylus long; pec, pecten of male present; bul, bulla of male next to cell bcu of wing; cer, ceromata present and oval (except in *Apodacus* in which they are narrow); wpd, wing of male with a dense area of microtrichia around cell bcu. *n*1 = number of species of known lure. *n*2 = number of species of known character state. Species for which males are known but for which no lure response data are available may be estimated as *n*2 – *n*1.

just be a byproduct of vibration essential to pheromone dispersal; as most of the work on sound production was carried out before the importance of wing vibration for pheromone dispersal was realized, that possibility does not appear to have been addressed. For example, Kanmiya (1988) makes no mention of playing back recordings of male "songs" to females in the manner reported by Aluja et al. (Chapter 15) for *Anastrepha*. Some workers have applied the term *stridulation* (e.g., Kanmiya 1988); however, there is no evidence of the sort of very specific sounds associated with the grinding together of special sound-producing organs in many Orthoptera and Coleoptera, although the presence of the pecten has been shown to increase the complexity of the sound produced in *B.* (*Zeugodacus*) *cucurbitae* (Kanmiya 1988). However, Kanmiya (1988) also showed that the surgical removal of the pecten had no effect on rate of copulation. Regrettably, differences between species have been little studied, but a comparison of the sounds produced by *B.* (*Bactrocera*) *dorsalis* and the Australian *B.* (*Bactrocera*) *opiliae* (Drew and Hardy 1981) showed that they differed by no more than did the sounds produced by two laboratory strains of *B.* (*Zeugodacus*) *cucurbitae*. Although this suggests that if sound is important in mate attraction, then it is of little help in recognizing conspecific mates from those of related species; it should be noted that no comparative studies of sympatric species have been carried out. Kanmiya (1988) did report a difference between male and female song signatures of *B.* (*Zeugodacus*) *cucurbitae* and the significance, if any, of that observation deserves study. He also reported that *B.* (*Paratridacus*) *garciniae* (Bezzi) [as *B. expandens* (Walker)], a species which has no pecten, has no song, although he did not give a source of that data or an indication of the number of replicates made, and so the possibility that species without a pecten lack songs (wing vibration) requires study. Clearly, the whole area of acoustics requires further investigation, and there is as yet little evidence of complex calling patterns such as those reported for *A. suspensa* (Loew) (Sivinski and Burk 1989). To test for interspecific differences, sympatric species will have to be compared.

## 20.4 MIMICRY

It is likely that most, perhaps all, Dacini are to some extent mimics of Hymenoptera Aculeata. However, very little work has been carried out to discover the "models" involved. The only specific mention known to the author is that Enderlein (1920) considered *B.* (*Tetradacus*) *minax* (Enderlein) to be a mimic of *Polistes marginatus* Fabricius, which was found in the same locality in Sikkim. *Bactrocera* (*Tetradacus*) *minax* and some other species of the subgenus *Tetradacus* have a slightly narrowed, but not markedly elongate, syntergite 1+2. That modification is carried further (and presumably independently) in *Dacus* subgenus *Callantra*, which have syntergite 1+2 longer than broad, and often constricted medially. This modification gives these flies a wasp-waisted (petiolate) appearance which is almost certainly "modeled" on that of a wasp. Furthermore, many of the markings typical of a great many dacine species are similar to the markings often seen on wasps. For example, postsutural yellow vittae (ltv, mdv in Table 20.1), notopleural wedges, and a yellow scutellum, contrasting with an otherwise dark scutum, is a common feature of both Dacini and many wasps (e.g., some *Polistes* spp.). On the abdomen, a pattern of dark lateral wedges and a dark midline on a light-colored, typically yellowish, ground are also typical of many wasps as well as most Dacini. In addition, many wasps have a dark anterior margin to the wing which may be the model for the costal band in the Dacini, and the anal streak of most dacines suggests a hind wing with a similar dark anterior feature. The long first flagellomere (third antennal segment; 1fg in Table 20.1) of *Callantra* spp. may also be part of the mimicry as it gives the antenna a similar appearance to that of many wasps, and the long ovipositor ("sting") of the family completes the disguise. Wasps have been shown to be the major predators of *Ceratitis capitata* (Hendrichs et al. 1994) but the main predators of adult dacines appear to be unrecorded. Furthermore, the author is unaware of any studies of dacine mimicry comparable with that carried out for *Rhagoletis zephyria* Snow, which appears to mimic jumping spiders in order to reduce spider predation (Mather and Roitberg 1987).

## 20.5 MATE RECOGNITION?

Most nondacine tephritids have complex wing patterns, and visual cues are of acknowledged importance in both their male–male and male–female encounters. Because most dacines lack complex wing patterns, if visual cues are of importance in their interactions, other morphological features may be involved. One character that has been overlooked in the taxonomy of Dacini is the pattern (distribution) of microtrichia on the integument. Microtrichia giving a silvery sheen when viewed from certain specific angles is a common feature of many Diptera and in many cases provides the only external character by which to distinguish closely related species. That in itself suggests that microtrichia patterning may often have a role in the recognition of conspecifics. In the Dacini microtrichia patterning has rarely been mentioned in descriptions, perhaps because most specimens have been trapped and this delicate structure can be damaged beyond recognition if the specimen becomes even slightly damp, as in an infrequently emptied trap. Furthermore, microtrichose patterns on the scutum tend to look quite different according to angle of view. White and Hancock (1997) did differentiate between two classes of microtrichia pattern in the *B.* (*Bactrocera*) *dorsalis* species complex, a group of over 60 known species. For example, *B.* (*Bactrocera*) *occipitalis* (Bezzi) and *B.* (*Bactrocera*) *philippinensis* Drew and Hancock are both very common species that occur sympatrically in the Philippines. When viewed with a low power microscope (or even the trained unaided eye) from above, both species appear to have a clear, nonmicrotrichose stripe down the middle of the scutum. However, when viewed from in front, the scutum of *B.* (*Bactrocera*) *philippinensis* has an unbroken microtrichose covering, whereas *B.* (*Bactrocera*) *occipitalis* has a wide, longitudinal medial stripe that lacks microtrichia. SEM studies of these microtrichia patterns by K. Mahmood (reported by White and Hancock 1997) showed that *B.* (*Bactrocera*) *philippinensis* has an unbroken covering of fine microtrichia, although of reduced density along the midline, while *B.* (*Bactrocera*) *occipitalis* has a wide, longitudinal medial stripe that totally lacks microtrichia. In the case of the parallel sympatric pair in Malaysia, *B.* (*Bactrocera*) *carambolae* and *B.* (*Bactrocera*) *papayae* both appear to have a complete covering of microtrichia on the scutum when examined under a light microscope from in front, but again have an apparent bare medial stripe when viewed from above. However, SEM study shows that in this pair of species only *B.* (*Bactrocera*) *carambolae* actually has a nonmicrotrichose medial stripe, similar, although narrower, to that of *B.* (*Bactrocera*) *occipitalis*.

All four species present the same striped pattern of microtrichia when viewed from above, and that pattern may serve to reinforce any mimicry provided by the lateral yellow vittae. That view from above may perhaps be considered the "predator's view." Conversely, each member of each sympatric pair has a very different appearance when viewed frontally, and it is tempting to suggest that such a view could help a fly visually recognize conspecifics from nonconspecifics, either in mating or competitive interactions. However, these flies do not face each other during courtship (R.A.I. Drew, personal communication), which for most species takes place at dusk on leaves, with each male defending an individual leaf (Fletcher 1987), so another explanation is needed. In considering this problem, R.A.I. Drew (personal communication) suggested that the differences in microtrichia pattern may be detected by the touch of the male's labella against the female during mating. However, the fly's perception of these patterns may be quite different from ours and R. J. Prokopy (personal communication) suggested that these patterns should be examined under ultraviolet light which may be an important part of their visual spectrum.

## 20.6 RESOURCE PARTITIONING

Another area of pure speculation arises from the fact that within each of the sympatric species pairs mentioned above, one species has a short ovipositor and the other a long ovipositor, and although the length of the aculeus (the hard apical segment of the ovipositor) does increase slightly with body size, the increase is not in proportion to body size. This is illustrated for *B.* (*Bactrocera*) *occipitalis* and *B.* (*Bactrocera*) *philippinensis* (Figure 20.1), but *B.* (*Bactrocera*) *carambolae* and *B.* (*Bactrocera*) *papayae* show an identical pattern (Table 20.3). Similarly, White and Marquardt (1989) showed that in *Chaetorellia* spp. (Tephritinae) the ratio of within-sample to between-sample variation in body size measures was an order of magnitude greater than in aculeus length. These observations suggest that regardless of body size, which is presumably determined by a combination

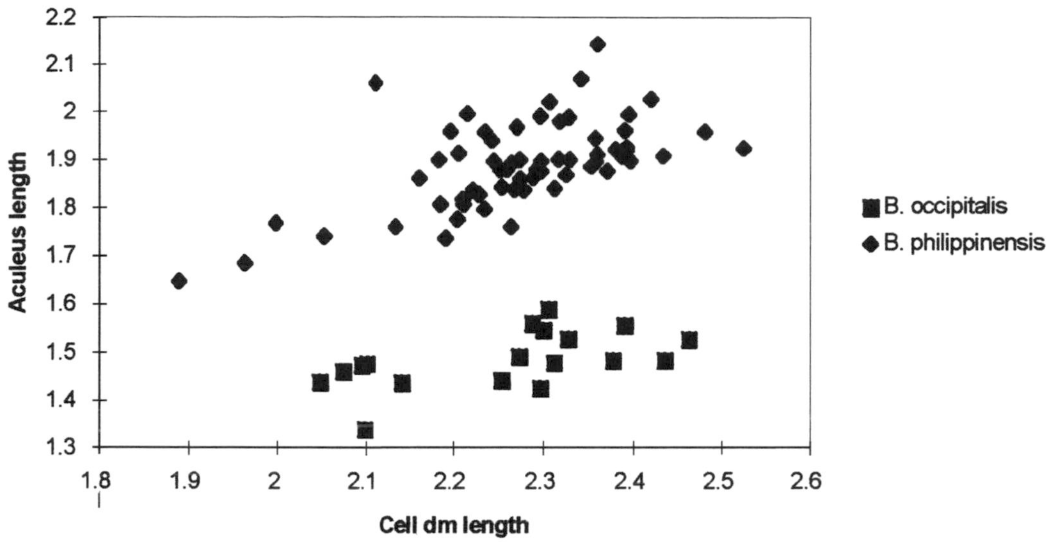


**FIGURE 20.1** Aculeus length of *B. philippinensis* and *B. occipitalis* plotted against a body size measure, namely, length of posterior edge of cell dm (chosen because it has well-defined end points and is rarely damaged even in cultured specimens with broken wing tips). Slopes are aculeus = 0.47 × cell dm length + 0.81 and aculeus = 0.26 × cell dm length + 0.91, respectively. Separation of aculeus length, and aculeus length/dm length ratio highly significant (2 tail, $t = 2.02$ $p < 0.001$ and $t = 2.04$ $p < 0.001$, respectively). (Data from unpublished studies, notably postgraduate work of K. Mahmood, 1993 to 1996.)

**TABLE 20.3**
**Aculeus Length, and Length Standardized for Body Size for Two Sympatric Species Pairs**

| Area | Species | Aculeus Length, mm | Aculeus/dm Ratio |
|---|---|---|---|
| SE Asian spp. | *B. carambolae* Drew and Hancock | 1.3–1.6 | 0.62–0.75 |
| | *B. papayae* Drew and Hancock | 1.5–2.1 | 0.77–0.99 |
| Philippines spp. | *B. occipitalis* (Bezzi) | 1.3–1.7 | 0.61–0.70 |
| | *B. philippinensis* Drew and Hancock | 1.6–2.1 | 0.76–0.98 |

*Note:* The body size index used was cell dm length measured along vein $Cu_1$. Data from White and Hancock (1997), based on pooled data from unpublished studies (notably postgraduate work of K. Mahmood, 1993 to 1996). 15 to 100+ specimens measured of each species.

of larval food quality and genetics, aculeus length is kept within very strict bounds. Such a developmental pattern would require a developmental control mechanism, and for that to have evolved, we may conclude that there is a strong selective reason for ensuring aculeus length is "correct" for the species, regardless of how big or small the rest of the fly becomes.

In a genus like *Chaetorellia*, where each species attacks only a very limited range of hosts (flowers of *Centaurea* spp., Asteraceae), aculeus length would be expected to be an adaptation to oviposition in each particular host. However, these four *Bactrocera* spp. are polyphagous and have most of their recorded hosts in common, so another explanation for such close control of aculeus length must be sought. It is known that some fruits are attacked at different stages than others, for example, *B.* (*Bactrocera*) *carambolae* attacks carambola fruits while they are very young, whereas *B.* (*Bactrocera*) *philippinensis* only attacks mango when it is ripening, and some other fruits may only be attacked when overripe or even after fall. Studies have been carried out to record what fruits are attacked by which flies and at which stage of fruit development, but, regrettably, those known to the author remain unpublished due to restrictions placed on commercially sensitive plant quarantine data. Further study may show that each of these species pairs (*B. occipitalis–B. philippinensis* and *B. carambolae–B. papayae*) in effect avoid competition (partition resources) by each species having the appropriate length of ovipositor to attack certain fruits at certain stages, in a manner that somehow differs from the other. Jones (1989) noted a negative correlation of dacine aculeus length and width vs. increasing host cuticle thickness. There is also evidence of a slight difference in habitat preference between the species in each "sympatric" pair. In the Philippines the longer ovipositor species (*B. philippinensis*) is dominant in orchards and backyard habitats while the shorter ovipositor species (*B. occipitalis*) is dominant in forests (observations by author). However, the converse is true in peninsular Malaysia where the shorter ovipositor species (*B. carambolae*) is dominant in orchards or disturbed areas while the longer ovipositor species (*B. papayae*) is dominant in forest areas (Ooi 1991).

Aculeus length provided by far the best means of discrimination between these species (Table 20.3). However, the greatest demand for identifications is for males collected in attractant traps. Casual observation by the author while doing dissections of the males and females of a wide range of Tephritidae is that the species with very long ovipositors are also the species with very long male phalli. This is presumably a simple functional relationship, because during copula the apex of the phallus must extend through the ovipositor to reach the appropriate place within the female abdomen; so if the ovipositor is long, the phallus in turn needs to be very long. However, the longer it is, perhaps the more likely it is to be damaged (broken pieces of phalli are sometimes found in dissected females). Thus there will be a selective advantage in the phallus not being any longer than is vital and so a correlation with ovipositor length would be expected. Table 20.4 lists the known ranges of aculeus and phallus length for each of the species for which both data were available, together with information on the aculeus tip shape (data on other species can be extracted from White and Hancock 1997). The correlation coefficient was found to be 0.66 when all 22 species were considered. It is clear that the measured specimens of *B.* (*Bactrocera*) *pedestris* (Bezzi) and *B.* (*Bactrocera*) *propinqua* (Hardy and Adachi) had a much longer phallus than would be predicted from aculeus length, although these species have unusual aculeus tip shapes and the samples were small ($n = 1$ or 2). When only the 16 species with a simple caudate aculeus tip were considered the correlation coefficient was 0.92, indicating a very strong association of phallus and aculeus length within a single category of aculeus tip shape. A similar analysis was carried out by Iwaizumi et al. (1997) who found a correlation of 0.97 for nine species sampled across the entire genus, eight of which were recorded by White and Hancock (1997) as having a pointed aculeus (although not necessarily caudate). It should be noted that the measurements given and discussed by Iwaizumi et al. (1997) as "aedeagus" (i.e., phallus) length are minus the glans. That is probably easier to measure accurately than the entire phallus as used by White and Hancock (1997), who also included the short glans which has a membranous (and so ill-defined) apex and can be bent at right angles to the phallus. Consequently, this difference of method should be noted if comparing the two sources of values. It should also be noted that phallus length gives a far less perfect separation of the species than aculeus length, probably due to a high error level incurred while trying to measure a normally coiled tube (White and Hancock (1997) explained how to roughly straighten the phallus).

Most Dacini for which data are available have a simple pointed aculeus tip shape, although trilobed and other more complex shapes are also known. The reason for the evolution of different aculeus tip shapes is unknown. One would expect a simple point to be most efficient if fruit piercing were the only factor (the sawlike shapes common in *Anastrepha* are very rare in the Dacini). Tip shape is also under strong genetic control. If the shape of one specimen is drawn at high magnification and then viewed against other specimens of the same species with the aid of a drawing tube, it will match exactly, even if the other specimens have a longer or shorter aculeus (i.e., the apex shape and size are even more controlled than overall length).

**TABLE 20.4**
**Aculeus Type and Length, and Phallus Length for Species in the Oriental Fruit Fly Species Complex**

| | Aculeus Tip Shape | Aculeus Length (mm) | Phallus Length (mm) |
|---|---|---|---|
| *B. arecae* (Hardy and Adachi) | Caudate | 1.0–1.2 | 2.6 |
| *B. cacuminata* (Hering) | Caudate | 1.2–1.4 | 2.8 |
| *B. carambolae* Drew and Hancock | Caudate | 1.3–1.6 | 2.5–3.2 |
| *B. caryeae* (Kapoor) | Caudate | 1.56–1.78 | 3.2 |
| *B. dorsalis* (Hendel) | Caudate | 1.3–1.80 | 2.8–3.5 |
| *B. endiandrae* (Perkins and May) | Caudate | 1.0–1.3 | 2.9 |
| *B. irvingiae* Drew and Hancock | Caudate | 1.8–2.0 | 3.4 |
| *B. kandiensis* Drew and Hancock | Caudate | 1.6–1.84 | 3.3 |
| *B. occipitalis* (Bezzi) | Caudate | 1.3–1.7 | 2.7–3.2 |
| *B. opiliae* (Drew and Hardy) | Caudate | 1.40 | 3 |
| *B. papayae* Drew and Hancock | Caudate | 1.5–2.1 | 3.0–3.6 |
| *B. philippinensis* Drew and Hancock | Caudate | 1.6–2.1 | 3.2–4 |
| *B. pyrifoliae* Drew and Hancock | Caudate | 1.96 | 3.3 |
| *B. raiensis* Drew and Hancock | Caudate | 1.87–2.3 | 3.9 |
| *B. trivialis* (Drew) | Caudate | 1.80 | 3.3 |
| *B. verbascifoliae* Drew and Hancock | Caudate | 1.2–1.4 | 2.9 |
| *B. melastomatos* Drew and Hancock | Blunt | 1.16 | 3.2 |
| *B. osbeckiae* Drew and Hancock | Blunt | 1.20 | 3.6 |
| *B. cognata* (Hardy and Adachi) | Tapered | 1.24–1.36 | 2.7–3.3 |
| *B. pedestris* (Bezzi) | Tapered | 1.83–1.94 | 4.8 |
| *B. propinqua* (Hardy and Adachi) | Trilobed | 1.84–2.1 | 4.5 |
| *B. quasipropinqua* Drew and Hancock | Trilobed | 1.58 | 3.4 |

*Note:* Data from White and Hancock (1997). Only those species for which both measurements were available are listed. For most species only a single or few specimens were measured, but for the pest species 30 to 100+ specimens were measured by K. Mahmood (unpublished postgraduate studies, 1993 to 1996).

Other factors that might influence tip shape should be considered. Eberhard and Pereira (1993) reported that during copula males of *Ceratitis capitata* used their surstyli to hold the aculeus tip, which tempts the suggestion that there would be selection pressure to evolve easily gripped aculeus tip shapes. However, work on the Tephritinae by Headrick and Goeden (Chapter 25) shows that the aculeus is gripped too far from the apex for that to be likely to have any influence on tip shape. Some genitalic structures in other insects are believed to convey species identity through touch against receptors in the mate (normally complex male structures rather than female, e.g., in some Coleoptera; Hammond 1972). During copula the aculeus tip comes to rest behind/below the epandrium (see work on Tephritinae, Headrick and Goeden 1994) and the touch of a complex aculeus tip just might be involved in some exchange of information between the sexes. The aculeus tip of a tephritid does bear a series of sensory structures, including preapical setulae arranged in grooves that have been shown to function both as gustatory and tactile sensilla (see Stoffolano 1989 and Rice 1989 for reviews). The need to hold the sensory structures in functionally useful positions, pierce the fruit, and perhaps communicate some information, may all be selected for in the development of aculeus tip shape.

The simple pointed aculeus tip shape is by far the most common in the Dacini. White and Hancock (1997) gathered available information on aculeus tip shape and an analysis of their data matrix shows that out of 204 species that have been studied, 76% have a simple point, 15% are trilobed, and just 9% have the six other tip shapes they defined. The distribution of the two main types differs between the two major subgenera, as follows: of 106 subgenus *Bactrocera* spp., 85% have simple points and 7% are trilobed; of 32 *B.* (*Zeugodacus*) spp., 50% have simple points and 44% are trilobed. Furthermore, all of the polyphagous pest species from all subgenera have a simple pointed aculeus tip. The distribution of preapical setulae (sensilla) may also differ among subgenera. White and Elson-Harris (1994) illustrated the aculei of the pest species by drawings of optical sections in which even very small setulae are visible (SEM illustrations of the tip in ventral view do not show the smaller structures as they are sometimes hidden in a groove). Most subgenus *Bactrocera* spp. have four pairs of setulae, close together and of similar length (~0.01 to 0.02 mm), an exception being *B.* (*Bactrocera*) *musae* which appears to have a very small fifth pair of setulae (~0.003 mm) placed proximally relative to other pairs. *Bactrocera* (*Notodacus*) *xanthodes* (Broun) also has this small structure. *Bactrocera* (*Tetradacus*) *minax* has five pairs of setulae but all of similar, short (~0.01 mm) length and grouped together. All of the pest species of subgenus *Bactrocera* (*Zeugodacus*), plus *Bactrocera* (*Hemigymnodacus*) *diversa* and *Bactrocera* (*Austrodacus*) *cucumis* (French), have four long pairs of setulae placed together, of which the two most apical pairs are often very long (~0.08 mm), plus a fifth pair, of smaller size (~0.01 mm), placed proximally relative to the others. All of the latter group infest Cucurbitaceae. The pest species of *Dacus*, which also infest Cucurbitaceae, all have a long, slender, pointed aculeus tip with three or four small pairs of setulae (~0.01 mm), in some cases arranged with the fourth pair separated some distance proximally from the remaining three. *Bactrocera* (*Daculus*) *oleae* (Rossi), which is restricted to attacking *Olea* spp., is exceptional in that it appears to lack any setulae (none can be seen using optical microscopy). These data suggest that the arrangement of setulae may differ between groups of subgenera and therefore may be of value in the formation of the generic and subgeneric classification.

## 20.7 OTHER FEATURES USED IN THE GENERIC AND SUBGENERIC CLASSIFICATION

The function of the remaining structures used in the generic and subgeneric classification of Dacini is uncertain. The separation of the genera *Dacus* and *Bactrocera* is based on the abdominal tergites (tgf in Table 20.1) being fused or unfused, respectively, although some *B.* (*Sinodacus*) spp. show at least partial fusion and their status requires more-detailed study — some species at present placed in *B.* (*Asidacus*) and *B.* (*Tetradacus*) present the same problem. Tergite fusion may have evolved as a modification for drier habitat types and it is certainly true that most *Bactrocera* spp. come from areas of the humid Indo-Australasian tropics (very few are African) while most *Dacus* spp. come from Africa, and very few of those are from rain forest areas. Another possibility is that fused terga may confer an advantage for oviposition into thick-skinned fruits such as the pods of Asclepiadaceae. It is likely that the African *Dacus* spp. are a monophyletic group that have evolved in relative isolation from the Indo-Australasian fauna, and the status of the few Indo-Australasian species with fused terga (currently in *Dacus*) and the few African species with unfused terga (currently in *Bactrocera*) requires further study with an open mind as to the possibility that tergite fusion may have evolved in separate lines, and perhaps even been secondarily lost in some cases.

The presence or absence of postsutural supra-alar, prescutellar acrostichal, and basal scutellar setae have been used to a considerable degree in the formation of the subgeneric classification (asp, prs, 4sc in Table 20.1). However, when a large enough sample of almost any species is studied, some or all of these characters prove to have some intraspecific variation. For example, *B. cucurbitae* belongs to subgenus *Zeugodacus* and should therefore have both postsutural supra-alar and prescutellar acrostichal setae, although it normally lacks the basal scutellar setae so typical of most *Zeugodacus* spp. When a long series ($n = 351$) of *B.* (*Zeugodacus*) *cucurbitae* was studied, the author found that 95.7% had postsutural supra-alar setae (not 100%), 98.9% had prescutellar acrostichal setae (not 100%), and 3.4% had basal scutellar setae (not 0%). The fact that species which normally lack a particular seta can sometimes "regain" it indicates that the genotypic description for the structure exists even in those species that normally lack the feature in their phenotype. Such characters are of little phylogenetic significance as they are likely to have "come and gone" almost at random through evolutionary time and subgenera defined by such homoplasious features of chaetotaxy are therefore of no real value. However, some general trends do occur. In particular, all species with fused terga lack prescutellar acrostichal setae and basal scutellar setae are a common feature of the *Zeugodacus* group of subgenera, in common with medial vittae and not usually responding to methyl eugenol.

The structure of the male terminalia in *Bactrocera* spp. can be divided into two distinct types, which implies that the details of copula may differ between the groups in which they occur. The posterior lobes of the male's surstylus are usually either very short or very long (sur in Table 20.1), and this distinction separates the *Bactrocera* and *Zeugodacus* groups of subgenera, respectively. Largely concordant with this distinction, species in the *Bactrocera* group of subgenera have a deep V-shaped emargination in the hind margin of sternite 5, whereas species of the *Zeugodacus* group of subgenera have only a shallow inward curve (against which the surstyli and other external parts of the terminalia normally rest) (S5V in Table 20.1). However, eight species "break the rules" by having long surstylus lobes combined with a deep emargination (subgenus *Queenslandacus*) or short surstylus lobes combined with a shallow emargination (subgenera *Hemigymnodacus*, *Hemisurstylus,* and *Melanodacus*). Drew (1989) called these the *Queenslandacus* and *Melanodacus* groups of subgenera, respectively, and the existence of these taxa indicates that the emargination of sternite 5 and the size of the posterior surstylus lobes can be regarded as independent characters for cladistic analysis.

## 20.8 PRELIMINARY CLADISTIC ANALYSIS

The large data matrix (507 species, 169 characters) that was developed for the identification of Indo-Australasian Dacini by White and Hancock (1997) constitutes a good starting point for phylogenetic study. Although a detailed study was beyond the scope of this chapter, a preliminary analysis of 51 species (all those categorized as pests by White and Elson-Harris 1994) was carried out, based largely on a subset of the characters used by White and Hancock (1997). That selection of species may be assumed to be independent of the present classification, as defined by White and Hancock (1997), who largely followed Drew (1989). Initially 53 characters were exported into Hennig86 data format (Farris 1988). Nonmorphological characters, primarily distribution, host, and lure response, were excluded, as were measurements. Ordered multistate characters which the White and Hancock (1997) system treated as "fuzzy" were either recoded (as present, absent, or variable) or removed (e.g., extent of anepisternal stripe and color of scutum). Characters only scored for a minority of taxa and complex unordered multistate characters, such as femora, scutellum, and tergite color characters, were also excluded. Many of these excluded characters would be important for elucidating relationships within subgenera and the assumption that they could be safely excluded for the purpose of this analysis was the only *a priori* assumption made. Characters describing the presence of a lamprine and tergite 6 reduction* were added, making 55 characters. The data were then "packed" to remove characters which did not vary among the chosen subset of taxa, or which constituted autapomorphies, leaving 38 characters (Tables 20.5 and 20.6). All remaining characters were given an initial weight of 1 and multistate characters were regarded as unordered so as not to make any *a priori* assumptions about the relationships of the states. Some species were identical to each other with respect to the remaining characters and in these cases only one species from each set of initially coded species was retained in the matrix, as follows: *B.* (*Bactrocera*) *dorsalis*,

* This character was taken to represent all fully correlated features of *Dacus* plus *Bactrocera*, e.g., deep cell bm and very long extension of cell bcu.

vs. *B.* (*Bactrocera*) *carambolae*, *B.* (*Bactrocera*) *occipitalis*, *B.* (*Bactrocera*) *papayae*, and *B.* (*Bactrocera*) *philippinensis*; *B.* (*Bactrocera*) *caryeae* vs. *B.* (*Bactrocera*) *kandiensis* Drew and Hancock; and *D.* (*Didacus*) *frontalis* Becker vs. *D.* (*Didacus*) *vertebratus* Bezzi.

The purpose of this preliminary analysis was to ascertain the extent to which a cladogram based on maximum parsimony did, or did not, support the major groupings defined by other recently proposed classifications of the Dacini (Drew 1989; Munro 1984) (see Figures 20.5 and 20.6). The taxonomic representation (9% of species) is insufficient to permit revision of the classification and the character set too restrictive to make for meaningful detail within groups of subgenera. A full review of earlier classifications is not given here (for that, see Drew 1989), although a few points are worthy of mention. Hendel (1914) used free and fused terga (character 25) but Bezzi (1915) wrote, "It is indeed very difficult to find a dividing line between the species with free and those with fixed abdominal segments." He then went on to say, "On the other hand, I have found a better character for dividing the Ethiopian [i.e., Afrotropical] species in the thoracic chaetotaxy" (characters 13, 14). These characters are still used as the criteria for the separation of many subgenera, together with male terminalia characters (31, 32), which were first used by Drew (1972). Some earlier classifications also used the presence or absence of a supernumerary lobe formed by an indent in the hind margin of the male wing (33) (see Hardy 1955 for discussion).

Candidate outgroups considered were *Monacrostichus citricola* Bezzi (to represent that genus of two species), *Ichneumonopsis burmensis* Hardy, and a member of the Ceratitidini (another tribe of Dacinae). It was decided that *Monacrostichus* was itself a taxon whose placement should be part of the study. Because the inclusion of *I. burmensis* within the Dacini is the subject of debate (see introduction) it was not chosen. Instead, *Ceratitis capitata* was chosen to represent the Ceratitidini whose close relationship to the Dacini is supported by some morphological evidence (Hancock 1986) and by a molecular cladogram (Han and McPheron 1997). Some characters were difficult to apply to *C. capitata* and the following decisions were taken: character 2, the postpronotal lobe of *C. capitata* is marked but not in a manner similar to any Dacini, so it was scored as unmarked; 18, the narrow part of cell br has microtrichia but not in the dense ridge formed in the dacine wing, so this was scored as no; 19, 23, 24, the wing is patterned in *C. capitata* but not in the manner seen in any Dacini so these patterns were assumed to be nonhomologous and scored absent; 33, the indent at the end of vein $A_1$+$Cu_2$ (the supernumerary lobe) is a specifically male feature in Dacini but is more or less present in both sexes of *C. capitata* and so was regarded as nonhomologous and scored absent.

The data were analyzed using NONA (Goloboff 1993), which is similar to Hennig86 (Farris 1988) except that it does not store multiple copies of ambiguously supported trees (I.J. Kitching, personal communication) (Figure 20.2). A second analysis was also carried out using PIWE (Parsimony Analysis with Implied Weights); Goloboff 1997), which applied successive approximations weighting, that is, characters which had a good fit to the initial cladogram were given high weight while those that showed a high level of homoplasy received low weight; the method repeats the reweighting process until a stable cladogram is produced (see Siebert 1992 for discussion) (Figure 20.3). Both analyses were carried out using the recommended options for making a series of heuristic cladograms based on randomly rearranged data matrices (hold 100; hold/20; mult*15;), followed by branch swapping (max*;) and the formation of a strict consensus tree (nelsen;); for PIWE the default concavity ($k = 3$) was used. In each case 100 minimum-length trees were stored and a consensus tree constructed. The length of each character on each cladogram, implied weights, and the manner in which previous classifications (Munro 1984; Drew 1989) used the characters, are listed in Table 20.7.

Both resulting cladograms supported the previous classifications (see Figures 20.5 and 20.6) in placing *Monacrostichus* as the sister group to *Bactrocera* plus *Dacus*, which together share some presumed apomorphies that are lacking in *Monacrostichus*, namely, the development of ceromata (character 29), the presence of a pecten and lamprine (30/36), and the female sixth tergite being hidden beneath tergite 5 (37). Both ceromata and the pecten/lamprine complex of structures are

## TABLE 20.5
## Characters Used for Cladistic Analysis

| | |
|---|---|
| 0 | Antenna: Pedicel + first flagellomere longer than ptilinal suture |
| — | Antenna: Arista plumose |
| 1 | Face with a dark spot in each antennal furrow |
| — | Face with an extra (upper) pair of dark spots |
| — | Face pattern (other than spot); 0 = none/spots only; 1 = line over mouth; 2 = black |
| — | Face with upper line |
| — | Dark mark between eye and antenna base |
| 2 | Postpronotal (= humeral) lobe color; 0 = yellow; 1 = part dark; 2 = dark |
| — | Yellow spot laterad of postpronotal lobe |
| 3 | Presutural part of scutum with lateral yellow/orange vittae (stripes) |
| 4 | Notopleuron colored |
| 5 | Notopleural suture with isolated wedge-shaped mark |
| 6 | Scutum with lateral postsutural vitta (yellow/orange stripe) |
| 7 | Lateral postsutural vitta extended anterior to suture |
| 8 | Scutum with a medial vitta or stripe |
| 9 | Scutellum extensively marked (patterned) |
| 10 | Scutellum with a deep basal band |
| — | Scutellum dark (concolorous with scutum) |
| — | Scutellum bilobed |
| 11 | Yellow marking on calli; 0 = both anatergite and katatergite; 1 = katatergite only |
| 12 | Postpronotal lobe (= humerus) with a seta |
| — | Notopleuron with anterior seta |
| 13 | Scutum with postsutural supra-alar seta |
| 14 | Scutum with prescutellar acrostichal seta |
| 15 | Scutellum with basal as well as apical setae |
| — | Vein $R_{2+3}$ sinuate near end of $R_1$ |
| — | Vein $R_{4+5}$ setulose |
| — | Cell bm tapered to base |
| — | Cell dm expanded apically |
| 16 | Cell bc with extensive covering of microtrichia |
| 17 | Cell c with extensive covering of microtrichia |
| 18 | Cell br (narrowed part) with extensive covering of microtrichia |
| 19 | Wing with a complete costal band |
| 20 | Wing with an anal streak |
| 21 | Cell bc colored |
| — | Cell c colored [same distribution as 21 for the selected taxa] |
| — | Transverse mark along crossvein BM-Cu |
| 22 | Transverse or sinuate mark covering both crossveins R-M and DM-Cu |
| 23 | Transverse mark over crossvein R-M (not direct to any on DM-Cu) |
| 24 | Transverse mark over crossvein DM-Cu (not direct to any on R-M) |
| — | Fore femur with stout ventral spines |
| 25 | Tergites fused |
| 26 | Abdomen wasp-waisted |
| 27 | Wasp waist — tergite 1 much longer than broad |
| 28 | Medial longitudinal stripe on tergite 4 |
| 29 | Ceromata present |
| — | Male wing with a bulla |
| 30 | Male tergite 3 with a pecten (setal comb) on each side |
| 31 | Male sternite 5 V-shaped |
| 32 | Lateral surstylus (male) with a long posterior lobe |
| 33 | Wing (male) with a deep indent at end of $A_1$+$Cu_2$ (supernumerary lobe) |
| 34 | Wing (male) with microtrichia area adjacent to cell bcu extension |
| 35 | Hind tibia (male) with a preapical pad |
| 36 | Lamprine or at least vestigial pecten/lamprine present |
| 37 | Female with tergite 6 hidden under tergite 5 |

*Note:* The 55 characters listed were initially used. The "pack data" option reduced that to 38 by discarding characters that were autapomorphies, identical to others, or not applicable to the taxa. Unless otherwise indicated, only the "present" state (coded 1) is listed.

**TABLE 20.6**
**Descriptions of 45 Pest Species of Dacini, plus *C. capitata*, in Terms of 38 Characters**

| | Character Number | | | | | | | |
|---|---|---|---|---|---|---|---|---|
| | | | 11111 | 11111 | 22222 | 22222 | 33333 | 333 |
| | 01234 | 56789 | 01234 | 56789 | 01234 | 56789 | 01234 | 567 |
| *C. capitata* | 00001 | 00000 | 00111 | 11100 | 00000 | 00000 | 00010 | 000 |
| *M. citricola* | 110?0 | ?1110 | 10010 | 01111 | 01000 | 01000 | 0?000 | 000 |
| *B. (Af.) jarvisi* | 01011 | 01000 | 00001 | 00011 | 10000 | 00011 | 11011 | 111 |
| *B. (Au.) cucumis* | 0100? | 01110 | 00000 | 10011 | 10000 | 00001 | 001?0 | 011 |
| *B. (B.) albistrigata* | 01101 | 0100? | 10011 | 0011? | 1?100 | 00011 | 11011 | 111 |
| *B. (B.) aquilonis* | 01001 | 01000 | 00011 | 00111 | 11000 | 000?1 | 11011 | 111 |
| *B. (B.) caryeae* | 01101 | 01000 | 00011 | 000?1 | 10000 | 00011 | 11011 | 111 |
| *B. (B.) correcta* | 0?001 | 01000 | 00011 | 00000 | 00000 | 00011 | 11011 | 111 |
| *B. (B.) curvipennis* | 01001 | 01000 | 00011 | 00111 | 11010 | 00001 | 11011 | 111 |
| *B. (B.) distincta* | 01001 | 0100? | 00011 | 00111 | 11100 | 00011 | 11011 | 111 |
| *B. (B.) dorsalis* | 01001 | 01000 | 00011 | 00011 | 10000 | 00011 | 11011 | 111 |
| *B. (B.) facialis* | 00001 | 0?00? | 10011 | 00011 | ?0000 | 00011 | 11011 | 111 |
| *B. (B.) frauenfeldi* | 01?01 | 01001 | 00011 | 0011? | 1?100 | 00011 | 11011 | 111 |
| *B. (B.) kirki* | 01?01 | 00001 | 00011 | 00?1? | 100?? | 00011 | 11011 | 111 |
| *B. (B.) latifrons* | 01001 | 01000 | 00011 | 00011 | 10000 | 000?1 | 11011 | 111 |
| *B. (B.) melanotus* | 0??00 | 00000 | 00011 | 00011 | 000?? | 00001 | 11011 | 111 |
| *B. (B.) musae* | 01001 | 01000 | 00011 | 00011 | 10000 | 000?1 | 11011 | 111 |
| *B. (B.) neohumeralis* | 01201 | 01000 | 00011 | 00111 | 11000 | 00011 | 11011 | 111 |
| *B. (B.) passiflorae* | 00201 | 00000 | ?0011 | 00011 | ?0000 | 000?1 | 11011 | 111 |
| *B. (B.) psidii* | 0??01 | 01001 | 00011 | 0001? | ?00?? | 00001 | 11011 | 111 |
| *B. (B.) trivialis* | 01001 | 01000 | 00011 | 00011 | 10000 | 000?1 | 11011 | 111 |
| *B. (B.) tryoni* | 01001 | 01000 | 00011 | 00111 | 11000 | 00011 | 11011 | 111 |
| *B. (B.) tuberculata* | 01001 | 01000 | 00011 | 00000 | 00000 | 00001 | 11011 | 111 |
| *B. (B.) umbrosa* | 01001 | 01000 | 00011 | 00?11 | 11100 | 000?1 | 11011 | 111 |
| *B. (B.) zonata* | 01001 | 01000 | 00011 | 00000 | 00000 | 000?1 | 11011 | 111 |
| *B. (D.) oleae* | 01001 | 0000? | 11000 | 00000 | 00000 | 00001 | 1?011 | 111 |
| *B. (H.) diversa* | 0?001 | 01110 | 00011 | ?0011 | 10000 | 000?1 | 00110 | 011 |
| *B. (N.) xanthodes* | 01?10 | 01111 | 01111 | 00011 | 00000 | 00001 | 11011 | 111 |
| *B. (Pr.) decipiens* | 01001 | 01110 | 00?10 | 10011 | 1?011 | 00001 | 10111 | ?11 |
| *B. (Pt.) atrisetosa* | 01001 | 01110 | 00011 | 10011 | 1?000 | 000?1 | 00110 | 011 |
| *B. (T.) minax* | 0101? | 01010 | 00000 | 00?11 | ??000 | 01011 | 11000 | 011 |
| *B. (T.) tsuneonis* | 01011 | 01010 | 00010 | 00?11 | 01000 | 01011 | 11000 | 011 |
| *B. (Z.) caudata* | 00001 | 01110 | 00011 | ?0011 | 10000 | 00011 | 10111 | 111 |
| *B. (Z.) cucurbitae* | 01001 | 01110 | 000?1 | ?0011 | 100?1 | 00011 | 10111 | 111 |
| *B. (Z.) depressa* | 01001 | 01110 | 00?1? | 10011 | 10000 | 00011 | 10111 | 111 |
| *B. (Z.) tau* | 01001 | 01110 | 00011 | 10011 | 10000 | 00011 | 10111 | 111 |
| *D. (C.) axanus* | 11001 | 1000? | 110?0 | 01111 | 11000 | 111?1 | 10001 | 011 |
| *D. (C.) longicornis* | 1??01 | 100?0 | ?1010 | 01111 | ?1000 | 11101 | 10001 | 011 |
| *D. (C.) solomonensis* | 1?001 | 10000 | 01010 | 01111 | 11000 | 11101 | 10000 | 011 |
| *D. (D.) bivittata* | 01?01 | ?1?10 | 00010 | 00111 | 1?000 | 1?001 | 10001 | 111 |
| *D. (D.) demmerezi* | 01001 | 10010 | 00010 | 00?11 | 10010 | 1?0?1 | 10001 | 111 |
| *D. (D.) punctatifrons* | 01001 | 01110 | 10010 | 00?11 | 100?0 | 100?1 | 10001 | 111 |
| *D. (D.) telfaireae* | 0?101 | 0?011 | 01010 | 00011 | 10010 | 1?001 | 10001 | 111 |
| *D. (Di.) ciliatus* | 01001 | ?0000 | 01000 | 00011 | ?0000 | 10001 | 10001 | 111 |
| *D. (Di.) frontalis* | 01001 | 10000 | ?0000 | 00011 | 10000 | 100?1 | 10001 | 111 |
| *D. (Di.) lounsburyii* | 01201 | 1101? | 11000 | 00?11 | 10000 | 110?1 | 10001 | 111 |

*Note:* Characters are grouped into blocks of five. ? = variable, intermediate, or unknown states. All characters were analyzed as unordered. Zero marks the absent state unless otherwise indicated in Table 20.5.

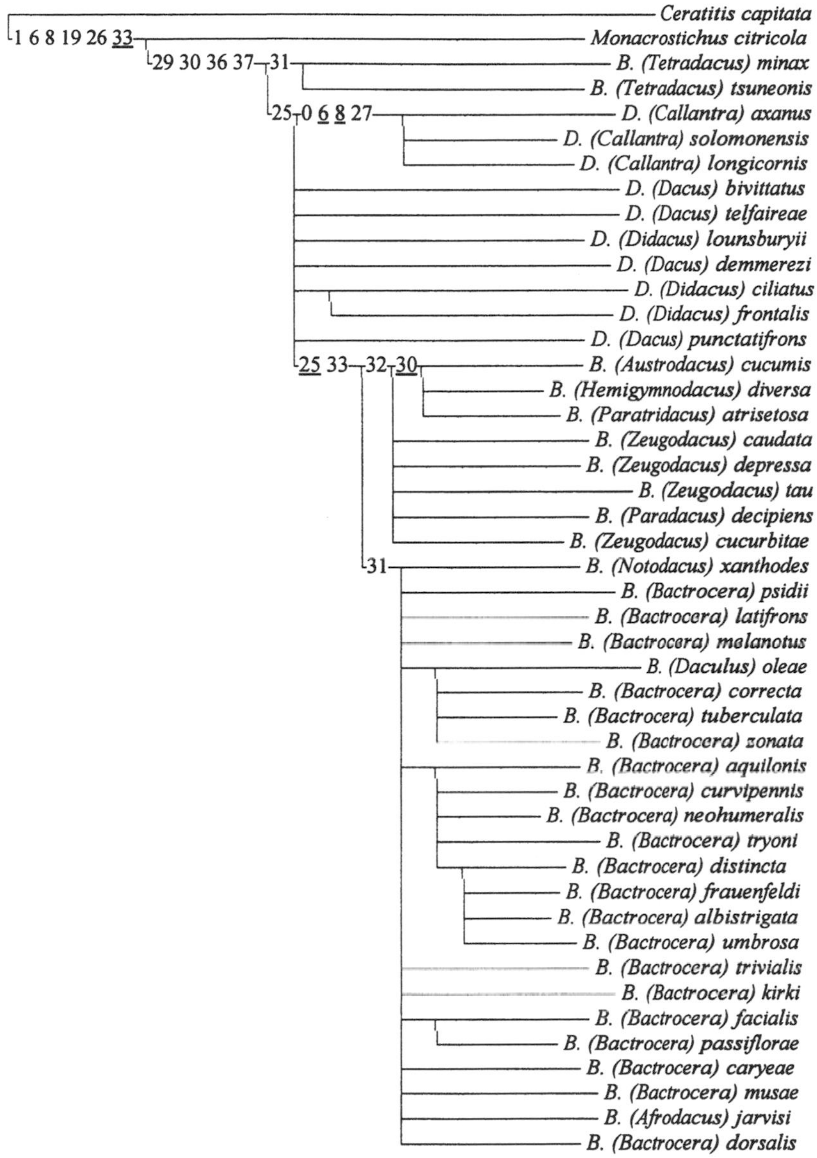


**FIGURE 20.2** Strict consensus cladogram produced using NONA (Goloboff 1993) to analyze the 51 pest species of Dacini, plus *C. capitata* as outgroup. (Data on Dacini based on White and Hancock 1997.) Length/Fit = 111; consistency index (ci) = 0.35; retention index (ri) = 0.70. Although it is not possible to plot character state changes properly on a consensus tree, where practical, characters of interest for comparison with the classifications of Munro (1984) and Drew (1989) are indicated using the numbering from Table 20.5; reversals underlined.

unique to this group of tephritids and the hiding of tergite 6 beneath tergite 5 is a feature that is rare and probably also unique to these genera.

The unweighted analysis placed *B.* (*Tetradacus*) spp. and *D.* (*Callantra*) spp. near the base of the cladogram. These subgenera share with *Monacrostichus* spp. the unusual feature of a wasp-waisted abdomen. Munro (1984) associated these two taxa (see Figure 20.5) but the current classification based on Drew (1989) places *B.* (*Tetradacus*) spp. within the *Bactrocera* group of subgenera (see Figure 20.6), and that is supported by the weighted analysis (see Figure 20.3).

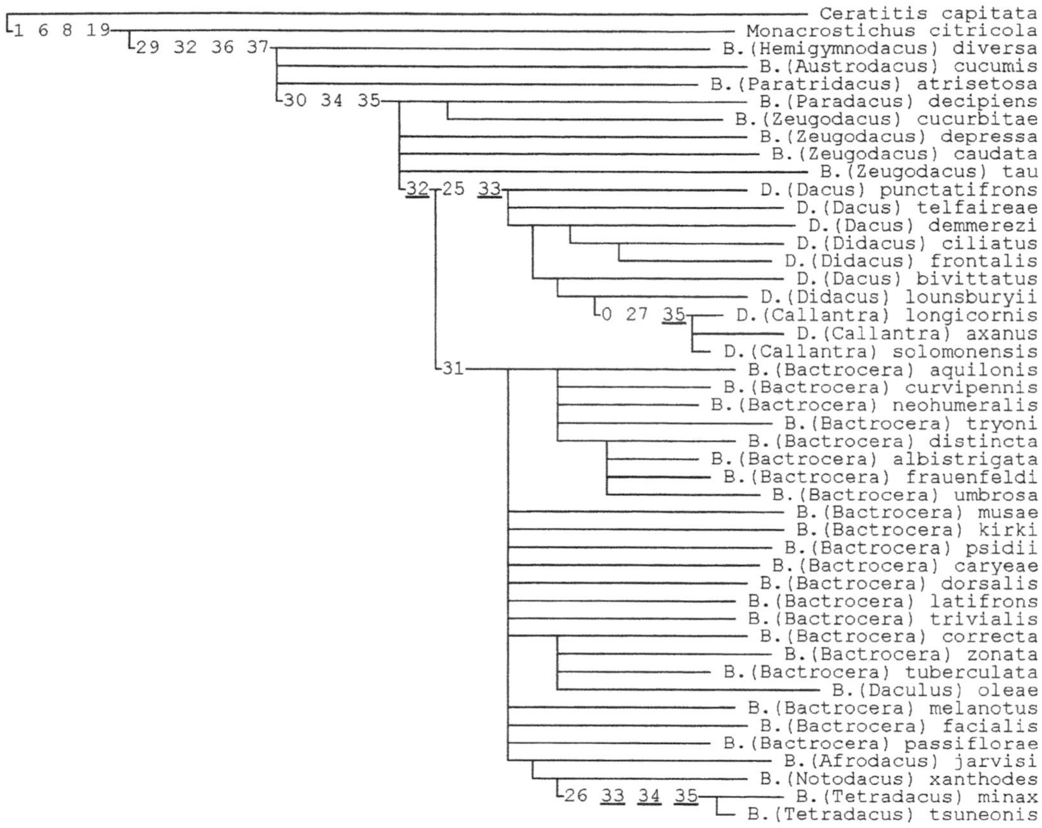


**FIGURE 20.3** Strict consensus cladogram produced using PIWE (Goloboff 1997) to analyze the 45 pest species of Dacini, plus *C. capitata* as outgroup. Data on Dacini based on White and Hancock (1997). Length = 114; Fit = 260.5. Although it is not possible to plot character state changes properly on a consensus tree, where practical, characters of interest for comparison with the classifications of Munro (1984) and Drew (1989) are indicated using the numbering from Table 20.5; reversals underlined.

Following these analyses the likelihood of the wasp-waist (character 26) in these taxa being homologous was reconsidered. To judge from drawings of the species here analyzed, such as those provided by White and Elson-Harris (1994), it would appear that in *D.* (*Callantra*) spp. the main point of constriction is at the join of tergites 1 and 2, and the basal "knobs" are well developed and prominent, while in *B.* (*Tetradacus*) spp. and *M. citricola,* the constriction and basal "knobs" are less distinct. However, examination of a wide range of species, not just those included in this analysis, indicated that the range of interspecific variation is considerable; all species have the "knobs," and on present evidence the structures have to be regarded as a single character for all these taxa. The unweighted analysis may also have been influenced by the fact that the three *D.* (*Callantra*) spp. all lack male hind tibial pads (character 35), in common with the included *B.* (*Tetradacus*) spp. However, some other (nonincluded) species in these groups have this feature. Thus, more detailed examination (dissection) of the wasp-waisted abdomen is needed to verify homology and a future analysis should also include more species from these subgenera.

The unweighted analysis showed *Dacus* as a paraphyletic group near the base of the cladogram (Figure 20.2); that is, tergite fusion has supposedly evolved and then later reversed. However, the weighted analysis (Figure 20.3) supported the presumed monophyly of *Dacus*, and hypothesizes that tergite fusion has evolved once and not been reversed. The genus *Dacus* is unique among the Tephritidae in having fused terga (character 25) and might therefore be expected to be a monophyletic

## TABLE 20.7
## The Comparative "Value" Accorded to Each Character in Each of the Compared Classifications

| Character | NONA Steps | PIWE Steps | PIWE Final Weight | Rank Order in Which Used in Classification | |
|---|---|---|---|---|---|
| | | | | Munro (1984) | Drew (1989) |
| 37 | 1 | 1 | 10.0 | 1 | 1 |
| 29 | 1 | 1 | 10.0 | 1 | 1 |
| 25 | 2 | 1 | 10.0 | 3 | 2 |
| 31 | 2 | 1 | 10.0 | | 3 |
| 30 | 2 | 1 | 10.0 | 4 | |
| 5 | 3 | 1 | 10.0 | | |
| 3 | 3 | 1 | 10.0 | | |
| 36 | 1 | 1 | 10.0 | | |
| 27 | 1 | 1 | 10.0 | | |
| 24 | 1 | 1 | 10.0 | | |
| 22 | 1 | 1 | 10.0 | | |
| 32 | 1 | 2 | 7.5 | | 3 |
| 0 | 2 | 2 | 7.5 | 2 | |
| 19 | 2 | 2 | 7.5 | | |
| 18 | 2 | 2 | 7.5 | | |
| 16 | 2 | 2 | 7.5 | | |
| 15 | 2 | 2 | 7.5 | | |
| 12 | 2 | 2 | 7.5 | | |
| 26 | 2 | 3 | 6.0 | 2 | |
| 34 | 3 | 3 | 6.0 | | |
| 20 | 3 | 3 | 6.0 | | |
| 4 | 3 | 3 | 6.0 | | |
| 1 | 3 | 3 | 6.0 | | |
| 35 | 2 | 3 | 6.0 | | |
| 33 | 2 | 3 | 6.0 | | |
| 17 | 2 | 3 | 6.0 | | |
| 9 | 5 | 4 | 5.0 | | |
| 23 | 4 | 4 | 5.0 | | |
| 21 | 3 | 4 | 5.0 | | |
| 7 | 3 | 4 | 5.0 | | |
| 2 | 6 | 6 | 4.2 | | |
| 11 | 5 | 5 | 4.2 | | |
| 8 | 4 | 5 | 4.2 | | |
| 13 | 6 | 6 | 3.7 | 4 | |
| 14 | 5 | 6 | 3.7 | 4 | |
| 6 | 6 | 6 | 3.7 | | |
| 10 | 7 | 7 | 3.3 | | |
| 28 | 6 | 8 | 3.0 | | |

*Note:* The table is ordered by decreasing magnitude of the final weights assigned to the characters following successive approximations weighting with PIWE. Column 1 = character number as listed in Table 20.5; column 2 = number of steps (changes) of the character on the cladogram produced by NONA; column 3 = number of steps (changes) of the character on the cladogram produced by PIWE; final weights assigned to each character following successive approximations weighting with PIWE; column 4 = rank order of use of each character by Munro (1984); column 5 = rank order of use of each character in the classification based on Drew (1989).

group, leaving *Bactrocera* as a possibly paraphyletic group delimited by the plesiomorphic state of the same character. However, as noted earlier in this chapter, tergite fusion is not as distinct as the present classifications based on Drew (1989) and Munro (1984) imply (see Figures 20.5 and 20.6), and that suggests that the character either needs better definition or the present classification is flawed. A study involving the dissection of the abdomen of a wide variety of species would therefore be advisable to redefine the tergite fusion character. Munro (1984) did carry out a detailed study of abdominal characters, but that was largely confined to African *Dacus* spp.

Drew (1989) divided *Bactrocera* spp. into four groups of subgenera based on all four possible permutations of two male characters, namely, shape of sternite 5 (character 31) and lateral surstylus shape (character 32). Most species (99%) belong to just two of these groups, namely, the *Bactrocera* group of subgenera (31 = 1, 32 = 0) or the *Zeugodacus* group of subgenera (31 = 0, 32 = 1), and both were represented in these analyses. The unweighted analysis (Figure 20.2) supported the *Zeugodacus* group (subgenera *Austrodacus*, *Hemigymnodacus*, *Paradacus*, *Paratridacus*, *Zeugodacus*) defined by the presumed apomorphy of the long posterior lobe of the surstylus (32 = 1). However, the weighted analysis (Figure 20.3) placed the *Zeugodacus* group of subgenera as a paraphyletic group near the base of the cladogram; that is, the long surstylus lobe must later have reversed. Both of the analyses detailed here supported the *Bactrocera* group of subgenera (i.e., *Afrodacus*, *Bactrocera*, *Daculus*, *Notodacus*, and *Tetradacus*), although *Tetradacus* was excluded by the unweighted analysis.

Although it might be expected that a parsimony-based analysis of the Dacini would to some extent support the genera and groups of subgenera defined by traditional approaches, it would not be expected to find support for the myriad of subgenera at present recognized. The classification of *Bactrocera* based on Drew (1989) now recognizes 29 subgenera, 28 of which are based on present/absent permutations of just five characters (13, 14, 30, 31, 32). Arithmetically there are 32 possible permutations of these characters and 88% of them are found in nature. Munro (1984) called what we refer to here as the genus *Dacus*, the supertribe "Dacina," having elevated what is at present called the tribe Dacini to family rank (Dacidae). Within his "Dacina" Munro (1984) recognized 36 "genera." These 36 "genera" were largely described in terms of present/absent permutations of six characters — 1, 6/8 (combined) 13, 30, 20, and narrow/broad wing pattern). Of the possible permutations 29 of 64 (45%) occur in nature. White and Elson-Harris (1994) proposed that those be reduced to just four subgenera based on the present/absent permutations of two characters (13, 30), plus *D.* (*Callantra*) (0, 26). This grouping of species by the use of most of the possible permutations of a few binary characters cannot possibly adhere to the principle of grouping by shared apomorphy, and it would be expected that a cladistic analysis would not support these subgenera, most of which should be placed in synonymy, leaving just the currently recognized "groups" of subgenera. This excessive use of permutations of a few characters, including those of chaetotaxy whose likelihood of homoplasy was discussed earlier in this chapter, derives from the work of Bezzi (1915), when a fraction of the currently recognized fauna was known and the resultant groupings were a practical aid to identification. In later years the use of such permutations led Munro (1984) to abandon dichotomous keys and attempt to produce a multiple-entry key on paper, despite that technique being almost inoperable unless computerized (as for the Indo-Australasian fauna, by White and Hancock 1997).

Two male secondary sexual characters not previously regarded as of importance in higher level classification had low levels of change in this analysis (see Table 20.7), namely, the presence of the hind tibial pad (35) and the lamprine (36). The distribution of the hind tibial pad has been discussed earlier in this chapter (Section 20.2 on pheromone dispersal). The lamprine character was added to indicate secondary loss of the pecten/lamprine complex of characters because all *Bactrocera* spp. included in these analyses that lacked a fully developed pecten were found to have fine hairs and at least a narrow lamprine, which were assumed to represent a vestigial (secondary loss) of the pecten. However, in many *Dacus* spp. that lack a pecten (the subgenera *Leptoxyda* and *Metidacus* were not represented in this analysis), no clear evidence of a vestigial pecten or lamprine could be detected. That may simply be a result of surface features being reduced in association

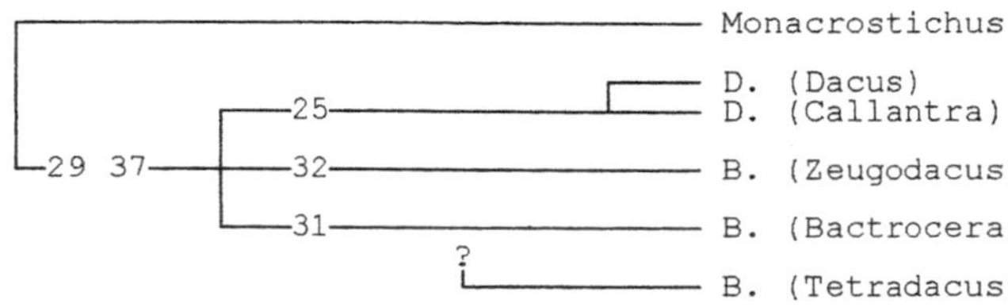


**FIGURE 20.4** Summary classification in the form of a tentative cladogram. Significant characters are numbered as in Table 20.5.

with tergite fusion, or it may be taken to support the possibility that at least some *Dacus* spp. derive from near the base of the phylogeny before the development of the pecten. A future analysis should investigate a method of improved character coding to associate these two characters together, since the apparent paraphyly of the *Zeugodacus* group in the weighted analysis may have been at least partly due to the apparent derivation of the lamprine (36) before the pecten (30), when the whole purpose of the lamprine character was to indicate secondary loss of the pecten (30).

In conclusion, a parsimony-based approach to the analysis of currently available data did largely support the classification of Drew (1989) (Figure 20.6), although the unweighted analysis supported the position of *B.* (*Tetradacus*) spp. as a relatively primitive group, as indicated by Munro (1984) (Figure 20.5). Furthermore, the characters used for the initial "splits" in the classification of Drew (1989) (see Table 20.7) are the characters accorded high weight (or a good fit to the cladograms). However, these analyses call into question the definition of numerous subgenera based on permutations of a few characters, many of which are highly homoplasious, and a high proportion of those subgenera must be defined by symplesiomorphy. Furthermore, many of these characters are present in the outgroup (e.g., postsutural supra-alar and prescutellar acrostichal setae, characters 13, 14) or are basal to the whole of *Dacus* plus *Bactrocera* (e.g., the pecten, character 30), and so the apomorphies for many subgenera are character reversals that might be expected to be highly homoplasious (Table 20.7 indicates the low weights given to characters 13, 14).

Although there is a clear need for a revision based on a more critical and broader-based analysis than those presented here, a summary classification is tentatively presented (Figure 20.4). This is shown as if it were a cladogram, although further study may either confirm that one of the three major groups is paraphyletic or that the apparent paraphyly indicated in the above analyses is an artifact caused by character coding problems. Note that the *B.* (*Bactrocera*) group and the *B.* (*Zeugodacus*) group are placed as a trichotomy with *Dacus*. That is because there is no apomorphy for the genus *Bactrocera,* defined here purely by the lack of fused terga (25). No attempt has been made to place the remaining two small groups of *Bactrocera*, namely *B.* (*Melanodacus*) group (31 = 0, 32 = 0) or *B.* (*Queenslandacus*) *exigua* (May) (31 = 1, 32 = 1). The former could be plesiomorphic for both characters, or represent a reversal of one or the other character; both hypotheses are equally parsimonious. However, *B.* (*Queenslandacus*) *exigua* poses a greater problem as one or the other character cannot after all be uniquely apomorphic, unless its long surstylus lobe in fact represents a slightly different structure to that found in the *B.* (*Zeugodacus*) group. As *B.* (*Queenslandacus*) *exigua*, along with most *B.* (*Zeugodacus*) group spp., have only been subjected to an external examination of this structure and not dissected, it is quite likely that more critical study would yield additional surstylus characters. Regrettably, only a single male of *B.* (*Queenslandacus*) *exigua* has ever been found so dissection would not be practical.

## 20.9 THE FIT OF OTHER DATA TO THE SUMMARY CLASSIFICATION

The summary classification (Figure 20.4) is insufficiently resolved to permit more than a superficial attempt at fitting behavioral or other features besides adult morphology to it. In fact, the only strictly behavioral data that can be discussed is male lure response. The attraction of males to ME is almost

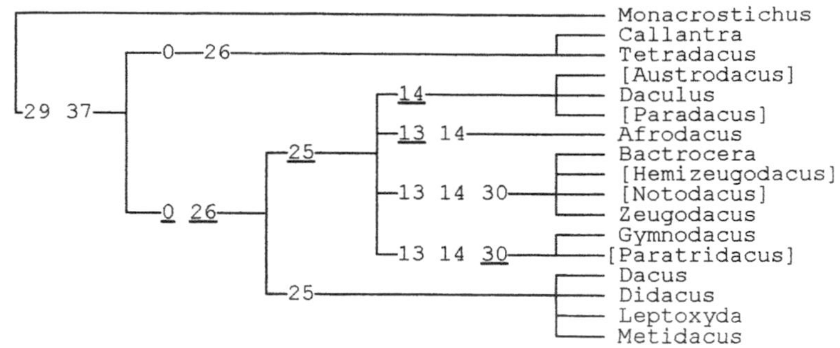


**FIGURE 20.5** Dendrogram summarizing the classification of Munro (1984). Nomenclature and ranking of groups modified to conform to current usage and to facilitate easy comparison. Some groups not specifically mentioned by Munro (1984) are shown in [ ]. Significant characters are numbered as in Table 20.5; absences are underlined.

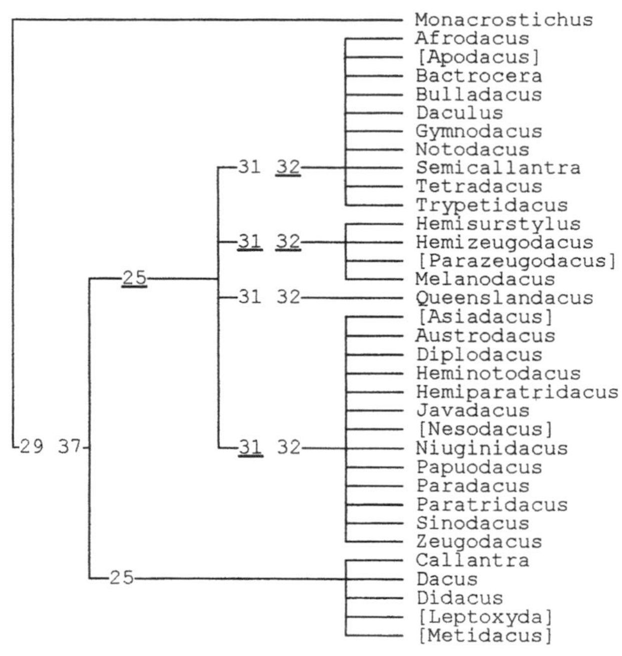


**FIGURE 20.6** Dendrogram summarizing the classification based on Drew (1989), as modified by Drew and Hancock (1994b; 1995) and White and Hancock (1997). Groups not specifically mentioned by Drew (1989) are shown in [ ]. Significant characters are numbered as in Table 20.5; absences are underlined.

exclusively confined to the *B.* (*Bactrocera*) group of subgenera. However, CL response is found in many *B.* (*Bactrocera*), *B.* (*Zeugodacus*), and *Dacus* spp. This suggests that CL response (or a precursor allowing its development) is plesiomorphic. However, a more resolved cladogram is needed before any conclusions can be drawn regarding ME response.

The larvae of most *B.* (*Zeugodacus*) group spp. and many *Dacus* spp. develop exclusively in the fruits or flowers of Cucurbitaceae, including all of the species included in the above analyses, but no *B.* (*Bactrocera*) group spp. are known to have a primary association with cucurbits. Without a more resolved cladogram it is not possible to say if this is a plesiomorphy lost in the *B.* (*Bactrocera*) group or an apomorphy supporting a common ancestry of the *B.* (*Zeugodacus*) group and *Dacus* spp., but the fact that the outgroup and other Ceratitidini, as well as *Monacrostichus* spp., are not normally associated with cucurbits, gives tentative support for the latter hypothesis. The association of *M. citricola* and the two included species of *B.* (*Tetradacus*) with *Citrus* spp. is noteworthy as it may account for the similarity of these two taxa, either due to a shared basal apomorphy that was later reversed (see unweighted cladogram, Figure 20.2) or due to convergence (see weighted cladogram, Figure 20.3).

M. M. Elson-Harris (in White and Elson-Harris 1994) described the third instar larvae of 22 pest species of Dacini. These included *D.* (*Callantra*) *axanus* (Hering) which has a large preapical tooth on each mouthhook; three species of the *B.* (*Zeugodacus*) group, each of which has a small preapical tooth; and 18 *B.* (*Bactrocera*) group spp., none of which have a preapical tooth. The mouthhook of the outgroup, *C. capitata*, also lacks a preapical tooth. However, the related *C. rosa* Karsch (Carroll 1998) and the second instar of all known *Bactrocera* spp. have a preapical tooth, that is, even the *B.* (*Bactrocera*) group spp. have the "teeth" in their genotype and their absence in the third instar is purely an ontogenetic "loss," although that could be interpreted as an apomorphy for that group.

Regarding geographic distribution, *Dacus* spp. are nearly all Afrotropical while most *Bactrocera* spp. are Indo-Australasian. A more resolved cladogram is needed before a species-area approach can be taken to study the zoogeography of the Dacini over a wide area, although that has been attempted for the limited South Pacific fauna (Michaux and White, in press).

## 20.10 CONCLUSIONS AND FUTURE RESEARCH NEEDS

The characters used for the identification of Dacini spp., from generic and subgeneric separation, down to species-level diagnosis, have been reviewed from the perspective of several different aspects of Dacini behavior. Work on the relationship of male lure and pheromone chemistry to mating behavior is in progress and will continue, given the great economic importance of this field, particularly with such new developments as the idea of feeding sterile flies on lure so that it increases their mating chances and decreases the likelihood of them being caught in male suppression traps (Shelly 1994). The possibility that lure gathering from natural plant sources may benefit the plant through pollination has been suggested by a few isolated observations and, if shown to be the case, would greatly expand our understanding of the biology of the flies. Furthermore, studies of acoustic signaling need to be broadened beyond the bounds of simply studying single species in mass-rearing if any real behavioral significance of sound production is to be confirmed and understood, particularly if any interspecific significance is to be discovered.

Many of the characters used for identification to species group or species level are color pattern differences likely to be involved in mimicry. While those of us who have observed Dacini in the field would generally agree that they are wasp mimics, no research has been carried out to identify the models. Such a program of work might lead to a greater understanding of many other aspects of Dacini behavior, genetics and evolution. This chapter also hypothesizes that scutal microtrichia pattern may be important in close encounters between some very closely related sympatric species in the Oriental fruit fly species complex that require recognition of conspecifics from nonconspecifics. Simple experiments in which the microtrichia are abraded may be possible to verify the importance of this feature to the choosing of the correct species in mating. Such work may be of importance in the proposed use of sterile insect technique against *B.* (*Bactrocera*) *philippinensis* in areas where *B.* (*Bactrocera*) *occipitalis* is also common. The differing aculeus length of these sympatric species may just be a coincidence, but the fact that the pattern is repeated in another pair of abundant sympatric species elsewhere starts to suggest a pattern that may be found in other sympatric species pairs. However, a real test of the possibility that abundant sympatric pairs have evolved a form of resource partitioning requires further study (and publication of existing but confidential data) of host preferences, including preferences for different fruits at different stages of fruit development.

Characters used in the subgeneric classification whose functional significance remains unknown were discussed. These included the tergite fusion of *Dacus* spp., which might have a function related to water conservation, but the only recent review of water relations in the Tephritidae (Meats 1989) did not consider this feature. The reason one species should have a certain seta on the scutum or scutellum, and another not, may be of interest for the geneticist but it is hard to conceive of a behavioral advantage to such features. However, the reason for the male terminalia differences that divide the four groups of *Bactrocera* subgenera suggests some differences in the mechanics of either terminalia retraction and storage or copula between these groups, and that may have some behavioral significance. At the very least it has phylogenetic significance and is worthy of study if we are to improve the classification of the group. A revised classification that gets away from the tradition of separating subgenera by such meaningless features as the presence or absence of postsutural supra-alar setae, would in turn be more predictive of nonmorphological features, and so reinforce our understanding of many of the economic aspects of the group.

Finally, the recent "traditional" classifications of Munro (1984) and Drew (1989) were compared with a cladistic analysis of the 51 pest species of Dacini, which were assumed to be a reasonably representative sampling of the tribe. It was found that the three major groups of subgenera recognized by Drew (1989), that is, *Dacus*, the *B.* (*Bactrocera*) group, and the *B.* (*Zeugodacus*) group, were tentatively supported by an unweighted or a weighted analysis, and a summary classification was produced (Figure 20.4). However, it was noted that the numerous subgenera (Figure 20.6) could not possibly be justified and it was concluded that before a revision of the subgenera could be carried out a more rigorous cladistic analysis of a larger sample of species should be undertaken. The resulting cladogram could then be tested against other independent data sources, namely, host, lure, and distribution data, plus larval morphology, in considerably more detail than was possible here.

## ACKNOWLEDGMENTS

Special thanks are due to Drs. Martín Aluja (Xalapa) and Allen Norrbom (USDA, Washington, D.C.) for kindly inviting me to the Xalapa meeting. I am also very grateful to Martín and his colleagues for arranging funds for the symposium and for my attendance. The following donors very kindly provided those funds: Campaña National contra las Moscas de la Fruta (Mexico), International Organization for Biological Control of Noxious Animals and Plants (IOBC), Instituto de Ecología, A.C. (Mexico), and the Consejo Nacional de Ciencia y Tecnología (Mexico). I also wish to thank Al Norrbom and Dick Drew (Griffith University, Brisbane) for encouraging me to attempt a cladistic analysis of some Dacini; the initial attempts were carried out at the meeting with the help of Al Norrbom and Bruce McPheron (Pennsylvania State University), and Ian Kitching (Natural History Museum) kindly helped with the final analyses and provided facilities for that work. I would also like to thank Martin Hall (Natural History Museum) for kindly reading an early draft of the behavioral sections, David Henshaw for advice on chemical nomenclature, all the other participants in the Xalapa meeting for their helpful discussion, and the many staff members of the Natural History Museum, London, who provided collection and library access.

## REFERENCES

Bezzi, M. 1915. On the Ethiopian fruit-flies of the genus *Dacus*. *Bull. Entomol. Res.* 6: 85–101.

Carroll, L.E. 1998. Description of the third instar larva of *Ceratitis rosa* Karsch (Diptera: Tephritidae). *Proc. Entomol. Soc. Wash.* 100: 88–94.

Chuah, C.H., H.S. Yong, and S.H. Goh. 1997. Methyl eugenol, a fruit-fly attractant, from the browning leaves of *Proiphys amboinensis* (Amaryllidaceae). *Biochem. Syst. Ecol.* 25: 391–393.

Cunningham, R.T. 1989. Parapheromones. In *Fruit Flies: Their Biology, Natural Enemies and Control* (A.S. Robinson and G. Hooper, eds.), pp. 221–230. In *World Crop Pests* (W. Helle, ed.), Vol. 3A. Elsevier Science Publishers, Amsterdam. 372 pp.

Drew, R.A.I. 1972. The generic and subgeneric classification of Dacini (Diptera: Tephritidae) from the South Pacific area. *J. Aust. Entomol. Soc.* 11: 1–22.

Drew, R.A.I. 1974. The responses of fruit fly species (Diptera: Tephritidae) in the South Pacific area to male attractants. *J. Aust. Entomol. Soc.* 13: 267–270.

Drew, R.A.I. 1987. Behavioral strategies of fruit flies of the genus *Dacus* (Diptera: Tephritidae) significant in mating and host-plant relationships. *Bull. Entomol. Res.* 77: 73–81.

Drew, R.A.I. 1989. The tropical fruit flies (Diptera: Tephritidae: Dacinae) of the Australasian and Oceanian Regions. *Mem. Queensl. Mus.* 26: 1–521.

Drew, R.A.I. and D.L. Hancock. 1994a. The *Bactrocera dorsalis* complex of fruit flies (Diptera: Tephritidae: Dacinae) in Asia. *Bull. Entomol. Res., Suppl.* 2, 68 pp.

Drew, R.A.I. and D.L. Hancock. 1994b. Revision of the tropical fruit flies (Diptera: Tephritidae: Dacinae) of South-east Asia. I. *Ichneumonopsis* Hardy and *Monacrostichus* Bezzi. *Invertebr. Taxon.* 8: 829–838.

Drew, R.A.I. and D.L. Hancock. 1995. New species, subgenus and records of *Bactrocera* Macquart from the South Pacific (Diptera: Tephritidae: Dacinae). *J. Aust. Entomol. Soc.* 34: 7–11.

Drew, R.A.I. and D.E. Hardy. 1981. *Dacus* (*Bactrocera*) *opiliae*, a new sibling species of the *dorsalis* complex of fruit flies from northern Australia (Diptera: Tephritidae). *J. Aust. Entomol. Soc.* 20: 131–137.

Drew, R.A.I., D.L. Hancock, and I.M. White. 1998. Revision of the tropical fruit flies (Diptera: Tephritidae: Dacinae) of South-east Asia. II. *Dacus* Fabricius. *Invertebr. Taxon.* 12: 567– 654.

Eberhard, W.G. and F. Pereira. 1993. Functions of the male genitalic surstyli in the Mediterranean fruit fly, *Ceratitis capitata* (Diptera: Tephritidae). *J. Kans. Entomol. Soc.* 66: 427–433.

Enderlein, G. 1920. Zur Kenntnis tropischer Frucht-Bohrfliegen. *Zool. Jarhb. Abt. Syst. Geogr. Biol. Tiere* 43: 336–360.

Farris, J.S. 1988. Hennig86, version 1.5. Program and manual. Privately distributed.

Fitt, G. 1981. The influence of age, nutrition and time of day on the responsiveness of male *Dacus opiliae* to the synthetic lure methyl eugenol. *Entomol. Exp. Appl.* 30: 83–90.

Fletcher, B.S. 1987. The biology of dacine fruit flies. *Annu. Rev. Entomol.* 32: 115–144.

Goloboff, P.A. 1993. *NONA (a bastard son of Pee-Wee), Version 1.8 (32 bit)*. American Museum of Natural History, New York.

Goloboff, P.A. 1997. *PIWE, Parsimony with Implied Weights, Version 2.8 (32 bit)*. American Museum of Natural History, New York.

Hammond, P.M. 1972. The micro-structure, distribution and possible function of peg-like setae in male Coleoptera. *Entomol. Scand.* 3: 40–54.

Han, H.-Y. and B.A. McPheron. 1997. Molecular phylogenetic study of Tephritidae (Insecta: Diptera) using partial sequences of the mitochondrial 16S ribosomal DNA. *Mol. Phylogenet. Evol.* 7: 17–32.

Hancock, D.L. 1986. Classification of the Trypetinae (Diptera: Tephritidae), with a discussion of the Afrotropical fauna. *J. Entomol. Soc. South. Afr.* 49: 275–305.

Hardy, D.E. 1955. A reclassification of the Dacini (Tephritidae — Diptera). *Ann. Entomol. Soc. Am.* 48: 425–437.

Hardy, D.E. 1973. The fruit flies (Tephritidae — Diptera) of Thailand and bordering countries. *Pac. Insects Monogr.* 31: 1–353.

Headrick, D.H. and R.D. Goeden. 1995. Reproductive behavior of California fruit flies and the classification and evolution of Tephritidae (Diptera) mating systems. *Stud. Dipterol.* 1: 194–252.

Hendel, F. 1914. Die Gattungen der Bohrfliegen (Analytische Übersichte aller bisher bekannten Gattungen der Tephritinae). *Wien. Entomol. Ztg.* 33: 73–98.

Hendrichs, J., B.I. Katsoyannos, V. Wornoayporn, and M.A. Hendrichs. 1994. Odor-mediated foraging by yellowjacket wasps (Hymenoptera: Vespidae): Predation on leks of pheromone-calling Mediterranean fruit fly males (Diptera: Tephritidae). *Oecologia* 99: 88–94.

Iwaizumi, R., M. Kaneda, and O. Iwahashi. 1997. Correlation of length of terminalia of males and females among nine species of *Bactrocera* (Diptera: Tephritidae) and differences among sympatric species of *B. dorsalis* complex. *Ann. Entomol. Soc. Am.* 90: 664–666.

Jones, S.R. 1989. Morphology and Evolution of the Aculei of True Fruit Flies (Diptera: Tephritidae) and Their Relationship to Host Anatomy. Ph.D. dissertation, Pennsylvania State University, University Park.

Kanmiya, K. 1988. Acoustic studies on the mechanism of sound production in the mating songs of the melon fly, *Dacus cucurbitae* Coquillett (Diptera: Tephritidae). *J. Ethol.* 6: 143–151.

Kanmiya, K., A. Tanaka, H. Kamiwada, K. Nakagawa, and T. Nishioka. 1987. Time-domain analysis of the male courtship songs produced by wild, mass-reared, and by irradiated melon flies, *Dacus cucurbitae* Coquillett (Diptera: Tephritidae). *Appl. Entomol. Zool.* 22: 181–194.

Kuba, H. and Y. Sokei. 1988 The production of pheromone clouds by spraying in the melon fly, *Dacus cucurbitae* Coquillett (Diptera: Tephritidae). *J. Ethol.* 6: 105–110.

Lewis, J.A., C.J. Moore, M.T. Fletcher, R.A.[I.] Drew, and W. Kitching. 1988. Volatile compounds from the flowers of *Spathiphyllum cannaefolium. Phytochemistry* 27: 2755–2757.

Mather, M.H. and B.D. Roitberg. 1987. A sheep in wolf's clothing: Tephritid flies mimic spider predators. *Science* 236(4799): 308–310.

Meats, A. 1989. Water relations of Tephritidae. In *Fruit Flies: Their Biology, Natural Enemies and Control* (A.S. Robinson and G. Hooper, eds.), pp. 241–247. In *World Crop Pests* (W. Helle, ed.), Vol. 3A. Elsevier Science Publishers, Amsterdam. 372 pp.

Michaux, B. and I.M. White. Systematics and biogeography of southwest Pacific *Bactrocera* (Diptera: Tephritidae: Dacini). *Palaeogeogr. Palaeoclimatol. Palaeoecol.*, 153: 337–351.

Munro, H.K. 1984. A taxonomic treatise on the Dacidae (Tephritoidea, Diptera) of Africa. *Entomol. Mem. S. Afr. Dept. Agric.* 61: 1–313.

Nishida, R., O. Iwahashi, and K.H. Tan. 1993. Accumulation of *Dendrobium superbum* (Orchidaceae) fragrance in the rectal glands by males of the melon fly, *Dacus cucurbitae. J. Chem. Ecol.* 19: 713–722.

Norrbom, A.L., L.E. Carroll, and A. Freidberg. 1999. Status of knowledge. In *Fruit Fly Expert Identification System and Systematic Information Database* (F.C. Thompson, ed.), pp. 9–47. *Myia* (1998) 9, 524 pp.

Ooi, C.S. 1991. Genetic variation in populations of two sympatric taxa in the *Dacus dorsalis* complex and their relative infestation levels in various fruit hosts. In *First International Symposium on Fruit Flies in the Tropics, Kuala Lumpur, 1988* (S. Vijaysegaran and A.G. Ibrahim, eds.), pp. 71–80. Malaysian Agricultural Research and Development Institute, Kuala Lumpur. 430 pp.

Prokopy, R.J., D.R. Papaj, J. Hendrichs, and T.T.Y. Wong. 1992. Behavioral responses of *Ceratitis capitata* flies to bait spray droplets and natural food. *Entomol. Exp. Appl.* 64: 247–257.

Rice, M.J. 1989. The sensory physiology of pest fruit flies: conspectus and prospectus. In *Fruit Flies: Their Biology, Natural Enemies and Control* (A.S. Robinson and G. Hooper, eds.), pp. 249–272. In *World Crop Pests* (W. Helle, ed.), Vol. 3A. Elsevier Science Publishers, Amsterdam. 372 pp.

Shah, A.H. and R.C. Patel. 1976. Role of tulsi plant (*Ocimum sanctum*) in control of mango fruitfly, *Dacus correctus* Bezzi (Tephritidae: Diptera). *Curr. Sci.* 45: 313–314.

Shelly, T.E. 1994. Consumption of methyl eugenol by male *Bactrocera dorsalis* (Diptera: Tephritidae): low incidence of repeat feeding. *Fla. Entomol.* 77: 201–208.

Shelly, T.E. and A.-L. M. Dewire. 1994. Chemically mediated mating success in male oriental fruit flies (Diptera: Tephritidae). *Ann. Entomol. Soc. Am.* 87: 375–382.

Shelly, T.E. and E.M. Villalobos. 1995. Cue lure and the mating behavior of male melon flies (Diptera: Tephritidae). *Fla. Entomol.* 78: 473–482.

Shelly, T., S. Resilva, M. Reyes, and H. Bignayan. 1996. Methyl eugenol and mating competitiveness of irradiated male *Bactrocera philippinensis* (Diptera: Tephritidae). *Fla. Entomol.* 79: 481–488.

Shiraki, T. 1968. Fruit flies of the Ryukyu Islands. *U.S. Nat. Mus. Bull.* 263: 1–104.

Siebert, D.J. 1992. Tree statistics; trees and "confidence"; consensus trees; alternatives to parsimony; character weighting; character conflict and its resolution. In *Cladistics; A Practical Course in Systematics* (P.L. Forey, C.J. Humphries, I.L. Kitching, R.W. Scotland, D.J. Siebert, and D.M. Williams, eds.), pp. 72–88. The Systematics Association Publication 10, 228 pp.

Sivinski, J. and T. Burk. 1989. Behavior; reproductive and mating behavior. In *Fruit Flies: Their Biology, Natural Enemies and Control* (A.S. Robinson and G. Hooper, eds.), pp. 343–351. In *World Crop Pests* (W. Helle, ed.), Vol. 3A. Elsevier Science Publishers, Amsterdam. 372 pp.

Stoffolano, J.G. 1989. Structure and function of the ovipositor of the tephritids. In *Fruit Flies of Economic Importance 87. Proceedings of the CEC/IOBC International Symposium, Rome, 1987* (R. Cavalloro, ed.), pp. 141–146. A.A. Balkema, Rotterdam. 626 pp.

Toong, Y.C. and K.H. Tan. 1994. Fruit fly attractants from local plants. In *Current Research on Tropical Fruit Flies and Their Management. Proceedings of the Symposium on Tropical Fruit Flies, Kuala Lumpur, 1992* (H.S. Yong and S.G. Khoo, eds.), p. 68. The Working Group on Malaysian Fruit Flies, Kuala Lumpur. 76 pp.

White, I.M. and M.M. Elson-Harris. 1994. *Fruit Flies of Economic Significance; Their Identification and Bionomics*, CAB INTERNATIONAL, Wallingford. Reprint with addendum. 601 pp.

White, I.M. and D.L. Hancock. 1997. *CABIKEY to the Dacini (Diptera, Tephritidae) of the Asia-Pacific-Australasian Regions*. CAB INTERNATIONAL, Wallingford. Windows CD-ROM.

White, I.M. and K. Marquardt. 1989. A revision of the genus *Chaetorellia* Hendel (Diptera: Tephritidae) including a new species associated with spotted knapweed, *Centaurea maculosa* Lam. (Asteraceae). *Bull. Entomol. Res.* 79: 453–487.

Yong, H.S. 1990. Flower of *Couroupita guianensis*: a male fruit-fly attractant of the methyl eugenol group. *Nat. Malays.* 15: 92–97.

Yong, H.S. 1993. Flowers of *Spathiphyllum cannaefolium* (Araceae): a male fruit fly attractant of the methyl eugenol type. *Nat. Malays.* 18: 60–63.

Yong, H.S. 1994. The Gnemon fruit fly. *Nat. Malays.* 19: 37–40.

## WORLD WIDE WEB SOURCES

*Acacia farnesiana* — http://sunsite.unc.edu/pub/academic/medicine/alternative-healthcare/herbal-medicine/SWSBM/Constituents/Acacia_farnesiana.txt

*Asarum* spp — http://balsam.methow.com/~gwooten/chem/hrbdata.htm

*Canella winterana* — http://www.uwcm.ac.uk/uwcm/dm/BoDD/BotDermFolder/BotDermC/CANE.html

*Dacrydium franklinii* — http://www.execpc.com/~goodscnt/data/rw10082913.html

*Hyssopus officinalis* — http://chili.rt66.com/hrbmoore/Constituents/Hyssopus_officinalis.txt

*Laurus nobilis* — http://bkfug.kfunigraz.ac.at/~katzer/engl/Laur_nob.html

*Ocimum* spp — http://www.hort.purdue.edu/newcrop/proceedings1990/v1-484.html

*Rosmarinus officinalis* — http://sunsite.unc.edu/london/orgfarm/Alternative- Healthcare/Southwest-School-of-Botanical-Medicine/Constituents/Rosmarinus_officinalis.txt

*Umbellularia californica* — http://chili.rt66.com/hrbmoore/Constituents/Umbellularia_californica.txt

# 21 The Biology and Behavior of Flies in the Tribe Dacini (Dacinae)

*Richard A.I. Drew and Meredith C. Romig*

## CONTENTS



## 21.1 INTRODUCTION

Extensive literature reviews of dacine ecology and biology have been published by Christenson and Foote (1960), Bateman (1972), Fletcher (1987b), Drew (1989), and Fletcher and Prokopy (1991). In keeping with the theme of this book, *Fruit Flies (Tephritidae): Phylogeny and Evolution of Behavior,* this chapter is not designed to be another such review but rather a discussion on those aspects of dacine behavior and biology that will help us to assess trends in the evolution of species and thus phylogenetic relationships. In particular, this chapter is designed to provide a thematic basis for the discussions in Drew and Hancock, Chapter 19, on the phylogenetic relationships in the Dacini.

Over the past four to five decades there has been considerable population ecology research conducted on a range of species of Dacini in Southeast Asia and the Pacific Region. The major pest species, *Bactrocera cucurbitae* (Coquillett) and *B. dorsalis* (Hendel) in Hawaii, and *B. tryoni* (Froggatt) in Australia, have been the primary focus and studies have included understanding the seasonal abundance of species and the influences on this of climatic factors such as light, temperature, moisture, and the availability of host plants. Much of this research has been on the species in locations into which they have been introduced and in which no dacine species have evolved, e.g., Hawaii and southeastern Australia.

In contrast, the research effort into dacine behavior and biology over the past few decades has been considerably smaller and is only a small fraction of that which has been undertaken on *Ceratitis,* particularly *C. capitata* (Wiedemann), and *Rhagoletis*, particularly *R. pomonella* (Walsh). We believe that a realistic understanding of processes of dacine speciation and phylogenetic relationships will only be achieved through a combination of extensive taxonomic studies plus in-

0-8493-1275-2/00/$0.00+$.50

depth biological and behavioral research into species in their endemic habitats. These habitats are tropical/subtropical rain forests in the case of the genus *Bactrocera* and the African savannah-type environments in the case of the genus *Dacus*. Very few such studies have ever been undertaken on species of Dacini. The few geographically inbetween species such as *B. oleae* (Rossi), *D. persicus* (Hendel), and the *Dacus* species of Southeast Asia and the South Pacific, will obviously form significant links in the overall picture.

Aspects of dacine behavior that have received most attention in research programs are adult fly feeding, courtship and mating, oviposition, and within-host foraging. Little information is available on responses to the host plant, behavior of individuals within the host plant, the micro-habitat of the host plant, and their influences on the insect behavioral patterns such as foraging, courtship, and mating. In addition, our understanding of the biology and behavior of species of Dacini would be greatly enhanced by a thorough knowledge of host plant characteristics such as plant surface leachates, physical and chemical components of fruit and leaves, larval feeding behavior, and vertebrate predation on immature stages.

Our thesis presented in this chapter is that the behavioral and biological interactions between fruit fly species and their host plants, especially within their endemic habitats, have influenced the evolution of fly species–host plant associations which in turn must have a strong influence on the evolution of the fly species and thus the process of speciation. Consequently, any assessment of phylogeny in the Dacini must include an analysis of their biology and behavior, especially within the host plant.

The behavioral strategies of the Dacinae are discussed in relation to the primary activities of adult fly feeding, courtship and mating, oviposition, larval behavior, dispersal, host recognition, and bacteria relationships. The role played by certain bacteria as mediators in the fruit fly–host plant associations is then discussed and emphasis placed on the importance of understanding these relationships in order to elucidate processes of speciation. In reading this chapter, it is essential to realize that each subfamily of Tephritidae possesses its own unique behavioral patterns and that what is presented in this chapter refers to the Dacini alone.

## 21.2 ADULT FLY FEEDING BEHAVIOR

This subject has been discussed in considerable detail by Drew and Yuval, Chapter 27. The basic nutrients required by adult Dacini have long been defined as protein (in the form of free amino acids), minerals, sugars, B-complex vitamins, and water (Hagen 1953). Tsitsipis (1989) provided a comprehensive review of adult and larval nutritional requirements and the understanding of these on the development of diets for mass production in artificial rearing systems. Most Dacinae have a high reproductive rate due to a large number of ovarioles per ovary (20 to 40), short life cycles allowing many generations per year (six to eight in the subtropics to tropics as long as host plants are available) and having multiple host species (about 50% being polyphagous). Protein in the female diet is essential to maintain this high rate of reproduction. There have been many theories regarding the source of this protein in nature, with the main candidates being honeydew, extrafloral plant exudates, nectar, pollen, bird feces, and bacteria. In the tropical/subtropical environments honeydew is rapidly colonized by sooty molds and rendered unavailable as a food source. It also lacks some amino acids and contains others in lower-than-required levels for fly development (Craig 1960; Tsitsipis 1989). Further, bird feces have been observed to repel flies while pollen grains could not enter through the pores in the labellae of *B. tryoni* (Vijaysegaran et al. 1997). Consequently, recent focus has been on certain bacteria growing on plant surface leachates and extrafloral exudates (Drew et al. 1983; Lloyd et al. 1986; Drew and Lloyd 1987; Prokopy et al. 1991; Vijaysegaran et al. 1997).

To investigate the roles of bacteria in the fly food chain, initial research utilized sterile dissection techniques to isolate and identify the bacteria species in the alimentary tract of four species of *Bactrocera* collected from host plants, namely, *B. cacuminata* (Hering), *B. musae* (Tryon), *B. neohumeralis* (Hardy), and *B. tryoni*. In addition, bacteria were isolated and identified from fly feces, host fruit surfaces, oviposition sites, and larvae-infested fruit tissue (Lloyd et al. 1986). The most common bacteria were the following species of Enterobacteriaceae: *Klebsiella oxytoca*, *Erwinia herbicola*, *Enterobacter cloacae*, *Citrobacter freundii*, *Providencia rettgeri*, and *Proteus* spp. In subsequent fruit fly studies these have been called "fruit fly type" bacteria.

Laboratory-based feeding and egg production studies showed that a diet of pure fruit fly type bacteria cells, fed to *B. tryoni* adults as the only protein source, was sufficient to enable a large egg production that exceeded that from normal yeast protein used in laboratory culturing and mass-rearing (Drew et al. 1983).

Additional research, using a labeled strain of *K. oxytoca*, showed that after ingesting the bacteria as the only protein source, *B. tryoni* adults distributed them onto the host fruit surfaces in caged trees by a combination of proboscis extension and exploring and regurgitation of droplets of fluid crop contents. The bacteria were subsequently inserted into the fruit during oviposition, produced the fruit rot in which the larvae fed, and were passed through all immature life cycle stages and into the next generation of adults (Lloyd 1991).

Research indicated that after inoculation onto the fruit surfaces, the bacteria grew and colonized the fruit surfaces and were then available as a protein food source for immature females that entered the tree (Drew and Lloyd 1987). Probably a combination of leachates and fruit fly type bacteria growing in them provide the flies with the required nutrients for survival and reproduction. It is important to realize that the significant food supply, especially for development of females to sexual maturity and thus for reproduction, is obtained within the fruiting host plant, when the fruit begin to reach a mature green stage susceptible to fruit fly oviposition (Drew and Lloyd 1987).

Tsiropoulos (1977) demonstrated that *B. oleae* adults could survive and reproduce on a diet of honeydew and pollen grains when supplemented with a supply of sucrose solution, but did not demonstrate that the pollen grains were ingested. Perhaps in the drier Mediterranean environment, bacteria are not a significant part of the plant surface microflora and thus less important as an adult fly food source. Also, if pollen grains are ingested by *B. oleae*, it would be valuable to study the functional morphology of the adult fly mouthparts in a way similar to that of Vijaysegaran et al. (1997) for *B. tryoni* which could not ingest pollen particles.

Koveos and Tzanakakis (1993) showed that adult female *B. oleae* developed to sexual maturity when they had access to olive fruits, yeast hydrolysate, or a bacterial culture from the fly's esophageal diverticulum, but the addition of streptomycin to the olive fruit diets prevented sexual development. It appears that olive fruit exudates may be a valuable adult fly food source for *B. oleae* (Fletcher et al. 1978; Fletcher and Kapatos 1983) as well as certain bacteria species from the fly's alimentary canal.

Although an instructive assessment of the evolution of feeding behavior in the Dacini can only be achieved through an analysis of feeding and nutrition in a wide range of species, combined with a study of the morphology of mouthparts, it appears that different nutritional requirements and modes of feeding have developed in association with the different habitat types. The *Dacus* species in Africa have not been studied, but *Dacus aequalis* (May) in Queensland is known to forage on and ingest substances from fruit surfaces (R. Drew, unpublished data). On the other hand, tropical/subtropical *Bactrocera* species appear to be adapted to feeding on bacteria cells, while *B. oleae* in the Mediterranean climate lies in between, utilizing both fruit-based liquids and bacteria.

## 21.3 COURTSHIP AND MATING BEHAVIOR

In Dacini, probably the most important behavioral characteristic that influences both survival of the species and the process of speciation is host plant mating. Zwölfer (1974) reported that 45 species of Tephritinae from 18 genera participated in courtship and mating behavior on their host plants. He noted that they exhibit characteristic rendezvous behavior where the males utilize the larval host plants as species-specific waiting places and territories for courtship and mating. Experiments with some species showed that if the males and females did not meet on their specific host plants they were not able to recognize their conspecific partners.

Drew and Lloyd (1987) recorded mating pairs of *B. tryoni* in a fruiting peach tree over a 14-day period. *Bactrocera cacuminata*, a monophagous species in eastern Australia, has been recorded mating in its host plant *Solanum mauritianum* Scop. (R. Drew, personal observation). In the peach tree study, Drew and Lloyd (1987) recorded a semipermanent population of *B. tryoni* with a range of behavioral patterns that were dependent on the host plant, namely, adult feeding, courtship and mating, oviposition, and larval feeding. Also, in some species of the genus *Dacus*, pupation occurs in the remnants of the host fruit after the larvae have utilized it.

In a study of the mating behavior of dung flies, *Scathophaga stercoraria* (L.), Parker (1979) stated that the insects' search for mates has much in common with their search for food and that mate-encounter sites have evolved in areas of resource value to females. In that study it was noted that males arrive at the oviposition site (dung pat) at different times to the females, probably as a result of response to different odor cues to the females. Dodson (1997) similarly noted that *Phytalmia* spp. also mated at the oviposition sites. The peach tree study (Drew and Lloyd 1987) recorded primarily sexually mature males and immature females arriving in the host tree and that on the basis of crop and midgut contents, the sexes fed on different substrates. It was proposed that the host plant, with fruit susceptible to oviposition, was a food attractant to the females and a sex attractant to males.

The production of pheromone and associated mating behavior by male *B. cucurbitae* has been well illustrated and described by Kuba and Koyama (1982; 1985). Chemical analyses of male pheromones have been reported for 24 species of *Bactrocera* (Fletcher and Kitching 1995). Some of these studies have shown widely different chemical compositions in apparently closely related species, e.g., *B. carambolae* Drew and Hancock and *B. papayae* Drew and Hancock (Drew and Hancock 1994). These two species mate in sympatric situations yet do not interbreed. The actual use of male pheromones in courtship and mating in *Bactrocera* has been well documented (Koyama 1989). For most species of *Bactrocera* where courtship and mating have been recorded, the males release the pheromones from a position on a leaf of the host plants. It would be interesting to investigate whether or not the host plant had an influence on pheromone release. In the Dacinae, mating pairs have rarely been recorded in nonhost plants and the published statement that mating does not necessarily occur on the host tree (Bateman 1972) has not been clearly supported by published field observations. In a study on *B. dorsalis* in host trees with and without fruit in Thailand (Prokopy et al. 1996), all sexual behavior of males and all mating occurred on trees with fruit and it was concluded that both sexes were strongly attracted to the odor of host fruit. Jang et al. (1997) recorded that *B. dorsalis* mated females were attracted to fresh whole leaves and solvent-water leaf extracts of a nonhost tree, *Polyscias guilfoylei* (W. Bull) L. H. Bailey. Presumably, the attractive chemical is one, or similar to one, that occurs in a host plant for this fly species. Occasional mating pairs were also observed in these plants but these observations were additional to those of Stark et al. (1994) who recorded *B. dorsalis* mating in guava trees only, a major host plant.

Differences in the chemical compositions of the male pheromones in sibling species have been used to assist in the determination of species in the *B. dorsalis* complex (Drew and Hancock 1994). This application of pheromone analyses in taxonomy will have increasing importance as we attempt to investigate more apparently closely related species groups. Also there is a real need to analyze the actual volatile emissions produced by male flies during courtship and not just the components of the glands and reservoir in the rectum. Definition of volatiles in pest species will have a high chance of success in the development of female attractants for both trapping and field control.

Studies on the feeding of methyl eugenol to adult *B. dorsalis*–complex species by Shelly and Dewire (1994), Nishida et al. (1997), and Hee and Tan (1998), demonstrated that males have increased precopulatory behavior, are more attractive to females, and possess enhanced mating competitiveness, especially in being able to mate more frequently, than non-methyl-eugenol-fed males. Similar responses resulted when *B. dorsalis* males fed on methyl eugenol–containing plants (Nishida et al. 1997). While these behavioral responses are significant findings, their relevance in nature awaits further research. In particular, no host or nonhost plant species in the endemic rain forest habitat have yet been found to contain methyl eugenol. If such plants can be discovered, then the above laboratory findings may be vital in understanding dacine field mating behavior.

Prokopy et al. (1996) suggested that the sexual behavior of polyphagous species, which still occurs in the host plant, could be an evolutionary remnant of behavior that existed when the host range of such species was less extensive. They also emphasized the need to study mating behavior in natural habitats containing native host species, and at a time when they are bearing fruit in optimal condition for egg laying.

There have been some studies that have indicated that some host-plant-based odors, especially from fruit, are rendezvous stimulants (Drew 1987a; Jang et al. 1997). Such attraction of both sexes to a central point for courtship and mating would seem to be a logical evolutionary result, especially in *Bactrocera* species which have individuals that disperse over very large distances. Long-distance dispersal causes a significant dilution in population densities, and rendezvous stimulants would seem essential to ensure mate encounters. Mark-recapture studies on *B. tryoni* in southeast Queensland (Courtice and Drew 1984) revealed that most released flies departed from localities in which fruit did not occur and moved large distances before aggregating in fruiting mango trees. Green et al. (1993), studying intratree foraging behavior of *B. dorsalis*, found flies departing trees without fruit and moving to trees with fruit and that increasing fruit density attracted larger numbers of individuals.

## 21.4 OVIPOSITION BEHAVIOR

In the Dacini, oviposition has been the least studied of all the behavioral strategies. The methods of exploration of the fruit surfaces by female flies prior to oviposition have been observed. Basically the flies undergo a pattern of proboscis examination and probing with the tip of the aculeus of the ovipositor before deciding whether or not to oviposit. However, little is known about the cues that are utilized to assess site acceptability and the actual process of penetration with the aculeus. Bateman (1972) defined what was thought to be some of the physical characteristics that define oviposition site choice; for example, there was an attraction to the leeward side of fruit, shade rather than sunlight, soft rather than hard areas, rough rather than smooth surfaces and a preference for cracks and broken fruit surfaces, and the oviposition holes of other flies. There has to be doubt over the choice of prior oviposition holes as some observations (R. Drew, personal observation) have indicated that females avoid such encounters. Green et al. (1993) stated that *B. dorsalis* females declined to oviposit in fruit containing conspecific larvae.

In *B. tryoni* and *B. jarvisi* (Tryon), Fitt (1984) showed that adult females discriminated between fruits with and without larvae and that this process was probably due to chemical changes in fruit associated with the presence of larvae. It was also suggested that this action of discrimination may be a more primitive state than the use of marking pheromones in some Trypetinae. In a study on the relationship of ovariole number and clutch size, Fitt (1990) suggested that generalist species produced smaller clutch sizes and specialists larger clutch sizes and that this may be a result of exploitation of a diverse range of wild fruits by generalist fly species. In contrast, Drew (unpublished data) has observed large numbers of small egg clutches of *B. cacuminata* (an oligophagous species) in the same pieces of *S. mauritianum* fruit and often in association with larvae at different stages of development. Fitt's (1990) suggestion that the clutch size and distribution may be related to host range is debatable. The high egg production and widespread distribution of oviposition sites may be a method of combating the high levels of vertebrate predation that occur in *Bactrocera* species within their endemic rain forest habitat (Drew 1987b).

Fletcher (1987a) showed that females of *B. tryoni* reabsorb their ovarian follicles and contents of eggs during the cold winter months and then at the end of winter feed on proteinaceous materials, mate, and produce fertile eggs. During winter, sperm also disappear from the spermathecae.

An area of research that would be useful in both phylogenetic analyses and in understanding oviposition behavior is a comparative analysis of ovipositor structure in relation to the physical characteristics of endemic host fruits such as the epicarp thickness. The suggestion by White (Chapter 20) that the shape of the apex of the aculeus is probably related to the male clasper structure, and an adaptation for mating, is unproved and most unlikely. For example, species of *Zeugodacus* and *Sinodacus* have similar long surstylus lobes in males, yet *Zeugodacus* species possess needle-shaped aculeus tips while in *Sinodacus* species they are trilobed.

## 21.5 LARVAL BEHAVIOR

The feeding activity and nutritional requirements of dacine larvae are poorly understood. In general, the larvae have fluid-feeding mouthpart structures and in *Bactrocera* species they inhabit the bacterial soup in the decaying areas of infested fruit. Studies by Drew and Lloyd (1987) using a labeled strain of *K. oxytoca* showed that the bacteria fed to flies as the only protein source were distributed onto fruit surfaces when the flies were released into a caged nectarine tree. The bacteria were the only microorganisms in oviposition holes, the fruit rot, alimentary tracts of larvae, pupae, and the next generation of adults. Whether or not the larvae ingest the bacteria is unproved and a thorough investigation of larval feeding behavior is needed.

The nutritional requirements of larvae have been reviewed by Tsitsipis (1989). This review discussed the reported host ranges of major pest species, the influences of various host fruits on larval developmental rates, pupal weights and pupal emergence rates, and the development of artificial larval diets for mass-rearing. Most larval diets contain hydrolyzed protein and some vitamins (in the form of yeast), sucrose as a carbohydrate source and antimicrobial agents. Modifications of this basic formula include the addition of certain nutrients to meet the requirements of specific species, particularly monophagous and oligophagous species such as *B. latifrons* (Hendel) and *B. oleae*. The pattern of emergence of mature larvae from fruit has also received little study in the Dacini. In some *Dacus*, especially the Asclepiadaceae feeders, the larvae pupate within the fruiting body. This characteristic is rare in the *Bactrocera* with the only known records being *B. oleae* which can pupate in fruit or soil (Koveos et al. 1993) and *B. melastomatos* Drew and Hancock which pupates in the fruiting body which is the base of the flower of *Melastomatos malabathricum*.

Host utilization by species, whether they be monophagous, oligophagous, or polyphagous, must depend on adult fly choice in terms of attraction to the host plant and fruit for oviposition, the adaptation by larvae to survive and develop in the specific regime of nutrients supplied by the fruit tissue and provision of a suitable pupation microhabitat. The development of such complex associations must have occurred through a system of evolution of the fly species in association with the endemic host plants. Through extensive host fruit surveys in endemic rain forest habitats throughout Southeast Asia and the South Pacific region since the mid-1980s, it has been shown that closely related fly species usually utilize closely related plant taxa that have evolved across adjacent geographic areas (Allwood et al., in press). For example, fruit fly species in the small subgenus *Bulladacus* Drew and Hancock, occur from India, through Southeast Asia to Fiji in the South Pacific. Most have been reared from different species of Gnetaceae across the distribution range of the subgenus. Future research into such fly species–host plant associations will be exciting and is much needed.

## 21.6 DISPERSAL BEHAVIOR

Dispersal of adult Dacini has been well researched and documented over the past four decades. General trapping surveys run continuously over climatic seasons and mark-recapture studies have provided considerable data. Bateman (1972) and Fletcher (1973) defined two basic types of movements, nondispersive or localized and dispersive or long distance. Examples of nondispersive movements were given as those for *Bactrocera curcurbitae* moving daily in and out of curcubit patches for oviposition and those associated with feeding, oviposition, and mating in species such as *B. tryoni* within their host plants. The dispersive movements were those long-distance ones undertaken by post-teneral flies and appear to have an inherent genetic base. MacFarlane et al. (1987) studied post-teneral dispersal of *B. tryoni* in northern Victoria and measured distances of flight up to 94 km. Also some tropical species have been recorded flying large distances after the disappearance of all fruit in their host trees and with the onset of warmer temperatures following a cold winter season (Fletcher 1973).

Basically Dacini are strong fliers capable of long-distance flight. It is important to understand this behavioral pattern to gain insights into how infestations develop in plantations and then to develop appropriate strategies for field control. Also, a knowledge of dispersal into and within tropical rain forests is significant to our understanding of host finding. In this, the endemic habitat of many dacines, host plants occur not in concentrated groups but mostly as individuals interspersed with many other plant species. Monophagous dacine species in particular would have to move over considerable distances before finding their specific host plant.

## 21.7 HOST RECOGNITION BEHAVIOR

Some individuals of dacine species exhibit an acute capacity to detect host plants when their fruits (or fruiting bodies) have developed to a stage that is susceptible to oviposition. Even host plants growing in isolated situations in both rain forest and open forest habitats are readily detected. There also appears to be a precise time period (or fruit development stage) when flies begin to occupy the tree, prior to which no flies can be observed (Drew and Lloyd, 1987). At the completion of the fruiting period, the flies depart and the tree remains free of flies until the next fruiting season.

In our studies in Queensland (Drew 1987a; Drew and Lloyd 1987), results have indicated that the host plant for dacine species is not just the oviposition and larval development site. We now have a more holistic view of the host plant and have coined the phrase that it is the "center of activity" for a species population (Drew 1987a). In a study on *B. tryoni*, it was shown to be the site of some adult fly feeding, development to sexual maturity for females, courtship and mating, oviposition, larval feeding, and pupation (Drew 1987a; Drew and Lloyd 1987). In this research it was shown that the greater proportion of flies that arrived in a fruiting host plant (peach tree) were sexually mature males and immature females (Drew and Lloyd 1987). Examination of crop contents of feeding flies showed that the females and males fed on different substrates and that the females fed mostly on the fruit surfaces (Courtice and Drew 1984; Drew 1987a; Drew and Lloyd 1987). Ingested food was utilized rapidly, passing through the alimentary tract in 2 to 5 h. Mature eggs were subsequently laid 3 days after feeding (Courtice and Drew 1984; Drew 1987a).

The results of studies on adult female fly feeding in the host plant, plus the observations on courtship and mating in the same trees, demonstrated the fact that mate encounter sites for a species are locations of resource value to females. The specialization of female activities in these host plant locales ensures that individuals carrying heavy egg loads only have to participate in the nondispersive (short distance) movements that were recorded by Sonleitner and Bateman (1963).

There are two important questions to ask with regard to host plant recognition:

1. In the endemic habitat of most *Bactrocera* species, the tropical/subtropical rain forests, individuals of plant species are distributed erratically throughout the forests and usually not in homogeneous clumps. How do flies find such host plants within an extremely large and variable biomass of forest canopy?
2. What are the cues used in host plant recognition particularly when considering that mostly sexually mature males and immature females are attracted?

In the endemic rain forest habitat, many dacine species are monophagous. For such species in this complex ecosystem, which has the richest species diversity of all forest types, efficient methods of host plant recognition are vital. The obvious cues that have been considered and researched are olfactory and visual. Basically, it is recognized that the Dacini olfactory responses operate over greater distances than do visual ones.

Drew (1987a) reported that fruit fly–type bacteria volatiles, and some of the known chemical components of these, attracted sexually mature males and immature females of *B. tryoni.* This substantiated the finding of Gow (1954) that volatiles produced by bacteria growing in soya meal attracted *B. dorsalis*, *B. cucurbitae,* and *C. capitata.* Further work by Drew and Fay (1988) and that of Gow (1954) showed that the bacteria volatiles were stronger attractants for dacine species than ammonia, a commonly used lure in earlier years. The responses of flies to the bacteria odors were similar to those reported by Drew and Lloyd (1987) for adult *B. tryoni* attracted to an isolated fruiting peach tree. In that study large numbers of sexually mature males and immature females arrived in the tree, after which time the females fed and developed to sexual maturity, mated and oviposited. Following the first 5 to 10 days of fly activity in the peach tree there was a major increase in the number of fruit fly–type bacteria on the fruit surfaces. Coinciding with this was a marked increase in numbers of flies arriving in the tree, indicating that fly activity and accompanying bacterial growth increased the attractancy of the host plant. The research of Prokopy et al. (1991) has been very significant in showing that (1) the fruit surfaces are initially inoculated with bacteria from sexually mature females that arrive in the tree; (2) bacterial odors attract protein-hungry but not protein-fed females; and (3) fruit odor attracts mature females but not immature females. Consequently, initial colonization appears to be due to responses to fruit odor.

The large-scale return of adult *B. oleae* to olive groves following the first abundant summer rains, led Scarpati et al. (1996) to investigate the cues that elicited such responses. They found that *B. oleae* were attracted to fruit maceration water and olive leaf leaching water and that the attractive components were styrene and ammonia, both products of metabolism of microbial flora that are present on olive fruit and leaf surfaces. Scarpati et al. (1993) showed that α-pinene, mostly emitted by olive tree leaves and half-ripe olives, was an oviposition stimulant.

While there appear to be other chemicals involved such as fruit and leaf volatiles and insect produced pheromones, evidence is mounting that bacterial volatiles are important factors in host recognition for *Bactrocera* species. The responses to these volatiles may also be the basis for fly attraction to protein bait sprays used for field control in orchards. In protein bait attractancy trials carried out over many years (R. Drew, personal observation), flies have been found to be attracted considerable distances. It is most likely that olfactory cues are very strong in host finding and that the host tree is a food attractant for female flies and sex attractant for males (as noted previously, females feed on fruit surfaces but males feed on different substrates, judging from the different color of crop contents).

With regard to visual responses, Prokopy et al. (1990) showed that adult *B. dorsalis* responded to visual stimuli and that responses were greater when fruit visual stimuli existed in combination with fruit odor stimuli. Vargas et al. (1991) captured most *B. dorsalis* on yellow and white spheres while Prokopy and Haniotakis (1975) attracted *B. oleae* to black, red, and yellow spheres. Similar studies on *B. tryoni* in Queensland revealed that this species responded most strongly to cobalt blue (R. Prokopy, personal communication). The visual responses of *Bactrocera* species appear to vary, different species responding more strongly to different colors. These visual responses probably operate over shorter distances than do olfactory cues and are probably utilized after the flies enter the host trees.

## 21.8 BACTERIA RELATIONSHIPS

In the above discussion it is evident that the fruit fly–type bacteria play a role in the life system of some *Bactrocera* species that is wider than just being a food substrate. It is possible that the bacteria are a significant linking factor in the relationship of fruit flies and their host plants. Krischik and Jones (1991) stated, "It is becoming increasingly clear that patterns of associations cannot be understood by considering the plant and the herbivore as only an isolated pair of interacting organisms," and, "Microorganisms play critical roles in plant–herbivore relationships." Jones (1984) noted that microorganisms have an extraordinary ability to concentrate, synthesize, convert, and make available nutrients to herbivores and this gives them considerable potential to mediate interactions between the plant and herbivorous insects. It was also noted (Krischik and Jones 1991) that although microbes were largely unseen, microbial mediation of plant–herbivore interactions is widespread and has diverse and significant impacts on plant–herbivore relationships.

The current understanding of fruit fly–type bacteria in the life system of species of *Bactrocera* indicates that they provide nutrients for adult females and possibly larvae as a food substrate, provide an olfactory cue to attract flies to the host plant, lure flies to the plant in a sex maturity ratio that ensures host plant courtship and mating, and may play a role in the fly defense mechanisms against bacterial pathogens such as *Serratia* species. Consequently, the fruit fly–type bacteria can be seen to mediate interactions between the fly and its host plant and that this association benefits both the insect and the microorganism. This association has led Krischik and Jones (1991) to define the bacteria associated with dacine fruit flies as insect mutualists, not symbionts as has long been accepted. They stated that the bacteria beneficially affect the capacity of the fly to explore the plant, and in turn the microorganism is affected by the insect-plant interaction. The long-held view that the bacteria associated with fruit flies were symbionts was first brought into question by Drew et al. (1983), Courtice and Drew (1984), and Drew and Lloyd (1987).

Krischik and Jones (1991) stated that microorganisms have a long evolutionary history of associations with plants and herbivores. A logical consequence of this is that the evolution of the association between fruit fly species and their host plants could have been strongly influenced by the bacteria associations and that the bacteria and host plant evolutionary changes may have been governing the evolution of fruit fly species. Certainly, it seems reasonable to accept that the host plant has a strong influence on the ecology of fruit fly species and patterns of speciation, at least, in the Dacini.

## 21.9 CONCLUSIONS AND FUTURE RESEARCH NEEDS

Current understanding of the biology and behavior of dacine species suggests that there is a close relationship between the fly species and their host plants. This relationship involves adult fly feeding, courtship and mating, oviposition, larval feeding, triggering mechanisms for aspects of dispersal when fruit disappears from plants, host plant recognition, and bacteria associations. The survival of the fly population appears strongly dependent upon its host, not just for oviposition and subsequent larval development but for most of the life system activities. Consequently, the host plant has a significant influence on the survival of the species and probably plays an important role in driving the processes of speciation.

The gene pool of a species is delimited by the mechanisms that function to bring about fertilization under natural conditions (Drew 1987a). In particular, the Specific Mate Recognition Systems (SMRS; Paterson 1985), which bring the sexes together and ensure fertilization, appear to occur within the mating environment of *Bactrocera* species, that is, the host plant. Thus, it could be expected that speciation changes in host plants would have influences on courtship and mating behavior of species, as well as on host plant detection and oviposition behavior.

Paterson (1993) stated that species were incidental consequences of adaptive evolution and that speciation was a consequence of small daughter populations of sexual organisms adapting to a habitat that was different from the normal habitat of the parental population. Restriction of a small population to a different habitat type will thus lead to destabilization of the fertilization system and then speciation.

If speciation occurs when small populations are restricted to new environmental conditions, within the endemic habitat, then the evolution of the *Bactrocera* species host plants could be expected to have a direct influence on the evolution of the fly species.

Because the SMRS are some of the behavioral aspects of the fly species, or are plant characteristics that influence fly behavior strategies, research into the behavior and biology of species will be extremely valuable to our understanding of speciation and phylogenetic relationships.

There are significant areas where we need to pursue further research, particularly plant influences upon courtship and mating, oviposition behavior, larval feeding behavior, host recognition cues, and bacteria relationships. Some of the excellent research carried out by Prokopy (1968; 1977), Prokopy and Bush (1972), Prokopy et al. (1987), Moericke et al. (1975) over many years on *R. pomonella* and *C. capitata* should be pursued in the Dacini. In particular, more information is needed on the responses of dacine species to visual cues, general host plant foraging, and prior learning in choices by adult flies of host plants and oviposition sites.

The areas of biology and behavior discussed in this chapter will continue to provide important foundations that will assist in the elucidation of phylogenetic relationships within the Dacinae. In particular the understanding of the strong fly species–host plant relationship provides confidence in the use of endemic host plant records discussed in Chapter 19 on the phylogeny of the Dacini.

## ACKNOWLEDGMENTS

We thank the following donors for partial funding: Campaña Nacional contra las Moscas de la Fruta (Mexico), International Organization for Biological Control of Noxious Animals and Plants (IOBC), Instituto de Ecología, A.C. (Mexico), and Consejo Nacional de Ciencia y Tecnología (Mexico).

## REFERENCES

Allwood, A.J., A. Chinajariyawong, R.A.I. Drew, E.L. Hamacek, D.L. Hancock, C. Hengsawad, J.C. Jipanin, M. Jirasurat, C. Kong Krong, S. Kritsaneepaiboon, C.T.S. Leong, and S. Vijaysegaran. Host plant records for fruit flies (Diptera: Tephritidae) in Southeast Asia. *Raffles Bull. Zool.,* Suppl., in press.

Bateman, M.A. 1972. The ecology of fruit flies. *Annu. Rev. Entomol.* 17: 493–518.

Christenson, L.D. and R.H. Foote. 1960. Biology of fruit flies. *Annu. Rev. Entomol.* 5: 171–192.

Courtice, A.C. and R.A.I. Drew. 1984. Bacterial regulation of abundance in tropical fruit flies (Diptera: Tephritidae). *Aust. Zool.* 21: 251–268.

Craig, R. 1960. The physiology of excretion in the insect. *Annu. Rev. Entomol.* 5: 53–68.

Dodson, G.N. 1997. Resource defense mating system in antlered flies, *Phytalmia* spp. (Diptera: Tephritidae). *Ann. Entomol. Soc. Am.* 90: 496–504.

Drew, R.A.I. 1987a. Behavioural strategies of fruit flies of the genus *Dacus* (Diptera: Tephritidae) significant in mating and host-plant relationships. *Bull. Entomol. Res.* 77: 73–81.

Drew, R.A.I. 1987b. Reduction in fruit fly (Tephritidae: Dacinae) populations in their endemic rain forest habitat by frugivorous vertebrates. *Aust. J. Zool.* 35: 283–288.

Drew, R.A.I. 1989. The tropical fruit flies (Diptera: Tephritidae: Dacinae) of the Australasian and Oceanian Regions. *Mem. Queensl. Mus.* 26: 1–521.

Drew, R.A.I. and H.A.C. Fay. 1988. Comparison of the roles of ammonia and bacteria in the attraction of *Dacus tryoni* (Froggatt) (Queensland fruit fly) to proteinaceous suspensions. *J. Plant Prot. Trop.* 5: 127–130.

Drew, R.A.I. and D.L. Hancock. 1994. The *Bactrocera dorsalis* complex of fruit flies (Diptera: Tephritidae: Dacinae) in Asia. *Bull. Entomol. Res.*, Suppl. 2, 68 pp.

Drew, R.A.I. and A.C. Lloyd. 1987. Relationship of fruit flies (Diptera: Tephritidae) and their bacteria to host plants. *Ann. Entomol. Soc. Am.* 80: 629–636.

Drew, R.A.I., A.C. Courtice, and D.S. Teakle. 1983. Bacteria as a natural source of food for adult fruit flies (Diptera: Tephritidae). *Oecologia* (Berlin) 60**:** 279–284.

Fitt, G.P. 1984. Oviposition behavior of two tephritid fruit flies, *Dacus tryoni* and *Dacus jarvisi*, as influenced by the presence of larvae in the host fruit. *Oecologia* 62: 37–46.

Fitt, G.P. 1990. Variation in ovariole number and egg size of species of *Dacus* (Diptera; Tephritidae) and their relation to host specialization. *Ecol. Entomol.* 15: 255–264.

Fletcher, B.S. 1973. The ecology of a natural population of the Queensland fruit fly, *Dacus tryoni*. IV. The immigration and emigration of adults. *Aust. J. Zool.* 21: 541–565.

Fletcher, B.S. 1987a. The overwintering strategy of the Queensland fruit fly, *Dacus tryoni*. In *Fruit Flies: Proceedings of the Second International Symposium, 16–21 September 1986, Colymbari, Crete, Greece* (A.P. Economopoulos, ed.), pp. 375–382. Elsevier Science Publishers, Amsterdam. 509 pp.

Fletcher, B.S. 1987b. The biology of dacine fruit flies. *Annu. Rev. Entomol.* 32: 115–144.

Fletcher, B.S. and E. Kapatos. 1983. The influence of temperature, diet and olive fruits on the maturation rates of female olive flies at different times of the year. *Entomol. Exp. Appl.* 33: 244–252.

Fletcher, B.S. and R.J. Prokopy. 1991. Host location and oviposition in tephritid fruit flies. In *Reproductive Behaviour of Insects. Individuals and Populations* (W.J. Bailey and J. Ridsdill-Smith, eds.), pp. 139–171. Chapman & Hall, New York. 339 pp.

Fletcher, B.S., S. Pappas, and E. Kapatos. 1978. Changes in the ovaries of olive flies (*Dacus oleae* [Gmelin]) during the summer, and their relationship to temperature, humidity and fruit availability. *Ecol. Entomol.* 3: 99–107.

Fletcher, M.T. and W. Kitching. 1995. Chemistry of fruit flies. *Chem. Rev.* 95: 789–828.

Gow, P.L. 1954. Proteinaceous bait for the Oriental fruit fly. *J. Econ. Entomol.* 47: 153–160.

Green, T.A., R.J. Prokopy, R.I. Vargas, D. Kanehisa, and C. Albrecht. 1993. Intra-tree foraging behaviour of *Dacus dorsalis* flies in relation to host fruit quantity, quality and type. *Entomol. Exp. Appl.* 66: 13–20.

Hagen, K.S. 1953. Influence of adult nutrition upon the reproduction of three fruit fly species. In *Third Special Report on the Control of the Oriental Fruit Fly (*Dacus dorsalis*) in the Hawaiian Islands*, pp. 72–76. The Senate of the State of California.

Hee, A.K. and K. Tan. 1998. Attraction of female and male *Bactrocera papayae* to conspecific males fed with methyl eugenol and attraction of females to male sex pheromone components. *J. Chem. Ecol.* 24: 753–764.

Jang, E.B., L.A. Carvalho, and J.D. Stark. 1997. Attraction of female Oriental fruit fly, *Bactrocera dorsalis*, to volatile semiochemicals from leaves and extracts of a nonhost plant, *Panax* (*Polyscias guilfoylei*) in laboratory and olfactometer assays. *J. Chem. Ecol.* 23: 1389–1401.

Jones, C. G. 1984. Microorganisms as mediators of plant resource exploitation by insect herbivores. In *A New Ecology: Novel Approaches to Interactive Systems* (P.W. Price, C.N. Slobodchikoff, and S.W. Gaud, eds.), pp. 53–99. John Wiley & Sons, New York.

Katsoyannos, B.I. 1989. Response to shape, size and color. In *Fruit Flies: Their Biology, Natural Enemies and Control* (A.S. Robinson and G. Hooper, eds.), pp. 307–324. In *World Crop Pests* (W. Helle, ed.), Vol. 3A. Elsevier Science Publishers, Amsterdam. 372 pp.

Koveos, D.S. and M.E. Tzanakakis. 1993. Diapause aversion in the adult olive fruit fly through effects of the host fruit, bacteria, and adult diet. *Ann. Entomol. Soc. Am.* 86: 668–673.

Koyama, J. 1989. Mating pheromones. In *Fruit Flies: Their Biology, Natural Enemies and Control* (A.S. Robinson and G. Hooper, eds.), pp. 165–168. In *World Crop Pests* (W. Helle, ed.), Vol. 3A. Elsevier Science Publishers, Amsterdam. 372 pp.

Krischik, V.A. and C.G. Jones. 1991. Prologue – Microorganisms: the unseen mediators. In *Microbial Mediation of Plant–Herbivore Interactions* (P. Barbosa, V.A. Krischik, and C.G. Jones eds.), pp. 1–6. John Wiley & Sons, New York.

Kuba, H. and J. Koyama. 1982. Mating behavior of the melon fly, *Dacus cucurbitae* Coquillett (Diptera: Tephritidae): comparative studies of one wild and two laboratory strains. *Appl. Entomol. Zool.* 17: 559–568.

Kuba, H. and J. Koyama. 1985. Mating behavior of wild melon flies, *Dacus cucurbitae* Coquillett (Diptera: Tephritidae) in a field cage: courtship behavior. *Appl. Entomol. Zool.* 20: 365–372.

Lloyd, A.C. 1991. Bacteria Associated with *Bactrocera* Species of Fruit Flies (Diptera: Tephritidae) and Their Host Trees in Queensland. Ph.D. dissertation, University of Queensland, Brisbane.

Lloyd, A.C., R.A.I. Drew, D.S. Teakle, and A.C. Hayward. 1986. Bacteria associated with some *Dacus* species (Diptera: Tephritidae) and their host fruit in Queensland. *Aust. J. Biol. Sci.* 39: 361–368.

MacFarlane, J.R., R.W. East, R.A.I. Drew, and G.A. Betlinski. 1987. Dispersal of irradiated Queensland fruit flies, *Dacus tryoni* (Froggatt) (Diptera: Tephritidae), in South-Eastern Australia. *Aust. J. Zool.* 35: 275–281

Moericke, V., R.J. Prokopy, S. Berlocher, and G.L. Bush. 1975. Visual stimuli eliciting attraction of *Rhagoletis pomonella* (Diptera: Tephritidae) flies to trees. *Entomol. Exp. Appl.* 18: 497–507.

Nishida, R., T.E. Shelly, and K.Y. Kaneshiro. 1997. Acquisition of female-attracting fragrance by males of Oriental fruit fly from a Hawaiian lei flower, *Fagraea berteriana*. *J. Chem. Ecol.* 23: 2275–2285.

Parker, G.A. 1979. Sex around the cow-pats. *N. Sci.* 82: 125–127.

Paterson, H.E.H. 1985. The recognition concept of species. In *Species and Speciation* (E.S. Vrba, ed.), pp. 21–29. Transvaal Museum Monograph No.4, Pretoria.

Paterson, H.E.H. 1993. Environment and species. In *Evolution and the Recognition Concept of Species, Collected Writings* (S.F. McEvey, ed.), pp. 158–167. Johns Hopkins University Press, Baltimore.

Prokopy, R.J. 1968. Visual responses of apple maggot flies, *Rhagoletis pomonella* (Diptera: Tephritidae): orchard studies. *Entomol. Exp. Appl.* 11: 403–422.

Prokopy, R.J. 1977. Stimuli influencing trophic relations in Tephritidae. *Colloq. Int. C.N.R.S.* 265: 305–336.

Prokopy, R.J. and G.L. Bush. 1972. Mating behaviour in *Rhagoletis pomonella* (Diptera: Tephritidae) III. Male aggregation in response to an arrestant. *Can. Entomol.* 104: 275–283.

Prokopy, R.J. and B.S. Fletcher. 1987. The role of adult learning in the acceptance of host fruit for egglaying by the Queensland fruit fly, *Dacus tryoni*. *Entomol. Exp. Appl.* 45: 259–263.

Prokopy, R.J. and G.E. Haniotakis. 1975. Reponses of wild and laboratory-cultured *Dacus oleae* to host plant color. *Ann. Entomol. Soc. Am.* 68: 73–77.

Prokopy, R.J., D.R. Papaj, S.B. Opp, and T.T.Y. Wong. 1987. Intra-tree foraging behaviour of *Ceratitis capitata* flies in relation to host fruit density and quality. *Entomol. Exp. Appl.* 45: 251–258.

Prokopy, R.J., T.A Green, and R.I. Vargas. 1990. *Bactrocera dorsalis* flies can learn to find and accept host fruit. *J. Insect Behav.* 3: 663–672.

Prokopy, R.J., R.A.I. Drew, B.N.E. Sabine, A.C. Lloyd, and E. Hamacek. 1991. Effect of physiological and experimental state of *Bactrocera tryoni* flies on intra-tree foraging behaviour for food (bacteria) and host fruit. *Oecologia* 87: 394–400.

Prokopy, R.J., R. Poramarcom, M. Sutantawong, R. Dokmaihom, and J. Hendrichs. 1996. Localization of mating behavior of released *Bactrocera dorsalis* flies on host fruit in an orchard. *J. Insect Behav.* 9: 133–142.

Scarpati, M.L., R.L. Scalzo, and G. Vita. 1993. *Olea europaea* volatiles attractive and repellent to the olive fruit fly (*Dacus oleae* Gmelin). *J. Chem. Ecol.* 19: 881–891.

Scarpati, M.L., R.L. Scalzo, G. Vita, and A. Gambacorta. 1996. Chemiotropic behaviour of female olive fly (*Bactrocera oleae* GMEL.) on *Olea europaea* L. *J. Chem. Ecol.* 22: 1027–1036.

Shelly, T.E. and A.M. Dewire. 1994. Chemically mediated mating success in male Oriental fruit flies (Diptera: Tephritidae). *Ann. Entomol. Soc. Am.* 87: 375–382.

Sonleitner, F.J. and M.A. Bateman. 1963. Mark-recapture analysis of a population of Queensland fruit fly, *Dacus tryoni* in an orchard. *J. Anim. Ecol.* 32: 259–269.

Stark, J.D., R.I. Vargas, and W.A. Walsh. 1994. Temporal synchrony and patterns in an exotic-host-parasitoid community. *Oecologia* 100: 196–199.

Tsiropoulos, G.J. 1977. Reproduction and survival of the adult *Dacus oleae* feeding on pollens and honeydews. *Environ. Entomol.* 6: 390–392.

Tsitsipis, J.A. 1989. Nutrition. In *Fruit Flies: Their Biology, Natural Enemies and Control* (A.S. Robinson and G. Hooper, eds.), pp. 103–119. In *World Crop Pests* (W. Helle, ed.), Vol. 3A. Elsevier Science Publishers, Amsterdam. 369 pp.

Vargas, R.I., J.D. Stark, R.J. Prokopy, and T.A. Green. 1991. Response of Oriental fruit fly (Diptera: Tephritidae) and associated parasitoids (Hymenoptera: Braconidae) to different-color spheres. *J. Econ. Entomol.* 84: 1503–1507.

Vijaysegaran, S., G.H. Walter, and R.A.I. Drew. 1997. Mouthpart structure, feeding mechanisms, and natural food sources of adult *Bactrocera* (Diptera: Tephritidae). *Ann. Entomol. Soc. Am.* 90: 184–201.

Zwölfer, H. 1974. Das Treffpunkt-Prinzip als Kommunikationsstrategie und Isolationsmechanismus bei Bohrfliegen (Diptera: Trypetidae). *Entomol. Ger.* 1: 11–20.

# *Section VI*

## *Subfamily Tephritinae*

# 22 Phylogeny of the Subfamily Tephritinae: Relationships of the Tribes and Subtribes

*Valery A. Korneyev*

## CONTENTS



## 22.1 INTRODUCTION

The subfamily Tephritinae is the most specialized subfamily of Tephritidae. The larvae of Tephritinae predominantly infest flowerheads of the Asteraceae, the largest and the most advanced and

0-8493-1275-2/00/$0.00+$.50

widespread family of the angiosperms. The tephritines are almost ubiquitous and have penetrated even to subarctic and mountain tundras and alpine and arid deserts.

With few exceptions, the tephritines are small or medium-sized flies, often with whitish, thickened postocular setae, dark wing pattern with hyaline spots, oval epandrium, and two spermathecae. As in other Tephritidae, these characters have a mosaic distribution among the Tephritinae, and it is a difficult task to provide a definition of the subfamily and an analysis of relationships among the taxa included.

The purpose of this chapter is to analyze the phylogenetic relationships among the tribes of the subfamily Tephritinae, which is hardly an accessible goal without testing each tribe for monophyly, studying the distribution of characters within the tribes, and reconstructing the ground plans of the subfamily and tribes. A number of genitalic and nongenitalic characters used in the analysis were examined in this study in 1986 to 1998 from materials deposited in the collections of Dr. B. Merz (Geneva), Instituto de Ecología, Xalapa; National Museum of Natural History, Washington, D.C.; Naturhistorisches Museum Wien; Schmalhausen Institute of Zoology, Kiev; Zoological Institute, St. Petersburg; and some other collections. A list of examined taxa is given in Appendix 22.1 (except for species mentioned in the text); for species known to me from the literature only, the reference is given in the text. The ground plan character states were determined by comparison with the outgroups (Gastrozonini, Trypetini, Zaceratini). In cases where this was impossible (for characters variable or not occurring in the outgroup), the character state common in the three more primitive tribes (Terelliini, Xyphosiini, and the *Tomoplagia* group of genera) was considered plesiomorphic (these characters were not used to analyze the relationships among these three tribes, so this is not circular reasoning). The cladistic analyses performed on the subfamily Tephritinae and on three of its tribes were conducted using the program Hennig86 (Farris 1988).

## 22.2 SUBFAMILY TEPHRITINAE

### 22.2.1 Monophyly and Ground Plan of Tephritinae

Foote et al. (1993) and Norrbom et al. (1999) suggested several characters as possible synapomorphies of the Tephritinae: the oval outline of epandrium and surstyli (characteristic of most genera of all the tribes); and the strongly dilated distal portion of the spermathecal ducts. The former, however, may be a synapomorphy also shared by the Zaceratini of the Trypetinae (see below).

Apparently, the best candidate to be the sister group of the Tephritinae is the tribe Zaceratini (= Plioreoceptini), as suggested by Korneyev (1996). The larvae of Zaceratini, known only for *Plioreocepta poeciloptera* (Schrank), and of Terelliini are known to have placoid ("cellular") structure of the mask, with numerous (up to 10 to 15) smooth-edged oral ridges. The oval epandrium and surstyli in both genera (*Zacerata* Coquillett and *Plioreocepta* Korneyev) of Zaceratini strongly resemble those of most Tephritinae, and the glans of *Zacerata* has both a sclerotized subapical lobe and paired lobes of the acrophallus, like in *Terellia* Robineau-Desvoidy. In addition, the sister group relationship of Zaceratini + Tephritinae was recently supported by DNA analysis (Han and McPheron 1997; and Chapter 5).

There are two papillose spermathecae in *Plioreocepta* Korneyev, but their ducts are slender. The spermathecae of *Zacerata* have not been reexamined since Hancock (1986) reported this genus to have three bare spherical spermathecae; later he wrote me (Hancock, personal communication) that this could have been a misinterpretation or artifact. In Gastrozonini and Ceratitidini there are also two spermathecae, and some species have the spermathecal ducts dilated (see Korneyev 1996, Figures 2-18, 2-19, 2-20) similar to Tephritinae. Both the presence of two spermathecae and of such a dilation are believed to appear in Tephritinae independently from Dacinae.

The whitish setulae covering the scutum are another probable synapomorphy. This character state is not known in Zaceratini, Trypetini, and Gastrozonini, but occurs in most Tephritinae, even in the generalized tribes, like Terelliini and Xyphosiini. In some tribes of the Tephritinae where there are only dark brown setulae covering the scutum, usually at least some genera or species possess whitish setulae. The only exception are Myopitini, where species with white lanceolate setulae are unknown.

For characters not occurring in Zaceratini, the reconstruction of the ground plan of the Tephritinae was based on the presumption that Terelliini, Xyphosiini, and the *Tomoplagia* group of genera are the most generalized tribes, as they include many species that share plesiomorphic states of several characters with some Trypetinae (Trypetini, Zaceratini) and Dacinae (Gastrozonini). Their members are moderately large flies with completely setulose $R_{4+5}$ (only in Xyphosiini and *Tomoplagia* group), mostly banded wing pattern, unicolorous black or yellow (never white) major setae, large phallus with the glans often almost as big as the epandrium, subapical lobe sclerotized (in Terelliini only), acrophallus bilobate (in Terelliini only), eversible membrane of ovipositor anteriorly with four moderately long and broad taeniae, on posterior portion evenly covered with blunt dentiform scales, aculeus apically blunt, rounded and rather broad, nonserrate, with conspicuously developed lateral setae (corresponding to the lateral setae of the cercal unit), and two elongate and papillose spermathecae. Females lay eggs mostly into open young flowerheads between florets rather than into flower buds, and do not pierce plant tissues. The larvae bore in the tissues of flowerhead, and often in stems, in many cases leaving the plant for pupation. Most of these characters are common to the ground plans of the tribes of the Trypetinae–Dacinae complex (for instance, in Gastrozonini, Ceratitidini, etc.) and probably to the ground plan of the Tephritinae, too.

Therefore, the following features are believed to belong to the subfamily ground plan:

1. Fronto-orbital plates: rather long, reaching from the anterior margin to the posterior half of frons;
2. Frontal setae: three concolorous, of equal length;
3. Vertical plates: rather short, reaching one-third of frons length;
4. Orbital setae: two reclinate, concolorous;
5. Frontal vitta (mesofrons) finely yellow to black setulose;
6. Lateral vertical seta concolorous with the medial vertical seta, longer than one-half the of length of the latter;
7. Row of postocular setae consists of long and concolorous setae only, without the shorter and darker setulae between them;
8. $R_{4+5}$ setulose over all its length, or at least to level of DM-Cu;
9. Wing pattern striate or banded;
10. No white lanceolate setae present;
11. Posterior notopleural seta concolorous with anterior one;
12. Scutum sparsely microtrichose;
13. Scutellum with four concolorous setae;
14. Scutellum finely setulose on posterior margin only;
15. Anepimeral seta yellow or black, concolorous with the dorsalmost anepisternal seta;
16. Katepisternal seta yellow or black, concolorous with the dorsalmost anepisternal seta;
17. Mediotergite sparsely microtrichose or shining;
18. Abdominal terga not narrowed, sparsely microtrichose;
19. Glans with membranous basal lobe;
20. Preglans area of phallus microtrichose or bare (not spinulose);
21. Glans with trumpet-like, sclerotized subapical lobe;
22. Acrophallus well developed, consisting of two semitubular lobes hidden within the praeputium;
23. Basiphallus not extended either posteriorly or ventrally, simple, ring-like;
24. Phallapodeme with both lateral and posterior arms narrow, sclerotized, and separate, not “delta-like”;
25. Abdominal sternites 5 and 6 of female each with rodlike apodeme anteriorly;

**TABLE 22.1**
**List of Characters Used in Analysis of the Relationships among the Tribes of Tephritinae**

1. Dummy character
2. Palpus: 0, of normal shape; 1, enlarged
3. Medial vertical seta: 0, brown to black; 1, yellow
4. Lateral vertical seta: 0, longer than anterior orbital seta; 1, shorter than anterior orbital seta
5. Lateral vertical seta: 0, concolorous with medial vertical seta; 1, white
6. Posterior orbital seta: 0, black, brown, or yellow, concolorous with anterior one; 1, white
7. Posterior orbital seta: 0, reclinate; 1, inclinate
8. Postocular setae: 0, of same length; 1, mixed long and short
9. Anepimeral seta: 0, black, brown, or yellow; 1, white
10. Scutal setulae: 0, black only; 1, mixed black and white; 2, white only
11. Posterior notopleural seta: 0, concolorous with anterior one; 1, white
12. Costa at Sc apex: 0, normal; 1, incised
13. Costal spur: 0, not more than 1.5 times as long as width of costa; 1, longer, approximately as long as flagellomere 1
14. Vein $R_1$ at Sc bend: 0, uniformly setulose; 1, with a bare gap
15. Cell bcu: 0, with triangular posterolateral lobe; 1, closed by incurved vein
16. Anterior surface of midtibia: 0, without erect setae; 1, with erect setae
17. Preglans area of phallus: 0, bare; 1, setulose; 2, with strong spinulae
18. Subapical lobe of glans: 0, sclerotized, trumpetlike; 1, neither sclerotized nor trumpetlike
19. Acrophallus: 0, with paired tubular sclerites (at least in taxon ground plan); 1, forming a single tube
20. Ventroapical portion of oviscape: 0, sparsely setulose; 1, densely microsetulose
21. Dorsal side of eversible membrane: 0, without medial groove; 1, with a medial bare groove between two rows of scales
22. Scales on ventral side of eversible membrane: 0, covering most of ventral surface more or less uniformly; 1, forming narrow isolated stripe
23. Apical portion of aculeus: 0, not very long and narrow nor needlelike; 1, very long and narrow, needlelike
24. Aculeus: 0, not basally constricted nor apically barbed; 1, basally constricted, apically barbed
25. Aculeus apically: 0, blunt; 1, pointed, acute
26. Apical portion of spermathecal duct: 0, not dilated; 1, dilated
27. Host-plants: 0, Asteraceae; 1, not Asteraceae (Lamiaceae, Acanthaceae, or Verbenaceae)
28. Host plants: 0, not Senecioneae; 1, Senecioneae (polarity unresolved)
29. Larvae: 0, not in stem or bud galls on Anthemideae; 1, forming stem or bud galls on Anthemideae (polarity unresolved)

*Note:* 0 = plesiomorphy; 1,2 = apomorphies.

26. Eversible membrane evenly covered by blunt triangular scales dorsally and ventrally;
27. Aculeus apex broad and stout (neither cutting, nor piercing);
28. Spermathecae drop shaped, elongate, papillose;
29. Larval mask with 10 to 15 oral ridges and placoid structures (usually smooth-edged) posterolaterally of them; and
30. Larva in unmodified plant tissues, not forming gall.

### 22.2.2 Cladistic Parsimony Analysis of Relationships among the Tribes of Tephritinae

Relationships among the tribes of Tephritinae were analyzed using a character state matrix that consisted of 33 terminal taxa and 29 characters (including dummy outgroup and character; see Tables 22.1 and 22.2). All characters that have three or more states were considered additive. Autapomorphies were not included. Several taxa were included as outgroups as were multiple taxa of most ingroups.

**TABLE 22.2**
**Character States of Taxa Used in Analysis of the Relationships among the Tribes of Tephritinae (character numbers refer to Table 22.1)**

| | Character numbers |
|---|---|
| | **00000000011111111112222222222** |
| | **12345678901234567890123456789** |
| Dummy outgroup | 000000000000000000000000000??0 |
| Zaceratini | 00000000000000000000000?0?00 |
| *Terellia* | 00000010020000000000000001000 |
| *Neaspilota* | 0000001002000000?000000?1000 |
| *Tomoplagia* | 00100000020000000110000001000 |
| *Xyphosia* | 00100010020000000110000001000 |
| *Polionota* | 0010000002000000001?0000001000 |
| *Gymnocarena* | 00100000020000000110000001000 |
| *Procecidochares* | 00000000120001000110000001000 |
| *Axiothauma* | 00000000000000001011000001010 |
| *Cryptophorellia* | 00000000020000001011000011010 |
| *Myopitora* | 0000000000000001001?0000001000 |
| *Myopites* | 000000000000000100??0000001000 |
| *Ensina* | 000000000200000001110001?1000 |
| *Jamesomyia* | 000000010100000001110001?1000 |
| *Ptiloedaspis* | 00011001010000000110110011001 |
| *Oedaspis* | 00011001010000000110110011001 |
| *Hendrella* | 00011001020001000110110011001 |
| *Dithryca* | 00011101010001000110110011001 |
| *Pliomelaena* | 00011001020001000110101011100 |
| *Aciura* | 00011001000001000110101011100 |
| *Acinia* | 00011101020011000110100011000 |
| *Merzomyia* | 01011101120011000110100011010 |
| "*Dictyotrypeta*" *longiseta* | 0101?001020011000110100011000 |
| *Paranoeeta* | 01011101020011000110100011000 |
| *Rhabdochaeta* | 01011001020101000110100011000 |
| *Schistopterum* | 01011?01020101000110100011000 |
| *Spathulina* | 00011101121001000110100011000 |
| *Tephritis* | 00011101121001000110100011000 |
| *Oxyna* | 00011101121001002110100011001 |
| *Campiglossa* | 00011101120001002110100011000 |
| *Parafreutreta* | 00011101120001001110100011010 |
| *Capitites* | 00011101120001000110100011000 |

A Nelson consensus tree was obtained from the trees resulting from the mhennig* option combined with bb* (branch swapping) option (Figure 22.1A). The tree length = 50, consistency index (ci) = 0.60, retention index (ri) = 0.86. The second tree (Figure 22.1B) is one of two trees obtained from the mhennig* option combined with a series of successive character weightings. The tree length = 239, ci = 0.86, ri = 0.96.

Both trees show certain important synapomorphies of the taxa within the subfamily Tephritinae. Due to the small sample of characters and taxa, the trees show only putative relationships and must be considered with precaution. Monophyly of several tribes and of the Higher Tephritinae is well supported by synapomorphies, as indicated on Figure 22.1.

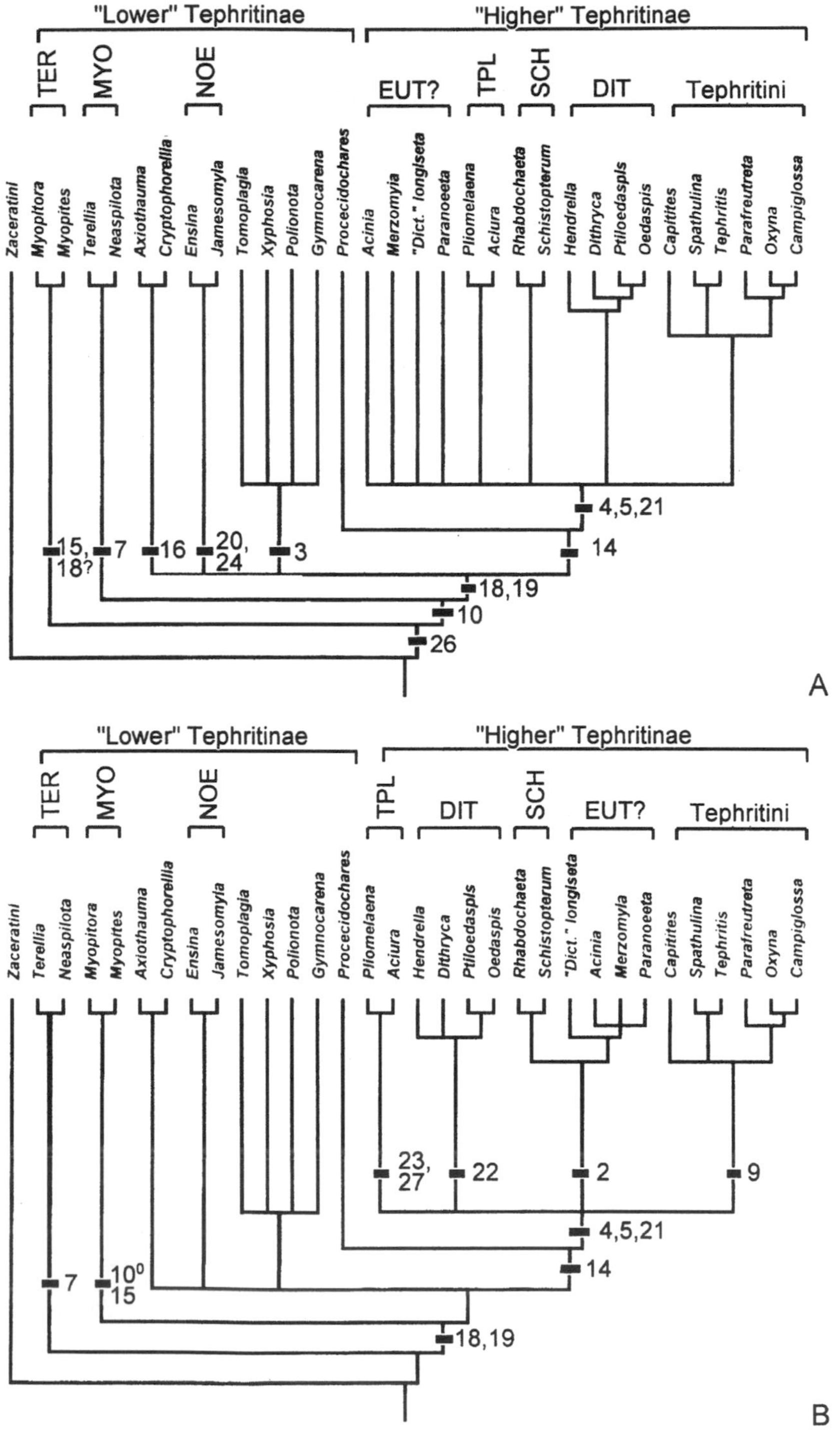


**FIGURE 22.1** Phylogenetic relationships of tribes of Tephritinae. (A) Nelson consensus tree from mhennig* bb* 100 trees (overflow); (B) mhennig tree obtained after series of successive weightings from the character matrix of Table 22.2. Character numbers at black bars represent state 1 unless followed by number of another state in superscript (see Table 22.1). DIT = Dithrycini, EUT = Eutretini, MYO = Myopitini, NOE = Noeetini, SCH = Schistopterini, TPL = Tephrellini.

### 22.2.3 The Lower Tephritinae

This aggregation of tribes retains many primitive characters, and may be paraphyletic, in contrast to the Higher Tephritinae, the larger group of tribes that share several advanced (apomorphic) characters, and therefore clearly comprise a monophyletic lineage.

#### 22.2.3.1 Terelliini

This tribe retains many characters in plesiomorphic condition and appears to be close to the ground plan of the subfamily, first of all, in phallus structure (the subapical lobe of glans trumpetlike, sclerotized; the sclerites of the acrophallus paired; the basal lobe of glans sclerotized). None of the species is known to form galls. The monophyly of the tribe is supported by the presence of the inclinate posterior orbital seta, the bare vein $R_{4+5}$, and the dark lyrate pattern of the scutum (synapomorphies). In addition to about 70 Old World and three New World species that belong here, the Nearctic genus *Neaspilota* Osten Sacken is usually classified in this tribe (Freidberg and Mathis 1986), but its position is less certain.

*Neaspilota* shares the synapomorphies of the Terelliini (the posterior orbital setae inclinate, bare $R_{4+5}$, the scutum with the dark lyrate pattern) (Freidberg and Mathis 1986), differing in the frontal vitta sometimes setulose (plesiomorphy), and in the absence of the subapical lobe of the glans (homoplasy with the remaining tribes of Tephritinae?). Some *Neaspilota* species possess the basal lobe of the glans and the paired sclerites of the acrophallus (see Freidberg and Mathis 1986: Figures 125, 162), sharing these plesiomorphies with the other Terelliini.

Currently, the tribe includes *Neaspilota*, the large and probably paraphyletic genus *Terellia* Robineau-Desvoidy, and several smaller, rather specialized, clearly monophyletic genera: *Orellia* Robineau-Desvoidy, *Craspedoxantha* Bezzi, *Chaetostomella* Hendel and *Chaetorellia* Hendel (Freidberg 1985; Korneyev 1985; Freidberg and Mathis 1986). The hypothesized phylogenetic relationships among the genera other than *Neaspilota* are shown in Figure 22.2, based on the cladistic analysis of the character matrix in Table 22.4.

This character state matrix consisted of 63 terminal taxa and 48 characters (including dummy outgroup and character; see Tables 22.3 and 22.4). Autapomorphies were not included. A Nelson consensus tree was obtained from the trees resulting from the mhennig* option combined with bb* (branch swapping) option (Figure 22.2A). The tree length = 152, ci = 0.38, ri = 0.80. The second tree (Figure 22.2B) was obtained from the mhennig* option combined with a series of successive character weightings. The tree length = 384, ci = 0.60, ri = 0.91.

Both trees show certain characters considered important synapomorphies of the taxa within the tribe Terelliini. Due to the small sample of characters and some restrictions of the techniques applied, the trees show only putative relationships, and must be considered with precaution. Several species groups and genera are supported as monophyletic by synapomorphies, as indicated on Figure 22.2.

The *quadratula–vilis* complex of *Terellia* is the most generalized group of species, in which the aculeus is usually blunt (plesiomorphy), subapical lobe of glans long (plesiomorphy), and acrophallus lobes broad. It includes eight Central Palearctic species (three undescribed) associated with rather generalized genera of the tribe Cardueae — *Echinops* L. and *Jurinea* Cass. Monophyly of this complex is putative, and is supported only by the absence of black setulae on the abdominal tergites, a character of uncertain polarity. This group corresponds to the *quadratula, deserta,* and *vilis* species groups (Korneyev 1985; 1988). If recognized as a genus, its valid name would be *Galada* Hering (type species: *G. vilis* Hering), now treated as a synonym of *Terellia.*

The next group, the genus *Orellia*, includes three Palearctic species associated with plants of the asteraceous tribe Cichorieae. They also have a blunt aculeus (plesiomorphy), long lobe of cell bcu, bulging glans with short subapical lobe (synapomorphies), and share the presence of black shining scutal spots on the transverse suture (apomorphic) with two undescribed species of the *vilis* complex associated with *Echinops*.

**TABLE 22.3**
**List of Characters Used in Analysis of the Relationships Species and Genera of Terelliini Exclusive of *Neaspilota***

1. Dummy character
2. Medial lobe of pedicel in male: 0, unmodified; 1, produced
3. Antennal first flagellomere: 0, elongate, rounded; 1, shortened, subquadrate; 2, shortened and acute apically
4. Palpus: 0, not long; 1, long, conspicuously extended over anterior margin of oral cavity
5. Vibrissal angle: 0, covered with short and thin setulae; 1, covered with short setulae and two to five long setae
6. Postocular setae: 0, all yellowish white; 1, black in lower portion of row; 2, mostly black
7. Lyrate pattern of mesonotum: 0, uniformly black; 1, reddish at center of scutum, the rest black; 2, completely reddish yellow
8. Presutural dorsocentral seta: 0, absent; 1, present
9. Transverse suture: 0, without bare spots; 1, with two bare, shining spots
10. Base of postsutural dorsocentral seta: 0, with round, shining black spot; 1, without black shining spot
11. Base of prescutellar acrostichal seta: 0, with round, shining black spot; 1, without black shining spot
12. Bases of postsutural anterior supra-alar seta: 0, without black shining spot; 1, with round, shining black spot
13. Scutellum: 0, with white setulae only; 1, with some black setulae
14. Bases of scutellar setae: 0, without black spots; 1, with round black spots
15. Scutellum apex: 0, without black spot; 1, with large black spot
16. Upper portions of anepisternum and katepisternum: 0, unicolorous with medial portion of anepisternum, pleura not striate; 1, lighter than medial portion of anepisternum,whitish yellow, grayish green, to bluish gray, pleura striate
17. Extension of cell bcu: 0, moderately short, not reaching level of crossvein BM-Cu; 1, very short, not exceeding level of its anterolateral corner; 2, very long, exceeding level of crossvein BM-Cu (*nonadditive*)
18. Wing pattern: 0, striate or spotted; 1, hyaline
19. Abdominal terga: 0, without spots; 1, spots small, hidden below margins of preceding terga; 2, normally developed (polarity unresolved)
20. Lateral spots on male abdominal tergum 5: 0, two pairs; 1, joined into triangular strip; 2, absent (*nonadditive*)
21. Spots on abdominal terga, when present: 0, form four separate rows; 1, medial pair fused; 2, only medial pair developed (*nonadditive*)
22. Abdominal terga: 0, with black setulae only; 1, terga 1 to 3 with whitish setulae; 2, terga 4 to 5 (to 6) at least laterally with whitish setae (polarity unresolved)
23. Male tergum 5: 0, shorter than terga 1 to 4 together; 1, not shorter than terga 1 to 4 together
24. Anterior part of hypandrium: 0, without sculptured pouch; 1, with pouch, sculptured with rounded structures
25. Lateral surstylus: 0, neither narrowed, nor elongate; 1, narrowed, elongate
26. Stipe of distiphallus: 0, bare; 1, spinulose
27. Subapical lobe of glans: 0, well developed; 1, reduced or absent
28. Subapical lobe of glans: 0, simple; 1, apically bilobate, fanlike compressed
29. Subapical lobe of glans apically: 0, caplike; 1, acute; 2, hooklike
30. Basal lobe of glans (ligula): 0, absent; 1, moderately developed; 2, very long
31. Semitubular sclerites of acrophallus: 0, separate at least at apex; 1, fused forming single tube
32. Semitubular sclerites of acrophallus, when separate: 0, not exceeding length of subapical lobe; 1, exceeding length of subapical lobe
33. Tubular basal part of acrophallus: 0, not very long, usually shorter or at most 1.5 times longer than semitubular sclerites; 1, very long, more than 2.5 times longer than semitubular sclerites
34. Medially directed folds on inner surface of praeputium: 0, absent; 1, elongate, covered with acute scales; two, forming 2 strongly incurved dentate lobes; 3, forming small, short papillose tubercle (*nonadditive*)
35. Praeputial sclerite: 0, without processes; 1, with two hooklike processes preapically
36. Praeputial sclerite: 0, not constricted; 1, constricted at middle of acrophallus length
37. Aculeus: 0, blunt apically; 1, narrowly pointed
38. Aculeus: 1, acute, but not narrowly pointed; 0, either pointed or narrowly pointed
39. Spermathecae: 0, well sclerotized; 1, weakly sclerotized, inconspicuous
40. Preapical portion of spermathecal ducts: 0, smooth; 1, with transverse striation
41. Preapical portion of spermathecal ducts: 0, small; 1, very large

**TABLE 22.3 (continued)**
**List of Characters Used in Analysis of the Relationships Species and Genera of Terelliini Exclusive of *Neaspilota***

42. Egg micropyle: 0, not on long neck; 1, on end of long tubular neck
43. Larva: 0, elongate, maggotlike, creamy white; 1, short ovoid, light orange
44. Last larval segment: 0, normal, not sclerotized; 1, heavily sclerotized, with single or double hornlike posterior protuberance
45. Host plants: 0, not Lactuceae; 1, Lactuceae (polarity unresolved)
46. Host plants: 0, not Cardueae; 1, Cardueae (polarity unresolved)
47. Host plants: 0, not Centaureinae; 1, Centaureinae (polarity unresolved)
48. Host plants: 0, not *Jurinea*; 1, *Jurinea* sp. (polarity unresolved)

*Note:* 1, 2 = apomorphies, 0 = plesiomorphy, if not indicated otherwise for the outgroup.

The following group, *Cerajocera* Rondani *sensu* Korneyev (1985; 1987), may be defined by the blunt (plesiomorphy), but very long aculeus (synapomorphy), as well as by the very characteristic, aberrant structure of the glans, with a large bifurcated or fan-shaped subapical lobe and the acrophallus completely reduced (synapomorphies). The larvae are borers in large flowerheads and stems of several Carduinae. No synapomorphies showing the sister group relationship of *Cerajocera* with other groups of species have been found.

Another complex includes the following species groups currently placed in *Terellia*: *blanda, megalopyge, colon, popovi,* and *virens* (Korneyev 1985), plus *T. amberboae* Korneyev and Merz, altogether a dozen Palearctic species whose broad, oval, reddish larvae breed in the flowerheads of *Serratula, Centaurea, Amberboa,* and *Carthamus,* belonging to the advanced subtribe Centaureinae of the tribe Cardueae. The valid generic name for this group is *Squamensina* Hering (type: *S. oasis* Hering, = ? *T. vectensis* Collin) (= *Whiterellia* Bassov and Nartshuk, type species: *T. virens* Loew). This complex is believed to be monophyletic because of similar larval body shape and coloration, but further study of larval characters is needed. Species of the *megalopyge* group, which breed in *Serratula,* and of the *blanda* group, whose larvae are not known, share peculiar long surstyli (synapomorphy), as well as the extremely long fifth tergite of the male (synapomorphy).

Two species of the *vilis* complex, *T. matrix* Korneyev and *T. orheana* Korneyev, share hyaline wings and white setulose abdominal terga with the *virens* group, and appear as members of the *colon–virens* complex in some cladograms when the successive weighting technique was used (Figure 22.2B). This may indicate homoplasy or that this is the true relationship of these species. It may mean that most or all of the *colon-virens* complex or perhaps even *Orellia* (see above) may be derived taxa of the *quadratula–vilis* complex, but the adult morphological data are insufficient to confirm this hypothesis.

Most of the previously enumerated groups have been formerly assigned either to *Orellia* (Hendel 1927) or to *Terellia* (Korneyev 1985), but actually share nothing with them but plesiomorphic (blunt aculeus, white setulae) or "negative" characters which are either of unresolved polarity or highly homoplastic (lack of spots, setae, wing pattern, *etc.*).

The next cluster includes the species now assigned to the genera *Craspedoxantha*, *Chaetostomella*, *Chaetorellia*, and the remaining species of *Terellia* (including its type species), and is clearly monophyletic. They all possess a granular sculptured hemispheric pouch in the anterior portion of the hypandrium (Figure 22.3) (synapomorphy not present in the other groups above), and a narrowly pointed aculeus (synapomorphy).

The genus *Terellia* (*s. str.*) includes the species of the *serratulae* and *ruficauda* groups, plus the Nearctic *T. occidentalis* (Snow) and *T. palposa* (Loew), all of which have long semitubular sclerites of the acrophallus (synapomorphy), and the paired flaps inside the glans sparsely covered with blunt spines (synapomorphy).

**TABLE 22.4**
**Character States of Taxa used in Preliminary Analysis of Relationships among the Species and Genera of Terelliini Exclusive of *Neaspilota* (character numbers refer to Table 22.3)**

| | Character Numbers |
|---|---|
| | **000000000111111111122222222223333333333444444444** |
| | **123456789012345678901234567890123456789012345678** |
| Dummy outgroup | 0000000000000000000???200000000100000000000000???? |
| *Orellia falcata* | 000000001001010020200000000001000000000110001000 |
| *O. scorzonerae* | 000000001000000000200000000001000000000000?001000 |
| *O. stictica* | 000000201001010020200000000001000000000110011000 |
| *Terellia vilis* | 0000000000000000001202 0?000001000001???????????? |
| *T.* sp. 1 from *Echinops* | 000000001000000000200200000001000001010 00???01?? |
| *T. quadratula* | 00000000000000000020200000011000000000000???0100 |
| *T. deserta* | 000000000000000001202000000010000010000 0?100101 |
| *T.* sp. 2 from *Jurinea* | 0000000000000000020020000000100000100000?100101 |
| *T. matrix* | 000000000110000001022?200000000100000 1???????????? |
| *T. orheana* | 000000000110000001 01?200000000100000100000?010101 |
| *T. colon* | 0000000000000000011200000000001000100000 10?100110 |
| *T.* sp. 3 from *Cousinia* | 00000000010000000112002000000001000300000 11?100100 |
| *T. odontolophi* | 000000000110000001210200000000100030000000?100110 |
| *T. popovi* | 000000000100100001100000001010101?0000??0???0110 |
| *T. dubia* | 000000000110000001211110100001011000000000?100110 |
| *T. megalopyge* | 00000000011000000121111010000101100000 0???100110 |
| *T. vectensis* | 000000000110000001211110100001011000000000?100110 |
| *T. blanda* | 000000010000000000210110100001001000 00?11??????? |
| *T. ermolenkoi* | 000000000000000000210110100001001000 00?00??????? |
| *T. uncinata* | 00000000000000000112002000000001001300000 10?100110 |
| *T. virens* | 00000000010000000112002000000001001300000 1??100110 |
| *Cerajocera armeniaca* | 01000000000000000120010000001010000000000????????? |
| *C. ceratocera* | 010000000000000000200100000010100000000100?011110 |
| *C. clarissima* | 000000000000000001200200000101000000000110?011101 |
| *C. gynaecochroma* | 000000000000000000200200000101000000000100?011100 |
| *C. nigronota* | 000000000000000000200100000101000000000100?000100 |
| *C. plagiata* | 010000000000000000200100000101000000000100?011110 |
| *C. setifera* | 000000000000000001200100000101000000000100?011101 |
| *C. tussilaginis* | 0000001000000000002001000001010000000001???000100 |
| *Terellia amberboae* | 002200000110000101222200000000100110001?00?100110 |
| *T. apicalis* | 00110000000000001002?020?00???1????00?0?????????? |
| *T. fuscicornis* | 00110000000000001012002010000010102001 00??????10? |
| *T. longicauda* | 001100000000000010120020100000101020010000 0010100 |
| *T. nigripalpis* | 00110000000000001012002010000010102 00?00???01?100 |
| *T. occidentalis* | 0011000000000000100200101000001010200100 00?0?0100 |
| *T. palposa* | 0011000000000000000200101000001010200100???0?0100 |
| *T. ruficauda* | 00110000000000001002001010000110102001000 0?010100 |
| *T. sabroskyi* | 00110000000000001012002010000110102001000 0?010100 |
| *T. serratulae* | 0011000000000000101200201000001010200100???010100 |
| *T. winthemi* | 0011000000000000000200101000001010200100???010100 |
| *Craspedoxantha bafut* | 000000200002100000200?010?0001000000010011???0?00 |
| *C. unimaculata* | 00000?000002111000200?010?0001000000010011???0?00 |
| *C. marginata* | 000000000002110000200001010002000000010011???0000 |
| *C. polyspila* | 00000?000002110000200?010?0001000000010011???0?00 |

**TABLE 22.4 (continued)**
**Character States of Taxa used in Preliminary Analysis of Relationships among the Species and Genera of Terelliini Exclusive of *Neaspilota* (character numbers refer to Table 22.3)**

| | Character Numbers |
|---|---|
| | **000000000111111111122222222223333333333444444444** |
| | **123456789012345678901234567890123456789012345678** |
| *C. octopunctata* | 00000?00000211000020000 10?0001000000010011???0000 |
| *C. manengubae* | 000001000002100000200201 0?000100000010 0111??0?00 |
| *Chaetorellia acrolophi* | 00000111000000100020000 10?00111000001000100?0110 |
| *C. australis* | 00000111000000100020000 10?0011100000100 01?0?0110 |
| *C. carthami* | 00000111000100100020000 10?001?100000100 01?0?0110 |
| *C. conjuncta* | 00000111000000100020000 10?00111000001000100?0110 |
| *C. hestia* | 00000101000000100020000 10?001?100000100 01?0?0110 |
| *C. isais* | 00000101000100100020000 10?0010100000100 01?0?0110 |
| *C. jaceae* | 00000111000000100020000 10?00111000001000 10000110 |
| *C. loricata* | 00000101000100100020000 10?00101000001000 11000110 |
| *Chaetostomella rossica* | 00001010000000101020010 10?00210000101000 00000101 |
| *C. succinea* | 00000101000100100020000 10?0010100000100 01?0?0110 |
| *C. cylindrica* | 00001200000000101020000 10100210000101000 0000110 |
| *C. steropea* | 000000?0000000101010 0?010?00210000101000 0?0?0100 |
| *C. stigmataspis* | 00001220000000101020000 10?00210000101000 0?000100 |
| *C. trimacula* | 00001010000000101020 0?010?00210000101000 0?0?0??0 |
| *C. undosa* | 00001200000000101020000 10?00210000101000 0?0?0100 |
| *C. vibrissata* | 00001210000000101020000 10?00210000101000 0?0?0100 |

The sister group of *Terellia* (*s. str.*) is *Craspedoxantha* + *Chaetostomella* + *Chaetorellia*. This clade is also monophyletic. Most of its members have black spots at the bases of the presutural supra-alar setae and on the scutellum (synapomorphies), but neither long semitubular sclerites of acrophallus, nor the spinose inner lobes of the glans (plesiomorphies). Each of these genera is well defined by several autapomorphies, but the relationships among them are not resolved. Freidberg and Mathis (1990) provided a comprehensive phylogenetic analysis of the genus *Craspedoxantha*.

These data show that the current concepts of the genera of the Terelliini serve diagnostic purposes, but do not mirror the phylogeny of the tribe. However, pending additional studies to confirm these results and resolve the remaining uncertain relationships, it seems preferable to retain the current classification rather than further splitting the non-monophyletic genus *Terellia* into smaller monophyletic genera based on male genitalic characters that would cause serious diagnostic difficulties.

### 22.2.3.2 Remaining Part of Tephritinae (Minus Terelliini)

To our knowledge, the Tephritinae, except Terelliini, do not have a basally sclerotized, apically trumpetlike, expanded subapical lobe of the glans, and if they possess one at all, it is a soft, membranous, tail-like process covered with fine microtrichia or spinules. In two cases (in *Neotaracia* Foote and *Stenopa* Loew) the basal part of the lobe is sclerotized like in Terelliini (symplesiomorphy), but apically it is long microtrichose (apparently a synapomorphy of the Tephritinae minus Terelliini). Some structures of the glans subapical "cap" in *Myopites* Blot are possibly remainders of the subapical lobe sclerotization. Very often, such a tail-like process is completely lacking in different distant genera, including *Neaspilota* (Terelliini?), *Urophora* Robineau-Desvoidy (Myopitini), *Oedaspis* Loew (Dithrycini), or *Tephritis* Latreille (Tephritini).

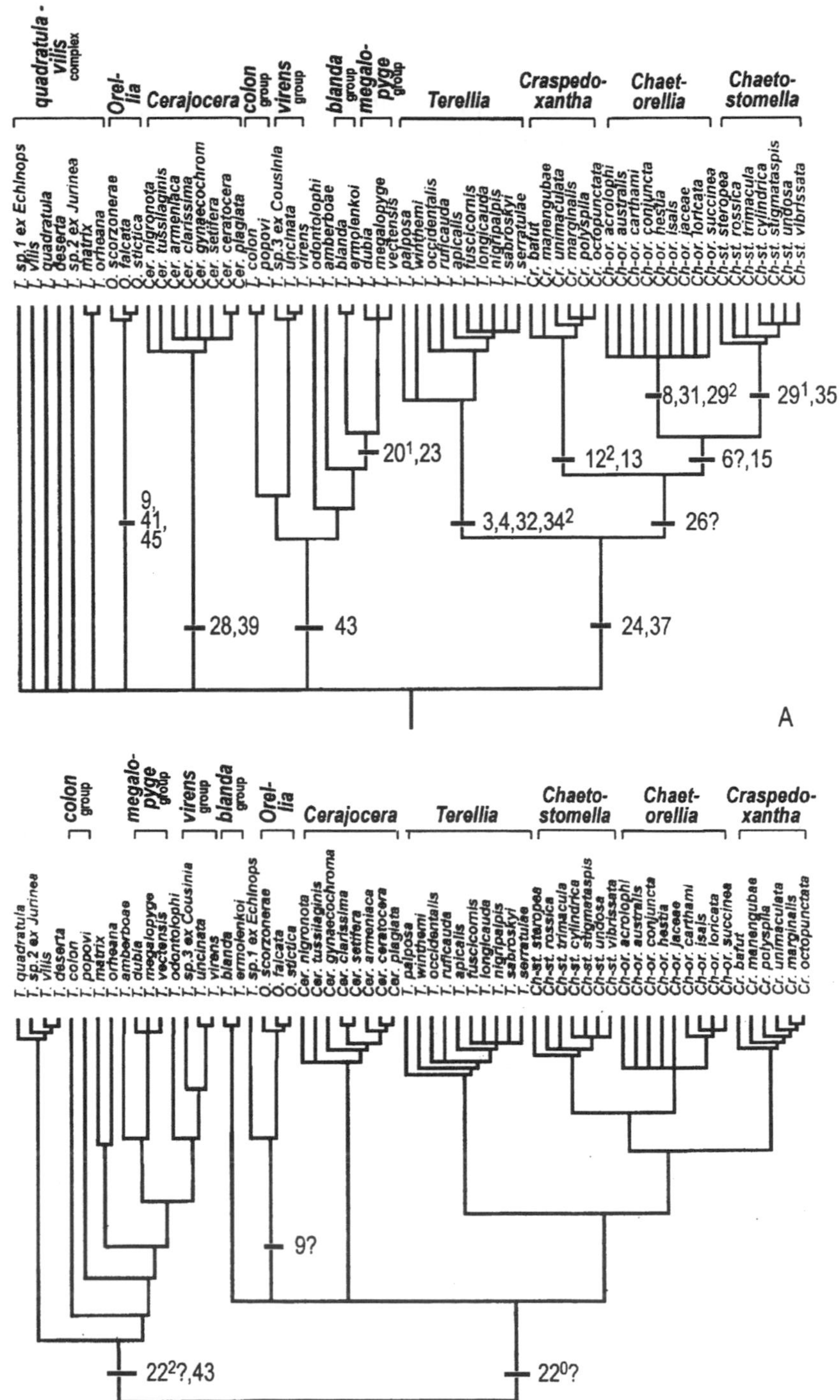


**FIGURE 22.2** Phylogenetic relationships of Terelliini exclusive of *Neaspilota*. (A) Nelson consensus tree from mhennig* bb* 100 trees (overflow); (B) mhennig tree obtained after a series of successive weightings from the character matrix of Table 22.4. Character numbers at black bars represent state 1 unless followed by number of another state in superscript (see Table 22.3).

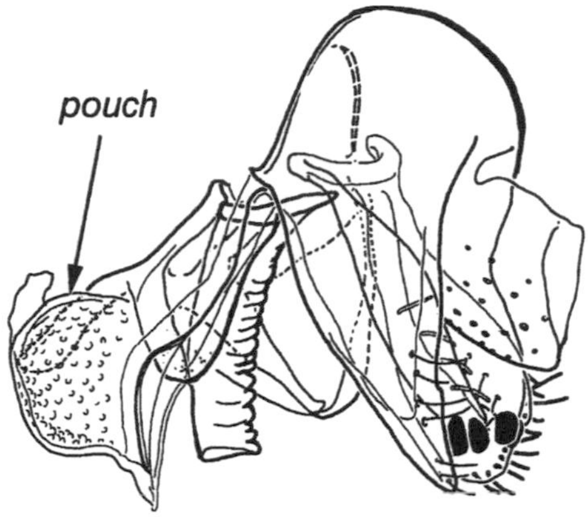


**FIGURE 22.3** Hypandrium and epandrium of *Chaetorellia loricata* (right lateroventral view).

The soft membranous subapical lobe of the glans with a microtrichose apex is believed to be a synapomorphy of the tribes of Tephritinae, except Terelliini. The reduction of the subapical lobe sclerotization appears to have taken place several times within the Tephritidae. In the other subfamilies there are species that have no subapical lobe, species that have the lobe, and species that have a tail-like process instead.

This large complex may be divided into two unequal groups. Although the members of each group are highly variable, and often differ considerably from each other or greatly resemble some members of the other group, they are rather consistent in respect to the following character.

The first group, which includes the *Tomoplagia* group of genera, Cecidocharini, Myopitini, Noeetini, and Xyphosiini, and probably, the *Axiothauma* group of genera, differs by the lateral vertical seta always concolorous with the medial vertical seta, at least two-thirds as long as the medial vertical seta, and somewhat longer than the anterior orbital seta. This character state is obviously plesiomorphic, as it occurs in all outgroups, including Terelliini.

The second group includes the Dithrycini, Eutretini, Schistopterini, Tephritini, and, with some reservations, Tephrellini. They share the lateral vertical seta white and shorter than the anterior orbital seta and less than half the length of the medial vertical seta. It is concolorous with the major white setae of the postoccular row that are inserted between the shorter, acute, and black setae. There are some exceptions, where all the white setae become black in some melanistic species of *Oedaspis*, as well as in some Tephrellini and Schistopterini, but otherwise this character remains quite stable to be accepted as a possible synapomorphy of this group of tribes.

The state of this character in the Afrotropical genera *Axiothauma* Munro, *Orthocanthoides* Freidberg and *Cryptophorellia* Freidberg and Hancock is intermediate. The lateral vertical seta is as long as the anterior orbital seta, and half as long as the medial vertical seta, and always black, concolorous with the uniformly black and rather long postocular setae. It may be either a symplesiomorphy with the Lower Tephritinae or a character reversal.

### 22.2.3.3 Group of Genera Related to *Tomoplagia* (Acrotaeniini, *partim*)

This group of genera was defined by Norrbom (1987; 1992) as a part of the tribe Acrotaeniini established by Foote et al. (1993) for the New World genera *Acrotaenia* Loew, *Acrotaeniacantha* Hering, *Baryplegma* Wulp, *Caenoriata* Foote, *Euarestopsis* Hering, *Neotaracia* Foote, *Polionota* Wulp, *Pseudopolionota* Lima, *Tetreuaresta* Hendel, and *Tomoplagia* Coquillett.

Reexamination of the genera included in Acrotaeniini shows that actually it is an artificial, heterogeneous aggregation. At least five genera (*Acrotaenia, Acrotaeniacantha, Baryplegma, Euarestopsis,* and *Tetreuaresta*) include species that have shorter black postocular setae between longer white or yellow ones, and short white lateral vertical seta, and which therefore belong to the Higher Tephritinae. Some species of *Acrotaenia* (*A. testudinea* (Loew), *A. tarsata* Doane) have uniformly yellow medial vertical seta, lateral vertical seta and postocular setae, and share vertical head with blunt fronto-facial angle, like in some *Tomoplagia*. *Acrotaenia otopappi* Doane has white short lateral vertical seta, short black postocular setae mixed with longer and white, and the fronto-facial angle close to 90°, showing strong similarity to *Acinia* Robineau-Desvoidy and some Eutretini of the Higher Tephritinae. Judging from the highly specific wing pattern and the structure of male terminalia, it is certainly congeneric with the type and other species of *Acrotaenia*. Otherwise, the lateral vertical seta in all the *Acrotaenia* species and in the allied genera is much shorter than the anterior orbital seta and half of the medial vertical seta length, so these genera apparently belong to the Higher Tephritinae.

*Tomoplagia, Polionota, Neotaracia,* and *Caenoriata* (and apparently *Pseudopolionota,* known to me from descriptions only) are not closely related to *Acrotaenia*, and must be retained in the Lower Tephritinae.

Examined species in this group of genera share the following characters: bare arista (autapomorphy or synapomorphy with *Gymnocarena* Hering?) (Norrbom 1992, Figure 2), bare frontal vitta (autapomorphy or synapomorphy with *Gymnocarena* and Terelliini *s. str.*?; highly homoplastic character); blunt fronto-facial angle (autapomorphy; homoplasy with some *Acrotaenia*); broad or very broad palpi (autapomorphy; apparently homoplasy with some *Acrotaenia* and Eutretini); posterior orbital seta reclinate, concolorous with the anterior one; lateral vertical seta longer than anterior orbital seta and longer than half the length of the medial vertical seta; postocular setae unicolorous (whitish-yellow to brownish-yellow) and uniformly long; costal spur at the apex of Sc never very long; vein $R_{4+5}$ setulose dorsally to level of $R_{2+3}$ apex, and ventrally to level of R-M; wing pattern striate, consisting of fused crossbands, or radiate, but not reticulate; scutum with more or less distinct scapular setae (plesiomorphies); glans with a very long tubular acrophallus free from posterodorsal wall of the praeputium (the sinus surrounding the base of the acrophallus); flaps of the latter supported by two proximal arms, often serrate (see Foote 1979, Figures 4 to 6) (synapomorphies with Terelliini, including *Neaspilota*?); acrophallus conspicuously asymmetrical, bowed to the right side (synapomorphy with *Neaspilota*?); a sclerotized basal lobe similar to that of Terelliini other than *Neaspilota* is found in some *Tomoplagia* species (plesiomorphy); aculeus broad and blunt in the ground plan of the group (plesiomorphy).

### 22.2.3.4 Xyphosiini

This tribe was proposed by Hendel (1927) to include some genera with yellow setae, a reticulate wing pattern and vein $R_{4+5}$ setulose. Recently, it was found to be heterogeneous (Korneyev 1995; and in preparation), and its concept must be restricted to two closely related Palearctic genera, *Xyphosia* Robineau-Desvoidy and *Ictericodes* Hering. They share the following characters: all setae uniformly yellow; no minute black postocular setae (plesiomorphies); posterior orbital seta inclinate (synapomorphy or homoplasy with Terelliini); $R_{4+5}$ setulose to DM-Cu (plesiomorphy); wing pattern reticulate (synapomorphy); aculeus broad and blunt; and larvae not forming galls (plesiomorphies). They share these characters plus yellow-colored setae with the *Tomoplagia* group (see above), differing from them by Palearctic distribution, frons finely setulose, and the reticulate wing pattern (the latter is probably the only true synapomorphy of the tribe). Another genus that may belong here is *Icterica* Loew (Korneyev 1995).

### 22.2.3.5 Noeetini

This group of genera was recognized to include *Jamesomyia* Quisenberry, *Acidogona* Loew, *Xenochaeta* Snow, *Noeeta* Robineau-Desvoidy, and *Paracanthella* Hendel and assigned (as the *Noeeta* group) to Eutretini by Foote et al. (1993). Later it was shown to have clear synapomorphies

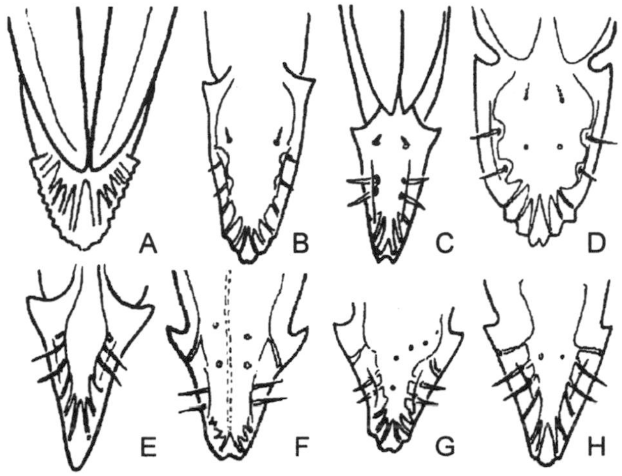


**FIGURE 22.4** Apex of aculeus of Noeetini (ventral view). (A) *Hypenidium graecum*; (B) *Ensina sonchi*; (C) *Jamesomyia geminata*; (D) *Paracanthella pavonina*; (E) *Hypenidium roborowskii*; (F) *Xenochaeta dichromata*; (G) *Acidogona melanura*; (H) *Noeeta pupillata.*

with *Ensina* Robineau-Desvoidy and *Hypenidium* Loew, and was defined as a separate tribe (Norrbom and Korneyev in Wang 1996; Norrbom and Korneyev in Norrbom et al. 1999). The distribution of this tribe is mainly Holarctic, with some species of *Ensina* Robineau-Desvoidy occurring in the Andes.

All the members of the tribe share the narrow aculeus with the barbed cercal unit (Figure 22.4; synapomorphy). Also, most of them share the oviscape densely microsetulose (synapomorphy), and larvae breeding in flowerheads of various Lactuceae plants, mostly *Lactuca* and *Hieracium*, often forming a nonlignified flower bud gall.

The Nearctic genus *Jamesomyia* seems to share general appearance with Xyphosiini and the *Tomoplagia* group in having yellow setae, setulose $R_{4+5}$, reticulate wing pattern, and vertical plates short (symplesiomorphies). The remaining genera of the tribe seem to have these ground plan characters variously modified.

A cladistic analysis of the relationships among the genera of Noeetini was conducted using the character state matrix shown in Table 22.6. It consisted of 30 terminal taxa and 44 characters (including dummy outgroup and character), which are listed in Table 22.5. Autapomorphies were not included. A Nelson consensus tree was obtained from the trees resulting from the mhennig* option combined with bb* (branch swapping) option (Figure 22.5A). The tree length = 120, ci = 0.42, ri = 0.75. The second tree (Figure 22.5B) was obtained from the mhennig* option combined with a series of successive character weightings. The tree length = 356, ci = 0.67, ri = 0.88.

Both trees show certain characters considered important synapomorphies of the taxa within the tribe Noeetini, together with different outgroups and some genera previously placed with *Noeeta* and *Paracanthella* in Dithrycini (Hendel 1927) or distantly related taxa with similar states of analyzed characters (for purposes of successive character weighting). The cladograms obtained by using the different techniques of analysis are consistent in respect to the monophyly of Noeetini and relationships of included genera, except for unresolved polytomies of *Noeeta* + *Acidogona* + *Xenochaeta* and of the species of *Hypenidium*, as indicated in Figure 22.5.

Initially, the Noeetini lineage bifurcates into two branches (Figure 22.5). The first one consists of the genus *Ensina* which differs by the head elongate and the posterior orbital seta lacking (autapomorphies), but the basiphallus and acrophallus short and unmodified (compared with the subfamily ground plan). The second branch includes the remaining genera, which share strongly elongate, narrowly pointed acrophallus and often long L-shaped modified basiphallus (synapomor-

## TABLE 22.5
## Character States of Taxa Used in Analysis of Relationships among the Genera of Noeetini

1. Dummy character
2. First flagellomere: 0, rounded apically; 1, slightly pointed and incised dorsoapically; 2, strongly pointed and incised dorsoapically
3. Anterior frontal seta: 0, in line with posterior frontal setae; 1, medial to line of posterior two or three setae
4. Number of long unicolorous frontal setae: 0, three; 1, four; 2, two (*nonadditive*)
5. Third frontal seta: 0, concolorous with other frontal seta; 1, whitish
6. Vertical plates: 0, restricted to posterior quarter or third of frons length; 1, long, reaching middle of frons length
7. Posterior orbital seta: 0, zero to one; 1, two
8. Posterior orbital seta: 0, yellow to brown, if whitish yellow, then unicolorous with anterior orbital and frontal setae; 1, whitish
9. Medial vertical seta: 0, dark brown; 1, yellow
10. Lateral vertical seta: 0, acuminate, black to yellow; 1, thickened and short, white
11. Number of longer and thickened white postocular setae: 0, none white, all black; 1, one to three; 2, more than three
12. Shorter setulae between longer setae of postocular row: 0, absent; 1, present
13. Eye: 0, rounded to moderately high; 1, vertical, 1.6 to 2.3 times higher than long
14. Fronto-orbital plate: 1, shining; 0, matte
15. Scutal setulae: 0, unicolorous yellow to yellowish brown; 1, mixed yellow and black; 2, black
16. Scutal setulae: 0, not in pattern; 1, in stripes or clusters
17. Posterior notopleural seta: 0, unicolorous with anterior seta; 1, white
18. Scutellum: 0, subshining or matte; 1, shining
19. Setulae between scutellar setae: 0, faint yellow, black or indistinct; 1, white, strong and erect
20. Anterior pair of white scutellar setae: 0, closer to margin or absent; 1, closer to middle of scutellum
21. Postnotum: 0, microtrichose; 1, shining medially
22. Costal spines: 0, very short (0.8 to 1.2 times as long as costa width); 1, moderately developed (1.2 to 1.5 times longer than costa width); 2, very strong (approximately as long as first flagellomere)
23. Costa at apex of subcosta: 0, not incised; 1, incised
24. Discal reddish bulla: 0, absent; 1, present
25. Posteroapical lobe of cell bcu: 0, moderately long; 1, short; 2, inconspicuous
26. Wing pattern: 0, striate or dark with hyaline wedges; 1, reticulate; 2, radiate
27. Abdominal terga 1 to 4: 0, shining; 1, subshining; 2, densely microtrichose
28. Abdominal tergum 5 of male and 6 of female: 0, subshining; 1, shining
29. Medial surstylus: 0, short, closely associated with lateral surstylus; 1, fingerlike, long, well separated from lateral surstylus
30. Preglans area of distiphallus: 0, finely setulose or bare; 1, spinulose
31. Subapical lobe of glans: 0, present, nail-like; 1, transformed into flagellum-like membranose appendix or absent; 2, short, thornlike, not transformed into flagellum
32. Acrophallus: 0, short, not exposed from the praeputium; 1, moderately long, exposed from the praeputium; 2, very long, subequal to basal part of glans
33. Apical portion of oviscape: 0, with sparse normal setulae; 1, densely covered with microscopic setulae
34. Ventral surface of eversible membrane: 0, with numerous, moderate to small, triangular scales; 1, with few strong hooklike scales in middle of its length
35. Dorsal surface of eversible membrane: 0, evenly covered with scales; 1, with shallow medial groove, devoid of scales
36. Ventral surface of eversible membrane: 0, with scales evenly covering most of surface; 1, with scales restricted to medial area
37. Aculeus: 0, slightly compressed dorsoventrally throughout its length ("swordlike"); 1, subconical in basal portion, sagittally compressed in medial part of its length (sagittate)
38. Aculeus: 0, gradually tapered; 1, with arrowlike, dorsoventrally compressed portion, conspicuously separate from its basal part
39. Spermathecae: 0, elongate; 1, spherical (polarity unresolved)
40. Host plant association: 0, not gall-forming species; 1, gall-formers in Anthemidae (polarity unresolved)
41. Host plant association: 0, larva not in flowerheads of Lactueae; 1, larva in flowerheads of Lactueae
42. Host plant association: 0, larva not in flowerheads of *Lactuca*; 1, larva in flowerheads of *Lactuca*
43. Host plant association: 0, larva not in flowerheads of *Hieracium*; 1, larva in flowerheads of *Hieracium*
44. Labella: 0, normal; 1, rudimentary

*Note:* 1, 2 = apomorphies, 0 = plesiomorphy, if not indicated otherwise for the outgroup.

**TABLE 22.6**
**Character States in Taxa Used in Preliminary Analysis of Relationships among the Genera of Noeetini and Other Genera Previously Assigned to Dithrycini (character numbers refer to Table 22.6)**

| | Character numbers<br>00000000011111111112222222222333333333344444<br>12345678901234567890123456789012345678901234 |
|---|---|
| Dummy outgroup | 000000000??00000000?10000?000000000000000000 |
| *Terellia* | 00000000002000000000010000100000000000000000 |
| *Polionota* | 0000000010200000000000100001000100000000??0000 |
| *Gymnocarena* | 000000001020000000000010000100010000000?00000 |
| *Xyphosia* | 00000000102000000000001000110001000000000000000 |
| *Ensina* | 0000000?0020000000000010010100 0?0110011001110 |
| *Jamesomyia* | 01110100002100100000001000110 00?2110011101100 |
| *Paracanthella* | 021101110021001011100100111110?2110011101100 |
| *Hypenidium graecum* | 0101000000001000000001000200100?2110011101100 |
| *H. roborowskii* | 0100000000000100000000100020010 0?1110011101?00 |
| *H. oculatum* | 0100000000000100000000100020010 0??1?00???01?00 |
| *Xenochaeta* | 011001000021001011001100111110?2000011101010 |
| *Acidogona* | 021001000021001011101100111110?2000011101010 |
| *Noeeta* | 021001110021001011101100022110?2000011101010 |
| *Oedaspis fissa* | 00000000000100200100110010110010001100010001 |
| *O. multifasciata* | 00010001012100000100010000200010001100010001 |
| *Dithryca* | 01000001012100000100010001210010001100010000 |
| *Rhabdochaeta* | 0212?11101110000001011111220101200?000000000 |
| *Schistopterum* | 0002?00?0111000100101111121100 1?01?000?00000 |
| *Brachiopterna* | 0002?00101110001000011112021001?00?000000000 |
| *Xanthomyia* (*s. str.*) | 0102?00101110000010001000121002000100000000 0 |
| *Xanthomyia* (*Paranoeeta*) | 02021011011100000000001000220002000100000000 0 |
| *Paracantha culta* | 00121101011101010011020002200010000000000000 |
| *P. ruficallosa* | 00021111011101010011021002200010000000000000 |
| *Strobelia* | 0002?00101110101000002001120001000000000000 0 |
| *Rachiptera* | 0002?0010111010100000200012000100000000000 00 |
| *Oxyna parietina* | 0002?001012100001000010001200 1?0001000010000 |
| *Campiglossa irrorata* | 0002?011012100000000001000120 01?1001000010000 |
| *C. achyrophori* | 0002?0010121000000000010001200 1?1001000010010 |
| *Tephritis oedipus* | 0002?00101210000100001000120001000100000 1100 |

phies), but nonelongate head. As far as is known, larvae of the members of the second cluster breed singly in flowerheads (soft flower bud galls).

This second branch may be divided into two separate clusters: (1) *Hypenidium* and (2) *Jamesomyia* + *Acidogona* + *Xenochaeta* + *Noeeta* + *Paracanthella*. Species of *Hypenidium* have the wing pattern mostly dark striate, hyaline round spots absent, eyes vertical, cell bcu with the posteroapical extension lacking, the epandrium very broad with short medially directed lateral surstyli (synapomorphies), but the frons narrow, medial surstyli short, and white setae not developed (plesiomorphies).

The second cluster differs by the presence of variously colored setae (white + black or yellow, or all three colors), both major and minor, and by the frons widened, with long vertical plates (synapomorphies), eyes never vertical, wing pattern reticulate, and cell bcu lobate (plesiomorphies).

With the exception of the monotypic genus *Jamesomyia* that retains the ground plan characters of the frons and male epandrium, in the remaining genera (*Acidogona* + *Xenochaeta* + *Noeeta* +

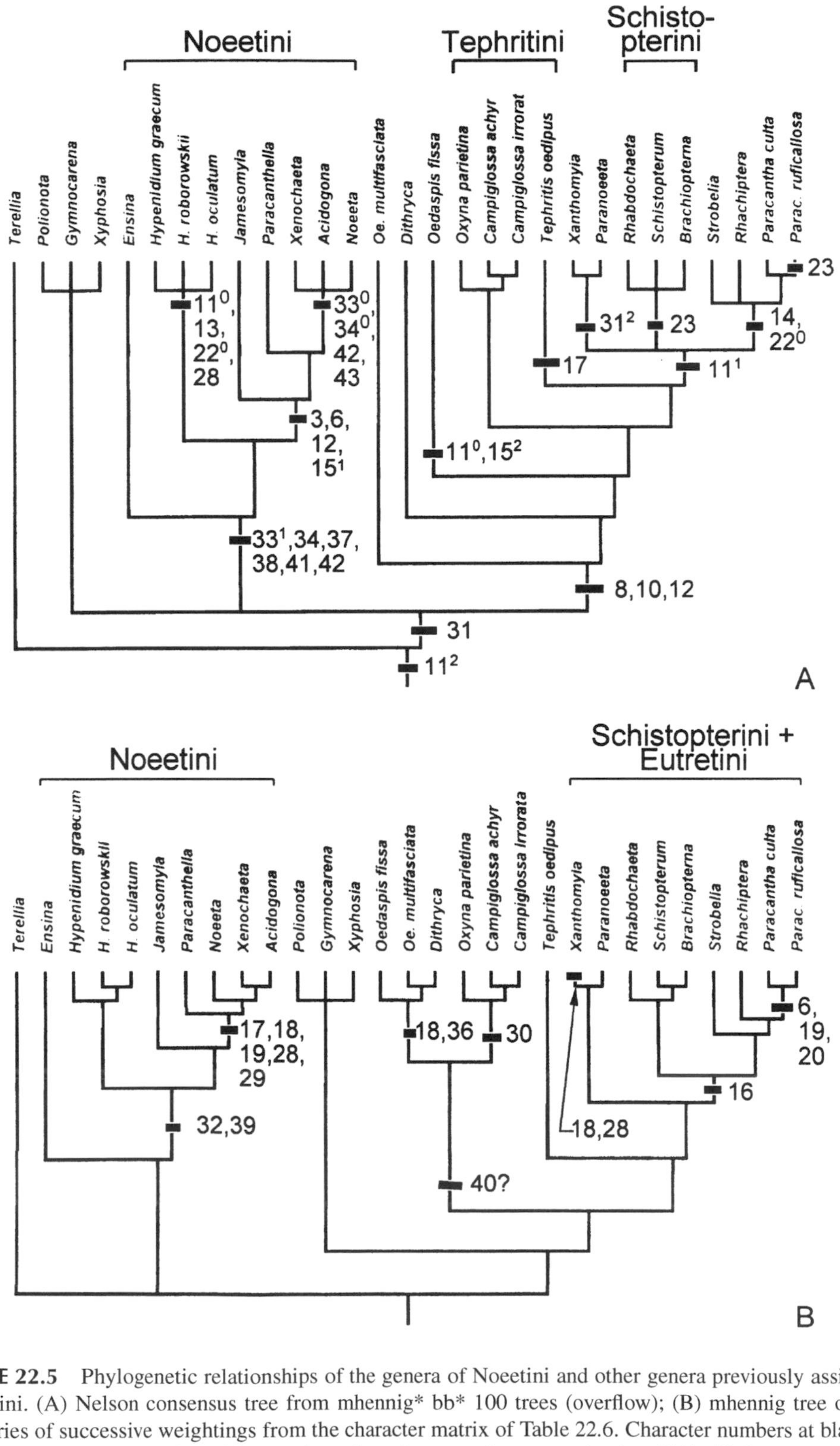


**FIGURE 22.5** Phylogenetic relationships of the genera of Noeetini and other genera previously assigned to Dithrycini. (A) Nelson consensus tree from mhennig* bb* 100 trees (overflow); (B) mhennig tree obtained after series of successive weightings from the character matrix of Table 22.6. Character numbers at black bars represent state 1 unless followed by number of another state in superscript (see Table 22.5).

*Paracanthella*) the frons is further widened, with the anterior frontal seta conspicuously displaced medially, the scutellum is convex and shining, with white erect minor setae between dark major scutellar setae, and the medial surstylus is separate from the lateral, and long and fingerlike (synapomorphies). The relationship among these genera is not completely resolved, as most characters used in the analysis have a mosaic distribution, and will be discussed in detail in the forthcoming revision of the tribe (Korneyev and Norrbom, in preparation).

### 22.2.3.6 Group of Genera Related to *Axiothauma*

This group includes three Afrotropical genera, *Axiothauma* Munro, *Orthocanthoides* Freidberg, and *Cryptophorellia* Freidberg and Hancock, which share having two (three in *Axiothauma*) frontal, two unicolorous orbital, and uniformly black and long postocular setae, and very similar structure of male terminalia. This group was initially established as part of the *Sphenella* group of genera of the Tephritini as defined by Freidberg (1987) and Freidberg and Hancock (1989). They based it on the common host plant tribe, Senecioneae (no known exceptions), although some other, unrelated tephritids utilize plants of the same tribe (e.g., *Trupanea* Schrank, *Campiglossa* Rondani, *Stemonocera* Rondani); *Axiothauma* spp. have not yet been reared, but are clearly associated with giant *Senecio* spp. (Freidberg, personal communication).

The presence of erect or suberect setae on the anterior surface of the midtibia certainly is an autapomorphy of this group of genera. Specimens of *A. nigrinitens* Munro and *Cryptophorellia flava* Freidberg and Hancock from Kenya examined in this study were found to have the lateral vertical seta as long as the anterior orbital seta, and half length of the medial vertical seta, and always black, concolorous with the uniformly black and rather long postocular setae.

Although the condition of the lateral vertical seta is intermediate between the Lower and the Higher Tephritinae, the absence of both the shorter black postocular setulae and the gap between the setulae of $R_1$ at the Sc bend level suggest that these genera belong to the Lower Tephritinae, rather than to the *Sphenella* group within the Higher Tephritinae (Figure 22.1). Species of *Orotava* Frey (Tephritini: *Sphenella* group) resemble some *Cryptophorellia* in having two frontal setae, setulose frontal vitta, lateral vertical seta rather long (but always white), black postocular setae and the mixed banded/maculose wing pattern. *Cryptophorellia* differs from *Orotava* by the absence of the following apomorphies found in all the Tephritini and most Higher Tephritinae: shorter setulae in the postocular row, white katepimeral seta, and a gap in the row of $R_1$ setulae.

Among the Lower Tephritinae, the *Axiothauma* group belongs to the complex of tribes that also includes Xyphosiini and Noeetini, and is characterized by the combination of setulose frontal vitta and $R_{4+5}$ (both plesiomorphic) and the tendency to have maculate wing patterns (synapomorphy?). In general appearance *C. phaeoptera* (Bezzi) is similar to *Xyphosia laticauda* (Meigen), but there is no further evidence of their close relationship.

### 22.2.3.7 Myopitini

The phylogenetic relationships among the genera of this tribe are discussed by Freidberg and Norrbom (Chapter 23), and those within the largest genus *Urophora* Robineau-Desvoidy by Korneyev and White (in press). Therefore, only the relationship of the Myopitini within the Tephritinae is considered here.

The monophyly of this tribe is supported by several autapomorphies: the two frontal and one orbital seta, cell bcu lacking posteroapical extension, vein $R_{4+5}$ bare, and the glans completely lacking acrophallic sclerites. The sister group relationship of the Myopitini is unclear, at least not from studies to date involving the morphology of the adults, although the presence of concolorous medial and lateral vertical setae shows that the tribe belongs to the Lower Tephritinae. In their ground plan, Myopitini have rather blunt aculei (as in *Eurasimona* Korneyev and White or *Myo-*

*pites*), although some advanced species of *Urophora s. str.* have long, narrow, and acute aculei, often with subapical steps, capable of piercing tissues between flowerhead bracts.

Members of the *Axiothauma* group resemble Myopitini in all the longer setae, including postoculars, black, but share no myopitine autapomorphies, except for having only two frontal setae (but in *Axiothauma* very often three to four), a character of questionable value.

Myopitini show affinities to some genera of Noeetini. *Hypenidium graecum* Loew has the head setae mostly black, the posterior orbital seta often (but not always!) lacking, the extension of cell bcu very short, and male forefemur spinulose, like in some *Urophora s. str.* These characters are believed to have appeared due to homoplasy, as the other species of this genus (*H. roborowskii* (Becker), *H. oculatum* (Becker)), and the species of *Ensina,* have the setae yellow, but the aculeus and the eversible membrane scales highly modified.

### 22.2.3.8 Cecidocharini

The Cecidocharini is a mainly Neotropical tribe, established by Hering (1947) and then redefined by Foote et al. (1993) and Norrbom et al. (1999) as the subtribe Cecidocharina of Dithrycini (= Oedaspidini). Recent study (Korneyev, in preparation) has shown that Cecidocharini and Dithrycini share only a striate wing pattern type (symplesiomorphy) and a swollen, shining scutellum (a character highly subject to homoplasy, found also in some genera of Ulidiidae, in Dacinae: Ceratitidini, Trypetinae: Carpomyini, Tephritinae: Noeetini). Cecidocharini do not possess either the short white lateral vertical seta, or the shorter dark setulae between longer postocular setae, or the highly specific pattern of the scales on the eversible membrane, and therefore do not share apomorphies of the Dithrycini.

The Cecidocharini is here redefined as a monophyletic group of genera to include the closely related genera *Cecidochares* Bezzi, *Neorhagoletis* Hendel, and *Procecidochares* Hendel that share approximated R-M and DM-Cu crossveins (synapomorphy), short costal spurs, and the scutellum and the area laterad of dorsocentral line bare and shining (autapomorphy). It apparently includes also *Procecidocharoides* Foote, which is known to me from descriptions only.

Male terminalia, based on examination of a few *Cecidochares* and *Procecidochares* species, have no posterior flanges (symplesiomorphy with other Lower Tephritinae?), the subapical lobe of the glans tail-like, setulose, nonsclerotized (synapomorphy with all the tribes except Terelliini and possibly Noeetini), bare sternum 8 (in *P. anthracina* (Doane)) (unique among examined Tephritidae; distribution of this character not examined) and praeputium longer than the preceding section of the glans (polarity unclear, also in Terelliini, *Tomoplagia* group and Xyphosiini); female ovipositor with the eversible membrane evenly covered with blunt and small scales both ventrally and dorsally (plesiomorphy).

At least two examined species of Cecidocharini have a gap in the row of setulae on $R_1$ opposite the Sc bend as do most Higher Tephritinae. This suggests that Cecidocharini may be the sister group of the Higher Tephritinae (Figure 22.1), but additional study of genitalic characters is needed to clarify this.

Several genera previously assigned to this tribe, including *Stenopa* Loew, *Cecidocharella* Hendel, *Dracontomyia* Becker, and *Ostracocoelia* Giglio-Tos, apparently belong to the Higher Tephritinae, and, very probably, to Eutretini (palpi broad and/or costal spurs very long), or elsewhere. *Gerrhoceras* Hering is transferred to Dithrycini herein, and the position of *Hetschkomyia* Hendel is still unknown to me.

Cecidocharini differ from other Tephritinae by the combination of the concolorous and long medial and lateral vertical setae (plesiomorphy), white postocular, postvertical, and postocellar setae, and stout white setae on the scutum and abdominal tergites (either a convergence or a synapomorphy with the Higher Tephritinae).

Cecidocharini share the setulose $R_{4+5}$, rather long extension of cell bcu (plesiomorphies, in the ground plan of the tribe) and some apomorphic features of the striate wing pattern with the

*Tomoplagia* group, and especially with the genus *Tomoplagia* (e.g., the second apical band, ending at the apex of M, and the oblique discal crossband from the pterostigma through the approximated R-M and DM-Cu, are well developed) which suggests these two groups may be closely related.

### 22.2.4 The Higher Tephritinae (Dithrycini + Eutretini + Acrotaeniini + Schistopterini + Tephritini + Tephrellini)

This complex is believed to be monophyletic. All its members have the lateral vertical seta shorter than the anterior orbital seta, 0.3 to 0.5 times as long as the medial vertical seta (synapomorphy), and usually white and stout. The posterior orbital seta is commonly white, and the row of setulae on the dorsal surface of vein $R_1$ has a bare space at the level of the subapical bend of Sc (except for the Dithrycini: Oedaspidina). The groups included herein are inferred to have the postocular setae and the mesonotal setulae white and thickened in the ground plan of each tribe (even if there are different conditions in some or many advanced genera of Tephrellini and Dithrycini).

#### 22.2.4.1 Dithrycini

This tribe was recently redefined by Foote et al. (1993) to include all the gall-forming Tephritinae with a sharply pointed aculeus and shining swollen scutellum previously placed in Oedaspidini and Cecidocharini, and a group of genera allied to *Eurosta* Loew recently excluded from the Dithrycini to form a new tribe, Eurostini Norrbom 1999. Freidberg and Kaplan (1992) limited the genus *Oedaspis* Bezzi to include only the species with a reduced proboscis.

Recent study (Korneyev, in preparation) has shown that the Dithrycini is apparently a monophyletic tribe, with at least one synapomorphy found in all examined genera (except for the Cecidocharini, whose removal from this group was discussed above): the scales of the eversible membrane on the ventral side are restricted to a narrow medial stripe, separated by bare areas from the two dorsal fields of scales that slightly extend onto the ventral surface. Redefined in this way, the tribe includes the genera allied to *Oedaspis* Loew (the subtribe Oedaspidina), the genera allied to *Hendrella* Munro, *Dithryca* Rondani (subtribe Dithrycina), and the genera close to *Eurosta* (subtribe Eurostina).

The ground plan of the tribe includes presence of three frontal and two (the dark and the white) orbital setae, white postocular, lateral vertical, postvertical, and postocellar setae, and four scutellar setae, frontal vitta finely setulose, $R_{4+5}$ setose to level of DM-Cu, mesonotum microtrichose, scutellum slightly swollen and microtrichose, mediotergite and abdomen microtrichose, epandrium rounded, with the surstyli of normal length and shape, slightly expanded, with smooth-edged flanges, glans with a tail-like, microtrichose flagellum and simple, short, tubular acrophallus, eversible membrane ventrally with the taeniae posteriorly approximated and continuing as a single narrow longitudinal stripe of scales (autapomorphy of the tribe), aculeus sharply pointed with nonserrate margins.

A preliminary cladistic analysis of the relationships among the genera of Dithrycini was conducted, but based on incomplete morphological data. The results (Figure 22.6) are therefore rather tentative and need additional data to be included. The character state matrix (Table 22.8) consisted of 39 terminal taxa and 38 characters (including dummy outgroup and character; Table 22.7). Autapomorphies were not included. A Nelson consensus tree was obtained from the trees resulting from the mhennig* option combined with bb* (branch swapping) option (Figure 22.6A). The tree length = 147, ci = 0.34, ri = 0.67. The second tree (Figure 22.6B) was obtained from the mhennig* option combined with a series of successive character weightings. The tree length = 266, ci = 0.70, ri = 0.90.

Both trees show certain characters considered important synapomorphies of the taxa within the tribe Dithrycini. The cladograms obtained by use of different techniques of analysis are not consistent with respect to the relationships of included genera and monophyly of the subtribes

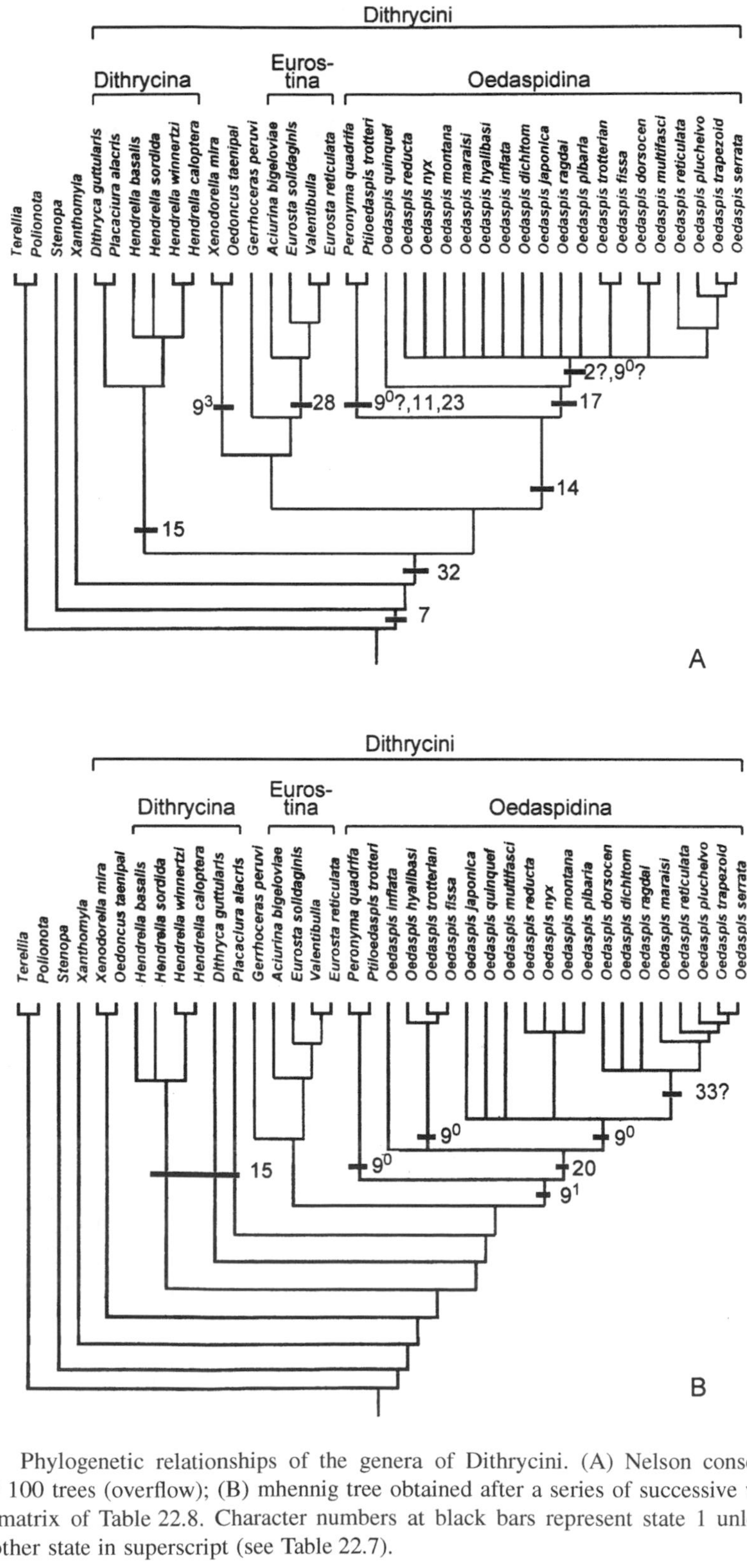


**FIGURE 22.6** Phylogenetic relationships of the genera of Dithrycini. (A) Nelson consensus tree from mhennig* bb* 100 trees (overflow); (B) mhennig tree obtained after a series of successive weightings from the character matrix of Table 22.8. Character numbers at black bars represent state 1 unless followed by number of another state in superscript (see Table 22.7).

**TABLE 22.7**
**Character States of Taxa Used in Analysis of Relationships among the Genera of Dithycini**

1. Dummy character
2. Number of unicolorous frontal setae: 0, three; 1, four; 2, two
3. Additional white setula before unicolorous frontal setae: 0, absent; 1, present
4. Frontal vita: 0, bare; 1, finely setulose
5. Vertical plates: 0, restricted to posterior quarter or third of frons length; 1, long, reaching middle of frons length
6. Lateral vertical seta: 0, black to yellow, concolorous with medial vertical seta; 1, white
7. Lateral vertical seta: 0, long; 1, short (see Table 22.1)
8. Posterior orbital seta: 0, yellow to brown; 1, whitish
9. Proboscis: 0, strongly reduced; 1, shortened; 2, normal, capitate; 3, long, geniculate
10. Shorter setulae between longer setae of postocular row present: 0, no; 1, yes
11. Presutural part of scutum: 0, of normal length; 1, shortened
12. Scutal setulae: 0, unicolorous yellow to white; 1, mixed yellow and black; 2, black
13. Tomentum of mesonotum: 0, dense; 1, sparse; 2, absent
14. Scutellum: 0, densely microtrichose; 1, sparsely microtrichose; 2, shining
15. Apical scutellar seta: 0, present; 1, absent
16. Presutural dorsocentral seta: 0, absent; 1, present
17. Scutellum: 0, flat or very slightly convex; 1, convex; 2, bilobate apically
18. Cell $r_1$: 0, without hyaline spots; 1, with one hyaline spot; 2, with two; 3, with three
19. Forefemur: 0, of normal shape; 1, thickened
20. Vein $R_{4+5}$ above: 0, setulose to level of crossvein DM-Cu; 1, at least with four to six setulae basally, at most setulose to R-M; 2, bare, or at most with one to three setulae basally
21. Wing pattern: 0, striate or dark with hyaline wedges; 1, reticulate; 2, radiate (*inactivated* character included for tracing changes of pattern type)
22. Wing: 0, without transverse hyaline crossband distal to DM-Cu; 1, with transverse hyaline crossband distal to DM-Cu
23. Dark stripe along vein Cu: 0, absent; 1, present
24. Cell $r_{2+3}$: 0, without hyaline spot posterior to apex of $R_{2+3}$; 1, with hyaline spot posterior to apex of $R_{2+3}$
25. Apex of cell $r_{4+5}$: 0, without hyaline spots; 1, with small hyaline spot; 2, mostly hyaline
26. Bulla in cell $r_{4+5}$: 0, absent; 1, present
27. Abdominal terga 1 to 4: 0, subshining; 1, densely microtrichose; 2, shining
28. Surstyli: 0, not narrow and elongate; 1, narrow and elongate
29. Dorsal lobe (flange) of lateral surstylus: 0, setulose; 1, densely microtrichose
30. Phallic glans: 0, with tail-like flagellum; 1, without tail-like flagellum
31. Dorsal surface of eversible membrane: 0, without medial groove, devoid of scales; 1, with shallow medial groove, devoid of scales
32. Ventral surface of eversible membrane: 0, with scales evenly covering most of surface; 1, with scales restricted to medial area
33. Aculeus: 0, gradually tapered; 1, with arrow-like, dorsoventrally compressed portion, conspicuously separate from basal part
34. Aculeus preapical portion: 0, smooth; 1, serrate
35. Aculeus: 0, not longer, than terga 4 to 6 together; 1, as long as abdomen
36. Host plant: 0, not Anthemideae; 1, Anthemideae
37. Host plant: 0, not Inuleae; 1, larva forming gall on Inuleae
38. Host plant: 0, not Heliantheae; 1, Heliantheae

Dithrycina and Oedaspidina, as indicated in Figure 22.6. Due to the small sample of characters and incompleteness of data, the trees show only putative relationships and must be considered with caution.

The largest portion of the Dithrycini includes the species with four scutellar setae and completely setulose $R_1$ (plesiomorphies), the scutellum nonmicrotrichose, the glans small and lacking both the apicodorsal tail-like appendage and the acrophallic tube. These characters were examined

**TABLE 22.8**
**Character States in Taxa Used in Preliminary Analysis of the Relationships among the Genera of Dithrycini (character numbers refer to Table 22.7)**

| | Character numbers |
|---|---|
| | **00000000011111111112222222222333333333** |
| | **12345678901234567890123456789012345678** |
| Dummy outgroup | 00010000200000000200000000000?00000000 |
| *Terellia* | 00000000200010000102000000000?00000000 |
| *Polionota* | 01000000200010010200000000000?00000000 |
| *Stenopa* | 00010011210002001202000000100?00000000 |
| *Xanthomyia alpestris* | 00010111210002001300100110100?10000000 |
| *Dithryca guttularis* | 02111111210102111301100110101011000100 |
| *Placaciura alacris* | 02010110110222101302000000201011000100 |
| *Peronyma quadrifasciata* | 0001011101111200210100100000??11000??? |
| *Ptiloedaspis trotteriana* | 00010111011112101201001100001011000100 |
| *Xenodorella mira* | 0001011131 0?000010010000000???1001??? |
| *Oedoncus taenipalpis* | 0001011121 0??00011010000000???1001??? |
| *Oedaspis trapezoidalis* | 0000011?010?02000202010120 00?1?1110010 |
| *O. serrata* | 0100011?010?02000202010120 00?1?1110010 |
| *O. plucheivora* | 0000011?010?12000202010110 00?1?1110010 |
| *O. reticulata* | 0000011?010?02000302100110 00?1?1110001 |
| *O. reducta* | 0101011?010?02002302000120 00?1?1?00001 |
| *O. nyx* | 010?011?010?12001202000110 00?1?1?00??? |
| *O. montana* | 010?011?010?12000202000110 00?1?1?00??? |
| *O. maraisi* | 0?0?011?010??200010200000000?1?1110000 |
| *O. hyalibasis* | 0100011001021210100200000000?1?1?????? |
| *O. inflata* | 010101110101020011 0?00000000?1?1?00010 |
| *O. dorsocentralis* | 01010111010002011002000000000?111100100 |
| *O. dichotoma* | 01010111010002001102000000000?111100100 |
| *O. japonica* | 0101011101000 20?1?0200000000?111?00100 |
| *O. quinqueifasciata* | 00010111110002001002000000000?111?00100 |
| *O. trotteri* | 01010110010222001102000120 00?111?00100 |
| *O. ragdai* | 01010111010002001102000000000?111100??? |
| *O. fissa* | 01010010010222001102000120 20?111?00?00 |
| *O. multifasciata* | 01010111010002011102000000 10?111?00100 |
| *O. pibaria* | 010?01110100020012020001100 0???10?0??? |
| *Hendrella basalis* | 02000111210001100201000000201011000100 |
| *H. sordida* | 02000111210001100201000000201011000100 |
| *H. winnertzi* | 02100111210001100202000000201011000100 |
| *H. caloptera* | 02100111210001100101000000201011000100 |
| *Valentibulla* | 02010111110022100312200121 21??11000100 |
| *Aciurina bigeloviae* | 00010111110000100212000001 01?111000100 |
| *Eurosta solidaginis* | 01010111110000100201110110 11?111000100 |
| *E. reticulata* | 02010111110000000301100120 11?111000100 |
| *Gerrhoceras peruviana* | 0101011111 0?00000100000000000101100???? |

only for Palearctic *Oedaspis* and the few Afrotropical species illustrated by Freidberg and Kaplan (1992), and they are considered synapomorphies only putatively.

The phylogeny of this branch, which corresponds to the subtribe Oedaspidina, is poorly understood. Most species have three frontal setae (plesiomorphy) or more. The genera included in this group have as yet no discovered synapomorphy, and the group is possibly paraphyletic.

According to Freidberg and Kaplan (1992), most species they assigned to *Oedaspis* have a reduced proboscis, but, as a matter of fact, *O. multifasciata* (Loew), the type species of the genus, *O. dichotoma* Loew, and *O. quinqueifasciata* Becker, have the labella rather conspicuous and apparently functional, as well as *Oedoncus* Speiser and *Xenodorella* Munro, whereas both *Peronyma* Loew and *Ptiloedaspis* Bezzi (see below) have rudimentary labella. It clearly shows that reduction of the labella took place independently in different lines of the Oedaspidina and that the mosaic distribution of all the diagnostic characters makes the definition of *Oedaspis*, even in the broadest concept, a hardly possible task without involving additional genital characters.

The position of *Oedoncus* and *Xenodorella* is rather controversial on the tree, and both these genera show no clear affinities with the other Dithrycini. They have the "*Oedaspis*-like" wing pattern (probably synapomorphy with *Oedaspis*), scutellum shining (synapomorphy or homoplasy with Oedaspidina?) with four setae, and $R_{4+5}$ setulose (plesiomorphies). Nevertheless, the arrangement of scales on the eversible membrane has never been examined in both *Oedoncus* and *Xenodorella*, and their inclusion into Oedaspidina is tentative.

Other genera that were found to possess the typical dithrycine state of the eversible membrane are *Peronyma* (southeastern part of the United States), *Ptiloedaspis* (Europe: Spain), and the Neotropical *Gerrhoceras* Hering.

The wing pattern of *Peronyma* (especially the stripe along Cu), the setulose $R_{4+5}$, the shortened presutural part of the mesonotum, the two large bare shining spots at the transverse suture, the labellum very short, nonfunctional, and mixed white and black setulae of the mesonotum show the strong affinity of this genus to the western Palearctic *Ptiloedaspis*. These genera certainly are sister groups, and their amphi-Atlantic distribution is uncommon for the Tephritidae. *Ptiloedaspis tavaresiana* Bezzi has the apical scutellar setae completely reduced (autapomorphy or synapomorphy with Dithrycina and Eurostina?), and *Peronyma quadrifasciata* (Macquart) has the apical pair of setae shortened.

*Gerrhoceras* has a microtrichose scutellum with four setae, setulose $R_{4+5}$, normal, mesally rounded surstyli, glans with the apicodorsal appendage (see Korytkowski 1976) (plesiomorphies), and the eversible membrane scales arranged typically for the Dithrycini (Norrbom and Korneyev, personal observation), and certainly belongs here. This genus appears to be the sister group to the rest of Dithrycini, as it has no other synapomorphies with the subtribes within the tribe.

The Nearctic genera *Aciurina* Curran, *Eurosta*, and *Valentibulla* Foote and Blanc form a monophyletic clade. Its monophyly is inferred from the very characteristic shape of surstyli (Steyskal and Foote 1977; Foote et al. 1993; Norrbom et al. 1999). Its phylogeny was analyzed by Ming (1989).

The group of Palearctic genera in which the apical scutellar setae are lacking may also be monophyletic (Figure 22.6A). It includes *Hendrella* Munro, *Placaciura* Hendel, and *Dithryca* Rondani. They share the simple, rounded epandrium shape (symplesiomorphy), and rounded, finely setulose flanges of the surstyli (synapomorphy?) and the same shape of the glans, with a long and thin microtrichose apicodorsal process (polarity unknown). Alternately, this group may be paraphyletic (Figure 22.6B), because *Hendrella* + *Placaciura* and the Eurostina share subshining, nonmicrotrichose abdominal terga and the presence of two frontal setae (at least in the ground plan of the genera), and possibly form a monophyletic branch.

The genera of the Eurostina and Dithrycina share the gap in the $R_1$ setulae at the level of the bend in Sc (synapomorphy with the other Higher Tephritinae), whereas in the Oedaspidina it is completely setulose (symplesiomorphy with the Lower Tephritinae or character reversal), which somewhat contradicts the hypothesis of monophyly of the Dithrycini.

### 22.2.4.2 *Acinia* Robineau-Desvoidy

This is a strictly Palearctic genus. The New World species attributed to *Acinia* are not congeners of the Old World species, differing by some genital structures, and apparently all belong to or fit near the genus *Baryplegma* Wulp (Eutretini?). Hendel (1927) assigned *Acinia* to Xyphosiini, but it certainly does not belong either to Xyphosiini or to the Lower Tephritinae, having posterior orbital, lateral vertical, postvertical, postocellar, and some postocular setae white and rather lanceolate, and the aculeus pointed.

Among the Higher Tephritinae, *Acinia* is rather close to the ground plan of the whole complex, because of having $R_{4+5}$ setulose to DM-Cu, and larvae forming a cocoonlike structure of the flower remains, like *Terellia, Xyphosia,* and *Ictericodes.* It shares long costal spines with the Eutretini (in *A. corniculata* (Zetterstedt), conspicuously longer than, and in *A. biflexa* (Loew), as long as width of first flagellomere), Tephritini and Tephrellini (synapomorphy). In *A. biflexa,* some specimens have a more or less conspicuous parafacial spot. Species of *Acinia* never have both enlarged palpi and/or white anepimeral setae, like Eutretini and Tephritini, nor a needlelike aculeus, like Tephrellini, and I consider this genus the sister group to the cluster of these three tribes. Compared with the hypothesized ground plan of these three tribes, *Acinia* has a maculose wing pattern (apomorphy) rather than one consisting of confluent crossbands.

### 22.2.4.3 Eutretini (including Acrotaeniini)

This tribe was proposed by Munro (1952) to include several Afrotropical species he believed to be closely related to the New World genus *Eutreta*, and recently was redefined by Foote et al. (1993) and Norrbom et al. (1999) to include the species usually with a parafacial spot, white scalelike setulae rather sparsely or unevenly spread over the mesonotum, multimaculate wing pattern, and the aculeus rather broad basally, and often barbed preapically.

Further characters that should be used to define the Eutretini are the moderately to enormously enlarged palpus (apomorphy; also in *Tomoplagia* group due to homoplasy), and long erect costal spines, exceeding the width of the first flagellomere (synapomorphy with *Acinia*; in Tephritini and Tephrellini the spurs are also long, but slightly shorter than the width of the first flagellomere). All the Eutretini possess also white posterior orbital seta, mixed black and white postocular seta, $R_1$ with the gap of setulae at the Sc bend level, serrate dorsal lobe or flange of the lateral surstylus, and the short praeputium, which are apomorphies of the Higher Tephritinae.

Norrbom et al. (1999) defined Eutretini to include *Afreutreta* Bezzi, *Cosmetothrix* Munro, *Cryptotreta* Blanc and Foote, *Dictyotrypeta* Hendel, *Eutreta* Loew, *Laksyetsa* Foote, *Paracantha* Coquillett, *Polymorphomyia* Snow, *Pseudeutreta* Hendel, *Rachiptera* Bigot, *Strobelia* Rondani, *Tarchonanthea* Freidberg and Kaplan, and *Xanthomyia* Phillips. Some additional genera should also be included.

*Stenopa* Loew has the consistent features of the Eutretini, including the broad wing, strong costal spurs, and large palpi (synapomorphies of the tribe), plus the surstylar flanges serrate and the acrophallus shorter than the previous section of the glans (synapomorphies of the Higher Tephritinae) and should be transferred here from the Cecidocharini, despite having no parafacial spot. Of the two species of *Stenopa*, one has white lateral vertical seta, and the second has this seta brownish-black (symplesiomorphy with the Lower Tephritinae or character reversal). Other genera that apparently belong here are *Cecidocharella* Hendel and *Ostracocoelia* Giglio-Tos, known to me from descriptions only (Aczél 1953, Figures 1 to 6).

The position of *Acrotaenia* Loew needs further clarification. At least one species of this genus, *A. otopappi* Doane, possesses the essential characters of the Eutretini, including the broad wing disk, strong costal spurs, large palpi, and parafacial spot. It is certainly congeneric with the type species of *Acrotaenia*, *A. testudinea* (Loew), and *A. tarsata* Wulp (the unique wing pattern, with several costal bullae and combination of maculate wing base and three apical crossbands, well support the monophyly of this genus). The latter two species lack the long costal spurs and the parafacial spot, and share the vertically elongate head shape with *Tomoplagia*. For this reason *Tomoplagia*, along with the closely related genera *Polionota, Caenoriata,* and *Neotaracia*, was included into Acrotaeniini (Foote et al. 1993; Norrbom et al. 1999) that comprises a heterogeneous conglomerate of genera. Relationships of the other genera formerly assigned to the Acrotaeniini, *Acrotaeniacantha* Hering, *Baryplegma* Wulp, *Euarestopsis* Hering, and *Tetreuaresta* Hendel, are still unclear.

The Palearctic *Merzomyia* Korneyev, recently assigned to Tephritini (*Sphenella* group) (Korneyev 1990, as *Orotava* Frey, *p. p.*; 1995), also has the long costal spur, the long and broad palpi, and $R_{4+5}$ setulose dorsally and ventrally, and rather generalized type of male and female genitalia, characteristic of many other genera of Eutretini, and apparently belongs here.

Norrbom et al. (1999) briefly discussed the phylogenetic relationships within the Eutretini. Although they were unsure if the tribe is monophyletic, they indicated three genus groups, and noted that *Xanthomyia* and allied genera belong here, and that the Paleotropical tribe Schistopterini is somehow related to them all.

Actually, all these taxa share the conspicuously enlarged palpi, the subcostal break more or less incised, the postvertical and some postocular setae thickened and scalelike, and the anal lobe and the alula dark. The Old World genera, as well as the Oriental *"Dictyotrypeta" longiseta* Hering and *"Xyphosia" malaisei* Hering, are closely related to the Nearctic *X. platyptera* (Loew) and eastern Palearctic *X. japonica* (Shiraki), but lack the characteristic apicodorsal glans sclerotization and the extremely pointed aculeus tip. The Afrotropical *Cosmetothrix* and *Tarchonanthea* (Freidberg and Kaplan 1993) are somehow associated with that group.

### 22.2.4.4 Schistopterini

The tribe Schistopterini possesses all the autapomorphies of Eutretini (large palpi, long costal spurs, *etc.*) and actually is a monophyletic group of small-sized Paleotropical genera within the Eutretini that has several additional derived features, like a deeply incised subcostal break, etc. In the present chapter I retain them as separate tribes, but with the reservation that in forthcoming taxonomic revision they could be synonymized.

This tribe is well defined by its numerous autapomorphies (the deeply broken costa, tiny size, etc.; see Hardy 1973; 1985). The examined species of *Rhabdochaeta* and *Schistopterum* share with *"Dictyotrypeta" longiseta* (Korneyev, personal observation) and *"Xyphosia" malaisei* (see Hardy 1973, Figure 167) the ventrally (not mesally) directed lateral surstyli, narrow, gradually tapered aculeus, and the long dentate scales of eversible membrane. The latter two taxa comprise the hypothetical sister group of Schistopterini within the Eutretini. Schistopterini are predominantly an Afrotropical group, with numerous undescribed species (Freidberg, personal communication). To my knowledge, none of them are gall-forming and many breed in Asteraceae with small flowerheads.

### 22.2.4.5 Tephritini

This tribe is the largest and most widespread of all tribes of the Tephritinae. Like Eutretini and Tephrellini, Tephritini share the long costal spurs with *Acinia*. The monophyly of the Tephritini is tentatively presumed from the following synapomorphies: (1) only two dark frontal setae (sometimes secondarily three to four or one to zero; this also occurs in some Oedaspidini and Schistopterini); (2) white anepimeral seta, also in some *Paracantha* (Eutretini) and *Hyaloctoides* (Tephrellini); (3) glans rather small, with the apicodorsal appendage weak; and (4) aculeus moderately broad, tapered before apex, commonly with one subapical step. Wing pattern usually reticulate, $R_{4+5}$ dorsally at most with four to five setulae before R-M.

Norrbom et al. (1999) recognized six main lineages (*Campiglossa* group, *Dyseuaresta* group, *Euarestoides* group, *Spathulina* group, *Sphenella* group, *Trupanea* group, and genera *incertae sedis*) inside the Tephritini. Merz (Chapter 24) included the *Trupanea* group and other genera in the *Tephritis* group, but I prefer to recognize the *Tephritis* group of genera in a broader sense, including also the *Spathulina* group of genera.

#### *22.2.4.5.1* Sphenella *Group of Genera*

This group was defined by Munro (1957b). It is characterized by the presence of the black posterior notopleural seta (plesiomorphy), long and acute flange of the lateral surstylus, and a multimaculate wing pattern with a crossband of hyaline confluent spots between R-M and DM-Cu (synapomorphies). The flange is secondarily lacking in some *Sphenella* that have glans structure exactly as in *Paratephritis*. This group was reviewed by Freidberg (1987), Freidberg and Hancock (1989), and Korneyev (1990). The group includes *Sphenella* Robineau-Desvoidy, *Parafreutreta* Munro, *Paratephritis* Shiraki, *Orotava* Frey, *Oedosphenella* Frey, *Telaletes* Munro and *Bevismyia* Munro. All the representatives of this group whose larvae are known are associated with the plants of the tribe Senecioneae. A detailed analysis of the relationships in the group is the subject of another paper (Korneyev, in preparation).

The genera related to *Axiothauma* formerly included in this group have a black and rather long lateral vertical seta, and belong to the Lower Tephritinae (see above). The Palearctic genus *Merzomyia* Korneyev, which previously was included here because of breeding in *Senecio* and having two frontal setae, is apparently associated with Eutretini (see above).

#### *22.2.4.5.2* Campiglossa *Group of Genera*

This group was preliminarily revised by Munro (1957a) for the Afrotropical fauna (as the *Paroxyna*-series), and then by Korneyev (1990) and Merz (1992) for the Palearctic fauna. Its monophyly is supported mostly by having the proboscis elongate and the phallic preglans area distinctly spinulose. *Scedella* Munro and some species of *Campiglossa* Rondani have the posterior notopleural seta black, while the other species of this genus, as well as those of *Oxyna* Robineau-Desvoidy and *Antoxya* Munro, have this seta white. As the latter two genera are the only ones that have a simple tubular acrophallus (modified in other taxa of the group), the white color of the posterior notopleural seta is presumed to be a plesiomorphic state in the genera allied to *Campiglossa* belonging to its ground plan.

#### *22.2.4.5.3* Tephritis *Group of Genera*

Analysis of this group is the subject of the chapter by Merz (Chapter 24). It is considered the sister group of the *Campiglossa* group, as they share at least one synapomorphy, the white posterior notopleural seta. This group, as currently defined, hardly can be separated from the other New World Tephritini (groups of genera related to *Euaresta* Loew, *Euarestoides* Benjamin, and *Dyseuaresta* Hendel) that also have white posterior notopleural seta. The analysis of their relationships is beyond the scope of the present work.

### 22.2.4.6 Tephrellini

This tribe includes mostly Paleotropical species, with a dozen species in the southern part of the Palearctic Region. Most examined species have a narrow aculeus with an apical portion that becomes needlelike and rather long posterior to the apex of sternum 8, and a shining or very lightly microtrichose abdomen (autapomorphies). Otherwise Munro (1947: 83, 89) gave a definition of "Group A" that corresponds to the current concept of the Tephrellini (= Aciurini, Platensinini) (Hancock 1990). The larvae of many Afrotropical and Palearctic species breed in seed capsules of nonasteraceous hosts (Acanthaceae, Lamiaceae, Verbenaceae), but very little is known about Oriental species assigned to *Platensina* Enderlein and *Pliomelaena* Bezzi.

Currently, subdivision of the tribe into the subtribes Tephrellina (= Aciurini Hering 1947) and Platensinina (= Tephrellini *sensu* Hering 1947, *p. p.*) is accepted (Norrbom et al. 1999), but the monophyly of these subtribes and the phylogenetic relationships among the included genera have never been tested.

Neither the sister group, nor the limits of this tribe are well established. I believe that the most generalized members are those possessing white head setae (lateral vertical and some postocular) and four scutellar setae, like most Higher Tephritinae.

Among Tephrellini, the Australasian species placed in *Protephritis* Shiraki (the genus recently considered a synonym of *Pliomelaena* Bezzi; e.g., Hardy 1977; 1988; Hardy and Foote 1989) seem to fit this definition, but it is questionable if they really belong to *Pliomelaena*. *Protephritis sauteri* (Enderlein), *P. sonani* from Taiwan, and "*Icterica*" *kashmerensis* Hendel from India have a head shape and chaetotaxy very similar to that of Afrotropical *Pliomelaena* species, but certainly differ from them by the abdomen more or less microtrichose, and the wing pattern including more small hyaline dots. *Protephritis sonani* was figured to have the aculeus needlelike (Shiraki 1933, Pl. XII, Figure 6), and this is the only evidence that these *Acinia*-like flies really belong to the Tephrellini. Nothing is known about their host plants.

*"Tephrella" variegata* Radhakrishnan from India was reared from galls on *Inula cappa* (DC) (Asteraceae) (Radhakrishnan 1984). From the description and figures, it looks to have a typically tephrelline, needlelike aculeus (contrary to *Hendrella* and other superficially similar Dithrycini), and male lateral surstylus deeply incised, like in *"Pliomelaena" callista* Hering (Hardy 1988, Figure 24). *"Tephrella" variegata* differs from them by the two scutellar setae and abdomen shining. It may fit near *Protephritis* and *Pliomelaena*.

Tephrellini are therefore hypothesized to be derived from the *Acinia*-like ancestral group that fed on Asteraceae, and secondarily changed larval feeding mode. These hypotheses of phylogenetic relationship and evolution of host usage of Tephrellini need further study that is beyond the scope of the present chapter.

## 22.3 CONCLUSIONS AND FUTURE RESEARCH NEEDS

The relationships among the genera of Tephritinae have been only preliminarily resolved by this study and are a rough phylogenetic scheme rather than a completely resolved phylogeny. Some tribes of Tephritinae (especially the *Tomoplagia* group and Eutretini) have their centers of diversity in the Neotropical and Nearctic Regions and many of their representatives were not available in this study. In some cases large gaps in data on male and female genitalic characters, and especially the absence or incompleteness of larval characters made a thorough phylogenetic analysis inapplicable.

In this study, the subfamily was found to consist of the Higher Tephritinae, a monophyletic complex of the advanced tribes Dithrycini, Eutretini, Schistopterini, Tephritini, and Tephrellini, and the Lower Tephritinae, comprising by Terelliini, *Tomoplagia* group of genera, Xyphosiini, Cecidocharini, Noeetini, Myopitini, and *Axiothauma* group of genera. Our data suggest that the latter complex is apparently paraphyletic, and that the Higher Tephritinae is a derived clade of the Lower Tephrititinae.

The tribe Terelliini is hypothesized to be the sister group of the remaining Tephritinae, but to prove this properly, the *Tomoplagia* group of genera (one of the most primitive among the other tephritines) must be thoroughly revised, with the description of genitalic and larval characters.

The tribes Dithrycini, Cecidocharini, and Acrotaeniini were found to be heterogeneous and needing further redefinition. The monophyly of some tribes is supported only tentatively, because the morphological data are incomplete for many included genera. Even for some tribes that are certainly proved to be monophyletic (e.g., Myopitini, Noeetini), the sister group relationships are uncertain. To perform a reliable analysis for the complete set of taxa, a large set of morphological characters must be involved, and, for this, much additional morphological study is necessary to provide unknown data, especially for many characters of the female and male genitalia, and the larval mask, spiracles, and cuticular structures for both the known species, and for the many undescribed species in all the tribes. This, together with the use of different techniques (e.g., successive weighting) would increase the resolution of the computer analysis.

Nevertheless, the current study shows that the level of homoplasy is very high and that even a very complete morphological data set would not necessarily guarantee that such an analysis would clearly resolve phylogenetic relationships within the subfamily. In many cases, such a result could be obtained only by various approaches of both molecular and morphological studies.

## ACKNOWLEDGMENTS

I am indebted to Allen Norrbom (USDA SEL, Washington, D.C.), Amnon Freidberg (Tel Aviv University), Bernhard Merz (ETH Zürich, currently Muséum d'Histoire Naturelle, Genève), Lynn Carroll (USDA SEL), Gary Steck, and David Headrick for fruitful discussions and dissemination of the new facts and data they discovered, as well as the specimens they shared with me. Their ideas and criticism were very beneficial for this work.

Papers recently written by R. A. I. Drew, A. Freidberg, D. L. Hancock, and A. L. Norrbom were essential sources of morphological information. I thank R. Contreras-Lichtenberg for the great opportunity to study specimens in the collection of the Naturhistorisches Museum Wien and for arranging a visitor's grant from the NHMW in 1995. Support generously provided by the Smithsonian Institution as a Visitor Grant in January 1993 greatly aided this investigation and is acknowledged. A. L. Norrbom and A. Freidberg offered valuable comments on the manuscript.

I wish to thank Martín Aluja, Instituto de Ecología, Xalapa, and A. L. Norrbom for their enormous efforts to organize the symposium and to invite me to attend this extraordinary meeting of the best experts in the taxonomy, phylogeny, and behavior of the Tephritidae. Financial support of Consejo Nacional de Ciencia y Tecnología (CONACyT), Subdirección de Asuntos Bilaterales is greatly appreciated. I acknowledge the support from Campaña Nacional contra las Moscas de la Fruta (SAGAR-IICA) and thank Lic. A. Meyenberg (CONACyT), Ing. J. Reyes (SAGAR), C. Leal (IE), V. Domínguez (IE), and E. Bombela (IE) for arranging my trip to Xalapa. My warmest and sincere thanks are due to B. Delfosse (IE) and the team members: L. Guillén, I. Jácome, R. Cervantes, and J. Piñero.

I also thank Elena and Severin for their patience. They inspired this work every minute of the day for many years.

## REFERENCES

Aczél, M.L. 1953. La familia Tephritidae en la region neotropical. I. *Acta Zool. Lilloana* 13: 97–200.

Farris, J.S. 1988. *Henning86, version 1.5*. Published by the author, Port Jefferson Station, New York.

Foote, R.H. 1979. A review of the Neotropical genus *Neotaracia* Foote (Diptera: Tephritidae). *J. Wash. Acad. Sci.* 69: 174–179.

Foote, R.H., F.L. Blanc, and A.L. Norrbom. 1993. *Handbook of the Fruit Flies (Diptera: Tephritidae) of America North of Mexico*. Comstock Publishing Associates, Ithaca. 571 pp.

Freidberg, A. 1985. Studies of Terelliinae (Diptera: Tephritidae): the genus *Craspedoxantha* Bezzi. *Ann. Natal Mus.* 27: 183–206.

Freidberg, A. 1987. *Orthocanthoides aristae*, a remarkable new genus and species of Tephritidae (Diptera) from Mount Kenya. *Ann. Natal Mus.* 28: 551–559.

Freidberg, A. and D.L. Hancock. 1989. *Cryptophorellia*, a remarkable new genus of Afrotropical Tephritinae (Diptera: Tephritidae). *Ann. Natal Mus.* 30: 15–52.

Freidberg, A. and F. Kaplan. 1992. Revision of Oedaspidini of the Afrotropical Region (Diptera: Tephritidae: Tephritinae). *Ann. Natal Mus.* 33: 51–98.

Freidberg, A. and F. Kaplan. 1993. A study of *Afreutreta* Bezzi and related genera (Diptera: Tephritidae: Tephritinae). *Afr. Entomol.* 1: 207–228.

Freidberg, A. and W.N. Mathis. 1986. Studies of Terelliinae (Diptera: Tephritidae): a revision of the genus *Neaspilota* Osten Sacken. *Smithson. Contrib. Zool.* 439, 75 pp.

Freidberg, A. and W.N. Mathis. 1990. A new species of *Craspedoxantha* and a revised phylogeny for the genus (Diptera: Tephritidae). *Proc. Entomol. Soc. Wash.* 92: 325–332.

Han, H.-Y. and B.A. McPheron. 1997. Molecular phylogenetic study of Tephritidae (Insecta: Diptera) using sequences of the mitochondrial 16S ribosomal DNA. *Mol. Phylogenet. Evol.* 7: 17–32.

Hancock, D.L. 1986. Classification of the Trypetinae (Diptera: Tephritidae), with a discussion of the Afrotropical fauna. *J. Entomol. Soc. South. Afr.* 49: 275–305.

Hancock, D.L. 1990. Notes on the Tephrellini-Aciurini (Diptera: Tephritidae), with a checklist of the Zimbabwe species. *Trans. Zimb. Sci. Assoc.* 64: 41–48.

Hardy, D.E. 1973. The fruit flies (Tephritidae — Diptera) of Thailand and bordering countries. *Pac. Insects Monogr.* 31: 1–353.

Hardy, D.E. 1977. Family Tephritidae (Trypetidae, Trupaneidae). In *A Catalog of the Diptera of the Oriental Region*. Vol. III, *Suborder Cyclorrhapha (excluding Division Aschiza)* (M.D. Delfinado and D.E. Hardy, eds.), pp. 44–134. University of Hawaii Press, Honolulu. 854 pp.

Hardy, D.E. 1985. The Schistopterinae of Indonesia and New Guinea. (Diptera: Tephritidae). *Proc. Hawaii. Entomol. Soc.* 25: 59–73.

Hardy, D.E. 1988. The Tephritinae of Indonesia, New Guinea, and the Bismarck and Solomon Islands (Diptera: Tephritidae). *Bishop Mus. Bull. Entomol.* 1, 92 pp.

Hardy, D.E. and R.H. Foote. 1989. Family Tephritidae. In *Catalog of the Diptera of the Australian and Oceanian Regions* (N.L. Evenhuis, ed.), pp. 502–531. Bishop Museum Special Publication 86, Bishop Museum Press and E.J. Brill, Honolulu.

Hendel, F. 1927. Trypetidae. In *Die Fliegen der Paläarktischen Region* (E. Lindner, ed.). E. Schweizerbart. Verlag (Erwin Nägele), Stuttgart: 5 (16–19): 221 pp.

Hering, E.M. 1947. Neue Arten and Gattungen der Fruchtfliegen. *Siruna Seva* 6: 1–16.

Korneyev, V.A. 1985. The fruit flies of the tribe Terelliini Hendel, 1927 (Diptera, Tephritidae). *Entomol. Obozr.* 64: 626–644 [in Russian].

Korneyev, V.A. 1987. A revision of the subgenus *Cerajocera* stat. n. of the genus *Terellia* (Diptera, Tephritidae) with description of a new species of fruit fly. *Zool. Zh.* 66: 237–243 [in Russian].

Korneyev, V.A. 1988. A new and some little known species of tephritid flies of the genus *Terellia* Robineau-Desvoidy (Diptera, Tephritidae) from Middle Asia and Transcaucasia. *Entomol. Obozr.* 67: 871–875 [in Russian].

Korneyev, V.A. 1990. A review of *Sphenella* and *Paroxyna* series of genera (Diptera: Tephritidae: Tephritinae) of the Eastern Palearctic. *Nasekomye Mongol.* (1989) 11: 395–470 [in Russian].

Korneyev, V.A. 1995. New records and synonymy in Xyphosiini and Tephritini (Diptera: Tephritidae: Tephritinae) from the Far East Russia. *Russ. Entomol. J.* 4(1–4): 3–10.

Korneyev, V.A. 1996. Reclassification of Palearctic Tephritidae (Diptera). Communication 3. *Vestn. Zool.* 1995 (5–6): 25–48.

Korneyev, V.A. and I.M. White. Fruit-flies of the genus *Urophora* R.-D. (Diptera, Tephritidae) of East Palaearctic. IV. Conclusion. *Entomol. Obozr.*, in press.

Korytkowski, C.A. 1976. El genero *Gerrhoceras* Hering, y descripcion de una nueve especie de habitos cecidogenos. *Rev. Bras. Biol.* 36: 411–417.

Merz, B. 1992. Revision der Westpaläarktischen Gattungen und Arten der *Paroxyna*-Gruppe und Revision der Fruchtfliegen der Schweiz (Diptera, Tephritidae). Dissertation, Eidgenössische Technische Hochschule-Zentrum, Zürich, Nr. 9902, 341 pp.

Ming, Y. 1989. A revision of the genus *Eurosta* Loew, with a scanning electron microscopy study of taxonomic characters (Diptera, Tephritidae). M.Sc. thesis, Washington State University, Pullman, 190 pp.

Munro, H.K. 1947. African Trypetidae (Diptera). A review of the transition genera between Tephritinae and Trypetinae, with a preliminary study of the male terminalia. *Mem. Entomol. Soc. South. Afr.* 1, 284 pp.

Munro, H.K. 1952. Les Trypetides, Dipteres cecidogenes de la serie *Eutreta–Oedaspis* a propos de deux nouvelles especies malgaches. *Mem. Inst. Sci. Madagascar Ser. E.* 1: 217–225.

Munro, H.K. 1957a. Trypetidae. *Ruwenzori Expedition 1934–1935. Br. Mus. Nat. Hist.* 2(9): 853–1054.

Munro, H. 1957b. *Sphenella* and some allied genera. *J. Entomol. Soc. South Afr.* 20: 14–57.

Norrbom, A. 1987. A revision of the Neotropical genus *Polionota* Wulp (Diptera: Tephritidae). *Folia Entomol. Mex.* 73: 101–123.

Norrbom, A.L. 1992. A revision of the Nearctic genus *Gymnocarena* Hering (Diptera: Tephritidae). *Proc. Entomol. Soc. Wash.* 93: 527–555.

Norrbom, A.L., L.E. Carroll, and A. Freidberg. 1999. Status of Knowledge. In *Fruit Fly Expert Identification System and Systematic Information Database* (F.C. Thompson, ed.), pp. 9–47. *Myia* (1998) 9, 524 pp.

Radhakrishnan, C. 1984. A new *Tephrella* (Diptera: Tephritidae) from Meghalaya, India. *Bull. Zool. Surv. India* 6(1–3): 27–29.

Shiraki, T. 1933. A systematic study of Trypetidae in the Japanese. *Mem. Fac. Sci. Agric. Taihoku Imp. University* 8 (Entomol. 2), 509 pp.

Steyskal, G.C. and R.H. Foote. 1977. Revisionary notes on North American Tephritidae (Diptera), with keys and descriptions of new species. *Proc. Entomol. Soc. Wash.* 79: 146–155.

Wang, X.-J. 1996. The fruit flies (Diptera: Tephritidae) of the East Asian Region. *Acta Zootaxon. Sin.*, Suppl., 338 pp.

## APPENDIX 22.1: TAXA EXAMINED

Terelliini: All Palearctic species; *Neaspilota alba* (Loew), *N. albidipennis* (Loew), *N. achilleae* Johnson, *N. punctistigma* Benjamin, *Craspedoxantha bafut* Freidberg and Mathis, *C. manengubae* Speiser, *C. marginalis* (Wiedemann), *C. vernoniae* Freidberg, *C. yaromi* Freidberg.

*Tomoplagia* group of species: *Gymnocarena diffusa* (Snow), *Neotaracia plaumanni* (Hering), *Tomoplagia* spp. (nine species).

Xyphosiini: All Palearctic species, *Icterica circinata* (Loew), *I. seriata* (Loew).

Noeetini: All Holarctic species.

*Axiothauma* group of species: *Axiothauma nigrinitens* Munro, *Cryptophorellia flava* Freidberg and Hancock.

Myopitini: All Palearctic species, *Stamnophora vernoniicola* (Bezzi), *Stamnophora* sp.

Cecidocharini: *Cecidochares* sp., *Procecidochares anthracina* (Doane).

Dithrycini: *Aciurina bigeloviae* (Cockerell), *Dithryca guttularis* (Meigen), *Hendrella* Munro (all species), *Oedaspis dichotoma* Loew, *O. dorsocentralis* Zia, *O. fissa* Loew, *O. japonica* Shiraki, *O. kaszabi* Richter, *O. multifasciata* (Loew), *O. quinquiefasciata* Becker, *O. ragdai* Hering, *O. trotteriana* Bezzi, *Peronyma quadrifasciata* (Macquart), *Placaciura alacris* (Loew), *Ptiloedaspis tavaresiana* Bezzi, *Valentibulla steyskali* Foote.

Eutretini: *Acrotaenia otopappi* Doane, *A. testudinea* (Loew), *A. tarsata* Wulp, *Eutreta diana* (Osten Sacken), *E. novaeboracensis* (Fitch), *E. simplex* Thomas, *Dictyotrypeta atacta* (Hendel), *D. strobelioides* (Hendel), *D. syssema* Hendel, "*Dictyotrypeta*" *longiseta* Hering, *Laksyetsa trinotata* Foote, *Merzomyia westermanni* (Meigen), *M. licenti* (Chen), *Paracantha culta* (Wiedemann), *P. cultaris* (Coquillett), *P. ruficallosa* Hering, *Rachiptera* sp. near *percnoptera* Hendel, *Rachiptera* sp. near *virginalis* Hering, *Stenopa affinis* Quisenberry, *S. vulnerata* (Loew), *Strobelia* spp., *Xanthomyia* (*s. str.*) *alpestris* (Pokorny), *X.* (*s. str.*) *nora* (Doane), *X.* (*s. str.*) *platyptera* (Loew), *X.* (*Paranoeeta*) *japonica* (Shiraki).

Schistopterini: *Brachiopterna katonae* Bezzi, *Rhabdochaeta asteria* Hendel, *Schistopterum moebiusi* Becker.

Tephritini: Most species of all Palearctic genera; *Dyseuaresta adelphica* (Hendel), *D. bilineata* (Foote), *Dyseuaresta* spp., *Parafreutreta regalis* Munro, *Telaletes ochraceus* (Loew).

Tephrellini: Most Palearctic species; *Afraciura quaternaria* (Bezzi), *A. zernyi* Hering, *Bezzina margaritifera* (Bezzi), *Brachyaciura limbata* (Bezzi), *Elaphromyia adatha* (Walker), *E. pterocallaeformis* (Bezzi), *Katonaia hemileoides* (Hering), *Metasphenisca negeviana* (Freidberg), *M. gracilipes* (Loew), *Munroella myopitina* Bezzi, *Platensina diaphasis* (Bigot), *P. zodiacalis* (Bezzi), *Pliomelaena zonogastra* Bezzi, *P. sauteri* Hendel, "*Icterica*" *cashmerensis* Hendel, *Psednometopum aldabrense* (Lamb), *Sphaeniscus atilius* (Walker), *S. filiolus* (Loew), *Stephanotrypeta nigrofemorata* Munro, *Tarchonanthea coleoptrata* Freidberg and Kaplan, *Ypsilomena compacta* (Bezzi).

# 23 A Generic Reclassification and Phylogeny of the Tribe Myopitini (Tephritinae)

*Amnon Freidberg and Allen L. Norrbom*

**CONTENTS**



## 23.1 INTRODUCTION

Despite the well-recognized economic importance of fruit flies, their higher classification has been unstable and their phylogeny is poorly resolved. One reason is that most tephritid taxonomists have treated the family on a regional basis and usually did not analyze large taxa comprehensively. Another is that there seems to be considerable homoplasy in some characters that have been used in tephritid higher classification, for example, chaetotaxy and wing pattern. By looking at a large number of characters, using the latest phylogenetic methodology, and studying a group on a worldwide basis, we hoped to overcome these problems and provide a stable, predictive classification for a fairly large tephritid group. We selected the Myopitini because it is a reasonably well defined group, and because of our longstanding interests in species of this tribe.

The tribe Myopitini (= Euribiinae, Urophorinae, Myopitininae) has been treated as a separate subfamily or as a tribe of either the Trypetinae or the Tephritinae. We recognize it as a tribe of Tephritinae (see Section 23.3).

The Myopitini include 130 currently recognized species, although we know of about 60 more that are undescribed (see checklist in Section 23.2). All species reared to date feed as larvae in Asteraceae (= Compositae), infesting a wide range of tribes and genera. The larvae of most species feed in flowerheads, on developing seeds or receptacle tissue, and many cause the formation of flowerhead galls. A few species breed in stem galls. Some species of the genus *Urophora* Robineau-Desvoidy are important agents for weed biocontrol.

In the most recent comprehensive treatment of the Myopitini, Steyskal (1979) included 11 genera. Five of these genera are no longer classified in this tribe: *Marriottella* Munro is now placed in the subfamily Tephritinae, tribe Tephritini; *Hypenidium* Loew and *Trigonochorium* Becker are now placed in the Tephritinae, tribe Noeetiini (Norrbom et al. 1999a); *Cycasia* Malloch (now a junior synonym of *Ornithoschema* Meijere) was assigned to the subfamily Trypetinae, tribe Rivelliomimini, by Hancock (1986a, b); and *Nitrariomyia* Rohdendorf has been placed in the Trypetinae, tribe Trypetini, subtribe Nitrariomyiina (Korneyev 1987; 1996a). Individual species, including *Urophora acuticornis* Steyskal and *U. sabroskyi* Steyskal, also have been removed from the tribe (Norrbom 1989), and *Urellia diluta* Enderlein, although included in *Urophora* by Steyskal (1979), is a *Trupanea* (Tephritini), not a myopitine (Hardy 1969; Norrbom, personal observation).

With the removal of the above small taxa, the tribe Myopitini is a well-defined, monophyletic group (see Section 23.3). Although the definition and limits of the tribe have been relatively well established for many years, the current generic classification is unsatisfactory. Prior to this study and publication of a world tephritid database (Norrbom et al. 1999b) in which a few species were reclassified based on our results, the large majority (110 of 130) of the currently recognized species were lumped in *Urophora*, which is polyphyletic as previously conceived (e.g., Steyskal 1979). Some authors have recognized that the non-Palearctic species probably do not belong there, but included them for lack of a thorough understanding of myopitine relationships. Steyskal (1979), for example, recognized this anomaly, but nevertheless included in *Urophora* 39 of the 40 New World species known to him. White and Korneyev (1989), Korneyev and White (1991; 1992; 1993; 1996), and Korneyev and Merz (1998) revised the Palearctic species of *Urophora* and partially clarified the problem. They clearly defined *Urophora* (*s. str.*) and recognized that several Palearctic species did not belong in it. They proposed the subgenera *Eurasimona* Korneyev and White, *Inuromaesa* Korneyev and White, *Myopitora* Korneyev and White, and *Promyopites* Korneyev and Merz for these species, but continued to include them under *Urophora* for lack of comprehensive understanding of myopitine relationships and in abeyance of our revision.

## 23.2 MATERIAL AND METHODS

We studied representatives of all genera of Myopitini recognized in this chapter. This included about 85% of the species, including all the described species, except some in *Myopites* Blot, *Neomyopites*, n. gen., and *Urophora* (see checklist below), and all the known undescribed species, except a few in *Urophora* and one in *Inuromaesa*. Most of this study was based on the extensive collections of the National Museum of Natural History, Washington, D.C. (USNM) and Zoological Museum, Tel Aviv University (TAU). We also borrowed many specimens from other collections that are listed in the Acknowledgments. Phylogenetic relationships were analyzed using Hennig86 software (Farris 1988; Fitzhugh 1989); see Section 23.7 for further explanation. For discussion of myopitine host plant relationships we follow the classification of Asteraceae of Bremer (1994).

### 23.2.1 Checklist of Genera and Species of Myopitini

(An asterisk indicates a species not seen by us.)

*Asimoneura* Czerny
- *indecora* (Loew)
- *pantomelas* (Bezzi)
- *petiolata* (Munro) (including ssp. *flava* Munro and *seminigra* Munro)
- *shirakii* (Munro)
- *stroblii* Czerny
- at least 25 undescribed Afrotropical and Oriental species

*Eurasimona* Korneyev and White, n. stat.
- *fedotovae* (Korneyev and White), n. comb
- *stigma* (Loew), n. comb

*Goedenia*, n. gen.
- *bajae* (Steyskal), n. comb.
- *caurina* (Doane), n. comb.
- *formosa* (Coquillett), n. comb.
- *grindeliae* (Coquillett), n. comb.
- *rufipes* (Curran), n. comb.
- *setosa* (Foote), n. comb.
- *stenoparia* (Steyskal), n. comb.
- *timberlakei* (Blanc and Foote), n. comb.

*Inuromaesa* Korneyev and White, n. stat. (= *Promyopites* Korneyev and Merz, n. syn.)
- *circumflava* (Korneyev and Merz), n. comb.
- *maura* (Frauenfeld), n. comb. (= *tecta* Hering)
- *sogdiana* (Korneyev and Merz), n. comb.
- *at least one undescribed Palearctic species

*Myopites* Blot (= *Nearomyia* Becker, n. syn.)
- *apicatus* Freidberg
- **boghariensis* Séguy
- *bonifaciae* Dirlbek
- *cypriacus* Hering (= *shiakidesi* Dirlbeck)
- *delottoi* Munro
- *flavovarius* (Becker), n. comb.
- *hemixanthus* (Munro)
- *inulaedyssentericae* Blot (= *blotti* Brebisson, *hebe* Newman, *septemmaculata* Macquart, *inulae* Roser, *jasoniae* Dufour, *damascenus* Rondani, *sardoa* Costa, *olivieri* Kieffer)
- **lelea* Dirlbek (= *lelae* Dirlbek)
- *limbardae* Schiner (= *stylata* Fabricius)
- *longirostris* (Loew) (= *mentharum* Robineau-Desvoidy, *frauenfeldi* Schiner, *eximia* Séguy)
- *nigrescens* Becker
- **olii* Dirlbek
- *orientalis* Korneyev
- *tenellus* Frauenfeld
- *variofasciatus* Becker
- *zernyi* Hering
- at least three undescribed Afrotropical species

*Myopitora* Korneyev and White, n. stat.
- *shatalkini* (Korneyev and White), n. comb.

*Neomyopites*, n. gen.

*acompsus* (Hendel), n. comb.
*adjacens* (Hering), n. comb.
*aereus* (Hering), n. comb.
*agnatus* (Hering), n. comb.
*chaetostomus* (Hering), n. comb.
*chimborazonis* (Steyskal), n. comb.
*claripennis* (Foote), n. comb.
*columbianus* (Hering), n. comb.
*cordilleranus* (Steyskal), n. comb.
*cubanus* (Dirlbek and Dirlbekova), n. comb.
*cuzconis* (Steyskal), n. comb.
*disjunctus* (Becker), n. comb.
*euryparius* (Steyskal), n. comb.
*eved* (Steyskal), n. comb.
**funebris* (Hering), n. comb.
*hodgesi* (Steyskal), n. comb.
*jamaicensis* (Steyskal), n. comb.
*mamarae* (Hendel), n. comb.
*melanops* (Steyskal), n. comb.
*mexicanus* (Steyskal), n. comb.
*morus* (Hering), n. comb.
*paulensis* (Steyskal), n. comb.
*regis* (Steyskal), n. comb.
*simplex* (Becker), n. comb.
**townsendi* (Bezzi), n. comb.
*tresmilius* (Steyskal), n. comb.
*trivirgulatus* (Foote), n. comb.
*unicus* (Becker), n. comb.
at least nine undescribed Neotropical species

*Rhynencina* Johnson
*dysphanes* (Steyskal)
*emphanes* (Steyskal)
*spilogaster* (Steyskal)
*longirostris* Johnson (= *alpha* Phillips)
*xanthogaster* (Steyskal)

*Spinicosta*, n. gen.
*agromyzella* (Bezzi), n. comb.
*cilipennis* (Bezzi), n. comb.
at least six undescribed Afrotropical species

*Stamnophora* Munro
*vernoniicola* (Bezzi)
at least ten undescribed Afrotropical species

*Urophora* Robineau-Desvoidy
*affinis* (Frauenfeld)
**anthropovi* Korneyev and White
*aprica* (Fallén) (? = *brunicornis* Robineau-Desvoidy, *scutellata* Rondani)
**bernhardi* Korneyev and White
*campestris* Ito
*cardui* (Linnaeus) (= *flexuosa* Ahrens, *reaumurii* Robineau-Desvoidy, ? = *sonchi* Robineau-Desvoidy)
**christophi* Loew

*congrua* Loew
*coronata* Bassov
*cuspidata* (Meigen)
**digna* Richter
*dzieduszyckii* Frauenfeld (= *erishischmidti* Hering, *syriaca* Hendel)
**egestata* (Hering) (= *ensata* Richter)
**formosana* (Shiraki)
*hermonis* Freidberg
*hispanica* Strobl
**iani* Korneyev and Merz
*impicta* (Hering)
*ivannikovi* Korneyev and White
*jaceana* (Hering) (= *conyzae* Hering)
*jaculata* Rondani
**kasachstanica* (Richter)
**korneyevi* White (= *arctii* Korneyev and White)
**longicauda* (Hendel)
*lopholomae* White and Korneyev
**mandschurica* (Hering) (= *chejudoensis* Kwon)
*mauritanica* Macquart (= *macrura* Loew, *lejura* Rondani, ? = *algira* Macquart, ? = *sejuncta* Becker)
**melanocera* Hering
*misakiana* (Matsumura) (= *bicoloricornis* Zia, *hoenei* Hering)
*neuenschwanderi* Freidberg
**nigricornis* Hendel
*pauperata* (Zaitzev) (= *calcitrapae* White and Korneyev)
*phaeocera* Hering
*phaleolepidis* Merz and White
*pontica* (Hering)
*quadrifasciata* (Meigen) (= *dejeanii* Robineau-Desvoidy; and including subspp. *algerica* (Hering) and *sjumorum* (Rohdendorf), ? = *armeniaca* Hering)
*repeteki* (Munro) (= *ligulipalpis* Hering)
*sachalinensis* (Shiraki)
**sciadocousiniae* Korneyev and White
**sinica* (Zia)
*sirunaseva* (Hering)
**solaris* Korneyev
*solstitialis* (Linnaeus) (= *arctii* De Geer, *dauci* Fabricius, *hastatus* Fabricius, *pugionata* Meigen, *veruata* Rondani, *sibynata* Rondani, *sonderupi* Hering, ? = *femoralis* Robineau-Desvoidy)
**spatiosa* (Becker) (= *angustifascia* Hering)
*spoliata* (Haliday)
*stalker* Korneyev (= *beikoi* Korneyev)
*stylata* (Fabricius) (= *cirsii* Schrank, *jacobeae* Panzer, *japonica* Shiraki, *leucacanthi* Schrank, *venabulata* Rondani, *vulcanica* Rondani, *pia* Hering)
*tengritavica* Korneyev and White
*tenuior* Hendel (= *tenuis* Hendel, *attingens* Munro, *heratensis* Dirlbek and Dirlbek)
*tenuis* Becker
*terebrans* (Loew) (= *eriolepidis* Loew, *manni* Hendel, *approximata* Hering, *satunini* Zaitzev)
**trinervii* Korneyev and White
**tsoii* Korneyev and White

*variabilis* Loew (= *kiritshenkoi* Zaitzev)
*vera* Korneyev and White
*volkovae* Korneyev
*xanthippe* (Munro)
at least five undescribed Palearctic and Afrotropical species
Unplaced species of Myopitini
*Trypeta conferta* Walker (probably belongs in *Neomyopites*)

## 23.3 RELATIONSHIPS AND DIAGNOSIS OF THE MYOPITINI

### 23.3.1 Relationships to Other Tephritinae

The subfamily Tephritinae, in the broad sense, including the Myopitini, Tephrellini, and other taxa that sometimes have been recognized as tribes of the Trypetinae or as separate subfamilies, is generally thought to be a monophyletic group, although defining it in terms of unique synapomorphies is difficult (Foote et al. 1993; Korneyev, Chapter 22). Most of the species (except the *Tephellini* and a few species in other tribes) have similar biologies; that is, their larvae live in flowerheads or stems of species of Asteraceae, often forming galls. Foote et al. (1993) and Norrbom et al. (1999a) suggested several possible synapomorphies: The shape of the male epandrium and surstyli, which are more or less oval in posterior view in most tephritine species; the shape of the spermathecal duct, in which the apical part (exceeding the length of the spermatheca) is expanded; and the type of sex determination method (heterogametic female). The former character actually appears to be a synapomorphy for Tephritinae + Zaceratini, and more study of the other characters is needed (Korneyev, Chapter 22).

Korneyev (Chapter 22) classifies the Myopitini within what he terms the "Lower Tephritinae," a probably paraphyletic group of tribes that seems to include the most basal lineages of the subfamily. Within this group, the exact relationship of the Myopitini to the rest of the Tephritinae is uncertain, and we do not know which other tribe or tribes form their sister group. Two character states, both apomorphic within the Tephritidae, have traditionally been used to diagnose the Myopitini: (1) a single orbital seta present; and (2) vein $Cu_2$ straight or convex (i.e., the basal cubital cell without a posteroapical lobe). Although each of these character states occurs occasionally in other tephritid genera, no nonmyopitines have this combination. A third possible synapomorphy involves the abdominal pleural membrane, which is opaque, granular, and usually black or darker than the sclerites. Several other characters are perhaps phylogenetically significant but difficult to evaluate.

Myopitini usually have distinct wing patterns, although many species have reduced patterns or entirely hyaline wings (Figures 23.4 and 23.5). The patterns are of two types. The *Myopites*-type (Figures 23.4C, F, H, and 23.5A, B, and E) generally has five transverse bands or rows of spots, including a small crossband in the middle of cell $r_1$, and the spots covering crossvein DM-Cu and the apex of vein $R_{2+3}$ are not aligned, or if forming a continuous band (Figure 23.4C), with a bend along vein M. The apical band is usually isolated. The *Urophora*-type (Figure 23.5G–H) generally has four transverse bands and is without a small crossband in cell $r_1$, but has a complete or nearly complete band covering DM-Cu that extends straight to the anterior wing margin. The apical band is usually connected to the band covering DM-Cu. Many Lower Tephritinae (*sensu* Korneyev) have banded wing patterns, suggesting that the *Urophora*-type pattern may be plesiomorphic for the Myopitini, but except for many Terelliini, the pattern is rather different in the other tribes, casting some doubt on this hypothesis of character polarity. Instead, the *Myopites*-type pattern, which is unique to the tribe, may be an autapomorphy present in the myopitine ground plan. Myopitini also have only dark, acuminate head and scutal setae and setulae, which is plesiomorphic within Tephritinae (most other tribes have some setae or setulae whitish and lanceolate, the apomorphic condition). However, the possibility that the absence of lanceolate setae and setulae in the Myopitini is due to reversal should not be overlooked (e.g., if the Myopitini is eventually shown to be the sister group of one of the tribes having lanceolate setulae), in which case their loss would be a synapomorphy of the tribe.

A phylogenetic analysis of all tephritid taxa is necessary before any definitive statements about character polarities can be made. Because the Myopitini as currently understood are relatively homogeneous, all consistent characters, whether apomorphic or plesiomorphic, are enumerated below in the diagnosis of the tribe.

## 23.3.2 Diagnosis

*Head* (Figures 23.1 and 23.2): One orbital seta; usually two frontal setae, rarely three or four (or supernumeraries sometimes present); frontal vitta bare; ocellar seta about as long as frontal setae; postocular setae acuminate and dark; eye usually small, oval, leaving high gena that is seldom narrower than antenna; parafacial often as wide as antenna; antenna shorter than face; first flagellomere rounded apically; arista pubescent, with hairs shorter than its basal width; proboscis usually spatulate to long geniculate, and usually with at least somewhat elongate haustellum, and often also rostrum; labella usually elongate, but sometimes short or appearing short, although proboscis rarely capitate (e.g., some *Neomyopites*).

*Thorax*: Scutum and scutellum often short and convex (Figure 23.3B), scutellum sometimes flat dorsally and more elongate (Figure 23.3A); one postpronotal, two notopleural, one presutural and one postsutural supra-alar, one intra-alar, and one postalar seta present; scapulars often distinguishable although not as clearly as in most Trypetinae; dorsocentral seta aligned from slightly anterior to postsutural supra-alar seta to about midway between it and acrostichal seta; two scutellar setae equal, or apical slightly (10%) longer than basal seta; setulae acuminate and dark; microtrichia dense or sparse, usually denser on scutum and not forming pattern, although rarely with narrow brownish dorsocentral stripes; pleura entirely microtrichose or sometimes with shiny area without microtrichia; yellow notopleural stripe usually present and conspicuous; scutellum entirely dark or yellow medially (Figure 23.3A-B); legs usually without overt features, although in species of the *melanops* group of *Neomyopites* (see Discussion for that genus) hind femur anteroventrally with long seta(e) (Figure 23.3E); wing veins straight and generally parallel (e.g., Figure 23.4B), but cell $r_{4+5}$ occasionally narrowed apically (Figure 23.4F-G); crossvein R-M typically opposite or near middle of cell dm, often proximal to middle, but rarely beyond 0.66 of length of cell; cell bcu without posteroapical lobe, with vein $Cu_2$ convex or straight; vein $R_1$ setulose dorsally without gap; no other setulae normally present on veins either dorsally or ventrally; wing banded, spotted, or hyaline.

*Abdomen*: Pleural membrane opaque, usually partially or entirely black or darker than sclerites (yellow in *Inuromaesa circumflava* and sometimes in *Rhynencina*); sclerites entirely black (Figure 23.6B) or partially to entirely yellow (Figure 23.6A); setulae acuminate and dark; microtrichia sparse or absent; terga often narrow in males, so pleural membrane visible in dorsal view (Figure 23.6A); epandrium and lateral surstylus rounded or oval in posterior view (Figure 23.7), often relatively long (along longitudinal axis); lateral surstylus short, often with lobes projected posterad; medial surstylus with prensisetae usually directed mesally; lateral sclerites asymmetric (right sclerite longer than left) (Figure 23.6H) or subequal, slender or fused flangelike to hypandrium (Figure 23.6F-G); arm of phallapodeme touching or connected to lateral sclerite; glans (Figures 23.9 and 23.10) short to very long, often with sclerotization weak, small, and/or restricted to basal area; female tergite 6 as long as or shorter than tergite 5; oviscape (Figure 23.11A) usually strongly conical at base and more distally cylindrical, often considerably longer than preabdomen; aculeus (Figure 23.11B-C) gradually tapered and pointed, or often slightly broadened preapically; tip sometimes (many *Urophora*) with minute preapical shoulders or truncate extreme apex; two spermathecae, usually not sclerotized and therefore difficult to detect in dissections of dried specimens (Figure 23.11D-E).

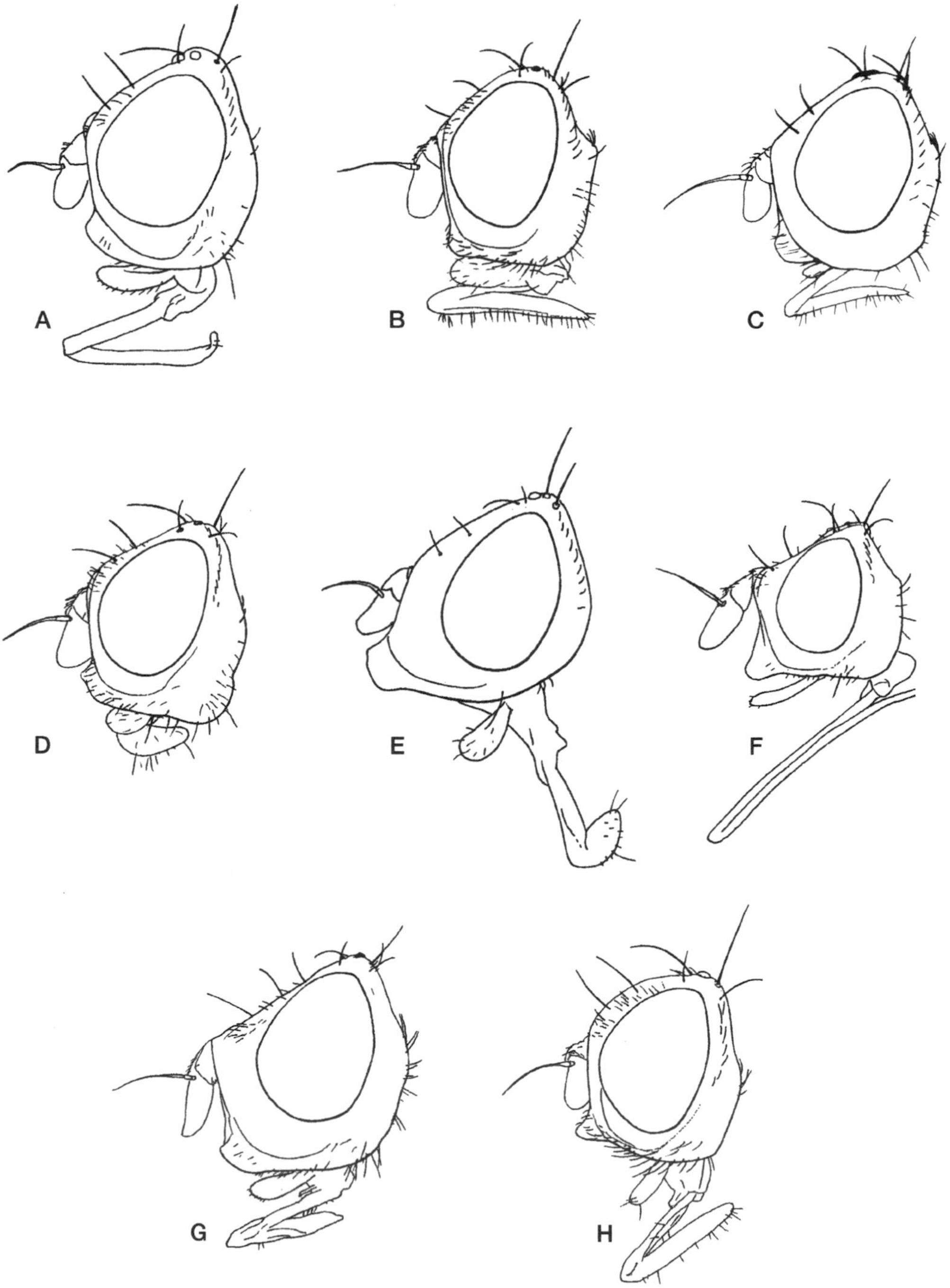


**FIGURE 23.1** Head, lateral view: (A) *Asimoneura stroblii*; (B) *Eurasimona stigma*; (C) *Goedenia rufipes*; (D) *Inuromaesa maura*; (E) *I. circumflava*; (F) *Myopites delottoi*; (G) *Myopites* sp. B; (H) *Myopitora shatalkini*.

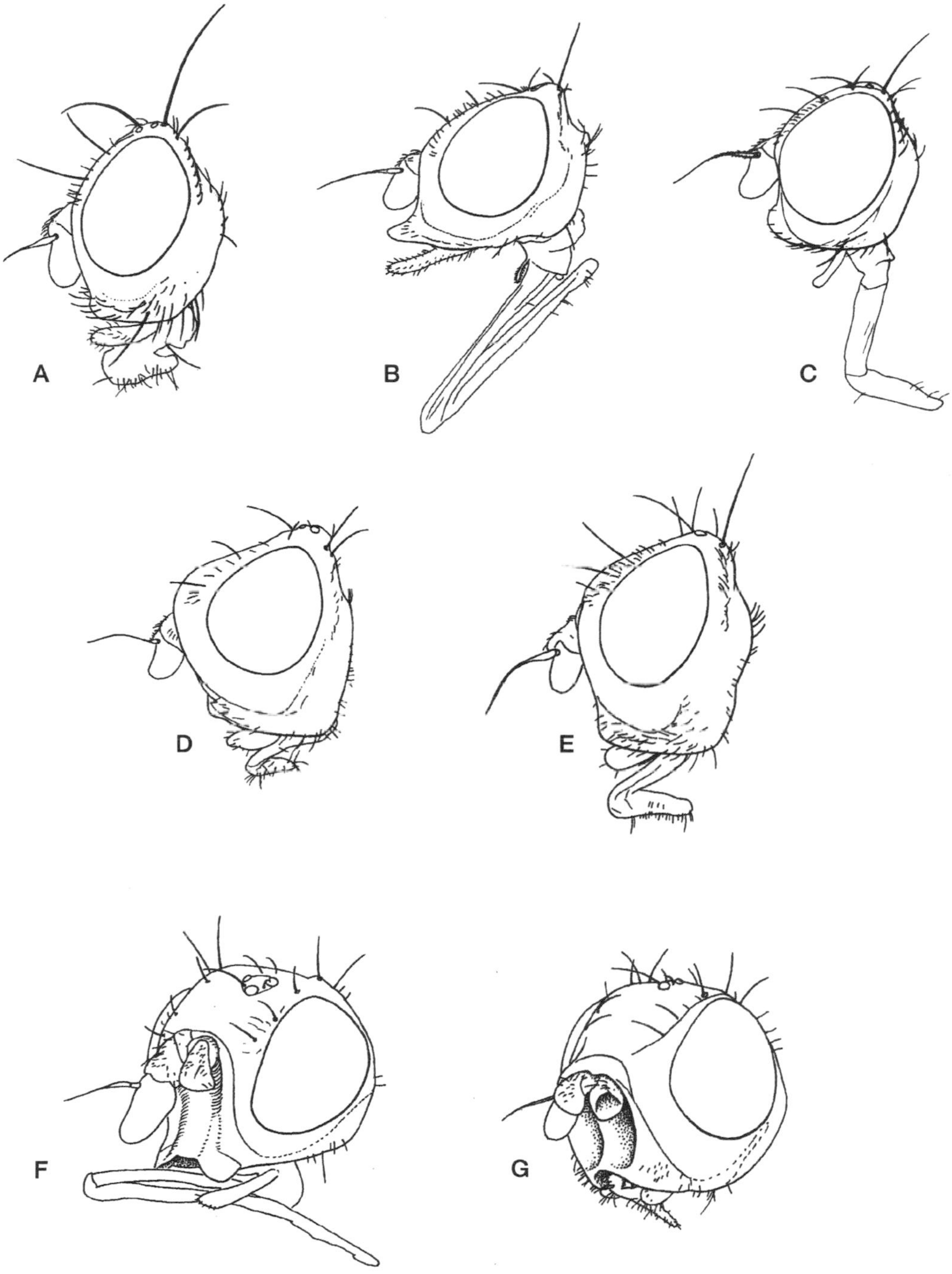


**FIGURE 23.2** Head: (A to E) lateral view; (F and G) anterolateral view, with left first flagellomere removed; (A) *Neomyopites* sp.; (B) *Rhynencina longirostris*; (C) *Spinicosta cilipennis*; (D) *Stamnophora vernoniicola*; (E) *Urophora cardui*; (F) *Myopites delottoi*; (G) *Goedenia rufipes*.

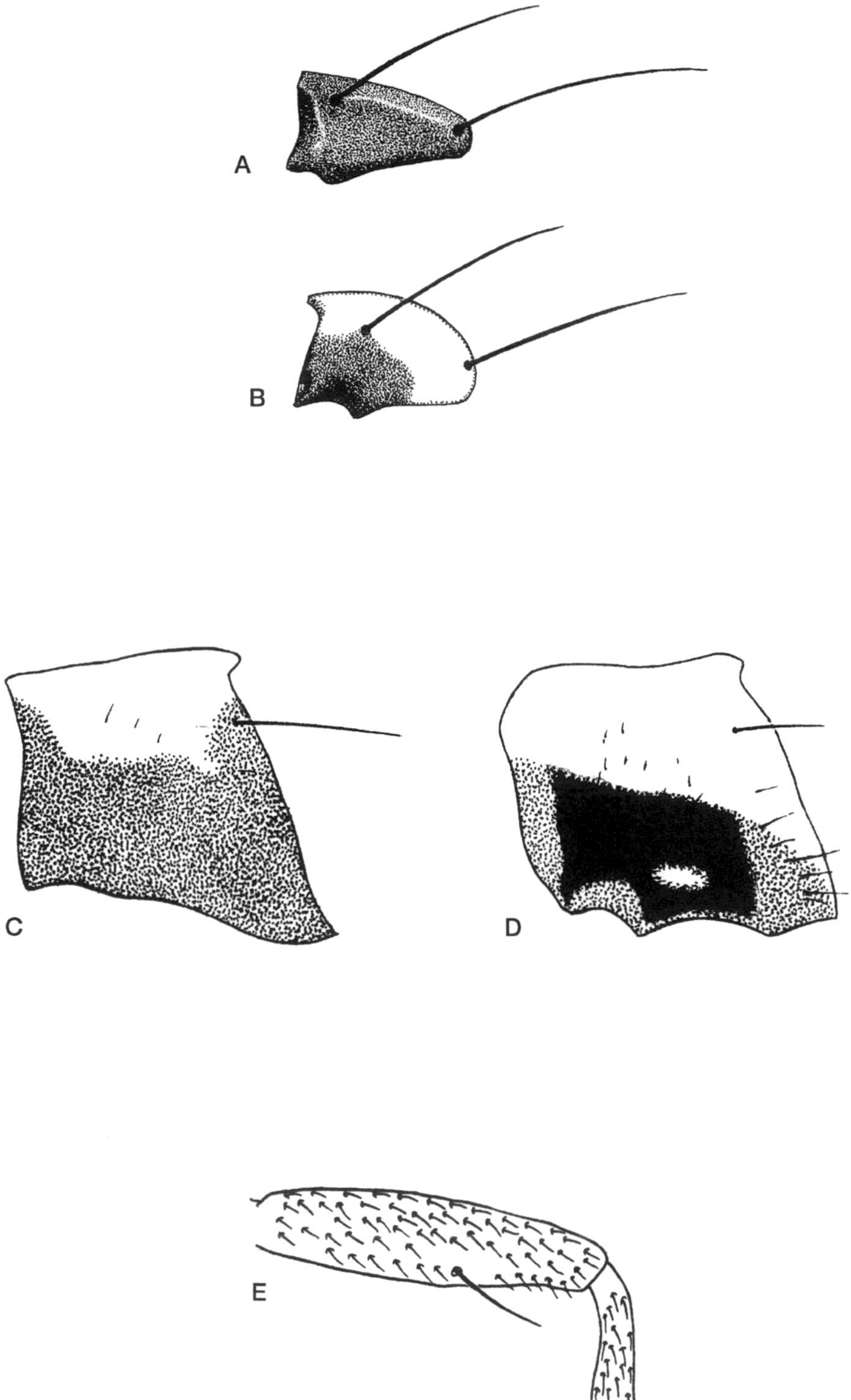


**FIGURE 23.3** (A and B) Scutellum, lateral view: (A) *Spinicosta cilipennis*; (B) *Urophora cardui*. (C and D) anepisternum: (C) *Rhynencina spilogaster*; (D) *Urophora hermonis*. (E) Hind femur and part of tibia, anterior view, *Neomyopites jamaicensis*.

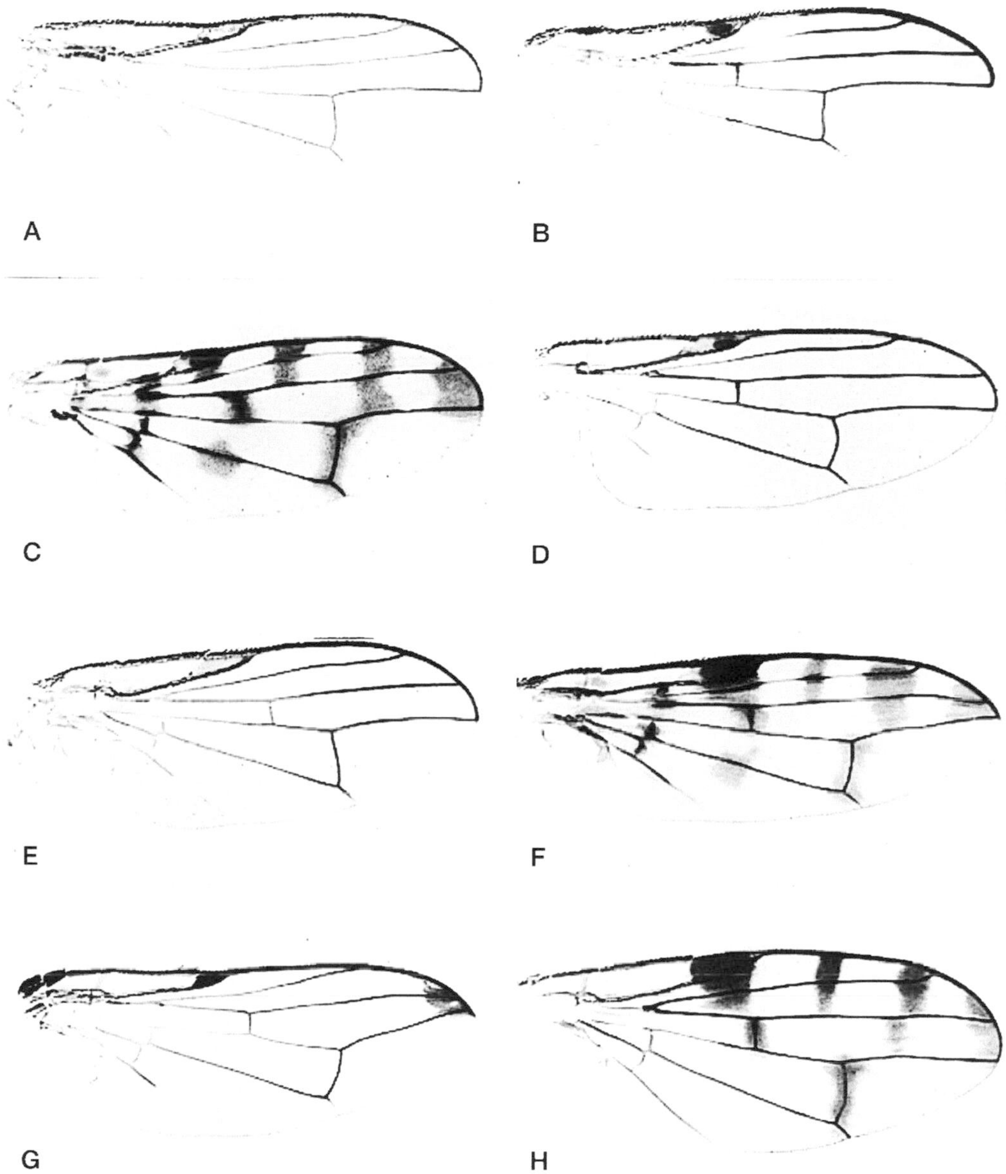

**FIGURE 23.4** Wing: (A) *Asimoneura stroblii*; (B) *Eurasimona stigma*; (C) *Goedenia formosa*; (D) *Goedenia timberlakei*; (E) *Inuromaesa maura*; (F) *Myopites* sp.; (G) *Myopites* sp. B; (H) *Myopitora shatalkini*.

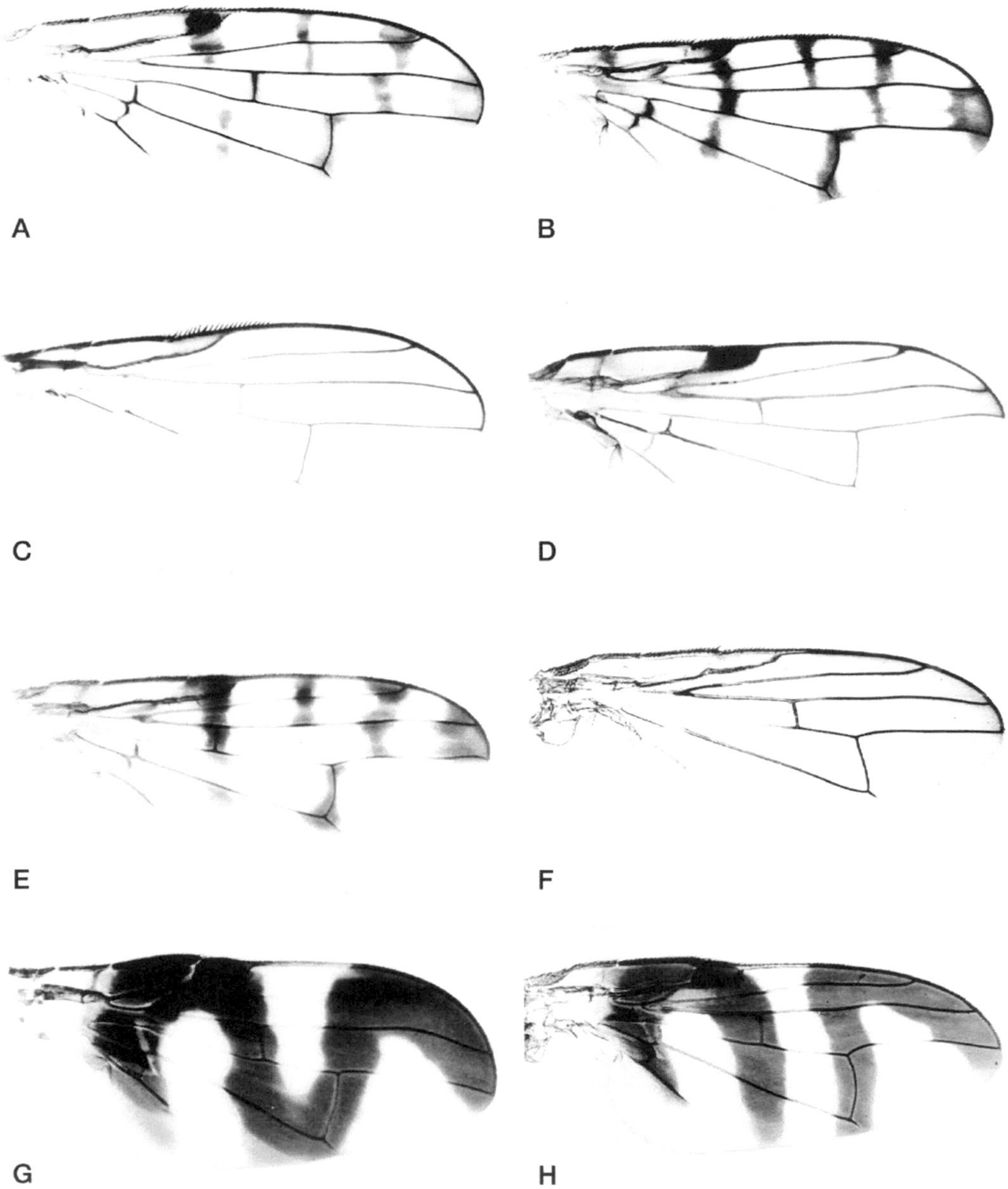


**FIGURE 23.5** Wing: (A) *Neomyopites* sp.; (B) *Rhynencina spilogaster*; (C) *Spinicosta cilipennis*; (D) *Stamnophora vernoniicola*; (E) *Stamnophora* sp.; (F) *Urophora hermonis*; G. *U. cardui*; (H) *U. quadrifasciata*.

## 23.4 KEY TO THE GENERA OF MYOPITINI

1. Anepisternum at least partially nonmicrotrichose and shiny (Figure 23.3D). Wing pattern, if not absent or reduced, often of *Urophora*-type, with band covering crossvein DM-Cu complete and extended straight to anterior wing margin (Figure 23.5G–H) . . . . . . . . . . . . . . . . . . . . . . . . . . . . . . . . . . . . . . . . . . . . . . . . 2

- Anepisternum entirely microtrichose (Figure 23.3C). Wing pattern variable, but if banded, rarely (one species of *Neomyopites*) of *Urophora*-type, usually with band covering crossvein DM-Cu incomplete or not extended straight to anterior wing margin (Figure 23.5B) . . . . . . . . . . . . . . . . . . . . . . . . . . . . . . . . . . . . . . . . . . . 6

2. Anepisternum, anepimeron, and katepimeron each at least partially bare and shiny, at most partly sparsely microtrichose. Dorsocentral seta well posterior to level of postsutural supra-alar seta, usually more than one-third distance from level of supra-alar seta to level of acrostichal seta. Face moderately to deeply concave (Figure 23.2G). Wing pattern, if not absent or reduced, of *Myopites*-type, with band covering crossvein DM-Cu incomplete or not extended straight to anterior wing margin (Figure 23.4C). Hosts: Astereae, subtribe Solidagininae. Western North America . . . . . . . . . . . . . . . . . . . . . . . . . . . . . . . . . . . . . . . *Goedenia*, n. gen.

• Anepimeron and katepimeron usually both entirely microtrichose, rarely anepimeron with bare area (*U. dzieduszyckii* group) or katepimeron with dorsal margin bare (some *Asimoneura*), but not both; anepisternum usually partially densely microtrichose. Dorsocentral seta more or less aligned with postsutural supra-alar seta, at most slightly posterior to it and less than one-third distance from level of supra-alar seta to level of acrostichal seta. Face slightly convex to moderately concave (Figure 23.2F). Wing pattern, if not absent or reduced, of *Urophora*-type, with band covering crossvein DM-Cu complete and extended straight to anterior wing margin (Figure 23.5G-H). . . . . . . . . . . . . . . . . . . . . . . 3

3. Proboscis long geniculate (Figure 23.1A, F); labella length:head length ratio 0.8 or greater; labella length:head height ratio 0.6 to 2.5, usually greater than 0.8. Wing pattern highly reduced or absent. Male abdominal tergites often narrowed (as in Figure 23.6A) . . . . . . . . . . . . . . . . . . . . . . . . . . . . . . . . . . . . . . . . . . . . . . . . . . 4

• Proboscis spatulate to short geniculate (Figure 23.2C, E); labella length:head length ratio 0.3 to 0.8; labella length:head height ratio 0.2 to 0.7. Wing pattern absent or with *Urophora*-type bands. Male abdominal tergites broad (Figure 23.6B) . . . . . . . . . . . . . . . . . . . . . . . . . . . . . . . . . . . . . . . . . . . . . . . . . . 5

4. Cell $r_{4+5}$ not strongly narrowed distally (Figure 23.4A). Abdominal tergites black. Hosts: Arctoteae, Anthemideae, Gnaphalieae. Old World, mostly South Africa . . . . . . . . . . . . . . . . . . . . . . . . . . . . . . . . . . . . . . . . . . *Asimoneura* Czerny, in part

• Cell $r_{4+5}$ strongly narrowed distally by anterior slant of vein M (Figure 23.4F-G). Abdominal tergites partially yellow. Hosts: Inuleae, Plucheeae, and Cardueae. Palearctic and Afrotropical.. . . . . . . . . . . . . . . . . . . *Myopites* Blot, in part (*M. delottoi* Munro, Ethiopia)

5. Costal setulae along pterostigma enlarged and erect (Figure 23.5C). Scutellum entirely black, flat (Figure 23.3A). Wing entirely hyaline (Figure 23.5C). Hosts: Arctoteae and Vernonieae. Afrotropical. . . . . . . . . . . . . . . . . . . . . . . . . . . . *Spinicosta*, n. gen.

• Costal setulae small, not erect (Figure 23.5G-H). Scutellum yellow medially, convex (Figure 23.3B). Wing usually with distinct bands (Figure 23.5G-H), sometimes reduced, rarely hyaline (Figure 23.5F). Hosts: Cardueae. Predominantly Palearctic, one sp. Afrotropical, two spp. Oriental, also introduced to North America, South Africa, and Australia . . . . . . . . . . . . . . . . . . . . *Urophora* Robineau-Desvoidy

6. Proboscis geniculate, long (Figure 23.1A–C, F, H); labella more than 0.6 times as long as head height, rarely less than 0.75 times head length, and usually more than six times as long as high (in lateral view); ventral part of face usually moderately to strongly protrudent. . . . . . . . . . . . . . . . . . . . . . . . . . . . . . . . . . . . . . . . . . 7

• Proboscis spatulate or rarely capitate (Figures 23.1D-E, 23.2A, D); labella less than 0.55 times as long as head height, less than six times as long as high; ventral part of face usually at most moderately protrudent, only in *Inuromaesa circumflava* (Figure 23.1E) strongly protrudent. . . . . . . . . . . . . . . . . . . . . . . . . 12

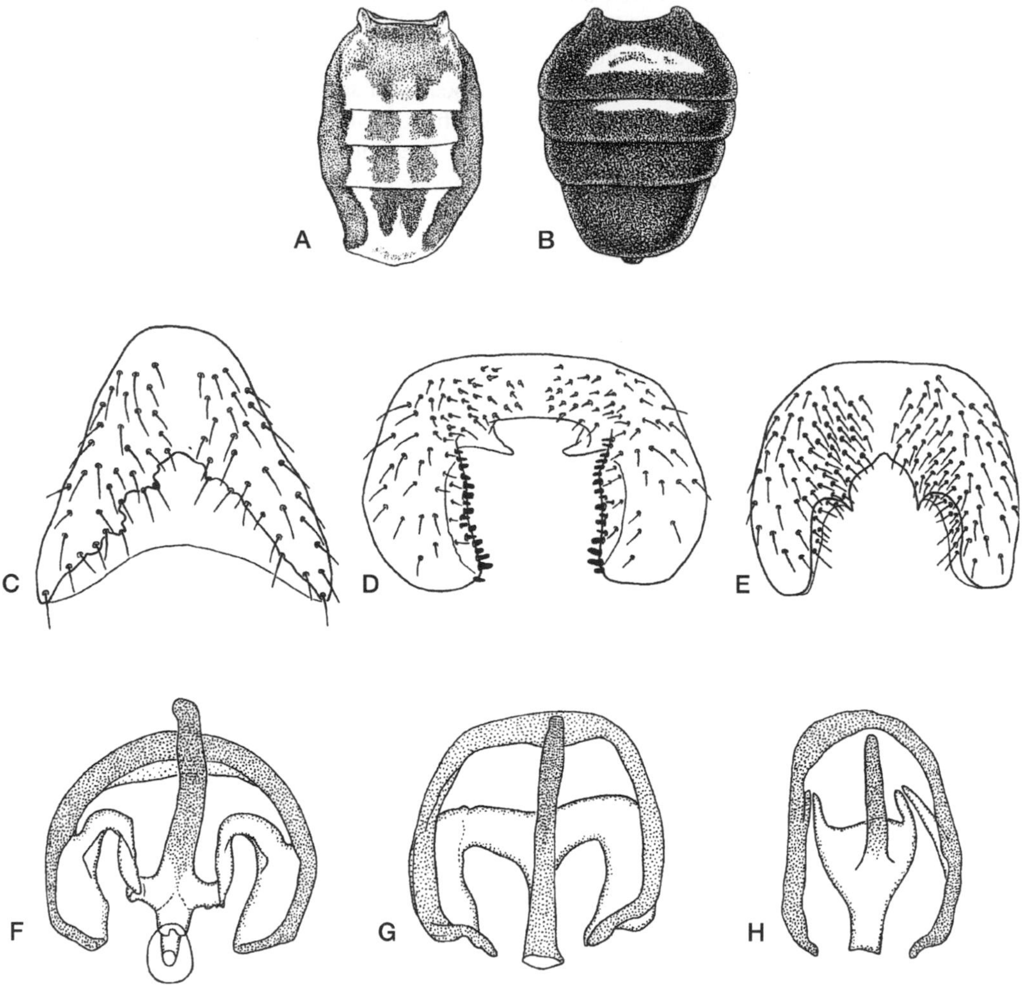


**FIGURE 23.6** Male abdomen: (A and B), dorsal view: (A) *Myopites* sp.; (B) *Urophora stylata*; (C to E) fifth sternite; (C) *Goedenia formosa*; (D) *Neomyopites melanops*; (E) *Neomyopites* sp.; (F and G) hypandrium and phallapodeme, dorsal view; (F) *Asimoneura stroblii*; (G) *Urophora cardui*; (H) *Goedenia rufipes*.

7. Tergites entirely black (similar to Figure 23.6B). Wing pattern usually reduced (Figure 23.4A-B), occasionally with extensive pattern (Figure 23.4H). Glans sclerotization strong and extensive (Figure 23.9B) or weak (Figures 23.9A, 23.10A) .......................................... 8
- Tergites at least partly yellow (Figure 23.6A). Wing usually with extensive pattern (Figures 23.4F, 23.5B). Glans with small, well differentiated basal sclerite (Figures 23.9E-F, 23.10C) .......................................... 10
8. Scutellum entirely black (Figure 23.3A). Wing without pattern (Figure 23.4A), except sometimes faintly along costa. Hosts: Anthemideae, Arctoteae, Gnaphalieae. Old World, mostly South Africa. .......................... *Asimoneura* Czerny, in part
- Scutellum yellow medially (Figure 23.3B). Wing mostly or entirely hyaline (Figure 23.4B) or with extensive pattern (Figure 23.4H) ...................... 9
9. Wing pattern reduced, at most with dark pterostigma and anteroapical spot (Figure 23.4B), wing length: wing width ratio about 2.7. Face moderately to deeply concave. Lateral surstylus apically strongly curved mesally (Figure 23.7B). Glans

extensively sclerotized (Figure 23.9B). Spermatheca weakly sclerotized. Hosts: Anthemideae. Europe, Central Asia . . . . . . . . . . . . . . . . . . . . . . . . . . *Eurasimona* Korneyev and White, in part (*E. stigma* (Loew))

- Wing with extensive pattern (Figure 23.4H), wing length: wing width ratio about 2.3. Face flat to moderately concave. Lateral surstylus apically not strongly curved mesally (Figure 23.7G). Glans with relatively little, weak sclerotization (Figure 23.10A). Spermatheca unknown. Hosts: Astereae. Eastern Asia . . . . . . . . . . . . . . . . . . . . . . . . . . . . . . . . . . . . . . . *Myopitora* Korneyev and White

10. Wing without pattern (similar to Figure 23.4A), cell $r_{4+5}$ not narrowed distally. Scutellum entirely black. Hosts: Anthemideae, Arctoteae, Gnaphalieae. Old World, mostly South Africa . . . . . . . . . . . . . . . . . . *Asimoneura*, in part (*A. shirakii* (Munro), Taiwan)

- Wing with extensive pattern of bands or spots (Figures 23.4F, 23.5B). Scutellum usually yellow medially, if entirely black (*Myopites nigrescens*), cell $r_{4+5}$ narrowed distally (as in Figure 23.4F) . . . . . . . . . . . . . . . . . . . . . . . . . . . . . . . . . . . . . . . . . . . . 11

11. Cell $r_{4+5}$ narrowed distally (slightly in *M. hemixanthus* and *apicatus*) (Figure 23.4F). Male abdominal tergites narrow, pleural membrane visible in dorsal view (Figure 23.6A). Glans usually very long, length-to-width ratio usually at least 10; acrophallus without striations (Figure 23.9E-F). Hosts: Inuleae, Plucheeae, and Cardueae. Palearctic and Afrotropical . . . . . . . . . . . . . . . . . . . . . . . . . *Myopites* Blot, in part

- Cell $r_{4+5}$ not narrowed distally (Figure 23.5B). Male abdominal tergites broad, pleural membrane not visible in dorsal view. Glans not extremely elongate, length-to-width ratio less than 10; acrophallus striate (Figure 23.10C). Hosts: Heliantheae. New World . . . . . . . . . . . . . . . . . . . . . . . . . . . . . . . . *Rhynencina* Johnson

12. Vein M ending anterior to wing apex (Figure 23.4G), cell $r_{4+5}$ slightly or strongly narrowed distally from level of crossvein DM-Cu . . . . . . . . . . . . . . . . . . . . . . . . . . . . . . . . 13

- Vein M ending at or posterior to wing apex (Figure 23.5A, D-E), cell $r_{4+5}$ not narrowed distally, or if slightly narrowed (some *Stamnophora*), then only in distal third . . . . . . . . . . . . . . . . . . . . . . . . . . . . . . . . . . . . . . . . . . . . . . . . . . . . . . . 14

13. Body entirely yellow. Wing hyaline, with cell $r_{4+5}$ slightly narrowed distally. Hosts: Inuleae and Cardueae. Europe, Central Asia . . . *Inuromaesa* Korneyev and White in part (*I. circumflava* (Korneyev and Merz), Central Asia)

- Body not entirely yellow. Wing with extensive bands or spots or at least a dark spot near apex in addition to dark pterostigma (Figure 23.4G); cell $r_{4+5}$ strongly narrowed distally. Hosts: Cardueae, Inuleae, and Plucheeae. Palearctic and Afrotropical . . . . . . . . . . . . . . . . . . . . . . . . . . . . . . . . . . . . . . . . . . . . . *Myopites* Blot, in part

14. Tergites at least partly yellow (mostly to entirely black in *S. vernoniicola*), in male narrow so that pleural membrane is visible in dorsal view (similar to Figure 23.6A). Wing pattern usually extensive (Figure 23.5E), occasionally reduced (Figure 23.5D). Lateral surstylus with large, broad posterolateral lobe (Figure 23.8K). Glans with small, well-differentiated basal sclerite (Figure 23.10E). Hosts: Vernonieae. Afrotropical . . . . . . . . . . . . . . . . . . *Stamnophora* Munro

- Tergites entirely black, in male usually broad (except in *Eurasimona*) so that pleural membrane not visible in dorsal view. Wing pattern variable. Lateral surstylus without posterolateral lobe (Figure 23.8B, D, H). Glans basal sclerotization strong and extensive (Figure 23.9B) or weak (Figures 23.9D, 23.10B) . . . . . . . . . . . . . . . . . . . . . . . . . . . . . . . . . . . . . . . . . . . . 15

15. Wing usually with extensive pattern of spots or bands (Figure 23.5A). Scutellum flat or at most slightly convex. Hind femur often with one to two outstanding anteroventral setae (Figure 23.3E). Puparium often with spines on posterior end (Figure 23.12E). Acrophallus of glans with dot-like marks (Figure 23.10B). Spermathecae and spermathecal ducts membranous. Hosts: Astereae,

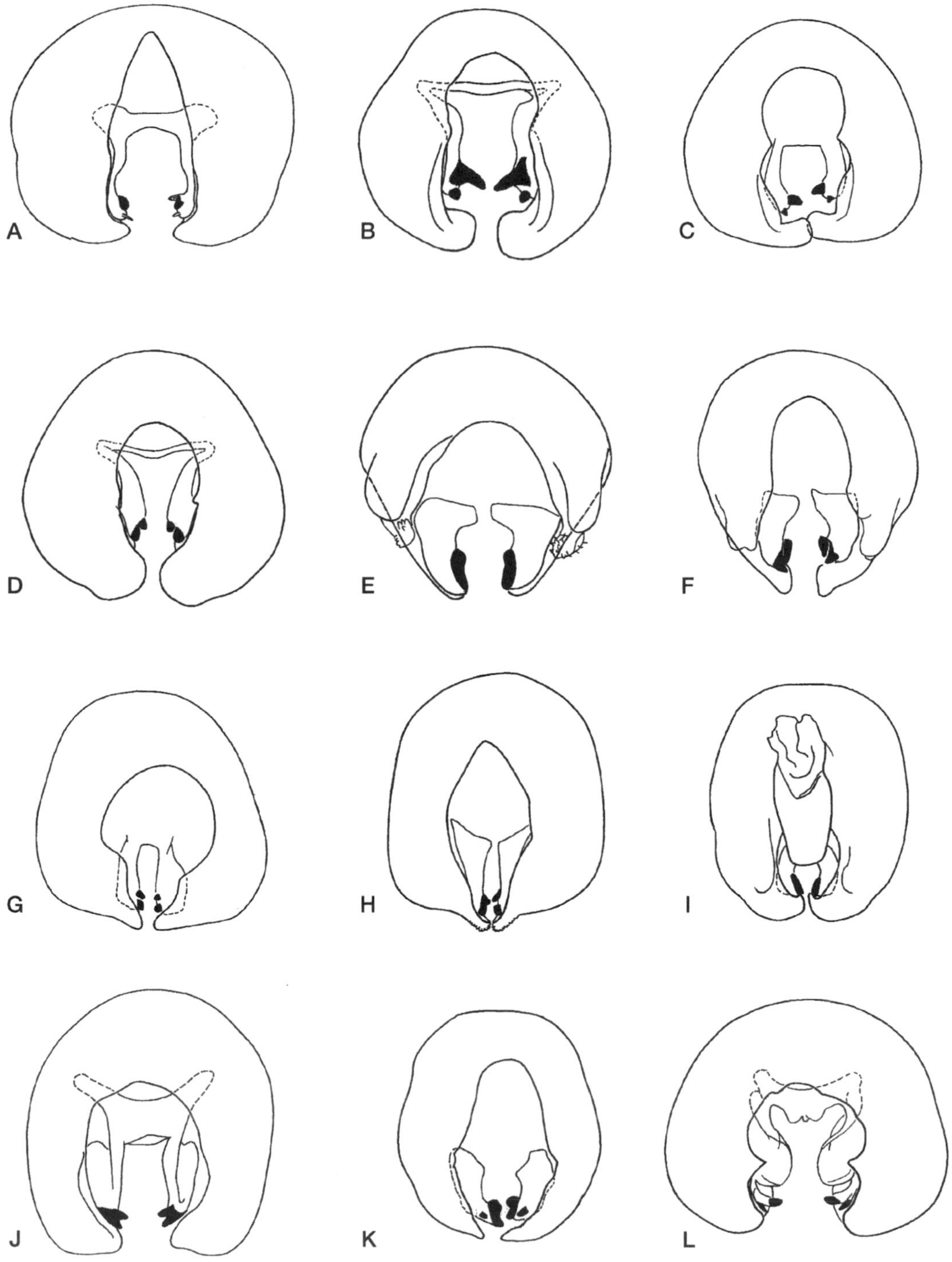


**FIGURE 23.7** Male genitalia, epandrium, and surstyli, posterior view: (A) *Asimoneura stroblii*; (B) *Eurasimona stigma*; (C) *Goedenia formosa*; (D) *Inuromaesa maura*; (E) *Myopites* sp. A; (F) *M. flavovarius*; (G) *Myopitora shatalkini*; (H) *Neomyopites melanops*; (I) *Rhynencina longirostris*; (J) *Spinicosta cilipennis*; (K) *Stamnophora vernoniicola*, (L) *Urophora cardui*.

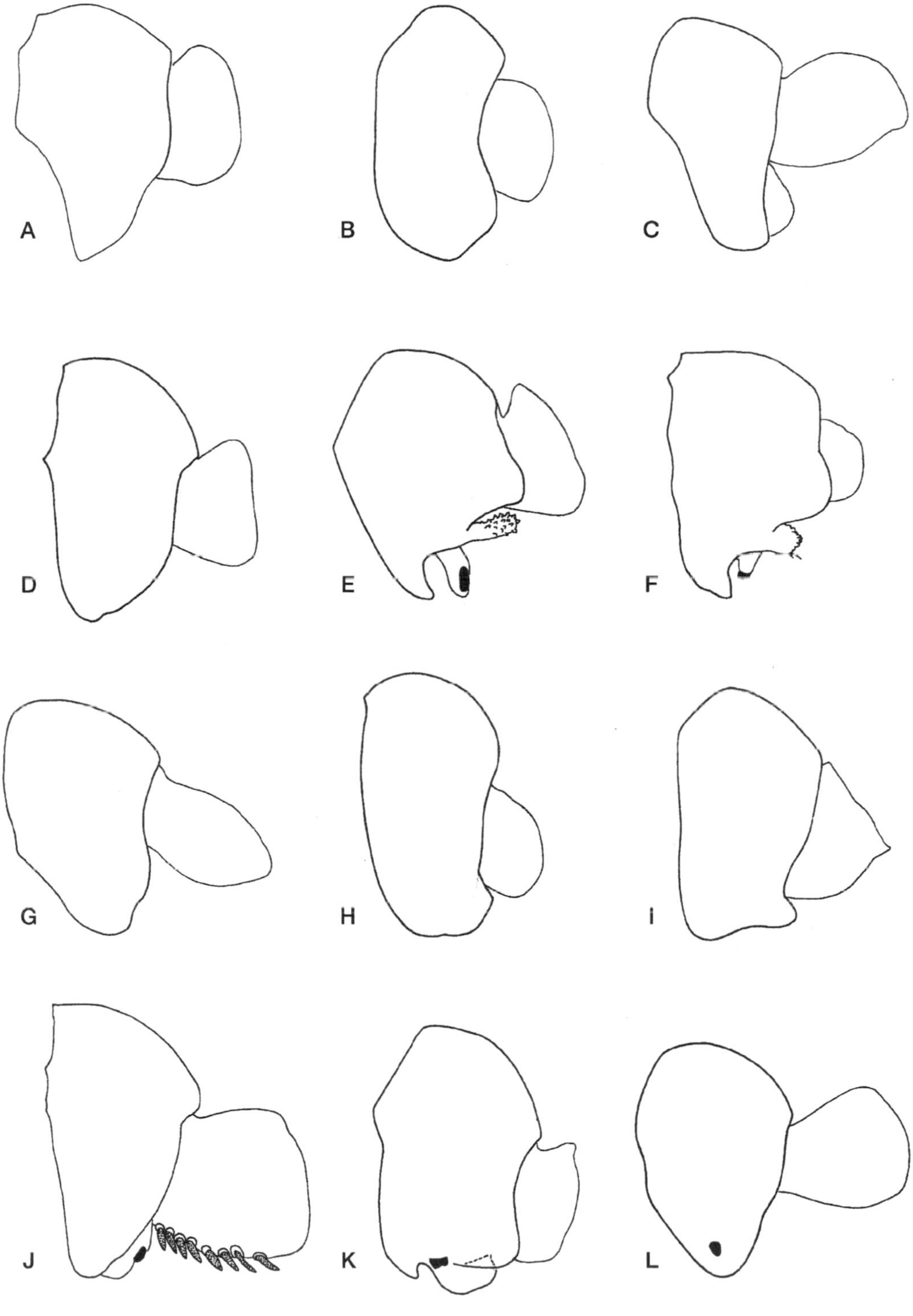


**FIGURE 23.8** Male genitalia, epandrium, surstyli, and proctiger, lateral view: (A) *Asimoneura stroblii*; (B) *Eurasimona stigma*; (C) *Goedenia formosa*; (D) *Inuromaesa maura*; (E) *Myopites* sp. A; (F) *M. flavovarius*; (G) *Myopitora shatalkini*; (H) *Neomyopites melanops*; (I) *Rhynencina longirostris*; (J) *Spinicosta cilipennis*; (K) *Stamnophora vernoniicola*; (L) *Urophora cardui*.

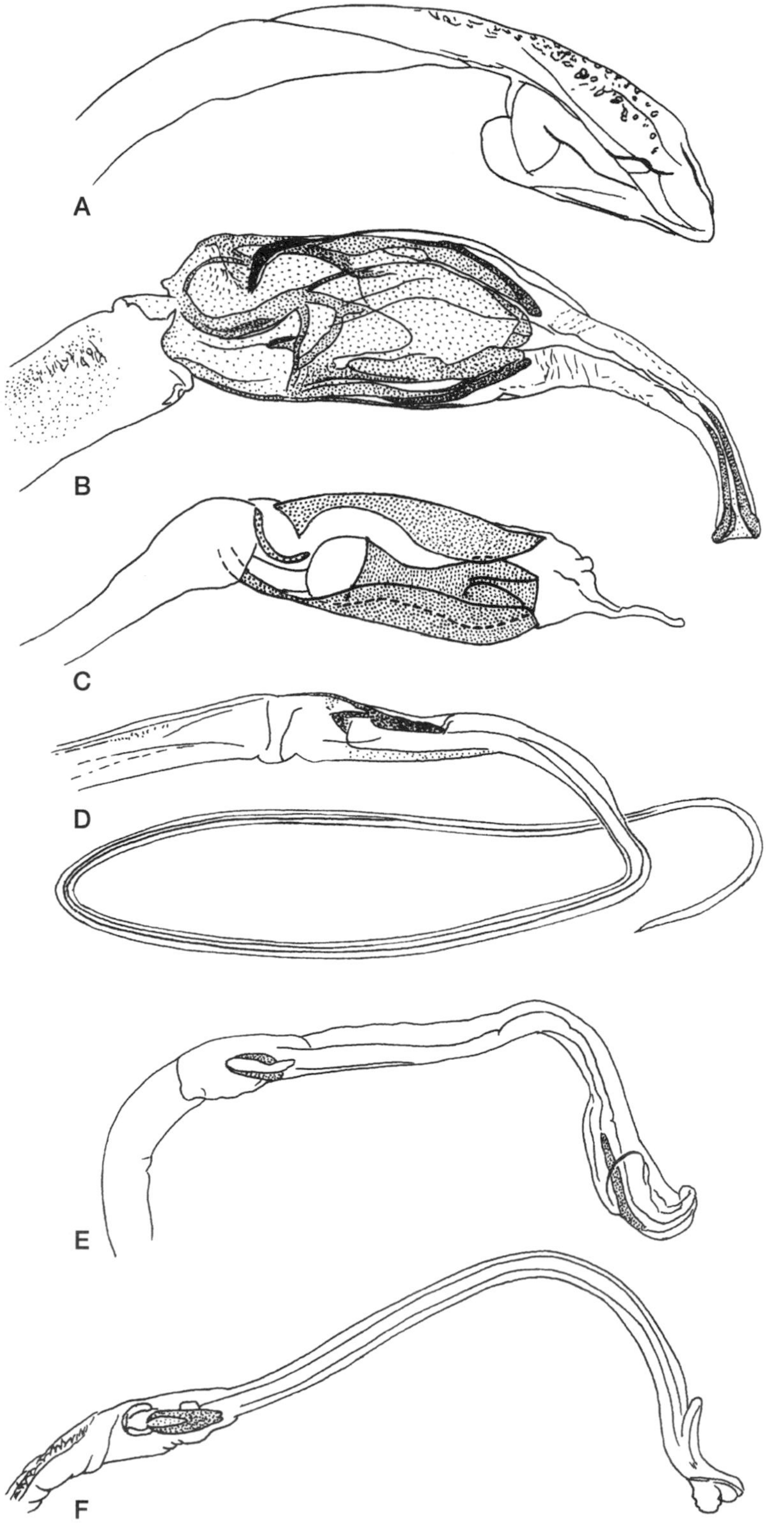


**FIGURE 23.9** Glans: (A) *Asimoneura stroblii*; (B) *Eurasimona stigma*; (C) *Goedenia formosa*; (D) *Inuromaesa maura*; (E) *Myopites* sp. A; (F) *M. flavovarius*.

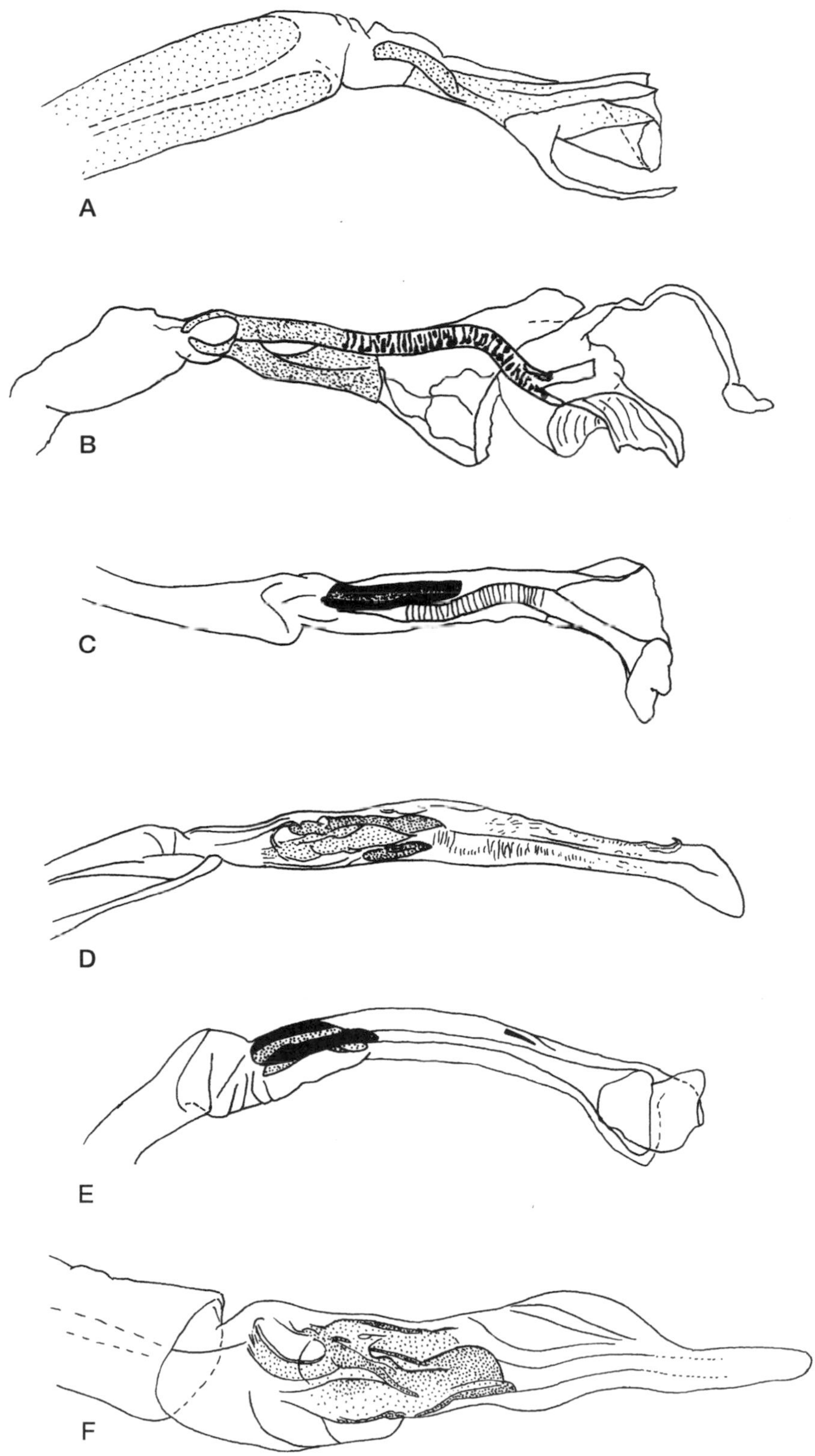


**FIGURE 23.10** Glans: (A) *Myopitora shatalkini*; (B) *Neomyopites melanops*; (C) *Rhynencina longirostris*; (D) *Spinicosta cilipennis*; (E) *Stamnophora vernoniicola*; (F) *Urophora cardui*.

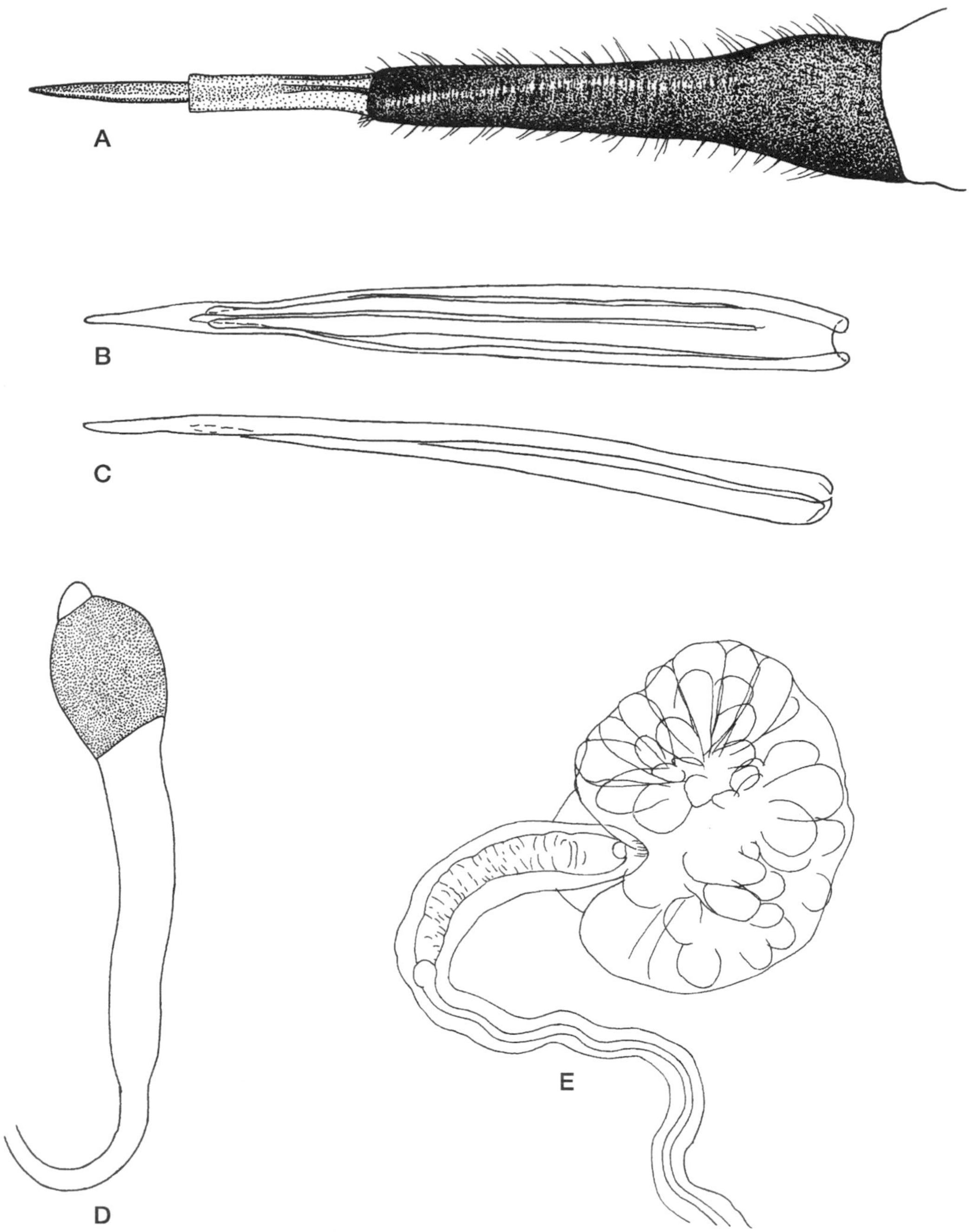


**FIGURE 23.11** Female terminalia: (A) ovipositor, dorsolateral view, *Urophora dzieduszyckii*; (B) aculeus, dorsal view, *Rhynencina longirostris*; (C) same, lateral view; (D and E) spermatheca; (D) *Inuromaesa maura*; (E) *Myopites stylatus* (drawn from a freshly killed specimen).

Eupatorieae, Heliantheae, Liabeae. Predominantly Neotropical, two spp. Nearctic. . . . . . . . . . . . . . . . . . . . . . . . . . . . . . . . . . . . . . . . . . . . . . . . *Neomyopites*, n. gen.

- Wing entirely or nearly entirely hyaline (Figure 23.4B, E). Scutellum strongly convex. Hind femur without outstanding anteroventral setae. Puparium without spines on posterior end (Figure 23.12A-D). Acrophallus of glans without dotlike marks (Figure 23.9B, D). Spermathecae or apex of spermathecal ducts at least weakly sclerotized (Figure 23.11D) ....................... 16

16. Crossvein R-M distal to middle of cell dm and aligned with or apical to apex of vein $R_1$ (Figure 23.4E). Male abdominal tergites broad, pleural membrane not visible in dorsal view. Glans extremely long and gradually tapered, weakly sclerotized basally (Figure 23.9D). Hosts: Inuleae and Cardueae. Europe, Central Asia ................................ *Inuromaesa* Korneyev and White, in part

• Crossvein R-M at or basal to middle of cell dm (Figure 23.4B); if crossvein DM-Cu indistinct (some specimens of *E. fedotovae*), then crossvein R-M aligned with basal half of pterostigma. Male abdominal tergites narrow, pleural membrane visible in dorsal view. Glans short, stout, extensively sclerotized basally (Figure 23.9B). Hosts: Anthemideae. Europe, Central Asia .................... *Eurasimona* Korneyev and White, in part (*E. fedotovae* (Korneyev and White), Central Asia)

## 23.5 SYNOPSES OF THE GENERA

### 23.5.1 *Asimoneura* Czerny

(Figures 23.1A, 23.4A, 23.6F, 23.7A, 23.8A, 23.9A)

*Asimoneura* Czerny, in Czerny and Strobl 1909: 253 (type species: *stroblii* Czerny, by monotypy).

*Diagnosis*: Face at most slightly concave, in lateral view ventral margin usually strongly produced, in a few species only slightly or not produced; proboscis long, geniculate, labella length:head length ratio 0.85 to 1.65, labella length:head height ratio 0.6 to 2.5, usually 0.8 to 1.3; dorsocentral seta usually aligned with supra-alar seta, occasionally slightly anterior or posterior to it, in *A. shirakii* and *A. stroblii* closer to acrostichal than to supra-alar; anepisternum entirely microtrichose in about half of species, the others with bare area or mostly bare; katepisternum entirely microtrichose or sometimes with dorsal margin bare; anepimeron entirely microtrichose; scutellum moderately convex, strongly so in several species (including *A. shirakii*), usually entirely black, but medially yellow in two undescribed Afrotropical species; fore femur posteroventral setae equal in male and female of about half of species, shorter in male of others; hind femur without outstanding anteroventral setae; costal setulae usually small, slightly enlarged along pterostigma in two species; crossvein R-M distal to midlength of cell dm except in two Afrotropical species, at midlength in *A. shirakii*; cell $r_{4+5}$ not narrowed distally; basal section of vein M complete, cells br and bm completely separated; wing usually without dark pattern, in two species some darkening present basally near costa; tergites entirely black except partly yellow in *shirakii*, extremely narrow in males of two-thirds of species, broad in others; male sternite 5 not modified; lateral surstylus without posterior lobe, apically not strongly curved mesally or ventrally; medial surstylus without basal lobe; anterior prensiseta usually not enlarged; arm of phallapodeme not fused or very narrowly fused to lateral sclerite of hypandrium; left and right lateral sclerites short, subequal, fused flangelike to hypandrium; glans with base not extensively sclerotized, rarely with a small, strong sclerite; acrophallus not striate; proctiger without stout setae; spermatheca not sclerotized; puparium (based on one undescribed Afrotropical species) posteriorly with black plate not including spiracles, without spines.

*Hosts*: Hosts are known for only two of the named species: *A. pantomelas* was reared from flowers of *Helichrysum foetidum* (Gnaphalieae) (Munro 1935); and *A. petiolata* from flowers of *Pentzia incana* (Anthemideae) (Munro 1931). Munro said that *A. petiolata* "larvae live and pupate in the thick bases" of the flowers. This is perhaps a gall of the *Myopites* type. The type species, *A. stroblii*, is associated with *Helichrysum* (B. Merz, personal communication) and probably develops in flowerheads of plants of this genus. In addition, two undescribed species have been reared from

*Berkheya* sp. (Arctoteae) and *Helichrysum* sp., respectively (Freidberg, personal observation). The statement of White and Korneyev (1989: 333) that many Afrotropical species are associated with Vernonieae appears to be incorrect.

*Distribution*: Although the type species, *A. stroblii*, is Palearctic (known from Spain and France) and *A. shirakii* is from Taiwan, most species are undescribed and are Afrotropical. Of these, about 20 are from South Africa and 5 are East African.

*Discussion*: Further study of this genus is needed, as its monophyly has not been demonstrated. There is considerable diversity among the included species in structure, coloration and vestiture; their exact relationships with other myopitines cannot be clarified before a revision of these species is completed. The generic placement of *A. shirakii* (Munro) is particularly uncertain and will probably remain in doubt until males are collected and their terminalia studied. It may possibly be related to the *Myopites* clade or *Inuromaesa*, as suggested by the yellow tergite color, rather than to the other species of *Asimoneura*. An undescribed species known from a female from Sumatra possibly belongs here. Hendel (1927) placed *Trypeta stigma* Loew here, but it was removed to *Eurasimona* by Korneyev and White (1991). The terminalia have been dissected and studied in only about a third of the species.

### 23.5.2 *Eurasimona* Korneyev and White, n. stat.

(Figures 23.1B, 23.4B, 23.7B, 23.8B, 23.9B)

*Urophora* subgenus *Eurasimona* Korneyev and White 1991: 217 (type species: *Trypeta stigma* Loew, by original designation).

*Diagnosis*: Face moderately to deeply concave, in lateral view ventral margin not or slightly produced; proboscis short, spatulate, or long, geniculate, labella length:head length ratio 0.52 to 1.20, labella length:head height ratio 0.4 to 0.8; dorsocentral seta midway between supra-alar and acrostichal setae; anepisternum, anepimeron, and katepisternum entirely microtrichose; scutellum strongly convex, medially yellow; fore femur posteroventral setae equal in male and female; hind femur without outstanding anteroventral setae; costal setulae small, not erect; crossvein R-M at 0.28 to 0.50 of length of cell dm; cell $r_{4+5}$ not narrowed distally; basal section of vein M complete, cells br and bm completely separated; crossvein DM-Cu sometimes pale to indistinct in *E. fedotovae*; wing pattern reduced, at most a pterostigmal spot and an infuscation anteroapically; tergites entirely black, narrow in male; male sternite 5 not modified; lateral surstylus apically strongly curved mesally and not curved ventrally, usually with small posterior lobe arising from mesal margin; medial surstylus without basal lobe; anterior prensiseta not enlarged; arm of phallapodeme narrowly fused to lateral sclerite of hypandrium; right lateral sclerite much longer than left, often narrowly connected to hypandrium; glans short, with at least basal half extensively sclerotized and stout, tapered to more membranous medioapical lobe; acrophallus not striate; proctiger without stout setae; spermatheca weakly sclerotized; puparium unknown.

*Hosts*: All known hosts are in the tribe Anthemideae, including *Anthemis*, *Achillea*, *Leucanthemum*, and *Tanacetum* species for *E. stigma* (White and Korneyev 1989; Merz 1994) and a *Tanacetum* species for *E. fedotovae* (Korneyev and White 1991).

*Distribution*: *Eurasimona stigma* is known from Europe, including most of the eastern part, Kyrghyzstan and Kazakstan; *E. fedotovae* only from Kazakstan.

*Discussion*: *Eurasimona fedotovae* has much shorter mouthparts than *E. stigma*, but otherwise closely resembles it, and these two species are clearly closely related.

### 23.5.3 *Goedenia*, n. gen.

(Figures 23.1C, 23.2G, 23.4C-D, 23.6C, 23.6H, 23.7C, 23.8C, 23.9C)

*Type species*: *Aleomyia rufipes* Curran (by present designation).

*Diagnosis*: Face deeply concave, in lateral view ventral margin slightly or occasionally strongly produced; proboscis moderately long to long, geniculate, labella length:head length ratio 0.7 to 1.1, labella length:head height ratio 0.64 to 0.97; dorsocentral seta distinctly posterior to supra-alar seta, usually one-third to one-half the distance from supra-alar to acrostichal seta; anepisternum, anepimeron, and katepisternum mostly shiny and bare, each at most partly sparsely microtrichose; scutellum moderately or usually strongly convex, usually medially yellow, sometimes entirely black; fore femur posteroventral setae equal in male and female; hind femur without outstanding anteroventral setae; costal setulae small, not erect; crossvein R-M at 0.4 to 0.55 of length of cell dm; cell $r_{4+5}$ not narrowed distally; basal section of vein M complete, cells br and bm completely separated; wing pattern *Myopites*-type, occasionally reduced to a few spots or hyaline; tergites entirely black, extremely narrow in male; male sternite 5 not modified; lateral surstylus apically strongly curved mesally and not curved ventrally, with posterior lobe arising from mesal margin; medial surstylus without basal lobe; anterior prensiseta not enlarged; arm of phallapodeme narrowly fused to lateral sclerite of hypandrium; right lateral sclerite much longer than left, often narrowly connected to hypandrium; glans short, with more than basal half extensively sclerotized and stout, tapered to membranous medioapical lobe; acrophallus not striate; proctiger without stout setae; spermatheca not sclerotized; puparium posteriorly with black plate including spiracles, without spines.

*Etymology*: This genus is named for Richard D. Goeden, whose extensive studies have greatly advanced the knowledge of the biology of the fruit flies of western North America. Gender feminine.

*Hosts*: Wasbauer (1972), Goeden (1987), and Goeden et al. (1995) listed the numerous host records for the species of this genus, which include species of *Acamptopappus*, *Amphipappus*, *Chrysothamnus*, *Grindelia*, *Gutierrezia*, and *Haplopappus*. As noted by Goeden (1987) and Goeden et al. (1995), all known hosts belong to the tribe Astereae, subtribe Solidagininae.

*Distribution*: Western North America.

*Discussion*: Species limits within this genus are poorly understood. The description of the puparium is based on two species, *G. timberlakei* (Goeden et al. 1995) and *G. grindeliae*.

### 23.5.4 *Inuromaesa* Korneyev and White, n. stat.

(Figures 23.1D-E, 23.4E, 23.7D, 23.8D, 23.9D, 23.11D, 23.12A-B)

*Urophora* subgenus *Inuromaesa* Korneyev and White 1991: 217 (type species: *Trypeta maura* Frauenfeld, by original designation).

*Urophora* subgenus *Promyopites* Korneyev and Merz 1998: 514 (type species: *Urophora circumflava* Korneyev and Merz, by original designation), n. syn.

*Diagnosis*: Face slightly to moderately concave, in lateral view ventral margin not produced, slightly produced, or strongly produced; proboscis short, spatulate, or (*I. circumflava*) with haustellum long, but labella short; labella length:head length ratio 0.40 to 0.55, labella length:head height ratio 0.3 to 0.4; dorsocentral seta slightly posterior to supra-alar seta to almost midway between supra-alar and acrostichal setae; anepisternum, anepimeron, and katepisternum entirely microtrichose; scutellum moderately to strongly convex, medially or entirely yellow; fore femur posteroventral setae equal in male and female; hind femur without outstanding anteroventral setae; costal setulae small, not erect; crossvein R-M at 0.6 to 0.7 of length of cell dm; cell $r_{4+5}$ not narrowed apically or (*I. circumflava*) slightly narrowed apically by distinctly anteriorly slanted distal section of vein M; basal section of vein M complete, cells br and bm completely separated; wing pattern absent; tergites entirely black or (*I. circumflava*) entirely yellow, broad in male; male sternite 5 not modified; lateral surstylus without posterior lobe, apically not strongly curved mesally or ventrally; medial surstylus without basal lobe; anterior prensiseta not enlarged; arm of phallapodeme narrowly or broadly fused to lateral sclerite of hypandrium; lateral sclerites subequal, broadly connected to hypandrium basally; glans with small basal sclerite, long to extremely long, gradually tapering, sometimes irregularly, to slender, nonsclerotized apex; acrophallus not striate; proctiger without

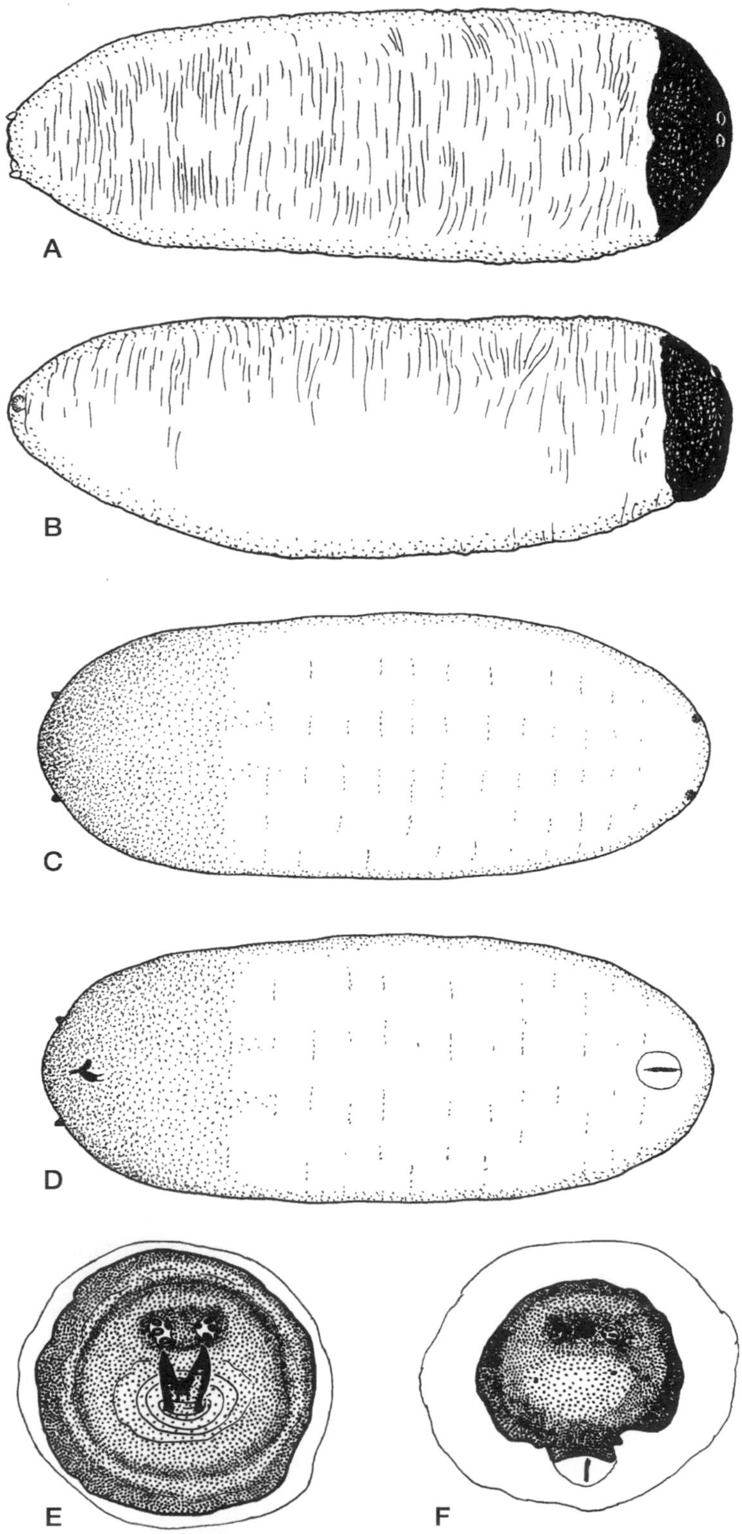


**FIGURE 23.12** Puparium: (A, C) dorsal view; (B) lateral view; (D) ventral view; (E and F) posterior view; (A and B) *Inuromaesa maura*; (C and D) *Stamnophora* sp.; (E) *Neomyopites melanops*; (F) *Neomyopites* sp.

stout setae; spermatheca membranous but apex of duct moderately sclerotized; puparium posteriorly with black plate including spiracles, without spines (spermathecal and puparial characters for *I. sogdiana* provided by V. A. Korneyev, personal communication).

*Hosts*: *Inuromaesa maura* infests several species of *Inula* (Inuleae) (White and Korneyev 1989), whereas *I. sogdiana* was reared from *Jurinea* (Cardueae) (Korneyev and Merz 1998). The host of *I. circumflava* is unknown.

*Distribution*: Central and southern Europe, Central Asia (Kyrghyzstan, Uzbekistan), China.

*Discussion*: *Inuromaesa maura* was traditionally placed in *Urophora*. Its distinctness from the other Palearctic species was recognized by Korneyev and White (1991), who proposed for it the subgenus *Inuromaesa*. Korneyev and Merz (1998) described a second species, *I. sogdiana*, and reported another, undescribed species from Central Asia. White and Korneyev (1989) suggested that *I. maura* might be closely related to *Myopites* based on the elongate glans and common host group (Inuleae; both genera are now known to also have host plants in the Cardueae), but our analysis does not support that hypothesis. In our cladistic analyses, the two species previously placed in *Inuromaesa* were grouped with *I. circumflava*, the type species of *Promyopites*. In one unweighted analysis the former two species were grouped together, but we did not discover any unambiguous synapomorphies for *Inuromaesa* as previously defined (i.e., without including *I. circumflava*) and therefore consider *Promyopites* a subjective junior synonym of *Inuromaesa*. In none of our analyses was *Inuromaesa* supported as the sister group of *Urophora*, and we thus remove this taxon from *Urophora* and elevate it to generic rank.

### 23.5.5 *Myopites* Blot

(Figures 23.1F-G, 23.2F, 23.4F-G, 23.6A, 23.7E-F, 23.8E-F, 23.9E-F, 23.11E)

*Myopites* Blot 1827: 102 (type species: *inulaedyssentericae* Blot, by monotypy).
*Nearomyia* Becker 1913: 646 (type species: *flavovaria* Becker, by monotypy). n. syn.

*Diagnosis*: Face usually slightly convex, flat, or slightly concave, moderately concave in undescribed spp. B and C, in lateral view ventral margin usually strongly produced, only slightly produced in *M. hemixanthus* and undescribed spp. B and C, and not produced in *M. flavovarius*; proboscis usually long and thin, geniculate, but only moderately long in undescribed spp. B and C, and short, spatulate in *M. flavovarius*; labella length:head length ratio usually greater than 1, rarely as low as 0.8, about 0.6 in undescribed spp. B and C, 0.25 in *M. flavovarius*; labella length:head height ratio usually 0.8 to 2, about 0.5 in undescribed spp. B and C, 0.18 in *M. flavovarius*; dorsocentral seta usually aligned with supra-alar seta or slightly anterior or posterior to it, in undescribed spp. B and C and *M. flavovarius* aligned halfway between supra-alar seta and acrostichal seta; anepisternum, anepimeron, and katepisternum entirely microtrichose except in *M. delottoi*, which has bare area on anepisternum; scutellum strongly convex, usually medially yellow, but entirely black sometimes in *M. lelea*, usually in *M. nigrescens*, and always in undescribed spp. B and C; fore femur posteroventral setae usually equal in male and female, in undescribed spp. B and C sexually dimorphic, shorter and thicker (more spinelike) in male than in female; hind femur without outstanding anteroventral setae; costal setulae small, not erect; crossvein R-M usually at 0.4 to 0.5 of length of cell dm, in undescribed spp. B and C at 0.55 to 0.6 of length of cell dm; cell $r_{4+5}$ slightly to strongly narrowed distally by anterior slant of vein M; basal section of vein M usually complete, but in undescribed spp. B and C incomplete and cells br and bm incompletely separated; wing pattern *Myopites*-type, highly reduced in *M. delottoi* and undescribed spp. B and C; tergites usually at least partly yellow, entirely black in undescribed spp. B and C, moderately to extremely narrow in male; male sternite 5 not modified; lateral surstylus apically not strongly curved mesally or ventrally, with large, often elongate lobe arising lateral to mesal margin, edge of epandrium dorsal to it often lobelike; medial surstylus without basal lobe, in undescribed spp. B and C strongly sclerotized, dark, with prensisetae not clearly demarcated, apically clawlike pointed, preapically broadened and rounded; in other species anterior prensiseta enlarged; arm of phallapodeme not fused to lateral sclerite of hypandrium; lateral sclerites short, subequal, fused flangelike to hypandrium; glans usually more than 15 times as long as wide, in *M. hemixanthus* and undescribed sp. A only about ten times as long as wide, always with small but strongly sclerotized, well-differentiated basal sclerite; acrophallus not striate; proctiger without stout setae; tip of aculeus often clearly distinguished from proximal part; spermatheca not sclerotized; puparium posteriorly without black plate or spines.

*Hosts*: The Palearctic and South African species of *Myopites* for which hosts are known form flowerhead galls in species of Inuleae, including *Geigeria*, *Inula*, *Phagnalon*, *Pulicaria*, and *Schizogyne* (Munro 1931; Freidberg 1980; 1984; Merz 1992). *Myopites delottoi* was reared from a species of *Sphaeranthus* (Plucheeae) (Munro 1955), although this record requires confirmation, and the two undescribed East African species (sp. B and sp. C) were reared from flowerheads of *Echinops* spp. (Cardueae).

*Distribution*: *Myopites* is mostly a circum-Mediterranean genus, with species extending to the Canary Islands, southern England, and Kazakstan. One species occurs in eastern Asia, and five species in the Afrotropical region, of which one (*M. delottoi*) is Ethiopian, two (undescribed spp. B and C) are East African, and two (*M. hemixanthus* and undescribed sp. A) are southern African.

*Discussion*: *Myopites* contains 17 currently valid species (15 Palearctic and 2 Afrotropical) plus at least three undescribed Afrotropical species. The two southern African species, *M. hemixanthus* and a similar undescribed species (sp. A), have the glans considerably shorter and lacking the distinctive apical sclerotization, and appear to be the sister group of the rest of the genus (see Section 23.7). The other two undescribed Afrotropical species (spp. B and C) are quite distinctive. They differ from the other species by a number of autapomorphies (forefemur setation, basal section of vein M incomplete, medial surstylus shape, large size), in their host associations (Cardueae vs. Inuleae or Plucheeae), and from all or most of the other species by several other character states that are hypothesized as homoplasy in our phylogenetic analysis. Despite their distinctiveness, they appear to be part of the clade of species exclusive of the two southern African species, and we therefore place them in *Myopites*. The type species of *Nearomyia*, *N. flavovaria*, which is known only from the unique male holotype from Iran, differs from typical *Myopites* species only by proboscis length, face shape, and location of the dorsocentral seta, characters weighted either 0 or 1 in our phlogenetic analysis. It also appears to belong to the clade of species exclusive of the two southern African species, so we therefore consider *Nearomyia* a subjective synonym of *Myopites*.

The name *Stomoxys stylata* Fabricius (1794: 396, not 353) has often been confused with *Musca stylata* Fabricius, currently a valid species of *Urophora*. For example, White and Korneyev (1989: 361) considered subsequent usage of it to be misidentifications of the latter name. Norrbom et al. (1999b) considered it the valid senior synonym of *Myopites limbardae* Schiner, but although it is an available name it is technically a secondary junior homonym that was replaced prior to 1961 (F. C. Thompson, personal communication), so it cannot be a valid name. Although Loew (1862: 66) placed the two *stylata* names in different genera (*Musca stylata* in *Urophora*, and *Stomoxys stylata* under *Myopites inlulae*) and was not fully certain of the specific identity of *S. stylata*, he considered it a junior homonym and replaced it with a junior synonym.

### 23.5.6 *Myopitora* Korneyev and White, n. stat.

(Figures 23.1H, 23.4H, 23.7G, 23.8G, 23.10A)

*Urophora* subgenus *Myopitora* Korneyev and White 1991: 217 (type species: *Urophora shatalkini* Korneyev and White, by original designation).

*Diagnosis*: Face moderately concave, in lateral view ventral margin strongly produced; proboscis moderately long, geniculate, labella length:head length ratio 0.78, labella length:head height ratio 0.68; dorsocentral seta slightly posterior to supra-alar seta, less than one-third the distance to acrostichal seta; anepisternum, anepimeron, and katepisternum entirely microtrichose; scutellum strongly convex, medially yellow; hind femur without outstanding anteroventral setae; costal setulae small, not erect; crossvein R-M at 0.49 of length of cell dm; basal section of vein M complete, cells br and bm completely separated; wing pattern *Myopites*-type; tergites entirely black, broad in male; male sternite 5 not modified; lateral surstylus apically not strongly mesally curved, slightly ventrally curved, with very small posterior lobe arising from mesal margin; medial surstylus without basal lobe; anterior prensiseta not enlarged; arms of phallapodeme and lateral sclerites of hypandrium not examined; glans short, base not extensively or strongly sclerotized, with slender membranous lateral lobe; acrophallus not striate nor with dotlike markings; proctiger without stout setae; female terminalia not studied; puparium unknown.

*Hosts*: V. A. Korneyev (personal communication) has examined specimens of *M. shatalkini* reared from *Erigeron acris* L. (Astereae).

*Distribution*: The single known species occurs in China (Freidberg, unpublished data) and the Russian Far East.

*Discussion*: Korneyev and White (1991) proposed *Myopitora* as a subgenus of *Urophora*, but it is not closely related to that taxon and we have therefore elevated it to genus rank. The shape of the epandrium and surstyli in posterior view resembles most *Neomyopites* species in the slight ventral curve of the apex of the lateral surstylus, and the glans also resembles some species of this genus, which we hypothesize to be the most closely related taxon.

## 23.5.7 *Neomyopites*, N. Gen.

(Figures 23.2A, 23.3E, 23.5A, 23.6D-E, 23.7H, 23.8H, 23.10B, 23.12E-F)

*Type species*: *Euribia aerea* Hering (by present designation).

*Diagnosis*: Face at most slightly concave, in lateral view ventral margin not or very slightly produced; proboscis usually short, capitate or spatulate, slightly longer in *N. cuzconis*, *mexicanus*, *regis*, and *paulensis*, labella length:head length ratio 0.36 to 0.68, labella length:head height ratio 0.27 to 0.51; dorsocentral seta slightly anterior to, aligned with, or slightly posterior to supra-alar seta; anepisternum, anepimeron, and katepisternum entirely microtrichose; scutellum flat to slightly convex, medially yellow or entirely black; fore femur posteroventral setae equal in male and female; hind femur with (*melanops* group (see Discussion)) or without outstanding anteroventral setae; costal setulae small, not erect; crossvein R-M at 0.4 to 0.55 of length of cell dm; cell $r_{4+5}$ not narrowed distally; basal section of vein M complete, cells br and bm completely separated; wing pattern *Myopites*-type or rarely *Urophora*-type (*N. trivirgulatus*); tergites entirely black, broad in male; male sternite 5 not modified, or (*melanops* group) broadly U- or V-shaped, concave medially, and usually with pair of small mesal lobes and/or modified setae; lateral surstylus apically usually curved ventrally, often with posterior lobe arising from mesal margin; medial surstylus without basal lobe; anterior prensiseta not enlarged; arm of phallapodeme fused, often broadly, to lateral sclerite of hypandrium; lateral sclerites usually asymmetric, broadly connected to hypandrium basally; glans usually short, with base not extensively sclerotized, although some species of *aereus* group (see Discussion) with small, strong sclerites; acrophallus usually with rows of dark dotlike marks; proctiger without stout setae; spermatheca not sclerotized; puparium posteriorly with black plate, including or not including spiracles, and with (*aereus* group) or without transverse row of large, stout spines.

*Etymology*: Derived from the Greek *"neos"* (new) and *"Myopites,"* in reference to the New World distribution of this taxon. The gender is masculine.

*Hosts*: Four species of the *aereus* group (see Discussion) have been reared from flowerheads of species in four closely related genera of the subtribe Verbesininae of the Heliantheae: *Espeletia*, *Espeletiopsis*, *Libanothamnus*, and *Ruilopezia*. In at least two of these species, the larvae feed on developing seeds and surrounding tissues, but they often burrow deeply into the receptacle. Other species of the *aereus* group have been reared from species of *Diplostephium* and *Gutierrezia* (Astereae), *Mikania* and *Koanophyllon* (Eupatorieae), *Viguiera* (Heliantheae: Helianthinae), and *Sinclairia* (Liabeae) (Steyskal 1979; Norrbom, personal observation). The larvae appear to feed on developing seeds inside the flowerhead without forming galls or modifying the receptacle; in all three species in which parts of the flowerhead were preserved with the specimens, the puparia were inside or among the remains of largely devoured achenes.

Two species of the *melanops* group (see Discussion), *N. melanops* and *tresmilius*, were reared from inside modified, greatly enlarged achenes of species of *Calea* and *Verbesina* (Heliantheae), respectively (Norrbom, personal observation). Several specimens of two other species have labels with the names of species of *Calea* or *Verbesina*, one of which also says "agallas" (galls). At least the latter presumably were reared.

*Distribution*: Widespread in the Neotropical Region, with two species occurring in the southern Nearctic, one as far north as Texas.

*Discussion*: Based on morphological characters, two monophyletic species groups can be recognized within *Neomyopites*, and the diversity of known host tribes also indicates that this genus may eventually need to be further divided. However, there are additional species of the genus lacking the autapomorphies of either of these two species groups and whose relationships are unclear, and we therefore have not recognized these species groups with subgeneric names. Species of the *melanops* group (including at least *N. acompsus, adjacens, agnatus, chaetostomus, chimborazonis, disjunctus, hodgesi, jamaicensis, mamarae, melanops, morus,* and *tresmilius*) have one to two outstanding anteroventral setae on the hind femur, and, in many, sternite 5 of the male has a pair of medial lobes and often has modified setae. Species of the *aereus* group (including at least *N. aereus, columbianus, cordilleranus, cuzconis, euryparius, eved, paulensis, regis, simplex,* and *unicus*) have spines on the posterior end of the puparium.

The wing pattern of *N. trivirgulatus* resembles those in *Urophora*, with a band extending straight from the costa to crossvein DM-Cu, but this species certainly seems to belong in *Neomyopites* based on other characters and the similarity with *Urophora* in wing pattern is apparently due to convergence. The figure of the wing of *N. cubanus* by Dirlbek and Dirlbekova (1973) also suggests that it has a *Urophora*-type pattern, but in the holotype female the band covering crossvein DM-Cu has a slight proximal bend along vein M. Thus, this band appears to be a fusion of that on DM-Cu with the usually small band in the middle of cell $r_1$ rather than with the spot on the apex of $R_{2+3}$. We suggest this is a modification of the *Myopites*-type pattern and is not the *Urophora*-type.

The description of the puparium is based on specimens of *N. aereus, claripennis, columbianus, cordilleranus, melanops, regis, tresmilius*, and four undescribed species. Most of the described species of this genus can be identified using the key of Steyskal (1979).

## 23.5.8 *Rhynencina* Johnson

(Figures 23.2B, 23.3C, 23.5B, 23.7I, 23.8I, 23.10C, 23.11B-C)

*Rhynencina* Johnson 1922: 24 (type species: *Rhynencina longirostris* Johnson, by original designation).

*Diagnosis*: Face at most slightly concave, in lateral view ventral margin slightly to strongly produced; proboscis moderately long (*R. emphanes*) to very long, labella length:head length ratio 0.68 to 1.64, labella length:head height ratio 0.64 to 1.77; dorsocentral seta slightly anterior to supra-alar seta to about midway between it and acrostichal seta; anepisternum, anepimeron, and katepisternum entirely microtrichose; scutellum moderately to strongly convex, medially yellow; fore femur posteroventral setae equal in male and female; hind femur without outstanding anteroventral setae; costal setulae small, not erect; crossvein R-M proximal to 0.5 of length of cell dm, except in *R. xanthogaster*; cell $r_{4+5}$ not narrowed distally; basal section of vein M complete, cells br and bm completely separated; wing pattern *Myopites*-type; tergites partly to entirely yellow, broad in male; male sternite 5 not modified; lateral surstylus apically not strongly curved mesally or ventrally, usually with small posterior lobe arising lateral to mesal margin (absent in *R. emphanes*); medial surstylus without basal lobe; anterior prensiseta enlarged; arm of phallapodeme not fused to lateral sclerite of hypandrium; lateral sclerites short, subequal, fused flangelike to hypandrium; glans moderately long, with small but strongly sclerotized, well-differentiated basal sclerite; acrophallus with evenly spaced weak transverse striations; proctiger without stout setae; spermatheca not sclerotized; puparium unknown.

*Hosts*: Only *R. emphanes* has been reared, from flowers of *Espeletia* (Norrbom, unpublished data), but *R. dysphanes*, *longirostris,* and *spilogaster* have been collected on *Polymnia* and *Smallanthus* species, and females of *R. spilogaster* have been observed ovipositing into flowers of the latter genus on several occasions (Norrbom, personal observation). All of these plants belong to the tribe Heliantheae.

*Distribution*: New World, including one species from eastern United States, one from Mesoamerica, and three from South America (Andean countries).

*Discussion*: *Rhynencina xanthogaster*, known only from the female holotype, is tentatively included here. To confirm this classification, males of this species are needed because most synapomorphics of this genus involve the male terminalia.

### 23.5.9 *Spinicosta*, n. gen.

(Figures 23.2C, 23.3A, 23.5C, 23.7J, 23.8J, 23.10D)

*Type species*: *Urophora cilipennis* Bezzi (by present designation).

*Diagnosis*: Face at most slightly concave, in lateral view ventral margin not or slightly produced; proboscis moderately long, spatulate, labella length:head length ratio 0.5 to 0.75, labella length:head height ratio 0.3 to 0.6; dorsocentral seta aligned with supra-alar seta; anepisternum with bare (nonmicrotrichose) area; anepimeron and katepisternum entirely microtrichose; scutellum flat, entirely black; apical seta slightly longer than basal seta; fore femur posteroventral setae equal in male and female; hind femur without outstanding anteroventral setae; costal setulae along pterostigma slightly enlarged and erect, especially in male; crossvein R-M at 0.5 to 0.6 of length of cell dm; veins $R_{4+5}$ and M usually parallel, cell $r_{4+5}$ at most slightly narrowed distally; basal section of vein M complete, cells br and bm completely separated; wing pattern entirely absent; tergites entirely black, broad in male; male sternite 5 not modified; lateral surstylus apically not strongly curved mesally or ventrally, without posterior lobe; medial surstylus without basal lobe; anterior prensiseta not enlarged; hypandrium with apical half slender; arm of phallapodeme fused, often broadly, to lateral sclerite of hypandrium; lateral sclerites asymmetric, broadly connected to hypandrium basally; glans short, with base not extensively or strongly sclerotized; acrophallus not striate; proctiger with stout ventral setae in at least some species; spermatheca not sclerotized; puparium unknown.

*Etymology*: Derived from the Latin "*spina*" and "*costa*" in reference to the long setulae on the costa. Gender feminine.

*Hosts*: At least three species have been reared from flowerheads of two *Vernonia* spp. and one *Ethulia* sp. (Vernonieae), respectively (Munro 1935: 38; Freidberg, personal observation), and at least two additional species were reared from flowerheads of *Berkheya* spp. (Arctoteae).

*Distribution*: Afrotropical. From at least Cameroon and Kenya to South Africa.

*Discussion*: Morphologically this is a very homogeneous genus. Based on known and suspected hosts as well as morphological characters, it may be divided into two species groups, one including the *Berkheya* (Arctoteae) breeders and the other the Vernonieae (*Vernonia* and *Ethulia*) breeders. Cogan and Munro (1980) considered *S. agromyzella* and *S. cilipennis* conspecific, but a study of their types showed that the latter species is valid (Freidberg, personal observation), and it is here removed from synonymy with *S. agromyzella*. *Spinicosta cilipennis* differs from *S. agromyzella* in having predominantly black tibiae (yellow in *S. cilipennis*) and a slightly longer proboscis. It breeds in *Berkheya*, whereas *S. agromyzella* breeds in *Vernonia*.

### 23.5.10 *Stamnophora* Munro

(Figures 23.2D, 23.5D-E, 23.7K, 23.8K, 23.10E, 23.12C-D)

*Stamnophora* Munro 1955: 415 (type species: *Tephritis vernoniicola* Bezzi, by original designation).

*Diagnosis*: Face slightly to moderately concave, in lateral view ventral margin usually not produced, rarely slightly produced; frons with two to four frontal setae; proboscis short to moderately long, spatulate, labella length:head length ratio 0.35 to 0.65, labella length:head height ratio 0.2 to 0.6; dorsocentral seta slightly posterior to supra-alar seta to about halfway between it and acrostichal seta; anepisternum, anepimeron, and katepisternum entirely microtrichose; scutellum strongly convex, medially yellow; fore femur posteroventral setae equal in male and female; hind femur without anteroventral setae; costal setulae small, not erect; crossvein R-M at 0.4 to 0.5 of length of cell dm; cell $r_{4+5}$ not or only slightly narrowed distally; basal section of vein M complete, cells br and bm completely separated; wing pattern *Myopites*-type, highly reduced in two species (including the type species); tergites usually partly yellow (mostly to entirely black in *S. vernoniicola*), moderately to extremely narrow in male; male sternite 5 not modified; lateral surstylus apically not strongly curved mesally or ventrally, with large, broad posterior lobe arising lateral to mesal margin, edge of epandrium dorsal to it often lobelike; medial surstylus without basal lobe; anterior prensiseta enlarged; arm of phallapodeme not fused to lateral sclerite of hypandrium; lateral sclerites short, subequal, fused flangelike to hypandrium; glans moderately long, with small but strongly sclerotized, well-differentiated basal sclerite; acrophallus not striate; proctiger without stout setae; spermatheca not sclerotized; puparium posteriorly without black plate or spines.

*Hosts*: *Stamnophora vernoniicola* has been reared from flowerhead galls on *Vernonia abyssinica* and *V. leptolepis* (Vernonieae) (Munro 1955), but also from conspicuous stem galls on other species of *Vernonia* (Munro 1955; Freidberg 1998). At least two additional species were reared from inconspicuous stem galls, and three other species were reared from flowerhead galls on various species of *Vernonia*.

*Distribution*: Afrotropical (see Discussion).

*Discussion*: This genus has hitherto been recorded from a single, widespread, Afrotropical species, *S. vernoniicola*. A second, similar, undescribed species occurs in West Africa. Both species have reduced wing patterns. The remaining species, all undescribed, have the *Myopites*-type wing pattern, and are divided more or less equally between Madagascar and mainland Africa. This geographic grouping may have phylogenetic significance as these two groups may be sister groups, but the relationships within *Stamnophora* need further analysis.

### 23.5.11 *Urophora* Robineau-Desvoidy

(Figures 23.2E, 23.3B, 23.3D, 23.5F-H, 23.6B, 23.6G, 23.7L, 23.8L, 23.10F, 23.11A)

*Urophora* Robineau-Desvoidy 1830: 769 (type species: *Musca cardui* Linnaeus, by designation of Westwood 1840:149).

*Diagnosis*: Face at most slightly concave, in lateral view ventral margin not or slightly produced; proboscis spatulate to short geniculate, labella length:head length ratio 0.3 to 0.8, labella length:head height ratio 0.2 to 0.7 (although greater than 0.5 only in several central Palearctic species related to *U. phaeocera*); dorsocentral seta aligned with supra-alar seta or slightly anterior to it (in *U. dzieduszyckii*, often more than one dorsocentral, with anteriormost near transverse suture and posteriormost aligned with supra-alar seta); anepisternum with bare (nonmicrotrichose) area; anepimeron usually entirely microtrichose, rarely (*dzieduszyckii* group) partially bare ventrally; katepisternum entirely microtrichose; scutellum moderately to strongly convex, medially yellow; fore femur posteroventral setae shorter in male than in female; hind femur without outstanding anteroventral setae; costal setulae small, not erect; crossvein R-M at 0.5 to 0.7 of length of cell dm; cell $r_{4+5}$ usually not narrowed distally, sometimes slightly narrowed by posterior curve of vein $R_{4+5}$; basal section of vein M complete, cells br and bm completely separated; wing pattern *Urophora*-type, occasionally reduced or absent; tergites entirely black, broad in male, except slightly narrower in *U. dzieduszyckii*; male sternite 5 not modified; lateral surstylus apically not strongly curved mesally or ventrally, without posterior lobe; medial surstylus usually with basal lobe, without lobe in *dzieduszyckii* group (see Discussion); anterior prensiseta not enlarged; hypandrium with anterior half slender; arm of phallapodeme fused, usually broadly, to lateral sclerite of hypandrium; lateral sclerites moderately long but usually subequal, broadly connected to hypandrium basally; glans short, with base not extensively or strongly sclerotized; acrophallus not striate; proctiger without stout setae; spermatheca not sclerotized; puparium posteriorly with black plate including spiracles, without spines.

*Hosts*: White and Korneyev (1989) gave a comprehensive host plant list for the western Palearctic species (also see Merz 1994). Much less is known of the hosts of the eastern Palearctic species (Korneyev and White 1992; 1993; 1996). All information indicates that associations are exclusively with plants of the tribe Cardueae (Korneyev, personal communication). Most species form galls in the flowerheads.

*Distribution*: As delimited here, *Urophora* is an essentially Palearctic genus, with two species native to the Oriental Region (one occurs in both the Palearctic and Oriental Regions) and an undescribed species in the Afrotropical Region. Six species have been introduced to North America for weed biocontrol, and another by accident. One of them also has been introduced to Australia and New Zealand for weed biocontrol, and another to India (Norrbom et al. 1999a).

*Discussion*: Even after the removal of almost all non-Palearctic species and all subgenera previously included in this genus, *Urophora* remains the largest genus of Myopitini, with 57 species currently recognized. It includes numerous species of economic importance as actual or potential biocontrol agents of weeds (White and Clement 1987). White and Korneyev (1989), Korneyev and White (1992; 1993; 1996), Wang (1996), and Korneyev and Merz (1998), and have revised most of the species and recognized a number of species groups within *Urophora*. The *dzieduszyckii* group (Korneyev and White 1992), which includes *U. dzieduszyckii, pontica, and solaris,* may be the sister group of the rest of the genus (see Section 23.7).

## 23.6 BIOLOGY, IMMATURE STAGES, AND ECONOMIC IMPORTANCE

Published information on the biology of Myopitini is scant except for some western Palearctic species of *Urophora* (e.g., Varley 1937; 1947; Sobhian and Zwölfer 1985; Zwölfer and Arnold-Rinehart 1993) and *Myopites* (Freidberg 1980) and the species of the western North American genus *Goedenia* (Goeden 1987, as *Urophora*). For other genera and regions, information is limited to a handful of host records (e.g., Munro 1931; 1935; 1955).

All available records indicate that members of the Myopitini infest plants of the family Asteraceae (Compositae) only, although within this family they attack various tribes, including Anthemideae, Arctoteae, Astereae, Cardueae, Eupatorieae, Gnaphalieae, Heliantheae, Inuleae, Liabeae, Plucheeae, and Vernonieae (Table 23.1). Most myopitine genera have a relatively narrow range of hosts, attacking plants in only one or two tribes of Asteraceae. For example, species of *Urophora*, the largest genus, attack only Cardueae, and those of *Goedenia* breed only in species of the subtribe Solidagininae of the Astereae. On the other hand, *Asimoneura, Myopites,* and *Neomyopites* have hosts in three or four tribes of Asteraceae.

Larvae of most species of Myopitini feed in the flowerheads of their host plants. Gall formation is apparently the rule for *Myopites, Stamnophora,* and *Urophora,* but may not be so for the rest of the genera (at least some species of *Neomyopites* do not form galls). Most species of *Myopites, Urophora,* and at least four species of *Stamnophora* (three undescribed) induce galls in the flowerheads

**TABLE 23.1**
**Number of Described and Undescribed Species and Summary of Host Plant and Distributional Data for Genera of Myopitini**

| Genus | # Spp. | Host Tribes | Distribution |
|---|---|---|---|
| *Asimoneura* | 5+25 | Anthemideae, Arctoteae, Gnaphalieae | Afrotropical, one sp. western Palearctic, two spp. Oriental |
| *Eurasimona* | 2 | Anthemideae | Western and central Palearctic |
| *Goedenia* | 8 | Astereae (Solidagininae) | Western Nearctic |
| *Inuromaesa* | 3+1 | Cardueae, Inuleae | Palearctic |
| *Myopites* | 17+3 | Cardueae, Inuleae, Plucheeae | Palearctic, Afrotropical |
| *Myopitora* | 1 | Astereae | Eastern Palearctic |
| *Neomyopites* | 28+9 | Astereae, Eupatorieae, Heliantheae, Liabeae | Neotropical, two spp. southern Nearctic |
| *Rhynencina* | 5 | Heliantheae | Neotropical, one sp. eastern Nearctic |
| *Spinicosta* | 2+6 | Arctoteae, Vernonieae | Afrotropical |
| *Stamnophora* | 1+10 | Vernonieae | Afrotropical |
| *Urophora* | 57+5 | Cardueae | Palearctic, two spp. Afrotropical, one sp. Oriental |

*Note:* The number of known undescribed species is indicated to the right of the + sign.

of their hosts, whereas at least three species of *Stamnophora* (two undescribed) and *U. cardui* cause the formation of stem galls (*S. vernoniicola* induces both flowerhead and stem galls) (Freidberg 1998). Galls are usually polythalamous, and those in the flowerheads comprise achenes, receptacles, or both.

Clear differences exist in the structure of the flowerhead galls of the three genera (Freidberg 1998). *Myopites* galls, which are often rather conspicuous, invariably comprise a swollen receptacle to which lignified achenes ("chimneys") are firmly attached. These chimneys indicate the probable site of egg deposition and the site where young maggots begin to feed. Pupariation, however, takes place in the receptacle (Freidberg 1984). *Urophora* flowerhead galls are essentially the heavily lignified achenes in which larval development and pupariation takes place; they are sometimes embedded in the receptacle (Zwölfer and Arnold-Rinehart 1993). Infestation cannot be observed unless the outer bracts of the flower head are removed. Adults of both *Myopites* and *Urophora* escape separately through their individual chimneys. *Stamnophora*-induced galls differ from those of *Myopites* and *Urophora* in having a common opening through which all adults escape, and by not including achenes. The stem galls of *U. cardui* and *S. vernoniicola* are similar in being conspicuous, discrete, spindle-shaped, or rounded swellings. The stem galls formed by the undescribed species of *Stamnophora* are entirely different (Freidberg 1998). Although they consist of numerous individual cells that open to a common longitudinal slit, through which the adults emerge, there is little swelling and they are inconspicuous externally and difficult to detect in the field.

For other genera little or no information is available about the mode of infestation. Munro (1931) stated that *Asimoneura petiolata* was reared from flowerheads of *Pentzia incana* that had thick bases, perhaps referring to receptacle galls. One species of *Neomyopites* possibly was reared from galls, but the record is poorly documented. *Neomyopites melanops* and *tresmilia* induce formation of a swollen, gall-like achene, within which the larva feeds, whereas several other species of *Neomyopites* feed within flowerheads without inducing gall formation (Norrbom, unpublished data). The remaining taxa with known hosts were reared from flowerheads and may not be gall formers, although special studies are needed to confirm these data in view of the existence of inconspicuous, thin-walled galls such as those formed by *U. quadrifasciata* (Harris and Myers 1984). Females of *Rhynencina spilogaster* oviposit into the sides of the flowerheads of their apparent host species of *Smallanthus*, orienting their ovipositors at an approximately right angle to the long axis of the flowerhead (Norrbom, personal observation).

The above summary of gall formation in the Myopitini indicates that Freidberg's (1984: 160) Figure 6 is an oversimplification of the evolutionary processes thought to have taken place in the Myopitini. Instead of assuming one linear process beginning with flowerheads lacking galls, continuing through flowerheads with galls, and ending in stem galls, we now assume three or more separate and more or less parallel processes, each represented by one of the genera *Myopites*, *Stamnophora*, or *Urophora*, and with two of these lineages resulting independently in stem galls.

The immature stages of Myopitini have been described in detail for only a few species of *Urophora* (e.g., Varley 1937; Persson 1963; Freidberg 1982), *Myopites* (Freidberg 1980), and *Goedenia* (Goeden et al. 1995). Much of this knowledge is summarized by Ferrar (1987). Two outstanding anatomical features of third-instar larvae are the small number (two to four) of digits, or lobes, on the anterior spiracle, and in *Myopites* the presence of only two slits on the posterior spiracle, the latter possibly a unique state in third-instar Cyclorrhapha. In some *Urophora* species, the first-instar larva molts to a second instar before hatching from the egg (Varley 1937; 1947; Persson 1963). The coloration of the larvae and puparia is usually pale (whitish to yellowish) or predominantly pale, although the anterior or posterior parts are sometimes darker. In the mature third-instar larva and the puparium of some genera, the posterior end has a black, sclerotized plate. In *Neomyopites*, this plate often bears a row of spinelike projections (Steyskal 1979; Norrbom, personal observation; Figure 23.12E). The puparia of *Inuromaesa maura* and *Stamnophora* sp. are illustrated here for the first time (Figure 23.12A–D).

Some *Urophora* species are actual or potential agents for the biological control of Asteraceae plants of Palearctic origin (e.g., *Carduus*, *Centaurea*, and *Cirsium* spp.) that have become noxious weeds elsewhere (White and Korneyev 1989). Six species have already been established in North America, and one each in Australia and New Zealand and in India (Norrbom et al. 1999a). White and Clement (1987) provided a key for the identification of those *Urophora* species of interest to North American weed biocontrol projects.

## 23.7 PHYLOGENETIC ANALYSIS AND CLASSIFICATION

We did not attempt to include in the cladistic analysis all of the approximately 160 species (described or undescribed) that were examined during this study. We chose instead representative species from the groups of similar species (putative genera) that we first recognized on a tentative basis. From each group, including all previously recognized supraspecific taxa, we chose sufficient species to represent the full range of variation in the group for the characters listed below. Because part of our purpose was to recognize monophyletic genera, at least two species were included in the matrix from each group, except for the monotypic taxa. Except for a few species, like *Rhynencina xanthogaster*, *Asimoneura shirakii*, and two Afrotropical species of *Asimoneura* with a medially yellow scutellum, for which we have incomplete data (male genitalia unknown), or as noted in the explanation of characters below (see characters 7, 9, and 24), the species not included in the matrix have the same set of character states as one of the included taxa. For example, we did not include *R. spilogaster* in the data matrix because it has the same states for the characters analyzed as *R. longirostris*. Table 23.2 lists the character state distributions for the terminal taxa used in the Hennig86 analyses. Character numbers are indicated by the "#" symbol hereafter.

Numerous morphological characters were studied during this analysis. The 29 characters listed below were found to have little or no intraspecific variation (see #7 and #9) and to vary among the preliminary groups of species. We did not include characters that varied only within single putative genera (although some of the final genera that we recognize include more than one of these preliminary groups, and some characters (e.g., #8) now vary only within one of them). In accordance with standard procedures for cladistic analysis, these characters were divided into states and polarized, as explained below. Korneyev (Chapter 22) places the Myopitini within the "Lower Tephritinae," but the exact relationship of the Myopitini to the rest of the Tephritinae is unknown, and we do not know which other tribe or tribes form its sister group. We therefore used a hypothetical

**TABLE 23.2**
**Character Matrix for Myopitini**

| Taxa | Characters<br>11111111112222222222<br>12345678901234567890123456789 |
|---|---|
| Outgroup | 000101?0000000000000000000000 |
| *Goedenia formosa* | 11122200101010001000000010210 |
| *G. timberlakei* | 21122220102010001000000010210 |
| *G. bajae* | 111321200010100010000000102?? |
| *Eurasimona stigma* | 111302001020100010000000101?? |
| *E. fedotovae* | 010302001020100010000000101?? |
| *Myopitora shatalkini* | 20120200001000000100101000??? |
| *Neomyopites aereus* | 00010010001000000100101002221 |
| *N.* sp. J | 00000110001000000100101002221 |
| *N. regis* | 10020010101000000100101002221 |
| *N. claripennis* | 00020110101000000100101002210 |
| *N. cordilleranus* | 00000000001000000100101002211 |
| *N. columbianus* | 00000000001000000100101002211 |
| *N. melanops* | 00000001001000000100101002210 |
| *N. tresmilius* | 00000001101000000100101002210 |
| *N. morus* | 000000110010000000100101002?? |
| *Asimoneura stroblii* | 201311100020001000000101000?? |
| *A. petiolata* | 001101101020101000011110002?? |
| *A.* sp. A | 201101100020001000000?010002?? |
| *A.* sp. B | 20101110002010100000101000220 |
| *A.* sp. C | 201112100020101000001020002?? |
| *Spinicosta agromyzella* | 000110100021001000001010002?? |
| *S. cilipennis* | 000110100021001000001010002?? |
| *Urophora cardui* | 00011200000000100010101000210 |
| *U. hermonis* | 00011200003000100010101000210 |
| *U. dzieduszyckii* | 00011200000000100000101000210 |
| *Inuromaesa maura* | 10030200002000000000101200310 |
| *I. sogdiana* | 00020200002000000000101200310 |
| *I. circumflava* | 211201000020010000001012003?? |
| *Rhynencina longirostris* | 201302001010010100011120012?? |
| *R. dysphanes* | 201201001010010100011120012?? |
| *R. emphanes* | 101001001010010000011120012?? |
| *Stamnophora* sp. A | 01030200101011220001102000200 |
| *S. vernoniicola* | 01030200102010220001102000200 |
| *S.* sp. B | 01020200101011220001102000200 |
| *Myopites apicatus* | 20110200111011220001102100200 |
| *M. delottoi* | 201112001120112200011021002?? |
| *M. nigrescens* | 201102101110112200011021002?? |
| *M. hemixanthus* | 201102001110112200011020002?? |
| *M. flavovarius* | 00030200111011220001102100??? |
| *M.* sp. B | 110302100120102200011021002?? |

outgroup, which was assigned the state for each character that occurs in the rest of the Tephritinae, or when more than one state occurs outside the Myopitini, the state occurring exclusively or more commonly in the Lower Tephritinae. In one case (#7) where the polarity was unclear, we coded the outgroup"?". For any characters in which the polarity is not straightforward, it is discussed below. Unless otherwise stated, state 0 is plesiomorphic.

1. Face shape, lateral view: 0, not produced (Figure 23.2D-E); 1, slightly produced (Figure 23.1B-D); 2, strongly produced (Figure 23.1E-H). This and #3 are somewhat correlated. Some Eutretini (e.g., *Paracantha*, some *Eutreta*) and some Schistopterini have a produced face, as do some Tephritini that have an elongate proboscis. However, the not-produced condition is much more widespread in Tephritinae in general and especially in the Lower Tephritinae and is almost certainly plesiomorphic in the Myopitini.

2. Face shape, anterior view: 0, convex to moderately concave, facial ridges not distinct (Figure 23.2F); 1, strongly concave, facial ridges distinct (Figure 23.2G). Although most taxa are clearly one state or the other, a few are intermediate and were difficult to code. We coded them as state 1 if there was any doubt.

3. Labella length:head height ratio: 0, less than 0.6 (Figure 23.2C-E); 1, more than 0.6 (Figure 23.1A-C, F). There is broad range of proboscis lengths among Myopitini, and we have undoubtedly underestimated the amount of evolution in this character, but our initial attempts to further divide it into additional states was limited by the almost continuous range of variation. For all taxa except *Urophora* (the latter mostly with ratio less than 0.6, but slightly higher in just a few species) there is a gap in the variation at the 0.6 ratio, which we used to recognize these two states. A few other Tephritinae have an elongate proboscis (e.g., various genera of the *Campiglossa, Sphenella,* and *Tephritis* groups in the Tephritini, *Hetschkomyia* in the Cecidocharini, and *Xenodorella* in the Dithrycini), but this state is so sporadic (and especially rare in the Lower Tephritinae) that state 0 is clearly plesiomorphic within the Myopitini. *Inuromaesa circumflava* is unusual in having relatively short labella, but the rest of the proboscis is elongate, so we coded it state 1.

4. Dorsocentral seta: 0, aligned anterior to postsutural supra-alar seta; 1, aligned with supra-alar seta; 2, aligned posterior to supra-alar but less than one-third the distance to acrostichal seta; 3, aligned one-third or more the distance from supra-alar to acrostichal seta. Most other Tephritinae have states 0 or 1, unless two pairs of dorsocentral setae are present. Assigning either as the plesiomorphic state has little effect on the resulting trees.

5. Thoracic pleura: 0, entirely microtrichose (Figure 23.3C); 1, matte (densely microtrichose) except distinct bare area on anepisternum (best viewed in oblique anterior view) (Figure 23.3D), and rarely a bare area on either anepimeron (some *Urophora*) or the extreme dorsal margin of the katepisternum (some *Asimoneura*); 2, mostly shiny, with anepisternum, anepimeron, and katepisternum each at most partly sparsely microtrichose. Other Tephritinae (except some Tephrellini and Cecidocharini) do not have a bare area, so state 0 is hypothesized as plesiomorphic for the Myopitini.

6. Scutellum shape: 0, flat (Figure 23.3A); 1, moderately convex; 2, strongly convex (Figure 23.3B). Most Tephritinae have an almost flat or slightly convex scutellum; in most Oedaspidina it is convex but of a different shape and probably the result of convergence. A few taxa were difficult to code for this character, as there is slight intergradation, but most taxa were easily coded as one of the three states.

7. Scutellum color: 0, yellow medially or entirely (Figure 23.3B); 1, entirely dark; 2, yellow area very narrow or intraspecifically variable. This character is variable in some *Goedenia* species, which were coded state 2. Scutellum color in other Tephitinae may be entirely dark to partially or entirely yellow. Many species are gradually paler apically. The partially yellow color pattern in myopitines is unique; it has distinct borders and extends to the base medially. We therefore coded the outgroup "?".

8. Hind femur: 0, without outstanding anteroventral setae; 1, with one to two outstanding anteroventral setae at two-thirds to three-quarters the length of femur (Figure 23.3E).

9. Location of crossvein R-M along length of cell dm: 0, at or distal to 0.5 (Figure 23.4E); 1, proximal to 0.5 (Figure 23.4A-B). Most genera are clearly one state or the other of this character. In some *Neomyopites* and *Goedenia* species the location of R-M varies intraspecifically, ranging from slightly above to slightly below 0.5 the length of cell dm. We ignored this variation in the analyses because, in *Goedenia* and all three preliminary groups of species now included in *Neomyopites*, both states occur exclusively in at least one species, and these are included in the matrix.

10. Cell $r_{4+5}$: 0, not distally narrowed by anterior slant of vein M; 1, distally narrowed by anterior slant of vein M (Figure 23.4F–G).

11. Wing pattern: 0, *Urophora*-type (Figure 23.5G-H), without small crossband in $r_1$, with band covering DM-Cu extended straight to anterior wing margin, apical band covering apex of vein $R_{2+3}$; 1, *Myopites*-type (Figures 23.4C, F, H, 23.5A-B, E), with small crossband in middle of cell $r_1$, spots covering crossvein DM-Cu and apex of vein $R_{2+3}$ not aligned, or if continuous band, with a bend along vein M, apical band usually isolated; 2, highly reduced (pattern type not recognizable) or entirely hyaline; 3, reduced or faint, derived from *Urophora*-type. Other Tephritinae have a great variety of wing patterns, including various types of banded, spotted, and reticulate patterns. Few taxa (some Terelliini, *Ensina*) have unpatterned wings, so the absence of wing pattern is probably apomorphic within the Myopitini, but it is difficult to determine whether state 0 or 1 is plesiomorphic without knowing the sister group of the Myopitini. We hypothesize that the *Urophora*-type pattern is plesiomorphic because it is most similar to the banded pattern in Terelliini, considered the most primative tribe of Tephritinae by Korneyev (Chapter 22), and because the *Myopites*-type pattern does not occur in any other Lower Tephritinae. A few *Urophora* species have highly reduced patterns; we coded *U. hermonis* differently from the species of other genera with reduced patterns because it still has faint traces of pattern similar to state 0.

12. Costal setulae: 0, setulae bordering pterostigma not erect nor enlarged; 1, setulae bordering pterostigma enlarged and erect (Figure 23.5C).

13. Male abdominal tergite width: 0, broad, pleural membrane not or occasionally only slightly visible in dorsal view (Figure 23.6B); 1, narrow, pleural membrane visible in dorsal view (Figure 23.6A).

14. Abdominal tergite coloration: 0, entirely dark (Figure 23.6B); 1, at least partially yellow (Figure 23.6A). Other Tephritinae vary considerably in tergite coloration. Many myopitine species with state 1 have a distinctive pattern with paired submedial black spots. Some other Tephritinae (e.g., some Terelliini, *Tomoplagia*) do have mostly yellow, dark-spotted abdomens, but this is not common, so we hypothesize the partially yellow state as apomorphic within Myopitini.

15. Epandrium and lateral surstylus shape in lateral view: 0, somewhat rectangular, rounded, or truncate ventrally (Figure 23.8B, D, G, H); 1, triangular, broad dorsally and evenly tapered ventrally, apex often acute (Figure 23.8A, J, L); 2, very broad dorsally, ventrally narrowed and lobe-like, with posterior margin usually concave (Figure 23.8E, F, K).

16. Posterolateral lobe of lateral surstylus: 0, absent; 1, small (Figure 23.8I); 2, large (Figure 23.8E, F, K). In the Myopitini with states 1 or 2, this lobe is more lateral than and probably not homologous with the typical dorsal lobe found in many Tephritinae, including some other Myopitini, which is formed by the projecting mesal edge of the lateral surstylus.

17. Lateral surstylus apex: 0, not strongly mesally curved; 1, strongly mesally curved (Figure 23.7B-C). Lateral surstylus shape varies considerably in other Tephritinae, but it is rarely so strongly mesally curved as in *Eurasimona* and *Goedenia*, and we therefore hypothesize this as apomorphic within the Myopitini.

18. Lateral surstylus apex: 0, not ventrally curved; 1, ventrally curved (Figure 23.7G-H). As stated above, there is considerable variation in lateral surstylus shape in other Tephritinae, but the apex is generally not ventrally curved.

19. Medial surstylus basal lobe: 0, absent; 1, present (Figure 23.7L).

20. Anterior prensiseta: 0, not enlarged; 1, greatly enlarged (broadened) (Figure 23.7E, F, I, K).

21. Lateral sclerites: 0, right sclerite much longer than left, both often slender (Figure 23.6H); 1, right sclerite medium length to short, sometimes subequal to left, both usually broadly fused to hypandrium basally (Figure 23.6F-G). This character has been little studied in other Tephritidae; there appears to be considerable variation across the family, but all other Tephritinae we examined have the right sclerite longer than the left.

22. Arms of phallapodeme: 0, slender to moderately broad; 1, extremely broad.

23. Glans sclerotization: 0, with extensive, strong sclerotization basally (Figure 23.9B-C); 1, with relatively little, mostly moderate or weak sclerotization basally (e.g., Figures 23.9D, 23.10A-B, F); 2, with small, but strong, well-differentiated basal sclerite shaped as in *Myopites* (Figures 23.9E-F, 23.10C, E).

24. Glans length: 0, not more than ten times as long as wide; 1, more than 12 times as long as wide, apex expanded with usually cup-shaped sclerite (Figure 23.9F); 2, more than ten times as long as wide, gradually tapered to slender apex (Figure 23.9D). *Neomyopites unicus* has an extremely long glans, more than 12 times as long as wide, but not tapered and with a paired apical sclerite. Because of its different glans structure and other character states indicating the relationships of this species, we consider its glans length to be independently evolved and that a short glans was the ground plan condition for *Neomyopites*. With *N. unicus* included in the matrix, the resulting trees were similar to those in Figures 23.13 through 23.15, but there was poorer resolution within *Neomyopites* because of missing data (#28 and #29) for *N. unicus*.

25. Glans shape: 0, not stout, not tapered to slender, medial lobe; 1, stout, tapered to slender medial lobe (Figure 23.9B-C). The shape of the glans varies tremendously in other Tephritinae. In many taxa it is stout, but we know of none in which it is tapered just as in *Eurasimona* and *Goedenia*, which we hypothesize as an apomorphic state within the Myopitini.

26. Acrophallus: 0, hyaline, smooth; 1, with striations (Figure 23.10C); 2, with dots or dashlike marks (Figure 23.10B).

27. Spermatheca: 0, strongly sclerotized; 1, weakly sclerotized; 2, membranous (Figure 23.11E); 3, membranous, but with apex of duct sclerotized (Figure 23.11D)

28. Puparium: 0, posterior end without unusual sclerotization (Figure 23.12C-D); 1, posterior end with large black, heavily sclerotized area, including spiracles (Figure 23.12A-B, E-F); 2, this black area not including spiracles. Some other Tephritinae have the posterior end of the puparium darkened (e.g., *Xanthaciura insecta* (Loew), at least some *Baryplegma* spp., many Schistopterini), but as this is uncommon, we hypothesize state 0 as plesiomorphic within the Myopitini.

29. Puparium: 0, posterior end without spines; 1, posterior end with spines (Figure 23.12E).

Two other characters, discussed below, at first appeared to be useful for phylogenetic analysis among the Myopitini. However, following further study, we found that they varied greatly within some of our tentatively recognized taxa, and, furthermore, the variation was too continuous to divide into character states. These characters thus were not used in the analysis. The arms of the phallapodeme are distinctly fused to the apices of the lateral sclerites in some Myopitini (Figure 23.6F-G; White and Korneyev 1989, Figure 6). However, the width of the arms varies greatly, and when the arms are narrow, it is difficult to determine if there is a narrow connection of sclerotization or only of membrane. The basiphallus also varies in size and shape, but the variation is continuous and is considerable within some genera.

Trees were calculated from the character matrix using the mhennig* and bb* options of Hennig86. One analysis (henceforth "unweighted analysis 1") was run in which all multistate characters were treated as nonadditive and no character weights were assigned. Another analysis (henceforth "unweighted analysis 2") was conducted in which characters #1, 4, 5, 6, 16, 23, and 28 were coded additive and characters #7, 11, 15, 24, 26, and 27 nonadditive; the former are those in which there are apparent trends in the transformation series, the latter those in which there is no obvious indication of the polarity of the derived states. In analysis 1, there was "overflow," meaning that there could be shorter trees or additional equally parsimonious trees not included in the set of trees produced. To minimize the likelihood of missing the shortest trees because of memory limitations, this analysis was rerun twice with the order of the taxa in the matrix randomly rearranged. Additional analyses were then conducted, started like each of those above, but followed by the successive weighting technique (Carpenter 1988) which was used to weight the characters and reanalyze the data set (by repeating the commands xs w; mh*; bb*; until there were no further changes in the results).

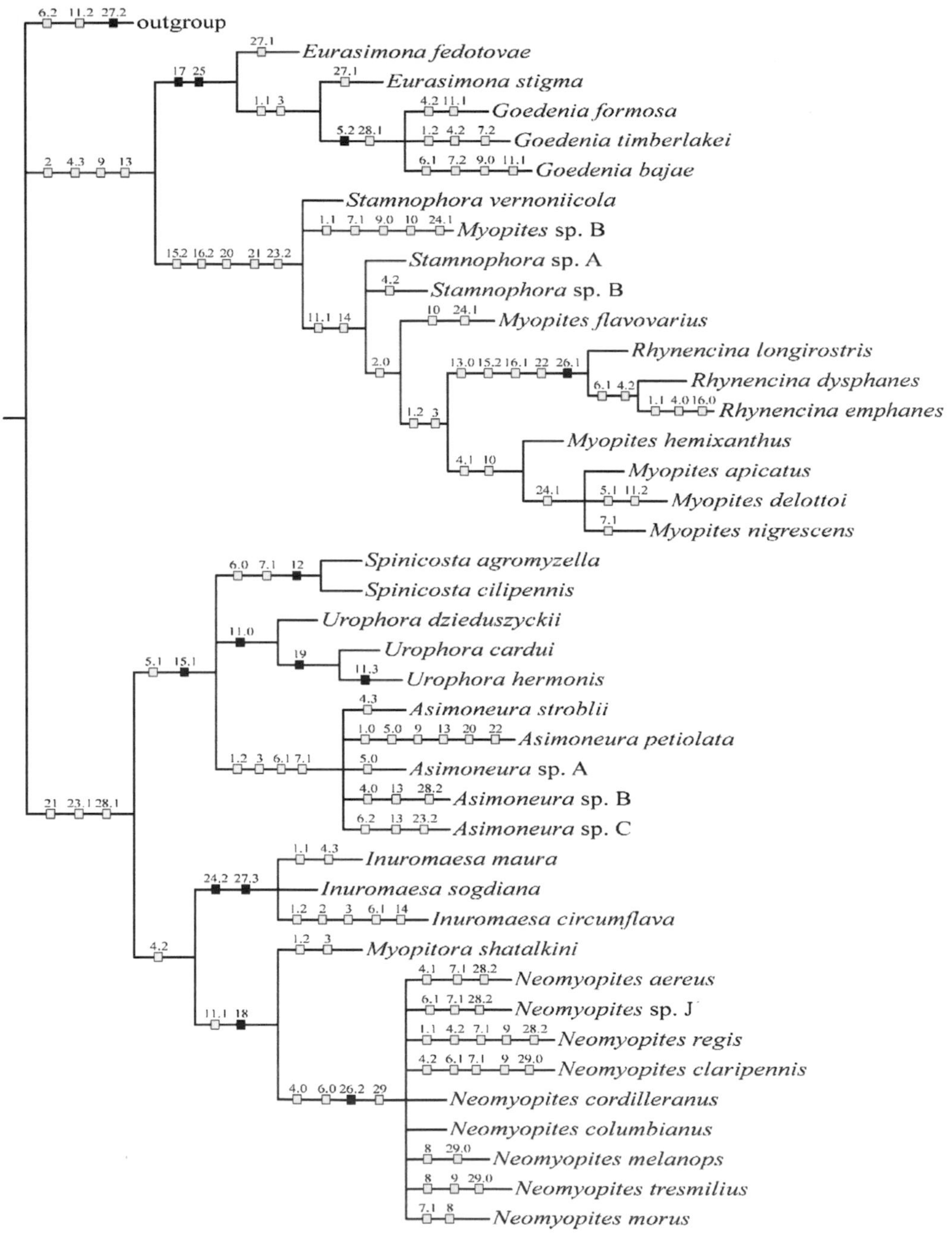


**FIGURE 23.13** Possible phylogenetic relationships of genera of Myopitini. Consensus tree based on analysis with all characters unweighted and all multistate characters nonadditive (unweighted analysis 1). Character state changes represented by black bars are unique; those by gray bars are homoplasious.

The results of the analyses of the matrix in Table 23.2 are as follows. In unweighted analysis 1, more than 2279 trees of equal, minimal length of 114 steps, consistency index (ci) = 0.39, retention index (ri) = 0.76, resulted. Because there was overflow, the matrix was reordered twice and rerun, but the strict (Nelson) consensus tree generated in all three cases was the same. This

**FIGURE 23.14** Possible phylogenetic relationships of genera of Myopitini. Consensus tree based on analysis with all characters unweighted and only characters 6, 7, 11, 15, 24, 26, and 27 nonadditive (unweighted analysis 2). Character state changes represented by black bars are unique; those by gray bars are homoplasious. Alternate plotting of characters 6, 11, and/or 13 is equally parsimonious.

tree (length 128, ci = 0.35, ri = 0.71) is shown in Figure 23.13. Of the genera we recognize, only *Eurasimona*, *Stamnophora*, and *Myopites* are not supported as monophyletic by this analysis. The *Goedenia* clade (*Goedenia* + *Eurasimona*), the *Myopites* clade (*Rhynencina* + *Stamnophora* + *Myopites*), the *Myopitora* clade (*Myopitora* + *Neomyopites*), and the *Urophora* clade (*Asimoneura*

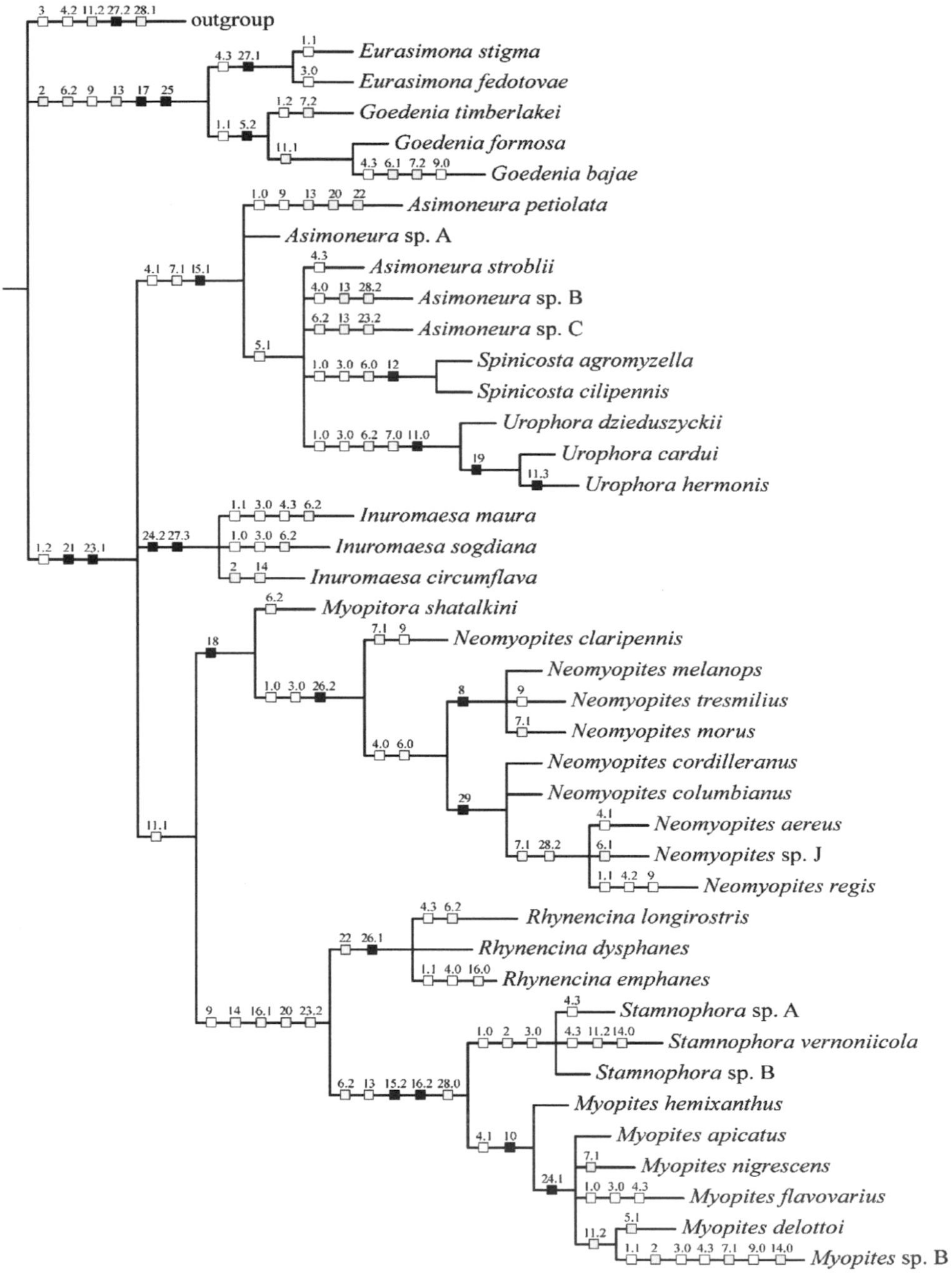


**FIGURE 23.15** Possible phylogenetic relationships of genera of Myopitini. Consensus tree based on analysis with only characters 6, 7, 11, 15, 24, 26, and 27 nonadditive followed by successive weighting. Character state changes represented by black bars are unique; those by gray bars are homoplasious; those by white bars were weighted 0. Alternate plotting of character 6 is equally parsimonious (characters 1, 3, and 4 can also be plotted differently).

+ *Urophora* + *Spinicosta*) also are supported. The *Goedenia* clade and the *Myopites* clade are hypothesized to form a monophyletic group, which is the sister group of the remaining taxa. The latter include the *Urophora* clade as the sister group of the *Myopitora* clade + *Inuromaesa*.

In unweighted analysis 2, 126 trees of 120 steps (ci = 0.37, ri = 0.77) resulted. The strict consensus tree (length 126, ci = 0.35, ri = 0.75) is shown in Figure 23.14. In this tree, all of the genera we recognize are supported as monophyletic except *Asimoneura*, *Stamnophora*, and *Myopites*. The *Goedenia* clade and the *Myopites* clade are supported, but not the *Myopitora* clade and the *Urophora* clade. *Neomyopites* is hypothesized as the sister group of the rest of the Myopitini.

The results of the two weighted analyses were very similar to each other. With some multistate characters additive (#1, 4, 5, 6, 16, 23, and 28; see above), 400 trees of 315 steps (ci = 0.76, ri = 0.93) resulted, from which the consensus tree (length 318, ci = 0.76, ri = 0.93) is shown in Figure 23.15. The characters were weighted as follows: 1, 3, 4, 9 = 0; 2, 6, 7, 14 = 1; 13 = 2; 11, 22, 28 = 3; 5, 20 = 4; 16, 23 = 6; all others = 10. With all of the multistate characters nonadditive, 132 trees of 305 steps (ci = 0.81, ri = 0.95) resulted. The consensus tree (length 307, ci = 0.81, ri = 0.95) is nearly identical in topology to Figure 23.15, differing only in that the three species of *Rhynencina* are resolved, with *R. emphanes* as the sister group of the other two species. The characters were assigned the same weights as in the other weighted analysis except the following: 6 = 0 and 5 = 5. In the consensus trees from these two weighted analyses, all of the genera we recognize were supported as monophyletic except for *Asimoneura*. The *Goedenia*, *Myopitora*, *Urophora*, and *Myopites* clades are supported, and the *Goedenia* clade is hypothesized as the sister group of a clade including all of the other myopitine genera.

In the weighted analyses, most of the genitalic characters were assigned high weights, whereas many external characters were assigned low or zero weight. The final weighted trees were not among the most parsimonious trees resulting from the unweighted analyses.

In the remainder of this section, we discuss the phylogenetic support for each of the genera and major clades of the Myopitini that we have recognized. Most of the genera are supported in both the weighted and unweighted analyses. The analyses differed considerably, however, concerning the relationships among the genera. Because of these differences, it is clear that further study of this aspect of the phylogeny of the Myopitini is needed. Although we have particular doubts about certain parts of the weighted trees, for example, the relationships within *Myopites*, we believe that the results of the weighted analyses (i.e., the hypotheses of relationships shown in Figure 23.15) are probably more reliable than the results of the unweighted analyses (Figures 23.13 and 23.14). We suggest this mainly because the results were similar in the weighted analyses, whereas they varied considerably in the unweighted analyses, especially regarding the lower nodes of the trees. This was true not only of the results presented here, but also in additional analyses that were done earlier in this project. During the course of this study, we ran numerous data sets that differed slightly from the matrix in Table 23.2 in terms of the included taxa or character data. For example, some earlier analyses did not include *Myopites* sp. B nor puparium data for *Neomyopites claripennis* and *Inuromaesa maura*, which were subsequently discovered, nor *I. sogdiana* and *circumflava*, which were recently described by Korneyev and Merz (1998). We also tried adding or substituting different species (e.g., *N. unicus*, which has a unique character state for glans length and shape but for which we have no puparium data, vs. *N. columbianus*). Other data sets varied in the states assigned to the outgroup for characters of uncertain polarity (e.g., #7, 11) or in the treatment of multistate characters. The results of the weighted analyses of all of these data sets were much more similar to each other than were the results of the unweighted analyses. Seemingly small differences in the data sets often produced large differences among the unweighted results, but this was much less so in the weighted analyses. At first glance, many of the hypothesized relationships among the genera on the unweighted trees (Figures 23.13 and 23.14) appear well supported by synapomorphies, but, in fact, most of the characters on these nodes are homoplasious (unique character state changes are indicated by black blocks, homoplasy by gray) and the basal nodes of these trees are unstable.

Based on the weighted analyses, we have grouped the genera of Myopitini in four clades (*Goedenia*, *Urophora*, *Myopitora*, and *Myopites* clades) plus *Inuromaesa*, whose sister group relationship is uncertain. The *Goedenia* clade (*Goedenia* + *Eurasimona*) is supported as monophyletic in all of the analyses by the following unequivocal synapomorphies: lateral surstylus strongly mesally curved (#17); and glans stout with slender apical lobe (#25). The face strongly concave (#2) is probably another synapomorphy (Figures 23.14 and 23.15). *Goedenia* is supported as monophyletic in all analyses by the reduced microtrichia of the thoracic pleura (#5.2). *Eurasimona* is supported in both weighted analyses (Figure 23.15) and in unweighted anlaysis 2 (Figure 23.14) by its weakly sclerotized spermatheca (#27.3). In unweighted analysis 1, *E. stigma* was grouped closer to *Goedenia* than to *E. fedotovae*, but this relationship is supported only by characters related to mouthpart length (#1 and 3) that appear to be homoplasious (given weight = 0 in the weighted analyses).

The *Urophora* clade (*Urophora* + *Spinicosta* + *Asimoneura*) is supported by the weighted analyses (Figure 23.15), although *Asimoneura* itself is not, and by unweighted anlaysis 1 (Figure 23.13). The triangular shape of the epandrium and lateral surstylus in lateral view (#15.1) is a consistent synapomorphy.

Both *Urophora* and *Spinicosta* are well supported as monophyletic groups in both the unweighted and weighted analyses. *Spinicosta* is supported by one unique apomorphy, the erect costal setulae (#12), and consistently by one homoplastic character, the flat scutellum (#6.0). *Urophora* is supported as monophyletic by its strongly convex scutellum (#6.2), medially yellow scutellum (#7.0), and/or by its wing pattern (#11.0). In the weighted analyses (Figure 23.15) and unweighted analysis 1 (Figure 23.13) the latter character is hypothesized as a synapomorphy, whereas in unweighted analysis 2 (Figure 23.14) it may or may not be (the evolution of this character is less certain; it may be plotted on the tree as shown in Figure 23.14, or state 2 may arise in the large clade (everything exclusive of *Neomyopites*) and reverse to state 0 in *Urophora*). In this *Urophora*-type pattern there is a straight band from crossvein DM-Cu to the anterior margin, which is unique within Myopitini (#11.0) except for one species of *Neomyopites*. It is possible that the ancestor of the *Urophora* clade or of *Urophora* + *Spinicosta* possessed this character state, but because *Asimoneura* and *Spinicosta* lack extensive wing patterns this cannot be evaluated at this time. Within *Urophora*, White and Korneyev (1989: 340) and Korneyev and White (1992) noted that *U. dzieduszyckii* (as *syriaca*; see Korneyev 1996b for synonymy of this species), *pontica,* and *solaris* share several probably apomorphic characters (broad palpus, black wing base, anepimeron partially bare, male fore coxa with spinelike setae, hosts *Echinops,* at least for the former two species) and form a distinctive species group, which they named the *dzieduszyckii* group. This group appears to be the sister group of the rest of *Urophora*; these species lack the basal lobe on the medial surstylus (#19) that is present in all other species we examined and which appears to be a unique synapomorphy of the rest of the genus.

The relationships of the species we have provisionally included in *Asimoneura*, most of which are undescribed, need to be further studied to test whether this genus is in fact a natural group. The species included in the matrix do not even include the full range of variation within the genus for the characters of the matrix; we did not include two undescribed Afrotropical species with a medially yellow scutellum nor *A. shirakii*, which varies in a number of character states, because of missing male genitalic character data for these species. The five *Asimoneura* species in the matrix were supported as a monophyletic group in unweighted analysis 1 (Figure 23.13) by several homoplastic characters, including the produced face (#1.2), elongate proboscis (#3), and slightly convex and entirely black scutellum (#6.1, 7). The latter character is shared with *Spinicosta* and, as noted above, varies in two undescribed Afrotropical species not included in the matrix. In the weighted analyses, *Asimoneura* is paraphyletic. Some species that have a bare area on the anepisternum (#5.1) are grouped with *Spinicosta* and *Urophora*, which also possess this character state, which otherwise occurs only in *M. delottoi*. A full revision of *Asimoneura* is needed to resolve the relationships of the species we have tentatively included here; the many undescribed species especially need more detailed study. Additional genera may need to be described for some of them, but such an action would be premature based on our current data.

The *Myopitora* clade includes *Myopitora* + *Neomyopites*, whose sister group relationship is supported in the weighted and some unweighted analyses (Figures 23.13 and 23.15) by the shape of the lateral surstylus (#18). This character is difficult to interpret as there is some variability in shape within *Neomyopites*, but it was easy to code in most species. *Myopitora* is monotypic. *Neomyopites* is supported as a monophyletic group in all analyses by the markings of the acrophallus (#26.2), a character state unique to this genus. Most of the species have a flat scutellum (#6.0). We originally thought that three genera might be recognizable for the species placed in *Neomyopites* based on scutellum color, leg chaetotaxy, and puparial characters. The analyses and recent new puparium data for several species indicate that although two large monophyletic species groups can be recognized within *Neomyopites*, there are additional species of the genus lacking the autapomorphies of both groups and whose relationships are unclear. We therefore have not recognized these species groups with subgeneric names. Species of the *melanops* group have one to two outstanding anteroventral setae on the hind femur (#8), and, in many, sternite 5 of the male has a pair of medial lobes and often has modified setae. Species of the *aereus* group have spines on the posterior end of the puparium (#29). All three characters are unique apomorphies within the Myopitini.

The monophyly of *Inuromaesa*, including *I. circumflava*, the type species of *Promyopites*, is supported in all of the analyses by the elongate, gradually tapered glans (#24.2) and sclerotized spermathecal duct (#27.2) of all three species (Korneyev and Merz 1998). The closer relationship of *I. maura* + *I. sodgiana*, the only two species previously included in *Inuromaesa*, is supported in unweighted analysis 2 (Figure 23.14) by face shape (#1.1), proboscis length (#3.0), and scutellum shape (#6.2), but not in unweighted anlaysis 1. In some of the most parsimonious trees resulting from the weighted analyses, scutellum shape also supports the relationship of these two species, but the evolution of this character is uncertain and it can be variously plotted on the consensus trees. The more evenly tapered glans might be interpreted as a synapomorphy for *I. maura* + *sogdiana,* but the unevenly tapered shape in *I. circumflava* could be derived from it. Thus, we have not discovered any unequivocal synapomorphies to indicate that these two species are more closely related to each other than to *I. circumflava*, which has a number of autapomorphies that readily distinguish it. For this reason we recognize only one genus for these three species and consider *Promyopites* a synonym of *Inuromaesa*.

The *Myopites* clade, which includes *Rhynencina*, *Myopites*, and *Stamnophora*, is supported in all of the analyses, although not all of the individual genera were supported in the unweighted analyses. The following characters were consistently hypothesized as synapomorphies for the *Myopites* clade: the lateral surstylus with a posterolateral lobe (#16.1), which is present in all species except *R. emphanes*; the anterior prensiseta enlarged (#20); and glans with a characteristically shaped basal sclerite (#23.2). In the weighted analyses (Figure 23.15) and unweighted analysis 2 (Figure 23.14) the abdominal tergites partially yellow (#14) is also hypothesized as a synapomorphy for this clade. This is a nearly unique character (otherwise found only in *I. circumflava*, which is almost entirely yellow, and *A. shirakii*), which is hyopthesized in these analyses to reverse in *Myopites* sp. B and *S. vernoniicola*.

*Rhynencina* is supported as a monophyletic group in all analyses by the broad arms of the phallapodeme (#22) and the striate acrophallus (#26.1). The latter character state is a unique apomorphy of this genus. *Rhynencina* appears to be the sister group of *Myopites* + *Stamnophora*, which are supported as a clade in the weighted analyses and in unweighted analysis 2 by the narrow male tergites (#13), the shape of the epandrium and lateral surstylus (#15.2), and the large size of the posterolateral lobe of the lateral surstylus (#16.2). The latter two character states are unique to these two genera.

*Myopites* is supported as monophyletic in the weighted analyses by one unique synapomorphy, the shape of cell $r_{4+5}$ (#10). Within *Myopites*, the species exclusive of *M. hemixanthus* and a very similar undescribed Afrotropical species not included in the matrix are grouped by their longer glans with the apical cup-shaped sclerite (#24.1). *Stamnophora* is supported as monophyletic in the weighted analyses, but only by a single homoplastic character, face shape (#2).

Relationships within *Myopites* + *Stamnophora* require special mention because of the status of *M. flavovarius*, n. comb., which is the type species of *Nearomyia*, here considered a junior synonym of *Myopites*. In the unweighted analyses, *Myopites* sp. B and *M. flavovarius* are either grouped within *Stamnophora* (e.g., Figure 23.14) based mainly on face shape (#1), proboscis length (#3.0), and dorsocentral seta position (#4.3), or are placed among the species of *Stamnophora*, which arise paraphyletically at the base of the clade (Figure 23.13). However, in the weighted analyses, they are both nested within *Myopites* (Figure 23.15), based mainly on the slanted vein M (#10) and glans shape (#24.1). We place both of these species in *Myopites* based on the weighted trees, in which characters #1, 3, and 4 have zero weight. These characters are assigned low weight in the successive weighting process because they are incongruent with other characters, which indicates more likelihood of homoplasy. *Myopites* sp. B and another very similar undescribed Afrotropical species that was not included in the analysis are difficult to evaluate. In addition to the above characters they share with *M. flavovarius*, they differ from other *Myopites* species in several additional characters, including face concave (#2), scutellum entirely black (#7.1; also in *M. nigrescens*), crossvein R-M beyond middle of cell dm (#9), and abdominal tergites entirely black (#14), all of which must be considered homoplastic if these species truly belong in *Myopites*. Despite these considerable differences, the weighted analyses nonetheless indicate that these two species are most closely related to the species of *Myopites* exclusive of *M. hemixanthus* and the undescribed Afrotropical species similar to it. We therefore treat them under *Myopites* because placing them in a separate genus would make *Myopites* paraphyletic.

The unweighted analyses support considerably different relationships among the genera of Myopitini than do the weighted analyses. In unweighted analysis 1 (Figure 23.13) the *Goedenia* clade + the *Myopites* clade is hypothesized as the sister group of a clade including all of the remaining taxa. In unweighted analysis 2 (Figure 23.14) *Neomyopites* is supported as the sister group of a clade including all other Myopitini. As stated above, however, we believe that neither of these hypotheses are as well supported as the relationships indicated by the weighted analyses (Figure 23.15). The weighted analyses support the hypothesis that the *Goedenia* clade is the sister group of the other genera. The remaining Myopitini exclusive of *Goedenia* + *Eurasimona* form a large clade supported mainly by lateral sclerite shape (#21) and reduced glans sclerotization (#23.1), which are unique apomorphies. This clade includes four large clades: the *Urophora* clade, the *Myopitora* clade, *Inuromaesa*, and the *Myopites* clade. The *Myopitora* clade and the *Myopites* clade were supported as sister taxa by a single synapomorphy, wing pattern *Myopites* type (#11.2), but this character is homoplastic and we thus do not consider any hypothesis of relationships among these four clades to be well supported.

## 23.8 CONCLUSIONS AND FUTURE RESEARCH NEEDS

In this chapter, the tribe Myopitini was redefined and demonstrated to be monophyletic, and a new generic classification was proposed, based on phylogenetic analysis using new and traditional characters. Eleven genera were recognized, of which three were described as new (*Goedenia, Neomyopites,* and *Spinicosta*), and an additional three were elevated from subgeneric status (*Eurasimona, Inuromaesa,* and *Myopitora*, all described as subgenera of *Urophora*). Two genus group names were synonymized: *Nearomyia* with *Myopites*, and *Promyopites* with *Inuromaesa*. A key to the genera, a checklist of included species, a diagnosis for each genus, and data about the hosts and distribution were provided. A short summary of the biology of the Myopitini and a detailed treatment of its phylogeny were given.

Most of the recognized genera are supported as monophyletic, although some questions remain about *Myopites*, *Stamnophora*, and especially *Asimoneura*. Although we could not fully resolve the relationships among the genera, the *Goedenia* clade (*Goedenia* + *Eurasimona*), the *Urophora* clade (*Urophora* + *Spinicosta* + *Asimoneura*), the *Myopitora* clade (*Myopitora* + *Neomyopites*), and the *Myopites* clade (*Rhynencina* + *Stamnophora* + *Myopites*) are strongly supported as monophyletic groups in the analyses using successive weighting (Figure 23.15), which we believe are a better estimate of the phylogeny of the Myopitini than the results of our unweighted analyses

(Figures 23.13 and 23.14). In these weighted analyses, the *Goedenia* clade was recognized as the sister group of the clade including all of the other genera. Within the *Myopites* clade, *Rhynencina* was indicated as the sister group of the other two genera.

As to our goal of providing a stable, predictive classification for the Myopitini by using a comprehensive, cladistic approach, only time will answer whether we were successful, but clearly without a worldwide study of this group it is doubtful that much progress could have been made. The relatively low consistency indices for the unweighted analyses indicate that there is considerable homoplasy in the characters used in the analysis. It is noteworthy that in the weighted analyses many of the external characters, such as proboscis length, position of the dorsocentral setae, location of crossvein R-M, and scutellum color, were assigned low weight compared with the genitalic characters, and thus appear to be highly homoplastic within the Myopitini.

Our experience with conducting a genus-level revision before the included genera themselves were revised has taught us that the reverse order may be preferable. We originally tried to restrict the study to a small number of selected species (especially type species), but we were continuously forced to add more and more species, ending up with partial revisions of most genera, which was not our original intention. Revisions of some of the largest genera (*Asimoneura, Goedenia, Myopites, Neomyopites, Rhynencina, Spinicosta,* and *Stamnophora*) are still needed (only *Urophora* has been reasonably revised). After these are completed, the generic classification of the tribe may need to be further revised. The majority of these genera are in the Neotropical and Afrotropical Regions, where many as yet undiscovered species likely exist. These two regions particularly require additional faunistic work.

## ACKNOWLEDGMENTS

We are grateful for the cooperation of Ian M. White (formerly CAB International Institute of Entomology) and Valery Korneyev (Ukrainian Academy of Sciences) in this project. They have revised much of the Palearctic Myopitini fauna and have generously shared data, discussed characters, and loaned specimens for study. We also thank Bernhard Merz (Muséum d'Histoire Naturelle, Genève) for his help, especially in supplying adults and immature stages, and for sharing information on host plants. We thank R. Contreras-Lichtenberg (Naturhistorisches Museum Wien), U. Kallweit (Staatliches Museum für Tierkunde, Dresden), V. Richter (Academy of Sciences, St. Petersburg), M. Mansell (Plant Protection Research Institute, Pretoria), B. Stuckenberg and J. Londt (Natal Museum, Pietermaritzburg), J. Ziegler (Deutsches Entomologisches Institut, Eberswalde), and X.-J. Wang (Academia Sinica, Beijing) for loan of study material, and to J. Kinkorova for carrying a loan from the Moravske Muzeum, Brno. We are grateful to S. Gaimari, W. N. Mathis (Smithsonian Institution), and D. Judd (Oregon State University) for assistance with the phylogenetic analysis, and to R. J. Gagné, B. Merz, W. N. Mathis and F. C. Thompson for reviewing the manuscript. W. Ferguson drew or inked the morphological illustrations, and L. Rodriguez prepared those of the phylogenetic trees.

We acknowledge the Campaña Nacional contra las Moscas de la Fruta (Mexico), International Organization for Biological Control of Noxious Animals and Plants (IOBC), and Instituto de Ecología, A.C. (Mexico) for their financial support of the Symposium, and USDA-ICD-RSED for funding ALN's travel.

## REFERENCES

Becker, T. 1913. Persische Dipteren von den Expeditionen des Herrn N. Zarudny 1898 und 1901. *Ezheg. Zool. Mus.* 17: 503–654.

Blot, F. 1827. Sur un nouveau genre et une nouvelle espèce de Diptère. *Mem. Soc. Linn. Normandie* 1826–1827: 101–106.

Bremer, K. 1994. Astraceae Cladistics and Classification. Timber Press, Inc., Portland. 752 p.

Carpenter, J.M. 1988. Choosing among multiple equally parsimonious cladograms. *Cladistics* 4: 291–296.

Cogan, B.H. and H.K. Munro. 1980. Family Tephritidae. In *Catalogue of the Diptera of the Afrotropical Region* (R.W. Crosskey, ed.), pp. 518–554. British Museum (Natural History), London. 1437 pp.

Czerny, L. and G. Strobl. 1909. Spanische Dipteren. III. Beitrag. *Verh. Zool. Bot. Ges. Wien* 59: 121–301.

Dirlbek, J. and O. Dirlbekova. 1973. Beitrag zur Kenntnis einiger Bohrfliegen (Diptera, Trypetidae) von Cuba. *Cas. Morav. Mus. Vedy. Prir.* 58: 121–129.

Fabricius, J.C. 1794. *Entomologia systematica emendata et aucta. Secundum classes, ordines, genera, species, adjectis synonimis, locis, observationibus, descriptionibus. Tome 4.* C.G. Proft, Hafniae (Copenhagen). 472 pp.

Farris, J.S. 1988. *Hennig86, version 1.5*. Published by the author, Port Jefferson Station, NY.

Ferrar, P. 1987. *A Guide to the Breeding Habits and Immature Stages of Diptera Cyclorrhapha*. Entomonograph 8. E.J. Brill/Scandinavian Science Press, Leiden. 2 vol., 907 pp.

Fitzhugh, K. 1989. Cladistics in the fast lane. *J. N.Y. Entomol. Soc.* 97: 234–241.

Foote, R.H., F.L. Blanc, and A.L. Norrbom. 1993. *Handbook of the Fruit Flies (Diptera: Tephritidae) of America North of Mexico*. Comstock Publishing Associates, Ithaca. 571 pp.

Freidberg, A. 1980. On the taxonomy and biology of the genus *Myopites* (Diptera: Tephritidae). *Isr. J. Entomol.* (1979) 13: 13–26.

Freidberg, A. 1982. *Urophora neuenschwanderi* and *Terellia sabroskyi* (Diptera: Tephritidae), two new species reared from *Ptilostemon gnaphaloides* in Crete. *Mem. Entomol. Soc. Wash.* 10: 56–64.

Freidberg, A. 1984. Gall Tephritidae (Diptera). In *Biology of Gall Insects* (T.N. Ananthakrishnan, ed.), pp. 129–167. Oxford and IBH Publishing Co., New Delhi.

Freidberg, A. 1998. Tephritid galls and gall Tephritidae revisited, with special emphasis on myopitine galls. In *The Biology of Gall-Inducing Arthropods* (G. Csóka, W.J. Mattson, G.N. Stone, and P.W. Price, eds.), pp. 36–43. *U.S. Dep. Agric. Forest Service, Gen. Tech. Rep.* NC-199, 329 pp.

Goeden, R.D. 1987. Host-plant relations of native *Urophora* spp. (Diptera: Tephritidae) in southern California. *Proc. Entomol. Soc. Wash.* 89: 269–274.

Goeden, R.D., D.H. Headrick, J.A. Teerink. 1995. Life history and description of immature stages of *Urophora timberlakei* Blanc and Foote (Diptera: Tephritidae) on native Asteraceae in southern California. *Proc. Entomol. Soc. Wash.* 97: 779–790.

Hancock, D.L. 1986a. Classification of the Trypetinae (Diptera: Tephritidae), with a discussion of the Afrotropical fauna. *J. Entomol. Soc. South. Afr.* 49: 275–305.

Hancock, D.L. 1986b. New genera and species of African Tephritinae (Diptera: Tephritidae), with comments on some currently unplaced or misplaced taxa and on classification. *Trans. Zimb. Sci. Assoc.* 63: 16–34.

Hardy, D.E. 1969. Lectotype designations for fruit flies (Diptera: Tephritidae). *Pac. Insects* 11: 477–481.

Harris, P. and J.H. Myers. 1984. *Centaurea diffusa* Lam. and *C. maculosa* Lam. s. lat., diffuse and spotted knapweed (Compositae). In *Biological Control Programmes against Insects and Weeds in Canada 1969–1980* (J.S. Kelleher and M.A. Hulme, eds.), pp. 127–137. Commonwealth Agricultural Bureau, Slough.

Hendel, F. 1927. Trypetidae. In *Die Fliegen der Paläarktischen Region* (E. Lindner, ed.), 5:1–221. Stuttgart.

Johnson, C.W. 1922. New genera and species of Diptera. *Occas. Pap. Boston Soc. Nat. Hist.* 5: 21–26.

Korneyev, V.A. 1987. The asparagus fly and its position in the systematics of the Tephritidae (Diptera). *Vestn. Zool.* 1987 (1): 39–44 [in Russian].

Korneyev, V.A. 1996a. Reclassification of Palaearctic Tephritidae (Diptera). Communication 3. *Vestn. Zool.* 1995 (5–6): 25–48.

Korneyev, V.A. 1996b. The status of *Urophora dzieduszyckii* Frauenfeld (Insecta: Diptera: Tephritidae). *Ann. Naturhist. Mus. Wien* 98B: 525–528.

Korneyev, V.A. and B. Merz. 1998. A supplement to revision of fruit-flies of the genus *Urophora* R.-D. (Diptera, Tephritidae) of Eastern Palaearctic. *Entomol. Obozr.* 77: 512–522.

Korneyev, V.A. and I.M. White. 1991. Fruit-flies of the genus *Urophora* R.-D. (Diptera, Tephritidae) of East Palaearctic. I. A key to subgenera and review of species (except for the subgenus *Urophora* s. str.). *Entomol. Obozr.* 70: 214–228 [in Russian].

Korneyev, V.A. and I.M. White. 1992. Fruitflies of the genus *Urophora* R.-D. (Diptera, Tephritidae) of East Palaearctic. II. Review of species of the subgenus *Urophora* s. str. (Communication 1). *Entomol. Obozr.* 71: 688–699 [in Russian].

Korneyev, V.A. and I.M. White. 1993. Fruit flies of the genus *Urophora* R.-D. (Diptera, Tephritidae) of East Palaearctic. II. Review of species of the subgenus *Urophora* s. str. (Communication 2). *Entomol. Obozr.* 72: 232–247 [in Russian].

Korneyev, V.A. and I.M. White. 1996. Fruit-flies of the genus *Urophora* R.-D. (Diptera, Tephritidae) of Eastern Palaearctic. II. Review of species of the subgenus *Urophora* s. str. (Communication 3). *Entomol. Obozr.* 75: 463–477 [in Russian].

Loew, H. 1862. *Die Europaischen Bohrfliegen (Trypetidae).* W. Junk, Wien [Vienna]. 128 pp.

Merz, B. 1992. The fruit flies of the Canary Islands (Diptera: Tephritidae). *Entomol. Scand.* 23: 215–231.

Merz, B. 1994. Diptera Tephritidae. *Insecta Helv. Fauna* 10, 198 pp.

Munro, H.K. 1931. New Trypetidae (Dipt.) from South Africa, II. *Bull. Entomol. Res.* 22: 115–126.

Munro, H.K. 1935. Biological and systematic notes and records of South African Trypetidae (fruit-flies, Diptera) with descriptions of new species. *Entomol. Mem. S. Afr. Dep. Agric.* 9: 18–59.

Munro, H.K. 1955. The influence of two Italian entomologists on the study of African Diptera and comments on the geographical distribution of some African Trypetidae. *Boll. Lab. Zool. Gen. Agrar. Portici* 33: 410–426.

Norrbom, A.L. 1989. The status of *Urophora acuticornis* and *U. sabroskyi* (Diptera: Tephritidae). *Entomol. News* 100: 59–66.

Norrbom, A.L., L.E. Carroll, and A. Freidberg. 1999a. Status of knowledge. In *Fruit Fly Expert Identification System and Systematic Information Database* (F.C. Thompson, ed.), pp. 9–47. *Myia* (1998) 9, 524 pp.

Norrbom, A.L., L.E. Carroll, F.C. Thompson, I.M. White, and A. Freidberg. 1999b. Systematic database of names. In *Fruit Fly Expert Identification System and Systematic Information Database* (F.C. Thompson, ed.), pp. 65–251. *Myia* (1998) 9, 524 pp.

Persson, P.I. 1963. Studies on the biology and morphology of some Trypetidae (Dipt.). *Opus. Entomol.* 28: 33–69.

Robineau-Desvoidy, J.B. 1830. Essai sur les Myodaires. *Mem. Pres. Div. Sav. Acad. R. Sci. Inst. Fr.* (ser. 2) 2: 1–813.

Sobhian, R. and H. Zwölfer. 1985. Phytophagous insect species associated with flower heads of yellow starthistle (*Centaurea solstitialis* L.). *Z. Angew. Entomol.* 99: 301–321.

Steyskal, G.C. 1979. *Taxonomic studies on fruit flies of the genus* Urophora *(Diptera: Tephritidae).* Entomological Society of Washington, Washington. 61 pp.

Varley, G.C. 1937. The life history of some trypetid flies with description of the early stages (Diptera). *Proc. R. Entomol. Soc. Lond. Ser. A Gen. Entomol.* 12: 109–122.

Varley, G.C. 1947. The natural control of population balance in the knapweed gall-fly (*Urophora jaceana*). *J. Anim. Ecol.* 16: 139–187.

Wang, X.-J. 1996. The fruit flies (Diptera: Tephritidae) of the East Asian Region. *Acta Zootaxon. Sin.* 21 (Suppl.): 338 + 82 pp.

Wasbauer, M.S. 1972. An annotated host catalog of the fruit flies of America north of Mexico (Diptera: Tephritidae). *Occas. Pap. Calif. Dept. Agric. Bur. Entomol.* 19: 1–172.

Westwood, J.O. 1840. Order XIII. Diptera Aristotle (Antliata Fabricius. Halteriptera Clairv.). In *An Introduction to the Modern Classification of Insects; Founded on the Natural Habits and Corresponding Organisation of the Different Families. Synopsis of the Genera of British Insects*, pp. 125–154. Orme, Brown, Green and Longmans, London. 158 pp.

White, I.M. and S.L. Clement. 1987. Systematic notes on *Urophora* (Diptera: Tephritidae) species associated with *Centaurea solstitialis* (Asteraceae, Cardueae) and other Palaearctic weeds adventive in North America. *Proc. Entomol. Soc. Wash.* 89: 571–580.

White, I.M. and V.A. Korneyev. 1989. A revision of the western Palaearctic species of *Urophora* Robineau-Desvoidy (Diptera: Tephritidae). *Syst. Entomol.* 14: 327–374.

Zwölfer, H. and J. Arnold-Rinehart. 1993. The evolution of interactions and diversity in plant–insect systems: the *Urophora-Eurytoma* food web in galls on Palearctic Cardueae. In *Biodiversity and Ecosystem Function* (E.D. Schulze and H.A. Mooney, eds.), pp. 211–233. (Ecological Studies 99). Springer-Verlag, Berlin.

# 24 Phylogeny of the Palearctic and Afrotropical Genera of the *Tephritis* Group (Tephritinae: Tephritini)

*Bernhard Merz*

## CONTENTS



0-8493-1275-2/00/$0.00+$.50

## 24.1 INTRODUCTION

Munro (1957a, b) first showed that the tribe Tephritini is rather heterogeneous and its genera may be clustered into groups. He named one of the groups the "*Trupanea-Tephritis* complex" (Munro 1957b), which he separated from the "*Paroxyna* group" and the "*Sphenella* group" based on differences in wing pattern. In the same year, Munro (1957a) proposed a slightly different classification of the Tephritinae. His "*Acanthiophilus* series" (including *Acanthiophilus* Becker, *Tephritomyia* Hendel, and *Pherothrinax* Munro), "*Euarestella* series" (including *Euarestella* Hendel and *Migmella* Munro) and "*Trupanea-Tephritis* series" (including *Trupanea* Schrank, *Dectodesis* Munro, *Goniurellia* Hendel, and *Tephritis* Latreille) share the characters of the "*Trupanea-Tephritis* complex" as defined by Munro (1957b). The three "series," however, were not characterized by descriptions. Therefore, it is not possible to recognize them by diagnostic characters nor to show relationships between the "series" or the genera included.

Later, Freidberg (1979) described two genera that he compared with *Tephritis*: *Hyalotephritis* Freidberg and *Tephritites* Freidberg. As part of a revision of the Tephritidae of Zimbabwe, Hancock (1986) clarified the status of some unplaced or misplaced species that belong in the same group of genera. He erected three new genera: *Brachydesis* Hancock, *Brachytrupanea* Hancock, and *Paradesis* Hancock, all of which he compared with genera of the "series" of Munro.

An attempt to clarify the classification of the American Tephritini was done by Foote et al. (1993), who characterized the genera allied to *Trupanea* by the structure of the male glans, which has a basal hooklike sclerite. In this restricted sense, *Tephritis* could not be placed in any known group of genera. It was outside the scope of their study to deal with non-Nearctic genera related to *Trupanea*, so that many Afrotropical and Palearctic genera treated by Munro (1957a, b) were not classified. Later, in their world classification of Tephritidae, Norrbom et al. (1999) copied the system of Foote et al. (1993), and left the other genera of the complex as genera *incertae sedis*.

Based on the present state of knowledge, it is clear that the higher classification of the Tephritini is largely unresolved, and sister group relationships have not been elucidated.

The aim of the present study is therefore to determine whether the genera of the "*Trupanea-Tephritis* complex," "*Acanthiophilus* series," and "*Euarestella* series" of Munro (1957a, b) and those later added by Freidberg (1979) and Hancock (1986), which together I call the *Tephritis* group, form a monophyletic group based on a cladistic analysis of adult characters. In a next step the monophyly of each described genus will be evaluated and new genera will be described for unplaced species or species groups. A key, diagnoses of the genera, and illustrations of important characters are provided to facilitate identification.

## 24.2 MATERIALS AND METHODS

### 24.2.1 Ingroup Species Selection

Prior to the cladistic analysis I checked over 100 species from the Palearctic and Afrotropical Regions that were placed in the genera of the *Tephritis* group or which seemed to belong to that group. For each species, 38 external characters were studied. I have excluded *Migmella* Munro, which according to Freidberg (personal communication) and personal observations, belongs to the

*Spathulina* group of genera. Those species that do not differ from the type species in any character used for the cladistic analysis have been excluded from further treatment. Therefore, the final selection for the cladistic analysis comprised 26 species, namely:

1. The type species of 16 described genera: *Acanthiophilus walkeri* (Wollaston), *Actinoptera discoidea* (Fallén), *Brachydesis rivularis* (Bezzi), *Brachytrupanea brachystigma* (Bezzi), *Capitites ramulosa* (Loew), *Dectodesis confluens* (Wiedemann), *Goniurellia tridens* (Hendel), *Hyalotephritis planiscutellata* (Becker), *Insizwa striatifrons* (Munro), *Paradesis auguralis* (Bezzi), *Pherothrinax redimitis* Munro, *Tephritis arnicae* (Linnaeus), *Tephritites australis* (Bezzi), *Tephritomyia lauta* (Loew), *Trupanea stellata* (Fuesslin), *Urelliosoma desertorum* (Efflatoun);
2. Two species of genera where only one specimen in bad quality of the type species was available: *Euarestella kugleri* Freidberg, *E. pninae* Freidberg; and
3. Nine additional species which differed from the species above in at least one character studied (original combination given here): *Campiglossa perspicillata* Bezzi, *Trypanea aurea* Bezzi, *T. dentiens* Bezzi, *T. goliath* Bezzi, *T. pulchella* Bezzi, *T. woodi* Bezzi, *Trypeta augur* Frauenfeld, and two undescribed species.

### 24.2.2 Outgroup Species Selection

As pointed out in the introduction, the phylogeny of the Tephritini is little known, and based on the present stage of knowledge it is not possible to recognize the sister group of the *Tephritis* group. The *Campiglossa, Sphenella,* and *Spathulina* groups may be most closely related to the *Tephritis* group, and two species of two of these groups (*C. producta* (Loew) and *S. sicula* Rondani) were selected as outgroup. At the least they are reasonable representatives of the rest of the Tephritini.

Freidberg (1987) and Korneyev (1989) gave some evidence that the *Campiglossa* and *Sphenella* groups may be monophyletic. From external characters (chaetotaxy, general appearance), these groups may be compared with the *Tephritis* group. However, both authors give some evidence that the *Campiglossa* and *Sphenella* groups form a clade based on synapomorphies in the male terminalia and their association with *Senecio* and related plant genera, which are not among the hosts of species of the *Tephritis* group. Moreover, no synapomorphy is known that would indicate a sister group relationship of these two groups with the *Tephritis* group.

Freidberg (personal communication) has informed me that *Spathulina* Rondani, *Elgonina* Munro, *Heringina* Aczél, and *Migmella* Munro form a well-defined, probably monophyletic group of genera, based on the shining abdominal tergites which may be considered an autapomorphy of this group. The *Spathulina* group has a similar external appearance and a similar host plant range as the *Tephritis* group, but the glans with its strong sclerotization exhibits still the plesiomorphic condition. However, no synapomorphy is yet known which would support a sister group relationship.

### 24.2.3 Techniques

The external characters of the specimens were studied with a "Leica M8" microscope. The terminalia were dissected with the method proposed by Merz (1994a). The drawings of the external characters were executed with a camera lucida usually directly from pinned specimens, including the wing illustrations of type specimens. Wings of non-type specimens were embedded in glycerol on a slide mount for drawing. Therefore, the proportions of some wings may differ slightly from reality. The terminalia were studied in glycerol under a Leica M20 compound microscope, which was also used for their illustrations.

For the phylogenetic analysis, PAUP, Version 3.0 (Swofford 1991) was used on a Macintosh PPC with the following heuristic search settings: simple and closest addition sequence, respectively; one tree held at each step during stepwise addition; tree bisection–reconnection branch swapping; mulpars option in effect. In a first run, all characters were treated as unordered and given the same weight. The resulting strict consensus tree did not sufficiently resolve the relationships of the ingroup taxa. Therefore, in a second run, character 4 (medial postocellar seta) was given double weight.

## 24.3 CHARACTER ANALYSIS

Of the 38 characters studied in total, I selected 20 for the cladistic analysis (Table 24.1). Excluded were those characters which were highly variable within species (e.g., setulae on frons, frontal stripe), those which could not be coded reasonably (e.g., degree of microtrichosity on different parts of the body), and those of uncertain homology (e.g., structures of the acrophallus). For each character used I discuss the plesiomorphic and apomorphic condition(s). This includes the range of states in other Tephritini, although decisions on character polarity were based on the state in the

**TABLE 24.1**
**Matrix of 29 Species and 20 Characters Used for the Cladistic Analysis of the *Tephritis* Group Presented in Figures 24.1 and 24.2**

| | Character No. |
|---|---|
| | **00000000011111111112** |
| | **12345678901234567890** |
| *Spathulina sicula* | 00000001000000000000 |
| *Campiglossa producta* | 00000000000000000000 |
| *Tephritis arnicae* | 10000000001000100000 |
| *Multireticula perspicillata* | 11000010001010100001 |
| *Dectodesis confluens* | 01000011011000100101 |
| *Dectocesis auguralis* | 01000010011000100101 |
| *Tephrodesis pulchella* | 11001010011000100001 |
| *Brachytrupanea brachystigma* | 1000001111????????01 |
| *Actinoptera discoidea* | 00100011111000100000 |
| *Brachydesis rivularis* | 01000011111000100000 |
| *Freidbergia mirabilis* | 01000010011000100011 |
| *Capitites augur* | 01001010011003102011 |
| *Capitites dentiens* | 000010000110031020?? |
| *Capitites ramulosa* | 01001000011002102011 |
| *Tephritomyia lauta* | 12001002000010103001 |
| *Acanthiophilus walkeri* | 12001000011111110001 |
| *Trupanea desertorum* | 12001011012011100000 |
| *Trupanea stellata* | 12001011012011100001 |
| *Trupanodesis aurea* | 12001002011000000011 |
| *Pherothrinax redimitis* | 12001000011000101011 |
| Gen. sp. 1 | 12011000001000101000 |
| *Stelladesis woodi* | 11011000011000101011 |
| *Insizwa goliath* | 01011000011000111011 |
| *Insizwa striatifrons* | 01011000001000121011 |
| *Hyalotephritis australis* | 10011000022000101011 |
| *Hyalotephritis planiscutellata* | 11011000022010101011 |
| *Goniurellia tridens* | 11011001011000101011 |
| *Euarestella kugleri* | 11011101011000111010 |
| *Euarestella pninae* | 11011101001010101010 |

outgroup taxa, except in one case (number of scutellar setae in *Spathulina*). Plesiomorphic states are coded 0, apomorphic states 1, 2, 3, respectively. The characters are listed in topographic order.

1. Mouthparts: 0, geniculate; 1, capitate. The shape of the mouthparts is highly variable among Tephritini, with very short labella in *Hendrella* Munro to extremely elongate mouthparts in *Capitites augur* (Frauenfeld), which makes it difficult to determine the polarity. Most genera of the *Sphenella, Campiglossa,* and *Spathulina* groups, however, have geniculate mouthparts with the labella at least twice as long as wide and clearly exceeding the mentum posteriorly (Figure 24.3D). The geniculate mouthparts are therefore regarded as plesiomorphic. Capitate mouthparts are widespread among genera of the *Tephritis* group, but geniculate mouthparts are also known for some genera.

2. Frontal setae: 0, two; 1, three, the anteriormost white, the others dark; 2, three to five, concolorous. The first condition is much more widespread in other Tephritini, as demonstrated in the two species chosen as outgroup. The presence of a small, usually rather lanceolate whitish seta in front is known only in *Noeeta* Robineau-Desvoidy and few other genera of the Tephritinae, but in none of the non-ingroup Tephritini studied. The presence of three or more concolorous frontal setae is rare in the Tephritini except in the taxa studied here and thus treated as the apomorphic condition.

3. Orbital setae: 0, two pairs; 1, one pair. The apomorphic condition does not occur in the outgroup taxa.

4. Medial postocellar seta: 0, absent; 1, present. A careful study of all Tephritinae available from the Palearctic and Afrotropical Regions showed that these setae are without exception present only in some genera of the *Tephritis* group. Only one single specimen of over 1000 specimens studied of species with this pair of setae exhibits them only on one side (the other setae of head and thorax are more variable within populations). Therefore, I assume that this character is very conservative and of great importance for the phylogeny of the group, and for this reason it was given double weight in the second cladistic analysis.

5. Postocular setae: 0, mixed dark and white; 1, only white. The apomorphic condition is not known in Tephritini outside the *Tephritis* group. As in the previous character, there is almost no variation within species. In almost all other families of acalyptrates and in the Trypetinae and many Tephritinae (Myopitini, Oedaspidina) of the Tephritidae the postocular setae are usually entirely dark. Most genera of the Tephritini have these setae mixed black and white. The white setae are usually lanceolate, thus differing from the acuminate dark setae.

6. Position of dorsocentral seta: 0, close to transverse suture, 1, at least half of its length posterior of suture. In all other Tephritini studied, the dorsocentral seta is situated just at the suture, which is regarded as the plesiomorphic condition.

7. Lower calypter: 0, convex on outer margin; 1, striplike, very narrow. The lower calypter of most Tephritini has a convex outer margin and it is almost as wide as the upper calypter. This state is treated here as plesiomorphic. In few genera of the *Tephritis* group the distal margin of the lower calypter is straight, and the surface is much reduced.

8. Apical scutellar seta: 0, present, at most 0.5 times as long as basal scutellar seta; 1, absent; 2, very long, at least 0.8 times as long as basal scutellar seta. The plesiomorphic condition is the most widespread within the Tephritini. Within the group studied, two directions of the development of the apical scutellar seta can be observed: some species and genera have lost this seta, but in others this seta is much enlarged, sometimes even longer than the basal scutellar seta. The apomorphic condition is less widespread in the Tephritinae. However, both states occur frequently in the Tephritini, which indicates that there may be frequent character state change.

9. Pterostigma: 0, about twice as long as high; 1, very short, about as long as high. The short pterostigma is known only from *Actinoptera, Brachydesis,* and *Brachytrupanea* within the Tephritini and therefore is regarded as apomorphic.

10. Wing pattern: 0, evenly reticulate in basal half; 1, stellate-shaped in basal four-fifths; 2, basal half of wing hyaline, at most inconspicuous spots present. The polarity of this character is uncertain because there are numerous different wing patterns in Tephritini and homoplasies occur in various groups (Freidberg and Kaplan 1992; Norrbom 1993). It may be assumed, however, that

in the ground plan, the wing of the Tephritini is usually mostly dark, interrupted by more or less regular hyaline spots, giving a "reticulate" appearance as in the *Campiglossa* and *Spathulina* groups which were chosen here as outgroup taxa. In contrast, in most species of the *Tephritis* group some of the hyaline spots are broadly connected. They leave only narrow dark rays from the central dark area to the wing margin, which results in a stellate appearance. There are two different types of this stellate pattern, the "elongate stellate-pattern" (Figure 24.7D) and the "narrow stellate-pattern" (Figure 24.8G). In the former, the dark area is restricted to the anterior and distal part of the wing, which is interrupted only in cell $r_1$ by one to two hyaline indentations, one hyaline, round spot at the tip of vein $R_{2+3,}$ and another just above crossvein DM-Cu in cell $r_{4+5}$. The apex of the wing, and cells m and dm contain four to six very narrow dark rays; otherwise the wing is more or less translucent. The "narrow stellate-pattern" is very similar, but the dark area proximad of crossvein R-M is reduced. In a few species the entire wing pattern is reduced (Figure 24.8F), which is regarded as a further evolutionary step in a transformation series. To delimit phylogenetic trends more precisely, two areas of the wing pattern were evaluated in the cladistic analysis: character 10 codes for the general appearance of the wing pattern, whereas character 11 codes for the wing tip distad of the tip of vein $R_{2+3}$. These two characters are partly linked to each other (an entirely hyaline wing, as in *Hyalotephritis*, is coded 2 in characters 10 and 11), but as seen in the matrix (Table 24.1), some species with a reticulate basal half of the wing may have either an evenly reticulate tip (*Campiglossa*) or a stellate tip (*Tephritis*). Conversely, in *Trupanea* the basal four-fifths of the wing appears stellate, but the wing tip is more or less hyaline.

11. Wing tip: 0, evenly reticulate; 1, apical fork present; 2, apex of wing hyaline. As I have discussed above, the stellate appearance is a feature almost exclusively found in the *Tephritis* group.

12. Surstyli: 0, normal; 1, thickened, partly separated from the epandrium by a seam. The apomorphic condition is only known from *Acanthiophilus* within the Tephritini.

The next six characters deal with the glans. This structure is very complexly sclerotized in most Tephritidae. It is difficult to homologize the different parts. Therefore, instead of coding each part of the sclerotization, the general appearance of the glans is evaluated, except for character 14, which involves a unique structure within the Tephritini.

13. Acrophallus: 0, tube-like; 1, reduced to some isolated sclerites which do not form a tube. It is assumed that in the ground plan of the Tephritini the acrophallus forms a well-sclerotized tube for the transfer of sperm. In some members of the *Tephritis* group the acrophallus is reduced to some isolated sclerites that do not form a tube. Because this type of character is not found in any other Tephritini studied, it is assumed to be the apomorphic condition.

14. Glans: 0, without hooks; 1, one hook present; 2, one modified hook present; 3, more than one hook present. While dissecting the phalli it was found that some species of the *Tephritis* group have well-developed, strongly sclerotized, movable hooks which are inserted on the outside of the cylindrical, membranous glans. They may be used to fix the glans in the female terminalia while transmitting the sperm. These hooks are not known in other Tephritini and are probably apomorphic. Three types of hooks with increasing complexity have been found: a simple hook (*Acanthiophilus, Trupanea*, Figure 24.12A), a Y-shaped hook (*Capitites ramulosa*, Figure 24.13A), and the presence of two or three hooks (other *Capitites*, Figure 24.13C). Whereas the simple hook in *Acanthiophilus* and *Trupanea* is short and not connected with the remaining acrophallus, there seems to be some connection between these two structures in *Capitites*. Therefore, the hooks in these groups may not be homologous.

15. Sclerotization of acrophallus: 0, strong; 1, weak. The *Spathulina* and *Campiglossa* groups and most other tribes of the Tephritinae have a well-sclerotized acrophallus. Except for *Trupanodesis aurea* (Figure 24.11C), the glans of the species of the *Tephritis* group is rather soft. The sclerotization appears weak, and even difficult to detect in some species. This condition is probably apomorphic.

16. Inner side of acrophallus: 0, smooth; 1, spiny; 2, with reticulated surface. In most Tephritini the acrophallus has a smooth inner surface. Some species of the *Tephritis* group, however, exhibit

some small spines on the inner side (Figure 24.14C) or they have a reticulated surface (Figure 24.14D), which may be bumpy, but this is difficult to see even with the strongest magnification of the compound microscope. These features are regarded here as apomorphic with respect to the outgroup.

17. Vesica: 0, small, rectangular; 1, long and thin, flattened; 2, voluminous; 3, long and tubular. In most species of Tephritini outside the *Tephritis* group the vesica is little developed, which is regarded here as plesiomorphic. Within the *Tephritis* group, different apomorphic conditions are exhibited: the long, but rather thin vesica (Figure 24.12E), the very voluminous vesica (Figure 24.13D), and the tubelike vesica (Figure 24.12C). States 1 and 3 appear externally similar, but they slightly differ in structure and may have evolved independently. Because the polarity is unknown, this character is treated unordered.

18. Tail on vesica: 0, absent; 1, present. A few species of the *Tephritis* group have one or more soft, narrow, hairy outgrows from the vesica which resemble tails (Figure 24.11E). Their function is unknown, but it may be speculated that the hairs have tactile functions inside the female terminalia. Because this character is not present in other Tephritini it is coded here as apomorphic.

19. Oviscape: 0, uniformly dark; 1, at least partly yellowish. In all Tephritini studied outside the *Tephritis* group, the oviscape is uniformly blackish, which is the plesiomorphic condition. Only within the *Tephritis* group, some genera have an oviscape that is partly or entirely yellowish.

20. Setulae on oviscape: 0, fine and dark; 1, whitish at base. The apomorphic condition is very rare outside the *Tephritis* group.

## 24.4 MONOPHYLY OF THE *TEPHRITIS* GROUP

In a first run, all characters were given the same weight, resulting in 90 most parsimonious trees of 64 branches with retention index (ri) = 0.71 and consistency index (ci) = 0.45, of which the consensus tree is presented in Figure 24.1. The monophyly of the *Tephritis* group is supported by three autapomorphies: Capitate mouthparts (character 1), apical fork in wing (character 11), and weakly sclerotized glans (character 15).

The apomorphic state for character 1 is present in all species except the *Actinoptera* clade, *Freidbergia*, *Dectodesis*, *Insizwa*, and *Capitites*, which do not form a monophyletic group. This result opens the question of whether capitate mouthparts is an autapomorphy of the *Tephritis* group with several later reversals, or whether the capitate mouthparts frequently evolved independently. Therefore, its support for the monophyly of the *Tephritis* group is weak.

Character 11 is present in apomorphic states 1 or 2 in all species of the *Tephritis* group, which at first appears to be strong evidence for its monophyly. But some species of *Spathulina* also have an apical fork (personal observation). It should be stressed that wing pattern is subject to high variation and therefore homoplasies are probably widespread in Tephritini (see Section 24.3), so this character also does not strongly support the monophyly of the *Tephritis* group.

The weak, simple sclerotization of the glans (character 15) gives the strongest evidence for the monophyly of the *Tephritis* group. Only one species, *Trupanodesis aurea*, has a strongly sclerotized acrophallus with a recurved sperm duct, which is, however, rather different in shape compared with *Campiglossa* or *Spathulina* and can therefore be regarded as independently derived. A similarly weak sclerotization of the glans has evolved, apparently independently, in the Oedaspidina (Freidberg and Kaplan 1992) and in some Myopitini (Freidberg and Norrbom, Chapter 23), which are not closely related to the *Tephritis* group.

## 24.5 DESCRIPTION OF THE *TEPHRITIS* GROUP

Species of the *Tephritis* group share the following set of characters. Body: Usually rather ash-gray dusted, notopleural region and hind margin of tergites often yellowish. Size ranging from

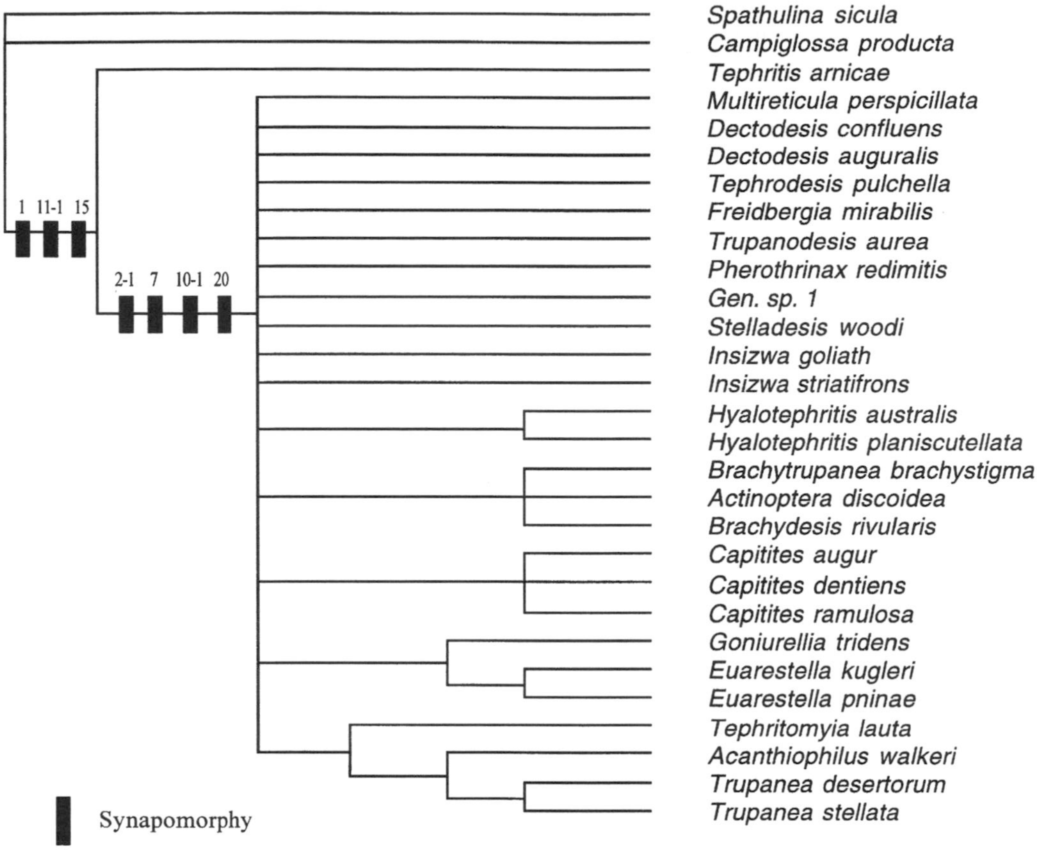


**FIGURE 24.1** Strict consensus tree for the 90 most parsimonious trees resulting from analysis of the matrix of Table 24.1 with *Campiglossa producta* and *Spathulina sicula* as outgroups. All characters are given the same weight (PAUP 3.0); Tree length: 64 steps; ri = 0.71; ci = 0.45.

2.0 to 6.0 mm (without oviscape). Head: In profile either almost square (Figure 24.5D) or distinctly higher than long (Figure 24.4G); therefore fronto-facial angle varying from 90° to 140°. Mouthparts capitate (Figure 24.4G) or short spatulate (Figure 24.3D) to extremely long geniculate (Figure 24.5D). Antenna normal; arista with short pubescence. Frons bare or setulose anteriorly; sometimes with conspicuous central stripe; ocellar triangle often setulose; occiput yellowish or dark, often with butterfly-like dark pattern on yellow ground color above occipital foramen. Chaetotaxy: two to four dark frontal setae, often with short, whitish seta in front at level of lunule (Figure 24.3F-G); one to two orbital setae; one ocellar seta, one medial and one lateral vertical seta, one postocellar seta; row of postocular setae either mixed white and dark or entirely white; some taxa have one to two medial postocellar setae between the postocellar setae. Thorax: As usual in the tribe Tephritini with the following variation: dorsocentral seta either at transverse suture or behind line of posterior notopleural seta; anepimeral seta whitish or dark; apical scutellar seta absent or when present varying from 0.2 to 1.1 times length of basal scutellar seta; lower calypter either very narrow and striplike or broad with convex margin. Legs as usual in the Tephritini, only forefemur in some genera with conspicuous spinules apicoventrally; foretarsus in males of some *Trupanea* with ornamentations. Wing: Vein $R_{4+5}$ ventrally bare or with some setulae; pterostigma very short (Figure 24.7J) or normal; wing either entirely hyaline or with dark pattern; the latter may be faint or strong; it covers only parts or whole wing surface, but usually not forming crossbands. Abdomen: Usually covered with dense, whitish setulae on tergites. Male terminalia: Epandrium either elliptical (Figure 24.10B) or more rounded (Figure 24.10D); dorsal lobe present in some taxa (Figure 24.9E-F); surstylus arched, fused with epandrium, or separated from it by faint constriction and then often modified (Figure 24.9A-D). Usually two pairs of prensisetae present, rarely one pair small (Figure 24.10D) or reduced. Basal part of distiphallus sometimes with appendages (Figure 24.11D). Glans usually with weakly sclerotized and irregularly shaped acrophallus that may be furnished with spines on inner side (Figure 24.14C) or with movable hooks (Figures 24.12A and 24.13C); rostrum present (Figure 24.12E) or absent (Figure 24.12F); vesica very long and narrow (Figure 24.14A), voluminous (Figure 24.13D), or almost absent (Figure 24.11C). Female terminalia: Oviscape black or to various degrees brownish; covered either with usual short, blackish setulae or with thick, whitish setulae at base. Eversible membrane uniform in all species. Aculeus gradually narrowed toward tip, either with apical indentation (Figure 24.10F-G) or pointed (Figure 24.10H-I), rarely serrated (Figure 24.10J-K).

## 24.6 PROPOSED CLASSIFICATION OF THE GENERA OF THE *TEPHRITIS* GROUP

The strict consensus tree of Figure 24.1 does not sufficiently resolve the relationships of the genera within the *Tephritis* group. Therefore, in a second run, character 4 (medial postocellar seta) was given double weight. The justification for this procedure is explained above (see Section 24.3). The cladistic analysis yielded 18 equally parsimonious trees of 65 steps (ri = 0.73, ci = 0.46) with the strict consensus tree shown in Figure 24.2.

As in the previous analysis with all characters given the same weight, *Tephritis* is the sister group of the remaining taxa. None of the analyzed characters is an autapomorphy for *Tephritis*, whereas the monophyly of the remaining taxa is based on four synapomorphies: three pairs of frontal setae (with two reversals, in *Actinoptera* and *Brachytrupanea*); striplike lower calypter (with one reversal in the *Trupanea* and *Goniurellia* clades); wing pattern strongly reticulate at base (with three reversals, in *Multireticula, E. pninae,* and Gen. sp. 1); and the oviscape with the setulae at base white (with four reversals, in *Actinoptera*, *Trupanea desertorum*, Gen. sp. 1, and *Euarestella*).

The relationships of the seven taxa of the next branch are still unresolved. This is due to the numerous homoplasies and reversals; in particular, the mouthparts, the number of scutellar setae, the shape of the acrophallus, and the color of the oviscape are highly variable and allow different similarly parsimonious trees. For instance, *Dectodesis auguralis* and *confluens* do not form a monophyletic unit in the cladogram (Figure 24.2), although they differ in the matrix (Table 24.1) only in the number of scutellar setae, which is subject to homoplasy. These two species share, however, a unique structure not found in other Tephritini, the "tail" which emerges from the glans and which is a synapomorphy for the species of *Dectodesis*. *Actinoptera*, *Brachytrupanea,* and *Brachystigma* may be a monophyletic unit based on one synapomorphy, the short pterostigma. The relationships within this group are uncertain, but may be resolved by study of the male terminalia of *B. brachystigma.* The monophyly of *Freidbergia* and *Capitites* receives little support, as it is based on one reversal (mouthparts) and one homoplasy (oviscape color). On the other hand, *Capitites* is a well-supported monophyletic group, as the three included species share two synapomorphies: shape of vesica and the presence of modified hooks on the glans. The *Trupanea* clade, *Trupanodesis*, *Pherothrinax,* and the *Goniurellia* clade are lumped together primarily based on one weak synapomorphy, the presence of three concolorous frontal setae, with one reversal within this clade (*Goniurellia* clade). Within the *Trupanea* clade, relationships are comparatively well resolved, because *Tephritomyia*, *Acanthiophilus,* and *Trupanea* each are supported by at least one strong synapomorphy. Good evidence is given for a sister group relationship between *Acanthiophilus* and *Trupanea*. Both genera have the same type of movable

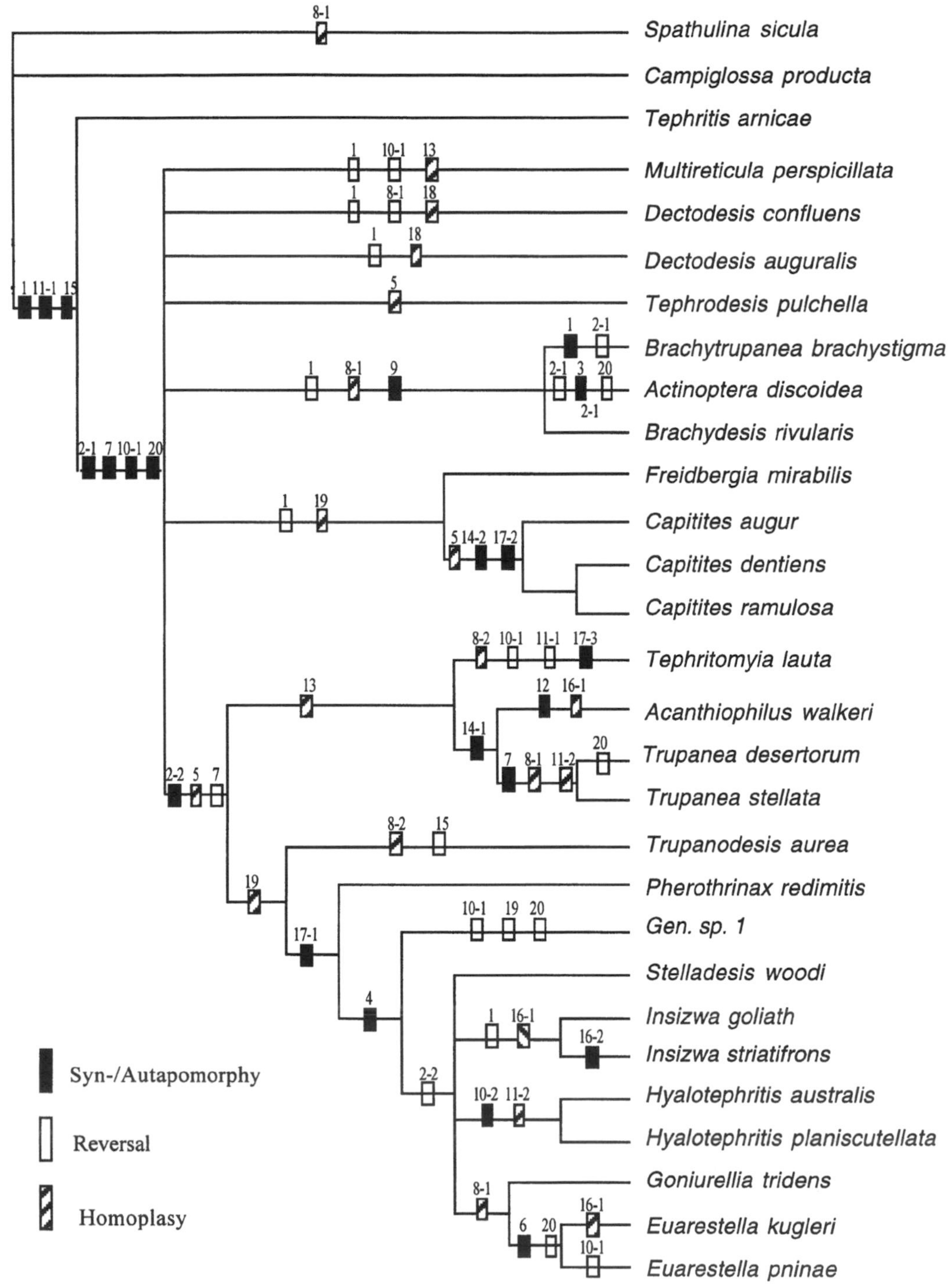


**FIGURE 24.2** Strict consensus tree for the 18 most parsimonious trees resulting from analysis of the matrix of Table 24.1 with *Campiglossa producta* and *Spathulina sicula* as outgroups. Character 4 is given double weight (PAUP 3.0); Tree length: 65 steps; ri = 0.73; ci = 0.46.

hook emerging from the glans, which is not known in other Tephritini. *Trupanodesis* is characterized by one reversal (glans sclerotization) and one homoplasy (scutellar setae length). Nevertheless, it seems justified to describe a new genus for the included species (*T. aurea* and an undescribed species), because they possess at least two autapomorphies (see Section 24.7.19). None of the characters analyzed is an autapomorphy for *Pherothrinax*; thus, the support to maintain it as a valid genus is weak. Other species previously placed in *Pherothrinax* (*woodi*, *pulchella*) belong according to this cladogram to different clades. The sister group of *P. redimitis* is the *Goniurellia* clade, which is a monophyletic unit well supported by character 4 (medial postocellar seta). Gen. sp. 1 represents an undescribed species that shows in relation to the other species of the *Goniurellia* clade a number of plesiomorphic characters (wing pattern reticulate at base, Figure 24.8H; oviscape black; and no white setulae at base of oviscape). The study of other characters did not yield any possible autapomorphy. Therefore, I prefer not to propose a new genus name for this species and leave it undescribed at this time. The other genera of the *Goniurellia* clade share a similar chaetotaxy of the head, but the evidence for the monophyly of this group is weak. Inside the group, relationships are little resolved. *Goniurellia* and *Euarestella* share the loss of the apical scutellar seta; this character, however, is difficult to interpret as many reversals and/or homoplasies involving it are hypothesized in the present cladogram.

Based on the cladistic analysis, the following generic classification is proposed for the *Tephritis* group (in parentheses the number of species known in the world; "+" indicates that undescribed species are known in these genera).

- *Tephritis* Latreille (168+ species)
- *Multireticula*, n. gen. (1 species)
- *Dectodesis* Munro (15+ species) = *Paradesis* Hancock, n. syn.
- *Tephrodesis*, n. gen. (3+ species)
- *Actinoptera* clade
    - *Actinoptera* Rondani (32+ species)
    - *Brachydesis* Hancock (1 species)
    - *Brachytrupanea* Hancock (2 species)
- *Capitites* clade
    - *Freidbergia*, n. gen. (1+ species)
    - *Capitites* Foote and Freidberg (4 species)
- *Trupanea* clade
    - *Trupanea* Schrank (217+ species) = *Urelliosoma* Hendel, n. syn.
    - *Acanthiophilus* Becker (10 species)
    - *Tephritomyia* Hendel (6 species)
- *Trupanodesis*, n. gen. (1+ species)
- *Pherothrinax* Munro (1 species)
- *Goniurellia* clade
    - *Stelladesis*, n. gen. (4+ species)
    - *Insizwa* Munro (5 species)
    - *Hyalotephritis* Freidberg (3 species) = *Tephritites* Freidberg, n. syn.
    - *Goniurellia* Hendel (9 species)
    - *Euarestella* Hendel (5 species)

## 24.7 THE GENERA OF THE *TEPHRITIS* GROUP

I do not give a full account of all taxa listed below, but references are indicated rather extensively when possible. However, special attention is drawn to the taxonomic changes that are a consequence of the phylogeny proposed in this chapter. The genera are listed in alphabetic order.

### 24.7.1 *Acanthiophilus* Becker, 1908

(Figures 24.9A-D and 24.12B)

*Type species*: *Tetanocera walkeri* Wollaston, 1858, by original designation.

*Recognition*: Hendel (1927), Freidberg and Kugler (1989), Merz (1994a).

*Discussion*: Except for the wing pattern, the species of this genus are rather uniform and probably belong to a monophyletic group characterized by the unique shape of the lateral surstylus (Figure 24.9A–D) and the spiny sclerotized bar in the glans (Figure 24.12B), both characters not known in other Tephritini.

*Biology*: All species with known biology breed in flowerheads of Cardueae (see Freidberg and Kugler 1989 and Merz 1994a for host plant lists).

### 24.7.2 *Actinoptera* Rondani, 1870

(Figure 24.3C-D)

*Type species*: *Trypeta aestiva* Meigen, 1826 (= *Tephritis discoidea* Fallén, 1814), by designation of Coquillett (1910).

*Recognition*: Hendel (1927), Munro (1957a).

*Discussion*: This genus, together with *Brachytrupanea* and *Brachydesis*, forms a well-founded monophyletic group based on the very short pterostigma, which is not known in other Tephritini. Species of *Actinoptera* are very similar; the loss of the posterior orbital seta (Figure 24.3C-D) supports the monophyly of the genus.

*Biology*: All species of known biology breed in flowerheads or stems of *Helichrysum*, *Filago,* or *Gnaphalium* (Asteraceae, Inuleae), often producing galls.

### 24.7.3 *Brachydesis* Hancock, 1986

(Figures 24.3F-G, 24.7I-J, 24.12F)

*Type species*: *Trypanea rivularis* Bezzi, 1924, by original designation.

*Recognition*: Hancock (1986). The original diagnosis of the genus is short, but sufficient. It should only be noted that the number of frontal setae is three, of which the anteriormost is short and whitish (Figure 24.3F-G) (not two as indicated).

*Discussion*: Contrary to the suggestion of Hancock (1986), *Brachydesis* appears very closely related to *Actinoptera* and *Brachytrupanea* and not *Migmella* Munro (which belongs to the *Spathulina* group; Freidberg, personal communication). It differs from its closest relatives mainly in the chaetotaxy of the head. In addition, the mouthparts are more strongly geniculate, but the phylogenetic value of this character is unknown. So far this South African genus is monotypic.

*Biology*: Unknown.

### 24.7.4 *Brachytrupanea* Hancock, 1986

(Figure 24.3E)

*Type species*: *Trypanea brachystigma* Bezzi, 1924, by original designation.

*Recognition*: Hancock (1986). His short diagnosis may be supplemented by the following additions. Head higher than long (Figure 24.3E), not almost square as in *Actinoptera* and *Brachydesis*; postocular setae white with some small black setulae; medial postocellar seta absent; labella fleshy and large, but apparently not spatulate; lower calypter very narrow, striplike; oviscape with some white setulae at base.

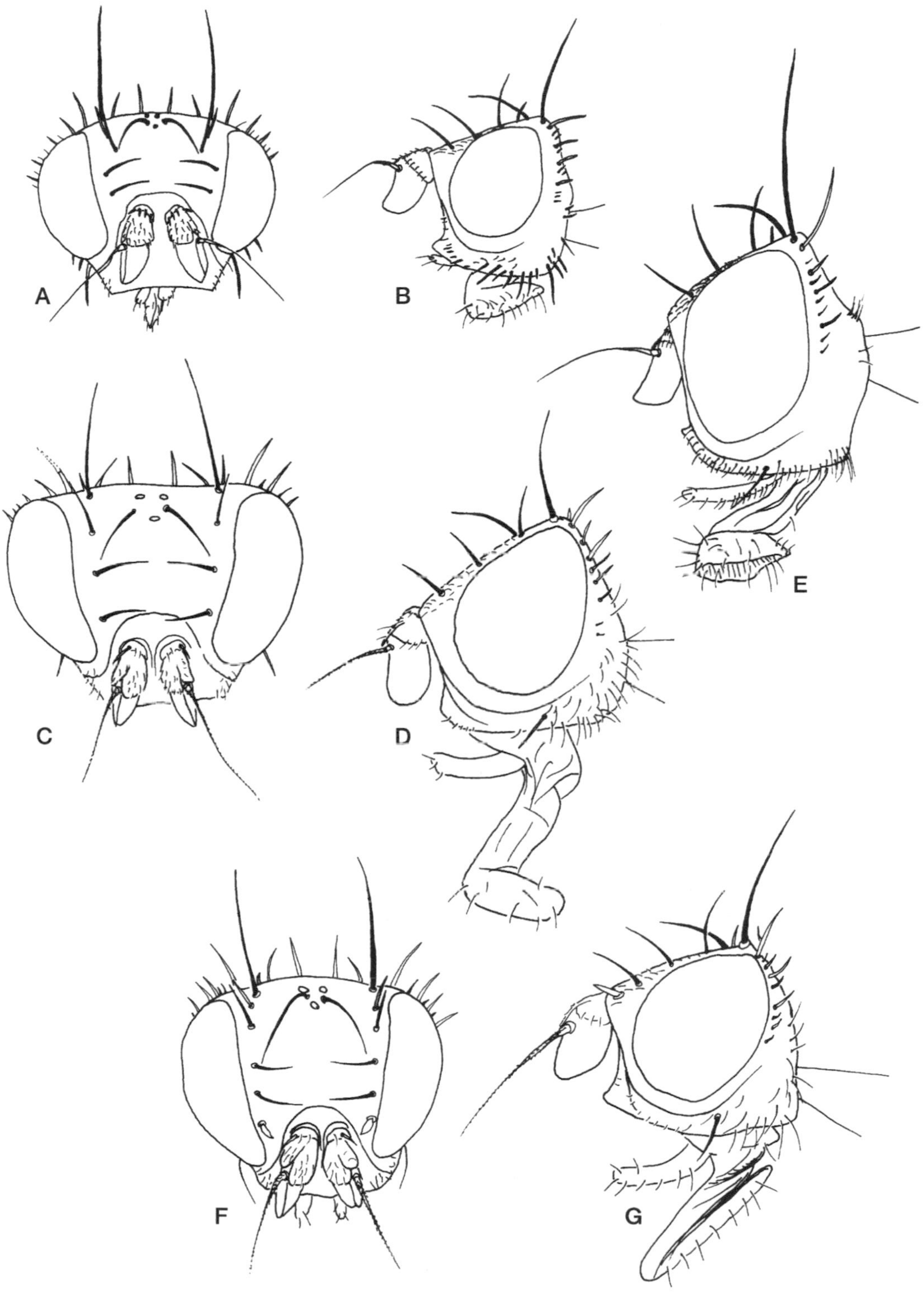


**FIGURE 24.3** Head, frontal and laterial views. (A and B) *Multireticula perspicillata*; (C and D) *Actinoptera discoidea*; (E) *Brachytrupanea brachystigma*; (F and G) *Brachydesis rivularis*.

*Discussion*: The present study reveals a close relationship of *Brachytrupanea* with *Actinoptera* and *Brachydesis* (see under these genera) and not with *Trupanea* and *Pherothrinax*, as suggested by Hancock (1986). Unfortunately, only a pair of syntypes of *B. brachystigma* was available for study and they were not dissected. It cannot be excluded, therefore, that a careful examination of the terminalia will change the phylogenetic placement of the genus. The second species placed in this genus by Hancock (1986), *B. semiatrata* (Hering), was not studied and its generic position is therefore tentative.

*Biology*: Unknown.

### 24.7.5 *Capitites* Foote and Freidberg, 1980

(Figures 24.5C-F, 24.10F-G, 24.13A, C-E)

*Type species*: *Trypeta ramulosa* Loew, 1844, by original designation.

*Recognition*: Foote and Freidberg (1980), Freidberg and Kugler (1989). Based on the phylogenetic analysis, the concept of this genus is redefined here.

*Diagnosis*: Head varying in shape from almost square (Figure 24.5D) to higher than long in profile (Figure 24.5F); mouthparts spatulate to geniculate; usually two dark, posterior and a small white, anterior frontal setae; medial postocellar seta absent; all postocular setae whitish; frons bare or setulose anteriorly, with or without central microtrichose stripe. Thorax ash-gray, notopleuron yellowish; two scutellar setae present, apical seta varying in size (0.2 to 0.9 times as long as basal scutellar seta); wing with elongate-stellate pattern; bulla present in all species studied; lower calypter narrow or well developed. Male terminalia with highly apomorphic glans consisting of one to three movable, sclerotized hooks and very large, voluminous vesica (Figure 24.13A, C-D). Female terminalia with oviscape at least partly reddish, basally with some white setulae; aculeus evenly acuminate or with apical indentation (Figure 24.10F-G).

*Discussion*: In the present concept, this genus comprises a number of species with a unique type of glans, which is clearly an apomorphic condition: the acrophallus is divided into a tubular basal part and some strongly modified hooks which are movable and extend from the membranous sack. Further, the species are all characterized by the very voluminous vesica. With this enlarged characterization of the genus, *C. augur* (Frauenfeld), n. comb., is transferred to *Capitites* (from *Dectodesis*), despite its aberrant shape of the proboscis and the short apical scutellar seta. On the other hand, the following species, which were transferred to *Capitites* by Hancock (1986), should be placed elsewhere: *Trupanodesis aurea* (Bezzi), *Insizwa goliath* (Bezzi), *I. kloofensis* (Munro), and *I. dicomala* (Munro). *Capitites albicans* (Munro) has not been examined and is retained tentatively in *Capitites*.

*Biology*: All species with known biology breed in flowerheads of Inuleae (Asteraceae): *Phagnalon* (*C. ramulosa*) and *Pulicaria* (*C. augur*).

### 24.7.6 *Dectodesis* Munro, 1957

(Figures 24.4A-B, 24.7C, 24.11E)

*Type species*: *Trypeta confluens* Wiedemann, 1830, by original designation.

*Paradesis* Hancock, 1986 (type species: *Urellia auguralis* Bezzi, 1908, by original designation), n. syn.

*Recognition*: Munro (1957a) gives a short but concise diagnosis. The description in Freidberg and Kugler (1989) includes also *Trupanea augur*, which is here placed in *Capitites*.

Discussion: The members of this genus are extremely uniform in external characters, in particular in head shape (Figure 24.4A-B) and wing pattern (Figure 24.7C). The hairy "tail" emerging from the vesica of the glans (Figure 24.11E) is a distinctive synapomorphy for the members of the genus. The species differ mainly in details of the male terminalia and scutal pattern. *Paradesis* was

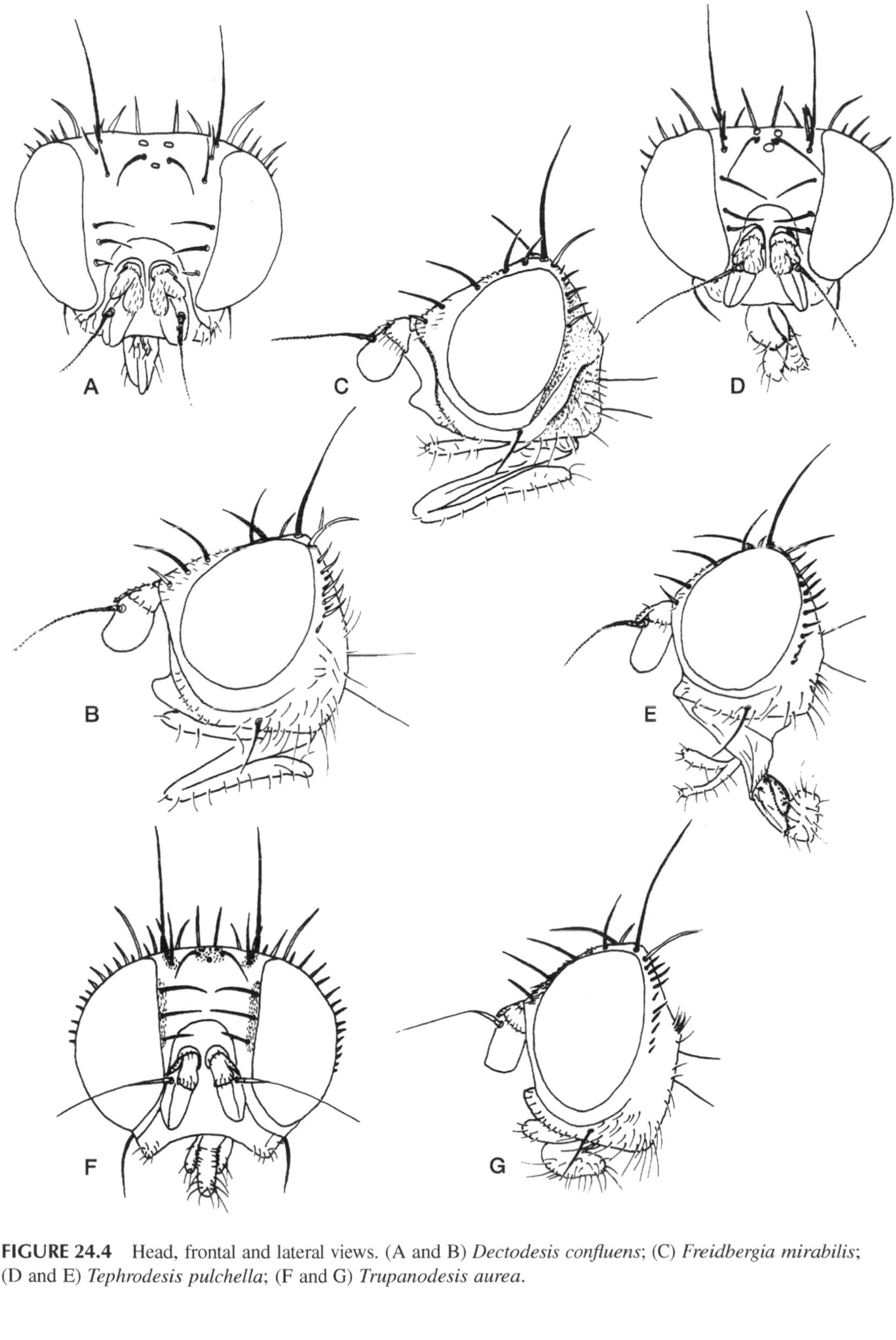


**FIGURE 24.4** Head, frontal and lateral views. (A and B) *Dectodesis confluens*; (C) *Freidbergia mirabilis*; (D and E) *Tephrodesis pulchella*; (F and G) *Trupanodesis aurea*.

proposed primarily based on the absence of apical scutellar setae. However, their length varies considerably throughout *Dectodesis*, from minute to well developed, so that this separation seems arbitrary. The two genera are synonymized here. The following species are therefore transferred from *Paradesis*: *Dectodesis bomolina* (Speiser), n. comb., *D. hexapoda* (Bezzi), n. comb., and *D. inundans* Munro, n. comb.

*Biology*: All species with known biology breed in flowerheads of *Helichrysum*, *Gnaphalium*, or *Erigeron* (Asteraceae, Inuleae).

## 24.7.7 *Euarestella* Hendel, 1927

(Figure 24.8C-E)

*Type species*: *Trypeta megacephala* Loew, 1846, by original designation.

*Recognition*: Freidberg and Kugler (1989).

*Discussion*: This genus is very heterogeneous in external characters, in particular in the wing pattern (Figure 24.8C–E). Its monophyly is supported by the position of the dorsocentral seta situated well behind the transverse suture. Freidberg and Kugler (1989) proposed that, based on terminalia and wing pattern, the four Palearctic species form two groups: one including *E. megacephala* (Loew) and *pninae* (Freidberg), and the other *E. kugleri* (Freidberg) and *iphionae* (Efflatoun). The monophyly of *E. megacephala* and *pninae* is corroborated by the small setulae on the ocellar triangle between the ocelli and by their biology (see below). The other two species are probably also a monophyletic group, because they share the same type of small spines on the inner side of the tube of the acrophallus (see Freidberg and Kugler 1989, Figure 87), which are known only in *Insizwa*, but this may be a convergent development. *Euarestella abyssinica* Hering was not studied, but according to its description this species belongs elsewhere. Pending the study of its type specimens it is tentatively left in *Euarestella*.

*Biology*: All species breed in Inuleae (*Pulicaria*, *Helichrysum*, *Iphiona*). Two species (*E. kugleri* and *iphionae*) live in flowerheads and the other two species (*E. megacephala* and *pninae*) produce stem galls (Hendel 1927; Freidberg and Kugler 1989).

## 24.7.8 *Freidbergia* Merz, n. gen.

(Figures 24.4C, 24.7D, 24.11D)

*Type species: Freidbergia mirabilis,* n. sp., by present designation.

*Diagnosis*: Head: Almost square in profile (Figure 24.4C), with very long, strongly geniculate mouthparts; labella and palpus markedly exceeding oral margin. Frons flat, bare, with faint frontal stripe; ocellar triangle without setulae. Face with two antennal grooves separated by distinct carina. Antenna a little shorter than face, first flagellomere only a little longer than wide, slightly pointed dorsoapically. Postgena with distinct groove. Chaetotaxy: two dark posterior and one white anterior frontal, one dark anterior and one white posterior orbital, one dark ocellar, one pale postocellar, one dark medial vertical and one white lateral vertical seta, postocular setae mixed white and dark, the dark setae smaller and often restricted to ventral part of row; no medial postocellar seta present. Thorax: Of yellow ground color, partly microtrichose. Chaetotaxy: one dark dorsocentral seta close to transverse suture, one dark postpronotal, one dark anterior and one white posterior notopleural setae, one dark presutual supra-alar, one dark prescutellar acrostichal, one dark postsutural supra-alar, one dark intra-alar, and one white postalar setae, one dark anepisternal, one dark katepisternal, and one white anepimeral setae. Basal scutellar seta well developed, dark; apical scutellar seta very short, white, barely longer than adjacent setulae. Legs normal. Wing with vein $R_{4+5}$ bare ventrally, pterostigma slightly more than twice as long as wide, pattern elongate-stellate. Lower calypter very narrow, striplike on outer side. Abdomen densely microtrichose. Male terminalia:

Epandrium and surstylus without peculiarities. Phallus (Figure 24.11D) with two distinct rows of laterally directed spines on basal part of distiphallus. Glans with weakly sclerotized acrophallus and short vesica. Female terminalia: Oviscape partly yellowish, densely covered with white setulae. Aculeus evenly acute.

*Discussion*: The cladistic analysis suggests the sister group relationship of *Freidbergia* with *Capitites*, but only on weak evidence. They share a reversal (mouthparts) and a homoplasy (oviscape color), but no unique synapomorphy is known. Nevertheless, I prefer to describe this new genus because of the following three autapomorphies (which are not included in the cladistic analysis): Unique arrangement of laterally directed spines on the basal part of the distiphallus, strong facial carina which separates the antennal grooves, and the conspicuous groove on the postgena, none of which are known from other Tephritini. Moreover, *Freidbergia* lacks the movable hooks of the glans, which is a synapomorphy of *Capitites*. As shown above, no unambiguous synapomorphy exists between the two genera, therefore, their lumping would make a diagnosis almost impossible.

Etymology: Named in honor of Amnon Freidberg, the outstanding authority of Afrotropical Tephritidae and an excellent friend whose support was substantially responsible for my interest in Tephritidae and who collected most of the specimens of the type species. The gender is feminine.

### 24.7.8.1 *Freidbergia mirabilis*, n. sp.

**Holotype** ♂, 10♂♂, 16♀♀ paratypes: KENYA: 1300 m, 2 km S Chemelil [35.11°E/0.11°S], 11.III.1993, leg. B. Merz. Additional paratypes: ETHIOPIA: 3♂♂, 1♀, Rd. Addis Ababa-Debre Zeyit, 20.XII.1989, A. Freidberg and F. Kaplan; 1♀, Mojo, 75 km SE Addis Ababa, 20.XII.1989, A. Freidberg and F. Kaplan. KENYA: 14♂♂, 8♀♀, Rt. A109, Athi River, 30.IV.1991, A. Freidberg and F. Kaplan, ex flowerhead *Pluchea ovalis* 1.V.1991; 1♂, Lake Bugoria, 26.VIII.1983, A. Freidberg; 1♀, same, but 29.XI.1989, A. Freidberg and F. Kaplan; 2♂♂, 3♀♀, Taru, 80 km NW Mombasa, 13.VIII.1983, A. Freidberg; 1♂, same, but ex flowerhead *P. dioscoridis* August 1983; 1♂, 2♀♀, Hunter's Lodge, 150 km SE Nairobi, 18.VIII.1983, A. Freidberg; 1♂, same, but ex flowerhead *P. ovalis* August 1983; 2♂♂, 8♀♀, 20 km W Mombasa, 15.VIII.1983, A. Freidberg; 2♂♂, 2♀♀, Rt. A1, Turkwe River, 26.XI.1986, A. Freidberg; 2♀♀, Sebit, Morun River, 26.XI.1986, I. Susman; 6♂♂, 4♀♀, N.W., Kainuk, on Morun River, 25.XI.1989, A. Freidberg and F. Kaplan; 4♂♂, 16♀♀, 50 km North Mombasa, 4.XII.1989, A. Freidberg and F. Kaplan; 14♂♂, 10♀♀, Lake Baringo, 25.VIII.1983, A. Freidberg. UGANDA: 1♂, S.W. Ruwenzori Mts., Kilembe, 1900 m, 31.I.1996, I. Yarom and A. Freidberg; 5♂♂, 6♀♀, S.W. Kasese, 1500 m, 3.I.1996, I. Yarom and A. Freidberg. TANZANIA: 1♂, Hedaru, Rt. B1, 16.IX.1992, A. Freidberg; 1♂, Mto. Wa Mbu nr. Lake Manyara, 900 m, 6.IX.1992, A. Freidberg; 1♂, nr. Buiko, Rt. B1, 9.IX.1992, A. Freidberg; 1♀, Usambara Mts., Rt. B124, 1000 m, 9.IX.1992, A. Freidberg; 3♀♀, Moshi, 30 km E, Rt. B1, 17.IX.1992, A. Freidberg. MALAWI: 3♂♂, 3♀♀, S. Lake Malawi, Nkopola, 25.X.1983, A. Freidberg.

The holotype is double-mounted on a minuten pin, in excellent condition and deposited in the MHNG, the paratypes in CBM, MHNG, TAU, and USNM.

*Description*: Pale, yellowish golden species, weakly microtrichose. Head: Yellow, only posterior part of occiput grayish microtrichose; scape with white, pedicel with black setulae. Bristles pale rather than brown. Thorax: Gray microtrichose only between dorsocentral setae, otherwise golden yellow. Wing as in Figure 24.7D: "elongate-star" pattern conspicuous, lower part of wing with faint reticulation. Male: Epandrium and surstylus yellow. Phallus as in Figure 24.11D. Female: Oviscape slightly longer than wide at base, about as long as last two tergites combined. Aculeus tip with small indentation (less than in Figure 24.10F); length: 0.65 mm ($n = 2$). Wing length: male: 2.31 ± 0.11 mm (2.20 to 2.56, $n = 10$); female: 2.32 ± 0.07 mm (2.20 to 2.44, $n = 17$).

*Biology*: Adults were reared in Kenya from flowerheads of *P. dioscoridis* and *ovalis*. The type specimens were swept on *P. dioscoridis* together with *Schistopterum moebiusi* (Becker).

*Etymology*: The name refers to the unique rows of setae on the basal half of the distiphallus.

*Remark*: A series of 2♂♂, 13♀♀ from Southern Madagascar (Berenty Res. and Fort Dauphin, deposited in TAU) differs from the specimens from mainland Africa by the absence of the hyaline spot at the tip of cell $r_1$. Further material is needed to establish the status of this population; therefore, it is not named here.

## 24.7.9 *Goniurellia* Hendel, 1927 (as subgenus of *Trypanea*)

(Figures 24.8J, 24.14A-B)

*Type species*: *Urellia tridens* Hendel, 1910, by designation of I.C.Z.N. 1208 (1982).

*Recognition*: Freidberg (1980), Freidberg and Kugler (1989).

*Discussion*: The members of this genus are very uniform in wing pattern (Figure 24.8J), except for *G. lacerata* (Becker), and in general body characters. The terminalia are rather similar as well (Figure 24.14A–B), but *G. spinifera* Freidberg exhibits very strong modifications in the shape of the surstyli, prensisetae, and glans (see Freidberg 1980, for details). The monophyly of the genus is supported by the structure of the male terminalia, especially the sclerotized tube of the glans, which bears two small toothlike processes (except in *G. spinifera*) and a very long vesica. A similarly long vesica is only present in *Euarestella pninae* and *E. megacephala* within the *Tephritis* group, but this apparently is convergence.

*Biology*: All species breed in flowerheads of Asteraceae, in particular in *Pulicaria* and *Pallenis* spp. (Inuleae).

## 24.7.10 *Hyalotephritis* Freidberg, 1979

(Figures 24.8F, 24.10C-E, 24.13B)

*Type species*: *Trypeta planiscutellata* Becker, 1903, by original designation.

*Tephritites* Freidberg, 1979 (type species: *Terellia planiscutellata* var. *australis* Bezzi, 1924, by original designation), n. syn.

*Recognition*: Freidberg (1979).

*Discussion*: The synonymy of *Hyalotephritis* and *Tephritites* proposed here is based on the following shared characters: frontal stripe absent, wing pattern reduced (absent or weak, Figure 24.8F), surstylus thickened (Figure 24.10C–E). The latter two characters are probably synapomorphies as they are not found elsewhere in the *Goniurellia* clade. The number of frontal setae in *H. planiscutellata* is somewhat variable with many specimens lacking the small whitish anterior one at least on one side. The small spinules on the apicoventral side of the forefemur, which are very characteristic for *H. australis*, are also present, but less conspicuous, in some specimens of *H. planiscutellata*.

*Biology*: All species breed in flowerheads of Inuleae: *H. planiscutellata* and *complanata* in *Pluchea dioscoridis; H. australis* in *Geigeria passerinoides* (Freidberg 1979).

## 24.7.11 *Insizwa* Munro, 1930

(Figures 24.6C-D, 24.8I, 24.10A-B, 24.14C-D)

*Type species*: *Euaresta striatifrons* Munro, 1929, by monotypy.

*Recognition*: Munro (1930). This genus was erected for one species and one subspecies with atypical wing pattern. The present study shows that at least three additional species should be placed here: *I. goliath* (Bezzi) (= *Trypanea haemorrhoa* Bezzi), n. comb., *I. kloofensis* (Munro), n. comb., and *I. dicomala* (Munro), n. comb., which are all transferred from *Capitites,* where they were classified by Hancock (1986). Therefore, the diagnosis given by Munro (1930) should be supplemented: Mouthparts capitate to short spatulate (Figure 24.6D); all postocular setae whitish;

medial postocellar seta present; frontal stripe well developed. Wing either with a reticulate (Figure 24.8I) or with an elongate-stellate pattern (like Figure 24.8J); vein $R_{4+5}$ ventrally bare or setulose. Glans with tubelike, well-sclerotized acrophallus, which exhibits modifications on the inner side of the tube (Figure 24.14C-D); rostrum well developed.

*Discussion*: The unusual modifications inside the acrophallus tube (otherwise only known in some *Euarestella*), the basal ringlike sclerotization of the glans, and the well-sclerotized rostrum indicate that *Insizwa* is a well-supported monophyletic genus. Its relationships to the other genera of the *Goniurellia* clade are little known.

*Biology*: *Insizwa striatifrons* was reared from flowerheads of *Gazania uniflora* (Munro 1930). *Insizwa goliath* and *I. kloofensis* emerged from flowerheads of *Dicoma zehyeri* and *D. speciosa,* and *I. dicomala* from flowerheads of *Dicoma anomala* (Munro 1935). These composites belong to the tribes Arctotideae and Mutisieae, respectively.

### 24.7.12 *Multireticula* Merz, n. gen.

(Figures 24.3A-B, 24.7B, 24.11B)

*Type species*: *Campiglossa perspicillata* Bezzi, 1924, by present designation.

*Diagnosis*: A very distinct genus which is easily recognized by the following set of characters: Head (Figure 24.3A-B): Almost square in profile; frons very wide, bare or with minute setulae anteriorly; frontal stripe present, but faint; labella short spatulate; palpus longer, slightly projecting; two frontal setae, sometimes a third, short, whitish one anteriorly; two orbital setae, one ocellar seta, one postocellar seta, one medial and one lateral vertical seta; postocular setae mixed white and black; medial postocellar seta absent. Scutum covered with white setulae; one dorsocentral seta situated at transverse suture; anepimeral and katepisternal seta white, anepisternal seta dark; apical scutellar seta about half as long as basal scutellar seta; lower calypter very narrow, striplike. Wing: Extremely broad (Figure 24.7B), with regular reticulate pattern, apical fork present; vein $R_{4+5}$ with setulae on ventral side; pterostigma twice as long as wide. Abdomen: With pairs of dark spots on tergites 3 to 6. Male terminalia: Glans (Figure 24.11B) with much reduced, weakly sclerotized acrophallus, and short, wide vesica, out of which some slender, membranous tail-like lobes emerge. Female terminalia: Oviscape black, with fine, dark setulae on dorsum and a few white, thick setulae on ventral side at base; aculeus very acute, pointed.

*Discussion: Multireticula* exhibits a large number of autapomorphic characters: head and wing shape, abdominal spots, structure of the glans. Many other characters are plesiomorphic (chaetotaxy of head, mouthparts, wing pattern extensive) which indicate its position near the base of the phylogeny of the *Tephritis* group, but its exact sister group relationship is not yet known. Superficially, this genus is similar to some *Campiglossa*, but the weak sclerotization of the glans, its wing pattern at the apex, and the host plant give good evidence for its inclusion in the *Tephritis* group.

*Biology*: The only known species produces terminal galls on *Helichrysum cymosum* (Inuleae) (Munro 1935).

*Etymology*: Named after the extensive reticulation of the wing. The gender is feminine.

### 24.7.13 *Pherothrinax* Munro, 1957

(Figures 24.5A-B, 24.12D)

*Type species*: *Pherothrinax redimitis* Munro, 1957, by original designation.

*Recognition*: The diagnosis of Munro (1957a) is very brief and his concept of the genus unclear. This may be the reason why Hancock (1986) included nine additional species in this genus, but merely on the basis of descriptions rather than examined specimens. The study of the types of all species, except *S. bistellata* (Bezzi), indicates that none of these species is congeneric with *P. redimitis*. The genus may be described as follows: Body dark gray; head higher than long in lateral view (Figure 24.5B); frons densely setulose anteriorly; frontal stripe present; ocellar triangle with some conspicuous white setulae in addition to the ocellar seta; three pairs of concolorous frontal setae; all postocular setae white; medial postocellar seta absent. Thorax with three darker stripes on scutum, one in middle, the other two along dorsocentral lines; anepimeral seta of same color as surrounding setulae; lower calypter convex, almost as large as upper calypter; apical scutellar seta slightly more than half as long as basal scutellar seta. Wing pattern reticulate, but elongate-stellate pattern is discernible. Male terminalia: Epandrium with pair of strong setae dorsally; glans (Figure 24.12D) with partly reduced, tubelike, weakly sclerotized acrophallus; no rostrum visible; vesica about six times as long as acrophallus. Female terminalia: Oviscape partly yellowish, covered at base with white setulae; aculeus pointed.

*Discussion*: As here delimited, *Pherothrinax* is monotypic. The cladogram (Figure 24.2) indicates a sister group relationship with the *Goniurellia* clade, based on the similar type of vesica and the reddish oviscape.

*Biology*: Virtually unknown, as all species formerly placed in *Pherothrinax* with known host plants have been transferred to other genera.

### 24.7.14 *Stelladesis* Merz, n. gen.

(Figures 24.6A-B, 24.8G, 24.12E)

*Type species*: *Trypanea woodi* Bezzi, 1924, by present designation.

*Included species*: *Stelladesis arrhiza* (Bezzi 1924), n. comb., *S. bistellata* (Bezzi 1924), n. comb., *S. lamborni* (Munro 1935), n. comb., *S. lutescens* (Bezzi 1924), n. comb., and *S. woodi* (Bezzi 1924), n. comb., all of which are transferred from *Pherothrinax* (Hancock 1986). There are at least five undescribed species known from the Afrotropical Region that also belong to this genus.

*Description*: Ash-gray to dove-gray species of 3.0 to 4.5 mm wing length. Head slightly higher than long in profile (Figure 24.6A-B), yellowish in ground color, but occiput with black pattern. Frons setulose anteriorly; frontal stripe weak, ocellar triangle setulose; mouthparts capitate, palpus short. The following setae present: one short, white anterior and two dark, larger posterior frontal, two orbital, one ocellar, one medial postocellar, one postocellar, one medial and one lateral vertical; all postocular setae whitish. Dorsocentral seta situated close to transverse suture; anepimeral seta of same color as adjacent setulae; lower calypter convex, well developed; apical scutellar seta 0.4 to 0.6 times as long as basal scutellar seta. Legs yellow, as usual in Tephritini. Wing (Figure 24.8G) with normal pterostigma; vein $R_{4+5}$ usually bare on ventral side; pattern of the elongate- or narrow-stellate type, apical fork present or partly reduced. Abdomen covered with dense whitish setulae. Male terminalia: Epandrium elliptical, on upper side with four to eight strong setae; surstylus fused with epandrium similar to Figure 24.10A-B, arched; glans (Figure 24.12E) with slightly reduced, tubelike acrophallus and long, parallel-sided vesica; well-sclerotized rostrum penetrating vesica. Female terminalia: Oviscape at least partly brownish; at base with numerous whitish, thick setulae; aculeus pointed.

*Discussion*: Although the cladistic analysis (Figure 24.2) failed to find a synapomorphy for the ten species of *Stelladesis* that have the same set of characters as *S. woodi*, a new genus is proposed because of their similarities in wing pattern and the structure of the glans, with the partly reduced acrophallus, the long rostrum, and the type of vesica. From external appearance, *Stelladesis* may be confused with *Pherothrinax*, but that genus lacks the medial postocellar seta, usually has three concolorous frontal setae (in *Stelladesis* the anteriormost is white), and differs in details of the male terminalia. Another externally similar genus is *Tephrodesis* (see below), which lacks the medial postocellar seta and differs in particular in the structure of the male terminalia.

*Biology*: As far as is known, all species are associated with *Vernonia* (Asteraceae, Vernonieae). At least two species were reared from flowerheads.

*Distribution*: The genus comprises five described and at least five undescribed species from throughout the Afrotropical Region, with the greatest diversity in eastern and southern Africa.

*Etymology*: Named after the stellate wing pattern of most species of the genus and its morphological similarity with *Dectodesis*. The gender is feminine.

## 24.7.15 *Tephritis* Latreille, 1804

(Figure 24.7A)

*Type species*: *Musca arnicae* Linnaeus, 1758, by designation of Cresson (1914).

*Recognition*: Freidberg and Kugler (1989), Merz (1994a).

*Discussion*: *Tephritis* is a very large genus with about 168 described (Norrbom et al. 1999) and dozens of undescribed species which are distributed mainly in the temperate parts of the Northern Hemisphere and in Australia (Hardy and Drew 1996). Phylogenetic trends within the genus were discussed by Merz (1994b). This genus is probably the sister group of the other taxa of the *Tephritis* group. In particular, the chaetotaxy of the head and the assumed ground plan of the wing pattern are still comparable to members of the *Spathulina* group. The apical fork of the wing and the weak sclerotization of the glans are synapomorphies shared with other taxa of the *Tephritis* group. However, more-detailed study is needed to demonstrate the monophyly of *Tephritis* clearly and to establish its phylogenetic position.

*Biology*: A wide range of Asteraceae (Asteroidea, Cichorioidea) are known as host plants (see Freidberg and Kugler 1989 and Merz 1994a for host plant lists). Most species live in flowerheads, but a few species have been reared from stem or even stem-base galls. Extensive studies on the biology, immature stages, and life cycles are known for some Nearctic (Goeden et al. 1993) and Palearctic species (Berube 1978; Romstöck-Völkl 1997).

## 24.7.16 *Tephritomyia* Hendel, 1927 (as subgenus of *Acanthiophilus*)

(Figures 24.8A and 24.12C)

*Type species*: *Oxyna lauta* Loew, 1869, by original designation.

*Recognition*: Freidberg and Kugler (1989).

*Discussion*: The monophyly of *Tephritomyia* is well supported by a number of autapomorphies (length of apical scutellar seta, irregular wing pattern, very long, tubelike vesica). The cladogram (Figure 24.2) suggests the possible relationship of this genus with *Acanthiophilus* and *Trupanea*, based on the reduced acrophallus. A detailed study of this structure, however, reveals that this may be homoplasy as they differ in details of the sclerotization. Further studies are needed to establish the exact position of the genus.

*Biology*: All species with known biology breed in flowerheads of *Echinops* (Asteraceae, Cardueae).

## 24.7.17 *Tephrodesis* Merz, n. gen.

(Figures 24.4D-E, 24.8B, 24.9E-F, 24.10H-I, 24.11A)

*Type species*: *Trypanea pulchella* Bezzi, 1924, by present designation.

*Included species: Tephrodesis mutila* (Bezzi 1924), n. comb., *T. pulchella* (Bezzi 1924), n. comb. (= *Trypanea subcompleta furcatella* Bezzi 1924, n. syn.), *T. subcompleta* (Bezzi 1920), n. comb., all of which are transferred from *Pherothrinax* (Hancock 1986). One undescribed species also belongs here.

*Description*: Externally similar to *Stelladesis*, with the following differences. Head (Figure 24.4D-E): Frons bare or with minute setulae anteriorly, frontal stripe absent; ocellar triangle bare; number of frontal setae variable, sometimes even within individual specimens; two to three dark and sometimes single small, white anterior seta present; all postocular setae white; medial postocellar seta absent. Lower calypter very narrow, striplike. Anepimeral seta varying in color depending on the illumination, but usually slightly darker than surrounding setulae; wing with stellate pattern, but posterior and proximal parts usually exhibit slight dark reticulation (Figure 24.8B). Male terminalia: Epandrium (Figure 24.9E-F) without conspicuous long setae dorsally; lateral surstylus with large dorsal lobe; glans (Figure 24.11A) with well-developed acrophallus and short vesica; rostrum absent. Female terminalia: Oviscape entirely black, with distinct whitish setulae at base.

*Discussion*: The species of this genus were previously placed in *Pherothrinax* (Hancock 1986), but they retain two plesiomorphic character states (the small vesica and the entirely black oviscape), indicating that they do not belong in that genus. The large dorsal lobe of the lateral surstylus may be an autapomorphy for the genus. The exact relationship of this genus among the other genera of the *Tephritis* group is still unknown.

*Biology*: *Tephrodesis pulchella* and an undescribed species of this genus were found in large numbers on *Tarchonanthus camphoratus* (Asteraceae, Inuleae) (Freidberg, personal communication).

*Distribution*: The genus contains three described and one undescribed species. They are widespread in eastern Africa from Kenya to South Africa.

*Etymology*: The name refers to the morphological similarity of this genus with *Tephritis* and *Dectodesis*. The gender is feminine.

## 24.7.18 *Trupanea* Schrank, 1795

(Figures 24.7E-H, 24.12A)

*Type species*: *Trupanea radiata* Schrank, 1795 (= *Musca stellata* Fuesslin, 1775), by monotypy.

*Urelliosoma* Hendel, 1927 (type species: *Tephritis desertorum* Efflatoun, 1924, by original designation), n. syn.

*Recognition*: Munro (1964), Freidberg and Kugler (1989), Merz (1994a).

*Discussion*: This is a very large genus of worldwide distribution. Further study is needed to show that all included species are congeneric. At least all species studied from the Holarctic and Afrotropical Regions are very similar in external characters, and they all have a unique type of glans (Figure 24.12A) with a hooklike process and otherwise only very weak sclerotization (apomorphic state). The characters used to separate *Urelliosoma* (only Mediterranean species studied, the Eastern Palaearctic species of subgenus *Allocraspeda* Richter excluded) from *Trupanea* do not reflect phylogenetic relationships: The lack of an apical fork in the wing (Figure 24.7E-H) and the absence of microtrichosity on the last tergites is known also from typical members of *Trupanea*, so that the two genera are herewith synonymized. The following additional species are therefore transferred from *Urelliosoma*: *Trupanea guimari* (Becker), n. comb., and *T. pulcherrimum* (Efflatoun), n. comb.

*Biology:* A wide range of Asteraceae (Asteroidea, Cichorioidea) are among the host plants. Most species breed in flowerheads, but some species are known to attack stems, sometimes inducing gall formation. The life cycles and behavior of some Nearctic species have been studied in detail by Goeden and Teerink (1997).

## 24.7.19 *Trupanodesis* Merz, n. gen.

(Figures 24.4F-G, 24.10J-K, 24.11C)

*Type species*: *Trypanea aurea* Bezzi, 1924, by present designation.

*Included species*: *Trupanodesis aurea* (Bezzi 1924), n. comb. (transferred from *Capitites*, Hancock 1986), and an undescribed species from East Africa.

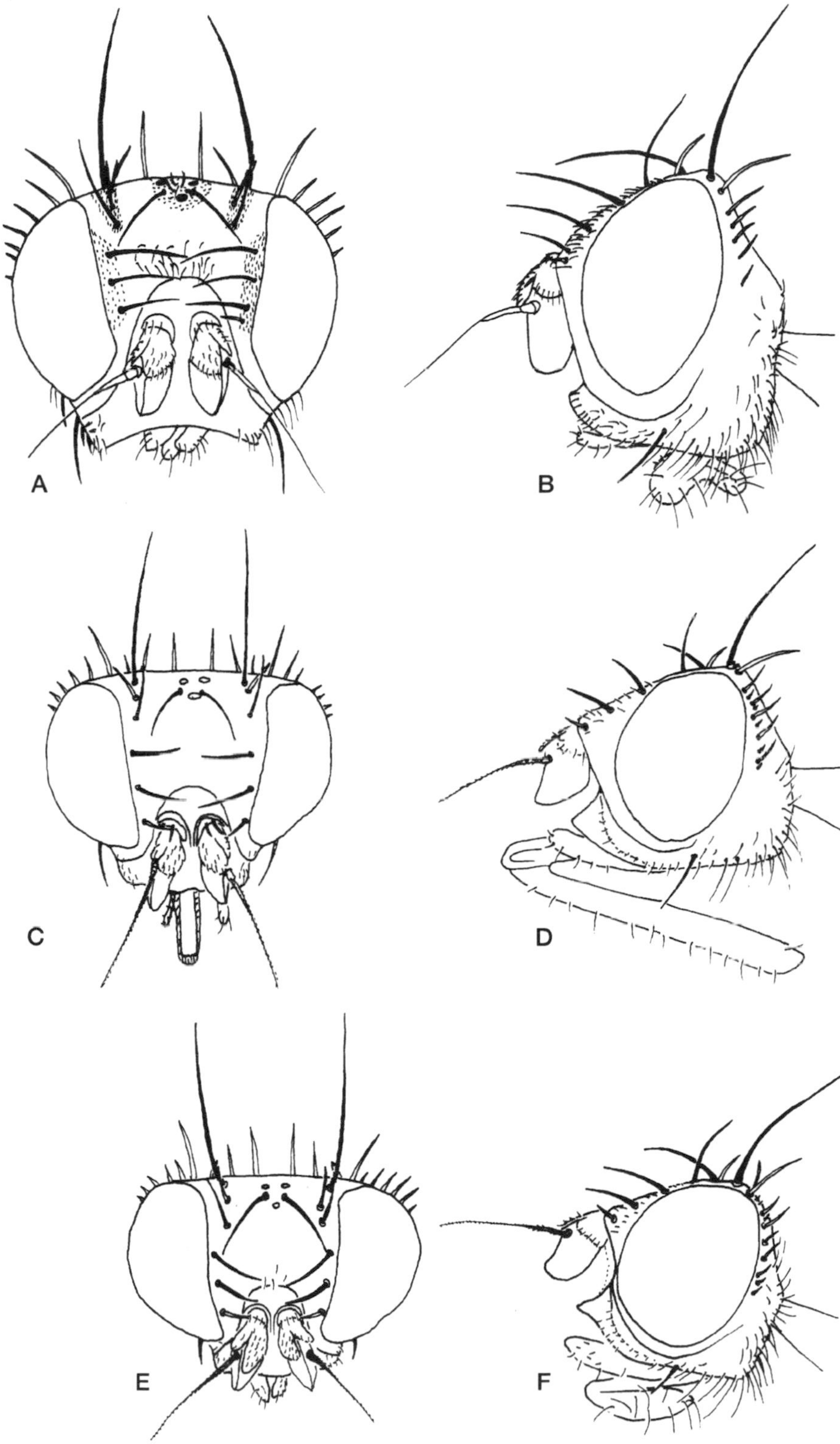


**FIGURE 24.5** Head, frontal and lateral views. (A and B) *Pherothrinax redimitis*; (C and D) *Capitites augur*; and (E and F) *Capitites ramulosa*.

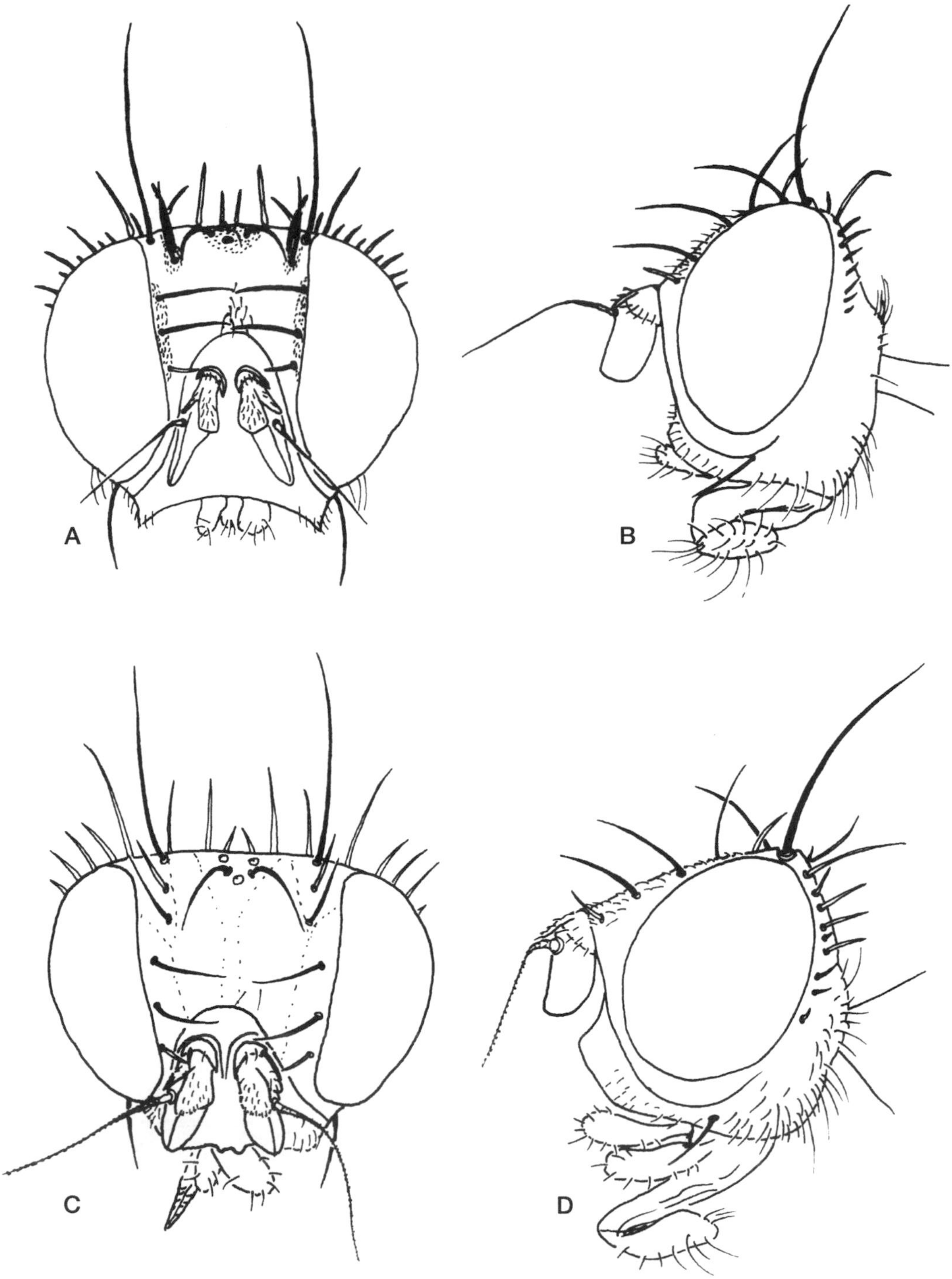


**FIGURE 24.6** Head, frontal and lateral views. (A and B) *Stelladesis* n. sp. nr. *lutescens*; (C and D) *Insizwa striatifrons*.

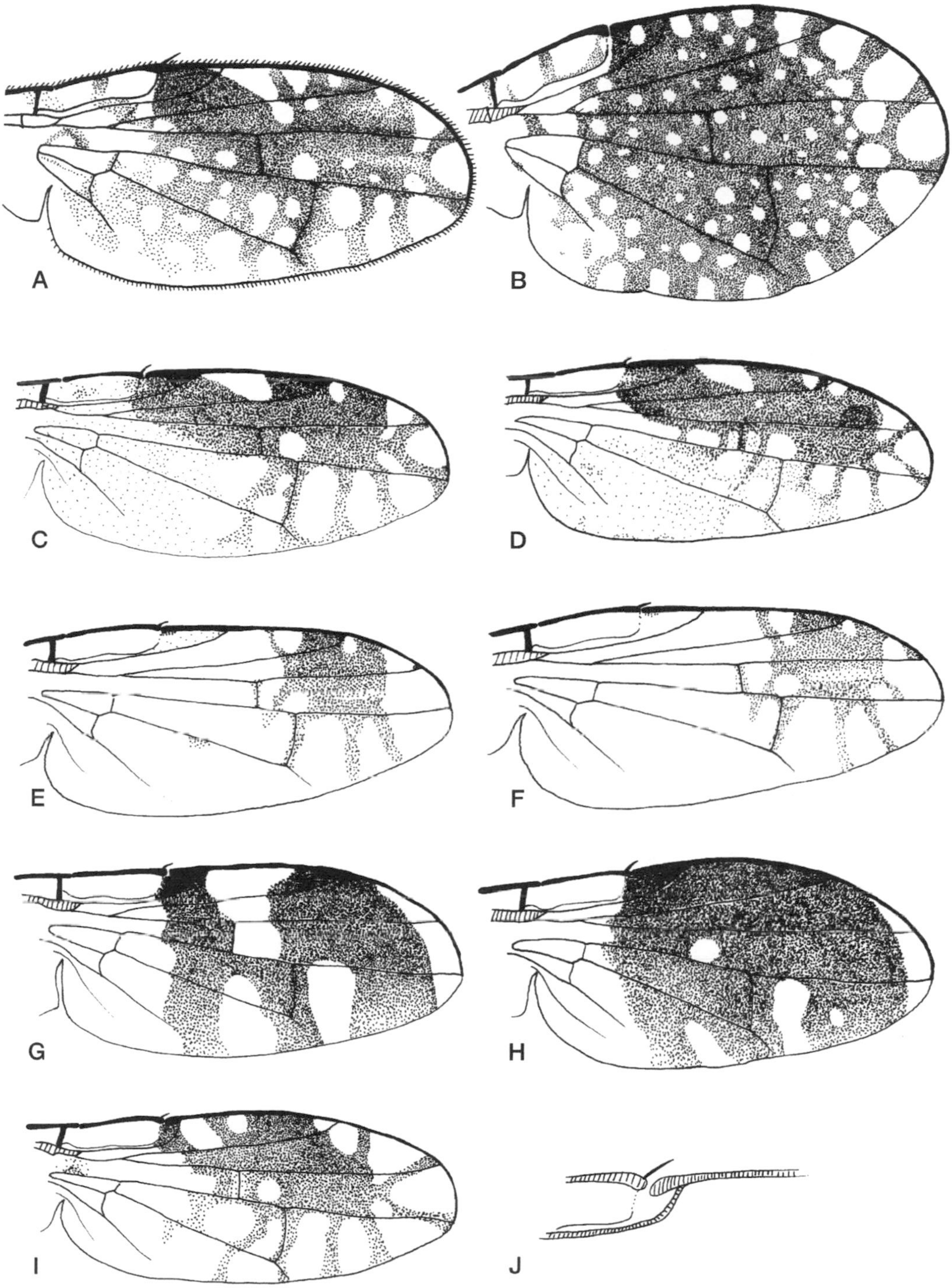


**FIGURE 24.7** Wing. (A) *Tephritis scorzonerae*; (B) *Multireticula perspicillata*; (C) *Dectodesis confluens*; (D) *Freidbergia mirabilis*; (E) *Trupanea stellata* ♂; (F) *T. stellata* ♀; (G) *Trupanea desertorum* ♂; (H) *T. desertorum* ♀; (I) *Brachydesis rivularis*; (J) *B. rivularis*, detail of pterostigma.

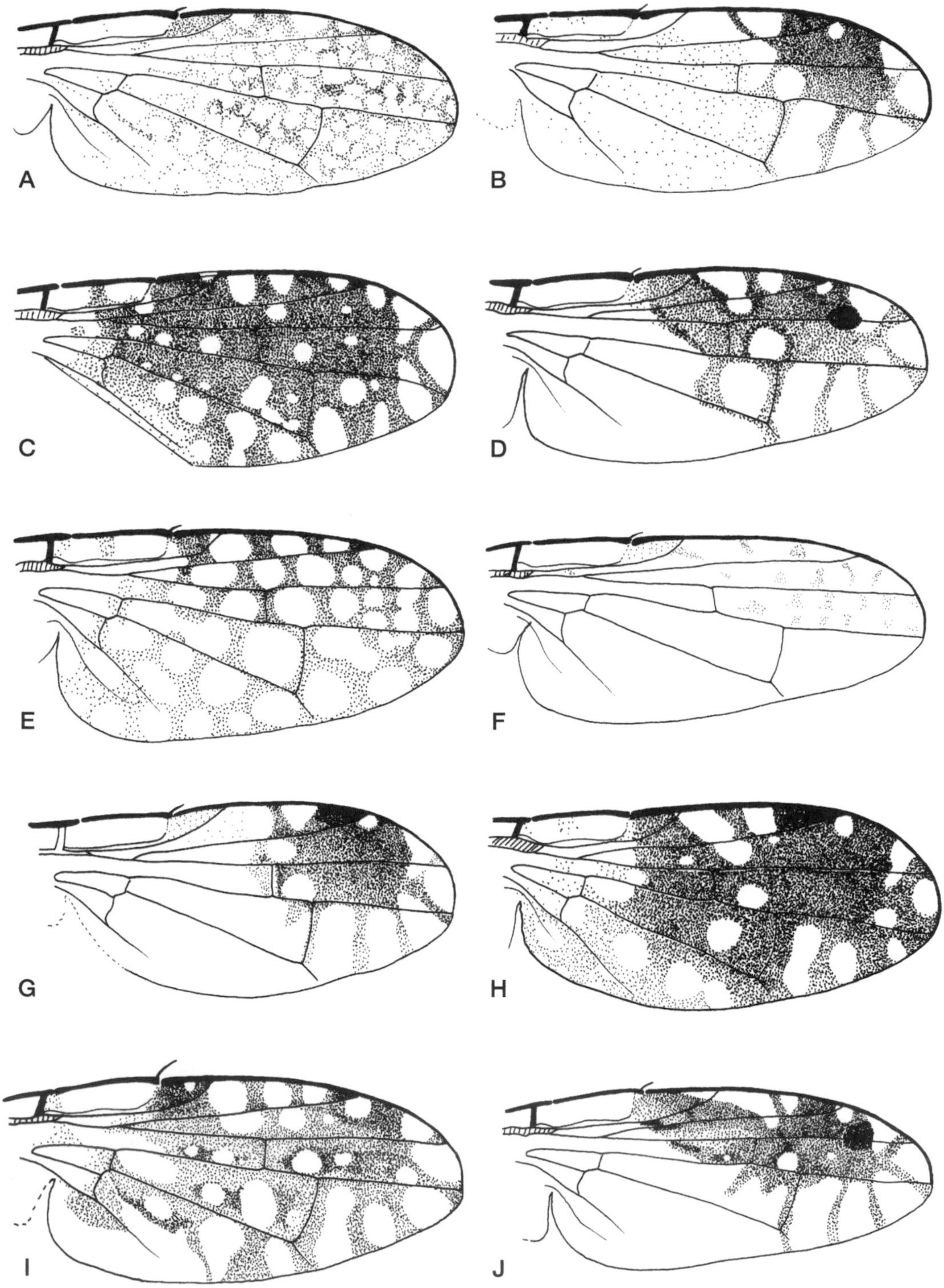


**FIGURE 24.8** Wing. (A) *Tephritomyia lauta*; (B) *Tephrodesis pulchella*; (C) *Euarestella megacephala*; (D) *Euarestella kugleri*; (E) *Euarestella pninae*; (F) *Hyalotephritis australis*; (G) *Stelladesis woodi*; (H) Gen. sp. 1; (I) *Insizwa striatifrons*; (J) *Goniurellia tridens*.

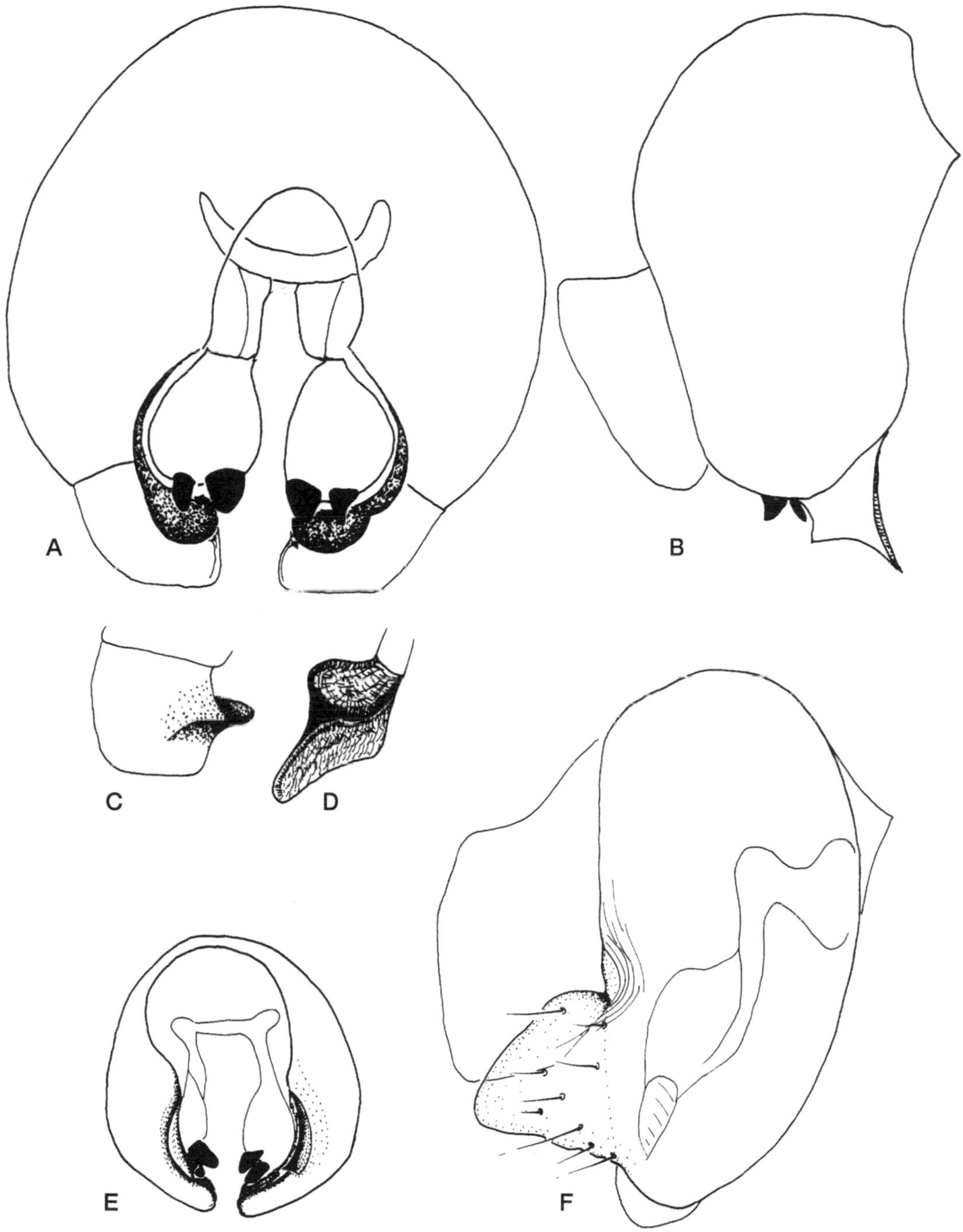


**FIGURE 24.9** Epandrium and surstyli. (A to D) *Acanthiophilus walkeri*; (E to F) *Tephrodesis pulchella*. (A and E) epandrium and surstyli in caudal view; (B and F) epandrium and surstyli in lateral view; (C and D) lateral surstylus in oblique view.

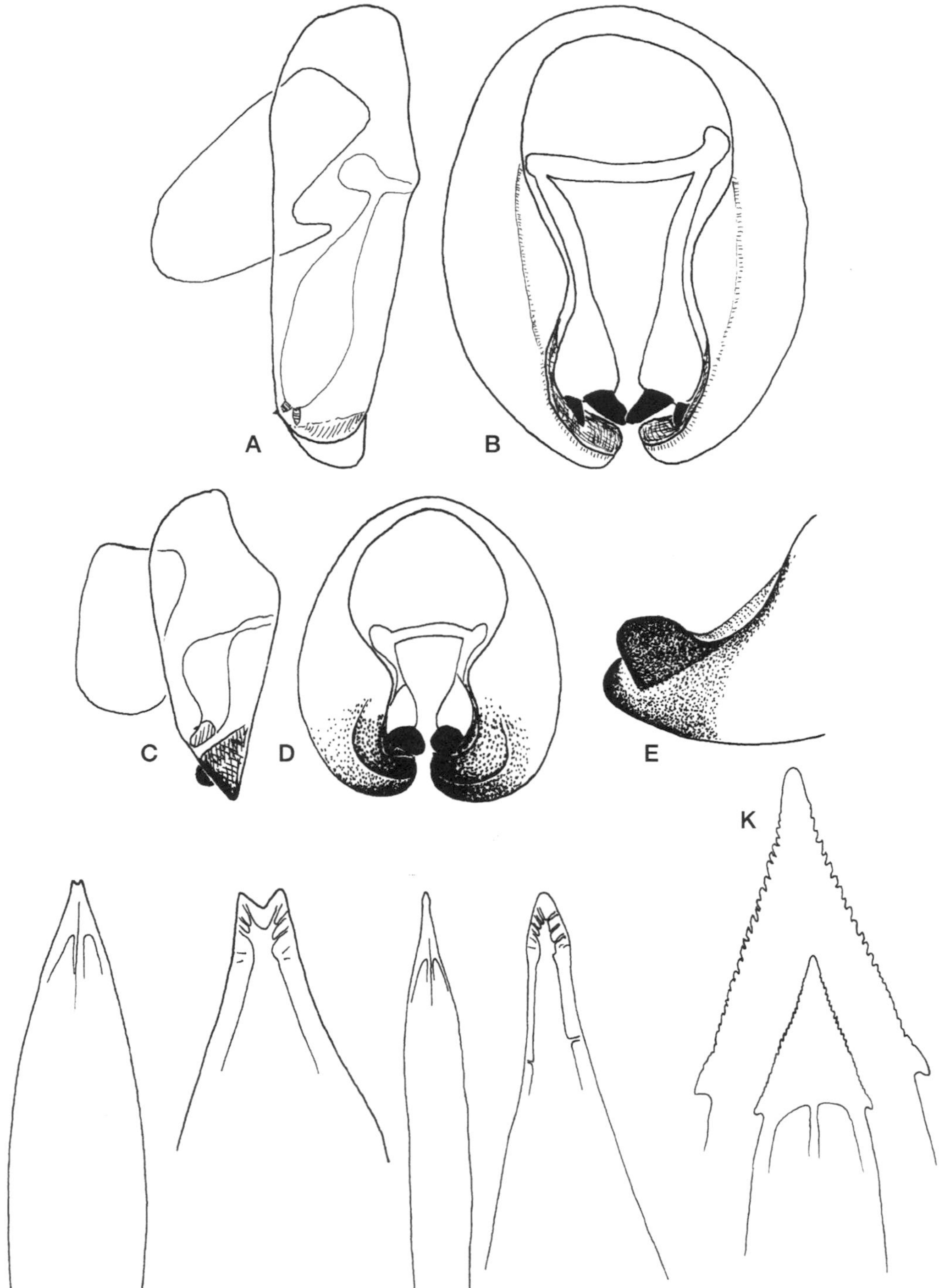


**FIGURE 24.10** Epandrium and surstylus (A to E) and aculeus (F to K). (A and B) *Insizwa striatifrons*; (C and D) *Hyalotephritis planiscutellata*; (E) *H. planiscutellata*, detail of surstylus in oblique view; (F and G) *Capitites augur*; (H and I) *Tephrodesis pulchella*; (J and K) *Trupanodesis aurea*.

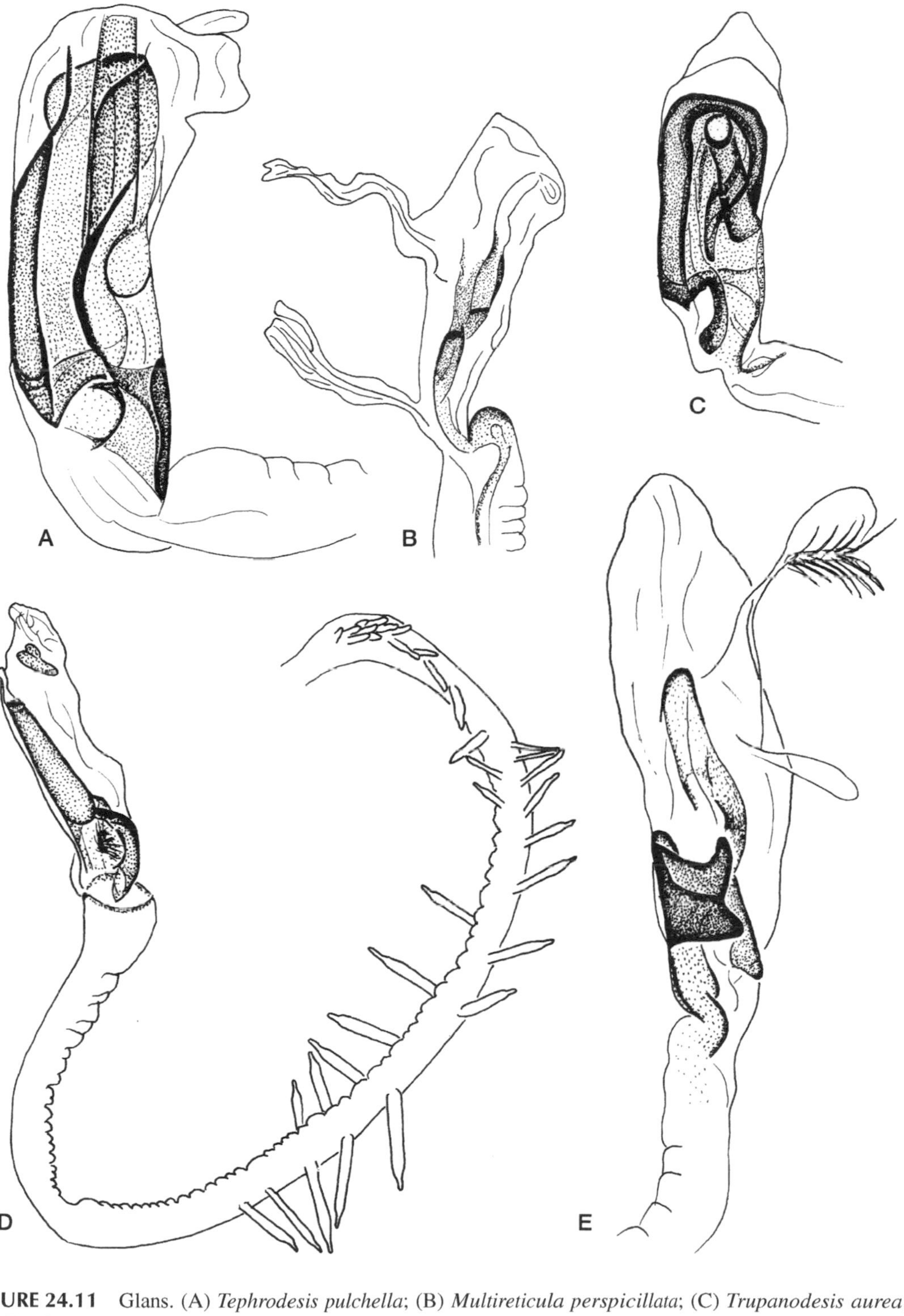


**FIGURE 24.11** Glans. (A) *Tephrodesis pulchella*; (B) *Multireticula perspicillata*; (C) *Trupanodesis aurea*; (D) *Freidbergia mirabilis*; (E) *Dectodesis auguralis*.

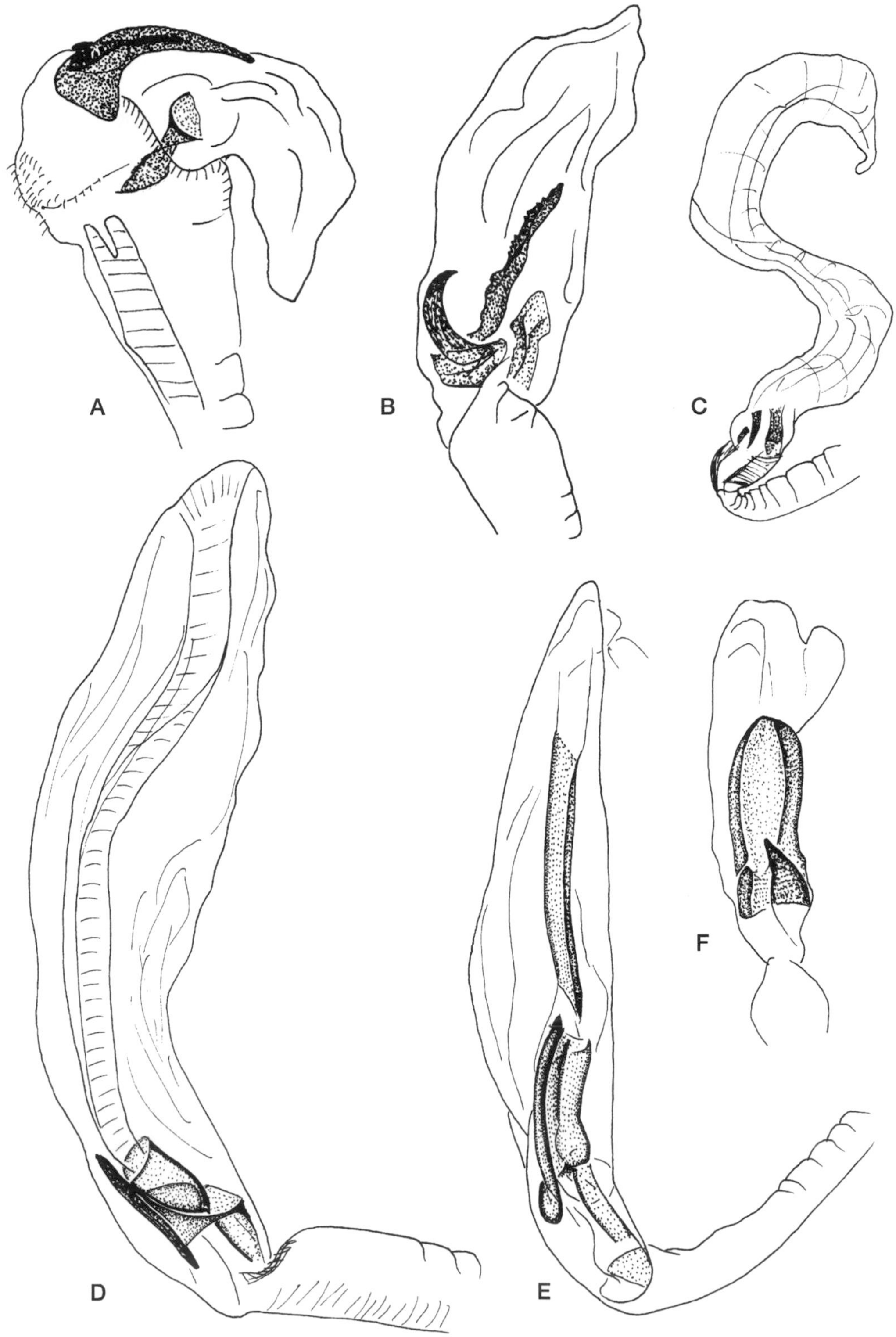


**FIGURE 24.12** Glans. (A) *Trupanea stellata*; (B) *Acanthiophilus walkeri*; (C) *Tephritomyia lauta*; (D) *Pherothrinax redimitis*; (E) *Stelladesis lamborni*; (F) *Brachydesis rivularis*.

*Description*: Head (Figure 24.4F-G) in profile higher than long; frons bare, except frontal plates densely setulose; frontal stripe absent; ocellar triangle with a few white setulae; mouthparts capitate, palpus short; three concolorous frontal setae; remaining chaetotaxy as in *Tephrodesis*. Thorax very densely covered with yellowish setulae; dorsocentral seta situated slightly distad of transverse suture; apical and basal scutellar setae almost equal in length; lower calypter larger than upper, strongly convex. Wing with elongate-stellate pattern (similar to Figure 24.7C); vein $R_{4+5}$ ventrally with some setulae; pterostigma twice as long as wide. Male terminalia: Epandrium fused with surstyli as in Figure 24.10A-B; glans (Figure 24.11C) with very strongly sclerotized acrophallus; vesica at most one third as long as acrophallus, unmodified. Female terminalia: Oviscape partly reddish, densely covered with white setulae; aculeus lanceolate, with serrate, pointed tip (Figure 24.10J-K).

*Discussion*: *Trupanodesis* has an unusual set of characters. The presence of only white postocular setae in combination with the absence of a medial postocellar seta suggests a close relationship with *Pherothrinax*, but the extremely short vesica is a plesiomorphic state. Synapomorphies for the two included species are the very long apical scutellar setae, the unusually dense setulae on the scutum, the strong sclerotization of the acrophallus, and the serrate tip of the aculeus.

*Biology*: The two known species (one yet undescribed) were swept and reared from *Vernonia* (Asteraceae, Vernonieae), where they live in the flowerheads.

*Etymology*: The name refers to the morphological similarity of this genus with *Trupanea* and *Dectodesis*. The gender is feminine.

## 24.8 KEY TO THE GENERA OF THE *TEPHRITIS* GROUP

1. Wing with pterostigma very short, about as long as wide (Figure 24.7J). Only basal scutellar seta present. Lower calypter very narrow, striplike . . . . . . . . . . 2
- Wing with pterostigma normal, at least twice as long as wide. 1-2 scutellar setae present. Lower calypter variable . . . . . . . . . . 4
2. Only one orbital seta (Figure 24.3C-D). . . . . . . . . . *Actinoptera*
- Two orbital setae, the posterior shorter (Figure 24.3E-G). . . . . . . . . . 3
3. Three frontal setae present, the anteriormost white; mouthparts strongly geniculate (Figure 24.3F-G). . . . . . . . . . *Brachydesis*
- Two concolorous frontal setae; mouthparts capitate or very slightly spatulate (Figure 24.3E) . . . . . . . . . . *Brachytrupanea*
4. Medial postocellar seta present (Figure 24.6A and C). All postocular setae white . . . . . . . . . . 5
- Medial postocellar seta absent (Figure 24.5A, C, and E). Postocular setae variable . . . . . . . . . . 9
5. Only basal scutellar seta present. Male: Glans without rostrum (Figure 24.14A) . . . . . . . . . . 6
- Two scutellar setae present. Male: Glans with well-developed rostrum (Figure 24.14C). . . . . . . . . . 7
6. Dorsocentral seta well behind transverse suture. Cell dm usually crossed by at least one complete broad dark ray (Figure 24.8C-E) . . . . . . . . . . *Euarestella*
- Dorsocentral seta situated almost on line of transverse suture. Cell dm at most with a narrow, transverse dark ray (Figure 24.8J) . . . . . . . . . . *Goniurellia*
7. Wing pattern strongly reduced (Figure 24.8F). Frontal stripe absent. Male: Surstylus thickened (Figure 24.10C-E) . . . . . . . . . . *Hyalotephritis*
- Wing pattern more extensive, with at least one entire hyaline spot (Figure 24.8G–I). Frontal stripe present, though sometimes weak. Male: Surstylus normal (Figure 24.10A-B) . . . . . . . . . . 8
8. Ocellar triangle without setulae (Figure 24.6C). Wing with a reticulate pattern (Figure 24.8I) or with elongate-stellate pattern and a bulla (as in Figure 24.8J). . . . . . *Insizwa*
- Ocellar triangle with some whitish setulae (Figure 24.6A). Wing with narrow-stellate pattern (Figure 24.8G). . . . . . . . . . *Stelladesis*

9. All postocular setae whitish (Figures 24.4D-G, 24.5, 24.6) .......................... 10
- At least some of the short postocular setae dark (Figures 24.3, 24.4A-C)............... 18

10. Only basal scutellar seta present. Male: Glans with hook (Figure 24.12A)........ *Trupanea*
- Two scutellar setae present. Male: Glans variable.................................. 11

11. Only two dark concolorous frontal setae present ..................................... 12
- Three frontal setae, the anteriormost seta sometimes white .......................... 13

12. Lower calypter with convex outer border (except for *T. praecox* and *T. luteipes*). Male: Glans with weakly sclerotized acrophallus and long vesica......... *Tephritis* (in part)
- Lower calypter very narrow, striplike, with straight outer border. Male: Glans with strongly sclerotized acrophallus and short vesica (Figure 24.11A) ..... *Tephrodesis* (in part)

13. Three frontal setae present, the anteriormost white. Mouthparts short spatulate (Figure 24.5F) or geniculate (Figure 24.5D). Wing with elongate-stellate pattern (as in Figure 24.8J). Male: Glans: with one or more complex, movable hooks and with voluminous vesica (Figure 24.13A, C-E)................................... *Capitites*
- Three concolorous dark frontal setae. Mouthparts capitate. Male: Glans at most with one simple, movable hook ...................................................... 14

14. Apical scutellar seta at least 0.8 times as long as basal scutellar seta. Female: Oviscape covered with strong whitish setulae at base ......................................... 15
- Apical scutellar seta at most half as long as basal scutellar seta. Female: Setulae on oviscape variable.................................................................. 16

15. Wing pattern elongate-stellate (as in Figure 24.7C). Frons bare, ocellar triangle with a few short setulae or bare. Male: Acrophallus strongly sclerotized, vesica short (Figure 24.11C). Female: Oviscape reddish at base; aculeus tip serrate (Figure 24.10J-K) ............................................... *Trupanodesis*
- Wing with faint reticulation (Figure 24.8A). Frons setulose, ocellar triangle always bare. Male: Acrophallus strongly reduced, vesica very long, narrow (Figure 24.12C). Female: Oviscape black; aculeus not serrate, evenly pointed................ *Tephritomyia*

16. Frons and ocellar triangle strongly setulose. Male: Glans with weak acrophallus and long, parallel-sided vesica (Figure 24.12D). Female: Oviscape reddish at base ........................................................ *Pherothrinax*
- Frons and ocellar triangle bare, rarely the former with a few setulae above lunule. Male: Shape of glans variable. Female: Oviscape uniformly black ................... 17

17. Lower calypter much larger than upper, convex on outer side. Male: Glans with movable hook and spiny bar (Figure 24.12B). Lateral surstylus very thick, separated from epandrium by a seam, without dorsal lobe (Figure 24.9A-D) .......... *Acanthiophilus*
- Lower calypter very narrow, striplike. Male: Glans with cylindrical acrophallus and very short vesica (Figure 24.11A). Lateral surstylus evenly narrowed apically, fused with epandrium, dorsal lobe large, conspicuous (Figure 24.9E-F)...... *Tephrodesis* (in part)

18. Wing very broad, strongly reticulate (Figure 24.7B). Abdomen with paired spots on tergites 3 to 5 ................................................ *Multireticula*
- Wing narrower (as in Figure 24.7A, C-D). Abdomen unspotted ...................... 19

19. Two concolorous frontal setae. Mouthparts capitate.................... *Tephritis* (in part)
- Three frontal setae, the anteriormost white. Mouthparts geniculate.................... 20

20. Face flat to smoothly concave (Figure 24.4B). Apical scutellar seta short or absent. Male: Glans with "tail" on vesica; distiphallus without spines (Figure 24.11E). Female: Oviscape uniformly dark ............................................ *Dectodesis*
- Face with distinct carina (Figure 24.4C). Apical scutellar seta short. Male: Glans without appendages but basal part of distiphallus with two rows of long spines (Figure 24.11D). Female: Oviscape reddish ............................ *Freidbergia*

## 24.9 BIOLOGY AND BIOGEOGRAPHY

Most species of the *Tephritis* group with known biology breed in plants of the tribes Cardueae, Inuleae, or Vernonieae of the family Asteraceae. At least ten genera attack the Inuleae, which is the dominant host plant tribe. Only some species of the large genera *Tephritis* and *Trupanea*, as well as all species of *Insizwa* breed in plants of other tribes (Table 24.2). However, the host plant(s) of three genera are still unknown. As far as known, the larvae of most species breed in the flowerheads without external damage. The larvae of these species usually feed on the seeds or the receptacle. The induction of galls has been recorded for some species of *Actinoptera, Euarestella, Multireticula, Tephritis,* and *Trupanea* (Freidberg 1984, but note that the species of *Acanthiophilus* in his Table 1 were transferred to *Afreutreta* Bezzi by Freidberg and Kaplan 1993). These species usually induce stem galls (except for some *Tephritis* that induce galls in flowerheads). These genera do not form a monophyletic unit (Figure 24.2); therefore, the induction of galls apparently evolved independently several times, unlike, for instance, in the Oedaspidina where this mode of life is shared by all species (Freidberg and Kaplan 1992). This hypothesis is corroborated by the observation that the type of gall produced varies among these genera. For example, the flowerhead galls of *Tephritis* are inconspicuous, simple and soft (Berube 1978), but some *Actinoptera* produce rather hard, well-differentiated stem galls (Freidberg 1984; Merz, personal observation).

The distributions of the genera of the *Tephritis* group are summarized in Table 24.2. It can be seen that the highest diversity of genera lies in the highlands of East Africa (Ethiopia, Kenya, Uganda), with only *Brachydesis* not recorded from this area, and *Euarestella* and *Tephritis* reaching this region only marginally. The steppe vegetation of the East African highlands is particularly rich in Inuleae and Vernonieae, which are the dominant host plant tribes. The number of genera (and in most cases also of species) of the *Tephritis* group declines toward the north and south (Table 24.2). Thus far, 13 genera have been recorded from southern Africa, but only 9 from the Mediterranean, 5 from the Oriental Region, 4 from the northern Palearctic Region, and 2 each from the New World and the Australasian Region (although unstudied genera from these regions may belong to the group). This pattern leads to the question of the origin of the *Tephritis* group: according to the cladistic analysis, *Tephritis* is the sister group of the other genera. The main center of diversity of *Tephritis* lies in the temperate areas of the Nearctic and Palearctic Regions with a marked impoverishment toward the tropics (Norrbom et al. 1999). It may be speculated that some ancient *Tephritis* invaded the East African region from the Palearctic Region through the Nile Valley or along the Indian Ocean where they found a wide range of suitable host plants that offered some niches not occupied by the resident fauna, followed by a strong radiation. On the other hand, the presumed sister group, the *Spathulina, Campiglossa,* and *Sphenella* groups of genera, have a major center of diversity in high altitudes of eastern and southern Africa (Munro 1938; 1957a, b; Freidberg 1987). Consequently, the *Tephritis* group could have originated in that region, and some genera, like *Tephritis, Trupanea,* or *Actinoptera*, penetrated into other biogeographic realms at a later time.

## 24.10 CONCLUSIONS AND FUTURE RESEARCH NEEDS

The present work deals with the species which had been assigned to *Trypanea* by Bezzi (1924a, b) and later to the *Trupanea–Tephritis* complex by Munro (1957a), supplemented by additional, subsequently described species apparently belonging to this group (Freidberg 1979; Hancock 1986). The main goal was the study of the phylogenetic relationships and the clarification of the limits of the genera described in this group of taxa. According to the cladistic analysis (Figures 24.1 and 24.2) they form the monophyletic *Tephritis* group, which is based on three autapomorphies. However, only the structure of the glans gives good support for its monophyly, whereas the other two characters (wing pattern, mouthparts) exhibit wide morphological variation and homoplasies are probably common. Within the *Tephritis* group, 19 described and one undescribed genera are

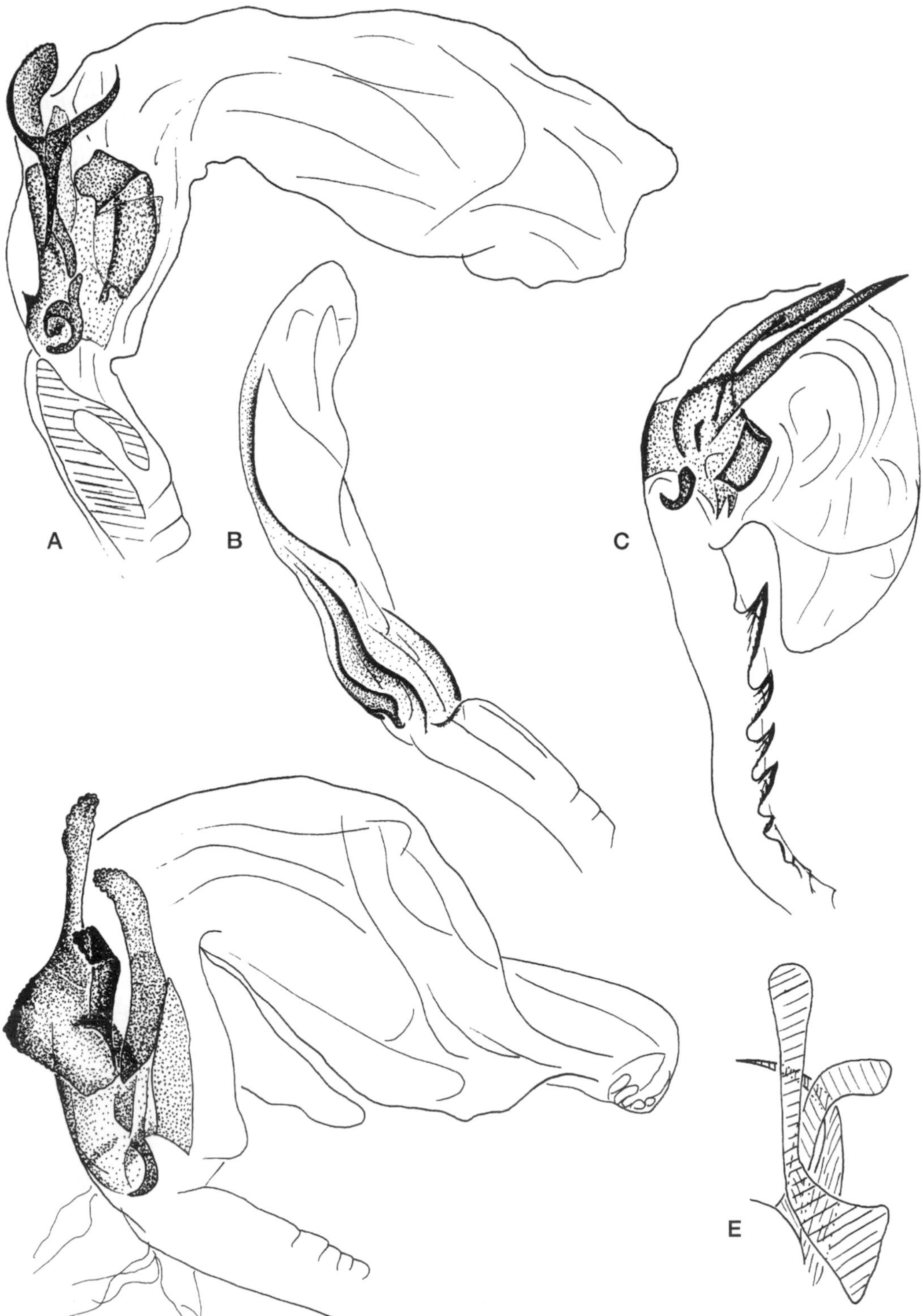


**FIGURE 24.13** Glans. (A) *Capitites ramulosa*; (B) *Hyalotephritis australis*; (C) *C. augur*; (D) *C. dentiens*; (E) *C. dentiens*, showing movable hooks from other direction.

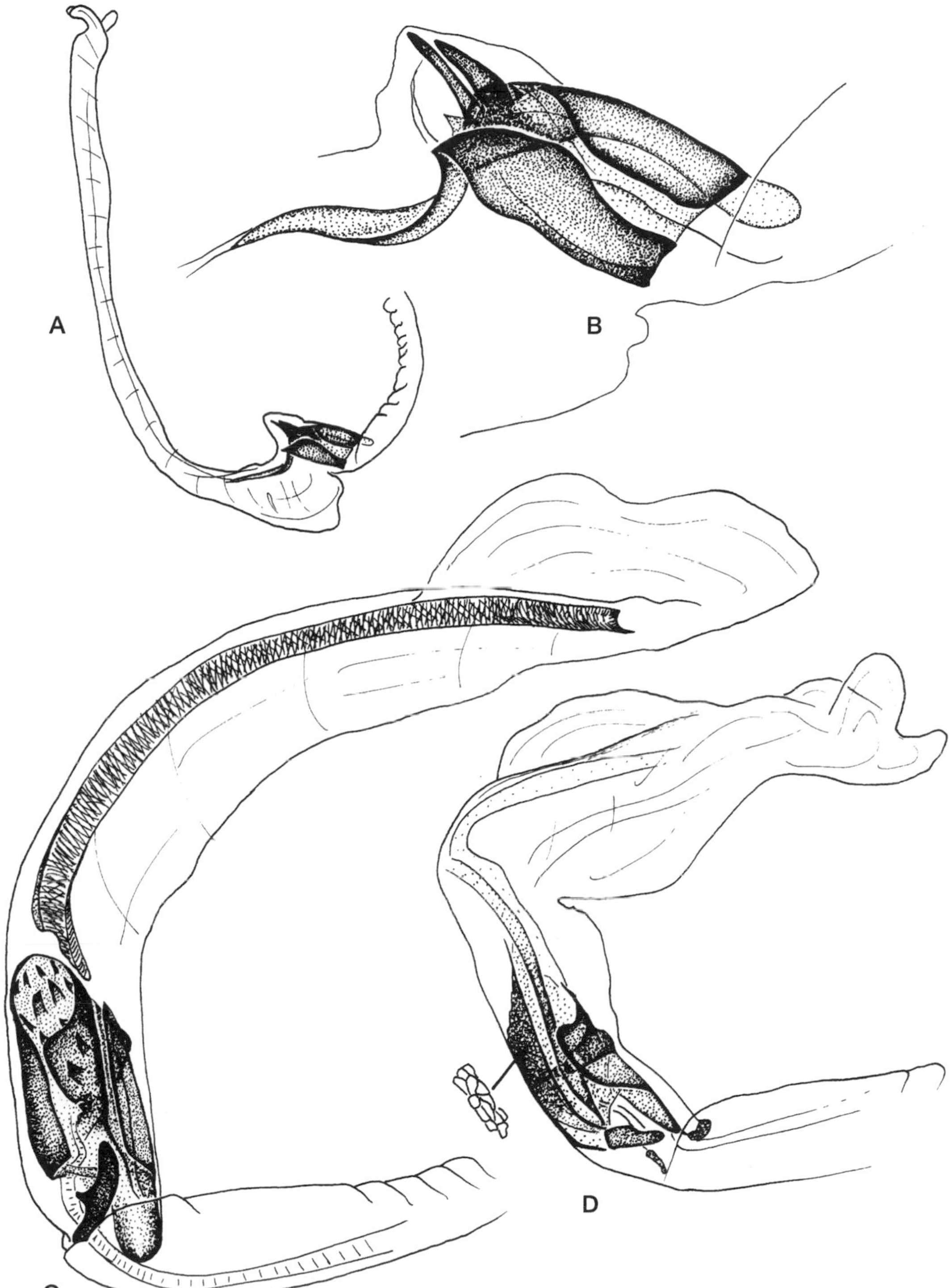


**FIGURE 24.14** Glans. (A) *Goniurellia tridens*; (B) *G. tridens*, stronger magnification of acrophallus; (C) *Insizwa goliath*; (D) *I. striatifrons*.

**TABLE 24.2**
**General Distributions and Host Plant Ranges of the Genera of the *Tephritis* Group**

| Genus | Central Europe | Mediterranean | Eastern Africa | Southern Africa | Other Regions | Host Plant Tribes |
|---|---|---|---|---|---|---|
| *Acanthiophilus* | (x) | x | x | | OR | Cardueae |
| *Actinoptera* | (x) | x | x | x | OR | Inuleae |
| *Brachydesis* | | | | x | | Unknown |
| *Brachytrupanea* | | | x | x | | Unknown |
| *Capitites* | | x | x | x | | Inuleae |
| *Dectodesis* | | | x | x | | Inuleae |
| *Euarestella* | | x | (x) | | | Inuleae |
| *Freidbergia* | | | x | | | Inuleae |
| *Goniurellia* | | x | x | x | OR | Inuleae |
| *Hyalotephritis* | | x | x | x | | Inuleae |
| *Insizwa* | | | x | x | | Arctotideae, Mutisicae |
| *Multireticula* | | | x | x | | Inuleae |
| *Pherothrinax* | | | x | | | Unknown |
| *Stelladesis* | | | x | x | | Vernonieae |
| *Tephritis* | x | x | (x) | | NE, NT, OR | Cardueae and other tribes |
| *Tephritomyia* | | x | x | | | Cardueae |
| *Tephrodesis* | | | x | x | | Inuleae |
| *Trupanea* | (x) | x | x | x | AU, NE, NT, OR | Inuleae and other tribes |
| *Trupanodesis* | | | x | x | | Vernonieae |
| Total | 1 (+3) | 9 | 16 (+2) | 13 | | |

AU = Australasian and Oceanian Region; NE = Nearctic Region; NT = Neotropical Region; OR = Oriental Region.

recognized. Some of them, however, are supported only by weak phylogenetic evidence. Therefore, further studies, and in particular more-detailed studies of the terminalia of males and females, are needed to understand their phylogenetic relationships fully. In addition, the present study is based on Afrotropical and Palearctic species only. It cannot be ruled out that the inclusion of species from other biogeographic regions will change the present classification.

Little is known about the biology of the species of the *Tephritis* group. For most species, at most only the host plant(s) and the part of the plant infested are known. Because they are of little economic importance, almost nothing is known about adult behavior (feeding, courtship, oviposition), larval feeding, and development except for some Palearctic (Berube 1978; Romstöck-Völkl 1997) and Nearctic (Goeden et al. 1993) species of *Tephritis*. As shown in other chapters of this volume, biological characters may change morphology-based phylogenetic models. As an example, Rotheray and Gilbert (1989) were able to resolve, by the study of larval, pupal, and feeding behavior characters, the phylogenetic position of a number of genera of Syrphidae, whose adult morphological data did not allow a well-founded phylogenetic placement.

Up to now, we can only speculate about the origin of the *Tephritis* group, or even the whole Tephritini. Much more work, including morphological and biogeographical data, is needed to understand better the phylogeny of the group.

## ACKNOWLEDGMENTS

It is my deepest wish to thank Allen L. Norrbom for his never-ending help in improving this manuscript, and him and Martín Aluja for the organization of the fruit fly symposium at Xalapa.

My sincere thanks go also to A. Freidberg (Tel Aviv), M. Hauser (Stuttgart), V. A. Korneyev (Kiev), M. W. Mansell (Pretoria), L. Matile (Paris), V. Michelsen (Copenhagen), and N. P. Wyatt and I. M. White (London) for the loan and/or donation of specimens. I wish to thank G. Bächli (Zürich) and G. Cuccodoro (Geneva) for their help in introducing me to the computer program PAUP. A. Freidberg and an anonymous referee are gratefully acknowledged for their very useful comments on an earlier version of the manuscript. The following institutions have contributed to this study through their financial support of the symposium: Campaña Nacional contra las Moscas de la Fruta (Mexico), International Organization for Biological Control of Noxious Animals and Plants (IOBC), and Insituto de Ecologia, A.C. (Mexico). I thank them all.

## REFERENCES

Berube, D.E. 1978. Larval descriptions and biology of *Tephritis dilacerata* (Dip.: Tephritidae), a candidate for the biocontrol of *Sonchus arvensis* in Canada. *Entomophaga* 23: 69–82.

Bezzi, M. 1924a. South African Trypaneid Diptera in the collection of the South African Museum. *Ann. S. Afr. Mus.* 19: 449–577 and 4 plates.

Bezzi, M. 1924b. Further notes on the Ethiopian fruit-flies, with keys to all the known genera and species (cont.). *Bull. Entomol. Res.* 15: 121–155.

Coquillett, D.W. 1910. The type-species of the North American genera of Diptera. *Proc. U.S. Natl. Mus.* 37: 499–647.

Cresson, E.T. 1914. Some nomenclatorial notes on the dipterous family Trypetidae. *Entomol. News* 25: 275–279.

Foote, R.H. and A. Freidberg. 1980. The taxonomy and nomenclature of some Palaearctic Tephritidae (Diptera). *J. Wash. Acad. Sci.* 70: 29–34.

Foote, R.H., F.L. Blanc, and A.L. Norrbom. 1993. *Handbook of the Fruit Flies (Diptera: Tephritidae) of America North of Mexico*. Comstock Publishing Associates, Ithaca. 571 pp.

Freidberg, A. 1979. The Afrotropical species assigned to *Terellia* R.D. (Diptera: Tephritidae). *J. Wash. Acad. Sci.* 69: 164–174.

Freidberg, A. 1980. A revision of the genus *Goniurellia* Hendel (Diptera: Tephritidae). *J. Entomol. Soc. South. Afr.* 43: 257–274.

Freidberg, A. 1984. Gall Tephritidae (Diptera). In *The Biology of Gall Insects* (T.N. Ananthakrishnan, ed.), pp. 129–167. Oxford and IBH Publishing Co., New Delhi. 362 pp.

Freidberg, A. 1987. *Orthocanthoides aristae*, a remarkable new genus and species of Tephritidae (Diptera) from Mount Kenya. *Ann. Natal Mus.* 28: 551–559.

Freidberg, A. and F. Kaplan. 1992. Revision of the Oedaspidini of the Afrotropical Region (Diptera: Tephritidae: Tephritinae). *Ann. Natal Mus.* 33: 51–94.

Freidberg, A. and F. Kaplan 1993. A study of *Afreutreta* Bezzi and related genera (Diptera: Tephritidae: Tephritinae). *Afr. Entomol.* 1: 207–228.

Freidberg, A. and J. Kugler. 1989. *Fauna Palaestina, Insecta IV, Diptera: Tephritidae*. The Israel Academy of Sciences and Humanities, Jesusalem. 212 pp. and 8 plates.

Goeden, R.D. and J.A. Teerink. 1997. Life history and description of immature stages of *Trupanea signata* Foote (Diptera: Tephritidae) on *Gnaphalium luteo-album* L. in southern California. *Proc. Entomol. Soc. Wash.* 99: 748–755.

Goeden, R.D., D.H. Headrick, and J.A. Teerink. 1993. Life history and descriptions of immature stages of *Tephritis arizonaensis* Quisenberry (Diptera: Tephritidae) on *Baccharis sarothroides* Gray in Southern California. *Proc. Entomol. Soc. Wash.* 95: 210–222.

Hancock, D.L. 1986. New genera and species of African Tephritinae (Diptera: Tephritidae), with comments on some currently unplaced or misplaced taxa and on classification. *Trans. Zimb. Sci. Assoc.* 63: 16–34.

Hardy, D.E. and R.A.I. Drew. 1996. Revision of the Australian Tephritini (Diptera: Tephritidae). *Invertebr. Taxon.* 10: 213–405.

Hendel, F. 1927. Trypetidae. In *Die Fliegen der Palaearktischen Region* (E. Lindner, ed.), 5: 1–221 + 17 plates.

I.C.Z.N. 1982. Opinion 1208. *Goniurellia* Hendel, 1927 (Insecta, Diptera): designation of type species. *Bull. Zool. Nom.* 39: 109–110.

Korneyev, V.A. 1989. A review of *Sphenella* and *Paroxyna* series of genera (Diptera, Tephritidae, Tephritinae) of Eastern Palaearctic. *Nasek. Mongolii* 11: 395–470 [in Russian].

Merz, B. 1994a. Diptera. Tephritidae. *Insecta Helv., Fauna* 10: 1–198.

Merz, B. 1994b. A revision of the western Palaearctic species of *Tephritis* (Diptera, Tephritidae). *Third International Congress of Dipterology, Abstr. Vol.*: 151–152.

Munro, H.K. 1930. New Trypetidae from South Africa. *Bull. Entomol. Res.* 20: 391–401.

Munro, H.K. 1935. Biological and systematic notes and records of South African Trypetidae (fruit-flies, Diptera) with descriptions of new species. *Entomol. Mem., S. Afr. Dep. Agric.* 9: 18–59.

Munro, H.K. 1938. A revision of the African species of the genus *Spathulina* Rond. (Diptera, Trypetidae). *Trans. R. Entomol. Soc. Lond.* 87: 417–429.

Munro, H.K. 1957a. Trypetidae. *Ruwenzori Expedition 1934–35*, Vol. 2 (9): 853–1054.

Munro, H.K. 1957b. *Sphenella* and some allied genera (Trypetidae, Diptera). *J. Entomol. Soc. South. Afr.* 20: 14–57.

Munro, H.K. 1964. The genus *Trupanea* in Africa. An analytical study in bio-taxonomy. *Entomol. Mem. S. Afr. Dep. Agric. Tech. Serv.* 8: 1–101.

Norrbom, A.L. 1993. New species and phylogenetic analysis of *Euaresta* Loew (Diptera: Tephritidae) with a key to the species from the Americas south of Mexico. *Proc. Entomol. Soc. Wash.* 95: 195–209.

Norrbom, A.L., L.E. Carroll, and A. Freidberg. 1999. Status of knowledge. In *Fruit Fly Expert Identification System and Systematic Information Database* (F.C. Thompson, ed.), pp. 9–47. *Myia* (1998) 9: 524 pp.

Romstöck-Völkl, M. 1997. Host race formation in *Tephritis conura*: determinants from three trophic levels. In *Vertical Food Web Interactions: Evolutionary Patterns and Driving Forces* (K. Dettner, G. Bauer, and W. Voelkl, eds.), pp. 21–38. Ecological Studies 130. Springer-Verlag, Berlin. 390 pp.

Rotheray, G. and F. Gilbert. 1989. The phylogeny and systematics of European predacious Syrphidae (Diptera) based on larval and puparial stages. *Zool. J. Linn. Soc. Lond.* 95: 29–70.

Swofford, D.L. 1991. *PAUP: Phylogenetic Analysis Using Parsimony, Version 3.0.* Illinois Natural History Survey, Champaign.

## APPENDIX 24.1: DEPOSITORIES OF SPECIMENS STUDIED

CBM: private collection B. Merz, Genève, Switzerland
ETHZ: Eidgenössische Technische Hochschule, Zürich, Switzerland
MNHN: Muséum National d'Histoire Naturelle, Paris, France
MHNG: Muséum d'Histoire Naturelle, Genève, Switzerland
NHML: Natural History Museum, London, U.K.
PPRI: Plant Protection Research Institute, Pretoria, South Africa
TAU: Tel Aviv University, Tel Aviv, Israel
USNM: National Museum of Natural History, Smithsonian Institution, Washington, D.C., U.S.A.
ZMUC: Zoological Museum, University of Copenhagen, Denmark

## APPENDIX 24.2: LIST OF SPECIMENS EXAMINED

Full data, as indicated on the labels, is given for those species which have been used in the cladistic analysis (marked with *) or for illustrations. Only the country is given for those species that were checked only for descriptions and keys. Additions are marked in [ ].

*Acanthiophilus brunneus* Munro: Kenya.

*Acanthiophilus ciconia* Munro: Kenya.

*Acanthiophilus helianthi* (Rossi): Austria, Crete, Netherlands, Israel, Kazakstan, Spain, Switzerland, Turkey.

**Acanthiophilus walkeri* (Wollaston): CANARY ISLANDS: 1♂, 1♀, Tenerife, Erjos, 28.IV.-5.V.1988 (CBM).

**Actinoptera discoidea* (Fallén): UKRAINE: 1♂, 1♀, SE Kherson, Black Sea Reserve, 1.-2.VII.1985, V. A. Korneyev (CBM).

*Actinoptera filaginis* (Loew): Italy.

*Actinoptera acculta* Munro: Kenya.

*Actinoptera contacta* Munro: Kenya.

*Actinoptera mamulae* (Frauenfeld): Crete.

*Actinoptera meigeni* Hendel: France, Spain.

**Brachydesis rivularis* (Bezzi): SOUTH AFRICA: 1♂, Jonkershoek, 22-23.X.1965 (Natal Museum); 1♂, E Transvaal, Vaalhoek, 6.II.1972, A. Freidberg (TAU); 1♀, Robbers' Pass, 29.XII.1994, A. Freidberg (TAU).

**Brachytrupanea brachystigma* (Bezzi): **Syntypes** 1♂, 1♀, NYASALAND [= Malawi], Cholo, XI.1919, R. C. Wood (NHML).

**Campiglossa producta* (Loew): ISRAEL: 1♂, 1♀, Herzliyya, 8.-10.VI.1996 (CBM).

**Capitites augur* (Frauenfeld): EGYPT: 2♂♂, 2♀♀, Near Taba, 12.IV.1992 (CBM, ETHZ). ISRAEL: 1♂, 1♀, Nahal Paran, 11.IV.1992 (CBM).

**Capitites dentiens* (Bezzi): ETHIOPIA: 1♂, Debre Libanos, 12.XII.1989, A. Freidberg (TAU). SOUTH AFRICA: 1♀, Natal, Empangeni distr., Enseleni N. Res., 8.X.1983, A. Freidberg (TAU); **Lectotype** ♂ (here designated) and **Paralectotype** ♂, S[OUTH] AFR[ICA], Pretoria, 4.I.1923, H.K.Munro (PPRI). 1♂, Pretoria, 28.IX.1923, H.K.Munro (NHML).

**Capitites ramulosa* (Loew): CANARY ISLANDS: 1♂, Gomera, Hermigua, 13.III.1990 (CBM); 1♀ Gomera, S. Sebastian, 12.III.1990 (CBM).

**Dectodesis auguralis* (Bezzi): KENYA: 1♂, 10 km NE Kericho, 10.III.1993 (CBM); 1♀, W-Nairobi, ILRAD, 8.III.1993 (CBM).

**Dectodesis confluens* (Wiedemann): KENYA: 1♂, Aberdares N.P., 3800 m, Mt. Lesatima, 7.III.1993 (CBM); 1♀, 2 km N Webuye, 14.III.1993 (CBM).

*Dectodesis inundans* Munro: Kenya.

*Dectodesis monticola* Munro: Kenya.

**Euarestella kugleri* Freidberg: EGYPT: 1♀, Sinai, near Nuweiba, 12.IV.1992 (CBM); ISRAEL: 2♂♂,1♀, Timna, 11.IV.1992 (CBM, ETHZ).

*Euarestella iphionae* (Efflatoun): Egypt, Israel.

*Euarestella megacephala* (Loew): MALTA: 1♀, Malta isl., Sct. Julian's, 4.-11.VI.1988, Stig Andersen (ZMUC).

**Euarestella pninae* Freidberg: EGYPT: 1♂, Sinai, near Nuweiba, 12.IV.1992 (CBM); ISRAEL: 1♀, Nahal Qumran, 10.IV.1992 (CBM).

**Freidbergia mirabilis* Merz, n. sp: see type material above.

*Goniurellia lacerata* (Becker): Iran.

*Goniurellia longicauda* Freidberg: Canary Islands, Israel.

*Goniurellia persignata* Freidberg: Israel.

*Goniurellia spinifera* Freidberg: Egypt, Israel.

**Goniurellia tridens* (Hendel): KAZAKSTAN: 1♀, Kara-Tau mts., 15 km N Atabaj, 10.V.1994 (CBM); UZBEKISTAN: 1♂, Fergana valley, Yazyavan-region, 18.V.1994 (CBM).

**Hyalotephritis australis* (Bezzi): SOUTH AFRICA: 2♂♂, 1♀, Pretoria, 13.II.1972, A. Freidberg (CBM); 1♀, E Transvaal, Vaalhoek, 6.II.1972, A. Freidberg (CBM).

*Hyalotephritis complanata* (Munro): KENYA: 2♂♂, 2♀♀, 50 km N Mombasa, 4.XII.1989, A. Freidberg (CBM).

**Hyalotephritis planiscutellata* (Becker): ISRAEL: 1♂, 1♀, Enot Zukim, 19.III.1995 (CBM); 1♂, 1♀, Tel Aviv, 9.-16.XI.1977 (CBM); En Gedi, 1♂, 2♀♀, 10.IV.1992 (CBM). KENYA: 2♂♂, 1♀, 15 km SE Hunters Lodge, 18.III.1993 (CBM); 1♂, Magadi Road, 3.III.1993 (CBM).

*Insizwa dicomala* (Munro): KENYA: 1♂, W-Nairobi, ILRAD, 16.III.1993 (CBM); 1♀, 15 km SE Hunters Lodge, 18.III.1993 (CBM).

**Insizwa goliath* (Bezzi): SOUTH AFRICA: 1♂, 1♀, Natal, Pietermaritzburg, Ukulinga Station, 3.X.1983, A. Freidberg (TAU).

**Insizwa striatifrons* (Munro): NAMIBIA: 1♀, 1500 m NN, 30 km W Okahandja, 2.-5.III.1997, MF [= Malaise Trap], O. Niehuis (CBM). SOUTH AFRICA: 1♂, Natal, South Coast, Umkoomas, 11.X.1983, A. Freidberg (CBM).

**Multireticula perspicillata* (Bezzi): KENYA: 1♀, 10 km SW Kitale, 14.III.1993 (CBM). 1♂, SOUTH AFRICA: W. Cape, Montagu Pass, 3322 Cd, between Comfer and George, 12.I.1983, A. Freidberg (TAU).

**Pherothrinax redimitis* Munro: **Paratypes** 1♂, 1♀, KENYA: Mt. Elgon, II.1935, B.M. E.Afr. Exp. (NHML); 1♀, 25 km NE Kericho, 19.-20.XI.1989 A. Freidberg (TAU). UGANDA: 1♂, 7 km NE Kabale, 23.XII.1995, A. Freidberg (TAU).

**Spathulina sicula* Rondani: CRETE: 1♂, Panagia, 17.IV.1991 (CBM); 1♀, Festos, 20.IV.1991 (CBM).

*Stelladesis arrhiza* (Bezzi) (described as *Trypanea woodi* var. *arrhiza*): **Lectotype** ♀ (here designated) [SOUTH AFRICA], Toise River, 3.XII.1923, H. K. Munro (PPRI); 1♂, Toise River, Jan. 25, H. K. Munro (NHML, not a type, collected after publication). KENYA: 1♂, 75 km SE Nairobi (A109), 18.III.1993 (CBM).

*Stelladesis lamborni* (Munro): **Paratypes** 2♂♂, 2♀♀, NYASALAND [= Malawi], Maiwale, 8.-16.V.1932, Dr. W. A. Lamborn (NHML).

*Stelladesis lutescens* (Bezzi): **Holotype** ♂ [SOUTH AFRICA], Pretoria, 11.XII.1916, H. K. Munro (PPRI).

*Stelladesis* n. sp. nr. *lutescens*: KENYA: 1♂, 1♀, 10 km NE Kericho, 9.III.1993 (CBM).

**Stelladesis woodi* (Bezzi): **Holotype** ♂, MOZAMBIQUE, Prov. du Gorongoza, Forêt d'Inhanconde, 350 m, 1907, G. Vasse (MNHNP). KENYA: 1♂, 15 km SE Nairobi (A109), 17.III.1993 (CBM), 1♀, W-Nairobi (ILRAD), 8.III.1993 (CBM).

**Tephritis arnicae* (Linnaeus): SWITZERLAND: 1♂, VS, 1520 m, Oberwald, coll. 12.VII.1989, ex flowerhead *Arnica montana* 1.VIII.1989 (CBM); 1♀, GR, 1800 m, Splügen, coll. 29.VII.1989, ex flowerhead *Arnica montana* 9.VIII.1989 (CBM).

*Tephritis scorzonerae* Merz: ITALY: 1♂, 1♀, Puglia, Mte. Gargano, S. Giovanni, 14.V.1990 (CBM).

*Tephritomyia grisea* Munro: Kenya.

**Tephritomyia lauta* (Loew): CRETE: 1♂, Festos, 20.IV.1991 (CBM); 1♀, Armeni, 19.IV.1991 (CBM); 1♀, Larani, 21.IV.1991 (CBM). ISRAEL: 1♂, Mt. Hermon, 8.IX.1971 (CBM).

**Tephrodesis pulchella* (Bezzi): **Lectotype** ♂ (here designated), **Paralectotype** ♀, [SOUTH AFRICA], Bloemfontein, 19.XI.1921, H. K. Munro (PPRI); **Paralectotype** ♂, Pretoria, 16.I.1923, H. K. Munro (PPRI). KENYA: 3♂♂, 3♀♀, Rt. 104, Gilgil, 8.V.1991, A. Freidberg (TAU, CBM).

*Tephrodesis furcatella* (Bezzi) (described as *Trypanea subcompleta* var. *furcatella*): **Lectotype** ♀ (here designated) [SOUTH AFRICA] East London, 12.VIII.1922, H. K. Munro (PPRI); one specimen **Paralectotype** (only one wing present on micropin), East London, 29.VII.1922, H.K. Munro (PPRI).

*Tephrodesis mutila* (Bezzi): **Holotype** ♀, [SOUTH AFRICA], East London, 20.VIII.1922, H. K. Munro (PPRI).

*Tephrodesis subcompleta* (Bezzi): **Holotype** ♀ [KENYA], Nairobi B. E. A., 27.IV.1911, T. J. Andersen (NHML).

*Trupanea amoena* (Frauenfeld): Canary Islands, Crete, Israel, Italy, Switzerland, Turkey.

**Trupanea desertorum* (Efflatoun): EGYPT: 1♂, Sinai, Wadi Rutia, 25.IX.1977, A. Freidberg (CBM); 1♀, Bin Znin, 8.IV.1976, A. Freidberg (CBM).

*Trupanea guimari* (Becker): Canary Islands.
*Trupanea pseudoamoena* Freidberg: Egypt, Israel.
*Trupanea pulcherrimum* (Efflatoun): Israel.
**Trupanea stellata* (Fuesslin): CRETE: 1♂, Panagia, 17.IV.1991 (CBM). SWITZERLAND: 1♀, TI, Biasca, 6.VII.1990 (CBM).
**Trupanodesis aurea* (Bezzi): KENYA: 2♂♂, 2♀♀, W Nairobi-ILRAD, 16.III.1993 (CBM).
*Undescribed genus and species: TANZANIA: 3♂♂, 3♀♀, Usambara Mts., Rt. B 124, Bumbuli, 14.IX.1992, A. Freidberg (TAU, CBM). KENYA: 1♀, Taita Hills, 4.V.1991, A. Freidberg (TAU).

# 25 Behavior of Flies in the Subfamily Tephritinae

*David H. Headrick and Richard D. Goeden*

## CONTENTS



0-8493-1275-2/00/$0.00+$.50

## 25.1 INTRODUCTION

The study of California tephritine fruit fly biology and behavior has been in progress for over 10 years. Our work has focused on the tephritines of southern California and currently we and co-workers have published detailed life histories on 42 species in 17 genera; we continue to publish behavioral descriptions for individual species as part of our ongoing studies. The behaviors of California tephritines were compiled and analyzed in a general overview and compared with other groups in the family (Headrick and Goeden 1994), but without reference to the phylogenetic relationships of the taxa involved.

The objective of this chapter is to summarize behaviors for each of the tephritine genera for which information is available, discuss the evolutionary trends within each genus, and then discuss some evolutionary trends for the tephritines as a whole. We should emphasize that each species newly studied by us provides new surprises and exceptions to most rules we attempt to make. Generalizing globally from the southern California fauna is done cautiously.

The genera herein are presented in a phylogenetic context as now understood for the subfamily (Korneyev, Chapter 22). Definitions for behavioral terms are to be found in the glossary (White et al., Chapter 33). The behaviors for each species are tabulated and presented in the chapter on behavioral evolution for comparative examination (Sivinski et al., Chapter 28). Each genus is briefly reviewed, the behaviors for the California species are described and analyzed, and these are compared with published records of other species within the genus in a section entitled "Comparison of Behaviors." Evolutionary trends are also discussed for most genera in this section. The last section, Conclusions and Future Research Needs, presents trends in the evolution of behavior for higher-level taxonomic groupings within the subfamily.

The majority of tephritine behaviors occurs in a reproductive context. These behaviors are the most frequently studied and, therefore, form the basis for most comparative examinations. There are other behavioral categories such as foraging, dispersal, diel patterns, and larval feeding. These types of behaviors have not been studied consistently among tephritine species. This, in part, reflects the rarity of many of these species. For several, we have only been offered brief glimpses at a few individuals; in the case of larval behaviors, they are concealed within plant tissues and are difficult to collect and observe. A review of these nonreproductive behaviors and their evolution has recently been published (Headrick and Goeden 1998).

## 25.2 BEHAVIORS OF THE CALIFORNIA TEPHRITINAE

### 25.2.1 BEHAVIORS OF *PROCECIDOCHARES* HENDEL

Subfamily Tephritinae; Tribe Cecidocharini

#### 25.2.1.1 Introduction

*Procecidochares* is strictly New World in distribution (except for species introduced elsewhere), with 13 described species occurring in North America (Foote et al. 1993), two of which were recently described from California (Goeden and Teerink 1997a). The taxonomic status of some of the remaining undescribed species occurring in California remains unclear (R. Goeden and D. Headrick, unpublished data). Green et al. (1993) clarified the biology and host associations of *P. stonei* Blanc and Foote in California. Silverman and Goeden (1980) described the biology and host associations of *P. kristineae* Goeden on *Ambrosia dumosa* (Gray) Payne in southern California. The life histories and immature stages also have been described for *P. anthracina* (Doane) (Goeden and Teerink 1997b), *P. flavipes* Aldrich (Goeden et al. 1994a), and *P. lisae* Goeden (Goeden and Teerink 1997a).

Other studies in North America have focused on the biology and host plant associations of the gall former, *P. minuta* Snow. This species has generally been recognized as a complex of allopatric or sympatric, cryptic sibling species widely distributed in western North America. Wangberg (1980) reported on the comparative biologies of three gall-forming, sympatric species in the *P. minuta* group on rabbitbrush in Idaho. Dodson (1987a) also reported on the life history, host plant associations, and reproductive behavior of a *Procecidochares* sp. nr. *minuta* in New Mexico. At present, the host plant associations, larval resource utilization, larval taxonomy, adult behavior, and morphology are being studied for the California *P. minuta* species group (R. Goeden and D. Headrick, unpublished data).

Reproductive behaviors of *Procecidochares* species have been described in greater detail in the recent literature (Headrick and Goeden 1994; Goeden et al. 1994; Goeden and Teerink 1997a, b); Dodson (1987a) is a notable exception, allowing for comparisons among *Procecidochares* species and among other tephritine genera.

### 25.2.1.2 Reproductive Behaviors

#### *25.2.1.2.1 Wing Displays*

Enantion is the most common wing display during all phases of adult behavior, exhibited either spontaneously or directed toward other individuals. Enantion displays are common to all banded-wing species (Headrick and Goeden 1994). Each species also exhibits unique wing displays, including hamation and male-agitation enantion, while in copula. *Procecidochares* species do not display asynchronous supination, commonly observed in *Trupanea* and *Tephritis* (Headrick and Goeden 1994).

#### *25.2.1.2.2 Courtship*

Males track females, typically from behind, with or without wing displays. The intensity of the wing display increases as a male nears a female. Males attempt to mount females either after tracking them or opportunistically. Males cease their wing displays before jumping at females. Males attempted to mount females from as far away as 10 cm in petri dish arenas (Headrick and Goeden 1994). Most mounting attempts are unsuccessful, as males miss physically contacting females completely, or are unable to hold onto them after mounting.

#### *25.2.1.2.3 Copulatory Induction Behavior*

Males grasp females around the abdomen near the thorax with their front legs, or with their foreclaws around the costal margins near the wing bases. The middle legs of the male grasp the female near the base of her ovipositor and the hind legs are bent underneath the ovipositor and abdomen. The hind legs are used to pull the aculeus apex to the epandrium and to drum asynchronously on the oviscape and posterior sternites of the abdomen. Females typically are receptive and do not struggle against a mounted male, and exert their aculeus within seconds after being mounted.

#### *25.2.1.2.4 Copulation*

Once the aculeus is exserted and held in position by the male terminalia, the apex is lifted to expose the eighth sternites. The phallus (aedeagus) uncoils from the right-hand side in all species observed and is inserted into the cloacal opening. In the final copulatory position, the front legs of males embrace the abdomen of females near the thorax, the middle legs grasp the base of the oviscape, and the hind legs generally rest on the substrate. A male may intermittently drum his hind legs against the venter of the abdomen of the female when agitated. The duration of copulation ranges from 0.5 to 5 h (see Sivinski et al., Chapter 28).

### 25.2.1.3 Comparison of Behaviors

#### *25.2.1.3.1 Wing Displays*

All *Procecidochares* species observed display enantion; enantion was first described for *P. stonei* (Green et al. 1993).

### 25.2.1.3.2 Courtship and Copulation

Males of *Procecidochares*, like those of *Aciurina*, displayed little courtship behavior in the laboratory or the field. There were no unique wing displays and males simply tracked females for mounting or mounted them as opportunity allowed. *Procecidochares* males did not display the abdominal pleural distension so commonly observed in males of most other tephritid species. Dodson (1987a) also reported male–female interactions for *P.* sp. nr. *minuta* in New Mexico similar to those reported here. However, Wangberg (1980) gave quite different accounts of three *P.* sp. nr. *minuta* populations in Idaho. Wangberg (1980) characterized the female as playing the dominant role by actively pursuing the male to elicit courtship behaviors. He also described frequent mounting attempts in the field by a male before successful copulation with a single female. His descriptions differed markedly from the behaviors observed by us and from behaviors previously reported by others for these tephritids. Populations from Idaho must be examined further and his results validated before comparisons can be made. Wangberg (1980) also concluded that *Procecidochares* spp. had courtship behaviors similar to *Valentibulla* spp. as described by Wangberg (1978). Again, the female in the latter genus was described as eliciting courtship with a male and usually with lengthy interaction between courting individuals (see our descriptions of *Valentibulla* below). Contrary to Wangberg, the interactions of *Procecidochares* individuals we observed were always brief, even when confined to arenas. A male of *P. minuta* typically had only one chance to mount a female, as females typically flew away from or otherwise avoided males.

Although Dodson (1987a) described similar behaviors for *P.* sp. nr. *minuta* on *Chrysothamnus nauseosus* (Pallas) in New Mexico to those reported for *P. minuta* by Headrick and Goeden (1994), some interesting differences are noted. Dodson (1987a) described all courtship and mating as specifically taking place away from the host plant on nearby plants of *Atriplex canescens* (Pursh) Nuttall and noted that the *P.* sp. nr. *minuta* population displayed a classical lek mating system. He concluded from release experiments that individuals of *P.* sp. nr. *minuta* were emerging from galls on *C. nauseosus* and aggregating on nearby *A. canescens* for courtship and copulation.

The *P.* sp. nr. *minuta* population on *C. nauseosus* in New Mexico had emerged from flowerhead galls in the fall (Sept.) according to Dodson (1987a). Flower head galls on *C. nauseosus* also occur in California from which *Procecidochares* sp. nr. *minuta* also emerge in the fall. These adults emerge at a time when their host plants are no longer in a stage suitable for oviposition. Unlike the *P. minuta* which emerges from stem galls in the spring and oviposits into newly developed leaf axils (Headrick and Goeden 1994), the flowerhead-infesting flies emerge and disperse to oviposit on alternate host plants. We currently believe that the two gall types on *C. nauseosus* (Pallas) Britton are produced by two distinct species of *Procecidochares* (R. Goeden and D. Headrick, unpublished data). Perhaps a similar situation explains the behavior Dodson (1987) reported for the *P.* nr. *minuta* species in New Mexico, as *Procecidochares* species typically are not long-lived (Green et al. 1993; Goeden and Teerink 1997a, b), and females of most gall-forming species emerge with a complement of full-size ovarian eggs (proovigenic). Females readily copulate upon emergence (Headrick and Goeden 1994) and oviposit shortly thereafter; *P. flavipes*, a non-gall former, is an exception to these generalizations (Goeden et al. 1994a).

The behavioral descriptions for *P. kristineae* (Headrick and Goeden 1994; Goeden and Teerink 1997a) augmented the initial report of Silverman and Goeden (1980). In the field, males of *P. kristineae* perched on and tracked females from the tops of new terminal growth, as was observed for *P. minuta* males (Headrick and Goeden 1994; Goeden and Teerink 1997a). These observations confirmed the observations of Silverman and Goeden (1980) that males contacted females atop terminal foliage and branches, where the males waited.

### 25.2.1.3.3 Cross-Matings

Several series of cross-matings were conducted with separate host plant–derived populations of *P. minuta* (D. Headrick and R. Goeden, unpublished data). Of special interest were 12 two-way, heterotypic crosses between adults that emerged from stem galls on *C. nauseosus* and adults from basal, axillary bud galls on *Gutierrizia sarothrae* (Pursh) Britton and Rusby. We believe the adults from *G. sarothrae* are part of a complex of species and probably represent a separate species from those that form galls in the flowerheads of *C. nauseosus*. The females from *C. nauseosus* never mated with males from *G. sarothrae*, but females from *G. sarothrae* readily mated with males from *C. nauseosus*. These results provided some evidence that these two species populations may be at least partially behaviorally isolated.

##### *25.2.1.3.4 Territoriality*

Dodson (1987a) reported that males of *P.* sp. nr. *minuta* interacted aggressively in laboratory arenas with head-butting and prolonged wrestling with their front legs. These types of aggressive encounters were not observed for the *P. minuta* we studied in the laboratory or the field (Headrick and Goeden 1994). However, the numbers of individuals confined in arenas or the numbers of individuals observed on host plants were not as large as those reported by Dodson (1987a). Therefore, relatively high population densities that increase the number of encounters between individuals may elicit this more aggressive behavior by males. Males of *P. anthracina* were reported to exhibit labellar touching when facing each other in laboratory cagings (Goeden and Teerink 1997a), indicative of at least one aspect of male–male encounters described for other tephritid species (Headrick and Goeden 1994).

### 25.2.2 Behaviors of *Stenopa* Loew

Subfamily Tephritinae; Tribe Eutretini

#### 25.2.2.1 Introduction

The genus *Stenopa* occurs only in North America and is represented by two species, *S. affinis* Quisenberry and *S. vulnerata* (Loew) (Foote et al. 1993). Korneyev (Chapter 22) has transferred this genus from the Cecidocharini to the Eutretini. The biology and behavior of *S. vulnerata*, the more common and widely distributed species, were described by Novak and Foote (1975). Goeden and Headrick (1990) reported the host association, distribution, life history, and immature stages of the much rarer and lesser-known species, *S. affinis*.

#### 25.2.2.2 Reproductive Behaviors

Goeden and Headrick (1990) were unable to examine courtship and copulation behaviors because of the rarity of adults.

#### 25.2.2.3 Comparison of Behaviors

The behavior of *S. vulnerata* appears complex, involving male production of membraneous "bubbles" from their mouth, production of froth masses, and use of "stylized wing movements" by receptive females and by males guarding froth masses. Field observations by Novak and Foote (1975) showed that males deposited these froth masses on leaves on the host plant and that copulation lasted from 5 to 20 min, was observed more frequently in the morning or afternoon hours, always took place on the host plant, and usually occurred on the upper surface of a leaf.

### 25.2.3 BEHAVIORS OF *ACIURINA* CURRAN

Subfamily Tephritinae; Tribe Dithrycini; Subtribe Eurostina

#### 25.2.3.1 Introduction

The genus *Aciurina* is strictly Nearctic and comprises 14 described species restricted to the western United States and Mexico (Foote et al. 1993; Hernández-Ortiz 1993; Goeden and Teerink 1996a, b, c; Headrick et al. 1997). Most California *Aciurina* species have dark wings with characteristic hyaline markings, green or gray thoraces, and large abdomens, which have brown sclerites and bright yellow pleura (Color Figure 19*). *Aciurina idahoensis* Steyskal is one exception; females have pale, brown stripes on hyaline wings, and males have a few small, light-brown markings on hyaline wings (Goeden and Teerink 1996b). Only one other North American *Aciurina* species has distinct sexually dimorphic wing patterns, *A. semilucida* (Bates), the female of which also has a striped pattern, but the male has wings that are fully infuscated (Goeden and Teerink 1996c). *Aciurina* species form galls on stems and branches of their host plants in the Asteraceae.

### 25.2.3.2 Reproductive Behavior

#### *25.2.3.2.1 Wing Displays*

Most *Aciurina* species have similar wing patterns and have common wing displays (Headrick and Goeden 1994). Both sexes display asynchronous extension of the wings to 90° with supination to 90°, = asynchronous supination (White et al., Chapter 33). During extension and supination, the wing blades are bent forward at the costal break. Adults display ancillary rotations (Headrick and Goeden 1994) of their wings during asynchronous supinations (except *A. idahoensis*). Ancillary rotations were also noted for *A. mexicana* (Aczél) by Jenkins (1990), who described wing extensions as being "smooth or (with) intermittent jerks," and further defined "jerks" as "slight rotational adjustments." Individuals of all species display abdominal flexures during asynchronous supinations. Both sexes raise and lower (= flex) their abdomens two to three times per wing extension. Individuals of all species also sway (White et al., Chapter 33) while facing other individuals and performing asynchronous wing displays. They also sway when turning to face a moving object or when startled by a sudden nearby motion. When individuals observe nearby movement, they respond by turning to face the object and display asynchronous supination, while swaying from one side to the other after each wing extension.

Adults also display wing hamation (White et al., Chapter 33). Hamation usually occurs when an individual finishes an asynchronous supination display. Males display hamation before mounting as they stand near females. Females also display wing hamation while in a defensive posture, for example, when startled by a sudden movement.

Enantion (White et al., Chapter 33) is also common to both sexes. Enantion is displayed during bouts of aggressive behavior while lunging at other individuals. Males also display enantion while *in copula* if other males come too close. Other wing displays, which differ slightly from those described above, were observed in the context of a particular behavior (i.e., courtship or aggression) and will be described for each species below.

#### *25.2.3.2.2 Courtship*

*Aciurina* male courtship displays comprise the following interaction displays: (1) synchronous wing supination; (2) abdominal pleural distension; (3) body swaying; (4) mouthpart extension; and (5) front leg waving. However, these elements are not always fully expressed by all species or individuals within a species. In arenas, males most often jump onto the dorsa of passing females trying to grasp their abdomens or an appendage without any prior displays. Occasionally, males displayed asynchronous supination or synchronous wing thrusts (see below) while facing a female. However, it was not discernible whether these were aggressive displays or part of courtship. Males only rarely distended their abdominal pleura. Our field observations of *A. trixa* Curran and laboratory observations of all *Aciurina* species studied indicated that males stalk females for mounting either passively or aggressively without any prior displays common to courting males in other genera (Headrick and Goeden 1994; Headrick et al. 1997).

#### *25.2.3.2.3 Copulation*

Copulation in *Aciurina* has two distinct elements, mounting and intromission, and they proceed similarly in all species. All *Aciurina* males have enlarged forefemora used to grasp and hold a

* Color Figures follow p. 204.

female for the purpose of mounting. All *Aciurina* females attempt to escape by walking rapidly and erratically once males have grasped them. Males primarily attempt to grasp a female's abdomen with their front legs, but occasionally they entangle a female's hind legs or wings. The important factor when a male mounts a female is to maintain his hold once she begins to walk rapidly. The male's middle and hind legs often dragged behind as the female walked, and only after he could pull himself up farther forward onto the female's abdomen were his other legs used to lift her abdomen to gain intromission.

##### 25.2.3.2.4 Aggression

Both sexes display aggression toward each other in laboratory arenas. Rapid, synchronous wing extensions to 90° with 90° supination are used in conjunction with lunging to startle or chase away an intruder. Swaying during wing displays occurs during visual inspection of nearby or distant moving objects. Females raise their front legs to approaching males and bat them if they approach too closely. Males typically retreat from aggressive females and do not display aggression in return.

##### 25.2.3.2.5 Territoriality

Field observations discussing male–male interactions for *A. trixa* are described by Headrick et al. (1997). Other *Aciurina* species were not observed in the field to substantiate territoriality or ritualized male–male combat.

#### 25.2.3.3 Comparison of Behaviors

##### 25.2.3.3.1 Wing Displays

All *Aciurina* species studied display asynchronous supination (Tauber and Tauber 1967; Wangberg 1981; Dodson 1987b; Jenkins 1990; Headrick and Goeden 1993; 1994; Goeden and Teerink 1996a, b, c; Headrick et al. 1997). Asynchronous supination is the most common wing display and is exhibited during all phases of adult behavior. It is distinct from asynchronous displays of other tephritid genera and species because the wing blades are bent and the blades are rotated during the display. These combined motions are also observed in *Blepharoneura* spp. (Condon and Norrbom, Chapter 7) and *Schistopterum moebiusi* Becker (Freidberg 1981). Jenkins (1990) diagrammed the wing blade of *A. mexicana* bent in the same manner during supination.

##### 25.2.3.3.2 Courtship

Discussion of courtship behaviors will be based on the following taxonomical hierarchy: behaviors common to the genus, behaviors shared by groups of species, behaviors displayed by each species, and ultimately by the individuals of that species. Six male courtship displays have been reported for five species of *Aciurina*: (1) synchronous wing supination; (2) abdominal pleural distension; (3) swaying; (4) front leg waving; (5) nuptial gift formation or trophallaxis; and (6) mouthpart extension. Each species exhibited its own subset of these displays. *Aciurina michaeli* Goeden displayed behaviors 1 and 6; *A. thoracica* Curran displayed 1, 2, and 3; *A. trixa* displayed 1, 4, and 6; *A. idahoensis* displayed 1. *Aciurina mexicana* was reported to display 1, 2, 3, and 5 (Jenkins 1990). Courtship displays are a function of life history strategy and population density (Headrick and Goeden 1994). There apparently is a correlation between life history strategy and the courtship behaviors displayed by a species, and a further correlation between population density and which displays an *individual* will exhibit. *Aciurina michaeli* and *A. trixa* are behaviorally similar because they have similar life history strategies; both are gall formers on *Chrysothamnus* spp., and both emerge as reproductively mature adults when their host plants are in a stage suitable for oviposition. Therefore, adults copulate and females oviposit shortly after emergence; indeed, teneral adults were observed *in copula* (Headrick and Goeden 1994). They do not aggregate as reproductive diapausing adults, but oversummer and overwinter in incipient galls as eggs and early instar larvae (Tauber and Tauber 1967; Dodson 1987b; Headrick and Goeden 1994; Goeden and Teerink 1996a, b, c). Gall growth resumes when winter rains stimulate plant growth and adults emerge after ~1 mo.

Neither of these two species exhibits a series of courtship displays requiring a female response. Rather, they stalk and pursue females either aggressively or passively. The wing displays and mouthpart extensions that *A. michaeli* and *A. trixa* share are part of their stalking behaviors, and females do not respond to either of those two displays. *Aciurina trixa* males did not distend their abdominal pleura in field observations, and only did so on two occasions in the laboratory when confined to cages with other males at what represented unusually high densities. Field observations on *A. trixa* showed that there were usually from one to ten individuals per plant and that encounter rates between individuals were infrequent (Headrick and Goeden 1994). However, it is known that *A. trixa* galls can be very numerous on individual host plants in other locations (Goeden 1988b). Perhaps, when densities are higher on host plants, behaviors such as abdominal pleura distension may serve some function not observed or accounted for when densities are low. Pheromone release has been linked with abdominal pleural distension, but it has been assumed to attract females as part of a lek mating system (Nation 1972; 1981; Arita and Kaneshiro 1986; Dodson 1987b). If abdominal pleural distension is only observed in nonaggregating species when population densities are high, then pheromone release may act as a dispersal agent and help maintain a more uniform distribution of males over a plant. *Aciurina michaeli* galls have not been observed to occur in high densities on individual plants, and thus males may not be able to invoke abdominal pleural distension as part of their interactive displays.

*Aciurina thoracica* and *A. mexicana* appear more similar to each other than either is to the other *Aciurina* species examined. They both form galls on *Baccharis sarothroides* Gray and both emerge as adults in early spring in the southwestern United States (Jenkins 1990; Headrick and Goeden 1993). Unfortunately, little is known about *A. mexicana*. *Aciurina thoracica* adults emerge reproductively immature, but like adults of *Eutreta diana* (Osten Sacken), another stenophagous gall former, ovigenesis is completed about 2 to 3 weeks after emergence (Goeden 1990a). Mating and oviposition for *A. thoracica* takes place on their host plants soon after egg maturation in February and March in southern California. Jenkins (1990) swept adults of *A. mexicana* from their host plants in February and March in New Mexico; thus, *A. thoracica* and *A. mexicana* may have similar life histories. Males of *A. thoracica* and *A. mexicana* exhibit typical courtship displays reported for other tephritid species: they face females, display their wings, sway, and distend their abdominal pleura. However, only *A. mexicana* has been observed to exhibit premating trophallaxis. Details of the behaviors associated with premating trophallaxis by *A. mexicana* reported by Jenkins (1990) compare well with other published accounts of tephritid premating trophallaxis (Stoltzfus and Foote 1965; Freidberg 1981; Headrick and Goeden 1990c; Goeden and Headrick 1992) with one exception: the exudate is not frothy (Jenkins 1990). *Aciurina mexicana* males produce a droplet on their labella which is then placed on the substrate. The reported reactions of other individuals in the same petri dish suggest that they are attracted to the droplet for feeding and that females were mounted for copulation while feeding on the droplet. Droplets were not produced in two of the seven trials in which copulations were observed (Jenkins 1990). Two males were also observed to add to their droplets when their mounting attempts were unsuccessful. Droplet formation of crop contents on the labellum is typical for all acalyptrate Diptera during their feeding behavior (Hendrichs et al. 1992; Headrick and Goeden 1994). During feeding, droplets are usually issued and reimbibed several times, then swallowed and shunted into the midgut for digestion.

To summarize, the male–female interactions of *Aciurina* form two distinct groupings: (1) nonaggregative (= circumnatal) species whose males do not exhibit female attraction or interaction displays, and (2) aggregative species whose males display female attraction and interaction displays. *Aciurina michaeli*, *A. trixa*, and *A. idahoensis* are circumnatal species, and *A. thoracica* and *A. mexicana* are aggregative species. The circumnatal species form galls on *Chrysothamnus* spp. and emerge reproductively mature (= proovigenic). When adults emerge they remain on or near their individual galled host plants. Males roam about the crowns of their plants, usually exploring the upper portions on the new growth. They stand on the apices of stems visually orienting to other individuals and there is little to no movement among plants. In other words, they do not aggregate on any particular individual plant. When a male encounters another male, they usually display their wings at each other, then move away. Territoriality displays or combat were not observed in the field or laboratory. When a male encounters a female, he does not display courtship behaviors, but, instead, immediately tries to mount her. In these attempts, males display a passive or aggressive approach, both observed equally as often. The aggressive approach was observed when males were lower on the crown of the plant moving from one area to another. In an aggressive mode, they chased females with their wings extended and lunged or jumped at females in an attempt to mount. Females flew or otherwise moved rapidly away from these aggressive male approaches. If a male managed to grasp a female and to hang on, copulation usually followed. The passive approach involved males perched on tops of stems. This is apparently one way males maximized their encounter rate with females, because of the females' stereotypical method of oviposition. When males encountered females on the apices of stems, they tracked them slowly. When males stopped and faced females from a distance, they often waved their front legs in the air; however, there was no observable reaction by females to this display. When males got close to females, they stood quietly with their wings held over their dorsa, and occasionally extended their mouthparts. Again, females did not respond. If females remained still long enough, males attempted to mount them and initiate copulation.

The aggregative group of species forms galls on *Baccharis* spp., primarily *B. sarothroides*, and at least *A. thoracica* adults emerge reproductively immature (= synovigenic), which is in keeping with a characteristic of other tephritid species that aggregate and display courtship (Jenkins 1990; Headrick and Goeden 1993; 1994).

#### *25.2.3.3.3 Copulation*

Tauber and Tauber (1967), who studied *A. michaeli* (as *A. ferruginea* (Doane)), reported that a male moved next to a female and pushed his head under a wing base, raising the wing blade, and simultaneously moved into a position beside her with his head still under her wing; from this position he climbed onto her abdomen ($n = 1$). Similar mountings were reported for *A. mexicana* (Jenkins 1990) and *A. trixa* (Dodson 1987b; Headrick et al. 1997). This scenario is not atypical, however, as males were observed by us to mount females from a variety of positions. Males positioning their heads under a wing of a female before mounting cannot be considered as a stereotypical courtship behavior for this genus because of the diverse nature of all observed mounting attempts.

The enlarged forefemora of *Aciurina* males may be adapted for maintaining a grip on the female during mounting. The forefemora of *A. trixa* populations from New Mexico were analyzed morphometrically by Dodson (1987c) and found to be sexually dimorphic. In 51 male–female interactions observed by Dodson (1987c), females were resistant to males 24 times, and were able to prevent mounting in 13 cases (25% compared with 14% observed by us). The higher percentage reported by Dodson (1987c) may be attributed to the natural host plant substrates on which his observations were made. Dodson (1987c) reported that males "leg-locked" females during mounting attempts in 24 of 36 (67%) observations. He observed that males used their front legs to grasp a female's abdomen and hind legs together, thus pinning her hind legs against her sides. The leg-lock behavior described by Dodson (1987b, c) for *A. trixa* in New Mexico was only observed by us on three occasions in laboratory arenas; however, each occasion led to a successful copulation. Dodson (1987c) hypothesized that females preferred mates better suited to capturing and subduing them; thus, selection would favor males with more robust front legs to aid in the capture process. However, comparisons between forefemora size of mated vs. single males showed no significant difference ($10.8 \pm 0.6$ vs. $10.6 \pm 0.7$ forefemur width; $34.1 \pm 2.0$ v. $34.0 \pm 1.5$ forefemur length in arbitrary units [1 mm = 30 units] for mated and single males, respectively, Dodson 1987c). Other tephritid species with enlarged forefemora do not display capture and leg-lock-type behaviors during courtship, but rather use them in defense of territories (Headrick et al. 1995).

Tauber and Tauber (1967) did not report leg-lock in their mating trials of *A. michaeli*, but we observed that *A. trixa* and *A. ferruginea* males occasionally grabbed one or both hind legs in their initial attempt at mounting. *Valentibulla californica* (Coquillett) and *Tephritis arizonaensis* Quisenberry males also have enlarged forefemora and their mounting and intromission methods are similar to those described above for *A. ferruginea* and *A. trixa* (Goeden et al. 1993; 1995; Goeden and Teerink 1996a; Headrick et al. 1997). However, both males and females of *T. arizonaensis* have enlarged forefemora, but a use has yet to be discerned for the female's (Goeden et al. 1993).

Tauber and Tauber (1967) reported mating durations of not less than 59 min for *A. michaeli*. Dodson (1987b) reported *A. trixa* copulations averaging 131 min, for 51 observations, and that repeated mating was frequent. Headrick and Goeden (1994) reported that nine copulations for *A. michaeli* averaged 2 h (range 1 to 4 h) and occurred throughout the day and at night under artificial lighting. A total of 47 copulations were observed for *A. trixa* averaging 2 h each in both the field and laboratory (Headrick et al. 1997). Therefore, copulatory duration appears similarly long for all *Aciurina* spp. (Headrick and Goeden 1994; Goeden and Teerink 1996a, b, c; Headrick et al. 1997).

#### *25.2.3.3.4 Territoriality*

Dodson (1987b) reported that males of *A. trixa* performed reciprocal wing displays followed by head-butting and boxing with their forelegs when they encountered each other on their host plants. No territoriality displays were observed for any *Aciurina* species we studied in either the laboratory or the field except for *A. semilucida*. Two newly emerged virgin males and a single female were caged together for observations. The two males displayed aggression toward each other only once. The aggressive display was exhibited by only one of the males and included a wing display and lunging. Again, under higher densities in the laboratory or the field such behaviors as male combat might be manifested in this and other *Aciurina* spp., as apparently was the case for *A. trixa* in New Mexico reported by Dodson (1987b).

## 25.2.4 Behaviors of *Valentibulla* Foote and Blanc

Subfamily Tephritinae; Tribe Dithrycini; Subtribe Eurostina

### 25.2.4.1 Introduction

*Valentibulla* is strictly Nearctic; no species have been recorded from south of the United States (Foote et al. 1993). Two of the three described Nearctic species occur in California, where the life history and immature stages of *V. californica* has been examined by Goeden et al. (1995a).

### 25.2.4.2 Reproductive Behaviors

Behaviors of *V. californica* are similar to *Aciurina* spp. *Valentibulla californica* and *A. trixa* are spatially and temporally sympatric on their sole host plant *Chrysothamnus nauseosus* in southern California.

#### *25.2.4.2.1 Wing Displays*

Adults display synchronous and asynchronous supination, with the wing blades bent and with ancillary rotations as described for *Aciurina* spp. Both sexes displayed abdominal flexure during asynchronous supination. A male in copula displayed lofting (White et al., Chapter 33) when agitated by the female's attempts to push him off her dorsum.

#### *25.2.4.2.2 Courtship and Copulation*

Males do not display aggregation behaviors, but rather stalked females over the host plant or nearby host plants from which they had recently emerged and jumped onto the dorsa of any nearby female. This behavior is consistent with species who exhibit a circumnatal life history strategy (Headrick and Goeden 1994; White et al., Chapter 33). Females walked quickly with males holding onto them with their forelegs. Copulatory induction behavior involved the males rubbing on the venter of the female abdomen with their hind legs. Receptive females exserted the aculeus and intromission was gained. Copulations lasted ~1 h.

### 25.2.4.3 Comparison of Behaviors

#### *25.2.4.3.1 Wing Displays*

Wangberg (1978) reported that adults of *Valentibulla* spp. on *C. nauseosus* in Idaho were active on host plants throughout the day and displayed their wings toward congeneric individuals, as did Dodson (1987b) for New Mexico populations of *V. dodsoni* Foote. Wing movements were the most frequently observed behaviors by both Wangberg (1978) and Dodson (1987b).

#### *25.2.4.3.2 Courtship*

Wangberg (1978) described females as vying for the attention of males, which then pursued females and followed them at similar distances. Males then continued to approach females and if a female remained motionless, the male moved behind her and tapped her with his front tarsi. This initial tracking and contact lasted from 1 to 2 s to 1 to 2 h. Receptive females remained motionless and males then mounted them. Wangberg (1978) incorrectly noted that males bent their abdomens ventrally in "an effort to penetrate (the female) with his genitalia." Headrick and Goeden (1994) showed that a female must first exsert her aculeus before a male can gain intromission. Dodson (1987b) reported that *V. dodsoni* males used the same leg-lock method as observed for *A. trixa* (Dodson 1987c; Headrick et al. 1997); however, males of *V. dodsoni* did not appear to distinguish between conspecific sexes when attempting to mount.

#### *25.2.4.3.3 Copulation*

Wangberg (1978) reported that copulations lasted 60 to 90 min. Dodson (1987b) reported that copulations for *V. dodsoni* averaged ~80 min. Disengagement behavior or repeated copulations by males were not reported. Repeated copulations by *V. californica* males that remain on females (Goeden et al. 1995a) resembled the mate guarding behavior of *D. picciola* (Bigot) (Headrick et al. 1996). However, further studies are needed to verify this behavior.

## 25.2.5 Behaviors of *Eutreta* Loew

Subfamily Tephritinae; Tribe Eutretini

### 25.2.5.1 Introduction

*Eutreta* is restricted to the New World, except for species introduced elsewhere, ranging from Canada to Brazil. Stoltzfus (1977) revised the known species. Little is known about the biology of *Eutreta*. Goeden (1990a, b) reported on the life history of *E. diana* (Osten Sacken) and *E. simplex* Thomas, gall formers on *Artemisia* spp. in California. Stoltzfus and Foote (1965) reported on the use of trophallaxis (or the formation of froth masses) during courtship by *E. novaeboracencis* (Fitch) (as *E. sparsa* (Wiedemann)), and Stoltzfus (1977) provided biological notes for North American species.

### 25.2.5.2 Reproductive Behaviors

#### *25.2.5.2.1 Wing Displays*

At rest, *Eutreta* held their wings slightly parted and arched. Arching of the wings is also observed in *Aciurina*, *Paracantha*, and *Valentibulla* (Headrick and Goeden 1994). There is no apparent rhythmicity to wing extensions and one wing was often repeatedly extended forward, while the other wing was held arched over the dorsum.

Both sexes of *E. angusta* Banks displayed asynchronous supinations. During these asynchronous supinations they also displayed abdominal flexures. Males displayed hamation, moving their wings over their dorsa, and also rotated them asynchronously while they were held over their dorsa.

Both sexes of *E. diana* displayed asynchronous supination spontaneously, and at other individuals, with ancillary rotations (pronation/supination) of the wing blades during extensions, as described for *Aciurina* species. Both sexes also displayed hamation at the end of asynchronous supination displays before returning their wings to a resting position. Males also displayed synchronous extensions without supination (= enantion) during male–male encounters. In these enantion displays the wings were rapidly and repeatedly extended from the resting position through ~10° fast enough that the wing blades were blurred.

### 25.2.5.2.2 Courtship

*Eutreta angusta* males did not display abdominal pleural distension, leg abduction, or trophallaxis during courtship; however, they did display a unique courtship dance. A male oriented to a female with his wings arched over his dorsa and with the costal margins parallel or slightly parted resting along the pleura. If a female remained still after a male faced her, he performed a step-by-step, side-to-side dance while slowly moving toward the female in a zigzag pattern. Each step was a single, rapid move to one side, pausing for ~1 s, then repeating the step to the opposite side. The male moved forward by moving one to three legs at a time with each side-to-side movement. A male also shifted his wings with each step across the dorsum toward the same side as the step taken, that is, incremental hamation. This wing shift is classified as hamation because the wings were moved together from one side to the other over the dorsum. Males also intermittently swayed during longer pauses in their dances. Each courtship display lasted 30 to 60 s ($n = 7$) and was repeated when males met females in their arenas. Females responded to this display by decamping or standing still.

Although *E. angusta* males displayed a courtship dance, females did not show receptivity to it and mounting by males was opportunistic ($n = 4$); that is, males jumped onto the dorsa of any nearby females, or after tracking them around the arena. Male mounting attempts were always initiated from behind a female.

Male *E. diana* courtship in laboratory arenas did not differ from displays observed in the field. Males typically held their wings arched over their dorsa. A male faced a female for courtship and synchronously abducted his middle legs perpendicular to his body through an arc of 60 to 70°, with respect to the substrate, at a rate of approximately one abduction per second. Middle leg abduction was a conspicuous display, as the femora were concolorous with the black body and the segments distad of the femora were yellow. Because males also held their wings slightly parted and arched over their dorsa, this provided a dark background, when viewed from the front, against which the light-colored segments of the middle legs were easily seen as they were abducted. Only males exhibited this display in *E. diana*. Only one other tephritid species, *Euaresta stigmatica* (Coquillett), is known to display middle leg abduction (Headrick et al. 1995). As a male *E. diana* moved forward toward a female he abducted one or both middle legs, but as he stood still in front of a female, he abducted both legs. The rate of abduction increased the nearer a male came to a female, as additionally, he pawed at the air toward the female with his front legs as was observed with *A. trixa* males (Headrick et al. 1997). Females moved away from males, stood still, or lunged at them with asynchronous supinations. A male attempted to mount a female only from behind. If a female remained still while facing away from a male, he then jumped onto her dorsum ($n = 4$). Males also opportunistically attempted to mount nearby females without any prior displays or tracking.

### 25.2.5.2.3 Copulatory Induction Behavior

After a male mounted a female, he moved into a copulatory position with his head just behind her scutellum, his front legs wrapped around her abdomen near her thorax, his middle legs wrapped around the base of her oviscape, and his hind legs bent beneath her abdomen. The male pressed his epandrium against the apex of the aculeus and drummed his hind tarsi asynchronously against her abdominal sternites.

Mounted males of *E. diana* moved into a copulatory position holding onto the abdomen of the female with his front legs near her thorax, his middle legs wrapped around the middle of her abdomen, and his hind legs held onto her oviscape. The hind legs of the male also were used to drum near the base of the oviscape. Drumming with the hind legs was asynchronous, in intermittent bursts that lasted 1 to 10 s. During drumming, a male curled the tip of his abdomen beneath him and pressed his epandrium against the apex of the oviscape. Males continued copulatory induction behavior for ~30 min and then dismounted if the female remained unreceptive.

##### *25.2.5.2.4 Copulation*

In the final copulatory position, *E. angusta* males were positioned with the head above the scutellum of the female, the front legs around her abdomen near the thorax, the middle legs around the middle of her abdomen, and his hind legs on the substrate. Four copulations were observed in the laboratory that averaged 1 h (range, 0.5 to 1.25 h). Adults reared from host plants did not copulate in the laboratory, only overwintered adults of *E. angusta* swept from field study sites, and then only rarely.

In the final copulatory position, the body of *E. diana* males were above the abdomen of the female, his head just behind her scutellum, his front legs were wrapped around her abdomen near the thorax, his middle legs were wrapped around the middle of her abdomen, and his hind legs rested on the substrate. The female had her wings forcibly parted and appressed to her pleura, their apices touching the substrate. The male's wings were slightly parted and arched over his dorsum. A total of 12 copulations were observed in the laboratory and 3 in the field. Copulations in the laboratory lasted ~2 h, differing only 5 to 10 min in duration, and occurred throughout the day. Both field copulations lasted ~2 h. This was the only species we studied in which copulation times were similar among trials in the laboratory and observations in the field. The only variation observed in the laboratory was in the time of day that copulation began. Pairs usually mated only once each day, but three pairs mated twice during the same day. Both field copulations ended before 12:00 h. Before disengaging from copulation, females became agitated, pushing at males with their hind legs and flexing their abdomens rapidly.

##### *25.2.5.2.5 Aggression*

In the field *E. diana* males displayed aggression ($n = 8$) toward each other; however, specific territories were apparent and fighting resulted in no observable advantage in access to females. When two males encountered each other, they both displayed middle leg abduction and rapid synchronous wing extensions without supination, but the wings were only extended ~10° from the midline of the body. Male aggression displays typically ended with a single lunge; then either male moved away.

#### 25.2.5.3 Comparison of Behaviors

In his revision, Stoltzfus (1977) partially described the reproductive behavior of five North American *Eutreta* species, including *E. diana*. Goeden (1990a) gave a partial description of copulation for *E. diana* and Goeden (1990b) described the reproductive behaviors of the very rare species, *E. simplex*; this was the most complete description of mating behavior for any *Eutreta* species.

##### *25.2.5.3.1 Wing Displays*

Except for asynchronous supinations, only one other behavior was common to more than one species. Males of both *E. angusta* and *E. caliptera* (Say) (Stoltzfus 1977) used rapid, lateral side steps when approaching females during courtship. The use of froth masses or nuptial gifts by *E. novaeboracensis* (Fitch) is unique in this genus, but as noted by Headrick and Goeden (1994), the use of nuptial gifts by a single species in a genus has precedence.

##### *25.2.5.3.2 Courtship*

Based on descriptions (Stoltzfus 1977; Goeden 1990a, b), courtship is abbreviated or nonexistent in *Eutreta* species. Stoltzfus (1977) described males of *E. caliptera*, *E. frontalis* Curran, *E. novaeboracensis*, *E. diana*, and *E. longicornis* Snow as being territorial; however, this territorial behavior of males was not distinguished from other behaviors on their host plants, including resting, feeding, and courtship. Stoltzfus (1977) described *E. diana* males as maintaining a territory on the upper half of plants, where suitable oviposition sites existed and reported that males employed "wing and body motions to chase away other insects or other males." Females approached male territories while in search of oviposition sites, and males courted females encountered by use of "stylized wing movements." The courtship displays described by Stoltzfus (1977) probably were asynchronous supinations (observed in all five of the above species). Stoltzfus (1977) did not report male–male combat, male tracking behavior, or the use of the middle leg display by combating or courting males as reported here for *E. diana*. Such disparity in observations reflects the need for more intensive fieldwork; however, such work is difficult, as *Eutreta* adults are encountered only rarely; Goeden (1990a), for instance, reported sighting only a single mated pair of *E. diana*.

##### *25.2.5.3.3 Copulation*

Both *E. angusta* and *E. diana* males only attempted to mount females from behind, as also was reported for *E. simplex* by Goeden (1990b). Stoltzfus (1977) reported that males of *E. caliptera*, *E. frontalis*, and *E. novaeboracensis* mounted females from the front after displaying asynchronous supination, but he did not observe mounting behavior in *E. diana*. The copulatory position appears the same for all *Eutreta* spp. (Stoltzfus 1977; Goeden 1990a, b; Headrick and Goeden 1994). Copulatory duration was ~1 h in *E. angusta* and ~2 h in *E. diana*. These times were shorter than the 4 h reported for *E. simplex* (Goeden 1990b), 4 h for *E. caliptera*, and the "several" hours for *E. novaeboracensis* reported by Stoltzfus (1977). Again, more work will be required before the common and unique elements of reproductive behavior can be defined for this genus.

### 25.2.6 BEHAVIORS OF *PARACANTHA* COQUILLETT

Subfamily Tephritinae; Tribe Eutretini

#### 25.2.6.1 Introduction

The genus *Paracantha* is distributed throughout North and South America from Canada to central and northern South America (Foote et al. 1993). The genera most closely related to *Paracantha* according to Foote (1980) are *Neorhabdochaeta* and *Laksyetsa*, neither of which is known to occur north of Mexico (Foote 1980). The genus most closely related to *Paracantha* in California is *Eutreta*, and they also share many behavioral attributes. The biology of *Paracantha* is well known for *P. gentilis* Hering (Headrick and Goeden 1990a, b, c), and partially known for *P. cultaris* (Coquillett) (Lamp and McCarty 1982; Cavender and Goeden 1984) and *P. culta* (Wiedemann) (Benjamin 1934; Phillips 1946). The reproductive behavior and reproductive morphology of *P. gentilis* was described by Headrick and Goeden (1990c). Adults of *P. cultaris* also were examined by us in laboratory arenas to confirm and add to the findings of Cavender and Goeden (1984).

#### 25.2.6.2 Reproductive Behaviors

##### *25.2.6.2.1 Wing Displays*

Headrick and Goeden (1990c) provided detailed descriptions of the adult behavior of *P. gentilis*. This section provides additional comparative information on *P. cultaris*. Both sexes of *P. cultaris* and *P. gentilis* displayed arching, enantion, hamation, and asynchronous supination. Adults of both species typically held their wings slightly spread over their dorsa and arched when at rest. Hamation was observed when individuals finished other wing displays before returning their wings to a resting position, or after a grooming episode. Enantion was less commonly observed. Males of both species displayed enantion during courtship by extending their wings forward from their resting position while facing or moving toward a female. Females of both species responded to males with enantion, or as a startle response by extending the wings through an arc of ~45° at a rate of about once per second with slight supination, then changing to asynchronous supination. Asynchronous supination was observed most commonly in both sexes throughout the day, either spontaneously or while facing other individuals. During asynchronous supination, the wings were extended forward, while remaining arched at 90° from the midline of the body, and while simultaneously being supinated 90° with respect to the substrate.

#### *25.2.6.2.2 Courtship*

Males of *P. cultaris* displayed courtship for ~2 h, typically between 20:00 and 22:00 h in laboratory arenas. Males always approached females for courtship and moved very slowly. As a male approached a female, he displayed abdominal pleural distension and synchronous supination. Walking sideways or making a zigzagging approach also occurred. With each side step both wings were extended forward once and supinated ~45° (= enantion). As the male neared the female, he extended his mouthparts fully. The female responded with asynchronous supination toward the male and turned and slowly walked away from him or remained still. When the male was within touching distance, the female often raised her front legs toward him. The male responded in kind and they wrestled with their front legs, although without raising up or stilting on their hind legs, as was observed with certain other tephritid species (Headrick and Goeden 1994; Headrick et al. 1995). When a female was receptive to a courting male, they stood facing each other and the male held his mouthparts fully extended. After a few seconds, he swayed his body rapidly from side to side. His anterior end passed through an arc of 25° as he pivoted on his hind legs. These side-to-side bursts lasted ~1 to 2 s and he continued until the female responded by extending her mouthparts forward. At the peak of his display, a male stopped swaying and began to vibrate the pseudotracheae of his labella rapidly while facing and nearly touching the female; then the vibrations ceased and he began rapid swaying once more. He continued intermittently with these displays until the female moved away or responded to his display. A receptive female exserted her mouthparts toward the displaying male, signaling her receptivity; the male then moved forward as they both opened their labella to expose the pseudotracheae and placed them together. Males kept their labella closed while swaying. Pairs held their mouthparts together for up to 20 min ($n = 10$). When pairs parted, strings of fluid were observed between their mouthparts indicating that some substance had been exchanged. Individual pairs continued coupling and uncoupling their mouthparts for up to 1 h. No copulation was observed with this species.

### 25.2.6.3 Comparison of Behaviors

#### *25.2.6.3.1 Wing Displays*

Cavender and Goeden (1984) described only asynchronous supination displays for *P. cultaris* adults. Headrick and Goeden (1990c) described wing displays for *P. gentilis* similar to these reported for the first time herein for *P. cultaris* (Headrick and Goeden 1990c; 1994). *Aciurina*, *Eutreta*, *Paracantha*, *Urophora,* and *Valentibulla* all had at least one species observed to display wing arching (Headrick and Goeden 1994).

#### *25.2.6.3.2 Courtship*

Cavender and Goeden (1984) reported that *P. cultaris* males approached females for courtship, displaying swaying and sidestepping, along with enantion and extended mouthparts as reported for *P. cultaris* herein. They also reported that males and females joined mouthparts after a courting male approached a female, and that they remained together for ~7 min, broke apart, then repeated copulation for another ~7 min. This trophallaxis-based display is similar to the display of *P. gentilis* reported by Headrick and Goeden (1990c), but with some differences. As observed for *P. cultaris* adults, Cavender and Goeden (1984) also reported that courtship was commonly observed in the laboratory, but that copulations did not follow.

*Paracantha* represents one of the few genera that have at least two species that display trophallaxis. Other tephritid genera each appear to have only one species which displays trophallaxis. *Aciurina* (Jenkins 1990; Headrick and Goeden 1994), *Eutreta* (Stoltzfus and Foote 1965), and *Neaspilota* (Goeden and Headrick 1992). The only other California species of *Paracantha* is *P. genalis* Malloch, an extremely rare species and not commonly collected, although the host is known (Goeden and Ricker 1987; Headrick and Goeden 1990c). Thus, other North American or perhaps Central or South American species will need to be studied to determine the commonality of behaviors such as trophallaxis within the genus.

## 25.2.7 Behaviors of *Goedenia* Freidberg and Norrbom

Subfamily Tephritinae; Tribe Myopitini

### 25.2.7.1 Introduction

The genus *Goedenia* is endemic to North America (Freidberg and Norrbom, Chapter 23). Eight species occur in the western United States and Mexico, along with perhaps several more undescribed species (Foote et al. 1993; Freidberg and Norrbom, Chapter 23). *Goedenia* has only recently been distinguished from the predominantly Palearctic genus *Urophora* (Freidberg and Norrbom, Chapter 23). Several European *Urophora* spp. are well known and have been introduced into North America for the biological control of knapweeds, *Centaurea* spp. However, little is known of the biology of *Goedenia* spp. Goeden (1987b) reported on the host associations of the six known California *Goedenia* spp. (as *Urophora*), and Goeden et al. (1995b) described the behaviors of adult *Goedenia* (as *Urophora*) *timberlakei* (Blanc and Foote), the only North America species studied in any detail to date.

### 25.2.7.2 Reproductive Behaviors

#### *25.2.7.2.1 Wing Displays*

All species of *Goedenia* exhibit hamation (Headrick and Goeden 1994; Goeden et al. 1995b). Both sexes exhibit this display throughout the day concurrent with other behaviors, such as grooming, resting, and feeding. Hamation in *Goedenia* is similar to that described for *Neaspilota viridescens* Quisenberry (Goeden and Headrick 1992). *Goedenia* spp. do not display asynchronous or synchronous supinations, except for males of *G.* sp. nr. *formosa* (D. Headrick and R. Goeden, unpublished data). Abdominal flexures during wing displays were observed and appear as described for other tephritid species studied (Headrick and Goeden 1994). Both sexes of all species of *Goedenia* occasionally displayed swaying during hamation while facing another individual. Individuals swayed less during hamation when no other individual fly was nearby. Both sexes also sidestepped during hamation while facing other individuals (Headrick and Goeden 1994; Goeden et al. 1995b).

#### *25.2.7.2.2 Courtship and Copulation*

Courtship and copulation were described for *G. timberlakei* (Goeden et al. 1995b). Males displayed abdominal pleural distension and attempted to mount females without any prior behavioral interactions. Males exhibited a unique side-to-side rapid shaking as part of copulatory induction behavior. Males also had to overcome the logistics of copulating with females with unusually long ovipositors, as compared with other California Tephritinae. Copulation durations were 1 and 1.5 h ($n = 2$) (Goeden et al. 1995b).

### 25.2.7.3 Comparison of Behaviors

There are no other published records for behavior in this genus.

### 25.2.8 BEHAVIORS OF *XENOCHAETA* SNOW

Subfamily Tephritinae; Tribe Noeetini

#### 25.2.8.1 Introduction

*Xenochaeta* is a rarely collected genus restricted in distribution to western North America. Goeden and Teerink (1997d) have published the only life history study for this genus.

#### 25.2.8.2 Reproductive Behaviors

The adults of *X. dichromata* (Snow) hold their wings arched and parted as observed for species of *Eutreta* (Headrick and Goeden 1994) and *Paracantha* (Headrick and Goeden 1990; 1994). Wing displays consist of synchronous supinations and, less commonly, asynchronous supination (Goeden and Teerink 1997d). Like many other tephritids, *X. dichromata* adults flex their abdomens during walking and wing displays (Goeden and Teerink 1997d). Courtship in this species is direct, without exhibition of many behaviors; however, males do display wing lofting embellished with rapid wing vibrations and sidestepping while tracking females before mounting attempts. Males were not observed to display abdominal pleural distention (Goeden and Teerink 1997d). Copulations lasted an average of ~90 min. (Goeden and Teerink 1997d).

#### 25.2.8.3 Comparison of Behaviors

There are no other published records for behavioral comparison.

### 25.2.9 BEHAVIORS OF *CAMPIGLOSSA* RONDANI

Subfamily Tephritinae; Tribe Tephritini; *Campiglossa* genus group

#### 25.2.9.1 Introduction

*Campiglossa* has a cosmopolitan distribution, with nearly 200 described species (Novak 1974; Norrbom et al. 1999). Foote (1980) reported that *Campiglossa* (as *Paroxyna*) was little studied and many of the Mexican and Neotropical species have not been described. They are generally small flies similar in habitus to *Tephritis*. *Campiglossa* and *Dioxyna* are morphologically distinguished from related genera, such as *Trupanea*, *Euaresta*, *Euarestoides*, *Tephritis*, and *Neotephritis*, by their geniculate mouthparts. *Goedenia* also has geniculate mouthparts. Goeden and Blanc (1986) synonomized *C. corpulenta* (Cresson) with *C. genalis* (Thomson) and reported on host associations for this species and *C. sabroskyi* (Novak).

The biologies and host associations of most *Campiglossa* species are unknown and many species are known only taxonomically. Novak and Foote (1968) reported on the biology of *C. albiceps* (Loew), and Goeden et al. (1994b) described the life history and immature stages of *C. genalis* (Thomson) in southern California.

#### 25.2.9.2 Reproductive Behaviors

The behaviors of the following species have been observed: *C. genalis* (Goeden et al. 1994b), *C. murina* (Doane), *C. sabroskyi*, *C. steyskali* (Novak), and *C. variablis* (Doane). *Campiglossa* species have common behavioral elements, some of which occur in other genera, for example, some wing displays; however, each species also displays unique behaviors. Adults are active from ~09:00 hours to dusk. Both males and females rest, groom, and feed while in arenas and when observed in the field. Mating occurs throughout the day and into the night in laboratory arenas under artificial lighting. Some matings continued in the dark in both the field and laboratory.

#### 25.2.9.2.1 Wing Displays

All species display wing lofting (White et al., Chapter 33), asynchronous supination, synchronous supination, and hamation. Wing lofting is the most common behavior; it is displayed by both sexes spontaneously and occurs with minor variations in each species. The angle between the wings varies considerably among species, as does the degree of supination and lofting. Another feature of lofting is the concomitant abdomen flexures, which mirror the same rate and degree of loft as the wings. The halteres also are simultaneously depressed when the wings and abdomen are raised. Individual species loft at various rates and some also hold their wings upright during displays and/or vibrate them. Both sexes exhibit spontaneous lofting at other individuals, or at moving objects, but lofting is always a part of male courtship displays.

Asynchronous supination is rarely observed. Females display asynchronous supination when visually orienting toward other individuals or moving objects. The movement is the same as described for other tephritid species. Hamation is also displayed in all species. Hamation follows other wing displays before the wings return to a resting position slightly parted over the dorsum. Males of *Campiglossa* spp. exhibit an agitation wing display while *in copula*. The wings are extended synchronously from a resting position over the dorsum to ~90° from the midline of the body with slight supination, and then are vibrated in a plane parallel to the blade for varying lengths of time. The wings are relaxed to ~60° from the midline of the body, then extended again perpendicular to the body to 90° and vibrated until the males are no longer agitated. Females of some species also display synchronous supination extensions (= enantion) with abdominal flexures as the wings are extended 45 to 90° from the midline of the body and supinated 45° with respect to the substrate. Swaying occurs in all species during lofting and asynchronous supinations, but not during synchronous supinations (Headrick and Goeden 1994). Males of *C. murina*, however, exhibit a unique swaying display as part of their courtship (Headrick and Goeden 1994).

#### 25.2.9.2.2 Courtship

*Campiglossa* males exhibit interest in females for periods of up to 8 h throughout the day in laboratory arenas, but each species has periods when displays are more prevalent (Headrick and Goeden 1994). During these periods, males visually track females while displaying wing lofting and abdominal pleural distension. Individual males always approach females for courtship, and if females remain and do not decamp, then other displays are exhibited. Males approach, move away from, and return to females during their display period and repeat this behavior up to 14 times during individual courtship displays, which last as long as ½ h. Four common elements in male courtship displays are: (1) labellar wagging; (2) abdominal pleural distension; (3) wing lofting; and (4) front leg waving.

Males loft their wings when they visually recognize a nearby female. Lofting is usually the first display exhibited during courtship and occurs throughout courtship until the female is mounted. Abdominal pleural distension characterizes displaying males and persists whether a female is approached or not. The abdominal pleura are distended from the second to fifth segment and deflate occasionally, especially during other behaviors like grooming and feeding. Males distend their abdominal pleura when approaching a female for courtship; they are only returned to normal size once a male mounts a female.

Another courtship display is labellar wagging. Both sexes of *Campiglossa* have geniculate mouthparts. Males always face females during labellar wagging and cease the display when a female turns away. Males extend their mouthparts while displaying to females 2 to 50 mm distant. The labella hang downward at ~90° from the rostrum. The entire mouthpart structure is moved from side to side through ~120° at a rate of approximately two wags per second. Episodes of lofting and wagging continue intermittently while males face females and appear to be mutually exclusive behaviors. Males of some species approach females and crouch before them, then begin displaying labellar wagging. Females do not extend their mouthparts toward males in response, as observed for females in other genera (e.g., *Euaresta*, *Eutreta*, and *Paracantha*; Headrick and Goeden 1994).

As courtship displays continue, males move closer to females. When a male stands facing and nearly touching a female and his display has reached a high intensity, the last of the four displays is exhibited. Males lift each front leg simultaneously above their heads 45° and the leg, including the tarsi, is fully extended in a straight line. Then each leg is waved synchronously and rapidly through 10°; the legs appear blurred when they are waved. After 1 to 2 s, the legs are simultaneously lowered to the substrate. Usually the male is close enough to the female that he touches her head when he lowers his legs. This display ends either with a female moving away from a displaying male or the male attempting to mount the female.

Females of three species give acceptance displays before being mounted by a courting male. Distinct, premounting acceptance displays are rare and have been observed only in two other species, *Paracantha gentilis* (Headrick and Goeden 1990c) and *Euaresta stigmatica* (Headrick et al. 1995). Receptive females give an acceptance display toward courting males, which consists of the females spreading their wings, raising their ovipositors, and lowering their anterior ends in a crouch. Males immediately cease displaying, lower their wings, deflate their abdominal pleura, and climb over the top of the females, then turn 180°, and grasp them with their legs. Males also mount females without prior displays.

Males of *Campiglossa* mount females farther anteriad for copulatory induction behavior (CIB) and copulation, than do members of other genera that we have studied. *Campiglossa* males usually hold their heads above or just behind that of the females. Thus mounted, a male grasps the female's humerus with his front legs, his middle legs wrap around her thorax behind the wing bases, and his hind legs grasp the base of the oviscape. There are slight differences in the mounting posture among species, which will be discussed below.

### *25.2.9.2.3 Copulatory Induction Behavior*

Males raise the abdomen of the female with their hind legs and begin CIB once mounted. Because a male is positioned farther forward on a female, the oviscape has to be flexed as much as 90° to reach the male's epandrium. CIB causes a receptive female to exsert her aculeus. A male rubs his hind legs on the oviscape of the mounted female, either down its sides or along its ventral aspect as it is raised to 90° with respect to the substrate. Rubbing is usually rapid, approximately five to six strokes per second, in short bursts. Sustained CIB occurs when a female is unreceptive and does not exsert her aculeus or when a female exserts her aculeus against the male during copulation (see below). A receptive female usually exserts her aculeus within 2 min after CIB is initiated. An unreceptive female does not exsert her aculeus and the male continues CIB for up to an hour before dismounting.

### *25.2.9.2.4 Copulation*

In the final copulatory position the oviscape is bent upward between 45 and 90°, depending on the species. The wings of the female are spread ~45° from the midline of the body and the male wings are slightly parted. The following activities occur during all copulations. Both sexes feed and groom throughout copulation. Females orient to moving objects and display asynchronous supinations toward them. If a female becomes agitated, she exerts hydrostatic pressure on her aculeus. Males respond with CIB. Exsertion of the aculeus by the female and CIB by the male are common throughout copulation. Males display synchronous wing extensions with or without vibrations, depending on the species involved, when they become agitated while *in copula* or if they observe nearby movement. This display continues until the stimulus is gone; then the male returns his wings to the resting position over his dorsum.

*Campiglossa genalis* display prolonged copulations during which males remain on the dorsa of females after removal of the phallus as a form of mate guarding (Goeden et al. 1994b). CIB is typically reinitiated and copulation again follows. Copulations are repeated from two to seven times in individual mating episodes. Otherwise, copulations ended with males turning 180°, walking off the dorsum of the mounted female, and pulling the phallus free as they moved away from each other (Headrick and Goeden 1994). Postcopulatory behavior was not observed on these occasions.

#### 25.2.9.2.5 Aggression

Females loft or spread their wings to 90° from the midline of the body, supinate their wings to 90° with respect to the substrate, and lunge forward at males while facing them. Males typically decamp. Males show aggression toward females by raising their front legs and lunging at them without any wing displays. Both sexes also raise their front legs at approaching individuals.

### 25.2.9.3 Comparison of Behaviors

The life history and reproductive behavior of *C.* (as *Paroxyna*) *albiceps* (Loew) was described in Ohio on *Aster* spp. (Novak and Foote 1968). *Campiglossa albiceps* overwintered as diapausing puparia, emerged throughout the summer from its *Aster* hosts and remained closely associated with its host plants after emergence. Adults were not observed mating until late summer, after a rather extensive premating/oviposition period (~60 days), during which females remained reproductively immature (Novak and Foote 1968). This life history scheme, however, is not commonly observed in California Tephritidae (Goeden and Headrick 1992).

#### 25.2.9.3.1 Wing Displays

*Campiglossa albiceps* adults were only observed to exhibit asynchronous supination, not lofting (Novak and Foote 1968). This would be a biologically significant deviation from other species in the genus, but again, without the aid of video-recording, similar wing displays may not have been easily observed.

#### 25.2.9.3.2 Courtship

Courtship for *C. albiceps*, as reported by Novak and Foote (1968), consisted of individuals of the opposite sexes facing each other at close range and raising their front legs to make "tarsal contact" with the other's head and antennae. This behavior most likely represented front leg waves by males guarding their territory and during female defensive reactions observed for *Campiglossa* spp. (Headrick and Goeden 1994). However, without the aid of video-recording equipment such behaviors may not have been seen in detail and tarsal contact would be a reasonable conclusion. Novak and Foote (1968) also reported that males and females approached each other, moved away, and reapproached several times after tarsal contact. This was consistent with observations for all other *Campiglossa* spp. observed in the present study.

#### 25.2.9.3.3 Copulation

Mounting or copulations were not observed by Novak and Foote (1968).

## 25.2.10 Behaviors of *Dioxyna* Frey

Subfamily Tephritinae; Tribe Tephritini; *Campiglossa* genus group

### 25.2.10.1 Introduction

The genus *Dioxyna* is widespread in the New World, but is represented by only two species in North America, *D. picciola* and *D. thomae* (Curran) (Foote et al. 1993). *Dioxyna* is closely related to *Campiglossa*. Headrick et al. (1996) provided the first detailed comparative analysis of behavior and host plant relationships. Novak (1974) keyed and discussed the U.S. species, and Goeden and Blanc (1986) discussed host associations for *D. picciola* in California.

### 25.2.10.2 Reproductive Behaviors

Headrick et al. (1996) provided detailed descriptions of the behaviors of *D. picciola*.

#### 25.2.10.3 Comparison of Behaviors

##### *25.2.10.3.1 Wing Displays*

*Dioxyna* is closely related to *Campiglossa* and these genera share many behaviors in common that differ from other related genera (Headrick and Goeden 1994). The primary wing display for both *Dioxyna* and *Campiglossa* is lofting, not asynchronous supination, although *Euaresta* males wing-loft in their courtship displays (Headrick and Goeden 1994; Headrick et al. 1995).

##### *25.2.10.3.2 Courtship*

Grewal and Kapoor (1984) described the floral disks of flowerheads of *Calendulla officianalis* L. as "sites of assembly" on which most male–female encounters took place. Grewal and Kapoor (1984) did not report male courtship displays, only that males rapidly mounted ovipositing females. Mountings were attempted in approximately half of all encounters and most were successful. Grewal and Kapoor (1984) also reported that pheromones were not involved in courtship or copulation and that conspecific recognition was by vision only. Males of *D. picciola* were not observed to display courtship in the field; however, courtship was observed in laboratory arenas (Headrick et al. 1996). The display of courtship behaviors probably is dependent on a number of factors, among which is local population densities. Even fly densities per single host plant are likely to be important (Headrick and Goeden 1994; Headrick et al. 1996).

##### *25.2.10.3.3 Copulation*

Grewal and Kapoor (1984) reported that males copulated with females an average of four times during a mating episode, and that between copulations, females laid two to five eggs singly. These results compared favorably with the three copulations per mating episode and one to three eggs laid between copulations reported by Headrick et al. (1996).

Grewal and Kapoor (1984) reported that males cleaned themselves after mounting and that male grooming produced a similar response in the mounted female. After a period of cleaning, males began pumping their proboscis, which repeatedly touched the vertical and postvertical bristles on the female's head. They reported that this action may have incited the female to raise her oviscape for the male to gain intromission. They did not report the CIB behavior described by Headrick et al. (1996), which involved mounted males rubbing the tops of females' abdomens with their hind tarsi. Copulation did not occur until CIB was initiated by mounted males (Headrick et al. 1996). In support of observations with *D. picciola*, most other tephritids observed have some form of CIB involving males rubbing the abdomen or oviscape of a mounted female with their hind legs (Headrick and Goeden 1990; 1991; 1994; Goeden and Headrick 1991).

Mate guarding behavior in the Tephritidae has rarely been reported. *Dioxyna picciola* males always remained on females after an initial copulation, and females were not observed to oviposit unaccompanied by a mounted male (Headrick et al. 1996). *Dioxyna* and *Campiglossa* share similar life history strategies, hosts, and copulatory behaviors, and they are the only genera known to exhibit mate guarding.

### 25.2.11 Behaviors of *Euarestoides* Benjamin

Subfamily Tephritinae; Tribe Tephritini; *Euarestoides* genus group

#### 25.2.11.1 Introduction

*Euarestoides* is commonly found in southern Canada, the United States, and to an undetermined extent in the Neotropics (Foote et al. 1993). *Euarestoides* is closely related to *Euaresta*, *Neotephritis*, *Campiglossa*, *Tephritis*, and *Trupanea*; however, little is known about the biology of most species. *Euarestoides* was first considered a subgenus of *Trupanea* (Foote et al. 1993). Only the biology of *E. acutangulus* (Thomson) is known among North American species (Piper 1976).

### 25.2.11.2 Reproductive Behaviors

#### *25.2.11.2.1 Wing Displays*

Both sexes exhibited asynchronous supination displays that were similar to *Trupanea* and *Tephritis*. These displays were spontaneous or induced while facing another individual. Asynchronous supination displays were typically followed by hamation. Hamation occurred as the wings were returned over the dorsum to their resting position. Both sexes exhibited swaying during asynchronous supinations while facing another individual.

#### *25.2.11.2.2 Courtship*

Courtship displays were commonly observed from approximately 12:00 to 16:00 hours for *E. flavus* (Adams). Males visually tracked and oriented toward females and displayed asynchronous supinations. Males approached females to within 5 mm and continued their wing displays. In all nine male–female interactions observed, females turned away from the displaying male and were immediately mounted from behind. Unreceptive females fought at males with their hind legs ($n$ = 6). Males tracked and attempted to mount unreceptive females for up to 3 h ($n$ = 3). Receptive females allowed males to remain mounted, which was followed by CIB.

#### *25.2.11.2.3 Copulatory Induction Behavior*

Mounted males used their hind legs to rub against the sides of the female's oviscape and simultaneously pressed the epandrium to the apex of the oviscape ($n$ = 3). A receptive female exserted her aculeus into the male's epandrium. CIB was continued after intromission, until the final copulatory position was gained.

#### *25.2.11.2.4 Copulation*

The final copulatory position was typical for most tephritid species. The male's front legs grasped the top of the abdomen of the female, his tarsi rested on the first tergite parallel with the long axis of her body, his middle legs wrapped around her abdomen near the base of the oviscape, and his hind legs held the oviscape during CIB, then were lowered to the substrate in the final position. During copulation both sexes groomed, formed droplets, and displayed wing extensions. Females oriented to movement and displayed swaying and wing extensions. The female began vigorous side-to-side swaying while exserting her aculeus and flexed her abdomen before disengagement. The male turned 180°, climbed down to the substrate, and walked away from the female while pulling his phallus out from her aculeus retracted into the oviscape. Both sexes immediately began grooming. Copulations lasted ~2 h ($n$ = 3) in laboratory arenas.

### 25.2.11.3 Comparison of Behaviors

Piper (1976) described the bionomics of *E. acutangulus* on *Ambrosia chamissonis* (Lessing) Greene, a native, coastal, ragweed species found on maritime sand dunes in southern and central California. *Euarestoides acutangulus* is bivoltine and overwinters as puparia. Adults emerge before flowering, after which mating and oviposition take place. $F_1$ adults emerged after ~1 month, and apparently this generation also mates and oviposits in *A. chamissonis*. The second generation overwintered as puparia in staminate flowerheads (Piper 1976). Adults were closely associated with their host plants and rarely were observed on nearby nonhost plants (Piper 1976). Adults were observed mating and ovipositing in June; $F_1$ adults emerged in late August (Piper 1976). The premating period for laboratory-reared adults averaged 11 days (Piper 1976). Piper (1976) reported that all matings in the field were observed on foliage and racemes. *Euarestoides acutangulus* also mated on, but was not confined to mating on, foliage and racemes of *A. dumosa* (Headrick and Goeden 1994).

#### *25.2.11.3.1 Wing Displays*

Piper (1976) reported that females displayed asynchronous supinations, and both sexes displayed synchronous supinations. However, the context in which these wing displays were used was not identified (Piper 1976).

#### *25.2.11.3.2 Courtship*

Courtship, as described by Piper (1976), consisted of males approaching females, exhibiting a wing display, and "walking around" a female. When they were within 15 mm, they turned to face each other and the male displayed side steps of ~10 mm and what are interpreted by us to be asynchronous supinations (Piper 1976).

#### *25.2.11.3.3 Copulation*

Piper (1976) reported that receptive females remained still and that males attempted to mount them from the rear. Headrick and Goeden (1994, and unpublished data) noted that *E. acutangulus* females were visually acutely aware of their surroundings and were never observed to "allow" males free access to move around them at such close range as described by Piper (1976). Instead, they decamped or showed aggression at the slightest movements of approaching males. *Euarestoides acutangulus* overwinters as adults in southern California (Goeden and Headrick, unpublished data). The populations studied by Headrick and Goeden (1994) were located at aggregation sites, but they did not express typical aggregation behavior. *Euarestoides acutangulus* males perched on tops of racemes, and tracked and mounted females without prior displays, similar to other aggregative species; however, they also exhibited territoriality, resource defense polygyny, and displays such as abdominal pleural distension. The seasonal history of *E. acutangulus* is complex, as potential hosts bloom nearly all year long (R. Goeden and D. Headrick, unpublished data). Thus, aggregative-type behaviors may be facultative and depend for expression on higher population densities at aggregation sites.

Laboratory copulations lasted from 1 to more than 3 h and were repeated frequently. Field copulations were not timed, but occurred throughout the day (Piper 1976). Laboratory copulations averaged 18 h ($n = 5$, range 6 to 24.5 h) and field copulations lasted for more than 24 h (Headrick and Goeden 1994, and unpublished data).

#### *25.2.11.3.4 Resource Guarding*

*Euarestoides acutangulus* adults fed regularly on cercopid spittle deposits (Headrick and Goeden 1994, and unpublished data). Males also defended these deposits against other males while allowing females to feed freely. This was the only case of what could be considered resource defense polygyny observed by Headrick and Goeden (1994). Male *E. acutangulus* guarding was infrequent and may be linked to one of two factors. First, the cercopid spittle may not be a critical resource for females; thus, the males who guarded these deposits were taking advantage of an opportunity provided by females frequenting spittle deposits. Opportunistic behavior, again, appears to be a hallmark of native California tephritids and the resource guarding behavior described may be part of a repertoire of guarding behaviors involving various resources (Headrick and Goeden 1994). Second, there may be a particular time of day when feeding takes place as field observations suggest (D. Headrick and R. Goeden, unpublished data). Thus, males may display territoriality at a resource that serves to draw both males and females and where there is a higher frequency of encounter between individuals.

### 25.2.12 Behaviors of *Trupanea* Schrank

Subfamily Tephritinae; Tribe Tephritini; *Tephritis* genus group

### 25.2.12.1 Introduction

The genus *Trupanea* has over 200 described species and occurs in all the major geographic regions of the world (Norrbom et al. 1999; Merz, Chapter 24). *Trupanea* is the largest and most commonly encountered genus of Tephritidae in California, where 16 species have been recorded thus far (Goeden 1992; Foote et al. 1993). What is known of the biology of *Trupanea* derives from life history studies of the California species *T. bisetosa* (Coquillett) (Cavender and Goeden 1982; Knio et al. 1996), *T. conjuncta* (Adams) (Goeden 1987), *T. imperfecta* (Coquillett) (Goeden 1988a), *T. californica* Malloch (Headrick and Goeden 1991), *T. nigricornis* (Coquillett) (Knio et al. 1996), *T. actinobola* (Loew) (Goeden et al. 1998), *T. signata* Foote (Goeden and Teerink 1997c), *T. jonesi* Curran (Goeden et al. 1998), *T. pseudovicina* Hering (Goeden and Teerink 1998), and *T. arizonensis* Malloch (Goeden and Teerink 1999). Goeden (1992) reported host plant associations for most of California's 16 species.

### 25.2.12.2 Reproductive Behaviors

#### *25.2.12.2.1 Wing Displays*

The asynchronous wing display is common to all species of *Trupanea* observed; it occurs throughout the day, during resting, grooming, and feeding at a frequency of about one wing extension per second. Each display consists of 1 to more than 20, alternating wing extensions in a given episode. One wing may also be extended more than once in succession. The wing movement only changes in frequency and synchrony when behavior changes, for example, when a male begins a courtship display or a female displays aggression. Thus, most of the unique wing movements of males are sexually dimorphic, typically involve courtship and mating, and will be described for each species below.

Both sexes of all species of *Trupanea* display hamation as defined by Headrick and Goeden (1991). All *Trupanea* species display abdominal flexures during asynchronous wing displays. This display is observed in other tephritid genera, but its function remains unknown. In *Trupanea*, the apex of the abdomen does not touch the substrate when it is flexed, so no deposition of material was suggested, as observed with *Neaspilota viridescens* Quisenberry males (Goeden and Headrick 1992). Both sexes also exhibit swaying during wing displays, or occasionally, with their wings held over the dorsum. Swaying in *Trupanea* also occurs during mating. Females swayed if they perceived nearby moving objects and extended their wings in a show of defense or aggression. The entire body is moved, but the head travels through the greatest arc. Swaying during a wing display occurs while one wing is held perpendicular to the body and supinated to 90°; however, the body is not swayed while the wings are in motion (Headrick and Goeden 1994).

#### *25.2.12.2.2 Courtship*

Both behavioral and anatomical changes identify a male *Trupanea* courtship display. Males distend their abdominal pleura and switch from an asynchronous to a synchronous wing display. This is observed with all *Trupanea* species and is probably the rule rather than the exception for the genus (Headrick and Goeden 1994).

Abdominal pleural distension is always accompanied by at least one unique wing display for each species. *Trupanea* is an exemplary genus for abdominal pleural distention, as this behavior is so widespread and, typically, very pronounced in all species (Headrick and Goeden 1994). Abdominal pleural distension occurs from the inception of a male's courtship display period and usually ceases when copulation begins or just before mounting.

#### *25.2.12.2.3 Copulatory Induction Behavior*

Just after mounting, a male grasps a female with all his legs in order to hold onto her, as most females resist being mounted (Headrick and Goeden 1994). The male then uses his hind legs to raise her abdomen and oviscape and press or drum his epandrium against its apex. Males of most species of *Trupanea* also rub their front legs asynchronously in 1-s bursts against the abdominal terga of females (Headrick and Goeden 1994). CIB is typically brief, as a receptive female exserts her aculeus soon after mounting and a male readily dismounts from an unreceptive female. Intromission is rapidly gained when the female exserts her aculeus.

##### *25.2.12.2.4 Copulation*

After intromission, both adults remain quiet for ~5 min until the male disengages from the female by turning and stepping down onto the substrate. The male walks away from the female while pulling his phallus from her aculeus. Typically, males remain active for brief periods after copulation, but ultimately both individuals move to opposite sides of the arena and groom.

##### *25.2.12.2.5 Aggression*

In some *Trupanea*, both sexes raise their front legs as a characteristic defensive posture to ward off an advancing intruder. If one adult is approached abruptly by another, the front legs immediately are raised in unison to a point where the tarsi are just above the head. If the intruder leaves, the legs are slowly lowered to the substrate. If the intruder does not leave and stays nearby, the legs remain above the head. If the intruder continues to move forward, the legs are vigorously brought down upon the intruder several times in a synchronous "pawing" fashion. If one adult approaches the side of another, instead of face forward, then only the leg on the side approached is raised in defense, and if the adult begins to paw at the intruder, only this one leg is used.

*Trupanea* males rarely show aggression toward females while in arenas. However, females often display aggression toward males. If a male approaches a female, she uses her front legs to drive him away. If the male persisted, the female thrusts and supinates both wings forward while lunging headfirst into the male. This is repeated and is effective in driving the male away.

#### 25.2.12.3 Comparison of Behaviors

##### *25.2.12.3.1 Wing Displays*

*Trupanea* is a behaviorally cohesive genus, only the wing displays by males during courtship vary significantly among species (Headrick and Goeden 1994). Both sexes of all species studied displayed asynchronous supinations as their primary wing display with little variation. Both sexes of all species also displayed hamation uniformly.

##### *25.2.12.3.2 Courtship*

In the species examined to date, males exhibited two distinct morphological changes in courtship: (1) abdominal pleural distension and (2) a change to synchronous wing extensions. Both of these behaviors continued throughout the courtship display period. Males were observed to display at particular times of the day and for consistent durations. Cavender and Goeden (1982) reported that males of *T. bisetosa* displayed early in the mornings and *T. nigricornis* males displayed in the late afternoon. *Trupanea californica* (Headrick and Goeden 1991), *T. nigricornis*, and *T. wheeleri* Curran (Headrick and Goeden 1994) were observed consistently to display courtship late in the afternoon, whereas *T. jonesi* (Goeden et al. 1998) and *T. signata* (Goeden and Teerink 1998) consistently displayed early in the mornings. However, *T. actinobola* (Goeden et al. 1998), *T. arizonensis* (Goeden and Teerink 1999), and *T. radifera* (Coquillet) (R. Goeden, J.A. Teerink, and D. Headrick, unpublished data) displayed courtship throughout the day. This suggested that the presence of females was not required for males to initiate their displays. Apparently, males responded to other environmental stimuli and their displays may serve to attract females from a distance.

The rate and range of synchronous wing extensions during courtship by males were typically species specific (Headrick and Goeden 1994). Most commonly, synchronous wing extensions by males during courtship were to 90° perpendicular to the body (Headrick and Goeden 1994).

Cavender and Goeden (1982) reported 90° extensions for *T. bisetosa* males; however, the males of a cryptic congener, *T. nigricornis*, extended their wings only to 45° (Headrick and Goeden 1994). Headrick and Goeden (1991) reported wing displays by *T. californica* males as 45°. *Trupanea radifera* males also extended their wings to ~45° during courtship (R. Goeden and D. Headrick, unpublished data).

##### *25.2.12.3.3 Copulation*

Although males of most *Trupanea* spp. readily displayed courtship behaviors, most have never been observed to copulate (Headrick and Goeden 1991; 1994; Knio et al. 1996; Goeden and Teerink 1997c; 1998). Copulation in *Trupanea* is typically brief, usually no more than 10 min (Headrick and Goeden 1994). This is on average the shortest duration for copulation of the genera discussed in the present chapter. Females remain quiet after disengagment and groom (Headrick and Goeden 1994). No female was ever observed to copulate more than once per day, and typically females mated only once during trials which lasted up to 1 month.

### 25.2.13 Behaviors of *Euaresta* Loew

Subfamily Tephritinae; Tribe Tephritini; *Incertae sedis*

#### 25.2.13.1 Introduction

*Euaresta* is endemic to the New World (two species have been introduced to the Old World), and is widespread in North America (Norrbom et al. 1999). Males in this genus have enlarged forefemora and distinctive external genitalia (Foote et al. 1993). Little is known about the biology of *Euaresta*. Batra (1979) described the biology and behavior of *E. bella* (Loew) and *E. festiva* (Loew), potential biological control agents for the control of ragweeds (*Ambrosia* spp.) in Eurasia. *Euaresta stigmatica* Coquillett was the only species examined for this genus in the behavior studies conducted by Headrick and Goeden (1994). *Euaresta stigmatica* infests the staminate flower heads and fruits of native *Ambrosia* spp. in southern California (Goeden and Ricker 1976a).

#### 25.2.13.2 Reproductive Behaviors

Headrick et al. (1995) described in detail the complex behaviors of adults of *E. stigmatica*.

#### 25.2.13.3 Comparison of Behaviors

Batra (1979) described the courtship and mating behavior of *E. bella* and *E. festiva* from *A. artemisiifolia* L. and *A. trifida* L., respectively. In Ohio *E. bella* and *E. festiva* are monophagous and univoltine on their respective host plants. Females of *E. bella* and *E. festiva* emerged and remained reproductively immature for ~4 weeks, after which reproduction (egg maturation through oviposition) occurred in a narrow window of ~4 weeks (Batra 1979). This compares favorably with the prereproductive period of ~3 weeks and 4 to 6 weeks, respectively, of reproductive behaviors for *E. stigmatica* observed by Headrick et al. (1995) at desert field sites. However, *E. stigmatica* $F_1$ adults emerged 2 to 3 weeks later, and overwintered as adults and not as puparia (Headrick et al. 1995). Reproduction in *E. bella* and *E. festiva* was contemporaneous with the flowering of their hosts, as with *E. stigmatica* (Headrick et al. 1995).

*Euaresta bella* and *E. festiva* adults were most active in the late afternoon, and both tended to remain on or near their host plants (Batra 1979). *Euaresta stigmatica* adults were active from approximately 09:00 to 15:00 hours, but females were generally more abundant and active after approximately 13:00 h (Headrick et al. 1995).

#### 25.2.13.3.1 Courtship

Courtship behaviors for *E. bella* and *E. festiva* were described by Batra (1979), and appear similar to those of *E. stigmatica*, except in terms of interpretation. Batra (1979) listed ten courtship behavior "patterns" observed for both *E. bella* and *E. festiva*, "(1) visual orientation; (2) alternate wing waving with vibration; (3) both wings extended horizontally; (4) both wings extended with proboscis extended; (5) both wings extended with head butting; (6) tapping with front feet; (7) rapid flicks of both wings simultaneously; (8) territoriality (lekking); (9) male following female, abdomen curved; (10) male following female, wings flattened against abdomen." Interpreting the courtship sequence for *E. bella* and *E. festiva* as described by Batra (1979) is difficult because other behaviors besides courtship were displayed simultaneously, such as aggression, male–male interactions, and female–female interactions. From the data presented by Batra (1979), during courtship, males approached females with a wing display other than lofting; mouthpart contact was made. Males of *E. festiva* followed females for courtship with their abdomens "curved," their wings flat over their dorsa, and male courtship took place while females were ovipositing; *E. bella* males did not court ovipositing females. Also according to Batra (1979), males of *E. bella* and *E. festiva* established territories, but of differing sizes. *Euaresta bella* males had a territory of one or two leaves, which males occupied for a few hours; however, *E. festiva* males had a territory of 1 $m^2$, which individual males occupied apparently for more than a few hours (Batra 1979). Copulation for *E. festiva* occurred late in the afternoon and lasted for ~1h ($n = 2$); copulation for *E. bella* occurred throughout the day and lasted between 20 to 60 min ($n = 15$) (Batra 1979).

The behavioral similarities of both of these species, and especially *E. festiva*, to *E. stigmatica* is an example of the many shared behavioral attributes of congeners. Behaviors apparently common to all three species include wing enantion or "rapid flicks of both wings simultaneously," asynchronous supination or "alternate wing waving with vibration" (although less common in *E. stigmatica*), mouthpart contact, and territoriality. *Euaresta stigmatica* and *E. festiva* also share two more features: males curling their abdomens underneath them and holding their wings flat over their dorsa while standing behind females. However, *E. stigmatica* did not court ovipositing females, and copulations only took place on the upper surfaces of leaves patrolled or defended by males. Other behaviors reported for *E. stigmatica* (Headrick et al. 1995) and not reported by Batra (1979) for *E. bella* and *E. festiva* include wing lofting, abdominal pleura distension, males raising their front legs as an initial greeting, male lofting/dance during courtship, a courting male's rapid sidestepping to get behind a female, a courting male raising his middle legs after initial rejection by an unreceptive female, and male ritual combat.

### 25.2.14 Behaviors of *Tephritis* Latreille and *Neotephritis* Hendel

Subfamily Tephritinae; Tribe Tephritini; *Tephritis* genus group

#### 25.2.14.1 Introduction

The genus *Tephritis* has 168 described species and is found in most zoogeographic regions (Norrbom et al. 1999; Merz, Chapter 24). North America contains 18 species (Goeden 1988b; 1993; Jenkins and Turner 1989; Goeden and Headrick 1991; Foote et al. 1993). *Tephritis* belongs to the same tribe as *Euaresta*, *Euarestoides*, *Neotephritis*, *Campiglossa,* and *Trupanea*, and species in these genera share similar wing patterns and behaviors.

The genus *Neotephritis* is restricted to the New World and Hawaii, with two species, *N. finalis* (Loew) and *N. inornata* (Coquillett), occurring north of Mexico (Goeden et al. 1987; Foote et al. 1993). *Neotephritis* is closely related to *Tephritis* and they share similar life history strategies and behaviors, thus behavioral descriptions of *N. finalis* will be included with *Tephritis*. The life history, host associations, and reproductive behavior of *N. finalis* were described by Goeden et al. (1987).

#### 25.2.14.2 Reproductive Behaviors

##### 25.2.14.2.1 *Wing Displays*

The most common wing display for both sexes of all species of *Tephritis* was asynchronous supination (Goeden et al. 1993; Headrick and Goeden 1994). Both sexes of all *Tephritis* species observed thus far displayed hamation (Goeden et al. 1993; Headrick and Goeden 1994, and unpublished data). Typically, hamation followed asynchronous supination. Males sometimes exhibited enantion when approaching females; this wing display was typical among *Tephritis* males (Goeden et al. 1993; Headrick and Goeden 1994).

##### 25.2.14.2.2 *Courtship*

Courtship displays were brief, if exhibited at all, as observed for *Trupanea* spp. (Headrick and Goeden 1994). The only consistent behaviors exhibited by courting males were tracking and wing displays, and a few displayed abdominal pleura distention (Headrick and Goeden 1994). Males rarely exhibited abdominal pleura distension characteristic of many tephritid species (Goeden et al. 1993; Headrick and Goeden 1994). Courting male wing displays consisted of asynchronous supination. Asynchronous supinations were often followed by synchronous supinations or hamation if the male tracked the female (Headrick and Goeden 1994, and unpublished data). Males of *T. stigmatica* displayed aggressive male–male interactions in laboratory arenas as described for other tephritids (Headrick and Goeden 1994).

##### 25.2.14.2.3 *Copulatory Induction Behavior*

Mounting and CIB in *Tephritis* were similar to *Trupanea*: males jumped on the dorsa of females without prior displays. A female often struggled after being mounted, using her hind legs to push at the male. CIB was initiated immediately after mounting. *Tephritis* males rubbed their hind legs ventrally near the oviscape base. Some males engage in CIB for up to 30 min before successfully gaining intromission or dismounting (Headrick and Goeden 1994, and unpublished data).

##### 25.2.14.2.4 *Copulation*

Copulation position, duration and activities during copulation were similar to those reported for *Trupanea* species (Headrick and Goeden 1994).

#### 25.2.14.3 Comparison of Behaviors

There are no other behavioral studies for comparison.

### 25.2.15 Behaviors of *Neaspilota* Osten Sacken

Subfamily Tephritinae; Tribe Terelliini

#### 25.2.15.1 Introduction

The genus *Neaspilota* is found entirely in the New World (Foote et al. 1993). Freidberg and Mathis (1986) revised *Neaspilota* and divided the 14 North American species into two subgenera, *Neaspilota sensu stricto* and *Neorellia* Freidberg and Mathis. All six species studied herein belong to *Neorellia*. Goeden (1989) reported host associations for 9 of the approximately 11 species of *Neaspilota* known to occur in California.

#### 25.2.15.2 Reproductive Behaviors

The biology of *Neaspilota* has been little studied because, until the revision by Freidberg and Mathis (1986), the genus was taxonomically poorly known. All California species infest the flowerheads of composites, mostly in the tribe Astereae (Goeden 1989). No biological studies had been reported for any species of *Neaspilota*, except for short notes on host associations, until Goeden and Headrick (1992) described the life history of *N. viridescens* Quisenberry in California.

### 25.2.15.3 Comparison of Behaviors

#### *25.2.15.3.1 Wing Displays*

*Neaspilota* species are unique because they display neither asynchronous or synchronous supinations (Headrick and Goeden 1994). Hamation is typically the primary wing display for courtship and copulation, but is unique in its periodicity (Goeden and Headrick 1992; Headrick and Goeden 1994). Mounted males also displayed rapid enantion when disturbed by other individuals (Goeden and Headrick 1992). The abdomen was flexed during wing displays as described for other tephritid species studied here (Headrick and Goeden 1994). Both sexes of *Neaspilota* spp. occasionally displayed swaying and sidestepping during hamation while interacting with other individuals (Headrick and Goeden 1994).

#### *25.2.15.3.2 Courtship*

Only two *Neaspilota* species have been observed during courtship and subsequently copulation: *N. achilleae* Johnson, which displayed no courtship, just males mounting females as opportunities arose in laboratory arenas (D. Headrick and R. Goeden, unpublished data) and *N. viridescens*, whose courtship involved wing displays, abdominal pleural distension, and nuptial gift formation (Goeden and Headrick 1992).

## 25.3 CONCLUSIONS AND FUTURE RESEARCH NEEDS

The sexual behaviors of fruit flies in the subfamily Tephritinae are highly diverse and typically overt in their presentation. The objectives in describing these behaviors include determining the types of behaviors exhibited, describing them in a scientific manner to promote comparison, and searching for patterns that will perhaps find utility in defining phylogenetic relationships. Headrick and Goeden (1994) provided an overview of tephritid behaviors for the subfamily Tephritinae and compared these to the family as a whole, but without reference to phylogeny. The current chapter allows for examination of the behaviors in the subfamily Tephritinae in a phylogenetic context.

We consider tephritine life history strategy to be a major factor in predicting the types of reproductive behaviors a species will exhibit. Following the patterns established by Bateman (1972) and Prokopy (1980), we categorized tephritines as either aggregative or circumnatal (Headrick and Goeden 1994). We find that the more primitive genera are more likely to exhibit circumnatal life history strategies, for example, *Procecidochares* and *Aciurina* (Sivinski et al., Chapter 28). Interestingly, species in these genera are typically gall formers, a trophic strategy thought of as highly evolved. The reproductive behaviors exhibited by circumnatal species are generally less interactive and involve few, if any, aggregative-type behaviors such as wing displays, abdominal pleural distention, and territoriality. The aggregative life history strategy apparently allows for the development of a greater variety of interactive behaviors and complex mating strategies such as mate guarding and resource defense.

We also find there is a distinct cohesiveness to the behaviors exhibited by species within a genus, but relatively little cohesiveness among related genera; for example, *Paracantha* and *Eutreta* or *Euaresta* and *Euarestoides*. Further, we find that species that exhibit long, involved, highly interactive aggregation, courtship, and mating do so in the absence of closely related congeners. A long-established theme in the evolution of complex behaviors is that they arise to aid in preventing mating mistakes between closely related, sympatric species. For species such as *P. gentilis* and *Euaresta stigmatica*, the absence of closely related species does little to help in determining the forces present to evolve such complex, interactive behaviors (Sivinski et al., Chapter 28).

Goeden and co-workers continue to publish detailed behavioral accounts for individual species studied in southern California. From this work, patterns have emerged allowing for some moderate predictability in what types of behaviors may be expected based on taxonomic affiliation.

Wing displays offer the opportunity to relate morphological characters and associated behaviors in an evolutionary context. All banded-wing genera display enantion; apparently this is the same wing display reported for such banded-wing genera as *Anastrepha* (Burk 1981), *Ceratitis* (Keiser et al. 1973), *Euleia* (Tauber and Toschi 1965), and *Zonosemata* (Greene et al. 1987). Banded-wing patterns and displaying enantion formed the basis for the hypothesis that banded-wing tephritids (studied experimentally with one species each of *Rhagoletis* and *Zonosemata*) elude salticid spider predators by mimicking their greeting dance (Greene et al. 1987; Mather and Roitberg 1987); however, size, gender, and age of the tephritid also play a significant role in this particular predator–prey system (Hasson 1995). There is probably a more prevalent selective force, as banded patterns are quite common in the family and many other families of Diptera (D. Headrick, unpublished data). Banding patterns on the body of an animal are indicative of disruptive coloration; bands on wings allows for the development of a greater range of expression in obscuring the outline figure of the body by changing wing position, thus providing greater protection from predation. For further analysis of this phenomenon, see Sivinski et al. (Chapter 28).

The study of *A. idahoensis* (Headrick and Goeden 1994; Goeden and Teerink 1996b) provided the opportunity to test one hypothesis about wing displays involving wing patterns and taxonomy. Tephritids have several distinct wing patterns that generally conform at the level of the genus, thus, most species in a genus share the same general wing pattern (Headrick and Goeden 1994). There are also groups of genera that have similar wing patterns, for example, *Procecidochares* and *Rhagoletis*; *Campiglossa*, *Dioxyna*, *Neotephritis,* and some *Tephritis*; other *Tephritis* and *Trupanea.* Species in these pattern groups also exhibit similar wing displays (Headrick and Goeden 1994). All the species with a striped pattern like *Procecidochares* display enantion (Headrick and Goeden 1994). However, *Aciurina* spp. are exceptional. The majority of species have a similar pattern, but *A. idahoensis* has a striped pattern similar to *Procecidochares*. The working hypothesis developed by Headrick and Goeden (1994) was that wing patterns and displays are closely linked because of their intimate association with courtship, copulation, territoriality, and aggression displays. Thus, the question raised by *A. idahoensis* was which wing display would it exhibit? In fact, *A. idahoensis* exhibited only one wing display — asynchronous supination, and this, very infrequently. The display was more like those displays observed with other *Aciurina* spp. rather than that typical of banded-winged species. Although the wing blades in *A. idhaoensis* were not bent and ancillary rotations were not observed, a distinction has been made by Headrick and Goeden (1994) between asynchronous supination and enantion. Thus, wing displays may conform more closely with taxonomic affinities, at least in *Aciurina.* This result does little to help clarify the long-assumed hypothesis that the primary function of wing patterns and displays is reproductive isolation or conspecific recognition (Tauber and Toschi 1965; Tauber and Tauber 1967; Burk 1981; Berube and Myers 1983; Dodson 1987b; Jenkins 1990; Headrick and Goeden 1994).

Another area that is given to comparison is mating strategies. Paternity assurance appears to be uniformly achieved in the tephritines by long durations for copulation that are followed directly by oviposition. The exception is the unusually short durations for copulation among *Trupanea* species. The mechanisms for having oviposition follow copulation remain unstudied, but probably include diel patterns and the transfer of behavior-modifying substances from the male to the female. Males increase their chance of paternity by remaining with females until they are ready to oviposit (Emlen and Oring 1977). This is facilitated by populations of tephritines being closely associated with their host plants at a time in which the oviposition substrate is suitable for larval development – such phenological timing is achieved by both aggregative and circumnatal species. *Euarestoides acutangulus* had the longest average copulation duration of any tephritid species examined by Headrick and Goeden (1994) and was the only species except *Dioxyna picciola* and *Tephritis baccharis* to copulate overnight (Goeden and Headrick 1991; Headrick et al. 1996). However,

*D. picciola* also displays mate guarding (Headrick et al. 1996), a classical example of paternity assurance (Thornhill and Alcock 1983). *Euarestoides acutangulus* males were never observed to remain with females while they oviposited (D. Headrick and R. Goeden, unpublished data). Again, see Sivinski et al. (Chapter 28) for more analyses of mating strategies in the family.

In summary, intimate ties with a host plant species and its seasonal development, again appears to be the driving force in the development of the various reproductive strategies exhibited by tephritine species.

Future research should target a continued drive toward meaningful terminology associated with tephritid behavior. Field studies of behavior will continue to be a challenge, but necessary to verify laboratory-based observations. Field studies are essential in accurately describing behavioral series, such as those observed during courtship. These behavioral series are a critical component in establishing behavioral homology relative to the study of phylogenetic relationships (see Sivinski et al., Chapter 28). Biology studies on tephritids should try and include at least observational data on behaviors as observed in the field or laboratory to expand the behavior database for the family. Tephritids also provide a unique opportunity to study evolution in sympatry and the role of behavior including oviposition behavior in speciation. Again, fieldwork coupled with an efficient terminology will facilitate such studies.

## ACKNOWLEDGMENTS

We would like to acknowledge the following for their generous support for this project. Partial funding was received from Campaña Nacional contra las Moscas de la Fruta (Mexico), International Organization for Biological Control of Noxious Animals and Plants, Instituto de Ecología, A. C. (Mexico), and USDA-ICD-RSED. We also gratefully acknowledge the help of two anonymous reviewers for their comments on early drafts of this chapter.

## REFERENCES

Arita, L.H. and K.Y. Kaneshiro. 1986. Structure and function of the rectal epithelium and anal glands during mating behavior in the Mediterranean fruit fly male. *Proc. Hawaii. Entomol. Soc.* 26: 27–30.

Bateman, M.A. 1972. The ecology of fruit flies. *Annu. Rev. Entomol.* 17: 493–518.

Batra, S.W.T. 1979. Reproductive behavior of *Euaresta bella* and *E. festiva* (Diptera: Tephritidae), potential agents for the biological control of adventive North American ragweeds (*Ambrosia* spp.) in Eurasia. *J. N.Y. Entomol. Soc.* 87: 118–125.

Benjamin, F.H. 1934. Descriptions of some native trypetid flies with notes on their habits. *U.S. Dep. Agric. Tech. Bull.* No. 401, 95 pp.

Berube, D.E. and J.H. Myers. 1983. Reproductive isolation between *Urophora affinis* and *U. quadrifasciata* (Diptera: Tephritidae) in British Columbia. *Can. J. Zool.* 61: 787–791.

Burk, T. 1981. Signaling and sex in acalyptrate flies. *Fla. Entomol.* 64: 30–43.

Cavender, G.L. and R.D. Goeden. 1982. Life history of *Trupanea bisetosa* (Diptera: Tephritidae) on wild sunflower in southern California. *Ann. Entomol. Soc. Am.* 75: 400–406.

Cavender, G.L. and R.D. Goeden. 1984. The life history of *Paracantha cultaris* (Coquillett) on wild sunflower, *Helianthus annuus* L. ssp. *lenticularis* (Douglas) Cockerell, in southern California (Diptera: Tephritidae). *Pan-Pac. Entomol.* 60: 213–218.

Dodson, G. 1987a. Host–plant records and life history notes on New Mexico Tephritidae (Diptera). *Proc. Entomol. Soc. Wash.* 89: 607–615.

Dodson, G. 1987b. Biological observations on *Aciurina trixa* and *Valentibulla dodsoni* (Diptera: Tephritidae) in New Mexico. *Ann. Entomol. Soc. Am.* 80: 494–500.

Dodson, G. 1987c. The significance of sexual dimorphism in the mating system of two species of tephritid flies (*Aciurina trixa* and *Valentibulla dodsoni*) (Diptera: Tephritidae). *Can. J. Zool.* 65: 194–198.

Emlen, S.T. and L.W. Oring. 1977. Ecology, sexual selection, and the evolution of mating systems. *Science* 197: 215–223.

Foote, R.H. 1980. Fruit fly genera south of the United States (Diptera: Tephritidae). *U.S. Dep. of Agric. Tech. Bull.* No. 1600, 79 pp.

Foote, R.H., F.L. Blanc, and A.L. Norrbom. 1993. *Handbook of the Fruit Flies (Diptera: Tephritidae) of America North of Mexico.* Comstock Publishing Associates, Ithaca, 571 pp.

Freidberg, A. 1981. Mating behavior of *Schistopterum moebiusi* Becker (Diptera: Tephritidae). *Isr. J. Entomol.* 15: 89–95.

Freidberg, A. and W.N. Mathis. 1986. Studies of Terelliinae (Diptera: Tephritidae): a revision of the genus *Neaspilota* Osten Sacken. *Smithson. Contrib. Zool.* No. 439, 75 pp.

Goeden, R.D. 1987a. Life history of *Trupanea conjuncta* (Adams) on *Trixis californica* Kellogg in southern California (Diptera: Tephritidae). *Pan-Pac. Entomol.* 63: 284–291.

Goeden, R.D. 1987b. Host-plant relations of native *Urophora* spp. (Diptera: Tephritidae) in southern California. *Proc. Entomol. Soc. Wash.* 89: 269–274.

Goeden, R.D. 1988a. Life history of *Trupanea imperfecta* (Coquillett) on *Bebbia juncea* (Bentham) Greene in the Colorado Desert of southern California (Diptera: Tephritidae). *Pan-Pac. Entomol.* 64: 345–351.

Goeden, R.D. 1988b. Gall formation by the capitulum-infesting fruit fly, *Tephritis stigmatica* (Diptera: Tephritidae). *Proc. Entomol. Soc. Wash.* 90: 37–43.

Goeden, R.D. 1989. Host-plants of *Neaspilota* in California (Diptera: Tephritidae). *Proc. Entomol. Soc. Wash.* 91: 164–168.

Goeden, R.D. 1990a. Life history of *Eutreta diana* (Osten Sacken) on *Artemisia tridentata* Nuttall in southern California. *Pan-Pac. Entomol.* 66: 24–32.

Goeden, R.D. 1990b. Notes on the life history of *Eutreta simplex* Thomas on *Artemisia ludoviciana* Nuttall in southern California (Diptera: Tephritidae). *Pan-Pac. Entomol.* 66: 33–38.

Goeden, R.D. 1992. Analysis of known and new host records for *Trupanea* from California (Diptera: Tephritidae). *Proc. Entomol. Soc. Wash.* 94: 107–118.

Goeden, R.D. 1993. Analysis of known and new host records of *Tephritis* from California and a description of a new species, *T. joanae* (Diptera: Tephritidae). *Proc. Entomol. Soc. Wash.* 95: 425–334.

Goeden, R.D. and F.L. Blanc. 1986. New synonymy, host, and California records in the genera *Dioxyna* and *Paroxyna* (Diptera: Tephritidae). *Pan-Pac. Entomol.* 62: 88–90.

Goeden, R.D. and D.H. Headrick. 1990. Notes on the biology and immature stages of *Stenopa affinis* Quisenberry (Diptera: Tephritidae). *Proc. Entomol. Soc. Wash.* 92: 641–648.

Goeden, R.D. and D.H. Headrick. 1991. Life history and descriptions of immature stages of *Tephritis baccharis* (Coquillett) on *Baccharis salicifolia* (Ruiz and Pavon) Persoon in southern California. *Pan-Pac. Entomol.* 67: 86–98.

Goeden, R.D. and D.H. Headrick. 1992. Life history and description of immature stages of *Neaspilota viridescens* Quisenberry (Diptera: Tephritidae) on native Asteraceae in southern California. *Proc. Entomol. Soc. Wash.* 94: 59–77.

Goeden, R.D. and D.W. Ricker. 1987. Phytophagous insect faunas of the native thistles *Cirsium brevistylum*, *Cirsium congdonii*, *Cirsium occidentale*, and *Cirsium tioganum*, in southern California. *Ann. Entomol. Soc. Am.* 80: 152–160.

Goeden, R.D. and J.A. Teerink. 1996a. Life histories and descriptions of adults and immature stages of two cryptic species, *Aciurina ferruginea* (Doane) and *A. michaeli*, new species (Diptera: Tephritidae), on *Chrysothamnus viscidiflorus* (Hooker) Nuttall in southern California. *Proc. Entomol. Soc. Wash.* 98: 415–438.

Goeden, R.D. and J.A. Teerink. 1996b. Life history and descriptions of adults and immature stages of *Aciurina idahoensis* Steyskal (Diptera: Tephritidae) on *Chrysothamnus viscidiflorus* (Hooker) Nuttall in southern California. *Proc. Entomol. Soc. Wash.* 98: 681–694.

Goeden, R.D. and J.A. Teerink. 1996c. Life history and descriptions of adults and immature stages of *Aciurina semilucida* (Bates) (Diptera: Tephritidae) on *Chrysothamnus viscidiflorus* (Hooker) Nuttall in southern California. *Proc. Entomol. Soc. Wash.* 98: 752–766.

Goeden, R.D. and J.A. Teerink. 1997a. Life histories and descriptions of adults and immature stages of *Procecidochares kristineae* and *P. lisae* new spp. (Diptera: Tephritidae) on *Ambrosia* spp. in southern California. *Proc. Entomol. Soc. Wash.* 99: 67–88.

Goeden, R.D. and J.A. Teerink. 1997b. Life history and description of immature stages of *Procecidochares anthracina* (Doane) (Diptera: Tephritidae) on *Solidago californica* Nuttall in southern California. *Proc. Entomol. Soc. Wash.* 99: 180–193.

Goeden, R.D. and J.A. Teerink. 1997c. Life history and description of immature stages of *Trupanea signata* Foote (Diptera: Tephritidae) on *Gnaphalium luteo-album* L. in southern California. *Proc. Entomol. Soc. Wash.* 99: 748–755.

Goeden, R.D. and J.A. Teerink. 1997d. Life history and description of immature stages of *Xenochaeta dichromata* Snow (Diptera: Tephritidae) on *Hieracium albiflorum* Hooker in central and southern California. *Proc. Entomol. Soc. Wash.* 99: 597–607.

Goeden, R.D. and J.A. Teerink. 1998. Life history and description of immature stages of *Trupanea pseudovicina* Hering (Diptera: Tephritidae) on *Porophyllum gracile* Bentham in southern California. *Proc. Entomol. Soc. Wash.* 100: 361–372.

Goeden, R.D. and J.A. Teerink. 1999. Life history and description of immature stages of *Trupanea arizonensis* Malloch (Diptera: Tephritidae) on *Trixis californica* Kellogg var. *californica* (Asteraceae) in southern California. *Proc. Entomol. Soc. Wash.* 101: 75–85.

Goeden, R.D., T.D. Cadatal, and G.L. Cavender. 1987. Life history of *Neotephritis finalis* (Loew) on native Asteraceae in southern California (Diptera: Tephritidae). *Proc. Entomol. Soc. Wash.* 89: 552–558.

Goeden, R.D., D.H. Headrick, and J.A. Teerink. 1993. Life history and descriptions of immature stages of *Tephritis arizonaensis* Quisenberry on *Baccharis sarothroides* Gray in southern California (Diptera: Tephritidae). *Proc. Entomol. Soc. Wash.* 95: 210–222.

Goeden, R.D., D.H. Headrick, and J.A. Teerink. 1994a. Life history and description of immature stages of *Procecidochares flavipes* Aldrich (Diptera: Tephritidae) on *Brickellia* spp. in southern California. *Proc. Entomol. Soc. Wash.* 96: 288–300.

Goeden, R.D., D.H. Headrick, and J.A. Teerink. 1994b. Life history and description of immature stages of *Paroxyna genalis* (Thomson) (Diptera: Tephritidae) on native Asteraceae in southern California. *Proc. Entomol. Soc. Wash.* 96: 612–629.

Goeden, R.D., D.H. Headrick, and J.A. Teerink. 1995a. Life history and description of immature stages of *Valentibulla californica* (Coquillett) (Diptera: Tephritidae) on *Chrysothamnus nauseosus* (Pallas) Britton in southern California. *Proc. Entomol. Soc. Wash.* 97: 548–560.

Goeden, R.D., D.H. Headrick, and J.A. Teerink. 1995b. Life history and description of immature stages of *Urophora timberlakei* Blanc and Foote (Diptera: Tephritidae) on native Asteraceae in southern California. *Proc. Entomol. Soc. Wash.* 97: 779–790.

Goeden, R.D., J.A. Teerink, and D.H. Headrick. 1998a. Life history and description of immature stages of *Trupanea jonesi* Curran (Diptera: Tephritidae) on native Asteraceae in southern California. *Proc. Entomol. Soc. Wash.* 100: 126–140.

Goeden, R.D., J.A. Teerink, and D.H. Headrick 1998b. Life history and description of immature stages of *Trupanea actinobola* (Loew) (Diptera: Tephritidae) on *Acamptopappus sphaerocephalus* (Harvey and Gray) Gray (Asteraceae) in southern California. *Proc. Entomol. Soc. Wash.* 100: 674–688.

Green, J.F., D.H. Headrick, and R.D. Goeden. 1993. Life history and descriptions of immature stages of *Procecidochares stonei* Blanc and Foote on *Viguiera* spp. in southern California (Diptera: Tephritidae). *Pan-Pac. Entomol.* 69: 18–32.

Greene, E., L.J. Orsak, and D.W. Whitman. 1987. A tephritid fly mimics the territorial displays of its jumping spider predators. *Science* 236: 310–312.

Grewal, J.S. and V.C. Kapoor. 1984. Courtship and mating behavior in the fruit fly *Dioxyna sororcula* (Wied.) (Diptera: Tephritidae). *Aust. J. Zool.* 1984. 32: 671–676.

Hasson, O. 1995. A fly in spider's clothing; what size the spider? *Proc. R. Soc. Lond. B* 261: 223–226.

Headrick, D.H. and R.D. Goeden. 1990a. Description of the immature stages of *Paracantha gentilis* Hering (Diptera: Tephritidae). *Ann. Entomol. Soc. Am.* 83: 220–229.

Headrick, D.H. and R.D. Goeden. 1990b. Resource utilization by larvae of *Paracantha gentilis* (Diptera: Tephritidae) in capitula of *Cirsium californicum* and *C. proteanum* (Asteraceae) in southern California. *Proc. Entomol. Soc. Wash.* 92: 512–520.

Headrick, D.H. and R.D. Goeden. 1990c. Life history of *Paracantha gentilis* (Diptera: Tephritidae). *Ann. Entomol. Soc. Am.* 83: 776–785.

Headrick, D.H. and R.D. Goeden. 1991. Life history of *Trupanea californica* Malloch (Diptera: Tephritidae) on *Gnaphalium* spp. (Asteraceae) in southern California. *Proc. Entomol. Soc. Wash.* 93: 559–570.

Headrick, D.H. and R.D. Goeden. 1993. Life history and description of immature stages of *Aciurina thoracica* Curran (Diptera: Tephritidae) on *Baccharis sarothroides* in southern California. *Ann. Entomol. Soc. Am.* 86: 68–79.

Headrick, D.H. and R.D. Goeden. 1994. Reproductive behavior of California fruit flies and the classification and evolution of Tephritidae (Diptera) mating systems. *Stud. Dipterol.* 1: 194–252.

Headrick, D.H. and R.D. Goeden. 1998. The biology of nonfrugivorous tephritid fruit flies. *Annu. Rev. Entomol.* 43: 217–241.

Headrick, D.H., R.D. Goeden, and J.A. Teerink. 1995. Life history and description of immature stages of *Euaresta stigmatica* (Diptera: Tephritidae) on *Ambrosia* spp. (Asteraceae) in southern California. *Ann. Entomol. Soc. Am.* 88: 58–71.

Headrick, D.H., R.D. Goeden, and J.A. Teerink. 1996. Life history and description of immature stages of *Dioxyna picciola* (Bigot) (Diptera: Tephritidae) on *Coreopsis* spp. (Asteraceae) in southern California. *Proc. Entomol. Soc. Wash.* 98: 332–349.

Headrick, D.H., R.D. Goeden, and J.A. Teerink. 1997. Taxonomy of *Aciurina trixa* Curran (Diptera: Tephritidae) and its life history on *Chrysothamnus nauseosus* (Pallas) Britton in southern California; with notes on *A. bigeloviae* (Cockerell). *Proc. Entomol. Soc. Wash.* 99: 415–428.

Hendrichs, J., S.S. Cooley, and R.J. Prokopy. 1992. Post-feeding bubbling behavior in fluid-feeding Diptera: concentration of crop contents by oral evaporation of excess water. *Physiol. Entomol.* 17: 153–161.

Hernández-Ortiz, V. 1993. A new species of *Aciurina* Curran (Diptera: Tephritidae) from Oaxaca, Mexico. *Folia Entomol. Mex.* 87: 49–54.

Jenkins, J. and W.J. Turner. 1989. Revision of the *Baccharis*-infesting (Asteraceae) fruit flies of the genus *Tephritis* (Diptera: Tephritidae) in North America. *Ann. Entomol. Soc. Am.* 82: 674–685.

Jenkins, J. 1990. Mating behavior of *Aciurina mexicana* (Aczél) (Diptera: Tephritidae). *Proc. Entomol. Soc. Wash.* 92: 66–75.

Keiser, I., R.M. Kobayashi, D.L. Chambers, and E.L. Schneider. 1973. Relation of sexual dimorphism in the wings, potential stridulation, and illumination to mating of Oriental fruit flies, melon flies, and Mediterranean fruit flies in Hawaii. *Ann. Entomol. Soc. Am.* 66: 413–417.

Knio, K.M., R.D. Goeden, and D.H. Headrick. 1996. Comparative biologies of the cryptic, sympatric species, *Trupanea bisetosa* and *T. nigricornis* (Diptera: Tephritidae) in southern California. *Ann. Entomol. Soc. Am.* 89: 252–260.

Lamp, W.O. and M.K. McCarty. 1982. Biology of predispersal seed predators of the Platte Thistle, *Cirsium canescens*. *J. Kans. Entomol. Soc.* 55: 305–316.

Mather, M.H. and B.D. Roitberg. 1987. A sheep in wolf's clothing: tephritid flies mimic spider predators. *Science* 236: 308–310.

Nation, J.L. 1972. Courtship behavior and evidence for a sex attractant in the male Caribbean fruit fly, *Anastrepha suspensa*. *Ann. Entomol. Soc. Am.* 65: 1364–1367.

Nation, J.L. 1981. Sex-specific glands in tephritid fruit flies of the genera *Anastrepha*, *Ceratitis*, *Dacus* and *Rhagoletis* (Diptera: Tephritidae). *Int. J. Insect Morphol. Embryol.* 10: 121–129.

Norrbom, A.L., L.E. Carroll, and A. Freidberg. 1999. Status of knowledge. In *Fruit Fly Expert Identification System and Systematic Information Database* (F.C. Thompson, ed.), pp. 9–47. *Myia* (1998) 9, 524 pp.

Novak, J.A. 1974. A taxonomic revision of *Dioxyna* and *Paroxyna* (Diptera: Tephritidae) for America north of Mexico. *Melanderia* 16: 1–53.

Novak, J.A. and B.A. Foote. 1968. Biology and immature stages of fruit flies: *Paroxyna albiceps* (Diptera: Tephritidae). *J. Kans. Entomol. Soc.* 41: 108–119.

Novak, J.A. and B.A. Foote. 1975. Biology and immature stages of fruit flies: the genus *Stenopa* (Diptera: Tephritidae). *J. Kans. Entomol. Soc.* 48: 42–52.

Phillips, V.T. 1946. The biology and identification of trypetid larvae (Diptera: Tephritidae). *Mem. Am. Entomol. Soc.* 12, 161 pp.

Piper, G.L. 1976. Bionomics of *Euarestoides acutangulus* (Diptera: Tephritidae). *Ann. Entomol. Soc. Am.* 69: 381–386.

Prokopy, R.J. 1980. Mating behavior of frugivorous Tephritidae in nature. In *Proceedings of the Symposium on Fruit Fly Problems* (J. Koyama, ed.), pp. 37–46. XVI Int. Cong. Entomol., Kyoto, Japan.

Silverman, J. and R.D. Goeden. 1980. Life history of a fruit fly, *Procecidochares* sp., on the ragweed, *Ambrosia dumosa* (Gray) Payne, in southern California (Diptera: Tephritidae). *Pan-Pac. Entomol.* 56: 283–288.

Stoltzfus, W.B. 1977. The taxonomy and biology of *Eutreta* (Diptera: Tephritidae). *Iowa State J. Res.* 51: 369–438.

Stoltzfus, W.B. and B.A. Foote. 1965. The use of froth masses in courtship of *Eutreta* (Diptera: Tephritidae). *Proc. Entomol. Soc. Wash.* 67: 262–264.

Thornhill, R. and J. Alcock. 1983. *The Evolution of Insect Mating Systems*. Harvard University Press, Cambridge. 547 pp.

Tauber, M.J. and C.A. Tauber. 1967. Reproductive behavior and biology of the gall-former *Aciurina ferruginea* (Doane) (Diptera: Tephritidae). *Can. J. Zool.* 45: 907–913.

Tauber, M.J. and C.A. Toschi. 1965. Bionomics of *Euleia fratria* (Loew) (Diptera: Tephritidae). I. Life history and mating behavior. *Can. J. Zool.* 43: 369–379.

Wangberg, J.K. 1978. Biology of gall-formers of the genus *Valentibulla* (Diptera: Tephritidae) on rabbitbrush in Idaho. *J. Kans. Entomol. Soc.* 51: 472–483.

Wangberg, J.K. 1980. Comparative biology of gall-formers in the genus *Procecidochares* (Diptera: Tephritidae) on rabbitbrush in Idaho. *J. Kans. Entomol. Soc.* 53: 401–420.

Wangberg, J.K. 1981. Gall forming habits of *Aciurina* spp. (Diptera: Tephritidae). *Proc. Entomol. Soc. Wash.* 86: 582–598.

# Section VII

## Evolution of Behavior

# 26 Genetic Population Structure in Tephritidae

*Stewart H. Berlocher*

## CONTENTS



## 26.1 INTRODUCTION

The study of genetic population structure can seem rather arcane and mathematically forbidding to new students of behavior, systematics, and ecology, but it is nonetheless highly relevant to these subjects. If one wants to understand why an insect species eats different host plants in different parts of its range, then one is studying population structure. If one wants to understand why sexual dimorphism is different in the north than the south of an insect's range, then one is studying population structure.

Genetic population structure is the term used to describe and understand genetic differences within, and especially among, different populations of a single species. Populations can be structured for molecular, morphological, ecological, or behavioral characters; the only requirement is that any character studied must be at least partially under genetic control.

This review has two parts. I first provide a broad overview of the entire area of population structure to provide a context in which to discuss studies in Tephritidae. I then review the available literature for tephritids and point out critical areas that must be resolved if progress is to be made in understanding tephritid population structure.

0-8493-1275-2/00/$0.00+$.50

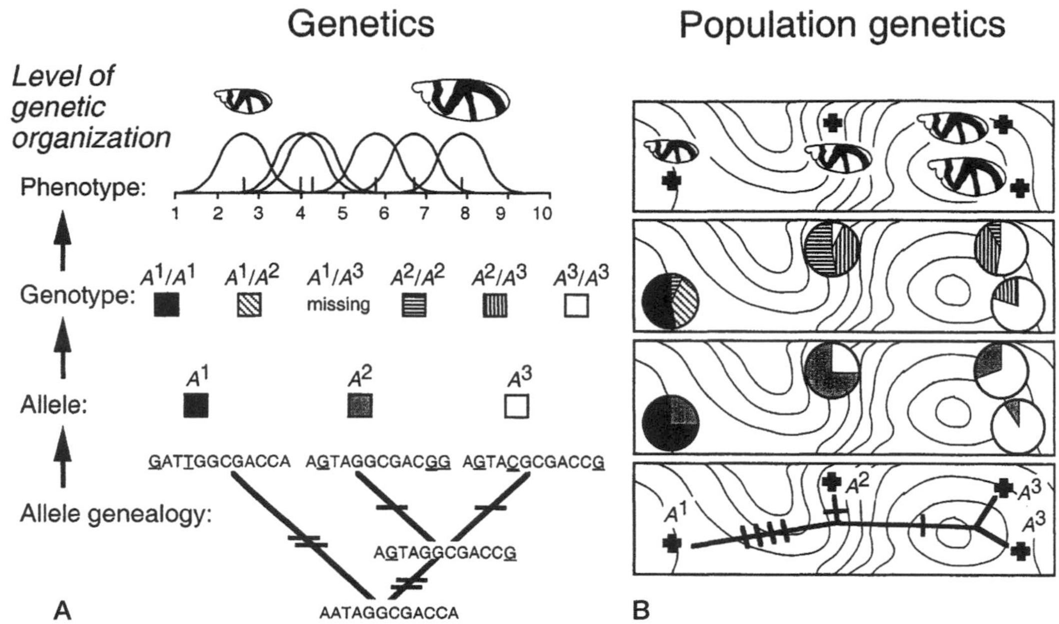


**FIGURE 26.1** Levels of genetic organization at which population structure can exist. (A) Basic genetic levels for a hypothetical gene *A*. (B) Hypothetical analysis of population structure at each of the genetic levels. Note that one genotype $A^1/A^3$ is labeled "missing" because it does not occur in any population. Also note that the mean size of each phenotype is graphed in the same order from left to right as shown in the row labeled "Genotype:"; the smallest (leftmost) genotype is $A^1/A^1$.

## 26.2 POPULATION STRUCTURE: AN OVERVIEW

In this overview I have attempted to distill the essence of a large literature, dating back to the 1930s. Consequently, I cite primarily review articles and books, although some original papers are cited either because of historical significance or recency. For readers who need an introduction to population genetics, a good place to start is Chapters 9 to 14 of Futuyma (1998) followed by the paperback text of Hartl (1988). An excellent summary of empirical studies of population structure from a systematic viewpoint is the recent book of Avise (1994). Weir (1990) has recently summarized the traditional allele frequency-based theory of population structure. A comparison of methods for estimating gene flow from genetic data has recently been made by Neigel (1997). Although the empirical focus is on humans, a good discussion of the uses of F-statistics and isolation-by-distance analysis can be found in Cavalli-Sforza et al. (1994: 111–125).

### 26.2.1 Levels of Genetic Organization

Genetic population structure (henceforth, just "population structure") is studied at all possible levels of genetic organization. In Figure 26.1A I have summarized the basic levels of genetic organization for a single gene. Complications involving multiple genes and chromosomes will be briefly discussed but not illustrated. I note that I follow the literature in using the terms gene and locus interchangeably in most applications, but locus is the broader term, applying to any variable site in the DNA, while a gene must encode a protein (or ribosomal or transfer RNA).

The most basal levels of organization (bottom of Figure 26.1A) are concerned with alleles. Although the exact definition of the term *allele* has been debated, I will here use the term in the broadest sense of being any sequence variation of a gene that is detectable with the technology at hand. Thus, if two enzyme mobility classes are observed with starch gel electrophoresis, then at least two alleles at the DNA level are indicated. If subsequent DNA sequencing reveals that there are more than two distinct sequences of the gene, then each of these is now an allele.

Throughout most of this century, the most basal level of genetic organization that could be studied is what I have labeled the "allele" level in Figure 26.1A. For example, in studies using allozymes, the most one can do is to count the total number of alleles; one cannot say anything about which alleles are most similar in terms of sequence divergence. However, with the PCR (polymerase chain reaction) revolution of the late 1980s, it became possible to obtain DNA sequences for alleles, and these sequences in turn made it possible to estimate allele genealogies (also referred to as allele phylogenies). Thus, a new level, what I am calling the "allele genealogy" level, was added below the allele level. Figure 26.1A shows the genealogy of a "gene" of 12 bases, gene $A$, that has evolved from an ancestral sequence to a total of three alleles, $A^1$, $A^2$, and $A^3$.

Genes do not occur in isolation, of course, but occur as segments of very long DNA molecules referred to as chromosomes. Despite this, single genes can often be studied as if they were isolated entities because recombination usually randomizes associations between alleles of different genes. In some cases, however, strong nonrandom associations between the alleles of different genes occur (say, if allele $A^1$ occurred exclusively or almost exclusively with allele $B^2$ of gene $B$ located close by on the chromosome). In such cases one needs to study a larger segment of DNA including at least the two genes. A convenient term for a length of DNA containing more than one gene is haplotype. (This is the original meaning of the term. However, "haplotype" has been broadened in the last two decades to describe any segment of DNA that can be isolated repeatably by some technique, and which includes nucleotide positions that can be shown by sequencing or by restriction endonuclease analysis to be variable, regardless of whether entire genes are included. When describing allozyme variation, which constitutes much of the tephritid work, the original concept is accurate.) Chromosome variation can be considered as an extreme case of haplotype variation, the difference being that the variable region of DNA being studied is large enough that it is observable under the light microscope.

The level of organization directly above the allele level is, of course, the level of diploid genotype, as has been universally understood since the rediscovery of Mendel in 1900. (As all tephritids are diploid, other ploidy levels need not be discussed.) One of the earliest insights of Mendelian genetics was the enormous number of genotype combinations possible from a modest number of alleles, following the relationship $g = a(a + 1)(1/2)$, where $g$ and $a$ are the number of genotypes and alleles, respectively. Thus, the three alleles in Figure 26.1A can be combined into 6 genotypes, while ten alleles can be combined into 55 genotypes. Haplotypes and chromosome variants, which are themselves at the same haploid level as alleles, also occur in various combinations at the diploid level.

Above the genotype level is the phenotype level. Relating the genotype and phenotype levels has been one of the central tasks of 20th-century genetics, one that is not yet complete (what I have called the phenotypic level in Figure 26.1A will need to be subdivided as more is learned of the molecular biology of development). However, one thing can be said with total certainty: many of the morphological and behavioral characteristics that vary within species do not fall into discrete phenotypic classes, but show a "continuous" distribution. Continuous distributions have two causes: (1) the existence of many slightly different genotypes (because of multiple alleles and polygeny); and (2) the "blurring" of phenotypic differences by environmental variation during development. In Figure 26.1A, I assume that gene $A$ has an effect on wing size and have shown the positions of the six genotypes on a phenotypic scale of wing sizes, along with normal curves showing the range of sizes ultimately produced by the environment acting on each genotype. Wings representing each end of the range are shown above the phenotypes. Depending on the frequencies of each allele, different populations will have different mean sizes, but sizes within each population will almost always be unimodally (and usually normally) distributed.

### 26.2.2 Concepts Used to Study Population Structure at Each Level

In Figure 26.1B I have outlined a hypothetical set of data for four populations. At the level of allele genealogy, the starting point for all analysis is a type of graph called a "tree" that shows the relationships of the alleles. The problem of estimating allele trees/genealogies has been approached in two different ways. The older way arises from systematics and applies to intraspecific alleles the same methods (maximum parsimony, distance methods, maximum likelihood, etc.) used for species and higher level taxa. This vast area has been masterfully reviewed by Swofford et al. (1996). The more recent approach, reviewed by Hudson (1990), is called coalescent theory and is a probabilistic method developed explicitly with alleles in mind. Haplotypes and chromosome variants have traditionally been analyzed using the older phylogenetic approach; the estimation of chromosome trees has a lengthy history, going back to Dobzhansky (1937).

Here I will restrict myself to the most widely used phylogenetic approach, maximum parsimony. If one had only the sequences of alleles $A^1$, $A^2$, and $A^3$, and the information that $A^1$ was the outgroup, the parsimony tree would be the same as the "known" evolutionary history of the alleles shown in Figure 26.1A. In reality, intraspecific trees are as subject to incorrect estimation due to homoplasy as are trees for species and higher-level taxa.

Intraspecific trees are most informative about population structure when they are mapped onto geographic space. This approach, termed intraspecific phylogeography, or just phylogeography (Avise et al. 1987; Avise 1994), is illustrated on the bottom of Figure 26.1B. For simplicity, an unrooted tree is shown. It is assumed that only a single individual has been sampled and sequenced from each of four samples, and that the most common allele in each sample was sampled (frequencies are in the panel immediately above). This very simple phylogeographic analysis shows that the three samples on the right are much more similar to each other at the sequence level than they are to the sample at the left, which differs by four base changes.

It is extremely important to stress that such a simple result is a perfect description of nature only when sites are totally "fixed" for unique alleles; if so, then samples of size one are in fact adequate. If one allele predominates in each population, as in Figure 26.1B, then samples of one individual are still a good representation of reality. However, if there is substantial variation in the populations, then larger samples must be taken, and the phylogeographic analysis modified accordingly. Avise (1994) discusses analyses in which multiple variable individuals per site are sampled.

The next level up, the allele level, is by far the most highly developed in terms of statistical methods for studying population structure. The fundamental unit here is the percentage or frequency of each allele in the sample. (Haplotypes and chromosome variants are also studied using frequencies.) In Figure 26.1B, frequencies of the three alleles are represented by the commonly used "pie" diagram.

The study of population structure began with Sewall Wright's (1931) work, and, since then, the theory of allele frequencies (or gene frequencies, as Wright always called them) has been very highly developed. Wright's $F$-statistics (Wright 1978, and references therein), in particular $F_{ST}$, are at the core of any analysis of allele frequencies. $F_{ST}$ has a remarkably large number of interpretations, but the simplest is that it is a standardized variance. If one has determined the frequency $p$ of an allele in a series of populations, then

$$F_{ST} = \frac{\sigma_p^2}{\bar{p}\bar{q}}$$

where $\sigma_p^2$ is the variance in $p$ across populations, $q = 1 - p$, and $\bar{p}$ and $\bar{q}$ are means of $p$ and $q$, respectively. I note that for algebraic simplicity Wright frequently used $n$ weighting rather than $n - 1$ weighting in expressions for variance, as I also have in the hypothetical examples to follow; in real data sets of any but the smallest size, the difference between $F_{ST}$ calculated with $n$ vs. $n - 1$

weighting is minuscule. The denominator of $F_{ST}$ is the maximum or limiting variance that a set of populations can have if an infinitely large set of populations, all with initial frequency $p_0$, drift to fixation. Thus $F_{ST}$ is the proportion of potential differentiation (expressed as variance) that is realized. In other words, $F_{ST}$ measures the magnitude of geographic differentiation, with 0.0 indicating no differentiation and 1.0 indicating maximum differentiation. The frequencies of allele $A^3$ in the four populations in Figure 26.1B are 0.00, 0.25, 0.67, and 0.92, so the variance in $p$ is 0.128, the mean is 0.460, the limiting variance is 0.248, and $F_{ST}$ for $A^3$ is thus 0.516 — about half of that possible if all populations were maximally differentiated. Wright (1978) weighted alleles by allele frequency to obtain a locus $F_{ST}$ value. In recent years much progress has been made in correcting $F_{ST}$ for biases relating to small number and size of samples; Cockerham and Weir's $\theta$ has become the most widely used estimator of $F_{ST}$ (Weir 1990).

$F_{ST}$ has an important advantage over genetic distances in describing population structure: Wright made $F$-statistics an intrinsic part of a broad theory unifying gene flow, population size, mutation rate, etc. A famous equation developed by Wright (1931) relates $F_{ST}$ to population size $N$ and gene flow $m$ (proportion of immigrants each generation) as

$$F_{ST} = \frac{1}{4Nm+1}$$

This equation is only strictly true for selectively neutral genes, small values of $m$, and an *island model* in which gene flow is equally likely between an infinite number of populations, but it provides a rough guide to what $F_{ST}$ to expect in nature. Note that the product $Nm$ (size of population times proportion of immigrants) can be thought of as the *number* of migrants per generation. If a set of populations all were of size $N = 100$ and the proportion of each population that came from another population each generation was $m = 0.01$, then $Nm = 1$ = one migrant per generation.

Most application of the equation is actually in the opposite direction. That is, one starts with $F_{ST}$ and solves for $Nm$. This is because in practice it is far easier to determine allele frequencies, calculate $F_{ST}$ and, from there, $Nm$, than it is to measure gene flow directly in the field. For the allele frequencies in Figure 26.1B, the number of migrants per generation is predicted to be 0.23 (or approximately one migrant every fourth generation). The use of genetic data to estimate gene flow has recently been reviewed (Neigel 1997).

Wright realized that his island model was not realistic for all species, as in many cases gene flow will be more likely between neighboring populations than distant ones. This led to his *stepping-stone model,* with its prediction of an effect called *isolation-by-distance* (IBD) (Wright 1943). Careful statistical analysis of allele frequencies for a set of populations can reveal whether the data lack IBD and thus are consistent with an island model, show IBD and are thus consistent with a stepping-stone model, or show some other pattern. A simple type of analysis that can reveal IBD starts with $F_{ST}$, but instead of computing one value for the entire set of populations, values are computed for just two populations at a time. A matrix of all possible such pair-wise $F_{ST}$ values is computed, an analogous matrix of pair-wise geographic distances is constructed, and then the correlation between corresponding entries in each matrix is calculated. An important consideration is that the test of significance must use a special method called a Mantel test (Mantel 1967) instead of conventional correlation methods; see papers cited below for examples of how this test can be implemented.

Hypothetical examples of geographic analysis of $F_{ST}$ are shown in Figure 26.2. If a result like Figure 26.2A is obtained, gene flow is not related to geographic distance, and there is no IBD. A recent example is the study of Schug et al. (1998) on the fish *Gambusia holbrooki* in the Bahamas. If a result like Figure 26.2B is obtained, gene flow is related to geographic proximity, and there is IBD; the populations then conform to Wright's later stepping-stone model of population structure. A recent, very informative study showing IBD is that of Patterson and Denno (1997) on winged

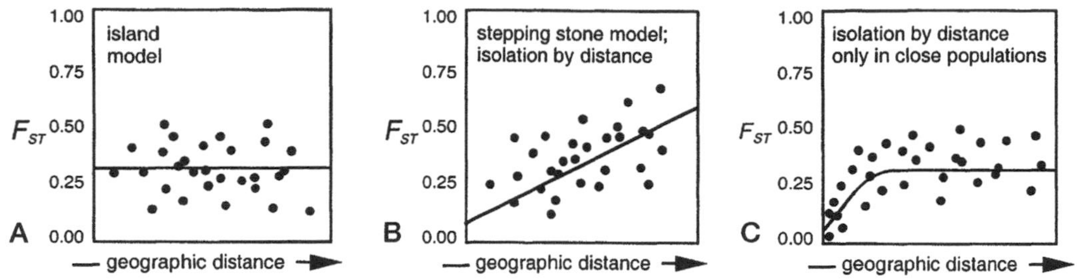


**FIGURE 26.2** Correlations of pair-wise $F_{ST}$ and geographic distance in different cases. (A) Island model, where gene flow is not correlated with geographic distance. (B) Stepping-stone model, where gene flow occurs only between adjacent populations, so isolation by distance is observed. (C) A model in which isolation by distance occurs only in geographically close populations.

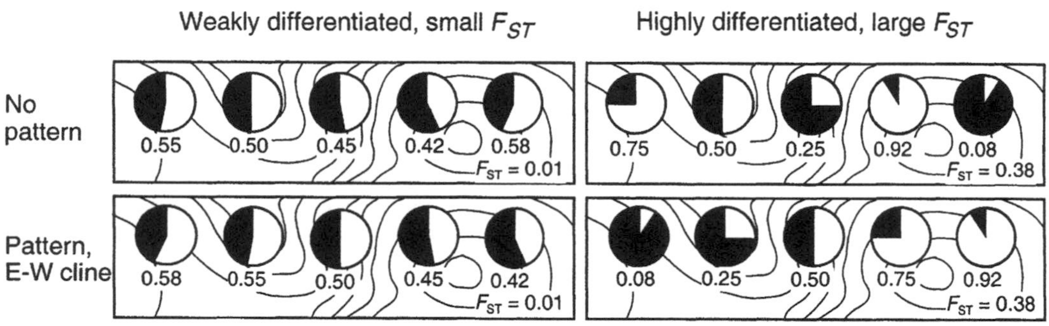


**FIGURE 26.3** Demonstration with hypothetical populations that magnitude of differentiation (as measured by $F_{ST}$) is not necessarily related to the existence of patterns such as clines. See text.

and wingless planthoppers. If a result like Figure 26.2C is obtained, only contiguous populations are similar, and there is a threshold of distance beyond which IBD is not seen. This pattern is illustrated by an informative study of rock outcrop beetles (King 1988). The study of geographic differentiation has developed enormously in recent years, with a variety of extensions of the approach described here and the proposal of several additional methods (private alleles, Slatkin 1985; Slatkin and Barton 1989; allele genealogy, Slatkin and Maddison 1990; variogram, Cavalli-Sforza et al. 1994: 111-125; spatial autocorrelation, Sokal and Wartenburg 1983).

The concept of isolation by distance leads to the general subject of patterns of geographic variation. Broad, wide-ranging patterns, such as clines, step clines, and races, occur in many species. Frequently, the presence of clines is associated with relatively large $F_{ST}$ values. But it is important to remember that the degree of patterning and the degree of differentiation (as measured by $F_{ST}$) can be independent of one another. This is illustrated in Figure 26.3, where six populations are sampled in a hypothetical geographic space. In the top two panels, no clinal structure is present, but the set of populations on the right have a much higher *magnitude* of differentiation. In the bottom two panels, populations showing the same degrees of differentiation are organized into east–west clines rather than being randomly patterned. The genetic analysis of clines has generated a large literature (Endler 1977), with one of the most important questions being whether a cline originated via primary differentiation or secondary contact.

As already noted, if alleles of different genes within haplotypes and chromosomes are randomly associated with one another, then the study of alleles and allele frequencies alone is sufficient. However, if nonrandom associations exist, then haplotype and chromosome, as well as allele, frequencies are needed to understand the population structure. When nonrandom associations between alleles of different genes occur in a population, *linkage disequilibrium* is said to exist. The best and most comprehensive review of linkage disequilibrium, including both a discussion of theory and a review of empirical studies, is unfortunately rather dated (Hedrick et al. 1978).

The next level above the allele level is the genotype level. For diploid sexual species such as tephritid flies, the key organizing principle is the well-known Hardy–Weinberg equilibrium (HWE). If there is random mating, then for two alleles $A^1$ and $A^2$ with frequencies $p$ and $q$, respectively, genotype frequencies for $A^1A^1$, $A^1A^2$, and $A^2A^2$ are $p^2$, $2pq$, and $q^2$, respectively. The great beauty of HWE is that, for species that do mate randomly within local populations, the genotype level often becomes unnecessary for studying geographic and other population structure. This is because in populations in HWE there is no information in genotype frequencies not also contained in allele frequencies; genotype frequencies can be reconstructed at any time from the allele frequencies. If this review was about, say, plants, then the statement above could not be made, as many plants have asexual populations at various ploidy levels, in which genotype frequencies contain unique information. With respect to tephritids, analysis at the genotype level is important primarily in the study of host races.

One commonly used genotype-level statistic, average heterozygosity $\overline{H}$, is often used to measure the total amount of variation in a population. $\overline{H}$ is just the mean over all loci of the HWE expected frequency of heterozygotes. $\overline{H}$ is of some use in comparing levels of variation within and between species, although it is often reported in allozyme studies in a rather ritualistic fashion, without any application to questions about population structure.

The study of population structure at the phenotype level has an old history, one intellectually independent of population genetics. Generally, such analysis has examined continuous and meristic morphological variation and has been referred to as morphometric analysis. The mathematical language of morphometrics, that of means, variances, and correlations, may at first seem unrelated to the methods used at lower levels of genetic organization, but as long as the traits are at least partially under genetic control, a morphological difference still implies a genetic difference. In most morphometric work, there is an implicit and untested assumption that the traits under study are under significant genetic control. To test this assumption, one should either estimate heritability (proportion of the variance that has a genetic basis) for each trait using parent–offspring regression (Hartl 1988, Chapter 4), or carry out a common garden experiment (raise samples of all populations one generation under the same environmental conditions before measuring traits). While it is sometimes not practical to control for environmental effects, an increasing number of morphometric studies do make the effort. For example, Kambhamphati and Mackauer (1988), in a morphometric study of parasitic wasps in the genus *Aphidius*, reared all samples one generation in the laboratory on the same host species to reduce host-related environmental effects.

### 26.2.3 Causes of Population Structure and Their Significance

The causes of population structure can be divided into those that increase differences between populations, and those that reduce differences (for reviews relevant to this section, see Hartl 1988; Avise 1994; Futuyma 1998). Differences are increased by genetic drift, by many forms of natural selection, and over the long run by mutation. Differences can be held constant by balancing selection acting in a similar way in different populations. Differences are reduced primarily by gene flow, but can also be decreased if directional selection begins operating to favor the same genotypes in different populations. In general, populations that are initially uniform will accumulate differences until an equilibrium between the forces that increase and decrease differences is reached. In this review I use the term *differentiation* to describe the process of populations becoming different over time. This term has not been formally defined, but is used informally in the literature with some frequency, and facilitates economy of description.

The significance of understanding the underlying causes of structure is that some affect all genes, while others affect only particular genes. Genetic drift and gene flow affect all genes equally (in the sense of having the same expectation, with a theoretically predictable variance in effect). Natural selection, on the other hand, acts differently on different loci. This distinction is critical for understanding problems like host race formation, where the relatively few loci that are under host-related divergent selection (or are close to those under selection) are expected to display a different level and pattern of differentiation from those that are not.

## 26.3 POPULATION STRUCTURE IN TEPHRITIDAE

In reviewing published work on population structure on tephritids, I have first organized studies by ecological and geographic characteristics, and then secondarily by the genetic levels just described. This is because the ecology of tephritids, as focused on the host plant, is the key organizing feature of tephritid evolution.

Before starting, some universal observations on the population genetics of typical tephritid populations can be made (and not repeated for each study). First, many tephritids are rich in genetic variation (at least as indicated by allozymes), with $\overline{H}$ in many species ranging as high as ~0.20 (e.g., *Rhagoletis pomonella* (Walsh); Berlocher and McPheron 1996). Second, local populations of tephritids conform well to HWE expectations in all of the studies discussed below, so that random mating can be reliably assumed within local populations of the same species (or host race, if applicable). Third, the same gene loci (for the enzymes aconitase and β-hydroxyacid dehydrogenase, for example) are consistently involved in host race differentiation, clinal and nonclinal geographic differentiation, and speciation (e.g., Feder et al. 1993; 1997a).

### 26.3.1 Tree-to-Tree Differentiation on the Same Host Species

Many tephritids infest hosts in which individual plants are discrete and have long life spans. For example, individual trees of *Juglans nigra* infested by the univoltine walnut husk fly *R. completa* Cresson may live for hundreds of fly generations. In such cases, the local (small geographic scale) population structure of the flies can be described as a set of large, discrete, relatively permanent "tree-centered" populations with some degree of gene flow between them. Unlike tephritids infesting herbaceous plants with transient populations, or those that have multiple generations on different tree hosts, univoltine tephritids on large, long-lived plants may be expected to develop local population structure that lasts more than one season.

This expectation is met in the two available studies of tree-to-tree population structure, both at the allele level (allozymes). The only study to focus entirely on such variation is that of McPheron et al. (1988a) on hawthorn (*Crataegus mollis*) populations of *R. pomonella*, in which significant tree-to-tree allele frequency differences were detected at six loci. The magnitude of differentiation was, however, very small, with $F_{ST} = 0.001$. Feder et al. (1990b) have also found similar significant, but very small, frequency differences between fly samples from individual hawthorn trees. *Rhagoletis pomonella* populations on a single hawthorn are often large, so that drift is an unlikely explanation for these differences (McPheron et al. 1988a), although this may not be the case for smaller host trees (e.g., small walnut species in the American Southwest; Berlocher, personal observation). Recent work on selection on allozymes in *R. pomonella* (Feder et al. 1997a, including three of the six loci studied by McPheron et al. 1988a; see also Tsakas and Zouros 1980; Cosmides et al. 1997) suggests the observed differences are the result of natural selection related to differences in fruit ripening times, although selection imposed by tree-to-tree differences in fruit chemistry should be examined. Large (>100) sample sizes are needed to study this level of population structure. IBD has not been studied at this level.

### 26.3.2 Differentiation on Different Host Species

If selection and drift can differentiate tephritid populations on the same host species, then it is reasonable to expect similar and probably larger differences when multiple host species are used. I stress that I am not discussing host races here (see later section), but simply populations of a species that use several host species, without such host use having any effect on reproductive isolation. For example, throughout the northeastern United States the black walnut fly *R. suavis* (Loew) infests two very distinctive species of walnuts, *J. nigra* and *J. cinerea* (Bush 1966); all flies can apparently mate and oviposit upon either of the two suitable hosts. Yet if some genotypes survive better on one host or the other, one may expect some host-related differentiation, which must be limited by gene flow, just as with tree-to-tree differentiation on the same host.

Differentiation related to different host species has been very poorly studied in tephritids; not a single study has had host species differentiation as its sole focus. Berlocher (1995), in an allozyme study focused primarily on geographic variation, examined the blueberry maggot *R. mendax* Curran from three host species at a site in New Jersey and four host species at a site in Maryland, and found no significant frequency differentiation. Sample sizes were small, however, and moreover the hosts were spatially intermixed, maximizing the potential for movement between host species. Malavasi and Morgante (1983) were attempting to find host races in their allozyme study of *Anastrepha fraterculus* (Wiedemann) in Brazil, but the lack of "host fidelity" (mating on the host; Feder et al. 1994) means that this species and most tropical generalists are unlikely to form host races. (At least based on available evidence, "*A. fraterculus*" seems to consist of many very poorly understood cryptic species (Steck 1991), some of which may in fact show host fidelity.) A few significant allele frequency differences were observed among the *A. fraterculus* samples from eight hosts growing in a single orchard, but the lack of control for temporal variation (see below) renders the cause of these problematical. The existence of tree-to-tree variation on different host species is an open question for tephritid population structure.

### 26.3.3 Temporal Variation

The fact that frequencies can cycle throughout the year was shown decades ago for chromosome inversions (Dobzhansky 1943). Given that some allozymes in *R. pomonella* are selected by temperature conditions at the pupal stage (Feder et al. 1997a, b), temporal variation is to be expected in tephritid populations. However, no studies have been published in this area. An ideal situation in which to seek temporal variation would be a multihost orchard population of a tropical generalist tephritid with multiple generations per year. To avoid statistical confounding with host effects (see preceding section), parallel studies of several orchards with different sets of host would be ideal.

### 26.3.4 Geographic Variation

Conventional geographic differentiation has been reasonably well studied in tephritids. All levels of genetic organization have been studied, albeit unequally.

Phylogeographic work on tephritids has just begun, with very few published studies. The focus of the study of Brown et al. (1996) was on the evolution of two apparent host races of *Eurosta solidaginis* (Fitch), with some insight into phylogeographic relationships among the races being gained. The data consists of partial sequences of the mitochondrial cytochrome oxidase (CO) I and II genes. The *Solidago altisima* race contains two haplotypes, a western one found in Michigan and adjoining states, and an eastern one in New England. The sequence difference between the two haplotypes is small at 0.8%. The *S. gigantea* race, in contrast, is essentially fixed for the eastern haplotype. With respect to *within*-region and *within*-race gene flow, the data are unfortunately uninformative, because of the almost complete genetic uniformity within regions and races. That is, with only one haplotype within an area, gene flow could be great or small and the population structure would be the same. Smith and Bush (1997) included a few widespread geographic samples of some species in the *R. pomonella* group in their phylogenetic study of the entire genus using the COII gene, but also found very little differentiation. Generalizations are difficult from so few studies.

The bulk of available data on geographic variation is at the allele level, the great majority of it using allozymes. In Table 26.1, I have compiled $F_{ST}$ values and other information such as the existence of clines, from both published and unpublished sources (theses). If $F_{ST}$ values had already been computed, I used them, regardless of the exact way in which $F_{ST}$ was calculated. For studies for which $F_{ST}$ values had not been published, I calculated traditional $F_{ST}$ (Wright 1978) if only allele frequencies were available, and Cockerham and Weir's $\theta$ (Weir 1990) if genotype numbers were available. For the former, I used BIOSYS-1 (Swofford and Selander 1981) and, for the latter, a version of BIOSYS-1 modified by W. Black (Department of Biology, University of Colorado, Boulder; Black's corrected version used).

I divided the studies into those in which the estimation of *Nm* from $F_{ST}$ appears to be reasonable, and those in which assumptions of the approach are violated. An important generalization to emerge from the 13 species in the first category is that $F_{ST}$ is relatively small ($0.083 \pm 0.089$), and the number of migrants *Nm* correspondingly large ($7.6 \pm 9.7$). The taxonomically complex *A. fraterculus* has been analyzed in two separate studies, and I have assumed that the "populations 1-9" of Malavasi and Morgante (1983) and "group I" of Steck (1991) represent the same single species. Overall, the *Nm* values from $F_{ST}$ accord well with the observation that tephritids can disperse substantial distances (McPheron et al. 1988a).

The eight species in the second group all violate at least one assumption of Wright's model. Latitudinal clines and strong regional differentiation exist in *R. pomonella* (six loci; Berlocher and McPheron 1996) and the undescribed flowering dogwood fly (two loci; Berlocher, in press), at a set of loci that cannot be treated as selectively neutral (Feder et al. 1997a, b). *Oxyna parietina* (Linnaeus) in Europe shows elevational clines (Eber et al. 1992). *Terellia palposa* (Loew) populations at high elevations in New Mexico have pronounced frequency differences at *Pgm* (Steck 1981), which are also probably due to selection. Introduced populations of *R. completa* (Berlocher 1984) and *Ceratitis capitata* (Wiedemann) (Gasperi et al. 1991) have almost certainly not reached an equilibrium between gene flow and drift. Finally, "group II" of *A. fraterculus* almost certainly contains several species, as discussed by Steck (1991). The species in my second group have a mean higher $F_{ST}$ ($0.158 \pm 0.096$) and smaller number of inferred migrants ($2.2 \pm 1.3$), as could be expected. The "take-home" lesson here is that estimation of gene flow from genetic data is only valid if assumptions are met.

Estimation of gene flow from sequences using other methods and nonallozyme data is very limited for tephritids. The "private alleles" method of Slatkin (1985) in tephritids has been used by Gasperi et al. (1991) for *C. capitata*, who found little agreement between estimates by the $F_{ST}$ and Slatkin methods. The only other application of this approach was by Eber and Brandl (1994), who reported a similar pattern of high gene flow in *O. parietina*, *Urophora cardui* (Linnaeus), and *R. alternata* (Fallén). Isolation by distance has been directly addressed in only one case. Eber and Brandl (1994) used spatial autocorrelation on allele frequency data for six alleles of *U. cardui*, but found no isolation by distance. Allele or haplotype frequencies at the DNA level that could potentially be used to calculate $F_{ST}$ are accumulating (Steck and Sheppard 1993; Haymer 1995; Gasparich et al. 1997), but are too few for reliable *Nm* estimation at present. The only study of intraspecific chromosomal variation in tephritids of which I am aware is on the composite feeder *U. cardui* (Pönisch and Brandl 1992), in which "B chromosomes" (small chromosomes composed of noncoding DNA) vary in frequency between populations in Germany. The difficulty of spreading polytene chromosomes in most tephritids has limited this area of study.

**TABLE 26.1**
**Geographic Population Structure in Tephritidae**

| Species | Loci | Sites | $F_{ST}$ | *Nm* | Statistic | Sampling area | References | Clines |
|---|---|---|---|---|---|---|---|---|
| **Taxa that are native lack strong clines or regional structure and do not contain more than one species** | | | | | | | | |
| *Rhagoletis mendax* | 16 | 26 | 0.015 | 16.4 | $\theta$ | Eastern United States | Berlocher 1995 | No |
| "Sparkleberry fly" | 16 | 12 | 0.018 | 13.6 | $\theta$ | Eastern United States | Payne and Berlocher 1995 | No |
| *R. completa* native | 4 | 8 | 0.007 | 35.5 | $F_{ST}$ | Midwestern United States | Berlocher 1984 | No |
| *Strauzia* "D" | 16 | 3 | 0.112 | 2.0 | $F_{ST}$ | Illinois | Lisowski 1979 | — |
| *Bactrocera umbrosa* | 3 | 3 | 0.290 | 0.6 | $\theta$ | Malaysia | Yong 1988 | — |
| *B. cucurbitae* | 5 | 4 | 0.017 | 15.0 | $F_{ST}$ | Selangor Malaysia | Yong 1992 | — |
| *Anastrepha obliqua* | 20 | 5 | 0.098 | 2.3 | $\theta$ | Costa Rica to Brazil | Steck 1991 | — |
| *A. distincta* | 19 | 4 | 0.062 | 3.8 | $\theta$ | Costa Rica to Brazil | Steck 1991 | — |
| *Terellia occidentalis* | 12 | 6 | 0.039 | 6.2 | $F_{ST}$ | Wyoming, New Mexico | Steck 1981 | — |
| *Anastrepha striata* | 14 | 6 | 0.207 | 1.0 | $\theta$ | Costa Rica to Brazil | Steck 1991 | — |
| *A. fraterculus* | 11 | 9 | 0.128 | 1.7 | $F_{ST}$ | Eastern South America | Malavasi and Morgante 1983 | — |
| *A. fraterculus* group 1 | 18 | 5 | 0.079 | 2.9 | $\theta$ | Mexico to S. Brazil | Steck 1991 | — |
| *Eurosta solidaginis* alt. | 6 | 9 | 0.079 | 2.9 | $F_{ST}$ | Northeastern United States | Waring et al. 1990 | — |
| Mean | | | 0.089 | 7.6 | | | | |
| S.D. | | | ±0.083 | ±9.7 | | | | |
| **Taxa that are (1) introduced, (2) possess strong clines or regional structure, or (3) may contain more than one species** | | | | | | | | |
| *Ceratitis capitata*[1] | 4 | 22 | 0.123 | 1.8 | $F_{ST}$ | Africa, Mediterranean | Gasperi et al. 1991 | — |
| *R. completa* introduced[1] | 4 | 8 | 0.054 | 4.4 | $F_{ST}$ | California | Berlocher 1984 | No |
| *R. pomonella*[2] | 17 | 22 | 0.148 | 1.4 | $\theta$ | Eastern United States | Berlocher and McPheron 1996 | Yes |
| "Dogwood fly"[2] | 17 | 21 | 0.084 | 2.7 | $\theta$ | Eastern United States | Berlocher, in press | Yes |
| *Terellia palposa*[2] | 12 | 6 | 0.217 | 2.9 | $F_{ST}$ | Texas to New Mexico | Steck 1981 | — |
| *E. solidaginis* alt.[2] | 6 | 9 | 0.219 | 0.9 | $F_{ST}$ | Northeastern United States | Waring et al. 1990 | ? |
| *Oxyna parietina*[2] | 8 | 10 | 0.079 | 2.9 | $\theta$ | Northern Europe | Eber et al. 1992 | Yes |
| *A. fraterculus* group 2[3] | 18 | 3 | 0.342 | 0.5 | $\theta$ | Peru to Brazil | Steck 1991 | — |
| Mean | | | 0.158 | 2.2 | | | | |
| S. D. | | | ±0.096 | ±1.3 | | | | |

*Note:* "Statistic" column explained in text; "—"indicates that no analysis for clines was performed. The "sparkleberry fly" and "dogwood fly" are undescribed species in the *Rhagoletis pomonella* species group.

A striking fact about tephritid population structure is that strong linkage disequilibrium occurs in at least one genus, *Rhagoletis*. Large disequilibria between autosomal allozyme loci occur in *R. pomonella* (Feder et al. 1990a; 1993; Berlocher and McPheron 1996) and *R. mendax* (Berlocher 1995). In both species the loci involved are either closely linked or are located in regions of reduced crossing-over (Roethele et al. 1997), and may be associated with genes that affect diapause timing and other life history traits. An especially intriguing type of disequilibrium in *Rhagoletis* involves sex chromosomes. Crossing-over is limited in *Rhagoletis* (Roethele et al. 1997), and in many species both sex chromosomes contain homologous regions containing many genes (Berlocher 1984; McPheron and Berlocher 1985); the result is that an allele of an allozyme locus can arise via mutation on a Y chromosome, and never cross over into the female side of the population. At least seven species have large disequilibria involving sex chromosomes and allozyme loci (Berlocher 1993). Moreover, in *R. completa* the pattern and strength of the disequilibria vary among populations, with introduced populations having greater disequilibria (Berlocher 1984).

At the phenotypic level, only a handful of morphometric and ecological studies of intraspecific variation have been carried out. In *O. parietina* both morphometrics and allozymes were analyzed, permitting a comparison of the two types of data (Eber et al. 1992). Unlike the clinal pattern seen for allozymes, no clinal variation was observed in wing measurements, and there was no association between allozyme and morphometric variation. In *R. cerasi* (Linnaeus), latitudinal clinal variation for postdiapause emergence timing has been observed (Boller and Bush 1974). Although both of these studies were otherwise well designed, neither controlled for environmental variation; future work in this area should employ the common garden approach of rearing samples through one generation under the same conditions on the same host, before measuring.

Turning to the issue of large-scale geographic patterns, one is struck by the number of clines that have been found. Out of seven allele-level studies in which clines have been sought, or have sufficiently dense sampling that they could be detected, clines have been observed in three (Table 26.1). The latitudinal clines in *R. pomonella* have received the most attention (Feder and Bush 1989b; 1991; Feder et al. 1990a; 1993; Berlocher and McPheron 1996), and the loci involved are almost certainly under selection relating to temperature and life history phenology (Feder et al. 1997a, b). Less extreme but similar clines are seen in the closely related flowering dogwood fly (Berlocher 1999). Shallow but significant latitudinal clines are also seen in *O. parietina* (Eber et al. 1992). Altitudinal clines occur in *U. cardui* (Eber and Brandl 1994), and may occur in *R. pomonella* as well (Berlocher and McPheron 1996).

As opposed to clines, well-defined geographic races or subspecies have not often been described in tephritids. No shortage of races and subspecies can be found in the older taxonomic literature, but many of these are the result of incomplete data or insufficient analysis. One possibly valid example of a pair of classical, morphologically defined subspecies occurs in *E. solidaginis* (Ming 1989, cited in Foote et al. 1993). The western and eastern subspecies differ in one feature of the wing pattern, and both wing patterns apparently occur in the area between the ranges of the two subspecies. The extent to which the wing pattern difference corresponds to genetic differences at the allelic or allele genealogical level (Waring et al. 1990; Brown et al. 1996) is unknown, as essentially only the eastern race has been studied.

Two special topics in the area of geographic structure need to be addressed. One is the subject of introductions. Because of the pest nature of many tephritids, and the current world-wide transport of fruits and vegetables, tephritid introductions are both common and noticeable. In terms of population structure, the pattern to be expected is loss of alleles during an introduction, if the founding population is sufficiently small. This expectation is met in *C. capitata*, which shows spectacular losses of variation for both allozymes (Kourti et al. 1992) and mitochondrial DNA restriction site polymorphisms (Gasparich et al. 1997). For a counterexample, see McPheron (1990). After the initial loss of variation, loci that remain variable may show greater differentiation in introduced than in native populations, as a result of secondary bottlenecks (Berlocher 1984).

The second special topic is reproductive incompatibility due to *Wolbachia*, bacteria that are restricted entirely to an intracellular existence in cells of the female reproductive tracts of insects (review in Werren 1997). *Wolbachia* can cause one-way sterility when an insect with *Wolbachia* mates with a conspecific lacking *Wolbachia*, or containing a different strain, and new strains can spread rapidly (Turelli and Hoffman 1991). In some cases two-way incompatibility due to two different strains can be established, as in Californian and Hawaiian *Drosophila simulans* Sturtevant (O'Neil and Karr 1990). In *R. cerasi*, one-way sterility exists; western European males crossed with eastern European females result in the production of sterile eggs, but the reciprocal cross is fertile (Boller and Bush 1974). While *R. cerasi* has not been tested for *Wolbachia*, the pattern is consistent with this explanation. It has been proposed that *Wolbachia* could initiate population structuring, and possibly even speciation, but such effects in *R. cerasi* have not been observed.

### 26.3.5 Host Races, with a "Key" to Their Recognition

Of the various aspects of population structuring in tephritids, by far the most attention has focused on host races. Host races are defined as "partially reproductively isolated populations specializing on different hosts" (Diehl and Bush 1984), and are of great theoretical interest because they are in theory intermediate stages in sympatric speciation (Bush 1969; 1994). Despite their importance (or perhaps because of it), the literature on host races is burdened with a fair number of putative examples that are not in fact host races. Diehl and Bush (1984) did an excellent job of discussing the variety of biological phenomena that can be confused with host races (e.g., polyphagous species, sibling species, *etc*.), and their account should be read by all those starting work in this area. Here I have tried to condense several papers (Diehl and Bush 1984; Berlocher 1989) into a "key" to help in correctly identifying host races.

I assume that one is studying a group of morphologically similar insect populations that are living on several different host plants, and that the goal is to determine whether host races exist. I make four other assumptions. One, I assume that insect samples reared from each host have been collected from an area of host sympatry, ideally in a state of "microsympatry." This term has not been formally defined by tephritid workers, but has been used to describe cases where individual host plants of putative host races occur in very close proximity (a few tens of meters at most), within the "cruising range" of individual insects. Such collections are necessary to avoid confounding variation due to host races with conventional geographic variation. An extreme example using *R. pomonella* illustrates how geographic variation could be confused for host race variation. *Rhagoletis pomonella* from the common northern host *Crataegus mollis* has very different allele frequencies from *R. pomonella* from the common southern host *C. branchyacantha*, but it is latitude, not host plants, that is controlling the allele frequencies (*R. pomonella* has very steep latitudinal clines in allele frequencies; Berlocher and McPheron 1996). In this case the geographic effect is so extreme that few workers would be misled, but in other cases with smaller geographic distances and complex (and often unstudied) geographic structuring, the potential for misunderstanding is quite real. Second, I assume that adult insects disperse far enough in their life span that they frequently must have the opportunity to land upon an alternative host, and thus to mate and oviposit there. If adults mate and oviposit primarily on the same tree on which they developed as larvae, then any genetic differences reflect only host-related selection, not host-related mating patterns. My third assumption is that large samples (>100) are available, because allele frequency differences between host races are not large (Feder 1998; Feder et al. 1988; McPheron et al. 1988b). Fourth, I assume that samples from each host are made at the same life stage, in a consistent manner. This is necessary because selection acting at different life stages may alter allele frequencies (Feder et al. 1997a, b).

#### 26.3.5.1 Key for Distinguishing Host Races from Other Host-Plant-Related Structuring

1. Samples reared from different hosts have significant allele frequency differences . . . . . . . . . . 2
• Samples reared from different hosts do not have significant frequency differences . . . . . . . . . . . *One polyphagous species with an "all-purpose genotype(s)."*
2. Genotypes of at least one locus mate preferentially on at least some hosts. . . . . . . . . . . . . . . . 3
• All genotypes mate randomly on all host . . . . . . . . . . . . . . *One randomly mating polyphagous species with structure due only to selection.*
3. No gene flow detected in typical ($N \sim 100$) studies . . . . . . . . . . . . . . . . . . . . . . . . . *One or more monophagous cryptic species.*
• Gene flow readily detected in typical ($N \sim 100$) studies . . . . . . . . . . . . . . . . . . . . . *One species composed of two or more host races.*

Each of the categories above requires comment. The concept of an "all-purpose genotype" is relatively old, but has received a fair amount of attention recently (e.g., Jaenike and Dombeck 1998). An all-purpose genotype (or genotypes) is said to occur in generalist species in which no genetic specialization relating to hosts occurs; one or more "jack-of-all-trades" genotype functions equally well on all hosts. I note that another more prosaic explanation for leaving the key at couplet 1 is simply sampling error — no loci that were directly involved in differentiation (e.g., under divergent selection in different hosts) happened to be sampled in the study at hand.

Couplet 2 is critical. If one finds statistically significant allele frequency differences in samples reared from different hosts, but no evidence for nonrandom mating of genotypes with respect to hosts, then what one has is one random mating population in which the different host plants are selecting for different genotypes in the larval stage. The case at hand cannot possibly represent host races if adults are mating completely randomly.

The process of determining whether there is host-related mating is absolutely critical for making the correct decision at couplet 2, so I describe this process in detail here. There are in principle three approaches. The first is not yet feasible for technical reasons. This is to study genotypes of loci that are *directly involved* in host selection behavior and determine whether genotypes that are attracted to different hosts are in fact mating preferentially on those hosts (as they should logically be doing). As more becomes known in the future about the molecular genetics of insect olfaction, diapause, visual discrimination, etc., this will become the approach of choice, but it is some years in the future.

The other two approaches are feasible at present. Approach two requires no use of genetic data (although the method is most powerful used in conjunction with genetic data). This is to use physical marking of flies emerging from different hosts (e.g., different paint dots for different hosts) to determine by direct behavioral observation if flies from different hosts are mating preferentially with each other, on the host from which they emerged. In practice this means carrying out a mark-release-recapture experiment at microsympatric sites. This is the approach Feder et al. (1994) used to estimate only 6% mating between the apple and hawthorn races of *R. pomonella* at the Grant, Michigan site. Note that while this is a direct estimate of gene flow, it is an indirect answer to the question in couplet 2.

Approach three uses genetic markers to determine whether different genotypes are mating, or are likely to be mating, preferentially with respect to hosts. Essentially one samples adult flies from each of the host plants, and then returns the flies to the laboratory to be analyzed for genetic differences. However, there are three possible levels of analysis possible for approach three, each requiring larger sample sizes, but providing more information, than the previous. The first ("all flies") level involves simply capturing adults from each host plant, regardless of how the fly is behaving at the time of capture — regardless of whether it is ovipositing or mating on the host. One may be restricted to the "all flies" level if adults are uncommon or are otherwise sufficiently difficult to obtain that one must pool flies across behaviors to obtain samples of, say, 100 from each host. At the "all flies" level one runs the risk of underestimating the amount of preferential mating that is occurring, because flies from one host race or species may conceivably visit nonhosts (for adult feeding, *etc.*) without consequently mating there. Nonetheless, if allele frequencies differ between adults captured from different hosts, preferential mating is very likely. Feder and Bush (1989a) provide a textbook example of the "all flies" approach, with species rather than host races, demonstrating that *R. pomonella* and *R. mendax* do not mate in the field.

The next level is the "ovipositing female" level. If one can obtain from each host samples of ~100 females while they are actually ovipositing, a more powerful test than at the "all flies" level is possible because one knows that the flies are not just "visiting" that host. Note that analysis at the "ovipositing female" level requires greater sampling effort than at the "all flies" level not for statistical reasons — the number of flies required for genetic and statistical analysis is the same — but because much larger amounts of time and effort may be required to sample enough ovipositing females. While the "ovipositing female" level of analysis provides more biological insight than the "all flies" level, male behavior is not being analyzed.

The ultimate level is the "mating pair" level. However, analysis at this level will require a great deal of work, for two reasons. First, mating pairs make up a small minority of flies observed on a typical host, so much sampling effort would be required to obtain, say, 100 pairs from each host. Second, to make optimum use of the information, analysis needs to be carried out at the genotype level. Of course, useful information is obtained by simply testing for allele frequency differences between samples of mating adults sampled from different hosts, but analysis at the genotype level can potentially provide information not only on whether genotypes are mating at random or not, but on the *amount* of nonrandom mating that is occurring. Using the basic allele frequency data, a null hypothesis of random mating genotype expectations can be constructed, which can then be tested with the observed frequencies of matings between different genotypes on each host. Furthermore, a maximum likelihood model could be constructed to allow estimation of the amount of mating occurring between host races (or species, if that is the case). I can find no evidence in the literature that this potentially very informative approach has been used, but future workers should give it serious consideration.

Couplet 3 presents a theoretical difficulty, which is that the amount of gene flow that characterizes a host race vs. a species is not clearly defined in the literature. The situation has grown more complicated in recent years with the realization that gene flow between animal species is not the extreme rarity that Mayr (1963) argued, but instead occurs with some frequency in some groups (Arnold 1992). Resolving this difficulty is beyond the scope of this review. Here I argue for a pragmatic decision: if one finds no evidence for gene flow (either from direct observation or inferred from the mating frequencies) in a study of typical size and statistical power (say, 100 pairs from each host), then the populations under study represent species rather than host races. If gene flow is unambiguously occurring, then one has host races. As loci and sample sizes in tephritid studies grow (hopefully) from the current tens of allozyme loci and few hundreds of individuals into much larger numbers, so that rare events can be detected, a more refined approach to couplet 3 will be needed.

By far the best-studied host races are the apple and hawthorn races of *R. pomonella* (Feder 1998). The apple race cannot be any older than about 450 years, the date of introduction of apples to the New World, but is probably only about 140 years old (Bush 1969). The two races differ in allozyme frequencies (Feder et al. 1988; 1990a, b; McPheron et al. 1988b), behavior (Prokopy et al. 1988), and diapause phenology (Smith 1988). Of critical importance, as already noted, mating between the two races is not random at microsympatric sites (Feder et al. 1994). Recognition that the apple and hawthorn populations are host races, not simply members of a single gene pool structured by host-related selection, has appeared in many influential textbooks (e.g., Futuyma 1998).

Other cases in tephritids generally lack sufficient data to make an informed decision about the existence of host races. Different oviposition locations of the European *Tephritis bardanae*

(Schrank) on two *Arctium* species may possibly indicate host races (Eber et al. 1991), but the morphological and allozyme samples from the two hosts were from different areas in Europe, so simple geographic variation may be a more parsimonious explanation. A more convincing case can be made for *T. conura* (Loew) on *Cirsium oleraceum* and *C. heterophyllum* (Seitz and Komma 1984). Although basic data such as allozyme frequencies were not included in the published study, Figure 26.9b of Seitz and Komma (1984) indicates that the frequencies of the *A* allele of the *hk2* locus were ~0.8 in flies reared from *C. heterophyllum* and ~0.2 in flies reared from *C. oleraceum*, at both single-host and two-host sites, which is a large difference to be maintained purely by selection in a random mating population; preferential mating is highly likely. More significantly, at two-host sites samples from both hosts analyzed singly fit HWE, whereas pooling of data from the two hosts at two-host sites produced highly significant departures from HWE (Seitz and Komma 1984: 153). These departures are in the form of heterozygote deficiency, as would be expected if two host races or species exist. Unfortunately, the data do not allow gene flow to be measured, so it is impossible to decide between these two possibilities.

The case of the *E. solidaginis* populations associated with *Solidago giganea* and *S. altissima* is much more thoroughly studied than that of the two *Tephtitis* species (see review by Abrahamson and Weis 1997). Flies from the two hosts do choose the host from which they were reared, and large, consistent allele and haplotype differences exist. However, from the standpoint of host races these differences are almost "too good"; the fixed mt-DNA haplotype differences in the western part of the range of the two races indicate that no gene flow is occurring, and that the races should be considered to be species. A notable feature of the *E. solidaginis* work is that the ancestral host can be inferred from phylogeographic analysis of the mt-DNA data to be *S. altissima* (Brown et al. 1996).

## 26.4 CONCLUSIONS AND FUTURE RESEARCH NEEDS

Several clear conclusions can be reached about tephritid population structure.

1. Local populations are inevitably in Hardy–Weinberg equilibrium; thus, local populations mate randomly (excluding the case of host races).
2. Geographic differentiation as measured by $F_{ST}$ and allozyme data is small, after cases with strong clines, possible cryptic species, or introductions are removed.
3. Gene flow inferred from $F_{ST}$ and allozyme data is high.
4. Clines in temperate species are common and may be related to temperature and diapause conditions.
5. Geographic races/subspecies (after removing cases from bad taxonomy) are not common.
6. Host races definitely exist in tephritids, as convincingly demonstrated by the *R. pomonella* apple and hawthorn host races, and may be common. However, great care must be taken to demonstrate for each putative case that host races, and not some simpler phenomenon, are involved.

As I have discussed, many areas of tephritid population structure need to be analyzed in much greater detail before we can generalize with great confidence about the forces that fashion population structure. Studies are particularly needed on host-related selection, temporal variation, and host races. However, even geographic structure, about which we have the most information, is inadequately studied in tephritids. My conclusion 5, that "classical" geographic races/subspecies are uncommon, could be a reflection of failure to study enough characters, or of a taxonomic tendency to describe all distinct allopatric populations as species rather than subspecies. The newer DNA markers, such as microsatellites, must be developed and applied to problems of population structure, although allozymes will remain a cost-effective and powerful way to analyze structure for some time into the future.

In particular, future studies should make the most of the strengths of the Tephritidae as research organisms. Among these strengths are simplicity of habitat and niche characterization, and relative ease of dispersal measurement. Some current workers feel that direct measurements of dispersal do not explain much about population structure and gene flow: "Empirical tests such as comparisons with mark-and-recapture estimates are at best ambiguous. Indirect methods estimate the cumulative effects of gene flow, acting over all temporal and spatial scales. In contrast, direct estimates of gene flow apply only to the interval of time and space over which observations are made" (Neigel 1997: 106). However, without direct empirical measurements, measurement of gene flow from genetic markers has the potential to become an exercise in circular reasoning. At least on a local scale, direct measurements of gene flow are both feasible and informative about the validity of indirect, genetic measurements (e.g., Pfenninger et al. 1996). For many problems in the Tephritidae, with host races being the premier case, both direct (behavioral) and indirect (genetic) gene flow measurements are absolutely essential.

## ACKNOWLEDGMENTS

Financial support for the symposium and the author's travel was generously provided by Campaña Nacional contra las Moscas del la Fruta (Mexico), International Organization for Biological Control of Noxious Animals and Plants, Instituto de Ecologia, A.C. (Mexico), and the USDA-ICD-RSED. The author's work has been supported at fitful intervals by the National Science Foundation.

## REFERENCES

Abrahamson, W.G. and A.E. Weis. 1997. *Evolutionary Ecology across Three Trophic Levels. Goldenrods, Gallmakers, and Natural Enemies.* Princeton University Press, Princeton. 456 pp.

Arnold, M.L. 1992. Natural hybridization as an evolutionary process. *Annu. Rev. Ecol. Syst.* 23: 237–261.

Avise, J.C. 1994. *Molecular Markers, Natural History, and Evolution.* Chapman & Hall, New York. 511 pp.

Avise, J.C., J. Arnold, R.M. Ball, E. Bermingham, T. Lamb, J.E. Neigel, C.A. Reeb, and N.C. Saunders. 1987. Intraspecific phylogeography: the mitochondrial DNA bridge between population genetics and systematics. *Annu. Rev. Ecol. Syst.* 18: 489–522.

Berlocher, S.H. 1984. Genetic changes coinciding with the colonization of California by the walnut husk fly, *Rhagoletis completa. Evolution* 38: 906–918.

Berlocher, S.H. 1989. The complexities of host races and some suggestions for their identification by enzyme electrophoresis. In *Electrophoretic Studies on Agricultural Pests* (H.D. Loxdale and J. den Hollander, eds.), pp. 51–68. Systematics Association special volume No. 39. Clarendon Press, Oxford.

Berlocher, S.H. 1993. Gametic disequilibrium between allozyme loci and sex chromosomes in the genus *Rhagoletis. J. Hered.* 84: 431–437.

Berlocher, S.H. 1995. Population structure of the blueberry maggot, *Rhagoletis mendax. Heredity* 74: 542–555.

Berlocher, S.H. Host race or species? Allozyme characterization of the "flowering dogwood fly," a member of the *Rhagoletis pomonella* complex. *Heredity,* in press.

Berlocher, S.H. and B.A. McPheron. 1996. Population structure of the apple maggot fly, *Rhagoletis pomonella. Heredity* 77: 83–99.

Boller, E.F. and G.L. Bush. 1974. Evidence for genetic variation in populations of the European cherry fruit fly, *Rhagoletis cerasi* (Diptera: Tephritidae), based on physiological parameters and hybridization experiments. *Entomol. Exp. Appl.* 17: 279–293.

Brown, J.M., W.G. Abrahamson, and P.A. Way. 1996. Mitochondrial DNA phylogeography of host races of the goldenrod ball gallmaker *Eurosta solidaginis* (Diptera: Tephritidae). *Evolution* 50: 777–786.

Bush, G.L. 1966. The taxonomy, cytology, and evolution of the genus Rhagoletis in North America (Diptera: Tephritidae). *Bull. Mus. Comp. Zool.* 134: 431–562.

Bush, G.L. 1969. Sympatric host race formation and speciation in frugivorous flies of the genus *Rhagoletis* (Diptera: Tephritidae). *Evolution* 23: 237–251.

Bush, G.L. 1994. Sympatric speciation in animals: new wine in old bottles. *Trends Ecol. Evol.* 9: 285–288.

Cavalli-Sforza, L.L., P. Menozzi, and A. Piazza. 1994. *The History and Geography of Human Genes.* Princeton University Press, Princeton.

Cosmides, N., M. Loukas, and E. Zouros. 1997. Differences in fitness components among alcohol dehydrogenase genotypes of the olive fruit fly (Diptera: Tephritidae) under artificial rearing. *Ann. Entomol. Soc. Am.* 90: 363–371.

Diehl, S.R. and G.L. Bush. 1984. An evolutionary and applied perspective of insect biotypes. *Annu. Rev. Entomol.* 29: 471–504.

Dobzhansky, Th. 1937. *Genetics and the Origin of Species.* Columbia University Press, New York.

Dobzhansky, Th. 1943. Genetics of natural populations: IX. Temporal changes in the composition of populations of *Drosophila pseudoobscura. Genetics* 24: 391–412.

Eber, S. and R. Brandl. 1994. Ecological and genetic spatial patterns of *Urophora cardui* (Diptera: Tephritidae) as evidence for population structure and biogeographical processes. *J. Anim. Ecol.* 63: 187–199.

Eber, S., P. Sturm, and R. Brandl. 1991 Genetic and morphological variation among biotypes of *Tephritis bardanae. Biochem. Syst. Ecol.* 19: 549–557.

Eber, S., R. Brandl, and S. Vidal. 1992. Genetic and morphological variation among populations of *Oxyna parietina* (Diptera: Tephritidae) across a European transect. *Can. J. Zool.* 70: 1120–1128.

Endler, J.A. 1977. *Geographic Variation, Speciation, and Clines.* Princeton University Press, Princeton.

Feder, J.L. 1998. The apple maggot fly, *Rhagoletis pomonella*: flies in the face of conventional wisdom about speciation? In *Endless Forms: Species and Speciation* (D.J. Howard and S.H. Berlocher, eds.), pp. 130–144. Oxford University Press, New York.

Feder, J.L. and G.L. Bush. 1989a. A field test of differential host usage between two sibling species of *Rhagoletis* fruit flies (Diptera: Tephritidae) and its consequences for sympatric models of speciation. *Evolution* 43: 1813–1819.

Feder, J.L. and G.L. Bush. 1989b. Gene frequency clines for host races of *Rhagoletis pomonella* (Diptera: Tephritidae) in the midwestern United States. *Heredity* 63: 245–266.

Feder, J.L. and G.L. Bush. 1991. Genetic variation among apple and hawthorn host races of *Rhagoletis pomonella* (Diptera: Tephritidae) across an ecological transition zone in the mid-western United States. *Entomol. Exp. Appl.* 59: 249–265.

Feder, J.L., C.A. Chilcote, and G.L. Bush. 1988. Genetic differentiation between sympatric host races of *Rhagoletis pomonella. Nature* 336: 61–64.

Feder, J.L., C.A. Chilcote, and G.L. Bush. 1990a. The geographic pattern of genetic differentiation between host associated populations of *Rhagoletis pomonella* (Diptera: Tephritidae) in the eastern United States and Canada. *Evolution* 44: 570–594.

Feder, J.L., C.A. Chilcote, and G.L. Bush. 1990b. Regional, local, and microgeographic allele frequency variation between apple and hawthorn populations of *Rhagoletis pomonella* in western Michigan. *Evolution* 44: 596–608.

Feder, J.L., T.A. Hunt, and G.L. Bush. 1993. The effect of climate, host plant phenology, and host fidelity on the genetics of apple and hawthorn-infesting races of *Rhagoletis pomonella. Entomol. Exp. Appl.* 69: 117–135.

Feder, J.L., S. Opp, B. Wlazlo, C. Reynolds, W. Go, and S. Spisak. 1994. The role of host fidelity in sympatric host race formation in the apple maggot fly, *Rhagoletis pomonella* (Diptera: Tephritidae) *Proc. Natl. Acad. Sci. U.S.A.* 91: 7990–7994.

Feder, J.L., J.B. Roethele, B. Wlazlo, and S.H. Berlocher. 1997a. Selective maintenance of allozyme differences between sympatric host races of the apple maggot fly. *Proc. Natl. Acad. Sci. U.S.A.* 94: 11417–11421.

Feder, J.L., U. Stolz, K.M. Lewis, W. Perry, J.B. Roethele, and A. Rogers. 1997b. The effects of winter length on the genetics of apple and hawthorn races of *Rhagoletis pomonella* (Diptera: Tephritidae). *Evolution* 51: 1862–1876.

Foote, R.H., F.L. Blanc, and A.L. Norrbom. 1993. *Handbook of the Fruit Flies (Diptera: Tephritidae) of America North of Mexico.* Comstock Publishing Associates, Ithaca.

Futuyma, D.J. 1998. *Evolutionary Biology.* 3rd ed. Sinauer Associates, Sunderland.

Gasparich, G.E., J.G. Silva, H.Y. Han, B.A. McPheron, G.J. Steck, and W.S. Sheppard. 1997. Population genetic structure of Mediterranean fruit fly (Diptera: Tephritidae) and implications for worldwide colonization patterns. *Ann. Entomol. Soc. Am.* 90: 790–797.

Gasperi, G., C.R. Gugleilmino, A.R. Malacrida, and R. Milani. 1991. Genetic variability and gene flow in geographical populations of *Ceratitis capitata* (Wied.) (Medfly). *Heredity* 67: 347–356.

Hartl, D.L. 1988. *A Primer of Population Genetics*. 2nd ed. Sinauer Associates, Sunderland.

Haymer, D.S. 1995. Genetic analysis of laboratory and wild strains of the melon fly (Diptera: Tephritidae) using random amplified polymorphic DNA-polymerase chain reaction. *Ann. Entomol. Soc. Am.* 88: 705–710.

Hedrick, P., S. Jain, and L. Holden. 1978. Multilocus systems in evolution. *Evol. Biol.* 11: 104–184.

Hudson, R.R. 1990. Gene genealogies and the coalescent process. *Oxford Sur. Evol. Biol.* 7.

Jaenike, J. and I. Dombeck. 1998. General-purpose genotypes for host species utilization in a nematode parasite of *Drosophila. Evolution* 52: 832–840.

Kambhamphati, S. and M. Makauer. 1988. Intra-and interspecific morphological variation in some *Aphidius* species (Hymenoptera: Aphidiidae) parasitic on the bean aphid in North America. *Ann. Entomol. Soc. Am.* 81: 1010–1016.

King, P.S. 1988. Distribution and genetic structure of two allopatric beetle (Coleoptera: Melyridae) species on rock outcrops in the Southeast. *Ann. Entomol. Soc. Am.* 81: 890–898.

Kourti, A., M. Loukas, and J. Sourdis. 1992. Dispersion pattern of the medfly from its geographic center of origin and genetic relationships of the medfly with two close relatives. *Entomol. Exp. Appl.* 63: 63–69.

Lisowski, E.A. 1979. Biochemical Systematics of the Genus *Strauzia* (Diptera: Tephritidae). M.S. thesis, University of Illinois at Urbana-Champaign, Urbana.

Malavasi, A. and J.S. Morgante. 1983. Population genetics of *Anastrepha fraterculus* (Diptera, Tephritidae) in different hosts: genetic differentiation and heterozygosity. *Genetica* (Dordrecht) 60: 207–212.

Mantel, N. 1967. The detection of disease clustering and a generalized regression approach. *Cancer Res.* 27: 209–220.

Mayr, E. 1963. *Animal Species and Evolution*. Belknap Press, Cambridge.

McPheron, B.A. 1990. Genetic structure of apple maggot fly (Diptera: Tephritidae) populations. *Ann. Entomol. Soc. Am.* 83: 568–577.

McPheron, B.A. and S.H. Berlocher. 1985. Segregation and linkage of allozymes of *Rhagoletis tabellaria. J. Hered.* 76: 218–219.

McPheron B.A., D.C. Smith, and S.H. Berlocher. 1988a. Microgeographic genetic variation in the apple maggot, *Rhagoletis pomonella. Genetics* 119: 445–451.

McPheron B.A., D.C. Smith, and S.H. Berlocher. 1988b. Genetic differences between host races of the apple maggot fly. *Nature* 336: 64–66.

Ming, Y. 1989. A Revision of the Genus *Eurosta* Loew, with a Scanning Electron Microscopic Study of Taxonomic Characters (Diptera: Tephritidae). M.S. thesis, Washington State University, Pullman.

Neigel, J.E. 1997. A comparison of alternative strategies for estimating gene flow from genetic markers. *Annu. Rev. Ecol. Syst.* 28: 105–128.

O'Neil, S.L. and T.L. Karr. 1990. Bidirectional incompatibility between conspecific populations of *Drosophila simulans. Nature* 348: 178–180.

Patterson, M.A. and R.F. Denno. 1997. The influence of intraspecific variation in dispersal strategies on the genetic structure of planthopper populations. *Evolution* 51: 1189–1206.

Payne, J.A. and S.H. Berlocher. 1995. Phenological and electrophoretic evidence for a new blueberry-infesting species in the *Rhagoletis pomonella* (Diptera: Tephritidae) sibling species complex. *Entomol. Exp. Appl.* 75: 183–187.

Pfenninger, M., A. Bahl, and B. Streit. 1996. Isolation by distance in a population of a land snail *Trochoidea geyeri*: evidence from direct and indirect methods. *Proc. R. Soc. London* 263: 1211–1217.

Pönisch, S. and R. Brandl. 1992. Cytogenetics and diversification of the phytophagous fly genus *Urophora* (Tephritidae). *Zool. Anz.* 228: 12–25.

Prokopy, R.J., S.R. Diehl, and S.S. Cooley. 1988. Behavioral evidence for host races in *Rhagoletis pomonella* flies. *Oecologia* 76: 138–147.

Roethele, J.R., J.R. Feder, S.H. Berlocher, and M.E. Kreitman. 1997. Toward the construction of a molecular genetic linkage map for the apple maggot fly, *Rhagoletis pomonella* (Diptera: Tephritidae): a comparison of alternative strategies. *Ann. Entomol. Soc. Am.* 90: 470–479.

Schug, M.D., J.F. Downhower, L.P. Brown, D.B. Sears, and P.A. Fuerst. 1998. Isolation and genetic diversity of *Gambusia hubbsi* (mosquitofish) populations in blueholes on Andros Island, Bahamas. *Heredity* 80: 336–346.

Seitz, A. and M. Komma. 1984. Genetic polymorphism and its ecological background in Tephritid populations (Diptera: Tephritidae). In *Population Biology and Evolution* (K. Wöhrmann and V. Loeschcke, eds.), pp. 141–158. Springer-Verlag, Berlin.

Slatkin, M. 1985. Rare alleles as indicators of gene flow. *Evolution* 39: 53–65.

Slatkin, M. and N. Barton. 1989. A comparison of three indirect methods for estimating average levels of gene flow. *Evolution* 43: 1349–1368.

Slatkin, M. and D. Maddison. 1990. Detecting isolation by distance using phylogenies of genes. *Genetics* 126: 249–260.

Smith, D.C. 1988. Heritable divergence of *Rhagoletis pomonella* host races by seasonal asynchrony. *Nature* 336: 66–68.

Smith, J.J. and G.L. Bush. 1997. Phylogeny of the genus *Rhagoletis* (Diptera: Tephritidae) inferred from DNA sequences of mitochondrial cytochrome oxidase II. *Mol. Phylog. Evol.* 7: 33–43.

Sokal, R.R. and D.E. Wartenburg. 1983. A test of spatial autocorrelation analysis using an isolation by distance model. *Genetics* 105: 219–237.

Steck, G.J. 1981. North American Terellinae (Diptera: Tephritidae): Biochemical Systematics and Evolution of Larval Feeding Niches and Adult Life History. Ph.D. dissertation, University of Texas at Austin, Austin.

Steck, G.J. 1991. Biochemical systematics and population genetic structure of *Anastrepha fraterculus* and related species (Diptera: Tephritidae). *Ann. Entomol. Soc. Am.* 84: 10–28.

Steck, G.J. and W.S. Sheppard. 1993. Mitochondrial DNA variation in *Anastrepha fraterculus*. In *Fruit Flies: Biology and Management* (M. Aluja and P. Liedo, eds.), pp. 9–14. Springer-Verlag, New York.

Swofford, D.S. and R.B. Selander. 1981. BIOSYS-1: a FORTRAN program for the comprehensive analysis of electrophoretic data in population genetics and systematics. *J. Hered.* 72: 281–283.

Swofford, D.L., G.J. Olsen, P.J. Wadell, and D.M. Hillis. 1996. Phylogenetic inference. In *Molecular Systematics,* 2nd ed. (D.M. Hillis, C. Moritz, and B.K. Mable, eds.), pp. 407–514. Sinauer Associates, Sunderland.

Tsakas, S.C. and E. Zouros. 1980. Genetic differences among natural and laboratory-reared populations of the olive fruit fly *Dacus oleae* (Diptera: Tephritidae). *Entomol. Exp. Appl.* 28: 268–276.

Turelli, M. and A.A. Hoffman. 1991. Rapid spread of an inherited incompatibility factor in California *Drosophila. Nature* 353: 440–442.

Waring, G.L., W.G. Abrahamson, and D.J. Howard. 1990. Genetic differentiation in the gallformer *Eurosta solidaginsis* (Diptera: Tephritidae) along host plant lines. *Evolution* 44: 1648–1655.

Weir, B.S. 1990. *Genetic Data Analysis*. Sinauer Associates, Sunderland. 377 pp.

Werren, J.H. 1997. Biology of *Wolbachia. Annu. Rev. Entomol.* 42: 587–609.

Wright, S. 1931. Evolution in Mendelian populations. *Genetics* 16: 97–159.

Wright, S. 1943. Isolation by distance. *Genetics* 28: 114–138.

Wright, S. 1978. *Evolution and the Genetics of Populations*. Vol. 4, *Variability in and among Natural Populations*. University of Chicago Press, Chicago. 580 pp.

Yong, H.S. 1988. Allozyme variation in the *Artocarpus* fruit fly *Dacus umbrosus* (Insecta: Tephritidae) from peninsular Malaysia. *Comp. Biochem. Physiol. B Comp. Biochem.* 91: 85–90.

Yong, H.S. 1992. Allozyme variation in the melon fly *Dacus cucurbitae* (Insecta: Diptera: Tephritidae) from peninsular Malaysia. *Comp. Biochem. Physiol. B Comp. Biochem.* 102: 367–370.

## Note Added in Proof:

I became aware of the recent paper of Eber and Brandl (Eber, S. and R. Brandl. 1997. Genetic differentiation of the tephritid fly *Urophora cardui* in Europe as evidence for its biogeographic history. *Mol. Ecol.* 6: 651–660) too late to include it in the text. Eber and Brandl used both Wright's (1978) F-statistics and the Slatkin (1985) private alleles approaches, and concluded that gene flow levels were high, and that gentic patterns predicted from Pleistocene refuge theory were not observed. However, the overall $F_{ST}$, 0.128, is among the higher values reported for tephritids, and frequency clines were inferred from a regression of the second principal component (from an analysis of Nei distances) against latitude. I would place this study in the second part of Table 26.1, cases in which inference of gene flow from genetic differentiation may not be valid.

# 27 The Evolution of Fruit Fly Feeding Behavior

*Richard A.I. Drew and Boaz Yuval*

## CONTENTS



## 27.1 INTRODUCTION

As immatures and often as adults, tephritid fruit flies eat to fuel daily survival and to acquire the resources critical for reproduction. Tephritid larvae spend most of their time feeding (they rarely do anything else), usually on the substrate or host they were placed on by their female parent. Hence, the evolution of larval feeding behavior is intimately related to the evolution of oviposition behavior. Most adult tephritids must find and consume fuel for their daily activities, and may forage for reproductive resources not acquired during the larval stage, thus forging a link between adult feeding and reproductive behavior.

Our basic assumption in this chapter is that the key to the biological success of the Tephritidae was the evolution of phytophagy, and that this was made possible primarily due to morphological and behavioral adaptations in the female, namely, ovipositors and specific oviposition preferences (see Zwölfer 1983; Sivinski, Chapter 2; Korneyev, Chapter 4; and Díaz-Fleischer et al., Chapter 30). The ability of the ancestral tephritid larva to develop and thrive in these novel environments was a critical preadaptation. Larvae in the families closely related to tephritids, indeed, most cyclorraphous larvae, are generally saprophagous, inhabiting and feeding on rotting and decaying organic material (see Zwölfer 1983 and Chapters 2, 4, and 30). The jump from saprophagy to phytophagy is not a major one, in terms of ingestive and digestive abilities (Terra 1990). It is reasonable to

0-8493-1275-2/00/$0.00+$.50

**TABLE 27.1**
**Larval Hosts in the Family Tephritidae**

| Subfamily | Larval Hosts |
|---|---|
| Tachiniscinae (18 species) | Parasitoids of Lepidoptera (biology known for only 1 sp.) |
| Blepharoneurinae (at least 33 species) | Specialize on flowers, seeds, fruit, and stems of Cucurbitaceae |
| Phytalmiinae (337 species in four tribes) | Mainly saprophagous, some species phytophagous. |
| Trypetinae (>1000 species in six tribes) | Extremely diverse — larvae of a few species predatory, others cover the whole phytophagous range — feed on fruits, seeds, flower buds, stems, or as leaf miners |
| Dacinae (>1000 species in three tribes) | Fruits, seed pods (rarely flowers) |
| Tephritinae (1847 species in 11 tribes) | Specialize on flowers of Asteraceae (some species form galls in stems, roots, or flowers) |

postulate that the sequence from saprophagy to frugivory and thence to feeding on seeds evolved independently in the various tephritid subfamilies (Table 27.1), as did specializations such as stem and leaf boring and, ultimately, gall forming. This point is best illustrated by the relationship of the Blepharoneurinae with cucurbit plant species and of the Tephritinae with species of Asteraceae. It would seem that independently in both these groups, larval feeding capabilities complemented shifts in female oviposition preferences.

Feeding strategies play a key role in the biology of adult tephritids because of their influences on other behavioral and physiological processes such as sexual maturation, courtship and mating, oviposition, and general responses to host plants. As we attempt to show below, adult feeding is dependent on reproductive demands, and the variety of specific patterns seen within the family depend in turn on larval nutrition.

This chapter is designed to update our knowledge of feeding behavior within the Tephritidae, help develop an understanding of the relationship of feeding behavior to other behavioral strategies, and identify gaps in our knowledge that need to be filled through future research. Some extremely valuable work has been done on feeding behavior of species across the family and is reviewed in this chapter. However, large gaps still exist in our understanding of what may be the least studied aspect of tephritid behavior.

## 27.2 LARVAL FEEDING BEHAVIOR

### 27.2.1 Nutritional Requirements

Compared with the quite massive amount of research done on oviposition behavior by tephritid females, very little has been done on the feeding of larvae. As stated above, it is frequently assumed that when larvae are capable of metabolically backing up female oviposition decisions, host ranges may be expanded. Whether monophagous species are in dead ends because of the inability of larvae to develop in other hosts is not known. Our understanding of nutritional requirements for larval stages of Tephritidae is currently limited to work done on the development of culture media for mass-rearing. The basic nutrient added to larval culture media is hydrolyzed protein in the form of brewer's yeast (Hagen et al. 1963; Mitchell et al. 1965; Tanaka et al. 1969; Schroeder et al. 1972). Probably this yeast also provides any required vitamins in the B-complex.

Research by Manoukas (1986) has shown that while a range of amino acids are essential for larval development, methionine and lysine were detrimental. Further, there are tolerance levels for the essential amino acids, above which adverse effects on larval development occur (Manoukas 1981). There is considerable variation in the growth, survival, and developmental times for larvae of tephritid species in different host fruits (Tsitsipis 1989). It may be that generalist species have a capacity to develop under a wider variation of nutrients, whereas monophagous species are more restricted and specific in their requirements. This possibility remains to be examined experimentally.

### 27.2.2 Feeding Mechanisms

Larvae in the subfamily Tephritinae have a pair of mouthhooks and a well-developed median oral lobe (Headrick and Goeden 1998). The latter appears to be characteristic of all nonfrugivorous Tephritidae and absent in the Trypetinae and Dacinae. The function of the median oral lobe appears to be to uptake fluids exuding from plant tissues that have been scraped by the mouthhooks. In the frugivorous species, the Trypetinae and Dacinae, a pair of mouthhooks and a simple median oral opening for intake of fluid are present.

It appears that the larvae of most tephritid species ingest liquid food and that the mouthhooks are used for maceration of food substrates and, in the case of fruit-infesting Tephritidae, cutting exit holes in fruit skins so that final instar larvae can emerge and enter the soil for pupation. Some dacines in the genus *Dacus* utilize fruit pods in the plant family Asclepiadaceae and pupate within the pods. This appears to be a mechanism that has evolved to ensure survival in hot, dry savannah environments where soil temperatures would be detrimental to the immature stages. In the Dacinae, the larvae exist and feed within a "bacterial soup" produced by the "fruit fly–type" bacteria rotting down the fruit tissue. In this situation the basic feeding mechanism appears to be the sucking in of fluid and the utilization of bacteria as a protein source.

### 27.2.3 Larval Environments

Resource exploitation by larval tephritids was discussed by Zwölfer (1983), who divided the family into three groups based on resource exploitation strategies. He distinguished between opportunistic broad-range exploiters of pulpy fruits, specialized exploiters of pulpy fruits, and specialized exploiters of vegetative structures and inflorescences. Indeed, tephritid larvae utilize a wide range of feeding sites. In the genus *Phytalmia*, larvae feed in rotting sapwood while species of Tephritinae utilize flowerheads of Asteraceae or form galls on stems and mines in leaves (Headrick and Goeden, Chapter 25). The various species of Blepharoneurinae specialize on flowers, seeds, fruit, and stems of cucurbits (Condon and Norrbom, Chapter 7). Many species of Trypetinae and Dacinae are primary fruit feeders while a few are secondary fruit invaders.

Some more specialized larval feeding activities are *Euphranta* species in mangrove fruit where the larvae pupate in parts of the fallen pods above the seawater line; *E. toxoneura* (Loew) feeding within galls produced by sawflies (White 1988); *Campiglossa misella* (Loew) within galls formed by a gall midge; *Oxyna palpalis* (Coquillett) on the host plant *Artemisia tridentata* (Sage); the leaf miner *Euleia heraclei* (L.) where the larvae are capable of movement and transfer between leaves; many Trypetinae species that feed in young shoots of giant bamboo, some of which have blue and orange colored larvae; *Chaetorellia australis* Hering that, within a breeding season, has two generations in star thistle and one in cornflower; *Toxotrypana curvicauda* Gerstaecker in papaya which produces up to 50 larvae per fruit, all of which emerge through the same exit hole on the fruit surface.

Although the larvae of most tephritid species are thought to be fluid feeders, some may not be. So little is known about the feeding activities of larvae of *Dacus* species in asclepiad pods, of gall formers in stems and flower heads, for example, that nonfluid food uptake may exist. Some of these species may ingest particulate matter or more viscous food.

### 27.2.4 Resource Partitioning

In the Tephritinae, the temporal and spatial utilization of resources by larvae is usually precise. For example, larvae of *Paracantha gentilis* feed centrally in thistle flowers and pupate in the same area in order to avoid competition with the lepidopteran flower feeder, *Rotruda mucidella*, which would prey upon the tephritid immature stages as well as utilize the larval food sources (Headrick and Goeden 1998). Similarly, of four species of *Aciurina* that utilize the same host plant in the same general locality, two species occupy the plant in spring at low altitudes while the other two breed in summer at higher altitudes. In coastal Queensland and offshore islands, a species of *Trupanea* and one species of *Rhabdochaeta* have been reared from the same *Wedelia* flowerheads at the same time (Hardy and Drew 1996). Although far from highly mobile, larvae of Mediterranean fruit fly, *Ceratitis capitata* (Wiedemann), can react to gradients of food quality and move within host fruits to sites of highest nutritional reward (e.g., Zucoloto 1991; Yuval and Hendrichs, Chapter 17).

In the Dacinae and Trypetinae, virtually nothing is known about the temporal and spatial differences in utilization of fruit by larvae. The ubiquity of host-marking pheromones in these species may be assumed to achieve considerable spacing of larvae (Díaz-Fleischer et al., Chapter 30). Nevertheless, some fruit species host more than one fly species at a time. *Bactrocera halfordiae* (Tryon) and *Dirioxa pornia* (Walker) utilize fruit of *Planchonella australis* at the same time in rain forests of southeast Queensland. In Malaysia, four species of *Bactrocera* have been bred from the same cucurbit host fruit in the endemic rain forest habitat.

In general, the protein and amino acid content of plant tissue is believed to be low (Burroughs 1970; Hansen 1970), and fruit is classified as a low-nitrogen product of little nutritional significance as a protein food. Tephritidae have evolved specific means of utilizing plant nutrients, and particularly proteins, in order to ensure adequate growth rates for the larval stages. Drew (1988) showed that fruit tissue containing tephritid larvae had considerably higher levels of amino acids than uninfested fruit tissue and that these increases were probably due to bacterial colonization in association with the larval activity. Another tactic is evident in tephritine galls. Headrick and Goeden (1998) reported that galls within flowerheads were metabolic sinks that drew nutrients from the host plant.

Adequate nutrition to ensure rapid development is essential in fruit-feeding tephritids in order to preempt predation. For example, Drew (1987b) showed that in the endemic rain forest habitat in southeast Queensland, fruit fly populations were reduced 70% by the activities of fruit-feeding vertebrates. Consequently, rapid larval growth (as well as burrowing toward the fruit center and rapid skipping activity of third instar larvae after leaving the fruit) appear to be functions that have evolved to ensure reproductive success in species subject to high levels of predation.

The fruit feeders appear to suffer more from predators and parasites than do the flower feeders and gall formers, and produce greater numbers of eggs than species using flowers. Some species of *Bactrocera* have at least 40 ovarioles per ovary compared with the tephritine species with fewer than ten. In order to produce such high egg loads, the fruit-feeding species require greater quantities of protein-rich food.

### 27.2.5 Bacteria Associations

Considerable research has been done on bacterial associations in fruit flies. The majority of the literature has been reviewed by Howard (1989) and Drew and Lloyd (1989). Most of the work that has been done reveals that certain bacteria are essential for larval growth and that this may be due to bacterial synthesis of nutrients from food substrates, rather than the microorganisms being ingested as a food source. However, Howard (1989) believes that the "premise of larval dependence on bacterial symbionts" is increasingly less tenable. Consequently, if the specific "fruit fly–type" bacteria are essential for larval development, at least in the fruit feeders, and symbiotic factors do not exist, then the utilization of these microorganisms as food cannot be dismissed and deserves further intensive research.

## 27.3 ADULT FLY FEEDING BEHAVIOR

### 27.3.1 Nutritional Requirements

Adult tephritids require a variety of nutrients in order to survive, fuel their various activities, and allow them to realize their reproductive potential (Tsitsipis 1989). In considering the subject of nutritional requirements and food sources in nature, one must be careful to assess these separately within the major supraspecific taxonomic groupings. For example, the Trypetinae (e.g., *Rhagoletis* species) and the Dacinae (e.g., *Bactrocera* species) may well have different nutritional and food requirements even though they are both primarily fruit infesters.

The first report on complete nutritional requirements of adult tephritids was by Hagen (1953), who found that both sexes of adult *B. cucurbitae* (Coquillett), *B. dorsalis* (Hendel), and *C. capitata* required a carbohydrate, protein in the form of free amino acids, minerals, B-complex vitamins, and water. These nutrients were essential for both reproduction and longevity.

The various contributions of the different nutrients have been well documented. Carbohydrates are utilized to fuel flight and foraging behavior of both sexes, as well as the often complex and lengthy courtship activities of males. Lipid reserves may be metabolized to support these needs, or added to when a surplus of carbohydrate is on hand (Jácome et al. 1995; Warburg and Yuval 1996). Proteins (free amino acids), minerals, and B-vitamins are essential for oogenesis (Hagen 1953; Christenson and Foote 1960; Bateman 1972; Webster and Stoffolano 1978; Drew 1987; Metcalf 1990) and certain elements of male reproductive behavior. In addition, water is essential for survival and reproduction may hinge on intake of a number of specific vitamins, minerals, and sterols (Tsitsipis 1989).

In general, the bulk of these nutrients are acquired by foraging during adult life (Drew et al. 1983; Hendrichs and Prokopy 1994). However, some may be carried over from the larval stage (Hagen 1953), synthesized *de novo* by the flies following ingestion of relevant precursors, or supplied by symbiotic organisms (Draser and Brandl 1992). Drew et al. (1983) questioned the existence and function of symbiotic microorganisms in the Dacinae and these bacteria have subsequently been classified as insect mutualists (Krischik and Jones 1991). Howard (1989) also stated that there was now substantial evidence to support the view "that species of *Rhagoletis* do not enter into symbiotic relationships with microorganisms."

### 27.3.2 Food Sources in Nature

Throughout the tephritid literature, many substrates that serve as food for adult flies in nature have been identified. These include fruit juices, extrafloral glandular exudates, nectar from flowers, pollen grains, honeydew, bird feces, yeasts, and bacteria (Hagen 1956; Christenson and Foote 1960; Bateman 1972; Prokopy 1976; Nishida 1980; Fitt and O'Brien 1985; Tsitsipis 1989; Hendrichs and Hendrichs 1990; Hendrichs et al. 1991; 1993a,b; Hendrichs and Prokopy 1994; Aluja 1994; Warburg and Yuval 1997b). There are contrasting opinions regarding the utilization of honeydew across the entire family, and this demands intensive research. The explosive evolution of the Diptera has been credited in part with the availability of honeydew produced by the homopterans who preceded them in evolutionary time (Downes and Dahlem 1987). How important this evolutionary association is for the Tephritidae is not clear.

Members of the various subfamilies feed on different substrates. A unique feeding behavior is that of some *Blepharoneura* species which grind and ingest leaf surface tissue (Condon and Norrbom, Chapter 7).

For the Dacinae, certain bacteria of the family Enterobacteriaceae, extrafloral exudates and plant surface leachates are the most probable food sources, whereas there is no evidence that nectar, pollen, honeydew, bird feces, and yeasts are fed on (Nishida 1958; Matsumoto and Nishida 1962; Drew et al. 1983; Lloyd et al. 1986; Drew and Lloyd 1987).

The chemical composition of plant leachates has been documented (Tukey 1971) and they contain most, if not all, inorganic and organic substances that occur within the plant. All of the amino acids, organic acids, and sugars found within plants have been detected in leachates. The richness of the chemical composition of fruit juices and leachates is in direct contrast to the paucity of nutrients in honeydew. Although containing a number of organic sugars (Gray and Fraenkel 1954) it is poor in amino acids (Craig 1960).

Nishida (1958) clearly demonstrated the utilization of extrafloral exudates by *B. cucurbitae,* and these fluids probably have a similar chemical and nutritional composition to plant surface leachates. Matsumoto (1962) found that honeydew was not attractive to *B. cucurbitae* and feeding on it gave rise to only a very slow rate of ovarian development. On the contrary, when this species was fed cut fruits of varying types, the fruit juices produced partial to complete ovarian development. Tsiropoulos (1977) showed that *B. oleae* (Rossi) adults could survive and reproduce on diets of honeydew and pollen grains supplementing sucrose but did not relate this to availability of food in the field.

There is evidence that flies (of several species) feed opportunistically on the juice of the ripe fruits that are available in their habitat. In an analysis of crop contents of field collected *B. dorsalis* in Hawaii (Chang et al. 1977), several distinctive crop content colors were observed, which probably correspond to the different colors of fruit juice fed on. In more recent studies (Courtice and Drew 1984; Drew 1987a) it was shown that such crop content colors were directly related to the colors of mature fruits upon which the flies were feeding. Similarly, Nishida (1980) recorded field-collected *B. dorsalis* in Hawaii with crop content colors that matched the colors of ripening fruits that were growing in the collection localities.

Natural sources of food in the subfamily Trypetinae are best known through studies on *Rhagoletis pomonella* (Walsh). Hagen (1956) summarized the early records of *Rhagoletis* species feeding on honeydew while Neilson and Wood (1966) conducted inconclusive experiments on *R. pomonella* utilization of honeydew. A detailed study on natural food sources of *R. pomonella* by Hendrichs et al. (1993b) showed that this species is not dependent on honeydew to satisfy its nutritional requirements. Further, this study revealed that the flies obtained carbohydrate from host foliage leachates and juice from ripe berries, while bird feces were the most useful source of protein to sustain a reasonable level of egg production. Recently, Lauzon et al. (1998), documented the highly specific association between this fly and bacteria of the family Enterobacteriacae. Foraging by adult *R. pomonella* for food is discussed in considerable detail by Prokopy and Papaj (Chapter 10).

Very little is known about adult fly feeding in the nonfrugivorous tephritids, subfamily Tephritinae. The knapweed gall fly, *Urophora jaceana* (Hering), was recorded feeding on honeydew by Varley (1947), while the adults of many other species are suspected not to eat at all (Steck 1984; personal communication).

### 27.3.3 Feeding Mechanisms

The morphological adaptions of tephritid adult mouthparts are related to the food substrates upon which they feed in nature.

In the Dacinae, most species are primarily fluid feeders. Tzanakakis et al. (1967) found that *B. oleae* adults fed diets of yeast hydrolysate and sugar in identical concentrations, except one was in liquid form and the other solid, produced significantly different egg loads. A higher percentage of females on the liquid diet oviposited and produced more eggs. Further, the regurgitation activity recorded by Drew et al. (1983) and Vijaysegaran (1995) is closely related to fluid food intake. Hendrichs et al. (1992; 1993c) also demonstrated in *R. pomonella* that the concentration of food significantly affected food-handling and -processing time and therefore foraging time. They showed that *R. pomonella* adults were primarily fluid feeders, most efficiently ingesting food in a liquid state, and that this species also performed bubbling and regurgitation activities after ingesting liquid meals. Vijaysegaran et al. (1997) demonstrated clearly that *B. tryoni* (Froggatt) adults were primarily fluid feeders and that the morphological structures of the labellar lobes were modified for ingesting particulate matter in liquid food. The strong robust labella on the proboscis of *B. tryoni* are modified to enable the ingestion of a fluid diet containing small bacteria cells while acting as a sieve for excluding all other larger particles that occur on the plant surface. The micropores on the pseudotracheae were so small that particles larger than the "fruit fly–type" bacteria could not be ingested. Thus, bacteria of the genera *Bacillus* and *Pseudomonas*, yeasts, fungal spores, and pollen grains could not be utilized in food for this and probably most species of *Bactrocera*. Although Howard (1989) stated that "the evidence that bacteria are an important source of nutrition remains unconvincing," the functional morphology studies of Vijaysegaran et al. (1997) provide strong arguments to the contrary (also see review by Drew and Lloyd 1989).

Similar information on species of Trypetinae and Tephritinae is not available. However, Condon and Norrbom (Chapter 7) have provided excellent descriptions of the toothed labellar lobes of *Blepharoneura* species which have developed to enable the fly to grind off the surface layers of leaves and other plant tissues for subsequent ingestion as the primary food source.

Some tephritine adults are nonfeeders (Steck 1984, and personal communication). An unusually small, almost vestigial, nonfunctional proboscis occurs in many species of the subtribe Oedaspidina (Tephritinae) (Freidberg and Kaplan 1992). Those are apparently nonfeeding adults and autogenous females. *Oedaspis reducta* Freidberg and Kaplan is an example. Other Tephritinae have long, slender geniculate probosces which are probably used for extracting nectar and other fluids from deep into flowerheads. The morphology of the mouthparts of these species has not been researched in detail and it would be valuable to compare with the known feeders.

### 27.3.4 Adult Feeding and Reproductive Success

A number of laboratory studies have investigated the relationship between feeding on artificial or natural diets, female reproductive success and longevity (Table 27.2). An interesting point becomes evident. In all frugivorous species, females are anautogenous – they need to feed on protein to realize their reproductive potential. Conversely, in species where larvae feed on seeds (e.g., *Toxotrypana* Gerstaecker and *Chaetostomella*) or galls (*U. affinis* (Frauenfeld) and *U. quadrifasciata* (Meigen)), females are autogenous and free of the burden of seeking protein for egg development.

In some tephritid species protein feeding is essential for males as well. A number of recent studies have revealed that male nutrition is linked with the expression of sexual behaviors, copulatory success, and, ultimately, reproductive success. Pheromone production and emission by male *Anastrepha suspensa* (Loew) is dependent on diet (Landolt and Sivinski 1992; Epsky and Heath 1993). Males with access to water only produce significantly less pheromone than males fed either sugar or sugar and protein. Protein-fed males exhibit a sharp increase in pheromone production, thus increasing their chances of gaining copulations (Sivinski et al. 1994). In addition, male diet affects the blend of components in pheromone emitted by males, protein-fed males producing the most attractive blend (Epsky and Heath 1993). In the closely related *A. obliqua* (Macquart), where male salivary glands are important in reproduction, protein feeding significantly increases the volume of these organs (Ferro and Zucoloto 1989). For the Mediterranean fruit fly, both laboratory (Landolt et al. 1992) and field studies (Yuval et al. 1998) indicate that nutritionally deprived males do not join leks. In addition, Blay and Yuval (1997) have shown that protein-fed males are significantly better than protein-deprived males at gaining copulations with virgin females and restricting the receptivity of these females to further copulations. Another example of diet affecting reproductive success is the case of *B. dorsalis*, where males that feed on methyl eugenol (a parapheromonal component) have a mating advantage over males that do not (Shelly 1994; Shelly and Dewire 1994; Nishida et al. 1997). All the above species exhibit a lek mating system which is associated with prodigious investments of time at the display site and energy in pheromone production, courtship displays and territory guarding (Shelly and Whittier 1997).

In studies on *B. tryoni*, Drew (1987a) found that protein feeding was not an essential prerequisite for males to copulate with females and fertilize their eggs. The relationship of diet to reproduction has also been studied in *R. pomonella*, a nonlekking species (Webster and Stoffolano 1977; Webster et al. 1979). Overall, males are less dependent on proteins than females. Females ingested greater amounts of protein and sucrose than males, and proteins are not essential for maturation of male accessory glands or spermatogenesis. Although additional information from other species is lacking at present, as is information on the effects of protein feeding on lifetime reproductive success in

**TABLE 27.2**
**Laboratory Studies on Diet, Survival, and Fecundity of Adult Tephritids**

| Species: | Variables Studied | Main Findings | Reference |
|---|---|---|---|
| *Urophora affinis* and *U. quadrifasciata* | Effect of sucrose and protein on longevity | Protein feeding *shortened* life, sugar feeding enhanced survival; flies not attracted to protein. | McCaffrey et al. 1994 |
| *Chaetostomella undosa* | Effect of diet on longevity | Flies survive for >20 days on water alone. | Steck 1984 |
| *Toxotrypana curvicauda* | Lifetime consumption of sucrose | Greater quantities of sucrose are ingested during the first 5 days of adult life; protein is not consumed (larvae are seed feeders). | Sharp and Landolt 1984; Landolt 1984 |
| *Bactrocera cucurbitae* | Effects of diet on survival and reproduction | Water and sugar essential for survival of adults; protein essential for oviposition. | Kaur and Srivastava 1995, and references therein |
| *B. tau* | Effect of diet on preoviposition period | Flies fed protein hydrolysate, vitamins, minerals, and carbohydrates had the shortest preoviposition period. | Bala 1987 |
| *B. oleae* | Effects of sucrose, protein, olives, amino acids, minerals, and sterols on survival and reproduction | Sucrose essential for survival and amino acids (or protein) for egg production (when minerals were present in the diets); vitamins important for both high egg production and egg hatch; cholesterol enhances fertility. | Economopoulos et al. 1976; Fletcher and Kapatos 1983; Girolami 1979; Tsiropoulos 1985 |
| *B. tryoni* | Effects of nutrition on fertility | Females fed on sugar and water required protein hydrolysate to produce eggs, but males were fertile with or without protein. | Drew 1987a |
| *Anastrepha ludens* | Male longevity on various diets | Males on a protein diet lived longer than males on a carbohydrate diet. | Martinez et al. 1987 |
| *A. suspensa* | Daily and lifetime patterns of sugar and protein feeding | Protein (yeast) consumption increases as flies age, peaks at day 6 and then falls off; sucrose consumption rises, peaks, and levels off. | Landolt and Davis-Hernández 1993 |
| *A. suspensa* | Lifetime feeding on sucrose and protein | Females feed more frequently than males; female protein feeding correlated to egg production. | Nigg et al. 1995 |

| | | | |
|---|---|---|---|
| *A. obliqua* | Effects of amino acid on fecundity | A diet containing 15% amino acids produced highest fertility. | Braga et al. 1981 |
| *Ceratitis capitata* | Influence of dietary protein on fecundity | Protein ingestion essential for egg production. | Cangussu and Zucoloto 1992; 1995 |
| *C. capitata* | Effect of natural food sources on longevity, fecundity, and fertility | Females required a substantial and varied diet to realize peak fecundity; grapes alone did not support any fecundity, contributing only to adult lifespan; figs sustained both longevity and egg production; bird feces alone supported neither egg production nor longevity; however, when added to a diet of figs, significantly increased fecundity. | Hendrichs et al. 1993a |
| *C. capitata* | Effect of diet on susceptibility to insecticide | Protein-fed flies were less susceptible to malathion than protein-deprived flies. | Vinuela and Arroyo 1983 |
| *C. capitata* | Effect of protein deprivation on longevity of males and females. | A surge in early female mortality under protein deprivation was found, and related to egglaying; male mortality and life expectancy were only mildly affected by protein deprivation. | Muller et al. 1997 |
| *Rhagoletis pomonella* | Effect of protein diet on ovarian maturation and patterns of feeding on protein and sugar | Protein essential for ovarian development; spermatogenesis independent of diet, and little protein was required for male accessory gland development. | Webster and Stoffolano 1978; Webster et al. 1979 |
| *R. pomonella* | Influence of natural and artificial food sources on fecundity and longevity | Carbohydrates sustain longevity; fecundity greatest on yeast hydrolysate; moderate fecundity achieved on bird droppings and honeydew, none on leaf surface bacteria, pollen, frass, or uric acid. | Hendrichs et al. 1993b |

the few species studied, we tentatively postulate that the relationship between male diet and reproductive success will be most pronounced in lekking species, where high levels of both protein and sugar will be mandatory. In species where the mating system is based on fruit guarding by males, other factors (such as male size or symmetry) may override nutritional status in its importance in determining reproductive success.

### 27.3.5 Temporal and Spatial Patterns of Feeding

In the Mediterranean fruit fly, feeding by males and particularly females may be seen throughout the day; however, a distinct peak of feeding occurs in the late afternoon. While females seem to feed opportunistically throughout the day, males have a more rigid timetable, lekking in the morning and midafternoon and feeding primarily in the late afternoon (Hendrichs and Hendrichs 1990; Hendrichs et al. 1991; Warburg and Yuval 1997a,b). It has been postulated that, while food in itself is not a limited resource in some habitats, time to consume it with minimum threat of predation is limited, resulting in a concentrated period of feeding after predator activity ceases (Warburg and Yuval 1997a). During the day, females move between oviposition and feeding sites (Hendrichs et al. 1991). Apparently, foraging for food by males is a factor in their dispersal: feeding sterile males prior to release reduces their tendency to disperse (Murtas et al. 1972). The search for food may enable females to colonize new habitats when this behavior mediates encounters with novel oviposition hosts (Prokopy et al. 1996).

Temporal and spatial patterns of feeding have been studied in several species of *Anastrepha*. Activity patterns of *A. obliqua* were studied in a habitat containing plum and mango trees (Aluja and Birke 1993). Behavioral activities are partitioned between the various hosts available. Thus, males lek on mango trees in the mornings and feed on plums in the afternoon. Similarly, females feed mainly on plum trees, with a small peak in the morning and a major one coinciding with male feeding in the afternoon. However, as in the Mediterranean fruit fly, females exhibited a less rigid pattern of behavior, feeding opportunistically on bird feces in mango trees when these were encountered. Similarly, *A. fraterculus* (Wiedemann) male lekking precedes the main feeding time of males, which occurs in midafternoon, although females feed throughout the day (Malavasi et al. 1983). In a greenhouse containing several species of host trees, *A. striata* Schiner exhibited a somewhat different pattern. Feeding on both protein and fruit occurred throughout the day, with a significant peak in the midmorning hours, preceding the peak of male sexual signaling (Aluja et al. 1993). Similarly, male *A. suspensa* feed mainly in the morning, devoting the afternoon to sexual activity (Burk 1983; Landolt and Davis-Hernández 1993). In reviewing these diverse patterns in *Anastrepha*, Aluja (1994) concluded that while the time of sexual activity was (within permissive environmental conditions) fixed, feeding behavior is much more plastic, responding to resource availability and local microhabitat conditions.

In a study of the activity patterns of *T. curvicauda* (whose larvae feed on seeds, see above) very few feeding events were observed, such that no pattern could be pinpointed (Aluja et al. 1997).

In the genus *Rhagoletis* several studies have examined the temporal behavior of flies in relation to food and oviposition resources. *Rhagoletis mendax* Curran feed mainly in the early morning and late afternoon, dedicating the interval to sexual behavior (Smith and Prokopy 1981). Averill and Prokopy (1993) found that the presence of both food and oviposition sites in a habitat prolonged the residence time of *R. pomonella* females. Subsequently, the comparative influence of varying fly physiological and experiential states on the location and duration of residence of females on apple trees, containing varying amounts of proteinaceous food and fruit, were investigated by Prokopy et al. (1994). Duration of residence was greater in patches where proteinaceous food and host fruit were present on all trees than in patches where either or both resources were absent from one or more trees, largely irrespective of fly physiological or experiential state. Flies that had no access to protein up to the day of release were sighted in greater numbers on proteinaceaous food than were flies that had continuous access to protein up to the day of release, whereas the reverse was true for flies sighted on host fruit. Flies that had partial access to protein up to the day of release were sighted in intermediate numbers on proteinaceous food and fruit. Mohammad and Aliniazee (1991) found that when protein is available near oviposition sites, oviposition is delayed, suggesting a higher priority for feeding over oviposition in this species. While feeding and oviposition are usually discrete events, separated temporally and spatially, this is not always the case. Female *R. berberis* Curran, who oviposit in the fruits of *Mahonia*, frequently combine oviposition and feeding. They feed on the juice of host fruit that oozes from punctures made with their ovipositors. A comparison of the incidence of feeding with successful and unsuccessful oviposition showed that 70% of unsuccessful ovipositions were followed by feeding compared with only 18% of successful ones (Mayes and Roitberg 1986).

In the Dacini, movements between oviposition and feeding sites are frequently associated with long-range displacements. There is evidence that large numbers of *B. tryoni*, *B. neohumeralis* (Hardy), and *B. endiandrae* (Perkins and May) migrate into urban areas in Queensland from other breeding areas up to 100 km away. Apparently, patches of tropical rain forest are important feeding areas for obtaining subsistence diets for these long-lived adult flies, even in the absence of larval food plants, and serve as stepping-stones between oviposition hosts that are usually available over a short time frame only (Drew et al. 1984).

In terms of resource positioning and partitioning in the Dacini, Drew (1987a) and Drew and Lloyd (1987) suggested that the host plant was the "center of activity" for a natural population of *Bactrocera* species. In studies on *B. tryoni* the host plant provided the primary resource for adult food, larval food, courtship and mating, and pupation in addition to oviposition. It is believed that outside of the host plant habitat, the species have little or no opportunity to obtain these resources. In addition, the host plant provides the essential timing opportunities whereby immature females arrive along with sexually mature males, feed on plant surface nutrients, go through ovarian development, courtship and mating and then oviposit.

### 27.3.6 Decision Making in Adult Fly Feeding

There is no evidence from field studies that protein and sugar feeding are discrete behaviors, occurring at different times of the day. However, specific hunger and age determine which segment of the population will be feeding on what substrate. Response to attractants and phagostimulants, in the several species where this question has been studied, is attenuated by feeding history and age. In the Mediterranean fruit fly, protein-deprived females respond more avidly to protein than do protein-fed flies (Prokopy et al. 1992; Cangussu and Zucoloto 1995). A similar response was seen in *R. pomonella* (Malavasi and Prokopy 1992; Prokopy et al. 1993) and *B. cucurbitae* (Liu et al. 1995). For *A. ludens* (Loew), the odor of fermenting chapote fruit was more attractive to hungry flies than that of yeast hydrolysate (Robacker et al. 1989). In this species, sugar or protein deprivation increased feeding preference for sugar or protein, respectively, whereas constant access to sugar or protein decreased preference for the respective nutrient. Sugar hunger, whether age dependent or deprivation induced, increased attraction of flies to a fruit-derived attractant, relative to proteinaceous lures. In addition, protein hunger dramatically increased attractiveness of bacterial volatiles (Robacker 1991; Robacker and García 1993; Robacker and Moreno 1995). Attraction to the odor of bacteria in these flies was attenuated by feeding on a relatively complete diet containing sugar, protein, fats, vitamins, and minerals. Attraction to bacterial odor decreased as the percentage of protein increased in a diet containing casein hydrolysate and sugar. However, dietary vitamins, minerals, fats, and percentage of protein as amino acids in the diet had no effect on the degree of attraction to bacterial volatiles (Robacker and Flath 1995).

Age-related patterns of feeding have been investigated in several species. In *A. suspensa* consumption of sucrose is low on the day following emergence, rises steadily until 1 week after emergence, and remains high thereafter. Protein feeding also increases during the first 5 days of adult life, but then drops off (Landolt and Davis-Hernández 1993). In *A. striata* protein is ingested mainly during the first

2 weeks of adult life, with a switch to fruit feeding thereafter (Aluja et al. 1993). Newly emerged *A. ludens* flies (up to 4 days old) prefer sugar to protein, whereas 5- to 9-day-old flies undergoing sexual maturation feed about equally on sugar and protein (Robacker 1991). In *R. pomonella* ingestion of protein and sucrose are highest during the first week of adult life, decrease slightly during weeks 2 and 3 and remain at a low level thereafter. Females consume greater amounts of protein and sucrose than males (Webster et al. 1979). Hendrichs et al. (1993c) demonstrated that food foraging time in *R. pomonella* was directly related to the quality and quantity of food consumed. Flies preferred solutions that minimize handling costs (such as liquid food) and maximize nutrient intake.

In analyzing feeding behavior in insects, Dethier (1966) noted that the factors underlying selection of food in nature were predominantly chemical and that the senses of taste and smell were the main means of detection. Although some of the stimuli used in food selection may not be the nutrients required by the insect, Dethier (1966) noted that sucrose was one of the nutrients of widespread importance to insects which was also a strong stimulant to feeding. In contrast, amino acids (also important in insect diets) were poorly stimulating. As amino acids and proteins do not produce volatiles and therefore alone could not attract fruit flies, the responses to protein baits must be due to other factors such as products of bacterial breakdown or fermentation.

Attraction to food sources and subsequent ingestion of food is mediated by volatiles derived from the food and phagostimulants contacted after landing on the food source. The attraction to olfactory cues associated with protein food is exploited in control operations by offering these cues as bait combined with poisons or in traps (Steiner 1952; 1955; Haniotakis et al. 1987; 1991; Buitendag and Naude 1994; Epsky et al. 1995; Heath et al. 1995; Katsoyannos and Hendrichs 1995). Ammonia, certain amines, certain fatty acids, and other unidentified volatiles (possibly of bacterial origin) seem to be the major factors responsible for attracting tephritids to protein (Morton and Bateman 1981). Drew and Fay (1988) showed that volatiles of bacterial origin were significantly more attractive than ammonia. Amino acids, although phagostimulatory, are not involved in long-range attraction. Furthermore, the water in a proteinaceous food bait could itself be a principal agent attractive to flies in low-rainfall climates (Cunningham et al. 1978).

Responses to attractants and phagostimulants appear to vary from species to species, although different results may stem from differences in experimental procedures (Galun et al. 1985). Studies on Mediterranean fruit fly showed that the optimum concentration of enzymatic casein hydrolysate for attracting adults of both sexes was 2.5%. However, of the individual amino acids within the hydrosylate, only arginine was found to be phagostimulatory (Galun et al. 1979). In *R. pomonella*, sucrose and fructose were the most stimulating of five sugars tested, whereas melezitose, glucose and maltose were less so. Flies did not respond to saccharine (which is sweet to humans). The amino acids phenylalanine, glutamine, leucine, methionine and arginine, when mixed with water at 0.1 *M* concentration were not phagostimulatory. When mixed with 4% sucrose in water, phenylalanine was significantly more phagostimulatory than sucrose alone, while the four other amino acids tested were not (Duan and Prokopy 1993). For *A. suspensa*, when amino acids were prepared as a 0.25% solution with 4 g of sucrose, 16 were phagostimulatory to males and 18 to females. Alanine, arginine, glutamic acid, glycine, isoleucine, and lysine were highly phagostimulatory to both sexes. However, cysteine and hydroxy-L-proline were highly inhibitory to females and proline was highly inhibitory to males (Sharp and Chambers 1984).

## 27.4 CONCLUSIONS AND FUTURE RESEARCH NEEDS

Feeding behavior is probably the least studied of all tephritid behaviors, yet valuable research has been carried out over the years. Although most of this research was motivated by, and used in, pest management programs, much of it may be subverted to the clarification of evolutionary patterns. Despite our modest efforts here, this remains to be done.

We have reviewed the variety of larval environments tephritids live in and feed upon. In evolutionary studies, the ability of the larvae to exploit host shifts caused by female oviposition choices is somehow taken for granted. Future studies should perhaps challenge this assumption, examining how larval adaptations support (or thwart) host shifts. Comparisons in larval feeding behavior between species along the gradient from feeding on fruit, seeds, leaf mining, and up to the highly specialized gall formers would reveal much about constraints and preadaptations. Furthermore, comparisons among the subfamilies that have arrived at similar trophic niches through specializing on one plant family – the Blepharoneurinae and Tephritinae – should be extremely rewarding.

With regard to adult feeding behavior, we have attempted to show how it is an extension of larval feeding, at least in terms of reproductive resources. The relationship between tephritid mating systems and oviposition resources has been studied extensively (review by Sivinski et al., Chapter 28). Although we have pointed here to some connections among mating systems, oviposition resources, and adult feeding, future studies should further explore these relationships.

Finally, the relationships of these flies, both as larvae and adults, with bacterial microorganisms should be investigated. Comparative phylogenies of the flies and their attendant microorganisms could provide a novel systematic tool, as well as reveal the true nature of the interactions between flies and their microbes.

In terms of evolutionary time, the Tachiniscinae, Blepharoneurinae and Phytalmiinae are considered to be the oldest branches of Tephritidae, followed by the Trypetinae and Dacinae (including Ceratitidini), with the Tephritinae being the youngest group (see Korneyev, Chapter 4). With regard to endemic habitats, a large number of Trypetinae and Dacinae appear to have originated in rain forests utilizing soft fleshy fruits while the Tephritinae in the drier savannah areas where they became established mostly in the Asteraceae. The Asteraceae, part of the mid-Tertiary flora, are considered to be younger than the rain forests which were well established by the early Tertiary period.

With regard to adult feeding behavior, no evolutionary trends or relationships are apparent in terms of tephritid subfamilies and habitat/host plants. With the larval feeding behavior, there has probably been an evolutionary transition from early fruit-feeding Trypetinae in two directions, one to the Dacinae and the other to the Tephritinae. The genus *Blepharoneura* and related genera all utilize stems, fruit, and flowers of Cucurbitaceae as larval hosts (Condon and Norrbom 1994). This is similar to some Southeast Asian *Bactrocera* (*Zeugodacus*) species. While such host utilization behavior could have evolved independently in different taxonomic groups, it may also represent some evolutionary connections through common ancestry.

In summary, such research will enable a clearer understanding of the evolution of feeding behavior within the Tephritidae and its role in relation to oviposition and sexual behavior.

## ACKNOWLEDGMENTS

We thank the community of tephritid biologists who, over the years, have produced a wealth of information on the feeding behavior of these flies. In our attempt to produce a synthesis of what is known and within constraints of time and space, we have referred to only part of this body of work, and may have dealt harshly with some of it. We thank R. Prokopy, M. Aluja, and two anonymous reviewers for many useful comments.

This chapter was prepared thanks to support from the Campaña Nacional contra las Moscas de la Fruta (Mexico), International Organization for Biological Control of Noxious Animals and Plants (IOBC), and Instituto de Ecología, A.C. (Mexico).

## REFERENCES

Aluja, M. 1994. Bionomics and management of *Anastrepha*. *Annu. Rev. Entomol.* 39: 155–178.

Aluja, M. and A. Birke. 1993. Habitat use by adults of *Anastrepha obliqua* (Diptera: Tephritidae) in a mixed mango and tropical plum orchard. *Ann. Entomol. Soc. Am.* 86: 799–812.

Aluja, M., I. Jacome, A. Birke, N. Lozada, and G. Quintero. 1993. Basic patterns of behavior in wild *Anastrepha striata* (Diptera: Tephritidae) flies under field-cage conditions. *Ann. Entomol. Soc. Am.* 86: 776–793.

Aluja, M., A. Jimenez, J. Piñero, M. Camino, L. Aldana, M.E. Valdes, V. Castrejon, I. Jácome, B. Dávila, and R. Figueroa. 1997. Daily activity patterns and within-field distribution of papaya fruit flies (Diptera: Tephritidae) in Morelos and Veracruz, Mexico. *Ann. Entomol. Soc. Am.* 90: 505–520.

Averill, A.L. and R.J. Prokopy. 1993. Foraging of *Rhagoletis pomonella* flies in relation to interactive food and fruit resources. *Entomol. Exp. Appl.* 66: 179–185.

Bala, A. 1987. Effect of nutritionally different diets on preoviposition period of the fruit fly, *Dacus tau* (Walker). *Ann. Agric. Res.* 8: 258–260.

Bateman, M.A. 1972. The ecology of fruit flies. *Annu. Rev. Entomol.* 17: 493–518.

Blay, S. and B. Yuval. 1997. Nutritional correlates to reproductive success of male Mediterranean fruit flies. *Anim. Behav.* 54: 59–66.

Braga, M.A.S., F.S. Zucoloto, and M.A. Simoes Braga. 1981. Studies on the best concentration of amino acids for adult flies of *Anastrepha obliqua* (Diptera, Tephritidae). *Rev. Bras. Biol.* 41: 75–79.

Buitendag, C.H. and W. Naude. 1994. Fruit-fly control: development of a new fruit-fly attractant and correct bait administration. *Citrus J.* 4: 22–25.

Burk, T. 1983. Behavioral ecology of mating in the Caribbean fruit fly, *Anastrepha suspensa* (Loew) (Diptera: Tephritidae). *Fla. Entomol.* 66: 330–344.

Burroughs, L.F. 1970. Amino acids. In *The Biochemistry of Fruits and Their Products* (A.C. Hume, ed.), pp. 119–146. Academic Press, London.

Cangussu, J.A. and F.S. Zucoloto. 1992. Nutritional value and selection of different diets by adult *Ceratitis capitata* flies (Diptera, Tephritidae). *J. Insect Physiol.* 38: 385–491.

Cangussu, J.A. and F.S. Zucoloto. 1995. Self-selection and perception threshold in adult females of *Ceratitis capitata* (Diptera, Tephritidae). *J. Insect Physiol.* 41: 223–227.

Chang, F., R.N. Winters, R.J. Vargas, S.L. Montgomery, and J.M. Takara. 1977. An analysis of crop sugars in the Oriental fruit fly, *Dacus dorsalis* Hendel (Diptera: Tephritidae) in Hawaii, and correlation with possible food sources. *Proc. Hawaii. Entomol. Soc.* 22: 461–468.

Christenson, L.D. and R.H. Foote. 1960. Biology of fruit flies. *Annu. Rev. Entomol.* 5: 171–192.

Condon, M.A. and A.L. Norrbom. 1994. Three sympatric species of *Blepharoneura* (Diptera: Tephritidae) on a single species of host (*Gurania spinulosa*, Cucurbitaceae): new species and taxonomic methods. *Syst. Entomol.* 19: 279–304.

Courtice, A.C. and R.A.I. Drew. 1984. Bacterial regulation of abundance in tropical fruit flies (Diptera: Tephritidae). *Aust. Zool.* 21: 251–268.

Craig, R. 1960. The physiology of excretion in the insect. *Annu. Rev. Entomol.* 5: 53–68.

Cunningham, R.T., S. Nakagawa, D.Y. Suda, and T. Urago. 1978. Tephritid fruit fly trapping: liquid food baits in high and low rainfall climates. *J. Econ. Entomol.* 71: 762–763.

Dethier, V.G. 1966. Feeding Behavior. *Symposia of the Royal Entomological Society of London,* No. 3. *Insect Behavior* (D.J. Haskell, ed.), Adlard and Son Limited, Bartholomew Press, London.

Downer, R.G.H. and J.R. Mathews. 1976. Patterns of lipid distribution and utilization in insects. *Am. Zool.* 16: 733–745.

Downes, W.L., Jr. and G.A. Dahlem. 1987. Keys to the evolution of Diptera: role of Homoptera. *Environ. Entomol.* 16: 847–854.

Draser, U. and R. Brandl. 1992. Microbial gut floras of eight species of tephritids. *Biol. J. Linn. Soc.* 45: 155–165.

Drew, R.A.I. 1987a. Behavioral strategies of fruit flies of the genus *Dacus* (Diptera: Tephritidae) significant in mating and host-plant relationships. *Bull. Entomol. Res.* 77: 73–81.

Drew, R.A.I. 1987b. Reduction in fruit fly (Tephritidae: Dacinae) populations in their endemic rain forest habitat by frugivorous vertebrates. *Aust. J. Zool.* 35: 283–288.

Drew, R.A.I. 1988. Amino acid increases in fruit infested by fruit flies of the family Tephritidae. *Zool. J. Linn. Soc.* 93: 107–112.

Drew, R.A.I. and H.A.C. Fay. 1988. Comparison of the roles of ammonia and bacteria in the attraction of *Dacus tryoni* (Froggatt) (Queensland fruit fly) to proteinaceous suspensions. *J. Plant Prot. Tropics.* 5: 127–130.

Drew, R.A.I. and A.C. Lloyd. 1987. Relationship of fruit flies (Diptera: Tephritidae) and their bacteria to host plants. *Ann. Entomol. Soc. Am.* 80: 629–636.

Drew, R.A.I. and A.C. Lloyd. 1989. Bacteria associated with fruit flies and their host plants. In *Fruit Flies: Their Biology, Natural Enemies and Control* (A.S. Robinson and G. Hooper, eds.), pp. 131–140. In *World Crop Pests* (W. Helle, ed.), Vol. 3A. Elsevier Science Publishers, Amsterdam.

Drew, R.A.I., A.C. Courtice, and D.S. Teakle. 1983. Bacteria as a natural source of food for adult fruit flies (Diptera: Tephritidae). *Oecologia* 60: 279–284.

Drew, R.A.I., M.P. Zalucki, and G.H.S. Hooper. 1984. Ecological studies of eastern Australian fruit flies (Diptera: Tephritidae) in their endemic habitat. I. Temporal variation in abundance. *Oecologia* 64: 267–272.

Duan, J.J. and R.J. Prokopy. 1993. Toward developing pesticide-treated spheres for controlling apple maggot flies, *Rhagoletis pomonella* (Walsh) (Dipt., Tephritidae). I. Carbohydrates and amino acids as feeding stimulants. *J. Appl. Entomol.* 115: 176–184.

Economopoulos, A.P., A.V. Voyadjoglou, and A. Giannakakis. 1976. Reproductive behavior and physiology of *Dacus oleae*: fecundity as affected by mating, adult diet and artificial rearing. *Ann. Entomol. Soc. Am.* 69: 725–729.

Epsky, N.D. and R.R. Heath. 1993. Food availability and pheromone production by males of *Anastrepha suspensa* (Diptera: Tephritidae). *Environ. Entomol.* 22: 942–947.

Epsky, N.D., R.R. Heath, A. Guzman, and W.L. Meyer. 1995. Visual cue and chemical cue interactions in a dry trap with food-based synthetic attractant for *Ceratitis capitata* and *Anastrepha ludens* (Diptera: Tephritidae). *Environ. Entomol.* 24: 1387–1395.

Ferro, M.I.T. and F.S. Zucoloto. 1989. Influence of protein nutrition on the salivary gland development in male *Anastrepha obliqua* Macquart, 1835 (Diptera, Tephritidae). *Cientifica* 17: 189–193.

Fitt, G.P. and R.W. O'Brien. 1985. Bacteria associated with four species of *Dacus* (Diptera: Tephritidae) and their role in the nutrition of the larvae. *Oecologia* 67: 447–454.

Fletcher, B.S. and E.T. Kapatos. 1983. The influence of temperature, diet and olive fruits on the maturation rates of female olive flies at different times of the year. *Entomol. Exp. App.* 33: 244–252.

Freidberg, A. and F. Kaplan. 1992. Revision of the Oedaspidini of the Afrotropical region (Diptera: Tephritidae: Tephritinae). *Ann. Natal. Mus.* 33: 51–94.

Galun, R., S. Gothilf, and S. Blondheim. 1979. The phago-stimulatory effects of protein hydrolysates and their role in the control of medflies. *Bull. SROP* 2: 81–90.

Galun, R., S. Gothilf, S. Blondheim, J.L. Sharp, M. Mazor, and A. Lachman. 1985. Comparison of aggregation and feeding responses by normal and irradiated fruit flies, *Ceratitis capitata* and *Anastrepha suspensa* (Diptera: Tephritidae). *Environ. Entomol.* 14: 726–732.

Girolami, V. 1979. Studies on the biology and population ecology of *Dacus oleae* (Gmelin). 1. Influence of environmental abiotic factors on the adult and on the immature stages. *Redia* 62: 147–191.

Gray, H.E. and G. Fraenkel. 1954. The carbohydrate components of honeydew. *Physiol. Zool.* 27: 56–65.

Hagen, K.S. 1953. Influence of adult nutrition upon the reproduction of three fruit fly species. In Third Special Report on the Control of the Oriental fruit fly (*Dacus dorsalis*) in the Hawaiian Islands, pp. 72–76. Senate of the State of California.

Hagen, K.S. 1958. Honeydew as an adult fruit fly diet affecting reproduction. In *Proceedings, 10th International Congress of Entomology* (E.C. Becker, ed.), pp. 25–30. Mortimer, Ottawa.

Hagen, K.S., L. Santos, and A. Tsekouras. 1963. A technique for culturing the olive fly *Dacus oleae* Gmelin on synthetic media under xenic conditions. In *Proceedings of the Symposium on Radiation and Radioisotopes Applied to Insects of Agricultural Importance,* Athens, Greece, 1963, International Atomic Energy Agency, Vienna, pp. 333–356.

Haniotakis, G.E., A. Vassiliou Waite, and A.V. Waite. 1987. Effect of combining food and sex attractants on the capture of *Dacus oleae* flies. *Entomol. Hellenica* 5: 27–33.

Haniotakis, G., M. Kozyrakis, T. Fitsakis, and A. Antonidaki. 1991. An effective mass trapping method for the control of *Dacus oleae* (Diptera: Tephritidae). *J. Econ. Entomol.* 84: 564–569.

Hansen, E. 1970. Proteins. In *The Biochemistry of Fruits and Their Products* (A.C. Hume, ed.), pp. 147–158. Academic Press, London.

Hardy, D.E. and R.A.I. Drew. 1996. Revision of the Australian Tephritini (Diptera: Tephritidae). *Invertebr. Taxon.* 10: 213–405.

Headrick, D.H. and R.D. Goeden. 1998. The biology of non frugivorous tephritid fruit flies. *Annu. Rev. Entomol.* 43: 217–241.

Heath, R.R., N.D. Epsky, A. Guzman, B. Dueben, A. Manukian, and W.L. Meyer. 1995. Development of a dry plastic insect trap with food-based synthetic attractant for the Mediterranean and Mexican fruit flies (Diptera: Tephritidae). *J. Econ. Entomol.* 88: 1307–1315.

Hendrichs, J. and M.A. Hendrichs. 1990. Mediterranean fruit fly (Diptera: Tephritidae) in nature: location and diel pattern of feeding and other activities on fruiting and nonfruiting hosts and nonhosts. *Ann. Entomol. Soc. Am.* 83: 632–641.

Hendrichs, J. and R.J. Prokopy, 1994. Food foraging behavior of frugivorous fruit flies. In *Fruit Flies and the Sterile Insect Technique* (C.O. Calkins, W. Klassen, and P. Liedo, eds.), pp. 37–55, CRC Press, Boca Raton.

Hendrichs, J., B.I. Katsoyannos, D.R. Papaj, and R.J. Prokopy. 1991. Sex differences in movement between natural feeding and mating sites and tradeoffs between food consumption, mating success and predator evasion in Mediterranean fruit flies (Diptera: Tephritidae). *Oecologia* 86: 223–231.

Hendrichs, J., S.S. Cooley, and R.J. Prokopy. 1992. Post-feeding bubbling behavior in fluid-feeding Diptera: concentration of crop contents by oral evaporation of excess water. *Physiol. Entomol.* 17: 153–161.

Hendrichs, J., B.I. Katsoyannos, and R.J. Prokopy. 1993a. Bird faeces in the nutrition of adult Mediterranean fruit flies *Ceratitis capitata* (Diptera: Tephritidae) in nature. *Mitt. Dtsch. Ges. Allg. Ang. Entomol.* 8: 4–6.

Hendrichs, J., C.R. Lauzon, S.S. Cooley, and R.J. Prokopy. 1993b. Contribution of natural food sources to adult longevity and fecundity of *Rhagoletis pomonella* (Diptera: Tephritidae). *Ann. Entomol. Soc. Am.* 86: 250–264.

Hendrichs, J., B.S. Fletcher, and R.J. Prokopy. 1993c. Feeding behavior of *Rhagoletis pomonella* flies (Diptera: Tephritidae): effect of initial food quantity and quality on food foraging, handling costs, and bubbling. *J. Insect Behav.* 6: 43–64.

Howard, D.J. 1989. The symbionts of *Rhagoletis*. In *Fruit Flies: Their Biology, Natural Enemies and Control* (A.S. Robinson and G. Hooper, eds.), pp. 121–129. In *World Crop Pests* (W. Helle ed.), Vol. 3A. Elsevier Science Publishers, Amsterdam.

Jácome, I., M. Aluja, P. Liedo, and D. Nestel. 1995. The influence of adult diet and age on lipid reserves in the tropical fruit fly *Anastrepha serpentina* (Diptera: Tephritidae). *J. Insect Physiol.* 41: 1079–1086.

Katsoyannos, B.I. and J. Hendrichs. 1995. Food bait enhancement of fruit mimics to attract Mediterranean fruit fly females. *J. Appl. Entomol.* 119: 211–213.

Kaur, S. and B.G. Srivastava. 1995. Longevity and reproduction of *Dacus cucurbitae* (Coquillett) adults in the absence of water, sucrose and yeast hydrolysate (enzymatic) individually. *Indian J. Entomol.* 57: 146–150.

Krischik, V.A. and C.G. Jones. 1991. Prologue — Microorganisms: the unseen mediators. In *Microbial Mediation of Plant–Herbivore Interactions* (P. Barbosa, V.A. Krischik, and C.G. Jones eds.), pp. 1–6. John Wiley & Sons, New York.

Landolt, P.J. 1984. Behavior of the papaya fruit fly *Toxotrypana curvicauda* Gerstaecker (Diptera: Tephritidae), in relation to its host plant, *Carica papaya* L. *Folia Entomol. Mex.* 61: 215–224.

Landolt, P.J. and K.M. Davis-Hernández. 1993. Temporal patterns of feeding by Caribbean fruit flies (Diptera: Tephritidae) on sucrose and hydrolyzed yeast. *Ann. Entomol. Soc. Am.* 86: 749–755.

Landolt, P.J. and J. Sivinski. 1992. Effects of time of day, adult food, and host fruit on incidence of calling by male Caribbean fruit flies (Diptera: Tephritidae). *Environ. Entomol.* 21: 382–387.

Landolt, P.J., R.R. Heath, and D.L. Chambers. 1992. Oriented flight responses of female Mediterranean fruit flies to calling males, odor of calling males and a synthetic pheromone blend. *Entomol. Exp. Appl.* 65: 259–266.

Lauzon, C.R., R.E. Sjogren, S.E. Wright, and R.J. Prokopy. 1998. Attraction of *Rhagoletis pomonella* (Diptera: Tephritidae) flies to odor of bacteria: apparent confinement to specialized members of enterobacteriaceae. *Environ. Entomol.* 27: 853–857.

Liu, Y., C. Chang, Y.C. Liu, and C.Y. Chang. 1995. Attraction of food attractants to melon fly, *Dacus cucurbitae* Coquillett. *Chin. J. Entomol.* 15: 69–80.

Lloyd, A.C., R.A.I. Drew, D.S. Teakle, and A.C. Hayward. 1986. Bacteria associated with some *Dacus* species (Diptera: Tephritidae) and their host fruit in Queensland. *Aust. J. Biol. Sci.* 39: 361–368.

Malavasi, A. and R.J. Prokopy. 1992. Effect of food deprivation on the foraging behavior of *Rhagoletis pomonella* (Diptera: Tephritidae) females for food and host hawthorn fruit. *J. Entomol. Sci.* 27: 185–193.

Malavasi, A., J.S. Morgante, and R.J. Prokopy. 1983. Distribution and activities of *Anastrepha fraterculus* (Diptera: Tephritidae) flies on host and nonhost trees. *Ann. Entomol. Soc. Am.* 76: 286–292.

Manoukas, A.G. 1981. Effect of excess levels of individual amino acids upon survival, growth and pupal yield of *Dacus oleae* (Gmel.) larvae. *Z. Angew. Entomol.* 91: 309–315.

Manoukas, A.G. 1986. Biological aspects of the olive fruit fly grown in different larval diets. In *Fruit Flies of Economic Importance 84* (R. Cavalloro, ed.), pp. 81–87. A.A. Balkema, Rotterdam.

Martinez, A.J., T.C. Holler, and J.N. Worley. 1987. A fructose and yeast hydrolysate diet for the irradiated Mexican fruit fly adult, *Anastrepha ludens* (Loew) and its effect on male longevity. *Southwest. Entomol.* 12: 317–320.

Matsumoto, B. and T. Nishida. 1962. Food preference and ovarian development of the melon fly, *Dacus cucurbitae* Coquillett, as influenced by diet. *Proc. Hawaii. Entomol. Soc.* 18: 137–144.

Mayes, C.F. and B.D. Roitberg. 1986. Host discrimination in *Rhagoletis berberis* (Diptera: Tephritidae). *J. Entomol. Soc. Br. C.* 83: 39–43.

McCaffrey, J.P., E.A. Vogt, and J.B. Johnson. 1994. Effects of adult diet on longevity of *Urophora affinis* and *U. quadrifasciata* (Diptera: Tephritidae). *Environ. Entomol.* 23: 623–628.

Metcalf, R.L. 1990. Chemical ecology of Dacinae fruit flies (Diptera: Tephritidae). *Ann. Entomol. Soc. Am.* 83: 1017–1030.

Mitchell, S., N. Tanaka, and L.F. Steiner. 1965. Methods of mass culturing melon flies and Oriental and Mediterranean fruit flies. USDA, ARS Series 33, 22 pp.

Mohammad, A.B. and M.T. Aliniazee. 1991. Influence of protein hydrolysate baits on oviposition in the apple maggot, *Rhagoletis pomonella. Entomol. Exp. Appl.* 59: 151–161.

Morton, T.C. and M.A. Bateman. 1981. Chemical studies on proteinacous attractants for fruit flies, including the identification of volatile constituents. *Aust. J. Agric. Res.* 32: 905–916.

Muller, H.G., J.L. Wang, W.B. Capra, P. Liedo, and J.R. Carey. 1997. Early mortality surge in protein-deprived females causes reversal of sex differential of life expectancy in Mediterranean fruit flies. *PNAS* 94: 2762–2765.

Murtas, I. De, U. Cirio, D. Enkerlin, and I.D. De Murtas. 1972. Dispersal of *Ceratitis capitata* Wied. on Procida Island. I. Distribution of sterilized Mediterranean fruit-fly on Procida and its relation to irradiation doses and feeding. *Bull. EPPO* 6: 63–68.

Neilson, W.T.A. and F.A. Wood. 1966. Natural source of food of the apple maggot. *J. Econ. Entomol.* 59: 997–998.

Nigg, N.H., S.E. Simpson, J.A. Attaway, S. Fraser, E. Burns, and R.C. Littell. 1995. Age-related response of *Anastrepha suspensa* (Diptera: Tephritidae) to protein hydrolysate and sucrose. *J. Econ. Entomol.* 88: 669–677.

Nishida, R., T.E. Shelly, and K.Y. Kaneshiro. 1997. Acquisition of female-attracting fragrance by males of Oriental fruit fly from a Hawaiian lei flower, *Fagraea berteriana. J. Chem. Ecol.* 23: 2275–2285.

Nishida, T. 1958. Extrafloral glandular secretions, a food source for certain insects. *Proc. Hawaii. Entomol. Soc.* 16: 379–386.

Nishida, T. 1980. Food system of tephritid fruit flies in Hawaii. *Proc. Hawaii. Entomol. Soc.* 23: 245–254.

Prokopy, R.J. 1976. Feeding, mating, and oviposition activities of *Rhagoletis fausta* flies in nature. *Ann. Entomol. Soc. Am.* 69: 899–904.

Prokopy, R.J., D.R. Papaj, J. Hendrichs, and T.T.Y. Wong. 1992. Behavioral responses of *Ceratitis capitata* flies to bait spray droplets and natural food. *Entomol. Exp. Appl.* 64: 247–257.

Prokopy, R.J., S.S. Cooley, L. Galarza, and C. Bergweiler. 1993. Bird droppings compete with bait sprays for *Rhagoletis pomonella* (Walsh) flies (Diptera: Tephritidae). *Canad. Entomol.* 125: 413–422.

Prokopy, R.J., S.S. Cooley, J.J. Prokopy, Q. Quan, and J.P. Buonaccorsi. 1994. Interactive effects of resource abundance and state of adults on residence of apple maggot (Diptera: Tephritidae) flies in host tree patches. *Environ. Entomol.* 23: 304–315.

Prokopy, R.J., J.J. Duan, and R.I. Vargas. 1996. Potential for host range expansion in *Ceratitis capitata* flies: impact of proximity of adult food to egg-laying sites. *Ecol. Entomol.* 21: 295–299.

Robacker, D.C. 1991. Specific hunger in *Anastrepha ludens* (Diptera: Tephritidae): effects on attractiveness of proteinaceous and fruit-derived lures. *Environ. Entomol.* 20: 1680–1686.

Robacker, D.C. and J.A. Garcia. 1993. Effects of age, time of day, feeding history, and gamma irradiation on attraction of Mexican fruit flies. (Diptera: Tephritidae), to bacterial odor in laboratory experiments. *Environ. Entomol.* 22: 1367–1374.

Robacker, D.C. and R.A. Flath. 1995. Attractants from *Staphylococcus aureus* cultures for Mexican fruit fly, *Anastrepha ludens. J. Chem. Ecol.* 21: 1861–1874.

Robacker, D.C. and D.S. Moreno. 1995. Protein feeding attenuates attraction of Mexican fruit flies (Diptera: Tephritidae) to volatile bacterial metabolites. *Fla. Entomol.* 78: 62–69.

Robacker, D.C., J.A. Garcia, and W.G. Hart. 1989. Attraction of a laboratory strain of *Anastrepha ludens* (Diptera: Tephritidae) to the odor of fermented chapote fruit and to pheromones in laboratory experiments. *Environ. Entomol.* 19: 403–408.

Schroeder, W.J., R.Y. Miyabara, and D.L. Chambers. 1972. Protein products for rearing three species of larval Tephritidae. *J. Econ. Entomol.* 65: 969–972.

Sharp, J.L. and D.L. Chambers. 1984. Consumption of carbohydrates, proteins, and amino acids by *Anastrepha suspensa* (Loew) (Diptera: Tephritidae) in the laboratory. *Environ. Entomol.* 13: 768–773.

Sharp, J.L. and P.J. Landolt. 1984. Gustory and olfactory behavior of the papaya fruit fly, *Toxotrypana curvicauda* Gerstaecker (Diptera: Tephritidae) in the laboratory with notes on longevity. *J. Ga. Entomol. Soc.* 19: 176–182.

Shelly, T.E. 1994. Consumption of methyl eugenol by male *Bactrocera dorsalis* (Diptera: Tephritidae): low incidence of repeat feeding. *Fla. Entomol.* 77: 201–208.

Shelly, T.E. and A.M. Dewire. 1994. Chemically mediated mating success in male Oriental fruit flies (Diptera: Tephritidae). *Ann. Entomol. Soc. Am.* 87: 375–382.

Shelly, T.E. and T.S. Whittier. 1997. Lek behavior of insects. In *Mating Systems in Insects and Arachnids* (J.C. Choe and B.J. Crespi, eds.), pp. 273–293. Cambridge University Press, Cambridge.

Sivinski, J.M., N.D. Epsky, and R.R. Heath. 1994. Pheromone deposition on leaf territories by male Caribbean fruit flies, *Anastrepha suspensa* (Loew) (Diptera: Tephritidae). *J. Insect Behav.* 7: 43–51.

Smith, D.C. and R.J. Prokopy. 1981. Seasonal and diurnal activity of *Rhagoletis mendax* flies in nature. *Ann. Entomol. Soc. Am.* 74: 462–466.

Steck, G.J. 1984. *Chaetostomella undosa* (Diptera: Tephritidae): biology, ecology and larval description. *Ann. Entomol. Soc. Am.* 77: 669–678.

Steiner, L.F. 1952. Fruit fly control in Hawaii with poison bait spray containing protein hydrolysate. *J. Econ. Entomol.* 45: 838–843.

Steiner, L.F. 1955. Bait sprays for fruit fly control. *Agric. Chem.* 10: 32–34.

Tanaka, N., L.F. Steiner, K. Ohinata, and R. Okamoto. 1969. Low-cost larval rearing medium for mass production of Oriental and Mediterranean fruit flies. *J. Econ. Entomol.* 62: 967–968.

Terra, W.R. 1990. Evolution of digestive systems of insects. *Annu. Rev. Entomol.* 35: 181–200.

Tsiropoulos, G. 1977. Reproduction and survival of the adult *Dacus oleae* feeding on pollen and honeydew. *Environ. Entomol.* 6: 390–392.

Tsiropoulos, G. 1985. The importance of dietary nitrogen and carbohydrate in the nutrition of the adult olive fruit fly, *Dacus oleae. Mitt. Dtsch. Ges. Allg. Angew. Entomol.* 4: 4–6.

Tsitsipis, J.A. 1989. Nutrition requirements. In *Fruit Flies: Their Biology, Natural Enemies and Control* (A.S. Robinson and G. Hooper, eds.), pp. 103–119. In *World Crop Pests* (W. Helle ed.), Vol. 3A. Elsevier Science Publishers, Amsterdam.

Tukey, H.B. 1971. Leaching of substances from plants. In *Ecology of Leaf Surface Microorganisms* (T.F. Preece and C.H. Dickinson, eds.), pp. 67–80. Academic Press, London.

Tzanakakis, M.E., J.A. Tsitsipis, and L.F. Steiner. 1967. Egg production of olive fruit fly fed solids or liquids containing protein hydrolysate. *J. Econ. Entomol.* 60: 352–354.

Varley, G.C. 1947. The natural control of *Urophora jaceana. J. Anim. Ecol.* 16: 139–187.

Vijaysegaran, S. 1995. Mouthpart Structure, Feeding Mechanisms and Natural Food Sources of Adult Fruit Flies in the Genus *Bactrocera* (Diptera: Tephritidae). Ph.D. dissertation, University of Queensland, Brisbane.

Vijaysegaran, S., G.H. Walter, and R.A.I. Drew. 1997. Mouthpart structure, feeding mechanisms, and natural food sources of adult *Bactrocera* (Diptera: Tephritidae). *Ann. Entomol. Soc. Am.* 90: 184–201.

Vinuela, E. and M. Arroyo. 1983. Effect of nutrition on the susceptibility of *Ceratitis capitata* Wied. (Dip.: Tephritidae) adults to malathion. Influence of adult food, physiological stage and age. *Acta Oecol. Oecol. Appl.* 4: 123–130.

Waldbauer, G.P. 1968. The consumption and utilization of food by insects. *Adv. Insect Physiol.* 5: 229–288.

Waldbauer, G.P. and S. Friedman. 1991. Self-selection of optimal diets by insects. *Annu. Rev. Entomol.* 36: 43–63.

Warburg, M.S. and B. Yuval. 1996. Effects of diet and activity on lipid levels of adult Mediterranean fruit flies. *Physiol. Entomol.* 21: 151–158.

Warburg, M.S. and B. Yuval. 1997a. Circadian patterns of feeding and reproductive activities of Mediterranean fruit flies (Diptera: Tephritidae) on various hosts in Israel. *Ann. Entomol. Soc. Am.* 90: 487–495.

Warburg, M.S. and B. Yuval. 1997b. Effects of energetic reserves on behavioral patterns of Mediterranean fruit flies (Diptera: Tephritidae). *Oecologia* 112: 314–319.

Webster, R.P. and J.G. Stoffolano, Jr. 1978. The influence of diet on the maturation of the reproductive system of the apple maggot, *Rhagoletis pomonella. Ann. Entomol. Soc. Am.* 71: 844–849.

Webster, R.P., J.G. Stoffolano, Jr., and R.J. Prokopy. 1979. Long-term intake of protein and sucrose in relation to reproductive behavior of wild and laboratory cultured *Rhagoletis pomonella. Ann. Entomol. Soc. Am.* 72: 41–46.

White, I.M. 1988. Tephritid flies, Diptera: Tephritidae. *Handb. Ident. Br. Insects* 10(5A): 134 pp.

Yuval, B., R. Kaspi, S. Shloush, and M. Warburg. 1998. Nutritional reserves regulate male participation in Mediterranean fruit fly leks. *Ecol. Entomol.* 23: 211–215.

Zucoloto, F.S. 1991. Effects of flavour and nutritional value on diet selection by *Ceratitis capitata* larvae (Diptera, Tephritidae). *J. Insect Physiol.* 37: 21–25.

Zwölfer, H. 1983. Life systems and strategies of resource exploitation in tephritids. In *Fruit Flies of Economic Importance* (R. Cavalloro, ed.), pp. 16–30, Proceedings of the CEC/IOBC International Symposium. A.A. Balkema, Rotterdam.

# 28 Topics in the Evolution of Sexual Behavior in the Tephritidae

*John Sivinski, Martín Aluja, Gary Dodson, Amnon Freidberg, David Headrick, Kenneth Kaneshiro, and Peter Landolt*

## CONTENTS

## 28.1 INTRODUCTION

The sex lives of the Tephritidae are wonderfully various. They range in complexity from males that couple after little preliminary courtship signaling to those that produce a repertoire of acoustic, pheromone, and visual displays, and from females that make few precopulatory mate choices to those that have information about potential mates broadcast to them via several different channels. There are instances of licking, transfer of regurgitants, bright coloration, feathered legs, and reflective setae. Beneath the often splendid surfaces are a variety of phallic structures, vaginas, and sperm storage organs, which might respectively represent organs of communication and the mechanisms of copulatory or postcopulatory mate choice. This wealth of diversity superimposed upon a common theme makes fruit flies ideal subjects for studies, particularly comparative studies, that attempt to illuminate the evolution of mate choice and sexual competition.

We are certainly not the first to appreciate the potential of the Tephritidae and a number of influential papers of general importance have centered on fruit fly sexual behavior (e.g., Prokopy 1981; Burk 1982). What we attempt here is to place into context the enormous amount of sexual information that the chapters of this volume contain, and in doing so we have made an effort to point out unsolved problems as well as currently attractive hypotheses for the explanation of reproductive behaviors.

The authorship of this chapter is somewhat unusual; it is a true communal effort and JS should be seen as an editor rather than the principal author. To make the authorship of a particular section clear to the reader, authors' names are placed after the titles in the Table of Contents. A consequence of this style of multiple authorship is an occasional difference of opinion among the various contributors. We trust that the reader will not be disturbed by the heterogeneity of perspective and interpretation, but rather see these disagreements as indicating unsettled areas in the understanding of tephritid biology and opportunities for further study. A second, perhaps inevitable, consequence is a certain amount of redundancy; one author may readdress a situation that has been previously discussed by another in order to better make a point. Again, we hope that the reader will be patient with this format and see this "chapter" for what it is — a series of short discussions on the evolution of sexual behaviors among Tephritidae.

The chapter is organized into four sections, an introduction, a description of various sexual behaviors, a phylogenetically organized table that lists the presence or absence of particular behaviors, and discussions of the interactions between phylogeny and sexual behavior. The descriptions of sexual behaviors begin with those that bring the sexes together in time and space, followed by male agonistic interactions. Courtship signals, visual, acoustic, and pheromonal, are considered next and, after these, discussions of the content of courtship signals, the role predation may have played in the evolution of signals, and what factors might have influenced copulation durations. The phylogenetic table will provide readers with an idea of the scope and distribution of sexual behaviors, and should be an invaluable starting point to those interested in comparative studies. Finally, two perspectives on the relationship of phylogeny and sexual behavior are considered: First, how sexual selection can result in divergence within populations and result in speciation and, second, how sexual behaviors can serve as characters in phylogenetic reconstructions.

## 28.2 A CATALOG OF SEXUAL AND AGONISTIC BEHAVIORS

### 28.2.1 Mating Sites and Rhythms

#### 28.2.1.1 The Influence of Resource Distribution on Mating Sites

Resource distributions, particularly those of breeding sites, influence where and when the sexes meet (Emlen and Oring 1977). In tephritids, arguments have been made by Prokopy (1980) and Burk (1981) that different host fruit distributions lead to either male defense of oviposition sites or to male signaling with pheromones and acoustic displays ("calling") away from oviposition sites. In brief, they proposed that when fruits are relatively rare (clumped), females can be predictably located at any particular oviposition site and that males that occupy such fruits can "force" females to mate. That is, it would be beneficial for females to mate with a resident male and then gain uninterrupted access to the resource rather than attempting to oviposit while being continually distracted by a courting male. Where there is little precopulatory female choice, there is little reason for males to invest in displays, and there is little if any courtship.

On the other hand, when resources are relatively abundant and homogeneous, females are not predictably located at any particular fruit, and males may either search among fruits or produce long-distance signals. Females have the freedom to leave fruits occupied by males and go to other, empty sites. This freedom of choice selects for males that advertise their qualities as mates and as a result courtship is complex. Males may aggregate either in good signaling sites, regions of female concentration, in the vicinity of unusually attractive males, or because of a female preference for grouped males (see Höglund and Alatalo 1995). The results are leks formed away from oviposition sites.

Both Prokopy and Burk argued that temperate species with narrow host ranges would be typified by fruit-guarding males with little courtship, but that polyphagous tropical species would form leks with complex displays. This is because highly polyphagous females presumably have so many choices of host that males can neither predict nor control access to oviposition sites.

Observations made over the intervening years have complicated the situation, and generalizations about temperate/tropic and monophagous/polyphagous distinctions are more difficult to make. For one thing, more examples have come to light of temperate and tropical monophagous species that form leks (e.g., Headrick and Goeden 1994). A peculiar tephrit*oid* example of mating system diversity on a common substrate is found in two related species of piophilids that lay their eggs on moose carcasses (see Sivinski, Chapter 2). Males of one species form mating aggregations on the antlers, while the other males of the other species are dispersed over the body and perform mate-guarding behaviors (Bonduriansky 1995). Even within a species some males may guard fruit while others participate in leks (e.g., *Ceratitis capitata* (Wiedemann), Warburg and Yuval 1995; *Anastrepha suspensa* (Loew), Burk 1983).

However, the core principle of the Emlen and Oring/Prokopy-Burk model, that the abundance of hosts relative to the number of female fruit flies influences male distributions, remains an attractive vehicle for the exploration of tephritid mating systems. For example, in the genus *Anastrepha,* males of one species guard fruits, other species call (see Section 28.2.3 on sexual signals) from host plant leaves, and many of these leaf-calling species form male mating aggregations. Although important information on host and *Anastrepha* densities are often unavailable at present, testable predictions of the "relative abundance model" can be put forward. For instance, the fruit guarding *A. bistrigata* Bezzi would be expected to be abundant relative to its guava (*Psidium* spp.) host fruits. Although *A. bistrigata* is thought to be stenophagous (Norrbom and Kim 1988), it is the relative scarcity of hosts and not monophagy or polyphagy per se, that is the critical factor in determining whether males find it profitable to wait for potential mates on a particular fruit, and that makes it unprofitable for females to search for unguarded oviposition sites.

In other apparently monophagous species, such as *A. hamata* (Loew), males call from host plant leaves rather than from fruits (M. Aluja, personal observation). The density model suggests that the fruits on these trees should typically be so abundant relative to females that a fruit-guarding male can no longer expect females to arrive at any particular fruit at a "profitable" rate. Furthermore, those females that alight on a male's fruit may be confident that there are unguarded and more easily exploited oviposition sites nearby.

While such males may now need to invest in more expensive advertisements, chemical and otherwise, to attract discriminating females, it is not clear why they abandoned fruit as a signaling platform. What advantages could leaves offer over fruit in these circumstances? It is possible that predators concentrate their foraging over fruits making residency dangerous (Hendrichs and Hendrichs 1990; see Landolt, Section 28.2.3.3). On the other hand, it may be that there are locations on host trees where fruits are not abundant, but which have high female densities because of favorable microhabitat. In *A. suspensa*, sexually inactive females and nonsignaling males accumulate in particular parts of host trees (Sivinski 1991). These same locations tend to be the sites occupied by sexually active males.

A particularly striking phenomenon among many species of leaf-calling males is their congregation into leks. These may occur in either polyphagous (e.g., *A. obliqua* (Macquart) and *A. suspensa*, Aluja and Birke 1993; Burk and Webb 1980) or monophagous species (e.g., *A. bezzii* Lima, M. Aluja, personal observation). Typically, several to many males signal with pheromones, acoustic signals, and perhaps visual displays from adjacent, or nearly adjacent, leaf territories which they defend from rival males (see Eberhard, Chapter 18, for discussion of lek definitions).

The significance of male fruit fly mating aggregations is not entirely understood. There are many hypotheses proposed for the evolution of insect leks (e.g., Shelly and Whittier 1997). Some of the more relevant ones are as follows.

1. Group calling amplifies the male signal and results in an average increase in the numbers of females encountered by the participants. However, the ranges of combined signals in most channels are at best additive, so there is no advantage to group signaling in terms of increased area covered. Pheromones are a possible exception to this rule (Bradbury 1981).
2. Females may prefer to choose mates from within groups of males, presumably because this facilitates accurate comparisons of potential mates or because mating inside a group protects females from predators.
3. Leks are an epiphenomenon. That is, they are simple accumulations of calling males in favorable microhabitats that might serve as resting sites for females, or in locations that are particularly suited to signaling, or in the neighborhoods of particularly attractive males.

Certain sites seem to be consistently used for lekking, suggesting that location plays an influential role in the formation of leks. As noted, *A. suspensa* leks form in the same areas that hold resting females, and these resting females often occur in clusters that have a structure that is superficially similar to that of a lek (Sivinski 1991). Such pockets of potentially high female concentration presumably constitute suitable sites for male calling. Lekking sites in *C. capitata* are stable over time. In a 60-day study by Shelly and Whittier (1995) only a few of the available trees consistently contained leks. There are a number of environmental factors, such as temperature, humidity, light, proximity to food sources, etc., that potentially influence the locations of leks (see Eberhard, Chapter 18). Males may also prefer to hold territories on sites where other males, or they themselves, had called previously. Pheromones are often deposited by males on leaf territory surfaces (see Section 28.2.3.1 on acoustic signals), and some of the components remain there until at least the following day. Female *A. suspensa* respond to these deposited chemicals, and perhaps leaf territories (and lek sites) acquire value with repeated use (Sivinski et al. 1994).

With a further decline in the numbers of females relative to oviposition sites, a point is reached where it is no longer profitable for males to forage for mates on host plants. In such cases they may turn to alternative sites for calling and lek formation. Such "encounter sites" may originally have been navigation markers that concentrated females as they moved through an area (Parker 1978; e.g., swarm markers or hilltops, Sivinski and Petersson 1997), or else a location that is particularly suited for signal broadcasting. There appear to be no examples of this sort of mating system in *Anastrepha,* although the peculiar leaf-based calling behavior of *A. robusta* Greene (Aluja 1994) may prove to be so when its larval host plant is finally discovered. Occasionally, leks form in trees neighboring host trees (e.g., in *A. obliqua*, Aluja and Birke 1993 and *A. fraterculus* (Wiedemann), Malavasi et al. 1983), but from the flies' perspective these are probably sensed as particularly favorable extensions of a host plant. However, there are tephritids in other genera that form leks on nonhost plants, for example, an undescribed *Blepharoneura* species from Costa Rica (Condon and Norrbom, Chapter 7) and *Procecidochares* sp. nr. *minuta* from the American southwest (Dodson 1987).

### 28.2.1.2 Species Isolation and the Timing of Sexual Activity

#### *28.2.1.2.1 Anastrepha*

Flies in the genus *Anastrepha* offer a unique opportunity to analyze the evolution of calling and mating rhythms. As reviewed by Aluja et al. (Chapter 15), there is an extremely wide range of calling rhythms in this genus, from species that call in the early morning hours (sunrise) to those that do so during the late afternoon (sunset). The most parsimonious interpretation of the evolution of such varied patterns is the gradual selection for sexual activity rhythms that limit interspecific interactions and hybridization. If so, the timing of sexual activities in sympatric species should diverge.

The Caribbean fruit fly (*A. suspensa*) and the West Indian fruit fly (*A. obliqua*) purportedly share a center of origin, and have daily patterns of calling that are almost perfectly reversed (Aluja et al., Chapter 15). While *A. obliqua* calls preferentially in the morning, *A. suspensa* calls during the afternoon hours. The calling patterns of three species living in sympatry in Mexico, *A. fraterculus, A. striata* Schiner, and *A. ludens* (Loew), also differ sharply among themselves (Aluja et al., Chapter 15).

An interesting phenomenon with respect to calling and mating rhythms is the appearance of differences in patterns among geographically distinct populations of the same species. For example, males of *A. serpentina* (Wiedemann) from the Pacific state of Chiapas in southwestern Mexico exhibit a bimodal pattern with peaks between 08:00 and 10:00 and 14:00 and 17:00 hours while those from the Gulf state of Veracruz only call from 11:00 to 17:00 hours (Aluja et al., Chapter 15). A similar geographic variation in calling rhythms is observed in *C. capitata.*

#### *28.2.1.2.2 Toxotrypana*

Information on this genus is restricted to one species, *T. curvicauda* Gerstaecker. Interestingly, reports vary sharply with respect to geographical origin. Individuals in populations from Florida call in the late afternoon hours (Aluja et al., Chapter 15). In sharp contrast to this, populations from the state of Morelos, Mexico call from late morning to early afternoon hours (depending on the time of year). It is possible that these differences reflect the presence of two biotypes or even two different species.

#### 28.2.1.2.3 *Bactrocera*

Flies in the genus *Bactrocera* (formerly *Dacus*) are a diverse group. Species such as *B. aglaiae* (Hardy), *aquilonis* (May), *cacuminata* (Hering), *cucumis* (French), *cucurbitae* (Coquillett), *diversa* (Coquillett), *decurtans* (May), *dorsalis* (Hendel), *endriandrae* (Perkins and May), *halfordiae* (Tryon), *jarvisi* (Tryon), *kraussi* (Hardy), *musae* (Tryon), *opiliae* (Drew and Hardy), *passiflorae* (Froggatt), *scutellaris* Bezzi, *tau* (Walker) (= *hageni* Meijere), *tryoni* (Froggatt), and *zonata* (Saunders) call and mate at dusk under low light intensity (see reviews by Fletcher 1987 and Smith 1989 for specific references). In contrast, there are some species, such as *B. expandens* (Walker), *oleae* (Rossi), and *neohumeralis* (Hardy), that mate during the middle of the photophase under high light intensities (Haniotakis 1974; Fletcher 1987). Finally, species such as *B. tenuifascia* (May) and *tsuneonis* (Miyake) initiate copulation during any time of the day. Why are there these differences? As previously suggested for species of *Anastrepha*, time of mating may be an effective barrier to hybridization between species. This can be illustrated in the case of the closely related species *B. tryoni* and *B. neohumeralis*. The former mates only at dusk, while the latter does so at midday. Perhaps the large number of species that mate at dusk do so in species-specific locations in their native forests.

#### 28.2.1.2.4 *Ceratitis*

The mating rhythms of only two species of this genus have been studied, the sympatric *C. capitata* and *C. rosa* Karsch. Caged, wild *C. rosa* on Reunion Island mate only in the late afternoon (peak between 18:30 and 19:00 hours), while sympatric *C. capitata* do so preferentially in early morning hours (peak between 07:00 and 09:00 hours with a smaller peak between 13:00 and 14 hours; a few individual *C. capitata* mate throughout the day; Quilici and Franck 1997).

*Ceratitis capitata* has become widespread around the world, and there appears to be a geographic pattern in *C. capitata* mating activity. In Guatemala, peak calling activity was reported between 11:00 and 14:00 hours. In Hawaii there were two peaks, a small one at 10:00 and a bigger one between midday and 14:00 hours (Whittier et al. 1992). Note that this is the reverse of the calling pattern on Reunion Island. In Egypt, the bimodal pattern resembles that reported for Reunion Island, a main peak in the morning (before the hottest part of the day) and a smaller one in the afternoon. Finally, in Chios (Greece) and Israel peak mating activity occurred during midday. In locations where interactions with close relatives are not possible (Greece, Israel, Guatemala, and Hawaii), mating activity occurs at times similar to the period of activity in *C. rosa* (i.e., afternoon hours). Thus, afternoon calling by *C. capitata* in regions of allopatry may represent a temporal shift into a block of time filled by the calling of *C. rosa* in areas of sympatry. If so, this further supports the hypothesis that varying mating rhythms evolved as a mechanism to restrict hybridization.

#### 28.2.1.2.5 *Rhagoletis*

Patterns of mating in *Rhagoletis* also differ according to species. This could be interpreted in a manner similar to the above (see Propkopy and Papaj, Chapter 10).

### 28.2.2 Agonistic Behavior

Defining agonistic interactions as "aggressive or defensive social interactions between conspecific individuals" does surprisingly little to elucidate what does or does not qualify as agonistic. A wide range of behaviors have been considered agonistic, and it is difficult to delineate the category clearly. While an extreme activity such as hostile, physical combat is readily viewed as aggressive, less pugilistic behaviors such as wing displays and other body postures are more difficult to classify. Noncontact interactions have often been described as agonistic in studies of tephritids and the justification seems to come mainly from assessment of the context of the encounter. For example, a particular wing movement by a male may be interpreted as a courtship signal when directed toward a female, but an agonistic display when directed intrasexually.

Most of us think of agonistic interactions as those that involve *fighting*. Indeed, Scott and Fredricson (1951) coined the term *agonistic* with specific reference to actions related to fighting. In the lay vernacular, fighting need not involve physical contact. Simply the threat of an action that would negatively impact the well-being (fitness) of another individual is often considered a fight (e.g., angry shouting by humans). Unfortunately, until neurological assessment of the motivation behind all nonhuman animal behaviors is possible, we will be forced to continue our speculation regarding the function of many potentially agonistic actions. Herein, we will focus on the more blatant examples of fighting in tephritids, but continue to consider ambiguous behaviors such as wing displays as likely components of agonistic activity. First, we review the theoretical framework that allows us to predict who should fight and when; and then we ask whether tephritids appear to follow predictions from this theory.

Ever since Darwin (1871) formalized the observation that males are more prone than females to exhibit overt intrasexual contests, there have been refinements of the theory attempting to explain why this is so. In a nutshell, females are expected to be the more discriminating sex in mate choice due to their overall lower variance in mating success and greater parental investment in offspring (Bateman 1948; Trivers 1972). Selection should therefore favor males that devote more of their total reproductive effort toward mating effort. Competition among conspecific males to attract, encounter, or monopolize potential mating partners leads to many contest situations.

This is not meant to suggest that females should never act aggressively. There is little doubt that females in most animal species routinely or occasionally direct aggression toward males when they are not receptive to mating attempts. Aggressive nonreceptivity can include charging, biting, striking, and other offensive behaviors. Since the message conveyed is likely to be unambiguous and males usually have little to gain by "fighting back," agonistic interactions of this type are not expected to be prolonged or highly ritualized. Even in the many described instances of "forced copulations" among tephritid species, the actions of males are most appropriately viewed as "coercive" rather than fighting. Female aggression toward males in tephritids is often observed by researchers (see taxon-specific chapters in this volume), but there are no reports of prolonged or escalated contests.

Aggressive, intrasexual competition is expected among females if a required resource (e.g., oviposition substrate or adult food source) is limited relative to the number of individuals attempting to use it. Tephritid females have been observed physically aggressing conspecific and heterospecific intruders while at food or oviposition sites (e.g., Pritchard 1969; Biggs 1972; Dodson 1982; and unpublished data; Papaj et al. 1989; Headrick and Goeden 1994). However, severe resource limitations are rare (see Dodson, Chapter 8, for the only documented case); therefore, prolonged combat is not expected from a cost/benefit perspective. If monopolizing a site requires a high energy expenditure or excessive time commitment, then seeking other sites may be the best option as long as they are accessible. Selection should favor less costly alternatives when competition for resources is not extreme. The well-documented examples of oviposition-deterring pheromones within tephritids (Propkopy and Papaj, Chapter 10) suggest that all parties gain from signals that verify the earlier presence of a competitor.

Male–male aggressiveness, on the other hand, is expected to be common. Male reproductive fitness is limited primarily by the number of matings acquired (Bateman 1948). Receptive females are therefore a limiting resource for males (except in the rare instances of a female-biased operational sex ratio, Emlen and Oring 1977). Animals such as male tephritids effectively compete for mates every day of their adult lives, excluding instances of reproductive diapause. It is thus not surprising that males routinely display agonistic behaviors whenever they encounter one another, even when no females or resources are currently present. However, the most intense combat is expected when females or female-required resources are distributed in such a way that individual males can guard them and monopolize access (Brown and Orians 1970), leading to female or resource defense mating systems (Emlen and Oring 1977). Habitual aggression is also expected when males are competing for space at display sites to which females come only for mating (lek or landmark mating systems). The least aggression should occur within species that exhibit a distribution of females and resources that are not clumped and cannot be monopolized, and in which males compete in isolation to find females rather than advertising and waiting for females at leks ("scramble competition" *sensu* Thornhill and Alcock 1983).

Examples of the scramble competition mating system and the predicted low-intensity agonistic interactions are represented by the gall-forming species included in Headrick and Goeden's (1994) circumnatal mating system category. Males of these species typically *stalk* (Headrick and Goeden 1994) females in isolation and only occasionally confront each other. Even when male–male encounters occur, there is little incentive for lengthy fighting since any females in the vicinity are likely to be moving on. Dodson (1987a, b) provided a detailed account of the mating and agonistic behaviors of a representative species (*Aciurina trixa* Curran).

Lek mating systems have been well documented within the family (Sivinski, Section 28.2.1.1) and virtually all accounts include descriptions of frequent male–male agonistic interactions. With some interesting exceptions, lekking flies are mainly polyphagous species (Sivinski and Burk 1989). Since a high proportion of the mating in these species occurs at the lek (most, if not all exhibit mating at oviposition sites as well), there is strong selection on males to acquire a position within the lek and perhaps even to compete intensely for the "best" positions therein (Sivinski and Burk 1989). These species represent a good test of the theoretical expectations of relative fighting intensity. In species where the highest proportion of mating success is achieved at leks (many species of *Anastrepha*, *Ceratitis capitata*?), fighting intensity should be greater at these display sites than at secondary mating sites. By contrast, when a resource-based mating strategy produces more matings within a species than the leklike alternative (some *Rhagoletis*?), fighting should be less intense at the leks.

It is a logical assumption that female-required resources will be scarce more often for monophagous than for polyphagous species. Indeed, of the reported cases of resource defense mating systems in tephritids, all involve species that are monophagous or oligophagous, and the resource being defended is typically an oviposition substrate — for example, in *Anastrepha bistrigata* (Morgante et al. 1993), *Dacus longistylus* (Wiedemann) (Hendrichs and Reyes 1987), *Phytalmia* spp. (Dodson 1997), *Rhagoletis boycei* Cresson (Papaj 1994), *R. completa* Cresson (Boyce 1934), *R. juglandis* Cresson (Papaj 1994), *Toxotrypana curvicauda* (Landolt and Hendrichs 1983). Other cases of resource defense involve nutritional resources for females that are either naturally occurring or male produced, for example, plant wounds guarded by *Paracantha gentilis* Hering (Headrick and Goeden 1990) or spittle masses guarded by *Euarestoides acutangulus* (Thomson) (Headrick and Goeden 1996), salivary secretions produced and guarded by male *Eutreta novaeboracensis* (Fitch) (Stoltzfus and Foote 1965) and *Neaspilota viridescens* Quisenberry (Goeden and Headrick 1992, and unpublished data). Males of several species are known to produce salivary secretions in the form of mounds on which females feed (Freidberg, Section 28.2.3.4). Although that involving *E. novaeboracensis* is the only published account of agonistic behavior expressly associated with guarding a mound, further observations are likely to reveal such behavior for some if not all of these species.

Having considered the ecological characteristics that predispose certain species to agonistic behavior, it is interesting to ask what kind of fighting is expected to occur. To this end, game theory models have greatly clarified our understanding of the nature of contest structure (Parker 1974; Maynard Smith 1982; Reichert 1998). The emergent consensus is that animal contests are designed to gain the maximum information about the quality, ability, and motivation of an opponent with as little cost as possible. All participants potentially benefit from the acquisition of information that reliably forecasts the outcome of a contest without expenditure of additional energy or the liability of greater risk resulting from further escalation. Actual fighting should be limited to instances in which opponents are so evenly matched (through combinations of fighting prowess, current condition, motivation, etc.) that the outcome is uncertain based on the information available early in a contest. In other words, escalation of contests is expected only when it is difficult for the participants to determine the superior opponent without extended assessment or when the value of the contested resource is especially high (Sigurjónsdóttir and Parker 1981).

Tephritids have little in the way of physical attributes that would lead us to expect injury risks in their agonistic encounters, and we are not aware of any report of damage inflicted on one male by another. Thus, their conflicts are best represented by the war of attrition models in which contest costs arise in the form of time and energy expenditures and winners are those that opt to persist longer (Maynard Smith 1974; Parker and Thompson 1980). Contests such as the stilting behavior described below may represent the most extreme method used by evenly matched tephritid combatants to test the persistence quotient of each opponent.

The extensively documented wing displays (Headrick and Goeden 1994) employed by males in almost all aggressive encounters may provide the most basic information about the size, vigor, and/or motivation of a contestant. Acoustic and pheromonal signals could play similar roles or possibly represent a second level of information if they are more costly to exhibit. Should these types of information fail to clarify the probable outcome of a contest sufficiently (i.e., neither contestant retreats), physical contact appears to be the final assessment mode. A perusal of the behavior chapters in this volume reveals that tephritid males make both gentle and forceful contact with each other wing-to-wing, wing-to-body, head-to-head, mouthpart-to-mouthpart, legs-to-legs, legs-to-wings, legs-to-body, and legs-to-head. We assume that all of these actions involve a tactile signal, and that at least some may also provide chemical cues. Contact with the legs has been described variously as boxing, sparring, pawing, striking, batting, grappling (and even "belaboring each other with their forelegs" [Brooks 1921]). They butt heads and push against each other with mouthparts, heads, and legs. These actions often appear frenetic, but can be very complex and seemingly choreographed in some species (e.g., Headrick and Goeden 1990; Headrick et al. 1995; Goeden et al. 1998).

Males mounting other males in a copulatory manner is a fairly common behavior observed by tephritid researchers and has been interpreted by most authors as evidence of poor gender discrimination by males. Such behavior could be a form of aggression rather than a "mistake" and closer examination of the activities leading up to this behavior has suggested this interpretation in one case. Iwahashi and Majima (1986) discovered distinctive behaviors preceding such mountings, indicating that males recognize the sex of the intruders prior to making contact with them. While it is reasonable to expect mistakes by these animals, this finding should serve as a reminder that extremely careful examination is required to detect such subtle details of motivation.

The most-escalated fights described thus far for tephritids have a similar composition. Males that have confronted each other head-to-head rise up on their middle and hind legs. Their unconstrained forelegs (and sometimes the middle legs, Boyce 1934) are then used to box at each other or are raised above their heads and held aloft, waved, or batted against the legs of their opponent. This form of fighting is illustrated in Figure 21 of Headrick and Goeden (1994) and is characteristic of *Euarestoides acutangulus, Euaresta stigmatica* Coquillett, *Tephritis stigmatica* (Coquillett), *Trupanea jonesi* Curran (Headrick and Goeden 1994), *Phytalmia* spp. (Dodson 1997), and the *suavis* group of *Rhagoletis* (Brooks 1921; Boyce 1934; Papaj, unpublished data). In most of these species, the mouthparts of the males are in contact during some or all of the time they are stilted. A variation is found in some *Phytalmia* species in which the epistomal margin of the face is used as a pushing surface in these upright contests (Dodson 1997). The intensive male–male fighting exhibited by the papaya specialist *T. curvicauda* does not include stilting, but the head and thorax are often elevated during contests at fruit territories (Landolt and Hendrichs 1983). Fighting similar to that of *T. curvicauda* was described in laboratory observations of *R. pomonella* (Walsh) (Biggs 1972).

Reinforcing the point made earlier, all of the species exhibiting this escalated fighting style are monophagous or stenophagous species and thus are known or presumed to be defending a limited resource with the prospect of increased mating encounters. The exceptional species is *T. jonesi*, which has the broadest known host range (104 species in eight tribes of Asteraceae) of any North American tephritid (Goeden et al. 1998). The intensive male–male fighting was observed in two instances in the laboratory (Goeden et al. 1998), and the location of this fighting in the field awaits observation. Perhaps this species defends a limited resource separate from an oviposition substrate, as does *E. acutangulus;* or consists of host races isolated on subsets of the known hosts. If *T. jonesi* is determined to be a polyphagous species fighting so intensely at a nonlimited resource, we will need to rethink our theory.

In sum, the theoretical models of animal conflict and agonistic behavior in tephritids seem to be in concert. Fighting is more frequent in males than females. Fighting is least intense in species exhibiting scramble competition (i.e., circumnatal) mating systems, more pronounced at leks, and escalated where required resources are economically defensible. A hierarchy of behaviors appears to be utilized, consistent with the expectation that contests will be settled as soon as sufficient information is obtained that reliably predicts the winner. Finally, the most energetically and temporally costly behaviors are used only when necessary in these fascinating tephritid wars of attrition.

## 28.2.3 Courtship

### 28.2.3.1 Acoustic Signals

Male, and occasionally female, fruit flies sometimes make rapid wing motions in sexual contexts (Tephritinae — e.g., Greene et al. 1987; Trypetinae — e.g., Burk and Webb 1983; Dacini — e.g., Kanmiya 1988). These movements could typically be characterized as hamation, evanation, or synchronic supination in the terminology of Headrick and Goeden (1994), and are often correlated to pheromone release and related behaviors, such as abdominal pleural distention and/or dabbing the substrate with the proctiger (e.g., Headrick and Goeden 1994; see Glossary, Chapter 33).

It has been supposed that the function of the movements is to waft pheromones into the airstream or toward an approaching female (e.g., Sivinski et al. 1994). In the genus *Anastrepha* all species evert the proctiger to expose pheromone-dampened anal membranes; however, primitive species such as *A. cordata* Aldrich and *A. aphelocentema* Stone do not simultaneously fan their wings while chemically signaling in the absence of females (Aluja et al., Chapter 15; see similar behavior in the sister genus *Toxotrypana*, Sivinski and Webb 1985b). More derived *Anastrepha* species wing-fan as they occupy leaf territories in leks (e.g., Burk 1983; Aluja and Birke 1993). Calling by male *A. cordata* and *A. aphelocentema* has not been observed in nature; perhaps they will be found to call, like *T. curvicauda*, on or near fruit whose additional female-attracting odors may allow them conserve the energy that would otherwise be spent in wing fanning. Alternatively, it may be that competition from nearby males in mating aggregations results in males expending considerable energy to establish their precise location. That is, females arriving at lek sites might find it easier to discriminate males with strong individual pheromone plumes generated by wing fanning.

While pheromone dispersal may have been the original purpose of wing fanning, the sounds produced by such movements, in at least some instances, have taken on a signaling significance of their own (e.g., Aluja et al., Chapter 15). Even additional sound-producing structures, such as the "pecten" (abdominal setae that are struck by the wing) of *Bactrocera cucurbitae* (Coquillett) (Kanmiya 1988), have evolved. In *A. suspensa* and *C. capitata* removal of the wing does not entirely mute the sound, suggesting that thoracic vibrations in addition to wing movement may be involved in sound production (Keiser et al. 1973; Sivinski and Webb 1985a).

There are several additional behaviors and structures that imply an acoustic communications role for wing movements. Female *C. capitata* perform high-speed wing movements identical to those of males (Sivinski and Webb 1989). The reason for these movements is obscure, although they may incite male courtship by mimicking a rival (Arita and Kaneshiro 1983). Since there is no evidence that females produce a pheromone (Heath et al. 1994), mimetic females are probably producing signals that have some acoustic (or visual) significance. This suggests that the original male wing motions, on which the female signals are modeled, are also likely to be acoustic displays rather than mere by-products of chemical dispersal behaviors. In another instance, *R. juglandis* males control access to individual walnuts and are not known to produce pheromones. However, they wing-fan as they approach females that have come onto their fruit and the low-frequency sound produced may serve as a courtship song (see Prokopy and Papaj, Chapter 10; alternatively, they may be directing host volatiles at females).

In the relatively well-studied case of *A. suspensa,* there are at least two forms of acoustic signals ("calling" in the vocabulary of Headrick and Goeden 1994). One takes place when the male expands its pleural glands and dabs pheromone unto the leaf territory surface ("calling song"), and the other, when the male mounts the female and attempts to engage her genitalia ("precopulatory song"; Sivinski et al. 1984). Calling songs elicit responses from virgin females and males, but not from mated females, and their rate of production and structure change with circumstance. For example, pulse trains (episodes of wing beating) increase in duration in the presence of males, and the interval between pulse trains decreases when females are nearby (Sivinski and Webb 1986). The precopulatory song is very energetic and its sound intensity was shown experimentally to be an important factor in determining whether females would allow singing males to copulate (Sivinski et al. 1984). As mentioned earlier, there is a third form of acoustic sexual signal, made only as males orient toward and court a nearby female. A relatively well-documented instance occurs in *T. curvicauda,* a member of the sister genus of *Anastrepha* (Sivinski and Webb 1985).

### 28.2.3.2 Visual Signals

#### *28.2.3.2.1 Color and Pattern*

As brief examinations of Foote et al. (1993), White and Elson-Harris (1992), and other taxonomic works will testify, many tephritid bodies and wings are marked with bands, spots, and blotches. In addition, the eyes are often brightly colored and sometimes banded or otherwise patterned (e.g., *Anastrepha, Ceratitis,* or *Phytalmia* spp.; Color Figure 20*; and see Moffett 1997). There is an understandable assumption that these colors, particularly when sexually dimorphic, serve communicative functions (e.g., Burk 1981). One plausible example are the violet pink and green-striped eyes of *Phytalmia megalotis* Gerstaecker that flank a pink face, which is itself extended laterally to form pink- and black-trimmed antlers (Wallace 1869). Such antlers have been shown to advertise size in agonistic communications among rival males (Dodson, Chapter 8), and it is likely that the facial colors and those of the eyes contribute to the display. Likewise, the sexually dimorphic eye colors and contrastingly colored capitate anterior orbital setae of *C. capitata* probably produce male visual signals (Eberhard, Chapter 18). Another display that seems almost certain to be a visual signal, and one that uses colors on a different part of the body, occurs in the tephritine *Eutreta diana* (Osten Sacken). In this species, the midfemora are black with yellow tips (Headrick and Goeden, Chapter 25), and are lifted (abducted) by the male during courtship. At the same time it holds its wings arched over its back, providing a dark background for the femoral display.

While there is a good deal of anecdotal natural history, there is little experimental evidence addressing intraspecific fruit fly visual communication. Female *C. capitata* will turn toward males separated from them by a pane of glass when the latter vibrate their wings (Féron 1962), and males of the same species are more likely to begin pheromone calling when kept in vials with mirrors (McDonald 1987). Females are more likely to oviposit when they perceive wing-waving by a female already present on a fruit (Prokopy and Duan 1998). However, the best documented cases for visual signaling are interspecific in nature and concern the "misleading" information transmitted by wing patterns of species such as *Zonostemata vittigera* (Coquillett) and *R. zephyria* Snow to jumping spiders (Greene et al. 1987; Mather and Roitberg 1987). When seen from behind, some pigmented bands create the illusion of a salticid seen face-on and the resemblance deters attacks.

Support for visual communication through color patterns on wings might be obtained by considering the behaviors, and the absence of certain behaviors, in taxa which lack wing patterns. The only tephritoid

---

* Color Figures follow p. 204.

family with typically unpatterned wings is the Lonchaeidae, and this is the only family in which aerial swarm-mating systems are common (Sivinski, Chapter 2). It may be that patterns are imperceptible in flying individuals and useless for communicating within swarms. Note also that fruit flies without distinctively patterned wings such as *T. curvicauda* and *Bactrocera* spp. do not have many of the stylized wing movements typical of other tephritids (Sivinski and Webb 1985b; Landolt, Chapter 14).

However, if wing patterns function in sexual communication, it seems surprising that sexual dimorphism appears to be rare, although there are instances of considerable differences; for example, in *Aciurina idahoensis* Steyskal, female wings are striped and males' spotted (Headrick and Goeden, Chapter 25), while in the related *A. semilucida* (Bates) females again have striped wings while those of the male are fully infuscated. In addition to being uncommon, those sexual dimorphisms that do occur are often not what is expected, with male wings being fainter than females' and/or bearing interrupted or missing markings (e.g., many *Trupanea* spp.; Foote et al. 1993). There has apparently been no thorough search for ultraviolet markings on the wings of fruit flies, and until such a survey has been completed it may be premature to discount the possibility of widespread sexual dimorphism.

#### *28.2.3.2.2 Ornaments*

Occasionally, male tephritids bear elaborate or novel structures that presumably have been sexually selected to function as signaling devices (e.g., Sivinski 1997). The above-mentioned antlers of *Phytalmia* species are an example. In this instance, the ornaments communicate male size to rivals during agonistic interactions for the control of oviposition sites (Dodson, Chapter 8). There is a variety of ornaments whose functions may vary as well. A few examples give a flavor of the range.

Head: While the rain forests of New Guinea and vicinity contain the antlered *Phytalmia*, even seemingly mundane locations such as the British Isles harbor several fruit flies with marvelously modified heads (White 1988). An impressive row of enlarged bristles projects from the lower face of the trypetine *Chetostoma curvinerve* Rondani. Forward-jutting projections on the frons of another trypetine, *Stemonocora cornuta* (Scopoli) also bear long, stout setae — in the United States similar setae arise from the upper portion of the head of the trypetine *Paramyiolia rhino* (Steyskal) (Foote et al. 1993). In the terelliine, *Cerajocera ceratocera* Hendel, it is the pedicel (second antennal segment) that sticks out like a bristled horn. See Han (Chapter 11) for additional examples of tephritids with projections or modified setae on the head.

Other examples of specialized head setae occur in *Ceratitis* subgenus *Ceratitis;* the reflective, paddlelike hairs are especially well developed in species such as *C. catoirii* Guérin-Méneville (White and Elson-Harris 1992) and *C. caetrata* Munro (Munro 1949; see Eberhard, Chapter 18).

Thorax: The front basitarsi of *Euphranta maculifemur* (Meijere) is broadened and concave (Hardy 1973). In several species of *Ceratitis* the midlegs are "feathered" with long setae on the tibia or tibia and femur (e.g., Freidberg 1991; White and Elson-Harris 1992; De Meyer, Chapter 16).

Abdomen: Male *Trupanea brunneipennis* Hardy have a mass of strong yellowish bristles along the fifth tergite of the abdomen (Hardy 1973). *Copiolepis quadrisquamosa* Enderlein from New Britain and New Guinea is dramatically attired with long bird-of-paradise-like "plumes" projecting from the abdomen (Enderlein 1920).

The functions of structures such as those listed above are often enigmatic, particularly when so few observations of courtship have been made, and they may or may not function in visual displays. For example, there is a large downward-pointing projection from the front femur of male *Ectopomyia baculigera* Hardy (Hardy 1973), but it is possible that such a structure is a mechanism used to secure mounted females — see projections on the legs of *Phytalmia* spp. (Dodson, Chapter 8), and the enlarged forefemora of *Aciurina* spp. (Headrick and Goeden, Chapter 25). Even if the ornaments are used in communication they may perform in a different, nonvisual, channel; for example, "feathered" legs might provide tactile stimuli.

### 28.2.3.2.3 What Do Displays Mean?

Courtship activities and structures are often hypothesized to have evolved in one of several contexts (e.g., Sivinski 1997; Endler and Basolo 1998). One of these general theories supposes that some information of importance to the receiver is contained in the signal, although selection may favor exaggeration in the display. MacAlpine's (1979) hypothetical account of the evolution of stalk eyes in Platystomatidae (Tephritoidea) is a particularly elegant example of this process, albeit one that describes the evolution of an agonistic rather than a courtship display (Sivinski, Chapter 2); tephritids such as *Pelmatops ichneumoneus* (Westwood) also have stalk eyes (Wilkinson and Dodson 1997; see also Dodson, Chapter 8). Suppose that flies entering a face-to-face agonistic interaction avoid costly combat by first estimating the size of their opponent and then decamping if the other fly is larger. If size is determined by the extent that the margins of the opponent's head (eyes) overlaps those of the observing fly, then an atypically broad head provides a psychological advantage. Of course, as broad heads become more common, then still greater head expansion is required to carry out a successful bluff against a typical opponent. This results in an "arms race" that pulls eyes farther and farther apart. The race concludes when the danger and expense of the ornament equals the competitive advantage it provides. Accordingly, while stalk eyes were originally deceitful, in the end they became honest advertisements since only the largest and most vigorous individuals could wield the largest ornaments (such ornaments may also give insight into the information-processing capacities of flies; it is difficult to imagine animals capable of synthesizing information from different perspectives, like cats or squirrels, evolving stalk eyes).

Honest signals, including stalk eyes, are of interest to mate-choosing females as well as rivals. In the diopsid *Cyrtodiopsis whitei* (Curran) females preferentially gather around males with long eye stalks (Burkhardt and Motte 1988).

Another perspective on "honest advertisements" supposes that during the evolution of ornaments and elaborate behaviors it is the expense of displaying that is selected from the very start (e.g., Zahavi and Zahavi 1997). Expensive and dangerous "handicaps" reflect the underlying qualities of the signaler. The fact that an animal has survived in spite of its display burden gives a potential mate greater insight into the signaler's abilities than what could be determined from the mere presence of an unornamented and untested rival.

Whether a signal has evolved through an "arms race" between emitters and receivers, or through competition among "reckless signalers" striving to guarantee their hardiness, there is the supposition that the messages, at least those directed at the opposite sex, contain evidence of genetic quality ("good genes"). What exactly might these advertised qualities be? There is an extensive literature on the "substance" of intersexual signals, one too large to address here. But, size and vigor (e.g., Burk and Webb 1981), low genetic loads (as expressed by high levels of bilateral symmetry; e.g., Thornhill 1992), and ability to resist pathogens and parasites (e.g., Hamilton and Zuk 1982) are all potentially heritable qualities that appear to interest females of certain species (see also Eberhard, Chapter 18). It is also possible for males to advertise a material rather than genetic quality (see Freidberg, Section 28.2.3.4).

At this time, there is little direct evidence for "good genes" content in male fruit fly displays directed toward females. Many structures and colors used in the displays of tephritids, as well as mammals and birds, are located on the head (e.g., capitate setae and bristled projections). The front of the head is likely to be the "signal platform" closest to the courted or threatened conspecific. Zahavi and Zahavi (1997) have argued that head extensions might clearly display the orientation of the head, and so serve as proof that the displaying animal is looking directly and uninterruptedly at the receiver. Since focused staring presumably makes the signaler vulnerable to predation, it may constitute a dangerous handicap and be an expression of "self-confidence"; that is, insects that behave in such a way, and are still alive, will tend to have superior senses and reflexes. If so, would female fruit fly orientations toward courting males be shorter in duration than male orientations to females, and do females "glance" about more often than males? Male *C. capitata* with more symmetrical anterior orbital bristles have greater mating success (Eberhard, Chapter 18). However, the orientation of the bristles during courtship may make them difficult to observe, and bristle symmetry might only reflect a general symmetry that is scrutinized by potential mates through other, more noticeable, structures. Expensive acoustic signals, such as the copulatory song of *A. suspensa,* have been hypothesized to be the result of discriminating females making increasing demands on males to display their vigor and energy reserves (Sivinski et al. 1984; Aluja et al., Chapter 15). Athleticism may be demonstrated by the males of some *Rhagoletis* spp. which initiate copulations either on leaf surfaces or in *midair* (e.g., Prokopy 1976).

A second explanation for the evolution of courtship signals supposes that there is no information embedded in male motions, structures, or colors that could independently corroborate a male's suitability as a sire. Rather, an arbitrary preference in females for the most extreme examples of a particular male trait can lead to "Fisherian runaway sexual selection"; that is, such a preference results in the presence of genes for both the extraordinary signal and a preference for the extraordinary signal in both males and females which can generate a sort of "chain reaction" self-selection for the increasingly extreme (see a lucid explanation in Dawkins 1986). There is debate over the likelihood of runaway selection in insects. Alexander et al. (1997) have argued that insects are typically ill-suited to this form of selection since it requires that females sample a range of males and then choose the extreme. They argue that most insects are too short lived to acquire broad experience of potential mates and lack the capacity to recall and compare the information they do obtain. This view, and the evidence supporting it, has been criticized by Eberhard (1997). Zahavi and Zahavi (1997) point out that extravagant signals are sometimes used both in sexual *and* agonistic encounters, and that runaway selection is unlikely to account for the evolution of an intrasexual signal. That is, males unburdened by ornamentation will presumably defeat those that are handicapped by an ostentatious display that had no correlation to size or vigor during the early stages of its evolution.

A third explanation, and one supported by a growing number of examples, is that signals evolve to exploit biases in their receivers. That is, a particular ornament or coloration evolved simply because females were predisposed to respond to the early stages of the signal. A particular bias might be a side effect of "other mate choice preferences, responses that evolved to locate prey or avoid predators, and limitations imposed by the more general operating principles of neural and cognitive systems" (Ryan 1998). For example, females of wolf spider species without ornamented males preferentially respond to the courtship displays of ornamented males of related species (McClintock and Uetz 1996). This sort of untapped female "preference" (or vulnerability to manipulation) has been found in amphibians, fish, and birds as well (Ryan 1998). Similar experiments might be attempted with the genus *Ceratitis*, some species of which have large capitate setae while others do not (see White and Elson-Harris 1992). Might tephritid wing motions and patterns, originally employed to distract or confound predators, have an effect on females as well and become secondarily useful in sexual communication (see Section 28.2.3.2.1)?

### 28.2.3.3 Predation and the Evolution of Sexual Behavior

#### *28.2.3.3.1 Introduction*

The sexual behavior of tephritid fruit fly species is quite varied, with many types of signals and strategies, and both intrasexual and intersexual interactions. The sexual behavioral repertoire evolved within a given species of Tephritidae may be a result in part of ecological parameters, such as host breadth, latitude, and climate (Emlen and Oring 1977; Prokopy 1980), as well as phylogenetic history. In many species of Tephritidae, selection pressures exerted by predation may also have had impacts on the signaling and strategies involved in the seeking and selecting of mates.

Monteith (1972) found that apple maggot flies in apple trees were not predated and Prokopy (1977) considered nonteneral adult tephritids to be generally free of predation. However, others report predation on adult fruit flies by wasps, odonates, mantids, and spiders (Brittain and Good 1917; Greene et al. 1987; Mather and Roitberg 1987; Van der Valk 1987; Whitman et al. 1988;

Hendrichs and Hendrichs 1990; Hendrichs et al. 1994). Fruit flies are probably eaten also by vertebrates such as frogs, lizards, and birds. Although most species of tephritids are agile fliers and able to evade predators, they are also generally defenseless, both physically and chemically.

Fruit flies may be particularly prone to predation when they are engaged in mating (Hendrichs and Hendrichs 1998) as well as other sexual behaviors such as courtship interactions. When engaged in such activities, they are likely to be more conspicuous or apparent to predators, may be less mobile (when in copula), and they also are likely to suffer a decrease in attentiveness. If an insect is displaying or signaling, it may place itself in a relatively open location, either to signal from a resource at that site or to increase its apparency to conspecifics. It would then also be more conspicuous to predators at such signaling sites. When signaling, whether visually, acoustically, or pheromonally, a fruit fly may also inadvertently advertise its presence to predators by those same signals. Additionally, it seems likely that male and female flies interacting in courtship, or males involved in territorial and related agonistic interactions, would be less apt to perceive danger from approaching predators, and might be more susceptible to predation. It is easy to imagine that predation pressures may have influence, via natural selection, on the sexual signaling and location of sexual interactions of tephritid fruit flies, and that specific morphological, physiological, and behavioral traits might be adaptive responses to such pressures, effectively reducing their susceptibility to predation.

### *28.2.3.3.2 Predation and Sexual Signals*

Sexual signaling in fruit flies includes visual signaling, chemical signaling, and acoustic signaling (Burk 1981). Fruit fly visual sexual signals may be used agonistically between competing males or as courtship signals between the sexes. These signals include species-specific patterns that are recognition signals, postures that may communicate size, and movements such as wing waving that communicate the species, sex, and physiological state of the signaler. Some fruit flies use chemical signals as sex pheromones to attract potential mates or in courtship interactions (Fletcher and Kitching 1995; Landolt and Averill 1999; Heath et al., Chapter 29). Acoustic signals of fruit flies may be attractive to the opposite sex and the same sex, may also be important to successful courtship of males, and may be agonistic (Sivinski 1988).

28.2.3.3.2.1 Visual Signals. There are no studies specifically addressing the question of whether or not fruit flies engaged in visual signaling, such as wing waving (hamation and supination), are more susceptible to predation. Such a study would be difficult in part because such activities are associated with other behaviors that may also affect fly susceptibility, such as pheromonal and acoustic calling, and focusing attention on another fly. Regardless, one may surmise that a visually oriented predator may more easily spot a moving, rather than a stationary, prey item. Vertebrate predators such as lizards and frogs do respond to movement and are less apt or unable to recognize stationary prey (e.g., Cott 1940).

28.2.3.3.2.2 Chemical Signals. Signaling with sex pheromones by insects may entail risks, by attracting predators (Vite and Williamson 1970) and parasitoids (Sternlicht 1973; Kennedy 1979; Aldrich et al. 1987; 1991). A well-documented example of this is that of Hendrichs et al. (1994) showing that German wasps respond to the male-produced pheromone of the Mediterranean fruit fly, locate pheromonally-calling males, and that calling males suffered a higher rate of predation than noncalling males and females.

Chemical communication is generally considered to be a fairly secure mode of signaling because of the typically small quantities released, hence the predominance of pheromonal calling by females in the Lepidoptera (Thornhill and Alcock 1983). However, the susceptibility of pheromone-releasing fruit flies to predation may be exacerbated relative to that of moths by two factors. First, the male-produced pheromones of tephritid fruit flies are generally released at rates that are orders of magnitude larger than that typical found for female-produced pheromones of nocturnal Lepidoptera, although the data sets for both groups are limited. For example, the female pheromone of the tobacco budworm moth *Helicoverpa virescens* (Fabricius) is released at about 1 n/h (Teal et al. 1986) while the male pheromone of the Mediterranean fruit fly is released at about 1 μg/h (Heath et al. 1991). A much greater amount of pheromone released may increase the likelihood of a predator being able to perceive the pheromone at a distance. Second, fruit flies are predominantly diurnal while moths are predominately nocturnal, making pheromone-releasing fruit flies more vulnerable to many groups of predators, such as predatory wasps and flies, Odonata, salticid spiders, lizards, and birds. Nocturnal predators might include bats, some web-spinning spiders, and frogs.

28.2.3.3.2.3 Acoustic Signals. There are as yet no indications that the acoustic signals of fruit flies, such as those of *A. suspensa* and *C. capitata*, are detectable by potential fruit fly predators or confer any additional risk on the signaler. There are examples, however, in other insects of vertebrate predator, and both invertebrate predator and parasitoid, utilization of prey acoustic signaling (Walker 1964; Cade 1975; Burk 1982). Perhaps this possibility should be kept in mind for future study of predator and parasite responses to fruit fly acoustic signals.

### *28.2.3.3.3 Predation and Encounter Sites*

Sexual rendezvous sites of fruit flies are generally the host fruit, foliage of host trees, or foliage of nonhost plants (see Sivinski, Section 28.2.1.1). When these mating sites are on fruit, they are considered to be resource-related, with the fruit being both an oviposition substrate for females and a source of food for adult fruit flies. The use of foliage away from fruit as a mating site is often considered a lek, whether on or off trees that contain host fruit suitable for oviposition or feeding.

28.2.3.3.3.1 Resource-related Station Taking. The use of fruit as a mating site clearly involves males searching for females at a resource (the fruit as oviposition substrates and adult food sources). Males may perch at and call from fruit, and defend fruit from other males as territories. Courtship interactions and mating may then take place on the fruit. For example, a mating strategy of *R. pomonella* is to encounter the opposite sex and mate on the host fruit. Fruit flies on fruit generally are more exposed and susceptible to predators, and fruit flies in foliage are more hidden from view and may be more difficult for certain types of predators to find. It also seems possible that the chemical odors of fruit, particularly fruit that is damaged, may attract predators, increasing the risk of using such sites for mating. For example, vespids, potential predators of fruit flies, are attracted to and feed on many types of fruit to obtain carbohydrates.

28.2.3.3.3.2 Leks. A number of polyphagous and pestiferous species of tephritids encounter potential mates in leks, which by definition are nonresource-based mating aggregations. Polyphagous tropical pest species generally use leks as mate-encounter sites. It was proposed by Hendrichs and Hendrichs (1990) that predation pressure on flies at host fruit may have driven flies to form mating aggregations, leks, on foliage and away from fruit. Although fruit flies in leks may be less exposed than flies on fruit and thus might be less vulnerable to predation, Mediterranean fruit fly males in leks are at times heavily preyed upon by German wasps that are attracted to their pheromone (Hendrichs et al. 1994). The greater amount of pheromone released by a number of males in a lek may increase their attractiveness to the wasps. Nevertheless, this theory has considerable merit for many species of fruit flies. Quantitative studies of predation pressure on flies on fruit vs. flies in leks would be illuminating.

28.2.3.3.3.3 Signaling. The use of particular mating strategies by some fruit fly species and not others suggests that they may be adaptations to predation pressure. Similarly, particular signal traits or characteristics used by some species and not others implies that they may be a result of predation pressure. These are discussed below.

The production and release of sex-attractant pheromones by males rather than females of Tephritidae, except in *B. oleae* (Haniotakis 1974), could be interpreted as an adaptation to predation risks involved in chemical signaling by flies during the daylight hours. It is assumed that males will be the signalers where signaling is risky or unduly expensive, or where males can control access to female-required resources, and that females will be the signalers where the risk and expenditure of resources is minimal (Thornhill and Alcock 1983). The predation suffered by pheromone-releasing male Mediterranean fruit flies from German wasps (Hendrichs et al. 1994) supports the contention that signaling can be very dangerous.

The nature of the chemicals produced by some fruit flies suggests chemical mimicry as a possible adaptation to predation on pheromonally calling fruit flies. Alkyl pyrazines are commonly found as exocrine products of Hymenoptera, including a number of species of ants and some wasps (Blum 1981). Similar compounds are produced by several species of Tephritidae; for example, 2-methyl-6-vinylpyrazine is the male pheromone of the papaya fruit fly (Chuman et al. 1987). Other pyrazines have been found in the glands or volatiles of species of *Bactrocera* (Metcalf 1990; Fletcher and Kitching 1995) and *Anastrepha* (Heath et al., Chapter 29). *N*-3-Methylbutylacetamide is an alarm pheromone of several species of *Vespula* wasps (Vespidae) (Heath and Landolt 1988) and is found in the male odors of a number of species of *Bactrocera* (Metcalf 1990). Another class of compounds, the spiroacetals, is found in a number of species of *Bactrocera* (see reviews by Metcalf 1990; and Fletcher and Kitching 1995) and is also present in the venoms of several social wasps (Francke et al. 1979; Aldiss 1983). Olean or (1,7)-dioxaspiro-(5,5) undecane, for example, is the major component of the female-produced pheromone of *B. oleae* (Baker et al. 1980), while the wasp compounds are alkyl-1,6-dioxaspiro[4,5]decanes. It is tempting to speculate that a fly smelling like a hymenopteran may be protected from some predators. Although suggestive, there are as yet no data indicating any deterrent or other protective effects of these compounds produced and emitted by fruit flies.

28.2.3.3.3.4 Mimicry. Some species of fruit flies may be visual mimics of other arthropods and may thus gain benefit from protection against some predators. Generally, it is suggested that some fruit flies may mimic spiders and others may mimic social wasps.

It was noted by Mather and Roitberg (1987) and Greene et al. (1987) that the tephritids *R. zephyria* and *Z. vittigera* resemble salticid jumping spiders. Further, it has been shown that salticid spiders avoid these flies and that the protection was derived principally from the patterns on the fly wings, which resemble spider legs when viewed from the front (Mather and Roitberg 1987; Greene et al. 1987; Whitman et al. 1988). It is concluded that these flies mimic jumping spiders and are not eaten by the spiders, which also do not prey on other salticids.

The papaya fruit fly appears to mimic social wasps and may gain some protection from predators by this mimicry. They mimic species of *Polistes* and *Mischocyttarus* in Florida and other species in Central America (Landolt 1984). They possess color patterns similar to these wasps, as well as dark shading to the fore part of the wing that is similar in appearance to the folded wing aspect of Vespidae. This species is highly exposed when sexually active and during oviposition, which occurs during daylight hours on the fruit of the tree. The fruit are located on the trunk below the foliage, making the fruit and flies particularly visible. Males perch on fruit when calling, and females are immobilized for extended periods of time while ovipositing in fruit (Landolt and Hendrichs 1983). The wasplike appearance of this fly may be of particular advantage because of its activity on exposed fruit and resultant visibility to predators. The long ovipositor of the female, an apparent adaptation to access the center of fruits for deposition of eggs (Landolt 1985), may be a preselection factor because it contributes much to the wasplike appearance of the female. However, papaya fruit flies are not immune to predation and are eaten by lizards, spiders, and other predators. Of note, is the death of a captive *Anolis* lizard that was fed female *T. curvicauda*, and subsequently died with a mass of fruit fly ovipositors lodged in and puncturing the intestine. There are also species of *Anastrepha* and *Bactrocera* with elongated ovipositors that permit the deep penetration and deposition of eggs within fruit. It remains to be determined whether these species also possess additional morphological and coloration traits that resemble those of stinging Hymenoptera.

**TABLE 28.1**
**Mating Trophallaxis in Tephritidae**

| Subfamily/Tribe | Species | Timing | Direct/ Indirect | Mating Trophallaxis Confirmed? | Ref. |
|---|---|---|---|---|---|
| **Phytalmiinae** | | | | | |
| Acanthonevrini | *Afrocneros mundus* (Loew) | Pre | Dir | Yes | Oldroyd 1964 |
| | *Dirioxa pornia* (Walker) | Pre+In | Ind | Yes | Pritchard 1967 |
| **Trypetinae** | | | | | |
| Toxotrypanini | *Anastrepha striata* Schiner | Pre | Dir | Yes | Aluja et al. 1993 |
| **Tephritinae** | | | | | |
| Eurostini | *Aciurina mexicana* (Aczel) | Pre | Ind | Yes | Jenkins 1990 |
| Eutretini | *Eutreta novaeboracensis* (Fitch) | Pre+In | Ind | Yes | Stoltzfus and Foote 1965 |
| | *Paracantha gentilis* Hering | Pre | Dir | Yes | Headrick and Goeden 1990 |
| | *Paracantha cultaris* (Coquillett) | ?Pre | Dir | No | Cavender and Goeden 1988 |
| | *Stenopa vulnerata* (Loew) | Pre+In | Ind | Yes | Novak and Foote 1975 |
| Schistopterini | *Schistopterum moebiusi* Becker | Pre+In | Ind | Yes | Freidberg 1981 |
| | *Schistopterum* sp. | Pre+In | Ind | Yes | Freidberg, unpublished |
| | *Eutretosoma* sp. | Pre+In | Ind | Yes | Freidberg, unpublished |
| Tephrellini | *Metasphenisca negeviana* (Freidberg) | Pre+Post | Dir | Yes | Freidberg 1997 |
| Tephritini | *Euaresta festiva* (Loew) | Pre | Dir | No | Batra 1979 |
| | *Spathulina sicula* Rondani | Post | Dir | Yes | Freidberg 1982 |
| Terelliini | *Chaetostomella undosa* (Coquillett) | Post | Dir | No | Steck 1984 |
| | *Chaetorellia carthami* Stackelberg | Post | Dir | Yes | Freidberg 1978 |
| | *Chaetorellia succinea* (Costa) | Post | Dir | No | Freidberg 1978 |
| | *Neaspilota pubescens* Freidberg and Mathis | Pre | Dir | ? | Headrick and Goeden 1994 and personal communication |
| | *Neaspilota viridescens* Quisenberry | Pre | Ind | Yes | Goeden and Headrick 1992 |
| | *Terellia quadratula* (Loew) | Post | Dir | No | Freidberg 1978 |
| Xyphosiini | *Icterica seriata* (Loew) | Pre+In | Ind | Yes | Foote 1967 |

*Abbreviations*: Dir = direct; In = in-mating; Ind = indirect; Post = postmating; Pre = premating.

### 28.2.3.4 Trophallaxis

#### *28.2.3.4.1 Introduction*

Mating trophallaxis constitutes an array of behaviors in which the males provide females with nuptial gifts which are then consumed. It is connected with copulation and may occur shortly before, during, or shortly after it (Freidberg 1981). Unlike the definition of trophallaxis in social insects, which is limited to the "exchange of alimentary liquid among colony members and guest organisms" (Wilson 1975), mating trophallaxis encompasses the exchange of both liquid and solid substances which may have originated from a variety of organs or even from outside the body of the donor. By this definition, cases of females cannibalizing their mates during or after insemination are instances of trophallaxis.

The study of mating trophallaxis confronts one theoretical and two practical problems. The theoretical problem focuses around the somewhat vague concept of "connected with copulation." For example, what is the longest interval of time between the two activities before the idea of association becomes invalid?

On the practical side, there are two kinds of difficulties, the first being the need to prove that a behavior that superficially appears to be mating trophallaxis does indeed involve the transfer of substances. As shown in Table 28.1, several studies describe a contact between the mouthparts of the mates (often termed a kiss), but fail to prove that a substance has been transferred during this contact. However, to provide a comprehensive treatment of the subject such ambiguous cases are nonetheless treated here as instances of mating trophallaxis.

The second practical difficulty has to do with experimental manipulation of mating trophallaxis. Because such trophallaxis is strongly associated with copulation, any experimental interference might significantly affect both behaviors. This is especially true in the apparently more widespread cases of premating trophallaxis (see below), in which any interference in the trophallactic sequence might prevent mating and thus also preclude interpretation of the significance of the trophallaxis.

Classification of the various phenomena grouped here under the general term *mating trophallaxis* is important both for the understanding of the similarities and differences in the various manifestations of the phenomenon, and for relating this behavior to evolution and phylogeny. However, readers should be aware that the suggested classification is somewhat artificial, and that categories are not always mutually exclusive.

*Premating trophallaxis* — Trophallaxis that is initiated before copulation, although sometimes continuing through part of or even the entire sexual process.

*In-mating trophallaxis* — Trophallaxis that occurs during copulation.

*Postmating trophallaxis* — Trophallaxis that occurs after copulation (= sperm transfer) has been completed.

*Direct trophallaxis* — Trophallaxis in which the trophallactic substance is directly transferred between the mates (without being placed on an intermediate substrate).

*Indirect trophallaxis* — Trophallaxis in which the trophallactic material is transferred from the donor to an intermediate substrate before being picked up by the recipient.

*Stomodeal trophallaxis* — Trophallaxis in which the donor secretes the trophallactic substance from its mouth.

*Proctodeal trophallaxis* — Trophallaxis in which the donor secretes the trophallactic substance from its anus.

Mating trophallaxis, at least in Tephritidae, appears to be a relatively rare phenomenon, and has been reported for only about 20 species. Knowledge about this phenomenon is, therefore, scarce, generally anecdotal, and incomplete. Presentation of the available knowledge and continued research on this topic are undoubtedly necessary, and should result in additional discoveries of this general phenomenon.

#### *28.2.3.4.2 Distribution in the Animal Kingdom*

Trophallaxis in vertebrates has been documented in a number of bird species (Johnston 1962). In invertebrates it is restricted to Arthropoda, and has been described in both spiders (Arachnida) and insects (Insecta). In the former, it is well known that female widow spiders devour their mates after sperm transfer is completed (Kaston 1970). In insects, the phenomenon has been reported in several orders, notably the Orthoptera (e.g., Wedell 1994), Dictyoptera (e.g., Roeder et al. 1960), Hymenoptera (e.g., Given 1953), Mecoptera (e.g., Thornhill 1976) and Diptera. In Diptera, as in Mecoptera, mating trophallaxis may occur in two distinct ways: (1) through the transfer of prey, as in Empididae (Kessel 1955); and (2) through secretions, as in Asteiidae (Freidberg 1984), Drosophilidae (Kaneshiro and Ohta 1982), Ephydridae (Mathis and Freidberg, unpublished data), Micropezidae (Wheeler 1924), Platystomatidae (Piersol 1907), Sciomyzidae (Green 1977; Berg and Valley 1985), and Tephritidae (see Table 28.1).

#### *28.2.3.4.3 Distribution in the Tephritidae*

Table 28.1 lists all the tephritid species in which mating trophallaxis has either been reported or is suspected to occur. Reports vary in length and depth, ranging from a few words or an illustration to entire articles devoted to the subject. Representative cases are described more fully below.

Of all the cases in Table 28.1, only a few have been described in enough detail to warrant summarizing here. Of these, the case of *Schistopterum moebiusi* Becker (Freidberg 1981) was selected as an example of premating trophallaxis because of the field observations of this species, which are generally difficult to obtain. *Spathulina sicula* Rondani (Freidberg 1982; as *S. tristis* (Loew)) was selected both because it is the only well-studied case of postmating trophallaxis, and because it is the only case combining observations with experimentation.

*Schistopterum moebiusi* is a tiny (length about 2 mm) but colorful, widespread species distributed from Israel to South Africa, whose sole known host plant is the shrub *Pluchea dioscoridis* (L.) DC. (Asteraceae). *Schistopterum moebiusi* exhibits premating, indirect, stomodeal trophallaxis. Males defend territories near inflorescences and engage in aggressive encounters with other male intruders. When a female approaches a territorial male, he receives her with the same agonistic behavior shown to a male intruder. An unreceptive female decamps by running or flying away. A receptive female remains on the same leaf, and the male continues his approach by scissoring (enantion; see Glossary, Chapter 33) his wings and partially circling the female several times, both clockwise and counterclockwise. During this activity the female either moves about or remains stationary, enanting and moving her wings slightly. This courtship behavior takes from 30 s to 2 min.

When a female stands motionless, the male extends his proboscis to the leaf surface and secretes a white, frothy material from his labella. This material builds up into a vertical pillar, with new material supplied to the top. In the last stage of secretion the male broadens the upper part of the pillar into a mushroomlike shape, and by applying pressure to it, tilts it to one side. The final height of the structure is about 1 mm, and its white color makes it conspicuous against the green background. During secretion of the froth, which lasts 10 to 50 s, the female faces the scene from a distance of usually less than 10 mm, mostly 1 to 2 mm. Her occasional attempts to approach the froth are stopped by enanting activity by the male. When secretion is completed, the male backs away a short distance. If the female is facing him, she immediately approaches, extrudes her proboscis, and feeds on the froth. While she is feeding, he mounts her. If the female is not facing the male, or stands farther away, the male circles her and, while enanting, he orients her toward the froth. In the latter case, either the female begins feeding, or the male first adds more froth on top of the pillar. If the female does not feed, the male may add froth a second or even a third time, until the female either begins to feed or decamps. A feeding female extends her aculeus, which is immediately grasped by the male's surstyli. The period from terminating secretion to establishing genital contact lasts only a few seconds. Copulation posture is generally similar to that of many other observed species, but with the aculeus greatly extruded, possibly its whole length. Froth feeding and copulation sometimes proceed uninterrupted, until the female has finished the froth or for several seconds thereafter. Such copulations last 3 to 5 min.

If the male dismounts from the female shortly before she has finished feeding, he usually stays close to her, or enanates nearby, but without trying to resume copulation. The female continues to feed on the froth until nothing is left (up to 1 more minute). In many instances the male dismounts while the female is still feeding, but behaves in a different manner: 2.5 to 4 min after onset of copulation the male dismounts backward, circles the female and what is left of the froth, and faces both, so that the froth is located between them. He then resumes secretion and reconstructs the froth, while the female, who has stopped feeding, continues to face him. The behavior of the couple during froth reconstruction is similar to that during the initial formation of the froth. This includes attempts at feeding by the female and preventative actions by the male. In one case, however, a female was observed feeding from one side of the froth head, while the male was busy reconstructing it at the other. After reconstruction the male mounts the female as described before, thus performing a "set" of sequential copulations. Such a "set" is composed of several alternating copulations and reconstructions. The male fully reconstructs the froth each time, or even enlarges it beyond its original dimensions, an action that requires 10 to 45 s. In one extreme case we observed five reconstructions in one "set" of copulations that lasted more than 24 min. Abandoned froth pillars and feeding on froth by flies other than the original female were occasionally observed.

Premating trophallaxis is the most widespread kind of mating trophallaxis in Tephritidae, and Freidberg (1981) compared the details of the behavior described above with those of four other species: *Stenopa vulnerata* (Loew), *Icterica seriata* (Loew), *Eutreta novaeboracensis* (Fitch) (as *E. sparsa* (Wiedemann)), and *Dirioxa pornia* (Walker), all exhibiting indirect, premating trophallaxis. However, additional cases of similar behaviors have since been reported (e.g., Jenkins 1990) or observed (in *Schistopterum* sp. and *Eutretosoma* sp.; Freidberg, unpublished data). It is important to stress that in all the species that exhibit indirect trophallaxis, copulation does not take place without mating trophallaxis.

Postmating trophallaxis has so far been reported for only a few species of Tephritidae, of which *Spathulina sicula* was the subject of the most-detailed study (Figure 10*). This species is a moderately small (3 to 4 mm long), blackish fly, with a reticulate wing pattern and a nearly circum-Mediterranean distribution. It induces the formation of terminal stem galls on three species of *Phagnalon* (Asteraceae) (Persson 1976). *Spathulina sicula* exhibits direct, stomodeal, postmating trophallaxis. Its reproductive behavior was primarily studied in the laboratory, where single males and females were placed together in petri dishes. Males initiate courtship and are usually successful in achieving copulation within 1 to 30 min after introduction, although coupling occasionally begins 2 h after introduction, or does not occur at all. The average duration of 105 copulations was 3.06 ± 0.05 h (range: 1:07 to 5:16). The male occasionally enanates (scissors) shortly before dismounting the female, then releases the female's aculeus, dismounts her by stepping slowly backward, and stands still very close to the female.

The association between the sexes was seldom immediately broken at dismounting. In more than 90% of the observed copulations this was only the beginning of postmating behavior. Sometimes one or both partners may engage in self-cleaning and grooming for several minutes, and they may also walk around. However, usually within a few seconds after the male dismounts, the female will turn and face him. Her proboscis is extended and moves as if the labella are "searching" for those of the male. Within a few seconds she finds his proboscis, and when matched in a "kissing" posture (Figure 10*), a milk-white fluid appears between the labella of the two flies. Body position during this stage is oblique to the substrate, the male somewhat more erect, so that the female's head is a little lower than the male's. During the "kiss" the female's proboscis is more active than that of the male, while strong constrictions can be observed in the male's abdomen. White fluid continuously appears between the labella, and when the behavior is observed in correct illumination under the stereoscopic microscope, this fluid seems to enter the food canal of the female's proboscis. The "kiss" may last several minutes (5 min 19 s ± 0 min 23 s, $n = 51$; range: 1 min 10 s to 13 min 36 s), and is often interrupted by one or several short intervals, during which the labella of the partners lose contact. The male seems to break the "kiss" more frequently, and males were often observed trying to withdraw while their labella were still attached to those of the female, in some cases by pushing at the female's head with their forelegs. Sometimes, they eventually succeeded in releasing themselves, but often they seemed to give up the struggle and continued "kissing." Initiation of grooming activity, particularly of the genitalia, may occur even before the labella of the partners finally detach.

Most other known cases of mating trophallaxis in tephritidae are generally rather similar to the two cases described above. There are, however, two notable exceptions. The first is the case of *N. viridescens* (Goeden and Headrick 1992), which is the only tephritid to exhibit proctodeal (in addition to stomodeal) trophallaxis. The second exceptional case is that of *Metasphenisca negeviana* (Freidberg), the only species to exhibit both premating and postmating trophallaxis (Freidberg 1997).

* Figure 10 follows p. 204.

#### 28.2.3.4.4 Anatomy, Ultrastrucure, and Biochemistry

It has been suggested that the male's salivary glands are the source of the trophallactic substance in at least some Tephritidae, although Jenkins (1990) deduced that contents of the crop contributed to the nuptial gift of *Aciurina mexicana* (Aczél). Freidberg (1982) described the sexually dimorphic salivary glands of *S. sicula*. In this species the sacs of the male glands are much larger than those of the female and contain a milky substance, whereas those of the female are translucent. Freidberg (1978) also described the strikingly sexually dimorphic salivary glands in several species of Terelliini, most of which were observed to perform postmating trophallaxis. However, a somewhat similar sexual dimorphism also occurs in *Anastrepha suspensa* (Nation 1974), a species that does not practice mating trophallaxis and in which the large male glands are associated with pheromone production. Pritchard (1967) showed different staining reaction in male and female salivary glands of *Dirioxa pornia*, and Freidberg (1982) gave circumstantial evidence that the milky substance produced and stored in the male salivary glands of *S. sicula* is transferred and ingested by the female during the "kiss." Freidberg (1978) showed that this "milk" is an emulsion containing tiny round particles (diameter about 1.5 μm), each with a concentric design. Preliminary biochemical tests of the "milk" detected high absorption at 280 to 290 μm. The molecular weight of most of the material was less than 10,000 D. Tests for amino acids and sugars were indecisive (Freidberg 1978), and additional tests, using more-modern techniques, should be employed to reevaluate these findings.

#### 28.2.3.4.5 Experiments

Experiments designed to reveal the potential benefits that might accrue to females receiving trophallactic substances should be based primarily on artificial prevention of trophallaxis. However, it may be difficult to draw conclusions from such experiments, especially in cases of premating trophallaxis, in which prevention of trophallaxis results in no copulation.

In contrast, experimentation on postmating trophallaxis is possible, and *S. sicula* provides a suitable model for such experiments. Freidberg (1982) used this species for testing the effect of trophallaxis on subsequent female sexual receptivity, longevity, and fecundity, and on female fertility and progeny success. The basic design of the experiments consisted of a comparison of a test group of females that copulated, but were separated from their mates before "kissing," with a control group of mated females that were allowed to "kiss." The females were not given food prior to the experiment.

After nine days neither "kissed" nor "unkissed" females mated a second time, so that any inhibitive factor was transferred/communicated during copulation, and not through trophallaxis. Differences in longevity and fecundity between the test and control group were not statistically significant. The effect of trophallaxis on fertility was only studied qualitatively, and offspring were recovered from developing galls induced by "unkissed" females.

#### 28.2.3.4.6 Evolutionary Implications and Phylogeny

Premating trophallaxis is a prerequisite for copulation, whereas postmating trophallaxis is not. Premating trophallaxis could be the result of sexual selection (see Darwin 1871); that is, males compete through their ability to produce and guard nuptial gifts (such as a mound of froth), and females use these behaviors and substances to choose mates from among their suitors. It is possible that these nuptial gifts are either valuable resources or chemical displays whose content is informational rather than nutritional. However, perhaps postmating trophallactic substances would be more likely to consist solely of male resources that enhance the fecundity or longevity of his mate (despite the present lack of experimental evidence of this male investment). After all, females have already copulated, and the opportunity to choose a mate has passed — or has it? An alternative view has been suggested by Eberhard (1994), who argued that courtship, including nuptial giving, may persist during, or even following, copulation. This entails "cryptic" (in the sense of internal, difficult to observe) female choice. If so, then it is possible that postmating trophallaxis somehow acts to influence females to employ the sperm of a particular male in the fertilization of her eggs. However, the absence of multiple matings in female *S. sicula*, the best-studied case of postcopulatory trophallaxis, makes it unlikely that there is an opportunity to perform cryptic mate choice in at least this instance.

All tephritid species practicing mating trophallaxis are listed in Table 28.1. The resulting inventory is obviously insufficient for cladistic analysis. Nevertheless, we would like to highlight two points: (1) mating trophallaxis has been definitely reported in the most primitive (Acanthonevrini) as well as the most derived (Tephritini) tribes of Tephritidae; (2) most cases have been reported in the Tephritinae, the most-derived subfamily, and these instances occur in no fewer than eight tribes. This may mean that the phenomenon is much more widespread than previously thought, and it would appear to have evolved independently many times within the family.

#### 28.2.3.5 Copulation Duration — Sperm Competition and Female Choice

The genus *Anastrepha* is one of the best-studied groups with respect to copulation duration. As described by Aluja et al. (Chapter 15), mean mating times vary from 24.3 ± 1.5 min in *A. bistrigata* to 350 ± 60 min in *A. hamata*. It is noteworthy that three of the species with the longest mating times are also among the larger species.

Why is there so much variability among species? Couplings that extend beyond what is required for sperm transfer in related species are often interpreted in terms of sperm competition avoidance (Parker 1970), protection from predators (Sivinski 1981), or "cryptic female choice," that is, courtship that continues through mating and influences the female to retain or utilize the sperm of the signaling male preferentially (Eberhard 1996; Belford and Jenkins 1998.) There are also instances that are seemingly inconsistent with mate guarding to avoid sperm competition. For example, *A. leptozona* Hendel form relatively long unions (>6 h), but these continue into the night, beyond the sexual signaling period, and presumably when no other males would be searching for mates. Couplings by *Euarestoides acutangulus* are even longer and those initiated at midday extend into the following afternoon (Headrick and Goeden 1994). This does not discount the possibility that males use this time to transfer materials that might induce refractory periods and thereby protect their ejaculates. However, in some species (e.g., *A. suspensa*), females appear to have considerable control over mating durations; because males have a difficult time maintaining their position when females become restless and move about. Males respond by producing what appear to be brief repetitions of "precopulatory song," an important acoustic courtship signal (see Aluja et al., Chapter 15). If long copulations are performed with the compliance of the female, it may be more likely that the male is either protecting its mate (and the mother of their offspring) from predators, or is continuing to provide the female with information she will use to make reproductive decisions (see Eberhard, Chapter 18; Belford and Jenkins 1998).

Materials other than refractory-period inducers, such as nutrients and defensive compounds, may also be transferred by males in their ejaculates, and the mechanics of these transfers might also influence copulation durations (e.g., Gwynne 1983). In several Diptera species males provide resources that are incorporated into female somatic tissue and developing ovaries (e.g., Markow and Ankney 1984). In *Drosophila pseudoobscura* Fralova multiple-mated females have greater fertility, suggesting that they have acquired multiple male "investments" (Turner and Anderson 1983). Radioactively labeled substances in the ejaculate of *A. suspensa* were later recovered in the unfertilized eggs and tissues of mated females (Sivinski and Smittle 1987). However, the amounts of these substances appear to be small, and are perhaps inconsequential as male investments.

## 28.3 THE PHYLETIC DISTRIBUTION OF SEXUAL BEHAVIORS

Information is presented in an annotated table (Table 28.2).

**TABLE 28.2**
**Phyletic Distribution of Sexual Behaviors**

| | Behavior | | | | | | | |
|---|---|---|---|---|---|---|---|---|
| | Life-History Strategy | | Wing Displays | | | | | |
| Taxa | Aggregative | Circumnatal | Hamation | Enantion | Asynchronous Supination | Synchronous Supination | Lofting | Pleural Distention |
| **Subfamily Blepharoneurinae** | | | | | | | | |
| *Blepharoneura* | | | | | | | | |
| *atomaria* | Y | N | Y | Y | Y | Y | Y | U |
| *manchesteri* | Y | N | Y | Y | Y | Y | U | U |
| *perkinsi* | Y | N | Y | Y | Y | Y | U | U |
| **Subfamily Phytalmiinae** | | | | | | | | |
| **Tribe Phytalmiini** | | | | | | | | |
| *Phytalmia* | | | | | | | | |
| *alcicornis* | Y | N | N | Y | N | U | Y | Y |
| *cervicornis* | Y | N | N | U | U | U | Y | Y |
| *mouldsi* | Y | N | N | Y | N | Y | Y | Y |
| **Subfamily Dacinae** | | | | | | | | |
| **Tribe Ceratitidini** | | | | | | | | |
| *Ceratitis* | | | | | | | | |
| *capitata* | Y | N | U | Y | U | Y | U | Y |
| **Tribe Dacini** | | | | | | | | |
| *Bactrocera* | | | | | | | | |
| *dorsalis* | Y | N | U | U | U | U | U | U |
| *oleae* | Y | N | U | U | U | U | U | Y |
| *tryoni* | Y | N | U | U | U | U | U | Y |
| **Subfamily Trypetinae** | | | | | | | | |
| **Tribe Carpomyini** | | | | | | | | |
| *Rhagoletis* | | | | | | | | |
| *cressoni* | Y | N | N | Y | N | N | Y | Y |
| *indifferens* | Y | N | U | Y | U | Y | Y | U |
| *pomonella* | Y | N | U | Y | U | U | U | Y |
| **Tribe Toxotrypanini** | | | | | | | | |
| *Anastrepha* | | | | | | | | |
| *bistrigata* | Y | N | U | U | U | Y | U | Y |
| *fraterculus* | Y | N | U | U | U | Y | U | Y |
| *ludens* | Y | N | U | U | U | Y | U | Y |
| *obliqua* | Y | N | U | U | U | Y | U | Y |
| *pseudoparallela* | Y | N | U | U | U | Y | U | Y |
| *sororcula* | Y | N | U | U | U | Y | U | Y |
| *suspensa* | Y | N | U | U | Y | Y | U | Y |
| *striata* | Y | N | U | Y | U | Y | Y | Y |
| *Toxotrypana* | | | | | | | | |
| *curvicauda* | Y | N | U | Y | U | U | U | Y |
| **Tribe Trypetini** | | | | | | | | |
| *Euleia* | | | | | | | | |
| *fratria* | U | U | Y | Y | U | Y | U | U |

| Behavior | | | | | | | | | | |
|---|---|---|---|---|---|---|---|---|---|---|
| Mouthpart Extension | Labellar Wagging | Trophallaxis | Foreleg Extension | Midleg Abduction | Side Step | Male Stalking | Mate Guarding | Resource Guarding | Multiple Matings | Copulation Duration |
| | | | | | | | | | | |
| Y | U | U | Y | U | Y | U | U | U | U | 0.3–2 h |
| U | U | U | U | U | U | U | U | U | U | — |
| U | U | U | U | U | U | U | U | U | U | — |
| | | | | | | | | | | |
| N | N | N | U | N | Y | N | Y | Y | Y | 2 min |
| N | N | N | U | N | Y | N | Y | Y | Y | U |
| N | N | N | U | N | Y | N | Y | Y | Y | 2 min |
| | | | | | | | | | | |
| U | U | U | U | U | U | Y | U | Y[a] | Y | U |
| | | | | | | | | | | |
| U | U | U | U | U | U | U | U | U | U | 2–12 h |
| U | U | U | U | U | U | U | U | U | U | U |
| U | U | U | U | U | U | U | U | U | U | 0.5 h |
| | | | | | | | | | | |
| Y | N | U | Y | U | Y | Y | U | Y | Y | 10 min |
| Y | U | U | Y | U | U | U | U | Y | U | U |
| Y | U | U | U | U | U | Y | U | Y | Y | 0.5 h |
| | | | | | | | | | | |
| U | U | U | U | U | U | Y | U | U | U | 24 min |
| U | U | U | U | U | U | U | U | U | U | 1–3 h |
| Y | U | U | U | U | U | U | U | U | U | U |
| U | U | U | U | U | U | Y | U | U | U | 1 h |
| U | U | U | U | U | U | Y | U | U | U | 110 s |
| U | U | U | U | U | U | Y | U | U | U | 1 |
| Y | U | U | U | U | U | U | U | U | U | 0.5 h |
| Y | U | U | U | U | Y | U | U | U | U | 0.5 h |
| U | U | U | U | U | U | U | U | Y | U | 91 min |
| | | | | | | | | | | |
| U | U | U | U | U | U | U | U | U | Y | 4 h |

**TABLE 28.2 (continued)**
**Phyletic Distribution of Sexual Behaviors**

| | Behavior | | | | | | | |
|---|---|---|---|---|---|---|---|---|
| | Life-History Strategy | | Wing Displays | | | | | |
| Taxa | Aggregative | Circumnatal | Hamation | Enantion | Asynchronous Supination | Synchronous Supination | Lofting | Pleural Distention |
| **Subfamily Tephritinae** | | | | | | | | |
| **Tribe Acrotaeniini** | | | | | | | | |
| *Tomoplagia* | | | | | | | | |
| *cressoni* | Y | N | U | Y | Y | U | U | Y |
| **Tribe Cecidocharini** | | | | | | | | |
| *Procecidochares* | | | | | | | | |
| *anthracina* | N | Y | N | Y | N | Y | N | N |
| *flavipes* | Y | N | Y | Y | N | U | N | N |
| *kristinae* | N | Y | N | Y | N | U | N | N |
| *lisae* | N | Y | N | Y | N | U | N | N |
| *minuta* | N | Y | N | Y | N | U | N | N |
| sp. nr. *minuta* (NM) | Y | N | U | Y | U | U | N | U |
| *stonei* | N | Y | N | Y | Y | U | N | N |
| **Tribe Eurostini** | | | | | | | | |
| *Aciurina* | | | | | | | | |
| *ferruginea* | N | Y | Y | U | Y | Y | N | U |
| *mexicana* | Y | N | Y | U | Y | Y | N | Y |
| *thoracica* | Y | N | Y | N | Y | Y | N | Y |
| *trixa* | N | Y | Y | N | Y | Y | N | U |
| *Eurosta* | | | | | | | | |
| *comma* | Y | N | U | Y | U | U | U | U |
| *Valentibulla* | | | | | | | | |
| *californica* | N | Y | U | U | Y | Y | U | U |
| *dodsoni* | N | Y | N | U | Y | Y | U | U |
| **Tribe Eutretini** | | | | | | | | |
| *Eutreta* | | | | | | | | |
| *angusta* | N | Y | Y | N | Y | Y | U | U |
| *diana* | N | Y | Y | N | Y | Y | U | U |
| *Paracantha* | | | | | | | | |
| *cultaris* | Y | N | Y | N | Y | Y | N | Y |
| *gentilis* | Y | N | Y | N | Y | Y | N | Y |
| *Stenopa* | | | | | | | | |
| *vulnerata* | Y | N | U | U | Y | U | U | U |
| **Tribe Myopitini** | | | | | | | | |
| *Goedenia* | | | | | | | | |
| *formosa* | Y | N | Y | N | N | N | U | Y |
| *timberlakei* | Y | N | Y | N | N | N | U | Y |
| species | Y | N | Y | N | N | Y | U | Y |
| *Urophora* | | | | | | | | |
| *affinis* | Y | N | U | Y | U | Y | U | U |
| *quadrifasciata* | Y | N | U | Y | U | Y | U | U |

| Behavior | | | | | | | | | | |
|---|---|---|---|---|---|---|---|---|---|---|
| Mouthpart Extension | Labellar Wagging | Trophallaxis | Foreleg Extension | Midleg Abduction | Side Step | Male Stalking | Mate Guarding | Resource Guarding | Multiple Matings | Copulation Duration |
| U | U | U | U | U | U | Y | U | U | | U |
| Y | U | U | U | U | U | Y | U | U | U | 1 h |
| U | U | U | U | U | U | Y | U | U | Y | 5 h |
| U | U | U | U | U | U | Y | U | U | U | 0.5 h |
| U | U | U | U | U | U | Y | U | U | U | 0.5 h |
| U | U | U | U | U | Y | Y | U | U | U | 1 h |
| U | U | N | U | U | U | Y | N | N | U | U |
| U | U | U | U | U | U | Y | U | U | N | 2 h |
| U | U | U | Y | U | U | Y | U | U | U | 1+ h |
| Y | U | Y | U | U | Y | Y | U | U | U | U |
| U | U | U | U | U | Y | Y | U | U | U | 1 h |
| Y | N | N | Y | N | Y | Y | N | N | Y | 1–2 h |
| U | U | U | U | U | U | U | U | U | U | 0.5–1 h |
| U | U | U | U | U | Y | Y | U | U | Y | 1 |
| N | N | N | U | N | U | Y | N | N | Y | 80 min |
| U | U | U | U | U | Y | Y | U | U | U | 1 h |
| U | U | U | U | Y | Y | Y | U | U | U | 2 h |
| Y | N | Y | Y | N | Y | N | U | U | U | U |
| Y | N | Y | Y | N | Y | N | N | N | Y | 2.5–4 h |
| Y | U | Y | U | U | U | U | U | Y | U | U |
| U | U | U | U | U | U | Y | U | U | U | 1.5 h |
| U | U | U | U | U | U | Y | U | U | U | 1 h |
| Y | U | U | U | U | U | Y | U | U | U | U |
| U | U | U | U | U | U | U | U | U | U | U |
| U | U | U | U | U | U | U | U | U | U | U |

**TABLE 28.2 (continued)**
**Phyletic Distribution of Sexual Behaviors**

| | Behavior | | | | | | | |
|---|---|---|---|---|---|---|---|---|
| | Life-History Strategy | | Wing Displays | | | | | |
| Taxa | Aggregative | Circumnatal | Hamation | Enantion | Asynchronous Supination | Synchronous Supination | Lofting | Pleural Distention |
| **Tribe Noeetini** | | | | | | | | |
| *Xenochaeta* | | | | | | | | |
| *dichromata* | Y | N | U | U | Y | Y | U | U |
| **Tribe Schistopterini** | | | | | | | | |
| *Schistopterum* | | | | | | | | |
| *moebiusi* | Y | N | Y | U | U | U | U | U |
| **Tribe Tephritini** | | | | | | | | |
| ***Campiglossa* genus group** | | | | | | | | |
| *Campiglossa* | | | | | | | | |
| *genalis* | Y | N | Y | N | Y | Y | Y | Y |
| *murina* | Y | N | Y | N | Y | Y | Y | Y |
| *sabroskyi* | Y | N | U | N | Y | U | Y | Y |
| *steyskali* | Y | N | Y | N | Y | U | Y | Y |
| *variabilis* | Y | N | Y | N | Y | Y | Y | Y |
| *Dioxyna* | | | | | | | | |
| *picciola* | Y | N | Y | N | Y | U | Y | Y |
| ***Euarestoides* genus group** | | | | | | | | |
| *Euarestoides* | | | | | | | | |
| *acutangulus* | Y | N | Y | Y | Y | U | U | Y |
| *flavus* | Y | N | Y | U | Y | U | U | U |
| ***Tephritis* genus group** | | | | | | | | |
| *Tephritis* | | | | | | | | |
| *araneosa* | Y | N | Y | Y | Y | Y | N | Y |
| *arizonaensis* | Y | N | Y | Y | Y | U | U | Y |
| *dilacerata* | Y | N | U | U | Y | U | U | U |
| *stigmatica* | Y | Y | Y | Y | Y | Y | Y | Y |
| *Neotephritis* | | | | | | | | |
| *finalis* | Y | N | Y | N | Y | Y | U | Y |
| *Trupanea* | | | | | | | | |
| *actinobola* | Y | N | Y | N | Y | Y | N | Y |
| *arizonaensis* | Y | N | Y | N | Y | Y | N | Y |
| *californica* | Y | N | Y | N | Y | Y | N | Y |
| *conjuncta* | Y | N | Y | N | Y | U | U | U |
| *imperfecta* | Y | N | Y | N | Y | U | U | U |
| *jonesi* | Y | N | Y | N | Y | Y | U | Y |
| *nigricornis* | Y | N | Y | N | Y | Y | U | Y |
| *radifera* | Y | N | Y | N | Y | Y | U | Y |
| *signata* | Y | N | Y | N | Y | Y | U | Y |
| *wheeleri* | Y | N | Y | N | Y | Y | U | Y |
| **Tephritini *Incertae Sedis*** | | | | | | | | |
| *Euaresta* | | | | | | | | |
| *bella* | Y | N | U | U | Y | Y | U | U |
| *festiva* | Y | N | U | U | Y | Y | U | U |
| *stigmatica* | Y | N | Y | Y | Y | Y | Y | Y |

| Behavior | | | | | | | | | | |
|---|---|---|---|---|---|---|---|---|---|---|
| Mouthpart Extension | Labellar Wagging | Trophallaxis | Foreleg Extension | Midleg Abduction | Side Step | Male Stalking | Mate Guarding | Resource Guarding | Multiple Matings | Copulation Duration |
| U | U | U | U | U | Y | Y | U | U | Y | 1.5 h |
| Y | U | Y | Y | U | Y | U | U | Y | U | U |
| Y | Y | U | Y | U | Y | N | Y | U | Y | 4 h |
| Y | Y | U | Y | U | U | N | U | U | U | 2 h |
| Y | Y | U | U | U | U | N | U | U | U | 5 h |
| Y | Y | U | Y | U | U | N | U | U | U | 5 h |
| Y | Y | U | Y | U | U | N | U | U | U | 4.6 h |
| Y | Y | U | Y | U | U | N | Y | U | Y | 1 (36) h |
| U | U | U | U | U | U | Y | Y | Y | Y | 24 h |
| U | U | U | U | U | Y | Y | U | U | U | 2 h |
| U | U | U | Y | U | U | Y | U | U | | |
| U | U | U | Y | U | U | Y | U | U | U | 2.5–8 h |
| U | U | U | U | U | U | Y | U | U | Y | >1 h |
| Y | U | U | Y | U | U | Y | U | U | U | 2–12.5 h |
| U | U | U | U | U | U | Y | U | U | U | 4 h |
| U | U | U | Y | U | U | Y | U | U | U | 0.08 h |
| U | U | U | U | U | U | Y | U | U | U | U |
| U | U | U | Y | U | U | Y | U | U | U | U |
| U | U | U | U | U | U | U | U | U | U | U |
| U | U | U | U | U | U | U | U | U | U | U |
| U | U | U | Y | U | U | Y | U | U | U | 0.17 h |
| U | U | U | Y | U | U | Y | U | U | U | 0.1 h |
| U | U | U | U | U | U | Y | U | U | U | U |
| U | U | U | U | U | U | Y | U | U | U | U |
| U | U | U | U | U | U | Y | U | U | U | 0.08 h |
| Y | U | U | U | U | U | U | U | U | U | 0.3–1 h |
| Y | U | U | U | U | U | U | U | U | U | 0.3–1 h |
| Y | Y | N | Y | Y | Y | N | N | U | U | 1 h |

**TABLE 28.2 (continued)**
**Phyletic Distribution of Sexual Behaviors**

| | Behavior | | | | | | | |
|---|---|---|---|---|---|---|---|---|
| | Life-History Strategy | | Wing Displays | | Asynchronous | Synchronous | | Pleural |
| Taxa | Aggregative | Circumnatal | Hamation | Enantion | Supination | Supination | Lofting | Distention |
| **Tribe Terelliini** | | | | | | | | |
| *Chaetostomella* | | | | | | | | |
| *undosa* | Y | N | U | Y | U | U | U | U |
| *Neaspilota* | | | | | | | | |
| *achilleae* | Y | N | Y | N | N | N | N | Y |
| *callistigma* | Y | N | Y | N | N | N | N | Y |
| *viridescens* | Y | N | Y | N | N | N | N | Y |
| **Tribe Xyphosiini** | | | | | | | | |
| *Icterica* | | | | | | | | |
| *circinata* | Y | N | U | U | Y | Y | U | U |
| *seriata* | Y | N | U | U | Y | Y | U | U |

*Key*: Y = confirmed; U = unobserved; N = does not occur.

## 28.4 PHYLOGENY AND BEHAVIOR

### 28.4.1 Sexual Selection and Speciation

In his book *Modes of Speciation*, White (1978) states that "the comparative study of speciation, in relation to the population structure and genetic architecture of living organisms, is assuming an increasing importance in evolutionary studies." In a volume titled *Mechanisms of Speciation*, Mayr (1982) contributed an article in which he stated, "Speciation … now appears as the key problem of evolution. It is remarkable how many problems of evolution cannot be fully understood until Speciation is understood."

Within the last decade there has been renewed interest in the process of speciation, as evidenced by two edited volumes on the topic (Otte and Endler 1989; Lambert and Spenser 1995). Otte and Endler (1989) in their preface state that the collection of papers "illustrates the inhomogeneity among diverse taxa in their patterns and processes of speciation" and "[w]e hope that this will encourage reassessment of both data and theory at all levels, and ultimately contribute to a new synthesis of evolutionary ideas." These thoughts by some of the leading researchers in speciation provide the primary thesis of this section, especially the hope that studies of speciation will lead to new ideas regarding evolutionary processes.

It is a generally accepted notion that the accumulation of genetic differences that result in reproductive isolation between daughter populations is the most important feature of the speciation process and, as such, has been the primary focus of attention in research on speciation. One school of thought is that isolation barriers arise as incidental by-products of natural selection during spatial isolation rather than as a direct result of selection for reproductive isolation (Muller 1942; Mayr 1963). Others believe that genetic barriers formed during allopatry are incomplete and that isolation is perfected following secondary contact of the daughter populations (Fisher 1930; Dobzhansky 1940). In the latter, it is suggested that some form of intrinsic barriers such as hybrid inferiority arise as a result of natural selection during allopatry and that selection acts against those parental genotypes that hybridize. Thus, hybridization actually strengthens interspecific isolation barriers, and premating barriers such as behavioral and ecological differences evolve as a response to natural selection against hybridization.

| Behavior | | | | | | | | | | |
|---|---|---|---|---|---|---|---|---|---|---|
| Mouthpart Extension | Labellar Wagging | Trophallaxis | Foreleg Extension | Midleg Abduction | Side Step | Male Stalking | Mate Guarding | Resource Guarding | Multiple Matings | Copulation Duration |
| Y | U | U | U | U | U | U | U | U | U | 2.5 h |
| U | U | U | U | U | Y | Y | U | U | U | 3 h |
| U | U | U | Y | U | Y | Y | U | U | U | U |
| Y | U | Y | Y | U | Y | Y | Y | Y | Y | 3 to 9 h |
| Y | U | U | U | U | U | U | U | U | U | U |
| U | U | Y | U | U | U | U | U | Y | Y | 1–3 h |

Based on the results of some of his earlier studies on the mating behavior of Hawaiian *Drosophila*, Kaneshiro (1989) concluded that "sexual selection may be a pivotal feature of the speciation process and may indeed play a prominent role in the origin of new species." For nearly a century, the role of sexual selection as an important factor in the speciation process has been largely ignored by evolutionary biologists. Even Darwin (1871), despite his strong convictions "of the power of sexual selection" noted that "sexual selection will also be dominated by natural selection." Only within the past two decades has there been renewed interest in investigating sexual selection and its influence on the "mutual adjustment of the sexes to what may be called the intraspecific sexual environments" (Carson 1978). Researchers began to focus on changes within the sexual environment as a major component of genetic adaptations during speciation (Lande 1981; 1982; Kirkpatrick 1982).

One of the classical theories of sexual selection is the notion that female choice and male character would coevolve very rapidly (i.e., "runaway selection") within an interbreeding population (Fisher 1930; O'Donald 1977; 1980). Lande (1981; 1982) developed polygenic models to confirm the runaway process of Fisher's original ideas. In all these models, it is assumed that two factors act to counterbalance the runaway process of sexual selection. On the one hand, female preference for a certain male trait acts to select for elaborate forms of that trait. On the other hand, natural selection acts to maintain the optimum male phenotype to survive in a particular environment. Thus, an essential component of the classical sexual selection model is the role of natural selection in checking the runaway process that results from the genetic correlation between male trait and female preference for that trait.

It is theorized that directional selection via female choice for males with an exaggerated secondary sexual character is counterbalanced by the forces of natural selection due to the reduced survivability of males with excessive adornments. Eventually, when the genetic variability for the upper limits of the exaggerated character is reduced to the point where selection can no longer produce males with structures detrimental to their survival, the optimum phenotype becomes fixed in the population. This is a paradox inherent in the runaway sexual selection model. The reduction in variability for a male trait means that there can no longer be selection for such traits. Does this mean that secondary sexual characters that appear to be used in some aspect of the mating system, whether intrasexual competition among males or in epigamic selection, are not under direct sexual selection?

To address the issue of reduced genetic variability inherent in the runaway model, Kaneshiro (1989) proposed an alternative sexual selection model. Based on mating studies of Hawaiian *Drosophila*, Kaneshiro (1976; 1980; 1983; 1987) suggested that there is a range of mating types segregating in the two sexes; that is, males that are highly successful in mating and females that are very choosy at one end of the distribution, with males that are not so successful and females that are not so choosy at the other end. Data from mate-choice experiments conducted in the laboratory, as well as observations of courtship encounters in the field, suggest that the most likely mating within an interbreeding population occurred between males with exceptional mating qualities and females that were nonchoosy. Observations of mating experiments in the laboratory indicate that in most cases, successful matings occurred very quickly, within a few seconds following initial encounter between a male and female. In many cases, although the male may perform courtship displays vigorously for a long period, the female will continue to reject the male's attempt to copulate. In the field, where numerous observations of courtship attempts have been observed (K. Kaneshiro and P. Conant, unpublished data), more than 90% result in the female rejecting and decamping from the male's territory. In the few cases where a courtship encounter resulted in successful copulation, the female appeared to accept the male after an extremely brief courtship display (i.e., within a few seconds). In all of the observations where courtship lasted for more than 15 s, the female inevitably decamped from the male's mating territory even if the male continued to court for several minutes. A possible explanation for these observations is that in cases where a female rejects the male even after a lengthy courtship display, either the female is very choosy or the male does not have the courtship ability to satisfy the requirements of most such females in the population. On the other hand, those cases that result in successful copulation after a very brief courtship display by the male would appear to be between males that are highly successful in mating ability and females that are not so choosy.

It was also hypothesized that there is a strong genetic correlation between male mating ability and female choosiness (Kaneshiro 1989). Some preliminary selection experiments for these two behavioral phenotypes in the two sexes provide some support for this hypothesis. Kaneshiro (1989) conducted selection experiments in a Hawaiian *Drosophila* species in which males assayed for high mating success were crossed with females assayed for high choosiness. Within a single generation of selection, the sons of such mating pairs displayed mating ability similar to their fathers and the daughters displayed choosiness similar to their mothers. Similarly, strains with the opposite phenotype, that is, poor male mating ability and nonchoosy females, could be selected with significant results even after a single generation of selection. The results of these experiments indicate that by selecting both behavioral phenotypes in the two sexes simultaneously, it was possible to obtain strains with males and females that resembled the mating behavioral qualities of their parents even within a single generation.

Thus, matings in the natural population between highly successful males and less choosy females and the strong genetic correlation between these behavioral phenotypes in the two sexes would maintain the entire range of mating types in both sexes. Consequently, with this model, in contrast to the runaway selection model, levels of genetic variability of any phenotypic trait involved in mating success would be maintained rather than reduced as predicted by the runaway model. Rather than natural selection acting to counterbalance the directional runaway selection as seen in the classical models of sexual selection, the model proposed by Kaneshiro views sexual selection acting by itself to maintain a balanced polymorphism of the mating system.

Kaneshiro (1989) further extended his ideas on sexual selection to its role in the speciation process. During periods when population size is small, such as might be expected when a subset of the parent population is isolated by some extrinsic (spatial) barrier, there is a strong selection for less choosy females in the population. Under these conditions, females that are choosy may never encounter males able to satisfy their courtship requirements. Within a few generations of small population size, there will be an increase in frequency of less choosy females in the population with a corresponding shift in gene frequencies toward the genotypes of these females. This is further accompanied by a destabilization of the coadapted genetic system resulting in the generation of novel genetic recombinants, some of which may be better adapted to the new habitat of the daughter population. Those recombinants that are closely linked or correlated with the genotypes of the less choosy females will be strongly selected and can spread quickly throughout the population. Thus, the dynamics of the sexual selection process permit the population to overcome any effects of such drastic reduction in population size and even to recover from the genetic disorganization that accompany such populational events. In rebuilding its coadapted genetic system as the population size increases, selection may result in a shift toward a new adaptive peak, which may include reproductive barriers that isolate the incipient population from the parental population. Thus, sexual selection is viewed as playing an extremely important role in the initial stages of species formation and providing a mechanism for generating novel genetic material with which the population can continue to respond to sexual as well as natural selection in completing the speciation process.

### 28.4.2 Mating Behavior and the Reconstruction of Phylogeny

Sexual behaviors in tephritids are many and diverse, and thus form a rich pool of candidate characters for phylogenetic analyses. However, their use will depend on the resolution of homology or homoplasy and the level of taxonomic analysis chosen. Pinto (1977), in his seminal work on meloid sexual behavior, stated that a rich diversity of behaviors does not translate into optimal taxonomic utility. For higher level classifications, the behaviors would have had to diverge early in the group's evolution and progress accordingly through time. Behaviors that achieve high levels of divergence in the group's recent history are thus restricted in phylogenetic value to lower-ranked taxa.

Sexual characters represent only one facet of an animal's behavior and may be on a par with any other generalized behavioral grouping such as oviposition, grooming, or feeding. Similarly, sexual behavior may also be on a par with other phenotypic expressions such as the products of behavior (nests, galls, spermatophores, and webs) or interspecific interactions (host finding and resource utilization) that have had previous phylogenetic utility (Wenzel 1992). The following will explore levels of taxonomic resolution and homology regarding tephritid sexual behavior.

#### 28.4.2.1 Homology and Homoplasy

Behavioral characters are rarely used in the development of phylogenetic reconstructions (Sanderson et al. 1993); however, based on a survey, Proctor (1996) argued convincingly that this rarity is due to behavioral characters not being readily available to the systematists rather than the pervasive notion that behavioral characters are too labial and homoplastic to be reliable. I am reminded of the words of William Sharp MacLeay (1829) in his paper first describing the Mediterranean fruit fly as an agricultural pest and providing its description. To paraphrase: naturalists are the historians of facts, some of which have obvious and immediate utility providing the discoverer with much reward, while others are more obscure. However, when the time comes to evaluate the "noblest branch of our science, the progression of natural affinities," we need all the data we can get — utilitarian or not.

Behaviors used as characters for phylogenetic reconstruction must conform to the criteria applicable to other types of characters used with phylogenetics. The first and foremost criterion is homology. Wenzel's (1992) review of behavioral homology stands as one of the best-organized and most enduring. He suggested that Remane's criteria are useful for postulating behavioral homology and can be translated into behavioral equivalents: morphological position equates to a behavioral sequence; special quality equates to a complex movement in a particular behavioral context; and linkage by intermediate forms is the same as it is for morphology (Wenzel 1992). Wenzel (1992) described many caveats and pitfalls in homologizing behaviors using Remane's criteria, one of which, in particular, is special quality. To use special quality for behaviors, the context in which the behavior occurs must be understood. A broadly defined attribute such as male courtship may be useful for analyzing higher-level taxa, whereas the individual components of male courtship may or may not be homologous depending on the taxonomic level under consideration. Headrick and Goeden (1994) described a situation that occurs in two distinctive species of tephritines, *Euaresta stigmatica* (Headrick et al. 1995) and *Paracantha gentilis* (Headrick and Goeden 1990). These two species exhibit some of the most complex courtship and mating behavior described for any species of animal, let alone tephritids. Many of the components of the male courtship display are remarkably similar. Are they homologous or homoplasious? If they are homologous, is it because they are borrowed from some other more basic behavior, such as wing displays being derived from flight movements? Even though some of the behavioral elements are different, they function in a similar manner. Thus, is the whole episode of courtship homologous based on function? The complexity of the courtship and mating has no clear adaptive value as neither of these two species cooccurs with any congeners or close relatives. Thus, reproductive isolation is not an adaptive consideration. Wenzel (1992) warned that traits based on function or adaptation should be avoided in phylogeny. Questions regarding homology arise for every behavior. Due to the innate complexity of many behaviors, including "motivation" and learning, systematists are required to develop postulates about homology that include many levels not usually encountered with morphological or molecular characters.

Further, establishing the polarity of behavioral characters is also difficult and compounded by a lack of understanding of the context in which an observed behavior occurs and uniformity of behavioral knowledge among variously allied taxa. There are many identical behaviors that occur in vastly different contexts. A particular wing display may be used by one tephritid species in a clearly defensive maneuver, and in another species as part of a courtship ritual. Again, are the movements (= behavior) homologous or homoplasious? Headrick and Goeden (1994) developed ideas regarding the polarity of larger behavioral groupings: wing displays, wing patterns; pheromone production; territorial displays; and courtship displays. Both pleisomorphic and apomorphic behaviors can be identified in most of these categories. Their rationales for polarity focused mainly on uniqueness in relation to an assumed primitive "root" behavior and the taxonomic ubiquity of the behavior, rather than comparisons with an outgroup. This first attempt is clearly that and requires further study.

On a positive note, Queiroz and Wimberger (1993) showed that behavioral characters are as "sound" as other characters, such as morphological and molecular, in developing phylogenies, or, in other words, they do not exhibit excessive homoplasy. Patterson et al. (1993) discovered that where morphological data and molecular data were available for the same taxa, neither had a greater resolving power over the other where there were highly branching topologies. In her review, Proctor (1996) noted that some authors had determined that behavioral characters produced more parsimonious trees than those generated by morphological data sets. Clearly, the most robust hypothesis regarding phylogeny is one where there is a high level of congruence among independent data sets.

Another point examined by Queiroz and Wimberger (1993) was that most behaviorally derived phylogenies occurred at the species or genus level, rather than for developing relations among higher taxa. Indeed, behaviors have had far more success in the area of taxonomic relationships at the opposite end of the scale — infraspecific taxa. Here behaviors are used to help distinguish between races, biotypes, subspecies, populations, etc. (Bush 1966; 1969; Gordh 1977). Thus, there is enough clarity in a behavioral repertoire to separate intraspecific groupings, but not enough to distinguish higher ranked taxa (Pinto 1977). Proctor (1996) suggested the reason for this pattern is that most behavioral studies focus on taxa that are easily observed and that most behavioral studies do not have taxonomic breadth and relationships as an objective due to logistical and/or time constraints. Headrick and Goeden (1994; and Chapter 25) set out to develop a broad behavioral database for tephritids for eventual use in a phylogenetic context. One of the main problems they encountered was how much behavioral description was needed and what the hierarchy was to be. A similar situation occurs with molecular data, in which the appropriate type of molecular data is matched with the level of taxonomic analysis. Slow changing molecular characters like rDNA work well for higher taxonomic levels; rapidly changing types like mtDNA work well for lower taxa

(Proctor 1996). We are still not sure what types of behavioral data are appropriate to which taxonomic levels in tephritids.

### 28.4.2.2 Level of Taxonomic Analysis

Behaviors are difficult to describe with accuracy and consistency. Consistency is the key for postulating homology (Headrick and Goeden 1994), and so is context (Wenzel 1992; Proctor 1996). The assumption arises, that once behaviors have been accurately described they can then be homologized. This assumption only scratches the surface of what it means to describe behaviors accurately and in what detail. Headrick and Goeden (1994) attempted to introduce a standardized terminology for tephritid behavioral descriptions with the hope of generating interest in describing behaviors in many different taxa and providing a language to facilitate comparisons — it is still a work in progress. Most behaviors occur as a continuum of movement, one into another, without a clear distinction. The wing displays of tephritids are a case in point. We can describe tephritid wing displays as "extensions," and under extensions we can have extending one wing at a time or together, and under extending one wing at a time we can have rotating the wing blade while extending the wing or keeping the blade parallel to the substrate. Now we have a hierarchy of behavior. The category of wing extension may be a useful binary character for some higher-level classification within the Tephritoidea, but is certainly far from sufficient to serve for any lower taxonomic evaluation. Farther down the descriptive line we run into problems of a different kind. Variations of a behavioral element do occur among individuals, leading to problems in determining what is a "root" behavior and what is individual embellishment by the performer of the behavior. Too much descriptive detail provides little or no resolving power for a behavioral character. Understanding *how* to describe tephritid behavior is a first step and will require examination of many more species. Headrick and Goeden (1994) have examined approximately 50 species, but they all occur in an evolutionarily advanced group. Examination of the behaviors of other groups, as is currently being done with the more ancestral genera *Blepharoneura* (see Condon and Norrbom, Chapter 7) and *Phytalmia* (Dodson, Chapter 8), is exciting as many behaviors observed in the higher tephritids also occur in these genera.

### 28.4.2.3 Conclusion

The use of behavioral data to develop new hypotheses of phylogenetic relationships, or to test existing hypotheses, is possible for tephritids. To achieve this we need an intersection, where well-known behavior, well know ecology, and at least a superficial understanding of phylogenetic relationships cross paths. Within our family this crossroads occurs for the genus *Rhagoletis* due to the work of Guy Bush, Ron Prokopy, Stewart Berlocher, Dan Papaj, Jim Smith, and their colleagues, and perhaps this is the best place to start. This genus can serve as a model system to help us determine how detailed our behavioral descriptions should be and at what taxonomic level they will provide the best results. There are many groups within the Tephritidae for which ecology, behavior, or morphology is relatively well known, and it is with these taxa that we can test the hypotheses built from studies on groups such as *Rhagoletis*.

## ACKNOWLEDGMENTS

We are grateful to James Lloyd, Denise Johanowiz, Kevina Vulinec, Naomi Paz, Ora Manheim, and Netta Dorchin for their comments on the manuscript. The writing of this chapter was made possible by the generous financial support of the International Organization for Biological Control of Noxious Animals and Plants (IOBC), the Campaña Nacional contra las Moscas de la Fruta (SAGAR-IICA, Mexico), and USDA-ICD-RSED.

## REFERENCES


Aldiss, J. 1983. Chemical Communication in British Social Wasps (Hymenoptera: Vespidae). Ph.D. dissertation, University of Southampton, Southampton. 252 pp.

Aldrich, J.R., J.E. Oliver, W.R. Lusby, J.P. Kochansky, and J.A. Lockwood. 1987. Pheromone strains of the cosmopolitan pest, *Nezara viridula*. *J. Exp. Zool.* 244: 171–175.

Aldrich, J.A., M.P. Hoffmann, J.P. Kochansky, W.R. Lusby, J.E. Eger, and J.A. Payne. 1991. Identification and attractiveness of a major pheromone component for nearctic *Euschistus* spp. stink bugs (Heteroptera: Pentatomidae). *Environ. Entomol.* 20: 477–483.

Alexander, R., D. Marshall, and J. Cooley. 1997. Evolutionary perspectives on insect mating. In *Mating Systems in Insects and Arachnids* (J.C. Crespi and B.J. Choe, eds.), pp. 4–41. Cambridge University Press, Cambridge. 387 pp.

Aluja, M. 1993. Unusual calling behavior of *Anastrepha robusta* (Diptera: Tephritidae) flies in nature. *Fla. Entomol.* 76: 391–395.

Aluja, M. and A. Birke. 1993. Habitat use by adults of *Anastrepha obliqua* (Diptera: Tephritidae) in a mixed mango and tropical plum orchard. *Ann. Entomol. Soc. Am.* 86: 799–812.

Aluja, M., M. Cabrera, and J. Hendrichs. 1983. Behavior and interactions between *Anastrepha ludens* (L.) and *A. obliqua* (M.) on a field caged mango tree. I. Lekking behavior and male territoriality. In *Fruit Flies of Economic Importance* (R. Cavalloro, ed.), pp. 122–133. A.A. Balkema, Rotterdam.

Aluja, M., I. Jácome, A. Birke, N. Lozada, and G. Quintero. 1993. Basic patterns of behavior in wild *Anastrepha striata* (Diptera: Tephritidae) flies under field-cage conditions. *Ann. Entomol. Soc. Am.* 86: 776–793.

Arakaki, N., H. Kuba, and H. Soemori. 1984. Mating behavior of the oriental fruit fly, *Dacus dorsalis* Hendel (Diptera: Tephritidae). *Appl. Entomol. Zool.* 19: 42–51.

Arita, L. and K. Kaneshiro. 1983. Pseudomale courtship behavior of the female Mediterranean fruit fly, *Ceratits capitata* (Wiedemann). *Proc. Hawaii. Entomol.* Soc. 24: 205–210.

Baker, R., R. Herbert, P.E. Howse, and O.T. Jones. 1980. Identification and synthesis of the major sex pheromone of the olive fly (*Dacus oleae*). *J. Chem. Soc. Chem. Commun.* 1: 52–53.

Barigozzi, C. 1982. *Mechanisms of Speciation.* Alan R. Liss, New York. 546 pp.

Bateman, A.J. 1948. Intra-sexual selection in Drosophila. *Heredity* 2: 349–368.

Batra, S.W.T. 1979. Reproductive behavior of *Euaresta bella* and *E. festiva* (Diptera: Tephritidae), potential agents for the biological control of adventive North American ragweeds (*Ambrosia* spp.) in Eurasia. *J. N.Y. Entomol. Soc.* 87: 118–125.

Belford, S.R. and M.D. Jenkins. 1998. Establishing cryptic female choice in animals. *Trends Ecol. Evol.* 13: 216–218.

Berg, C.O. and K. Valley. 1985. Nuptial feeding in *Sepedon* spp. (Diptera: Sciomyzidae). *Proc. Entomol. Soc. Wash.* 87: 622–633.

Biggs, J.D. 1972. Aggressive behavior in the adult apple maggot. *Can. Entomol.* 140: 349–353.

Blum, M.S. 1981. *Chemical Defenses of Arthropods.* Academic Press, New York. 562 pp.

Bonduriansky, R. 1995. A new Nearctic species of *Protopiophila* Duda (Diptera: Piophilidae) with notes on its behavior and comparison with *P. latipes* (Meigen). *Can. Entomol.* 127: 859–863.

Boyce, A.M. 1934. Bionomics of the walnut husk fly *Rhagoletis completa*. *Hilgardia* 8: 363–579.

Bradbury, J.W. 1981. The evolution of leks. In *Natural Selection and Social Behavior* (R.D. Alexander and D.W. Tinkle, eds.), pp. 138–169. Chiron Press, New York. 532 pp.

Brittain, W.H. and C.A. Good. 1917. The apple maggot in Nova Scotia. *Bull. Nova Scotia Dep. Agric.* 9: 1–70.

Brooks, F.E. 1921. Walnut husk fly. *U.S. Dep. Agric. Bull.* No. 992.

Brown, J.L. and G.H. Orians. 1970. Spacing patterns in mobile animals. *Annu. Rev. Ecol. Syst.* 1: 239–262.

Burk, T. 1981. Signaling and sex in acalyptrate flies. *Fla. Entomol.* 64: 30–43.

Burk, T. 1982. Evolutionary significance of predation on sexually signaling males. *Fla. Entomol.* 65: 90–104.

Burk, T. 1983. Behavioral ecology of mating in the Caribbean fruit fly, *Anastrepha suspensa* (Loew). *Fla. Entomol.* 66: 330–344.

Burk, T. and J.C. Webb. 1983. Effect of male size on calling propensity, song parameters, and mating success in the Caribbean fruit fly, *Anastrepha suspensa* (Loew) (Diptera: Tephritidae). *Ann Entomol. Soc. Am.* 76: 678–682.

Burkhardt, D. and I. de la Motte. 1988. Big 'antlers' are favoured: female choice in stalk-eyed flies (Diptera, Insecta), field collected harems and laboratory experiments. *J. Comp. Physiol. A Sensory Neural Behav. Physiol.* 162: 649–652.

Bush, G.L. 1966. The taxonomy, cytology, and evolution of the genus *Rhagoletis* in North America (Diptera: Tephritidae). *Bull. Mus. Comp. Zool.* 134: 431–562.

Bush, G.L. 1969. Mating behavior, host specificity, and the ecological significance of sibling species in frugivorous flies of the genus *Rhagoletis* (Diptera: Tephritidae). *Am. Nat.* 103: 669–672.

Cade, W. 1975. Acoustically orienting parasitoids: fly phonotaxis to cricket song. *Science* 190: 1312–1313.

Carson, H.L. 1978. Speciation and sexual selection in Hawaiian *Drosophila*. In *Ecological Genetics: The Interface* (P.F. Brussard, ed.), pp. 93–107. Springer-Verlag, New York. 247 pp.

Cavender, G.L. and R.D. Goeden. 1984. The life history of *Paracantha cultaris* (Coquillett) on wild sunflower, *Helianthus annus* L. ssp. *lenticularis* (Douglas) Cockerell, in southern California (Diptera: Tephritidae). *Pan-Pac. Entomol.* 60: 213–218.

Chuman, T., P.J. Landolt, R.R. Heath, and J.H. Tumlinson. 1987. Isolation, identification, and synthesis of male-produced sex pheromone of papaya fruit fly, *Toxotrypana curvicauda* Gerstaecker (Diptera: Tephritidae). *J. Chem. Ecol.* 13: 1979–1992.

Condon, M.A. and A.L. Norrbom. 1994. Three sympatric species of *Blepharoneura* (Diptera: Tephritidae) on a single species of host (*Gurania spinulosa*, Cucurbitaceae): new species and new taxonomic methods. *Syst. Entomol.* 19: 279–304.

Cott, H.B. 1940. *Adaptive Coloration in Animals.* Methuen and Co. Ltd., London. 508 pp.

Darwin, C. 1871. *The Descent of Man and Selection in Relation to Sex* (reprinted). Modern Library, New York. 1000 pp.

Dawkins, R. 1986. *The Blind Watchmaker.* W.W. Norton, New York. 332 pp.

Dobzhansky, T. 1940. Speciation as a stage in evolutionary divergence. *Am. Nat.* 74: 312–321.

Dodson, G.N. 1982. Mating and territoriality in wild *Anastrepha suspensa* (Diptera: Tephritidae) in field cages. *J. Ga. Entomol. Soc.* 17: 189–200.

Dodson, G.N. 1985. Lek mating system and large male aggressive advantage in a gall-forming tephritid fly (Diptera: Tephritidae). *Ethology* 72: 99–108.

Dodson, G.N. 1987a. The significance of sexual dimorphism in the mating systems of two species of tephritid flies, *Aciurina bigeloviae* and *Valentibulla dodsoni* (Diptera: Tephritidae). *Can. J. Zool.* 65: 194–198

Dodson, G.N. 1987b. Biological observations on *Aciurina trixa* and *Valentibulla dodsoni* (Diptera: Tephritidae) in New Mexico. *Ann. Entomol. Soc. Am.* 80: 494–500.

Dodson, G.N. 1997. Resource defense mating system in antlered flies, *Phytalmia* spp. (Diptera: Tephritidae). *Ann. Entomol. Soc. Am.* 90: 496–504.

Downes, J.A. 1969. The swarming and mating flight of Diptera. *Annu. Rev. Entomol.* 14: 271–298.

Eberhard, W.G. 1994. Evidence for widespread courtship during copulation in 131 species of insects and spiders, and implications for cryptic female choice. *Evolution* 48: 711–733.

Eberhard, W.G. 1996. *Female Control: Sexual Selection by Cryptic Female Choice.* Princeton University Press, Princeton. 504 pp.

Eberhard, W.G. 1997. Sexual selection by cryptic female choice in insects and arachnids. In *The Evolution of Mating Systems in Insects and Arachnids* (J.C. Choe and B.J. Crespi, eds.), pp. 32–57. Cambridge University Press, Cambridge. 387 pp.

Emlen, S.T. and LW. Oring. 1977. Ecology, sexual selection, and the evolution of mating systems. *Science* 197: 215–223.

Enderlein, G. 1920. Zur Kenntis tropischer Frucht-Bohrfliegen. *Zool. Jahrb. Abt. Syst. Geogr. Biol. Tiere* 43: 336–360.

Endler, J.A. and A.L. Basolo. 1998. Sensory ecology, receiver biases, and sexual selection. *Trends Ecol. Evol.* 13: 415–420.

Féron, M. 1962. *Le Comportement de reproduction chez la mouche méditerranéenne des fruits, Ceratitis capitata Wied (Dipt. Trypetidae): Comportement sexuel, comportement de ponte.* Université de Paris, Paris. 131 pp.

Fisher, R.A. 1930. *The Genetical Theory of Natural Selection.* Claredon Press, Oxford. 272 pp.

Fletcher, B.S. 1987. The biology of dacine fruit flies. *Annu. Rev. Entomol.* 32: 115–144.

Fletcher, B.S. and W. Kitching.1995. Chemistry of fruit flies. *Chem. Rev.* 95: 789–828.

Foote, B.A. 1967. Biology and immature stages of fruit flies: the genus *Icterica* (Diptera, Tephritidae). *Ann. Entomol. Soc. Am.* 60: 1295–1305.

Foote, B.A., F.L. Blanc, and A.L. Norrbom. 1993. *Handbook of the Fruit Flies (Diptera: Tephritidae) of America North of Mexico.* Comstock Publishing Associates, Ithaca. 571 pp.

Franke, W., G. Hindorf, and W. Reith. 1979. Alkyl-1,6-dioxaspiro[4,5]decanes. A new class of pheromones. *Naturwissenschaften* 66: 618–619.

Freidberg, A. 1978. Reproductive behaviour of Fruit Flies. Ph.D. dissertation, Tel Aviv University, Tel Aviv.

Freidberg, A. 1981. Mating behaviour of *Schistopterum moebiusi* Becker (Diptera: Tephritidae). *Isr. J. Entomol.* 15: 89–95.

Freidberg, A. 1982. Courtship and post-mating behaviour of the fleabane gall fly, *Spathulina tristis* (Diptera: Tephritidae). *Entomol. Gen.* 7: 273–285.

Freidberg, A. 1984. The mating behavior of *Asteia elegantula* with biological notes on some other Asteiidae (Diptera). *Entomol. Gen.* 9: 217–224.

Freidberg, A. 1991. A new species of *Ceratitis* (*Ceratitis*) (Diptera: Tephritidae), key to species of subgenera *Ceratitis* and *Pterandrus*, and record of *Pterandrus* fossil. *Bishop Mus. Occas. Pap.* 31: 166–173.

Freidberg, A. 1997. Mating trophallaxis in *Metasphenisca negeviana* (Freidberg) (Diptera: Tephritidae). *Isr. J. Entomol.* 31: 199–203.

Given, B.B. 1953. Evolutionary trends in the Thynninae (Hymenoptera: Tiphiidae) with special reference to feeding habits of Australian species. *Trans. R. Entomol. Soc. Lond.* 105: 1–10.

Goeden, R.D. and D.H. Headrick. 1992. Life history and description of immature stages of *Neaspilota viridescens* Quisenberry (Diptera: Tephritidae) on native *Asteraceae* in southern California. *Proc. Entomol. Soc. Wash.* 94: 59–77.

Goeden, R.D., J.A. Teerink, and D.H. Headrick. 1998. Life history and description of immature stages of *Trupanea jonesi* Curran (Diptera: Tephritidae) on native Asteraceae in Southern California. *Proc. Entomol. Soc. Wash.* 100: 126–140.

Gordh, G. 1977. Biosystematics of natural enemies. In *Biological Control by Augmentation of Natural Enemies* (R.L. Ridgway and S.B. Vinson, eds.), pp. 125–150. Plenum Press, New York.

Green, T. 1977. A man's obsession reveals the riches of a hidden world. *Smithsonian* 8: 80–86.

Greene, E., L.J. Orsack, and D.W. Whitman. 1987. A tephritid fly mimics the territorial displays of its jumping spider predators. *Science* 236: 310–312.

Gwynne, D.T. 1983. Male nutritional investment and the evaluation of sexual differences in Tettigoniidae and other Orthoptera. In *Orthoptera Mating Systems* (D. Gwynne and G. Morris, eds.), pp. 337–366. Westview Press, Boulder.

Hamilton, W.D. and M. Zuk. 1982. Heritable true fitness and bright birds: a role for parasites? *Science* 218: 384–387.

Haniotakis, G.E. 1974. Sexual attraction in the olive fruit fly, *Dacus oleae* (Gmelin). *Environ. Entomol.* 3: 82–86.

Hardy, D.E. 1973. The Fruit Flies (Tephritidae — Diptera) of Thailand and Bordering Countries. *Pac. Insects Monogr.* 31: 353 pp.

Headrick, D.H. and R.D. Goeden. 1990. Life history of *Paracantha gentilis* (Diptera: Tephritidae). *Ann. Entomol. Soc. Am.* 83: 776–785.

Headrick, D.H. and R.D. Goeden. 1994. Reproductive behavior of California fruit flies and the classification and evolution of Tephritidae (Diptera) mating systems. *Stud. Dipterol.* 1: 194–252.

Headrick, D.H. and R.D. Goeden. 1996. Issues concerning the eradication or establishment and biological control of the Mediterranean fruit fly, *Ceratitis capitata* (Wiedemann) (Diptera: Tephritidae), in California. *Biol. Control* 6: 412–421.

Headrick, D.H., R.D. Goeden, and J.A. Teerink. 1995. Life history and description of immature stages of *Euaresta stigmatica* (Diptera: Tephritidae) on *Ambrosia* spp. (Asteraceae) in southern California. *Ann. Entomol. Soc. Am.* 88: 58–71.

Heath, R.R. and P.J. Landolt. 1988. The isolation, identification, and synthesis of the alarm pheromone of *Vespula squamosa* (Drury) (Hymenoptera: Vespidae) and associated behavior. *Experientia* 44: 82–83.

Heath, R.R., P.J. Landolt, J.H. Tumlinson, D.L. Chambers, R.E. Murphy, R.E. Doolittle, B.D. Dueben, J. Sivinski, and C.O. Calkins. 1991. Analysis, synthesis, formulation, and field testing of three major components of male Mediterranean fruit fly pheromone. *J. Chem. Ecol.* 17: 1925–1940.

Heath, R.R., N. Epsky, B. Dueben, A. Guzman, and L.E. Andrade. 1994. Gamma radiation effects on production of four pheromonal components of male Mediterranean fruit fly (Diptera: Tephritidae). *J. Chem. Ecol.* 17: 1925–1940.

Hendrichs, J. and M.A. Hendrichs. 1990. Mediterranean fruit flies (Diptera: Tephritidae) in nature: location and diel pattern of feeding and other activities on fruiting and nonfruiting hosts and nonhosts. *Ann. Entomol. Soc. Am.* 83: 632–641.

Hendrichs, J. and J. Reyes. 1987. Reproductive behavior and post-mating female guarding in the monophagous multivoltine *Dacus longistylus* (Wied.) (Diptera: Tephritidae) in southern Egypt. In *Fruit Flies: Proceedings of the Second International Synposium, 16–21 September 1986, Colymbari, Crete, Greece* (A.P. Economopoulos, ed.), pp. 303–313. Elsevier Science Publishers, Amsterdam.

Hendrichs, J., Katsoyannos, D. Papaj, and R. Prokopy. 1991. Sex differences in movement between feeding sites and mating sites and tradeoffs between food consumption, mating success, and predatory evasion in Mediterranean fruit flies (Diptera: Tephritidae). *Oecologia* 86: 223–231.

Hendrichs, J., B.I. Katsoyannos, V. Wornoayporn, and M.A. Hendrichs. 1994. Odour-mediated foraging by yellowjacket wasps (Hymenoptera: Vespidae): predation on leks of pheromone-calling Mediterranean fruit fly males (Diptera: Tephritidae). *Oecologia* 99: 88–94.

Hendrichs, M.A. and J. Hendrichs. 1998. Perfumed to be killed: interception of Mediterranean fruit fly (Diptera: Tephritidae) sexual signaling by predatory foraging wasps (Hymenoptera: Vespidae). *Ann. Entomol. Soc. Am.* 91: 228–234.

Höglund, J. and R.V. Alatalo. 1995. *Leks.* Monographs in Behavior and Ecology Series. Princeton University Press, Princeton.

Iwahashi, O. and T. Majima. 1986. Lek formation and male–male competition in the melon fly, *Dacus cucurbitae* Coquillett (Diptera: Tephritidae). *Appl. Entomol. Zool.* 21: 70–75.

Jenkins. J. 1990. Mating behavior of *Aciurina mexicana* (Aczel)(Diptera: Tephritidae). *Proc. Entomol. Soc. Wash.* 92: 66–75.

Johnston, R.F. 1962. A review of courtship feeding in birds. *Bull. Kans. Ornithol. Soc.* 13: 25–32.

Kaneshiro, K.Y. 1976. Ethological isolation and phylogeny in the *planitiba* subgroup of Hawaiian *Drosophila. Evolution* 30: 740–745.

Kaneshiro, K.Y. 1980. Sexual selection, speciation, and the direction of evolution. *Evolution* 34: 437–444.

Kaneshiro, K.Y. 1983. Sexual selection, and direction of evolution in the biosystematics of Hawaiian Drosophilidae. *Annu. Rev. Entomol.* 28: 161–178.

Kaneshiro, K.Y. 1987. The dynamics of sexual selection and its pleiotropic effects. *Behav. Genetics* 17: 559–569.

Kaneshiro, K.Y. 1989. The dynamics of sexual selection and founder effects in species formation. In *Genetics, Speciation, and the Founder Principle* (L.V. Giddings, K.Y. Kaneshiro, and W.W. Anderson, eds.), pp. 279–296. Oxford University Press, Oxford. 373 pp.

Kaneshiro, K.Y. and A.T. Ohta. 1982. The flies fan out. *Nat. Hist.* 91: 54–58.

Kanmiya, K. 1988. Acoustic studies on the mechanism of sound production in the mating songs of the melon fly, *Dacus cucurbitae* Coquillett (Diptera: Tephritidae). *J. Ethol.* 6: 143–151.

Kaston, B.J. 1970. Comparative biology of American black widow spiders. *Trans. San Diego Soc. Nat. Hist.* 16: 34–82.

Keiser, I., R.M. Kabayashi, D.L. Chambers, and E.L. Schneider. 1973. Relation of sexual dimorphism in the wings, potential stridulation, and illumination to mating of Oriental fruit flies, melon flies, and Mediterranean fruit flies in Hawaii. *Ann. Entomol. Soc. Am.* 66: 937–941.

Kennedy, B.H. 1979. The effect of multilure on parasites of the European elm bark beetle, *Scolytus multistriatus. Bull. Entomol. Soc. Am.* 25: 116–118.

Kessel, E.L. 1955. The mating activities of balloon flies. *Syst. Zool.* 4: 97–104.

Kirkpatrick, M. 1982. Sexual selection and the evolution of female choice. *Evolution* 36: 1–2.

Lambert, D.M. and H.G. Spenser. 1995. *Speciation and the Recognition Concept.* Johns Hopkins University Press, Baltimore. 502 pp.

Lande, R. 1981. Models of speciation by selection on polyphyletic traits. *Proc. Natl. Acad. Sci. U.S.A.* 78: 3721–3725.

Lande, R. 1982. Rapid origin of sexual isolation and character divergence in a cline. *Evolution* 36: 213–223.

Landolt, P.J. 1984. Behavior of the papaya fruit fly *Toxotrypana curvicauda* Gerstaecker (Diptera: Tephritidae), in relation to its host plant, *Carica papaya* L. *Folia Entomol. Mex.* 61: 215–224.

Landolt, P.J. 1985. Papaya fruit fly eggs and larvae (Diptera: Tephritidae) in field-collected papaya fruit. *Fla. Entomol.* 68: 354–356.

Landolt, P.J. and J. Hendrichs. 1983. Reproductive behavior of the papaya fruit fly, *Toxotrypana curvicauda* Gerstaecker (Diptera: Tephritidae). *Ann. Entomol. Soc. Am.* 76: 413–417.

MacLeay, W.S. 1829. Notice of *Ceratitis citriperda*, an insect very destructive to orange. *Zool. J.* 4: 475–482.

Malavasi, A., J.S. Morgante, and R.J. Prokopy. 1983. Distribution and activities of *Anastrepha fraterculus* (Diptera: Tephritidae) flies on host and nonhost trees. *Ann. Entomol. Soc. Am.* 76: 286–292.

Markow, T.M. and P.F. Ankney. 1984. *Drosophila* males contribute to oogenesis in a multiple mating species. *Science* 224: 302–303.

Mather, M.H. and B.D: Roitberg. 1987. A sheep in wolf's clothing: tephritid flies mimic spider predators. *Science* 236: 308–310.

Maynard Smith, J. 1974. The theory of games and the evolution of animal conflict. *J. Theor. Biol.* 47: 209–222.

Maynard Smith, J. 1982. *Evolution and the Theory of Games.* Cambridge University Press, Cambridge.

Mayr, E. 1982. Processes of speciation in animals. In *Mechanisms of Speciation* (C. Barigozzi, ed.), pp. 1–19. A.R. Liss, New York. 546 pp.

Mayr, E. 1963. *Animal Species and Evolution.* Belknap Press of Harvard University, Cambridge. 797 pp.

McAlpine, D.K. 1979. Agonistic behavior in *Achias australis* (Diptera: Platystomatidae) and the significance of eye stalks. In *Sexual Selection and Reproductive Competition in Insects* (M.S. Blum and N.A. Blum, eds.), pp. 221–230. Academic Press, New York. 463 pp.

McClintock, W.J. and G.W. Uetz. 1996. Female choice and pre-existing bias: Visual cues during courtship in two *Schizocosa* wolf spiders (Araneae: Lycosidae). *Anim. Behav.* 52: 167–181.

Metcalf, R.L. 1990. Chemical Ecology of Dacinae fruit flies (Diptera: Tephritidae). *Ann. Entomol. Soc. Am.* 83: 1017–1030.

Moffett, M.W. 1997. Flies that fight. *Nat. Geogr.* 192: 68–77.

Monteith, L.G. 1972. Status of predators of the adult apple maggot, *Rhagoletis pomonella* in Ontario. *Can. Entomol.* 104: 257–262.

Morgante, J.S., D. Selivon, V.N. Solferini, S.R. Matioli. 1993. Evolutionary patterns in specialist and generalist species of *Anastrepha.* In *Fruit Flies: Biology and Management* (M. Aluja and P. Liedo, eds.), pp. 15–20. Springer-Verlag, New York.

Muller, H.J. 1942. Isolating mechanisms, evolution and temperature. *Biol. Symp.* 6: 71–125.

Munro, H.K. 1949. A remarkable new species of trypetid fly of the genus *Ceratitis* (sensu stricto) from east Africa in the collection of the United States National Museum. *Proc. U.S. Natl. Mus.* 99: 499–501.

Nation, J.L. 1974. The structure and development of two sex specific glands in the male Caribbean fruit flies. *Ann. Entomol. Soc. Am.* 67: 731–734.

Norrbom, A.L. and K.C. Kim. 1988. A List of the Reported Host Plants of the Species of *Anastrepha* (Diptera: Tephritidae). USDA-APHIS 81–52.

Novak, J.A. and B.A. Foote. 1975. Biology and immature stages of fruit flies: the genus *Stenopa* (Diptera, Tephritidae). *J. Kans. Entomol. Soc.* 48: 42–52.

O'Donald, P. 1977. Theoretical aspects of sexual selection. *Theor. Popul. Biol.* 12: 298–334.

O'Donald, P. 1980. *Genetic Models of Sexual Selection.* Cambridge University Press, Cambridge. 250p.

Oldroyd, H. 1964. *The Natural History of Flies.* Weidenfeld and Nicolson, London. 324 pp.

Otte, D. and J.A. Endler. 1989. *Speciation and Its Consequences.* Sinauer Associates Inc., Sunderland. 679 pp.

Papaj, D.R. 1994. Oviposition site guarding by male walnut flies and its possible consequences for mating success. *Behav. Ecol. Sociobiol.* 34: 187–195

Papaj, D.R., J. Hendrichs, and B.E. Katsovannos. 1989. Use of fruit wounds in oviposition by the Mediterranean fruit fly. *Entomol. Exp. Appl.* 53: 203–209

Parker, G.A. 1970. Sperm competition and its evolutionary consequences in the insects. *Cambridge Phil. Soc. Biol. Rev.* 45: 525–567.

Parker, G.A. 1974. Assessment strategy and the evolution of fighting behaviour. *J. Theor. Biol.* 47: 223–243.

Parker, G.A. 1978. Evolution of competitive mate searching [insects]. *Annu. Rev. Entomol.* 23: 173–196.

Parker, G.A. and E.A. Thompson. 1980. Dung fly struggles: a test of the war of attrition. *Behav. Ecol. Sociobiol.* 7: 37–44.

Persson, P.I. 1976. Description of third instar larval characters in *Spathulina trisitis* (Loew) from Crete (Diptera: Tephritidae). *Entomol. Scand.* 7: 307–308.

Piersol, W.H. 1907. The curious mating habit of the fly *Rivellia boscii. Am. Nat.* 41: 465–467.

Pinto, J.D. 1977. Comparative sexual behavior in blister beetles of the subtribe Eupomphina (Coleoptera: Meloidae), and an evaluation of its taxonomic significance. *Ann. Entomol. Soc. Am.* 70: 937–951.

Pritchard, G. 1967. Laboratory observations on the mating behaviour of the island fruit fly *Rioxa pornia* (Diptera: Tephritidae). *J. Aust. Entomol. Soc.* 6: 127–132.

Pritchard, G. 1969. The ecology of a natural population of Queensland fruit fly *Dacus tryoni* II. The distribution of eggs and its relation to behaviour. *Aust. J. Zool.* 17: 293–311.

Proctor, H.C. 1996. Behavioral characters and homoplasy: perception vs. practice. In *Homoplasy: The Recurrence of Similarity in Evolution* (M.J. Sanderson and L. Hufford, eds.), pp. 131–152. Academic Press, San Diego.

Prokopy, R.J. 1976. Feeding, mating, and oviposition activities of *Rhagoletis fausta* flies in nature. *Ann. Entomol. Soc. Am.* 69: 899–904.

Prokopy, R.J. 1977. Stimuli influencing trophic relations in Tephritidae. *Coll. Int. C.N.R.S.* 265: 305–336.

Prokopy, R.J. 1980. Mating behavior of frugivorous Tephritidae in nature. *Proc. Symp. Fruit Fly Problems, XVI International Cong. Entomol.*, Kyoto, pp. 37–46.

Prokopy, R.J. and G.L. Bush. 1973. Mating behavior of *Rhagoletis pomonella* (Diptera: Tephritidae) IV. Courtship. *Can. Entomol.* 105: 873–891.

Prokopy, R.J. and J.J. Duan. 1998. Socially facilitated egglaying behavior in Mediterranean fruit flies. *Behav. Ecol. Sociobiol.* 42: 117–122.

Quieroz, A. de and P.H. Wimberger. 1993. The usefulness of behavior for phylogeny estimation: Levels of homoplasy in behavioral and morphological characters. *Evolution* 47: 46–60.

Quilici, S. and A. Franck. 1997. Field cage studies on mating behavior of *Ceratitis* spp. (Diptera: Tephritidae) in Reunion Island. 3rd FAO/IAEA Research Coordination meeting on *Medfly Mating Behavior Studies Under Field Cage Conditions*. Tel Aviv, Israel. 11 pp.

Reichert, S.E. 1998. Game theory and animal contests. In *Game Theory and Animal Behavior* (L.A. Dugatkin and H.K. Reeve, eds.), pp. 64–93. Oxford University Press, New York. 320 pp.

Roeder, K.D., L. Tozian, and E.A. Weinst. 1960. Endogenous nerve activity and behavior in the mantis and cockroach. *J. Insect Physiol.* 4: 45–62.

Ryan, M.J. 1998. Sexual selection, receiver biases, and the evolution of sex differences. *Science* 281: 1999–2003.

Sanderson, M.J., B.G. Baldwin, G. Bharathan, C.S. Campbell, C. Dohlen, D. Ferguson, J.M. Porter, M.F. Wojciechowski, and M.J. Donoghue. 1993. The growth of phylogenetic information and the need for a phylogenetic database. *Syst. Biol.* 42: 562–568.

Scott, J.P. and E. Fredericson. 1951. The causes of fighting in mice and rats. *Physiol. Zool.* 24: 273–309.

Shelly, T.E. and T.S. Whittier. 1995. Lek distribution in the Mediterranean fruit fly: influence of tree size, foliage density and neighborhood. *Proc. Hawaii. Entomol. Soc.* 32: 113–121.

Sigurjónsdóttir, H. and G.A. Parker. 1981. Dung fly struggles: evidence for assessment strategy. *Behav. Ecol. Sociobiol.* 8: 219–230.

Sivinski, J. 1981. The effects of mating on predation in the stick insect *Diapheromera veliei. Ann. Entomol. Soc. Am.* 73: 553–556.

Sivinski, J. 1989. Lekking and the small-scale distribution of the sexes in the Caribbean fruit fly, *Anastrepha suspensa* (Loew). *J. Insect Behav.* 2: 3–13.

Sivinski, J. 1997. Ornaments in the Diptera. *Fla. Entomol.* 80: 142–164.

Sivinski, J. and T. Burk. 1989. Reproductive and mating behavior. In *Fruit Flies: Their Biology, Natural Enemies and Control* (A.S. Robinson and G. Hooper, eds.), pp. 343–351. In *World Crop Pests* (W. Helle, ed.), Vol. 3A. Elsevier Science Publishers, Amsterdam.

Sivinski, J. and G.N. Dodson. 1992. Sexual dimorphism in *Anastrepha suspensa* (Loew) and other tephritid flies: possible roles of developmental rate, fecundity, and dispersal. *J. Insect Behav.* 5: 491–506.

Sivinski, J.M. and E. Petersson. 1997. Mate choice and species isolation in swarming insects. In *The Evolution of Mating Systems in Insects and Arachnids* (J.C. Choe and B.J. Crespi, eds.), pp. 294–309. Cambridge University Press, Cambridge. 387 pp.

Sivinski, J. and B. Smittle. 1987. Male transfer of materials to mates in the Caribbean fruit fly, *Anastrepha suspensa* (Loew) (Diptera: Tephritidae). *Fla. Entomol.* 70: 233–238.

Sivinski, J. and J.C. Webb. 1985a. Sound production and reception in the caribfly, *Anastrepha suspensa* (Loew) (Diptera: Tephritidae). *Fla. Entomol.* 68: 273–278.

Sivinski, J. and J.C. Webb. 1985b. The form and function of acoustic courtship of the papaya fruit fly, *Toxotrypana curvicauda* Gerstacker (Tephritidae). *Fla. Entomol.* 68: 634–641.

Sivinski, J. and J.C. Webb. 1986. Changes in a Caribbean fruit fly acoustic signal with social situation (Diptera: Tephritidae). *Ann. Entomol. Soc. Am.* 79: 146–149.

Sivinski, J. and J.C. Webb. 1989. Comparisons of acoustic courtship signals in wild and laboratory reared Mediterranean fruit flies, *Ceratits capitata. Fla Entomol.* 72: 212–214.

Sivinski, J., T. Burk, and J.C. Webb. 1984. Acoustic courtship signals in the Caribbean fruit fly, *Anastrepha suspensa* (Loew). *Anim. Behav.* 32: 1011–1016.

Sivinski, J., N. Epsky, and R. Heath. 1994. Pheromone deposition on leaf territories by male Caribbean fruit flies, *Anastrepha suspensa* (Loew) (Diptera: Tephritidae). *J. Insect Behav.* 7: 43–51.

Smith, D.C. 1985a. General activity and reproductive behavior of *Rhagoletis cornivora* (Diptera: Tephritidae) flies in nature. *J. N.Y. Entomol. Soc.* 93: 1052–1056.

Smith, D.C. 1985b. General activity and reproductive behavior of *Rhagoletis tabellaria* (Diptera: Tephritidae) flies in nature. *J. Kans. Entomol. Soc.* 58: 737–739.

Smith, P.H. 1989. Behavioral partitioning of the day and circadian rhythmicity. In *Fruit Flies: Their Biology, Natural Enemies and Control* (A.S. Robinson and G. Hooper, eds.), pp. 325–341. In *World Crop Pests* (W. Helle, ed.), Vol. 3A. Elsevier Science Publishers, Amsterdam.

Steck, G.L. 1984. *Chaetostomella undosa* (Diptera: Tephritidae): biology, ecology, and larval description. *Ann. Entomol. Soc. Am.* 77: 669–678.

Sternlicht, M. 1973. Parasitic wasps attracted by the sex pheromone of their coccid host. *Entomophaga* 18: 339–342.

Stoltzfus, W.B. and B.A. Foote. 1965. The use of froth masses in courtship of *Eutreta* (Diptera: Tephritidae). *Proc. Entomol. Soc. Wash.* 67: 263–264.

Teal, P.E.A., J.H. Tumlinson, and R.R. Heath. 1986. Chemical and behavioral analyses of volatile sex pheromone components released by calling *Heliothis virescens* (F.) females (Lepidoptera: Noctuidae). *J. Chem. Ecol.* 12: 107–126.

Thornhill, R. 1976. Sexual selection and nuptial feeding behavior in *Bittacus apicalis* (Insecta: Mecoptera). *Am. Nat.* 110: 529–548.

Thornhill, R. 1992. Female preference for the pheromone of males with low fluctuating asymmetry in the Japanese scorpionfly (*Panorpa japonica*: Mecoptera). *Behav. Ecol.* 3: 277–283.

Thornhill, R. and J. Alcock. 1983. *The Evolution of Insect Mating Systems.* Harvard University Press, Cambridge. 547 pp.

Trivers, R.L. 1972. Parental investment and sexual selection. In *Sexual Selection and the Descent of Man* (B. Campbell, ed.), pp. 1871–1971. Aldine, Chicago.

Turner, M.E. and W.W. Anderson. 1983. Multiple mating and female fitness in *Drosophila pseudoobscura. Evolution* 37: 714–723.

Van der Valk, H. 1987. Spatial and Temporal Dynamics of Mediterranean Fruit Fly, *Ceratitis capitata* Wied. on a Host Tree in the Field. M.S. thesis, Wageningen Agricultural University, The Netherlands.

Vite, J.P. and D.L. Williamson. 1970. *Thanasimus dubius*: prey perception. *J. Insect Physiol.* 16: 233–239.

Walker, T.J. 1964. Experimental demonstration of a cat locating orthopteran prey by the prey's calling song. *Fla. Entomol.* 102: 204–208.

Wallace, A.F. 1869. *The Malay Archipelago.* Dover Press, New York.

Warburg, M.S. and B. Yuval. 1997. Circadian patterns of feeding and reproductive activities of Mediterranean fruit flies (Diptera: Tephritidae) on various hosts in Israel. *Ann. Entomol. Soc. Am.* 90: 487–495.

Webb, J.C., J. Sivinski, and C. Litzkow. 1984. Acoustical behavior and sexual success in the Caribbean fruit fly, *Anastrepha suspensa* (Loew) (Diptera: Tephritidae). *Environ. Entomol.* 13: 650–656.

Wedell, N. 1994. Variation in nuptial gift quality in bush crickets (Orthoptera: Tettigoniidae). *Behav. Ecol.* 5: 418–425.

Wenzel, J.W. 1992. Behavioral homology and phylogeny. *Annu. Rev. Ecol. Syst.* 23: 361–382.

Wheeler, W.M. 1924. Courtship of the Calobatas. *J. Hered.* 15: 485–495.

White, I.M. 1988. Tephritid Flies (Diptera: Tephritidae). *Handb. Identif. Br. Insects* 10(5a): 134 pp.

White, I.M. and M.M. Elson-Harris. 1992. *Fruit Flies of Economic Significance: Their Identification and Bionomics.* CAB. International, Wallingford. 601 pp.

White, M.J.D. 1978. *Modes of Speciation.* Freeman, San Francisco. 455 pp.

Whitman, D.W., L. Orsack, and E. Greene. 1988. Spider mimicry in fruit flies (Diptera: Tephritidae): further experiments on the deterrence of jumping spiders (Araneae: Salticidae) by *Zonosemata vittigera* (Coquillet). *Ann. Entomol. Soc. Am.* 81: 532–536.

Whittier, T.S., K. Kaneshiro, and L.D. Prescott. 1992. Mating behavior of Mediterranean fruit flies (Diptera: Tephritidae) in a natural environment. *Ann. Entomol. Soc. Am.* 85: 214–218.

Wilkinson. G.S. and G. Dodson. 1997. Function and evolution of antlers and eye stalks in flies. In *The Evolution of Mating Systems in Insects and Arachnids* (J.C. Choe and B.J. Crespi, eds), pp. 310–328. Cambridge University Press, Cambridge. 387 pp.

Wilson, O.E. 1975. *Sociobiology.* Belknap, Cambridge.

Zahavi, A. and A. Zahavi. 1997. *The Handicap Principle.* Oxford University Press, Oxford. 286 pp.

# 29 Sexual Pheromones of Tephritid Flies: Clues to Unravel Phylogeny and Behavior

*Robert R. Heath, Peter J. Landolt, David C. Robacker, Barbara D. Dueben, and Nancy D. Epsky*

**CONTENTS**



## 29.1 INTRODUCTION

Knowledge of pheromones used within a group of insect species may provide insight into the phylogenies or patterns of relatedness of those different species. Attempts have been made with other insect groups to use the structures of sex pheromones to confirm or elucidate patterns of relatedness. For example, Roelofs and Brown (1982) compared female sex pheromones of a number of species of Tortricidae moths. He used pheromonal components as indicators of primitive and advanced character states, and to delineate relatedness. Chemicals used by Noctuidae moths were evaluated as taxonomic characters indicating relatedness at the subfamily and family levels by Renou et al. (1988). Often sex pheromones are highly, although not purely, species specific in their chemical structures and blend compositions, making readily identifiable characters. Also, the patterns of use of sex pheromones by fruit flies may provide useful behavioral characters for study. Examples might be the temporal patterns of pheromone release or the nature of calling sites by flies that are releasing pheromone.

Obvious limitations to any attempts to conduct such a study with tephritid fruit flies are the comparatively small number of species for which putative pheromone chemicals have been isolated and identified and the much smaller subset of species for which sex pheromones are chemically known and demonstrated behaviorally. If some or many of these putative pheromone chemicals are not in sex attraction or courtship interactions, their usefulness as characters for consideration of phylogenies would be compromised. However, speculation on how pheromone chemistry and related behavior varies within and among taxa may still provide interesting ideas on tephritid evolution and should stimulate more research on fruit fly sex pheromones.

In this chapter, we explore the possible sex pheromone systems of several fruit fly species, including chemicals involved with sex attraction or courtship signaling. We discuss chemical structures and how environmental and artificial factors may affect the release of these chemicals by fruit flies. The discussion will focus on the pheromonal systems of *Ceratitis capitata* (Wiedemann), *Anastrepha ludens* (Loew), *A. suspensa* (Loew), *A. obliqua* (Macquart), *A. fraterculus* (Wiedemann), *A. serpentina* (Wiedemann), *A. striata* Schiner, and *Bactrocera oleae* (Rossi).

## 29.2 MEDITERRANEAN FRUIT FLY, *CERATITIS CAPITATA*

### 29.2.1 Chemical Identification

The first report of a putative *C. capitata* pheromone was published by Jacobson et al. (1973). Subsequent research did not support their findings and the report was probably incorrect because of the limited analytical methodology available at the time. Baker et al. (1985) identified nine volatile compounds emitted and/or extracted from sexually mature male *C. capitata*. The identified components included ethyl-(*E*)-3-octenoate; geranyl acetate; (3*E*,6*E*)-1,3,6,10-dodecatetraene (*E*,*E*-α-farnesene); 3,4-dihydro-2H-pyrole (delta$^1$-pyrroline); *E*-2-hexenoic acid; dihydro-3-methylfuran-2(3,H)-one; 2-ethyl-3,5-dimethylpyrazine; linalool; and ethyl acetate. It was claimed that the delta$^1$-pyrroline was active but no data were reported to support pheromonal activity of this or any of the other compounds. The amounts and ratio of the nine components were not reported. Jang et al. (1989) detected 69 compounds using gas chromatographic analysis of collections of headspace from sexually mature male laboratory-reared *C. capitata* and identified 56 of these compounds. Six of these compounds, i.e., ethyl-(*E*)-3-octenoate; geranyl acetate; *E*,*E*-α-farnesene; delta$^1$-pyrroline; linalool; and ethyl acetate, showed significant activity when tested individually and as a blend compared with blanks. Baker et al. (1990) in tests with sterile *C. capitata* tested racemic linalool; 2,3-dimethylpyrazine; 2,5-dimethylpyrazine, and geranyl acetate individually and in combination, and found that yellow delta traps and yellow square traps baited with large amounts of the compounds singularly and in combination were more attractive to released female *C. capitata* than traps without chemicals.

It is apparent that a considerable number of chemicals produced by male *C. capitata* may function as pheromonal components. A principal problem in interpreting the results of early research was the lack of quantitative data in analysis of male-produced volatiles and the lack of control over chemical release rate and ratios in tests of compounds for pheromonal activity. Heath et al. (1991) found that ethyl-(*E*)-3-octenoate, geranyl acetate, and *E*,*E*-α-farnesene were the three major components released by male *C. capitata*, and they found differences in component ratios of volatiles emitted by wild males vs. laboratory-reared males. A formulation method was developed that released these three compounds in the same ratio as that released by wild males (Heath et al. 1991). Field tests with a feral population in Guatemala demonstrated that formulated blends that released 1.6 and 6.4 male per hour equivalent (MHe) were more attractive than blends that released 0, 0.3, and 3.2 MHe. Thus, *C. capitata* females were highly responsive to dose, which demonstrated that a precise control of release rate of pheromone is needed for optimal performance. Optimal response to two concentrations of pheromone, with lesser capture at an intermediate concentration, observed in these tests with *C. capitata* was similar to that observed for the papaya fruit fly, *Toxotrypana curvicauda* Gerstaecker, attraction to its pheromone 2-methyl-6-vinylpyrazine (Landolt and Heath 1988; 1990). The apparently bimodal nature of the numbers of papaya fruit fly females captured was likely due to responses of mated females vs. virgin females to the pheromone-baited traps (Landolt and Heath 1990). Female *C. capitata* captured in our tests were not preserved and dissected to determine their mating status. However, subsequent research (R.R. Heath et al., unpublished data) suggested that the bimodal response was due to preferential response of mated and unmated female flies to different release rates. In laboratory bioassay, female response to the three component synthetic pheromone blend was similar to response to natural male odor (Landolt et al. 1992).

Originally, research on the biological activity of the delta$^1$-pyrroline was hampered by lack of information regarding its equilibrium and stability properties and also by the lack of a quantitative method for analysis. Research conducted in 1992 resolved these problems and an analytical method was developed for delta$^1$-pyrroline (Baker et al. 1992). A typical reconstructed chromatogram obtained from analysis of the *C. capitata* male-produced pheromone is shown in Figure 29.1. In experiments using a flight tunnel bioassay system, it was shown that the addition of the delta$^1$-pyrroline to the three component synthetic pheromone blend resulted in a significantly improved lure for female *C. capitata* (Heath and Epsky 1993). In laboratory tests of both feral and factory-reared female *C. capitata*, 90% of the tested females were attracted to five caged males and 10% to the previously field-tested three-component blend containing ethyl-(*E*)-3-octenoate, geranyl acetate, and E,*E*-α-farnesene. The addition of delta$^1$-pyrroline resulted in approximately 30% response to the synthetic four-component lure. In field tests, however, few females were captured in response to either the three-component or the four-component blend or to live caged males (R.R. Heath et al., unpublished data). Jang et al. (1994) tested a five-component blend in which ethyl acetate was added to the four components listed above. In laboratory tests, similar numbers of females responded to the synthetic blend or to the live males tested against clean air blanks (~19 and ~24, respectively); however, only one-third responded to the synthetic blend when tested against the live males (~ 6 and ~17, respectively).

### 29.2.2 Effect of Irradiation on Pheromone Production

Studies were conducted to determine if gamma radiation affected pheromone production of *C. capitata*. Pheromone production among fruit-reared, factory-reared fertile, and factory-reared sterile male *C. capitata* was compared in tests conducted in Guatemala (Heath et al. 1994). There were no significant differences in pheromone production (nanograms per male per hour) from 06:00 to 14:00 hours (Table 29.1). However, in collections made from 14:00 to 17:00 hours, factory-reared fertile males produced significantly more of the three major terpene components (geranyl acetate, ethyl-(*E*)-3-octenoate, *E*,*E*-α-farnesene), while the factory-reared sterile males produced significantly more of the four-component blend (the three terpenes plus delta$^1$-pyrroline) than fruit-reared males. Sterile males produced a significantly higher percentage of ethyl-(*E*)-3-octenoate, based on the four-component pheromone blend, during the 10:00 to 14:00 hour collections. Thus, the primary difference in pheromone production among the tested flies was that the fruit-reared males produced pheromone over a shorter time period during the day. Gamma radiation did not adversely affect the total amount of pheromone produced, but did affect component ratios in the pheromone blend.

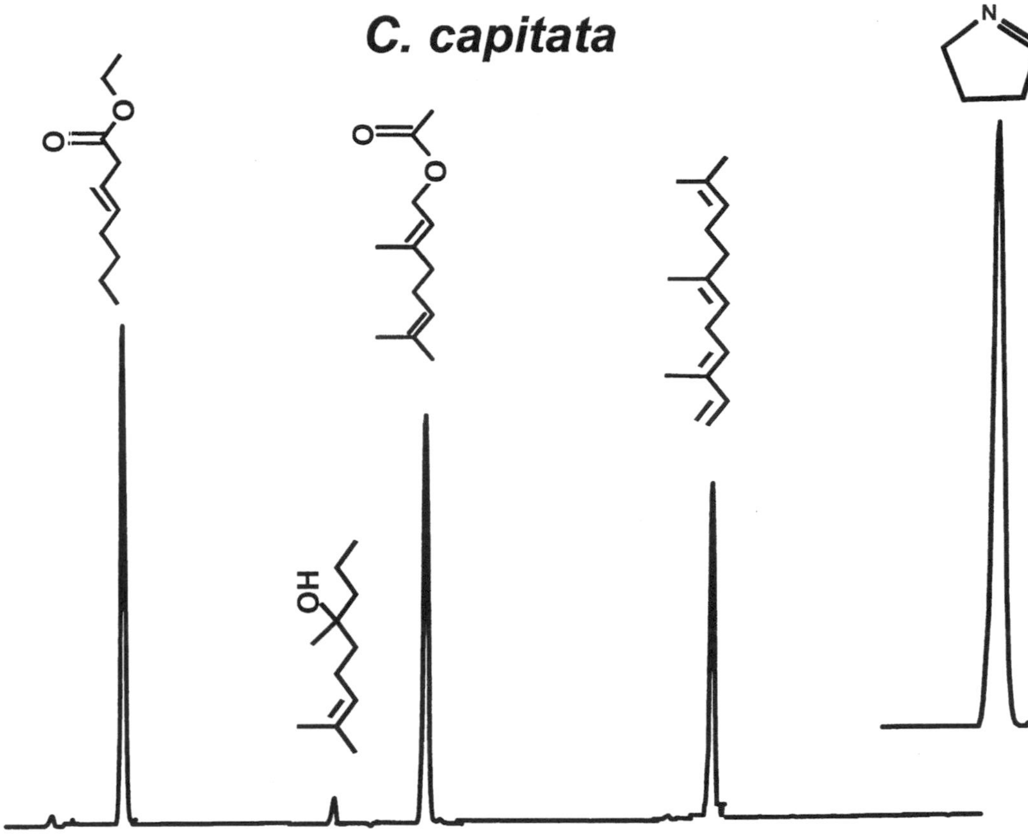


**FIGURE 29.1** Typical chromatogram of pheromonal volatiles emitted by male *C. capitata* obtained using a nonpolar gas chromatographic column. Structures from left to right are ethyl-(*E*)-3-octenoate, linalool, geranyl acetate, and *E,E*-α-farnesene. The 3,4-dihydro-2H-pyrole (delta[1]-pyrroline) is inserted on right.

## 29.3 CARIBBEAN FRUIT FLY, *ANASTREPHA SUSPENSA*

### 29.3.1 Chemical Identification

The most current information on the pheromone components can be found in Rocca et al. (1992) and references therein. Initial studies on the chemical nature of the pheromone extracted from abdomens of sexually mature male *A. suspensa* resulted in the identification of (*Z*)-3-nonenol and (*Z,Z*)-3,6-nonadienol (Nation 1983). Subsequent investigations of abdominal extracts of male flies resulted in the identification of two additional components, the lactones anastrephin (*trans*-hexahydro-*trans*-4,7α-dimethyl-4-vinyl-2-(3H)-benzofuranone) and epianastrephin (*trans*-hexahydro-*cis*-4,7α-dimethyl-4-vinyl-2-(3H)-benzofuranone) (Battiste et al. 1983). Laboratory bioassays of these compounds showed that all were individually attractive to females, but a blend of all four components was the most attractive to females (Nation 1975). A synthetic mix of the four components, however, failed to attract flies in field trials (Nation 1989). A fifth component, a macrolide (*E,E*)-4,8-dimethyl-3,8-decadien-10-olide was identified, synthesized, and named suspensolide (Chuman et al. 1988). Additionally, β-bisabolene, ocimene and *E,E*-α-farnesene have been reported as volatiles emitted by *A. suspensa* (Rocca et al. 1992). The structures of identified pheromone components are shown in Figure 29.2. The chemistry of the male *A. suspensa* pheromone is complex. Analysis of some of the pheromonal components is further complicated because of their thermal lability. For example, the identification of suspensolide was not completed until 1988 and the

**TABLE 29.1**
**Average Percentage (SD) of Individual Pheromone Components in the Total Blend Produced by Male Mediterranean Fruit Flies That Were Fruit-Reared, Factory-Reared Fertile, or Factory-Reared Sterile (Irradiated)**

| Collection Period | Source of Males | Ethyl-(*E*)-3-Octenoate | Geranyl Acetate | *E,E*-α-Farnesene | Delta¹-Pyrroline |
|---|---|---|---|---|---|
| 06:00 to 10:00 | Fruit | 18.6a[a] (7.1) | 22.2a (5.4) | 24.9a (7.8) | 33.8a (15.0) |
| | Fertile | 15.4a (8.5) | 27.5a (9.0) | 24.7a (8.0) | 32.8a (19.4) |
| | Sterile | 18.1a (8.6) | 24.2a (7.8) | 22.2a (7.5) | 35.5a (20.8) |
| 10:00 to 14:00 | Fruit | 18.2a (4.4) | 24.2a (7.4) | 29.5a (10.2) | 28.1a (17.4) |
| | Fertile | 17.8b (8.9) | 29.4a (7.6) | 32.0a (9.0) | 20.8a (18.8) |
| | Sterile | 25.2b (7.4) | 27.5a (7.4) | 28.1a (6.4) | 19.2a (16.6) |
| 14:00 to 17:00 | Fruit | 15.5a (9.1) | 22.6a (11.0) | 31.8a (11.9) | 30.2a (26.6) |
| | Fertile | 15.8a (7.9) | 26.9a (7.0) | 32.2a (10.6) | 25.3a (17.9) |
| | Sterile | 21.8a (5.7) | 25.9a (8.3) | 31.3a (6.8) | 21.0 (17.4) |

Collections were made under natural light conditions in Guatemala City, Guatemala with photophase extending from 0600 to 1800 hours.

[a] Means within a collection period for each component followed by the same letter are not significantly different (LSD, $P = 0.05$).

identification of *E,E*-α-farnesene not until 1992. Elucidation of these components in earlier research was hampered by the lack of appropriate analytical methodologies now available to researchers. Exact determination of the stereochemistry of the lactones was accomplished in 1993 (Baker and Heath 1993).

### 29.3.2 Environmental Factors Affecting Pheromone Production

Burk (1983) documented the start of the afternoon calling period by wild males in host trees. Males were inactive from 10:00 hours until the initiation of calling around 15:00 hours, and this activity peaked at 17:00 to 18:00 hours. Hendrichs (1986) observed a minor calling period in the early morning by a mixed sample of laboratory and wild males. This activity occurred high in the canopy of a host tree in a field cage, and thus may not have been observed by earlier studies. Landolt and Sivinski (1992) also documented an early-morning peak calling period in laboratory-reared *A. suspensa* when tested in a greenhouse, but no such peak was observed in flies tested in a laboratory. They speculated that lack of the early-morning calling period in flies in the laboratory was due to the absence of low light intensities that normally occur at dawn. Hendrichs (1986) noted that calling in field cages was affected by temperature and light intensity, and that the brief morning calling period corresponded to the short period of time in which the temperature and light conditions fell within the range appropriate for male sexual activity.

Pheromone release from male flies held under artificial light (laboratory) and natural light (greenhouse) has been compared (Epsky and Heath 1993a). The pheromone release during peak production under laboratory conditions (i.e., 14:30 to 16:30 hours) was 1.46 μg/male/h, with 1.9% β-ocimene, 5.4% nonenols, 18.5% suspensolide, 5.6% *E,E*-α-farnesene, 28.4% β-bisabolene, 6.7% anastrephin, and 33.5% epianastrephin. These percentages were equivalent to those obtained from the total (16-h) pheromone release. The pheromone release during peak production under greenhouse conditions (i.e., 17:00 to 18:00 hours) was 1.79 μg/male/h, with 2.6% β-ocimene, 5.1% nonenols, 13.1% suspensolide, 3.9% *E,E*-α-farnesene, 24.5% β-bisabolene, 8.3% anastrephin, and 42.5% epianastrephin. Although there was little difference in the total amount of pheromone released by flies during the two peak periods, the differences in percent anastrephin and percent

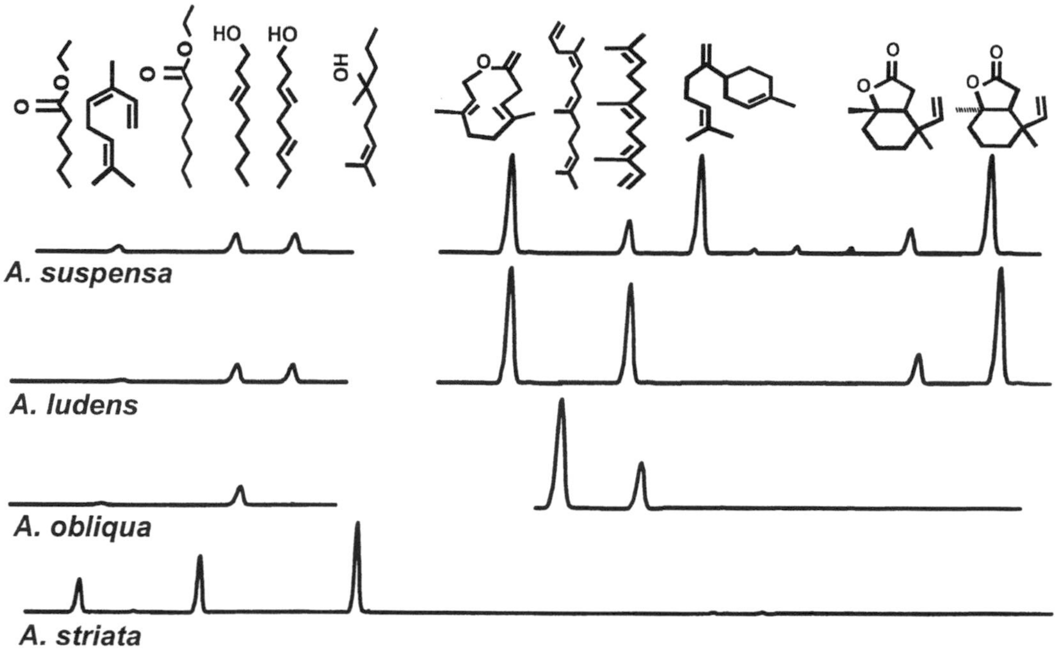


**FIGURE 29.2** Comparisons of pheromone components released by *A. suspensa, A. ludens, A. obliqua, and A. striata.* Chromatographic peaks indicate a hypothetical illustration of separation obtained based on the volatility of the compounds. Peak heights represent relative amounts of the pheromone component. Structures from left to right are ethyl hexanoate, ocimene; ethyl octanoate, nonenol; nonadienol; linalool; suspensolide; (Z,E)-α-farnesene; (*E,E*)-α-farnesene; β-bisabolene; anastrephin; and epianastrephin.

epianastrephin released during the peak period were significant. Some calling activity was observed throughout the day in our study. It has been noted that calling activity is not always indicative of quantity of pheromone released (Nation 1991), and our data support this observation. Broad late-afternoon peaks in calling activity were observed in flies under both laboratory and greenhouse conditions (Landolt and Sivinski 1992). Hendrichs (1986) observed that cloudy conditions early in the day stimulated calling activity which ceased quickly when the sun reappeared. These observations may explain the differences in the periodicity of pheromone release observed in our study. Under artificial light, the abrupt off/on light cycle may trigger a brief flurry of male calling that is quickly terminated or dampened, thus resulting in a sharp early-morning peak under artificial light. Some calling activity continues and, since the light intensity in the laboratory is apparently too low to inhibit pheromone release, pheromone release increases steadily until a peak late in the afternoon. High light intensity and low relative humidity appear to act to inhibit pheromone release by *A. suspensa* males and, if this is true, environmental conditions act to modify the endogenous bimodal periodicity to produce a broad minor early-morning calling period and a sharp major afternoon calling period in flies in the field. Environmental variables in laboratory tests may not be at inhibitory levels and, under these conditions, the afternoon calling period is solely dependent on endogenous circadian rhythms and calling becomes much less synchronized within the tested group of males.

Periodicity of pheromone production by males under simulated natural environmental conditions mirrored the periodicity of calling activity that has been observed under field conditions. The nonenols, anastrephin and epianastrephin, were found to be the major female attractants in laboratory bioassays (Nation 1991). Thus, these differences in the amounts of anastrephin and epianastrephin may have a strong effect on female response. The results of this study underline the importance of obtaining chemical data under environmental regimes that approximate natural conditions.

**TABLE 29.2**
**Total Mean (SD) Pheromone (μg/male/h) Produced during 2-h Volatile Collections from Male *A. suspensa* ($n = 5$) That Were Held Overnight with Water Only (water-only), with Water and Sugar (sugar-only), or with Water, Sugar, and Protein (fully fed)**

| | Food Treatments | | |
|---|---|---|---|
| Collection Period (hours) | Water-Only | Sugar-Only | Fully Fed |
| 12:30 to 14:30 | 0.019aA (0.007) | 0.071aA (0.042) | 0.020aA (0.008) |
| 14:30 to 16:30 | 0.091aA (0.052) | 0.223aAB (0.118) | 0.067aAB (0.023) |
| 16:30 to 18:30 | 0.116aA (0.050) | 0.576bAB (0.136) | 0.271abBC (0.121) |
| 19:00 to 21:00 | 0.048aA (0.018) | 0.457bB (0.142) | 0.491bC (0.203) |

Means within a row followed by the same small letter or within a column followed by the same capital letter are not significantly different (Tukey's mean separation test on transformed [$\log x + 1$] data, $P = 0.05$; untransformed means presented).

Bioassays that linked female response with chemical analysis of pheromone produced by live males were conducted in a greenhouse under natural light conditions (Heath et al. 1993). There was no correlation between number of females captured per hour in response to pheromone from live males and the amount of any of the individual components or the sum of all components produced per hour. There was, however, a correlation between proportion of epianastrephin in the pheromone and number of females captured; and this was most apparent early in the calling period (Heath et al. 1993).

### 29.3.3 Effect of Food Availability on Pheromone Production

Studies (Epsky and Heath 1993b) were conducted to determine the release of pheromone from males that were given protein, sugar, and water (fully fed), sugar and water (sugar-only), or water (water-only) (Table 29.2). There were significant differences in pheromone production among males on the different food treatments during the 2-h collection periods of peak pheromone production under greenhouse conditions. Water-only males produced significantly less pheromone than sugar-only males during 16:30 to 18:30 hours, and significantly less than both sugar-only and fully fed males during 19:00 to 21:00 hours. Time period significantly affected both sugar-only and fully fed males. Pheromone production by water-only males was suppressed during all collection times and was not affected by the collection time period.

Food regime had an effect on component ratios in the pheromone blend. There were significant differences in the percentages of suspensolide and epianastrephin during 16:30 to 18:30 hours, and in the percentage of suspensolide during 19:00 to 21:00 hours. For both time periods, percentage epianastrephin was highest and percentage suspensolide was lowest in water-only males. A potential link between these components has been proposed, with suspensolide as a possible precursor to both anastrephin and epianastrephin or with a common precursor for all of these components (Chuman et al. 1988). The peaks in production of these two pheromone components are separated in time, with the suspensolide peak occurring earlier in photophase than the anastrephin and epianastrephin peaks (Nation 1990). This same pattern was observed for males from all food regimes. The percentage of suspensolide was highest during 14:00 to 16:30 hours, while the percentage of epianastrephin was highest during 19:00 to 21:00 hours for all food regimes.

## 29.4 MEXICAN FRUIT FLY, *ANASTREPHA LUDENS*

Abdominal extracts of sexually mature male *A. ludens* yielded (*Z*)-3-nonenol and (*Z,Z*)-3,6-nonadienol (Esponda-Gaxiola 1977; Nation 1983), and two lactones, anastrephin and epianastrephin (Battiste et al. 1983; Stokes et al. 1983). The stereochemistry of these chiral lactones and their release as volatiles were not determined in these studies. Ensuing investigations by Robacker and Hart (1985) of volatiles released by sexually mature, 8- to 26-day-old male *A. ludens* indicated that (*Z*)-3-nonenol, (*Z,Z*)-3,6-nonadienol, anastrephin, and epianastrephin were released at an average rate of 0, 40, 60, and 300 ng/male, respectively, based on 3-h collections made during active calling periods. Concurrent with the volatile collections, abdominal extracts from sexually mature males were obtained during this same time period and yielded 100, 40, 200, and 700 ng/male, respectively. Based on bioassays of individual and blends of synthetic chemicals, Robacker (1988) concluded that treatments containing (*Z*)-3-nonenol, (*Z,Z*)-3,6-nonadienol, and (*S,S*)-(–)-epianastrephin elicited strong behavioral response by virgin female *A. ludens*. The addition of the other antipodes of anastrephin and epianastrephin to the three-component blend did not result in increased attraction of female fruit flies. Recent efforts to reevaluate the pheromonal components emitted by male *A. ludens* resulted in the identification of *E,E*-α-farnesene, the 11-member macrolide suspensolide, and also trace amounts of limonene (Rocca et al. 1992).

As part of a research program to identify male-produced pheromones of Tephritidae and to develop better attractants for *A. ludens*, periodicity of volatiles emitted by sexually mature *A. ludens* males was analyzed chemically (R.R. Heath et al., unpublished data). Flies were from a laboratory culture maintained in Weslaco, TX, for approximately 25 generations on a laboratory diet containing carrot and corncob in addition to protein and sugar sources. The strain originated from yellow chapote (*Sargentia greggii* Wats.) fruit field-collected in Nuevo Leon, Mexico. Laboratory rearing conditions were 22 ± 2°C, 60 ± 10% relative humidity, and photophase from 06:30 to 20:00 hours under fluorescent lights. Conditions for the collection of emitted volatiles were similar except lighting was provided by a combination of fluorescent and natural light. Volatiles were collected from 12- to 19-day-old males in groups of nine to ten flies per chamber. Collections were made from each group of flies for consecutive 2- to 2.25-h time periods from 07:00 to 20:00 hours and a 1-h time period from 20:30 to 21:30 hours. Two groups were used on each of 3 days. Additionally, emissions from 15- and 16-day-old virgin females were collected from 07:00 to 21:30 hours on one day. The description of the system used to collect volatiles has been described (Heath and Manukian 1992; Heath et al. 1993).

Volatiles from male *A. ludens* were first detected beginning at 11:00 hours. At the onset of volatile emission only small amounts of (*Z*)-3-nonenol, (*Z,Z*)-3,6-nonadienol, and suspensolide were detected. Pheromone emission began to increase at 13:45 hours and the greatest amounts of pheromone were obtained in collections made from 16:00 to 18:00 hours. Analyses of collections from throughout the afternoon indicated that the amount of suspensolide released from 13:45 to 18:00 hours was relatively constant. Average (± SD) amounts were 157.3 ± 51.2, 229.3 ± 53.7, and 226.3 ± 90.0 ng/male/h from collections made 13:45 to 16:00, 16:00 to 18:00, and 18:00 to 20:00 hours, respectively. During these time periods the amount of nonenol, nonadienol, farnesene, anastrephin, and epianastrephin released increased (Figure 29.3). None of the male-produced compounds were detected in collections from virgin female *A. ludens* made throughout the photophase. Assignment of the stereochemistry of synthetic chiral lactones anastrephin and the two epianastrephins was done by comparison of the analyses of natural and synthetic materials using chiral capillary columns. Male *A. ludens* released only (*S,S*)-(–)-anastrephin and (*R,R*)-(+) and (*S,S*)-(–)-epianastrephin in a ratio of approximately 1:10.

*Anastrepha ludens* is not sympatric with *A. suspensa*. It is interesting to note that there is considerable overlap in the chemicals released by both species (see Figure 29.2). (*Z*)-3-Nonenol, (*Z,Z*)-3,6-nonenol, anastrephin, epianastrephin, and suspensolide are common to both species. (*E,E*)-α-Farnesene is also a volatile emitted by *A. ludens* (R.R. Heath et al., unpublished data). Thus, it appears that the major difference in the pheromones of the two species is the release of

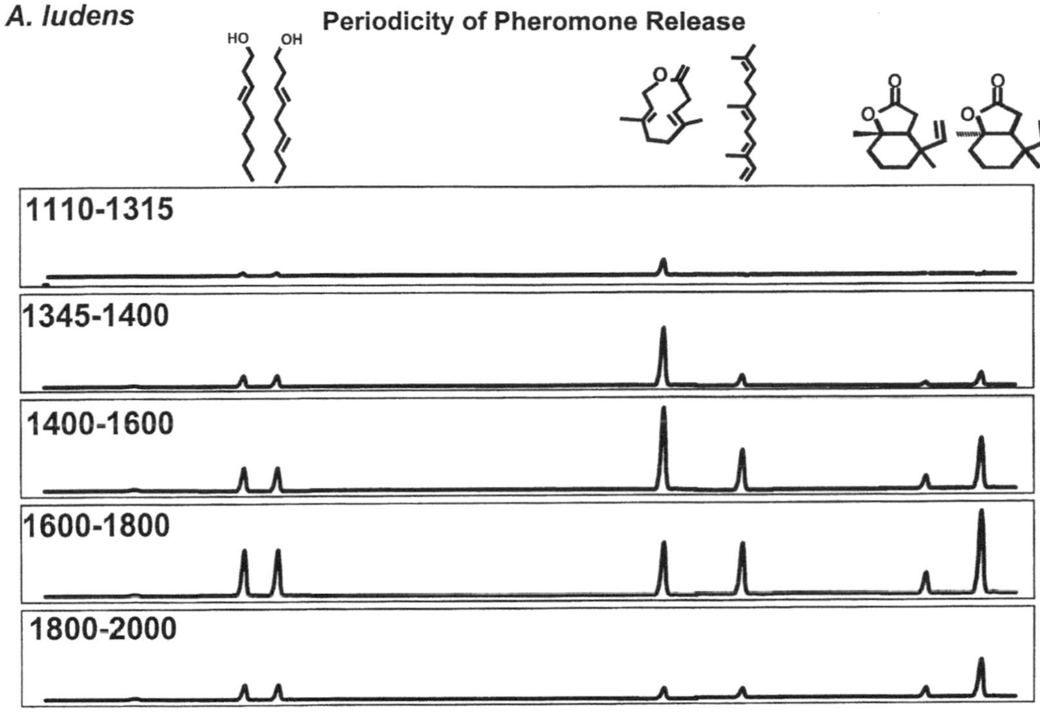


**FIGURE 29.3** Periodicity of release of pheromone components of *A. ludens* during periods of the day when pheromone is released. Chromatograms illustrate separation obtained based on the volatility of the compounds. Peak heights represent relative amounts of the pheromone component. Structures from left to right are nonenol; nonadienol; suspensolide; (*E,E*)-α-farnesene; anastrephin, and epianastrephin.

β-bisabolene and ocimene by *A. suspensa* but not by *A. ludens*. In addition, the enantiomer composition for anastrephin and epianastrephin in *A. suspensa* as reported by Battiste et al. (1983) is 55 ± 3% (–) enantiomer to 45 ± 3% (+) enantiomer, respectively, which is different from that found in volatiles released by *A. ludens*.

## 29.5 WEST INDIAN FRUIT FLY, *ANASTREPHA OBLIQUA*

Volatiles emitted by sexually mature males were collected principally from laboratory-cultured flies in Weslaco, TX that originated from field-collected mangos in Haiti and had been reared on artificial diet for approximately 35 generations. Rearing diet contained papaya fruit in addition to torula yeast, wheat germ, sugar, and agar. Sexually immature adults were sexed when 2- to 3-day-old and the sexes were held separately in wood-framed, screen cages (30 cm per side) containing water and a mixture of sugar and yeast hydrolysate. Laboratory holding conditions were 22 ± 2°C, 60 ± 10% relative humidity, and photophase from 06:30 to 20:00 hours under fluorescent lights. Test conditions were similar except lighting was provided by a combination of fluorescent and natural light. Volatiles were collected from 14- to 26-day-old males in groups of nine to ten flies per chamber. Collections were made from each group of flies for consecutive 2- to 2.25-h periods from 06:30 to 21:30 hours. Two groups were used on each of 3 days. Emissions also were collected from four groups of males during three consecutive 1- to 1.25-h periods beginning at 05:30 hours on another day. Finally, emissions from 21-day-old virgin females in two groups of ten each were collected from 06:30 to 21:30 hours on one day.

Two compounds, (*Z,E*)-α-farnesene and (*E,E*)-α-farnesene, were identified as the major volatiles and the likely pheromone components emitted by laboratory-reared male *A. obliqua*. Emission of these volatiles occurred at the onset of light, decreased during midday, and increased prior to darkness. Although the release rate changed during the day, the ratio of (*Z,E*)-α-farnesene to (*E,E*)-α-farnesene was fixed at 5.5:1. (*Z*)-3-Nonenol was also identified in the volatiles collected. (*Z,Z*)-3,6-Nonadienol, β-bisabolene, suspensolide, anastrephin, and epianastrephin, identified as pheromonal components released by male *A. suspensa* and *A. ludens,* were not detected in volatiles from *A. obliqua.*

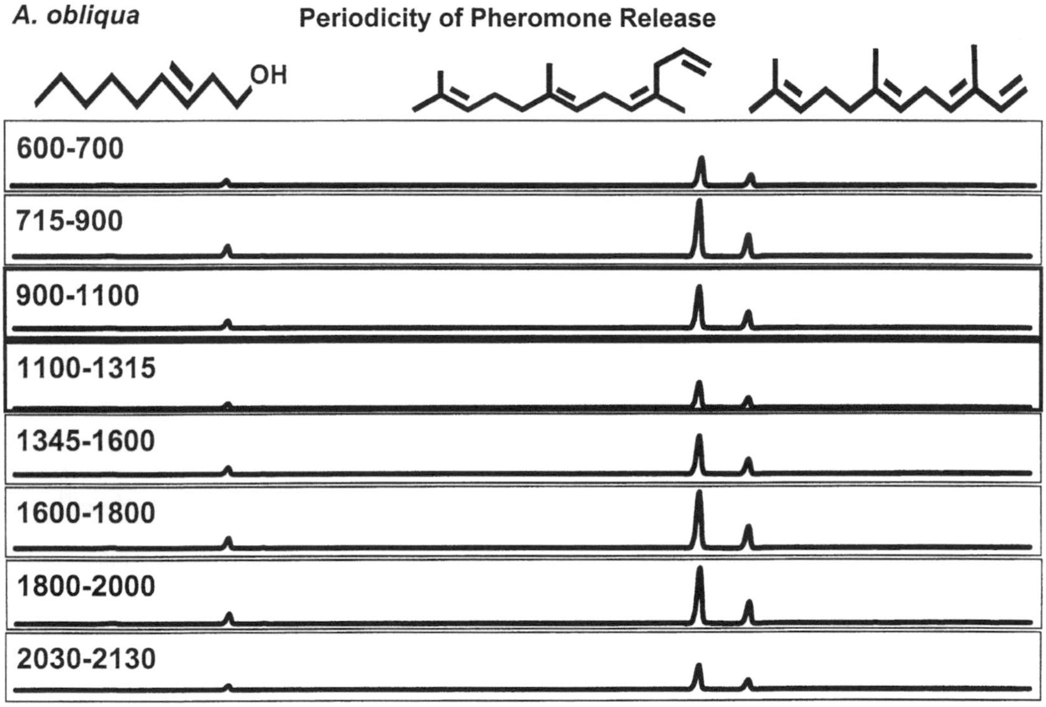


**FIGURE 29.4** Periodicity of release of pheromone components of *A. obliqua* during periods of the day when pheromone is released. Chromatograms illustrate separation obtained based on the volatility of the compounds and peak heights represent relative amounts of the pheromone component. Structure from left to right are nonenol; (Z,E)-α-farnesene and (*E,E*)-α-farnesene.

The results of the analyses of volatiles collected on Porapak-Q® from groups of ten male *A. obliqua* for 2-h collection periods ($n$ = 6 for each 2-h period) from 06:00 to 21:30 hours is shown in Figure 29.4. Greatest mean (±SD) amount (866 ± 218 ng/male/h) of pheromone emission occurred at 07:15 to 09:00 hours, with less material released during midday, and with increased release occurring at 16:00 to18:00 hours (766 ± 338 ng/male/h) and 18:00 to 20:00 hours (798 ± 237 ng/male/h). Analyses of collections made from 05:30 to 06:30 hours resulted in <20 ± 15 ng/male/h ($n$ = 4) of pheromone detected. No farnesenes were detected in collections from virgin female *A. obliqua* for 2-h periods throughout the photophase ($n$ = 4 for each 2-h time period). Analyses of the ratio of (*E,E*)-α-farnesene and (*Z,E*)-α-farnesene released during the seven time periods from 06:00 to 21:30 hours resulted in an average (± SD) of 84.8% ± 1.1 ($n$ = 48) of the (*Z,E*) isomer to 15.2% ± 1.0 of the (*E,E*)-α-farnesene.

In general, periodicity of volatile emissions corresponded with sexual behavior of males. Calling behavior, similar to that reported for *A. suspensa* (Nation 1972) and for *A. ludens* (Robacker et al. 1985), occurred during most of the day. However, no decline in sexual behavior was observed between 10:00 and 13:15 hours when volatile emission declined to a midday low. Aluja et al. (1983), observing wild *A. obliqua* on a field-caged host tree, also reported males calling throughout most of the day. In addition, they observed two peaks of lekking behavior at 08:30 to 10:30 hours and at 13:30 to 14:30 hours with a distinct decline at 12:30 hours. The timing of the morning lekking activity coincides well with the midmorning peak in volatiles, but the afternoon period was not as prolonged in the field as the coinciding period of volatile release in the laboratory (see Figure 29.4). The two major compounds emitted by virgin male *A. obliqua* are released in a precise ratio that is independent of time.

## 29.6 *ANASTREPHA STRIATA, A. FRATERCULUS* COMPLEX, AND *A. SERPENTINA*

A complete description of the pheromones of these species is not available. Collections from male *A. striata* have been obtained from insects maintained in colonies in Guatemala, Costa Rica, and Texas and then analyzed in Gainesville, FL. Based on limited spectroscopic data, it appears that the males release linalool, ethyl hexanoate, and ethyl octanoate (Figure 29.2) and that the period of maximum release of pheromone is late in the afternoon. We did not detect β-bisabolene, suspensolide, anastrephin, or epianastrephin.

Analysis of volatiles from *A. fraterculus* complex males that were laboratory reared in Guatemala and Costa Rica indicated that they release nonenol(s) and large amounts of anastrephin and epianastrephin. Because of the limited number of samples, periods of maximum release of pheromone were not determined. Other compounds emitted by *A. suspensa* and *A. ludens* may be present in small amounts, but their identification is uncertain. De Lima et al. (1996) found that extracts of salivary glands from Brazilian *A. fraterculus* complex males act as short range attractants and arrestants for females. The major components in the salivary glands were (*Z,E*)-α-farnesene, (*E,Z*)-α-farnesene, and (*E,E*)-suspensolide, in addition to several pyrazines. The Brazilian *A. fraterculus* complex flies may be a separate species from the Central American flies (see Norrbom et al., Chapter 12); thus it is not known if the observed variation is interspecific, intraspecific, or due to differences in chemical extracted from salivary glands or released as volatiles during calling.

Analysis of a small number of volatile collections from *A. serpentina* did not reveal β-bisabolene, suspensolide, anastrephin, or epianastrephin. *Anastrepha serpentina* may release chemicals similar to those of *A. striata* (R.R. Heath et al., unpublished data).

## 29.7 DISCUSSION OF *ANASTREPHA* PHEROMONES

A comparison of the pheromone components of *A. suspensa, A ludens, A. obliqua,* and *A. striata* (see Figure 29.2) indicates that the patterns of chemicals used by these species are not entirely consistent with the phylogenies suggested by Norrbom et al. (Chapter 12) and by McPheron et al. (Chapter 13). Mitochondrial ribosomal DNA analysis indicates that the *A. fraterculus* complex is in the species cluster with *A. suspensa, A. ludens,* and *A. obliqua* (McPheron et al., Chapter 13) and thus may be expected to use similar pheromone components. There is a large amount of duplication in the pheromone components of *A. suspensa* and *A. ludens,* some duplication in the components used by *A. fraterculus* complex, but considerable divergence from the pheromonal compounds of *A. obliqua* and *A. striata.* The major pheromone components emitted by *A. suspensa* and *A. ludens* are complex macro cyclic lactones with minor terpenes and sequiterpenes components. A close genetic similarity of these two species is indicated by mitochondrial ribosomal DNA analysis (McPheron et al., Chapter 13). There are differences in the pheromones of these two species including the presence of β-bisabolene and ocimene in *A. suspensa*, the absence of these components in *A. ludens,* and the difference in enantiomer compositions of anastrephin and epianastrephin. Although studies are incomplete, the presence of anastrephin, epianastrephin, nonenols, isomers of alpha farnesene, and suspensolide are indeed indicated in analyses of glands and headspace of male *A. fraterculus* (R.R. Heath et al., unpublished data). The presence of unidentified pyrazines in male *A fraterculus* volatiles (De Lima et al. 1996), however, is puzzling. Most of the compounds found in *A. suspensa*, *A. ludens*, and *A. fraterculus* complex are not found in *A. obliqua*. However, the presence of (Z)-3-nonenol and isomers of alpha farnesene still indicates an alliance of sorts of *A. obliqua* with the other species in this cluster.

The pheromone of *A. striata* is completely different from the known pheromones of *A. suspensa*, *A. ludens*, *A. fraterculus* complex, and *A. obliqua*, although there may be similarities to *A. serpentina* pheromone. This pattern is noteworthy in that *A. striata* and *A. serpentina* belong to species groups that may be closely related to each other (see Norrbom et al., Chapter 12). Information on the volatiles of males of *A. striata* and A. *serpentina*, however, are too incomplete to warrant speculation on their relatedness.

## 29.8 PAPAYA FRUIT FLY, TOXOTRYPANA CURVICAUDA

The sexual behavior, including the sex pheromone, of the papaya fruit fly is reviewed by Landolt (Chapter 14). A brief summary of our knowledge of the pheromone is included here also.

The sex pheromone of the papaya fruit fly was first isolated from airflow over calling males by Landolt et al. (1985) using Poropak filters extracted with hexane. This isolate was shown to be behaviorally active in an arena-type assay. Chuman et al. (1987) subsequently identified 2-methyl-6-vinylpyrazine as the principal volatile compound in air passed over calling males and demonstrated its attractiveness to female papaya fruit flies in a flight tunnel assay. Volatile collections made from individual calling males indicated a pheromone release rate of $63.2 \pm 33.2$ ng/h (Chuman et al. 1987). In further studies of papaya fruit fly responses to 2-methyl-6-vinylpyrazine, using a laboratory flight tunnel, Landolt and Heath (1988) found that female responsiveness to pheromone coincided with egg maturation and that female attraction to the pheromone remained high following mating.

The original objective of the studies of papaya fruit fly pheromone was to develop an attractant and trapping system that could be used to monitor flies in papaya plantings. This objective was subsequently pursued — to develop controlled release systems for the pheromone, to develop effective traps, and to assess efficacy as a monitoring method. By using a glass capillary formulation for controlled release of 2-methyl-6-vinylpyrazine and a green sphere coated with an adhesive as a visual target, a system was first developed (Landolt et al. 1988) that was effective in attracting and capturing both mated and mature virgin female papaya fruit flies in papaya groves. This system was also tested in Guatemala and Costa Rica and was found to be highly effective in removing both sexes of papaya fruit flies from papaya plantings (Landolt et al. 1991). The glass capillary method of providing controlled release of pheromone and the green sphere coated with tanglefoot were found to be problematic in their manufacture and their use in the field. Further improvements in a lure-and-trap system made it less expensive, longer lasting, and easier to use (Heath et al. 1996). This system comprises a membrane-based formulation for controlled release of 2-methyl-6-vinylpyrazine and a green cylindrical trap with a replaceable sticky covering.

## 29.9 OLIVE FRUIT FLY, *BACTROCERA OLEAE*

*Bactrocera oleae* is unusual among tephritids in that the female produces a pheromone that is attractive to males. Studies in the laboratory and field demonstrated that female flies release a volatile mixture that attracts male flies (Economopoulos et al. 1971; Schultz and Boush 1971). Subsequent work by Haniotakis (1974) demonstrated male olive fruit fly attraction to females in the laboratory and noted that female attractiveness to males was correlated with the age at which they mated.

The first report of female olive fruit fly sex pheromone components was by Baker et al. (1980). The major component was identified as (1,7)-dioxaspiro-[5,5]undecane (olean), a spiroacetal, and was isolated from extracts of female rectal glands. Field tests conducted in Granada with the spiroacetal demonstrated that the pheromone did attract predominantly male olive flies. Mazomenos and Pomonis (1983) reported the identification of four chemicals from cold-trap condensation of female-produced volatiles, only two of which were present in rectal gland secretions. Subsequently, it was determined by Haniotakis et al. (1986) that the R-(–)-enantiomer attracted males, while the S-(+)-enantiomer attracted females in the laboratory, although not in the field. Three other components of the female pheromone, α-pinene, *n*-nonanol, and ethyl dodecanoate, were identified by Mazomenos and Haniotakis (1985). The maximum response to the spiroacetal in laboratory experiments occurred at a dosage of 10 µg of a racemic mixture applied to filter paper (Haniotakis and Pittara 1994). The spiroacetal is an effective lure for males in the field and can be used as part of an integrated approach to controlling populations of olive fruit fly in olive groves. Haniotakis et al. (1983) showed that yellow sticky panels baited with a sex pheromone lure comprising of a racemic mixture of the enantiomers of the spiroacetal reduced fly populations and fruit infestation. Subsequently, Haniotakis et al. (1991) developed a mass-trapping approach to control *B. oleae* that included a food attractant (ammonium bicarbonate), sex attractant (racemic olean), phagostimulant (sugar), visual atttractant (yellow panel), and pesticide.

## 29.10 *BACTROCERA DORSALIS, B. CUCURBITAE,* AND *B. TRYONI*

Sexual pheromones produced by males that are attractive to females are indicated in the Oriental fruit fly, *B. dorsalis* (Hendel), the melon fly, *B. cucurbitae* (Coquillett) (Kobayashi et al. 1978; Poramarcom 1988), and the Queensland fruit fly, *B. tryoni* (Froggatt) (Fletcher 1969; Fletcher and Giannakakis 1973). Short-range orientation of females to live males and to rectal glands of males was demonstrated for all three species and several chemicals have been identified from rectal gland secretions (Fletcher and Kitching 1995). However, demonstration of long-range attraction to sex pheromone, either upwind-oriented flights in laboratory assays or responses in the field, are lacking for these species (Koyama 1989).

## 29.11 CONCLUSIONS AND FUTURE RESEARCH NEEDS

The pheromone systems used by the tephritid fruit flies are very complex. Not only are numerous chemical compounds released from calling males, but the amounts and ratios may vary over time or among different populations of flies of the same species. In addition, the range in volatilities of compounds produced by male fruit flies has increased the difficulties in formulating synthetic blends that mimic the release rates and ratios of pheromones from live males. Thus, it is not known if lack of field efficacy of synthetic compounds that have been tested is due to absence of biological activity or to inadequate formulation. The pheromonal system used by male *C. capitata* has been studied relatively intensively. Numerous volatile compounds have been identified (e.g., Baker et al. 1985; Jang et al. 1989). Female response to formulations that contain three, four, and five of the synthetic components of male-produced *C. capitata* pheromone has been demonstrated in laboratory and field trials; however, response is low in comparison with authentic pheromone released from live males in wind-tunnel bioassays (Heath and Epsky 1993; Jang et al. 1994) or to other types of lures, such as trimedlure or food-based lures, in field trials (Heath et al. 1991, and unpublished data). Similarly, several volatile chemicals have been identified from *A. suspensa* and *A. ludens* males (Chuman et al. 1988; Nation 1990; Rocca et al. 1992). Glass capillaries of different inside diameters and lengths have been used to formulate the more volatile compounds, and rubber septa to formulate the less volatile compounds of *C. capitata* (Heath et al. 1991) and *A. suspensa* (R.R. Heath et al., unpublished data) pheromones. An eight-component synthetic blend of *A. suspensa* pheromone was as effective as live males in laboratory tests with factory-reared flies when compared with a clean air blank (Heath et al., unpublished data). Effectiveness in the field, however, has not been tested. More studies on biological activity are needed to confirm that the volatile compounds collected from the other fruit fly species are actually components of their pheromones.

Although there are very minor overlaps in pheromone chemistries of fruit flies in different genera, they are quite different overall with no indications yet of any similarities among taxa above the species level. Similarities include the presence of α-farnesenes and linalool in *C. capitata* and in some *Anastrepha* species, and 2,6-methylvinyl pyrazine of the papaya fruit fly and the possible presence of pyrazines in the pheromone of *A. fraterculus*. At this time, the data are tenuous at best and should be verified. The types of chemicals isolated and identified from species of *Bactrocera* are generally unique to that group.

As previously acknowledged, the information on pheromone chemistry of tephritid flies is quite limited, permitting only rudimentary comparisons among species and higher taxa. It is hoped that additional strides in the near future on both the chemistries of additional species and their respective roles in fruit fly mate-finding and mating behavior will permit additional elucidation of their relatedness. An additional problem in attempting to use pheromone chemistry as characters to suggest phylogeny is the potential importance of the ecological relationships between fruit fly pheromone chemistry and plants or other insects. Such relationships may obscure true phylogenetic relationships among fruit fly species. For example, there are numerous similarities among volatile chemicals from fruit fly males, plant foliage, and fruits. These include α-farnesene isomers, linalool, geranyl acetate, and others. Other fruit fly pheromone components, such as delta$^1$-pyrroline and ethyl acetate found in the Mediterranean fruit fly, overlap with food odors. A third theme is the similarity between the pheromone chemistries of some species of fruit flies and stinging Hymenoptera. These include various alkyl pyrazines that are reported from several species of fruit flies and are found in gland analyses of some ants and wasps (Chuman et al. 1987), and also *N*–3-methylbutylacetamide, a *Bactrocera* chemical that is an alarm pheromone of species of *Vespula* spp. (Vespidae) (Heath and Landolt 1988). While all of these examples may be coincidental, we should consider the possibility that fruit fly males may release pheromone components that include food or host cues, and that some similarities among pheromone chemistries may relate better to host plants or food sources.

Improvements in techniques for chemical identification, quantification, and formulation have facilitated evaluation of fruit fly pheromone systems. Additional research should be directed at full evaluation of the species reported herein and expansion of this type of research to other fruit fly species. Patterns of types of pheromonal components used and periodicity of pheromonal release will provide important insight into the evolution of behavior in Tephritidae and may contribute to the control of pest species.

## ACKNOWLEDGMENTS

The authors would like to thank Adron T. Proveaux (USDA/ARS, CMAVE, Gainesville, FL) for mass spectroscopic analysis, the staff at the USDA, APHIS-International Services, U.S. Embassy, Guatemala and the staff of the Cooperative Moscamed Program in Guatemala for providing insects and technical assistance, the California Department of Food and Agriculture for funding aspects of this research, and the following donors for partial funding in support of this conference — Campaña Nacional contra las Moscas de la Fruta (Mexico), International Organization for Biological Control of Noxious Animals and Plants (IOBC), Instituto de Ecología, A.C. (Mexico), Consejo Nacional de Ciencia y Tecnología (Mexico) and USDA-ICD-RSED. This chapter reports the results of research only. Mention of a proprietary product does not constitute an endorsement or recommendation for its use by the USDA.

## REFERENCES

Aluja, M., J. Hendrichs, and M. Cabrera. 1983. Behavior and interactions between *Anastrepha ludens* (L) and *A. oblicua* [sic] (M) on a field-caged mango tree — lekking behavior and male territoriality. In *Fruit Flies of Economic Importance* (R. Cavalloro, ed.), pp. 122–133. A.A. Balkema, Rotterdam.

Baker, J.D. and R.R. Heath. 1993. NMR spectral assignment of lactone pheromone components emitted by Caribbean and Mexican fruit flies. *J. Chem. Ecol.* 19: 1511–1519.

Baker, J.D., R.R. Heath, and J.G. Millar. 1992. An equilibrium and stability study of delta$^1$-pyrroline. *J. Chem. Ecol.* 18: 1595–1602.

Baker, P.S., P.E. Howse, R.N. Ondarza, and J. Reyes. 1990. Field trials of synthetic sex pheromone components of the male Mediterranean fruit fly (Diptera: Tephritidae) in southern Mexico. *J. Econ. Entomol.* 83: 2235–2245.

Baker, R., R. Herbert, P.E. Howse, and O.T. Jones. 1980. Identification and synthesis of the major sex pheromone of the olive fly (*Dacus oleae*). *J. Chem. Soc. Chem. Commun.* 1: 52–53.

Baker, R., R.H. Herbert, and G.G. Grant. 1985. Isolation and identification of the sex pheromone of the Mediterranean fruit fly, *Ceratitis capitata* (Wied.). *J. Chem. Soc. Chem. Commun.* 12: 824–825.

Battiste, M.A., L. Strekowski, D.P. Vanderbilt, M. Visnick, R.W. King, and J.L. Nation. 1983. Anastrephin and epianastrephin, novel lactone components isolated from the sex pheromone blend of male Caribbean and Mexican fruit flies. *Tetrahedron Lett.* 24: 2611–2614.

Burk, T. 1983. Behavioral ecology of mating in the Caribbean fruit fly, *Anastrepha suspensa* (Loew) (Diptera: Tephritidae). *Fla. Entomol.* 66: 330–344.

Chuman, T., P.J. Landolt, R.R. Heath, and J.H. Tumlinson. 1987. Isolation, identification, and synthesis of male-produced sex pheromone of papaya fruit fly, *Toxotrypana curvicauda* Gerstaecker (Diptera: Tephritidae). *J. Chem. Ecol.* 13: 1979–1992.

Chuman, T., J. Sivinski, R.R. Heath, C.O. Calkins, J.H. Tumlinson, M.A. Battiste, R.L. Wydra, L. Strekowski, and J.L. Nation. 1988. Suspensolide, a new macrolide component of male Caribbean fruit fly (*Anastrepha suspensa* [Loew]) volatiles. *Tetrahedron Lett.* 29: 6561–6564.

De Lima, I.S., P.E. Howse, and I.D.R. Stevens. 1996. Volatile components from the salivary glands of calling males of the South American fruit fly, *Anastrepha fraterculus*: partial identification and behavioural activity. In *Fruit Fly Pests: A World Assessment of Their Biology and Management* (B.A. McPheron and G.J. Steck, eds.), pp. 107–113. St. Lucie Press, Delray Beach, FL.

Economopoulos, A.P., A. Gianakakis, M.E. Tzanakakis, and A. Voystzoglou. 1971. Reproductive behavior and physiology of the olive fruit fly. *Ann. Entomol. Soc. Am.* 64: 1112–1116.

Epsky, N.D. and R.R. Heath. 1993a. Pheromone production by males of *Anastrepha suspensa* (Diptera: Tephritidae) under natural light cycles in greenhouse studies. *Environ. Entomol.* 22: 464–469.

Epsky, N.D. and R.R. Heath. 1993b. Food availability on pheromone production by males of *Anastrepha suspensa* (Diptera: Tephritidae). *Environ. Entomol.* 22: 942–947.

Esponda-Gaxiola, R.E. 1977. Contribución al Estudio Químico del Atrayente Sexual de la Mosca Mexicana de la Fruta, *Anastrepha ludens* (Loew). Thesis, Instituto Tecnológico y de Estudios Superiores Monterrey, Monterrey, N.L., Mexico.

Fletcher, B.S. 1969. The structure and function of the sex pheromone glands of the Queensland fruit fly *Dacus tryoni*. *J. Insect Physiol.* 13: 1309–1322.

Fletcher, B.S. and A. Giannakakis. 1973. Factors limiting the response of females of the Queensland fruit fly, *Dacus tryoni*, to the sex pheromone of the male. *J. Insect Physiol.* 19: 1147–1155.

Fletcher, M.T. and W. Kitching. 1995. The chemistry of fruit-flies. *Chem. Rev.* 95: 789–828.

Haniotakis, G.E. 1974. Male olive fly attraction to virgin females in the field. *Ann. Zool. Ecol. Anim.* 9: 273–276.

Haniotakis, G.E. and I.S. Pittara. 1994. Response of *Bactrocera* (*Dacus*) *oleae* males (Diptera: Tephritidae) to pheromones as affected by concentration, insect age, time of day, and previous exposure. *Environ. Entomol.* 23: 726–731.

Haniotakis, G.E., M. Kozyrakis, and I. Hardakis. 1983. Applications of pheromones for the control of the olive fruit fly. In *Proc. Internat. Conf. Int. Pl. Prot.* Vol. 4, pp. 164–171. Budapest, Hungary, July 4–9, 1983.

Haniotakis, G.E., W. Francke, K. Mori, H. Redlich, and V. Schurig. 1986. Sex-specific activity of (R)-(–)- and (S)-(–)-1,7-dioxaspiro[5,5]undecane, the major pheromone of *Dacus oleae*. *J. Chem. Ecol.* 12: 1559–1568.

Haniotakis, G.E., M. Kozyrakis, T. Fitsakis, and A. Antonidaki. 1991. An effective mass trapping method for the control of *Dacus oleae* (Diptera: Tephritidae). *J. Econ. Entomol.* 84: 564–569.

Heath, R.R. and N.D. Epsky. 1993. Recent progress in the development of attractants for monitoring the Mediterranean fruit fly and several *Anastrepha* species. In *Management of Insect Pests: Nuclear and Related Molecular and Genetic Techniques,* pp. 463–472. Int. Symp. Manage. of Insect Pests, IAEA, Vienna.

Heath, R.R. and P.J. Landolt. 1988. The isolation, identification, and synthesis of the alarm pheromone of *Vespula squamosa* (Drury) (Hymenoptera: Vespidae) and associated behavior. *Experientia* 44: 82–88.

Heath, R.R. and A. Manukian. 1992. Development and evaluation of systems to collect volatile semiochemicals from insects and plants using a charcoal-infused medium for air purification. *J. Chem. Ecol.* 18: 1209–1226

Heath, R.R., P.J. Landolt, J.H. Tumlinson, D.L. Chambers, R.E. Murphy, R.E. Doolittle, B.D. Dueben, J. Sivinski, and C.O. Calkins. 1991. Analysis, synthesis, formulation, and field testing of three major components of male Mediterranean fruit fly pheromone. *J. Chem. Ecol.* 17: 1925–1940.

Heath, R.R., A. Manukian, N.D. Epsky, J. Sivinski, C.O. Calkins, and P.J. Landolt. 1993. A bioassay system for collecting volatiles while simultaneously attracting tephritid fruit flies. *J. Chem. Ecol.* 19: 2395–2410.

Heath, R.R., N.D. Epsky, B.D. Dueben, A. Guzman, and L.E. Amdrada. 1994. Gamma radiation effect on production of four pheromonal components of male Mediterranean fruit flies (Diptera: Tephritidae). *J. Econ. Entomol.* 87: 904–909.

Heath, R.R., N.D. Epsky, A. Jimenez, B.D. Dueben, P.J. Landolt, W.L. Meyer, M. Aluja, J. Rizzo, M. Camino, F. Jeronimo, and R.M. Baranowski. 1996. Improved pheromone-based trapping systems to monitor *Toxotrypana curvicauda* (Diptera: Tephritidae). *Fla. Entomol.* 79: 37–48.

Hendrichs, J.P. 1986. Sexual Selection in Wild and Sterile Caribbean Fruit Flies, *Anastrepha suspensa* (Loew) (Diptera, Tephritidae). M.S. thesis, University of Florida, Gainesville.

Jacobson, J., K. Ohinata, D.L. Chambers, W.A. Jones, and M.S. Fujimoto. 1973. Insect sex attractants. 13. Isolation, identification and synthesis of sex pheromones of the male Mediterranean fruit fly. *J. Med. Chem.* 16: 248–251.

Jang, E.G., D.M. Light, R.A. Flath, J.T. Nagata, and T.R. Mon. 1989. Electroantennogram responses of the Mediterranean fruit fly, *Ceratitis capitata* to identified volatile constituents from calling males. *Entomol. Exp. Appl.* 50: 7–19, 248–251.

Jang, E.G., D.M. Light, R.G. Binder, R.A. Flath, and L.A. Carvalho. 1994. Attraction of female Mediterranean fruit flies to the five major components of male-produced pheromone in a laboratory flight tunnel. *J. Chem. Ecol.* 20: 9–20.

Kobayashi, R.M., K. Ohinata, D.L. Chambers, and M.S. Fujimoto. 1978. Sex pheromones of the oriental fruit fly and the melon fly: mating behavior, bioassay method, and attraction of females by live males and by suspected pheromone. *Environ. Entomol.* 7: 107–112.

Koyama, J. 1989. Tropical Dacines. In *Fruit Flies: Their Biology, Natural Enemies and Control* (A.S. Robinson and G. Hooper, eds.), pp. 165–168. In *World Crop Pests* (W. Helle, ed.), Vol. 3A. Elsevier Science Publishers, Amsterdam.

Landolt, P.J. and R.R. Heath. 1988. Effects of age, mating, and time of day on behavioral responses of female papaya fruit fly, *Toxotrypana curvicauda* Gerstaecker (Diptera: Tephritidae), to synthetic sex pheromone. *Environ. Entomol.* 17: 47–51.

Landolt, P.J. and R.R. Heath. 1990. Effects of pheromone release rate and time of day on catches of male and female papaya fruit flies (Diptera: Tephritidae) on fruit model traps baited with pheromone. *J. Econ. Entomol.* 83: 2040–2043.

Landolt, P.J. and J. Sivinski. 1992. Effects of time of day, adult food, and host fruit on incidence of calling by male Caribbean fruit flies (Diptera: Tephritidae). *Environ. Entomol.* 21: 382–387.

Landolt, P.J., R.R. Heath, and J.R. King. 1985. Behavioral responses of female papaya fruit flies, *Toxotrypana curvicauda* Gerstaecker (Diptera: Tephritidae), to male-produced sex pheromones. *Ann. Entomol. Soc. Am.* 78: 751–755.

Landolt, P.J., R.R. Heath, H.R. Agee, J.H. Tumlinson, and C.O. Calkins. 1988. Sex pheromone-based trapping system for papaya fruit fly (Diptera: Tephritidae). *J. Econ. Entomol.* 81: 1163–1169.

Landolt, P.J., M. Gonzalez, D.L. Chambers, and R.R. Heath. 1991. Comparison of field observations and trapping of papaya fruit fly in papaya plantings in Central America and Florida. *Fla. Entomol.* 74: 408–414.

Landolt, P.J., R.R. Heath, and D.L. Chambers. 1992. Oriented flight responses of female Mediterranean fruit flies to calling males, odor of calling males, and a synthetic pheromone blend. *Entomol. Exp. Appl.* 65: 259–266.

Mazomenos, B.E. and G.E. Haniotakis. 1985. Male olive fruit fly attraction to synthetic sex pheromone components in laboratory and field tests. *J. Chem. Ecol.* 11: 397–405.

Mazomenos, B.E. and J.G. Pomonis. 1983. Male olive fruit fly pheromone: isolation, identification, and lab bioassays. CEC/IOBC Symp., Athens, Nov. 1982. pp. 96–103.

Nation, J.L. 1972. Courtship behavior and evidence for a sex attractant in the male Caribbean fruit fly, *Anastrepha suspensa*. *Ann. Entomol. Soc. Am.* 65: 1364–1367.

Nation, J.L. 1975. The sex pheromone blend of Caribbean fruit fly males: isolation, biological activity, and partial chemical characterization. *Environ. Entomol.* 4: 27–30.

Nation, J.L. 1983. Sex pheromone of the Caribbean fruit fly: chemistry and field ecology. In *IUPAAC Pesticide Chemistry, Human Welfare and the Economy, Vol. 2* (J. Miyamoto and P.C. Kearney, eds.), pp. 109–110. Pergamon Press, New York.

Nation, J.L. 1989. The role of pheromones in the mating system of *Anastrepha* fruit flies. In *Fruit Flies: Their Biology, Natural Enemies and Control* (A.S. Robinson and G. Hooper, eds.), pp. 189–205. In *World Crop Pests* (W. Helle, ed.), Vol. 3A. Elsevier Science Publishers, Amsterdam.

Nation, J.L. 1990. Biology of pheromone release by male Caribbean fruit flies, *Anastrepha suspensa* (Diptera: Tephritidae). *J. Chem. Ecol.* 16: 553–572.

Nation, J.L. 1991. Sex pheromone components of *Anastrepha suspensa* and their role in mating behavior. In *Proceedings of the International Symposium on the Biology and Control of Fruit Flies* (K. Kawasaki, O. Iwahashi, and K.Y. Kaneshiro, eds.), pp. 224–236. Ginowan, Okinawa.

Poramarcom, R. 1988. Sexual Communication in the Oriental Fruit Fly, *Dacus dorsalis* Hendel (Diptera: Tephritidae). Ph.D. dissertation, University of Hawaii, Honolulu. 126 pp.

Renou, W.L., B. Lalanne-Cassou, D. Michelot, G. Gordon, and J.C. Dore. 1988. Multivariate analysis of the correlation between Noctuidae subfamilies and their sex pheromones or male attractants. *J. Chem. Ecol.* 14: 1187–1215.

Robacker, D.C. 1988. Behavioral responses of female Mexican fruit flies, *Anastrepha ludens*, to components of male-produced sex pheromone. *J. Chem. Ecol.* 14: 1715–1726.

Robacker, D.C. and W.G. Hart. 1985. (*Z*)-3-Nonenol, (*Z,Z*)-3,6-nonadienol and (*S,S*)-(−)-epianestrephin: male produced pheromones of the Mexican fruit fly. *Entomol. Exp. Appl.* 39: 103–108.

Robacker, D.C., S.J. Ingle, and W.G. Hart. 1985. Mating frequency and response to male-produced pheromone by virgin and mated females of the Mexican fruit fly. *Southwest. Entomol.* 10: 215–221.

Rocca, J.R., J.L. Nation, L. Strekowski, and M.A. Battiste. 1992. Comparison of volatiles emitted by male Caribbean and Mexican fruit flies. *J. Chem. Ecol.* 18: 223–244.

Roelofs, W.L. and R.L. Brown. 1982. Pheromones and evolutionary relationships of Tortricidae. *Annu. Rev. Ecol. Syst.* 3: 395–422.

Schultz, C.A. and G.M. Boush. 1971. Suspected sex pheromone glands in three economically important species of *Dacus*. *J. Econ. Entomol.* 84: 564–569.

Stokes, J.B., E.C. Uebel, J.D. Warthen, Jr., M. Jacobson, J.L. Flippen-Anderson, R. Gilardi, L.M. Spishakoff, and K.R. Wilzer. 1983. Isolation and identification of novel lactones from male Mexican fruit flies. *J. Agric. Food. Chem.* 31: 1162–1167.

# 30 Evolution of Fruit Fly Oviposition Behavior

*Francisco Díaz-Fleischer, Daniel R. Papaj, Ronald J. Prokopy, Allen L. Norrbom, and Martín Aluja*

## CONTENTS

## 30.1 INTRODUCTION: OPPORTUNISM AND INNOVATION AS FACTORS IN THE EVOLUTION OF TEPHRITID OVIPOSITION BEHAVIOR

Two general concepts in evolutionary biology have fundamental application to our understanding of the evolution of oviposition behavior in tephritid flies, namely, the concepts of opportunism and key innovation. The first, *opportunism*, refers to the opportunistic nature of natural selection. Natural selection does not necessarily move organisms along the path that leads to the peak of highest fitness on the adaptive landscape. Rather, natural selection often chooses a relatively expedient path, even if that path does not lead to a theoretically maximum fitness. There are two types of opportunism that we will address in this chapter. The first involves opportunistic use of what the animal itself has available to be modified to serve a particular function. The conversion of the second pair of wings in the Diptera to the halteres, structures that serve a gyroscopic function, is a classic example of morphological opportunism in natural selection.

Opportunism can take a less commonly recognized form, involving opportunistic use of what the environment offers to the animal in terms of available niches. For example, with respect to host specialization in insects, a truly spectacular example relates to the evolution of mistletoe feeding in weevils whose relatives feed on the host plants of the parasitic mistletoe (Anderson 1994). This transition from exploitation of a resource to exploitation of something that exploits the resource is a kind of ecological opportunism. Here again, the path that natural selection has taken in modifying the animal may be more expedient than the path to some perhaps more fitness-lucrative, but ecologically less accessible niche. With respect to either form of opportunism, an animal's phylogenetic history is critical to understanding the evolution of its behavior. In this chapter, we will address the extent to which opportunism figured in the evolution of oviposition behavior within the family Tephritidae.

A second concept in evolutionary biology relevant to the evolution of oviposition behavior as well as the diversification of the family Tephritidae is the concept of *key innovation*. Key innovations are traits which, once evolved, increase the rate of cladogenesis within a lineage. Examples of key innovations in the diversification of phytophagous insects are thought to abound with respect to adaptations for dealing with the secondary plant compounds found in prospective host plants. That key innovations might account for the diversification of the Tephritidae is consistent with the ideas of Southwood (1973). Southwood argued that life on higher plants presents a formidable evolutionary hurdle that most groups of insects have conspicuously failed to overcome. Once the hurdle is cleared (via one or more key innovations), radiation may be dramatic. In accordance with this view, the Tephritidae are both the most diverse family within the Tephritoidea (more than 4200 species in 471 genera are known from all parts of the world; Norrbom et al. 1999). At the same time, the Tephritidae is the only family in that superfamily whose members uniformly have phytophagous larvae, with the exception of Tachiniscidae that are parasitic and Phytalmiinae whose members are saprophagous (Dodson, Chapter 8). It appears that, with respect to phytophagy within the Tephritidae, an evolutionary hurdle was cleared that led to an impressive radiation of species. Important questions to be addressed in this chapter include what, in terms of oviposition behavior, that evolutionary hurdle was and how it was cleared. What key innovations related to oviposition permitted tephritids to exploit living plants and thus diversify to a greater extent than allied groups?

In this chapter, we will explore the role of opportunism and key innovation in the evolution of oviposition behavior and the diversification of the Tephritidae. Toward that end, we will adopt a comparative perspective. In most cases, we will forgo rigorous phylogenetic tests of hypotheses and simply put forward hypotheses along with the usually meager evidence that bears on these hypotheses. It is the aim of this chapter to point the way for those interested in gathering information of value in testing these hypotheses.

## 30.2 EVOLUTION OF HOST USE

In this section, we summarize patterns of host use within the superfamily Tephritoidea. This summary is followed by an interpretation of those patterns with respect to oviposition behavior, making reference to the concepts of opportunism and key innovation where appropriate.

### 30.2.1 PATTERNS WITHIN THE SUPERFAMILY TEPHRITOIDEA

#### 30.2.1.1 Overview of Host Use in the Tephritoidea

As reported earlier by Korneyev (Chapter 1), the superfamily Tephritoidea consists of eight families: Lonchaeidae, Pallopteridae, Piophilidae, Platystomatidae, Pyrgotidae, Richardiidae, Ulidiidae, and Tephritidae. Within the Tephritoidea, the relationships among the families are not fully resolved, but the Lonchaeidae appear to be the sister group to the rest of the superfamily. The Tephritidae belong to a monophyletic clade, termed the Higher Tephritoidea by Korneyev, that also includes the Ulidiidae, Platystomatidae, and Pyrgotidae. Within that clade, the Ulidiidae are the sister group of the other three families, and the Pyrgotidae is probably the most closely related family to the Tephritidae.

The adult breeding and larval feeding habits in most families of Tephritoidea are incompletely known, and detailed studies of this and other aspects of their biology are rare or completely lacking for many genera, tribes, or even subfamilies within the superfamily. Although our knowledge of their biology is biased by several factors, such as economic importance and distribution (temperate taxa and those including pest species are generally better known), enough information is available to give a probably accurate general picture of the breeding habits of the families of Tephritoidea. The following summary is largely based on the impressive data set compiled by Ferrar (1987); distribution and species diversity data are taken from McAlpine (1989) unless otherwise indicated.

##### *30.2.1.1.1 Lonchaeidae*

This family includes approximately 700 species, almost half of which remain undescribed. They occur in all biogeographic regions, with the least diversity in the Australasian Region. The great majority of the Lonchaeidae are scavengers, although a few species feed on plants or are predators. The saprophagous species breed in a broad range of decaying organic materials, such as dung, rotting vegetation, fruits, vegetables, or fungi, dead insects, frass, and damaged areas of living plants. The exact feeding mode of the larvae of most species is unknown. Many species have been reared from under the bark of fallen or damaged trees; they breed in the damaged plant tissues, on frass of other insects, and/or on dead beetle larvae or facultatively as predators. Many species of

*Silba* and *Neosilba* have been reared from fruits and vegetables, but most are thought to be capable of breeding only in damaged tissues; they often breed in fruits previously attacked by Tephritidae or other insects. True phytophagy has been reported in various *Earomyia* species that attack conifer seeds and *Neosilba perezi* (Romero and Ruppel) which mines *Manihot* stems. Some, probably most, species of *Dasiops* are saprophagous, but a few species are primary invaders in *Passiflora* flowers or fruit (Norrbom and McAlpine 1997), and several others in cacti. *Dasiops alveofrons* McAlpine has been reported as a primary invader in apricots, and to oviposit into them in a way similar to that of tephritids (Moffitt and Yaruss 1961). However, it also has been reared from walnut husks, currants, and sunflower stems. In Arizona, where it has been observed on walnuts, females oviposit into oviposition cavities produced in the husk by two tephritid specialists on walnuts, *Rhagoletis juglandis* Cresson and *R. boycei* Cresson (D.R. Papaj, personal observation). *Dasiops alveofrons* is likely to be either a saprophage or a secondary invader.

#### *30.2.1.1.2 Pallopteridae*

This small family contains only 54 species, distributed in the Americas, and the Palearctic and Australasian Regions (Pitkin 1989). Their feeding habits are poorly understood. The larvae of some species have been found in stems of herbaceous plants or in flowerheads galled by tephritids, but they may be secondary invaders. Other species have been found under tree bark, frequently in association with beetles. Predation has been observed, but may be opportunistic. In the absence of detailed studies, Ferrar (1987) assumed that larvae of Pallopteridae can survive on a saprophagous diet, but further assumed that some species are facultatively predaceous on insect larvae, and some are facultatively phytophagous. That these nonsaprophagous modes of nutrition are relatively recent developments is suggested by the lack of specialization of the head skeleton, which is of normal saprophagous form with well-developed ventral pharyngeal ridges.

#### *30.2.1.1.3 Piophilidae*

This small family includes only about 70 recognized species. It occurs in all regions of the world, but with greatest diversity in the Holarctic Region. The majority of species appear to be scavengers as larvae, although some feed on fungi, and at least one species feeds on the blood of nestling birds and perhaps sometimes on dead nestling carcasses. The saprophagous species appear to be specialized on high-protein, often dry, substances. Many species have been reared from carrion, often in an advanced state of decay, and others have been reared from animal bones or skins, mink and human dung, or animal products. The cheese skipper, *Piophila casei* (L.), is a pest in preserved foods. Its larvae feed deeply inside relatively dry, nutritious animal material, particularly in preserved meats, ham, bacon, dried fish, and cheese (Oldroyd 1964).

#### *30.2.1.1.4 Richardiidae*

This family includes approximately 170 currently recognized species that are restricted to the Americas, and for the most part, to the Neotropical Region. This is the least biologically known family of Tephritoidea. One species has been reared from rotten cactus, another from a diseased coconut palm, and a third from flowers of *Heliconia* spp., where the larvae feed on nectar, petals, and other flower parts.

#### *30.2.1.1.5 Ulidiidae*

The Ulidiidae (= Otitidae) includes at least 800 recognized species, although there are many undescribed species, particularly in the Neotropical Region, where the diversity of the family appears to be the greatest. The breeding habits of this family were discussed by Allen and Foote (1992) in addition to Ferrar (1987). Most species that have been reared are saprophagous, although a few species are phytophagous. The saprophagous species use a broad range of decaying organic materials for breeding, although many genera are specific in the type of decaying material. Various genera, such as *Physiphora*, are generalists capable of breeding in decaying vegetation, dung, or even carrion. Many other genera attack rotting or damaged tissues, such as stems, inflorescences, or fruits, in certain types of plants. Several genera are specialized on rotting cacti, and others on grasses or other monocots. A few genera breed in decaying cambium and other tissues in trees. They are somewhat specific in the type of tree used, not unlike some Phytalmiinae in the Tephritidae. The exact role of many species that have been bred from plant tissues is unclear; whether they are primary or secondary invaders needs to be determined. The phytophagous nature of a few species is well documented, including two *Tritoxa* species that attack *Allium* bulbs, *Tetanops myopaeformis* (Roeder) which breeds in taproots of several genera of Chenopodiaceae, and *Eumetopiella rufipes* (Macquart) which breeds in grass inflorescences. Plant feeding also appears to have evolved independently within the genera *Chaetopsis* and *Euxesta*, each genus including some species that breed in rotting vegetation, some species that are secondary invaders in damaged plants, and one or more species that are phytophagous.

##### *30.2.1.1.6 Platystomatidae*

This family includes more than 1000 currently recognized species, with the greatest diversity in the Paleotropics. Only one of the five subfamilies occurs in the Americas. The biology of this family is very poorly known. The majority of species within the group appear to be saprophagous, having been bred from decaying or damaged plant parts, such as rotting logs, coconuts, or bulbs, or dead hearts of rice or sugar cane. Two species have been reared from fruits, but without indication whether the fruits were living or rotting when attacked (Coquillett 1904; McAlpine 1973). A few species are predators, including one that attacks locust eggs, and some may breed in dung or carrion, or at least are attracted to it. Other species have been reared from beetle galleries or damage or have been found feeding on dead insects, although it is not clear if they are scavengers or predators. At least some species of *Rivellia* are true phytophages, attacking root nodules of Fabaceae.

##### *30.2.1.1.7 Pyrgotidae*

This family includes about 330 species, which occur in all biogeographic regions, but with greatest diversity in the tropics and south temperate areas (Steyskal 1987). The biology of very few species of Pyrgotidae has been studied, but all that have been reared are parasitoids of adult scarabaeoid beetles or Hymenoptera. The crepuscular habits known for many species (pyrgotids are commonly taken at light traps) and the modified female genitalia of most species also suggest that they are also parasitoids. The biology of a few primitive Neotropical genera is unknown and if different could shed light upon the evolution of parasitism in this family. Some species capture their host in flight and oviposit into the host's abdomen, usually through the soft dorsum, whereas species like *Maenomenus ensifer* Bezzi oviposit in the anus of feeding scarab hosts (citations in Ferrar 1987).

##### *30.2.1.1.8 Tephritidae*

This family is by far the largest within the Tephritoidea. More than 4200 species in 471 genera are currently recognized (Norrbom et al. 1999). The vast majority of species are phytophagous, although the Tachiniscinae are parasitoids and some, possibly most, Phytalmiinae are saprophagous. The larvae of most phytophagous species develop in the seed-bearing organs of their host plants, although some form galls or mine stems, roots, or leaves. About 35% of the species attack fruits, and another 40% (most of the subfamily Tephritinae) breed in flowers or galls of Asteraceae (White and Elson-Harris 1992). Diversity data for subfamilies listed below are based on Norrbom et al. (1999).

#### **30.2.1.2 Overview of Host Use within the Tephritidae**

##### *30.2.1.2.1 Tachiniscinae*

This group, previously ranked as a small family of three monotypic genera, is recognized as a subfamily of Tephritidae by Korneyev (Chapters 1 and 4), who expanded it to include the tribe Ortalotrypetini (5 genera, 15 species), formerly placed in the Trypetinae. The group occurs in the

Neotropical, Afrotropical, Palearctic and Oriental Regions. The only known host record is for an undescribed species of *Bibundia* reared by Roberts (1969) from a caterpillar of a saturniid moth.

#### 30.2.1.2.2 Blepharoneurinae

The relationships of the five genera of this subfamily are analyzed by Norrbom and Condon (Chapter 6) and their host plant associations are reviewed by Condon and Norrbom (Chapter 7). The group includes 33 extant species, although many more undescribed species of *Blepharoneura* are known. All species reared to date are phytophagous, although the biology of the genera *Ceratodacus* and *Problepharoneura,* whose relationships are somewhat uncertain, remains unknown. Most of the reported host records, all from the Cucurbitaceae, are for the Neotropical genus *Blepharoneura,* but the two other genera, the Afrotropical *Baryglossa* and Oriental/eastern Palearctic *Hexaptilona,* also appear to be associated with this plant family. *Blepharoneura* species are highly host and tissue specific, and different species can use different tissues of a single host plant, including male flowers, female flowers, seeds, fruits, or stems.

#### 30.2.1.2.3 Phytalmiinae

This subfamily includes four tribes: the Acanthonevrini (76 genera, 282 species), Epacrocerini (4 genera, 7 species), Phascini (6 genera, 14 species), and Phytalmiini (13 genera, 34 species). This group is restricted to the Old World and mainly to the Paleotropics; only three genera reach the Palearctic Region. Most species of Phytalmiinae that have been reared appear to be saprophagous (see Dodson, Chapter 8), although Hardy (1986) and Permkam and Hancock (1995b) reported that species of *Clusiosoma, Clusiosomina, Cheesmanomyia,* and *Rabaulia* are known to infest fruit of *Ficus* spp. *Dirioxa pornia* (Walker) has been bred from fruits of a wide variety of plants, but normally only in damaged or decaying fruits; it also has been reared from fallen *Araucaria* cones (Permkam and Hancock 1995b). Species of *Acanthonevra, Felderimyia, Polyara,* and *Ptilona* have been reared from damaged bamboo or internodal spaces in the shoots, and species of various other genera have been collected on bamboo and may breed in it (Hardy 1986; Hancock and Drew 1995a, b). Species of *Austronevra, Dacopsis, Diarrhegma, Diarrhegmoides, Lumirioxa,* and *Phytalmia* breed in decomposing tree trunks or rotting parts of trees, although individual species of at least *Phytalmia* are fairly specific to certain types of trees (Hardy 1986; Dodson and Daniels 1988; Permkam and Hancock 1995b; Dodson, Chapter 8). Some *Afrocneros* and *Ocnerioxa* have been found under the bark of living trees (Munro 1967). *Termitorioxa termitoxena* (Bezzi) breeds in termite galleries in tree trunks (Hill 1921).

#### 30.2.1.2.4 Trypetinae

This mainly phytophagous and frugivorous subfamily, which may be paraphyletic, includes nearly 1000 species divided among six tribes, including Adramini (26 genera, 181 species), Carpomyini (12 genera, 115 species), Rivelliomimini (3 genera, 6 species), Toxotrypanini (3 genera, 216 species), Trypetini (44 genera, 385 species), and Zaceratini (2 genera, 2 species), plus an additional 24 genera (with 54 species) which are unplaced. The Trypetinae is the most diverse subfamily in terms of larval feeding habits. The Adramini (= Euphrantini), which are restricted to the Old World except for two species of *Euphranta* from North America, breed in fruits, seeds, flower buds, or stems (Hardy 1983; 1986b; White and Elson-Harris 1992; Hancock and Drew 1994), although one species, *E. toxoneura* (Loew), is a predator of sawfly larvae within their galls (Kopelke 1984). The Carpomyini, which occur mainly in the Holarctic and Neotropical Regions, breed exclusively in fruits, and most species are highly host specific (White and Elson-Harris 1992; Smith and Bush, Chapter 9). The Toxotrypanini, which are restricted to the Americas and mainly to the Neotropical Region, also breed almost exclusively in fruits or in the seeds within; some species are generalists, but many are specialized on specific plant groups such as *Passiflora* or latex-bearing families such as Sapotaceae, Moraceae, Apocynaceae, Caricaceae, or Asclepiadaceae (Norrbom et al., Chapter 12). The Rivelliomimini, which occur only in tropical Africa, Asia, and the Pacific, are poorly known biologically; only one species appears to have been reared, from *Cycas* spp. (White and Elson-Harris 1992). The Trypetini is most diverse in the Oriental and Palearctic Regions. Species of this tribe are mostly leaf or stem miners or fruit breeders (Han, Chapter 11), although the small subtribe Acidoxanthina which is sometimes included in the Trypetini includes flower breeders. The two species of Zaceratini, one Afrotropical and one Palearctic, are stem borers (White and Elson-Harris 1992).

##### *30.2.1.2.5 Dacinae*

This subfamily of more than 1000 species includes three tribes: the Ceratitidini (20 genera, 198 species), Dacini (3 genera, 723 species), and Gastrozonini (19 genera, 112 species). The group is restricted to the Old World except for species spread by humans. Most species of Dacini and Ceratitidini breed in fruits, seed pods, or seeds, although a few dacine species breed in flowers and the ceratitidine genera *Capparimyia* and *Paraceratitella* attack flower buds (Hancock 1987; White and Elson-Harris 1992; Permkam and Hancock 1995a). The few species of Gastrozonini of known biology attack living stems of Poaceae, mainly bamboo (Hardy 1988; Hancock and Drew 1995a).

##### *30.2.1.2.6 Tephritinae*

This is the largest subfamily of Tephritidae, with 203 genera and 1847 currently recognized species divided among 11 tribes. This group is most diverse in the Holarctic Region and temperate or high-altitude areas of the Afrotropical and Neotropical Regions. Except for the tribe Tephrellini and a few species of *Eutreta* and *Oedaspis*, the subfamily is associated exclusively with the plant family Asteraceae. The majority of species breed in flowerheads, although a good number form galls in stems, roots, or flowers (White and Elson-Harris 1992).

#### 30.2.1.3 Interpretation of Patterns of Resource Use in the Tephritoidea

##### *30.2.1.3.1 General themes in host use in the Tephritoidea*

The preceding summary of resource use in the Tephritoidea reveals several themes of interest in understanding the evolution of tephritid oviposition behavior. First, the families of flies most closely related to Tephritidae tend to feed as larvae in rotting substrates, including decaying vegetation. In several of these related families, predation and parasitism (often of creatures that inhabit sites of decay) and/or even phytophagy is known, but the dominant theme is one of life on rot and decay.

Second, as one proceeds from more-basal to more-derived groups within the Tephritoidea (see phylogenies by Korneyev, Chapter 1), there is a progressive tendency for proportionately more of the rotting and decaying substrates utilized in breeding by a particular group to be predominantly plant derived. Thus, the most basal group in the Tephritoidea, the Loncheidae, use the broadest possible range of substrates, including dung, insects, fungi, and plants. Similarly, another basal group, the Piophilidae, range widely in terms of substrates, using carrion, fungi, and even blood. In contrast, the more-derived Platystomatidae, while also predominantly saprophagous, use mainly plant-derived rot and decay as breeding substrates. The Ulidiidae, which are basal to the rest of the higher tephritoids, show a mixed pattern in type of substrates. Some genera range broadly in terms of substrates of decay, whereas others predominantly use plant-derived substrates. Finally, the progressive use of substrates of decay of plant origin is observed in a basal group of the Tephritidae, the Phytalmiinae. This group is relatively unusual in the family for their saprophagous lifestyle; however, their breeding substrates are entirely plant derived.

Third, based on current knowledge of the phylogeny of the Tephritoidea and Tephritidae (Korneyev, Chapters 1 and 4), the transition to feeding upon living plant tissue has occurred multiple times within the superfamily, including several times each within the Lonchaeidae and Ulidiidae, at least once within the Platystomatidae, and probably at least twice within the Tephritidae, in the Blepharoneurinae and in the Trypetinae + Dacinae + Tephritinae clade (perhaps also within the Acanthonevrini). Only in the Tephritidae has this transition led to a rich diversity of extant taxa.

### *30.2.1.3.2 Opportunism in relation to host use*

Based on the first two phylogenetic trends noted above, we propose that the transition from a saprophagous to a phytophagous lifestyle within the Tephritoidea is an example of opportunism. Phytophagy arose when the ecological niche exploited by saprophagous ancestors provided an opportunity for phytophagy, specifically when saprophagous groups began to utilize rotting substrates of plant origin. Exactly what facilitated the transition from plant-based saprophagy to phytophagy is largely a matter of conjecture. Phenology may be a factor; even in the tropics, plant decay probably has a seasonal component and phenological matching to a seasonal pattern of a plant's decay in a saprophagous ancestor could conceivably have facilitated a transition onto the living tissue of that plant. Similar arguments may be made with respect to ovipositional or larval feeding stimulants as facilitative factors in the transition to living tissue.

### *30.2.1.3.3 The tephritid ovipositor as a key innovation*

If saprophagy on plants provided the opportunity for the origin of phytophagy, what key innovations facilitated the shift to living plant tissue and the rapid diversification of phytophagous groups? There are two likely candidates, one involving changes to the ovipositor and one involving changes that permitted larvae to develop on a diet of plant tissue.

One key innovation probably involved deposition of eggs into plant parts. This innovation involved specializations of the ovipositor (as well as the mechanical means and behavior pattern associated with its deployment) which permitted the tough surfaces of fruits or other plant parts to be penetrated (see also White, Chapter 20). The evolution of the tephritid ovipositor can be viewed as occurring in two stages, the first of which occurred early in the diversification of the Tephritoidea. All members of the superfamily have a specialized ovipositor, characterized by a well-sclerotized, tubular or conical oviscape formed by the tergite and sternite of segment 7. Except in some Piophilidae and Pallopteridae, these sclerites are completely fused. The tephritoid ovipositor also includes an elongate eversible membrane and elongate, slender aculeus, which telescope, at least partially, into the oviscape when at rest. This type of ovipositor presumably allowed its saprophagous owners to lay their eggs deeper into decaying substrates, or into crevices or cracks (e.g., in tree bark or damaged plant parts).

One factor favoring the insertion of eggs into decaying substrates with a specialized ovipositor may have been protection of eggs from desiccation. Supporting this idea is the absence of any obvious adaptations in the eggs of Tephritoidea for reducing risk of desiccation. Chorionic modifications for reducing desiccation are found, for example, in the eggs of many Lepidoptera that deposit eggs on exposed plant surfaces (Zeh et al. 1989, and references within). What prevents such modifications within the Tephritoidea is not known; however, such a constraint, if present, would have generated an advantage to inserting eggs into cracks, crevices, or substrates that offer a relatively moist microclimate for developing eggs.*

A second factor favoring the insertion of eggs into decaying substrates with a specialized ovipositor may have been protection from predators and parasites. Yet, a third factor may have been an escape from competition afforded by use of novel substrates. In this regard, the primitive tephritoid ovipositor presumably facilitated use of drier types of decaying substrates than substrates such as dung or carrion used by saprophagous members of the Lower Tephritoidea. These drier substrates include the bark of dead or injured trees, which probably could not be used unless eggs were laid on the undersurface, and damaged fruit, which probably could not be used unless eggs were inserted below the fruit surface. Such substrates are used by many members of the Loncheidae (Lower Tephritoidea) and Ulidiidae (Higher Tephritoidea) as well as most members of the Phytalminae (a relatively basal group within the Tephritidae).

* With insertion into a substrate comes a potential problem: eggs may suffer from reduced availability of oxygen. The eggs of some tephritid species have specialized lobes which probably function to increase the availability of oxygen to eggs placed within moist plant tissues.

An important second stage in the evolution of the tephritid ovipositor involved strengthening of the aculeus to permit a more effective piercing of the tissues of living plants. The aculeus includes the tergite and the pair of sternites of segment 8, plus an apical cercal unit, formed from cerci, which are usually at least partially fused, and probably the hypoproct (Korneyev, Chapter 1). In saprophagous members of the Tephritoidea, the cercal unit is not fused to the eighth tergite and is thus movable. By contrast, in many, and possibly most, phytophagous species of Tephritoidea, the cercal unit is fused to the eighth tergite, forming a stronger, perhaps more rigid aculeus capable of piercing the tissues of living plants. This modification occurs in all phytophagous Tephritidae, in all phytophagous species of the genus *Dasiops* (Lonchaeidae) that breed in *Passiflora* and at least some species that breed in cacti (Norrbom and McAlpine 1997; A. Norrbom, personal observation). In *Chaetopsis* (Ulidiidae), some species of which are phytophagous and some of which are saprophagous, the cercal unit may be free or fused to varying extents with the rest of the aculeus (Norrbom, personal observation). In *Rivellia* (Platystomatidae), at least some of which are phytophagous, the cercal unit is movable (e.g., see Hara 1989). However, these flies do not oviposit into the plant, but instead place eggs in detritus on the soil surface; larvae then migrate down through the soil and attack root nodules (Foote 1985). The Pyrgotidae and Tachiniscinae (Tephritidae), which are parasitoids that deposit their eggs inside living insects, also do not possess a separate cercal unit; that unit is presumably fused with the rest of the aculeus in these taxa. Additional modifications of the oviscape (e.g., the shape of the base and oviscape length) and eversible membrane (e.g., taeniae closer together, various patterns and shapes of denticles) occur among the phytophagous Tephritidae, at least some of which are probably related to oviposition.

We view the tephritid-type ovipositor, together with the mechanical means and behavior pattern associated with its deployment, as a key innovation facilitating the tephritoid transition to life on living plant tissue. Further secondary modifications of the aculeus may have facilitated the diversification of tephritid species via specialization. Jones (1989) reported that variation in aculeus morphology among tephritids that use living plant tissues as oviposition substrate is related to characteristics of the host. Tephritids whose hosts possess thick cuticles tend to have aculei with smaller tip widths (e.g., more pointed) and more acute tip angles. Both attributes should facilitate host penetration. Presumably, small tip widths and acute tip angles have some cost; otherwise, the aculei of all tephritid species would possess these features.

Among polyphagous *Anastrepha* species, the relationship between aculeus width and host cuticle thickness is not strong. Possibly, the large body size typical of *Anastrepha* (Jones 1989) enables them to penetrate thick-skinned fruits despite possession of long, relatively blunt-tipped aculei. Alternatively, the serrations present on the tip of the aculeus in many *Anastrepha* species may serve the same function as slender, sharp aculeus tips (A. Norrbom, personal observation). There may exist some as-yet-undefined cost of slender, sharp-tipped aculei that is particularly high in *Anastrepha* species.

#### *30.2.1.3.4 Oviposition-mediated bacterial transfer as a key innovation*

A second key innovation promoting the shift to living plant tissue presumably involved traits permitting larvae that fed ancestrally on rotting substrates to feed on living plant tissue. On this subject, weighty statements are hard to make, owing to the dearth of detailed information on larval nutrition in tephritids and related groups. Some tentative points are worth raising. First, the larvae of many frugivorous forms, although inserted as eggs into fresh fruit flesh, are nevertheless feeding on rotting flesh. In gregarious forms such as *Rhagoletis* species in the *suavis* species group and *Anastrepha* species that lay eggs in clutches, the larvae themselves appear to generate an advancing front of rot along which they feed (C. Nufio and D. Papaj, personal observation; M. Aluja and colleagues, personal observation). For at least some groups (e.g., many frugivorous tephritids), transition to life on living plant tissues may have involved the evolution of traits involved in transferring bacteria to fruit. Such bacteria began or enhanced a process of fruit degradation. Transfer of such bacteria may be mediated during oviposition (Howard et al. 1985).

Innovations in transfer of bacteria from fly to fruit may have been mediated by an intermediate stage, such as that found in lonchaeids and phytalmiines, in which females of most species lay eggs in small areas of rot within the intact resource. It is easy to imagine a sequence of evolutionary transitions from "self-rotting" substrates to substrates in which rot is fly enhanced to substrates in which rot is wholly fly mediated, and finally to substrates used in the absence of rot. During oviposition, females of a lonchaeid-like or phytalmiine-like ancestor might passively transfer bacteria from rotting areas to areas that were accessible for oviposition yet not rotting. Bacterial transfer could have facilitated development of the rot, oviposition into such rot, and survival of young. From this rudimentary form of transfer may have evolved traits in both females and bacteria that favored harboring of bacteria within females, transfer of those bacteria during oviposition, and promotion of bacterially mediated tissue degradation in the vicinity of deposited eggs. Such traits may in turn have favored the shift onto living plant tissue in which rot was wholly or largely fly mediated.

Once the transition to living plant tissue was made, via the exploitation of rot within that tissue, the way was paved for use of living plant tissue without the requirement of tissue decay. In a number of relatively derived tephritid species, including *Toxotrypana* and some *Anastrepha* spp. that feed on seeds or associated tissues, larval trails do not appear to be associated with visible rot (Baker et al. 1944; M. Aluja, personal communication; P. Landolt, personal communication). It is possible that biochemical changes invisible to the naked eye are taking place as larvae feed, but it is also possible that larvae in these species are capable of digesting living tissue without degrading that tissue prior to ingestion. Most Tephritinae and leaf-mining trypetine spp. do not seem to require rotting material and, at least in some species, larvae even feed directly on plant sap. In nonfrugivorous tephritids, the aculeus often pierces plant tissues and eggs are laid within those tissues. Occasionally, necrosis is generated at the site egg deposition (compare Goeden et al. 1994) but, more usually, little necrosis of invaded plant tissues is obvious (Headrick and Goeden 1998).

To evaluate the importance of oviposition-related innovations in bacterial transfer related to larval use of living plant tissue, more knowledge about larval nutrition in both the tephritids or allied groups is needed (but see Drew and Yuval, Chapter 27). We need to determine more definitively whether or not the bacteria transferred by adult flies to fruit are truly promoters of larval performance in at least some members of the Tephritidae and Higher Tephritoidea. We also need to know more about the process of transfer of bacteria by ovipositing females, again from a comparative perspective. There are two alternative hypotheses that such information (along with more robust family-level phylogenies) might permit us to distinguish. First, it is possible that the importance of bacterial decay, induced either passively or actively, in exploitation of living plant tissues declines gradually over evolutionary time. This hypothesis might predict that more-derived groups within the family depend less on bacterial transfer during oviposition than more-basal groups. Alternatively, it is possible that the importance of bacterial decay depends less on phylogenetic position and more on the ecology of particular groups, and especially on the kinds of hosts or host parts that are used for larval development. These alternatives are not mutually exclusive and their level of importance may depend on the level of phylogenetic resolution considered. The first hypothesis may receive more support at the level of the superfamily; the second hypothesis may receive more support at the level of tribe or even genus.

#### *30.2.1.3.5 Key innovations, host specialization, and diversification*

We argued above that innovations involving the ovipositor and oviposition-mediated patterns of bacterial transfer facilitated a shift onto living plant tissue. However, such innovations can be characterized as "key innovations" only if they increase rates of cladogenesis. It is our position that these innovations promoted diversification, essentially by paving the way for resource specialization. In other groups of insects, the transition to life on plants appears to have been associated with high rates of cladogenesis (Mitter et al. 1988). Any innovation that facilitates use of living plant tissue can be viewed as an innovation that facilitates host specialization and thereby promotes diversification. In our view, host specialization is the driving force behind speciation and ultimately the rich pattern of diversification observed in the family Tephritidae, but key innovations in ovipositor morphology, oviposition behavior, and patterns of bacterial transfer set the stage for specialization. The relationship between host specialization and diversification in the tephritids is treated elsewhere in this volume.

## 30.3 EVOLUTION OF HOST SELECTION BEHAVIOR

We argued above that key innovations in ovipositor morphology and oviposition behavior facilitated a shift onto living plant tissue and, in so doing, paved the way for host specialization. Host specialization is in part a function of host selection behavior. In this section, we address aspects of oviposition behavior such as host finding and host acceptance that contribute directly to host specialization in tephritid flies.

### 30.3.1 Host Preference — Performance Trade-Offs

As discussed by Fletcher (1989), most species of tephritid flies, irrespective of subfamily or tribal affiliation, are either monophagous or stenophagous. One might therefore expect that, in most tephritid species, behavior associated with host plant finding and acceptance would consist of quite specific responses to chemical and/or physical cues from host plants. This expectation appears to hold true for the limited number of monophagous or stenophagous species for which sufficient information is available. However, polyphagous tephritid species, while fewer in number, have received most of the attention with respect to host selection behavior. While this disparity in research emphasis is not surprising in view of the major pest status of certain polyphagous species, it means that comparative data are rarely robust enough to define evolutionary trends in host selection behavior of relevance to host specialization.

When searching for comparative trends in the evolution of oviposition behavior, one point of focus for any phytophagous insect is a possible linkage between preference for host oviposition sites and performance of larvae in hosts. Oviposition site selection can be regarded as a form of maternal investment in which females invest time and energy to select sites in which offspring are more likely to survive. Particularly under conditions of egg limitation (wherein females tend to deplete their egg supply before opportunities to oviposit are exhausted), females should be selective, placing eggs on those hosts that tend to be associated with relatively high juvenile growth and survival.

Traditional within-species evaluations of the host preference–performance relationship conducted on various species of *Bactrocera* (Bower 1977; Fitt 1986), *Rhagoletis pomonella* (Walsh) (Neilson 1967; Reissig et al. 1990), and *Eurosta solidaginis* (Fitch) (Horner and Abrahamson 1992) show little correlation between hosts preferred for oviposition and hosts that best support larval growth and development. Still, a failure to find preference–performance correlations within a given species should not deter us from examining such correlations on an among-species basis. It is well known that correlations among traits assessed within a species can differ in magnitude and even in sign from those assessed on an among-species basis.

Preference–performance correlations can be evaluated with available phylogenetic techniques. An investigator might determine which of a range of potential hosts were preferred for oviposition by various fly species and which of the same range of hosts supported larval development in those species. Host preference measurements could be mapped onto a phylogenetic tree of the group under study for evaluation of trends in host preference; similarly, larval performance measurements could be mapped onto a tree of the same form and trends again evaluated. Finally, the two trees could be compared side by side to determine if, at the level of the entire phylogenetic tree, there was evidence of a match between host preference and larval performance.

Unfortunately, such analyses have not been performed on tephritid flies to date. In lieu of comparative studies of this sort, we will devote the remainder of our discussion of host selection behavior to a review of comparative data on host-finding and host-acceptance behavior. With respect to host finding and acceptance per se, little or no experimental information exists for members of the subfamilies Tachiniscinae, Blepharoneurinae, and Phytalmiinae. Hence, members of these subfamilies, considered to be the oldest lineages of the Tephritidae, will not be considered here. Focus will be primarily on members of two more-derived subfamilies (Trypetinae and Dacinae) and to a lesser extent on the most-derived subfamily (Tephritinae). The approach we will take involves exposition of various components of host finding and acceptance behavior, with special attention to species differences in behavior that appear to relate to differences in level of host specialization.

### 30.3.2 Host Finding

Arrival of females on host plants may sometimes result from activities not specifically directed toward oviposition site foraging, and involve responses to stimuli from food, mates, or shelter sites. As suggested by Drew and Romig (Chapter 21), evidence is mounting that at least in some species of *Bactrocera*, volatiles from bacteria associated with feeding sites on host plants may also draw oviposition-site-seeking females to hosts. Here, we will confine our discussion to those stimuli that appear to be related exclusively to foraging for egg-laying sites.

Females of several species of tephritids have been shown to move upwind toward sources of odor emitted from potential oviposition sites (Jang and Light 1996). Odors from nonhost plants are infrequently attractive to females seeking egg-laying sites (Jang and Light 1996). From available information, it appears that monophagous or stenophagous species respond positively to a narrower set of host plant volatiles than do polyphagous species. For example, substantial attraction of stenophagous *R. pomonella* flies to volatiles from fresh, ripe fruit of native hosts (hawthorns) and recently acquired agricultural hosts (apples) is limited to five closely related esters emanating from each of these host types (Carle et al. 1987; C. Linn and W. Roelofs, unpublished data). Also, substantial attraction of monophagous *Bactrocera oleae* (Rossi) flies to volatiles from partially ripe host olive fruit seems to be limited to two related terpenes (Scarpati et al. 1993). On the other hand, studies on *Anastrepha suspensa* (Loew) (Nigg et al. 1994), *Ceratitis capitata* (Wiedemann) (Light et al. 1988; 1992; Hernández et al. 1996; Warthen et al. 1997), *B. dorsalis* (Hendel) (Light and Jang 1987), and *B. tryoni* (Froggatt) (Eismann and Rice 1992), all of which are polyphagous, suggest positive response to the odor of a broad range of unrelated compounds emitted by ripening host fruit. Unfortunately, no convincing studies have been conducted on possible response to host plant odor by members of the Tephritinae, a subfamily composed mainly of specialists. Conceivably, breadth of odor response in terms of numbers of compounds is directly related to breadth of host range.

For frugivorous tephritids, plant or foliar color, shape, or size may provide comparatively short-range generalized visual cues helping to guide odor-responsive females to fruit-bearing host plants (Moericke et al. 1975; Meats 1983; Green et al. 1994). All tephritids studied to date are most sensitive to light reflected at about 500 nm (Agee et al. 1982), approaching that part of the spectrum where green leaves are maximally reflective (550 nm) within the visual spectrum of tephritids (350 to 600 nm). There is no evidence that any tephritid is able to distinguish a specific host plant by the reflectance composition of the foliage, even in species such as *B. oleae*, whose host olive trees bear leaves that differ markedly in light reflectance (having a greater amount of ultra-violet light reflected from lower leaf surfaces) from typical foliage (Prokopy and Haniotakis 1975). There is also no evidence that any frugivorous tephritid responds selectively to the particular form or size of its host plant. There is some evidence to suggest, however, that *Plioreocepta poeciloptera* (Schrank) (Zaceratini; probably closely related to tephritine flies), a specialist stem borer in asparagus, uses the vertical aspect of shoots of asparagus as a visual cue (Eckstein 1931).

Once a tephritid female closely approaches or arrives on a host plant, both olfactory and visual cues emanating from potential oviposition sites may be used to locate such sites. Although insufficient for drawing firm conclusions, limited evidence suggests that monophagous and stenophagous tephritids rely more on visual cues for locating egg-laying sites than polyphagous ones. Studies on the fruit-foraging behavior of the specialist species *R. pomonella* in field-caged host trees indicate that when host fruit are abundant and visually conspicuous, fruit visual stimuli alone elicit alighting; fruit odor stimuli enhance alighting on fruit only when fruit are scarce or visually inconspicuous, and even then to only a slight degree (Aluja and Prokopy 1993). Similarly, field-cage studies on the specialist *B. oleae* suggest that when fruit are visually conspicuous, visual cues alone are sufficient to elicit alighting (Prokopy and Haniotakis 1976). On the other hand, odor of fruit proved as important as or even more important than visual stimuli of fruit in eliciting alighting of generalist *B. dorsalis* and *B. tryoni* flies on visually conspicuous fruit in field-caged host trees (Prokopy et al. 1990; 1991).

Among the physical oviposition site properties of shape, size, and color, shape elicits the strongest and most specific positive responses in alightings of tephritid flies. Response to size appears to be less specific and, to color, least specific. Analogous to the aforementioned pattern of tephritid responses to host odor, it appears that specialist tephritids respond to a narrower profile of visual stimuli than do generalist tephritids. For example, with respect to form of oviposition sites, females of the specialist frugivores *R. pomonella*, *R. cerasi* (L.), and *B. oleae*, as well as the composite-infesting tephritine specialists *Urophora sirunaseva* (Hering) and *Chaetorellia australis* Hering, are highly attracted toward visual mimics whose form closely resembles that of prospective oviposition sites (Prokopy 1968; 1969; Zwölfer 1969; Prokopy and Haniotakis 1976; Pittara and Katsoyannos 1992). The form of oviposition site also plays a role in attraction of females of the generalist frugivores *Anastrepha ludens* (Loew), *C. capitata*, *B. tryoni*, and *B. dorsalis* (Nakagawa et al. 1978; Hill and Hooper 1984; Robacker 1992; Cornelius et al., in press), but responses to host form in these species are less specific than those by specialist tephritids. With respect to attraction of tephritid flies toward different sizes of oviposition sites, response patterns are confounded by the phenomenon that fly perception of site size varies according to distance of the fly from the site (Roitberg 1985). In cases in which tephritid fly responses toward different-size models of oviposition sites have been evaluated (reviewed by Katsoyannos 1989), females tend to be more attracted to models substantially larger than natural oviposition sites than to the sites themselves. Too few species have been examined in sufficient detail to reveal possible differences in responses. Interpretation of findings on attraction of tephritids to different colors of oviposition sites is somewhat complicated by the fact that response to color varies according to the visual background of the responding fly (reviewed by Katsoyannos 1989). Nonetheless, females of specialist tephritids may be attracted to a narrower range of oviposition site colors than are females of generalist tephritids. In frugivorous tephritids, for example, consider the degree of difference in attractiveness between the most attractive color of fruit-mimicking spheres and a white sphere. Few species of ripening fruit are white, and white may be considered to represent a rather neutral surface, reflecting a full range of incoming wavelengths of sky light within the visible spectrum of tephritids, except possibly ultraviolet light. Females of the specialist species *R. pomonella*, *R. cerasi,* and *B. oleae* strongly prefer dark-colored spheres over white ones (Prokopy 1968; 1969; Prokopy and Haniotakis 1976). In contrast, females of the generalist species *A. fraterculus* (Wiedemann), *A. ludens*, *C. capitata*, *B. dorsalis,* and *B. tryoni* show little or no discrimination between white spheres and spheres of other colors (Nakagawa et al. 1978; Cytrynowicz et al. 1982; Katsoyannos 1987; Vargas et al. 1991; Robacker 1992; Cornelius et al. 1998; Drew and Prokopy, unpublished data).

### 30.3.3 Host Acceptance

After arrival at a prospective oviposition site, tephritid females use a variety of cues to determine whether the site is acceptable for oviposition. Stimuli include chemicals in surface waxes, various exterior physical characteristics such as shape, size, and color, and the chemical composition and physical structure of the interior (reviewed by Katsoyannos 1989; Fletcher and Prokopy 1991;

Prokopy and Papaj, Chapter 10; Aluja et al., Chapter 15; Yuval and Hendrichs, Chapter 17). Chemical stimuli associated either with the surface or interior of oviposition sites stimulate oviposition in the specialist species *E. solidaginis*, *R. pomonella*, *R. mendax* Curran, *R. cerasi*, and *B. oleae* (Haisch and Levinson 1980; Girolami et al. 1983; Abrahamson et al. 1989; Bierbaum and Bush 1990; Kombargi et al. 1998), as well as in the generalist species *A. suspensa*, *C. capitata*, and *B. tryoni* (Pritchard 1969; Szentesi et al. 1979; Freeman and Carey 1990; Eismann and Rice 1992). However, too few detailed comparisons of responses of ovipositing females to chemical profiles associated with hosts vs. nonhosts have been made to permit meaningful speculation on degree of specificity of response in relation to degree of breadth of host range.

With respect to physical stimuli, there appears to be a trend toward decreasing specificity in the physical characteristics of hosts that elicit egg laying as host range expands from monophagous and stenophagous species to polyphagous species. Many species in the composite-infesting tephritine genera *Urophora*, *Chaetorellia*, *Eurosta*, *Tephritis*, and *Cerajocera* are essentially monophagous and respond to a narrow range of oviposition site physical properties, notably size (diameter), shape (or shape of bracts associated with the site), and, to a lesser extent, color (Zwölfer 1969; 1972a, b; Abrahamson et al. 1989; Straw 1989b, c; Pittara and Katsoyannos 1992). Similarly, monophagous or stenophagous frugivorous tephritids such as *R. cerasi*, *R. pomonella*, *R. mendax*, *R. zephyria* Snow, *R. cornivora* Bush, *R. completa* Cresson, *R. indifferens* Curran, and *B. oleae* respond positively to a narrow range of shapes and sizes after alighting, although somewhat more broadly or variably to a range of colors (Wiesmann 1937; Prokopy 1966; Prokopy and Boller 1971; Cirio 1972; Prokopy and Bush 1973; Haisch and Levinson 1980; Katsoyannos and Pittara 1983; Katsoyannos et al. 1985; Messina 1990). Among these frugivorous species, those having the narrowest host ranges respond to a narrower set of fruit sizes than those having broader host ranges. Although solid information on ovipositional responses of polyphagous tephritids to a range of fruit-mimicking physical stimuli is confined almost exclusively to *C. capitata*, positive responses in this species at least appear not to be as canalized as in monophagous tephritids (Féron 1962; Sanders 1962; Katsoyannos et al. 1986; Freeman and Carey 1990).

Once a tephritid female has accepted a potential oviposition site and begun to bore, compounds in fruit such as glucose, fructose, sodium chloride, and calcium chloride may stimulate or inhibit egg deposition, depending on concentration (Tsiropoulos and Hagen 1979; Girolami et al. 1986; Eismann and Rice 1985). Another factor of possible major importance to egg deposition is the length and shape of the aculeus in relation to characteristics of a potential ovipositional site. A striking example of the relevance of aculeus length to egg deposition is given by Zwölfer (1983), who showed a strong positive correlation between sizes of flowerheads and lengths of aculei of host-specific *Urophora* females ovipositing into them. Species-typical characteristics of aculeus morphology, in addition to changes in morphology resulting from aculeus wear (Jones and Kim 1994), may constrain the potential host range of a tephritid.

### 30.3.4 Related Themes

In sum, comparative evidence suggests that monophagous and stenophagous tephritids tend to be chemical, visual, and tactile specialists when searching for and evaluating oviposition sites, whereas polyphagous tephritids tend toward being odor, visual, and tactile generalists. Below we address some themes worthy of additional attention.

#### 30.3.4.1 Evolutionary Trends in Level of Specialization

It is commonly believed that specialists arise from generalist ancestors and not the other way around (but see Kaneshiro, Chapter 32). To date, the great majority of studies on host finding and acceptance behavior in tephritids have been carried out on species in the subfamilies Trypetinae and Dacinae. Because relationships among members of these two subfamilies remain incompletely resolved (with even the monophyletic status of the Trypetinae in some doubt), it is not possible at this time to determine whether a generalist or a specialist pattern is ancestral in these subfamilies. Within the Toxotrypanini, polyphagy, or at least broad polyphagy, appears to be a derived trait, apparently associated with the type of fruit tissue attacked (pulp vs. developing seeds) (Norrbom et al., Chapter 12). Additionally, there is evidence that, in historical times at least, host range has tended to expand rather than contract. Range expansion was facilitated by the increased availability of new types of host fruit as a consequence of agricultural breeding and production, beginning ten or more millennia ago. Initial expansion of host range probably involved shifts onto agriculturally developed or agriculturally introduced species whose chemical and physical cues used in host finding and acceptance bore close resemblance to native hosts. Such a shift occurred when the stenophagous species *R. pomonella* expanded its host range, adding introduced apple fruit to a range formerly including only the native hawthorn fruit. As host range expanded, species probably broadened their response pattern to include an ever-greater range of host cues.

Patterns of egg maturation in dacine flies suggest that, in terms of oviposition behavior at least, expansion of host range is more likely for a generalist than a specialist. Egg maturation in three Australian dacine specialist species, *B. cacuminata* (Hering), *B. cucumis* (French), and *B. jarvisi* (Tryon), was retarded when females were deprived of hosts (Fitt 1986). Egg maturation in the highly polyphagous *B. tryoni*, in contrast, continued unabated. Eggs thus accumulated with continued deprivation, a pattern accompanied by a decline in host selectivity (Fitt 1986). Accumulation of eggs by a generalist and the decline in selectivity associated with egg load could promote deposition of eggs into fruit not previously used as hosts. The generalist dacine is, in a way, "better prepared" to shift onto novel hosts than is the specialist. Such differences in response to host deprivation could have the effect of promoting continued specialization in the specialist, while facilitating expansion of host range in the generalist.

#### 30.3.4.2 Specialization and Suites of Correlated Traits

Like any trait, oviposition behavior does not evolve in a phenotypic vacuum. Variation in ovipositional responses is undoubtedly accompanied by correlated variation in other aspects of reproductive morphology and physiology. For example, ovipositors, particularly aculei, of polyphagous tephritids may be better adapted for exploiting a broad range of skin thicknesses and hardnesses of host fruit than are ovipositors of nonpolyphagous tephritids. A specialist's ovipositor, by contrast, may be better adapted to penetration of its host's surface than is a generalist ovipositor. In other words, ovipositors should be adapted for either specialist or generalist roles.

Similar arguments may be made with respect to ovarian development. The dacine species differences in pattern of ovarian maturation discussed above (Fitt 1986) could reflect an adaptive difference in the way that generalists and specialists respond to host deprivation. A specialist faced with a shortage of hosts can either disperse to find hosts or wait out the period of scarcity. In the face of a dispersal-reproduction trade-off, reducing egg production might improve dispersal to areas with hosts. A generalist faced with a shortage of a host species, by contrast, can resort to use of alternative species.

## 30.4 EVOLUTION OF HOST-MARKING BEHAVIOR

### 30.4.1 Basic Description and Effects of Host-Marking Behavior

Host-marking behavior is a conspicuous and well-studied aspect of the oviposition behavior of many tephritids, particularly frugivorous species. In tephritids, host-marking behavior generally involves dragging the aculeus about the fruit or other plant part, resulting in deposition of a substance termed a host-marking pheromone (HMP) (Color Figure 15* and Figure 16). HMPs exert numerous and complex effects on behavior in female and male flies, including suppression of oviposition

---

* Color Figure 15 and Figure 16 follow p. 204.

attempts (see review by Averill and Prokopy 1989b; Papaj et al. 1992), interruption of oviposition once initiated (Papaj et al. 1989b), stimulation of emigration from areas of high infestation (Roitberg et al. 1982; Roitberg and Prokopy 1984; Papaj et al. 1989b), reduction of clutch size and number of clutches laid per visit (Papaj et al. 1989b; 1990) and, where the mating system is resource based, arrestment of males on fruit (Prokopy and Bush 1972; Katsoyannos 1975).

### 30.4.2 Taxonomic Distribution of Host Marking in Fruit Flies

Marking of an oviposition site has been reported in many frugivorous species of the genera *Anastrepha*, *Ceratitis*, *Paraceratitella* and *Rhagoletis* (also see reviews in this volume by Prokopy and Papaj, Chapter 10; Yuval and Hendrichs, Chapter 17; Headrick and Goeden, Chapter 25; and Aluja et al., Chapter 15): *A. bistrigata* Bezzii (Selivon 1991), *A. fraterculus* (Prokopy et al. 1982b), *A. grandis* (Macquart) (Silva 1991), *A. ludens* (Papaj and Aluja 1993), *A. obliqua* (Macquart) (Aluja and Díaz-Fleischer, unpublished data), *A. pseudoparallela* (Loew) (Polloni and Silva 1987), *A. serpentina* (Wiedemann) (Aluja and Díaz-Fleischer, unpublished data), *A. sororcula* Zucchi (Simoes et al. 1985), *A. striata* Schiner (Aluja et al. 1993), *A. suspensa* (Prokopy et al. 1977), *C. capitata* (Prokopy et al. 1978), *Paraceratitella eurycephala* Hardy (Fitt 1981), *R. completa* (Cirio 1972), *R. pomonella* (Prokopy 1972), *R. cingulata* (Loew) (Prokopy et al. 1976), *R. cerasi* (Katsoyannos 1975), *R. fausta* (Osten Sacken) (Prokopy 1975), *R. cornivora* (Prokopy et al. 1976), *R. indifferens* (Prokopy et al. 1976), *R. mendax* (Prokopy et al. 1976), *R. tabellaria* (Fitch) (Prokopy et al. 1976), *R. basiola* (Osten Sacken) (Averill and Prokopy 1981), and *R. zephyria* (Averill and Prokopy 1982). There is further evidence that *A. alveata* Stone, *A. leptozona* Hendel (M. Aluja and I. Jácome, unpublished data), *A. acris* Stone (M. Aluja and C. Miguel, unpublished data), and *A. spatulata* Stone (M. Aluja and M. López, unpublished data) also exhibit host-marking behavior. Interestingly, no host marking has been reported in *Toxotrypana curvicauda* Gerstaecker, which is closely related to *Anastrepha,* or in *Anastrepha* species that feed on seeds. In the case of the genus *Bactrocera*, aculeus dragging without HMP deposition has been reported in *B. cucurbitae* (Coquillett) by Prokopy and Koyama (1982), *B. dorsalis* by Prokopy et al. (1989), and *B. tryoni* and *B. jarvisi* (Tryon) by Fitt (1984).

There are comparatively few reports of aculeus-dragging behavior with concomitant deposition of HMP in nonfrugivorous tephritids. The few known cases of host-marking behavior include those of *Tephritis bardanae* (Schrank) (Straw 1989a), *Chaetorellia australis* (Pitarra and Katsoyannos 1990), and *Terellia ruficauda* (Fabricius) (Lalonde and Roitberg 1992). Nonfrugivorous tephritids have been studied less intensively than frugivorous ones. Thus, whether the apparent association between frugivory and host marking is real or a consequence of disparity in research effort remains to be evaluated.

### 30.4.3 Ecological Factors Influencing Host-Marking Behavior

#### 30.4.3.1 Competition as a Key Factor in the Evolution of Host-Marking Behavior

The central function of host-marking phermone communication is a reduction of competition incurred by a female's progeny. The fruit or plant parts used by tephritids are limited resources. By reducing investment of progeny in already-infested hosts, females presumably increase the reproductive success of their offspring. A theoretical model of the evolution of host marking (Roitberg and Mangel 1988) predicted that host-marking behavior can evolve if the mark enables a female to avoid a second oviposition in a host previously used by the same female. Host marking can also evolve even if second ovipositions by the same female are unlikely (see review by Godfray 1994). Even in this case, the payoff to host marking is a reduction in levels of larval competition.

In principle, use of HMP should be linked to host characteristics that tend to increase levels of competition: (1) small size of plant part on which larvae feed; (2) ephemeral nature of plant part on which larvae feed; (3) limited number of feeding and/or resting sites on the host plant;

(4) narrow host breadth; (5) long lifetime of host plants (such that insect populations build up on individual plants over many generations); and (6) patchy distribution of hosts in time and space (Prokopy 1981; Fitt 1984; Roitberg and Prokopy 1987; Averill and Prokopy 1989a). Use of HMPs should also be linked to life history characteristics of the insects themselves, such as limited mobility of parents or progeny (Roitberg and Prokopy 1987) and cannibalistic potential of larvae.

#### 30.4.3.2 Comparative Data on Role of Competition in Host-Marking Behavior

To date, a comparative approach has not been rigorously applied to assessment of the competition hypothesis. One candidate taxon in which such an approach may be useful is the mainly temperate genus *Rhagoletis*. The occurrence of host-marking pheromone in this genus is perhaps better studied than in any other insect taxon. As reviewed by Prokopy and Papaj (Chapter 10), most members of most species groups within *Rhagoletis* mark consistently. The *suavis* group, however, stands as a striking exception. In the *suavis* group, virtually all species have been observed to mark at least occasionally. However, the consistency of marking behavior is variable across species. Host marking in both *R. boycei* and *R. completa* has been observed but, in both species, is sporadic in occurrence (*R. boycei*: Papaj 1994; *R. completa*: Cirio 1972; C. Nufio and D. Papaj, unpublished data; R. Lalonde, personal communication). Only in *R. juglandis* within the *suavis* group do females engage vigorously and consistently in aculeus-dragging behavior after oviposition. Even in this case, data are lacking as to the effect of the mark on oviposition.

The species group differences in expression of host-marking behavior is correlated with a difference in the size of native host fruit. Members of *Rhagoletis* species groups that specialize on small hosts (for example, the *alternata*, *indifferens*, and *pomonella* groups) tend to engage in host-marking behavior (Prokopy and Papaj, Chapter 10). *Rhagoletis pomonella*, for example, uses hawthorn fruit as its native host. Hawthorn fruit are small, supporting no more than three larvae per fruit (usually only one) (Averill and Prokopy 1987) and *R. pomonella* host-marks consistently. By contrast, all members of the *suavis* group use walnut (*Juglans* spp.), a host not used by any other North American members of the genus. Walnuts tend to be larger and walnut husks offer more food for larvae than host fruit of members of other species groups. A walnut husk will typically yield dozens of *R. completa* or *R. boycei* pupae (Lalonde and Mangel 1994; C. Nufio and D. Papaj, unpublished data). Perhaps not surprisingly, members of the *suavis* group frequently add clutches to already-infested fruit (Lalonde and Mangel 1994; Papaj 1993; 1994). In short, consistency of host marking appears to be correlated with host size, which in turn is directly related to levels of larval competition.

Appealing though this interpretation may be, as a test of the hypothesis that use of HMP depends on fruit size, comparative data are weak. Because all members within a given species group are more closely related to one another than they are to members of other species groups, data on different species within a species group cannot be treated as independent. In effect, each species group constitutes one data point. Unfortunately, this means that we essentially have just a single data point, the *suavis* group, with respect to use of HMP on large fruit. Worse still, that particular data point is somewhat ambiguous: one species in the group marks vigorously, others mark, but inconsistently. Finally, it is not clear that species using large fruit, even in theory, should benefit less from use of HMP than species using small fruit. Papaj et al. (1992) and Papaj (1993) proposed that, in species using large fruit, HMP could act as an indicator of level of larval competition. HMP deposited after deposition of the first or even second clutch in a fruit may have little effect on subsequent oviposition. However, as more and more clutches were deposited in a fruit, accumulating levels of HMP might finally deter females from depositing more clutches in a fruit. A pattern of dosage dependency in response to HMP constitutes a mechanism by which females could respond to rising levels of competition in large host fruit.

Such a mechanism was proposed to account for dynamics in fruit use by the Mediterranean fruit fly, *C. capitata*, which multiply infests large fruit (Papaj et al. 1989a, b; 1990; 1992; Papaj

1993). *Ceratitis capitata* females mark hosts and HMP is unambiguously deterrent (Prokopy et al. 1978; Papaj et al. 1989b; 1992). Moreover, females adjust use of an infested fruit in response to host size exactly as one would expect if larvae were competing for resources; females are more likely to reinfest large fruit than small fruit (Papaj and Messing 1996). In short, species that use large fruit can potentially benefit from deployment of an HMP.

Multiple oviposition is also observed in the walnut-infesting *Rhagoletis*. Why then do species such as *R. boycei* and *R. completa* exhibit inconsistent patterns in host-marking behavior? A possible alternative explanation for inconsistency in host marking within the *suavis* species group concerns the occurrence of *male* host-marking behavior. Recently (Papaj et al. 1996), it was discovered that *R. boycei* males frequently touch the host fruit with their proctigers, leaving a clear viscous substance on the fruit in the process. Females prefer to attempt oviposition in the vicinity of the putative mark vs. a control area of the fruit devoid of mark. Papaj et al. (1996) hypothesized that, since on-fruit mating in *R. boycei* (as most *Rhagoletis* species generally) takes place as females attempt to oviposit, male marking functions to increase a male's mating success. Of relevance here is the occurrence of male-marking in light of the erratic pattern in host-marking behavior noted in female *R. boycei*. Is it possible that, from a female's perspective, the male's host mark substitutes in function for the female's own mark and has led to the loss or reduction of a female's mark? The laboratory of one of us (D. Papaj) is evaluating this hypothesis by surveying the occurrence of male and female marking among members of the genus *Rhagoletis*. Thus far, male marking has been observed in two members of the *suavis* group, *R. boycei* and *R. suavis* (Loew) (D. Papaj, unpublished data), and has been confirmed to be absent in another member, *R. juglandis*. The latter species marks vigorously, whereas the former two species mark inconsistently at best. There are no reports of male marking in the remaining two members of the *suavis* group, *R. completa* and *R. zoqui* Bush. There also are no reports in any member of any other species group within the genus, including species that have been well characterized with respect to use of HMP (Prokopy and Papaj, Chapter 10).

#### 30.4.3.3 Factors in Host-Marking Behavior Other Than Competition

Variation among species in use of HMPs depends on factors other than competition. Roitberg and Prokopy (1987) proposed that, aside from the requirement of larval competition, host marking is favored under two conditions: (1) when conspecifics are inconspicuous and (2) when other signals associated with oviposition are weak. In tephritids, eggs are almost always deposited within the plant part and thus visually concealed, satisfying condition (1). However, with respect to condition (2), oviposition may generate other signals, such as damage to the fruit surface. That fruit wounds can signal egg infestation is dramatically illustrated by the behavior of *B. oleae* females, who mark fruit in a fundamentally different way than most tephritids. In this species, females use their labella to distribute juices that flow from the oviposition wound over the surface of the fruit (Cirio 1971; Girolami et al. 1981).

The interplay between fruit damage and HMPs as signals of egg infestation is illustrated by variation in deployment of HMPs among species in the closely related genera *Anastrepha* and *Toxotrypana*. Two species, *A. cordata* Aldrich and *A. hamata* (Loew), are exceptions to a genus-typical pattern of host marking. The papaya fruit fly, *T. curvicauda*, also does not mark hosts. Hosts of all three species release latex after being punctured; possibly, the released latex is a sufficient signal to another conspecific female that a fruit is infested and preempts use of an HMP. Consistent with this inference is the observation that two *Anastrepha* species, *A. sagittata* (Stone) and *A. serpentina*, vary in their tendency to mark according to the amount of latex that each releases during oviposition in fruit of a host species shared between them. Larvae of *A. sagittata* feed on seeds of the yellow sapote, *Pouteria campechiana* (Kunth); larvae of *A. serpentina* feed on pulp of the same species. Larvae of both species can be found in a single fruit (Baker et al. 1944). Yet *A. serpentina* females engage in host-marking behavior, whereas *A. sagittata* females do not.

Possibly, the mark of fruit latex left by *A. sagittata* females ovipositing deep in the pulp to reach the young seed is a sufficient signal of the occurrence of eggs in a fruit. In contrast, *A. serpentina* females lay eggs less deeply in fruit of a more mature stage; relatively less latex is released during oviposition by *A. serpentina* females. A host-marking pheromone might thus be required to signal previous occupation. While plausible, this last observation has an alternative explanation related to levels of competition. Seeds in an optimal stage for oviposition are highly ephemeral; a female's young may escape competition, regardless of the presence of HMP, simply because another female is less likely to find and use the fruit. In contrast, pulp is less ephemeral and hence another female relatively more likely to find the fruit and lay eggs in it. A difference in the ephemerality of the resource thus sets up a difference in the expected benefit of an HMP. These alternative explanations are not mutually exclusive and may jointly explain the species differences in host-marking behavior.

### 30.4.4 Opportunism and the Origin of Host-Marking Behavior

The origin of host-marking behavior within the Tephritidae may reflect two instances of opportunism, one involving production of HMP and one involving dispersion of HMP over the fruit surface. In *Rhagoletis,* HMP is produced in the midgut (Prokopy et al. 1982). We propose that HMP production represents an example of opportunism in which gut products were commandeered and used as an indicator of the presence of conspecific eggs in fruit. Consistent with this proposition is the observation that *Rhagoletis* and *Anastrepha* feces deter oviposition, albeit to a lesser extent than HMP extracts (Hurter et al. 1976; Aluja and Díaz-Fleischer, unpublished data). Parallel examples of opportunism in phermone production in relation to digestive products are known in the Hymenoptera (Wilson 1971; Holldobler and Wilson 1990). Those cases differ from the general tephritid case in that, in many Hymenoptera, glands have been found which actively produce pheromone. To date, no specific gland has been identified as a site of HMP production in tephritids (Prokopy et al. 1982a).

The motor pattern involved in dispersion of HMP over the fruit surface may likewise constitute an example of opportunism. Fitt (1984) pointed out that, in some *Bactrocera* species which do not host-mark, females circle the fruit and occasionally drag and clean the aculeus. This behavior evidently removes bits of fruit skin that stuck to the aculeus during oviposition. Similarly, *A. cordata* females clean their ovipositor immediately after removing it from the fruit. Females generally groom the aculeus tip with their legs but sometimes drag the tip against the fruit surface. Aculeus-cleaning behavior of this type may have been commandeered opportunistically as the motor pattern used to distribute HMP over the fruit surface (Fitt 1984).

The relatively poorly resolved relationships among tribes within the Tephritidae make it impossible to conduct a meaningful phylogenetic analysis of the origin of host marking at the family level, especially as regards the origin of the aculeus-dragging behavior, on one hand, and the origin of HMP, on the other. The fact that some *Bactrocera* species engage in aculeus-dragging behavior without deposition of HMP, but that no species of which we are aware deposits HMP without such behavior, raises the possibility that the origin of aculeus-dragging behavior predated the origin of HMP. It is certainly intriguing to consider that the behavior used to disperse HMP evolved before the HMP itself. However, it would be equally parsimonious to suggest that aculeus-dragging behavior and HMP arose simultaneously early in the diversification of the Tephritidae, but that HMP deterrency (though not aculeus-dragging behavior) was lost in *Bactrocera* (compare Prokopy and Koyama 1982). Alternatively, the fact that feces of nonmarking *Anastrepha* and *Toxotrypana* species have deterrent effects on oviposition (Aluja et al. 1998) raises the possibility that the HMP arose before aculeus-dragging behavior. Which of these hypotheses is correct must await the construction of better-resolved phylogenetic trees among groups within the family Tephritidae. A similar consideration of origins may be made of the production vs. recognition of HMP. It too must await construction of better phylogenetic trees.

### 30.4.5 Phylogenetic Patterns in Interspecific Mark Recognition

#### 30.4.5.1 *Rhagoletis* Flies as a Case Study

As reported by Prokopy and Papaj (Chapter 10), clear phylogenetic patterns of recognition of HMP have emerged from cross-recognition assays (Prokopy et al. 1976). First, overall, species respond more strongly to their own HMP than to that of other species. Second, cross-deterrence is greater among species within the same species group than for species in different species groups. Third, cross-deterrence between a given pair of species is not always perfectly symmetrical. In assays on members of the *pomonella* species group, for example, *R. pomonella* showed greater response to *R. mendax*, *R. cornivora,* and *R. zephyria* HMP than any of the latter showed to *R. pomonella* HMP.

The second pattern, greater cross-recognition among members of the same species group, could be a consequence of either shared ancestry or some ecological factor shared in common among members of a given group. For example, if species within a given species group overlapped more in host use than did members of different groups, cross-recognition of HMP might be expected to be greater among members of the same group than among members of different groups.

In fact, members within each of the species groups for which data have been collected (the *pomonella*, *cingulata*, *alternata*, and *tabellaria* groups) share virtually no hosts in common. Synchronization in emergence with the fruiting phenology of the host plant greatly influences host availability and preference, and, as a result, overlap in host use is minimal (Messina and Jones 1990; Messina and Subler 1995). We thus believe that the phylogenetic pattern in cross-recognition reflects the effects of shared ancestry.

In any of these assays, less than complete cross-recognition of one species HMP by another may reflect a difference in what is produced by one species, what is recognized by each species, or (more likely) both. These assays cannot by themselves inform us as to evolutionary change in sending vs. receiving the HMP message. An obvious first step in determining where phylogenetic change is occurring is identification of HMP components. Economic motives for more-limited sampling notwithstanding, we advocate that such identification be carried out across a broad range of taxa, in a way that is phylogenetically informative.

More information on more species with respect to HMP effects is warranted as well. In particular, the ideas of Bush (1966; 1969) with respect to mode of speciation among the North American *Rhagoletis* groups suggest an intriguing hypothesis with respect to HMP cross-recognition. Bush (1966) argued that, whereas these species groups mainly speciated sympatrically by shifting host species and becoming allochronically isolated, one group, the *suavis* species group, probably speciated allopatrically (see also Berlocher and Bush 1982). Members of the *suavis* group, he pointed out, use species of the same genus of host, *Juglans*, precluding a sympatric shift of the kind postulated for members of other species groups. Moreover, members of the *suavis* species group probably came repeatedly into secondary contact over the course of speciation. Bush (1969) argued that such secondary contact should promote greater expression of reproductive isolating mechanisms, such as courtship signaling, than in the sympatrically speciating clades. Evidence for such biases in signaling is emerging (Smith and Bush, Chapter 9). Bush's arguments suggest an analogous hypothesis with respect to responses to host-marking pheromone, namely, that members of the *suavis* group should show greater cross-recognition of each other's pheromone than is shown among members of sympatrically speciating groups. One might further hypothesize that cross-recognition should be greater for the member of a given species pair that is the relatively weaker competitor in the larval stage.

As mentioned before, host-marking behavior has been observed in three members of the *suavis* group (*R. boycei*, *R. completa,* and *R. juglandis*). However, only for *R. completa* is there evidence that the pheromone has any effect on conspecific behavior, and there is no evidence regarding heterospecific responses. It would be particularly intriguing if those species that are erratic in their host marking nevertheless show strong responses to the HMP of species that mark consistently. It would also be useful to assay responses among populations inside and outside zones of current secondary contact.

#### 30.4.5.2 *Anastrepha* Flies as a Case Study

In contrast to *Rhagoletis* in which interspecific recognition of HMP is generally limited to species of the same species group, cross-recognition in the limited number of *Anastrepha* species tested to date is absolute and independent of phylogenetic relatedness (Aluja et al. 1998). Cross-recognition probably relates to use by various species of common host resources. Whereas the genus *Rhagoletis* comprises mainly stenophagous or even monophagous species, the genus *Anastrepha* comprises mainly stenophagous to polyphagous species that exhibit broad overlap in host use and thus potentially compete for the same resource. Such is the case for *A. striata*, *A. fraterculus,* and *A. obliqua*, all of which use guavas (*Psidium guajava* L.). On occasion, larvae of *A. striata* and *A. fraterculus* species are found in a single fruit (López et al. 1999). Such is also the case for *A. serpentina* and *A. leptozona* with respect to infestation of "baricoco" (*Micropholis mexicana* (Gilly)) (Aluja and Díaz-Fleischer, unpublished data). Under such circumstances, interspecific recognition of host-marking pheromone would be favored among at least some unrelated *Anastrepha* species. In *Rhagoletis* species, by contrast, host overlap, as discussed above, is minimal. One exception is found with respect to *R. fausta, R. cingulata,* and *R. pomonella*, all of which infest sour cherry, but show no interspecific cross-recognition of HMP (Prokopy et al. 1976). However, sour cherry is not a native host for any of these species, and the overlap in host use is thus relatively recent in origin. Another exception is found among members of the *suavis* group which commonly share native host species, even being found in the same tree at the same time (Bush 1966; Papaj 1994). However, here the occurrence of host-marking behavior is spotty and information as to cross-recognition, as discussed above, is lacking.

### 30.4.6 Evolution of Oviposition Site Reuse

Despite the commonness of host-marking behavior in several tribes of Tephritidae, flies sometimes not only do not avoid oviposition into infested fruit but actually reuse fruit and even existing oviposition punctures established by other flies. Active reuse of oviposition punctures occurs in at least two families within the Tephritoidea, the Lonchaeidae and the Tephritidae. In the Tephritidae, reuse occurs in *C. capitata* (Papaj et al. 1989a, b; 1990; 1992; Papaj 1993), *B. dorsalis* (Prokopy et al. 1989), *B. tryoni* (Pritchard 1969), and probably all members of the *R. suavis* species group (*R. completa, R. boycei, R. suavis,* and *R. juglandis*) (Papaj 1993; 1994; Lalonde and Mangel 1994; Papaj and Alonso-Pimentel 1997). In many species of Lonchaeidae, reuse is obligatory and interspecific in nature. For example, as noted above, *Dasiops alveofrons* oviposits in oviposition cavities established by walnut-infesting *Rhagoletis* flies (D.R. Papaj, unpublished data). While many species have not been examined for the occurrence of this behavior and any evaluation of origin of the trait is therefore tentative, it would appear at present that the trait has arisen independently at least three times within the Tephritoidea, once in the Loncheidae, once in the genus *Rhagoletis*, and once in the lineage including *Ceratitis* and *Bactrocera*.

With three putatively independent origins in hand, it should be possible to determine what selective forces in common at the points of origin, if any, favor repeated use of the same oviposition site. Reuse among these groups could conceivably reflect commonality of benefits, commonality of costs, both, or neither. As too-often stated or implied in this volume, comparative data are at present insufficient to address this issue in detail.

Benefits include direct female benefits such as time saving and reduction of aculeus wear, and indirect benefits such as improvement of larval performance (Papaj 1993). At present, the only clear-cut evidence of a benefit of reuse of existing sites has been made with respect to time saved in egg deposition. Time savings in reuse have been demonstrated for *C. capitata*, *R. juglandis,* and *R. boycei* (Papaj et al. 1989a; Papaj 1993; 1994; Papaj and Messing 1996; Papaj and Alonso-Pimentel 1997). In *R. juglandis,* where it has been best studied, females save time in terms of the time required to excavate a cavity in the walnut husk (Papaj and Alonso-Pimentel 1997).

Time savings may translate into a higher rate of egg laying, if females are time limited, or a reduction of predation risk in situations where oviposition is a relatively risky activity (Papaj 1993). Time saving has even been demonstrated in a hymenopterous parasitoid which engages in reuse of sites on host eggs (Takasu and Hirose 1991) and thus seems likely to be a general advantage of reuse. In fact, time saving is almost too general an advantage: few are the fruits and plant parts used by tephritids which would not afford some kind of savings in conjunction with reuse of sites.

The same point applies to aculeus wear, which ought to be a factor in use of many of the hosts of many, if not most, tephritid species. The significance of aculeus wear to reuse remains to be demonstrated. While aculeus wear occurs in walnut flies (D.R. Papaj and H. Alonso-Pimentel, unpublished data), no one to date has demonstrated, for any species, that females that reuse sites suffer less aculeus wear than females that do not. Neither has anyone demonstrated that females with worn aculei tend subsequently to reuses sites more than females with intact aculei. Finally, despite repeated suggestions that larvae enjoy a direct benefit in having clutches clustered within a fruit, supporting data are wholly lacking.

The only proposed function for site reuse that has not received even nominal support to date is improvement in larval performance. All evidence to date instead indicates that reuse of oviposition sites exacts a cost to larvae in terms of increased competition (Papaj et al. 1989b; C. Nufio and D. Papaj, unpublished data). Consistent with this evidence is the observation that *C. capitata* females adjust reuse in response to host size as one would expect if larvae were competing for resources; females are more likely to reuse sites on large fruit than on small fruit (Papaj and Messing 1996). Also consistent with a finding of larval competition are the observations that *C. capitata* engage in host-marking behavior and that the mark is invariably deterrent (Prokopy et al. 1978; Papaj et al. 1989b; 1990; 1992). Members of the *R. suavis* group also mark, although sometimes inconsistently. In other words, in terms of host-marking behavior, females behave as though their larvae are competing with one another. As noted above, HMP communication in species that reuse sites may signal the level of competition a female is likely to incur by laying another clutch in a multiply infested fruit.

The occurrence of larval competition in species that reuse oviposition sites does not mean that competition is not a factor in taxonomic patterns of reuse. Tephritid species which reuse sites may do so not only because they enjoy special benefits, for example, in terms of saving time, but also because they suffer reduced costs in terms of larval competition. The host ranges of species that reuse sites includes relatively large fruit capable of supporting the survival to pupation of a number of individuals that exceeds (sometimes exceeds greatly) the size of individual clutches. Native hosts of members of the *suavis* group, for example, support many more larvae to pupation than do the hosts of members of other *Rhagoletis* species groups in which reuse has never been observed.

Given that time savings and reduction of aculeus wear would seem to offer a benefit to almost any tephritid species, the taxonomic distribution of reuse perhaps relates more to variation in the cost of competition (as influenced by host size) than variation in the benefit of time saved. Finally, in considering the role of competition in reuse, it is critical to consider the options that are available to the ovipositing female. Arguing that improvement of larval performance is not a factor in site reuse because larvae compete with one another implies that the female has an option of placing her larvae somewhere where they will not compete with other larvae (for example, an uninfested fruit). However, in many situations, uninfested fruit are not available. In a walnut grove in southern Arizona, for example, 100% of all fruit on a majority of trees were infested within a week of the appearance of the first infested fruit (C. Nufio and D.R. Papaj, unpublished data). Thus, in a very short time, females may confront a situation in which they can either lay eggs in infested fruit or not lay eggs at all. At this point, a female fly must, teleologically speaking, weigh the cost of competition for a clutch laid in an existing oviposition site against the cost of competition for a clutch at a new site elsewhere on the same fruit. It is not clear that the former cost will be higher than the latter cost. If, for example, a benefit in terms of bacterial decay initiated at an existing site offsets the cost of being in close proximity to other larvae (e.g., exploitation competition or possible cannibalism), then it may be to the advantage of a female to reuse an oviposition site despite an overall pattern of larval competition (C. Nufio and D.R. Papaj, personal communication). The female would, in this instance, be "making the best of a bad job."

## 30.5 CONCLUSIONS AND FUTURE RESEARCH NEEDS

*Evolution of Host Use.* The use of fruit and other plant parts by tephritid larvae represents a shift from a general saprophagous habit of feeding on rotting fruit and vegetation to a habit of feeding on intact plant material, including stems, shoots, leaves, seeds, and fruit. We propose that this shift involved two key innovations: (1) an aculeus robust enough to penetrate intact plant material and (2) transfer of bacteria during oviposition that degrade plant material. The first proposition would be bolstered by an analysis of oviposition from a biomechanical perspective. Despite the uniqueness and conspicuousness of tephritid ovipositors, what we know about the precise mechanism by which the surface of fruit and other plant parts are penetrated could be fit, well, on the tip of an aculeus. The relationship between oviposition behavior and aculeus morphology could be a particularly fruitful area of future research. A phylogenetic approach would help to elucidate the evolutionary interplay between behavior and morphology. The second proposition concerning bacterial transfer would be bolstered by efforts to describe the microbial ecology of a diversity of members of the Tephritoidea (compare Drew and Yuval, Chapter 27).

*Evolution of Host Finding and Acceptance.* The vast majority of tephritids is either monophagous or stenophagous, but some species, confined to clades ovipositing into the flesh of developing fruit, are polyphagous. We propose that monophagous/stenophagous species tend to be chemical, visual, and tactile specialists in terms of the nature of host cues to which they respond when searching for and accepting oviposition sites, whereas polyphagous species tend to be chemical, visual, and tactile generalists. Evaluation of this proposition will be stymied until studies of host selection behavior adopt a phylogenetic perspective and use standardized protocols to study the behavior of various species. At present, differences in methodologies make it difficult to compare the behavior of different species.

We further propose that, in historical times, species have tended to expand onto abundant agricultural hosts whose chemical and physical cues at least initially resembled cues of native hosts. Future research is needed to assess the degree of chemical and physical resemblance of newly acquired hosts to native hosts as a clue to future host range expansion. We also need to define the extent to which range expansion in polyphages actually involves evolutionary change. How different, in terms of host preference, is the Mediterranean fruit fly that has circumnavigated the globe from the presumably more-specialized fly that originated in Africa? Population comparisons of fly preference in concert with population-level phylogenies, perhaps constructed with molecular methods, might furnish answers.

Finally, host selection behavior does not evolve in a phenotypic vacuum, but in concert with physiological traits associated with larval performance on hosts. We propose that present-day polyphagous tephritids, for example, have capitalized on the broad availability of toxin-free flesh for larval development associated with the advent of agricultural hosts. The pattern of rapid and widespread acquisition of new hosts by current polyphagous members of the Trypetinae and Dacinae is as much a consequence of the suitability of fruit for larval survival as it is the breadth of response to stimuli in host selection behavior or the pattern of egg maturation.

*Evolution of Host-Marking Behavior.* Comparative data offer limited support for the hypothesis that variation in patterns of host-marking behavior are a result of variation in levels of larval competition incurred by re-using hosts. However, variation in other factors, most notably the occurrence of fruit-derived signals of egg infestation, may also account for variation in host-marking behavior.

Use of gut products as host-marking pheromones (HMP) and of aculeus-cleaning behavior as a method of HMP dispersal are proposed as examples of opportunism in natural selection. The relative timing of these opportunistic events in evolutionary history awaits construction of better-resolved phylogenetic trees at the family and subfamily levels.

Finally, a strong phylogenetic component to cross-recognition due to shared ancestry was noted among members of the genus *Rhagoletis.* In contrast, no such component has been noted among members of the genus *Anastrepha* where cross-recognition is absolute in all species pairs examined to date and appears to relate to host overlap. Prokopy and colleague's studies on cross-recognition in the genus *Rhagoletis* are nearly 20 years old and were not made from a phylogenetic perspective. In light of improvements in phylogenetic techniques and generally strong interest in phylogenetic biology, the relative contribution of host overlap and phylogenetic ancestry to patterns of cross-recognition deserves renewed attention. HMP communication in tephritids could prove to be a model system for studies of the evolution of communication.

*Evolution of Oviposition Site Reuse.* The evolutionary interplay between reuse and HMP communication is complex and constitutes a fascinating direction for future research. Also worthy of attention is the role of bacterial decay in reuse. Reuse of oviposition sites means that females oviposit directly into sites of rot on a fruit, albeit sites associated with prior egg deposition. Given the tendency of groups allied to the Tephritidae to exploit decaying matter, orientation to rot may represent atavistic behavior. Does it? If so, what factor(s) disposes some species toward such atavism and what constrains others from expressing it? Does phylogenetic ancestry or properties of host tissue or some combination of factors account for the taxonomic distribution of reuse? The apparent multiple origin of reuse within the Tephritoidea offers opportunities for comparative analyses that bear on these and other issues. Costs and benefits of reuse need to be assessed at each of its putative points of origin, and compared with the potential costs and benefits in allied groups that do not reuse.

## ACKNOWLEDGMENTS

We thank Bruno Patrian (Swiss Federal Research Station, Wädenswil), Gina Posey (U.S. Department of Agriculture, Gainesville, FL), and Juan Rull (University of Massachusetts) for sending us badly needed reprints and for doing literature searches for us. This chapter was written thanks to the financial support provided by the Campaña Nacional contra las Moscas de la Fruta (SAGAR-IICA, Mexico), the International Organization for Biological Control of Noxious Animals and Plants (IOBC), the Instituto de Ecología, A.C. (Mexico), and USDA-ICD-RSED.

## REFERENCES

Abrahamson, W.G., K.D. McCrea, and S.S. Anderson. 1989. Host preference and recognition by the goldenrod ball gallmaker *Eurosta solidaginis. Am. Midland Nat.* 121: 322–330

Agee, H.R., E. Boller, U. Remund, J.C. Davies, and D.L. Chambers. 1982. Spectral sensitivities and visual attractant studies on the Mediterranean fruit fly, *Ceratitis capitata*, olive fly, *Dacus oleae*, and the European cherry fly, *Rhagoletis cerasi. J. Appl. Entomol.* 93: 403–412.

Allen, E.J. and B.A. Foote. 1992. Biology and immature stages of *Chaetopsis massyla* (Diptera: Otitidae), a secondary invader of herbaceous stems of wetland monocots. *Proc. Entomol. Soc. Wash.* 94: 320–328.

Aluja, M. 1994. Bionomics and management of *Anastrepha. Annu. Rev. Entomol.* 39: 155–178.

Aluja, M. and E.F. Boller. 1992. Host marking pheromone of *Rhagoletis cerasi:* foraging behavior in response to synthetic pheromonal isomers. *J. Chem. Ecol.* 18: 1299–1311.

Aluja, M. and R.J. Prokopy. 1993. Host odor and visual stimuli interaction during intratree host finding behavior of *Rhagoletis pomonella* flies. *J. Chem. Ecol.* 19: 2671–2686.

Aluja, M., I. Jácome, A. Birke, N. Lozada, and G. Quintero. 1993. Basic patterns of behavior in wild *Anastrepha striata* (Diptera: Tephritidae) flies under field cage conditions. *Ann. Entomol. Soc. Am.* 86: 766–783

Aluja, M., F. Díaz-Fleischer, A.J.F. Edmunds, and L. Hagmann. 1998. Aislamiento, Determinación Estructural, Síntesis, Actividad Biológica y Aplicación como Agente de Control de la Feromona Marcadora de Hospedero (y sus Derivados) de las Moscas de la Fruta del Género *Anastrepha* (Diptera: Tephritidae). Pending Patent No. 988732. Instituto Mexicano de la Propiedad Industrial, México D.F., México.

Anderson, R.S. 1994. A review of New-World weevils associated with Viscaceaeae (mistletoes [in part]) including descriptions of new genera and new species (Coleoptera, Curculionidae). *J. Nat. Hist.* 28: 435–492.

Averill, A.L. and R.J. Prokopy. 1981. Oviposition-deterring fruit marking pheromone in *Rhagoletis basiola. Fla. Entomol.* 64: 221–226.

Averill, A.L. and R.J. Prokopy. 1982. Oviposition-deterring fruit marking pheromone in *Rhagoletis zephyria. J. Ga. Entomol. Soc.* 17: 315–319.

Averill, A.L. and R.J. Prokopy. 1987. Intraspecific competition in the tephritid fruit fly *Rhagoletis pomonella. Ecology* 68: 878–886.

Averill, A.L. and R.J. Prokopy. 1989a. Distribution patterns of *Rhagoletis pomonella* (Diptera: Tephritidae) eggs in hawthorn. *Ann. Entomol. Soc. Am.* 82: 38–44.

Averill, A.L. and R.J. Prokopy. 1989b. Host-marking pheromones. In *Fruit Flies: Their Biology, Natural Enemies and Control* (A.S. Robinson and G. Hooper, eds.), pp. 207–219. In *World Crop Pests* (W. Helle, ed.), Vol. 3A. Elsevier Science Publishers, Amsterdam.

Baker, A.C., W.E. Stone, C.C. Plummer, and M. McPhail. 1944. A review of studies on the Mexican fruitfly and related Mexican species. *U.S. Dep. Agric. Misc. Publ.* No. 531, 155 pp.

Berlocher, S.H. and G.L. Bush. 1982. An electrophoretic analysis of *Rhagoletis* (Diptera: Tephritidae) phylogeny. *Syst. Zool.* 31: 136–155.

Bierbaum, T.J. and G.L. Bush. 1990. Host fruit chemical stimuli eliciting distinct ovipositional responses from sibling species of *Rhagoletis* fruit flies. *Entomol. Exp. Appl.* 56: 165–177.

Bower, C.C. 1977. Inhibition of larval growth of the Queensland fruit fly, *Dacus tryoni*, in apples. *Ann. Entomol. Soc. Am.* 70: 97–100.

Bush, G.L. 1966. The taxonomy, cytology and evolution of the genus *Rhagoletis* in North America (Diptera: Tephritidae). *Bull. Mus. Comp. Zool.* 134: 431–562.

Bush, G.L. 1969. Sympatric host race formation and speciation in frugivorous flies of the genus *Rhagoletis* (Diptera: Tephritidae). *Evolution* 23: 237–251.

Carey, J.R. 1984. Host specific demographic studies of the Mediterranean fruit fly, *Ceratitis capitata. Ecol. Entomol.* 9: 261–270.

Carle, S.A., A.L. Averill, G.S. Rule, W.H. Reissig, and W.L. Roelofs. 1987. Variation in host fruit volatiles attractive to the apple maggot fly. *J. Chem. Ecol.* 13: 795–805.

Christenson, L.D. and R.H. Foote. 1960. Biology of fruit flies. *Annu. Rev. Entomol.* 5: 171–192.

Cirio, U. 1972. Observazioni sul comportamento di ovideposizione della *Rhagoletis completa* (Diptera: Trypetidae) in laboratorio. *Atti Congr. Naz. Ital. Entomol.* 9: 99–117.

Coquillett, D.W. 1904. New Diptera from India and Australia. *Proc. Entomol. Soc. Wash.* 6: 137–140.

Cornelius, M.L., J.J. Duan, and R.H. Messing. Response of female Oriental fruit flies and associated parasitoids to the color, shape and size of fruit-mimicking visual traps. *J. Econ. Entomol.*, in press.

Cytrynowicz, M., J.S. Morgante, and H.M.L. DeSouza. 1982. Visual responses of South American fruit flies, *Anastrepha fraterculus*, and Mediterranean fruit flies, *Ceratitis capitata*, to colored rectangles and spheres. *Environ. Entomol.* 11: 1202–1210.

Diehl, S.R., R.J. Prokopy, and S. Henderson. 1986. The role of stimuli associated with branches and foliage in host selection by *Rhagoletis pomonella*. In *Fruit Flies of Economic Importance 84* (R. Cavalloro, ed.), pp. 191–196. A. A. Balkema, Rotterdam.

Dodson, G. and G. Daniels. 1988. Diptera reared from *Dysoxylum gaudichaudianum* (Juss.) Miq. at Iron Range, northern Queensland. *Aust. Entomol. Mag.* 15: 77–79.

Eckstein, F. 1931. Über die Verwendung von Kodermitteln zur Schadlingsbekämpfung. *Z. Angew. Entomol.* 18: 726–743.

Eismann, C.H. and M.J. Rice. 1985. Oviposition behavior of *Dacus tryoni*: the effect of some sugars and salt. *Entomol. Exp. Appl* 39: 61–71.

Eismann, C.H. and M.J. Rice. 1992. Behavioral evidence for hygro-and mechanoreception by ovipositor sencilla of *Dacus tryoni. Physiol. Entomol.* 14: 273–277.

Féron, M. 1962. L'instinct de réproduction chez la mouche méditerranéenne des fruits *Ceratitis capitata* Wied. (Dipt. Trypetidae). Comportement sexuel. Comportement de ponte. *Rev. Pathol. Veg. Entomol. Agric. Fr.* 41(1–2): 3–129.

Ferrar, P. 1987. *A Guide to the Breeding Habits and Immature Stages of Diptera Cyclorrapha.* Entomonograph, Vol. 8, Parts 1–2. E.J. Brill/Scandinavica Science Press, Leiden. 2 vol., 907 pp.

Fitt, G.P. 1981. Observations on the biology and behaviour of *Paraceratitella eurycephala* (Diptera: Tephritidae) in Northern Australia. *J. Aust. Entomol. Soc.* 20: 1–7.

Fitt, G.P. 1984. Oviposition behaviour of two tephritid fruit flies, *Dacus tryoni* and *Dacus jarvisi*, as influenced by the presence of larvae in the host fruit. *Oecologia* 62: 37–46.

Fitt, G.P. 1986. The influence of shortage of hosts on the specificity of oviposition behaviour in species of *Dacus* (Diptera: Tephritidae). *Physiol. Entomol.* 11: 133–143.

Fitt, G.P. 1990. Variation in ovariole number and egg size of species of *Dacus* (Diptera: Tephritidae) and their relation to host specialization. *Ecol. Entomol.* 15: 255–264.

Fletcher, B.S. 1989. Life history strategies of tephritid fruit flies. In *Fruit Flies: Their Biology, Natural Enemies and Control* (A.S. Robinson and G. Hooper, eds.), pp. 195–208. In *World Crop Pests* (W. Helle, ed.), Vol. 3B. Elsevier Science Publishers, Amsterdam.

Fletcher, B.S. and R.J. Prokopy. 1991. Host location and oviposition in tephritid fruit flies. In *Reproductive Behavior of Insects: Individuals and Populations* (W.J. Bailey and J. Ridsdill-Smith, eds.), pp. 138–171. Chapman & Hall, New York.

Foote, B.A. 1967. Biology and immature stages of fruit flies: the genus *Icterica*. *Ann. Entomol. Soc. Am.* 60: 1295–1305.

Foote, B.A. 1985. Biology of *Rivellia pallida* (Diptera: Platystomatidae) a consumer of the notrogen-fixing root nodules of *Amphicarpa bracteata* (Leguminosae). *J. Kans. Entomol. Soc.* 58: 27–35.

Foote, R.H., F.L. Blanc, and A.L. Norrbom. 1993. *Handbook of the Fruit Flies (Diptera: Tephritidae) of America North of Mexico.* Comstock Publishing Associates, Ithaca. 571 pp.

Freeman, R. and J.R. Carey. 1990. Interaction of host stimuli in the ovipositional response of the Mediterranean fruit fly. *Environ. Entomol.* 19: 1075–1080.

Freese, G. and H. Zwölfer. 1996. The problem of the optimal clutch size in a tritrophic system: the oviposition strategy of the thistle gallfly *Urophora cardui* (Diptera: Tephritidae). *Oecologia* 108: 293–302.

Girolami, V.A., A. Vianello, A. Strapazzon, E. Ragazzi, and E. Veronese. 1981. Ovipositional deterrents in *Dacus oleae*. *Entomol. Exp. Appl.* 29: 177–188.

Girolami, V., A. Strapazzon, and P.F. De Gerloni. 1983. Insect/plant relationship in olive flies: general aspects and new findings. In *Fruit Flies of Economic Importance* (R. Cavalloro, ed.), pp. 258–267. A.A. Balkema, Rotterdam.

Girolami, V.A., A. Strapazzon, R. Crnjar, A.M. Anjioy, P. Pietra, J.G. Stoffolano, and R.J. Prokopy. 1986. Behavior and sensory physiology of *Rhagoletis pomonella* in relation to oviposition stimulants and deterrents in fruit. In *Fruit Flies of Economic Importance 84* (R. Cavalloro, ed.), pp. 183–190. A.A. Balkema, Rotterdam.

Godfray, H.C.J. 1994. *Parasitoids: Behavioral and Evolutionary Ecology.* Princeton University Press, Princeton. 473 pp.

Goeden, R.D., D.H. Headrick and J.A. Teerink. 1994. Life history and description of immature stages of *Paroxyna genalis* (Thomson) (Diptera: Tephritidae) on native Asteraceae in southern California. *Proc. Entomol. Soc. Wash.* 96: 612–629.

Green, T.A., R.J. Prokopy, and D.W. Hosmer. 1994. Distance of response to host tree models by female apple maggot flies, *Rhagoletis pomonella*: interaction of visual and olfactory stimuli. *J. Chem. Ecol.* 20: 2393–2413.

Haisch, A. and H.Z. Levinson. 1980. Influences of fruit volatiles and coloration on oviposition of the cherry fruit fly. *Naturwissenschaften* 67: 44–45.

Hancock, D.L. 1987. Notes on some African Ceratitinae (Diptera: Tephritidae), with special reference to the Zimbabwean fauna. *Trans. Zimb. Sci. Assoc.* 63: 47–57.

Hancock, D.L. and R.A.I. Drew. 1994. New species and records of Asian Trypetinae (Diptera: Tephritidae). *Raffles Bull. Zool.* 42: 555–591.

Hancock, D.L. and R.A.I. Drew. 1995a. Observations on the genus *Acanthonevra* Macquart in Thailand and Malaysia (Diptera: Tephritidae: Trypetinae). *Entomologist* 114: 99–103.

Hancock, D.L. and R.A.I. Drew. 1995b. New genus, species and synonyms of Asian Trypetinae (Diptera: Tephritidae). *Malays. J. Sci.* 16A: 45–59.

Hara, H. 1989. Identity of *Rivellia fusca* (Thomson) and description of a new allied species, with special reference to the structure of abdomen of *Rivellia* (Diptera, Platystomatidae). *Jpn. J. Entomol.* 57: 793–802.

Hardy, D.E. 1986a. Fruit flies of the subtribe Acanthonevrini of Indonesia, New Guinea and the Bismarck and Solomon Islands (Diptera: Tephritidae: Trypetidae: Acanthonevrini). *Pac. Insects Monogr.* 42: 1–191.

Hardy, D.E. 1986b. The Adramini of Indonesia, New Guinea and adjacent islands of the Bismarcks and Solomons (Diptera: Tephritidae: Trypetinae). *Entomography* 5: 247–373.

Headrick, D.H. and R.D. Goeden. 1998. The biology of nonfrugivorous tephritid fruit flies. *Annu. Rev. Entomol.* 43: 217–241.

Hernández, M.M., I. Sanz, M. Adelantado, S. Ballach, and E. Primo. 1996. Electroantennogram activity from antennae of *Ceratitis capitata* to fresh orange airborne volatiles. *J. Chem. Ecol.* 22: 1607–1619.

Hill, A.R. and G.H.S. Hooper. 1984. Attractiveness of various colors to Australian tephritid flies in the field. *Entomol. Exp. Appl.* 356: 119–128.

Hill, G.F. 1921. Notes on some Diptera found in association with termites. *Proc. Linn. Soc. N.S. W.* 46: 216–220.

Holldobler, B. and E.O. Wilson. 1990. *The Ants.* Harvard University Press, Cambridge.

Horner, J.D. and W.G. Abrahamson. 1992. Influence of plant genotype and environment on oviposition preference and offspring survival in a gallmaking herbivore. *Oecologia* 90: 323–332.

Howard, D.J., G.L. Bush, and J.A. Breznak. 1985. The evolutionary significance of bacteria associated with *Rhagoletis. Evolution* 39: 405–417.

Hurter, J., B. Katsoyannos, E.F. Boller, and P. Wirz. 1976. Beitrag zur Anreicherung und teilweisen Reinigung des eiablageverhindernden Pheromons der Kirschenfliege, *Rhagoletis cerasi* L. (Dipt., Trypetidae). *Z. Angew. Entomol.* 80: 50–56

Jang, E.B. and D.M. Light. 1996. Olfactory semiochemicals of tephritids. In *Fruit Fly Pests: A World Assessment of Their Biology and Management* (B.A. McPheron and G.J. Steck, eds.), pp. 73–90. St. Lucie Press, Delray Beach.

Jones, S.R. 1989. Morphology and Evolution of the Aculei of True Fruit Flies (Diptera; Tephritidae) and Their Relationship to Host Anatomy. Ph.D. thesis, Pennsylvania State University, University Park. 320 pp.

Jones, S.R. and K.C. Kim. 1994. Aculeus wear and oviposition in four species of Tephritidae. *Ann. Entomol. Soc. Am.* 87: 104–107.

Katsoyannos, B.I. 1975. Oviposition-deterring, male-arresting, fruit-marking pheromone in *Rhagoletis cerasi. Environ. Entomol.* 4: 801–807.

Katsoyannos, B.I. 1987. Field responses of Mediterranean fruit flies to colored spheres suspended in fig, citrus and olive trees. In *Insects–Plants* (V. Labeyrie, G. Fabres, and D. Lachaise, eds.), pp. 167–172. Junk, Dordrecht.

Katsoyannos, B.I. 1989. Response to shape, size and color. In *Fruit Flies: Their Biology, Natural Enemies and Control* (A.S. Robinson and G. Hooper, eds.), pp. 307–324. In *World Crop Pests* (W. Helle, ed.), Vol. 3A. Elsevier Science Publishers, Amsterdam.

Katsoyannos, B.I. and J.S. Pittara. 1983. Effect of size of artificial oviposition substrates and presence of natural host fruits on the selection of oviposition site by *Dacus oleae. Entomol. Exp. Appl.* 34: 326–332.

Katsoyannos, B.I., G. Patsouras, and M. Vrehoussi. 1985. Effect of color hue and brightness of artificial oviposition substrates on the selection of oviposition site by *Dacus oleae. Entomol. Exp. Appl.* 38: 205–214.

Katsoyannos, B.I., K. Panagiotidou, and I. Kechagia. 1986. Effect of color properties on the selection of oviposition site by *Ceratitis capitata. Entomol. Exp. Appl.* 42: 187–193.

Kombargi, W.S., S.E. Michelakis, and C.A. Petrakis. 1998. Effect of olive surface waxes on oviposition by *Bactrocera oleae. J. Econ. Entomol.* 91: 993–998.

Kopelke, J.P. 1984. Der erste Nachweis eines Brutparasiten unter den Bohrfliegen. *Nat. Mus. Frankfurt* 114: 1–28.

Kothe, J.K. and J.W. Van Duyn. 1988. Influence of soil surface conditions and host plants on soybean nodule fly *Rivellia quadrifasciata* (Macquart) (Diptera: Platystomatidae) oviposition. *J. Entomol. Sci.* 23: 251–256.

Lalonde, R.G. and M. Mangel. 1994. Seasonal effects on superparasitism by *Rhagoletis completa. J. Anim. Ecol.* 63: 583–588.

Lalonde, R.G. and B.D. Roitberg. 1992. Host selection behavior of the thistle-feeding fly: choices and consequences. *Oecologia* 90: 534–539.

Lalonde, R.G. and B.D. Roitberg. 1994. Pollen availability, seed production and seed predator clutch size in tephritid-thistle system. *Evol. Ecol.* 8: 188–195.

Light, D.M. and E.B. Jang. 1987. Electroantennogram responses of the oriental fruit fly, *Dacus dorsalis*, to a spectrum of alcohol and aldehyde plant volatiles. *Entomol. Exp. Appl.* 45: 55–64.

Light, D.M., E.B. Jang, and J.C. Dickens. 1988. Electroantennogram responses of the Mediterranean fruit fly, *Ceratitis capitata*, to a spectrum of plant volatiles. *J. Chem. Ecol.* 14: 159–180.

Light, D.M., E.B. Jang, and R.A. Flath. 1992. Electroantennogram responses of the Mediterranean fruit fly, *Ceratitis capitata*, to the volatile constituents of nectarines. *Entomol. Exp. Appl.* 63: 13–26.

López, M., M. Aluja, and J. Sivinski. 1999. Hymenopterous larval-pupal and pupal parasitoids of *Anastrepha* flies (Diptera: Tephritidae) in Mexico. *Biol. Control* 15: 119–129.

McAlpine, D.K. 1973. The Australian Platystomatidae (Diptera: Schizofora) with a revision of five genera. *Mem. Aust. Mus.* 15: 1–256

McAlpine, J.F. 1989. Phylogeny and classification of the Muscomorpha. In *Manual of Neartic Diptera* Vol. 3 (J.F. McAlpine, ed.), pp. 1397–1518. Monograph of the Biosystematics Research Centre, No. 32. Agriculture Canada, Ottawa.

McMichael, B., B.A. Foote, and B. Bowker. 1990. Biology of *Rivellia melliginis* (Diptera: Platystomatidae): a consumer of the nitrogen-fixing root nodules of black locust (Leguminosae). *Ann. Entomol. Soc. Am.* 83: 967–974.

Meats, A. 1983. The response of the Queensland fruit fly, *Dacus tryoni*, to tree models. In *Fruit Flies of Economic Importance* (R. Cavalloro, ed.), pp. 285–289. A.A. Balkema, Rotterdam.

Messina, F.J. 1990. Components of host choice by two *Rhagoletis* species in Utah. *J. Kans. Entomol. Soc.* 63: 80–87.

Messina, F.J. and V.P. Jones. 1990. Relationship between fruit phenology and infestation by the apple maggot (Diptera: Tephritidae) in Utah. *Ann. Entomol. Soc. Am.* 83: 742–752.

Messina, F.J. and J.K. Subler. 1995. Conspecific and heterospecific interactions of male *Rhagoletis* flies (Diptera: Tephritidae) on a shared host. *J. Kans. Entomol. Soc.* 68: 206–213.

Mitter, C., B. Farrell, and B. Wiegmann. 1988. The phylogenetic study of adaptive zones: has phytophagy promoted insect diversification? *Am. Nat.* 132: 107–128.

Moericke, V., R.J. Prokopy, S. Berlocher, and G.L. Bush. 1975. Visual stimuli eliciting attraction of *Rhagoletis pomonella* flies to trees. *Entomol. Exp. Appl.* 18: 497–507.

Moffitt, H. and F. Yaruss. 1961. *Dasiops alveofrons*, a new pest of apricots in California. *J. Econ. Entomol.* 54: 504–505.

Munro, H.K. 1967. Fruit flies allied to species of *Afrocneros* and *Ocnerioxa* that infest *Cussonia*, the umbrella tree or kiepersol (Araliaceae) (Diptera: Trypetidae). *Ann. Natal Mus.* 18: 571–594 .

Nakagawa, S., R.J. Prokopy, T.T.Y. Wong, J.R. Zeigler, S.M. Mitchell, T. Urago, and E.J. Harris. 1978. Visual orientation of *Ceratitis capitata* flies to fruit models. *Entomol. Exp. Appl.* 24: 193–198.

Neilson, W.T.A. 1967. Development and mortality of the apple maggot, *Rhagoletis pomonella*, in crab apples. *Can. Entomol.* 99: 217–219.

Nigg, H.N., L.L. Mallory, S.E. Simpson, S.B. Callahan, J.P. Toth, S. Fraser, M. Klim, S. Nagy, J.L. Nation, and J.A. Attaway. 1994. Caribbean fruit fly, *Anastrepha suspensa*, attraction to host fruit and host kairomones. *J. Chem. Ecol.* 20: 727–743.

Norrbom, A.L. 1994. New genera of Tephritidae (Diptera) from Brazil and Dominican amber, with phylogenetic analysis of the tribe Ortalotrypetini. *Insecta Mundi* 8: 1–16.

Norrbom, A.L. and J.F. McAlpine. 1997. A revision of the neotropical species of *Dasiops* Rondani (Diptera: Lonchaeidae) attacking *Passiflora* (Passifloraceae). *Mem. Entomol. Soc. Wash.* 18: 189–211.

Norrbom, A.L., L.E. Carroll, and A. Freidberg. 1999. Status of knowledge. In *Fruit Fly Expert Identification System and Systematic Information Database* (F.C. Thomson, ed.), pp. 9–47. *Myia* (1998) 9, 524 pp.

Oldroyd, H. 1964. *The Natural History of Flies.* W.W. Norton, New York. 324 pp.

Papaj, D.R. 1993. Use and avoidance of occupied host as a dynamic process in tephritid flies. In *Insects–Plants Interactions,* Vol. V (E.A. Bernays, ed.), pp. 25–46. CRC Press, Boca Raton.

Papaj, D.R. 1994. Oviposition site guarding by male walnut flies and its possible consequences for mating success. *Behav. Ecol. Sociobiol.* 34: 187–195.

Papaj, D.R. and H. Alonso-Pimentel. 1997. Why walnut flies superparasitize: time savings as a possible explanation. *Oecologia* 109: 166–174.

Papaj, D.R. and M. Aluja. 1993. Temporal dynamics of host-marking in the tropical tephritid fly *Anastrepha ludens*. *Physiol. Entomol.* 18: 279–284.

Papaj, D.R. and R.H. Messing. 1996. Functional shifts in the use of parasitized host by a tephritid fly: the role of host quality. *Behav. Ecol.* 7: 235–242.

Papaj, D.R., B.D.Katsoyannos, and J.G. Hendrichs. 1989a. Use of fruit wounds in oviposition by the Mediterranean fruit fly. *Entomol. Exp. Appl.* 53: 203–209.

Papaj, D.R., B.D. Roitberg, and S.B. Opp. 1989b. Serial effects of host infestation on egg allocation by the Mediterranean fruit fly: a rule of a thumb and its functional significance. *J. Anim. Ecol.* 58: 955–970.

Papaj, D.R., B.D. Roitberg, S.B. Opp, M. Aluja, R.J. Prokopy, and T.T.Y. Wong. 1990. Effect of marking pheromone on clutch size in the Mediterranean fruit fly. *Physiol. Entomol.* 15: 463–468.

Papaj, D.R., A.L. Averill, R.J. Prokopy, and T.T.Y. Wong. 1992. Host-marking pheromone and use of previously established oviposition sites by the Mediterranean fruit fly (Diptera: Tephritidae). *J. Insect Behav.* 5: 583–598.

Papaj, D.R., J.M. Garcia, and H. Alonso-Pimentel. 1996. Marking of host fruit by male *Rhagoletis boycei* Cresson flies (Diptera: Tephritidae) and its effect on egg-laying. *J. Insect Behav.* 9: 585–598.

Paramonov, S.J. 1958. A review of the Australian Pyrgotidae (Diptera). *Aust. J. Zool.* 6: 89–137.

Permkam, S. and D.L. Hancock. 1995a. Australian Ceratitinae (Diptera: Tephritidae). *Invertebr. Taxon.* (1994) 8: 1325–1341.

Permkam, S. and D.L. Hancock. 1995b. Australian Trypetinae (Diptera: Tephritidae). *Invertebr. Taxon.* 9: 1047–1209.

Pierce, N.E. 1985. Lycaenid butterflies and ants selection for nitrogen-fixing and other protein-rich food plants. *Am. Nat.* 125: 888–895.

Pitarra, K. and B.I. Katsoyannos. 1990. Evidence for a host-marking pheromone in *Chaetorellia australis*. *Entomol. Exp. Appl.* 54: 287–295.

Pittara, I.S. and B.I. Katsoyannos. 1992. Effect of shape, size and color on selection of oviposition sites by *Chaetorellia australis*. *Entomol. Exp. Appl.* 63: 105–113.

Pitkin, B.R. 1989. Family Pallopteridae. In *Catalog of the Diptera of the Australasian and Oceanian Regions* (N.L. Evenhuis, ed.), p. 532. Bishop Museum Press, Honolulu, and E.J. Brill, Leiden.

Polloni, Y.J. and M.T. da Silva. 1987. Considerations on the reproductive behavior of *Anastrepha pseudoparallela* Loew 1873 (Diptera Tephritidae). In *Fruit Flies: Proceedings of the Second International Symposium, 11–21 September 1986, Colymbari, Crete, Greece* (A.P. Economopoulos, ed.), pp. 295–301. Elsevier Science Publishers, Amsterdam.

Pritchard, G. 1969. The ecology of a natural population of Queensland fruit fly *Dacus tryoni*. II. The distribution of eggs and its relation to behavior. *Aust. J. Zool.* 15: 303–335.

Prokopy, R.J. 1968. Visual responses of apple maggot flies, *Rhagoletis pomonella*: orchard studies. *Entomol. Exp. Appl.* 11: 403–422.

Prokopy, R.J. 1969. Visual responses of European cherry fruit flies, *Rhagoletis cerasi*. *Pol. Pismo Entomol.* 39: 539–566.

Prokopy, R.J. 1972. Evidence for a marking pheromone deterring repeated oviposition in apple maggot flies. *Environ. Entomol.* 1: 326–332.

Prokopy, R.J. 1975. Oviposition-deterring fruit marking pheromone in *Rhagoletis fausta*. *Environ. Entomol.* 4: 298–300.

Prokopy, R.J. 1996. Artificial oviposition devices for apple maggot. *J. Econ. Entomol.* 59: 384–387.

Prokopy, R.J. and E.F. Boller. 1971. Stimuli eliciting oviposition of *Rhagoletis cerasi* (Diptera: Tephritidae) into inanimate objects. *Entomol. Exp. Appl.* 14: 1–14.

Prokopy, R.J. and G.L. Bush. 1973a. Mating behavior in *Rhagoletis pomonella* (Diptera: Tephritidae). IV. Courtship. *Can. Entomol.* 105: 873–891.

Prokopy, R.J. and G.L. Bush. 1973b. Ovipositional responses to different sizes of artificial fruit by flies of *Rhagoletis pomonella* species groups. *Ann. Entomol. Soc. Am.* 66: 927–929.

Prokopy, R.J. and G.E. Haniotakis. 1975. Responses of wild and laboratory cultured *Dacus oleae* to host plant odor. *Ann. Entomol. Soc. Am.* 68: 73–77.

Prokopy, R.J. and G.E. Haniotakis. 1976. Host detection by wild and laboratory-cultured olive flies. In *The Host Plant in Relation to Insect Behavior and Reproduction* (T. Jermy, ed.), pp. 209–214, Budapest.

Prokopy, R.J. and J. Koyama. 1982. Oviposition site partitioning in *Dacus cucurbitae*. *Entomol. Exp. Appl.* 31: 428–432.

Prokopy, R.J., W.H. Reissig, and V. Moericke. 1976. Marking pheromone deterring repeated oviposition in *Rhagoletis* flies. *Entomol. Exp. Appl.* 20: 170–178.

Prokopy, R.J., P.D. Greany, and D.L. Chambers. 1977. Oviposition-deterring pheromone in *Anastrepha suspensa*. *Environ. Entomol.* 6: 463–465.

Prokopy, R.J., J.R. Ziegler, and T.T.Y. Wong. 1978. Deterrence of repeated oviposition by fruit marking pheromone in *Ceratitis capitata* (Diptera: Tephritidae). *J. Chem. Ecol.* 4: 55–63.

Prokopy, R.J., A.N. Averill, C.M. Bardinelli, E.S. Bowdan, S.S. Cooley, R.M. Crnjar, E.A. Dundulis, C.A. Roitberg, P.J. Spatcher, J.H. Tumlinson, and B.L. Weeks. 1982a. Site of production of an oviposition deterring pheromone component in *Rhagoletis* flies. *J. Insect Physiol.* 28: 1–10

Prokopy, R.J., A. Malavasi, and J.S. Morgante. 1982b. Oviposition-deterring pheromone in *Anastrepha fraterculus* flies. *J. Chem. Ecol.* 8: 763–771.

Prokopy, R.J., D.R. Papaj, and T.T.Y. Wong. 1986. Fruit foraging behavior of Mediterranean fruit fly females on host and nonhost plants. *Fla. Entomol.* 69: 651–657.

Prokopy, R.J., T.A. Green, W.A. Olson, R.F. Vargas, D. Kaneshisa, and T.Y. Wong. 1989. Discrimination by *Dacus dorsalis* females (Diptera: Tephritidae) against larval-infested fruit. *Fla. Entomol.* 72: 319–323.

Prokopy, R.J.,T.A. Green, and R.I. Vargas. 1990. *Dacus dorsalis* flies can learn to find and accept host fruit. *J. Insect Behav.* 3: 663–672.

Prokopy, R.J., R.A.I. Drew, B.N.E. Sabine, A.C. Lloyd, and E. Hamacek. 1991. Effect of physiological and experiential state of *Bactrocera tryoni* flies on intra-tree foraging behavior for food (bacteria) and host fruit. *Oecologia* 87: 394–400.

Reissig, W.H., S.K. Brown, R.C. Lamb, and J.N. Cummins. 1990. Laboratory and field studies of resistance of crab apple clones to *Rhagoletis pomonella*. *Environ. Entomol.* 19: 565–572.

Robacker, D.C. 1992. Effects of shape and size of colored traps on attractiveness to irradiated, laboratory-strain Mexican fruit flies. *Fla. Entomol.* 75: 230–241.

Roberts, H. 1969. Forest insects in Nigeria with notes of their biology and distribution. *Inst. Pap. Commonw. For. Inst.* 44: 1–206.

Roitberg, B.D. and M. Mangel. 1988. On the evolutionary ecology of marking pheromones. *Evol. Ecol.* 2: 289–315.

Roitberg, B.D. 1985. Search dynamics in fruit parasitic insects. *J. Insect Physiol.* 31: 865–872.

Roitberg, B.D. and R.J. Prokopy. 1981. Experience required for pheromone recognition in the apple maggot fly. *Nature* 292: 540–541.

Roitberg, B.D. and R.J. Prokopy. 1982. Influence of inter-tree distance on foraging behaviour of *Rhagoletis pomonella* in the field. *Ecol. Entomol.* 7: 437–442.

Roitberg, B.D. and R.J. Prokopy. 1983. Host deprivation influence on response of *Rhagoletis pomonella* to its oviposition deterring pheromone. *Physiol. Entomol.* 8: 69–72.

Roitberg, B.D. and R.J. Prokopy. 1984. Host visitation as a determinant of search persistence in fruit parasitic tephritid flies. *Oecologia* 62: 7–12.

Roitberg, B.D. and R.J. Prokopy. 1987. Insects that mark host plants. *Bioscience* 37: 400–406.

Roitberg, B.D., J.C. Van Lenteren, J.J.M. Van Alphen, F. Galis, and R.J. Prokopy. 1982. Foraging behaviour of *Rhagoletis pomonella,* a parasite of hawthorn (*Crataegus viridis*), in nature. *J. Anim. Ecol.* 51: 307–325.

Sanders, W. 1962. Das Verhalten der Mittelmeerfruchtfliege *Ceratitis capitata* Wied. bei der Eiablage. *Z. Tierpsychol.* 19: 1–28

Scarpati, M.L., R.L. Scalzo, and G. Vita. 1993. *Oleae europaea* volatiles attractive and repellent to the olive fruit fly (*Dacus oleae*). *J. Chem. Ecol.* 19: 881–891.

Selivon, D. 1991. Alguns Aspectos do Comportamento de *Anastrepha striata* Schiner e *Anastrepha bistrigata* Bezzii (Diptera: Tephritidae). M.Sc. thesis, Instituto de Biociencias da Universidade de São Paulo, São Paulo.

Silva, J.G. 1991. Biologia e comportamento de *Anastrepha grandis* (Macquart, 1846) (Diptera: Tephritidae). M.Sc. thesis, Instituto de Biociencias da Universidade de São Paulo, São Paulo. 135 pp.

Simoes, M.H., Y.J. Polloni, and M.A. Paludetti. 1978. Biologia de algunas especies de *Anastrepha* (Diptera: Tephritidae) en laboratorio. Third Latin-American Entomology Congress, Iheus, Bahia, Brazil.

Sivinski, J. 1987. Acoustical oviposition cues in the Caribbean fruit fly, *Anastrepha suspensa* (Diptera: Tephritidae). *Fla. Entomol.* 70: 171–172.

Southwood, T.R.E. 1973. The insect/plant relationship — an evolutionary perspective. *Symp. R. Entomol. Soc. Lond.* 6: 3–30

Steyskal, G.C. 1979. Biological, anatomical, and distribution notes on the genus *Callopistromyia* (Diptera: Otitidae). *Proc. Entomol. Soc. Wash.* 81: 450–455.

Steyskal, G.C. 1986. Taxonomy of the adults of the genus *Strauzia* Robineau-Desvoidy (Diptera: Tephritidae). *Insecta Mundi* 1: 101–117.

Steyskal, G.C. 1987. Pyrgotidae. In *Manual of Nearctic Diptera*, Vol. 2 (J.F. McAlpine, ed.), pp. 813–816. Monograph of the Biosystematics Research Centre, No. 28. Agriculture Canada, Ottawa.

Straw, N.A. 1989a. Evidence for an oviposition-deterring pheromone in *Tephritis bardanae* (Schrank) (Diptera: Tephritidae). *Oecologia* 78: 121–130.

Straw, N.A. 1989b. The timing of oviposition and larval growth by two tephritid fly species in relation to host plant development. *Ecol. Entomol.* 14: 443–454.

Straw, N.A. 1989c. Taxonomy, attack strategies and host relations of flowerhead Tephritinae: a review. *Ecol. Entomol.* 14: 455–462.

Szentesi, A., P.D. Greany, and D.L. Chambers. 1979. Oviposition behavior of laboratory-reared and wild Caribbean fruit flies (*Anastrepha suspensa*) to selected chemical influences. *Entomol. Exp. Appl.* 26: 227–238.

Takasu, K. and Y. Hirose. 1991. The parasitoid *Oencyrtus nezarae* (Hymenoptera: Encyrtidae) prefers hosts parasitized by conspecifics over unparasitized hosts. *Oecologia* 87: 319–323.

Tsiropoulos, G.J. and K.S. Hagen. 1979. Oviposition behavior of laboratory-reared and wild Caribbean fruit flies: selected chemical influences. *Z. Angew. Entomol.* 89: 547–550.

Vargas, R.I., J.D. Stark, R.J. Prokopy, and T.A. Green. 1991. Response of Oriental fruit fly and associated parasitoids to different color spheres. *J. Econ. Entomol.* 84: 1503–1507.

Warthen, J.D., C.J. Lee, E.B. Jang, D.R. Lance, and D.O. McInnis. 1997. Volatile potential attractants from ripe coffee fruit for female Mediterranean fruit fly. *J. Chem. Ecol.* 23: 1891–1900.

White, I.M. and M.M. Elson-Harris. 1992. *Fruit Flies of Economic Significance: Their Identification and Bionomics*. CAB International, Wallingford. 601 pp.

Wiesmann, R. 1937. Die Orientierung der Kirschenfliege *Rhagoletis cerasi* L. *Landwirtschaft. Jahrb. Schweiz* 51: 1080–1109.

Wilson, E.O. 1971. *The Insect Societies*. Belknap Press, Cambridge.

Zeh, D.W., J.E. Zeh, and R.L. Smith. 1989. Ovipositors, amnions and eggshell architecture in the diversification of terrestrial arthropods. *Q. Rev. Biol.* 64: 147–174.

Zwölfer, H. 1969. *Urophora siruna-seva* (Hg.) (Diptera: Trypetidae), a potential insect for the biological control of *Centaurea solstitialis* L. in California. *Tech Bull. Commonw. Inst. Biol. Control* 11: 105–154.

Zwölfer, H. 1972a. Investigation of *Chaetorellia* species associated with *Centaurea solstitialis*. Commonwealth Institute of Biological Control of Weeds Project. University of California Report 7, 1–21.

Zwölfer, H. 1972b. Investigation on *Urophora stylata* Fabr., a possible agent for the biological control of *Cirsium vulgare* in Canada. Weed Projects for Canada Progress Report. *Commonw. Inst. Biol. Control Rep.* 29: 20 pp.

Zwölfer, H. 1983. Life systems and strategies of resource exploitation in tephritids. In *Fruit Flies of Economic Importance* (R. Cavalloro, ed.), pp. 16–30. A.A. Balkema, Rotterdam.

# 31 Changes in Sexual Behavior and Life History Traits of Tephritid Species Caused by Mass-Rearing Processes

*Jean-Pierre Cayol*

**CONTENTS**



## 31.1 INTRODUCTION

In the past decades, the shift to biological control methods for insect pests, such as the Sterile Insect Technique (Knipling 1953; 1984), has resulted in expansion in the number of facilities rearing insects on a large scale. This procedure involves the colonization of insects from natural populations into artificial habitats. For biological control programs of fruit fly species, large numbers of insects are necessary (Kuba et al. 1996; Penrose 1996; Rull Gabayet et al. 1996) and managers of rearing facilities have to focus, among other goals, on the quantity and quality of the product with production efficiency often the first priority (Miyatake 1996). However, the success of these methods of pest population management requires considerable interaction between wild populations and artificially

0-8493-1275-2/00/$0.00+$.50

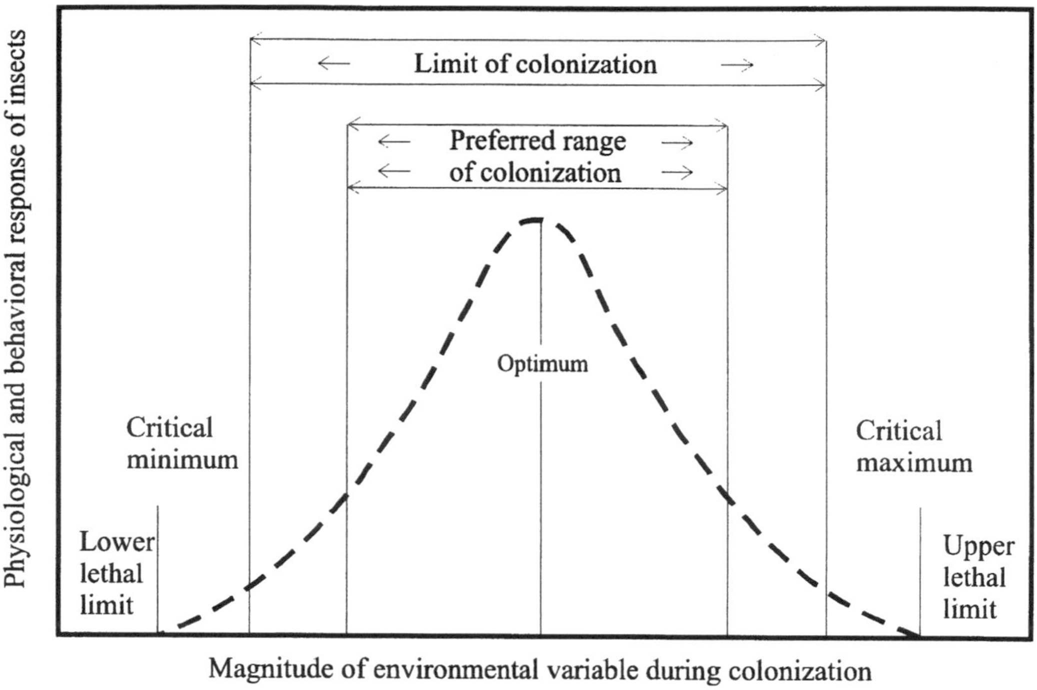


**FIGURE 31.1** Responses of insects to environmental variables during colonization showing the extreme limits and preferred range of environmental variables within which colonization is possible. (Modified with permission from Ochieng-Odero 1994.)

colonized insects that often have been reared in the laboratory for many generations (Mason et al. 1987). The difficult task of fruit fly-rearing managers is to balance the benefits of intensive insect production with the behavioral quality of the end product.

The process by which an insect population transfers from the natural environment to the laboratory is often erroneously referred to as as "adaptation process." However, according to the Darwinian definition, adaptation is a product of genetic change, which may or may not occur under these conditions (Ochieng-Odero 1994). True adaptation must therefore be minimized in rearing systems where mass-reared insects should resemble as closely as possible the wild populations. Ochieng-Odero (1994) defined the term laboratory *colonization* of insect species as consisting of three processes.

First, the *acclimatization* process which, in successive generations, gradually shifts the optimal performance of the insects from the extreme limits to the preferred range of colonization (see Figure 31.1). The limits of colonization bracket the original broad range of conditions within which the wild population was able to maintain itself successfully. Near the limits of colonization, and despite the fact that the population can still be maintained, the performance of the population is drastically reduced. Outside of this limit, the population cannot be maintained successfully and the extreme limits are called lower and upper lethal limits. The preferred range of colonization is relatively narrow and represents the magnitude of environmental variables within which the population shows optimal performance. Through acclimatization, a gain in performance is noticed by restricting the magnitude of environmental variables to those of the preferred range of colonization. Acclimatization occurs within variable time ranges, which are very much shorter than the evolutionary timescale.

Second, the *selection* process is involved when insects are forced directly into mass production. The optimum performance represents the equilibrium ("center") of the bell-shaped curve shown in Figure 31.1. By applying directed and unrelenting selection pressure on successive generations, the

**TABLE 31.1**
**Divergent Outcomes Sought through the Life History Strategies Adopted by Wild and by Mass-Reared Tephritid Populations**

| Wild Population | Mass-Reared Strain |
|---|---|
| 1. To ensure the adaptation to various environmental conditions | 1. To ensure a rapid turnover of successive generations |
| 2. To promote the quality of the progeny | 2. To promote egg productivity |

optimum peak of performance can be displaced within the range of the limits of colonization. Selection changes the characteristics of the population, which then shows a maximum performance under conditions which differ from those of the original wild population.

The third process in colonization is *domestication*. Domestication is defined as the gradual acceptance by insect species of new constraints which do not exist in the wild, and the ability to perform despite them (Ochieng-Odero 1994). This process can be due to conditioning or changes in behavior that are not determined by alterations of the genotype (Boller 1972, as cited in Ochieng-Odero 1994).

Both wild and laboratory populations have to pass through the process of colonization of their respective ecosystems. In nature, tephritid populations face changing environmental conditions and predation, whereas in mass-rearing facilities, they face a constant environment focused toward a rapid turnover of the successive generations (Table 31.1). During a new colonization in the wild by a tephritid population, its behavioral traits and life history strategy are shaped by natural selection. However, during a new colonization of a tephritid population in the laboratory, its behavioral traits and life history strategy are wrenched in a new direction by acclimatization, artificial selection, and domestication processes. Therefore, laboratory strains often come to possess behavioral and physiological traits that diverge from those required for success by wild populations.

The major findings, which document the divergent consequences of natural selection and laboratory colonization on behavioral traits and patterns and on life history strategies, are reviewed in this chapter.

## 31.2 DIVERGENT COLONIZATION STRATEGIES OF WILD AND MASS-REARED POPULATIONS: HIGH ADAPTABILITY VS. HIGH SPECIFICITY

The vastly different requirements for biological success of colonization in the wild vs. the laboratory lead to divergent life history strategies and behaviors.

Thus, during field colonization, selection pressure is focused predominantly on the egg/larval stages, but also tends to increase the adult prematuration period and to promote high adaptability to various environmental conditions. By contrast, during laboratory colonization, selection pressure is focused predominantly on the adult stage and induces a short adult life span and high specificity for performance only in the standard laboratory environment.

### 31.2.1 Differences in Larval Stage Strategies

In nature, there are high selection pressures on the egg and larval stages. For frugivorous tephritid species, fruits often represent poor substrates and short-lived resources (Hendrichs and Prokopy 1994) (Table 31.2). Extended egg and larval stages increase the risk of parasitism and predation by vertebrate seed dispensers (frugivorous mammals, birds) and humans. Consequently, there is selection for rapid larval development. In addition, wild tephritid populations can adjust their larval developmental period to the type and status of the host (size, ripening stage, etc.), which can result in a doubling of this period as shown in the Mediterranean fruit fly, *Ceratitis capitata* (Wiedemann) (Féron and Sacantanis 1955), and in the Natal fruit fly, *C. rosa* Karsch (Etienne 1973; see Table 31.2).

**TABLE 31.2**
**Divergent Nature of Selection Factors and Outcomes That Impinge on the Larval Stage in Wild and Mass-Reared Tephritid Populations**

| Wild Population | Mass-Reared Strain |
|---|---|
| 1. Strongly selected by natural enemies, especially high risk of predation | 1. No selection occurs unless applied deliberately |
| 2. Larval diets (fruits) are short-lived, variable resources, and have low protein content | 2. Larval diets (artificial) are long-lived resources, and have high protein content |
| 3. Variability in larval developmental periods is promoted | 3. Homogeneity in larval developmental periods is promoted |

In rearing facilities, larvae develop in artificial diets. When compared with natural hosts of tephritid species, these artificial diets represent richer substrates and relatively long-lived resources, and there is no risk of predation (see Table 31.2). Artificial larval diets have a higher proportion of protein than do natural larval hosts (Nakamori and Kakinohana 1980), up to about 8% in Mediterranean fruit fly diet (Bruzzone et al. 1990). Larvae in these diets mature faster than larvae in fruits. This has been observed in *C. rosa* (Etienne 1973), *C. capitata* (Vargas and Carey 1989; Dridi 1990), Caribbean fruit fly (*Anastrepha suspensa* (Loew)) (Chambers 1977), olive fly (*Bactrocera oleae* (Rossi)) (Tsitsipis 1983), and melon fly (*B. cucurbitae* (Coquillett)) (Miyatake 1995). In addition, larvae on artificial diets are selected to develop synchronously (Miyatake 1993).

The shortening of the larval developmental period through laboratory colonization is accomplished within a limited number of generations — three in Mediterranean fruit fly (Dridi 1990), five in Caribbean fruit fly (Chambers 1977). Different larval-rearing practices in the normal acclimatization process can sometimes lead to very different changes. Thus, in one facility in Japan, the melon fly strain developed a long larval period, while in a second facility with different rearing practices, the strain developed a short larval period (Suenaga et al. 1992; Miyatake 1995; Miyatake 1996).

### 31.2.2 Differences in Adult Stage Strategies

Adult strategies are somehow influenced by the characteristics of the larval stage. In nature, some flies might delay development to avoid competition. By having a large range of differently aged flies, a population can bridge the ripening period of successive host varieties (Hendrichs and Prokopy 1994; Buyckx 1994; Cayol 1996). However, the limited nutritive value of natural larval food results in low energy reserves carried over from the larval stage (Hendrichs and Prokopy 1994). Wild adult tephritids have therefore to build up their reserves before being able to initiate mating (Table 31.3). In the laboratory, constant larval developmental periods are encouraged so that most flies emerge from the pupae within a few days, and their nearly uniform chronological and physiological ages ensure a rapid turnover of generations. Artificial larval diets have high nutritive levels and adults emerge with high energy reserves and consequently are sexually mature earlier than wild insects. This was demonstrated in melon fly (Suzuki and Koyama 1980; Soemori 1980), olive fly (Zervas 1983), Mediterranean fruit fly (Rössler 1975; Dridi 1990), and Mexican fruit fly, *A. ludens* (Loew) (Liedo et al. 1996; see Table 31.3). Through the laboratory colonization process of a wild Mediterranean fruit fly population, it was shown that, after only three to four generations under rearing, the presexual maturity period of the flies was significantly shortened when compared with that of the original wild population (Dridi 1990).

In wild populations, priority is given to a long adult life span as it represents a greater chance for individuals to mate and find suitable hosts for oviposition (see Table 31.3). In the laboratory, flies are held for only a short time, which coincides with their period of greatest oviposition, and are then discarded (Calkins 1989a, b; 1991). Since the main goal of some managers in mass-rearing

**TABLE 31.3**
**Divergent Nature of Requirements for Survival and of the Selection Factors and Outcome That Impinge on the Adult Stage in Wild and Mass-Reared Tephritid Populations**

| Wild Population | Mass-Reared Strain |
|---|---|
| 1. Extended adult survival is the key to survival | 1. Early adult mating performance and egg production are the keys to survival |
| 2. Long presexual maturity period, caused by low energy reserves carried over from larval stage | 2. Short presexual maturity period because of high energy reserves accumulated during larval stage |
| 3. Variability in adult ages and sizes in the same ecosystem | 3. Homogeneity in adult ages and sizes in the same rearing cage |
| 4. Adults are selected for high irritability that is needed to escape from predators | 4. Adults accustomed to crowded conditions and lose much of their irritability |
| 5. Wide range of adaptations to various changing environmental conditions | 5. Maximum productivity obtained under optimum controlled conditions |

is to ensure a rapid succession of generations, extended life spans are viewed as detrimental. Thus, the acclimatization process tends to shorten the adult life span (see Table 31.3).

In nature, juice from wounded fruits, fluids deposited on leaves by aphids, and bird feces (Hendrichs and Hendrichs 1990) represent scattered feeding resources and provide considerable energy reserves. By contrast in the laboratory, while food is available in quantity in the rearing cages, getting it may be difficult for the flies in the current form it is presented. In mass-rearing cages, adult feeding stations are often limited in surface relative to the high population density (e.g., in Mediterranean fruit fly mass-rearing cages, 1 $cm^2$ of feeding surface is available for about 14 adult flies, J.P. Cayol, unpublished data). This induces stress and often results in a high mortality shortly after the peak of adult emergence (J.P. Cayol, unpublished data).

In nature, the survival of adult tephritids also depends on their ability to escape from predators such as wasps and birds. Such reactions are often referred to as "irritability" or "startle behavior" (Boller et al. 1981; see Table 31.3). In nature, low irritability predisposes flies to a high risk of predation (Hendrichs et al. 1996), whereas, in the laboratory, with high adult densities, frequent interactions lead to a decreased irritability level (Boller and Calkins 1984; Table 31.3). It was shown in the Mediterranean fruit fly that the initially high irritability of wild flies disappears after only six generations of laboratory rearing (Boller and Calkins 1984). The negative effects of low irritability on the survival of mass-reared tephritids was demonstrated in the Mediterranean fruit fly (Hendrichs et al. 1993; 1994). The authors found that mass-reared adult flies were about three to four times as likely as wild flies to be captured by yellow jacket wasps. This might be one reason why, in SIT programs, only a few mass-reared flies can be found in the field 1 week after the last release (Shelly et al. 1994).

The activity of tephritid flies is regulated mainly by the environmental conditions and they can tolerate a range of environmental factors. The constant temperatures and abrupt shifts between photophases that prevail in the laboratory do not require colonized flies to cope with variability in environmental changes. However, in the wild, the fluctuating temperatures and effects of dawn and dusk require wild flies to have a much higher tolerance of variability than that of mass-reared flies. Obviously, the wild population as a whole must be able to "react" appropriately to changing conditions (Ochieng-Odero 1994). Indeed, wild tephritid populations adjust their behavioral pattern to fluctuating environmental conditions (Prokopy and Hendrichs 1979; Hendrichs and Hendrichs 1990; see Table 31.3) within the critical limits of colonization described by Ochieng-Odero (1994). In the laboratory, tephritid flies are kept under constant optimal conditions of light, temperature, and relative humidity. The "sieve" of the colonization processes may retain only those individuals that are better adapted to these specific and constant conditions (see Table 31.3). The regulation of circadian activity rhythms has been described for the melon fly (Back and Pemberton 1917; Miyatake 1997), the Mexican fruit fly (Baker et al. 1944), the Queensland fruit fly (*Bactrocera tryoni* (Froggatt)) (Myers 1952), the olive fly (Sacantanis 1953; Féron 1960; Causse et al. 1966), the oriental fruit fly (*Bactrocera dorsalis* (Hendel)) (Roan et al. 1954), and the Mediterranean fruit fly (Causse et al. 1966). Under very high midday temperature conditions, wild Mediterranean fruit fly populations exhibit a bimodal activity rhythm with an activity peak during the first and the last hours of daylight, respectively (Hendrichs and Hendrichs 1990; Cayol 1996). However, under temperate conditions, the Mediterranean fruit fly displays a unimodal activity rhythm centered on the warmer temperatures at midday (Prokopy and Hendrichs 1979). Because of the acclimatization to constant conditions, laboratory-reared insects under natural conditions do not exhibit the same activity pattern as their wild counterparts. Thus, on cloudy days, mass-reared Mediterranean fruit fly males are not competitive with wild males for mates (Zapién et al. 1983). Seasonal behavioral patterns exist in wild melon fly population, which lead to considerable modifications in the age of sexual maturity at different times of the year (Miyatake and Iwahashi 1994). However, under the constant conditions in rearing facilities, mass-reared strains do not exhibit corresponding seasonal changes in sexual maturity.

## 31.3 SEXUAL BEHAVIOR: PROLONGED AND INTRICATE VS. BRIEF AND SIMPLIFIED

Tephritid flies in nature have developed an intricate multiple-step sequence of precopulatory courtship, which culminates in the acceptance of males by females. In the laboratory, this behavior is modified according to the conditions prevailing in the artificial environment.

### 31.3.1 Changes in Daily Sexual Activity

Mass-reared tephritids reach sexual maturity earlier than their wild counterparts, and their sexual activity is not regulated by changes in the environment. By contrast, in nature the initiation of daily sexual activity is regulated by a change in the environmental conditions (light and/or temperature). Daily sexual activity starts at dawn in the South American fruit fly, *A. fraterculus* (Wiedemann) (Malavasi et al. 1983), in the late afternoon in the Caribbean fruit fly (Burk 1983), and at dusk in the Mexican fruit fly (Aluja et al. 1983), the melon fly (Iwahashi and Majima 1986), and the Queensland fruit fly (Tychsen 1977). Often the main factor regulating the initiation of sexual activity in these species is an increase or decrease of light intensity. In some species, such as the Mediterranean fruit fly (Hendrichs and Hendrichs 1990), the West Indian fruit fly, *A. obliqua* (Macquart), and the papaya fruit fly, *Toxotrypana curvicauda* Gerstaecker (Landolt and Hendrichs 1983), the main limiting factor is temperature, so that the sexual activity of these species is confined largely to the warmer hours of the day. In rearing facilities, temperature and relative humidity are constant and the transition between light and dark phases is abrupt. In addition, because of the high adult densities, individual tephritid flies must initiate sexual activities speedily in order to obtain a mate (Calkins 1989b). Consequently, mass-reared tephritid flies may initiate sexual activity earlier during the day than their wild counterparts. This has been observed in the melon fly (Koyama et al. 1986), the Mexican fruit fly (Moreno et al. 1991), and in the Mediterranean fruit fly in some cases (Guerra et al. 1986). Interestingly, in Mediterranean fruit fly, Blay and Yuval (1997) showed that protein-fed males start their mating activity earlier in the day than protein-deprived males. In rearing facilities, proteins represent as much as one-fourth of the adult Mediterranean fruit fly diet (one volume of yeast hydrolysate for three volumes of sugar; K. Fisher, personal communication). In the wild, although proteins are crucial for lekking and mating success (Yuval et al. 1998; Yuval and Hendrichs, Chapter 17), adult Mediterranean fruit fly diet is apparently relatively poor in protein (Hendrichs and Hendrichs 1990; Hendrichs et al. 1991; Warburg and Yuval 1997a, b). This difference in adult diet may also partially explain the fact that mass-reared Mediterranean fruit fly males initiate their sexual activity earlier than wild males.

**TABLE 31.4**
**Relative Importance of Chemical and Physical Cues in Wild and Mass-Reared Tephritid Populations**

| Wild Population | Mass-Reared Strain |
|---|---|
| 1. Emission of quality male pheromone plays a major role in attracting the females to mate | 1. Pheromone production plays a limited role since rearing cages are filled with pheromone |
| 2. In lekking species, males that join the best-located leks have greatest chances to mate | 2. Crowded conditions preclude the formation of leks |

### 31.3.2 Relative Importance of Chemical and Physical Cues

In nature, tephritid populations engage in complex behavioral patterns involving frequent movements within a three-dimensional ecosystem, which result in the location of a feeding source, a mating partner, or an oviposition site (Hendrichs and Hendrichs 1990; Hendrichs and Prokopy 1994). The emission of male pheromone plays an important role in attracting females and in the female's selection of a sexual partner from a lek (in lekking species) (Prokopy and Hendrichs 1979; Arita and Kaneshiro 1989; Whittier et al. 1992; Sivinski et al. 1994) (Table 31.4). In mass-rearing cages, crowding precludes the formation of leks. Also male-quality cues such as pheromone production and pheromone component ratios cannot be judged as measures of male quality because the cage is filled with pheromone and females have difficulty establishing from which male a given pheromone plume is emitted (Calkins 1989b; see Table 31.4). In addition, in some species such as the Caribbean fruit fly, food availability may impact male pheromone production qualitatively and quantitatively (Epsky and Heath 1993).

In lekking tephritid species such as Queensland fruit fly (Tychsen 1977), Mediterranean fruit fly (Prokopy and Hendrichs 1979), melon fly (Iwahashi and Majima 1986), Mexican fruit fly (Moreno et al. 1991), and Caribbean fruit fly (Sivinski et al. 1994), meeting of males and females is mostly restricted to lek sites where highly discriminating females choose the best partner from among males (Hendrichs et al. 1991). In rearing cages where no leks are formed, each male has an equal chance to be approached by a female in any part of the cage, while in nature a male who is included in the best-located lek has an increased chance of being selected for mating. In the field, wild Mediterranean fruit fly males compete to join leks on parts of the tree where most of the mating with wild females occurs (Calkins 1989b), but mass-reared males often do not select a specific location (Cayol et al. 1999).

### 31.3.3 Divergent Characteristics of Male Courtship Behavior

In nature, males emit pheromone during the period when females are receptive and visit leks to select mates (Cayol et al. 1999). However, in rearing cages males are able to court females at any time and they modify their pattern of calling. When mass-reared males are released in the field, they often call throughout most of the day, regardless of female receptivity (Hendrichs et al. 1996; Cayol et al. 1999).

The transition between the different steps of male courtship in nature is modulated by and integrated with the responses of the females, and impetuous males which try to shortcut part of the courtship are rejected by wild females (Lux et al. 1997) (Table 31.5). In rearing cages, the courtship of tephritid males often is simplified and males do not systematically interpret the unresponsiveness of females as a signal to interrupt courtship (see Table 31.5). Such a departure from the behavior of wild tephritid males was documented in an old strain of the Mexican fruit fly that had been reared for more than 250 generations (Mangan 1997) and in the Mediterranean fruit fly (Shelly 1996; Liimatainen et al. 1997). In extreme cases of deviation of courtship behavior, as

**TABLE 31.5**
**Difference in Male Courtship Behaviors in Wild and Mass-Reared Tephritid Populations**

| Wild Population | Mass Reared Strain |
|---|---|
| 1. Males call only during the period when females are receptive | 1. Males call throughout the day, regardless of female receptivity |
| 2. Males which try to shortcut part of the courtship are rejected by wild females | 2. Rapidity in copulating under crowded conditions is required to assure that a male will reproduce |
| 3. Only males that perform the long intricate courtship succeed in mating | 3. Courting time is minimized to preclude interruption by other insects |

documented in the Mexican fruit fly (Moreno et al. 1991), mass-reared males partially substituted courtship with what the authors described as "mating aggression" and, as a consequence, were successful mainly with mass-reared females.

In nature, a female tephritid selects a mate on the basis of long and detailed courtship (Eberhard and Briceño 1996). However, through colonization, mass-reared tephritid males have evolved briefer courtships to decrease the risk of frequent interruptions which occur in rearing cages (Eberhard and Briceño 1996); Table 31.5). Evidence of the shortening of courtship in mass-reared males was documented in the melon fly (Kanmiya et al. 1987; Hibino and Iwahashi 1991) and in the Mediterranean fruit fly (Eberhard and Briceño 1996). In the Mediterranean fruit fly, a correlation was demonstrated between the degree of shortening of courtship and the number of generations under mass-rearing (Eberhard and Briceño 1996). Yet in mass-reared Mediterranean fruit flies, even though the total duration of the courtship has been reduced, the duration of some of the male acoustic signals (calling buzzes) has been increased (Briceño and Eberhard 1997). It is believed that the prolonged sounds overcome the background noise which pervades the rearing cages and that prolonged songs are interpreted by females as an indicator of male quality.

### 31.3.4 Mating Behavior: The Relative Importance of Discrimination by the Female in Selection of a Mate

The selection of a mate in nature is made by the female on the basis of chemical, physical, and visual cues (Eberhard, Chapter 18) (Table 31.6). The interaction of males and females is limited in time to the period of the day when females are receptive and visit the sexually mature males. In the field, females can select mating partners among males of different ages (age is an indicator of sexual maturity) and sizes. The size of the male plays an important role in female choice, and it has been shown that larger males produce stronger wing vibrations indicative of male quality. Thus, larger males have been found to be more likely to be selected as sexual partners in Caribbean fruit flies (Burk and Webb 1983) and in Mediterranean fruit flies (Orozco and López 1993; Blay and Yuval 1997). In mass-rearing cages, the choices offered to females are drastically limited since all of the flies in the same cage are the same age, of similar size, and they reach sexual maturity at about the same time (Calkins 1991). In nature, the female tends to choose the mates that are most fit to assure the quality of the progeny; in mass-rearing, the main goal is egg productivity in a limited period of time and females that mate earliest usually have the most progeny. In the laboratory, quantity overcomes quality. Under such circumstances, discriminating (selective) females tend to be unsuccessful in reproducing (Calkins 1989b; see Table 31.6). Moreover, unlike in nature where unreceptive or immature females can avoid being courted by avoiding male territories (Mangan 1997), the crowded conditions in mass-rearing cages might force immature and unreceptive females, which would not mate under natural conditions, to mate and produce offspring (Calkins 1989b; see Table 31.6).

**TABLE 31.6**
**Difference in Selection of a Mate by Female in Wild and Mass-Reared Tephritid Populations**

| Wild Population | Mass-Reared Strain |
|---|---|
| 1. Females are highly selective in mating; selection of a male is based on various chemical, physical, and visual cues, among males of different size (fitness) and different age (sexual maturity) | 1. Because of factory schedule, only those eggs laid in the first days are used to maintain the colony; the importance of original cues is limited, and all males are of the age and similar size; rapidity of copulation is the key; less "choosy" females mate earlier and give offspring |
| 2. Impetuous males are rejected by wild females | 2. Due to crowded conditions, females cannot avoid meeting males; females might be forced to mate and give offspring |

In brief, the key role of the female in the wild is to select the mate best able to father robust and successful progeny, but this role is compromised severely under conditions of mass-rearing.

### 31.3.5 Female Postmating Behavior: Attraction to Oviposition Sites

In tephritids, the postmating activities of females are very different from premating activities; hence, mating induces a major switch in the female's behavioral pattern. Upon reaching sexual maturity, females are strongly attracted to male pheromone. However, once mated, the females switch imperatives from mating to oviposition. In the Mediterranean fruit fly, during copulation the male transfers substances from the accessory glands to the female. These accessory gland substances result in the female switching from preferential attraction to male-produced pheromone to compelling attraction to host fruit odors emitted from potential oviposition sites (Jang 1995). In nature, such a behavior may preclude remating by females after a successful copulation.

Under mass-rearing conditions, where no fruit odor is present, do mass-reared females retain this postmating behavior and do mass-reared males retain this ability to induce this behavioral switch in the females with whom they mate? Jang et al. (1998) found that both wild and mass-reared virgin Mediterranean fruit fly females were significantly more attracted by the male pheromone than by fruit (guava) odor. Moreover, both types of females when mated to a mass-reared or to a wild male, switched their behavior from attraction to the pheromone to attraction to fruit odor. This suggests strongly that the rearing procedures had not impaired the effectiveness of the males to induce a switch in female behavior. Of utmost importance for SIT programs is the fact that mass-reared males are also able to induce the same switch in wild female behavior, and so to inhibit or delay remating. Thus, with little remating, there is no need to worry about the impact of sperm displacement, or precedence (Yuval and Hendrichs, Chapter 17), on the effectiveness of SIT programs.

### 31.3.6 Female Oviposition Behavior: Relative Importance of the Host-Marking Pheromone

After laying eggs in host fruit, females of some tephritid species mark the oviposition site by circling it with the extended aculeus of the ovipositor from which a substance known as host-marking pheromone (HMP) is deposited (Prokopy 1972). The HMP deters oviposition by other females of the same species on the respective fruit. As a consequence, this behavior indirectly contributes to a better distribution of eggs (and larvae) in nature, and a better use of larval food resources by the wild population. Evidence of an HMP has been reported in the apple maggot fly, *Rhagoletis pomonella* (Walsh) (Prokopy 1972), the cherry fruit fly, *R. cerasi* (L.) (Katsoyannos 1975), the Caribbean fruit fly (Prokopy et al. 1977), and the Mediterranean fruit fly (Prokopy et al. 1978).

In mass-rearing facilities, artificial oviposition sites are used such as textile egging nets in the case of Mediterranean fruit fly or gel-covered textile panels in the case of *Anastrepha* Schiner species. In both cases, mass-reared females must share their oviposition sites with many other females. Under such circumstances, high sensitivity to an HMP would limit egg production. Indeed in Mediterranean fruit fly, this problem is encountered often during the laboratory colonization process of a wild population, and three to four generations of selection can be required for the colony to be fully adapted to the new oviposition situation (Dridi 1990). To overcome this oviposition bottleneck, Mediterranean fruit fly genetic sexing strains are founded on a female population already adapted to artificial oviposition devices (G. Franz, personal communication), the wild genetic material being introduced through the male lineage. As shown in the Mediterranean fruit fly, females from a strain with a long history of being mass-reared demonstrate a reduced ability to discriminate HMP when compared with wild females (Averill and Prokopy 1989; Boller et al. 1994). In addition, wild Mediterranean fruit fly females spend more than 60 s dragging their aculei while in females from long-term mass-reared strains this period has become shortened to 2 to 3 s (Boller et al. 1994).

In conclusion, the imperative to maintain high egg productivity under mass-rearing conditions tends to change oviposition behavior by decreasing the female's sensitivity to the HMP or perhaps to the quantity of HMP released by the female while dragging her aculeus.

## 31.4 IMPLICATIONS OF CHANGES IN MASS-REARED STRAINS FOR SIT PROGRAMS

For SIT programs, it is of major importance that the sterile insects released exhibit a behavioral pattern similar to that of the wild individuals so that they will compete successfully for wild insect mates. Taking into account the changes which can occur in tephritid strains under mass-rearing conditions, various tests have been developed to measure differences between wild and mass-reared flies (Fried 1971; Chambers 1975; Boller and Chambers 1977; Boller et al. 1981; Chambers et al. 1983). More recently, the measurement of the behavioral quality of mass-reared strains has been developed further. Improved tests were defined to measure the mating competitiveness of mass-reared flies and the sexual compatibility of mass-reared and wild Mediterranean fruit fly populations (McInnis et al. 1996; Cayol et al. 1999). Through the experience gained in these quality control tests and in SIT action programs against tephritid species, various implications of the changes in behavior induced by mass-rearing in tephritids were identified, and these are described in the following sections.

### 31.4.1 Reduced Mating Competitiveness in the Field

Because mass-reared tephritids mate under high density conditions, the aspects of mating competitiveness measured depend mainly on the conditions of the tests. When mass-reared tephritid adults are tested under high fly densities and laboratory conditions, their relative mating competitiveness is often higher than that of wild flies. This outcome has occurred in tests on the melon fly (Kakinohana 1980; Soemori et al. 1980). Under high-fly-density conditions in the laboratory, the mating competitiveness of mass-reared flies is at least equal to that of wild flies. This was the case in the melon fly (Soemori et al. 1980; Nakamori 1987; as cited in Miyatake and Haraguchi 1996), the oriental fruit fly (Shelly 1995), and the cherry fruit fly (Boller et al. 1977).

Under field-cage or open-field conditions, which have greater relevance to SIT programs, the competitiveness of mass-reared tephritids is reduced (Shelly and Whittier 1996). In field-cage tests of Mediterranean fruit fly, with a 1-to-1 wild-to-mass-reared male ratio, mass-reared and wild males engage about one-third and two-thirds of the wild female mates, respectively (Cayol, in press). However, the reduced competitiveness of mass-reared males is largely overcome when mass-reared males outnumber wild males, as in the case of SIT.

The modification of sexual behavior which occurs following long-term rearing was shown to be the main reason for deterioration in mating competitiveness (Iwahashi et al. 1983; Kanmiya et al. 1987; Hibino and Iwahashi 1991; Shelly 1996; Eberhard and Briceño 1996). Taking this phenomenon into account, the difficult task of tephritid mass-rearing facility managers is to balance the benefits derived from long-term rearing against the progressive loss in mating efficiency under field conditions. In Metapa, Mexico, the management of the Mediterranean fruit fly–rearing facility replaces the strain when there is a significant decrease in mating competitiveness as assessed in annual field-cage tests (C. Garcia, personal communication).

### 31.4.2 The Extreme Case of "Behavioral Resistance"

Behavioral resistance of a wild population to mating with a mass-reared tephritid strain would constitute the ultimate shift in the sexual behavior of the mass-reared strain. Behavioral resistance has been reported twice in tephritid species: in melon fly on Okinawa Island, Japan (Hibino and Iwahashi 1991) and in Mediterranean fruit fly on Kauai Island, Hawaii (McInnis et al. 1996).

In these two cases, the wild females were able to identify mass-reared males and to reject them as mating partners. Nevertheless, in Kauai wild Mediterranean fruit fly males did mate with mass-reared females (McInnis et al. 1996). In the case of the melon fly, the reduced courtship duration of mass-reared males probably was the cause of "resistance" (Hibino and Iwahashi 1991). In the case of the Mediterranean fruit fly, the strain used for field release had been mass-reared for about 38 years (McInnis et al. 1996), and the behavioral quality of the sterile flies probably had deteriorated so much that these flies had diverged behaviorally from their wild counterparts. The selection pressure of the prolonged releases had induced the capacity of wild females to recognize released males and led to the noncompetitiveness of these flies for mating with their wild counterparts. However, after the release program had been terminated for several years, the wild females were found to have lost their ability to recognize released males as such (D. McInnis, personal communication).

The problem posed by the behavioral resistance can be solved either by (1) increasing the number of flies released to overcome the poor level of interbreeding between the two populations, provided that some cross-matings still occur, or (2) colonizing a new strain based on wild pupae collected in the target area. In Okinawa, where the first solution was applied, the melon fly was successfully eradicated in 1993 (Itô and Kakinohana 1995).

The possibility of development of behavioral resistance strongly underscores the importance of the practice of regular strain replacement for use in SIT programs.

## 31.5 CONCLUSIONS AND FUTURE RESEARCH NEEDS

The widely dissimilar life history strategies developed during colonization in the wild as compared with the laboratory result in a drastic divergence in the behavior of the mass-reared and wild tephritids. In nature, natural selection tends to shorten the high risk larval stage, to increase the adult life span, and to promote diversity (various fly sizes and longevities, variety of hosts utilized, and adaptations to various environmental conditions). Sexual behavior in wild populations is based on female choice after the male has presented her with a long series of intricate chemical, visual, acoustical, and physical cues (Eberhard, Chapter 18). However, mass-rearing tends to shorten each phase of the life cycle, to reduce differences (flies come to have nearly the same size and longevity), and to select for high egg production in young females. Sexual behavior in laboratory strains seems to be characterized by fast mating of less selective females with impetuous and sometimes aggressive males. These altered behaviors can arise because, in the laboratory, the flies are exposed to different levels of chemical, visual, and physical cues than are experienced by wild flies.

During the 1960s, it appears that most managers of rearing facilities considered production efficiency to be the major priority (Finney and Fisher 1964), leading to the extreme view that long-term artificial rearing was beneficial in that it improved performance significantly, e.g., shortened the preoviposition period and increased egg production. The unavoidable selection of a different set of behavioral attributes can be expected to lead to reduced effectiveness of the mass-reared insects in the field. In Mediterranean fruit fly, some strains were artificially selected for "fast mating" (Harris et al. 1983), but the impetuous males were rejected by wild females. Recently, Fisher and Caceres (in press) developed an innovative system to rear tephritid species that reduces the selection pressure on adult flies. This new rearing philosophy, called Filter Rearing System (FRS), is based on the maintenance of a small colony in which adult flies are maintained under relaxed conditions (low fly density, sufficient feeding surfaces available). The subsequent steps in colony production are unidirectional in that no insects are returned to the original small colony. This prevents the accumulation of unsuitable traits. The FRS is currently under evaluation and development. By using the FRS, various aspects of improvement of tephritid behavior could be investigated: (1) mixing flies of a range of ages and sizes in the same cage; (2) increasing the total surface available for the flies to limit the effect of crowding, without affecting the productivity; (3) designing new rearing cages taking into account the behavior of tephritid species in the field (in the wild, feeding, male courtship, mating, and oviposition often occur in different locations, which is not possible in mass-rearing cages); (4) questioning whether or not a cage is even required to rear the initial small-scale colony — could not it be done in a small greenhouse containing host trees?

The FRS is a major breakthrough in the control of some of the negative effects of mass-rearing on tephritid behavior but has highlighted the needs for better quality control tests to assess the behavior of mass-reared tephritid flies relative to that of wild individuals. Initially, laboratory tests were developed that measured mating ability, sperm competitiveness (Haisch 1970), assessed overall quality (Boller and Chambers 1977; Boller et al. 1981) and overall competitiveness (Fried 1971). In laboratory "mating propensity test," the sexual competitiveness of a mass-reared strain was measured by mating speed in a Plexiglas cage (30 × 30 × 40cm). Subsequently, Zapién et al. (1983) and Chambers et al. (1983) used field-cage tests as a quality control test to monitor the competitiveness of Mediterranean fruit fly laboratory strains. Despite this, the standard quality control manual still recommended that mating propensity of mass-reared tephritid flies be assessed under laboratory conditions (Brazzel et al. 1986). These laboratory tests were inadequate to assess the behavioral repertoire of mass-produced insects and field-cage tests became essential (McInnis et al. 1996; Cayol et al. 1999; Cayol, in press). The current quality control manual for sterile tephritid fruit flies release programs (Anonymous 1999) recommends that periodic mating compatibility tests be carried out in field cages. This is a step forward in the appropriate evaluation of the behavioral quality of mass-reared tephritid strains.

Do field-cage mating tests answer all the questions related to the behavioral quality of mass-reared tephritid flies? The answer is obviously no, as (1) these tests can only be used to evaluate the relative quality of a mass-reared strain but cannot be used to predict field results; (2) the size of the standard field cage usually allows a single tree to be placed inside (this may not be representative of an open-field situation, especially in the case of lekking species); (3) the flies tested are virgin and often of the same age; (4) the same flies are tested for one day only and remating response is not measured; (5) the egg fertility is not measured; (6) the tree often needs to be pruned to facilitate the observations, which may affect the response of the flies to various stimuli (e.g., light intensity). Future research on the field evaluation of the behavioral quality of mass-reared tephritid strains could address the following questions:

1. How can field cage tests be improved to simulate open field conditions better (e.g., by increasing the size of the cage, keeping the flies for consecutive days, releasing flies of different ages, virgin and already mated)?
2. Can a model be designed to predict the field performance of mass-reared insects on the basis of a relatively simple test?

Mass-reared tephritid flies will always be inferior to wild flies. However, it is the responsibility of the rearing facility managers to minimize this inferiority by developing strategies that maintain a certain defined level of quality. One simple option is to replace the strain on a regular basis. It is easier to include such preventive actions than to face disastrous field results, and their consequences.

## ACKNOWLEDGMENTS

I would like to acknowledge the following donors — Campaña Nacional contra las Moscas de la Fruta (Mexico), the International Organization for Biological Control of Noxious Animals and Plants (IOBC), and the Instituto de Ecología, A.C. (Mexico) — for partial funding of my participation to the symposium. My best thanks go to M. Aluja and his team and to A. Norrbom for the organization of such a beautiful symposium, and for their warm hospitality. I am grateful to A.S. Robinson and J. Hendrichs for their useful suggestions at the early stage of the manuscript. I also thank five anonymous reviewers for their comments, and B. Eberhard, B. Yuval, D. Papaj, J. Sivinski, and K. Kaneshiro for the final revision of the manuscript.

## REFERENCES

Aluja, M., J. Hendrichs, and M. Cabrera. 1983. Behavior and interactions between *Anastrepha ludens* (L.) and *A. obliqua* (M.) on a field caged mango tree. I. Lekking behavior and male territoriality. In *Fruit Flies of Economic Importance* (R. Cavalloro, ed.), pp. 122–133. A.A. Balkema, Rotterdam.

Anonymous. 1999. Product quality control, irradiation and shipping procedures for mass-reared tephritid fruit flies for sterile insect release programs. International Atomic Energy Agency, Vienna.

Arita, L.H. and K.Y. Kaneshiro. 1989. Sexual selection and lek behavior in the Mediterranean fruit fly, *Ceratitis capitata* (Diptera: Tephritidae). *Pac. Sci.* 43: 135–143.

Averill, A.L. and R.J. Prokopy. 1989. Host marking pheromones. In *Fruit Flies: Their Biology, Natural Enemies and Control* (A.S. Robinson and G. Hooper, eds.), pp. 207–219. In *World Crop Pests* (W. Helle, ed.), Vol. 3A. Elsevier Science Publishers, Amsterdam.

Back, E.A. and C.E. Pemberton. 1917. The melon fly in Hawaii. *U.S. Dep. Agric. Bull.* 491, 1–64.

Baker, A.C., W.E. Stone, C.C. Plummer, and M. McPhail. 1944. A review of studies on the Mexican fruit fly and related Mexican species. *U.S. Dep. Agric. Misc. Publ.* 531, 155 pp.

Blay, S. and B. Yuval. 1997. Nutritional correlates to reproductive success of male Mediterranean fruit flies. *Anim. Behav.* 54: 59–66.

Boller, E.F. 1972. Behavioral aspects of mass-rearing insects. *Entomophaga* 17: 9–25.

Boller, E.F. and C.O. Calkins. 1984. Measuring, monitoring and improving the quality of mass-reared Mediterranean fruit flies, *Ceratitis capitata* (Wied.) 3. Improvement of quality by selection. *Z. Angew. Entomol.* 98: 1–15.

Boller, E.F. and D.L. Chambers. 1977. Quality control: an idea book for fruit fly workers. *IOBC/WPRS Bull.* 162 pp.

Boller, E.F., U. Remund, B.I. Katsoyannos, and W. Berchtold. 1977. Quality control in European cherry fruit fly: evaluation of mating activity in laboratory and field cage tests. *Z. Angew. Entomol.* 83: 183–201.

Boller, E.F., B.I. Katsoyannos, U. Remund, and D.L. Chambers. 1981. Measuring, monitoring and improving the quality of mass-reared Mediterranean fruit flies, *Ceratitis capitata* (Wied.) 1. The RAPID quality control system for early warning. *Z. Angew. Entomol.* 92: 67–83.

Boller, E.F., C. Hippe, R.J. Prokopy, W. Enkerlin, B.I. Katsoyannos, J.S. Morgante, S. Quilici, D. Crespo De Stilinovic, and M. Zapater. 1994. Response of wild and laboratory-reared *Ceratitis capitata* Wied. (Dipt., Tephritidae) flies from different geographic origins to a standard host marking pheromone solution. *J. Appl. Entomol.* 118: 84–91.

Brazzel, J.R., C.O. Calkins, D.L. Chambers, and D.B. Gates. 1986. Required Quality Control Tests, Quality Specifications, and Shipping Procedures for Laboratory Produced Mediterranean Fruit Flies for Sterile Insect Control Programs. USDA Manual. *APHIS* 81–51, 28 pp.

Briceño, R.D. and B. Eberhard. 1997. Comparisons of the songs of successful and unsuccessful males of different genetic sexing strains of medfly, *Ceratitis capitata* (Diptera: Tephritidae). In *Medfly Mating Behaviour Studies under Field Cage Conditions: Report of the Third Research Co-ordination Meeting*. International Atomic Energy Agency, Vienna. 34 pp.

Bruzzone, N.D., A.P. Economopoulos, and H.S. Wang. 1990. Mass rearing *Ceratitis capitata*: reuse of the finisher larval diet. *Entomol. Exp. Appl.* 56: 103–106.

Burk, T. 1983. Behavioral ecology of mating in the Caribbean fruit fly, *Anastrepha suspensa* (Loew) (Diptera: Tephritidae). *Fla. Entomol.* 66: 330–344.

Burk, T. and J.C. Webb. 1983. Effect of male size on calling propensity, song parameters, and mating success in Caribbean fruit flies, *Anastrepha suspensa* (Loew) (Diptera: Tephritidae). *Ann. Entomol. Soc. Am.* 76: 678–682.

Buyckx, E.J. 1994. Unfecundated dates, host of the Mediterranean fruit fly (Diptera: Tephritidae) in the oases of Tozeur, Tunisia. *IOBC/WPRS Bull.* 17: 25–37.

Calkins, C.O. 1989a. Quality control. In *Fruit Flies: Their Biology, Natural Enemies and Control.* (A.S. Robinson and G. Hooper, eds.), pp. 153–165. In *World Crop Pests* (W. Helle, ed.), Vol. 3B. Elsevier Science Publishers, Amsterdam. 447 pp.

Calkins, C.O. 1989b. Lekking behaviour in fruit flies and its implications for quality assessments. In *Fruit Flies of Economic Importance 87* (R. Cavalloro, ed.), pp. 135–139. A.A. Balkema, Rotterdam.

Calkins, C.O. 1991. The effect of mass-rearing on mating behavior of Mediterranean fruit flies. In *Proceedings of the International Symposium on the Biology and Control of Fruit Flies* (K. Kawasaki, O. Iwahashi, and K.Y. Kaneshiro, eds.), pp. 153–160. Ginowan, Okinawa.

Causse, R., M. Féron, and M.M. Serment. 1966. Rythmes nycthéméraux d'activité sexuelle inverses l'un de l'autre chez deux diptères Trypetidae: *Dacus oleae* Gmelin et *Ceratitis capitata* Wiedeman. *C.R. Acad. Sci.* 262: 1558–1560.

Cayol, J.P. 1996. Box thorn, key early season host of the Mediterranean fruit fly. *Int. J. Pest Manage.* 42: 325–329.

Cayol, J.P. World-wide sexual compatibility in medfly, *Ceratitis capitata* Wied., and its implications for SIT. In *Area-Wide Control of Fruit Flies and Other Major Insect Pests* (K.T. Hon, ed.). Universiti Sains Malaysia, Penang, Malaysia, in press.

Cayol, J.P., J. Vilardi, E. Rial, and M.T. Vera. 1999. New indices to measure the sexual compatibility and mating performance of medfly (Diptera: Tephritidae) laboratory reared strains under field cage conditions. *J. Econ. Entomol.* 92: 140–145.

Chambers, D.L. 1975. Quality in mass-produced insects: definition and evaluation. In *Controlling Fruit Flies by the Sterile Insect Technique*, pp. 19–32. International Atomic Energy Agency, Vienna.

Chambers, D.L. 1977. Quality control in mass-rearing. *Annu. Rev. Entomol.* 22: 289–308.

Chambers, D.L., C.O. Calkins, E.F. Boller, Y. Itô, and R.T. Cunningham. 1983. Measuring, monitoring, and improving the quality of mass-reared Mediterranean fruit flies, *Ceratitis capitata* Wied. 2. Field tests for confirming and extending laboratory results. *Z. Angew. Entomol.* 95: 285–303.

Dridi, B. 1990. Etude de Quelques Aspects de la Biologie de la Mouche Méditerranéenne des Fruits: *Ceratitis capitata* Wiedemann (Diptère, Trypetidae). Différenciation entre Souche d'Élevage et Population Sauvage Provenant d'Algérie. Ph.D. thesis, Université Aix-Marseille 3, Marseille.

Eberhard, B. and R.D. Briceño. 1996. Genetic assimilation of the effects of crowding on the duration of premounting courtship behaviour of male medflies (Diptera: Tephritidae). In *Medfly Mating Behaviour Studies under Field Cage Conditions: Report of the Second Research Co-ordination Meeting*. International Atomic Energy Agency, Vienna. 53 pp.

Epsky, N.D. and R.R. Heath. 1993. Food availability and pheromone production by males of *Anastrepha suspensa* (Diptera: Tephritidae). *Environ. Entomol.* 22: 942–947.

Etienne, J. 1973. Conditions artificielles nécessaires à l'élevage massif de *Ceratitis rosa* (Diptera: Trypetidae). *Entomol. Exp. Appl.* 16: 380–388.

Féron, M. 1960. L'appel sonore du mâle dans le comportement sexuel de *Dacus oleae* Gmel. *Bull. Soc. Entomol. Fr.* 65: 139–143.

Féron, M. and K. Sacantanis. 1955. L'élevage permanent de *Ceratitis capitata* Wied. au laboratoire. *Ann. Epiphyties* 2: 201–214.

Finney, G.L. and T.W. Fisher. 1964. Culture of entomophagous insects and their hosts. In *Biological Control of Insect Pests and Weeds* (P. DeBach and E.T. Schlinger, eds.), pp. 328–355. Chapman & Hall, London.

Fisher, K. and C. Caceres. A filter rearing system for mass-reared medfly. In *Area-Wide Control of Fruit Flies and Other Major Insect Pests* (K.H. Tan, ed.). Universiti Sains Malaysia, Penang, Malaysia, in press.

Fried, M. 1971. Determination of sterile-insect competitiveness. *J. Econ. Entomol.* 64: 869–872.

Guerra, M., D. Orozco, A. Schwarz, and P. Liedo. 1986. Mating competitiveness of mass-reared and sterilized medflies compared with wild flies. In *Fruit Flies of Economic Importance 84* (R. Cavalloro, ed.), pp. 113–119. A.A. Balkema, Rotterdam.

Haisch, A. 1970. Some observations on decreased vitality of irradiated Mediterranean fruit flies. In *Sterile-Male Technique for Control of Fruit Flies*, pp. 71–75. International Atomic Energy Agency, Vienna.

Harris, D.J., R.J. Wood, and S.E.R. Bailey. 1983. Studies on mating in the Mediterranean fruit fly, *Ceratitis capitata* Wied. In *Fruit Flies of Economic Importance* (R. Cavalloro, ed.), pp. 197–202. A.A. Balkema, Rotterdam.

Hendrichs, J. and M.A. Hendrichs. 1990. Mediterranean fruit fly (Diptera: Tephritidae) in nature: location and diel pattern of feeding and other activities on fruiting and nonfruiting hosts and nonhosts. *Ann. Entomol. Soc. Am.* 88: 632–641.

Hendrichs, J. and R.J. Prokopy. 1994. Food foraging behavior of frugivorous fruit flies. In *Fruit Flies and the Sterile Insect Technique* (C.O. Calkins, W. Klassen, and P. Liedo, eds.), pp. 37–55. CRC Press, Boca Raton.

Hendrichs, J., B.I. Katsoyannos, D.R. Papaj, and R.J. Prokopy. 1991. Sex differences in movement between natural feeding and mating sites and trade-offs between food consumption, mating success and predator evasion in Mediterranean fruit flies (Diptera: Tephritidae). *Oecologia* 86: 223–231.

Hendrichs, J., V. Wornoayporn, B.I. Katsoyannos, and K. Gaggl. 1993. First field assessment of the dispersal and survival of mass-reared sterile Mediterranean fruit fly males of an embryonal, temperature sensitive genetic sexing strain. In *Management of Insect Pests: Nuclear and Related Molecular and Genetic Techniques,* pp. 453–462. International Atomic Energy Agency, Vienna.

Hendrichs, J., B.I. Katsoyannos, V. Wornoayporn, and M.A. Hendrichs. 1994. Odour-mediated foraging by yellowjacket wasps (Hymenoptera: Vespidae): predation on leks of pheromone-calling Mediterranean fruit fly males (Diptera: Tephritidae). *Oecologia* 99: 88–94.

Hendrichs, J., B. Katsoyannos, K. Gaggl, and V. Wornoayporn. 1996. Competitive behavior of males of Mediterranean fruit fly, *Ceratitis capitata*, genetic sexing strain Vienna-42. In *Fruit Fly Pests: a World Assessment of their Biology and Management* (B.A. McPheron and G.J. Steck, eds.), pp. 405–414. St. Lucie Press, Delray Beach.

Hibino, Y. and O. Iwahashi. 1991. Appearance of wild females unreceptive to sterilized males on Okinawa Is. in the eradication program of the melon fly, *Dacus cucurbitae* Coquillett (Diptera: Tephritidae). *Appl. Entomol. Zool.* 26: 265–270.

Itô, Y. and H. Kakinohana. 1995. Eradication of the melon fly (Diptera: Tephritidae) from the Ryukyu Archipelago with the sterile insect technique: possible reasons for its success. In *The Mediterranean Fruit Fly in California: Defining Critical Research* (J.G. Morse, R.L. Metcalf, J.R. Carey, and R.V. Dawell, eds.), pp. 215–229. University of California, Riverside.

Iwahashi, O. and T. Majima. 1986. Lek formation and male–male competition in the melon fly, *Dacus cucurbitae* Coquillett (Diptera: Tephritidae). *Appl. Entomol. Zool.* 21: 70–75.

Iwahashi, O., Y. Itô, and M. Shiyomi. 1983. A field evaluation of the sexual competitiveness of sterile melon flies, *Dacus* (*Zeugodacus*) *cucurbitae*. *Ecol. Entomol.* 8: 43–48.

Jang, E.B. 1995. Effects of mating and accessory gland injections on olfactory mediated behavior in the female Mediterranean fruit fly, *Ceratitis capitata*. *J. Insect Physiol.* 41: 705–710.

Jang, E.B., D.O. McInnis, D.R. Lance, and L.A. Carvalho. 1998. Mating-induced changes in olfactory-mediated behavior of laboratory-reared normal, sterile, and wild female Mediterranean fruit flies (Diptera: Tephritidae) mated to conspecific males. *Ann. Entomol. Soc. Am.* 91: 139–144.

Kakinohana, H. 1980. Qualitative change in the mass-reared melon fly, *Dacus cucurbitae* Coq. In *Proceedings of a Symposium on Fruit Fly Problems,* pp. 27–36. NIAS, Ibaraki, Japan.

Kanmiya, K., A. Tanaka, H. Kamiwada, K. Nakagawa, and T. Nishioka. 1987. Time-domain analysis of the male courtship songs produced by wild, mass-reared, and by irradiated melon flies, *Dacus cucurbitae* Coquillett (Diptera: Tephritidae). *Appl. Entomol. Zool.* 22: 181–194.

Katsoyannos, B.I. 1975. Oviposition-deterring, male-arresting fruit marking pheromone in *Rhagoletis cerasi*. *Environ. Entomol.* 4: 801–807.

Knipling, E.F. 1953. Possibilities of insect control or eradication through the use of sexually sterile males. *J. Econ. Entomol.* 48: 459–462.

Knipling, E.F. 1984. What colonization of insects means to research and pest management. In *Advances and Challenges in Insect Rearing* (E.G. King and N.C. Leppla, eds.), pp. 4–5. USDA ARS Publishers, Washington, D.C.

Koyama, J., H. Nakamori, and H. Kuba. 1986. Mating behavior of wild and mass-reared strains of the melon fly, *Dacus cucurbitae* Coquillett (Diptera: Tephritidae), in a field cage. *Appl. Entomol. Zool.* 21: 203–209.

Kuba, H., T. Kohama, H. Kakinohana, M. Yamagishi, K. Kinjo, Y. Sokei, T. Nakasone, and Y. Nakamoto. 1996. The successful eradication programs of the melon fly in Okinawa. In *Fruit Fly Pests: A World Assessment of Their Biology and Management* (B.A. McPheron and G.J. Steck, eds.), pp. 543–550. St Lucie Press, Delray Beach.

Landolt, P.J. and J. Hendrichs. 1983. Reproductive behavior of the papaya fruit fly, *Toxotrypana curvicauda* Gerstaecker (Diptera: Tephritidae). *Ann. Entomol. Soc. Am.* 76: 413–417.

Liedo, P., J. Toledo, and M.I. Barrios. 1996. Comparación de los parámetros demográficos de una cepa silvestre y una de laboratorio de *Anastrepha ludens* (Diptera: Tephritidae). In *Proceedings: 2nd Meeting of the Working Group on Fruit Flies of the Western Hemisphere, Viña del Mar, Chile, Nov. 3–8 1996*, pp. 33–34.

Liimatainen, J.O., A. Hoikkala, and T.E. Shelly. 1997. Courtship behavior in the Mediterranean fruit fly, *Ceratitis capitata* (Diptera: Tephritidae): a comparison of wild and mass-reared males. *Ann. Entomol. Soc. Am.* 90: 836–843.

Lux, S.A., J. Vilardi, and F. Muniri. 1997. Comparison between two Argentinean strains of medfly: the mass-reared in Mendoza and wild ones from Patagonia. In *Medfly Mating Behaviour Studies under Field Cage Conditions: Report of the Third Research Co-ordination Meeting*. International Atomic Energy Agency, Vienna. 34 pp.

Malavasi, A., J.S. Morgante, and R.J. Prokopy. 1983. Distribution and mating activities of *Anastrepha fraterculus* (Diptera: Tephritidae) flies on host and nonhost trees. *Ann. Entomol. Soc. Am.* 76: 286–292.

Mangan, R.L. 1997. Effects of strain and access to males on female longevity, lifetime oviposition rate, and egg fertility of the Mexican fruit fly (Diptera: Tephritidae). *J. Econ. Entomol.* 90: 945–954.

Mason, L.J., D.P. Pashley, and S.J. Johnson. 1987. The laboratory as an altered habitat: phenotypic and genetic consequences of colonization. *Fla. Entomol.* 70: 49–58.

McInnis, D.O., D.R. Lance, and C.G. Jackson. 1996. Behavioral resistance to the Sterile Insect Technique by Mediterranean fruit fly (Diptera: Tephritidae) in Hawaii. *Ann. Entomol. Soc. Am.* 89: 739–744.

Miyatake, T. 1993. Difference in the larval and pupal periods between mass-reared and wild strains of the melon fly, *Bactrocera cucurbitae* (Diptera: Tephritidae). *Appl. Entomol. Zool.* 28: 577–581.

Miyatake, T. 1995. Two-way artificial selection for developmental period in *Bactrocera cucurbitae* (Diptera: Tephritidae). *Ann. Entomol. Soc. Am.* 88: 848–855.

Miyatake, T. 1996. Comparison of adult life history traits in lines artificially selected for long and short larval and pupal developmental periods in the melon fly, *Bactrocera cucurbitae* (Diptera: Tephritidae). *Appl. Entomol. Zool.* 31: 335–343.

Miyatake, T. 1997. Correlated responses to selection for developmental period in *Bactrocera cucurbitae* (Diptera: Tephritidae): time of mating and daily activity rhythms. *Behav. Genet.* 27: 489–498.

Miyatake, T. and D. Haraguchi. 1996. Mating success in *Bactrocera cucurbitae* (Diptera: Tephritidae) under different rearing densities. *Ann. Entomol. Soc. Am.* 89: 284–289.

Miyatake, T. and O. Iwahashi. 1994. Delayed and asynchronous sexual maturity during the winter season in wild melon fly females *Bactrocera cucurbitae* (Coquillett) (Diptera: Tephritidae). *Biol. Mag. Okinawa* 32: 1–5.

Moreno, D.S., M. Sanchez, D.C. Robacker, and J. Worley. 1991. Mating competitiveness of irradiated Mexican fruit fly (Diptera: Tephritidae). *J. Econ. Entomol.* 84: 1227–1234.

Myers, K. 1952. Oviposition and mating behaviour of the Queensland fruit fly *Dacus* (*Strumela*) *tryoni* Frogg. and the solanum fruit fly *Dacus* (*Strumela*) *cacuminatus* Hering. *Aust. J. Sci. Res.* 5: 457–474.

Nakamori, H. 1987. Variation of reproductive characters in wild and mass-reared melon flies, *Dacus cucurbitae* Coquillett (Diptera: Tephritidae). *Appl. Entomol. Zool.* 31: 309–314.

Nakamori, H. and H. Kakinohana. 1980. Mass-production of the melon fly, *Dacus cucurbitae* Coquillett, in Okinawa, Japan. *Rev. Plant. Prot. Res.* 13: 37–53.

Ochieng-Odero, J.P.R. 1994. Does adaptation occur in insect rearing systems, or is it a case of selection, acclimatization and domestication? *Insect Sci. Appl.* 15: 1–7.

Orozco, D. and R.O. López. 1993. Mating competitiveness of wild and laboratory mass-reared medflies: effect of male size. In *Fruit Flies: Biology and Management* (M. Aluja and P. Liedo, eds.), pp. 185–188. Springer-Verlag, New York.

Penrose, D. 1996. California's 1993/1994 Mediterranean fruit fly eradication program. In *Fruit Fly Pests: A World Assessment of Their Biology and Management* (B.A. McPheron and G.J. Steck, eds.), pp. 551–554. St. Lucie Press, Delray Beach.

Prokopy, R.J. 1972. Evidence for a marking pheromone deterring repeated oviposition in apple maggot flies. *Environ. Entomol.* 1: 326–332.

Prokopy, R.J. and J. Hendrichs. 1979. Mating behavior of *Ceratitis capitata* on a field caged host tree. *Ann. Entomol. Soc. Am.* 72: 642–648.

Prokopy, R.J., P.D. Greany, and D.L. Chambers. 1977. Oviposition-deterring pheromone in *Anastrepha suspensa. Environ. Entomol.* 6: 463–465.

Prokopy, R.J., J.R. Ziegler, and T.T.Y. Wong. 1978. Deterrence of repeated oviposition by fruit-marking pheromone in *Ceratitis capitata. J. Chem. Ecol.* 4: 55–63.

Roan, C.C., N.E. Flitters, and C.J. Davis. 1954. Light intensity and temperature as factors limiting the mating of the oriental fruit fly *Dacus dorsalis* Hendel. *Ann. Entomol. Soc. Am.* 47: 593–594.

Rössler, Y. 1975. Reproductive differences between laboratory-reared and field-collected populations of the Mediterranean fruit fly, *Ceratitis capitata. Ann. Entomol. Soc. Am.* 68: 987–991.

Rull Gabayet, J.A., J. Reyes Flores, and W. Enkerlin Hoeflich. 1996. The Mexican national fruit fly eradication campaign: largest fruit fly industrial complex in the world. In *Fruit Fly Pests: A World Assessment of Their Biology and Management* (B.A. McPheron and G.J. Steck, eds.), pp. 561–563. St. Lucie Press, Delray Beach.

Sacantanis, K. 1953. Méthode d'élevage au laboratoire de la mouche des olives *Dacus oleae* Gmel. *Rev. Pathol. Veg. Entomol. Agric. Fr.* 32: 247–257.

Shelly, T.E. 1995. Methyl eugenol and the mating competitiveness of irradiated male *Bactrocera dorsalis* (Diptera: Tephritidae). *Ann. Entomol. Soc. Am.* 88: 883–886.

Shelly, T.E. 1996. Courtship in the Mediterranean fruit fly — a comparison of wild and sterile males. In *Medfly Mating Behaviour Studies under Field Cage Conditions: Report of the Second Research Co-ordination Meeting*. International Atomic Energy Agency, Vienna. 53 pp.

Shelly, T.E. and T.S. Whittier. 1996. Mating competitiveness of sterile male Mediterranean fruit flies (Diptera: Tephritidae) in male only releases. *Ann. Entomol. Soc. Am.* 89: 754–758.

Shelly, T.E., T.S. Whittier, and K.Y. Kaneshiro. 1994. Sterile insect release and the natural mating system of the Mediterranean fruit fly, *Ceratitis capitata* (Diptera: Tephritidae). *Ann. Entomol. Soc. Am.* 87: 470–481.

Sivinski, J.M., N. Epsky, and R.R. Heath. 1994. Pheromone deposition on leaf territories by male Caribbean fruit flies, *Anastrepha suspensa* (Loew) (Diptera: Tephritidae). *J. Insect Behav.* 7: 43–51.

Soemori, H. 1980. Difference in mating occurrence between mass-reared and wild strains of the melon fly, *Dacus cucurbitae* Coquillett. *Bull. Okinawa Agric. Exp. Stn.* 5: 69–71.

Soemori, H., S. Tsukaguchi, and H. Nakamori. 1980. Comparison of mating ability and mating competitiveness between mass-reared and wild strains of the melon fly, *Dacus cucurbitae* Coquillett (Diptera: Tephritidae). *Appl. Entomol. Zool.* 24: 246–250.

Suenaga, H., H. Kamiwada, A. Tanaka, and N. Chisaki. 1992. Difference in the timing of larval jumping behavior of mass-reared and newly-colonized strains of the melon fly, *Dacus cucurbitae* Coquillett (Diptera: Tephritidae). *Appl. Entomol. Zool.* 27: 177–183.

Suzuki, Y. and J. Koyama. 1980. Temporal aspects of mating behavior of the melon fly, *Dacus cucurbitae* Coquillett (Diptera: Tephritidae): a comparison between laboratory and wild strains. *Appl. Entomol. Zool.* 15: 215–224.

Tsitsipis, J.A. 1983. Changes of a wild ecotype of the olive fruit fly during adaptation to lab rearing. In *Fruit Flies of Economic Importance* (R. Cavalloro, ed.), pp. 416–421. A.A. Balkema, Rotterdam.

Tychsen, P.H. 1977. Mating behavior of the Queensland fruit fly, *Dacus tryoni*, in field cages. *J. Aust. Entomol. Soc.* 16: 459–465.

Vargas, R.I. and J.R. Carey. 1989. Comparison of demographic parameters for wild and laboratory-adapted Mediterranean fruit fly (Diptera: Tephritidae). *Ann. Entomol. Soc. Am.* 82: 55–59.

Warburg, M. and B. Yuval. 1997a. Circadian patterns of feeding and reproductive activities of Mediterranean fruit flies (Diptera: Tephritidae) on various hosts in Israel. *Ann. Entomol. Soc. Am.* 90: 487–495.

Warburg, M. and B. Yuval. 1997b. Effects of energetic reserves on behavioral patterns of Mediterranean fruit flies (Diptera: Tephritidae). *Oecologia* 113: 314–319.

Whittier, T.S., K.Y. Kaneshiro, and L.D. Prescott. 1992. Mating behavior of Mediterranean fruit flies (Diptera: Tephritidae) in a natural environment. *Ann. Entomol. Soc. Am.* 85: 214–218.

Yuval, B., R. Kaspi, S. Shloush, and M. Warburg. 1998. Nutritional reserves regulate male participation in Mediterranean fruit fly leks. *Ecol. Entomol.* 23: 211–215.

Zapién, G., J. Hendrichs, P. Liedo, and A. Cisneros. 1983. Comparative mating behavior of wild and mass-reared medfly *Ceratitis capitata* (Wied.) on a field cage host tree. II. Female mate choice. In *Fruit Flies of Economic Importance* (R. Cavalloro, ed.), pp. 397–409. A.A. Balkema, Rotterdam.

Zervas, G.A. 1983. Sexual and reproductive maturation in wild and lab cultured olive fruit flies *Dacus oleae* (Gmelin) (Diptera: Tephritidae). In *Fruit Flies of Economic Importance* (R. Cavalloro, ed.), pp. 429–438. A.A. Balkema, Rotterdam.

# 32 Sexual Selection and Speciation in Hawaiian *Drosophila* (Drosophilidae): A Model System for Research in Tephritidae

*Kenneth Y. Kaneshiro*

## CONTENTS



## 32.1 INTRODUCTION

Species in the dipteran family Drosophilidae, especially *Drosophila melanogaster* Meigen, have been subjects of intense genetic, evolutionary, developmental, molecular, and behavioral research. Tremendous numbers of papers in all aspects of biology have been published on this group of flies. According to Powell (1997), more than 60,000 papers have been written on some aspect of the biology of drosophilids, and the community of researchers investigating this group continues to grow exponentially. In his book entitled *Progress and Prospects in Evolutionary Biology: The Drosophila Model*, Powell (1997: 3) makes the statement:

> Of the millions of species that inhabit the earth, biological researchers tend to concentrate on relatively few organisms that subsequently become "model systems." The reason is obvious: Research builds on past research. To advance the forefront of knowledge, the system one studies must be known up to that forefront.... Many organisms have been studied, and by a process not unlike natural selection, certain organisms come to the fore as popular models. Examples include the house mouse, yeast, *Escherichia coli*, corn and *Drosophila*.... Of all these models, it is arguable that none has received as much attention as has *Drosophila*.

0-8493-1275-2/00/$0.00+$.50

Certainly, results of research on *Drosophila* have had significant influence on research dealing with all aspects of the biology of species in the family Tephritidae. In this chapter, I want to first present a discussion of research we have conducted on the ecology, behavior, and sexual selection in a group of *Drosophila* species endemic to the Hawaiian Islands. The primary thesis of this section is the role of sexual selection in the speciation of Hawaiian *Drosophila* but also its role as a "driver for genetic change" (Carson 1997). Second, I will present a brief discussion about how the results of the research on the Hawaiian *Drosophila* have been applied to the pest tephritid species we have been studying in Hawaii. Where relevant, I will also make brief references to some of the other groups of tephritid species discussed in this volume.

## 32.2 THE HAWAIIAN DROSOPHILIDAE

### 32.2.1 Natural History of the Hawaiian Archipelago

The Hawaiian Islands have been recognized as one of the best places in the world for conducting evolutionary research. Special geologic features of the Hawaiian Archipelago present evolutionary biologists a unique opportunity not only to test classical concepts of evolutionary theory, but, more important, to formulate new ideas on the processes of evolution. First, the islands are the most isolated landmass occupying a fixed position near the middle of the vast Pacific Plate (Dalrymple et al. 1973). The evidence indicates that the islands have always been isolated by more than 3500 km of ocean from any continental or island group. Thus, founder events to the Hawaiian Islands are relatively rare events, and the lack of many groups of plants and animals in Hawaii reflects the rare incidences of successful dispersal from the landmasses around the Pacific Rim that encircle the Hawaiian Islands. Second, these islands were formed sequentially over a fixed "hot-spot" in the Earth's mantle below the Pacific plate (Clague and Dalrymple 1987). As the plate moves in a northwesterly direction at a rate of 9 cm/year, a plume of lava periodically punches through the ocean floor that emerges above the ocean surface as oceanic islands that form the single-file formation of the Hawaiian chain. Thus, it is possible to document, without doubt, the sequential ages of the islands, which can be used as a tool for predicting the ages of populations that occur on each island (Carson and Clague 1995). Presumably, in most cases, the most ancestral populations occur on the oldest islands, and vice versa, although occasional back-migrations from a younger to an older island have been ascertained based on genetic analyses. Third, in most instances, each island is isolated from the adjacent island by a deep channel with no indication that the islands were connected by land bridges within the distant or recent geologic history. Only the complex of islands called the Maui Nui (i.e., Maui, Molokai, and Lanai) is known to have been connected and separated more than twice during the Pleistocene period. Indeed, for many groups of organisms, Maui Nui is considered a single biological island with species common to more than one of the five volcanoes that comprise Maui Nui. And fourth, present-day volcanic activity, where fresh lava flows have dissected old-stand forests into smaller patches of forests called "kipukas," provides researchers with an opportunity to investigate the genetic consequences of severe population bottlenecks and to test classical theories of population genetics especially in regard to reductions of genetic variability as a consequence of severely reduced population size.

Climatically, the Hawaiian Islands are also somewhat unique. Within relatively short distances, extreme conditions range from dry, desertlike conditions with as little as 65 cm of rainfall in an average year to a deep, rain forest situation where rainfall can average up to 800 cm/year. Hawaii even claims to have the "wettest spot on earth" at Mt. Waialeale, where rainfall can average upward of 1300 cm/year. These conditions provide the natural ingredients for the tremendous diversity of habitats and ecosystems found in the Hawaiian Islands.

Another important component of Hawaii's position as one of the best places in the world for evolutionary research is the diversity of the fauna and flora that have evolved there. Because of the long-distance dispersal required to reach these isolated islands, many groups of organisms are not represented here and the fauna and flora of Hawaii are often considered "depauperate" (Zimmerman 1948). However, those groups that were successful in surviving the long-distance trans-Pacific crossing and landed within the Hawaiian Archipelago, with a few exceptions, speciated profusely. For example, while Hawaii's tropical and subtropical rain forests are similar to habitats found in Central and South America or Southeast Asia where hundreds of butterfly species have evolved, only two endemic butterflies are found in the Hawaiian Islands. On the other hand, there are nearly a thousand species of moths that have evolved here. Similarly, for the coleopteran groups, those that made the long-distance dispersal to Hawaii are represented by a large number of species that evolved from a few founders. Perhaps the most striking and best-studied group of insects is the Hawaiian Drosophilidae, which will be the major topic of discussion for this chapter.

### 32.2.2 Systematics

Currently, there are 511 species of flies in the family Drosophilidae that have been named and described as being endemic to the Hawaiian Islands (i.e., found nowhere else in the world) (Kaneshiro 1997). However, it is estimated that this number represents only about 50% of the total number of species that exists in the Hawaiian fauna. Another 250 to 300 new and undescribed species are already in the collection at the University of Hawaii awaiting taxonomic treatment. In addition, as fieldwork on this group continues, new species continue to be discovered. A total of 1000 species in this family in the native Hawaiian fauna may be a conservative estimate. Yet, the evidence based on morphological, developmental, behavioral, and genetic (including DNA sequence) data indicate that the group is monophyletic and that the entire group arose from a single founder that arrived in the Hawaiian Islands "X" million years ago (the timescale on the origin of the Hawaiian Drosophilidae is controversial and will not be addressed here). While the total landmass of the islands represents less than 0.2% of the land area in the United States, about one quarter of the total number of species in the family Drosophilidae inhabit the Hawaiian Islands, an incredible example of explosive adaptive radiation.

One of the most conspicuous features of the endemic *Drosophila* fauna is the remarkable morphological diversity among the species. Some of the bizarre morphological structures observed in this group led earlier taxonomists (Bryan 1934; 1938; Grimshaw 1901–1902; Malloch 1938; Wheeler 1952; Hardy 1965) to recognize nine genera in the endemic fauna. With the combined effort of a multidisciplinary team of investigators, it is now clear that such a taxonomic treatment of the endemic Drosophilidae based wholly on morphological relationships may invite a misleading interpretation of evolutionary divergence. Thus, Kaneshiro (1976a), by pooling corroborating evidence from mating behavior studies (Spieth 1966; 1968), cytology (Carson et al. 1967; Yoon et al. 1972), internal anatomy (Throckmorton 1966), ecology (Heed 1968; 1971), and especially a comparative study of the external male genitalia (Takada 1966; Kaneshiro 1969), presented evidence for the existence of only two major lineages in the evolution of the Hawaiian Drosophilidae. It was clearly demonstrated that the "key" characters previously used to differentiate drosophiloid species into generic groups are not "good" generic characters. The tremendous morphological diversity observed in the Hawaiian *Drosophila* is a manifestation of the elaborate, species-specific courtship displays, which resulted in sexual selection for ornate secondary sexual structures in the males. Females, on the other hand, including those of species that were previously described in separate genera, retain characteristics typical of the genus *Drosophila*.

### 32.2.3 Ecology

The drosophilids of Hawaii inhabit an extremely diverse range of ecosystems from dry-land forests to wet rain forests. For the most part, the larvae are saprophytic in decaying parts of native plants (Heed 1968; 1971; Montgomery 1975). Some species are specialized on the decaying leaves while others breed only in the decaying bark of native plants. About 60 species are host-specific on fungi and mushrooms (Heed 1968; and Spieth and K. Kaneshiro, unpublished data) while a few species have become predaceous on the egg masses of a native spider (Wirth 1952). A few species utilize flowers such as morning glory blossoms or the flower heads of endemic composit species, the Hawaiian silversword, *Argyroxiphium sanwicense* subsp. *macrocephalum* (A. Gray) Meyrat. One species may have invaded the aquatic habitat with the larvae breeding on the green algae below the surface in a freshwater stream on the island of Hawaii (K. Kaneshiro, unpublished data). The natural breeding substrates of about 50% of the described Hawaiian drosophilids are known (Heed 1971; Montgomery 1975) and, in most cases, they are monophagous (i.e., utilize a single host plant as the larval breeding substrate). Interestingly, while many of these species require that we provide the natural substrate to stimulate oviposition in the laboratory, once the eggs hatch, the larvae are able to complete development in other decaying vegetation as well as our standard laboratory medium. Thus, it would appear that ovipositional behavior of the females rather than nutritional requirements of the larvae may be the primary mechanism for host selectivity and that nutritional requirements provided by a specific host plant may not be the critical factor in determining host specificity.

In a recent article, Kaneshiro and Kambysellis (1999) described a new species of Hawaiian *Drosophila,* which is morphologically essentially identical with a related species. *Drosophila craddockae* Kaneshiro and Kambysellis from the islands of Kauai and Oahu cannot be distinguished from *D. grimshawi* Oldenberg from the islands of Maui, Molokai, and Lanai based on classical taxonomic characters. However, there are clear genetic and ecological differences, which readily differentiate these two species as a clear case of allopatric sibling species. More interestingly, *D. craddockae*, which is generally accepted to be the ancestral species based on geologic, cytological, behavioral, and molecular genetic data, is an ecological specialist, breeding in the decaying stems of a single native plant species. *Drosophila grimshawi*, on the other hand, is certainly derived from either the Kauai or Oahu population of *D. craddockae*, but has been reared from the decaying parts of ten different families of plants, and is a generalist (Montgomery 1975). Thus, while longstanding ecological theory states that specialization is a derived condition, the biological and genetic evidence indicate that specialism in *D. craddockae* is the ancestral condition and that generalism evolved in *D. grimshawi* as a derived trait.

While the ecological diversity of the Hawaiian *Drosophila* is not the primary thesis of this chapter, it is still important to understand that these species as a group provide an excellent example of adaptive radiation into a diverse range of macro- and microniches. The tremendous diversity of macro- and microniches occupied by the Hawaiian drosophilids represents a microcosm of habitats and niches that can be observed in the Tephritidae (see other chapters in this volume) and the results of the research on the Hawaiian Drosophilidae can provide insights into a better understanding of the ecology of tephritids globally.

The evolutionary research conducted on the Hawaiian *Drosophila* indicate that natural selection, while an important force in the evolution of the group, has not played the most important role in the formation of new species (Kaneshiro 1993a). Rather, it has been suggested that sexual selection and shifts within the sexual environment have played a critical role as the entering wedge of new species formation (Kaneshiro 1989). This will be the main topic of focus for the remainder of this chapter.

### 32.2.4 The Role of Sexual Selection in Species Formation

Carson (1997), in presenting the Wilhelmine Key Distinguished Lecture during the 50th Anniversary of the Society for the Study of Evolution in 1996, spoke on the topic, "Sexual Selection: A Driver of Genetic Change in Hawaiian *Drosophila*." In his paper, Carson presents a discussion of how sexual selection in "local genetically rich populations can either maintain a cohesive status quo or forge a disruptive evolutionary novelty." He suggests that the latter, that is, the evolution of novelty, is even more evident when the population is subjected to bottlenecks. These ideas are based on earlier studies on the behavior of the Hawaiian *Drosophila* which I will discuss briefly here.

It was Spieth (1966; 1968) who first suggested that elaborate, oftentimes bizarre, secondary sexual structures found in the males of most Hawaiian *Drosophila* species are manifestations of the complex courtship behavior displayed by the males. Extreme modifications of the mouthparts, front legs, wing venation and maculations, and even head shape can be found among males of Hawaiian *Drosophila*. Some of these modifications occurred in traits that usually define generic groupings in the family Drosophilidae and led earlier taxonomists to group species that shared these bizarre structures into separate genera. However, studies of the male genitalia as well as behavioral, ecological, and genetic studies provided evidence that these characters were nothing more than species group differences and that these structures were used primarily as part of the complex mating behavior displayed by the males. It became clear, more than 30 years ago, during the early years of the research on the Hawaiian *Drosophila* that sexual selection was an extremely important part of the evolutionary process in the Hawaiian fauna.

Spieth (1966; 1968; 1974a; 1974b; 1982) was also the first to describe lek mating behavior in the Hawaiian *Drosophila*. Unlike in the Tephritidae where Sivinski and Burk (1989) observed that lekking tephritid species are mainly polyphagous species, most of the Hawaiian drosophilids that display lek behavior are monophagous. While Spieth's descriptions of the physical and environmental characteristics of the lek (mating arena) sites were qualitative, they were the most important observations that enabled future researchers to find and collect many of the species which were not readily attracted to the standard baiting techniques. He described the aggressive behavior displayed by the males against other males that arrived at territories within a lek site. He also carefully described other behaviors displayed by males that occupied lek territories. In one group of species, the *D. adiastola* Hardy group, males advertise their presence by raising the tips of their abdomens and extruding or pulsating an anal droplet. He concluded that such displays "...serve to release a pheromone which acts as a stimulus to sexually receptive females and allows her to orient upon and move to the lek site" (Spieth 1982). In one of the species of the *adiastola* group, *D. clavisetae* Hardy, the males have evolved a row of specialized bristles that form a fanlike arrangement on the ventral surface of the last abdominal segment. During courtship, the male raises his abdomen up between his wings and over his thorax, superficially resembling a scorpion-like behavior. A bubble, presumably a sex pheromone, is produced from the anal papillae and is pointed toward the female. The row of abdominal bristles, which are broadened apically, is now projected on the "upper surface" of the abdomen which is then vibrated in a back-and-forth motion apparently "fanning" the pheromone toward the female (K. Kaneshiro, unpublished data). Thus, in this species, an elaborate secondary sexual character for which the species is named, the row of clavate bristles, evolved to focus and concentrate the sex pheromone toward the female.

While Spieth continued his work on describing the courtship pattern of representative species of the Hawaiian *Drosophila* until the early 1980s, it was not until the mid-1970s when a better understanding of the role of sexual selection in the speciation process began to unfold (Kaneshiro 1976b; 1980; 1983; Ohta 1978; Kaneshiro and Kurihara 1981). First, it was observed that when mate preference experiments between allopatric species pairs were conducted, the outcome was usually asymmetrical sexual isolation (Kaneshiro 1976b). In most cases, females of species from a geologically older island discriminated against males of species from a younger island. Reciprocally, females of the species from the younger island mated readily with males from the older island. When results of mate preference tests among other groups of Hawaiian *Drosophila* indicated similar direction of asymmetry which appear to be correlated with other genetic criteria (e.g., analyses of the banding pattern of the giant polytene chromosomes), the species on the older island was, in most cases, deemed to be ancestral to the species on the younger island (e.g., see Ohta 1978). Thus, it appeared that it was possible to determine the "direction of evolution" of a group of Hawaiian *Drosophila* species by analyzing data obtained from mate preference studies.

The hypothesis that females of ancestral species discriminated against males of more-derived species but that, conversely, females of derived species mated readily with males from an ancestral population became the topic of numerous studies outside of the Hawaiian *Drosophila* group, including other *Drosophila* and insect groups, but also vertebrate species groups. The results of these studies were split; some appeared to validate the hypothesis, whereas others provided alternative explanations. So, the value of mate preference experiments for predicting the direction of evolution among related species remained controversial for a number of years since the hypothesis was first formulated (Kaneshiro 1976b). Subsequent review papers addressing the controversy appear to provide further arguments in favor of the Kaneshiro hypothesis. Giddings and Templeton (1983), in an invited review article for *Science*, conclude, "The Kaneshiro hypothesis has been extensively tested.... All these results are compatible with the Kaneshiro hypothesis ..." and "... the application of this model in the future should expand our ability to resolve phylogenetic relationships." However, the controversy continued when three invited articles in the prominent book series *Evolutionary Biology* (Volume 21, 1987) focused on the asymmetry hypothesis. In the first article, Ehrman and Wasserman (1987) concluded that "the direction of asymmetrical isolation, taken by itself, is an unreliable indicator of the direction of evolution." In the second article, DeSalle and Templeton (1987) state that "the central thesis of the preceding chapter by Ehrman and Wasserman is that there is more than one mechanism for yielding asymmetrical isolation, and hence mating asymmetry alone cannot be used to infer the direction of evolution without qualification. We are in complete agreement with this central thesis. Although we agree with the central thesis of Ehrman and Wasserman, we do disagree on other issues." DeSalle and Templeton conclude their chapter by stating, "we feel that these recent molecular studies confirm the validity of the Kaneshiro model when its assumptions are satisfied" and that "these molecular studies offer strong support for the conclusions of Giddings and Templeton (1983) that the conditions that they made explicit are necessary conditions for the applicability of the Kaneshiro model." In the third chapter, Kaneshiro and Giddings (1987) conclude that "the challenge is not just to determine whether mating asymmetries exist within the group of organisms being studied and whether the direction of evolution predicted by the various asymmetry models points to the correct direction based on other evidence. Rather, we hope that investigators will ask the question of why such asymmetries exist and how they arose."

Kaneshiro (1989), in response to the challenge issued in the Kaneshiro and Giddings (1987) chapter, expanded on his asymmetry model (Kaneshiro 1976b; 1980) this time with an emphasis on the "why" and "how" such asymmetrical sexual isolation between allopatric pairs of species arise. Based on the results of mate preference experiments, it was suggested that there is a range of mating types segregating in both sexes. That is, some males are more successful in mating while others are just not able to satisfy the courtship requirements of most females in the populations. Similarly, among females, some are very choosy in selecting a mate while others are less choosy. It was also suggested that within a single interbreeding population, the most likely matings occur between males with high mating success rate and those females that are less choosy. Observations of a relatively large number (i.e., >100) of courtship encounters among several different species of Hawaiian *Drosophila* in the field (K. Kaneshiro, unpublished data; Conant 1978) indicate that successful matings, that is, courtships that culminate in copulation, occur relatively quickly, within a few seconds after the male initiates courtship. Courtship encounters that continue for more than a few seconds (e.g., 15 s or more) inevitably result in the female decamping from the mating territory. These observations seem to provide evidence that matings in the natural population occur between males with high mating ability and females that are not so choosy, thus the extremely short courtship period leading to copulation. In the more than 95% of the courtship encounters which are prolonged and result in the female rejecting the male's courtship repertoire, the females are either highly discriminant in mate choice or the males are not very good at courtship.

Kaneshiro (1989) suggested that the "differential selection" for opposite ends of the mating distribution in the two sexes, that is, high mating males and less choosy females, and the genetic correlation between the mating phenotypes in the two sexes is what maintains the range of mating types in the two sexes generation after generation. That is, sexual selection itself serves as a stabilizing mechanism in maintaining a balanced polymorphism in the mating distribution. The differential selection hypothesis is a significant departure from the classical runaway sexual selection model, which states that there is strong genetic correlation between a male sexual trait (either behavioral or morphological) and female preference for that trait. Coevolution of male characters and female preference for that trait then results in the elaboration of secondary sexual structures among the males. Mayr (1972) states that "natural selection will surely come into play as soon as sexual selection leads to the production of excesses that significantly lower the fitness of the species." That is, natural selection exerts its forces to maintain the optimum male phenotype that is able to survive within a particular environment. The runaway sexual selection model, on the other hand, requires that natural selection act as a stabilizing force to maintain the "optimum" phenotype within the interbreeding population, implying a reduction in phenotypic variability for that character. The differential selection model infers that sexual selection itself maintains a balanced polymorphism for sexual selection in both sexes without having to rely on natural selection as a stabilizing mechanism.

Given the range of mating types segregating within an interbreeding population, Kaneshiro (1989) suggested that the differential sexual selection model may be applied to species formation via what has been referred to as "founder event speciation" (Carson 1968; 1971; Mayr 1972) in the evolutionary biology of Hawaiian *Drosophila*. It was proposed that, when a single fertilized female is carried across the ocean channels that separate the islands of the Hawaiian Archipelago and is able to locate suitable substrate into which the female can oviposit, the $F_1$ and at least for a few generations beyond will likely be faced with small population size. Under these conditions, it is suggested that there would be strong selection against females that are too choosy in mate choice since they may never encounter males that would be able to satisfy their courtship requirements. On the other hand, there would be strong selection for the less choosy females in the population and that within a few generations there could be a significant increase in frequency of less choosy females in the population. As a consequence of such a shift in the distribution of mating types toward an increased frequency of less choosy females, there would be a corresponding shift in gene frequencies in the population, which may result in a destabilized genetic condition such that novel genetic recombinants may be generated. Conventional population genetic theory holds that population bottlenecks, a severe form of which can be ascribed to founder events, cause a significant loss of genetic variability. While infrequent alleles in the parent population may be lost due to drift, recent data demonstrate that genetic variance may actually increase following a single population bottleneck (Bryant et al. 1986; Carson and Wisotzkey 1989; Carson 1990). It is suggested that the genetic destabilization that accompanies the demographic consequences of founder events results in the conversion of balanced epistatic genetic variance to additive variance that can respond to selective pressures exerted by biotic and abiotic factors in the environment. In other words, coadapted blocks of genes that are held in a balanced state by the fitness of such genotypes are now, as a result of the destabilized condition, allowed to recombine, generating novel genetic recombinants. Some of the newly generated additive genetic variants may be preadapted to the new habitat (in the case of a founder population) or to the environmental stress (in the case where the bottleneck is a result of some catastrophic environmental event). Thus, the dynamics of sexual selection and its potential influence on the stability of the balanced genetic composition of the population can play an extremely important role not only in maintaining levels of variability but also in generating an increased level of additive genetic variance immediately following the bottleneck event.

### 32.2.5 Sexual Selection and Natural Hybridization

An aspect of the biology of populations not usually considered when developing control programs for pest species is that of natural hybridization between related sympatric or parapatric species pairs. The results of research on natural hybrids discovered in Hawaiian *Drosophila* are relevant to the thesis of this symposium because of the role of sexual selection in "permitting" natural hybridization to occur and also because of the potential impact natural hybridization might have when implementing a control program on a pest fruit fly species.

In a paper that included a brief review of natural hybridization between sympatric *Drosophila* species, Kaneshiro (1990) suggested that the dynamics of sexual selection may actually permit natural hybridization to occur under certain conditions. Citing studies of natural hybridization in *Drosophila* where genetic information is available, it was shown that gene exchange between two sympatric populations was asymmetrical. That is, genetic material from one species appears to leak across into the gene pool of a related sympatric species, but not vice versa. Based on the observations of asymmetrical mating preference between closely related species when mating experiments are conducted in the laboratory, Kaneshiro (1990) proposed an intuitive model to explain the asymmetrical natural hybridization between sympatric species pairs. It was suggested that a drastic reduction in population size due to environmental stress conditions could induce strong selection for less choosy females in the population. Continued selection for less choosy females over even a few generations will result in a significant shift in the distribution of mating types toward an increase in frequency of less choosy females segregating in the population. Under these conditions, the females from the bottlenecked population may occasionally accept males of a related species, which has been less susceptible to the environmental stress conditions. While it is not implied that these conditions lead to significant "gene flow" between the two populations, it is suggested that there is "leakage" of genetic material across the species barrier, but certainly not enough to destroy the integrity of the separate gene pools.

Kaneshiro (1990) suggested that such a mechanism whereby natural hybridization is "permitted" by the dynamics of sexual selection may have evolved as a mechanism which will enable populations to "replenish" genetic variability that might be lost due to drift as a result of the population bottleneck. Sexual selection is described as a density-dependent process that enables a population to overcome harsh environmental conditions and drastic reduction in population size. Furthermore, as more and more cases are reported in the literature, it is becoming clear that natural hybridization between sympatric species pairs is not an uncommon phenomenon and that sexual selection may be the underlying mechanism which permits interspecific mating to occur rather than just a mechanism to "reinforce" reproductive isolation.

Natural hybridization between animal species is a more widespread phenomenon than reported in the earlier literature. With the development of molecular tools (e.g., analyses of the maternally inherited mitochondrial DNA) to be able to assay even single individuals for genetic introgression (e.g., using the PCR technique), it has been possible to document cases of natural hybridization between closely related species. Paradoxically, while molecular techniques have, over the past three decades, become more and more relied upon for elucidating evolutionary phylogenies of groups of species (e.g., see Avise 1989), the occurrence of interspecific natural hybridization may create a problem for interpreting genetic similarities between taxa. That is, it may be difficult to interpret phylogenetic relationships between pairs of sympatric species especially since it is difficult, if not impossible, to differentiate genetic similarities as homologies due to common ancestry or due to natural hybridization. For example, among the Hawaiian *Drosophila*, there are two documented cases of natural hybridization between pairs of sympatric species (Kaneshiro 1990). In both cases, the levels of genetic similarities as evidenced both by chromosomal banding patterns and by molecular markers infer homologies by common ancestry. However, morphological and behavioral evidence indicates that for each of the two pairs, there were two separate founders from two different adjacent islands that gave rise to each of the species. In both cases then, the high level of genetic similarity is probably a result of natural hybridization rather than a result of common ancestry as indicated by the molecular phylogenies. This is an issue that needs to be addressed in developing phylogenies for all groups of organisms, including those of the Tephritidae.

## 32.3 PARALLELS TO TEPHRITIDAE

The results of research on the mating behavior of Hawaiian *Drosophila* provide some important insights into the sexual selection system in tephritid fruit fly species which can be useful for understanding the evolution of behavior, but also in the development of more effective control/eradication protocols of pest species. For example, our studies of the ecology of lek behavior in Hawaiian *Drosophila* enabled Arita and Kaneshiro (1985) to discover and describe the lek mating system in the natural population of the Mediterranean fruit fly, *Ceratitis capitata* (Wiedemann). Although Prokopy and Hendrichs (1979) first described the notion of a lek mating system in the Mediterranean fruit fly (referred to as the medfly, henceforth), their study had been conducted in a field-cage situation. Conducting their studies under totally natural conditions, Arita and Kaneshiro (1985) described some of the environmental parameters essential in lek formation and mating success. It was shown that the lek system is an intensification of the sexual selection process which strongly influenced the effective mating population. I want to summarize briefly some of the findings on the mating behavior of medfly that have been studied under totally natural conditions, that is, in contrast to laboratory or field-caged conditions.

### 32.3.1 The Medfly

In the medfly, there are two essential components to successful mating in the natural population (at least in Hawaii). First, males must be able to locate and participate in lek formation to even encounter sexually receptive females. It should be mentioned here that in all our observations of lek behavior in wild medfly populations in Hawaii, we have never observed males defend oviposition substrates for the purposes of mating as described by Yuval and Hendrichs (Chapter 17). Second, males must be able to perform complex courtship displays in order to satisfy the mating requirements of the females. Leks are aggregations of males that defend territories to which females are attracted for the sole purpose of mating (Bradbury 1981). The individual male's territory within these aggregations does not contain any other resources vital to the fitness of the female, and, therefore, in most lek species female choice is based on male traits, that is, morphological, acoustical, pheromonal, visual, etc. features of the complex mating behavior, which comprise the sexual selection system. Males interact with other males as they jockey for the opportunity to occupy a territory within a lek system (intrasexual selection) and thus the opportunity to encounter receptive females. Male mating success is often highly skewed in species that display lek behavior with a few males accounting for the majority of the matings (Bradbury and Gibson 1983). However, in most cases, a male must also be able to perform courtship behavior adequately before a female will accept him as a mate (intersexual or epigamic selection).

It turns out that because of the relatively well-defined lek system observed in the medfly, it has been possible to conduct field studies that have provided some important sights into mate choice in this species (Prokopy and Hendrichs 1979; Arita 1983; Arita and Kaneshiro 1989). Arita (1983) first described the various environmental parameters that constitute a medfly lek system. She observed that individual males occupied the underside of single leaves of host plants and defended the leaf as a mating territory. The leaves that were used as territories were always at least partially exposed to sunlight, while those in total shade were never occupied as territories. Laboratory experiments with light penetrating through the leaf showed that light is a critical component of the lek system. It was shown that males rely on the silhouette produced by the male or female that alights on the topside of the leaf to orient and direct pheromone calling in the direction of the individual on the top of the leaf. Arita and Kaneshiro (1989) reported that the lekking male on the underside of the leaf oriented toward the silhouette of the individual on the top of the leaf and began wing vibrations, which inevitably resulted in the individual walking to the edge of the leaf toward which the male on the underside was facing. When the individual on the top of the leaf walked to the underside of the leaf, it would always be facing the lekking male sitting on the underside of the leaf. Either male–male aggression will occur if a male had arrived at the territory or courtship displays would ensue if the individual turned out to be a female.

Another important environmental parameter that constitutes the lek system is wind direction. Arita and Kaneshiro (1989) reported that leks are always formed on the upwind side of a tree rather than the downwind side. It was postulated that this served to disperse chemical cues (pheromones) released by the lekking males through the crown of the tree rather than being carried away from the tree. Lekking males take up a "pheromone calling" posture on the underside of leaves with the terminalia bent in an dorsal position and with the rectal epithelium extruded into a balloonlike structure which is coated by a putative sex pheromone secreted by the anal glands found within the rectal epithelium (Arita and Kaneshiro 1986). During the pheromone calling display, the pheromone is carried passively by wind currents as a medium-distance attractant. The upwind orientation to wind direction would not only serve to attract females from within the tree but may also serve to attract females to the specific trees on which leks are formed. It is hypothesized that the formation of leks (i.e., aggregations of males) serves to increase the concentration of the pheromone relative to what might be dispersed by a single male. Being able to attract females to specific trees may be extremely important since lek sites are apparently not randomly distributed among a group of host trees. In both of their study sites on the Island of Maui, Arita and Kaneshiro (1989) reported that lek sites were established on the same four or five trees at each study site year after year over a period of more than 10 years. Whittier et al. (1992) did a census of all the flies observed on 118 host trees within their study site (which was the same as one of the sites studied by Arita and Kaneshiro) and reported that more than 80% of the flies observed were found on 10 of the trees. Furthermore, 73% of all the mating occurred on just three of the 118 trees at this study site. These data indicate that receptive females are cueing into certain trees where leks are being established and it is likely that pheromonal cues may be playing an important role in attracting females to these trees.

On average, three to five males can be observed occupying territories within a sphere of about 35 to 40 cm diameter although the number of males within a lek system can vary from 2 to as many as 12 males (Arita and Kaneshiro 1989). In Hawaii, we have observed that lek formation starts at about midmorning (i.e., approximately 09:00 hours) depending on environmental conditions, with the approximate size of the lek (i.e., number of males participating) established within about 10 to 15 min after the first male arrives at the lek site. During this period and for a short period following lek formation, males aggressively attempt to displace other males from their territories in apparent competition for territories, although there was little evidence for what might be referred to as a "preferred" territory. Whittier et al. (1992) showed that nearly 70% of the male–male encounters observed in the field resulted in the intruder displacing the resident male. Furthermore, it was observed that out of 71 total matings observed only two occurred on the same leaf, which further suggests that females were not cueing into particular territories within the lek system.

Even though males are able to occupy and perhaps defend a suitable territory within the lek system, they still must be able to perform a complex series of courtship behavior to be successful in mating with females. Once the female arrives at the territory of a male, the male's entire behavioral action shifts from passive to active dispersal of the sex pheromone. As indicated earlier, once the lekking male is visually stimulated by the arrival of a female on the topside of his territory, he immediately orients to the silhouette, tucks his abdomen ventrally (from the dorsal orientation during pheromone calling display), and begins to vibrate his wings. Inevitably, the female walks to the edge of the leaf in the exact direction that the male is facing so that once she walks to the underside of the leaf, she is automatically facing the male. As the female continues to walk toward the male, he continues vigorous wing vibrations until the female is within about 1 or 2 cm from the male. At this point, the male initiates a series of head movements, oscillating his head in both directions at about 45° from the normal resting position. Almost simultaneously, the male also begins a second series of wing movements referred to as wing-fanning which is superimposed on the wing vibrations. It is believed that the entire courtship display performed by the male medfly is aimed at dispersing and concentrating the pheromone in front of the female. During wing-fanning, the female's wing reacts in an up-and-down motion which corresponds to the forward and backward movement of the male's wings. The one-to-one correlation of the female's wing motion with that of the male's wing-fanning indicates that some kind of airflow is being generated by the male toward the female. In addition, the specialized secondary sexual structures found in the males (i.e., the paddle-shaped anterior orbital bristle on the head and the dense brush of long yellow bristles found on the ventral surface of the femur of the males) may have evolved as part of sexual selection for concentrating the pheromone in front of the female. Initially, it was speculated that the anterior orbital bristle served as a visual cue to the female as the male oscillated his head back and forth. However, videotaped recordings of this behavior indicate that the flattened surface of the bristle is not oriented toward the female. Rather, the flattened surface of the "fan-shaped" bristle faces almost downward toward the front of the female as the female stands facing the male. As described for the behavior of *D. clavisetae* above, the specialized bristles may have evolved to enhance the dispersal of the sex pheromone toward the female. Thus, it is postulated that the oscillating motion of the head together with the wing vibration and wing-fanning may serve to form some kind of a convection current which stirs the "pheromone cloud" in front of the female where the chemical receptors are likely located. Furthermore, the dense brush of yellow bristles on the ventral surface of the femur of the male's forelegs may also serve to "collect" molecules of the pheromone and further concentrate the chemical stimulus in front of the female. It is suggested then that every aspect of the courtship sequence of the medfly male appears to be focused on delivering the pheromone to the female and that secondary sexual structures such as the specialized anterior orbital bristle on the head and bristles on the forelegs evolved to enhance the ability of the male to satisfy the mating requirements of the female.

As with the studies of mating behavior in the Hawaiian *Drosophila*, it is suggested that the most likely mating within the natural population of the medfly would be between the highly successful males and the less choosy females at least under normal population conditions. Again, the evidence for this hypothesis is the fact that females inevitably reject the males' courtship overtures and decamps from the mating territory when the courtship display continues for more than a few seconds. In more than 100 observations of courtship encounters where the female was observed arriving on the territory of a lekking male, successful copulation occurs only when the female appear to accept an abbreviated courtship display. On the other hand, courtship encounters that proceed for more than 10 or 15 s inevitably result in the female terminating the courtship even if the male continues to court for several minutes. Also, we have observed many more copulating pairs within the lek sites although we did not observe the courtship display probably because copulation occurred so rapidly in these instances.

Clearly, sexual selection may be playing a powerful role in maintaining a balanced polymorphism in the mating system of the medfly. That is, the genetic correlation between the two behavioral phenotypes in the two sexes is what maintains the entire range of mating types segregating in the population generation after generation under normal conditions. Under stressful environmental conditions, or during control programs when the population is faced with reduced size, there may be a shift in the mating distribution toward an increase in frequency of less choosy females. Females that are choosy in mate choice may never encounter males that are able to satisfy their courtship requirements especially under severely reduced population size, and, even after a few generations, there may be a significant increase in frequency of less choosy females in the populations. As discussed above, these conditions provide the ingredients for a possible destabilization of the genome and the generation of new genetic recombinants. Selection for genotypes better adapted to the stress conditions (whether due to environmental conditions or to the effects of control programs) is then followed by the reorganization of the gene pool until a new balanced genetic system is attained. The resulting genetic architecture of the population is further strengthened by its correlation with less choosy females in the population. Genotypes that render the population more resistant or adaptable to the stress conditions that forced the population through a bottleneck will spread quickly through the population especially if such genotypes are linked or correlated to those of less-choosy females. Even resistance to insecticides can evolve very quickly through the dynamics of the sexual selection system discussed above.

Thus, when control or eradication programs for fruit flies or other pest species are being developed, it is important to consider the role of sexual selection in not only maintaining levels of genetic variability in the population but also the possible evolution of resistance to the control technology. Kaneshiro (1993b) suggested that the sexual selection model described above might have played a crucial role in the presence of medfly populations in the Los Angeles basin each year for several years beginning in 1986. In fact, wild medfly captures increased from a single fly in 1986, to 45 in 1987, to 54 in 1988, and to more than 300 wild flies in 1989 to 1990 with wild flies being trapped periodically through the winter months during this latter period. It was postulated that the reduction of population size resulting from the eradication program might have inadvertently selected for a resident medfly population that became better adapted to the Los Angeles region. Thus, there was an actual increase in numbers of wild flies trapped with each subsequent year. On the other hand, the sexual selection model described above could also be applied to increase the effectiveness of the control/eradication technologies such as the Sterile Insect Technique (SIT). Kaneshiro (1993b) suggested that when the wild population has been reduced to such a low level (below that of detection), there would have been selection for less-choosy females in the wild population, at which point the wild population would be most susceptible to the SIT. That is, the less-choosy females in the wild population would more readily accept the sterile males and by continuing to release sterile flies for a few more generations (rather than stopping the releases at three generations beyond the last wild fly captured as called for in the earlier SIT protocol), it would be possible to eradicate totally any residual populations remaining. In fact, this new protocol has been implemented during the 2-year period 1994 to 1995 (Dowell and Penrose 1995), and while the results are still circumstantial, it appears that medflies may indeed have been totally eradicated from the Los Angeles basin.

We have some recent data (K. Kaneshiro and Kennelly, unpublished data) which are still preliminary at the time of this publication but which may be an extremely important aspect of the medfly population biology. As discussed in the section on Hawaiian *Drosophila*, the differential sexual selection model suggests that sexual selection itself serves as a stabilizing mechanisms for maintaining a balanced polymorphism of the mating system. The genetic correlation between male (high mating ability) and female (low discrimination) mating phenotypes is what maintains the entire range of mating types in the two sexes generation after generation. This mechanism also applies to the medfly; however, there may be further insurance to maintain the balanced polymorphism in this species. Arita and Kaneshiro (1983) observed that females that were held as virgins beyond the optimum age during which the females should have mated display behavior that resembled male courtship behavior. Females kept separate from males for a few weeks posteclosion begin to display "pseudomale" behavior, including pheromone calling posture, wing-vibrations, wing-fanning, head oscillations, and even attempts to mount and copulate with another individual, male or female. It was also observed that such females, when approached by a sexually mature male, would cease pseudomale behavior and almost immediately accept his courtship rituals and allow him to copulate. It seems that a physiological change triggers the pseudomale behavior in females if they have not mated by a certain age but also lowers the threshold of receptivity in these females. For the medfly then, choosy females that may not be able to encounter males that are able to satisfy their courtship requirements may experience a physiological change which results in lowering their threshold of receptivity. Such females may mate with the first male they encounter, which, especially under conditions of small population size, is likely to be a male that is less successful in mating. Because of the correlation between male and female behavioral phenotypes, when there is an increase in frequency of less-choosy females, there is a corresponding increase in frequency of "dud" males. The matings between genetically choosy females and "dud" males as a result of physiological changes in the females also result in generation of the entire range of mating types in the subsequent generations. Thus, for the medfly, there appears to be an additional mechanism by which a balanced polymorphism for the entire range of mating types can be maintained in the population. This notion may be a possible explanation for the colonizing ability of this species throughout many parts of the world without having undergone much genetic change. We are pursuing this aspect of the biology of the medfly and will attempt to substantiate whether the phenomenon of pseudomale behavior does indeed occur in natural populations under certain conditions.

## 32.4 THE BIOLOGICAL SIGNIFICANCE OF SYNTHETIC LURES AND NATURAL PLANT KAIROMONES

The observations of courtship behavior in the Hawaiian *Drosophila adiastola* species group, especially that of *D. clavisetae,* had significant influence in our research on the role of sex pheromones in sexual selection. Nishida et al. (1988; 1990; 1993; 1997) showed that wild males of *Bactrocera dorsalis* (Hendel) (Oriental fruit fly, OFF) and *B. cucurbitae* (Coquillett) (melon fly, MF) contain chemical components in their rectal gland which are missing in laboratory-reared males. However, when laboratory-reared males are exposed to plants that secrete analogs of methyl eugenol for OFF males and cuelure for MF males, it was shown that the missing components were recovered in the extracts obtained from the rectal glands. It was shown that the males that were allowed to feed on these plants were able to sequester the chemical components that comprise the putative sex pheromone found in the wild males. It is clear that the compounds secreted by these plants provide important precursors for the sex pheromone emitted by the males during their courtship displays. Kuba and Sokei (1988), using special photographic techniques, showed unequivocally that MF males produce a "cloud" of pheromone which totally engulfs the female during courtship encounters, attesting to the importance of the sex pheromone in the mating system of these species. Furthermore, mating experiments indicate that males of both OFF and MF that have been allowed to feed on these natural plant compounds are significantly more successful in mating with females compared with control males (Shelly and Dewire 1994; Shelly 1995; Shelly and Villalobos 1995). So, for the *Bactrocera* species, at least, the biological significance of males being strongly attracted to synthetic lures such as methyl eugenol and cuelure used for monitoring populations and to plant kairomones found in nature is to sequester important components of their sex pheromone.

Similar studies of the effects of trimedlure and other natural compounds on the mating success of the medfly have only just begun (Shelly et al. 1996; and T.E. Shelly and K. Kaneshiro, unpublished data). Unlike the reaction of *Bactrocera* males to their respective lures where the males are observed to feed on these compounds, medfly males typically do not feed on trimedlure. Rather, they simply approach and perch on substrates nearby the trimedlure source, in some instances displaying what might be normally observed as lek behavior (e.g., pheromone calling posture). However, the absence of feeding does not rule out the possible beneficial effects of trimedlure, and experiments have been carried out to investigate the effects of the lure on the sexual behavior of male medflies. Similar to the results obtained for the studies involving *Bactrocera* species, the studies with the medfly also indicate that exposure to trimedlure had a significant effect on the mating success of the treated males. However, unlike the results of the *Bactrocera* studies, the mating success of treated medfly males had a very short-term effect lasting for 24 h or less. Similar experiments involving other compounds known to be attractive to medfly males such as α-copaene, angelica seed oil, and ginger root oil also displayed varying degrees of mating success among treated males. Ginger root oil, especially, appears to be a promising compound for further studies since the mating success rate is very high and the effects are longer lasting than those observed with trimedlure and the other compounds. The preliminary studies with α-copaene are also interesting in that females exposed to this compound display pseudomale behavior (discussed above) and may serve to reduce the threshold of receptivity of wild females, which could be developed as a technique to increase the effectiveness of the SIT.

## 32.5 CONCLUSIONS AND FUTURE RESEARCH NEEDS

The research on the Hawaiian *Drosophila*, especially on the complex mating system, has certainly served as model for our studies of the mating behavior of tephritid species. An understanding of the processes of sexual selection and its role in speciation and in the maintenance of genetic variability in Hawaiian *Drosophila* has provided important insights into our understanding of population dynamics in the fruit flies. Clearly, an evolutionary approach to investigations of the biology of pest species such as the medfly has had significant influence on the development of more effective control technologies, and the potential for further improvement as we learn more about the basic biology of these species is substantial. The discussions that have taken place at this symposium and the resulting chapters in this volume have been some of the most important in terms of the ecology and behavior of tephritid species. It is suggested that further research into the evolutionary biology of the Tephritidae be conducted in order to understand better phylogenetic relationships among species and higher-level taxa. As has been seen among Hawaiian *Drosophila* species, natural hybridization among related sympatric species is certainly a possibility, and the "leakage" of genetic material between such pairs of species may give a misleading impression of phylogenetic relationships. While molecular techniques have proved to be an important tool for analyzing phylogeny, without knowing whether natural hybridization may have occurred between sympatric species pairs, it may be difficult to differentiate homology due to common ancestry vs. that due to hybridization. It is hoped that more research on the basic evolutionary biology of the tephritids, whether they are pest species or not, will be encouraged and adequately funded so that our understanding of the phylogeny of the group can be enhanced. For those species that are important agricultural pests, more effective control techniques can be developed as we gain further insights into the behavioral ecology of these species.

## ACKNOWLEDGMENTS

I want to thank Drs. Martín Aluja and Allen Norrbom for their efforts in organizing the symposium which resulted in this volume. I also wish to acknowledge partial funding from the following: Campaña Nacional contra las Moscas de la Fruta (Mexico), International Organization for Biological Control of Noxious Animals and Plants (IOBC), Instituto de Ecología, A.C. (Mexico), and the USDA-ICD-RSED, without which I would not have been able to participate in the symposium.

## REFERENCES

Arita, L.H. 1983. The Mating Behavior of the Mediterranean Fruit Fly, *Ceratitis capitata* (Wiedemann). Ph.D. dissertation, University of Hawaii, Honolulu.

Arita, L.H. and K.Y. Kaneshiro. 1983. Pseudomale courtship behavior of the Mediterranean fruit fly, *Ceratitis capitata* (Wied.). *Proc. Hawaii. Entomol. Soc.* 24: 205–210.

Arita, L.H. and K.Y. Kaneshiro. 1985. The dynamics of the lek system and mating success in males of the Mediterranean fruit fly, *Ceratitis capitata* (Wiedemann). *Proc. Hawaii. Entomol. Soc.* 25: 205–210.

Arita, L.H. and K.Y. Kaneshiro. 1986. The structure and function of the anal glands during mating behavior of the Mediterranean fruit fly, *Ceratitis capitata* (Wiedemann). *Proc. Hawaii. Entomol. Soc.* 26: 27–30.

Arita, L.H. and K.Y. Kaneshiro. 1989. Sexual selection and lek behavior in the Mediterranean fruit fly, *Ceratitis capitata* (Diptera: Tephritidae). *Pac. Sci.* 43: 135–143.

Avise, J.R. 1989. Gene trees and organismal histories: a phylogenetic approach to population biology. *Evolution* 43: 1192–1208.

Bradbury, J.W. 1981. The evolution of leks. In *Natural Selection and Social Behavior* (R.D. Alexander and D.W. Tinkle, eds.), pp. 138–169. Chiron Press, New York. 532 pp.

Bradbury, J. W and R.M. Gibson. 1983. Leks and mate choice. In *Mate Choice* (P. Bateson, ed.), pp. 109–138. Cambridge University Press, Cambridge, U.K. 462 pp.

Bryan, E.H. 1934. A review of the Hawaiian Diptera, with descriptions of new species. *Proc. Hawaii. Entomol. Soc.* 8: 434–440, 456–457.
Bryan, E.H. 1938. Key to the Hawaiian Drosophilidae and descriptions of new species. *Proc. Hawaii. Entomol. Soc.* 10: 25–42.
Bryant, E.H., S.A. McCommas, and L.M. Combs. 1986. The effect of an experimental bottleneck upon quantitative genetic variation in the housefly. *Genetics* 114: 1191–1211.
Carson, H.L. 1968. The population flush and its genetic consequences. In *Population Biology and Evolution* (R.C. Lewontin, ed.), pp. 123–137. Syracuse University Press, Syracuse. 205 pp.
Carson, H.L. 1971. Speciation and the Founder Principle. University of Missouri, *Stadler Symp.* 3: 51–70.
Carson, H.L. 1990. Evolutionary process as studied in population genetics: clues from phylogeny. *Oxford Surv. Evol. Biol.* 7: 129–156.
Carson, H.L. 1997. Sexual selection: a driver of genetic change in Hawaiian *Drosophila*. *J. Hered.* 88: 343–352.
Carson, H.L. and D.A. Clague. 1995. Geology and biogeography of the Hawaiian Islands. In *Hawaiian Biogeography* (W.L. Wagner and V.A. Funk, eds.), pp. 14–29. Smithsonian Institution Press, Washington, D.C. 467 pp.
Carson, H.L. and R.G. Wisotzkey. 1989. Increase in genetic variance following a population bottleneck. *Am. Nat.* 134: 668–673.
Carson, H.L., F.E. Clayton, and H.D. Stalker. 1967. Karyotypic stability and speciation in Hawaiian *Drosophila*. *Proc. Natl. Acad. Sci. U.S.A.* 57: 1280–1285.
Clague, D.A. and G.B. Dalrymple. 1987. The Hawaiian-Emperor volcanic chain. In *Volcanism in Hawaii* (R.W. Decker, T.L. Wright, and P.H. Stauffler, eds.), pp. 1–54. U.S. Geological Survey Professional Paper 1350. U.S. Government Printing Office, Washington, D.C. 1667 pp.
Conant, P. 1978. Lek Behavior and Ecology of Two Sympatric Homosequential Hawaiian *Drosophila*: *Drosophila heteroneura* and *Drosophila silvestris*. M.S. thesis, University of Hawaii, Honolulu.
Dalrymple, G.B., E.A. Silver, and E.D. Jackson. 1973. Origin of the Hawaiian Islands. *Am. Sci.* 61: 294–308.
DeSalle, R. and A.R. Templeton. 1987. Comments on the "Significance of asymmetrical sexual isolation," In *Evolutionary Biology,* Vol. 21 (M.K. Hecht, B. Wallace, and G.T. Prance, eds.), pp. 21–28. Plenum Press, New York. 434 pp.
Dowell, R.V. and R. Penrose. 1995. Mediterranean fruit fly eradication in California 1994–1995. In *The Mediterranean Fruit Fly in California: Defining Critical Research* (J.G. Morse, R.L. Metcalf, J.R. Carey, and R.V. Dowell, eds.), pp. 161–185. College of Natural and Agricultural Sciences, University of California, Riverside. 318 pp.
Ehrman, E. and M. Wasserman. 1987. The significance of asymmetrical sexual isolation. In *Evolutionary Biology,* Vol. 21 (M.K. Hecht, B. Wallace, and G.T. Prance, eds.), pp. 1–20. Plenum Press, New York. 434 pp.
Giddings, L.V. and A.R. Templeton. 1983. Behavioral phylogenies and the direction of evolution. *Science* 220: 372–377.
Grimshaw, P.H. 1901–1902. *Fauna Hawaii.* 3: 51–73, 86.
Hardy, D.E. 1965. Diptera: Cyclorrhapha II, Series Schizophora, Section Acalyptratae I. Family Drosophilidae. *Insects of Hawaii,* Vol. 12. University of Hawaii Press, Honolulu. 814 pp.
Heed, W.B. 1968. Ecology of the Hawaiian Drosophilidae. *Univ. Texas Publ.* 6818, pp. 387–419.
Heed, W.B. 1971. Host plant specificity and speciation in Hawaiian *Drosophila*. *Taxon* 20: 115–121.
Kaneshiro, K.Y. 1969. A study of the relationships of Hawaiian *Drosophila* species based on external male genitalia. *Univ. Texas Publ.* 6918: 55–70.
Kaneshiro, K.Y. 1976a. A revision of generic concepts in the biosystematics of Hawaiian Drosophilidae. *Proc. Hawaii. Entomol. Soc.* 22: 255–278.
Kaneshiro, K.Y. 1976b. Ethological isolation and phylogeny in the *planitibia* subgroup of Hawaiian *Drosophila*. *Evolution* 30: 740–745.
Kaneshiro, K.Y. 1980. Sexual selection, speciation, and the direction of evolution. *Evolution* 34: 437–444.
Kaneshiro, K.Y. 1983. Sexual selection, direction of evolution and the biosystematics of Hawaiian Drosophilidae. *Annu. Rev. Entomol.* 28: 161–178.
Kaneshiro, K.Y. 1989. The dynamics of sexual selection and founder effects in species formation. In *Genetics, Speciation, and the Founder Principle* (L.V. Giddings, K.Y. Kaneshiro, and W.W. Anderson, eds.), pp. 279–296. Oxford University Press, Oxford. 373 pp.
Kaneshiro, K.Y. 1990. Natural hybridization in *Drosophila* with special reference to species from Hawaii. *Can. J. Zool.* 68: 1800–1805.

Kaneshiro, K.Y. 1993a. Habitat-related variation and evolution by sexual selection. In *Evolution of Insect Pests* (K.C. Kim and B.A. McPheron, eds.), pp. 89–101. John Wiley & Sons, New York. 479 pp.

Kaneshiro, K.Y. 1993b. Introduction, colonization, and establishment of exotic insect populations: fruit flies in Hawaii and California. *Am. Entomol.* 39: 23–29.

Kaneshiro, K.Y. 1997. R.C.L. Perkins' legacy to evolutionary research on Hawaiian Drosophilidae (Diptera). *Pac. Sci.* 51: 450–461.

Kaneshiro, K.Y. and L.V. Giddings. 1987. The significance of asymmetrical sexual isolation and the formation of new species. In *Evolutionary Biology,* Vol. 21 (M.K. Hecht, B. Wallace, and G.T. Prance, eds.), pp. 29–44. Plenum Press, New York. 434 pp.

Kaneshiro, K.Y. and M.P. Kambysellis. 1999. Description of a new allopatric sibling species of Hawaiian picture-winged Drosophila. *Pac. Sci.* 53: 208–213.

Kaneshiro, K.Y. and J.S. Kurihara. 1981. Sequential differentiation of sexual behavior in populations of *Drosophila silvestris. Pac. Sci.* 35: 177–183.

Kuba, H. and Y. Sokei. 1988. The production of pheromone clouds by spraying in the melon fly, *Dacus cucurbitae* Coquillett (Diptera: Tephritidae). *J. Ethol.* 6: 105–110.

Malloch, J.R. 1938. Two genera of Hawaiian Drosophilidae (Diptera). *Proc. Hawaii. Entomol. Soc.* 10: 53–55.

Mayr, E. 1972. Sexual selection and natural selection. In *Sexual Selection and the Descent of Man, 1871–1971* (B. Campbell, ed.), pp. 87–104. Aldine, Chicago. 378 pp.

Montgomery, S.L. 1975. Comparative breeding site ecology and the adaptive radiation of picture-winged *Drosophila. Proc. Hawaii. Entomol. Soc.* 22: 65–102.

Nishida, R., K.H. Tan, M. Serit, N.H. Lajis, A.M. Sukari, S. Takahashi, and H. Fukami. 1988. Accumulation of phenylpropanoids in the rectal glands of males of the Oriental fruit fly, *Dacus dorsalis. Experientia* 44: 534–536.

Nishida, R., K.H. Tan, S. Takahashi, and H. Fukami. 1990. Volatile compounds of male rectal glands of the melon fly, *Dacus cucurbitae* Coquillett (Diptera: Tephritidae). *Appl. Entomol. Zool.* 25: 105–112.

Nishida, R., O. Iwahashi, and K.H. Tan. 1993. Accumulation of *Dendrobium superbum* (Orchidaceae) fragrance in the rectal glands by males of the melon fly, *Dacus cucurbitae. J. Chem. Ecol.* 19: 713–722.

Nishida, R., T.E. Shelly, and K.Y. Kaneshiro. 1997. Acquisition of female-attracting fragrance by males of Oriental fruit fly from a Hawaiian lei flower, *Fagracea berteriana. J. Chem. Ecol.* 23: 2275–2285.

Ohta, A.T. 1978. Ethological isolation and phylogeny in the grimshawi species complex of Hawaiian *Drosophila. Evolution* 32: 485–492.

Powell, J.R. 1997. *Progress and Prospects in Evolutionary Biology: The Drosophila Model.* Oxford University Press, New York. 562 pp.

Prokopy, R.J. and J. Hendrichs. 1979. Mating behavior of *Ceratitis capitata* on a field-caged tree. *Ann. Entomol. Soc. Am.* 72: 642–648.

Shelly, T.E. 1995. Methyl eugenol and the mating competitiveness of irradiated male *Bactrocera dorsalis* (Diptera: Tephritidae). *Ann. Entomol. Soc. Am.* 88: 883–886.

Shelly, T.E. and A.L. M. Dewire. 1994. Chemically mediated mating success in male Oriental fruit flies (Diptera: Tephritidae). *Ann. Entomol. Soc. Am.* 87: 375–382.

Shelly, T.E. and E.M. Villalobos. 1995. Cuelure and the mating behavior of male melon flies, *Bactrocera cucurbitae* (Diptera: Tephritidae). *Fla. Entomol.* 78: 473–482

Shelly, T.E., T.S. Whittier, and E.M. Villalobos. 1996. Trimedlure affects mating success and mate attraction in male Mediterranean fruit flies. *Entomol. Exp. App.* 78: 181–185.

Sivinski, J. and T. Burk. 1989. Reproductive and mating behaviour. In *Fruit Flies: Their Biology, Natural Enemies and Control* (A.S. Robinson and G. Hooper, eds.), pp. 343–351. In *World Crop Pests* (W. Helle, ed.), Vol. 3A. Elsevier Science Publishers, Amsterdam. 372 pp.

Spieth, H.T. 1966. Courtship behavior of endemic Hawaiian *Drosophila. Univ. Texas Publ.* 6615, pp. 245–313.

Spieth, H.T. 1968. Evolutionary implications of the mating behavior of the species of *Antopocerus* (Drosophilidae) in Hawaii. *Univ. Texas Publ.* 6818, pp. 319–333.

Spieth, H.T. 1974a. Mating behavior and evolution of the Hawaiian *Drosophila.* In *Genetic Mechanisms of Speciation in Insects* (M.J.D. White, ed.), pp. 94–101. Australia and New Zealand Book Co., Sydney. 170 pp.

Spieth, H.T. 1974b. Courtship behavior in *Drosophila. Annu. Rev. Entomol.* 19: 385–405.

Spieth, H.T. 1982. Behavioral biology and evolution of the Hawaiian picture-winged species group of *Drosophila*. In *Evolutionary Biology,* Vol. 14 (M.K. Hecht, B. Wallace, and G.T. Prance, eds.), pp. 351–437. Plenum, New York. 445 pp.

Takada, H. 1966. Males genitalia of some Hawaiian Drosophilidae. *Univ. Texas Publ.* 6615, pp. 315–333.

Throckmorton, L.H. 1966. The relationships of the endemic Hawaiian Drosophilidae. *Univ. Texas. Publ.* 6615, pp. 335–396.

Wheeler, M.R. 1952. A key to the Drosophilidae of the Pacific Islands (Diptera). *Proc. Hawaii. Entomol. Soc.* 14: 421–423.

Whittier, T.S., K.Y. Kaneshiro, and L.D. Prescott. 1992. Mating behavior of Mediterranean fruit flies (Diptera: Tephritidae) in a natural environment. *Ann. Entomol. Soc. Am.* 85: 214–218.

Wirth, W.W. 1952. Two new spider egg predators from the Hawaiian Islands (Diptera: Drosophilidae). *Proc. Hawaii. Entomol. Soc.* 14: 415–417.

Yoon, J.S., K. Resch, and M.R. Wheeler. 1972. Intergeneric chromosomal homology in the family Drosophilidae. *Genetics* 71: 477–480.

Zimmerman, E.C. 1948. Introduction. *Insects of Hawaii,* Vol. 1. University of Hawaii Press, Honolulu.

# Section VIII

## Glossary

# 33 Glossary

*Ian M. White, David H. Headrick, Allen L. Norrbom, and Lynn E. Carroll*

## CONTENTS



## 33.1 INTRODUCTION

The following glossary is primarily a collection of definitions of specialized terms used in the preceding chapters of this book. However, to make it more generally useful, it also includes and cross references a wide selection of outdated or less preferred morphological and behavioral terms that might be encountered in the broader tephritid literature. Most of the included terms can be categorized as one of the following: terms to describe fruit fly behavior or ecology; names of morphological parts of adult or larval fruit flies; nomenclatural terms dealing with rules for correctly naming fruit fly species, genera, or other taxa; or terms pertaining to analysis of fruit fly phylogeny. If not otherwise annotated (followed by (**larva**) or (**behavior**), etc.), entries refer to adult morphology.

The glossary of White (1988), which was later expanded by White and Elson-Harris (1994), was used as the basis for the morphological and nomenclatural terms, although most definitions were extensively revised by us or contributing colleagues (see Acknowledgments). Simple terms that can be found in any general introduction to entomology and terms that are easily explained by the figures are omitted. For additional morphological terms or explanations, we highly recommend the excellent treatments of general Diptera morphology by McAlpine (1981) and Teskey (1981), from whom we have copied liberally. For additional nomenclatural terms and definitions see the glossary and text of the International Code of Zoological Nomenclature (ICZN 1985), the source of many of the entries we have included. Most of the behavioral terms are taken from Headrick and Goeden (1994), based on the behaviors observed in 49 species of Tephritidae from southern California. Some terms used by various other authors to describe behavior in other tephritid species have also been added with cross-references, but some terms could not be understood clearly because the previous lack of a standardized terminology for tephritid behavior.

Entries for obsolete or less preferred terms are given in *italics* and, unless easily explained by the figures, a cross reference is given to the preferred term (underlined). Many of these outdated or nonpreferred terms are also listed under the corresponding entry of the preferred term. Other cross references are also underlined. We have not attempted to include every outdated term, and emphasis has been placed on selecting those most likely to be encountered in the tephritid literature. No attempt was made to include non-English language terms; however, an appendix has been added to give equivalents between old German (Hendel 1927) wing vein/cell terms and those now used.

## 33.2 TERMS

**Abdominal pleural distension (behavior)** See male display behavior.

**Accessory costal band** or **crossband** See crossband.

**Accessory plates (larva)** (Figure 33.7A, ac pl) Small plates, often toothed, immediately lateral to the oral ridges, often poorly differentiated from them.

**Acuminate seta** A slender, tapered, acute seta (Freidberg and Kugler 1989). Also see lanceolate seta.

***Aciura*-type pattern** (Figure 33.3F) A predominantly dark brown or black wing pattern, with several anterior and posterior hyaline marginal incisions (Freidberg and Kugler 1989).

**Acrophallus** (Figure 33.5, acroph) The sclerites guarding the gonopore in the glans of the male genitalia. It is comprised by two, rarely three, semitubular lobes (in most Tephritinae fused to form one tube). It has also been called the *endophallus*. It may contain a chamber, the praeputium.

**Acrostichal seta** (Figure 33.2, acr s) In Diptera, the row or rows of setae nearest to the midline of the scutum. Tephritidae have at most one pair of these setae, placed just anterior to the scuto-scutellar suture. Consequently, some authors have called them the *prescutellar setae*. They are occasionally absent (e.g., in *Dacus* spp.).

**Aculeus** (Figure 33.4, acul) In Tephritoidea, the piercing part of the female ovipositor, which is normally retracted inside the oviscape (it often must be dissected to be examined). It consists of an elongate eighth tergite, a pair of elongate eighth sternites (st8), which have also been called *egg guides*, *genital flaps*, *valves*, *ventral flaps,* or *ventral sclerites*, and an apical cercal unit, a diamond-shaped or triangular part probably derived from the cerci and the subanal plate (hypoproct) (V.A. Korneyev, personal observation). In most Tephritidae the cercal unit is completely fused to the eighth tergite, but in most Phytalminae it is free or there are sutures indicating the limits of these two sclerites. The aculeus has also been called the ovipositor, *oviscapt*, *apical part of the ovipositor*, *piercer*, *ovipositor blade,* or *gynium* (see Norrbom and Kim 1988). Its ultrastructure and function were discussed by Stoffolano and Yin (1987).

**Aculeus tip** (Figure 33.4, acul t) The apical part of the aculeus. By convention, especially in *Anastrepha*, its length is measured on the ventral side from the inner margin of the sclerotized area, which has been erroneously called the apex of the oviduct or *apparent genital opening* (also see cloaca).

**Adventive (ecology)** Adjective meaning accidental or not native, as in a species introduced to a new geographic region as a result of human activity.

***Aedeagal apodeme*** See phallapodeme.

***Aedeagal glans*** See glans.

***Aedeagus*** See phallus.

**Aggregative life history (behavior)** A life history strategy of some temperate tephritid species that are univoltine to multivoltine and the adults synovigenic (Headrick and Goeden 1994). Adults are usually long-lived, up to a year, and may remain in a reproductive diapause until their host plants are in a stage suitable for oviposition. These species are characterized by having highly interactive courtship and copulation behavior, and have a variety of mating strategies such as mate guarding and resource defense (Headrick and Goeden 1994, 1998). These species also exhibit many different types of host plant utilization including ectophagous and endophagous strategies. Also see circumnatal life history.

**Agonistic interactions (behavior)** Interactions between two or more individuals involved in aggressive behavior such as threat displays, chases, or physical interactions.

**Allochronic (ecology)** Adjective describing two or more species or populations whose potential to interbreed is reduced by a temporal separation, either in terms of time of year or due to mating at different times of day. See also allopatric, sympatric and synchronic.

**Allopatric (ecology)** Adjective describing two or more species or populations that are geographically isolated, thus preventing them from interbreeding. Two species that evolved from one, in geographical isolation, are said to have evolved allopatrically (see Bush 1966; 1969b). Also see allochronic, sympatric and synchronic.

**Allotype (nomenclature)** See type specimen.

**Anal cells/anal lobe** The cells posterior to veins $A_1$ and $A_2$ are the anal cells, but they are not closed in Tephritidae and the term anal lobe (Figure 33.4A) is more often used for the entire area posterior to vein $A_1$. The term "anal cell" was often incorrectly used in the past for the basal cubital cell.

***Anal cell extension*** See basal cubital cell.

**Anal elevation (larva)** (Figure 33.7, an elev) The area surrounding the anal opening.

**Anal lobe (larva)** (Figures 33.6 and 33.7, an lb) A paired lobe flanking the anal opening.

**Anal streak** (Figure 33.3E, AS) A diagonal marking that covers cell bcu and part of cell $cu_1$ in most species with a wasp mimicry pattern (e.g., most Dacini, *Toxotrypana*). It is sometimes called the *anal stripe*.

***Anal stripe*** See anal streak.

***Anastrepha*-type pattern** (Figure 33.3D) A wing pattern with a short costal band ending at the apex of vein $R_1$, a strongly oblique radial-medial band and an anterior apical band joined to form an S-band, and subapical and posterior apical bands often joined to form a V-band (Lima 1934; Stone 1942). Also see crossband.

**Anatergite** (Figure 33.2, anatg) In lateral view, this sclerite is anterior to the mediotergite and above the haltere. The Adramini differ from other tephritids by having long, fine, pale-colored setulae on the anatergite. The anatergite and katatergite together form the laterotergite, which is the *pleurotergite* or *hypopleural callus* of many authors.

**Anepimeron** (Figure 33.2, anepm) The lateral thoracic sclerite that is below the wing base. It has also been called the *pteropleuron*. In Tephritidae it bears an anepimeral seta (Figure 33.2, anepm s).

**Anepisternal phragma** (Figure 33.2, anepst phgm) In Tephritidoidea the anepisternum has a vertical phragma just anterior to the series of setae near the posterior edge. This phragma, which has also been called the *mesopleural suture*, is usually visible in the Phytalmiinae and Trypetinae, but is obscured by dense microtrichia in most Tephritinae (in the latter it can usually be seen if the anepisternum is wetted with a drop of alcohol).

**Anepisternal stripe** A yellow stripe (of the xanthine type) which covers the posterior part of the anepisternum and the anterior portion of the anepimeron. It is common in *Bactrocera* and *Dacus* spp., in which it usually extends from the notopleural callus down and onto the upper part of the katepisternum. In some species the upper end extends anteriorly to the postpronotal lobe. It is the *mesopleural stripe* of many authors.

**Anepisternum** (Figure 33.2, anepst) The large pleural sclerite of the thorax between the anterior spiracle and the wing base. It has also been called the *mesopleuron*. In Tephritidae, near the posterior margin, it bears one or a row of anepisternal setae (Figure 33.2, anepst s) or setulae which decrease in size ventrally.

**Antenna (larva)** (Figure 33.7, ant) The more dorsal of the two sensory organs on the anterior end of the head. It consists of one to three apparent segments, including two sclerotized segments and a conical to flattened tip. It has also been termed *anterior sense organ* (e.g., Snodgrass 1924; Exley 1955; Phillips 1946), *antennal sensory organ* (White and Elson-Harris), or *dorsal sense organ* (e.g., Bolwig 1946; Snodgrass 1953; Novak and Foote 1968; Chu and Axtell 1971).

***Antennal sensory organ (larva)*** See antenna.

**Anterior apical band** or **crossband** See crossband.

***Anterior lobes (larva)*** See maxillary sense organ.

**Anterior sclerite (larva)** (Figure 33.6, a scl) A sclerite on either side of the pharyngeal sclerite projecting anteriorly from just below the dorsal bridge.

***Anterior sense organ (larva)*** See antenna.

**Anterior spiracle (larva)** (Figure 33.6, a spr) There are two functional pairs of spiracles in tephritid larvae. The anterior spiracle projects laterally from the prothoracic segment. It is fan-shaped or bimodal, with 2 to 53 tubules with openings along the outer edge.

***Anterior supra-alar seta*** See supra-alar setae.

**Apical band** or **crossband** See crossband.

***Apicodorsal rod*** See subapical lobe.

**Apomorphy (phylogenetic analysis)** A derived trait or character state, that is, of two (or more) states of a character, one is plesiomorphic, or ancestral, and the other(s) is apomorphic.

***Apparent genital opening*** A term used by White and Elson-Harris (1992) for the inner margin of the sclerotization on the ventral side of the aculeus tip, which is used as a point of reference for measuring the aculeus tip.

**Arching (behavior)** See wing displays.

**Argent** (Figure 33.3J, arg) In the wing pattern, a silvery or white area, usually a spot, which changes color or appearance when the wing is viewed at different angles (Munro 1947).

**Arista** (Figure 33.1, ar) In Tephritidae, the flagellum is modified into a large first flagellomere and 3 very slender ones that arise dorsally near the base of the first, and form the style-like or seta-like arista. Most tephritids have a micropubescent arista, that is, it is covered in a microscopic downy pile, but some (e.g., Gastrozonina, many Acanthonevrini and Adramini) have a plumose or pectinate arista, and in others (e.g., *Baryglossa*, *Gymnocarena*) it is mostly or entirely bare.

**Autapomorphy (phylogenetic analysis)** A derived trait possessed by only one of a group of taxa whose relationships are being analyzed. Although useful to diagnose or to support the monophyly of the taxon that possesses it, an autapomorphy holds no information regarding that taxon's relationship to the other taxa, that is, it does not resolve relationships among the larger group of taxa and is therefore phylogenetically uniformative at that level.

**Basal cells** A general term for the basal radial, basal medial and basal cubital cells (br, bm and bcu). The old terms *1st and 2nd basal cells* meant cells br and bm, respectively.

**Basal cubital cell** (Figure 33.4A, bcu) The basal wing cell bounded anteriorly by the base of vein Cu, apically by vein $Cu_2$, and posteriorly by vein $A_1$. In most Tephritidae, vein $Cu_2$ is concave or has a distinct bend, forming an acute posteroapical extension on cell bcu. This cell was long commonly known as the "anal cell" (e.g., Munro 1947) or the *1st anal cell*, which is incorrect because it is anterior to vein $A_1$. McAlpine (1981) called it the *posterior cubital cell* based on the untracheated structure (called vein CuP by McAlpine) that incompletely crosses the cell's anterior third. But at most, only the part posterior to "CuP" should be called cell cu*p*, but even that is questionable because the homology of the structure is uncertain (Steyskal 1984) and it does not reach vein $Cu_2$ to form a complete cell. This cell has also been called the *cubital cell* or abbreviated as *cell cu* (e.g., Hardy 1973).

**Basal medial cell** (Figure 33.4A, bm) The basal wing cell bounded anteriorly by vein M, apically by crossvein BM-Cu, and posteriorly by vein Cu. It has also been called *cell M* (e.g., Hardy 1973) or the *2nd basal cell.*

**Basal medial-cubital crossvein** (Figure 33.4A, BM-Cu) The more basal of the two transverse veins connecting veins M and $Cu_1$, in Tephritidae located just distal to the fork in the cubital vein. This crossvein has also been called $M_3$ or the *anterior basal crossvein.*

**Basal radial cell** (Figure 33.4A, br) The elongate, basalmost radial cell, bordered anteriorly by the base of veins R and Rs and by $R_{4+5}$, apically by R-M, and posteriorly by vein M. It has also been termed or abbreviated as *cell R* or the *1st basal cell.*

***Basicostal band*** A term used by White and Elson-Harris (1994) for the short costal band in most *Anastrepha* spp.

**Basiphallus** (Figure 33.5, bph) The very short, well-sclerotized, basal part of the male phallus. It has also been called the *phallobase* (Munro 1947). This term has also been used for all of the phallus except the glans (e.g., McAlpine 1981), but that homology is incorrect (V.A. Korneyev, personal observation).

**Bivoltine (ecology)** Having two generations per year.

**Body swaying (behavior)** See male display behavior.

**Bootstrapping (phylogenetic analysis)** A technique whereby a loose form of confidence limit is assigned to clades within a tree. Felsenstein (1985) recommends randomly sampling the characters in a data matrix (with replacement, but maintaining association of character states with taxa) to build a hypothetical data matrix of the same size as the original one. This matrix is analyzed in the same way as the original set, and this procedure is repeated at least 100 times (often 1000 times for molecular systematics data sets). Within the obtained analyses of the simulated data sets, certain clades found in the original analysis will recur with varying percentages. The percentage of occurrences will give an indication of support to these clades, but this should not be confused with confidence limits as used in formal statistics.

**Bristle** An alternative term for a large macrotrichium, or seta.

***Bubbling*** **(behavior)** See droplet formation.

**Bulla** (plural: **bullae**) This term has been given a variety of uses. Freidberg and Kugler (1989) used it to refer to a "specialized" area of wing pattern that is usually black or brown and oval as in some genera of Tephritini (e.g., *Goniurellia* spp.) that have a dark oval mark on $R_{4+5}$. However, in other groups the bullae are more distinct structures, rather than just a darker marking. In *Schistopterum* spp. (Schistopterini) the wing has raised areas in cells $r_{2+3}$, dm, and m (see Figure 33.3G, bul). Drew and Hancock (1995) applied the term to a swelling adjacent to cell bcu found in species of the subgenus *Bulladacus* of *Bactrocera*. Permkam and Hancock (1995) applied the term to a pair of raised black bulbous areas on the last visible tergite of *Ornithoschema* spp. (Rivelliomimini) (tergite 5 in male and apparent tergite 6 in female).

***Button (larva)*** See ecdysial scar.

**Capitulum** (plural: **capitula) (botanical term)** The compound flower of a plant belonging to the family Asteraceae (= Compositae), often referred to as a flowerhead or seedhead.

**Calling (behavior)** A general term applied to all male courtship displays. See male calling under male display behavior. Future usage should limit it to sound production in male courtship.

***Carina*** See facial carina.

**Caudal ridge (larva)** (Figure 33.7, caud rdg) A narrow, transverse ridge of thickened cuticle running dorsad of the I1a, I1b, and I2 sensilla on the intermediate area of the caudal segment in species of Dacinae.

**Caudal segment (larva)** (Figures 33.6 and 33.7, caud sg) Abdominal segment 8.

**Cell bcu** See basal cubital cell.

**Cell bm** See basal medial cell.

***Cell cu*** See basal cubital cell.

***Cell cup*** See basal cubital cell.

**Cell m** See medial cells and basal medial cell.

**Cell sc** See subcostal cell.

**Cephalic segment (larva)** (Figure 33.6, ceph sg) The outer, membranous part of the head in Cyclorrhapha. It has also been called the *gnathocephalon*.

***Ceratitis*-type pattern** (Figure 33.3C) A wing pattern with discal, subapical, anterior apical, and sometimes posterior apical bands, and with dark spots or streaks in or near the costal and basal cells.

***Ceromae*** See ceromata.

**Ceromata** A pair of slightly depressed shiny areas on tergite 5 of *Bactrocera* and *Dacus* spp. They were called *ceromae* by Munro (1984) and *shining spots* or *tergal glands* by others. These areas are covered with wax glands (Munro 1984).

**Character weighting (phylogenetic analysis)** A method used in cladistic analysis to give greater importance to certain characters by assigning them weights, usually on a 1 to 3 or 1 to 10 scale. This may be done subjectively by the investigator; for example, complex characters are often considered less likely to be homoplasious than simple ones or those involving loss of an attribute and therefore assigned a higher weight. Most commonly, however, a method known as successive weighting or successive approximation is used, in which each character is weighted based on its congruence with the other characters in the data matrix. Successive analyses are conducted in which the characters are reweighted based on their average consistency indices in previous analyses, followed by reanalysis of the data matrix until the resulting trees no longer change. See Carpenter (1994) and included references.

**Chorion (egg)** The outer surface of the egg, which may appear smooth or reticulate.

***Ciliate*** See pecten.

**Circumnatal life history (ecology, behavior)** A life history strategy of some temperate tephritid species that are univoltine, with adults that are proovigenic and typically emerge when their host plants are in a stage suitable for oviposition (Headrick and Goeden 1994). Adults are usually short-lived and proceed to copulate with no aggregation behavior and very little courtship behavior. The longest stages are the early larval instars. These species usually form galls on their host plants. Also see aggregative life history.

**Clade (phylogenetic analysis)** An alternate term for monophyletic group, usually used in referring to part of a cladogram, for example, "the clade including taxa a-us, b-us, and c-us" or "the a-us clade."

**Cladistics (phylogenetic analysis)** A method used to analyze phylogenetic relationships among a group of organisms. Specifically, it attempts to determine the historical branching pattern or recency of common ancestry among these taxa based on the derived character states they share. Also see parsimony.

**Cladogram (phylogenetic analysis)** A branching diagram, or "tree," produced from a cladistic analysis that shows the hypothesized relationships of common ancestry (the historical branching sequence) among a group of taxa.

**Cloaca/cloacal opening** (Figure 33.4, cl op) In most Diptera the genital and alimentary canals have separate openings. The genital opening is between segments 8 and 9 ventrally, and the anus is on the apical segment below the cerci (McAlpine 1981). In Tephritidae, the genital and alimentary canals join internally to form a cloaca, which opens on the aculeus between or just beyond the apices of the eighth sternites (Dean 1935; Stoffolano and Yin 1987; Valdez and Prado 1991). This opening in Tephritidae has been misnamed the genital opening (e.g., Norrbom and Kim 1988), which in these flies is at the base of the cloaca, or *apex of oviduct* (e.g., Stone 1942). Both eggs and waste are passed through the cloaca, and its opening is also the point of insertion for the male's phallus during copulation.

***Clypeal ridge*** See facial carina.

**Combination (nomenclature)** The association of a generic name and a specific name to form the full scientific name of a species, or of those names and a subspecific name to form the name of a subspecies. The original combination is that used when a species is first described. If the species is transferred to another genus, the use of the new generic name and the previously published species name is a subsequent combination. The first use of a non-original generic name with a previously published species name is called a new combination.

***Complex*** See species complex.

**Consensus tree (phylogenetic analysis)** In cladistic analysis, this term is commonly used to mean a strict consensus tree, which is a tree containing only the branches common to all of the most parsimonious (i.e., shortest) trees resulting from an analysis. For example, an analysis may indicate that there are three equally parsimonious cladograms for a group of taxa that differ only in the relationships among three of the taxa. In the consensus tree, those three taxa will be an unresolved polytomy, but the tree will indicate the relationships of the other taxa that are consistently supported. This is most useful in analyses of large numbers of taxa where many equally parsimonious trees may result.

**Consistency index (phylogenetic analysis) (ci)** A measure, on a scale from 0 to 1, of how much homoplasy there is in a cladogram or a character (how well that character "fits" a particular cladogram). Those with little homoplasy have high values (1 means no homoplasy), and those with more homoplasy have low values. This is most commonly reported for the most parsimonious tree or set of trees resulting from a maximum parsimony analysis. See Farris (1989).

**Convergence (phylogenetic analysis)** See homoplasy.

**Convergent seta** A seta that is inclinate, that is, leans toward the midline of the fly. The Terelliini, for example, have the posterior pair of orbital setae convergent.

**Copulation (behavior)** Copulation includes the acts of intromission of the male phallus and subsequent sperm transfer. Intromission, the insertion of the male phallus, can begin only after the female has exserted her aculeus and the male has grasped it with his surstyli to expose the cloacal opening. Sperm transfer occurs sometime after the phallus is fully inserted, and is followed (not necessarily immediately) by retraction of the phallus. The term "mating" should not function as a synonym for copulation; the latter is one aspect of the process of mating. The term "in copula" is sometimes used to describe a copulating pair. See Eberhard and Pereira (1993, 1998) and Headrick and Goeden (1994) for further details.

**Copulatory induction behavior (behavior)** The highly interactive sequence of behaviors between a male and a female after mounting, following courtship. While mounted, the male performs a series of tactile and wing movements to induce the female to exsert her aculeus. Exsertion of the aculeus is the response that marks the acceptance of the male by the female for intromission. Copulatory induction behavior and exsertion of the aculeus were observed in all species studied by Headrick and Goeden (1994).

**Cornu** (plural: **cornua) (larva)** See dorsal cornu and ventral cornu.

***Costagium*** A term used by Munro (1984) for the basicosta.

**Costal band** (Figure 33.3D-E, CB) A band along the anterior margin of the wing, typical of species presumed to be wasp mimics. It may vary in extent, for example, it extends the entire wing length in *Toxotrypana*, from cell sc to the wing apex in most *Bactrocera* and *Dacus* spp., or from the wing base to the apex of vein $R_1$ in most *Anastrepha* spp. It is probably independently derived within various unrelated tephritid lineages.

**Costal cells** (Figure 33.4A) A collective term for the basal costal cell (bc) and costal cell (c). *1st and 2nd costal cells* (or *1st C* and *2nd C*) are old terms for these two cells, respectively.

**Courtship (behavior)** A series of behavioral events between two conspecific individuals of opposite sex that may result in the mounting of the female by the male. In Tephritidae, courtship involves ritualistic male display behavior, sometimes followed by a female's response that may eventually lead to mounting. Courtship ends at mounting, as the male and female begin to interact differently during ensuing copulatory induction behavior.

Courtship typically takes place on host plants, which usually are in a suitable stage for oviposition. Courtship away from the host usually involves a lek. Male display behaviors (i.e., displays not directed toward any particular individual) are often different from courtship displays which are directed toward solitary females.

**Creeping welt (larva)** (Figure 33.6, cr wlt) A locomotory structure on the ventral surface of an abdominal segment consisting of a membranous, transverse, swollen ridge bearing rows of small spinules or rounded projections.

**Crop** A blind sac of the alimentary system whose opening is near the foregut-midgut interface, extending posteriorly as a small tube through the thorax, and expanding into an enlarged sac within the abdomen. Used for the initial storage of liquid food.

**Crossband** (Figure 33.3) A transverse wing band. Various types of wing patterns that have crossbands occur in Tephritidae, for example, the *Anastrepha*-type, *Ceratitis*-type, and *Rhagoletis*-type patterns. Jenkins (1996) attempted to homologize the elements of banded patterns found in Carpomyini and some other Trypetinae. He noted the usefulness of the positions of the three campaniform sensilla on vein $R_{4+5}$ for determining homologies of some bands, but correctly stated that phylogenetic analysis is needed to resolve homologies within any group. As far as possible, use of his concepts for other tephritids is recommended, although the traditional names for the bands proposed by Foote (1981), Bush (1966), and Steyskal (1979) for *Rhagoletis* and *Urophora* are retained below. These names are applicable in many tephritid genera (e.g., *Cryptophorellia* Freidberg and Hancock 1989), although the bands may not be strictly homologous in unrelated taxa.

Jenkins (1996) recognized a number of pattern elements, which may be variously present or combined in patterns with crossbands. They are located more or less on the following areas: on the humeral crossvein; on the basal cells; on the pterostigma; on the discal medial cell, often including crossvein R-M; in cells $r_1$ and $r_{2+3}$, sometimes including R-M; on crossvein DM-Cu; along the apical costal margin; and in cells $r_{4+5}$ and m. Of the bands listed below, not all are present in any one species; some of them are formed from different combinations of the same elements (e.g., the mark on the basal cells can be part of the subbasal band if joined to the humeral band, or part of the subcostal band if joined to the mark on the subcostal cell).

**Humeral band** (Figure 33.3B, HB) A band on the humeral crossvein. It is often part of the subbasal band.

**Subbasal band** (Figure 33.3A, SBB) A band crossing the humeral crossvein and the basal cells (br, bm, bcu) (e.g., in the *Rhagoletis*-type pattern). It may be divided into a humeral band, and a posterior part on the basal cells that may fuse with a spot or band on the subcostal cell (usually on the pterostigma) to form a subcostal band.

**Subcostal band** (Figure 33.3B, SCB) A band over the basal cells and subcostal cell (usually on the pterostigma). It was called the *proximal subcostal band* by Jenkins (1996).

**Discal band** (Figure 33.3A, C, DB) A band crossing cell dm, typically starting on the pterostigma and usually covering crossvein R-M (e.g., in the *Rhagoletis*-type pattern). It has also been called the *medial band* or *distal subcostal band* (Bush 1966; Jenkins 1996).

**Accessory costal band** (Figure 33.3A, ACB) A short crossband between the discal and subapical bands in the radial cells (e.g., in the *Rhagoletis*-type pattern). It has also been called the *intercalary band* (Bush 1966; Foote et al. 1993).

**Radial-medial band** (Figure 33.3B, RMB) A band covering crossvein R-M and not the pterostigma. It may be formed by the accessory costal band and the posterior part of the discal band.

**Subapical band** (Figure 33.3A-D, SAB) A band covering crossvein DM-Cu (e.g., in the *Rhagoletis*-type pattern). It has also been called the *cubital band* (e.g., Freidberg 1991), *discal medial-cubital band* (Jenkins 1996), or *preapical crossband*.

**Apical** or **anterior apical band** (Figure 33.3A-D, AAB) A band on or near the wing margin on the apical part of the wing, typically running from cell $r_1$ to the wing apex (e.g., in the *Rhagoletis*-type pattern). In *Ceratitis* it has also been called the *marginal band* (e.g., Freidberg 1991).

**Posterior apical band** (Figure 33.3A, C-D, PAB) A band between the subapical and anterior apical bands, usually in cells $r_{4+5}$ and m (e.g., in the *Rhagoletis*-type pattern). It is often joined to the subapical band or to the junction of the subapical and anterior apical bands. In most *Anastrepha* spp. it forms part of the V-band. In *Ceratitis*, it has sometimes been called the *medial band* (e.g., Freidberg 1991).

**Crossvein DM-Cu** or **dm-cu** See discal medial-cubital crossvein.

***Crossvein i-m*** An old name for the radial-medial crossvein.

**Crossvein R-M** or **r-m** See radial-medial crossvein.

***Crossvein t-p*** An old name for the discal medial-cubital crossvein.

**Cryptic behavior (behavior)** A behavior that is known to be exhibited in a species, but is not used by certain individuals of that species. Tephritids exhibit a wide range of behaviors, some of which occur sequentially (e.g., courtship). However, not every individual will necessarily display every behavior in a given sequence. This is especially true of newly emerged or otherwise virginal individuals. Typically, within a species some individuals never display a complete courtship and copulation sequence, others do so occasionally, and yet others do so regularly. In some species many individuals must be observed over extended periods before all of their behaviors can be cataloged (Headrick and Goeden 1994).

**Cubital cells** (Figure 33.4A) The cells posterior to the cubital vein; in Tephritidae the basal cubital cell (bcu) and the first cubital cell ($cu_1$). The terms *third posterior cell, anterior cubital cell, apical cubital cell,* and $cua_1$ have been used for $cu_1$.

**Cubital vein/cubitus** (Figure 33.4A, Cu) According to McAlpine (1981), both the anterior and posterior branches of the cubitus are present in Diptera, but the homology of the untracheated structure he called the posterior cubitus (CuP) was questioned by Steyskal (1984), who considered it merely a sclerotized fold. The well-developed part of the cubitus, technically CuA, has two branches, here abbreviated for simplicity as $Cu_1$ and $Cu_2$. Vein $Cu_1$ has been termed or abbreviated as the cubitus, *Cu*, $CuA_1$, *fifth longitudinal vein*, or $M_3+Cu$. Vein $Cu_2$ has been called the *posterior basal crossvein, cu-an* or $CuA_2$. In Tephritidae, it fuses with $A_1$ to close the basal cubital cell. This combined vein has been termed the *sixth longitudinal vein* or $Cu_2+2nd\ A$.

**Dental sclerites (larva)** (Figure 33.6, den scl) A pair of small sclerites lying close to the ventral margin of the mouthhooks. They are common in the Dacini but absent or inconspicuous in other groups.

**Dimidiate pattern** A wing pattern in which the wing is divided more or less evenly into an anterior dark area and a posterior hyaline area.

**Discal band** or **crossband** See crossband.

***Discal cell*** An old term for the discal medial cell (dm).

**Discal medial cell** (Figure 33.4A, dm) The cell near the middle of the wing bounded anteriorly by vein M, basally by crossvein BM-Cu, posteriorly by vein $Cu_1$, and apically by crossvein DM-Cu. It has also been called cell *1st* $M_2$ (e.g., Hardy 1973), the *discal cell*, or the *discoidal cell.*

***Discal medial-cubital band*** Another name for the subapical band. See crossband.

**Discal medial-cubital crossvein** (Figure 33.4A, DM-Cu) The more distal of the two transverse veins connecting veins M and $Cu_1$, in Tephritidae usually located at about two-thirds of the wing length. It has also been called the *hind posterior crossvein, lower crossvein, M crossvein, median crossvein, posterior crossvein, posterior transverse vein, vein im, vein m*, or *vein tp.*

***Discoidal cell*** An old term for the discal medial cell (dm).

**Distiphallus** (Figure 33.5, distph) The main part of the male phallus in Tephritidae. It includes an elongate basal part and an expanded, apical glans. The basal part, which coils at rest, has a pair of weak, elongate sclerites on the outer (posterior) side and numerous membranous folds on the inner (anterior) side. The term "distiphallus" has also been used to mean only the glans, but that homology is incorrect (V.A. Korneyev, personal observation).

**Dorsal area (larva)** (Figure 33.7, d area) The area above the posterior spiracle on the caudal segment.

**Dorsal bridge (larva)** (Figure 33.6, d brg) Part of the pharyngeal sclerite that anteriorly joins the dorsal cornua.

**Dorsal cornu** (plural: **dorsal cornua) (larva)** (Figure 33.6, d corn) A paired, dorsal, wing-like portion of the pharyngeal sclerite. It is frequently sclerotized and often cleft on the outer margin.

***Dorsal sense organ (larva)*** See antenna.

**Dorsal sensilla (larva)** (Figure 33.7) Two pairs of sensilla, each often associated with a papilla, located on the dorsal area of the caudal segment. The individual sensilla have been termed D1 and D2.

**Dorsocentral setae** (Figure 33.2) In Diptera, a series of paired setae between the acrostichal and intra-alar series. Tephritids have at most two pairs. There is usually one postsutural dorsocentral seta, usually termed simply the dorsocentral seta (dc s). It is rarely absent (e.g., *Bactrocera* and *Dacus* spp.). A few tephritids (e.g., *Chaetorellia* spp.) have a presutural dorsocentral seta (presut dc s) in addition to the postsutural seta. The relative position of the postsutural dorsocentral seta compared with the postsutural supra-alar seta is of some use in the higher classification of tephritids; in Tephritini the dorsocentrals are usually placed in front of an imaginary line between the supra-alar setae; in the Trypetinae and Terelliini they are usually on or behind that line. The dorsocentral seta should not be confused with the acrostichal seta.

**Dorsolateral group of sensilla (larva)** This term was introduced by Singh and Singh (1984) to refer to the three sensilla of the maxillary sense organ external to the sclerotized ring of the maxillary palp; their innervation suggests derivation from both antenna and maxillary elements. These sensilla have also been referred to as the *lateral sense organ* (e.g., Headrick and Goeden 1990a).

**Droplet formation (behavior)** Regurgitation of a small droplet of the crop contents, usually a clear fluid, held on the rostrum. Evaporation of water and subsequent concentration of nutrients is the likely purpose. This process may also result in evaporative cooling of the head. *Bubbling* is used as a synonym in much of the tephritid literature.

**Ecdysial scar (larva)** (Figure 33.7, ecdys sc) A mark on the posterior spiracle which marks the area occupied by the spiracle of the previous instar.

***Emplexis*** See microtrichia.

**Enantion (behavior)** See wing displays.

**Epandrium** (Figure 33.5, epand) The male abdominal tergite 9, which in Teprhitidae has a broad, inverted-U shape and bears paired surstyli ventrally or apicoventrally.

**Epipharyngeal sclerite (larva)** A minute, horizontal, triangular sclerite located dorsal to the hypopharyngeal sclerite.

***Epistome*** See lower facial margin.

**Eversible membrane** (Figure 33.4, ev memb) The membranous part of the ovipositor between the oviscape and the aculeus. It is derived anteriorly from segment 7, and posteriorly from the intersegmental membrane between segments 7 and 8 (Foote and Steyskal 1987). It and the aculeus are normally retracted inside the oviscape, but they evert, at least partially, during oviposition and copulation. Basally the eversible membrane usually bears a dorsal and ventral pair of short sclerites, termed taeniae. The eversible membrane also bears minute, toothlike scales or denticles, which may be simple, multidentate, or comblike. In *Anastrepha* and *Toxotrypana*, a group of dorsobasal scales, varying in number and arrangement, are greatly enlarged. Stone (1942) used the term *rasper* for this group of teeth, but their function for rasping is unproven and use of this term has been largely abandoned. Foote and Steyskal (1987) used the term more broadly for all of the denticles of the eversible membrane. The eversible membrane has also been called the *eversible ovipositor sheath*, *inversion membrane*, *ovipositubus*, or *segment 8* (see Norrbom and Kim 1988).

***Eversible ovipositor sheath*** See eversible membrane.

**Eye stalk** A lateral enlargement of the part of the head that bears the eye. A few tropical genera of Tephritidae have eye stalks, but their antennae are still close together, unlike in Diopsidae in which the antennae are near the ends of the stalks. In *Pelmatops* (Adramini) the stalks are very long in the male and short in the female. In males of *Themara* (Acanthonevrini) stalk size varies among species, from little more than a broadened head to well-developed stalks, each longer than the width of the vertex.

**Facial carina** A keel-like medial protrusion of the face. It has erroneously been called the *clypeal ridge* (e.g., Stone 1942).

***Facial mask (larva)*** See mask.

**Female reproductive system** The female reproductive system includes the external parts comprising the ovipositor, and internal parts, including paired ovaries and lateral oviducts, a common oviduct, 2-4 spermathecae and their ducts, accessory glands, and the genital chamber which bears the ventral receptacle and opens into the cloaca. See Dean (1935), Hanna (1938), Drew (1969), Dodson (1978), or De Carlo et al. (1994) for further details.

***Female terminalia*** See ovipositor.

**Fertilization (behavior)** The fusion of a sperm and the egg pronucleus. Fertilization does not usually occur immediately after sperm transfer. These actions are not part of the same behavior sequence because the female may store sperm in her spermathecae, and the eggs are not fertilized until she is ready to oviposit. Typically, sperm is stored several hours or days, but presumably it can be stored for several months (Myers et al. 1976).

***Fifth vein*** An old term that usually meant vein $Cu_1$.

**First flagellomere** (Figure 33.1, flgm 1) In Tephritidae, the apparent third segment of the antenna (see arista). It has also been called the *postpedicel*. It may be apically pointed (e.g., in many Carpomyini and Gastrozonini) or elongate (e.g., in most Dacini).

***First posterior cell*** An old name for cell $r_{4+5}$ (see radial cells).

***First vein*** An old term that usually meant vein $R_1$.

**Flagellomere** See first flagellomere.

***Fourth vein*** An old term that usually meant vein M.

**Frons** (Figure 33.1, fr) The anterodorsal area of the head, bounded laterally by the eyes, posteriorly by the vertex, and anteriorly by the antennae.

**Frontal setae** (Figure 33.1, fr s) The row of setae next to each eye on the lower part of the frons. In Tephritidae they are usually all inclinate, and most species have between one and five pairs (usually two to three), that is, one to five setae next to each eye. They have also been called *inferior* or *lower fronto-orbital setae*.

**Frugivore/frugivorous (behavior)** A species whose larvae feed on fruit. They may attack the fleshy part of the fruit, the seeds at various stages of development, or both. Most, although not all, of the economically important species of Tephritidae are frugivorous.

**Gena** (plural: **genae**) (Figure 33.1, gn) The area ventral to the eye, posterior to the parafacial and facial ridge, and anterior to the postgena.

**Genal seta** (Figure 33.1, gn s) In general, any setae on the gena. In Tephritidae there is usually one seta larger than the surrounding setulae that is called the genal seta. It should not be confused with the well-developed subvibrissal setae along the anteroventral margin in genera such as *Chetostoma*.

**Genital opening** In Tephritidae this occurs where the genital chamber opens into the cloaca. This term has often been misused for the cloacal opening in Tephritidae.

**Glans** (plural: **glandes**) (Figure 33.5, gls) The apical, expanded part of the male distiphallus. It usually contains complex internal sclerotization (see acrophallus and praeputium) and may have a membranous vesica and a subapical lobe of various shapes. In many Trypetinae the glans has a membranous basal lobe (the *preglans lobe* of Korneyev 1996), usually covered with minute spicules, which in *Ceratitis capitata* expands and flexes during copulation, apparently to help insert the phallus by moving the glans through the female cloaca and genital chamber (Eberhard and Pereira 1998). The glans has also been called the *aedeagal glans* or the distiphallus, but it is only part of the latter.

***Gnathocepalon* (*larva*)** See cephalic segment.

**Greater ampulla** (Figure 33.2, gr amp) A small, dome-shaped area on the anterodorsal part of the anepimeron in front of the wing base. It is variably produced, but present, in most Tephritidae.

**Ground plan (phylogenetic analysis)** The set of character states hypothesized to have been present in the most recent common ancestor of a monophyletic group.

***Gynium*** See aculeus.

**Hamation (behavior)** See wing displays.

**Hennig86 (phylogenetic analysis)** A DOS software program for cladistic analysis written by J.S. Farris.

**Holotype (nomenclature)** See type specimen.

**Homonym (nomenclature)** If the same name has been used for more than one taxon (e.g., species, genus or family) it is a homonym. For example, *Dacus humeralis*, from Africa, was described in 1915 by Bezzi; *Chaetodacus humeralis*, from Australia, was described in 1934 by Perkins. When transferred to *Dacus* the younger Australian species name became a junior homonym of the more senior African name. Consequently, the Australian species had to be given a new name; Hardy (1951) gave it the name *D. neohumeralis,* which is still the valid name even though it is now called *Bactrocera neohumeralis*.

**Homoplasy (phylogenetic analysis)** In cladistic analysis, when different characters support conflicting hypotheses of relationship among a group of taxa, at least one of these characters is homoplasious. This is the result of convergent evolution of similar attributes (i.e., two independently evolved character states are erroneously perceived as one and coded the same), or due to loss or reversal to the plesiomorphic state in one or more taxa. Homoplasy can only be detected after the analysis, that is, a particular character is considered homoplasious only if one or more of its states evolve more than once or reverse on the most parsimonious tree derived from the entire character matrix.

**Host plant (behavior)** A plant that is used as a food source, in Tephritidae usually only by the larvae. It should be noted that some fruit fly species have been forcibly reared under artificial conditions from plants that should not be considered natural hosts. Furthermore, any species of plant upon which an adult merely is collected or observed is not necessarily a host plant, and no plant should be called a host without evidence of larval feeding. Many aspects of the environment factor into the behavior of adult tephritids, but none so much as their host plants. The importance of the host plant in tephritid reproductive behavior was emphasized by Bush (1966; 1969a), who hypothesized the role of plant chemicals as providing the secondary sexual compounds used by the sexes for indentification of conspecifics.

**Humeral band** or **crossband** See crossband.

***Humeral callus*** or ***lobe*** See postpronotal lobe.

***Humeral seta*** An old term for the postpronotal seta.

***Humeralis*** A term used by Munro (1984) for the section of costa between the costagial and humeral breaks.

**Hyaline** Clear, as in an unpatterned part of a wing. A hyaline wing means one with no markings.

**Hypandrium** (Figure 33.5, hypd) The male abdominal sternite 9, which in Tephritidae is a slender, semicircular sclerite closely associated with the phallapodeme. It usually bears an anterior hypandrial apodeme and basolateral processes which have been termed lateral sclerites (possibly derivatives of the pregonites; V.A. Korneyev, personal communication). The latter are not always fused to the hypandrium (e.g., in *Anastrepha*). The hypandrium has also been called the *genital ring*.

**Hypopharyngeal sclerite (larva)** (Figure 33.6, hyphar scl) An H-shaped sclerite consisting of two elongate, sclerotized, posteriorly directed plates connected by a small crossbar. It articulates anteriorly with the mouthhooks.

***Hypopleural calli*** See anatergite.

***Hypostomal sclerite (larva)*** See hypopharyngeal sclerite.

**Incertae sedis (nomenclature)** Of uncertain taxonomic position. For example, a genus belonging to the subfamily Tephritinae whose tribal classification is unclear would be treated under "Tephritinae, incertae sedis."

***Inner vertical seta*** See vertical setae.

***Inferior fronto-orbital setae*** See frontal setae.

***Inferior orbital setae*** See frontal setae.

***Inner surstylus*** See medial surstylus.

***Intercalary band*** Another name for the accessory costal crossband. See crossband.

**Intermediate area (larva)** (Figure 33.7, i area) An area on the caudal segment, slightly lateral to the midline, between the posterior spiracle and the ventral area; often protuberant and in some species (e.g., *Bactrocera cucurbitae*), almost linked by a pigmented transverse band.

**Intermediate sensilla (larva)** (Figure 33.7) Four pairs of sensilla located on the intermediate area of the caudal segment, often associated with papillae and/or tubercles. The individual sensilla have been termed I1a, I1b, I2, and I3.

**Intra-alar setae** (Figure 33.2, ial s) A series of setae between the dorsocentral and supra-alar series; tephritids have only one pair of intra-alars, placed near the level of the acrostichal setae.

**Intrapostalar seta** (Figure 33.2, ipal s) A seta near the posterior margin of the scutum, very slightly lateral to the dorsocentral line.

**Intromission (behavior)** The insertion of the phallus by the male into the cloaca and then the genital chamber of the female. This process was described by Eberhard and Pereira (1998). Also see copulation.

***Inversion membrane*** See eversible membrane.

***Juxta*** See vesica.

**Katatergite** (Figure 33.2, ktg) The lateral thoracic sclerite anteroventral to the anatergite and between the wing base and the posterior spiracle. In Tephritidae it is more produced than surrounding sclerites. Also see anatergite.

**Katepisternum** (Figure 33.2, kepst) The triangular sclerite between the coxae of the fore- and midlegs. Most species have a well-developed katepisternal seta (Figure 33.2, kepst s) near the posterodorsal corner, but it may be absent (e.g., in many Adramini and Dacina). This sclerite has also been called the *sternopleuron*.

**Keilin's organ (larva)** One of a pair of trifid sensilla on the ventral part of each thoracic segment, sometimes represented externally only by a pit.

***Labial lobe (larva)*** A term used by Belcari (1989), as "lobo labiale," for the labium and for the median oral lobe.

**Labial sclerites (larva)** (Figure 33.6, lab scl) Two sclerites forming a V-shape in the floor of the mouth between the hypopharyngeal sclerites and the mouthhooks.

**Labium (larva)** (Figure 33.7, lab) A large triangular fleshy lobe on the ventral margin of the mouth.

**Lamprine** Munro (1984) used this term to refer to the smooth area between the pecten and the margin of tergite 3 in *Bactrocera* and *Dacus* spp.

**Lanceolate seta** In many Tephritinae, a flattened or spindle-shaped seta that is wider at its middle than its base (Freidberg and Kugler 1989). Also see acuminate seta.

**Lateral area (larva)** (Figure 33.7, l area) The area lateral to the posterior spiracles on the caudal segment.

**Lateral sclerite** (Figure 33.5, l scl) A slender, paired sclerite of the male genitalia, possibly derived from the pregonite (V.A. Korneyev, personal communication). It is usually basally fused to the hypandrium, but may be separate (e.g., in *Anastrepha*). Apically each lateral sclerite articulates with the apex of a lateral arm of the phallapodeme. Usually the right lateral sclerite is longer than the left.

***Lateral sense organ (larva)*** See dorsolateral group of sensilla.

**Lateral sensillum (larva)** (Figure 33.7, L) A sensillum located on the lateral region of the caudal segment, often associated with a papilla.

**Lateral surstylus** (plural: **lateral surstyli**) (Figure 33.5, l sur) The more lateral of the two paired surstyli in the male genitalia (see surstylus). In Tephritidae the lateral surstylus is fused to the epandrium, sometimes to such an extent that the limits of these sclerites are unclear. In many Tephritidae the lateral surstylus has two lobes: an anterior lobe, sometimes called the *mesal lobe* (e.g., Norrbom 1994), which is recognizable by the presence of denticles and usually one or more sensilla; and a posterior lobe (Jenkins 1990). Occasionally it has a third, medial lobe. In Tephritinae it often has a posteriorly directed dorsal lobe basally, sometimes called the *flange* (Munro 1957).

**Lateral vertical seta** (Figure 33.1, l vt s) See vertical setae.

**Lectotype (nomenclature)** See type specimen.

**Leg-lock (behavior)** Males of some species grasp the female with their legs until she becomes quiescent for later mounting (Dodson 1987a, b).

**Lek (behavior)** A group of males defending territories for mating purposes. Lek has also been used to describe a type of mating system or strategy. Wilson (1975) defined a lek as a "communal display area where males congregate for the sole purpose of attracting and courting females and to which females come for mating." In a discussion of acalyptrate mating systems, Burk (1981) defined a lek as "all aggregations of displaying males away from female-required resources." These definitions contain components that are not yet clearly defined and thus uninterpretable. The term lek has been broadly applied in the literature to a vast array of taxa. Any attempt to use the term lek relative to tephritids should take into account the role of the host plant in distinguishing a communal display area for the sole purpose of reproduction away from a female-required resource. In each published description of a male defending a territory, of which there are only a few complete reports in the literature, the territory's boundaries were undefined in space and time. Yet, however nebulous the boundaries of these territories, some male tephritids do defend areas within the context of reproduction. Nonfrugivorous tephritid males are not known to exhibit communal displays (Headrick and Goeden 1998) and thus the term lek has not been used to describe the area where they occur or the mating system employed. Some tropical species (some *Anastrepha* and *Ceratitis*) have been shown to exhibit lek mating systems. Thus, the application of the term lek should be made on a case-by-case basis with a clear statement pertaining to its use (i.e., an area or a mating system).

**Lofting (behavior)** See wing displays.

**Lower facial margin** The lower anterior part of the head, below the face and above or in front of the mouth opening. The lower facial margin is the *epistome* of many authors.

***Lower fronto-orbital setae*** See frontal setae.

***Lower orbital setae*** See frontal setae.

**Lunule** (Figure 33.1, lun) The semicircular plate above the antennal bases and below the ptilinal fissure.

***M crossvein*** An old term for the discal medial-cubital crossvein (DM-Cu) (e.g., Hardy 1973).

$M_{3+4}$ An old abbreviation for the basal medial-cubital crossvein plus the first cubital vein ($Cu_1$) (e.g., Hardy 1973).

**Majority rule consensus tree (phylogenetic analysis)** A consensus tree including clades found in more than 50% of the most parsimonious trees (Margush and McMorris 1981) resulting from an analysis. Each clade is usually marked with the percentage of the trees in which it is found; only those marked 100% occur in all of the most parsimonious trees. Use of these trees is controversial and they are often misinterpreted.

**Male display behavior (behavior)** Male displays involve the production of visual, olfactory, and/or auditory stimuli to attract conspecific individuals. Aggregation displays attract both sexes, whereas courtship displays attract or influence only females. Some specific display behaviors may be used for both purposes, but others are used only in courtship.

A. **Male aggregation displays** Behaviors used by a male to attract conspecific males and females. This also includes territoriality (see territory). The following are terms for each type of sensory stimulus produced or used by the male during aggregation behavior. These displays also may be used in courtship in some species.

1. **Male calling** This term has been used for the entire sequence of male display behaviors (Burk 1981), but as used here refers only to the production of auditory stimuli for the purpose of attracting mates. Only a few tephritid species have been shown to use sound as part of their courtship ritual (Keiser et al. 1973; Burk and Webb 1983; Kanmiya 1988). Male calling does not include stridulatory noises that are part of defensive behavior used by both sexes in conjunction with wing displays and lunging. The relationship between defensive buzzing and stridulation has not been established (Nation 1972; Keiser et al. 1973; Tyschen 1977; Burk and Webb 1983). See precopulatory song.

2. **Olfactory stimuli** Production of sexual attractants or pheromones has been suspected to occur in tephritids, yet few reports of experimental evidence of pheromones are available (see Heath et al., Chapter 29).

a. **Abdominal pleural distension** One site suspected for pheromone release is the distended abdominal pleura of displaying males. The pleura of the segments that distend (segments 3 to 5) are morphologically distinct from those of the other abdominal segments (Pritchard 1967; Nation 1981; Headrick and Goeden 1994). They are two to three times thicker and are filled with cavernous spaces giving them a spongy appearance. These intercellular areas become filled with hemolymph, which causes the pleural walls to distend. The function of abdominal pleural distension remains unknown. It may provide a larger surface area for the volatilization of a pheromone or help in the release of pheromone from glandular cells within the pleural wall. This behavior has also been called *abdominal inflation* (Jenkins 1990).

b. Other reports have implicated enlarged sacs located in the abdomen and rectum of males and salivary glands as producing pheromones (Lhoste and Roche 1960; Fletcher 1969; Economopoulos et al. 1971; Arita and Kaneshiro 1986; Little and Cunningham 1987). Glands in the head, with external pores near the base of the rostrum, have also been identified as possible pheromone sources (Headrick and Goeden 1990).

3. **Visual display** The use of body parts by a male in a characteristic or stylized manner to which the female responds during courtship. Male visual displays include the use or movement of the legs, wings, abdomen, or entire body. The most conspicuous and ubiquitous visual displays involve the wings. However, wing displays also are used in most other aspects of tephritid behavior and only certain species use unique wing displays during courtship (see wing displays).

B. **Male courtship displays** Behaviors used by a male to attract or influence a conspecific female. These displays often include a continuation of one or more of those described under aggregation displays and additionally one or more of the following behaviors.

1. **Middle leg abduction** The middle legs are abducted in an extended position in a plane perpendicular to the long axis of the body through an arc of ~60%. The periodicity of abduction depends on the species.
2. **Body swaying** Movement of the body from side-to-side over the legs in a plane parallel to the substrate (i.e., the tarsi of each leg remain in one position), while the body moves from one side to the other. The entire body can be swayed, or the anterior end may pass through an arc while the posterior end remains stationary.
3. **Mating trophallaxis** The regurgitation and exchange of a fluid, or nuptial gift, between a male and a female during reproduction (Freidberg 1981). Pre- and post-mating trophallaxis involve the exchange of fluid during different parts of the mating sequence. Typically, a nuptial gift is part of the male courtship display, and is deposited on the substrate by a courting male and consumed by the female (i.e., indirect trophallaxis), or it is exchanged directly through labellar contact between males and females (Stoltzfus and Foote 1965; Freidberg 1981; Headrick and Goeden 1994). See Sivinski et al., Chapter 28.

**Male genitalia/male terminalia** (Figure 33.5) In male Tephritidae, the genitalia are ventroapical and they may be hidden from above by the elongate tergite 5. The external parts include a broad epandrium, a slender hypandrium and associated lateral sclerites, an apical baglike proctiger, a paired lateral surstylus fused to the epandrium, a subepandrial sclerite joining a pair of medial surstyli that are closely associated with the lateral surstyli, an elongate phallus, and a phallapodeme with a pair of lateral arms. In Ulidiidae, some Platystomatidae and Pyrgotidae there are small, buttonlike, rudimentary gonostyli (= parameres of McAlpine 1981), each with four to five sensilla. In Tephritidae, they are usually completely absent, but are present in *Tachinisca* (V.A. Korneyev, personal observation).

**Male reproductive system** The male reproductive system includes the external male genitalia and the following internal parts: paired testes and vas deferens, several accessory glands, and an ejaculatory apodeme, sperm sac, and ejaculatory duct which connects to the phallus. See Hanna (1938), Drew (1969), Dodson (1978), or De Marzo et al. (1976) for further details.

***Mandible (larva)*** See mouthhook.

**Manuscript name (nomenclature)** A scientific name which has not been properly published. Such names are not valid nor even available names for taxa and their creation should be avoided.

***Marginal cell*** An old name for cell $r_1$ (see radial cells).

**Mask (larva)** (Figure 33.7) The area on the head surrounding the antenna, maxillary sense organ, and part of the mouth (Kandybina 1977).

**Mating (behavior)** All behaviors performed by males and females to acquire mates for the purpose of reproduction.

**Mating mistakes (behavior)** Any attempt by a tephritid to mate with any insect other than a conspecific individual of the opposite sex. Males may mount and attempt to mate with conspecific males instead of females (Prokopy and Bush 1973). This has been called homosexual behavior (Tauber and Toschi 1965), but that term implies that the males actively sought other males for sexual purposes. True homosexuality has never been demonstrated in Tephritidae, only various levels and abilities of discrimination by males.

**Maxillary palp (larva)** The part of the maxillary sense organ contained within a sclerotized ring and including 11 sensilla. Also termed *posterior sense organ* (e.g., Snodgrass 1924; Phillips 1946; Exley 1955), *ventral sense organ* (e.g., Snodgrass 1953; Novak and Foote 1968), *terminal sense organ* (e.g., Bolwig 1946; Chu-Wang and Axtell 1972), or *maxillary (terminal) sense organ* (Singh and Singh 1984) (see Carroll 1992).

**Maxillary sense organ (larva)** (Figure 33.7, mx sen org) The maxillary palp plus the dorsolateral group of three sensilla. This organ has also been called the *anterior lobes* (Headrick and Goeden 1990a). In early works this term may refer to the maxillary palp.

**Media/medial vein** (Figure 33.4A, M) Technically the posterior medial vein, but in Diptera abbreviated for simplicity as vein M because the anterior branch of the media (MA) is very small. Although it is unbranched in most Cyclorrhapha, vein M may have up to three branches in other Diptera, and the single vein in Tephritidae has therefore sometimes been named $M_1$ or $M_{1+2}$, but that convention is not followed here. This vein has also been called the *discoidal vein* or *4th longitudinal vein*.

**Medial cells** (Figure 33.4A) There are three cells posterior to vein M: the basal medial cell (bm), discal medial cell (dm), and medial cell (m). The latter has also been called the *apical medial cell, cell am, 2nd* $M_2$, or the *second posterior cell.*

**Medial surstylus** (plural: **medial surstyli**) (Figure 33.5, m sur) In the male genitalia, the slender lobe connected basally to the subepandrial sclerite and usually closely associated with the lateral surstylus. The limits and homology of the medial surstylus and subepandrial sclerite are not well understood, but at least the former appears to be derived from the single surstylus of other acalyptrate flies (V.A. Korneyev, personal observation). Subapically each medial surstylus bears a pair of prensisetae. It has also been called the *inner surstylus*.

**Medial vertical seta** (Figure 33.1, m vt s) See vertical setae.

**Median oral lobe (larva)** An elongate, partially sclerotized lobe between the mouthhooks in Tephritinae (Headrick and Goeden 1990a), and apparently also some Trypetinae (e.g., Trypetini). It consists of a sclerotized dorsal rib that may be homologous with the epipharyngeal sclerite, and a ventral lobe of uncertain homology, perhaps derived partially from the labium, with which it is closely associated.

**Mediotergite** (Figure 33.2, mtg) The sclerite below the scutellum and subscutellum. Some authors have called this the *metanotum* (e.g., Stone 1942), *mesophragma*, or the postnotum, but it is only a part of the latter.

**MEGA (phylogenetic analysis)** Molecular Evolutionary Genetics Analysis, a software program for distance analysis of systematic and population genetics data sets written by S. Kumar, K. Tamura, and M. Nei (DOS version).

**Mesonotum** (Figure 33.2) In flies most of the thorax is derived from the mesothorax, and the mesonotum, which includes the scutum, scutellum, and postnotum, forms most of the dorsum. In Tephritidae, the postpronotal lobes are the only dorsally visible parts of the thorax that are not part of the mesonotum. The length of the mesonotum in dorsal view is that of the scutum + scutellum (the latter projects over the postnotum).

***Mesopleural stripe*** See anepisternal stripe.

***Mesopleuron*** See anepisternum.

***Metanotum*** See mediotergite.

**Micropyle (egg)** A small, often nipplelike, structure at the anterior end of the egg where the spermatozoan enters.

**Microtrichia/microtrichose** (Figure 33.2C) Usually minute projections of the cuticle that lack alveoli. Microtrichia are usually hairlike or scalelike under compound or scanning electron microscopy. They may cover parts of the wing membrane, making them look darker, or parts of the body. On the body, depending upon their shape and density, they may produce a subshining, silvery or other colored, or dull, matt appearance. Many species have patterns, especially on the scutum, of bare and microtrichose areas, or due to variation in microtrichia density or shape. The appearance of these patterns may change depending on the angle of view, and they are usually not visible in specimens in fluid or may be obscured in specimens dried from fluids. Males of most Dacini have an extra dense area of microtrichia around the end of vein $A_1$+$Cu_2$ that Munro (1984) called the *emplexis*, although that term has not been generally accepted. The terms *pollinose*, *pruinose*, and *microtomentose* have also been used to describe microtrichose areas. Also see seta. The scutal microtrichia should not be confused with the scutal setulae.

**Middle leg abduction (behavior)** See male display behavior.

**Monophagous (ecology)** Adjective meaning to feed on only one plant species; in Tephritidae meaning a species that breeds in only one species of host plant. See also oligophagous, polyphagous, and stenophagous.

**Monophyletic group (phylogenetic analysis)** A taxon including a common ancestor and all of its decendants (e.g., Mammalia). It is based upon synapomorphies. See paraphyletic group and polyphyletic group.

**Monotypy (nomenclature)** See type species.

***Morphospecies*** See species complex.

***Morula gland*** See ventral receptacle.

**Mounting (behavior)** A male positioning his body upon a female's body to attempt copulation.

**Mouthhook (larva)** (Figures 33.6 and 33.7, mh) A paired, curved, strongly sclerotized, hooklike sclerite found in Cyclorrhapha larvae. It articulates posteriorly with the hypopharyngeal sclerite and may have an additional preapical tooth. The mouthhook is often called the mandible (Teskey 1981) and may be homologous with that sclerite in other Diptera.

**Multilocular gall (botanical term)** A gall which has many chambers; not to be confused with capitula containing several unilocular galls that have fused.

**Multivoltine (ecology)** Having more than one generation per year.

**Neighbor joining (phylogenetic analysis)** A distance-based method of reconstructing evolutionary relationships that attempts to minimize the sum of all branch lengths on a bifurcating tree. It is not a parsimony-based method.

**Neotype (nomenclature)** See type species.

**New combination (nomenclature)** See combination.

**Node (phylogenetic analysis)** A branching point on a cladogram.

**Nominal species (nomenclature)** A species name. It may be the valid name for a biological species or an invalid name (a synonym or homonym).

***Notopleural suture*** See transverse suture.

**Notopleuron** (Figure 33.2, npl) A lateral sclerite on the mesonotum, derived from the scutum. In Tephritidae, it bears two setae, the anterior and posterior notopleural setae (npl s). The color of the notopleuron is sometimes used as a character (e.g., in *Bactrocera*), and the color of the posterior notopleural seta is a useful character within Tephritinae.

**Nuptial gift (behavior)** The fluid passed from the male to the female during mating trophallaxis (Thornhill and Alcock 1983) (see male display behavior). The contents of the clear and viscous fluid have not been identified. It is often churned into a froth by the action of the pseudotracheae of the labella of the male (Stoltzfus and Foote 1965; Headrick and Goeden 1994).

**Ocellar seta** (Figure 33.1, oc s) A paired seta in front of the posterior ocellus. It may be absent (e.g., in many *Bactrocera* and *Dacus* spp.), and is often reduced (e.g., *Anastrepha*, *Trypeta*).

**Ocellar triangle** (Figure 33.1, oc tr) The subtriangular area that encloses the three ocelli, which are themselves arranged as a triangle.

**Oligophagous (ecology)** Adjective meaning to feed only on closely related plant species; in Tephritidae meaning a species that breeds in a limited range of closely related host plant species (e.g., all of its hosts are members of a single family). This is broader than stenophagous. See also monophagous and polyphagous.

**Oral ridges (larva)** (Figure 33.7, or rg) Several rows of ridges on each side of the mouth opening which may be entire (unserrated) or toothed on their lower (posterior) edge.

***Oral setae*** See subvibrissal setae.

**Orbital setae** (Figure 33.1, orb s) Setae on the upper, lateral part of the frons. In Tephritidae there are usually two, although the posterior pair are sometimes absent (e.g., in Myopitini), and there are rarely three (e.g., *Paracantha*) or none. They are usually reclinate, although the posterior seta is sometimes inclinate (e.g., Terelliini), and the anterior seta is proclinate in males of most species of *Ceratitis* (*Ceratitis*). These setae have also been called *superior* or *upper fronto-orbital setae*.

**Original designation (nomenclature)** See type species.

***Outer surstylus*** See lateral surstylus.

***Outer vertical seta*** See vertical setae.

**Outgroup comparison (phylogenetic analysis)** A method used in cladistics to assign character polarity, that is, which state is plesiomorphic and which state (or states if there are more than two) is apomorphic. The outgroup is a related taxon, ideally the sister group of the taxon being studied (ingroup). Of two or more states of a character occurring within the ingroup, that found in the outgroup is hypothesized to be plesiomorphic.

**Ovipositor** (Figure 33.4) The parts of the female abdomen including and apical to segment 7, which are the main parts used in oviposition. In Tephritidae these are highly modified into three main parts: a tubular or conical oviscape; an elongate, membranous eversible membrane; and a needlelike or bladelike aculeus. The eversible membrane and aculeus are normally retracted, telescope-like, within the oviscape (Figure 33.4D). Berube and Zacharuk (1983) and Stoffolano and Yin (1987) discussed the structure and operation of the ovipositor. The term ovipositor has been used by many American authors to mean only the aculeus. See Norrbom and Kim (1988) for discussion.

***Ovipositor piercer*** See aculeus.

***Ovipositor sheath*** See oviscape.

***Ovipositubus*** See eversible membrane.

**Oviscape** (Figure 33.4, ovscp) The basal, tubular or conical segment of the ovipositor, which is formed by the fusion of tergite 7 and sternite 7. Technically it is syntergosternite 7 (Norrbom and Kim 1988), but the shorter term oviscape has gained wider usage. It bears spiracle 7 basolaterally and large internal phragmata basally. It has also been called the *basal segment of the ovipositor* or, especially by American workers, the *ovipositor sheath*.

***Oviscapt*** A term which has been variously used in Tephritidae for the ovipositor, oviscape, or aculeus.

**Papilla** (plural: **papillae) (larva)** A small tubercle.

**Paralectotype (nomenclature)** See type specimen.

**Paraphyletic group (phylogenetic analysis)** An artificial taxon including a common ancestor and some but not all of its decendants (usually one or more highly derived taxa are excluded), for example, Reptilia, from which Aves (birds) was excluded although they are most closely related to some groups of dinosaurs. Paraphyletic groups are based upon symplesiomorphies or the lack of certain synapomorphies.

**Parastomal bars (larva)** (Figure 33.6, pastm b) Two long rod-shaped sclerites lying dorsally, parallel to the hypopharyngeal sclerite.

**Paratype (nomenclature)** See type specimen.

**Parsimony (phylogenetic analysis)** The principle that a simple hypothesis, requiring few if any assumptions, is preferable to a complicated one. In cladistic analysis, shorter trees requiring fewer steps, or character state changes, are said to be more parsimonious than, and preferable to, trees requiring more steps.

**PAUP (phylogenetic analysis)** Phylogenetic Analysis Using Parsimony, a software program for cladistic analysis and other systematic applications written by D.L. Swofford (Macintosh and DOS versions).

**Pecten** (Figure 33.5, pect) The row of setae on each side near the posterior margin of tergite 3 of the males of most *Bactrocera* and *Dacus* spp. Some authors call this a *comb*, whereas tergite 3 is said to be *ciliate* by some other authors. The pecten is separated from the margin by a narrow flat area called the lamprine.

**Pedicel** (Figure 33.1, ped) The second segment of the antenna.

***Penis*** See phallus.

***Peristomal hairs*** See subvibrissal setae.

**Peritreme (larva)** (Figure 33.7, perit) The plate surrounding the three rimae of the posterior spiracle. In almost all Tephritidae it is unsclerotized.

**Phallapodeme** (Figure 33.5, phapod) The sclerite which articulates with the base of the phallus (Cumming et al. 1995). In Tephritidae it bears, usually on its middle third, a pair of lateral arms or vanes that articulate apically with the lateral sclerites, which are usually fused to the base of the hypandrium. The lateral arms are sometimes fused basally to form a Y-shaped structure. The phallapodeme has also been called the *aedeagal apodeme* (McAlpine 1981) or *fultella* (Munro 1947).

**Phallus** (Figure 33.5, ph) The male intromittent organ in Cyclorrhapha (Cumming et al. 1995), which in Tephritidae consists of a short basiphallus and a usually very elongate distiphallus. The length of the phallus is usually correlated with that of the female oviscape. At rest it is coiled and stored in a pocket above the postabdomen and below tergite 5. It has also been called the *aedeagus* (e.g., Foote and Steyskal 1987).

**Pharyngeal sclerite (larva)** (Figure 33.6, phr scl) The largest and posteriormost sclerite of the hypopharyngeal skeleton. It includes paired dorsal and ventral cornua connected medially by the tentorial phragma. It has also been called the *tentoropharyngeal sclerite*.

**Phylogeny (phylogenetic analysis)** The evolutionary history of a group of organisms, often illustrated in the form of a branching diagram or tree.

***Piercer*** See aculeus.

***Pilidium*** A term used by Munro (1984) for the dorsal side of the abdomen.

**Plesiomorphy (phylogenetic analysis)** An ancestral or primitive trait or character state (see apomorphy).

**Pleuron** (plural: **pleura**) The lateral side of a body part or segment, as in the thoracic pleuron.

***Pleurotergite*** See anatergite.

***Pochette*** Munro (1984) used this term for a gland formed in a pocket-like fold between sternites 1 and 2 in the Dacini.

***Pollinosity/pollinose*** See microtrichia.

**Polychotomy/polytomy (phylogenetic analysis)** A monophyletic group of three or more taxa whose relationships are unresolved (on a cladogram represented as three or more branches meeting at the same node).

**Polyphagous (ecology)** Adjective meaning to feed on many plant species; in Tephritidae, meaning a species that breeds in a broad range of host plants belonging to unrelated groups (e.g., several plant families). See also monophagous, oligophagous, and stenophagous.

**Polyphyletic group (phylogenetic analysis)** An artificial group including two or more taxa that are not very closely related. It is based upon convergent characters (similar, but independently evolved, traits).

**Postalar seta** (Figure 33.2, pal s) The posterolateralmost seta on the scutum (Foote 1981). In some Diptera one or more postalar setae occur on a distinct postalar callus, but that callus is not differentiated in Tephritidae and so the homology of this seta is uncertain. Some authors have interpreted it as a posterior supra-alar (e.g., Drew 1989; White and Elson-Harris 1992), but it is more lateral than the supra-alar line (this is more obvious in lateral view) and in the typical position, on the posterolateral corner of the scutum, of a postalar seta.

**Posterior apical band** or **crossband** See crossband.

***Posterior cubital cell* (cup)** See basal cubital cell.

***Posterior sense organ* (*larva*)** See maxillary palp.

**Posterior spiracle (larva)** (Figures 33.6A and 33.7C-D, p spr) There are two functional pairs of spiracles in tephritid larvae. The pair of posterior spiracles is positioned on dorsal half of the caudal segment. Each spiracle has two to three spiracular openings (two in first and second instars, usually three in third instars), frequently arranged almost parallel to each other.

***Posterior supra-alar seta*** See postalar seta and supra-alar setae.

**Postocellar setae** (Figure 33.1, poc s) One to two pairs of setae behind the ocellar triangle.

**Postocular setae** (Figure 33.1, pocl s) The row of small setae behind each eye. These are usually acuminate and unicolorous yellow to black, but in many Tephritinae at least some of these setae are white and lanceolate.

**Postpronotal lobe** (Figure 33.2, pprn lb) The anterolateral "shoulder" of the thorax known in earlier works as the *humeral lobe* or *humeral callus*. In most Tephritidae it bears one postpronotal seta (Figure 33.2, pprn s), which has also been called the *humeral seta*.

***Postsutural vittae*** See scutal vittae.

**Praeputium** (Figure 33.5, prput) A chamber or sinus formed within the complex sclerotization of the glans by the two lateral flaps of the acrophallus (Korneyev 1996). Its inner surface is often sculptured (e.g., in the Trypetini; see Han, Chapter 11).

***Preapical crossband*** Another name for the subapical band. See crossband.

**Preapical tooth (larva)** (Figure 33.7, prap th) An additional tooth on the ventral surface of the mouthhook. More than one tooth may be present in some species.

**Precopulatory song (behavior)** Sound production by courting males generated by fanning or vibrating the wings, and in most Dacini, rubbing them against the pecten, a series of stiff setae on the abdomen. See male calling under male display behavior.

***Pre-mating period*** See reproductive period.

**Prensisetae** (Figure 33.5, prens) Highly modified, short, stout setae on the medial surstylus. In Tephritidae there are usually two on each medial surstylus, usually located subapically.

**Preoral lobes (larva)** (Figure 33.7, pror lb) A small series of lobes or ridges, with entire or serrated edges, one of which, the primary lobe, bears the preoral organ.

**Preoral organ (larva)** (Figure 33.7, pror org) A minute sensory organ at the anterolateral corner of the mouth usually bearing several sensilla (Kandybina 1977). It is usually borne on a small round or quadrate lobe (the primary preoral lobe). It has also been referred to as the *stomal organ*, *stomal sensory organ*, *secondary* [*maxillary*] *sense organ* (Teskey 1981), or *maxillary* (*ventral*) *sense organ* (Singh and Singh 1984).

***Preoral sense organ* (*larva*)** See preoral organ.

**Preoral teeth (larva)** (Figure 33.7, pror th) Small teeth or finger-like projections at the base of the preoral lobes found in most Carpomyini spp. (e.g., *Carpomya* and most *Rhagoletis*). They have also been called *stomal guards*.

***Prescutellar acrostichal seta*** See acrostichal seta.

***Prescutellar seta*** See acrostichal seta.

***Prestigma*** A term used by Munro (1984) for the section of the costa between the humeral and subcostal breaks.

***Presutural setae*** See supra-alar setae.

**Proclinate seta** Any seta that lean forward.

**Proctiger** (Figure 33.5, proct) In the male, the baglike structure attached posteriorly to the epandrium and subepandrial sclerite and bearing the anus. It is partly membranous, but ventrally and laterally it is weakly to moderately sclerotized, setulose and often microtrichose. It is sometimes involved in dispersion of pheromones produced by the rectum (see De Marzo et al. 1978). It has sometimes been called the cerci, and its sclerotized areas may be derived from them.

**Profile (phylogenetic analysis)** A set of trees with the same terminal nodes (usually the result of an analysis through a computer program, e.g., the profile of most parsimonious trees obtained after running Hennig86).

**Proovigenic (behavior)** Adjective describing a female that emerges with a complement of mature eggs ready for fertilization. Proovigenic females emerge and copulate at a time when their host plant is in a suitable stage for oviposition, thus their developmental history is tied to that of their host plant species (Bush 1966; Zwölfer 1974; Headrick and Goeden 1994; 1998). Also see synovigenic.

***Proximal subcostal band*** Another name for the subcostal band. See crossband.

***Pruinescence/pruinosity/pruinose*** See microtrichia. The term pruinose was used by McAlpine (1981) to describe body surfaces covered by microtrichia, but as noted by Sabrosky (1983), pruinose means covered with a white powdery substance and is no more appropriate than other terms such as *pollinose*.

***Pteropleuron*** An old term for the anepimeron.

**Pterostigma** (Figure 33.4A) A sinus in the apical part of the subcostal cell that is often more opaque than surrounding areas. In Tephritidae it is the part of the cell distal to the bend of vein Sc. It has also been called the *stigma* or *mediastinal cell.*

**Ptilinal fissure** (Figure 33.2, ptil fis) The inverted U-shaped slit which runs above the antennal bases, ending in the genal grooves. It marks the edge of the sclerite including the face and lunule that is pushed forwards when the ptilinum is expanded.

**Ptilinum** A sacklike structure which is inflated by the adult as a mechanism for bursting the puparium. It folds back inside the head soon after the fly emerges.

**Pupariation (behavior)** The formation of the puparium.

**Puparium (larva/pupa)** The hardened skin of the last larval instar within which pupation takes place.

**Radial cells** (Figure 33.4A) In Tephritidae, there are four cells defined by the radial veins: the basal radial cell (br), and cells $r_1$, $r_{2+3}$ and $r_{4+5}$. The latter two abbreviations are sometimes shortened to $r_3$ and $r_5$, respectively. Cell $r_1$ has also been called the *marginal cell* or subcostal cell, cell $r_{2+3}$ has been called the *submarginal cell*, and cell $r_{4+5}$ has been called the *1st posterior cell.*

**Radial vein/radius** (Figure 33.4A, R) In Diptera, this vein may have up to five branches, but in Tephritidae only three are present. In Tephritidae it first divides into $R_1$ (technically RA) and the radial sector (Rs), which divides into veins $R_{2+3}$ and $R_{4+5}$. The setation of the radial vein is taxonomically useful within the Tephritidae. $R_1$ is finely setulose on the entire dorsal and sometimes part of the ventral side, and $R_{2+3}$ and/or $R_{4+5}$ may have setulae on either side. $R_1$, $R_{2+3}$ and $R_{4+5}$ have also been called the *1st*, *2nd,* and *3rd longitudinal* *veins*, respectively, and when unbranced as in Tephritidae, $R_{2+3}$ and $R_{4+5}$ are sometimes shortened to $R_3$ and $R_5$.

**Radial-medial band** or **crossband** See crossband.

**Radial-medial crossvein** (Figure 33.4A, R-M) The transverse vein connecting veins $R_{4+5}$ and M. It has also been called the *anterior crossvein, anterior transverse vein, median crossvein*, or *upper crossvein* or abbreviated as *i-m* or *ta*.

**Radiate pattern** (Figure 33.3G) A wing pattern with the middle largely patterned, with rays extending to the margins.

***Rasper*** See eversible membrane.

**Ray** (Figure 33.3G-H, r) On a wing with a radiate or stellate pattern, a usually short band running from the main part of the pattern to or toward the wing margin.

**Receptacle (botanical term)** In Asteraceae, the solid, sometimes fleshy, basal part of the capitulum, beneath where the seeds form.

**Reclinate seta** Any seta that leans backward.

**Reproductive behavior (behavior)** All behaviors performed by males and females for progeny production. This includes mating behaviors as well as oviposition and associated behaviors (e.g., defense of oviposition sites). Reproductive behavior is sometimes applied to only a part of a sequence (e.g., courtship, copulation) and should remain as a broad conceptual term rather than applied to specific behaviors or sequences.

**Reproductive isolation (behavior)** A condition in which interbeeding between two or more populations is prevented by intrinsic factors, called reproductive isolating mechanisms. It is the basis for the biological species concept. This term has several synonyms, for example, ethological isolation and prevention of gene flow. Reproductive isolation is a populational phenomenon and results from the sequential process of individuals determining the correct conspecific sexes for reproduction. Thus, species recognition and mating barriers contribute to reproductive isolation. Wing patterns have been assumed to be the primary conspecific recognition factor in the Tephritidae (Bush 1966; 1969a; Tauber and Toschi 1965; Tauber and Tauber 1967).

**Reproductive period (behavior)** Reproduction can be divided into discrete periods based on the life-history strategy of the tephritid species:

1. **Pre-coitus** The period from adult eclosion to first mating; it is typically applied to the female. Precoitus is used to avoid confusion with the term premating, which has been used as an equivalent term for courtship, for any period before the act of mating, and for any period of time between subsequent matings. Precoitus is an especially useful term with tropical species, multivoltine species, or species that have a host plant in a suitable stage for oviposition at the time of eclosion. Precoitus distinguishes the period up to the first mating from the period(s) between subsequent matings. Female tephritids can mate multiple times (Prokopy and Hendrichs 1979; Headrick and Goeden 1994). Precoitus may be a confusing term because synovigenic tephritids may enter a state of reproductive diapause in which the ovaries of the female will not develop or the eggs will be resorbed if ovipositional sites are unavailable (Goeden 1990a, b). This is usually due to environmental factors, as adults of many species will either estivate over the long summers, as in southern California, and/or diapause over winter until their host plants have reached the proper stage for oviposition (Headrick and Goeden 1994). In these cases, the precoitus period could last several months until the flies return to their host plants to begin reproduction. Precoitus refers to the period from adult eclosion to the time of the first mating, irrespective of duration.
2. **Postcoitus** The period from the end of a mating episode to oviposition. Again, depending on the seasonality of a given species or the state of development of the ovaries, this can represent a few hours to several months.

3. **Postovipositional period** The period between oviposition and the next mating episode. Multiple matings are important to the reproductive potential of several tropical species, including *Ceratitis capitata* (Prokopy and Hendrichs 1979; Myers et al. 1976), because the female will not mate for up to 2 weeks, depending on the species, until she is again ready for or requires insemination.
4. **Intercopulatory period** The period between subsequent matings and before oviposition. It includes the periods between repeated copulations in a single mating episode. An intercopulatory period may be a subunit of the postcoitus period.

**Retention index (phylogenetic analysis) (ri)** A measure, on a scale from 0 to 1, of the amount of homoplasy (specifically the amount of apparent vs. actual synapomorphy) in a character or cladogram. Those with little homoplasy have high values (1 means no homoplasy), and those with more homoplasy have low values. The retention index differs from the consistency index in that it accounts for how many taxa have the derived state. It is most commonly reported for the most parsimonious tree or set of trees resulting from an analysis. See Farris (1989).

**Reticulate pattern** (Figure 33.3I-J) A wing pattern that is netlike, with many hyaline spots against a dark background, or more hyaline with numerous, often connected, dark spots.

***Rhagoletis*-type pattern** (Figure 33.3A) A wing pattern with many or all of the following crossbands: subbasal, discal, accessory costal, subapical, and anterior and posterior apical (Foote 1981). See crossband.

**Rima (larva)** (Figure 33.7, rm) The marginal supporting sclerotization of each spiracular opening.

**S-band** (Figure 33.3D, SB) A band in most *Anastrepha* spp. formed from a strongly oblique radial-medial band and the anterior apical band. It runs from cell bcu, diagonally across crossvein R-M to join the costa in the apical part of cell $r_1$, and then follows the edge of the wing to the wing apex. It may be joined to the costal band and/or the V-band.

**Scapular setae** (Figure 33.2, scap s) Small setae, only slightly larger than the scutal setulae, near the anterior margin of the scutum (Munro 1947). There are often one to two pairs in Tephritidae, although they are poorly differentiated or absent in most Tephritinae.

**Scutal stripes** or **vittae** Diverse scutal color patterns occur across the Tephritidae, and dark stripes are not uncommon. Many Tephritidae (e.g., *Rhagoletotrypeta*, most *Bactrocera*, *Dacus* and *Anastrepha* spp.) have one to three pale or sometimes bright yellow or white stripes or vittae whose color appears to be determined by internal tissues and often changes with death (see xanthine). These may include a paired sublateral vitta (sometimes called the lateral vitta) that extends from or sometimes partially along the transverse suture back to or toward the intra-alar seta. Many species have an unpaired medial vitta in addition to or independent of the sublateral vittae. Frequently there are only postsutural vittae, but sometimes the vittae, especially the medial vitta, extend anteriorly beyond the transverse suture. Also see vitta.

**Scutellum** (Figure 33.2, sctl) The triangular or semicircular sclerite posterior to the scutum. In Tephritidae it bears one to two, or rarely more, large marginal setae and may have additional setulae. The basalmost marginal seta is called the basal scutellar seta (Figure 33.2, b sctl s) or *anterior scutellar* or *lateral scutellar seta*, and the apicalmost is termed the apical scutellar seta (Figure 33.2, ap sctl s) or *posterior scutellar seta*.

**Scutum** (Figure 33.2, sct) In Tephritidae, most of the thorax visible in dorsal view is the scutum. Only the postpronotal lobe, the scutellum, and the postnotum are not part of it. It includes pre- and postsutural areas incompletely divided by the transverse suture.

***Second posterior cell*** An old name for the medial cell.

***Second vein*** An old term that usually meant vein $R_{2+3}$.

**Sensillum** (plural: **sensilla**) Simple sense organs, for example, on the larva or on the aculeus tip.

**Seta/setula** (plural: **setae/setulae**) (Figure 33.2) Hairlike surface structures that are articulated (having alveoli or sockets) are macrotrichia. A relatively large one is called a seta (e.g., the acrostichal and dorsocentral setae on the scutum), and a smaller one is called a setula (e.g., the scutal setulae that cover much of the scutum in many species). The setulae should not be confused with the much smaller microtrichia that are often present on the scutum and other parts of the body.

***Shining spots*** See ceromata.

***Sibling species*** See species complex.

**Sister group (phylogenetic analysis)** The most closely related taxon to the group being studied, for example, McAlpine (1989) hypothesized that Lonchaeidae is the sister group of the clade including the other families of Tephritoidea.

***Sixth vein*** An old term that usually meant vein $A_1$+$Cu_2$.

**Species complex (nomenclature)** A group of species that cannot be distinguished from each other using the morphological criteria normally used to identify related species. For example, most *Rhagoletis* spp. can be identified using simple combinations of color and wing pattern characters, but some species that are genetically distinct are difficult or impossible to distinguish morphologically. The individual species within a species complex are called cryptic or sibling species. The term morphospecies is sometimes used to refer to a group of cryptic species, and the term species complex is used by some authors (e.g., Drew 1989) in a wider sense, to refer to any group of species, that is, what is usually called a species group. Conversely, Bush (1966) used the term species group to mean species complex.

**Spermatheca** (plural: **spermathecae**) A usually sclerotized female internal organ used to store sperm. Tachiniscinae, Blepharoneurinae, Phytalmiinae, and many Trypetinae have three spermathecae, whereas Dacini, Tephritinae, and some Trypetinae have only two, and *Oedicarena* have four. They are normally found within abdominal segments 3 to 6, and are connected to the genital chamber by spermathecal ducts. The shape of the spermathecae varies considerably within Tephritidae, and their surface may be smooth, wrinkled, or covered by various papillose or dentate structures.

**Spermathecal ducts** Most of the Tephritoidea have two spermathecal ducts, of which the right apically bifurcates and bears two spermathecae. In Pyrgotidae and Tephritidae there are two separate spermathecal ducts on the right side (three total) that independently connect to the genital chamber. The apical portion of the spermathecal ducts is sometimes dilated (e.g., in Tephritinae), or may be sclerotized, giving the spermathecae the appearance of a figure 8 (e.g., *Enicoptera, Celidodacus,* some Trypetini).

**Spinule (larva)** (Figure 33.7, spn) A small spinelike projection.

**Spiracular hairs (larva)** (Figure 33.7, spr h) Translucent hairs arranged in four groups or bundles around the outer edge of the posterior spiracle.

**Spiracular openings** or **slits (larva)** (Figure 33.7, spr op) The external openings of the spiracular chamber. On the posterior spiracle they are variable in shape and length. There are normally three on each posterior spiracle in third instars of Tephritidae, and two on first and second instars.

***Star-shaped pattern*** See stellate pattern.

**Stellate pattern** (Figure 33.3H) A wing pattern with a somewhat star-shaped subapical area, with rays extending to or toward the wing margins, often including an apical fork (rays running to the apices of veins $R_{4+5}$ and M). This pattern is found in some genera of Tephritinae (e.g., most *Trupanea* spp.; see Merz, Chapter 24).

**Stenophagous (ecology)** Adjective meaning to feed only on very closely related plant species; in Tephritidae meaning a species that breeds in a very narrow range of closely related host plant species (e.g., all of its hosts are members of a single genus). This is narrower than oligophagous. See also monophagous and polyphagous.

***Sternite 10* (*male*)** See subepandrial sclerite.

**Sternites** (Figures 33.4C and 33.5A, st) The ventral abdominal sclerites. The shape of sternite 5 of the male is sometimes a useful character (e.g., see Drew 1989).

***Sternopleuron*** See katepisternum.

***Stigma*** See pterostigma.

***Stigmacosta*** A Munro (1984) term for vein C between the subcostal break and the end of vein $R_1$.

***Stomal guards* (*larva*)** See preoral teeth.

***Stomal organ/stomal sensory organ* (*larva*)** See preoral organ.

**Subapical band** or **crossband** See crossband.

**Subapical lobe** (Figure 33.5, sbap l) An unpaired lobe or bar of varying shape and sclerotization is present near the apex of the vesica of the glans in many Tephritidae (Munro 1984). Whether all of these lobes are homologous is uncertain. The subapical lobe may be simple, "T-shaped" (e.g., in some Toxotrypanini), "trumpet-shaped" (e.g., in Dacinae and some Trypetinae), or variously lobed or haired, and partially sclerotized or membranous. It has been called the *apicodorsal rod* (Munro 1984), the *T-shaped sclerite* (Norrbom 1985), or the *accessory sclerite, apical sclerite, juxta, tubular structure*, or *subapical distiphallic lobe* (see Jenkins 1996).

**Subbasal band** or **crossband** See crossband.

**Subcosta/subcostal vein** (Figure 33.4A, Sc) A major diagnostic feature of the Tephritidae is that the subcosta is abruptly bent forward subapically and is weak beyond the bend. This vein has also been called the *auxiliary vein* or *mediastinal vein.*

**Subcostal band** or **crossband** See crossband.

**Subcostal cell** (Figure 33.4A, sc) The cell bounded anteriorly by the subcosta and costa, and posteriorly and distally by vein $R_1$. Also see pterostigma.

**Subepandrial sclerite** (Figure 33.5, sbepand scl) In Cyclorrhapha, a sclerite between the epandrium and the phallus (Cumming et al. 1995). In Tephritidae, it is the sclerite connecting the basal ends of the medial surstyli, although the limits and homology of these sclerites are not well understood. The subepandrial sclerite has two anteriorly projecting lobes and one or two transverse connections or bridges. It has also been called *sternite 10* (McAlpine 1981) or the *interparameral sclerite* (e.g., Norrbom and Kim 1988).

***Subhypostomium* or *subhypostomal sclerite* (*larva*)** See labial sclerite.

***Submarginal cell*** An old name for cell $r_{2+3}$ (see radial cells).

**Subscutellum** (Figure 33.2, sbsctl) The small sclerite, best seen in posterior view of the thorax, below the scutellum and above the mediotergite. It has also been called the *postscutellum.*

**Subsequent designation (nomenclature)** See type species.

**Subvibrissal setae** (Figure 33.1, sbvb s) Setae on the anteroventral margin of the gena. They have also been called *oral setae* or *peristomal hairs.*

***Superior fronto-orbital setae*** See orbital setae.

***Superior orbital setae*** See orbital setae.

**Supernumerary lobe** Males of most *Bactrocera* spp. have a slight indentation of the hind margin of the wing at the end of vein $A_1$+$Cu_2$. This forms a shallow lobe between the end of that vein and vein $Cu_1$.

**Supination (behavior)** See wing displays.

**Supra-alar setae** (Figure 33.2, spal s) Tephritids have up to three pairs of supra-alar setae in a longitudinal row between the intra-alar and postalar setae. The presutural supra-alar seta (presut spal s), often called simply the presutural seta, is occasionally absent (e.g., most Dacina, some Adramini). There is usually one postsutural supra-alar seta (psut spal s), usually called the supra-alar seta (or the anterior supra-alar), but a second one is rarely present (e.g., *Ceratodacus*, *Ortalotrypeta*, *Tachinisca,* and a few genera of Acanthonevrini). The postalar seta has been regarded as a posterior supra-alar seta by some authors.

**Surstylus** (plural: **surstyli**) (Figure 33.5) In Cyclorrhapha, a paired clasping organ that articulates with the epandrium (Cumming et al. 1995). In Tephritidae it is divided into two parts that are usually very closely associated and difficult to distinguish. The lateral surstylus is fused to the epandrium, and the medial surstylus is fused to the subepandrial sclerite. During copulation they hold the female aculeus (see Eberhard and Pereira 1993; Headrick and Goeden 1994). The surstyli have also been called *claspers* (e.g., Stone 1942).

**Sympatric (ecology)** Adjective describing two or more species or populations found in the same place and having the potential to interbreed. Two species which evolved from one, without geographical isolation, are said to have evolved sympatrically (see Bush 1966; 1969b). Also see allopatric, allochronic and synchronic.

**Symplesiomorphy (phylogenetic analysis)** An ancestral character state shared by two or more taxa. A symplesiomorphy does not indicate recency of common ancestry. It does not support the hypothesis that the taxa which possess it are more closely related than taxa that do not possess it. See synapomorphy.

**Synapomorphy (phylogenetic analysis)** A derived character state shared by two or more taxa. A synapomorphy is indicative of close phylogenetic relationship. It supports the hypothesis that the taxa which possess it are more closely related than taxa that do not possess it.

**Synchronic (ecology)** Adjective describing two or more species or populations whose potential to interbreed is not reduced by a temporal separation. See also allochronic, allopatric and sympatric.

**Synonym (nomenclature)** Different names that have been used to describe the same biological taxon (genus, species, etc.) are synonyms. Only the first, or senior, name is valid; subsequent names are junior synonyms. An exception occurs when the senior name is a junior homonym. In that case the next-to-oldest name is used (if not also a junior homonym) or a new name must be published.

**Synovigenic (behavior)** Adjective describing a female that emerges reproductively immature, with her ovaries small. The stimulus that causes females to produce eggs is unknown, but is undoubtedly tied to the phenology of their host-plant species (Bush 1966; Zwölfer 1974; Drew 1987; Headrick and Goeden 1994). Females of some species require a protein source to begin egg development after emergence (Drew 1987; Headrick and Goeden 1994). Also see proovigenic.

**Syntergosternite 7** See oviscape.

**Syntype (nomenclature)** See type specimen.

***T-shaped sclerite*** See subapical lobe.

**Taenia** (plural: **taeniae**) (Figure 33.4, tae) A paired, striplike, basal sclerite of the eversible membrane (Steyskal 1984).

**Taxon** (plural: **taxa) (nomenclature)** A group of organisms at some taxonomic level (e.g., a species, genus, or family).

**Teneral** A freshly emerged adult with a soft, pale-colored body and poorly formed wing markings, and often with the ptilinum exposed. Reared specimens should always be kept alive for a few days to allow their bodies to harden and colors to develop before being killed.

***Tentoropharyngeal sclerite* (*larva*)** See pharyngeal sclerite.

***Tergal glands*** See ceromata.

**Tergites** (Figures 33.4B–C and 33.5A, tg) In the higher Diptera the first visible dorsal abdominal sclerite is formed by the fusion of tergites 1 and 2, and is called syntergite 1+2. In Tephritidae, the preabdomen of the male thus has four large tergites (1+2, 3, 4, and 5) (Figure 33.5A). Females have five (Figure 33.4B-C); however, tergite 6 may be short or be hidden under tergite 5 so that it is not visible from above. In *Dacus* spp. all the tergites are fused, but shiny lines or depressions marking the tergite borders can still be seen.

***Terminal sense organ (larva)*** See maxillary palp.

***Terminalia*** See male genitalia or ovipositor (for female).

**Territory (behavior)** A defended area (Noble 1939). Specific to fruit flies: A particular site or area on a plant where males exhibit defensive behavior and may display to attract females. No criterion as yet exists to identify the components of a territory objectively, and the term has been loosely applied to areas in which tephritid male defensive behavior, courtship, and copulation are observed. Many tephritid males display behaviors presumed to attract females within a particular site or area. Males of some species defend an area from intrusion by other males by highly ritualized combat, but a male also may exhibit these aggressive or combative behaviors toward other males where there is no clearly definable territory and the outcome of disputes is ambiguous. Males defend territories such as leaves, fruits, flowerheads, and stems on their host plant, or on a nonhost in a lek, but these territories are temporary. Two related words, territorial and territoriality, also are commonly used in describing tephritid behavior, but have differing implications. A male may be said to be territorial or exhibiting territoriality; however, this may refer to a particular "attitude" adopted by the male based on subjective interpretation of behavior one male may display toward another male. Thus, the latter two terms refer to the aggressive or competitive behaviors exhibited by a male toward other male(s) while in the course of mate aquisition and not to the territory specifically. The behaviors exhibited by males while exhibiting defensive or aggressive behaviors toward other arthropods, moving objects, or conspecifics may be similar to those exhibited when males also are trying to attain mates. Clear distinctions among these behaviors have not been made for any tephritid species. Also, not all encounters that appear aggressive should be described as territorial, exhibiting territoriality or defending a territory. Headrick and Goeden (1994) observed seven species in the field under natural conditions. The males of some of these species defended territories while exhibiting courtship displays. Territorial behaviors also were observed in laboratory arenas. These males were referred to as displaying territoriality, because their behaviors suggested that within their spheres of influence, they were attempting to exlude other males from acquiring mates. Also see lek.

***Third posterior cell*** An old name for cell $cu_1$.

***Third vein*** An old term that usually meant vein $R_{4+5}$.

***Tomentose/tomentosity/tomentum*** See microtrichia.

**Transverse suture** (Figure 33.2, trn sut) Calyptrate Diptera have a complete suture across the scutum between the notopleura. In the Acalyptratae, including the Tephritidae, the central part of this suture is absent but the lateral parts are distinct and extend mesally from each notopleuron. This partial suture divides the presutural and postsutural parts of the scutum. It has also been called the *notopleural suture*.

**Trophallaxis (behavior)** See male display behavior.

**Tubercle (larva)** A small raised area on the caudal segment often forming a base for a sensillum. Also see papilla.

**Tubule (larva)** A small tubular or fingerlike process bearing a spiracular opening on the outer edge of the anterior spiracle.

**Type genus (nomenclature)** The nominal genus that is the name-bearing type of a nominal family-group taxon (e.g., *Trypeta* is the type genus of Trypetinae).

**Type locality (nomenclature)** The geographic place of capture or collection of the primary type specimen of a nominal species or subspecies.

**Type species (nomenclature)** Every genus has one species, called its type species, whose function is to identify which species rightfully hold an existing generic name when generic (or subgeneric) limits are revised. Present-day taxonomists are obliged to choose the type species at the time of description, which is called original designation, but in the past the rules were less strict. If a genus was originally described for a single species, that species is the type by monotypy, but if a genus was originally described for several species, the type species must be chosen from among the originally included species. This is called subsequent designation. Through any changes of generic limits, the name of the genus always stays with the type species, even if that means placing it in synonymy with another genus or retaining the name of a large, well-known genus for a single species. For example, almost all Dacini used to be called *Dacus*, but when it was decided that the group was really two distinct genera (Drew 1989), only the minority of species could retain the name *Dacus* because the type species (*D. armatus*) belonged to the smaller of the two groups; most species, including most pests, were transferred to *Bactrocera*.

**Type specimen (nomenclature)** When a new (previously unknown) species is described, the taxonomist is obliged to designate a type specimen whose function is to identify which biological species holds that name if there is need for further revision. For example, when Drew (1991) discovered that the Oriental fruit fly was a complex of several species that had previously all been known as *B. dorsalis* (Hendel), he examined Hendel's original specimens from Taiwan to establish which one of the many members of the complex rightfully held that name. Modern taxonomists select a single specimen called a holotype and name any other specimens that they use for describing the species as paratypes; only the holotype matters if the subsequent identity of the species is disputed. Authors such as Hendel, working in the early part of this century, called all the specimens they used to describe a species "types." Modern authors call the specimens in a series which lack a single holotype syntypes, and to guard against future dispute they may select a single specimen as a lectotype (in effect, a primary type by subsequent designation); all remaining specimens in the series are paralectotypes, which like paratypes have no real nomenclatural significance. The term allotype may be used to designate a paratype of the opposite sex to the holotype, but such specimens have no more significance than other paratypes. In cases where the original type specimen is known to have been destroyed, a neotype may be designated as the primary type. A neotype is to be designated only in exceptional circumstances when it is necessary in the interests of stability of nomenclature.

**Univoltine (ecology)** Having one generation per year.

***Upper fronto-orbital setae*** See orbital setae.

***Upper orbital setae*** See orbital setae.

**V-band** (Figure 33.3D, VB) An inverted V-shaped wing band in most *Anastrepha* spp., formed from the subapical band (the "proximal arm" of the V-band), which covers cross-vein DM-Cu, and the posterior apical band (the "distal arm"), which crosses cell m. They are usually joined in cell $r_{4+5}$, but may be separated or the distal arm may be reduced or absent. In some species the V-band is joined to the S-band. Also see crossband.

**Vein M** See media/medial vein.

***Vein $M_{1+2}$*** See media/medial vein.

***Vein $M_{3+4}$*** An old name for vein $Cu_1$. See cubital vein/cubitus.

**Vein $R_1$** See radial vein/radius.

**Vein $R_{4+5}$** See radial vein/radius

**Vein Sc** See subcosta/subcostal vein.

***Veins 1/2/3/4/5/6*** See first, second, third, fouth, fifth, and sixth veins, respectively.

**Ventral area (larva)** (Figure 33.7, v area) On the caudal segment, the area between the intermediate area and the anal elevation.

**Ventral cornu** (plural: **ventral cornua) (larva)** (Figure 33.6, v corn) The paired, ventral, winglike portion of the pharyngeal sclerite. It may have a clear or unsclerotized area called a window.

***Ventral lobe (larva)*** See medial oral lobe.

**Ventral receptacle** In Tephritidae, a small, sclerotized, multichambered structure on the venral side of the genital chamber. This has also been called the *morula gland.*

***Ventral sense organ (larva)*** See maxillary palp and preoral organ.

**Ventral sensilla (larva)** (Figure 33.7) Three pairs of sensilla located on the ventral area of the caudal segment. The individual sensilla have been termed V1, V2, and V3.

**Vertex** (Figure 33.1, vrt) The uppermost part of the head, between the eyes and around the ocellar triangle.

**Vertical setae** (Figure 33.1) There are two pairs of vertical setae on or slightly posterior to the dorsalmost part of the head, or vertex, near the margin of the eye. The medial vertical seta (m vt s) is large, reclinate and/or inclinate. The lateral vertical seta (l vt s) is usually lateroclinate; it is usually large, but in the Tephritinae it is sometimes smaller, white, and lanceolate, and may be difficult to distinguish from the postocular setae. McAlpine (1981) termed these setae the *inner vertical seta* and *outer vertical seta*, respectively, but as they are external in both cases the adjectives medial and lateral are more appropriate (A. Freidberg, personal communication).

**Vesica** (Figure 33.5, ves) The membranous apical part of the glans (Bush 1966). It has also been called the *juxta* (Korneyev 1985).

***Vibrissal setae*** See genal setae.

**Vitta** (plural: **vittae)** The Latin term for stripe, which is a longitudinal color marking (as opposed to a band, which is transverse). Some authors have used the term vittae exclusively for the bright yellow or orange stripes on the scutum of most *Bactrocera* and *Dacus* spp. (see scutal stripes and xanthine).

**Wasp mimicry pattern** (Figure 33.3E) A wing pattern with a costal band and often an anal streak, which is presumed to mimic the appearance of vespid wasps or other stinging Hymenoptera.

**Window (larva)** See ventral cornu.

**Wing displays (behavior)** The array of wing movements for purposes other than flight. Few families of Diptera have such stylized wing movements as the Tephritidae, and none are more elaborate or complex. Tephritids move their wings in characteristic series or sequences. One difficulty in describing these motions is based on the anatomy of their joints. The wings are attached to the body by a membranous area that contains small articular sclerites or pteralia. These articulations have not been described or comparatively studied in tephritids, and it is unknown if they differ from other Diptera, which complicates the definition of terms to describe tephritid wing movements.

Wing movements are usually observed in conjunction with other activities. Some wing movements are specific to an activity such as courtship and thus are not seen at other times. Other wing displays are not specific to any activity, can be observed at any time, and are often spontaneous. The following terms are used to describe specific wing movements:

1. **Arching** This type of display has rarely been reported, probably due to its specialized nature. As described for *Paracantha gentilis* Hering (Headrick and Goeden 1990b), the wings are held over the dorsum, slightly spread, and arched from the base to the apex such that the tips nearly touch the substrate. This action must require musculature in the wings, but this needs further study.
2. **Enantion** Enantion, derived from the Greek preposition meaning “against” or “opposite” (Green et al. 1993), is the extension of both wings away from the body simulta-

neously. The wings are parted, slightly supinated and held along the pleura at rest. From this position each wing blade is simultaneously extended forward up to 90° away from the midline of the body. Enantion was the third most common wing display observed by Headrick and Goeden (1994) (in 32 of 49 species). It was most commonly observed in species with a banded wing pattern (e.g., *Rhagoletis* and *Procecidochares*). This display occurs during all activities such as feeding, grooming, and courtship. Enantion is distinct from synchronous supinations because the wing blades are not supinated during extension. It has also been called *scissoring*.

3. **Hamation** Hamation, derived from the Greek preposition meaning "together with" (Headrick and Goeden 1991), is the movement of the wings together over the dorsum or while they are extended away from the body. The wings are moved synchronously, side-to-side over the dorsum, and parallel with the substrate with or without supination. This was the most common wing display observed by Headrick and Goeden (1994) (in 42 of 49 species).
4. **Lofting** A display in which both wings are extended upward to 90° above the substrate and supinated up to 90°. Thus, when the wings are lofted at their maximum, the wing blades are parallel to each other above the dorsum and the costal margins are perpendicular to the substrate. Males of some *Campiglossa* species hold their wings raised 45° above their dorsa and then loft them to 90° as described above. Other tephritid species commonly only loft their wings from a resting position to ~45° above their dorsa. In this latter display, the wing are only supinated ~45°, and are not parallel to each other, but form a "V" of ~45°. In the 49 species studied by Headrick and Goeden (1994), lofting was uncommon and observed in only nine species.
5. **Supination** This display consists of bringing the wing forward perpendicular to the long axis of the body while the ventral surface of the wing is turned to face anterior such that the costal margin of the wing is dorsal. The rotation of the wing (or any appendage) in such a manner in anatomical terms is defined as supination; the opposite movement is pronation. Many terms have been used for this wing movement (e.g., *wing waving*, *flicking*, and *flexing*). It may involve one or both wings being extended forward (i.e., synchronous or asynchronous supination). It has not been observed in other families of Diptera, but is common to most species of tephritids for which adult behavior has been described (in 34 of 49 species studied by Headrick and Goeden 1994). Asynchronous supination was the most common supination display. Supination is embellished in many species and has been observed during aggression, male displays, courtship, mating, and oviposition.

**Wing fanning (behavior)** A general term used to describe the full array of stylized movements of the wings during mating. See wing displays.

**Xanthine** A term used by Munro (1984) for vittae and other areas of the body on Dacini that are bright yellow or orange and appear to be formed from a body of fat which is visible through a "window" in the integument. This term has not been widely used but it could be an appropriate general term for these structures. Specific terms were as follows: *humeral xanthines* for yellow postpronotal lobes; *postsutural xanthines* for lateral and medial scutal vittae; *pleurosutural xanthines* for anepisternal stripe, and the "wedge" of yellow that may be present at the ends of the transverse suture; and *postalar xanthines* for the yellow on the anatergite and katatergite.

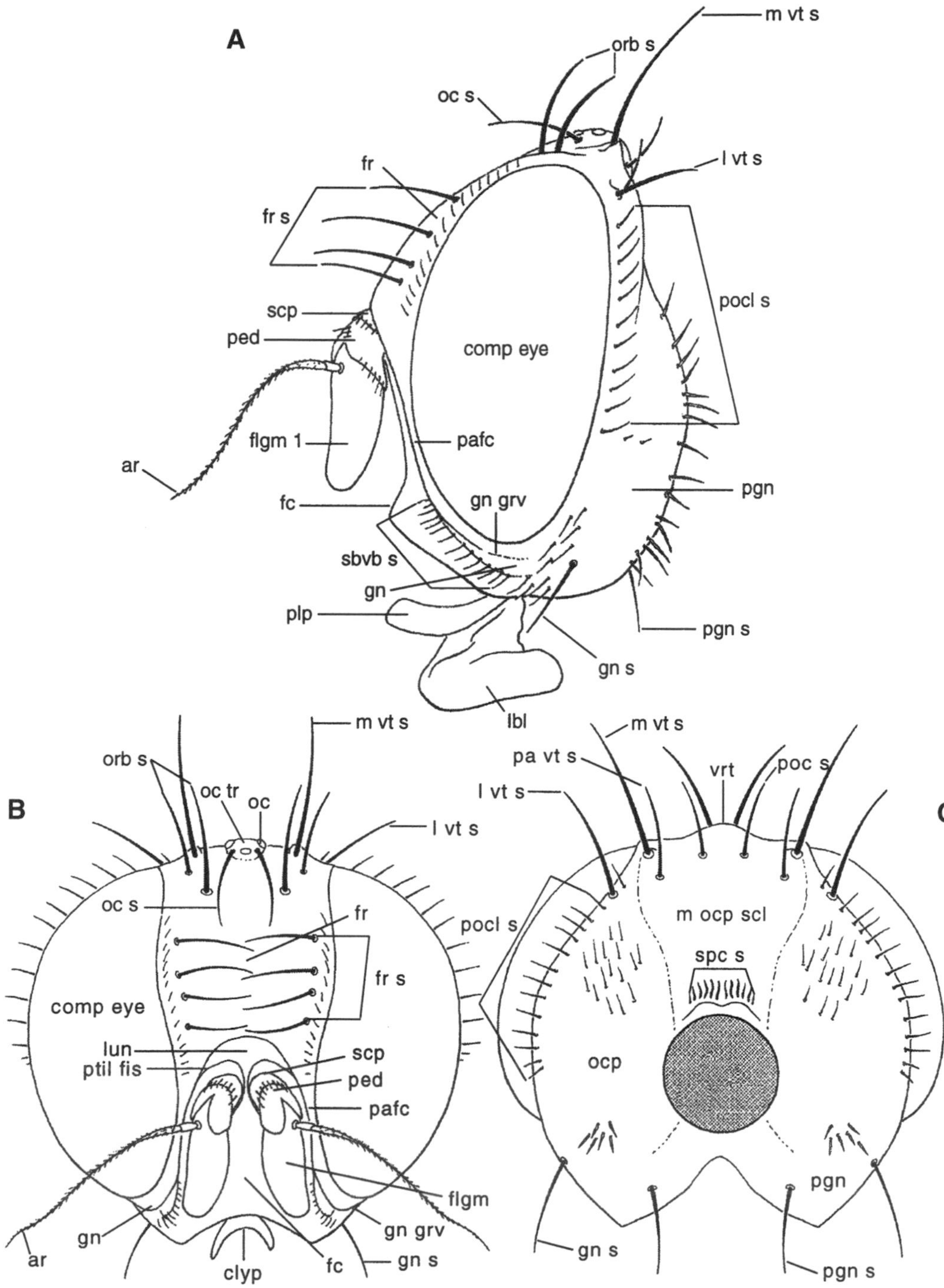


**FIGURE 33.1** Adult head of generalized tephritid. (A) lateral view; (B) anterior view; (C) posterior view. Abbreviations are explained in Table 33.1. (Modified from figures based on *Anastrepha ludens* (Loew), from Foote R.H. et al., *Handbook of the Fruit Flies (Diptera: Tephritidae) of America North of Mexico,* Comstock Publishing Associates, Ithaca, NY, 1993. With permission.)

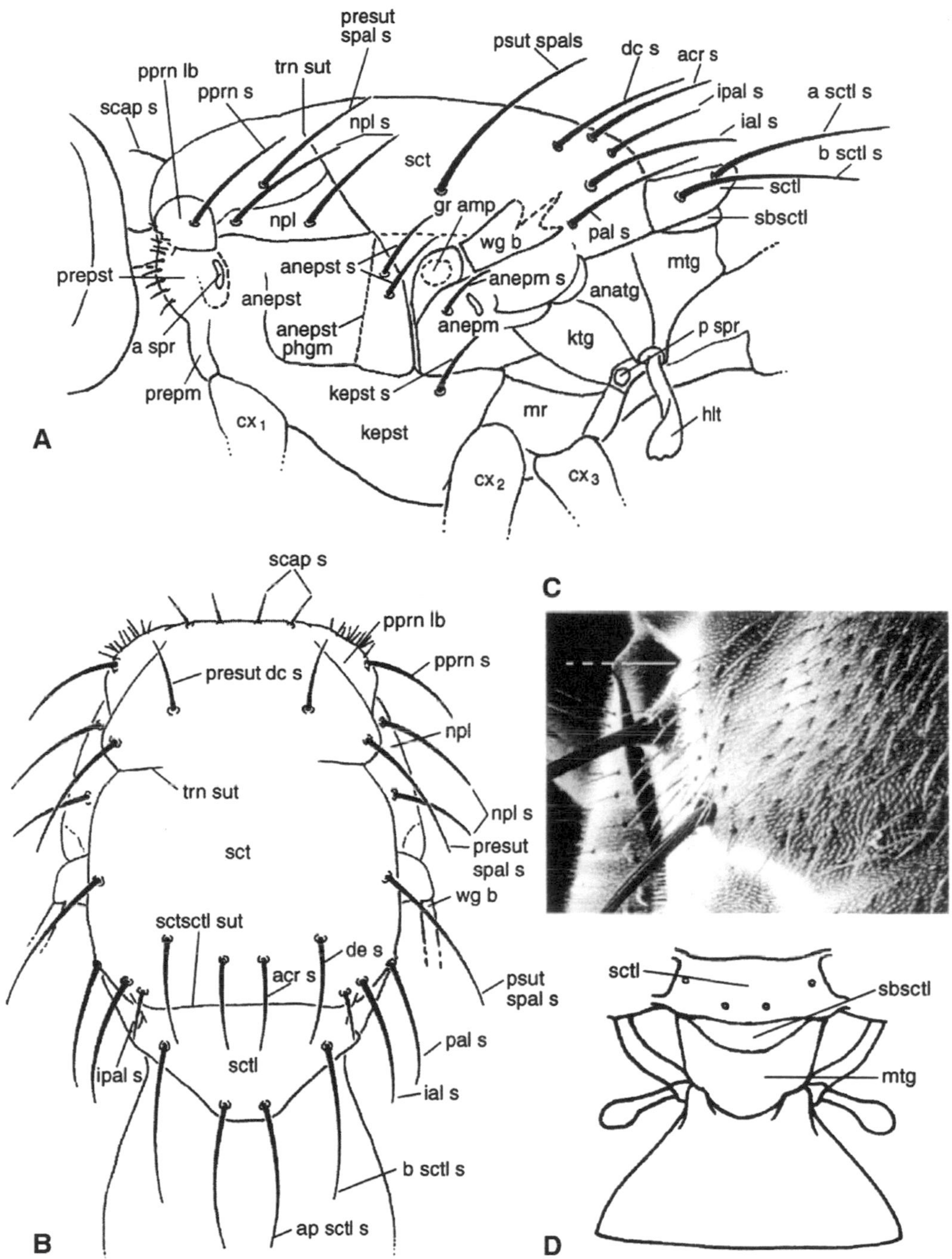


**FIGURE 33.2** (A to C), Thorax of generalized tephritid. (A) Lateral view; (B) Dorsal view; (D) Posterior view. (C) Scanning electron micrograph of scutum near left anterior notopleural and presutural supra-alar setae of *Anastrepha fraterculus* (Wiedemann) showing numerous scutal setulae and microtrichia. Abbreviations are explained in Table 33.1. (A and B, modified from figures based on *Anastrepha ludens* (Loew) from Foote R.H. et al., *Handbook of the Fruit Flies (Diptera: Tephritidae) of America North of Mexico,* Comstock Publishing Associates, Ithaca, NY, 1993. With permission of Cornell University Press; D, from Norrbom, A.L., Dissertation, Pennsylvania State University, University Park, 1985. With permission.)

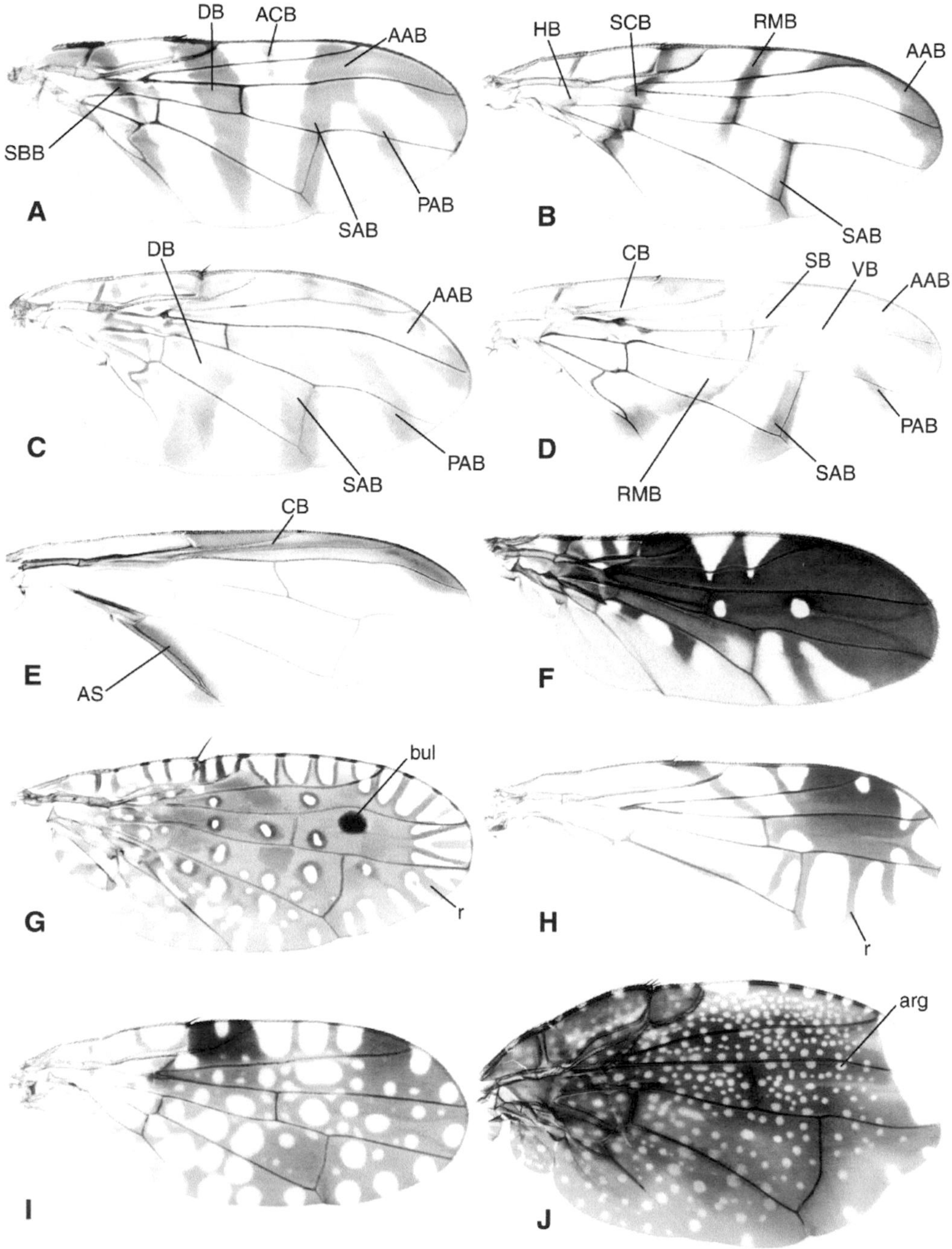


**FIGURE 33.3** Wing patterns. (A) *Rhagoletis psalida* Hendel, *Rhagoletis*-type pattern; (B) *Chetostoma californicum* Blanc; (C) *Ceratitis venusta* (Munro), *Ceratitis*-type pattern; (D) *Anastrepha ludens* (Loew), *Anastrepha*-type pattern; (E) *Dacus humeralis* (Bezzi); (F) *Xanthaciura mallochi* Aczél, *Aciura*-type pattern; (G) *Paracantha genalis* Malloch, radiate pattern; (H) *Trupanea wheeleri* Curran, stellate pattern; (I) *Campiglossa albiceps* (Loew), reticulate pattern; J, *Eutreta distincta* (Schiner), reticulate pattern. Abbreviations are explained in Table 33.1.

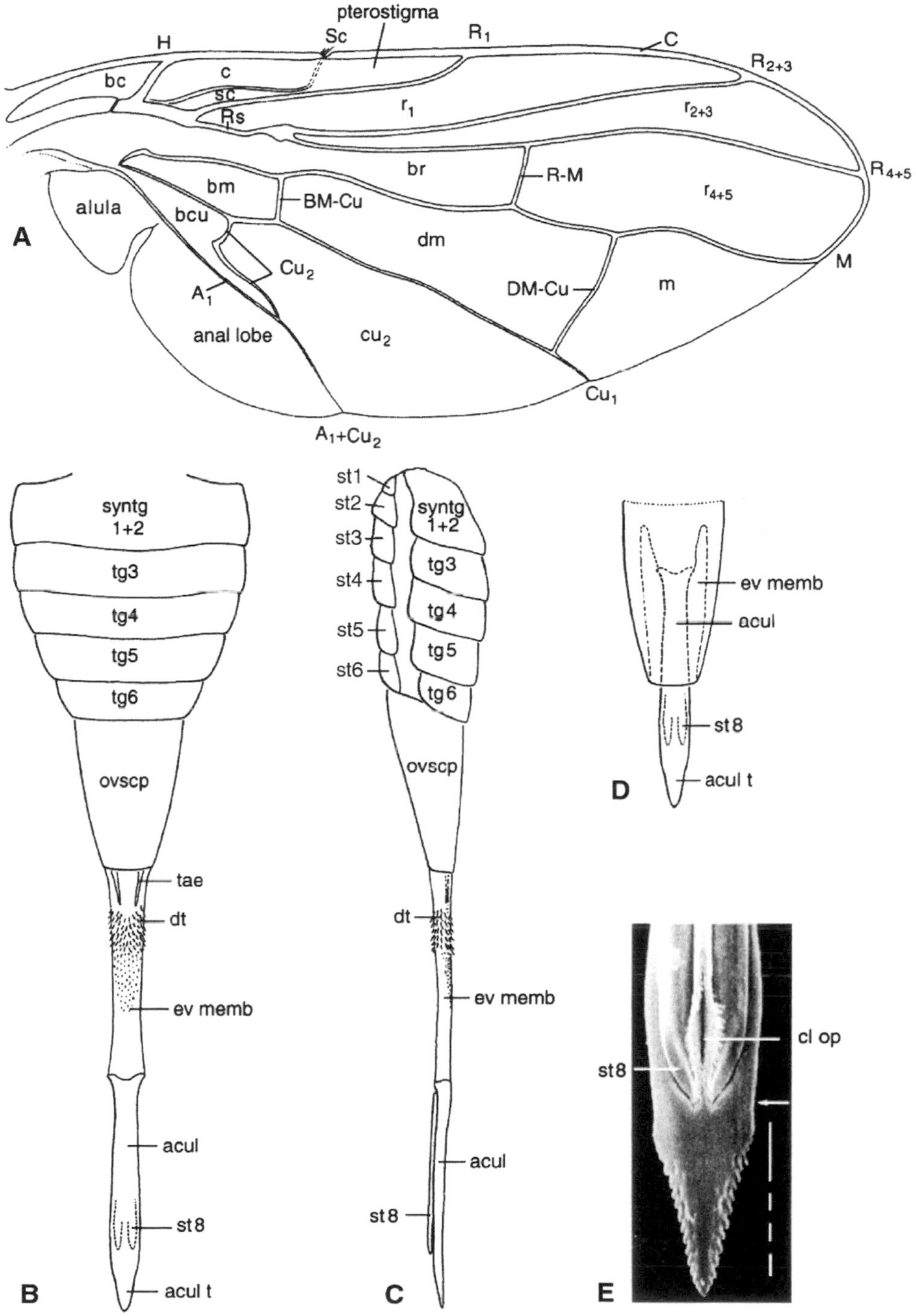


**FIGURE 33.4** (A) Right wing of *Anastrepha ludens* (Loew); (B) female abdomen of generalized tephritid with ovipositor fully extended, dorsal view; (C) same, lateral view; (D) same, ovipositor with aculeus partially retracted; E, scanning electron micrograph of aculeus tip of *A. obliqua* (Macquart). Abbreviations are explained in Table 33.1. (A to D, modified from Foote R.H. et al., *Handbook of the Fruit Flies (Diptera: Tephritidae) of America North of Mexico,* Comstock Publishing Associates, Ithaca, NY, 1993. E, modified from Norrbom, A.L., Ph.D. dissertation, Pennsylvania State University, University Park, 1985. With permission).

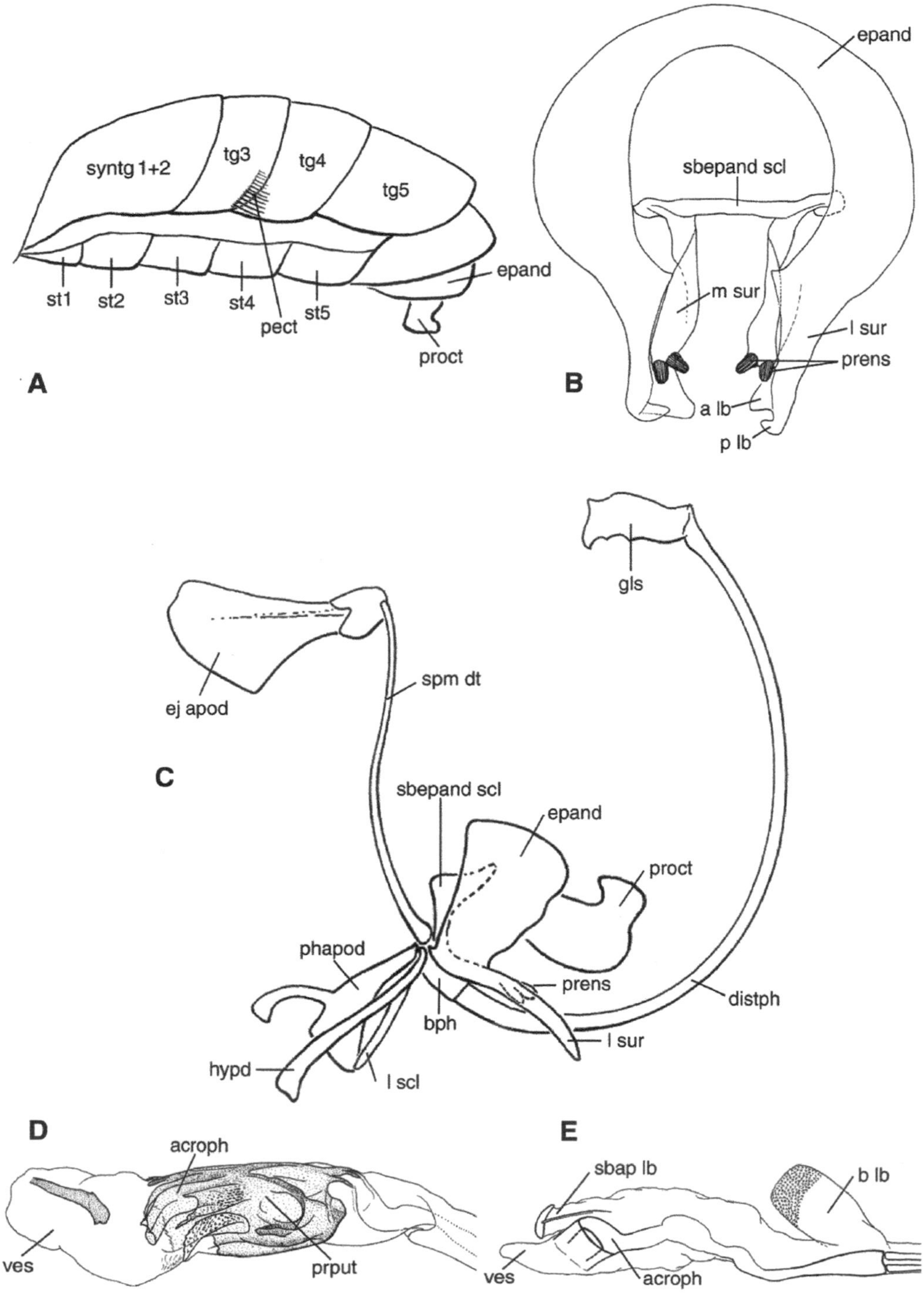


**FIGURE 33.5** (A) Male abdomen of generalized tephritid, lateral view; (B) same, genitalia, lateral view; (C) same, epandrium and surstyli, posterior view (proctiger omitted); (D, E) glans enlarged. Abbreviations are explained in Table 33.1. (A, B, modified from Foote R.H. et al., *Handbook of the Fruit Flies (Diptera: Tephritidae) of America North of Mexico,* Comstock Publishing Associates, Ithaca, NY, 1993. C, E, from Freidberg, A. and Kugler, J., *Fauna Palaestina, Insecta IV. Diptera: Tephritidae,* Israel Academy of Sciences & Humanities, Jerusalem, 1989. With permission.)

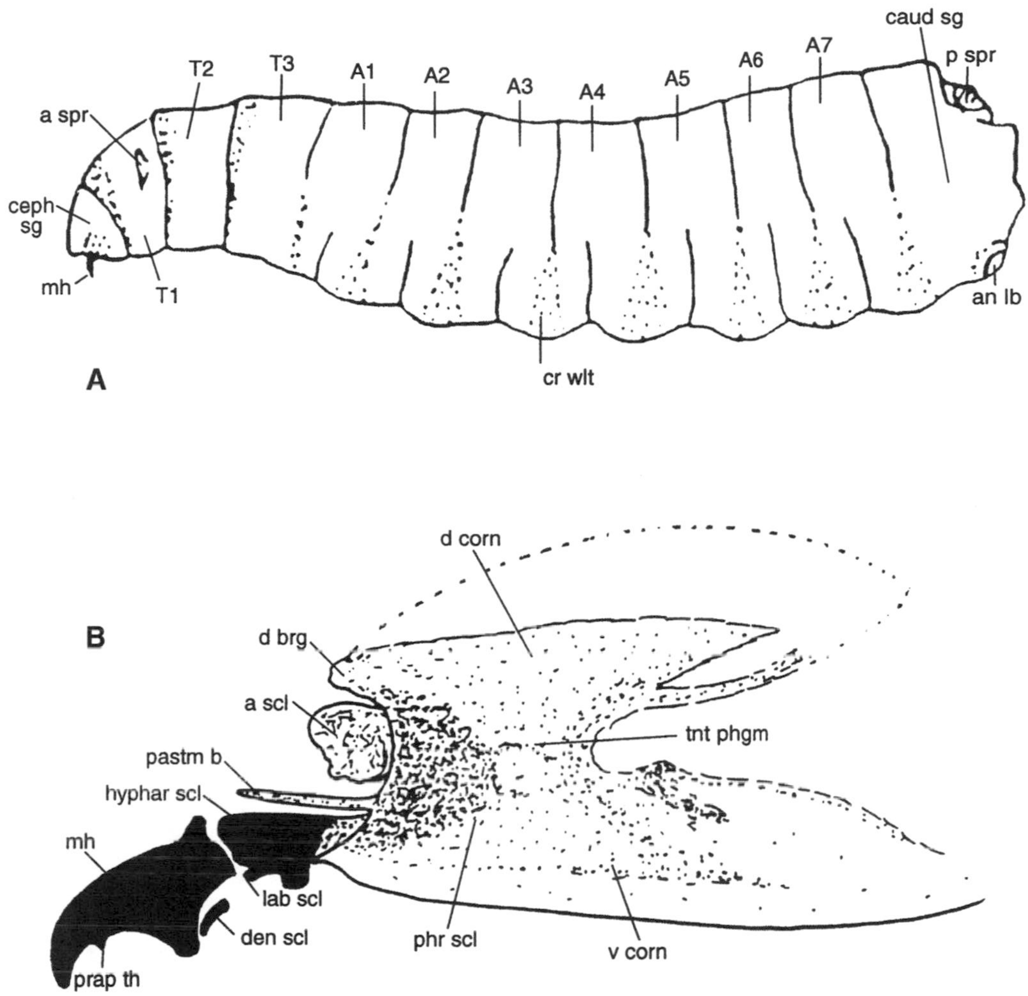


**FIGURE 33.6** *Bactrocera* sp., third instar larva. (A) lateral view; (B) cephalopharyngeal skeleton, lateral view. Abbreviations are explained in Table 33.1. (From White, I.M. and Elson-Harris, M.M., *Fruit Flies of Economic Significance: Their Identification and Bionomics,* CAB INTERNATIONAL, Wallingford, 1994. With permission.)

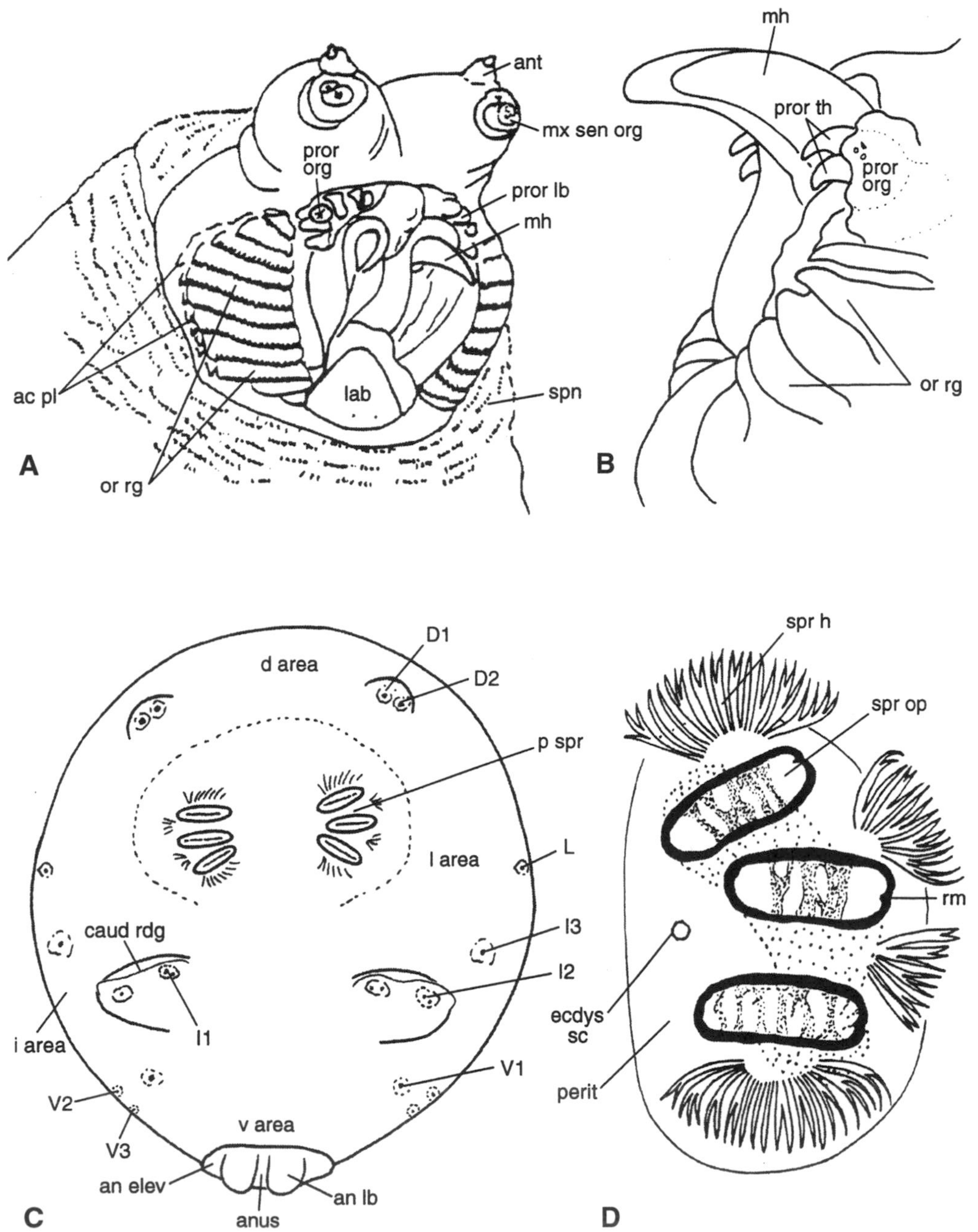


**FIGURE 33.7** Third instar larva. (A, C, D) *Bactrocera* sp.; (B) *Rhagoletis* sp. (A) Head, ventral view; (B) oral area, lateral view; (C) caudal segment, posterior view; (D) posterior spiracle. Abbreviations explained in Table 33.1. (From White, I.M. and Elson-Harris, M.M., *Fruit Flies of Economic Significance: Their Identification and Bionomics,* CAB INTERNATIONAL, Wallingford, 1994. With permission.)

## TABLE 33.1
## Figure Abbreviations

### Head

ar = arista; clyp = clypeus; comp eye = compound eye; fc = face; flgm 1 = first flagellomere; fr = frons; fr s = frontal setae; gn = gena; gn grv = genal groove; gn s = genal seta; lbl = labella; l vt s = lateral vertical seta; lun = lunule; m ocp scl = median occipital sclerite; m vt s = medial vertical seta; oc = ocellus; oc s = ocellar seta; oc tr = ocellar triangle; ocp = occiput; orb s = orbital setae; pafc = parafacial area; ped = pedicel; pgn = postgena; pgn s = postgenal seta; plp = palpus; poc s = postocellar seta, medial and lateral; pocl s = postocular setae; ptil fis = ptilinal fissure; pavt s = paravertical seta; sbvb s = subvibrissal setulae; scp = scape; spc s = supracervical setae; vrt = vertex.

### Thorax

a spr = anterior spiracle; acr s = acrostichal seta; anatg = anatergite; anepm = anepimeron; anepm s = anepimeral seta; anepst = anepisternum; anepst phgm = anepisternal phragma; anepst s = anepisternal setae; ap sctl s = apical scutellar seta; b sctl s = basal scutellar seta; cx1, cx2, cx3 = coxae of fore, mid and hind legs; dc s = dorsocentral seta; hlt = halter; gr amp = greater ampulla; ial s = intra-alar seta; ipal s = intrapostalar seta; kepst = katepisternum; kepst s = katepisternal seta; ktg = katatergite; mtg = mediotergite; npl = notopleuron; npl s = notopleural setae; pal s = postalar seta; pprn lb = postpronotal lobe; pprn s = postpronotal seta; prepim = proepimeron; prepst = proepisternum; presut dc s = presutural dorsocentral seta; presut spal s = presutural supra-alar seta; psut spal s = postsutural supra-alar seta; sbsctl = subscutellum; scap s = scapular setae; sct = scutum; sctl = scutellum; sctsctl sut = scuto-scutellar suture; trn sut = transverse suture; wg b = wing base

### Wing venation (cells in lowercase letters, veins in capitals)

$A_1$ = first anal vein; $A_1$+$Cu_2$ = second branch of cubital vein plus first anal vein; bc = basal costal cell; bcu = basal cubital cell; bm = basal medial cell; BM-Cu = basal medial-cubital crossvein; br = basal radial cell; C = costa (costal vein); c = costal cell; $Cu_1$ = first branch of cubital vein; $cu_1$- anterior cubital cell; $Cu_2$ = second branch of cubital vein; dm = discal medial cell; DM-Cu = discal medial-cubital crossvein; H = humeral crossvein (basal crossvein); M = medial vein; m = medial cell; R-M = radial-medial crossvein; $R_1$- first branch of radial vein; $r_1$- first (anterior) radial cell; $R_{2+3}$ = first branch of radial sector vein; $r_{2+3}$ = third radial cell; $R_{4+5}$ = last branch of radial vein; $r_{4+5}$ = fifth radial cell; Rs = sectoral branch of radial vein; Sc = subcostal vein; sc = subcostal cell.

### Wing Pattern

AAB = anterior apical band; ACB = accessory costal band; arg = argent; AS = anal streak; bul = bulla; CB = costal band; DB = discal band; HB = humeral band; PAB = posterior apical band; r = ray; RMB = radial-medial band; SAB = subapical band; SB = S-band; SBB = subbasal band; SCB = subcostal band; VB = V-band.

### Female Abdomen

acul = aculeus; acul t = aculeus tip; cl op = cloacal opening; dt = denticles; ev memb = eversible membrane; ovscp = oviscape; st1 to st6 = sternites of abdominal segments 1 to 6; st8 = 8th sternite; syntg1+2 = syntergite of abdominal segments 1 and 2; tae = taenia; tg3 to tg6 = tergites of abdominal segments 3 to 6

### Male Abdomen

acroph = acrophallus; a lb = anterior lobe of lateral surstylus; b lb = basal lobe; bph = basiphallus; distph = distiphallus; ej apod = ejaculatory apodeme; epand = epandrium; gls = glans; hypd = hypandrium; l scl = lateral sclerite; l sur = lateral surstylus; m sur = medial surstylus; pect = pecten; phapod = phallapodeme; p lb = posterior lobe of lateral surstylus; prens = prensisetae; proct = proctiger; sbap lb = subapical lobe; sbepand scl = subepandrial sclerite; st1 to st5 = sternites of abdominal segments 1 to 5; spm dt = sperm duct; syntg1+2 = syntergite of abdominal segments 1 and 2; tg3 to tg5 = tergites of abdominal segments 3 to 5; ves = vesica.

### Larva

A1 = A7 = abdominal segments 1 to 7; ac pl = accessory plates; an elev = anal elevation; an lb = anal lobe; ant = antenna; a scl = anterior sclerite; a spr = anterior spiracle; ceph sg = cephalic segment; caud rdg = caudal ridge; caud sg = caudal segment; cr wlt = creeping welt; D1, D2 = dorsal sensilla 1 and 2; d area = dorsal area; d brg = dorsal bridge; d corn = dorsal cornu; den scl = dental sclerite; ecdys sc = ecdysial scar; hyphar scl = hypopharyngeal sclerite; I1, I2, I3 = intermediate sensilla 1, 2, and 3; i area = intermediate area; L = lateral sensillum; lab = labium; lab scl = labial sclerite; l area = lateral area; mh = mouthhook; mx sen org = maxillary sense organ; or rg = oral ridges; pastm b = parastomal bar; perit = peritreme;

**TABLE 33.1 (continued)**
**Figure Abbreviations**

phr scl = pharyngeal sclerite; prap th = preapical tooth of mouthhook; pror lb = preoral lobes; pror org = preoral organ; pror th = preoral teeth; p spr = posterior spiracle; rm = rima; spn = spinules; spr h = spiracular hairs; spr op = spiracular opening; T1, T2, T3 = pro-, meso-, and metathorax; tnt phgm = tentorial phragma; V1, V2, V3 = ventral sensilla 1, 2, and 3; v area = ventral area; v corn = ventral cornu

## ACKNOWLEDGMENTS

Compiling this glossary required the help of many of the participants. Those who contributed or revised entries included: Valery Korneyev (male terminalia terms), Ho-Yeon Han (nomenclatural terms), Bruce McPheron and Jim Smith (cladistic terms), and Martín Aluja and Boaz Yuval (behavior terms). Others who contributed important comments included Amnon Freidberg, Marc De Meyer and Ken Kaneshiro. Marian Kotrba, Stuart McKamey, and Ed Barrows and kindly reviewed the manuscript. Permission for the use of most figures of the adults was kindly granted by Cornell University Press and for those of the larvae by CAB INTERNATIONAL and Marlene Elson-Harris.

## REFERENCES

Arita, L.H. and K.Y. Kaneshiro. 1986. Structure and function of the rectal epithelium and anal glands during mating behavior in the Mediterranean fruit fly male. *Proc. Hawaii. Entomol. Soc.* 26: 27–30.

Belcari, A. 1989. Contributi alla conoscenza dei ditteri tefritidi. IV. Descrizione della larva di terza eta' di *Acanthiophilus helianthi* (Rossi), *Dacus oleae* (Gmel.), *Ceratitis capitata* (Wied.), *Acidia cognata* Wied. e considerazioni preliminari sulle differenziazioni morfologiche legate al diverso trofismo. *Frustula Entomol.* (N.S.). (1987) 10: 82–125.

Berube, D.E. and R.Y. Zacharuk. 1983. The abdominal musculature associated with oviposition in two gall-forming tephritid fruit flies in the genus *Urophora*. *Can. J. Zool.* 61: 1805–1814.

Bolwig, N. 1946. Senses and sense organs of the anterior end of the housefly larva. *Vidensk. Medd. Dan. Naturhist. Foren.* 109: 81–217.

Burk, T. 1981. Signaling and sex in acalypterate flies. *Fla. Entomol.* 64: 30–43.

Burk, T. and J.C. Webb. 1983. Effect of male size on calling propensity, song parameters, and mating success in Caribbean fruit flies, *Anastrepha suspensa* (Loew) (Diptera: Tephritidae). *Ann. Entomol. Soc. Am.* 76: 678–682.

Bush, G.L. 1966. The taxonomy, cytology and evolution of the genus *Rhagoletis* in North America. *Bull. Mus. Comp. Zool.* 134: 431–562.

Bush, G.L. 1969a. Mating behavior, host specificity, and the ecological significance of sibling species in frugivorous flies of the genus *Rhagoletis* (Diptera: Tephritidae). *Am. Nat.* 103: 669–672.

Bush, G.L. 1969b. Sympatric host formation and speciation in frugivorous flies of the genus *Rhagoletis*. *Evolution* 23: 237–251.

Carpenter, J.M. 1994. Successive weighting, reliability and evidence. *Cladistics* 10: 215–220.

Carroll, L.E. 1992. Systematics of Fruit Fly Larvae (Diptera: Tephritidae). Ph.D. dissertation, Texas A&M University, College Station. 341 pp.

Chu, I.-W. and R.C. Axtell. 1971. Fine structure of the dorsal organ of the house fly larva, *Musca domestica* L. *Z. Zellforch.* 117: 17–34.

Chu-Wang, I.-W. and R.C. Axtell. 1972. Fine structure of the terminal organ of the house fly larva, *Musca domestica* L. *Z. Zellforch. Mikrosk. Anat.* 127: 287–305.

Cumming, J.M., B.J. Sinclair, and D.M. Wood. 1995. Homology and phylogenetic implications of male genitalia in Diptera — Eremoneura. *Entomol. Scand.* 26: 120–151.

Dean, R.W. 1935. Anatomy and postpupal development of the female reproductive system in the apple maggot fly, *Rhagoletis pomonella* Walsh. *N.Y. Agric. Exp. Stn. Tech. Bull.* 229: 1–31.

De Carlo, J.M., G.N. Pellerano, and L.I. Martinez. 1994. Saco del oviducto medio de *Ceratitis capitata* Wied. (Diptera, Tephritidae): consideraciones histo-funcionales. *Physis* (B. Aires) (1991) 49: 19–25.

De Marzo, L., G. Nuzzaci, and M. Solinas. 1976. Aspetti anatomici, strutturali, ultrastrutturali e fisiologici delle ghiandole genitali accessorie del maschio di *Dacus oleae* Gmel. in relazione alla maturità ed all'attività sessuale. *Entomologica* (Bari) 12: 213–240.

De Marzo, L., G. Nuzzaci, and M. Solinas. 1978. Studio anatomico, istologico, ultrastrutturale e fisiologico del retto ed osservazione etologiche in relazione alla possibile produzione di feromoni sessuali nel maschio di *Dacus oleae* Gmel. *Entomologica* (Bari) 14: 203–266.

Dodson, G. 1978. Morphology of the reproductive system in *Anastrepha suspensa* (Loew) and notes on related species. *Fla. Entomol.* 61: 231–240.

Dodson, G. 1982. Mating and territoriality in wild *Anastrepha suspensa* (Diptera: Tephritidae) in field cages. *J. Ga. Entomol. Soc.* 17: 189–200.

Dodson, G. 1987a. Biological observations on *Aciurina trixa* and *Valentibulla dodsoni* (Diptera: Tephritidae) in New Mexico. *Ann. Entomol. Soc. Am.* 80: 494–500.

Dodson, G. 1987b. The significance of sexual dimorphism in the mating system of two species of tephritid flies (*Aciurina trixa* and *Valentibulla dodsoni*) (Diptera: Tephritidae). *Can. J. Zool.* 65: 194–198.

Drew, R.A.I. 1969. Morphology of the reproductive system of *Strumeta tryoni* (Froggatt) (Diptera: Trypetidae) with a method of distinguishing sexually mature adult males. *J. Aust. Entomol. Soc.* 8: 21–32.

Drew, R.A.I. 1987. Behavioral strategies of fruit flies of the genus Dacus (Diptera: Tephritidae) significant in mating and host-plant relationships. *Bull. Entomol. Res.* 77: 73–81.

Drew, R.A.I. 1989. The tropical fruit flies (Diptera: Tephritidae: Dacinae) of the Australasian and Oceanian Regions. *Mem. Queensl. Mus.* 26**:** 1–521.

Drew, R.A.I. 1991. Taxonomic studies on Oriental fruit fly. In *First International Symposium on Fruit Flies in the Tropics, Kuala Lumpur, 1988* (S. Vijaysegaran and A.G. Ibrahim, eds), pp. 63–66. Malaysian Agricultural Research and Development Institute, Kuala Lumpur.

Drew, R.A.I. and D.L. Hancock. 1995. New species, subgenus and records of *Bactrocera* Macquart from the south Pacific (Diptera: Tephritidae: Dacinae). *J. Aust. Entomol. Soc.* 34: 7–11.

Eberhard, W.G. and F. Pereira. 1993. Functions of the male genitalic surstyli in the Mediterranean fruit fly, *Ceratitis capitata* (Diptera: Tephritidae). *J. Kans. Entomol. Soc.* 66: 427–433.

Eberhard, W.G. and F. Pereira. 1998. The process of intromission in the Mediterranean fruit fly, *Ceratitis capitata* (Diptera: Tephritidae). *Psyche* (1995) 102: 99–120.

Economopoulos, A.P., A. Giannakakis, M.E. Tzanakakis, and A.V. Voyadjoglou. 1971. Reproductive behavior and physiology of the olive fruit fly. 1. Anatomy of the adult rectum and odors emitted by adults. *Ann. Entomol. Soc. Am.* 64: 1112–1116.

Exley, E.M. 1955. Comparative morphological studies of the larvae of some Queensland Dacinae (Trypetidae, Diptera). *Queensl. J. Agric. Sci.* 12: 119–150.

Farris, J.S. 1989. The retention index and rescaled consistency index. *Cladistics* 5: 417–419.

Felsenstein, J. 1985. Confidence limits on phylogenies: an approach using the bootstrap. *Evolution* 39: 783–791.

Fletcher, B.S. 1969. The structure and function of the sex pheromone glands of the male Queensland fruit fly, *Dacus tryoni*. *J. Insect Physiol.* 15: 1309–1322.

Foote, R.H. 1981. The genus *Rhagoletis* Loew south of the United States (Diptera: Tephritidae). *U.S. Dep. Agric. Tech. Bull.* 1607: iv + 75 p.

Foote, R.H. and G.C. Steyskal. 1987. Tephritidae. In *Manual of Nearctic Diptera, Vol. 2* (J.F. McAlpine, ed.), pp. 817–831. Monograph of the Biosystematics Research Centre, No. 28. Agriculture Canada, Ottawa.

Foote, R.H., F.L. Blanc, and A.L. Norrbom. 1993. *Handbook of the Fruit Flies (Diptera: Tephritidae) of America North of Mexico*. Comstock Publishing Associates, Ithaca. 571 pp.

Freidberg, A. 1981. Mating behavior of *Schistopterum moebiusi* Becker (Diptera: Tephritidae). *Isr. J. Entomol.* 15: 89–95.

Freidberg, A. 1991. A new species of *Ceratitis* (*Ceratitis*) (Diptera: Tephritidae), key to species of subgenera *Ceratitis* and *Pterandrus*, and record of *Pterandrus* fossil. *Bishop Mus. Occas. Pap.* 31: 166–173.

Freidberg, A. and D.L. Hancock. 1989. *Cryptophorellia*, a remarkable new genus of Afrotropical Tephritinae (Diptera: Tephritidae). *Ann. Natal Mus.* 30: 15–32.

Freidberg, A. and J. Kugler. 1989. *Fauna Palaestina, Insecta IV. Diptera: Tephritidae*. Israel Academy of Sciences and Humanities, Jerusalem. 212 pp.

Goeden, R.D. 1990a. Life history of *Eutreta diana* (Osten Sacken) on *Artemisia tridentata* Nuttall in southern California. *Pan-Pac. Entomol.* 66: 24–32.

Goeden, R.D. 1990b. Notes on the life history of *Eutreta simplex* Thomas on *Artemisia ludoviciana* Nuttall in southern California (Diptera: Tephritidae). *Pan-Pac. Entomol.* 66: 33–38.

Hanna, A.D. 1938. Studies on the Mediterranean fruit-fly: *Ceratitis capitata* Wied. 1. The structure and operation of the reproductive organs. *Bull. Soc. Entomol. Egypte* 22: 39–60.

Hardy, D.E. 1951. The Krauss collection of Australian fruit flies (Tephritidae-Diptera). *Pac. Sci.* 5: 115–189.

Hardy, D.E. 1973. The fruit flies (Tephritidae — Diptera) of Thailand and bordering countries. *Pac. Insects Monogr.* 31, 353 p.

Headrick, D.H. and R.D. Goeden. 1990a. Description of the immature stages of *Paracantha gentilis* (Diptera: Tephritidae). *Ann. Entomol. Soc. Am.* 83: 220–229.

Headrick, D.H. and R.D. Goeden. 1990b. Life history of *Paracantha gentilis* (Diptera: Tephritidae). *Ann. Entomol. Soc. Am.* 83: 776–785.

Headrick, D.H. and R.D. Goeden. 1991. Life history of *Trupanea californica* Malloch (Diptera: Tephritidae) on *Gnaphalium* spp. (Asteraceae) in southern California. *Proc. Entomol. Soc. Wash.* 93: 559–570.

Headrick, D.H. and R.D. Goeden. 1994. Reproductive behavior of California fruit flies and the classification and evolution of Tephritidae (Diptera) mating systems. *Stud. Dipterol.* 1: 195–252.

Headrick, D.H. and R.D. Goeden. 1998. The biology of non-frugivorous tephritid fruit flies. *Annu. Rev. Entomol.* 43: 217–241.

Hendel, F. 1927. Trypetidae. In *Die Fliegen der Palaearktischen Region* (E. Lindner, ed.) 5 (1): 1–221. Stuttgart.

International Commission on Zoological Nomenclature. 1985. *International Code of Zoological Nomenclature, 3rd ed.* International Trust for Zoological Nomenclature, London. 338 pp.

Jenkins, J. 1990. Mating behavior of *Aciurina mexicana* (Aczél) (Diptera: Tephritidae). *Proc. Entomol. Soc. Wash.* 92: 66–75.

Jenkins, J. 1996. Systematic Studies of *Rhagoletis* and Related Genera (Diptera: Tephritidae). Dissertation, Michigan State University, Ann Arbor. 184 pp.

Kandybina, M.N. 1977. Larvae of fruit-infesting fruit flies (Diptera, Tephritidae). *Opred. Faune SSSR* No. 114: 1–210 [in Russian: unpublished English translation, 1987, produced by National Agricultural Library, Beltsville, MD].

Kanmiya, K. 1988. Acoustic studies on the mechanism of sound production in the mating songs of the melon fly, *Dacus curcurbitae* Coquillett (Diptera: Tephritidae). *J. Ethol.* 6: 143–151.

Keiser, I., R.M. Kobayashi, D.L. Chambers, and E.L. Schneider. 1973. Relation of sexual dimorphism in the wings, potential stridulation, and illumination to mating of Oriental fruit flies, melon flies, and Mediterranean fruit flies in Hawaii. *Ann. Entomol. Soc. Am.* 66: 937–941.

Korneyev, V.A. 1985. Fruit flies of the tribe Terelliini Hendel, 1927 (Diptera, Tephritidae) of the fauna of the USSR. *Entomol. Obozr.* 64: 626–644.

Korneyev, V.A. 1996. Reclassification of Palaearctic Tephritidae (Diptera). Communication 3. *Vestn. Zool.* 1995 (5–6): 25–48.

Lhoste, J. and A. Roche. 1960. Organes odoriférants des mâles de *Ceratitis capitata* (Diptera: Trypetidae). *Bull. Soc. Entomol. Fr.* 65: 206–210.

Lima, A.M. da Costa. 1934. Moscas de frutas do genero *Anastrepha* Schiner, 1868 (Diptera: Trypetidae). *Mem. Inst. Oswaldo Cruz Rio de J.* 28: 487–575.

Little, H.F. and R.T. Cunningham. 1987. Sexual dimorphism and presumed pheromone gland in the rectum of *Dacus latifrons* (Diptera: Tephritidae). *Ann. Entomol. Soc. Am.* 80: 765–767.

Margush, T. and F.R. McMorris. 1981. Consensus n-trees. *Bull. Math. Biol.* 43: 239–244.

McAlpine, J.F. 1981. Morphology and terminology-adults. In *Manual of Nearctic Diptera,* Vol. 1 (J.F. McAlpine, B.V. Peterson, G.E. Shewell, H.J. Teskey, J.R. Vockeroth, and D.M. Wood, eds.), pp. 9–63. Monograph of the Biosystematics Research Institute, No. 27. Agriculture Canada, Ottawa.

McAlpine, J.F. 1989. Phylogeny and classification of the Muscomorpha. In *Manual of Nearctic Diptera,* Vol. 3 (J.F. McAlpine, ed.), pp. 1397–1518. Monograph of the Biosystematics Research Centre, No. 32. Agriculture Canada, Ottawa.

Munro, H.K. 1947. African Trypetidae (Diptera); a review of the transition genera between Tephritinae and Trypetinae, with a preliminary study of the male terminalia. *Mem. Entomol. Soc. South. Afr.* 1: 1–284.

Munro, H.K. 1957. *Sphenella* and some allied genera (Trypetidae, Diptera). *J. Entomol. Soc. South. Afr.* 20: 14–57.

Munro, H.K. 1984. A taxonomic treatise on the Dacidae (Tephritoidea, Diptera) of Africa. *Entomol. Mem. S. Afr. Dep. Agric.* 61, 313 pp.

Myers, H.S., B.D. Barry, J.A. Burnside, and R.H. Rhode. 1976. Sperm precedence in female apple maggots alternately mated to normal and irradiated males. *Ann. Entomol. Soc. Am.* 69: 39–41.

Nation, J.L. 1972. Courtship behavior and evidence for a sex attractant in the male Caribbean fruit fly, *Anastrepha suspensa*. *Ann. Entomol. Soc. Am.* 65: 1364–1367.

Nation, J.L. 1981. Sex-specific glands in tephritid fruit flies of the genera *Anastrepha*, *Ceratitis*, *Dacus* and *Rhagoletis* (Diptera: Tephritidae). *Int. J. Insect Morphol. Embryol.* 10: 121–129.

Noble, G.K. 1939. The role of dominance in the social life of birds. *The Auk* 56: 263–273.

Norrbom, A.L. 1985. Phylogenetic Analysis and Taxonomy of the *cryptostrepha*, *daciformis*, *robusta*, and *schausi* Species Groups of *Anastrepha* Schiner (Diptera: Tephritidae). Ph.D. dissertation, Pennsylvania State University, University Park. 354 pp.

Norrbom, A.L. and K.C. Kim 1988. Revision of the *schausi* group of *Anastrepha* Schiner (Diptera: Tephritidae), with a discussion of the terminology of the female terminalia in the Tephritoidea. *Ann. Entomol. Soc. Am.* 81: 164–173.

Novak, J.A. and B.A. Foote. 1968. Biology and immature stages of fruit flies: *Paroxyna albiceps* (Diptera: Tephritidae). *J. Kans. Entomol. Soc.* 41: 108–119.

Phillips, V.T. 1946. The biology and identification of trypetid larvae. *Mem. Am. Entomol. Soc.* 12: 1–161.

Pritchard, G. 1967. Laboratory observations on the mating behavior of the Island fruit fly *Rioxa pornia* (Diptera: Tephritidae). *J. Aust. Entomol. Soc.* 6: 127–132.

Prokopy, R.J. and G.L. Bush. 1973. Mating behavior of *Rhagoletis pomonella* (Diptera: Tephritidae). IV. Courtship. *Can. Entomol.* 105: 873–891.

Prokopy, R.J. and J. Hendrichs. 1979. Mating behavior of *Ceratitis capitata* on a field-caged host tree. *Ann. Entomol. Soc. Am.* 72: 642–648.

Sabrosky, C.W. 1983. A synopsis of the world species of *Desmometopa* Loew (Diptera, Milichiidae). *Contrib. Am. Entomol. Inst.* 19 (8), 69 pp.

Singh, R.N. and K. Singh. 1984. Fine structure of the sensory organs of *Drosophila melanogaster* Meigen larva (Diptera: Drosophilidae). *Int. J. Insect Morphol. Embryol.* 13: 255–273.

Snodgrass, R.E. 1924. Anatomy and metamorphosis of the apple maggot. *J. Agric. Res.* 28: 1–36.

Snodgrass, R.E. 1953. The metamorphosis of a fly's head. *Smithson. Misc. Collect.* 122 (3): 1–25.

Steyskal, G.C. 1979. *Taxonomic Studies on Fruit Flies of the Genus Urophora (Diptera: Tephritidae)*. Entomological Society of Washington, Washington, D.C. 61 pp.

Steyskal, G.C. 1984. A synoptic revision of the genus *Aciurina* Curran, 1932 (Diptera: Tephritidae). *Proc. Entomol. Soc. Wash.* 86: 582–598.

Stoffolano, J.G., Jr. and L.R.S. Yin. 1987. Structure and function of the ovipositor and associated sensilla of the apple maggot, *Rhagoletis pomonella* (Walsh) (Diptera: Tephritidae). *Int. J. Insect Morphol. Embryol.* 16: 41–69.

Stoltzfus, W.B. and B.A. Foote. 1965. The use of froth masses in courtship of *Eutreta* (Diptera: Tephritidae). *Proc. Entomol. Soc. Wash.* 67: 262–264.

Stone, A. 1942. The fruitflies of the genus *Anastrepha*. *U.S. Dep. Agric. Misc. Publ.* 439, 112 pp.

Tauber, M.J. and C.A. Tauber. 1967. Reproductive behavior and biology of the gall-former *Aciurina ferruginea* (Doane) (Diptera: Tephritidae). *Can. J. Zool.* 45: 907–913.

Tauber, M.J. and C.A. Toschi. 1965. Bionomics of *Euleia fratria* (Loew) (Diptera: Tephritidae). I. Life history and mating behavior. *Can. J. Zool.* 43: 369–379.

Teskey, H.J. 1981. Morphology and terminology — larvae. In *Manual of Nearctic Diptera, Vol. 1* (J.F. McAlpine, B.V. Peterson, G.E. Shewell, H.J. Teskey, J.R. Vockeroth, and D.M. Wood, eds.), pp. 65–88. Monograph of the Biosystematics Research Institute No. 27. Agriculture Canada, Ottawa.

Thornhill, R. and J. Alcock. 1983. *The Evolution of Insect Mating Systems*. Harvard University Press, Cambridge. 547 pp.

Tyschen, P.H. 1977. Mating behaviour of the Queensland fruit fly, *Dacus tryoni* (Diptera: Tephritidae), in field cages. *J. Aust. Entomol. Soc.* 16: 459–465.

Valdez-Carrasco, J. and E. Prado-Beltran. 1991. Esqueleto y musculatura de la mosca del Mediterraneo, *Ceratitis capitata* (Wiedemann) (Diptera: Tephritidae). *Folia Entomol. Mex.* (1990) 80: 59–225.

White, I.M. 1988. Tephritid flies (Diptera: Tephritidae). *Handb. Identif. Br. Insects* 10 (5a): 1–134.
White, I.M. and M.M. Elson-Harris. 1994. *Fruit Flies of Economic Significance: Their Identification and Bionomics*. CAB INTERNATIONAL, Wallingford. Reprint with addendum (first published 1992).
Wilson, E.O. 1975. *Sociobiology, The New Synthesis*. Harvard University Press, Cambridge. 697 pp.
Zwölfer, H. 1974. Innerartliche Kommunikations-systeme bei Bohrfliegen. *Biol. Unserer Zeit* 4: 146–153.

## APPENDIX 33.1: HENDEL (1927) WING VEIN AND CELL TERMS

Cc (Kostalzelle) = refers to both cells bc and c
Csc (Subkostalzelle, Pterostigma) = cell sc
Cm (Marginalzell) = cell $r_1$
Csm (Submarginalzell) = cell $r_{2+3}$
$Cb_1$ (1. Basalzelle) = cell br
$Cb_2$ (2. Basalzelle) = cell bm
Cd (Diskalzelle) = cell dm
$Cp_1$ (1. Hinterrandzelle) = cell $r_{4+5}$
$Cp_2$ (2. Hinterrandzelle) = cell m
$Cp_3$ (3. Hinterrandzelle) = cell $cu_1$
Can (Analzelle) = cell bcu

# Index

## A

## B

## C

## D

## E

## G



## H

## I

## N

# O

## P

## Q



## R

## S

## T

## U



## V

## W



## X



## Y



## Z

*9780367399108*

T - #0450 - 071024 - C988 - 254/178/43 - PB - 9780367399108 - Gloss Lamination